Ullmann's Encyclopedia of Industrial Chemistry

Volume B 1

Fundamentals of
Chemical Engineering

Ullmann's Encyclopedia of Industrial Chemistry

Volumes A 1 – A 28: alphabetically arranged articles
Volumes B 1 – B 8: basic knowledge

Editorial Advisory Board

Hans-Jürgen Arpe
Hoechst Aktiengesellschaft,
Federal Republic of Germany

Ernst Biekert
Knoll AG, Federal Republic of Germany

H. Ted Davis
University of Minnesota, USA

Wolfgang Gerhartz
Degussa AG, Federal Republic of Germany

Heinz Gerrens
BASF Aktiengesellschaft,
Federal Republic of Germany

Wilhelm Keim
RWTH Aachen, Federal Republic of Germany

John L. McGuire
Johnson and Johnson, USA

Akio Mitsutani
Nippon Chemtech Consulting, Japan

Howard Pilat
Hoechst Celanese Corp., USA

Sir Charles Reece
Imperial Chemical Industries, UK

David P. Sheetz
The Dow Chemical Co., USA

Howard E. Simmons
E. I. Du Pont de Nemours, USA

Eberhard Weise
Bayer AG, Federal Republic of Germany

Rudolf Wirtz
Hoechst Aktiengesellschaft, Federal Republic of Germany

Hans-Rudolf Wüthrich
Metallgesellschaft AG, Federal Republic of Germany

Ullmann's Encyclopedia of Industrial Chemistry

Fifth, Completely Revised Edition

Volume B1:
Fundamentals of Chemical Engineering

Editors: Barbara Elvers, Stephen Hawkins, Gail Schulz

Volume Editor: Hanns Hofmann,
Universität Erlangen-Nürnberg

Numerical data, descriptions of methods or equipment, and other information presented in this book have been carefully checked for accuracy. Nevertheless, authors and publishers do not assume any liability for misprints, faulty statements, or other kinds of errors. Persons intending to handle chemicals or to work according to information derived from this book are advised to consult the original sources as well as relevant regulations in order to avoid possible hazards.

Production Director: Maximilian Montkowski
Production Manager: Myriam Nothacker

Library of Congress Card No. 84-25-829

Deutsche Bibliothek, Cataloguing-in-Publication Data:

Ullmann's encyclopedia of industrial chemistry / ed.: Barbara Elvers ... [Ed. advisory board Hans-Jürgen Arpe ...]. — Weinheim ; Basel (Switzerland) ; Cambridge ; New York, NY : VCH
  Teilw. executive ed.: Wolfgang Gerhartz.
  Bis 4. Aufl. u. d. T.: Ullmanns Encyklopädie der technischen Chemie
NE: Gerhartz, Wolfgang [Hrsg.]; Elvers, Barbara [Hrsg.]; Encyclopedia of industrial chemistry
Vol. B. Basic knowledge.
  1. Fundamentals of Chemical Engineering. — 1. — 5., completely rev. ed. — 1990
  ISBN 3-527-20131-9 (Weinheim ...) Pp.
  ISBN 0-89573-536-9 (New York) Pp.

British Library Cataloguing in Publication Data
Ullmann's encyclopedia of industrial chemistry.
  — 5th completely rev. ed.
  Vol. B1. Fundamentals of Chemical Engineering
  I. Ullmann, Fritz, **1875–1939** II. Gerhartz, Wolfgang III. [Ullmanns Enzyklopädie der technischen Chemie]. **English**
661′.003′21

ISBN 3-527-20131-9

© VCH Verlagsgesellschaft mbH, D-6940 Weinheim (Federal Republic of Germany), 1990.
Printed on acid-free paper

Distribution
VCH Verlagsgesellschaft, P.O. Box 10 11 61. D-6940 Weinheim (Federal Republic of Germany)
Switzerland: VCH Verlags-AG, P.O. Box, CH-4020 Basel (Switzerland)
Great Britain and Ireland: VCH Publishers (UK) Ltd., 8 Wellington Court, Wellington Street, Cambridge
            CB1 1HZ (Great Britain)
USA and Canada: VCH Publishers, Suite 909, 220 East 23rd Street, New York NY 10010-4606 (USA)

All rights reserved (including those of translation into other languages). No part of this book may be reproduced in any form — by photoprint, microfilm, or any other means — transmitted or translated into a machine language without written permission from the publishers.
Authorization to photocopy items for internal or personal use, or the internal or personal use of specific clients, is granted for libraries and other users registered with the Copyright Clearance Center (CCC) Transactional Reporting Service, provided that the base fee of $ 1.00 per copy, plus $ 0.25 per page is paid directly to CCC, 27 Congress Street, Salem, MA 01970. 0740-9451/85 $ 1.00 + 0.25.
Registered names, trademarks, etc. used in this book and not specifically marked as such are not to be considered unprotected.

Cover design: Wolfgang Schmidt
Composition, printing, and bookbinding: Graphischer Betrieb Konrad Triltsch, D-8700 Würzburg
Printed in the Federal Republic of Germany

Foreword

The fifth completely revised edition of Ullmann's Encyclopedia has been designed to help researchers, scientists, engineers, and managers keep up with the broad spectrum of chemical technology. In order to do this the 36 volumes of the Encyclopedia have been divided into two series. The 28 "A" volumes contain alphabetically arranged articles on chemicals, product groups, processes, and technological concepts. The 8 "B" volumes are compendia of basic knowledge and are structured as follows:

B1 Fundamentals of Chemical Engineering
B2 Unit Operations I
B3 Unit Operations II
B4 Chemical Reaction Engineering and Plant Design
B5 Analytical Methods I
B6 Analytical Methods II
B7 Environmental Protection and Plant Safety I
B8 Environmental Protection and Plant Safety II

The first volume of the B series is devoted to important fundamentals of chemical engineering. The articles are divided into three main groups: basic theory; materials of construction with emphasis on properties and testing; and information.

Basic Theory. This section starts with an article on *mathematics in chemical engineering* that gives a comprehensive survey of the mathematical "tools" needed by chemical engineers. One of the most important applications of mathematics in chemical engineering is the *modeling* of unit operations, processes, and reactors. *Dimensional analysis* is another commonly used technique for solving engineering problems, particularly scale-up and treatment of variable physical properties. The analysis of *transport phenomena* that occur as a result of driving forces in momentum, mass, and heat also finds a variety of applications, for example in process design and analysis. Chemical engineers should also be equipped with a good knowledge of *fluid mechanics* to deal with the large variety of liquids and gases they are likely to encounter. Chemical process design and operation also requires methods for the *estimation of physical properties* in cases where their values are not known.

Materials of Construction. Heavy demands are placed on *construction materials* used in the chemical industry. Their properties and testing are thus of crucial importance. One of the major causes of damage is *corrosion*, and corrosion resistance is one of the most important criteria for selecting materials. Mechanical *abrasion and erosion* may, however, in some cases also be a crucial factor. The success of a material in use depends on proper selection; this requires satisfactory *mechanical testing*. Many faults of equipment in operating plants can, however, only be detected by *nondestructive testing*.

Information. Industrial chemists and chemical engineers rely on the rapidly expanding resources of scientific information that are now available both in printed and electronic form. The handling of *information and documentation* as well as *patents* are therefore of prime importance.

The international team of authors contributing to volume B1 have been selected from academia and industry as renowned experts in their respective fields. We gratefully acknowledge their efforts and cooperation. Our thanks also go to our advisors and reviewers, who have helped make Ullmann's Encyclopedia the unparalleled international authority on industrial chemistry.

The Editors

Contents

1. Mathematics in Chemical Engineering
2. Mathematical Modeling
3. Dimensional Analysis
4. Transport Phenomena
5. Fluid Mechanics
6. Estimation of Physical Properties
7. Construction Materials in Chemical Industry
8. Corrosion
9. Abrasion and Erosion
10. Mechanical Properties and Testing of Metallic Materials
11. Nondestructive Testing
12. Information and Documentation
13. Patents

Symbols and Units

Symbols and units agree with SI standards (for conversion factors see pp. VIII–IX). The following list gives the most important symbols used in the encyclopedia. Articles with many specific units and symbols have a similar list as front matter.

Symbol	Unit	Physical Quantity
a_B		activity of substance B
A_r		relative atomic mass (atomic weight)
A	m²	area
c_B	mol/m³, mol/L (M)	concentration of substance B
C	C/V	electric capacity
c_p, c_v	J kg⁻¹K⁻¹	specific heat capacity
d	cm, m	diameter
d		relative density (ϱ/ϱ_{water})
D	m²/s	diffusion coefficient
D	Gy (= J/kg)	absorbed dose
e		elementary charge
E	J	energy
E	V/m	electric field strength
E	V	electromotive force
E_A	J	activation energy
f		activity coefficient
F	C/mol	Faraday constant
F	N	force
g	m/s²	acceleration due to gravity
G	J	Gibbs free energy
h	m	height
h	W·s²	Planck constant
H	J	enthalpy
I	A	electric current
I	cd	luminous intensity
k	(variable)	rate constant of a chemical reaction
k	J/K	Boltzmann constant
K	(variable)	equilibrium constant
l	m	length
m	g, kg, t	mass
M_r		relative molecular mass (molecular weight)
n_D^{20}		refractive index (sodium D-light, 20°C)
n	mol	amount of substance
N_A	mol⁻¹	Avogadro constant (6.023×10^{23} mol⁻¹)
p	Pa; bar*	pressure
Q	J	quantity of heat
r	m	radius
R	J K⁻¹mol⁻¹	gas constant
R	Ω	electric resistance
S	J/K	entropy
t	s, min, h, d, month, a	time
t	°C	temperature
T	K	absolute temperature
u	m/s	velocity

* The official unit of pressure is the pascal (Pa).

Symbols and units (continued from p. VII)

Symbol	Unit	Physical Quantity
U	V	electric potential
U	J	internal energy
V	m^3, L, mL	volume
w		mass fraction
W	J	work
x_B		mole fraction of substance B
α		cubic expansion coefficient
α	$W\,m^{-2}K^{-1}$	heat-transfer coefficient (heat-transfer number)
α		degree of dissociation of electrolyte
$[\alpha]$	$10^{-2} \deg\, cm^2 g^{-1}$	specific rotation
η	Pa · s	dynamic viscosity
θ	°C	temperature
$\varkappa$		c_p/c_v
λ	$W\,m^{-1}K^{-1}$	thermal conductivity
λ	nm, m	wavelength
μ		chemical potential
ν	Hz; s^{-1}	frequency
ν	m^2/s	kinematic viscosity (η/ϱ)
π	Pa	osmotic pressure
ϱ	g/cm^3	density
σ	N/m	surface tension
τ	Pa (N/m^2)	shear stress
φ		volume fraction
χ	Pa^{-1} (m^2/N)	compressibility

Conversion Factors

SI unit	Non-SI unit	From SI to non-SI multiply by
Mass		
kg	pound (avoirdupois)	2.205
kg	ton (long)	9.842×10^{-4}
kg	ton (short)	1.102×10^{-3}
Volume		
m^3	cubic inch	6.102×10^4
m^3	cubic foot	35.315
m^3	gallon (U.S., liquid)	2.642×10^2
m^3	gallon (Imperial)	2.200×10^2
Temperature		
°C	°F	°C × 1.8 + 32
Force		
N	dyne	1.0×10^5

Conversion factors (continued from p. VIII)

SI unit	Non-SI unit	From SI to non-SI multiply by
Energy, Work		
J	Btu (int.)	9.480×10^{-4}
J	cal (int.)	2.389×10^{-1}
J	eV	6.242×10^{18}
J	erg	1.0×10^{7}
J	kW · h	2.778×10^{-7}
J	kp · m	1.020×10^{-1}
Pressure		
MPa	at	10.20
MPa	atm	9.869
MPa	bar	10
kPa	mbar	10
kPa	mm Hg	7.502
kPa	psi	0.145
kPa	torr	7.502

Prefixes for Powers of Ten

T (tera) 10^{12} k (kilo) 10^{3} d (deci) 10^{-1} μ (micro) 10^{-6}
G (giga) 10^{9} h (hecto) 10^{2} c (centi) 10^{-2} n (nano) 10^{-9}
M (mega) 10^{6} m (milli) 10^{-3} p (pico) 10^{-12}

Abbreviations

The following is a list of the abbreviations used in the text. Common terms, the names of publications and institutions, and legal agreements are included along with their full identities. Other abbreviations will be defined wherever they first occur in an article. For further abbreviations, see p. VII (Symbols and Units), p. XIV (Companies and Country Codes in Patent References). The names of periodical publications are abbreviated exactly as done by Chemical Abstracts Service.

abs.	absolute
a.c.	alternating current
ACGIH	American Conference of Governmental Industrial Hygienists
ACS	American Chemical Society
ADI	acceptable daily intake
ADN	accord européen relatif au transport international des marchandises dangereuses par voie de navigation interieure (European agreement concerning the international transportation of dangerous goods by inland waterways)
ADNR	ADN par le Rhin (regulation concerning the transportation of dangerous goods on the Rhine and all national waterways of the countries concerned)
ADP	adenosine 5'-diphosphate
ADR	accord européen relatif au transport international des marchandises dangereuses par route (European agreement concerning the international transportation of dangerous goods by road)
AEC	Atomic Energy Commission (United States)
AG	Aktiengesellschaft
AIChE	American Institute of Chemical Engineers

Abbreviations

AIME	American Institute of Mining, Metallurgical, and Petroleum Engineers	DAB 9	Deutsches Arzneibuch, 9th ed., Deutscher Apotheker-Verlag, Stuttgart 1986
AMP	adenosine 5'-monophosphate	d.c.	direct current
APhA	American Pharmaceutical Association	decomp.	decompose, decomposition
API	American Petroleum Institute	DFG	Deutsche Forschungsgemeinschaft (German Science Foundation)
ASTM	American Society for Testing and Materials	dil.	dilute, diluted
ATP	adenosine 5'-triphosphate	DIN	Deutsche Industrie Norm (Federal Republic of Germany)
BAM	Bundesanstalt für Materialprüfung (Federal Republic of Germany)	DMF	dimethylformamide
		DNA	deoxyribonucleic acid
BAT	Biologischer Arbeitsstoff-Toleranzwert (biological tolerance value for a working material, established by MAK commission, see MAK)	DOE	Department of Energy (United States)
		DOT	Department of Transportation – Materials Transportation Bureau (United States)
Beilstein	Beilstein's Handbook of Organic Chemistry, Springer, Berlin–Heidelberg–New York	DTA	differential thermal analysis
		ed.	editor, editors, edition, edited
BET	Brunauer–Emmett–Teller	EEC	European Economic Community
BGBl.	Bundesgesetzblatt (Federal Republic of Germany)	e.g.	for example
		emf	electromotive force
BIOS	British Intelligence Objectives Subcommitee Report (see also FIAT)	EPA	Environmental Protection Agency (United States)
BOD	biological oxygen demand	EPR	electron paramagnetic resonance
bp	boiling point	Eq.	equation
B.P.	British Pharmacopeia	ESCA	electron spectroscopy for chemical analysis
BS	British Standard		
ca.	circa	esp.	especially
calcd.	calculated	ESR	electron spin resonance
CAS	Chemical Abstracts Service	Et	ethyl substituent $(-C_2H_5)$
cat.	catalyst; catalyzed	et al.	and others
cf.	compare	etc.	et cetera
CFR	Code of Federal Regulations (United States)	EVO	Eisenbahnverkehrsordnung (Federal Republic of Germany)
Chap.	chapter	exp (…)	$e^{(…)}$, mathematical exponent
ChemG	Chemikaliengesetz (Federal Republic of Germany)	FAO	Food and Agriculture Organization (United Nations)
C.I.	Colour Index	FDA	Food and Drug Administration (United States)
CIOS	Combined Intelligence Objectives Subcommitee Report (see also FIAT)	FD & C	Food, Drug and Cosmetic Act (United States)
		FHSA	Federal Hazardous Substances Act (United States)
CNS	central nervous system	FIAT	Field Information Agency, Technical (United States reports on the chemical industry in Germany, 1945)
Co.	Company		
COD	chemical oxygen demand		
conc.	concentrated		
const.	constant	Fig.	figure
Corp.	Corporation	*fp*	freezing point
crit.	critical	Friedländer	P. Friedländer, Fortschritte der Teerfarbenfabrikation und verwandter Industriezweige, Vol. 1–25, Springer, Berlin 1888–1942
CTFA	The Cosmetic, Toiletry and Fragrance Association (United States)		
		FT	Fourier transform

(g)	gas, gaseous	IMO	Inter-Governmental Maritime Consultive Organization (in the past: IMCO)
GC	gas chromatography		
GefStoffV	Gefahrstoffverordnung (regulations in the Federal Republic of Germany concerning hazardous substances)	Inst.	Institute
		i.p.	intraperitoneal
		IR	infrared
		ISO	International Organization for Standardization
GGVE	Verordnung in der Bundesrepublik Deutschland über die Beförderung gefährlicher Güter mit der Eisenbahn (regulation in the Federal Republic of Germany concerning the transportation of dangerous goods by rail)	IUPAC	International Union of Pure and Applied Chemistry
		i.v.	intravenous
		Kirk-Othmer	Encyclopedia of Chemical Technology, 3rd ed., J. Wiley & Sons, New York–Chichester–Brisbane–Toronto 1978–1984
		(l)	liquid
GGVS	Verordnung in der Bundesrepublik Deutschland über die Beförderung gefährlicher Güter auf der Straße (regulation in the Federal Republic of Germany concerning the transportation of dangerous goods by road)	Landolt-Börnstein	Zahlenwerte u. Funktionen aus Physik, Chemie, Astronomie, Geophysik u. Technik, Springer, Heidelberg 1950–1980; Zahlenwerte und Funktionen aus Naturwissenschaften und Technik, Neue Serie, Springer, Heidelberg, since 1961
GGVSee	Verordnung in der Bundesrepublik Deutschland über die Beförderung gefährlicher Güter mit Seeschiffen (regulation in the Federal Republic of Germany concerning the transportation of dangerous goods by sea-going vessels)	LC_{50}	lethal concentration
		LCLo	lowest published lethal concentration
		LD_{50}	lethal dose
		LDLo	lowest published lethal dose
		ln	logarithm (base e)
GLC	gas-liquid chromatography	LNG	liquefied natural gas
Gmelin	Gmelin's Handbook of Inorganic Chemistry, 8th ed., Springer, Berlin–Heidelberg–New York	log	logarithm (base 10)
		LPG	liquefied petroleum gas
		M	mol/L
GRAS	generally recognized as safe	M	metal (in chemical formulas)
Hal	halogen substituent ($-F$, $-Cl$, $-Br$, $-I$)	MAK	Maximale Arbeitsplatz-Konzentration (maximum concentration at the workplace in the Federal Republic of Germany); cf. Deutsche Forschungsgemeinschaft (ed.): Maximale Arbeitsplatzkonzentrationen (MAK) und Biologische Arbeitsstoff-Toleranz-Werte (BAT), VCH Verlagsgesellschaft, Weinheim (published annually)
Houben-Weyl	Methoden der organischen Chemie, 4th ed., Georg Thieme Verlag, Stuttgart		
HPLC	high performance liquid chromatography		
IARC	International Agency for Research on Cancer, Lyon, France		
IAEA	International Atomic Energy Agency		
IATA-DGR	International Air Transport Association, Dangerous Goods Regulations		
		max.	maximum
		MCA	Manufacturing Chemists Association (United States)
ICAO	International Civil Aviation Organization		
i.e.	that is	Me	methyl substituent ($-CH_3$)
i.m.	intramuscular	Methodicum Chimicum	Methodicum Chimicum, Georg Thieme Verlag, Stuttgart
IMDG	International Maritime Dangerous Goods Code		

Abbreviations

MIK	maximale Immissionskonzentration (maximum immission concentration)
min.	minimum
mp	melting point
MS	mass spectrum, mass spectrometry
NAS	National Academy of Sciences (United States)
NASA	National Aeronautics and Space Administration (United States)
NBS	National Bureau of Standards (United States)
NCTC	National Collection of Type Cultures (United States)
NIH	National Institutes of Health (United States)
NIOSH	National Institute for Occupational Safety and Health (United States)
NMR	nuclear magnetic resonance
no.	number
NRC	Nuclear Regulatory Commission (United States)
NRDC	National Research Development Corporation (United States)
NSC	National Service Center (United States)
NSF	National Science Foundation (United States)
NTSB	National Transportation Safety Board (United States)
OECD	Organization for Economic Cooperation and Development
OSHA	Occupational Safety and Health Administration (United States)
p., pp.	page, pages
Patty	G. D. Clayton, F. E. Clayton (eds.): Patty's Industrial Hygiene and Toxicology, 3rd ed., Wiley Interscience, New York
PB report	Publication Board Report (U.S. Department of Commerce, Scientific and Industrial Reports)
PEL	permitted exposure limit
Ph	phenyl substituent ($-C_6H_5$)
Ph. Eur.	European Pharmacopoeia, 2nd. ed., Council of Europe, Strasbourg 1981 –
phr	part per hundred rubber (resin)
PNS	peripheral nervous system
q. v.	which see (quod vide)
ref.	refer, reference
resp.	respectively
R_f	retention factor (TLC)
R. H.	relative humidity
RID	règlement international concernant le transport des marchandises dangereuses par chemin de fer (international convention concerning the transportation of dangerous goods by rail)
RNA	ribonucleic acid
rpm	revolutions per minute
RTECS	Registry of Toxic Effects of Chemical Substances, edited by the National Institute of Occupational Safety and Health (United States)
(s)	solid
SAE	Society of Automotive Engineers (United States)
s.c.	subcutaneous
SI	International System of Units
SIMS	secondary ion mass spectrometry
STEL	Short Term Exposure Limit (see TLV)
STP	standard temperature and pressure (0° C, 101.325 kPa)
T_g	glass transition temperature
TA Luft	Technische Anleitung zur Reinhaltung der Luft (clean air regulation in Federal Republic of Germany)
TA Lärm	Technische Anleitung zum Schutz gegen Lärm (low noise regulation in Federal Republic of Germany)
TDLo	lowest published toxic dose
THF	tetrahydrofuran
TLC	thin layer chromatography
TLV	Threshold Limit Value (TWA and STEL); published annually by the American Conference of Governmental Industrial Hygienists (ACGIH), Cincinnati, Ohio
TOD	total oxygen demand
TRK	Technische Richtkonzentration (lowest technically feasible level)
TSCA	Toxic Substances Control Act (United States)
TÜV	Technischer Überwachungsverein (Technical Control Board of the Federal Republic of Germany)
TWA	Time Weighted Average
Ullmann	Ullmanns Encyklopädie der Technischen Chemie, 4th ed., Verlag Chemie, Weinheim 1972–1984; 3rd ed., Urban und Schwarzenberg, München 1951–1970

USAEC	United States Atomic Energy Commission		for storage, filling, and transportation of flammable liquids; classification according to the flash point of liquids, recently in accordance with the classification in the United States)
USAN	United States Adopted Names		
USD	United States Dispensatory		
USDA	United States Department of Agriculture		
U.S.P.	United States Pharmacopeia	VDE	Verband Deutscher Elektroingenieure (Federal Republic of Germany)
UV	ultraviolet		
UVV	Unfallverhütungsvorschriften der Berufsgenossenschaft (workplace safety regulations in the Federal Republic of Germany)	VDI	Verein Deutscher Ingenieure (Federal Republic of Germany)
		vol	volume
VbF	Verordnung in der Bundesrepublik Deutschland über die Errichtung und den Betrieb von Anlagen zur Lagerung, Abfüllung und Beförderung brennbarer Flüssigkeiten (regulation in the Federal Republic of Germany concerning the construction and operation of plants	vol.	volume (of a series of books)
		vs.	versus
		WHO	World Health Organization (United Nations)
		Winnacker-Küchler	Chemische Technologie, Carl Hanser Verlag, München
		wt	weight
		$	U.S. dollar, unless otherwise stated

1. Mathematics in Chemical Engineering

BRUCE A. FINLAYSON, Department of Chemical Engineering, University of Washington, Seattle, Washington 98195, United States (Chaps. 1–9, 11, 12)

LORENZ T. BIEGLER, IGNACIO E. GROSSMANN, ARTHUR W. WESTERBERG, Carnegie Mellon University, Pittsburgh, Pennsylvania 15231, United States (Chap. 10)

1.	Solution of Equations	1-3
1.1.	Linear Algebraic Equations	1-4
1.2.	Nonlinear Algebraic Equations	1-9
1.3.	Linear Difference Equations	1-11
1.4.	Eigenvalues	1-12
2.	Approximation and Integration	1-13
2.1.	Introduction	1-13
2.2.	Global Polynomial Approximation	1-13
2.3.	Piecewise Approximation	1-15
2.4.	Quadrature	1-18
2.5.	Linear Least Squares	1-20
2.6.	Nonlinear Least Squares	1-22
2.7.	Fourier Transforms of Discrete Data	1-23
2.8.	Two-Dimensional Interpolation and Quadrature	1-25
3.	Complex Variables	1-26
3.1.	Introduction to the Complex Plane	1-26
3.2.	Elementary Functions	1-27
3.3.	Analytic Functions of a Complex Variable	1-28
3.4.	Integration in the Complex Plane	1-29
3.5.	Other Results	1-32
4.	Integral Transforms	1-33
4.1.	Fourier Transforms	1-33
4.2.	Laplace Transforms	1-37
4.3.	Solution of Partial Differential Equations by Using Transforms	1-42
5.	Vector Analysis	1-44
6.	Ordinary Differential Equations as Initial Value Problems	1-53
6.1.	Solution by Quadrature	1-54
6.2.	Explicit Methods	1-55
6.3.	Implicit Methods	1-58
6.4.	Stiffness	1-60
6.5.	Differential-Algebraic Systems	1-61
6.6.	Computer Software	1-62
6.7.	Stability, Bifurcations, Limit Cycles	1-63
6.8.	Sensitivity Analysis	1-65
6.9.	Eigenvalues and Roots by Initial Value Techniques	1-66
7.	Ordinary Differential Equations as Boundary Value Problems	1-67
7.1.	Solution by Quadrature	1-67
7.2.	Shooting Methods	1-68
7.3.	Finite Difference Method	1-70
7.4.	Orthogonal Collocation	1-73
7.5.	Orthogonal Collocation on Finite Elements	1-76
7.6.	Galerkin Finite Element Method	1-78
7.7.	Cubic B-Splines	1-80
7.8.	Adaptive Mesh Strategies	1-81
7.9.	Comparison	1-82
7.10.	Singular Problems and Infinite Domains	1-83
8.	Partial Differential Equations	1-83
8.1.	Classification of Equations	1-84
8.2.	Hyperbolic Equations	1-86
8.3.	Parabolic Equations in One Dimension	1-88
8.4.	Elliptic Equations	1-94
8.5.	Parabolic Equations in Two or Three Dimensions	1-97
9.	Integral Equations	1-97
9.1.	Classification	1-98
9.2.	Numerical Methods for Volterra Equations of the Second Kind	1-99

9.3.	Numerical Methods for Fredholm, Urysohn, and Hammerstein Equations of the Second Kind	1-102	10.6.	Mixed-Integer Programming 1-120
9.4.	Numerical Methods for Eigenvalue Problems	1-103	10.7.	Solution of Dynamic Optimization Problems ... 1-123
9.5.	Green's Functions	1-103	11.	Probability and Statistics 1-128
9.6.	Boundary Integral Equations and Boundary Element Method	1-105	11.1.	Concepts ... 1-128
10.	Optimization	1-106	11.2.	Sampling and Statistical Decisions . 1-131
10.1.	Introduction	1-106	11.3.	Error Analysis in Experiments 1-135
10.2.	Conditions for Optimality	1-107	11.4.	Factorial Design of Experiments and Analysis of Variance ... 1-135
10.3.	Strategies of Optimization	1-110	12.	Multivariable Calculus Applied to Thermodynamics ... 1-138
10.4.	Successive Quadratic Programming (SQP)	1-114	12.1.	State Functions ... 1-138
10.5.	Linear Programming	1-117	12.2.	Applications to Thermodynamics ... 1-138
10.5.1.	Basic Properties	1-118	12.3.	Partial Derivatives of All Thermodynamic Functions ... 1-140
10.5.2.	Simplex Algorithm	1-119	13.	References ... 1-141

In addition to the standard symbols defined in the front matter of this volume, the following symbols are used:

Variables

a scalar constant in quadratic approximation for F, the objective function
A $m \times n$ matrix of constant coefficients for equality constraints in linear programming model
b vector of constants premultiplying the r independent variables u in the quadratic approximation for F, the objective function. Also vector of m right-hand-sides for equality constraints in linear programming problem
B approximation of the $n \times n$ Hessian matrix for the Lagrange function for the successive quadratic programming algorithm. Also $m \times m$ non-singular matrix corresponding to the basis variables in a linear programming problem. Also coefficient matrix for binary variables y in the set of equality constraints for a mixed integer programming problem
Bi Biot number
c vector of n constant cost coefficients for all variables z in linear programming problem
Co Courant number
d search direction in the space of all n variables z (both dependent and independent variables)
D diffusion constant
Da Damköhler number
f nonlinear scalar contribution to objective function which is a function only of the continuous variables, y, for a mixed integer programming problem
F scalar objective function for an optimization problem
g set of p inequality constraints for an optimization problem
h set of m equality constraints for an optimization problem
H the inverse of matrix Q. Also the scalar Hamiltonian function for a dynamic optimization problem
I identity matrix
k iteration matrix
L scalar Lagrange function
m number of equality constraints for an optimization problem
n number of total variables in an optimization problem
N $(n-m) \times m$ matrix corresponding to the non-basis variables in a linear programming problem
p number of inequality constraints for an optimization problem. Also vector of parameters (do not vary with time) for dynamic optimization problem
Pe Peclet number
Q an $r \times r$ matrix of constants used in defining the quadratic term for a quadratic approximation for F, the objective function
r number of independent variables for an optimization problem ($r = n-m$)
Re the space of real numbers
R matrix defined in the quadratic programming algorithm. Also the vector of residual equations in the finite element approach to solving dynamic optimization problems
Re^n real number vector space of dimension n
s change in the independent variables u (a vector with r elements in it). Also sensitivity of the functions f with respect to the parameter p for a dynamic optimization problem
Sh Sherwood number
u vector of r independent variables for an optimization problem. For a time-varying problem, the independent time-varying control variables
W $r \times r$ weighting matrix used in computing Forbenius norm
x vector of m dependent variables for an optimization problem (the state variables for a dynamic optimization problem)

y	variables used in linear programming problem. Also binary variables in the formulation of a mixed integer programming problem
Y	$n \times m$ matrix whose columns are vectors that span the range space of the linearized constraints (see section on successive quadratic programming algorithm)
z	vector of all n variables—both the m dependent variables x and the r independent variables u—for an optimization problem
$\bar{z}$	interior point for linear programming problem
Z	$n \times (n-m)$ matrix whose columns are vectors that span the null space of the linearized constraints (see section on successive quadratic programming algorithm)

Greek symbols

α	scalar for setting the step size in a line search algorithm for the successive quadratic programming algorithm. Also a vector of coefficients to form a convex combination of extreme points for a linear programming problem. Also scalar variable used as the objective for the mixed integer programming problem in equation 76. Also length of an element in the finite element approach to solving dynamic optimization problems
γ	change in gradient of the objective function in moving from point k to point $k+1$. γ is a vector with r elements in it
δ	Kronecker delta
Δ	sampling rate
ε	small scalar
ϕ	vector of polynomial approximation functions for the state variables for finite element method for dynamic optimization problem, also Thiele modulus
κ	condition number
λ	vector of m Lagrange multipliers for an optimization problem. Also for a dynamic optimization problem, the Lagrange multiplier for the state variable constraints.
μ	vector of p Kuhn–Tucker multipliers
ν	vector of Lagrange multipliers (vary with time) for the algebraic equality constraints for a dynamic optimization problem
τ	rescaled time so it lies in range [0, 1]
Y	vector of polynomial approximation function for the control variables for finite element method for dynamic optimization problem

Special symbols

|	subject to
:	mapping. For example, $h: R^n \to R^m$, states that functions h map real numbers into m real numbers. There are m functions h written in terms of n variables
$\in$	member of
$\to$	maps into

Subscripts

A	g_A are the active constraints (i.e., precisely equal to zero) among the inequality constraints
B	basis variables in a linear programming problem
f	value at the final time for a dynamic optimization problem
i	i-th element of a vector of variables
N	non-basis variables in a linear programming problem
NE	number of elements in a finite element model
OA	outer approximation
0	value at the initial time for a dynamic optimization problem

Superscript

k	iteration index
L	lower bound
T	transpose of a vector or matrix
U	upper bound
$\wedge$	base point for a Taylor series expansion

1. Solution of Equations

Mathematical models of chemical engineering systems can take many forms: they can be sets of algebraic equations, differential equations, or integral equations. Mass and energy balances of chemical processes typically lead to large sets of *algebraic* equations:

$$a_{11}x_1 + a_{12}x_2 = b_1$$

$$a_{21}x_1 + a_{22}x_2 = b_2$$

Mass balances of stirred tank reactors may lead to ordinary *differential* equations:

$$\frac{dy}{dt} = f[y(t)]$$

Radiative heat transfer may lead to *integral* equations:

$$y(x) = g(x) + \lambda \int_0^1 K(x,s) f(s) \, ds$$

Even when the model is a differential equation or integral equation, the most basic step in the algorithm is the solution of sets of algebraic equations. The solution of sets of algebraic equations is the focus of Chapter 1.

A single linear equation is easy to solve for either x or y:

$$y = ax + b$$

If the equation is nonlinear,

$$f(x) = 0$$

it may be more difficult to find the x satisfying this equation. These problems are compounded when there are more unknowns, leading to simultaneous equations. If the unknowns appear in a linear fashion, then an important consideration is the structure of the matrix representing the equations; special methods are presented here for special structures. They are useful because they increase the speed of solution. If the unknowns appear in a nonlinear fashion, the problem is much more difficult. Iterative techniques must be used (i.e., make a guess of the solution and try to improve the guess). An important question then is whether such an iterative scheme converges. Other important types of equations are linear difference equations and eigenvalue problems, which are also discussed.

1.1. Linear Algebraic Equations

Consider the $n \times n$ linear system

$$a_{11} x_1 + a_{12} x_2 + \ldots + a_{1n} x_n = f_1$$

$$a_{21} x_1 + a_{22} x_2 + \ldots + a_{2n} x_n = f_2$$

$$a_{n1} x_1 + a_{n2} x_2 + \ldots + a_{nn} x_n = f_n$$

In this equation $a_{11}, \ldots, a_{nn}$ are known parameters, $f_1, \ldots, f_n$ are known, and the unknowns are $x_1, \ldots, x_n$. The values of all unknowns that satisfy every equation must be found. This set of equations can be represented as follows:

$$\sum_{j=1}^{n} a_{ij} x_j = f_j \quad \text{or} \quad A x = f$$

The most efficient method for solving a set of linear algebraic equations is to perform a lower-upper (LU) decomposition of the corresponding matrix A. This decomposition is essentially a Gaussian elimination, arranged for maximum efficiency.

The LU decomposition is done by calculating in turn

```
for i = 1, n do
    for j = 1, n do
        a_{ij}^{(1)} = a_{ij}
    enddo
enddo
for k = 2, n do
    for i = k + 1, n do
        for j = k + 1, n do

            a_{ij}^{(k)} = a_{ij}^{(k-1)} - \frac{a_{i,k-1}^{(k-1)}}{a_{k-1,k-1}^{(k-1)}} a_{k-1,j}^{(k-1)}

        enddo
    enddo
enddo
```

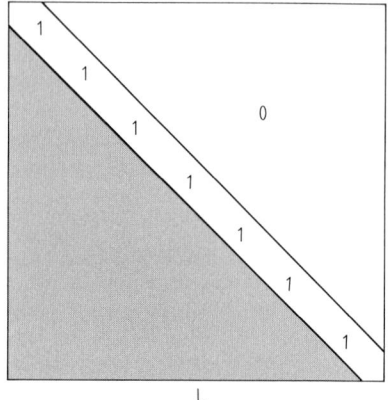
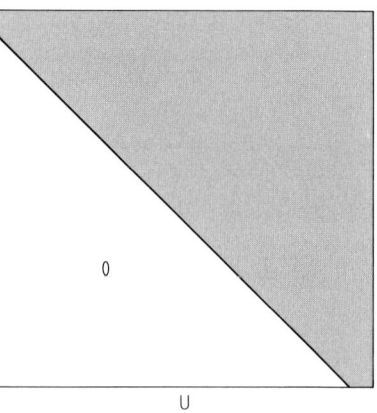

Figure 1. Structure of L and U matrices

Then two matrices are formed. The upper matrix $U = A^{(n)}$ is defined as

$$U_{ij} = A_{ij}^{(n)} = \begin{cases} 0 & i > j \\ a_{ij}^{(i)} & i \leq j, \quad 1 \leq j \leq n \end{cases}$$

The lower matrix L is defined as

$$L_{ij} = \begin{cases} 0 & i < j \\ 1 & i = j \\ \dfrac{a_{ij}^{(j)}}{a_{jj}^{(j)}} & i > j \end{cases}$$

The U is upper triangular; it has zero elements below the main diagonal and possibly nonzero values along the main diagonal and above it (see Fig. 1). The L is lower triangular. It has the value 1 in each element of the main diagonal, nonzero values below the diagonal, and zero values above the diagonal (see Fig. 1). Thus,

$$A = LU$$

The original problem can be solved in two steps:

$$Ly = f, \ Ux = y \text{ solves } Ax = LUx = f$$

Each of these steps is straightforward because the matrices are upper triangular or lower triangular. The solution is performed using the equations

for $i = 1, n$ do
  $f_i^{(1)} = f_i$
enddo
for $k = 2, n$ do
  for $i = k+1, n$ do
    $f_i^{(k)} = f_i^{(k)} - \dfrac{a_{i,k-1}^{(k-1)}}{a_{k-1,k-1}^{(k-1)}} f_{k-1}^{(k-1)}$
  enddo
enddo

$x_n = f_n^{(n)}/a_{nn}^{(n)}$

for $i = n-1, 1$ do
  $x_i = \dfrac{\left[f_i^{(i)} - \sum_{j=i+1}^{n} a_{ij}^{(i)} x_j\right]}{a_{ii}^{(i)}}$
enddo

When f is changed, the last steps can be done without recomputing the LU decomposition. Thus, multiple right-hand sides can be computed efficiently. The number of multiplications and divisions necessary to solve for m right-hand sides is:

$$\text{Operation count} = \frac{1}{3}n^3 - \frac{1}{3}n + mn^2$$

The actual algorithm used will be different for parallel computers.

The *determinant* is given by

$$\text{Det } A = \prod_{i=1}^{n} a_{ii}^{(i)}$$

This should be calculated as the LU decomposition is performed. If the value of the determinant is a very large or very small number, it can be divided or multiplied by 10 to retain accuracy in the computer; the scale factor is then accumulated separately. The condition number κ can be defined in terms of the singular value decomposition as the ratio of the largest w_i to the smallest w_i (see below). It can also be expressed in terms of the norm of the matrix:

$$\kappa(A) = \|A\| \, \|A^{-1}\|$$

where the norm is defined as

$$\|A\| \equiv \sup_{x \neq 0} \frac{\|Ax\|}{\|x\|} = \max_k \sum_{j=1}^{n} |a_{jk}|$$

If this number is infinite, the set of equations is singular. If the number is too large, the matrix is said to be ill-conditioned. The definition of "large" refers to the accuracy of the computer being used; the criterion is the inverse of the floating point precision. If the machine's floating point precision is 10^{-6} then 10^6 is large; if double precision is used then the precision is 10^{-12} and 10^{12} is large. Calculation of the condition number can be lengthy so another criterion is also useful. Compute the ratio of the largest to the smallest pivot and make judgments on the ill-conditioning based on that.

$$\text{Ratio} = \frac{\max_k |a_{kk}^{(k)}|}{\min_k |a_{kk}^{(k)}|}$$

Another empirical test is the quantity V; when V is small the matrix is ill-conditioned.

$$V = \frac{|\det A|}{\alpha_1 \alpha_2 \dots \alpha_n}, \ \alpha_i = (a_{i1}^2 + a_{i2}^2 + \dots + a_{in}^2)^{1/2}$$

When a matrix is ill-conditioned the LU decomposition must be performed by using pivoting (or the singular value decomposition described below). With pivoting, the order of the elimination is rearranged. At each stage, one looks for the largest element (in magnitude); the next stages if the elimination are on the row and column containing that largest element. The largest element can be obtained from only the diagonal entries (partial pivoting) or from all the remaining entries. If the matrix is nonsingular, Gaussian elimination (or LU decomposition) could fail if a zero value were to occur along the diagonal and were to be a pivot. With full pivoting, however, the Gaussian elimination (or LU decomposition) cannot fail because the matrix is nonsingular.

A matrix is *symmetric* if

$$a_{ij} = a_{ji}$$

and it is *positive definite* if

$$x^T A x = \sum_{i=1}^{n} \sum_{j=1}^{n} a_{ij} x_i x_j \geq 0$$

for all x and the equality holds only if $x = 0$ (where x^T is the transpose of the matrix, see p. 1-45)

The *Cholesky decomposition* can be used for real, symmetric, positive definite matrices. This algorithm saves on storage (divide by about 2) and reduces the number of multiplications (divide by 2), but adds n square roots.

The *inverse of a matrix* can also be used to solve sets of linear equations. The inverse is a matrix such that when A is multiplied by its inverse the result is the identity matrix, a matrix with 1.0 along the main diagonal and zero elsewhere.

$$A A^{-1} = I$$

Then the linear equations are solved by

$$x = A^{-1} f$$

Generally, the inverse is not used in this way because it requires three times more operations than solving with an LU decomposition. However, if an inverse is desired, it is calculated most efficiently by using the LU decomposition and then solving

$$A x^{(i)} = b^{(i)}$$

$$b_j^{(i)} = \begin{cases} 0 & j \neq i \\ 1 & j = i \end{cases}$$

Then set

$$A^{-1} = (x^{(1)} | x^{(2)} | x^{(3)} | \ldots | x^{(n)})$$

Generally software packages are available to solve this problem, and the user must submit the elements of the matrix in proper fashion. Remembering that the information passed from calling program to subroutine is a linear array of numbers, and that actually only the location of the first number is passed to the computer subroutine, is useful for interpretating user instructions. If the matrix is an $n \times n$ matrix stored in an $m \times m$ array, then what is passed is the location of a_{11} and the numbers are expected to be in the arrangement

$a_{11}, a_{21}, \ldots, a_{n1}, -,-, a_{12}, a_{22}, \ldots, a_{n2}, -,-,\ldots$

Entry 1, 2, ..., n, -,-, m, $m+1$, $m+2$, ..., $m+n$, -,-, $2n$, ...

Solutions of Special Matrices. Special matrices can be handled even more efficiently. A *tridiagonal matrix* is one with nonzero entries

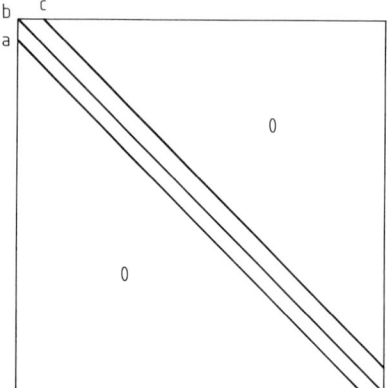

Figure 2. Structure of tridiagonal matrices

along the main diagonal, and one diagonal above and below the main one (see Fig. 2). The corresponding set of equations can then be written as

$$a_i x_{i-1} + b_i x_i + c_i x_{i+1} = d_i$$

The LU decomposition algorithm for solving this set is

$b'_1 = b_1$
for $k = 2, n$ do

$$a'_k = \frac{a_k}{b'_{k-1}}, \quad b'_k = b_k - \frac{a_k}{b'_{k-1}} c_{k-1}$$

enddo
$d'_1 = d_1$
for $k = 2, n$ do

$$d'_k = d_k - a'_k d'_{k-1}$$

enddo

$$x_n = d'_n / b'_n$$

for $k = n-1, 1$ do

$$x_k = \frac{d'_k - c_k x_{k+1}}{b'_k}$$

enddo

The number of multiplications and divisions for a problem with n unknowns and m right-hand

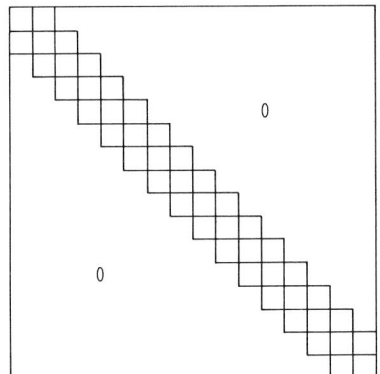

Figure 3. Structure of block tridiagonal matrix

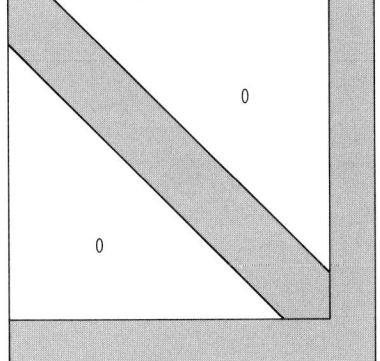

Figure 4. Structure of arrow matrix

sides is

Operation count $= 2(n-1) + m(3n-2)$

If

$|b_i| > |a_i| + |c_i|$

no pivoting is necessary. For solving two-point boundary value problems and partial differential equations this is often the case.

A *block tridiagonal matrix* (see Fig. 3) frequently arises in solving multiple differential equations, either two-point boundary value problems or partial differential equations. When solving several simultaneous equations the unknowns must be arranged in a particular way to achieve the block tridiagonal form of the matrix and the resulting efficiencies. Suppose we are solving for two unknowns, for example, temperature T and concentration c, and a value for each exists at each node $i = 1$ to n, the unknowns are $\{c_1, c_2, \ldots, c_n, T_1, T_2, \ldots, T_n\}$. If the unknowns are ordered in this way, however, the resulting linear equations to be solved, when solving two-point boundary value problems or partial differential equations, will be dense, that is, the nonzero entries of the matrix A will have no special pattern. However, if we arrange the unknowns in the pattern $\{c_1, T_1, c_2, T_2, \ldots, c_n, T_n\}$, then special patterns emerge. If the finite difference method is used, the block tridiagonal matrix arises. For other methods, other special patterns arise. Solution algorithms are most efficient if these patterns are taken into account in the LU decomposition. Clearly, a trade-off occurs between the programming time (needed to exploit any special structure) and the value received from a more efficient solution. If the problem is being solved only a few times or the time is extremely small, the special structure need not be exploited and usual software packages are suitable. If the problem is being solved thousands of times or the computer time is large (when n is large, e.g., 1000), the special structure must be exploited. Suppose the block tridiagonal matrix is composed of blocks that are $n_b \times n_b$, with n columns of blocks (along the diagonal), then the number of multiplications and divisions is (for large n and n_b) [1.1, p. 61]

$$\text{Operation count} = \frac{5}{3} n n_b^3$$

If $n = 100$ and $n_b = 3$, the block tridiagonal technique requires only 4500 operations, whereas the dense matrix technique requires 9×10^6 operations. Thus, the block tridiagonal technique is 2000 times faster.

Another matrix with special structure is the *arrow matrix* (see Fig. 4). This matrix has a large, banded portion, with a much smaller number of columns at one side and rows at the bottom. This matrix arises when solving sets of fluid flow problems with free surfaces, when using continuation methods or parameter estimation methods. The matrix problem is subdivided as

$$\begin{bmatrix} A & B \\ C & D \end{bmatrix} \begin{bmatrix} u \\ v \end{bmatrix} = \begin{bmatrix} f \\ g \end{bmatrix}$$

and

$Ax = f$

is solved by using the same LU decomposition of A,

$$Ay = B$$

These problems are sparse and the special methods are applied to them. Next, the dense matrix problem

$$(D - Cy)v = Cx + g$$

is solved and

$$u = x - yv$$

is evaluated, the solution is then (u^T, v^T).

Sparse matrices are ones in which the majority of elements are zero. If the zero entries occur in special patterns, efficient techniques can be used to exploit the structure, as was done above for tridiagonal matrices, block tridiagonal matrices, arrow matrices, etc. These structures typically arise from numerical analysis applied to solve differential equations. Other problems, such as modeling chemical processes, lead to sparse matrices but without such a neatly defined structure—just a lot of zeros in the matrix. For matrices such as these, special techniques must be employed: efficient codes are available [1.2]. These codes usually employ a symbolic factorization, which must be repeated only once for each structure of the matrix. Then an LU factorization is performed, followed by a solution step using the triangular matrices. The symbolic factorization step has significant overhead, but this is rendered small and insignificant if matrices with exactly the same structure are to be used over and over [1.3]–[1.7].

The efficiency of a technique for solving sets of linear equations obviously depends greatly on the arrangement of the equations and unknowns because an efficient arrangement can reduce the bandwidth, for example. Techniques for renumbering the equations and unknowns arising from elliptic partial differential equations are available for finite difference methods [1.8] and for finite element methods [1.9].

Solutions with Iterative Methods. Sets of linear equations can also be solved by using iterative methods; these methods have a rich historical background. Some of them are discussed in Chapter 8 and include Jacobi, Gauss–Seidel, and overrelaxation methods. As the speed of computers increases, direct methods become preferable for the general case, and those are the methods presented here. One iterative method deserves special mention, however, because of its importance in solving partial differential equations.

The *conjugate gradient method* is an iterative method that can solve a set of n linear equations in n iterations. The method primarily requires multiplying a matrix by a vector, which can be done very efficiently on parallel computers: for sparse matrices this is a viable method. The original method was devised by HESTENES and STIEFEL [1.10], however, current implementations use a *preconditioned* conjugate gradient method because it converges faster, provided a good "preconditioner" can be found. An efficient implementation of such a method is described by EISENSTAT [1.11]. The system of n linear equations

$$Ax = f$$

where A is symmetric and positive definite, is to be solved. A preconditioning matrix M is defined in such a way that the problem

$$Mt = r$$

is easy to solve exactly (M might be diagonal, for example). Then the preconditioned conjugate gradient method is

Guess x_0
Calculate $r_0 = f - Ax_0$
Solve $Mt_0 = r_0$, and set $p_0 = t_0$
for $k = 1, n$ (or until convergence)

$$a_k = \frac{r_k^T t_k}{p_k^T A p_k}$$

$$x_{k+1} = x_k + a_k p_k$$

$$r_{k+1} = r_k - a_k A p_k$$

Solve $M t_{k+1} = r_{k+1}$

$$b_k = \frac{r_{k+1}^T t_{k+1}}{r_k^T t_k}$$

$$p_{k+1} = t_{k+1} + b_k p_k$$

test for convergence
enddo

Additional information is available in [1.12] and [1.13].

The singular value decomposition is useful when the matrix is singular or nearly singular, or when the system of equations is overdetermined. An $m \times n$ matrix A can be represented by

$$A = UWV^T, \quad W = \begin{pmatrix} w_1 & & & \\ & w_2 & & \\ & & \ddots & \\ & & & w_n \end{pmatrix}$$

where the matrices U and V are orthogonal in the following sense:

$$\sum_{i=1}^{m} U_{ik} U_{ij} = \delta_{kj}, \quad 1 \le k \le n$$

$$\sum_{i=1}^{n} V_{ik} V_{ij} = \delta_{kj}, \quad 1 \le j \le n$$

In addition

$$VV^T = I$$

This decomposition can always be performed, even for singular matrices. The condition number is the ratio of the largest w_j to the smallest w_j. The inverse of A is

$$A^{-1} = VW^{-1}U^T$$

The rank r of a matrix is a value such that all $r+1 \times r+1$ determinants are zero. If an $n \times n$ matrix A is singular, the rank of the matrix is $r < n$. The columns of U whose same-numbered elements w_j are nonzero are an orthonormal set of basis vectors that span the range. The columns of V whose same-numbered w_j are zero provide an orthonormal basis for the nullspace. The solution to the problem $Ax = f$ is

$$x = VW^{-1}U^T f$$

In this Equation if $w_j = 0$, $1/w_j$ is replaced with zero [1.14]. This is also the least-squares solution to a set of overdetermined equations (i.e., A is an $m \times n$ matrix, x is an $n \times 1$ vector, and f is an $m \times 1$ vector).

In dimensional analysis if the dimensions of each physical variable P_j (there are n of them) are expressed in terms of fundamental measurement units m_j (such as time, length, mass; there are m of them):

$$[P_j] = m_1^{\alpha_{1j}} m_2^{\alpha_{2j}} \dots m_m^{\alpha_{mj}}$$

then a matrix can be formed from the α_{ij}. If the rank of this matrix is r, $n - r$ independent dimensionless groups govern that phenomenon. In chemical reaction engineering the chemical reaction stoichiometry can be written as

$$\sum_{i=1}^{n} \alpha_{ij} C_i = 0, \quad j = 1, 2, \dots, m$$

where there are n species and m reactions. Then if a matrix is formed from the coefficients α_{ij}, which is an $n \times m$ matrix, and the rank of the matrix is r, there are r independent chemical reactions. The other $n - r$ reactions can be deduced from those r reactions.

1.2. Nonlinear Algebraic Equations

In considering a single nonlinear equation in one unknown,

$$f(x) = 0$$

one quick method for finding a root is to program a calculator to evaluate f for a given x and then supply various x until $f(x) = 0$. Another alternative is to program a microcomputer to make a table of x, $f(x)$ and look for the root in the table, refining the table where needed. These methods can be employed when the problem is not too hard and it need be solved only once or twice. Even experienced programmers may do this to save time in some problems.

Iterative methods are also employed to solve the equation, and the k-th iteration is denoted as x^k. The *successive substitution* method is written as

$$x^{k+1} = x^k + \beta f(x^k) \equiv g(x^k)$$

If the constant β is chosen correctly these iterations will converge to a solution. The conditions are that [1.15]

$$\left| \frac{dg}{dx}(x) \right| \le \mu \quad \text{for} \quad |x - \alpha| < h$$

where μ and h are constants, and α is the unknown solution. In other words, if the derivative can be bounded for all x then the method converges. The convergence is linear and may be slow, requiring many iterations, but the method is easy to program. This method is difficult to apply to systems of equations, but it is applied in the solution of ordinary differential equations.

In that case, the constant β is proportional to Δt, and Δt is reduced until the method converges.

The *Newton–Raphson method* is based on a Taylor series of the equation about the k-th iterate:

$$f(x^{k+1}) = f(x^k) + \left.\frac{df}{dx}\right|_{x^k}(x^{k+1} - x^k) + \left.\frac{d^2 f}{dx^2}\right|_{x^k}$$

$$\tfrac{1}{2}(x^{k+1} - x^k)^2 + \ldots$$

The second and higher-order terms are neglected and $f(x^{k+1}) = 0$. Rearrangement gives

$$x^{k+1} = x^k - \frac{f(x^k)}{df/dx(x^k)}$$

This method converges if the following inequalities are satisfied [1.1, p. 115].

$$\left.\frac{df}{dx}\right|_{x^0} > 0, \quad |x^1 - x^0| = \left.\frac{f(x^0)}{df/dx(x^0)}\right|_{x^0} \leq b,$$

$$\text{and} \quad \left|\frac{d^2 f}{dx^2}\right| \leq c$$

where b and c are bounded constants. In practice the method may not converge unless the initial guess is good, or it may converge for some parameters and not others. Unfortunately, when the method is nonconvergent the results look as though a mistake occurred in the computer programming; distinguishing between these situations is difficult, so careful programming and testing are required. If the method converges the difference between successive iterates is something like 0.1, 0.01, 0.0001, 10^{-8}. The error (when it is known) goes the same way; the method is said to be quadratically convergent when it converges. The method is not robust because it can fail for poor initial guesses. If the derivative is difficult to calculate a numerical approximation may be used.

$$\left.\frac{df}{dx}\right|_{x^k} = \frac{f(x^k + \varepsilon) - f(x^k)}{\varepsilon}$$

In the *secant method* the same formula is used as for the Newton–Raphson method, except that the derivative is approximated by using the values from the last two iterates:

$$\left.\frac{df}{dx}\right|_{x^k} = \frac{f(x^k) - f(x^{k-1})}{x^k - x^{k-1}}$$

This is equivalent to drawing a straight line through the last two iterate values on a plot of $f(x)$ versus x. The Newton–Raphson method is equivalent to drawing a straight line tangent to the curve at the last x. In the *method of false position* (or regula falsi), the secant method is used to obtain x^{k+1}, but the previous value is taken as either x^{k-1} or x^k. The choice is made so that the function evaluated for that choice has the opposite sign to $f(x^{k+1})$. This method is slower than the secant method, but it is more robust and keeps the root between two points at all times. In all these methods, appropriate strategies are required for bounds on the function or when $df/dx = 0$. Brent's method combines bracketing, bisection, and an inverse quadratic interpolation to provide a method that is fast and guaranteed to converge, if the root can be bracketed initially [1.14, p. 251].

In the *method of bisection*, if a root lies between x_1 and x_2 because $f(x_1) < 0$ and $f(x_2) > 0$, then the function is evaluated at the center, $x_c = 0.5(x_1 + x_2)$. If $f(x_c) > 0$, the root lies between x_1 and x_c. If $f(x_c) < 0$, the root lies between x_c and x_2. The process is then repeated. If $f(x_c) = 0$, the root is x_c. If $f(x_1) > 0$ and $f(x_2) > 0$, more than one root may exist between x_1 and x_2 (or no roots).

For systems of equations the Newton–Raphson method is widely used, especially for equations arising from the solution of differential equations.

$$f_i(\{x_j\}) = 0, \quad 1 \leq i, j \leq n,$$
$$\text{where } \{x_j\} = (x_1, x_2, \ldots, x_n) = x$$

Then, an expansion in several variables occurs:

$$f_i(x^{k+1}) = f_i(x^k) + \sum_{j=1}^{n} \left.\frac{\partial f_i}{\partial x_j}\right|_{x^k}(x_j^{k+1} - x_j^k) + \ldots$$

The Jacobian matrix is defined as

$$J_{ij}^k = \left.\frac{\partial f_i}{\partial x_j}\right|_{x^k}$$

and the Newton–Raphson method is

$$\sum_{j=1}^{n} J_{ij}^k (x^{k+1} - x^k) = -f_i(x^k)$$

For convergence, the norm of the inverse of the Jacobian must be bounded, the norm of the function evaluated at the initial guess must be bounded, and the second derivative must be bounded [1.1, p. 115], [1.15, p. 12].

A review of the usefulness of solution methods for nonlinear equations is available [1.16].

This review concludes that the Newton–Raphson method may not be the most efficient. Broyden's method approximates the inverse to the Jacobian and is a good all-purpose method, but a good initial approximation to the Jacobian matrix is required. Furthermore, the rate of convergence deteriorates for large problems, for which the Newton–Raphson method is better. Brown's method [1.16] is very attractive, whereas Brent's is not worth the extra storage and computation.

Homotopy methods can be used to ensure finding the solution when the problem is especially complicated. Suppose an attempt is made to solve $f(x) = 0$, and it fails; however, $g(x) = 0$ can be solved easily, where $g(x)$ is some function, perhaps a simplification of $f(x)$. Then, the two functions can be embedded in a homotopy by taking

$$h(x, t) = t f(x) + (1 - t) g(x)$$

In this equation, h can be a $n \times n$ matrix for problems involving n variables; then x is a vector of length n. Then $h(x, t) = 0$ can be solved for $t = 0$ and t gradually changes until at $t = 1$, $h(x, 1) = f(x)$. If the Jacobian of h with respect to x is nonsingular on the homotopy path (as t varies), the method is guaranteed to work. In classical methods, the interval from $t = 0$ to $t = 1$ is broken up into N subdivisions. Set $\Delta t = 1/N$ and solve for $t = 0$, which is easy by the choice of $g(x)$. Then set $t = \Delta t$ and use the Newton–Raphson method to solve for x. Since the initial guess is presumably pretty good, this has a high chance of being successful. That solution is then used as the initial guess for $t = 2\Delta t$ and the process is repeated by moving stepwise to $t = 1$. If the Newton–Raphson method does not converge, then Δt must be reduced and a new solution attempted.

Another way of using homotopy is to create an ordinary differential equation by differentiating the homotopy equation along the path (where $h = 0$).

$$\frac{d h[x(t), t]}{dt} = \frac{\partial h}{\partial x} \frac{dx}{dt} + \frac{\partial h}{\partial t} = 0$$

This can be expressed as an ordinary differential equation for $x(t)$:

$$\frac{\partial h}{\partial x} \frac{dx}{dt} = -\frac{\partial h}{\partial t}$$

If Euler's method is used to solve this equation, a value x^0 is used, and dx/dt from the above equation is solved for. Then

$$x^{1,0} = x^0 + \Delta t \frac{dx}{dt}$$

is used as the initial guess and the homotopy equation is solved for x^1.

$$\frac{\partial h}{\partial x}(x^{1,k+1} - x^{1,k}) = -h(x^{1,k}, t)$$

Then t is increased by Δt and the process is repeated.

In arc-length parameterization, both x and t are considered parameterized by a parameter s, which is thought of as the arc length along a curve. Then the homotopy equation is written along with the arc-length equation.

$$\frac{\partial h}{\partial x} \frac{dx}{ds} + \frac{\partial h}{\partial t} \frac{dt}{ds} = 0$$

$$\frac{dx^T}{ds} \frac{dx}{dx} + \left(\frac{dt}{ds}\right)^2 = 1$$

The initial conditions are

$$x(0) = x^0$$
$$t(0) = 0$$

The advantage of this approach is that it works even when the Jacobian of h becomes singular because the full matrix is rarely singular. Illustrations applied to chemical engineering are available [1.11].

1.3. Linear Difference Equations

Difference equations arise in chemical engineering from staged operations, such as distillation or extraction, as well as from differential equations modeling adsorption and chemical reactors. The value of a variable in the n-th stage is noted by a subscript n. For example, if $y_{n,i}$ denotes the mole fraction of the i-th species in the vapor phase on the n-th stage of a distillation column, $x_{n,i}$ is the corresponding liquid mole fraction, R the reflux ratio (ratio of liquid returned to the column to product removed from the condenser), and $K_{n,i}$ the equilibrium constant, then the mass balances about the top of the column give

$$y_{n+1,i} = \frac{R}{R+1}x_{n,i} + \frac{1}{R+1}x_{0,i}$$

and the equilibrium equation gives

$$y_{n,i} = K_{n,i} x_{n,i}$$

If these are combined,

$$K_{n+1,i} x_{n+1,i} = \frac{R}{R+1}x_{n,i} + \frac{1}{R+1}x_{0,i}$$

is obtained, which is a linear difference equation. This particular problem is quite complicated, and the interested reader is referred to [1.18, Chap. 6]. However, the form of the difference equation is clear. Several examples are given here for solving difference equations. More complete information is available in [1.19].

An equation in the form

$$x_{n+1} - x_n = f_n$$

can be solved by

$$x_n = \sum_{i=1}^{n} f_i$$

Usually, difference equations are solved analytically only for linear problems. When the coefficients are constant and the equation is linear and homogeneous, a trial solution of the form

$$x_n = \varphi^n$$

is attempted; φ is raised to the power n. For example, the difference equation

$$c x_{n-1} + b x_n + a x_{n+1} = 0$$

coupled with the trial solution would lead to the equation

$$a\varphi^2 + b\varphi + c = 0$$

This gives

$$\varphi_{1,2} = \frac{-b \pm \sqrt{b^2 - 4ac}}{2a}$$

and the solution to the difference equation is

$$x_n = A\varphi_1^n + B\varphi_2^n$$

where A and B are constants that must be specified by boundary conditions of some kind.

When the equation is nonhomogeneous, the solution is represented by the sum of a particular solution and a general solution to the homogeneous equation.

$$x_n = x_{n,P} + x_{n,H}$$

The general solution is the one found for the homogeneous equation, and the particular solution is any solution to the nonhomogeneous difference equation. This can be found by methods analogous to those used to solve differential equations: the method of undetermined coefficients and the method of variation of parameters. The last method applies to equations with variable coefficients, too. For a problem such as

$$x_{n+1} - f_n x_n = 0$$
$$x_0 = c$$

the general solution is

$$x_n = c \prod_{i=1}^{n} f_i$$

This can then be used in the method of variation of parameters to solve the equation

$$x_{n+1} - f_n x_n = g_n$$

1.4. Eigenvalues

The $n \times n$ matrix A has n eigenvalues λ_i, $i = 1, \ldots, n$, which satisfy

$$\det(A - \lambda_i I) = 0$$

If this equation is expanded, it can be represented as

$$P_n(\lambda) = (-\lambda)^n + a_1(-\lambda)^{n-1} + a_2(-\lambda)^{n-2} + \ldots + a_{n-1}(-\lambda) + a_n = 0$$

If the matrix A has real entries then a_i are real numbers, and the eigenvalues either are real numbers or occur in pairs as complex numbers with their complex conjugates (for definition of complex numbers, see Chap. 3). The Hamilton–Cayley theorem [1.18, p. 127] states that the matrix A satisfies its own characteristic equation.

$$P_n(A) = (-A)^n + a_1(-A)^{n-1} + a_2(-A)^{n-2} + \ldots + a_{n-1}(-A) + a_n I = 0$$

A laborious way to find the eigenvalues of a matrix is to solve the n-th order polynomial for the λ_i—far too time consuming. Instead the matrix is transformed into another form whose eigenvalues are easier to find. In the *Givens method* and the *Housholder method* the matrix is transformed into the tridiagonal form; then, in a fixed number of calculations the eigenvalues can be found [1.14]. The Givens method requires $4n^3/3$ operations to transform a real symmetric matrix to tridiagonal form, whereas the Householder method requires half that number [1.1]. Once the tridiagonal form is found, a Sturm sequence is applied to determine the eigenvalues. These methods are especially useful when only a few eigenvalues of the matrix are desired.

If all the eigenvalues are needed, the QR algorithm is preferred. In the QR algorithm [1.20] a modified Householder transformation is applied to A, transforming it to the form

$$A = QR$$

where Q is orthogonal and R is upper triangular. If A is banded, then Q and R are banded.

The eigenvalues of a certain tridiagonal matrix can be found analytically. If A is a tridiagonal matrix with

$$a_{ii} = p,\ a_{i,i+1} = q,\ a_{i+1,i} = r,\ qr > 0$$

then the eigenvalues of A are [1.21]

$$\lambda_i = p + 2(qr)^{1/2}\cos\frac{i\pi}{n+1} \quad i = 1, 2, \ldots, n$$

This result is useful when finite difference methods are applied to the diffusion equation.

2. Approximation and Integration

2.1. Introduction

Two types of problems arise frequently:

1) A *function* is known exactly at a set of points and an interpolating function is desired. The interpolant may be exact at the set of points, or it may be a "best fit" in some sense. Alternatively it may be desired to represent a function in some other way.
2) *Experimental data* must be fit with a mathematical model. The data have experimental error, so some uncertainty exists. The parameters in the model as well as the uncertainty in the determination of those parameters is desired.

These problems are addressed in this chapter. Section 2.2 gives the properties of polynomials defined over the whole domain and Section 2.3 of polynomials defined on segments of the domain. In Section 2.4, quadrature methods are given for evaluating an integral. Least-squares methods for parameter estimation for both linear and nonlinear models are given in Sections 2.5 and 2.6. Fourier transforms to represent discrete data are described in Section 2.7. The chapter closes with extensions to two-dimensional representations.

2.2. Global Polynomial Approximation

A *global* polynomial $P_m(x)$ is defined over the entire region of space

$$P_m(x) = \sum_{j=0}^{m} c_j x^j$$

This polynomial is of degree m (highest power is x^m) and order $m + 1$ ($m + 1$ parameters $\{c_j\}$). If a set of $m + 1$ points is given,

$$y_1 = f(x_1),\ y_2 = f(x_2), \ldots, y_{m+1} = f(x_{m+1})$$

then Lagrange's formula yields a polynomial of degree m that goes through the $m + 1$ points:

$$P_m(x) = \frac{(x-x_2)(x-x_3)\ldots(x-x_{m+1})}{(x_1-x_2)(x_1-x_3)\ldots(x_1-x_{m+1})}y_1 +$$

$$\frac{(x-x_1)(x-x_3)\ldots(x-x_{m+1})}{(x_2-x_1)(x_2-x_3)\ldots(x_2-x_{m+1})}y_2 + \ldots +$$

$$\frac{(x-x_1)(x-x_2)\ldots(x-x_m)}{(x_{m+1}-x_1)(x_{m+1}-x_2)\ldots x_{m+1}-x_m}y_{m+1}$$

Note that each coefficient of y_j is a polynomial of degree m that vanishes at the points $\{x_j\}$ (except for one value of j) and takes the value of 1.0 at that point, i.e.,

$$P_m(x_j) = y_j \quad j = 1, 2, \ldots, m+1$$

If the function $f(x)$ is known, the error in the approximation is [2.1]

$$|\text{error}(x)| \leq \frac{|x_{m+1} - x_1|^{m+1}}{(m+2)!}$$

$$\max_{x_1 \leq x \leq x_{m+1}} |f^{(m+2)}(x)|$$

The evaluation of $P_m(x)$ at a point other than the defining points can be made with Neville's algorithm [2.2]. Let P_1 be the value at x of the unique function passing through the point (x_1, y_1); i.e., $P_1 = y_1$. Let P_{12} be the value at x of the unique polynomial passing through the points x_1 and x_2. Likewise, $P_{ijk...r}$ is the unique polynomial passing through the points x_i, x_j, $x_k, \ldots, x_r$. The following scheme is used:

$$\begin{array}{l} x_1 \; y_1 = P_1 \\ \quad\quad\quad P_{12} \\ x_2 \; y_2 = P_2 \quad\quad P_{123} \\ \quad\quad\quad P_{23} \quad\quad\quad P_{1234} \\ x_3 \; y_3 = P_3 \quad\quad P_{234} \\ \quad\quad\quad P_{34} \\ x_4 \; y_4 = P_4 \end{array}$$

These entries are defined by using

$$P_{i(i+1)\ldots(i+m)} = $$

$$\frac{(x - x_{i+m})\, P_{i(i+1)\ldots(i+m-1)} + (x_i - x)\, P_{(i+1)(i+2)\ldots(i+m)}}{x_i - x_{i+m}}$$

Consider P_{1234}: the terms on the right-hand side of the equation involve P_{123} and P_{234}. The "parents," P_{123} and P_{234}, already agree at points 2 and 3. Here $i = 1$, $m = 3$; thus, the parents agree at $x_{i+1}, \ldots, x_{i+m-1}$ already. The formula makes $P_{i(i+1)\ldots(i+m)}$ agree with the function at the additional points x_{i+m} and x_i. Thus, $P_{i(i+1)\ldots(i+m)}$ agrees with the function at all the points $\{x_i, x_{i+1}, \ldots, x_{i+m}\}$.

Orthogonal Polynomials. Another form of the polynomials is obtained by defining them so that they are orthogonal. It is required that $P_m(x)$ be orthogonal to $P_k(x)$ for all $k = 0, \ldots, m-1$.

$$\int_a^b W(x) P_k(x) P_m(x)\, dx = 0 \quad k = 0, 1, 2, \ldots, m-1$$

The orthogonality includes a nonnegative weight function, $W(x) \geq 0$ for all $a \leq x \leq b$. This procedure specifies the set of polynomials to within multiplicative constants, which can be set either by requiring the leading coefficient to be one or by requiring the norm to be one.

$$\int_a^b W(x) P_m^2(x)\, dx = 1$$

The polynomial $P_m(x)$ has m roots in the closed interval a to b.

The polynomial

$$p(x) = c_0 P_0(x) + c_1 P_1(x) + \ldots c_m P_m(x)$$

minimizes

$$I = \int_a^b W(x) [f(x) - p(x)]^2\, dx$$

when

$$c_j = \frac{\int_a^b W(x) f(x) P_j(x)\, dx}{W_j},$$

$$W_j = \int_a^b W(x) P_j^2(x)\, dx$$

Note that each c_j is independent of m, the number of terms retained in the series. The minimum value of I is

$$I_{\min} = \int_a^b W(x) f^2(x)\, dx - \sum_{j=0}^m W_j c_j^2$$

Such functions are useful for continuous data, i.e., when $f(x)$ is known for all x.

Typical orthogonal polynomials are given in Table 1. Chebyshev polynomials are used in spectral methods (see Chap. 6). The last two rows of Table 1 are widely used in the orthogonal collocation method in chemical engineering. The last entry (the shifted Legendre polynomial as a

Table 1. Orthogonal polynomials [2.21]

a	b	$W(x)$	Name	Recursion relation
-1	1	1	Legendre	$(i+1) P_{i+1} = (2i+1) \times P_i - i P_{i-1}$
-1	1	$\dfrac{1}{\sqrt{1-x^2}}$	Chebyshev	$T_{i+1} = 2 \times T_i - T_{i-1}$
0	1	$x^{q-1}(1-x)^{p-q}$	Jacobi (p, q)	
$-\infty$	∞	e^{-x^2}	Hermite	$H_{i+1} = 2 \times H_i - 2i H_{i-1}$
0	∞	$x^c e^{-x}$	Laguerre (c)	$(i+1) L_{i+1}^c = (-x + 2i + c + 1) L_i^c = -(i+c) L_{i-1}^c$
0	1	1	shifted Legendre	
0	1	1	shifted Legendre, function of x^2	

function of x^2) is defined by

$$\int_0^1 W(x^2) P_k(x^2) P_m(x^2) x^{a-1} dx = 0$$

$$k = 0, 1, \ldots, m-1$$

where $a = 1$ is for planar, $a = 2$ for cylindrical, and $a = 3$ for spherical geometry. These functions are useful if the solution can be proved to be an even function of x.

Rational Polynomials. Rational polynomials are ratios of polynomials. A rational polynomial $R_{i(i+1)\ldots(i+m)}$ passing through $m+1$ points

$$y_i = f(x_i), \quad i = 1, \ldots, m+1$$

is

$$R_{i(i+1)\ldots(i+m)} = \frac{P_\mu(x)}{Q_\nu(x)} = \frac{p_0 + p_1 x + \ldots + p_\mu x^\mu}{q_0 + q_1 x + \ldots + q_\nu x^\nu},$$

$$m + 1 = \mu + \nu + 1$$

An alternative condition is to make the rational polynomial agree with the first $m+1$ terms in the power series, giving a Padé approximation, i.e.,

$$\frac{d^k R_{i(i+1)\ldots(i+m)}}{dx^k} = \frac{d^k f(x)}{dx^k} \quad k = 0, \ldots, m$$

The Bulirsch–Stoer recursion algorithm can be used to evaluate the polynomial:

$$R_{i(i+1)\ldots(i+m)} = R_{(i+1)\ldots(i+m)}$$
$$+ \frac{R_{(i+1)\ldots(i+m)} - R_{i(i+1)\ldots(i+m-1)}}{\text{Den}}$$

$$\text{Den} = \left(\frac{x - x_i}{x - x_{i+m}}\right)$$
$$\left(1 - \frac{R_{(i+1)\ldots(i+m)} - R_{i(i+1)\ldots(i+m-1)}}{R_{(i+1)\ldots(i+m)} - R_{i(i+1)\ldots(i+m-1)}}\right) - 1$$

Rational polynomials are useful for approximating functions with poles and singularities, which occur in Laplace transforms (see Section 4.2).

Fourier series are discussed in Section 4.1. Representation by sums of exponentials is also possible [2.3].

In summary, for discrete data, Legendre polynomials and rational polynomials are used. For continuous data a variety of orthogonal polynomials and rational polynomials are used. When the number of conditions (discrete data points) exceeds the number of parameters, then see Section 2.5.

2.3. Piecewise Approximation

Piecewise approximations can be developed from difference formulas [2.4]. Consider a case in which the data points are equally spaced

$$x_{n+1} - x_n = \Delta x$$
$$y_n = y(x_n)$$

forward differences are defined by

$$\Delta y_n = y_{n+1} - y_n$$
$$\Delta^2 y_n = \Delta y_{n+1} - \Delta y_n = y_{n+2} - 2y_{n+1} + y_n$$

Then, a new variable is defined

$$\alpha = \frac{x_\alpha - x_0}{\Delta x}$$

and the finite interpolation formula through the points $y_0, y_1, \ldots, y_n$ is written as follows:

$$y_\alpha = y_0 + \alpha \Delta y_0 + \frac{\alpha(\alpha - 1)}{2!} \Delta^2 y_0 + \ldots +$$
$$\frac{\alpha(\alpha - 1)\ldots(\alpha - n + 1)}{n!} \Delta^n y_0 \quad (1)$$

Keeping only the first two terms gives a straight line through $(x_0, y_0) - (x_1, y_1)$; keeping the first three terms gives a quadratic function of position going through those points plus (x_2, y_2). The value $\alpha = 0$ gives $x = x_0$; $\alpha = 1$ gives $x = x_1$, etc.

Backward differences are defined by

$$\nabla y_n = y_n - y_{n-1}$$
$$\nabla^2 y_n = \nabla y_n - \nabla y_{n-1} = y_n - 2y_{n-1} + y_{n-2}$$

The interpolation polynomial of order n through the points $y_0, y_{-1}, y_{-2}, \ldots$ is

$$y_\alpha = y_0 + \alpha \nabla y_0 + \frac{\alpha(\alpha + 1)}{2!} \nabla^2 y_0 + \ldots +$$
$$\frac{\alpha(\alpha + 1)\ldots(\alpha + n - 1)}{n!} \nabla^n y_0$$

The value $\alpha = 0$ gives $x = x_0$; $\alpha = -1$ gives $x = x_{-1}$. Alternatively, the interpolation polynomial of order n through the points $y_1, y_0, y_{-1}, \ldots$ is

$$y_\alpha = y_1 + (\alpha - 1) \nabla y_1 + \frac{\alpha(\alpha - 1)}{2!} \nabla^2 y_1 + \ldots +$$
$$\frac{(\alpha - 1)\alpha(\alpha + 1)\ldots(\alpha + n - 2)}{n!} \nabla^n y_1$$

Now $\alpha = 1$ gives $x = x_1$; $\alpha = 0$ gives $x = x_0$.

The *finite element method* can be used for piecewise approximations [2.5]. In the finite element method the domain $a \le x \le b$ is divided into elements as shown in Figure 5. Each function $N_i(x)$ is zero at all nodes except x_i; $N_i(x_i) = 1$. Thus, the approximation is

$$y(x) = \sum_{i=1}^{NT} c_i N_i(x) = \sum_{i=1}^{NT} y(x_i) N_i(x)$$

where $c_i = y(x_i)$. For convenience, the trial functions are defined within an element by using new coordinates:

$$u = \frac{x - x_i}{\Delta x_i}$$

The Δx_i need not be the same from element to element. The trial functions are defined as $N_i(x)$ (Fig. 5A) in the global coordinate system and $N_I(u)$ (Fig. 5B) in the local coordinate system (which also requires specification of the element). For $x_i < x < x_{i+1}$

$$y(x) = \sum_{i=1}^{NT} c_i N_i(x) = c_i N_i(x) + c_{i+1} N_{i+1}(x)$$

because all the other trial functions are zero there. Thus

$$y(x) = c_i N_{I=1}(u) + c_{i+1} N_{I=2}(u),$$
$$x_i < x < x_{i+1}, \quad 0 < u < 1$$

Then

$$N_{I=1} = 1 - u, \quad N_{I=2} = u$$

and the expansion as is rewritten as

$$y(x) = \sum_{I=1}^{2} c_I^e N_I(u) \tag{2}$$

x in e-th element and $c_i = c_I^e$ within the element e. Thus, given a set of points (x_i, y_i), a finite element approximation can be made to go through them.

Quadratic approximations can also be used within the element (see Fig. 6). Now the trial functions are

$$N_{I=1} = 2(u - 1)(u - \tfrac{1}{2})$$
$$N_{I=2} = 4u(1 - u) \tag{3}$$
$$N_{I=3} = 2u(u - \tfrac{1}{2})$$

The approximation going through an odd number of points (x_i, y_i) is then

$$y(x) = \sum_{I=1}^{3} c_I^e N_I(u) \quad x \text{ in } e\text{-th element}$$

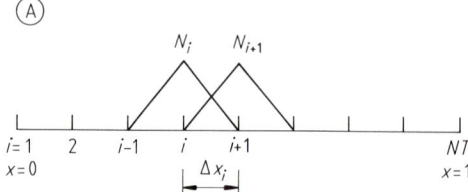

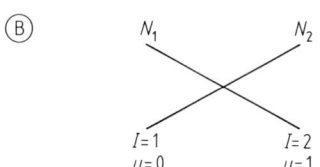

Figure 5. Galerkin finite element method – linear functions
A) Global numbering system; B) Local numbering system

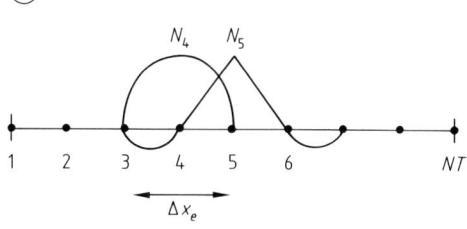

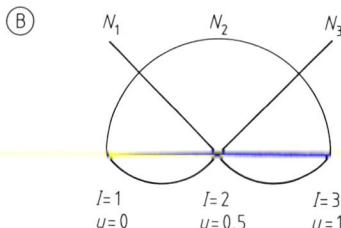

Figure 6. Finite elements approximation – quadratic elements
A) Global numbering system; B) Local numbering system

with $c_I^e = y(x_i), \quad i = (e - 1)2 + I$
in the e-th element

Hermite cubic polynomials H which have continuous first derivatives can be used for the N_I [2.5]. Now, write

$$y(x) = \sum_{I=1}^{4} c_I^e H_I(u) \quad x \text{ in } e\text{-th element}$$

where

$$H_1 = (1 - u)^2 (1 + 2u)$$
$$H_2 = u(1 - u)^2 \Delta x_e$$
$$H_3 = u^2 (3 - 2u)$$
$$H_4 = u^2 (u - 1) \Delta x_e$$

Identify the points within the e-th element as
$x_i : u = 0$; $x_{i+1} : u = 1$. Then

$$c_1^e = y(x_i), \quad c_3^e = y(x_{i+1})$$
$$c_2^e = \frac{dy}{dx}(x_i)\Delta x_e, \quad c_4^e = \frac{dy}{dx}(x_{i+1})\Delta x_e$$

Thus, at the points

$$x_1, x_2, \ldots, x_{NT}$$

both the function and first derivative are necessary:

$$y(x_1), \quad y(x_2), \ldots, y(x_{NT})$$
$$\frac{dy}{dx}(x_1), \quad \frac{dy}{dx}(x_2), \ldots, \frac{dy}{dx}(x_{NT})$$

Splines. Splines are functions that match given values at the points $x_1, \ldots, x_{NT}$, shown in Figure 7, and have continuous derivatives up to some order at the knots, or the points $x_2, \ldots, x_{NT-1}$. Cubic splines are most common. In this case the function is represented by a cubic polynomial within each interval and has continuous first and second derivatives at the knots.

Consider the points shown in Figure 7 A. The notation for each interval is shown in Figure 7 B. Within each interval the function is represented as a cubic polynomial.

$$C_i(x) = a_{0i} + a_{1i}x + a_{2i}x^2 + a_{3i}x^3$$

The interpolating function takes on specified values at the knots.

$$C_{i-1}(x_i) = C_i(x_i) = f(x_i)$$

Given the set of values $\{x_i, f(x_i)\}$, the objective is to pass a smooth curve through those points, and the curve should have continuous first and second derivatives at the knots.

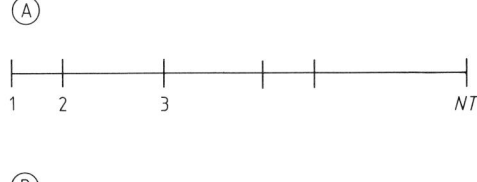

Figure 7. Finite elements for cubic splines
A) Notation for spline knots. B) Notation for one element

$$C'_{i-1}(x_i) = C'_i(x_i)$$
$$C''_{i-1}(x_i) = C''_i(x_i)$$

The formulas for the cubic spline are derived as follows for one region. Since the function is a cubic function the third derivative is constant and the second derivative is linear in x. This is written as

$$C''_i(x) = C''_i(x_i) + [C''_i(x_{i+1}) - C''_i(x_i)]\frac{x - x_i}{\Delta x_i}$$

and integrated once to give

$$C'_i(x) = C'_i(x_i) + C''_i(x_i)(x - x_i) + [C''_i(x_{i+1})$$
$$- C''_i(x_i)]\frac{(x - x_i)^2}{2\Delta x_i}$$

and once more to give

$$C_i(x) = C_i(x_i) + C'_i(x_i)(x - x_i) + C''_i(x_i)$$
$$\frac{(x - x_i)^2}{2\Delta x_i} + [C''_i(x_{i+1}) - C''_i(x_i)]\frac{(x - x_i)^3}{6\Delta x_i}$$

Now

$$y_i = C_i(x_i), \quad y'_i = C'_i(x_i), \quad y''_i = C''_i(x_i)$$

is defined so that

$$C_i(x) = y_i + y'_i(x - x_i) + \tfrac{1}{2}y''_i(x - x_i)^2$$
$$+ \frac{1}{6\Delta x_i}(y''_{i+1} - y''_i)(x - x_i)^3$$

A number of algebraic steps make the interpolation easy. These formulas are written for the i-th element as well as the $i - 1$-th element. Then the continuity conditions are applied for the first and second derivatives, and the values y'_i and y'_{i-1} are eliminated [2.6, p. 163]. The result is

$$y''_{i-1}\Delta x_{i-1} + y''_i 2(\Delta x_{i-1} + \Delta x_i) + y''_{i+1}\Delta x_i =$$
$$6\left(\frac{y_i - y_{i-1}}{\Delta x_{i-1}} - \frac{y_{i+1} - y_i}{\Delta x_i}\right)$$

This is a tridiagonal system for the set of $\{y''_i\}$ in terms of the set of $\{y_i\}$. Since the continuity conditions apply only for $i = 2, \ldots, NT - 1$, only $NT - 2$ conditions exist for the NT values of y''_i. Two additional conditions are needed, and these are usually taken as the value of the second derivative at each end of the domain, y''_1, y''_{NT}. If these values are zero, the *natural cubic splines* are obtained; they can also be set to achieve some other purpose, such as making the first derivative match some desired condition at the two

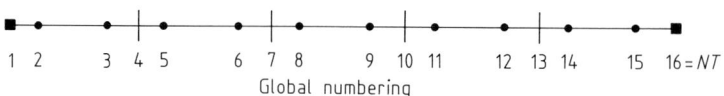

Figure 8. Notation for orthogonal collocation on finite elements
● Residual condition; ■ Boundary conditions; | Element boundary, continuity
NE = total no. of elements.
$NT = (NCOL + 1)NE + 1$

ends. With these values taken as zero, in the natural cubic spline, an $NT - 2$ system of tridiagonal equations exists, which is easily solved. Once the second derivatives are known at each of the knots, the first derivatives are given by

$$y'_i = \frac{y_{i+1} - y_i}{\Delta x_i} - y''_i \frac{\Delta x_i}{3} - y''_{i+1} \frac{\Delta x_i}{6}$$

The function itself is then known within each element

Orthogonal Collocation on Finite Elements. In the method of orthogonal collocation on finite elements the solution is expanded in a polynomial of order $NP = NCOL + 2$ within each element [2.5]. The choice $NCOL = 1$ corresponds to using quadratic polynomials, whereas $NCOL = 2$ gives cubic polynomials. The notation is shown in Figure 8. Set the function to a known value at the two endpoints

$$y_1 = y(x_1)$$
$$y_{NT} = y(x_{NT})$$

and then at the $NCOL$ interior points to each element

$$y^e_I = y_i = y(x_i), \quad i = (NCOL + 1)e + I$$

The actual points x_i are taken as the roots of the orthogonal polynomial.

$$P_{NCOL}(u) = 0 \text{ gives } u_1, u_2, \ldots, u_{NCOL}$$

and then

$$x_i = x_{(e)} + \Delta x_e u_I \equiv x^e_I$$

The first derivatives must be continuous at the element boundaries:

$$\left.\frac{dy}{dx}\right|_{x=x_{(2)-}} = \left.\frac{dy}{dx}\right|_{x=x_{(2)+}}$$

Within each element the interpolation is a polynomial of degree $NCOL + 1$. Overall the function is continuous with continuous first derivatives. With the choice $NCOL = 2$, the same approximation is achieved as with Hermite cubic polynomials.

2.4. Quadrature

To calculate the value of an integral, the function can be approximated by using each of the methods described in Section 2.3. Using the first three terms in Equation (1) gives

$$\int_{x_0}^{x_0+h} y(x)\,dx = \int_0^1 y_\alpha h\,d\alpha$$

$$= \frac{h}{2}(y_0 + y_1) - \frac{1}{12}h^3 y''_0(\xi), \quad x_0 \leq \xi \leq x_0 + h$$

This corresponds to passing a straight line through the points (x_0, y_0), (x_1, y_1) and integrating under the interpolant. For equally spaced points at $a = x_0$, $a + \Delta x = x_1$, $a + 2\Delta x = x_2, \ldots, a + N\Delta x = x_N$, $a + (N+1)\Delta x = b = x_{n+1}$, the trapezoid rule is obtained.

Trapezoid Rule.

$$\int_a^b y(x)\,dx = \frac{h}{2}(y_0 + 2y_1 + 2y_2 + \ldots + 2y_N + y_{N+1}) + O(h^3)$$

The first five terms in Equation (1) are retained and integrated over two intervals.

$$\int_{x_0}^{x_0+2h} y(x)\,dx = \int_0^2 y_\alpha h\,d\alpha = \frac{h}{3}(y_0 + 4y_1 + y_2)$$

$$- \frac{h^5}{90} y_0^{(IV)}(\xi), \quad x_0 \leq \xi \leq x_0 + 2h$$

This corresponds to passing a quadratic function through three points and integrating. For an even number of intervals and an odd number of points, $2N+1$, with $a = x_0$, $a + \Delta x = x_1$, $a + 2\Delta x = x_2, \ldots, a + 2N\Delta x = b$, Simpson's rule is obtained.

Simpson's Rule.

$$\int_a^b y(x)\,dx = \frac{h}{3}(y_0 + 4y_1 + 2y_2 + 4y_3 + 2y_4 + \ldots + 2y_{2N-1} + 4y_{2N} + y_{2N+1}) + O(h^5)$$

Within each pair of intervals the interpolant is continuous with continuous derivatives, but only the function is continuous from one pair to another.

If the *finite element representation* is used (Eq. 2), the integral is

$$\int_{x_i}^{x_{i+1}} y(x)\,dx = \int_0^1 \sum_{I=1}^{2} c_I^e N_I(u)(x_{i+1} - x_i)\,du$$

$$= \Delta x_i \sum_{I=1}^{2} c_I^e \int_0^1 N_I(u)\,du = \Delta x_i (c_1^e \tfrac{1}{2} + c_2^e \tfrac{1}{2})$$

$$= \frac{\Delta x_i}{2}(y_i + y_{i+1})$$

Since $c_1^e = y_i$ and $c_2^e = y_{i+1}$, the result is the same as the trapezoid rule. These formulas can be added together to give *linear elements*:

$$\int_a^b y(x)\,dx = \sum_e \frac{\Delta x_e}{2}(y_1^e + y_2^e)$$

If the *quadratic* expansion is used (Eq. 3), the endpoints of the element are x_i and x_{i+2}, and x_{i+1} is the midpoint, here assumed to be equally spaced between the ends of the element:

$$\int_{x_i}^{x_{i+2}} y(x)\,dx = \int_0^1 \sum_{I=1}^{3} c_I^e N_I(u)(x_{i+2} - x_i)\,du$$

$$= \Delta x_i \sum_{I=1}^{3} c_I^e \int_0^1 N_I(u)\,du = \Delta x_e (c_1^e \tfrac{1}{6} + c_2^e \tfrac{2}{3} + c_3^e \tfrac{1}{6})$$

For many elements, with different Δx^e, *quadratic elements*:

$$\int_a^b y(x) = \sum_e \frac{\Delta x_e}{6}(y_1^e + 4y_2^e + y_3^e)$$

If the element sizes are all the same this gives Simpson's rule.

For Hermite cubic functions the quadrature rule is

$$\int_a^b y(x)\,dx = \sum_e \Delta x_e \left[\frac{1}{2}(c_1^e + c_3^e) + \frac{\Delta x_e}{12}(c_2^e - c_4^e)\right]$$

For cubic splines the quadrature rule within one element is

$$\int_{x_i}^{x_{i+1}} C_i(x)\,dx = \tfrac{1}{2}\Delta x_i(y_i + y_{i+1})$$

$$- \tfrac{1}{24}\Delta x_i^3(y_i'' + y_{i+1}'')$$

For the entire interval the quadrature formula is

$$\int_{x_1}^{x_{NT}} y(x)\,dx = \frac{1}{2}\sum_{i=1}^{NT-1} \Delta x_i(y_i + y_{i+1})$$

$$- \frac{1}{24}\sum_{i=1}^{NT-1} \Delta x_i^3(y_i'' + y_{i+1}'')$$

with $y_1'' = 0$, $y_{NT}'' = 0$ for natural cubic splines.

When orthogonal polynomials are used, as in Equation (1), the m roots to $P_m(x) = 0$ are chosen as quadrature points and called points $\{x_j\}$. Then the quadrature is *Gaussian*:

$$\int_0^1 y(x)\,dx = \sum_{j=1}^{m} W_j y(x_j)$$

The quadrature is exact when y is a polynomial of degree $2m - 1$ in x. The m weights and m Gauss points result in $2m$ parameters, chosen to exactly represent a polynomial of degree $2m - 1$, which has $2m$ parameters. The Gauss points and weights are given in Table 2. The weights can be defined with $W(x)$ in the integrand as well.

For orthogonal collocation on finite elements the quadrature formula is

$$\int_0^1 y(x)\,dx = \sum_e \Delta x_e \sum_{j=1}^{NP} W_j y(x_j^e)$$

Table 2. Gaussian quadrature points and weights*

N	x_i	W_i
1	0.5000000000	0.6666666667
2	0.2113248654	0.5000000000
	0.7886751346	0.5000000000
3	0.1127016654	0.2777777778
	0.5000000000	0.4444444445
	0.8872983346	0.2777777778
4	0.0694318442	0.1739274226
	0.3300094783	0.3260725774
	0.6699905218	0.3260725774
	0.9305681558	0.1739274226
5	0.0469100771	0.1184634425
	0.2307653450	0.2393143353
	0.5000000000	0.2844444444
	0.7692346551	0.2393143353
	0.9530899230	0.1184634425

* For a given N the quadrature points $x_2, x_3, \ldots, x_{NP-1}$ are given above. $x_1 = 0$, $x_{NP} = 1$. For $N = 1$, $W_1 = W_3 = 1/6$ and for $N \geq 2$, $W_1 = W_{NP} = 0$.

Each special polynomial has its own quadrature formula. For example, Gauss–Legendre polynomials give the quadrature formula

$$\int_0^\infty e^{-x} y(x)\, dx = \sum_{i=1}^n W_i y(x_i)$$

(points and weights are available in mathematical tables) [2.1].

For Gauss–Hermite polynomials the quadrature formula is

$$\int_{-\infty}^\infty e^{-x^2} y(x)\, dx = \sum_{i=1}^n W_i y(x_i)$$

(points and weights are available in mathematical tables) [2.1].

Romberg's method uses extrapolation techniques to improve the answer [2.2]. If I_1 is the value of the integral obtained by using interval size $h = \Delta x$, I_2 the value of I obtained by using interval size $h/2$, and I_0 the true value of I, then the error in a method is approximately h^m, or

$$I_1 \approx I_0 + c h^m$$

$$I_2 \approx I_0 + c \left(\frac{h}{2}\right)^m$$

Replacing the $\approx$ by an equality (an approximation) and solving for c and I_0 give

$$I_0 = \frac{2^m I_2 - I_1}{2^m - 1}$$

This process can also be used to obtain I_1, $I_2, \ldots$, by halving h each time, calculating new estimates from each pair, and calling them $J_1, J_2, \ldots$ (i.e., in the formula above, I_0 is replaced with J_1). The formulas are reapplied for each pair of J's to obtain $K_1, K_2, \ldots$. The process continues until the required tolerance is obtained.

$$I_1\ I_2\ I_3\ I_4$$
$$J_1\ J_2\ J_3$$
$$K_1\ K_2$$
$$L_1$$

Romberg's method is most useful for a low-order method (small m) because significant improvement is then possible.

When the integrand has singularities, a variety of techniques can be tried. The integral may be divided into one part that can be integrated analytically near the singularity and another part that is integrated numerically. Sometimes a change of argument allows analytical integration. Series expansion might be helpful, too. When the domain is infinite, Gauss–Legendre or Gauss–Hermite quadrature can be used. Also a transformation can be made [2.2]. For example, let $u = 1/x$ and then

$$\int_a^b f(x)\, dx = \int_{1/b}^{1/a} \frac{1}{u^2} f\left(\frac{1}{u}\right) du \quad a, b > 0$$

2.5. Linear Least Squares

The following description of maximum likelihood applies to both linear and nonlinear least squares [2.2]. If each measurement point y_i has a measurement error Δy_i that is independently random and distributed with a normal distribution about the true model $y(x)$ with standard deviation σ_i, then the probability of a data set is

$$P = \prod_{i=1}^N \left\{ \exp\left[-\frac{1}{2}\left(\frac{y_i - y(x_i)}{\sigma_i}\right)^2\right] \Delta y \right\}$$

(For definition of probability, normal distribution, and standard deviation, see Chap. 11.) Here y_i is the measured value, σ_i is the standard deviation of the i-th measurement, and Δy is needed in order that a measured value $\pm \Delta y$ has a certain probability. Given a set of parameters (maximizing this function), the probability that this data set (plus or minus Δy) could have occurred is P. This probability is maximized (giving the maximum likelihood) if the negative of the logarithm is minimized.

$$-\log P = \sum_{i=1}^{N} \left(\frac{y_i - y(x_i)}{\sqrt{2}\,\sigma_i}\right)^2 - N \log \Delta y$$

Because N, σ_i, and Δy are constants, this is the same as minimizing χ^2:

$$\chi^2 = \sum_{i=1}^{N} \left[\frac{y_i - y(x_i; a_1, \ldots, a_M)}{\sigma_i}\right]^2 \quad (4)$$

with respect to the parameters $\{a_j\}$. Note that the standard deviations $\{\sigma_i\}$ of the measurements are expected to be known. The goodness of fit is related to the number of degrees of freedom, $v = N - M$. The probability P that χ^2 would exceed a particular value $(\chi_0)^2$ is

$$P = 1 - P\left(\frac{v}{2}, \frac{1}{2}\chi_0^2\right)$$

where $P(a, x)$ is the incomplete gamma function

$$P(a, x) = \frac{1}{\Gamma(a)} \int_0^x e^{-t} t^{a-1} \, dt \quad (a > 0)$$

and $\Gamma(a)$ is the gamma function

$$\Gamma(a) = \int_0^\infty t^{a-1} e^{-t} \, dt$$

Both functions are tabulated in mathematical handbooks. The function P gives the goodness of fit. After Equation (4) is minimized, let χ_0^2 be the value of χ^2 at the minimum. Then $P > 0.1$ represents a believable fit; $P > 0.001$ might be an acceptable fit; smaller values of P indicate that the model may be in error (or the σ_i are really larger). A "typical" value of χ^2 for a moderately good fit is $\chi^2 \sim v$. Asymptotically for large v, the statistic χ^2 becomes normally distributed with a mean v and a standard deviation $\sqrt{2v}$.

If values σ_i are not known in advance, assume $\sigma_i = \sigma$ (then its value does not affect the minimization of χ^2). Find the parameters by minimizing χ^2 and compute

$$\sigma^2 = \sum_{i=1}^{N} \frac{[y_i - y(x_i)]^2}{N}$$

This gives some information about the errors (i.e., the variance and standard deviation of each data point), although the goodness of fit P cannot be calculated.

The minimization of χ^2 requires

$$\sum_{i=1}^{N} \left[\frac{y_i - y(x_i)}{\sigma_i^2}\right] \frac{\partial y(x_i; a_1, \ldots, a_M)}{\partial a_k} = 0,$$

$$k = 1, \ldots, M \quad (5)$$

When the model is a straight line

$$\chi^2(a, b) = \sum_{i=1}^{N} \left[\frac{y_i - a - b x_i}{\sigma_i}\right]^2$$

Define

$$S = \sum_{i=1}^{N} \frac{1}{\sigma_i^2}, \quad S_x = \sum_{i=1}^{N} \frac{x_i}{\sigma_i^2}, \quad S_y = \sum_{i=1}^{N} \frac{y_i}{\sigma_i^2}$$

$$S_{xx} = \sum_{i=1}^{N} \frac{x_i^2}{\sigma_i^2}, \quad S_{xy} = \sum_{i=1}^{N} \frac{x_i y_i}{\sigma_i^2},$$

$$t_i = \frac{1}{\sigma_i}\left(x_i - \frac{S_x}{S}\right), \quad S_{tt} = \sum_{i=1}^{N} t_i^2$$

Then

$$b = \frac{1}{S_{tt}} \sum_{i=1}^{N} \frac{t_i y_i}{\sigma_i}, \quad a = \frac{S_y - S_x b}{S},$$

$$\sigma_a^2 = \frac{1}{S}\left(1 + \frac{S_x^2}{S S_{tt}}\right), \quad \sigma_b^2 = \frac{1}{S_{tt}}$$

$$\text{Cov}(a, b) = -\frac{S_x}{S S_{tt}}, \quad r_{ab} = \frac{\text{Cov}(a, b)}{\sigma_a \sigma_b}$$

Thus, the values of a and b with maximum likelihood are obtained: the variances of a and b. By using the value of χ^2 for this a and b, the goodness of fit P can also be calculated. In addition, the linear correlation coefficient r is related by

$$\chi^2 = (1 - r^2) \sum_{i=1}^{N} (y_i - \bar{y})^2$$

Here

$$r = \frac{\sum_{i=1}^{N} \frac{(x_i - \bar{x})(y_i - \bar{y})}{\sigma_i^2}}{\sqrt{\sum_{i=1}^{N} \frac{(x_i - \bar{x})^2}{\sigma_i^2}} \sqrt{\sum_{i=1}^{N} \frac{(y_i - \bar{y})^2}{\sigma_i^2}}}$$

Values of r near 1 indicate a positive correlation, r near -1 means a negative correlation, and r near zero means no correlation.

A general linear model is one expressed as

$$y(x) = \sum_{k=1}^{M} a_k X_k(x)$$

where the parameters are $\{a_k\}$, and the expression is linear with respect to them, and $X_k(x)$ can be any (nonlinear) functions of x, not depending on the parameters $\{a_k\}$. Equation (5) is then

$$\sum_{i=1}^{N} \frac{1}{\sigma_i^2} \left[y_i - \sum_{j=1}^{M} a_j X_j(x_i) \right] X_k(x_i) = 0,$$

$$k = 1, \ldots, M$$

which is rewritten as

$$\sum_{j=1}^{M} \left[\sum_{i=1}^{N} \frac{1}{\sigma_i^2} X_j(x_i) X_k(x_i) \right] a_j = \sum_{i=1}^{N} \frac{y_i}{\sigma_i^2} X_k(x_i)$$

or as

$$\sum_{j=1}^{M} \alpha_{kj} a_j = \beta_k \tag{6}$$

Solving this set of equations gives the parameters $\{a_j\}$, which maximize the likelihood. The variance of a_j is

$$\sigma^2(a_j) = C_{jj}$$

where $C_{jk} = \alpha_{jk}^{-1}$, i.e., C is the inverse of α. The covariance of a_j and a_k is given by C_{jk}. If round-off errors affect the result, try to make the functions orthogonal. For example, using

$$X_k(x) = x^{k-1}$$

will cause round-off errors for a smaller M than

$$X_k(x) = P_{k-1}(x) \tag{7}$$

where P_{k-1} are orthogonal polynomials. If necessary, a singular value decomposition can be used rather than solving Equation (6) directly.

Various global and piecewise polynomials can be used to fit the data. Some of the approximations only make sense for $N = M$, leading to a perfect fit of the data but perhaps a highly oscillatory model. These include the Lagrangian polynomial, and the forward and backward differences. However, the other approximations can be used with $M < N$. Consider orthogonal polynomials as expressed in Equation (7). Simply evaluate the polynomials at the points $\{x_i\}$, and solve Equation (6) for the coefficients. When piecewise polynomials are used, write

$$y(x) = \sum_{k=1}^{M} a_k N_k(x) = \sum_{e} \sum_{J=1}^{NP} a_J^e N_J(u)$$

and then minimize

$$\chi^2 = \sum_{i=1}^{N} \left[\frac{y_i - \sum_{e} \sum_{J=1}^{NP} a_J^e N_J(u_I)}{\sigma_i^2} \right]^2$$

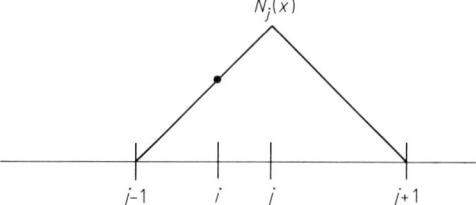

Figure 9. Fitting discrete data using linear elements

Equation (6) is still solved, but with $X_j(x_i)$ given by $N_j(x_i)$, as shown in Figure 9 for $NP = 2$.

Sometimes, particularly in two-dimensional problems, elements may have no, or too few, data points in them. Then the matrix α_{kj} would be singular. In that case a smoothing term can be added to the criterion and the equation minimized:

$$I = \alpha_1 \chi^2 + \alpha_2 \int_a^b \left(\frac{dy}{dx} \right)^2 dx =$$

$$\alpha_1 \chi^2 + \alpha_2 \sum_{j=1}^{N} \sum_{i=1}^{N} a_j a_i \int_a^b \frac{dN_i}{dx} \frac{dN_j}{dx} dx$$

Generally, small values of α_2 are used. The format for the last term, in terms of finite elements (see Section 7.5), is

$$\int_a^b \frac{dN_i}{dx} \frac{dN_j}{dx} dx = \sum_e \frac{1}{\Delta x_e} \int_0^1 \frac{dN_I}{du} \frac{dN_J}{du} du$$

The effect of this smoothing on the statistics for $\{a_k\}$ is unknown.

2.6. Nonlinear Least Squares

The Levenberg–Marquardt method is used when the parameters of the model appear nonlinearly. Define

$$\chi^2(\boldsymbol{a}) = \sum_{i=1}^{N} \left[\frac{y_i - y(x_i; \boldsymbol{a})}{\sigma_i^2} \right]^2$$

and, near the optimum, represent χ^2 by

$$\chi^2(\boldsymbol{a}) = \chi_0^2 - \boldsymbol{d}^T \cdot \boldsymbol{a} + \tfrac{1}{2} \boldsymbol{a}^T \cdot \boldsymbol{D} \cdot \boldsymbol{a}$$

where $\boldsymbol{d}$ is an $M \times 1$ vector and $\boldsymbol{D}$ is an $M \times M$ matrix. Then calculate iteratively

$$\boldsymbol{D} \cdot (\boldsymbol{a}^{k+1} - \boldsymbol{a}^k) = - \nabla \chi^2(\boldsymbol{a}^k) \tag{8}$$

The notation a_l^k means the l-th component of $\boldsymbol{a}$ evaluated on the k-th iteration. If $\boldsymbol{a}^k$ is a poor approximation to the optimum, steepest descent might be used instead

$$a^{k+1} - a^k = - \text{constant} \times \nabla \chi^2(a^k) \qquad (9)$$

and the constant chosen somehow to decrease χ^2 as much as possible. The gradient of χ^2 is

$$\frac{\partial \chi^2}{\partial a_k} = -2 \sum_{i=1}^{N} \frac{y_i - y(x_i; a)}{\sigma_i^2} \frac{\partial y(x_i; a)}{\partial a_k}$$

$$k = 1, 2, \ldots, M$$

The second derivative (in D) is

$$\frac{\partial^2 \chi^2}{\partial a_k \partial a_l} = 2 \sum_{i=1}^{N} \frac{1}{\sigma_i^2} \left\{ \frac{\partial y(x_i; a)}{\partial a_k} \frac{\partial y(x_i; a)}{\partial a_l} \right.$$

$$\left. - [y_i - y(x_i; a)] \frac{\partial^2 y(x_i; a)}{\partial a_k \partial a_l} \right\}$$

Equations (8) and (9) are included in

$$\sum_{l=1}^{M} \alpha'_{kl} (a_l^{k+1} - a_l^k) = \beta_k \qquad (10)$$

where

$$a'_{kl} = \sum_{i=1}^{N} \frac{1}{\sigma_i^2} \frac{\partial y(x_i; a)}{\partial a_k} \frac{\partial y(x_i; a)}{\partial a_l} \quad k \neq l$$

$$a'_{kk} = \sum_{i=1}^{N} \frac{1}{\sigma_i^2} \left[\frac{\partial y(x_i; a)}{\partial a_k} \right]^2 (1 + \lambda)$$

$$\beta_k = \sum_{i=1}^{N} \frac{y_i - y(x_i; a)}{\sigma_i^2} \frac{\partial y(x_i; a)}{\partial a_k}$$

The second term in the second derivative is dropped because it is usually small [remember that y_i will be close to $y(x_i, a)$]. The Levenberg–Marquardt method then iterates as follows:

1) Choose a and calculate $\chi^2(a)$.
2) Choose λ, say $\lambda = 0.001$.
3) Solve Equation (10) for a^{k+1} and evaluate $\chi^2(a^{k+1})$.
4) If $\chi^2(a^{k+1}) \geq \chi^2(a^k)$, increase λ by a factor of 10, for example, and go back to step 3. This makes the step more like a steepest descent.
5) If $\chi^2(a^{k+1}) < \chi^2(a^k)$, then update a (i.e., use $a = a^{k+1}$), decrease λ by a factor of 10, and go back to step 3.
6) Stop the iteration when the decrease in χ^2 from one step to another is not statistically meaningful (i.e., less than 0.1 or 0.01 or 0.001).

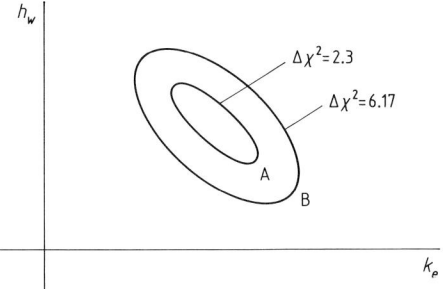

Figure 10. Parameter estimation for heat transfer

7) Set $\lambda = 0$ and compute the estimated covariance matrix.

$$C = \alpha^{-1}$$

This gives the standard errors in the fitted parameters a.

For normally distributed errors the parameter region in which $\chi^2 = \text{constant}$ can give boundaries of the confidence limits. The value of a obtained in the Marquardt method gives the minimum $\chi^2_{\min}$. Setting $\chi^2 = \chi^2_{\min} + \Delta\chi^2$ for some $\Delta\chi^2$ and looking at contours in parameter space where $\chi_1^2 = \text{constant}$ give confidence boundaries at the probability associated with χ_1^2. For example, in a chemical reactor with radial dispersion, the heat-transfer coefficient and radial effective heat conductivity are closely connected: decreasing one and increasing the other can still give a good fit. Thus, the confidence boundaries may look something like Figure 10. The ellipse defined by $\Delta\chi^2 = 2.3$ contains 68.3% of the normally distributed data. The curve defined by $\Delta\chi^2 = 6.17$ contains 95.4% of the data.

2.7. Fourier Transforms of Discrete Data [2.2]

Suppose a signal $y(t)$ is sampled at equal intervals

$$y_n = y(n\Delta), \quad n = \ldots, -2, -1, 0, 1, 2, \ldots$$
Δ = sampling rate (e.g., number of samples per second)

The Fourier transform and inverse transform are

$$Y(\omega) = \int_{-\infty}^{\infty} y(t) e^{i\omega t} dt$$

$$y(t) = \frac{1}{2\pi} \int_{-\infty}^{\infty} Y(\omega) e^{-i\omega t} dt$$

(For definition of i, see Chap. 3.) The Nyquist critical frequency or critical angular frequency is

$$f_c = \frac{1}{2\Delta}, \quad \omega_c = \frac{\pi}{\Delta}$$

If a function $y(t)$ is bandwidth limited to frequencies smaller than f_c, i.e.,

$$Y(\omega) = 0 \text{ for } \omega > \omega_c$$

then the function is completely determined by its samples y_n. Thus, the entire information content of a signal can be recorded by sampling at a rate $\Delta^{-1} = 2f_c$. If the function is not bandwidth limited, then aliasing occurs. Once a sample rate Δ is chosen, information corresponding to frequencies greater than f_c is simply aliased into that range. The way to detect this in a Fourier transform is to see if the transform approaches zero at $\pm f_c$; if not, aliasing has occurred and a higher sampling rate is needed.

Next, for N samples, where N is even

$$y_k = y(t_k), \quad t_k = k\Delta, \quad k = 0, 1, 2, \ldots, N-1$$

and the sampling rate is Δ; with only N values $\{y_k\}$ the complete Fourier transform $Y(\omega)$ cannot be determined. Calculate the value $Y(\omega_n)$ at the discrete points

$$\omega_n = \frac{2\pi n}{N\Delta}, \quad n = -\frac{N}{2}, \ldots, 0, \ldots, \frac{N}{2}$$

$$Y_n = \sum_{k=0}^{N-1} y_k e^{2\pi i k n/N}$$

$$Y(\omega_n) = \Delta Y_n$$

The discrete inverse Fourier transform is

$$y_k = \frac{1}{N} \sum_{n=0}^{N-1} Y_n e^{-2\pi i k n/N}$$

The *fast Fourier transform* (*FFT*) is used to calculate the Fourier transform as well as the inverse Fourier transform. A discrete Fourier transform of length N can be written as the sum of two discrete Fourier transforms, each of length $N/2$.

$$Y_k = Y_k^e + W^k Y_k^o$$

Here Y_k is the k-th component of the Fourier transform of y, and Y_k^e is the k-th component of the Fourier transform of the even components of $\{y_j\}$ and is of length $N/2$; similarly Y_k^o is the k-th component of the Fourier transform of the odd components of $\{y_j\}$ and is of length $N/2$; and W is a constant, taken to the k-th power.

$$W = e^{2\pi i/N}$$

Because Y_k has N components, whereas Y_k^e and Y_k^o have $N/2$ components, Y_k^e and Y_k^o are repeated once to give N components in the calculation of Y_k. This decomposition can be used recursively. Thus, Y_k^e is split into even and odd terms of length $N/4$.

$$Y_k^e = Y_k^{ee} + W^k Y_k^{eo}$$
$$Y_k^o = Y_k^{oe} + W^k Y_k^{oo}$$

This process is continued until only one component remains. For this reason the number N is taken as a power of 2. The vector $\{y_j\}$ is filled with zeroes, if need be, to make $N = 2^p$ for some p. For the computer program, see [2.2, p. 381]. The standard Fourier transform takes N^2 operations to calculate, whereas the fast Fourier transform takes only $N \log_2 N$. For large N the difference is significant; at $N = 100$ it is a factor of 15, but for $N = 1000$ it is a factor of 100.

The discrete Fourier transform can also be used for *differentiating* a function; this is used in the spectral method for solving differential equations. Consider a grid of equidistant points:

$$x_n = n\Delta x, \quad n = 0, 1, 2, \ldots, 2N-1, \quad \Delta x = \frac{L}{2N}$$

the solution is known at each of these grid points $\{Y(x_n)\}$. First, the Fourier transform is taken:

$$y_k = \frac{1}{2N} \sum_{n=0}^{2N-1} Y(x_n) e^{-2ik\pi x_n/L}$$

The inverse transformation is

$$Y(x) = \frac{1}{L} \sum_{k=-N}^{N} y_k e^{2ik\pi x/L}$$

which is differentiated to obtain

$$\frac{dY}{dx} = \frac{1}{L} \sum_{k=-N}^{N} y_k \frac{2\pi i k}{L} e^{2ik\pi x/L}$$

Thus, at the grid points

$$\left.\frac{dY}{dx}\right|_n = \frac{1}{L} \sum_{k=-N}^{N} y_k \frac{2\pi i k}{L} e^{2ik\pi x_n/L}$$

The process works as follows. From the solution at all grid points the Fourier transform is obtained by using FFT $\{y_k\}$. This is multiplied by $2\pi i k/L$ to obtain the Fourier transform of the derivative:

$$y'_k = y_k \frac{2\pi i k}{L}$$

The inverse Fourier transform is then taken by using FFT, to give the value of the derivative at each of the grid points:

$$\left.\frac{dY}{dx}\right|_n = \frac{1}{L} \sum_{k=-N}^{N} y'_k e^{2ik\pi x_n/L}$$

2.8. Two-Dimensional Interpolation and Quadrature

If the domain is square, $a \le x \le b$, $c \le y \le d$, then the approximation can be made by using tensor products of orthogonal polynomials.

$$z(x,y) = \sum_{i=0}^{m1} \sum_{j=0}^{m2} c_{ij} P_i(x) P_j(y)$$

The coefficients are chosen to minimize I.

$$I = \int_a^b \int_c^d W(x) W(y) \left[z(x,y) - \sum_{i=0}^{m1} \sum_{j=0}^{m2} c_{ij} P_i(x) P_j(y) \right]^2 dx\,dy$$

$$c_{kl} = \int_a^b \int_c^d W(x) W(y) z(x,y) P_k(x) P_l(x) dx\,dy /$$

$$\int_a^b W(x) P_k^2(x) dx \int_c^d W(y) P_l^2(y) dy$$

Bicubic splines can be used to interpolate a set of values on a regular array, $f(x_i, y_j)$. Suppose NX points occur in the x direction and NY points occur in the y direction. Press et al. [2.2] suggest computing NY different cubic splines of size NX along lines of constant y, for example, and storing the derivative information. To obtain the value of f at some point x, y, evaluate each of these splines for that x. Then do one spline of size NY in the y direction, doing both the determination and the evaluation.

If the points are distributed randomly within the domain, the finite element method can be used, which is especially useful if the domain is irregular. The expansion is

$$z(x,y) = \sum_{l=1}^{M} \tilde{z}_l N_l(x,y)$$

Let the data points be $z_i = z(x_i, y_i)$, $i = 1, \ldots, N$. The maximum likelihood of the parameter fit is

$$\chi^2 = \sum_{i=1}^{N} \left[\frac{z_i - \sum_{l=1}^{M} \tilde{z}_l N_l(x_i; y_i)}{\sigma_i^2} \right]^2$$

and this is minimized with respect to z_k.

$$\sum_{i=1}^{N} \frac{1}{\sigma_i^2} \left[z_i - \sum_{l=1}^{M} \tilde{z}_l N_l(x_i, y_i) \right] N_k(x_i; y_i) = 0$$

$$k = 1, \ldots, M$$

With two-dimensional domains, ensuring enough data points with each element is difficult. Thus, minimize

$$I = \alpha_1 \chi^2 + \alpha_2 \int_\Omega \nabla z \cdot \nabla z \, dx\,dy$$

The minimum is achieved for

$$\sum_{l=1}^{M} \left\{ \alpha_1 \sum_{i=1}^{N} \frac{1}{\sigma_i^2} \frac{1}{\sigma_k^2} N_l(x_i, y_i) N_k(x_i, y_i) \right.$$

$$\left. + \alpha_2 \int_\Omega \nabla N_l \cdot \nabla N_k \, dx\,dy \right\} \tilde{z}_l = \alpha_1 \sum_{i=1}^{N} \frac{\tilde{z}_i}{\sigma_i^2} \frac{N_k(x_i, y_i)}{\sigma_k^2}$$

The finite element integrals are calculated by [2.5]

$$\int_\Omega \nabla N_l \cdot \nabla N_k \, dx\,dy = \sum_e \int_{\Omega_e} \nabla N_L \cdot \nabla N_K \, dx\,dy$$

If linear elements on triangles are used and the nodes are (x_J, y_J), (x_K, y_K), and (x_L, y_L), the integral is

$$\int_{\Omega_e} \nabla N_L \cdot \nabla N_K \, dx\,dy = \frac{1}{4\Delta} (b_L b_K + c_L c_K)$$

where

$$\Delta = \det \begin{bmatrix} 1 & x_J & y_J \\ 1 & x_K & y_K \\ 1 & x_L & y_L \end{bmatrix},$$

$$b_K = y_K - y_J, \text{ and } c_K = x_J - x_K$$

plus permutations on J, K, L that are arranged in a counterclockwise orientation.

Quadrature follows directly from the approximation. If orthogonal polynomials are used, then

$$\int_a^b \int_c^d z(x,y) dx\,dy = \sum_{j=1}^{m1-1} \sum_{k=1}^{m2-1} W_{xj} W_{jk} z(x_j, y_k)$$

and the limits of integration must be transformed to coincide with those of the defining polynomial. If finite elements are used, then

$$\int_\Omega z(x,y)\,dx\,dy = \sum_{l=1}^{M} \tilde{z}_l \int_\Omega N_l(x,y)\,dx\,dy$$

$$= \sum_e \sum_{L=1}^{NP} \tilde{z}_L^e \int_{\Omega_e} N_L(x,y)\,dx\,dy$$

For the same linear elements on triangles

$$\int_{\Omega_e} N_L(x,y)\,dx\,dy = \frac{\Delta}{3}$$

Multidimensional integrals can also be broken down into one-dimensional integrals. For example,

$$\int_a^b \int_{f_1(x)}^{f_2(x)} z(x,y)\,dx\,dy = \int_a^b G(x)\,dx;$$

$$G(x) = \int_{f_1(x)}^{f_2(x)} z(x,y)\,dx$$

3. Complex Variables

3.1. Introduction to the Complex Plane

A complex number is an ordered pair of real numbers, x and y, that is written as

$$z = x + iy$$

The variable i is the imaginary unit which has the property

$$i^2 = -1$$

The real and imaginary parts of a complex number are often referred to:

$$Re(z) = x, \quad Im(z) = y$$

A complex number can also be represented graphically in the complex plane, where the real part of the complex number is the abscissa and the imaginary part of the complex number is the ordinate (see Fig. 11).

Another representation of a complex number is the *polar form*, where r is the magnitude and θ is the argument.

$$r = |x + iy| = \sqrt{x^2 + y^2}, \quad \theta = \arg(x + iy)$$

Write

$$z = x + iy = r(\cos\theta + i\sin\theta)$$

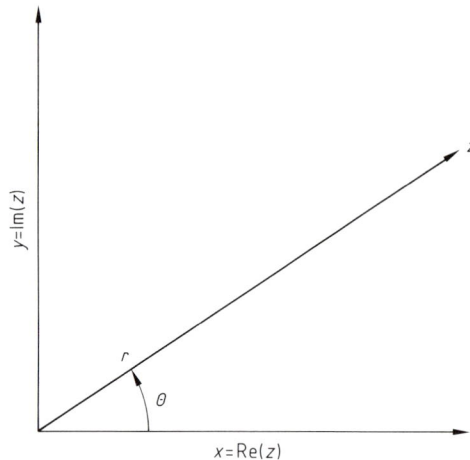

Figure 11. The complex plane

so that

$$x = r\cos\theta, \quad y = r\sin\theta$$

and

$$\theta = \arctan\frac{y}{x}$$

Since the arctangent repeats itself in multiples of π rather than 2π, the argument must be defined carefully. For example, the θ given above could also be the argument of $-(x+iy)$. The function $r = \cos\theta + i\sin\theta$ obeys $|\cos\theta + i\sin\theta| = 1$.

The rules of equality, addition, and multiplication are

$$z_1 = x_1 + iy_1, \quad z_2 = x_2 + iy_2$$
Equality: $z_1 = z_2$ if and only if $x_1 = x_2$ and $y_1 = y_2$
Addition: $z_1 + z_2 = (x_1 + x_2) + i(y_1 + y_2)$
Multiplication: $z_1 z_2 = (x_1 x_2 - y_1 y_2) + i(x_1 y_2 + x_2 y_1)$

The last rule can be remembered by using the standard rules for multiplication, keeping the imaginary parts separate, and using $i^2 = -1$. In the complex plane, addition is illustrated in Figure 12. In polar form, multiplication is

$$z_1 z_2 = r_1 r_2 [\cos(\theta_1 + \theta_2) + i\sin(\theta_1 + \theta_2)]$$

The magnitude of $z_1 + z_2$ is bounded by

$$|z_1 \pm z_2| \le |z_1| + |z_2| \text{ and } |z_1| - |z_2| \le |z_1 \pm z_2|$$

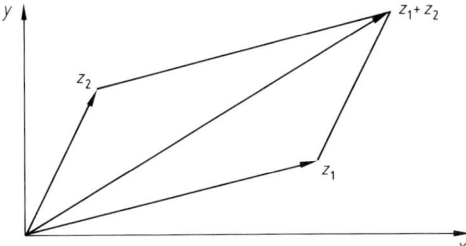

Figure 12. Addition in the complex plane

as can be seen in Figure 12. The magnitude and arguments in multiplication obey

$$|z_1 z_2| = |z_1||z_2|, \quad \arg(z_1 z_2) = \arg z_1 + \arg z_2$$

The *complex conjugate* is $z^* = x - iy$ when $z = x + iy$ and $|z^*| = |z|$, $\arg z^* = -\arg z$
For complex conjugates then

$$z^* z = |z|^2$$

The reciprocal is

$$\frac{1}{z} = \frac{z^*}{|z|^2} = \frac{1}{r}(\cos\theta - i\sin\theta), \quad \arg\left(\frac{1}{z}\right) = -\arg z$$

Then

$$\frac{z_1}{z_2} = \frac{x_1 + iy_1}{x_2 + iy_2} = (x_1 + iy_1)\frac{x_2 - iy_2}{x_2^2 + y_2^2}$$

$$= \frac{x_1 x_2 + y_1 y_2}{x_2^2 + y_2^2} + i\frac{x_2 y_1 - x_1 y_2}{x_2^2 + y_2^2}$$

and

$$\frac{z_1}{z_2} = \frac{r_1}{r_2}[\cos(\theta_1 - \theta_2) + i\sin(\theta_1 - \theta_2)]$$

3.2. Elementary Functions

Properties of elementary functions of complex variables are discussed here [3.1]. When the polar form is used, the argument must be specified because the same physical angle can be achieved with arguments differing by 2π. A complex number taken to a real power obeys

$$u = z^n, \quad |z^n| = |z|^n, \quad \arg(z^n) = n \arg z \pmod{2\pi}$$
$$u = z^n = r^n(\cos n\theta + i \sin n\theta)$$

Roots of a complex number are complicated by careful accounting of the argument

$$z = w^{1/n} \text{ with } w = R(\cos\Theta + i\sin\Theta),$$
$$0 \leq \Theta \leq 2\pi$$

then

$$z_k = R^{1/n}\left\{\cos\left[\frac{\Theta}{n} + (k-1)\frac{2\pi}{n}\right] + i\sin\left[\frac{\Theta}{n} + (k-1)\frac{2\pi}{n}\right]\right\}$$

such that

$(z_k)^n = w$ for every k
$z = r(\cos\theta + i\sin\theta)$
$r^n = R, \quad n\theta = \Theta \pmod{2\pi}$

The exponential function is

$$e^z = e^x(\cos\theta + i\sin\theta)$$

Thus,

$$z = r(\cos\theta + i\sin\theta)$$

can be written

$$z = re^{i\theta}$$

and

$$|e^z| = e^x, \quad \arg e^z = y \pmod{2\pi}$$

The exponential obeys

$$e^z \neq 0 \text{ for every finite } z$$

and is periodic with period 2π:

$$e^{z + 2\pi i} = e^z$$

Trigonometric functions can be defined by using

$$e^{iy} = \cos\theta + i\sin\theta, \text{ and } e^{-iy} = \cos\theta - i\sin\theta$$

Thus,

$$\cos y = \frac{e^{iy} + e^{-iy}}{2} = \cosh iy$$

$$\sin y = \frac{e^{iy} + e^{-iy}}{2i} = -i \sinh iy$$

The second equation follows from the definitions

$$\cosh z \equiv \frac{e^z + e^{-z}}{2}, \quad \sinh z \equiv \frac{e^z - e^{-z}}{2}$$

The remaining hyperbolic functions are

$$\tanh z \equiv \frac{\sinh z}{\cosh z}, \quad \coth z \equiv \frac{1}{\tanh z}$$

$$\text{sech } z \equiv \frac{1}{\cosh z}, \quad \text{csch } z \equiv \frac{1}{\sinh z}$$

The circular functions with complex arguments are defined

$$\cos z = \frac{e^{iz} + e^{-iz}}{2}, \quad \sin z = \frac{e^{iz} - e^{-iz}}{2},$$

$$\tan z = \frac{\sin z}{\cos z}$$

and satisfy

$$\sin(-z) = -\sin z, \quad \cos(-z) = \cos z$$
$$\sin(iz) = i \sinh z, \quad \cos(iz) = \cosh z$$

All trigonometric identities for real, circular functions with real arguments can be extended without change to complex functions of complex arguments. For example,

$$\sin^2 z + \cos^2 z = 1,$$
$$\sin(z_1 + z_2) = \sin z_1 \cos z_2 + \cos z_1 \sin z_2$$

The same is true of hyperbolic functions. The absolute boundaries of $\sin z$ and $\cos z$ are not bounded for all z.

Trigonometric identities can be defined by using

$$e^{i\theta} = \cos\theta + i\sin\theta$$

For example,
$$\begin{aligned}e^{i(\alpha+\beta)} &= \cos(\alpha+\beta) + i\sin(\alpha+\beta)\\ &= e^{i\alpha}e^{i\beta} = (\cos\alpha + i\sin\alpha)\\ &\quad(\cos\beta + i\sin\beta)\\ &= \cos\alpha\cos\beta - \sin\alpha\sin\beta\\ &\quad + i(\cos\alpha\sin\beta + \cos\beta\sin\alpha)\end{aligned}$$

Equating real and imaginary parts gives

$$\cos(\alpha+\beta) = \cos\alpha\cos\beta - \sin\alpha\sin\beta$$
$$\sin(\alpha+\beta) = \cos\alpha\sin\beta + \cos\beta\sin\alpha$$

The logarithm is defined as

$$\ln z = \ln z + i\arg z$$

and the various determinations differ by multiples of $2\pi i$. Then,

$$e^{\ln z} = z$$
$$\ln(e^z) - z \equiv 0 \pmod{2\pi i}$$

Also,

$$\ln(z_1 z_2) - \ln z_1 - \ln z_2 \equiv 0 \pmod{2\pi i}$$

is always true, but

$$\ln(z_1 z_2) = \ln z_1 + \ln z_2$$

holds only for some determinations of the logarithms. The principal determination of the argument can be defined as $-\pi < \arg \leq \pi$.

3.3. Analytic Functions of a Complex Variable

Let $f(z)$ be a single-valued continuous function of z in a domain D. The function $f(z)$ is differentiable at the point z_0 in D if

$$\lim_{h \to 0} \frac{f(z_0 + h) - f(z_0)}{h}$$

exists as a finite (complex) number and is independent of the direction in which h tends to zero. The limit is called the derivative, $f'(z_0)$. The derivative now can be calculated with h approaching zero in the complex plane, i.e., anywhere in a circular region about z_0. The function $f(z)$ is differentiable in D if it is differentiable at all points of D; then $f(z)$ is said to be an analytic function of z in D. Also, $f(z)$ is analytic at z_0 if it is analytic in some ε neighborhood of z_0. The word analytic is sometimes replaced by holomorphic or regular [3.1].

The *Cauchy–Riemann equations* can be used to decide if a function is analytic. Set

$$f(z) = f(x + iy) = u(x, y) + iv(x, y)$$

Theorem [3.1]. Suppose that $f(z)$ is defined and continuous in some neighborhood of $z = z_0$. A necessary condition for the existence of $f'(z_0)$ is that $u(x, y)$ and $v(x, y)$ have first-order partials and that the Cauchy–Riemann conditions (see below) hold.

$$\frac{\partial u}{\partial x} = \frac{\partial v}{\partial y} \quad \text{and} \quad \frac{\partial u}{\partial y} = -\frac{\partial v}{\partial x} \quad \text{at } z_0$$

Theorem [3.2]. The function $f(z)$ is analytic in a domain D if and only if u and v are continuously differentiable and satisfy the Cauchy–Riemann conditions there.

If $f_1(z)$ and $f_2(z)$ are analytic in domain D, then $\alpha_1 f_1(z) + \alpha_2 f_2(z)$ is analytic in D for any (complex) constants α_1, α_2.

$f_1(z) + f_2(z)$ is analytic in D
$f_1(z)/f_2(z)$ is analytic in D except where $f_2(z) = 0$

An analytic function of an analytic function is analytic. If $f(z)$ is analytic, $f'(z) \neq 0$ in D,

$f(z_1) \neq f(z_2)$ for $z_1 \neq z_2$, then the inverse function $g(w)$ is also analytic and

$$g'(w) = \frac{1}{f'(z)} \quad \text{where} \quad w = f(z),$$

$$g(w) = g[f(z)] = z$$

Analyticity implies continuity but the converse is not true: $z^* = x - iy$ is continuous but, because the Cauchy–Riemann conditions are not satisfied, it is not analytic. An entire function is one that is analytic for all finite values of z. Every polynomial is an entire function. Because the polynomials are analytic, a ratio of polynomials is analytic except when the denominator vanishes. The function $f(z) = |z^2|$ is continuous for all z but satisfies the Cauchy–Riemann conditions only at $z = 0$. Hence, $f'(z)$ exists only at the origin, and $|z|^2$ is nowhere analytic. The function $f(z) = 1/z$ is analytic except at $z = 0$. Its derivative is $-1/z^2$, where $z \neq 0$. If $\ln|z| = \ln z + i \arg z$ in the cut domain $-\pi < \arg z \leq \pi$, then $f(z) = 1/\ln z$ is analytic in the same cut domain, except at $z = 1$, where $\log z = 0$. Because e^z is analytic and $\pm iz$ are analytic, $e^{\pm iz}$ is analytic and linear combinations are analytic. Thus, the sine and cosine and hyperbolic sine and cosine are analytic. The other functions are analytic except when the denominator vanishes.

The derivatives of the elementary functions are

$$\frac{d}{dz} e^z = e^z, \quad \frac{d}{dz} z^n = n z^{n-1}$$

$$\frac{d}{dz} (\ln z) = \frac{1}{z}, \quad \frac{d}{dz} \sin z = \cos z,$$

$$\frac{d}{dz} \cos z = -\sin z$$

In addition,

$$\frac{d}{dz}(fg) = f \frac{dg}{dz} + g \frac{df}{dz}$$

$$\frac{d}{dz} f[g(z)] = \frac{df}{dg} \frac{dg}{dz}$$

$$\frac{d}{dz} \sin w = \cos w \frac{dw}{dz}, \quad \frac{d}{dz} \cos w = -\sin w \frac{dw}{dz}$$

Define $z^a = e^{a \ln z}$ for complex constant a. If the determination is $-\pi < \arg z \leq \pi$, then z^a is analytic on the complex plane with a cut on the negative real axis. If a is an integer n, then $e^{2\pi i n} = 1$ and z^n has the same limits approaching the cut from either side. The function can be made continuous across the cut and the function is analytic there, too. If $a = 1/n$ where n is an integer, then

$$z^{1/n} = e^{(\ln z)/n} = |z|^{1/n} e^{i(\arg z)/n}$$

So $w = z^{1/n}$ has n values, depending on the choice of argument.

Laplace Equation. If $f(z)$ is analytic, where

$$f(z) = u(x, y) + iv(x, y)$$

the Cauchy–Riemann equations are satisfied. Differentiating the Cauchy–Riemann equations gives the Laplace equation:

$$\frac{\partial^2 u}{\partial x^2} = \frac{\partial^2 v}{\partial x \partial y} = \frac{\partial^2 v}{\partial y \partial x} = -\frac{\partial^2 u}{\partial y^2} \quad \text{or} \quad \frac{\partial^2 u}{\partial x^2} + \frac{\partial^2 u}{\partial y^2} = 0$$

Similarly,

$$\frac{\partial^2 v}{\partial x^2} + \frac{\partial^2 v}{\partial y^2} = 0$$

Thus, general solutions to the Laplace equation can be obtained from analytic functions [3.1, p. 83], [3.2, p. 223]. For example,

$$\ln \frac{1}{|z - z_0|}$$

is analytic so that a solution to the Laplace equation is

$$\ln [(x - a)^2 + (y - b)^2]^{-1/2}$$

A solution to the Laplace equation is called a harmonic function. A function is harmonic if, and only if, it is the real part of an analytic function. The imaginary part is also harmonic. Given any harmonic function u, a conjugate harmonic function v can be constructed such that $f = u + iv$ is locally analytic [3.1, p. 85].

Maximum Principle. If $f(z)$ is analytic in a domain D and continuous in the set consisting of D and its boundary C, and if $|f(z)| \leq M$ on C, then $|f(z)| < M$ in D unless $f(z)$ is a constant [3.2].

3.4. Integration in the Complex Plane

Let C be a rectifiable curve in the complex plane

$$C: z = z(t), \quad 0 \leq t \leq 1$$

where $z(t)$ is a continuous function of bounded variation; C is oriented such that $z_1 = z(t_1)$ precedes the point $z_2 = z(t_2)$ on C if and only if $t_1 < t_2$. Define

$$\int_C f(z)\,dz = \int_0^1 f[z(t)]\,dz(t)$$

The integral is linear with respect to the integrand:

$$\int_C [\alpha_1 f_1(z) + \alpha_2 f_2(z)]\,dz$$

$$= \alpha_1 \int_C f_1(z)\,dz + \alpha_2 \int_C f_2(z)\,dz$$

The integral is additive with respect to the path. Let curve C_2 begin where curve C_1 ends and $C_1 + C_2$ be the path of C_1 followed by C_2. Then,

$$\int_{C_1+C_2} f(z)\,dz = \int_{C_1} f(z)\,dz + \int_{C_2} f(z)\,dz$$

Reversing the orientation of the path replaces the integral by its negative:

$$\int_{-C} f(z)\,dz = -\int_C f(z)\,dz$$

If the path of integration consists of a finite number of arcs along which $z(t)$ has a continuous derivative, then

$$\int_C f(z)\,dz = \int_0^1 f[z(t)]\,z'(t)\,dt$$

Also if $s(t)$ is the arc length on C and $l(C)$ is the length of C

$$\left|\int_C f(z)\,dz\right| \leq \max_{z \in C} |f(z)|\,l(C)$$

and

$$\left|\int_C f(z)\,dz\right| \leq \int_C |f(z)|\,|dz| = \int_0^1 |f[z(t)]|\,ds(t)$$

Cauchy's Theorem [3.1]. Suppose $f(z)$ is an analytic function in a domain D and C is a simple, closed, rectifiable curve in D such that $f(z)$ is analytic inside and on C. Then

$$\oint_C f(z)\,dz = 0 \qquad (11)$$

If D is simply connected, then Equation (11) holds for every simple, closed, rectifiable curve C in D. If D is simply connected and if a and b are any two points in D, then

$$\int_a^b f(z)\,dz$$

is independent of the rectifiable path joining a and b in D.

Cauchy's Integral. If C is a closed contour such that $f(z)$ is analytic inside and on C, z_0 is a point inside C, and z traverses C in the counterclockwise direction,

$$f(z_0) = \frac{1}{2\pi i} \oint_C \frac{f(z)}{z - z_0}\,dz$$

$$f'(z_0) = \frac{1}{2\pi i} \oint_C \frac{f(z)}{(z - z_0)^2}\,dz$$

Under further restriction on the domain [3.1, p. 178],

$$f^{(m)}(z_0) = \frac{1}{2\pi i} \oint_C \frac{f(z)}{(z - z_0)^{m+1}}\,dz$$

Power Series. If $f(z)$ is analytic interior to a circle $|z - z_0| < r_0$, then at each point inside the circle the series

$$f(z) = f(z_0) + \sum_{n=1}^{\infty} \frac{f^{(n)}(z_0)}{n!}(z - z_0)^n$$

converges to $f(z)$. This result follows from Cauchy's integral. As an example, e^z is an entire function (analytic everywhere) so that the MacLaurin series

$$e^z = 1 + \sum_{n=1}^{\infty} \frac{z^n}{n!}$$

represents the function for all z.

Another result of Cauchy's integral formula is that if $f(z)$ is analytic in an annulus R, $r_1 < |z - z_0| < r_2$, it is represented in R by the Laurent series

$$f(z) = \sum_{n=-\infty}^{\infty} A_n (z - z_0)^n, \quad r_1 < |z - z_0| \leq r_2$$

where

$$A_n = \frac{1}{2\pi i} \int_C \frac{f(z)}{(z-z_0)^{n+1}} dz,$$

$n = 0, \pm 1, \pm 2, \ldots,$

and C is a closed curve counterclockwise in R.

Singular Points and Residues [3.3, p. 159]. If a function in analytic in every neighborhood of z_0, but not at z_0 itself, then z_0 is called an isolated singular point of the function. About an isolated singular point, the function can be represented by a Laurent series.

$$f(z) = \ldots + \frac{A_{-2}}{(z-z_0)^2} + \frac{A_{-1}}{z-z_0} + A_0$$
$$+ A_1(z-z_0) + \ldots 0 < |z-z_0| \le r_0 \quad (12)$$

In particular,

$$A_{-1} = \frac{1}{2\pi i} \oint_C f(z) dz$$

where the curve C is a closed, counterclockwise curve containing z_0 and is within the neighborhood where $f(z)$ is analytic. The complex number A_{-1} is the residue of $f(z)$ at the isolated singular point z_0; $2\pi i A_{-1}$ is the value of the integral in the positive direction around a path containing no other singular points.

If $f(z)$ is defined and analytic in the exterior $|z-z_0| > R$ of a circle, and if

$$g(\zeta) = f\left(z_0 + \frac{1}{\zeta}\right) \text{ obtained by } \zeta = \frac{1}{z-z_0}$$

has a removable singularity at $\zeta = 0$, $f(z)$ is analytic at infinity. It can then be represented by a Laurent series with nonpositive powers of $z - z_0$.

If C is a closed curve within which and on which $f(z)$ is analytic except for a finite number of singular points $z_1, z_2, \ldots, z_n$ interior to the region bounded by C, then the residue theorem states

$$\oint_C f(z) dz = 2\pi i (\varrho_1 + \varrho_2 + \ldots \varrho_n)$$

where ϱ_n denotes the residue of $f(z)$ at z_n.

The series of negative powers in Equation (12) is called the principal part of $f(z)$. If the principal part has an infinite number of nonvanishing terms, the point z_0 is an essential singularity. If $A_{-m} \ne 0$, $A_{-n} = 0$ for all $m < n$, then z_0 is called a pole of order m. It is a simple pole if $m = 1$. In such a case,

$$f(z) = \frac{A_{-1}}{z-z_0} + \sum_{n=0}^{\infty} A_n (z-z_0)^n$$

If a function is not analytic at z_0 but can be made so by assigning a suitable value, then z_0 is a removable singular point.

When $f(z)$ has a pole of order m at z_0,

$$\varphi(z) = (z-z_0)^m f(z), \quad 0 < |z-z_0| < r_0$$

has a removable singularity at z_0. If $\varphi(z_0) = A_{-m}$, then $\varphi(z)$ is analytic at z_0. For a simple pole,

$$A_{-1} = \varphi(z_0) = \lim_{z \to z_0} (z-z_0) f(z)$$

Also $|f(z)| \to \infty$ as $z \to z_0$ when z_0 is a pole. Let the function $p(z)$ and $q(z)$ be analytic at z_0, where $p(z_0) \ne 0$. Then

$$f(z) = \frac{p(z)}{q(z)}$$

has a simple pole at z_0 if, and only if, $q(z_0) = 0$ and $q'(z_0) \ne 0$. The residue of $f(z)$ at the simple pole is

$$A_{-1} = \frac{p(z_0)}{q'(z_0)}$$

If $q^{(i-1)}(z_0) = 0$, $i = 1, \ldots, m$, then z_0 is a pole of $f(z)$ of order m.

Branch [3.3, p. 163]. A branch of a multiple-valued function $f(z)$ is a single-valued function that is analytic in some region and whose value at each point there coincides with the value of $f(z)$ at the point. A branch cut is a boundary that is needed to define the branch in the greatest possible region. The function $f(z)$ is singular along a branch cut, and the singularity is not isolated. For example,

$$z^{1/2} = f_1(z) = \sqrt{r}\left(\cos\frac{\theta}{2} + i \sin\frac{\theta}{2}\right)$$

$-\pi < \theta < \pi, \quad r > 0$

is double valued along the negative real axis. The function tends to $\sqrt{r}i$ when $\theta \to \pi$ and to $-\sqrt{r}i$ when $\theta \to -\pi$; the function has no limit as $z \to -r$ ($r > 0$). The ray $\theta = \pi$ is a branch cut.

Analytic Continuation [3.3, p. 165]. If $f_1(z)$ is analytic in a domain D_1 and domain D contains D_1, then an analytic function $f(z)$ may exist that equals $f_1(z)$ in D_1. This function is the analytic continuation of $f_1(z)$ onto D, and it is unique. For example,

$$f_1(z) = \sum_{n=0}^{\infty} z^n, \quad |z| < 1$$

is analytic in the domain $D_1 : |z| < 1$. The series diverges for other z. Yet the function is the MacLaurin series in the domain

$$f_1(z) = \frac{1}{1-z}, \quad |z| < 1$$

Thus,

$$f_1(z) = \frac{1}{1-z}$$

is the analytic continuation onto the entire z plane except for $|z| = 1$.

An extension of the Cauchy integral formula is useful with Laplace transforms. Let the curve C be a straight line parallel to the imaginary axis and z_0 be any point to the right of that (see Fig. 13). A function $f(z)$ is of order z^k as $|z| \to \infty$ if positive numbers M and r_0 exist such that

$$|z^{-k} f(z)| < M \quad \text{when} \quad |z| > r_0, \quad \text{i.e.,}$$
$$|f(z)| < M |z|^k \quad \text{for} \quad |z| \text{ sufficiently large}$$

Theorem [3.3, p. 167]. Let $f(z)$ be analytic when $R(z) \geq \gamma$ and $O(z^{-k})$ as $|z| \to \infty$ in that half-plane, where γ and k are real constants and $k > 0$. Then for any z_0 such that $R(z_0) > \gamma$

$$f(z_0) = -\frac{1}{2\pi i} \lim_{\beta \to \infty} \int_{\gamma - i\beta}^{\gamma + i\beta} \frac{f(z)}{z - z_0} dz,$$

i.e., integration takes place along the line $x = \gamma$.

3.5. Other Results

Theorem [3.1, p. 84]. Let $P(z)$ be a polynomial of degree n having the zeroes $z_1, z_2, \ldots, z_n$ and let π be the least convex polygon containing the zeroes. Then $P'(z)$ cannot vanish anywhere in the exterior of π.

If a polynomial has real coefficients, the roots are either real or form pairs of complex conjugate numbers.

The radius of convergence R of the Taylor series of $f(z)$ about z_0 is equal to the distance from z_0 to the nearest singularity of $f(z)$.

Conformal Mapping. Let $u(x, y)$ be a harmonic function. Introduce the coordinate transformation

$$x = \hat{x}(\xi, \eta), \quad y = \hat{y}(\xi, \eta)$$

It is desired that

$$U(\xi, \eta) = u[\hat{x}(\xi, \eta), \hat{y}(\xi, \eta)]$$

be a harmonic function of ξ and η.

Theorem [3.2, p. 237]. The transformation

$$z = f(\zeta) \tag{13}$$

takes all harmonic functions of x and y into harmonic functions of ξ and η if and only if either $f(\zeta)$ or $f^*(\zeta)$ is an analytic function of $\zeta = \xi + i\eta$.

Equation (13) is a restriction on the transformation which ensures that

$$\text{if} \quad \frac{\partial^2 u}{\partial x^2} + \frac{\partial^2 u}{\partial y^2} = 0 \quad \text{then} \quad \frac{\partial^2 U}{\partial \xi^2} + \frac{\partial^2 U}{\partial \eta^2} = 0$$

Such a mapping with $f(\zeta)$ analytic and $f'(\zeta) \neq 0$ is a conformal mapping.

If Laplace's equation is to be solved in the region exterior to a closed curve, then the point at infinity is in the domain D. For flow in a long channel (governed by the Laplace equation) the

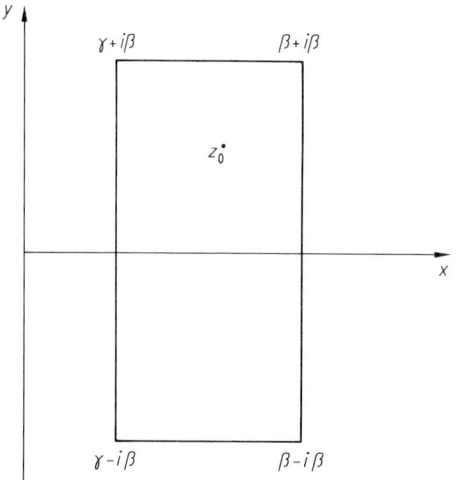

Figure 13. Integration in the complex plane

inlet and outlet are at infinity. In both cases the transformation

$$\zeta = \frac{az+b}{z-z_0}$$

takes z_0 into infinity and hence maps D into a bounded domain D*.

4. Integral Transforms

4.1. Fourier Transforms

Fourier Series [4.1]. Let $f(x)$ be a function that is periodic on $-\pi < x < \pi$. It can be expanded in a Fourier series

$$f(x) = \frac{a_0}{2} + \sum_{n=1}^{\infty} (a_n \cos nx + b_n \sin nx)$$

where

$$a_0 = \frac{1}{\pi} \int_{-\pi}^{\pi} f(x)\,dx, \quad a_n = \frac{1}{\pi} \int_{-\pi}^{\pi} f(x) \cos nx\,dx,$$

$$b_n = \frac{1}{\pi} \int_{-\pi}^{\pi} f(x) \sin nx\,dx$$

The values $\{a_n\}$ and $\{b_n\}$ are called the finite cosine and sine transform of f, respectively. Because

$$\cos nx = \frac{1}{2}(e^{inx} + e^{-inx})$$

and $\sin nx = \frac{1}{2i}(e^{inx} - e^{-inx})$

the Fourier series can be written as

$$f(x) = \sum_{n=-\infty}^{\infty} c_n e^{-inx}$$

where

$$c_n = \begin{cases} \frac{1}{2}(a_n + ib_n) & \text{for } n \geq 0 \\ \frac{1}{2}(a_{-n} - ib_{-n}) & \text{for } n < 0 \end{cases}$$

and

$$c_n = \frac{1}{2\pi} \int_{-\pi}^{\pi} f(x) e^{inx}\,dx$$

If f is real

$$c_{-n} = c_n^*.$$

If f is continuous and piecewise continuously differentiable

$$f'(x) = \sum_{-\infty}^{\infty} (-in) c_n e^{-inx}$$

If f is twice continuously differentiable

$$f''(x) = \sum_{-\infty}^{\infty} (-n^2) c_n e^{-inx}$$

Inversion. The Fourier series can be used to solve linear partial differential equations with constant coefficients. For example, in the problem

$$\frac{\partial T}{\partial t} = \frac{\partial^2 T}{\partial x^2}$$

$$T(x, 0) = f(x)$$

$$T(-\pi, t) = T(\pi, t)$$

Let

$$T = \sum_{-\infty}^{\infty} c_n(t) e^{-inx}$$

Then,

$$\sum_{-\infty}^{\infty} \frac{dc_n}{dt} e^{-inx} = \sum_{-\infty}^{\infty} c_n(t)(-n^2) e^{-inx}$$

Thus, $c_n(t)$ satisfies

$$\frac{dc_n}{dt} = -n^2 c_n, \quad \text{or } c_n = c_n(0) e^{-n^2 t}$$

Let $c_n(0)$ be the Fourier coefficients of the initial conditions:

$$f(x) = \sum_{-\infty}^{\infty} c_n(0) e^{-inx}$$

The formal solution to the problem is

$$T = \sum_{-\infty}^{\infty} c_n(0) e^{-n^2 t} e^{-inx}$$

Fourier Transform [4.1]. When the function $f(x)$ is defined on the entire real line, the Fourier transform is defined as

$$F[f] \equiv \hat{f}(\omega) = \int_{-\infty}^{\infty} f(x) e^{i\omega x}\,dx$$

This integral converges if

$$\int_{-\infty}^{\infty} |f(x)|\,dx$$

does. The inverse transformation is

$$f(x) = \frac{1}{2\pi} \int_{-\infty}^{\infty} \hat{f}(\omega) e^{-i\omega x} d\omega$$

If $f(x)$ is continuous and piecewise continuously differentiable,

$$\int_{-\infty}^{\infty} f(x) e^{i\omega x} dx$$

converges for each ω, and

$$\lim_{x \to \pm \infty} f(x) = 0$$

then

$$F\left[\frac{df}{dx}\right] = -i\omega F[f]$$

If f is real $F[f(-\omega)] = F[f(\omega)^*]$. The real part is an even function of ω and the imaginary part is an odd function of ω.

A function $f(x)$ is *absolutely integrable* if the improper integral

$$\int_{-\infty}^{\infty} |f(x)| dx$$

has a finite value. Then the improper integral

$$\int_{-\infty}^{\infty} f(x) dx$$

converges. The function is *square integrable* if

$$\int_{-\infty}^{\infty} |f(x)|^2 dx$$

has a finite value. If $f(x)$ and $g(x)$ are square integrable, the product $f(x)g(x)$ is absolutely integrable and satisfies the Schwarz inequality:

$$|\int_{-\infty}^{\infty} f(x) g(x) dx|^2 \leq \int_{-\infty}^{\infty} |f(x)|^2 dx \int_{-\infty}^{\infty} |g(x)|^2 dx$$

The triangle inequality is also satisfied:

$$\left\{\int_{-\infty}^{\infty} |f+g|^2 dx\right\}^{1/2}$$

$$\leq \left\{\int_{-\infty}^{\infty} |f|^2 dx\right\}^{1/2} \left\{\int_{-\infty}^{\infty} |g|^2 dx\right\}^{1/2}$$

A sequence of square integrable functions $f_n(x)$ converges in the mean to a square integrable function $f(x)$ if

$$\lim_{n \to \infty} \int_{-\infty}^{\infty} |f(x) - f_n(x)|^2 dx = 0$$

The sequence also satisfies the Cauchy criterion

$$\lim_{\substack{n \to \infty \\ m \to \infty}} \int_{-\infty}^{\infty} |f_n - f_m|^2 dx = 0$$

Theorem [4.1, p. 307]. If a sequence of square integrable functions $f_n(x)$ converges to a function $f(x)$ uniformly on every finite interval $a \leq x \leq b$, and if it satisfies Cauchy's criterion, then $f(x)$ is square integrable and $f_n(x)$ converges to $f(x)$ in the mean.

Theorem (Riesz–Fischer) [4.1, p. 308]. To every sequence of square integrable functions $f_n(x)$ that satisfy Cauchy's criterion, there corresponds a square integrable function $f(x)$ such that $f_n(x)$ converges to $f(x)$ in the mean. Thus, the limit in the mean of a sequence of functions is defined to within a null function.

Square integrable functions satisfy the Parseval equation.

$$\int_{-\infty}^{\infty} |\hat{f}(\omega)|^2 d\omega = 2\pi \int_{-\infty}^{\infty} |f(x)|^2 dx$$

This is also the total power in a signal, which can be computed in either the time or the frequency domain. Also

$$\int_{-\infty}^{\infty} \hat{f}(\omega) \hat{g}*(\omega) d\omega = 2\pi \int_{-\infty}^{\infty} f(x) g*(x) dx$$

Fourier transforms can be used to solve differential equations too. Then it is necessary to find the inverse transformation. If $f(x)$ is square integrable, the Fourier transform of its Fourier transform is $2\pi f(-x)$, or

$$f(x) = F[\hat{f}(\omega)] = \frac{1}{2\pi} \int_{-\infty}^{\infty} \hat{f}(\omega) e^{-i\omega x} d\omega$$

$$= \frac{1}{2\pi} \int_{-\infty}^{\infty} \int_{-\infty}^{\infty} f(x) e^{i\omega x} dx e^{-i\omega x} d\omega$$

$$f(x) = \frac{1}{2\pi} F[Ff(-x)] \text{ or } f(-x) = \frac{1}{2\pi} F[F[f]]$$

Properties of Fourier Transforms [4.1, p. 324], [4.2].

$$F\left[\frac{df}{dx}\right] = -i\omega F[f] = i\omega \hat{f}$$

$$F[ixf(x)] = \frac{d}{d\omega} F[f] = \frac{d}{d\omega} \hat{f}$$

$$F[f(ax-b)] = \frac{1}{|a|} e^{i\omega b/a} \hat{f}\left(\frac{\omega}{a}\right)$$

$$F[e^{icx} f(x)] = \hat{f}(\omega + c)$$

$$F[\cos \omega_0 x f(x)] = \frac{1}{2}[\hat{f}(\omega + \omega_0) + \hat{f}(\omega - \omega_0)]$$

$$F[\sin \omega_0 x f(x)] = \frac{1}{2i}[\hat{f}(\omega + \omega_0) - \hat{f}(\omega - \omega_0)]$$

$$F[e^{-i\omega_0 x} f(x)] = \hat{f}(\omega - \omega_0)$$

If $f(x)$ is real, then $f(-\omega) = \hat{f}*(\omega)$.
If $f(x)$ is imaginary, then $\hat{f}(-\omega) = -\hat{f}*(\omega)$.
If $f(x)$ is even, then $\hat{f}(\omega)$ is even.
If $f(x)$ is odd, then $\hat{f}(\omega)$ is odd.
If $f(x)$ is real and even, then $\hat{f}(\omega)$ is real and even.
If $f(x)$ is real and odd, then $\hat{f}(\omega)$ is imaginary and odd.
If $f(x)$ is imaginary and even, then $\hat{f}(\omega)$ is imaginary and even.
If $f(x)$ is imaginary and odd, then $\hat{f}(\omega)$ is real and odd.

Convolution [4.1, p. 326].

$$f * h(x_0) \equiv \int_{-\infty}^{\infty} f(x_0 - x) h(x) \, dx$$

$$= \frac{1}{2\pi} \int_{-\infty}^{\infty} e^{i\omega x_0} \hat{f}(-\omega) \hat{h}(-\omega) \, d\omega$$

Theorem. The product

$$\hat{f}(\omega) \hat{h}(\omega)$$

is the Fourier transform of the convolution product $f * h$. The convolution permits finding inverse transformations when solving differential equations. To solve

$$\frac{\partial T}{\partial t} = \frac{\partial^2 T}{\partial x^2}$$

$$T(x, 0) = f(x), \quad -\infty < x < \infty$$

T bounded

take the Fourier transform

$$\frac{d\hat{T}}{dt} + \omega^2 \hat{T} = 0$$

$$\hat{T}(\omega, 0) = \hat{f}(\omega)$$

The solution is

$$\hat{T}(\omega, t) = \hat{f}(\omega) e^{-\omega^2 t}$$

The inverse transformation is

$$T(x, t) = \frac{1}{2\pi} \int_{-\infty}^{\infty} e^{-i\omega x} \hat{f}(\omega) e^{-\omega^2 t} \, d\omega$$

Because

$$e^{-\omega^2 t} = F\left[\frac{1}{\sqrt{4\pi t}} e^{-x^2/4t}\right]$$

the convolution integral can be used to write the solution as

$$T(x, t) = \frac{1}{\sqrt{4\pi t}} \int_{-\infty}^{\infty} f(y) e^{-(x-y)^2/4t} \, dy$$

Finite Fourier Sine and Cosine Transform [4.3]. In analogy with finite Fourier transforms (on $-\pi$ to π) and Fourier transforms (on $-\infty$ to $+\infty$), finite Fourier sine and cosine transforms (0 to π) and Fourier sine and cosine transforms (on 0 to $+\infty$) can be defined.

The finite Fourier sine and cosine transforms are

$$f_s(n) = F_s^n[f] = \frac{2}{\pi} \int_0^{\pi} f(x) \sin nx \, dx,$$

$$n = 1, 2, \ldots,$$

$$f_c(n) = F_c^n[f] = \frac{2}{\pi} \int_0^{\pi} f(x) \cos nx \, dx$$

$$n = 0, 1, 2, \ldots$$

$$f(x) = \sum_{n=1}^{\infty} f_s(n) \sin nx,$$

$$f(x) = \tfrac{1}{2} f_c(0) + \sum_{n=1}^{\infty} f_c(n) \cos nx$$

They obey the operational properties

$$F_s^n\left[\frac{d^2 f}{dx^2}\right] = -n^2 F_s^n[f]$$

$$+ \frac{2n}{\pi}[f(0) - (-1)^n f(\pi)]$$

f, f' are continuous, f' is piecewise continuous on $0 \leq x \leq \pi$.

$$f_s(n) \cos nk$$
$$= F_s^n[\tfrac{1}{2} f_1(x-k) + \tfrac{1}{2} f_1(x+k)]$$
$$f_s(n) (-1)^{n+1} = F_s^n[f(\pi - x)]$$

f is piecewise continuous on $0 \leq x \leq \pi$

and f_1 is the extension of f, k is a constant

$$\left.\begin{array}{l} f_1(x) = f(x) \quad 0 < x < \pi \\ f_1(-x) = -f_1(x) \\ f_1(x + 2\pi) = f_1(x) \end{array}\right\} -\infty < x < \infty$$

Also,

$$F_c^n\left[\frac{d^2 f}{dx^2}\right] = -n^2 F_c^n[f] - \frac{2}{\pi}\frac{df}{dx}(0)$$
$$+ (-1)^n \frac{2}{\pi}\frac{df}{dx}(\pi)$$

$$f_c(n) \cos nk = F_c^n[\tfrac{1}{2}f_2(x-k) + \tfrac{1}{2}f_2(x+k)]$$
$$f_c(n)(-1)^n = F_c^n[f(\pi - x)]$$

and f_2 is the extension of f.

$$\left.\begin{aligned}f_2(x) &= f(x) \quad 0 < x < \pi\\ f_2(-x) &= -f_2(x)\\ f_2(x + 2\pi) &= f_2(x)\end{aligned}\right\} -\infty < x < \infty$$

Also,

$$F_s^n\left[\frac{df}{dx}\right] = -n F_c^n[f]$$

$$F_c^n\left[\frac{df}{dx}\right] = n F_s^n[f] - \frac{2}{\pi}f(0) + (-1)^n \frac{2}{\pi}f(\pi)$$

When two functions $F(x)$ and $G(x)$ are defined on the interval $-2\pi < x < 2\pi$, the function

$$F(x) * G(x) = \int_{-\pi}^{\pi} f(x-y)g(y)\,dy$$

is the convolution on $-\pi < x < \pi$. If F and G are both even or both odd, the convolution is even; it is odd if one function is even and the other odd. If F and G are piecewise continuous on $0 \le x \le \pi$, then

$$f_s(n)g_s(n) = F_c^n[-\tfrac{1}{2}F_1(x) * G_1(x)]$$
$$f_s(n)g_c(n) = F_s^n[f][\tfrac{1}{2}F_1(x) * G_2(x)]$$
$$f_c(n)g_c(n) = F_c^n[\tfrac{1}{2}F_2(x) * G_2(x)]$$

where F_1 and G_1 are odd extensions of F and G, respectively, and F_2 and G_2 are even extensions of F and G, respectively. Finite sine and cosine transforms are listed in Tables 3 and 4.

On the semi-infinite domain, $0 < x < \infty$, the Fourier sine and cosine transforms are

$$F_s^\omega[f] \equiv \int_0^\infty f(x)\sin\omega x\,dx,$$

$$F_c^\omega[f] \equiv \int_0^\infty f(x)\cos\omega x\,dx \quad \text{and}$$

$$f(x) = \frac{2}{\pi}F_s^{\omega'}[F_s^\omega[f]], \quad f(x) = \frac{2}{\pi}F_c^{\omega'}[F_c^\omega[f]]$$

Table 3. Finite sine transforms [4.3]

$f_s(n) = \int_0^\pi F(x)\sin nx\,dx \ (n = 1,2,\ldots)$	$F(x)\ (0 < x < \pi)$
$(-1)^{n+1} f_s(n)$	$F(\pi - x)$
$\dfrac{1}{n}$	$\dfrac{\pi - x}{\pi}$
$\dfrac{(-1)^{n+1}}{n}$	$\dfrac{x}{\pi}$
$\dfrac{1-(-1)^n}{n}$	1
$\dfrac{\pi}{n^2}\sin nc\ (0 < c < \pi)$	$\begin{cases}(\pi-c)x & (x \le c)\\ c(\pi-x) & (x \ge c)\end{cases}$
$\dfrac{\pi}{n}\cos nc\ (0 \le c \le \pi)$	$\begin{cases}-x & (x < c)\\ \pi - x & (x > c)\end{cases}$
$\dfrac{\pi^2(-1)^{n-1}}{n} - \dfrac{2[1-(-1)^n]}{n^3}$	x^2
$\pi(-1)^n\left(\dfrac{6}{n^3} - \dfrac{\pi^2}{n}\right)$	x^3
$\dfrac{n}{n^2 + c^2}[1 - (-1)^n e^{c\pi}]$	e^{cx}
$\dfrac{n}{n^2 + c^2}$	$\dfrac{\sinh c(\pi - x)}{\sinh c\pi}$
$\dfrac{n}{n^2 - k^2}\ (\|k\| \ne 0,1,2,\ldots)$	$\dfrac{\sin k(\pi - x)}{\sin k\pi}$
$0(n \ne m);\ f_s(m) = \dfrac{\pi}{2}$	$\sin mx\ (m = 1,2,\ldots)$
$\dfrac{n}{n^2 - k^2}[1 - (-1)^n \cos kx]$	$\cos kx\ (\|k\| \ne 1,2,\ldots)$
$\dfrac{n}{n^2 - m^2}[1 - (-1)^{n+m}],\ (n \ne m)$	$\cos mx\ (m = 1,2,\ldots)$

The material is reproduced with permission of McGraw-Hill, Inc.

The sine transform is an odd function of ω, whereas the cosine function is an even function of ω. Also,

$$F_s^\omega\left[\frac{d^2 f}{dx^2}\right] = f(0)\omega - \omega^2 F_s^\omega[f]$$

$$F_c^\omega\left[\frac{d^2 f}{dx^2}\right] = -\frac{df}{dx}(0) - \omega^2 F_c^\omega[f]$$

provided $f(x)$ and $f'(x) \to 0$ as $x \to \infty$. Thus, the sine transform is useful when $f(0)$ is known and the cosine transform is useful when $f'(0)$ is known.

Table 4. Finite cosine transforms [4.3]

$f_c(n) = \int_0^\pi F(x) \cos nx \, dx \; (n=0,1,\ldots)$	$F(x) \; (0 < x < \pi)$
$(-1)^n f_c(n)$	$F(\pi - x)$
0 when $n = 1,2,\ldots; f_c(0) = \pi$	1
$\dfrac{2}{n} \sin nc; f_c(0) = 2c - \pi$	$\begin{cases} 1 & (0 < x < c) \\ -1 & (c < x < \pi) \end{cases}$
$-\dfrac{1-(-1)^n}{n^2}; f_c(0) = \dfrac{\pi^2}{2}$	x
$\dfrac{(-1)^n}{n^2}; f_c(0) = \dfrac{\pi^2}{6}$	$\dfrac{x^2}{2\pi}$
$\dfrac{1}{n^2}; f_c(0) = 0$	$\dfrac{(\pi - x)^2}{2\pi} - \dfrac{\pi}{6}$
$\dfrac{(-1)^n e^{cx} - 1}{n^2 + c^2}$	$\dfrac{1}{c} e^{cx}$
$\dfrac{1}{n^2 + c^2}$	$\dfrac{\cosh c(\pi - x)}{c \sinh c\pi}$
$\dfrac{(-1)^n \cos k\pi - 1}{n^2 - k^2} \; (\lvert k\rvert \ne 0,1,\ldots)$	$\dfrac{1}{k} \sin kx$
$\dfrac{(-1)^{n+m} - 1}{n^2 - m^2}; f_c(m) = 0 \; (m = 0,1,\ldots)$	$\dfrac{1}{m} \sin mx$
$\dfrac{1}{n^2 - k^2} (\lvert k\rvert \ne 0,1,\ldots)$	$-\dfrac{\cos k(\pi - x)}{k \sin kx}$
$o(n \ne m); f_c(m) = \dfrac{\pi}{2} \; (m = 1,2,\ldots)$	$\cos mx$

The material is reproduced with permission of McGraw-Hill, Inc.

Hsu and Dranoff [4.4] solved a chemical engineering problem by applying finite Fourier transforms and then using the fast Fourier transform (see Chap. 2).

4.2. Laplace Transforms

Consider a function $F(t)$ defined for $t > 0$. The Laplace transform of $F(t)$ is [4.3]

$$L[F] = f(s) = \int_0^\infty e^{-st} F(t) \, dt$$

The Laplace transformation is linear, that is,

$$L[F + G] = L[F] + L[G]$$

Thus, the techniques described herein can be applied only to linear problems. Generally, the assumptions made below are that $F(t)$ is at least piecewise continuous, that it is continuous in each finite interval within $0 < t < \infty$, and that it may take a jump between intervals. It is also of exponential order, meaning $e^{-\alpha t}\lvert F(t)\rvert$ is bounded for all $t > T$, for some finite T.

The unit step function is

$$S_k(t) = \begin{cases} 0 & 0 \le t < k \\ 1 & t > k \end{cases}$$

and its Laplace transform is

$$L[S_k(t)] = \dfrac{e^{-ks}}{s}$$

In particular, if $k = 0$ then

$$L[1] = \dfrac{1}{s}$$

The Laplace transforms of the first and second derivatives of $F(t)$ are

$$L\left[\dfrac{dF}{dt}\right] = s f(s) - F(0)$$

$$L\left[\dfrac{d^2 F}{dt^2}\right] = s^2 f(s) - s F(0) - \dfrac{dF}{dt}(0)$$

More generally,

$$L\left[\dfrac{d^n F}{dt^n}\right] = s^n f(s) - s^{n-1} F(0)$$
$$- s^{n-2} \dfrac{dF}{dt}(0) - \ldots - \dfrac{d^{n-1} F}{dt^{n-1}}(0)$$

The inverse Laplace transformation is

$$F(t) = L^{-1}[f(s)] \text{ where } f(s) = L[F]$$

The inverse Laplace transformation is not unique because functions that are identical except for isolated points have the same Laplace transform. They are unique to within a null function. Thus, if

$$L[F_1] = f(s) \text{ and } L[F_2] = f(s)$$

it must be that

$$F_2 = F_1 + N(t)$$

where $\int_0^T N(t) \, dt = 0 \;$ for every T

Laplace transforms can be inverted by using Table 5, but knowledge of several rules is helpful.

Table 5. Laplace transforms [see 4.5 for a more complete list]

$L[F]$	$F(t)$
$\dfrac{1}{s}$	1
$\dfrac{1}{s^2}$	t
$\dfrac{1}{s^n}$	$\dfrac{t^{n-1}}{(n-1)!}$
$\dfrac{1}{\sqrt{s}}$	$\dfrac{1}{\sqrt{\pi t}}$
$s^{-3/2}$	$2\sqrt{t/\pi}$
$\dfrac{\Gamma(k)}{s^k}\ (k>0)$	t^{k-1}
$\dfrac{1}{s-a}$	e^{at}
$\dfrac{1}{(s-a)^n}\ (n=1,2,\ldots)$	$\dfrac{1}{(n-1)!}t^{n-1}e^{at}$
$\dfrac{\Gamma(k)}{(s-a)^k}\ (k>0)$	$t^{k-1}e^{at}$
$\dfrac{1}{(s-a)(s-b)}$	$\dfrac{1}{a-b}(e^{at}-e^{bt})$
$\dfrac{s}{(s-a)(s-b)}$	$\dfrac{1}{a-b}(ae^{at}-be^{bt})$
$\dfrac{1}{s^2+a^2}$	$\dfrac{1}{a}\sin at$
$\dfrac{s}{s^2+a^2}$	$\cos at$
$\dfrac{1}{s^2-a^2}$	$\dfrac{1}{a}\sinh at$
$\dfrac{s}{s^2-a^2}$	$\cosh at$
$\dfrac{s}{(s^2+a^2)^2}$	$\dfrac{t}{2a}\sin at$
$\dfrac{s^2-a^2}{(s^2+a^2)^2}$	$t\cos at$
$\dfrac{1}{(s-a)^2+b^2}$	$\dfrac{1}{b}e^{at}\sin bt$
$\dfrac{s-a}{(s-a)^2+b^2}$	$e^{at}\cos bt$

Substitution.

$$f(s-a) = L[e^{at}F(t)]$$

This can be used with polynomials. Suppose

$$f(s) = \frac{1}{s} + \frac{1}{s+3} = \frac{2s+3}{s(s+3)}$$

Because

$$L[1] = \frac{1}{s}$$

Then

$$F(t) = 1 + e^{-3t}, \quad t \geq 0$$

More generally, translation gives the following.

Translation.

$$f(as-b) = f\left[a\left(s-\frac{b}{a}\right)\right] = L\left[\frac{1}{a}e^{bt/a}F\left(\frac{t}{a}\right)\right], \quad a > 0$$

The step function

$$S(t) = \begin{cases} 0 & 0 \leq t < \dfrac{1}{h} \\ 1 & \dfrac{1}{h} \leq t < \dfrac{2}{h} \\ 2 & \dfrac{2}{h} \leq t < \dfrac{3}{h} \end{cases}$$

has the Laplace transform

$$L[S(t)] = \frac{1}{s}\frac{1}{1-e^{-hs}}$$

The Dirac delta function $\delta(t-t_0)$ ($\to$ 2. Mathematical Modeling, Eq. 116) has the property

$$\int_0^\infty \delta(t-t_0)F(t)\,dt = F(t_0)$$

Its Laplace transform is

$$L[\delta(t-t_0)] = e^{-st_0}, \quad t_0 \geq 0, \quad s > 0$$

The square wave function illustrated in Figure 14 has Laplace transform

$$L[F_c(t)] = \frac{1}{s}\tanh\frac{cs}{2}$$

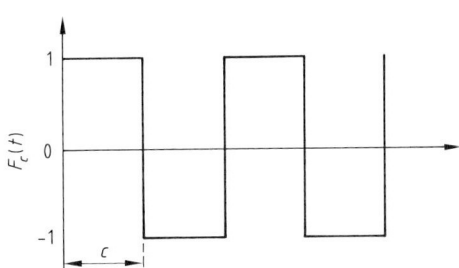

Figure 14. Square wave function

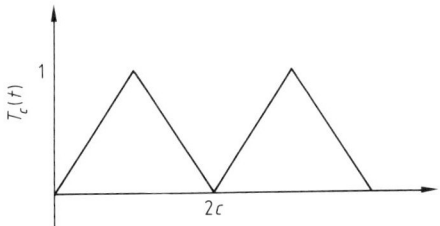

Figure 15. Triangular wave function

The triangular wave function illustrated in Figure 15 has Laplace transform

$$L[T_c(t)] = \frac{1}{s^2}\tanh\frac{cs}{2}$$

Other Laplace transforms are listed in Table 5.

Convolution properties are also satisfied:

$$F(t) * G(t) = \int_0^t F(\tau)G(t-\tau)d\tau$$

and

$$f(s)g(s) = L[F(t)*G(t)]$$

Derivatives of Laplace Transforms. The Laplace integrals $L[F(t)], L[tF(t)], L[t^2 F(t)], \ldots$ are uniformly convergent for $s \geq \alpha_1 > \alpha$ and

$$\lim_{s \to \infty} f(s) = 0, \quad \lim_{s \to \infty} L[t^n F(t)] = 0, \quad n = 1, 2, \ldots$$

and

$$\frac{d^n f}{ds^n} = L[(-t)^n F(t)]$$

Integration of Laplace Transforms.

$$\int_s^\infty f(\xi)d\xi = L\left[\frac{F(t)}{t}\right]$$

If $F(t)$ is a periodic function, $F(t) = F(t+a)$, then

$$f(s) = \frac{1}{1-e^{-as}}\int_0^a e^{-st}F(t)dt,$$

where $F(t) = F(t+a)$

Partial Fractions [4.6, p. 229]. Suppose $q(s)$ has m factors

$$q(s) = (s-a_1)(s-a_2)\ldots(s-a_m)$$

All the factors are linear, none are repeated, and the a_n are all distinct. If $p(s)$ has a smaller degree than $q(s)$, the Heaviside expansion can be used to evaluate the inverse transformation:

$$L^{-1}\left[\frac{p(s)}{q(s)}\right] = \sum_{i=1}^m \frac{p(a_i)}{q'(a_i)}e^{a_i t}$$

If the factor $(s-a)$ is repeated $n+1$ times, then

$$f(s) = \frac{p(s)}{q(s)} = \frac{A_m}{(s-a)^m} + \frac{A_{m-1}}{(s-a)^{m-1}} + \ldots + \frac{A_1}{s-a} + h(s)$$

where

$$\varphi(s) \equiv \frac{(s-a)^m p(s)}{q(s)}$$

$$A_m = \varphi(a), \quad A_k = \frac{1}{(m-k)!}\frac{d^{m-k}\varphi(s)}{ds^{m-k}}\bigg|_a,$$

$$k = 1, \ldots, m-1$$

The term $h(s)$ denotes the sum of partial fractions not under consideration. The inverse transformation is then

$$F(t) = e^{at}\left(A_m \frac{t^{m-1}}{(m-1)!} + A_{m-1}\frac{t^{m-2}}{(m-2)!} + \ldots + A_2 \frac{t}{1!} + A_1\right) + H(t)$$

The term in $F(t)$ corresponding to

$s-a$ in $q(s)$ is $\varphi(a)e^{at}$
$(s-a)^2$ in $q(s)$ is $[\varphi'(a) + \varphi(a)t]e^{at}$
$(s-a)^3$ in $q(s)$ is $\frac{1}{2}[\varphi''(a) + 2\varphi'(a)t + \varphi(a)t^2]e^{at}$

For example, let

$$f(s) = \frac{1}{(s-2)(s-1)^2}$$

For the factor $s - 2$,

$$\varphi(s) = \frac{1}{(s-1)^2}, \quad \varphi(2) = 1$$

For the factor $(s - 1)^2$,

$$\varphi(s) = \frac{1}{s-2}, \quad \varphi'(s) = -\frac{1}{(s-2)^2}$$

$$\varphi(1) = -1, \quad \varphi'(1) = -1$$

The inverse Laplace transform is then

$$F(t) = e^{2t} + [-1 - t]e^t$$

Quadratic Factors. Let $p(s)$ and $q(s)$ have real coefficients, and $q(s)$ have the factor

$$(s - a)^2 + b^2, \quad b > 0$$

where a and b are real numbers. Then define $\varphi(s)$ and $h(s)$ and real constants A and B such that

$$f(s) = \frac{p(s)}{q(s)} = \frac{\varphi(s)}{(s-a)^2 + b^2}$$

$$= \frac{As + B}{(s-a)^2 + b^2} + h(s)$$

Let φ_1 and φ_2 be the real and imaginary parts of the complex number $\varphi(a + ib)$.

$$\varphi(a + ib) \equiv \varphi_1 + i\varphi_2$$

Then

$$f(s) = \frac{1}{b} \frac{(s-a)\varphi_2 + b\varphi_1}{(s-a)^2 + b^2} + h(s)$$

$$F(t) = \frac{1}{b} e^{at}(\varphi_2 \cos bt + \varphi_1 \sin bt) + H(t)$$

To solve ordinary differential equations by using these results:

$$Y''(t) - 2Y'(t) + Y(t) = e^{2t}$$
$$Y(0) = 0, \; Y'(0) = 0$$

Taking Laplace transforms

$$L[Y''(t)] - 2L[Y'(t)] + L[Y(t)] = \frac{1}{s-2}$$

using the rules

$$s^2 y(s) - sY(0) - Y'(0) - 2[sy(s) - Y(0)] + y(s) = \frac{1}{s-2}$$

and combining terms

$$(s^2 - 2s + 1)y(s) = \frac{1}{s-2}$$

$$y(s) = \frac{1}{(s-2)(s-1)^2}$$

lead to

$$Y(t) = e^{2t} - (1 + t)e^t$$

To solve an integral equation:

$$Y(t) = a + 2\int_0^t Y(\tau)\cos(t - \tau)\,d\tau$$

it is written as

$$Y(t) = a + Y(t) * \cos t$$

Then the Laplace transform is used to obtain

$$y(s) = \frac{a}{s} + 2y(s)\frac{s}{s^2 + 1}$$

or $y(s) = \dfrac{a(s^2 + 1)}{s(s-1)^2}$

Taking the inverse transformation gives

$$Y(t) = s[1 + 2te^t]$$

Next, let the variable s in the Laplace transform be complex. $F(t)$ is still a real-valued function of the positive real variable t. The properties given above are still valid for s complex, but additional techniques are available for evaluating the integrals. The real-valued function is $O[\exp(x_0 t)]$:

$$|F(t)| < Me^{x_0 t}, \quad z_0 = x_0 + iy_0$$

The Laplace transform

$$f(s) = \int_0^\infty e^{-st} F(t)\,dt$$

is an analytic function of s in the half-plane $x > x_0$ and is absolutely convergent there; it is uniformly convergent on $x \geq x_1 > x_0$.

$$\frac{d^n f}{ds^n} = L[(-t)^n F(t)] \quad n = 1, 2, \ldots, \; x > x_0$$

and $f^*(s) = f(s^*)$

The functions $|f(s)|$ and $|xf(s)|$ are bounded in the half-plane $x \geq x_1 > x_0$ and $f(s) \to 0$ as $|y| \to \infty$ for each fixed x. Thus,

$$|f(x + iy)| < M, \quad |xf(x + iy)| < M,$$
$$x \geq x_1 > x_0$$

$$\lim_{y \to \pm\infty} f(x + iy) = 0, \quad x > x_0$$

If $F(t)$ is continuous, $F'(t)$ is piecewise continuous, and both functions are $O[\exp(x_0 t)]$, then $|f(s)|$ is $O(1/s)$ in each half-plane $x \geq x_1 > x_0$.

$$|sf(s)| < M$$

If $F(t)$ and $F'(t)$ are continuous, $F''(t)$ is piecewise continuous, and all three functions are $O[\exp(x_0 t)]$, then

$$|s^2 f(s) - s F(0)| < M, \quad x \geq x_1 > x_0$$

The additional constraint $F(0) = 0$ is necessary and sufficient for $|f(s)|$ to be $O(1/s^2)$.

Inversion Integral [4.3]. Cauchy's integral formula for $f(s)$ analytic and $O(s^{-k})$ in a half-plane $x \geq y, k > 0$, is

$$f(s) = \frac{1}{2\pi i} \lim_{\beta \to \infty} \int_{\gamma - i\beta}^{\gamma + i\beta} \frac{f(z)}{s - z} dz, \quad \mathrm{Re}(s) > \gamma$$

Applying the inverse Laplace transformation on either side of this equation gives

$$F(t) = \frac{1}{2\pi i} \lim_{\beta \to \infty} \int_{\gamma - i\beta}^{\gamma + i\beta} e^{zt} f(z) dz$$

If $F(t)$ is of order $O[\exp(x_0 t)]$ and $F(t)$ and $F'(t)$ are piecewise continuous, the inversion integral exists. At any point t_0, where $F(t)$ is discontinuous, the inversion integral represents the mean value

$$F(t_0) = \lim_{\varepsilon \to \infty} \tfrac{1}{2}[F(t_0 + \varepsilon) + F(t_0 - \varepsilon)]$$

When $t = 0$ the inversion integral represents $0.5 F(O +)$ and when $t < 0$, it has the value zero.

If $f(s)$ is a function of the complex variable s that is analytic and of order $O(s^{-k-m})$ on $R(s) \geq x_0$, where $k > 1$ and m is a positive integer, then the inversion integral converges to $F(t)$ and

$$\frac{d^n F}{dt^n} = \frac{1}{2\pi i} \lim_{\beta \to \infty} \int_{\gamma - i\beta}^{\gamma + i\beta} e^{zt} z^n f(z) dz,$$

$$n = 1, 2, \ldots, m$$

Also $F(t)$ and its n derivatives are continuous functions of t of order $O[\exp(x_0 t)]$ and they vanish at $t = 0$.

$$F(0) = F'(0) = \ldots F^{(m)}(0) = 0$$

Series of Residues [4.3]. Let $f(s)$ be an analytic function except for a set of isolated singular points. An isolated singular point is one for which $f(z)$ is analytic for $0 < |z - z_0| < \varrho$ but z_0 is a singularity of $f(z)$. An isolated singular point is either a pole, a removable singularity, or an essential singularity. If $f(z)$ is not defined in the neighborhood of z_0 but can be made analytic at z_0 simply by defining it at some additional points, then z_0 is a *removable singularity*. The function $f(z)$ has a *pole* of order $k \geq 1$ at z_0 if $(z - z_0)^k f(z)$ has a removable singularity at z_0 whereas $(z - z_0)^{k-1} f(z)$ has an unremovable isolated singularity at z_0. Any isolated singularity that is not a pole or a removable singularity is an *essential* singularity.

Let the function $f(z)$ be analytic except for the isolated singular point $s_1, s_2, \ldots, s_n$. Let $\varrho_n(t)$ be the residue of $e^{zt} f(z)$ at $z = s_n$ (for definition of residue, see Section 3.4). Then

$$F(t) = \sum_{n=1}^{\infty} \varrho_n(t)$$

When s_n is a simple pole

$$\varrho_n(t) = \lim_{z \to s_n} (z - s_n) e^{zt} f(z)$$

$$= e^{s_n t} \lim_{z \to s_n} (z - s_n) f(z)$$

When

$$f(z) = \frac{p(z)}{q(z)}$$

where $p(z)$ and $q(z)$ are analytic at $z = s_n$, $p(s_n) \neq 0$, then

$$\varrho_n(t) = \frac{p(s_n)}{q'(s_n)} e^{s_n t}$$

If s_n is a removable pole of $f(s)$, of order m, then

$$\varphi_n(z) = (z - s_n)^m f(z)$$

is analytic at s_n and the residue is

$$\varrho_n(t) = \frac{\Phi_n(s_n)}{(m-1)!} \quad \text{where} \quad \Phi_n(z) = \frac{\partial^{m-1}}{\partial z^{m-1}}[\varphi_n(z) e^{zt}]$$

An important inversion integral is when

$$f(s) = \frac{1}{s} \exp(-s^{1/2})$$

The inverse transform is

$$F(t) = 1 - \mathrm{erf}\left(\frac{1}{2\sqrt{t}}\right) = \mathrm{erfc}\left(\frac{1}{2\sqrt{t}}\right)$$

where erf is the error function and erfc the complementary error function.

4.3. Solution of Partial Differential Equations by Using Transforms

A common problem facing chemical engineers is to solve the heat conduction equation or diffusion equation

$$\varrho C_p \frac{\partial T}{\partial t} = k \frac{\partial^2 T}{\partial x^2} \quad \text{or} \quad \frac{\partial c}{\partial t} = D \frac{\partial^2 c}{\partial x^2}$$

The equations can be solved on an infinite domain $-\infty < x < \infty$, a semi-infinite domain $0 \leq x < \infty$, or a finite domain $0 \leq x \leq L$. At a boundary, the conditions can be a fixed temperature $T(0, t) = T_0$ (boundary condition of the first kind, or Dirichlet condition), or a fixed flux

$$-k \frac{\partial T}{\partial x}(0, t) = q_0 \quad \text{(boundary condition of the}$$

second kind, or Neumann condition), or a combination

$$-k \frac{\partial T}{\partial x}(0, t) = h[T(0, t) - T_0]$$

(boundary condition of the third kind, or Robin condition).

The functions T_0 and q_0 can be functions of time. All properties are constant (ϱ, C_p, k, D, h), so that the problem remains linear. Solutions are presented on all domains with various boundary conditions for the heat conduction problem.

$$\frac{\partial T}{\partial t} = \alpha \frac{\partial^2 T}{\partial x^2}, \quad \alpha = \frac{k}{\varrho C_p}$$

Problem 1. Infinite domain, on $-\infty < x < \infty$.

$T(x, 0) = f(x)$, initial conditions
$T(x, t)$ bounded

Solution is via Fourier transforms

$$\hat{T}(\omega, t) = \int_{-\infty}^{\infty} T(x, t) e^{i \omega x} dx$$

Applied to the differential equation

$$F\left[\frac{\partial^2 T}{\partial x^2}\right] = -\omega^2 \alpha F[T]$$

$$\frac{\partial \hat{T}}{\partial t} + \omega^2 \alpha \hat{T} = 0, \quad \hat{T}(\omega, 0) = \hat{f}(\omega)$$

By solving

$$\hat{T}(\omega, t) = \hat{f}(\omega) e^{-\omega^2 \alpha t}$$

the inverse transformation gives [4.1, p. 328], [4.7, p. 58]

$$T(x, t) = \frac{1}{2\pi} \lim_{L \to \infty} \int_{-L}^{L} e^{-i\omega x} \hat{f}(\omega) e^{-\omega^2 \alpha t} d\omega$$

Another solution is via Laplace transforms; take the Laplace transform of the original differential equation.

$$s t(s, x) - f(x) = \alpha \frac{\partial^2 t}{\partial x^2}$$

This equation can be solved with Fourier transforms [4.1, p. 355]

$$t(s, x) = \frac{1}{2\sqrt{s\alpha}} \int_{-\infty}^{\infty} e\left\{-\sqrt{\frac{s}{\alpha}}|x - y|\right\} f(y) dy$$

The inverse transformation is [4.1, p. 357], [4.7, p. 53]

$$T(x, t) = \frac{1}{2\sqrt{\pi \alpha t}} \int_{-\infty}^{\infty} e^{-(x-y)^2/4\alpha t} f(y) dy$$

Problem 2. Semi-infinite domain, boundary condition of the first kind, on $0 \leq x \leq \infty$

$T(x, 0) = T_0 = \text{constant}$
$T(0, t) = T_1 = \text{constant}$

The solution is

$T(x, t) = T_0 + [T_1 - T_0][1 - \text{erf}(x/\sqrt{4\alpha t})]$
or $T(x, t) = T_0 + (T_1 - T_0)\text{erfc}(x/\sqrt{4\alpha t})$

Problem 3. Semi-infinite domain, boundary condition of the first kind, on $0 \leq x < \infty$

$T(x, 0) = f(x)$
$T(0, t) = g(t)$

The solution is written as

$$T(x, t) = T_1(x, t) + T_2(x, t)$$

where

$T_1(x, 0) = f(x)$, $T_2(x, 0) = 0$
$T_1(0, t) = 0$, $T_2(0, t) = g(t)$

Then T_1 is solved by taking the sine transform

$$U_1 = F_s^\omega[T_1]$$

$$\frac{\partial U_1}{\partial t} = -\omega^2 \alpha U_1$$

$$U_1(\omega, 0) = F_s^\omega[f]$$

Thus,

$$U_1(\omega, t) = F_s^\omega[f] e^{-\omega^2 \alpha t}$$

and [4.1, p. 322]

$$T_1(x, t) = \frac{2}{\pi} \int_0^\infty F_s^\omega[f] e^{-\omega^2 \alpha t} \sin \omega x \, d\omega$$

Solve for T_2 by taking the sine transform

$$U_2 = F_s^\omega[T_2]$$

$$\frac{\partial U_2}{\partial t} = -\omega^2 \alpha U_2 + \alpha g(t)\omega$$

$$U_2(\omega, 0) = 0$$

Thus,

$$U_2(\omega, t) = \int_0^t e^{-\omega^2\alpha(t-\tau)} \sqrt{\alpha} g(\tau) d\tau$$

and [4.1, p. 435]

$$T_2(x, t) = \frac{2\alpha}{\pi} \int_0^\infty \omega \sin \omega x \int_0^t e^{-\omega^2\alpha(t-\tau)} g(\tau) d\tau d\omega$$

The solution for T_1 can also be obtained by Laplace transforms.

$$t_1 = L[T_1]$$

Applying this to the differential equation

$$st_1 - f(x) = \alpha \frac{\partial^2 t_1}{\partial x^2}, \quad t_1(0, s) = 0$$

and solving gives

$$t_1 = \frac{1}{\sqrt{s\alpha}} \int_0^x e^{-\sqrt{s/\alpha}(x'-x)} f(x') dx'$$

and the inverse transformation is [4.1, p. 437], [4.7, p. 59]

$$T_1(x, t) = \frac{1}{\sqrt{4\pi\alpha t}} \int_0^\infty [e^{-(x-\xi)^2/4\alpha t} - e^{-(x+\xi)^2/4\alpha t}] f(\xi) d\xi$$

Problem 4. Semi-infinite domain, boundary conditions of the second kind, on $0 \leq x < \infty$.

$$T(x, 0) = 0$$

$$-k\frac{\partial T}{\partial x}(0, t) = q_0 = \text{constant}$$

Take the Laplace transform

$$t(x, s) = L[T(x, t)]$$

$$st = \alpha \frac{\partial^2 t}{\partial x^2}$$

$$-k\frac{\partial t}{\partial x} = \frac{q_0}{s}$$

The solution is

$$t(x, s) = \frac{q_0 \sqrt{\alpha}}{k s^{3/2}} e^{-x\sqrt{s/\alpha}}$$

The inverse transformation is [4.3, p. 131], [4.7, p. 75]

$$T(x, t) = \frac{q_0}{k}\left[2\sqrt{\frac{\alpha t}{\pi}} e^{-x^2/4\alpha t} - x \operatorname{erfc}\left(\frac{x}{\sqrt{4\alpha t}}\right)\right]$$

Problem 5. Semi-infinite domain, boundary conditions of the third kind, on $0 \leq x < \infty$

$$T(x, 0) = f(x)$$

$$k\frac{\partial T}{\partial x}(0, t) = hT(0, t)$$

Take the Laplace transform

$$st - f(x) = \alpha \frac{\partial^2 t}{\partial x^2}$$

$$k\frac{\partial t}{\partial x}(0, s) = ht(0, s)$$

The solution is

$$t(x, s) = \int_0^\infty f(\xi) g(x, \xi, s) d\xi$$

where [4.3, p. 227]

$$2\sqrt{s}g(x, \xi, s) = \exp(-|x - \xi|\sqrt{s/\alpha}) + \frac{\sqrt{s} - h\alpha/k}{\sqrt{s} + h\alpha/k}$$
$$\exp[-(x + \xi)\sqrt{s/\alpha}]$$

One form of the inverse transformation is [4.2, p. 228]

$$T(x, t) = \frac{2}{\pi} \int_0^\infty e^{-\beta^2 t} \cos[\beta x - \mu(\beta)] \int_0^\infty f(\xi)$$
$$\cos[\beta\xi - \mu(\beta)] d\xi d\beta$$

$$\mu(\beta) = \arg(\beta + ih\sqrt{\alpha/k})$$

Another form of the inverse transformation when $f = T_0 = $ constant is [4.3, p. 231], [4.7, p. 71]

$$T(x, t) = T_0\left[\operatorname{erf}\left(\frac{x}{\sqrt{4\alpha t}}\right) + e^{hx/k} e^{h^2\alpha t/k^2}\right]$$

$$\operatorname{erfc}\left(\frac{h\sqrt{\alpha t}}{k} + \frac{x}{\sqrt{4\alpha t}}\right)$$

Problem 6. Finite domain, boundary condition of the first kind

$$T(x, 0) = T_0 = \text{constant}$$
$$T(0, t) = T(L, t) = 0$$

Take the Laplace transform

$$st(x, s) - T_0 = \alpha \frac{\partial^2 t}{\partial x^2}$$

$$t(0, s) = t(L, s) = 0$$

The solution is

$$t(x, s) = -\frac{T_0}{s} \frac{\sinh\sqrt{\frac{s}{\alpha}}x}{\sinh\sqrt{\frac{s}{\alpha}}L} - \frac{T_0}{s}\frac{\sinh\sqrt{\frac{s}{\alpha}}(L-x)}{\sinh\sqrt{\frac{s}{\alpha}}L} + \frac{T_0}{s}$$

The inverse transformation is [4.3, p. 220], [4.7, p. 96]

$$T(x, t) = \frac{2}{\pi} T_0 \sum_{n=1,3,5,\ldots} \frac{2}{n} e^{-n^2\pi^2\alpha t/L^2} \sin\frac{n\pi x}{L}$$

or (depending on the inversion technique) [4.1, pp. 362, 438]

$$T(x, t) = \frac{T_0}{\sqrt{4\pi\alpha t}} \int_0^L \sum_{n=-\infty}^{\infty} [e^{-[(x-\xi)+2nL]^2/4\alpha t}$$
$$- e^{-[(x+\xi)+2nL]^2/4\alpha t}] d\xi$$

Problem 7. Finite domain, boundary condition of the first kind

$T(x, 0) = 0$
$T(0, t) = 0$
$T(L, 0) = T_0 = \text{constant}$

Take the Laplace transform

$$st(x, s) = \alpha \frac{\partial^2 t}{\partial x^2}$$

$$t(0, s) = 0, \quad t(L, t) = \frac{T_0}{s}$$

The solution is

$$t(x, s) = \frac{T_0}{s} \frac{\sinh \frac{x}{L}\sqrt{s}}{\sinh \sqrt{s}}$$

and the inverse transformation is [4.3, p. 201], [4.7, p. 313]

$$T(x, t) = T_0 \left[\frac{x}{L} + \frac{2}{\pi} \sum_{n=1}^{\infty} \frac{(-1)^n}{n} e^{-n^2 \pi^2 \alpha t / L^2} \sin \frac{n \pi x}{L} \right]$$

An alternate transformation is [4.3, p. 139], [4.7, p. 310]

$$T(x, t) = T_0 \sum_{n=0}^{\infty} \left[\operatorname{erf}\left(\frac{(2n+1)L + x}{\sqrt{4 \alpha t}}\right) \right.$$
$$\left. - \operatorname{erf}\left(\frac{(2n+1)L - x}{\sqrt{4 \alpha t}}\right) \right]$$

Problem 8. Finite domain, boundary condition of the second kind

$T(x, 0) = T_0$
$\frac{\partial T}{\partial x}(0, t) = 0, \quad T(L, t) = 0$

Take the Laplace transform

$$st(x, s) - T_0 = \alpha \frac{\partial^2 t}{\partial x^2}$$

$$\frac{\partial t}{\partial x}(0, s) = 0, \quad t(L, s) = 0$$

The solution is

$$t(x, s) = \frac{T_0}{s}\left[1 - \frac{\cosh x \sqrt{s/\alpha}}{\cosh L \sqrt{s/\alpha}} \right]$$

Its inverse is [4.3, p. 138]

$$T(x, t) = T_0 - T_0 \sum_{n=0}^{\infty} (-1)^n \left[\operatorname{erfc}\left(\frac{(2n+1)L - x}{\sqrt{4 \alpha t}}\right) \right.$$
$$\left. + \operatorname{erfc}\left(\frac{(2n+1)L + x}{\sqrt{4 \alpha t}}\right) \right]$$

5. Vector Analysis

Notation. A *scalar* is a quantity having magnitude but no direction (e.g., mass, length, time, temperature, and concentration). A *vector* is a quantity having both magnitude and direction (e.g., displacement, velocity, acceleration, force). A second-order *dyadic* has magnitude and two directions associated with it, as defined precisely below. The most common examples are the stress dyadic (or tensor) and the velocity gradient (in fluid flow). Vectors are printed in boldface type and identified in this chapter by lower-case Latin letters. Second-order dyadics are printed in bold-face type and are identified in this chapter by capital or Greek letters. Higher order dyadics are not discussed here. Dyadics can be formed from the components of tensors including their directions, and some of the identities for dyadics are more easily proved by using tensor analysis, which is not presented here (see also, → 4. Transport Phenomena, pp. 4-9–4-10). Vectors are also first-order dyadics.

Vectors. Two vectors **u** and **v** are equal if they have the same magnitude and direction. If they have the same magnitude but the opposite direction, then **u** = −**v**. The sum of two vectors is identified geometrically by placing the vector **v** at the end of the vector **u**, as shown in Figure 16. The product of a scalar m and a vector **u** is a vector $m\mathbf{u}$ in the same direction as **u** but with a magnitude that equals the magnitude of **u** times the scalar m. Vectors obey commutative and associative laws.

$\mathbf{u} + \mathbf{v} = \mathbf{v} + \mathbf{u}$	Commutative law for addition
$\mathbf{u} + (\mathbf{v} + \mathbf{w}) = (\mathbf{u} + \mathbf{v}) + \mathbf{w}$	Associative law for addition
$m\mathbf{u} = \mathbf{u}m$	Commutative law for scalar multiplication
$m(n\mathbf{u}) = (mn)\mathbf{u}$	Associative law for scalar multiplication
$(m + n)\mathbf{u} = m\mathbf{u} + n\mathbf{u}$	Distributive law
$m(\mathbf{u} + \mathbf{v}) = m\mathbf{u} + m\mathbf{v}$	Distributive law

The same laws are obeyed by dyadics, as well.

A unit vector is a vector with magnitude 1.0 and some direction. If a vector has some magni-

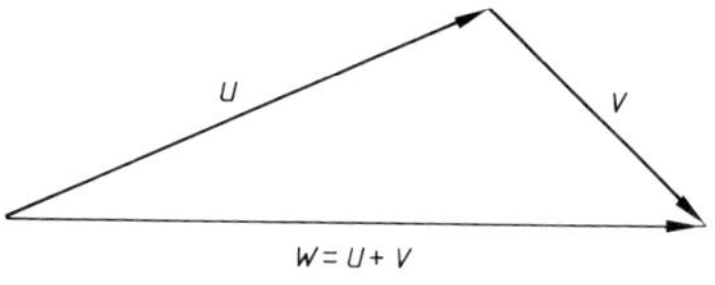

Figure 16. Addition of vectors

tude (i.e., not zero magnitude), a unit vector e_u can be formed by

$$e_u = \frac{u}{|u|}$$

The original vector can be represented by the product of the magnitude and the unit vector.

$$u = |u| e_u$$

In a cartesian coordinate system the three principle, orthogonal directions are customarily represented by *unit vectors*, such as $\{e_x, e_y, e_z\}$ or $\{i, j, k\}$. Here, the first notation is used (see Fig. 17). The coordinate system is right handed; that is, if a right-threaded screw rotated from the x to the y direction, it would advance in the z direction. A vector can be represented in terms of its components in these directions, as illustrated in Figure 18. The vector is then written as

$$u = u_x e_x + u_y e_y + u_z e_z$$

The magnitude is

$$u = |u| = \sqrt{u_x^2 + u_y^2 + u_z^2}$$

The position vector is

$$r = x e_x + y e_y + z e_z$$

with magnitude

$$r = |r| = \sqrt{x^2 + y^2 + z^2}$$

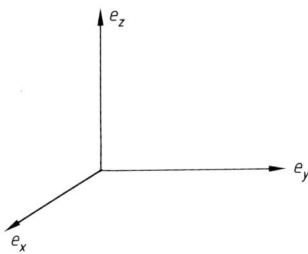

Figure 17. Cartesian coordinate system

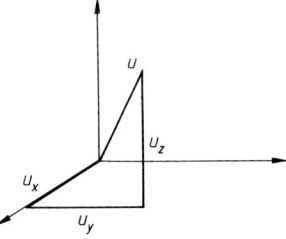

Figure 18. Vector components

Dyadics. The dyadic A is written in component form in cartesian coordinates as

$$\begin{aligned} A = &A_{xx} e_x e_x + A_{xy} e_x e_y + A_{xz} e_x e_z \\ &+ A_{yx} e_y e_x + A_{yy} e_y e_y + A_{yz} e_y e_z \\ &+ A_{zx} e_z e_x + A_{zy} e_z e_y + A_{zz} e_z e_z \end{aligned}$$

Quantities such as $e_x e_y$ are called *unit dyadics*. They are second-order dyadics and have two directions associated with them, e_x and e_y; the order of the pair is important. The components $A_{xx}, \ldots, A_{zz}$ are the components of the tensor A_{ij} which here is a 3×3 matrix of numbers that are transformed in a certain way when variables undergo a linear transformation. The yx momentum flux can be defined as the flux of x momentum across an area with unit normal in the y direction ($\to$ 5. Fluid Mechanics, p. **5**-7; $\to$ 4. Transport Phenomena, p. **4**-29). Since two directions are involved, a second-order dyadic (or tensor) is needed to represent it, and because the y momentum across an area with unit normal in the x direction may not be the same thing, the order of the indices must be kept straight. The dyadic A is said to be symmetric if

$$A_{ij} = A_{ji}$$

Here, the indices i and j can take the values x, y, or z; sometimes $(x, 1)$, $(y, 2)$, $(x, 3)$ are identified and the indices take the values 1, 2, or 3. The dyadic A is said to be antisymmetric if

$$A_{ij} = -A_{ji}$$

The transpose of A is

$$A_{ij}^T = A_{ji}$$

Any dyadic can be represented as the sum of a symmetric portion and an antisymmetric portion.

$$A_{ij} = B_{ij} + C_{ij}, \; B_{ij} \equiv \tfrac{1}{2}(A_{ij} + A_{ji}),$$
$$C_{ij} \equiv \tfrac{1}{2}(A_{ij} - A_{ji})$$

An ordered pair of vectors is a second-order dyadic.

$$u v = \sum_i \sum_j e_i e_j u_i v_j$$

The transpose of this is

$$(u v)^T = v u$$

but

$$uv \neq vu$$

The Kronecker delta is defined as

$$\delta_{ij} = \begin{cases} 1 \text{ if } i = j \\ 0 \text{ if } i \neq j \end{cases}$$

and the unit dyadic is defined as

$$\delta = \sum_i \sum_j e_i e_j \delta_{ij}$$

Operations. The *dot* or *scalar product* of two vectors is defined as

$$u \cdot v = |u||v| \cos \theta, \quad 0 \leq \theta \leq \pi$$

where θ is the angle between u and v. The scalar product of two vectors is a scalar, not a vector. It is the magnitude of u multiplied by the projection of v on u, or vice versa. The scalar product of u with itself is just the square of the magnitude of u.

$$u \cdot u = |u^2| = u^2$$

The following laws are valid for scalar products

$$u \cdot v = v \cdot u \quad \text{Commutative law for scalar products}$$
$$u \cdot (v + w) = u \cdot v + u \cdot w \quad \text{Distributive law for scalar products}$$

$$e_x \cdot e_x = e_y \cdot e_y = e_z \cdot e_z = 1$$
$$e_x \cdot e_y = e_x \cdot e_z = e_y \cdot e_z = 0$$

If the two vectors u and v are written in component notation, the scalar product is

$$u \cdot v = u_x v_x + u_y v_y + u_z v_z$$

If $u \cdot v = 0$ and u and v are not null vectors, then u and v are perpendicular to each other and $\theta = \pi/2$.

The *single dot product* of two dyadics is

$$A \cdot B = \sum_i \sum_j e_i e_j \left(\sum_k A_{ik} B_{kj} \right)$$

The *double dot product* of two dyadics is

$$A : B = \sum_i \sum_j A_{ij} B_{ji}$$

Because the dyadics may not be symmetric, the order of indices and which indices are summed are important. The order is made clearer when the dyadics are made from vectors.

$$(uv) \cdot (wx) = u(v \cdot w)x = ux(v \cdot w)$$
$$(uv) : (wx) = (u \cdot x)(v \cdot w)$$

The dot product of a dyadic and a vector is

$$A \cdot u = \sum_i e_i \left(\sum_j A_{ij} u_j \right)$$

The *cross* or *vector product* is defined by

$$c = u \times v = a|u||v| \sin \theta, \quad 0 \leq \theta \leq \pi$$

where a is a unit vector in the direction of $u \times v$. The direction of c is perpendicular to the plane of u and v such that u, v, and c form a right-handed system. If $u = v$, or u is parallel to v, then $\theta = 0$ and $u \times v = 0$. The following laws are valid for cross products.

$$u \times v = -v \times u \quad \text{Commutative law fails for vector product}$$

$$u \times (v \times w) \neq (u \times v) \times w \quad \text{Associative law fails for vector product}$$

$$u \times (v + w) = u \times v + u \times w \quad \text{Distributive law for vector product}$$

$$e_x \times e_x = e_y \times e_y = e_z \times e_z = 0$$
$$e_x \times e_y = e_z, \; e_y \times e_z = e_x, \; e_z \times e_x = e_y$$

$$u \times v = \det \begin{bmatrix} e_x & e_y & e_z \\ u_x & u_y & u_z \\ v_x & v_y & v_z \end{bmatrix} = e_x(u_y v_z - v_y u_z) \\ + e_y(u_z v_x - u_x v_z) \\ + e_z(u_x v_y - u_y v_x)$$

This can also be written as

$$u \times v = \sum_i \sum_j \varepsilon_{kij} u_i v_j e_k$$

where

$$\varepsilon_{ijk} = \begin{cases} 1 \text{ if } i, j, k \text{ is an even permutation of 123} \\ -1 \text{ if } i, j, k \text{ is an odd permutation of 123} \\ 0 \text{ if any two of } i, j, k \text{ are equal} \end{cases}$$

Thus $\varepsilon_{123} = 1$, $\varepsilon_{132} = -1$, $\varepsilon_{312} = 1$, $\varepsilon_{112} = 0$, for example.

The magnitude of $u \times v$ is the same as the area of a parallelogram with sides u and v. If $u \times v = 0$ and u and v are not null vectors, then u and v are parallel. Certain triple products are useful.

$$(u \cdot v)w \neq u(v \cdot w)$$
$$u \cdot (v \times w) = v \cdot (w \times u) = w \cdot (u \times v)$$
$$u \times (v \times w) = (u \cdot w)v - (u \cdot v)w$$
$$(u \times v) \times w = (u \cdot w)v - (v \cdot w)u$$

The cross product of a dyadic and a vector is defined as

$$A \times u = \sum_i \sum_j e_i e_j \left(\sum_k \sum_l \varepsilon_{klj} A_{ik} u_l \right)$$

The magnitude of a dyadic is

$$|A| = A = \sqrt{\tfrac{1}{2}(A:A^T)} = \sqrt{\tfrac{1}{2} \sum_i \sum_j A_{ij}^2}$$

There are three *invariants* of a dyadic. They are called invariants because they take the same value in any coordinate system and are thus an intrinsic property of the dyadic. They are the trace of A, A^2, A^3 [5.1].

$$I = \operatorname{tr} A = \sum_i A_{ii}$$

$$II = \operatorname{tr} A^2 = \sum_i \sum_j A_{ij} A_{ji}$$

$$III = \operatorname{tr} A^3 = \sum_i \sum_j \sum_k A_{ij} A_{jk} A_{ki}$$

The invariants can also be expressed as

$$I_1 = I$$
$$I_2 = \tfrac{1}{2}(I^2 - II)$$
$$I_3 = \tfrac{1}{6}(I^3 - 3 I \cdot II + 2 III) = \det A$$

Invariants of two dyadics are available [5.2]. Because a second-order dyadic has nine components, the *characteristic equation*

$$\det(\lambda \delta - A) = 0$$

can be formed where λ is an eigenvalue. This expression is

$$\lambda^3 - I_1 \lambda^2 + I_2 \lambda - I_3 = 0$$

An important theorem of HAMILTON and CAYLEY [5.3] is that a second-order dyadic satisfies its own characteristic equation.

$$A^3 - I_1 A^2 + I_2 A - I_3 \delta = 0 \qquad (14)$$

Thus A^3 can be expressed in terms of δ, A, and A^2. Similarly, higher powers of A can be expressed in terms of δ, A, and A^2. Decomposition of a dyadic into a symmetric and an antisymmetric part was shown above. The antisymmetric part has zero trace. The symmetric part can be decomposed into a part with a trace (the isotropic part) and a part with zero trace (the deviatoric part).

$$A = \underbrace{\tfrac{1}{3}\delta\delta:A}_{\text{Isotropic}} + \underbrace{\tfrac{1}{2}[A + A^T] - \tfrac{2}{3}\delta\delta:A]}_{\text{Deviatoric}} + \underbrace{\tfrac{1}{2}[A - A^T]}_{\text{Antisymmetric}}$$

Differentiation. The derivative of a vector is defined in the same way as the derivative of a scalar. Suppose the vector u depends on t. Then

$$\frac{du}{dt} = \lim_{\Delta t \to 0} \frac{u(t + \Delta t) - u(t)}{\Delta t}$$

If the vector is the position vector $r(t)$, then the difference expression is a vector in the direction of Δr (see Fig. 19). The derivative

$$\frac{dr}{dt} = \lim_{\Delta t \to 0} \frac{\Delta r}{\Delta t} = \lim_{\Delta t \to 0} \frac{r(t + \Delta t) - r(t)}{\Delta t}$$

is the velocity. The derivative operation obeys the following laws.

$$\frac{d}{dt}(u + v) = \frac{du}{dt} + \frac{dv}{dt}$$

$$\frac{d}{dt}(u \cdot v) = \frac{du}{dt} \cdot v + u \cdot \frac{dv}{dt}$$

$$\frac{d}{dt}(u \times v) = \frac{du}{dt} \times v + u \times \frac{dv}{dt}$$

$$\frac{d}{dt}(\varphi u) = \frac{d\varphi}{dt} u + \varphi \frac{du}{dt}$$

If the vector u depends on more than one variable, such as x, y and z, partial derivatives are defined in the usual way. For example,

if $u(x, y, z)$, then

$$\frac{\partial u}{\partial x} = \lim_{\Delta t \to 0} \frac{u(x + \Delta x, y, z) - u(x, y, z)}{\Delta x}$$

Rules for differentiation of scalar and vector products are

$$\frac{\partial}{\partial x}(u \cdot v) = \frac{\partial u}{\partial x} \cdot v + u \cdot \frac{\partial v}{\partial x}$$

$$\frac{\partial}{\partial x}(u \times v) = \frac{\partial u}{\partial x} \times v + u \times \frac{\partial v}{\partial x}$$

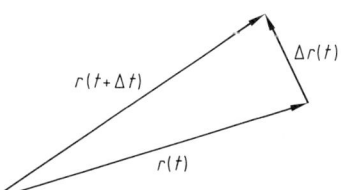

Figure 19. Vector differentiation

Differentials of vectors are

$$d\mathbf{u} = du_x \mathbf{e}_x + du_y \mathbf{e}_y + du_z \mathbf{e}_z$$
$$d(\mathbf{u} \cdot \mathbf{v}) = d\mathbf{u} \cdot \mathbf{v} + \mathbf{u} \cdot d\mathbf{v}$$
$$d(\mathbf{u} \times \mathbf{v}) = d\mathbf{u} \times \mathbf{v} + \mathbf{u} \times d\mathbf{v}$$
$$d\mathbf{u} = \frac{\partial \mathbf{u}}{\partial x} dx + \frac{\partial \mathbf{u}}{\partial y} dy + \frac{\partial \mathbf{u}}{\partial z} dz$$

If a curve is given by $\mathbf{r}(t)$, the length of the curve is [5.4, p. 373]

$$L = \int_a^b \sqrt{\frac{d\mathbf{r}}{dt} \cdot \frac{d\mathbf{r}}{dt}}\, dt$$

The arc-length function can also be defined:

$$s(t) = \int^a \sqrt{\frac{d\mathbf{r}}{dt^*} \cdot \frac{d\mathbf{r}}{dt^*}}\, dt^*$$

This gives

$$\left(\frac{ds}{dt}\right)^2 = \frac{d\mathbf{r}}{dt} \cdot \frac{d\mathbf{r}}{dt} = \left(\frac{dx}{dt}\right)^2 + \left(\frac{dy}{dt}\right)^2 + \left(\frac{dz}{dt}\right)^2$$

Because

$$d\mathbf{r} = dx\,\mathbf{e}_x + dy\,\mathbf{e}_y + dz\,\mathbf{e}_z$$

then

$$ds^2 = d\mathbf{r} \cdot d\mathbf{r} = dx^2 + dy^2 + dz^2$$

The derivative $d\mathbf{r}/dt$ is tangent to the curve in the direction of motion

$$\mathbf{u} = \frac{\frac{d\mathbf{r}}{dt}}{\left|\frac{d\mathbf{r}}{dt}\right|}$$

Also,

$$\mathbf{u} = \frac{d\mathbf{r}}{ds}$$

Differential Operators (see also, → 4. Transport Phenomena, pp. 4-9–4-10). The vector differential operator (del operator) ∇ is defined in cartesian coordinates by

$$\nabla = \mathbf{e}_x \frac{\partial}{\partial x} + \mathbf{e}_y \frac{\partial}{\partial y} + \mathbf{e}_z \frac{\partial}{\partial z}$$

The *gradient* of a scalar function is defined

$$\nabla \varphi = \mathbf{e}_x \frac{\partial \varphi}{\partial x} + \mathbf{e}_y \frac{\partial \varphi}{\partial y} + \mathbf{e}_z \frac{\partial \varphi}{\partial z}$$

and is a vector. If φ is height or elevation, the gradient is a vector pointing in the uphill direction. The steeper the hill, the larger is the magnitude of the gradient.

The *divergence* of a vector is defined by

$$\nabla \cdot \mathbf{u} = \frac{\partial u_x}{\partial x} + \frac{\partial u_y}{\partial y} + \frac{\partial u_z}{\partial z}$$

and is a scalar. For a volume element ΔV, the net outflow of a vector $\mathbf{u}$ over the surface of the element is

$$\int_{\Delta S} \mathbf{u} \cdot \mathbf{n}\, dS$$

This is related to the divergence by [5.5, p. 411]

$$\nabla \cdot \mathbf{u} = \lim_{\Delta V \to 0} \frac{1}{\Delta V} \int_{\Delta S} \mathbf{u} \cdot \mathbf{n}\, dS$$

Thus, the divergence is the net outflow per unit volume.

The *curl* of a vector is defined by

$$\nabla \times \mathbf{u} = \left(\mathbf{e}_x \frac{\partial}{\partial x} + \mathbf{e}_y \frac{\partial}{\partial y} + \mathbf{e}_z \frac{\partial}{\partial z}\right) \times (\mathbf{e}_x u_x + \mathbf{e}_y u_y + \mathbf{e}_z u_z)$$

$$= \mathbf{e}_x \left(\frac{\partial u_z}{\partial y} - \frac{\partial u_y}{\partial z}\right) + \mathbf{e}_y \left(\frac{\partial u_x}{\partial z} - \frac{\partial u_z}{\partial x}\right)$$
$$+ \mathbf{e}_z \left(\frac{\partial u_y}{\partial x} - \frac{\partial u_x}{\partial y}\right)$$

and is a vector. It is related to the integral

$$\oint \mathbf{u} \cdot d\mathbf{s} = \int_C \mathbf{u}_s\, ds$$

which is called the circulation of $\mathbf{u}$ around path C. This integral depends on the vector and the contour C, in general. If the circulation does not depend on the contour C, the vector is said to be irrotational; if it does, it is rotational. The relationship with the curl is [5.5, p. 419]

$$\mathbf{n} \cdot (\nabla \times \mathbf{u}) = \lim_{\Delta S \to 0} \frac{1}{\Delta S} \int_C \mathbf{u} \cdot d\mathbf{s}$$

Thus, the normal component of the curl equals the net circulation per unit area enclosed by the contour C.

The gradient, divergence, and curl obey a distributive law but not a commutative or associative law.

$$\nabla(\varphi + \psi) = \nabla \varphi + \nabla \psi$$
$$\nabla \cdot (\mathbf{u} + \mathbf{v}) = \nabla \cdot \mathbf{u} + \nabla \cdot \mathbf{v}$$
$$\nabla \times (\mathbf{u} + \mathbf{v}) = \nabla \times \mathbf{u} + \nabla \times \mathbf{v}$$
$$\nabla \cdot \varphi \neq \varphi \nabla$$
$$\nabla \cdot \mathbf{u} \neq \mathbf{u} \cdot \nabla$$

Useful formulas are [5.6]

$$\nabla \cdot (\varphi \boldsymbol{u}) = \nabla\varphi \cdot \boldsymbol{u} + \varphi \nabla \cdot \boldsymbol{u}$$
$$\nabla \times (\varphi \boldsymbol{u}) = \nabla\varphi \times \boldsymbol{u} + \varphi \nabla \times \boldsymbol{u}$$
$$\nabla \cdot (\boldsymbol{u} \times \boldsymbol{v}) = \boldsymbol{v} \cdot (\nabla \times \boldsymbol{u}) - \boldsymbol{u} \cdot (\nabla \times \boldsymbol{v})$$
$$\nabla \times (\boldsymbol{u} \times \boldsymbol{v}) = \boldsymbol{v} \cdot \nabla \boldsymbol{u} - \boldsymbol{v}(\nabla \cdot \boldsymbol{u}) - \boldsymbol{u} \cdot \nabla \boldsymbol{v} + \boldsymbol{u}(\nabla \cdot \boldsymbol{v})$$
$$\nabla \times (\nabla \times \boldsymbol{u}) = \nabla(\nabla \cdot \boldsymbol{u}) - \nabla^2 \boldsymbol{u}$$
$$\nabla \cdot (\nabla \varphi) = \nabla^2 \varphi = \frac{\partial^2 \varphi}{\partial x^2} + \frac{\partial^2 \varphi}{\partial y^2} + \frac{\partial^2 \varphi}{\partial z^2}, \text{ where } \nabla^2$$

is called the Laplacian operator $\nabla \times (\nabla \varphi) = 0$. The curl of the gradient of φ is zero. $\nabla \cdot (\nabla \times \boldsymbol{u}) = 0$. The divergence of the curl of $\boldsymbol{u}$ is zero.

Formulas useful in fluid mechanics are

$$\nabla \cdot (\nabla \boldsymbol{v})^T = \nabla(\nabla \cdot \boldsymbol{v})$$
$$\nabla \cdot (\boldsymbol{\tau} \cdot \boldsymbol{v}) = \boldsymbol{v} \cdot (\nabla \cdot \boldsymbol{\tau}) + \boldsymbol{\tau} : \nabla \boldsymbol{v}$$
$$\boldsymbol{v} \cdot \nabla \boldsymbol{v} = \tfrac{1}{2} \nabla(\boldsymbol{v} \cdot \boldsymbol{v}) - \boldsymbol{v} \times (\nabla \times \boldsymbol{v})$$

If a coordinate system is transformed by a rotation and translation, the coordinates in the new system (denoted by primes) are given by

$$\begin{pmatrix} x' \\ y' \\ z' \end{pmatrix} = \begin{pmatrix} l_{11} & l_{12} & l_{13} \\ l_{21} & l_{22} & l_{23} \\ l_{31} & l_{32} & l_{33} \end{pmatrix} \begin{pmatrix} x \\ y \\ z \end{pmatrix} + \begin{pmatrix} a_1 \\ a_2 \\ a_3 \end{pmatrix}$$

Any function that has the same value in all coordinate systems is an invariant. The gradient of an invariant scalar field is invariant; the same is true for the divergence and curl of invariant vectors fields.

The gradient of a vector field is required in fluid mechanics because the velocity gradient is used. It is defined as

$$\nabla \boldsymbol{v} = \sum_i \sum_j \boldsymbol{e}_i \boldsymbol{e}_j \frac{\partial v_j}{\partial x_i} \text{ and } (\nabla \boldsymbol{v})^T = \sum_i \sum_j \boldsymbol{e}_i \boldsymbol{e}_j \frac{\partial v_i}{\partial x_j}$$

The divergence of dyadics is defined

$$\nabla \cdot \boldsymbol{\tau} = \sum_i \boldsymbol{e}_i \left(\sum_j \frac{\partial \tau_{ji}}{\partial x_j} \right) \text{ and}$$
$$\nabla \cdot (\varphi \boldsymbol{u} \boldsymbol{v}) = \sum_i \boldsymbol{e}_i \left[\sum_j \frac{\partial}{\partial x_j} (\varphi u_j v_i) \right]$$

where $\boldsymbol{\tau}$ is any second-order dyadic.
Useful relations involving dyadics are

$$(\varphi \boldsymbol{\delta} : \nabla \boldsymbol{v}) = \varphi (\nabla \cdot \boldsymbol{v})$$
$$\nabla \cdot (\varphi \boldsymbol{\delta}) = \nabla \varphi$$
$$\nabla \cdot (\varphi \boldsymbol{\tau}) = \nabla \varphi \cdot \boldsymbol{\tau} + \varphi \nabla \cdot \boldsymbol{\tau}$$
$$\boldsymbol{n} \boldsymbol{t} : \boldsymbol{\tau} = \boldsymbol{t} \cdot \boldsymbol{\tau} \cdot \boldsymbol{n} = \boldsymbol{\tau} : \boldsymbol{n} \boldsymbol{t}$$

A surface can be represented in the form

$$f(x, y, z) = c = \text{constant}$$

The normal to the surface is given by

$$\boldsymbol{n} = \frac{\nabla f}{|\nabla f|}$$

provided the gradient is not zero. Operations can be performed entirely within the surface. Define [5.7]

$$\boldsymbol{\delta}_{\text{II}} \equiv \boldsymbol{\delta} - \boldsymbol{n}\boldsymbol{n}, \quad \nabla_{\text{II}} \equiv \boldsymbol{\delta}_{\text{II}} \cdot \nabla, \quad \frac{\partial}{\partial n} \equiv \boldsymbol{n} \cdot \nabla$$

$$\boldsymbol{v}_{\text{II}} \equiv \boldsymbol{\delta}_{\text{II}} \cdot \boldsymbol{v}, \quad v_n \equiv \boldsymbol{n} \cdot \boldsymbol{v}$$

Then a vector and del operator can be decomposed into

$$\boldsymbol{v} = \boldsymbol{v}_{\text{II}} + \boldsymbol{n} v_n, \quad \nabla = \nabla_{\text{II}} + \boldsymbol{n} \frac{\partial}{\partial n}$$

The velocity gradient can be decomposed into

$$\nabla \boldsymbol{v} = \nabla_{\text{II}} \boldsymbol{v}_{\text{II}} + (\nabla_{\text{II}} \boldsymbol{n}) v_n + \boldsymbol{n} \nabla_{\text{II}} v_n + \boldsymbol{n} \frac{\partial \boldsymbol{v}_{\text{II}}}{\partial n} + \boldsymbol{n}\boldsymbol{n} \frac{\partial v_n}{\partial n}$$

The surface gradient of the normal is the negative of the curvature dyadic of the surface.

$$\nabla_{\text{II}} \boldsymbol{n} = -\boldsymbol{B}$$

The surface divergence is then

$$\nabla_{\text{II}} \cdot \boldsymbol{v} = \boldsymbol{\delta}_{\text{II}} : \nabla \boldsymbol{v} = \nabla_{\text{II}} \cdot \boldsymbol{v}_{\text{II}} - 2 H v_n$$

where H is the mean curvature.

$$H = \tfrac{1}{2} \boldsymbol{\delta}_{\text{II}} : \boldsymbol{B}$$

The surface curl can be a scalar

$$\nabla_{\text{II}} \times_s \boldsymbol{v} = -\boldsymbol{\varepsilon}_{\text{II}} : \nabla \boldsymbol{v} = -\boldsymbol{\varepsilon}_{\text{II}} : \nabla_{\text{II}} \boldsymbol{v}_{\text{II}} = -\boldsymbol{n} \cdot (\nabla \times \boldsymbol{v}),$$
$$\boldsymbol{\varepsilon}_{\text{II}} = \boldsymbol{n} \cdot \boldsymbol{\varepsilon}$$

or a vector

$$\nabla_{\text{II}} \times \boldsymbol{v} \equiv \nabla_{\text{II}} \times \boldsymbol{v}_{\text{II}} = \boldsymbol{n} \nabla_{\text{II}} \times_s \boldsymbol{v} = \boldsymbol{n}\boldsymbol{n} \cdot \nabla \times \boldsymbol{v}$$

Vector Integration [5.5, pp. 206–212]. If $\boldsymbol{u}$ is a vector, then its integral is also a vector.

$$\int \boldsymbol{u}(t) \, dt = \boldsymbol{e}_x \int u_x(t) \, dt + \boldsymbol{e}_y \int u_y(t) \, dt + \boldsymbol{e}_z \int u_z(t) \, dt$$

If the vector $\boldsymbol{u}$ is the derivative of another vector, then

$$\boldsymbol{u} = \frac{d\boldsymbol{v}}{dt}, \quad \int \boldsymbol{u}(t) \, dt = \int \frac{d\boldsymbol{v}}{dt} dt = \boldsymbol{v} + \text{constant}$$

If $r(t)$ is a position vector that defines a curve C, the line integral is defined by

$$\int_C u \cdot dr = \int_C (u_x dx + u_y dy + u_z dz)$$

Theorems about this line integral can be written in various forms.

Theorem [5.4, p. 460]. If the functions appearing in the line integral are continuous in a domain D, then the line integral is independent of the path C if and only if the line integral is zero on every simple closed path in D.

Theorem [5.4, p. 461]. If $u = \nabla \varphi$ where φ is single-valued and has continuous derivatives in D, then the line integral is independent of the path C and the line integral is zero for any closed curve in D.

Theorem [5.4, p. 460]. If f, g, and h are continuous functions of x, y, and z, and have continuous first derivatives in a simply connected domain D, then the line integral

$$\int_C (f dx + g dy + h dz)$$

is independent of the path if and only if

$$\frac{\partial h}{\partial y} = \frac{\partial g}{\partial z}, \quad \frac{\partial f}{\partial z} = \frac{\partial h}{\partial x}, \quad \frac{\partial g}{\partial x} = \frac{\partial f}{\partial y}$$

or if f, g, and h are regarded as the x, y, and z components of a vector v:

$$\nabla \times v = 0$$

Consequently, the line integral is independent of the path (and the value is zero for a closed contour) if the three components in it are regarded as the three components of a vector and the vector is derivable from a potential (or zero curl). The conditions for a vector to be derivable from a potential are just those in the third theorem. In two dimensions this reduces to the more usual theorem.

Theorem [5.5, p. 207]. If M and N are continuous functions of x and y that have continuous first partial derivatives in a simply connected domain D, then the necessary and sufficient condition for the line integral

$$\int_C (M dx + N dy)$$

to be zero around every closed curve C in D is

$$\frac{\partial M}{\partial y} = \frac{\partial N}{\partial x}$$

If a vector is integrated over a surface with incremental area dS and normal to the surface n, then the surface integral can be written as

$$\iint_S u \cdot dS = \iint_S u \cdot n dS$$

If u is the velocity then this integral represents the flow rate past the surface S.

Divergence Theorem [5.5], [5.6]. If V is a volume bounded by a closed surface S and u is a vector function of position with continuous derivatives, then

$$\int_V \nabla \cdot u dV = \int_S n \cdot u dS = \int_S u \cdot n dS = \int_S u \cdot dS$$

where n is the normal pointing outward to S. The normal can be written as

$$n = e_x \cos(x, n) + e_y \cos(y, n) + e_z \cos(z, n)$$

where, for example, $\cos(x, n)$ is the cosine of the angle between the normal n and the x axis. Then the divergence theorem in component form is

$$\int_V \left(\frac{\partial u_x}{\partial x} + \frac{\partial u_y}{\partial y} + \frac{\partial u_z}{\partial z} \right) dx dy dz = \int_S [u_x \cos(x, n) + u_y \cos(y, n) + u_z \cos(z, n)] dS$$

If the divergence theorem is written for an incremental volume

$$\nabla \cdot u = \lim_{\Delta V \to 0} \frac{1}{\Delta V} \int_{\Delta S} u_n dS$$

the divergence of a vector can be called the integral of that quantity over the area of a closed volume, divided by the volume. If the vector represents the flow of energy and the divergence is positive at a point P, then either a source of energy is present at P or energy is leaving the region around P so that its temperature is decreasing. If the vector represents the flow of mass and the divergence is positive at a point P, then either a source of mass exists at P or the density is decreasing at the point P. For an incompressible fluid the divergence is zero and the rate at which fluid is introduced into a volume must equal the rate at which it is removed.

Various theorems follow from the divergence theorem.

Theorem. If φ is a solution to Laplace's equation

$$\nabla^2 \varphi = 0$$

in a domain D, and the second partial derivatives of φ are continuous in D, then the integral of the normal derivative of φ over any piecewise smooth closed orientable surface S in D is zero. Suppose $\boldsymbol{u} = \varphi \nabla \psi$ satisfies the conditions of the divergence theorem: then *Green's theorem* results from use of the divergence theorem [5.4, p. 451].

$$\int_V (\varphi \nabla^2 \psi + \nabla \varphi \cdot \nabla \psi) \, dV = \int_S \varphi \frac{\partial \psi}{\partial n} \, dS$$

and

$$\int_V (\varphi \nabla^2 \psi - \psi \nabla^2 \varphi) \, dV = \int_S \left(\varphi \frac{\partial \psi}{\partial n} - \psi \frac{\partial \varphi}{\partial n} \right) dS$$

Also if φ satisfies the conditions of the theorem and is zero on S then φ is zero throughout D. If two functions φ and ψ both satisfy the Laplace equation in domain D, and both take the same values on the bounding curve C, then $\varphi = \psi$; i.e., the solution to the Laplace equation is unique.

The divergence theorem for dyadics is

$$\int_V \nabla \cdot \tau \, dV = \int_S \boldsymbol{n} \cdot \tau \, dS$$

Stokes Theorem [5.5], [5.6, p. 106]. Stokes theorem says that if S is a surface bounded by a closed, nonintersecting curve C, and if $\boldsymbol{u}$ has continuous derivatives then

$$\oint_C \boldsymbol{u} \cdot d\boldsymbol{r} = \iint_S (\nabla \times \boldsymbol{u}) \cdot \boldsymbol{n} \, dS = \iint_S (\nabla \times \boldsymbol{u}) \cdot d\boldsymbol{S}$$

The integral around the curve is followed in the counterclockwise direction. In component notation, this is

$$\oint_C [u_x \cos(x, s) + u_y \cos(y, s) + u_z \cos(z, s)] \, ds =$$

$$\iint_S \left[\left(\frac{\partial u_z}{\partial y} - \frac{\partial u_y}{\partial z} \right) \cos(x, n) + \left(\frac{\partial u_x}{\partial z} - \frac{\partial u_z}{\partial x} \right) \cos(y, n) + \left(\frac{\partial u_y}{\partial x} - \frac{\partial u_x}{\partial y} \right) \cos(z, n) \right] dS$$

Applied in two dimensions, this results in Green's theorem in the plane:

$$\oint_C (M \, dx + N \, dy) = \iint_S \left(\frac{\partial N}{\partial x} - \frac{\partial M}{\partial y} \right) dx \, dy$$

The formula for dyadics is

$$\iint_S \boldsymbol{n} \cdot (\nabla \times \tau) \, dS = \oint_C \tau^T \cdot d\boldsymbol{r}$$

Representation. Two theorems give information about how to represent vectors that obey certain properties.

Theorem [5.5, p. 422]. The necessary and sufficient condition that the curl of a vector vanish identically is that the vector be the gradient of some function.

Theorem [5.5, p. 423]. The necessary and sufficient condition that the divergence of a vector vanish identically is that the vector is the curl of some other vector.

Leibniz Formula. In fluid mechanics and transport phenomena, an important result is the derivative of an integral whose limits of integration are moving. Suppose the region $V(t)$ is moving with velocity $\boldsymbol{v}_s$. Then Leibniz's rule holds:

$$\frac{d}{dt} \iiint_{V(t)} \varphi \, dV = \iiint_{V(t)} \frac{\partial \varphi}{\partial t} \, dV + \iint_S \varphi \boldsymbol{v}_s \cdot \boldsymbol{n} \, dS$$

Curvilinear Coordinates. Many of the relations given above are proved most easily by using tensor analysis rather than dyadics. Once proven, however, the relations are perfectly general in any coordinate system. Displayed here are the specific results for cylindrical and spherical geometries. Results are available for a few other geometries: parabolic cylindrical, paraboloidal, elliptic cylindrical, prolate spheroidal, oblate spheroidal, ellipsoidal, and bipolar coordinates [5.1], [5.8].

For *cylindrical coordinates,* the geometry is shown in Figure 20. The coordinates are related to cartesian coordinates by

$$x = r \cos \theta \qquad r = \sqrt{x^2 + y^2}$$

$$y = r \sin \theta \qquad \theta = \arctan\left(\frac{y}{x}\right)$$

$$z = z \qquad z = z$$

The unit vectors are related by

$$\boldsymbol{e}_r = \cos \theta \, \boldsymbol{e}_x + \sin \theta \, \boldsymbol{e}_y \quad \boldsymbol{e}_x = \cos \theta \, \boldsymbol{e}_r - \sin \theta \, \boldsymbol{e}_\theta$$

$$\boldsymbol{e}_\theta = -\sin \theta \, \boldsymbol{e}_x + \cos \theta \, \boldsymbol{e}_y \quad \boldsymbol{e}_y = \sin \theta \, \boldsymbol{e}_r + \cos \theta \, \boldsymbol{e}_\theta$$

$$\boldsymbol{e}_z = \boldsymbol{e}_z \qquad \boldsymbol{e}_z = \boldsymbol{e}_z$$

Derivatives of the unit vectors are

$$d\boldsymbol{e}_\theta = -\boldsymbol{e}_r \, d\theta, \quad d\boldsymbol{e}_r = \boldsymbol{e}_\theta \, d\theta, \quad d\boldsymbol{e}_z = 0$$

$$\nabla \cdot v = \frac{1}{r}\frac{\partial}{\partial r}(rv_r) + \frac{1}{r}\frac{\partial v_\theta}{\partial \theta} + \frac{\partial v_z}{\partial z}$$

$$\nabla \times v = e_r\left(\frac{1}{r}\frac{\partial v_z}{\partial \theta} - \frac{\partial v_\theta}{\partial z}\right) + e_\theta\left(\frac{\partial v_r}{\partial z} - \frac{\partial v_z}{\partial r}\right)$$
$$+ e_z\left[\frac{1}{r}\frac{\partial}{\partial r}(rv_\theta) - \frac{1}{r}\frac{\partial v_r}{\partial \theta}\right]$$

$$\nabla \cdot \tau = e_r\left[\frac{1}{r}\frac{\partial}{\partial r}(r\tau_{rr}) + \frac{1}{r}\frac{\partial \tau_{\theta r}}{\partial \theta} + \frac{\partial \tau_{zr}}{\partial z} - \frac{\tau_{\theta\theta}}{r}\right] +$$
$$+ e_\theta\left[\frac{1}{r^2}\frac{\partial}{\partial r}(r^2\tau_{r\theta}) + \frac{1}{r}\frac{\partial \tau_{\theta\theta}}{\partial \theta} + \frac{\partial \tau_{z\theta}}{\partial z} + \frac{\tau_{\theta r} - \tau_{r\theta}}{r}\right] +$$
$$+ e_z\left[\frac{1}{r}\frac{\partial}{\partial r}(r\tau_{rz}) + \frac{1}{r}\frac{\partial \tau_{\theta z}}{\partial \theta} + \frac{\partial \tau_{zz}}{\partial z}\right]$$

$$\nabla v = e_r e_r \frac{\partial v_r}{\partial r} + e_r e_\theta \frac{\partial v_\theta}{\partial r} + e_r e_z \frac{\partial v_z}{\partial r} +$$
$$+ e_\theta e_r\left(\frac{1}{r}\frac{\partial v_r}{\partial \theta} - \frac{v_\theta}{r}\right) + e_\theta e_\theta\left(\frac{1}{r}\frac{\partial v_\theta}{\partial \theta} + \frac{v_r}{r}\right) +$$
$$+ e_\theta e_z \frac{1}{r}\frac{\partial v_z}{\partial \theta} + e_z e_r \frac{\partial v_r}{\partial z} + e_z e_\theta \frac{\partial v_\theta}{\partial z} + e_z e_z \frac{\partial v_z}{\partial z}$$

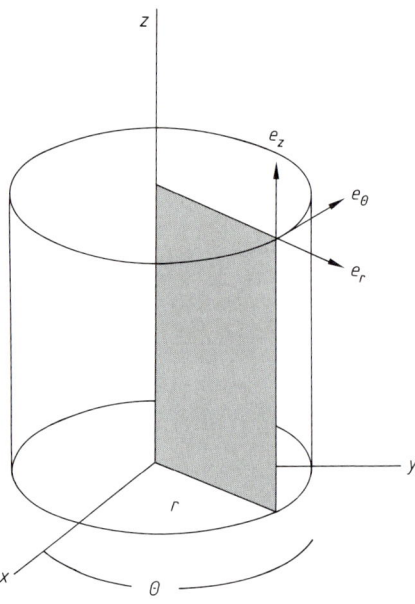

Figure 20. Cylindrical coordinate system

$$\nabla^2 v = e_r\left[\frac{\partial}{\partial r}\left(\frac{1}{r}\frac{\partial}{\partial r}(rv_r)\right)\right.$$
$$\left. + \frac{1}{r^2}\frac{\partial^2 v_r}{\partial \theta^2} + \frac{\partial^2 v_r}{\partial z^2} - \frac{2}{r^2}\frac{\partial v_\theta}{\partial \theta}\right] +$$
$$+ e_\theta\left[\frac{\partial}{\partial r}\left(\frac{1}{r}\frac{\partial}{\partial r}(rv_\theta)\right)\right.$$
$$\left. + \frac{1}{r^2}\frac{\partial^2 v_\theta}{\partial \theta^2} + \frac{\partial^2 v_\theta}{\partial z^2} + \frac{2}{r^2}\frac{\partial v_r}{\partial \theta}\right] +$$
$$+ e_z\left[\frac{1}{r}\frac{\partial}{\partial r}\left(r\frac{\partial v_z}{\partial r}\right) + \frac{1}{r^2}\frac{\partial^2 v_z}{\partial \theta^2} + \frac{\partial^2 v_z}{\partial z^2}\right]$$

For *spherical coordinates*, the geometry is shown in Figure 21. The coordinates are related to cartesian coordinates by

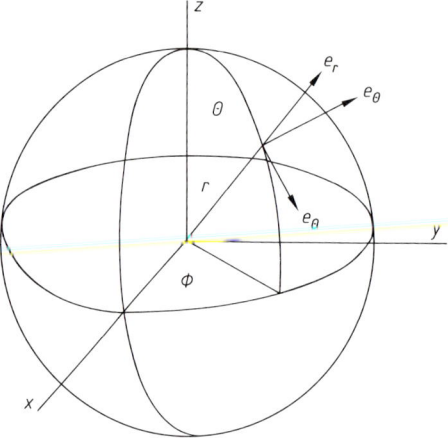

Figure 21. Spherical coordinate system

$$x = r\sin\theta\cos\varphi \qquad r = \sqrt{x^2 + y^2 + z^2}$$
$$y = r\sin\theta\sin\varphi \qquad \theta = \arctan(\sqrt{x^2 + y^2}/z)$$
$$z = r\cos\theta \qquad \varphi = \arctan\left(\frac{y}{x}\right)$$

Differential operators are given by [5.1]

$$\nabla = e_r\frac{\partial}{\partial r} + \frac{e_\theta}{r}\frac{\partial}{\partial \theta} + e_z\frac{\partial}{\partial z},$$

$$\nabla\varphi = e_r\frac{\partial \varphi}{\partial r} + \frac{e_\theta}{r}\frac{\partial \varphi}{\partial \theta} + e_z\frac{\partial \varphi}{\partial z}$$

$$\nabla^2\varphi = \frac{1}{r}\frac{\partial}{\partial r}\left(r\frac{\partial \varphi}{\partial r}\right) + \frac{1}{r^2}\frac{\partial^2 \varphi}{\partial \theta^2} + \frac{\partial^2 \varphi}{\partial z^2}$$
$$= \frac{\partial^2 \varphi}{\partial r^2} + \frac{1}{r}\frac{\partial \varphi}{\partial r} + \frac{1}{r^2}\frac{\partial^2 \varphi}{\partial \theta^2} + \frac{\partial^2 \varphi}{\partial z^2}$$

The unit vectors are related by

$$e_r = \sin\theta\cos\varphi\, e_x + \sin\theta\sin\varphi\, e_y + \cos\theta\, e_z$$
$$e_\theta = \cos\theta\cos\varphi\, e_x + \cos\theta\sin\varphi\, e_y - \sin\theta\, e_z$$
$$e_\varphi = -\sin\varphi\, e_x + \cos\varphi\, e_y$$
$$e_x = \sin\theta\cos\varphi\, e_r + \cos\theta\cos\varphi\, e_\theta - \sin\varphi\, e_\varphi$$
$$e_y = \sin\theta\sin\varphi\, e_r + \cos\theta\sin\varphi\, e_\theta + \cos\varphi\, e_\varphi$$
$$e_z = \cos\theta\, e_r - \sin\theta\, e_\theta$$

Derivatives of the unit vectors are

$$\frac{\partial e_r}{\partial \theta} = e_\theta, \quad \frac{\partial e_\theta}{\partial \theta} = -e_r$$

$$\frac{\partial e_r}{\partial \varphi} = e_\phi \sin \theta, \quad \frac{\partial e_\theta}{\partial \varphi} = e_\phi \cos \theta,$$

$$\frac{\partial e_\phi}{\partial \varphi} = -e_r \sin \theta - e_\theta \cos \theta$$

Others 0

Differential operators are given by [5.1]

$$\nabla = e_r \frac{\partial}{\partial r} + e_\theta \frac{1}{r} \frac{\partial}{\partial \theta} + e_\phi \frac{1}{r \sin \theta} \frac{\partial}{\partial \varphi},$$

$$\nabla \psi = e_r \frac{\partial \psi}{\partial r} + e_\theta \frac{1}{r} \frac{\partial \psi}{\partial \theta} + e_\phi \frac{1}{r \sin \theta} \frac{\partial \psi}{\partial \varphi}$$

$$\nabla^2 \psi = \frac{1}{r^2} \frac{\partial}{\partial r}\left(r^2 \frac{\partial \psi}{\partial r}\right) + \frac{1}{r^2 \sin \theta} \frac{\partial}{\partial \theta}\left(\sin \theta \frac{\partial \psi}{\partial \theta}\right) + \frac{1}{r^2 \sin^2 \theta} \frac{\partial^2 \psi}{\partial \varphi^2}$$

$$\nabla \cdot v = \frac{1}{r^2} \frac{\partial}{\partial r}(r^2 v_r) + \frac{1}{r \sin \theta} \frac{\partial}{\partial \theta}(v_\theta \sin \theta) + \frac{1}{r \sin \theta} \frac{\partial v_\phi}{\partial \varphi}$$

$$\nabla \times v = e_r \left[\frac{1}{r \sin \theta} \frac{\partial}{\partial \theta}(v_\phi \sin \theta) - \frac{1}{r \sin \theta} \frac{\partial v_\theta}{\partial \varphi}\right] + e_\theta \left[\frac{1}{r \sin \theta} \frac{\partial v_r}{\partial \varphi} - \frac{1}{r} \frac{\partial}{\partial r}(r v_\phi)\right] + e_\phi \left[\frac{1}{r} \frac{\partial}{\partial r}(r v_\theta) - \frac{1}{r} \frac{\partial v_r}{\partial \theta}\right]$$

$$\nabla \cdot \tau = e_r \left[\frac{1}{r^2} \frac{\partial}{\partial r}(r^2 \tau_{rr}) + \frac{1}{r \sin \theta} \frac{\partial}{\partial \theta}(\tau_{\theta r} \sin \theta) + \frac{1}{r \sin \theta} \frac{\partial \tau_{\phi r}}{\partial \varphi} - \frac{\tau_{\theta\theta} + \tau_{\phi\phi}}{r}\right] + e_\theta \left[\frac{1}{r^3} \frac{\partial}{\partial r}(r^3 \tau_{r\theta}) + \frac{1}{r \sin \theta} \frac{\partial}{\partial \theta}(\tau_{\theta\theta} \sin \theta) + \frac{1}{r \sin \theta} \frac{\partial \tau_{\phi\theta}}{\partial \varphi} + \frac{\tau_{\theta r} - \tau_{r\theta} - \tau_{\phi\phi} \cot \theta}{r}\right] + e_\phi \left[\frac{1}{r^3} \frac{\partial}{\partial r}(r^3 \tau_{r\phi}) + \frac{1}{r \sin \theta} \frac{\partial}{\partial \theta}(\tau_{\theta\phi} \sin \theta) + \frac{1}{r \sin \theta} \frac{\partial \tau_{\phi\phi}}{\partial \varphi} + \frac{\tau_{\phi r} - \tau_{r\phi} + \tau_{\phi\theta} \cot \theta}{r}\right]$$

$$\nabla v = e_r e_r \frac{\partial v_r}{\partial r} + e_r e_\theta \frac{\partial v_\theta}{\partial r} + e_r e_\phi \frac{\partial v_\phi}{\partial r} + e_\theta e_r \left(\frac{1}{r} \frac{\partial v_r}{\partial \theta} - \frac{v_\theta}{r}\right) + e_\theta e_\theta \left(\frac{1}{r} \frac{\partial v_\theta}{\partial \theta} + \frac{v_r}{r}\right) + e_\theta e_\phi \frac{1}{r} \frac{\partial v_\phi}{\partial \theta} + e_\phi e_r \left(\frac{1}{r \sin \theta} \frac{\partial v_r}{\partial \varphi} - \frac{v_\phi}{r}\right) + e_\phi e_\theta \left(\frac{1}{r \sin \theta} \frac{\partial v_\theta}{\partial \varphi} - \frac{v_\phi}{r} \cot \theta\right) + e_\phi e_\phi \left(\frac{1}{r \sin \theta} \frac{\partial v_\phi}{\partial \varphi} + \frac{v_r}{r} + \frac{v_\theta}{r} \cot \theta\right)$$

$$\nabla^2 v = e_r \left[\frac{\partial}{\partial r}\left(\frac{1}{r^2} \frac{\partial}{\partial r}(r^2 v_r)\right) + \frac{1}{r^2 \sin \theta} \frac{\partial}{\partial \theta}\left(\sin \theta \frac{\partial v_r}{\partial \theta}\right) + \frac{1}{r^2 \sin^2 \theta} \frac{\partial^2 v_r}{\partial \varphi^2} - \frac{2}{r^2 \sin \theta} \frac{\partial}{\partial \theta}(v_\theta \sin \theta) - \frac{2}{r^2 \sin \theta} \frac{\partial v_\phi}{\partial \varphi}\right] + e_\theta \left[\frac{1}{r^2} \frac{\partial}{\partial r}\left(r^2 \frac{\partial v_\theta}{\partial r}\right) + \frac{1}{r^2} \frac{\partial}{\partial \theta}\left(\frac{1}{\sin \theta} \frac{\partial}{\partial \theta}(v_\theta \sin \theta)\right) + \frac{1}{r^2 \sin^2 \theta} \frac{\partial^2 v_\theta}{\partial \varphi^2} + \frac{2}{r^2} \frac{\partial v_r}{\partial \theta} - \frac{2 \cot \theta}{r^2 \sin \theta} \frac{\partial v_\phi}{\partial \varphi}\right] + e_\phi \left[\frac{1}{r^2} \frac{\partial}{\partial r}\left(r^2 \frac{\partial v_\phi}{\partial r}\right) + \frac{1}{r^2} \frac{\partial}{\partial \theta}\left(\frac{1}{\sin \theta} \frac{\partial}{\partial \theta}(v_\phi \sin \theta)\right) + \frac{1}{r^2 \sin^2 \theta} \frac{\partial^2 v_\phi}{\partial \varphi^2} + \frac{2}{r^2 \sin \theta} \frac{\partial v_r}{\partial \varphi} + \frac{2 \cot \theta}{r^2 \sin \theta} \frac{\partial v_\theta}{\partial \varphi}\right]$$

6. Ordinary Differential Equations as Initial Value Problems

A differential equation for a function that depends on only one variable (often time) is called an ordinary differential equation. The general solution to the differential equation includes many possibilities; the boundary or initial conditions are required to specify which of those are desired. If all conditions are at one point, the problem is an initial value problem and can be integrated from that point on. If some of the conditions are available at one point and others at another point, the ordinary differential equations become two-point boundary value problems, which are treated in Chapter 7. Initial value

6.1. Solution by Quadrature

When only one equation exists, even if it is nonlinear, solving it by quadrature may be possible. For

$$\frac{dy}{dt} = f(y)$$

$$y(0) = y_0$$

the problem can be separated

$$\frac{dy}{f(y)} = dt$$

and integrated:

$$\int_{y_0}^{y} \frac{dy'}{f(y')} = \int_0^t dt = t$$

If the quadrature can be performed analytically then the exact solution has been found. For example, consider the kinetics problem with a second-order reaction.

$$\frac{dc}{dt} = -kc^2, \quad c(0) = c_0$$

To find the function of the concentration versus time, the variables can be separated and integrated.

$$\frac{dc}{c^2} = -k\,dt,$$

$$-\frac{1}{c} = -kt + D$$

Application of the initial conditions gives the solution:

$$\frac{1}{c} = kt + \frac{1}{c_0}$$

For other ordinary differential equations an integrating factor is useful. Consider the problem governing a stirred tank with entering fluid having concentration c_{in} and flow rate F, as shown in Figure 22. The flow rate out is also F

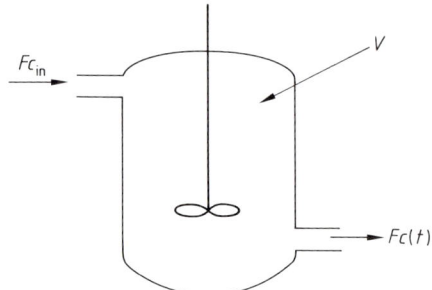

Figure 22. Stirred tank

and the volume of the tank is V. If the tank is completely mixed, the concentration in the tank is c and the concentration of the fluid leaving the tank is also c. The differential equation is then

$$V\frac{dc}{dt} = F(c_{in} - c), \quad c(0) = c_0$$

Upon rearrangement,

$$\frac{dc}{dt} + \frac{F}{V}c = \frac{F}{V}c_{in}$$

is obtained. An integrating factor is used to solve this equation. The integrating factor is a function that can be used to turn the left-hand side into an exact differential and can be found by using Fréchet differentials [6.1]. In this case,

$$\exp\left(\frac{Ft}{V}\right)\left[\frac{dc}{dt} + \frac{F}{V}c\right] = \frac{d}{dt}\left[\exp\left(\frac{Ft}{V}\right)c\right]$$

Thus, the differential equation can be written as

$$\frac{d}{dt}\left[\exp\left(\frac{Ft}{V}\right)c\right] = \exp\left(\frac{Ft}{V}\right)\left[\frac{F}{V}c_{in}\right]$$

This can be integrated once to give

$$\exp\left(\frac{Ft}{V}\right)c = c(0) + \frac{F}{V}\int_0^t \exp\left(\frac{Ft'}{V}\right)c_{in}(t')\,dt'$$

or

$$c(t) = \exp\left(-\frac{Ft}{V}\right)c_0$$

$$+ \frac{F}{V}\int_0^t \exp\left(-\frac{F(t-t')}{V}\right)c_{in}(t')\,dt'$$

If the integral on the right-hand side can be calculated, the solution can be obtained analytically. If not, the numerical methods described in the next sections can be used. Laplace transforms

6.2. Explicit Methods

Consider the ordinary differential equation

$$\frac{dy}{dt} = f(y)$$

Multiple equations that are still initial value problems can be handled by using the same techniques discussed here. A higher order differential equation

$$y^{(n)} + F(y^{(n-1)}, y^{(n-2)}, \ldots, y', y) = 0$$

with initial conditions

$$G_i(y^{(n-1)}(0), y^{(n-2)}(0), \ldots, y'(0), y(0)) = 0$$
$$i = 1, \ldots, n$$

can be converted into a set of first-order equations. By using

$$y_i \equiv y^{(i-1)} = \frac{d^{(i-1)}y}{dt^{(i-1)}} = \frac{d}{dt} y^{(i-2)} = \frac{dy_{i-1}}{dt}$$

the higher order equation can be written as a set of first-order equations:

$$\frac{dy_1}{dt} = y_2$$

$$\frac{dy_2}{dt} = y_3$$

$$\frac{dy_3}{dt} = y_4$$

$$\ldots$$

$$\frac{dy_n}{dt} = -F(y_{n-1}, y_{n-2}, \ldots, y_2, y_1)$$

The initial conditions would have to be specified for variables $y_1(0), \ldots, y_n(0)$, or equivalently $y(0), \ldots, y^{(n-1)}(0)$. The set of equations is then written as

$$\frac{dy}{dt} = f(y, t)$$

All the methods in this chapter are described for a single equation; the methods apply to the multiple equations as well. Taking the single equation in the form

$$\frac{dy}{dt} = f(y)$$

multiplying by dt, and integrating once yields

$$\int_{t_n}^{t_{n+1}} \frac{dy}{dt'} dt' = \int_{t_n}^{t_{n+1}} f(y(t')) dt'$$

This is

$$y^{n+1} = y^n + \int_{t_n}^{t_{n+1}} \frac{dy}{dt'} dt'$$

The last substitution gives a basis for the various methods. Different interpolation schemes for $y(t)$ provide different integration schemes; using low-order interpolation gives low-order integration schemes [6.1], [6.2].

Euler's method is first order

$$y^{n+1} = y^n + \Delta t f(y^n)$$

Adams – Bashforth Methods. The second-order Adams – Bashforth method is

$$y^{n+1} = y^n + \frac{\Delta t}{2} [3 f(y^n) - f(y^{n-1})]$$

The *fourth-order* Adams – Bashforth method is

$$y^{n+1} = y^n + \frac{\Delta t}{24} [55 f(y^n) - 59 f(y^{n-1}) + 37 f(y^{n-2}) - 9 f(y^{n-3})]$$

Notice that the higher order explicit methods require knowing the solution (or the right-hand side) evaluated at times in the past. Because these were calculated to get to the current time, this presents no problem except for starting the evaluation. Then, Euler's method may have to be used with a very small step size for several steps to generate starting values at a succession of time points. The error terms, order of the method, function evaluations per step, and stability limitations are listed in Table 6. The advantage of the fourth-order Adams – Bashforth method is that it uses only one function evaluation per step and yet achieves high-order accuracy. The disadvantage is the necessity of using another method to start.

Runge – Kutta Methods. Runge – Kutta methods are explicit methods that use several function evaluations for each time step. The general form

Table 6. Properties of integration methods for ordinary differential equations

Method	Error term	Order	Final evaluation step	Stability limit, $\lambda \Delta t \leq$
Explicit methods				
Euler	$\dfrac{h^2}{2} y''$	1	1	2.0
Second-order Adams–Bashforth	$\dfrac{5}{12} h^3 y'''$	2	1	
Fourth-order Adams–Bashforth	$\dfrac{251}{720} h^5 y^{(5)}$	4	1	0.3
Second-order Runge–Kutta	$h^3 y''$	2	2	
Second-order Runge–Kutta (midpoint)		2	2	2.0
Runge–Kutta–Gill		4	4	2.8
Runge–Kutta–Feldberg	$y^{n+1} - z^{n+1}$	5	6	2.1
Predictor–corrector methods				
Second-order Runge–Kutta		2	2	2.0
Adams, fourth-order		2	2	1.3
Implicit methods, stability limit ∞				
Backward Euler		1	many, iterative	∞*
Trapezoid rule		2	many, iterative	2*
Fourth-order Adams–Moulton		4	many, iterative	3*

* Oscillation limit, $\lambda \Delta t \leq$.

of the methods is

$$y^{n+1} = y^n + \sum_{i=1}^{v} w_i k_i$$

with

$$k_i = \Delta t f\left(t^n + c_i \Delta t, y^n + \sum_{j=1}^{i-1} a_{ij} k_j\right)$$

Runge–Kutta methods traditionally have been writen for $f(t, y)$ and that is done here, too. If these equations are expanded and compared with a Taylor series, restrictions can be placed on the parameters of the method to make it first order, second order, etc. Even so, additional parameters can be chosen. A *second-order* Runge–Kutta method is

$$y^{n+1} = y^n + \frac{\Delta t}{2}[f^n + f(t^n + \Delta t, y^n + \Delta t f^n)]$$

The midpoint scheme is another second-order Runge–Kutta method:

$$y^{n+1} = y^n + \Delta t f\left(t^n + \frac{\Delta t}{2}, y^n + \frac{\Delta t}{2} f^n\right)$$

A popular *fourth-order* method is the Runge–Kutta–Gill method with the formulas

$$k_1 = \Delta t f(t^n, y^n)$$

$$k_2 = \Delta t f\left(t^n + \frac{\Delta t}{2}, y^n + \frac{k_1}{2}\right)$$

$$k_3 = \Delta t f\left(t^n + \frac{\Delta t}{2}, y^n + a k_1 + b k_2\right)$$

$$k_4 = \Delta t f\left(t^n + \frac{\Delta t}{2}, y^n + c k_2 + d k_3\right)$$

$$y^{n+1} = y^n + \tfrac{1}{6}(k_1 + k_4) + \tfrac{1}{3}(b k_2 + d k_3)$$

$$a = \frac{\sqrt{2}-1}{2}, \quad b = \frac{2-\sqrt{2}}{2},$$

$$c = -\frac{\sqrt{2}}{2}, \quad d = 1 + \frac{\sqrt{2}}{2}$$

Another fourth-order Runge–Kutta method is given by the Runge–Kutta–Feldberg formulas [6.3]; although the method is fourth-order, it achieves fifth-order accuracy. The popular integration package RKF45 is based on this method.

$$k_1 = \Delta t f(t^n, y^n)$$

$$k_2 = \Delta t f\left(t^n + \frac{\Delta t}{4}, y^n + \frac{k_1}{4}\right)$$

$$k_3 = \Delta t f\left(t^n + \frac{3}{8}\Delta t, y^n + \frac{3}{32}k_1 + \frac{9}{32}k_2\right)$$

$$k_4 = \Delta t\, f\left(t^n + \frac{12}{13}\Delta t,\, y^n + \frac{1932}{2197}k_1\right.$$
$$\left. - \frac{7200}{2197}k_2 + \frac{7296}{2197}k_3\right)$$

$$k_5 = \Delta t\, f\left(t^n + \Delta t,\, y^n + \frac{439}{216}k_1\right.$$
$$\left. - 8k_2 + \frac{3680}{513}k_3 - \frac{845}{4104}k_4\right)$$

$$k_6 = \Delta t\, f\left(t^n + \frac{\Delta t}{2},\, y^n - \frac{8}{27}k_1 + 2k_2\right.$$
$$\left. - \frac{3544}{2565}k_3 + \frac{1859}{4104}k_4 - \frac{11}{40}k_5\right)$$

$$y^{n+1} = y^n + \frac{25}{216}k_1 + \frac{1408}{2565}k_3 + \frac{2197}{4104}k_4 - \frac{1}{5}k_5$$

$$z^{n+1} = y^n + \frac{16}{135}k_1 + \frac{6656}{12825}k_3$$
$$+ \frac{28561}{56430}k_4 - \frac{9}{50}k_5 + \frac{2}{55}k_6$$

The value of $y^{n+1} - z^{n+1}$ is an estimate of the error in y^{n+1} and can be used in step-size control schemes.

Generally, a high-order method should be used to achieve high accuracy. The Runge–Kutta–Gill method is popular because it is high order and does not require a starting method (as does the fourth-order Adams–Bashforth method). However, it requires four function evaluations per time step, or four times as many as the Adams–Bashforth method. For problems in which the function evaluations are a significant portion of the calculation time this might be important. Given the speed of computers and the widespread availability of microcomputers, the efficiency of a method is most important only for very large problems that are going to be solved many times. For other problems the most important criterion for choosing a method is probably the time the user spends setting up the problem.

The stability of an integration method is best estimated by determining the rational polynomial corresponding to the method. Apply this method to the equation

$$\frac{dy}{dt} = -\lambda y,\quad y(0) = 1$$

and determine the formula for r_{mn}:

$$y^{k+1} = r_{mn}(\lambda \Delta t)\, y^k$$

The rational polynomial is defined as

$$r_{mn}(z) = \frac{p_n(z)}{q_m(z)} \approx e^{-z}$$

and is an approximation to $\exp(-z)$, called a Padé approximation. The stability limits are the largest positive z for which

$$|r_{mn}(z)| \le 1$$

The method is A acceptable if the inequality holds for $\mathrm{Re}\, z > 0$. It is $A(0)$ acceptable if the inequality holds for z real, $z > 0$ [6.11]. The method will not induce oscillations about the true solution provided

$$r_{mn}(z) > 0$$

A method is L acceptable if it is A acceptable and

$$\lim_{z \to \infty} r_{mn}(z) = 0$$

For example, Euler's method gives

$$y^{n+1} = y^n - \lambda \Delta t\, y^n \text{ or } y^{n+1} = (1 - \lambda \Delta t)\, y^n$$
or $\quad r_{mn} = 1 - \lambda \Delta t$

The stability limit is then

$$\lambda \Delta t \le 2$$

The Euler method will not oscillate provided

$$\lambda \Delta t \le 1$$

The stability limits listed in Table 6 are obtained in this fashion. The limit for the Euler method is 2.0; for the Runge–Kutta–Gill method it is 2.785; for the Runge–Kutta–Feldberg method it is 2.057. The rational polynomials for the various explicit methods are illustrated in Figure 23. As can be seen, the methods approximate the exact solution well as $\lambda \Delta t$ approaches zero, and the higher order methods give a better approximation at high values of $\lambda \Delta t$.

In solving sets of equations

$$\frac{d\mathbf{y}}{dt} = A\mathbf{y} + \mathbf{f},\quad \mathbf{y}(0) = \mathbf{y}_0$$

all the eigenvalues of the matrix A must be examined. FINLAYSON [6.1] and AMUNDSON [6.4, p. 197–199] both show how to transform these equations into an orthogonal form so that each equation becomes one equation in one unknown, for which single equation analysis applies. For

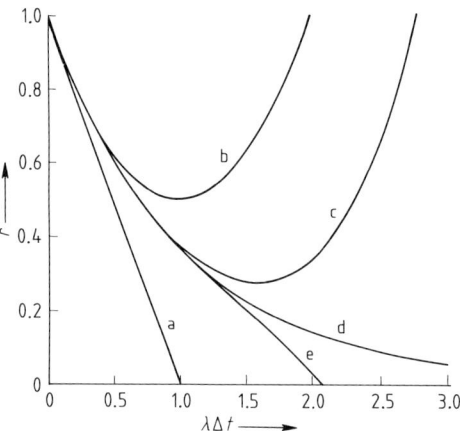

Figure 23. Rational approximations for explicit methods
a) Euler; b) Runge–Kutta–2; c) Runge–Kutta–Gill; d) Exact curve; e) Runge–Kutta–Feldberg

linear problems the eigenvalues do not change, so the stability and oscillation limits must be satisfied for every eigenvalue of the matrix A. When solving nonlinear problems the equations are linearized about the solution at the local time, and the analysis applies for small changes in time, after which a new analysis about the new solution must be made. Thus, for nonlinear problems the eigenvalues keep changing.

Richardson extrapolation can be used to improve the accuracy of a method. Step forward one step Δt with a p-th order method. Then redo the problem, this time stepping forward from the same initial point but in two steps of length $\Delta t/2$, thus ending at the same point. Call the solution of the one-step calculation y_1 and the solution of the two-step calculation y_2. Then an improved solution at the new time is given by

$$y = \frac{2^p y_2 - y_1}{2^p - 1}$$

This gives a good estimate provided Δt is small enough that the method is truly convergent with order p. This process can also be repeated in the same way Romberg's method was used for quadrature (see Section 2.4).

The accuracy of a numerical calculation depends on the step size used, and this is chosen automatically by efficient codes. For example, in the Euler method the local truncation error LTE is

$$\text{LTE} = \frac{\Delta t^2}{2} y_n''$$

Yet the second derivative can be evaluated by using the difference formulas as

$$y_n'' = \nabla(\Delta t\, y_n') = \Delta t\,(y_n' - y_{n-1}') = \Delta t\,(f_n - f_{n-1})$$

Thus, by monitoring the difference between the right-hand side from one time step to another, an estimate of the truncation error is obtained. This error can be reduced by reducing Δt. If the user specifies a criterion for the largest local error estimate, then Δt is reduced to meet that criterion. Also, Δt is increased to as large a value as possible, because this shortens computation time. If the local truncation error has been achieved (and estimated) by using a step size Δt_1

$$\text{LTE} = c\,\Delta t_1^p$$

and the desired error is ε, to be achieved using a step size Δt_2

$$\varepsilon = c\,\Delta t_2^p$$

then the next step size Δt_2 is taken from

$$\frac{\text{LTE}}{\varepsilon} = \left(\frac{\Delta t_1}{\Delta t_2}\right)^p$$

Generally, things should not be changed too often or too drastically. Thus one may choose not to increase Δt by more than a factor (such as 2) or to increase Δt more than once every so many steps (such as 5) [6 5]. In the most sophisticated codes the alternative exists to change the order of the method as well. In this case, the truncation error of the orders one higher and one lower than the current one are estimated, and a choice is made depending on the expected step size and work.

6.3. Implicit Methods

By using different interpolation formulas, involving y^{n+1}, implicit integration methods can be derived. Implicit methods result in a nonlinear equation to be solved for y^{n+1} so that iterative methods must be used. The backward Euler method is a first-order method:

$$y^{n+1} = y^n + \Delta t\, f(y^{n+1})$$

The trapezoid rule (see Section 2.4) is a second-order method:

$$y^{n+1} = y^n + \frac{\Delta t}{2}[f(y^n) + f(y^{n+1})]$$

When the trapezoid rule is used with the finite difference method for solving partial differential equations it is called the Crank–Nicolson method. Adams methods exist as well, and the fourth-order Adams–Moulton method is

$$y^{n+1} = y^n + \frac{\Delta t}{24}[9f(y^{n+1}) + 19f(y^n) - 5f(y^{n-1}) + f(y^{n-2})]$$

The properties of these methods are given in Table 6. The implicit methods are stable for any step size but do require the solution of a set of nonlinear equations, which must be solved iteratively. An application to dynamic distillation problems is given in [6.6].

All these methods can be written in the form

$$y^{n+1} = \sum_{i=1}^{k} \alpha_i y^{n+1-i} + \Delta t \sum_{i=0}^{k} \beta_i f(y^{n+1-i})$$

or

$$y^{n+1} = \Delta t \beta_0 f(y^{n+1}) + w^n$$

where w^n represents known information. This equation (or set of equations for more than one differential equation) can be solved by using successive substitution:

$$y^{n+1,k+1} = \Delta t \beta_0 f(y^{n+1,k}) + w^n$$

Here, the superscript k refers to an iteration counter. The successive substitution method is guaranteed to converge, provided the first derivative of the function is bounded and a small enough time step is chosen. Thus, if it has not converged within a few iterations, Δt can be reduced and the iterations begun again. The Newton–Raphson method (see Section 1.2) would solve the problem as

$$y^{n+1,k+1} = \Delta t \beta_0 f(y^{n+1,k}) + \Delta t \beta_0 \left.\frac{\partial f}{\partial y}\right|_{y^{n+1,k}}$$
$$(y^{n+1,k+1} - y^{n+1,k}) + w^n$$

or

$$\left(I - \Delta t \beta_0 \left.\frac{\partial f}{\partial y}\right|_{y^{n+1,k}}\right)(y^{n+1,k+1} - y^{n+1,k}) =$$
$$= \Delta t \beta_0 f(y^{n+1,k}) + w^n - y^{n+1,k}$$

For multiple equations, I is the Dirac Kronecker function δ_{ij}, and $\partial f/\partial y$ is the Jacobian $\partial f_i/\partial y_j$. As Δt becomes smaller the conditions for convergence are more likely to be satisfied, but if the solution does not converge, Δt can be reduced and the process repeated. In many computer codes, iteration is allowed to proceed only a fixed number of times (e.g., 3) before Δt is reduced. Because a good history of the function is available from previous time steps, a good initial guess is usually possible.

The best software packages for stiff equations (see Section 6.4) use Gear's backward difference formulas. The formulas of various orders are [6.7], [6.8, p. 263]

1: $y^{n+1} = y^n + \Delta t f(y^{n+1})$

2: $y^{n+1} = \frac{4}{3} y^n - \frac{1}{3} y^{n-1} + \frac{2}{3} \Delta t f(y^{n+1})$

3: $y^{n+1} = \frac{18}{11} y^n - \frac{9}{11} y^{n-1} + \frac{2}{11} y^{n-2}$
$+ \frac{6}{11} \Delta t f(y^{n+1})$

4: $y^{n+1} = \frac{48}{25} y^n - \frac{36}{25} y^{n-1} + \frac{16}{25} y^{n-2} - \frac{3}{25} y^{n-3}$
$+ \frac{12}{25} \Delta t f(y^{n+1})$

5: $y^{n+1} = \frac{300}{137} y^n - \frac{300}{137} y^{n-1} + \frac{200}{137} y^{n-2}$
$- \frac{75}{137} y^{n-3} + \frac{12}{137} y^{n-4}$
$+ \frac{60}{137} \Delta t f(y^{n+1})$

The stability properties of these methods are determined in the same way as explicit methods. They are always expected to be stable, no matter what the value of Δt is, and this is confirmed in Figure 24.

Predictor–corrector methods can be employed in which an explicit method is used to predict the value of y^{n+1}. This value is then used in an implicit method to evaluate $f(y^{n+1})$. The first-order method is merely the Runge–Kutta method described above.

$$\bar{y}^{n+1} = y^n + \Delta t f(y^n)$$

$$y^{n+1} = y^n + \frac{\Delta t}{2}[f(\bar{y}^{n+1}) + f(y^n)]$$

Note we do not iterate on y^{n+1}, as we would in implicit methods. The fourth-order Adams predictor–corrector method uses the Adams–Bashforth method to provide a predicted value:

$$\bar{y}^{n+1} = y^n + \frac{\Delta t}{24}[55 f(y^n) + \ldots]$$

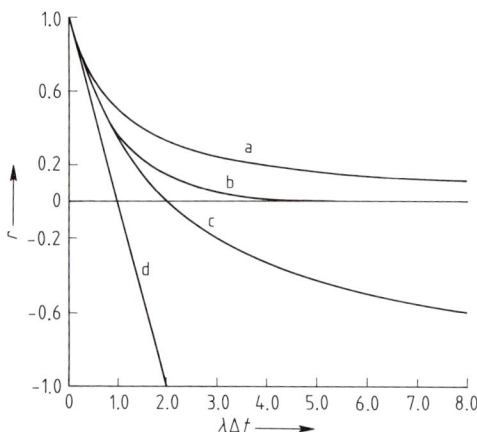

Figure 24. Rational approximations for implicit methods
a) Backward Euler; b) Exact curve; c) Trapezoid; d) Euler

and the Adams–Moulton method to correct it:

$$y^{n+1} = y^n + \frac{\Delta t}{24}[9 f(\bar{y}^{n+1}) + 19 f(y^n) + \ldots]$$

The last step can be applied over and over, if desired. The stability properties of these methods are listed in Table 6. Step-size control can also be employed for these methods.

6.4. Stiffness

Why is it desirable to use implicit methods that lead to sets of algebraic equations that must be solved iteratively whereas explicit methods lead to a direct calculation? The reason lies in the stability limits; to understand their impact, the concept of stiffness is necessary. When modeling a physical situation, the time constants governing different phenomena should be examined. Consider flow through a packed bed, as illustrated in Figure 25.

The superficial velocity u is given by

$$u = \frac{Q}{A\varphi}$$

where Q is the volumetric flow rate, A is the cross-sectional area, and φ is the void fraction. A time constant for flow through the device is then

$$t_{flow} = \frac{L}{u} = \frac{\varphi A L}{Q}$$

where L is the length of the packed bed. If a chemical reaction occurs, with a reaction rate

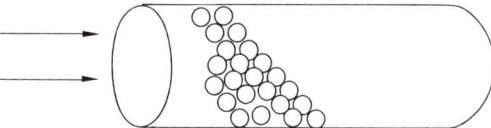

Figure 25. Flow through packed bed

given by

$$\frac{\text{Moles}}{\text{Volume time}} = -kc$$

where k is the rate constant (time^{-1}) and c is the concentration (moles/volume), the characteristic time for the reaction is

$$t_{rxn} = \frac{1}{k}$$

If diffusion occurs inside the catalyst, the time constant is

$$t_{internal\ diffusion} = \frac{\varepsilon R^2}{D_e}$$

where ε is the porosity of the catalyst, R is the catalyst radius, and D_e is the effective diffusion coefficient inside the catalyst. The time constant for heat transfer is

$$t_{internal\ heat\ transfer} = \frac{R^2}{\alpha} = \frac{\varrho_s C_s R^2}{k_e}$$

where ϱ_s is the catalyst density, C_s is the catalyst heat capacity per unit mass, k_e is the effective thermal conductivity of the catalyst, and α is the thermal diffusivity. The time constants for diffusion of mass and heat through a boundary layer surrounding the catalyst are

$$t_{external\ diffusion} = \frac{R}{k_g}$$

$$t_{external\ heat\ transfer} = \frac{\varrho_s C_s R}{h_p}$$

where k_g and h_p are the mass-transfer and heat-transfer coefficients, respectively. The importance of examining these time constants comes from realization that their orders of magnitude differ greatly. For example, in the model of an automobile catalytic converter [6.9] the time constant for internal diffusion was 0.3 s, for internal heat transfer 21 s, and for flow through the device 0.003 s. Flow through the device is so fast that it might as well be instantaneous. Thus, the time derivatives could be dropped from the mass balance equations for the flow, leading to a set of differential-algebraic equations (see below). If the original equations had to be solved, the eigenvalues would be roughly proportional to the inverse of the time constants. The time interval over which to integrate would be a small number (e.g., five) multiplied by the longest time constant. Yet the explicit stability limitation applies to all the eigenvalues, and the largest eigenvalue would determine the largest permissible time step. If that eigenvalue was very large, very small time steps would have to be used, but a long integration would be required to reach steady state. Such problems are termed stiff, and implicit methods are very useful for them. In that case the stable time constant is not of any interest, because any time step is stable. What is of interest is the largest step for which a solution can be found. If a time step larger than the smallest time constant is used, then any phenomena represented by that smallest time constant will be overlooked—at least transients in it will be smeared over. However, the method will still be stable. Thus, if the very rapid transients of part of the model are not of interest, they can be ignored and an implicit method used.

The idea of stiffness is best explained by considering a system of linear equations:

$$\frac{dy}{dt} = Ay$$

Let λ_i be the eigenvalues of the matrix A. This system can be converted into a system of n equations, each of them having only one unknown; the eigenvalues of the new system are the same as the eigenvalues of the original system [6.1, pp. 39–42], [6.4, pp 197–199], [6.10]. Then the stiffness ratio SR is defined as [6.11, p. 32]

$$SR = \frac{\max_i |Re(\lambda_i)|}{\min_i |Re(\lambda_i)|}$$

SR = 20 is not stiff, SR = 10^3 is stiff, and SR = 10^6 is very stiff. If the problem is nonlinear, the solution is expanded about the current state:

$$\frac{dy_i}{dt} = f_i[y(t^n)] + \sum_{j=1}^{n} \frac{\partial f_i}{\partial y_j}[y_j - y_j(t^n)]$$

The question of stiffness then depends on the eigenvalue of the Jacobian at the current time. Consequently, for nonlinear problems the problem can be stiff during one time period and not stiff during another. Packages have been developed for problems such as these. Although the chemical engineer may not actually calculate the eigenvalues, knowing that they determine the stability and accuracy of the numerical scheme, as well as the step size employed, is useful.

6.5. Differential–Algebraic Systems

Sometimes models involve ordinary differential equations subject to some algebraic constraints. For example, the equations governing one equilibrium stage (as in a distillation column) are

$$M \frac{dx^n}{dt} = V^{n+1} y^{n+1} - L^n x^n - V^n y^n + L^{n-1} x^{n-1}$$

$$x^{n-1} - x^n = E^n(x^{n-1} - x^{*,n})$$

$$\sum_{i=1}^{N} x_i = 1$$

where x and y are the mole fractions in the liquid and vapor, respectively; L and V are liquid and vapor flow rates, respectively; M is the holdup; and the superscript n is the stage number. The efficiency is E, and the concentration in equilibrium with the vapor is x^*. The first equation is an ordinary differential equation for the mass of one component on the stage, whereas the third equation represents a constraint that the mass fractions add to one. As a second example, the following kinetics problem can be considered:

$$\frac{dc_1}{dt} = f(c_1, c_2)$$

$$\frac{dc_2}{dt} = k_1 c_1 - k_2 c_2^2$$

The first equation could be the equation for a stirred tank reactor, for example. Suppose both k_1 and k_2 are large. The problem is then stiff, but the second equation could be taken at equilibrium. If

$$c_1 \rightleftharpoons 2c_2$$

The equilibrium condition is then

$$\frac{c_2^2}{c_1} = \frac{k_1}{k_2} \equiv K$$

Under these conditions the problem becomes

$$\frac{dc_1}{dt} = f(c_1, c_2)$$

$$0 = k_1 c_1 - k_2 c_2^2$$

Thus, a differential-algebraic system of equations is obtained. In this case, the second equation can be solved and substituted into the first to obtain differential equations, but in the general case that is not possible.

Differential-algebraic equations can be written in the general notation

$$F\left(t, y, \frac{dy}{dt}\right) = 0$$

or the variables and equations may be separated according to whether they come primarily from differential or algebraic equations:

$$\frac{dy}{dt} = f(t, y, x), \quad g(t, y, x) = 0$$

Another form is not strictly a differential-algebraic set of equations, but the same principles apply; this form arises frequently when the Galerkin finite element is applied:

$$A \frac{dy}{dt} = f(y)$$

Suppose the general problem is to be solved by using the backward Euler method. Then, the nonlinear differential equation is replaced by the nonlinear algebraic equation for one step:

$$F\left(t, y^{n+1}, \frac{y^{n+1} - y^n}{\Delta t}\right) = 0$$

This equation must be solved for y^{n+1}. The Newton–Raphson method can be used, and if convergence is not achieved within a few iterations, the time step can be reduced and the step repeated. In actuality, higher order backward-difference Gear methods are used in the computer program DASSL [6.12].

Differential-algebraic systems are more complicated than differential systems because the solution may not always be defined. PONTELIDES et al. [6.13] introduced the term "index" to identify possible problems. The index is defined as the minimum number of times the equations must be differentiated with respect to time to convert the system to a set of ordinary differential equations. These higher derivatives may not exist, and the process places limits on which variables can be given initial values. Sometimes the initial values must be constrained by the algebraic equations [6.13]. For a differential-algebraic system modeling a distillation tower, the index depends on the specification of pressure for the column [6.13]. Several chemical engineering examples of differential-algebraic systems and a solution for one involving two-phase flow are given in [6.14].

6.6. Computer Software

Efficient software packages are widely available for solving ordinary differential equations as initial value problems. Three of them—RKF45, LSODE, and EPISODE—are discussed here. In each of the packages the user specifies the differential equation to be solved and a desired error criterion. The package then integrates in time and adjusts the step size to achieve the error criterion, within the limitations imposed by stability.

A popular explicit Runge–Kutta package is RKF45, developed by FORSYTHE et al. [6.3, Chap. 6]. The method is based on the Runge–Kutta–Feldberg formulas (see Section 6.2). An estimate of the truncation error at each step is available. Then the step size can be reduced until this estimate is below the user-specified tolerance. The method is thus automatic, and the user is assured of the results. Note, however, that the tolerance is set on the local truncation error, namely, from one step to another, whereas the user is generally interested in the global trunction error, i.e., the error after several steps. The global error is generally made smaller by making the tolerance smaller, but the absolute accuracy is not the same as the tolerance. If the problem is stiff, then very small step sizes are used and the computation becomes very lengthy. The RKF45 code discovers this and returns control to the user with a message indicating the problem is too hard to solve with RKF45.

A popular implicit package is LSODE, a version of Gear's method [6.7] written by ALAN HINDMARSH at Lawrence Livermore Laboratory [6.15]. Earlier versions of this were GEAR and GEARB [6.16]. In this package, the user specifies the differential equation to be solved and the tolerance desired. Now the method is implicit and, therefore, stable for any step size. The accuracy may not be acceptable, however, and sets of nonlinear equations must be solved. Thus, in practice, the step size is limited but not nearly so much as in the Runge–Kutta methods. In these packages both the step size and the order of the method are adjusted by the package itself. Suppose a k-th order method is being used. The truncation error is determined by the $(k+1)$-th order derivative. This is estimated by using difference formulas and the values of the right-hand sides at previous times. An estimate is also made for the k-th and $(k+2)$-th derivative. Then, the errors in a $(k-1)$-th order method, a k-th order method, and a $(k+1)$-th order method can be estimated. Furthermore, the step size required to satisfy the tolerance with each of these methods can be determined. Then the method and step size for the next step that achieves the biggest step can be chosen, with appropriate adjustments due to the different work required for each order. The package generally starts with a very small step size and a first-order method—the backward Euler method. Then it integrates along, adjusting the order up (and later down) depending on the error estimates. The user is thus assured that the local truncation error meets the tolerance. A further difficulty arises because the set of nonlinear equations must be solved. Usually a good guess of the solution is available, because the solution is evolving in time and past history can be extrapolated. Thus, the Newton–Raphson method will usually converge. The package protects itself, though, by only doing a few (i.e., three) iterations. If convergence is not reached within these iterations, the step size is reduced and the calculation is redone for that time step. The convergence theorem for the Newton–Raphson method (Chap. 1) indicates that the method will converge if the step size is small enough. Thus, the method is guaranteed to work. Further economies are possible. The Jacobian needed in the Newton–Raphson method can be fixed over several time steps. Then if the iteration does not converge, the Jacobian can be reevaluated at the current time step. If the iteration still does not converge, then the step size is reduced and a new Jacobian is evaluated. The successive substitution method can also be used—which is even faster, except that it may not converge. However, it too will converge if the time step is small enough.

Comparisons of the methods and additional details are provided for chemical engineering problems by FINLAYSON [6.1, pp. 54–56] and CARNAHAN and WILKES [6.8]. Generally the Runge–Kutta methods give extremely good accuracy, especially when the step size is kept small for the stability reason. When the computation time is comparable for LSODE and RKF45, the RKF45 package generally gives much more accurate results. The RKF45 package is unsuitable, however, for many chemical reactor problems because these problems are so stiff. Other comparisons are available by HULL et al. [6.17], SHAMPINE et al. [6.18], BYRNE et al. [6.19], and JUNCE et al. [6.20]. Other packages are available through IMSL etc; see [6.5, pp. 291–292] and [6.21, pp. 439, 451]. Generally, standard packages must have a high-order explicit method (usually a version of Runge–Kutta) and a multistep, implicit method (usually a version of GEAR, EPISODE, or LSODE). The package DASSL [6.12] uses similar principles to solve the differential-algebraic systems. The packages LSODA and LSODAR are sister programs to LSODE. They have the feature of switching between stiff and nonstiff methods automatically; LSODAR also has a stopping feature when some condition is satisfied, which can be used for finding roots.

[The software described here is available by electronic mail over BITNET. Sending the message:

"mail netlib@ornl.gov", "send index"

will retrieve an index and descriptions of how to obtain the software. The packages DASSL and ODEPACK (containing LSODE...LSODAR) can be obtained for a nominal fee on a computer tape from Professor W. E. Schiesser, Department of Chemical Engineering, Lehigh University, Bethlehem, Pennsylvania 18015.]

6.7. Stability, Bifurcations, Limit Cycles

In this section, bifurcation theory is discussed in a general way. Some aspects of this subject involve the solution of nonlinear equations; other aspects involve the integration of ordinary differential equations; applications include chaos and fractals as well as unusual operation of some chemical engineering equipment. An excellent introduction to the subject and details needed to apply the methods are given in [6.22]. For more details of the algorithms described below and a concise survey with some chemical engineering examples, see [6.23] and [6.24]. Bifurcation results are closely connected with stability of the steady states, which is essentially a transient phenomenon.

Consider the problem

$$\frac{\partial u}{\partial t} = F(u, \lambda)$$

The variable u can be a vector, which makes F a vector, too. Here, F represents a set of equations that can be solved for the steady state:

$$F(u, \lambda) = 0$$

If the Newton–Raphson method is applied,

$$F_u^s \delta u^s = -F(u^s, \lambda)$$
$$u^{s+1} = u^s + \delta u^s$$

is obtained, where

$$F_u^s = \frac{\partial F}{\partial u}(u^s)$$

is the Jacobian. Look at some property of the solution, perhaps the value at a certain point or the maximum value or an integral of the solution. This property is plotted versus the parameter λ; typical plots are shown in Figure 26. At the point shown in Figure 26A, the determinant of the Jacobian is zero:

$$\det F_u = 0$$

For the limit point,

$$\frac{\partial F}{\partial \lambda} \neq 0$$

whereas for the bifurcation-limit point

$$\frac{\partial F}{\partial \lambda} = 0$$

The stability of the steady solutions is also of interest. Suppose a steady solution u_{ss}; the function u is written as the sum of the known steady state and a perturbation u':

$$u = u_{ss} + u'$$

This expression is substituted into the original equation and linearized about the steady-state

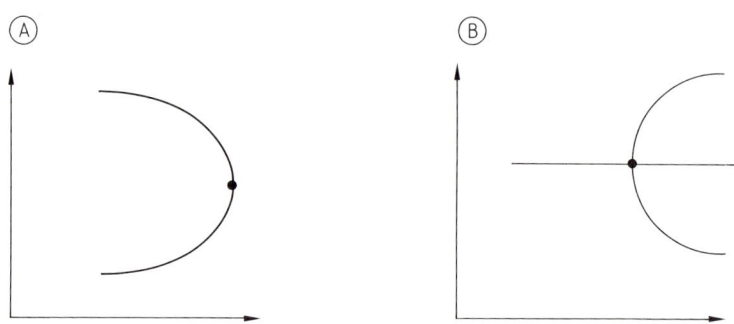

Figure 26. Limit points and bifurcation–limit points
A) Limit point (or turning point); B) Bifurcation-limit point (or singular turning point or bifurcation point)

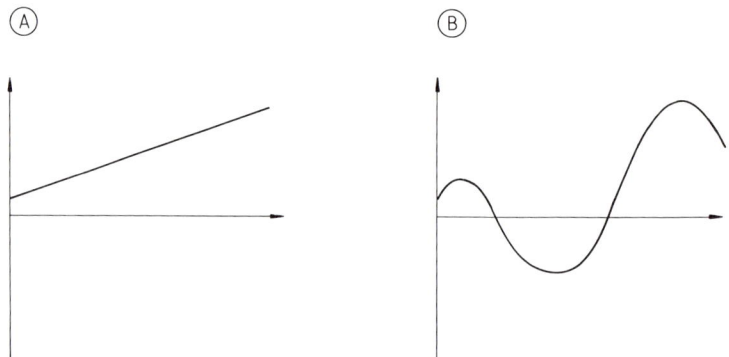

Figure 27. Stationary and oscillatory instability
A) Stationary instability; B) Oscillatory instability

value:

$$\frac{\partial u_{ss}}{\partial t} + \frac{\partial u'}{\partial t} = F(u_{ss} + u', \lambda)$$

$$\approx F(u_{ss}, \lambda) + \left.\frac{\partial F}{\partial u}\right|_{u_{ss}} u' + \ldots$$

The result is

$$\frac{\partial u'}{\partial t} = F_u^{ss} u'$$

A solution of the form

$$u'(x, t) = e^{\sigma t} X(x)$$

gives

$$\sigma e^{\sigma t} X = F_u^{ss} e^{\sigma t} X$$

The exponential term can be factored out and

$$(F_u^{ss} - \sigma \delta) X = 0$$

A solution exists for X if and only if

$$\det |F_u^{ss} - \sigma \delta| = 0$$

The σ are the eigenvalues of the Jacobian. Now clearly if $\text{Re}(\sigma) > 0$ then u' grows with time, and the steady solution u_{ss} is said to be unstable to small disturbances. If $\text{Im}(\sigma) = 0$ it is called stationary instability, and the disturbance would grow monotonically, as indicated in Figure 27 A. If $\text{Im}(\sigma) \ne 0$ then the disturbance grows in an oscillatory fashion, as shown in Figure 27 B, and is called oscillatory instability. The case in which $\text{Re}(\sigma) = 0$ is the dividing point between stability and instability. If $\text{Re}(\sigma) = 0$ and $\text{Im}(\sigma) = 0$ — the point governing the onset of stationary instability — then $\sigma = 0$. However, this means that $\sigma = 0$ is an eigenvalue of the Jacobian, and the determinant of the Jacobian is zero. Thus, the points at which the determinant of the Jacobian is zero (for limit points and bifurcation-limit points) are the points governing the onset of stationary instability. When $\mathrm{Re}(\sigma) = 0$ but $\mathrm{Im}(\sigma) \neq 0$, which is the onset of oscillatory instability, an even number of eigenvalues pass from the left-hand complex plane to the right-hand complex plane. The eigenvalues are complex conjugates of each other (a result of the original equations being real, with no complex numbers), and this is called a Hopf bifurcation. Numerical methods to study Hopf bifurcation are very computationally intensive and are not discussed here [6.22].

To return to the problem of solving for the steady-state solution: near the limit point or bifurcation-limit point two solutions exist that are very close to each other. In solving sets of equations with thousands of unknowns, the difficulties in convergence are obvious. For some dependent variables the approximation may be converging to one solution, whereas for another set of dependent variables it may be converging to the other solution; or the two solutions may all be mixed up. Thus, solution is difficult near a bifurcation point, and special methods are required. These methods are discussed in [6.23].

The first approach is to use *natural continuation* (also known as Euler–Newton continuation). Suppose a solution exists for some parameter λ. Call the value of the parameter λ_0 and the corresponding solution u_0. Then

$$F(u_0, \lambda_0) = 0$$

Also, compute u_λ as the solution to

$$F_u^{ss} u_\lambda = -F_\lambda$$

at this point $[\lambda_0, u_0]$. Then predict the starting guess for another λ using

$$u^0 = u_0 + u_\lambda (\lambda - \lambda_0)$$

and apply Newton–Raphson with this initial guess and the new value of λ. This will be a much better guess of the new solution than just u_0 by itself.

Even this method has difficulties, however. Near a limit point the determinant of the Jacobian may be zero and the Newton method may fail. Perhaps no solutions exist at all for the chosen parameter λ near a limit point. Also, the ability to switch from one solution path to another at a bifurcation-limit point is necessary. Thus, other methods are needed as well: arc-length continuation and pseudo-arc-length continuation

[6.23]. The latter method can use the arrow matrix LU decomposition described in Chapter 1.

6.8. Sensitivity Analysis

Often, when solving differential equations, the solution as well as the sensitivity of the solution to the value of a parameter must be known. Such information is useful in doing parameter estimation (to find the best set of parameters for a model) and in deciding whether a parameter needs to be measured accurately. The differential equation for $y(t, \alpha)$ where α is a parameter, is

$$\frac{dy}{dt} = f(y, \alpha), \quad y(0) = y_0$$

If this equation is differentiated with respect to α, then because y is a function of t and α

$$\frac{\partial}{\partial \alpha}\left(\frac{dy}{dt}\right) = \frac{\partial f}{\partial y}\frac{\partial y}{\partial \alpha} + \frac{\partial f}{\partial \alpha}$$

Exchanging the order of differentiation in the first term leads to the ordinary differential equation

$$\frac{d}{dt}\left(\frac{\partial y}{\partial \alpha}\right) = \frac{\partial f}{\partial y}\frac{\partial y}{\partial \alpha} + \frac{\partial f}{\partial \alpha}$$

The initial conditions on $\partial y/\partial \alpha$ are obtained by differentiating the initial conditions

$$\frac{\partial}{\partial \alpha}[y(0, \alpha) = y_0], \text{ or } \frac{\partial y}{\partial \alpha}(0) = 0$$

Next, let

$$y_1 = y, \quad y_2 = \frac{\partial y}{\partial \alpha}$$

and solve the set of ordinary differential equations

$$\frac{dy_1}{dt} = f(y_1, \alpha) \quad y_1(0) = y_0$$

$$\frac{dy_2}{dt} = \frac{\partial f}{\partial y}(y_1, \alpha) y_2 + \frac{\partial f}{\partial \alpha} \quad y_2(0) = 0$$

Thus, the solution $y(t, \alpha)$ and the derivative with respect to α are obtained. To project the impact of α, the solution for $\alpha = \alpha_1$ can be used:

$$y(t, \alpha) = y_1(t, \alpha_1) + \frac{\partial y}{\partial \alpha}(t, \alpha_1)(\alpha - \alpha_1) + \ldots$$

$$= y_1(t, \alpha_1) + y_2(t, \alpha_1)(\alpha - \alpha_1) + \ldots$$

This is a convenient way to determine the sensitivity of the solution to parameters in the problem.

6.9. Eigenvalues and Roots by Initial Value Techniques

Initial value methods can also be used to find roots of equations. Doing this requires a code that stops when the dependent variable (e.g., y) takes a certain value, as opposed to stopping when the independent variable takes a certain value (e.g., t). The RKF45 code has been modified in this way [6.25], [6.26], as has the LSODAR code. To find the values of t that satisfy

$$f(t) = 0$$

set

$$y = f(t)$$

and differentiate to get

$$\frac{dy}{dt} = \frac{df}{dt}$$

Now integrate the problem

$$\frac{dy}{dt} = \frac{df}{dt} \equiv g(t), \quad y(0) = y_0 \text{ (arbitrary)}$$

until $y = 0$. The t for which this occurs gives the root to the equation. Continued integration will give multiple roots.

This technique can also be used to solve certain eigenvalue problems. If separation of variables is applied to unsteady heat conduction

$$\frac{\partial T}{\partial t} = \frac{1}{r}\frac{\partial}{\partial r}\left(r\frac{\partial T}{\partial r}\right)$$

$$T(1, t) = 0, \quad \frac{\partial T}{\partial r}(0, t) = 0$$

the following is obtained:

$$T(r, t) = T(t) R(r)$$

$$\frac{1}{T}\frac{dT}{dt} = \frac{1}{rR}\frac{d}{dr}\left(r\frac{dR}{dr}\right) = \text{constant} = -\alpha^2$$

Then the function $R(r)$ must satisfy

$$\frac{1}{r}\frac{d}{dr}\left(r\frac{dR}{dr}\right) + \alpha^2 R(r) = 0$$

or

$$\frac{d^2R}{dr^2} + \frac{1}{r}\frac{dR}{dr} + \alpha^2 R(r) = 0, \quad R(1) = 0, \quad \frac{dR}{dr}(0) = 0$$

The solution is a zero-order Bessel function.

$$R(r) = J_0(\alpha_k r) \text{ because } J_0(\alpha_k) = 0$$

Now a new variable is defined

$$z = \alpha r$$

and the equation rearranged for R:

$$\frac{d^2R}{dz^2} + \frac{1}{z}\frac{dR}{dz} + R = 0, \quad \frac{dR}{dz}(0) = 0, \quad R(0) = 1$$

(arbitrary, take = 1)

This equation can be solved as an initial value problem. The solution is continued until

$$R = 0$$

Suppose this occurs for z_k. To have

$$R = 0$$

when

$$r = 1$$

choose

$$\alpha_k = z_k$$

The function

$$R(\alpha_k r) = R(z)$$

then satisfies the correct differential equation and the appropriate boundary conditions. This gives a method of finding the Bessel functions of order zero.

When the Graetz problem for heat transfer to a fluid flowing in a tube is considered,

$$(1 - r^2)\frac{\partial T}{\partial t} = \frac{1}{r}\frac{\partial}{\partial r}\left(r\frac{\partial T}{\partial r}\right)$$

$$T(1, t) = 0, \quad \frac{\partial T}{\partial r}(0, t) = 0$$

the corresponding eigenvalue problem is

$$\frac{d^2R}{dr^2} + \frac{1}{r}\frac{dR}{dr} + \alpha^2(1 - r^2) R(r) = 0,$$

$$R(1) = 0, \quad \frac{dR}{dr}(0) = 0$$

Now the initial value methods apparently cannot be used because the transformation is not possible.

7. Ordinary Differential Equations as Boundary Value Problems

Diffusion problems in one dimension lead to boundary value problems. The boundary conditions are applied at two different spatial locations: at one side the concentration may be fixed and at the other side the flux may be fixed. Because the conditions are specified at two different locations the problems are not initial value in character. To begin at one position and integrate directly is impossible because at least one of the conditions is specified somewhere else and not enough conditions are available to begin the calculation. Thus, methods have been developed especially for boundary value problems.

7.1. Solution by Quadrature

When only one equation exists, even if it is nonlinear, it may possibly be solved by quadrature. For

$$\frac{dy}{dt} = f(y)$$

$$y(0) = y_0$$

the problem can be separated

$$\frac{dy}{f(y)} = dt$$

and integrated

$$\int_{y_0}^{y} \frac{dy'}{f(y')} = \int_0^t dt = t$$

If the quadrature can be performed analytically, the exact solution has been found.

As an example, consider the flow of a non-Newtonian fluid in a pipe, as illustrated in Figure 28. The governing differential equation is [7.1]

$$\frac{1}{r}\frac{d}{dr}(r\tau) = -\frac{\Delta p}{L}$$

where r is the radial position from the center of the pipe, τ is the shear stress, Δp is the pressure drop along the pipe, and L is the length over which the pressure drop occurs. The variables are separated once

$$d(r\tau) = -\frac{\Delta p}{L} r \, dr$$

and then integrated to give

$$r\tau = -\frac{\Delta p}{L}\frac{r^2}{2} + c_1$$

Proceeding further requires choosing a constitutive relation relating the shear stress and the velocity gradient as well as a condition specifying the constant. For a Newtonian fluid

$$\tau = -\eta \frac{dv}{dr}$$

where v is the velocity and η the viscosity. Then the variables can be separated again and the result integrated to give

$$-\eta v = -\frac{\Delta p}{L}\frac{r^2}{4} + c_1 \ln r + c_2$$

Now the two unknowns must be specified from the boundary conditions. This problem is a two-point boundary value problem because one of the conditions is usually specified at $r = 0$ and the other at $r = R$, the tube radius. However, the technique of separating variables and integrating works quite well.

When the fluid is non-Newtonian, it may not be possible to do the second step analytically. For example, for the Bird–Carreau fluid [7.2, p. 171], stress and velocity are related by

$$\tau = \frac{\eta_0}{\left[1 + \lambda\left(\frac{dv}{dr}\right)^2\right]^{(1-n)/2}}$$

where η_0 is the viscosity at $v = 0$ and λ the time constant.

Putting this value into the equation for stress as a function of r gives

$$\frac{\eta_0}{\left[1 + \lambda\left(\frac{dv}{dr}\right)^2\right]^{(1-n)/2}} = -\frac{\Delta p}{L}\frac{r}{2} + \frac{c_1}{r}$$

This equation cannot be solved analytically for dv/dr, except for special values of n. For problems, such as this, numerical methods must be used.

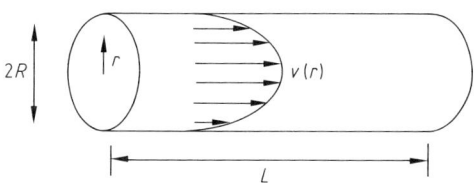

Figure 28. Flow in pipe

7.2. Shooting Methods

A shooting method is one that utilizes the techniques for initial value problems but allows for an iterative calculation to satisfy all the boundary conditions. Suppose the nonlinear boundary value problem

$$\frac{d^2y}{dx^2} = f\left(x, y, \frac{dy}{dx}\right)$$

with the boundary conditions

$$a_0 y(0) - a_1 \frac{dy}{dx}(0) = \alpha, \quad a_i \geq 0$$

$$b_0 y(1) + b_1 \frac{dy}{dx}(1) = \beta, \quad b_i \geq 0$$

Convert this second-order equation into two first-order equations along with the boundary conditions written to include a parameter s.

$$\frac{du}{dx} = v$$

$$\frac{dv}{dx} = f(x, u, v)$$

$$u(0) = a_1 s - c_1 \alpha$$

$$v(0) = a_0 s - c_0 \alpha$$

The parameters c_0 and c_1 are specified by the analyst such that

$$a_1 c_0 - a_0 c_1 = 1$$

This ensures that the first boundary condition is satisfied for any value of parameter s. If the proper value for s is known, $u(0)$ and $u'(0)$ can be evaluated and the equation integrated as an initial value problem. The parameter s should be chosen so that the last boundary condition is satisfied. Define the function

$$\chi(s) = b_0 u(1, s) + b_1 \frac{du}{dx}(1, s) - \beta$$

$$= b_0 u(1, s) + b_1 v(1, s) - \beta$$

which should preferably be zero. Note that the solution depends on s.

$$\chi(s) = 0$$

Iteration on s is required to find the solution to the problem. The condition at $x = 0$ is satisfied for any s; the differential equation is satisfied by the integration routine; and all that must be ensured is that the last boundary condition is satisfied. A successive substitution method would use

$$s^{k+1} = s^k - m\chi(s^k)$$

KELLER [7.3] showed that if

$$\frac{\partial f}{\partial y} \leq N$$

for some N and $0 < m < 2\Gamma$, where Γ increases as N increases, then the iteration scheme converges as k approaches infinity.

Newton's method of iteration would use

$$s^{k+1} = s^k - \frac{\chi(s^k)}{\frac{d\chi}{ds}(s^k)}$$

The function $d\chi/ds$ is determined by integrating two more equations obtained by differentiating the original equations with respect to s. If

$$\zeta = \frac{\partial u(x, s)}{\partial s}, \quad \eta = \frac{\partial v(x, s)}{\partial s}$$

the additional differential equations are

$$\frac{d\zeta}{dx} = \eta, \quad \frac{d\eta}{dx} = \frac{\partial f}{\partial u}\zeta + \frac{\partial f}{\partial v}\eta$$

These equations are integrated along with the initial conditions

$$\zeta(0) = a_1, \quad \eta(0) = a_0$$

Then the derivative is

$$\frac{d\chi}{ds} = b_0 \zeta(1, s) + b_1 \eta(1, s)$$

The process can be used when multiple conditions must be satisfied, such as might arise when more than one dependent variable exists. Suppose two equations must be satisfied at one boundary

$$\chi_1(s_1, s_2) = 0, \quad \chi_2(s_1, s_2) = 0$$

Then two variables s_1 and s_2 are introduced, and the Newton–Raphson method is applied in the form

$$\begin{bmatrix} \frac{\partial \chi_1}{\partial s_1} & \frac{\partial \chi_1}{\partial s_2} \\ \frac{\partial \chi_2}{\partial s_1} & \frac{\partial \chi_2}{\partial s_2} \end{bmatrix} \begin{bmatrix} s_1^{k+1} - s_1^k \\ s_2^{k+1} - s_2^k \end{bmatrix} = -\begin{bmatrix} \chi_1(s_1^k, s_2^k) \\ \chi_2(s_1^k, s_2^k) \end{bmatrix}$$

The problem for *reaction and diffusion in a catalyst pellet* is

$$\frac{1}{r^{a-1}}\frac{d}{dr}\left(r^{a-1}\frac{dc}{dr}\right) = \varphi^2 R(c) \qquad (15)$$

$$\frac{dc}{dr}(0) = 0, \quad c(1) = 1$$

where φ is the Thiele modulus. WEISZ and HICKS [7.4] showed how to transform this problem into one that can be solved by using initial value methods. The problem can be written as

$$\frac{1}{z^{a-1}}\frac{d}{dz}\left(z^{a-1}\frac{dc}{dz}\right) = d^2 R(c)$$

$$\frac{dc}{dz}(0) = 0, \quad c(bz = 1) = 1$$

by using $r = bz$ and $d = b\varphi$. For any d, choose an arbitrary $c(0) = c_0$ and integrate until the concentration reaches 1.0. Suppose that occurs at $z = z_1$. Then, let

$$b = \frac{1}{z_1}, \quad \varphi = \frac{d}{b} = dz_1$$

and the solution for the problem is obtained without iteration. We do not know in advance what φ the problem has been solved for, so this technique is especially useful to solve the problem for a range of φ. Employing this method requires a code that stops when the dependent variable (here c) equals some value, rather than one that stops when the independent variable (here z) equals some value.

The model for a *chemical reactor with axial diffusion* is

$$\frac{1}{Pe}\frac{dc^2}{dz^2} - \frac{dc}{dz} = Da\,R(c),$$

$$-\frac{1}{Pe}\frac{dc}{dz}(0) + c(0) = c_{in}, \quad \frac{dc}{dz}(1) = 0$$

where Pe is the Péclet number and Da the Damköhler number.

The boundary conditions are due to DANCKWERTS [7.5] and to WEHNER and WILHELM [7.6]. This problem can be treated by using initial value methods also, but the method is highly sensitive to the choice of the parameter s, as outlined above. Starting at $z = 0$ and making small changes in s will cause large changes in the solution at the exit, and the boundary condition at the exit may be impossible to satisfy. By starting at $z = 1$, however, and integrating backwards, the process works and an iterative scheme converges in many cases [7.7]. However, if the problem is extremely nonlinear the iterations may not converge. In such cases, the methods for boundary value problems described below must be used.

Computer software exists for solving two-point boundary value problems: for example, the IMSL program DTPTB used DVERK, which employs Runge–Kutta integration to integrate the ordinary differential equations [7.8, p. 301].

Initial value methods can also be used when the sensitivity of the solution to parameters must be determined. The sensitivity might be desired for studying the stability of the equations, especially when the solution bifurcates into more than one solution, or it might be required for parameter estimation. Knowing the sensitivity gives clues as to what experimental measurements are most important for verifying a mathematical model. As an example, suppose the sensitivity of the solution to Equation (15) with respect to the Thiele modulus squared φ^2 must be known. The variables are defined

$$u(r) = c(r), \quad v(r) = \frac{du}{dr}$$

and the differential equation is written as

$$\frac{du}{dr} = v$$

$$\frac{dv}{dr} = \varphi^2 R(u) - \frac{a-1}{r}v$$

Next, the equation is differentiated with respect to φ^2. Then the following variables and equations are added:

$$\zeta = \frac{\partial u}{\partial \varphi^2}, \quad \eta = \frac{\partial v}{\partial \varphi^2}$$

$$\frac{d\zeta}{dr} = \eta$$

$$\frac{d\eta}{dr} = R(u) + \varphi^2 \frac{dR}{du}\zeta - \frac{a-1}{r}\eta$$

These can be solved along with the original problem (on each iteration) or only after the proper s is found from the iterations. The result of this calculation is that we have $c(x, \varphi^2)$ for some φ^2 and we also know $\partial c/\partial \varphi^2$ at the same value of φ^2. Then,

$$c(r, \varphi_2^2) \approx c(r, \varphi_1^2) + \zeta(r, \varphi_1^2)(\varphi_2^2 - \varphi_1^2)$$

can be used.

7.3. Finite Difference Method

To apply the finite difference method, we first spread grid points through the domain. Figure 29 shows a uniform mesh of n points (nonuniform meshes are described below). The unknown, here $c(x)$, at a grid point x_i is assigned the symbol $c_i = c(x_i)$. The finite difference method can be derived easily by using a Taylor expansion of the solution about this point.

$$c_{i+1} = c_i + \left.\frac{dc}{dx}\right|_i \Delta x + \left.\frac{d^2c}{dx^2}\right|_i \frac{\Delta x^2}{2} + \ldots$$

$$c_{i-1} = c_i - \left.\frac{dc}{dx}\right|_i \Delta x + \left.\frac{d^2c}{dx^2}\right|_i \frac{\Delta x^2}{2} - \ldots \quad (16)$$

These formulas can be rearranged and divided by Δx to give

$$\left.\frac{dc}{dx}\right|_i = \frac{c_{i+1} - c_i}{\Delta x} - \left.\frac{d^2c}{dx^2}\right|_i \frac{\Delta x}{2} + \ldots \quad (17)$$

$$\left.\frac{dc}{dx}\right|_i = \frac{c_i - c_{i-1}}{\Delta x} + \left.\frac{d^2c}{dx^2}\right|_i \frac{\Delta x}{2} + \ldots \quad (18)$$

which are representations of the first derivative. Alternatively the two equations can be subtracted from each other, rearranged and divided by Δx to give

$$\left.\frac{dc}{dx}\right|_i = \frac{c_{i+1} - c_{i-1}}{2\Delta x} - \left.\frac{d^3c}{dx^3}\right|_i \frac{\Delta x^2}{3!} \quad (19)$$

If the terms multiplied by Δx or Δx^2 are neglected, three representations of the first derivative are possible. In comparison with the Taylor series, the truncation error in the first two expressions is proportional to Δx, and the methods are said to be first order. The truncation error in the last expression is proportional to Δx^2, and the method is said to be second order. Usually, the last equation is chosen to ensure the best accuracy.

The finite difference representation of the second derivative can be obtained by adding the two expressions in Equation (16). Rearrangement and division by Δx^2 give

$$\left.\frac{d^2c}{dx^2}\right|_i = \frac{c_{i+1} - 2c_i + c_{i-1}}{\Delta x^2} - \left.\frac{d^4c}{dx^4}\right|_i \frac{2\Delta x^2}{4!} + \ldots \quad (20)$$

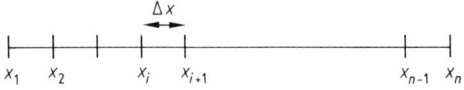

Figure 29. Finite difference mesh; Δx uniform

The truncation error is proportional to Δx^2.

To see how to solve a differential equation, consider the equation for convection, diffusion, and reaction in a tubular reactor:

$$\frac{1}{Pe}\frac{d^2c}{dx^2} - \frac{dc}{dx} = Da\,R(c)$$

To evaluate the differential equation at the i-th grid point, the finite difference representations of the first and second derivatives can be used to give

$$\frac{1}{Pe}\frac{c_{i+1} - 2c_i + c_{i-1}}{\Delta x^2} - \frac{c_{i+1} - c_{i-1}}{2\Delta x} = Da\,R(c_i) \quad (21)$$

This equation is written for $i = 2$ to $n - 1$ (i.e., the internal points). The equations would then be coupled but would involve the values of c_1 and c_n, as well. These are determined from the boundary conditions.

If the boundary condition involves a derivative, the finite difference representation of it must be carefully selected; here, three possibilities can be written. Consider a derivative needed at the point $i = 1$. First, Equation (17) could be used to write

$$\left.\frac{dc}{dx}\right|_1 = \frac{c_2 - c_1}{\Delta x} \quad (22)$$

Then a second-order expression is obtained that is one-sided. The Taylor series for the point c_{i+2} is written:

$$c_{i+2} = c_i + \left.\frac{dc}{dx}\right|_i 2\Delta x + \left.\frac{d^2c}{dx^2}\right|_i \frac{4\Delta x^2}{2!}$$
$$+ \left.\frac{d^3c}{dx^3}\right|_i \frac{8\Delta x^3}{3!} + \ldots$$

Four times Equation (16) minus this equation, with rearrangement, gives

$$\left.\frac{dc}{dx}\right|_i = \frac{-3c_i + 4c_{i+1} - c_{i+2}}{2\Delta x} + O(\Delta x^2)$$

Thus, for the first derivative at point $i = 1$

$$\left.\frac{dc}{dx}\right|_i = \frac{-3c_1 + 4c_2 - c_3}{2\Delta x} \quad (23)$$

This one-sided difference expression uses only the points already introduced into the domain. The third alternative is to add a false point, outside the domain, as $c_0 = c(x = -\Delta x)$. Then the centered first derivative, Equation (18), can be used:

$$\left.\frac{dc}{dx}\right|_1 = \frac{c_2 - c_0}{2\Delta x}$$

Because this equation introduces a new variable, another equation is required. This is obtained by also writing the differential equation (Eq. 21), for $i = 1$.

The same approach can be taken at the other end. As a boundary condition, any of three choices can be used:

$$\left.\frac{dc}{dx}\right|_n = \frac{c_n - c_{n-1}}{\Delta x}$$

$$\left.\frac{dc}{dx}\right|_n = \frac{c_{n-2} - 4c_{n-1} + 3c_n}{2\Delta x}$$

$$\left.\frac{dc}{dx}\right|_n = \frac{c_{n+1} - c_{n-1}}{2\Delta x}$$

The last two are of order Δx^2 and the last one would require writing the differential equation (Eq. 21) for $i = n$, too.

Generally, the first-order expression for the boundary condition is not used because the error in the solution would decrease only as Δx, and the higher truncation error of the differential equation (Δx^2) would be lost. For this problem the boundary conditions are

$$-\frac{1}{Pe}\frac{dc}{dx}(0) + c(0) = c_{in}$$

$$\frac{dc}{dx}(1) = 0$$

Thus, the three formulations would give first order in Δx

$$-\frac{1}{Pe}\frac{c_2 - c_1}{\Delta x} + c_1 = c_{in}$$

$$\frac{c_n - c_{n-1}}{\Delta x} = 0$$

plus Equation (21) at points $i = 2$ through $n - 1$; second order in Δx, by using a three-point one-sided derivative

$$-\frac{1}{Pe}\frac{-3c_1 + 4c_2 - c_3}{2\Delta x} + c_1 = c_{in}$$

$$\frac{c_{n-2} - 4c_{n-1} + 3c_n}{2\Delta x} = 0$$

plus Equation (21) at points $i = 2$ through $n - 1$; second order in Δx, by using a false boundary point

$$-\frac{1}{Pe}\frac{c_2 - c_0}{2\Delta x} + c_1 = c_{in}$$

$$\frac{c_{n+1} - c_{n-1}}{2\Delta x} = 0$$

plus Equation (21) at points $i = 1$ through n.

The sets of equations can be solved by using the Newton–Raphson method, as outlined in Section 1.2. The form of the equations is important, however, because the equations give a tridiagonal structure. For the first-order method the equations have the form shown in Figure 30 A and the standard routines for solving tridiagonal equations suffice. For the second-order method with a one-sided three-point derivative the form of the equations is shown in Figure 30 B. Now the tridiagonal structure is broken. However, if the second equation is multiplied by a constant and added to the first one, a zero can be obtained in the third column of the first row. Then the standard routines for tridiagonal matrices can be called. Similar manipulations are necessary in the last row of the matrix to obtain the tridiagonal form of the equations. For the second-order method with a false boundary point, the form of the equations is shown in Figure 30 C. Now, the two additional equations are almost tridiagonal. With the manipulations described above, they can be put into a tridiagonal form and the standard routines can be called.

Frequently, the transport coefficients (e.g., diffusion coefficient D or thermal conductivity) depend on the dependent variable (concentration or temperature, respectively). Then the differential equation might look like

$$\frac{d}{dx}\left(D(c)\frac{dc}{dx}\right) = 0$$

This could be written as

$$-\frac{dJ}{dx} = 0 \qquad (24)$$

in terms of the mass flux J, where the mass flux is given by

$$J = -D(c)\frac{dc}{dx}$$

Because the coefficient depends on c the equations are more complicated. A finite difference method can be written in terms of the fluxes at

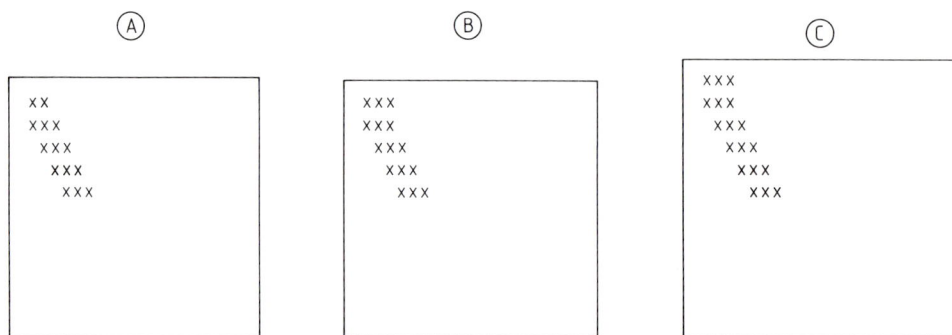

Figure 30. Equation structure for different boundary conditions
A) First-order method; B) Second-order method with one-sided three-point derivative; C) Second-order method using false boundary point

the midpoints, $i + 1/2$. Thus,

$$-\frac{J_{i+1/2} - J_{i-1/2}}{\Delta x} = 0$$

Then the constitutive equation for the mass flux can be written as

$$J_{i+1/2} = -D(c_{i+1/2})\frac{c_{i+1} - c_i}{\Delta x}$$

If these are combined,

$$\frac{D(c_{i+1/2})(c_{i+1} - c_i) - D(c_{i-1/2})(c_i - c_{i-1})}{\Delta x^2} = 0$$

This represents a set of nonlinear algebraic equations that can be solved with the Newton–Raphson method. However, in this case a viable iterative strategy is to evaluate the transport coefficients at the last value and then solve

$$\frac{D(c^k_{i+1/2})(c^{k+1}_{i+1} - c^{k+1}_i) - D(c^k_{i-1/2})(c^{k+1}_i - c^{k+1}_{i-1})}{\Delta x^2} = 0$$

The advantage of this approach is that it is easier to program than a full Newton–Raphson method. If the transport coefficients do not vary radically, the method converges. If the method does not converge, use of the full Newton–Raphson method may be necessary.

Three ways are commonly used to evaluate the transport coefficient at the midpoint. The first one employs the transport coefficient evaluated at the average value of the solutions on either side:

$$D(c_{i+1/2}) \approx D[\tfrac{1}{2}(c_{i+1} + c_i)]$$

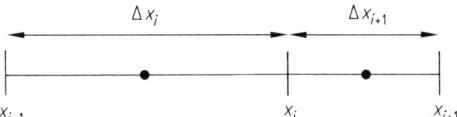

Figure 31. Finite difference grid with variable spacing

The truncation error of this approach is Δx^2 [7.9, Chap. 14]. The second approach uses the average of the transport coefficients on either side:

$$D(c_{i+1/2}) \approx \tfrac{1}{2}[D(c_{i+1}) + D(c_i)] \qquad (25)$$

The truncation error of this approach is also Δx^2 [7.9, Chap. 14], [7.10, p. 215]. The third approach employs an "upstream" transport coefficient.

$$D(c_{i+1/2}) \approx D(c_{i+1}), \text{ when } D(c_{i+1}) > D(c_i)$$
$$D(c_{i+1/2}) \approx D(c_i), \text{ when } D(c_{i+1}) < D(c_i)$$

This approach is used when the transport coefficients vary over several orders of magnitude and the "upstream" direction is defined as the one in which the transport coefficient is larger. The truncation error of this approach is only Δx [7.9, Chap. 14], [7.10, p. 253], but this approach is useful if the numerical solutions show unrealistic oscillations [7.9], [7.10].

If the grid spacing is not uniform the formulas must be revised. The notation is shown in Figure 31. The finite difference form of Equation (24) is then

$$-\frac{J_{i+1/2} - J_{i-1/2}}{\tfrac{1}{2}(\Delta x_i + \Delta x_{i+1})} = 0$$

and the constitutive relations are

$$J_{i+1/2} = -D_{i+1/2} \frac{c_{i+1} - c_i}{\Delta x_{i+1}},$$

$$J_{i-1/2} = -D_{i-1/2} \frac{c_i - c_{i-1}}{\Delta x_i},$$

Combined with Equation (25), this gives the following representation for Equation (24)

$$\frac{1}{\Delta x_{i+1} + \Delta x_i} \left[\frac{1}{\Delta x_{i+1}} (D_{i+1} + D_i)(c_{i+1} - c_i) - \frac{1}{\Delta x_i} (D_i + D_{i-1})(c_i - c_{i-1}) \right] = 0$$

Rigorous error bounds for linear ordinary differential equations solved with the finite difference method are dicussed by ISAACSON and KELLER [7.11, p. 431]. Computer software exists to solve two-point boundary value problems. The IMSL routine DVCPR uses the finite difference method, with a variable step size [7.8, p. 301]; FDRXN is given for reaction problems in [7.10, p. 335].

7.4. Orthogonal Collocation

The orthogonal collocation method has found widespread application in chemical engineering, particularly for chemical reaction engineering. In the collocation method [7.10], the dependent variable is expanded in a series.

$$y(x) = \sum_{i=1}^{N+2} a_i y_i(x) \tag{26}$$

Suppose the differential equation is

$$N[y] = 0$$

Then the expansion is put into the differential equation to form the residual:

$$\text{Residual} = N\left[\sum_{i=1}^{N+2} a_i y_i(x) \right]$$

In the collocation method, the residual is set to zero at a set of points called collocation points:

$$N\left[\sum_{i=1}^{N+2} a_i y_i(x_j) \right] = 0, \quad j = 2, \ldots, N+1$$

This provides N equations; two more equations come from the boundary conditions, giving $N + 2$ equations for $N + 2$ unknowns. This procedure is especially useful when the expansion is in a series of orthogonal polynomials, and when the collocation points are the roots to an orthogonal polynomial, as first used by LANCZOS [7.12], [7.13]. A major improvement was the proposal by VILLADSEN and STEWART [1.14] that the entire solution process be done in terms of the solution at the collocation points rather than the coefficients in the expansion. Thus, Equation (26) would be evaluated at the collocation points

$$y(x_j) = \sum_{i=1}^{N+2} a_i y_i(x_j), \quad j = 1, \ldots, N+2$$

and solved for the coefficients in terms of the solution at the collocation points:

$$a_i = \sum_{j=1}^{N+2} [y_i(x_j)]^{-1} y(x_j), \quad i = 1, \ldots, N+2$$

Furthermore, if Equation (26) is differentiated once and evaluated at all collocation points, the first derivative can be written in terms of the values at the collocation points:

$$\frac{dy}{dx}(x_j) = \sum_{i=1}^{N+2} a_i \frac{dy_i}{dx}(x_j), \quad j = 1, \ldots, N+2$$

This can be expressed as

$$\frac{dy}{dx}(x_j) = \sum_{i,k=1}^{N+2} [y_i(x_k)]^{-1} y(x_k) \frac{dy_i}{dx}(x_j),$$
$$j = 1, \ldots, N+2$$

or shortened to

$$\frac{dy}{dx}(x_j) = \sum_{k=1}^{N+2} A_{jk} y(x_k),$$

$$A_{jk} = \sum_{i=1}^{N+2} [y_i(x_k)]^{-1} \frac{dy_i}{dx}(x_j)$$

Similar steps can be applied to the second derivative to obtain

$$\frac{d^2 y}{dx^2}(x_j) = \sum_{k=1}^{N+2} B_{jk} y(x_k),$$

$$B_{jk} = \sum_{i=1}^{N+2} [y_i(x_k)]^{-1} \frac{d^2 y_i}{dx^2}(x_j)$$

This method is next applied to the differential equation for reaction in a tubular reactor, after the equation has been made nondimensional so that the dimensionless length is 1.0.

$$\frac{1}{Pe} \frac{d^2 c}{dx^2} - \frac{dc}{dx} = Da\, R(c),$$

$$-\frac{dc}{dx}(0) = Pe\,[c(0) - c_{in}], \quad \frac{dc}{dx}(1) = 0 \tag{27}$$

The differential equation at the collocation points is

$$\frac{1}{Pe} \sum_{k=1}^{N+2} B_{jk} c(x_k) - \sum_{k=1}^{N+2} A_{jk} c(x_k) = Da\, R(c_j) \quad (28)$$

and the two boundary conditions are

$$-\sum_{k=1}^{N+2} A_{1k} c(x_k) = Pe(c_1 - c_{in}),$$

$$\sum_{k=1}^{N+2} A_{N+2,k} c(x_k) = 0 \quad (29)$$

Note that 1 is the first collocation point ($x = 0$) and $N+2$ is the last one ($x = 1$). To apply the method, the matrices A_{ij} and B_{ij} must be found and the set of algebraic equations solved, perhaps with the Newton–Raphson method. If orthogonal polynomials are used and the collocation points are the roots to one of the orthogonal polynomials, the orthogonal collocation method results.

In the orthogonal collocation method the solution is expanded in a series involving orthogonal polynomials, where the polynomials $P_{i-1}(x)$ are defined in Section 2.2.

$$y = a + bx + x(1-x) \sum_{i=1}^{N} a_i P_{i-1}(x)$$

$$= \sum_{i=1}^{N+2} b_i P_{i-1}(x) \quad (30)$$

which is also

$$y = \sum_{i=1}^{N+2} d_i x^{i-1}$$

The collocation points are shown in Figure 32. There are N interior points plus one at each end, and the domain is always transformed to lie on 0 to 1. To define the matrices A_{ij} and B_{ij} this expression is evaluated at the collocation points; it is also differentiated and the result is evaluated at the collocation points.

$$y(x_j) = \sum_{i=1}^{N+2} d_i x_j^{i-1}$$

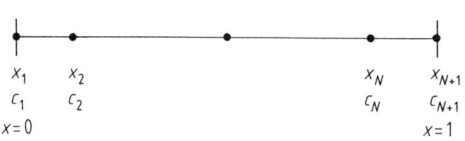

Figure 32. Orthogonal collocation points

$$\frac{dy}{dx}(x_j) = \sum_{i=1}^{N+2} d_i (i-1) x_j^{i-2}$$

$$\frac{d^2 y}{dx^2}(x_j) = \sum_{i=1}^{N+2} d_i (i-1)(i-2) x_j^{i-3}$$

These formulas are put in matrix notation, where Q, C, and D are $N+2$ by $N+2$ matrices.

$$y = Q d, \quad \frac{dy}{dx} = C d, \quad \frac{d^2 y}{dx^2} = D d$$

$$Q_{ji} = x_j^{i-1}, \quad C_{ji} = (i-1) x_j^{i-2},$$
$$D_{ji} = (i-1)(i-2) x_j^{i-3}$$

In solving the first equation for d, the first and second derivatives can be written as

$$d = Q^{-1} y, \quad \frac{dy}{dx} = CQ^{-1} y = Ay,$$

$$\frac{d^2 y}{dx^2} = DQ^{-1} y = By \quad (31)$$

Thus the derivative at any collocation point can be determined in terms of the solution at the collocation points. The same property is enjoyed by the finite difference method (and the finite element method described below), and this property accounts for some of the popularity of the orthogonal collocation method. In applying the method to Equation (27), the same result is obtained; Equations (28) and (29), with the matrices defined in Equation (31). To find the solution at a point that is not a collocation point, Equation (30) is used; once the solution is known at all collocation points, d can be found; and once d is known, the solution for any x can be found.

To use the orthogonal collocation method, the matrices are required. They can be calculated as shown above for small N ($N < 8$) and by using more rigorous techniques, for higher N (see Chap. 2). However, having the matrices listed explicitly for $N = 1$ and 2 is useful; this is shown in Table 7.

For some reaction diffusion problems, the solution can be an even function of x. For example, for the problem

$$\frac{d^2 c}{dx^2} = kc, \quad \frac{dc}{dx}(0) = 0, \quad c(1) = 1 \quad (32)$$

The solution can be proved to involve only even powers of x. In such cases, an orthogonal collocation method, which takes this feature into account, is convenient. This can easily be done by using expansions that only involve even powers

Table 7. Matrices for orthogonal collocation

$$N = 1, a = 0.1666666667$$

$$x_j = \begin{pmatrix} 0 \\ 0.5 \\ 1 \end{pmatrix}, W_j = \begin{pmatrix} a \\ 2a \\ a \end{pmatrix}, A_{ji} = \begin{pmatrix} -3 & 4 & -1 \\ -1 & 0 & 1 \\ 1 & -4 & 3 \end{pmatrix}, B_{ji} = \begin{pmatrix} 4 & -8 & 4 \\ 4 & -8 & 4 \\ 4 & -8 & 4 \end{pmatrix}, Q_{ji}^{-1} = \begin{pmatrix} 1 & 0 & 0 \\ -3 & 4 & -1 \\ 2 & -4 & 2 \end{pmatrix}$$

$$N = 2$$

$$x_j = \begin{pmatrix} 0 \\ e \\ 1-e \\ 1 \end{pmatrix}, W_j = \begin{pmatrix} 0 \\ 0.5 \\ 0.5 \\ 0 \end{pmatrix}, Q_{ji}^{-1} = \begin{pmatrix} 1 & 0 & 0 & 0 \\ -7 & 7+b & -1-b & 1 \\ 12 & -18-f & 12+f & -6 \\ -6 & 10+d & -10-d & 6 \end{pmatrix}$$

$$A_{ji} = \begin{pmatrix} -7 & 7+b & -1-b & 1 \\ -1-a & a & a & 1-a \\ -1+a & -a & -a & 1+a \\ -1 & 1+b & -7-b & 7 \end{pmatrix}, B_{ji} = \begin{pmatrix} 24 & -37-c & 25+c & -12 \\ 16+d & -24 & 12 & -4-d \\ -4-d & 12 & -24 & 16+d \\ -12 & 25+c & -37-c & 24 \end{pmatrix}$$

where $a = 1.732050808$, $b = 1.196152423$, $c = 0.17691454$, $d = 0.392304846$
$e = 0.2113248654$, $f = 0.58845727$

---

of x. Thus, the expansion

$$y(x^2) = y(1) + (1-x^2)\sum_{i=1}^{N} a_i P_{i-1}(x^2)$$

is equivalent to

$$y(x^2) = \sum_{i=1}^{N+1} b_i P_{i-1}(x^2) = \sum_{i=1}^{N+1} d_i x^{2i-2}$$

The polynomials are defined to be orthogonal with the weighting function $W(x^2)$.

$$\int_0^1 W(x^2) P_k(x^2) P_m(x^2) x^{a-1} dx = 0,$$

$$k \leq m-1 \tag{33}$$

where the power on x^{a-1} defines the geometry as planar or cartesian ($a = 1$), cylindrical ($a = 2$), and spherical ($a = 3$). An analogous development is used to obtain the $(N+1) \times (N+1)$ matrices

$$y(x_j) = \sum_{i=1}^{N+1} d_i x_j^{2i-2}$$

$$\frac{dy}{dx}(x_j) = \sum_{i=1}^{N+1} d_i (2i-2) x_j^{2i-3}$$

$$\nabla^2 y(x_j) = \sum_{i=1}^{N+1} d_i \nabla^2 (x^{2i-2})|_{x_j} d_j$$

$$y = Qd, \quad \frac{dy}{dx} = Cd, \quad \frac{d^2y}{dx^2} = Dd$$

$$Q_{ji} = x_j^{2i-2}, \quad C_{ji} = (2i-2) x_j^{2i-3},$$

$$D_{ji} = \nabla^2 (x^{2i-2})|_{x_j}$$

$$d = Q^{-1}y, \quad \frac{dy}{dx} = CQ^{-1}y = Ay,$$

$$\frac{d^2y}{dx^2} = DQ^{-1}y = By$$

In addition, the quadrature formula is

$$WQ = f, \quad B = fQ^{-1}$$

where

$$\int_0^1 x^{2i-2} x^{a-1} dx = \sum_{j=1}^{N+1} W_j x_j^{2i-2}$$

$$= \frac{1}{2i-2+a} \equiv f_i$$

As an example, for the problem

$$\frac{1}{x^{a-1}} \frac{d}{dx}\left(x^{a-1} \frac{dc}{dx}\right) = \varphi^2 R(c)$$

$$\frac{dc}{dx}(0) = 0, \quad c(1) = 1$$

orthogonal collocation is applied at the interior points

$$\sum_{i=1}^{N+1} B_{ji} c_i = \varphi^2 R(c_j), \quad j = 1, \ldots, N$$

and the boundary condition solved for is

$$c_{N+1} = 1$$

Table 8. Collocation points for orthogonal collocation with symmetric polynomials and $W = 1$

N	Planar	Cylindrical	Spherical
1	0.5773502692	0.7071067812	0.7745966692
2	0.3399810436	0.4597008434	0.5384693101
	0.8611363116	0.8880738340	0.9061793459
3	0.2386191861	0.3357106870	0.4058451514
	0.6612093865	0.7071067812	0.7415311856
	0.9324695142	0.9419651451	0.9491079123
4	0.1834346425	0.2634992300	0.3242534234
	0.5255324099	0.5744645143	0.6133714327
	0.7966664774	0.8185294874	0.8360311073
	0.9602898565	0.9646596062	0.9681602395
5	0.1488743390	0.2165873427	0.2695431560
	0.4333953941	0.4803804169	0.5190961292
	0.6794095683	0.7071067812	0.7301520056
	0.8650633667	0.8770602346	0.8870625998
	0.9739065285	0.9762632447	0.9782286581

The boundary condition at $x = 0$ is satisfied automatically by the trial function. After the solution has been obtained, the effectiveness factor η is obtained by calculating

$$\eta \equiv \frac{\int_0^1 R[c(x)] x^{a-1} dx}{\int_0^1 R[c(1)] x^{a-1} dx} = \frac{\sum_{i=1}^{N+1} W_j R(c_j)}{\sum_{i=1}^{N+1} W_j R(1)}$$

Note that the effectiveness factor is the average reaction rate divided by the reaction rate evaluated at the external conditions. Error bounds have been given for linear problems [7.15, p. 356]. For planar geometry the error is

$$\text{Error in } \eta = \frac{\varphi^{2(2N+1)}}{(2N+1)!(2N+2)!}$$

This method is very accurate for small N (and small φ^2); note that for finite difference methods the error goes as $1/N^2$, which does not decrease as rapidly with N. If the solution is desired at the center (a frequent situation because the center concentration can be the most extreme one), it is given by

$$c(0) = d_1 = \sum_{i=1}^{N+1} [Q^{-1}]_{1i} y_i$$

The collocation points are listed in Table 8. For small N the results are usually more accurate when the weighting function in Equation (33) is $1 - x^2$. The matrices for $N = 1$ and $N = 2$ are given in Table 9 for the three geometries. Computer programs to generate matrices and a program to solve reaction diffusion problems, OCRXN, are available [7.10, p. 325, p. 331].

Orthogonal collocation can be applied to distillation problems. STEWART et al. [7.16], [7.17] developed a method using Hahn polynomials that retains the discrete nature of a plate-to-plate distillation column. Other work treats problems with multiple liquid phases [7.18].

7.5. Orthogonal Collocation on Finite Elements

In the method of orthogonal collocation on finite elements, the domain is first divided into elements, and then within each element orthogonal collocation is applied. Figure 33 shows the domain being divided into NE elements, with $NCOL$ interior collocation points within each element, and $NP = NCOL + 2$ total points per element, giving $NT = NE * (NCOL + 1) + 1$ total number of points. Within each element a local coordinate is defined

$$u = \frac{x - x_{(k)}}{\Delta x_k}, \quad \Delta x_k = x_{(k+1)} - x_{(k)}$$

The reaction–diffusion equation is written as

$$\frac{1}{x^{a-1}} \frac{d}{dx}\left(x^{a-1} \frac{dc}{dx}\right) = \frac{d^2 c}{dx^2} + \frac{a-1}{x} \frac{dc}{dx} = \varphi^2 R(c)$$

and transformed to give

$$\frac{1}{\Delta x_k^2} \frac{d^2 c}{du^2} + \frac{a-1}{x_{(k)} + u \Delta x_k} \frac{1}{\Delta x_k} \frac{dc}{du} = \varphi^2 R(c)$$

The boundary conditions are typically

$$\frac{dc}{dx}(0) = 0, \quad -\frac{dc}{dx}(1) = Bi_m [c(1) - c_B]$$

where Bi_m is the Biot number for mass transfer. These become

$$\frac{1}{\Delta x_1} \frac{dc}{du}(u = 0) = 0, \quad \text{in the first element;}$$

$$-\frac{1}{\Delta x_{NE}} \frac{dc}{du}(u = 1) = Bi_m [c(u = 1) - c_B],$$

in the last element.

Table 9. Matrices for orthogonal collocation with symmetric polynomials and $W = 1 - x^2$

Planar geometry, $a = 1$
$N = 1$
$$x_j = \begin{pmatrix} 0.447214 \\ 1.000000 \end{pmatrix}, A_{ji} = \begin{pmatrix} -1.118034 & 1.118034 \\ -2.500000 & 2.500000 \end{pmatrix}$$

$$W_j = \begin{pmatrix} 0.833333 \\ 0.166667 \end{pmatrix}, B_{ji} = \begin{pmatrix} -2.5 & 2.5 \\ -2.5 & 2.5 \end{pmatrix}, Q_{ji}^{-1} = \begin{pmatrix} 1.25 & -0.25 \\ -1.25 & 1.25 \end{pmatrix}$$

$N = 2$
$$x_j = \begin{pmatrix} 0.285232 \\ 0.765055 \\ 1.000000 \end{pmatrix}, W_j = \begin{pmatrix} 0.554858 \\ 0.378475 \\ 0.066667 \end{pmatrix}, A_{ji} = \begin{pmatrix} -1.752962 & 2.507614 & -0.754652 \\ -1.370599 & -0.653547 & 2.024146 \\ 1.791503 & -8.791503 & 7.000000 \end{pmatrix}$$

$$B_{ji} = \begin{pmatrix} -4.73987 & 5.67713 & -0.93725 \\ 8.32288 & -23.26013 & 14.93725 \\ 19.07189 & -47.07190 & 28.00000 \end{pmatrix}, Q_{ji}^{-1} = \begin{pmatrix} 1.26430 & -0.38930 & 0.125 \\ -3.42435 & 5.17435 & -1.750 \\ 2.16005 & -4.78505 & 2.625 \end{pmatrix}$$

Cylindrical geometry, $a = 2$
$N = 1$
$$x_j = \begin{pmatrix} 0.577350 \\ 1.000000 \end{pmatrix}, W_j = \begin{pmatrix} 0.375 \\ 0.125 \end{pmatrix}, A_{ji} = \begin{pmatrix} -1.732051 & 1.732051 \\ -3.000000 & 3.000000 \end{pmatrix}$$

$$B_{ji} = \begin{pmatrix} -6 & 6 \\ -6 & 6 \end{pmatrix}, Q_{ji}^{-1} = \begin{pmatrix} 1.5 & -0.5 \\ -1.5 & 1.5 \end{pmatrix}$$

$N = 2$
$$x_j = \begin{pmatrix} 0.39377 \\ 0.80309 \\ 1.000000 \end{pmatrix}, W_j = \begin{pmatrix} 0.18820 \\ 0.25624 \\ 0.05555 \end{pmatrix}, A_{ji} = \begin{pmatrix} -2.53958 & 3.82562 & -1.28603 \\ -1.37768 & -1.24519 & 2.62287 \\ 1.71548 & -9.71548 & 8.00000 \end{pmatrix}$$

$$B_{ji} = \begin{pmatrix} -9.90238 & 12.29966 & -2.39728 \\ 9.03367 & -32.76429 & 23.73061 \\ 22.7575 & -65.42415 & 42.66667 \end{pmatrix}, Q_{ji}^{-1} = \begin{pmatrix} 1.58808 & -0.89141 & 0.33333 \\ -3.97389 & 6.64056 & -2.66667 \\ 2.41582 & -5.74914 & 3.33333 \end{pmatrix}$$

Spherical geometry, $a = 3$
$N = 1$
$$x_j = \begin{pmatrix} 0.654654 \\ 1.000000 \end{pmatrix}, W_j = \begin{pmatrix} 0.233333 \\ 0.100000 \end{pmatrix}, A_{ji} = \begin{pmatrix} -2.291288 & 2.291288 \\ -3.500000 & 3.500000 \end{pmatrix}$$

$$B_{ji} = \begin{pmatrix} -10.5 & 10.5 \\ -10.5 & 10.5 \end{pmatrix}, Q_{ji}^{-1} = \begin{pmatrix} 1.75 & -0.75 \\ -1.75 & 1.75 \end{pmatrix}$$

$N = 2$
$$x_j = \begin{pmatrix} 0.46885 \\ 0.83022 \\ 1.00000 \end{pmatrix}, W_j = \begin{pmatrix} 0.09491 \\ 0.19081 \\ 0.04762 \end{pmatrix}, A_{ji} = \begin{pmatrix} -3.19933 & 5.01517 & -1.81584 \\ -1.40870 & -1.80674 & 3.21544 \\ 1.69677 & -10.69677 & 9.00000 \end{pmatrix}$$

$$B_{ji} = \begin{pmatrix} -15.66996 & 20.03488 & -4.36492 \\ 9.96512 & -44.33004 & 34.36492 \\ 26.93285 & -86.93229 & 60.00000 \end{pmatrix}, Q_{ji}^{-1} = \begin{pmatrix} 1.88193 & -1.50693 & 0.625 \\ -4.61225 & 8.36225 & -3.750 \\ 2.73032 & -6.85532 & 4.125 \end{pmatrix}$$

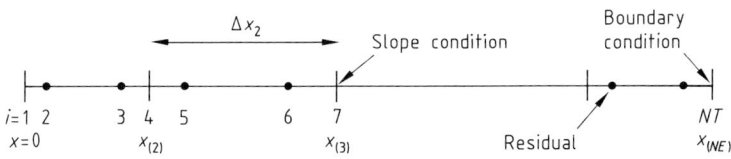

Figure 33. Grid for orthogonal collocation on finite elements

The orthogonal collocation method is applied at each interior collocation point.

$$\frac{1}{\Delta x_k^2} \sum_{J=1}^{NP} B_{IJ} c_J + \frac{a-1}{x_{(k)} + u_I \Delta x_k} \frac{1}{\Delta x_k} \sum_{J=1}^{NP} A_{IJ} c_J =$$

$$= \varphi^2 R(c_I), \quad I = 2, \ldots, NP - 1$$

The local points $i = 2, \ldots, NP - 1$ represent the interior collocation points. Continuity of the function and the first derivative between elements is achieved by taking

$$\frac{1}{\Delta x_{k-1}} \sum_{J=1}^{NP} A_{NP,J} c_J \bigg|_{\text{element } k-1}$$

$$= \frac{1}{\Delta x_k} \sum_{J=1}^{NP} A_{1,J} c_J \bigg|_{\text{element } k}$$

at the points between elements. Naturally, the computer code has only one symbol for the solution at a point shared between elements, but the derivative condition must be imposed. Finally, the boundary conditions at $x = 0$ and $x = 1$ are applied:

$$\frac{1}{\Delta x_k} \sum_{J=1}^{NP} A_{1,J} c_J = 0, \quad \text{in the first element}$$

$$-\frac{1}{\Delta x_{NE}} \sum_{J=1}^{NP} A_{NP,J} c_J = Bi_m [c_{NP} - c_B],$$

in the last element.

These equations can be assembled into an overall matrix problem

$$A A c = f$$

The form of these equations is special and is discussed by FINLAYSON [7.10, p. 116], who also gives the computer code to solve linear equations arising in such problems. Reaction–diffusion problems are solved by the program OCFERXN [7.10, p. 337]. See also the program COLSYS described below.

Another way to approach the same problem is to use the Hermite polynomials (see Section 2.3). Then the continuity of the first derivative between elements is already part of the basis set, leading to a smaller set of equations. If cubic Hermite polynomials are used, the solution is identical to that obtained above with the orthogonal collocation method when using cubic polynomials ($N_{COL} = 2$). This is not usually done in chemical engineering; the interested reader is referred to [7.10, p. 121].

The error bounds of DEBOOR [7.19] give the following results for second-order problems solved with cubic trial functions on finite elements with continuous first derivatives. The error at all positions is bounded by

$$\left\| \frac{d^i}{dx^i} (y - y_{\text{exact}}) \right\|_\infty \leq \text{constant } |\Delta x|^2$$

The error at the collocation points is more accurate, giving what is known as superconvergence.

$$\left| \frac{d^i}{dx^i} (y - y_{\text{exact}}) \right|_{\text{collocation points}} \leq \text{constant } |\Delta x|^4$$

7.6. Galerkin Finite Element Method

In the finite element method the domain is divided into elements and an expansion is made for the solution on each finite element. In the Galerkin finite element method an additional idea is introduced: the Galerkin method is used to solve the equation. The Galerkin method is explained before the finite element basis set is introduced.

To solve the problem

$$\frac{1}{x^{a-1}} \frac{d}{dx} \left(x^{a-1} \frac{dc}{dx} \right) = \varphi^2 R(c)$$

$$\frac{dc}{dx}(0) = 0, \quad -\frac{dc}{dx}(1) = Bi_m [c(1) - c_B]$$

the unknown solution is expanded in a series of known functions $\{b_i(x)\}$, with unknown coefficients $\{a_i\}$.

$$c(x) = \sum_{i=1}^{NT} a_i b_i(x)$$

The series (the trial solution) is inserted into the differential equation to obtain the residual:

$$\text{Residual} = \sum_{i=1}^{NT} a_i \frac{1}{x^{a-1}} \frac{d}{dx} \left(x^{a-1} \frac{db_i}{dx} \right)$$

$$- \varphi^2 R \left[\sum_{i=1}^{NT} a_i b_i(x) \right]$$

The residual is then made orthogonal to the set of basis functions.

$$\int_0^1 b_j(x) \left\{ \sum_{i=1}^{NT} a_i \frac{1}{x^{a-1}} \frac{d}{dx} \left(x^{a-1} \frac{db_i}{dx} \right) \right.$$

$$\left. - \varphi^2 R \left[\sum_{i=1}^{NT} a_i b_i(x) \right] \right\} x^{a-1} dx = 0 \quad (34)$$

$$j = 1, \ldots, NT$$

This process makes the method a Galerkin method. The basis for the orthogonality condition is that a function that is made orthogonal to each member of a complete set is then zero. The residual is being made orthogonal, and if the basis functions are complete, and an infinite number of them are used, then the residual is zero. Once the residual is zero the problem is solved. It is necessary also to allow for the boundary conditions. This is done by integrating the first term of Equation (34) by parts and then inserting the boundary conditions:

$$\int_0^1 b_j(x) \frac{1}{x^{a-1}} \frac{d}{dx}\left(x^{a-1} \frac{db_i}{dx}\right) x^{a-1} dx$$

$$= \int_0^1 \frac{d}{dx}\left[b_j(x) x^{a-1} \frac{db_i}{dx}\right] dx - \int_0^1 \frac{db_j}{dx} \frac{db_i}{dx} x^{a-1} dx$$

$$= \left[b_j(x) x^{a-1} \frac{db_i}{dx}\right]_0^1 - \int_0^1 \frac{db_j}{dx} \frac{db_i}{dx} x^{a-1} dx$$

$$= -\int_0^1 \frac{db_j}{dx} \frac{db_i}{dx} x^{a-1} dx - Bi_m b_j(1) [b_i(1) - c_B]$$

(35)

Combining this with Equation (34) gives

$$-\sum_{i=1}^{NT} \int_0^1 \frac{db_j}{dx} \frac{db_i}{dx} x^{a-1} dx \, a_i$$

$$- Bi_m b_j(1) \left[\sum_{i=1}^{NT} a_i b_i(1) - c_B\right]$$

$$= \varphi^2 \int_0^1 b_j(x) R \left[\sum_{i=1}^{NT} a_i b_i(x)\right] x^{a-1} dx$$

$$j = 1, \ldots, NT \qquad (36)$$

This equation defines the Galerkin method, and a solution that satisfies this equation (for all $j = 1, \ldots, \infty$) is called a weak solution. For an approximate solution the equation is written once for each member of the trial function, $j = 1, \ldots, NT$. If the boundary condition is

$$c(1) = c_B$$

then the boundary condition is used (instead of Eq. 36) for $j = NT$,

$$\sum_{i=1}^{NT} a_i b_i(1) = c_B$$

The Galerkin finite element method results when the Galerkin method is combined with a finite element trial function. Both linear and quadratic finite element approximations are described in Chapter 2. The trial functions $b_i(x)$ are then generally written as $N_i(x)$.

$$c(x) = \sum_{i=1}^{NT} c_i N_i(x)$$

Each $N_i(x)$ takes the value 1 at the point x_i and zero at all other grid points (Chap. 2). Thus c_i are the nodal values, $c(x_i) = c_i$. The first derivative must be transformed to the local coordinate system, $u = 0$ to 1 when x goes from x_i to $x_i + \Delta x$.

$$\frac{dN_J}{dx} = \frac{1}{\Delta x_e} \frac{dN_J}{du}, \quad dx = \Delta x_e du, \quad \text{in } e\text{-th element}$$

Then the Galerkin method is

$$-\sum_e \frac{1}{\Delta x_e} \sum_{I=1}^{NP} \int_0^1 \frac{dN_J}{du} \frac{dN_I}{du} (x_e + u\Delta x_e)^{a-1} du \, c_I^e$$

$$- Bi_m \sum_e N_J(1) \left[\sum_{I=1}^{NP} c_I^e N_I(1) - c_1\right]$$

$$= \varphi^2 \sum_e \Delta x_e \int_0^1 N_J(u) R\left[\sum_{I=1}^{NP} c_I^e N_I(u)\right]$$

$$(x_e + u\Delta x_e)^{a-1} du \qquad (37)$$

The element integrals are defined as

$$B_{JI}^e = -\frac{1}{\Delta x_e} \int_0^1 \frac{dN_J}{du} \frac{dN_I}{du} (x_e + u\Delta x_e)^{a-1} du,$$

$$F_J^e = \varphi^2 \Delta x_e \int_0^1 N_J(u) R\left[\sum_{I=1}^{NP} c_I^e N_I(u)\right]$$

$$(x_e + u\Delta x_e)^{a-1} du$$

whereas the boundary element integrals are

$$BB_{JI}^e = -Bi_m N_J(1) N_I(1),$$
$$FF_J^e = -Bi_m N_J(1) c_1$$

Then the entire method can be written in the compact notation

$$\sum_e B_{JI}^e c_I^e + \sum_e BB_{JI}^e c_I^e = \sum_e F_J^e + \sum_e FF_J^e$$

The matrices for various terms are given in Table 10. This equation can also be written in the form

$$AAc = f$$

where the matrix AA is sparse. If linear elements are used the matrix is tridiagonal. If quadratic elements are used the matrix is pentadiagonal. Naturally the linear algebra is most efficiently carried out if the sparse structure is taken into account. Once the solution is found the solution

Table 10. Element matrices for Galerkin method

Linear shape functions

$$N_1 = 1 - u, \quad N_2 = u, \quad \frac{dN_1}{du} = -1, \quad \frac{dN_2}{du} = 1$$

Quadratic shape functions

$$N_1 = 2(u-1)\left(u - \frac{1}{2}\right), \quad N_2 = 4u(1-u), \quad N_3 = 2u\left(u - \frac{1}{2}\right),$$

$$\frac{dN_1}{du} = 4u - 3, \quad \frac{dN_2}{du} = 4 - 8u, \quad \frac{dN_3}{du} = 4u - 1$$

$$\int_0^1 \frac{dN_J}{du}\frac{dN_I}{du} du = \begin{pmatrix} 1 & -1 \\ -1 & 1 \end{pmatrix}, \quad \int_0^1 N_J \frac{dN_I}{du} du = \begin{pmatrix} -\frac{1}{2} & \frac{1}{2} \\ -\frac{1}{2} & \frac{1}{2} \end{pmatrix}$$

$$\int_0^1 \frac{dN_J}{du}\frac{dN_I}{du} du = \frac{1}{3}\begin{pmatrix} 7 & -8 & 1 \\ -8 & 16 & -8 \\ 1 & -8 & 7 \end{pmatrix}, \quad \int_0^1 N_J \frac{dN_I}{du} du = \frac{1}{6}\begin{pmatrix} -3 & 4 & -1 \\ -4 & 0 & 4 \\ 1 & -4 & 3 \end{pmatrix}$$

$$\int_0^1 N_J N_I du = \begin{pmatrix} \frac{1}{3} & \frac{1}{6} \\ \frac{1}{6} & \frac{1}{3} \end{pmatrix}, \quad \int_0^1 N_J du = \begin{pmatrix} \frac{1}{2} \\ \frac{1}{2} \end{pmatrix}, \quad \int_0^1 N_J u\, du = \begin{pmatrix} \frac{1}{6} \\ \frac{1}{3} \end{pmatrix}$$

$$\int_0^1 N_J N_I du = \frac{1}{30}\begin{pmatrix} 4 & 2 & -1 \\ 2 & 16 & 2 \\ -1 & 2 & 4 \end{pmatrix}, \quad \int_0^1 N_J du = \frac{1}{6}\begin{pmatrix} 1 \\ 4 \\ 1 \end{pmatrix}, \quad \int_0^1 N_J u\, du = \frac{1}{6}\begin{pmatrix} 0 \\ 2 \\ 1 \end{pmatrix}$$

at any point can be recovered from

$$c^e(u) = c^e_{I=1}(1-u) + c^e_{I=2}u$$

for linear elements

$$c^e(u) = c^e_{I=1} 2(u-1)\left(u - \tfrac{1}{2}\right) + c^e_{I=2} 4u(1-u) + c^e_{I=2} 2u\left(u - \tfrac{1}{2}\right)$$

for quadratic elements

Because the integrals in Equation (36) may be complicated, they are usually formed by using Gaussian quadrature. If NG Gauss points are used, a typical term would be

$$\int_0^1 N_J(u) R \left[\sum_{I=1}^{NP} c^e_I N_I(u) \right] (x_e + u\Delta x_e)^{a-1} du$$

$$= \sum_{k=1}^{NG} W_k N_J(u_k) R \left[\sum_{I=1}^{NP} c^e_I N_I(u_k) \right] (x_e + u_k \Delta x_e)^{a-1}$$

For an application of the finite element method in fluid mechanics, see → 5. Fluid Mechanics, pp. 5-46–5-49.

7.7. Cubic B-Splines

Cubic B-splines have cubic approximations within each element, but first and second derivatives continuous between elements. The functions are the same ones discussed in Chapter 2, but the treatment for differential equations follows SINCOVEC [7.20]. The trial function is taken as

$$y = \sum_{i=1}^{NT+1} a_i S_i$$

Table 11. Function and derivative values at knots of cubic B-spline

	x_{i-2}	x_{i-1}	x_i	x_{i+1}	x_{i+2}
S_i	0	$\frac{1}{4}$	1	$\frac{1}{4}$	0
S'_i	0	$\frac{3}{4h}$	0	$-\frac{3}{4h}$	0
S''_i	0	$\frac{3}{2h^2}$	$-\frac{3}{h^2}$	$\frac{3}{2h^2}$	0

The values of the function and the first and second derivatives at the knots are listed in Table 11. The knots are the points between elements. If a differential equation

$$\frac{d^2 y}{dx^2} = f\left(x, y, \frac{dy}{dx}\right)$$

must be solved, the derivatives are approximated by

$$\left.\frac{d^2 y}{dx^2}\right|_i = \frac{3}{2h^2} a_{i-1} - \frac{3}{h^2} a_i + \frac{3}{2h^2} a_{i+1}$$

$$= \frac{3}{2h^2}(a_{i-1} - 2a_i + a_{i+1})$$

The differential equation is then satisfied at each knot.

$$a_{i-1} - 2a_i + a_{i+1} = \frac{2h^2}{3} f\left(x_i, \frac{1}{4}a_{i-1} + a_i + \frac{1}{4}a_{i+1}, \frac{3}{4h}(a_{i+1} - a_{i-1})\right)$$

$$i = 1, \ldots, NT$$

The boundary conditions must also be applied. If the boundary condition at the left is

$$y(x_1) = y_{\text{left}}$$

it is satisfied by

$$y(x_1) = \sum_{i=0}^{NT+1} a_i S(x_1) = a_0 S_0(x_1) + a_1 S_1(x_1)$$
$$+ a_1 S_1(x_1) = a_0 \tfrac{1}{4} + a_1 + a_1 \tfrac{1}{4} = y_{\text{left}}$$

Similar considerations apply at the right-hand side. To preserve the tridiagonal nature of the equations, the manipulations discussed in Section 7.3 must be used.

7.8. Adaptive Mesh Strategies

In many two-point boundary value problems, the difficulty in the problem is the formation of a boundary layer region, or a region in which the solution changes very dramatically. In such cases small mesh spacing should be used there, either with the finite difference method or the finite element method. If the region is known a priori, small mesh spacings can be assumed at the boundary layer. If the region is not known though, other techniques must be used. These techniques are known as adaptive mesh techniques. A simple technique that has proven useful [7.9, Chap. 7] is presented first, and more complicated (and more robust) techniques that have been implemented are then described.

The adaptive mesh technique requires some criteria for deciding whether to add or remove points. The policy is taken as

$$E_j < 0.1\, E_a \quad \text{remove node } j$$
$$0.1\, E_a \le E_j \le 10\, E_a \quad \text{keep node } j$$
$$10\, E_a < E_j \le 100\, E_a \quad \text{add 1 node}$$
$$100\, E_a < E_j \le 1000\, E_a \quad \text{add 2 nodes}$$
$$1000\, E_a < E_j \quad \text{add 3 nodes}$$

$$E_a = \sum_{j=2}^{n-1} E_j \Big/ (n-2)$$

FINLAYSON [7.9, Chap. 7] tried several criteria: the residual, the first derivative, the second derivative, etc. The first and second derivative worked best and are the easiest so they are described. If a finite difference method or a linear finite element method is used, the truncation error in the method is [7.21]:

$$\text{Error}_j = C\Delta x_j^2 \left\| \frac{d^2 c}{dx^2} \right\|_j$$

Thus, the pointwise value of the second derivative is used as the criterion. Because the grid points or mesh points are spaced at irregular intervals, an expression must be used for the second derivative that accounts for the irregularity. For the notation shown in Figure 30, the criterion is then

$$E_j = \Delta x_j^2 \frac{d^2 c}{dx^2} = \Delta x_j^2 \frac{\dfrac{c_{j+1}-c_j}{x_{j+1}-x_j} - \dfrac{c_j-c_{j-1}}{x_j-x_{j-1}}}{\tfrac{1}{2}(x_{j+1}-x_{j-1})}$$

and this should be made uniform throughout the domain. The first derivative for the irregular mesh is

$$\left.\frac{dc}{dx}\right|_j = \frac{c_{j+1}-c_j}{x_{j+1}-x_j}$$

However, the mean square derivative can also be used.

$$E_j = \left[\int_{x_{j-1}}^{x_j} \left(\frac{dc}{dx}\right)^2 dx\right]^{1/2}$$
$$= \left[\Delta x_{j-1}\left(\frac{c_j-c_{j-1}}{\Delta x_{j-1}}\right)^2\right]^{1/2} = \frac{|c_j-c_{j-1}|}{\sqrt{\Delta x_{j-1}}}$$

These adaptive mesh methods work as follows. An initial mesh is assumed. This might be a uniform mesh with only a few points, or it could have some features of the solution built into it. Then the problem is solved on this mesh. One of the criteria is then applied to decide if more points should be added or if points should be removed. Once an entire new mesh is found, the points can be smoothed somewhat by using [7.22]

$$x_k = \tfrac{1}{2}(x_k + x_{k+1})$$

This ensures that the mesh does not change too drastically. Then the old solution is interpolated onto the new mesh and the problem is resolved. The process is continued until the solution is good enough, which might be defined as making the second (or first) derivative smaller than some fixed number over the entire domain.

The adaptive mesh strategy was employed by ASCHER et al. [7.23] and by RUSSELL and CHRISTIANSEN [7.24]. For a second-order differential equation and cubic trial functions on finite elements, the error in the i-th element is given by

$$\|\text{Error}\|_i = C \Delta x_i^4 \|u^{(4)}\|_i$$

Because cubic elements do not have a nonzero fourth derivative, the third derivative in adjacent elements is used [7.10, p. 166]:

$$a_i = \frac{1}{\Delta x_i^3} \frac{d^3 c^i}{du^3}, \quad a_{i+1} = \frac{1}{\Delta x_{i+1}^3} \frac{d^3 c^{i+1}}{du^3}$$

$$\|u^{(4)}\|_i \approx \frac{1}{2} \left[\frac{a_i - a_{i-1}}{\frac{1}{2}(x_{i+1} - x_{i-1})} + \frac{a_{i+1} - a_i}{\frac{1}{2}(x_{i+2} - x_i)} \right]$$

Element sizes are then chosen so that the following error bounds are satisfied

$$C \Delta x_i^4 \|u^{(4)}\|_i \leq \varepsilon \text{ for all } i$$

These features are built into the code COLSYS.

The error expected from a method one order higher and one order lower can also be defined. Then a decision about whether to increase or decrease the order of the method can be made by taking into account the relative work of the different orders. This provides a method of adjusting both the mesh spacing (Δx, sometimes called h) and the degree of polynomial (p). Such methods are called $h-p$ methods.

7.9. Comparison

What method should be used for any given problem? Obviously the error decreases with some power of Δx, and the power is higher for the higher order methods, which suggests that the error is less. For example, with linear elements the error is

$$y(\Delta x) = y_{\text{exact}} + c_2 \Delta x^2$$

for small enough (and uniform) Δx. A computer code should be run for varying Δx to confirm this. For quadratic elements, the error is

$$y(\Delta x) = y_{\text{exact}} + c_3 \Delta x^3$$

If orthogonal collocation on finite elements is used with cubic polynomials, then

$$y(\Delta x) = y_{\text{exact}} + c_4 \Delta x^4$$

However, the global methods, not using finite elements, converge even faster [7.25], for example,

$$y(N) = y_{\text{exact}} + c_N \left(\frac{1}{NCOL}\right)^{NCOL}$$

Yet the workload of the methods is also different. These considerations are discussed in [7.10]. Here, only sweeping generalizations are given.

If the problem has a relatively smooth solution, then the orthogonal collocation method is preferred. It gives a very accurate solution, and N can be quite small so the work is small. If the problem has a steep front in it, the finite difference method or finite element method is indicated, and adaptive mesh techniques should probably be employed. Consider the reaction–diffusion problem: as the Thiele modulus φ increases from a small value with no diffusion limitations to a large value with significant diffusion limitations, the solution changes as shown in Figure 34. The orthogonal collocation method is initially the method of choice. For intermediate values of φ, $N = 3-6$ must be used, but orthogonal collocation still works well (for η down to approximately 0.01). For large φ, use of the finite difference method, the finite element method, or an asymptotic expansion for large φ is better. The decision depends entirely on the type of solution that is obtained. For steep fronts the finite difference method and finite element method with adaptive mesh are indicated.

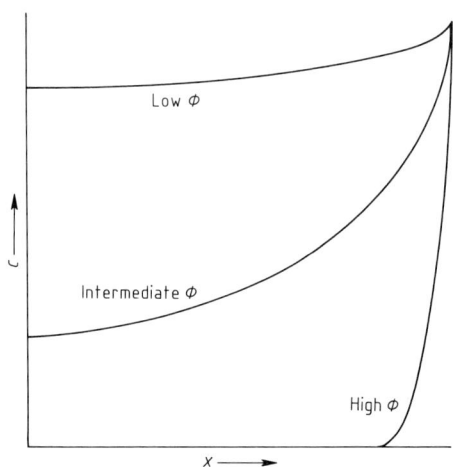

Figure 34. Concentration solution for different values of Thiele modulus

7.10. Singular Problems and Infinite Domains

If the solution being sought has a singularity, a good numerical solution may be hard to find. Sometimes even the location of the singularity may not be known [7.26, pp. 230–238]. One method of solving such problems is to refine the mesh near the singularity, by relying on the better approximation due to a smaller Δx. Another approach is to incorporate the singular trial function into the approximation. Thus, if the solution approaches $f(x)$ as x goes to zero, and $f(x)$ becomes infinite, an approximation may be taken as

$$y(x) = f(x) + \sum_{i=1}^{N} a_i y_i(x)$$

This function is substituted into the differential equation, which is solved for a_i. Essentially, a new differential equation is being solved for a new variable:

$$u(x) \equiv y(x) - f(x)$$

The differential equation is more complicated but has a better solution near the singularity (see [7.27, pp. 189–192], [7.28, p. 611]).

Sometimes the domain is infinite. Boundary layer flow past a flat plate is governed by the Blasius equation for stream function [7.29, p. 117].

$$2\frac{d^3f}{d\eta^3} + f\frac{d^2f}{d\eta^2} = 0$$

$$f = \frac{df}{d\eta} = 0 \text{ at } \eta = 0$$

$$\frac{df}{d\eta} = 1 \text{ at } \eta \to \infty$$

Because one boundary is at infinity using a mesh with a constant size is difficult! One approach is to transform the domain. For example, let

$$z = e^{-\eta}$$

Then $\eta = 0$ becomes $z = 1$ and $\eta = \infty$ becomes $z = 0$. The derivatives are

$$\frac{dz}{d\eta} = -e^{-\eta} = -z, \quad \frac{d^2z}{d\eta^2} = e^{-\eta} = z$$

$$\frac{df}{d\eta} = \frac{df}{dz}\frac{dz}{d\eta} = -z\frac{df}{dz}$$

$$\frac{d^2f}{d\eta^2} = \frac{d^2f}{dz^2}\left(\frac{dz}{d\eta}\right)^2 + \frac{df}{dz}\frac{d^2z}{d\eta^2} = z^2\frac{d^2f}{dz^2} + z\frac{df}{dz}$$

The Blasius equation becomes

$$2\left[-z^3\frac{d^3f}{dz^3} - 3z^2\frac{d^2f}{dz^2} - z\frac{df}{dz}\right]$$
$$+ f\left[z^2\frac{d^2f}{dz^2} + z\frac{df}{dz}\right] = 0$$

The differential equation now has variable coefficients, but these are no more difficult to handle than the original nonlinearities.

Another approach is to use a variable mesh, perhaps with the same transformation. For example, use $z = e^{-\eta}(-\eta)$ and a constant mesh size in z. Then with 101 points distributed uniformly from $z = 0$ to $z = 1$, the following are the nodal points:

$z = 0., 0.01, 0.02, \ldots, 0.99, 1.0$
$\eta = \infty, 4.605, 3.912, \ldots, 0.010, 0$
$\Delta\eta = \infty, 0.693, \ldots, 0.01$

Still another approach is to solve on a finite mesh in which the last point is far enough away that its location does not influence the solution [7.30]. A location that is far enough away must be found by trial and error.

8. Partial Differential Equations

Partial differential equations are differential equations in which the dependent variable is a function of two or more independent variables. These can be time and one space dimension, or time and two or more space dimensions, or two or more space dimensions alone. Problems involving time are generally either hyperbolic or parabolic, whereas those involving spatial dimensions only are often elliptic. Because the methods applied to each type of equation are very different, the equation must first be classified as to its type. Then the special methods applicable to each type of equation are described. For a discussion of all methods, see [8.1]; for a discussion oriented more toward chemical engineering applications, see [8.2]. Many chemical engineering examples that use some of the methods described below are given in [8.3].

8.1. Classification of Equations

A set of differential equations may be hyperbolic, elliptic, or parabolic, or it may be of mixed type. The type may change for different parameters or in different regions of the flow. This can happen in the case of nonlinear problems; an example is a compressible flow problem with both subsonic and supersonic regions. *Characteristic curves* are curves along which a discontinuity can propagate. For a given set of equations, it is necessary to determine if characteristics exist or not, because that determines whether the equations are hyperbolic, elliptic, or parabolic.

Linear Problems. For linear problems, the theory summarized by JOSEPH et al. [8.4] can be used.

$$\frac{\partial}{\partial t}, \frac{\partial}{\partial x_i}, \ldots, \frac{\partial}{\partial x_n}$$

is replaced with the Fourier variables

$$i\xi_0, i\xi_1, \ldots, i\xi_n$$

If the m-th order differential equation is

$$P = \sum_{|\alpha|=m} a_\alpha \partial^\alpha + \sum_{|\alpha|<m} b_\alpha \partial^\alpha$$

where

$$\alpha = (\alpha_0, \alpha_1, \ldots, \alpha_n), \quad |\alpha| = \sum_{i=0}^{n} \alpha_i$$

$$\partial^\alpha = \frac{\partial^{|\alpha|}}{\partial t^{\alpha_0} \partial x_1^{\alpha_1} \ldots \partial x_n^{\alpha_n}}$$

the characteristic equation for P is

$$\sum_{|\alpha|=m} a_\alpha \sigma^\alpha = 0, \quad \sigma = (\sigma_0, \sigma_1, \ldots, \sigma_n) \quad (38)$$

$$\sigma^\alpha = \sigma_0^{\alpha_0} \sigma_1^{\alpha_1} \ldots \sigma_n^{\alpha_n}$$

where σ represents coordinates. Thus only the highest derivatives are used to determine the type. The surface is defined by this equation plus a normalization condition:

$$\sum_{k=0}^{n} \sigma_k^2 = 1$$

The shape of the surface defined by Equation (38) is also related to the type: elliptic equations give rise to ellipses; parabolic equations give rise to parabolas; and hyperbolic equations give rise to hyperbolas.

$$\frac{\sigma_1^2}{a^2} + \frac{\sigma_2^2}{b^2} = 1, \quad \sigma_0 = a\sigma_1^2, \quad \sigma_0^2 - a\sigma_1^2 = 0$$

Ellipse   Parabola   Hyperbola

If Equation (38) has no nontrivial real zeroes then the equation is called elliptic. If all the roots are real and distinct (excluding zero) then the operator is hyperbolic.

This formalism is applied to three basic types of equations. First consider the equation arising from steady diffusion in two dimensions:

$$\frac{\partial^2 c}{\partial x^2} + \frac{\partial^2 c}{\partial y^2} = 0$$

This gives

$$-\xi_1^2 - \xi_2^2 = -(\xi_2^1 + \xi_2^2) = 0$$

Thus,

$$\sigma_1^2 + \sigma_2^2 = 1 \text{ (normalization)}$$
$$\sigma_1^2 + \sigma_2^2 = 0 \text{ (equation)}$$

These cannot both be satisfied so the problem is elliptic. When the equation is

$$\frac{\partial^2 u}{\partial t^2} - \frac{\partial^2 u}{\partial x^2} = 0$$

then

$$-\xi_0^2 + \xi_1^2 = 0$$

Now real ξ_0 can be solved and the equation is hyperbolic

$$\sigma_0^2 + \sigma_1^2 = 1 \text{ (normalization)}$$
$$-\sigma_0^2 + \sigma_1^2 = 0 \text{ (equation)}$$

When the equation is

$$\frac{\partial c}{\partial t} = D \left(\frac{\partial^2 c}{\partial x^2} + \frac{\partial^2 c}{\partial y^2} \right)$$

then

$$\sigma_0^2 + \sigma_1^2 + \sigma_2^2 = 1 \text{ (normalization)}$$
$$\sigma_1^2 + \sigma_2^2 = 0 \text{ (equation)}$$

thus we get

$$\sigma_0^2 = 1 \text{ (for normalization)}$$

and the characteristic surfaces are hyperplanes with t = constant. This is a parabolic case.

Consider next the telegrapher's equation:

$$\frac{\partial T}{\partial t} + \beta \frac{\partial^2 T}{\partial t^2} = \frac{\partial^2 T}{\partial x^2}$$

Replacing the derivatives with the Fourier variables gives

$$i\xi_0 - \beta\xi_0^2 + \xi_1^2 = 0$$

The equation is thus second order and the type is determined by

$$-\beta\sigma_0^2 + \sigma_1^2 = 0$$

The normalization condition

$$\sigma_0^2 + \sigma_1^2 = 1$$

is required. Combining these gives

$$1 - (1 + \beta)\sigma_0^2 = 0$$

The roots are real and the equation is hyperbolic. When $\beta = 0$

$$\xi_1^2 = 0$$

and the equation is parabolic.

First-order quasi-linear problems are written in the form

$$\sum_{l=0}^{n} A_l \frac{\partial \boldsymbol{u}}{\partial x_l} = \boldsymbol{f}, \quad \boldsymbol{x} = (t, x_1, \ldots, x_n) \quad (39)$$

$$\boldsymbol{u} = (u_1, u_2, \ldots, u_k)$$

The matrix entries A_l is a $k \times k$ matrix whose entries depend on $\boldsymbol{u}$ but not on derivatives of $\boldsymbol{u}$. Equation (39) is hyperbolic if

$$A = A_\mu$$

is nonsingular and for any choice of real λ_l, $l = 0, \ldots, n$, $l \neq \mu$ the roots α_k of

$$\det\left(\alpha A - \sum_{\substack{l=0 \\ l \neq \mu}}^{n} \lambda_l A_l\right) = 0$$

are real. If the roots are complex the equation is elliptic; if some roots are real and some are complex the equation is of mixed type.

Apply these ideas to the convection equation

$$\frac{\partial u}{\partial t} + F(u)\frac{\partial u}{\partial x} = 0$$

Thus,

$$\det(\alpha A_0 - \lambda_1 A_1) = 0 \text{ or } \det(\alpha A_1 - \lambda_0 A_0) = 0$$

In this case,

$$n = 1, A_0 = 1, A_1 = F(u)$$

Using the first of the above equations gives

$$\det(\alpha - \lambda_1 F(u)) = 0, \text{ or } \alpha = \lambda_1 F(u)$$

Thus, the roots are real and the equation is hyperbolic.

The final example is the heat conduction problem written as

$$\varrho C_p \frac{\partial T}{\partial t} = -\frac{\partial q}{\partial x}, \quad q = -k\frac{\partial T}{\partial x}$$

In this formulation the constitutive equation for heat flux is separated out; the resulting set of equations is first order and written as

$$\varrho C_p \frac{\partial T}{\partial t} + \frac{\partial q}{\partial x} = 0$$

$$k\frac{\partial T}{\partial x} = -q$$

In matrix notation this is

$$\begin{bmatrix} \varrho C_p & 0 \\ 0 & 0 \end{bmatrix} \begin{bmatrix} \frac{\partial T}{\partial t} \\ \frac{\partial q}{\partial t} \end{bmatrix} + \begin{bmatrix} 0 & 1 \\ k & 0 \end{bmatrix} \begin{bmatrix} \frac{\partial T}{\partial x} \\ \frac{\partial q}{\partial x} \end{bmatrix} = \begin{bmatrix} 0 \\ -q \end{bmatrix}$$

This compares with

$$A_0 \frac{\partial \boldsymbol{u}}{\partial x_0} + A_1 \frac{\partial \boldsymbol{u}}{\partial x_1} = \boldsymbol{f}$$

In this case A_0 is singular whereas A_1 is nonsingular. Thus,

$$\det(\alpha A_1 - \lambda_0 A_0) = 0$$

is considered for any real λ_0. This gives

$$\begin{vmatrix} -\varrho C_p \lambda_0 \alpha & \alpha \\ k\alpha & 0 \end{vmatrix} = 0$$

or

$$\alpha^2 k = 0$$

Thus the α is real, but zero, and the equation is parabolic.

8.2. Hyperbolic Equations

The most common situation yielding hyperbolic equations involves unsteady phenomena with convection. A prototype equation is

$$\frac{\partial c}{\partial t} + \frac{\partial F(c)}{\partial x} = 0$$

Depending on the interpretation of c and $F(c)$, this can represent accumulation of mass and convection. With $F(c) = uc$, where u is the velocity, the equation represents a mass balance on concentration. If diffusive phenomenon are important, the equation is changed to

$$\frac{\partial c}{\partial t} + \frac{\partial F(c)}{\partial x} = D\frac{\partial^2 c}{\partial x^2} \quad (40)$$

where D is a diffusion coefficient. Special cases are the *convective diffusive equation*

$$\frac{\partial c}{\partial t} + u\frac{\partial c}{\partial x} = D\frac{\partial^2 c}{\partial x^2} \quad (41)$$

and *Burger's viscosity equation*

$$\frac{\partial u}{\partial t} + u\frac{\partial u}{\partial x} = v\frac{\partial^2 u}{\partial x^2} \quad (42)$$

where u is the velocity and v is the kinematic viscosity. This is a prototype equation for the Navier–Stokes equations (→ 5. Fluid Mechanics, p. 5-11) in a shock. For adsorption phenomena [8.5, p. 202],

$$\varphi\frac{\partial c}{\partial t} + \varphi u\frac{\partial c}{\partial x} + (1-\varphi)\frac{df}{dc}\frac{\partial c}{\partial t} = 0 \quad (43)$$

where φ is the void fraction and $f(c)$ gives the equilibrium relation between the concentrations in the fluid and in the solid phase. In these examples, if the diffusion coefficient D or the kinematic viscosity v is zero, the equations are hyperbolic. If D and v are small, the phenomenon may be essentially hyperbolic even though the equations are parabolic. Thus the numerical methods for hyperbolic equations may be useful even for parabolic equations.

Equations for several methods are given here, as taken from [8.6]. If the convective term is treated with a centered difference expression the solution exhibits oscillations from node to node, and these vanish only if a very fine grid is used. The simplest way to avoid the oscillations with a hyperbolic equation is to use upstream derivatives. If the flow is from left to right, this would give the following for Equations (40):

$$\frac{dc_i}{dt} + \frac{F(c_i) - F(c_{i-1})}{\Delta x} = D\frac{c_{i+1} - 2c_i + c_{i-1}}{\Delta x^2}$$

for Equation (42):

$$\frac{du_i}{dt} + u_i\frac{u_i - u_{i-1}}{\Delta x} = v\frac{u_{i+1} - 2u_i + u_{i-1}}{\Delta x^2}$$

and for Equation (43):

$$\varphi\frac{dc_i}{dt} + \varphi u_i\frac{c_i - c_{i-1}}{\Delta x} + (1-\varphi)\left.\frac{df}{dc}\right|_i \frac{dc_i}{dt} = 0$$

If the flow were from right to left, then the formula would be

$$\frac{dc_i}{dt} + \frac{F(c_{i+1}) - F(c_i)}{\Delta x} = D\frac{c_{i+1} - 2c_i + c_{i-1}}{\Delta x^2}$$

If the flow could be in either direction, a local determination must be made at each node i and the appropriate formula used. The effect of using upstream derivatives is to add artificial or numerical diffusion to the model. This can be ascertained by taking the finite difference form of the convective diffusion equation

$$\frac{dc_i}{dt} + u\frac{c_i - c_{i-1}}{\Delta x} = D\frac{c_{i+1} - 2c_i + c_{i-1}}{\Delta x^2}$$

and rearranging

$$\frac{dc_i}{dt} + u\frac{c_{i+1} - c_{i-1}}{2\Delta x}$$
$$= \left(D + \frac{u\Delta x}{2}\right)\frac{c_{i+1} - 2c_i + c_{i-1}}{\Delta x^2}$$

Thus the diffusion coefficient has been changed from

$$D \text{ to } D + \frac{u\Delta x}{2}$$

Another method often used for hyperbolic equations is the *MacCormack method*. This method has two steps; it is written here for Equation (41).

$$c_i^{*n+1} = c_i^n - \frac{u\Delta t}{\Delta x}(c_{i+1}^n - c_i^n)$$
$$+ \frac{\Delta t\, D}{\Delta x^2}(c_{i+1}^n - 2c_i^n + c_{i-1}^n)$$

$$c_i^{n+1} = \frac{1}{2}(c_i^n + c_i^{*n+1}) - \frac{Pe\,\Delta t}{2\,\Delta x}(c_i^{*n+1} - c_{i-1}^{*n+1})$$

$$+ \frac{\Delta t\,D}{2\,\Delta x^2}(c_{i+1}^{*n+1} - 2c_i^{*n+1} + c_{i-1}^{*n+1})$$

The concentration profile is steeper for the MacCormack method than for the upstream derivatives, but oscillations can still be present. The flux-corrected transport method can be added to the MacCormack method. A solution is obtained both with the upstream algorithm and the MacCormack method; then they are combined to add just enough diffusion to eliminate the oscillations without smoothing the solution too much. The algorithm is complicated and lengthy but well worth the effort [8.6]–[8.8].

If finite element methods are used, an explicit Taylor–Galerkin method is appropriate. For the convective diffusion equation the method is

$$\tfrac{1}{6}(c_{i+1}^{n+1} - c_{i+1}^n) + \tfrac{2}{3}(c_i^{n+1} - c_i^n) + \tfrac{1}{6}(c_{i-1}^{n+1} - c_{i-1}^n)$$

$$= -\frac{u\,\Delta t}{2\,\Delta x}(c_{i+1}^n - c_{i-1}^n)$$

$$+ \left(\frac{\Delta t\,D}{\Delta x^2} + \frac{u^2\,\Delta t^2}{2\,\Delta x^2}\right)(c_{i+1}^n - 2c_i^n + c_{i-1}^n)$$

Leaving out the $u^2\,\Delta t^2$ terms gives the Galerkin method. Replacing the left-hand side with

$$c_i^{n+1} - c_i^n$$

gives the Taylor finite difference method, and dropping the $u^2\,\Delta t^2$ terms in that gives the centered finite difference method. This method might require a small time step if reaction phenomena are important. Then the implicit Galerkin method (without the Taylor terms) is appropriate

$$\tfrac{1}{6}(c_{i+1}^{n+1} - c_{i+1}^n) + \tfrac{2}{3}(c_i^{n+1} - c_i^n) + \tfrac{1}{6}(c_{i-1}^{n+1} - c_{i-1}^n)$$

$$= -\frac{u\,\Delta t}{2\,\Delta x}[1 - \theta][c_{i+1}^n - c_{i-1}^n] - \frac{u\,\Delta t}{2\,\Delta x}$$

$$\theta[c_{i+1}^{n+1} - c_{i-1}^{n+1}] + \frac{\Delta t\,D}{\Delta x^2}[1-\theta][c_{i+1}^n - 2c_i^n$$

$$+ c_{i-1}^n] + \frac{\Delta t\,D}{\Delta x^2}\theta[c_{i+1}^{n+1} - 2c_i^{n+1} + c_{i-1}^{n+1}]$$

The Taylor terms are not required because the implicit time step provides the same effect as diffusion.

For the nonlinear Equation (40) the Taylor–Galerkin method is

$$\frac{1}{6}\frac{c_{i+1}^{n+1} - c_{i+1}^n}{\Delta t} + \frac{2}{3}\frac{c_i^{n+1} - c_i^n}{\Delta t} + \frac{1}{6}\frac{c_{i-1}^{n+1} - c_{i-1}^n}{\Delta t}$$

$$= \frac{1}{2\,\Delta x}[-F_{i+1}^n + F_{i-1}^n] + \frac{\Delta t}{4\,\Delta x^2}$$

$$\left\{\left[\left(\frac{dF}{dc}\right)_{i+1}^2 + \left(\frac{dF}{dc}\right)_i^2\right][c_{i+1}^n - c_i^n]\right.$$

$$\left. - \left[\left(\frac{dF}{dc}\right)_i^2 + \left(\frac{dF}{dc}\right)_{i-1}^2\right][c_i^n - c_{i-1}^n]\right\}$$

$$+ \frac{D}{\Delta x^2}(c_{i+1}^n - 2c_i^n + c_{i-1}^n)$$

A stability diagram for the explicit methods applied to the convective diffusion equation is shown in Figure 35. Notice that all the methods require

$$Co = \frac{u\,\Delta t}{\Delta x} \leq 1$$

where Co is the Courant number. How much Co should be less than one depends on the method and on $r = D\,\Delta t/\Delta x^2$, as given in Figure 35. The MacCormack method with flux correction requires a smaller time step than the MacCormack method alone (curve a), and the implicit Galerkin method (curve e) is stable for all values of Co and r shown in Figure 35 (as well as even larger values).

Each of these methods tries to avoid oscillations that would disappear if the mesh were fine

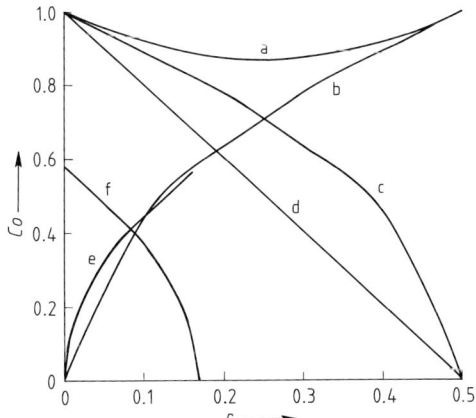

Figure 35. Stability diagram for convective diffusion equation (stable below curve)
a) MacCormack; b) Centered finite difference; c) Taylor finite difference; d) Upstream; e) Galerkin; f) Taylor–Galerkin

enough. For the steady convective diffusion equation these oscillations do not occur provided

$$\frac{u\Delta x}{2D} \leq 1 \qquad (44)$$

For large u, Δx must be small to meet this condition. An alternative is to use a small Δx in regions where the solution changes drastically. Because these regions change in time, the elements or grid points must move. The criteria to move the grid points can be quite complicated, and typical methods are reviewed in [8.6]. The criteria include moving the mesh in a known way (when the movement is known a priori), moving the mesh to keep some property (e.g., first- or second-derivative measures) uniform over the domain, using a Galerkin or weighted residual criterion to move the mesh, and Euler–Lagrange methods which move part of the solution exactly by convection and then add on some diffusion after that.

The final illustration is for adsorption in a packed bed, or chromatography. Equation (43) can be solved when the adsorption phenomenon is governed by a Langmuir isotherm.

$$f(c) = \frac{\alpha c}{1 + Kc}$$

Similar numerical considerations apply and similar methods are available [8.6].

8.3. Parabolic Equations in One Dimension

In this section several methods are applied to parabolic equations in one dimension: separation of variables, combination of variables, finite difference method, finite element method, and the orthogonal collocation method. Separation of variables is successful for linear problems, whereas the other methods work for linear or nonlinear problems. The finite difference, finite element, and the orthogonal collocation methods are numerical, whereas the separation or combination of variables can lead to analytical solutions.

Analytical Solutions. Consider the diffusion equation

$$\frac{\partial c}{\partial t} = D\frac{\partial^2 c}{\partial x^2}$$

with boundary and initial conditions

$$c(x, 0) = 0$$
$$c(0, t) = 1, \quad c(1, t) = 0$$

A solution of the form

$$c(x, t) = T(t)X(x)$$

is attempted and substituted into the equation, with the terms separated to give

$$\frac{1}{DT}\frac{dT}{dt} = \frac{1}{X}\frac{d^2X}{dx^2}$$

One side of this equation is a function of x alone, whereas the other side is a function of t alone. Thus, both sides must be a constant. Otherwise, if x is changed one side changes, but the other cannot because it depends on t. Call the constant $-\lambda$ and write the separate equations

$$\frac{dT}{dt} = -\lambda DT, \qquad \frac{d^2X}{dx^2} = -\lambda X$$

The first equation is solved easily

$$T(t) = T(0)e^{-\lambda Dt}$$

and the second equation is written in the form

$$\frac{d^2X}{dx^2} + \lambda X = 0$$

Next consider the boundary conditions. If they are written as

$$c(1, t) = 1 = T(t)X(1)$$
$$c(0, t) = 0 = T(t)X(0)$$

the boundary conditions are difficult to satisfy because they are not homogeneous, with a zero right-hand side. Thus, the problem must be transformed to make the boundary conditions homogeneous. The solution is written as the sum of two functions, one of which satisfies the nonhomogeneous boundary conditions, whereas the other satisfies the homogeneous boundary conditions.

$$c(x, t) = f(x) + u(x, t)$$
$$u(0, t) = 0$$
$$u(1, t) = 0$$

Thus, $f(0) = 1$ and $f(1) = 0$ are necessary.

Now the combined function satisfies the boundary conditions. In this case the function $f(x)$ can be taken as

$$f(x) = 1 - x$$

The equation for u is found by substituting for c in the original equation and noting that the $f(x)$ drops out for this case; it need not disappear in the general case:

$$\frac{\partial u}{\partial t} = D \frac{\partial^2 u}{\partial x^2}$$

The boundary conditions for u are

$$u(0, t) = 0$$
$$u(1, t) = 0$$

The initial conditions for u are found from the initial condition

$$u(x, 0) = c(x, 0) - f(x) = x - 1$$

Separation of variables is now applied to this equation by writing

$$u(x, t) = T(t) X(x)$$

The same equation for $T(t)$ and $X(x)$ is obtained, but with $X(0) = X(1) = 0$.

$$\frac{d^2 X}{dx^2} + \lambda X = 0$$
$$X(0) = X(1) = 0$$

Next $X(x)$ is solved for. The equation is an eigenvalue problem. The general solution is obtained by using e^{mx} and finding that $m^2 + \lambda = 0$; thus $m = \pm i\sqrt{\lambda}$. The exponential term

$$e^{\pm i\sqrt{\lambda}x}$$

is written in terms of sines and cosines, so that the general solution is

$$X = B \cos\sqrt{\lambda}\,x + E \sin\sqrt{\lambda}\,x$$

The boundary conditions are

$$X(1) = B \cos\sqrt{\lambda} + E \sin\sqrt{\lambda} = 0$$
$$X(0) = B = 0$$

If $B = 0$, then $E \neq 0$ is required to have any solution at all. Thus, λ must satisfy

$$\sin\sqrt{\lambda} = 0$$

This is true for certain values of λ, called eigenvalues or characteristic values. Here, they are

$$\lambda_n = n^2 \pi^2$$

Each eigenvalue has a corresponding eigenfunction

$$X_n(x) = E \sin n\pi x$$

The composite solution is then

$$X_n(x) T_n(t) = E A \sin n\pi x \, e^{-\lambda_n D t}$$

This function satisfies the boundary conditions and differential equation but not the initial condition. To make the function satisfy the initial condition, several of these solutions are added up, each with a different eigenfunction, and EA is replaced by A_n.

$$u(x, t) = \sum_{n=1}^{\infty} A_n \sin n\pi x \, e^{-n^2 \pi^2 D t}$$

The constants A_n are chosen by making $u(x, t)$ satisfy the initial condition.

$$u(x, 0) = \sum_{n=1}^{\infty} A_n \sin n\pi x = x - 1$$

The residual $R(x)$ is defined as the error in the initial condition:

$$R(x) = x - 1 - \sum_{n=1}^{\infty} A_n \sin n\pi x$$

Next, the Galerkin method is applied, and the residual is made orthogonal to a complete set of functions, which are the eigenfunctions.

$$\int_0^1 (x - 1) \sin m\pi x \, dx$$

$$= \sum_{n=1}^{\infty} A_n \int_0^1 \sin m\pi x \sin n\pi x \, dx = \frac{A_m}{2}$$

The Galerkin criterion for finding A_n is the same as the least-squares criterion [8.9, p. 183]. The solution is then

$$c(x, t) = 1 - x + \sum_{n=1}^{\infty} A_n \sin n\pi x \, e^{-n^2 \pi^2 D t}$$

This is an "exact" solution to the linear problem. It can be evaluated to any desired accuracy by taking more and more terms, but if a finite number of terms are used, some error always occurs.

For large times a single term is adequate, whereas for small times many terms are needed. For small times the Laplace transform method is also useful, because it leads to solutions that converge with fewer terms. For small times, the method of combination of variables may be used as well. For nonlinear problems, the method of separation of variables fails and one of the other methods must be used.

The method of combination of variables is useful, particularly when the problem is posed in a semi-infinite domain. Here, only one example is provided; more detail is given in [8.9]–[8.11]. The method is applied here to the nonlinear problem

$$\frac{\partial c}{\partial t} = \frac{\partial}{\partial x}\left[D(c)\frac{\partial c}{\partial x}\right] = D(c)\frac{\partial^2 c}{\partial x^2} + \frac{dD(c)}{dc}\left(\frac{\partial c}{\partial x}\right)^2$$

with boundary and initial conditions

$$c(x, 0) = 0$$
$$c(0, t) = 1, \quad c(\infty, t) = 0$$

The transformation combines two variables into one

$$c(x, t) = f(\eta) \quad \text{where} \quad \eta = \frac{x}{\sqrt{4 D_0 t}}$$

The use of the 4 and D_0 makes the analysis below simpler. The equation for $c(x, t)$ is transformed into an equation for $f(\eta)$

$$\frac{\partial c}{\partial t} = \frac{df}{d\eta}\frac{\partial \eta}{\partial t}, \quad \frac{\partial c}{\partial x} = \frac{df}{d\eta}\frac{\partial \eta}{\partial x}$$

$$\frac{\partial^2 c}{\partial x^2} = \frac{d^2 f}{d\eta^2}\left(\frac{\partial \eta}{\partial x}\right)^2 + \frac{df}{d\eta}\frac{\partial^2 \eta}{\partial x^2}$$

$$\frac{\partial \eta}{\partial t} = -\frac{x/2}{\sqrt{4 D_0 t^3}}, \quad \frac{\partial \eta}{\partial x} = \frac{1}{\sqrt{4 D_0 t}}, \quad \frac{\partial^2 \eta}{\partial x^2} = 0$$

The result is

$$\frac{d}{d\eta}\left[K(c)\frac{df}{d\eta}\right] + 2\eta\frac{df}{d\eta} = 0$$

$$K(c) = D(c)/D_0$$

The boundary conditions must also combine. In this case the variable η is infinite when either x is infinite or t is zero. Note that the boundary conditions on $c(x, t)$ are both zero at those points. Thus, the boundary conditions can be combined to give

$$f(\infty) = 0$$

The other boundary condition is for $x = 0$ or $\eta = 0$,

$$f(0) = 1$$

Thus, an ordinary differential equation must be solved rather than a partial differential equation. When the diffusivity is constant the solution is the well-known complementary error function:

$$c(x, t) = 1 - \text{erf } \eta = \text{erfc } \eta$$

$$\text{erf } \eta = \frac{\int_0^\eta e^{-\xi^2} d\xi}{\int_0^\infty e^{-\xi^2} d\xi}$$

This is a tabulated function [8.12].

Numerical Methods. Numerical methods are applicable to both linear and nonlinear problems on finite and semi-infinite domains. The *finite difference method* is applied by using the method of lines [8.13]. In this method the same equations are used for the spatial variations of the function, but the function at a grid point can vary with time. Thus the linear diffusion problem is written as

$$\frac{dc_i}{dt} = D\frac{c_{i+1} - 2c_i + c_{i-1}}{\Delta x^2} \qquad (45)$$

This can be written in the general form

$$\frac{dc}{dt} = A A c$$

This set of ordinary differential equations can be solved by using any of the standard methods. The stability of explicit schemes is deduced from the theory presented in Chapter 6. The equations are written as

$$\frac{dc_i}{dt} = D\frac{c_{i+1} - 2c_i + c_{i-1}}{\Delta x^2} = \frac{D}{\Delta x^2}\sum_{j=1}^{n+1} B_{ij} c_j$$

where the matrix $\mathbf{B}$ is tridiagonal. The stability of the integration of these equations is governed by the largest eigenvalue of $\mathbf{B}$. If Euler's method is used for integration,

$$\Delta t \frac{D}{\Delta x^2} \leq \frac{2}{|\lambda|_{\max}}$$

The largest eigenvalue of $\mathbf{B}$ is bounded by the Gerschgorin theorem [8.14, p. 135].

$$|\lambda|_{max} \leq \max_{2 < j < n} \sum_{i=2}^{n} |B_{ji}| = 4$$

This gives the well-known stability limit

$$\Delta t \frac{D}{\Delta x^2} \leq \frac{1}{2}$$

If other methods are used to integrate in time, then the stability limit changes according to the method. It is interesting to note that the eigenvalues of Equation (45) range from $D\pi^2/L^2$ (smallest) to $4D/\Delta x^2$ (largest), depending on the boundary conditions. Thus the problem becomes stiff as Δx approaches zero [8.9, p. 263].

Another way of studying the stability of explicit equations is to use the positivity theorem. If the Euler method is applied, the equations can be written in the form

$$\frac{c_i^{n+1} - c_i^n}{\Delta t} = D \frac{c_{i+1}^n - 2c_i^n + c_{i-1}^n}{\Delta x^2}$$

where

$$c_i^n = c(x_i, t^n)$$

Then the new value is given by

$$c_i^{n+1} = \frac{D\Delta t}{\Delta x^2} c_{i+1}^n + \left(1 - 2\frac{D\Delta t}{\Delta x^2}\right) c_i^n + \frac{D\Delta t}{\Delta x^2} c_{i-1}^n$$

Theorem. If

$$c^{n+1} = A c_{i+1}^n + B c_i^n + C c_{i-1}^n$$

and A, B, and C are positive, and $A + B + C \leq 1$, then the scheme is stable and the errors die out.

Here the theorem requires

$$\left(1 - 2\frac{D\Delta t}{\Delta x^2}\right) > 0$$

which gives the same stability condition [8.9, p. 217].

Implicit methods can also be used. Write a finite difference form for the time derivative and average the right-hand sides, evaluated at the old and new times:

$$\frac{c_i^{n+1} - c_i^n}{\Delta t} = D(1-\theta) \frac{c_{i+1}^n - 2c_i^n + c_{i-1}^n}{\Delta x^2}$$
$$+ D\theta \frac{c_{i+1}^{n+1} - 2c_i^{n+1} + c_{i-1}^{n+1}}{\Delta x^2}$$

Now the equations are of the form

$$-\frac{D\Delta t\,\theta}{\Delta x^2} c_{i+1}^{n+1} + \left[1 + 2\frac{D\Delta t\,\theta}{\Delta x^2}\right] c_i^{n+1} - \frac{D\Delta t\,\theta}{\Delta x^2} c_{i-1}^{n+1}$$
$$= c_i^n + \frac{D\Delta t(1-\theta)}{\Delta x^2}(c_{i+1}^n - 2c_i^n + c_{i-1}^n)$$

and require solving a set of simultaneous equations, which have a tridiagonal structure. Using $\theta = 0$ gives the Euler method (as above); $\theta = 0.5$ gives the Crank–Nicolson method; $\theta = 1$ gives the backward Euler method. The stability limit is given by

$$\frac{D\Delta t}{\Delta x^2} \leq \frac{0.5}{1 - 2\theta}$$

whereas the oscillation limit is given by

$$\frac{D\Delta t}{\Delta x^2} \leq \frac{0.25}{1 - \theta}$$

If a time step is chosen between the oscillation limit and stability limit, the solution will oscillate around the exact solution, but the oscillations remain bounded. For further discussion, see [8.9, p. 218].

The *finite element method* is handled in a similar fashion, as an extension of two-point boundary value problems by letting the solution at the nodes depend on time. For the diffusion equation the finite element method gives

$$\sum_e \sum_I C_{JI}^e \frac{dc_I^e}{dt} = \sum_e \sum_I B_{JI}^e c_I^e$$

with the mass matrix defined by

$$C_{JI}^e = \Delta x_e \int_0^1 N_J(u) N_I(u)\, du$$

This set of equations can be written in matrix form

$$CC \frac{dc}{dt} = AAc$$

Now the matrix CC is not diagonal, so that a set of equations must be solved for each time step, even when the right-hand side is evaluated explicitly. This is not as time-consuming as it seems, however. The explicit scheme is written as

$$CC_{ji} \frac{c_i^{n+1} - c_i^n}{\Delta t} = AA_{ji} c_i^n$$

and rearranged to give

$$CC_{ji}(c_i^{n+1} - c_i^n) = \Delta t \, AA_{ji} c_i^n \quad \text{or}$$

$$CC(c^{n+1} - c^n) = \Delta t \, AA \, c$$

This is solved with an LU decomposition (see Section 1.1) that retains the structure of the mass matrix CC. Thus,

$$CC = LU$$

At each step, calculate

$$c^{n+1} - c^n = \Delta t \, U^{-1} L^{-1} AA \, c^n$$

This is quick and easy to do because the inverse of L and U are simple. Thus the problem is reduced to solving one full matrix problem and then evaluating the solution for several right-hand sides. For implicit methods the same approach yields

$$CC_{ji} \frac{c_i^{n+1} - c_i^n}{\Delta t} = AA_{ji}[(1-\theta)c_i^n + \theta c_i^n]$$

which can be rearranged to give

$$(CC_{ji} - \Delta t \, \theta \, AA_{ji}) c_i^{n+1}$$
$$= CC_{ji} c_i^n + \Delta t (1-\theta) AA_{ji} c_i^n \equiv f_j$$

This is again solved as

$$c^{n+1} = U^{-1} L^{-1} f$$

In both cases the LU decomposition need be redone only when the time step size is changed.

The *method of orthogonal collocation* uses a similar extension: the same polynomial of x is used but now the coefficients depend on time.

$$c(x,t) = a(t) + b(t)x + x(1-x) \sum_{i=1}^{N} a_i(t) P_{i-1}(x)$$

The same spatial derivatives evaluated at the collocation points can be derived:

$$c(x,t) = \sum_{i=1}^{N+2} d_i(t) x^{i-1};$$

$$c(x_j, t) = \sum_{i=1}^{N+2} d_i(t) x_j^{i-1}; \quad c(t) = Q d(t)$$

$$\left.\frac{\partial c}{\partial x}\right|_{x_j} = \sum_{i=1}^{N+2} d_i(t)(i-1) x^{i-2}\bigg|_{x_j};$$

$$\frac{\partial c}{\partial x} = C d(t)$$

$$\frac{\partial c}{\partial x} = CQ^{-1} c \equiv Ac$$

Now both c and $\partial c/\partial x$ are functions of time, but the matrix A is constant in time. For the time derivatives,

$$\left.\frac{\partial c}{\partial t}\right|_{x_j} = \frac{dc(x_j, t)}{dt} = \frac{dc_j}{dt}$$

Thus, for diffusion problems

$$\frac{dc_j}{dt} = \sum_{i=1}^{N+2} B_{ji} c_j, \quad j = 2, \ldots, N+1$$

This can be integrated by using the standard methods for ordinary differential equations as initial value problems. Stability limits for explicit methods are available [8.9, p. 204].

The method of orthogonal collocation on finite elements can also be used, and details are provided elsewhere [8.9, pp. 228–230]. An application to chemical reactors, where the radial direction is handled by using the method of orthogonal collocation on finite elements and the axial direction is handled by using the method of lines (PDECOL), is given by PIRKLE et al. [8.16].

Spectral methods employ Chebyshev polynomials and the fast Fourier transform, and are quite useful for hyperbolic or parabolic problems on rectangular domains [8.16]. The Chebyshev polynomial of degree n is defined by

$$T_n(\cos \theta) = \cos n\theta$$

The first few polynomials are

$$T_0(x) = 1, \quad T_1(x) = x, \quad T_2(x) = 2x^2 - 1,$$

$$T_3(x) = 4x^3 - 3x, \quad T_4(x) = 8x^4 - 8x^2 + 1$$

They satisfy the orthogonality condition

$$\int_{-1}^{1} \frac{1}{\sqrt{1-x^2}} T_n(x) T_m(x) \, dx = \frac{\pi c_n}{2} \delta_{nm}$$

where the constants are given by

$$c_0 = 2, \quad c_n = 0 \, (n < 0), \quad c_n = 1 \, (n > 0)$$

In solving a differential equation for $f(x,t)$, for example, the function f is expanded in Chebyshev polynomials

$$f = \sum_{n=0}^{\infty} a_n T_n, \quad |x| \leq 1$$

Various derivatives are also expanded in Chebyshev polynomials. For differential operator L, formally

$$Lf = \sum_{n=0}^{\infty} b_n T_n, \quad |x| \leq 1$$

For specific cases [8.16, p. 160] the relations between the coefficients of the function (a_n) and the derivatives (b_n) are given by

$$Lf = \frac{df}{dx}, \quad c_n b_n = 2 \sum_{\substack{p=n+1 \\ p+n \text{ odd}}}^{\infty} p a_p$$

$$Lf = \frac{d^2 f}{dx^2}, \quad c_n b_n = \sum_{\substack{p=n+2 \\ p+n \text{ even}}}^{\infty} p(p^2 - n^2) a_p$$

Derivatives can be evaluated even more efficiently by using a recursion relation. When

$$S_n = \sum_{\substack{p=n+1 \\ p+n \text{ odd}}}^{\infty} p a_p$$

the following can be used [8.16, p. 117]:

$$S_n = S_{n+2} + (n+1) a_{n+1}, \quad 0 \leq n \leq N-1$$
$$S_N = 0, \quad S_{N+1} = 0$$

This recurrence relation is assured by

$$2 T_n = \frac{1}{n+1} \frac{dT_{n+1}}{dx} - \frac{1}{n-1} \frac{dT_{n-1}}{dx}, \quad n > 1$$

In the Chebyshev collocation method, $N+1$ collocation points are used

$$x_j = \cos \frac{\pi j}{N}, \quad j = 0, 1, \ldots, N$$

As an example, consider the equation

$$\frac{\partial u}{\partial t} + f(u) \frac{\partial u}{\partial x} = 0$$

An explicit method in time can be used

$$\frac{u^{n+1} - u^n}{\Delta t} + f(u^n) \left.\frac{\partial u}{\partial x}\right|^n = 0$$

and evaluated at each collocation point

$$\frac{u_j^{n+1} - u_j^n}{\Delta t} + f(u_j^n) \left.\frac{\partial u}{\partial x}\right|_j^n = 0$$

The trial function is taken as

$$u_j(t) = \sum_{p=0}^{N} a_p(t) \cos \frac{\pi p j}{N}, \quad u_j^n = u_j(t^n) \quad (46)$$

Assume that the values u_j^n exist at some time. Then invert Equation (46) using the fast Fourier transform to obtain $\{a_p\}$ for $p = 0, 1, \ldots, N$; then calculate S_p

$$S_p = S_{p+2} + (p+1) a_{p+1}, \quad 0 \leq p \leq N-1$$

$$S_N = 0, \quad S_{N+1} = 0$$

and finally

$$a_p^{(1)} = \frac{2 S_p}{c_p}$$

Thus, the first derivative is given by

$$\left.\frac{\partial u}{\partial x}\right|_j = \sum_{p=0}^{N} a_p^{(1)}(t) \cos \frac{\pi p j}{N}$$

This is evaluated at the set of collocation points by using the fast Fourier transform again. Once the function and the derivative are known at each collocation point the solution can be advanced forward to the $n+1$-th time level.

The advantage of the spectral method is that it is very fast and can be adapted quite well to parallel computers. It is, however, restricted in the geometries that can be handled.

Software packages exist that use various discretizations in the spatial direction and an integration routine in the time variable: PDECOL uses B-splines for the spatial direction and various GEAR methods in time [8.17, p. 346]; PDEPACK and DSS [8.17, p. 351] use finite differences in the spatial direction and GEARB in time [8.18, p. 163]; and REACOL [8.9, p. 191] use orthogonal collocation in the radial direction and LSODE in the axial direction, whereas REACFD uses finite difference in the radial direction (both codes are restricted to modeling chemical reactors).

The maximum eigenvalue for all the methods is given by

$$|\lambda|_{max} = \frac{LB}{\Delta x^2} \quad (47)$$

where the values of LB are as follows:

Finite difference	4
Galerkin, linear elements, lumped	4
Galerkin, linear elements	36
Galerkin, quadratic elements	60
Orthogonal collocation on finite elements, cubic	36

8.4. Elliptic Equations

Elliptic equations can be solved with both finite difference and finite element methods. One-dimensional elliptic problems are two-point boundary value problems and are covered in Chapter 7. Two- and three-dimensional elliptic problems are often solved with iterative methods when the finite difference method is used and with direct methods when the finite element method is used. Thus, two aspects must be considered: how the equations are discretized to form sets of algebraic equations and how the algebraic equations are then solved.

The prototype elliptic problem is steady-state heat conduction or diffusion,

$$k\left(\frac{\partial^2 T}{\partial x^2} + \frac{\partial^2 T}{\partial y^2}\right) = Q$$

possibly with a heat generation term per unit volume, Q. The boundary conditions can be

Dirichlet or 1st kind: $T = T_1$ on boundary S_1

Neumann or 2nd kind: $k\dfrac{\partial T}{\partial n} = q_2$ on boundary S_2

Robin, mixed, or 3rd kind: $-k\dfrac{\partial T}{\partial n} = h(T - T_3)$ on boundary S_3

Illustrations are given for constant physical properties k, h, while T_1, q_2, T_3 are known functions on the boundary and Q is a known function of position. The finite difference formulation is given by using the following nomenclature

$$T_{i,j} = T(i\Delta x, j\Delta y)$$

The finite difference formulation is then

$$\frac{T_{i+1,j} - 2T_{i,j} + T_{i-1,j}}{\Delta x^2} + \frac{T_{i,j+1} - 2T_{i,j} + T_{i,j-1}}{\Delta y^2} = Q_{i,j} \quad (48)$$

$T_{i,j} = T_1$ for i,j on boundary S_1

$k\dfrac{\partial T}{\partial n}\bigg|_{i,j} = q_2$ for i,j on boundary S_2

$-k\dfrac{\partial T}{\partial n}\bigg|_{i,j} = h(T_{i,j} - T_3)$ for i,j on boundary S_3

If the boundary is parallel to a coordinate axis the boundary slope is evaluated as in Chapter 7, by using either a one-sided, centered difference or a false boundary. If the boundary is more irregular and not parallel to a coordinate line, more complicated expressions are needed and the finite element method may be the better method.

Equation (48) is rewritten in the form

$$2\left(1 + \frac{\Delta x^2}{\Delta y^2}\right)T_{i,j} = T_{i+1,j} + T_{i-1,j} + \frac{\Delta x^2}{\Delta y^2}(T_{i,j+1} + T_{i,j-1}) - \Delta x^2 \frac{Q_{i,j}}{k}$$

The Jacobi method is

$$2\left(1 + \frac{\Delta x^2}{\Delta y^2}\right)T_{i,j}^{s+1} = T_{i+1,j}^s + T_{i-1,j}^s + \frac{\Delta x^2}{\Delta y^2}(T_{i,j+1}^s + T_{i,j-1}^s) - \Delta x^2 \frac{Q_{i,j}}{k}$$

If the points are located in a regular order, and the calculations proceed from low to high i, then from low to high j, the Gauss–Seidel method can be used:

$$2\left(1 + \frac{\Delta x^2}{\Delta y^2}\right)T_{i,j}^{s+1} = T_{i+1,j}^s + T_{i-1,j}^{s+1} + \frac{\Delta x^2}{\Delta y^2}(T_{i,j+1}^s + T_{i,j-1}^{s+1}) - \Delta x^2 \frac{Q_{i,j}}{k}$$

This method converges twice as fast as the Jacobi method. The relaxation method uses

$$2\left(1 + \frac{\Delta x^2}{\Delta y^2}\right)T_{i,j}^* = T_{i+1,j}^s + T_{i-1,j}^{s+1} + \frac{\Delta x^2}{\Delta y^2}(T_{i,j+1}^s + T_{i,j-1}^{s+1}) - \Delta x^2 \frac{Q_{i,j}}{k}$$

$$T_{i,j}^{s+1} = T_{i,j}^s + \beta(T_{i,j}^* - T_{i,j}^s)$$

If $\beta = 1$, this is the Gauss–Seidel method. If $\beta > 1$, it is overrelaxation; if $\beta < 1$, it is underrelaxation. The value of β may be chosen empirically, $0 < \beta < 2$, but it can be selected theoretically for simple problems like this [8.19, p. 100], [8.9, p. 282]. In particular, the optimal value of the iteration parameter is given by

$$-\ln(\beta_{\text{opt}} - 1) \approx R$$

and the error (in solving the algebraic equation) is decreased by the factor $(1 - R)^N$ for every N iterations. For the heat conduction problem and Dirichlet boundary conditions,

$$R = \frac{\pi^2}{2n^2}$$

(when there are n points in both x and y directions). For Neumann boundary conditions, the value is

$$R = \frac{\pi^2}{2n^2} \frac{1}{1 + \max[\Delta x^2/\Delta y^2, \Delta y^2/\Delta x^2]}$$

These iterative methods are point iterative methods. Line iterative methods [8.19, p. 113] use

$$2\left(1 + \frac{\Delta x^2}{\Delta y^2}\right) T_{i,j}^* = T_{i+1,j}^* + T_{i-1,j}^*$$
$$+ \frac{\Delta x^2}{\Delta y^2}(T_{i,j+1}^s + T_{i,j-1}^s) - \Delta x^2 \frac{Q_{i,j}}{k}$$
$$T_{i,j}^{s+1} = T_{i,j}^s + \beta(T_{i,j}^* - T_{i,j}^s)$$

The last equation is applied after the solution for the entire line (current value of i). Now the value of R is

$$R = \frac{2\pi}{n}\left(1 + \frac{\Delta y^2}{\Delta x^2}\right)^{1/2} \quad \text{for Dirichlet boundary conditions}$$

$$R = \frac{2\pi}{n} \quad \text{for Neumann boundary conditions}$$

Line Jacobi converges twice as fast as point Jacobi (when $\Delta x = \Delta y$), and line successive overrelaxation converges twice as fast as point successive overrelaxation [8.19, pp. 110, 114].

The alternating direction method can be employed for elliptic problems by using sequences of iteration parameters [8.9, pp. 283–286], [8.19, pp. 120–127]. The method is well suited to transient problems.

These are the classical iterative techniques. Recently, preconditioned conjugate gradient methods have been developed (see Chap. 1). In these methods a series of matrix multiplications are done iteration by iteration; and the steps lend themselves to the efficiency available in parallel computers. In the multigrid method the problem is solved on several grids, each more refined than the previous one. In iterating between the solutions on different grids, one converges to the so-lution of the algebraic equations. A chemical engineering application is given in [8.20].

Another way to solve the elliptic equations is to convert them to parabolic equations and integrate them to "steady state". One method of doing this, operator splitting, is illustrated for the Navier–Stokes equations in Section 8.5.

The Galerkin finite element method (FEM) is useful for solving elliptic problems and is particularly effective when the domain or geometry is irregular. As an example, cover the domain with triangles and define a trial function on each tringle. The trial function takes the value 1.0 at one corner and 0.0 at the other corners, and is linear in between (see Fig. 36). These trial functions on each triangle are pieced together to give a trial function on the whole domain. General treatments of the finite element method are available [8.21]–[8.23]. For the heat conduction problem the method gives [8.9]

$$\sum_e \sum_j A_{IJ}^e T_J^e = \sum_e \sum_J F_I^e \quad (49)$$

where

$$A_{IJ}^e = -\int k \nabla N_I \cdot \nabla N_J \, dA - \int_{C_3} h_3 N_I N_J \, dC$$

$$F_I^e = \int N_I Q \, dA + \int_{C_2} N_I q_2 \, dC - \int_{C_3} N_I h_3 T_3 \, dC$$

Also, a necessary condition is that

$$T_i = T_1 \text{ on } C_1$$

In these equations I and J refer to the nodes of the triangle forming element e and the summa-

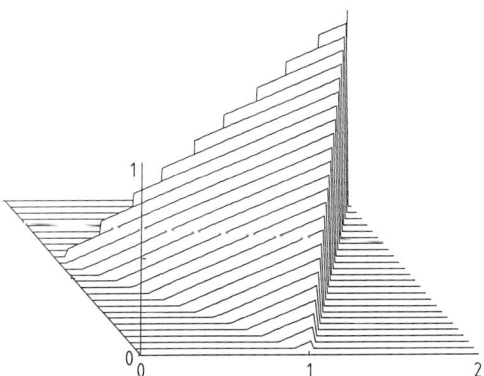

Figure 36. Finite elements trial function: linear polynomials on triangles

tion is made over all elements. These equations represent a large set of linear equations. They are solved simultaneously, with only one iteration, by using either the frontal solution method [8.24] or various sparse matrix techniques (Chap. 1).

If the problem is nonlinear, e.g., with k a function of temperature,

$$\frac{\partial}{\partial x}\left[k(T)\frac{\partial T}{\partial x}\right] + \frac{\partial}{\partial y}\left[k(T)\frac{\partial T}{\partial y}\right] = Q(T)$$

then the equations must be solved iteratively. Let T_I^s be the known temperature at node I. Then Equation (49) can be used with

$$A_{IJ}^e(T^e) = -\int k(T^e)\nabla N_I \cdot \nabla N_J dA$$

$$- \int_{C_3} h_3 N_I N_J dC$$

$$F_I^e(T^e) = \int N_I Q(T^e) dA + \int_{C_2} N_I q_2 dC$$

$$- \int_{C_3} N_I h_3 T_3 dC$$

or

$$\sum_e \sum_J A_{IJ}^e(T^{e,s}) T_J^{e,s+1} = \sum_e \sum_J F_J^e(T^{e,s})$$

This is a successive substitution method. The Newton–Raphson method can also be used in the form

$$\sum_e \sum_J A_{IJ}^e(T^{e,s}) T_J^{e,s+1}$$

$$+ \sum_e \sum_K \frac{dA_{IJ}^e}{dT_K}(T^{e,s})(T_K^{e,s+1} - T_K^{e,s})$$

$$= \sum_e \left[F_I^e(T^{e,s}) + \sum_K \frac{dF_I^e}{dT_K}(T^{e,s})(T_K^{e,s+1} - T_K^{e,s})\right]$$

The integrals in these formulas can be calculated analytically when the physical properties are constant. The results are given for a triangle with nodes I, J, and K in counterclockwise order. Within an element,

$$T = N_I(x, y) T_I + N_J(x, y) T_J + N_K(x, y) T_K$$

$$N_I = \frac{a_I + b_I x + c_I y}{2\Delta}$$

$$a_I = x_J y_K - x_K y_J$$
$$b_I = y_I - y_K \quad \text{plus permutation on } I, K, J$$
$$c_I = x_K - x_J$$

$$2\Delta = \det\begin{bmatrix} 1 & x_I & y_I \\ 1 & x_J & y_J \\ 1 & x_K & y_K \end{bmatrix} = 2 \text{ (area of triangle)}$$

$$a_I + a_J + a_K = 1$$
$$b_I + b_J + b_K = 0$$
$$c_I + c_J + c_K = 0$$

$$A_{IJ}^e = -\frac{k}{4\Delta}(b_I b_J + c_I c_J)$$

$$F_{IJ}^e = \frac{Q}{2}(a_I + b_I \bar{x} + c_I \bar{y}) = \frac{Q\Delta}{3}$$

$$\bar{x} = \frac{x_I + x_J + x_K}{3}, \quad \bar{y} = \frac{y_I + y_J + y_K}{3},$$

$$a_I + b_I \bar{x} + c_I \bar{y} = \frac{2}{3}\Delta$$

One nice feature of the finite element method is the use of natural boundary conditions. In this problem the natural boundary conditions are the Neumann or Robin conditions. When using Equation (49), the problem can be solved on a domain that is shorter than needed to reach some limiting condition (such as at an outflow boundary). The externally applied flux is still applied at the shorter domain, and the solution inside the truncated domain is still valid. Examples are given in [8.6] and [8.25]. The effect of this is to allow solutions in domains that are smaller, thus saving computation time and permitting the solution in semi-infinite domains. Use of the collocation method on finite elements is discussed in [8.1, pp. 330–339].

The methods can also be combined by using one method for one direction and another method for the other direction. KLEIN [8.26, p. 603] treated two-dimensional reactors using orthogonal collocation in the radial direction and a shooting method in the axial direction. PIRKLE et al. [8.27] were able to derive an equivalent one-dimensional model of a chemical reactor to avoid solving the two-dimensional problem. DE-GREVE et al. [8.28] used an adaptive mesh calculation for combustion problems (see also the examples in [8.6]).

A general-purpose package for general two-dimensional domains and rectangular three-dimensional domains is ELLPACK [8.11, p. 348]. This package allows choice of a variety of methods: finite difference, Hermite collocation, spline Galerkin, collocation, and others. Comparisons of the various methods are available [8.29]. The program FISHPAK solves the Helmholtz equation in multiple dimensions when the domain is separable (because fast methods like FFT are used) [8.17, p. 346].

8.5. Parabolic Equations in Two or Three Dimensions

Computations become much more lengthy with two or more spatial dimensions, for example, the unsteady heat conduction equation

$$\varrho C_p \frac{\partial T}{\partial t} = k\left(\frac{\partial^2 T}{\partial x^2} + \frac{\partial^2 T}{\partial y^2}\right) - Q$$

or the unsteady diffusion equation

$$\frac{\partial c}{\partial t} = D\left(\frac{\partial^2 c}{\partial x^2} + \frac{\partial^2 c}{\partial y^2}\right) - R(c)$$

In the finite difference method an explicit technique would evaluate the right-hand side at the n-th time level:

$$\varrho C_p \frac{T_{i,j}^{n+1} - T_{i,j}^n}{\Delta t}$$

$$= \frac{k}{\Delta x^2}(T_{i+1,j}^n - 2 T_{i-1,j}^n + T_{i-1,j}^n)$$

$$+ \frac{k}{\Delta y^2}(T_{i,j+1}^n - 2 T_{i,j}^n + T_{i,j-1}^n) - Q$$

When $Q = 0$ and $\Delta x = \Delta y$, the time step is limited by

$$\Delta t \leq \frac{\Delta x^2 \varrho C_p}{4 k} \quad \text{or} \quad \frac{\Delta x^2}{4 D}$$

These time steps are smaller than for one-dimensional problems. For these dimensions, the limit is

$$\Delta t \leq \frac{\Delta x^2}{6 D}$$

To avoid such small time steps, which become smaller as Δx decreases, an implicit method could be used. This leads to large sparse matrices, rather than convenient tridiagonal matrices. These can be solved, but the alternating direction method is also useful [8.19, pp. 57–63]. This reduces a problem on an $n \times n$ grid to a series of $2n$ one-dimensional problems on an n-grid. Here, step forward $\Delta t/2$ by using the Crank–Nicolson method

$$\varrho C_p \frac{T_{i,j}^{n+1/2} - T_{i,j}^n}{\Delta t/2} = \frac{k}{\Delta x^2}(T_{i+1,j}^{n+1/2} - 2 T_{i,j}^{n+1/2}$$

$$+ T_{i-1,j}^{n+1/2}) + \frac{k}{\Delta y^2}(T_{i,j+1}^n - 2 T_{i,j}^n + T_{i,j-1}^n) - Q$$

and then another $\Delta t/2$ using an implicit method

$$\varrho C_p \frac{T_{i,j}^{n+1} - T_{i,j}^{n+1/2}}{\Delta t/2} = \frac{k}{\Delta x^2}(T_{i+1,j}^{n+1/2} - 2 T_{i,j}^{n+1/2}$$

$$+ T_{i-1,j}^{n+1/2}) + \frac{k}{\Delta y^2}(T_{i,j+1}^{n+1} - 2 T_{i,j}^{n+1} + T_{i,j-1}^{n+1}) - Q$$

This method is second order in Δt errors $[O(\Delta t^2)]$ and reduces the approximate computational burden from n^4 to $6n^2$.

The method of operator splitting is also useful when different terms in the equation are best evaluated by using different methods or as a technique for reducing a larger problem to a series of smaller problems. Here the method is illustrated by using the Navier–Stokes equations. In vector notation the equations are

$$\varrho \frac{\partial \boldsymbol{u}}{\partial t} + \varrho \boldsymbol{u} \cdot \nabla \boldsymbol{u} = \varrho \boldsymbol{f} = \varrho \boldsymbol{f} - \nabla p + \mu \nabla^2 \boldsymbol{u}$$

The equation is solved in the following steps

$$\varrho \frac{\boldsymbol{u}^* - \boldsymbol{u}^n}{\Delta t} = -\varrho \boldsymbol{u}^n \cdot \nabla \boldsymbol{u}^n + \varrho \boldsymbol{f} + \mu \nabla^2 \boldsymbol{u}^n$$

$$\nabla^2 p^{n+1} = \frac{1}{\Delta t} \nabla \cdot \boldsymbol{u}^*$$

$$\varrho \frac{\boldsymbol{u}^{n+1} - \boldsymbol{u}^*}{\Delta t} = -\nabla p$$

This can be done by using the finite difference [8.30, p. 162] or the finite element method [8.31]. In the finite element method, the need to solve large sets of equations simultaneously must be eliminated. To do this, the matrices multiplying time derivatives are approximated by moving all terms to the diagonal (called lumping). The effects of this have been examined in detail in [8.32].

9. Integral Equations [9.1]–[9.4]

If the dependent variable appears under an integral sign an equation is called an integral equation; if derivatives of the dependent variable appear elsewhere in the equation it is called an integrodifferential equation. This chapter describes the various classes of equations, gives information concerning Green's functions, and presents numerical methods for solving integral equations.

9.1. Classification

Volterra integral equations have an integral with a variable limit, whereas Fredholm integral equations have a fixed limit. Volterra equations are usually associated with initial value or evolutionary problems, whereas Fredholm equations are analogous to boundary value problems. The terms in the integral can be unbounded, but still yield bounded integrals, and these equations are said to be weakly singular. A *Volterra equation of the second kind* is

$$y(t) = g(t) + \lambda \int_a^t K(t, s) y(s) \, ds \qquad (50)$$

whereas a *Volterra equation of the first kind* is

$$y(t) = \lambda \int_a^t K(t, s) y(s) \, ds$$

Equations of the first kind are very sensitive to solution errors so that they present severe numerical problems.

An example of a problem giving rise to a Volterra equation of the second kind is the following heat conduction problem:

$$\varrho C_p \frac{\partial T}{\partial t} = k \frac{\partial^2 T}{\partial x^2}, \quad 0 \le x < \infty, \quad t > 0$$

$$T(x, 0) = 0, \quad \frac{\partial T}{\partial x}(0, t) = -g(t),$$

$$\lim_{x \to \infty} T(x, t) = 0, \quad \lim_{x \to \infty} \frac{\partial T}{\partial x} = 0$$

If this is solved by using Fourier transforms the solution is

$$T(x, t) = \frac{1}{\sqrt{\pi}} \int_0^t g(s) \frac{1}{\sqrt{t - s}} e^{-x^2/4(t-s)} \, ds$$

Suppose the problem is generalized so that the boundary condition is one involving the solution T, which might occur with a radiation boundary condition or heat-transfer coefficient. Then the boundary condition is written as

$$\frac{\partial T}{\partial x} = -G(T, t), \quad x = 0, \quad t > 0$$

The solution to this problem is

$$T(x, t) = \frac{1}{\sqrt{\pi}} \int_0^t G(T(0, s), s) \frac{1}{\sqrt{t - s}} e^{-x^2/4(t-s)} \, ds$$

If $T(t)$ is used to represent $T(0, t)$, then

$$T(t) = \frac{1}{\sqrt{\pi}} \int_0^t G(T(s), s) \frac{1}{\sqrt{t - s}} \, ds$$

Thus the behavior of the solution at the boundary is governed by an integral equation. NAGEL and KLUGE [9.5] use a similar approach to solve for adsorption in a porous catalyst.

The existence and uniqueness of the solution can be proved. One example is the following:

Theorem [9.3, p. 30]. If $K(t, s)$ is continuous in $0 \le s \le t \le \alpha$ and $g(t)$ is continuous in $0 \le t \le \alpha$, then the integral equation (Eq. 50) possesses a unique continuous solution for $0 \le t \le \alpha$. More general theorems are also available [9.3, p. 32]. The solution to the integral equation can be obtained by using a Picard or successive substitution method (this is how the existence theorem is proved).

$$y_n(t) = g(t) + \lambda \int_a^t K(t, s) y_{n-1}(s) \, ds$$

Sometimes the kernel is of the form

$$K(t, s) = K(t - s)$$

Equations of this form are called convolution equations and can be solved by taking the Laplace transform. For the integral equation

$$Y(t) = G(t) + \lambda \int_0^t K(t - \tau) Y(\tau) \, d\tau$$

$$K(t) * Y(t) \equiv \int_0^t K(t - \tau) Y(\tau) \, d\tau$$

the Laplace transform is

$$y(s) = g(s) + k(s) y(s)$$
$$k(s) y(s) = L [K(t) * Y(t)]$$

Solving this for $y(s)$ gives

$$y(s) = \frac{g(s)}{1 - k(s)}$$

If the inverse transform can be found, the integral equation is solved.

A *Fredholm equation of the second kind* is

$$y(x) = g(x) + \lambda \int_a^b K(x, s) y(s) \, ds \qquad (51)$$

whereas a *Fredholm equation of the first kind* is

$$\int_a^b K(x, s) y(s) \, ds = g(x)$$

The limits of integration are fixed, and these problems are analogous to boundary value problems. An eigenvalue problem is a homogeneous equation of the second kind.

$$y(x) = \lambda \int_a^b K(x, s) y(s) \, ds \qquad (52)$$

Solutions to this problem occur only for specific values of λ, the eigenvalues. Usually the Fredholm equation of the second or first kind is solved for values of λ different from these, which are called regular values.

Nonlinear Volterra equations arise naturally from initial value problems. For the initial value problem

$$\frac{dy}{dt} = F(t, y)(t))$$

both sides can be integrated from 0 to t to obtain

$$y(t) = y(0) + \int_0^t F(s, y(s)) \, ds$$

which is a nonlinear Volterra equation. The general nonlinear Volterra equation is

$$y(t) = g(t) + \int_0^t K(t, s, y(s)) \, ds \qquad (53)$$

A successive substitution method for its solution is

$$y_{n+1}(t) = g(t) + \int_0^t K[t, s, y_n(s)] \, ds$$

Theorem [9.3, p. 55]. If $g(t)$ is continuous, the kernel $K(t, s, y)$ is continuous in all variables and satisfies a Lipschitz condition

$$|K(t, s, y) - K(t, s, z)| \le L|y - z|$$

then the nonlinear Volterra equation has a unique continuous solution.

Nonlinear Fredholm equations have special names. The equation

$$f(x) = \int_0^1 K[x, y, f(y)] \, dy$$

is called the Urysohn equation [9.2 p. 208]. The special equation

$$f(x) = \int_0^1 K[x, y] F[y, f(y)] \, dy$$

is called the Hammerstein equation [9.2, p. 209]. Iterative methods can be used to solve these equations, and these methods are closely tied to fixed point problems. A fixed point problem is

$$x = F(x)$$

and a successive substitution method is

$$x_{n+1} = F(x_n)$$

Local convergence theorems prove the process convergent if the solution is close enough to the answer, whereas global convergence theorems are valid for any initial guess [9.2, p. 229–231]. The successive substitution method for nonlinear Fredholm equations is

$$y_{n+1}(x) = \int_0^1 K[x, s, y_n(s)] \, ds$$

Typical conditions for convergence include that the function satisfies a Lipschitz condition.

9.2. Numerical Methods for Volterra Equations of the Second Kind

Volterra equations of the second kind are analogous to initial value problems. An initial value problem can be written as a Volterra equation of the second kind, although not all Volterra equations can be written as initial value problems [9.3, p. 7]. Here the general nonlinear Volterra equation of the second kind is treated (Eq. 53). The simplest numerical method involves replacing the integral by a quadrature using the trapezoid rule

$$y_n \equiv y(t_n) = g(t_n) + \Delta t \left\{ \tfrac{1}{2} K(t_n, t_0, y_0) \right.$$

$$\left. + \sum_{i=1}^{n-1} K(t_n, t_i, y_i) + \tfrac{1}{2} K(t_n, t_n, y_n) \right\}$$

This equation is a nonlinear algebraic equation for y_n. Since y_0 is known it can be applied to solve for $y_1, y_2, \ldots$ in succession. For a single integral equation, at each step one must solve a single nonlinear algebraic equation for y_n. Typically, the error in the solution to the integral equation is proportional to Δt^μ, and the power μ is the same as the power in the quadrature error [9.3, p. 97].

The stability of the method [9.3, p. 111] can be examined by considering the equation

$$y(t) = 1 - \lambda \int_0^t y(s)\,ds$$

whose solution is

$$y(t) = e^{-\lambda t}$$

Since the integral equation can be differentiated to obtain the initial value problem

$$\frac{dy}{dt} = -\lambda y, \quad y(0) = 1$$

the stability results are identical to those for initial value methods. In particular, using the trapezoid rule for integral equations is identical to using this rule for initial value problems. The method is A-stable.

The next highest quadrature method is Simpson's rule. When expressed for three points, each Δt apart, this is

$$\int_{t_i}^{t_{i+2}} y(s)\,ds = \frac{\Delta t}{3}[y(t_i) + 4 y(t_{i+1}) + y(t_{i+2})]$$

$$t_{i+1} = t_i + \Delta t, \quad t_{i+2} = t_i + 2\Delta t$$

When applied to an integral equation in the first two intervals,

$$y_{i+2} = g(t_{i+2}) + \frac{\Delta t}{3}\{K(t_{i+2}, t_i, y_i)$$
$$+ 4K(t_{i+2}, t_{i+1}, y_{i+1}) + K(t_{i+2}, t_{i+2}, y_{i+2})\}$$

This is the equation for y_{i+2}, but now there is no equation for y_{i+1}. This is obtained by subdividing the region from t_i to t_{i+1} with another point at $t_{i+1/2}$. Then an integral equation can be written as

$$y_{i+1} = g(t_{i+1}) + \frac{\Delta t}{3}\{K(t_{i+1}, t_i, y_i)$$
$$+ 4K(t_{i+1}, t_{i+1/2}, y_{i+1/2})$$
$$+ K(t_{i+1}, t_{i+1}, y_{i+1})\}$$

To obtain $y_{i+1/2}$ a quadratic interpolation is used through the original three points:

$$y_{i+1/2} = \tfrac{3}{8} y_i + \tfrac{3}{4} y_{i+1} - \tfrac{1}{8} y_{i+2}$$

Now two equations must be solved for the two points y_{i+1} and y_{i+2}. In this block-by-block method [9.3, p. 114], the solution proceeds to solve for two unknowns at a time, block by block.

Explicit Runge–Kutta methods have an analogous formula for integral equations. When solving nonstiff differential equations, Runge–Kutta methods are fast and efficient until the time step is very small to meet the stability requirements. Then implicit methods are used, even though the set of simultaneous algebraic equations must be solved. This time-consuming step can be justified only for stiff problems. When solving integral equations, however, the time-consuming part of the calculation is the repeated approximation of the integrals; thus the effort needed to solve the algebraic equations is not as large a fraction of the total effort for integral methods. Thus, implicit methods, like the trapezoid rule given above, tend to be preferred for integral equations [9.3, p. 124].

Predictor-corrector methods can be used, however. Using fourth-order Adams–Moulton and Adams–Bashforth methods gives

$$\bar{y}_{n+1} = g_{n+1} + \frac{\Delta t}{24}[55 K(t_{n+1}, t_n, y_n)$$
$$- 59 K(t_{n+1}, t_{n-1}, y_{n-1})$$
$$+ 37 K(t_{n+1}, t_{n-2}, y_{n-2})$$
$$- 9 K(t_{n+1}, t_{n-3}, y_{n-3})]$$

and

$$y_{n+1} = g_{n+1} + \frac{\Delta t}{24}[9 K(t_{n+1}, t_{n+1}, \bar{y}_{n+1})$$
$$+ 19 K(t_{n+1}, t_n, y_n)$$
$$- 5 K(t_{n+1}, t_{n-1}, y_{n-1})$$
$$+ K(t_{n+1}, t_{n-2}, y_{n-2})]$$

The next situation that must be addressed is how to solve problems when the kernel is infinite at certain points, i.e., when the problem has a weak singularity. Clearly the formulas given above would be unsuitable when the kernel is infinite at the quadrature point. To handle problems like this, product equations are employed. The problem is written as

$$y(t) = g(t) + \int_0^t p(t, s) K[t, s, y(s)]\,ds$$

where the kernel K is Lipschitz continuous and the singular behavior is contained in p. Then, the functional form of p is included in the quadrature rule. For example, the integral

$$I = \int_a^b p(s)\psi(s)\,ds$$

where p is singular at some points. Each quadrature method corresponds to some method of interpolating the solution. That interpolation is written as

$$\psi(t) = \sum a_i \psi_i(t)$$

Integrals of the form

$$\int_a^b p(s)\psi_i(s)\,ds$$

must be integrated exactly. Then methods can be constructed to handle weakly singular problems.

The product trapezoid rule and product Simpson's rule are contained in the following algorithm. Subdivide the interval into n equal subintervals of length Δt, having points at $t_0, t_1, \ldots, t_n$. Within each subinterval, introduce a further subdivision with points $t_i \leq t_{i0} < t_{i1} < \ldots t_{im} \leq t_{i+1}$. Approximate the function with a polynomial of degree m in each subinterval using a Lagrangian interpolation (see Section 2.2).

$$\psi(t) = \sum_{i=0}^m l_{ij}(t)\psi(t_{ij}), \quad t_{i0} \leq t \leq t_{im}$$

$$l_{ij} = \frac{(t - t_{i0})\ldots(t - t_{i,j-1})(t - t_{i,j+1})\ldots(t - t_{im})}{(t_{ij} - t_{i0})\ldots(t_{ij} - t_{i,j-1})(t_{ij} - t_{i,j+1})\ldots(t_{ij} - t_{im})}$$

Inserting this approximation into the integral gives

$$I = \sum_{i=0}^{n-1}\sum_{j=0}^m w_{ij}\psi(t_{ij})$$

where

$$w_{ij} = \int_{t_i}^{t_{i+2}} p(s)l_{ij}(s)\,ds$$

If the following integrals can be calculated

$$\int s^\mu p(s)\,ds, \quad \mu = 0, 1, \ldots, m$$

then the weights can be found explicitly.

An example is the function as piecewise constant [9.3, p. 132]. Then the integrals are

$$\int_{t_i}^{t_{i+1}} K[t_n, s, y(s)]\,p(s)\,ds =$$

$$K[t_n, t_i, y(t_i)]\int_{t_i}^{t_{i+1}} p(s)\,ds$$

In using this expression the numerical form of the integral equation is

$$y_n = g_n + \sum_{i=0}^{n-1} w_{ni} K[t_n, t_i, y_i], \quad n = 1, 2, \ldots$$

where the weights are

$$w_{ni} = \int_{t_i}^{t_{i+1}} p(s)\,ds$$

When the function $p(s, t) = (s - t)^{-1/2}$, the weights are

$$w_{ni} = 2[\sqrt{t_n - t_i} - \sqrt{t_n - t_{i+1}}]$$

By starting with $y_0 = g(0)$ successive values of y_1, y_2, etc. can be calculated. This corresponds to the Euler method of integration and is only accurate to $O(\Delta t)$. To improve the accuracy, the Richardson extrapolation techniques described in Section 6.2 can be used.

Next, consider the product trapezoid rule [9.3, p. 135]. The kernel is approximated by piecewise linear functions

$$K[t, s, y(s)] = \frac{s - t_i}{\Delta t} K[t, t_{i+1}, y(t_{i+1})]$$

$$+ \frac{t_{i+1} - s}{\Delta t} K[t, t_i, y(t_i)], \quad t_i \leq s \leq t_{i+1}$$

Then the integration formula is

$$\int_0^{t_n} p(t_n, s) K[t_n, s, y(s)]\,ds = \alpha_{n1} K[t_n, t_0, y(t_0)]$$

$$+ \sum_{i=1}^{N-1}(\alpha_{n,i+1} + \beta_{ni}) K[t_n, t_i, y(t_i)]$$

$$+ \beta_{nn} K[t_n, t_n, y(t_n)]$$

where

$$\alpha_{n,i+1} = \frac{1}{\Delta t}\int_{t_i}^{t_{i+1}}(t_{i+1} - s) p(t_n, s)\,ds,$$

$$\beta_{n,i+1} = \frac{1}{\Delta t}\int_{t_i}^{t_{i+1}}(s - t_i) p(t_n, s)\,ds$$

The numerical version of the integral equation is

$$y_n = g_n + \alpha_n K[t_n, t_0, y(t_0)]$$
$$+ \sum_{i=1}^{N-1} (\alpha_{n,i+1} + \beta_{ni}) K[t_n, t_i, y(t_i)]$$
$$+ \beta_{nn} K[t_n, t_n, y(t_n)]$$

Higher order methods can be solved block by block in a similar fashion.

Volterra equations of the first kind are not well posed, and small errors in the solution can have disastrous consequences [9.3, p. 71]. Only lowest order methods have been recommended for problems of the first kind [9.3, p. 151].

9.3. Numerical Methods for Fredholm, Urysohn, and Hammerstein Equations of the Second Kind

Whereas Volterra equations could be solved from one position to the next, like initial value differential equations, Fredholm equations must be solved over the entire domain, like boundary value differential equations. Thus, large sets of equations will be solved and the notation is designed to emphasize that.

The methods are also based on quadrature formulas. For the integral

$$I(\varphi) = \int_a^b \varphi(y) \, dy$$

a quadrature formula is written:

$$I(\varphi) = \sum_{i=0}^{n} w_i \varphi(y_i)$$

Then the integral Fredholm equation can be rewritten as

$$f(x) - \lambda \sum_{i=0}^{n} w_i K(x, y_i) f(y_i) = g(x), \quad a \le x \le b \tag{54}$$

If this equation is evaluated at the points $x = y_j$,

$$f(y_j) - \lambda \sum_{i=0}^{n} w_i K(y_j, y_i) f(y_i) = g(y_i)$$

is obtained, which is a set of linear equations to be solved for $\{f(y_j)\}$. The solution at any point is then given by Equation (54).

Because the quadrature method is often defined on subintervals, the equation is rewritten in terms of information obtained on the subintervals. The finite element notation is a natural notation. On a subinterval with points $y_{I=1}$, $y_{I=2}, \ldots$ the quadrature formula is

$$\int_{x_i}^{x_{i+2}} \varphi(y) \, dy = \Delta x \int_0^1 \varphi(u) \, du = \sum_{I=1}^{NP} w_I \varphi(y_I)$$

An element diagonal matrix is then constructed:

$$D = \begin{bmatrix} w_1 & 0 & . & 0 \\ 0 & w_2 & . & 0 \\ . & . & . & . \\ 0 & 0 & . & w_n \end{bmatrix}$$

and the element kernel

$$K_{iJ}^e = K(x_i, x^e + \Delta x^e u_J)$$

The numerical version of the integral equation is then

$$f_i - \lambda \sum_{e} \sum_{J=1}^{NP} K_{iJ}^e D_{JJ}^e f_J^e = g_i$$

In matrix notation this can be written as

$$(I - \lambda K D) f = g$$

The structure of this matrix is typically dense, because the kernel is nonzero even when the points x and y are in different elements. Thus, the economies obtained by using sparse matrix routines to solve differential equations do not always hold for integral equations.

A common type of integral equation has a singular kernel along $x = y$. This can be transformed to a less severe singularity by writing

$$\int_a^b K(x, y) f(y) \, dy = \int_a^b K(x, y) [f(y) - f(x)] \, dy$$
$$+ \int_a^b K(x, y) f(x) \, dy = \int_a^b K(x, y) [f(y)$$
$$- f(x)] \, dy + f(x) H(x)$$

where

$$H(x) = \int_a^b K(x, y) f(x) \, dy$$

is a known function. The integral equation is

then replaced by

$$f(x) = g(x) + \sum_{i=0}^{n} w_i K(x, y_i)[f(y_i) - f(x)] + f(x) H(x)$$

In matrix notation this is

$$(I - \lambda M)f = g$$

and the matrix M is the same as KD except for the diagonal elements, which are

$$M_{ii} = H_i - \sum K_{ik} w_k$$

The product quadrature can be used to handle weak singularities, as with Volterra equations (see [9.1, p. 540] for an example).

Collocation methods can be applied as well [9.1, p. 396]. To solve integral Equation (51) expand f in the function

$$f = \sum_{i=0}^{n} a_i \varphi_i(x)$$

Substitute f into the equation to form the residual

$$\sum_{i=0}^{n} a_i \varphi_i(x) - \lambda \sum_{i=0}^{n} a_i \int_a^b K(x, y) \varphi_i(y) \, dy = g(x)$$

Evaluate the residual at the collocation points

$$\sum_{i=0}^{n} a_i \varphi_i(x_j) - \lambda \sum_{i=0}^{n} a_i \int_a^b K(x_j, y) \varphi_i(y) \, dy = g(x_j)$$

The expansion can be in piecewise polynomials, leading to a collocation finite element method, or global polynomials, leading to a global approximation. If orthogonal polynomials are used then the quadratures can make use of the accurate Gaussian quadrature points to calculate the integrals. Galerkin methods are also possible [9.1, p. 406]. MILLS et al. [9.6] consider reaction–diffusion problems and say the choice of technique cannot be made in general because it is highly dependent on the kernel.

When the integral equation is nonlinear, iterative methods must be used to solve it. Convergence proofs are available, based on Banach's contractive mapping principle. Consider the Urysohn equation, with $g(x) = 0$ without loss of generality:

$$f(x) = \int_a^b F[x, y, f(y)] \, dy$$

The kernel satisfies the Lipschitz condition

$$\max_{a \leq x, y \leq b} |F[x, y, f(y)] - F[x, z, f(z)]| \leq K|y - z|$$

Theorem [9.2, p. 214]. If the constant K is < 1 and certain other conditions hold, the successive substitution method

$$f_{n+1}(x) = \int_a^b F[x, y, f_n(y)] \, dy, \quad n = 0, 1, \ldots$$

converges to the solution of the integral equations.

9.4. Numerical Methods for Eigenvalue Problems

Eigenvalue problems are treated similarly to Fredholm equations, except that the final equation is a matrix eigenvalue problem instead of a set of simultaneous equations. For example,

$$\sum_{i=1}^{n} w_i K(y_i, y_j) f(y_i) = \lambda f(y_j), \quad i = 0, 1, \ldots, n$$

leads to the matrix eigenvalue problem

$$KDf = \lambda f$$

9.5. Green's Functions

Integral equations can arise from the formulation of a problem by using Green's function. For example, the equation governing heat conduction with a variable heat generation rate is represented in differential forms as

$$\frac{d^2 T}{dx^2} = \frac{Q(x)}{k}, \quad T(0) = T(1) = 0$$

In integral form the same problem is [9.1, pp. 57–60]

$$T(x) = \frac{1}{k} \int_0^1 G(x, y) Q(y) \, dy$$

$$G(x, y) = \begin{cases} -x(1-y) & x \leq y \\ -y(1-x) & y \leq x \end{cases}$$

Green's functions for typical operators are given below.

For the Poisson equation with solution decaying to zero at infinity

$$\nabla^2 \psi = -4\pi\varrho$$

the formulation as an integral equation is

$$\psi(r) = \int_V \varrho(r_0) G(r, r_0) dV_0$$

where Green's function is [9.7, p. 891]

$$G(r, r_0) = \frac{1}{r} \text{ in three dimensions}$$

$$= -2 \ln r \text{ in two dimensions}$$

where $r = \sqrt{(x-x_0)^2 + (y-y_0)^2 + (z-z_0)^2}$ in three dimensions

and $r = \sqrt{(x-x_0)^2 + (y-y_0)^2}$ in two dimensions

For the problem

$$\frac{\partial u}{\partial t} = D\nabla^2 u, \quad u = 0 \text{ on } S,$$

with a point source at x_0, y_0, z_0

Green's function is [9.8, p. 355]

$$u = \frac{1}{8[\pi D(t-\tau)]^{3/2}}$$

$$e^{-[(x-x_0)^2 + (y-y_0)^2 + (z-z_0)^2]/4D(t-\tau)}$$

When the problem is

$$\frac{\partial c}{\partial t} = D\nabla^2 c$$

$c = f(x, y, z)$ in S at $t = 0$
$c = \varphi(x, y, z)$ on $S, t > 0$

the solution can be represented as [9.8, p. 353]

$$c = \iiint (u)_{\tau=0} f(x, y, z) \, dx \, dy \, dz$$

$$+ D \int_0^t \iint \varphi(x, y, z, \tau) \frac{\partial u}{\partial n} dS \, dt$$

When the problem is two dimensional,

$$u = \frac{1}{\sqrt{4\pi D(t-\tau)}} e^{-[(x-x_0)^2 + (y-y_0)^2]/4D(t-\tau)}$$

$$c = \iint (u)_{\tau=0} f(x, y) \, dx \, dy$$

$$+ D \int_0^t \int \varphi(x, y, \tau) \frac{\partial u}{\partial n} dC \, dt$$

For the following differential equation and boundary conditions

$$\frac{1}{x^{a-1}} \frac{d}{dx}\left(x^{a-1} \frac{dc}{dx}\right) = f[x, c(x)],$$

$$\frac{dc}{dx}(0) = 0, \quad \frac{2}{Sh} \frac{dc}{dx}(1) + c(1) = g$$

where Sh is the Sherwood number, the problem can be written as a Hammerstein integral equation:

$$c(x) = g - \int_0^1 G(x, y, Sh) f[y, c(y)] y^{a-1} dy$$

Green's function for the differential operators are [9.9]

$a = 1$
$$G(x, y, Sh) = \begin{cases} 1 + \dfrac{2}{Sh} - x, & y \leq x \\ 1 + \dfrac{2}{Sh} - y, & x < y \end{cases}$$

$a = 2$
$$G(x, y, Sh) = \begin{cases} \dfrac{2}{Sh} - \ln x, & y \leq x \\ \dfrac{2}{Sh} - \ln y, & x < y \end{cases}$$

$a = 3$
$$G(x, y, Sh) = \begin{cases} \dfrac{2}{Sh} + \dfrac{1}{x} - 1, & y \leq x \\ \dfrac{2}{Sh} + \dfrac{1}{y} - 1, & x < y \end{cases}$$

Green's functions for the reaction diffusion problem were used to provide computable error bounds by FERGUSON and FINLAYSON [9.9]. If Green's function has the form

$$K(x, y) = \begin{cases} u(x)v(y) & 0 \leq y \leq x \\ u(y)v(x) & x \leq y \leq 1 \end{cases}$$

the problem

$$f(x) = \int_0^1 K(x, y) F[y, f(y)] dy$$

may be written as

$$f(x) - \int_0^x [u(x)v(y) - u(y)v(x)] F[y, f(y)] dy$$

$$= \alpha v(x)$$

where

$$\alpha = \int_0^1 u(y) F[y, f(y)] \, dy$$

Thus, the problem ends up as one directly formulated as a fixed point problem:

$$f = \Phi(f)$$

When the problem is the diffusion–reaction one, the form is

$$c(x) = g - \int_0^x [u(x) v(y) \\
- u(y) v(x)] f[y, c(y)] y^{a-1} \, dy - \alpha v(x)$$

$$\alpha = \int_0^1 u(y) f[y, c(y)] y^{a-1} \, dy$$

DIXIT and TAULARIDIS [9.10] solved problems involving Fischer–Tropsch synthesis reactions in a catalyst pellet using a similar method.

9.6. Boundary Integral Equations and Boundary Element Method

The boundary element method utilizes Green's theorem and integral equations. Here, the method is described briefly for the following boundary value problem in two or three dimensions

$$\nabla^2 \varphi = 0, \quad \varphi = f_1 \text{ on } S_1, \quad \frac{\partial \varphi}{\partial n} = f_2 \text{ on } S_2$$

Green's theorem (see p. 1-51) says that for any functions sufficiently smooth

$$\int_V (\varphi \nabla^2 \psi - \psi \nabla^2 \varphi) \, dV = \int_S \left(\varphi \frac{\partial \psi}{\partial n} - \psi \frac{\partial \varphi}{\partial n} \right) dS$$

Suppose the function ψ satisfies the equation

$$\nabla^2 \psi = 0$$

In two and three dimensions, such a function is

$$\psi = \ln r, \quad r = \sqrt{(x - x_0)^2 + (y - y_0)^2}$$
in two dimensions

$$\psi = \frac{1}{r}, \quad r = \sqrt{(x - x_0)^2 + (y - y_0)^2 + (z - z_0)^2}$$
in three dimensions

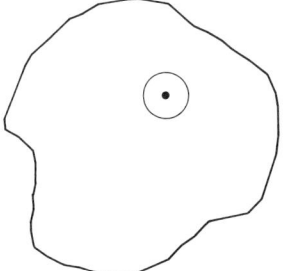

Figure 37. Domain with singularity at P

where $\{x_0, y_0\}$ or $\{x_0, y_0, z_0\}$ is a point in the domain. The solution φ also satisfies

$$\nabla^2 \varphi = 0$$

so that

$$\int_S \left(\varphi \frac{\partial \psi}{\partial n} - \psi \frac{\partial \varphi}{\partial n} \right) dS = 0$$

Consider the two-dimensional case. Since the function ψ is singular at a point, the integrals must be carefully evaluated. For the region shown in Figure 37, the domain is $S = S_1 + S_2$; a small circle of radius r_0 is placed around the point P at x_0, y_0. Then the full integral is

$$\int_S \left(\varphi \frac{\partial \ln r}{\partial n} - \ln r \frac{\partial \varphi}{\partial n} \right) dS$$

$$+ \int_{\theta=0}^{\theta=2\pi} \left(\varphi \frac{\partial \ln r_0}{\partial n} - \ln r_0 \frac{\partial \varphi}{\partial n} \right) r_0 \, d\theta = 0$$

As r_0 approaches 0,

$$\lim_{r \to \infty} r_0 \ln r_0 = 0$$

and

$$\lim_{r_0 \to \infty} \int_{\theta=0}^{\theta=2\pi} \varphi \frac{\partial \ln r_0}{\partial n} r_0 \, d\theta = -\varphi(P) 2\pi$$

Thus for an internal point,

$$\varphi(P) = \frac{1}{2\pi} \int_S \left(\varphi \frac{\partial \ln r}{\partial n} - \ln r \frac{\partial \varphi}{\partial n} \right) dS \quad (55)$$

If P is on the boundary, the result is [9.11, p. 464]

$$\varphi(P) = \frac{1}{\pi} \int_S \left(\varphi \frac{\partial \ln r}{\partial n} - \ln r \frac{\partial \varphi}{\partial n} \right) dS$$

Putting in the boundary conditions gives

$$\pi \varphi (P) = \int_{S_1} \left(f_1 \frac{\partial \ln r}{\partial n} - \ln r \frac{\partial \varphi}{\partial n} \right) dS$$
$$+ \int_{S_2} \left(\varphi \frac{\partial \ln r}{\partial n} - f_2 \ln r \right) dS \quad (56)$$

This is an integral equation for φ on the boundary. Note that the order is one less than the original differential equation. However, the integral equation leads to matrices that are dense rather than banded or sparse, so some of the advantage of lower dimension is lost. Once this integral equation (Eq. 56) is solved to find φ on the boundary, it can be substituted in Equation (55) to find φ anywhere in the domain.

In the boundary finite element method, both the function and its normal derivative along the boundary are approximated.

$$\varphi = \sum_{j=1}^{N} \varphi_j N_j(\xi), \quad \frac{\partial \varphi}{\partial n} = \sum_{j=1}^{N} \left(\frac{\partial \varphi}{\partial n} \right)_j N_j(\xi)$$

One choice of trial functions can be the piecewise constant functions shown in Figure 38. The integral equation then becomes

$$\pi \varphi_i = \sum_{j=1}^{N} \left[\varphi_j \int_{S_j} \frac{\partial \ln r_i}{\partial n} dS - \left(\frac{\partial \varphi_j}{\partial n} \right) \int_{S_j} \ln r_i \, ds \right]$$

The function φ_j is of course known along s_1, whereas the derivative $\partial \varphi_j/\partial n$ is known along s_2. This set of equations is then solved for φ_i and $\partial \varphi_i/\partial n$ along the boundary. This constitutes the boundary integral method applied to the Laplace equation.

If the problem is Poisson's equation

$$\nabla^2 \varphi = g(x, y)$$

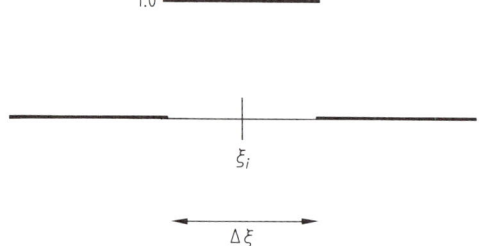

Figure 38. Trial function on boundary for boundary finite element method

Green's theorem gives

$$\int_S \left(\varphi \frac{\partial \ln r}{\partial n} - \ln r \frac{\partial \varphi}{\partial n} \right) dS + \int_A g \ln r \, dA = 0$$

Thus, for an internal point,

$$2 \pi \varphi (P) = \int_S \left(\varphi \frac{\partial \ln r}{\partial n} - \ln r \frac{\partial \varphi}{\partial n} \right) dS$$
$$+ \int_A g \ln r \, dA \quad (57)$$

and for a boundary point,

$$\pi \varphi (P) = \int_S \left(\varphi \frac{\partial \ln r}{\partial n} - \ln r \frac{\partial \varphi}{\partial n} \right) dS + \int_A g \ln r \, dA \quad (58)$$

If the region is nonhomogeneous this method can be used [9.11, p. 475], and it has been applied to heat conduction by HSIEH and SHANG [9.12]. The finite element method can be applied in one region and the boundary finite element method in another region, with appropriate matching conditions [9.11, p. 478]. If the problem is nonlinear, then it is more difficult. For example, consider an equation such as Poisson's in which the function depends on the solution as well

$$\nabla^2 \varphi = g(x, y, \varphi)$$

Then the integral appearing in Equation (58) must be evaluated over the entire domain, and the solution in the interior is given by Equation (57). For further applications, see [9.13] and [9.14].

10. Optimization

10.1. Introduction

The design and operation of chemical processes involve making decisions that commonly have a large economic impact. In most cases, selecting the appropriate decisions is a complex process. Chemical processes tend to exhibit strong interactions among units as well as many different types of trade-offs. Furthermore, how to select decisions that satisfy design or product specifications is not always obvious. A clear need exists for the use of systematic techniques to model the decision making in chemical processes. The methods of optimization, or more precisely, mathematical programming, are concerned with this objective.

In its most general form, an optimization problem can be modeled as follows:

$$\min F = F(z)$$
$$\text{such that } \boldsymbol{h}(z) = 0 \quad (59)$$
$$\boldsymbol{g}(z) \leq 0$$

where F represents a specified objective function that is to be minimized. The case of maximization can be modeled by minimizing the negative of the objective. The functions $\boldsymbol{h}$ and $\boldsymbol{g}$ represent equality and inequality constraints that must be satisfied as the minimum of the function F is determined. These constraints define the feasible region for the allowable choices of variables z. Also, any inequality constraint can be rearranged in the above form.

In Equation (59) the variables z are often continuous in nature because they are used mainly to model flow, pressure, temperature, and size. However, in a number of cases, some of these variables can be restricted to take only discrete values. This arises when in modeling choices of standard sizes or logical choices that are represented by Boolean variables 0–1. The objective function F is assumed to be a scalar function typically representing total cost or net present value. When several objective functions are specified (e.g., minimize cost, maximize reliability), these objectives are commonly combined into one function, or else one is selected for optimization whereas the others are specified as constraints by stating values they are to attain. The equations $\boldsymbol{h}(z) = 0$ are commonly linear or nonlinear algebraic equations when steady-state processes are being modeled. However, in dealing with distributed or dynamic models, ordinary or partial differential equations might be involved. The inequalities $\boldsymbol{g}(z) \leq 0$, correspond to algebraic functions, many of which are simply lower and upper bounds on the variables. Finally, the objective function and the constraints may be given explicitly as equations, or they may be given as implicit functions when computed through procedures (e.g., by calling a subroutine or by running a process simulator).

Different types of optimization problems arise depending on the nature of the functions involved in Equation (59). The simplest case, when no constraints are involved, gives rise to unconstrained nonlinear optimization problems. An example in chemical engineering of this problem type is the fitting of parameters for thermodynamic correlations. When both equality and inequality constraints are involved, a nonlinear programming (NLP) problem must be solved. These problems are used to model the optimization of sizes and operating conditions of flow sheets operating at steady state. The particular case when all the functions and constraints are linear gives rise to linear programming (LP) problems. Linear programming problems are used in the planning and scheduling of refinery operations. At the other extreme, when differential equations are involved in the model, a variational or optimal control problem exists. Such problems are encountered in the determination of optimal temperature profiles in batch reactors. Finally, when a subset of the variables is restricted to take discrete (e.g., 0–1) values, a mixed-integer optimization problem exists. Such formulations are used to synthesize the configuration of a process flow sheet.

This chapter presents on overview of the forementioned optimization problems. Section 10.2 presents the optimality conditions for nonlinear unconstrained and constrained continuous variable optimization problems, introducing constrained derivatives and the closely related Lagrange theory. The next section presents both basic and state-of-the-art algorithms for these problems. Because of their importance as problem classes, individual sections covering linear programming, mixed-integer programming, and the optimization of variational problems complete this chapter.

10.2. Conditions for Optimality

In this section, conditions are presented by which a candidate point for the solution of the optimization problem (Eq. 59) can be tested to see if it is indeed optimal. Generally, local tests are developed, which can tell only if the candidate point is a minimum point for the objective function among all points in a small but finite neighborhood of points around it.

To prove that a point is the global optimum requires stating something about the structure of the overall problem. A sufficient condition is that the problem is *convex*, a property that can be proved for certain classes of problems (e.g., linear and some geometric programming problems). For a convex problem, all local optimum points (if many exist) are connected to each other (form a flat spot) and must be the global optimum.

An intuitive feel for the concept of a convex problem can be gained by visualizing an objective function $F(z_1, z_2)$ plotted above two variables, z_1 and z_2. Constraints of the form $g(z_1, z_2) \le 0$ and $h(z_1, z_2) = 0$ limit the allowed values of z_1 and z_2. A three-dimensional volume is defined by joining vertically all allowed values of z_1 and z_2 to their respective values of F. If any two points in this three-dimensional volume are joined by a straight line and that line always stays within the volume, the problem is convex; otherwise, it is not. A nonlinear equality constraint in z_1 and z_2 makes the problem nonconvex.

Local Minimum Point for Unconstrained Problems. The first problem for which optimality conditions are presented is for the minimization of a scalar objective function F of r independent variables u_i, $i = 1, 2, \ldots r$, where the variables u are unconstrained, i.e.,

$$\text{Min}\{F(u) | u \in Re^r\}$$

If F is sufficiently smooth (is continuous and has continuous first and second derivatives), a *necessary* condition for optimality is that F is stationary with respect to all variations in the independent variables u at the candidate point $\hat{u}$, i.e.,

$$\frac{\partial F}{\partial u_i} = 0, \quad i = 1, 2, \ldots r \quad \text{or} \quad \mathbf{V}_u^T F = \mathbf{0}^T \text{ at } u = \hat{u}$$

The point $\hat{u}$ may, however, be a minimum, maximum, or saddle point.

Sufficient conditions are harder to state. They require that any local move away from the optimal point $\hat{u}$ must give rise to an increase in the objective function. Here, F is expanded in a Taylor series locally around the candidate point $\hat{u}$ up to second-order terms:

$$F(u) = F(\hat{u}) + \mathbf{V}_u^T F|_{\hat{u}} (u - \hat{u}) + \tfrac{1}{2}(u - \hat{u})^T \mathbf{V}_{uu} F|_{\hat{u}} (u - \hat{u}) + \ldots$$

In the neighborhood of $\hat{u}$ where $\Delta u = u - \hat{u}$ is sufficiently small that all higher order terms are negligible, $F(u)$ will increase if the matrix of second partial derivatives $\mathbf{V}_{uu} F$ evaluated at $\hat{u}$ is positive definite. Because $\mathbf{V}_{uu} F$ is symmetric, all of its eigenvalues are real numbers, and it will be positive definite if and only if all of its eigenvalues are greater than zero. Thus a set of sufficient conditions that F has a local minimum at $\hat{u}$ is that $\mathbf{V}_u^T F$ is zero and that $\mathbf{V}_{uu} F$ is positive definite at $\hat{u}$.

Equality Constrained Problems—Constrained Derivatives. The necessary conditions for minimizing a scalar objective function F written in terms of n variables z and subject to m equality constraints $h(z) = 0$ are more complex. A formal definition of the problem is

$$\text{Min}\{F(z) | h(z) = 0, z \in Re^n, h: Re^n \to Re^m\}$$

Assume the candidate point $\hat{z}$ is to be tested to see if it could be an optimum point. Linearizing the m equality constraints around $\hat{z}$ yields

$$h(\hat{z} + \Delta z) = h(\hat{z}) + \mathbf{V}_z^T h|_{\hat{z}} \Delta z$$

where $\Delta z = z - \hat{z}$. Require that the move Δz be selected in such a way that to a first-order approximation, the equality constraints remain at zero, i.e., $\Delta h = h(\hat{z} + \Delta z) - h(\hat{z}) = 0$ (this is constraining the allowed move). Therefore, m constraints have been written which must be satisfied for any move away from $\hat{z}$.

Next partition the variables Δz into a set of m dependent variables Δx and $n - m = r$ independent variables Δu, and rewrite the previous equations in terms of these partitioned variables.

$$\Delta h = \mathbf{V}_x^T h|_{\hat{z}} \Delta x + \mathbf{V}_u^T h|_{\hat{z}} \Delta u = 0$$

These equations are solved for the dependent variables Δx in terms of the independent variables Δu:

$$\Delta x = -[\mathbf{V}_x^T h|_{\hat{z}}]^{-1} \mathbf{V}_u^T h|_{\hat{z}} \Delta u = 0 \qquad (60)$$

Because the inverse of the Jacobian matrix $\mathbf{V}_x^T h|_{\hat{z}}$ must be formed, we are constrained to partition the variables Δz into variables Δx and Δu in such a way that this Jacobian matrix will be nonsingular. Only the values of the independent variables Δu can be changed arbitrarily because Equation (60) provides us with the corresponding values for the dependent variables Δx.

The objective function $F(z)$ can also be linearized in terms of the partitioned variables

$$\Delta F = \mathbf{V}_x^T F|_{\hat{z}} \Delta x + \mathbf{V}_u^T F|_{\hat{z}} \Delta u$$

and Δx can be replaced by using Equation (60):

$$\Delta F = \{\mathbf{V}_u^T F|_{\hat{z}} - \mathbf{V}_x^T F|_{\hat{z}} [\mathbf{V}_x^T h|_{\hat{z}}]^{-1} \mathbf{V}_u^T h\} \Delta u$$

$$= \left(\frac{dF}{du}\right)^T_{\Delta h = 0} \Delta u = \sum_{i=1}^{r} \left(\frac{dF}{du_i}\right)_{\Delta h = 0} \Delta u_i \qquad (61)$$

The terms in the braces {} are in the form of a row vector, one term for each variable Δu_i. Each tells how the objective function F will change if an infinitesimal change is made in the corresponding independent variable Δu_i while corresponding changes in the variables x are being computed such that the equality constraints remain satisfied. These terms are called constrained derivatives.

Necessary conditions for optimality are now written directly: F should be stationary with respect to any infinitesimal move in the independent variables u. The coefficients in these equations must be zero, i.e.,

$$\left(\frac{dF}{du_i}\right)_{\Delta h = 0} = 0, \quad i = 1, 2, \ldots r$$

Partitioning the variables z into a legal set of dependent and independent variables, and at the same time computing these constrained derivatives by performing the forward elimination step of a Gauss elimination for solving linear equations, are rather easy. The following $m + 1$ equations are written in the $m = n + r$ variables z. The Jacobian should be evaluated at $\hat{z}$, i.e., it should be an array of numbers:

$$\mathbf{V}_z^T h|_{\hat{z}} \Delta z = 0 \quad m \text{ rows}$$

$$\mathbf{V}_z^T F|_{\hat{z}} \Delta z = 0 \quad 1 \text{ row}$$

Next a forward Gauss elimination is performed on this Jacobian. The eliminations are done in the last row ($\mathbf{V}_z^T F \Delta z = 0$) without pivoting within that row. Pivot using any of the variables z, and rearrange the variables and equations so they are in the order in which they were pivoted. Figure 39 shows the structure of the result. The pivoted variables become the dependent variables x. Because one could pivot with these variables, the Jacobian matrix $\mathbf{V}_x^T h|_{\hat{z}}$ will be nonsingular, as required. The remaining unpivoted variables are selected to be the independent variables u. Finally, and most conveniently, the nonzero portion of the last row beneath the variables u contains the numerical evaluation for the constrained derivatives given in Equation (61). The numbers here are exact, as if the matrix operations in Equation (61) had been carried out.

Equality Constrained Problems—Lagrange Multipliers. Look at exactly the same optimization problem given in the last section, again with a point $\hat{z}$, and consider what conditions are necessary for $\hat{z}$ to be an optimum point. A scalar function is formed by adding to the objective function each of the equality constraints multiplied by an arbitrary multiplier and partitioning the variables as shown above.

$$L(x, u, \lambda) = F(x, u) + \sum_{i=1}^{m} \lambda_i h_i(x, u)$$

$$= F(x, u) + \lambda^T h(x, u)$$

This function is termed the Lagrange function and the multipliers are called Lagrange multipliers. At any point where the functions $h(z)$ are zero, the Lagrange function equals the objective function.

Next, the stationarity conditions for L are written with respect to all the variables of which it is a function, x, u, and λ.

$$\mathbf{V}_x^T L|_{\hat{z}} = \mathbf{V}_x^T F|_{\hat{z}} + \lambda^T \mathbf{V} h_x^T|_{\hat{z}} = 0$$
$$\mathbf{V}_u^T L|_{\hat{z}} = \mathbf{V}_u^T F|_{\hat{z}} + \lambda^T \mathbf{V} h_u^T|_{\hat{z}} = 0$$
$$\mathbf{V}_\lambda^T L|_{\hat{z}} = h^T(x, u) \qquad\qquad = 0$$

The first equation is solved for the Lagrange multipliers

$$\lambda^T = -\mathbf{V}_x^T F [\mathbf{V} h_x^T]^{-1} \tag{62}$$

and this result is used to eliminate the multipliers from the second equation.

$$\mathbf{V}_u^T L = \mathbf{V}_u^T F - \mathbf{V}_x^T F [\mathbf{V} h_x^T] \mathbf{V} h_u^T = 0 \tag{63}$$

The results are examined. The third stationarity equation for the Lagrange function with respect to the Lagrange multipliers simply states that the equality constraints must be zero. Equation (63)

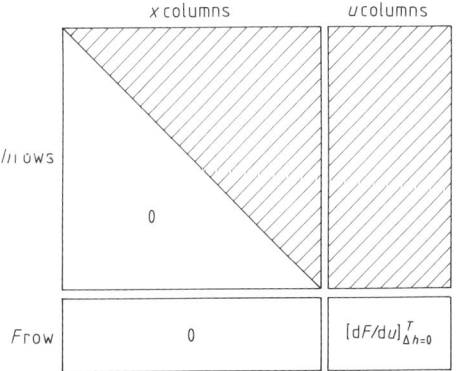

Figure 39. Computing constrained derivatives by Gaussian elimination

states that $\nabla_u L$ is equal to the constrained derivatives for the problem, which can be seen by comparing them to Equation (61). These should be zero. Thus the necessary conditions for optimality, derived in the last section, have been reproduced rather nicely.

These Lagrange multipliers are very interesting variables. They are often referred to as shadow prices, adjoint variables, or dual variables, depending on the context. The notion of a shadow price is that the multiplier λ_i tells one the marginal cost of requiring the constraint h_i to be held at zero. If h_i is moved slightly away from zero at a point that is currently the solution for the optimization problem, and reoptimization is performed (which will result in all the other constraints being held at zero), consider how the objective function will change. The perturbation in the Lagrange function L is given by

$$\Delta L = \Delta F + \lambda_i \Delta h_i = 0$$

which is zero because, as above, L (and not the objective Function F) is at a stationary point at the optimum. Solving for the change in the objective function yields

$$\Delta F = -\lambda_i \Delta h_i = -\lambda_i h_i \qquad (64)$$

Thus, with the multiplier, an estimation can be made of how the objective function will change for a small change in the value of a constraint.

For example, suppose a chemical process flow sheet has been optimized, subject to a constraint that held the sulfur dioxide concentration leaving in the stack gas at 1 ppm. In solving with the Lagrange formulation, the multiplier would have been evaluated for that constraint. Then an estimate could be made of the cost of decreasing the concentration to 0.95 ppm by simply applying Equation (64) rather than by resolving. (If this change could be tied to projected human mortality rates downstream of the emission, our multiplier would allow calculation of how many dollars would have to be spent to save another life per year.)

Equality and Inequality Constrained Problems — Kuhn–Tucker Multipliers. In this section, the following problem is considered:

Min $\{F(z) | h(z) = 0, \quad g(z) \leq 0$,
where $z \in Re^n$, $F: Re^n \rightarrow Re^1$,
$h: Re^n \rightarrow Re^m$, $g: Re^n \rightarrow Re^p\}$

Now, p inequality constraints $g(z) \leq 0$ are being added to the equality constrained problem considered in the last two sections. A Lagrange function can be written, as before:

$$L(z, \lambda, \mu) \equiv F(z) + \lambda^T h(z) + \mu^T g(z)$$

Here, to the Lagrange function is added each of the inequality constraints $g_i(z)$, multiplied by a so-called Kuhn–Tucker multiplier μ_i. The necessary conditions for optimality are then stated and interpreted. These conditions are called the (Karush-)Kuhn–Tucker conditions for inequality-constrained optimization problems.

$$\nabla_z^T L|_{\hat{z}} = \nabla_z^T F|_{\hat{z}} + \nabla_z^T h|_{\hat{z}} \lambda + \nabla_z^T g|_{\hat{z}} \mu = 0 \qquad (65\,\text{a})$$
$$\nabla_\lambda^T L = h(z) = 0 \qquad (65\,\text{b})$$
$$g(z) \leq 0 \qquad (65\,\text{c})$$
$$\mu_i g_i(z) = 0, \quad i = 1, 2, \ldots p \qquad (65\,\text{d})$$
$$\mu_i \geq 0, \quad i = 1, 2, \ldots p \qquad (65\,\text{e})$$

The conditions in Equation (65d) are called complementary slackness conditions. Each one says that either the constraint $g_i(z) = 0$ or its corresponding multiplier μ_i is zero (or both can be zero). Intuitively, this makes sense. If the constraint $g_i(z) = 0$, it is behaving like an equality constraint, and its Kuhn–Tucker multiplier μ_i can be thought of as a Lagrange multiplier that can be nonzero. In this case the inequality is being treated exactly like an equality constraint. If, on the other hand, the constraint is away from zero, it is not a part of the problem. It can be removed from the Lagrange function by simply setting its corresponding Kuhn–Tucker multiplier to zero.

If the objective function is to be minimized, then KUHN and TUCKER [10.1] showed that the multipliers must be positive, conditions (Eq. 65e). Because a Kuhn–Tucker multiplier is a Lagrange multiplier for an inequality constraint that is being treated as an equality constraint, the interpretation that it is a shadow price can be used to understand this condition. If $g_i(z)$ is currently zero, then a feasible change would be for it to become negative ($g_i(z) \leq 0$). From Equation (64), if the multiplier for the constraint is negative, this change will increase the objective function F which says the constraint should be kept at zero. If the multiplier is negative, this change will decrease the objective function F. The constraint should be released; the point $\hat{z}$ is almost certainly not a minimum point.

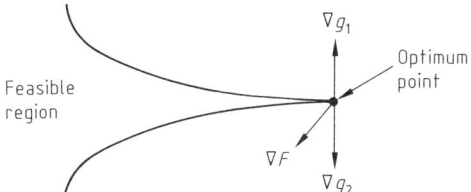

Figure 40. Special case where necessary conditions do not hold

Constraint Qualifications. In special cases the necessary conditions above do not hold. For example, if two nonlinear inequality constraints form a cusp as shown in Figure 40 and the optimum is exactly at that point as shown, then (1) both constraints are needed to define the optimum, but (2) they are locally colinear at that point. Thus the equation

$$\nabla F + \mu_1 \nabla g_1 + \mu_2 \nabla g_2 = 0$$

which states that the gradient of the objective function can be written as a linear combination of the gradients of the constraints, can be untrue at this point, as Figure 40 illustrates. By proving that the constraints have to be independent at the solution, one can be assured that this problem will not occur (but proving this condition will hold for a general nonlinear problem is not easy).

Kuhn–Tucker Sufficiency Conditions. Sufficiency conditions can be stated which will assure that an optimum that meets the Kuhn–Tucker conditions is indeed a local minimum point. The matrix of second derivatives of the Lagrange function with respect to all the variables z evaluated at $\hat{z}$ must be generated and used in a test to guarantee that all moves away from $\hat{z}$ which remain feasible with respect to the constraints will lead to an increase in the objective function. This test is formidable and is seldom done.

10.3. Strategies of Optimization

The theory covered in Section 10.2 explains how to decide if a point is an optimum point—or better, if a point cannot be an optimum point. This section explores how to find such a candidate point. This is rather like being blindfolded and starting a walk to find the lowest point in a valley below. One can imagine being fooled by local optima (multiple low points); running into a fence that is not allowed to be crossed (inequality constraint) but that might be hung on to; being required to find the solution along a trail (equality constraint) that can perhaps be found only by groping, bumping into a cliff (discontinuities), and so forth.

The simplest strategy to finding the minimum of a function is to place a grid of points throughout the feasible space and evaluate the objective function at every grid point. The point yielding the highest value for the objective function would be chosen as the solution. Although this is simple, with 20 variables over which to search and a rather coarse grid of 10 points in each direction, 10^{20} function evaluations would be required to solve the problem. At 1 µs per evaluation, more than four million years would be necessary to carry out these evaluations. Clearly, nothing but the smallest problem can be solved with this approach.

Most strategies limit themselves to finding a local minimum point in the vicinity of the starting point for the search. This would be like the person with a blindfold moving only in the downhill direction until reaching a spot that appeared locally level. Such a strategy will find the global optimum if the problem is convex. It will work if the roughness that makes the problem nonconvex does not lead to local minimum points along the search path.

Pattern Search. Occasionally a problem is confronted for which no mathematical model is available for assessing the objective function value. For example, in trying to improve a recipe where the objective is that the product looks and smells nice: the recipe can be varied by adjusting the relative amounts of materials used in making it, by changing the temperature at which it is processed, and so forth. Manufacturing several examples and handing them to a panel of judges may lead to a preference ordering only. For this type of problem, pattern search methods can be used to find the better conditions for manufacturing the product. Only the ideas behind this approach are described here. Details of one approach to implement it can be found in [10.2].

One pattern search method called the complex method is as follows. First, a "complex" of at least $r + 1$ different points is formed, at which to manufacture the product by picking a range of suitable values for the r independent variables in manufacturing process. The product is manufactured at each of these points. Next judges are asked to decide which product is the worst. For each independent variable, form the average value at which it was run in the complex. Draw a line from the coordinates of the worst point through the average point and continue on that line a distance twice that between these two points. Choose that point as the next test point if it is feasible and if it is better than the worst point (the product will have to be manufactured at that point and judged). If not feasible and better, back off on the distance past the average point until a feasible, better point is located (each trial point requires the product to be manufactured and judged). When a successful point is located, throw out the worst point and include the new one in the complex of points. Repeat the cycle until the points in the complex are so close together that the differences among them are uninteresting. A series of steps that successfully uses the doubled step size starts to take quite large steps in a generally downhill direction. When the step size must be reduced for a series of successive steps, the complex of points will come quite close together and signal that the optimum is near.

The method works adequately but, as might be expected, takes an enormous number of steps relative to methods discussed next that work when a reasonably behaved mathematical model is available. It also collapses into searching only in a subspace if it encounters any sort of ridge along the way. This method should thus be restarted frequently by setting up a new complex of points around the current best point.

Quadratic Fit—Unconstrained Case. An often effective method for optimization is to convert the problem into an unconstrained one and solve it by using a sequence of quadratic approximations to it. Suppose a computer program is available that can be run to give the value of the objective function F, if it is supplied with values for a set of independent variables u.

Assume that F can be approximated by the quadratic function in the variables u

$$F \approx a + b^T u + \tfrac{1}{2} u^T Q u$$

where a is a scalar, b a vector, and Q an $r \times r$ symmetric positive definite matrix. If a, b, and Q are known then the minimum could be found by finding a stationary point for this equation:

$$\nabla_u^T F = b + Q u = 0$$

or

$$u = - Q^{-1} b \qquad (66)$$

However, a, b, and Q are not known. They contain $q = 1 + r + r(r+1)/2$ unknown coefficients in a, b, and Q. If the computer program were run q times, q equations could be generated in these unknown coefficients which could then be solved to estimate their values from the following set of linear equations (note that the values for F^i and u^i are known, the coefficients are to be computed):

$$F^1 = a + b^T u^1 + \tfrac{1}{2} u^{1T} Q u^1$$
$$F^2 = a + b^T u^2 + \tfrac{1}{2} u^{2T} Q u^2$$
$$F^q = a + b^T u^q + \tfrac{1}{2} u^{qT} Q u^q$$

The points at which to run the code must be selected to ensure that these equations are not singular. Once an estimate for a, b, and Q is available, Equation (66) gives the next point at which to evaluate the variables u. The newest point is then included in the set and the oldest removed, giving again q points at which to evaluate a, b, and Q. If removal of the oldest point leads to a singular set of equations, a different point must be selected for removal. An approach to this is to keep all the older equations, with the new ones added to the top of the list. Then pivoting can be done by proceeding down the list until a nonsingular set of q equations has been found. The older equations will not be used unless they have to be. In addition, clever schemes, to be considered in the section on quasi-Newton methods, can be used to minimize the effort to carry out the pivoting from one step to the next because only a single equation has been changed for each step.

If gradients as well as objective function values are provided by the computer model, the process is even more efficient. The following equations are written a sufficient number of times to allow one to estimate b and Q:

$$\nabla_u^T F^1 = b + Q u^1$$
$$\nabla_u^T F^2 = b + Q u^2 \qquad (67)$$

Again, past equations are kept on a stack, and newer ones are placed at the top. Then, pivoting from the top down until a nonsingular set of equations exists provides a scheme for estimating b and Q and for deleting older equations as the iterations proceed.

For a problem with three independent variables u, $3 + 3(4)/2 = 9$ unknown coefficients exist. Each numerical experiment would supply three equations. Thus, the program would have to be run at least three times to estimate the coefficients b and Q.

Quadratic Fit—Equality-Constrained Case. To solve the following equality-constrainted optimization problem

$$\text{Min } \{F(z) | h(z) = 0, z \in Re^n, h : R^n \to R^m\}$$

proceed as follows:

1) Partition the variables z into dependent variables x and independent variables u.
2) Guess values for the variables u^k, $k = 0$, where k is the iteration counter.
3) Solve equation $h(x, u)$ for x^k given the values guessed for u^k. These will be m equations in m unknowns. Solving in general will require an iterative procedure (e.g., the Newton–Raphson method).
4) Use Equation (62) to solve for the Lagrange multipliers λ^k. If the Newton–Raphson method is used to solve the equations, the Jacobian matrix $\nabla_x^T h|_{\hat{z}}$ will already have been generated at the point $z^k = x^k, u^k$ found in previous step.
5) Substitute λ^k to form Equation (63), which in general will not be zero; $\nabla_u L^k$ computed will be the constrained derivatives of F with respect to the independent variables u^k.

This effectively creates a method for computing $F(u)$ and $\nabla_u^T F(u)$ if u is given. Now directly apply the ideas in the last section for using the quadratic fit method to solve unconstrained optimization problems for the case in which derivatives are available. In other words, generate a table of u values at which to compute $\nabla_u^T F(u)$ sufficient in number that b and Q can be estimated for a quadratic fit. These can be estimated by using Equation (67). Then the next value is estimated for the independent variables u by using Equation (66), and so forth.

This method is termed a *reduced gradient method* and is one of the more effective approaches for solving equality-constrained optimization problems.

Quasi-Newton Updates. The above approach on quadratic fit for finding an optimum point is a secant method which is a form of quasi-Newton method. Approaches based on the Newton method have a number of advantages. Most important is that convergence near the solution is very fast (i.e., at a superlinear or, for the Newton method itself, quadratic rate). If the matrix of second partial derivatives of the objective function ∇F_{uu} (termed the Hessian matrix) remains positive definite, the Newton method can be made globally convergent by choosing a step-size α between 0 and 1 along the Newton direction that will guarantee a decrease of the objective function and movement toward the optimum. However, if the Hessian matrix becomes singular or indefinite, a modification of Newton's method must be considered. Moreover, Newton's method itself requires the evaluation of second derivatives for the Hessian matrix and the solution of a set of linear equations similar to Equation (66):

$$\nabla_{uu} F(u^k) d = -\nabla_u F(u^k)$$

where $d = u^{k+1} - u^k$ at each iteration k.

Now develop efficient update formulas for the approach where gradients are available. Assume that the next point to test as the optimum point is u^{k+1} by solving equations similar to those above for d (or by Eq. 66). Evaluate the gradient of F with respect to u, $\nabla F_u(u^{k+1})$ at that point by running the model. The goal is to update the estimate of the Hessian matrix at u^{k+1}. Proceed as follows:

Second derivative information may be approximated by application of a secant relationship (also known as the quasi-Newton condition):

$$Q^{k+1} s = \gamma$$

where $s = u^{k+1} - u^k$, $\gamma = \nabla F^{k+1} - \nabla F^k$, and Q is the same matrix as in the above section on quadratic fit (see Eq. 66). In one dimension, this condition is sufficient for construction of Q. For multiple dimensions, additional conditions must be specified, and here conditions are imposed that are desirable for the optimization problem, such as symmetry and positive definiteness of Q.

Although detailed derivation of quasi-Newton update formulas is beyond the scope of this chapter, considering two of the most popular quasi-Newton methods and the concepts underlying their derivation is constructive. The older update formula (the *DFP formula*) was developed by DAVIDON [10.3] and analyzed by

FLETCHER and POWELL [10.4]; this rank 2 formula can be stated as follows:

$$Q^{k+1} = s^T Q^k s \left(I - \frac{\gamma s^T}{\gamma^T s} \right) Q^k \left(I - \frac{s \gamma^T}{\gamma^T s} \right) + \frac{\gamma \gamma^T}{\gamma^T s}$$

That Q^{k+1} satisfies the secant relationship and is symmetric as long as Q^k is can be shown, as can the fact that Q^{k+1} is positive definite as long as Q^k is positive definite and $\gamma^T s > 0$. This formula can be derived conceptually by using the following variational problem [10.5].

$$\text{Min } \| W^{1/2} (Q^{k+1} - Q^k) W^{1/2} \|_F^2$$
$$\text{such that } Q^{k+1} = (Q^{k+1})^T$$
$$Q^{k+1} s = \gamma$$

Here, the weighting matrix W must satisfy the inverse secant relationship $W\gamma = s$. Also, the Frobenius norm squared ($\| \cdot \|_F^2$) is the sum of squares of the matrix elements. The DFP formula can also be represented by an inverse relationship where $H^k = (Q^k)^{-1}$ and

$$H^{k+1} = H^k + \frac{s s^T}{s^T \gamma} - \frac{H^k \gamma \gamma^T H^k}{\gamma^T H^k \gamma}$$

In this way the quasi-Newton search direction $d^k = - H^k \nabla_u F(u^k)$ is evaluated directly by a simple matrix multiplication, rather than by solving a system of linear equations.

An alternative to the DFP update was discovered in 1970 independently by BROYDEN, FLETCHER, GOLDFARB, and SHANNO (BFGS) [10.6]–[10.9]. Matrix and inverse formulas for this BFGS update are given as

$$Q^{k+1} = Q^k - \frac{Q^k s s^T Q^k}{s^T Q^k s} + \frac{\gamma \gamma^T}{\gamma^T s}$$

$$H^{k+1} = \left(I - \frac{s \gamma^T}{\gamma^T s} \right) H^k \left(I - \frac{\gamma s^T}{s^T \gamma} \right) + \frac{s s^T}{s^T \gamma}$$

As with the DFP formula, symmetry conditions and the secant relation can easily be verified. Similarly, these updates remain positive definite as long as $\gamma^T s > 0$. Conceptually, this update can be derived by the following variational problem:

$$\text{Min } \| W^{1/2} (H^{k+1} - H^k) W^{1/2} \|_F^2$$
$$H^{k+1} = (H^{k+1})^T$$
$$H^{k+1} \gamma = s$$

where the weighting matrix W satisfies $Ws = \gamma$. Note the similarities in the BFGS and DFP formulas. From the variational formulations and the update formulas themselves the two updates are seen to complement each other. The BFGS update, in fact, is often referred to as the complementary DFP formula.

Once either Q^{k+1} or H^{k+1} has been updated, d can be computed for the next move. Again, at this new point, the model is run to obtain the gradient of F with respect to u, update Q^{k+2} or H^{k+2}, etc.

Both quasi-Newton methods exhibit superlinear convergence on unconstrained optimization problems and have been very successful in solving moderate-sized (e.g., $n < 100$ variables) optimization problems. However, extensive numerical testing has shown that the BFGS update is consistently more reliable and efficient for these problems. More information about theoretical properties, numerical experience, and implementation details of these quasi-Newton methods can be found in [10.10] and [10.11].

Inequality-Constrained Problems. To solve inequality-constrained problems, we must develop a strategy that can decide which of the inequality constraints should be treated as equalities. The problem often is split into two phases: phase 1 where the objective is to find a point that is feasible with respect to the equality and inequality constraints, and phase 2 where the optimum is sought while feasibility is maintained. Many optimization strategies are based on this idea and are called feasible path approaches. The sequential quadratic programming approach discussed in Section 10.4, in contrast, does not worry about any of the constraints being feasible except after the final iteration. It is termed an infeasible path approach and is one of the most effective approaches in terms of computational efficiency for large problems.

10.4. Successive Quadratic Programming (SQP)

Because the solution of the nonlinear programming problem is a stationary point represented by a set of optimality conditions (i.e., the Kuhn–Tucker conditions), a straightforward alternative to the reduced gradient approach can be derived by considering a Newton method applied directly to these conditions. The advantage of this formulation is that, as with any Newton-type method, convergence is rapid near the solution. However, advantage must be taken of the special structure of the optimality conditions to ensure convergence from a poor starting point.

A concise derivation of this method can be developed by considering the Kuhn–Tucker conditions as written below:

$$\nabla_z L(z, \mu, \lambda) = \nabla F + \nabla g_A \mu + \nabla h \lambda = 0$$
$$g_A(z) = 0$$
$$h(z) = 0$$

Assume, for the moment, that the active inequalities g_A are known at the solution and that their corresponding multipliers are positive. Applying Newton's method to these conditions from an initial guess (x^i, μ^i, λ^i) leads to the solution of the following system of linear equations to determine the search direction:

$$\begin{bmatrix} \nabla_{zz} L(z^i, u^i, \lambda^i) & \nabla g_A(z^i) & \nabla h(z^i) \\ \nabla g_A(z^i)^T & 0 & 0 \\ \nabla h(z^i)^T & 0 & 0 \end{bmatrix} \begin{bmatrix} z - z^i \\ \mu - \mu^i \\ \lambda - \lambda^i \end{bmatrix}$$
$$= - \begin{bmatrix} \nabla_{zz} L(z^i, u^i, \lambda^i) \\ g_A(z^i) \\ \nabla h(z^i) \end{bmatrix}$$

An equivalent linear system written only in terms of the search direction for z and the multipliers is given by

$$\begin{bmatrix} \nabla_{zz} L(z^i, u^i, \lambda^i) & \nabla g_A(z^i) & \nabla h(z^i) \\ \nabla g_A(z^i)^T & 0 & 0 \\ \nabla h(z^i)^T & 0 & 0 \end{bmatrix} \begin{bmatrix} d \\ \mu \\ \lambda \end{bmatrix}$$
$$= - \begin{bmatrix} \nabla_z F(z^i) \\ g_A(z^i) \\ h(z^i) \end{bmatrix}$$

This system of linear equations, however, may not have a solution. A sufficient condition for a unique Newton direction is that the matrix of constraint derivatives is of full rank (linear independence of constraints) and the Hessian matrix of the Lagrange function ($\nabla_{zz} L(z, \mu, \lambda)$) projected into the space of the linearized constraints is positive definite. Moreover, under these conditions the linearized system actually represents the solution of the following quadratic programming problem:

$$\text{Min } \nabla F(z^i)^T d + \tfrac{1}{2} d^T \nabla_{zz} L(z^i, \mu^i, \lambda^i) d$$
$$\text{such that } g_A(z^i) + \nabla g_A(z^i)^T d = 0$$
$$h(z^i) + \nabla h(z^i)^T d = 0$$

Note that the quadratic program still requires second derivatives for construction of the Hessian matrix.

The question of choosing the active set of inequalities, mentioned above, was resolved by WILSON in 1963 simply by adding all of the inequality constraints to the quadratic program. Because a quadratic program is solved at every iteration, the active set that minimizes the quadratic program is updated and therefore becomes the active set at the solution to the nonlinear program. Also, a solution of zero to the linear system for the search direction implies satisfaction of the Kuhn–Tucker conditions for the nonlinear program.

Problems with calculating second derivatives as well as maintaining positive definiteness of the Hessian matrix can be avoided by approximating this matrix by B^i using a positive definite quasi-Newton formula (such as BFGS). Here, gradients of the Lagrange function are used to calculate γ in the update formula [10.12], [10.13]. The resulting quadratic program, which generates the search direction at each iteration i, therefore becomes

$$\text{Min } \nabla F(z^i)^T d + \tfrac{1}{2} d^T B^i d$$
$$\text{such that } g(z^i) + \nabla g(z^i)^T d \leq 0$$
$$h(z^i) + \nabla h(z^i)^T d = 0$$

Because this problem has linear constraints and a strictly convex objective function, it has a unique solution. Moreover, quadratic programs can be solved quite efficiently with numerous finite step algorithms including primal projection methods [10.14], dual projection methods [10.15], or modified simplex methods [10.16], [10.17].

Finally, to ensure convergence of this algorithm from poor starting points, a step size α is chosen along the search direction so that the point at the next iteration ($z^{i+1} = z^i + \alpha d$) is closer to the solution of the NLP.

Line search criteria for determining α usually consist of obtaining a sufficient decrease for some merit function that consists of the objective function and the violated constraints. Examples of these merit functions include the exact penalty function [10.2], augmented Lagrange functions [10.18] and the *watchdog hybrid function* [10.19]. For many problems, performance of these line search methods is similar and generally quite good. However, all of these line search methods can experience difficulties, and much current research focuses on their development and improvement.

In numerous numerical comparisons (e.g., [10.18] and [10.20]) successive quadratic programming requires consistently fewer function evaluations than other NLP algorithms, including reduced gradient methods described above. However, for larger problems the cost of creating and storing a large, dense Hessian matrix and of solving large quadratic programs can be prohibitive.

To adapt SQP to deal with larger problems, two approaches are available. On one hand, quadratic programming algorithms can be developed that take advantage of problem sparsity, especially in the matrix of constraint derivatives. Also, the structure of the Hessian matrix must be exploited, either by providing second-derivative information directly or by implementing structured quasi-Newton updating formulas. These approaches have met with some success on process optimization problems [10.21]. However, several problems still remain, including loss of positive definiteness in the Hessian matrix and the determination of a systematic step-size adjustment strategy.

Many process problems, however, consist of large complex models (i.e., many nonlinear equality constraints) and relatively few independent variables. Moreover, for these problems the approximation of the sparse, indefinite Hessian matrix by a large, dense positive definite matrix may not be appropriate. Also, from the second-order optimality conditions, the small Hessian matrix projected into the constraint space (the constrained second derivatives for the independent variables) is generally dense and positive definite at the solution. Consequently, BERNA et al. [10.22] proposed the use of decomposition algorithms to exploit these problem characteristics. Problems with a few thousands of dependent and a few tens of independent variables were solved. The more recent algorithms that exploit these problem characteristics are presented here.

Consider the linear equations that represent the optimality conditions of an equality-constrained quadratic programming subproblem:

$$\begin{bmatrix} B^i & \nabla h \\ \nabla h^T & 0 \end{bmatrix} \begin{bmatrix} d \\ \lambda \end{bmatrix} = - \begin{bmatrix} \nabla F(z^i) \\ h(z^i) \end{bmatrix}$$

To simplify the notation the inequality constraints are omitted (if the active ones are known, they can also be combined within the set of equalities). By performing an orthonormal factorization of the m columns of ∇h, an $n \times m$ matrix Y is obtained that spans the range space of the linearized constraints. Orthogonal to the columns of Y, an orthonormal $n \times (n - m)$ matrix Z is developed, whose columns span the null space of the linearized constraints. These orthonormal matrices have the following properties [10.23]:

$$Y^T Y = I_m \qquad Y^T \nabla h = R$$
$$Z^T Z = I_{n-m} \qquad Y^T Z = 0 \qquad \nabla h^T Z = 0$$

Also, the following orthonormal matrix Q:

$$Q = \begin{bmatrix} [Y\ Z] & 0 \\ 0 & I_m \end{bmatrix} \qquad QQ^T = I_{n+m}$$

is constructed, and the QP optimality conditions are multiplied through by Q to transform this system to

$$Q^T \begin{bmatrix} B^i & \nabla h \\ \nabla h^T & 0 \end{bmatrix} QQ^T \begin{bmatrix} d \\ \lambda \end{bmatrix} = - Q^T \begin{bmatrix} \nabla F \\ h \end{bmatrix}$$

which upon multiplication becomes

$$\begin{bmatrix} Y^T B^i Y & Y^T B^i Z & R \\ Z^T B^i Y & Z^T B^i Z & 0 \\ R^T & 0 & 0 \end{bmatrix} \begin{bmatrix} d_Y \\ d_Z \\ \lambda \end{bmatrix} = - \begin{bmatrix} Y^T \nabla F \\ Z^T \nabla F \\ h \end{bmatrix}$$

Note that the range space direction d_Y can now be determined directly from the bottom row. Moreover, the null space direction d_Z is independent of the multipliers λ. Consequently, the multipliers can be estimated by $R\lambda = - Y^T \nabla F$ because the other terms become unimportant as both d_Z and d_Y approach zero. As suggested initially by MURRAY and WRIGHT [10.24], the $Y^T BZ$ term in the second row could be ignored because it becomes unimportant for small d_Y. With these simplifications, the linear system of equations becomes uncoupled, and null steps, range steps, and multipliers can be determined independently as follows:

$$R^T d_Y = - h$$
$$(Z^T B^i Z) = - Z^T \nabla F$$
$$R\lambda = - Y^T \nabla F$$
$$d = Y d_Y + Z d_Z$$

NOCEDAL and OVERTON [10.25] proved that the convergence rate of this procedure is superlinear at every second iteration; this represents a slight deterioration from the "ideal" convergence rate of the original SQP method. However, testing on a set of small problems shows no loss in performance with this approach. Further testing on small problems as well as evaluation of a number of strategies for handling inequality constraints were recently reported by GURWITZ and OVERTON [10.26].

VASANTHARAJAN and BIEGLER [10.27] extended this approach to deal with general inequality constraints. Here, the null space step is determined by a small quadratic program, which represents a slight modification of the uncoupled linear equations given above. Multipliers for the inequality constraints are also determined from the quadratic program, and the equation for λ is modified slightly. They also developed a factorization of Y and Z that allows the application of sparse matrix methods. Although columns of Y and Z are no longer orthonormal, the overall search direction predicted by the range and null space algorithm still remains the same. The range and null space decomposition of SQP has been very efficient and reliable in solving large nonlinear programming problems. VASANTHARAJAN et al. [10.28] demonstrated this approach on process optimization problems of up to 1000 variables. Their results showed it to be competitive with recent implementations of MINOS and very reliable for difficult, nonlinear problems.

10.5. Linear Programming

A very important special case of the nonlinear programming problem arises when both the objective function and the constraints are linear. This case gives rise to a linear programming problem which, due to its special properties, can be solved much more efficiently than general nonlinear programs.

To provide some insight into the nature of this problem, consider the following linear programming optimization problem:

$\min F = -z_1 - 3z_2$
such that $z_1 + z_2 \leq 4$
$z_1 + 2z_2 \leq 6$
$z_1 \geq 0, \quad z_2 \geq 0$

The feasible region and the contours for this problem are shown in Figure 41. The feasible region is convex because the inequalities define a polyhedral set. Also, because the objective function is linear, it corresponds to a convex function. The unique minimum value of the objective function of $F = -9$ is located at the point $z_1 = 0, z_2 = 3$ (point B). This point corresponds to a vertex in the feasible region given by the intersection of the constraints $z_1 + 2z_2 = 6$ and $z_1 = 0$. The two inequality constraints to which

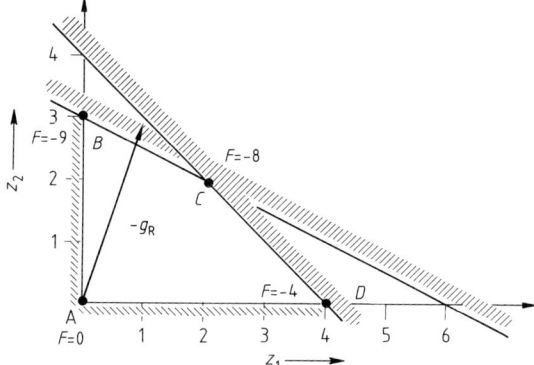

Figure 41. Feasible region for linear programming problem

these correspond are said to be active at this point.

The observations in the above example on the nature of the optimal solution for linear programs can be generalized. Consider first the general form of a linear program as given by

$\min F = c^T z$
such that $Az = b$ (68)
$z \geq 0$

where z is an n vector of continuous variables, c is an n vector of coefficients, A is an $m \times n$ matrix ($m < n$), and b is an m vector of right-hand sides.

Note that any linear program can be written in the form of Equation (68). As an example, consider the above problem. The inequalities can be converted to equalities by introducing slack variables and rewriting the constraints as

$z_1 + z_2 + z_3 = 4$
$z_1 + 2z_2 + z_4 = 6$

where z_3 und z_4 are nonnegative slack variables. Thus, the example can be expressed in the form of Equation (68):

$$\min F = [-1 \ -3 \ 0 \ 0] \begin{bmatrix} z_1 \\ z_2 \\ z_3 \\ z_4 \end{bmatrix}$$

$$\begin{bmatrix} 1 & 1 & 1 & 0 \\ 1 & 2 & 0 & 1 \end{bmatrix} \begin{bmatrix} z_1 \\ z_2 \\ z_3 \\ z_4 \end{bmatrix} = \begin{bmatrix} 4 \\ 6 \end{bmatrix} \quad (69)$$

$z_1, z_2, z_3, z_4 \geq 0$

Inequalities that are greater or equal to zero can similarly be converted into equations by the use of nonnegative slack variables. For example, $6z_1 - 3z_2 \geq 9$, can be converted into $6z_1 - 3z_2 - z_3 = 0$, where z_3 is a slack variable.

10.5.1. Basic Properties

The linear programming problem (Eq. 68) has the following basic properties, which are stated without proof (see [10.29], [10.30], for details).

Property 1. The linear program (Eq. 68) has a unique minimum value for the objective function. Otherwise the problem is either unbounded or infeasible.

As mentioned previously, the uniqueness of the optimal objective function value follows from the fact that Equation (68) corresponds to a convex programming problem. Note, however, that although the minimum value of the objective function is unique, the values of the optimum variables may not be. For illustration, assume that the objective function in the example is replaced by $F = -z_1 - 2z_2$. From Figure 41, it can easily be verified that the minimum $F = -6$ occurs at the two vertices $z_1 = 0, z_2 = 3$ and at $z_1 = z_2 = 2$, as well as at any point that is a linear combination of these two points (i.e., boundary of $z_1 + 2z_2 = 2 \leq 6$ in the feasible region).

Note also that if Equation (68) has no finite solution, it is either unbounded or infeasible. An example of the former arises if the two inequalities in the problem are deleted. An example of the latter arises if the constraint $z_1 + z_2 \geq 5$ is added to the problem.

Property 2. A vertex or extreme solution in Equation (68) corresponds to a partition of those equations which has the form

$$Bz_B + Nz_N = b \qquad (70)$$

with $z_B \geq 0$ and $z_N = 0$, where B is an $m \times m$ nonsingular matrix denoted as the basis, N is an $(n-m) \times m$ matrix, z_B is an m vector of basic variables, z_N is an $n-m$ vector of nonbasic variables.

Again for illustration purposes, consider vertex A in Figure 41 for the problem where $z_1 = z_2 = 0$. If these two variables are selected as nonbasic, then from Equations (69) and (70),

$$Bz_B = b$$

or

$$\begin{bmatrix} 1 & 0 \\ 0 & 1 \end{bmatrix} \begin{bmatrix} z_1 \\ z_2 \end{bmatrix} = \begin{bmatrix} 4 \\ 6 \end{bmatrix}$$

In this way the basic variables $z_3 = 4, z_4 = 6$ correspond precisely to nonzero slacks for the two inequalities that are inactive at that point.

As another example, consider vertex B for which $z_1 = 0, z_2 = 3$, and for which the second inequality is active (i.e., slack $z_4 = 0$). In this case, $z_2 = z_4 = 0$ are the nonbasic variables and thus the basic equations are given by

$$\begin{bmatrix} 1 & 1 \\ 2 & 0 \end{bmatrix} \begin{bmatrix} z_1 \\ z_2 \end{bmatrix} = \begin{bmatrix} 4 \\ 6 \end{bmatrix}$$

for which $z_2 = 3, z_3 = 1$. Note that at vertex $B, z_2 = 3$, and the first constraint has a slack of 1.

Property 3. The minimum of Equation (68) is located at an extreme point of the feasible space. Furthermore, at that extreme point the following inequality is satisfied:

$$-c_B^T B^{-1} N + c_N^T \geq 0^T \qquad (71)$$

where c_B and c_N are coefficients of the objective for the basic and nonbasic variables, respectively.

As was illustrated with the example problem, the optimum was indeed located at the vertex point which is also predicted by property 2. The reason why an interior point $\bar{z}$ (nonvertex) will generally be nonoptimal is that such a point can always be expressed as a linear combination of extreme points, that is,

$$\bar{z} = \sum_i \alpha_i z^i \qquad (72)$$

where z^i is an extreme point, and α_i is a scalar between 0 and 1 for which $\sum \alpha_i = 1$. Therefore,

$$c^T \bar{z} = c^T \sum_i \alpha_i z^i$$

This equation can clearly not be satisfied for $0 < \alpha_i < 1$ if $c^T z < c^T z^i$. Thus, the optimum will always correspond to an extreme point.

As for Equation (71), recall that the reduced gradient g_R in a linearly constrained optimization problem is given by

$$g_R^T = -\nabla_x F^T [\nabla_x h]^{-1} \nabla_u h + \nabla_u F^T \qquad (73)$$

Because the dependent variables x correspond to the basic variables z_B, $\nabla_x F^T = c_B^T, \nabla_x h = B$. Also, because the independent variables correspond to the nonbasic variables z_N, $\nabla_u F^T = c_N^T, \nabla_u h = N$. Thus, the reduced gradient in Equation (68) is given by

$$g_R^T = -c_B^T B^{-1} N + c_N^T$$

Thus, Equation (68) can be expressed in the space of the non-basic (reduced) variables z_N as

$$\min F = [-c_B^T B^{-1} N + c_N^T] z_N$$
$$\text{such that } -z_N \leq 0$$

From the Kuhn–Tucker conditions presented earlier,

$$-c_B^T B^{-1} N + c_N^T - \mu^T = 0^T$$
$$-z_N \leq 0, \quad \mu \geq 0, \quad \mu^T z_N = 0$$

Thus, because the multipliers $\mu \geq 0$, Equation (71) holds.

10.5.2. Simplex Algorithm

The simplex algorithm [10.31] exploits the basic properties described in the previous section by searching only among the vertex points in the feasible space. It verifies whether Equation (73) holds to identify the optimum solution. Rather than enumerating all possible vertices, the algorithm proceeds by moving sequentially from one vertex to another, ensuring that the objective function value decreases.

To explain the essence of the simplex algorithm, rewrite the objective function of the equation

$$F - c^T z = 0$$

Then, for a given selection of the basis, the problem can be presented through the following tableau [see the above equations and Equation (72)]:

(T 1)

	F	z_B^T	z_N^T	Right-hand side
Objective row	1	$-c_B^T$	$-c_N^T$	0
Constraint row	0	B	N	b

To update this, premultiply the constraint rows by B^{-1} and add to the objective row the product of c_B^T times the new constraint; that is,

(T 2)

	F	z_B^T	z_N^T	Right-hand side
Objective row	1	0	$c_B^T B^{-1} N - c_N^T$	$c_B^{T\bar{b}}$
Constraint row	0	I	$B^{-1} N$	$\bar{b}$

where $\bar{b} = N^{-1} b$. Note that the entry in the objective row for z_N^T displays the information of the negative of the reduced gradient. Thus, at the optimum each of these components must be nonpositive.

In general, the simplex algorithm proceeds as follows. First, an initial basis is selected. This is trivial for the case in which the constraints are of the form $Az \leq b$ where $b > 0$, because then the initial basis to be selected can be the slack variables for which $B = I$. Otherwise, one has to resort to the use of artificial variables and a two-phase procedure [10.30]. Second, the optimality condition is verified by checking if

$$c_B^T B^{-1} n_i - c_{N_i}^T \leq 0$$

for each nonbasic variable z_{N_i}. If this condition is not satisfied, move to a new vertex which requires the selection of a new basis. This is accomplished as follows:

1) Introduce in the basis the nonbasic variable z_{Nj} with largest positive value for the above optimality condition. This ensures a move to a neighboring vertex that will lead to the largest decrease in the objective per unit of distance moved.

2) Remove from the basis the basic variable z_{Br} that will become zero as a step is taken along the j-th component of the reduced gradient. The r-th basic variable can be determined by computing the smallest step size a from among all the basic variables z_{Bi}:

$$a = \min \{b_i/y_{ij}\} = b_r/y_{rj} \geq 0$$

where y_{ij} is the i-th in row in the vector $B^{-1} n_j$ which corresponds to the j-th column of the nonbasic variables.

3) Pivot on the r-th row and the j-th column of the tableau to obtain the identity matrix for the basic variables.

To illustrate the application of the simplex algorithm, construct the tableau for the example problem. From Equation (69),

F	z_1	z_2	z_3	z_4	Right-hand side
1	1	3	0	0	0
0	1	1	1	0	4
0	1	2	0	1	6

If the initial basis z_3 and z_4 is chosen, the point is feasible because $z_1 = z_2 = 0$ (nonbasic) and $z_3 = 4$, $z_4 = 6$ (vertex A in Fig. 41). Furthermore, because the matrix in the basis corresponds to the identity matrix, the tableau has the form of (T 2). The entries in the first row under z_1 and z_2 correspond to the negative of the reduced gradient. Select z_2 as the variable to enter the basis because it has the largest negative reduced gradient, -3. Now compute the step size a:

$$a = \min \{4/1, 6/2\} = 6/2 = 3$$

Because this corresponds to the right-hand side element that is the value of z_4, select z_4 as the variable to leave the basis. The second constraint will become active as its slack variable z_4 is set to zero when it becomes nonbasic. Notice from Figure 41 that $a = 3$ along the direction (0, 1) corresponds to the boundary of the second constraint; $a = 4$ would lead to the boundary of the first constraint, which would then violate the second constraint.

By selecting z_2 as the variable to enter the basis and z_4 as the one to leave it, one pivots on the entry with the value of "2" in the column beneath z_2 in the bottom row above. This then leads to the following new tableau:

F	z_1	z_2	z_3	z_4	Right-hand side
-9	$-1/2$	0	0	$-3/2$	
0	1/2	0	1	$-1/2$	1
0	1/2	1	0	1/2	3

Then, because the elements of the negative of the reduced gradient $(-1/2, -3/2)$ (variables z_1 and z_4 are the nonbasic variables) are negative, stop. Thus, the optimal solution is given at $z_1 = 0$, $z_2 = 3$, and $z_3 = 1$ (slack for first constraint), $z_4 = 0$ (second constraint is active) with $F = -9$.

Finally, note that most current commercial codes for LP (MPSX, SCICONIC, MINOS) are based on extensions of the simplex algorithm and can typically handle problems with up to 15 000 constraints. The new interior point methods (see [10.8] for a review) seem to offer the potential of solving larger problems more efficiently.

10.6. Mixed-Integer Programming

Up to this point, this chapter has assumed that only continuous variables are involved in the optimization problems. A number of important applications, however, require that all or a subset of the variables be constrained to take only integer or discrete values. Simple examples occur in modeling the number of batches to be produced or discrete sizes to be selected for a piece of equipment. Another example is modeling the selection of process units in a flow sheet (i.e., yes or no decisions).

When an optimization problem involves both discrete and continuous variables, it is called a mixed-integer programming problem. This section considers the case in which the objective function and constraints are linear (MILP) first and then the case in which nonlinearities are involved (MINLP). Also, for convenience, assume that all the discrete variables are of the 0–1 type.

Mixed-Integer Linear Programming (MILP). Assume a linear programming problem in which a subset of the variables y is restricted to take only 0 or 1 values. This then gives rise to the MILP problem:

$$\min F = c^T z + b^T y$$
$$\text{such that } Az + By \leq d \qquad (73)$$
$$z \geq 0, \; y \in \{0, 1\}$$

A first approach to solve this MILP problem is to solve a linear programming problem for every combination of 0–1 variables and pick as the solution the 0–1 combination with the lowest value for the objective function. The major drawback with such an approach is that the number of 0–1 combinations can be very large. For example, an MILP problem with ten 0–1 variables would require the solution of $2^{10} = 1024$ linear programs, whereas a problem with fifty 0–1 variables would require the solution of $2^{50} = 1.13 \times 10^{15}$ programs. This approach is, in general, computationally infeasible.

A second alternative is to relax the 0–1 requirements and treat the variables y as continuous with bounds $0 \leq y \leq 1$. The problem with such an approach, however, is that except for few special cases (e.g., assignment problems), there is no guarantee that the variables y will take integer values at the relaxed LP solution. As an example, consider the pure integer program

$$\min F = -1.2 y_1 - y_2$$
$$\text{such that } 1.2 y_1 + 0.5 y_2 \leq 1$$
$$y_1, y_2 = 0, 1$$

By relaxing y_1 and y_2 to be continuous the solution yields $y_1 = 0.715$, $y_2 = 0.285$, $F = 1.148$. It might then be tempting to simply round the variables to the nearest integer value, namely $y_1 = 1$, $y_2 = 0$. This point is an infeasible solution because it violates the first constraint. In fact, the optimal solution is $y_1 = 0$, $y_2 = 1$, $F = -1$. Thus, solving the MILP problem by relaxation of the y variables will in general not lead to the correct solution. Note, however, that the relaxed LP has the property that its optimal objective value provides a lower bound to the integer solution.

To obtain the solution to Equation (73), the most common approach is the branch and bound method [10.32], where the basic objective is to perform an enumeration without having to examine all the 0–1 combinations. The basic idea is to represent all the 0–1 combinations through a binary tree such as the example shown in Figure 42. At each node of the tree the solution of the linear program is considered subject to integer constraints for the subset of the y variables that are fixed in previous branches. For example, in node A the root of the tree involves the solution of the relaxed LP, whereas node B involves the solution of the LP with fixed $y_1 = 0$, $y_2 = 1$ and with $0 \le y_3 \le 1$.

To avoid enumeration of all the nodes in the binary tree, the following basic properties can be exploited. Let k denote a descendent node of another node l in the tree (e.g., $k = B, l = A$) and let (P^k) and (P^l) denote the corresponding LP subproblems. Then the following properties can be easily established:

1) If (P^l) is infeasible then (P^k) is also infeasible.
2) If (P^k) is feasible than (P^l) is also feasible, and $(F^l)^* \le (F^k)^*$. That is, the optimal value of the objective of subproblem (P^l) corresponds to a lower bound of the optimal value of the objective for subproblem (P^k).
3) If the optimal solution of subproblem (P^k) is such that all $y = 0$ or 1, then $(F^k)^* \ge F^*$. That is, the optimal objective of subproblem (P^k) corresponds to an upper bound of F^*, the optimal MILP solution.

These properties can be used to fathom nodes in the tree within an enumeration procedure. The question of how actually to enumerate the tree involves the use of branching nodes. It is not necessary to follow the order of the index of the variables y for branching as might be implied in Figure 42. A simple alternative is to branch instead on the 0–1 variable that is closest to 0.5.

Alternatively, a priority ordering for the 0–1 variables can be specified, or else a more sophisticated scheme can be used, such as the penalties described in [10.33]. After solving the LP at a node in the tree, decide which node to examine next. Here the two primary alternatives are to use a depth-first (last in–first out) or a breadth-first (best second rule) enumeration. In the former case, one of the branches of the most recent node is expanded first: if all of them have been examined, backtrack to another node. In the latter case, the two branches of the node with the lowest bound are expanded successively; in this case, no backtracking is required. Although the depth-first enumeration requires less storage, the breadth-first enumeration generally requires examination of fewer nodes.

In summary, the branch and bound method consists of first solving the relaxed LP problem. If all the integer variables y take integer values, stop. Otherwise, proceed to enumerate the nodes in the tree according to some prespecified branching rules. At each node the corresponding LP subproblem is solved. (Typically the dual problem is updated because it requires less work. The dual problem for an LP is not discussed in this chapter.) By making use of the properties stated above, either the node is fathomed (i.e., terminate looking at it and all nodes emanating from it) if the LP for it is infeasible or if its lower bound is greater or equal to its upper bound) or it is kept open for further examination.

As an example, consider the following MILP problem involving one continuous variable and

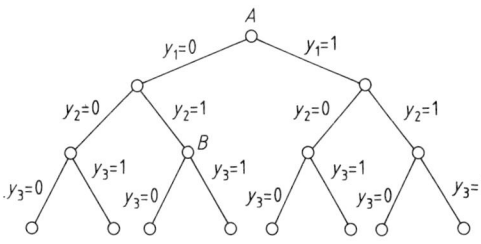

Figure 42. Binary tree for three 0–1 variables

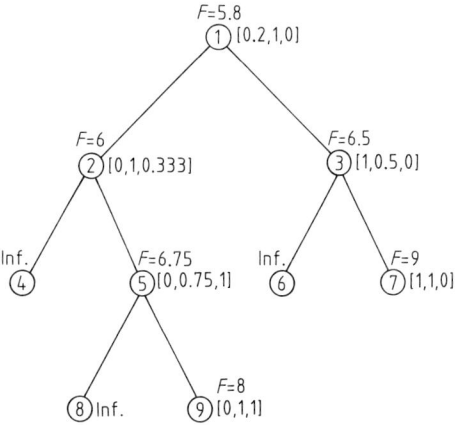

Figure 43. Branch and bound search

three 0–1 variables:

$$\min F = z + y_1 + 3y_2 + 2y_3$$
such that $-z + 3y_1 + 2y_2 + y_3 \leq 0$
$-5y_1 - 8y_1 - 3y_3 \leq -9$
$z \geq 0, y_1, y_2, y_3 = 0, 1$

The branch and bound enumeration using a breadth-first enumeration is shown in Figure 43, where the number in the circles represents the order in which 9 out of the 15 nodes in the tree are examined to find the optimum. Note that the relaxed solution (node 1) has a lower bound of $F = 5.8$ and that the optimum is found in node 9 where $F = 8$, $y_1 = 0$, $y_2 = y_3 = 1$, and $z = 3$.

Mixed-Integer Nonlinear Programming (MINLP). Consider now the case of the nonlinear optimization problem involving both continuous and 0–1 variables (MINLP, Eq. 74):

$$\min F = F(z, y)$$
such that $h(z, y) = 0$
$g(z, y) \leq 0$ (74)
$z \in Re^n \quad y \in \{0, 1\}^m$

This mixed-integer nonlinear program can in principle also be solved with the branch and bound method presented in the previous section. The major difference here would be that the examination of each node would require the solution of a nonlinear program rather than the solution of an LP. Provided the solution of each NLP subproblem is unique, similar properties as in the case of the MILP would hold with which the rigorous global solution of the MINLP can be guaranteed.

An important drawback of the branch and bound method for MINLP is that solution of the NLP subproblems is much more expensive, and they cannot be updated readily as in the case of the MILP. Therefore, to reduce the computational expense involved in solving many NLP subproblems, two other methods can be used: generalized Benders decomposition [10.34] and outer-approximation [10.35].

In both these methods the MINLP (Eq. 74) is assumed to be linear in the 0–1 variables and nonlinear in the continuous variables z; that is, MINLP (Eq. 75)

$$\min F = f(z) + c^T y$$
such that $h(z) = 0$
$g(z) + By \leq 0$ (75)
$z \in Re^n, \quad y \in \{0, 1\}^m$

The basic idea in both methods is to solve an alternating sequence of NLP subproblems and MILP master problems. The NLP subproblems are solved by optimizing the continuous variables z for a given fixed value of y, and their solution yields an upper bound to the optimal solution of Equation (75). The MILP master problems consist of a linear approximation that is refined as iterations proceed, and they have the objective of predicting new values of the binary variables y as well as a lower bound on the optimal solution. The alternate sequence of NLP subproblems and MILP master problems is continued up to the point where the predicted lower bound of the MILP master is greater than or equal to the best upper bound obtained from the NLP subproblems.

The MILP master problem in generalized Benders decomposition is given at any iteration K by Equation (76):

$$F_{GB}^K = \min \alpha$$
such that
$\alpha \geq f(z^k) + c^T y + (\mu^k)^T [g(z^k) + By]$ (76)
$k = 1, 2, \ldots K$
$\alpha \in Re^1, y \in \{0, 1\}^m$

where α is the largest Lagrangian approximation obtained from the solution of the K NLP subproblems; τ^k and μ^k correspond to the optimal solution and multiplier of the k-th NLP subproblem; F_{GB}^K corresponds to the predicted lower bound at iteration K.

In the case of the outer-approximation method, the MILP master problem is given by Equation (77)

$$F_{OA}^K = \min \alpha$$
such that
$\left.\begin{array}{l}\alpha \geq f(z^k) + \nabla f(z^k)^T (z - z^k) + c^T y \\ T^k \nabla h(z^k)^T (z - z^k) \leq 0 \\ g(z^k) + \nabla g(z^k)^T (z - z^k) + By \leq 0\end{array}\right\} k = 1, 2, \ldots K$
$\alpha \in Re^1, \quad Z \in Re^n, \quad y \in \{0, 1\}^m$ (77)

where α is the largest linear approximation of the objective function subject to linear approximations of the feasible region obtained from the solution of the K NLP subproblems; T^k is a diagonal matrix whose entries $t_{ii}^k = sign(\lambda_i^k)$, where λ_i^k is the Lagrange multiplier of equation h_i at iteration k and is used to relax the equations in the form of inequalities [10.36].

Note that in both master problems the predicted linear bounds F_{GB}^K, and F_{OA}^K increase monotonically as iterations K proceed, because the linear approximations are refined by accumulating the Lagrangian (in Eq. 76) or linearizations (in Eq. 77) of previous iterations. Also, in both cases, rigorous lower bounds can only be ensured when certain convexity conditions hold [10.34], [10.35].

In comparing the two methods, the lower bounds predicted by the outer-approximation method are always greater or equal to the lower bounds predicted by generalized Benders decomposition. Hence, the outer-approximation method requires the solution of fewer NLP subproblems and MILP master problems. On the other hand, the MILP master in the outer-approximation method is more expensive to solve, so the generalized Benders method may require less time if the NLP subproblems are inexpensive to solve. For a more extensive discussion and computational experience in chemical engineering applications, see [10.36]–[10.38].

10.7. Solution of Dynamic Optimization Problems

This section deals with the solution of optimization problems that include differential equations as well as algebraic constraints. In addition to parametric decision variables, they can include control profiles that are functions of time. Differential–algebraic models appear in all aspects of process engineering. Optimization problems that include these models are extremely common, covering problems as fundamental as the design of a single catalyst pellet to the optimal design and operation of an entire chemical plant. However, solving even the simplest and smallest of these optimization problems is typically difficult and time-consuming. Consequently, the development of efficient methods for these problems represents an interesting research area with a wealth of important and challenging applications.

The general optimization problem under consideration can be represented by Equation (78):

$$\min_{p, x(t), u(t)} F(p, x(t_f))$$

such that
$$h(p, u(t), x(t)) = 0$$
$$h_f(p, x(t_f)) = 0$$
$$g(p, u(t), x(t)) \leq 0$$
$$g_f(p, x(t_f)) \leq 0 \quad (78)$$
$$\dot{x}(t) = f(p, u(t), x(t))$$
$$x(0) = x_0$$
$$p^L \leq p \leq p^U$$
$$u^L \leq u(t) \leq u^U$$
$$x^L \leq x(t) \leq x^U$$

where F is the objective function; g, g_f represent the design inequality constraint vector; h, h_f the design equality constraint vector; p is a parameter, decision variable vector; $x(t)$ the state profile vector; $u(t)$ the control profile vector; p^L, p^U are the parameter variable bounds; x^L, x^U the state profile bounds; and u^L, u^U the control profile bounds. Note that algebraic constraints have been classified into conditions at final time t_f as well as constraints that must be enforced over the time domain. Also, although the model is given as an initial value problem for convenience, general ordinary differential equations can be handled in a straightforward manner. Finally, note the distinction between continuous variables p (parameters that do not vary with time) and $u(t)$ (time-varying control profiles) as decision variables.

Optimality conditions for the above problem formulation can be derived in a manner similar to nonlinear programming problems. Here, a Lagrange function is formed that consists of the objective function and the weighted sum of the constraints evaluated at each point in time. This weighted sum is expressed more concisely as an integral and the resulting function can be written as

$$L(x, u, p, \mu, v, \lambda) = F(x(t_f)) + \int_0^{t_f} \{\lambda^T (f(x, u) - \dot{x})$$
$$+ \mu^T g(x, u) + v^T h(x, u)\} dt$$
$$+ \mu_f^T g_f(x(t_f)) + v_f^T h_f(x(t_f))$$

To express this function entirely in terms of parameters and state and control variables, the following transformation is applied:

$$\int_0^{t_f} \lambda^T \dot{x} \, dt = \lambda(t_f)^T x(t_f) - \lambda(0)^T x(0) - \int_0^{t_f} \dot{\lambda}^T x \, dt$$

and the Lagrange function becomes

$$L(x, u, p, \mu, v, \lambda) = [F(x(t_f)) - \lambda(t_f)^T x(t_f)\\
+ \mu_f^T g_f(x(t_f)) + v_f^T h_f(x(t_f))]\\
+ \lambda(0)^T x(0) + \int_0^{t_f} \{\lambda^T f(x, u)\\
+ \lambda^T x + \mu^T g(x) + v^T h(x)\} dt$$

Note that multipliers λ, μ, and v have been introduced, which correspond to $f(x, u, p) - \dot{x}$, $g(x, u, p)$, and $h(x, u, p)$, and that all are functions of t. Because the multipliers λ perform a special function for the sensitivity of the differential equations, they are denoted as adjoint variables. On the other hand, multipliers μ_f and v_f, which correspond to the final time conditions, do not vary with time and function in the same manner as multipliers in nonlinear programming.

Stationary conditions of this Lagrange function can be given as follows with respect to $x(t)$, $u(t)$, and p. In addition, complementarity and nonnegativity conditions relating to the inequality constraints and feasibility conditions make up the balance of the optimality conditions given below:

$$x(t): \dot{\lambda} + \nabla_x f \lambda + \nabla_x g \mu(t) + \nabla_x h v(t) = 0$$
$$x(t_f): \nabla_{x_f} F - \lambda_f + \nabla_{x_f} g_f \mu_f + \nabla_{x_f} h_f v_f = 0$$
$$u(t): \nabla_u f \lambda + \nabla_u g \mu(t) + \nabla_u h v(t) = 0$$

$$p: [\nabla_p F + \nabla_p g_f^T \mu_f + \nabla_p h^T v_f] + \int_0^{t_f} [\nabla_p f \lambda\\
+ \nabla_p g \mu + \nabla_p h v] dt = 0$$
$$\mu_f^T g_f = 0 \quad \mu, \mu_f \geq 0 \quad g, g_f \leq 0$$
$$\mu^T g = 0 \quad h, h_f = 0 \quad \dot{x} = f(x, u), x(0) = x_0$$

The derivation of these conditions is very similar to the Kuhn–Tucker conditions for the nonlinear programming problem. The additional differential equations encountered for λ are known as adjoint equations and, in this case, have final conditions associated with them. In the same manner as for nonlinear programs, second-order conditions can also be derived, based on constrained projections of the second variations of the Lagrange functions. A detailed derivation of these conditions is given in [10.39].

Moreover, several straightforward modifications can be made to extend this problem. For example, if some or all initial conditions are not specified, they can be represented by parameters p_0 that are determined as part of the optimization. In this case, an additional term $v_0^T(x(0) - p_0)$ is appended to the Lagrange function, and the corresponding stationary condition with respect to p_0 becomes

$$x_0: v_0 - \lambda_0 = 0$$
$$p_0: v_0 = 0$$
$$\lambda_0 = 0, \text{ an additional boundary condition for the adjoint equations}$$

In addition, if final time t_f is not specified for this problem, t can be normalized by $\tau = t/p_f$ between 0 and 1. Note that a new parameter p_f, whose value is determined as part of the optimization, is introduced to represent final time. By writing the problem with τ substituted for t, differential equations are merely rewritten as

$$\frac{dx}{d\tau} = p_f f(x, u); \quad x(0) = x_0$$

and final conditions are evaluated at $\tau = 1$.

Finally, a large body of literature appears for this problem in which the algebraic constraints H and the profile bounds are deleted. For this case, the Hamiltonian is defined as $\lambda^T f$ and the resulting simplified optimality conditions are denoted as the Euler–Lagrange equations:

$$\frac{\partial H}{\partial u} = \frac{\partial f}{\partial u} \lambda = 0$$
$$\dot{\lambda} = -\frac{\partial H}{\partial x} = \nabla_{x_f} f \lambda$$
$$\lambda(t_f) = -\nabla_{x_f} F$$

To illustrate these optimality conditions, the simple example given below is considered:

Min $x_2(t_f)$
such that $\dot{x}_1 = -2x_1 + u^2 - u$
$\dot{x}_2 = x_1$
$x_1(0) = 1$
$x_2(0) = 1 \quad t_f = 1$

Here, the optimal control profile over a 1-h period must be determined for this dynamic system. This problem has no algebraic side conditions and no parameters. The Lagrange function for this system is given by

$$L = x_2(t_f) + \int_0^{t_f} \lambda_1 (u^2 - u - 2x_1 - \dot{x}_1)$$
$$+ \lambda_2 (x_1 - \dot{x}_2) dt = x_2(t_f) + \lambda_1(t_f) x_1(t_f)$$
$$+ \lambda_2(t_f) x_2(t_f) - \lambda_1(0) x_1(0) - \lambda_2(0) x_2(0)$$
$$+ \int_0^{t_f} \{\lambda_1 (u^2 - u - 2x_1)$$
$$+ \dot{\lambda}_1 x_1 + \lambda_2 x_1 + \dot{\lambda}_2 x_2\} dt$$

and the stationary conditions with respect to $x(t)$, $x(1.0)$, and $u(t)$ are given by:

x_1: $\dot{\lambda}_1 = 2\lambda_1 - \lambda_2$ $\lambda_1(1) = 0$
x_2: $\dot{\lambda}_2 = 0$ $\lambda_2(1) = 1$
u: $\lambda_1(2u - 1) = 0$

The adjoint variables λ_1 and λ_2 can be determined directly from the adjoint equations and the associated final conditions. These are

$\lambda_1(t) = \{1 - \exp(2t - 2)\}/2$
$\lambda_2(t) = 1$

Because λ_1 is zero only at $t = 1$, the stationary condition for $u(t)$ is satisfied only when $u = 0.5$. Consequently, it can be verified that $u(t) = 0.5$ minimizes $x_2(1)$.

Solution of the differential–algebraic optimization problem becomes especially difficult in the presence of state variable inequality and equality constraints. Because many process problems are constrained, optimization problems of this type are not considered frequently. Methods for tackling these problems can be divided into three basic types:

1) iterative methods based on variational conditions;
2) feasible path nonlinear programming methods; and
3) simultaneous nonlinear programming methods.

The analytical approach to solving small design problems, such as the above example, naturally leads to iterative algorithms based on variational conditions. Based on the optimality conditions derived above, control vector iteration algorithms were proposed that involve the solution of model equations forward in time, adjoint equations backward in time, and intermittent updating of the control profile to minimize the Hamiltonian function, $\lambda^T f(x, u)$ (or which make $dH/du = 0$). Methods that are analogous to those for solving unconstrained optimization problems can be applied here. For example, using the gradient of the Hamiltonian with respect to $u(t)$, BRYSON and DENHAM [10.40] proposed a steepest descent method. LASDON, MITTER, and WARREN [10.41] and LASDON [10.42] accelerated this approach by proposing conjugate gradient and variable metric methods, respectively, for these types of problems. Finally, a Newton-type extension to optimal control problems is presented in [10.43].

These methods work best on problems with initial value ordinary differential equation models without state variable and final time constraints. For these simple problems, solutions have been reported that require from several dozen to several hundred model (and adjoint equation) evaluations [10.44], [10.45]. Moreover, any additional constraints in this problem require the search for appropriate multiplier values, μ and ν, which often requires an outer loop in the solution algorithm and can easily lead to a prohibitive number of model evaluations, even for small systems. Consequently, the control vector iteration methods are limited to the simplest optimal control problems.

Instead, the optimal control problem can be approached as a nonlinear programming problem which results after the control variable is discretized. Here, the ordinary differential equation model is solved repeatedly in an inner loop while parameters representing $u(t)$ are updated on the outside. Initially, the updating was performed by direct search or "hill climbing" algorithms and could become costly for large problems. Use of more sophisticated methods, on the other hand, must lead to a consideration of how gradients can be calculated efficiently from the differential equation model.

Using this nonlinear programming approach (also termed the feasible path approach), denote u_p as the vector of parameters representing $u(t)$. For example, if $u(t)$ is assumed piecewise constant over a variable distance, include u_i and t_i in u_p. The original optimization problem then becomes the following nonlinear programming problem:

$$\min_{u_p, p} F(u_p, p)$$
such that $h(u_p, p) = 0$
$g(u_p, p) \leq 0$
$p^L \leq p \leq p^U$
$u_p^L \leq u_p \leq u_p^U$

and the ordinary differential equation model:

$$\dot{x}(t) = f(p, u_p, x(t))$$
$$x(0) = x_0$$
$$x_{K+1}(t_{i-1,f}) = x_{i0} \quad i = 2, NE$$
$$x_{K+1}(t_{NE,f}) = x_f$$
$$x^L \le x_{il} \le x^U$$
$$u^L \le u_{il} \le u^U$$
$$p^L \le p \le p^U$$

is solved.

Note that the independent variable, time, disappears from this problem, and constraints imposed at final time appear naturally in the above problem; other constraints that must be enforced over time have to be treated in a more complex manner. For example, they can be converted to final time constraints by integrating the square of the constraint violations and forcing these to be less than a tolerance at the final time.

To solve the nonlinear program, gradients can be calculated in a number of ways with respect to u_p from the ordinary differential equation model. The easiest, but least efficient and accurate, way is simply to resolve the model for each perturbation of the parameters. Sensitivity information can also be obtained by solving the adjoint equations

$$\dot{\lambda}_i = \nabla_x f(x, u) \lambda_i, \quad i = 0, m$$
$$\lambda_0 = \nabla_{x_f} F; \qquad \lambda_i = \nabla_{x_f} g_{f_i} \text{ or } \nabla_{x_f} h_{f_i}$$

and evaluating parameteric sensitivities by using the following relation:

$$\frac{\partial F}{\partial p} = \int_0^{t_f} [\nabla_p f^T \lambda_0] dt$$

$$\frac{\partial g_{f_i}}{\partial p} \text{ or } \frac{\partial h_{f_i}}{\partial p} = \int_0^{t_f} \nabla_p f^T \lambda_i \, dt$$

These equations are very similar to the optimality conditions derived above.

Gradients with respect to u_p can also be calculated through sensitivity equations derived from the ordinary differential equation model. Here, for continuous state variable profiles, the sensitivity equations with respect to parameters p, for example, are

$$\frac{\partial}{\partial p}\{\dot{x} = f(x, u)\} \to \frac{\partial}{\partial p}\left(\frac{dx}{dt}\right) = \frac{\partial f}{\partial p} + \left(\frac{\partial x}{\partial p}\right)\left(\frac{\partial f}{\partial x}\right)$$

which, upon changing orders of differentiation and defining $s = \partial x/\partial p$, leads to

$$\dot{s} = \partial f/\partial p + s(\partial f/\partial x)$$

Either approach results in gradient calculations with costs proportional to problem size; effort for evaluating gradients with adjoint approaches is proportional to the number of (objective and constraint) functions evaluated at final time, whereas effort for sensitivity equations is proportional to the number of parameters u_p and p. Consequently, the choice of gradient calculation approach depends on the structure of the particular feasible path problem. Nevertheless, both CARCOTSIOS and STEWART [10.46] and SARGENT and SULLIVAN [10.47] demonstrated that gradient calculation approaches can be accelerated considerably by tailoring the ordinary differential equation solver to include sensitivity or adjoint equations.

The feasible path approach has been very successful in solving large process problems. However, this approach still requires repeated solution of the process model (and sensitivities). For large processes or for processes that require the solution of rigorous underlying procedures, this approach can become expensive. Moreover, for stiff or otherwise difficult systems, this approach is only as reliable as the ordinary differential equation solver. The feasible path approach also offers only indirect ways of handling time-dependent constraints. Finally, the optimal solution with this approach is only as good as its control-variable parameterization, which often can only be improved by a priori information about the specific problem. Consequently, a simultaneous nonlinear programming approach is also considered as an alternative solution method.

Instead of parameterizing the control profile and solving the system as a nonlinear program, the simultaneous approach begins with a parameterization of both the control and the state variable profiles, and solves a mathematical programming problem consisting of algebraic equations. However, early application of the simultaneous approach suffered from two drawbacks. First, simultaneous approaches lead to much larger nonlinear programs than feasible path approaches. Consequently, nonlinear programming methods must be very efficient to compete with smaller feasible path formulations. Here, a trade-off occurs between the expense of repeated ordinary differential equation model solution and solution of a larger nonlinear program. Second, care must be taken in the formulation to yield an accurate algebraic representation of the differential equations.

Recently, the SQP algorithm, described in a previous section, and orthogonal collocation have been applied successfully to a number of optimal control problems; this approach shows considerably better performance than control vector iteration methods. Moreover, through appropriate discretization of the differential equations, this simultaneous approach can handle constraints on state and control profiles directly. Moreover, by using the range and null space decomposition technique for SQP, very efficient solutions can be achieved with the simultaneous approach.

Because their approximation and stability properties are well studied, various ordinary differential equation solution methods can be applied directly to discretize the differential equations to algebraic constraints. Here, implicit Runge–Kutta methods are considered, which coincidentally also include the method of orthogonal collocation. By defining state and control profiles at $\tau_l \varepsilon [0, 1]$ [e.g., the shifted roots of orthogonal (Legendre) polynomials], these profiles are parameterized as Lagrange-form polynomials over $\tau \varepsilon [0, 1]$:

$$x_{K+1}(\tau) = \sum_{l=0}^{K} x_l \varphi_l(\tau)$$

$$u_K(\tau) = \sum_{l=1}^{K} u_l \psi_l(\tau)$$

$$\varphi_l(\tau) = \prod_{k=0, l}^{K} \frac{(\tau - \tau_k)}{(\tau_l - \tau_k)}, \quad \psi_l(\tau) = \prod_{k=1, l}^{K} \frac{(\tau - \tau_k)}{(\tau_l - \tau_k)}$$

where $x_{K+1}>(\tau)$ and u_K are $(K+1)$-th order (degree $< K+1$) and K-th order polynomials, respectively. (Here the notation $k = 0, l$ refers to the index k starting at zero but not equal to l.)

The state variables are one order higher than the controls because they have explicit interpolation coefficients defined at the beginning of each element. With this representation of $x(t)$ and $u(t)$ this approach is extended to piecewise polynomials and orthogonal collocation on NE finite elements (of length $\Delta \alpha_i$) is applied. The differential equations are now represented by the following nonlinear algebraic equations:

$$R(t_{il}, \Delta \alpha_i) = \sum_{j=0}^{K} x_{ij} \varphi_j(\tau_l) - \Delta \alpha_i f(p, u_{il}) = 0$$

$$l = 1, \ldots K, \quad i = 1, \ldots NE$$

with

$$x_{i,o} = x_o \quad \text{and} \quad x_{K+1}(t_{i-1, f}) = x_{io} \quad i = 2, NE$$

where t_{il} represent shifted roots of Legendre polynomials over the i-th element length and $t_{i-1, f}$ is the time at the end of the $(i-1)$-th element. Note that state profiles satisfy continuity conditions across finite elements $\Delta \alpha_i$, whereas this property is not enforced for control profiles. Substituting this representation of the ordinary differential equation model into the dynamic optimization problem with $\Delta \alpha_i$ fixed, leads to the following nonlinear programming problem:

$$\min_{p, x_{il}, u_{il}} F(p, x_f)$$

such that $h(p, u_{il}, x_{il}) = 0$

$$g(p, u_{il}, x_{il}) \leq 0$$

$$R(t_{il}, \Delta \alpha_i) = 0$$

$$i = 1, \ldots NE, \quad l = 1, \ldots K$$

$$x_{io} = x_o$$

By taking advantage of the orthogonal properties of the polynomial representation, the Kuhn–Tucker conditions of this nonlinear program can be shown to be parameterizations of the optimality conditions of the dynamic problem. Therefore, the only requirement is that $\Delta \alpha$ be chosen appropriately (i.e., be sufficiently small for accurate approximation of the differential equations) and that the breakpoints for the optimal control profile can be located.

Choosing the element lengths to render an accurate discretization of the differential equations can also be performed automatically by the nonlinear programming problem. For example, the element lengths $\Delta \alpha_i$ could be made decision variables, and a measure of the approximation error could be included as an inequality constraint. Here, a suitable approximation measure is the residual of the differential equation evaluated at a noncollocation point $\bar{t}$ within each element, i.e., $|R(\bar{t}, \Delta \alpha_i)| \leq \varepsilon$ for some ε tolerance.

For dynamic optimization problems without control profiles and with a sufficiently large number of elements, this simultaneous nonlinear programming approach can be shown to yield accurate solutions to difficult dynamic optimization problems. Therefore, the choice of approach for parameter optimization problems depends on the difficulty of solving the ordinary differential equation model. If the model and its sensitivity information can be determined quickly, then a feasible path approach is probably more efficient than a simultaneous strategy. On the other hand, if the model is expensive and difficult to solve at intermediate points, and if state variable constraints must be enforced over time, then a simultaneous approach with an efficient nonlinear programming strategy should be considered. Here, the advantage of this approach is that the ordinary differential equation model is solved only once and state variable constraints can be enforced directly.

For problems with control profiles, on the other hand, one must be especially careful about the stability properties of the ordinary differential equation discretization. Without control profiles, stability properties of the ordinary differential equation discretization are determined by properties of the corresponding ordinary differential equation solver (e.g., implicit Runge–Kutta methods). However, for dynamic optimization problems with control profiles, the optimality conditions form a set of differential–algebraic equations that may lead to different approximation and stability properties for the same discretization method.

Difficulties in solving differential–algebraic equation systems normally occur when the discretized system of algebraic equations becomes singular. This can occur, for example, if time-dependent equalities (h) and active inequalities (g) are not functions of $u(t)$. Although solutions to the differential–algebraic equation system (and the optimization problem) do exist, instabilities and loss of accuracy may result unless the discretization is of reasonably high order and has very strong stability properties. BRENAN and PETZOLD [10.48] and LOGSDON and BIEGLER [10.49] mention that collocation formulas can deal with these difficult systems, but the orthogonal collocation method may have to be modified before it can be applied.

Consequently, to deal with dynamic optimization problems with control profiles, it is necessary to determine the difficulty of the system (a straightforward analysis can be found in [10.48] and [10.50]) to see if an appropriately accurate and stable discretization (e.g., orthogonal collocation, implicit Runge–Kutta, etc.) is available. If one cannot be found, then the only recourse may be to parameterize the control profile and settle for a suboptimal solution. The advantage of this, however, is that the methods described above for parameter problems can be applied directly without any difficulties due to discretization of differential–algebraic equation systems.

11. Probability and Statistics

[11.1]–[11.4]

The models treated thus far have been deterministic, that is, if the parameters are known the outcome is determined. In many situations, all the factors cannot be controlled and the outcome may vary randomly about some average value. Then a range of outcomes has a certain probability of occurring, and statistical methods must be used. This is especially true in quality control of production processes and experimental measurements. This chapter presents standard statistical concepts, sampling theory and statistical decisions, and factorial design of experiments or analysis of variances. Multivariant linear and nonlinear regression is treated in Chapter 2.

11.1. Concepts

Suppose N values of a variable y, called $y_1, y_2, \ldots, y_N$, might represent N measurements of the same quantity. The *arithmetic mean* $E(y)$ is

$$E(y) = \frac{\sum_{i=1}^{N} y_i}{N}$$

The *median* is the middle value (or average of the two middle values) when the set of numbers is arranged in increasing (or decreasing) order. The *geometric mean* $\bar{y}_G$ is

$$\bar{y}_G = (y_1 y_2 \ldots y_N)^{1/N}$$

The *root-mean-square* or quadratic mean is

$$\text{Root-mean-square} = \sqrt{E(y^2)} = \sqrt{\sum_{i=1}^{N} y_i^2 / N}$$

The *range* of a set of numbers is the difference between the largest and the smallest members in the set. The *mean deviation* is the mean of the deviation from the mean.

$$\text{Mean-deviation} = \frac{\sum_{i=1}^{N} |y_i - E(y)|}{N}$$

The *variance* is

$$\text{var}(y) = \sigma^2 = \frac{\sum_{i=1}^{N} (y_i - E(y))^2}{N}$$

and the *standard deviation* σ is the square root of the variance.

$$\sigma = \sqrt{\frac{\sum_{i=1}^{N}(y_i - E(y))^2}{N}}$$

If the set of numbers $\{y_i\}$ is a small sample from a larger set, then the *sample average*

$$\bar{y} = \frac{\sum_{i=1}^{n} y_i}{n}$$

is used in calculating the *sample variance*

$$s^2 = \frac{\sum_{i=1}^{n}(y_i - \bar{y})^2}{n-1}$$

and the sample standard deviation

$$s = \sqrt{\frac{\sum_{i=1}^{n}(y_i - \bar{y})^2}{n-1}}$$

The value $n-1$ is used in the denominator because the deviations from the sample average must total zero:

$$\sum_{i=1}^{n}(y_i - \bar{y}) = 0$$

Thus, knowing $n-1$ values of $y_i - \bar{y}$ and the fact that there are n values automatically gives the n-th value. Thus, only $n-1$ degrees of freedom v exist. This occurs because the unknown mean $E(y)$ is replaced by the sample mean $\bar{y}$ derived from the data.

If data are taken consecutively, running totals can be kept to permit calculation of the mean and variance without retaining all the data:

$$\sum_{i=1}^{n}(y_i - \bar{y})^2 = \sum_{i=1}^{n} y_i^2 - 2\bar{y}\sum_{i=1}^{n} y_i + (\bar{y})^2$$

$$\bar{y} = \sum_{i=1}^{n} y_i/n$$

Thus,

$$n, \quad \sum_{i=1}^{n} y_i^2, \quad \text{and} \quad \sum_{i=1}^{n} y_i$$

are retained, and the mean and variance are computed when needed.

Repeated observations that differ because of experimental error often vary about some central value in a roughly symmetrical distribution in which small deviations occur more frequently than large deviations. In plotting the number of times a discrete event occurs, a typical curve is obtained, which is shown in Figure 44. Then the probability p of an event (score) occurring can be thought of as the ratio of the number of times it was observed divided by the total number of events. A continuous representation of this probability density function is given by the *normal distribution*

$$p(y) = \frac{1}{\sigma\sqrt{2\pi}} e^{-[y-E(y)]^2/2\sigma^2}. \tag{79}$$

This is called a normal probability distribution function. It is important because many results are insensitive to deviations from a normal distribution. Also, the central limit theorem says that if an overall error is a linear combination of component errors, then the distribution of errors tends to be normal as the number of components increases, almost regardless of the distribution of the component errors (i.e., they need not be normally distributed). Naturally, several sources of error must be present and one error cannot predominate (unless it is normally distributed). The normal distribution function is calculated easily; of more value are integrals of the function, which are given in Table 12; the region of interest is illustrated in Figure 45.

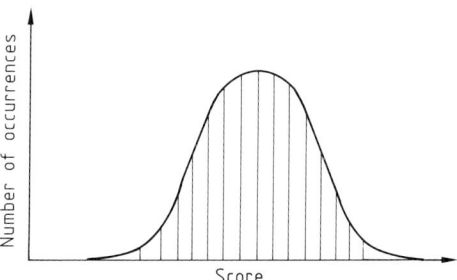

Figure 44. Frequency of occurrence of different scores

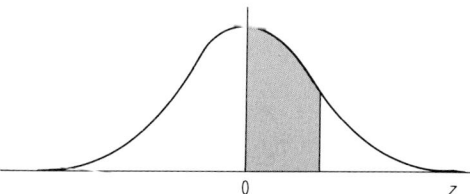

Figure 45. Area under normal curve

Table 12. Area under normal curve*

$$F(z) = \frac{1}{\sqrt{2\pi}} \int_0^z e^{-z^2/2}\, dz$$

z	F(z)"	z	F(z)"
0.0	0.0000	1.5	0.4332
0.1	0.0398	1.6	0.4452
0.2	0.0793	1.7	0.4554
0.3	0.1179	1.8	0.4641
0.4	0.1554	1.9	0.4713
0.5	0.1915	2.0	0.4772
0.6	0.2257	2.1	0.4821
0.7	0.2580	2.2	0.4861
0.8	0.2881	2.3	0.4893
0.9	0.3159	2.4	0.4918
1.0	0.3413	2.5	0.4938
1.1	0.3643	2.7	0.4965
1.2	0.3849	3.0	0.4987
1.3	0.4032	4.0	0.499968
1.4	0.4192	5.0	0.4999997

* Table gives the probability F that a random variable will fall in the shaded region of Figure 45. For a more complete table (in slightly different form), see [11.6, Table 26.1].

Table 13. Percentage points of area under Students t-distribution*

v	$\alpha = 0.10$	$\alpha = 0.05$	$\alpha = 0.01$	$\alpha = 0.001$
1	6.314	12.706	63.657	636.619
2	2.920	4.303	9.925	31.598
3	2.353	3.182	5.841	12.941
4	2.132	2.776	4.604	8.610
5	2.015	2.571	4.032	6.859
6	1.943	2.447	3.707	5.959
7	1.895	2.365	3.499	5.405
8	1.860	2.306	3.355	5.041
9	1.833	2.262	3.250	4.781
10	1.812	2.228	3.169	4.587
15	1.753	2.131	2.947	4.073
20	1.725	2.086	2.845	3.850
25	1.708	2.060	2.787	3.725
30	1.697	2.042	2.750	3.646
∞	1.645	1.960	2.576	3.291

* Table gives t values such that a random variable will fall in the shaded region of Figure 47 with probability α. For a one-sided test the confidence limits are obtained for $\alpha/2$. For a more complet table (in slightly different form), see [11.6, Table 26.10].

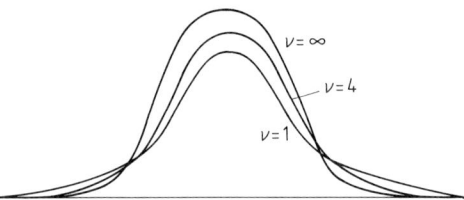

Figure 46. Student's t-distribution. For explanation of v see text

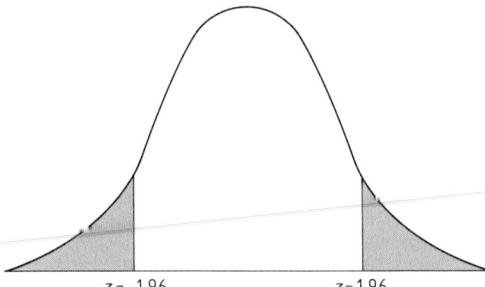

Figure 47. Percentage points of area under Student's t-distribution

For a small sample, the variance can only be estimated with the sample variance s^2. Thus, the normal distribution cannot be used because σ is not known. In such cases *Student's t-distribution*, shown in Figure 46 [11.1, p. 70], is used:

$$p(y) = \frac{y_0}{\left(1 + \frac{t^2}{n-1}\right)^{n/2}}, \quad t = \frac{\bar{y} - E(y)}{s/\sqrt{n}}$$

and y_0 is chosen such that the area under the curve is one. The number $v = n - 1$ is the degrees of freedom, and as v increases, Student's t-distribution approaches the normal distribution. The normal distribution is adequate (rather than the t-distribution) when $v > 15$, except for the tails of the curve which require larger v. Integrals of the t-distribution are given in Table 13, the region of interest is shown in Figure 47.

Other probability distribution functions are useful. Any distribution function must satisfy the following conditions:

$0 \leq F(x) \leq 1$
$F(-\infty) = 0,\ F(+\infty) = 1$
$F(x) \leq F(y)$ when $x \leq y$

The *probability density function* is

$$p(x) = \frac{dF(x)}{dx}$$

where

$$dF = p\, dx$$

is the probability of x being between x and $x + dx$. The probability density function satisfies

$$p(x) \geq 0$$

$$\int_{-\infty}^{\infty} p(x)\,dx = 1$$

The *Bernoulli distribution* applies when the outcome can take only two values, such as heads or tails, or 0 or 1. The probability distribution function is

$$p(x = k) = p^k(1-p)^{1-k}, \quad k = 0 \text{ or } 1$$

and the mean of a function $g(x)$ depending on x is

$$E[g(x)] = g(1)p + g(0)(1-p)$$

The *binomial distribution function* applies when there are n trials of a Bernoulli event; it gives the probability of k occurrences of the event, which occurs with probability p on each trial

$$p(x=k) = \frac{n!}{k!(n-k)!} p^k (1-p)^{n-k}$$

The mean and variance are

$$E(x) = np$$
$$\text{var}(x) = np(1-p)$$

The *hypergeometric distribution function* applies when there are N objects, of which M are of one kind and $N-M$ are of another kind. Then the objects are drawn one by one, without replacing the last draw. If the last draw had been replaced the distribution would be the binomial distribution. If x is the number of objects of type M drawn in a sample of size n, then the probability of $x = k$ is

$$p(x=k)$$
$$= \frac{M!(N-M)!\,n!(N-n)!}{k!(M-k)!(n-k)!(N-M-n+k)!\,N!}$$

The mean and variance are

$$E(x) = \frac{nM}{N}$$

$$\text{var}(x) = np(1-p)\frac{N-n}{N-1}$$

The *Poisson distribution* is

$$p(x=k) = e^{-\lambda}\frac{\lambda^k}{k!}$$

with a parameter λ. The mean and variance are

$$E(x) = \lambda$$
$$\text{var}(x) = \lambda$$

The simplest continuous distribution is the uniform distribution. The probability density function is

$$p = \begin{cases} \dfrac{1}{b-a} & a < x < b \\ 0 & x < a, \quad x > b \end{cases}$$

and the probability distribution function is

$$F(x) = \begin{cases} 0 & x < a \\ \dfrac{x-a}{b-a} & a < x < b \\ 1 & b < x \end{cases}$$

The mean and variance are

$$E(x) = \frac{a+b}{2}$$

$$\text{var}(x) = \frac{(b-a)^2}{12}$$

The *normal distribution* is given by Equation (79), p. 130, with variance σ^2.

The *log normal probability density* function is

$$p(x) = \frac{1}{x\sigma\sqrt{2\pi}} \exp\left[-\frac{(\log x - \mu)^2}{2\sigma^2}\right]$$

and the mean and variance are [11.5, p. 89]

$$E(x) = \exp\left(\mu + \frac{\sigma^2}{2}\right)$$

$$\text{var}(x) = \exp(\sigma^2 - 1)\exp(2\mu + \sigma^2)$$

11.2. Sampling and Statistical Decisions

Two variables can be statistically dependent or independent. For example, the height and diameter of all distillation towers are statistically dependent, because the distribution of diameters of all columns 10 m high is different from that of columns 30 m high. If y_B is the diameter and y_A the height, the distribution is written as

$p(y_B|y_A = \text{constant})$, or here
$p(y_B|y_A = 10) \neq p(y_B|y_A = 30)$

A third variable y_C, could be the age of the operator on the third shift. This variable is probably unrelated to the diameter of the column, and for the distribution of ages is

$$p(y_C y_A) = p(y_C)$$

Thus, variables y_A and y_C are distributed independently. The joint distribution for two variables is

$$p(y_A, y_B) = p(y_A) p(y_B | y_A)$$

if they are statistically dependent, and

$$p(y_A, y_B) = p(y_A) p(y_B)$$

if they are statistically independent. If a set of variables $y_A, y_B, \ldots$ is independent and identically distributed,

$$p(y_A, y_B, \ldots) = p(y_A) p(y_B) \ldots$$

Conditional probabilities are used in hazard analysis of chemical plants.

A measure of the linear dependence between variables is given by the *covariance*

$$Cov(y_A, y_B) = E\{[y_A - E(y_A)][y_B - E(y_B)]\}$$

$$= \frac{\sum_{i=1}^{N} [y_{Ai} - E(y_A)][y_{Bi} - E(y_B)]}{N}$$

The correlation coefficient ϱ is

$$\varrho(y_A, y_B) = \frac{Cov(y_A, y_B)}{\sigma_A \sigma_B}$$

If y_A and y_B are independent, then $Cov(y_A, y_B) = 0$. If y_A tends to increase when y_B decreases then $Cov(y_A, y_B) < 0$. The sample correlation coefficient is [11.4, p. 484]

$$r(y_A, y_B) = \frac{\sum_{i=1}^{n} (y_{Ai} - \bar{y}_A)(y_{Bi} - \bar{y}_B)}{(n-1) s_A s_B}$$

If measurements are for independent, identically distributed observations, the errors are independent and uncorrelated. Then $\bar{y}$ varies about $E(y)$ with variance σ^2/n, where n is the number of observations in $\bar{y}$. Thus if something is measured several times today and every day, and the measurements have the same distribution, the variance of the means decreases with the number of samples in each day's measurement n. Of course, other factors (weather, weekends) may cause the observations on different days to be distributed nonidentically.

Suppose Y, which is the sum or difference of two variables, is of interest:

$$Y = y_A \pm y_B$$

Then the mean value of Y is

$$E(Y) = E(y_A) \pm E(y_B)$$

and the variance of Y is

$$\sigma^2(Y) = \sigma^2(y_A) + \sigma^2(y_B)$$

More generally, consider the random variables $y_1, y_2, \ldots$ with means $E(y_1), E(y_2), \ldots$ and variances $\sigma^2(y_1), \sigma^2(y_2), \ldots$ and correlation coefficients ϱ_{ij}. The variable

$$Y = \alpha_1 y_1 + \alpha_2 y_2 + \ldots$$

has a mean

$$E(Y) = \alpha_1 E(y_1) + \alpha_2 E(y_2) + \ldots$$

and variance [11.1, p. 87]

$$\sigma^2(Y) = \sum_{i=1}^{n} \alpha_i^2 \sigma^2(y_i)$$

$$+ 2 \sum_{i=1}^{n} \sum_{j=i+1}^{n} \alpha_i \alpha_j \sigma(y_i) \sigma(y_j) \varrho_{ij}$$

or

$$\sigma^2(Y) = \sum_{i=1}^{n} \alpha_i^2 \sigma^2(y_i)$$

$$+ 2 \sum_{i=1}^{n} \sum_{j=i+1}^{n} \alpha_i \alpha_j Cov(y_i, y_j) \quad (80)$$

If the variables are uncorrelated and have the same variance, then

$$\sigma^2(Y) = \left(\sum_{i=1}^{n} \alpha_i^2\right) \sigma^2$$

This fact can be used to obtain more accurate cost estimates for the purchased cost of a chemical plant than is true for any one piece of equipment. Suppose the plant is composed of a number of heat exchangers, pumps, towers, etc., and that the cost estimate of each device is $\pm 40\%$ of its cost (the sample standard deviation is 20% of its cost). In this case the α_i are the numbers of each type of unit. Under special conditions, such as equal numbers of all types of units and comparable cost, the standard deviation of the plant costs is

$$\sigma(Y) = \frac{\sigma}{\sqrt{n}}$$

and is then $\pm (40/\sqrt{n})\%$. Thus the standard deviation of the cost for the entire plant is the standard deviation of each piece of equipment divided by the square root of the number of units. Under less restrictive conditions the actual numbers change according to the above equations, but the principle is the same.

Suppose modifications are introduced into the manufacturing process. To determine if the modification causes a significant change, the mean of some property could be measured before and after the change; if these differ, does it mean the process modification caused it, or could the change have happened by chance? This is a statistical decision. A hypothesis H_0 is defined; if it is true, action A must be taken. The reverse hypothesis is H_1; if this is true, action B must be taken. A correct decision is made if action A is taken when H_0 is true or action B is taken when H_1 is true. Taking action B when H_0 is true is called a type I error, whereas taking action A when H_1 is true is called a type II error.

The following test of hypothesis or test of significance must be defined to determine if the hypothesis is true. The level of significance is the maximum probability that an error would be accepted in the decision (i.e., rejecting the hypothesis when it is actually true). Common levels of significance are 0.05 and 0.01, and the test of significance can be either one or two sided. If a sampled distribution is normal, then the probability that the z score

$$z = \frac{y - \bar{y}}{s_y}$$

is in the unshaded region is 0.95. The value given in Table 12 for $F = 0.475$ is $z = 1.96$. Because a two-sided test is desired, $F = 0.95/2 = 0.475$. If the test was one-sided, at the 5% level of significance, $F = 0.45$ or $z = 1.645$. In the two-sided test (see Fig. 48), if a single sample is chosen and $z < -1.96$ or $z > 1.96$, then this could happen with probability 0.05 if the hypothesis were true. This z would be significantly different from the expected value (based on the chosen level of sig-

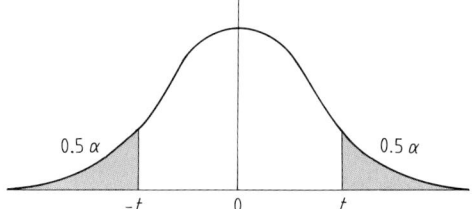

Figure 48. Two-sided statistical decision

nificance) and the tendency would be to reject the hypothesis. If the value of z was between -1.96 and 1.96, the hypothesis would be accepted.

The same type of decisions can be made for other distributions. Consider Student's t-distribution. At a 95% level of confidence, with $v = 10$ degrees of freedom, the t values are ± 2.228. Thus, the sample mean would be expected to be between

$$\bar{y} \pm t_c \frac{s}{\sqrt{n}}$$

with 95% confidence. If the mean were outside this interval, the hypothesis would be rejected.

The *chi-square distribution* is useful for examining the variance or standard deviation. The statistic is defined as

$$\chi^2 = \frac{ns^2}{\sigma^2}$$
$$= \frac{(y_1 - \bar{y})^2 + (y_2 - \bar{y})^2 + \ldots + (y_n - \bar{y})^2}{\sigma^2}$$

and the chi-square distribution is

$$p(y) = y_0 \chi^{v-2} e^{-\chi^2/2}$$

$v = n - 1$ is the number of degrees of freedom and y_0 is chosen so that the integral of $p(y)$ over all y is 1. The probability of a deviation larger than χ^2 is given in Table 14; the area in question, in Figure 49. For example, for 10 degrees of freedom and a 95% confidence level, the critical values of χ^2 are 0.025 and 0.975. Then

$$\frac{s\sqrt{n}}{\chi_{0.975}} < \sigma < \frac{s\sqrt{n}}{\chi_{0.025}}$$

or

$$\frac{s\sqrt{n}}{20.5} < \sigma < \frac{s\sqrt{n}}{3.25}$$

Table 14. Percentage points of area under chi-square distribution with v degrees of freedom *

v	$\alpha=0.995$	$\alpha=0.99$	$\alpha=0.975$	$\alpha=0.95$	$\alpha=0.5$	$\alpha=0.05$	$\alpha=0.025$	$\alpha=0.01$	$\alpha=0.005$
1	7.88	6.63	5.02	3.84	0.455	0.0039	0.0010	0.0002	0.0000
2	10.6	9.21	7.38	5.99	1.39	0.103	0.0506	0.0201	0.0100
3	12.8	11.3	9.35	7.81	2.37	0.352	0.216	0.115	0.072
4	14.9	13.3	11.1	9.49	3.36	0.711	0.484	0.297	0.207
5	16.7	15.1	12.8	11.1	4.35	1.15	0.831	0.554	0.412
6	18.5	16.8	14.4	12.6	5.35	1.64	1.24	0.872	0.676
7	20.3	18.5	16.0	14.1	6.35	2.17	1.69	1.24	0.989
8	22.0	20.1	17.5	15.5	7.34	2.73	2.18	1.65	1.34
9	23.6	21.7	19.0	16.9	8.34	3.33	2.70	2.09	1.73
10	25.2	23.2	20.5	18.3	9.34	3.94	3.25	2.56	2.16
12	28.3	26.2	23.3	21.0	11.3	5.23	4.40	3.57	3.07
15	32.8	30.6	27.5	25.0	14.3	7.26	6.26	5.23	4.60
17	35.7	33.4	30.2	27.6	16.3	8.67	7.56	6.41	5.70
20	40.0	37.6	34.2	31.4	19.3	10.9	9.59	8.26	7.43
25	46.9	44.3	40.6	37.7	24.3	14.6	13.1	11.5	10.5
30	53.7	50.9	47.0	43.8	29.3	18.5	16.8	15.0	13.8
40	66.8	63.7	59.3	55.8	39.3	26.5	24.4	22.2	20.7
50	79.5	76.2	71.4	67.5	49.3	34.8	32.4	29.7	28.0
60	92.0	88.4	83.3	79.1	59.3	43.2	40.5	37.5	35.5
70	104.2	100.4	95.0	90.5	69.3	51.7	48.8	45.4	43.3
80	116.3	112.3	106.6	101.9	79.3	60.4	57.2	53.5	51.2
90	128.3	124.1	118.1	113.1	89.3	69.1	65.6	61.8	59.2
100	140.2	135.8	129.6	124.3	99.3	77.9	74.2	70.1	67.3

* Table value is χ_α^2; $\chi^2 < \chi_\alpha^2$ with probability α. For a more complete table (in slightly different form), see [11.6, Table 26.8].

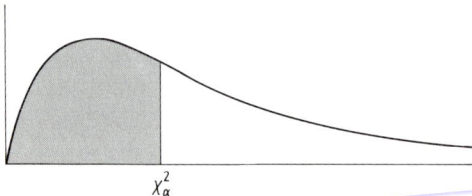

Figure 49. Percentage points of area under chi-squared distribution with v degrees of freedom

with 95 % confidence.

Tests are available to decide if two distributions that have the same variance have different means [11.4, p. 465]. Let one distribution be called x, with N_1 samples, and the other be called y, with N_2 samples. First, compute the standard error of the difference of the means:

$$s_D = \sqrt{\frac{\sum_{i=1}^{N_1}(x_i - \bar{x})^2 + \sum_{i=1}^{N_2}(y_i - \bar{y})^2}{N_1 + N_2 - 2}\left(\frac{1}{N_1} + \frac{1}{N_2}\right)}$$

Next, compute the value of t

$$t = \frac{\bar{x} - \bar{y}}{s_D}$$

and evaluate the significance of t using Student's t-distribution for $N_1 + N_2 - 2$ degrees of freedom.

If the samples have different variances, the relevant statistic for the t-test is

$$t = \frac{\bar{x} - \bar{y}}{\sqrt{var(x)/N_1 + var(y)/N_2}}$$

The number of degrees of freedom is now taken aproximately as

$$v = \frac{\left(\frac{var(x)}{N_1} + \frac{var(y)}{N_2}\right)^2}{\frac{[var(x)/N_1]^2}{N_1 - 1} + \frac{[var(y)/N_2]^2}{N_2 - 1}}$$

There is also an F-test to decide if two distributions have significantly different variances. In this case, the ratio of variances is calculated:

$$F = \frac{var(x)}{var(y)}$$

where the variance of x is assumed to be larger. Then, a table of values is used to determine the significance of the ratio. The table [11.6, Table 26.9] is derived from the formula [11.4, p. 169]

$$Q(F|v_1, v_2) = I_{v_2/(v_2 + v_1 F)}\left(\frac{v_2}{2}, \frac{v_1}{2}\right)$$

where the right-hand side is an incomplete beta function.

11.3. Error Analysis in Experiments

Suppose a measurement of several quantities is made and a formula or mathematical model is used to deduce some property of interest. For example, to measure the thermal conductivity of a solid k, the heat flux q, the thickness of the sample d, and the temperature difference across the sample ΔT must be measured. Each measurement has some error. The heat flux q may be the rate of electrical heat input $\dot{Q}$ divided by the area A, and both quantities are measured to some tolerance. The thickness of the sample is measured with some accuracy, and the temperatures are probably measured with a thermocouple, to some accuracy. These measurements are combined, however, to obtain the thermal conductivity, and the error in the thermal conductivity must be determined. The formula is

$$k = \frac{d}{A \Delta T} \dot{Q}$$

If each measured quantity has some variance, what is the variance in the thermal conductivity?

Suppose a model for Y depends on various measurable quantities, $y_1, y_2, \ldots$ Suppose several measurements are made of $y_1, y_2, \ldots$ under seemingly identical conditions and several different values are obtained, with means $E(y_1), E(y_2), \ldots$ and variances $\sigma_1^2, \sigma_2^2, \ldots$ Next suppose the errors are small and independent of one another. Then a change in Y is related to changes in y_i by

$$dY = \frac{\partial Y}{\partial y_1} dy_1 + \frac{\partial Y}{\partial y_2} dy_2 + \ldots$$

If the changes are indeed small, the partial derivatives are constant among all the samples. Then the expected value of the change is

$$E(dY) = \sum_{i=1}^{N} \left(\frac{\partial Y}{\partial y_i} \right) E(dy_i)$$

Naturally $E(dy_i) = 0$ by definition so that $E(dY) = 0$, too. However, since the errors are independent of each other and the partial derivatives are assumed constant because the errors are small, the variances are given by Equation (80)

$$\sigma^2(dY) = \sum_{i=1}^{N} \left(\frac{\partial Y}{\partial y_i} \right)^2 \sigma_i^2$$

Thus, the variance of the desired quantity Y can be found. This gives an independent estimate of the errors in measuring the quantity Y from the errors in measuring each variable it depends upon.

11.4. Factorial Design of Experiments and Analysis of Variance

Statistically designed experiments consider, of course, the effect of primary variables, but they also consider the effect of extraneous variables, the interactions among variables, and a measure of the random error. Primary variables are those whose effect must be determined. These variables can be quantitative or qualitative. Quantitative variables are ones that may be fit to a model to determine the model parameters. Curve fitting of this type is discused in Chapter 2. Qualitative variables are ones whose effect needs to be known; no attempt is made to quantify that effect other than to assign possible errors or magnitudes. Qualitative variables can be further subdivided into type I variables, whose effect is determined directly, and type II variables, which contribute to performance variability, and whose effect is averaged out. For example, in studying the effect of several catalysts on yield in a chemical reactor, each different type of catalyst would be a type I variable, because its effect should be known. However, each time the catalyst is prepared, the results are slightly different, because of random variations; thus, several batches may exist of what purports to be the same catalyst. The variability between batches is a type II variable. Because the ultimate use will require using different batches, the overall effect including that variation should be known, because knowing the results from one batch of one catalyst precisely might not be representative of the results obtained from all batches of the same catalyst. A randomized block design, incomplete block design, or Latin square design, for example, all keep the effect of experimental error in the blocked variables from influencing the effect of the primary variables. Other uncontrolled variables are accounted for by introducing randomization in parts of the experimental design. To study all variables and their interaction requires a factorial design, involving all possible combinations of each variable, or a fractional factorial design, involving only a selected set. Statistical techniques are then used to determine the important variables, the important interactions and the error in estimating these effects. The discussion here is a brief overview of [11.1].

If only two methods exist for preparing some product, to see which treatment is best, the sampling analysis discussed in Section 11.2 can be used to deduce if the means of the two treatments differ significantly. With more treatments, the analysis is more detailed. Suppose the experimental results are arranged as shown in Table 15, i.e., several measurements for each treatment. The objective is to see if the treatments differ significantly from each other, that is, whether their means are different. The samples are assumed to have the same variance. The hypothesis is that the treatments are all the same, and the null hypothesis is that they are different. Deducing the statistical validity of the hypothesis is done by an analysis of variance.

The data for $k = 4$ treatments are arranged in Table 15. Each treatment has n_t experiments, and the outcome of the i-th experiment with treatment t is called y_{ti}. The treatment average is

$$\bar{y}_t = \frac{\sum_{i=1}^{n_t} y_{ti}}{n_t}$$

and the grand average is

$$\bar{y} = \frac{\sum_{t=1}^{k} n_t \bar{y}_t}{N}, \quad N = \sum_{t=1}^{k} n_t$$

Next, the sum of squares of deviations is computed from the average within the t-th treatment

$$S_t = \sum_{i=1}^{n_t} (y_{ti} - \bar{y}_t)^2$$

Since each treatment has n_t experiments, the number of degrees of freedom is $n_t - 1$. Then the sample variances are

$$s_t^2 = \frac{S_t}{n_t - 1}$$

The within-treatment sum of squares is

$$S_R = \sum_{t=1}^{k} S_t$$

and the within-treatment sample variance is

$$s_R^2 = \frac{S_R}{N - k}$$

Now, if no difference exists between treatments, a second estimate of σ^2 could be obtained by calculating the variation of the treatment averages about the grand average. Thus, the between-treatment mean square is computed:

$$s_T^2 = \frac{S_T}{k - 1}, \quad S_T = \sum_{t=1}^{k} n_t (\bar{y}_t - \bar{y})^2$$

Basically the test for whether the hypothesis is true or not hinges on a comparison between the within-treatment estimate s_R^2 (with $v_R = N - k$ degrees of freedom) and the between-treatment estimate s_T^2 (with $v_T = k - 1$ degrees of freedom). The test is made based on the F distribution for v_R and v_T degrees of freedom [11.6, Table 26.9], [11.1, p. 636].

Next consider the case in which *randomized blocking* is used to eliminate the effect of some variable whose effect is of no interest, such as the batch-to-batch variation of the catalysts in the chemical reactor example. With k treatments and n experiments in each treatment, the results from nk experiments can be arranged as shown in Table 16; within each block, various treatments are applied in a random order. The block average, the treatment average, and the grand average are computed as before. The following

Table 15. Estimating the effect of four treatments

Treatment	1	2	3	4
	—	—	—	—
	—	—	—	—
	—	—	—	—
	—	—	—	—
Treatment average	—	—	—	—
Grand average		—		

Table 16. Block design with four treatments and five blocks

Treatment	1	2	3	4	Block average
Block 1	—	—	—	—	—
Block 2	—	—	—	—	—
Block 3	—	—	—	—	—
Block 4	—	—	—	—	—
Block 5	—	—	—	—	—
Treatment average	—	—	—	—	grand average

quantities are also computed for the analysis of variance table:

Name	Formula	Degrees of freedom
Average	$S_A = nk\bar{\bar{y}}^2$	1
Blocks	$S_B = k \sum_{i=1}^{n} (\bar{y}_i - \bar{\bar{y}})^2$	$n - 1$
Treatments	$S_T = n \sum_{t=1}^{k} (\bar{y}_t - \bar{\bar{y}})^2$	$k - 1$
Residuals	$S_R = \sum_{t=1}^{t} \sum_{i=1}^{n} (y_{ti} - \bar{y}_i - \bar{y}_t + \bar{\bar{y}})^2$	$(n-1)(k-1)$
Total	$S = \sum_{t=1}^{t} \sum_{i=1}^{n} y_{ti}^2$	$N = nk$

The key test is again a statistical one, based on the value of

$$s_T^2/s_R^2, \text{ where } s_T^2 = \frac{S_T}{k-1}$$

$$\text{and} \quad s_R^2 = \frac{S_R}{(n-1)(k-1)}$$

and the F distribution for v_R and v_T degrees of freedom [11.1, p. 636]. The assumption behind the analysis is that the variations are linear [11.1, p. 218]. Ways to test this assumption as well as transformations to make if it is not true are provided in [11.1], where an example is given of how the observations are broken down into a grand average, a block deviation, a treatment deviation, and a residual. For two-way factorial design, in which the second variable is a real one rather than one you would like to block out, see [11.1, p. 228].

To measure the effects of variables on a single outcome, a *factorial design* is appropriate. In a two-level factorial design, each variable is considered at two levels only, a high and low value, often designated as a + and a −. The two-level factorial design is useful for indicating trends and showing interactions; it is also the basis for a fractional factorial design. As an example, consider a 2^3 factorial design, with 3 variables and 2 levels for each. The experiments are indicated in Table 17. The main effects are calculated by determining the difference between results from all high values of a variable and all low values of a variable; the result is divided by the number of experiments at each level. For example, for the first variable, calculate

Effect of variable 1 = $[(y_2 + y_4 + y_6 + y_8)$
$- (y_1 + y_3 + y_5 + y_7)]/4$

Table 17. Two-level factorial design with three variables

Run	Variable 1	Variable 2	Variable 3
1	−	−	−
2	+	−	−
3	−	+	−
4	+	+	−
5	−	−	+
6	+	−	+
7	−	+	+
8	+	+	+

Note that all observations are being used to supply information on each of the main effects and each effect is determined with the precision of a fourfold replicated difference. The advantage of a one-at-a-time experiment is the gain in precision if the variables are additive and the measure of nonadditivity if it occurs [11.1, p. 313].

Interaction effects between variables 1 and 2 are obtained by comparing the difference between the results obtained with the high and low value of 1 at the low value of 2 with the difference between the results obtained with the high and low value 1 at the high value of 2. The 12-interaction is

12 interaction = $[(y_4 - y_3 + y_8 - y_7)$
$- [(y_2 - y_1 + y_6 - y_5)]/2$

The key step is to determine the errors associated with the effect of each variable and each interaction so that the significance can be determined. Thus, standard errors need to be assigned. This can be done by repeating the experiments, but it can also be done by using higher order interactions (such as 123 interactions in a 2^4 factorial design). These are assumed negligible in their effect on the mean but can be used to estimate the standard error [11.1, pp. 319–328]. Then calculated effects that are large compared to the standard error are considered important, whereas those that are small compared to the standard error are considered due to random variations and are unimportant.

In a fractional factorial design, only part of the possible experiments is performed. With k variables, a factorial design requires 2^k experiments. When k is large, the number of experiments can be large; for $k = 5$, $2^5 = 32$. For k this large, Box et al. [11.1, p. 376] do a fractional factorial design. In the fractional factorial design with $k = 5$, only 16 experiments are chosen. CROPLEY [11.7] gives an example of how to combine heuristics and statistical arguments in application to kinetics mechanisms in chemical engineering.

12. Multivariable Calculus Applied to Thermodynamics

Many of the functional relationships required in thermodynamics are direct applications of the rules of multivariable calculus. In this short chapter, those rules are reviewed in the context of the needs of thermodynamics. These ideas were expounded in one of the classic books on chemical engineering thermodynamics [12.1].

12.1. State Functions

State functions depend only on the state of the system, not on its past history or how one got there. If z is a function of two variables x and y, then $z(x, y)$ is a state function, because z is known once x and y are specified. The differential of z is

$$dz = M\,dx + N\,dy$$

By the theorem on p. 1-32 the line integral

$$\int_C (M\,dx + N\,dy)$$

is independent of the path in x–y space if and only if

$$\frac{\partial M}{\partial y} = \frac{\partial N}{\partial x} \tag{81}$$

Because the total differential can be written as

$$dz = \left(\frac{\partial z}{\partial x}\right)_y dx + \left(\frac{\partial z}{\partial y}\right)_x dy \tag{82}$$

for path independence

$$\frac{\partial}{\partial y}\left(\frac{\partial z}{\partial x}\right)_y = \frac{\partial}{\partial x}\left(\frac{\partial z}{\partial y}\right)_x$$

or

$$\frac{\partial^2 z}{\partial y\,\partial x} = \frac{\partial^2 z}{\partial x\,\partial y} \tag{83}$$

is needed.

Various relationships can be derived from Equation (82). If z is constant,

$$\left[0 = \left(\frac{\partial z}{\partial x}\right)_y dx + \left(\frac{\partial z}{\partial y}\right)_x dy\right]_z$$

Rearrangement gives

$$\left(\frac{\partial z}{\partial x}\right)_y = -\left(\frac{\partial y}{\partial x}\right)_z \left(\frac{\partial z}{\partial y}\right)_x = -\frac{(\partial y/\partial x)_z}{(\partial y/\partial z)_x} \tag{84}$$

Alternatively, if Equation (82) is divided by dy while some other variable w is held constant,

$$\left(\frac{\partial z}{\partial y}\right)_w = \left(\frac{\partial z}{\partial x}\right)_y \left(\frac{\partial x}{\partial y}\right)_w + \left(\frac{\partial z}{\partial y}\right)_x \tag{85}$$

Dividing both the numerator and the denominator of a partial derivative by dw while holding a variable y constant yields

$$\left(\frac{\partial z}{\partial x}\right)_y = \frac{(\partial z/\partial w)_y}{(\partial x/\partial w)_y} = \left(\frac{\partial z}{\partial w}\right)_y \left(\frac{\partial w}{\partial x}\right)_y \tag{86}$$

In thermodynamics the state functions include the internal energy U, the enthalpy H, and the Helmholtz and Gibbs free energies A and G, respectively, which are defined as follows:

$$H = U + pV$$
$$A = U - TS$$
$$G = H - TS = U + pV - TS = A + pV$$

where S is the entropy, T the absolute temperature, p the pressure, and V the volume. These are also state functions, in that the entropy is specified once two variables (e.g., T and p) are specified. Likewise V is specified once T and p are specified, and so forth.

12.2. Applications to Thermodynamics

All of the following applictions are for closed systems with constant mass. If a process is reversible and only p–V work is done, one form of the first law states that changes in the internal energy are given by the following expression

$$dU = T\,dS - p\,dV \tag{87}$$

If the internal energy is considered a function of S and V, then

$$dU = \left(\frac{\partial U}{\partial S}\right)_V dS + \left(\frac{\partial U}{\partial V}\right)_S dV$$

This is the equivalent of Equation (82) and

$$T = \left(\frac{\partial U}{\partial S}\right)_V, \quad p = -\left(\frac{\partial U}{\partial V}\right)_S$$

Because the internal energy is a state function, Equation (83) is required:

$$\frac{\partial^2 U}{\partial V \partial S} = \frac{\partial^2 U}{\partial S \partial V}$$

which here is

$$\left(\frac{\partial T}{\partial V}\right)_S = -\left(\frac{\partial p}{\partial S}\right)_V \quad (88)$$

This is one of the Maxwell relations and is merely an expression of Equation (83).

The differentials of the other energies are

$$dH = T dS + V dp \quad (89)$$
$$dA = -S dT - p dV \quad (90)$$
$$dG = -S dT + V dp \quad (91)$$

From these differentials, other Maxwell relations can be derived in a similar fashion by applying Equation (83).

$$\left(\frac{\partial T}{\partial p}\right)_S = \left(\frac{\partial V}{\partial S}\right)_p \quad (92)$$

$$\left(\frac{\partial S}{\partial V}\right)_T = \left(\frac{\partial p}{\partial T}\right)_V \quad (93)$$

$$\left(\frac{\partial S}{\partial p}\right)_T = -\left(\frac{\partial V}{\partial T}\right)_p \quad (94)$$

The heat capacity at constant pressure is defined as

$$C_p = \left(\frac{\partial H}{\partial T}\right)_p$$

If entropy and enthalpy are taken as functions of T and p, the total differentials are

$$dS = \left(\frac{\partial S}{\partial T}\right)_p dT + \left(\frac{\partial S}{\partial p}\right)_T dp$$

$$dH = \left(\frac{\partial H}{\partial T}\right)_p dT + \left(\frac{\partial H}{\partial p}\right)_T dp$$

$$= C_p dT + \left(\frac{\partial H}{\partial p}\right)_T dp$$

If the pressure is constant,

$$dS = \left(\frac{\partial S}{\partial T}\right)_p dT \text{ and } dH = C_p dT$$

When enthalpy is considered a function of S and p, the total differential is

$$dH = T dS + V dp$$

When the pressure is constant, this is

$$dH = T dS$$

Thus, at constant pressure

$$dH = C_p dT = T dS = T\left(\frac{\partial S}{\partial T}\right)_p dT$$

which gives

$$\left(\frac{\partial S}{\partial T}\right)_p = \frac{C_p}{T}$$

When p is not constant, using the last Maxwell relation gives

$$dS = \frac{C_p}{T} dT - \left(\frac{\partial V}{\partial T}\right)_p dp \quad (95)$$

Then the total differential for H is

$$dH = T dS + V dp = C_p dT - T\left(\frac{\partial V}{\partial T}\right)_p dp + V dp$$

Rearranging this, when $H(T, p)$, yields

$$dH = C_p dT + \left[V - T\left(\frac{\partial V}{\partial T}\right)_p\right] dp \quad (96)$$

This equation can be used to evaluate enthalpy differences by using information on the equation of state and the heat capacity:

$$H(T_2, p_2) - H(T_1, p_1) = \int_{T_1}^{T_2} C_p(T, p_1) dT$$
$$+ \int_{p_1}^{p_2} \left[V - T\left(\frac{\partial V}{\partial T}\right)_p\right]_{T_2, p} dp \quad (97)$$

The same manipulations can be done for internal energy:

$$\left(\frac{\partial S}{\partial T}\right)_V = \frac{C_v}{T} \quad (98)$$

$$dS = -\left[\frac{(\partial V/\partial T)_p}{(\partial V/\partial p)_T}\right] dV + \frac{C_v}{T} dT \quad (99)$$

$$dU = C_v dT - \left[p + T\frac{(\partial V/\partial T)_p}{(\partial V/\partial p)_T}\right] dV$$

12.3. Partial Derivatives of All Thermodynamic Functions

The various partial derivatives of the thermodynamic functions can be classified into six groups. In the general formulas below, the variables U, H, A, G, or S are denoted by Greek letters, whereas the variables V, T, or p are denoted by Latin letters.

Type 1 (3 possibilities plus reciprocals).

General: $\left(\dfrac{\partial a}{\partial b}\right)_c$, Specific: $\left(\dfrac{\partial p}{\partial T}\right)_V$

Equation (84) yields

$$\left(\frac{\partial p}{\partial T}\right)_V = -\left(\frac{\partial V}{\partial T}\right)_p \left(\frac{\partial p}{\partial V}\right)_T = -\frac{(\partial V/\partial T)_p}{(\partial V/\partial p)_T} \quad (100)$$

This relates all three partial derivatives of this type.

Type 2 (30 possibilities).

General: $\left(\dfrac{\partial \alpha}{\partial b}\right)_c$, Specific: $\left(\dfrac{\partial G}{\partial T}\right)_V$

Using Equation (91) gives

$$\left(\frac{\partial G}{\partial T}\right)_V = -S + V\left(\frac{\partial p}{\partial T}\right)_V$$

Using the other equations for U, H, A, or S gives the other possibilities.

Type 3 (15 possibilities plus reciprocals).

General: $\left(\dfrac{\partial a}{\partial b}\right)_\alpha$, Specific: $\left(\dfrac{\partial V}{\partial T}\right)_S$

First the derivative is expanded by using Equation (84), which is called expansion without introducing a new variable:

$$\left(\frac{\partial V}{\partial T}\right)_S = -\left(\frac{\partial S}{\partial T}\right)_V \left(\frac{\partial V}{\partial S}\right)_T = -\frac{(\partial S/\partial T)_V}{(\partial S/\partial V)_T}$$

Then the numerator and denominator are evaluated as type 2 derivatives, or by using Equations (98) and (99):

$$\left(\frac{\partial V}{\partial T}\right)_S = -\frac{C_v/T}{-(\partial V/\partial T)_p (\partial p/\partial V)_T} = \frac{C_v}{T}\frac{\left(\dfrac{\partial V}{\partial p}\right)_T}{\left(\dfrac{\partial V}{\partial T}\right)_p} \quad (101)$$

These derivatives are important for reversible, adiabatic processes (e.g., in an ideal turbine or compressor) because the entropy is constant. Similar derivatives can be obtained for isenthalpic processes, such as a pressure reduction at a valve. In that case, the Joule–Thomson coefficient is obtained:

$$\left(\frac{\partial T}{\partial p}\right)_H = \frac{1}{C_p}\left[-V + T\left(\frac{\partial V}{\partial T}\right)_p\right]$$

Type 4 (30 possibilities plus reciprocals).

General: $\left(\dfrac{\partial \alpha}{\partial \beta}\right)_c$, Specific: $\left(\dfrac{\partial G}{\partial A}\right)_p$

Now, expand through the introduction of a new variable using Equation (86):

$$\left(\frac{\partial G}{\partial A}\right)_p = \left(\frac{\partial G}{\partial T}\right)_p \left(\frac{\partial T}{\partial A}\right)_p = \frac{(\partial G/\partial T)_p}{(\partial A/\partial T)_p}$$

This operation has created two type 2 derivatives. Substitution yields

$$\left(\frac{\partial G}{\partial A}\right)_p = \frac{S}{S + p(\partial V/\partial T)_p}$$

Type 5 (60 possibilities).

General: $\left(\dfrac{\partial \alpha}{\partial b}\right)_\beta$, Specific: $\left(\dfrac{\partial G}{\partial p}\right)_A$

Starting from Equation (91) for dG gives

$$\left(\frac{\partial G}{\partial p}\right)_A = -S\left(\frac{\partial T}{\partial p}\right)_A + V$$

The derivative is a type 3 derivative and can be evaluated by using Equation (84).

$$\left(\frac{\partial G}{\partial p}\right)_A = S\frac{(\partial A/\partial p)_T}{(\partial A/\partial T)_p} + V$$

The two type 2 derivatives are then evaluated:

$$\left(\frac{\partial G}{\partial p}\right)_A = \frac{Sp(\partial V/\partial p)_T}{S + p(\partial V/\partial T)_p} + V$$

These derivatives are also of interest for free expansions or isentropic changes.

Type 6 (30 possibilities plus reciprocals).

General: $\left(\dfrac{\partial \alpha}{\partial \beta}\right)_\gamma$, Specific: $\left(\dfrac{\partial G}{\partial A}\right)_H$

Equation (86) is used to obtain two type 5 derivatives.

$$\left(\frac{\partial G}{\partial A}\right)_H = \frac{(\partial G/\partial T)_H}{(\partial A/\partial T)_H}$$

These can then be evaluated by using the procedures for type 5 derivatives.

The difference in molar heat capacities $(C_p - C_v)$ can be derived in similar fashion. Using Equation (97) for C_v yields

$$C_v = T\left(\frac{\partial S}{\partial T}\right)_V$$

To evaluate the derivative, Equation (85) is used to express dS in terms of p and T:

$$\left(\frac{\partial S}{\partial T}\right)_V = -\left(\frac{\partial V}{\partial T}\right)_p \left(\frac{\partial p}{\partial T}\right)_V + \frac{C_p}{T}$$

Substitution for $(\partial p/\partial T)_v$ and rearrangement give

$$C_p - C_v = T\left(\frac{\partial V}{\partial T}\right)_p \left(\frac{\partial p}{\partial T}\right)_V = -T\left(\frac{\partial V}{\partial T}\right)_p^2 \left(\frac{\partial p}{\partial V}\right)_T$$

Use of this equation permits the rearrangement of Equation (101) into

$$\left(\frac{\partial V}{\partial T}\right)_S = \frac{(\partial V/\partial T)_p^2 + \dfrac{C_p}{T}(\partial V/\partial p)_T}{(\partial V/\partial T)_p}$$

The ratio of heat capacities is

$$\frac{C_p}{C_v} = \frac{T(\partial S/\partial T)_p}{T(\partial S/\partial T)_v}$$

Expansion by using Equation (84) gives

$$\frac{C_p}{C_v} = \frac{-(\partial p/\partial T)_S (\partial S/\partial p)_T}{-(\partial V/\partial T)_S (\partial S/\partial V)_T}$$

and the ratios are then

$$\frac{C_p}{C_v} = \left(\frac{\partial p}{\partial V}\right)_S \left(\frac{\partial V}{\partial p}\right)_T$$

Using Equation (84) gives

$$\frac{C_p}{C_v} = -\left(\frac{\partial p}{\partial V}\right)_S \left(\frac{\partial T}{\partial p}\right)_V \left(\frac{\partial V}{\partial T}\right)_p$$

Entropy is a variable in at least one of the partial derivatives.

13. References

References for Chapter 1

General references

    J. J. Dongarra, C. B. Moler, J. R. Bunch, G. W. Stewart: *Linpack User's Guide*, SIAM, Philadelphia 1979.
    G. Forsyth, C. B. Moler: *Computer Solution of Lineart Algebraic Systems*, Prentice-Hall, Englewood Cliffs 1967.

Specific references

[1.1] E. Isaacson, H. B. Keller, *Analysis of Numerical Methods*, J. Wiley and Sons, New York 1966.
[1.2] S. C. Eisenstat, M. H. Schultz, A. H. Sherman: "Algorithms and Data Structures for Sparse Symmetric Gaussian Elimination," *SIAM J. Sci. Stat. Comput.* **2** (1981) 225–237.
[1.3] V. A. Barker (ed.): *Sparse Matrix Techniques – Copenhagen 1976*, Lecture Notes in Mathematics 572, Springer Verlag, Berlin – Heidelberg – New York – Tokyo 1977.
[1.4] J. R. Bunch, D. J. Rose (ed.): *Sparse Matrix Computations*, Academic Press, New York 1976.
[1.5] I. S. Duff (ed.): *Sparse Matrices and Their Uses*, Academic Press, New York 1981.
[1.6] I. S. Duff: *Direct Methods for Sparse Matrices*, Charendon Press, Oxford 1986.
[1.7] J. R. Rice: *Matrix Computations and Mathematical Software*, McGraw-Hill, New York 1983.
[1.8] H. S. Price, K. H. Coats, "Direct Methods in Reservoir Simulation," *Soc. Pet. Eng. J.* **14** (1974) 295–308.
[1.9] A. Bykat: "A Note on an Element Re-Ordering Scheme," *Int. J. Num. Methods Egn.* **11** (1977) 194–198.
[1.10] M. R. Hestness, E. Stiefel: "Methods of conjugate gradients for solving linear systems," *J. Res. Nat. Bur. Stand* **29** (1952) 409–439.
[1.11] S. C. Eisenstat: "Efficient Implementation of a Class of Preconditioned Conjugate Gradient Methods," *SIAM J. Sci. Stat. Comput.* **2** (1981) 1–4.
[1.12] M. R. Hestenes: *Conjugate Gradient Methods in Optimization*, Springer-Verlag, Berlin – Heidelberg – New York – Tokyo 1980.
[1.13] A. McIntosh: *Fitting Linear Models: an Application of Conjugate Graduent Algorithms*, Springer Verlag, Berlin – Heidelberg – New York – Tokyo 1982.
[1.14] W. H. Press, B. P. Flannery, S. A. Teukolsky, W. T. Vetterling: *Numerical Recipes*, Cambridge University Press, Cambridge 1986.
[1.15] B. A. Finlayson: *Nonlinear Analysis in Chemical Engineering*, McGraw-Hill, New York 1980.
[1.16] R. W. H. Sargent: "A Review of Methods for Solving Non-linear Algebraic Equations," in R. S. H. Mah, W. D. Seider (eds.): *Foundations of Computer-Aided Chemical Process Design*, American Institute of Chemical Engineers, New York 1981.
[1.17] J. D. Seader: "Computer Modeling of Chemical Processes," *AIChE Monogr. Ser.* **81** (1985) no. 15.
[1.18] N. R. Amundson: *Mathematical Methods in Chemical Engineering*, Prentice-Hall, Englewood Cliffs, N.J. 1966.
[1.19] J. H. Perry: *Chemical Engineers' Handbook*, 6th ed., McGraw-Hill, New York 1984.
[1.20] D. S. Watkins: "Understanding the QR Algorithm," *SIAM Rev.* **24** (1982) 427–440.
[1.21] G. F. Carey, K. Sepehrnoori: "Gershgorin Theory for Stiffness and Stability of Evolution Systems and Convection-Diffusion," *Comp. Meth. Appl. Mech.* **22** (1980) 23–48.

References for Chapter 2

[2.1] M. Abranowitz, I. A. Stegun: *Handbook of Mathematical Functions*, National Bureau of Standards, Washington, D.C. 1964.
[2.2] W. H. Press, B. P. Flannery, S. A. Teukolsky, W. T. Vetterling: *Numerical Recipes*, Cambridge University Press, Cambridge 1986.
[2.3] J. C. Daubisse: "Some Results about Approximation of Functions of One or Two Variables by Sums of Exponentials," *Int. J. Num. Meth. Eng.* **23** (1986) 1959–1967.
[2.4] L. Lapidus, J. H. Seinfeld: *Numerical Solution of Ordinary Differential Equations*, Academic Press, New York 1971.
[2.5] B. A. Finlayson: *Nonlinear Analysis in Chemical Engineering*, McGraw-Hill, New York 1980.
[2.6] J. M. Ortega, W G. Poole Jr.: *An Introduction to Numerical Methods for Differential Equations*, Pitman, Marshfield, Mass. 1981.
[2.7] R. Courant, D. Hilbert: *Methods of Mathematical Physics*, vol. 1, Wiley-Interscience, New-York 1953.

References for Chapter 3

[3.1] E. Hille: *Analytic Function Theory*, Ginn and Co., Boston 1959.
[3.2] H. F. Weinberger: *A First Course in Partical Differential Equations*, Blaisdell, Waltham, Mass. 1965.
[3.3] R. V. Churchill: *Operational Mathematics*, McGraw-Hill, New York 1958.

References for Chapter 4

[4.1] H. F. Weinberger: *A First Course in Partical Differential Equations*, Blaisdell, Waltham, Mass. 1965.

[4.2] W. H. Press, B. P. Flanner, S. A. Teukolsky, W. T. Vetterling: *Numerical Recipes*, Cambridge Univ. Press, Cambridge 1986.
[4.3] R. V. Churchill: *Operational Mathematics*, McGraw-Hill, New York 1958.
[4.4] J. T. Hsu, J. S. Dranoff: "Numerical Inversion of Certain Laplace Transforms by the Direct Application of Fast Fourier Transform (FFT) Algorithm," *Comput. Chem. Eng.* **11** (1987) 101–110.
[4.5] M. Abramowitz, I. A. Stegun: *Handbook of Mathematical Functions*, National Bureau of Standards, Washington, D.C. 1964.
[4.6] E. Kreyszig: *Advanced Engineering Mathematics*, 4th ed., J. Wiley and Sons, New York 1979.
[4.7] H. S. Carslaw, J. C. Jaeger: *Conduction of Heat in Solids*, 2nd ed., Clarendon Press, Oxford – London 1959.

References for Chapter 5

[5.1] R. B. Bird, R. C. Armstrong, O. Hassager: *Dynamics of Polymeric Liquids*, 2nd ed., Appendix A, Wiley-Interscience, New York 1987.
[5.2] R. S. Rivlin, *J. Rat. Mech. Anal.* **4** (1955) 681–702.
[5.3] N. R. Amundson: *Mathematical Methods in Chemical Engineering; Matrices and Their Application*, Prentice-Hall, Englewood Cliffs, N.J. 1966.
[5.4] E. Kreyszig: *Advanced Engineering Mathematics*, 4th ed., J. Wiley and Sons, New York 1979.
[5.5] I. S. Sokolnikoff, E. S. Sokolnikoff: *Higher Mathematics for Engineers and Physicists*, McGraw-Hill, New York 1941.
[5.6] M. R. Spiegel: *Vector Analysis*, Schaum Publishing, New York 1959.
[5.7] L. E. Scriven: personal communication.
[5.8] P. M. Morse, H. Feshbach: *Methods of Theoretical Physics*, McGraw-Hill, New York 1953.

References of Chapter 6

[6.1] B. A. Finlayson: *Nonlinear Analysis in Chemical Engineering*, McGraw Hill, New York 1980.
[6.2] L. Lapidus, J. H. Seinfeld: *Numerical Solution of Ordinary Differential Equations*, Academic Press, New York 1971.
[6.3] G. Forsythe, M. Malcolm, C. Moler: *Computer Methods for Mathematical Computation*, Prentice-Hall, Englewood Cliffs, N.J. 1977.
[6.4] N. R. Amundson: *Mathematical Methods in Chemical Engineering*, Prentice-Hall, Englewood Cliffs, N.J. 1966.
[6.5] J. R. Rice: *Numerical Methods, Software, and Analysis*, McGraw-Hill, New York 1983.
[6.6] M. B. Bogacki, K. Alejski, J. Szymanowski: "The Fast Method of the Solution of Reacting Distillation Problem," *Comput. Chem. Eng.* **13** (1989) 1081–1085.
[6.7] C. W. Gear: *Numerical Initial-Value Problems in Ordinary Differential Equations*, Prentice-Hall, Englewood Cliffs, N.J. 1971.
[6.8] B. Carnahan, J. O. Wilkes: "Numerical Solution of Differential Equations – An Overview," in *Foundations of Computer-Aided Chemical Process Design*, American Institute of Chemical Engineers, New York 1981, pp. 225–340.
[6.9] N. B. Ferguson, B. A. Finlayson: "Transient Modeling of a Catalytic Converter to Reduce Nitric Oxide in Automobile Exhaust," *AIChE J* **20** (1974) 539–550.
[6.10] I. M. Smith, J. L. Siemienivich, I. Gladwell: "A Comparison of Old and New Methods for Large Systems of Ordinary Differential Equations Arising from Parabolic Partial Differential Equations," *Num. Anal. Rep.* no. 13, Department of Engineering, University of Manchester, 1975.
[6.11] J. D. Lambert: *Computational Methods in Ordinary Differential Equations*, J. Wiley and Sons, New York 1973.
[6.12] L. R. Petzold: "A description of DASSL: a Differential-Algebraic System Solver," *Sandia Nat. Lab. Rech. Rep.* **SAND 82-8637**; also in R. S. Stepleman et al. (eds.): *IMACS Trans. on Scientific Computing*, vol. 1, North Holland Publ. Co, Amsterdam 1983, pp. 65–68.
[6.13] C. C. Pontelides, D. Gritsis, K. R. Morison, R. W. H. Sargent: "The Mathematical Modelling of Transient Systems Using Differential-Algebraic Equations," *Comput. Chem. Eng.* **12** (1988) 449–454.
[6.14] G. A. Byrne, P. R. Ponzi: "Differential-Algebraic Systems, Their Applications and Solutions," *Comput. Chem. Eng.* **12** (1988) 377–382.
[6.15] A. C. Hindmarsh: "LSODE and LSODI, Two New Initial Value Ordinary Differential Equations Solvers," *ACM SIGNUM Newsletter* **15** (1980) 10–11.
[6.16] A. C. Hindmarsh: "GEARB: Solution of Ordinary Differential Equations Having Banded Jacobian," *Computer Documentation*, UCID-30059, Rev. 1, Lawrence Livermore Laboratory, University of California, 1975.
[6.17] T. E. Hull, W. H. Enright, B. M. Fellen, A. E. Sedgwick: "Comparing Numerical Methods for Ordinary Differential Equations," *SIAM J. Num. Anal.* **9** (1972) 603–637.
[6.18] L. F. Shampine, H. A. Watts, S. M. Davenport: "Solving Nonstiff Ordinary Differential Equations – The State of the Art," *SIAM Rev.* **18** (1976) 376–411.
[6.19] G. Byrne, A. Hindmarsh, K. R. Jackson, H. G. Brown: "A Comparison of Two ODE Codes: GEAR and EPISODE," *Comput. Chem. Eng.* **1** (1977) 133–147.
[6.20] G. Junce, V. Lauric, R. Mihail: "A Comparison Between the Insensitive Runge–Kutta, The Semi-Implicit Extrapolation and the Backward Differencing Methods in Solving the ODE Systems Which Describe the Radical Hydrocarbon Pyrolysis," *Comput. Chem. Eng.* **13** (1989) 1075–1079.
[6.21] W. F. Ramirez: *Computational Methods for Process Simulations*, Butterworths, Boston 1989.
[6.22] M. Kubicek, M. Marek: *Computational Methods in Bifurcation Theory and Dissipative Structures*, Springer Verlag, Berlin – Heidelberg – New York – Tokyo 1983.
[6.23] T. F. C. Chan, H. B. Keller: "Arc-Length Continuation and Multi-Grid Techniques for Nonlinear Elliptic Eigenvalue Problems," *SIAM J. Sci. Stat. Comput.* **3** (1982) 173–194.
[6.24] M. F. Doherty, J. M. Ottino: "Chaos in Deterministic Systems: Strange Attractors, Turbulence and Applications in Chemical Engineering," *Chem. Eng. Sci.* **43** (1988) 139–183.
[6.25] K. L. Hiebert, L. F. Shampine: "Implicity Defined Output Points for Solutions of Ordinard Differential Equations," *Sandia Nat. Lab. Tech. Rep.*

SAND 80-180 (1980); SAND 84-0812; SAND 83-2560.

[6.26] H. Stern, B. A. Finlayson: *On Terminating the Integration of ODE's Based on the Value of the Dependent Variable,* Unpublished report, April, 10, 1985, University of Washington.

References for Chapter 7

[7.1] R. B. Bird, W. E. Stewart, E. N. Lightfoot: *Transport Phenomena,* J. Wiley and Sons, New York 1960.
[7.2] R. B. Bird, R. C. Armstrong, O. Hassager: *Dynamics of Polymeric Liquids,* 2nd ed., Wiley-Interscience, New York 1987.
[7.3] H. B. Keller: *Numerical Methods for Two-Point Boundary-Value Problems,* Blaisdell, New York 1972.
[7.4] P. B. Weisz, J. S. Hicks: "The Behavior of Porous Catalyst Particles in View of Internal Mass and Heat Diffusion Effects," *Chem. Eng. Sci.* **17** (1962) 265–275.
[7.5] P. V. Danckwerts: "Continuous Flow Systems," *Chem. Eng. Sci.* **2** (1953) 1–13.
[7.6] J. F. Wehner, R. Wilhelm: "Boundary Conditions of Flow Reactor," *Chem. Eng. Sci.* **6** (1956) 89–93.
[7.7] V. Hlaváček, H. Hofmann: "Modeling of Chemical Reactors-XVI-Steady-State Axial Heat and Mass Transfer in Tubular Reactors. An Analysis of the Uniqueness of Solutions," *Chem. Eng. Sci.* **25** (1970) 173–185.
[7.8] J. R. Rice: *Numerical Methods, Software, and Analysis,* McGraw-Hill, New York 1983.
[7.9] B. A. Finlayson: *Numerical Methods for Problems with Moving Fronts,* to be published, 1990.
[7.10] B. A. Finlayson: *Nonlinear Analysis in Chemical Engineering,* McGraw-Hill, New York 1980.
[7.11] E. Isaacson, H. B. Keller: *Analysis of Numerical Methods,* J. Wiley and Sons, New York 1966.
[7.12] C. Lanczos: "Trigonometric Interpolation of Empirical and Analytical Functions," *J. Math. Phys. (Cambridge Mass.)* **17** (1938) 123–199.
[7.13] C. Lanczos: *Applied Analysis,* Prentice-Hall, Englewood Cliffs, N.J. 1956.
[7.14] J. Villadsen, W. E. Stewart: "Solution of Boundary-Value Problems by Orthogonal Collocation," *Chem. Eng. Sci.* **22** (1967) 1483–1501.
[7.15] M. L. Michelsen, J. Villadsen: "Polynomial Solution of Differential Equations" pp. 341–368 in R. S. H. Mah, W. D. Seider (eds.): *Foundations of Computer-Aided Chemical Process Design,* Engineering Foundation, New York 1981.
[7.16] W. E. Stewart, K. L. Levien, M. Morari: "Collocation Methods in Distillation," in A. W. Westerber, H. H. Chien (eds.): *Proceedings of the Second Int. Conf. on Computer-Aided Process Design,* Computer Aids for Chemical Engineering Education (CACHE), Austin Texas, 1984, pp. 535–569.
[7.17] W. E. Stewart, K. L. Levien, M. Morari: "Simulation of Fractionation by Orthogonal Collocation," *Chem. Eng. Sci.* **40** (1985) 409–421.
[7.18] C. L. E. Swartz, W. E. Stewart: "Finite-Element Steady State Simulation of Multiphase Distillation," *AIChE J.* **33** (1987) 1977–1985.
[7.19] C. deBoor, B. Swartz: "Collocation at Gaussian Points," *SIAM J. Num. Anal.* **10** (1973) 582–606.
[7.20] R. F. Sincovec: "On the Solution of the Equations Arising From Collocation With Cubic B-Splines," *Math. Comp.* **26** (1972) 893–895.
[7.21] P. M. Prenter, *Splines and Variational Methods,* J. Wiley and Sons, New York 1975.
[7.22] C. E. Pearson: "On a Differential Equation of Boundary Layer-Type," *J. Math. Phys. (Cambridge Mass.)* **47** (1968) 351–358.
[7.23] U. Ascher, J. Christiansen, R. D. Russell: "A Collocation Solver for Mixed-Order Systems of Boundary-Value Problems," *Math. Comp.* **33** (1979) 659–679.
[7.24] R. D. Russell, J. Christiansen: "Adaptive Mesh Selection Strategies for Solving Boundary-Value Problems," *SIAM J. Num. Anal.* **15** (1978) 59–80.
[7.25] P. G. Ciarlet, M. H. Schultz, R. S. Varga: "Nonlinear Boundary-Value Problems I. One Dimensional Problem," *Num. Math.* **9** (1967) 394–430.
[7.26] W. F. Ames: *Numerical Methods for Partial Differential Equations,* 2nd ed., Academic Press, New York 1977.
[7.27] J. F. Botha, G. F. Pinder: *Fundamental Concepts in The Numerical Solution of Differential Equations,* Wiley-Interscience, New York 1983.
[7.28] W. H. Press, B. P. Flanner, S. A. Teukolsky, W. T. Vetterling: *Numerical Recipes,* Cambridge Univ. Press, Cambridge 1986.
[7.29] H. Schlichting, *Boundary Layer Theory,* 4th ed. McGraw-Hill, New York 1960.
[7.30] B. Carnahan, H. A. Luther, J. O. Wilkes: *Applied Numerical Methods,* J. Wiley and Sons, New York 1969.

References for Chapter 8

[8.1] L. Lapidus, G. F. Pinter: *Numerical Solution of Partial Differential Equations in Science and Engineering,* Wiley-Interscience, New York 1982.
[8.2] B. Carnahan, J. O. Wilkes: "Numerical Solution of Differential Equations – An Overview" in *Foundations of Computer-Aided Chemical Process Design,* vol. 1, Engineering Foundation, New York 1981, pp. 225–340.
[8.3] W. F. Ramirez: *Computational Methods for Process Simulation,* Butterworths, Boston 1989.
[8.4] D. D. Joseph, M. Renardy, J. C. Saut: "Hyperbolicity and Change of Type in the Flow of Viscoelastic Fluids," *Arch. Rational Mech. Anal.* **87** (1985) 213–251.
[8.5] H. K. Rhee, R. Aris, N. R. Amundson: *First-Order Partial Differential Equations,* Prentice-Hall, Englewood Cliffs, N.J. 1986.
[8.6] B. A. Finlayson: *Numerical Methods for Problems with Moving Fronts,* to be published, 1990.
[8.7] D. L. Book: *Finite-Difference Techniques for Vectorized Fluid Dynamics Calculations,* Springer Verlag, Berlin – Heidelberg – New York – Tokyo 1981.
[8.8] G. A. Sod: *Numerical Methods in Fluid Dynamics,* Cambridge University Press, Cambridge 1985.
[8.9] B. A. Finlayson: *Nonlinear Analysis in Chemical Engineering,* McGraw-Hill, New York 1980.
[8.10] W. F. Ames: "Recent Developments in the Nonlinear Equations of Transport Processes," *Ind. Eng. Chem. Fundam.* **8** (1969) 522–536.
[8.11] W. F. Ames: *Nonlinear Partial Differential Equations in Engineering,* Academic Press, New York 1965.

[8.12] M. Abramowitz, I. A. Stegun: *Handbook of Mathematical Functions*, National Bureau of Standards, Washington, D.C. 1964.
[8.13] M. B. Carver: "Methods of Lines Solution of Differential Equations. Fundamental Principles and Recent Extensions" in *Foundations of Computer-Aided Chemical Process Design*, vol. 1, Engineering Foundation, New York 1981, pp. 369–402.
[8.14] E. Isaacson, H. B. Keller: *Analysis of Numerical Methods*, J. Wiley and Sons, New York 1966.
[8.15] J. C. Pirkle, I. E. Wachs, J. E. Sobel: "Numerical Methods for Simulation of Fixed-Bed Reactors for Complex Exothermic Reactions" in *Foundations of Computer-Aided Chemical Process Design*, vol. 2, Engineering Foundation, New York 1981, pp. 401–429.
[8.16] D. Gottlieb, S. A. Orszag: *Numerical Analysis of Spectral Methods: Theory and Applications*, SIAM, Philadelphia, PA 1977.
[8.17] J. R. Rice: *Numerical Methods, Software, and Analysis*, McGraw-Hill, New York 1983.
[8.18] M. E. Davis: *Numerical Methods and Modeling for Chemical Engineers*, J. Wiley and Sons, New York 1984.
[8.19] D. W. Peaceman: *Fundamentals of Numerical Reservoir Simulation*, Elsevier, Amsterdam 1977.
[8.20] G. Juncu, R. Mihail: "Multigrid Solution of the Diffusion-Convection-Reaction Equations which Describe the Mass and/or Heat Transfer from or to a Spherical Particle," *Comput. Chem. Eng.* **13** (1989) 259–270.
[8.21] G. Strang, G. J. Fix: *An Analysis of the Finite Element Method*, Prentice-Hall, Englewood Cliffs, N.J. 1973.
[8.22] E. B. Becker, G. F. Carey, J. T. Oden: *Finite Elements. An Introduction*, vol. 1, Prentice-Hall, Englewood Cliffs, N.J. 1981.
[8.23] J. E. Akin: *Finite Element Analysis for Undergraduates*, Academic Press, New York 1986.
[8.24] P. Hood, "Frontal Solution Program for Unsymmetric Matrices," *Int. J. Num. Methods Eng.* **10** (1976) 379–399; **11** (1977) 1202.
[8.25] M. W. Chang, B. A. Finlayson: "On the Proper Boundary Condition for the Thermal Entry Problem," *Int. J. Num. Methods Eng.* **15** (1980) 935–942.
[8.26] H. H. Klein: "Computer Modeling of Chemical Process Reactors," in A. W. Westerberg, H. H. Chien (eds.): *Proceedings of the Second International Conference on Foundations of Computer-Aided Process Design*, Computer Aids for Chemical Engineering Education (CACHE), Austin, Texas 1984, pp. 571–661.
[8.27] J. C. Pirkl Jr., S. C. Reyes, P. S. Hagan, H. Kheshgi, W. E. Schiesser: "Solution of Dynamic Distributed Parameter Model of Nonadiabatic, Fixed-Bed Reactor," *Comput. Chem. Eng.* **11** (1987) 737–747.
[8.28] J. Degreve, P. Dimitriou, J. Puszynski, V. Hlavacek: "Use of 2-D-Adaptive Mesh in Simulation of Combustion Front Phenomena," *Comput. Chem. Eng.* **11** (1987) 749–755.
[8.29] E. N. Houstis, W. F. Mitchell, T. S. Papatheodoros: "Performance Evaluation of Algorithms for Mildly Nonlinear Elliptic Problems," *Int. J. Num. Methods Eng.* **19** (1983) 665–709.
[8.30] R. Peyret, T. D. Taylor: *Computational Methods for Fluid Flow*, Springer Verlag, Berlin – Heidelberg – New York – Tokyo 1983.
[8.31] P. M. Gresho, S. T. Chan, C. Upson, R. L. Lee: "A Modified Finite Element Method for Solving the Time-Dependent, Incompressible Navier–Stokes Equations," *Int. J. Num. Method. Fluids* **4** 'Part 1. Theory", pp. 557–589; "Part 2: Applications," pp. 619–640 (1984).
[8.32] P. M. Gresho, S. Chan: "A New Semi-Implicit Method for Solving the Time-Dependent Conservation Equations for Incompressible Flow," *Proc. Num. Methods in Laminar and Turbulent Flow*, Pineridge Press, Ltd., Swansea, U.K. 1985.

References for Chapter 9

[9.1] C. T. H. Baker: *The Numerical Treatment of Integral Equations*, Clarendon Press, Oxford 1977.
[9.2] L. M. Delves, J. Walsh (eds.): *Numerical Solution of Integral Equations*, Clarendon Press, Oxford 1974.
[9.3] P. Linz: *Analytical and Numerical Methods for Volterra Equations*, SIAM Publications, Philadelphia 1985.
[9.4] M. A. Golberg (ed.): *Numerical Solution of Integral Equations*, Plenum Press, New York 1990.
[9.5] G. Nagel, G. Kluge: "Non-Isothermal Multicomponent Adsorption Processes and Their Numerical Treatment by Means of Integro-Differential Equations," *Comput. Chem. Eng.* **13** (1989) 1025–1030.
[9.6] P. L. Mills, S. Lai, M. P. Duduković, P. A. Ramachandran: "A Numerical Study of Approximation Methods for Solution of Linear and Nonlinear Diffusion-Reaction Equations with Discontinuous Boundary Conditions," *Comput. Chem. Eng.* **12** (1988) 37–53.
[9.7] P. M. Morse, H. Feshbach: *Methods of Theoretical Physics*, vol. I, McGraw-Hill, New York 1953.
[9.8] H. S. Carslaw, J. C. Jaeger: *Conduction of Heat in Solids*, 2nd ed., Clarendon Press, Oxford 1959.
[9.9] N. D. Ferguson, B. A. Finlayson: "Error Bounds for Approximate Solutions to Nonlinear Ordinary Differential Equations," *AIChE J.* **18** (1972) 1053–1059.
[9.10] R. S. Dixit, L. L. Taularidis: "Integral Method of Analysis of Fischer–Tropsch Synthesis Reactions in a Catalyst Pellet," *Chem. Eng. Sci.* **37** (1982) 539–544.
[9.11] L. Lapidus, G. F. Pinder: *Numerical Solution of Partial Differential Equations in Science and Engineering*, Wiley-Interscience, New York 1982.
[9.12] C. K. Hsieh, H. Shang: "A Boundary Condition Dissection Method for the Solution of Boundary-Value Heat Conduction Problems with Position-Dependent Convective Coefficients," *Num. Heat Trans. Part B: Fund.* **16** (1989) 245–255.
[9.13] C. A. Brebbia, J. Dominguez: *Boundary Elements – An Introductory Course*, Computational Mechanics Publications, Southhampton 1988.
[9.14] J. Mackerle, C. A. Brebbia (eds.): *Boundary Element Reference Book*, Springer Verlag, Berlin – Heidelberg – New York – Tokyo 1988.

References for Chapter 10

[10.1] H. W. Kuhn, A. W. Tucker: "Nonlinear Programming" in J. Neyman (ed.): *Proc. Second Berkeley Symp. Mathematical Statistics and Probability*,

Univ. California Press, Berkeley, CA, 1951, pp. 402–411.

[10.2] T. Umeda, A. Ichikawa, *Ind. Eng. Chem. Process Des. Dev.* **10** (1971) 229.

[10.3] W. C. Davidon: "Variable Metric Methods for Minimization," *AEC R & D Report ANL-5990*, (rev.) Argonne, Illinois 1959.

[10.4] R. Fletcher, M. J. D. Powell: "A Rapidly Converging Descent Method for Minimization," *Comput. J.* **6** (1963) 163.

[10.5] R. Fletcher: *Practical Methods of Optimization*, Wiley-Interscience, New York 1987.

[10.6] C. G. Broyden: "The Convergence of a Class of Double Rank Minimization Algorithms," *J. Inst. Math. Its. Appl.* **6** (1970) 76.

[10.7] R. Fletcher: "A new approach to variable metric algorithms," *Comput. J.* **13** (1970) 317.

[10.8] D. Goldfarb: "A Family of Variable Metric Methods Derived by Variational Means," *Math. Comp.* **24** (1970) 23.

[10.9] D. Goldfarb, M. J. Todd: "Linear Programming," chap. II, in G. L. Nemhauser, A. H. G. Rinnoy Kan, M. J. Todd (eds.): *Optimization*, North Holland Publ. Co, Amsterdam 1989.

[10.10] D. F. Shanno: "Conditioning of Quasi-Newton Methods for Function Minimization," *Math. Comp. Co* **24** (1970) 647.

[10.11] J. E. Dennis, J. J. More: "Quasi-Newton Methods, Motivation and Theory," *SIAM Rev.* **21** (1977) 443.

[10.12] S.-P. Han: "A Globally Convergent Method for Nonlinear Programming," *J. Opt. Theo. Applics.*, **22** (1977) 297.

[10.13] M. J. D. Powell: "A Fast Algorithm for Nonlinearly Constrained Optimization Calculations," *Lecture Notes in Mathematics 630*, Springer Verlag, Berlin – Heidelberg– New York – Tokyo 1977.

[10.14] P. Gill, W. Muray: "Numerically Stable Methods for Quadratic Programming," *Math. Prog.*, **14** (1978) 349.

[10.15] D. Goldfarb, A. Idnani: "A Numerically Stable Dual Method for Solving Strictly Convex Quadratic Programs," *Math. Prog.* **27** (1983) 1.

[10.16] P. Wolfe: "The Simplex Method for Quadratic Programming," *Econometrica* **27** (1959) no. 3, 382.

[10.17] C. E. Lemke: "A Method of Solution for Quadratic Programs," *Manag. Science* **8** (1962) 442.

[10.18] K. Schittkowski: "The Nonlinear Programming Algorithm of Wilson, Han and Powell with an Augmented Lagrangian Type Line Search Function," *Num. Math.* **38** (1982) 83.

[10.19] R. M. Chamberlain, C. Lemarechal, H. C. Pedersen, M. J. D. Powell: "The Watchdog Technique for Forcing Convergence in Algorithms for Constrained Optimization," *Math. Prog. Study* **16** (1982).

[10.20] W. Hock, K. Schittkowski: "Test Examples for Nonlinear Programming Codes," *Lecture Notes in Economics 187*, Springer Verlag, Berlin – Heidelberg – New York – Tokyo 1981.

[10.21] A. Lucia, A. Kumar: "Distillation Optimization," *Comput. Chem. Engl.* **12** (1988) no. 12, 1263.

[10.22] T. J. Berna, M. H. Locke, A. W. Westerberg: "A New Approach to Optimization of Chemical Processes," *AIChE J.* **26** (1980) 37–43.

[10.23] P. E. Gill, W. Murray, M. Wright: *Practical Optimization*, Academic Press, New York 1981.

[10.24] W. Murray, M. Wright: "Projected Lagrangian Methods Based on the Trajectories of Penalty and Barrier Functions," *SOL Report 78-23*, Stanford University 1978.

[10.25] J. Nocedal, M. L. Overton: "Projected Hessian Updating Algorithms for Nonlinearly Constrained Optimization," *SIAM J. Numer. Anal.* **22** (1985) 821.

[10.26] C. B. Gurwitz, M. L. Overton: "SQP Methods Based on Approximating a Projected Hessian Matrix," *SIAM J. Sci. Stat. Comput.* **10** (1989) no. 4, 631.

[10.27] A. Vasantharajan, L. T. Biegler: "Large-Scale Decomposition Strategies for Successive Quadratic Programming," *Comput. Chem. Eng.* **12** (1988) no. 11, 1087.

[10.28] S. Vasantharajan, J. Viswanathan, L. T. Biegler: "Large Scale Development of Reduced Successive Quadratic Programming," *CORS/TIMS/ORSA Meeting*, Vancouver, BC, May, 1989.

[10.29] V. Chvatal: *Linear Programming*, Freeman, New York 1983.

[10.30] K. G. Murty: *Linear Programming*, J. Wiley & Sons, New York 1983.

[10.31] G. Dantzig: *Linear Programming and Extensions*, Princeton University Press, Princeton 1963.

[10.32] G. L. Nemhauser, L. A. Wolsey: *Integer and Combinatorial Optimization*, J. Wiley & Sons, New York 1988.

[10.33] R. S. Garfinkel, G. L. Nemhauser: *Integer Programing*, J. Wiley & Sons, New York 1972.

[10.34] A. M. Geoffrion: "Generalized Benders Decomposition," *J. Optim. Theory Appl.* **10** (1972) 237.

[10.35] M. A. Duran, I. E. Grossmann: "An Outer-Approximation Algorith for a Class of Mixed-Integer Nonlinear Programs," *Math. Prog.* **36** (1986) 307.

[10.36] G. R. Kocis, I. E. Grossmann: "Relaxation Strategy for the Structural Optimization of Process Flowsheets," *Ind. Eng. Chem. Prod. Res. Dev.* **27** (1987) no. 8, 1869.

[10.37] I. E. Grossmann: "MINLP Optimization Strategies and Algorithms for Process Synthesis", in J. J. Siirola, I. E. Grossmann, G. Stephanopoulos (eds.): *Proceedings of FOCAPD '89 Conference*, Elsevier, Amsterdam 1989.

[10.38] J. Viswanathan, I. E. Grossmann: "Combined Penalty Function and Outer-Approximation Method for MINLP Optimization," *Paper MD 18.5, CORS/TIMS/ORSA Meeting*, Vancouver 1989.

[10.39] A. E. Bryson, Y.-C. Ho: *Applied Optimal Control*, Hemisphere Publishing, Washington, D.C. 1975.

[10.40] A. E. Bryson, W. F. Denham, *J. Appl. Mech.* **29** (1962) 247.

[10.41] L. S. Lasdon, S. K. Mitter, A. D. Warren, *IEEE Trans. Autom. Control* **AC-12** (1967) 132.

[10.42] L. S. Lasdon, *IEEE Trans. Autom. Control* **AC-15** (1970) 268.

[10.43] L. Lapidus, R. Luus: *Optimal Control of Engineering Processes*, Blaisdell, Waltham, MA 1967.

[10.44] D. I. Jones, J. W. Finch: "Comparison of Optimization Algorithms," *Int. J. Control* **40** (1984) 747.

[10.45] S. N. Rao, R. Luus: "Evaluation and Improvement of Control Vector Iteration Procedures," *Can. J. Chem. Eng.* **50** (1972) 777.

[10.46] M. Caracotsios, W. E. Stewart: "Sensitivity Analysis of Initial Value Problems with Mixed ODE's and Algebraic Equations," *Comput. Chem. Eng.* **9** (1985) 359–365.

[10.47] R. W. H. Sargent, G. R. Sullivan: "The Development of an Efficient Optimal Control Package," *Proceedings of the 8th IFIP Conference on Optimization Techniques Pt. 2*, Würzburg 1977.

[10.48] K. E. Brenan, L. R. Petzold: *The Numerical Solution of Higher Index Differential/Algebraic Equations by Implicit Runge–Kutta Methods*, UCRL-95905, preprint, Lawrence Livermore National Laboratories, Livermore, Ca. 1986.

[10.49] J. S. Logsdon, L. T. Biegler: "Accurate Solution of Differential-Algebraic Optimization Problems," *Ind. Eng. Chem. Res.* **28** (1989) 1628.

[10.50] L. R. Petzold: "Differential/Algebraic Equations Are Not ODEs," *SIAM J. Sci. Stat. Comput.* **3** (1982) 367–385.

References for Chapter 11

[11.1] G. E. P. Box, W. G. Hunter, J. S. Hunter: *Statistics for Experimenters*, J. Wiley and Sons, New York 1978.

[11.2] E. Kreyszig: *Advanced Engineering Mathematics*, 4th ed., J. Wiley and Sons, New York 1979.

[11.3] J. L. Devore: *Probability and Statistics for Engineering and the Sciences*, Brooks/Cole, Monterery, California 1987.

[11.4] W. H. Press, B. P. Flannery, S. A. Teukolsky, W. T. Vetterling: *Numerical Recipes. The Art of Scientific Computing*, Cambridge University Press, Cambridge 1986.

[11.5] B. W. Lindgren: *Statistical Theory*, Macmillan, New York 1962.

[11.6] M. Abramowitz, I. A. Stegun: *Handbook of Mathematical Functions*, U.S. Government Printing Office, Washington, D.C. 1964.

[11.7] J. B. Cropley: "Heuristic Approach to Comples Kinetics," *ACS Symp. Ser.* **65** (1978) 292–302.

References for Chapter 12

[12.1] O. A. Hougen, K. M. Watson, R. A. Ragatz: *Chemical Process Principles*, 2nd ed., part II, "Thermodynamics," J. Wiley and Sons, New York 1959.

2. Mathematical Modeling

HENNING BOCKHORN, Technische Hochschule, Darmstadt, Federal Republic of Germany

1.	Introduction	2-4
1.1.	Terminology	2-4
1.2.	Application Areas of Mathematical Modeling in Industrial Chemistry and Chemical Engineering	2-4
1.3.	Limitations of Mathematical Models	2-5
2.	Construction and Classification of Mathematical Models	2-6
2.1.	Construction of Mathematical Models	2-6
2.2.	Classification of Mathematical Models	2-7
3.	Empirical Models	2-10
3.1.	Linear Empirical Models	2-11
3.1.1.	Linear Empirical Models with One Variable	2-11
3.1.1.1.	Parameter Estimation (Linear Regression)	2-11
3.1.1.2.	Assessment of Estimated Parameter Values	2-12
3.1.1.3.	Sensitivity Analysis	2-13
3.1.1.4.	Example of a Linear Model	2-14
3.1.1.5.	Concluding Remarks	2-16
3.1.2.	Linear Empirical Models with Several Variables	2-16
3.1.2.1.	Parameter Estimation	2-16
3.1.2.2.	Assessment of Estimated Parameter Values	2-18
3.1.2.3.	Concluding Remarks	2-18
3.2.	Nonlinear Empirical Models	2-19
3.2.1.	Parameter Estimation (Nonlinear Regression)	2-19
3.2.1.1.	Transformation of the Model into Linear Form	2-20
3.2.1.2.	Direct Search Methods	2-20
3.2.1.3.	Gradient Methods	2-21
3.2.2.	Assessment of Estimated Parameter Values	2-24
3.2.3.	Example of a Nonlinear Model	2-24
3.2.4.	Concluding Remarks	2-25
3.3.	Further Calculation Methods and Model Types	2-25
3.3.1.	Further Methods of Parameter Estimation	2-25
3.3.2.	Other Types of Models	2-26
3.3.2.1.	Models with Constraints on the Parameters	2-26
3.3.2.2.	Models Based on Differential Equations	2-27
4.	Models Based on Transport Equations for Probability Density Functions	2-28
4.1.	Terminology	2-29
4.1.1.	One-Dimensional Distribution and Probability Density Functions	2-29
4.1.2.	Multidimensional Distribution and Probability Density Functions	2-32
4.1.3.	Conditional Probability Density Functions	2-33
4.2.	Transport Equations for Single-Point Probability Density Functions	2-34
4.2.1.	General Form of the Transport Equations for Probability Density Functions	2-35
4.2.2.	Limitations of Single-Point Probability Density Functions	2-38
4.3.	Examples of Calculating Probability Density Functions	2-39
4.3.1.	Solutions for Deterministic Systems	2-39
4.3.1.1.	Age (Residence Time) Distributions in Chemical Reactors	2-39
4.3.1.2.	Crystal Size Distribution in Continuously Operating Crystallizers	2-43
4.3.2.	Solutions for Statistical Systems	2-43
4.3.2.1.	Closure of the Transport Equation for Probability Density Functions	2-44
4.3.2.2.	Solution Methods of the Transport Equation for Probability Density Functions	2-47
4.3.2.3.	Example: Combustion of Propane in a Turbulent Diffusion Flame	2-48
5.	Models Based on Physicochemical Principles (Transport Phenomena)	2-50
5.1.	Applications of the Principle of Conservation of Momentum	2-51
5.1.1.	Laminar Tube Flow	2-51
5.1.2.	Turbulent Nonreactive Free Jets	2-52
5.1.2.1.	Models for Reynolds Stresses	2-53

5.1.2.2.	Solution Method of the Resulting System of Partial Differential Equations	2-55	5.3.1.3.	Examples of Simultaneous Mass and Heat Transfer: Dynamic Models ... 2-65
5.1.2.3.	Example: Turbulent Flow of Nitrogen in Air	2-58	5.3.1.4.	Examples of Simultaneous Mass and Heat Transfer: Static Models 2-67
5.2.	Applications of the Principle of Conservation of Enthalpy	2-59	5.3.2.	Mass Transfer with Chemical Reactions ... 2-71
5.2.1.	Heat Conduction	2-59	5.3.3.	Chemical Reactions in the Homogeneous Phase ... 2-73
5.2.2.	Heat Transfer	2-60	5.3.3.1.	Isothermal Reactors with Frictionless Flow, Constant Density, and Reactions Without Volume Changes ... 2-73
5.2.2.1.	Exact Solution for a Boundary Layer Problem	2-60		
5.2.2.2.	General Principles of Modeling Heat Transfer	2-62	5.3.3.1.1.	Stability of Isothermal Reactors . 2-77
			5.3.3.1.2.	Sensitivity Analysis ... 2-82
5.3.	Applications of the Principle of Conservation of Mass	2-63	5.3.3.2.	Nonisothermal Reactors ... 2-83
5.3.1.	Mass Transfer without Chemical Reaction	2-64	5.3.3.2.1.	Heterogeneous Catalytic Reactions ... 2-83
5.3.1.1.	Exact Solution for a Boundary Layer Problem	2-64	5.3.3.2.2.	Stability Analysis of Nonisothermal Reactors ... 2-90
5.3.1.2.	General Principles for Modeling Mass Transfer	2-65	5.3.3.2.3.	Use on Statistical Processes ... 2-95
			6.	References ... 2-99

In addition to the standard symbols defined in the front matter of this volume, the following symbols are used:

a	general coefficient, temperature conductivity
$\boldsymbol{a}$	general matrix
A	chemical component
A	general coefficient, source term in momentum conservation equation
b	general coefficient, general model parameter (estimate)
$\boldsymbol{b}$	general model parameter in vector notation (estimate)
B	chemical component
B	general coefficient
$\boldsymbol{B}$	general matrix
Bi	Biot number
Bo	Bodenstein number
c	concentration, general coefficient
$\boldsymbol{c}$	general matrix
c_p	specific heat capacity
C	chemical component
C	model parameter in turbulence models
C_1, C_2, C_3, C_4	integration constants
D	chemical component
D	diameter, diffusion coefficient
D_R	tube diameter
Da_I, Da_{II}	Damköhler number (of 1st and 2nd kind)
E	statistical event
$\boldsymbol{E}$	statistical error in vector notation
E_a	activation energy
f	dimensionless flow function, mixture fraction (normalized mass fraction of an element in a mixture)
F	cross-sectional area, feed flow of a liquid, value of the F-distribution
F	distribution function of a statistical variable
g	acceleration due to gravity
G	mass flow of gas
h	enthalpy
$\boldsymbol{h}$	step size in vector notation
Δh_v	specific latent heat of vaporization
H	Heaviside step function, height
Ha	Hatta number
HTU	height of one transfer unit
$\boldsymbol{I}$	unit matrix
I_u, I_Φ, I_p	class of integrals
j	total flux of a scalar quantity
$\boldsymbol{j}$	flux of scalar quantities per unit area
$\boldsymbol{J}$	Jacobi matrix
k	general parameter, rate coefficient, total heat-transfer coefficient, turbulence energy
K	"adsorption" coefficient
K_p	class of integrals
L	length, likelihood function, mass flow of liquid, number of grid nodes in the x-direction, size, transport operator
Le	Lewis number
m	deterministic moment, mass fraction, number of occurrences of an event
M	number of grid nodes in the y- or r-direction
M_i	molecular mass
n	number of data sets, number of units
Nu	Nusselt number
NTU	number of transfer units
O	order of magnitude
p	general parameter, pressure, statistical weight
$\boldsymbol{p}$	statistical weight in vector notation
P	chemical component
P	probability density function of a statistical variable

Pe	Péclet number
Pr	Prandtl number
q	general parameter, number of independent variables in multivariable models, volumetric flow rate
Q	general function
r	general parameter, radial distance, rate of reaction
R	gas constant, radius, ratio of enthalpies of vaporization
Re	Reynolds number
s	specific surface area, general parameter, estimate for standard deviation
S	sink flow of a liquid, source term in conservation equation, surface area
s_r^2	mean square deviation about regression
s_ξ^2	mean square deviation of data set
$s_{Y_i}^2$	mean square deviation about $\bar{Y}_i$
Sc	Schmidt number
Sh	Sherwood number
t	time, value of student t-variable
T	temperature
u	velocity
$\mathbf{u}$	velocity in vector notation, i.e., $\mathbf{u} = (u_1, u_2, u_3)$; $\mathbf{u} = (u, v, w)$
U	velocity in a turbulent (statistical) flow
$\mathbf{U}$	velocity in a turbulent (statistical) flow in vector notation, i.e., $\mathbf{U} = (U_1, U_2, U_3)$; $\mathbf{U} = (U, V, W)$
V	volume
w	residence time distribution, velocity
W	residence time distribution function, probability, velocity in a turbulent (statistical) flow
x	distance, general independent variable
$\mathbf{x}$	distance in vector notation, i.e., $\mathbf{x} = (x_1, x_2, x_3)$; $\mathbf{x} = (x, y, z)$, general independent variable in vector notation
X	partial derivatives of expected values with respect to model parameters in matrix notation
y	general dependent variable
$\mathbf{y}$	general dependent variable in vector notation
Y	general statistical variable
$\mathbf{Y}$	general statistical variable in vector notation

Greek Symbols

α	age, general coefficient, heat-transfer coefficient, normalized mass fraction, significance level
β	general coefficient, general model parameter, mass-transfer coefficient, normalized mass fraction
$\boldsymbol{\beta}$	general model parameter in vector notation
β_r	acceleration factor
β_H	Prater number
γ	Arrhenius number, correlation coefficient, scaling factor
γ_r	reciprocal Arrhenius number
δ	Dirac delta function
δ	film thickness, Kronecker symbol
ε	dissipation rate, intergranular volume, normalized rate coefficient, statistical error
ζ	phase shift
η	effectiveness factor, normalized distance, similarity coordinate
$\boldsymbol{\eta}$	expected values in vector notation
θ	general variable, normalized age, normalized residence time, normalized temperature
Θ	general statistical variable, source term in scalar quantities equation
κ	permeability, normalized rate coefficient
λ	eigenvalues, Lagrangian multipliers, mean free path, scaling factor, thermal conductivity, wavelength
μ	dynamic viscosity, normalized mass fraction
μ_t	apparent turbulent viscosity
μ_n	nth moment of a probability density function
ν	frequency, kinematic viscosity, stoichiometric coefficient
ξ	general variable
Ξ	general statistical variable
π	normalized mass fraction
ϱ	density, scaling factor
σ	standard deviation
σ_φ	turbulent Schmidt/Prandtl numbers
τ	normalized time, residence time, turbulent time scale, viscous stress
τ_N	cooling time
φ	general variable, normalized mass fraction
Φ	general variable, general statistical variable
$\boldsymbol{\Phi}$	general variable and general statistical variable in vector notation, sum of squares of deviations
Φ_K	Thiele modulus
ψ	general variable
Ψ	general variable, general statistical variable, stream function
$\boldsymbol{\Psi}$	general variable and general statistical variable in vector notation
Ω	general function, general operator

Subscripts

a, b	reference states a and b
A	ambient, chemical component
ad	adiabatic
B	chemical component
Bf	bifurcation
c	central
C	cooling
Cat	catalyst
diff	diffusion
e	east, end, expansion
eff	effective
ext	external
G	gas phase
h	enthalpy
i	component of a vector, chemical component
int	internal
j	component of a vector, chemical component
k	component of a vector, chemical component, contraction, turbulence energy, summation according to the Einstein summation convention

l	number of elements in a difference equation	α	a component
		ε	dissipation rate
L	liquid phase	φ	variable
L	total length	ω	chemical component, summation according to the Einstein summation convention
m	mean		
m	mass	0	initial, reference state
n	north		
o	upper limit		
p	pressure, pressure-related	**Superscripts**	
Ph	phase boundary		
qs	quasi-steady-state	*	transformed or normalized variable or parameter
r	reaction, reflected		
ref	reference	′	scaled variable
s	south	~	weighted or normalized variable
S	surface	–	mean
t	turbulent	^	calculated model response
T	temperature	A, B, C, D	chemical component
tot	total	eq	equilibrium
u	lower limit	f	fluctuating
u	velocity	G	gas phase
w	west	h	enthalpy
W	wall	L	liquid phase
Wi	inner wall	m	mass
Wo	outer wall	(n)	number of iterations
x	local	T	transpose of a vector or matrix

1. Introduction

1.1. Terminology

Models are used in almost every branch of science and technology. It is difficult to encompass the many meanings of the term "model" with one single definition. The classical definition is "An object M (an object, physical or ideal system, or process) is a model if analogies exist between M and another object O that permit conclusions to be made about O" [1]. This definition describes the representation of the object O and presupposes its isolation from the whole of reality. Thus a model represents reality or a part of it. Since only certain conclusions are required, the model is reduced to parts or individual aspects of reality. The restriction to analogies between model M and object O results in limited conformity of their function, structure, and behavior. The representation of the object O by the model M may have limited accuracy and the model M can be based on a different scale than the object O. The model is satisfactory as long as important variables and phenomena are correctly represented for the specific context or investigation.

An adequate model according to the above definition can rarely explain reality in the form of a theory. Information that has been lost in the representation cannot be retrieved from the model. Attempts at theoretically interpreting the object O with the help of model M must therefore be discussed according to the aspect, scale, and accuracy of the representation.

Analogies between model M and object O can be established in the form of mathematical equations. A mathematical model therefore represents a set of algebraic and/or differential and/or integral equations which are used to describe the behavior of the object O. The terms variables and parameters are used in this article in accordance with their definitions given in mathematical relationships, and not as defined in process control, see for example [2].

1.2. Application Areas of Mathematical Modeling in Industrial Chemistry and Chemical Engineering

The two major tasks facing engineers and scientists in industrial chemistry are (1) the operation and optimization of existing processes and (2) the design of new or improved ones. In this article the term process means a sequence of operations or treatments and transformations performed on specific materials.

For the first task engineers and scientists must obtain exact information about the particular process under consideration. Qualitative aspects and criteria must be quantified. The basic variables and parameters of the relationships which describe the individual parts of the process must be worked out. The individual parts of the process must be combined. Mathematical models are very important in this respect.

For the second task mathematical models help in applying existing processes to new or modified plants and in the definition of safer, more economically viable operating conditions. Data for the construction of new plants cannot be obtained from an operating process by running it to its technical limits; this entails a high degree of risk. In contrast, a mathematical model of a process is easy to manipulate. Unusual operating conditions can easily be simulated. The plant can even be modeled under hazardous conditions to define the limits of operating parameters or risk areas.

The main applications of mathematical models in industrial chemistry are summarized as follows:

1) *Simple Experimentation.* By means of mathematical models, existing processes can be investigated more quickly, economically, and thoroughly than in operating industrial plants. Mathematical models can simulate the plant in time lapse or slow motion. The results from the model are therefore available more quickly or have a higher time resolution.
2) *Repetition of Experiments.* Simulation of processes with mathematical models allows the influence of variations in variables or parameters to be investigated. Errors can easily be implanted into or removed from the model. In a running industrial process, this type of empirical sensitivity analysis has the disadvantage that it requires a high degree of technical effort and entails a high degree of risk.
3) *Sensitivity Analysis.* With a mathematical model, sensitivity analysis does not have to be carried out on an empirical basis by repetition of tests with systematic changes of the variables and parameters. Sensitivity analysis as a mathematical tool allows calculation of the gradients of dependent variables with respect to the model parameters.
4) *Control and Operation.* Mathematical models can be used as an economical way of estimating the stability of industrial processes or subsystems as a prerequisite for effective control or operation. Again, use of the industrial process for direct experimentation must be ruled out because of the high risk involved.
5) *Optimization.* Description of a complex system in terms of a mathematical model allows optimization objectives for the process to be developed. The optimization objectives can easily be adjusted in accordance with changing requirements. Optimum values of operating variables or parameters can also be determined easily.
6) *Extrapolation.* A suitable mathematical model can be used to test extreme operating conditions that are not possible or practical in the running process. Surfaces of operating variables and parameters can then be constructed and the optimum conditions for running the plant can be extrapolated. Extrapolability must, however, always be discussed in the light of the limits of the model and the part of the object O being modeled.

1.3. Limitations of Mathematical Models

Mathematical models have some significant limitations. The first is the type, quantity, and accuracy of the available data. The success and results of mathematical models depend largely on the information about the object O being modeled. A mathematical model cannot be better than the physical or chemical data on which it is based. In many cases available data are not sufficient and engineers and scientists have to obtain sufficient data for a model by experimental analysis. The alternative is to simplify the model adequately. The two following examples illustrate the associated problems.

The number of theoretical plates necessary for solving a separation problem (e.g., the number of theoretical plates of a rectification column) can usually be calculated more accurately than the effectiveness factor of a separation plate in a real piece of equipment. The latter value depends on the fluid dynamics of the system and on the operating conditions of the column. However, the inaccuracy in the estimate of the effectiveness factor strongly affects the commercial viability of a separation process with a large number of theoretical plates.

Another example concerns the kinetic data for industrial processes that are mainly derived from laboratory experiments. On a laboratory scale, side reactions may be much less noticeable than in an industrial plant. Lack of information about side reactions can lead to rate data that are unsuitable for industrial process design. A further problem is the impurities of the reactants used in an industrial plant; impurities can lead to changes in the reaction rate that cannot be predicted with the pure chemicals used in laboratory experiments.

The accuracy required for the individual parameters of the model depends on the sensitivity with which the results of the model respond to changes in these parameters (see Sections 3.1.1.3 and 5.3). Those parameters which exert the greatest influence on the results of the model must be determined with the greatest accuracy.

A second restriction in the use of mathematical models lies in the mathematical tools that are available for solving the resulting system of equations. Complex structures can often be formulated mathematically. However, the mathematical and numerical methods required to solve the resulting system of equations are either not yet developed or are beyond the capacity of available computers. The rapid development of supercomputers in recent years has enabled increasingly complex problems to be solved.

A further restriction concerns the interpretation of the results of the model and the extrapolability of the representation. This limitation is illustrated in the simple example in Figure 1. The solid curve shows the dynamics of a physicochemical quantity. The object could be, for example, an intermediate in consecutive chemical reactions, φ represents its concentration:

$$\varphi = \frac{k_1}{k_2 - k_1} \psi_0 [\exp(-k_1 t) - \exp(-k_2 t)] \quad (1)$$

where ψ_0 is the initial concentration of starting compound from which φ is formed, t is the time, and k_1 and k_2 are the rate coefficients of two consecutive first-order reactions, $k_1 \neq k_2$. The available data cover the range $0 < t < t_1$ and are shown in Figure 1 as squares. An obvious empirical model based on these data is $\varphi = kt$ and is given by the dashed line. In fact the object is represented by this model in the range $k_1 t \ll 1$ and $k_2 t \ll 1$. If the exponential functions in Equation (1) are developed for the above condition, it follows that:

$$\varphi \approx \frac{k_1}{k_2 - k_1} \psi_0 [(k_2 - k_1) t] = k_1 \psi_0 t \quad (2)$$

It can be seen that for the range covered by experimental data $(0 < t < t_1)$, an analogy exists between the model and the object and the phenomenon of interest is also adequately reproduced within this region. The representation has a coarser scale, i.e., $\varphi = kt$ instead of $\varphi = k_1 \psi_0 t$. However, the model does not explain the physicochemical characteristics of the object—an intermediate in a system of consecutive chemical reactions. Finally, extrapolation of this model beyond the range covered by the available data leads to a false representation.

2. Construction and Classification of Mathematical Models

2.1. Construction of Mathematical Models

Based on the terminology used in Section 1.1, the construction of a mathematical model consists of setting up a consistent set of mathematical relationships. In the ideal case these would be identical to the relationships between the process variables. Often, only certain phenomena of particular interest can be incorporated in the model due to the complexity of the real process.

The strategy of analysis used for developing mathematical models generally contains the following steps (see also [3]–[10]):

1) Formulation of the problem and compilation of objectives and decision criteria
2) Investigation of the process and its classification with the aim of splitting it up into subsystems (process elements)
3) Determination of the relationships between the subsystems

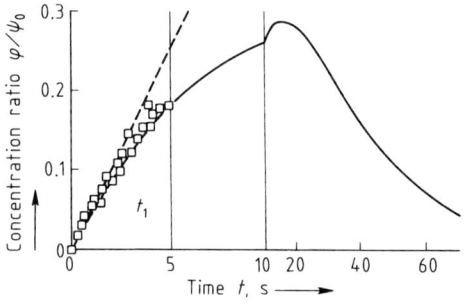

Figure 1. Relationship between a model and object with limited data

4) Analysis of the variables and relationships between the variables for the individual process elements
5) Setting up of mathematical relationships with variables and parameters; data acquisition
6) Investigation of the representation of the process by the model; comparison of the simulation with real process data
7) Installation of the model; interpretation and collation of the results

For further evaluation of the model the above steps can be complemented by:

8) Sensitivity analysis of the model: identification of the parameters with strong and weak influences on the response of the model
9) Model simplification

If the above steps are compiled in a flow diagram, then steps 4–9 must be repeated, possibly several times, until the interpretation of the results of the model still makes sense within the framework of the formulated objectives and expected solutions to the problem.

Industrial processes are very complex. The success of mathematical models in describing them often depends on whether they can be subdivided into process elements. The entire process can then be built up from the individual elements if the relationships between them have been worked out. Subdivision into process elements is not necessarily based on the physical principles of the entire process. For example, a nonideal tubular reactor can be simulated by a series of ideal stirred tank reactors although such units may not exist in the industrial process. A packed absorption column can be represented by a number of theoretical plates even though mass transfer occurs continuously and does not result in stepwise attainment of phase equilibrium.

By correct treatment and adjustment of the process elements, modeling strives to represent the entire process as accurately as possible on the basis of simple, established principles of the individual elements.

Industrial processes are not only very complex, they can also consist of a number of qualitatively different process elements. A chemical process may, for example, encompass process elements for the preparation of the reactants, the chemical reactor, heat exchangers, and finally process elements for separation and possibly processing of the product. The different qualities of the process elements require different mathematical models depending on the physical or physicochemical principles on which they are based. An industrial heat exchanger is represented by a different type of model from that required for a rectification column, the latter has a different mathematical description from a chemical reactor, and so on. Due to the underlying physicochemical principles, mathematical modeling and the construction of mathematical models employ a wide variety of mathematical methods and methods of solution. For example, the design of distillation columns with models based on theoretical plates leads to a system of linear equations which is often solved graphically. The representation of a nonideal reactor by a series of ideal stirred tank reactors leads to a system of algebraic equations which can be highly nonlinear depending on the chemical reaction under consideration. Finally, a reactor model for a homogeneous, turbulent-flow reactor based on the modeled Navier–Stokes equations leads to a system of coupled partial differential equations that can be solved by various methods.

Mathematical modeling is used to formulate widely differing physical and physicochemical phenomena: transfer of heat, mass, and momentum, as well as chemical reactions in homogeneous and heterogeneous systems. Mathematical modeling is thus used in the design of mass-transfer operations, calculation of heat exchangers, chemical reaction engineering, and finally process control. A wide range of methods are used to formulate models and to solve the resulting systems of equations.

For this reason further formalization of the steps for building mathematical models and a further division into individual steps will be avoided. The following section examines the classification of models based on the physical background of the model, the type of system of equations, and the necessary methods for their solution.

2.2. Classification of Mathematical Models

Mathematical models can be classified according to their physical background, the type of systems of equations, and the corresponding methods of solution. Alternatives are discussed in [3], [8]–[12]. Figure 2 shows three classes of

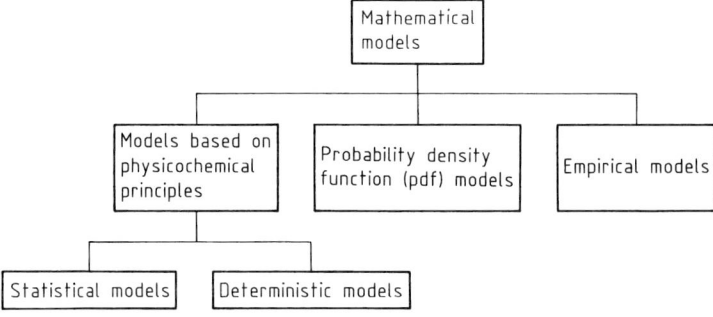

Figure 2. Classification of mathematical models according to their physical principles

models with different physical backgrounds: models based on physicochemical principles, probability density function (pdf) models, and empirical models.

The first category comprises models based mainly on the mathematical formulation of transport phenomena. Use of this principle requires that the process can be subdivided into process elements that can be described by the laws governing the transport of mass, momentum, and energy, i.e., their conservation principles.

Such models are subdivided into deterministic and statistical models. Deterministic models or model elements have a determined value or set of values for each variable or parameter for any given set of conditions. In contrast, the variables and parameters used in statistical models (or elements of statistical models) are statistical quantities. They can only be given with a certain probability or in terms of moments of probability density functions. If, for example, the probability density function $P(Y)$ holds for the statistical variable Y, then $P(Y)\,\mathrm{d}Y$ is the probability for the variable lying in the range $\mathrm{d}Y$ around Y. Normally the full statistical information is not necessary for technical purposes; therefore statistical variables are often described by moments of the probability density functions. These moments are defined as

$$\mu_n \equiv \int_{-\infty}^{+\infty} Y^n P(Y)\,\mathrm{d}Y$$

For example the first moment

$$\mu_1 \equiv \int_{-\infty}^{+\infty} Y P(Y)\,\mathrm{d}Y$$

is the mean value of the statistical variable Y, the second moment about the mean value is its variance $\mathrm{Var}(Y)$:

$$\mu_2' \equiv \int_{-\infty}^{+\infty} (Y - \mu_1)^2 P(Y)\,\mathrm{d}Y, \text{ etc.}$$

A statistical mathematical model is the mathematical description of a process, e.g., in terms of the moments defined above. Figure 3 shows some possibilities for statistical processes. Nor-

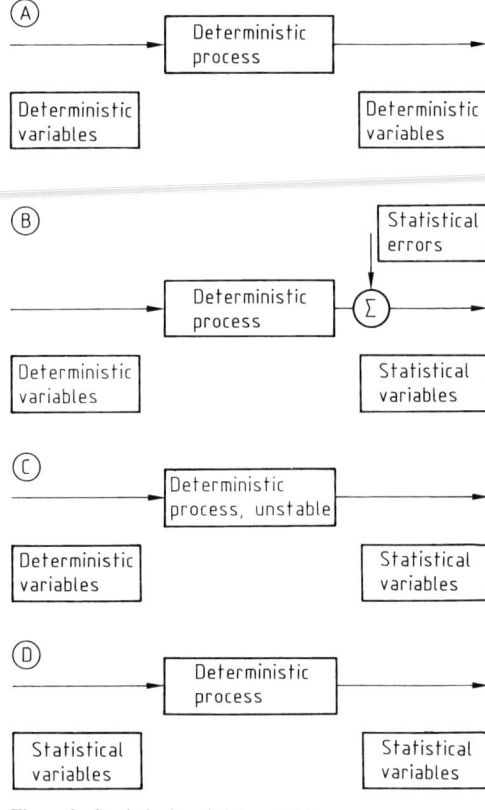

Figure 3. Statistical and deterministic processes

mally a deterministic process leads to deterministic results (Fig. 3A). If a statistical error is added to the process results (Fig. 3B), it leads to a statistical result. The same result is obtained if the deterministic process becomes unstable (Fig. 3C); a simple example for this is buoyancy-driven turbulence. If the input process variables are statistical variables then the results of the process are also statistical variables (Fig. 3D). Figure 3 shows that the macroscopic phenomena in statistical processes change but that the physicochemical background of the process and hence of the process model remains the same. The principles of the conservation of mass, momentum, and energy are also valid for statistical variables; the mathematical overhead for handling these cases is, however, more complicated. The transport equations used to describe a statistical process by means of statistical moments contain, for example, a number of statistical hypotheses and model assumptions for their closure but the character of the model equations does not change. For these reasons statistical models are classified as shown in Figure 3. Methods for representing statistical processes will be discussed in Chapter 5.

A further subdivision of models based on physicochemical principles according to the type of model equations is shown in Figure 4. The complexity of the method of solution decreases from right to left. The division is based on the most commonly used types of models but is by no means definitive. Thus, multidimensional models can be formulated in the form of algebraic relationships; models which can be described by algebraic relationships do not have to be steady-state models. Steady-state models describe processes in which the accumulation terms (the changes in the variables over time) disappear. Nonsteady-state models include changes in the variables over time. In models with constant variables, the properties and state are not functions of space so that the system is homogeneous. Description with locally distributed variables considers local changes in the dependent variables. (The nomenclature of process analysis does not make the distinction between variables and parameters introduced in Section 1.1; the latter two types of models are then classified as models with constant or concentrated parameters or as models with distributed parameters [10], [13]–[15].) Figure 4 also shows that the description of processes by differential equations and difference equations is frequently equivalent. Differential equations must normally be solved by numerical methods. In order to achieve this they are transformed into difference equations. These methods, which are really mathematical tools, have their physical analogy in the description of processes with

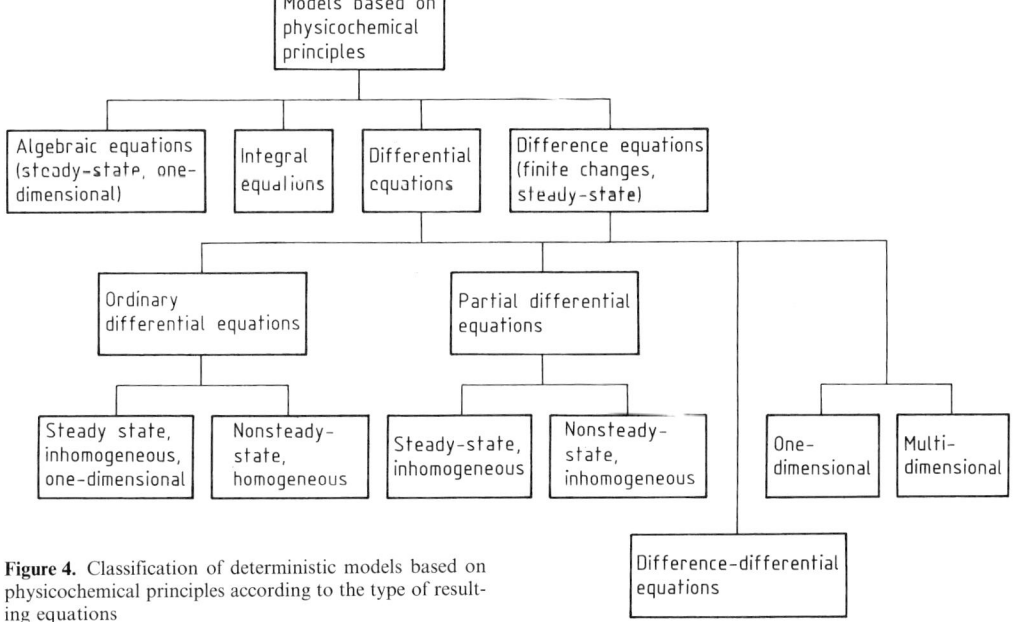

Figure 4. Classification of deterministic models based on physicochemical principles according to the type of resulting equations

models based on finite changes (see Sections 4.3.1 and 5.3.3.1).

Models based on transport equations for probability density functions generally give their results in the form of functionals $P(\Phi_1, \ldots, \Phi_n)$. These are defined in such a way that the probability of finding the dependent variables $(\Phi_1, \ldots, \Phi_n)$ in the small range $d\Phi_1, \ldots, d\Phi_n$ around the functions $\Phi_1(x, t), \ldots, \Phi_n(x, t)$ is $P(\Phi_1, \ldots, \Phi_n) d\Phi_1, \ldots, d\Phi_n$. Probability density functions $P(\Phi_1, \ldots, \Phi_n; x, t)$ that are defined analogously for given values of the independent variables x, t are more common. They give the probability that the dependent variables at a fixed distance $x = (x_1, x_2, x_3)$ and time t lie in the range $d\Phi_1, \ldots, d\Phi_n$ around the values $\Phi_1, \ldots, \Phi_n$ as $P(\Phi_1, \ldots, \Phi_n; x, t) d\Phi_1, \ldots, d\Phi_n$. These models provide the complete statistical information for statistical processes or give distribution functions of particular process variables. They are closely related to the description of statistical processes by the statistical moments discussed on p. 2–8, the difference is that the transport equations for the probability density functions are formulated.

Classical examples for models based on transport equations for probability density functions are to be found in statistical mechanics and kinetic gas theory. This concept was introduced in industrial systems to describe flow and mixing in nonideally mixed reactors [16]. One-dimensional models which represent the "macromixing" in a chemical reactor in terms of age (residence time) distributions are often sufficient to allow estimation of its behavior. Other simple pdf models are used in other areas, for example to model the crystal size distribution in crystallization, the activity distribution of catalyst pellets, or the age and size distribution of microbiological cultures [17]. Recently this concept has been used to describe reacting turbulent flow [18], [19].

Many industrial processes cannot be described with pdf models or models based on physicochemical principles because of their complexity. In these cases empirical models are used. In such cases the response of the process to variation of one or more process variables is known and the model correlates the process results. The simplest example of this is the fitting of a polynomial to experimental data. Another example in process control is the description of a process response in the form of transfer functions in the time or frequency domain. Empirical models are statistical (Fig. 3B) because the necessary data have to be obtained experimentally and contain statistical errors. Empirical models have only limited value for describing processes or process elements, e.g., if predictions are to be made outside the range covered by experimental data or if a theoretical approach is to be verified. In spite of this empirical mathematical models are still preferred in many cases.

Discussion on mathematical modeling in the following sections is based on the classification of mathematical models shown in Figures 3 and 4. Empirical models will be discussed first, followed by models based on the transport equations for probability density functions, and finally models based on physicochemical principles (transport phenomena). Presentation of the important principles of mathematical modeling will be given priority over a complete description.

Nonstandard numerical and mathematical methods required for solving the resulting set of equations for the developed mathematical models are discussed in more depth. More detailed classifications of industrial processes or process variables can be found in the literature [2], [10], [13]–[15]. Relevant aspects of such classifications will be discussed where necessary.

Mathematical modeling of physical or physicochemical phenomena or industrial processes is a concept that is applied to many other disciplines. Aspects of mathematical modeling related to other topics of engineering are described elsewhere in the Encyclopedia: → Chemical Reaction Engineering, **B4**; → Unit Operations, **B2, B3**; → 4. Transport Phenomena, this volume; → 5. Fluid Mechanics, this volume; → Process Control Engineering, **B6**.

3. Empirical Models

The fundamental variables of a process are known, for example, from a series of experiments. If, however, the complexity of the process prevents the formulation of a model on the basis of physicochemical principles, the process can be described by empirical models. The fitting of a polynomial or a similar function to a set of measured values is the simplest empirical model. An empirical model represents a general relationship:

$$y = f(x, \beta) \qquad (3)$$

where y is the vector of the dependent variables (model responses, process responses), x is the vector of the independent variables (model variables, process variables), and β is the vector of the parameters from which the functional relationship is constructed. This relationship is empirical and must be found on the basis of existing experimental data. In the following sections empirical models will be introduced through several examples.

3.1. Linear Empirical Models

If Equation (3) can be represented by the relation

$$y = \beta x \qquad (4)$$

the model is linear. Equation (4) is linear with respect to both its parameters and its independent variables. This implies that the first derivatives with respect to the components of the parameter vector and the vector of the independent variables disappear. The simplest relationship of type (4) is a linear model with one dependent variable:

$$y = \beta_0 + \beta_1 x \qquad (5)$$

where the estimates b_0 and b_1 of the parameters β_0 and β_1 have to be determined from experimental data. This will be outlined briefly in the following. Only the essential steps of the calculation will be presented. Detailed derivations and proofs are given in [2], [4, [5]], [7], [20].

3.1.1. Linear Empirical Models with One Variable

The determination of the estimates of the parameters β_0 and β_1 will be discussed in detail for a linear model with one dependent variable. This model is based on the following assumptions:

1) The dependent variable is a random variable so that the model in Equation (5) can be given in the following form:

$$\langle \bar{Y}_i \rangle = \bar{Y}_i - \varepsilon_i = \beta'_0 + \beta_1 (x_i - \bar{x}) \qquad (6)$$

where $\bar{Y}_i$ is the mean value of the dependent variables of n data sets with repeated measurements at x_i; ε_i is the random error in $\bar{Y}_i$.

Hence $(\bar{Y}_i - \langle \bar{Y}_i \rangle) = \varepsilon_i$ has an expected value of zero and a constant variance. The angular brackets denote the expected values or mathematical expectation.

2) The variance of Y_i is also constant, has a normal distribution, and is equal to the variance of ε_i.
3) The observations of Y are statistically independent so that the random errors ε_i are also statistically independent.
4) The variable x is not a random variable.

A process based on these assumptions is shown in Figure 5 and is discussed as an example in Section 3.1.1.4.

3.1.1.1. Parameter Estimation (Linear Regression)

Estimates b'_0 and b_1 of the parameters β'_0 and β_1 are found by minimizing the sum of the squares of the deviations between the observed values Y_i and the expected values of $\bar{Y}_i, \langle \bar{Y}_i \rangle$, i.e., the variance

$$\sum_{i=1}^{n} (\bar{Y}_i - \langle \bar{Y}_i \rangle)^2 \stackrel{!}{=} \mathrm{Min} \qquad (7)$$

This is known as least squares estimation.

The problem can also be posed in other ways. For the overdetermined system of equations for β'_0 and β_1

$$\bar{Y}_1 - \beta'_0 - \beta_1 (x_1 - \bar{x}) = 0 \qquad (8\,\mathrm{a})$$
$$\bar{Y}_2 - \beta'_0 - \beta_1 (x_2 - \bar{x}) = 0 \qquad (8\,\mathrm{b})$$
$$\vdots$$
$$\bar{Y}_n - \beta'_0 - \beta_1 (x_n - \bar{x}) = 0 \qquad (8\,n)$$

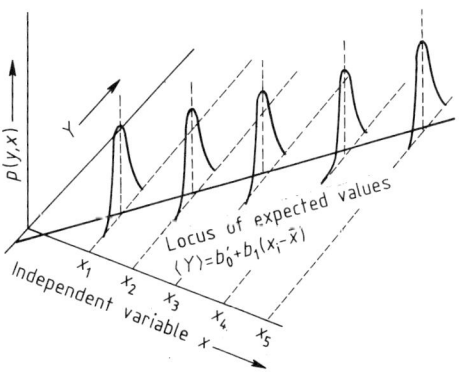

Figure 5. Representation of the statistical process described in Section 3.1.1

the estimated values b_0 and b_1 have to be found that reduce the residuals on the left-hand sides to as close to zero as possible.

The problem is the minimization of the sum:

$$\Phi = \sum_{i=1}^{n} p_i(\bar{Y}_i - \langle \bar{Y}_i \rangle)^2$$

$$= \sum_{i=1}^{n} p_i[\bar{Y}_i - \beta'_0 - \beta_1(x_i - \bar{x})]^2 \qquad (9)$$

where p_i are statistical weights determined by the number of measurements made for each data set, i.e., the number of measured values of the dependent variables at each value of the independent variable x. If Equation (9) is partially differentiated with respect to the parameters β'_0 and β_1 and the derivatives set to zero, rearrangement gives the normal equations

$$\sum_{i=1}^{n} p_i \bar{Y}_i = b'_0 \sum_{i=1}^{n} p_i + b_1 \sum_{i=1}^{n} p_i(x_i - \bar{x}) \qquad (10\,\text{a})$$

and

$$\sum_{i=1}^{n} p_i \bar{Y}_i(x_i - \bar{x}) = b'_0 \sum_{i=1}^{n} p_i(x_i - \bar{x})$$
$$+ b_1 \sum_{i=1}^{n} p_i(x_i - \bar{x})^2 \qquad (10\,\text{b})$$

in which the parameters β_0 and β_1 are replaced by their estimated values b'_0 and b_1. Since

$$\sum_{i=1}^{n} p_i(x_i - \bar{x}) \equiv 0 \qquad (11)$$

the coupling between Equations (10a) and (10b) disappears and they can be solved separately for b'_0 and b_1:

$$b'_0 = \frac{\sum_{i=1}^{n} p_i \bar{Y}_i}{\sum_{i=1}^{n} p_i} = \bar{Y} \qquad (12\,\text{a})$$

$$b_1 = \frac{\sum_{i=1}^{n} p_i \bar{Y}_i(x_i - \bar{x})}{\sum_{i=1}^{n} p_i(x_i - \bar{x})^2} \qquad (12\,\text{b})$$

By forming the second partial derivatives it can easily be shown that Equations (10a) and (10b) actually represent a minimum of the sum Φ. The simple uncoupled form only arises using the scaled form given in Equation (6).

3.1.1.2. Assessment of Estimated Parameter Values

Although the estimation of the parameter values is important, the quality of the empirical model is of particular interest (see points 6 and 7, p. 2-7). The first assessment of the empirical model is the test of the hypothesis $\langle \bar{Y}_i \rangle = \beta'_0 + \beta_1(x_i - \bar{x})$. For this the mean square deviations

$$s_r^2 = \frac{1}{n-2} \sum_{i=1}^{n} p_i(\bar{Y}_i - \hat{Y}_i)^2 \qquad (13)$$

and

$$s_\varepsilon^2 = \frac{\sum_{i=1}^{n} \sum_{j=1}^{p_i} (Y_{ij} - \bar{Y}_i)^2}{\sum_{i=1}^{n} p_i - n} \qquad (14)$$

are formed. Equation (13) represents the mean square of the residuals, i.e., the mean square deviations from the regression line. Equation (14) represents the mean square deviations within the data sets; it is a measure of the random error which occurs in the experiments at the points x_i.

The hypothesis $\langle Y_i \rangle = \beta'_0 + \beta_1(x_i - \bar{x})$ is confirmed if the mean squares of the residuals are significantly smaller than the mean deviations within the data sets. The hypothesis can be verified by means of the F-test [4], [5], [20]–[23]. If, for a preset confidence number $(1-\alpha)$, for example, 0.95

$$\frac{s_r^2}{s_\varepsilon^2} > F_{(1-\alpha)}^{(n-2),(\Sigma p_i - n)} \qquad (15)$$

then the mean squares of the residuals are significantly greater than the mean square deviations within one data set, and the hypothesis is not valid. In Equation (15) $F_{(1-\alpha)}^{(n-2),(\Sigma p_i - n)}$ is the value of the F-distribution for the confidence number $(1-\alpha)$ and the degrees of freedom $(n-2)$, $(\sum_i p_i - n)$. In the positive case both s_r^2 and s_ε^2 are good estimates for $\sigma_{\bar{Y}_i}^2$, the variance in $\bar{Y}_i$. If the hypothesis is valid, the variance of $\bar{Y}_i$ is required for the calculation of the variances of the estimated values of the model parameters and their confidence intervals. The variances as given by their definitions are

$$\text{Var}(b'_0) = \langle (b'_0 - \beta'_0)^2 \rangle = \frac{\sigma_{\bar{Y}_i}^2}{\sum_i p_i} \qquad (16)$$

(cf. Eq. 12a) and

$$\text{Var}(b_1) = \langle (b_1 - \beta_1)^2 \rangle = \frac{\sigma_{\bar{Y}_i}^2}{\sum_i p_i (x_i - \bar{x})^2} \quad (17)$$

(cf. Eq. 12b). The variance of the model response $\hat{Y}_i$ calculated from the linear model is given by

$$\text{Var}(\hat{Y}_i) = \text{Var}(b_0') + (x_i - \bar{x})^2 \text{Var}(b_1)$$

$$= \sigma_{\bar{Y}_i}^2 \left[\frac{1}{\sum_i p_i} + \frac{(x_i - \bar{x})^2}{\sum_i p_i (x_i - \bar{x})^2} \right] \quad (18)$$

since b_0' and b_1 are statistically independent. Equation (18) shows that the variance of Y_i is dependent on x and shows a minimum at $\bar{x}$. A first assessment of the linear model defined by the estimated parameter values b_0' and b_1 is achieved by means of the calculated variances.

In addition to the variances, which are a measure of the scatter of the estimated values, the confidence limits of the estimated values are used and can be calculated from the variances for further assessment. Since b_0' and b_1 are linear combinations of Y_i (cf. Eq. 12), they also have a normal distribution about β_0' and β_1 respectively because of the assumed normal distribution of Y_i. Under these conditions the Student t-variable can be formed [4], [5], [21]–[23]:

$$t = \frac{b_0' - \beta_0'}{s_{b_0'}} = \frac{b_0' - \beta_0'}{\sigma_{\bar{Y}_i}/(\sum_i p_i)^{1/2}} \quad (19)$$

This has a t-distribution with $\sum_i p_i - 2$ degrees of freedom. Estimates of the confidence interval can be obtained using the t-distribution in the following form:

$$b_0' - t_{(1-\alpha/2)}^{(\sum_i p_i - 2)} s_{b_0'} \leq \beta_0' \leq b_0' + t_{(1-\alpha/2)}^{(\sum_i p_i - 2)} s_{b_0'} \quad (20)$$

With known $\sigma_{\bar{Y}_i}^2$ the interval in which the expected value β_0' lies is given. Values of the t-distribution can be found in statistics books [4], [5], [21]–[23].

An analogous procedure is adopted for β_1. The Student t-variable is then

$$t = \frac{b_1 - \beta_1}{s_{b_1}} = \frac{b_1 - \beta_1}{\sigma_{\bar{Y}_i}^2/[\sum_i p_i (x_i - \bar{x})^2]^{1/2}} \quad (21)$$

This quantity also has a t-distribution with $\sum_i p_i - 2$ degrees of freedom. As before, the confidence interval for β_1 becomes:

$$b_1 - t_{(1-\alpha/2)}^{(\sum_i p_i - 2)} s_{b_1} \leq \beta_1 \leq b_1 + t_{(1-\alpha/2)}^{(\sum_i p_i - 2)} s_{b_1} \quad (22)$$

Confidence intervals for the expected values $\langle \bar{Y}_i \rangle$ are defined in a similar fashion. Since the $\bar{Y}_i$ values are assumed to have a normal distribution around $\langle \bar{Y}_i \rangle$, the t-variable can be expressed in the form

$$t = \frac{\bar{Y}_i - \langle \bar{Y}_i \rangle}{s_{\bar{Y}_i}} = \frac{\bar{Y}_i - \langle \bar{Y}_i \rangle}{\sigma_{\bar{Y}_i}\left/\left[\frac{1}{\sum_i p_i} + \frac{(x_i - \bar{x})^2}{\sum_i p_i (x_i - \bar{x})^2}\right]^{1/2}\right.} \quad (23)$$

which again has a t-distribution with $\sum_i p_i - 2$ degrees of freedom. In analogy with Equations (20) and (22) the confidence interval for $\langle \bar{Y}_i \rangle$ is as follows:

$$\hat{Y}_i - t_{(1-\alpha/2)}^{(\sum_i p_i - 2)} s_{\bar{Y}_i} \leq \langle \bar{Y}_i \rangle \leq \hat{Y}_i + t_{(1-\alpha/2)}^{(\sum_i p_i - 2)} s_{\bar{Y}_i} \quad (24)$$

Further discussion and illustration of the method of parameter estimation and assessment are given in Section 3.1.1.4.

3.1.1.3. Sensitivity Analysis

In addition to the analysis of the variances of β_0', β_1, and $\bar{Y}_i$, the sensitivity of the linear empirical model $\langle \bar{Y}_i \rangle = \beta_0' + \beta_1(x_i - \bar{x})$ can help in interpreting the results obtained from the model as well as in the design of experiments for establishing an empirical model. The sensitivity of the model may be described in various ways. In order to select experiments and methods of measurement for verification of the fundamental variables of the process, it is useful to have a measure of the sensitivity of the sum of the squares of the deviations and of the estimated parameter values with respect to Y_i. Sensitivity can be regarded as the relative change in the sum of squares of the deviations $\partial \Phi_{min}/\Phi_{min}$ or the relative change in the estimated parameter values $\partial b_i/b_i$ for a relative change $\partial \bar{Y}_i/\bar{Y}_i$ in $\bar{Y}_i$. Since

$$\Phi_{min} = \sum_{i=1}^{n} p_i (\bar{Y}_i - \hat{Y}_i)^2 \quad (25)$$

(cf. Section 3.1.1.1), the sensitivity of Φ_{min} with

respect to $\bar{Y}_i$ is:

$$\frac{\partial \Phi_{min}}{\Phi_{min}} \bigg/ \frac{\partial \bar{Y}_i}{\bar{Y}_i} = \frac{2\bar{Y}_i p_i(\bar{Y}_i - \hat{Y}_i)}{\sum_{i=1}^{n} p_i(\bar{Y}_i - \hat{Y}_i)^2} \quad (26)$$

Hence, relative change in a value $\bar{Y}_i$ causes a change in the minimum of the sum of squares, which increases as the deviation between $\bar{Y}_i$ and the model solution $\hat{Y}_i$ becomes larger and as the minimum becomes smaller.

The sensitivity of the estimated parameter values b'_0 and b_1 to changes in $\bar{Y}_i$ can be obtained simply and directly from Equations (12a) and (12b):

$$\frac{\partial b'_0}{b'_0} \bigg/ \frac{\partial \bar{Y}_i}{\bar{Y}_i} = \frac{1}{\sum_i p_i} \frac{p_i \bar{Y}_i}{b'_0} \quad (27)$$

and

$$\frac{\partial b_1}{b_1} \bigg/ \frac{\partial \bar{Y}_i}{\bar{Y}_i} = \frac{1}{\sum_{i=1}^{n} p_i(x_i - \bar{x})^2} \frac{p_i \bar{Y}_i(x_i - \bar{x})}{b_1} \quad (28)$$

The sensitivity obtained from Equations (27) and (28) is equally simple to interpret. The more reliable the measured value is, i.e., the larger p_i becomes in comparison to $\sum_i p_i$, the more rapidly b'_0 changes with $\bar{Y}_i$. The same applies to b_1, here the interval ($x_i - \bar{x}$) at which the values are measured has an additional weight.

Finally, an estimate of the sensitivity of the model with respect to its parameters is a sensible measure to evaluate the experimental effort required in the model assessment. This sensitivity is defined by the relative change in the expected value $\langle\bar{Y}_i\rangle$, $\partial\langle\bar{Y}_i\rangle/\langle\bar{Y}_i\rangle$, caused by a change of $\partial\beta_i/\beta_i$ in the model parameter. For the linear model described here formulation of the problem is trivial since the solutions

$$\frac{\partial\langle\bar{Y}_i\rangle}{\langle\bar{Y}_i\rangle} \bigg/ \frac{\partial\beta'_0}{\beta'_0} = \frac{\beta'_0}{\langle\bar{Y}_i\rangle} \quad (29)$$

and

$$\frac{\partial\langle\bar{Y}_i\rangle}{\langle\bar{Y}_i\rangle} \bigg/ \frac{\partial\beta_1}{\beta_1} = \frac{\beta_1}{\langle\bar{Y}_i\rangle}(x_i - \bar{x}) \quad (30)$$

are obvious and can be derived directly from Equation (6). For more complex models there are other more subtle relationships for deciding the degree of complexity of the experiments required for the model development. This type of sensitivity analysis will be discussed in detail later (Section 5.3.3.1).

3.1.1.4. Example of a Linear Model

The determination of the model parameters will now be illustrated using a simple example of a linear model with one dependent variable, as defined in Section 3.1.1.1.

As an example we will consider the correlation of the mean heat-transfer coefficients for hot air at the wall of a pipe with turbulent flow (Table 1). The gas velocity (indicated by x in Table 1) is varied and can be precisely adjusted so that it can be regarded as a deterministic variable. The measured values (the mean heat-transfer coefficients denoted by Y) are statistical quantities which satisfy the assumptions presented on p. 2–11. Table 1 also contains all conversions necessary for linear regression.

According to Equations (12a) and (12b), the estimated values for the parameters of the linear model $\langle\bar{Y}_i\rangle = \beta'_0 + \beta_1(x_i - \bar{x})$ are $b'_0 = 15.002$ and $b = 1.282$. To test the hypothesis, the mean square deviations s_r^2 and s_ε^2 are first obtained from Equations (13) and (14): $s_r^2 = 7.749/3 = 2.583$ and $s_\varepsilon^2 = 7.613/9 = 0.846$. The ratio of the mean square deviations is then 3.05. The associated F-value [4] is $F_{0.95}^{3,9} = 3.86$, so that the hypothesis is valid for a significance level of 0.05. To obtain an estimated value $s_{\bar{Y}_i}^2$ for $\sigma_{\bar{Y}_i}^2$ it is better to use the combined square deviations from s_r^2 and s_ε^2:

$$s_{\bar{Y}_i}^2 = \frac{\sum_{i=1}^{n}\sum_{j=1}^{p_i}(Y_{ij} - \hat{Y}_i)^2}{\sum_{i=1}^{n} p_i - 2} = \frac{7.613 + 7.746}{9 + 3} = 1.28$$

The variances of the estimated values of the model parameters can be obtained using Equations (16) and (17):

$$\text{Var}(b'_0) = \frac{1.28}{14} = 9.14 \times 10^{-2}$$

$$\text{Var}(b_1) = \frac{1.28}{206.25} = 6.20 \times 10^{-3}$$

and finally the x_i-dependent variance of the model responses $\hat{Y}_i$ are given by Equation (18).

Table 1. Example of linear regression, model with one independent variable

Data set	p_i	x_i	Y	$\bar{Y}_i$	$p_i(x_i-\bar{x})\bar{Y}_i$	$p_i(x_i-\bar{x})^2$	$(Y-\bar{Y}_i)^2$	$(\bar{Y}_i-\hat{Y}_i)^2$	$(x_i-\bar{x})$
1	3	8.00	8.08	9.09	−136.35	75.00	1.020	0.248	−5.00
			9.95				0.739		
			9.26				0.029		
2	3	10.50	11.21	12.18	− 91.35	18.75	0.941	0.146	−2.50
			12.36				0.032		
			12.98				0.640		
3	2	13.00	14.12	15.04	0.00	0.00	0.846	0.001	0.00
			15.96				0.846		
4	2	15.50	16.80	17.71	88.55	12.50	0.828	0.247	2.50
			18.62				0.828		
5	4	18.00	19.60	20.18	403.60	100.00	0.336	1.517	5.00
			20.08				0.010		
			20.90				0.518		
			20.16				0.004		
$\sum n = 5$	$\sum_i p_i = 14$	$\bar{x} = 13.00$	$\bar{Y} = 15.002$		$\sum_i p_i \bar{Y}_i(x_i-\bar{x})$ $= 264.45$	$\sum_i p_i(x_i-\bar{x})^2$ $= 206.25$	$\sum_i(Y-\bar{Y}_i)^2$ $= 7.613$	$\sum_i p_i(\bar{Y}_i-\hat{Y}_i)^2$ $= 7.746$	

$$\text{Var}(\hat{Y}_i) = 1.28\left[\frac{1}{14} + \frac{1}{206.25}(x_i - \bar{x})^2\right]$$

$$= 9.14 \times 10^{-2} + 6.20 \times 10^{-3}(x_i - \bar{x})^2$$

The individual confidence intervals of the expected values of these three quantities can be calculated for a significance level of 0.05 using Equations (20), (22), and (24) ($t_{0.975}^{(\sum p_i - 2)} = 2.179$, [4]):

$$b'_0 - 0.658 \leq \beta'_0 \leq b'_0 + 0.658$$

and

$$b_1 - 0.171 \leq \beta_1 \leq b_1 + 0.171$$

The individual confidence intervals for $\langle \bar{Y}_i \rangle$ are also dependent on x_i because of the dependence of the variance on x_i:

$$\hat{Y}_i - [1.99 \times 10^{-1} + 1.35 \times 10^{-2}(x_i - \bar{x})^2]^{1/2}$$
$$\leq \langle \bar{Y}_i \rangle \leq \hat{Y}_i + [1.99 \times 10^{-1} + 1.35 \times 10^{-2}(x_i - \bar{x})^2]^{1/2}$$

The results of the regression analysis are illustrated in Figure 6: the regression line, the confidence intervals of the parameters, and the confidence intervals for the expected values of the model responses are shown. In spite of the verification of the hypothesis, the results of the linear regression are only moderately satisfactory due to the relatively large scatter of the measurements. Sensitivity analysis emphasizes this. For example, for data set number 5 from Table 1

$$\frac{\partial \Phi_{\min}}{\Phi_{\min}} \bigg/ \frac{\partial \bar{Y}_5}{\bar{Y}_5} = \frac{2 \times 20.28 \times 4 \times 1.231}{7.746} = 25.66$$

A change in the measured value $\bar{Y}_5$ of 1% causes a change of 25.55% in $\Phi_{\min}$. Similarly high sensitivities arise with

$$\frac{\partial b'_0}{b'_0} \bigg/ \frac{\partial \bar{Y}_5}{\bar{Y}_5} = \frac{1}{14} \cdot \frac{4 \times 20.18}{15.002} = 0.384$$

and

$$\frac{\partial b_1}{b_1} \bigg/ \frac{\partial \bar{Y}_5}{\bar{Y}_5} = \frac{1}{206.25} \cdot \frac{4 \times 20.18 \times 5}{1.282} = 1.526$$

so that the experimenter concerned with this task would be well advised to repeat the experiments with a more precise measurement technique. In fact other experiments for this example give the

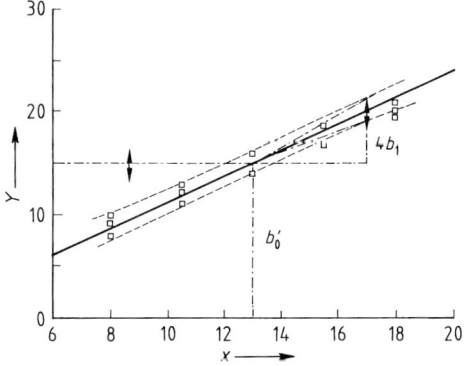

Figure 6. Results of linear regression for the example discussed in Section 3.1.1.4
Arrows indicate the confidence intervals of the parameters, square symbols denote measured values, and the dashed lines the confidence intervals of the expected values.

relationship

$$Nu = \frac{\xi/8\,(Re - 1000)\,Pr}{1 + 12.7\sqrt{\xi/8}\,(Pr^{2/3} - 1)} \left[1 + \left(\frac{D_R}{L}\right)^{2/3}\right]$$

where $\xi = (1.82 \log Re - 1.64)^{-2}$ [24], [25]: this relationship does not produce a linear dependency $Nu = f(Re)$ and hence a linear model appears to be inadequate.

3.1.1.5. Concluding Remarks

The assumptions stated on p. 2–11 are not always appropriate; for example, the random error of the measurements may be a function of the independent variables. In addition, the random errors may be statistically dependent on each other and, in a given process, both the dependent and the independent variables can be statistical quantities. Estimation of the model parameters by linear regression in such cases does not differ in principle from that presented so far: more complex calculations may be required and further hypotheses or statistical rules may be needed. For discussion of these special cases, the reader is referred to the literature [4], [5], [7], [20].

Standardized computer programs are available for linear regression by least squares estimations for models with one dependent variable [26]. The solution of linear equations required for linear regression is also available through program libraries for numerical problems or for problems in linear algebra. Most of these libraries are readily available [26]–[32], so that special-purpose computer programs can easily be assembled for particular requirements.

3.1.2. Linear Empirical Models with Several Variables

In the case of linear empirical models with several independent variables, similar problems to those in Section 3.1.1 arise. Thus in the following, only the generalized notation will be discussed.

The linear relationship for a model with several independent variables is given by

$$y = \boldsymbol{\beta}\,\boldsymbol{x} \tag{31}$$

and in extended notation

$$y = \beta_0 + \beta_1 x_1 + \beta_2 x_2 + \ldots + \beta_q x_q \tag{32}$$

so that the problem is extended to q variables and $(q + 1)$ parameters. The same assumptions hold as in Section 3.1.1, so that the model given by Equation (31) can be written in the form

$$\langle \bar{Y}_i \rangle = \bar{Y}_i - \varepsilon_i = \beta_0' + \sum_{j=1}^{q} \beta_j (x_{ij} - \bar{x}_j) \tag{33}$$

The $(q+1)$ model parameters are calculated from the measured values defined at the n data points in the same way as in Section 3.1.1.

In matrix notation Equation (33) reads

$$\langle \bar{\boldsymbol{Y}} \rangle = \bar{\boldsymbol{Y}} - \boldsymbol{E} = \boldsymbol{\beta}\,\boldsymbol{x}' \tag{34}$$

with the abbreviations

$$\langle \bar{\boldsymbol{Y}} \rangle = \begin{pmatrix} \langle \bar{Y}_1 \rangle \\ \langle \bar{Y}_2 \rangle \\ \vdots \\ \langle \bar{Y}_n \rangle \end{pmatrix}, \quad \boldsymbol{\beta} = \begin{pmatrix} \beta_0' \\ \beta_1 \\ \vdots \\ \beta_q \end{pmatrix},$$

and $\boldsymbol{x}' = \begin{pmatrix} 1 & (x_{11} - \bar{x}_1) & \cdots & (x_{1q} - \bar{x}_q) \\ 1 & (x_{21} - \bar{x}_1) & \cdots & (x_{2q} - \bar{x}_q) \\ \vdots & \vdots & \ddots & \vdots \\ 1 & (x_{n1} - \bar{x}_1) & \cdots & (x_{nq} - \bar{x}_q) \end{pmatrix}\,p$

The first step is the same as for models with only one independent variable—the estimation of model parameters $\boldsymbol{\beta}$.

3.1.2.1. Parameter Estimation

As in Section 3.1.1.1, the estimates of the parameters $\boldsymbol{\beta}$ are found by minimizing the sum of squares of the deviations between the observed values $\bar{Y}_i$ and the expected values $\langle \bar{Y}_i \rangle$ (i.e., the sum of the variances):

$$(\bar{\boldsymbol{Y}} - \langle \bar{\boldsymbol{Y}} \rangle)^T \boldsymbol{p}\,(\bar{\boldsymbol{Y}} - \langle \bar{\boldsymbol{Y}} \rangle) \stackrel{!}{=} \text{Min} \tag{35}$$

By using Equation (34), the sum of squares of the deviations can be formulated as

$$\Phi = (\bar{\boldsymbol{Y}} - \boldsymbol{\beta}\,\boldsymbol{x}')^T \boldsymbol{p}\,(\bar{\boldsymbol{Y}} - \boldsymbol{\beta}\,\boldsymbol{x}') \tag{36}$$

which has to be differentiated with respect to all the elements of vector $\boldsymbol{\beta}$ and then set to zero.

In Equation (36) p once again represents the number of repeated measurements or, in general terms, the statistical weights. In matrix notation

$$p = \begin{pmatrix} p_1 & 0 & \cdots & 0 \\ 0 & p_2 & \cdots & 0 \\ \vdots & \vdots & \ddots & \vdots \\ 0 & 0 & \cdots & p_n \end{pmatrix}$$

Differentiation yields

$$\frac{\partial \Phi}{\partial \beta} = \begin{pmatrix} \frac{\partial \Phi}{\partial \beta_0} \\ \frac{\partial \Phi}{\partial \beta_1} \\ \vdots \\ \frac{\partial \Phi}{\partial \beta_q} \end{pmatrix} = \frac{2[\partial(\bar{Y} - \beta x')^T] p (\bar{Y} - \beta x')}{\partial \beta} \quad (37)$$

so that for a minimum to occur

$$(x')^T p (\bar{Y} - \beta x') = 0 \quad (38)$$

Further differentiation shows that Equation (38) actually represents a minimum. After rearrangement the normal equation

$$(x')^T p x' b = (x')^T p \bar{Y} \quad (39)$$

is obtained.

In Equation (39) the parameters β_i are replaced by their estimated values. Equation (39) is the matrix equivalent of Equations (9a, b). Solution of Equation (39) for the parameters gives

$$b = [(x')^T p x']^{-1} (x')^T p \bar{Y} \quad (40)$$

Inversion of the matrix $(x')^T p x \equiv a$ is required for the solution. A condition for this is that the determinant is not equal to zero, or in other words that matrix a is nonsingular. Numerical problems related to ill-conditioned matrices a can be avoided by suitably scaling or normalizing the independent variables or by suitable experimental design. This can be illustrated by a simple example. From measurements at

$$x = \begin{pmatrix} 9.5 \\ 10.0 \\ 10.5 \end{pmatrix},$$

a relationship of the form $y = \beta_0 + \beta_1 x$ will be investigated. Estimation of the model parameters involves inversion of the matrix

$$a \equiv x^T p x = \begin{pmatrix} 3.0 & 30.0 \\ 30.0 & 300.5 \end{pmatrix}$$

As can be easily verified,

$$\text{Det}(a) = 3.0 \times 300.5 - 30.0 \times 30.0$$
$$= 901.5 - 900.0 = 1.5$$

However, because of the spread of the measured values, the matrix is ill-conditioned for inversion since $\text{Det}(a) = 0$ if only the first two digits are significant. If it is scaled instead and we use $y = \beta'_0 + \beta_1(x - \bar{x})$ then

$$a \equiv (x')^T p x' = \begin{pmatrix} 3.0 & 0 \\ 0 & 0.5 \end{pmatrix}$$

or $\text{Det}(a) = 3.0 \times 0.5 - 0 = 1.5$

The result is the same. The problem of subtraction of large numbers which arises from the design of the experiments is thereby avoided. As can be seen from the numbers given above, the problem of handling differences between large numbers is aggravated if the empirical model is in the form of a polynomial, i.e.,

$$y = \beta_0 + \beta_1 x_1 + \beta_2 x_2$$

with $x_1 = x$, $x_2 = x^2$ etc.

Numerical problems with the inversion of a are also avoided if the main diagonal of a can be occupied predominantly through proper normalization of the independent variables. This technique can be demonstrated using the simple example of measured values at

$$x = \begin{pmatrix} 5.0 & 10.0 \\ 5.0 & 20.0 \\ 10.0 & 10.0 \\ 10.0 & 20.0 \end{pmatrix}$$

If the variables are normalized in the form

$$\tilde{x}_{i1} = \frac{x_{i1} - 7.5}{2.5}$$

and $\tilde{x}_{i2} = \frac{x_{i2} - 15.0}{5.0}$ then

$$\tilde{x}' = \begin{pmatrix} 1.0 & -1.0 & -1.0 \\ 1.0 & -1.0 & 1.0 \\ 1.0 & 1.0 & -1.0 \\ 1.0 & 1.0 & 1.0 \end{pmatrix}$$

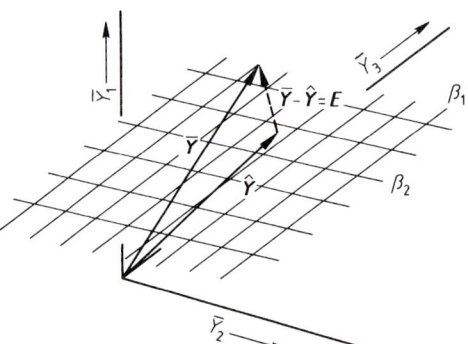

Figure 7. Geometrical illustration of the normal equations for a linear model

With $a \equiv (\tilde{x}')^T p \tilde{x}'$ this leads to the occupation of the main diagonal only with

$$a = \begin{pmatrix} 4.0 & 0 & 0 \\ 0 & 0.4 & 0 \\ 0 & 0 & 4.0 \end{pmatrix}$$

so that inversion does not cause any problems. The variables normalized in this form are orthogonal since $\sum x_0 x_{i1} = \sum x_0 x_{i2} = \sum x_{i1} x_{i2} = 0$. Ill-conditioning of a can therefore be avoided by planning the experiment in a manner which allows simple orthogonalization of the dependent variables.

A brief geometric explanation of the normal equations will now be given. Figure 7 shows the vector $Y = (\bar{Y}_1, \bar{Y}_2, \bar{Y}_3)$ in the space of measured values. Since the relationship $y = \beta_0 + \beta_1(x - \bar{x})$ is linear in its parameters, a plane surface for the calculated model solutions that depends on the estimated values b_0' and b_1 results.

The set of estimated values b_0' and b_1 that gives the shortest distance between the end points of the vectors $\bar{Y}$ and $\hat{Y}$ now has to be found. Obviously the shortest distance $\bar{Y} - \hat{Y}$ is given by the normal from the end of $\bar{Y}$ onto the surface of the calculated model solution. The normal Equations (10) and (39) fix b such that the vector of the residuals $\bar{Y} - \hat{Y}$ passes through the end point of $\bar{Y}$ and is perpendicular to the surface of values calculated from the model.

3.1.2.2. Assessment of Estimated Parameter Values

The approach used to assess the estimated parameter values is the same as that in Section 3.1.1.2, and so will only be briefly discussed. To hypothesis $\langle \bar{Y} \rangle = \beta x'$, the mean square deviations from the regression line must be obtained. If $\bar{Y} - \langle \bar{Y} \rangle$ is represented by E then

$$s_r^2 = \frac{(E^T p E)}{n - q - 1} \tag{41}$$

where s_r^2 is the sum of squares of the residuals divided by the number of degrees of freedom. The number of degrees of freedom is the number of data points minus the number of conditions constraining the minimization, in this case the number of parameters. The mean square deviations for the individual data points are calculated from Equation (19), so that the F-test can be conducted using the ratio s_r^2/s_ε^2. The variances of the parameters and the model solutions can be calculated by means of the estimated values $s_{\bar{Y}_i}^2$ of $\sigma_{\bar{Y}_i}^2$ which are obtained from s_r^2 and/or s_ε^2 in the case of a positive F-test. The covariances of the parameters are

$$\text{Cov}(b) = \langle (b - \beta)^T (b - \beta) \rangle = \sigma_{\bar{Y}_i}^2 a^{-1} \tag{42}$$

If a^{-1} is denoted as c, then the variances of the estimated parameter values can be obtained from the diagonal elements of c with estimated values $s_{\bar{Y}_i}^2$ for $\sigma_{\bar{Y}_i}^2$:

$$s_{b_k}^2 = s_{\bar{Y}_i}^2 c_{kk} \tag{43}$$

With

$$\text{Var}(\hat{Y}_i) = \text{Var}(x' b) = x' \text{Var}(b)(x')^T$$
$$= \sigma_{\bar{Y}_i}^2 x' c (x')^T \tag{44}$$

and $s_{\hat{Y}_i}^2$ for $\sigma_{\hat{Y}_i}^2$, the following is obtained:

$$s_{\hat{Y}_i} = s_{\bar{Y}_i} \sqrt{x' c (x')^T} \tag{45}$$

so that, as in Section 3.1.1.2 (Eqs. 20, 22, and 24), the confidence intervals for b and $\hat{Y}_i$ are obtained.

More detailed methods for the assessment of the estimated parameter values for models with several independent variables are treated in [4], [5], [20].

3.1.2.3. Concluding Remarks

The sensitivity analysis for empirical models with several independent variables will not be discussed. The relationships presented in Section 3.1.1.3 can be used, and so need not be repeated.

The basic rules of linear regression were illustrated using an example with one independent variable, a further example for a model with several independent variables will not be given. Cases which do not comply with the assumptions presented here are described in the literature; for example, processes with random errors which are not normally distributed or with random errors which depend on x are discussed in [4], [7]. Standardized computer programs are available for the treatment of empirical linear models with several independent variables or for multiple linear or polynomial regression [26], [33]. Programs can be easily constructed to fit users' requirements with the aid of modules from extensive program libraries for matrix and linear algebra [27]–[32].

3.2. Nonlinear Empirical Models

In nonlinear models the relationship from Equation (3) cannot be represented by Equation (4). The derivatives of the model equation with respect to the parameters are themselves functions of the parameters. The nonlinear model $y = f(x, \beta)$ is once again based on several assumptions:

1) The dependent variable is a random variable for which n data sets of measured values $\bar{Y}_i$ are available. The model can therefore be represented in the form

$$\langle \bar{Y} \rangle = \bar{Y} - E = f(x, \beta) \qquad (46)$$

where E once again represents the random error in $\bar{Y}_i$, so that $\bar{Y} - \langle \bar{Y} \rangle = E$. The expected value of the random error is zero and its variance is constant (independent of x).
2) The random variable Y has a normal distribution about $\bar{Y}_i$ and has the same variance as ε_i.
3) Both the random variable and the random error are statistically independent.
4) The variables x are not random variables.

The empirical nonlinear statistical model can be represented analogously to that in Figure 5, whereby the relationship is nonlinear and is characterized by several independent variables. As with the linear model, the first task consists of estimating the model parameters using the experimental process data.

3.2.1. Parameter Estimation (Nonlinear Regression)

Estimation of the model parameters β involves the now familiar task of defining the estimated values b such that the sum of squares of the deviations between the values $\bar{Y}_i$ observed at the n data points and the estimated values $\langle \bar{Y}_i \rangle$ is a minimum. [In the following, the equations are expressed in a simplified form, without repeated measurements, i.e., with unit statistical weights p, Det $(p) = 1$]:

$$(\bar{Y} - \langle \bar{Y} \rangle)^T (\bar{Y} - \langle \bar{Y} \rangle) \stackrel{!}{=} \text{Min} \qquad (47)$$

A geometric representation of this task is shown in Figure 8. The end point of the vector $\bar{Y} = (\bar{Y}_1, \bar{Y}_2, \bar{Y}_3)$ in the space of measured values should be linked by the shortest possible distance with the end point of the vector $\hat{Y}$. This is once again given by the normal to the surface of model responses in the parameter space, which in this case is curved because of the nonlinearity of the relationship $y = f(x, \beta)$.

Partial differentiation of the sum of squares of the deviations

$$\Phi = [\bar{Y} - f(x, \beta)]^T [\bar{Y} - f(x, \beta)] \qquad (48)$$

with respect to the parameters and then setting the derivatives to zero produces a system of non-

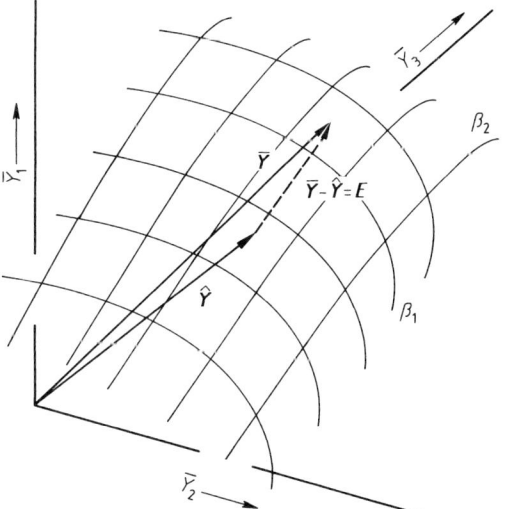

Figure 8. Geometrical illustration of the normal equation for a nonlinear model

linear normal equations. This is the main difference from linear regression in which a system of linear equations in $\boldsymbol{\beta}$ is obtained for the normal equations. This can be easily verified by analyzing the sums of squares (Eqs. 9, 36, and 48).

Nonlinear regression is therefore the solution of a system of nonlinear equations for $\boldsymbol{\beta}$.

3.2.1.1. Transformation of the Model into Linear Form

The simplest method of estimating parameters in nonlinear regression is transformation of the model equations into linear form to avoid the problem of solving the nonlinear equations described above. Certain types of nonlinear models can be transformed into linear form but this may involve several problems; this will now be briefly illustrated.

The relationship

$$y = \beta_0 x_1^{\beta_1} x_2^{\beta_2} \tag{49}$$

e.g., the rate of a complex chemical reaction, can be transformed into linear form by taking logarithms:

$$\log y = \log \beta_0 + \beta_1 \log x_1 + \beta_2 \log x_2 \tag{50}$$

The transformed model is therefore

$$\log \bar{Y}_i - \varepsilon_i = \log \beta_0 + \beta_1 \log x_{i1} + \beta_2 \log x_{i2} \tag{51}$$

The previously presented methods of regression analysis for testing hypotheses and calculating the confidence intervals of the parameters or model solutions presume that the additive errors ε_i and random variables have normal distributions. For the nonlinear model of Equation (49), the additive error must be multiplicative:

$$\bar{Y}_i = \beta_0 x_{i1}^{\beta_1} x_{i2}^{\beta_2} \varepsilon_i' \tag{52}$$

and, like the random variable Y_i, must have a log-normal distribution to satisfy the above-mentioned presumptions. This is not always the case and must be experimentally verified.

Another example is the relationship of the form

$$y = \frac{\beta_1 \beta_2 x}{(1 + \beta_2 x)^2} \tag{53}$$

which represents, for example, the rate of a heterogeneous catalytic reaction (Hougen–Watson–Langmuir–Hinshelwood kinetics). Transformation into a linear model gives

$$\left(\frac{x}{y}\right)^{1/2} = \frac{1}{(\beta_1 \beta_2)^{1/2}} + \frac{\beta_2}{(\beta_1 \beta_2)^{1/2}} x \tag{54}$$

The additive statistical error ε_i in the linear model for the estimation of the parameters (Eq. 54) is not identical to the random error of the nonlinear model (Eq. 53). Inverse transformation of Equation (54) gives:

$$y = \frac{\beta_1 \beta_2 x}{(1 + \beta_2 x)^2} + \frac{\beta_1 \beta_2}{(1 + \beta_2 x)^2} (y \varepsilon'^2 - 2\sqrt{y x \varepsilon'}) \tag{55}$$

which contains an additive random error that is dependent on x, y, β_1, and β_2.

3.2.1.2. Direct Search Methods

Another way of circumventing the problem of solving nonlinear equations is to use direct search methods. For this the normal equations are not derived from the sum of squares of the deviations (Eq. 48), instead the minimum of the function Φ is sought directly. Vectorial search methods [34], Simplex methods [35], and Simplex methods with variable Simplex geometry [36] are the most commonly used direct search methods. The structure of these methods is presented schematically for a model with two parameters in Figure 9.

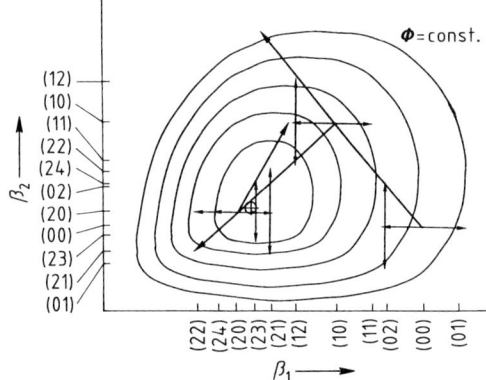

Figure 9. Representation of direct search methods
The first digits in the parentheses indicate the number of iteration. The second digits denote the number of successive variation of β_i. For example, (02) on the β_1 axis means the initial guess of β_1 and the second variation

Vectorial search methods start from estimated values $b_i^{(0)}$ for the parameters β_i. The direction of the search vector is established from successive calculations of Φ by varying the estimated values b_i. The direction of the vector leading to a reduction in Φ is used to calculate improved estimates of b_i. If there is no further reduction in Φ along the search vector, the direction is redefined and the minimum of Φ is located by reducing the step size (Fig. 9).

Simplex methods are more effective, especially with large numbers of parameters. For a two-parameter problem (Fig. 10), a regular Simplex is an equilateral triangle which, starting from estimated values $b_i^{(0)}$, is placed at a point on the surface of Φ in the parameter space. During the search, Φ is evaluated at each of the three vertices, and the triangle is reflected about the side opposite the largest value of Φ. If there is no further reduction in Φ after reflection, then the Simplex is reduced in size to allow more accurate localization of the minimum.

Simplexes with variable geometry are obtained by expanding after successful reflection or contracting after unsuccessful reflection. An example is given in Figure 11 for a two-parameter problem. The reflection algorithm is

$$b_{ir} = (1 + \gamma_r) b_{ic} - \gamma_r b_{i\,\text{max}}$$

where b_{ir}, $b_{i\,\text{max}}$, and b_{ic} are the coordinates of the reflected Simplex point, of the Simplex point with the largest sum of squares of deviations, and of the midpoint of the side lying opposite it, respectively; γ_r is the reflection factor. The expansion algorithm is

$$b_{ie} = \gamma_e b_{ir} + (1 - \gamma_e) b_{ic}$$

where b_{ie} are the coordinates of the expanded Simplex point and γ_e is the expansion factor. After unsuccessful expansion or reflection, contraction is achieved with the algorithm

$$b_{ik} = (1 - \gamma_k) b_{ic} + \gamma_k b_{i\,\text{max}}$$

where b_{ik} and γ_k have analogous meanings to those in reflection or expansion.

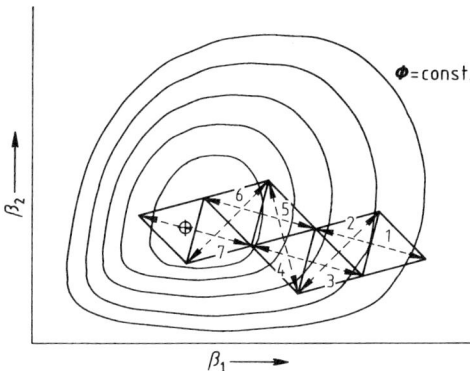

Figure 10. Representation of the Simplex method
The numbers represent the number of reflection

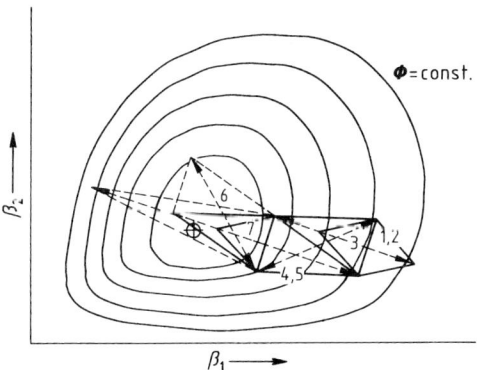

Figure 11. Representation of the Simplex method with variable geometry
The numbers represent the number of reflection, expansion, or contraction

3.2.1.3. Gradient Methods

Discussion of the methods for solving nonlinear systems of equations arising from the minimum conditions for Φ will be limited to two methods. Further methods are described in the literature on nonlinear optimization [5], [7], [20], [37]–[39]; see also → 1. Mathematics in Chemical Engineering, pp. 1-106 – 1-128.

Through use of the abbreviation

$$\langle \bar{Y} \rangle \equiv \eta = f(x, \beta) \tag{56}$$

the sum of the squares of the deviations (Eq. 48) can be written as

$$\Phi = (\bar{Y} - \eta)^T (\bar{Y} - \eta) \tag{57}$$

The condition for the minimum with respect to

the parameters $\boldsymbol{\beta}$ then becomes:

$$\frac{\partial \boldsymbol{\Phi}}{\partial \boldsymbol{\beta}} = \begin{pmatrix} \dfrac{\partial \boldsymbol{\Phi}}{\partial \beta_1} \\ \dfrac{\partial \boldsymbol{\Phi}}{\partial \beta_2} \\ \vdots \\ \dfrac{\partial \boldsymbol{\Phi}}{\partial \beta_q} \end{pmatrix} = \frac{2[\partial(\bar{\boldsymbol{Y}} - \boldsymbol{\eta})^T](\bar{\boldsymbol{Y}} - \boldsymbol{\eta})}{\partial \boldsymbol{\beta}} = 0 \quad (58)$$

The first method discussed here is based on the Newton–Raphson method for the solution of nonlinear systems of equations. Here the nonlinear Equation (58) is expanded around an estimated solution into a Taylor series, which is truncated after the first term. For Equation (58) this leads to the expansion of $\boldsymbol{\eta}$ about $\boldsymbol{\eta}^{(0)}$ at $\boldsymbol{b}^{(0)}$, hence

$$\boldsymbol{\eta} = \boldsymbol{\eta}^{(0)} + \left(\frac{\partial \boldsymbol{\eta}}{\partial \boldsymbol{\beta}}\right)^{(0)} \Delta \boldsymbol{b}^{(0)} \quad (59)$$

with $\Delta \boldsymbol{b}^{(0)} = \boldsymbol{\beta} - \boldsymbol{b}^{(0)}$. A system of linear equations for the new variable $\Delta \boldsymbol{b}^{(0)}$ then arises, solution of which leads to an improved estimated value for the parameters:

$$\boldsymbol{b}^{(1)} = \boldsymbol{b}^{(0)} + \Delta \boldsymbol{b}^{(0)} \quad (60)$$

If this procedure is applied to Equation (58), substitution of Equation (59) into Equation (58) gives

$$\frac{2\left[\partial\left(\bar{\boldsymbol{Y}} - \boldsymbol{\eta}^{(0)} - \left(\dfrac{\partial \boldsymbol{\eta}}{\partial \boldsymbol{\beta}}\right)^{(0)} \Delta \boldsymbol{b}^{(0)}\right)^T\right]\left[\left(\dfrac{\partial \boldsymbol{\eta}}{\partial \boldsymbol{\beta}}\right)^{(0)} \Delta \boldsymbol{b}^{(0)}\right]}{\partial \Delta \boldsymbol{b}^{(0)}} = 0 \quad (61)$$

This gives the following system of equations for $\Delta \boldsymbol{b}^{(0)}$:

$$(\boldsymbol{X}^T \boldsymbol{X})^{(0)} \boldsymbol{B}^{(0)} = (\boldsymbol{X}^T \boldsymbol{E})^{(0)} \quad (62)$$

which employs the abbreviations

$$\boldsymbol{X} \equiv \frac{\partial \boldsymbol{\eta}}{\partial \boldsymbol{\beta}}, \quad \boldsymbol{E} \equiv (\bar{\boldsymbol{Y}} - \boldsymbol{\eta}), \text{ and } \boldsymbol{B} \equiv \Delta \boldsymbol{b}$$

Equation (62) is similar to Equation (39). In Equation (62) $\boldsymbol{B}^{(0)}$ is the dependent variable, whereas Equation (39) represents a system of equations for the parameters themselves. (Since there are no repeated measurements according to the assumptions in Section 3.2.1 the statistical weights for all measured points are equal, p is replaced by the unit matrix). The solution of Equation (62) for $\boldsymbol{B}^{(0)}$ gives

$$\boldsymbol{B}^{(0)} = [(\boldsymbol{X}^T \boldsymbol{X})^{(0)}]^{-1}(\boldsymbol{X}^T \boldsymbol{E})^{(0)} \quad (63)$$

By using Equation (60) an improved set of parameters is obtained and the method is iterated until a previously defined convergence condition of $\boldsymbol{B}^{(n)}$ or $\boldsymbol{\Phi}$ is satisfied.

With the second method for the estimation of parameters from nonlinear empirical models, the sum of squares itself is linearized. Expansion of Equation (48) around $\boldsymbol{b}^{(0)}$ into a Taylor series truncated after the first term gives

$$\boldsymbol{\Phi} \approx (\boldsymbol{\Phi})^{(0)} + \sum_{j=1}^{q} \left(\frac{\partial \boldsymbol{\Phi}}{\partial \beta_j}\right)^{(0)} (\beta_j - b_j^{(0)}) \quad (64)$$

The direction of search used to calculate improved estimated values from the relation

$$\boldsymbol{b}^{(n+1)} = \boldsymbol{b}^{(n)} + \boldsymbol{h}^T \Delta \boldsymbol{b}^{(n)} \quad (65)$$

($\boldsymbol{h}$ is a vector of the step size) can be defined by the gradient of $\boldsymbol{\Phi}$; grad $\boldsymbol{\Phi}$ is a vector that is normal to the surface of $\boldsymbol{\Phi}$ in the parameter space and indicates the direction of the steepest ascent in $\boldsymbol{\Phi}$ at the point $\boldsymbol{b}$. Conversely, the vector $-\text{grad } \boldsymbol{\Phi}$ shows the direction of the steepest descent of $\boldsymbol{\Phi}$. For parameter estimation the components of the vectors at $\boldsymbol{b}^{(0)}$

$$-\text{grad } \boldsymbol{\Phi} = -\left(\frac{\partial \boldsymbol{\Phi}}{\partial \beta_1}\right)^{(0)} \delta \beta_1 - \left(\frac{\partial \boldsymbol{\Phi}}{\partial \beta_2}\right)^{(0)} \delta \beta_2 - \ldots - \left(\frac{\partial \boldsymbol{\Phi}}{\partial \beta_q}\right)^{(0)} \delta \beta_q \quad (66)$$

are calculated. For this the vector $-\text{grad } \boldsymbol{\Phi}$, normalized by its absolute value is used:

$$\frac{-\text{grad } \boldsymbol{\Phi}}{\|-\text{grad } \boldsymbol{\Phi}\|} = \frac{-\left(\dfrac{\partial \boldsymbol{\Phi}}{\partial \beta_1}\right)^{(0)} \delta \beta_1 - \left(\dfrac{\partial \boldsymbol{\Phi}}{\partial \beta_2}\right)^{(0)} \delta \beta_2 - \ldots - \left(\dfrac{\partial \boldsymbol{\Phi}}{\partial \beta_q}\right)^{(0)} \delta \beta_q}{\sqrt{\left(-\dfrac{\partial \boldsymbol{\Phi}}{\partial \beta_1}\right)^{(0)2} + \left(-\dfrac{\partial \boldsymbol{\Phi}}{\partial \beta_2}\right)^{(0)2} + \ldots + \left(-\dfrac{\partial \boldsymbol{\Phi}}{\partial \beta_q}\right)^{(0)2}}} \quad (67)$$

In Equations (66) and (67), $\delta \beta_i$ denotes the unit vectors in the direction of β_i. The components of the normalized vector (Eq. 67) produce the term Δb in Equation (64) from which the iteration formula (Eq. 65) can proceed.

Both methods (solution of the system of nonlinear normal equations and the method of steepest descent) can, under certain conditions of the sum of squares Φ, lead to unfavorable trajectories over the surface of Φ in the parameter space [40]. To deal with this, according to MARQUARDT a diagonal matrix is added to the coefficient matrix $(X^T X)$ in Equation (62):

$$(X^T X + \lambda I) B = X^T E \quad \lambda \geq 0 \quad (68)$$

where I is the unit matrix and λ is a scaling factor. When $\lambda = 0$ Equation (62) is obtained. As $\lambda \to \infty$, then $B \approx X^T E/\lambda$ which, with the identity

$$\lambda = \frac{\|-\operatorname{grad} \Phi^{(n)}\|}{h^{(n)}}$$

leads to the method of steepest descent for calculating B. By fitting λ using Marquardt's method, the essential characteristics of both methods can be combined, [4], [7], [40].

Numerical Problems Connected with Gradient Methods.

Iterative solution of Equation (62) or the method of steepest descent can give rise to problems, one of them being the initial guesses $b^{(0)}$. Under certain conditions, the selection of $b^{(0)}$ leads to a neighboring minimum in Φ, or is so unfavorable that iteration does not lead to convergence. The first of these problems requires a more precise investigation of the sum of the squares of the deviations or repetition of the calculation with various initial values for $b^{(0)}$. In order to avoid divergence of the iterative method (especially with poorly chosen $b^{(0)}$), only the direction of the vector $B^{(0)}$ should be used; new estimated values for β are then calculated using Equation (65) with a reduced step size in place of Equation (60). In Equation (65), h has the character of a user-selected damping factor, which can be adjusted during each iteration in accordance with the evolution of Φ. With favorable conditioning of the nonlinear equation system, this factor can also be used to accelerate convergence.

Another problem lies in the mathematical form of the model. For example, for a model consistent with Equation (53) it is easy to show that η, as well as the partial derivatives $\partial \eta / \partial \beta_1$ and $\partial \eta / \partial \beta_2$, are unbounded if $\beta_2 x = -1$. The only way to circumvent this problem is to constrain the range of the estimated values of the parameters on the basis of the physicochemical characteristics of the process.

The sum of the squares of the deviations can contain terms of differing orders of magnitude, which lead to very different gradients in the direction of the individual parameters. In such cases scaling of the parameters is always advisable, whereby the individual components in Equation (66) are brought to the same order of magnitude by linear transformation.

Transformations of the independent variables are necessary if interactions exist between individual parameters. This is the case, for example, with a simple Arrhenius type of rate coefficient of a chemical reaction

$$k = k_0 e^{-E_a/RT}$$

where E_a is the activation energy, R the universal gas constant, and T the temperature.

If the exponential function is expanded so that

$$k \approx k_0 - k_0 \frac{E_a}{RT} + \frac{1}{2} k_0 \left(\frac{E_a}{RT}\right)^2 - \cdots$$

it becomes clear that k_0 has a multiplicative effect on the other parameter E_a. Minima of the sum of squares can thus be found for various combinations of k_0 and E_a that do not make sense in physical terms. In such cases either the range of allowable estimated values of the parameters is restricted on the basis of the physicochemical process characteristics, or the matrix $(X^T X)$ is brought into a form in which the diagonal elements dominate through transformation of the independent variables. The smaller the off-diagonal elements of the matrix $(X^T X)$ are compared to the diagonal elements, the smaller the interaction between the parameters becomes. This can be demonstrated for the above example of the simple Arrhenius type of rate coefficient in which T is the independent variable. Since

$$\frac{\partial k}{\partial k_0} = e^{-\frac{E_a}{RT}} = \frac{k}{k_0} \quad \text{and} \quad \frac{\partial k}{\partial (E_a/R)} = -\frac{k_0}{T} e^{-\frac{E_a}{RT}} = \frac{k}{T}$$

then $(X^T X) = \begin{pmatrix} \frac{1}{k_0^2} \sum_i k_i^2 & -\frac{1}{k_0} \sum_i \frac{k_i^2}{T_i} \\ -\frac{1}{k_0} \sum_i \frac{k_i^2}{T_i} & \sum_i \left(\frac{k_i^2}{T_i}\right)^2 \end{pmatrix}$

Therefore

$$\text{Det}(X^T X) = \frac{1}{k_0^2}\left[\sum_i k_i^2 \sum_i k_i^2 \left(\frac{k_i^2}{T_i}\right)^2 - \left(\sum_i \left(\frac{k_i^2}{T_i}\right)\right)^2\right]$$

which can become singular under certain conditions, since all terms in the square brackets are positive for all regions of the measured k_i and T_i.

If the transformation $T^* = (T - \bar{T})/T$ is used, we obtain $k = k_0^* e^{E^* T^*}$ with $k_0^* = k_0/e^{E^*}$ and $E^* = E_a/R\bar{T}$. For the transformed model

$$\frac{\partial k}{\partial k_0^*} = e^{E^* T^*} = \frac{k}{k_0^*}$$

and

$$\frac{\partial k}{\partial E^*} = -k_0^* T^* e^{E^* T^*} = kT^*$$

Hence

$$(X^T X) = \begin{pmatrix} \dfrac{1}{k_0^{*2}} \sum_i k_i^2 & -\dfrac{1}{k_0^*} \sum_i k_i^2 T_i^* \\ -\dfrac{1}{k_0^*} \sum_i k_i^2 T_i^* & \sum_i (k_i T_i^*)^2 \end{pmatrix}$$

Since T_i^* can be positive or negative, the diagonal elements of the matrix dominate, and in

$$\text{Det}(X^T X) = \frac{1}{k_0^{*2}}\left[\sum_i k_i^2 \sum_i (k_i T_i^*)^2 - \left(\sum_i k_i^2 T_i^*\right)^2\right]$$

the first term in the square brackets always dominates. The danger of the matrix $(X^T X)$ being singular (which would prevent solution of Eq. 63) is thus overcome.

3.2.2. Assessment of Estimated Parameter Values

Several methods have been developed to assess the estimated parameter values from nonlinear empirical models [20], [41], [42]. A method which is similar to the approach used in Sections 3.1.1.2 and 3.1.2.2 will be presented here.

The nonlinear model is linearized around the estimated values b in the parameter space, i.e., is expanded into a Taylor series truncated after the first term. The approximate variances and covariances of the parameters are then given by (cf. Eqs. 42 and 59):

$$\text{Cov}(b) \approx (X^T X)^{-1} \sigma_{\bar{Y}_i}^2 \tag{69}$$

Each element of X is calculated either analytically or numerically from the value of b indicated by the minimum. If the hypothesis is valid, s_r^2 can once again be used as an estimated value for $\sigma_{\bar{Y}_i}^2$ (Eq. 41). If repeated measurements are available, the hypothesis $\langle \bar{Y}_i \rangle = f(x, \beta)$ can be tested beforehand. The approximate variances of the estimated values of the parameters are obtained from the diagonal elements of $(X^T X)^{-1}$:

$$s_{b_k}^2 \approx s_{\bar{Y}_i}^2 \cdot c_{kk} \tag{70}$$

using the abbreviation $c = (X^T X)^{-1}$. The approximate confidence intervals can be estimated from this as described in Section 3.1.1.2.

The variances of the model solutions are obtained from the linearized model according to

$$s_{\hat{Y}_i}^2 \approx \sum_{j=1}^q \left(\frac{\partial \hat{Y}}{\partial b_j}\right) s_{b_j}^2$$

$$+ \sum_{i=1}^q \sum_{j=1, i \neq j}^q \left(\frac{\partial \hat{Y}}{\partial b_i}\right)\left(\frac{\partial \hat{Y}}{\partial b_j}\right) \text{Cov}(b_i b_j)$$

$$\approx s_{\bar{Y}_i} \sum_{i=1}^q \sum_{j=1, i \neq j}^q \left(\frac{\partial \hat{Y}}{\partial b_i}\right)\left(\frac{\partial \hat{Y}}{\partial b_j}\right) c_{ij} \tag{71}$$

so that the approximate confidence intervals of the expected values can also be estimated.

3.2.3. Example of a Nonlinear Model

The estimation of the parameters of a nonlinear empirical model will be demonstrated by further discussion of the process presented in Section 3.1.1.4. Because of the relatively large scatter of the measured values within the individual data sets, the result of the linear regression is merely tolerable in spite of a positive test of the hypothesis $\langle \bar{Y}_i \rangle = \beta_0' + \beta_1 (x_i - \bar{x})$. The assessment of the estimated parameters gives large individual confidence intervals for the parameters and the expected values. Furthermore, the sum of squares and the estimated parameter values show a high sensitivity to the measured values $\bar{Y}_i$.

The same process will now be tested with a second approximation using a model of the form

$$\langle \bar{Y}_i \rangle = \beta_0 x_i^{\beta_1} \tag{72}$$

The data for nonlinear regression are given in Table 1 (p. 2-15). The Newton–Raphson method is used to solve the system of nonlinear equations (Eq. 58). The results from the application of the linear model $\langle \bar{Y}_i \rangle =

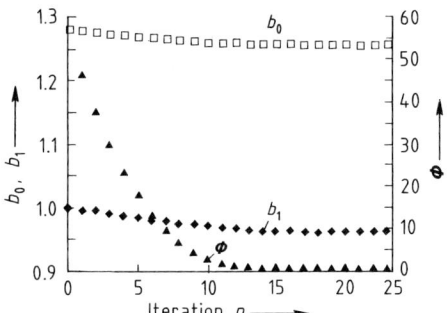

Figure 12. Results of nonlinear regression for the example discussed in Section 3.2.3

$\beta'_0 + \beta_1(x_i - \bar{x})$ are used as an initial guess for b_0 and b_1, hence $b_0^{(0)} = 1.282$, $b_1^{(0)} = 1$. The step size $h_i^{(n)}$ (Eq. 65) is adjusted automatically according to the improvement in Φ with each iteration. Figure 12 shows the estimated values $b_0^{(n)}$ and $b_1^{(n)}$ and the sum of squares of the deviations $\Phi^{(n)}$ plotted against the number of iterations n. The method is regarded as convergent when $\Phi^{(n-1)} - \Phi^{(n)} \leq 10^{-4}$.

For the region of the experimentally adjusted gas velocities, the calculated results $b_0 = 1.257$ and $b_1 = 0.9626$ produce a smaller minimum in Φ, $\Phi_{min} = 0.332$, than the linear regression with $\Phi_{min} = 7.746$ (Fig. 6). The improved quality of the regression is obvious from the value of the mean square deviations about the regression function ($s_r^2 = 0.1105$) so that the F-test is fulfilled to a higher level of significance. If the combined squared deviations s_r^2 and s_e^2 are used to estimate $\sigma_{\bar{Y}_i}^2$, $s_{\bar{Y}_i}^2 = 0.996$ is obtained; the improvement seen in Figure 12 also results in a smaller variance in the estimated parameter values b_0 and b_1 and in the model solution $\bar{Y}_i$. From Equations (70) and (71) the model linearized in the minimum of Φ gives

$$s_{b_0}^2 \approx 0.996 \times 5.21 \times 10^{-2} = 5.03 \times 10^{-2}$$

$$s_{b_1}^2 \approx 0.996 \times 4.51 \times 10^{-3} = 4.3 \times 10^{-3}$$

$$s_{\bar{Y}_i}^2 \approx 0.996\,[(x_i^{0.9626})^2\,(5.21 \times 10^{-2} - 3.6 \ln x_i + 7.13 \times 10^{-3} (\ln x_i)^2)]$$

The relation $\langle \bar{Y}_i \rangle = \beta_0 x_i^{\beta_1}$ with $b_1 \leq 1$ reflects the true dependency better (see Section 3.1.1.4).

3.2.4. Concluding Remarks

The treatment of nonlinear empirical models is conducted as in Sections 3.1.1. and 3.1.2 using a relatively simple model to obtain the simplest formulation possible. The principles of nonlinear regression and some of the resulting problems can be clarified using this simple model. For cases in which the assumptions listed on p. 2–19 do not hold, more advanced literature should be consulted, [4], [5], [7], [20]. Many computer programs are available for routine problems of parameter estimation with nonlinear empirical models (e.g., for gradient methods [27], for Marquardt's method [43], or for direct search methods [44]). The individual steps of the Newton–Raphson method (see Sections 3.1.1.5 and 3.1.2.3) can be obtained as modules from readily available program libraries [26]–[32].

3.3. Further Calculation Methods and Model Types

The discussion on the quantitive definition of empirical models (Sections 3.1 and 3.2) referred to the minimization of the sum of squares of the deviations as well as to simple models based on Equations (5) and (46). The section on empirical models will now be concluded with a short discussion of other methods of parameter estimation for empirical models and other types of models.

3.3.1. Further Methods of Parameter Estimation

Further methods of defining the parameters of empirical models include the maximum likelihood method.

A statistical empirical model can be described with the use of the probability density function $P(x, y)$ (Section 2.2 and Fig. 5). If a model for the process can be represented in the form $\langle \bar{Y}_i \rangle = \beta'_0 + \beta_1(x_i - \bar{x})$ (Section 3.1.1), the probability density function is defined by the parameters β'_0, β_1, and $\sigma_{\bar{Y}_i}^2$, hence $P(\bar{Y}, x; \beta'_0, \beta_1, \sigma_{\bar{Y}_i}^2)$. The problem of estimating the parameters now consists of defining β'_0, β_1, and $\sigma_{\bar{Y}_i}^2$ such that the likelihood function becomes a maximum. With one observation the likelihood function L for the parameters is the probability density function in which the variables $\bar{Y}$ and x are viewed as parameters and the parameters β'_0, β_1, and $\sigma_{\bar{Y}_i}^2$ are viewed as variables.

$$L(\beta'_0, \beta_1, \sigma_{\bar{Y}_i}^2; \bar{Y}_1, x_1) = P(\bar{Y}_1, x_1; \beta'_0, \beta_1, \sigma_{\bar{Y}_i}^2) \tag{73}$$

With n statistically independent observations the likelihood function is the product of the individual functions:

$$L(\beta'_0, \beta_1, \sigma^2_{\bar{Y}_i}; \bar{Y}_1, x_1 \ldots \bar{Y}_n, \bar{x}_n)$$

$$= \prod_{i=1}^{n} L(\beta'_0, \beta_1, \sigma^2_{\bar{Y}_i}; \bar{Y}_i, x_i)$$

$$= P(\bar{Y}_1, x_1; \beta'_0, \beta_1, \sigma^2_{\bar{Y}_i}) \cdot P(\bar{Y}_2, x_2; \beta'_0, \beta_1, \sigma^2_{\bar{Y}_i})$$
$$\ldots P(\bar{Y}_n, x_n; \beta'_0, \beta_1, \sigma^2_{\bar{Y}_i}) \quad (74)$$

To define β'_0, β_1, and $\sigma^2_{\bar{Y}_i}$, L must now be maximized. The logarithms of Equation (74) are taken giving the following relations for the parameters:

$$\frac{\partial \ln L}{\partial \beta'_0} = \frac{\partial \sum_{i=1}^{n} P(\bar{Y}_i, x_i; \beta'_0, \beta_1, \sigma^2_{\bar{Y}_i})}{\partial \beta'_0} = 0 \quad (75a)$$

$$\frac{\partial \ln L}{\partial \beta_1} = \frac{\partial \sum_{i=1}^{n} P(\bar{Y}_i, x_i; \beta'_0, \beta_1, \sigma^2_{\bar{Y}_i})}{\partial \beta_1} = 0 \quad (75b)$$

$$\frac{\partial \ln L}{\partial \sigma^2_{\bar{Y}_i}} = \frac{\partial \sum_{i=1}^{n} P(\bar{Y}_i, x_i; \beta'_0, \beta_1, \sigma^2_{\bar{Y}_i})}{\partial \sigma^2_{\bar{Y}_i}} = 0 \quad (75c)$$

For the linear problem $\langle \bar{Y}_i \rangle = \beta'_0 + \beta_1(x_i - \bar{x})$ where the random variable Y is assumed to have a normal distribution, it can easily be shown that Equations (9) for defining β'_0 and β are obtained from Equation (75) since

$$P(\bar{Y}_i, x_i; \beta'_0, \beta_1, \sigma^2_{\bar{Y}_i})$$
$$= \frac{1}{\sqrt{2\pi \sigma^2_{\bar{Y}_i}}} \cdot \exp\left[-\frac{1}{2\sigma^2_{\bar{Y}_i}} (\bar{Y}_i - \langle \bar{Y}_i \rangle)^2 p_i\right] \quad (76)$$

Detailed discussion of this method can be found in [4], [22].

Further search algorithms are discussed in more advanced literature on optimization, → 1. Mathematics in Chemical Engineering, pp. 1-106–1-128. [5], [7], [38], [39].

3.3.2. Other Types of Models

In addition to the models described by Equations (5) and (46), empirical models in which underlying conditions constrain the parameters are frequently encountered in industrial chemistry. Furthermore the functional relationship may not be in the simple form of algebraic equations (Eqs. 5, 32, and 46) but in the form of differential equations or transfer functions.

3.3.2.1. Models with Constraints on the Parameters

Constraints on parameters are found in many models used in industrial chemistry. A simple example is the reaction rate of a heterogeneously catalyzed reaction according to Equation (53), which only makes sense when

$$\beta_i \geq 0 \quad (77a)$$

Equally common are constraints of the type

$$\beta_i \geq k_i, \text{ where } k_i \text{ is positive} \quad (77b)$$

$$0 \leq \beta_i \leq 1 \quad (77c)$$

$$k_i^* \leq \beta_i \leq k_i \quad (77d)$$

Explicit constraints of this type can easily be removed with appropriate transformations. For example, the general transformation rule for Equation (77b) is

$$\beta_i = k_i + e^{\beta_i^*} \quad (78a)$$

in which the unknown β_i^* is not affected by any constraints. For Equation (77d) the general transformation rule

$$\beta_i = k_i^* + (k_i - k_i^*) \sin^2 \beta_i^* \quad (78b)$$

can be used, whereas for Equation (77a)

$$\beta_i = \beta_i^{*2} \text{ or } \beta_i = e^{\beta_i^*} \quad (78c, d)$$

are suitable transformations.

The general constraint

$$g_i(Y, x, \beta) \geq 0, \quad i = 1, \ldots q \quad (79)$$

for which transformations are not applicable, can be added as a penalty function to the sum of squares of the deviations:

$$\Phi + \sum_{i=1}^{q} \lambda_i (g_i)^{\varrho} = \Phi^* \stackrel{!}{=} \text{Min} \quad (80)$$

For constraints given in Equation (77b) the penalty function is $g_i = 0$ for $b_j \geq k_j$ and $g_i = (b_j - k_j)^2$ for $b_j \leq k_j$. Here the scaling factors λ and ϱ are set to 1 and 2, respectively.

Similarly constraints in the form

$$g_i(Y, x, \beta) = 0, \quad i = 1, \ldots q \quad (81)$$

can be handled in the form of penalty functions which are added to the sum of squares of the deviations. They disappear when the associated conditions (Eq. 81) are satisfied. Alternatively they are weighted according to the magnitude of the deviation and added to Φ.

A more rigorous method for the treatment of associated constraints in the form of equations is based on their geometrical interpretation. In Figure 13 the sum of squares of the deviations for a two-parameter problem is shown in the parameter space. A constraint of the form

$$\beta_1 + a\beta_2 = c \tag{82}$$

can be represented as a straight line which intersects the lines of constant Φ. The minimum of Φ is now searched for along this straight line, so that the two-dimensional problem is reduced to a one-dimensional one. This can be easily demonstrated for a linear model of the form according to Equation (5) with the constraint given by Equation (82) because one parameter is directly eliminated by substitution.

In problems involving many parameters or nonlinear models, reduction of a multidimensional problem with one or more constraints by substitution is either impossible or very difficult. In such cases the method of Lagrangian multipliers is used. Figure 13 illustrates that for the constrained minimum, the slopes $\partial \beta_1 / \partial \beta_2$ of the two curves $\Phi = \text{const.}$ and $\beta_1 + a\beta_2 - c = 0$ are equal. In general these conditions can be expressed as

$$\frac{\partial \Phi / \partial \beta_1}{\partial \Phi / \partial \beta_2} - \frac{\partial g / \partial \beta_1}{\partial g / \partial \beta_2} = 0 \tag{83}$$

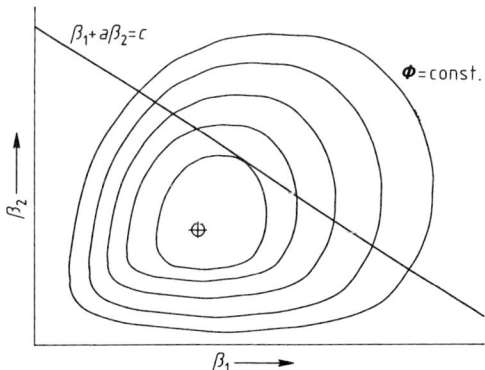

Figure 13. Geometric representation of least squares estimations with constraints on the parameters

or as the gradient ratio

$$\frac{\partial \Phi / \partial \beta_1}{\partial g / \partial \beta_1} = \frac{\partial \Phi / \partial \beta_2}{\partial g / \partial \beta_2} = \lambda \tag{84}$$

From Equation (84) the system of equations

$$\frac{\partial \Phi}{\partial \beta_1} - \lambda \frac{\partial g}{\partial \beta_1} = 0 \tag{85a}$$

and

$$\frac{\partial \Phi}{\partial \beta_2} - \lambda \frac{\partial g}{\partial \beta_2} = 0 \tag{85b}$$

is obtained. Equations (85a), (85b), and the constraint given by Equation (82) may be derived from the extended sum of squares

$$\Phi^* = \Phi + \lambda(\beta_1 + a\beta_2 - c) \tag{86}$$

by partial differentiation with respect to the parameters and the Lagrangian multipliers λ. In general, therefore, when the associated constraints are in the form of Equation (81), the minimum of the extended sum of squares

$$\Phi + \sum_{i=1}^{q} \lambda_i g_i(Y, x, \beta) = \Phi^* \stackrel{!}{=} \text{Min} \tag{87}$$

must be found by varying the parameters β and the Lagrange multipliers λ_i. This can be carried out using the methods discussed previously (see Section 3.2.1).

3.3.2.2. Models Based on Differential Equations

Chemical processes often give rise to models in the form of differential equations. Examples are models of reactors based on transport equations or the rates of chemical reactions. For differential equations that are simple to integrate the parameters can be determined according to the principles presented in Section 3.2, for example, as with the common first-order differential equation

$$\frac{dy(t)}{dt} = \alpha y(t) + x(t), \quad \text{where} \quad y_{t=0} = y_0 \tag{88}$$

The general solution of Equation (88) is [45]

$$y(t) = y_0 e^{\alpha t} + \int_0^t x(\tau) e^{\alpha(t-\tau)} d\tau \tag{89}$$

If discrete measured values $Y(t_i)$ are available, the condition for the minimum with respect to α and y_0 is given by

$$\Phi = \sum_{i=1}^{n} [Y(t_i) - y_0 e^{\alpha t_i} + \frac{x_0}{\alpha}(1 - e^{-\alpha t_i})]^2 \qquad (90)$$

This leads to the system of nonlinear equations

$$\sum_{i=1}^{n} [Y(t_i) - \hat{y}_0 e^{\hat{\alpha} t_i} + \frac{x_0}{\hat{\alpha}}(1 - e^{-\hat{\alpha} t_i})] e^{\hat{\alpha} t_i} = 0 \qquad (91\,a)$$

and

$$\sum_{i=1}^{n} [Y(t_i) - \hat{y}_0 e^{\hat{\alpha} t_i} + \frac{x_0}{\hat{\alpha}}(1 - e^{-\hat{\alpha} t_i})]$$
$$\cdot [t_i \hat{y}_0 e^{\hat{\alpha} t_i} - \frac{x_0}{\hat{\alpha}^2}(e^{\hat{\alpha} t_i}(1 - \hat{\alpha} t_i) - 1)] = 0 \qquad (91\,b)$$

which can be solved for $\hat{y}_0$ and $\hat{\alpha}$ by an appropriate method.

In the majority of cases, the differential equations of the model are more complex than Equation (88), so that no analytical solution is available; numerical solution of the differential equations is required to evaluate $\langle Y_i \rangle$. New problems then arise depending on the type of differential equation. Especially with systems of coupled partial differential equations used to model chemical reactors on the basis of transport equations (see Chap. 5), the solution of the differential equations is the real problem. In such cases other methods are often used to determine the model parameters. Changing certain process variables produces a characteristic behavior in some model solutions which is studied in suitably designed experiments.

An example is the spread of the concentration profile of a tracer added in the form of a Dirac function into reactors with axial and/or radial dispersion. By measurement of the concentration at two points in the reactor, the Péclet number Pe can be derived in a relatively simple manner. Further examples are discussed in [4], [46]. A feature of these methods is that an analytical or numerical solution of the differential equations is often not necessary because the model parameters are comparatively simple functions of the deterministic moments of the measured variables. Deterministic moments are defined analogously to the moments of probability density functions of statistical variables (Section 2.2) and are related to the distribution functions of the deterministic variables. If, for example, $w(t)$ is the residence time distribution for a chemical reactor, then the first two deterministic moments m_1 and m'_2 are defined as

$$m_1 = \bar{t} = \int_0^\infty t \cdot w(t)\,dt$$

$$m'_2 = \int_0^\infty (t - m_1)^2 w(t)\,dt$$

An example is the determination of the Péclet number for a chemical reactor with back mixing on the basis of measurements of the residence-time distribution [46], [47].

Finally, it should be noted that for processes which can be described in the form of systems of common linear differential equations analogous to Equation (88), i.e.,

$$\frac{dy}{dt} + \beta y - x(t), \quad y_{t=0} = 0 \qquad (92)$$

the parameters can be determined with the Laplace transforms of the transfer functions. The transfer functions can be given in the physical time coordinates as well as in the coordinates of the Laplace transformation. The methods are discussed in detail in [4], [9], [13]–[15].

4. Models Based on Transport Equations for Probability Density Functions

Probability density functions and the probabilities which can be calculated from them are the most adequate way of describing statistical variables. Examples of statistical variables can be found at the molecular level in the molecular velocity resulting from random thermal movements; the corresponding probability density function is the Maxwell–Boltzmann velocity distribution [48]. The use of this concept in kinetic gas theory and statistical mechanics has been referred to in Section 2.2.

Statistical variables can also be found at the macroscopic level. In the example discussed in Chapter 3, statistical process variables with presumed distribution or probability density functions were examined and their expected values, variances, and functional relationships to other variables were determined.

Another example is the jet stirred fluid-phase reactor with turbulent flow presented schematically in Figure 14. The surrounding fluid entrains the jet by exchange of momentum. The intake volume of the jet is greater than the volume of the surrounding fluid. This leads to recirculation of material from further downstream and mixing of the reactor contents. The fluid in the jet exhibits turbulent flow. The velocity of the fluid particles is so high that they can no longer be held in regular streamlines by the viscous forces, and irregular vortices are formed. Vortices of different sizes with different velocities and different macroscopic properties pass a fixed observer. Therefore, the fixed observer measures statistical, time-dependent variations in the physical quantity Φ.

In Figure 14 this is shown at two positions in the jet. At the edge of the nozzle (A) either the properties of the jet or the surroundings are measured. Further downstream (B) the macroscopic properties are balanced due to turbulent mixing. All physical parameters are thus characterized by position- and time-dependent distribution or probability density functions. The objective of models based on transport equations for probability density functions is the calculation of functions $P(\Phi_1, \ldots, \Phi_n)$ in which $\Phi_1, \ldots, \Phi_n$ is the vector of the dependent statistical variables. This has to be specified for each problem. Models based on transport equations for probability density functions generally give a differential equation for $P(\Phi_1, \ldots, \Phi_n)$.

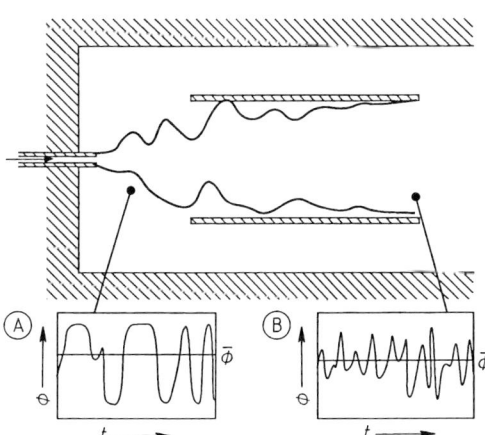

Figure 14. Example of a process with statistical variables: a jet stirred reactor with turbulent flow

4.1. Terminology

4.1.1. One-Dimensional Distribution and Probability Density Functions

The example shown in Figure 14 will be used to explain terminology. The temperature of the fluid is treated as a macroscopic physical quantity. Since the temperature varies between two limits $T_b > T > T_a$, it is better to use a normalized variable so that the statistical variable Φ is defined as

$$\Phi = \frac{T - T_b}{T_a - T_b} \tag{93}$$

The sample space for Φ (see Fig. 15) is a line on which only values between 0 and 1 can be realized. An event is defined as $E \equiv \varphi_u \leq \Phi \leq \varphi_o$. The probability of the occurrence of a particular event is between 0 and 1. When

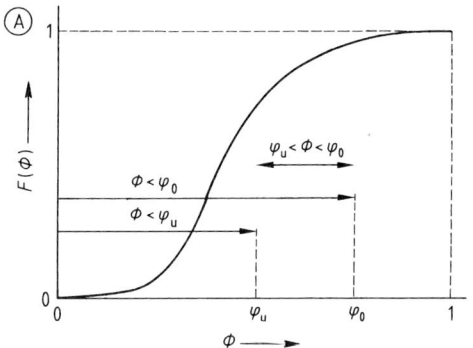

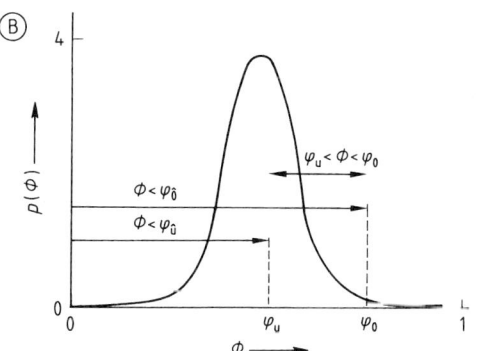

Figure 15. Representation of the distribution function (A) and the probability density function (B) of a statistical variable
The φ axis represents the sample space for the statistical variable Φ. The values φ_o and φ_u confine a finite region within the sample space.

$W(E) = 0$, $\varphi_u \leq \Phi \leq \Phi_o$ never occurs and when $W(E) = 1$, $\varphi_u < \Phi < \varphi_o$ always occurs. Probability defined in this way does not predict E for the next experiment but instead the fraction of occurences of E over a large number of experiments. Since each event is linked to a defined area of the sample space, the probability of an event is the probability that a random variable lies in this area. The probability of an event can be defined from the distribution function of the random variables:

$$F(\Phi) \equiv W(\Phi \leq \varphi) \tag{94}$$

Since the normalized temperatures defined in Equation (93) can only have values between $0 \leq \Phi \leq 1$, then

$$W(\Phi \leq 0) = F(0) = 0 \quad \text{and} \tag{95a}$$

$$W(\Phi \leq 1) = F(1) = 1 \tag{95b}$$

Any event, i.e., any interval $E \equiv \varphi_u \leq \Phi \leq \varphi_o$ in the sample space can be obtained by subtraction (see Fig. 15):

$$E = (\Phi \leq \varphi_o) - (\Phi \leq \varphi_u) \tag{96}$$

Thus

$$W(E) = W(\varphi_u \leq \Phi \leq \varphi_o) = W(\Phi \leq \varphi_o)$$
$$- W(\Phi \leq \varphi_u) = F(\varphi_o) - F(\varphi_u) \tag{97}$$

The fundamental characteristics of the distribution function are easily derived by similar reasoning. Since in the general case $W(\Phi \leq -\infty) = F(-\infty) = 0$ and $W(\Phi \leq +\infty) = F(+\infty) = 1$, then

$$W(E) = F(\varphi_o) - F(\varphi_u) \geq 0 \tag{98}$$

and

$$F(\varphi_o) \geq F(\varphi_u) \quad \text{for} \quad \varphi_o \geq \varphi_u \tag{99}$$

The distribution function $F(\Phi)$ is therefore a nondecaying function which grows from 0 at $\varphi = -\infty$ to 1 at $\varphi = +\infty$ (Fig. 15A).

The probability density function of the random variable Φ is the derivative of the distribution function with respect to the random variable (Fig. 15B).

$$P(\Phi) = \frac{d}{d\Phi}[F(\Phi)] \tag{100}$$

Using this result then

$$\int_{\varphi_u}^{\varphi_o} P(\Phi) d\Phi = F(\varphi_o) - F(\varphi_u)$$
$$= W(\varphi_u \leq \Phi \leq \varphi_o) \tag{101}$$

According to Equation (101) the probability that a random variable falls in a particular area is equal to the integral of the probability density function over that area. For an infinitely small area $d\varphi$ this then becomes

$$W(\varphi \leq \Phi \leq \varphi + d\varphi) = P(\Phi) d\Phi \tag{102}$$

(cf. Section 2.2). The fundamental characteristics of $P(\Phi)$ can be derived from Equation (100). Since $F(\Phi)$ is a nondecaying function this gives

$$P(\Phi) \geq 0 \tag{103}$$

Further

$$\int_{-\infty}^{+\infty} P(\Phi) d\Phi = 1 \tag{104}$$

since according to Equation (101) the probability of the occurrence of the random variable in the area from $-\infty$ to $+\infty$ in the sample space must be 1. The third fundamental characteristic of the probability density function follows from the monotonic approximation of the distribution function to 0 or 1 when $|\Phi| \to \infty$. Thus

$$\frac{d}{d\Phi}[F(\Phi)]_{-\infty} = P(\Phi)_{-\infty}$$
$$= \frac{d}{d\Phi}[F(\Phi)]_{+\infty} = P(\Phi)_{+\infty} = 0 \tag{105}$$

For discontinuous distribution functions with statistical variables which can only take discrete values (see Chap. 3) the above relationships can be given with additional derivation steps [18], [19], [22]. Continuity of the distribution function is not a necessary condition for the above relationships.

The relationships for transformations of distribution functions and probability density functions are easy to see from Figure 15 and Equations (98)–(100). For example if the relationship $\psi = f(\varphi)$ holds for a given transformation of Φ, then the events $\Phi \leq \varphi$ and $\Psi \leq \psi = f(\varphi)$ are identical. From this

$F(\Phi) = F(\Psi)$ and according to Equation (100)

$$P(\Psi) = \frac{d}{d\Psi} F(\Psi) = \frac{d}{d\Phi} F(\Phi) \cdot \frac{d\Phi}{d\Psi} = P(\Phi) \cdot \frac{d\Phi}{d\Psi} \quad (106)$$

Figure 14 shows that the distribution functions and probability density functions can be dependent on position and are thus not always homogeneous. For the area at the edge of the jet close to the nozzle exit an almost bimodal probability density function is produced due to intermittencies of the jet and the surrounding fluid. Further downstream in the middle of the turbulent jet these intermittencies no longer affect the form of the probability density function. Thus probability density functions hold for a given position at a given time; they will be written as $P(\Phi; x, t)$ from now on.

For engineering applications the probability density functions themselves are not very illustrative. Statistical variables are usually described in terms of expected values (see Section 2.2 and Chap. 3). These are easier to handle but have a lower information content. The expected values of statistical variables (synonyms are mean values, mathematical expectations, and first moment) are given by

$$\langle \Phi(x, t) \rangle = \int_{-\infty}^{+\infty} \Phi P(\Phi; x, t) d\Phi \quad (107)$$

For statistical steady-state processes or for $d\langle\Phi\rangle/dt \ll d\Phi/dt$ the equivalent form of the averaging is

$$\langle \Phi(x, t) \rangle = \frac{1}{\Delta \tau} \int_{t}^{t+\Delta\tau} \Phi(x, t) dt \quad (108)$$

For this a continuous function of Φ versus time is necessary as indicated in the plot of Φ against time given in Figure 14.

Functions of statistical variables are themselves statistical values, so that the operations related to Equation (107) for deriving expected values of functions can be used

$$\langle Q(\Phi) \rangle = \int_{-\infty}^{+\infty} Q(\Phi) P(\Phi; x, t) d\Phi \quad (109)$$

The definitions of Equations (107) and (109) respectively are analogously applicable to higher moments

$$\mu_n \equiv \langle \Phi(x, t)^n \rangle = \int_{-\infty}^{+\infty} \Phi^n P(\Phi; x, t) d\Phi \quad (110)$$

For example, the second moment around the mean (the variance) is also a characteristic of statistical quantities which is often used in engineering applications (see Chap. 3):

$$\mathrm{Var}(\Phi) \equiv \mu'_2 \equiv \langle (\Phi - \langle \Phi \rangle)^2 \rangle$$
$$= \int_{-\infty}^{+\infty} (\Phi - \langle \Phi \rangle)^2 P(\Phi; x, t) d\Phi \quad (111)$$

All of these quantities can be determined from the probability density function. However, the probability density function can only be given approximately from a finite number of moments which are relatively easy to measure [49].

If a statistical quantity is split up into the expected value and a fluctuating component Φ^f

$$\Phi = \langle \Phi \rangle + \Phi^f \quad (112)$$

then from Equation (111)

$$\mathrm{Var}(\Phi) = \langle (\Phi^f)^2 \rangle \quad \text{and} \quad \langle \Phi^f \rangle = 0 \quad (113\,a, b)$$

Rules for measuring the statistical moments of statistical steady-state processes are derived from Equation (108). The continuous function $\Phi(x, t)$ is necessary for this. However measurements of statistical variables are mostly available in the form of discrete values rather than in continuous form. Therefore all values referred to here must be derived from discrete data. This has been done for the empirical models discussed in Chapter 3 where discrete values of the statistical variables were available.

For n independent repetitions of an experiment under constant conditions the ensemble mean value

$$\langle \Phi \rangle_n = \frac{1}{n} \sum_{i=1}^{n} \varphi_i \quad (114)$$

gives a good approximation of $\langle \Phi \rangle$ when $n \to \infty$.

In the examples discussed in Chapter 3 the number of occurrences is so small that the mean values from the individual data sets do not represent the expected values but are statistical values. Similarly to Equation (114) the means of functions $Q(\Phi)$ or higher moments are estimated, c.f. Chapter 3 where the sums of squares of the deviations were used as estimates for the variances. The probability density function is measurable through the number of occurrences of φ_i in a given area $\Delta\varphi$ of the sample space. If this number is denoted as $m(\Phi, \Delta\Phi)$, the normalized density of occurrences in the sample space provides a good estimate for the probability density function:

$$P_n(\Phi) = \frac{m(\Phi, \Delta\Phi)}{n\Delta\Phi} \approx P(\Phi; x, t) \quad \text{for} \quad n \to \infty \tag{115a}$$

The discrete form of the probability density function can also be expressed as

$$P_n(\Phi) = \frac{1}{n}\sum_{i=1}^{n} \delta(\Phi - \varphi_i) \tag{115b}$$

Here $\delta(a - b)$ is the impulse or Dirac delta function which is defined as the derivative of the Heaviside or step function:

$$\delta(a - b) = \frac{dH(a - b)}{da} \tag{116}$$

with $H(a - b) = 0$ for $a \le b$ and $H(a - b) = 1$ for $a > b$.

According to Equation (115b) the discrete form of the probability density function in the sample space for Φ is the number impulses of discrete occurrences normalized with respect to the total count. The discrete form of the distribution function

$$F_n(\Phi) = \frac{1}{n}\sum_{i=1}^{n} H(\Phi - \varphi_i) \tag{115c}$$

becomes a stepped curve as shown in Figure 16.

4.1.2. Multidimensional Distribution and Probability Density Functions

For the example shown in Figure 14 the normalized temperature alone is clearly not sufficient to describe the state at a point x and time t. Other variables such as the axial velocity U can be used which is expressed in normalized form as

$$\Psi = \frac{U - U_b}{U_a - U_b} \tag{117}$$

For the statistical variable Ψ the same definitions hold as previously for Φ so that

$$W(\varphi \le \Phi \le \varphi + d\varphi) = P(\Phi; x, t)\,d\Phi$$

and

$$W(\psi \le \Psi \le \psi + d\psi) = P(\Psi; x, t)\,d\Psi$$

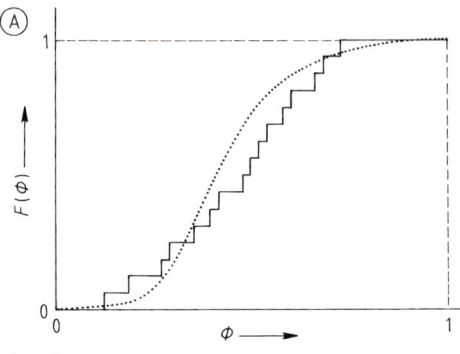

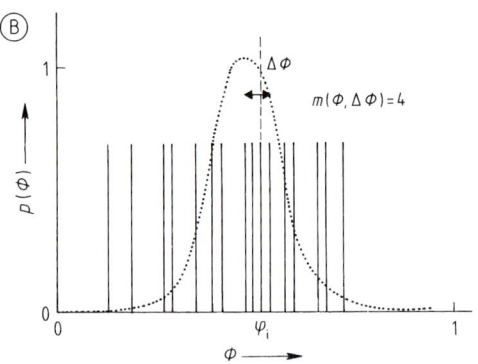

Figure 16. Discrete forms of the distribution function (A) and the probability density function (B)
The continuous functions are shown as dotted lines.

However, this information is not sufficient. More information is required about the probability of the joint events

$$E_\varphi = (\varphi \le \Phi \le \varphi + d\varphi)$$

and

$$E_\psi = (\psi \le \Psi \le \psi + d\psi)$$

This information can be derived from the joint distribution function $F(\Phi, \Psi; x, t)$ which has characteristics analogous to those of the simple distribution function:

$$0 \le F(\Phi, \Psi) \le 1, \quad \text{nondecaying,} \tag{118a}$$
$$F(-\infty, \Psi) = F(\Phi, -\infty) = 0, \tag{118b}$$
$$F(\infty, \Psi) = F(\Psi), \quad \text{and} \tag{118c}$$
$$F(\Phi, \infty) = F(\Phi) \tag{118d}$$

The joint probability density function becomes

$$P(\Phi, \Psi; x, t) = \frac{\partial^2}{\partial \Phi \partial \Psi}[F(\Phi, \Psi; x, t)] \quad (119)$$

which possesses the same fundamental characteristics as the simple probability density function:

$$P(\Phi, \Psi; x, t) \geq 0 \quad (120\,a)$$

$$\iint P(\Phi, \Psi; x, t)\,d\Psi\,d\Phi = 1 \quad (120\,b)$$

The simple probability density function or marginal distributions (the reduced information) can be derived from the joint probability density function:

$$\int P(\Phi, \Psi; x, t)\,d\Psi = P(\Phi; x, t) \quad (121\,a)$$

$$\int P(\Phi, \Psi; x, t)\,d\Phi = P(\Psi; x, t) \quad (121\,b)$$

Conversely the joint probability density functions cannot normally be derived from the marginal distributions. Arguments analogous to those above may be used for the functions $Q(\Phi, \Psi)$.

Covariances are another quality of joint probability density functions:

$$\mathrm{Cov}(\Phi, \Psi) = \langle \Phi^f \Psi^f \rangle$$

$$= \int_{-\infty}^{+\infty} \int_{-\infty}^{+\infty} (\Phi - \langle \Phi \rangle)(\Psi - \langle \Psi \rangle)$$

$$\cdot P(\Phi, \Psi; x, t)\,d\Phi\,d\Psi \quad (122)$$

They are mostly given in normalized form as correlation coefficients:

$$\gamma = \frac{\mathrm{Cov}(\Phi, \Psi)}{[\mathrm{Var}(\Phi)\,\mathrm{Var}(\Psi)]^{1/2}} \quad (123)$$

With the help of Equation (122) it is easy to show that $-1 \leq \gamma \leq 1$. If a linear relationship of the form $\psi = \alpha + \beta \varphi$ exists, then $|\gamma| = 1$.

For joint multidimensional probability density functions, transformation rules are also valid. Using the same reasoning as with one-dimensional probability density functions and assuming that $\xi = \xi(\varphi, \psi)$ and $\theta = \theta(\varphi, \psi)$

$$P(\Xi, \Theta) = P(\Phi, \Psi) \cdot |J|^{-1} \quad (124)$$

where J is the Jacobi matrix of the system of transformation equations.

4.1.3. Conditional Probability Density Functions

Conditional probability is a prerequisite for the derivation of transport equations for probabiliby density functions. It denotes the probability of an event, for example $\Psi \leq \psi$, subject to the condition that the event $\varphi \leq \Phi \leq \varphi + d\varphi$ occurs. It is given by the probability for a general event divided by the probability of the condition:

$$W[(\Psi \leq \psi)|(\varphi \leq \Phi \leq \varphi + d\varphi)]$$

$$= \frac{W(\Psi \leq \psi)(\varphi \leq \Phi \leq \varphi + d\varphi)}{W(\varphi \leq \Phi \leq \varphi + d\varphi)} \quad (125)$$

Consequently

$$W[(\Psi \leq \psi)|(\varphi \leq \Phi \leq \varphi + d\varphi)]$$

$$= \frac{\int_{-\infty}^{\psi} \int_{\varphi}^{\varphi+d\varphi} P(\Phi, \Psi)\,d\Phi\,d\Psi}{\int_{\varphi}^{\varphi+d\varphi} P(\Phi)\,d\Phi} \quad (126)$$

The conditional distribution function is the limit of the conditional probability for $d\varphi \to 0$ so that

$$F(\Psi|\varphi) = P[(\Psi \leq \psi)|(\Phi = \varphi)]$$

$$= \int_{-\infty}^{\psi} \frac{P(\Psi, \Phi)\,d\Psi}{P(\Phi)} \quad (127)$$

The usual definition of probability density function then gives the following:

$$P(\Psi|\varphi) = \frac{d}{d\Psi} F(\Psi|\varphi) = \frac{P(\Psi, \Phi)}{P(\Phi)} \quad (128)$$

It is easy to show that the conditional distribution functions and probability density functions exhibit the same characteristics as unconditional ones. By using the same operations as above, the conditional expected value of a function $Q(\Psi, \Phi)$ can be derived

$$\langle Q(\Psi, \Phi)|\Phi = \varphi \rangle = \int_{-\infty}^{+\infty} Q(\Psi, \Phi)P(\Psi|\varphi)\,d\Psi \quad (129)$$

From Equation (128) it can be seen that the joint probability density function $P(\Psi, \Phi)$ can be obtained from the conditional probability density function $P(\Psi|\varphi)$ and the marginal distribution $P(\Phi)$. This important result is used in the determination of expected values and leads to a further important result:

$$\langle Q(\Psi, \Phi) \rangle = \int_{-\infty}^{+\infty} \int_{-\infty}^{+\infty} Q(\Psi, \Phi) P(\Psi, \Phi) \, d\Psi \, d\Phi$$

$$= \int_{-\infty}^{+\infty} P(\Phi) \left\{ \int_{-\infty}^{+\infty} Q(\Psi, \Phi) P(\Psi|\varphi) \, d\Psi \right\} d\Phi$$

$$= \int_{-\infty}^{+\infty} P(\Phi) < Q(\Psi, \Phi) | \Phi = \varphi > d\Phi \qquad (130)$$

4.2. Transport Equations for Single-Point Probability Density Functions

Before going into some examples, the following sections will briefly discuss transport equations for single-point probability density functions starting from the general conservation equations. Only the important relationships will be described. A more detailed discussion of the general conservation equations is given in [48].

The state of a reacting fluid mixture (e.g., the contents of the reactor shown in Fig. 14) is accurately defined by the velocity $U(x, t)$, the pressure $p(x, t)$, the mass fraction of the chemical species present in the mixture $m_i(x, t)$, $i = 1, \ldots, N$, and the specific enthalpy $h(x, t)$. A complete statistical single-point description of the mixture is given by the joint probability density function $P(U, \Psi; x, t)$ where $U = (U_1, U_2, U_3)$ and Φ is a vector of length $N + 2$ which includes the rest of the scalar quantities. Thus the required function is a surface in the $(N + 3 + 2)$-dimensional Euclidean space.

The relationship between the instantaneous values of the variables stated above is given by the laws of conservation of total mass, momentum, and mass of the individual chemical species, and by the enthalpy. These can be given in the form [48], [50], [51]:

Accumulation + Change due to convective transport = Change due to molecular transport + Source

This leads to

$$\frac{\partial \varrho}{\partial t} + \frac{\partial}{\partial x_k}(\varrho U_k) = 0 \qquad (131)$$

for the total mass,

$$\varrho \frac{DU_i}{Dt} = \frac{\partial \tau_{ik}}{\partial x_k} - \frac{\partial p}{\partial x_i} + \varrho g_i, \quad i = 1, 2, 3 \qquad (132)$$

for the component of the momentum,

$$\varrho \frac{Dm_i}{Dt} = -\frac{\partial J_k^{m_i}}{\partial x_k} + \varrho S_i, \quad i = 1, \ldots, N-1 \qquad (133)$$

for the mass of the individual chemical species, and finally

$$\varrho \frac{Dh}{Dt} = -\frac{\partial J_k^h}{\partial x_k} + \varrho S_h \qquad (134)$$

for the enthalpy.

In Equations (132)–(134) the operator D denotes the substantial derivative:

$$\frac{D}{Dt} = \frac{\partial}{\partial t} - U_k \frac{\partial}{\partial x_k} \qquad (135)$$

The Einstein summation convention is also used, for example:

$$\frac{\partial}{\partial x_k}(\varrho U_k) \equiv \frac{\partial}{\partial x_1}(\varrho U_1) + \frac{\partial}{\partial x_2}(\varrho U_2) + \frac{\partial}{\partial x_3}(\varrho U_3)$$

The formulation of Equations (132)–(134) implies the continuity equation; ϱ is the density; m_i is the mass fraction of component i; τ_{ij} is the viscous stress tensor; ϱg_i is the specific mass force in the x_i direction (e.g., ϱg_1, where g is the acceleration due to gravity); S_i is the reaction rate and S_h the rate of enthalpy conversion produced by pressure changes, viscosity effects etc.; J^m is the mass flux of the chemical species and J^h is the enthalpy flux, based on molecular transport mechanisms, like τ_{ij} they are functions of local physical quantities and their gradients.

Equations (131)–(134) can be closed by a thermal equation of state

$$\varrho = \varrho(m_i, h, p) \qquad (136)$$

The expressions for the sources S_i and S_h are also generally functions of the scalar quantities:

$$S_i = S_i(m_i, h, p) \qquad (137\,\text{a})$$

$$S_h = S_h\left(m_i, h, p, \frac{\partial p}{\partial t}\right) \qquad (137\,\text{b})$$

The assumptions and conditions included in this system of equations are discussed fully in [48], [50]–[52], see also Chapter 5. The system of Equations (131)–(137) is deterministic, this means that a determined solution exists for each set of determined boundary conditions.

The solution of this system of equations for deterministic variables is illustrated with some examples in Chapter 5. Problems arise for statistical variables [52] which can only be solved for simple cases [53], [54] or by introducing far-reaching assumptions (see Chap. 5). The system of conservation equations serves as a starting point for the derivation of the transport equations for the probability density functions which are used to overcome the principle problems for solving this set of equations for statistical variables.

4.2.1. General form of the Transport Equations for Probability Density Functions

Equations (131)–(137) show that the state of a chemically reacting system at a position $x = (x_1, x_2, x_3)$ and time t is determined by the velocity vector $U = (U_1, U_2, U_3)$ in the sample space of velocity (Section 3.1.2.1 and Fig. 7) and the vector of scalar quantities $\Phi = (m_1, \ldots, m_N, h)$ of length σ in the sample space of scalar quantities. U and Φ are vectors of random variables, i.e., random vectors. This formulation holds for low Mach numbers.

The state of the reacting system represents a measured point in the three-dimensional velocity space of the vector U and likewise a measured point in the Euclidean σ-dimensional space of the vector Φ.

For a homogeneous mixture at rest with constant enthalpy, Equation (133) gives

$$\frac{dm_i}{dt} = S_i \qquad (138)$$

A chemical reaction can be represented in the composition space as a trajectory between two measured points. A system with two chemical species is shown in Figure 17. The reaction rate is the velocity in the composition space, and the reaction rates of the individual chemical species are given by the tangents to the trajectory in the direction of each species. Since the state of the system can be described by a vector with its end point in the plane of the required probability density function, changes of state due to chemical reactions are therefore trajectories in this plane. Chemical reactions that constitute the source terms in the conservation equations of the chemical species (Eq. 133) are contained in the transport equations for the probability density functions merely as transport in the composition space.

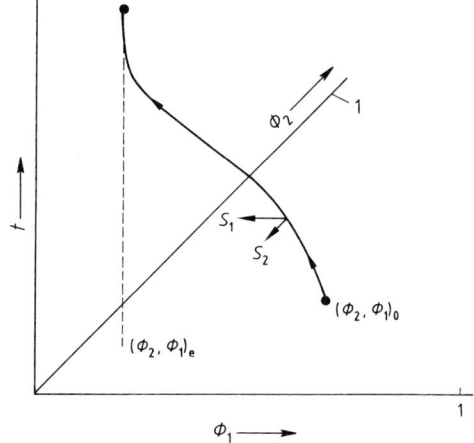

Figure 17. Chemical reactions for a system with two chemical species represented as transport in the sample space of the concentration (Φ_1, Φ_2) of the species
S_1 and S_2 are the partial derivatives of the trajectory in the Φ_1 and Φ_2 direction, the subscripts 0 and e denote initial and final states, respectively.

Similarly, forces which are the source terms in the momentum conservation equations are expressed in the transport equations for probability density functions as transport in the velocity space. These are the main differences between the simple conservation equations and the transport equations for the probability density functions. The latter do not include forces or chemical reactions as sources.

Both U and Φ are necessary to complete the information. The complete information at position $x = (x_1, x_2, x_3)$ and time t is represented by a measured point in Euclidean $(\sigma + 3)$-dimensional space. Consequently, the objective of the complete description is a $(\sigma + 3)$-dimensional distribution function $F(U, \Phi; x, t)$ or the corresponding probability density function:

$$\frac{\partial^{\sigma+3}}{\partial U_1 \partial U_2 \ldots \partial \Phi_{N+1}}[F(U, \Phi; x, t)]$$
$$= P(U, \Phi; x, t) \qquad (139)$$

The characteristics described in Sections 4.1.1–4.1.3 apply to these functions. In the terminology of probabilities these functions predict the fraction of occurrences of discrete values u and φ for a large number of occurrences.

A transport equation for the joint velocity–composition probability density function can be derived with help of Equations (132)–(134) and the terminology described in Section 4.1. This method is one possibility for deriving the transport equation for probability density functions. This will be used in preference to other methods [18], [55]–[58] because of its simplicity.

Equations (132) and (133) are rewritten as Equations (140) and (141), respectively:

$$\frac{DU_i}{Dt} = A_i \qquad (140)$$

$$\frac{D\Phi_\alpha}{Dt} = \Theta_\alpha \qquad (141)$$

where

$$\varrho A_i(x, t) = \frac{\partial \tau_{ik}}{\partial x_k} - \frac{\partial p}{\partial x_i} + \varrho g_i \quad \text{and} \qquad (142\text{a})$$

$$\varrho \Theta_\alpha(x, t) = -\frac{\partial J_k^\alpha}{\partial x_k} + \varrho S_\alpha \qquad (142\text{b})$$

For any function $Q(U, \Phi)$ averaging according to Equation (109) gives

$$\left\langle \varrho \frac{DQ(U, \Phi)}{Dt} \right\rangle = \frac{\partial}{\partial t} \iint \varrho(\Phi) Q(U, \Phi) P \, dU \, d\Phi$$

$$+ \frac{\partial}{\partial x_k} \iint \varrho(\Phi) U_k Q(U, \Phi) P \, dU \, d\Phi$$

$$= \iint Q(U, \Phi) \left\{ \varrho(\Phi) \frac{\partial P}{\partial t} + \varrho(\Phi) U_k \frac{\partial P}{\partial x_k} \right\} dU \, d\Phi \qquad (143)$$

In Equation (143) and in the following discussions, P will be used to represent $P(U, \Phi; x, t)$. In addition, dU will be used to describe the vector dU_1, dU_2, dU_3 and $d\Phi$ for $d\Phi_1, \ldots, d\Phi_{N+1}$. The integral sign in Equation (143) will also be interpreted in this way. The continuity equation (Eq. 131) is used to derive the expression $\left\langle \varrho \frac{DQ(U, \Phi)}{Dt} \right\rangle$ in Equation (143).

A second independent expression for $\left\langle \varrho \frac{DQ(U, \Phi)}{Dt} \right\rangle$ is obtained if $Q(U, \Phi)$ is differentiated with respect to U and Φ (the Einstein summation convention is used for ω in the same way as for k in Eqs. 131–135).

$$\frac{DQ(U, \Phi)}{Dt} = \frac{\partial Q(U, \Phi)}{\partial U_k} \cdot \frac{DU_k}{Dt} + \frac{\partial Q(U, \Phi)}{\partial \Phi_\omega} \cdot \frac{D\Phi_\omega}{Dt} \qquad (144)$$

If Equations (140) and (141) are used, the substantial derivatives can be substituted with the respective right-hand sides. Thus,

$$\left\langle \varrho \frac{DQ(U, \Phi)}{Dt} \right\rangle$$

$$= \left\langle \varrho \frac{\partial Q(U, \Phi)}{\partial U_k} A_k \right\rangle + \left\langle \varrho \frac{\partial Q(U, \Phi)}{\partial \Phi_\omega} \Theta_\omega \right\rangle \qquad (145)$$

The terms on the right-hand side of Equation (145) can be rewritten as conditional expected values by using Equation (130). For the first term of the right-hand side of Equation (145) this gives

$$\left\langle \varrho \frac{\partial Q(U, \Phi)}{\partial U_k} A_k \right\rangle$$

$$= \iint \left\langle \varrho \frac{\partial Q(U, \Phi)}{\partial U_k} A_k \middle| u, \varphi \right\rangle P \, dU \, d\Phi \qquad (146)$$

Since the differential quotient $\partial Q(U, \Phi)/\partial U_k$ is a known function of u and φ for a given u and φ then from Equation (146)

$$\left\langle \varrho \frac{\partial Q(U, \Phi)}{\partial U_k} A_k \right\rangle$$

$$= \iint \varrho \frac{\partial Q(U, \Phi)}{\partial U_k} \langle A_k | u, \varphi \rangle P \, dU \, d\Phi \qquad (147)$$

If Equation (147) is partially integrated, then

$$\left\langle \varrho \frac{\partial Q(U, \Phi)}{\partial U_k} A_k \right\rangle$$

$$= I_U - \iint Q(U, \Phi) \frac{\partial}{\partial U_k} (\varrho \langle A_k | u, \varphi \rangle P) \, dU \, d\Phi \qquad (148)$$

where

$$I_U = \iint \frac{\partial}{\partial U_k} [\varrho(\Phi) Q(U, \Phi) \langle A_k | u, \varphi \rangle P] \, dU \, d\Phi$$

An analogous procedure for the second term on the right-hand side of Equation (145) gives

$$\left\langle \varrho \frac{\partial Q(U, \Phi)}{\partial \Phi_\omega} \Theta_\omega \right\rangle$$

$$= I_\Phi - \iint Q(U, \Phi) \frac{\partial}{\partial \Phi_\omega} (\varrho \langle \Theta_\omega | u, \varphi \rangle P) \, dU \, d\Phi \qquad (149)$$

where

$$I_\Phi = \iint \frac{\partial}{\partial \Phi_\omega} [\varrho(\Phi) Q(U, \Phi) \langle \Theta_\omega | u, \varphi \rangle P] dU d\Phi$$

It can be shown that the integrals I_U and I_Φ disappear for a wide class of functions $Q(U, \Phi)$. The general condition for this is that the functions are bounded and continuous [19]. With this a second independent expression for $\left\langle \varrho \dfrac{DQ(U, \Phi)}{Dt} \right\rangle$ is obtained from Equation (145):

$$\left\langle \varrho \frac{DQ(U, \Phi)}{Dt} \right\rangle$$

$$= -\iint Q(U, \Phi) \frac{\partial}{\partial U_k} (\varrho \langle A_k | u, \varphi \rangle P) dU d\Phi$$

$$- \iint Q(U, \Phi) \frac{\partial}{\partial \Phi_\omega} (\varrho \langle \Theta_\omega | u, \varphi \rangle P) dU d\Phi \quad (150)$$

Subtraction of Equation (150) from Equation (143) gives

$$\iint Q(U, \Phi) \Bigg\{ \varrho(\Phi) \frac{\partial P}{\partial t} + \varrho(\Phi) U_k \frac{\partial P}{\partial x_k}$$

$$+ \frac{\partial}{\partial U_k} [\varrho(\Phi) \langle A_k | u, \varphi \rangle P]$$

$$+ \frac{\partial}{\partial \Phi_\omega} [\varrho(\Phi) \langle \Theta_\omega | u, \varphi \rangle P] \Bigg\} dU d\Phi = 0 \quad (151)$$

All the terms in the curved brackets are independent of $Q(U, \Phi)$. If $Q(U, \Phi)$ is the type of function where the integrals I_u and I_Φ disappear, then a sufficient condition to fulfil Equation (151) is that the sums in the curved brackets are zero. With this the transport equation for the joint probability density function $P(U, \Phi; x, t)$ is

$$\varrho(\Phi) \frac{\partial P}{\partial t} + \varrho(\Phi) U_k \frac{\partial P}{\partial x_k}$$

$$= -\frac{\partial}{\partial U_k} [\varrho(\Phi) \langle A_k | u, \varphi \rangle P] - \frac{\partial}{\partial \Phi_\omega}$$

$$\cdot [\varrho(\Phi) \langle \Theta_\omega | u, \varphi \rangle P] \quad (152)$$

For further discussion of this transport equation, A_i and Θ_α are replaced by their respective definition functions:

$$\varrho(\Phi) \langle \Theta_\alpha | u, \varphi \rangle = -\left\langle \frac{\partial J_k^\alpha}{\partial x_k} + \varrho(\Phi) S_\alpha(\Phi) | u, \varphi \right\rangle$$

$$= -\left\langle \frac{\partial J_k^\alpha}{\partial x_k} | u, \varphi \right\rangle + \varrho(\Phi) S_\alpha(\Phi) \quad (153)$$

Using the decomposition $p^f \equiv p - \langle p \rangle$ gives

$$\varrho(\Phi) \langle A_i | u, \varphi \rangle$$

$$= \left\langle \frac{\partial \tau_{ik}}{\partial x_k} - \frac{\partial \langle p \rangle}{\partial x_i} - \frac{\partial p^f}{\partial x_i} + \varrho g_i | u, \varphi \right\rangle$$

$$= \left\langle \frac{\partial \tau_{ik}}{\partial x_k} | u, \varphi \right\rangle - \frac{\partial \langle p \rangle}{\partial x_i} - \left\langle \frac{\partial p^f}{\partial x_i} | u, \varphi \right\rangle$$

$$+ \langle \varrho g_i | u, \varphi \rangle \quad (154)$$

Substitution of these expressions into Equation (152) gives

$$\varrho(\Phi) \frac{\partial P}{\partial t} + \varrho(\Phi) U_k \frac{\partial P}{\partial x_k} + \left[\varrho(\Phi) g_k - \frac{\partial \langle p \rangle}{\partial x_k} \right] \frac{\partial P}{\partial U_k}$$

$$+ \frac{\partial}{\partial \Phi_\omega} [\varrho(\Phi) S_\omega(\Phi) P]$$

$$= -\frac{\partial}{\partial U_k} \left[\left\langle \frac{\partial \tau_{lk}}{\partial x_l} + \frac{\partial p^f}{\partial x_l} | u, \varphi \right\rangle P \right]$$

$$+ \frac{\partial}{\partial \Phi_\omega} \left[\left\langle \frac{\partial J_k^\omega}{\partial x_k} | u, \varphi \right\rangle P \right] \quad (155)$$

This is the general form of the transport equation for the probability density functions in the combined velocity–composition space (in the first term of the right-hand side k and l must be summed according to the Einstein summation convention). Equation (155) can be solved if the conditional expected values on the right-hand side are known. All physical processes represented by the terms on the left-hand side are described exactly in the context of the method of description given here. These processes are transport in physical space (first and second terms), transport in the velocity space due to gravitational forces and average pressure gradient (third term), and transport in the composition space by chemical reactions (fourth term). All of these processes can be described without any model assumptions.

The expressions on the right-hand side of Equation (155) describe transport in the velocity space due to momentum transfer at the molecular level as well as transport in the composition space due to molecular diffusion. For a turbulent flow the decay of turbulent fluctuations is caused by these mechanisms. Viscous forces are responsible for the dissipation of turbulent (statistical) fluctuations in velocity. Molecular diffusion is responsible for the dissipation of turbulent fluctuations in scalar quantities.

Before Equation (155) can be solved, these terms must be defined or approximated using models. The following sections will describe with some examples how this can be done.

4.2.2. Limitations of Single-Point Probability Density Functions

Solution of Equation (155) gives the joint probability density function $P(U, \Phi; x, t)$ of the random vectors U and Φ at position $x = (x_1, x_2, x_3)$ and time t. This result contains the complete statistical information at a defined position for a given time. Even this information includes only a partial view of the statistical process being considered. The functionals $P(U, \Phi)$ described in Section 2.2 or multipoint probability density functions, such as the two-point probability density function $P(U_1, \Phi_1; x_1, t_1, U_2, \Phi_2; x_2, t_2)$ which can be derived from them, provide more comprehensive information. The meaning of these multipoint probability density functions can be illustrated by the example shown in Figure 14. The structure of the vortices in the turbulent flow of the jet stirred reactor is dependent on position and time. The "lifetime" and "tracks" of the vortices being transported in the flow are crucial for the chemical reaction. They give an indication of the degree of mixing and the "segregation" of the flow. (Segregation denotes the state of a fluid in which individual fluid parcels maintain their identities when passing through the reactor and do not exchange properties with the surrounding fluid [24], [46].) Second-order moments (the correlation functions) are used to describe mixing and segregation:

$$R_{\Phi, \Psi} = \langle (\Phi(x_1, t_1) - \langle \Phi(x_1, t_1) \rangle)(\Psi(x_2, t_2) - \langle \Psi(x_2, t_2) \rangle)\rangle$$
$$= \iint (\Phi(x_1, t_1) - \langle \Phi(x_1, t_1) \rangle)(\Psi(x_2, t_2) - \langle \Psi(x_2, t_2) \rangle) \cdot P(U_1, \Phi_1; x_1, t_1, U_2, \Phi_2; x_2, t_2) d\Phi d\Psi$$

For $\Phi = \Psi$ the autocorrelation functions are obtained which provide data about the vortex structure for a physical property of the fluid e.g., the velocity. Time and length scales for the turbulent exchange of the considered property can be derived from the autocorrelation functions [49], [59]–[61]; these play an important role in the modeling of turbulent processes.

The dimensionality of the transport equations for multipoint probability density functions is proportional to the number of points (Eq. 155). The problem soon becomes unmanageable, even with a small number of scalar quantities and points.

Single-point probability density functions can be derived from probability density functionals or multipoint probability density functions:

$$P(\Phi_1; x, t) = \int_{-\infty}^{+\infty} P(\Phi_1; x_1, t_1, \Phi_2; x_2, t_2) d\Phi_2 \quad (157)$$

Single-point probability density functions only give limited information about the field of random variables. They give complete statistical information at a given site and time but not joint information about two sites and/or two times. For the example shown in Figure 14 the single-point functions therefore contain no information about the length or time scales of the statistical (turbulent) fluctuations.

This deficiency can be illustrated by a simple example. Suppose that a turbulent fluctuation of a physical variable is in the form of a sine wave of frequency v and wavelength λ. The statistical character is due to a statistical phase shift ζ which is evenly distributed between 0 and 1. The amplitude of the physical variable is given by

$$\Phi(x, t) = \tfrac{1}{2} + \tfrac{1}{2} \sin(2\pi v t + x_1 \lambda + \zeta) \quad (158)$$

If the amplitude is measured sufficiently often for a given x, t then the probability density function can be obtained from the density of measured points in the sample space (Eq. 115a). The density of the measured points is given by

$$\left(\frac{\Delta \Phi}{\Delta \zeta}\right)_{x,t} \approx \left(\frac{\partial \Phi}{\partial \zeta}\right)_{x,t} = \frac{1}{\pi}[1 - (2\Phi - 1)^2]^{-1/2}$$

The single-point probability density function is thus independent of the frequency v and the wavelength λ. In addition, information about values such as $\langle (\partial \Phi/\partial x)^2 \rangle$ cannot be derived from the single-point probability density function. This value is, however, dependent on the wavelength, $\langle (\partial \Phi/\partial x)^2 \rangle = \tfrac{1}{2}(\pi^2/\lambda^2)$, compare Equation (158). The measured single-point probability density function gives no information about the spatial and time propagation of the disturbances.

In spite of these limitations, single-point probability density functions are improvements over the handling of statistical variables with the moments of probability density functions.

4.3. Examples of Calculating Probability Density Functions

For further discussions Equation (155) is rewritten as

$$\varrho(\Phi)\frac{\partial P(U, \Phi; x, t)}{\partial t} + \varrho(\Phi) U_k \frac{\partial P(U, \Phi; x, t)}{\partial x_k}$$
$$+ \frac{\partial}{\partial U_k}[\varrho(\Phi)\langle A_k|u, \varphi\rangle P(U, \Phi; x, t)]$$
$$+ \frac{\partial}{\partial \Phi_\omega}[\varrho(\Phi)\langle \Theta_\omega|u, \varphi\rangle P(U, \Phi; x, t)] = 0 \quad (159)$$

where A_i and Θ_α are given by Equations (142a, b). Equation (159) is deterministic, i.e., determined solutions exist for given initial and boundary conditions. The conditional expected values $\langle A_i|u, \varphi\rangle$ and $\langle \Theta_\alpha|u, \varphi\rangle$ are not statistical values. In the general case they are unknown functions of the variables u and φ. For the general case (i.e., for fields of statistical variables) model assumptions must be introduced so that Equation (155) or (159) can be solved. For deterministic systems Equations (155) and (159) can be treated without further assumptions because $\langle A_i|u, \varphi\rangle = A_i(u, \varphi)$ and $\langle \Theta_\alpha|u, \varphi\rangle = \Theta_\alpha(u, \varphi)$, respectively.

4.3.1. Solutions for Deterministic Systems

4.3.1.1. Age (Residence Time) Distributions in Chemical Reactors

The individual types of chemical reactors are normally treated as deterministic systems although under industrial conditions flow is not laminar. This simplification is justified if the turbulent time scales for the reactions occurring in the reactor lie several orders of magnitude above that for the turbulent mass or momentum exchange. The reactor contents have then reached the state of molecular mixing, without noticeable chemical conversion having taken place. Distribution functions or probability density functions may be given for the residence time, for example, and can thus indicate the probability for the residence time (the age α) of a volume element in an interval dα around α; residence time distribution functions are easily derived from Equation (159).

Ideally Mixed Stirred-Tank Reactor. In the mathematical sense an ideally mixed reactor is a one-dimensional control volume with homogeneous internal properties. The upstream characteristics are transported into the control volume while the characteristics at the downstream boundary surface relate to the internal properties (Fig. 18). Viscosity, pressure, and gravitational forces and other transport processes (with the exception of convection) are ignored.

For expediency Equation (159) is integrated with respect to the volume:

$$\int_V \varrho(\Phi)\frac{\partial P}{\partial t}dV + \int_V \varrho(\Phi) u \frac{\partial P}{\partial x}dV$$
$$+ \int_V \frac{\partial}{\partial \Phi_\omega}[\varrho(\Phi) S_\omega(u, \Phi) P]dV = 0 \quad (160)$$

Using the Gauss–Ostrogradski integrals, Equation (160) can be rewritten as

$$\frac{\partial}{\partial t}\int_V \varrho(\Phi) P\,dV + \oint_F \varrho(\Phi) u P\,dF$$
$$+ \int_V \frac{\partial}{\partial \Phi_\omega}[\varrho(\Phi) S_\omega(u, \Phi) P]dV = 0 \quad (161)$$

where integration has to be carried out for the surfaces perpendicular to the direction of the flow.

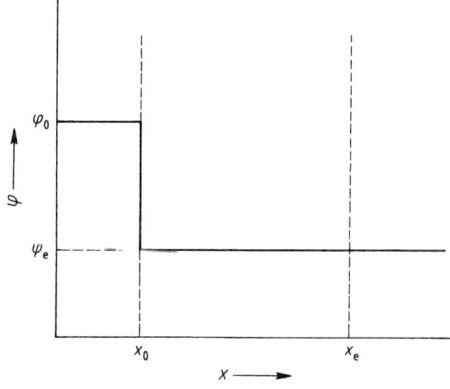

Figure 18. Mathematical definition of an ideally mixed reactor
The subscripts 0 and e denote inlet and outlet, respectively.

If the age (residence time) distribution is considered, the variables u, Φ in $P(u, \Phi)$ and $S(u, \Phi)$ must be substituted by the age α. Since intervals in the age space are time intervals, the velocity in the age space is equal to $\partial \alpha / \partial t = S(\alpha) = 1$. With this result and for constant density the following expression can be derived from Equation (161):

$$\frac{\partial}{\partial t} P(\alpha) dV + u P(\alpha) dz dy - u P_0(\alpha) dz dy$$

$$+ \frac{\partial}{\partial \alpha} P(\alpha) dV = 0 \qquad (162)$$

In the steady state, Equation (162) is easy to solve since the first term on the left-hand side is zero:

$$\frac{\partial}{\partial \alpha} [P(\alpha)] = \frac{1}{\tau} [P_0(\alpha) - P(\alpha)] \qquad (163\,a)$$

where $\tau = u dz dy / dV$ is the mean hydrodynamic residence time. With the boundary condition $P_0(\alpha) = \delta(\alpha - 0)$ and the condition $\int_0^{+\infty} P(\alpha) d\alpha = 1$ (cf. Eq. 104) then

$$P(\alpha) = \frac{1}{\tau} \exp\left(-\frac{\alpha}{\tau}\right) \qquad (164)$$

$\delta(\alpha - 0)$ denotes the impulse function or Dirac delta function (Section 4.1.1).

$$\int_0^{+\infty} \alpha \delta(\alpha - b) g(\alpha) d\alpha = a g(b); \quad b \geq 0$$

is used to integrate Equation (163a). Equation (164) is the familiar residence time distribution function for ideally mixed reactors (denoted as $w(t)$ in Section 3.3.2.2). If it is transformed to normalized age ($\theta = \alpha / \tau$) then

$$P(\theta) \equiv w(\theta) = P(\alpha) \frac{\partial \alpha}{\partial \theta} = \frac{1}{\tau} \exp\left(-\frac{\alpha}{\tau}\right) \tau$$

$$= \exp(-\theta) \qquad (165)$$

In Equation (165) the distinction between age α and time t is no longer maintained. The integral over the residence time distribution

$$W(t) = \int_0^{+\infty} w(t) dt \quad \text{or}$$

$$W(\theta) = \int_0^{+\infty} w(\theta) d\theta$$

is denoted as distribution function. Figure 19 shows the residence time distribution and the distribution function for an ideal stirred-tank reactor.

Alternative boundary conditions lead to alternative solutions of Equation (163a). For example, if n ideally mixed reactors are linked in a cascade, then $P_{n-1}(\alpha) = P_{0n}(\alpha)$ etc. With $P_{01}(\alpha) = \delta(\alpha - 0)$ and $\tau_1 = \tau_2 = \cdots \tau_n = \tau/n$, the solution of Equation (163a) then becomes

$$w_n(t) = \frac{1}{(n-1)!} n^n \frac{t^{(n-1)}}{\tau^n} \exp\left(-n \frac{t}{\tau}\right) \qquad (166)$$

Figure 20 shows residence time distribution functions and distribution functions for a cascade of n ideally mixed reactors for various values of n. It is easy to prove from Equation (166) that simple relationships exist between the deterministic moments

$$m_1 = \int_0^{+\infty} t w(t) dt = \bar{t} \quad \text{and}$$

$$m_2' = \int_0^{+\infty} (t - m_1)^2 w(t) dt$$

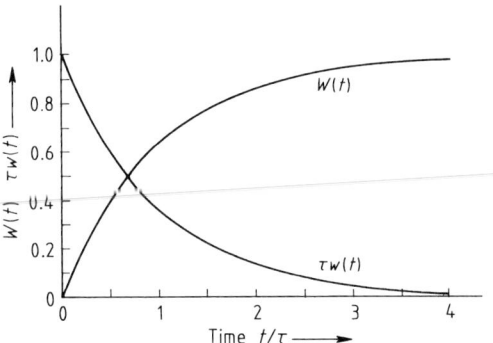

Figure 19. Residence time distribution $w(t)$ and distribution function $W(t)$ for the ideal stirred-tank reactor

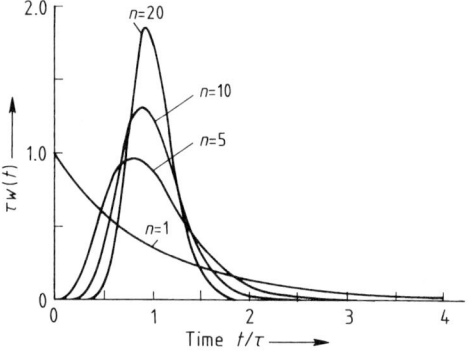

Figure 20. Residence time distribution for a cascade of n ideal stirred-tank reactors

(see Section 3.3.2.2) and the number of elements in the cascade. Thus

$$m_1 = n\tau_i = \tau \quad \text{and} \quad m_2' = n\tau_i^2 = \frac{\tau^2}{n} \quad (167\,\text{a, b})$$

Given that a reactor can be represented by a cascade of ideal stirred-tank reactors, the number of elements required to represent the reactor can easily be determined with Equation (167) from the first two deterministic moments of the measured residence time distribution, namely from the mean residence time and the variance.

For the nonsteady case, Equation (162) becomes:

$$\frac{\partial}{\partial t}[P(\alpha;t)] + \frac{\partial}{\partial \alpha}[P(\alpha;t)] = \frac{1}{\tau}[P_0(\alpha;t) - P(\alpha;t)] \quad (163\,\text{b})$$

The general solution for Equation (163 b) can easily be found by using Laplace or Fourier transforms and applying the concept of generalized functions [62], [63]:

$$P(\alpha;t) = \exp\left(-\frac{t}{\tau}\right) P_0(\alpha - t)$$

$$+ \frac{1}{\tau} \exp\left(-\frac{\alpha}{\tau}\right)[H(\alpha) - H(\alpha - t)] \quad (168)$$

The solution holds for the initial conditions $P(\alpha;t)_{t=0} = P(\alpha)$ and the same boundary conditions as above; $H(\alpha)$ denotes the Heaviside or step function (cf. Eq. 116). It can easily be shown that Equation (168) becomes Equation (165) for $t \to \infty$.

Ideally Nonmixed Reactor (Ideal Tubular Reactor). An ideally nonmixed reactor is a one-dimensional flow system with inhomogeneous internal properties. Analogously to the ideally mixed reactor, in this one-dimensional model viscosity, pressure, gravitational forces, and other transport processes (excluding convection) are neglected. The changes in the internal properties occur as a result of physical or chemical processes.

The age (residence time) distribution of this model is obtained from Equation (159):

$$\frac{\partial}{\partial t}[P(\alpha;x,t)] + u\frac{\partial}{\partial x}[P(\alpha;x,t)]$$

$$+ \frac{\partial}{\partial \alpha}[P(\alpha;x,t)] = 0 \quad (169\,\text{a})$$

In the steady state, Equation (169 a) becomes

$$u\frac{\partial}{\partial x}[P(\alpha;x)] + \frac{\partial}{\partial \alpha}[P(\alpha;x)] = 0 \quad (169\,\text{b})$$

The partial differential Equation (169 b) can be solved in a number of ways. The calculation domain (the tubular reactor) can be divided into a number of finite control volumes, for each of which Equation (169 b) holds. The individual control volumes have the same characteristics as the ideally mixed reactor. The same operations as discussed above can be carried out for each of the finite control volumes and an equation of type (163 b) is obtained for each control volume. Physically this means that an ideally nonmixed reactor is represented by a cascade with an infinite number of elements. Numerically this means that the partial differential Equation (169 b) is solved with an up-wind difference scheme. The solution for the boundary condition $P_0(\alpha) = \delta(\alpha - 0)$ is Equation (166) with $n \to \infty$, i.e., an impulse function at $t/n\tau_i = 1$.

Equation (169 b) can also be solved analytically, the general solution is

$$P(\alpha;x) = \Omega\left(\alpha - \frac{1}{u}x\right) \quad (170)$$

where Ω is any differentiable function [64]. Equation (170) clarifies the behavior of ideally nonmixed reactors. A fluid flowing into the reactor with a given age distribution retains the form of the age distribution. The age of the individual fluid elements increases uniformly by the time necessary to pass the distance x, $t = x/u$. For example, at the reactor inlet (i.e., $x = 0$)

$$\Omega(\alpha - 0) = \delta(\alpha - 0)$$

at the reactor outlet (i.e., $x = L$)

$$P(\alpha;x) = \delta\left(\alpha - \frac{L}{u}\right) = \delta(\alpha - \tau)$$

This is identical to Equation (166) for $n \to \infty$.

If an ideally mixed reactor with a mean residence time τ_1 and an ideally nonmixed reactor with a mean residence time τ_2 are connected together, then $\Omega(\alpha - 0) = H(\alpha - 0)\exp(-\alpha/\tau_1)$, and from Equation (170)

$$P(\alpha) = H(\alpha - \tau_2) \cdot \frac{1}{\tau_1} \exp\left[-\frac{1}{\tau_1}(\alpha - \tau_2)\right]$$

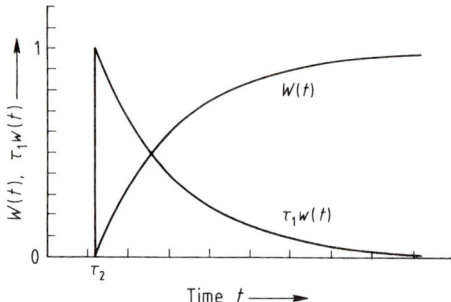

Figure 21. Residence time distribution $w(t)$ and distribution function $W(t)$ for an ideal stirred-tank reactor connected in series with an ideal tubular reactor

It is seen that the age distribution of the ideally mixed reactor is shifted by the mean residence time τ_2 of the nonmixed reactor. Figure 21 shows the residence time distribution and the distribution function for this case.

Reactors with Finite Mixing. Reactors with finite mixing can be thought of as the transport mechanisms discussed on pp. 2-39–2-41 combined with another transport mechanism. Usually this is treated in the form of a "diffusion" mechanism in which the relevant diffusion coefficient has the character of a model parameter (dispersion model).

Diffusion is expressed in the transport equation for the probability density function as an additional transport term in the sample space of scalar values (see Eq. 155, last term on the right-hand side and Eq. 159). The basic treatment of these terms is discussed in [19], [65]–[67], see also Section 4.3.2.1. Here, a heuristic method will be used to find a transport equation for the age distribution in reactors where diffusion is superimposed on convective transport.

The problem consists of deriving a term for the additional transport in the age space for a superimposed diffusion process. In chemical reactors with finite mixing, fluid parcels are convected by turbulent motion, i.e., this process consists of the diffusion of fluid elements which can be handled in the same way as molecular diffusion.

The fluid parcels experience a displacement due to the "diffusion" in addition to the displacement resulting from convective transport with velocity u. Analogous to molecular motion, the probability density function for the displacement by diffusion is a normal distribution [68]:

$$P(x)_{\text{diff}} = \frac{1}{2\sqrt{\pi D t}} \exp\left(-\frac{x^2}{4Dt}\right) \quad (171)$$

where D is the diffusion coefficient. The probability density function for the displacement x of the fluid elements due to diffusion can be rewritten as the probability density function for the age with the help of the transformation rule given by Equation (106) and $\alpha = x/u$:

$$P(\alpha)_{\text{diff}} = P(x)_{\text{diff}} \cdot \frac{dx}{d\alpha} = \frac{u}{2\sqrt{\pi D \alpha}} \exp\left(-\frac{x^2}{4D\alpha}\right) \quad (172)$$

Equation (172) represents the age distribution for a diffusion process that is analogous to molecular diffusion. The term required for the transport equation is the differential of this distribution in the age space, i.e., $\frac{\partial}{\partial \alpha}[P(\alpha)_{\text{diff}}]$. Equation (169b) is now supplemented by the term $\frac{\partial}{\partial \alpha}[P(\alpha)_{\text{diff}}]$ so that the transport equation for the age distribution in the steady-state case takes the form

$$\frac{\partial}{\partial \alpha}[P(\alpha;x)] + u\frac{\partial}{\partial x}[P(\alpha;x)] + D\frac{\partial^2}{\partial x^2}[P(\alpha)] = 0 \quad (173)$$

Equation (173) can be solved analytically for simple boundary conditions. For the boundary condition

$$\frac{\partial}{\partial x}[P(\alpha;x)] = 0 \quad \text{for} \quad x = 0 \quad \text{and} \quad x = L$$

this gives

$$P(\alpha) = \frac{1}{2}\sqrt{\frac{Pe}{\pi \alpha}} \exp\left(-\frac{(1-\alpha/\tau)^2 Pe}{4\alpha/\tau}\right) \quad (174)$$

where Pe is the axial Péclet number, $Pe = uL/D$ (Section 3.3.2.2), it is often called the Bodenstein number, Bo.

Figure 22 shows the residence time distributions (Eq. 174) for different Bodenstein numbers. For large Bodenstein numbers the figure resembles the distributions for the cascade of ideally mixed reactors (Fig. 20). This similarity is inevitable, because the cascade of ideally mixed reactors represents the numerical solution of the transport Equation (169b) as $n \to \infty$ and Equation (173) will be identical to this for $D \to 0$. The deterministic moments m_1 and m'_2 can be calculated from Equation (174):

$$m_1 = \left(1 + \frac{2}{Bo}\right)\tau \quad \text{and} \quad m'_2 = \left(\frac{2}{Bo} + \frac{8}{Bo^2}\right)\tau^2$$

(175 a, b)

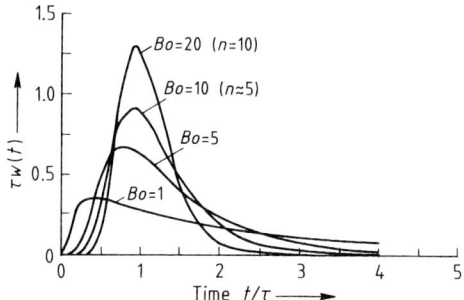

Figure 22. Residence time distribution $\tau w(t)$ for a reactor with axial mixing
Bo = Bodenstein number; n = number of elements of a cascade of ideally mixed reactors to represent the reactor with the corresponding Bodenstein number

For $Bo \to \infty$, the first moment corresponds to the mean hydrodynamic residence time (Eq. 167a). If Equation (175b) is compared with Equation (167b) on the basis of the similarity of the solutions of the age distribution for large Bodenstein numbers, then the relationship

$$\frac{1}{n} = \frac{2}{Bo} + \frac{8}{Bo^2}$$

can be derived. For large Bodenstein numbers (i.e., little axial dispersion), $n \approx Bo/2$.

This discussion shows that the definition or experimental determination of model parameters can be relatively simple even for complex models. In the above case the formal diffusion coefficient, the axial mixing coefficient, or the Bodenstein number can be simply determined from the second deterministic moment of the age (residence time) distribution. Furthermore, complex processes may be built up from a number of simple elements. In the case described above, a reactor with finite axial mixing is constructed from a cascade of ideally mixed reactors; a simple relationship exists between the number of elements and the model parameters of this complex one-dimensional process. Further examples for the representation of complex processes by simple elements are found in [4], [11], [12], [24], [46], and text books on chemical reaction engineering, see also Chapter 5.

4.3.1.2. Size Distribution in Continuously Operating Crystallizers

The model used for the representation of a continuously operating crystallizer is an ideally mixed reactor. The supersaturated solution is fed into the crystallizer and the substrate grows on the existing crystals. Using the operations discussed on pp. 2-39–2-40 this gives the transport equation

$$\frac{\partial}{\partial \Phi}[S(\Phi) P(\Phi)] = \frac{1}{\tau}[P_0(\Phi) - P(\Phi)] \tag{176}$$

The property under consideration is the crystal size L. The rate of crystal growth can be described empirically by $S(L) = S(L, c)$, i.e., it is a function of the crystal size and the concentration of the substrate c. Often $S(L) = S_0(c)L^b$ is used so that Equation (176) becomes

$$\frac{\partial}{\partial L}[S_0(c) L^b P(L)] = \frac{1}{\tau}[P_0(L) - P(L)] \tag{177}$$

If the feed does not contain solids, $P_0(L) = \delta(L - 0)$. Assuming that the substrate concentration is approximately constant, the solution of Equation (177) is

$$P(L) = C_1 L^{-b} \exp\left[-\frac{L^{1-b}}{S_0(c)\tau(1-b)}\right] \tag{178}$$

where C_1 is an integration constant. For $b = 0$ (McCabe ΔL law) Equation (178) becomes

$$P(L) = P(L_0) \exp\left(-\frac{L}{S_0(c)\tau}\right) \tag{179}$$

where C_1 is substituted by the initial condition $P(L_0)$, i.e., the number density of the nuclei in the crystallizer. Equation (179) represents the number density distribution of crystal sizes for this simple case ($b = 0$).

4.3.2. Solutions for Statistical Systems

The discussion in Section 4.3.1 was confined to deterministic systems for which either the physical quantities are not subject to turbulent (statistical) fluctuations or for which the time scales of the individual physicochemical processes are of vastly different orders of magnitude. In the ideally mixed reactor the time scale for the chemical reactions is so large in comparison with that of the turbulent mass exchange that the reactor contents can be considered as being mixed at the molecular level without having reacted. The point of this discussion was to bring the transport equations for the probability density functions (Eqs. 155, 159) into a simple form which is closed without any model assumptions and is also easy to solve analytically in simple cases.

These simplifications do not apply to statistical systems, such as the chemical reactions in a turbulent single-phase flow shown in Figure 14. The transport equations for the probability density function in the combined velocity–composition space have advantages for describing such processes because, in contrast to the simple equations for turbulent reacting flows (see Section 5.1.2), the most important processes (convection and chemical reaction) can be handled without model assumptions. Assumptions must be introduced for the dissipation of turbulent velocity fluctuations or fluctuations of scalar quantities by molecular mixing. These are manifested in the transport equations for probability density functions as transport in the velocity or composition space.

A method of solving the transport equations must also be found. Analytical solutions can only be obtained for very simple flows, simple models, or deterministic systems under simple conditions (Section 4.3.1). In general, Equation (155) must be solved numerically with the relevant terms on the right-hand side being modeled. Methods using finite differences (see Section 4.3.1.1) can be ruled out for this since $P(U, \Phi; x, t)$ is a function of $(\sigma + 6)$ independent variables and time. If, for example, a case was investigated which required 20 finite volumes in the direction of each of these variables, then the number of grid points for a statistical two-dimensional flow with only two chemical species under isothermal conditions would be $20^7 \approx 10^9$. Other methods must therefore be used.

4.3.2.1. Closure of the Transport Equation for Probability Density Functions

The equation for the derivation of the joint probability density function for a turbulent reacting flow is given by

$$\varrho(\Phi)\frac{\partial P}{\partial t} + \varrho(\Phi) U_k \frac{\partial P}{\partial x_k} + \left[\varrho(\Phi) g_k - \frac{\partial \langle p \rangle}{\partial x_k}\right]\frac{\partial P}{\partial U_k}$$
$$+ \frac{\partial}{\partial \Phi_\omega}[(\varrho(\Phi) S_\omega(\Phi) P]$$
$$= -\frac{\partial}{\partial U_k}\left[\left\langle \frac{\partial \tau_{lk}}{\partial x_l} + \frac{\partial p^f}{\partial x_l} \bigg| u, \varphi \right\rangle P\right]$$
$$+ \frac{\partial}{\partial \Phi_\omega}\left[\left\langle \frac{\partial J_k^\omega}{\partial x_k} \bigg| u, \varphi \right\rangle P\right] \quad (155)$$

All the terms on the right-hand side of the equation (convection in physical space, transport in the velocity space due to gravitational forces, forces due to the gradient of the average pressure, and transport in the composition space due to chemical reactions) are closed and do not require any model assumptions.

The term $\varrho(\Phi)$ is a known function (e.g., a form of thermal equation of state, Eq. 136) and $S_\alpha(\Phi)$ is given by Equation (137a). For example, for a bimolecular reaction $A + B \to C + D$ this relationship is $S(A) = r_A = -k m_A m_B$ where the rate coefficient k can often be represented by three parameters in the form $k = k_0 T^\alpha \exp(-E_a/RT)$. If the equations for the reaction rate and the rate coefficients are not considered to be a model, then the representation of the reaction rates in Equation (155) is free of models.

No explicit equation for the mean pressure is contained in the system of Equations (131)–(137). The momentum conservation equations (Eq. 132) contain only the pressure gradient in each of the directions under consideration. Equations (131)–(137) assume a known pressure field which is given implicitly by the conservation equations for the components of momentum and the continuity equation. The mean pressure can be determined from a Poisson equation which is derived from the conservation equations for the momentum components and the continuity equation. Equation (132) can be rewritten as

$$\frac{\partial \varrho U_i}{\partial t} + \frac{\partial}{\partial x_k}(\varrho U_i U_k) = -\frac{\partial \langle p \rangle}{\partial x_i} + \varrho A_i' \quad (180)$$

with

$$A_i' \equiv \frac{\partial \tau_{ik}}{\partial x_k} - \frac{\partial p^f}{\partial x_i} + \varrho g_i$$

If Equation (180) is averaged and differentiated with respect to x_k and the continuity equation differentiated with respect to t

$$\frac{\partial}{\partial x_k}\frac{\partial}{\partial t}(\langle \varrho U_k \rangle) = -\frac{\partial^2 \langle \varrho \rangle}{\partial t^2} \quad (181)$$

is subtracted, then the Poisson equation for the mean pressure is obtained which has to be solved simultaneously to the equation for the probability density function:

$$\frac{\partial^2 \langle p \rangle}{\partial x_k \partial x_k} = \frac{\partial^2 \langle \varrho \rangle}{\partial t^2} - \frac{\partial^2 \langle \varrho U_k U_k \rangle}{\partial x_k x_k} + \frac{\partial \langle \varrho A_k' \rangle}{\partial x_k} \quad (182)$$

In the next step the model assumptions for the conditional expected values on the right-hand side of Equation (155) must be discussed, in particular for

$$\frac{\partial}{\partial U_k}\left[\left\langle\frac{\partial \tau_{ik}}{\partial x_k}+\frac{\partial p^{\mathrm{f}}}{\partial x_k}\bigg|\boldsymbol{u},\varphi\right\rangle P\right] \text{ and}$$

$$\frac{\partial}{\partial \Phi_\alpha}\left[\left\langle\frac{\partial J_k^\alpha}{\partial x_k}\bigg|\boldsymbol{u},\varphi\right\rangle P\right]$$

Modeling of these expected conditional values is the main problem in developing mathematical models for chemically reacting turbulent flows on the basis of transport equations for probability density functions. The principles used in these methods will be discussed here, for further details see [18], [19], [70], [71], [79], [80].

The probability density functions in the velocity space and the composition space will be treated separately. They can be calculated by integrating Equation (155) over the velocity space and composition space or by using the methods shown in Section 4.2.1 for the momentum components or the scalar quantities. For further simplification a turbulent system is considered with constant density, homogeneous turbulence, and no gradients of the mean velocity.

A uniform fluid flow which flows through a turbulence-generating grid represents a good approximation to these simplified assumptions. If the coordinate system moves with the flow at the mean flow velocity, then no relative convection occurs. The probability density function becomes

$$\frac{\partial P(\boldsymbol{U};t)}{\partial t}$$
$$= -\frac{\partial}{\partial U_k}\left[\left\langle\left(\mu\frac{\partial^2 U_k}{\partial x_k^2}-\frac{\partial p^{\mathrm{f}}}{\partial x_k}\right)\bigg|\boldsymbol{u}\right\rangle P(\boldsymbol{U};t)\right]$$
(183)

Equation (183) represents the time development of the probability density function (μ is the dynamic viscosity). The only term to be modeled is

$$\frac{\partial}{\partial U_k}\left[\left\langle\mu\frac{\partial^2 U_k}{\partial x_k^2}-\frac{\partial p^{\mathrm{f}}}{\partial x_k}\bigg|\boldsymbol{u}\right\rangle P(\boldsymbol{U};t)\right]$$

Experimental observations in a turbulent flow of this type show that the turbulent velocity fluctuations decay with time. Further, if the eddy dissipation model outlined on p. 2-28 is used, it can be shown that the turbulent momentum exchange is an isotropic diffusion process of the turbulent eddies [49] that is rate determining for the dissipation of the velocity fluctuations at the molecular level. Similarly to the diffusion process examined on p. 2-42, this model gives a normal distribution for the velocity (a three-dimensional distribution in this case).

A normal distribution is precisely defined by the mean value and the variances of the quantities concerned. Since the normal distribution being discussed is three dimensional, the variances and covariances of the velocities are important and, using Equations (112) and (113), can be expressed as $\langle U_i^{\mathrm{f}} U_j^{\mathrm{f}}\rangle$ (Reynolds stresses). The problem now is to find models for the Reynolds stresses and their decay over time which will guarantee the required normal distribution as the solution for Equation (183).

A frequently used model for the Reynolds stresses (see also Section 5.1.2.1) which satisfies these conditions, is that developed by LUMLEY [72] and LAUNDER [73] and modified by ROTTA [49], [74]:

$$\frac{\partial}{\partial t}\langle U_i^{\mathrm{f}} U_j^{\mathrm{f}}\rangle = [-\langle U_i^{\mathrm{f}} U_j^{\mathrm{f}}\rangle$$
$$- C_{\langle U_i^{\mathrm{f}} U_j^{\mathrm{f}}\rangle}(\langle U_i^{\mathrm{f}} U_j^{\mathrm{f}}\rangle - \tfrac{2}{3}k)\delta_{ij}]/\tau$$
(184)

where δ_{ij} is the Kronecker delta, $\delta_{ij}=1$ for $i=j$, $\delta_{ij}=0$ for $i\neq j$; k is the specific kinetic energy of the turbulent velocity fluctuations, $k=\tfrac{1}{2}\langle U_k^{\mathrm{f}} U_k^{\mathrm{f}}\rangle$; and τ is the turbulent time scale, which has to be defined from model assumptions.

The first term on the right-hand side of Equation (184) causes exponential decay of the Reynolds stresses. The second term is responsible for the isotropy of the "diffusion" of the kinetic energy of the turbulent fluctuations.

The second problem for the solution of Equation (183) is the modeling of the conditional expected values

$$\left\langle\left(\frac{\partial p^{\mathrm{f}}}{\partial x_i}\right)\bigg|\boldsymbol{u}\right\rangle$$

A Poisson equation for the pressure variations can be derived from the conservation equations for the momentum components and the continuity equation in a similar manner to the mean pressure:

$$\frac{\partial^2 p^{\mathrm{f}}}{\partial x_k \partial x_k}=-2\varrho\frac{\partial\langle U_k\rangle}{\partial x_l}\cdot\frac{\partial U_l^{\mathrm{f}}}{\partial x_k}-\varrho\frac{\partial^2 U_k^{\mathrm{f}} U_l^{\mathrm{f}}}{\partial x_k \partial x_l} \quad (185)$$

Thus the pressure fluctuations have two sources. The first is the interactions of the turbulent fluctuations themselves (second term of the right-hand side) and the interactions of the turbulent fluctuations with the gradients of the mean velocities. The second term is described by the equation for the Reynolds stressses (Eq. 184); the first term ("rapid pressure") must be modeled, this can also be carried out with the help of the Reynolds stresses [19], [66].

The third problem is the description of the conditional expected values

$$\frac{\partial}{\partial \Phi_\alpha}\left[\left\langle\frac{\partial J_k^\alpha}{\partial x_k}|u, \varphi\right\rangle P\right]$$

These terms can be modeled by using the simplifications described above. For the same homogeneous system at constant density with isotropic turbulence and a moving coordinate system, the change in the probability density function over time for a passive scalar quantity, e.g., the mass fraction of a nonreacting chemical species is given by:

$$\frac{\partial P(\Phi; t)}{\partial t} = \frac{\partial}{\partial \Phi}\left[\left\langle\frac{\partial J_k^\Phi}{\partial x_k}|\varphi\right\rangle P(\Phi; t)\right]$$

$$= \frac{\partial}{\partial \Phi}\left[\left\langle\frac{D}{\varrho}\frac{\partial^2 \Phi}{\partial x_k^2}|\varphi\right\rangle P(\Phi; t)\right] \quad (186)$$

where D is the diffusion coefficient. Equation (186) is obtained for the nonreacting scalar Φ from the general transport equation by integrating over the velocity space or by using the method given in Section 4.2.1. As before with the dissipation of velocity fluctuations, the eddy transport model allows the turbulent exchange of mass to be considered as "eddy diffusion". This "diffusion" is rate determining for the dissipation of the turbulent fluctuations of the scalar at the molecular level. The term

$$\frac{\partial}{\partial \Phi}\left[\left\langle\frac{D}{\varrho}\frac{\partial^2 \Phi}{\partial x_k^2}|\varphi\right\rangle P(\Phi; t)\right]$$

must be modeled in such a way that the solution of Equation (186) is a normal distribution which decays over time. Thus, the problem consists of modeling the second moment $\langle\Phi^{f2}\rangle$ (the variance of the scalar quantities).

A simple deterministic model [75] which satisfies the above requirements is

$$\left\langle\left(\frac{D}{\varrho}\frac{\partial^2 \Phi}{\partial x_k^2}\right)|\varphi\right\rangle = -\frac{1}{2}C_\Phi(\Phi - \langle\Phi\rangle)\frac{1}{\tau} \quad (187)$$

in which τ is the turbulent time scale $\tau/C_\Phi = \tau_\Phi$. Since diffusion transport is in the composition space, Equation (187) can be interpreted as transport velocity in the composition space which is proportional to the deviation from the mean value. However, model Equation (187) does not yield a normal distribution from all initial conditions. Instead, the initial conditions must contain information about the form of the distribution [19] as in the applications described in [57], [76], [77]. A general equation can be derived with a statistical particle interaction model, for details refer to [19].

The variances of Φ, $\langle\Phi^{f2}\rangle$, according to Equation (187) are described by an equation similar to Equation (184):

$$\frac{\partial}{\partial t}\langle\Phi^{f2}\rangle = -C_{\langle\Phi^{f2}\rangle}\langle\Phi^{f2}\rangle/\tau \quad (188)$$

For the combined velocity–composition space of the homogeneous system with isotropic turbulence the same line of reasoning can be applied. The joint probability density function $P(U, \Phi; t)$ is normally distributed so that it is defined by the mean values and the second moments $\langle U_i^f U_j^f\rangle$, $\langle\Phi_\alpha^f\Phi_\beta^f\rangle$, and $\langle U_i^f\Phi_\alpha^f\rangle$. The latter quantities (the Reynolds fluxes) are usually defined by a model equation analogous to Equations (188) and (184):

$$\frac{\partial}{\partial t}\langle U_i^f\Phi_\alpha^f\rangle = -C_{\langle U_i^f\Phi_\alpha^f\rangle}\langle U_i^f\Phi_\alpha^f\rangle/\tau \quad (189)$$

which does not change the character of the solution for the joint probability density function. In the combined velocity–composition space, the diffusion of the Reynolds stresses $\langle U_i^f U_j^f\rangle$, the fluctuations of scalar quantities $\langle\Phi^{f2}\rangle$, and the Reynolds fluxes $\langle U_i^f\Phi_\alpha^f\rangle$ are all superimposed on one another. The remaining problem is then the modeling of the turbulent time scales for these "diffusion" processes and the treatment of the "rapid pressure". Some ideas for modeling these are discussed in Sections 5.1 and 5.3.

Under industrial conditions, chemical reactions in turbulent flows occur in nonhomogeneous systems that usually involve gradients of the mean velocities and nonisotropic turbulence.

The extension of the simple model described above to such cases is still being developed [18], [19]. The general principle is that the main properties of the model for the second moments are retained. The additional effects are treated as being superimposed on the undisturbed "diffusion". Often "hybrid" models are also used where the expected values of the velocities are computed by other methods (cf. Section 5.1.2.3) and only the joint probability density function for the composition space is evaluated (cf. Section 4.3.2.3)

4.3.2.2. Solution Methods of the Transport Equation for Probability Density Functions

Equation (155) (Section 4.2.1)

$$\varrho(\Phi)\frac{\partial P}{\partial t} + \varrho(\Phi) U_k \frac{\partial P}{\partial x_k} + \left[\varrho(\Phi) g_k - \frac{\partial \langle p \rangle}{\partial x_k}\right]\frac{\partial P}{\partial U_k}$$
$$+ \frac{\partial}{\partial \Phi_\omega}[\varrho(\Phi) S_\omega(\Phi) P]$$
$$= -\frac{\partial}{\partial U_k}\left[\left\langle\left(\frac{\partial \tau_{lk}}{\partial x_l} + \frac{\partial p^f}{\partial x_l}\right)\Big| u, \varphi\right\rangle P\right]$$
$$+ \frac{\partial}{\partial \Phi_\omega}\left[\left\langle\frac{\partial J_k^\omega}{\partial x_k}\Big| u, \varphi\right\rangle P\right] \quad (155)$$

is a partial differential equation in $(\sigma + 6)$ independent variables and time. Finite difference methods for solving this equation fail due to the large number of variables if drastic simplifications cannot be applied. Successful methods of solution have been based on Monte Carlo simulations of the discrete form of the probability density functions in the velocity–composition space [19]. These methods will be described briefly before an example is discussed.

Equation (155) can be rewritten as

$$\frac{\partial P}{\partial t} = (\Omega_1 + \Omega_2 + \Omega_3) P \quad (190)$$

in which Ω_1, Ω_2, and Ω_3 are operators that are obtained by comparison with Equation (155):

$$\Omega_1 \equiv -g_k \frac{\partial}{\partial U_k} - S_\omega \frac{\partial}{\partial \Phi_\omega} - \frac{\partial S_\omega}{\partial \Phi_\omega} I \quad (191\,\text{a})$$

$$\Omega_2 \equiv [\ldots] \quad (191\,\text{b})$$

$$\Omega_3 \equiv U_k \frac{\partial}{\partial x_k} + \frac{1}{\varrho(\Phi)}\frac{\partial \langle p \rangle}{\partial x_k}\frac{\partial}{\partial U_k} \quad (191\,\text{c})$$

In Equation (190) P denotes $P(U, \Phi; x, t)$ and in Equation (191 a) I is the identity operator. The terms represented by dots in Equation (191 b) symbolize the models used for each of the "diffusion" processes, i.e., the types of models to describe the conditional expected values on the right-hand side of Equation (155). Equation (190) indicates that all three operators influence the probability density function simultaneously. However, for a solution the operators are separated.

A first-order approximation $P^{(1)}(t)$ for $P(t)$ starting from given initial conditions $P(t_0)$ is estimated by calculating the separate effects of the three operators in the following way:

$$P_1^{(1)}(t) = (I + \Delta t\,\Omega_1) P^{(1)}(t) \quad (192\,\text{a})$$

$$P_2^{(1)}(t) = (I + \Delta t\,\Omega_2) P_1^{(1)}(t) \quad (192\,\text{b})$$

$$P^{(1)}(t + \Delta t) = (I + \Delta t\,\Omega_3) P_2^{(1)}(t) \quad (192\,\text{c})$$

If the fractional changes $P_1^{(1)}(t)$ and $P_2^{(2)}(t)$ are eliminated backwards, then $P^{(1)}(t + \Delta t)$ can be expressed by $P^{(1)}(t)$:

$$P^{(1)}(t + \Delta t)$$
$$= (I + \Delta t\,\Omega_3)(I + \Delta t\,\Omega_2)(I + \Delta t\,\Omega_1) P^{(1)}(t)$$
$$= P^{(1)}(t) + \Delta t\,(\Omega_1 + \Omega_2 + \Omega_3) P^{(1)}(t) + O(\Delta t^2)$$
$$\quad (193\,\text{a})$$

On the other hand if $P^{(1)}(t)$ is developed as a Taylor series, then

$$P^{(1)}(t + \Delta t) = P^{(1)}(t) + \frac{\partial P^{(1)}(t)}{\partial t}\Delta t$$
$$+ \frac{\partial^2 P^{(1)}(t)}{\partial t^2}\frac{\Delta t^2}{2!} + \ldots = P^{(1)}(t)$$
$$+ \frac{\partial P^{(1)}(t)}{\partial t}\Delta t + O(\Delta t^2) \quad (193\,\text{b})$$

If Equations (193 a) and (193 b) are set equal to each other and divided by Δt then

$$\frac{\partial P^{(1)}(t)}{\partial t} = (\Omega_1 + \Omega_2 + \Omega_3) P^{(1)}(t) + O(\Delta t^2) \quad (194)$$

Equation (194) is identical to Equation (190) apart from the development errors of the order $O(\Delta t^2)$. Thus this procedure of separating the operators gives a first-order approximation for $P(t)$.

The Monte Carlo simulation of the development of the probability density function uses the discrete form of the probability density function with n statistical fluid particles:

$$P_n(U, \Phi; x, t) = \frac{1}{n} \sum_{i=1}^{n} \delta(U - u_i) \delta(\Phi - \varphi_i) \delta(x - x_i) \quad (195)$$

(cf. Eq. 115b, Section 4.1.1). The discrete form of the probability density function converges with $\sqrt{n}$ towards $P(U, \Phi; x, t)$. The numerical problem is to determine the movement of n statistical fluid particles in the physical, velocity, and composition spaces consistent with the operators Ω_1, Ω_2, and Ω_3. Stated in another way: starting from a given initial distribution $P_n(U, \Phi; x, t = 0)$, n fluid particles with the properties $U_i(t_0)$, $\Phi_i(t_0)$, and $x_i(t_0)$ are chosen at random. The fractional changes of P_n are then calculated which lead to $P_n(U, \Phi; x, t = t + \Delta t)$ with the particle properties $U_i(t = t_0 + \Delta t)$, $\Phi_i(t = t_0 + \Delta t)$, and $x_i(t = t_0 + \Delta t)$. This gives an algorithm for calculating the development of $P_n(U, \Phi; x, t)$ over time by using forward differences with respect to time. The movement equations consistent with the operators Ω_1, Ω_2, and Ω_3 are for the third fractional change Ω_3

$$\frac{dU}{dt} = -\frac{1}{\varrho(\Phi(t))} \left[\frac{\partial \langle p \rangle}{\partial x_k}\right]_{x(t)}, \quad \frac{d\Phi_\omega}{dt} = 0, \quad \text{and}$$

$$\frac{dx}{dt} = U(t) \quad (196\,a, b, c)$$

According to Equation (191a), for the operator Ω_1 and thus for the first fractional change:

$$\frac{dU}{dt} = g_k + [\ldots], \quad \frac{d\Phi_\omega}{dt} = S_\omega[\Phi(t)], \quad \text{and}$$

$$\frac{dx}{dt} = 0 \quad (197\,a, b, c)$$

In Equation (197a) the contents of the square brackets indicated by dots must be substituted by the model being used for describing the dissipation of the turbulent velocity fluctuations. The second operator only causes a fractional change in the composition space so that the movement equations for the second fractional change are

$$\frac{dU}{dt} = 0, \quad \frac{d\Phi_\omega}{dt} = [\ldots], \quad \text{and} \quad \frac{dx}{dt} = 0$$

$$(198\,a, b, c)$$

The movement of the statistical fluid particles in the composition space (Eq. 198b) must again be obtained from the model being used.

After i time steps, each with three fractional changes $P_n(U, \Phi; x, t = t + i\Delta t)$ is obtained in discrete form and the remaining task is to calculate the mean values, variances, covariances, and all values of interest from the discrete probability density function. This may be achieved by the averaging procedure according to Equation (114) and the analogous rules for higher moments. However, for discrete probability density functions, special interpolation routines are recommended for calculating the moments [19].

4.3.2.3. Example: Combustion of Propane in a Turbulent Diffusion Flame

The solution of Equation (155) for turbulent reacting flows gives the joint probability density function $P(U, \Phi; x, t)$ in discrete form according to the methods outlined in Section 4.3.2.2. All the information required for engineering applications can be obtained from this complete statistical description.

To solve Equation (155) a number of model parameters must be determined (see Eqs. 184, 188, or 189 for simple systems) as well as the turbulent time scale τ. Equation (155) contains $(\sigma + 6)$ independent variables so a finite difference method of solution is ruled out. Even if the method of solution shown in Section 4.3.2.2 overcomes these problems, a reduction in the number of variables by a simplification of Equation (155) is worthwhile. In most applications this is achieved by solving Equation (155) for the composition space only [57], [70], [71], [78]–[80]. The result of this is that, besides the turbulent time scale τ, the expected values of the velocities must also be calculated by other models (Section 5.1.2).

The combustion of propane with air in a turbulent diffusion flame will be described as an example. The propane flows from a simple jet assembly with a Reynolds number of 40 000 related to the nozzle outlet. The gas is mixed with the surrounding air and burns in a turbulent diffusion flame. This example may resemble an industrial flue gas flare.

The numerical calculation is confined to the probability density function in the composition space [70]. Integration of Equation (155) over the velocity space according to Equation (121) gives

$$\frac{\partial[\varrho(\boldsymbol{\Phi})P(\boldsymbol{\Phi};\boldsymbol{x},t)]}{\partial t}+\frac{\partial}{\partial x_k}[\varrho(\boldsymbol{\Phi})\langle U_k|\varphi\rangle$$

$$\cdot P(\boldsymbol{\Phi};\boldsymbol{x},t)]+\frac{\partial}{\partial \Phi_\omega}[(\varrho(\boldsymbol{\Phi})S_\omega(\boldsymbol{\Phi})P(\boldsymbol{\Phi};\boldsymbol{x},t)]$$

$$=\frac{\partial}{\partial \Phi_\omega}\left[\left\langle\frac{\partial J_k^\omega}{\partial x_k}\bigg|u,\varphi\right\rangle P(\boldsymbol{\Phi};\boldsymbol{x},t)\right] \quad (199)$$

By using density-weighted variables (indicated by a tilde) [81]

$$\tilde{P}(\boldsymbol{\Phi};\boldsymbol{x},t)=\frac{1}{\langle\varrho\rangle}\int\varrho(\boldsymbol{\Phi})P(\boldsymbol{\Phi};\boldsymbol{x},t)\,d\varrho \quad (200)$$

and decomposing the velocities according to Equation (112)

$$U_i|\varphi=\langle\tilde{U}_i\rangle+\langle\tilde{U}_i^f|\varphi\rangle \quad (201)$$

the following is obtained:

$$\langle\varrho\rangle\frac{\partial\tilde{P}(\boldsymbol{\Phi};\boldsymbol{x},t)}{\partial t}+\frac{\partial}{\partial x_k}[\langle\varrho\rangle\langle\tilde{U}_k\rangle$$

$$\cdot P(\boldsymbol{\Phi};\boldsymbol{x},t)]+\frac{\partial}{\partial \Phi_\omega}[(\langle\varrho\rangle S_\omega\tilde{P}(\boldsymbol{\Phi};\boldsymbol{x},t)]$$

$$=\frac{\partial}{\partial \Phi_\omega}\left[\left\langle\frac{1}{\varrho}\frac{\partial J_k^\omega}{\partial x_k}\bigg|\varphi\right\rangle\tilde{P}(\boldsymbol{\Phi};\boldsymbol{x},t)\right]$$

$$-\frac{\partial}{\partial x_k}[\langle\tilde{U}_k^f|\varphi\rangle\langle\varrho\rangle\tilde{P}(\boldsymbol{\Phi};\boldsymbol{x},t)] \quad (202)$$

The problem in using Equation (202) is the unknown velocity $\langle U\rangle$ and the modeling of the terms for the dissipation of turbulent fluctuations of the scalar quantities at the molecular level (first term on the right-hand side of Eq. 202) and transport in the physical space by turbulent velocity fluctuations (second term on the right-hand side of Eq. 202). For the latter term a gradient transport approach is used [70] that describes the transport of higher moments analogously to diffusion. The background of such equations is briefly treated in Section 5.1. In the special case of propane combustion:

$$\langle\tilde{U}_k^f|\varphi\rangle\langle\varrho\rangle\tilde{P}(\boldsymbol{\Phi};\boldsymbol{x},t)$$

$$\approx-C_s\langle\varrho\rangle\frac{\tilde{k}}{\tilde{\varepsilon}}\langle U_k^fU_l^f\rangle\frac{\partial}{\partial x_l}\tilde{P}(\boldsymbol{\Phi};\boldsymbol{x},t) \quad (203)$$

where C_s is a model parameter; $\tilde{k}$ is the turbulence energy (see Section 4.3.2.1); $\tilde{\varepsilon}=-d\tilde{k}/dt$ such that $\tilde{k}/\tilde{\varepsilon}\equiv\tau$ the turbulent time scale; and finally $\langle U_i^f U_j^f\rangle$ are the Reynolds stresses. The last three quantities are unknown and must be provided before the solution of Equation (202). In [70] this occurs together with the calculation of the expected velocity values using a "second-order closure model". In this procedure the conservation equations are averaged and the correlations which arise from averaging (the Reynold stresses and Reynolds fluxes) are modeled directly (see also Section 5.1). The remaining problem is the modeling of the dissipation of the turbulent fluctuations of the scalar quantities at the molecular level, this is taken care of with information about the form of the distribution using the method described in Section 4.3.2.1 [57], [71], [76], [78].

The combustion of propane with air in a diffusion flame, comprises ca. 300 elementary reactions between ca. 45 chemical species [82]. The composition space is therefore approximately 45-dimensional and numeric simulation using the Monte Carlo method is prohibitive due to the high dimensionality of the problem. The reaction mechanism for the combustion of propane may be reduced to [70]

$$C_3H_8+\tfrac{3}{2}O_2\rightarrow 3CO+4H_2 \quad (i)$$

$$CO+OH\rightleftharpoons CO_2+H \quad (ii)$$

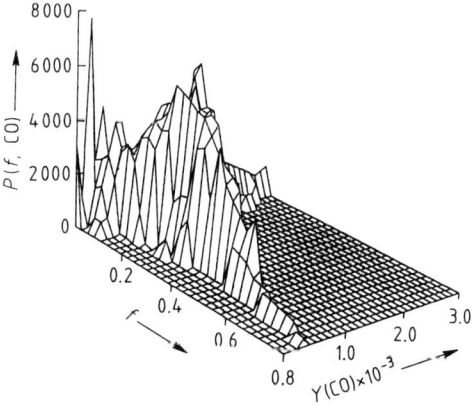

Figure 23. Joint probability density function for the normalized elemental mass fraction (mixture fraction) and mass fraction of CO according to [70] for a propane air diffusion flame at 10 nozzle diameters above the burner nozzle
f = mixture fraction

and the recombination reactions of the H_2-O_2 system

$$H + O_2 + M \rightleftharpoons HO_2 + M \quad \text{(iii)}$$

$$H + H + M \rightleftharpoons H_2 + M \quad \text{(iv)}$$

$$H + OH + M \rightleftharpoons H_2O + M \quad \text{(v)}$$

$$H + O + M \rightleftharpoons OH + M \quad \text{(vi)}$$

where M denotes a third body collision partner.

The chemical system is thus described by four scalar variables; the normalized elemental mass fraction of carbon $\Phi_1 \equiv f$, the mixture fraction; the total molar concentration $\Phi_2 \equiv c_{tot}$; the molar concentration of propane $\Phi_3 \equiv c_{C_3H_8}$; and the molar concentration of carbon monoxide $\Phi_4 \equiv c_{CO}$. This manipulation reduces the number of variables in Equation (202) to four. The reaction rates of these four variables are derived from the reduced mechanism:

$$S(\Phi_1) = 0$$
$$S(\Phi_2) = 4\tfrac{1}{2}r_{(i)} - r_{(ii)} - r_{(iv)} - r_{(v)} - r_{(vi)}$$
$$S(\Phi_3) = -r_{(i)}$$
$$S(\Phi_4) = 3r_{(i)} - r_{(iv)}$$

Figures 23 and 24 show the results of the numeric simulation according to [70]. The results given in Figure 23 are the joint probability density function of f at a distance of 10 jet diameters from the fuel nozzle. There is a peak near the beginning of the coordinates which means that in the region near the fuel nozzle, the air from the surrounding fluid entrains the flow. Figure 24 A gives radial profiles of the mean values of the propane mole fraction and the mixture fraction at a distance of 40 mm from the burner nozzle. These results are important for setting up an exhaust gas flare and can be derived from the general probability density function. There is relatively good correlation between the calculated and measured values. The correlation between the combustion products (CO and CO_2) is not as good (Fig. 24 B). This is due to the models contained in the transport equation for the joint probability density function and the chemical model which contains drastic assumptions about the concentrations of radicals formed during propane combustion (Eqs. i–vi).

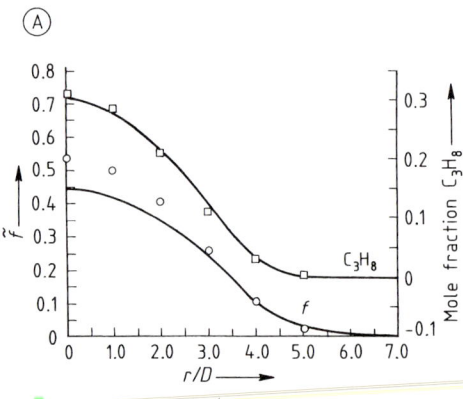

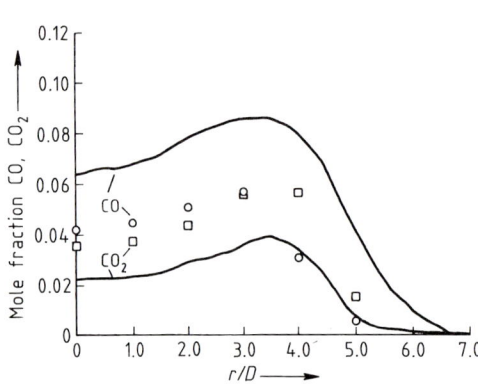

Figure 24. Simulation of a propane air diffusion flame 40 mm above the burner nozzle according to [70]
A) Mean values for the normalized elemental mass fraction (mixture fraction) and the mole fraction of propane
B) Mean values for the mole fraction of carbon monoxide and carbon dioxide
The open squares and circles denote measured values, the curves denote calculated values; r = radial distance; D = nozzle diameter.

5. Models Based on Physicochemical Principles (Transport Phenomena)

See also → 4. Transport Phenomena and → 5. Fluid Mechanics.

Models based on physicochemical principles apply the principles of the transport of momentum, mass, and energy to chemical engineering processes. The representation of processes by this type of model consists of the formulation of the conservation equations for the total mass, the mass of the individual chemical species, components of momentum, and the enthalpy for the system or subsystem under consideration. The conservation equations can be given in the form shown in Section 4.2:

Accumulation + Change due to convective transport = Change due to molecular transport + Source

The required number, combination, and formulation of the conservation equations and any other physicochemical relationships necessary for closing the set of equations depend on the kind of process and the physical boundary conditions for the system under consideration.

In the following sections, applications of the principles of conservation of mass, momentum, and energy will be discussed. Wherever possible the classification according to Figure 4 will be used.

5.1. Application of the Principle of Conservation of Momentum

(The notation used in this chapter is the same as that employed in Chapter 4.)

The detailed notation for the equation for the conservation of momentum (Eq. 132) is

$$\varrho \frac{\partial u_i}{\partial t} + \varrho u_k \frac{\partial u_i}{\partial x_k} = \frac{\partial \tau_{ik}}{\partial x_k} - \frac{\partial p}{\partial x_i} + \varrho g_i$$

$$i = 1, 2, 3 \quad (204)$$

Physically this equation represents the change in momentum flux in a differential control volume caused by volume-specific forces: viscous stress (the first term on the right-hand side of Eq. 204), pressure (the second term on the right-hand side), and body forces, in this case the gravitational forces (the third term on the right-hand side). The assumptions made in the formulation of Equation (204) are discussed in [48], [50], [51].

For Newtonian fluids a gradient expression relates the viscous stresses to the flow velocities. In accordance with Newton's law of friction the shear stress tensor component τ_{ij} can be written as

$$\tau_{ij} = \mu \left(\frac{\partial u_i}{\partial x_j} + \frac{\partial u_j}{\partial x_i} \right) - \frac{2}{3} \mu \frac{\partial u_k}{\partial x_k} \delta_{ij} \quad (205)$$

where δ_{ij} is the Kronecker delta: $\delta_{ij} = 1$ for $i = j$ and $\delta_{ij} = 0$ for $i \neq j$. The assumptions used in the formulation of Equation (205) are discussed in [48], [50], [51]. Except for the pressure gradients, gravitational forces, density ϱ, and the dynamic viscosity of the material μ, Equation (204) is closed. In the case of gases, for example, μ can be determined from the kinetic theory of gases; other methods of calculating the dynamic viscosity are given in [48], [50]. For systems with constant density (e.g., incompressible fluids) and a known pressure field, Equation (204) is sufficient for calculating the velocity field. This will be demonstrated for the simple example of laminar flow in a tube.

5.1.1. Laminar Tube Flow

Laminar tube flow is illustrated in Figure 25. The flow is assumed to be fully developed and steady state, the accumulation term and the gradients in the direction of flow therefore disappear. There are no components of velocity transverse to the direction of flow. The problem is thus reduced to a one-dimensional problem with one dependent variable.

With the definitions specified in Figure 25 and the appropriate formulation in cylindrical coordinates, Equation (204) can easily be arranged as follows

$$0 = \frac{1}{r} \frac{\partial}{\partial r} (r \tau_{rz}) - \frac{\partial p}{\partial z} + \varrho g_z \quad (206)$$

If the pressure gradient in Equation (206) is replaced by the pressure difference over length L, which is taken as known, Equation (207) is obtained:

$$\frac{\partial}{\partial r} (r \tau_{rz}) = - \left(\frac{p_0 - p_L}{L} + \varrho g \right) r \quad (207)$$

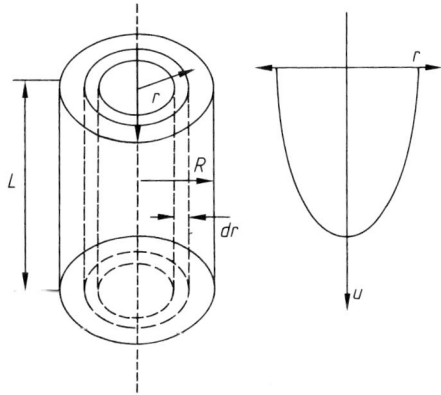

Figure 25. Laminar tube flow and parabolic velocity profile
L = length; R = radius; r = radial distance; u = velocity in the axial direction

Integration of Equation (207) gives

$$r\tau_{rz} = -\left(\frac{p_0 - p_L}{L} + \varrho g\right)\frac{r^2}{2} + C_1 \qquad (208)$$

where C_1 is an integration constant. If the term from Newton's law of friction is substituted for the laminar viscous shear stress τ_{rz}, then Equation (208) becomes

$$\frac{du}{dr} = -\left(\frac{p_0 - p_L}{L} + \varrho g\right)\frac{r}{2\mu} + \frac{C_1}{\mu r} \qquad (209)$$

Integration of Equation (209) results in

$$u = -\left(\frac{p_0 - p_L}{L} + \varrho g\right)\frac{r^2}{4\mu} + \frac{C_1}{\mu}\ln r + C_2 \qquad (210)$$

Since for $r = 0$ the velocity is finite, C_1 must be zero. The constant C_2 can be calculated from a further boundary condition, $u = 0$ for $r = R$:

$$C_2 = \left(\frac{p_0 - p_L}{L} + \varrho g\right)\frac{R^2}{4\mu}$$

Equation (210) thus takes the form

$$u = -\left(\frac{p_0 - p_L}{L} + \varrho g\right)\frac{R^2}{4\mu}\left[1 - \left(\frac{r}{R}\right)^2\right] \qquad (211)$$

This is the well-known parabolic velocity profile for laminar tube flow (Hagen–Poiseuille flow). Further integration of Equation (211) with the relationship $dq = u\,2\pi r\,dr$ provides the Hagen–Poiseuille relationship for the volumetric flow rate q in laminar tube flow

$$q = \left(\frac{p_0 - p_L}{L} + \varrho g\right)\frac{\pi R^4}{8\mu} \qquad (212)$$

Apart from the assumptions and simplifications associated with the momentum balance itself, the example of laminar flow does not contain any troublesome modeling hypotheses. This is not the case for turbulent flow.

5.1.2. Turbulent, Nonreactive Free Jets

Turbulent flows were described in Chapter 4 as flows with high kinetic energy so that the damping viscous forces can no longer hold the fluid particles on controlled paths due to their high inertial forces. This leads to the development of irregular statistical fluctuations. Turbulent flows are therefore basically unsteady and three-dimensional. This does not, however, represent the principal problem in mathematical modeling. As indicated in Section 2.2, compliance with physicochemical laws does not lose its validity even in statistical processes, such as turbulent flows. Direct numerical simulation of such flows in the nonsteady state with three-dimensionally formulated conservation equations is possible in principle.

However, systematic difficulties arise which can only be solved for simple cases [49], [53], [54]. These problems include the small measurements of characteristic turbulent length scales (Kolmogorov length) and the extreme sensitivity of the nonsteady, three-dimensional conservation equations to changes in the boundary conditions. Other options for describing statistical processes are the use of probability density functions (Sections 4.2 and 4.3.2) or the moments of probability density functions. Although the latter do not provide complete statistical information, time- or ensemble-averaged values are often adequate for engineering purposes.

The most important features of models of statistical processes in the form of conservation equations for the moments of probability density functions will be outlined below for turbulent nonreactive flow. The aim is to derive transport equations for the mean values of the velocities starting from the conservation Equation (204) for the compounds of momentum which form a system of differential equations for the instantaneous values of the velocity components.

The example used is based on a jet stirred reactor, similar to that depicted in Figure 14.

The first step is easily carried out by using the decomposition Equation (112) and is demonstrated on a convective term in Equation (204). Equation (204) is first converted into

$$\frac{\partial}{\partial t}(\varrho U_i) + \frac{\partial}{\partial x_k}(\varrho U_k U_i) = \frac{\partial \tau_{ik}}{\partial x_k} - \frac{\partial p}{\partial x_i} + \varrho g_i$$

$$i = 1, 2, 3 \qquad (213)$$

using the continuity Equation (131). From the convective term in one direction and Equation (112) the following is obtained:

$$\varrho U_i U_j = (\langle\varrho\rangle + \varrho^f)(\langle U_i\rangle + U_i^f)(\langle U_j\rangle + U_j^f)$$
$$= \langle\varrho\rangle\langle U_i\rangle\langle U_j\rangle + \varrho\langle U_i\rangle U_j^f$$
$$+ \langle\varrho\rangle U_i^f\langle U_j\rangle + \langle\varrho\rangle U_i^f U_j^f$$
$$+ \varrho^f\langle U_i\rangle\langle U_j\rangle + \varrho^f\langle U_i\rangle U_j^f$$
$$+ \varrho^f U_i^f\langle U_j\rangle + \varrho^f U_i^f U_j^f \qquad (214)$$

If Equation (214) is averaged, then by taking Equation (113b) into account we get

$$\langle \varrho U_i U_j \rangle = \langle \varrho \rangle \langle U_i \rangle \langle U_j \rangle + \langle \varrho \rangle \langle U_i^f U_j^f \rangle$$
$$+ \langle U_i \rangle \langle \varrho^f U_j^f \rangle + \langle U_j \rangle \langle \varrho^f U_i^f \rangle$$
$$+ \langle \varrho^f U_i^f U_j^f \rangle \quad (215)$$

This shows the basic problem in the derivation of conservation equations for the expected values: in addition to the expected values, the convective terms contain the covariances of the velocity components and density–velocity covariances. These terms must be modeled or replaced by assumptions. This is the major difference to dealing with the problem in terms of probability density functions, where convection is not modeled in the transport equations (cf. Section 4.2).

If Equation (215) is substituted into Equation (213), taking the average of the other terms and neglecting fluctuations in density, then

$$\frac{\partial}{\partial t}(\langle \varrho \rangle \langle U_i \rangle) + \frac{\partial}{\partial x_k}(\langle \varrho \rangle \langle U_k \rangle \langle U_i \rangle)$$
$$= \frac{\partial}{\partial x_k}(\langle \tau_{ik} \rangle - \langle \varrho \rangle \langle U_i^f U_k^f \rangle)$$
$$- \frac{\partial \langle p \rangle}{\partial x_i} + \langle \varrho \rangle \langle g_i \rangle, \quad i = 1, 2, 3 \quad (216)$$

Neglection of the density fluctuations in Equation (216) represents an important modeling assumption. Further model formation concerns the terms $\langle \varrho \rangle \langle U_i^f U_j^f \rangle$ which have the units of stresses and are thus called Reynolds stresses.

5.1.2.1. Models for Reynolds Stresses

The modeling of Reynolds stresses has already been referred to in Section 4.3.2.1. Discussion is continued here for the closure of the averaged transport equations for the momentum components. More detailed information is given in [49], [51], [52], [83]–[86].

The most commonly used model in engineering is based on the analogy between laminar and turbulent transfer of momentum (Chap. 4), established by BOUSSINESQ [87] and expanded by PRANDTL [88], [89]. Analogous to the gradient expression for laminar viscous shear stresses (Eq. 205) the apparent turbulent viscous stresses (Reynolds stresses) are coupled with the gradients of the mean velocity.

A tensor for the turbulent shear stresses is obtained:

$$\tau_{t_{ij}} = \mu_t \left(\frac{\partial \langle U_i \rangle}{\partial x_j} + \frac{\partial \langle U_j \rangle}{\partial x_i} \right) - \frac{2}{3} \mu_t \frac{\partial \langle U_k \rangle}{\partial x_k} \delta_{ij} \quad (217)$$

in which δ_{ij} is the Kronecker delta and μ_t is the apparent turbulent viscosity.

The problem is now to determine the apparent turbulent viscosity, which is not (as in the laminar case) a propensity of the material, but a characteristic generated by the flow. When determining this characteristic the analogy between molecular and turbulent transfer of momentum is retained.

In the laminar case the viscosity can be given as

$$\mu = \tfrac{1}{2} \varrho \lambda \bar{u} \quad (218)$$

from the kinetic gas theory, where $\bar{u}$ is the mean velocity and λ the mean free path of the molecules. For turbulent transfer of momentum the same argument leads to

$$\mu_t \sim \bar{\varrho} L^* u^* \quad (219)$$

where L^* is the analogous term for the mean free path of the molecules and u^* the analogue for the mean velocity of the molecules. This corresponds with Prandtl's theory of mixing lengths [60]. In the turbulent case, not molecules but turbulent eddies are the elements which transfer momentum. Both of the quantities L^* and u^* representing the eddy viscosity hypothesis must therefore be determined by plausible arguments [84]. The characteristic velocity for the turbulent eddies is proportional to the fluctuations of the velocities $u^* \sim (\tfrac{2}{3} \langle U_k^f U_k^f \rangle)^{1/2}$ and the characteristic length is the distance which the turbulent eddies travel perpendicularly to the direction of flow, before they lose their identity through dissipation at a molecular level.

This can be expressed as $L^* \sim u^* \tau$, where τ is the turbulent time scale; using the approximation $\tau = \langle k \rangle / \langle \varepsilon \rangle$ (Section 4.3.2.3) gives $L^* \sim \langle U_k^f U_k^f \rangle^{3/2} / \langle \varepsilon \rangle$. The eddy viscosity hypothesis is thus

$$\mu_t = C_\mu \langle \varrho \rangle \frac{\langle k \rangle^2}{\langle \varepsilon \rangle} \quad (220)$$

where C_μ is a modeling parameter. As a result of measurements on turbulent boundary layers its value is given as $C_\mu = 0.09$ [60].

Many of the problems involved in applying the eddy viscosity hypothesis (e.g., the physical requirements for gradient transport approaches) are discussed in the literature [84]–[86]. A mathematical model of this simplicity entails a series of other problems which are physically awkward to deal with. For the model with gradient transport approach and the eddy viscosity hypothesis ($k-\varepsilon$ turbulence model) these problems lie in the determination of the turbulence energy $\langle k \rangle$ and its dissipation rate $\langle \varepsilon \rangle$ and require the derivation of corresponding transport equations [84].

The description of turbulent flows in terms of the mean values of the velocities thus requires the calculation of second moments, in this case of the variances $\langle U_i^f U_j^f \rangle$, from which the turbulence energy and its dissipation rate are derived. An equation for the turbulence energy may be derived by multiplying Equation (213) formulated for the i-direction by U_j^f and formulated for the j-direction by U_i^f. Addition of the two equations followed by use of Equation (112) for the components of the velocity, rearrangement of differential quotients, consideration of the continuity equation, and averaging finally give

$$\frac{\partial}{\partial t}(\langle \varrho \rangle \langle U_i^f U_j^f \rangle) + \frac{\partial}{\partial x_k}(\langle \varrho \rangle \langle U_k \rangle \langle U_i^f U_j^f \rangle)$$

$$= -\frac{\partial}{\partial x_k}(\langle \varrho \rangle \langle U_k^f U_i^f U_j^f \rangle)$$

$$- \langle \varrho \rangle \langle U_i^f U_k^f \rangle \frac{\partial \langle U_j \rangle}{\partial x_k}$$

$$- \langle \varrho \rangle \langle U_j^f U_k^f \rangle \frac{\partial \langle U_i \rangle}{\partial x_k}$$

$$+ \left\langle U_i^f \frac{\partial \tau_{jk}}{\partial x_j} \right\rangle + \left\langle U_j^f \frac{\partial \tau_{ik}}{\partial x_i} \right\rangle$$

$$- \left\langle U_i^f \frac{\partial p}{\partial x_j} \right\rangle - \left\langle U_j^f \frac{\partial p}{\partial x_i} \right\rangle \quad (221)$$

gravitational forces being ignored. Summation over the diagonal elements from Equation (221) gives half the trace of the tensor for the Reynolds stresses $\langle U_i^f U_l^f \rangle$, $\langle k \rangle = \frac{1}{2}\langle U_l^f U_l^f \rangle$, and hence the equation for turbulence energy

$$\frac{\partial}{\partial t}(\langle \varrho \rangle \langle k \rangle) + \frac{\partial}{\partial x_k}(\langle \varrho \rangle \langle U_k \rangle \langle k \rangle)$$

$$= -\frac{\partial}{\partial x_k}(\langle \varrho \rangle \langle U_k^f k \rangle) - \langle \varrho \rangle \langle U_l^f U_k^f \rangle \frac{\partial \langle U_l \rangle}{\partial x_k}$$

$$+ \left\langle U_l^f \frac{\partial \tau_{lk}}{\partial x_l} \right\rangle - \left\langle U_l^f \frac{\partial p}{\partial x_l} \right\rangle \quad (222)$$

The description of statistical flows using mean values (first moments) additionally requires the calculation of second moments, in this case those of the velocity variances in the form of turbulence energy. Transport equations can be simply derived for this (see Eq. 222), but they contain terms with moments of the next higher order so that this procedure never leads to a closed equation system. The first expression on the right-hand side of Equation (222), for example, contains moments of the third order. To close the transport equations at a certain order, model assumptions have to be introduced for the higher order moments. These are often based on the same assumptions as the simple gradient transport assumptions and the eddy viscosity hypothesis. The physical basis and the larger number of model parameters arising in Reynolds stress models are the reasons that further development of these [83], [90], [91] is only gradually gaining acceptance in engineering.

The generally accepted form of the transport equation for turbulence energy (Eq. 223) is obtained from Equation (222) by using the definition for turbulent shear stresses, gradient transport approaches for third-order moments, the dissipation hypothesis [49], and other model assumptions [92]:

$$\frac{\partial}{\partial t}(\langle \varrho \rangle \langle k \rangle) + \frac{\partial}{\partial x_k}(\langle \varrho \rangle \langle U_k \rangle \langle k \rangle)$$

$$= \frac{\partial}{\partial x_k}\left[\frac{\mu_t}{\sigma_{k\,\text{eff}}} \frac{\partial \langle k \rangle}{\partial x_k}\right] + G_k - \langle \varrho \rangle \langle \varepsilon \rangle \quad (223)$$

with the abbreviation

$$G_k = \mu_t \left[\left(\frac{\partial \langle U_l \rangle}{\partial x_k} + \frac{\partial \langle U_k \rangle}{\partial x_l}\right) - \frac{2}{3}\left(\frac{\partial \langle U_k \rangle}{\partial x_k} + \langle \varrho \rangle \langle k \rangle\right)\delta_{ik}\right]\frac{\partial \langle U_k \rangle}{\partial x_k} \quad (224)$$

The individual steps in model formation for Equation (123) are described in [48], [51], [86], [92] and Section 4.3.2.1. For the simplifying homogeneous cases described in Section 4.3.2.1, Equation (223) becomes Equation (184) where $\langle \varepsilon \rangle = \langle k \rangle / \tau$.

For complete definition one more equation is needed for the dissipation rate $\langle \varepsilon \rangle = \mathrm{d}\langle k \rangle / \mathrm{d}t$ which occurs both in Equation (222) and also in the eddy viscosity hypothesis (Eq. 220). The derivation takes a similar path to that for the equation for turbulence energy and is described in [49], [92]. The accepted form is

$$\frac{\partial}{\partial t}(\langle\varrho\rangle\langle\varepsilon\rangle) + \frac{\partial}{\partial x_k}(\langle\varrho\rangle\langle U_k\rangle\langle\varepsilon\rangle)$$

$$= \frac{\partial}{\partial x_k}\left[\frac{\mu_t}{\sigma_{\varepsilon\,\text{eff}}}\frac{\partial\langle\varepsilon\rangle}{\partial x_k}\right]$$

$$+ [C_{\varepsilon 1} G_k - 2 C_{\varepsilon 2}\langle\varrho\rangle\langle\varepsilon\rangle]\frac{\langle\varepsilon\rangle}{\langle k\rangle} \quad (225)$$

In both Equations (223) and (225) $\sigma_{\varphi\,\text{eff}}$ is the turbulent Schmidt number, which compares the mixing length for the velocity fluctuations to the mixing length for the Reynolds stresses or the dissipation rate respectively; $C_{\varepsilon 1}$ and $C_{\varepsilon 2}$ are further model parameters. From the solution of Equation (225), the turbulent time scale can be determined; this is a precondition for the closure of the models discussed in Sections 4.3.2.1 and 4.3.2.3.

Describing a nonreactive, isothermal turbulent flow with the model discussed above in terms of the expected velocity values leads to the averaged conservation equations of momentum (Eq. 216), as well as the equations for turbulence energy (Eq. 223) and the dissipation rate (Eq. 225). This system of simultaneous partial differential equations has to be solved by a suitable numerical procedure (see Section 5.1.2.2).

5.1.2.2. Solution Method of the Resulting System of Partial Differential Equations

The partial, simultaneous nonlinear differential Equations (216), (223), and (225) can be put into the form

$$\text{div } J_\varphi = \langle S_\varphi\rangle$$
$$\langle \Phi\rangle = \langle U\rangle, \langle V\rangle, \langle W\rangle, \langle k\rangle, \langle\varepsilon\rangle \quad (226)$$

for the steady-state problem being considered here. In Equation (226) J_φ is

$$J_\varphi = \langle\varrho\rangle\langle\Phi\rangle\langle U_k\rangle - \frac{\mu_t}{\sigma_{\varphi\,\text{eff}}}\frac{\partial\langle\Phi\rangle}{\partial x_k} \quad (227)$$

and $\langle S_\varphi\rangle$ accomodates all other expressions not included by Equation (227). For the momentum equation $\langle S_\varphi\rangle$ also contains specific terms for the turbulent shear stress tensor from Equation (217) and the right-hand sides of the equation for $\langle k\rangle$ and $\langle\varepsilon\rangle$ can be seen from Equations (223) and (225).

The laminar transport of momentum is generally ignored compared with the turbulent shear stresses. To develop a numerical solution procedure, Equation (226) is integrated over a control volume (Section 4.3.1.1, pp. 2-39–2-40). Using the Gauss–Ostrogradski integration principle this gives

$$\int_V \text{div } J_\varphi = \oint_F J_{\varphi k} = \int_V \langle S_\varphi\rangle \quad (228)$$

The integration has to be carried out over the surfaces of the control volume which are perpendicular to the components $J_{\varphi i}$, so that for a cartesian system of coordinates the relationship

$$\Delta J_{\varphi k}\,\text{d}x_i\,\text{d}x_{j,i,j\neq k} = \int_V \langle S_\varphi\rangle$$
$$\langle\Phi\rangle = \langle U\rangle, \langle V\rangle, \langle W\rangle, \langle k\rangle, \langle\varepsilon\rangle \quad (229)$$

results, in which the operator Δ denotes the difference at two opposite surfaces of the control volume. Equation (229) is the basis for setting up difference equations which are substituted for the continuous differential operators in Equations (216), (223), and (225) to allow numerical integration.

The conservation equations for the momentum components contain the simple pressure gradients as sources; a reduction in pressure in the direction of the ith coordinate thus leads to an increase of momentum in this direction. Equations (216), (223), and (225) do not, however, contain a pressure term so that the pressure $\langle p\rangle$ or the pressure gradients $\partial\langle p\rangle/\partial x_i$ cannot be directly calculated from them. The same problem occurs in closing the transport equation for the probability density functions (Section 4.3.2.1). Solution of these equations assumes a known pressure field, which is implicitly given through the conservation equations for the momentum components and the continuity equation. If an estimate for the pressure field is assumed for a numerical solution, then the continuity equation (Eq. 131) is usually not complied with. Consequently the numerical solution in an iterative procedure may get further and further away from physically meaningful results because the equation system does not contain correction factors to ensure compliance with the continuity equation.

In the numerical simulation of turbulent steady flow or when using the Navier–Stokes equations, this problem is frequently solved by replacing the primitive variables $\langle U \rangle$ and $\langle p \rangle$ with stream functions and vorticity [94], [95] through the von Mises transformation [93]. The stream function is defined by the von Mises transformation such that the continuity equation is complied with (see also Section 5.2.2.1). In the procedure outlined below another possibility is suggested, in which the pressure correction is obtained from the continuity equation and a simplified form of the momentum balance (the SIMPLE algorithm) [96], [97].

For the example considered here of a turbulent free jet the formulation in cylinder coordinates is used as in Section 5.1.1. Due to the rotational symmetry Equation (226) can be expressed as

$$L(\Phi) = \langle S_\varphi \rangle, \quad \Phi = \langle U \rangle, \langle V \rangle, \langle k \rangle, \langle \varepsilon \rangle \quad (230)$$

The turbulent transport operator $L(\Phi)$ is given by

$$L(\Phi) \equiv \frac{\partial(\langle \varrho \rangle \langle U \rangle \langle \Phi \rangle)}{\partial x} + \frac{1}{r}\frac{\partial(r \langle \varrho \rangle \langle V \rangle \langle \Phi \rangle)}{\partial r}$$

$$- \frac{\partial}{\partial x}\left(\frac{\mu_t}{\sigma_{\varphi\,\text{eff}}}\frac{\partial \langle \Phi \rangle}{\partial x}\right) - \frac{1}{r}\frac{\partial}{\partial r}\left(r\frac{\mu_t}{\sigma_{\varphi\,\text{eff}}}\frac{\partial \langle \Phi \rangle}{\partial r}\right)$$

$$\langle \Phi \rangle = \langle U \rangle, \langle V \rangle, \langle k \rangle, \langle \varepsilon \rangle \quad (231)$$

Since an axially symmetrical system is invariant with respect to rotation around the axis of symmetry, then a two-dimensional grid with L nodal points in the x-direction and M nodal points in the r-direction is used for replacing the differential operators in Equation (231) by differences (Fig. 26). The spatial structure of the grid arises from rotation about the axis of symmetry, and the control volumes are defined by revolving the grid through the unit angle. The control volume is indicated in Figure 26 by a shaded area, the relevant flows, for example $J_{\varphi 1w}$, can be expressed as

$$J_{\varphi 1w} = -\left(\text{Max}\left\{+ (\langle \varrho \rangle \langle U \rangle)_w, 0\right\} \langle \Phi \rangle_{l-1}\right.$$

$$\left. - \text{Max}\left\{-(\langle \varrho \rangle \langle U \rangle)_w, 0\right\} \langle \Phi \rangle_l\right.$$

$$\left. - \frac{\mu_t}{\sigma_{\varphi\,\text{eff}}}\frac{\langle \Phi \rangle_l - \langle \Phi \rangle_{l-1}}{x_l - x_{l-1}}\right) F_w \quad (232)$$

The formulation given in Equation (232) contains an "up-wind" difference scheme [Max$\{+(\langle \varrho \rangle \langle U \rangle)_w, 0\} \langle \Phi \rangle_{l-1}$ − Max$\{-(\langle \varrho \rangle \langle U \rangle)_w, 0\} \langle \Phi \rangle_l$] for the convective terms (analogous to the cascade of ideally mixed stirred reactors discussed in Section 4.3.1.1) and a central difference scheme for the diffusive terms. As with the model for ideally mixed reactors, the up-wind difference scheme is based on the assumption that the upstream value of a physical quantity is transported over the control surfaces.

The profiles of the variables brought into discrete form in the x-direction in accordance with the "up-wind" difference scheme are also shown in Figure 26. Problems associated with this difference scheme (e.g., "numerical diffusion") and measures to limit them are described in [97]–[101].

The grid with L nodal points in the x-direction and M nodal points in the r-direction and the difference scheme can be used to develop the difference equations from Equation (229); the fluxes through the remaining surfaces of the control volume are defined analogously to Equation (232). Summation of the fluxes gives

$$-\left(\text{Max}\left\{+ (\langle \varrho \rangle \langle U \rangle)_w, 0\right\} \langle \Phi \rangle_{l-1}\right.$$

$$\left. - \text{Max}\left\{-(\langle \varrho \rangle \langle U \rangle)_w, 0\right\} \langle \Phi \rangle_l\right.$$

$$\left. - \frac{\mu_t}{\sigma_{\varphi\,\text{eff}}}\frac{\langle \Phi \rangle_l - \langle \Phi \rangle_{l-1}}{x_l - x_{l-1}}\right) F_w$$

$$+\left(\text{Max}\left\{+ (\langle \varrho \rangle \langle U \rangle)_e, 0\right\} \langle \Phi \rangle_l\right.$$

$$\left. - \text{Max}\left\{-(\langle \varrho \rangle \langle U \rangle)_e, 0\right\} \langle \Phi \rangle_{l+1}\right.$$

$$\left. - \frac{\mu_t}{\sigma_{\varphi\,\text{eff}}}\frac{\langle \Phi \rangle_{l+1} - \langle \Phi \rangle_l}{x_{l+1} - x_l}\right) F_e$$

$$-\left(\text{Max}\left\{+ (\langle \varrho \rangle \langle V \rangle)_s, 0\right\} \langle \Phi \rangle_{l-L}\right.$$

$$\left. - \text{Max}\left\{-(\langle \varrho \rangle \langle V \rangle)_s, 0\right\} \langle \Phi \rangle_l\right.$$

$$\left. - \frac{\mu_t}{\sigma_{\varphi\,\text{eff}}}\frac{\langle \Phi \rangle_l - \langle \Phi \rangle_{l-L}}{x_l - x_{l-L}}\right) F_s$$

$$+\left(\text{Max}\left\{+ (\langle \varrho \rangle \langle V \rangle)_n, 0\right\} \langle \Phi \rangle_l\right.$$

$$\left. - \text{Max}\left\{-(\langle \varrho \rangle \langle V \rangle)_n, 0\right\} \langle \Phi \rangle_{l+L}\right.$$

$$\left. - \frac{\mu_t}{\sigma_{\varphi\,\text{eff}}}\frac{\langle \Phi \rangle_{l+L} - \langle \Phi \rangle_l}{x_{l+L} - x_l}\right) F_n = \langle S \rangle_{\varphi l}$$

$$(233)$$

In Equation (233) $\langle S \rangle_{\varphi l}$ represents the source terms from Equation (230) integrated over the

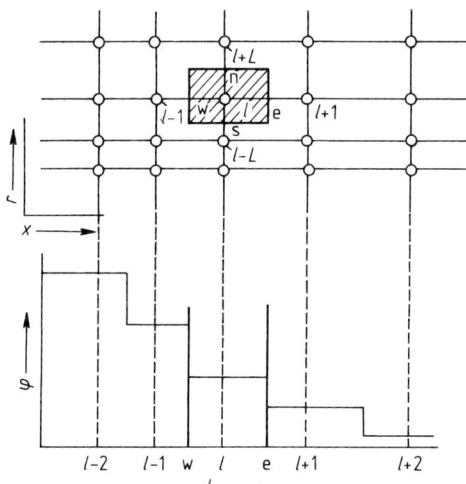

Figure 26. Diagram illustrating the "up-wind" difference scheme for elliptical differential equations
l = number of grid point; L = number of grid points in n direction; n = north; e = east; s = south; w = west; φ = any variable

control volume. If the continuity equation in the corresponding discrete form is taken into account

$$(\langle\varrho\rangle\langle U\rangle)_w F_w - (\langle\varrho\rangle\langle U\rangle)_e F_e + (\langle\varrho\rangle\langle V\rangle)_s F_s - (\langle\varrho\rangle\langle V\rangle)_n F_n = 0 \quad (234)$$

then Equation (233), after addition of Equation (234) multiplied by $\langle\Phi\rangle_l$, can be transformed into

$$-a_{l-L}\langle\Phi\rangle_{l-L} - a_{l-1}\langle\Phi\rangle_{l-1} + b_l\langle\Phi\rangle_l$$
$$- a_{l+1}\langle\Phi\rangle_{l+1} - a_{l+L}\langle\Phi\rangle_{l+L} - \langle S\rangle_{\varphi l} = 0 \quad (235)$$

In accordance with Equation (233) the coefficients for Equation (235) are

$$a_{l-L} = \left[\text{Max}\left\{+(\langle\varrho\rangle\langle V\rangle)_s, 0\right\}\right.$$
$$\left. + \frac{\mu_t}{\sigma_{\varphi\,\text{eff}}} \frac{1}{x_l - x_{l-L}}\right] F_s \quad (236\,\text{a})$$

$$a_{l-1} = \left[\text{Max}\left\{+(\langle\varrho\rangle\langle U\rangle)_w, 0\right\}\right.$$
$$\left. + \frac{\mu_t}{\sigma_{\varphi\,\text{eff}}} \frac{1}{x_l - x_{l-1}}\right] F_w \quad (236\,\text{b})$$

$$a_{l+1} = \left[\text{Max}\left\{-(\langle\varrho\rangle\langle U\rangle)_e, 0\right\}\right.$$
$$\left. + \frac{\mu_t}{\sigma_{\varphi\,\text{eff}}} \frac{1}{x_{l+1} - x_l}\right] F_e \quad (236\,\text{c})$$

$$a_{l+L} = \left[\text{Max}\left\{-(\langle\varrho\rangle\langle V\rangle)_n, 0\right\}\right.$$
$$\left. + \frac{\mu_t}{\sigma_{\varphi\,\text{eff}}} \frac{1}{x_{l+L} - x_l}\right] F_n \quad (236\,\text{d})$$

$$b_l = a_{l-L} + a_{l-1} + a_{l+1} + a_{l+L} \quad (236\,\text{e})$$

Difference equations corresponding to Equation (235) can be formulated for the $(L-2)$, $(M-2)$ inner grid points of the calculation domain for all variables $\langle\Phi\rangle = \langle U\rangle, \langle V\rangle, \langle k\rangle, \langle\varepsilon\rangle$. Similar relationships are derived for the grid points lying on the boundaries of the calculation domain. The resulting system of equations is nonlinear and coupled via $\langle S\rangle_{\varphi l}$ and the coefficients a and b, which are not constant but depend on the components of the velocity and turbulent viscosity. Because the numerical procedure for solving Equation (235) is an iterative one these dependencies are not resolved.

An analogous equation for pressure corrections can be obtained with the procedure suggested in [96], [97]:

$$-a_{pl-L}\Delta\langle p\rangle_{l-L} - a_{pl-1}\Delta\langle p\rangle_{l-1} + b_{pl}\Delta\langle p\rangle_l$$
$$- a_{pl+1}\Delta\langle p\rangle_{l+1} - a_{pl+L}\Delta\langle p\rangle_{l+L} - Rs_l = 0 \quad (237)$$

where $\Delta\langle p\rangle$ represents the corrections for the given pressure field, and Rs_l the "continuity error", which is obtained from the residuals of the continuity equation with the calculated velocity field. The coefficients of Equation (237) are obtained from a simplified form of the momentum balance [96], [97]. To avoid numerical problems which originate from the structure of the averaged momentum equations, Equations (235) and (237) are best solved with a "staggered grid" arrangement for the velocities [98].

Equations (235) and (237) constitute a system of nonlinear equations with $5(L-2)(M-2)$ unknowns for the example considered. The values of the variables at the boundaries of the calculation domain are given by plausible boundary conditions. The system matrix for this system of nonlinear equations has a block pentadiagonal structure due to the elliptical formulation of the problem (Eq. 231), and the difference scheme used. The system of equations can now be solved, for example with the Newton–Raphson procedure outlined in Section 3.2.1.3 or other methods.

The model for the Reynolds stresses discussed in Section 5.1.2.1 as well as the numerical procedure given in this section are examples from a rapidly developing field of engineering. Other models for the Reynolds stresses represent the phenomena observed in turbulent flows better than the $k-\varepsilon$ turbulence model [52], [84]–[86], [90]–[92]. Difference schemes other than the combination of "up-wind" and central differences are also used for partial elliptical differential equations and other solution procedures [101]–[104] are suggested.

5.1.2.3. Example: Turbulent Flow of Nitrogen in Air

In the example discussed here nitrogen is discharged at room temperature from a tube of 10-mm diameter with an exit velocity of 85 m/s into air which is almost at rest (0.3 m/s) (Fig. 14). The system of nonlinear Equations (235) and (237) is solved for a tensor product grid by an alternating direction implicit (ADI) procedure [101]. In this procedure the block structure is first resolved by solving the system of equations one after the other for the individual variables. The estimated values from the previous iteration are used for the variables contained in the couplings. The system is converted into a quasi-one-dimensional one by including the elements of the two outer diagonals in the right-hand sides.

The system of equations is solved alternately along a line of the grid in Figure 26 in the x- or r-direction. The coefficients situated on the neighboring lines are calculated with the estimated values from the previous iteration.

To increase the velocity of convergence, linear relaxation is used (Section 3.2.1.3) and the source terms are linearized in the form $\langle S_\varphi \rangle = \langle S_{\text{lin }\varphi} \rangle + \langle S_{\text{prop }\varphi} \rangle \langle \Phi \rangle$. The latter measure is particularly useful for stiff differential equation systems, for which the source terms are

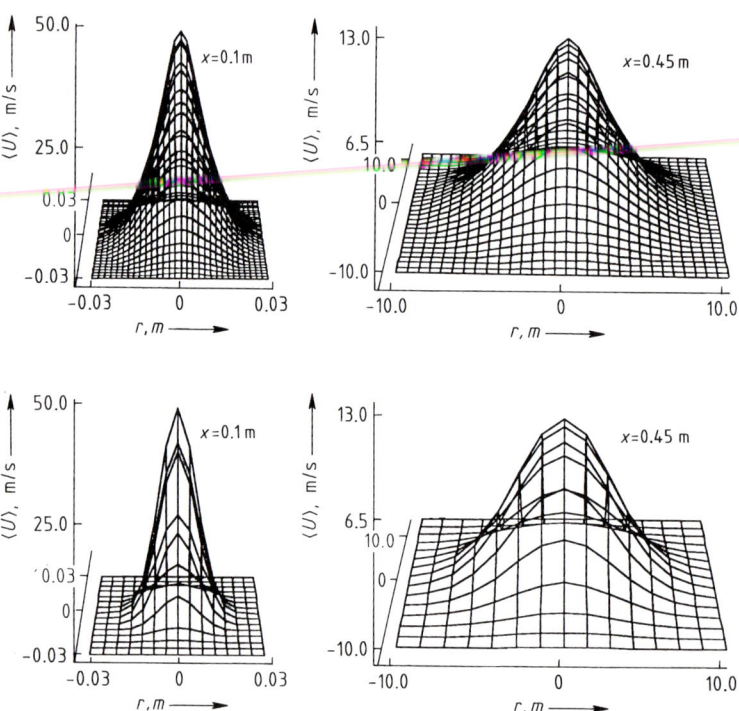

Figure 27. Profiles of the expected values for velocities in the x-direction of a turbulent inert flow from a free turbulent jet. The upper diagrams show numerical simulations, the lower diagrams show similarity solutions. The radial profiles at two distances from the nozzle are given.

highly nonlinear and coupled. The modeling parameters [91] used for calculation are $C_\mu = 0.09$, $C_{\varepsilon 1} = 1.4$, $C_{\varepsilon 2} = 0.925$, $\sigma_{k\,\text{eff}} = 1.0$, and $\sigma_{\varepsilon\,\text{eff}} = 1.3$. Figure 27 shows the predicted values for the mean velocity in the x-direction for the turbulent flow of nitrogen at two distances from the nozzle outlet in comparison with the similarity solutions from [105]. The latter give the flow velocity as

$$\langle U \rangle_m = \langle U \rangle_0 \Big/ \left(0.16 \frac{x}{D_R} - 1.5\right) \tag{238}$$

and

$$\langle U \rangle = \langle U \rangle_m \exp\left[-C_u \left(\frac{r}{x}\right)^2\right] \tag{239}$$

where $C_u = 90$.

5.2. Applications of the Principle of Conservation of Enthalpy

The conservation equation of enthalpy (Eq. 134) is written in detail as

$$\varrho \frac{\partial h}{\partial t} + \varrho u_k \frac{\partial h}{\partial x_k} = -\frac{\partial J_k^h}{\partial x_k} + \varrho S_h \tag{240a}$$

In accordance with Equation (240a) enthalpy transport occurs due to convection and molecular transport. The term S_h contains all the mechanisms for conversion of enthalpy (e.g., volumetric work due to a change in pressure, work against physical forces or radiation).

Fourier's simple gradient transport approach can be used to represent molecular transport:

$$J_i^h = -\lambda \frac{\partial T}{\partial x_i} \tag{241}$$

where λ is the thermal conductivity. If a Lewis number, $Le = Pr/Sc$, of one is assumed and other transport effects are neglected, the simple form of the conservation of energy equation is obtained

$$\varrho \frac{\partial h}{\partial t} + \varrho u_k \frac{\partial h}{\partial x_k} = \frac{\partial}{\partial x_k} \lambda \frac{\partial T}{\partial x_k} + \varrho S_h \tag{240b}$$

Detailed discussion of the assumptions and simplifications for this formulation are found in [48], [50], [51].

The conservation equation of enthalpy also contains the density and the components of the velocity. The enthalpy and momentum equations are thus coupled and have to be solved simultaneously. To avoid this a simple heat conduction problem is discussed first.

5.2.1. Heat Conduction

A simple heat conduction problem is shown in Figure 28. A pipe consisting of two different materials (pipe and insulation) has an internal temperature T_{Wi}, the external surface has a temperature of T_{Wo}. If only molecular conduction is considered, Equation (240a) is reduced to

$$-\frac{1}{r}\frac{d(r J_r^h)}{dr} = 0 \tag{242}$$

Integration gives

$$-r J_r^h = C_1 \tag{243a}$$

Applying Equation (241) gives

$$r \lambda \frac{dT}{dr} = C_1 \tag{243b}$$

which after integration results in the relationship

$$T = \frac{C_1}{\lambda} \ln r + C_2 \tag{244a}$$

for the temperature profile. For the inner pipe surface, the boundary condition $T = T_{Wi}$ applies for $r = r_{Wi}$. Thus $C_2 = T_{Wi} - (C_1/\lambda_i) \ln r_{Wi}$ and

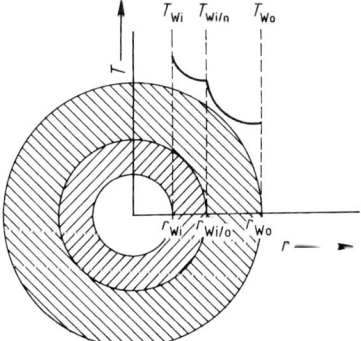

Figure 28. Arrangement and temperature profile for a simple heat conduction problem
The subscripts Wi and Wo denote inner wall and outer wall, respectively.

Equation (244a) becomes

$$T = T_{wi} - \frac{C_1}{\lambda_i} \ln \frac{r}{r_{wi}} \qquad (244b)$$

Correspondingly for the outer pipe surface the boundary condition is

$$T_{wi/o} = T_{wi} - \frac{C_1}{\lambda_i} \ln \frac{r_{wi/o}}{r_{wi}}$$

and

$$T = T_{wi/o} - \frac{C_1}{\lambda_o} \ln \frac{r}{r_{wi/o}} \qquad (244c)$$

In Equations (244a–c), C_1 is given by the total heat flux (Eq. 243a).

5.2.2. Heat Transfer

5.2.2.1. Exact Solution for a Boundary Layer Problem

An exact solution for the heat transfer from a fluid to a solid wall is possible for laminar flow over a flat plate (Fig. 29). A fluid flows over a flat plate of infinitely small thickness and of length L at a velocity u_∞. Due to the viscous stresses at the wall the fluid particles slow down and a boundary layer of increasing thickness forms along the plate in which the velocity increases from zero to u_∞. This gives the boundary conditions $u = 0$ for $y = 0$ and $u_{\delta_u}(x) = u_\infty$. For the sake of simplicity, constant pressure is assumed, $p_{\delta_u}(x) = p_\infty = $ const. The flow is steady-state and the plate extends infinitely in one direction so that the problem can be formulated two-dimensionally. The boundary conditions for

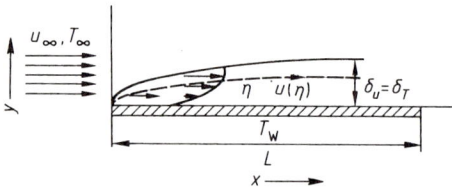

Figure 29. Development of a velocity (δ_u) and temperature (δ_T) boundary layer for laminar incompressible flow over a flat plate
$Pr = 1$; $u_\infty =$ velocity in the x-direction outside the boundary layer; $T_\infty =$ temperature outside the boundary layer; $\eta =$ similarity coordinate; $T_w =$ wall temperature; $L =$ plate length

temperature are $T = T_w$ for $y = 0$ and $T_{\delta_T}(x) = T_\infty$. An equation for the temperature can be obtained from Equation (240b) by neglecting sources such as the conversion of mechanical energy or viscous dissipation and by using the equation for the definition of enthalpy $dh = c_p dT$ and the usual boundary layer approximations [59], [60]:

$$c_p \varrho u \frac{\partial T}{\partial x} + c_p \varrho v \frac{\partial T}{\partial y} = \lambda \frac{\partial^2 T}{\partial y^2} \qquad (245a)$$

At constant density and specific heat

$$u \frac{\partial T}{\partial x} + v \frac{\partial T}{\partial y} = a \frac{\partial^2 T}{\partial y^2} \qquad (245b)$$

where $a = \lambda/\varrho c_p$.

The differential equation for the temperature still contains the unknown velocities u and v. Equations for these variables are obtained from the continuity equation (Eq. 131) and the momentum equation (Eq. 204) in the x-direction

$$\frac{\partial u}{\partial x} + \frac{\partial v}{\partial y} = 0 \qquad (246)$$

$$u \frac{\partial u}{\partial x} + v \frac{\partial u}{\partial y} = v \frac{\partial^2 u}{\partial y^2} \qquad (247)$$

where $v = \mu/\varrho$ is the kinematic viscosity. For $Pr = v/a = 1$ and hence for $v = a$, the enthalpy equation is identical with the momentum equation, so that the solution of the latter produces a solution to the enthalpy equation. Furthermore, the coupling of the momentum and enthalpy equations are removed by these assumptions.

The general expression for temperature is

$$\frac{T}{T_\infty} = A \frac{u}{u_\infty} + B$$

where $A = \dfrac{T_\infty - T_w}{T_\infty}$ and $B = \dfrac{T_w}{T_\infty}$ (248)

The heat flux to the wall is then

$$J_w^h = -\lambda \left(\frac{\partial T}{\partial y}\right)_w = -\lambda \frac{T_\infty - T_w}{u_\infty} \left(\frac{\partial u}{\partial y}\right)_w \qquad (249)$$

Determination of the heat flux to the wall thus depends on determination of the velocity gradient at the wall, which is a function of x and can be obtained from the solution of Equations (246) and (247).

For the solution, the stream function Ψ is used to eliminate the velocities u and v [59], [60]:

$$u = \frac{\partial \Psi}{\partial y} \quad \text{and} \quad v = -\frac{\partial \Psi}{\partial x} \quad (250\,a, b)$$

The stream function invariably satisfies the continuity equation, as can be seen by suitable differentiation. With this transformation, Equation (247) becomes a partial differential equation of the third order

$$\Psi_y \Psi_{yx} - \Psi_x \Psi_{yy} = v \Psi_{yyy} \quad (251)$$

in which Ψ_x and the appropriate symbols for each of the differential operators are used.

In the boundary layer formed in laminar flow locations of constant velocity u lie on a parabola $\eta \sim y/\sqrt{x}$ (Fig. 29). If the equation system is therefore transformed to this similarity coordinate η, then Equations (247) and (251) can be reduced to ordinary differential equations. The exact transformation is

$$\eta(x, y) = y\sqrt{\frac{u_\infty}{v x}} \quad (252)$$

in which u_∞/v represents a scale factor that makes η dimensionless. As u is only a function of η, the general expression $u/u_\infty = g(\eta)$ can be used for making the stream function dimensionless:

$$\Psi(x, y) = \int u\, dy = u_\infty \int g(\eta)\, d\eta$$

$$= u_\infty \sqrt{\frac{u_\infty}{v}} \int g(\eta)\, d\eta \quad (253)$$

where $\int g(\eta)\, d\eta \equiv f(\eta)$ is called the dimensionless stream function. Equations (250a, b), (252), and (253) provide all the transformations and conversions that are needed to convert the partial differential Equation (251) into an ordinary differential equation of dimensionless quantities.

$$\Psi_y = u_\infty f_\eta, \quad \Psi_x = \frac{1}{2}\sqrt{\frac{u_\infty}{v x}}\,[\eta f_\eta - f(\eta)],$$

$$\Psi_{yx} = -\frac{1}{2}\frac{u_\infty}{x}\eta f_{\eta\eta},$$

$$\Psi_{yy} = u_\infty \sqrt{\frac{u_\infty}{v x}} f_{\eta\eta}, \quad \Psi_{yyy} = u_\infty \frac{u_\infty}{v x} f_{\eta\eta\eta}$$

(254 a, b, c, d, e)

In Equations (254) f_η and the corresponding symbols denote the derivatives of f with respect to η. The ordinary differential equation

$$f_{\eta\eta\eta} + \tfrac{1}{2} f \cdot f_{\eta\eta} = 0 \quad (255)$$

is thus obtained from Equation (251) which has to be solved for the boundary conditions $f(\eta) = 0$, $f_\eta = 0$ for $y = 0$ or $\eta = 0$ and also for $f_\eta = 1$ for $y \to \infty$ or $\eta \to \infty$.

Equation (255) can be solved numerically but not analytically. The essential quantities for the boundary layer problem, $f(\eta) = \int (u/u_\infty)\, d\eta$, $f_\eta = u/u_\infty$, and $f_{\eta\eta} = u_\infty^{-1}(\partial u/\partial \eta)$, can be found in the literature, tabulated as a function of η [60], [106]. According to [106], $u/u_\infty = 0.99$ for $\eta = 5.0$ and $f_{\eta\eta} = 0.332$ for $\eta = 0$.

The boundary layer thickness is calculated from the first of these two values and is defined here as the distance δ_u at which the velocity reaches 99% of the flow velocity outside the boundary layer. It then follows from Equation (252) that

$$\delta_u = 5\sqrt{\frac{v x}{u_\infty}} \quad \text{or} \quad \frac{\delta_u}{x} = \frac{5}{\sqrt{Re_x}} \quad (256\,a, b)$$

The essential information provided by Equation (256) is that the development of the boundary layer thickness for laminar flow over a flat plate is proportional to $\sqrt{x}$ and $1/\sqrt{u_\infty}$, and inversely proportional to the square root of the Reynolds number $Re_x = u_\infty x/v$. The solutions of the momentum and enthalpy equations are similar because $Pr = 1$; as a result $\delta_u = \delta_T$. A detailed discussion of further implications of the solution of the boundary layer equations is given in [59], [60].

Using the value

$$0.332 = f_{\eta\eta} = \frac{1}{u_\infty}\frac{\partial u}{\partial \eta} = \frac{1}{u_\infty}\frac{\partial u}{\partial y}\frac{\partial y}{\partial \eta}$$

$$= \frac{1}{u_\infty}\frac{1}{\sqrt{u_\infty/v x}}\frac{\partial u}{\partial y}$$

the following relation is obtained:

$$\left(\frac{\partial u}{\partial y}\right)_w = 0.332\, u_\infty \sqrt{\frac{u_\infty}{v x}} \quad (257)$$

Substitution of this term in Equation (249) gives

$$J_w^h = -0.332\, \lambda \sqrt{\frac{u_\infty}{v x}}(T_\infty - T_w) \quad (258)$$

If the definition for the heat-transfer coefficient $J_w^h \equiv \alpha \Delta T$ where $\Delta T = (T_w - T_\infty)$ is now substituted in Equation (258), then

$$\alpha = 0.332 \lambda \sqrt{\frac{u_\infty}{\nu x}} = 0.332 \lambda \frac{1}{x} \sqrt{Re_x} \qquad (259)$$

Rearrangement of this equation in dimensionless groups using $Nu_x = \alpha x/\lambda$ yields the solution for the heat-transfer problem for laminar flow over a flat plate:

$$\frac{Nu_x}{\sqrt{Re_x}} = 0.332 \qquad (260)$$

The physical background of Equations (259) and (260) is that the local heat-transfer coefficient, like the boundary layer thickness, is inversely proportional to the distance in the x-direction and the velocity of flow. This is easy to understand because the temperature and velocity gradients level out at the wall with the formation of the boundary layer.

In technical problems the local heat-transfer coefficient is often not as important as its mean value α_m from the beginning of the plate to a distance x or the total length L. Equation (259) can be written as $\alpha = \text{const.} \cdot x^{-1/2}$:

$$\alpha_m = \frac{1}{L}\int_0^L \alpha(x)\,dx = \frac{\text{const.}}{L}\int_0^L \frac{dx}{\sqrt{x}}$$

$$= 2\frac{\text{const.}}{L}\sqrt{x}\Big|_0^L = 2\alpha_L \qquad (261\,\text{a})$$

The mean heat-transfer coefficient for a plate with length L is thus double the local heat-transfer coefficient at the distance L. This can also be formulated with Equation (260):

$$Nu_m = 0.664\sqrt{Re_L} = 2\,Nu_L \qquad (261\,\text{b})$$

The analysis described here only applies under the highly simplified assumptions of incompressible flow and constant density and also $Pr = \nu/a = 1$. Nevertheless this allows the general principles of modeling for heat transfer to be described.

5.2.2.2. General Principles of Modeling Heat Transfer

Under industrial conditions the simplifications introduced in the previous section become void. The consequence of $Pr \neq 1$ and variable density is that Equations (245), (213), and (131) are coupled through the density. Moreover, the solutions for the momentum balances and the enthalpy balances are no longer identical. The dimensions of the temperature boundary layer δ_T and the hydrodynamic boundary layer δ_u differ. In addition, flow is mostly turbulent sometimes with very complicated geometries.

As the heat transfer under such conditions can still be defined as

$$J_W^h = -\lambda\left(\frac{\partial T}{\partial y}\right)_W \equiv \alpha \Delta T \qquad (262)$$

or from the appropriate mean value over a surface, the problem is thus to determine the temperature gradients at the wall. Heat transfer at the wall is primarily a heat conduction problem. The temperature profile in the vicinity of the wall is however determined by simultaneous transfer of momentum, heat, and mass. Analytical or numerical solutions for the system of Equations (131), (213), and (245) will only succeed in a few cases with a justifiable amount of effort [25], [59], [107], [108], see also the solution of the momentum equations for the turbulent inert jet of gas in Section 5.1.2.2.

In most engineering applications semi-empirical methods are therefore used which assume a linearized temperature profile at the wall. Figure 30 shows the temperature profile for laminar flow over a plate.

If the temperature is developed at the wall, $y = 0$ in a Taylor series that is truncated after the first member, then

$$T = T_W + \left(\frac{\partial T}{\partial y}\right)_W dy \text{ or } T_\infty = T_W + \left(\frac{\partial T}{\partial y}\right)_W \delta \qquad (263)$$

If this is substituted into Equation (262) then

$$\lambda\frac{T_\infty - T_W}{\delta} = \alpha(T_\infty - T_W) \text{ or } \alpha = \frac{\lambda}{\delta} \qquad (264\,\text{a, b})$$

Here δ is the film thickness produced by linearization of the temperature profile as shown in Figure 30. This is not identical with the boundary layer thickness δ_u or δ_T. As can be seen from the solutions for the laminar boundary layer

$$\left(\frac{\partial T}{\partial y}\right)_W = 0.332(T_\infty - T_W)\sqrt{Re_x}/x$$

whereas

$$\left(\frac{\partial T}{\partial y}\right)_W^* = \frac{1}{5}(T_\infty - T_W)\sqrt{Re_x}/x$$

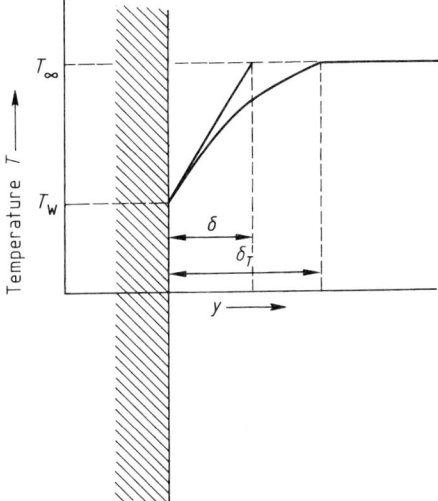

Figure 30. Representation of a general model for heat transfer
δ = film thickness; δ_T = thickness of temperature boundary layer; T_w = wall temperature; T_∞ = temperature outside the boundary layer.

when the thickness δ_T of the temperature boundary layer is used for the difference quotients. This latter quantity is thus only ca. 60 % of the true temperature gradient. The determination of heat transfer thus depends on determination of the film thickness as in Equation (264). This problem has been solved experimentally in most technical applications, where the mean heat-transfer number depending on the various variables is represented in the form $Nu = f(Re, Pr, L/D...)$, see also Equations (260) and (261).

These relationships for industrial heat exchangers are not treated here. They are described in [25], see also → 2. Heat Exchange, B3. An example of such a relationship for the transfer of heat for turbulent flow was discussed in Sections 3.1.1.4 and 3.2.3.

With the establishment of the mean heat-transfer coefficients α and the relevant temperature difference, heat transfer (Eq. 262) is clearly defined. Arrangements such as that shown in Figure 28 can be calculated completely with regard to temperature profiles or heat losses. Summation of the temperature differences for the individual "phases" of the arrangement allows heat transfer to be related to an "overall resistance"

$$\frac{1}{k} = \frac{1}{\alpha_{w_i}} + \sum \frac{1}{\lambda_i/(r_{w_o} - r_{w_i})} + \frac{1}{\alpha_{w_o}}$$

and to the total temperature difference. The equation for heat transfer analogous to Equation (262) will then be used depending on the statement of the question. The calculation procedure will be shown in an analogous problem of mass transfer in Section 5.3.1.3.

5.3. Applications of the Law of Conservation of Mass

The law of conservation of mass is formulated through the continuity equation (Eq. 131) and the conservation equations for the mass of the individual chemical species, m_i

$$\varrho \frac{\partial m_i}{\partial t} + \varrho u_k \frac{\partial m_i}{\partial x_k} = -\frac{\partial J_k^{m_i}}{\partial x_k} + \varrho S_i$$
$$i = 1, \ldots, N-1 \quad (265\,\mathrm{a})$$

A further equation results from the condition $\sum_{1}^{N} m_i = 1$. Analogous to the formulations for the conservation of momentum and enthalpy, the transport of chemical species results from convection and molecular transport. The term S_i contains sources or sinks based on all possible forms of transformation, for example due to chemical reactions or due to phase changes in multiphase systems. If molecular transport is considered to be adequately described by Fick's law (gradient transport approach)

$$J_j^{m_i} = -\varrho D_i \frac{\partial m_i}{\partial x_j} \quad (241\,\mathrm{a})$$

then a simple form of the conservation equations is obtained that is analogous to Equation (240b)

$$\varrho \frac{\partial m_i}{\partial t} + \varrho u_k \frac{\partial m_i}{\partial x_k} = \frac{\partial}{\partial x_k}\left[\varrho D_i \frac{\partial m_i}{\partial x_k}\right] + \varrho S_i$$
$$i = 1, \ldots, N-1 \quad (265\,\mathrm{b})$$

In Equation (265b) thermal and pressure diffusion are neglected. Furthermore in general $S_i = S_i(m_i, h, p)$ (Eq. 137a) is to be set in general for the transformation term so that Equation (265b) is coupled with the conservation equation of momentum through the density and velocities and coupled with the conservation equation of enthalpy through the transformation term. The coupled system of equations given in Section 5.2 is thus expanded by one equation.

5.3.1. Mass Transfer without Chemical Reaction

5.3.1.1. Exact Solution for a Boundary Layer Problem

The flat plate in Figure 29 is replaced by a water surface (Fig. 31) from which the water evaporates into the higher temperature air flow. Under the same assumptions as in Section 5.2.2.1 and with the boundary layer approximations for Equation (265 b), the following system of equations is obtained for this steady-state problem:

$$\frac{\partial u}{\partial x} + \frac{\partial v}{\partial y} = 0 \tag{246}$$

$$u\frac{\partial u}{\partial x} + v\frac{\partial u}{\partial y} = v\frac{\partial^2 u}{\partial y^2}, \quad v = \frac{\mu}{\varrho} \tag{247}$$

$$u\frac{\partial T}{\partial x} + v\frac{\partial T}{\partial y} = a\frac{\partial^2 T}{\partial y^2}, \quad a = \frac{\lambda}{\varrho c_p} \tag{245b}$$

$$u\frac{\partial m}{\partial x} + v\frac{\partial m}{\partial y} = D\frac{\partial^2 m}{\partial y^2} \tag{266}$$

where $m = m_{H_2O}$. The equation system is to be solved for the boundary conditions $u = 0$, $v = v_W$, $T = T_W$, and also $m = m_W$ for $y = 0$ and $u = u_\infty$, and $m = m_\infty$ for $y \to \infty$.

Since the inclusion of the conservation equation of mass has not altered the nature of the equations, it is again logical to transform the equations into the similarity coordinates

$$\eta(x, y) = y\sqrt{\frac{u_\infty}{v x}} = \frac{y}{x}\sqrt{Re_x}$$

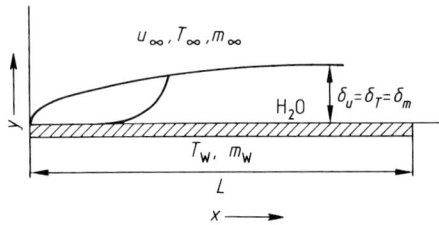

Figure 31. Development of a velocity (δ_u), temperature (δ_T), and concentration (δ_m) boundary layer for laminar incompressible flow over a flat water surface
$Pr = Sc = 1$
The subscripts ∞ and W indicate values outside the boundary layer and at the wall, respectively.

(see Eq. 252), and to express them in the dimensionless streams function $f(\eta) = \int (u/u_\infty)\, d\eta$. Due to the different boundary conditions for temperature and mass fraction and in analogy with $f_\eta = u/u_\infty$, the term $\theta = (T - T_W)/(T_\infty - T_W)$ is used for the temperature and $\varphi = (m - m_W)/(m_\infty - m_W)$ for the mass fraction. With the aid of the relationships given in Section 5.2.2.1, these transformations yield the equation system

$$f_{\eta\eta\eta} + \tfrac{1}{2} f f_{\eta\eta} = 0 \tag{255}$$

$$\theta_{\eta\eta} + \tfrac{1}{2} Pr f \theta_\eta = 0 \tag{267}$$

$$\varphi_{\eta\eta} + \tfrac{1}{2} Sc f \varphi_\eta = 0 \quad \text{where} \quad Sc = v/D \tag{268}$$

that has to be solved simultaneously.

The boundary conditions with the transformed variables are $f_\eta = 0$, $\theta = 0$, $\varphi = 0$ and also $(v_W/u_\infty)\sqrt{Re_x} = \text{const.}$ for $\eta = 0$ and $f_\eta = 1$, $\theta = 1$, $\varphi = 1$ for $\eta \to \infty$.

The system of Equations (255), (267), (268) cannot be solved analytically, but for $Le = Pr/Sc = 1$ all profiles are identical and thus result in identical values for δ_u, δ_T, and δ_m. All transport mechanisms exhibit the same development of the boundary layer. For $Pr = Sc = 1$ and $Pr = Sc = 0.7$ and also some values of $(v_W/u_\infty)\sqrt{Re_x}$, Figure 32 shows the profiles $f_\eta = u/u_\infty$, θ, and φ [59]. The curve for

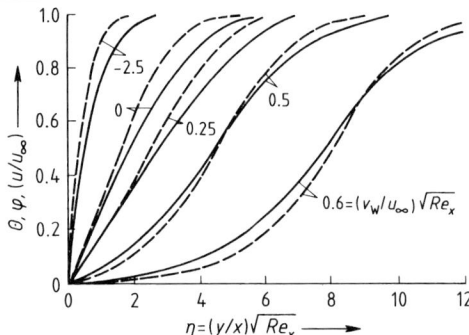

Figure 32. Concentration and temperature distribution with mass transfer for laminar flow over a flat surface
The curves for $Pr = Sc = 1$ indicated by the dashed lines also give the velocity profiles. The curves for $Pr = Sc = 0.7$ indicated by the solid lines show temperature and concentration distributions.
$\theta = (T - T_W)/(T_\infty - T_W)$; $\varphi = (m - m_W)/(m_\infty - m_W)$; $\eta = $ similarity coordinate $= (y/x)\sqrt{Re_x}$; $v = $ velocity in y-direction; $u = $ velocity in x-direction; $Re_x = $ Reynolds number defined with u_∞ and axial distance x; the subscripts W and ∞ indicate values at the wall and outside the boundary layer, respectively.

$Pr = Sc = 1$ and $(v_w/u_\infty)\sqrt{Re_x} = 0$ corresponds to the profile discussed in the previous sections and is called the Blasius profile. The value of $\eta = 5$ for $\theta = \varphi = u/u_\infty \approx 0.99$ can be read off from the curve for $(v_w/u_\infty)\sqrt{Re_x} = 0$. Figure 32 shows that, depending on the magnitude of the "blow out parameter" $(v_w/u_\infty)\sqrt{Re_x}$, the profiles of the flat plate will be either blown off or sucked on. For many technical applications this is taken advantage of for stabilization or removal of boundary layers.

In the case of the evaporation example shown in Figure 31, $(v_w/u_\infty)\sqrt{Re_x}$ can be neglected so that the mass transfer, which is to be defined analogously to the heat transfer, is written in the form

$$J_w^m = -\varrho D \left(\frac{\partial m}{\partial y}\right)_w$$
$$= -\varrho D(m_\infty - m_w)\sqrt{\frac{u_\infty}{vx}}\left(\frac{\partial \varphi}{\partial \eta}\right)_w \quad (269)$$

If the Blasius solution $f_{\eta\eta} = \varphi_\eta = \theta_\eta = 0.332$ is used for $Pr = Sc = 1$ and, analogous to the heat-transfer coefficient, a mass-transfer coefficient is defined as $J_w^m = \varrho \beta \Delta m$ where $\Delta m = (m_w - m_\infty)$, it follows that

$$\beta = 0.332\, D \sqrt{\frac{u_\infty}{vx}} = 0.332\, D \frac{1}{x}\sqrt{Re_x} \quad (270)$$

In order to complete the analogy with heat transfer the Sherwood number $Sh_x = \beta x/D$ should be substituted. Thus for $Pr = Sc = 1$

$$\frac{Sh_x}{\sqrt{Re_x}} = 0.332 \quad (271)$$

Further treatment of mass-transfer coefficients—for example averaging over the length—follows the scheme given in Section 5.2.2.1 and need not be repeated here.

5.3.1.2. General Principles for Modeling Mass Transfer

As mentioned in Section 5.2.2.2 for heat transfer, only in very few cases do the Prandtl or Schmidt numbers have the value one; in addition the geometric or hydrodynamic assumptions listed in Section 5.3.1.1 seldom apply. Even if the system of Equations (204), (240b), and (265b) can be solved under such conditions, the resulting solutions deviate from the complete analogy.

For engineering applications mass transfer is therefore also treated by semi-empirical models which assume a linearization of the profile of the concentrations at the phase boundary. Because of the analogy with heat transfer (Fig. 30) discussion of this topic can be kept short. If the mass fraction profile is linearized at the phase boundary

$$m = m_w + \left(\frac{\partial m}{\partial y}\right)_w dy \quad \text{or} \quad m_\infty = m_w + \left(\frac{\partial m}{\partial y}\right)_w \delta \quad (272)$$

the equation defining the mass-transfer coefficient results in

$$D\frac{m_\infty - m_w}{\delta} = \beta(m_\infty - m_w) \quad \text{or} \quad \beta = \frac{D}{\delta} \quad (273\,\text{a, b})$$

where δ again represents the film thickness produced by linearization of the mass fraction profile. Determination of the mass-transfer coefficient thus depends on the determination of the film thickness, which is carried out experimentally for many engineering systems. The results for the mean mass-transfer coefficients in the form of the relationships $Sh = f(Re, Sc, L/D)$ are described in [25], [59], [109]. For evaluation of industrial mass transfer equipment there remains the problem, analogous to heat transfer, of the addition of the "individual resistances" for mass transfer and reference to the total concentration difference.

5.3.1.3. Examples of Simultaneous Mass and Heat Transfer: Dynamic Models

The system used to illustrate the general principles of the modeling of simultaneous heat and mass transfer (Sections 5.2.2.2 and 5.3.1.2) is based on the boundary layer example shown in Figure 31. Figure 33 shows an evaporation cooler which is used in an air-conditioning device. A liquid (water) trickles in a thin film over suitable packings and comes into contact with a gas (air) flowing in the same direction. The liquid evaporates and is thus cooled, so that heat flow occurs from gas to liquid thus lowering the gas temperature.

The final steady-state temperature of the gas must be determined for a given area, $S = sFH$, where s is the specific transfer area. All terms are

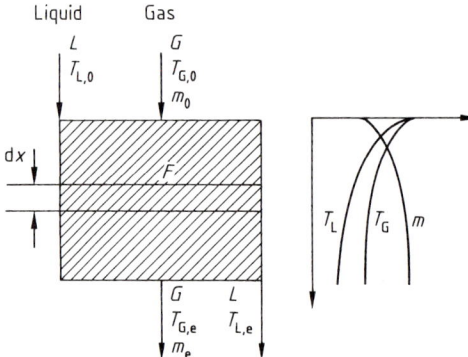

Figure 33. Example of simultaneous mass and heat transfer in an evaporation cooler
L = mass flow of liquid; G = mass flow of gas; F = cross sectional area; x = axial distance
The subscripts L, G, 0, and e denote liquid, gas, initial, and final, respectively.

defined in Figure 33. The thickness of the layer of liquid flowing over the packings is assumed to be so small that the temperature in the liquid phase is uniform. The mean heat- and mass-transfer coefficients for this problem are known. The temperature changes lie in a range which allows the material values to be regarded as constants. The functions for the Nusselt and Sherwood numbers generally contain terms which are functions of the properties of the material. The assumption of approximately constant material properties avoids recalculation of the example with improved Nusselt and Sherwood numbers. The mass transfer is so small compared with the mass flow of gas and liquid (G and L, respectively) that the latter can likewise be regarded as nearly constant. The required quantities only change in the x-direction so that the problem can be formulated one-dimensionally.

The heat balance for the gas phase gives the temperature change in the gas based on the heat transfer to the liquid

$$\frac{dT_G}{dx} = -\alpha \frac{sF}{Gc_{pG}}(T_G - T_L) \tag{274}$$

where c_p is the specific heat. The mass transfer to the gas phase gives the change in mass fraction in the gas phase

$$\frac{dm}{dx} = \beta \frac{\varrho s F}{G}(m_{Ph} - m) \tag{275}$$

where β is the mass-transfer coefficient and m_{Ph} the mass fraction of liquid in the phase boundary. The temperature change in the liquid phase is finally obtained from the heat balance for the liquid phase

$$\frac{dT_L}{dx} = \alpha \frac{sF}{Lc_{pL}}(T_G - T_L) - \beta \frac{\Delta h_v \varrho s F}{Lc_{pL}}(m_{Ph} - m) \tag{276}$$

where Δh_v is the specific latent heat of vaporization of the liquid.

Equations (274)–(276) represent a system of simultaneous ordinary differential equations for T_G, T_L, and m. In order to close these equations a relationship has to be found for m_{Ph}. This can be obtained from the Clausius–Clapeyron equation for vapor pressure $\ln p = -A'/T + B'$. Since the changes are assumed to be very small, a linearized expression can be used in the temperature interval being considered, so that $m_{Ph} = AT_L + B$ or $dm_{Ph}/dx = A(dT_L/dx)$. By substituting the transformations $\theta = T_G - T_L$ and $\varphi = m_{Ph} - m$, the number of equations can be reduced by one:

$$\frac{d\theta}{dx} = -\alpha \frac{sF}{Gc_{pG}Lc_{pL}}(Gc_{pG} + Lc_{pL})\theta$$
$$+ \beta \frac{\Delta h_v \varrho s F}{Lc_{pL}}\varphi \tag{277a}$$

$$\frac{d\varphi}{dx} = \alpha \frac{AsF}{Lc_{pL}}\theta + \left[\beta \frac{A\Delta h_v \varrho s F}{Lc_{pL}} + \beta \frac{\varrho s F}{G}\right]\varphi \tag{277b}$$

Equations (277a) and (277b) can also be abbreviated as

$$\theta_x = a_1\theta + b_1\varphi \tag{278a}$$

$$\varphi_x = a_2\theta - b_2\varphi \tag{278b}$$

where θ_x and φ_x are the derivatives of θ and φ with respect to x. Differentiation of Equation (278a) and elimination of φ_x and φ give the ordinary homogeneous second-order differential equation

$$\theta_{xx} + (a_1 + b_2)\theta_x + (a_1 b_2 - a_2 b_1)\theta = 0 \tag{279}$$

to which the system of coupled equations is now reduced. Since $(a_1 + b_2)^2 > 4(a_1 b_2 - a_2 b_1)$, the characteristic equation of differential Equation (279) $\lambda^2 + (a_1 + b_2)\lambda + (a_1 b_2 - a_2 b_1) = 0$ has two real roots. The general solution for Equation

(279) is thus

$$\theta = C_1 e^{\lambda_1 x} + C_2 e^{\lambda_2 x} \qquad (280\,\text{a})$$

The integration constants C_1 and C_2 are determined from the boundary conditions, which are simplified here as $T_{L0} = T_{G0}$, since water and air are both assumed to have the same (room) temperature. This leads to $\theta = 0$ and $\theta_x = b_1 \varphi_0$ for $x = 0$. These conditions give $C_1 = -C_2$ and $\lambda_1 C_1 + \lambda_2 C_2 = b_1 \varphi_0$. Thus the solution of Equation (279) is

$$\theta = \frac{b_1 \varphi_0}{\lambda_1 - \lambda_2} [e^{\lambda_1 x} - e^{\lambda_2 x}] \qquad (280\,\text{b})$$

where φ_0 is to be determined from the known initial temperature of the water and the initial mass fraction of water in the air; λ_1 and λ_2 have to be calculated from the characteristic equation. Equation (280 b) shows that cooling is initiated entirely by mass transfer into the gas phase. The gas temperature can now be determined with the known solution for θ by integration of Equation (274):

$$T_G = T_{G0} - a_3 \frac{b_1 \varphi_0}{(\lambda_1 - \lambda_2)} \left[\frac{1}{\lambda_1} e^{\lambda_1 x} - \frac{1}{\lambda_2} e^{\lambda_2 x} \right] \qquad (281)$$

where $a_3 = \alpha s F / G c_{pG}$.

5.3.1.4. Examples of Simultaneous Mass and Heat Transfer: Static Models

In the examples outlined in Section 5.3.1.3, the changes in temperature and mass fractions were calculated from the rates of transport of enthalpy and mass between the individual phases. The concept used in this model is shown in Figure 34 A for the mass transfer of a component from the gas phase into a counterflowing liquid phase. At the phase boundary the mass fraction profile shows a jump caused by the relevant phase equilibrium which may be expressed by the relationship e.g., $m_{PhG} = A m_{PhL}$. In contrast to the cases discussed in Sections 5.3.1.1 and 5.3.1.3 the component being transferred is dissolved in the liquid phase. The mass fraction differences between the bulk phases and the phase boundary (Fig. 34 A) can be expressed as

$$(m_G - m_{PhG}) = \frac{j^m}{\varrho_G \beta_G s F \, dx} \qquad (282\,\text{a})$$

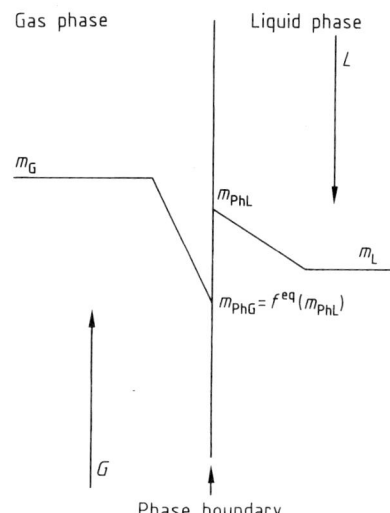

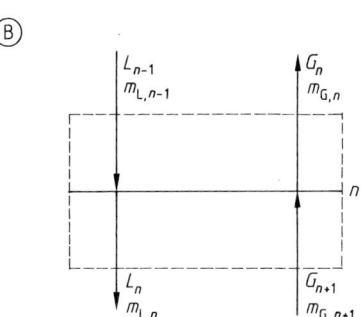

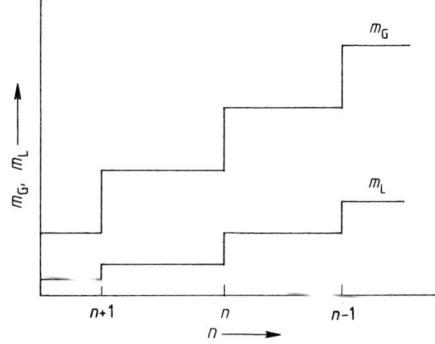

Figure 34. Representation of the dynamic (A) and static (B) models for mass transfer
L = mass flow of liquid; G = mass flow of gas
The subscripts L, G, Ph, and n denote liquid, gas, phase boundary, and the number of theoretical plates, respectively. The superscript eq denotes equilibrium.

and

$$(m_{\text{PhL}} - m_{\text{L}}) = \frac{j^m}{\varrho_{\text{L}} \beta_{\text{L}} s F \, dx} \quad (282\,\text{b})$$

where j^m is the mass flux of the transferred component. If Equations (282a) and (282b) are added and the relationship for the phase equilibrium is used to eliminate m_{PhL}, then the following equation is obtained for the steady-state case, taking into consideration the mass balance for the gas phase:

$$j^m = \frac{sF}{\dfrac{1}{\varrho_{\text{G}} \beta_{\text{G}}} + \dfrac{A}{\varrho_{\text{L}} \beta_{\text{L}}}} (m_{\text{G}} - A m_{\text{L}}) dx = - dm_{\text{G}} G \quad (283)$$

Rearrangement gives

$$-\frac{dm_{\text{G}}}{(m_{\text{G}} - A m_{\text{L}})} = \frac{sF}{\left(\dfrac{1}{\varrho_{\text{G}} \beta_{\text{G}}} + \dfrac{A}{\varrho_{\text{L}} \beta_{\text{L}}}\right) G} dx \quad (284)$$

and after integration

$$\int_{m_{\text{Ge}}}^{m_{\text{Go}}} \frac{dm_{\text{G}}}{(m_{\text{G}} - A m_{\text{L}})} = \frac{sF}{\left(\dfrac{1}{\varrho_{\text{G}} \beta_{\text{G}}} + \dfrac{A}{\varrho_{\text{L}} \beta_{\text{L}}}\right) G} \int_0^H dx \quad (285\,\text{a})$$

which can be abbreviated as

$$NTU = \frac{H}{HTU} \quad (285\,\text{b})$$

where NTU denotes the number of transfer units and HTU the height of one transfer unit.

The notation and assumptions for the derivation of Equation (285b) are the same as those in Section 5.3.1.3. The same procedure as indicated in Section 5.2.2.2 for the heat transfer leads to Equation (285b). A similar relationship is often used for total heat transfer in heat exchangers [108].

Equation (285b) is used to determine the height of a separation column with a known specific area and cross-sectional area or conversely to determine the concentration difference for a given height of the separation column. The quantity $A m_{\text{L}}$ contained in the integral NTU still has to be expressed in terms of m_{G} and the boundary conditions $m_{\text{G},0}$ and $m_{\text{L},0}$ or $m_{\text{L},e}$. All quantities in the HTU expression are known.

A completely different model for mass and heat transfer in mass separation processes is shown in Figure 34 B. Here the mass transfer device consists of a series of separation stages, in which phase equilibrium occurs between the counterflowing phases. Mass transfer is assumed to occur infinitely quickly and the separation stage is homogeneous with regard to the variables of state. The mass balance in the steady-state case for the components transferred is then

$$L m_{\text{L},n-1} + G m_{\text{G},n+1} = L m_{\text{L},n} + G m_{\text{G},n} \quad (286)$$

Since phase equilibrium should exist in the separation stage, then $m_{\text{G},n}$ can be expressed by $m_{\text{L},n}$. Due to the modeling assumptions, the mass balance does not contain any transport terms. If the relationship for the phase equilbrium is expressed in the form $m_{\text{G}} = A m_{\text{L}} + B$, Equation (286) becomes

$$L m_{\text{L},n-1} + G(A m_{\text{L},n+1} + B)$$
$$= L m_{\text{L},n} + G(A m_{\text{L},n} + B) \quad (287\,\text{a})$$

or

$$L m_{\text{L},n-1} - (L + AG) m_{\text{L},n} + AG m_{\text{L},n+1} = 0 \quad (287\,\text{b})$$

For a separation column with N separation stages (Fig. 35) for given mass fractions $m_{\text{G},0}$ and

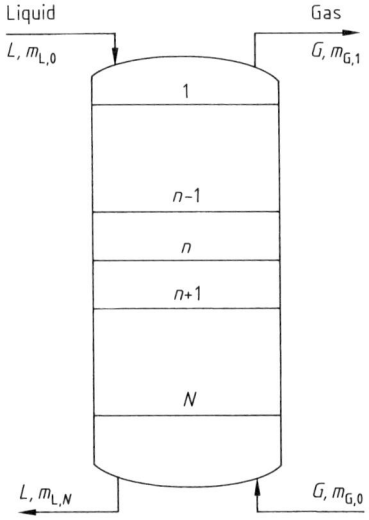

Figure 35. Schematic representation of a counterflow separation column
For explanation of symbols see Figure 34.

$m_{L,0}$, the relevant linear equation system is

$$-(L+AG)m_{L,1} + AGm_{L,2} = -Lm_{L,0} \quad (288, 1)$$
$$\vdots$$
$$Lm_{L,n-1} - (L+AG)m_{L,n} + AGm_{L,n+1} = 0 \quad (288, n)$$
$$\vdots$$
$$Lm_{L,N-1} - (L+AG)m_{L,N} = -G(m_{G,0} - B) \quad (288, N)$$

In matrix form this can be written as

$$\begin{pmatrix} -(L+AG) & AG & & \cdots \\ L & -(L+AG) & AG & \cdots \\ \vdots & & \ddots & \vdots \\ \cdots & & L & -(L+AG) \end{pmatrix} \begin{pmatrix} m_{L,1} \\ \vdots \\ \vdots \\ m_{L,N} \end{pmatrix}$$

$$= \begin{pmatrix} -Lm_{L,0} \\ \vdots \\ -G(m_{G,0} - B) \end{pmatrix} \quad (288\,a)$$

or

$$Bm = c \quad (288\,b)$$

The solution of this equation with one of the methods discussed in Section 3.1.2 is

$$m = B^{-1}c \quad (289)$$

and provides the mass fraction profile in discrete form for the component being transferred to the liquid phase over the entire length of the column. Using the appropriate relationships for phase equilibrium the corresponding profile can easily be determined for the gas phase. The system of equations could equally well be derived for the mass fractions in the gas phase.

The above problem can easily be solved in another way. Transformation of Equation (286) gives

$$m_{L,1} - m_{L,0} = \frac{G}{L}(m_{G,2} - m_{G,1}) \quad \text{or} \quad (290\,a)$$

$$\Delta m_L = \frac{G}{L}\Delta m_G \quad (290\,b)$$

for the first separation stage in the column shown in Figure 35. The geometric interpretation of this equation is to find the point with coordinates $(m_{L,1}, m_{G,2})$ on a straight line of

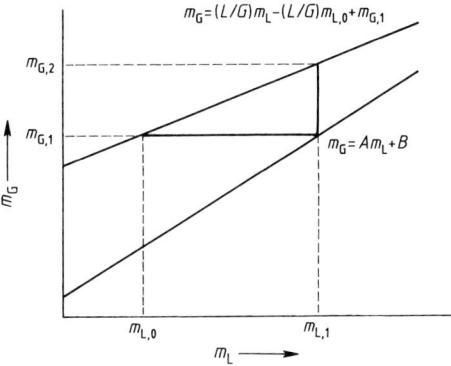

Figure 36. Graphical solution of the equation system (288) for a counterflow separation column: McCabe–Thiele diagram
A and B are coefficients of the relation for thermodynamic phase equilibrium $m_G = f^{eq}(m_L)$. For explanation of other symbols see Figure 34.

slope (L/G) starting from the point with the coordinates $(m_{L,0}, m_{G,1})$. As the first step, a point $(m_{L,1}, m_{G,1})$ has to be found that satisfies the phase equilibrium relationship $m_G = Am_L + B$. This is done by inserting $m_{G,1}$ into the phase equilibrium relationship as shown by the horizontal line in Figure 36. Working from this point, point $(m_{L,1}, m_{G,2})$ can be found by inserting $m_{L,1}$ into the mass balance equation (Eq. 290a); this is shown in Figure 36 by the vertical line. In this way Equation (288) can easily be solved graphically by completing the construction for all separation stages step by step. The graphical solution method is the basis for the McCabe–Thiele method and similar procedures for calculating mass separation equipment ($\rightarrow$ 4. Distillation and Rectification, **B3**, pp. 4-22–4-24).

Static models can also be used for simultaneous heat and mass transfer in mass separation processes. This will be described briefly for the separation stage of the rectification column shown in Figure 37. Phase equilibrium is assumed in the separation stage and the separation stage is regarded as being homogeneous with respect to the variables of state. Phase equilibria relationships exist both for the mass fractions and the enthalpies. With these preconditions, balances for the total mass, the mass of the individual components of the mixture, and the total enthalpy can be given.

For the sake of simplicity the mixture is taken to be a binary mixture of components A and B. The balances are considered for the steady-state case. (The mass and enthalpy balances can, how-

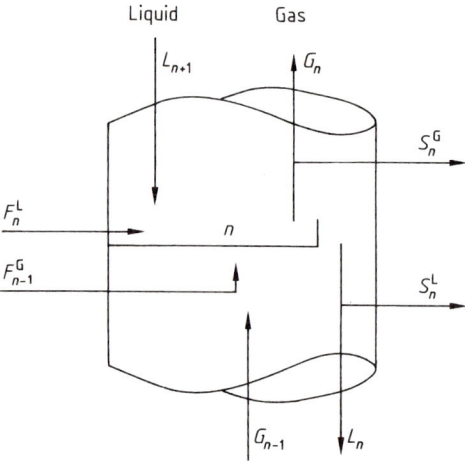

Figure 37. Mass flows in a separation stage of a rectification column
F = liquid or gaseous feed; S = liquid or gaseous sink
The superscripts L and G denote liquid and gas phase, respectively. For explanation of other symbols see Figure 34.

ever, also be formulated without difficulty for a multicomponent mixture or for the nonsteady state, for details see [3], [8], [9].) The equation for the total mass is (notation is defined in Fig. 37)

$$L_{n+1} + G_{n-1} + F_n^L + F_{n-1}^G$$
$$= L_n + G_n + S_n^G + S_n^L \quad (291)$$

Analogously, the equation for the mass of a component is

$$L_{n+1} m_{L,n+1} + G_{n-1} m_{G,n-1} + F_n^L m_{L,n}$$
$$+ F_{n-1}^G m_{G,n-1}$$
$$= L_n m_{L,n} + G_n m_{G,n} + S_n^G m_{G,n} + S_n^L m_{L,n} \quad (292)$$

and finally that for the total enthalpy is

$$h_{L,n+1} + h_{G,n-1} + h_{L,F_n^L} + h_{G,F_{n-1}^G}$$
$$= h_{L,n} + h_{G,n} + h_{G,S_n^G} + h_{L,S_n^L} \quad (293)$$

For a rectification column of N separation stages, for example, the mass fraction in the liquid phase, the mass flows of the liquid phase, and the temperature have to be calculated. To do this Equations (291)–(293) are simplified by setting all flows to and from the external surroundings to zero (see also Fig. 37). This simplification does not change the character of the resulting equations.

The expansions of the equations for inputs or outputs at the individual separation stages, and likewise the formulations for the condensor at the top of the column and the reboiler at the bottom of the column, offer no major difficulties. To reduce Equations (291)–(293) to a system of equations in L, m_L, and T_L further relationships are needed to eliminate m_G and G. Relationships for the phase equilibrium are used for this: $m_G = f^{eq}(m_L)$, and $G_n/G_{n-1} = R_n$.

Here f^{eq} is the relation for phase equilibrium and R_n is the ratio of the evaporation enthalpies $R_n = \Delta h_{v,n}/\Delta h_{v,n-1}$. If mixing effects are neglected the enthalpies are additive, so that for example $h_L = (h_L^A - h_L^B)m_L + h_L^B$. Finally, neglecting the mixing effects, $p_{tot} = [p^A(T) - p^B(T)]m_L + p^B(T)$ is used to obtain a correlation between the temperature and the composition of the liquid phase. Using $h = h_0 + c_p(T - T_0)$, the enthalpies can finally be eliminated. The resulting system of equations is

$$[p^A(T_n) - p^B(T_n)]m_{L,n} + p^B(T_n) = p_{tot} \quad (294\,a)$$

$$\left[m_{L,n} - \left(\frac{1}{1+R_n}\right) f^{eq}(m_{L,n}) \right.$$
$$+ \left(\frac{R_n}{1-R_n}\right) f^{eq}(m_{L,n-1}) \right] L_n$$
$$+ \left[\left(\frac{1}{1+R_n}\right) f^{eq}(m_{L,n}) - \left(\frac{R_n}{1-R_n}\right) f^{eq}(m_{L,n-1}) \right.$$
$$\left. - m_{L,n-1} \right] L_{n+1} = 0 \quad (294\,b)$$

$$\left[\left(\frac{R_n}{1-R_n}\right)(L_{n+1} - L_n) \right]$$
$$\cdot \{[(h_L^A + \Delta h_v^A) - (h_L^B + \Delta h_v^B)]m_{L,n-1}$$
$$+ (h_L^B + \Delta h_v^B)\} - L_n[(h_L^A - h_L^B)m_{L,n} + h_L^B]$$
$$- \left[\left(\frac{1}{1-R_n}\right)(L_{n+1} - L_n) \right]$$
$$\cdot \{[(h_L^A + \Delta h_v^A) - (h_L^B + \Delta h_v^B)]m_{L,n}$$
$$+ (h_L^B + \Delta h_v^B)\}$$
$$+ L_{n+1}[(h_L^A - h_L^B)m_{L,n+1} + h_L^B] = 0 \quad (294\,c)$$

In Equations (294) the relationships for $m_G = f^{eq}(m_L)$, $p^i(T)$, $h_L^i(T)$, and R_n are not explicitly formulated. In this form Equation (294) should be adequate to point to the similarity to Equation (288). A system of N equations is obtained whose system matrix exhibits a distorted tridiagonal structure. Each element consists of a block with 3×3 elements. In contrast to Equation (288) the sytem is nonlinear and coupled.

Consideration of output and input at the individual separation stages or at the top and bottom of the column does not alter the structure of this system of equations; further terms are simply added to the right-hand side and the dimensions of the blocks of the system matrix are increased on account of consideration of a multicomponent mixture. Equation (294) is easy to solve because of the conditioning of the system matrix. Equation (294a) contains entries only in the main diagonals, and Equation (294b) on two diagonals. In most applications the system is not therefore solved simultaneously, but is solved successively starting with Equation (294a) and predefined boundary conditions. Examples of computer programs for the solution of these problems with multicomponent formulations and nonsteady-state conditions are found in [8], [9].

5.3.2. Mass Transfer with Chemical Reactions

In the dynamic and static models for calculating simultaneous mass and heat transfer, chemical reactions have not yet been considered. In many industrial mass separation processes, mass transfer is associated with chemical reactions. The effect of chemical reactions on mass transfer is now discussed using a simple example (Fig. 38) based on the simple boundary layer problem given in Sections 5.2.1.1 and 5.3.1.1. The plate is replaced by a soluble component A which passes into the fluid phase where it reacts with a second component B. The reaction rate is first order with respect to the individual components, the reaction rate is thus

$$\frac{dm_A}{dt} = -k'(T) m_A m_B \qquad (295)$$

The component B is assumed to be present in excess such that $m_B \approx$ const. If $m = m_A$, then Equation (295) can be simplified as

$$\frac{dm}{dt} = -k(T, m_B) m \qquad (296)$$

Using the same assumptions as in Sections 5.2.2.1 and 5.3.1.1 (the boundary layer approximations and negligible reaction enthalpy), the following system of equations is obtained

$$\frac{\partial u}{\partial x} + \frac{\partial v}{\partial y} = 0 \qquad (246)$$

$$u \frac{\partial u}{\partial x} + v \frac{\partial u}{\partial y} = v \frac{\partial^2 u}{\partial y^2}, \qquad (247)$$

$$u \frac{\partial T}{\partial x} + v \frac{\partial T}{\partial y} = a \frac{\partial^2 T}{\partial y^2} \qquad (245b)$$

From the addition of a source term S_A to Equation (266) and with $S_A = dm/dt$:

$$u \frac{\partial m}{\partial x} + v \frac{\partial m}{\partial y} = D \frac{\partial^2 m}{\partial y^2} - k(T, m_B) m \qquad (297)$$

Due to the simplification used in the formulation of Equation (296), the conservation equations for the chemical species are no longer coupled.

Equation system (246), (247), (245b), and (297) cannot be solved analytically. Numerical solutions for this problem will not be discussed here. However, also for $Pr = Sc = 1$ the similarity of the equations is lost due to the source term $-k(T, m_B) m$ in Equation (297) originating from the chemical reaction. Thus the boundary layer δ_m develops differently from δ_u and δ_T. Since the source term in Equation (297) is negative, a larger gradient of the mass fraction of A occurs and this leads to a thinner boundary layer δ_m in comparison with δ_u, δ_T. From the definition of the mass-transfer number (p. 2-65) it can be seen that the local mass-transfer coefficient increases due to the larger gradients. Mass transfer is accelerated by the chemical reaction.

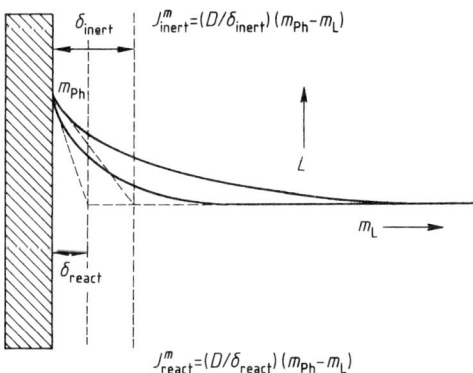

Figure 38. Acceleration of mass transfer by chemical reactions
δ = film thickness; J^m = mass fluxes; D = diffusion coefficient
For explanation of other symbols see Figure 34.

For a more thorough discussion of this acceleration effect, the boundary layer problem is further simplified by considering the situation at a large distance from the edge of the plate. Since $\delta_u \sim \sqrt{x}$, the gradients of u, T, and m in the x-direction become extremely small, and v approaches zero. Thus for $x \to \infty$ Equation (297) is simplified to

$$D \frac{d^2 m}{dy^2} - k(T, m_B) m \approx 0 \tag{298}$$

If y is transformed into $\eta = y/\delta_m$, then

$$D \frac{d^2 m}{d\eta^2} - Ha^2 m = 0 \tag{299}$$

where

$$Ha = \delta_m \sqrt{\frac{k}{D}} = \frac{\delta_m}{D}\sqrt{Dk} = \frac{1}{\beta}\sqrt{Dk}$$

is the Hatta number. The boundary conditions given in Figure 38 are $m = m_{Ph}$ for $\eta = 0$ and $m = m_L$ for $\eta = 1$. The characteristic equation of the ordinary homogenous differential Equation (299), $\lambda^2 - Ha^2 = 0$ has the solutions $\lambda_1 = Ha$ and $\lambda_2 = -Ha$ and thus the general solution for Equation (299) is

$$m = C_1 e^{Ha\eta} + C_2 e^{-Ha\eta} \tag{300}$$

From the above boundary conditions

$$C_1 = \frac{1}{e^{Ha} - e^{-Ha}}[m_L - m_{Ph} e^{-Ha}]$$

and

$$C_2 = -C_1 = \frac{1}{e^{Ha} - e^{-Ha}}[m_{Ph} e^{-Ha} - m_L]$$

so that the solution becomes

$$m = \frac{1}{\sinh Ha}[m_{Ph} \sinh(Ha(1-\eta)) + m_L \sinh(Ha\eta)] \tag{301}$$

Mass transfer is defined as usual by the condition

$$J_W^m = -\varrho D \left(\frac{dm}{dy}\right)_{y=0} = -\varrho D \frac{1}{\delta_m}\left(\frac{dm}{d\eta}\right)_{\eta=0}$$

From Equation (301) this condition gives

$$(J_W^m)_{react} = \varrho \frac{D}{\delta_m} \frac{Ha}{\sinh Ha}[m_{Ph} \cosh Ha - m_L] \tag{302}$$

If the mass-transfer rates from Equation (302) are compared with the condition without reaction (indicated by the subscript inert)

$$(J_W^m)_{inert} = \varrho \frac{D}{\delta_m}(m_{Ph} - m_L) \tag{303}$$

the accelerating effect of the chemical reaction is immediately obvious. The ratio of Equations (302) and (303) gives the "acceleration factor" β_r, i.e., the factor by which mass transfer is accelerated as a result of the chemical reaction:

$$\beta_r = \frac{Ha}{\tanh Ha} \frac{m_{Ph}}{m_{Ph} - m_L}\left[1 - \frac{m_L}{m_{Ph} \cosh Ha}\right] \tag{304a}$$

For $m_L = 0$

$$\beta_r = \frac{Ha}{\tanh Ha} \tag{304b}$$

which in turn, for $Ha > 3$, gives

$$\beta_r = Ha \tag{304c}$$

The acceleration of mass transfer is therefore proportional to the square root of the rate coefficient of the reaction rate. This only applies, however, if the assumptions made in deriving the equation for the expression of the reaction rate (Eq. 296) are fulfilled. For very fast reactions, i.e., for very high Hatta numbers, this is no longer the case.

The concentration of component B is then no longer "unlimited". On the contrary, transport of B into the boundary layer becomes rate-determining. The simultaneous solution for the conservation equations for A and B produces a result which is given in Figure 39 [110]. For $Ha \gg 3$

$$\beta_r = 1 + \frac{v_A m_{BL} D_B}{v_B m_{APh} D_A}$$

The accelerating effect of the reaction is controlled by the limited availability of B.

Introduction of the acceleration factor β_r means that the effect of the chemical reactions is incorporated into the mass transfer; the chemical reactions are no longer coupled with the mass

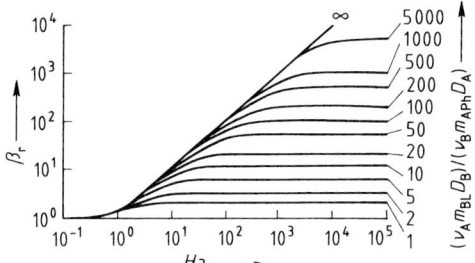

Figure 39. Acceleration factor β_r as a function of the Hatta number Ha for various stoichiometric characteristic values [110]
v = stoichiometric coefficient; D = diffusion coefficient
The subscripts A and B denote components, L and Ph denote liquid and phase boundary, respectively.

transfer. Further aspects of mass transfer with chemical reactions are discussed in Section 5.3.3.1.

5.3.3. Chemical Reactions in the Homogeneous Phase

Many chemical reactors are single phase. The term single phase should be understood here to mean that mass transfer from one phase to the other is not rate determining. Many heterogeneous systems also come within the term "single phase", for example heterogeneous catalytic reactions, which are dealt with as quasi-single phase. The mathematical treatment of single-phase reactors differs from the problems discussed in the earlier chapters due to the boundary conditions and the models for the processes occurring at and in the phase boundary layers. Definition of the state of the reactive mixture in chemically reactive flows with the fields $u(x, t)$, $p(x, t)$, $m_i(x, t)$, $i = 1, \ldots, N$ and $h(x, t)$ requires the solution of the conservation equations for the total mass, the components of the momentum, the mass of the individual chemical species, and the enthalpy. For closure of the system of equations other fundamental equations may have to be incorporated. With the simplifications introduced in the earlier sections for the molecular transport processes this system of equations can be given in the form

$$\frac{\partial \varrho}{\partial t} + \frac{\partial}{\partial x_k}(\varrho u_k) = 0 \tag{131}$$

for the total mass,

$$\varrho \frac{\partial u_i}{\partial t} + \varrho u_k \frac{\partial u_i}{\partial x_k} = \frac{\partial \tau_{ik}}{\partial x_k} - \frac{\partial p}{\partial x_i} + \varrho g_i$$
$$i = 1, 2, 3 \tag{204}$$

for the components of the momentum,

$$\varrho \frac{\partial h}{\partial t} + \varrho u_k \frac{\partial h}{\partial x_k} = \frac{\partial}{\partial x_k} \lambda \frac{\partial T}{\partial x_k} + \varrho S_h \tag{240b}$$

for the enthalpy, and

$$\varrho \frac{\partial m_i}{\partial t} + \varrho u_k \frac{\partial m_i}{\partial x_k} = \frac{\partial}{\partial x_k}\left[\varrho D_i \frac{\partial m_i}{\partial x_k}\right] + \varrho S_i$$
$$i = 1, \ldots, N - 1 \tag{265b}$$

for the chemical species. Since the system of Equations (131), (204), (240b), and (265b) is coupled, simultaneous solution is necessary. This is usually done numerically.

Before discussing a numerical procedure, some fundamental characteristics of the modeling of reactors will be shown by a simplified version of the system of Equations (131), (204), (240b), and (265b). The first simplification is the assumption of constant density, isothermal conditions, chemical reactions that do not involve volume changes, and frictionless flow.

5.3.3.1. Isothermal Reactors with Frictionless Flow, Constant Density, and Reactions Without Volume Changes

Under the above conditions it is easy to show that the equation system

$$\varrho \frac{\partial m_i}{\partial t} + \varrho u \frac{\partial m_i}{\partial x} = \frac{\partial}{\partial x}\left[\varrho D_i \frac{\partial m_i}{\partial x}\right] + \varrho S_i$$
$$i = 1, \ldots, N - 1 \tag{305}$$

is adequate for the definition of the problem provided that pressure effects and physical forces are neglected and the assumption is made that $v = w = 0$ (one-dimensional formulation).

If only two chemical species A and B are considered which react in a first-order reaction $A + B \rightarrow$ products, Equation (305) for the steady-state case and with $S_A = -k m_A$ is

$$\varrho u \frac{dm}{dx} - \varrho D \frac{d^2 m}{dx^2} + \varrho k m = 0 \tag{306}$$

In Equation (306) $m = m_A$, and thus also $S_A = S = -km$. Analytical solutions can be found for this equation for various boundary conditions. If the equation is, for example, normalized after dividing by the density ϱ in the form $\eta = x/L$, then

$$\frac{d^2 m}{d\eta^2} - Pe \frac{dm}{d\eta} - Da_1 Pe\, m = 0$$

Here $Pe = uL/D$, and Da_1 is the Damköhler number of the first kind, $Da_1 = k\tau$, where $\tau = L/u$ is the mean hydrodynamic residence time. For $Da_1 < Pe$, the characteristic equation of the above differential equation has two real roots

$$\lambda_{1,2} = \frac{1}{2} Pe \pm \sqrt{\frac{Pe^2}{4} + Da_1 Pe}$$

The general solution is thus

$$m = C_1 e^{\lambda_1 x} + C_2 e^{\lambda_2 x}$$

where C_1 and C_2 are found from the boundary conditions $m = m_0$ for $x = 0$ and $dm/dx = 0$ for $x = L$ as

$$C_1 = -m_0 \frac{\lambda_2 e^{\lambda_2 L}}{\lambda_1 e^{\lambda_1 L} - \lambda_2 e^{\lambda_2 L}} \quad \text{and}$$

$$C_2 = m_0 - C_1 = m_0 \left[1 + \frac{\lambda_2 e^{\lambda_2 L}}{\lambda_1 e^{\lambda_1 L} - \lambda_2 e^{\lambda_2 L}}\right]$$

Here however the numerical procedure analogous to Section 5.1.2.2 is discussed in order to compare various concepts of mathematical modeling.

To convert the differential operators into differences, Equation (306) is subjected to the same manipulations as Equation (226) and brought into discrete form with an "up-wind" difference scheme. For a one-dimensional problem a line along the x-direction of the grid in Figure 26 is adequate (Fig. 40). In Figure 40 the profile for m is also shown. When the operations carried out in Section 5.1.2.2 are applied to Equation (306) with the grid given in Figure 40, then the system of equations

$$-a_{l-1} m_{l-1} + b_l m_l - a_{l+1} m_{l+1} - \varrho S \Delta V_l = 0 \tag{307}$$

is obtained. The coefficients of this tridiagonal system of equations, with the notation from Fig-

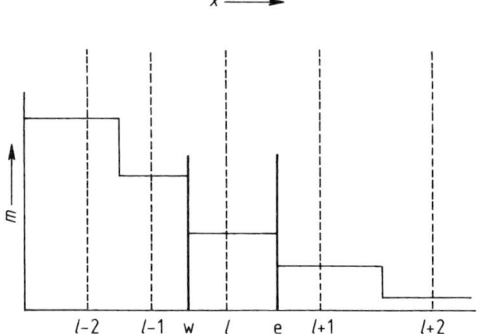

Figure 40. "Up-wind" difference scheme for a second-order differential equation
For explanation of symbols see Figure 26.

ure 40 are given by

$$a_{l-1} = \left[+ (\varrho u)_w + \varrho D \frac{1}{x_l - x_{l-1}} \right] F_w \tag{308 a}$$

$$a_{l+1} = \left[+ \varrho D \frac{1}{x_{l+1} - x_l} \right] F_e \tag{308 b}$$

$$b_l = a_{l-1} + a_{l+1} \tag{308 c}$$

Under the conditions given above $u > 0$. Assuming that $(uF) = \Delta V_l / \tau_l$ where τ_l is the hydrodynamic residence time over a distance l, and Pe_l is a local Péclet number $Pe_l = u(x_l - x_{l-1})/D$, the coefficients can be converted into

$$a_{l-1} = \frac{\varrho}{\tau_l} \Delta V_l + \frac{\varrho}{\tau_l Pe_l} \Delta V_l \tag{308 d}$$

$$a_{l+1} = \frac{\varrho}{\tau_l Pe_{l+1}} \Delta V_l \tag{308 e}$$

$$b_l = \frac{\varrho}{\tau_l} \Delta V_l + \frac{\varrho}{\tau_l Pe_l} \Delta V_l + \frac{\varrho}{\tau_l Pe_{l+1}} \Delta V_l \tag{308 f}$$

From Equations (307) and (308 d–f) the following tridiagonal equation system

$$\frac{\varrho}{\tau_l}\left[1 + \frac{1}{Pe_l}\right] m_{l-1} - \frac{\varrho}{\tau_l}\left[1 + \frac{1}{Pe_l} + \frac{1}{Pe_{l+1}}\right] m_l$$

$$+ \frac{\varrho}{\tau_l} \frac{1}{Pe_{l+1}} m_{l+1} + \varrho S(m_l) = 0 \tag{309}$$

is obtained. If the case of convectively dominated flows is considered where $Pe \to \infty$, Equation (309) becomes

$$\frac{\varrho}{\tau_l}(m_{l-1} - m_l) + \varrho S(m_l) = 0 \qquad (310)$$

This is the mass balance for a cascade of ideally mixed reactors. This analysis again shows that reactors with very large Bodenstein or Péclet numbers (ideally unmixed reactors) can be represented by a cascade of ideally mixed reactors (see Fig. 41 and Section 4.3.1.1).

Numerically this physical model finds its equivalent in the discrete form of differential Equation (306) using an "up-wind" difference scheme for the convectively dominated case. Equation (310) can easily be solved either by elimination or graphically by plotting both terms of the equation against m_l. The graphical solution is shown in Figure 41 for a nonequidistant grid, i.e., for ideally mixed reactors with different hydrodynamic residence times. It is important that there is no feedback from elements further downstream.

If the flow becomes predominantly diffusive, then Pe is very small and Equation (309) is written as

$$\frac{\varrho}{\tau_l}\frac{1}{Pe_l}m_{l-1} - \frac{\varrho}{\tau_l}\left[\frac{1}{Pe_l} + \frac{1}{Pe_{l+1}}\right]m_l$$
$$+ \frac{\varrho}{\tau_l}\frac{1}{Pe_{l+1}}m_{l+1} + \varrho S(m_l) = 0 \qquad (311)$$

Equation (311) has the same structure as Equation (310) but the feed for the ideally mixed reactor with the number l consists of a flow with the composition m_{l-1} and a flow with the composition m_{l+1} (Fig. 42). There is therefore a feedback from the elements further downstream, and the volumetric mass flows are increased by the factor $1/Pe_l$ etc., as compared with the convective mass flow $\varrho m_l / \tau_l$. Algorithms for the solution of the tridiagonal Equation (311) on the basis of LU decomposition of the system matrix can be found in [29]–[32].

The general case where $1/Pe_l = O(1)$ has already been formulated with Equation (309). Feedback dependent on the local Péclet number exists from the element $l + 1$ and the volumetric mass flows are increased as against the convective flows by the factor $[1 + (1/Pe_l)]$ etc. (Fig. 43). For solution, the modeling parameter Pe is required, which contains the molecular dif-

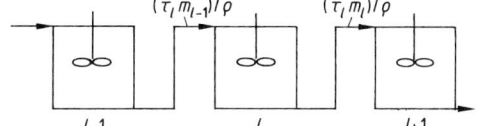

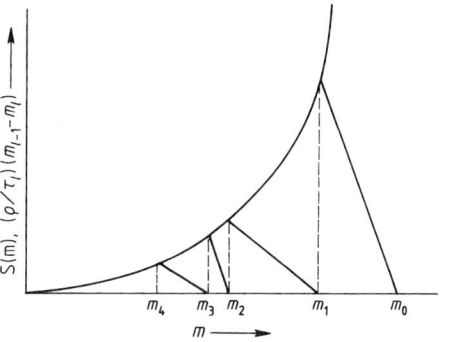

Figure 41. Cascade of ideally mixed reactors equivalent to an ideal tubular reactor
τ = residence time; ϱ = density; l = number of reactor.

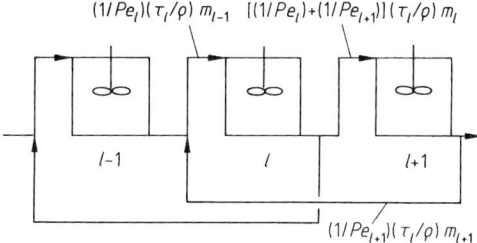

Figure 42. Cascade of ideally mixed reactors with feedback, equivalent to a tubular reactor with strong back-mixing
Pe = Péclet number. For explanation of other symbols see Figure 41.

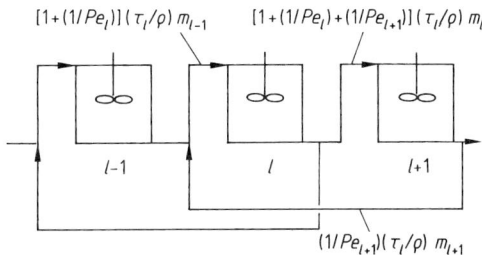

Figure 43. Cascade of ideally mixed reactors with feedback, equivalent to a tubular reactor with back-mixing, general case
Pe = Péclet number. For explanation of symbols see Figure 41.

fusion coefficients or experimentally determined effective mixing coefficients (see pp. 2-42–2-43).

In the foregoing analysis real reactors were described by numerical solution of differential Equation (306). This is equivalent to a representation by elements of ideally mixed reactors. Alternative structures of matrices of elements of ideal reactors can be derived by physical analysis of complex reactors [4], [11], [12], [24], [111], [112]. An example which is of some importance in chemical engineering will be described here.

Example of Combinations of Elements of Ideal Reactors: the Loop Reactor. The layout of an ideal plug flow reactor with recirculation (loop reactor) is shown in Figure 44, part of the product flow is added to the feed. For analysis the same assumptions apply as in Section 5.3.3.1.

With the definitions given in Figure 44, the mass fraction m_e at the outlet of the reactor can be obtained from Equation (310)

$$m_e = m'_0 \left[\frac{1}{(1 + k\tau_l)^n} \right] \quad (312)$$

For the sake of simplicity the ideal plug flow reactor in n equidistant sections so that $\tau_l = \tau = \tau_{tot}/n$. Thus from Equation (312)

$$m_e = m'_0 \left[\frac{1}{\left(1 + k\tau_{tot} + \frac{n-1}{n2!}(k\tau_{tot})^2 + \frac{(n-1)(n-2)}{n^2 3!}(k\tau_{tot})^3 + \ldots\right)} \right] \quad (313)$$

For $n \gg 1$ and $k\tau_{tot} \ll 1$

$$m_e \approx m'_0 \left[\frac{1}{\left(1 + k\tau_{tot} + \frac{1}{2!}(k\tau_{tot})^2 + \frac{1}{3!}(k\tau_{tot})^3 + \ldots\right)} \right] = m'_0 e^{-k\tau_{tot}} \quad (314)$$

where m'_0 and τ_{tot} are unknown and depend on the magnitude of the recirculated volume q_r.

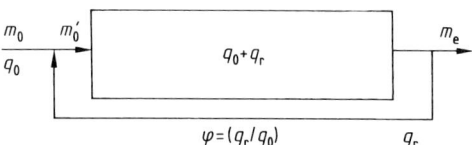

Figure 44. Schematic representation of a loop reactor
q = flow rate; m = mass fraction
The subscripts 0, r, and e denote inlet, loop (recirculated), and final, respectively; the superscript (') denotes conditions after mixing of inlet and backflow.

The term m'_0 results from the combined mass fractions of the flows entering the plug flow reactor and τ_{tot} from the volumes of the flow reactor and the total volumetric flow. If $\varphi = q_r/q_0$ (the recirculation ratio), then

$$m'_0 = m_0 \frac{1}{1 + \varphi} + m_e \frac{\varphi}{1 + \varphi} \quad (315\,a)$$

and

$$\tau_{tot} = \tau_0 \frac{1}{1 + \varphi} \quad (315\,b)$$

where τ_0 is the hydrodynamic residence time without recirculation. Substitution into Equation (314) gives

$$m_e = m_0 \frac{e^{-k\tau_0/(1+\varphi)}}{1 + \varphi(1 + e^{-k\tau_0/(1+\varphi)})} \quad (316)$$

For $\varphi \to 0$ the loop reactor takes on the character of the ideal plug flow reactor and $m_e = m_0 e^{-k\tau_0}$. For $k\tau_0 = Da_1 = 0.1$, $m_e/m_0 \approx 90\%$. For $\varphi \to \infty$, $e^{-k\tau_0/\varphi} = 1 - (k\tau_0/\varphi)$ and $m_e = m_0/(1 + k\tau_0)$. The loop reactor is similar to an ideally stirred reactor with correspondingly lower conversions.

The adjustable recirculation ratio of the loop reactor is fully exploited in important engineering applications. When measuring kinetic data in laboratory reactors the lowest possible conversion is required in the reactor so that the measured reaction rates can be assigned to specific temperatures and concentrations. Thus $Da_1 \ll 1$. On the other hand the concentration differences measured for evaluating the reaction rates should be as large as possible to minimize statistical errors. As may be calculated from

Equations (314) and (316), for $Da_1 = 0.1$ and $\varphi = 10$ the conversion in the plug flow reactor is $1 - (m_e/m_0') \approx 1\%$. The measured difference in the mass fractions is, however, $m_e/m_0 \approx 90\%$.

For reactions with positive reaction order relative to the reactants, the reaction rate falls off as the concentration decreases. Recirculation of the product lowers the concentration in the feed due to dilution. There is however a class of reactions in which the reaction rate first increases with decreasing concentration and then decreases after passing through a maximum (e.g., in heterogeneous catalysis, enzymatic catalysis, and autocatalysis). Recirculation has a positive effect on conversion in these cases.

5.3.3.1.1. Stability of Isothermal Reactors

Steady-State Cases. The importance of the stability of chemical reactors for safety and economic reasons is obvious. The description of stability analysis given here refers to the description of processes in physical space, location $x = (x_1, x_2, x_3)$ and time t. Stability analysis and theories often refer to the description of chemical processes in the form of transfer functions in the space of Laplace-transformed variables or to the frequency behavior. Detailed discussion of these relationships can be found in [3], [9], [113]–[116].

To illustrate stability analysis in isothermal, steady-state reactors one element of the ideal plug flow reactor will be used (Fig. 40).

This can be regarded as an ideally mixed reactor where the essentials of stability analysis can be shown. Three types of chemical reactions will be considered:

1) Normal reactions with declining reaction rate

   $A + B \rightarrow$ products

   where $dm_A/dt = -k_2 m_A m_B$ and
   $m_B = m_{B_0} - (m_{A_0} - m_A)$

2) Autocatalysis

   $A + 2B \rightarrow 3B$

   where $dm_A/dt = -k_3 m_A m_B^2$ and
   $m_B = m_{B_0} + (m_{A_0} - m_A)$

3) Self poisoning

   $A \rightarrow B$

   where $dm_A/dt = -k_1 m_A/(1 + K m_A)^2$ (Section 3.2.1.1, Eq. 53).

Along with the assumptions from Section 5.3.3.1, equal molar masses of A and B are also assumed. Using the abbreviations $\alpha = m_A/m_{A_0}$ and $\beta_0 = m_{B_0}/m_{A_0}$ and Equation (310), the following equations are obtained for the three cases

$$\frac{1 - \alpha}{\tau k_2 m_{A_0}} = \alpha^2 + (\beta_0 - 1)\alpha \tag{317a}$$

$$\frac{1 - \alpha}{\tau k_3 m_{A_0}^2} = \alpha[(\beta_0 + 1) - \alpha]^2 \tag{317b}$$

$$\frac{1 - \alpha}{\tau k_1} = \alpha \left[\frac{1}{1 + \alpha m_{A_0} K(2 + \alpha m_{A_0} K)}\right] \tag{317c}$$

where $1 - \alpha$ is the conversion. The term $\tau k_2 m_{A_0}$, $\tau k_3 m_{A_0}^2$, or τk_1 can be regarded as the ratio of the hydrodynamic residence time to a characteristic chemical time and is denoted as the Damköhler number of the first kind Da_1. Equations (317) can conveniently be solved graphically by plotting both sides against $1 - \alpha$. Figures 45–47 show such plots for the three cases above. For the normal case (Fig. 45 A), decreasing reaction rate with decreasing reactant concentration,

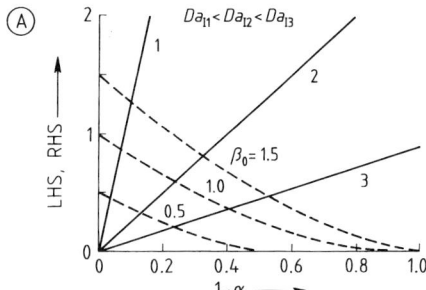

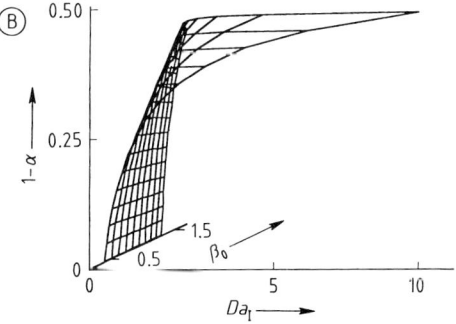

Figure 45. Mono-steady-state conversion points for "normal" chemical reactions (A) and bifurcation diagram (B) for isothermal conditions in a well-stirred reactor
LHS = left-hand side (Eq. 317a); RHS = right-hand side (Eq. 317a); $1 - \alpha$ = conversion; Da_1 = Damköhler number of first kind; $\beta_0 = m_{B_0}/m_{A_0}$

an unambiguous steady-state solution exists for each Damköhler number. With increasing Damköhler number (i.e., increasing hydrodynamic residence time), the steady-state solutions give higher conversions. All steady-state solutions, which are plotted against the Damköhler number in Figure 45 B, are stable. If a small disturbance occurs to the left or right on the conversion axis in the steady-state solution, opposing changes are produced in the right-hand side (RHS) or left-hand side (LHS) respectively of Equation (317a).

The "convective term" on the LHS decreases (increases), whilst the "reaction term" on the RHS increases (decreases). A lesser (greater) availability of A due to convection therefore counterbalances a greater (lesser) consumption of A due to reaction, so that the conversion is shifted to the right or to the left. The general condition for stability can be given in the terminology of Equation (317) as

$$\frac{d}{d(1-\alpha)}[\text{LHS}] > \frac{d}{d(1-\alpha)}[\text{RHS}]$$

at LHS = RHS   (318a)

For Equation (317a), $\frac{d}{d(1-\alpha)}[\text{LHS}] = 1/Da_1$ and $\frac{d}{d(1-\alpha)}[\text{RHS}] = 2(1-\alpha) - \beta - 1$; this satisfies the condition for stability for all possible values of $1-\alpha$ based on the inlet condition β_0.

A completely different picture results for autocatalysis (Fig. 46 A) or self poisoning (Fig. 47 A). In these cases multiple steady-state solutions occur for a particular range of Damköhler numbers. Some of the solutions do not comply with the conditions for stability (Eq. 318). For Equation (317b) $\frac{d}{d(1-\alpha)}$ [RHS] = $[\beta_0 + (1-\alpha)][2 - 3(1-\alpha) - \beta_0]$, so that the stability criterion is not met for small values of $(1-\alpha)$.

In this range a small disturbance of $(1-\alpha)$ to the right leads to a larger increase in the "reaction term" than in the availability due to convection; conversion therefore increases. The reaction system thus always heads for the nearest adjacent stable operating point. In the bifurcation diagrams shown in Figures 46 B and 47 B, these unstable steady-state solutions lie on an S-shaped branch. With a continuous increase in the Damköhler number (residence time) however

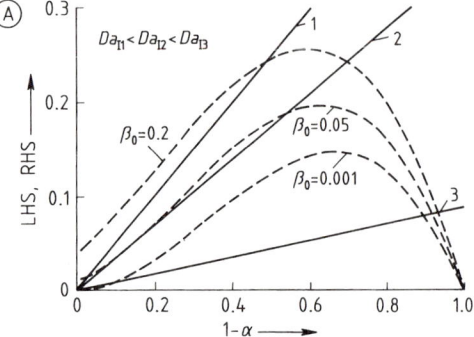

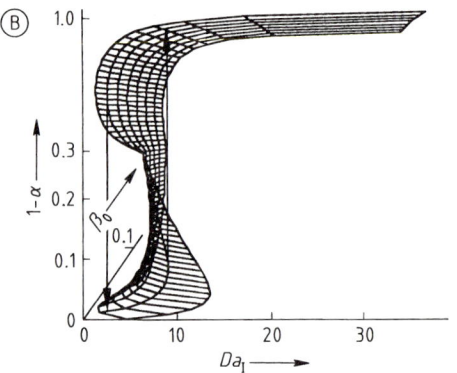

Figure 46. Multiple steady-state conversion points for autocatalytic chemical reactions (A) and bifurcation diagram (B) for isothermal conditions in a well-stirred reactor
LHS = left-hand side (Eq. 317b); RHS = right-hand side (Eq. 317b)
For explanation of other symbols see Figure 45.

the system cannot pass through this S-shaped branch, but jumps from a condition with low conversion to one with high conversion. The bifurcation diagrams show that in both cases with multiple steady-state solutions the Damköhler number for "igniting" the reaction is higher than that for "extinguishing" it. Thus hysteresis occurs. The jumping points are given by the condition for the coalescence of an unstable with a stable steady-state solution:

$$\frac{d}{d(1-\alpha)}[\text{LHS}] = \frac{d}{d(1-\alpha)}[\text{RHS}]$$

at LHS = RHS   (318b)

In the case of autocatalysis this condition occurs for

$$\frac{8}{Da_1} = 1 + 20\beta_0 - 8\beta_0^2 \pm (1 - 4\beta_0)\sqrt{(1 - 8\beta_0)}$$

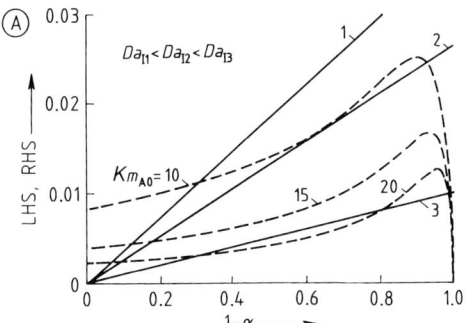

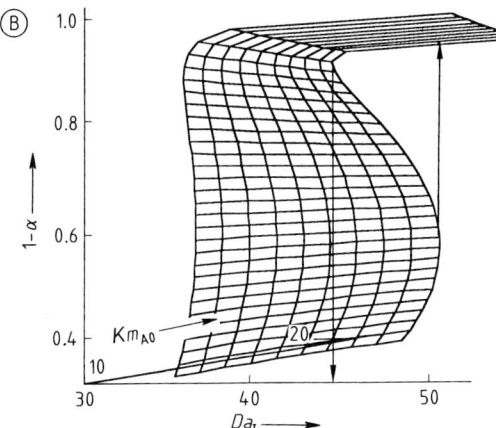

Figure 47. Multiple steady-state conversion points for chemical reactions with self poisoning (A) and bifurcation diagram (B) for isothermal conditions in a well-stirred reactor
K = adsorption coefficient; m_{A_0} = inlet mass fraction of component A; LHS = left-hand side (Eq. 317c), RHS = right-hand side (Eq. 317c)
For explanation of other symbols see Figure 45.

The positions of the two jumping points move closer together and, with an increase in the inlet concentration of the catalyst β_0, the S-shaped curve unfolds. Increasing the concentration of the catalyst at the reactor inlet results in stable reactor behavior, shown by the disappearance of multiple steady-state solutions.

An analogous picture results for self poisoning. The appropriate conditions can easily be calculated from Equation (317c) and the stability condition. Here the parameter controlling the nonambiguity of the solutions is Km_{A_0} which can be regarded as a measure of the poisoning. A decrease in this quantity (lower pressure or lower inlet concentration of A in heterogeneous catalytic reactions) leads to a range of conversions that exhibits unambiguous, stable, steady-state solutions (compare Figs. 45 B–47 B).

Nonsteady-State Cases. The stability analysis of steady-state, isothermal, ideally mixed reactors has shown that the phenomenon of multiple steady states with hysteresis always occurs when there is feedback. A feedback system will now be considered for the stability analysis of nonsteady-state reactions. The chemical reaction is formulated as

$$P \xrightarrow{k_0} A \xrightarrow{k_u} \overset{k_1}{\underset{}{B}} \xrightarrow{k_2} C$$

where the step A → B is a noncatalytic first-order reaction and a cubic (third-order) reaction catalyzed with B. The feedback mechanism in this reaction is clear: the more B is formed, the faster A is consumed. Since the supply of A from P remains limited, at high concentrations of catalyst B the reaction A → B breaks down until adequate A is again supplied.

Analysis will be performed on a batch reactor so that the system of Equations (305) can be written in the form

$$\frac{dm_P}{dt} = -k_0 m_P \qquad (319\,a)$$

$$\frac{dm_A}{dt} = k_0 m_P - k_1 m_A m_B^2 - k_u m_A \qquad (319\,b)$$

$$\frac{dm_B}{dt} = k_1 m_A m_B^2 + k_u m_A - k_2 m_B \qquad (319\,c)$$

A condition for m_C results from the mass balance of the system. The quantity $(k_2/k_1)^{1/2}$ is used to normalize the mass fraction, and the normalization of the time is achieved through $(1/k_2)$. The system of equations can then be written as

$$\frac{d\pi}{d\tau} = -\varepsilon\pi \qquad (320\,a)$$

where $\pi = m_P(k_2/k_1)^{1/2}$, $\varepsilon = (k_0/k_2)$, and $\tau = t/k_2$

$$\frac{d\alpha}{d\tau} = \varepsilon\pi - \alpha\beta^2 - \kappa\alpha \qquad (320\,b)$$

where $\alpha = m_A(k_2/k_1)^{1/2}$, $\beta = m_B(k_2k_1)^{1/2}$, and $\kappa = (k_u/k_2)$, and finally

$$\frac{d\beta}{d\tau} = \alpha\beta^2 + \kappa\alpha - \beta \qquad (320\,c)$$

The initial conditions are $\pi = \pi_0$, $\alpha = \beta = 0$ for $\tau = 0$. Thus Equation (320a) can easily be solved. The result

$$\pi = \pi_0 e^{-\varepsilon\tau} \tag{321}$$

is used in the system of Equations (320) which thus become a system of coupled first-order differential equations

$$\frac{d\alpha}{d\tau} = \mu - \alpha\beta^2 - \kappa\alpha = f(\alpha, \beta) \tag{322a}$$

$$\frac{d\beta}{d\tau} = \alpha\beta^2 + \kappa\alpha - \beta = g(\alpha, \beta) \tag{322b}$$

where the abbreviation μ denotes $\mu = \varepsilon\pi_0 e^{-\varepsilon\tau} = \mu_0 e^{-\varepsilon\tau}$. For $k_0 \ll k_1$, k_u, k_2 the principle of quasi-steady state is applicable for the mass fractions α and β. Thus Equations (322 a, b) are solved without integration; approximate solutions for α and β can be obtained from the conditions

$$\frac{d\alpha}{d\tau} = \frac{d\beta}{d\tau} = 0$$

From

$$\mu - \alpha\beta^2 - \kappa\alpha = 0 \tag{323a}$$

and

$$\alpha\beta^2 + \kappa\alpha - \beta = 0 \tag{323b}$$

the quasi-steady-state mass fractions α_{qs}, β_{qs} are given by

$$\alpha_{qs} = \mu/(\mu^2 + \kappa) \text{ and } \beta_{qs} = \mu \tag{324a, b}$$

For the stability analysis α and β are regarded as being subject to small disturbances $\Delta\alpha$ and $\Delta\beta$, respectively. The time evolution of these disturbances are then considered. The perturbed quantities thus become

$$\alpha = \alpha_0 + \Delta\alpha, \; f(\alpha, \beta) = f(\alpha, \beta)_0 + \left(\frac{\partial f(\alpha, \beta)}{\partial \alpha}\right)_0 \Delta\alpha$$

$$+ \left(\frac{\partial f(\alpha, \beta)}{\partial \beta}\right)_0 \Delta\beta + \ldots \tag{325a}$$

and

$$\beta = \beta_0 + \Delta\beta, \; g(\alpha, \beta) = g(\alpha, \beta)_0 + \left(\frac{\partial g(\alpha, \beta)}{\partial \alpha}\right)_0 \Delta\alpha$$

$$+ \left(\frac{\partial g(\alpha, \beta)}{\partial \beta}\right)_0 \Delta\beta + \ldots \tag{325b}$$

The functions $f(\alpha, \beta)$ and $g(\alpha, \beta)$ in Equations (325) are developed in Taylor series at α_0 and β_0 that are truncated after the first term. This is a reasonable approximation for small disturbances. If Equation (325) is substituted into Equation (322), the time evolution of the disturbances can be expressed as

$$\frac{\partial\alpha_0}{\partial\tau} + \frac{\partial\Delta\alpha}{\partial\tau} = f(\alpha, \beta)_0 + \left(\frac{\partial f(\alpha, \beta)}{\partial\alpha}\right)_0 \Delta\alpha$$

$$+ \left(\frac{\partial f(\alpha, \beta)}{\partial\beta}\right)_0 \Delta\beta \tag{326a}$$

and

$$\frac{\partial\beta_0}{\partial\tau} + \frac{\partial\Delta\beta}{\partial\tau} = g(\alpha, \beta)_0 + \left(\frac{\partial g(\alpha, \beta)}{\partial\alpha}\right)_0 \Delta\alpha$$

$$+ \left(\frac{\partial g(\alpha, \beta)}{\partial\beta}\right)_0 \Delta\beta \tag{326b}$$

If the quasi-steady-state solutions are considered, then Equation (326) becomes the system of coupled ordinary differential equations

$$\begin{pmatrix} \frac{d\Delta\alpha}{d\tau} \\ \frac{d\Delta\beta}{d\tau} \end{pmatrix} - J_{qs}\begin{pmatrix} \Delta\alpha \\ \Delta\beta \end{pmatrix} = 0 \tag{327a}$$

where J_{qs} is the Jacobi matrix of Equation (323)

$$J_{qs} = \begin{pmatrix} \frac{\partial f(\alpha, \beta)}{\partial\alpha} & \frac{\partial f(\alpha, \beta)}{\partial\beta} \\ \frac{\partial g(\alpha, \beta)}{\partial\alpha} & \frac{\partial g(\alpha, \beta)}{\partial\beta} \end{pmatrix}_{qs} \tag{327b}$$

Equation (327) can easily be solved analytically after being transformed into a second-order differential equation (see Section 5.3.1.3). The general solution is

$$\Delta\alpha(\tau) = C_1 e^{\lambda_1 \tau} + C_2 e^{\lambda_2 \tau} \tag{328a}$$

$$\Delta\beta(\tau) = C_3 e^{\lambda_1 \tau} + C_4 e^{\lambda_2 \tau} \tag{328b}$$

where the eigenvalues λ_1 and λ_2 are determined from the characteristic equation

$$\lambda^2 - \text{Tr}(J_{qs})\lambda + \text{Det}(J_{qs}) = 0 \tag{329}$$

This expression has already been used in Sections 5.3.1.3, 5.3.2, and 5.3.3.1. The trace of the Jacobi matrix is the sum of the diagonal elements

$$\text{Tr}(J_{qs}) = \left(\frac{\partial f(\alpha, \beta)}{\partial\alpha}\right)_{qs} + \left(\frac{\partial g(\alpha, \beta)}{\partial\beta}\right)_{qs}$$

and its determinant is

$$\mathrm{Det}(J_{qs}) = \left(\frac{\partial f(\alpha,\beta)}{\partial \alpha}\right)_{qs}\left(\frac{\partial g(\alpha,\beta)}{\partial \beta}\right)_{qs}$$
$$- \left(\frac{\partial g(\alpha,\beta)}{\partial \alpha}\right)_{qs}\left(\frac{\partial f(\alpha,\beta)}{\partial \beta}\right)_{qs}$$

so that the solution of the characteristic equation is

$$\lambda_{1,2} = \tfrac{1}{2}[\mathrm{Tr}(J_{qs}) \pm \{\mathrm{Tr}(J_{qs})^2 - 4\,\mathrm{Det}(J_{qs})\}^{1/2}] \tag{330}$$

The time evolution of the disturbances $\Delta\alpha$ and $\Delta\beta$ is affected by the eigenvalues of $\lambda_{1,2}$, which depend on the values of $\mathrm{Tr}(J_{qs})$ and $\mathrm{Det}(J_{qs})$. The different domains of the solutions of Equations (327) are shown in Figure 48 in a stability diagram.

To demonstrate some of the possible cases the simplifying assumption is made that the first-order noncatalytic reaction $(A \to B)$ proceeds very slowly, i.e., $\kappa \ll 1$. Equations (323) and (324) can then be further simplified

$$\mathrm{Tr}(J_{qs}) = (1 - \mu^2) \tag{331a}$$

$$\mathrm{Det}(J_{qs}) = \mu^2 \tag{331b}$$

$$\mathrm{Tr}(J_{qs})^2 - 4\,\mathrm{Det}(J_{qs}) = \mu^4 - 6\mu^2 + 1 \tag{331c}$$

As μ is a function of time, it is to be expected that during the course of the reaction with the depletion of the reservoir P, the discriminants, Equation (331c), as well as $\mathrm{Tr}(J_{qs})$ and $\mathrm{Det}(J_{qs})$ will change their values and signs, so that $\lambda_{1,2}$ take all possible combinations

1) Initial condition, $\mu > 1 + \sqrt{2}$

   Initially only P is present so μ is very large. Then $\mathrm{Tr}(J_{qs})$ is negative, $\mathrm{Det}(J_{qs})$ is positive, and the discriminant is also positive. Thus $\lambda_{1,2}$ become both real and negative. Small disturbances $\Delta\alpha$ and $\Delta\beta$ decay monotonically, and the quasi-steady-state mass fractions represent a stable node of the system. The terms node (and focus) are derived from the phase diagrams of $\Delta\alpha$ and $\Delta\beta$, see [114]. In Figure 49 this range is shown in the plot of α_{qs} and β_{qs} against μ.

2) Progressive reaction, $1 < \mu < 1 + \sqrt{2}$

   In this range of conversion of P, $\mathrm{Tr}(J_{qs})$ is negative and $\mathrm{Det}(J_{qs})$ positive. The discriminant has now changed its sign however. Due

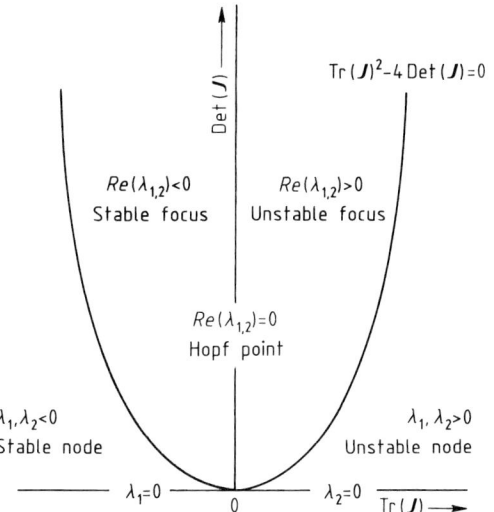

Figure 48. Stability diagram and nature of the steady-state solutions for the example in Section 5.3.3.1 $\lambda_{1,2}$ = eigenvalues of Equations (328).

to this $\lambda_{1,2}$ become conjugate complex with a negative real part. In this case the disturbances $\Delta\alpha$ and $\Delta\beta$ behave as damped oscillations and the quasi-steady-state solutions represent a stable focus of the system.

3) Further progress of the reaction, $\sqrt{2} - 1 < \mu < 1$

   In this concentration range of P, $\mathrm{Tr}(J_{qs})$ is positive and the discriminant negative. Here too $\lambda_{1,2}$ are conjugate complex but with a positive real part. The disturbances $\Delta\alpha$ and $\Delta\beta$ increase in the form of divergent oscillations. The quasi-steady-state solutions are an unstable focus of the system.

4) Depletion of P, $0 < \mu < \sqrt{2} - 1$

   Towards the end of the reaction the trace of the Jacobi matrix is always positive; the discriminant has changed its sign and is also positive again. Now $\lambda_{1,2}$ are both real and positive so that the disturbances grow exponentially. The state of the system is described as an unstable node.

Between the conditions (2) and (3), a state exists at which $\mu = 1$. At this point $\mathrm{Tr}(J_{qs}) = 0$ and $\mathrm{Det}(J_{qs}) > 0$ so that $\lambda_{1,2}$ become imaginary. Here the disturbances $\Delta\alpha$ and $\Delta\beta$ behave as undamped sinusoidal oscillations and the system

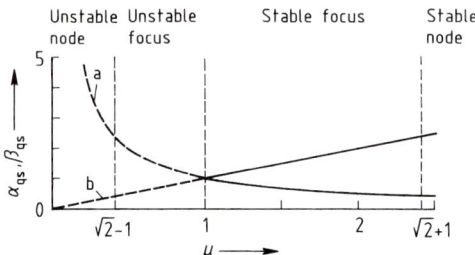

Figure 49. Quasi-steady-state solutions (a) $\alpha_{qs}(\mu)$ and (b) $\beta_{qs}(\mu)$ for an isothermal nonsteady-state system with cubic autocatalysis
For further explanation of symbols see text.

forms stable oscillations around the quasi-steady-state solutions. This point is designated the Hopf bifurcation point.

For the simplified system without the noncatalytic reaction, the trace of the Jacobi matrix is given by Equation (331 a). Thus the condition for the change of sign, i.e., $\text{Tr}(\mathbf{J}_{qs}) = 0$ and $d\text{Tr}(\mathbf{J}_{qs})/d\mu \neq 0$, only occurs when $\mu = 1$. If the noncatalytic reaction is also considered, then

$$\text{Tr}(\mathbf{J}_{qs}) = \frac{\mu^4 - (1 - 2\kappa)\mu^2 + \kappa(1 + \kappa)}{\mu^2 + \kappa}$$

(see also Eq. 323). Two Hopf bifurcation points are then obtained from

$$(\mu_{1,2\,\text{Bf}})^2 = \tfrac{1}{2}[(1 - 2\kappa) \pm (1 - 8\kappa)^{1/2}]$$

For $\kappa \ll 1$ the upper equation may be expanded into a series and the two bifurcation points are

$$\mu_{1\,\text{Bf}} = 1 - \tfrac{3}{2}\kappa + \ldots \quad \text{and}$$

$$\mu_{2\,\text{Bf}} = \kappa^{1/2}(1 + 2\kappa) + \ldots$$

Since $\text{Det}(\mathbf{J}_{qs})$ is always positive, then with consumption of P the system passes through two Hopf bifurcation points between which solutions for $\lambda_{1,2}$ occur with a predominantly imaginary part. In this range instabilities can develop in the form of stable oscillations, the amplitude and frequency of which alter with the decreasing concentration of P [113].

In chemical engineering practice instabilities can only be avoided in systems with the properties quoted above for the range $\mu > 1$. Details of the Hopf bifurcation analysis and the detailed calculation of the oscillation properties (frequency, amplitude, and stability) are given in [113]–[116], see also → 1. Mathematics in Chemical Engineering, pp. 1-64–1-65.

5.3.3.1.2. Sensitivity Analysis

In Section 3.1.1.3 sensitivity analysis was described as the change in the dependent variables of a model with a change of its modeling parameters. If the dependent variables $\boldsymbol{\Phi}$ are developed as a Taylor series at a specific solution $\boldsymbol{\Phi}_0$, the change is given by

$$\boldsymbol{\Phi} = \boldsymbol{\Phi}_0 + \left(\frac{\partial \boldsymbol{\Phi}}{\partial \boldsymbol{\beta}}\right)_0 \Delta \boldsymbol{\beta} + \left(\frac{\partial^2 \boldsymbol{\Phi}}{\partial \boldsymbol{\beta}^2}\right)_0 \frac{\Delta \boldsymbol{\beta}^2}{2!} + \ldots \quad (332)$$

where $\boldsymbol{\beta}$ is the vector of the parameters.

The simplest measure of the system's sensitivity are the gradients $(\partial \boldsymbol{\Phi}/\partial \boldsymbol{\beta})$ with which the change in the dependent variables is calculated with the aid of a Taylor series truncated after the first term. The gradients $(\partial \boldsymbol{\Phi}/\partial \boldsymbol{\beta})$ are denoted as a first-order sensitivity coefficients, the curve and the higher derivatives of Equation (332) as sensitivity coefficients of the second and correspondingly higher orders.

If the solution $\boldsymbol{\Phi}$ is available in explicit form (e.g., Eq. 323), calculation of the sensitivity coefficients is simple. For very many models of chemical engineering processes the differential equations which constitute the model cannot, however, be solved analytically or with the help of simple assumptions. The sensitivity coefficients must then be determined from the differential equations or from the numerical solutions [117], [118]. The latter method will now be described.

Numerical solutions provide the dependent variables not in the form of continuous functions but in discrete form (discrete values), calculation of the gradients is therefore not trivial. The numerical solution of a differential equation is (e.g., for the example in Section 5.3.3.1) given implicitly through Equation (309) in the form of a system of equations.

$$F(\boldsymbol{\Phi}) = 0 \qquad (333)$$

The first-order sensitivity coefficients are obtained from Equation (333) by the differentiation rule

$$\frac{\partial F(\boldsymbol{\Phi})}{\partial (\boldsymbol{\Phi})} \frac{\partial (\boldsymbol{\Phi})}{\partial \boldsymbol{\beta}} + \frac{\partial F(\boldsymbol{\Phi})}{\partial \boldsymbol{\beta}} = 0 \qquad (334)$$

Thus the gradients are given by

$$\frac{\partial (\boldsymbol{\Phi})}{\partial \boldsymbol{\beta}} = -\mathbf{J}^{-1} \frac{\partial F(\boldsymbol{\Phi})}{\partial \boldsymbol{\beta}} \qquad (335)$$

In the example from Section 5.3.3.1 the corresponding relations are easily verified: the Jacobi matrix is given by the coefficients a_{l-1}, $-[b_l - \varrho(\partial S(m_l)/\partial m_l)]$, and a_{l+1} and is identical to the coefficient matrix for the equation system (Eq. 309). For $\beta = k$ the derivatives with respect to the parameters are simply $\varrho(\partial S(m_l)/\partial k)$. For a first-order reaction $\varrho(\partial S(m_l)/\partial k) = -\varrho m_l$. If we simplify in Equation (309) $Pe_l = Pe_{l+1} = Pe$ then

$$\frac{\partial F}{\partial Pe} = -\frac{\varrho}{\tau_l Pe^2}(m_{l-1} - 2m_l + m_{l+1})$$

so for the sensitivity coefficients the equation system (336) has to be solved. Equation system (336) for the gradients $\partial(\Phi)/\partial\beta$ has the same structure as the original equation system (Eq. 309) and can consequently be solved by similar methods.

$$\begin{pmatrix} \frac{\partial m_1}{\partial k} & \frac{\partial m_1}{\partial Pe} \\ \vdots & \vdots \\ \frac{\partial m_n}{\partial k} & \frac{\partial m_n}{\partial Pe} \end{pmatrix} = -\begin{pmatrix} -(b_1+\varrho k) & a_2 & & \\ a_1 & -(b_2+\varrho k) & a_3 & \\ & & \ddots & \\ & & a_{n-1} & -(b_n+\varrho k) \end{pmatrix}^{-1} \quad (336)$$

$$\cdot \begin{pmatrix} -\varrho m_1 & -\frac{\varrho}{\tau_1 Pe^2}(m_0 - 2m_1 + m_2) \\ \vdots & \vdots \\ -\varrho m_n & -\frac{\varrho}{\tau_n Pe^2}(m_{n-1} - 2m_n) \end{pmatrix}$$

For the calculation and discussion of the sensitivity coefficients of higher order, reference should be made to the literature [117], [118]. An example of calculating first-order sensitivity coefficients for a complex model follows in Section 5.3.3.2.

5.3.3.2. Nonisothermal Reactors

In nonisothermal reactors the density does not remain constant. The couplings in the system of Equations (131), (204), (240b), and (265b) therefore cannot be resolved and the equations must be solved simultaneously.

5.3.3.2.1. Heterogeneous Catalytic Reactions

Example: Heterogeneous Catalytic Dehydration of Ethylbenzene to Styrene. The heterogeneous catalytic dehydration of ethylbenzene to styrene is outlined in Figure 50. Ethylbenzene is preheated and mixed with superheated steam. The mixture is passed through a cylindrical reactor packed with an iron oxide catalyst. Dehydration is endothermic. The steam, used as a heat carrier, causes some secondary reactions. A suitable reaction scheme is [11], [119], [120]:

$$C_6H_5-C_2H_5 \text{ (1)} \rightarrow C_6H_5-C_2H_3 \text{ (2)} + H_2 \text{ (3)} \quad \text{(i)}$$

$$C_6H_5-C_2H_5 \rightarrow C_6H_6 \text{ (4)} + C_2H_4 \text{ (7)} \quad \text{(ii)}$$

$$C_6H_5-C_2H_5 + H_2 \rightarrow C_6H_5-CH_3 \text{ (5)} + CH_4 \text{ (6)} \quad \text{(iii)}$$

$$\tfrac{1}{2}C_2H_4 + H_2O \text{ (10)} \rightarrow CO \text{ (8)} + 2H_2 \quad \text{(iv)}$$

$$CH_4 + H_2O \rightarrow CO + 3H_2 \quad \text{(v)}$$

$$CO + H_2O \rightarrow CO_2 \text{ (9)} + H_2 \quad \text{(vi)}$$

The formulation of this set of reactions is necessary because they all have an appreciable reaction enthalpy so that conversion into the various byproducts also affects the temperature profile. Furthermore the reaction rate of reaction (i) is given by $r_{(i)} = k_1(m_1 - m_2m_3/K)$ [8], [108] which indicates poisoning due to the products styrene and hydrogen, and thus the mass fraction m_1 is coupled with m_2, m_3.

The appropriate formulation of Equations (131), (204), (240b), and (265b) is in cylindrical coordinates for this axially symmetric problem. The continuity equation in this form is

$$\frac{\partial \varrho}{\partial t} + \varrho \frac{\partial}{\partial x}(u) + \varrho \frac{1}{r}\frac{\partial}{\partial r}(rv) = 0 \quad (337a)$$

If we use the normalized variables $u^* = u/u_0$, $v^* = v/u_0$, $\varrho^* = \varrho/\varrho_0$, $t^* = t/\tau$, $x^* = x/L$, and $r^* = r/R$, then from Equation (337a) it follows

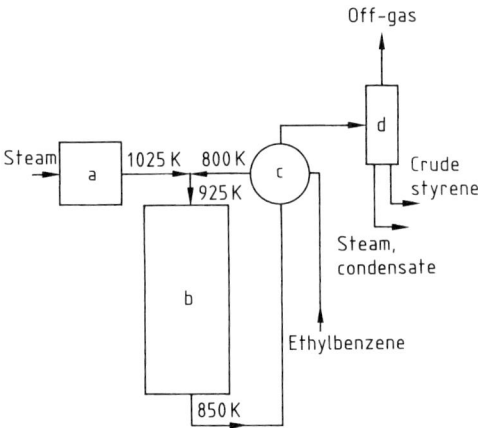

Figure 50. Schematic of a pilot plant for catalytic dehydration of ethylbenzene
a) Superheater; b) Catalytic reactor; c) Heat exchanger; d) Separator.

that

$$\frac{\varrho_0}{\tau}\frac{\partial \varrho^*}{\partial t^*} + \frac{\varrho_0 u_0}{L}\varrho^*\frac{\partial}{\partial x^*}(u^*)$$

$$+ \frac{\varrho_0 u_0}{R}\varrho^*\frac{1}{r^*}\frac{\partial}{\partial r^*}(r^* v^*) = 0 \qquad (337\,\text{b})$$

This can be transformed into

$$\frac{\partial \varrho^*}{\partial t^*} + \varrho^*\frac{\partial}{\partial x^*}(u^*) + \varrho^*\frac{1}{r^*}\frac{\partial}{\partial r^*}\left(r^* v^* \frac{R}{L}\right) = 0 \qquad (337\,\text{c})$$

Since u^*, x^*, and r^* are all of the order of magnitude $O(1)$, then $v^*(L/R) = O(1)$ or $v^* \approx (R/L)$, and thus for $(R/L) \ll 1$ the third term in Equation (337c) can be neglected. In dimensional quantities the continuity equation for the steady-state case thus has the form

$$\varrho \frac{du}{dx} = 0 \qquad (337\,\text{d})$$

The balance of momentum (Eq. 204) with the above result in cylindrical coordinates is

$$\varrho \frac{\partial u}{\partial t} + \varrho u \frac{\partial u}{\partial x} = \left[\frac{\partial \tau_{xx}}{\partial x} + \frac{1}{r}\frac{\partial (r\tau_{rx})}{\partial r} + \frac{1}{r}\frac{\partial \tau_{\theta x}}{\partial \theta}\right] - \frac{\partial p}{\partial x} \qquad (338\,\text{a})$$

For flows through packed catalyst beds the viscous stress term can be expressed by the permeability κ according to Darcy's Law [8], [121]. The term in the square brackets in Equation (338a) then becomes

$$\left[\frac{\partial \tau_{xx}}{\partial x} + \frac{1}{r}\frac{\partial (r\tau_{rx})}{\partial r} + \frac{1}{r}\frac{\partial \tau_{\theta x}}{\partial \theta}\right] = \frac{\mu u}{\kappa}$$

If the same normalized variables are used as in Equation (337) together with the normalized pressure $p^* = p/p_0$, Equation (338a) becomes

$$\frac{R}{\tau u_0}\frac{\partial u^*}{\partial t^*} + \frac{R}{L}u^*\frac{\partial u^*}{\partial x^*}$$

$$= -\frac{\mu R}{\kappa u_0 \varrho_0}\frac{u^*}{\varrho^*} - \frac{R p_0}{L \varrho_0 u_0^2}\frac{1}{\varrho^*}\frac{\partial p^*}{\partial x^*} \qquad (338\,\text{b})$$

Estimates of the orders of magnitude of the factors in front of the differential quotients in Equation (338b) under pilot plant conditions ($\tau \approx 1$ s, $u_0 \approx 0.2$ m/s, $R/L \approx 5 \times 10^{-2}$, $\varrho_0 \approx 0.5$ kg/m³, $p_0 \approx 1.5 \times 10^5$ Pa, $\mu \approx 10^{-5}$ kg m^{-1} s^{-1}, and finally $\kappa \approx 1.5 \times 10^{-8}$ m²) leads to

$$\frac{R}{L} < \frac{R}{\tau u_0} < 1 < \frac{\mu R}{\kappa u_0 \varrho_0} \ll \frac{R p_0}{L \varrho_0 u_0^2}$$

Thus, for the given conditions, the pressure term predominates in the steady-state balance of momentum which in dimensional quantities simplifies to

$$\frac{dp}{dx} \approx 0 \quad \text{or} \quad p \approx \text{const.} \qquad (338\,\text{c})$$

Consequently the continuity equation and the momentum equation are no longer coupled with the other transport equations and pressure p and velocity u are simple to estimate.

Using the definition equation for enthalpy, the conservation equation for the enthalpy (Eq. 240b) can be transformed into a differential equation for temperature (see Section 5.2.2.1). For a multicomponent system the conservation equations for the mass of the chemical species must be used for this. The result, also in cylindrical coordinates, is

$$\varrho c_p \frac{\partial T}{\partial t} + \varrho c_p u \frac{\partial T}{\partial x} + \lambda \left[\frac{\partial^2 T}{\partial x^2} + \frac{1}{r}\frac{\partial}{\partial r}\left(r \frac{\partial T}{\partial r}\right)\right]$$

$$= \Sigma r_i(-\Delta h_{r_i}) + S_C \qquad (339\,\text{a})$$

Equation (339a) contains the assumptions introduced on p. 2-59. In addition since $p \approx$ const., S_C only includes a term for heat transfer from the gas phase to the catalyst. Due to the transformation of the conservation equation of enthalpy into a differential equation for temperature, the term $\sum r_i(-\Delta h_{r_i})$ occurs which describes the release of heat due to chemical reactions. As in Equations (337) and (338), all quantities are normalized to the conditions at the inlet of the reactor. For Equation (339a) this results in

$$\frac{L}{\tau u_0}\varrho^* c_p^* \frac{\partial T^*}{\partial t^*} + \varrho^* c_p^* u^* \frac{\partial T^*}{\partial x^*}$$

$$= \frac{\lambda_0}{L\varrho_0 c_{p0} u_0}\lambda^* \frac{\partial^2 T^*}{\partial x^{*2}}$$

$$+ \frac{\lambda_0 L}{R^2 \varrho_0 c_{p0} u_0}\frac{\lambda^*}{r^*}\frac{\partial}{\partial r^*}\left(r^* \frac{\partial T^*}{\partial r^*}\right)$$

$$+ \frac{L\alpha s}{\varepsilon \varrho_0 c_{p0} u_0}(T_{\text{Cat}}^* - T^*)$$

$$+ \frac{L}{u_0 T_0 \varrho^0 c_{p0}}\sum r_i(-\Delta h_{r_i}) \qquad (339\,\text{b})$$

where α is the heat-transfer coefficient, s the specific surface area, ε the intergranular volume, and T_{Cat} the catalyst temperature. Estimation of the factors preceding the differential quotients in Equation (339b) ($c_{p0} \approx 2.5\,\text{kJ kg}^{-1}\,\text{K}^{-1}$, $\lambda_0 \approx 4 \times 10^{-5}\,\text{kJ m}^{-1}\,\text{K}^{-1}\,\text{s}^{-1}$, $\alpha \approx 4 \times 10^{-2}\,\text{kW} \cdot \text{m}^{-2}\,\text{K}^{-1}$, $\varepsilon \approx 0.36$, and $s \approx 1.5 \times 10^3\,\text{m}^{-1}$) results in

$$\frac{\lambda_0 L}{R^2 \varrho_0 c_{p0} u_0} < \frac{\lambda_0}{L\varrho_0 c_{p0} u_0} \ll \frac{L}{\tau u_0} \approx 1 \ll \frac{L\alpha s}{\varepsilon \varrho_0 c_{p0} u_0}$$

The fundamental mechanisms for energy transport are therefore expressed in the one-dimensional form of the temperature equation, which in steady-state formulation with dimensional quantities becomes

$$\varrho c_p u \frac{dT}{dx} = \frac{\alpha s}{\varepsilon}(T_{\text{Cat}} - T) + \sum r_i(-\Delta h_{r_i}) \quad (339\,\text{c})$$

The heat-transfer term in Equation (339c) couples the temperature of the gas phase T with the catalyst temperature T_{Cat}. This coupling requires an equation for determining the catalyst temperature, which is derived from the heat balance for the catalyst bed:

$$\varrho_{\text{Cat}} c_{p\text{Cat}} \frac{\partial T_{\text{Cat}}}{\partial t} = \lambda_{\text{eff}} \frac{\partial^2 T_{\text{Cat}}}{\partial x^2} + \frac{\lambda_{\text{eff}}}{r}\frac{\partial}{\partial r}\left(r\frac{\partial T_{\text{Cat}}}{\partial r}\right)$$

$$- \left(\frac{1}{1-\varepsilon}\right)\alpha s(T_{\text{Cat}} - T) \qquad (340\,\text{a})$$

Normalizing the equation in similar fashion to Equations (337), (338), or (339) results in

$$\varrho_{\text{Cat}} c_{p\text{Cat}}\left(\frac{T_0}{\tau}\right)\frac{\partial T_{\text{Cat}}^*}{\partial t^*} = \frac{\lambda_{\text{eff}} T_{\text{Cat}\,0}}{L^2}\frac{\partial^2 T_{\text{Cat}}^*}{\partial x^{*2}}$$

$$+ \lambda_{\text{eff}}\frac{T_{\text{Cat}\,0}}{R^2}\frac{1}{r^*}\frac{\partial}{\partial r^*}\left(r^*\frac{\partial T_{\text{Cat}}^*}{\partial r^*}\right)$$

$$- \left(\frac{1}{1-\varepsilon}\right)\alpha s\, T_{\text{Cat}\,0}(T_{\text{Cat}}^* - T^*) \qquad (340\,\text{b})$$

Estimation of the orders of magnitude of the factors preceding the differential quotients gives

$$\frac{\lambda_{\text{eff}} T_{\text{Cat}\,0}}{L^2} \ll \lambda_{\text{eff}}\frac{T_{\text{Cat}\,0}}{R^2}$$

and thus the fundamental mechanism for the transport of enthalpy in the catalyst bed in the steady state is represented by an equation which is one-dimensional in the radial direction

$$0 = \frac{\lambda_{\text{eff}}}{r}\frac{d}{dr}\left(r\frac{dT_{\text{Cat}}}{dr}\right) - \left(\frac{1}{1-\varepsilon}\right)\alpha s(T_{\text{Cat}} - T) \qquad (340\,\text{c})$$

The remaining problem is the expression of the conservation equations for the chemical species by means of similar estimates of orders of magnitudes. The conservation equation for the chemical species with the simplified continuity equation (Eq. 337d)

$$\varrho \frac{\partial m_i}{\partial t} + \varrho u \frac{\partial m_i}{\partial x} = \frac{\partial}{\partial x}\left[\varrho D_i \frac{\partial m_i}{\partial x}\right]$$

$$+ \frac{1}{r}\frac{\partial}{\partial r}\left[r\varrho D_i \frac{\partial m_i}{\partial r}\right] + \varrho S_i$$

$$i = 1,\ldots, 8 \qquad (341\,\text{a})$$

is first formulated with the variables normalized to the state at the reactor inlet.

$$\frac{\varrho_0 m_{i0}}{\tau}\varrho^* \frac{\partial m_i^*}{\partial t^*} + \frac{\varrho_0 m_{i0} u_0}{L}\varrho^* u^* \frac{\partial m_i^*}{\partial x^*}$$

$$= \frac{\varrho_0 D_{i0} m_{i0}}{L^2}\varrho^* D_i^* \frac{\partial^2 m_i^*}{\partial x^{*2}}$$

$$+ \frac{\varrho_0 D_{i0} m_{i0}}{R^2}\frac{1}{r^*}\frac{\partial}{\partial r^*}\left(r^*\varrho^* D_i^* \frac{\partial m_i^*}{\partial r^*}\right)$$

$$+ \varrho_0 \varrho^* S_i, \quad i = 1,\ldots, 8 \qquad (341\,\text{b})$$

With $D_{i0} \approx 10^{-3}\,\text{m}^2\,\text{s}^{-1}$

$$\frac{\varrho_0 D_{i0} m_{i0}}{L^2} \ll \frac{\varrho_0 D_{i0} m_{i0}}{R^2} \ll \frac{\varrho_0 m_{i0} u_0}{L}$$

and for the pilot plant conditions assumed here, obviously convective transport in the axial direction is dominant. The following one-dimensional equation results for the steady-state case

$$\varrho u \frac{dm_i}{dx} = \varrho S_i = \varrho r_i, \quad i = 1, \ldots, 8 \quad (341\,c)$$

The above estimates have resulted in a reduction of the model for the fixed bed catalyst reactor: convective transport of mass in the axial direction is the main transport mechanism for the gas phase, whilst for the catalyst phase conductive transport of heat in the radial direction along with heat transfer predominates. Consequently the couplings in Equations (131), (204), (240b), and (265b) are largely reduced.

To solve the system of Equations (339c), (340c), and (341c), Equation (340c) for the catalyst temperature is transformed into normalized variables with the transformations

$$r^* = \frac{r}{R}, \quad \theta = \frac{T_{\text{Cat}} - T}{T_{\text{W}} - T}$$

(T_{W} is the wall temperature) and

$$\beta^2 = \frac{R^2 \alpha s}{\lambda_{\text{eff}} (1 - \varepsilon)}$$

This gives

$$\frac{d^2\theta}{dr^{*2}} + \frac{1}{r^*}\frac{d\theta}{dr^*} - \beta^2 \theta = 0 \quad (342)$$

with the boundary condition $(d\theta/dr^*) = 0$ at $r^* = 0$. This is the modified Bessel differential equation with variable coefficients, the general solution of which is given by [122]:

$$\theta = C_1 I_0(\beta r^*) + C_2 K_0(\beta r^*) \quad (343)$$

where

$$I_p(\beta r^*) = \sum_{n=0}^{\infty} \left(\frac{\beta r^*}{2}\right)^{2n+p} \frac{1}{n!(n+p)!} \quad \text{and}$$

$$K_p(\beta r^*) = I_p(\beta r^*) \int \frac{d(\beta r^*)}{\beta r^* I_p^2(\beta r^*)}$$

From the above boundary condition and the properties of the Bessel functions $dI_0/d(\beta r^*) = I_{-1}(\beta r^*)$ and $dK_0/d(\beta r^*) = K_{-1}(\beta r^*)$, then C_2 is zero. With a further boundary condition $\theta = 1$ for $r^* = 1$, it follows that $C_1 = 1/I_0(\beta)$, so the general solution changes to

$$\theta = \frac{I_0(\beta r^*)}{I_0(\beta)} \quad \text{or} \quad (T_{\text{C}} - T) = \frac{I_0(\beta r^*)}{I_0(\beta)}(T_{\text{W}} - T) \quad (344)$$

Equations (339c) and (341c) can only be solved numerically. Equation (339c) has to be transformed into a difference equation in the axial direction. With the solution of Equation (344) for the catalyst temperature the cross-section-averaged heat flow to the catalyst bed can be calculated for each control volume. For this Equation (339c) with solution Equation (344) is integrated over the cross section of the reactor:

$$0 = -\int_0^1 \varepsilon 2\pi r^* \varrho c_p u \frac{dT}{dx} dr^*$$

$$+ \int_0^1 \varepsilon 2\pi r^* \frac{\alpha s}{\varepsilon} (T_{\text{C}} - T) dr^*$$

$$+ \int_0^1 2\pi r^* \sum r_i (-\Delta h_{r_i}) dr^* \quad (345)$$

The solution of the integral $\int_0^1 (T_{\text{C}} - T) r^* dr^*$ also results from the general properties of the Bessel functions

$$\int_0^1 (T_{\text{C}} - T) r^* dr^* = \frac{R^{3/2}}{\sqrt{\beta}} \frac{I_1(\beta)}{I_0(\beta)} (T_{\text{W}} - T) \quad (346)$$

This expression can be inserted into Equation (339c) which is then solved numerically and simultaneously with Equation (341c).

Numerical solutions for pilot plant conditions with integration routines from [28] are taken from [119] and shown in Figure 51. Kinetic data for reactions (i)–(vi) (p. 2-83) are given in [8], [11], [119]. A drop in temperature of ca. 75 K and a continuous increase in conversion occur over the length of the reactor. At higher conversions the reactions forming the byproducts gain greater importance.

The estimates used to obtain the model in the form of Equations (339c), (340c), and (341c) are undoubtedly drastic. They should therefore be regarded as an example of how a two-dimensional problem (a reactor model with axial and radial dispersion) can be reduced to a one-dimensional problem and of how the system of model differential equations is decoupled. Nevertheless the model successfully reproduces experimental temperature and concentration profiles measured in

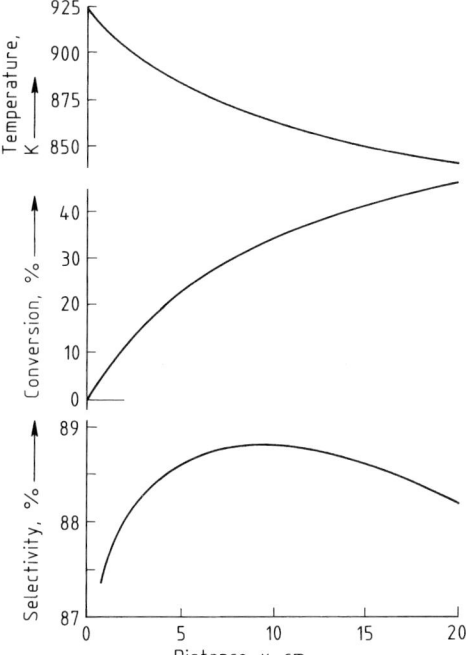

Figure 51. Calculated temperature, conversion, and selectivity profiles for the catalytic dehydration of ethylbenzene [119]

a pilot plant [11]. Another special feature of the above model is the treatment of the heterogeneous catalytic reactions. Mass transport processes in the catalyst phase were assumed to have no effect on the reaction rates. The heterogeneous system is thereby regarded as being quasi-homogeneous and all couplings between mass fractions in the gas phase and catalyst phase are neglected. This assumption is not always true. The effect of transport processes on the reaction rates and the repercussion on models for reactors with heterogeneous catalysis are discussed below.

Some Particulars of Heterogeneous Catalytic Reactions. The effect of transport processes on reaction rates in heterogeneous catalysis can be treated analogously to the acceleration of mass transport by chemical reactions described in Section 5.3.2 (Fig. 38). The core of the liquid phase is replaced by a porous catalyst sphere that is one-dimensional in spherical coordinates (Fig. 52). Furthermore if a simple reaction A → products is considered, the heterogeneous reaction rate of which is assumed to follow a formal first-order rate law $r = -k'm_A = -k'm$,

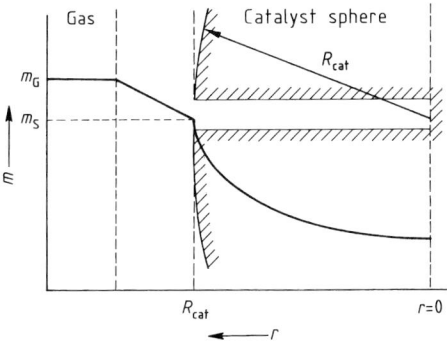

Figure 52. The effect of mass transport in heterogeneous catalytic reactions
m = mass fraction; r = radial distance; R_{cat} = radius of catalyst sphere
The subscripts G and S denote gas phase and surface of catalyst, respectively.

where k' is a rate coefficient related to the surface area of the catalyst.

The steady-state mass flux to the surface of the catalyst (see also Section 5.3.1) is given by

$$j^m = \varrho \beta s (m_G - m_S) \qquad (347)$$

This flow is compensated for by the consumption of component A by the chemical reaction. If an "effective" reaction rate $r_S = -k's m_S$ with the mass fraction at the surface of the catalyst is formulated, then the mass fraction m_S for steady-state conditions is given by

$$m_S = \frac{m_G}{\left(1 + \dfrac{k}{\beta s}\right)} \qquad (348)$$

where $k = k's$. In Equation (347) m_S can now be eliminated so that

$$j^m = \varrho \beta s m_G \left(1 - \left(1 + \frac{k}{\beta s}\right)^{-1}\right) \qquad (349)$$

Equation (349) also gives the effective reaction rate for steady-state conditions. As in Section 5.2.3, Equation (349) illustrates the acceleration of mass transport with an increasing ratio $Da_{II} = k/\beta s$, i.e., with increasing reaction rate where Da_{II} is the Damköhler number of the second kind. The degree of external utilization (i.e., effectiveness) of the catalyst η_{ext} is defined as the ratio of the surface reaction rate r_S to the reaction rate for $Da_{II} \to 0$, i.e., $m_S \to m_G$. It thus follows that (cf. Eq. 348)

$$\eta_{ext} = \frac{k m_G}{k m_G \left(1 + \frac{k}{\beta s}\right)} = \frac{1}{1 + Da_{II}} \quad (350)$$

For nonisothermal conditions the temperature dependence of the reaction rate must be considered. If Da_{II} is defined with the conditions in the gas phase and constant density is assumed, it follows from Equation (350) that

$$\eta_{ext} = \frac{k(T_S)}{k(T_G)} \left(1 + \frac{k(T_S)}{k(T_G)} Da_{II}\right)^{-1} \quad (351)$$

Given the temperature dependence of the rate coefficient in the form of an Arrhenius equation $k = k_0 e^{-E_a/RT}$, the temperature at the surface T_S must be eliminated from Equation (351) to calculate η_{ext}. The temperature at the surface T_S may be calculated in a similar way as m_S. In the steady-state case

$$j^h = \alpha s (T_S - T_G) \quad (352)$$

The heat transferred to the surroundings is released by chemical reactions occurring at the catalyst. Thus from the steady-state condition $j^h = r_S(-\Delta h_r)$

$$\frac{T_G}{T_S} = 1 + \left[\frac{\varrho(-\Delta h_r) m_G \, \eta_{ext} k(T_G)}{\frac{\lambda}{D_{eff}} Le \, T_G} \cdot \frac{1}{\beta s}\right]^{-1} \quad (353)$$

To derive Equation (353) full analogy for heat and mass transfer, i.e., $Pr = Le \cdot Sc$, is assumed. Using the abbreviations $\gamma = E_a/RT_G$ (Arrhenius number), and

$$\beta_H = \frac{\varrho(-\Delta h_r) m_G}{\frac{\lambda}{D_{eff}} Le \, T_G} \quad \text{(Prater number)}$$

Equations (351) and (353) are combined to give

$$\frac{\eta_{ext}}{\exp[-\gamma(1 + \beta_H \eta_{ext} Da_{II})^{-1}]} = \{1 + Da_{II} \exp[-\gamma(1 + \beta_H \eta_{ext} Da_{II})^{-1}]\}^{-1} \quad (354)$$

η_{ext} can also take values greater than 1 in the nonisothermal case if $\beta_H > 0$. In this case the reaction at the surface of the catalyst occurs at a higher temperature than the gas-phase temperature T_G. Since all quantities in Equation (354) are related to T_G, then values of $\eta_{ext} > 1$ are also consistent with the definition of η_{ext}.

Conditions inside the catalyst can be illustrated in a similar fashion. For spherical geometry (Fig. 52) diffusive mass flux inside the catalyst is given by

$$j^m = -\varrho D_{eff} 4\pi r^2 \frac{dm}{dr} \quad (355)$$

Changes in the diffusive mass flux are caused by chemical reactions at the internal surface of the catalyst, so that under steady-state conditions the mass balance is

$$\frac{dj^m}{dr} = \varrho k' 4\pi r^2 m \quad (356)$$

Using Equation (355)

$$\frac{d^2 \varphi}{d\eta^2} + \frac{2}{\eta} \frac{d\varphi}{d\eta} - \Phi_K^2 \varphi = 0 \quad (357)$$

is obtained where the normalized quantities $\varphi = m/m_S$ and $\eta = r/R_{Cat}$ and the abbreviation $\Phi_K = R_{Cat} \sqrt{k's/D_{eff}}$ are used (R_{Cat} is the radius of the catalyst sphere). The solution of Equation (357) with the boundary conditions $\varphi = 1$ for $\eta = 1$ and $d\varphi/d\eta = 0$ for $\eta = 0$ is

$$\varphi = \frac{\sinh(\Phi_K \eta)}{\eta \sinh(\Phi_K)} \quad (358)$$

Completely analogous to the external mass transfer, a degree of internal utilization of the catalyst η_{int} (internal effectiveness factor) can be defined as the ratio of the mean reaction rate in the catalyst,

$$r_{eff} = \int_0^{R_{Cat}} k' s m 4\pi r^2 \, dr$$

and reaction rate for $m \to m_S$. Integration of Equation (358) gives

$$\eta_{int} = \frac{3}{\Phi_K} \left(\tanh \Phi_K - \frac{1}{\Phi_K}\right)^{-1} \quad (359)$$

For $\Phi_K > 3$, $\eta_{int} \approx 3/\Phi_K$, for $\Phi_K < 0.3$, $\eta_{int} \approx 1$. The result is similar to the acceleration of mass transfer by chemical reactions, cf. Section 3.3.2. The calculation of η_{int} for other catalyst geometries including more realistic pore structures and other classes of reaction are described in [24], [123].

For nonisothermal conditions the temperature dependence of the reaction rate must once again be taken into consideration. If the simple Arrhenius expression is substituted for k' in Equation (356), then assuming constant density and neglecting the temperature dependence of the effective diffusion coefficient, the result is

$$\frac{d^2m}{dr^2} + \frac{2}{r}\frac{dm}{dr} = \frac{k'_0}{D_{eff}} s \exp\left(-\frac{E_a}{RT}\right) m \quad (360)$$

Inside the catalyst heat is transported by conduction. The conductive heat flow is given by

$$j^h = -\lambda 4\pi r^2 \frac{dT}{dr} \quad (361)$$

Changes in heat flow are caused by heat release due to chemical reactions. Under steady-state conditions

$$\frac{dj^h}{dr} = \varrho k' s 4\pi r^2 m(-\Delta h_r) \quad (362)$$

From Equations (361) and (362) a differential equation for the temperature is obtained

$$\frac{d^2T}{dr^2} + \frac{2}{r}\frac{dT}{dr} = \frac{\varrho(-\Delta h_r) k'_0 s \exp\left(-\frac{E_a}{RT}\right)}{D_{eff}\frac{\lambda_{eff}}{D_{eff}}} m \quad (363)$$

Subtracting Equation (363) from Equation (360) gives

$$\frac{d^2m}{dr^2} + \frac{2}{r}\frac{dm}{dr} - \frac{\frac{\lambda_{eff}}{D_{eff}}}{\varrho(-\Delta h_r)}\left[\frac{d^2T}{dr^2} + \frac{2}{r}\frac{dT}{dr}\right] = 0 \quad (364)$$

which, after integrating twice with the boundary conditions $m = m_S$ and $T = T_S$ for $r = R_{Cat}$, and $dm/dr = 0$, and $dT/dr = 0$ for $r = 0$, leads to the result

$$T - T_S = \frac{\varrho D_{eff}(-\Delta h_r)}{\lambda_{eff}}(m_S - m) \quad (365)$$

With this the temperature can now be eliminated from Equation (360). If the Prater number, Arrhenius number, and Thiele modulus Φ_K are defined with the conditions at the outer surface of the catalyst, i.e.,

$$\beta_H = \frac{\varrho(-\Delta h_r) m_S}{\frac{\lambda}{D_{eff}} T_S}, \quad \gamma = \frac{E_a}{RT_S}, \quad \text{and}$$

$$\Phi_K^2 = R_{Cat}^2 \frac{k'_0 s}{D_{eff}} \exp\left(-\frac{E_a}{RT_S}\right)$$

Equation (364) becomes

$$\frac{d^2\varphi}{d\eta^2} + \frac{2}{\eta}\frac{d\varphi}{d\eta}$$

$$- \Phi_K^2 \varphi \exp\left[\gamma \beta_H \frac{(1-\varphi)}{1+\beta_H(1-\varphi)}\right] = 0 \quad (366)$$

Equation (366) cannot be solved analytically, numerical solutions are given in [124]. If the internal effectiveness factor η_{int} is calculated from the numerical solutions [124] the results shown in Figure 53 are obtained. For $\gamma = 20$, multiple solutions exist for η_{int} at high values of β_H and low values of the Thiele modulus. The middle solutions are unstable so that ignition and extinction of the exothermic chemical reaction in the catalyst can occur for certain values of Thiele moduli. As all parameters are related to the state

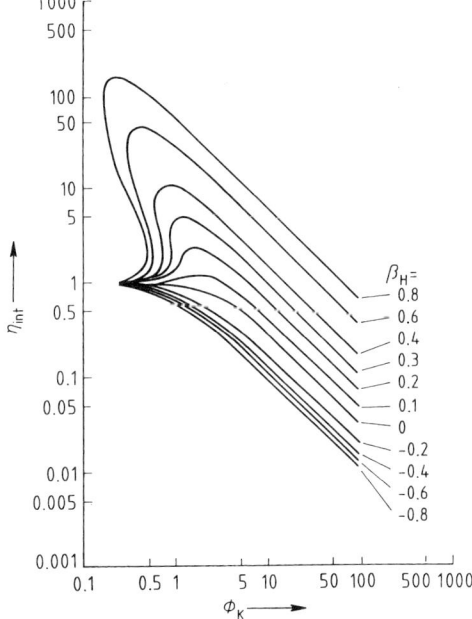

Figure 53. Effectiveness factor of a catalyst as a function of the Thiele modulus Φ_K and the Prater number β_H for spherical catalyst geometry (Arrhenius number $\gamma = 20$) [124] γ = Arrhenius number; $\eta_{int.}$ = internal effectiveness factor.

at the outer surface of the catalyst, values of $\eta_{int} > 1$ are consistent with the definition of the internal effectiveness factor. Due to the exothermic reaction the temperature inside the catalyst can be higher than temperature on the outer surface.

By introducing an effective reaction rate $r_{eff} = \eta_{eff} r_G$, the rate of heterogeneous catalytic reaction is related to the conditions in the gas phase and is thus no longer coupled with the internal transport in the catalyst. This permits the description of reactors with heterogeneous catalysis as quasi-single-phase systems. The effectiveness factors η_{int} or η_{ext} can be calculated from the variables of state at the surface of the catalyst or in the gas phase, respectively. For the isothermal case and spherical geometry

$$\eta_{eff} = \left(\frac{1}{\eta_{int}} + \frac{\Phi_K^2}{3\,Bi} \right)^{-1}$$

where the Biot number $Bi = \beta R_{Cat}/D_{eff}$.

Nonisothermal conditions and other geometries are discussed in [24], [123], [124].

The discussion in this section was based on first-order reactions with no changes in the mole number. Very few reactions in heterogeneous catalysis can be described with these expressions, higher order reactions are much more frequent. Furthermore the catalysts may be "poisoned" by adsorption of the reactants or products (see p. 2-83) yielding nonintegral orders of reaction. Inhibition by mass transport influences the order of the catalytic reaction. In the extreme case $r_{eff} = \beta s (m_G - m_S)$. This rate expression is of the first order with respect to the mass fraction of A. Inhibition by mass transport shifts the order of reaction in the direction of first order. Selectivity in parallel or secondary reactions will be similarly affected.

The temperature dependence of the reaction rate is also influenced by inhibition by mass transport. If the expression $r_{eff} = -\eta_{eff} k m_G = -k^* m_G$ is used for a first-order reaction, and the temperature dependence is given as usual by $k^* = k_0^* e^{-E_a^*/RT}$, then

$$\ln k^* = \ln k_0^* - \frac{E_a^*}{RT} \qquad (367)$$

For $\eta_{eff} = 1$ there is chemical control, $k_0^* = k_0$ and $E_a^* = E_a$. If diffusion in the catalyst pores is rate controlling, $\eta \approx 3/\Phi_K$. Then $\ln k^* = \ln k_0 + \ln 3 - \ln \Phi_K - E_a/RT$. Substitution of $\Phi_K = R_{Cat}\sqrt{k/D_{eff}}$ gives

$$\ln k^* = \frac{1}{2} \ln k_0 + \ln 3 + \frac{1}{2} \ln D_{eff} - \frac{1}{2}\frac{E_a}{RT}$$

thus $k_0^* = 3\sqrt{k_0 D_{eff}}$ and $E_a^* = \frac{1}{2} E_a$. If the temperature dependence of the effective diffusion coefficients is neglected, the activation energy for pore diffusion-controlled reaction is about half that for chemical control. If control is exclusively due to mass transfer, then $r_{eff} \approx -\beta s m_G$ and $k^* = \beta s \sim D_{eff}$. Taking into account the temperature dependence of D_{eff}, then $\ln k^* = \ln k_0^* + \frac{3}{2} \ln T$ which leads to an activation energy E_a^* of ca. 5 kJ/mol for this case. Thus the experimental determination of the effective formal activation energy for a heterogeneous catalytic reaction clarifies the extent of control by different transport processes. For further discussion of these relationships see [24], [123], [124].

In addition to the formulation of conservation equations for reactor models for heterogeneous catalytic reactions, combinations of simple reactor elements are frequently used for modeling these systems. For further information, see [12], [125]. Other complicated reactors (e.g., fluidized bed reactors) for heterogeneous catalytic reactions can be similarly dealt with [24], [123].

5.3.3.2.2. Stability Analysis of Nonisothermal Reactors

Steady-State Cases. The stability analysis of isothermal reactors discussed in Section 5.3.3.1 demonstrated the close connection of the phenomenon of multiple steady states in steady-state systems and the instability of nonsteady-state systems to a feedback mechanism. For the isothermal cases a chemical feedback mechanism was discussed in the form of autocatalysis and self-poisoning. Another feedback mechanism is the heat release from a chemical reaction. This will be discussed for a simple first-order reaction, $A \rightarrow B +$ heat, the temperature dependence of the rate coefficient is of the Arrhenius form, $k = k_0 e^{-E_a/RT}$. With increasing conversion of A heat is released. The released heat increases the temperature of the reaction mixture and the reaction rate. This in turn increases the rate of heat release. The reaction rate is an exponential function of the temperature. Therefore the feedback mechanism is nonlinear in temperature.

Stability analysis for nonisothermal steady-state reactors will again be carried out for a well-stirred reactor (one element of a reactor cascade of ideally mixed reactors which is used to represent a flow reactor with convectively dominated flow). In analogy with Equation (310), the mass balance for A gives

$$\frac{\varrho}{\tau}(m_0 - m) - \varrho m k_0 e^{-E_a/RT} = 0 \tag{368}$$

From the enthalpy balance and transformation of Equation (240b) into an equation for temperature, the following equation is obtained:

$$\frac{c_p \varrho}{\tau}(T_0 - T) + (-\Delta h_r)\varrho m k_0 e^{-E_a/RT}$$
$$- \alpha s(T - T_A) = 0 \tag{369}$$

where the specific surface area s is the effective surface for heat transfer to the surroundings divided by the total volume of the ideally mixed reactor. The source term in Equation (240b) is thus represented by a heat-transfer term. As in Section 5.3.3.1, Equations (368) and (369) are transformed into normalized variables: $\alpha = m_A/m_{A_0}$, $\theta = (T - T_0) E_a/RT_0^2$, $\theta_C = (T_A - T_0) E_a/RT_0^2$, and $Da_I = \tau k_0$. The Damköhler number (the ratio of the mean residence time to a characteristic chemical time, the time required until $m/m_0 = e^{-1}$) is defined with the inlet conditions. If the adiabatic temperature difference $\Delta T_{ad} = (-\Delta h_r)m_0/c_p$ is also defined with E_a/RT_0^2, then $\theta_{ad} = (-\Delta h_r)m_0 E_a/c_p RT_0^2$. Finally a "cooling time"

$$\tau_N = \frac{c_p \varrho}{\alpha s t_{chem}} = \frac{c_p \varrho k_0}{\alpha s}$$

can be defined from the heat transfer. A reciprocal Arrhenius number $\gamma_r = RT_A/E_a$ is also introduced. With these abbreviations Equations (368) and (369) become

$$\frac{1-\alpha}{Da_I} - \alpha \exp\left(\frac{\theta}{1+\gamma_r \theta}\right) = 0 \tag{370a}$$

$$\theta_{ad} \alpha \exp\left(\frac{\theta}{1+\gamma_r \theta}\right) - \left(\frac{1}{Da_I} + \frac{1}{\tau_N}\right)\theta + \frac{\theta_C}{\tau_N} = 0 \tag{370b}$$

Equations (370a, b) contain the variables α and θ as a function of six parameters. To reduce the number of parameters, the case of very high activation energy, $\gamma_r \ll 1$, $T_A = T_0$, and $\theta_C = 0$ is considered. With these assumptions the following simplified equations are obtained

$$\frac{1-\alpha}{Da_I} - \alpha e^\theta = 0 \tag{371a}$$

$$\theta_{ad} \alpha e^\theta - \left(\frac{1}{Da_I} + \frac{1}{\tau_N}\right)\theta = 0 \tag{371b}$$

In order to remain as close as possible to the notation used in Section 5.3.3.2, θ in Equation (371a) is replaced by Equation (371b) so that

$$\frac{1-\alpha}{Da_I} = \alpha \exp\left[\frac{(1-\alpha)\theta_{ad}}{1+\frac{Da_I}{\tau_N}}\right] \tag{372a}$$

The temperature increase can easily be attributed to α from Equations (371a, b):

$$\theta = \left[\frac{(1-\alpha)\theta_{ad}}{1+\frac{Da_I}{\tau_N}}\right] \tag{372b}$$

Equation (372a) can be solved graphically by plotting both sides against the progress of the reaction $1 - \alpha$ (conversion). In order to further reduce the number of parameters, ideal adiabatic conditions are assumed, i.e., $\tau_N \to \infty$. Thus the left-hand side (LHS) of Equation (372a) is only dependent on Da_I, whilst the right-hand side (RHS) is only dependent on the adiabatic temperature increase. The solutions for the case discussed here (Fig. 54) are qualitatively similar to the solutions for the isothermal case with self poisoning (see p. 2-79). Here too for a given residence time (Damköhler number) multiple steady states occur for particular areas of the adiabatic temperature increase θ_{ad}. Using the same argumentation as in Section 5.3.3.2, the stability condition Equation (318a)

$$\frac{d}{d(1-\alpha)}(\text{LHS}) > \frac{d}{d(1-\alpha)}(\text{RHS})$$

at LHS = RHS (318a)

is not complied with for the middle one of the multiple solutions. In this state a small perturbation of the conversion causes an increase or decrease in the "convective term" (LHS). The "re-

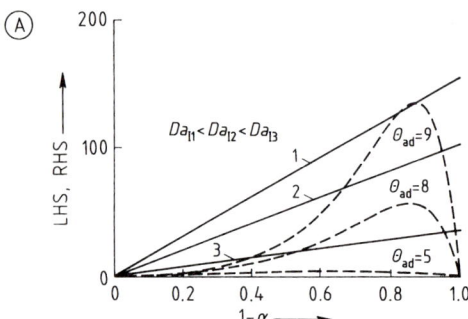

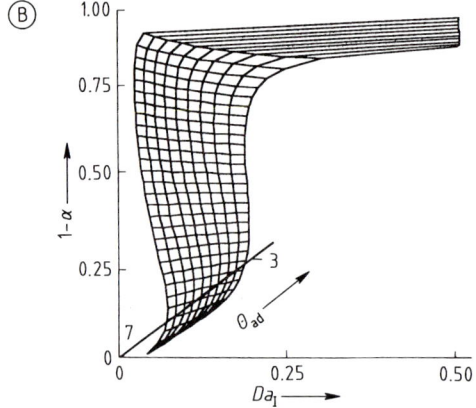

Figure 54. Multiple steady-state conversion points for "normal" exothermic chemical reactions (A) and bifurcation diagram (B) for nonisothermal conditions in a well-stirred reactor

θ_{ad} = adiabatic temperature increase; for explanation of other symbols see Figure 15.

action term" (RHS) however reacts in the same way so that a greater (smaller) availability of A due to convection is followed by a greater (smaller) consumption due to reaction. The system therefore tends towards one of the external stable steady-state solutions. As can be seen from Figure 54 A, the shape of the curve depends on the adiabatic temperature increase θ_{ad}. Figure 54 B shows that as θ_{ad} becomes smaller the S-shaped profile vanishes and the surface of solutions unfolds. Figure 54 B also demonstrates that the unstable branch of the solutions is not passed. The system jumps from low conversion to high conversion with increasing residence time Da_1. On igniting and extinguishing the reaction, hysteresis occurs and the jumping points can be calculated from Equation (318 b)

$$\frac{d}{d(1-\alpha)}(\text{LHS}) = \frac{d}{d(1-\alpha)}(\text{RHS})$$

at LHS = RHS (318 b)

For the example discussed here the jumping points are given by

$$(1-\alpha) = \frac{1}{2}\left[1 \pm \sqrt{\left(1 - \frac{4}{\theta_{ad}}\right)}\right] \quad (373)$$

Since θ_{ad} is always positive for exothermic reactions, then for the point at which the discriminant in Equation (373) disappears, $\theta_{ad} = 4$. If $\theta_{ad} > 4$, multiple steady states are observed.

If the simplification $\gamma_r \to 0$ is discarded, the following equation is obtained for the steady-state solutions

$$\frac{1-\alpha}{Da_1} = \alpha \exp\left[\frac{(1-\alpha)\theta_{ad}}{1 + \gamma_r \theta_{ad}(1-\alpha)}\right] \quad (374\text{a})$$

The right-hand side of Equation (374 a) now depends on two parameters. In general it can be written in the form

$$F(x, y; p, q, r, s, \ldots) = 0 \quad (374\text{b})$$

where x is the variable for describing the steady state, y the bifurcation parameter which controls the multiple steady state, and $p, q, r, s, \ldots$ are additional parameters which produce folding or unfolding of the surface of x in space y and $p, q, r, s, \ldots$

One or more steady-state solutions exist for the condition given by Equation (374 b). The steady-state solution is stable for

$$F(x, y; p, q, r, s, \ldots) = 0 \text{ and } F_x > 0 \quad (374\text{c})$$

For the unfolding of the hysteresis loop the following applies

$$F(x, y; p, q, r, s, \ldots) = 0, \quad F_x > 0,$$
$$\text{and} \quad F_{xx} = 0 \quad (374\text{d})$$

and "islands" finally occur in the bifurcation diagram which grow into "mushrooms" when

$$F(x, y; p, q, r, s, \ldots) = 0, \quad F_x = 0, \quad F_y = 0,$$
$$F_{xy} \neq 0, \quad F_{xx} \neq 0, \quad \text{and} \quad F_{yy} \neq 0 \quad (374\text{e})$$

Depending on the process under consideration these conditions apply for various parameter combinations. For Equation (374 a) the condition for unfolding occurs when $\theta_{ad}(4\gamma_r - 1) + 4 = 0$. In the system of Equations (374), F_x, F_y, etc. denote the derivatives according to the variable in the index.

Nonsteady-State Cases. As in the stability analysis of isothermal reactors, the stability of nonsteady-state nonisothermal reactors is investigated for a consecutive reaction $P \to A \to B$ where A is produced from a large reservoir of P. The second reaction $A \to B$ is exothermic and its rate coefficient k_1 is temperature dependent. For the reaction sequence $P \to A \to B$ where $k_1 = k_{01} e^{-E_a/RT}$, the mass and enthalpy balances for a batchwise operating ideally mixed reactor are given by

$$\frac{dm_P}{dt} = -k_0 m_P \tag{375}$$

$$\frac{dm_A}{dt} = k_0 m_P - k_1 m_A \tag{376a}$$

$$\varrho c_p \frac{dT}{dt} = (-\Delta h_r) k_1 m_A - \alpha s (T - T_A) \tag{376b}$$

(see p. 2-79). The feedback in this system results from the temperature dependence of the second reaction rate which increases the reaction rate with increasing conversion. As previously, normalized variables are used: $\alpha = m_A/m_{ref}$, $\pi = m_P/m_{ref}$, $\theta = (T - T_A) E_a/RT_A^2$ and $\gamma_r = RT_A/E_a$. For the temperature-dependent reaction rate coefficient it thus follows that

$$k_1 = k_{01} e^{-E_a/RT_A} \exp[\theta/(1 + \gamma_r \theta)]$$

The cooling time is not related to a characteristic chemical time $\tau_N = \varrho c_p/\alpha s$, so that a normalized time can be given as $\tau = t/\tau_N$. Thus the rate coefficients k_0 and k_1 are normalized so that $\varepsilon = k_0 \tau_N$ and $\kappa = \tau_N k_{01} e^{-E_a/RT_A}$. Finally m_{ref} is established from

$$m_{ref} = \frac{\alpha s RT_A^2}{\varrho E_a (-\Delta h_r) k_{01} e^{-E_a/RT_A}}$$

There is no feedback to reservoir P thus Equation (375) can be integrated separately:

$$m_P = m_{0P} e^{-k_0 t} \quad \text{or} \quad \pi = \pi_0 e^{-\varepsilon \tau} \tag{377}$$

Using this Equations (376a) and (377b) give

$$\frac{d\alpha}{d\tau} = \varepsilon \pi_0 e^{-\varepsilon \tau} - \kappa \alpha \exp\left[\frac{\theta}{(1 + \gamma_r \theta)}\right]$$

$$= \mu - \kappa \alpha \exp\left[\frac{\theta}{(1 + \gamma_r \theta)}\right] \tag{378a}$$

$$\frac{d\theta}{d\tau} = \alpha \exp\left[\frac{\theta}{(1 + \gamma_r \theta)}\right] - \theta \tag{378b}$$

where $\mu = \varepsilon \pi_0 e^{-\varepsilon \tau} = \mu_0 e^{-\varepsilon \tau}$. For the sake of simplicity the case of high activation energy (i.e., $\gamma_r \ll 1$) will be investigated. Equations (378a, b) are thus again simplified into

$$\frac{d\alpha}{d\tau} = \mu - \kappa \alpha e^{\theta} \tag{379a}$$

$$\frac{d\theta}{d\tau} = \alpha e^{\theta} - \theta \tag{379b}$$

For very small values of ε (i.e., for a very much quicker time scale for the changes in α and θ than for the consumption of P) the principle of quasi-steady-state can again be used, and hence Equations (379a, b) do not have to be integrated. The quasi-steady-state solutions for α and θ result from $d\alpha/d\tau = 0$ and $d\theta/d\tau = 0$, hence

$$\mu - \kappa \alpha e^{\theta} = 0 \tag{380a}$$

$$\alpha e^{\theta} - \theta = 0 \tag{380b}$$

giving

$$\theta_{qs} = \frac{\mu}{\kappa} \tag{381a}$$

$$\alpha_{qs} = \frac{\mu}{\kappa} e^{-\mu/\kappa} \tag{381b}$$

The approximate solutions α_{qs} and θ_{qs} are a function of the ratio μ/κ, and also of time since μ is a function of time. With these operations and simplifications the quasi-steady-state solutions α_{qs} and θ_{qs} in Equation (381) have the same form as for the isothermal case (Eq. 324). The stability analysis which now follows is carried out in the same way as for the isothermal case (Section 5.3.3.1).

The development of Equation (380a) into a Taylor series around the quasi-steady-state solutions provides a system of coupled differential equations for the disturbances $\Delta \alpha$ and $\Delta \theta$, the general solution of which is given by

$$\Delta \alpha (\tau) = C_1 e^{\lambda_1 \tau} + C_2 e^{\lambda_2 \tau} \tag{382a}$$

$$\Delta \theta (\tau) = C_3 e^{\lambda_1 \tau} + C_4 e^{\lambda_2 \tau} \tag{382b}$$

The eigenvalues λ_1, λ_2 once again determine the characteristics of the time evolution of the disturbances. They are given by the solutions to the characteristic Equation (329)

$$\lambda^2 - \text{Tr}(J_{qs}) \lambda + \text{Det}(J_{qs}) = 0 \tag{329}$$

i.e.,

$$\lambda_{1,2} = \tfrac{1}{2}[\mathrm{Tr}(J_{qs}) \pm \{\mathrm{Tr}(J_{qs})^2 - 4\mathrm{Det}(J_{qs})\}^{1/2}] \tag{330}$$

In this case the Jacobi matrix takes the form

$$J_{qs} = \begin{pmatrix} \dfrac{\partial f(\alpha,\theta)}{\partial \alpha} & \dfrac{\partial f(\alpha,\theta)}{\partial \theta} \\ \dfrac{\partial g(\alpha,\theta)}{\partial \alpha} & \dfrac{\partial g(\alpha,\theta)}{\partial \theta} \end{pmatrix}_{qs} \tag{383}$$

The three quantities which decide the nature of the solutions are the discriminant of the characteristic equation $\mathrm{Tr}(J_{qs})^2 - 4\mathrm{Det}(J_{qs})$, the trace of the Jacobi matrix

$$\mathrm{Tr}(J_{qs}) = \left(\frac{\mu}{\kappa}\right) - 1 - \kappa e^{\mu/\kappa} \tag{384a}$$

and also their determinant

$$\mathrm{Det}(J_{qs}) = \kappa e^{\mu/\kappa} \tag{384b}$$

As μ/κ is always positive, not all the combinations of λ_1, λ_2 given in the stability diagram (Fig. 48) will be passed through with increasing conversion of P.

Two states will now be discussed that are of practical importance. The first case is the transition of the solutions from a node of the system into a focus. The condition for this is that λ_1, λ_2 become conjugate complex, the discriminant thus passes through zero. If μ in Equations (384a, b) is replaced by Equation (381a), the result is the following equation for the discriminant

$$\kappa^2 e^{2\theta_{qs}} - 2\kappa(1+\theta_{qs})e^{\theta_{qs}} + (\theta_{qs}-1)^2 = 0 \tag{385}$$

For each positive value of θ_{qs}, there are thus two values of κ

$$\kappa_{1,2} = (\sqrt{\theta_{qs}} \pm 1)^2 e^{-\theta_{qs}} \tag{386a}$$

which satisfy the above condition for the discriminant to be zero. The appropriate values for μ are

$$\mu_{1,2} = \kappa_{1,2}\theta_{qs} \tag{386b}$$

The stability diagram is best displayed in the form given in Figure 55 with the two parameters μ and κ. Two closed loops are obtained, the outer of which corresponds to the larger root from Equation (386a) and the inner to the smaller.

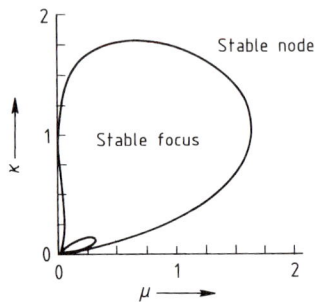

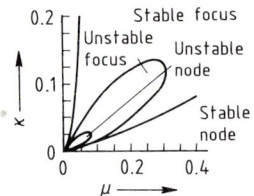

Figure 55. Stability diagram for exothermic consecutive reactions for quasi-steady-state nonisothermal conditions [113]
For explanation of symbols see Figure 49 and text.

The area outside the outer curve represents stable nodes, stable or unstable foci lie between the two curves, and the area within the inner curve gives unstable nodes.

The second case is the loss of stability of the system given by the condition $\mathrm{Tr}(J_{qs}) = 0$. This is also a condition for Hopf bifurcation. If μ in Equation (384a) is replaced by Equation (381a), this condition is given as

$$\kappa = (\theta_{qs} - 1)e^{-\theta_{qs}} \quad \text{and} \quad \mu = \kappa\theta_{qs} \tag{387a, b}$$

The pertinent curve in the stability diagram is a closed loop which lies between the two loops for the transition from nodes to foci. The region within this curve denotes unstable foci, see the enlarged section of Figure 55 in the area of the origin. The maximum of this curve is at e^{-2}.

Stability analysis for nonisothermal nonsteady-state batchwise operating models exhibits the following important results: for each combination of the experimental conditions μ and κ there is an unambiguous solution for the quasi-steady-state concentration of the intermediate α_{qs} and the normalized temperature increase θ_{qs}. If the normalized reaction rate coefficient is $\kappa > e^{-2}$, then the quasi-steady-state solutions are always stable. For $\kappa \approx 1.78$ transition from stable nodes to stable foci occurs. For $\kappa < e^{-2}$

the quasi-steady-state solutions are unstable. At the location of the Hopf bifurcations two solutions occur (Fig. 55). Between the two bifurcation points for α_{qs} and θ_{qs} as functions of the conversion (i.e., as functions of μ), oscillations of varying frequency and amplitude occur.

For engineering purposes it is important that, for the simplified case discussed here, stability can be controlled through the parameter κ. For $\kappa > e^{-2}$ the system remains stable. If the properties of the chemical reaction are given by $k_1 = k_{01} e^{-E_a/RT}$, stability can only be achieved by altering the cooling time by means of increased heat transfer to the surroundings. Detailed discussions on the stability of nonisothermal reactors can be found in [113], [114].

5.3.3.2.3. Use on Statistical Processes

The description of statistical processes by averaging the conservation equations of momentum is discussed in Section 5.1.2 for isothermal inert systems. The averaged equations however contain terms with higher moments which must be modeled; in Equation (216) these are the Reynolds stresses $\langle \varrho \rangle \langle U_i^f U_j^f \rangle$.

A model for the Reynolds stresses which shows them in analogy to the laminar viscous stresses as an isotropic tensor of turbulent viscous stresses

$$\tau_{t_{ij}} = \mu_t \left(\frac{\partial \langle U_i \rangle}{\partial x_j} + \frac{\partial \langle U_j \rangle}{\partial x_i} \right) - \frac{2}{3} \mu_t \frac{\partial \langle U_k \rangle}{\partial x_k} \delta_{ij}$$

is given in Section 5.1.2.1. The averaged equations then take the form

$$\frac{\partial}{\partial t}(\langle \varrho \rangle \langle U_i \rangle) + \frac{\partial}{\partial x_k}(\langle \varrho \rangle \langle U_k \rangle \langle U_i \rangle)$$

$$= \frac{\partial}{\partial x_k}(\langle \tau_{t_{ik}} \rangle) - \frac{\partial \langle p \rangle}{\partial x_i} + \langle \varrho \rangle \langle g_i \rangle, \quad i = 1, 2, 3$$

(388)

which is a transport equation for the first moment of the velocities. The accompanying problem of describing the turbulent viscosity was solved using the eddy viscosity hypothesis, $\mu_t = C_\mu \langle \varrho \rangle (\langle k \rangle^2 / \langle \varepsilon \rangle)$.

As well as the solution of Equation (388), the model requires the solution of the equations for the turbulence energy, $\langle k \rangle$ (Eq. 222), and its dissipation rate, $\langle \varepsilon \rangle$ (Eq. 225), thereby defining a turbulent time scale according to $\tau = \langle k \rangle / \langle \varepsilon \rangle$.

A numerical solution that takes into account the special nature of the averaged conservation equations of momentum is demonstrated in Section 5.1.2.2.

For nonisothermal reacting systems this concept has to be expanded to the equations for the conservation of enthalpy (Eq. 134 or 240a) and for the conservation of mass of the chemical species (Eq. 133 or 265a). Averaging Equations (240a) and (265a) using the same assumptions as in Section 5.1.2.1 results in the equations

$$\frac{\partial}{\partial t}(\langle \varrho \rangle \langle h \rangle) + \frac{\partial}{\partial x_k}(\langle \varrho \rangle \langle U_k \rangle \langle h \rangle)$$

$$= -\frac{\partial}{\partial x_k}(\langle J^h \rangle + \langle \varrho \rangle \langle h^f U_k^f \rangle) + \langle \varrho S_h \rangle \quad (389)$$

and

$$\frac{\partial}{\partial t}(\langle \varrho \rangle \langle m_i \rangle) + \frac{\partial}{\partial x_k}(\langle \varrho \rangle \langle U_k \rangle \langle m_i \rangle)$$

$$= -\frac{\partial}{\partial x_k}(\langle J^{m_i} \rangle + \langle \varrho \rangle \langle m_i^f U_k^f \rangle) + \langle \varrho S_i \rangle$$

$$i = 1, \ldots, N-1 \quad (390)$$

The problem now lies in modeling the Reynolds fluxes $\langle \varrho \rangle \langle h^f U_j^f \rangle$ and $\langle \varrho \rangle \langle m_i^f U_j^f \rangle$, and also the expected values of the source term $\langle \varrho S_i \rangle$ and $\langle \varrho S_h \rangle$.

Modeling the Reynolds Fluxes. The problem of modeling the Reynolds fluxes $\langle \varrho \rangle \langle h^f U_j^f \rangle$ and $\langle \varrho \rangle \langle m_i^f U_j^f \rangle$ can be solved analogously to the modeling of the Reynolds stresses. As with the transport equations for the Reynolds stresses (see Section 5.1.2.1, Eq. 221), the analogous transport equation for the Reynolds fluxes is not closed however. The same reasons as were discussed in Section 5.1.2.1 for the Reynolds stresses, lie behind the fact that transport equations for Reynolds fluxes are usually not solved in engineering but are modeled directly.

The direct modeling of Reynolds fluxes is based on the Boussinesq hypothesis and assumes that there is a similarity between the turbulent transport of momentum, and enthalpy, and mass. This principle has already been used in Sections 5.2.2.1 and 5.3.1.1 for deterministic systems. The turbulent flux of a scalar quantity is thus related to the gradients of the expected values of the scalar. The proportionality number is a turbulent transport coefficient which is obtained from the turbulent viscosity:

$$-\langle\varrho\rangle\langle\Phi^{\rm f} U_k^{\rm f}\rangle = \frac{\mu_t}{\sigma_\varphi}\frac{\partial\langle\Phi\rangle}{\partial x_k} \tag{391}$$

This expression has already been used in the equation for turbulence energy (Eq. 223); σ_φ is a turbulent Prandtl or Schmidt number, which relates the turbulent transfer of the scalar in question and the turbulent transfer of momentum. With the help of the eddy viscosity hypothesis, σ_φ can be interpreted as the ratio of the mixing length of the velocity fluctuations to the mixing length of the fluctuations in the scalar.

With this hypothesis and neglecting the proportion of laminar flux to turbulent flux, Equations (389) and (390) become

$$\frac{\partial}{\partial t}(\langle\varrho\rangle\langle h\rangle) + \frac{\partial}{\partial x_k}(\langle\varrho\rangle\langle U_k\rangle\langle h\rangle)$$

$$= \frac{\partial}{\partial x_k}\frac{\mu_t}{\sigma_h}\frac{\partial\langle h\rangle}{\partial x_k} + \langle\varrho S_h\rangle \tag{392}$$

and

$$\frac{\partial}{\partial t}(\langle\varrho\rangle\langle m_i\rangle) + \frac{\partial}{\partial x_k}(\langle\varrho\rangle\langle U_k\rangle\langle m_i\rangle)$$

$$= \frac{\partial}{\partial x_k}\frac{\mu_t}{\sigma_m}\frac{\partial\langle m_i\rangle}{\partial x_k} + \langle\varrho S_i\rangle$$

$$i = 1,\ldots, N-1 \tag{393}$$

The remaining problem is now the modeling of the terms $\langle\varrho S_h\rangle$ and $\langle\varrho S_i\rangle$. If the change in enthalpy due to mechanical work and radiation are neglected in Equation (392), then the term $\langle\varrho S_h\rangle$ vanishes. A modeling expression then only needs to be found for $\langle\varrho S_i\rangle$.

Modeling of the Mean Reaction Rates. The discussion of models for the mean reaction rates in turbulent (statistical) flows is of similar importance in engineering to the modeling of Reynolds stresses and Reynolds fluxes. Reviews are given in [126]–[128].

The reaction rates are generally a function of mass fractions and temperature (Eq. 137a). The expected values $\langle\varrho S_i\rangle$ are obtained from Equation (109) since the averaging rule (Eq. 107) is applicable to functions. If changes in density are neglected, then Equation (137a) gives

$$\langle S_i\rangle = \int_{T_u}^{T_o 1}\int_0^1\ldots\int_0^1 P(m_1,\ldots, m_o, T)$$
$$\cdot S_i(m_1,\ldots, m_o, T)\,dm_1\ldots dm_o\,dT \tag{394}$$

In Equation (394) $P(m_1,\ldots, m_o, T)$ is a joint probability density function of temperature and the mass fractions of the chemical species that are contained in the expression for the reaction rates. $S_i(m_1,\ldots, m_o, T)$ are usually known functions and hence to determine $\langle S_i\rangle$, the joint probability density functions have to be established. The example shown in Figure 14 demonstrates that a priori determinations for this are difficult; there are intermittencies in the region at the jet boundary so that $P(m_1,\ldots, m_o, T)$ becomes "bimodal". The bimodal structure is not dominant further downstream and within the turbulent jet. An acceptable assumption for the form of $P(m_1,\ldots, m_o, T)$ is a multidimensional normal distribution, which must be suitably clipped due to the restricted domain of definition of temperature and mass fractions. Although this form does not cover the intermittencies at the boundary of the turbulent jet, it is a good approximation for the inside of the turbulent jet and further downstream. For many applications the deficiencies in representation by a multidimensional normal distribution is of no consequence because the mean reaction rates in the areas with strong intermittencies are mostly small compared with those inside the turbulent jet. If the approximation of $P(m_1,\ldots, m_o, T)$ by means of a multidimensional normal distribution is adequate, the form of $P(m_1,\ldots, m_o, T)$ must be established quantitatively. As in the one-dimensional case, this is achieved through the expected values and the second moments. Consequently in order to determine $P(m_1,\ldots, m_o, T)$ calculation of $\langle m_1\rangle,\ldots, \langle m_o\rangle, \langle T\rangle, \langle m_i^{\rm f} m_j^{\rm f}\rangle$, $i, j = l, o$, and $\langle m_i^{\rm f} T^{\rm f}\rangle$, $i = l, o$ is necessary.

Equation (394) can be presented in the form

$$\langle S_i\rangle = S_i(\langle m_1\rangle,\ldots, \langle m_o\rangle, \langle T\rangle)$$
$$\cdot F(\langle m_i^{\rm f} m_j^{\rm f}\rangle, \langle m_i^{\rm f} T^{\rm f}\rangle, \quad i, j = l, o) \tag{395}$$

in which $F(\langle m_i^{\rm f} m_j^{\rm f}\rangle, \langle m_i^{\rm f} T^{\rm f}\rangle, i, j = l, o)$ are correction functions for reproducing the effect of turbulent fluctuations in temperature and mass fractions in the averaging procedure of weighting with the joint probability density functions.

These correction functions can be given as polynomials of the second moments $\langle m_i^{\rm f} m_j^{\rm f}\rangle$, $\langle m_i^{\rm f} T^{\rm f}\rangle$ (or more precisely of the turbulence intensities and correlation coefficients) and of the activation energies for the reactions in question [129], [130]. The coefficients of these polynomials can be obtained by applying the averaging procedure (Eq. 394) for a number of predefined values of $\langle m_i^f m_j^f \rangle$, $\langle m_i^f T^f \rangle$ and E_a; $\langle S_i \rangle$ is then represented as a function of these quantities by an empirical polynomial expression [129]. Thus integration of Equation (394) is unnecessary for the numerical solution of the resulting system of equations.

Calculation of the expected values $\langle S_i \rangle$ of the reaction rates from Equation (395) assumes knowledge of the second moments $\langle m_i^f m_j^f \rangle$, $i, j = 1, \ldots, N$ and $\langle m_i^f T^f \rangle$, $i = 1, \ldots, N$. In this case too closure of the equation system at the level of the expected values necessitates calculation of the second moments. The relevant transport equations can be derived by similar considerations and modeling assumptions used for deriving the equation for other second moments, see Section 5.1.2.1 [52], [126]–[128], [130].

The numerical work involved in calculating a statistical reacting nonisothermal flow is considerable, despite the relatively simple structure of the modeling components. In addition to the solution of the balances of momentum, direct modeling of Reynolds stresses requires the solution of two further equations for turbulence energy and the turbulent time scale. The Reynolds fluxes are modeled on the same basis as the Reynolds stresses, so that apart from the enthalpy balance (Eq. 392) and the transport equations for the mass of the chemical species (Eq. 393), no other equations originate from this part of the model. Modeling of the mean chemical reaction rates through the simple expression (Eq. 394) with presumed shape joint probability density functions requires the solution of equations for the variances and covariances $\langle m_i^f m_j^f \rangle$, $i, j = 1, \ldots, N$, and $\langle m_i^f T^f \rangle$, $i = 1, \ldots, N$. In addition the density has to be calculated from the averaged form of the thermal equation of state

$$\langle \varrho \rangle = \frac{\langle p \rangle}{R} \sum_{i=1}^{N} \frac{M_i}{\langle T \rangle \langle m_i \rangle + \langle m_i^f T^f \rangle} \quad (396)$$

which also contains the second moments $\langle m_i^f T^f \rangle$, $i = 1, \ldots, N$. Equation (396) can easily be derived from the ideal gas equation. The temperature must be determined from the expected value of the enthalpy of the mixture through the definition of the enthalpy:

$$\langle h \rangle = \sum_{i=1}^{N} \langle m_i h_i \rangle = \sum_{i=1}^{N} \langle m_i (h_{0i} + \int c_{pi} dT) \rangle \quad (397)$$

Equation (397) is not solved here for the temperature, for further details see [129]–[131].

Example: Combustion of Hydrogen in a Turbulent Diffusion Flame. The simulation of a turbulent nonisothermal reacting flow by the system of equations outlined on pp. **2**-95–**2**-97 will be demonstrated briefly on an example. A circular jet of hydrogen with a Reynolds number of 12 000 relative to the nozzle outlet is injected into virtually stationary air and is burned in a horizontal combustion chamber. In the experiments described in [132]–[134], velocities, temperatures, and concentrations of stable chemical species and of OH radicals are measured downstream of the burner nozzle. The system is axially symmetrical.

The numerical solution contains the solutions of Equations (223), (225), (388), (392), (393) and also of the equations for the variances and covariances $\langle m_i^f m_j^f \rangle$, $i, j = 1, \ldots, N$ and $\langle m_i^f T^f \rangle$, $i = 1, \ldots, N$ in axially symmetrical formulation. The numerical procedure is the same as that outlined for the isothermal case in Section 5.1.2.3. The chemical reactions are described by 44 elementary reactions between eleven chemical species: H_2, O_2, H_2O, N_2, H, O, OH, HO_2, H_2O_2 and also N and NO [135]. The reaction rates of each of these components consists of the summation of reaction rates of all the elementary reactions j in which the component in question i occurs, i.e., $r_i = \sum_{j=1}^{R} r_{ij}$. The rate coefficients of the reactions are expressed in the form $k_j = k_{0j} T^{\alpha j} e^{-E_a/RT}$ and are thus not freely selectable modeling parameters. For details of the reaction rate coefficients of the H_2–O_2 system see [130], [135]. The modeling parameters used are the coefficients $C_\mu = 0.09$, $C_{\varepsilon 1} = 1.4$, $C_{\varepsilon 2} = 0.925$, $\sigma_{k\,\mathrm{eff}} = 1.0$, $\sigma_{\varepsilon\,\mathrm{eff}} = 1.3$; the turbulent Prandtl numbers in Equations (392) and (393); and also the equations for the second moments which are defined as $\sigma_\varphi = 0.7$ [91].

Figure 56 shows the result of the simulations from the model compared with the measured results given in [132]–[134]. The correspondence between the measured results and the model is satisfactory. In particular the penetration of oxygen to the axis of the turbulent jet close to the nozzle is well predicted. Along with the calculation of the OH radical concentrations, this is a particular potential of the model that is based on an elementary reaction scheme for describing the chemical reactions for the H_2–O_2–N_2 system.

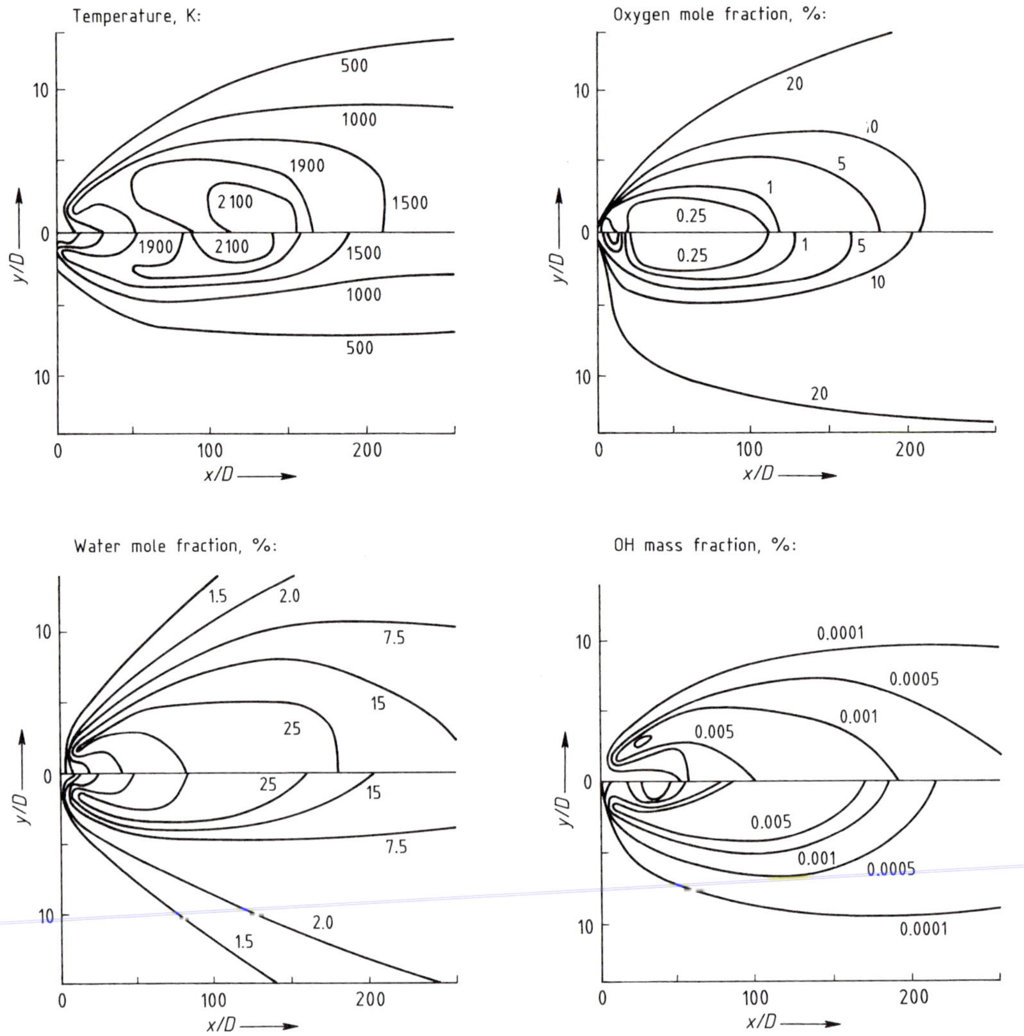

Figure 56. Measured (upper section) and calculated (lower section) profiles for temperature, mole fraction of oxygen and water, and mass fraction of OH radicals for a turbulent hydrogen–air diffusion flame [130], [131]. x = axial distance; y = radial distance; D = nozzle diameter

Another aspect of the model presented here can be shown with the results of sensitivity analysis. The first-order sensitivity coefficients are calculated as described on pp. 2-82–2-83. Due to the elliptical form of the conservation equations and the difference form which follows from this (see p. 2-56), the resulting system of equations has a block pentadiagonal structure and is solved by an ADI procedure.

Figure 57 A shows the results in the form of the gradients of the axial velocity $\langle U \rangle$ with respect to the model parameter C_μ, which controls the turbulent viscosity. The sensitivity coefficients are given in relative form $(\partial \langle U \rangle / \langle U \rangle)/(\partial \langle C_\mu \rangle / C_\mu)$ as in Equations (29) and (30). The sensitivity with respect to a change in the parameter C_μ is particularly large in areas with high velocity gradients. The sensitivity analysis thus reflects the physical background and the limitations of the model used: since the turbulent stresses are described by an approach which relates them to the gradients of the mean velocities,

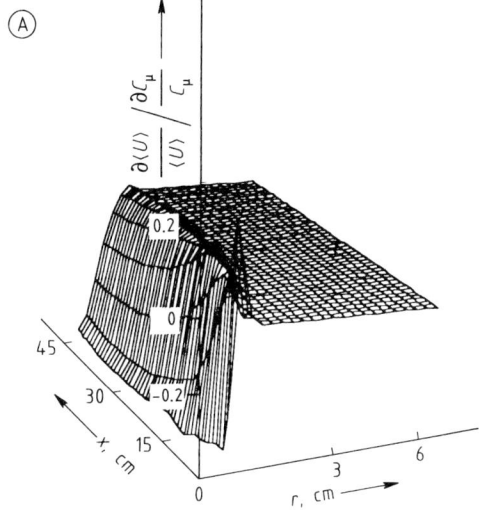

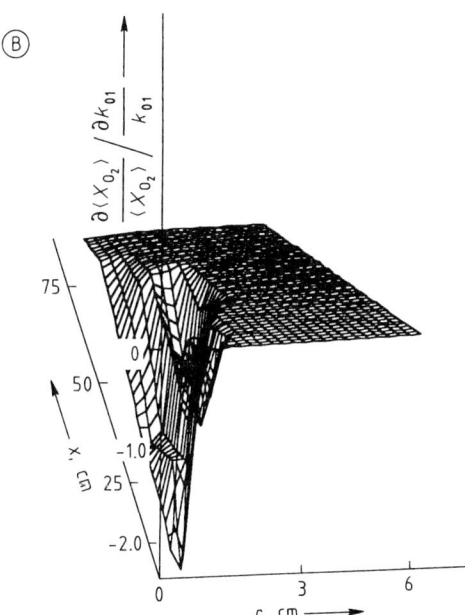

Figure 57. Relative sensitivity coefficient for a H_2–O_2 diffusion flame
A) Coefficients for the axial velocity $\langle U \rangle$ with respect to the modeling parameter C_μ; B) Coefficients for the mole fraction of oxygen $\langle X_{O_2} \rangle$ with respect to the rate coefficient k_{01}
x = axial distance; r = radial distance.

large sensitivities inevitably result in response to changes in the apparent turbulent viscosity in areas with high velocity gradients.

Figure 57 B shows the gradients of the molar fractions for oxygen with respect to the rate coefficient for the reaction $O_2 + H \rightarrow OH + O$, this is one of the most important oxygen-consuming reactions. Here to, the sensitivity coefficients are given in relative form $(\partial \langle X_{O_2} \rangle / \langle X_{O_2} \rangle)/(\partial k_{01}/k_{01})$. Sensitivity is only obvious in the main reaction zones of the turbulent hydrogen flame. Thus, the molar fraction of oxygen only reacts sensitively to the change in these rate coefficients in the main reaction zone of the flame.

The result of averaging the reaction rates with Equations (394) and (395) can be interpreted as the change in the rate coefficients due to the correction functions. The unsatisfactory representation of the joint probability density function $P(m_1 \ldots, m_o, T)$ by a multidimensional normal distribution used to calculate these correction functions is therefore no longer of importance at the edge of the jet for the prediction of the oxygen molar fraction. This is due to the low reaction rates for the consumption of oxygen in this region.

For further discussion of sensitivity analysis, especially for discussion of the possibility of reduction of the chemical models, see [131].

6. References

[1] G. Klaus: *Wörterbuch der Kybernetik*, Fischer Verlag, Frankfurt 1969.
[2] A. I. Bojarinow, W. W. Kafarow: *Optimierungsmethoden in der chemischen Technologie*, Verlag Chemie, Weinheim 1972.
[3] D. M. Himmelblau, K. B. Bischoff: *Process Analysis and Simulation: Deterministic Systems*, J. Wiley and Sons, New York 1968.
[4] D. M. Himmelblau: *Process Analysis by Statistical Methods*, J. Wiley and Sons, New York 1970.
[5] U. Hoffmann, H. Hofmann: *Einführung in die Optimierung mit Beispielen aus dem Chemie-Ingenieur-Wesen*, Verlag Chemie, Weinheim 1971.
[6] D. M. Himmelblau in A. Bisio, R. Kabel (eds.): *Scaleup of Chemical Processes*, J. Wiley and Sons, New York 1985.
[7] T. E. Edgar, D. M. Himmelblau: *Optimization of Chemical Processes*, McGraw-Hill, New York 1988.
[8] F. W. Ramirez: *Computational Methods for Process Simulation*, Butterworth Publishers, Stoneham 1989.
[9] W. F. Luyben: *Process Modeling, Simulation and Control for Chemical Engineers*, McGraw-Hill, New York 1990.
[10] *Ullmann*, 4th ed., **4**, 451.
[11] G. Emig, *Ber. Bunsenges. Phys. Chem.* **90** (1986) 986.

[12] H. Hofmann: "Future Trends in Chemical Engineering Modelling," *Proceedings of the 18th EFCE Congress*, Apr. 26–30, 1987, p. 579.
[13] W. W. Kafarow: *Kybernetische Methoden in der Chemie und chemischen Technologie*, Verlag Chemie, Weinheim 1971.
[14] L. A. Gould: *Chemical Process Control: Theory and Applications*, Addison-Wesley Publishing Company, Reading 1969.
[15] P. S. Buckley: *Techniques of Process Control*, Robert E. Krieger Publishing Company, Huntington 1979.
[16] P. V. Danckwerts, *Chem. Eng. Sci.* **2** (1953) 1.
[17] A. D. Randolph, *Can. J. Chem. Eng.* **42** (1964) 280.
[18] E. E. O'Brien in P. A. Libby, F. A. Williams (eds.): *Turbulent Reacting Flows*, Springer Verlag, Berlin-Heidelberg 1980, p. 191.
[19] S. B. Pope, *Prog. Energy Combust. Sci.* **11** (1985) 119.
[20] G. Emig, U. Hoffmann, H. Hofmann: *DECHEMA-Kurs: Planung und Auswertung von Versuchen zur Erstellung mathematischer Modelle*, parts I and II, DECHEMA, Frankfurt-Main 1974.
[21] E. Kreyszig: *Statistische Methoden und ihre Anwendung*, Verlag Vandenhoeck und Ruprecht, Göttingen 1972.
[22] M. Fisz: *Wahrscheinlichkeitsrechnung und mathematische Statistik*, VEB Deutscher Verlag der Wissenschaften, Berlin 1980.
[23] A. Lindner: *Statistische Methoden*, Birkhäuser, Stuttgart 1964.
[24] M. Baerns, H. Hofmann, A. Renken: *Chemische Reaktionstechnik, Lehrbuch der Technischen Chemie 1*, Thieme Verlag, Stuttgart 1987.
[25] VDI-Gesellschaft Verfahrenstechnik und Chemieingenieurwesen: "Berechnungsblätter für den Wärmeübergang," *VDI-Wärmeatlas*, VDI-Verlag, Düsseldorf 1984.
[26] IBM – Scientific Subroutine Package – System 360, IBM Corporation, White Plains 1972.
[27] NAG Fortran Library-Mark 8, Numerical Algorithms Group Inc., Downers Grove 1981.
[28] IMSL Library Reference Manual, IMSL Inc., Houston 1980.
[29] W. H. Press, B. B. Flannery, S. A. Teukolsky, W. T. Vetterling: *Numerical Recipes. The Art of Scientific Computing (Fortran Version)*, Cambridge University Press, Cambridge 1989 (also available in Pascal and C).
[30] W. H. Press, B. B. Flannery, S. A. Teukolsky, W. T. Vetterling: *Numerical Recipes. Example Book (Fortran)*, Cambridge University Press, Cambridge 1985 (also available in Pascal and C).
[31] G. Engeln-Müllges, F. Reutter: *Numerische Mathematik für Ingenieure*, B.-I. Wissenschaftsverlag, Mannheim 1987.
[32] G. Engeln-Müllges, F. Reutter: *Formelsammlung zur Numerischen Mathematik mit Standard-Fortran 77-Programmen*, B.-I. Wissenschaftsverlag, Mannheim 1988.
[33] J. Hildenrath et al.: "OMNITAB, A Computer Program for Statistical and Numerical Analysis," *Handbook 101*, Nat. Bur. of Standards," U.S. Government Printing Office, Washington, D.C., 1968.
[34] R. Hooke, T. A. Jeeves, *J. Assoc. Compt. Mach.* **8** (1961) 212.
[35] N. Spendley, G. R. Hext, F. R. Himsworth, *Technometrics* **4** (1962) 441.
[36] J. A. Nelder, R. Mead, *Comput. J.* **7** (1965) 308.
[37] H. H. Rosenbrock, C. Storey: *Computational Techniques for Chemical Engineers*, Pergamon Press, Oxford 1966.
[38] C. S. Beightler, D. T. Phillips, D. G. Wilde: *Foundations of Optimization*, Prentice Hall, Englewood Cliffs, N.Y. 1979.
[39] P. E. Gill, W. Murray, M. H. Wright: *Practical Optimization*, Academic Press, New York 1981.
[40] D. W. Marquardt, *J. Soc. Ind. Appl. Math.* **11** (1963) 431.
[41] H. O. Hartley, A. Booker, *Ann. Math. Stat.* **36** (1965) 638.
[42] E. J. Williams, *J. Royal Stat. Soc. B* **24** (1962) 125.
[43] J. J. More in G. A. Watson (ed.): *Numerical Analysis, Lecture Notes in Mathematics*, vol. 630, Springer Verlag, Berlin 1977, p. 105.
[44] Z. E. Beisinger, S. Bell: H2 SAND MIN, Sandia Corporation.
[45] I. N. Bronstein, K. H. Semendjajew: *Taschenbuch der Mathematik*, Verlag Harry Deutsch, Thun 1983.
[46] O. Levenspiel: *Chemical Reaction Engineering*, J. Wiley and Sons, New York 1972.
[47] O. Levenspiel, W. K. Smith, *Chem. Eng. Sci.* **6** (1975) 227.
[48] J. O. Hirschfelder, C. F. Curtiss, R. B. Bird: *Molecular Theory of Gases and Liquids*, J. Wiley and Sons, New York 1954.
[49] J. C. Rotta: *Turbulente Strömungen*, B. G. Teubner, Stuttgart 1972.
[50] R. B. Bird, W. E. Stewart, E. N. Lightfoot: *Transport Phenomena*, J. Wiley and Sons, New York 1960.
[51] F. A. Williams: *Combustion Theory*, The Benjamin/Cummings Publishing Company, Menlo Park, Calif. 1985.
[52] P. A. Libby, F. A. Wiliams in P. A. Libby, F. A. Williams (eds.): *Turbulent Reacting Flows*, Springer Verlag, Berlin – Heidelberg 1980, p. 1.
[53] P. Givi, *Prog. Energy Combust. Sci.* **15** (1989) 121.
[54] M. Lesieur: *Turbulence in Fluids*, Martinus Nijhoff Publishers, Dordrecht 1987.
[55] T. S. Lundgreen, *Phys. Fluids* **12** (1969) 485.
[56] S. B. Pope, *Combust. Flame* **27** (1976) 294.
[57] J. Janicka, W. Kolbe, W. Kollmann in C. T. Crowe, W. L. Grosshandler (eds.): *Proc. of the 1978 Heat Transf. Fluid Mech. Inst.*, Standford University Press, Stanford 1978, p. 296.
[58] S. B. Pope, *Philos. Trans. R. Soc. Lond. Ser. A* **291** (1979) 529.
[59] M. Jischa: *Konvektiver Impuls-, Wärme- und Stoffaustausch*, Vieweg und Sohn, Braunschweig 1982.
[60] H. Schlichting: *Grenzschichttheorie*, Verlag G. Braun, Karlsruhe 1982.
[61] N. Peters in E. S. Oran, J. P. Boris (eds.): *Approaches to Combustion Modeling*, in print.
[62] M. J. Lighthill: *Introduction to Fourier Analysis and Generalized Functions*, Cambridge University Press, Cambridge 1958.
[63] H. M. Hulburt, S. Katz, *Chem. Eng. Sci.* **19** (1964) 555.
[64] E. Kamke: *Differentialgleichungen, Lösungsmethoden und Lösungen II*, B. G. Teubner Verlag, Stuttgart 1979.
[65] P. Langevin, *C. R. Acad. Sci. Paris* **146** (1908) 530.
[66] S. B. Pope, *Phys. Fluids* **24** (1981) 588.
[67] L. Arnold: *Stochastic Differential Equations: Theory and Applications*, J. Wiley and Sons, New York 1974.

[68] W. Jost: *Diffusion in Solids, Liquids and Gases*, Academic Press, New York 1952.
[69] O. Levenspiel, K. B. Bischoff: *Adv. Chem. Eng.* **4** (1963) 95.
[70] J.-Y. Chen, W. Kollmann: *Twenty-Second Symposium (International) on Combustion*, The Combustion Institute, Pittsburgh 1989, p. 645.
[71] J.-Y. Chen, W. Kollmann, *Combust. Flame* **79** (1990) 75.
[72] J. L. Lumley, *Adv. Appl. Mech.* **18** (1978) 123.
[73] B. E. Launder, C. J. Reece, W. Rodi, *J. Fluid. Mech.* **68** (1975) 573.
[74] J. C. Rotta, *Z. Phys.* **129** (1951) 547.
[75] C. Dopazo, *Phys. Fluids* **18** (1975) 397.
[76] J. Janicka, W. Kolbe, W. Kollman, *J. Non-Equilib. Thermodyn.* **4** (1978) 47.
[77] C. Dopazo, *Phys. Fluids* **22** (1979) 20.
[78] J. Janicka, W. Kollman in L. J. S. Bradbury et al. (eds.): *Turbulent Shear Flows 4*, Selected Papers from the Fourth International Symposium on Turbulent Shear Flows, Springer Verlag, Berlin – Heidelberg 1985, p. 73.
[79] S. B. Pope, *Combust. Sci. Technol.* **25** (1981) 159.
[80] T. V. Nguyen, S. B. Pope, *Combust. Sci. Technol.* **42** (1984) 13.
[81] A. Favre in Society for Industrial and Applied Mathematics (ed.): *Problems of Hydrodynamics and Continuous Mechanics*, Philadelphia 1969, p. 231.
[82] G. Stahl, J. Warnatz, B. Rogg in AIAA (ed.): *Progr. Aeronaut. Astronaut.*, vol. 113, Washington 1988, p. 195.
[83] P. Bradshaw in P. Bradshaw (ed.): *Turbulence*, Springer Verlag, Berlin – Heidelberg 1978, p. 1.
[84] B. E. Launder, D. B. Spalding: *Lectures in Mathematical Models of Turbulence*, Academic Press, London – New York 1972.
[85] K. K. Kuo: *Principles of Combustion*, J. Wiley and Sons, New York 1986.
[86] J. O. Hinze: *Turbulence*, McGraw-Hill, New York 1975.
[87] J. Boussinesq, *Mém. prés. Acad. Sci.* XXIII, 46, Paris 1877.
[88] L. Prandtl, *Z. Angew. Math. Mech.* **22** (1942) 241.
[89] L. Prandtl, *Nachr. Akad. Wiss. Göttingen, Math. Phys. Kl.* **2** (1945) 6–19.
[90] J. Janicka: *Twenty-First Symposium (International) on Combustion*, The Combustion Institute, Pittsburgh 1988, p 1409.
[91] S. M. Correa, W. Shyy, *Prog. Energy Combust. Sci.* **13** (1987) 249.
[92] H. Tennekes, J. L. Lumley: *A First Course in Turbulence*, MIT-Press, Boston 1972.
[93] W. F. Ames: *Nonlinear Partial Differential Equations in Engineering*, Academic Press, New York – London 1965.
[94] A. D. Gosman et al.: *Heat and Mass Transfer in Recirculating Flows*, Academic Press, London 1969.
[95] M. D. Smooke, A. A. Turnbull, R. E. Mitchell, D. E. Keyes in C.-M. Brauner, C. Schmidt-Lainé (eds.): *Mathematical Modeling in Combustion and Related Topics*, Martinus Nijhoff Publishers, Dordrecht 1988, p. 261.
[96] D. B. Spalding: *Basic Equations of Fluid Mechanics and Heat and Mass Transfer*, Imperial College London, Mechanical Engineering Department, Report HTS/76/6, London 1976.
[97] S. V. Patankar: *Numerical Heat Transfer and Fluid Flow*, McGraw-Hill, New York 1980.
[98] F. H. Harlow, J. E. Welch, *Phys. Fluids* **8** (1965) 2182.
[99] A. O. Demurren, *Comput. Fluids* **13** (1985) 411.
[100] P. J. Roach: *Computational Fluid Dynamics*, Hermosa Publishers, Albuquerque 1976.
[101] E. S. Oran, J. P. Boris: *Numerical Simulation of Reactive Flow*, Elsevier Science Publishing Company, New York 1987.
[102] C. Hirsch: *Numerical Computation of Internal and External Flows*, vol. 1, J. Wiley and Sons, New York 1988.
[103] C. Hirsch: *Numerical Computation of Internal and External Flows*, vol. 2, J. Wiley and Sons, New York 1990.
[104] A. J. Baker: *Finite Element Computational Fluid Mechanics*, McGraw-Hill, New York 1985.
[105] J. M. Beer, N. A. Chigier: *Combustion Aerodynamics*, Applied Science Publishers Ltd., London 1972.
[106] L. Horvarth, *Proc. R. Soc. London Ser. A* **164** (1938) 547.
[107] E. R. G. Eckert, R. M. Drake: *Analysis of Heat and Mass Transfer*, McGraw-Hill, New York 1972.
[108] M. N. Özisik: *Basic Heat Transfer*, McGraw-Hill, New York 1985.
[109] H. Brauer: *Stoffaustausch einschließlich chemischer Reaktionen*, Verlag Sauerländer, Aarau 1971.
[110] P. Trambouze in A. E. Rodriguez, J. M. Calo, N. H. S. Sweed (eds.): *Multiphase Chemical Reactors, Fundamentals* (vol. I), Sijthoff and Noordhoff, Alphen 1981.
[111] J. A. van der Vusse, *Chem. Eng. Sci.* **17** (1962) 507.
[112] L. Cloutier, *Can. J. Chem. Eng.* **37** (1959) 105.
[113] P. Gray, S. K. Scott: *Chemical Oscillations and Instabilities: Nonlinear Chemical Kinetics*, Clarendon Press, Oxford 1990.
[114] G. Iooss, D. D. Joseph: *Elementary Stability and Bifurcation Theory*, Springer Verlag, New York 1980.
[115] J. H. Merkin, D. J. Needham, S. K. Scott, *Proc. R. Soc. London Ser. A* **406** (1986) 299.
[116] B. F. Gray, M. J. Roberts, *Proc. R. Soc. London Ser. A* **416** (1988) 391.
[117] P. M. Frank: *Introduction to System Sensitivity Theory*, Academic Press, New York 1978.
[118] T. P. Coffee, J. M. Heimerl, *Combust. Flame* **50** (1983) 323.
[119] D. E. Clough, W. F. Ramirez, *AIChE J.* **22** (1976) no. 4, 1097.
[120] J. C. P. Sheel, C. M. Crowe, *Can. J. Chem. Eng.* **47** (1969) 183.
[121] R. A. Greenkorn: *Flow Phenomena in Porous Media*, Marcel Dekker, New York 1983.
[122] E. Kreyszig: *Advanced Engineering Mathematics*, J. Wiley and Sons, New York 1988.
[123] G. F. Fromment, K. B. Bishoff: *Chemical Reactor Analysis and Design*, J. Wiley and Sons, New York 1990.
[124] P. B. Weisz, J. S. Hicks, *Chem. Eng. Sci.* **17** (1962) 265.
[125] J. Ganoulis, F. Durst: "Finite Elements in Water Resources," *Proceedings VI Int. Conf.*, Lisbon 1986, p. 655.
[126] S. N. B. Murthy (ed.): *Turbulent Mixing in Nonreactive and Reactive Flows*, Plenum Press, New York – London 1975.
[127] R. W. Bilger in P. A. Libby, F. A. Williams (eds.): *Turbulent Reacting Flows*, Springer Verlag, Berlin – Heidelberg 1980, p. 65.
[128] R. Borghi, *Prog. Energy Combust. Sci.* **14** (1988) 245.

[129] H. Bockhorn in C.-M. Brauner, C. Schmidt-Lainé (eds.): *Mathematical Modeling in Combustion and Related Topics,* Martinus Nijhoff Publishers, Dordrecht 1988, p. 411.
[130] H. Bockhorn: *Twenty-Second Symposium (International) on Combustion,* The Combustion Institute, Pittsburgh 1988, p. 665.
[131] H. Bockhorn: *Twenty-Third Symposium (International) on Combustion,* The Combustion Institute, Pittsburgh, in print.
[132] M. C. Drake et al.: *Twentieth Symposium (International) on Combustion,* The Combustion Institute, Pittsburgh 1984, p. 327.
[133] M. C. Drake, R. W. Pitz, M. Lapp: *AIAA Pap.* **84-0544** (1984).
[134] G. M. Faeth, G. S. Samuelson, *Prog. Energy Combust. Sci.* **12** (1986) 305.
[135] J. Warnatz in W. C. Gardiner, Jr. (ed.): *Combustion Chemistry,* Springer Verlag, New York 1984, p. 196.

3. Dimensional Analysis

MARKO ZLOKARNIK, Bayer AG, Leverkusen, Federal Republic of Germany

1.	Introduction	3-2	3.3. The Π Relationship	3-14
2.	Dimensional Analysis	3-3	3.4. Reduction of Matrix Size	3-16
2.1.	Historical Survey	3-3	3.5. Change of Dimensional Systems	3-17
2.2.	Introduction	3-4	4. Similarity and Scale-Up	3-17
2.3.	Fundamentals	3-6	4.1. Basic Principles of Scale-Up	3-17
2.3.1.	Physical Quantities and Relationships Between Them	3-6	4.2. Experimental Methods for Scale-Up	3-18
2.3.2.	Consistency of Secondary Units and Invariance of Physical Relationships	3-7	4.3. Scale-Up Under Conditions of Partial Similarity	3-19
2.3.3.	Physical Dimensions, Systems of Dimensions, and Dimensional Constants	3-7	5. Treatment of Variable Physical Properties by Dimensional Analysis	3-22
2.3.4.	The Dimensional Matrix and Its Linear Dependence	3-8	5.1. Dimensionless Representation of the Material Function	3-22
2.3.5.	The Π Theorem	3-9	5.2. The Π Set for Variable Physical Properties	3-23
3.	Description of a Physical Process with a Full Set of Dimensionless Numbers	3-10	5.3. Treatment of Non-Newtonian Liquids by Dimensional Analysis	3-24
3.1.	The Relevance List for a Problem	3-10	5.4. Treatment of Viscoelastic Liquids by Dimensional Analysis	3-25
3.2.	Determination of a Complete Set of Dimensionless Numbers	3-13	6. References	3-26

A comprehensive list of dimensionless numbers and their definitions is given in Table 3, p. 3-17. Symbols used in this article are as follows:

- a volume-related interface surface area
- A cross-sectional area
- c concentration
- d diameter
- D diffusivity, rate of shear
- Eu Euler number
- F drag resistance
- Fr Froude number
- g acceleration due to gravity
- G gas absorption rate, gravitational constant
- H liquid height in column, viscosity constant
- l length
- m mass
- n rotational speed of stirrer
- N_1 normal stress
- Ne Newton number
- Nu Nusselt number
- p pressure
- P power
- Pr Prandtl number
- q volume throughput
- R universal gas constant
- Re Reynolds number
- t time, period of oscillation
- T temperature
- v (superficial) velocity
- V volume
- Wi Weissenberg number
- α amplitude
- β temperature coefficient of density
- γ temperature coefficient of viscosity
- μ scale-up factor
- ν kinematic viscosity
- ϱ density

σ surface tension
τ shear stress
θ mixing time, time constant
φ volume ratio
η dynamic viscosity

Subscripts

crit critical
f   form
F   full-scale
L   liquid
M   model
p   polymer
r   friction
w   wall
0   reference value

1. Introduction

The process engineer is generally concerned with the industrial implementation of processes in which chemical or microbiological conversion of material takes place in conjunction with the transfer of mass, heat, and momentum. These processes are scale dependent; i.e., they behave differently on a small scale (in laboratories or pilot plants) and on a large scale (in production). They include heterogeneous chemical reactions and most unit operations (e.g., mixing, screening, sifting, filtration, centrifugation, grinding, drying, and combustion processes). Understandably, process engineers want to simulate these processes in models to gain insights that will assist them in designing new industrial plants. Occasionally, they are faced with the same problem for another reason: an industrial facility already exists but does not function properly, and suitable measurements must be carried out to determine the cause of the difficulties and to provide a solution.

Irrespective of whether the model involved represents "scaling up" or "scaling down," important questions always apply:

1) How small can the model be? Is one model sufficient or should tests be carried out on models of different sizes?
2) When must or can physical properties differ; when must measurements on the model be carried out with the original system of materials?
3) What rules govern the adaptation of process parameters in the model measurements to those of the full-scale plant?
4) Can complete similarity be achieved between processes in the model and those in its full-scale counterpart?

These questions touch on fundamental issues of the theory of models, a theory derived from the theory of similarity which is, in turn, based on dimensional analysis. The following is a theoretical discussion of these questions.

In today's world of technological progress, it is increasingly important to optimize processes both as a whole and in their individual steps with regard to both economy and environmental protection. The days of generous estimates and consequent "safety margins"—described by GRIGULL as "ignorance margins" in the early 1970s—are over. "All-round" plants, capable of manufacturing one product one day and another the next, and ultimately (after being turned to the relevant process on site by the "black-box method") having a production capacity many times their intended output, are a thing of the past.

Attempting to optimize a process means looking more deeply into its physical and technical elements. The effects of each parameter that influences a technical process must be examined carefully. This involves answering increasingly complicated questions, for which mathematical solutions often do not exist. The researcher becomes more and more dependent on model experiments, at a time when qualified laboratory personnel are scarce. Thus, problems of process technology have to be solved with a minimum of assistance as regards both finances and staff. Future research expenditure will probably be reduced still further while gleaning a maximum of information from the work performed.

In such a situation the theory of similarity may be of invaluable service. This theory facilitates the sensible planning and simple execution of experiments and evaluation of the resulting data so as to produce reliable information on the size and process parameters of the large-scale plant; this assumes, of course, that the method is applied correctly and as early as possible.

The present article is divided into five chapters:

Chapter 1 serves as an introduction.

Chapter 2 deals with the *fundamentals of dimensional analysis* whose importance lies in the fact that, by way of the Π theorem, they offer the only means of dealing with problems that cannot be formulated mathematically.

Chapter 3 shows how a physical process can be described with a complete set of relevant physical quantities. This comprehensive pool of information serves as the source of a *set of dimensionless numbers*, which is usually much smaller than the set of physical quantities but describes the problem just as completely.

Chapter 4 discusses *similarity and scale-up*. It explains why reliable scale-up is possible only within the framework of a complete set of dimensionless numbers and investigates the problems of partial similarity. Reliable scale-up is shown to be possible even when complete similarity is impossible because of the unavailability of small-scale substitutes for the physical properties of a full-scale system.

Chapter 5 considers specific questions that arise if *substances with variable physical properties* (e.g., temperature dependent or non-Newtonian) are to be treated by means of dimensional analysis.

When dimensional analysis is applied to process engineering, compilation of a complete list of problem-related physical quantities always causes the most difficulties. The reader should refer to the literature for a series of *practical examples* covering the fields of mechanical, thermal, and chemical process engineering [18]–[39] (→ 25. Stirring, **B2**).

The following *advantages* result from the correct use of dimensional analysis:

1) *Reduction of the number of parameters required to define the problem.* The Π theorem states that a physical problem can always be described in dimensionless terms. This means that the number of dimensionless groups required to fully describe it is much smaller than the number of physical quantities (and generally equals the number of physical quantities minus the number of basic units contained in them).
2) *Reliable scale-up of the desired operating conditions from the model to the full-scale plant.* According to the theory of models, two processes may be considered similar to one another if they occur under geometrically similar conditions and if all dimensionless numbers that describe the process have the same numerical value.
3) *A deeper insight into the physical nature of the process.* By presenting experimental data in a dimensionless form, distinct physical states can be isolated from one another (e.g., turbulent or laminar flow region), and the effect of individual physical variables can be identified.
4) *Flexibility in the choice of parameters and their reliable extrapolation within the range covered by the dimensionless numbers.* These advantages become clear in examples like the Reynolds number, $Re = vl/v$, which can be varied by altering the characteristic velocity v, characteristic length l, or kinematic viscosity v. By choosing appropriate model fluids the viscosity can easily be altered by several orders of magnitude. Once the effect of the Reynolds number is known, extrapolation of both v and l is allowed within the examined range of Re.

2. Dimensional Analysis

2.1. Historical Survey

The end result of dimensional analysis is a complete set of dimensionless numbers that describe a physical process and outline the conditions under which this process behaves "similarly" in the model and its full-sized counterpart; dimensional analysis is the basis of scale-up methods. Lord RAYLEIGH was aware of this when he referred to studies in which he employed dimensional analysis as "the study of similitude." Let us take this reference as our starting point in a historical survey of dimensional analysis which begins with the first attempts to scale up a model, attempts made at a time when the very concept of dimensions was unknown.

The desire to obtain information about a full-scale process by first carrying out tests on models has existed for centuries. LEONARDO DA VINCI wrote [40]

"VITRUVIUS says that small models are of no avail for ascertaining the effects of large ones; and I here propose to prove that this conclusion is a false one."

GALILEI later investigated the strength of mechanical parts of machines [41], and NEWTON clearly defined the concept of "mechanical similarity" [42]. BERTRAND formulated these rules unequivocally in 1847 and expressed Newton's general law of similarity as a constraint that must be satisfied by the four ratios of length, time, force, and mass [43]. However, the technical breakthrough in the field of similarity did not occur until 1869–1870, when FROUDE determined the drag on a ship by using model experiments [44], and 1883, when REYNOLDS published the results of his model-based experiments on the flow of liquids through pipes [45]. The basic works of PRANDTL, NUSSELT, GRÖBER, and many others followed. The "science of models" was born. The significance of this method is expressed by the words of BAEKELAND [46]:

"Commit your blunders on a small scale and make your profits on a large scale."

How was this done? The "general physical similarity" of basic quantities (geometrical, temporal, thermal, and other forms of similarity) was asserted and used to formulate "scale-up rules" for secondary quantities such as velocity and rate of acceleration. The comparison of forces, energies, etc., led to the definition of the first dimensionless groups, which were later referred to as

Newton's, Reynolds', or Froude's "model laws." WEBER described these techniques of deriving dimensionless numbers in two fundamental papers in 1919 and 1930 [1]–[3, Chap. 3].

These numbers can also be derived from the Navier–Stokes equations of hydrodynamics [3, Chap. 5], [4]. Application of this method is limited because most of the processes of current interest do not lend themselves to mathematical description. Differential equations do, however, provide valuable information on the dimensionless numbers describing a process even if they cannot actually be solved (cf. the work of DAMKÖHLER [47]).

Based on the dimensional homogeneity of physical relationships, dimensional analysis provides a far more elegant approach to deriving dimensionless groups. In retrospect the development of this basic method has been corrected and supplemented since GÖRTLER's studies in 1975 [48]. These changes are incorporated into this article.

In 1822 FOURIER [49] coined the term *physical dimension* and emphasized that physical equations must satisfy the property now described as dimensional homogeneity in relation to the chosen system of units. HELMHOLTZ examined the dimensionless groups governing hydrodynamics [50] and came up with the "Reynolds number" ten years before the Reynolds' publication appeared. Lord RAYLEIGH was particularly influential in encouraging the use of dimensional analysis [51].

In 1890 VASCHY [52] attempted to formulate the Π theorem as a consequence of the dimensional homogeneity of physical equations. However, in 1911 FEDERMANN [53] proved the Π theorem in a mathematical analysis of partial differential equations. FEDERMANN's work was not known to BUCKINGHAM who credited RIABOUCHINSKY with discovery of the Π theorem [54]. Posterity, on the other hand, gave all the credit to BUCKINGHAM because his paper of 1914 [55], introducing the term Π theorem, finally aroused scientific interest in this remarkable method. The theory states that a physical problem can always be described in dimensionless form, with the advantage that the number of dimensionless groups required is equal to the number of dimensional parameters minus the number of basic dimensions involved. BRIDGMAN later demonstrated (1922) that this "rule of thumb" does not always apply [5]. The rank of the dimensional matrix is the more correct consideration, rather than the number of basic units.

The excellent textbooks of BRIDGMAN (1922, 1931) [5] and LANGHAAR (1951) [6] helped popularize dimensional analysis and the theory of models derived from it. PANKHURST's book [7] is another important English publication, and attention must be drawn to SEDOV's Russian textbook which enjoys a good reputation [8]. Of all contributions in German, that of PAWLOWSKI is most worthy of mention [9]. This book deals with a series of questions related to the theory of similarity in an exact, mathematical fashion and is the only one to consider the treatment of variable physical properties (e.g., non-Newtonian viscosity) by dimensional analysis. It also provides extremely simple guidelines for using matrix calculations to derive dimensionless numbers. PAWLOWSKI offers his own proof of the Π theorem. The book by GÖRTLER [10] includes a comprehensive, mathematical treatment of dimensional analysis, and compares and comments on the lines of proof offered by several other authors.

2.2. Introduction

How should a physical problem be approached? Three increasingly complex problems are considered here:

1) The period of oscillation of a pendulum
2) The period of oscillation of small drops of liquid under the influence of their own surface tension
3) The pressure drop of a fluid in a straight, smooth pipe

1) The Period of Oscillation of a Pendulum. This problem is used in several treatises on dimensionless analysis (e.g., [5]) as a simple but elegant means of demonstration. To establish what determines the period of oscillation of a pendulum, a list of all the variables that may be relevant to it (i.e., *the relevance list*) is necessary. The dimensions (L length, M mass, T time) of these variables are then introduced. These variables should form a "physical relationship" that is independent of the system of dimensions chosen to measure them, a condition called *dimensional homogeneity*.

The period of oscillation of a pendulum is assumed to depend on the length and mass of the pendulum, the gravitational acceleration, and the amplitude of swing:

Physical quantity	Symbol	Dimension
Period of oscillation	t	T
Length of pendulum	l	L
Mass of pendulum	m	M
Gravitational acceleration	g	LT^{-2}
Amplitude (angle)	α	

The aim is to express t as a function of $l, m, g,$ and α:

$$t = f(l, m, g, \alpha)$$

The functional relationship f must remain independent (invariant) of the choice of system of units. This self-evident requirement affects the structure of the relationship in a significant way. The numerical value given for time depends only on the size of the basic unit of time being used and is therefore always the same even if the basic units used for mass and length change. To ensure that the function remains unchanged even when the basic units for mass and length are changed, the corresponding values (containing M and L) in the function arguments must be combined in such a way as to remain unaffected (i.e., these values have to be made dimensionless with regard to M and L).

The basic unit for mass M occurs only in the mass m itself. Changing this basic unit (e.g., from kilograms to pounds) changes the numerical value of the function, which is not acceptable. Either the list should include another variable containing M, or mass is not a relevant variable. If the latter is assumed, the above relationship is reduced to

$$t = f(l, g, \alpha)$$

Both l and g incorporate the basic unit of length. When combined as a ratio (l/g), they become dimensionless with regard to L and thus independent of changes in the basic unit of length:

$$t = f(l/g, \alpha)$$

Because the angle α has no dimension, the dimension T remains on the left-hand side of the equation and T^2 on the right. To remedy this, $\sqrt{l/g}$ must be written as

$$t = \sqrt{l/g}\, f(\alpha)$$

This is the only equation that dimensional analysis can offer in this case. It is not capable of producing information on the form of f. Integration of Newton's equation of motion for small amplitudes leads to $f = 2\pi$ and is independent of α [10]. The relationship can now be expressed as

$$t\sqrt{g/l} = 2\pi$$

The expression on the left is a *dimensionless number* with a numerical value of 2π.

The elegant solution of this first example should not tempt the reader to believe that dimensional analysis can be used to solve every problem that arises. To treat this example by dimensional analysis, the law of free fall had to be known. This knowledge was gained empirically by GALILEI [41] in 1604 through his experiments with the inclined plane. BRIDGMAN's comment is particularly appropriate [5]:

> "The problem cannot be solved by the philosopher in his armchair, but the knowledge involved was gathered only by someone at some time soiling his hands with direct contact."

2) The Period of Oscillation of Small Drops of Liquid under the Influence of their own Surface Tension. This problem is slightly more complex than the period of oscillation of a pendulum. The drops should be unaffected by gravity, and the oscillation should cause no more than periodic deformation (sphere to ellipsoid). The period of oscillation then depends only on the surface tension of the drops, their density, and diameter.

Physical quantity	Symbol	Dimension
Period of oscillation	t	T
Surface tension	σ	MT^{-2}
Density of the liquid	ϱ	ML^{-3}
Diameter of the drop	d	L

The aim is to determine the relationship

$$t = f(\sigma, \varrho, d)$$

which must be such that the numerical value for t is independent of the choice of the system of dimensions. By proceeding as in the first example, the equation can be made dimensionless with regard to M by using $\sigma/\varrho\,[L^3T^{-2}]$ and with regard to L via $\sigma/\varrho d^3\,[T^{-2}]$. Dimensional equality with T can be achieved via $\sqrt{\varrho d^3/\sigma}\,[T]$. The result

$$t = \text{const.}\,\sqrt{\varrho d^3/\sigma} \text{ or } t\sqrt{\sigma/\varrho d^3} = \text{const.}$$

can be confirmed experimentally.

The question arises of how t was known to depend only on the physical properties ϱ and σ and not on other factors such as the viscosity or compressibility of the liquid. The only reason is that prior experiments have shown that the role of viscosity is insignificant in nonviscous liquids, that compressibility is irrelevant when tiny drops are concerned, and that the above conclusion is valid only under these conditions. BRIDGMAN has a provocative question to ask on this subject:

"What use is all this dimensional analysis if we still need to resort to extensive preliminary experiments to solve the problem?"

The answer is simple. It is indeed true that in compiling a reliable relevance list, comprehensive preliminary experiments are often required. However, the two above examples show that dimensional analysis allows discovery of physical laws that would otherwise have required further systematic experiments.

The dimensional analysis presented in Examples 1 and 2 can be performed in this simple manner only when the result is a single dimensionless group. Problems nowadays, however, are much more complex and require a more systematic approach. Example 3 introduces a method that is still presented in most textbooks as the standard method.

3) Pressure Drop of a Homogeneous Fluid in a Straight Smooth Pipe.

In the flow of a homogeneous fluid (e.g., water or air) through a straight smooth pipe, the pressure drop Δp must be a function of the pipe geometry (diameter d and length l), the physical properties of the fluid (density ϱ and viscosity η), and its velocity v. The relevance list thus becomes

$$\{\Delta p; d, l; \varrho, \eta; v\}$$

For the relationship between these quantities

$$f(\Delta p, d, l, \varrho, \eta, v) = 0 \qquad (1)$$

to be dimensionally homogeneous, one must be able to form a *dimensionless product* Π ($\Pi = 1$) with them:

$$\Pi = \Delta p^\alpha d^\beta l^\gamma \varrho^\delta \eta^\varepsilon v^\zeta \qquad (2)$$
$$\Pi = [ML^{-1}T^{-2}]^\alpha [L]^\beta [L]^\gamma [ML^{-3}]^\delta$$
$$\cdot [ML^{-1}T^{-1}]^\varepsilon [LT^{-1}]^\zeta = 1 \qquad (3)$$

For the three basic dimensions (mass M, length L, time T),

for M: $\quad \alpha + \delta + \varepsilon = 0$ (4)
for L: $-\alpha + \beta + \gamma - 3\delta - \varepsilon + \zeta = 0$ (5)
for T: $\quad -2\alpha - \varepsilon - \zeta = 0$ (6)

Since only three equations with six unknowns are available, the number of possible Π numbers is infinite. For example, if α, β, and γ are considered as given, the remainder can be derived as follows. By solving Equation (4) for ε and substituting this in Equations (5) and (6),

$$\beta + \gamma - 2\delta + \zeta = 0 \qquad (5a)$$
$$-\alpha + \delta - \zeta = 0 \qquad (6a)$$

Elimination of ζ from these equations leads to

$$\delta = -\alpha + \beta + \gamma$$
$$\varepsilon = -\beta - \gamma$$
$$\zeta = -2\alpha + \beta + \gamma$$

Equation (2) can now be rewritten as follows:

$$\Pi = \Delta p^\alpha d^\beta l^\gamma \varrho^{-\alpha+\beta+\gamma} \eta^{-\beta-\gamma} v^{-2\alpha+\beta+\gamma}$$

or

$$\Pi = \left(\frac{\Delta p}{\varrho v^2}\right)^\alpha \left(\frac{d\varrho v}{\eta}\right)^\beta \left(\frac{l\varrho v}{\eta}\right)^\gamma$$

The resulting dimensionless product Π thus consists of three dimensionless groups (numbers). The first is the Euler number $Eu = \Delta p/(\varrho v^2)$; the second, the Reynolds number $Re = d\varrho v/\eta$; and the third, when recombined with the second, the aspect ratio l/d. The pressure drop of a fluid in a straight, smooth pipe can therefore be expressed in dimensionless form with the relationship (cf. p. 3-13 and Fig. 3):

$$Eu = f(Re, l/d)$$

The above method originates from Lord RAYLEIGH [56]. LANGHAAR [6] points out that BUCKINGHAM's method [55] differs only superficially from that of RAYLEIGH. The various methods are not discussed further because they are presented in detail elsewhere (e.g., [6], [10]). Note, however, that the method which employs the dimensional matrix (Chap. 3) is based on the same algebraic steps employed in Example 3.

These three examples clearly show how dimensional analysis deals with specific problems and what conclusions it allows. Lord RAYLEIGH's sarcastic comment should now be easier to understand [51]:

> "I have often been impressed by the scanty attention paid even by original workers in physics to the great principle of similitude. It happens not infrequently that results in the form of "laws" are put forward as novelties on the basis of elaborate experiments, which might have been predicted a priori after a few minutes' consideration."

2.3. Fundamentals

2.3.1. Physical Quantities and Relationships Between Them

In studying physical processes, one strives to obtain quantitative relationships between physical quantities that may or may not be of the same type. (Quantities of different types may, however, have the same dimensions; e.g., rotational speed, frequency, and shear rate; kinematic viscosity and diffusivity.) Types of physical quantities or *entities* (e.g., lengths, masses, and times) are regarded as purely qualitative, and are described by their method of measurement. A *physical quantity*, on the other hand, is a quantitatively described object (e.g., a mass of 5 kg). Therefore, one must discover what conditions allow physical appearances to be quantified and treated as mathematical objects for which functional relationships can then be formulated.

These questions touch on the principles of dimensional theory, which in turn deals with the rules of the formal language used to express physical statements. They are based on two important postulates:

1) *Any two physical quantities of the same type can be compared with one another by suitable means of measurement, and a positive real number can then be assigned to this pair.* If, for example, two bodies are weighed, the amount (numerical value) by which one body is lighter or heavier than the other can be determined. *Basic (or primary) quantities*, such as mass, length, or time, are thus quantified through comparison with their corresponding "standards." Once a standard (the basic unit) has been arbitrarily chosen, every quantity can be associated with a positive real number called its numerical value with respect to the chosen basic unit.

$$\left\{ \begin{matrix} \text{Physical} \\ \text{quantity} \end{matrix} \right\} = \text{Numerical value} * \text{Basic unit}$$

Example: a mass of 5 kg = 5 ∗ kg. (The symbol ∗ indicates that the combination is the result of a physical comparison; 5 ∗ kg is abbreviated as 5 kg for practical purposes.)

2) *Every physical interrelation can be represented as a relationship between the physical quantities involved.* For example, the acceleration of a body can be described by a relationship $R(F, m, a)$, which states that a specific force F is required to impart an acceleration a to a mass m. According to Newton's second law of motion, this relationship has the form $F = ma$. Numerical calculations with this equation, however, are not possible until numerical values are assigned to these quantities. This can only be accomplished through a choice of arbitrary basic units and quantification of the physical quantities involved by comparison as outlined in postulate 1.

The second example includes two physical quantities—force F and acceleration a—which are termed *secondary or derived quantities* because they consist of a number of basic ones. The line separating primary and secondary quantities is largely arbitrary. Until recently, a system of dimensions (the so-called technical system) was used in which force was a primary dimension instead of mass.

Secondary quantities are not quantified by comparison with standards of the same type. Instead, relevant basic quantities ae measured, and the results are combined according to specific rules. (Example: To obtain a numerical value for velocity, the distance and time required to cover it are measured. The first of these numerical values is then divided by the second.)

2.3.2. Consistency of Secondary Units and Invariance of Physical Relationships

For physical relationships to be independent (invariant) of the choice of system of units, two requirements must be met: (1) all secondary units must be consistent with the basic units, and (2) the relationships must be dimensionally homogeneous. The first requirement means that in the SI system with the basic units {kg, m, s, . . } the velocity, for example, cannot be expressed in kilometers per hour but rather in meters per second. The second requirement means that all additive terms in a relationship must have the same dimension and any function arguments (e.g., sin, exp) must be dimensionless.

2.3.3. Physical Dimensions, Systems of Dimensions, and Dimensional Constants

Physical laws define relationships for secondary or derived quantities (e.g., velocity = length/time) and thus for secondary units. For example, Newton's second law states that the force F is given by $F = m[\text{kg}]\, a[\text{m s}^{-2}]$. From this, one can conclude that force has the unit of $[\text{kg m s}^{-2}]$ and thus the dimension $[MLT^{-2}]$. In this way dimensions can be assigned to all physical entities, which then form their own *system of dimensions*. The currently used international system of dimensions (SI, Système International

Table 1. Common physical quantities and their dimensions according to the current SI system (left) and the dimensional system (right) in which heat is a basic quantity (dimension H)

Physical quantity	SI [MLTΘ]	[MLTΘH]
Mass	M	
Length	L	
Time	T	
Temperature	Θ	
Heat		H
Area	L^2	
Volume	L^3	
Angular velocity frequency shear rate	T^{-1}	
Velocity	LT^{-1}	
Acceleration	LT^{-2}	
Kinematic viscosity diffusion coefficient heat diffusivity	L^2T^{-1}	
Density	ML^{-3}	
Moment of inertia	ML^2	
Surface tension	MT^{-2}	
Dynamic viscosity	$ML^{-1}T^{-1}$	
Momentum	MLT^{-1}	
Force	MLT^{-2}	
Pressure, stress	$ML^{-1}T^{-2}$	
Energy, torque	ML^2T^{-2}	
Angular momentum	ML^2T^{-1}	
Power	ML^2T^{-3}	
Mechanical equivalent of heat, J	1	$ML^2T^{-2}H^{-1}$
Specific heat capacity	$L^2T^{-2}\Theta^{-1}$	$M^{-1}\Theta^{-1}H$
Thermal conductivity	$MLT^{-3}\Theta^{-1}$	$L^{-1}T^{-1}\Theta^{-1}H$
Heat-transfer coefficient	$MT^{-3}\Theta^{-1}$	$L^{-2}T^{-1}\Theta^{-1}H$
Stefan–Boltzmann constant	$MT^{-3}\Theta^{-4}$	$L^{-2}T^{-1}\Theta^{-4}H$

Table 2. The SI dimensions and units of primary quantities

Primary quantity	Primary dimension	Primary unit
Length	L	m
Mass	M	kg
Time	T	s
Thermodynamic temperature	Θ	K
Amount of substance	mol	mol
Electric current	I	A
Luminous intensity	I_v	cd

d'Unités; Table 1), prescribed by the Geneva Convention in 1954, is based on the primary quantities and their corresponding primary units given in Table 2 [57].

Some important secondary units have been named after famous scientists:

Force expressed in newtons: $N = kg\,m\,s^{-2}$
Energy expressed in joules: $J = N\,m = kg\,m^2\,s^{-2}$
Power expressed in watts: $W = J\,s^{-1} = N\,m\,s^{-1}$
$= kg\,m^2\,s^{-3}$

In building a system of dimensions, *dimensional constants* must sometimes be introduced to preserve the consistency of the system.

Example 1: According to Newton's second law, force is expressed as $F \sim ma$ with the dimension $[MLT^{-2}]$. According to Newton's third law, however, force is also given by $F \sim (m_1 m_2)/r^2$, which would lead to the dimension $[M^2L^{-2}]$. This inconsistency is resolved by agreeing on $F = ma$ and postulating a dimensional ("gravitational") constant G, defined by the third law: $F = G(m_1 m_2)/r^2$ with the dimension for $G[M^{-1}L^3T^{-2}]$. This example shows that the dimensional homogeneity of physical relationships and the invariance of their functional representations can always be preserved through the introduction of dimensional constants.

Example 2: Mechanical energy is defined by the relationship energy = force × distance, leading to the dimension $[ML^2T^{-2}]$. On the other hand, thermal energy is given as $E = JQ$. Here, heat Q is a basic quantity with the primary dimension $[H]$ and the primary unit calories or BTU, respectively. The dimensional constant J (Joule's mechanical equivalent of heat) has the dimension $[ML^2T^{-2}H^{-1}]$ and the value 4200 $kg\,m^2\,s^{-2}\,kcal^{-1}$.

An existing system of dimensions can be altered by making a dimensional constant dimensionless. Example: If the mechanical equivalent of heat J is made equal to a dimensionless constant with the value of one, heat Q has the dimension of energy $[ML^2T^{-2}]$. The system of dimensions {MLTΘH} is thus reduced to the SI system {MLTΘ}, in which calorific quantities such as c, λ, k, and α now have new dimensions (see Table 1). This illustrates that dimensions are not inherent properties of physical quantities but merely expressions of the laws that define them.

2.3.4. The Dimensional Matrix and Its Linear Dependence

Practical application of dimensional analysis leads to the solution of homogeneous, linear equations of the sort already discussed on p. 3-6. Mathematical calculations with determinants (quadratic arrangements of numbers) or with matrices (right-angled arrangements of numbers) can be used for this purpose.

An example is a problem involving velocity v, length l, mass m, density ϱ, viscosity η, and gravitational acceleration g. A dimensional matrix is assembled in which the dimensions of these physical quantities are arranged as in a table (the dimension of g is $[M^0L^1T^{-2}]$):

	v	l	m	ϱ	η	g
M	0	0	1	1	1	0
L	1	1	0	-3	-1	1
T	-1	0	0	0	-1	-2

This matrix consists of three "rows" and six "columns".

For the next step we need to know whether the rank r of this matrix is really three and whether the rows of the matrix are linearly independent of each other. If one of the rows is simply the sum of the other two, it can be described as a linear combination of these two and is dependent on them. In this case the rank of the matrix is lower than the number of rows it contains. The rank of the matrix does not change if the rows that constitute the linear combination are eliminated. The rank of the matrix is normally given by the order of determinants; this can be found easily with the aid of the Gaussian algorithm (as suggested by PAWLOWSKI [9]). The dimensional matrix must exhibit a zero-free main diagonal, and all the places below it must consist exclusively of zeros. The number of elements of the main diagonal that do not disappear but form a continuous sequence is the rank r of the matrix.

In practice the structural elements of the dimensional matrix (here the dimensions of the physical quantities) are arranged in such a way as to fulfill this requirement. If necessary, the columns can also change places (as in this example):

	m	l	v	ϱ	η	g
M	1	0	0	1	1	0
L	0	1	1	-3	-1	1
T	0	0	-1	0	-1	-2

If this method fails, additional *equivalence transformations* can be used, in which single rows or linear combinations of rows are added together. This simple method can be illustrated with an example. The three physical quantities—density ϱ, dynamic viscosity η, and kinematic viscosity v—must be proved to depend linearly on one another because they are bound together by the definition $v \equiv \eta/\varrho$. The rank of the dimensional matrix cannot be three but must be two. The proof is as follows:

The structure of the middle matrix shows that elimination of -1 in the second column of the T-row automatically eliminates -1 in the third column too. The last row then consists exclusively of zeros (see the right matrix). This proves that the rank of this dimensional matrix is only two.

2.3.5. The Π Theorem

The Π theorem forms the basis for a discussion of physical relationships within the framework of the theory of similarity; it reads

Every physical relationship between n physical quantities can be reduced to a relationship between $m = n - r$ mutually independent dimensionless groups, whereby r stands for the rank of the dimensional matrix, made up of the physical quantities in question and generally equal to the number of the basic quantities contained in them.

The Π theorem can be proved in a number of different ways (e.g., see [10]). However, since PAWLOWSKI's method [9] of deriving complete sets of dimensionless groups by a matrix calculation has been chosen, his proof of the Π theorem, which is based on the same procedure, will be presented here.

Let us examine a physical relationship involving x_k ($k = 1, 2, \ldots, n$) physical properties of any type. These quantities can be separated into two sets: the i set and the j set. The i set contains only x_i quantities with dimensions that are linearly independent of one another; their units may be either primary or secondary. The j set consists of all the other quantities x_j, the dimensions of which are linearly dependent on the i set. Since every physical relationship consists of at least one quantity with a dimension that can be expressed in terms of the dimensions of the other quantities (a consequence of the dimensional homogeneity), each set must contain at least one element; it follows for $n \geq 2$ that neither r nor m is zero.

This classification of the physical quantities involved in the process being considered is illustrated by the dimensional matrix shown in Figure 1. This matrix consists of a quadratic *unity matrix* and a *residual matrix*. The unity matrix

	ϱ	η	v
M	1	1	0
L	-3	-1	2
T	0	-1	-1

M
L 1/2(3 M + L) + T
T

	ϱ	η	v
M	1	1	0
L	0	1	1
T	0	-1	-1

M
L 1/2(3 M + L) + T
T 1/2(3 M + L) + 2 T

	ϱ	η	v
M	1	1	0
L	0	1	1
T	0	0	0

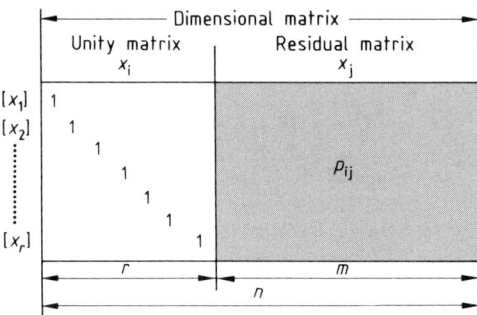

Figure 1. Dimensional matrix showing classification of physical quantities with dimensions that are linearly dependent or independent of one another

contains only x_i quantities which are linearly independent of one another; its main diagonal is formed solely of ones and the remaining elements are all zero. The x_j quantities form the residual matrix. Its elements p_{ij} depend on the dimensions of the x_j quantities with respect to x_i ones.

The physical relationship under consideration

$$f_1(x_k) = 0 \quad \text{with } k = 1, 2, \ldots, n$$

can thus be expressed by the equation

$$f_2(x_i, x_j) = 0$$

Replacing the original primary units with new ones, which differ from them by a factor of $(1/a_i)$, results in both a change of the numerical values from x_i to $a_i x_i$ and substitution of the x_j quantities by the expression:

$$\left(\prod_{i=1}^{r} a_i^{p_{ij}}\right) x_j$$

Thus,

$$f_2 = \left(a_i x_i, x_j \prod_{i=1}^{r} a_i^{p_{ij}}\right) = 0$$

This equation is valid for all positive values of a_i. If a_i values are chosen so as to obtain $a_i x_i = 1$, the equation changes to

$$f_2 = \left(1, x_j \prod_{i=1}^{r} x_i^{-p_{ij}}\right) = 0$$

Obviously, the relationship has now shrunk from its original n arguments to only m arguments,

whereby $m = n - r$. This being the case,

$$f_2 = \left(x_j \prod_{i=1}^{r} x_i^{-p_{ij}}\right) = \Phi(\Pi_j) = 0$$

These new variables

$$\Pi_j = x_j \prod_{i=1}^{r} x_i^{-p_{ij}}$$

are often called Π variables (because of the product symbol Π) or dimensionless numbers or groups. In German technical literature the term "Kennzahl" (criterion) is commonly used. It emphasizes the fact that the dimensionless numbers characterize specific conditions (e.g., different ranges in hydro- or thermodynamics). From here on the term *dimensionless number or group* is used in preference to Π variable.

The relationships between dimensionless numbers are called *characteristics* when they characterize specific properties of a piece of apparatus or a process [e.g., power characteristic $Ne(Re)$ of a stirrer, or heat transfer characteristic $Nu(Re, Pr)$ of a mixing vessel, or the pressure drop characteristic $Eul/d(Re)$ of a fluid in a pipe].

3. Description of a Physical Process with a Full Set of Dimensionless Numbers

3.1. The Relevance List for a Problem

All the essential ("relevant") physical quantities (variables, parameters) that describe a physical or technological interrelation must be known before the process can be described with a full set of dimensionless numbers. This demands a thorough and critical appraisal of the process being examined.

LANGHAAR [6] points out that this first step of naming process parameters may often require a "philosophical insight" into natural phenomena. BRIDGMAN [5] goes further still: in discussing the example of the period of oscillation of a pendulum (see p. 3-4), he remarked that, to clearly define physical interdependence, preliminary tests must sometimes be carried out "by someone at some time soiling his hands with direct contact."

The application of dimensional analysis is indeed heavily dependent on available knowledge.

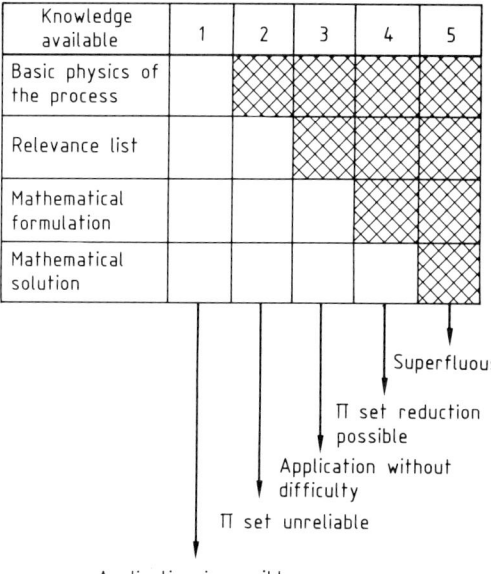

Figure 2. Ease of application of dimensional analysis depending on the degree of knowledge available about a particular problem [58]

PAWLOWSKI outlines the following five cases (Fig. 2) [58]:

1) The physics of the basic phenomenon is unknown:
   → Dimensional analysis cannot be applied
2) Enough is known about the physics of the basic phenomenon to compile a first, tentative relevance list:
   → The resultant Π set is unreliable
3) All the relevant physical variables describing the problem are known:
   → The application of dimensional analysis is unproblematic
4) The problem can be expressed in terms of a mathematical equation:
   → A closer insight into the Π relationship is feasible and may facilitate reduction of the set of dimensionless numbers
5) A mathematical solution of the problem exists:
   → Dimensional analysis is superfluous

Approaching a problem from the point of view of dimensional analysis remains useful even if all the variables relevant to the problem are not known (case 2). The timely application of dimensional analysis may often lead to the discovery of forgotten variables or the exclusion of fakes (see example of pendulum, p. 3-4).

The relevant physical variables comprise a single target quantity [the only dependent variable; this can also be a target function (e.g., residence time distribution) or even a target field (e.g., a temperature field)] and a series of parameters that influence it. These parameters can be divided into three categories:

1) Geometric variables
2) Material parameters (physical properties)
3) Process-related variables

The relevance list may also contain universal physical constants and intermediate quantities.

Geometric Variables. The geometric variables are all measurements of length, radii of curvature, angles, etc., that define the geometry of the problem under examination. In using dimensional analysis, however, only one "characteristic" measurement of length l need be entered in the relevance list. This serves for the dimensionless expression of all other geometric parameters: $\Pi_{geom} = (l_i/l, r_j/l, \alpha_z)$. (Angles are dimensionless geometrical parameters because they are defined as the ratios of two lengths.) The characteristic length chosen should have a particular significance for the process.

When tests are carried out on a single piece of apparatus to which no geometric changes are made (e.g., on a model of an existing full-scale device), the relevance list need include only the characteristic length, but Π_{geom} = idem (= identical value) must be recorded because the Π relationship for this particular piece of apparatus is functionally dependent on Π_{geom}. If, however, the intention is to vary some of the important geometric variables during testing, these must be included in the relevance list. In this case, only the remainder of the Π_{geom} parameters that are not varied in the test have to be listed separately.

Material Parameters. Material parameters include not only the physical properties of the system (e.g., viscosity, density, concentration) but also pore volume ε or mass or volume fraction φ_i of the phases. The last two variables are, by definition, dimensionless. Variables such as kinematic viscosity ν must not be included in the problem relevance list if dynamic viscosity η and density ϱ have already been listed, because ν is linked to these variables by the definition relation $\nu \equiv \eta/\varrho$ (see example in Section 2.3.4). It is, however, completely irrelevant which of the combinations ϱ and η, ϱ and ν, or η and ν is chosen. Other examples of definition relations between physical properties are $\gamma \equiv \varrho g$ for gravity and $a \equiv \lambda/\varrho c$ for heat diffusivity.

Process-Related Parameters. The process-related parameters chosen should be measured directly and not derived (e.g., the throughput q of a liquid, instead of its flow velocity $v \sim q/A$; revolutions per minute n and not tip velocity $u \sim nd$). Any process parameter to be included in the relevance list must be measurable.

Example: The relative velocities of solid particles in a dissolution process in a stirring vessel or of the dispersed phases in a countercurrent extraction column are not measurable. If these relative velocities were accessible to measurement, they could be interpreted as "intermediate quantities" and be usefully incorporated into the relevance list because this would reduce the number of parameters, see below.

Universal Physical Constants. The relevance list must also include universal physical constants such as the universal gas constant R, the speed of light in a vacuum c, or even the acceleration of a gravitational field (on Earth the acceleration due to gravity g) if these constants influence the process involved. The fact that a relevant physical quantity is a constant cannot be a reason for excluding it from the relevance list. By failing to consider the relevance of gravitational acceleration g, a process engineer may make a serious error. This problem is clearly not new. Lord RAYLEIGH [51] complained,

"I refer to the manner in which gravity is treated. When the question under consideration depends essentially upon gravity, the symbol of gravity (g) makes no appearance, but when gravity does not enter the question at all, g obtrudes itself conspicuously."

This is all the more surprising in view of the fact that the relevance of this quantity is easy enough to recognize if one asks the following question: *Would the process function differently if it took place on the moon instead of on Earth?* If the answer to this question is yes, g is a relevant variable.

The gravitational acceleration can be effective only in connection with the density ϱ, i.e., in the form of the gravity ϱg. When inertial forces play a role, the density also has to be listed. Thus it follows that:

1) In cases involving the ballistic (Galilean) movement of bodies, the formation of vortices in stirring, the bow wave of a ship, the movement of a pendulum, and other oscillation processes affected by Earth's gravity, the relevance list comprises ϱg and ϱ.
2) Creeping flow in a gravitational field is only governed by gravity. Only ϱg has to be listed.
3) In heterogeneous material systems with differences in density (sedimentation or buoyancy movements), the difference in specific gravity $g\Delta\varrho$ as well as ϱ are effective.

Intermediate Quantities. In some cases a closer look at the problem (or previous experience) facilitates reduction of the number of physical quantities in the relevance list. This is the case when some relevant variables affect the process by way of an *intermediate quantity*. If this intermediate variable can be measured experimentally, it should be included in the problem relevance list when it facilitates removal of more than one variable from the list.

Example 1: Dispersion processes (e.g., emulsification or aeration) in heterogeneous, fluid material systems (e.g., liquid–liquid and gas–liquid in stirred vessels). In the turbulent flow range (η is irrelevant) the volume-related interphase surface $a \equiv A/V$ depends on the diameter of the stirrer d, the physical properties of the bulk phase (density ϱ and surface tension σ), and the rotational speed n of the stirrer:

$$a = f(d, \varrho, \sigma, n)$$

In the range of fully developed turbulence, a quasi-homogeneous material system exists. Here, the variables n and d can be replaced by P/V (mixing power P per volume of liquid V), and thus the number of variables can be reduced by one:

$$a = f(P/V, \varrho, \sigma)$$

Example 2: Homogenization of liquid mixtures with different densities and viscosities. The mixing time θ needed for complete mixing of two Newtonian liquids, one resting in a layer on top of the other, in a vessel of known geometry (characteristic measurement of length is the stirrer diameter d) depends on the rotational speed of the stirrer n, the physical properties of the two liquids ($\varrho_1, \varrho_2, v_1, v_2$), their volume ratio φ, and the gravity difference $g\Delta\varrho$:

$$\theta = f(d, \varphi; \varrho_1, \varrho_2, v_1, v_2; n, g\Delta\varrho)$$

As mixing progresses, premixing of the two liquids is seen to depend essentially on $g\Delta\varrho$, whereas final homogenization occurs within a material system to which the physical properties of the uniform mixture apply: $v^* = f(v_1, v_2, \varphi)$ and $\varrho^* = f(\varrho_1, \varrho_2, \varphi)$. Introduction of the intermediate physical quantities v^* and ϱ^* allows deletion of three variables from the relevance list:

$$\theta = f(d; \varrho^*, v^*; n, g\Delta\varrho).$$

(See [18] for further details.)

Example 3: Dissolved air flotation makes use of the fact that tiny gas bubbles float the hydrophobic solid particles they have entrapped during desorption. The relevant physical properties are only partially known and are not easy to measure. The problem can be circumvented by introducing the "floating" (surfacing) velocity of the particles as a "lumped parameter." This can be utilized as an intermediate quantity in dimensional analysis provided it is determined separately in the same material system. (See [19] for further details.)

3.2. Determination of a Complete Set of Dimensionless Numbers

The relevance list represents the starting point for the determination of a complete set of dimensionless numbers. The process uses matrix calculation and consists of the following steps [9]:

1) construction of the *dimensional matrix*;
2) application of the *Gaussian algorithm* to determine the rank r of the matrix, which may reduce matrix size (the Gaussian algorithm reveals whether or not physical quantities in the core matrix are linearly independent of each other);
3) formation of the *unity matrix*;
4) formation of the *dimensionless numbers*; and
5) possible *transformation* of these dimensionless numbers to provide more common (usually named) expressions, or dimensionless groups that are more suitable for handling or describing the problem.

The procedure for *constructing the dimensional matrix and determining the complete Π set* is demonstrated by the pressure drop of volume flow in a straight, smooth pipe (inlet effects are ignored). Here, the relevance list consists of

1) the *target quantity*: pressure drop Δp;
2) the *geometric variables*: diameter d and length l of the pipe;
3) the *physical properties*: density ϱ and viscosity v of the fluid; and
4) the *process-related parameter*: volume throughput q:

$\{\Delta p; d, l; \varrho, v; q\}$

When combined with the SI dimensional system, the dimensional matrix takes shape as follows:

	Δp	q	d	l	ϱ	v
M	1	0	0	0	1	0
L	−1	3	1	1	−3	2
T	−2	−1	0	0	0	−1
	Core matrix			Residual matrix		

The nature of the steps of the subsequent process makes this dimensional matrix less than ideal because one has to know that each individual element of the residual matrix appears in only one of the dimensionless numbers, whereas the elements of the core matrix may appear as "fillers" in the denominators of all of them. The residual matrix should therefore be loaded with essential variables like the target quantity and the most important physical properties and process-related parameters. Variables with an as yet uncertain influence on the process must also be included in this group. Should such variables later prove irrelevant, only the dimensionless number concerned will have to be deleted, whereas others may remain unaltered.

Since the core matrix must be transformed into a unity matrix, the "fillers" should be arranged in such a way as to facilitate a minimum of linear transformations. The following reorganization of the above dimensional matrix achieves both of these aims:

	ϱ	d	v	Δp	q	l
M	1	0	0	1	0	0
L	−3	1	2	−1	3	1
T	0	0	−1	−2	−1	0
	Core matrix			Residual matrix		

The next step is the *application of the Gaussian algorithm* (zero-free main diagonal, only zeros below):

	ϱ	d	v	Δp	q	l
Z_1	1	0	0	1	0	0
Z_2	0	1	2	2	3	1
Z_3	0	0	1	2	1	0

$Z_1 = M$
$Z_2 = 3M + L$
$Z_3 = -T$

The rank of the matrix is three ($r = 3$). Only one more linear transformation of the rows is required for *transformation of the core matrix into a unity matrix*.

	ϱ	d	v	Δp	q	l
Z'_1	1	0	0	1	0	0
Z'_2	0	1	0	−2	1	1
Z'_3	0	0	1	2	1	0
	Unity matrix			Residual matrix		

$Z'_1 = Z_1$
$Z'_2 = Z_2 - 2Z_3$
$Z'_3 = Z_3$

In *generating dimensionless numbers*, each element of the residual matrix forms the numerator of a fraction whose denominator consists of the fillers from the uniform matrix with the exponents indicated in the residual matrix:

$$\Pi_1 \equiv \frac{\Delta p}{\varrho^1 d^{-2} v^2} = \frac{\Delta p\, d^2}{\varrho v^2}; \quad \Pi_2 \equiv \frac{q}{\varrho^0 d^1 v^1} = \frac{q}{d v};$$

$$\Pi_3 \equiv \frac{l}{\varrho^0 d^1 v^0} = \frac{l}{d}$$

The dimensionless number Π_1 does not usually occur as a target number for Δp. Unfortunately, it contains the essential physical property v which is already contained in the process number (where it belongs). This disadvantage can easily be overcome by appropriately combining the dimensionless numbers Π_1 and Π_2, which results in the well-known Euler number $Eu \equiv \Pi_1 \Pi_2^{-2} = \Delta p d^4 / \varrho q^2$; this is often combined with $\Pi_3 = l/d$ to obtain an intensively formulated target number. (If one had considered earlier that—by neglecting the inlet effects—Δp is proportional to l, $\Delta p/l$ could have been entered into the relevance list straight away):

$$Eu\, d/l \equiv \Pi_1 \Pi_2^{-2} \Pi_3^{-1} = \frac{\Delta p\, d^4}{\varrho q^2} d/l$$

These *transformations* show how the dimensionless numbers obtained can be appropriately combined into forms that correspond to common or named dimensionless groups or are particularly suitable for evaluating or describing test results.

The structure of dimensionless numbers depends on the variables contained in the core matrix. The Euler number, obtained by combining Π_1 and Π_2 in the example above, would have been obtained automatically if v and q had been exchanged in the core matrix:

	ϱ	d	q	Δp	v	l
Z_1	1	0	0	1	0	0
Z_2	0	1	0	-4	-1	1
Z_3	0	0	1	2	1	0

$Z_1 = M$
$Z_2 = 3M + L + 3T$
$Z_3 = -T$

$$\Pi'_1 \equiv \frac{\Delta p\, d^4}{\varrho q^2} \equiv Eu \quad \Pi'_2 \equiv \frac{v d}{q} \quad \Pi'_3 \equiv \frac{l}{d}$$

All Π sets obtained from one and the same relevance list are equivalent to each other and can be mutually transformed at will.

The dimensionless number Π_2 is in fact the Reynolds number Re. The Reynolds number is defined as any dimensionless number combining a characteristic velocity v and a characteristic measurement of length l with the kinematic viscosity of the fluid $v \equiv \eta/\varrho$. The following dimensionless numbers are equally capable of meeting these requirements:

$$Re \equiv v\frac{d}{v}\left(= \frac{4q}{\pi d^2}\frac{d}{v}\right) \sim \frac{q}{d v} \quad \text{and}$$

$$Re \equiv n d \frac{d}{v} = \frac{n d^2}{v}$$

This method of compiling a complete set of dimensionless numbers shows clearly that the numbers formed in this way can contain neither numerical values nor any other constant. These appear in dimensionless groups only when they are established and interpreted as ratios on the basis of known physical interrelations. For example, $Re \equiv \pi n d^2/v$, where $\pi n d$ is the tip speed, or $Eu = \Delta p/(v^2 \varrho/2)$, where $v^2 \varrho/2$ is the kinetic energy. Since such expressions have the same value as analytically derived ones, the definition must always be given.

In the case of the pressure drop of volume flow in a straight pipe treated above, this method of compiling a complete set of dimensionless numbers produces the relationship

$$f(Eu\, d/l, Re) = 0$$

The information contained in this relationship is the maximum that dimensional analysis can offer on the basis of a relevance list, which is assumed to be complete. Dimensional analysis cannot provide any information about the form of the function f (i.e., the sort of Π relationship) involved. This information can only be obtained experimentally.

3.3. The Π Relationship

STANTON and PANNELL evaluated the function f of the interrelation $f(Eu\, d/l, Re) = 0$ by measurements [20]. Figure 3 shows the result of their work, which demonstrates the significance of the Reynolds number for pipe flow; λ represents the "friction" or "resistance" coefficient, which is defined as

$$\lambda = \frac{\Delta p}{(\varrho/2) v^2} \frac{d}{l} = 2(\pi/4)^2 \frac{\Delta p\, d^4}{\varrho q^2} \frac{d}{l} = 1.24\, Eu\, d/l$$

Referring to this figure, ECK remarks:

"If one represented—as it was once usual—λ as a function of velocity v, one would obtain not a curve but a galaxy. Here, Reynolds' law must strike even a beginner with an absolute clarity" [59].

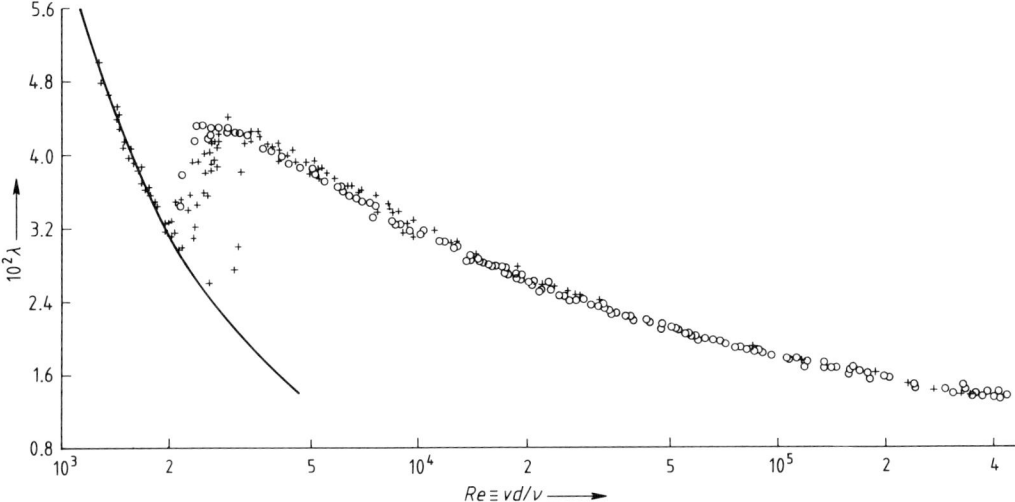

Figure 3. The relationship $\lambda(Re)$ for a smooth pipe with water ($\bigcirc$) and air ($+$) according to [20]
Tube diameters ranged from 0.361 to 2.855 cm (water) and from 0.361 to 12.63 cm (air). The solid curve indicates the theoretical curve for streamline motion.

The drawn-in curve is valid for the laminar flow range ($Re < 2300$) where the Π relationship is $\lambda = 64\,Re^{-1}$ or $\lambda\,Re = 64$. (This connection could have been clearly demonstrated had the authors chosen to present their test results in a double-logarithmic plot; this would have produced a straight line with a gradient of -1 in this range of Re.)

Thus, $\lambda\,Re$ can be viewed as a new dimensionless number that does not include the physical property density ϱ. Only with the Π relationship do the relevance to the problem and the operational range of individual variables become clear.

This example also shows that the Π set compiled on the basis of the relevance list does no more than define the maximum Π space, which may well shrink in a Π relationship like the one cited above.

The following Π relationships are valid for the turbulent flow region ($Re > 2300$):

$\lambda = 0.3164\,Re^{-0.25}$ $Re \leq 8 \times 10^4$
(Blasius)

$\lambda = 0.0054 + 0.396\,Re^{-0.3}$ $Re \leq 1.5 \times 10^6$
(Hermann)

Figure 4 shows the same interrelation in a double-logarithmic plot. (The two axes do not have the same scale: the ordinate has been stretched against the abscissa by a factor of two.) In this case an additional dimensionless geometric parameter, wall roughness ($\delta/d =$ grain diameter/pipe diameter), is added to the "pressure drop characteristic" $\lambda(Re)$. (Measurements are taken from [21].) This variable was not included in the relevance list compiled on p. **3-13** for two reasons: (1) its exclusion allowed us to begin with the results of STANTON and PANNELL, who carried out their tests in smooth pipes; (2) it showed that any mistakes or omissions can be corrected when test results are available.

The fact that analytical presentations of the Π relationships encountered in engineering literature often take the shape of power products does not stem from certain laws inherent in dimensional analysis. It can be explained simply by the engineering preference for depicting test results in double-logarithmic plots. Those sections of the curves that can be approximated as straight lines are then analytically expressed as power products. When this is difficult, the engineer is often satisfied with the curves alone (Figs. 3 and 4).

The "benefits" of dimensional analysis are often discussed. The above example provides a welcome opportunity to make a few comments. The five-parameter dimensional relationship

$\{\Delta p/l;\ d;\ \varrho,\ v;\ q\}$

can be represented by means of dimensional analysis as $\lambda(Re)$ and plotted as a single curve (Fig. 3). To represent this relationship in a di-

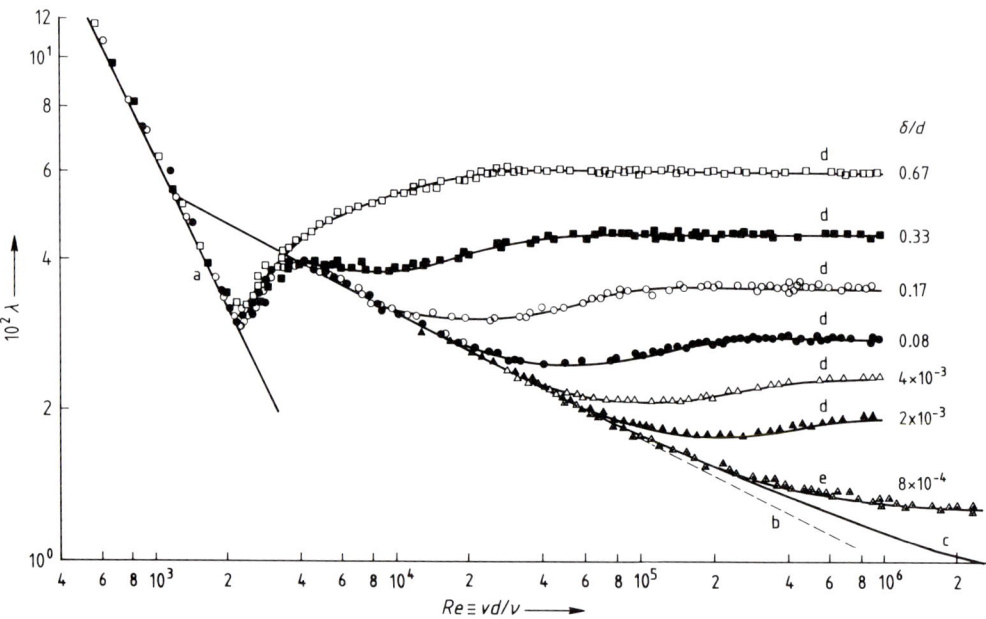

Figure 4. The relationship $\lambda(Re, \delta/d)$ for smooth, industrially rough, and artificially roughened pipes (δ = grain diameter, d = pipe diameter) [59]
a) Laminar flow, $\lambda = 64\, Re^{-1}$; b) Turbulent flow, $\lambda = 0.316\, Re^{0.25}$ (BLASIUS); c) Smooth pipe (STANTON and PANNELL);
d) Artificially roughened pipes (NIKURADZE) [21]; e) Industrially roughened pipe (GALAVICS)

mensional way and avoid creating a "galaxy" at the same time [59], 25 diagrams with five curves in each would be required. If only five measurements per curve are assumed to be sufficient, graphic representation of this problem would still require 625 measurements. The enormous savings in time and energy made possible by the application of dimensional analysis are thus easy to appreciate [6].

3.4. Reduction of Matrix Size

In Section 3.1, mention was made of BUCKINGHAM's assumption that the quantity of Π numbers required to create a complete picture of a problem can be derived from the number of dimensional variables minus the number of basic units of measurement included in them [55]. BRIDGMAN [5] corrected this pointing out that the "rule of thumb" does not always work and that reference to the rank of the matrix, rather than the number of basic units of measurement, is preferable.

The above statement can be demonstrated with a simple example. If two mutually miscible liquids with the same density and viscosity are mixed in a stirring vessel until the mixture is completely homogeneous, the mixing time θ depends on a characteristic measurement of length (stirrer diameter d), the physical properties of the mixture (density ϱ, diffusivity D, and viscosity v), and the process parameter, the rotational speed of the stirrer n. The relevance list is therefore $\{\theta; d; \varrho, D, v; n\}$.

The corresponding dimensional matrix is

	ϱ	d	n	θ	v	D
M	1	0	0	0	0	0
L	−3	1	0	0	2	2
T	0	0	−1	1	−1	−1

The basic unit of measurement for mass is present only in the density ϱ. Therefore, column ϱ and row M must be deleted. The resultant matrix has the rank 2:

	d	n	θ	v	D
L	1	0	0	2	2
−T	0	1	−1	1	1

The derivative Π set is

$\{n\theta,\ Re \equiv nd^2/v,\ Sc \equiv v/D\}$

Table 3. Selection of important dimensionless numbers [a]

Name	Symbol	Group	Remarks
Mechanical process engineering			
Archimedes	Ar	$g\Delta\varrho l^3/(\varrho v^2)$	$\equiv Ga(\varrho/\Delta\varrho)$
Euler	Eu	$p/(\varrho v^2)$	
Froude	Fr	$v^2/(lg)$	
	Fr^*	$v^2\varrho/(lg\Delta\varrho)$	$\equiv Fr(\varrho/\Delta\varrho)$
Galilei	Ga	gl^3/v^2	$\equiv Re^2/Fr$
Knudsen	Kn	λ_m/l	
Mach	Ma	v/v_s	
Newton	Ne	$F/(\varrho v^2 l^2)$	
		$P/(\varrho v^3 l^2)$	
Reynolds	Re	vl/v	$v \equiv \eta/\varrho$
Weber	We	$\varrho v^2 l/\sigma$	
Thermal process engineering (heat transfer)			
Fourier	Fo	at/l^2	
Grashof	Gr	$\beta\Delta Tgl^3/v^2$	$\equiv \beta\Delta T Ga$
Nusselt	Nu	$\alpha l/\lambda$	
Péclet	Pe	vl/a	$\equiv Re\, Pr$
Prandtl	Pr	v/a	$a \equiv \lambda/(\varrho c_p)$
Rayleigh	Ra	$\beta\Delta Tgl^3/(av)$	$\equiv Gr\, Pr$
Stanton	St	$\alpha/(v\varrho c_p)$	$\equiv Nu/(Re\, Pr)$
Thermal process engineering (mass transfer)			
Bodenstein	Bo	vl/D	$\equiv Re\, Sc$
Lewis	Le	a/D	$\equiv Sc/Pr$
Schmidt	Sc	v/D	
Sherwood	Sh	kl/D	
Stanton	St	k/v	$\equiv Sh/(Re\, Sc)$
Chemical process engineering			
Arrhenius	Arr	$E/(RT)$	
Damköhler	Da_I	$k_1 \tau$	
	Da_{II}	$k_1 l^2/D$	$\equiv Da_I Bo = Da_I Re\, Sc$
	Da_{III}	$k_1 \tau \left(\dfrac{c\Delta H_R}{c_p \varrho T_0}\right)$	$\equiv Da_I \left(\dfrac{c\Delta H_R}{c_p \varrho T_0}\right)$
	Da_{IV}	$\dfrac{k_1 c\Delta H_R l^2}{\lambda T_0}$	$\equiv Da_I Re\, Pr \left(\dfrac{c\Delta H_R}{c_p \varrho T_0}\right)$
Hatta	Hat	$(k_1 D)^{1/2} k_L$	
		$(k_2 c_2 D)^{1/2} k_L$	

[a] Definitions of symbols:
a — thermal diffusivity ($\equiv \lambda/\varrho c$)
c, c_1, c_2 — concentration
c_p — heat capacity at constant pressure
D — diffusivity
E — activation energy
F — force
g — gravitational constant
ΔH_R — heat of reaction
k, k_L — mass-transfer coefficient (index L — liquid side)
k_1, k_2 — reaction rate constant (index = reaction order)
l — length
$p, \Delta p$ — pressure, pressure difference
P — power
R — universal gas constant
t — time
$T, \Delta t$ — temperature, temperature difference
v — velocity
v_s — velocity of sound
α — heat-transfer coefficient
β — temperature coefficient of density
λ — thermal conductivity
λ_m — molecular free path length
η — dynamic viscosity
v — kinematic viscosity
$\varrho, \Delta\varrho$ — density, density difference
σ — surface tension
τ — residence time

Of course, this does not mean that density is irrelevant in this problem; rather it is already represented by the kinematic viscosity $v \equiv \eta/\varrho$. (Exchanging kinematic viscosity v for dynamic viscosity $\eta\,[M L^{-1} T^{-1}]$ would result in the same three-parameter Π set. In that case, however, the matrix would retain its rank $r = 3$.)

3.5. Change of Dimensional Systems

A change of dimensional system leading to an increase in the number of basic dimensions (e.g., adding the basic quantity heat H in thermodynamic problems, cf. both dimensional systems in Table 1) may at first seem tempting to reduce the number of Π variables for a given number of variables in the relevance list. However, expanding (or reducing) a dimensional system means that the corresponding relevance list must also be expanded (or reduced) by the appropriate dimensional constant. This procedure, therefore, has no effect on the resulting number of Π variables. If, however, the dimensional constant is foreseen to be irrelevant to the problem, it need not be added to the relevance list and the quantity of dimensionless numbers can be reduced by one. This often occurs in thermodynamic problems (cooling, heating, steady-state heat transfer) when mechanical heat generation is negligible and Joule's mechanical heat equivalent J is therefore irrelevant.

A selection of important dimensionless numbers and their definitions are given in Table 3.

4. Similarity and Scale-Up

4.1. Basic Principles of Scale-Up

Chapter 1 points out that using dimensional analysis to handle a physical problem, and thus to present it in the framework of a complete set of dimensionless numbers, is a sure way to obtain a simple and reliable scale-up from a small-scale model to the full-scale technical plant. The theory of models states that

Two processes may be considered completely similar if they take place in similar geometrical space and if all the dimensionless numbers necessary to describe them have the same numerical value ($\Pi_i = $ idem).

This statement is supported by the results shown in Figure 3. The researchers carried out their measurements in smooth pipes with diameters d in the range 0.36–12.63 cm. The physical properties of the fluid tested (water or air) varied widely. Nevertheless, every numerical value of Re still corresponds to a specific numerical value of λ. The Π relationship presented is thus valid not only for the laboratory devices examined but also for any other geometrically similar arrangement:

Every point in a Π framework, determined by the Π relationship, corresponds to an infinite number of possible implementations.

This characteristic of Π representation forms the basis of the dimensionally analytical concept of similarity:

Processes that are described by the same Π relationship are considered similar to each other if they correspond to the same point in the Π space.

Two realizations of the same physical interrelation are considered similar (complete similarity) when $m - 1$ dimensionless numbers of the m-dimensional Π framework have the same numerical value ($\Pi_i = $ idem) because the mth Π number will then automatically also have the same numerical value.

Scale-up with the simple example of Figure 3 can be attempted by imagining a pipeline several hundred kilometers long, through which a given fluid (natural gas or crude oil) is to be transported with a given throughput. The aim is to determine the pressure drop of the fluid flow in the pipeline in order to design pumps and compressors.

A geometrically similar small-scale model of the technical pipeline is first constructed. The physical properties of the fluid, its throughput, and the dimensions of the technical plant are known, as is therefore the numerical value of Re in operation. This value can be kept constant in the test apparatus by the correct choice of conveying device (pump, compressor, etc.) and model fluid. The pressure drop measured under these conditions allows calculation of the Euler number in the model. In this case the condition $Re = $ idem automatically implies $Eu = $ idem. The numerical value of the measured Euler number therefore corresponds to that of the full-scale plant. This then allows determination of the numerical value of Δp in the technical plant from the numerical value of Eu in the model and the given operational parameters.

The concept of complete similarity does not guarantee that a process is the same in the model and the full-scale version in every respect; it is only the same with regard to the particular aspect under examination which has been described by the appropriate Π relationship. To demonstrate this fact with the help of the above example, remember that the flow conditions in two smooth pipes of different scales should be considered similar when $Re = $ idem and, according to the pressure drop characteristics, therefore have the same numerical value of $Eu\,d/l$. This, however, does not mean that heat-transfer conditions in the two pipes are the same; for that to be the case, the relevant Π relationship $Nu = f(Re, Pr)$ requires that both the Reynolds number and the Prandtl number have the same numerical value (temperature-independent physical properties of the medium are supposed).

The more comprehensive the similarity demanded between model and full-scale device and the greater the

scale-up factor $\mu \equiv l(\text{model})/l(\text{technical plant})$

the harder it is to perform the scale-up. It can even fail completely if a material system with the physical properties required for model experiments cannot be obtained (cf. Section 4.3). A further difficulty is that scale-up involving large changes of scale may cause changes to the Π space. An example is the case of forced nonisothermal flow, in which progression in scale results in free convection and thus in the Grashof number, $Gr \equiv \beta \Delta T\, l^3 g/\nu^2$, becoming relevant to the problem (β is the temperature coefficient of density).

4.2. Experimental Methods for Scale-Up

Chapter 1 posed a number of questions often asked in connection with model experiments [60]:

1) *How small can a model be?* The size of a model depends on the scale factor μ and on the accuracy of experimental measurements. When $\mu = 1:10$, a 10% margin of error may already be excessive. A larger scale for the model must therefore be chosen to reduce μ.
2) *Is one model scale sufficient or should tests be carried out on models of different sizes?* One model scale is sufficient if the relevant numerical values of the dimensionless numbers necessary to describe the problem (the "process point" in the Π space describing the operational condition of the technical plant) can be adjusted by choosing the appropriate process parameters or physical properties of the material model system. If this is impossible, the process characteristics must be determined in a series of models of different sizes, or the process point must be extrapolated from experiments in technical plants of different sizes (cf. Section 4.3).
3) *When must model experiments be carried out exclusively with the original material system?* If the material model system is unavailable (e.g., in the case of non-Newtonian fluids) or the relevant physical properties are unknown (e.g., foams, sludges, slimes), model experiments must be carried out with the original material system. In this case, measurements must be performed in models of various size. Problems entailed in the nonavailability of a model material system can occasionally limit the applicability of the theory of similarity. It would, however, be wrong to refer to "limits of the theory of similarity."

4.3. Scale-Up Under Conditions of Partial Similarity

When appropriate substances are not available for model experiments, accurate simulation of the working conditions in an industrial plant on a laboratory scale may be impossible. Experiments with equipment of different sizes are then customarily used before extrapolation of the results to full-scale operating conditions. Sometimes this expensive, unreliable procedure can be replaced by a well-planned experimental strategy, in which the process is divided into parts that are then investigated separately (e.g., prediction of the drag resistance of a ship's hull by using FROUDE's approach; see Example 1 in this section) or by deliberately abandoning certain similarity criteria and checking the effect on the entire process (e.g., combined mass and heat transfer in a catalytic pipe reactor via DAMKÖHLER's approach; see [47]).

Several "rules of thumb" for dimensioning different types of process equipment are, in fact, scale-up rules based unknowingly on partial similarity. These rules include the so-called *volume-related mixing power* P/V, widely used for dimensioning mixing vessels, and the *superficial velocity* $v = q/A$, which is normally used to scale-up bubble columns. Some remarks on both these rules are given in Examples 2 and 3 at the end of this section (see also [61]).

Example 1: Drag Resistance of a Ship's Hull. This problem represents the birth and breakthrough of scale-up rules and is closely linked to the name of WILLIAM FROUDE (1810–1879). FROUDE solved this significant scale-up problem with a clear physical concept based on carefully executed experiments.

This problem is first treated here by using dimensional analysis. The drag resistance F of a ship's hull of a given geometry (characteristic length l and displacement volume V) depends on the speed v of the ship, the density ϱ and kinematic viscosity ν of the water and, because of bow wave formation, also on g, the acceleration due to gravity. The list of relevant quantities is thus

$$\{F;\ l,\ V;\ \varrho,\ \nu;\ v,\ g\}$$

The dimensional matrix

	ϱ	l	v	F	ν	g	V
M	1	0	0	1	0	0	0
L	−3	1	1	1	2	1	3
T	0	0	1	−2	−1	−2	0

leads, after only two linear transformations, to the following unity matrix (with rank $r = 3$) and the residual matrix

	ϱ	l	v	F	ν	g	V
M_1	1	0	0	1	0	0	0
L_1	0	1	0	2	1	−1	3
T_1	0	0	1	2	1	2	0

$M_1 = M$
$L_1 = 3M + L + T$
$T_1 = -T$

The dimensionless numbers are as follows:

$\Pi_1 = F/(\rho l^2 v^2) \equiv Ne$ (Newton number)
$\Pi_2 = \nu/lv \equiv Re^{-1}$ (Reynolds number)
$\Pi_3 = gl/v^2 \equiv Fr^{-1}$ (Froude number)
$\Pi_4 = V/l^3$ dimensionless displacement volume

The problem is thus completely defined by the Π set

$\{Ne, Re, Fr, V/l^3\}$

While maintaining geometric similarity (V/l^3 = const.), experiments on the scale of $\mu = 1:100$ should be carried out to obtain Ne. However, the same Re and Fr cannot be set simultaneously, because when using the same liquid (water) the requirement $Re =$ idem demands that vl be constant, whereas $Fr =$ idem demands that v^2/l be constant.

If the speed v of the model (subscript M) is specified by maintaining the same Fr value as in the full-scale application (subscript F):

$Fr =$ idem: $(v^2/l)_M = (v^2/l)_F \rightarrow v_M = v_F \mu^{1/2}$

then the same Re value in the model experiment must be attained by means of the appropriate kinematic viscosity ν:

$Re =$ idem: $(vl/\nu)_M = (vl/\nu)_F \rightarrow \nu_M = \nu_F \mu^{3/2}$

For the model liquid at $\mu = 1:100$, therefore, $\nu_M = 10^{-3} \nu_F$. However, no fluid satisfies the condition $\nu = 10^{-3} \nu_{water}$.

If the scale-up factor were not necessarily so small ($\mu = 1:100$) and models were not so expensive to build, experiments could be conducted in water at the same value of Fr with models of various sizes, and the results extrapolated to $Ne(Re_F)$. In view of the powerful model reduction and the resulting extreme differences in the Reynolds number,

$\mu = 1; 0.1; 0.01 \rightarrow Re_M/Re_F = 1; 0.032; 0.001$

extrapolation appears risky, particularly when the cost of the motor used in the full-scale application is considered.

Naturally, these results of dimensional analysis and their consequences were unknown to ship builders of the 19th century. Since the time of RANKINE, the total drag resistance of the ship has been divided into three parts: surface friction, stern vortex, and bow wave. However, the concept of Newtonian mechanical similarity, known at that time, stated only that for mechanically similar processes the forces vary as $F \sim \rho l^2 v^2$; scale-up was not considered for assessing the effect of gravity.

FROUDE observed that the resistance due to the stern vortex was relatively small compared to the other two resistances and decided to combine it with the bow wave resistance to obtain the *form drag* F_f. By careful investigations and theoretical considerations, he concluded that the wave formation of the ship could be simulated by using scale models, and he arrived at the *law of appropriate velocities*:

The wave formations at the ship and the model are (geometrically) similar, if the velocities are in the ratio of the square root of the linear dimensions.

FROUDE also found that for similar wave formations, the hull drag (*friction drag* F_r) behaves not as $F_r \sim v^2 l^2 \rho$, but as $F_r \sim v^{1.825} A \rho$ (A = surface area); he developed computational methods for scaling down models and ships by length and the type of wetted surface. Thus, he was able to calculate the form drag F_f from the total drag after subtracting the predictable friction drag. He found:

"If we adhere to the law of the appropriate velocities in scaling up the ship, the form resistances will correspond to the cubes of their dimensions (that is to say, their displacement volumes)" [44].

In summary,

1) $F_{total} - F_r = F_f$
2) If $v^2 \sim l$, then $F_f \sim l^3$

Dividing the functional dependence (2) by $\rho l^2 v^2$ in order to transform F_f into the Newton number of the form drag requires that

$$\frac{F_f}{\rho l^2 v^2} \sim \frac{l^3}{\rho l^2 v^2} = \frac{l}{\rho v^2} = \text{const.}$$

This means that

$Ne_f =$ idem at $Fr =$ idem with $Fr \equiv v^2/lg$

To verify these experimental results, the corvette *Greyhound* was towed by the corvette *Active* under the command of FROUDE, and the drag force in the tow rope was measured. Observed deviations from the model predictions were in the range 7–10 % [62].

WEBER points out that this procedure is not entirely correct and can never provide real proof, because complete similarity between the model and its full-scale counterpart cannot be achieved [1]. The procedure can therefore represent nothing more than an excellent approximation of reality. He says

"The fact that FROUDE was able to achieve his goal with such a large measure of success despite all the difficulties, lies in his ingenuity which enabled him to itemize and to assess all the practical and theoretical details of drag resistance and finally to trace a clear picture of this intricate phenomenon."

FROUDE's performance cannot be judged too highly, especially in view of the measuring techniques available to him.

PAWLOWSKI discussed an interesting alternative experimental approach to this scale-up problem [58]. He also started by splitting the drag resistance into friction, depending only on Re, and bow wave resistance, depending only on Fr (index F indicates full-scale device):

$$Ne_F = f_1(Re_F) + f_2(Fr_F)$$

However, he proposed a different strategy from that of FROUDE. In the first experiment, measurements at $Fr_1 = Fr_F$ are made with the model ship in water, with $Re_1 = Re_F \mu^{-3/2}$ (i.e., measurement is carried out at a correct value of Fr and a false value of Re). As a result, a value of Ne_1 is obtained from the relationship

$$Ne_1 = f_1(Re_1) + f_2(Fr_F)$$

Two additional experiments are carried out, not with the model ship but with a totally immersed form (Fig. 5) whose shape is given by reflecting the immersed portion of a ship's hull at the water line (at $V/l^3 = $ idem). In these experiments the Froude number is irrelevant; the friction corresponding to the surface area of the model must be divided by 2.

The measurement in water is carried out at Re_1 and Re_F to obtain $Ne_2 = f_1(Re_F)$ and $Ne_3 = f_1(Re_1)$. The desired Ne_F can now be calculated:

$$Ne_F = f_1(Re_F) + f_2(Fr_F) = Ne_1 - Ne_3 + Ne_2$$

Example 2: Mixing Power per Unit Volume as a Criterion for Scale-Up of Mixing Vessels. Scaling up stirred tanks by using the criterion of constant impeller power per unit volume, P/V, is an important example of scale-up under conditions of partial similarity in chemical engineering. Obviously, the complicated fluid mechanical processes that govern mass and heat transfer in

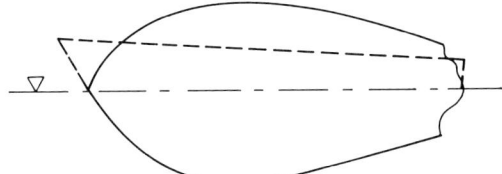

Figure 5. Sketch of a completely submerged streamlined body

stirred tanks cannot generally be described adequately with the help of such a simple criterion.

The scale-up criterion P/V is only adequate in gas–liquid contacting and in liquid–liquid dispersion processes, when the impeller power is as uniformly dissipated as possible in the tank volume (microscale mixing, isotropic turbulence; see also → 25. Mixing, **B2**).

The most important type of mixing operation (i.e., homogenization of liquid mixtures) depends on the scale of the convective bulk transport (macromixing). Measurements result in the relationship $n\theta = f(Re)$ which depends heavily on the type of stirrer and the vessel geometry [22].

Convective bulk transport (which is also responsible for the flow pattern at the tank bottom) is extremely important in a suspension of solids in a stirred tank; P/V cannot be used as a scale-up criterion in this process either. Measurements have shown that the minimum rotational speed n_{crit} of the stirrer which is necessary for the suspension (whirling up) of particles in the turbulent regime is given by the appropriate Froude number:

$$Fr_{crit} \equiv n_{crit}^2 d\varrho/g\Delta\varrho$$

What is the connection between P/V and the Froude number, the latter being the scale-up criterion? The answer follows.

Because the Froude number is the scale-up rule here, one begins with $Fr \sim n^2 d$ (the same material system is assumed) and expresses it in terms of $P/V \sim n^3 d^2$ [$P \sim n^3 d^5$ (turbulent region) and $V \sim d^3$]:

$$Fr \sim n^2 d \sim (n^2 d)^{3/2} = (P/V)d^{-1/2} = \text{idem}$$

From this is obtained (the scale factor being defined as $\mu = d_M/d_F$),

$$[(P/V)d^{-1/2}]_F = [(P/V)d^{-1/2}]_M$$
$$Fr = \text{idem} \to (P/V)_F = (P/V)_M \mu^{-1/2}$$

Example 3: Superficial Velocity as a Criterion for Scale-Up of Bubble Columns. Bubble columns are often designed on the basis of the *superficial velocity* $v \equiv q/A$ (q = gas throughput, A = cross-sectional area). Many authors have found that gas–liquid mass transfer in bubble columns is indeed governed by this quantity:

$$k_L a \sim v \to k_L a/v = \text{const.}$$

where k_L is the liquid-side mass-transfer coefficient and a is the interfacial area per unit volume. This is only understandable when one considers the interdependence of the volume-related mass-transfer coefficient $k_L a = G/V\Delta c$ and the superficial velocity $v = q/A$, as well the fact that volume $V = HA$ (H = liquid height in the column):

$$\frac{k_L a}{v} = \frac{G}{V \Delta c} \cdot \frac{A}{q} = \frac{G}{Hq\Delta c} = \text{const.}$$

Thus, the gas absorption rate $G[MT^{-1}]$ is proportional to the liquid height H, the gas throughput q, and the concentration difference Δc [24].

However, even for homogenization of the liquid content in the column by rising gas bubbles, v is not the only relevant parameter because, analogous to the stirred tank (Example 2), liquid bulk transport (back-mixing) must take place over the entire height of the liquid column. Experiments performed on different scales gave the following expression for the mixing time θ in a bubble column of given geometry [25]:

$$\theta (g/d)^{1/2} \sim Fr^{-1/4}$$

where $Fr \equiv v^2/dg$ and d is column diameter. Thus,

$$\theta \sim v^{-1/2} d^{3/4} \text{ or } \theta = \text{idem} \to v^{-1/2} d^{3/4} = \text{idem}$$

This leads to the conclusion that

$$v_F = v_M \mu^{-3/2}$$

Thus, in a bubble column geometrically scaled up by a factor of 10 the same mixing time as in the model is obtained only when the superficial velocity is increased by a factor of $10^{3/2} = 32$. Hence, v is not a scale-up criterion here.

Examples 2 and 3 show that a particular scale-up criterion, which is valid in a given type of apparatus for a particular process, is not necessarily applicable to other processes occurring in the same device.

5. Treatment of Variable Physical Properties by Dimensional Analysis

When using dimensional analysis to tackle engineering problems, the physical properties of the material system are generally assumed to remain unaltered in the course of the process. Relationships such as the heat-transfer characteristic of a technical device (e.g., vessel, pipe), $Nu = f(Re, Pr)$, are valid for any material system with Newtonian viscosity and for any constant process temperature (i.e., for constant physical properties). However, constancy of physical properties cannot be assumed in every physical process: a temperature field may well generate a viscosity field or even a density field; in non-Newtonian (structurally viscous or viscoelastic) liquids, a shear rate can also produce a viscosity field.

Although most physical properties (e.g., viscosity, density, heat conductivity and capacity, surface tension) must be regarded as variable, only the viscosity can vary by many orders of magnitude under certain process conditions. In the following, dimensional analysis is applied solely to describe the variability of viscosity. However, the approach can be adapted to other physical properties.

5.1. Dimensionless Representation of the Material Function

Similar behavior of a certain physical property common to different material systems can be visualized only by dimensionless representation of the material function of that property. Furthermore this function should be formulated as uniformly as possible. This can be achieved by "standard representation" of the material function [49] in which a standardizing transformation of the material function $\eta(T)$ is defined, such that the expression produced

$$\eta/\eta_0 = f\{-\gamma_0(T - T_0)\}$$

meets the requirement: $f(0) = f'(0) = 1$ where

$$\gamma_0 \equiv \left(\frac{1}{\eta} \cdot \frac{\delta \eta}{\delta T}\right)_{T_0}$$

is the temperature coefficient of the viscosity and $\eta_0 \equiv \eta(T_0)$, where T_0 is a reference temperature. Figures 6A and 6B depict the dramatic effect of this standard transformation.

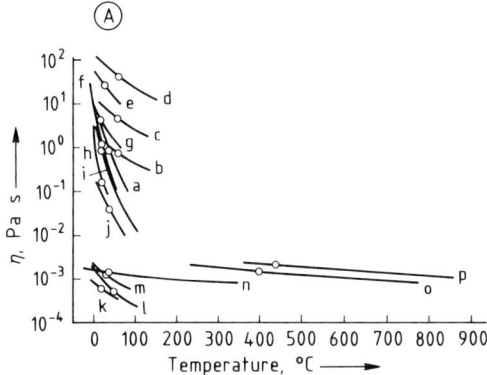

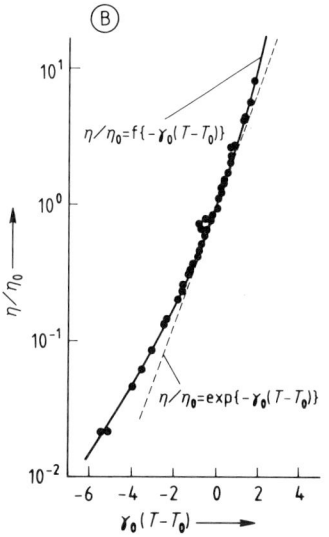

Figure 6. Temperature dependence of the viscosity $\eta(T)$ (A) and its standard representation (B) which leads to a reference-invariant approximation of the material function [13]
a) Baysilon resin; b) Baysilon M 1000; c) Baysilon M 10.000; d) Baysilon M 100.000; e) 25% Perbunan in benzene; f) Glycerol; g) 25% Levapren in benzene; h) Rapeseed oil; i) Castor oil; j) Olive oil; k) Methanol; l) Water; m) Turpentine oil; n) Mercury; o) Molten tin; p) Molten lead
η = viscosity at temperature T; η_0 = viscosity at reference temperature T_0; γ_0 = temperature coefficient of viscosity

Engineers prefer the representation

$$\eta/\eta_0 = \exp\{-\gamma_0(T - T_0)\}$$

although this is not the best possible approximation (see dashed line in Fig. 6 B). A slightly better approximation of the material function $\eta(T)$ is provided by the well-known Arrhenius relationship (cf. [13, Fig. 1.3.3]):

$$\frac{\eta}{\eta_0} = \exp\left\{-\frac{E_0}{RT_0}\left(\frac{T_0}{T} - 1\right)\right\}$$

To describe the process by dimensional analysis, the reference temperature T_0 should be formulated in a process-related manner by using a characteristic such as mean process temperature as reference. A class of functions exists, however, which are independent (invariant) of the reference point T_0. The solid line in Figure 6 B shows a function of this type.

5.2. The Π Set for Variable Physical Properties

The type of dimensionless representation of the material function affects the (extended) Π set within which the process relationship is formulated. When the standard representation is used, the relevance list must include the reference viscosity η_0, instead of η, and incorporate two additional parameters γ_0 and T_0. This leads to two additional dimensionless numbers in the process characteristic. With regard to the heat-transfer characteristic of a device, $Nu = f(Re, Pr)$, it now follows that

$$Nu = f(Re_0, Pr_0, \gamma_0 \Delta T, \Delta T/T_0)$$

where the index 0 in Re and Pr denotes that these two dimensionless numbers are to be formed with η_0 (which is the numerical value of η at T_0).

Because the standard transformation of the material function can be expressed invariantly with regard to the reference temperature T_0 (Fig. 6 B), the relevance list is extended by only one additional parameter γ_0. This, in turn, leads to only one additional dimensionless number. For the above problem then,

$$Nu = f(Re_0, Pr_0, \gamma_0 \Delta T)$$

In process characteristics for heat transfer, the temperature dependence of the density is normally not taken into account by the dimensionless number $\beta \Delta T$ (β is the temperature coefficient of the density, defined in a way similar to γ). Instead the Grashof number Gr is normally employed thus taking into account the fact that the density differences can only be effective in the presence of gravity:

$Gr = \beta \Delta T\, Re^2\, Fr^{-1} = g\beta \Delta T l^3 v^{-2}$

For more in-depth information on Sections 5.1 and 5.2, see [9, Chap. 3], [13], [63], [64].

5.3. Treatment of Non-Newtonian Liquids by Dimensional Analysis

The main characteristic of Newtonian liquids is that simple shear flow (e.g., Couette flow) generates shear stress τ which is proportional to the rate of shear D. For a full description of flow in non-Newtonian fluids, see → 5. Fluid Mechanics, pp. 24–40. The proportionality constant, the dynamic viscosity η, is the only material constant in the law of flow:

$\tau = \eta D$

η depends only on pressure and temperature.

In non-Newtonian liquids, η depends on D as well. These liquids can be classified in various categories of materials depending on their flow behavior (cf. DIN 1342/1, 1342/2, and 13 342). The graphic representation of flow behavior by using $D(\tau)$ is called a flow curve; when $\eta(D)$ or $\eta(\tau)$ is used, it is called a viscosity curve.

Figure 7 depicts the viscosity curve for a mixture of steam engine cylinder oil and ca. 7% aluminum stearate. The asymptote a corresponds to the so-called yield point $\tau_0 = 50$ N/m²; at a high rate of shear, $\eta_\infty = 9.7$ Pa·s = const. is reached (asymptote b). Figure 8 shows a dimensionless representation of the viscosity curves of a synthetic and a biological polymer (rheological material functions).

PAWLOWSKI points out that the rheological properties of many non-Newtonian liquids can be described by material parameters whose di-

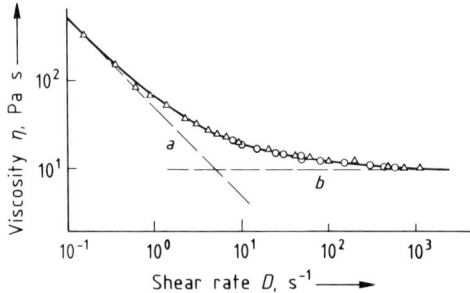

Figure 7. Viscosity curve for a mixture of steam engine cylinder oil and ca. 7% aluminum stearate [9]

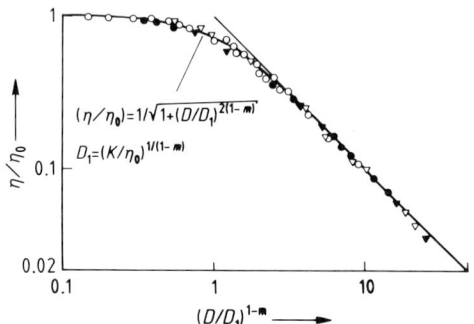

Figure 8. Dimensionless representation of polymer viscosity curves for a synthetic polymer (carboxymethyl cellulose) and a biological polymer (xanthan) [64]
η = viscosity; η_0 = viscosity at reference temperature; c_p = polymer concentration in water; D = shear rate; D_1 = characteristic shear rate; K = consistency index at $D_1 = 1$; m = flow exponent

mensional matrix has a rank $r = 2$ [9, Chap. 5], [65]. These physical properties can be usefully linked to produce two dimensional constants, a characteristic viscosity constant H and a characteristic time constant θ, and possibly a set of dimensionless material numbers Π_{rheol}. The relevance list is thus

$\{H, \theta, \Pi_{\text{rheol}}\}$

(In Fig. 7, $H = b$ and $\theta = H/\tau_0 = 0.194$ s.)

In changing from a Newtonian to a non-Newtonian liquid, the above interrelation leads to following consequences with regard to the complete Π set:

1) every dimensionless number incorporating η must now be formulated with the dimensionally equivalent H (e.g., η_∞);
2) a single process-related number containing θ is added; and
3) the pure material numbers are increased by Π_{rheol}.

These rules can be demonstrated on the heat-transfer characteristic of a smooth, straight pipe

(A represents temperature-independent, and B temperature-dependent, viscosity):

	Newtonian liquid	non-Newtonian liquid
A	Nu, Re, Pr	$Nu, Re_H, Pr_H, v\theta/l$
B	$Nu, Re, Pr, \gamma_0 \Delta T$	$Nu, Re_H, Pr_H, v\theta/l,$ $\gamma_H \Delta T, \gamma_\theta/\gamma_H$

In B, $\gamma_H \equiv \delta \ln H/\delta T$ and $\gamma_\theta = \delta \ln \theta/\delta T$ must be added; ΔT denotes the temperature difference between the bulk of the liquid and the wall. The conventionally used expression η/η_w (index w = wall) and $\gamma \Delta T$ are related by

$$\eta/\eta_w = \exp[-\gamma(T - T_w)]$$

Because little is known about the rheological properties of material systems, model experiments must be carried out with the same substance to be used in the full-scale plant. Since $\Pi_{material}$ (here Pr_H) = idem and Π_{rheol} = idem, the process takes place within a Π range with only one more dimensionless number ($v\theta/l$) than in the case of Newtonian liquids. However, when scaling up from the model to the full-scale plant, complete similarity cannot be attained by using the same material system. In the above example, keeping ϱ, H, θ = idem makes it impossible to ensure that $Re_H \equiv \varrho v l/H$ and $v\theta/l$ also remain identical. Therefore, the same substance should be retained but the scale of the model should be altered in the experiments (Section 4.2). Pure fluid mechanical processes involving creeping (ϱ being irrelevant), steady-state, and isothermal flow are exceptions. In these cases, mechanical similarity can be obtained despite the constancy of physical properties.

Non-Newtonian liquids with flow curves obeying the so-called power law $\tau = KD^m$ are known as *Ostwald–de Waele fluids*. The dimension of K depends on the numerical value of the exponent m and is therefore not a consistent physical quantity. To produce a material constant X having the dimension of viscosity, K must be combined with a characteristic velocity v and a characteristic length l:

$$X \equiv K(v/l)^{m-1}$$

With this sort of liquid, the extension of the Π set has the following consequences: (1) X replaces η in every dimensionless number in which it appears; and (2) pure material numbers are extended by m (m belongs to Π_{rheol}).

In the case of an Ostwald–de Waele fluid the heat-transfer characteristic of a smooth straight pipe is

$$Nu = f(Re_X, Pr_X, m)$$

where $Re_X = \varrho v^{2-m} L^m/X$. (To learn more about the questionability of this Π space, which originates from the inconsistency of K, see [14], [49].)

5.4. Treatment of Viscoelastic Liquids by Dimensional Analysis

Almost every biological solution of low viscosity (as well as viscous biopolymers such as xanthan, Fig. 8) and dilute solutions of long-chain polymers [carboxymethyl cellulose (CMC, Fig. 8), polyacrylamide, polyacrylonitrile, etc.] display not only viscous but also viscoelastic flow behavior. These liquids are capable of storing part of the deformation energy elastically and reversibly. They evade mechanical stress by contracting like rubber bands. This behavior causes secondary flow that often runs contrary to the flow produced by mass forces (e.g., the liquid "climbs" the shaft of a stirrer, the so-called Weissenberg effect).

Elastic behavior of liquids is characterized mainly by the ratio of first differences in normal stress N_1 to the shear stress τ. This relation, the Weissenberg number $Wi = N_1/\tau$, is usually represented as a function of the rate of shear D.

HENZLER takes an approach similar to that of the Ostwald–de Waele law [64]:

$$Wi = AD^a \quad \text{or preferably} \quad Wi = AD^a + BD^b$$

To transform these material functions into dimensionless forms, a reference Weissenberg number Wi_0 is chosen. This leads to the generation of the characteristic rates of shear D_1 and D_2. Thus,

$$Wi/Wi_0 = (D/D_1)^a + (D/D_2)^b$$

In this case, the following constituent parts of the dimensionless material function must be incorporated in the relevance list:

$$\{Wi_0, D_1, D_2, a, b\}$$

In the case of a viscoelastic liquid the material functions of both the viscous and the elastic behavior must be considered. If the viscous behavior of the liquid concerned is described as in Figure 8 and the above Wi/Wi_0 function is valid for its elastic behavior, the Π set must be modified as follows:

1) η_0 replaces η in every dimensionless number in which η appears
2) The pure material-related numbers are extended by the following six purely rheological numbers:

$$\Pi_{\text{rheol}} = Wi_0, \ D/D_1, \ D/D_2, \ m, \ a, \ b$$

(See [66] for rheological characteristic of fermentation broths.)

6. References

General References

[1] M. Weber: "Die Grundlagen der Ähnlichkeitsmechanik und ihre Verwertung bei Modellversuchen," *Jahrb. Schiffbautech. Ges.* **20** (1919) 355–477.
[2] M. Weber: "Das allgemeine Ähnlichkeitsprinzip der Physik und sein Zusammenhang mit der Dimensionslehre und der Modellwissenschaft," *Jahrb. Schiffbautech. Ges.* **31** (1930) 274–388.
[3] R. E. Johnstone, M. W. Thring: *Pilot Plants, Models, and Scale-up Methods in Chemical Engineering*, McGraw-Hill, New York 1957.
[4] Gröber/Erk/Grigull: *Die Grundgesetze der Wärmeübertragung*, Springer Verlag, Berlin – Göttingen – Heidelberg 1963, pp. 159 ff.
[5] P. W. Bridgman: *Dimensional Analysis*, Yale University Press, New Haven 1922, 1931, 1951; reprinted by AMS Press, New York 1978. German transl. by H. Holl: *Theorie der physikalischen Dimensionen – Ähnlichkeitsbetrachtungen in der Physik*, Verlag Teubner, Leipzig – Berlin 1932.
[6] H. L. Langhaar: *Dimensional Analysis and Theory of Models*, John Wiley & Sons, New York 1951; reprinted by R. E. Krieger Publ., Huntingdon, N.Y., 1980.
[7] R. C. Pankhurst: *Dimensional Analysis and Scale Factors*, Chapman and Hall, London 1964.
[8] L. I. Sedov: *Similarity and Dimensional Methods in Mechanics*, Moscow 1943; Engl. transl. by Academic Press, New York 1959.
[9] J. Pawlowski: *Die Ähnlichkeitstheorie in der physikalisch-technischen Forschung – Grundlagen und Anwendungen*, Springer Verlag, Berlin – Heidelberg – New York 1971.
[10] H. Görtler: *Dimensionsanalyse – Theorie der physikalischen Dimensionen mit Anwendungen*, Springer Verlag, Berlin – Heidelberg – New York 1975.
[11] W. Haeder, E. Gärtner: *Die gesetzlichen Einheiten in der Technik*, issued by the German Standards Committee (Deutscher Normenausschuß DNA) Berlin, Beuth-Vertrieb, Berlin – Köln – Frankfurt.
[12] W. Matz: *Anwendungstechnik des Ähnlichkeitsgrundsatzes in der Verfahrenstechnik*, Springer Verlag, Berlin – Göttingen – Heidelberg 1954.
[13] J. Pawlowski: *Veränderliche Stoffgrößen in der Ähnlichkeitstheorie*, Verlag H. R. Sauerländer, Aarau 1991.
[14] J. Pawlowski: *Einwellen-Schnecken; Förder-, Homogenisier- und Wärmeaustausch-Verhalten*, Verlag H. R. Sauerländer, Aarau 1990.
[15] J. Zierep: *Ähnlichkeitsgesetze und Modellregeln der Strömungslehre*, G. Braun (Wissenschaft + Technik), Karlsruhe 1982.
[16] H. Klingspor: *Experimentelle und theoretische Untersuchungen von Rektifiziervorgängen im Lichte der Ähnlichkeitstheorie*, Dechema Monographien, **37** (1960) 171–271.
[17] H. Wetzler: *Kennzahlen der Verfahrenstechnik*, Hüthig Verlag, Heidelberg 1985.

Specific References

[18] M. Zlokarnik: "Einfluß der Dichte- und Zähigkeitsunterschiede auf die Mischzeit beim Homogenisieren von Flüssigkeitsgemischen," *Chem.-Ing.-Tech.* **42** (1970) no. 15, 1009–1011.
[19] M. Zlokarnik: "Neue Flotationstechniken zur Abtrennung und Eindickung von Klärschlamm bei der biologischen Abwasserreinigung, part 2: Entgasungsflotation," *Korresp. Abwasser* **32** (1985) no. 7, 598–603.
[20] T. E. Stanton, J. R. Pannell: "Similarity of Motion in Relation to the Surface Friction of Fluids," *Philos. Trans. R. Soc. London A* **214** (1914) 199–225.
[21] J. Nikuradze: "Strömungsgesetze in rauhen Rohren," *VDI-Forschungsh.* **361** (1933) July/August.
[22] M. Zlokarnik: "Eignung von Rührern zum Homogenisieren von Flüssigkeitsgemischen," *Chem.-Ing.-Tech.* **39** (1967) nos. 9/10, 539–548.
[23] M. Zlokarnik: "Rührleistung in begaster Flüssigkeit," *Chem.-Ing.-Tech.* **45** (1973) no. 10 a, 689–692.
[24] M. Zlokarnik: "Sorptions-Charakteristiken des Schlitzstrahlers in Abhängigkeit von Koaleszenzbedingungen des Systems," *Chem.-Ing.-Tech.* **50** (1978) no. 9, 715. "Sorption Characteristics of Slot Injectors and Their Dependency on the Coalescence Behaviour of the System," *Chem. Eng. Sci.* **34** (1970) no. 10, 1265–1271.
[25] M. Zlokarnik: "Homogenisieren von Flüssigkeiten durch aufsteigende Gasblasen," *Chem.-Ing.-Tech.* **40** (1968) no. 15, 765–768.
[26] M. Zlokarnik: "Auslegung von Hohlrührern zur Flüssigkeitsbegasung," *Chem.-Ing.-Tech.* **38** (1966) no. 3, 357–366.
[27] W. Müller, H. Rumpf: "Das Mischen von Pulvern in Mischern mit axialer Mischbewegung," *Chem.-Ing.-Tech.* **39** (1967) nos. 5/6, 365–373.
[28] M. Zlokarnik: "Eignung von Einlochböden als Gasverteiler in Blasensäulen," *Chem.-Ing.-Tech.* **43** (1971) no. 6, 329–335.
[29] M. Zlokarnik: "Auslegung und Dimensionierung eines mechanischen Schaumzerstörers," *Chem.-Ing.-Tech.* **56** (1984) no. 11, 839–844. "Design and Scale-up of Mechanical Foam Breakers," *Ger. Chem. Eng.* **9** (1986) no. 5, 314–320.
[30] M. Zlokarnik: "Wärmeübergang an der Wand eines Rührbehälters beim Kühlen und Heizen im Bereich $10^{\circ} < Re < 10^5$," *Chem.-Ing.-Tech.* **41** (1969) no. 22, 1195–1202.
[31] J. Pawlowski, M. Zlokarnik: "Optimieren von Rührern für eine maximale Ableitung von Reaktionswärme," *Chem.-Ing.-Tech.* **44** (1972) no. 16, 982–986.

[32] W. Kast: "Untersuchungen zum Wärmeübergang in Blasensäulen," *Chem.-Ing.-Tech.* **35** (1963) no. 11, 785–788.

[33] M. Zlokarnik: "Auslegung von Hohlrührern zur Flüssigkeitsbegasung," *Chem.-Ing.-Tech.* **38** (1966) no. 7, 717–723.

[34] M. Zlokarnik: "Sorption Characteristics for Gas–Liquid Contacting in Mixing Vessels," *Adv. Biochem. Eng.* **8** (1978) 133–151.

[35] H. Judat: "Stoffaustausch Gas/Flüssigkeit im Rührkessel – eine kritische Bestandsaufnahme," *Chem.-Ing.-Tech.* **54** (1982) no. 7, 520–521. "Gas/Liquid Mass Transfer in Stirred Vessels—A Critical Review," *Ger. Chem. Eng.* **5** (1982) no. 6, 357–363.

[36] M. Zlokarnik: "Eignung und Leistungsfähigkeit von Oberflächenbelüftern für biologische Abwasserreinigungsanlagen," *Korresp. Abwasser* **27** (1980) no. 7, 14–21.

[38] M. Zlokarnik: "Die Leistungsfähigkeit optimierter Injektoren für den Sauerstoffeintrag in biologischen Abwasserreinigungsanlagen," *VT Verfahrenst.* **13** (1978) nos. 7/8, 601–604.

[39] M. Zlokarnik: "Tower-Shaped Reactors for Aerobic Biological Waste Water Treatment" in H.-J. Rehm, G. Reed (eds.): *Biotechnology*, vol. 2, VCH Verlagsgesellschaft, Weinheim 1985, pp. 537–569.

[40] Leonardo da Vinci: *Notebooks*, around 1500; from [3].

[41] Galileo Galilei: *Discorsi*, 1638; German in Ostwalds Klassiker, vol. 11, pp. 106–109.

[42] Isaac Newton: *Principia*, 1687, liber II, sectio VII, propositio 32.

[43] R. Bertrand, *C. R.* **25** (1847) 163; *J. Ec. Polytech.* (*Paris*) **1848**, 32, 189.

[44] C. W. Merrifield: "The Experiments Recently Proposed on the Resistance of Ships," *Trans. Inst. Naval Arch.* (*London*) **11** (1870) 80–93.

[45] O. Reynolds: "An Experimental Investigation of the Circumstances which Determine Whether the Motion of Water Shall be Direct or Sinuous, and of the Law of Resistance in Parallel Channels," *Philos. Trans. R. Soc. London* **174** (1883) 935–982.

[46] L. H. Baekeland: "Practical Life as a Complement to University Education – Medal Address," *J. Ind. Eng. Chem.* **8** (1916) 184–190.

[47] G. Damköhler: "Einflüsse der Strömung, Diffusion und des Wärmeüberganges auf die Leistung von Reaktionsöfen," *Z. Elektrochem.* **42** (1936) 846–862.

[48] H. Görtler: "Zur Geschichte des Π-Theorems," *Z. Angew. Math. Mech.* **55** (1975) 3–8.

[49] J. B. J. Fourier: *Théorie analytique de la chaleur*, Paris 1822.

[50] H. v. Helmholtz: "Über ein Theorem, geometrisch ähnliche Bewegungen flüssiger Körper betreffend, nebst Anwendung auf das Problem, Luftballons zu lenken," *Monatsber. Kgl. Preuß. Akad. Wiss. Berlin* **1873**, 501–514.

[51] Lord Rayleigh: "The Principle of Similitude," *Nature* (*London*) **95** (1915) no. 2368, 66–68.

[52] A. Vaschy: *Traité d'électricité et de magnétisme*, vol. I, Baudry et Cie., Paris 1890.

[53] A. Federmann: "Über einige allgemeine Integrationsmethoden der partiellen Differentialgleichungen erster Ordnung," *Ann. Polytech. Inst. Peter der Große, St. Petersburg* **16** (1911) 97–154.

[54] D. Riabouchinsky: "Méthode des variables de dimension zéro," *L'aérophile* **19** (1911) 407–408.

[55] E. Buckingham: "On Physically Similar Systems; Illustrations of the Use of Dimensional Equations," *Phys. Rev.* **4** (1914) no. 4, 345–376.

[56] Lord Rayleigh: "On the Viscosity of Argon as Affected by Temperature," *Proc. R. Soc. London* **66** (1899/1900) 68–74.

[57] *SI Units and Recommendations for the Use of Their Multiples and Certain Other Units*, ISO/DIS, January 1972.

[58] J. Pawlowski, *Seminar on the Theory of Similarity*, Bayer AG, Leverkusen 1967.

[59] E. Beck: *Technische Strömungslehre*, Springer Verlag, Berlin–Göttingen–Heidelberg 1961, p. 123.

[60] M. Zlokarnik: "Modellübertragung in der Verfahrenstechnik," *Chem.-Ing.-Tech.* **55** (1982) no. 5, 363–372; "Scale-up in Process Engineering," *Ger. Chem. Eng.* **7** (1984) 150–159.

[61] M. Zlokarnik: "Modellübertragung bei partieller Ähnlichkeit," *Chem.-Ing.-Tech.* **57** (1985) no. 5, 410–416; "Scale-Up Under Conditions of Partial Similarity," *Int. Chem. Eng.* **27** (1987) no. 1, 1–9.

[62] W. Froude: "On Experiments with H.M.S. "Greyhound"," *Trans. Inst. Naval Arch.* (*London*) **15** (1874) 36–73.

[63] J. Pawlowski: "Modellversuche an Newtonschen Flüssigkeiten mit temperaturabhängiger Viskosität," *Verfahrenstechnik* **8** (1974) no. 9, 269–272.

[64] H.-J. Henzler: "Rheologische Stoffeigenschaften – Erklärung, Messung, Erfassung und Bedeutung," *Chem.-Ing.-Tech.* **60** (1988) no. 1, 1–8.

[65] J. Pawlowski: "Zur Theorie der Ähnlichkeitsübertragung bei Transportvorgängen in nicht-Newtonschen Stoffen," *Rheol. Acta* **6** (1967) no. 1, 54–61. "Relationship Between Process Equations for Processes in Connection with Newtonian and Non-Newtonian Substances," *AIChE J.* **15** (1969) no. 2, 303–305.

[66] H.-J. Henzler, E. E. Schäfer: "Viskose und elastische Eigenschaften von Fermentationslösungen," *Chem.-Ing.-Tech.* **59** (1987) 940–944.

4. Transport Phenomena

RAYMOND W. FLUMERFELT, CHARLES J. GLOVER, Department of Chemical Engineering, Texas A & M University, College Station, Texas 77843, United States

1.	**Foundations**	4-6	2.1.5.	Composite Systems ... 4-25
			2.1.6.	Steady Conduction with Heat Generation and in Multidimensions . 4-26
1.1.	**Mathematical Preliminaries**	4-7		
1.1.1.	Coordinate Systems	4-7	2.1.7.	Transient Heat Conduction—Lumped Capacity Systems ... 4-26
1.1.2.	Vector and Tensor Operations	4-7		
1.1.3.	The Jacobian	4-8	2.1.8.	Transient Heat Conduction—More General Solutions ... 4-26
1.1.4.	Calculus of Vectors and Tensors	4-9		
1.1.5.	Divergence Theorem	4-10	**2.2.**	**Steady, One-Dimensional Flows** ... 4-29
1.1.6.	Kinematic Relations	4-10	2.2.1.	Generalized Couette Flow ... 4-30
1.1.7.	Partial and Total Derivatives	4-11	2.2.2.	One-Dimensional Poiseuille Flows ... 4-32
1.1.8.	Relation Between Different Time Derivatives	4-11	2.2.3.	Flow in Channels of Arbitrary Cross Section ... 4-33
			2.2.4.	Poiseuille Flow of Non-Newtonian Fluids ... 4-33
1.2.	**Basic Equations for Compositionally Homogeneous Systems**	4-11	2.2.5.	Two-Phase Concentric Flow in a Tube—Segregated Flow ... 4-35
1.2.1.	The Reynolds Transport Theorem	4-11		
1.2.2.	Conservation of Total Mass	4-12	**2.3.**	**Multidimensional Momentum Transfer** 4-36
1.2.3.	Conservation of Linear Momentum	4-12	2.3.1.	Two-Dimensional Flows—Stream Function Equations ... 4-36
1.2.4.	Condition of Local Stress Equilibrium	4-13	2.3.2.	Creeping Flow Around a Sphere and Other Bodies of Revolution ... 4-37
1.2.5.	Stress Tensor	4-13	2.3.3.	Flow in Channels with Varying Cross Sections—Lubrication Analysis ... 4-40
1.2.6.	Equation of Motion	4-13		
1.2.7.	Conservation of Angular Momentum	4-13	**2.4.**	**Coupled Momentum and Energy Transfer** ... 4-42
1.2.8.	Mechanical Energy Accounting Equation	4-13	2.4.1.	Heat Transfer in Laminar Tube Flow ... 4-43
1.2.9.	Conservation of Total Energy	4-14		
1.2.10.	Thermal Energy Accounting Equation	4-14	2.4.2.	Momentum and Heat Transfer in Laminar Boundary Layers ... 4-45
1.2.11.	Forms of the Governing Equations	4-14	2.4.3.	Free Convection on a Vertical Plate . 4-49
1.2.12.	Entropy Inequality	4-15		
1.2.13.	Linear Transport Fluxes and Relations	4-15	**2.5.**	**Turbulent Momentum and Energy Transfer** ... 4-50
1.2.14.	Transport Properties from Molecular Theories	4-16	2.5.1.	Physical Characteristics ... 4-50
			2.5.2.	Time-Smoothed Momentum and Energy Balances for Turbulent Flow 4-53
1.2.15.	Non-Newtonian Fluids	4-17		
1.3.	**Summary of Basic Equations**	4-19	2.5.3.	Mixing Length Theories ... 4-54
1.4.	**Boundary Conditions**	4-19	2.5.4.	Turbulent Heat Transfer in Tubes—Reynolds, Prandtl, and von Karman Analogies ... 4-55
1.5.	**Solution Philosophy**	4-19		
1.6.	**Dimensionless Equations of Change**	4-20	2.5.5.	Turbulent Momentum and Heat Transfer on a Flat Plate ... 4-57
2.	**Transport in Compositionally Homogeneous Systems**	4-22	**2.6.**	**Summary of Heat-Transfer Relations** 4-58
			3.	**Transport in Multicomponent Systems** ... 4-58
2.1.	**Heat Conduction in Solids**	4-22		
2.1.1.	Heat Conduction Equations	4-22		
2.1.2.	Initial and Boundary Conditions	4-23	3.1.	**Diffusive Mass Transfer in Binary Systems** ... 4-58
2.1.3.	Steady, One-Dimensional Conduction	4-23		
2.1.4.	Convective Boundary Conditions	4-25	3.1.1.	Species Mass Balances ... 4-58

3.1.2.	Species Diffusion Fluxes	4-62	3.3.4.	Relations for High Mass-Transfer Rates ... 4-79
3.1.3.	Fick's First Law of Diffusion	4-62		
3.1.4.	Special Cases of Diffusion	4-63	**4.**	**Macroscopic Systems** ... **4-81**
3.1.5.	Diffusivities of Gases and Liquids	4-64	**4.1.**	**Conservation of Total Mass** ... **4-82**
3.1.6.	Theoretical Foundations of Steady-State Measurement of Diffusion	4-65	**4.2.**	**Conservation of Linear Momentum** .. **4-83**
3.1.7.	Diffusion with Homogeneous Chemical Reaction	4-67	**4.3.**	**Conservation of Angular Momentum** . **4-83**
3.1.8.	Diffusion with Heterogeneous Reaction	4-69	**4.4.**	**An Accounting of Mechanical Energy** **4-83**
			4.5.	**Conservation of Total Energy** ... **4-84**
3.1.9.	Perspective	4-70	**4.6.**	**An Accounting of Thermal Energy** .. **4-85**
3.2.	**Convective Mass Transport**	**4-70**	**4.7.**	**The Second Law of Thermodynamics** **4-86**
3.2.1.	Gas Absorption in Falling Film with Reaction	4-70	**4.8.**	**Tabular Summary of the Macroscopic Equations** ... **4-87**
3.2.2.	Mass Transfer in Laminar and Turbulent Boundary Layers	4-73	**4.9.**	**Some Examples** ... **4-87**
			4.9.1.	Pressure Drops and Temperature Changes for Closed Channel Flow .. 4-87
3.3.	**Mass Transfer Across Interfaces**	**4-75**		
3.3.1.	Mass-Transfer Coefficients	4-75	4.9.2.	Compressible Flow in a Tube ... 4-88
3.3.2.	Functional Forms of Mass-Transfer Coefficient Relations	4-78	4.9.3.	Heterogeneous Reaction in a Fluidized-Bed Reactor ... 4-89
3.3.3.	Relations for Low Mass-Transfer Rates	4-79	**5.**	**References** ... **4-91**

In addition to the standard symbols defined in the front matter of this volume the following symbols are used:

a, b	major and minor axes of spheroids
a	acceleration
a, b, c	arbitrary vectors
A	area
A	area vector (magnitude and direction)
A, B	arbitrary second-order tensors
A_c	cross-sectional area
A_n	numerical parameters defined by Equation (2.174)
A_w	wall surface area
A_ξ	area perpendicular to ξ-direction
$\hat{A}$	Helmholtz free energy per unit mass
ATSB	A through stagnant B diffusion situation
b	flow geometry height
b^*	dimensionless flow channel height, defined below Equation (2.174)
b_0, b_L	flow channel vertical dimensions at $x = 0$ and $x = L$, Figure 13
B	applied load, Figure 13
Bi	Biot number; Equation (2.19) and definition below Equation (2.27)
Br	Brinkman number, Equation (1.118a)
c	molar concentration, moles per volume
c_p	constant pressure heat capacity per unit mass
c_v	constant volume heat capacity per unit mass
c_A, c_B, c_i	moles of A, B, i in a mixture per unit volume of mixture
c_A^*	dimensionless molar concentration of species A
$\bar{c}_A$	time-averaged value of c_A
c_{A0}	molar concentration at interface
$c_{A\infty}$	molar concentration of A in free stream, or in bulk fluid in contact with surface
C_D	drag coefficient defined by Equation (2.155)
C_{Dx}	local drag coefficient defined in Equation (2.235)
C_{DL}	overall drag coefficient defined by Equation (2.236)
C_k	coefficients in Equation (2.49)
$C_1, C_2, \hat{C}_1, \hat{C}_2$	integrating constants
d	diameter
d/dt	time derivative seen by an observer moving at arbitrary velocity; the total time derivative of a function
D/Dt	time derivative seen by observer maintaining constant material coordinates (following the motion of an element, Eq. 1.50)
D_{AB}	diffusivity of species A in B
D_{AB}°	diffusivity of infinitely dilute A in solvent B
Da_δ	Damköhler number based on film thickness, Equation (3.100)
Da_L	Damköhler number based on film length
e	height of wall roughness (protuberances)
e_{ijk}	permutation symbol, Equation (1.12)
E^2	differential operator, Equation (2.137)
$E(k)$	complete elliptic integral with argument k
ECD	equimolar counter diffusion situation
f	in Section 2.4.1, function of r defined by Equation (2.197); in Section 2.3.2, function of r defined by Equations (2.133) and (2.134); in Section 2.3.3, a function defined below Equation (2.162); friction factor defined by Equation (2.79)
	a function of time (1.56)
F	force, a vector
F_B	buoyancy force

F_D	drag force		in Section 2.5.3, proportionality constant in Equation (2.299); reaction rate constant, Equation (4.29)
$\boldsymbol{F}_{ext}$	external force		
F_{in}, F_{out}	volumetric flow rate entering and leaving a system	k'	reaction rate constant, Equation (3.98)
F_x, F_y, F_z	force components	k'_y	drift factor-corrected gas-phase mass-transfer coefficient
F_x^*, F_y^*, F_z^*	dimensionless force components		
$\mathscr{F}$	in Chapter 1, an arbitrary scalar, vector, or tensor function of time and position; in Section 2.5.2, any linear time or spatial derivative operator	k_c	single-phase mass-transfer coefficient based on molar concentrations
		k_{cL}	average mass-transfer coefficient defined by Equation (3.112)
Fr	Froude number, Equation (1.114)	k_G	gas-phase mass-transfer coefficient based on gas-phase partial pressures, Equation (3.132)
$\hat{F}$	conversion of mechanical energy to thermal energy by viscous dissipation, per unit mass (fluid friction losses)		
		k_x	liquid-phase mass-transfer coefficient based on liquid-phase mole fractions
$\boldsymbol{g}$	body force per unit mass or gravitational acceleration	k_y	gas-phase mass-transfer coefficient based on gas-phase mole fractions
$\boldsymbol{g}_i, \boldsymbol{g}_j, \boldsymbol{g}_k$	i, j, k base vectors in a curvilinear coordinate system	k_1	first-order reaction rate constant
		K	pressure gradient defined by Equation (2.68); overall reaction rate constant, Equation (4.31)
g_x, g_y, g_z	components of gravitational force per unit mass, in x, y, z directions		
G	scalar function in Equation (2.125)		
Gr_x	local Grashoff number, Equation (2.273)	K^*	dimensionless pressure gradient, defined by Equation (2.73)
Gr_L	Grashoff number associated with surface of length L		
		K_c	overall two-phase mass-transfer coefficient based on liquid-phase molar concentrations
$\hat{G}$	Gibbs free energy per unit mass		
h	heat-transfer coefficient, Equation (2.6)		
h_k	heat-transfer coefficient associated with surface at position x_k	K_G	overall two-phase mass-transfer coefficient based on gas-phase partial pressures, Equation (3.134)
h_{lm}	log mean heat-transfer coefficient, Equation (2.206)		
		K_x	overall two-phase mass-transfer coefficient based on liquid-phase mole fractions
$h_{x, lam}$	local heat-transfer coefficient, laminar boundary layer		
		K_y	overall two-phase mass-transfer coefficient based on gas-phase mole fractions
$h_{x, turb}$	local heat-transfer coefficient, turbulent boundary layer		
		$\widehat{KE}$	kinetic energy per unit mass
h_L	overall heat-transfer coefficient over surface of length L, Equation (2.254)	l	characteristic length, Equation (2.82); in Section 2.5.3, the Prandtl mixing length
$\bar{h}$	average heat-transfer coefficient over surface A	$\boldsymbol{L}$	angular momentum
		L	characteristic length for a process or flow, Chapter 1; length of body, surface, or flow channel
$\hat{H}$	enthalpy per unit mass		
$\boldsymbol{i}_i, \boldsymbol{i}_j, \boldsymbol{i}_k$	base vectors in a rectangular cartesian coordinate system		
		m, n	material parameters for power-law non-Newtonian fluid, Equation (2.97)
$\boldsymbol{i}_x, \boldsymbol{i}_y, \boldsymbol{i}_z$	unit base vectors in x, y, z directions		
$\boldsymbol{i}_r, \boldsymbol{i}_\theta, \boldsymbol{i}_z$	unit base vectors in cylindrical coordinates	n	coordinate in direction normal to surface, Equation (2.6)
$\boldsymbol{i}_r, \boldsymbol{i}_\theta, \boldsymbol{i}_\phi$	unit base vectors in spherical coordinates	$\boldsymbol{n}$	outwardly directed unit normal on surface enclosing volume V
$\boldsymbol{I}$	identity tensor		
$I_\Delta, II_\Delta, III_\Delta$	first, second, and third scalar invariants of second-order tensor Δ, defined by Equations (1.107)–(1.109)	$\boldsymbol{N}_A$	molar flux (vector) of species A
		N_{Ay}	mole flux of A in y-direction
		$\bar{N}_{Ay}$	time-averaged value of N_{Ay}
I_u	intensity of turbulence, Equation (2.280)	Nu_{AB}	mass-transfer Nusselt number, $Nu_{AB} = k_x d / c D_{AB}$
I_ϕ	a measure of the fluctuating part of ϕ in a turbulent flow, Equation (2.279)		
		Nu_d	Nusselt number based on diameter, Equation (2.202)
$\boldsymbol{j}_A$	mass diffusion flux of species A relative to $\boldsymbol{u}$, Equation (3.16)		
		$Nu_{d, lm}$	log mean Nusselt number, Equations (2.205) and (2.206)
$\tilde{\boldsymbol{j}}_A$	mass diffusion flux of species A relative to $\tilde{\boldsymbol{u}}$, Equation (3.19)		
		Nu_L	overall Nusselt number associated with surface of length L, Equation (2.255)
J	Jacobian relating differential volume elements in two coordinate systems and defined by Equations (1.23) and (1.24)		
		Nu_x	local heat-transfer Nusselt number, Equation (2.251)
$J_n( )$	Bessel functions of the first kind with argument $( )$ and $n = 0, 1, \ldots$		
		$O( )$	order of magnitude of $( )$
		$\boldsymbol{p}$	linear momentum
$\boldsymbol{J}_A$	molar diffusion flux of species A, relative to $\boldsymbol{u}$, Equation (3.18)	p	pressure, the isotropic stress
		p_0	reference pressure
$\tilde{\boldsymbol{J}}_A$	molar diffusion flux of species A, relative to $\tilde{\boldsymbol{u}}$, Equation (3.14)	p_A, p_B	partial pressure of species A, B
		p_H	hydrostatic pressure, Equation (2.63)
k	arbitrary constant, Equation (1.4); in Section 2.2.3, defined below Equation (2.94);	$\bar{p}$	time-averaged pressure
		P	dynamic pressure, Equation (2.62)

P_0	reference dynamic pressure	r, θ, z	coordinates in cylindrical coordinate system
ΔP	pressure drop across flow channel in direction of flow	r, θ, ϕ	coordinates in spherical coordinate system
Pe	Péclet number, Equation (1.117)	s	defined below Equation (2.98)
Pe_L	mass-transfer Péclet number based on film length	$\mathbf{s}$	displacement vector
Pe_δ	mass-transfer Péclet number based on film thickness, Equation (3.83)	S	control volume surface
		Sc	Schmidt number ($\equiv \nu/D_{AB}$)
Pr	Prandtl number ($\equiv \nu/\alpha$)	Sh	Sherwood number ($\equiv k_x d/cD_{AB}$)
$\widehat{PE}$	potential energy per unit mass	Sh_L	Sherwood number based on length L, Equation (3.95)
q	a dimensionless parameter defined by Equation (3.103)	Sh_x	local Sherwood number defined below Equation (3.110)
$\mathbf{q}$	heat flux vector (heat flow rate per unit area)	St	Stanton number, Equation (2.311)
$\bar{\mathbf{q}}$	time-averaged heat flux	$\hat{S}$	entropy per unit mass
q_n	heat flux component normal to surface, Equation (2.6)	$\dot{S}_{gen}$	entropy generation rate
		$\dot{S}_{in}$	rate of entropy addition
q_r, q_θ, q_z	heat flux components in cylindrical coordinates	t	time
		t_{exp}	diffusion penetration theory exposure time, Equations (3.136) and (3.137)
q_w	wall heat flux		
q_x, q_y, q_z	heat flux components in x-, y-, and z-directions	$\mathbf{t}_{(n)}$	traction vector acting on surface A, Equation (1.67)
$q_y^{(t)}$	turbulent heat flux component in y-direction, Equation (2.301)	t_0	in Section 2.5.1, time over which quantities are averaged; elsewhere, a reference time
q_ξ	heat flux in ξ-direction		
Q	energy transferred as heat, Chapter 4; volumetric flow rate, Chapter 2	t^*	dimensionless time, defined below Equation (2.27)
$Q, \hat{Q}$	volumetric flow rates in inner and outer regions, respectively, Section 2.2.5	tr	trace operation of a second-order tensor, Equation (1.8)
Q^*	defined by Equation (2.110)	T	absolute temperature
$\tilde{Q}$	heat transfer per unit mass	$\mathbf{T}$	total stress tensor, a symmetric second-order tensor
$\dot{Q}$	heat transfer rate		
r	molecular separation, Lennard–Jones function	T^*	dimensionless temperature, Equation (2.34)
$\mathbf{r}$	position vector	T_b	bulk average temperature, Equation (2.195)
r^*	dimensionless radial position		
r_A, r_B, r_i	mass rate of production of species A, B, i per unit volume due to reaction ($r_A > 0$ for generation of A; $r_A < 0$ for consumption of A)	T_f	film temperature [$\equiv (T_w + T_\infty)/2$] in Section 2.5.2.3 and Table 7; bulk fluid temperature
		T_{fk}	temperature of bulk fluid next to surface at x_k
r_c	radial position of interface between inner and outer phases	T_i	initial temperature
		T_k	temperature at position x_k
r_c^*	dimensionless radial position, defined below Equation (2.107)	T_0	characteristic reference temperature for a process
r_H	hydraulic radius defined below Equation (2.96)	T_w	wall temperature
		T_{w0}, T_{wL}	wall temperatures at positions $z = 0$ and $z = L$
r_Y	radial position of yield surface		
r_0, r_1	radii associated with cylinder or spherical surfaces	T_x, T_y, T_z, T_r	solutions defined below Equations (2.54)–(2.56)
R	ideal gas constant; in Section 2.3.3, the flow channel radius	T_ξ	time-averaged temperature at $y = \xi$
		T_∞	temperature in free stream
R_A, R_B, R_i	molar rate of production of species A, B, i per unit volume of mixture due to reaction ($R_A > 0$ for production of A; $R_A < 0$ for consumption of A)	T_0, T_1, T_2	temperatures at positions x_0, x_1, and x_2
		Δ_T	ratio of momentum and thermal boundary layer thicknesses ($\equiv \delta_m/\delta_T$)
R_{cond}	conductive resistance, Equation (2.14)	ΔT_{lm}	log mean temperature difference, Equation (2.207)
$R_{conv, i}$	convective heat-transfer resistance, Equation (2.24)	$\bar{T}$	average spatial temperature in a region, Section 2.1.7; time-averaged temperature, Section 2.5.2
Re	Reynolds number, Equations (1.113) and (2.81)		
Re_{cr}	critical Reynolds number for laminar to turbulent transition	$\mathbf{u}, u$	velocity vector (momentum per unit mass) of magnitude u; in a mixture: a mass average velocity, Equation (3.8)
Re_d	Reynolds number based on diameter d, defined below Equation (2.127)		
Re_L	Reynolds number based on length L	$\mathbf{u}'$	fluctuating part of velocity in turbulent field
Re_x	local Reynolds number, Equation (2.231)	u'_x, u'_y, u'_z	fluctuating velocity components
Re_δ	Reynolds number based on length scale δ	u_{avg}	average velocity

u_c	characteristic velocity, Equation (2.268)	y_A	mole fraction of species A in a gas mixture
u_i	velocity of species i in a mixture	y_{Ae}	mole fraction of species A in the gas phase required for equilibrium with a given bulk liquid-phase mole fraction, Figure 29
$u_{\text{interface}}$	fluid velocity at a gas–liquid interface		
$u_{\max}$	maximum velocity		
u_x, u_y, u_z	components of velocity in rectangular cartesian system	y_{Ai}, y_{Ab}	mole fraction of species A at the interface in the gas phase and in the bulk gas phase, Figure 29
u_0	characteristic velocity, Equation (2.159)		
u_∞	free stream velocity; see Figures 10, 11, 14–16	y_i, y_j, y_k	i, j, k components of a vector y or position vector in a rectangular cartesian coordinate system
u_*	friction velocity, Equation (2.286)		
u^*	dimensionless velocity	y'	y coordinate measured from channel centerline, defined above Equation (2.76)
u^+	dimensionless velocity, Equation (2.284)		
$\tilde{u}$	mixture molar average velocity, Equation (3.12)	y^*	dimensionless y coordinate
		y^+	dimensionless distance from wall, Equation (2.285)
$\bar{u}$	time-averaged velocity in turbulent field defined by Equation (2.277)		
		z_e	thermal entry length defined above Equation (2.204)
$\bar{u}_x, \bar{u}_y, \bar{u}_z$	time-averaged velocity components		
$u_z, \hat{u}_z$	velocity components in inner and outer regions, respectively, Section 2.2.5	z_1, z_2	elevation at points 1 and 2 in a flow conduit
$\bar{u}_\xi$	time-averaged velocity at $y = \xi$	z^*	dimensionless z coordinate
U	characteristic velocity for a process or flow; velocity of moving surface, Figure 5; velocity of lower plate, Figure 13	**Greek Symbols**	
		α	thermal diffusivity ($\equiv \lambda/\varrho\, c_p$)
$\hat{U}$	internal energy per unit mass	α_{KE}	factor representing deviation from a flat velocity profile, Equation (4.6)
V	volume		
V_A	molar volume of species A, volume per mole	α_{mom}	factor representing deviation from a flat velocity profile, Equation (4.4)
V_0	material volume in the reference coordinates	β	dimensionless quantity defined by Equation (2.43); isobaric coefficient of thermal expansion, Equation (2.185)
$\hat{V}$	volume per unit mass ($\equiv 1/\varrho$)		
w	arbitrary velocity of an observer, Equation (1.55)	β_n	eigenvalues given by Equations (2.48), (2.51), or (2.53) with $n = 1, 2,...$
w	flow geometry width	β_∞	coefficient of thermal expansion at T_∞
w_A	mass fraction of species A in a mixture: $w_A = \varrho_A/\varrho$	γ	defined below Equation (3.69)
		$\dot{\gamma}$	shear rate, Equation (1.110c)
W	film width (in x direction), Figure 25; work, Equation (4.8)	Δ	rate of deformation tensor, Equation (1.105)
$\dot{\mathscr{W}}_A$	rate of mass transfer, moles/time	δ	film thickness
$\dot{W}$	work rate (power)	δ, δ_{ij}	Kronecker delta defined by Equations (1.10) and (1.11)
$\dot{W}_{es}$	work done on the fluid at the entrance and exit points by extra stress, per unit mass	δ_c	boundary layer thickness for mass transfer analogous to Equations (2.211) and (2.237)
$\dot{W}_{np}$	nonpressure work rate, Section 4.5		
$\dot{W}_s$	shaft work per unit mass	δ_m	momentum boundary layer thickness, Equation (2.211)
x, x^i, x^j, x^k	i, j, k components of a position vector x in a curvilinear coordinate system		
		δ_T	thermal boundary layer thickness, Equation (2.237)
x_A	mole fraction of species A in a mixture: $x_A = c_A/c$		
		ε	fractional penetration of a gas into a film, used for nondimensionalizing the penetration, Equation (3.81a); smallness parameter, Equation (2.158); in Section 2.5.2.1, the eddy diffusivity; in Section 3.2.1, a smallness parameter defined by Equation (3.84); Lennard–Jones potential function energy parameter, Chapter 1
x_{Ae}	mole fraction of species A in the liquid phase required for equilibrium with a given bulk gas-phase mole fraction, Figure 29		
x_{Ai}, x_{Ab}	mole fraction of species A at the interface in the liquid phase and in the bulk liquid phase, Figure 29		
x_{cr}	position on surface where boundary layer becomes turbulent, Figure 14	$\varepsilon_A, \varepsilon_B, \varepsilon_{AB}$	Lennard–Jones potential function energy parameters for species A and B and for the mixture, $\varepsilon_{AB} = (\varepsilon_A \varepsilon_B)^{1/2}$
x^*	dimensionless coordinate position		
x_0, y_0	principal axes of elliptic cross section; defined above Equation (2.92)	$\varepsilon_c, \varepsilon_m, \varepsilon_T$	eddy mass, momentum, and thermal diffusivities, Equations (3.113), (2.296), and (2.297)
x_0, y_0, z_0	linear half-thickness associated with rectangular parallelepiped or rectangular bar, Section 2.1.8; reference position, Section 2.2	$\varepsilon_\mathscr{M}$	eddy diffusivity associated with transfer of quantity $\mathscr{M}$ by turbulent fluctuations
		ε_s	surface shear viscosity
Δx_k	thickness of kth slab ($x_k - x_{k-1}$)	ξ	dependent variable, Equation (2.35)
X_i, X_j, X_k	material coordinates (y_i, y_j, y_k) at a reference time) of an element of mass	η	viscosity; catalyst effectiveness factor, Equation (4.28)

$\eta, \hat{\eta}$	in Section 2.2.5, viscosities of core fluid and outer fluid, respectively; in Section 2.3.1, viscosities of continuous and drop phases, respectively
η_B	viscosity of solvent B, Equation (3.29)
η_p	plastic viscosity in Bingham plastic model
η_∞, η_w	viscosities evaluated at bulk and wall temperatures, respectively
θ	angle subtended by two vectors, Equation (1.5)
κ	heat capacity ratio, c_p/c_v; equal to b_L/b_0 in Section 2.3.3
Λ	interference parameter associated with hindered settling (see Eq. 2.157)
λ	thermal conductivity
$\bar{\lambda}$	average thermal conductivity similar to Equation (2.10) with T replacing T_1
$\bar{\lambda}_1$	average thermal conductivity, Equation (2.10)
$\bar{\lambda}_k$	average thermal conductivity, Equation (2.10) with T_k replacing T_1
ν	kinematic viscosity ($\equiv \mu/\varrho$)
ξ	coordinate direction; dummy integration variable; y position associated with edge of sublayer in turbulent flow
ϱ	density, mass per unit volume
$\varrho_A, \varrho_B, \varrho_i$	mass per unit volume of A, B, i in a mixture
ϱ_s	density of sphere
ϱ_∞	density at T_∞
σ	hard-sphere diameter, Equation (1.99); Lennard–Jones potential function separation parameter, Equation (1.100)
$\sigma_A, \sigma_B, \sigma_{AB}$	Lennard–Jones hard-sphere diameters for species A and B and for the mixture, $\sigma_{AB} = (\sigma_A + \sigma_B)/2$
τ	extra stress second-order tensor, Equation (1.84)
τ_{ij}	stress components
$\tau_{rz}, \hat{\tau}_{rz}$	stress components in inner and outer regions, respectively, Section 2.2.5
τ_w	magnitude of wall shear stress
τ_Y	yield stress in Bingham plastic model
$\tau_{yx}^{(t)}$	turbulent stress component, Equation (2.300)
τ_0	shear stress at $y = 0$
Φ	arbitrary flux vector or tensor, Equation (1.39); in Equations (2.289) and (2.290), represents a scalar, vector, or tensor function; rate of heat production and rate of radiation energy absorption, per unit volume
Φ_v	rate of conversion of mechanical energy to thermal energy by viscous dissipation, per unit mass (Eq. 1.97) for a constant-density Newtonian material
Φ_0, Φ_1, Φ_2	coefficients for the general viscous fluid, Equation (1.106)
ϕ	function described by Equations (2.90) and (2.91); association factor, Equation (3.29); fractional flow, Equation (2.109)
$\phi(y^*, z^*)$	in Section 3.2.1, the c_A solution from Equation (3.91)
χ	ratio of concentric cylinder inner and outer radii
ψ	stream function defined by Equation (2.125); in Section 2.4.1, function defined by Equation (2.210); in Equations (2.289) and (2.290), represents a scalar, vector, or tensor function
Ω_D	collision integral correction factor for intermolecular interactions, Equation (3.28)
Ω_v	collision integral correction factor for intermolecular interactions, Equations (1.101) and (1.102)
$\partial/\partial t$	time derivative seen by an observer maintaining constant spatial coordinates (position)

Others

∇	Del operator for spatial differentiation, defined by Equation (1.29) for rectangular cartesian coordinate systems
∇^2	Laplacian operator, $\nabla^2 = \nabla \cdot \nabla$, Equation (1.38)
$\sim$	order of magnitude estimate

1. Foundations

The objective of this article is to provide the foundations for understanding and analyzing phenomena that occur in continua as a result of driving forces in momentum, mass, and heat transfer. The descriptions can be concerned with continua on either differential or macroscopic scales; the results can be used for process analysis or for design to improve the efficiency of existing processes and to develop new processes.

The analysis of transport processes relies on a small set of fundamental laws. These laws, together with specific information and data about the properties and behavior of the process materials and about the constitutive relations between driving forces and their resulting fluxes, are used to provide a mathematical description of the behavior of the process or phenomenon. This mathematical description, together with process constraints such as geometry and boundary conditions, defines the mathematical problem. Its solution, then, is approached by using a variety of mathematical and engineering strategies for solutions to algebraic, ordinary, and partial differential equations.

The governing laws that establish the behavior of continua are the conservation laws for mass, energy, and linear and angular momentum (and their subset accounting equations for species mass, mechanical energy, and thermal energy) plus the second law of thermodynamics. These laws require a knowledge of the properties and behavior of materials in the form of (1) volumetric (pressure–volume–temperature) property relations and data; (2) thermodynamic property relations and data (such as internal energy, enthalpy, and entropy); and (3) transport flux relations and data (for heat, mass, and momentum fluxes in terms of their driving forces: temperature gradient, mole fraction gradient, and velocity gradient).

The foundations establish conventions and identities that are used throughout this article: (1) kinematic relationships used in describing the motion of continua, (2) development of the conservation laws and their corollary equations related to mechanical and thermal energy balances, (3) linear transport relations for heat and momentum fluxes, and (4) discussion of how this material is combined in a methodology establishing quantitative models for describing the behavior of continua. Additionally, the molecular foundations for transport properties of materials are discussed. Theories for these properties, used to quantify the relations between the transport fluxes and their driving forces, work reasonably well for gases to calculate properties such as thermal conductivity, viscosity, and diffusivity. However, for more dense phases in which intermolecular interactions play a significant role the theories serve primarily as a guiding framework for empirical relations. Finally, a brief discussion of non-Newtonian fluids and their constitutive relations is presented.

After development of the basic equations, boundary conditions and a solution philosophy for attacking problems are discussed briefly. Also, the equations of change in dimensionless form are presented as the basis for dimensionless group correlations and analogies between different modes of transport.

Chapter 2 is limited to compositionally homogeneous systems, whereas transport in multicomponent systems is discussed in Chapter 3.

1.1. Mathematical Preliminaries

1.1.1. Coordinate Systems

The study of transport processes is concerned with scalars, which are represented by a single number (their magnitude), and with vectors and tensors, which have directions in space as well as magnitude associated with them.

To describe vectors and tensors a coordinate system is defined for representing positions in three-dimensional space (direction and magnitude). A variety of systems are possible and can be used. Each coordinate system has defined base vectors, and all other vectors can be described as linear combinations of these base vectors. Any three noncoplanar vectors can be used as base vectors. However, orthogonal vectors (mutually perpendicular) that form a right-handed system are normally used. Base vectors are cartesian if they are of unit length. If they are both orthogonal and cartesian, then they are orthonormal. Base vectors are rectangular, cartesian if they are orthonormal and everywhere the same (in both magnitude and direction). In general, the base vectors for a coordinate system are not independent of position. For example, the base vectors for a cylindrical or spherical system vary with position (in a cylindrical system, the radius vector changes direction as the angular coordinate varies and the z coordinate is held constant). A coordinate system is referred to as curvilinear if it is not rectangular.

1.1.2. Vector and Tensor Operations

Useful references for vector and tensor notation, and their operation are given in [1.1] and [1.2]. The relations presented below summarize operations for *rectangular cartesian systems* for the purpose of setting notation and conventions.

Accordingly, any vector a may be written as a linear combination of its components and base vectors in the following way:

$$a = a_i i_i \equiv \sum_{i=1}^{3} a_i i_i \qquad (1.1)$$

A second-order tensor A is written in terms of its base vectors and components in the following way:

$$A = A_{ij} i_i i_j \qquad (1.2)$$

Here summation occurs over both indices i and j. The addition of two vectors is a vector that is obtained by adding corresponding components:

$$a + b = \sum_{i=1}^{3} (a_i + b_i) i_i \qquad (1.3)$$

Multiplication of a vector or tensor by a scalar k is performed by multiplying each component by that scalar:

$$k a = k(a_i i_i) = (k a_i) i_i \qquad (1.4)$$

The *dot product* of two vectors (which is a scalar) by definition (→ 1. Mathematics in Chemical Engineering, pp. 1-45–1-46) is the product of their magnitudes and the cosine of their subtended angle, which implies a sum of the products of the corresponding components (for an orthonormal system):

$$\boldsymbol{a} \cdot \boldsymbol{b} = |\boldsymbol{a}||\boldsymbol{b}| \cos \theta = a_i b_i \qquad (1.5)$$

The dot product of a vector and tensor is calculated by a similar sum over adjacent indices according to

$$\boldsymbol{a} \cdot \boldsymbol{B} = a_i B_{ij} \boldsymbol{i}_j$$
$$\boldsymbol{B} \cdot \boldsymbol{a} = B_{ij} a_j \boldsymbol{i}_i \qquad (1.6)$$

The components thus obtained are analogous to the elements of a vector obtained from the matrix multiplication of a vector and a square matrix. A different result is obtained for pre- versus postmultiplication. The dot product of two second-order tensors is a second-order tensor given by

$$\boldsymbol{A} \cdot \boldsymbol{B} = A_{ij} B_{jk} \boldsymbol{i}_i \boldsymbol{i}_k \qquad (1.7)$$

The *trace* of a second-order tensor is a scalar given by the sum of its diagonal terms

$$\mathrm{tr}(\boldsymbol{A}) = A_{ii} \qquad (1.8)$$

and the trace of the dot product of two second-order tensors is a double sum performed in the following way:

$$\mathrm{tr}(\boldsymbol{A} \cdot \boldsymbol{B}) = A_{ij} B_{ji} \qquad (1.9)$$

The identity second-order tensor is defined such that

$$\boldsymbol{\delta} = \delta_{ij} \boldsymbol{i}_i \boldsymbol{i}_j \qquad (1.10)$$

where δ_{ij} is the Kronecker delta defined by

$$\delta_{ij} = \begin{cases} 1 & \text{for } i = j \\ 0 & \text{for } i \neq j \end{cases} \qquad (1.11)$$

This identity tensor, either pre- or post-dotted with a vector, returns the original vector. Likewise, when it is dotted with a second-order tensor, the original second-order tensor is obtained.

The *permutation symbol* e_{ijk} is defined such that

$$e_{ijk} = \begin{cases} 1 & \text{for an even permutation of 123} \\ 0 & \text{when any two indices are repeated} \\ -1 & \text{for an odd permutation of 123} \end{cases}$$

$$\qquad (1.12)$$

The *cross product* of two vectors is then defined by

$$\boldsymbol{a} \times \boldsymbol{b} = a_i b_j e_{ijk} \boldsymbol{i}_k \qquad (1.13)$$

and the cross product of a vector and a second-order tensor is similarly given by

$$\boldsymbol{a} \times \boldsymbol{B} = a_i B_{jm} e_{ijk} \boldsymbol{i}_k \boldsymbol{i}_m \qquad (1.14)$$

Two convenient identities between the permutation symbol and the Kronecker delta are

$$e_{ijk} e_{mnk} = (\delta_{im} \delta_{jn} - \delta_{in} \delta_{jm}) \qquad (1.15)$$

and

$$e_{ijk} e_{mjk} = 2 \delta_{im} \qquad (1.16)$$

The *dyadic* or *tensor product* of two vectors is a second-order tensor whose dot product with a vector is given by

$$(\boldsymbol{a}\boldsymbol{b}) \cdot \boldsymbol{c} = \boldsymbol{a}(\boldsymbol{b} \cdot \boldsymbol{c}) \qquad (1.17)$$

In terms of index notation the tensor product $\boldsymbol{a}\boldsymbol{b}$ is written

$$\boldsymbol{a}\boldsymbol{b} = a_i b_j \boldsymbol{i}_i \boldsymbol{j}_j \qquad (1.18)$$

The *transpose* $\boldsymbol{A}^T$ of a second-order tensor is defined so that the ij component of $\boldsymbol{A}$ is the ji component of the second-order tensor $\boldsymbol{A}^T$. Accordingly,

$$\boldsymbol{A}^T = A_{ij} \boldsymbol{i}_j \boldsymbol{i}_i \qquad (1.19)$$

1.1.3. The Jacobian

The Jacobian is the ratio of differential volume elements represented in two coordinate systems. It has special value when one of the coordinate systems involves material coordinates and the other involves the spatial coordinate system discussed below in kinematics. In rectangular cartesian coordinates, the differential volume element can be written as

$$dV = dy_1 \, dy_2 \, dy_3 \qquad (1.20)$$

which can be expressed equivalently in terms of the triple scalar product:

$$dV = dy_1 \boldsymbol{i}_1 \cdot (dy_2 \boldsymbol{i}_2 \times dy_3 \boldsymbol{i}_3)$$

A position vector in space can be represented in

terms of a rectangular coordinate system or in terms of an alternate system according to

$$r = y_i i_i = x_i g_i \tag{1.21}$$

where x_i are the components of the position vector in the alternate system with base vectors g_i. Each of the cartesian coordinates can be expressed as a function of the other system coordinates, $y_i = y_i(x_1, x_2, x_3)$. Then, the differential volume element can be expressed in terms of the triple scalar product of three vectors written in terms of the derivatives with respect to each of the new coordinate system's three coordinates:

$$dV = \frac{\partial y}{\partial x_1} dx_1 \cdot \left(\frac{\partial y}{\partial x_2} dx_2 \times \frac{\partial y}{\partial x_3} dx_3 \right) \tag{1.22}$$

or

$$dy_1 dy_2 dy_3 = J dx_1 dx_2 dx_3 \tag{1.23}$$

and the Jacobian is the determinant of the partial derivatives relating changes in coordinates between the two systems:

$$J \equiv \begin{vmatrix} \frac{\partial y_1}{\partial x_1} & \frac{\partial y_2}{\partial x_1} & \frac{\partial y_3}{\partial x_1} \\ \frac{\partial y_1}{\partial x_2} & \frac{\partial y_2}{\partial x_2} & \frac{\partial y_3}{\partial x_2} \\ \frac{\partial y_1}{\partial x_3} & \frac{\partial y_2}{\partial x_3} & \frac{\partial y_3}{\partial x_3} \end{vmatrix} \tag{1.24}$$

The Jacobian only exists if there is a unique transformation between the two coordinate systems.

1.1.4. Calculus of Vectors and Tensors

In transport phenomena generally, two kinds of vector and tensor derivatives exist with respect to time and with respect to position.

Derivatives with Respect to Time. The derivative of a vector that is a function of time is a vector; a derivative of a second-order tensor with respect to time is a second-order tensor. Derivatives are obtained straightforwardly by differentiating each of their components with respect to time:

$$\frac{d\boldsymbol{a}}{dt} = \frac{da_i}{dt} i_i \tag{1.25}$$

$$\frac{d\boldsymbol{A}}{dt} = \frac{dA_{ij}}{dt} i_i i_j \tag{1.26}$$

Differentiating products of vectors is similar to differentiating products of scalars:

$$\frac{d}{dt}(\boldsymbol{a} \cdot \boldsymbol{b}) = \frac{d\boldsymbol{a}}{dt} \cdot \boldsymbol{b} + \boldsymbol{a} \cdot \frac{d\boldsymbol{b}}{dt} \tag{1.27}$$

$$\frac{d}{dt}(\boldsymbol{a} \times \boldsymbol{b}) = \frac{d\boldsymbol{a}}{dt} \times \boldsymbol{b} + \boldsymbol{a} \times \frac{d\boldsymbol{b}}{dt} \tag{1.28}$$

Derivatives with Respect to Position. The *del operator* ∇ is defined to differentiate quantities that are functions of position. In rectangular cartesian coordinate systems the del operator is

$$\nabla \equiv i_i \frac{\partial}{\partial y_i} \tag{1.29}$$

The del operator can operate in dyadic, dot, or cross product forms on scalars, vectors, or tensors. As a dyadic operation, *grad* (φ) is

$$\nabla \varphi = \frac{\partial \varphi}{\partial y_i} i_i \tag{1.30}$$

and grad (**a**)

$$\nabla \boldsymbol{a} = \frac{\partial a_j}{\partial y_i} i_i i_j \tag{1.31}$$

For grad (**a**), note the order of the index corresponding to the del operator with respect to that corresponding to the vector. The opposite order is sometimes used in the literature, and care must be taken to follow a consistent convention throughout a given work.

The *dot product* of the del operator with a vector **a** or tensor **A** is given by

$$\nabla \cdot \boldsymbol{a} = \frac{\partial a_i}{\partial y_i} \tag{1.32}$$

$$\nabla \cdot \boldsymbol{A} = \frac{\partial A_{ij}}{\partial x_i} i_j \tag{1.33}$$

and is called the *divergence* of the vector or tensor. Again, note the order of the indices corresponding to the del operator and the second-order tensor. The divergence of a vector is a scalar, so the order is immaterial. The del operator can also be crossed with a vector or tensor, and these operations are given by

$$\nabla \times \boldsymbol{a} = \frac{\partial a_j}{\partial y_i} e_{ijk} i_k \tag{1.34}$$

$$\nabla \times A = \frac{\partial}{\partial y_i} A_{jm} e_{ijk} i_k i_m \quad (1.35)$$

and called the *curl* of the vector or tensor. Here, only one definition with respect to the order of the indices seems to exist in the literature.

Because of its dual vector–operator role, the del operator may be dotted with one entity and operate on (differentiate) another. For example, in the thermal energy equation in Section 1.2 (Eq. 1.96), the term

$$u \cdot \nabla F \quad (1.36)$$

appears, in which the del operator is pre-dotted with the velocity vector u but operates (by differentiation) on the temperature (see also Eq. 1.53):

$$u \cdot \nabla T = u_i \frac{\partial T}{\partial y_i} \quad (1.37)$$

The *Laplacian* is another operator, a scalar operator, defined by $\nabla \cdot \nabla$. For rectangular cartesian systems it is given by

$$\nabla^2 \varphi = \nabla \cdot \nabla \varphi = \frac{\partial^2 \varphi}{\partial y_i \partial y_i} \quad (1.38)$$

1.1.5. Divergence Theorem

The divergence theorem is extremely important in deriving and understanding the basic transport relations. This theorem equates an integral over the entire volume of a region of space enclosed by a closed surface to an integral over that surface:

$$\int_V \nabla \cdot \boldsymbol{\Phi} \, dV = \int_A \boldsymbol{n} \cdot \boldsymbol{\Phi} \, dA \quad (1.39)$$

In this equation, $\boldsymbol{\Phi}$ can be a vector or a second-order tensor. If it is an isotropic second-order tensor ($p\boldsymbol{I}$, for example), then a form involving a scalar function is obtained:

$$\int_V \nabla p \, dV = \int_A p \boldsymbol{n} \, dA \quad (1.40)$$

In transport phenomena, $\boldsymbol{\Phi}$ represents the flux of an extensive property and the surface integral represents the rate at which the property (as written here using the outward normal vector $\boldsymbol{n}$) is leaving the volume element across the surface. With the del operator defined as above, the surface integral must be written with the surface outward normal vector pre-dotting the quantity $\boldsymbol{\Phi}$.

1.1.6. Kinematic Relations

A fundamental concept of kinematics, the description of motion without reference to its cause, is that of material coordinates. Given a fluid or material undergoing motion (translation, rotation, deformation), each particle of that material has a current position y

$$y = y_i i_i \quad (1.41)$$

Also, each element of the material can be identified with the position it had at some reference time ($t = 0$, for example):

$$X = X_i i_i \quad (1.42)$$

The *spatial coordinates* y_i are used to identify a position in space, and the *material coordinates* X_i are used to identify elements of the material. Furthermore, the position in space can be defined in terms of the element of the material that resides there (identified by its position at time $t = 0$, its reference position) at time t:

$$y = y(X, t) \quad (1.43)$$

Inversely, the material coordinates X can be defined in terms of the spatial coordinates and time t; each element of mass can be defined in terms of its location at time t

$$X = X(y, t) \quad (1.44)$$

This inversion depends on $0 < J < \infty$ where J is the Jacobian.

Every quantity expressed in terms of spatial coordinates and time can just as well be expressed in terms of material coordinates and time. For example, for a flowing fluid the density, velocity, etc., vary with position

$$\varrho = \varrho(y, t) \quad (1.45)$$

$$u = u(y, t) \quad (1.46)$$

or, equivalently, vary from material element to material element:

$$\varrho = \varrho(X, t) \quad (1.47)$$

$$u = u(X, t) \quad (1.48)$$

1.1.7. Partial and Total Derivatives

The partial derivative with respect to time [of a function $\mathfrak{F}(x, t)$], if position is held constant, is written as

$$\frac{\partial \mathfrak{F}(x, t)}{\partial t} \equiv \left(\frac{\partial \mathfrak{F}}{\partial t}\right)_x \quad (1.49)$$

and represents changes in the variable of interest with time at a fixed point in space.

The partial derivative with respect to time [of a function $\mathfrak{F}(X, t)$], if the material coordinates are kept constant, is called the *material derivative* or *substantial derivative*, and represents changes observed while following the motion of the material:

$$\frac{D\mathfrak{F}(X, t)}{Dt} \equiv \left(\frac{\partial \mathfrak{F}}{\partial t}\right)_X \quad (1.50)$$

Note that if $\mathfrak{F}$ is the spatial coordinate of an element of fluid, then $D\mathfrak{F}/Dt = Dx/Dt$ is the change in position with time of a material element, i.e., the fluid velocity:

$$u \equiv \left(\frac{\partial x}{\partial t}\right)_X = \frac{Dx}{Dt} \quad (1.51)$$

The *total derivative* represents changes that would be observed while moving around at an arbitrary velocity w; it is represented by $d\mathfrak{F}/dt$. This velocity w then represents the total derivative of the position vector:

$$w = dx/dt$$

1.1.8. Relation Between Different Time Derivatives

A property $\mathfrak{F}$ written in terms of x and t has the total differential

$$d\mathfrak{F} = \left(\frac{\partial \mathfrak{F}}{\partial t}\right)_x dt + dx_i \left(\frac{\partial \mathfrak{F}}{\partial x_i}\right)_t \quad (1.52)$$

from which the relation between the substantial and partial derivatives and u may be determined:

$$\left(\frac{\partial \mathfrak{F}}{\partial t}\right)_X = \frac{D\mathfrak{F}}{Dt} = \left(\frac{\partial \mathfrak{F}}{\partial t}\right)_x + u \cdot \nabla \mathfrak{F} \quad (1.53)$$

In an analogous way, the total derivative is related to the partial and substantial derivatives and to w by

$$\frac{d\mathfrak{F}}{dt} = \left(\frac{\partial \mathfrak{F}}{\partial t}\right)_x + \frac{dx}{dt} \cdot \nabla \mathfrak{F} \quad (1.54)$$

or

$$\frac{d\mathfrak{F}}{dt} = \left(\frac{\partial \mathfrak{F}}{\partial t}\right)_x + w \cdot \nabla \mathfrak{F} \quad (1.55)$$

A physical interpretation of the three derivatives follows [1.3]. An observer interested in determining the changes in the concentration of fish in a river c_f can do so while staying at a fixed position in the river $\partial c_f/\partial t$, while drifting in a canoe at the velocity of the river Dc_f/Dt, or while moving around in a motorboat with velocity w (dc_f/dt).

1.2. Basic Equations for Compositionally Homogeneous Systems [1.3]–[1.7]

1.2.1. The Reynolds Transport Theorem

In deriving the conservation equations of mass, energy, and momentum for a continuum, a volume (system) that always encloses the same elements of mass should be defined. This volume must then move and deform with the fluid, which is convenient because no convected input or output term need be considered. This concept, and the substantial derivative and Jacobian from kinematics, are very useful in deriving the desired conservation equations.

If $\mathfrak{F}(x, t)$ is a function of position and time (scalar, vector, or tensor), representing the amount of the desired property per unit volume, and $V(t)$ is the closed volume moving with the fluid, then

$$F(t) \equiv \int_V \mathfrak{F}(x, t) dV \quad (1.56)$$

represents the total amount of the desired property contained within the system $V(t)$. For the conservation equation, then, the accumulation rate of the property within the system is required $DF(t)/Dt$. Here the substantial derivative must be used because it is the derivative "following the motion." Then

$$\frac{DF}{Dt} = \frac{D}{Dt} \int_V \mathfrak{F}(x, t) dV \quad (1.57)$$

However, the derivative cannot be taken across the integral in this form because the volume is a function of time. This difficulty is circumvented by transforming the system to the reference coordinates, differentiating, and then transforming back. In the reference coordinates, the volume element is a constant. Now,

$$dV = J\,dV_0 \tag{1.58}$$

So

$$\frac{DF}{Dt} = \frac{D}{Dt}\int_{V_0} \mathfrak{F}[x(X,t),t]\,J\,dV_0$$

$$= \int_{V_0}\left(\frac{D\mathfrak{F}}{Dt}J + \mathfrak{F}\frac{DJ}{Dt}\right)dV_0 \tag{1.59}$$

$$\frac{DJ}{Dt} = J\nabla \cdot \boldsymbol{u} \quad \text{so}$$

$$\frac{DF}{Dt} = \int_{V_0}\left(\frac{D\mathfrak{F}}{Dt} + \mathfrak{F}\nabla \cdot \boldsymbol{u}\right)J\,dV_0 \tag{1.60}$$

Finally, converting back to spatial coordinates gives

$$\frac{DF}{Dt} = \int_{V}\left(\frac{D\mathfrak{F}}{Dt} + \mathfrak{F}\nabla \cdot \boldsymbol{u}\right)dV \tag{1.61}$$

which is known as the *Reynolds transport theorem* (RTT).

1.2.2. Conservation of Total Mass

In the absence of nuclear conversions, total mass is conserved; no generation or consumption occurs. (The term balance is commonly used to apply to both conservation and accounting equations. In this section, and in Chapter 4 on macroscopic systems, the term "accounting" is used for properties that may be generated or consumed and the term "conservation" for properties that may not be generated or consumed.) Now, for the volume that moves and deforms with the flowing fluid, as described above, no mass enters or leaves the system, so the mass conservation equation is

$$\frac{D}{Dt}\int_V \varrho\,dV = 0 \tag{1.62}$$

By the Reynolds transport theorem then,

$$\int_V \left(\frac{D\varrho}{Dt} + \varrho\nabla \cdot \boldsymbol{u}\right)dV = 0 \tag{1.63}$$

The volume chosen as the system is completely arbitrary, so this result implies

$$\frac{D\varrho}{Dt} = -\varrho\nabla \cdot \boldsymbol{u} \tag{1.64}$$

or, in terms of the partial derivative,

$$\frac{\partial \varrho}{\partial t} = -\nabla \cdot \varrho\boldsymbol{u} \tag{1.65}$$

This equation (in either form) is called the continuity equation (see also, → 5. Fluid Mechanics, pp. 5-4–5-7). Note that if $\mathfrak{F}$ in Equation (1.61) is proportional to ϱ (i.e., $\mathfrak{F} = \varrho G$), then by continuity

$$\frac{D}{Dt}\int_V \varrho G\,dV = \int_V \varrho\frac{DG}{Dt}\,dV \tag{1.66}$$

1.2.3. Conservation of Linear Momentum

An equation for the conservation of momentum or energy for a continuum must be able to express the forces that are exerted on an element in the continuum. These forces are of two kinds:

1) *Body forces* act upon the entire volume (reach inside the volume and act upon each element) and may be gravitational, electromagnetic, etc.
2) *Contact (or internal) forces* act at the surface of the element of the continuum.

Cauchy's stress principle states that if $\boldsymbol{t}_{(n)}$ is the force per unit area (*stress vector* or *traction*) at a point on the surface of a system volume (surface orientation indicated by $\boldsymbol{n}$), then $\boldsymbol{t}_{(n)}$ is a function of $\boldsymbol{x}$, t, and the orientation of the surface. The stress vector at a point $\boldsymbol{x}$ at a particular time is not sufficient; the surface upon which the stress vector acts must also be defined.

Note that $\boldsymbol{t}_{(n)}$ is a vector force per area (the stress vector or traction vector) and that $\boldsymbol{t}_{(n)}\,dA$ is the force acting from outside the surface. For a volume (with a surface) the total force acting on this volume due to these contact forces is the integral over the entire surface A:

$$\int_A \boldsymbol{t}_{(n)}\,dA \tag{1.67}$$

Body forces can be summed by a volume integral (g is force per unit mass)

$$\int_V \varrho g \, dV \tag{1.68}$$

and the conservation of linear momentum (for a system that moves with the material) can be expressed as

$$\frac{D}{Dt} \int_V \varrho u \, dV = \int_V \varrho g \, dV + \int_A t_{(n)} \, dA \tag{1.69}$$

This is known as Euler's first law.

1.2.4. Condition of Local Stress Equilibrium

As a direct consequence of the conservation of linear momentum, the stress vector (traction) acting at a point on a surface is matched by an opposite and equal reaction force on the surroundings. That is, if momentum is conserved, the transfer of momentum between the system and its surroundings due to forces is mutual (*condition of local stress equilibrium*).

1.2.5. Stress Tensor

The condition of local stress equilibrium leads to an important decoupling of stress from the surface orientation. Consideration of the stresses acting on the surfaces of a tetrahedron shows that

$$t_{(n)} = n \cdot T \tag{1.70}$$

where T is the second-order *stress tensor* (→ 5. Fluid Mechanics, pp. 5-9–5-10), which may be a function of position and time but not of surface orientation.

The state of stress in a continuum at a point is the totality of all possible pairs of traction vectors $t_{(n)}$ and surface orientations n (i.e., an infinite number). However, the nine-component stress tensor can be used to obtain $t_{(n)}$ if n is given.

1.2.6. Equation of Motion

Now,

$$\frac{D}{Dt} \int_V \varrho u \, dV = \int_V \varrho g \, dV + \int_A n \cdot T \, dA \tag{1.71}$$

or, by the divergence theorem and Reynolds transport theorem,

$$\int_V \varrho \frac{Du}{Dt} \, dV = \int_V \varrho g \, dV + \int_V \nabla \cdot T \, dV \tag{1.72}$$

so (again, V is arbitrary)

$$\varrho \frac{Du}{Dt} = \varrho g + \nabla \cdot T \tag{1.73}$$

or

$$\varrho a = \varrho g + \nabla \cdot T \tag{1.74}$$

where $a = Du/Dt$, the acceleration. This is known as Cauchy's first law, the conservation of linear momentum, or simply the *equation of motion* for a continuum. For further details, see → 5. Fluid Mechanics, pp. 5-7–5-8.

1.2.7. Conservation of Angular Momentum

If angular momentum is conserved locally in a continuum (the usual situation),

$$\frac{D}{Dt} \int_V \varrho (u \times x) \, dV = \int_V \varrho (g \times x) \, dV + \int_A (t_{(n)} \times x) \, dA \tag{1.75}$$

which is Euler's second law. By writing $t_{(n)}$ in terms of T (Eq. 1.70) and using the divergence theorem, Reynold's transport theorem, and vector–tensor identities, this statement of the conservation of angular momentum reduces to the result that the stress tensor is symmetric. That is, a necessary and sufficient condition for satisfying the conservation of angular momentum is:

$$T = T^T \tag{1.76}$$

However, this is restricted to fluids that have no couple stresses (i.e., torques on the fluid are the result only of forces) [1.5], [1.6].

Also, if T is a symmetric second-order tensor (nondiagonal, in general), a rotated coordinate system exists in which T is diagonal. In this system, the coordinate axes are called the principal axes of stress and the three normal stresses (eigenvalues) are called the principal stresses. In mechanics, analysis of the principal axes of stress provides the basis for Mohr's stress circle technique [1.6], [1.7].

1.2.8. Mechanical Energy Accounting Equation

Now, because of the conservation of linear momentum (Eq. 1.74),

$$(\varrho \boldsymbol{a} - \varrho \boldsymbol{g} - \nabla \cdot \boldsymbol{T}) \cdot \boldsymbol{u} = 0 \tag{1.77}$$

or

$$\varrho \frac{D}{Dt}\left(\frac{1}{2}\boldsymbol{u} \cdot \boldsymbol{u}\right) - \varrho \boldsymbol{g} \cdot \boldsymbol{u} - (\nabla \cdot \boldsymbol{T}) \cdot \boldsymbol{u} = 0 \tag{1.78}$$

$$\left[\frac{D(\boldsymbol{u} \cdot \boldsymbol{u})}{Dt} = \frac{D\boldsymbol{u}}{Dt} \cdot \boldsymbol{u} + \boldsymbol{u} \cdot \frac{D\boldsymbol{u}}{Dt}\right.$$
$$\left. = \boldsymbol{a} \cdot \boldsymbol{u} + \boldsymbol{u} \cdot \boldsymbol{a} = 2\boldsymbol{a} \cdot \boldsymbol{u}\right]$$

Now, as an identity, for $\boldsymbol{T}$ symmetric

$$\nabla \cdot (\boldsymbol{T} \cdot \boldsymbol{u}) = (\nabla \cdot \boldsymbol{T}) \cdot \boldsymbol{u} + \text{tr}\,(\boldsymbol{T} \cdot \nabla \boldsymbol{u}) \tag{1.79}$$

so that

$$\varrho \frac{D}{Dt}\left(\frac{1}{2}\boldsymbol{u} \cdot \boldsymbol{u}\right) = \varrho \boldsymbol{g} \cdot \boldsymbol{u} + \nabla \cdot (\boldsymbol{T} \cdot \boldsymbol{u}) - \text{tr}\,(\boldsymbol{T} \cdot \nabla \boldsymbol{u}) \tag{1.80}$$

This scalar equation is a rate accounting equation of mechanical energy in the form of kinetic energy (the left-hand side), work done by body forces ($\varrho \boldsymbol{g} \cdot \boldsymbol{u}$), work done by traction forces [$\nabla \cdot (\boldsymbol{T} \cdot \boldsymbol{u})$], and the conversion of mechanical energy to thermal energy due to traction through both fluid compression and viscous dissipation [tr$(\boldsymbol{T} \cdot \nabla \boldsymbol{u})$], all per unit volume. Obviously, this result is not independent of the equation of motion and, in fact, as a scalar equation must contain less information than the vector equation of motion.

1.2.9. Conservation of Total Energy

If $\hat{U}$ is the internal energy per unit mass, then a total energy balance is

$$\frac{D}{Dt}\int_V \varrho\left(\frac{1}{2}u^2 + \hat{U}\right)dV \tag{1.81}$$
$$= \int_V \varrho \boldsymbol{g} \cdot \boldsymbol{u}\,dV + \int_A \boldsymbol{t}_{(n)} \cdot \boldsymbol{u}\,dA - \int_A \boldsymbol{n} \cdot \boldsymbol{q}\,dA + \int_V \Phi\,dV$$

where $\boldsymbol{q}$ is the heat flux at the system boundary. The left side of this equation represents changes in kinetic and internal energy, and the terms on the right side have the following meaning:

$\int_V \varrho \boldsymbol{g} \cdot \boldsymbol{u}\,dV$ rate at which work is done by body forces,

$\int_A \boldsymbol{t}_{(n)} \cdot \boldsymbol{u}\,dA$ rate at which work is done due to traction,

$-\int_A \boldsymbol{n} \cdot \boldsymbol{q}\,dA$ rate of energy transfer due to $\boldsymbol{q}$, and

$\int_V \Phi\,dV$ rate of radiant energy absorption.

This last term may also be used to represent heat generation (conversion of energy from other forms to thermal energy) by mechanisms such as electrical heating and chemical or nuclear reactions instead of accounting for these conversions through changes in, e.g., changes in internal energy from one form to another within the material volume. From Equation (1.81) the continuum statement of total energy conservation is

$$\varrho \frac{D}{Dt}\left(\frac{1}{2}u^2 + \hat{U}\right)$$
$$= \varrho \boldsymbol{g} \cdot \boldsymbol{u} + \nabla \cdot (\boldsymbol{T} \cdot \boldsymbol{u}) - \nabla \cdot \boldsymbol{q} + \Phi \tag{1.82}$$

1.2.10. Thermal Energy Accounting Equation

Subtracting the mechanical energy accounting equation from the total energy conservation equation gives an accounting of thermal energy:

$$\varrho \frac{D\hat{U}}{Dt} = -\nabla \cdot \boldsymbol{q} + \text{tr}\,(\boldsymbol{T} \cdot \nabla \boldsymbol{u}) + \Phi \tag{1.83}$$

In this result, tr$(\boldsymbol{T} \cdot \nabla \boldsymbol{u})$ represents the conversion of mechanical energy to thermal energy. Equation (1.83) is not an independent equation but has the advantage of not involving mechanical energy terms. It can be used in place of the total energy equation.

1.2.11. Forms of the Governing Equations

Many possible forms of the continuity, motion, and energy equations exist, depending, for example, on whether partial or substantial derivatives are used and whether heat capacities (c_p or c_v) are used instead of internal energy. A convenient table for compositionally homogeneous continua summarizing these forms appears in [1.3]. Further summary of the governing equations is given in Section 1.4.

In these forms, the stress tensor is commonly separated into *isotropic stress* $p\boldsymbol{I}$ and *extra stress* τ ($\rightarrow$ 5. Fluid Mechanics, pp. **5**-9–**5**-10). Following the convention of BIRD,

$$\boldsymbol{T} = -\tau - p\boldsymbol{I} \tag{1.84}$$

where $p \equiv -\text{tr}(\boldsymbol{T})/3$. As a result $\nabla \cdot \boldsymbol{T}$ becomes $-\nabla \cdot \tau - \nabla p$ and tr$(\boldsymbol{T} \cdot \nabla \boldsymbol{u})$ becomes $-\text{tr}\,(\tau \cdot \nabla \boldsymbol{u}) - p\nabla \cdot \boldsymbol{u}$.

1.2.12. Entropy Inequality

If $T = -\tau - p\mathbf{I}$ for a compressible fluid, then the thermal energy equation is

$$\varrho \frac{D\hat{U}}{Dt} = -\nabla \cdot \mathbf{q} - p\nabla \cdot \mathbf{u} - \text{tr}(\tau \cdot \nabla \mathbf{u}) + \Phi \quad (1.85)$$

Now if $\hat{U} = \hat{U}(\hat{S}, \hat{V})$, where $\hat{S}$ is the entropy per unit mass and $\hat{V}$ the volume per unit mass, then

$$d\hat{U} = Td\hat{S} - pd\hat{V} \quad (1.86)$$

or by the equation of continuity

$$\varrho \frac{D\hat{U}}{Dt} = \varrho T \frac{D\hat{S}}{Dt} - \varrho p \frac{D\hat{V}}{Dt} = \varrho T \frac{D\hat{S}}{Dt} - p\nabla \cdot \mathbf{u} \quad (1.87)$$

This gives for the thermal energy equation

$$\varrho T \frac{D\hat{S}}{Dt} = -\nabla \cdot \mathbf{q} - \text{tr}(\tau \cdot \nabla \mathbf{u}) + \Phi \quad (1.88)$$

or

$$\varrho \frac{D\hat{S}}{Dt} = -\nabla \cdot \left(\frac{\mathbf{q}}{T}\right) - \mathbf{q} \cdot \frac{\nabla T}{T^2} - \frac{1}{T}\text{tr}(\tau \cdot \nabla \mathbf{u}) + \frac{\Phi}{T} \quad (1.89)$$

Now, from experience, $\mathbf{q}$ is known to be in the opposite direction of ∇T, so $-\mathbf{q} \cdot \nabla T \geq 0$. Furthermore, and also from experience, $-\text{tr}(\tau \cdot \nabla \mathbf{u}) \geq 0$. Consequently, an inequality results that must hold for all continuum processes:

$$\varrho \frac{D\hat{S}}{Dt} + \nabla \cdot \left(\frac{\mathbf{q}}{T}\right) - \frac{\Phi}{T} \geq 0 \quad (1.90)$$

Alternatively, $-\mathbf{q} \cdot \nabla T \geq 0$ and $-\text{tr}(\tau \cdot \nabla v) \geq 0$ may be viewed as constraints on allowable constitutive equations for $\mathbf{q}$ and $\mathbf{T}$.

1.2.13. Linear Transport Fluxes and Relations

The above transport equations describe the conservation of mass, linear momentum, and total energy, and include terms representing heat and momentum fluxes. The flux relations that are normally adopted follow the observation that the flux is proportional to a gradient driving force; i.e., the heat flux is proportional to the temperature gradient and the momentum flux is proportional to the velocity gradient (actually, the symmetric part of the velocity gradient tensor).

Fourier's Law. For *heat conduction*,

$$\mathbf{q} = -\lambda \nabla T \quad (1.91)$$

where λ is the thermal conductivity and, in general, may be a function of location in the material and temperature. (To be more general, the heat flux can be written in terms of an anisotropic thermal conductivity tensor to allow for directionality in the thermal conductivity.) In rectangular cartesian coordinates the components of the heat flux are

$$q_x = -\lambda \left(\frac{\partial T}{\partial x}\right)_{y,z}$$

$$q_y = -\lambda \left(\frac{\partial T}{\partial y}\right)_{x,z} \quad (1.92)$$

$$q_z = -\lambda \left(\frac{\partial T}{\partial z}\right)_{x,y}$$

Momentum Flux Relations. For viscous materials the most common momentum flux relation is Newton's law of viscosity, which for an incompressible material is written in terms of the *rate of deformation* tensor Δ as

$$\tau = -\eta \Delta \quad (1.93)$$

where η is the viscosity. For Newtonian materials the viscosity is constant at constant temperature (i.e., it is independent of shear rate), resulting in this linear flux relation. For one-dimensional shear flow with only one velocity gradient component, this result becomes

$$\tau_{yx} = -\eta \frac{du_x}{dy} \quad (1.94)$$

These linear relations can be used together with the conservation equations to model the behavior of continua. For a Newtonian fluid of constant viscosity and density (or for isochoric flow for which $\nabla \cdot \mathbf{u} = 0$), the equation of motion becomes the Navier–Stokes equation (for further details, see → 5. Fluid Mechanics, pp. 5-11 – 5-15)

$$\varrho \left(\frac{\partial \mathbf{u}}{\partial t} + \mathbf{u} \cdot \nabla \mathbf{u}\right) = \varrho \mathbf{g} - \nabla p + \eta \nabla^2 \mathbf{u} \quad (1.95)$$

Also, for a Newtonian fluid of constant density, thermal conductivity, and viscosity, the thermal energy equation becomes

$$\varrho c_p \left(\frac{\partial T}{\partial t} + \boldsymbol{u} \cdot \nabla T \right) = \lambda \nabla^2 T + \Phi_v \tag{1.96}$$

where the distinction between constant pressure and constant volume heat capacities vanishes for true incompressibility. In this result, Φ_v accounts for the conversion of mechanical energy to thermal energy through viscous dissipation and is given by

$$\Phi_v = \eta \left\{ \text{tr}\,[\nabla \boldsymbol{u} \cdot \nabla \boldsymbol{u}] + \text{tr}\,[(\nabla \boldsymbol{u})^T \cdot \nabla \boldsymbol{u}] \right\} \tag{1.97}$$

Further simplifications are often appropriate, such as for a solid material whose velocity terms in the thermal energy equation become insignificant, giving

$$\varrho c_p \left(\frac{\partial T}{\partial t} \right) = \lambda \nabla^2 T \tag{1.98}$$

Understanding material behavior with respect to thermal conductivity and viscosity is an important factor in understanding transport phenomena. Theoretical foundations for calculating these properties for low-density gases and non-Newtonian stress relations are given in the following paragraphs.

1.2.14. Transport Properties from Molecular Theories

The calculation of transport properties by using molecular transport theories with parameters estimated from experimental data is quite reliable for dilute (low-density) gases. This is a direct result of the fact that at low densities, intermolecular collisions and forces are minimized. Details of the theories are available in the original references or in summaries such as [1.8]. In this section, theories for viscosity and thermal conductivity are considered. Section 3.1.5 discusses theories for diffusivities.

Viscosity. Based on two-body, hard-sphere collisions the viscosity of a gas (in micropascal seconds, i.e., 10^{-6} Pa · s) theoretically is given by

$$\eta = \frac{\frac{5}{16}(\pi M_r R T)^{1/2}}{\pi \sigma^2} = 26.69 \frac{\sqrt{M_r T}}{\sigma^2} \tag{1.99}$$

where T is in kelvin and the hard-sphere diameter σ is in ångströms (i.e., 10^{-10} m); M_r is the molecular mass of the gas.

To adjust for non-hard-sphere interactions, CHAPMAN and ENSKOG introduced the use of spherically symmetric molecular potential functions [1.9], [1.10]. The most commonly used function is the Lennard–Jones 12–6 potential

$$\psi(r) = 4\varepsilon \left[\left(\frac{\sigma}{r} \right)^{12} - \left(\frac{\sigma}{r} \right)^6 \right] \tag{1.100}$$

from which can be calculated the interaction forces between molecules as a result of their separation r and model parameters σ and ε. These forces are then used to adjust the viscosity calculations through the dimensionless collision integral Ω_v, which is a function of kT/ε where k is the Boltzmann constant. Tabulations of this quantity are readily found in the literature, and a convenient and accurate computational form for Ω_v is given by NEUFELD et al. [1.8], [1.11]. The value of Ω_v is unity for the ideal hard sphere; thus, deviations are a measure of the departure from this ideal model. The Chapman–Enskog relation for viscosity, then, is

$$\eta = \frac{\frac{5}{16}(\pi M_r R T)^{1/2}}{(\pi \sigma^2)\Omega_v} = 26.69 \frac{\sqrt{M_r T}}{\sigma^2 \Omega_v} \tag{1.101}$$

where again η is in 1 micropascal seconds, T in kelvin, and σ in ångströms.

Other potential functions have been used. The Stockmayer potential adjusts for polarity and should be used instead of the Lennard–Jones potential for polar molecules. Calculation of the Stockmayer potential by using the analytical correction to the Lennard–Jones potential provided by BROKAW is recommended [1.8], [1.12].

These potential energy functions rely on molecular parameters that must be determined from data. In practice, tabulations of these parameters, which are available for individual materials, are derived from experimental data for viscosity by using the appropriate potential function with the Chapman–Enskog theory. Using the parameters, then, along with the potential function (collision integral) to calculate viscosity, amounts to adjusting the known values to the desired temperature by using the square-root temperature dependence of the molecular theory. As long as the form of the theory is correct, the calculations should be very good. In fact, when tabulated potential function parameters are used for a dilute gas, calculated viscosity values are correct to within ca. 1 %. Methods also exist for estimating the parameters; these too give quite reasonable viscosity values, with errors from 1 to 3 % [1.8].

Thermal Conductivity. The molecular theory for thermal conductivity is complicated by the fact that internal energy is stored in molecules in internal vibrational and rotational modes as well as in the translational mode, unlike viscosity (momentum) which is manifest only in translation. Consequently, although the basic molecular theory parallels that for viscosity, an additional term for these internal energy modes is necessary for polyatomic molecules. For monatomic gases (which do not have the internal modes) the Chapman–Enskog thermal conductivity relation is

$$\lambda = \frac{\frac{25}{32}(\pi M_r R T)^{1/2} c_v}{M_r (\pi \sigma^2) \Omega_v} = 0.0833 \frac{\sqrt{T/M_r}}{\sigma^2 \Omega_v} \quad (1.102)$$

where T is expressed in K, σ in Å, and λ in $J\,m^{-1}\,s^{-1}\,K^{-1}$. In this relation a value of $c_v = 3\,k/2$ (monatomic ideal gas) was used. By using the viscosity relation (Eq. 1.101), the thermal conductivity can be expressed in terms of the viscosity and constant volume heat capacity as

$$\frac{\lambda M_r}{\eta c_v} = 2.5 \quad (1.103)$$

This result works quite well for monatomic gases. For polyatomic gases it must be adjusted to account for the additional degrees of freedom. The Eucken equation

$$\frac{\lambda M_r}{\eta} = c_v + 1.872 \times 10^4 \quad (1.104)$$

is a simple, yet reasonably accurate method for this correction. Other methods are available and are discussed and evaluated by REID et al. [1.8].

1.2.15. Non-Newtonian Fluids
(see also → 5. Fluid Mechanics, pp. **5**-24–**5**-44)

In the momentum and energy balances, q and τ represent heat and momentum fluxes. Before problems can be solved, constitutive relations must be selected for these fluxes in terms of appropriate driving forces (e.g., ∇T or ∇u).

The specific relation for τ that is appropriate for a particular material is the subject of rheology. Specific rules dictate the form of constitutive equations or set constraints on them. Useful references on non-Newtonian rheology are [1.13]–[1.15].

The rheological constitutive equation that is applicable in a given circumstance can be viewed as depending on either the type of material or on the specific flow situation. That is, the material may behave in a certain way because it behaves according to that model in all circumstances (i.e., it is that kind of material) or because all materials behave in that way for a particular flow situation.

From a materials viewpoint, behavior is normally classified as viscous, elastic, or viscoelastic. Complicating this classification by material type is the presence of history or memory effects (→ 5. Fluid Mechanics, p. **5**–28).

One very common type of constant stretch history flow is *viscometric flow*. The importance of viscometric flows with respect to the characterization of materials is that all materials, whether they are viscous, elastic, or viscoelastic, can be characterized by the same set of viscometric functions. The functions are referred to as the shear stress (or, equivalently, the viscosity) and the first and second normal stress differences (for further details, see → 5. Fluid Mechanics, pp. **5**-35–**5**-36).

Other kinds of flows are *elongational, slow, and small deformation oscillatory flows*. The last are periodic and yield a set of dynamic viscoelastic functions such as the dynamic elasticity and the dynamic viscosity (also referred to as the storage and loss moduli, respectively).

A Newtonian fluid has a constant viscosity independent of shear rate (at constant temperature). Other materials may show a decrease (pseudoplastic) or an increase (dilatant) in viscosity with increasing shear rate or exhibit a yield stress (e.g., the Bingham plastic). Complicating this picture is the fact that time-dependent behavior may also be observed and such fluids are referred to as thixotropic (decrease in viscosity with time at fixed shear stress) or rheopectic (increase).

In addition to shear stress, normal stresses may be generated by steady shear flow. These are stresses that are normal to the flow direction and are additional to the isotropic stress (pressure). The generation of normal stresses can have some rather dramatic effects on the flow of fluids and may have to be considered in process design. Experimentally, the only materials that have been observed to exhibit normal stresses are viscoelastic, such as polymer melts and polymer solutions.

More quantitatively, a viscous fluid is one whose stress tensor is a function only of the rate of deformation tensor Δ (a function of strain rate and not of strain), where

$$\Delta \equiv \nabla u + (\nabla u)^T \quad (1.105)$$

A completely general form for a purely viscous fluid is

$$T = \Phi_0 I + \Phi_1 \Delta + \Phi_2 \Delta^2 \quad (1.106)$$

where the coefficients Φ_0, Φ_1, and Φ_2 are functions of the three scalar invariants of Δ (I_Δ, II_Δ, and III_Δ) and are defined in the following way:

$$I_\Delta = \text{tr}(\Delta) \qquad (1.107)$$

$$II_\Delta = \tfrac{1}{2}[(\text{tr}(\Delta))^2 - \text{tr}(\Delta^2)] \qquad (1.108)$$

$$III_\Delta = \det \Delta \qquad (1.109)$$

Note that $I_\Delta = \text{tr}(\Delta) = 2\nabla \cdot \boldsymbol{u}$, which is zero for an incompressible fluid or for isochoric flow. For incompressible flow then, the general viscous equation reduces to

$$\boldsymbol{T} = -p\boldsymbol{I} + \Phi_1 \Delta + \Phi_2 \Delta^2 \qquad (1.110)$$

which is called the Reiner–Rivlin equation. In this case, Φ_1 and Φ_2 are functions only of II_Δ and III_Δ because $I_\Delta = 0$. Apparently, actual fluids of this type have not been documented (i.e., fluids with a nonzero second-order term and which are purely viscous have not been observed). Nevertheless, it may be a useful model because the second-order term provides a mechanism for generating normal stress differences in steady shear flow. However, because normal stress differences are not generally observed with viscous fluids, the second-order term is usually dropped ($\Phi_2 = 0$), and the model then reduces to the form generally used for incompressible Newtonian and non-Newtonian (non-constant-viscosity) fluids:

$$\boldsymbol{T} = -p\boldsymbol{I} - \eta \Delta \qquad (1.110\,\text{a})$$

or, in terms of the extra stress τ,

$$\tau = -\eta \Delta \qquad (1.110\,\text{b})$$

where $\eta = \eta(II_\Delta, III_\Delta)$ is the non-Newtonian viscosity. For shear flows, $[u_1 = u_1(x_2)]$, III_Δ is zero so that η is only a function of the second invariant which can be related to the shear rate $\dot\gamma$ by

$$\dot\gamma \equiv |du_1/dx_2| = (\tfrac{1}{2}|II_\Delta|)^{1/2} \qquad (1.110\,\text{c})$$

Equation (1.110b) is commonly used with $\eta = \eta(\dot\gamma)$ for other kinds of flow (nonshear), even though to do so is not well-founded theoretically.

Table 1. Summary of the basic transport relations

Equation name	Conservation law	Equation	Equation number
Continuity	total mass	$\dfrac{D\varrho}{Dt} = -\varrho \nabla \cdot \boldsymbol{u}$	(1.64)
	total mass	$\dfrac{\partial \varrho}{\partial t} = -\nabla \cdot (\varrho \boldsymbol{u})$	(1.65)
Motion	linear momentum	$\varrho \boldsymbol{a} = \varrho \boldsymbol{g} + \nabla \cdot \boldsymbol{T}$	(1.74)
	linear momentum	$\varrho\left(\dfrac{\partial \boldsymbol{u}}{\partial t} + \boldsymbol{u} \cdot \nabla \boldsymbol{u}\right) = \varrho \boldsymbol{g} + \nabla \cdot \boldsymbol{T}$	
	linear momentum	$\varrho\left(\dfrac{\partial \boldsymbol{u}}{\partial t} + \boldsymbol{u} \cdot \nabla \boldsymbol{u}\right) = \varrho \boldsymbol{g} - \nabla \cdot \tau - \nabla p$	
	angular momentum	$\boldsymbol{T} = \boldsymbol{T}^{\mathrm{T}}$	(1.76)
	total energy	$\varrho \dfrac{D}{Dt}(\tfrac{1}{2}u^2 + \hat{U}) = \varrho \boldsymbol{g} \cdot \boldsymbol{u} + \nabla \cdot (\boldsymbol{T} \cdot \boldsymbol{u}) - \nabla \cdot \boldsymbol{q} + \Phi$	(1.82)
Thermal energy accounting		$\varrho \dfrac{D\hat{u}}{Dt} = -\nabla \cdot \boldsymbol{q} + \text{tr}(\boldsymbol{T} \cdot \nabla \boldsymbol{u}) + \Phi$	(1.83)
		$\varrho c_v \dfrac{DT}{Dt} = -\nabla \cdot \boldsymbol{q} + T\left(\dfrac{\partial p}{\partial T}\right)_\varrho (\nabla \cdot \boldsymbol{u}) - \text{tr}(\tau \cdot \nabla \boldsymbol{u}) + \Phi$	
		$\varrho c_p \dfrac{DT}{Dt} = -\nabla \cdot \boldsymbol{q} + \left(\dfrac{\partial \ln \hat{V}}{\partial \ln T}\right)_p \dfrac{Dp}{Dt} - \text{tr}(\tau \cdot \nabla \boldsymbol{u}) + \Phi$	
Entropy inequality		$\varrho \dfrac{D\hat{S}}{Dt} + \nabla \cdot \left(\dfrac{\boldsymbol{q}}{T}\right) - \dfrac{\varrho \Phi}{T} \geq 0$	(1.90)

1.3. Summary of Basic Equations

A handful of conservation laws govern physical behavior and, as such, provide the basics of transport phenomena: total mass, linear momentum, angular momentum, and total energy. In addition, subsets of energy give useful accounting relations, and the entropy inequality (second law of thermodynamics) provides constraints on the conservation laws. These relations are summarized in Table 1 along with some forms not presented in the text.

1.4. Boundary Conditions

The transport conservation relations (mass, total energy — or an accounting of thermal energy — and linear and angular momentum), along with the second law of thermodynamics and relations that describe material behavior, are not completely adequate for describing transport processes. With the addition of appropriate initial and boundary conditions the transport partial differential equations can be solved to obtain future behavior of the process over the required region of space. Specific examples appear in later sections. The principal types of conditions that are encountered and applied are outlined briefly in this section.

The conditions apply to the dependent variables of the transport equations (such as temperature or velocity), their derivatives, or other aspects of their behavior. The following are the types of conditions encountered:

1) A dependent variable is known throughout the region of space of interest at time zero; values at subsequent times result from the behavior of the system as destined through transport principles. For example, the temperature of a fluid over a region of space may be initially set to a prescribed value or function of position. Likewise, the velocity of the fluid may be known initially over the region of interest (e.g., if the fluid initially is stagnant, then $u = 0$ at $t = 0$).
2) The dependent variable at a boundary of the region of space of interest is known; it may be fixed at a constant value over time or it may be a known variable as a function of time. For example, in fluid mechanics a fluid in contact with a solid usually meets the no-slip condition: the velocity of the fluid at this solid boundary is equal to the velocity of the solid surface. Similarly, the temperature at a boundary could be a known value or function of time.
3) The transport flux at a boundary of the region of interest is known as a function of time. This is not a condition on a dependent variable directly; however, it is an indirect condition. For example, in Fourier's law the heat flux is proportional to the temperature gradient; thus, specifying the heat flux at the boundary places a condition on the behavior of the temperature at the boundary. If the component of the heat flux normal to the surface is known, a condition on $n \cdot q$ is established. For momentum transport, a known traction (stress vector) at the boundary is stated in terms of $n \cdot \tau$. As an example, at gas–liquid interfaces the stresses supported by the gas phase are very low (low viscosity), the momentum flux (traction) due to the extra stress at the interface is therefore zero: $n \cdot \tau = 0$.
4) The continuity of properties across the surface can also serve as a boundary condition. For example, at a surface (boundary) velocity, temperature, traction, and heat flux are normally continuous. The continuity of heat transfer across the interface holds as long as no thermal energy accumulates at the surface of interest. This, however, does not imply that temperature gradients are continuous; thermal conductivities may change at a boundary. Likewise, the continuity of traction is necessitated by the condition of local stress equilibrium (i.e., by the conservation of linear momentum).
5) The stress in fluids is finite, that is, fluids cannot sustain stresses without deforming so as to reduce those stresses. This condition is applicable, for example, in pipe flow problems where, if such a condition is not imposed a priori, solutions to the equation of motion could yield an infinite stress at the center of the pipe.

1.5. Solution Philosophy

We now have a total of five equations and one inequality. The equations are (1) continuity (mass conservation), (2) linear momentum conservation (three equations, one for each spatial direction), and (3) total energy conservation.

Additionally, the conservation of angular momentum constrains T to be symmetric, and the entropy inequality places further constraints on q and T. Also, linear momentum conservation can be used to obtain an accounting of mechanical energy (not an independent equation), which can be employed with total energy conservation to give an accounting of thermal energy that can be used in place of the total energy conservation law. The above equations have six independent variables: ϱ, u_x, u_y, u_z (or $x, y,$ and z), p, and T. [The stress T and heat flux q are expressed in terms of position derivatives and temperature gradients, respectively. Furthermore, the thermodynamic properties ($\hat{U}, \hat{H}$, and $\hat{S}$) are expressed as functions of T and pressure (or ϱ) through heat capacity and $p-\varrho-T$ data or equation of state.] The five equations plus one volumetric equation of state [$p = p(\varrho, T)$] along with appropriate boundary or initial conditions dictate the subsequent behavior of the system.

If temperature is constant, the number of variables is reduced by one, as is the number of equations (energy conservation reduces to the mechanical energy accounting equation, which is not an independent equation).

If density is constant, the number of variables and equations (volumetric equation of state) both decrease by one.

1.6. Dimensionless Equations of Change
(→ 3. Dimensional Analysis)

As discussed above, for a given flow situation involving momentum and heat transfer, a scalar equation of continuity, a vector equation of motion, and a scalar equation of energy (either total energy or a thermal energy accounting equation) are available. These laws govern the behavior of the process at hand. In these equations, each term has a specific physical meaning or origin important to the current situation. Consequently, each term carries with it dimensional characteristics that result from those physical meanings.

Frequently, from both computational and conceptual perspectives, equations are usefully recast into a dimensionless form by means of dimensions or parameters that are characteristic of the process. These dimensionless forms have several advantages. First, grouping the characteristic parameters or dimensions into a reduced number of parameters allows parametric calculations for a reduced number of cases. Second, the dimensionless equations allow comparison of processes of different size, which otherwise are the same; the dimensionless equations clarify the adjustments that must be made between the two processes to have them behave the same dynamically. In this way, a small-scale process can give an accurate prediction of a large-scale process for design purposes. Third, correlations of experimental data can be made very efficiently by using dimensionless groups and correlations. Correlations developed for one transport phenomenon can even be used for another one when the situations (process, apparatus, and boundary conditions) are analogous (e.g., heat-transfer results can be used to predict mass-transfer behavior). This is the very important *Reynolds analogy*, which rests upon the similarity of conservation laws (momentum, mass, and energy) and flux relations (Fourier's law, Newton's law of viscosity, and Fick's first law of diffusion) for certain situations. Finally, dimensionless forms of the transport laws have considerable value in understanding the relative importance of the various contributions to process behavior (e.g., convective versus conductive heat transfer). This order-of-magnitude scaling is discussed further in Chapter 2.

The objective of dimensional analysis using the governing equations, then, is to convert the equation to a dimensionless form, thereby extracting dimensionless groups. First, the variables of the equations are expressed in terms of dimensionless quantities (herein denoted by *) by dividing each of them by a characteristic parameter of the same dimensions. For example, a temperature-dependent variable would be divided by a reference characteristic temperature or, as is more frequently used, a temperature difference ($T_1 - T_0$); velocity, by a velocity U that is characteristic of the process; variables involving distance, such as a derivative with respect to a distance coordinate, by a characteristic length L; time, either by a characteristic process time (e.g., t_c, a relaxation time) or by the ratio of the characteristic distance to the characteristic velocity L/U; mass, by a characteristic mass or, if the material density is constant, by the density (M/L^3); and pressure, by ϱU^2. Then, when all the variables have been nondimensionalized by appropriate constants, the resulting coefficients of the various terms can be grouped into a smaller number of coefficients. When this is carried out, some terms in a given equation will consist solely of the nondimensional variables of the problem, and other terms will consist of quantities involving nondimensional variables multiplied by nondimensional groups of the constants and characteristic parameters of the equation.

The selection of specific constants for the nondimensionalizing process is not necessarily unique. In comparing two processes of geometrical similarity and comparable boundary conditions, however, the same physical dimension must be selected for the two processes (e.g., the diameter of a process tank for both situations). Also, for order-of-magnitude scaling, the characteristic dimensions chosen must represent the physical phenomena in a quantitatively meaningful way.

Nondimensionalizing the equation of motion for an incompressible, Newtonian material provides a basis for comparing the relative importance of viscous forces, inertial forces, body forces, etc. The Navier–Stokes equation (Eq. 1.95),

$$\varrho\left(\frac{\partial \boldsymbol{u}}{\partial t} + \boldsymbol{u} \cdot \nabla \boldsymbol{u}\right) = \varrho \boldsymbol{g} - \nabla p + \eta \nabla^2 \boldsymbol{u} \quad (1.111)$$

can be nondimensionalized by the three characteristic dimensions L, U, and ϱ (time is nondimensionalized by L/U) to give

$$\frac{\partial \boldsymbol{u}^*}{\partial t^*} + \boldsymbol{u}^* \cdot \nabla \boldsymbol{u}^*$$
$$= \left(\frac{Lg}{U^2}\right)\left(\frac{\boldsymbol{g}^*}{g}\right) - \nabla^* p^* + \left(\frac{\eta}{LU\varrho}\right) \nabla^{*2} \boldsymbol{u}^* \quad (1.112)$$

where in Equation (1.111) the $\varrho \boldsymbol{u} \cdot \nabla \boldsymbol{u}$ term represents inertial forces and the viscosity term represents viscous forces. Body forces such as gravity are accounted for by $\varrho \boldsymbol{g}$. In obtaining the dimensionless form (Eq. 1.112) the coefficient of the inertial term has been set to unity by division so that the remaining coefficients represent a ratio to the inertial forces. For example, the coefficient of the viscous term in Equation (1.112) is a dimensionless group representing the ratio of the viscous force to the inertial force characteristic quantities. The inverse of this group is the Reynolds number

$$Re = \frac{LU\varrho}{\eta} \quad (1.113)$$

The coefficient of the body force term represents the ratio of the body forces to the inertial forces and is the inverse of the *Froude number*:

$$Fr = \frac{U^2}{gL} \quad (1.114)$$

These dimensionless groups represent the relative importance of these terms in the equation of motion. A large Reynolds number signifies large inertial forces compared to viscous forces, and a large Froude number signifies a greater importance of inertial forces relative to the body forces (e.g., forced convection versus natural convection).

The equation of energy can be nondimensionalized in a similar way to obtain analogous dimensionless groups. In dimensional form the equation of thermal energy for an incompressible Newtonian fluid with Fourier's law of heat conduction is

$$\varrho c_p \left(\frac{\partial T}{\partial t} + \boldsymbol{u} \cdot \nabla T\right) = \lambda \nabla^2 T + \Phi_v \quad (1.115)$$

where Φ_v is the viscous dissipation defined by Equation (1.97). This energy equation can be nondimensionalized to give

$$\frac{\partial T^*}{\partial t^*} + \boldsymbol{u}^* \cdot \nabla^* T^* = \frac{1}{Pe} \nabla^{*2} T^* + \frac{Br}{Pe} \Phi_v^* \quad (1.116)$$

In Equation (1.115) the $\boldsymbol{u} \cdot \nabla T$ term is the transfer of heat as a result of fluid convection, $\lambda \nabla^2 T$ represents heat transfer by conduction, and Φ_v is the irreversible conversion of mechanical energy to thermal energy by viscous dissipation. When these terms are nondimensionalized and the coefficient of the convection heat-transfer term is made equal to unity, the inverse of the coefficient of the conduction term represents the relative importance of convention to conduction and is termed the *Péclet number*:

$$Pe = \frac{LU\varrho c_p}{\lambda} \quad (1.117)$$

It may also be viewed as the product of the Reynolds number and another dimensionless group termed the *Prandtl number*, given by

$$Pr = \frac{c_p \eta}{\lambda} \quad (1.118)$$

Evidently, the Prandtl number may be thought of as a conversion factor relating momentum transport (the ratio of transport by convection to that by molecular processes, i.e., stresses) to heat transport (the ratio of transport by convection to that by molecular processes, i.e., conduction) and expressed alternatively as the ratio of momentum diffusivity (kinematic viscosity) $v(=\mu/\varrho)$ to thermal diffusivity, $\alpha(=\lambda/\varrho c_p)$. The coefficient of the viscous dissipation term represents the relative importance of viscous dissipation to convection and is normally expressed as the ratio of the Brinkman to the Péclet number, where the *Brinkman number* is

$$Br = \frac{\eta U^2}{\lambda T_0} \qquad (1.118\,\text{a})$$

and T_0 is a reference temperature or temperature difference.

Use of dimensionless forms of the governing equations provides an excellent basis for comparing the same process at different scales. If two processes are the same, that is, if they accomplish the same task in geometrically similar apparatuses and with comparable boundary conditions, then they are mathematically identical with respect to these governing equations. For further details, see → 3. Dimensional Analysis.

2. Transport in Compositionally Homogeneous Systems

2.1. Heat Conduction in Solids

Heat transfer by conduction arises from spatial variations of temperature in a body. If the body is a stationary, opaque solid, conduction is the only mechanism for heat transfer. In this section the basic equations for heat conduction are summarized and specific results are presented that illustrate the effects of geometry, thermal properties, initial state, and boundary conditions. Good general references for this material are [2.1, Chaps. 2–4], [2.2, Chaps. 9–11], [2.3, Chaps. 1–5], and [2.4].

2.1.1. Heat Conduction Equations

The general energy balance was given in Equation (1.83). For heat conduction in solids it is simplified in several ways: In the case of stationary materials where $\boldsymbol{u} = \boldsymbol{0}$, only the accumulation, conduction, and generation terms are retained. Also, the continuity equation (Eq. 1.65) implies that the density ϱ is time-invariant. Further, local equilibrium is assumed, and the specific internal energy $\hat{U}$ is represented in terms of the density and temperature. Then,

$$\frac{\partial}{\partial t}\hat{U}(\varrho, T) = c_v \frac{\partial T}{\partial t} \qquad (2.1)$$

For solids, the specific heat capacity at constant volume $c_v \equiv (\partial \hat{U}/\partial T)_\varrho$ is approximately equal to the specific heat capacity at constant pressure $c_p = (\partial \hat{H}/\partial T)_p$. The latter quantity is usually measured and found in data tables.

With these simplifications, the energy balance (Eq. 1.83) reduces to

$$\varrho c_p \frac{\partial T}{\partial t} = -\nabla \cdot \boldsymbol{q} + \boldsymbol{\Phi} \qquad (2.2)$$

This equation is the differential form of the heat conduction equation. The term on the left is the accumulation term, and the terms on the right are the conductive flux and thermal energy generation and radiation terms, respectively.

To use the above equations to determine temperature fields and heat-transfer rates, the heat flux vector $\boldsymbol{q}$ must be related to the temperature gradient through Fourier's law, i.e.,

$$\boldsymbol{q} = -\lambda \nabla T \qquad (2.3)$$

which is written here for isotropic materials. In general, the thermal conductivity λ depends on temperature, pressure, and composition. For homogeneous stationary solids, only the temperature dependence is important.

Substituting Fourier's law into the heat conduction equation gives

$$\varrho c_p \frac{\partial T}{\partial t} = \nabla \cdot (\lambda \nabla T) + \boldsymbol{\Phi} \qquad (2.4)$$

which, with the specifications of $\lambda(T)$, $c_p(T)$, and $\boldsymbol{\Phi}$, as well as the necessary boundary and initial conditions, can be used to determine the temperature field. Once the temperature field is known, the directional heat-transfer rates can be determined from Fourier's law.

When a temperature-dependent thermal conductivity is used, the conduction equation is nonlinear and advanced analytical or numerical methods are generally required for solution [2.4, pp. 10–12]. For many materials, the dependence of λ on T is weak and λ can be taken as constant. Under these conditions, the heat conduction equation reduces to the nonhomogeneous form of the classical heat conduction (or diffusion) equation:

$$\frac{\partial T}{\partial t} = \alpha \nabla^2 T + \frac{\Phi}{\varrho c_p} \quad (2.5)$$

The quantity $\alpha \equiv \lambda/\varrho c_p$ is the *thermal diffusivity*, which plays the same role in heat transfer that mass diffusivity plays in mass transfer; it also has the same units (m²/s).

Typical values of the thermal properties of various solid materials are given in Table 2. Properties can vary widely depending on the material, composition, and structural form.

2.1.2. Initial and Boundary Conditions

The initial condition required in the solution of the conduction equation involves the specification of temperature as a function of position at some initial time, say $t = 0$.

The boundary conditions can take several forms:

1) The temperature on the bounding surface of the solid can be specified as a function of position and time. Although this is the most convenient boundary condition from a solution standpoint, surface temperatures are very difficult to measure. More commonly, the temperature of the fluid in contact with the surface is known (see condition 3 below).

Table 2. Thermal properties of various solid materials* at $\theta = 20\,°C$

Material	λ, W m⁻¹ K⁻¹	c_p, J kg⁻¹ K⁻¹	ϱ, kg/m³	$\alpha \times 10^6$, m²/s
Aluminum	237	905	2 707	96.7
Copper	398	384	8 954	115.7
Cast iron (4% C)	52	420	7 272	17.0
Stainless steel (304)	13.8	400	8 000	4.0
Lead	35	130	11 373	23.4
Silver	427	236	10 524	171.9
Common red brick	0.69	840	1 600	0.51
Concrete	1	800	2 100	0.6
Glass	0.76	800	2 690	0.35
Plaster	0.48	800	1 400	0.43
Wood, pine	0.1	2 500	600	0.07
Asbestos	0.2		580	
Glass wool	0.038		24	
Corkboard	0.043		160	

* Adapted from [2.1, p. 543] and [2.6, p. 592]. The values for construction materials can vary significantly depending on the specific composition and packing of the material.

Table 3. Order of magnitude of heat-transfer coefficients

Physical situation	h, W m⁻² K⁻¹
Free convection, air	4–50
Forced convection, air	10–300
Forced convection, oil	60–1800
Forced convection, water	250–15 000
Boiling water	300–100 000
Condensing steam	4000–160 000

2) The heat flux normal to the bounding surface is specified as a function of position and time. A special case arises with a perfectly insulated surface where the normal heat flux component is zero.

3) For a fluid–solid boundary, the normal component of the conductive heat flux on the solid side is equated to the "convective" flux away from the surface on the fluid side. This is expressed as

$$\text{On } A: \quad q_n = -\lambda \frac{\partial T}{\partial n} = h(T - T_f) \quad (2.6)$$

where the conductive flux has been represented by Fourier's law and the "convective" flux by Newton's law of cooling. The latter relation represents the "convective" flux in terms of the temperature difference between the surface and the fluid (T_f) and the quantity h, which is called the *heat-transfer coefficient*. Actually, the flux on the fluid side arises from complex, coupled, conduction, and convection mechanisms that depend on the flow field in the neighborhood of the surface. Such effects are not represented explicitly in this relation but are embedded in the heat-transfer coefficient h, which is an effective thermal conductance that depends on fluid properties, geometry, and fluid motion. Large values of h are associated with efficient fluid mixing and efficient convective–conductive heat transfer in the neighborhood of the surface (e.g., turbulent transport in high-conductivity liquids). Small values of h are associated with nearly stagnant fluids of low conductivity. Typical values of h for various physical situations are given in Table 3.

2.1.3. Steady, One-Dimensional Conduction

In many problems, the temperature field is time-invariant and depends on only one spatial variable. Consider first the plane slab case illustrated in Figure 1A where no heat is generated, conduction is in the x direction, and the temper-

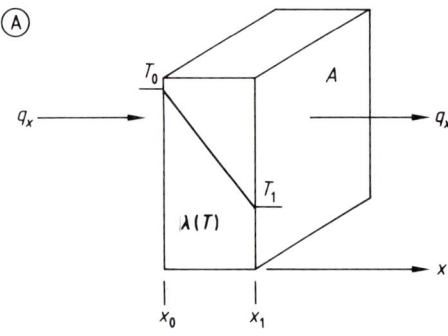

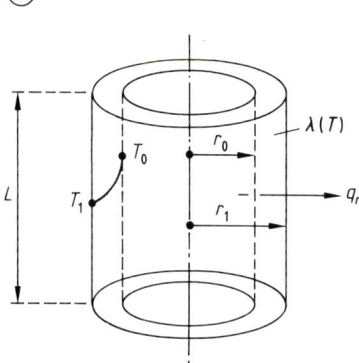

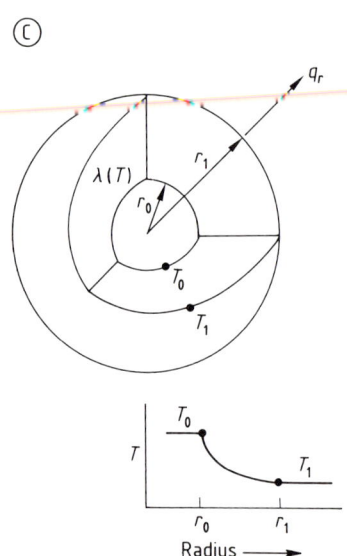

Figure 1. Steady one-dimensional heat conduction in various geometries
A) Conduction in a plane wall; B) Radial conduction in a cylindrical shell; C) Radial conduction in a spherical shell Conductive resistance R_{cond} obtained from Equation (2.14); heat-transfer rate $\dot{Q}$ from Equation (2.13).

ature dependence is $T(x)$. In this case, Equation (2.2) reduces to

$$\frac{dq_x}{dx} = 0 \tag{2.7}$$

or

$$q_x = \frac{\dot{Q}}{A} = \text{constant} \tag{2.8}$$

where $\dot{Q}$ is the heat-transfer rate in joules per second across the slab face of area A. If the temperatures T_0 and T_1 are specified at x_0 and x_1, and Fourier's law is used for the heat flux, the heat-transfer rate $\dot{Q}$ can be obtained by direct integration:

$$\dot{Q} = \frac{\bar{\lambda}_1 A (T_0 - T_1)}{\Delta x_1} \tag{2.9}$$

where $\Delta x_1 \equiv x_1 - x_0$ and $\bar{\lambda}_1$ is the average thermal conductivity defined by

$$\bar{\lambda}_1 \equiv \frac{1}{(T_1 - T_0)} \int_{T_0}^{T_1} \lambda \, dT \tag{2.10}$$

The temperature field is obtained similarly except that the range of integration is from x_0 to any position x:

$$T = T_0 - \frac{\dot{Q}}{\bar{\lambda} A}(x - x_0) \tag{2.11}$$

where $\bar{\lambda} = \lambda(T)$ is the average conductivity in the range from T_0 to T and is defined by an equation similar to Equation (2.10) except that T_1 is replaced everywhere by T. If Equation (2.9) is substituted into Equation (2.11), then

$$T = T_0 - \frac{\bar{\lambda}_1 (T_0 - T_1)}{\bar{\lambda} \Delta x_1}(x - x_0) \tag{2.12}$$

If the dependence of λ on T is sufficiently weak and $\bar{\lambda} \approx \bar{\lambda}_1$, a linear temperature profile results. If λ is a stronger function of temperature, nonlinear profiles develop.

These results can be extended to more general geometries with a few modifications. If ξ is the direction of the nonzero heat flux component, the heat-transfer rate can be expressed as

$$\dot{Q} = q_\xi A_\xi = \frac{(T_0 - T_1)}{R_{\text{cond}}} \tag{2.13}$$

with the conductive resistance R_{cond} given by

$$R_{\text{cond}} = \frac{1}{\bar{\lambda}_1} \int_{\xi_0}^{\xi_1} \frac{d\xi}{A_\xi} \tag{2.14}$$

Here A_ξ is the cross-sectional area perpendicular to the ξ direction. For radial conduction in cylindrical shells (see Fig. 1 B), ξ corresponds to r and A_ξ is $2\pi r L$; for radial conduction in spherical shells (see Fig. 1 C), ξ corresponds to r and A_ξ is $4\pi r^2$. The temperature profile is obtained in the same way as for the plane slab:

$$T = T_0 - \frac{\dot{Q}}{\lambda(T)} \int_{\xi_0}^{\xi} \frac{d\xi}{A_\xi} \tag{2.15}$$

If the dependence of thermal conductivity on temperature is sufficiently weak, where $\lambda(T) \approx \lambda(T_1)$, using Equations 2.13 and 2.14 gives

$$T = T_0 - \frac{(T_0 - T_1)}{\int_{\xi_0}^{\xi_1} \frac{d\xi}{A_\xi}} \int_{\xi_0}^{\xi} \frac{d\xi}{A_\xi} \tag{2.16}$$

For the cylindrical and spherical geometries illustrated in Figure 1, the conductive resistances and the heat-transfer rates are given by:

(Cylindrical) $\quad R_{\text{cond}} = \dfrac{\ln(r_1, r_0)}{2\pi L \bar{\lambda}_1}$ \hfill (2.16a)

$$\dot{Q} = \frac{2\pi \bar{\lambda}_1 (T_0 - T_1)}{\ln(r_1/r_0)} \tag{2.16b}$$

(Spherical) $\quad R_{\text{cond}} = \dfrac{(r_1 - r_0)}{4\pi r_0 r_1 \bar{\lambda}_1}$ \hfill (2.16c)

$$\dot{Q} = \frac{4\pi r_0 r_1 \bar{\lambda}_1 (T_0 - T_1)}{(r_1 - r_0)} \tag{2.16d}$$

2.1.4. Convective Boundary Conditions

As noted previously, the temperature of the fluid in contact with a surface is more commonly known than the surface temperature. Consider again the plane slab case in Figure 1 A, but now assume that the left face is in contact with a fluid of temperature T_{f0}, then from Newton's law of cooling,

$$T_{f0} - T_0 = \frac{q_x}{h_0} = \frac{\dot{Q}}{h_0 A} \tag{2.17}$$

Eliminating T_0 between Equations (2.9) and (2.17) gives

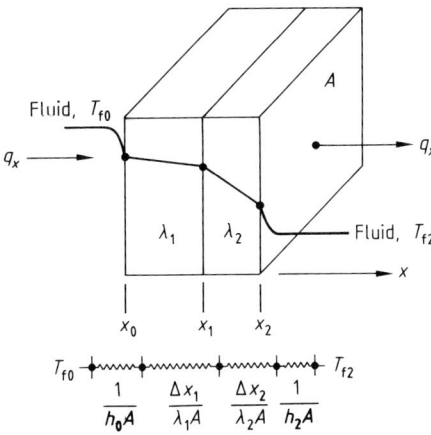

Figure 2. Conduction through a composite two-layer slab Left face is exposed to a fluid at temperature T_{f0}; right face is exposed to a fluid of temperature T_{f2}. The overall driving force for the heat transfer and the conductive resistances are shown in the equivalent electronic circuit.

$$\dot{Q} = \frac{(T_{f0} - T_1)}{\left(\dfrac{1}{h_0 A} + \dfrac{\Delta x_1}{\bar{\lambda}_1 A}\right)} \tag{2.18}$$

where $\Delta x_1 = x_1 - x_0$. This equation gives the total heat-transfer rate in terms of the temperature driving force $T_{f0} - T_1$ and the series of convective and conductive resistances. The ratio of the conductive ($\Delta x_1 / \bar{\lambda}_1 A$) and the convective resistances ($1/h_0 A$) is the dimensionless Biot number:

$$Bi \equiv \frac{h_0 \Delta x_1}{\bar{\lambda}_1} \tag{2.19}$$

When the Biot number is large, conduction is the controlling heat-transfer mechanism and $T_0 \approx T_{f0}$. When the Biot number is small, convection is controlling and $T_0 \approx T_1$.

2.1.5. Composite Systems

The results of Sections 2.1.1–2.1.4 can be extended to composite or multilayered systems by simply applying equations similar to Equation (2.9) across each solid layer and Newton's law of cooling at each fluid–solid boundary. For the two-layer slab shown in Figure 2

$$\dot{Q} = \frac{\bar{\lambda}_1 (T_0 - T_1)}{\Delta x_1} = \frac{\bar{\lambda}_2 (T_1 - T_2)}{\Delta x_2} \tag{2.20}$$
$$= \alpha_0 (T_{f0} - T_0) = \alpha_2 (T_2 - T_{f2})$$

where $\Delta x_k = x_k - x_{k-1}$. Using these equations to eliminate T_0, T_1, and T_2 yields

$$\dot{Q} = \frac{(T_{f0} - T_{f2})}{\left(\dfrac{1}{\alpha_0 A} + \dfrac{\Delta x_1}{\bar{\lambda} A} + \dfrac{\Delta x_2}{\bar{\lambda}_2 A} + \dfrac{1}{\alpha_2 A}\right)} \quad (2.21)$$

or

$$\dot{Q} = \frac{(T_{f0} - T_{f2})}{\sum R_{\text{conv}} + \sum R_{\text{cond}}} \quad (2.22)$$

Generalizing to N layers gives

$$\dot{Q} = \frac{(T_{f0} - T_{fN})}{\left[\dfrac{1}{\alpha_0 A} + \sum_{i=1}^{N}\left(\dfrac{\Delta x_i}{\bar{\lambda}_i A}\right) + \dfrac{1}{\alpha_N A}\right]} \quad (2.23)$$

Similar expressions can be written for composite systems in other geometries (cylinders, spheres, etc.). In these cases, the heat-transfer rate is given by Equation (2.22), with the $R_{\text{cond},i}$ contributions for each layer obtained from Equation (2.14) and the $R_{\text{conv},i}$ contributions from

$$R_{\text{conv},i} = \frac{1}{h_i A_i} \quad (2.24)$$

where h_i is the heat-transfer coefficient at the "ith" fluid–solid surface and A_i the corresponding surface area.

2.1.6. Steady Conduction with Heat Generation and in Multidimensions

Heat is commonly generated in solids by electrical heating as well as by chemical and nuclear reactions. If heat generation is uniform and conduction is one-dimensional, the temperature field can be determined by direct integration. For more complex problems involving multidimensional conduction and nonuniform generation, advanced mathematical methods are required. These are outside the scope of this article but are discussed in [2.4].

2.1.7. Transient Heat Conduction—Lumped Capacity Systems

When a solid body with an initial temperature T_i is exposed to a fluid of temperature T_f, the temperature field in the body changes with time to decrease the magnitude of the temperature difference. Under the conditions of small Biot numbers, the determination of the resulting transient temperature change can be analyzed by using a lumped parameter approach. In particular, if Newton's law of cooling describes the conditions at the fluid–solid boundary, an order-of-magnitude estimate (symbolized by "$\sim$") gives

$$\Delta T \sim \left(\frac{hl}{\lambda}\right)(T_0 - T_f) \quad (2.25)$$

where l is the characteristic length scale of the body and ΔT is the approximate temperature change occurring over this length scale. Clearly, when $Bi \equiv (hl/\lambda) \ll 1$, the spatial temperature change ΔT in the body is much smaller than the temperature changes $T_0 - T_f$ across the fluid layer adjacent to the bounding face. Under such conditions, the surface temperature of the body (T_0) can be assumed to be approximately equal to the average body temperature ($\bar{T}$). The energy balance then becomes

$$\varrho c_p V \frac{d\bar{T}}{dt} = -\bar{h} A (\bar{T} - T_f) \quad (2.26)$$

where $\bar{h}$ and $\bar{T}$ represent average values over the surface and volume V of the body, respectively. If the material properties are taken to be constant, this equation can be integrated to give

$$\bar{T} = T_f + (T_i - T_f) \exp\{-Bi\,t^*\} \quad (2.27)$$

where, in this case, the characteristic length is $l \equiv V/A$, the Biot number is $Bi \equiv l h/\lambda$, and the dimensionless time is $t^* \equiv \alpha t/l^2$. The corresponding rate of heat transfer to the fluid is given by

$$\dot{Q} = \bar{h} A (T_i - T_f) \exp\{-Bi\,t^*\} \quad (2.28)$$

The significance of these results for $Bi \ll 1$ is that the rate of heat transfer given by Equation (2.28) represents an upper bound; i.e., the cooling or heating rates found at higher Biot numbers will be slower and the characteristic time for temperature accommodation of the body with the fluid will be longer.

2.1.8. Transient Heat Conduction—More General Solutions

At higher Biot numbers, the spatial variations of temperature cannot be neglected and the problem is mathematically more complex than that given in the previous section. In this section, solutions for the semi-infinite body, the plane slab, the infinite cylinder, and the sphere are summarized. Mathematical details are presented in [2.4, pp. 50–73 (semi-infinite solid), pp. 119–127 (plane slab), pp. 201–203 (infinite cylinder), and pp. 237–238 (sphere)].

Consider first the transient heating (or cooling) of a body bounded by the plane $x = 0$ and extending to infinity in the positive x direction. The body is initially at temperature T_i and its face at $x = 0$ is held at temperature T_f. If constant properties are assumed, the governing equations for $T(x, t)$ are

$$\frac{\partial T}{\partial t} = \alpha \frac{\partial^2 T}{\partial x^2} \tag{2.29}$$

$$T(x, 0) = T_i, \quad x \geq 0 \tag{2.30}$$

$$T(0, t) = T_f, \quad t > 0 \tag{2.31}$$

$$T(\infty, t) = T_i, \quad t \geq 0 \tag{2.32}$$

These equations suggest a solution of the form

$$T - T_f = f(x, t, \alpha, T_i - T_f) \tag{2.33}$$

If dimensional analysis is used, the dimensionless temperature can be expressed in terms of a single independent variable, i.e.,

$$T^* \equiv \frac{T - T_f}{T_i - T_f} = T^*(\zeta) \tag{2.34}$$

where

$$\zeta = \frac{x}{\sqrt{4 \alpha t}} \tag{2.35}$$

When Equation (2.34) is substituted into Equations (2.29)–(2.32), the problem is reduced to the solution of an ordinary differential equation

$$\frac{d^2 T^*}{d\zeta^2} + 2\zeta \frac{dT^*}{d\zeta} = 0 \tag{2.36}$$

subject to the boundary conditions

$$T^*(0) = 0 \tag{2.37}$$

$$T^*(\infty) = 1 \tag{2.38}$$

Integrating Equation (2.36) first for $dT^*/d\zeta$ and then again for $T^*(\zeta)$ and using the boundary conditions to determine the two integrating constants give

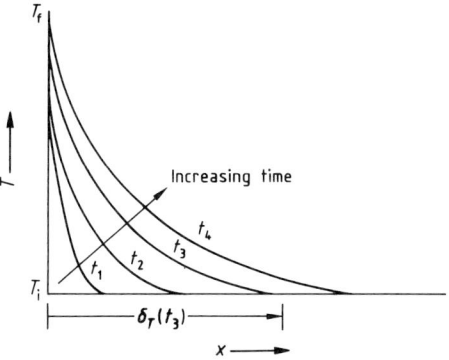

Figure 3. Time evolution of the temperature profile during heating of a semi-infinite slab ($x > 0$). The x position where $T = T_f - 0.99\,(T_f - T_i)$ defines the thermal boundary layer δ_T which according to Equation (2.40) increases in proportion to $\sqrt{\alpha t}$.

$$T^* = \frac{2}{\sqrt{\pi}} \int_0^{\zeta} e^{-\xi^2} d\xi \equiv \mathrm{erf}(\zeta) \tag{2.39}$$

The middle term is called the error function [erf(ζ)]. Tabulated values can be found in various sources [2.4, pp. 485–487]. Of particular interest here is the fact that the error function (as well as T^*) takes on a value of approximately 0.99 when its argument ζ takes on a value of 1.8. Physically, this implies that the temperature change $(T - T_f)$ reaches 99% of the maximum change $(T_i - T_f)$ within a thermal boundary layer defined by

$$\delta_T = 1.8 \sqrt{4 \alpha t} = 3.6 \sqrt{\alpha t} \tag{2.40}$$

This is illustrated in Figure 3 where the time evolution of the temperature profile is plotted for the transient heating of a semi-infinite body.

If the temperature boundary condition at $x = 0$ (Eq. 2.31) is replaced by a convective flux condition, i.e.,

$$\lambda \frac{\partial T}{\partial x}(0, t) = h[T(0, t) - T_f] \tag{2.41}$$

the solution takes the form

$$T^* = \mathrm{erf}(\zeta) + \exp(\zeta \beta + \tfrac{1}{4} \beta^2) [\mathrm{erfc}(\zeta + \tfrac{1}{2} \beta)] \tag{2.42}$$

where

$$\beta \equiv \frac{2\sqrt{\alpha t}\, h}{\lambda} \tag{2.43}$$

and $\mathrm{erfc}(y) = 1 - \mathrm{erf}(y)$ is the complementary error function.

Although these results have been developed in the context of a semi-infinite body, they are also valid for a body of finite thickness (e.g., a slab of thickness $2l$) for all times when $\delta_T \leq l$. For times longer than this, the boundary layer exceeds the half thickness and a semi-infinite solution is no longer valid. Then solutions for finite bodies must be used. Solutions of the latter type are now presented for the plane slab, the infinite cylinder, and the sphere.

Consider one-dimensional heat conduction in a plane slab bounded by the surfaces at $x = \pm l$. For constant thermal properties the governing energy equation is Equation (2.29); the initial and boundary conditions take the form

$$T(x, 0) = T_i, \quad -l \leq x \leq l \quad (2.44)$$

$$\frac{\partial T}{\partial x}(0, t) = 0, \quad \text{all } t \quad (2.45)$$

$$-\lambda \frac{\partial T}{\partial x}(l, t) = h[T(l, t) - T_f], \quad t > 0 \quad (2.46)$$

The second equation arises from the fact that the midplane ($x = 0$) is a plane of symmetry and the flux across it is zero.

The solution to this problem in terms of the dimensionless position $x^* \equiv x/l$ and the dimensionless time $t^* \equiv \alpha t/l^2$ is obtained as a Fourier series:

$$T^* \equiv \frac{T - T_f}{T_i - T_f} \quad (2.47)$$

$$= 2 \sum_{n=1}^{\infty} \frac{\sin \beta_n \cos(\beta_n x^*)}{\beta_n + \sin \beta_n \cos \beta_n} \exp\{-\beta_n^2 t^*\}$$

where the eigenvalues β_n are obtained from

$$\beta_n \tan \beta_n = Bi \quad (2.48)$$

For short times the convergence of Equation (2.47) is slow and many terms must be taken before an accurate solution can be obtained. However, for $t^* > 0.25$, only the first term of this equation is needed to obtain results with errors less than 1%. Even in these cases the solution is not convenient because β_1 must be determined from an implicit relation (Eq. 2.48). Fortunately, the roots of the latter equation can be accurately fitted to an expression of the form

$$\beta_1 = \sum_{k=1}^{5} C_k (\log Bi)^{k-1} \quad \text{for } 0.02 \leq Bi \leq 8 \quad (2.49)$$

For the plane slab case, the coefficients C_k are given in the first column of Table 4. When this equation is used with the first term of Equation (2.47), temperatures can be predicted to within 5% and, in most cases, depending on the Biot number, to within 1%. When $Bi > 8$, tabulated values of β_n can be used [2.4, p. 491]. Such approaches are preferred to the time–temperature charts used by many investigators [2.7]–[2.10] because these charts are often difficult to read and to interpolate accurately.

Similar types of solutions are available for the infinite cylinder (radius r_0) and the sphere (also radius r_0). Here again, the body is assumed to be at an initial temperature T_i and at $t = 0$ is immersed in a fluid at temperature T_f. For the cylinder, the solution in terms of the first- and second-order Bessel functions of the first kind is

$$T^* = 2 \sum_{n=1}^{\infty} \frac{J_1(\beta_n) J_0(\beta_n r^*)}{\beta_n [J_0^2(\beta_n) + J_1^2(\beta_n)]} \exp\{-\beta_n^2 t^*\} \quad (2.50)$$

where T^* is defined as in Equation (2.47), $r^* \equiv r/r_0$ and $t^* \equiv \alpha t/r_0^2$. The eigenvalues β_n are obtained from

$$\beta_n \frac{J_1(\beta_n)}{J_0(\beta_n)} = Bi \quad (2.51)$$

As before, for $t^* > 0.25$, only the first term of Equation (2.50) is important. Also, the first root of Equation (2.51) can be obtained from Equation (2.49) with the coefficients C_k given in the second column of Table 4.

For a sphere, the solution is

$$T^* = \frac{2 Bi}{r^*} \sum_{n=1}^{\infty} \frac{\sin \beta_n \sin(\beta_n r^*)}{\beta_n [\beta_n - \sin \beta_n \cos \beta_n]} \exp\{-\beta_n^2 t^*\} \quad (2.52)$$

with the eigenvalues given by

$$\beta_n \cot \beta_n = 1 - Bi \quad (2.53)$$

The one-term solution is again valid for $t^* > 0.25$, and the first eigenvalue can be obtained from Equation (2.49) by using the coefficients in the third column of Table 4.

Table 4. Coefficients for Equation (2.49)

Coefficient	Plane slab	Cylinder	Sphere
C_1	0.859777	1.256701	1.573305
C_2	0.702996	1.112492	1.450377
C_3	0.050636	0.148517	0.251563
C_4	−0.144345	−0.260841	−0.306763
C_5	−0.044631	−0.111339	−0.139485

The solutions of these one-dimensional unsteady-state problems can be used in simple ways to obtain solutions to multidimensional, unsteady-state problems. These are now summarized for several cases. Consider first the case of an *infinite rectangular bar* bounded by the surfaces at $x = \pm x_0$ and $y = \pm y_0$. The solution at any position (x, y) at time t is given by

$$T(x, y, t) = T_x(x, t) T_y(y, t) \tag{2.54}$$

Here $T_x(x, t)$ is obtained from Equation (2.47) with x_0 replacing l as the length scale. Similarly, $T_y(y, t)$ is obtained from Equation (2.47) with y replacing x in that equation and y_0 replacing l.

For a *rectangular parallelepiped* with bounding surfaces at $x = \pm x_0$, $y = \pm y_0$, and $z = \pm z_0$, the solution is given by

$$T(x, y, z, t) = T_x(x, t) T_y(y, t) T_z(z, t) \tag{2.55}$$

The T_x and T_y solutions are obtained in the manner just described for the infinite rectangular bar. The T_z solution is obtained from Equation (2.47) with z replacing x in that equation and z_0 replacing l.

Finally, for a *finite cylinder* bounded by the surfaces $r = r_0$ and $z = \pm l$, the solution is

$$T(r, z, t) = T_r(r, t) T_z(z, t) \tag{2.56}$$

Here T_r is obtained from Equation (2.50) and T_z from Equation (2.47) (z replaces x).

Examples of the use of Equations (2.54)–(2.56) are given in many texts [2.11, pp. 307–310], [2.12, pp. 174–177]. The approach suggested here is different in that the one-dimensional solutions are obtained from one-term analytical solutions instead of time–temperature charts.

2.2. Steady, One-Dimensional Flows
(→ 5. Fluid Mechanics, pp. **5**-11–**5**-13)

As in the sections above where heat transfer arose from the existence of temperature gradients, momentum transfer arises from fluid-deforming velocity gradients. In particular, for the one-dimensional unidirectional flow

$$\boldsymbol{u} \equiv (u_x, u_y, u_z) = [0, 0, u_z(y)] \tag{2.57}$$

the resulting momentum transfer involves z-momentum being transferred in the y-direction. This

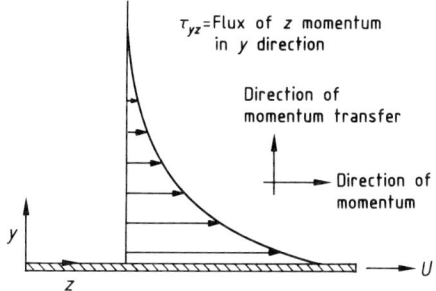

Figure 4. Flow resulting from the movement of a surface
In this case, z-momentum is being transferred from layer to layer in the y-direction. The stress component τ_{yz} can be interpreted as a momentum flux (z momentum transferred in the y direction).

is illustrated in Figure 4 where faster moving fluid layers transfer momentum to adjacent slower moving layers. The nonzero stress components associated with this velocity field can be interpreted as momentum fluxes [2.2, Chap. 2], and if the fluid is Newtonian, these components can be expressed as

$$\tau_{yz} = \tau_{zy} = -\eta \frac{du_z}{dy} \tag{2.58}$$

which is analogous to Fourier's law for one-dimensional heat conduction in the y-direction.

The flow above is a special case of the general class of unidirectional flows. Unidirectional flows commonly arise with internal flows in straight channels where the cross-sectional area is invariant in the flow direction. Under such conditions, the velocity u_z does not vary in the z-direction and the most general form of the velocity field is $\boldsymbol{u} = [0, 0, u_z(x, y)]$. The latter two-dimensional problem arises with systems of arbitrary cross section. For simpler systems such as flow between parallel plates or flow in cylindrical tubes, the problem becomes one-dimensional and u_z is dependent on only one spatial variable.

The driving force for unidirectional flows can be pressure gradients, gravity, or moving boundaries. The pressure gradient and gravity effects appear directly in the equation of motion and can be combined into an overall driving force term $\boldsymbol{\nabla}P$:

$$\boldsymbol{\nabla}P = \boldsymbol{\nabla}p - \varrho\boldsymbol{g} \tag{2.59}$$

To achieve a unidirectional flow without moving boundaries, $\boldsymbol{\nabla}P$ must be nonzero and the flow direction must be coincident with the directions of ∇P. For flow in the z-direction where $P(z)$, the only nonzero component of Equation (2.59) is

$$\frac{dP}{dz} = \frac{dp}{dz} - \varrho g_z \tag{2.60}$$

where g_z is the component of gravity in the z-direction. Integration of this equation along the flow channel from $z = z_0$ to any point z gives

$$P - P_0 = p - p_0 - \varrho g_z(z - z_0) \tag{2.61}$$

That is, the driving force for flow arises from variations of pressure and potential energy along the flow direction. The quantity P is sometimes referred to as the *dynamic pressure* because it can be written as the difference between the pressure existing during flow and that arising under no flow or hydrostatic conditions:

$$P = p - p_H \tag{2.62}$$

Here, p_H is the *hydrostatic pressure* which is obtained from

$$\nabla p_H - \varrho \mathbf{g} = 0 \tag{2.63}$$

In Equation (2.61) the hydrostatic pressure part is $p_0 + \varrho g_z(z - z_0)$.

Another characteristic of steady unidirectional flows is that these are noninertial flows. Specifically, a fluid element in this flow does not accelerate along its path, and the nonlinear inertial term $\mathbf{u} \cdot \nabla \mathbf{u}$ in the equation of motion is zero. The z-component of the equation of motion then takes the form

$$(\nabla \cdot \boldsymbol{\tau}) \cdot \mathbf{i}_z + \frac{dP}{dz} = 0 \tag{2.64}$$

where $\mathbf{i}_z$ is the unit base vector in the z direction. If the fluid is an incompressible Newtonian fluid, this equation can be expressed as

$$\eta \nabla^2 u_z - \frac{dP}{dz} = 0 \tag{2.65}$$

With respect to Newton's laws of motion, this equation can be interpreted as a balance of forces per unit volume on any fluid element in the flow; specifically, the driving force for the flow is $(-dP/dz)$ and the viscous resisting force is $\eta \nabla^2 u_z$.

The only material parameter that arises in the treatment of steady, unidirectional flows is the viscosity η. In more general flows, the density ϱ also arises. In Table 5, typical values of density and viscosity for a number of fluids are presented. Values for the heat capacity c_p, the thermal conductivity λ, and the Prandtl number $Pr \equiv \nu/\alpha$ are also provided. All of these quantities arise later in the treatment of convective heat transfer.

2.2.1. Generalized Couette Flow

A flow pattern arising in many different applications (polymer extrusion, screw pumping and expression, journal-bearing lubrication, etc.) is the unidirectional plane parallel flow illustrated in Figure 5. The flow is driven by both the movement of the upper plate at velocity U and the application of a pressure gradient $(P_0 - P_L)/L$ in the direction of flow. The flow geometry is assumed to be of thickness b in the y-direction, width w in the x-direction, and length L in the z-direction. The thickness-to-width ratio (b/w) and the thickness-to-length ra-

Table 5. Thermophysical properties of various gases and liquids*

Fluid	θ, °C	ϱ, kg/m³	η, Pa s	λ, W m⁻¹ K⁻¹	c_p, J kg⁻¹ K⁻¹	Pr
Gases (101.3 kPa)						
Air	27	1.183	1.853×10^{-5}	0.02614	1.003×10^3	0.711
Carbon dioxide	27	1.7973	1.4958×10^{-5}	0.016572	0.871×10^3	0.770
Helium	−18	0.1906	1.817×10^{-5}	0.1357	5.200×10^3	0.70
Hydrogen	27	0.08185	0.8963×10^{-5}	0.182	14.314×10^3	0.706
Nitrogen	27	1.1233	1.784×10^{-5}	0.0259	1.0408×10^3	0.715
Oxygen	27	1.3007	2.063×10^{-5}	0.02676	0.09203×10^3	0.709
Liquids (saturated)						
Dichlorodifluoromethane	27	1305	0.2545×10^{-3}	0.0690	980	3.62
Glycerol	20	1261	1412×10^{-3}	0.285	2350	11 630
Mercury	27	13 611	1.633×10^{-3}	8.34	139.1	0.027
Light machine oil	16	907	145.1×10^{-3}			
Water	27	996.6	0.8232×10^{-3}	0.6084	4177	5.65

* Adapted from [2.6, pp. 595–601].

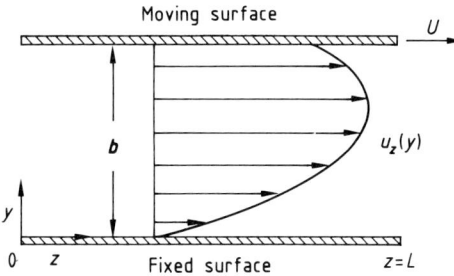

Figure 5. Generalized Couette flow between parallel surfaces
The upper surface moves with velocity U; the lower surface is stationary. The nature of the flow depends on the magnitude and sign of the applied pressure gradient dP/dz. In the case above, $DP/Dz > 0$.

tio (b/L) are assumed to be sufficiently small that edge effects and entrance or exit effects can be neglected, and the velocity field can be represented by $\boldsymbol{u} = (u_x, u_y, u_z) = [0, 0, u_z(y)]$.

The nonzero stress components for this flow are τ_{yz} and τ_{zy}, and Equation (2.64) reduces to

$$\frac{d\tau_{yz}}{dy} + \frac{dP}{dz} = 0 \qquad (2.66)$$

Since the first term on the left-hand side is at most a function of y and the second term at most a function of z, this equation is satisfied only if both terms are equal to a constant. Integration then gives

$$\tau_{yz} = K y + \tau_0 \qquad (2.67)$$

$$K \equiv \frac{(P_0 - P_L)}{L} \qquad (2.68)$$

where τ_0 is the shear stress at $y = 0$. At this point, the quantity τ_0 is unknown. It is determined by using the velocity boundary conditions.

If the fluid is an incompressible Newtonian fluid, Equation (2.58) can be combined with Equation (2.67) and integrated to give the velocity profile

$$u_z = \frac{b^2 \Delta P}{2 \eta L}\left(\frac{y}{b}\right)\left(1 - \frac{y}{b}\right) + U\left(\frac{y}{b}\right) \qquad (2.69)$$

where the quantity τ_0 and the additional constant that arises from the integration are evaluated by using the boundary conditions

$$u_z(0) = 0 \qquad (2.70)$$

$$u_z(b) = U \qquad (2.71)$$

In dimensionless form, the velocity is

$$\frac{u_z}{U} = K^*\left(\frac{y}{b}\right)\left(1 - \frac{y}{b}\right) + \frac{y}{b} \qquad (2.72)$$

where

$$K^* = \frac{b^2 \Delta P}{2 \eta U L} \qquad (2.73)$$

is the dimensionless pressure gradient.

Depending on the magnitude and sign of K^*, the velocity profile can take various forms (see Fig. 6). When $K^* = 0$, the profile is linear and corresponds to simple Couette flow:

$$u_z = U\left(\frac{y}{b}\right) \qquad (2.74)$$

When $K^* > 0$, the pressure gradient and the moving upper boundary both contribute to flow in the positive z direction, and if K^* is sufficiently large a maximum occurs in the velocity profile (see the $K^* \geq 2$ cases in Fig. 6). When $K^* < 0$, the pressure gradient causes a reverse flow. If the magnitude of this adverse gradient is sufficiently large, the fluid moves in the positive z-direction in the upper part of the flow channel (due to motion of the upper boundary) and in the negative z-direction in the lower part of the channel (see the $K^* = -4$ case in Fig. 6).

The volumetric flow rate Q and the average velocity u_{avg} associated with this flow are given by

$$Q = u_{\text{avg}} b w = \int_0^b u_z w \, dy = \frac{b^3 w \Delta P}{12 \eta L} + \tfrac{1}{2} b w U \qquad (2.75)$$

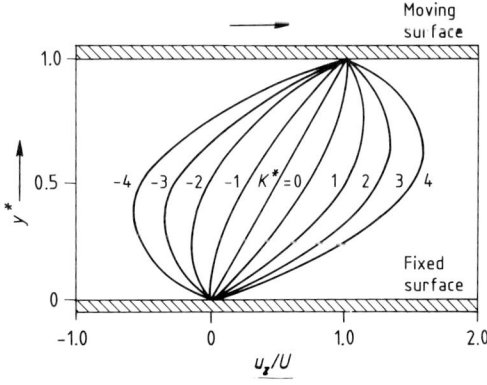

Figure 6. Dimensionless velocity profiles for generalized Couette flow

The first term on the right represents the flow contribution arising from the pressure gradient and the second that from the moving boundary. When $K^* = -3$, the volumetric flow rate associated with the moving boundary is balanced by the flow rate arising from the adverse pressure gradient and $Q = 0$.

If the fluid is non-Newtonian, significantly more complex flow equations are obtained for generalized Couette flow. For the case of power-law, viscous non-Newtonian behavior, see [2.13], [2.14] (see also → 5. Fluid Mechanics, pp. **5**-26 – **5**-27).

2.2.2. One-Dimensional Poiseuille Flows
(→ 5. Fluid Mechanics)

When $U = 0$, the flow illustrated in Figure 5 becomes a pressure-driven Poiseuille flow. Under such conditions, the velocity profile becomes parabolic with a plane of symmetry at $y = b/2$. Specifically, in terms of a y coordinate measured from the channel centerline, i.e., $y' = y - b/2$, Equation (2.69) takes the form

$$u_z = \left(\frac{b^2 \Delta P}{8 \eta L}\right)\left[1 - 4\left(\frac{y'}{b}\right)^2\right] \qquad (2.76)$$

and the volumetric flow rate reduces to a Hagen – Poiseuille relation where Q is proportional to $(P_0 - P_L)/L$, i.e.,

$$Q = u_{avg} b w = \frac{b^3 w \Delta P}{12 \eta L} \qquad (2.77)$$

In the case of one-dimensional Poiseuille flows in channels with circular and concentric annulus cross-sectional geometries, analogous results can be obtained. In all cases the volumetric flow rate is proportional to the pressure drop (for further details, see → 5. Fluid Mechanics, pp. **5**-11 – **5**-12).

An alternate representation of flow rate – pressure drop results for unidirectional flows is in terms of the friction factor relation. In particular, the wall shear stress $\tau_w \equiv -\eta |du_z/dy'|_w$ is assumed to be proportional to the kinetic energy per unit volume expressed in terms of the average velocity u_{avg}:

$$\tau_w = f\left(\tfrac{1}{2} \varrho u_{avg}^2\right) \qquad (2.78)$$

Here, the constant of proportionality is the friction factor f. Since the pressure force must be balanced by the drag force on the walls

$$\Delta P A_c = \tau_w A_w = f A_w \left(\tfrac{1}{2} \varrho u_{avg}^2\right) \qquad (2.79)$$

Here A_c is the cross-sectional area and A_w is the wall surface area. The friction factor can then be expressed as

$$f Re = C \equiv \frac{l^2 (\Delta P/L)}{2 \eta u_{avg}} \qquad (2.80)$$

where the Reynolds number Re and the characteristic length l are defined by

$$Re = \frac{\varrho u_{avg} l}{\eta} \qquad (2.81)$$

and

$$l = \frac{4 A_c L}{A_w} \qquad (2.82)$$

For plane Poiseuille flow where $l = 2b$, Equation (2.77) can be used in Equation (2.80) to obtain

$$f Re = C = 24 \qquad (2.83)$$

where $Re \equiv 2 \varrho u_{avg} b/\eta$.

Another flow that arises in many industrial applications is Poiseuille flow in a tube. If the tube has radius r_1 and length L, the stress and velocity fields for incompressible Newtonian fluids are

$$\tau_{rz} = \tau_{zr} = \frac{1}{2}\left(\frac{\Delta P}{L}\right) r \qquad (2.84)$$

$$u_z = \frac{r_1^2}{4\eta}\left(\frac{\Delta P}{L}\right)\left[1 - \left(\frac{r}{r_1}\right)^2\right] \qquad (2.85)$$

where r is the radial coordinate. The volumetric flow rate is then

$$Q = u_{avg} \pi r_1^2 = 2\pi \int_0^{r_1} u_z r \, dr = \frac{\pi r_1^4}{8\eta}\left(\frac{\Delta P}{L}\right) \qquad (2.86)$$

If the latter result is used in Equation (2.80), then

$$f Re = 16 \qquad (2.87)$$

where the characteristic length l is $2 r_1$, and the Reynolds number is defined by $Re \equiv 2 \varrho u_{avg} r_1/\eta$.

Similar results can be obtained for axial flow in the annular region between concentric cylinders of radii r_0 and r_1 where $r_1 > r_0$. In this case, the characteristic length is $l = 2(r_1 - r_0)$, and the value of C in Equation (2.80) varies with the ratio $\chi \equiv r_0/r_1$. In the limit of small values of r_0/r_1, C takes on the value of 16; it approaches an upper limit of 24 as r_0/r_1 approaches 1.

2.2.3. Flow in Channels of Arbitrary Cross Section

For flows in channels with more complex cross sections (e.g., flow in rectangular, elliptic, or eccentric annular channels) or flow in channels with irregular cross sections, the velocity u_z is a function of two spatial variables, i.e.,

$$u_z = u_z(x, y) \tag{2.88}$$

For incompressible Newtonian fluids, Equation (2.65) is applicable, and the solution to this equation is given by

$$u_z = \frac{1}{4\eta}(x^2 + y^2)\left(\frac{dP}{dz}\right) + \varphi(x, y) \tag{2.89}$$

where the function $\varphi(x, y)$ is a solution of Laplace's equation

$$\frac{\partial^2 \varphi}{\partial x^2} + \frac{\partial^2 \varphi}{\partial y^2} = 0 \tag{2.90}$$

If a no-slip velocity condition is assumed (i.e., $u_{z,s} = 0$), then

$$\varphi = \varphi_s = \left[-\frac{1}{4\eta}(x^2 + y^2)\left(\frac{dP}{dz}\right)\right]_s \tag{2.91}$$

on the bounding surfaces.

For a channel with an *elliptic cross section* defined by $(x/x_0)^2 + (y/y_0)^2 = 1$, the velocity is given by [2.15, p. 38]

$$u_z = \frac{x_0^2 y_0^2 \Delta P}{2\eta(x_0^2 + y_0^2)L}\left[1 - \left(\frac{x}{x_0}\right)^2 - \left(\frac{y}{y_0}\right)^2\right] \tag{2.92}$$

and the volumetric flow rate by

$$Q = u_{avg}\pi x_0 y_0 = \frac{\pi x_0^3 y_0^3 \Delta P}{4\eta(x_0^2 + y_0^2)L} \tag{2.93}$$

In terms of the friction factor, this result can be written

$$fRe = \frac{2\pi^2}{[E(k)]^2}\left[1 + \left(\frac{y_0}{x_0}\right)^2\right] = \left\{\text{function of } \frac{y_0}{x_0}\right\} \tag{2.94}$$

where $E(k)$ is the complete elliptic integral with argument $k = [1 - (y_0/x_0)^2]$ [2.16, p. 509]. When $y_0/x_0 \to 0$, $fRe \to 19.74$, and when $y_0/x_0 = 1$, $fRe = 16$ (the cylindrical tube result).

For flow in channels with *square and equilateral-triangle cross sections* with sides a, the flow rate–pressure drop relations are given by

Square: $fRe = 14.2$ (2.95)

Triangular: $fRe = 13.3$ (2.96)

Results for other cross sections (rectangular and eccentric annuli) are given in [2.15, pp. 33–39]. For even more complex and less regular cross-sectional geometries, estimates (sometimes crude) can be obtained by using a hydraulic radius approach where an effective radius is defined for the channel, i.e., $r_H \equiv l/2 = 2A_c L/A_w$, and C in Equation (2.80) is taken to be 16, the value corresponding to Poiseuille flow in a tube. However, this approach is approximate and can lead to sizable errors if the actual C for the geometry differs significantly from 16. For flow in channels with elliptic cross sections, the maximum error would be ca. 19%; for square channels, the error would be 12%.

All of these results are for laminar flow. When the Reynolds number reaches sufficiently high values, the flow becomes unstable and turbulent flow arises. For flow in circular tubes, turbulent flow is generally observed when $Re \equiv 2\rho u_{avg} r_1/\eta > 2100$. For other geometries, the transition values are different and must be determined in each individual case.

2.2.4. Poiseuille Flow of Non-Newtonian Fluids
(→ 5. Fluid Mechanics, pp. 5-28–5-33)

The flow characteristics of non-Newtonian fluids in tubes and planar channels can be quite different from those observed with Newtonian fluids. This is illustrated by considering the steady flow of a power-law, non-Newtonian fluid in a tube of radius r_0. As in the Newtonian case, the only nonzero stress components are $\tau_{rz} = \tau_{zr}$, and these are described by Equation (2.84). Because $du_z/dr < 0$ for this flow, the shear stress can be written as

$$\tau_{rz} = m\left(-\frac{du_z}{dr}\right)^n \tag{2.97}$$

Combining this equation with Equation (2.84), integrating, and using the no-slip boundary condition $u_z = 0$ at the tube wall give

$$u_z = \left(\frac{\tau_w}{m}\right)^s \frac{r_1}{s+1}\left[1 - \left(\frac{r}{r_1}\right)^{s+1}\right] \tag{2.98}$$

where $s \equiv 1/n$, and τ_w is the wall shear stress given by

$$\tau_w \equiv |\tau_{rz}|_{r=r_1} = \frac{1}{2}\left(\frac{\Delta P}{L}\right)r_1 \quad (2.99)$$

Because the maximum velocity occurs at $r = 0$, Equation (2.98) can also be written as

$$u_z = u_{max}\left[1 - \left(\frac{r}{r_1}\right)^{s+1}\right] \quad (2.100)$$

with u_{max} given by

$$u_{max} = \left(\frac{\tau_w}{m}\right)^s \frac{r_1}{(s+1)} \quad (2.101)$$

For $s = 1$ (or $n = 1$), the behavior is Newtonian and the profile is parabolic. For pseudoplastic fluids where $s > 1$ (or $n < 1$), the profile becomes more blunted and pluglike as s takes on larger and larger values (see Fig. 7). For dilatant fluids where $s < 1$ (or $n > 1$), the velocity profile becomes more and more pointed as s decreases.

Comparison of the average velocity with the maximum velocity gives

$$\frac{u_{avg}}{u_{max}} = \frac{s+1}{s+3} \quad (2.102)$$

For Newtonian fluids ($s = 1$), this reduces to $u_{avg}/u_{max} = 1/2$. For pseudoplastic fluids ($s > 1$), this ratio increases with s and approaches the limit of 1 (plug flow) as s gets very large ($n \to 0$). For dilatant fluids, u_{avg}/u_{max} ranges from 1/2 to 1/3 as s decreases (or n increases). These results are illustrated in Figure 7.

The flow rate–pressure drop expression for this flow is obtained directly from $Q = \pi r_1^2 u_{avg}$. For values of s different from 1, this relation is nonlinear, with $d(\Delta P/L)/dQ$ increasing with Q for dilatant fluids and $d(\Delta P/L)/dQ$ decreasing with Q for pseudoplastic fluids.

In the case of a Bingham plastic with a yield stress τ_Y and a plastic viscosity η_p, the nature of the flow depends on the radial position of the yield surface, i.e., the position $r = r_Y$ where $|\tau_{rz}| = \tau_Y$. From Equation (2.84)

$$r_Y = \frac{\tau_Y r_1}{\tau_w} = \frac{2\tau_Y}{(\Delta P/L)} \quad (2.103)$$

At radial positions outside the yield surface ($r > r_Y$), the magnitude of the shear stress exceeds the yield stress and a shear flow [$u_z = u_z(r)$]

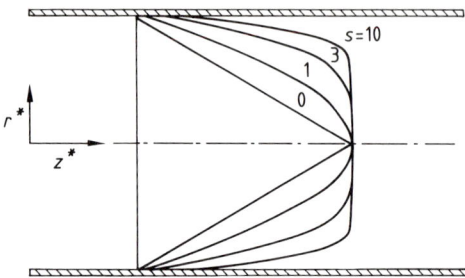

Figure 7. Velocity profiles for the flow of a power-law fluid in a tube
The parameter s is the reciprocal of the power-law slope parameter n. When $s = 1$, Newtonian behavior is observed. When $s > 0$, the behavior is dilatant.

results. At radial positions inside the yield surface ($r < r_Y$), the magnitude of the shear stress is less than τ_Y and a uniform plug flow occurs with velocity $u_z(r_Y)$. As $(\Delta P/L)$ decreases, the radial position of the yield surface moves closer and closer to the tube wall. At a value of $(\Delta P/L) = 2\tau_Y/r_1$, the position of the yield surface coincides with the wall and flow ceases. Hence, application of a finite pressure gradient does not ensure flow in the case of materials with plastic behavior.

Flow rate results for Newtonian fluids, power-law fluids, and Bingham plastics may be summarized as follows:

Flow geometry: Circular tube radius r_1, length L
Newtonian fluid (viscosity η)

$$Q = \frac{\pi r_1^4}{8\eta}\frac{\Delta P}{L}$$

Power-law fluid (parameters m, n)

$$Q = \frac{\pi r_1^3}{(s+3)}\left(\frac{\Delta P r_1}{2Lm}\right)^s, \quad s \equiv \frac{1}{n}$$

Bingham plastic (parameters τ_Y, η_p)

$$Q = \frac{\pi r_1^3 \tau_w}{4\eta_p}\left[1 - \frac{4}{3}\left(\frac{\tau_Y}{\tau_w}\right) + \frac{1}{3}\left(\frac{\tau_Y}{\tau_w}\right)^4\right], \quad \text{when } \tau_w > \tau_Y$$

where

$$\tau_w \equiv \frac{\Delta P r_1}{2L}$$

$$Q = 0, \quad \text{when } \tau_w \leq \tau_Y$$

Expressions for other types of non-Newtonian fluids can be found in various sources. These developments are described and the results summarized for various non-Newtonian fluid models [2.17, Chap. 5], [2.18]–[2.20]. Further details are given in $\to$ 5. Fluid Mechanics, pp. **5**-29–**5**-30.

2.2.5. Two-Phase Concentric Flow in a Tube—Segregated Flow

An interesting example of viscous, one-dimensional laminar flow in a tube takes place when two immiscible fluids are injected into a tube of radius r_1 such that one fluid occupies the core region $r < r_c$ and the other fluid the outer region $r > r_c$ (see Fig. 8). The core fluid has viscosity η and the outer fluid viscosity $\hat{\eta}$.

Such problems are often encountered in pipelines where the flow of a viscous fluid is facilitated by the injection of a second low-viscosity immiscible phase. The interest here is in the flow rate enhancement that results from the injection of this second fluid.

The analysis of flow in the separate fluid regions is similar to that for Poiseuille flow in cylindrical tubes and annuli. In particular, for Newtonian fluids, Equations (2.64) and (2.65) can be integrated to give the stress and velocity fields for the inner fluid region as

$$\tau_{rz} = \frac{1}{2}\left(\frac{\Delta P}{L}\right)r + \frac{C_1}{r} \tag{2.104}$$

$$u_z = -\frac{1}{4\eta}\left(\frac{\Delta P}{L}\right)r^2 - \frac{C_1}{\eta}\ln r + C_2 \tag{2.105}$$

where C_1 and C_2 are the integrating constants. Identical expressions are obtained for the outer fluid region except that τ_{rz}, u_z, η, C_1, and C_2 are replaced by $\hat{\tau}_{rz}$, $\hat{u}_z$, $\hat{\eta}$, $\hat{C}_1$, and $\hat{C}_2$. It writing these expressions, the axial pressure distributions in the two phases are assumed to be equal.

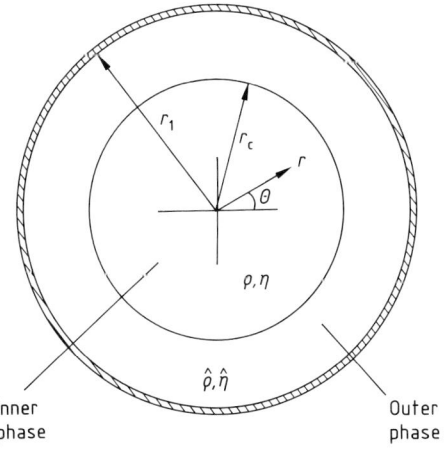

Figure 8. Segregated two-phase flow in a tube

The integrating constants can be evaluated by using the conditions (1) that the stress τ_{rz} is finite at $r = 0$, (2) that the velocity u_z is zero at the tube wall, and (3) that the stress and velocity are continuous at the boundary $r = r_c$ between the phases [$\tau_{rz}(r_c) = \hat{\tau}_{rz}(r_c)$ and $u_z(r_c) = \hat{u}_z(r_c)$]. Integrating the respective velocity profiles over the inner and outer fluid regions according to Equation (2.86) gives the corresponding volumetric flow rates

$$Q = \frac{\pi}{8\hat{\eta}}\left(\frac{\Delta P}{L}\right)r_1^4\left[\frac{\hat{\eta}}{\eta}r_c^{*4} + 2(1 - r_c^{*2})r_c^{*2}\right] \tag{2.106}$$

$$\hat{Q} = \frac{\pi}{8\hat{\eta}}\left(\frac{\Delta P}{L}\right)r_1^4(1 - r_c^{*2})^2 \tag{2.107}$$

Here, $r_c^* \equiv r_c/r_1$ is the dimensionless radial position of the boundary between the inner and outer fluid regions.

Eliminating $\Delta P/L$ between these equations gives an implicit expression for r_c^*:

$$(1 - \varphi)(1 - r_c^{*2})^2 - \varphi r_c^{*2}\left[\frac{\hat{\eta}}{\eta}r_c^{*2} + 2(1 - r_c^{*2})\right] = 0 \tag{2.108}$$

where

$$\varphi = \frac{\hat{Q}}{\hat{Q} + Q} \tag{2.109}$$

is the fractional flow of the outer fluid. Specification of φ and the viscosity ratio $\hat{\eta}/\eta$ then allows the determination of r_c^* by using simple numerical root finding methods. The flow rate enhancement for the inner fluid is

$$Q^* \equiv \frac{Q}{(Q)_{r_c^* = 1}} = r_c^{*4} + 2\left(\frac{\eta}{\hat{\eta}}\right)(1 - r_c^{*2})r_c^{*2} \tag{2.110}$$

Here $(Q)_{r_c^* = 1}$ is the volumetric flow rate when the inner fluid occupies the entire flow region (no outer fluid present).

For a specified viscosity ratio $\hat{\eta}/\eta$, Equations (2.108)–(2.110) can be used to determine the flow rate enhancement in terms of the fractional flow of the outer fluid and the viscosity ratio. The numerical results for several values of viscosity ratio are given in Figure 9. In particular, for the case in which the outer fluid is 100 times less viscous than the inner fluid ($\hat{\eta}/\eta = 0.01$) and the fractional flow is 10%, the flow rate of the more viscous inner fluid is enhanced by a factor of 30. For higher fractional flows and

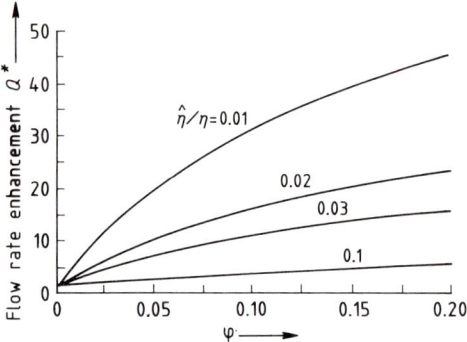

Figure 9. Flow rate enhancement associated with the injection of an immiscible low-viscosity fluid into a transport line

lower viscosity ratios, even greater enhancement factors are obtained. Clearly, when such segregated flows are possible, significant increases in throughput can be achieved per unit of applied pressure drop. In practice, the application of such approaches is limited only by the ability to create and maintain a stable outer concentric region of the lower viscosity fluid.

2.3. Multidimensional Momentum Transfer

2.3.1. Two-Dimensional Flows—Stream Function Equations

The flow examples considered to this point have involved only one nonvanishing velocity component. More complex flows are now considered, particularly two-dimensional incompressible flows of Newtonian fluids having two nonvanishing velocity components that depend on two spatial variables and time. The governing equations are the continuity equation and the two nonvanishing components of the equation of motion. Rather than solving these equations directly, the equations are more conveniently transformed into a single equation in terms of a stream function ψ ($\rightarrow$ 5. Fluid Mechanics, pp. **5**-5 – **5**-7). Because the stream function is defined in terms of the nonvanishing velocity components, solution of this equation then provides the velocity field. Physically, lines of constant ψ represent streamlines of the flow.

To illustrate this more explicitly, consider the flow

$$u_x = u_x(x, y, t), \quad u_y = u_y(x, y, t), \quad u_z = 0 \quad (2.111)$$

where the governing equations of continuity and motion are

$$\frac{\partial u_x}{\partial x} + \frac{\partial u_y}{\partial y} = 0 \quad (2.112)$$

and

$$\varrho \left(\frac{\partial u_x}{\partial t} + u_x \frac{\partial u_x}{\partial x} + u_y \frac{\partial u_x}{\partial y} \right)$$
$$= -\frac{\partial P}{\partial x} + \eta \left(\frac{\partial^2 u_x}{\partial x^2} + \frac{\partial^2 u_x}{\partial y^2} \right) \quad (2.113)$$

$$\varrho \left(\frac{\partial u_y}{\partial t} + u_x \frac{\partial u_y}{\partial x} + u_y \frac{\partial u_y}{\partial y} \right)$$
$$= -\frac{\partial P}{\partial y} + \eta \left(\frac{\partial^2 u_y}{\partial x^2} + \frac{\partial^2 u_y}{\partial y^2} \right) \quad (2.114)$$

If the stream function ψ is defined in terms of the velocity components by

$$u_x = -\frac{\partial \psi}{\partial y} \quad (2.115)$$

$$u_y = \frac{\partial \psi}{\partial x} \quad (2.116)$$

the continuity equation (Eq. 2.112) is satisfied identically. To obtain the desired stream function equation, the motion equations are combined in the following way. First operate on Equation (2.113) by $\partial/\partial y$ and Equation (2.114) by $\partial/\partial x$; then subtract the resulting equations to eliminate the pressure terms; finally, replace u_x and u_y with Equations (2.115) and (2.116). After simplification, the resulting equation is

$$\frac{\partial (\nabla^2 \psi)}{\partial t} + \frac{\partial (\psi, \nabla^2 \psi)}{\partial (x, y)} = v \nabla^4 \psi \quad (2.117)$$

where $v \equiv \eta/\varrho$ and

$$\nabla^2 \equiv \frac{\partial^2}{\partial x^2} + \frac{\partial^2}{\partial y^2}, \quad \nabla^4 \equiv \nabla^2 \nabla^2 \quad (2.118)$$

Also,

$$\frac{\partial (f, g)}{\partial (x, y)} \equiv \begin{vmatrix} \frac{\partial f}{\partial x} & \frac{\partial f}{\partial y} \\ \frac{\partial g}{\partial x} & \frac{\partial g}{\partial y} \end{vmatrix} \quad (2.119)$$

where a determinant operation is implied by the straight brackets.

Solution of Equation (2.117) gives the stream function $\psi(x, y, t)$. The velocity components are then obtained from Equations (2.115) and (2.116), and the pressure P is obtained from these velocity components and Equations (2.113) and (2.114).

A similar approach can be used in other two-dimensional flows, but the velocity–stream function relations and the stream function equation may take different forms depending on flow characteristics. For example, for axisymmetric flow around a sphere (see Fig. 10) where

$$u_r = u_r(r, \theta, t), \quad u_\theta = u_\theta(r, \theta, t), \quad u_\varphi = 0 \quad (2.120)$$

the velocity–stream function relations are

$$u_r = \frac{1}{r^2 \sin \theta} \frac{\partial \psi}{\partial \theta} \quad (2.121)$$

$$u_\theta = -\frac{1}{r \sin \theta} \frac{\partial \psi}{\partial r} \quad (2.122)$$

and the stream function equation is

$$\frac{\partial (E^2 \psi)}{\partial t} - \frac{1}{r^2 \sin \theta} \frac{\partial (\psi, E^2 \psi)}{\partial (r, \theta)} + \frac{2 E^2 \psi}{r^2 \sin^2 \theta}$$
$$\cdot \left(\frac{\partial \psi}{\partial r} \cos \theta - \frac{1}{r} \frac{\partial \psi}{\partial \theta} \sin \theta \right) = \nu E^2 (E^2 \psi) \quad (2.123)$$

where E^2 is an operator defined by

$$E^2 \equiv \frac{\partial^2}{\partial r^2} + \frac{\sin \theta}{r^2} \frac{\partial}{\partial \theta} \left(\frac{1}{\sin \theta} \frac{\partial}{\partial \theta} \right) \quad (2.124)$$

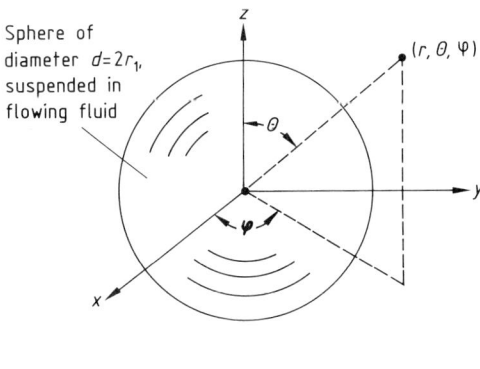

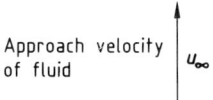

Figure 10. Axisymmetric flow past a sphere

Similar relations for other flows are given in [2.2, p. 131], [2.21, pp. 114–115]. Specification of the velocity in terms of the stream function ψ is given by

$$\boldsymbol{u} = \nabla \psi \times \nabla G \quad (2.125)$$

where G is a scalar function. This form satisfies the continuity equation ($\nabla \cdot \boldsymbol{u} = 0$) for incompressible flow [2.5, pp. 227–228]. For axisymmetric flow, the scalar function G is simply the coordinate ξ corresponding to the azimuthal angle about the axis of symmetry. For axisymmetric flow around a sphere, $G = \varphi$ and the resulting components of Equation (2.125) are the same as those of Equations (2.121) and (2.122). For two-dimensional axisymmetric flow in a cylindrical geometry with $u_\theta = 0$, $G = \theta$ is used, and the velocity–stream function relations take the forms

$$u_r = -\frac{1}{r} \frac{\partial \psi}{\partial z}, \quad u_z = \frac{1}{r} \frac{\partial \psi}{\partial r} \quad (2.126)$$

If Equation (2.123) is put in dimensionless form by scaling t, r, and ψ using d/u_∞, d, and $u_\infty d^2$, respectively, where d is the sphere diameter and u_∞ the free stream velocity, then

$$Re_d \left\{ \begin{array}{c} \text{left-hand side} \\ \text{terms} \end{array} \right\} = E^{*4} \psi^* \quad (2.127)$$

where $Re_d \equiv \varrho u_\infty d/\eta$ is the Reynolds number. Since all terms except Re_d are scaled to order 1 by the selections of d and u_∞, when $Re_d \ll 1$ the left-hand side can be set to zero and a *creeping flow* solution can be obtained. The latter is generally associated with small spheres, viscous fluids, and low velocities. When $Re_d \gg 1$, the right-hand side can be set to zero and a nonviscous flow solution can be obtained.

2.3.2. Creeping Flow Around a Sphere and Other Bodies of Revolution

Consider the steady, axisymmetrical flow of an incompressible Newtonian fluid past a sphere as illustrated in Figure 10. If the coordinate system is attached to the sphere, this representation is valid either for flow past a stationary sphere as illustrated or for the steady fall of a sphere (at terminal velocity u_∞) in a fluid at rest. The velocity and pressure fields under creeping flow conditions ($Re_d \ll 1$) are of particular interest, along with the drag force associated with viscous and pressure effects.

Under creeping flow conditions, Equation (2.123) reduces to

$$E^4 \psi = 0 \tag{2.128}$$

with boundary conditions

$$\text{at } r = r_1, \quad u_r = \frac{1}{r^2 \sin\theta} \frac{\partial \psi}{\partial \theta} = 0 \tag{2.129}$$

$$\text{at } r = r_1, \quad u_\theta = -\frac{1}{r \sin\theta} \frac{\partial \psi}{\partial r} = 0 \tag{2.130}$$

$$\text{as } r \to \infty, \quad u_z = \frac{u_r}{\cos\theta} = -\frac{u_\theta}{\sin\theta} \to u_\infty \tag{2.131}$$

By using Equations (2.121) and (2.122), the latter condition can be written as

$$r \to \infty, \quad \psi \to \tfrac{1}{2} u_\infty r^2 \sin^2\theta \tag{2.132}$$

Because Equation (2.132) is valid at all θ, the θ dependence at smaller r values might be expected to be the same. This suggests a solution to Equation (2.128) of the form

$$\psi = f(r) \sin^2\theta \tag{2.133}$$

Substituting this expression into Equation (2.128) and using Equations (2.129), (2.130), and (2.132) give

$$f = -\frac{1}{4} u_\infty \left(\frac{r_1^3}{r} + 3 r_1 r - 2 r^2 \right) \tag{2.134}$$

Then, from Equations (2.133), (2.121), and (2.122),

$$u_r = -u_\infty \left[1 - \frac{3}{2}\left(\frac{r_1}{r}\right) + \frac{1}{2}\left(\frac{r_1}{r}\right)^3 \right] \cos\theta \tag{2.135}$$

$$u_\theta = -u_\infty \left[1 - \frac{3}{4}\left(\frac{r_1}{r}\right) - \frac{1}{4}\left(\frac{r_1}{r}\right)^3 \right] \sin\theta \tag{2.136}$$

The dynamic pressure P is obtained from a line integration between the position $r \to \infty$ and $\theta = \pi/2$ (where $P = 0$) and any position r, θ [where $P = P(r, \theta)$], i.e.,

$$P = \int_0^{P(r,\theta)} dP = \int_\infty^r \left(\frac{\partial P}{\partial r}\right)_{\theta=\pi/2} dr + \int_{\pi/2}^\theta \left(\frac{\partial P}{\partial \theta}\right)_r d\theta \tag{2.137}$$

If $(\partial P/\partial r)$ and $(\partial P/\partial \theta)$ are obtained from the r and θ components of the equation of motion (creeping flow conditions) by using Equations (2.135) and (2.136), then

$$P = -\frac{3 \eta u_\infty}{2 r_1} \left(\frac{r_1}{r}\right)^2 \cos\theta \tag{2.138}$$

Also, the hydrostatic pressure is

$$p_H = p_0 - \varrho g z \tag{2.139}$$

where p_0 is the pressure at $r \to \infty$ and $\theta = \pi/2$. Combining the dynamic and hydrostatic results (Eq. 2.62) gives the pressure field

$$p = p_0 - \varrho g z - \frac{3 \eta u_\infty}{2 r_1} \left(\frac{r_1}{r}\right)^2 \cos\theta \tag{2.140}$$

The drag force of the fluid on the sphere is then obtained from

$$F = \int_S \mathbf{n} \cdot (-p\mathbf{I} - \boldsymbol{\tau}) \, dS \tag{2.141}$$

where $\mathbf{n} = \mathbf{i}_r$ is the outwardly directed unit normal to the sphere surface. Expanding Equation (2.141) and finding the component in the z-direction give

$$F_z = -\int_0^{2\pi}\int_0^\pi [(p + \tau_{rr})\cos\theta - \tau_{r\theta}\sin\theta]_{r_1} r_1^2 \sin^2\theta \, d\theta \, d\varphi \tag{2.142}$$

Then, by using Equations (2.135) and (2.136), the stress components and the pressure are evaluated at r_1 with the results

$$\tau_{rr}\Big|_{r_1} = -2\eta \frac{\partial u_r}{\partial r}\Big|_{r_1} = 0 \tag{2.143}$$

$$\tau_{r\theta}\Big|_{r_1} = -\eta \left[r \frac{\partial}{\partial r}\left(\frac{u_r}{r}\right) + \frac{1}{r}\frac{\partial u_r}{\partial \theta} \right]_{r_1}$$

$$= \frac{3 \eta u_\infty}{r_1} \sin\theta \tag{2.144}$$

and

$$p\Big|_{r_1} = p_0 - \varrho g r_1 \cos\theta - \frac{3 \eta u_\infty}{2 r_1} \cos\theta \tag{2.145}$$

Thus,

$$F_z = \frac{4}{3}\pi r_1^3 \varrho g + 2\pi\eta r_1 u_\infty + 4\pi\eta r_1 u_\infty \tag{2.146}$$

$$\quad\text{Buoyancy} \quad \text{Form drag} \quad \text{Viscous drag}$$
$$\quad\text{from } p_H \quad\quad \text{from } P \quad\quad\,\text{from } \tau_{r\theta}$$

The first term arises from hydrostatic pressure effects around the sphere (pressure is higher as $\theta \to \pi$ causing a net upward force); the second term, from dynamic pressure effects (again from higher pressure on the sphere bottom); and the third, from viscous effects associated with $\tau_{r\theta}$ on the sphere surface. If this result is represented in terms of a buoyancy force F_B (static effect) and a drag force F_D (dynamic effect), i.e.,

$$F_z = F_B + F_D \tag{2.147}$$

then F_B is just the first term of Equation (2.146) and F_D the sum of the other two terms, i.e.,

$$F_D = 6\pi\eta r_1 u_\infty \tag{2.148}$$

This equation is known as *Stokes law* and indicates that the drag force on a solid sphere increases in direct proportion to viscosity, radius, and terminal (or free stream) velocity.

If a sphere of density ϱ_s settles under its own weight in a fluid, the terminal velocity can be determined by equating the sum of F_B and F_D to the gravitational force on the sphere ($\frac{4}{3}\pi r_1^3 \varrho_s g$). The result is

$$u_\infty = \frac{2(\varrho_s - \varrho)r_1^2 g}{9\eta} \tag{2.149}$$

This result indicates that for particle settling in viscous fluids, the terminal velocity is most sensitive to variations in particle size.

Finally, the drag force results for several variations of this problem are noted. In particular, for creeping flow past a *fluid sphere* (or droplet) of viscosity $\hat{\eta}$ [2.15, pp. 127–129]

$$F_D = 6\pi\eta r_1 u_\infty \left(\frac{3\hat{\eta} + 2\eta}{3\hat{\eta} + 3\eta}\right) \tag{2.150}$$

For a *very viscous droplet* or *solid sphere* where $\hat{\eta} \gg \eta$, this result reduces to Stokes law. For *gas bubbles* where $\hat{\eta} \ll \eta$

$$F_D = 4\pi\eta r_1 u_\infty \tag{2.151}$$

In this case, the drag is two-thirds that of a rigid sphere.

If *surface-active agents* are present that give rise to a surface shear viscosity ε_s, the drag force for a fluid sphere is [2.15, pp. 129]:

$$F_D = 6\pi\eta r_1 u_\infty \left(\frac{(\varepsilon_s/r_1) + 3\hat{\eta} + 2\eta}{(\varepsilon_s/r_1) + 3\hat{\eta} + 3\eta}\right) \tag{2.152}$$

The important point here is that if r_1 is sufficiently small, surface viscosity effects dominate to give rigid sphere results, even when $\hat{\eta}$ is small (gas bubble case). Small bubbles often behave as rigid spheres when contaminant amounts of surface-active agents are present.

For *nonspherical* solid bodies, somewhat different drag force results are obtained. For oblate and prolate spheroids, the results are summarized in Figure 11. Note that the drag force for the prolate spheroid is particularly sensitive to the ratio of the minor and major axes. In the limit as b/a goes to zero, the oblate spheroid reduces to a circular flat disk with the drag force given by

$$F_D = 5.093\pi\eta a u_\infty \tag{2.153}$$

where a is the radius of the flat disk. In the limit as b/a goes to zero for the prolate spheroid, the results for an elongated rod of radius b and length $2a$ are

$$F_D = \frac{4\pi\eta a u_\infty}{\ln(a/b) + 0.1932} \tag{2.154}$$

Additional drag force results for creeping flow around bodies are given in [2.15].

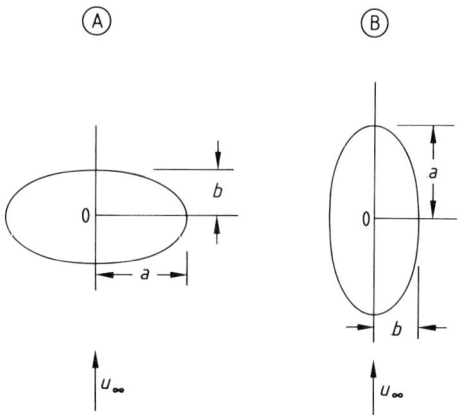

Figure 11. Drag force for creeping flow around oblate and prolate spheroids
A) Oblate spheroid, drag force $F_D = 6\pi\eta a u_\infty K$; B) Prolate spheroid, drag force $F_D = 6\pi\eta b u_\infty K$

b/a	K (oblate)	K (prolate)
0.0	0.8488	∞
0.1	0.8525	2.6471
0.5	0.9053	1.2039
1.0	1.0000	1.0000

An alternate way of representing drag force results around bodies is in terms of a drag coefficient C_D ($\rightarrow$ 5. Fluid Mechanics, pp. **5**-16–**5**-17):

$$F_D = C_D \binom{\text{Characteristic}}{\text{area of body}} \binom{\text{Characteristic kinetic}}{\text{energy of fluid per unit volume}} \quad (2.155)$$

For flow past a sphere

$$F_D = C_D \left(\frac{\pi}{4} d^2\right)\left(\tfrac{1}{2}\varrho u_\infty\right)^2 \quad (2.156)$$

$$C_D = \frac{24}{Re_d} \quad (2.156\,a)$$

For the settling of multiple bodies, interaction effects can be quite important, even over fairly large separation distances. In the case of two identical spheres of diameter d settling either parallel or perpendicular to their line of centers, the drag coefficient on each sphere is

$$C_D = \frac{24\,\Lambda}{Re_d} \quad (2.157)$$

where Λ is an interference parameter which is a function of the ratio l/d of the distance between the particles and the particle diameter. When $l/d \rightarrow \infty$, $\Lambda \rightarrow 1$. For two spheres settling side by side, Λ is found to be < 1 for finite values of l/d. As a result, two interfering spheres fall faster than isolated spheres.

Even though two spheres may fall faster than an isolated sphere, a suspension of many spheres usually experiences a higher drag than the isolated sphere case. As a result, when a suspension settles in a beaker, a sharp line is observed separating the clear fluid from the particle-laden fluid. This hindered settling phenomenon arises because particles in the low-concentration regions near the top of the settling suspension tend to overtake the more slowly settling particles in the underlying high-concentration regions.

2.3.3. Flow in Channels with Varying Cross Sections—Lubrication Analysis

In many problems the flow geometry is sufficiently complex that exact solutions of the motion equations are not possible, and approximate or numerical methods are required to obtain the solutions desired. In this section an approximate analysis is described, called lubrication theory. (The name lubrication theory arose from early applications of this approximate analysis to thin-film lubrica-

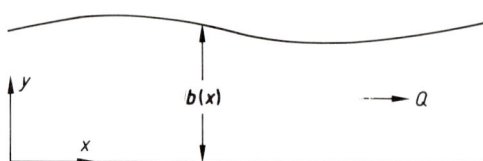

Figure 12. Flow in a two-dimensional channel with varying cross section

Flow can arise from an applied pressure gradient or from moving wall(s).

tion problems). This method is applicable when the change in flow channel geometry in the direction of flow is weak.

Consider steady two-dimensional flow $\boldsymbol{u} = [u_x(x, y), u_y(x, y), 0]$ in a channel with a slowly changing cross-sectional area as shown in Figure 12. The flow channel gap b varies with x, but the magnitude of this change at any x is small, i.e.,

$$\varepsilon \equiv \frac{db}{dx} \ll 1 \quad (2.158)$$

The equation of continuity and the x and y components of the equation of motion are the same as Equations (2.112)–(2.114), but without the time derivative terms. To scale these equations and determine the most dominant terms, the velocity component u_x must first be scaled to the characteristic average velocity u_0 in the channel:

$$u_x \sim u_{\text{avg}}(x) = \frac{Q}{b(x)w} \sim \frac{Q}{b_0 w} \equiv u_0 \quad (2.159)$$

Here Q is the volumetric flow rate, w the channel width, b_0 the channel gap at $x = 0$, and the symbol $\sim$ indicates an order-of-magnitude estimate. From this result,

$$\frac{\partial u_x}{\partial x} \sim \frac{Q}{b^2 w}\frac{db}{dx} \sim \frac{u_0}{b_0}\varepsilon \quad (2.160)$$

and from the equation of continuity,

$$\frac{\partial u_y}{\partial y} = -\frac{\partial u_x}{\partial x} \sim \frac{u_0}{b_0}\varepsilon \quad (2.161)$$

or

$$u_y \sim u_0\varepsilon, \quad \frac{\partial f}{\partial x} \sim \frac{\varepsilon}{b_0}f, \quad \frac{\partial f}{\partial y} \sim \frac{1}{b_0}f \quad (2.162)$$

where f may be u_x or u_y, or higher derivatives of these variables.

Based upon these scalings, the magnitudes of the various terms in the x component of the equation of motion can now be compared:

$$\varrho\left(u_x\frac{\partial u_x}{\partial x} + u_y\frac{\partial u_x}{\partial y}\right) = -\frac{\partial P}{\partial x} + \eta\frac{\partial^2 u_x}{\partial x^2} + \eta\frac{\partial^2 u_x}{\partial y^2} \quad (2.163)$$

$$O\left(\frac{\varrho u_0^2}{b_0}\varepsilon\right) \qquad O\left(\frac{\eta u_0}{b_0^2}\varepsilon^2\right) \; O\left(\frac{\eta u_0}{b_0^2}\right)$$

The inertial term will be negligible compared to the y derivative viscous term if

$$\frac{\varrho u_0^2}{b_0}\varepsilon \ll \frac{\eta u_0}{b_0^2} \quad (2.164)$$

$$\frac{\varrho u_0 b_0}{\eta}\varepsilon \ll 1 \quad (2.164\,a)$$

Similarly, the x-derivative viscous term will be negligible compared to the y-derivative viscous term if

$$\frac{\eta u_0}{b_0^2}\varepsilon^2 \ll \frac{\eta u_0}{b_0^2} \quad (2.165)$$

or

$$\varepsilon^2 \ll 1 \quad (2.165\text{a})$$

Equation (2.163) then reduces to

$$0 = -\frac{\partial P}{\partial x} + \eta \frac{\partial^2 u_x}{\partial y^2} \quad (2.166)$$

In a similar way, the y-component of the equation of motion can be simplified to

$$0 = -\frac{\partial P}{\partial y} + \eta \frac{\partial^2 u_y}{\partial y^2} \quad (2.167)$$

From these equations,

$$\frac{\partial P}{\partial y} \sim \frac{\eta u_0}{b_0^2}\varepsilon \sim \frac{\partial P}{\partial x}\varepsilon \quad (2.168)$$

which indicates that when $\varepsilon \ll 1$, pressure changes in the y-direction are negligible compared to those in the x-direction. Hence, changes of P in the y-direction can be neglected and the assumption made that $P = P(x)$.

The solution of Equation (2.166) with the boundary conditions

at $y = 0$, $\quad u_x = U \quad (2.169)$

at $y = h(x)$, $\quad u_x = 0 \quad (2.170)$

is

$$u_x = U\left(1 - \frac{y}{b}\right) - \frac{b^2}{2\eta}\left(\frac{dP}{dx}\right)\left[\frac{y}{b} - \left(\frac{y}{b}\right)^2\right] \quad (2.171)$$

with the volumetric flow rate being

$$Q = \int_0^b u_x w\,dy = \frac{1}{2}Ubw - \frac{b^3 w}{12\eta}\left(\frac{dP}{dx}\right) \quad (2.172)$$

If this equation is solved for (dP/dx) and then integrated, the pressure drop between $x = 0$ and $x = L$ is

$$\Delta P \equiv \frac{12 A_3 \eta Q L}{w h_0^3} - \frac{6 A_2 \eta U L}{h_0^2} \quad (2.173)$$

Here A_2 and A_3 are numerical parameters that depend on the geometry of the flow channel [$h(x)$]; they are given by

$$A_n = \int_0^1 \frac{dx^*}{h^{*n}}, \quad n = 2, 3 \quad (2.174)$$

where $x^* \equiv x/L$ and $b^* \equiv b/b_0$. Equation (2.173) is the pressure drop–flow rate relation for flow in a two-dimensional channel with a weakly varying cross section.

Similar relations can be obtained for axisymmetric flows where the cross section varies along the axis of flow [2.24]. This includes flows in sinusoidal tubes, and converging or diverging ducts. Lubrication theory is applicable in such problems as long as $\varepsilon \equiv dR/dz$ is sufficiently small, where $R(Z)$ is the local channel radius. Good engineering approximations are often possible even for values of ε as high as 0.2.

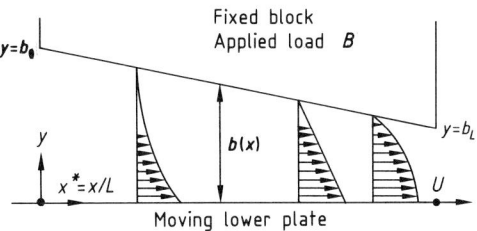

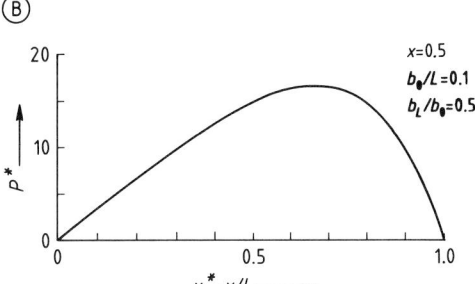

Figure 13. Lubrication flow between block and moving surface
A) Geometry; B) Pressure distribution

To conclude this section an application of lubrication theory is presented, which involves the analysis of slipper block performance. Consider the slipper block shown in Figure 13, which could represent a wiping block or ring on a moving piston. In some applications, the surface moves and the block is stationary (see Fig. 13); in others, the block moves and the surface is stationary. The thickness of the film gap between the block and the moving surface, as well as the drag force on the block, depends on the applied load B, the speed U, the viscosity of the film fluid η, the width w, the length L, and the variation of the gap thickness with x. Lubrication theory is used to develop the appropriate relations between these variables.

Begin by using Equation (2.172):

$$\frac{dP}{dx} = \frac{12\eta}{b^3}\left(\frac{Ub}{2} - \frac{Q}{w}\right) \quad (2.175)$$

In this problem, Q is not known and must be determined from known pressure conditions. In particular, at $x = 0$ and $x = L$ the pressure is assumed to be zero. Integration of Equation (2.175) between these points then yields

$$\int_0^L dP = 0 = \frac{12\eta L}{h_0^2}\left(\frac{1}{2}A_2 U - \frac{A_3 Q}{h_0 w}\right) \quad (2.176)$$

where the A_n are defined by Equation (2.174). The volumetric flow rate is then

$$Q = \frac{A_2 w b_0 U}{2 A_3} \quad (2.177)$$

If the gap geometry is described by the linear relation

$$b = b_0 + (b_L - b_0)\frac{x}{L} \quad (2.178)$$

the dimensionless pressure distribution can be obtained from Equations (2.175) and (2.177) with the result

$$P^* \equiv \frac{(P - P_0)L}{6\eta U} = \frac{(b^* - 1)(b^* - \kappa)}{(b_0/L)^2(\kappa^2 - 1)} \quad (2.179)$$

where $\kappa \equiv b_L/b_0$. As depicted in Figure 13, the pressure is positive in the gap and exhibits a maximum toward the narrow gap end. The magnitude of the pressure increases with U and η, and decreases with (b_0/L). The pressure also causes a *vertical force* F_y on the *block*, which is obtained by integrating P over the block length. The result is

$$F_y^* \equiv \frac{F_y}{6\eta U w} = \frac{1}{(b_0/L)^2 (\kappa - 1)^2} \left[\frac{2(\kappa - 1)}{\kappa + 1} - \ln \kappa \right] \quad (2.180)$$

For a given b_0/L, the maximum vertical force is obtained for a block geometry where $\kappa = 2.19$. Also, because the vertical pressure force must be equal to the load B, the resulting thickness b_0^* is

$$b_0^* \equiv \frac{b_0}{L} = \left\{ \frac{6\eta U w}{B(\kappa - 1)^2} \left[\frac{2(\kappa - 1)}{\kappa + 1} - \ln \kappa \right] \right\}^{1/2} \quad (2.181)$$

Here, for a given κ, the gap thickness increases with increasing U and η, and decreases with increasing load B, and each case has a square-root dependence.

The drag force F_x on the *moving plate is* obtained by integrating τ_{yx} over this surface:

$$F_x^* \equiv \frac{F_x}{6\eta w U} = \frac{1}{(b_0/L)(\kappa - 1)} \left[\frac{(\kappa - 1)}{\kappa + 1} - \frac{2}{3} \ln \kappa \right] \quad (2.182)$$

Finally, if Equation (2.181) is used to eliminate b_0/L,

$$F_x = \left(\frac{2}{3} \eta U w B \right)^{1/2} \left\{ \frac{3 \left(\frac{\kappa - 1}{\kappa + 1} \right) - 2 \ln \kappa}{2 \left(\frac{\kappa - 1}{\kappa + 1} \right) - \ln \kappa} \right\} \quad (2.183)$$

which shows that the drag force is proportional to the square root of the applied load. This half power dependence when a lubricating film is present is in contrast to the first-power dependence generally assumed between sliding solid surfaces with no lubricant. Also, the effective coefficient of friction for lubricated surfaces is orders of magnitude smaller than that associated with direct–contact solid surfaces (→ Lubricants and Related Products, **A15**, pp. 424–428).

2.4. Coupled Momentum and Energy Transfer

This discussion has been concerned with examples that involve only heat or momentum transfer, but not both. In this section, examples are considered in which heat and momentum are transferred simultaneously and, specifically, heat is being transferred in a flowing fluid. When such heat transfer results directly from fluid movement, the transfer is termed *heat convection*. If the flow arises from applied pressure gradients or moving boundaries, the heat transfer is called *forced convection*. When the flow arises from buoyancy effects associated with density variations in nonisothermal fluids, it is called *free or natural convection*.

In analyzing heat transfer with flow, the energy balance is expressed in the form [2.1, pp. 215–216]:

$$\varrho c_p \left(\frac{\partial T}{\partial t} + \boldsymbol{u} \cdot \boldsymbol{\nabla} T \right) = \boldsymbol{\nabla} \cdot (\lambda \boldsymbol{\nabla} T)$$

$$+ T\beta \left(\frac{\partial p}{\partial t} + \boldsymbol{u} \cdot \boldsymbol{\nabla} p \right) + \mathrm{tr}\,(\boldsymbol{\tau} \cdot \boldsymbol{\nabla} \boldsymbol{u}) + \Phi \quad (2.184)$$

which is equivalent to Equation (1.85) with Fourier's law (Eq. 1.91) used for the conductive heat flux $\boldsymbol{q}$; β is the coefficient of thermal expansion defined by

$$\beta \equiv -\frac{1}{\varrho} \left(\frac{\partial \varrho}{\partial T} \right)_p \quad (2.185)$$

In Equation (2.184) the second term on the left, which involves velocity, accounts for convective heat transfer, and the terms on the right are the conductive, reversible work (fluid expansion–contraction), viscous dissipation, and thermal generation terms, respectively.

Consider now the case of steady flow of a Newtonian fluid in a heated tube where the temperature of the fluid varies in the radial and axial directions because of heat exchange with the walls, viscous heat generation, and reversible work effects. If free convection, generation, and radiation effects are neglected, and the velocity field is assumed to be $\boldsymbol{u} = [0, 0, u_z(r)]$, Equation (2.184) takes the form

$$\varrho c_p u_z \frac{\partial T}{\partial z} = \boldsymbol{\nabla} \cdot (\lambda \boldsymbol{\nabla} T) + T\beta u_z \frac{dp}{dz} + 2\eta \left(\frac{du_z}{dr} \right)^2$$

$$(2.186)$$

The contribution of *viscous dissipation* to the temperature gradient in the z-direction can be estimated from

$$\left(\frac{\partial T}{\partial z} \right)_{\text{viscous dissipation}} = \frac{2\eta (du_z/dr)^2}{\varrho c_p u_z} \sim O\left(Re_\mathrm{d} \frac{v^2}{c_p d^3} \right) \quad (2.187)$$

where $Re_\mathrm{d} \equiv (u_{\mathrm{avg}} d/v)$ is the Reynolds number; $v \equiv \eta/\varrho$, the momentum diffusivity, d the tube diameter; and u_{avg} the average velocity in the tube. For air at room temperature ($v \approx 2 \times 10^{-5}\,\mathrm{m^2/s}$, $c_p \approx 10^3\,\mathrm{J\,kg^{-1}\,K^{-1}}$) flowing in a

0.025-m tube at a Reynolds number $Re_d = 10^3$, the temperature gradient due to viscous effects is on the order of 3×10^{-5} K/m. For water under the same conditions ($v \approx 8 \times 10^{-7}$ m²/s, $c_p \approx 4 \times 10^3$ J kg^{-1} K^{-1}), the gradient is on the order of 1×10^{-8} K/m; for an oil with properties $v \approx 9 \times 10^{-4}$ m²/s, $c_p \sim 2 \times 10^3$ J kg^{-1} K^{-1}, the gradient is ca. 3×10^{-2} K/m. In most heat-exchange applications, such effects are negligible compared to the conduction contributions (first term on right-hand side, Eq. 2.186) which can be on the order of several kelvin per meter. Therefore, viscous dissipation effects are generally neglected in most heat-exchange applications. However, viscous dissipation can be important in lubrication problems where viscous oils are being sheared at high rates between moving surfaces.

If a similar order-of-magnitude analysis is carried out for the contribution of the *reversible work term* to the axial temperature gradient:

$$\left(\frac{\partial T}{\partial z}\right)_{\text{rev. work}} = \frac{T\beta(dp/dz)}{\varrho c_p} \sim O\left(32\, Re_d \frac{T\beta v^2}{c_p d^3}\right) \quad (2.188)$$

Here, the pressure gradient is related to velocity through the Hagen–Poiseuille equation (Eq. 2.86). Comparing Equations (2.187) and (2.188) gives

$$\left(\frac{\partial T}{\partial z}\right)_{\text{rev. work}} \sim \pm 32\, T\beta \left(\frac{\partial T}{\partial z}\right)_{\substack{\text{viscous} \\ \text{dissipation}}} \quad (2.189)$$

where the sign depends on whether dp/dz in Equation (2.188) is positive or negative. When dp/d$z < 0$, the reversible work results in a decrease in temperature in the direction of flow; in contrast, the viscous dissipation causes an increase in temperature. For air at room temperature where $\beta = -1/T$ (ideal gas), the temperature gradient contribution associated with reversible work effects is on the order of 30 times that associated with viscous dissipation effects. For liquids where $\beta \sim 10^{-3}$ K^{-1}, the reversible work contribution to the temperature gradient is on the order of 10 times that arising from viscous dissipation. Still, in most applications, these effects are small compared to conduction or convection effects and can be neglected.

The other effect that is not commonly considered in convection analyses is the *temperature dependence of thermal conductivity*. For most fluids, the magnitude of $(1/\lambda)(\partial\lambda/\partial T)$ is on the order of 10^{-3} K^{-1} and

$$\mathbf{V}\cdot(\lambda\mathbf{V}T) = \lambda\left[\nabla^2 T + \frac{1}{\lambda}\left(\frac{\partial\lambda}{\partial T}\right)|\mathbf{V}T|^2\right] \sim \lambda\nabla^2 T \quad (2.190)$$

Hence, if viscous dissipation and reversible work effects are neglected, the heat generation term is zero, and the conductivity is assumed to be temperature-independent, the energy balance (Eq. 2.184) reduces to

$$\frac{\partial T}{\partial t} + \mathbf{u}\cdot\mathbf{V}T = a\,\nabla^2 T \quad (2.191)$$

where $\alpha \equiv \lambda/\varrho c_p$ is the thermal diffusivity. This is the equation used in the forced convection examples considered here.

2.4.1. Heat Transfer in Laminar Tube Flow

Consider the heat exchange that occurs when a fluid is passed through a heated or cooled tube under steady laminar flow conditions. If a Newtonian fluid, constant physical properties, negligible axial conduction, and fully developed velocity and temperature profiles are assumed, the energy balance can be written as

$$u_z\frac{\partial T}{\partial z} = \alpha\frac{1}{r}\frac{\partial}{\partial r}\left(r\frac{\partial T}{\partial r}\right) \quad (2.192)$$

where

$$u_z = \tfrac{1}{2}u_{\text{avg}}\left[1-\left(\frac{r}{r_1}\right)^2\right] \quad (2.193)$$

with r_1 being the tube radius and u_{avg} the average velocity. Depending on the boundary conditions, various solutions are possible. When the heat flux at the wall q_w is constant, a particularly simple solution is possible. In particular, if Equation (2.192) is multiplied by r and integrated with respect to r, then

$$\frac{dT_b}{dz} = \frac{2\alpha\left(\frac{\partial T}{\partial r}\right)_{r_1}}{r_1 u_{\text{avg}}} = \frac{2\alpha q_w}{r_1 u_{\text{avg}}\lambda} \quad (2.194)$$

where T_b is the *bulk average temperature* defined by

$$T_b = \frac{1}{\pi r_1^2 u_{\text{avg}}}\int_0^{r_1} u_z T\, 2\pi r\, dr \quad (2.195)$$

The bulk average temperature is simply the "cup-mixing temperature" that would be measured if the tube were chopped off at z and the fluid issuing forth were collected and thoroughly mixed in a container.

Equation (2.194) implies that T_b varies linearly with z. The achievement of a fully developed temperature profile requires that

$$\frac{\partial}{\partial z}\left(\frac{T_w - T}{T_w - T_b}\right) = 0 \qquad (2.196)$$

or

$$\frac{T_w - T}{T_w - T_b} = f(r) \qquad (2.197)$$

From Newton's law of cooling,

$$h = -\frac{\lambda\left(\frac{\partial T}{\partial r}\right)_{r_1}}{(T_w - T_b)} = \lambda\left[\frac{\partial}{\partial r}\left(\frac{T_w - T}{T_w - T_b}\right)\right]_{r_1}$$

$$= \lambda f(r_1) = \text{const.} \qquad (2.198)$$

or

$$T_w - T_b = \frac{q_w}{h} = \text{const.} \qquad (2.199)$$

By using this equation with Equations (2.194) and (2.197),

$$\frac{dT_w}{dz} = \frac{dT_b}{dz} = \frac{\partial T}{\partial z} = \frac{dT}{dz} = \frac{2q_w \alpha}{r_1 u_{avg} \lambda} \qquad (2.200)$$

Substituting this result into Equation (2.192) and integrating over r give

$$T_w - T_b = \frac{11}{24}\frac{q_w r_1}{\lambda} = \frac{11}{48}\frac{q_w d}{\lambda} \qquad (2.201)$$

The latter result indicates that when the fluid is heated (or cooled) under *constant wall heat flux* conditions, the temperature difference between the wall and the bulk fluid is constant.

If Equations (2.201) and (2.199) are combined,

$$Nu_d \equiv \frac{hd}{\lambda} = 4.364 \quad (q_w = \text{const., large } z) \qquad (2.202)$$

where Nu_d is the dimensionless Nusselt number based upon the tube diameter d. This result provides a simple expression for estimating h under constant wall heat flux conditions and in zones (large z) where the profiles are fully developed. For the case of *constant wall temperature*, the solution is somewhat more involved [2.23]; however, Nu_d is still constant but equal to a different value

$$Nu_d = 3.658 \quad (T_w = \text{const., large } z) \qquad (2.203)$$

Use of this result is limited to z values greater than the thermal entry length z_e which is the position from the tube entrance where Equation (2.196) is approximately satisfied, say $\partial(T_w - T)/\partial z \sim 0.01$. For laminar flow, this entry length can be estimated from

$$\frac{z_e}{d} \approx 0.05 \, Re_d \, Pr \qquad (2.204)$$

In particular, for water at room temperature ($Pr \equiv \nu/\alpha \sim 10$) flowing at a Reynolds number of 10^3, the entry length would be 500 diameters; for oils with $Pr \sim 10^4$, the entry lengths are very long and fully developed thermal profiles are seldom achieved. In the latter cases, more complete solutions must be used to describe heat transfer in the developing regions, and these are considerably more complicated than the fully developed solutions presented above [2.23]–[2.27]. The classical Graetz solution [2.23], which involves the solution of Equation (2.192) for parabolic and flat velocity profiles and constant wall temperatures are the best known among these solutions.

A useful empirical expression is given by SIEDER and TATE [2.28]; it is an empirical modification of the Graetz solution, (parabolic velocity) i.e.,

$$Nu_{d,\text{lm}} \equiv \frac{h_{\text{lm}} d}{\lambda}$$

$$= 1.86 \, Re_d^{1/3} \, Pr^{1/3} (L/d)^{-1/3} (\eta_{\text{bm}}/\eta_w)^{0.14} \qquad (2.205)$$

Here, the physical properties of the fluid are evaluated at the mean bulk fluid temperature, $T_{\text{bm}} \equiv (T_{\text{b0}} + T_{\text{bL}})/2$, where T_{b0} and T_{bL} are the entering and exit bulk temperatures for a tube section of length L. The viscosity ratio (η_{bm}/η_w) is the ratio of the viscosities evaluated at the mean bulk temperature and at the mean wall temperature; the quantity h_{lm} is the log-mean film heat-transfer coefficient defined by

$$h_{\text{lm}} = \frac{\dot{Q}}{\lambda L d \, \Delta T_{\text{lm}}} = -\frac{\int_0^L (q_r)_{r_1} dz}{L \Delta T_{\text{lm}}} \qquad (2.206)$$

where

$$\Delta T_{lm} \equiv \left[\frac{(T_{wL} - T_{bL}) - (T_{w0} - T_{b0})}{\ln\left(\frac{T_{wL} - T_{bL}}{T_{w0} - T_{b0}}\right)} \right] \quad (2.207)$$

is the log-mean temperature difference. The heat-transfer rate $\dot{Q}$ across the wall can be expressed in terms of the difference between the convected energy input and output to the heated (or cooled) section by

$$\dot{Q} = \frac{\pi d^2}{4} u_{avg} [(\varrho c_p T)_L - (\varrho c_p T)_0] \quad (2.208)$$

For large values of L/d, Equation (2.205) predicts that the log-mean Nusselt number goes to zero, which is not consistent with results for the fully developed regions. WHITAKER [2.1, pp. 324–325] suggests the use of the following expression, which is based on the work of HAUSEN [2.27]:

$$Nu_{d,lm} = 3.66 + \frac{0.0745 \, \psi^3}{1 + 0.04 \, \psi^2} \quad (2.209)$$

where

$$\psi \equiv (Re \, Pr)^{1/3} \, (L/d)^{-1/3} \, (\eta_{bm}/\eta_w)^{0.14} \quad (2.210)$$

This result is accurate to ca. ± 10% for situations in which the wall temperature is constant or nearly constant. In the case of constant wall heat flux, variations of T_w can be large; however, the local Nusselt numbers are only ca. 20% higher than those for constant-temperature cases. As a result, Equation (2.209) is used even for the case of constant heat flux and surface temperature variations are ignored.

When the entering fluid temperature and the wall temperature are known, the exit temperature T_L and the overall heat transfer rate $\dot{Q}$ can be calculated by using Equations (2.206)–(2.210). Since the evaluation of the properties requires a knowledge of T_L, an iterative procedure is required in which T_L is assumed and h_{lm} is calculated from Equation (2.209). Then this value of the heat-transfer coefficient is used with Equations (2.207) and (2.208) to determine T_L and $\dot{Q}$. The new value of T_L is then employed to update the value of h_{lm}, and the procedure is repeated until convergence is obtained.

2.4.2. Momentum and Heat Transfer in Laminar Boundary Layers

A characteristic of real fluids moving around solid bodies is the adherence of the fluid to the solid surface. This no-slip condition causes fluid-deforming velocity gradients that tend to be large in the neighborhood of the body and to diminish far away from it. If the Reynolds number based on the characteristic length of the body is sufficiently high, the resulting velocity gradient is restricted largely to a thin region in the neighborhood of the boundary. Under such conditions, a boundary layer approach is possible where the flow field is divided into two regions: (1) a thin boundary layer in which both viscous and inertial effects are important, and (2) an external region in which viscous effects (and fluid-deforming velocity gradients) are negligible. When applicable, such an approach simplifies the solution considerably, and approximate analytical solutions are possible in many cases.

Momentum Transfer in a Laminar Boundary Layer. To illustrate this approach, consider first the momentum transfer associated with flow past the surface shown in Figure 14. In this case, the flow is two dimensional where $\mathbf{u} = [u_x(x, y), u_y(x, y), 0]$, with the x-coordinate measured along the surface and the y-coordinate measured normal to the surface. As the fluid moves over the surface with velocity $\mathbf{u}$, a boundary layer is formed that encompasses most of the velocity changes resulting from interactions of the fluid with the surface. Although many different ways of defining the boundary layer thickness exist [2.29], in this article thickness is defined as the y-position at which the velocity is 99% of the mainstream velocity u_∞, i.e.,

$$\delta_m(x) \equiv \left\{ \begin{array}{l} \text{The } y\text{-position at each } x \\ \text{where } u_x = 0.99 \, u_\infty \end{array} \right\} \quad (2.211)$$

The subscript m indicates the boundary layer thickness for momentum transfer; boundary lay-

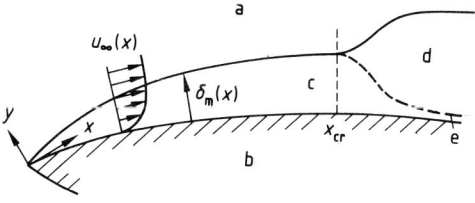

Figure 14. Momentum boundary layer on a surface
a) Fluid; b) Solid body; c) Laminar boundary layer; d) Turbulent boundary layer; e) Laminar sublayer

er thicknesses for heat- and mass-transfer processes are defined later and denoted by δ_T and δ_c, respectively.

As illustrated in Figure 14, the boundary layer thickness increases with distance x along the surface. For a certain distance $x < x_{cr}$, flow in the boundary layer is laminar. At the position x_{cr}, which can be estimated from

$$Re_{cr} \equiv \frac{u_\infty x_{cr}}{v} \approx 2 \times 10^5 \qquad (2.212)$$

inertial effects become sufficiently large compared to viscous effects to destabilize the flow and except for a very thin laminar sublayer next to the surface of the body, the flow exhibits sporadic vortex-like instabilities that eventually grow into fully developed turbulence. Physically, this is observed in flow visualization when the boundary layer thickness increases rather abruptly as shown in Figure 14 (in graphically depicting a boundary layer as in Fig. 14, the thickness is exaggerated and not to scale). Generally, x_{cr} may be on the order of meters and δ_m on the order of millimeters.

The velocity field in the laminar portion of the boundary layer, and the associated drag force on the surface are now determined. If the radius of curvature of the surface is sufficiently large compared to the boundary layer thickness, the flow can be represented as that along a flat plate as shown in Figure 15. If the scalings

$$u_x \sim u_\infty \qquad (2.213)$$

$$\frac{\partial u_x}{\partial x} \sim \frac{u_\infty}{L} \qquad (2.214)$$

are used, the equation of continuity

$$\frac{\partial u_x}{\partial x} + \frac{\partial u_y}{\partial y} = 0 \qquad (2.215)$$

yields

$$\frac{\partial u_y}{\partial y} \sim \frac{u_\infty}{L} \qquad (2.216)$$

$$u_y \sim u_\infty \frac{\delta_m}{L} \qquad (2.217)$$

By using these scalings, the x- and y-components of the equation of motion can be simplified. In particular, if $\delta_m/L \ll 1$, the x component takes

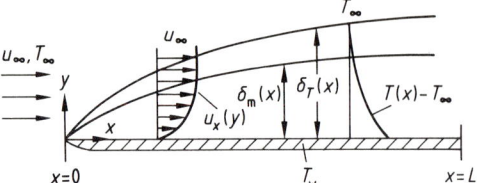

Figure 15. Laminar momentum and thermal boundary layers on a flat plate

the form [2.29, p. 131]

$$u_x \frac{\partial u_x}{\partial x} + u_y \frac{\partial u_x}{\partial y} = -\frac{1}{\varrho} \frac{dp}{dx} + v \frac{\partial^2 u_x}{\partial y^2} + g_x \qquad (2.218)$$

The y-component simply reduces to $\partial p/\partial y = 0$, which indicates that the pressure is independent of y and hence is simply the pressure $p(x)$ outside the boundary layer. If the external flow is assumed to be nonviscous the pressure field can be determined from:

$$p = -\tfrac{1}{2}\varrho u_\infty^2 - \varrho g_x x + \text{constant} \qquad (2.219)$$

using this equation in Equation (2.218) gives

$$u_x \frac{\partial u_x}{\partial x} + u_y \frac{\partial u_x}{\partial y} = u_\infty \frac{du_\infty}{dx} + v \frac{\partial^2 u_x}{\partial y^2} \qquad (2.220)$$

In principle, if u_∞ is known (obtained from the solution of the nonviscous equations for the flow geometry [2.30]), Equations (2.215) and (2.220) can be solved for the velocity components u_x and u_y. Although these equations can be solved numerically [2.29], [2.31], here the focus is on the use of approximate integral methods.

In particular, note that

$$u_y \frac{\partial u_x}{\partial y} = \frac{\partial}{\partial y}(u_y u_x) - \frac{\partial u_y}{\partial y} u_x \qquad (2.221)$$

If $(\partial u_y/\partial y)$ is replaced with $-(\partial u_x/\partial x)$ by using the equation of continuity, Equation (2.220) can then be written as

$$\frac{\partial}{\partial x}(u_x^2) + \frac{\partial}{\partial y}(u_y u_x) = \frac{1}{2}\frac{d}{dx}(u_\infty^2) + v \frac{\partial^2 u_x}{\partial y^2} \qquad (2.222)$$

When this equation is integrated over y from $y = 0$ to $y = \infty$, and the conditions

$$u_x = 0, \quad u_y = 0 \quad \text{at } y = 0 \qquad (2.223)$$

$$u_x = u_\infty, \quad \frac{\partial u_x}{\partial y} = 0 \quad \text{at } y \geq \delta_m \qquad (2.224)$$

along with (from Eq. 2.215)

$$u_y(x, \infty) = -\int_0^{\delta_m} \frac{\partial u_x}{\partial x} dy \quad (2.225)$$

are used, the result is

$$\frac{d}{dx}\int_0^{\delta_m} u_x(u_\infty - u_x) dy + \frac{du_\infty}{dx}\int_0^{\delta_m} (u_\infty - u_x) dy = \frac{\tau_w}{\varrho} \quad (2.226)$$

where τ_w is the wall shear stress given by

$$\tau_w = \eta \left(\frac{\partial u_x}{\partial y}\right)_{y=0} \quad (2.227)$$

Equation (2.226) is the *von Karman momentum integral equation*. For a given external flow $u_\infty(x)$, its solution requires a knowledge of the velocity u_x as a function of y. Even with crude approximations for $u_x(y)$, reasonably accurate estimates of the drag force can be obtained.

As an example, consider the case of a flat plate in a free stream flow where u_∞ = constant. If a cubic relation is assumed for $u_x(y)$, i.e.,

$$u_x = a + by + cy^2 + dy^3 \quad (2.228)$$

and the boundary conditions of Equations (2.223) and (2.224) are used, along with $(\partial^2 u_x/\partial y^2) = 0$ at $y = \delta_m$ to evaluate the constants, then

$$u_x = u_\infty \left[\frac{3}{2}\frac{y}{\delta_m} - \frac{1}{2}\left(\frac{y}{\delta_m}\right)^3\right] \quad (2.229)$$

If this assumed profile is substituted in the momentum integral balance (Eq. 2.226) and the necessary integrations are performed, the following differential equation for boundary layer thickness is obtained:

$$\frac{d\delta_m}{dx} = \frac{140 \nu}{13 u_\infty} \frac{1}{\delta_m} \quad (2.230)$$

Integration with the condition $\delta_m = 0$ at $y = 0$ gives the boundary layer thickness

$$\frac{\delta_m}{x} = \frac{4.64}{Re_x^{1/2}} \quad (2.231)$$

where $Re_x \equiv u_\infty x/\nu$ is the local Reynolds number. The corresponding drag coefficient is given by

$$C_{Dx} \equiv \frac{\tau_w}{\frac{1}{2}\varrho u_\infty^2} = \frac{2}{\varrho u_\infty^2}\frac{3\eta u_\infty}{2\delta_m} = \frac{0.646}{Re_x^{1/2}} \quad (2.232)$$

where the subscript x on C_D indicates the coefficient is a local value that varies with x. The drag force per unit width is then

$$\frac{F_D}{w} = \int_0^L \tau_w dx = C_{DL} L (\tfrac{1}{2}\varrho u_\infty^2) \quad (2.233)$$

where the overall film coefficient C_{DL} is given by

$$C_{DL} \equiv \frac{1}{L}\int_0^L C_{Dx} dx = \frac{1.292}{Re_L^{1/2}} \quad (2.234)$$

and $Re_L \equiv u_\infty L/\nu$.

The *exact* solution for a laminar boundary layer on a flat plate with a uniform free stream velocity was obtained by BLASIUS [2.29]. The corresponding local and overall drag coefficient results are

$$C_{Dx} = \frac{0.664}{Re_x^{1/2}} \quad (2.235)$$

$$C_{DL} = \frac{1.328}{Re_L^{1/2}} \quad (2.236)$$

The approximate results of Equations (2.232) and (2.234) compare favorably with the exact results. This demonstrates the utility of the integral method. Even more accurate results could have been obtained if higher order approximations (beyond the cubic) were used for the velocity profile. In general, such integral approaches are valuable tools in engineering analysis and provide excellent approximations with comparable ease relative to the numerical methods required to obtain exact results.

Heat Transfer in a Laminar Boundary Layer. Just as the momentum boundary layer thickness was defined as the region in which most of the fluid-deforming velocity gradients occur, a thermal boundary layer thickness δ_T can be defined as that region next to a heated (or cooled) surface in which most of the temperature gradients occur. Specifically, for heat transfer from a flat plate at temperature T_w to a flowing fluid with free stream temperature T_∞, the thermal boundary layer is defined as

$$\delta_T \equiv \left\{\begin{array}{l}\text{The } y\text{-position where} \\ T - T_w = 0.99 (T_\infty - T_w)\end{array}\right\} \quad (2.237)$$

In general, as shown in Figure 15, the thermal boundary layer thickness is not equal to the momentum boundary layer thickness. However, the relation between the thermal and momentum boundary layers in laminar flow can be approximated by [2.32]

$$\frac{\delta_m}{\delta_T} \equiv \Delta_T = \text{const.} \tag{2.238}$$

Equation (2.238) implies that the dependence of δ_T on x is the same as that of δ_m on x.

For heat transfer in a laminar boundary layer on a flat plate where $\delta_T/L \ll 1$, the energy balance (Eq. 2.191) can be simplified to

$$u_x \frac{\partial T}{\partial x} + u_y \frac{\partial T}{\partial y} = \alpha \frac{\partial^2 T}{\partial y^2} \tag{2.239}$$

By integrating over y from 0 to ∞ and using the equation of continuity to replace u_y (the steps involved are analogous to Eqs. 2.221–2.226), the energy balance can be put into the integral form

$$\frac{d}{dx} \int_0^{\delta_T} u_x (T - T_\infty) \, dy = \frac{q_w}{\rho c_p} \tag{2.240}$$

where q_w is the wall heat flux given by

$$q_w = -\lambda \left. \frac{\partial T}{\partial y} \right|_{y=0} \tag{2.241}$$

Equation (2.240) is analogous to the momentum integral equation (Eq. 2.226). To obtain a solution to this integral energy balance, a temperature profile must be assumed that satisfies the conditions

at $y = 0$, $T = T_w$ (2.242)

at $y = \delta_T$, $T = T_\infty$ (2.243)

at $y = \delta_T$, $\dfrac{\partial^n T}{\partial y^n} = 0$ for $n = 1, 2, \ldots$ (2.244)

If a cubic power relation is assumed for the temperature profile

$$T = a + by + cy^2 + dy^3 \tag{2.245}$$

the boundary conditions of Equations (2.242)–(2.244) can be used to determine the coefficients, with the result

$$T = T_w + (T_\infty - T_w) \left[\frac{3}{2} \left(\frac{y}{\delta_T} \right) - \frac{1}{2} \left(\frac{y}{\delta_T} \right)^3 \right] \tag{2.246}$$

When this profile is substituted into the integral form of the energy balance, Equation (2.229) is used for u_x, and Equation (2.238) is used for δ_m, then

$$\delta_T \frac{d\delta_T}{dx} = \frac{10\alpha}{u_\infty} \left(\frac{14\Delta_T^3}{14\Delta_T^2 - 1} \right) \tag{2.247}$$

where $\Delta_T \equiv \delta_m/\delta_T$ has been assumed to be constant. By integrating this equation with the condition $\delta_T = 0$ at $y = 0$, the thermal boundary layer thickness is found to be

$$\frac{\delta_T}{x} = \frac{4.47 \Delta_T^{3/2}}{Re_x^{1/2} Pr^{1/2}} \left(\frac{14}{14\Delta_T^2 - 1} \right)^{1/2} \tag{2.248}$$

where $Pr = \nu/\alpha$ is the Prandtl number, the ratio of the momentum and thermal diffusivities. Then, from Equations (2.231), (2.238), and (2.248),

$$\Delta_T \equiv \frac{\delta_m}{\delta_T} \left[\frac{14}{13} \left(1 - \frac{1}{14\Delta_T^2} \right) \right]^{1/3} Pr^{1/3} \approx Pr^{1/3} \tag{2.249}$$

and

$$\frac{\delta_T}{x} = 4.47 \, Re_x^{-1/2} \, Pr^{1/3} \tag{2.250}$$

where it has been assumed

$$14\Delta_T^2 \gg 1$$

Based upon these results, the momentum boundary layer thickness increases relative to the thermal boundary layer as the momentum diffusivity ν increases relative to the thermal diffusivity α (or Pr increases). For gases, the Prandtl number is on the order of 1 and the boundary layer thickness δ_m and δ_T have similar magnitudes. For liquids, the Prandtl number can be much higher than 1 and δ_m can be larger than δ_T. For liquid metals, $Pr \ll 1$ and the momentum boundary layer thickness is much smaller than the thermal boundary layer thickness.

Thus, from Equations (2.241), (2.246), and Newton's law of cooling,

$$Nu_x \equiv \frac{h_x \chi}{\lambda} = 0.36 \, Re_x^{1/2} \, Pr^{1/3} \tag{2.251}$$

where Nu_x is the dimensionless *local Nusselt number*. This result is ca. 8% higher than the exact result from numerical solutions, which is given by

$$Nu_x = 0.332\, Re_x^{1/2}\, Pr^{1/3} \qquad (2.252)$$

Heat transfer from the plate can then be determined from

$$\dot{Q} = \int_0^L h(T_w - T_\infty) w\, dx = h_L w L (T_w - T_\infty) \qquad (2.253)$$

where

$$h_L \equiv \frac{1}{L} \int_0^L h\, dx \qquad (2.254)$$

Using Equation (2.252) yields

$$Nu_L \equiv \frac{h_L L}{\lambda} = 0.664\, Re_L^{1/2}\, Pr^{1/3} \qquad (2.255)$$

where $Re_L \equiv (u_\infty L/\nu)$. Hence, if the properties are determined from the average film temperature, $1/2(T_w + T_\infty)$, the total heat-transfer rate from the plate can be determined from Equation (2.253), with h_L estimated from Equation (2.255).

For the case of *constant heat flux* at the plate surface, the local Nusselt number is given by

$$Nu_x = 0.453\, Re_x^{1/2}\, Pr^{1/3} \qquad (2.256)$$

In this case, wall temperature varies with position x on the plate according to

$$T_w = T_\infty + \frac{q_w}{h_x} \qquad (2.257)$$

Although the above relations have been derived for flat surfaces, they are also useful as approximations for curved surfaces if the characteristic radius of curvature for the surface is much larger than the boundary layer thickness.

2.4.3. Free Convection on a Vertical Plate

When a stagnant fluid is in contact with a heated surface, the fluid layers near the surface are at higher temperatures than the fluid layers further removed from the surface. Because of their higher temperatures, the layers near the surface have lower densities and will rise relative to the stagnant bulk fluid. Such free convection processes can be analyzed in much the same way as forced convection on a flat plate was analyzed by using boundary layer methods.

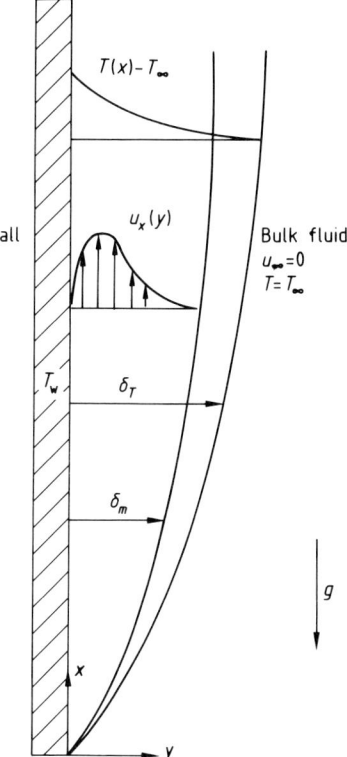

Figure 16. Free convection from a vertical heated plate, development of momentum, and thermal boundary layers

The velocity and temperature field inside the momentum and thermal boundary layers are illustrated in Figure 16. The boundary layer equations, Equations (2.218) and (2.239), describe the behavior of these variables. Outside the boundary layer, the pressure gradient is simply the hydrostatic gradient

$$\frac{dp}{dx} = -\varrho_\infty g \qquad (2.258)$$

where ϱ_∞ is the density of the fluid at T_∞. Inside the thermal boundary layer, the density can be related to the temperature by

$$\varrho = \varrho_\infty + \left(\frac{\partial \varrho}{\partial T}\right)_{T_\infty} (T - T_\infty)$$

$$= \varrho_\infty [1 - \beta_\infty (T - T_\infty)] \qquad (2.259)$$

Substituting Equations (2.258) and (2.259) into Equation (2.218) gives the momentum equation for free convection, i.e.,

$$u_x \frac{\partial u_x}{\partial x} + u_y \frac{\partial u_x}{\partial y} = \nu \frac{\partial^2 u_x}{\partial y^2} + g\beta_\infty (T - T_\infty) \quad (2.260)$$

In integral form, this equation becomes

$$\frac{d}{dx} \int_0^{\delta_m} u_x^2 \, dy = -\nu \left.\frac{\partial u_x}{\partial y}\right|_{y=0} + g\beta_\infty \int_0^{\delta_T} (T - T_\infty) \, dy \quad (2.261)$$

The integral form of the energy balance is the same as Equation (2.240).

In selecting appropriate velocity and temperature expressions to use in these equations, the following boundary conditions are applicable:

at $y = 0$, $u_x = 0$ (2.262)

$T = T_w$ (2.263)

at $y = \delta_m$, $u_x = 0$ (2.264)

$\dfrac{\partial^n u_x}{\partial y^n} = 0$, $n = 1, 2, \ldots$ (2.265)

at $y = \delta_T$, $T = T_\infty$ (2.266)

$\dfrac{\partial^n T}{\partial y^n} = 0$, $n = 1, 2, \ldots$ (2.267)

In particular, if a cubic expression is used for $u_x(y)$ and a quadratic expression for $T(y)$, the resulting profiles that are consistent with the boundary conditions are

$$u_x = u_c(x) \frac{y}{\delta_m}\left(1 - \frac{y}{\delta_m}\right)^2 \quad (2.268)$$

$$T = T_\infty + (T_w - T_\infty)\left(1 - \frac{y}{\delta_T}\right)^2 \quad (2.269)$$

where u_c is a characteristic velocity that varies with x. If $u_c = A x^m$ and $\delta_m = \delta_T = B x^n$, and these relations are substituted into Equations (2.261) and (2.240), the boundary layer thickness δ_m and the local and overall heat-transfer coefficients can be determined [2.6, pp. 371–376]. The results are

$$\frac{\delta}{x} = 3.936 \left(\frac{0.952 + Pr}{Pr^2}\right)^{1/4} \frac{1}{Gr_x^{1/4}} \quad (2.270)$$

$$Nu_x \equiv \frac{h_x x}{\lambda}$$

$$= 0.508 \, (Pr \, Gr_x)^{1/4} \left(\frac{Pr}{0.952 + Pr}\right)^{1/4} \quad (2.271)$$

and

$$Nu_L \equiv \frac{h_L L}{\lambda}$$

$$= 0.678 \, (Pr \, Gr_L)^{1/4} \left(\frac{Pr}{0.952 + Pr}\right)^{1/4} \quad (2.272)$$

Here Gr_x is the local Grashof number given by

$$Gr_x = \frac{g\beta_\infty (T_w - T_\infty) x^3}{\nu} \quad (2.273)$$

and $Pr \, Gr_x$ is the Rayleigh number given by

$$Ra_x \equiv Pr \, Gr_x = \frac{g\beta (T_w - T_\infty) x^3}{\nu \alpha} \quad (2.273\,a)$$

The assumption of $\delta_m = \delta_T$ would seem to limit the results to Prandtl numbers on the order of 1; however, because the effect of the velocity profile on the temperature profile is small as Pr increases, the result is also useful at higher Prandtl numbers.

Empirically, CHURCHILL and CHU [2.32] have found that free convection data on vertical surfaces correlate quite accurately with

$$Nu_L = 0.68 + 0.67 (Pr \, Gr_L)^{1/4}$$

$$\cdot \left[1 + \left(\frac{0.492}{Pr}\right)^{9/16}\right]^{-4/9} \quad (2.274)$$

In using this correlation, the thermal properties should be evaluated at the mean boundary layer temperature, except for β_∞ which should be evaluated at T_∞ if the fluid is a gas. Equation (2.272) is quite accurate at $Pr \approx 1$ or greater; however, it is not valid at low Prandtl numbers (liquid metals), and Equation (2.274) must be used instead.

2.5 Turbulent Momentum and Energy Transfer

2.5.1. Physical Characteristics

Much flow of industrial importance is turbulent flow. Momentum and energy transport in turbulent flow are characterized by time-varying velocity and temperature fields. In particular, for turbulent flow in a tube under constant flow rate conditions, the axial velocity component at a specific position in the flow appears as in Figure 17A with the velocity fluctuating around a time-averaged value. The same fluctuating behavior is

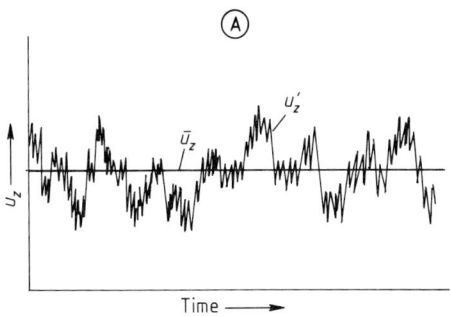

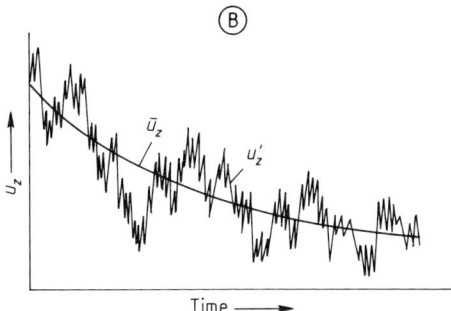

Figure 17. Illustration of velocity fluctuations in turbulent flow
The velocity consists of a time-average part ($\bar{u}_z$) and a fluctuating part (u'_z).
A) Steady turbulent flow, $\bar{u}_z$ constant with time; B) Unsteady turbulent flow, $\bar{u}_z$ varying with time

also observed for the other velocity components. Also, if heat is convected between the wall and the fluid, the temperature at each spatial position in the flow also fluctuates.

In describing this behavior, the velocity and temperature variables are conveniently separated into time-averaged values plus fluctuating parts, i.e.,

$$\boldsymbol{u} = \bar{\boldsymbol{u}} + \boldsymbol{u}' \quad (2.275)$$

and

$$T = \bar{T} + T' \quad (2.276)$$

where $\bar{\boldsymbol{u}}$ and $\bar{T}$ are the time-averaged values and $\boldsymbol{u}'$ and T' are the fluctuating parts. The time-averaged value at time t of a variable φ is defined by

$$\bar{\varphi} \equiv \frac{1}{t_0} \int_t^{t+t_0} \varphi \, dt \quad (2.277)$$

where φ can correspond to $\boldsymbol{u}$, T, p, or any other dependent variable. The time t_0 over which the average is taken is long compared to the reciprocal frequency of the fluctuations, but short compared to the time of macroscopic changes in pressure drop, wall temperature, and other externally controlled variables.

Although turbulence is inherently unsteady, steady conditions in turbulent flow can be considered if

$$\frac{\partial \bar{\varphi}}{\partial t} = 0 \quad (2.278)$$

In Figure 17 A, steady turbulent flow is illustrated; in Figure 17 B, unsteady turbulent flow. The latter might arise in a flow field if the applied pressure drop decreases with time.

To obtain a measure of the relative magnitude of the fluctuating part of a turbulent flow variable, the quantity

$$I_\varphi \equiv \frac{\sqrt{\overline{(\varphi')^2}}}{\bar{\varphi}} \quad (2.279)$$

can be defined. In particular, for a shear flow [$\bar{u}_x = \bar{u}_x(y)$] the intensity of turbulence is defined by

$$I_u \equiv \frac{\sqrt{\overline{(u'_x)^2} + \overline{(u'_y)^2} + \overline{(u'_z)^2}}}{\bar{u}} \quad (2.280)$$

In Figure 18, the intensities of individual velocity components are shown for a turbulent boundary layer on a flat plate as observed by KLEBANOFF [2.1, p. 275], [2.34]. As can be seen,

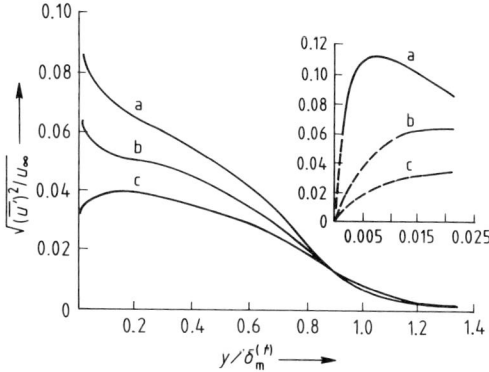

Figure 18. Distribution of intensities of the different velocity components in a turbulent boundary layer

$$a = \frac{\sqrt{\overline{(u'_x)^2}}}{u_\infty}, \quad b = \frac{\sqrt{\overline{(u'_z)^2}}}{u_\infty}, \quad c = \frac{\sqrt{\overline{(u'_y)^2}}}{u_\infty}$$

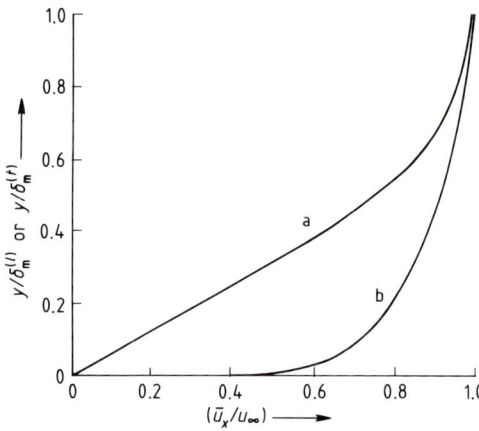

Figure 19. Dimensionless laminar and turbulent velocity profiles for flow on a flat plate at a local Reynolds number $Re_x = 10^6$
Here, $\delta_m^{(l)}$ and $\delta_m^{(t)}$ denote the momentum boundary layer thickness for laminar (a) and turbulent flow, (b) respectively; $\delta_m^{(t)}/\delta_m^{(l)} \approx 4.8$.

At distances sufficiently close to the wall, the turbulent fluctuations diminish and a *laminar sublayer* is observed. In this region, momentum or energy transfer occurs largely from conventional molecular and laminar convection mechanisms. At distances sufficiently remote from the wall, a fully developed *turbulent core* is present where the momentum- and energy-transfer mechanisms are largely a result of convective mixing of the turbulent eddies. In the *buffer zone* between the laminar sublayer and the turbulent core, the flow is neither laminar nor fully developed turbulent flow. In this region, both types of molecular and turbulent transport mechanisms are present.

The velocity distribution across these three regions can be described by the semiempirical relations

Laminar sublayer:

$$u^+ = y^+ \quad \text{for} \quad y^+ < 5 \qquad (2.281)$$

Buffer zone:

$$u^+ = -3.05 + 5 \ln y^+ \quad \text{for} \quad 5 \leq y^+ < 30 \qquad (2.282)$$

Turbulent core:

$$u^+ = 5.5 + 2.5 \ln y^+ \quad \text{for} \quad 30 \leq y^+ \qquad (2.283)$$

Here, u^+ and y^+ are defined by

$$u^+ \equiv u_x/u_* \qquad (2.284)$$

$$y^+ \equiv y u_*/\nu, \qquad (2.285)$$

and u_* is the friction velocity defined by

$$u_* \equiv \sqrt{\tau_w/\varrho} \qquad (2.286)$$

where τ_w is the wall shear stress. The velocity distribution represented by Equations (2.281)–(2.283) is commonly referred to as the *universal velocity distribution*.

Another form of the velocity distribution that is commonly used to describe turbulent flow in tubes is

$$\frac{\bar{u}_x}{u_{max}} = \left(\frac{y}{r_1}\right)^{1/n} \qquad (2.287)$$

where y is the distance measured from the wall and r_1 is the tube radius. NIKURADSE [2.35] obtained excellent agreement with experimental observations when n was allowed to vary with

the largest component intensity is that associated with the contribution in the direction of flow (u_x), with a maximum value of 0.11 which occurs very close to the wall ($y/\delta_m^{(t)} \approx 0.007$). The smallest component intensity is that associated with u_y, the velocity component normal to the wall, where a value of 0.04 is observed at $y/\delta_m^{(t)} = 0.15$. Although the fluctuating intensities may appear small, these quantities are important in the heat- and mass-transfer rates observed in turbulent flow.

In Figure 19, the normalized velocity profile $\bar{u}_x/u_\infty$ for a turbulent boundary layer on a flat plate (b) is compared with that for a laminar boundary layer (a), both corresponding to a local Reynolds number Re_x of 10^6. The turbulent boundary layer thickness is nearly five times that of the laminar boundary layer, and in turbulent flow the velocity profile is more uniform with a higher ratio of average velocity to maximum velocity. Also, the wall velocity gradient and the corresponding wall friction drag are higher in the turbulent case. The latter might suggest that a laminar boundary layer would be more desirable in minimizing frictional drag on surfaces; however, in most cases a turbulent boundary layer is preferred because it is more effective in resisting boundary layer separation and increased form drag.

For turbulent flow in tubes and along flat surfaces, various flow regions can be identified.

Table 6. Characteristics of velocity profile for turbulent flow in a tube [2.35]

Re_d	n in Equation (2.287)	u_{avg}/u_{max}
4×10^3	6	0.791
1.1×10^5	7	0.817
3.24×10^6	10	0.865

Reynolds number (see Table 6). For Reynolds numbers of ca. 10^5, a value of $n = 7$ is commonly used. Note that Equation (2.287) gives an infinite velocity gradient at the tube wall. As a result, drag force values cannot be predicted directly from velocity gradients calculations at the wall.

Although Equation (2.281)–(2.283) and (2.287) were developed originally for tube flow, they are also valid for turbulent boundary layers on flat plates. In this case, u_{max} in these equations is replaced by u_∞ and r_1 by $\delta_m^{(t)}$. Also, because $\tau_w = f \varrho u_{avg}^2 / 2$ for tubes, an expression for u_{avg}/u_{max} must be used to apply the results to flat plates. As illustrated in Table 6, this ratio for tube flows varies with Reynolds number. For Reynolds numbers of ca. 10^5, a value of $u_{avg}/u_{max} \approx 0.8$ is commonly used.

2.5.2. Time-Smoothed Momentum and Energy Balances for Turbulent Flow

The equations of continuity and motion given previously are perfectly valid for describing turbulent flow; however, because of the unsteady, three-dimensional fluctuations of velocity and other dependent variables, the analysis requires simultaneous solution of the full three-dimensional equations of motion along with the equation of continuity. Even if appropriate boundary and initial conditions could be specified, such solutions are beyond current mathematical capabilities. A less ambitious approach is to focus on the time-smoothed quantities and to describe these variables as a function of position and time by using time-smoothed forms of the equations of continuity, motion, and energy.

The specific forms of the *continuity* and *motion equations* can be obtained by time-smoothing Equations (1.65) and (1.73) term by term. Before carrying out these operations, Equation (1.73) can be put in the equivalent form

$$\frac{\partial}{\partial t}(\varrho \boldsymbol{u}) + \nabla \cdot (\varrho \boldsymbol{u}\boldsymbol{u}) = -\nabla p - \nabla \cdot \boldsymbol{\tau} + \varrho \boldsymbol{g} \quad (2.288)$$

where the equation of continuity has been used on the left side of Equation (1.73) and Equation (1.84) on the right side. The key time-smoothing identities are

$$\overline{\mathscr{F}\varphi} = \mathscr{F}\overline{\varphi} \quad (2.289)$$

$$\overline{\mathscr{F}\varphi\psi} = \mathscr{F}\overline{\varphi}\,\overline{\psi} + \mathscr{F}\overline{\varphi'\psi'} \quad (2.290)$$

where $\mathscr{F}$ is any linear time or spatial derivative operator (e.g., $\partial/\partial t$, ∇, $\nabla \cdot$) and φ and ψ represent scalar, vector, or tensor functions, or combinations thereof. By using these identities, the following time-smoothed equations are obtained (incompressible fluids):

$$\nabla \cdot \bar{\boldsymbol{u}} = 0 \quad (2.291)$$

$$\varrho \left(\frac{\partial \bar{\boldsymbol{u}}}{\partial t} + \bar{\boldsymbol{u}} \cdot \nabla \bar{\boldsymbol{u}} \right) = -\nabla \bar{p} - \nabla \cdot (\bar{\boldsymbol{\tau}} - \varrho \overline{\boldsymbol{u}'\boldsymbol{u}'}) + \varrho \boldsymbol{g}$$

$$(2.292)$$

If this equation is compared with the form of Equation (2.288) for incompressible fluids, i.e.,

$$\varrho \left(\frac{\partial \boldsymbol{u}}{\partial t} + \boldsymbol{u} \cdot \nabla \boldsymbol{u} \right) = -\nabla p - \nabla \cdot \boldsymbol{\tau} + \varrho \boldsymbol{g} \quad (2.293)$$

the time-smoothed equation is seen to have the same form, except that $\bar{\boldsymbol{u}}$ and $\bar{p}$ replace $\boldsymbol{u}$ and p, and $\boldsymbol{\tau}^{(t)} \equiv \bar{\boldsymbol{\tau}} - \varrho \overline{\boldsymbol{u}'\boldsymbol{u}'}$ replaces $\boldsymbol{\tau}$. The components of $\varrho \overline{\boldsymbol{u}'\boldsymbol{u}'}$ are called the *Reynolds stresses*. Physically, these represent the convective momentum flux components associated with the fluctuating part of the velocity.

One could expect that just as $\boldsymbol{\tau}$ depends on the deformation rate, the Reynolds stresses might be expressible in terms of the mean velocity and its spatial derivatives. Unfortunately, this does not appear to be possible, and when attempted, the coefficients associated with the mean velocity terms and its derivatives are spatially dependent and specific to each flow field.

In general, a proper description of the $\varrho \overline{\boldsymbol{u}'\boldsymbol{u}'}$ contributions requires a more complete statistical analysis of turbulence. Although progress is being made in this area, an approach that can be used in engineering analyses is not yet available. As a result, only forms of the Reynolds stresses that are valid for specific flows can be obtained. This is the approach taken here in restricting attention to flow in tubes and along flat surfaces.

Finally, note that the time-smoothed form of the *energy balance* Equation (2.191) is

$$\varrho c_p \left(\frac{\partial \bar{T}}{\partial t} + \bar{u} \cdot \nabla \bar{T} \right) = -\nabla \cdot (\bar{q} + \varrho c_p \overline{u' T'}) \quad (2.294)$$

where $\bar{q} = -\lambda \nabla \bar{T}$. This equation is of the same form as Equation (2.199), except that $q^{(t)} \equiv \bar{q} + \varrho c_p \overline{u' T'}$ replaces $q = -\lambda \nabla T$. Also, this equation neglects viscous heat dissipation, reversible work, and heat generation. The quantity $\varrho c_p \overline{u' T'}$ is the heat flux associated with turbulent fluctuations. As with Reynolds stresses, $\varrho c_p \overline{u' T'}$ must be expressed in terms of the time-smoothed quantities and their derivatives before solutions to Equation (2.294) can be obtained. Similar to Reynolds stresses, these relations are not universal and the coefficients in such expressions are spatially dependent and specific to each flow.

2.5.3. Mixing Length Theories

One of the common approaches used in relating $\varrho \overline{u' u'}$ and $\varrho c_p \overline{u' T'}$ to time-smoothed quantities is through a mixing length interpretation of turbulent transport. In particular, this approach assumes a mechanism for turbulent transport which involves irregular movement of discrete fluid elements in the flow in a manner analogous to the kinetic behavior of molecules in a gas.

For the case of unidirectional turbulent flow along a surface as illustrated in Figure 20, where the time-smoothed entity $\mathcal{M}$ (momentum, energy, mass, etc.) varies with distance y from the surface, the flux arising from the turbulent fluctuations is given by

$$\text{Flux of } \mathcal{M} \text{ in } y \text{ direction} = -\varepsilon_\mathcal{M} \frac{d \bar{\mathcal{M}}}{dy} \quad (2.295)$$

Here $\varepsilon_\mathcal{M}$ is the *eddy diffusivity* associated with the transfer of $\mathcal{M}$ by the turbulent fluctuations. Basically, the fluctuations give rise to the "diffusion" of discrete fluid elements that produce a flux of $\mathcal{M}$ down the gradient of decreasing $\bar{\mathcal{M}}$.

For the exchange of x-momentum between different y layers moving at different velocities $\bar{u}_x$:

$$\varrho \overline{u'_y u'_x} = \varepsilon_m \frac{d}{dy}(\varrho \bar{u}_x) = \varrho \varepsilon_m \frac{d \bar{u}_x}{dy} \quad (2.296)$$

where ε_m is the *eddy momentum diffusivity*.

Similarly, for the heat flux between different y layers at different temperatures $\bar{T}$:

$$\varrho c_p \overline{u'_y T'} = \varepsilon_T \frac{d}{dy}(\varrho c_p \bar{T}) = \varrho c_p \varepsilon_T \frac{d \bar{T}}{dy} \quad (2.297)$$

where ε_T is the *eddy thermal diffusivity*.

From largely physical arguments, PRANDTL [2.31], [2.36] was able to relate ε_m and ε_T to the motion of discrete fluid elements between layers with different mean velocities to obtain

$$\varepsilon_m = \varepsilon_T \equiv \varepsilon = l^2 \left| \frac{d \bar{u}_x}{dy} \right| \quad (2.298)$$

where l is the *Prandtl mixing length*. The mixing length is a measure of the distance (in the y-direction) that the discrete fluid elements move before they accommodate to the mean velocity of the surroundings. This quantity is somewhat analogous to the mean free path in the kinetic theory of gases. For one-dimensional flow in tubes and along surfaces (e.g., Fig. 20), the mixing length has been empirically found to increase

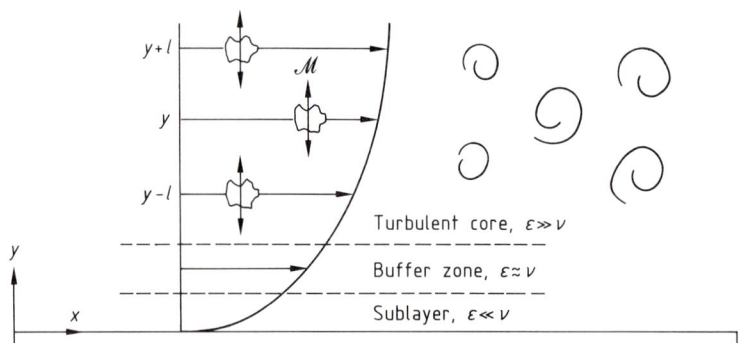

Figure 20. Turbulent transport in unidirectional flow along a surface

with distance from the wall. Specifically, the relation

$$l = ky \tag{2.299}$$

appears to represent experimental observations with the constant k being ca. 0.4.

In summary, for turbulent transport in shear flows where $\bar{u}_x = \bar{u}_x(y)$, the total x-momentum flux in the y-direction can be given by

$$\tau_{yx}^{(t)} = -\varrho(v+\varepsilon)\frac{d\bar{u}_x}{dy} \tag{2.300}$$

and the total heat flux $q_y^{(t)}$ in the y-direction by

$$q_y^{(t)} = -\varrho c_p(\alpha+\varepsilon)\frac{d\bar{T}}{dy} \tag{2.301}$$

where v and α are the molecular momentum and thermal diffusivities, respectively, and ε is the eddy diffusivity. At positions close to the wall in the laminar sublayer, the fluctuations have negligible importance, and $v \gg \varepsilon$ and $\alpha \gg \varepsilon$; whereas in the bulk turbulent field away from the wall, $\varepsilon \gg v$ and $\varepsilon \gg \alpha$ for high-intensity turbulent flow. In the buffer region between the laminar sublayer and the turbulent core, the molecular and eddy diffusivities are generally of the same order of magnitude.

2.5.4. Turbulent Heat Transfer in Tubes—Reynolds, Prandtl, and von Karman Analogies

Results are now presented for estimating heat-transfer coefficients for convective heat transfer in tubes under turbulent flow conditions. The heat transfer is assumed to arise from temperature differences between the heated (or cooled) bounding surface and the bulk fluid. The analysis follows that of PRANDTL [2.36] and is aimed at developing an expression involving the heat-transfer coefficient h, the friction factor f, the average velocity u_{avg}, and the properties of the fluid.

The key assumptions are (1) the turbulent flow field is divided into two zones, a laminar sublayer that extends to $y^+ = 5$ and a turbulent core that occupies the region beyond this point; (2) the velocity profile in the laminar sublayer is given by Equation (2.281); and (3) the ratio $\bar{q}_y/\bar{\tau}_{yx}$ of the heat and momentum fluxes is constant in both the sublayer and the turbulent core. The last assumption implies that the temperature and velocity profiles are functionally similar.

In the *laminar sublayer* where turbulent eddy transport is negligible,

$$\frac{\bar{q}_y}{\bar{\tau}_{yx}} = \frac{-\varrho c_p \alpha \dfrac{d\bar{T}}{dy}}{-\varrho v \dfrac{d\bar{u}_x}{dy}} = \frac{c_p \alpha}{v}\frac{d\bar{T}}{d\bar{u}_x} \tag{2.302}$$

or, because $\bar{q}_y/\bar{\tau}_{yx}$ is assumed to be constant,

$$\frac{\bar{q}_y}{\bar{\tau}_{yx}} = \frac{q_w}{(-\tau_w)} = \frac{c_p}{Pr}\frac{d\bar{T}}{d\bar{u}_x} \tag{2.303}$$

Integration of this equation over the laminar sublayer gives

$$\bar{T}_\xi - T_w = -\frac{Pr}{c_p}\left(\frac{q_w}{\tau_w}\right)\bar{u}_\xi \tag{2.304}$$

where $Pr = v/\alpha$. Here the edge of the sublayer is denoted by $y = \xi$, and $\bar{T}_\xi$ and $\bar{u}_\xi$ are the temperature and velocity at this position.

In the *turbulent core* the eddy momentum and energy transport dominate the molecular transport. As a result,

$$\frac{\bar{q}_y}{\bar{\tau}_{yx}} = -\frac{q_w}{\tau_w} \approx \frac{-\varrho c_p \varepsilon_T \dfrac{d\bar{T}}{dy}}{-\varrho \varepsilon_m \dfrac{d\bar{u}_x}{dy}} \tag{2.305}$$

or

$$-\frac{q_w}{\tau_w} = c_p \frac{d\bar{T}}{d\bar{u}_x} \tag{2.306}$$

where $\varepsilon_T = \varepsilon_m = \varepsilon$ is assumed as in the Prandtl mixing length theory. Integrating this equation from $y = \xi$, where $\bar{u}_x = \bar{u}_\xi$ and $\bar{T} = \bar{T}_\xi$, to the bulk of the fluid, where $\bar{u}_x = u_{avg}$ and $\bar{T} = T_\infty$, gives

$$T_\infty - \bar{T}_\xi = -\frac{1}{c_p}\left(\frac{q_w}{\tau_w}\right)(\bar{u}_{avg} - \bar{u}_\xi) \tag{2.307}$$

Adding Equations (2.304) and (2.307) and using

$$q_w = h(T_w - T_\infty) \tag{2.308}$$

$$\tau_w = f\tfrac{1}{2}\varrho u_{avg}^2 \tag{2.309}$$

and,

$$\bar{u}_\xi = 5\sqrt{\tau_w/\varrho} = 5 u_{avg}\sqrt{f/2}, \qquad (2.310)$$

give

$$St \equiv \frac{Nu_d}{Re_d Pr} = \frac{h}{\varrho c_p u_{avg}} = \frac{f/2}{1 + 5\sqrt{f/2}\,(Pr - 1)} \qquad (2.311)$$

where St is the Stanton number. This equation is called the *Prandtl analogy* and relates the heat-transfer coefficient to the friction factor (or drag coefficient). The quantities Nu_d and Re_d are the Nusselt and Reynolds numbers based on the tube diameter d and the average velocity, i.e., hd/λ and $u_{avg}d/\nu$, respectively.

For Prandtl numbers of 1 (applicable for many gases), Equation (2.311) reduces to

$$\frac{h}{\varrho c_p u_{avg}} = \frac{f}{2} \qquad (2.312)$$

which is the *Reynolds analogy*. This result can be obtained directly from the above analysis by neglecting the sublayer and allowing the turbulent core to occupy the entire flow region. Hence, $u_\xi = 0$ and $T_\xi = T_w$ in Equation (2.307), and the Reynolds analogy follows directly from Equations (2.308) and (2.309).

VON KARMAN [2.37, pp. 223–227] conducted a more complete analysis including all three zones in the turbulent field (i.e., laminar sublayer, buffer zone, and turbulent core). The final result was

$$St \equiv \frac{h}{\varrho c_p u_{avg}} \qquad (2.313)$$

$$= \frac{f/2}{1 + 5\sqrt{f/2}\,\{Pr - 1 + \ln[1 + \tfrac{5}{6}(Pr - 1)]\}}$$

This analogy includes the Reynolds and Prandtl analogies as special cases.

To use the above analogies to estimate heat-transfer coefficients, an expression is needed for the friction factor f in terms of the Reynolds number and appropriate roughness parameters.

For smooth tubes, a useful empirical expression for f in the range $4 \times 10^3 \leq Re_d \leq 10^5$ is the Blasius formula [2.38] ($\to$ 5. Fluid Mechanics, pp. 5-12):

$$f \equiv \frac{\tau_w}{\tfrac{1}{2}\varrho u_{avg}^2} = \frac{0.0791}{Re_d^{1/4}} \qquad (2.314)$$

At higher Reynolds numbers ($> 10^5$), the universal velocity profile can be used to develop an expression for the friction factor. In particular, if the turbulent core is assumed to extend to the wall, Equation (2.283) can be integrated over the tube cross section to obtain the average velocity:

$$u_{avg} = u_* \left[1.75 + 2.5 \ln\left(\frac{u_* r_1}{\nu}\right)\right] \qquad (2.315)$$

Note that

$$u_* \equiv \sqrt{\tau_w/\varrho} = u_{avg}\sqrt{f/2} \qquad (2.316)$$

thus, Equation (2.315) can be simplified to the von Karman friction factor relation, i.e.,

$$\frac{1}{\sqrt{f}} = 4.07 \log(Re\sqrt{f}) - 0.601 \qquad (2.317)$$

This result is quite close to the empirical result of NIKURADSE [2.35] given by

$$\frac{1}{\sqrt{f}} = 4.0 \log(Re\sqrt{f}) - 0.40 \qquad (2.318)$$

Both of these results are for smooth tubes and are generally used at elevated Reynolds numbers ($> 10^5$).

For *rough tubes,* the effects of relative roughness e/d ($e \approx$ height of wall protuberances) must be included. For Reynolds numbers above the value $Re\sqrt{f} = 0.01\, d/e$, the Colebrook equation can be used to determine f [2.39]:

$$\frac{1}{\sqrt{f}} = -4 \log\left(\frac{4.67}{Re\sqrt{f}} + \frac{e}{d}\right) + 2.28 \qquad (2.319)$$

The Colebrook equation reduces to the Nikuradse result when $e/d = 0$.

The degree of roughness of a tube is not always easy to determine accurately. As a result, pressure drops predicted by substituting the above results into Equation (2.80) can often involve considerable error because of uncertainties in estimating e/d. This is particularly true for tubing susceptible to scaling and corrosion and tubing that has been in use for some time.

In addition to the above analogies, another result can be obtained from the boundary layer results of Equations (2.235) and (2.255). These latter equations can be combined to give

$$St\, Pr^{2/3} = f/2 \qquad (2.320)$$

where f has been substituted for C_{Dx}. This equation is identical with the Reynolds analogy (Eq. 2.312) when $Pr = 1$. Based on this agreement at $Pr = 1$, COLBURN suggested the use of Equation (2.320) at higher Prandtl numbers even for turbulent flow (Colburn analogy). Although it is strictly valid only for laminar flow, this simple relation has been found to be a useful approximation for turbulent flow applications particularly, where there is no form drag and where $0.5 < Pr < 50$.

The Colburn analogy can be put in another form:

$$j_H = \frac{f}{2} \qquad (2.320\,a)$$

where

$$j_H \equiv St\, Pr^{2/3} \qquad (2.320\,b)$$

is the *Colburn j factor* for heat transfer. The mass transfer analogy of j_H will be discussed later.

2.5.5. Turbulent Momentum and Heat Transfer on a Flat Plate

For a turbulent boundary layer on a flat plate, the wall shear stress can be related to the free stream velocity u_∞ and the turbulent boundary layer thickness $\delta_m^{(t)}$ by the Blasius expression:

$$\tau_w = 0.0225\, \varrho\, u_\infty^2 \left(\frac{v}{u_\infty \delta_m^{(t)}}\right)^{1/4} \qquad (2.321)$$

This equation follows from Equation (2.314) if $d = 2r_1 \to 2\delta_m^{(t)}$, $u_{max} \to u_\infty$, and $u_{avg}/u_{max} \approx 0.8$. It is valid for flat plate Reynolds numbers up to 10^7. In terms of the drag coefficient C_{Dx}, this equation can be written as

$$C_{Dx} \equiv \frac{\tau_w}{\frac{1}{2}\varrho u_\infty^2} = \frac{0.0450}{Re_x^{1/4}} \left(\frac{x}{\delta_m^{(t)}}\right)^{1/4} \qquad (2.322)$$

To use Equations (2.321) and (2.322), the turbulent boundary layer thickness $\delta_m^{(t)}$ must be determined. If Equation (2.287) is used with $n = 7$, $r_1 \to \delta_m^{(t)}$, and $u_{max} \to u_\infty$ in the time-smoothed momentum integral balance (the same as Eq. 2.226, but with the dependent variables replaced by time-smoothed quantities),

$$\frac{\delta_m^{(t)}}{x} = \frac{0.371}{Re_x^{1/5}} \qquad (2.323)$$

and

$$C_{Dx} = \frac{0.0577}{Re_x^{1/5}} \qquad (2.324)$$

which are valid up to a local Reynolds number of 10^7. In determining $\delta_m^{(t)}$, the turbulent boundary layer has been assumed to extend to the leading edge of the plate ($\delta_m^{(t)} = 0$ at $x = 0$). Obviously, this can lead to considerable error, particularly because the laminar boundary layer that exists up to $x_{cr} = 2 \times 10^5$ (v/u_0) has a square-root dependence on x compared to the 4/5 power dependence that results from Equation (2.323). Regardless, if Equation (2.324) is used in the Colburn analogy (with C_{Dx} replacing f and St_x replacing St)

$$Nu_x = 0.0289\, Re_x^{4/5}\, Pr^{1/3} \qquad (2.325)$$

is obtained. Given the restrictions on Equation (2.322), this equation is limited to $Re_x < 10^7$.

Based on the experimental work of ZHUKAUSKAS and AMBRAZYAVICHYUS [2.40], WHITAKER [2.1, pp. 333–336] suggests the use of

$$Nu_x = 0.029\, Re_x^{4/5}\, Pr^{0.43} \left(\frac{\eta_\infty}{\eta_w}\right)^{1/4} \qquad (2.326)$$

instead of Equation (2.325), where all properties are evaluated at the free stream temperature T_∞. This equation is similar to Equation (2.325), except for the power dependence on the Prandtl number and the addition of a temperature-dependent viscosity effect. Remember that the origin of the 1/3 power dependence on Pr in Equation (2.325) was from laminar boundary theory. In extending this result to turbulent boundary layers, some modifications would be expected. The 1/4 power dependence on (η_∞/η_w) is different from that for tube flow where a 0.14 dependence was found. Although such differences are justified on the basis of available data, relatively small experimental errors could cause these differences. As a result, some uncertainty must be associated with the power dependences on Pr and (η_∞/η_w).

For a plate of length L, the overall heat-transfer coefficient can be determined from

$$h_L = \frac{1}{L}\left\{\int_0^{x_{cr}} h_{x,\,lam}\, dx + \int_{x_{cr}}^L h_{x,\,turb}\, dx\right\} \qquad (2.327)$$

Using a modified form of Equation (2.252) for $\alpha_{x,\text{lam}}$ [Eq. 2.252 with $(\eta_\infty/\eta_w)^{1/4}$ as a multiplier], and Equation (2.326) for $h_{x,\text{turb}}$ gives

$$Nu_L = 0.0361\, Re_L^{4/5}\, Pr^{0.43} \left(\frac{\eta_\infty}{\eta_w}\right)^{1/4}$$

$$\cdot \left\{1 - 17\,400\, Re_L^{-4/5}\left(1 - \frac{0.473}{Pr^{0.1}}\right)\right\} \quad (2.328)$$

This equation assumes that the transition Reynolds number is given by $Re_{\text{cr}} \equiv (u_\infty x_{\text{cr}}/\nu) = 2 \times 10^5$. For sufficiently large values of Re_L where $Re_{\text{cr}}/Re_L \ll 1$, Equation (2.328) can be simplified to

$$Nu_L = 0.0361\, Re_L^{4/5}\, Pr^{0.43} \left(\frac{\eta_\infty}{\eta_w}\right)^{1/4} \quad (2.329)$$

As with Equation (2.326), all properties in Equations (2.328) and (2.329) are evaluated at the free stream conditions. These equations are valid for $Re_L < 10^7$, Pr between 0.7 and 380, and viscosity ratios between 0.26 and 3.5.

WHITAKER found that the use of Equation (2.328) versus Equation (2.329) depends on the degree of turbulence in the free stream. When the latter is high, the transition Reynolds number is less than 2×10^5 and Equation (2.329) tends to be valid. When the free stream turbulence is low, Equation (2.328) represents experimental observations more accurately. In addition to this effect of free stream turbulence, the roughness of the plate also has a significant effect on the transition Reynolds number. In particular, very smooth plates can exhibit transition values considerably above 2×10^5. Because such effects are difficult to characterize in many applications, some degree of uncertainty is associated with the use of Equations (2.328) or (2.329).

If the fluid properties are evaluated at the film temperature $T_f \equiv (T_w + T_\infty)/2$ instead of T_∞, the (η_∞/η_w) dependence in Equations (2.328) and (2.329) can be dropped. Such differences are noted in Table 7.

To calculate the overall drag coefficient on a plate of length L that extends into the turbulent boundary layer region, we use

$$C_{DL} = \frac{1}{L}\left\{\int_0^{x_{\text{cr}}} C_{Dx,\text{lam}}\, dx + \int_{x_{\text{cr}}}^L C_{Dx,\text{turb}}\, dx\right\} \quad (2.330)$$

Using Equations (2.235) and (2.324) and assuming that $Re_{\text{cr}} = 2 \times 10^5$ give

$$C_{DL} = 0.072\, Re_L^{-1/5}(1 - 9200\, Re_L^{-4/5}) \quad (2.331)$$

This equation is valid for $Re_L < 10^7$. At higher Reynolds numbers, the result of PRANDTL can be used [2.41]:

$$C_{DL} = \frac{0.455}{[\log Re_L]^{2.58}} \quad (2.332)$$

which is valid for Re_L from 10^6 to 10^9.

2.6. Summary of Heat-Transfer Relations

In Table 7, heat-transfer and friction (or drag) coefficient results are summarized for various geometries. They include the flat plate and tube flow results presented here, as well as results for heat transfer around cylinders and spheres. In addition, free convection results are provided for several common situations.

3. Transport in Multicomponent Systems

Multicomponent transport processes require consideration of the fact that to the extent mixture compositions are not uniform, diffusion (the motion of species of the mixture relative to each other) will be nonzero. In binary systems, diffusion processes can be analogous to heat transfer. Multicomponent systems are considerably more complex, however. Useful references on diffusion and general transport references are given in [3.1]–[3.5]. An extensive review of methods for obtaining diffusion coefficients appears in [3.6].

3.1. Diffusive Mass Transfer in Binary Systems

3.1.1. Species Mass Balances

In considerations of single-component (compositionally homogeneous) systems the equation of continuity (conservation of total mass)

$$\frac{\partial \varrho}{\partial t} = -\nabla \cdot (\varrho\, \mathbf{u}) \quad (3.1)$$

plays a vital role in establishing relationships between components of velocity. In this form,

Table 7. Summary of heat-transfer and friction coefficient results (all properties evaluated at $T_f \equiv (T_w - T_\infty)/2$ unless otherwise noted)

Situation	Equation	Restrictions	Equation number () or reference []
Flow over Flat Plates			
	Heat-transfer coefficients		
Laminar, local	$Nu_x = 0.332\, Re_x^{1/2}\, Pr^{1/3}$	constant wall temperature; $Re_x < 2 \times 10^5$, $0.6 < Pr < 50$	(2.252)
Laminar, local	$Nu_x = 0.453\, Re_x^{1/2}\, Pr^{1/3}$	constant wall heat flux; $Re_x < 2 \times 10^5$, $0.6 < Pr < 50$	(2.256)
Laminar, overall	$Nu_L = 0.664\, Re_L^{1/2}\, Pr^{1/3}$	constant wall temperature; $Re_L < 2 \times 10^5$, $0.6 < Pr < 50$	(2.255)
Turbulent, local	$Nu_x = 0.029\, Re_x^{4/5}\, Pr^{0.43}\, (\eta_\infty/\eta_w)^{1/4}$	constant wall temperature; $2 \times 10^5 < Re_x < 10^7$, $0.7 < Pr < 380$; properties evaluated at T_∞	
Turbulent, overall	$Nu_L = 0.0361\, Re_L^{4/5}\, Pr^{0.43}\, (\eta_\infty/\eta_w)^{1/4}$ $\times \left[1 - 17400\, Re_L^{-4/5} \left(1 - \dfrac{0.473}{Pr^{0.1}} \right) \right]$	constant wall temperature; $2 \times 10^5 < Re_L < 10^7$, $0.7 \leq Pr \leq 380$; relatively low free stream turbulence; properties evaluated at T_∞	(2.328)
Turbulent, overall	$Nu_L = 0.0361\, Re_L^{4/5}\, Pr^{0.43}\, (\eta_\infty/\eta_w)^{1/4}$	same as above except relatively high free stream turbulence	(2.329)
	Drag coefficients		
Laminar, local	$C_{Dx} = 0.664\, Re_x^{-1/2}$	$Re_x < 2 \times 10^5$	(2.235)
Laminar, overall	$C_{DL} = 1.328\, Re_L^{-1/2}$	$Re_L < 2 \times 10^5$	(2.236)
Turbulent, local	$C_{Dx} = 0.0577\, Re_x^{-1/5}$	$2 \times 10^5 \leq Re_x < 10^7$	(2.324)
Turbulent, overall	$C_{DL} = 0.072\, Re_L^{-1/5}\, (1 - 9200\, Re_L^{-4/5})$	$2 \times 10^5 \leq Re_L < 10^7$	(2.331)
Flow in Tubes			
(Properties evaluated at average bulk mean temperature between $z = 0$ and $z = L$ unless otherwise noted)			
	Heat-transfer coefficients		
Fully developed laminar	$Nu_d = 3.658$	$T_w = \text{const.}$; $Re_d < 2100$; $z > 0.05\, d\, Re_d\, Pr$	(2.203)
Fully developed laminar	$Nu_d = 4.364$	$q_w = \text{const.}$; $Re_d < 2400$; $z > 0.05\, d\, Re_d\, Pr$	(2.202)
Developing laminar	$Nu_{d,\text{lm}} = 1.86\, Re_d^{1/3}\, Pr^{1/3}\, (L/d)^{-1/3}\, (\eta_\infty/\eta_w)^{0.14}$	$T_w = \text{const.}$; $Re_d < 2400$; intermediate-length tubes	(2.205)
Developing laminar	$Nu_{d,\text{lm}} = 3.66 + \dfrac{0.0745\, \psi^3}{1 + 0.04\, \psi^2}$ where $\psi \equiv (Re_d\, Pr)^{1/3}\, (L/d)^{-1/3}\, (\eta_{\text{bm}}/\eta_w)^{0.14}$	$Re_d < 2400$	(2.209) (2.210)
Turbulent	$St \equiv \dfrac{Nu_d}{Re_d\, Pr} = \dfrac{f/2}{1 + 5\sqrt{f/2}\, \{Pr - 1 + \ln[1 + \tfrac{5}{6}(Pr - 1)]\}}$	$Re_d > 2400$	(2.313)
Turbulent	$Nu_{d,\text{lm}} = 0.015\, Re_d^{0.83}\, Pr^{0.42}\, (\eta_{\text{bm}}/\eta_w)^{0.14}$	$Re_d > 4000$; $0.4 < Pr < 600$	[2.42]
	Friction coefficients		
Fully developed laminar	$f = 16/Re_d$	$Re_d < 2100$; $z > 0.05\, Re_d\, d$	(2.87)
Turbulent	$f = 0.0791\, Re_d^{-1/4}$	$2100 \leq Re_d \leq 10^5$; smooth tubes	(2.314)
Turbulent	$\dfrac{1}{\sqrt{f}} = -4 \log\left(\dfrac{4.67}{Re_d\, \sqrt{f}} + \dfrac{e}{d} \right) + 2.28$	$Re_d > 10^5$; smooth ($e = 0$) and rough tubes	(2.319)

Table 7. (continued)

Situation	Equation	Restrictions	Equation number () or reference []
Flow Around Bodies	*Heat-transfer coefficients*		
Flow normal to cylinder	$Nu_d = 0.3 + \dfrac{0.62\, Re_d^{1/2}\, Pr^{1/3}}{[1 + (0.4/Pr)^{2/3}]^{3/4}} \cdot \left[1 + \left(\dfrac{Re_d}{28200}\right)^{5/8}\right]^{4/5}$	$10^2 < Re_d < 4 \times 10^6$; $Re_d \equiv u_\infty d/\nu$; properties evaluated at film temperature	[2.43]
Flow past a sphere	$Nu_d = 2 + (0.4\, Re_d^{1/2} + 0.06\, Re_d^{2/3}) \cdot Pr^{0.4} (\eta_\infty/\eta_w)^{1/4}$	$3.5 < Re_d < 8 \times 10^4$; $0.7 < Pr < 380$; properties evaluated at T_∞	[2.44]
Free Convection from Various Geometries			
Vertical surface, overall	$Nu_L = 0.68 + \dfrac{0.67\, Ra_L^{1/4}}{\left[1 + \left(\dfrac{0.492}{Pr}\right)^{9/16}\right]^{4/9}}$	$Ra_L < 10^9$	[2.45]
Vertical surface, overall	$Nu_L = \left\{0.825 + \dfrac{0.387\, Ra_L^{1/6}}{[1 + (0.492/Pr)^{9/16}]^{8/27}}\right\}^2$	$T_w = $ const.; $10^{-1} < Ra_L < 10^{12}$	[2.45]
Vertical surface, laminar, local	$Nu_x = 0.60\, Ra_x^{1/5}$	$q_w = $ const.; $10^5 < Gr_x Nu_x < 10^{11}$	[2.46]
Vertical surface, laminar, local	$Nu_x = 0.17\, Ra_x^{1/4}$	$q_w = $ const.; $2 \times 10^{13} < Gr_x Nu_x Pr < 10^{16}$	[2.46]
Horizontal cylinder, laminar, overall	$Nu_d = 0.36 + \dfrac{0.518\, Ra_d^{1/4}}{[1 + (0.559/Pr)^{9/16}]^{4/9}}$	$T_w = $ const.; $10^{-6} \leq Ra_d \leq 10^9$	[2.46]
Horizontal cylinder, overall	$Nu_d = \left\{0.60 + 0.387\left[\dfrac{Ra_d}{[1 + (0.559/Pr)^{9/16}]^{16/9}}\right]^{1/6}\right\}^2$	$10^{-5} \leq Ra_d \leq 10^{12}$	[2.46]
Horizontal plate, downward facing hot plates or upward facing cold plates; overall	$Nu_L = 0.58\, Ra_L^{1/3}$	$T_w = $ const.; $10^6 < Ra_L < 10^{11}$	[2.47]
Horizontal plate, upward facing hot plates or	$Nu_L = 0.13\, Ra_L^{1/3}$	$T_w = $ const.; $Ra_L < 2 \times 10^8$	[2.47]
downward facing cold plates; overall	$Nu_L = 0.16\, Ra_L^{1/3}$	$T_w = $ const.; $2 \times 10^8 \leq Ra_L < 10^{11}$	[2.47]

$(\partial \varrho/\partial t)_x$ physically represents the accumulation rate per unit volume of mass; the expression $(-\nabla \cdot \varrho \boldsymbol{u})$ represents the net input rate per unit volume to the volume fixed in space as a result of the mass flux $\varrho \boldsymbol{u}$.

An analogous result can be obtained for an individual species (e.g., species A), however, since individual chemical species are not conserved (except for the elements), the net production of A due to reaction must be considered. Accordingly, an equation accounting for the amount of species A is given by

$$\frac{\partial \varrho_A}{\partial t} = -\nabla \cdot (\varrho_A \boldsymbol{u}_A) + r_A \qquad (3.2)$$

where r_A is the rate of production of species A per unit volume in the mixture, ϱ_A is the mass of species A per unit volume of the mixture, and $\boldsymbol{u}_A$ is the average velocity of the molecules of species A in the mixture.

This equation can also be expressed in terms of molar concentration by dividing by molecular mass to give

$$\frac{\partial c_A}{\partial t} = -\nabla \cdot (c_A \boldsymbol{u}_A) + R_A \qquad (3.3)$$

where c_A is the number of moles per volume of species A in the mixture and R_A is the molar rate of production of species A by reaction per unit volume of the mixture. The velocity in this equation is the same as in the preceding equation.

An equation of this sort in either mass or molar form can be written for each species in the mixture. When all the equations for all the species of an n-component mixture are added together, the following result is obtained:

$$\sum_{i=1}^{n} \frac{\partial \varrho_i}{\partial t} = -\sum_{i=1}^{n} \nabla \cdot (\varrho_i \boldsymbol{u}_i) + \sum_{i=1}^{n} r_i \qquad (3.4)$$

which, with interchange of the summations and derivatives, is

$$\frac{\partial \sum_{i=1}^{n} \varrho_i}{\partial t} = -\nabla \cdot \left(\sum_{i=1}^{n} \varrho_i \boldsymbol{u}_i \right) + \sum_{i=1}^{n} r_i \qquad (3.5)$$

Now, by definition, the sum of the individual species' densities (mass per unit volume of the mixture) is the total mass per unit volume of the mixture, which is customarily defined as ϱ to be consistent with the notation for compositionally homogeneous systems. Furthermore, the sum of all the individual species production rates by reaction must be zero, when expressed in mass units, because total mass is conserved (i.e., no net generation of total mass occurs). Consequently, the equation for the total mixture becomes

$$\frac{\partial \varrho}{\partial t} = -\nabla \cdot (\sum \varrho_i \boldsymbol{u}_i) \qquad (3.6)$$

Because of the similarity of this equation to the total mass conservation equation, the sum of the individual species fluxes is commonly defined as the product of the total mass density times an average velocity for the mixture. In fact, this becomes a defining relation for what is referred to as the mixture *mass average velocity* $\boldsymbol{u}$

$$\varrho \boldsymbol{u} \equiv \sum_{i=1}^{n} \varrho_i \boldsymbol{u}_i \qquad (3.7)$$

or

$$\boldsymbol{u} \equiv \frac{\sum_{i=1}^{n} \varrho_i \boldsymbol{u}_i}{\sum_{i=1}^{n} \varrho_i} \qquad (3.8)$$

Thus, in a mixture, different species may move at different velocities and the mixture velocity is defined as a weighted average of these individual velocities.

A parallel set of equations exists for molar concentrations. Accordingly, the sum of the molar concentrations of the individual species gives

$$\sum_{i=1}^{n} \frac{\partial c_i}{\partial t} = -\sum_{i=1}^{n} \nabla \cdot (c_i \boldsymbol{u}_i) + \sum_{i=1}^{n} R_i \qquad (3.9)$$

Again the summation and derivatives can be interchanged and the sum of the individual species concentrations (number of moles of species i per unit volume of the mixture) gives the total number of moles of the mixture per unit volume, customarily denoted as c. Accordingly, an accounting equation for the total number of moles in the mixture is

$$\frac{\partial c}{\partial t} = -\nabla \cdot \sum c_i \boldsymbol{u}_i + \sum R_i \qquad (3.10)$$

In terms of moles, however, the net number of moles produced by reaction is not necessarily zero; the total number of moles due to reaction is not conserved. Consequently, this total molar production rate must be included in the total molar accounting equation for the mixture. However, for convenience, a mixture average velocity is defined such that the summation of the individual species fluxes represents a mixture average flux in terms of the total molar concentration and the mixture *molar average velocity*. Accordingly,

$$c\tilde{\boldsymbol{u}} \equiv \sum_{i=1}^{n} c_i \boldsymbol{u}_i \qquad (3.11)$$

or, equivalently,

$$\tilde{\boldsymbol{u}} \equiv \frac{\sum_{i=1}^{n} c_i \boldsymbol{u}_i}{\sum_{i=1}^{n} c_i} \qquad (3.12)$$

This molar average velocity is not the same as the mass average velocity (Eqs. 3.7 and 3.8) because the individual species are not necessarily of the same molecular mass. The definition of a molar average velocity for the mixture allows the molar accounting equation to be written in the form

$$\frac{\partial c}{\partial t} = -\nabla \cdot (c\tilde{\boldsymbol{u}}) + \sum_{i=1}^{n} R_i \qquad (3.13)$$

Again, the net number of moles produced by reaction is not necessarily zero as it was for mass, and as a result, this term must remain in the equation. Mass and molar accounting equations are thus obtained for the individual species, along with statements of the conservation of total mass for the mixture. Now diffusion must be defined for the individual species and methods for calculating diffusion fluxes must be introduced.

3.1.2. Species Diffusion Fluxes

The preceding discussion recognized that each species may be characterized by its own velocity which, in general, is different from the average velocity of the mixture. Furthermore, the mixture average velocity may be defined in terms of mass or in terms of number of moles, and the two velocities in general are different. As a result, each species has a relative motion with respect to the mixture average motion.

This difference in fluxes between each species relative to the mixture average is referred to as diffusion. Because two mixture average velocities were defined, one based on number of moles and one based on mass, two different diffusion fluxes exist for each species, one with respect to the mixture molar average velocity and one with respect to its mass average velocity.

Accordingly, the molar diffusion flux $\tilde{J}_A$ of species A can be defined as the difference between the total flux of species A and that due to the mixture motion on the average [3.1]:

$$\tilde{J}_A \equiv c_A u_A - x_A(c\tilde{u}) = c_A u_A - c_A \tilde{u} \qquad (3.14)$$

where x_A is the mole fraction of species A. Using this definition of a diffusion flux in the accounting equation for species A gives

$$\frac{\partial c_A}{\partial t} = -\nabla \cdot (c_A \tilde{u}) - \nabla \cdot \tilde{J}_A + R_A \qquad (3.15)$$

Here, the accumulation rate of the number of moles of A results from the molar flux of A due to the average flow of the mixture, plus the molar diffusion flux of species A above and beyond this mixture molar average flow, plus the molar rate of production of A per unit volume.

Similarly, a diffusion flux can be defined in terms of mass fluxes, i.e., the difference between the total mass flux of species A and the mass flux of species A that results from the mixture moving at its mass average velocity:

$$j_A \equiv \varrho_A u_A - \varrho_A u \qquad (3.16)$$

This mass diffusion flux used in the mass accounting equation for species A gives

$$\frac{\partial \varrho_A}{\partial t} = -\nabla \cdot (\varrho_A u) - \nabla \cdot j_A + r_A \qquad (3.17)$$

Again the accumulation rate of species A at a point in space results from the mass flux of species A due to the bulk flow of the fluid (the mass average velocity of the mixture) plus the diffusion flux of species A plus the mass rate of production (per unit volume) of A by reaction.

In addition to the molar diffusion flux of A and the mass diffusion flux of A as defined above, a molar diffusion flux of A could be defined relative to the mass average velocity according to the equation

$$J_A \equiv c_A u_A - c_A u \qquad (3.18)$$

or a mass diffusion flux of A could be defined relative to the molar average velocity:

$$\tilde{j}_A \equiv \varrho_A u_A - \varrho_A \tilde{u} \qquad (3.19)$$

Neither of these is as commonly used as the first two, so they are not considered further here. The point of this discussion is that when talking about diffusion and diffusion coefficients the velocity used to characterize the mixture must be specified. If the molar average velocity is used, one value for diffusion is appropriate; if the mass average velocity is used, another (different) diffusion is described.

3.1.3. Fick's First Law of Diffusion

Until now the transport equations presented in this section have been truly general for multicomponent (as opposed to binary) systems. The species accounting equations and the mixture mass conservation equations are correct, independent of whether binary or multicomponent systems are involved. Likewise, the definitions of diffusion flux are still valid for multicomponent systems.

However, at this point a relationship is defined between diffusion driving forces and species fluxes. The relationship to be discussed is restricted to binary systems and is known as Fick's first law of diffusion. This law is a linear transport equation relating mass transfer or diffusion driving forces to the diffusion flux. Accordingly, in a binary system the driving force for diffusion (i.e., the reason different species move with different velocities) is the compositional inhomogeneity (the existence of mole fraction gradients). (Pressure and temperature gradients may also induce diffusion [3.1]. These effects are normally small and are not considered further here.) In the presence of such gradients, species within the mixture move in such a way as to remove these mole fraction gradients, a motion that occurs spontaneously (implying an increase in entropy of the mixture).

In accordance with a linear relation between this driving force and diffusion flux, *Fick's first law of diffusion* is written as

$$\tilde{J} = -cD_{AB}\nabla x_A \quad (3.20)$$

In this equation, ∇x_A is the driving force, and the product of the total mixture molar concentration and the diffusivity is the proportionality factor required to convert this driving force to a molar flux. The minus sign indicates that the diffusion flux is in the direction opposite to the direction of increasing mole fraction. The diffusivity is a parameter that describes how easily species A moves through the rest of the mixture, species B—consequently, the notation D_{AB}.

Similarly, the mass diffusion flux of species A can be written in terms of a mass fraction gradient with the proportionality factor being the product of the mixture mass density and the same diffusion coefficient as for the molar diffusion flux. A priori, the same diffusivity might not be expected to appear in both the molar and mass diffusion flux equations. However, this must be the case for binary systems. Furthermore $D_{AB} = D_{BA}$ in a binary system, because $x_A = 1 - x_B$ and $\nabla x_A = -\nabla x_B$, and the sum of the two species diffusion fluxes must be zero. Consequently, for binary mixtures a single diffusivity needs to be reported for any given mixture composition of A and B.

3.1.4. Special Cases of Diffusion

Using the relations for diffusion molar and mass fluxes (Fick's first law, Section 3.1.3) and the accounting equations presented in Section 3.1.2 yields the general results that account for species A in either moles or mass. Accordingly, in molar units

$$\frac{\partial c_A}{\partial t} = -\nabla \cdot (c_A \tilde{u}) + \nabla \cdot (cD_{AB}\nabla x_A) + R_A \quad (3.21)$$

or in mass units

$$\frac{\partial \varrho_A}{\partial t} = -\nabla \cdot (\varrho_A u) + \nabla \cdot (\varrho D_{AB}\nabla w_A) + r_A \quad (3.22)$$

where w_A is the mass fraction of A in the mixture. These equations are quite general within the restrictions of Fick's first law. They do not assume constancy of total molar concentration or of diffusivities.

Three special cases are of particular interest [3.1].

Case 1. If both the mixture mass density ϱ and D_{AB} are constant, then the following result is obtained:

$$\frac{\partial \varrho_A}{\partial t} = -u \cdot \nabla \varrho_A + \varrho D_{AB}\nabla^2 w_A + r_A \quad (3.23)$$

which may be divided by molecular mass to yield a molar accounting equation

$$\frac{\partial c_A}{\partial t} = -u \cdot \nabla c_A + cD_{AB}\nabla^2 x_A + R_A \quad (3.24)$$

This result, which assumes that the divergence of the velocity is zero (the mixture mass density is constant), is appropriate for dilute liquid solutions at constant temperature and pressure. This particular form of the equation is also expressed in terms of mixed quantities; the concentration is given in terms of molar concentration, but the mixture average velocity is the mass average velocity.

Case 2. If the mixture total molar concentration c and D_{AB} are constant, the mixture continuity equation gives

$$\nabla \cdot \tilde{u} = \frac{R_A + R_B}{c} \quad (3.25)$$

and the accounting equation for species A becomes

$$\frac{\partial c_A}{\partial t} = -\tilde{u} \cdot \nabla c_A + cD_{AB}\nabla^2 x_A$$
$$+ R_A(1 - x_A) - x_A R_B \quad (3.26)$$

This result is appropriate for low-density gases at constant temperature and pressure.

Case 3. Furthermore, if no chemical reaction occurs, if (for liquids) the mixture mass density is constant, and if the mixture mass average velocity is zero or (for gases) the mixture molar concentration is constant and the mixture molar average velocity is zero (i.e., no bulk motion of the fluid occurs), then Equations (3.25) and (3.26) reduce to the form

$$\frac{\partial c_A}{\partial t} = D_{AB} \nabla^2 c_A \tag{3.27}$$

This result is known as *Fick's second law of diffusion* and states that the accumulation of species A at a point in space results solely from the diffusion flux of species A; no convective transfer occurs, nor does any generation or consumption as a result of chemical reactions. This result is also called the diffusion equation and is identical to the partial differential equations (1) that are obtained for heat conduction in solids of constant heat capacity, thermal conductivity, and density; and (2) that result from the equation of motion for a fluid of constant density and viscosity near a wall suddenly set in motion. This type of agreement between different modes of transport provides the basis for analogous correlations between certain heat-, mass-, and momentum-transfer problems. That is, the same mathematical equations can govern these three different modes of transport. Then, with analogous geometrical systems and boundary conditions, exactly the same mathematical problem exists, resulting in indentical mathematical solutions. Consequently, the results of many diffusion problems are available in the heat-transfer literature, e.g., [3.5], [3.7].

3.1.5. Diffusivities of Gases and Liquids

Gases. The diffusivities of gases can be estimated quite well by using molecular theory (similar to viscosity). Diffusion calculations are not complicated by multiple internal energy modes as in the case of thermal conductivity, although the presence of two different molecules must be taken into account. Accordingly, the Chapman–Enskog equation for diffusivity (in square centimeters per second) in a binary mixture of gases is

$$D_{AB} = 1.883 \times 10^{-3} T^{3/2} \frac{[(M_{r,A} + M_{r,B})/M_{r,A} M_{r,B}]^{1/2}}{p \sigma_{AB}^2 \Omega_D} \tag{3.28}$$

where p is in bar, σ_{AB} is in ångströms, and T is in kelvin. This result incorporates a molecular potential energy function such as the Lennard–Jones 12–6 function, which may be used for mixtures of nonpolar species. For a binary mixture the characteristic separation distance σ_{AB} is an arithmetic average of the values for the two pure species. Also, the collision integral Ω_D is a function of kT/ε_{AB}, where ε_{AB} is the geometric mean of the values for the two species. As for viscosity, values of the collision integral are tabulated or may be calculated by using a relation given by NEUFELD et al. [3.8] and presented by REID et al. [3.6]. However, the values for calculating diffusivities are different from those used for viscosity. For binary mixtures with at least one polar species, a different potential function may be used (along with its own values for the molecular parameters) to calculate a different collision integral. Alternatively, the Lennard–Jones collision integral may be modified by an additive contribution due to molecular polarity as manifested by the dipole moment [3.9].

Diffusivities for low-density gases can also be estimated by using empirical relations built around the basic form of the Chapman–Enskog relation. The relation of FULLER et al. [3.10] is applicable to both polar and nonpolar molecules and is similar to a group contribution method wherein the different structural units of a molecule are assigned specific diffusion volume increments. The total diffusion volume for each molecule is calculated and used to determine the binary diffusivity. Finally, corresponding states methods also are available for estimating gas diffusivities.

Liquids. Theories for estimating liquid diffusivities are not nearly as successful as for gases because of the much smaller separation distances between molecules, which result in multibody interactions and magnified interaction forces. The Wilke–Chang method is probably the most widely used relation for estimating the diffusivity (in square centimeters per second) of a solute A at infinite dilution in a solvent B [3.11]:

$$D_{AB}^0 = 7.4 \times 10^{-11} \frac{(\varphi M_{r,B})^{1/2} T}{\eta_B V_A^{0.6}} \tag{3.29}$$

Accordingly, $M_{r,B}$ is the molecular mass of the solvent; η_B the solvent viscosity, in pascal seconds, and D_{AB}^0 the diffusivity when the solute is at infinite dilution in the solvent. Also, an association factor φ accounts for the degree of association exhibited by the solvent. It is 1.0 for nonassociated solvents and increases with association: 1.5 for ethanol, 1.9 for methanol, and 2.6 for water. Other methods are available and are summarized and compared in [3.6]; see also → 6. Estimation of Physical Properties, pp. 6-50 – 6-52.

3.1.6. Theoretical Foundations of Steady-State Measurement of Diffusion

To understand the techniques for measurement of diffusion, two situations are addressed with respect to gas-phase diffusion that occur as idealized limiting cases. The first is equimolar counterdiffusion (ECD) in which the two species of a binary mixture have exactly equal and opposite molar fluxes. The second is the diffusion of one species A through a stagnant film of a second species B (ATSB); the flux of one species is not zero whereas the other is exactly zero. Both situations are considered to be occurring at steady state. Since they are gas-phase diffusion problems at constant pressure the equations are given in molar form. In each case analysis involves using a continuity equation for each species (Eq. 3.3), a diffusion flux equation giving the total flux of a species as the sum of that due to the mixture molar average flux plus that due to diffusion (Eq. 3.14), and finally Fick's first law of diffusion (Eq. 3.20).

Equimolar Counterdiffusion. The term equimolar counterdiffusion means that the two species in the binary mixture diffuse at exactly the same molar flow rates per unit area but in opposite directions. A diffusion cell schematic that approximates ECD is depicted in Figure 21.

This is a one-dimensional problem for which the mixture molar average velocity (Eq. 3.11) is zero; hence, the flux of either species occurs only because of its diffusion. If N_A is defined as the total molar flux of species A, from Equation (3.14)

$$N_A = c_A u_A = c_A \tilde{u} + \tilde{J}_A = \tilde{J}_A \quad (3.30)$$

or in one dimension using Fick's first law

$$N_{Az} = -c D_{AB} \frac{dy_A}{dz} \quad (3.31)$$

Now, an accounting of species A (the continuity equation, Eq. 3.3) with no reactions and at steady state gives

$$\nabla \cdot N_A = 0 \quad (3.32)$$

which in one-dimensional form is

$$\frac{dN_{Az}}{dz} = 0 \quad (3.33)$$

and from which the molar flux of species A (and similarly of species B) is seen to be invariant with respect to position in the direction of flow.

Consequently, Equation (3.31) can be integrated directly (for constant total concentration and diffusivity) to give a linear relation for the concentration of species A (and hence also of species B) with respect to position in the direction of flow

$$y_A(z) = y_{A0} + \left(\frac{y_{AL} - y_{A0}}{L}\right) z \quad (3.34)$$

where y_{A0} and y_{AL} are the mole fractions of A at $z = 0$ and $z = L$, respectively. The molar flux in terms of diffusivity and concentrations at the ends of the diffusion tube is

$$N_{Az} = \frac{-c D_{AB}}{L} (y_{AL} - y_{A0}) \quad (3.35)$$

This result then can be used to measure diffusivities if the tube-end concentrations are known as a function of time. The above derivation assumes a steady state, i.e., the two end-cell concentrations do not change with time. The result can still be used if the changes in concentration are slow compared to the establishment of the concentration profiles in the diffusion cell. By using this pseudosteady-state assumption the changes in concentration at the two cell ends can be related to the flux through the cell by writing a mass accounting balance for species A in each cell. Accordingly,

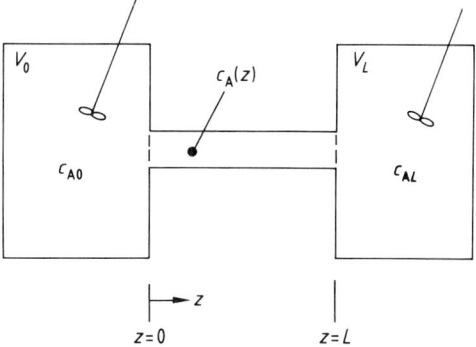

Figure 21. Schematic of an equimolar counterdiffusion (ECD) cell

$$\frac{d}{dt}(c_{A0}V_0) = N_{Az}A_c = -\frac{A_c D_{AB}}{L}(c_{AL} - c_{A0}) \quad (3.36)$$

where A_c is the cross-sectional area through which diffusion occurs and

$$\frac{d}{dt}(c_{AL}V_L) = -N_{Az}A_c = \frac{A_c D_{AB}}{L}(c_{AL} - c_{A0}) \quad (3.37)$$

from which

$$\frac{d}{dt}(c_{A0} - c_{AL}) = -\frac{A_c D_{AB}}{L}(c_{AL} - c_{A0})\left[\frac{1}{V_0} + \frac{1}{V_L}\right] \quad (3.38)$$

so that

$$[c_{AL}(t) - c_{A0}(t)] = [c_{AL}(t_0) - c_{A0}(t_0)]$$
$$\cdot \exp\left\{-\frac{A_c D_{AB}}{L}\left(\frac{1}{V_0} + \frac{1}{V_L}\right)(t - t_0)\right\} \quad (3.39)$$

Measuring the concentration difference between cells as a function of time gives D_{AB}. The diffusion cell itself can consist of a glass frit or even filter paper. In the latter case the cell length is ill-defined because of tortuosity and the coefficient $A_c(1/V_0 + 1/V_L)/L$ is determined by calibration with a material of known diffusivity [3.3].

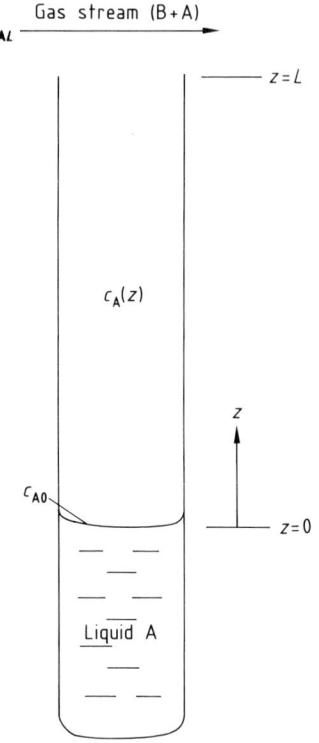

Figure 22. Schematic of "A through stagnant B" (ATSB) diffusion process

ATSB Diffusion. For the diffusion of species A through a stagnant film of B (ATSB), the situation is again taken to be constant total pressure and steady state. Diffusion occurs through a tube of uniform diameter and length L. In this case, however, the motion of one of the species is constrained to be zero as in the evaporation of a liquid A from a tube open at one end (Fig. 22). The second gas-phase component is not soluble in A (or at least A is saturated with B), so its flux at the interface of liquid A with the gas phase is zero. The liquid evaporates at the interface and diffuses to the open end of the tube where its concentration is maintained at a reduced level by a steady flow of fresh gas phase across the open end of the tube.

For steady state with no reactions in the gas phase, the continuity of species A in the gas phase requires (as in the equimolar counterdiffusion problem) that

$$0 = -\nabla \cdot N_A = -\frac{dN_{Az}}{dz} \quad (3.40)$$

Again, the molar flux of A along the length of the tube is constant. Likewise, a similar result is obtained for species B. Because liquid A is saturated with B, the flux of species B into the liquid at the interface must be zero. Then, because this flux is constant along the entire length of the tube, it must be zero everywhere; species B is stagnant. Now, by writing $N_A (= c_A u_A)$ and $N_B (= c_B u_B)$ as the molar fluxes of species A and B, respectively, the combined molar flux for the mixture is given by (see Eq. 3.11)

$$c\tilde{u} = c_A u_A + c_B u_B = N_A + N_B \quad (3.41)$$

which, combined with Equation (3.14), yields

$$N_A = y_A(N_A + N_B) + \tilde{J}_A \quad (3.42)$$

i.e., the total flux of A is that fraction due to the total mixture average flow plus the flow due to the diffusion of species A. With species B stagnant, $N_B = 0$ and

$$N_A = \frac{\tilde{J}_A}{(1 - y_A)} \qquad (3.43)$$

which in one-dimensional form is

$$N_{Az} dz = -c D_{AB} \frac{dy_A}{(1 - y_A)} \qquad (3.44)$$

Because of the constancy of N_{Az} along the length of the tube, this result can be integrated directly to give either a result for the molar flux of A (in terms of tube length, total pressure, diffusivity, and concentrations at the end of the tube)

$$N_{Az} = \frac{c D_{AB}}{L} \ln\left(\frac{1 - y_{AL}}{1 - y_{A0}}\right) \qquad (3.45)$$

or a result for the concentration profile of species A along the length of the tube

$$y_A(z) = 1 - (1 - y_{A0})\left(\frac{1 - y_{AL}}{1 - y_{A0}}\right)^{z/L} \qquad (3.46)$$

Note that, contrary to equimolar counterdiffusion, the concentration profile is nonlinear even though the flux of species A is still constant along the tube length.

Comparison of ATSB and ECD Results. Comparing the ATSB and ECD results shows that for the same mixture concentration c, diffusivity D_{AB}, and driving force for diffusion dy_A/dz, the flux for ATSB is greater than that for ECD due to the fact that the mole fraction of species A is less than unity and hence the driving force is divided by $(1 - y_A)$. This enhancement is perhaps best understood by looking at Equation (3.42) for the total molar flux of A in terms of the bulk flow of the mixture and the diffusion flux. The bulk flow of the mixture is not zero for ATSB, whereas it is zero for ECD. Enhancement occurs in the flux of species A out of the tube as a result of this nonzero bulk flow. Note also that even though B is stagnant, its diffusion flux, through Fick's law, is nonzero because its mole fraction gradient is nonzero. With species B, however, the diffusion flux down the tube toward the liquid in the bottom of the tube is exactly counterbalanced by its flow up the tube because of the bulk flow of the gas mixture. The net effect is that species B is stagnant ($N_B = 0$) even though it is in fact diffusing.

In principle, this experimental situation provides another method for measuring gas-phase diffusivities. In terms of changes in the amount of evaporating liquid in the tube,

$$N_{Az}(t) = \frac{\varrho}{M_r} \frac{dL}{dt} \qquad (3.47)$$

so that Equation (3.45) becomes

$$\frac{\varrho}{M_r} \frac{dL}{dt} = \frac{c D_{AB}}{L} \ln\left(\frac{1 - y_{AL}}{1 - y_{A0}}\right) \qquad (3.48)$$

which can be integrated to give

$$L^2 = \frac{2 c D_{AB} M_r}{\varrho} \left[\ln\left(\frac{1 - y_{AL}}{1 - y_{A0}}\right)\right](t - t_0) + L_0^2 \qquad (3.49)$$

Measuring L as a function of time (by weight) can then be used to determine D_{AB}. This result, like the ECD method, rests upon the pseudosteady-state assumption, in this case that the concentration profile and the flux adjust rapidly to changes in L. Further complications in this technique involve maintaining constant temperature in the presence of evaporation and convection due to differences in the molecular mass of the two gases [3.1].

3.1.7. Diffusion with Homogeneous Chemical Reaction

A situation is now considered that involves chemical reaction in addition to diffusion: the transfer of a component from a gas phase to a liquid film followed by diffusion away from the gas–liquid interface and reaction within the film. At a distance δ from the gas–liquid interface, the film meets an impermeable solid boundary (Fig. 23). As a boundary condition it is assumed that diffusion of the gas to the liquid

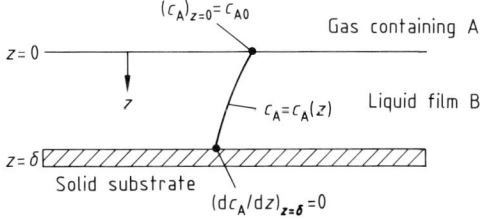

Figure 23. Schematic of diffusion and chemical reaction in a liquid film

film is not limiting so that the concentration of the gaseous material in the liquid film at the gas–liquid interface is that established by equilibrium with the bulk gas-phase concentration. Species A then diffuses away from the gas–liquid interface in an attempt to increase its concentration within the liquid film. However, the chemical reaction serves to deplete the concentration and a nonlinear steady-state concentration profile is achieved. At the solid–liquid interface, the concentration profile must be flat; because of the impermeability of the solid surface, the flux at that surface is zero.

The objective of the analysis is to obtain a relation for the molar flow rate of A within the film and the concentration profile of A across the film. For reactions of species A with the liquid (species B), consider a situation with the reaction given by

$$A + B \rightarrow C \tag{3.50}$$

which is characterized by a rate constant k_1 such that the rate of production of species A is given by the kinetic first-order equation

$$R_A = -k_1 c_A \tag{3.51}$$

($R_A < 0$ means that A is being consumed)

This reaction occurs homogeneously throughout the liquid film in the sense that it takes place everywhere in the film that species A and B exist together. Reaction rate varies with position because of variation of the concentration of A with position z through the film.

As usual, we begin with the equation of continuity for A which is

$$\frac{\partial c_A}{\partial t} = -\nabla \cdot \mathbf{N}_A + R_A \tag{3.52}$$

For the steady-state condition and for the rate of production of A given by chemical kinetics, this equation becomes

$$\frac{dN_{Az}}{dz} = -k_1 c_A(z) \tag{3.53}$$

This situation differs in a fundamental way from the previously described situations in that the flux of A in the diffusion direction is not constant because of the depletion of A by the chemical reaction. The concentration of A is a function of position z so that the result cannot be integrated to obtain an immediate conclusion about the flux of A. Instead, the flux equation of A is used in terms of bulk flow in the solution and diffusion:

$$\mathbf{N}_A = y_A(\mathbf{N}_A + \mathbf{N}_B + \mathbf{N}_C) - c D_{AB} \nabla y_A \tag{3.54}$$

Because species A reacts as it diffuses into the thin film, a reasonable assumption (depending on the reaction kinetics versus the diffusion rate) may be that the concentration of A in the liquid film is everywhere small so that the bulk flow (convection) term that contributes to the total flux of species A can be neglected relative to the diffusion flux term. In this case and for one dimension, the equation becomes

$$N_{Az} = -c D_{AB} \frac{dy_A}{dz} \tag{3.55}$$

Combining this with the continuity equation (Eq. 3.53) gives a second-order ordinary differential equation for the concentration of species A

$$\frac{d^2 y_A}{dz^2} - \frac{k_1}{D_{AB}} y_A = 0 \tag{3.56}$$

This equation is subject to the boundary conditions at the gas–liquid film interface where the concentration in the liquid film is established by thermodynamic equilibrium ($y_A = y_{A0}$) and at the solid surface ($z = \delta$) where $N_{Az} = 0$ and $dy_A/dz = 0$. Imposing these conditions allows integration of the equation to give

$$y_A(z) = y_{A0} [\cosh(z\sqrt{k_1/D_{AB}}) \\ - \tanh(\delta\sqrt{k_1/D_{AB}}) \sinh(z\sqrt{k_1/D_{AB}})] \tag{3.57}$$

The flux of A is a function of position and is given by

$$N_{Az}(z) = c y_{A0} \sqrt{k_1 D_{AB}} \cdot [-\sinh(z\sqrt{k_1/D_{AB}}) \\ + \tanh(\delta\sqrt{k_1/D_{AB}}) \cosh(z\sqrt{k_1/D_{AB}})] \tag{3.58}$$

Note that the concentration profile through the liquid film is exponential in form through the cosh and sinh functions and that the degree of decay depends on the ratio (relative magnitude) of the reaction rate to the diffusion rate. The larger the reaction rate, the more steeply the profile falls as the distance into the film away from the gas interface increases. Conversely, the faster the diffusion (i.e., the higher the diffusion rate relative to the reaction rate), the more slowly the (steady-state) concentration falls or decreases with increasing distance from the interface.

A similar problem for an infinitely thick liquid film is approached in a similar way except that instead of the no-flux boundary condition at the solid surface the condition $c_A \to 0$ as $z \to \infty$ is used. Such a solution is appropriate even for liquid films of finite thickness if the reaction rate is large enough compared to the diffusion rate so that the diffusing species A in fact never "sees" the solid surface, i.e., it is depleted by reaction before it reaches the solid surface.

3.1.8. Diffusion with Heterogeneous Reaction

The next problem addresses diffusion through a gas film to a surface at which a reaction occurs and from which the reaction product must diffuse back through the gas phase. This is the situation in many catalytic conversion reactions where a catalyst pellet may be considered to be surrounded by a thin stagnant film through which a species must diffuse. Because the film is stagnant, diffusion rather than reaction kinetics may be the controlling phenomenon.

The situation is depicted schematically in Figure 24 for a spherical catalyst particle of radius R surrounded by a boundary layer gaseous film of thickness δ. Reactant A diffuses through the film to the surface where the reaction $A + A \to A_2$ is assumed to occur very quickly compared to the diffusion processes. As a result, at the catalyst surface the concentration of reacting species is assumed to be zero; the reacting species is depleted by reaction as quickly as it is delivered to the surface by diffusion. This is a binary system with A and A_2 as the two components. Because the reaction occurs at the solid surface rather than in the gas phase, it is not

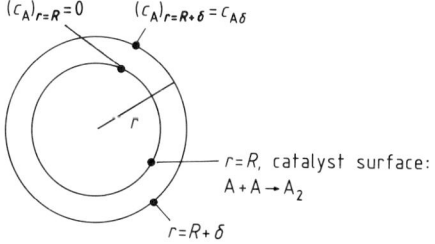

Figure 24. Schematic of diffusion and heterogeneous chemical reaction at a catalyst surface

relevant to the continuity equation of species A within the gas phase. The role of the reaction stoichiometry and kinetics is to provide a boundary condition on species fluxes at the catalyst surface.

As for the preceding problems, the objective is to establish quantitative relations for the flux of reactant A through the stagnant thin film and for the concentration profile of A across the film. As before, the continuity equation for reacting species A is employed:

$$\frac{\partial c_A}{\partial t} = -\nabla \cdot \mathbf{N}_A + R_A \quad (3.59)$$

For steady state with no homogeneous reaction, this gives

$$\nabla \cdot \mathbf{N}_A = 0 \quad (3.60)$$

which in spherical coordinates is

$$\frac{1}{r^2}\frac{d}{dr}(r^2 N_{Ar}) = 0 \quad (3.61)$$

which is integrated to give

$$r^2 N_{Ar} = \text{constant} \quad (3.62)$$

In this spherical geometry, the flux of a species is not constant but is proportional to $1/r^2$. Now because at any position r the spherical surface area is proportional to r^2, the molar flow rate of A rather than the flux is constant, independent of position, according to

$$r^2 N_{Ar} = \text{constant} = \frac{r^2 N_A}{4\pi r^2} = \frac{N_A}{4\pi} \quad (3.63)$$

Here N_A is the number of moles of species A per unit time passing through the entire spherical shell at radius r. This molar flow rate is a constant, independent of radial position. As before, the result for the molar flux of species A is written in terms of the convective bulk flow and diffusion:

$$\mathbf{N}_A = y_A = (\mathbf{N}_A + \mathbf{N}_{A_2}) - cD_{AB}\nabla y_A \quad (3.64)$$

which, in one-dimensional form, is

$$N_{Ar} = y_A(N_{Ar} + N_{A_2r}) - cD_{AB}\frac{dy_A}{dr} \quad (3.65)$$

Because of the reaction stoichiometry, the molar flow rate of species A can be related to that of product A_2 according to

$$N_A = -2 N_{A_2} \tag{3.66}$$

These are molar flow rates integrated over the entire surface of a spherical shell and, as shown above, because of continuity they are independent of radial position. The fluxes of each of the species can be written in terms of these molar flow rates so that Equation (3.65) becomes

$$\frac{N_A}{r^2} = y_A \left(\frac{N_A}{r^2} + \frac{N_{A_2}}{r^2} \right) - 4\pi c D_{AB} \frac{dy_A}{dr} \tag{3.67}$$

which, because of the relation between the molar flow rates of A and A_2, gives

$$N_A \frac{dr}{r^2} = \frac{4\pi c D_{AB} dy_A}{(1 - \frac{1}{2} y_A)} \tag{3.68}$$

As the molar flow rate is a constant, this result can be integrated directly to give a relation for the concentration profile through the film. In doing so, the boundary conditions are that the concentration of reactant A is zero at the surface of the pellet because it is instantaneously depleted by the reaction, and the concentration at the outer extent of the film (i.e., at $r = R + \delta$) is a known value (the concentration in the bulk gas phase). The resulting concentration profile is

$$y_A(r) = 2[1 - (1 - \tfrac{1}{2} y_{A\delta})^{\left(\frac{\gamma}{R+\gamma}\right)\left(\frac{R+\delta}{\delta}\right)}] \tag{3.69}$$

where $\gamma \equiv r - R$ and

$$N_A = 8\pi c D_{AB} \frac{R(R+\delta)}{\delta} \ln(1 - \tfrac{1}{2} y_{A\delta}) \tag{3.70}$$

Now this result is written in terms of an unknown film thickness δ. It can be evaluated, however, for two limiting cases to obtain limiting or boundary concentration profiles and molar flow rates. If gas flow across the pellet is slow or essentially stagnant, then the film thickness goes to infinity and the results obtained are

$$y_A(r) = 2[1 - (1 - \tfrac{1}{2} y_{A\delta})^{(1 - \frac{R}{r})}] \tag{3.71}$$

and

$$N_A = 8\pi c D_{AB} R \ln(1 - \tfrac{1}{2} y_{A\delta}) \tag{3.72}$$

Alternatively, if the gas velocity is high enough, the stagnant film becomes very thin ($\delta \ll R$) and results obtained are

$$y_A(r) = 2[1 - (1 - \tfrac{1}{2} y_{A\delta})^{\left(\frac{\gamma}{\delta}\right)}] \tag{3.73}$$

and

$$N_A = 8\pi c D_{AB} \frac{R^2}{\delta} \ln(1 - \tfrac{1}{2} y_{A\delta}) \tag{3.74}$$

In these latter equations, the film thickness remains as a parameter, whereas in the stagnant film situation the velocity profile and molar flow rate can be calculated directly; as the gas film becomes very thick, the molar flow rate and concentration profile become independent of the film thickness. At the other extreme, for a very thin film, the molar flow rate becomes inversely proportional to the film thickness; a doubling of film thickness halves the molar flow rate.

3.1.9. Perspective

The aforementioned results are useful in the context of this article; not so much for the specific problems addressed as for the types of approaches that are made to solving diffusion problems. In these problems, steady-state situations have involved either thin films or infinitely thick films, and slowly changing processes have been observed which can be treated in a pseudosteady-state manner by first solving the steady-state problem and then integrating over time to obtain the results as a function of time. Flat and spherical geometries have also been observed, although curved geometries, mathematically, look exactly like flat geometries when the film thickness is small in relation to the radius of curvature. Therefore, flat geometry calculations take on an importance above and beyond their actual application to specific physical problems. Finally, both homogeneous and heterogeneous chemical reactions have been addressed; homogeneous chemical reactions occur throughout the phase through which the diffusion or motion of molecules is occurring, whereas heterogeneous reactions occur at a boundary or surface associated with the diffusion process. As a result, homogeneous reaction kinetics appear directly in the continuity equations for the reacting species within the diffusion phase, whereas heterogeneous reaction kinetics play a role in the fluxes that occur at interfaces or boundaries of the diffusion region.

3.2. Convective Mass Transport

In the previous examples, motion of the individual species resulted from the diffusive fluxes that occur because of concentration gradients. In this section, examples in which species transport arises from bulk flow (or convection), as well as diffusion, are considered.

3.2.1. Gas Absorption in a Falling Film with Reaction

A problem encountered in many industrial scrubbers, fixed-bed gas–liquid reactors, and other mass-transfer devices, is the exchange of a gas species A between a gas mixture and a liquid film that moves on a solid surface. A simple example of such a situation is shown in Figure 25, where the gas species A is exchanged between a bulk gas phase and a falling liquid film of component B. This problem is analyzed first as steady-state *simple absorption with no chemical reaction*. Then, this solution is used to analyze the more general case in which *simultaneous absorption and reaction* occur.

In the analysis, A is assumed to be only slightly soluble in Newtonian liquid B, so that the viscosity and density of the liquid film are not changed appreciably. Under such conditions, the film velocity profile is given by (end effects neglected)

$$u_z = u_{max}\left[1 - \left(\frac{y}{\delta}\right)^2\right] \quad (3.75)$$

The concentration of A in the film varies with both y and z, and the species mass balance (Eq. 3.24) can be written as

$$u_z \frac{\partial c_A}{\partial z} = D_{AB}\left(\frac{\partial^2 c_A}{\partial y^2} + \frac{\partial^2 c_A}{\partial z^2}\right) \quad (3.76)$$

with the boundary conditions

at $z = 0$, $c_A = 0$ (3.77)

at $z \to \infty$, $c_A \to c_{A0}$ (3.78)

at $y = 0$, $c_A = c_{A0}$ (3.79)

at $y = \delta$, $\dfrac{\partial c_A}{\partial y} = 0$ (3.80)

The first boundary condition expresses the fact that pure B enters at $z = 0$; the second, that the film is saturated with A after a sufficiently long exposure distance. The third condition gives the concentration c_{A0} at the gas–liquid boundary. Generally, unless significant resistance to mass transfer occurs on the gas side of the interface, the concentration c_{A0} is the concentration in the liquid that is in equilibrium with the gas. The fourth condition simply indicates that no mass flux exists normal to the solid wall; hence, the concentration gradient $\partial c_A/\partial y$ is zero at the wall.

By assuming that over some film length $z \sim \delta$ the gas penetrates only a fraction of the distance into the film (e.g., $y \sim \varepsilon\delta$), the y and z variables can be scaled as

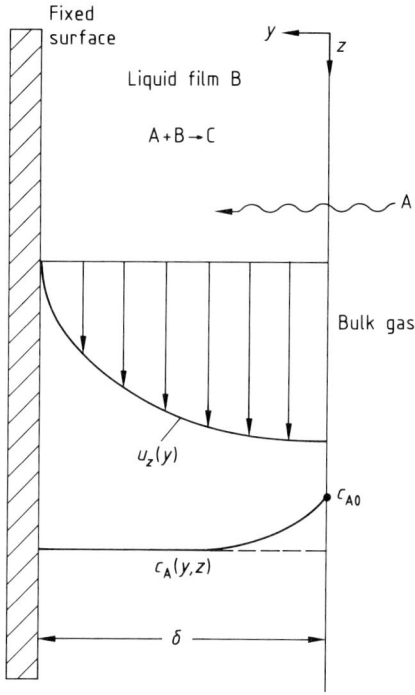

Figure 25. Absorption and reaction in a falling liquid film

$$y \equiv \varepsilon\delta y^* \quad (3.81\,\text{a})$$

$$z \equiv \delta z^* \quad (3.81\,\text{b})$$

where y^* and z^* are dimensionless quantities of order 1. Further, if c_A and u_z are scaled by

$$c_A = c_{A0} c_A^* \quad (3.81\,\text{c})$$

and

$$u_z = u_{max} u_z^* \quad (3.81\,\text{d})$$

Equation (3.76) can be written as

$$(1 - \varepsilon^2 y^{*2})\frac{\partial c_A^*}{\partial z^*} = \frac{1}{Pe_\delta}\left(\frac{1}{\varepsilon^2}\frac{\partial^2 c_A^*}{\partial y^{*2}} + \frac{\partial^2 c_A^*}{\partial z^{*2}}\right) \quad (3.82)$$

where

$$Pe_\delta \equiv \frac{u_{max}\delta}{D_{AB}} \quad (3.83)$$

is the mass-transfer Péclet number.

If convection is important, the left-hand side of Equation (3.82) must be of the same order as the largest diffusion term on the right-hand side (the y diffusion term). In particular, this requires that

$$\varepsilon \equiv \frac{1}{Pe_\delta^{1/2}} \tag{3.84}$$

Equation (3.82) can then be written as

$$\left(1 - \frac{1}{Pe_\delta} y^{*2}\right) \frac{\partial c_A^*}{\partial z^*} = \frac{\partial^2 c_A^*}{\partial y^{*2}} + \frac{1}{Pe_\delta} \frac{\partial^2 c_A^*}{\partial z^{*2}} \tag{3.85}$$

Equation (3.84) indicates that short penetration distances ($\varepsilon \ll 1$) are associated with large Péclet numbers. For the specific case in which $u_{max} \approx 0.1$ cm/s, $\delta \approx 0.1$ cm, and $D_{AB} \approx 10^{-5}$ cm^2/s, the value of Pe_δ is 10^3 (or $\varepsilon \approx 0.03$). Such high Péclet number, short penetration distance conditions are typical of those encountered in many gas–liquid contact processes. Under such conditions, Equation (3.82) can be reduced to

$$\frac{\partial c_A^*}{\partial z^*} = \frac{\partial^2 c_A^*}{\partial y^{*2}} \tag{3.86}$$

with the boundary conditions

at $z^* = 0$, $c_A^* = 0$ (3.87)

at $y^* = 0$, $c_A^* = 1$ (3.88)

at $y^* \to \infty$, $c_A^* \to 0$ (3.89)

The solution of Equation (3.86) subject to these boundary conditions is [3.1, p. 339]

$$c_A^* = \frac{c_A}{c_{A0}} = 1 - \frac{2}{\sqrt{\pi}} \int_0^{\frac{y}{\sqrt{4D_{AB} z/u_{max}}}} e^{-\zeta^2} d\zeta$$

$$= 1 - \mathrm{erf}\left(\frac{y}{\sqrt{4D_{AB} z/u_{max}}}\right) \tag{3.90}$$

or

$$\frac{c_A}{c_{A0}} = \mathrm{erfc}\left(\frac{y}{\sqrt{4D_{AB} z/u_{max}}}\right) \tag{3.91}$$

where erf and erfc are the error and complementary error functions, respectively. For a contact film of length L and width W, the rate of mass transfer (number of moles of A per unit time) to the film is given by

$$\mathscr{W}_A = \int_0^L N_{Ay}|_{y=0} W\,dz = \int_0^L \left(-D_{AB} \frac{\partial c_A}{\partial y}\right)\bigg|_{y=0} W\,dz \tag{3.92}$$

or, if Equation (3.91) is used

$$\mathscr{W}_A = LWc_{A0}\sqrt{\frac{4D_{AB} u_{max}}{\pi L}} \tag{3.93}$$

If the mass-transfer coefficient k_c is defined by

$$\mathscr{W}_A = k_c LW c_{A0} \tag{3.94}$$

then

$$Sh_L \equiv \frac{k_c L}{D_{AB}} = \frac{2}{\sqrt{\pi}} Re_L^{1/2} Sc^{1/2} = \frac{2}{\sqrt{\pi}} Pe_L^{1/2} \tag{3.95}$$

where Sh_L is the Sherwood number, and Re_L and Sc are the Reynolds and Schmidt numbers, respectively, which are defined by

$$Re_L \equiv \frac{u_{max} L}{\nu}, \quad Sc \equiv \frac{\nu}{D_{AB}} \tag{3.96}$$

Hence, for short penetration distances ε (or short contact times or high Péclet numbers), the dimensionless form of the mass-transfer coefficient (Sh_L) varies as the square root of the product of Re_L and Sc.

If the *chemical reaction*

$$A + B \to C \tag{3.97}$$

in the film is now considered and a sufficient excess of B is assumed so that a pseudo-first-order reaction expression is valid, i.e.,

$$R_A = k' c_A c_B \approx k' c_{B0} c_A = k_1 c_A \tag{3.98}$$

the mass balance takes the form

$$\frac{\partial c_A^*}{\partial z^*} = \frac{\partial^2 c_A^*}{\partial y^{*2}} - \frac{Da_\delta c_A^*}{Re_\delta Sc} \tag{3.99}$$

where Da_δ is the Damköhler number defined by

$$Da_\delta \equiv \frac{k_1 \delta^2}{D_{AB}} \tag{3.100}$$

Here again, in writing Equation (3.99), the penetration distance is assumed to be small (i.e., $Pe_\delta \equiv Re_\delta Sc$ is large). Also, the boundary conditions are assumed to be the same as given by Equations (3.87)–(3.89).

The solution to this problem can be expressed in terms of the solution for the nonreacting case (Eq. 3.91) by [3.7, pp. 32–33]

$$c_A^* = \frac{c_A}{c_{A0}} = \varphi(y^*, z^*) \exp\left(\frac{Da_\delta}{Re_\delta Sc} z^*\right)$$

$$+ \frac{Da_\delta}{Re_\delta Sc} \int_0^{z^*} \varphi(y^*, \zeta) \exp\left(\frac{Da_\delta}{Re_\delta Sc} \zeta\right) d\zeta \quad (3.101)$$

where $\varphi(y^*, z^*)$ is the c_A solution from Equation (3.91). Using Equation (3.92) to determine the total gas exchange over a surface of length L and width W gives

$$\dot{W}_A = W c_{A0} \frac{u_{max}}{\sqrt{k_1/D_{AB}}}$$

$$\cdot \left[\left(\frac{1}{2} + q\right) \operatorname{erf} \sqrt{q} + \sqrt{\frac{q}{\pi}} \cdot \exp(-q)\right] \quad (3.102)$$

where

$$q \equiv \frac{Da_L}{Re_L Sc} \quad (3.103)$$

In terms of $Sh_L \equiv k_c L/D_{AB}$, this result can be written as

$$Sh_L = Re_L^{1/2} Sc^{1/2}$$

$$\cdot \left[\frac{(1 + 2q)}{2\sqrt{q}} \operatorname{erf} \sqrt{q} + \frac{1}{\sqrt{\pi}} \exp(-q)\right] \quad (3.104)$$

If $q \to 0$ (no chemical reaction), this equation reduces to Equation (3.95); if $q \to \infty$,

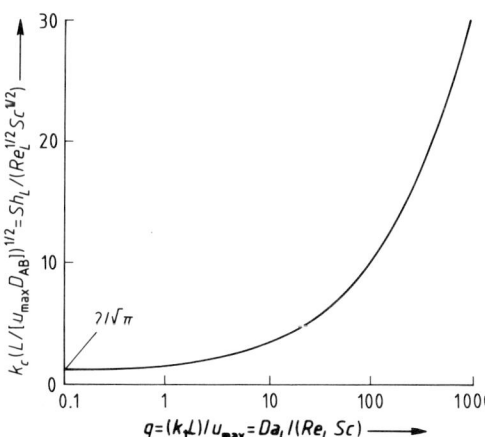

Figure 26. Enhancement of mass transfer with chemical reaction in a falling liquid film

$$Sh_L = Da_L^{1/2} = (q Re_L Sc)^{1/2} \quad (3.105)$$

In Figure 26, the results are given for a full range of q values. As q increases (higher reaction rates), mass transfer is significantly enhanced (i.e., higher mass-transfer coefficients) compared to the nonreacting case ($q = 0$). Many scrubbers and other absorption processes utilize absorption with reaction to obtain the most efficient removal of dilute gases from bulk gas streams ($\to$ 8. Absorption, **B3**).

3.2.2. Mass Transfer in Laminar and Turbulent Boundary Layers

Consider the convective mass transfer of species A from a flat or nearly flat surface into a free stream of species B. Assume that concentration changes of A are restricted to a concentration boundary layer of thickness δ_c. Conceptually, this is analogous to the problem illustrated in Figure 15, except that δ_c replaces δ_T, $c_{A\infty}$ replaces T_∞, and c_{A0} replaces T_w.

For this problem where $c_A = c_A(x, y)$, the species mass balance takes the form

$$u_x \frac{\partial c_A}{\partial x} + u_y \frac{\partial c_A}{\partial y} = D_{AB} \frac{\partial^2 c_A}{\partial y^2} \quad (3.106)$$

where the x diffusion term $[D_{AB}(\partial^2 c_A/\partial x^2)]$ has been neglected compared to the y diffusion term (right-hand side of above equation). This equation is analogous to Equation (2.239) for heat transfer in thermal boundary layers and can be integrated over the boundary layer thickness to obtain

$$\frac{d}{dx} \int_0^{\delta_c} u_x (c_A - c_{A\infty}) dy - N_{Ay}|_{y=0}$$

$$= -D_{AB} \frac{\partial c_A}{\partial y}\bigg|_{y=0} \quad (3.107)$$

The solution of this equation for $\delta_c(x)$ follows in the same way δ_T was obtained from the integral form of the energy balance (Eq. 2.240). If a cubic expression is used for c_A in terms of y and the constants a, b, c, and d are evaluated with the boundary conditions

at $y = 0$, $\quad c_A = c_{A0} \quad (3.108)$

at $y = \delta_c$, $\quad c_A = c_{A\infty} \quad (3.109)$

at $y = \delta_c$, $\dfrac{\partial^n c_A}{\partial y^n} = 0$, $n = 1, 2, \ldots$ \hfill (3.110)

the same results can be obtained as for Equations (2.250) and (2.251), with δ_c replacing δ_T, $Sc \equiv \nu/D_{AB}$ replacing Pr, and $Sh_x \equiv k_c x/D_{AB}$ replacing Nu_x. The Blasius solutions (Eqs. 2.252 and 2.255) also have the same forms for the mass-transfer results if the same replacements are made. The mass-transfer rate (number of moles A per unit time) from the surface is then

$$W_A = \int_0^L k_c (c_{A0} - c_{A\infty}) W \, dx$$

$$= k_{cL} W L (c_{A0} - c_{A\infty}) \hfill (3.111)$$

where

$$k_{cL} \equiv \frac{1}{L} \int_0^L k_c \, dx \hfill (3.112)$$

is the average mass-transfer coefficient over the plate length L.

These results are valid for the laminar part of the boundary layer. In the turbulent part ($x \geq x_{cr} = 2 \times 10^5 \, \nu/u_\infty$), the molar flux in the y-direction can be written as

$$\bar{N}_{Ay} = -(D_{AB} + \varepsilon_c) \frac{d\bar{c}_A}{dy} \hfill (3.113)$$

where $\bar{N}_{Ay}$ and $\bar{c}_A$ are time-smoothed quantities and ε_c is the eddy mass diffusivity. If a Prandtl mixing length expression is assumed for ε_c, then

$$\varepsilon_c = \varepsilon_m = \varepsilon_T = \varepsilon = l^2 \left| \frac{d\bar{u}_x}{dy} \right| \hfill (3.114)$$

with the mixing length l given by Equation (2.299). Here again, as with energy transfer in turbulent flow, $D_{AB} \gg \varepsilon$ for the laminar sublayer next to the wall, $D_{AB} \approx \varepsilon$ in the buffer zone, and $D_{AB} \ll \varepsilon$ in the turbulent core.

Carrying out an analysis analogous to that of Section 2.5.4 gives the mass-transfer forms of the various analogies:

Reynolds analogy: $\dfrac{Sh_x}{Re_x Sc} = \dfrac{k_c}{u_\infty} = C_{Dx}/2$ \hfill (3.115)

Prandtl analogy: $\dfrac{Sh_x}{Re_x Sc} = \dfrac{C_{Dx}/2}{1 + 5\sqrt{C_{Dx}/2}\,(Sc - 1)}$ \hfill (3.116)

Von Karman analogy:

$$\frac{Sh_x}{Re_x Sc} = \frac{C_{Dx}/2}{1 + 5\sqrt{C_{Dx}/2}\,\{Sc - 1 + \ln\left[(1 + 5Sc)/6\right]\}} \hfill (3.117)$$

Further, the Colburn analogy can be written as

$$\frac{Sh_x}{Re_x Sc} = \frac{k_c}{u_\infty} = \frac{C_{Dx}/2}{Sc^{2/3}} \hfill (3.118)$$

The last result is obtained directly from laminar boundary layer results; however, as noted in Section 2.5.4 it can be used as a first approximation to the Reynolds analogy for turbulent transport at Schmidt numbers different from 1. All of these analogies can be used for turbulent mass transport in tubes if u_∞ is replaced by u_{avg}.

Equation (3.118) can also be written as

$$j_D = \frac{C_{Dx}}{2} \hfill (3.118a)$$

where

$$j_D \equiv \frac{k_c Sc^{2/3}}{2 u_\infty} \hfill (3.118b)$$

is the j factor for mass transfer. When this result is combined with Equation (2.320a) with C_{Dx} replacing f, we obtain

$$j_D = j_H = \frac{C_{Dx}}{2} \hfill (3.118c)$$

This result is called the Chilton–Colburn analogy and connects the mass transfer, heat transfer, and momentum transfer processes in a given flow geometry. Although Equation (3.118c) is obtained directly from analysis of transport on laminar boundary layers, it can be used as an approximation for other geometries and for turbulent flows as long as there is no form drag. If form drag is present, the first part of Equation (3.118c), i.e., $j_D = j_H$, can still be used to relate mass transfer coefficients to heat transfer coefficients if $0.6 < Sc < 2500$ and $0.6 < Pr < 100$.

Finally, analogous to Equation (2.325),

$$Sh_x = 0.0289 \, Re_x^{4/5} Sc^{1/3} \hfill (3.119)$$

for mass transfer in turbulent boundary layers. Combining this with the mass-transfer equivalent of Equation (2.252) for the laminar part of the boundary layer, i.e.,

$$Sh_x = 0.332 \, Re_x^{1/2} \, Sc^{1/3} \tag{3.120}$$

yields

$$Sh_L = 0.0361 \, Re_L^{4/5} \, Sc^{1/3} (1 - 9170 \, Re_L^{-4/5}) \tag{3.121}$$

for the overall mass-transfer coefficient relation over a plate of length L where $L \geq x_{cr}$. Equation (3.121) is the mass-transfer equivalent of Equation (2.328).

3.3. Mass Transfer Across Interfaces

In the situations described in the preceding sections, the rate of mass transfer within a single phase is calculated by using fundamental transport relations (mass balances and diffusion flux relations) and transport properties (diffusivities). Such calculations are possible for well-defined geometries and flow situations of sufficient simplicity that can be modeled and calculated a priori and without empiricism.

This not always the case, however, because a great many more situations occur in which the flow of the fluid phases or a complex geometry prevents exact modeling or calculation. An example is forced convection transfer across an interface. In this case, the situation is described by using *mass-transfer coefficients* instead of diffusivities. These coefficients play a role similar to diffusivities in that they describe the transport rate of mass that occurs primarily because of molecular motion, but they also allow for other effects such as the enhancement of mass transfer because of forced or natural convection. Once these coefficients have been determined experimentally for a number of flow and mass-transfer situations, and correlated by dimensionless groups, the coefficients for analogous situations can be estimated and used for process design.

The calculations are complicated, however, by the dependence of mass-transfer coefficients on mass flux. For example, in Section 3.1.6 the diffusion of species A through a stagnant film B was enhanced, above and beyond pure diffusion, by the net bulk flow of the mixture. Similarly, mass-transfer coefficients that apply to low mass-transfer rates are affected by the species mass fluxes and should be adjusted accordingly for the most accurate design calculations at high mass-transfer rates.

Figure 27 shows how mass-transfer coefficients that are calculated for low mass-transfer rates by using diffusion coefficients or dimensionless group correlations can be adjusted for high mass-transfer rates. Then the coefficients for mass transfer through films on both sides of an interface can be combined (a process requiring thermodynamic equilibrium between the two phases) to give an overall coefficient for transfer between phases.

The commonly used definitions of mass-transfer coefficients for each side of an interface and for the overall coefficients calculated from them are given in Section 3.3.1; their functional forms based upon theoretical considerations for low mass-transfer rates or equimolar counterdiffusion are presented in Section 3.3.2; and correlations for calculating mass-transfer coefficients for low mass-transfer rates and the adjustments that may be made for high mass-transfer rates are considered in Sections 3.3.3 and 3.3.4, respectively. General references providing more detailed discussion of mass-transfer coefficient definitions and calculations include texts in the areas of transport phenomena [3.1], [3.12] and chemical engineering unit operations [3.13]–[3.17].

3.3.1. Mass-Transfer Coefficients

Mass transfer occurs because of an imbalance of concentrations, a departure from equilibrium. This imbalance provides a driving force for mass transfer. Uniformity of composition is the equilibrium state in a single phase; if mole fractions are not uniform, then a nonequilibrium condition exists and diffusion occurs until uniformity is reached. Two-phase thermodynamic equilibrium is the equilibrium state across an interface; to the extent that the two phases on opposite sides of an interface are not in equilibrium, mass transfer tends to occur in such a way as to move the system toward equilibrium.

The degree of departure from equilibrium directly affects the rate of mass transfer. In a single phase, the degree of departure from equilibrium is represented by the mole fraction (or mass fraction) gradient, and Fick's first law of diffusion (the most commonly used flux relation) models the diffusion flux as proportional to this driving force. The proportionality factor defines the diffusivity.

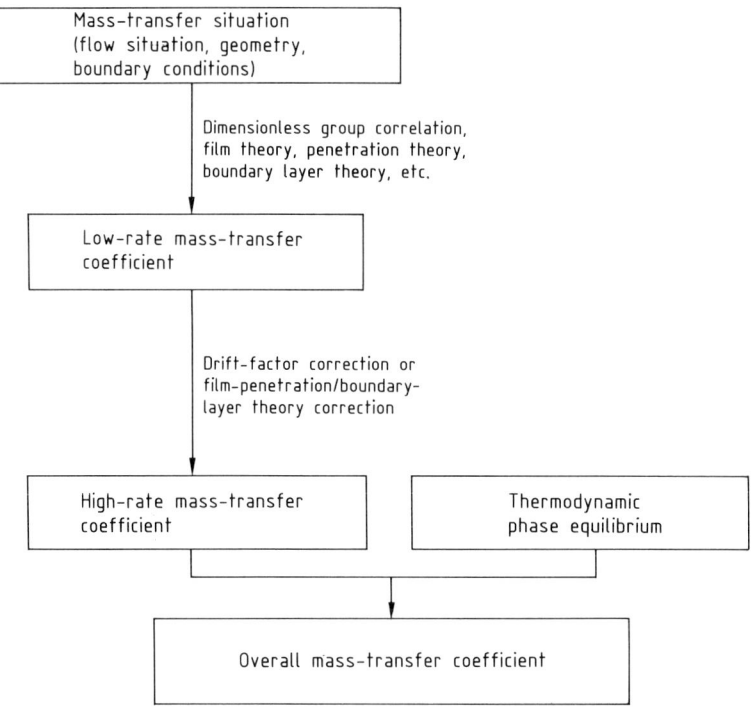

Figure 27. Representation of the calculation of overall mass-transfer coefficients showing corrections of the single-phase coefficients for high mass-transfer rates

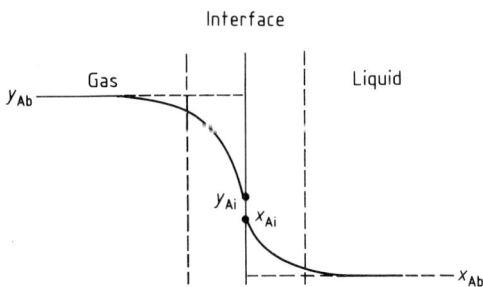

Figure 28. Hypothetical concentration profiles across the gas–liquid interface region with the transfer of A from the gas to the liquid
y_{Ab} = mole fraction of A in bulk gas phase; y_{Ai} = gas-phase mole fraction of A at interface; x_{Ai} = liquid-phase mole fraction of A at interface; x_{Ab} = mole fraction of A in bulk liquid phase

For mass transfer across interfaces or films, an analogous relationship is normally used to define mass-transfer coefficients. The mass-transfer flux of a species at an interface is modeled as proportional to the driving force (concentration difference) which exists for that transfer, through a thin film next to the interface. This situation is depicted in Figure 28. At the interface the two phases are normally assumed to be in thermodynamic equilibrium. Away from the interface, however, the bulk concentrations of the two phases are not necessarily at equilibrium with each other, and possible concentration or mole fraction profiles are shown as a function of distance from the interface. The majority of the concentration change is modeled to occur over a laminar film region near the interface. The actual concentrations and film depths are not known, however, which makes the definitions quite empirical and dependent on parameters such as fluid flow and turbulence. In Figure 28, concentration profiles are shown in both phases and, for simplicity, one phase is called a gas phase and the other a liquid phase, although this is not a limitation or constraint on the situation. The discussion could just as well be for two liquid phases or for a fluid and a solid phase. The model also normally assumes that concentrations at the interface are at steady state; flux to the interface through one phase equals that away from the interface through the other.

Mass-transfer coefficients, then, are defined for each of the two phases. The definition of a *liquid-phase mass-transfer coefficient* (based on a liquid-phase mole fraction driving force) is

Flux of A $= k_x(x_{Ai} - x_{Ab})$ (3.122)

Likewise, the defining relation for the *gas-phase mass-transfer coefficient* for species A based on the gas-phase mole fractions is

Flux of A $= k_y(y_{Ab} - y_{Ai})$ (3.123)

In each of these equations, a departure from equilibrium exists that represents the extent to which the interface mole fraction (x_{Ai} or y_{Ai}) differs from that in the bulk fluid (x_{Ab} or y_{Ab}) of the same phase.

Whereas the above relations define mass-transfer coefficients for a driving force within a single phase at an interface, interphase mass-transfer coefficients are also defined according to concentration or mole fraction differences that exist across the two phases, where the average or bulk concentrations are used for each phase. In this case the mass-transfer coefficients K_x and K_y are defined according to the relations

Flux of A $= K_x(x_{Ae} - x_{Ab})$ (3.124)

Flux of A $= K_y(y_{Ab} - y_{Ae})$ (3.125)

and are called *overall mass-transfer coefficients*. They describe the flux in terms of mole fractions in the bulk phases.

Here, instead of defining a driving force that exists within one phase or the other, a driving force that spans the two phases is defined. The mole fractions and driving forces are shown relative to a typical interfacial equilibrium curve in Figure 29. For a mass-transfer coefficient based on liquid-phase mole fractions, the driving force that is used is the difference between the actual mole fraction of A in the bulk liquid phase (x_{Ab}) and the mole fraction of A that would exist (x_{Ae}) if the liquid phase were in equilibrium with the mole fraction of A in the bulk gas phase. Likewise, in terms of gas-phase concentrations, mass transfer of A occurs to the extent that the bulk gas-phase mole fraction (y_{Ab}) differs from the value that would exist (y_{Ae}) if the gas phase were in equilibrium with the actual bulk liquid-phase mole fraction.

The slopes of lines that represent the *ratios of mass-transfer coefficients* are also shown in Figure 29. If species A does not accumulate at the interface, the liquid- and gas-phase relationships

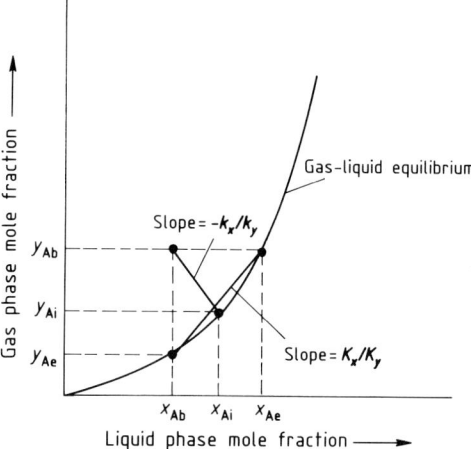

Figure 29. Relationships among interface, bulk, and equilibrium concentrations used in mass-transfer rate equations

for flux in terms of mass-transfer coefficients must be equal. Accordingly,

$k_x(x_{Ai} - x_{Ab}) = k_y(y_{Ab} - y_{Ai})$ (3.126)

which gives

$$-\frac{k_y}{k_x} = \frac{x_{Ab} - x_{Ai}}{y_{Ab} - y_{Ai}}$$ (3.127)

and the ratio of the interphase mass-transfer coefficients is the slope of a tieline connecting the point with composition coordinates equal to the liquid- and gas-phase bulk concentrations to a point with coordinates equal to the equilibrium interface liquid- and gas-phase concentrations. Similarly, a ratio can be obtained for the overall transfer coefficients:

$$\frac{K_x}{K_y} = \frac{y_{Ab} - y_{Ae}}{x_{Ae} - x_{Ab}}$$ (3.128)

In the limit of small driving forces or for a linear isotherm this ratio is the slope m of a tangent to the equilibrium curve in the concentration region of interest.

From the definition of the mass-transfer coefficients and for a locally linear isotherm (slope $= m$),

$$\frac{1}{K_x} = \frac{1}{k_x} + \frac{1}{mk_y}$$ (3.129)

and

$$\frac{1}{K_y} = \frac{1}{k_y} + \frac{m}{k_x} \quad (3.130)$$

Hence, the overall or combined resistance to mass transfer through the two phases ($1/K_x$ or $1/K_y$) is equal to the sum of the resistances through each of the phases individually. Before summing, however, one of the individual phase coefficients must be scaled by using the (local) slope of the equilibrium curve in order to be consistent with the resistance offered by the other mass-transfer coefficient. Note that if $k_x/m \gg k_y$, then the gas-phase mass transfer is limiting and $K_y \approx k_y$.

Dimensions of Mass-Transfer Coefficients. Because the flux of A is the number of moles of A per time per (cross-sectional) area, the mass-transfer coefficients as defined by these relations must also have the dimensions of number of moles per time per area. Other definitions using different driving force concentration units are employed, however, and the dimensions of the mass-transfer coefficient vary accordingly. For example, number of moles per volume is frequently used for liquid-phase concentrations and partial pressure for gas-phase concentration. In these situations, mass-transfer coefficients may be defined according to

Flux of A $= k_c(c_{Ai} - c_{Ab})$ (3.131)

Flux of A $= k_G(p_{Ab} - p_{Ai})$ (3.132)

Flux of A $= K_c(c_{Ae} - c_{Ab})$ (3.133)

Flux of A $= K_G(p_{Ab} - p_{Ae})$ (3.134)

Here, k_c and K_c have the dimensions of volume per time per area (length per time), and k_G and K_G have the dimension of number of moles per time per area per unit pressure.

3.3.2. Functional Forms of Mass-Transfer Coefficient Relations

In the preceding discussion, the mass flux through a film is assumed to be proportional to the driving force (the concentration difference) across a film, and this proportionality factor defines the mass-transfer coefficient. Combining this definition with theoretical models for mass transfer in terms of diffusion coefficients allows calculation of the dependence of mass-transfer coefficients on diffusivity. These are only estimates, however, because the situations are too complex for exact modeling, which is why mass-transfer coefficients were defined in the first place. A complete discussion of several models appears in [3.13].

Two-Film Theory. A particularly simple but useful model for steady-state transfer between two phases, which is frequently used to describe mass-transfer coefficients in terms of diffusion coefficients, is the two-film theory of WHITMAN [3.18]. The concentration profile is taken to be flat (independent of position) in the bulk fluid and then to vary linearly across a thin film of thickness δ up to the interface (Fig. 30). For steady-state mass transfer at low mass-transfer rates the results of this model are the same as for equimolar counterdiffusion (see Section 3.1.6); the molar flux is directly proportional to the diffusivity, and the mass-transfer coefficient, in terms of film thickness and diffusivity, is

$$k_x = \frac{c D_{Ab}}{\delta} \quad (3.135)$$

Hence, this film theory predicts a mass-transfer coefficient that is directly proportional to diffusivity through the film.

Other Theories. Other theories relating the coefficient to diffusion may also be appropriate. Examples are the *penetration theory* [3.19] and its adaptations [3.20], [3.21]. The penetration theory is applicable to a species that is being transferred from a gas to a liquid phase; it gives an estimate of the mass-transfer coefficient in the liquid phase based upon a nonsteady-state situa-

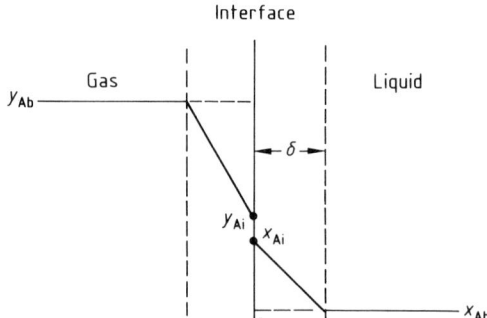

Figure 30. Simplified concentration profiles for the two-film theory

tion. The liquid phase is assumed to be a falling film and the velocity profile of the falling film at the gas–liquid interface is essentially flat, i.e., the velocity of the falling liquid is nearly independent of penetration depth from the gas into the liquid as long as the penetration depth is low. This is true if the time that the gas is in contact with the liquid is short. In this situation, any one packet of fluid is in contact with the gas for a fixed amount of exposure time; therefore, the diffusion of a species from the gas to the liquid—from the viewpoint of this packet of fluid—is a nonsteady-state process as opposed to the steady-state diffusion of the film theory. This different type of model gives rise to a mass-transfer coefficient that is proportional to the square root of diffusivity:

$$k_x = 2c \sqrt{\frac{D_{AB}}{\pi t_{exp}}} \tag{3.136}$$

Here, t_{exp} is the exposure time of a given packet or element of fluid to the gas stream and k_x is an average value over this time. Exposure time can be calculated in terms of the fluid velocity at the gas film interface and the distance of contact of the gas and liquid L:

$$t_{exp} = \frac{L}{u_{interface}} \tag{3.137}$$

A third theory, actually a combination of both, is the *film-penetration theory* [3.22]. For some flow situations, the film theory result is recovered, whereas for others, the penetration theory's $D^{1/2}$ dependence holds, suggesting that different dependencies might be expected in different flow situations.

3.3.3. Relations for Low Mass-Transfer Rates

By using a somewhat different approach, dimensionless group correlations can be postulated to exist for mass-transfer coefficients. For example, the Sherwood number ($Sh_{AB} = k_x d/c D_{AB}$) can be expressed in terms of a function of the Reynolds number ($Re = du\varrho/\eta$), Schmidt number ($Sc = \eta/\varrho D_{AB}$), and L/d [3.1]:

$$Sh_{AB} = f(Re, Sc, L/d) \tag{3.138}$$

The specific functional form depends on the actual flow situation.

Furthermore, as discussed in Section 3.2, analogous correlations exist for heat, mass, and momentum transfer because of similarities in the conservation equations that govern all these situations and in the corresponding flux relations. Consequently, correlations exist for mass-transfer coefficients that are very similar to those for heat or momentum transfer (friction loss) in analogous situations of geometry and boundary conditions. Results for some specific flow and mass-transfer situations are summarized in Table 8. Note that in these relations, the dependence of the mass-transfer coefficient on diffusivity ranges from about $D_{AB}^{0.6}$ to D_{AB}, corresponding to the $D_{AB}^{0.5}$ to D_{AB} predicted by the penetration and film theories, respectively.

3.3.4. Relations for High Mass-Transfer Rates

In Section 3.1.6, two important limiting cases of binary diffusion are considered for a one-dimensional diffusion process. In the case of equimolar counterdiffusion the flux of each species is the result of diffusion only and is calculated by using Fick's first law of diffusion; for each species the flux is directly proportional to the mole fraction gradient of that species. For the case of one species diffusing through a stagnant film of the other, however, the diffusion of each species is the result of the combination of Fick's law of diffusion and a bulk flow that is induced by the diffusion process. For the species that was defined to be stagnant, the diffusion flux exactly balances its flux due to the mixture bulk flow. For the other species, however, the total flux is that due to Fick's law of diffusion enhanced by the bulk flow, which results in the Fick's law flux being divided by $(1 - x_A)$.

In the latter case, if the nonstagnant species is present at low molar concentration (infinite dilution or approaching infinite dilution), then obviously the same result for the flux of this species is obtained as for the equimolar counterdiffusion situation but for a different reason. The other species is still stagnant, and the bulk flow is not reduced by the low concentration but rather by transport of the nonstagnant species that occurs because of reduced bulk flow.

These two limiting cases (ECD and ATSB) provide a basis for understanding the corrections to mass-transfer coefficients that are made for high mass-transfer rates. The mass-transfer coefficients described in the correlations of Table 8

Table 8. Mass-transfer coefficients[a]

Situation	Equation	Restrictions	Equation ( ), reference [ ]
Transfer from a Cylindrical Tube Wall to a Fluid			
[Tube diameter = d, tube length = L, average fluid velocity = u_{avg}, $Sh = k_c d/D_{AB}$, $Re = du\varrho/\mu$, $Sc = \eta/(\varrho D_{AB})$]			
Coefficient at distance L downstream with fully developed laminar velocity profile, constant wall concentration	$Sh = 3.66 + \dfrac{0.0668\,[(d/L)\,Re_d\,Sc]}{1 + 0.04\,[(d/L)\,Re_d\,Sc]^{2/3}}$	$Re < 2000$	[3.23]
Same as above but with constant wall mass flux	$Sh = 4.36 + \dfrac{0.023\,[(d/L)\,Re_d\,Sc]}{1 + 0.0012\,[(d/L)\,Re_d\,Sc]}$	$Re < 2000$	[3.23]
Turbulent liquid flow, $L/d > 6$, transfer of solid from the wall[b]	$Sh = 0.023\,Re^{0.83}\,Sc^{1/3}$	$2000 < Re < 70\,000$ $0.6 < Sc < 3000$	[3.24]
Turbulent gas flow, $L/d > 6$, transfer from liquid film at the wall (wetted-wall column)[b]	$Sh = 0.023\,Re^{0.83}\,Sc^{0.44}$	$2000 < Re < 35\,000$ $0.6 < Sc < 2.5$	[3.25]
Turbulent liquid flow, high Re, high Sc numbers	$Sh = 0.0096\,Re^{0.913}\,Sc^{0.346}$	$10\,000 < Re < 100\,000$ $430 < Sc < 100\,000$	[3.26]
Transfer from a Flat Plate Immersed in a Fluid Parallel to the Flow			
[$Sh_x = k_c x/D_{AB}$, $Sh_L = k_{cL} L/D_{AB}$, $Re_x = \varrho u_\infty x/\eta$, $Re_L = \varrho u_\infty L/\eta$, $Sc = \eta/(\varrho D_{AB})$, u_∞ = fluid velocity far from the plate]			
Laminar flow, coefficient at distance x from the plate leading edge	$Sh_x = 0.332\,Re_x^{0.5}\,Sc^{1/3}$	$Re_x < 2 \times 10^5$ $0.6 < Sc < 2500$	(3.120), [3.27]
Laminar flow, coefficient averaged over the plate from the leading edge to distance L	$Sh_L = 0.664\,Re_L^{0.5}\,Sc^{1/3}$	$Re_L < 2 \times 10^5$ $0.6 < Sc < 2500$	[3.17], [3.27]
Turbulent flow	$Sh_L = 0.036\,Re_L^{0.8}\,Sc^{1/3}$	$2 \times 10^5 < Re_L < 10^7$ $0.6 < Sc < 2500$	[3.27]
Transfer from Single Cylinders Perpendicular to the Flow			
[Cylinder diameter = d, $Sh = k_c d/D_{AB}$, $Re = du_\infty \varrho/\eta$, $Sc = \eta/(\varrho D_{AB})$]			
Flow of gases	$Sh = 0.281\,Re^{0.6}\,Sc^{0.44}$	$400 < Re < 25\,000$ $0.6 < Sc < 2.6$	[3.16], [3.28]
Transfer in Packed or Fluidized Beds			
[Particle diameter = d_p, superficial velocity = u', void fraction = ε, $Re = d_p u' \varrho/\eta$, $Re' = d_p u' \varrho/\eta(1-\varepsilon)$, $Sc = \eta/(\varrho D_{AB})$, $j_D = k_c Sc^{2/3}/u'$]			
Packed or fluidized beds; preliminary design only unless gases for $Re < 10$	$Sh = (0.765\,Re^{0.18} + 0.365\,Re^{0.614})\,\dfrac{Sc^{1/3}}{\varepsilon}$	$0.01 < Re < 15\,000$	[3.33]
Gas or liquid fluidized bed with spherical particles	$Sh = \left[0.01 + \dfrac{0.863}{Re^{0.58} - 0.483}\right]\dfrac{Re\,Sc^{1/3}}{\varepsilon}$	$10 < Re < 2000$	[3.34]
Packed or fluidized bed, gases or liquids, spherical or regular cylindrical pellets[d]	$j_D = 5.7\,Re'^{-0.78}$	$0 < Re' < 30$ $0.6 < Sc < 1400$	[3.35]
	$j_D = 177\,Re'^{-0.44}$	$30 < Re' < 10\,000$ $0.6 < Sc < 1400$	[3.35]

[a] Unless otherwise noted, the coefficients are average values and solute concentrations are constant at the interface. [b] The difference between mass transfer from a liquid film and transfer from a solid is attributed to ripples in the liquid film. [c] Also reported are correction factors for cylinders [for $L/d = 1$, $\varepsilon j_D/(\varepsilon j_D)_{sphere} = 0.79$] and cubes [$\varepsilon j_D/(\varepsilon j_D)_{cube} = 0.71$] where $j_D = k_c Sc^{2/3}/u'$. [d] Correction factors for irregularly shaped particles—Raschig rings, Berl saddles, and flakes—are also reported.

are defined for equimolar counterdiffusion-like situations; that is, they apply to true equimolar counterdiffusion situations or to low transfer rate calculations. Evidently, then, if they are to be used for ATSB situations, they must be corrected in a way similar to the comparison between equimolar counterdiffusion and ATSB calculations; the presence of a stagnant film and high mass-transfer rates provides an induced bulk flow that enhances the mass-transfer coefficient. If the low mass-transfer coefficient is used in a high mass-transfer stagnant film situation, the mass-transfer rates may be significantly underestimated.

A common correction to the low mass-transfer rate–ECD coefficients is to introduce a drift factor $\overline{(1 - y_A)}_L$ analogous to the adjustment of ECD to ATSB. A common, although not unique, notation is to distinguish between these two mass-transfer coefficients with a prime ('); however, the notation may be switched with k' indicating ECD rather than ATSB as is done here:

$$k'_y = \frac{k_y}{\overline{(1 - y_A)}_L} \tag{3.139}$$

where $\overline{(1 - y_{Ai})}_L$ is the nonlinear, log-mean average of $(1 - y_{Ai})$ and $(1 - y_{Ab})$:

$$\overline{(1 - y_A)}_L = \frac{(1 - y_{Ai}) - (1 - y_{Ab})}{\ln \frac{(1 - y_{Ai})}{(1 - y_A)}} \tag{3.140}$$

This correction is commonly used for ATSB-like situations and appears in design calculations such as packed gas-absorption towers where one species in the gas phase is absorbed and the others are not.

Instead of a simple ATSB correction to the mass-transfer coefficient, however, less restrictive theories for flux ratios may be used to make the adjustment. These are (1) the film theory, (2) the penetration theory, and (3) the boundary layer theory. Each of these theories provides a means of calculating a correction factor to the mass-transfer coefficient.

The situation of ATSB diffusion is actually a specific limiting case of the more complete film theory summarized by BIRD et al. [3.1]. They use the film theory to adjust EDC mass-transfer coefficients for the enhancements which result from diffusion-induced bulk flow. The model allows for relative fluxes of the two species (A and B) that differ from the stagnant B situation. For a given flux ratio, adjustments to the low mass-transfer rate coefficient can be calculated (or, equivalently, the ECD mass-transfer coefficient), which would be calculated or estimated from the correlations of the previous section. Again, ATSB situation falls within this less restrictive film theory.

The penetration theory of HIGBIE [3.19] has also been used to extend the low mass-transfer rate calculations to high mass-transfer rates due to diffusion-induced enhancement. The results obtained are presented in a similar way to those for the film theory, and correction factors are given for a boundary layer theory approach [3.1]. Qualitatively, the corrections calculated by these three theories are very similar although the quantitative results are somewhat different. The penetration and boundary layer theories predict an even larger effect of mass transfer on the coefficients than does the film theory.

4. Macroscopic Systems

The development of conservation and accounting equations for macroscopic systems parallels the development of those for differential systems that led to the partial differential equations of Table 1. Each law or statement is written for a single extensive property (a property that can be counted, for which the whole is equal to the sum of the parts) for a specified region (system) inside a closed-surface boundary and for a specified time period (either finite or differential) (Fig. 31). In the case of macroscopic (finite-size) systems, either process rate equations are obtained, which are ordinary differential equations, or for a finite time period, an algebraic equation is obtained. For conserved properties, the generation (or consumption) is always zero.

The conservation and accounting equations and their development are summarized below for compositionally homogeneous systems. The purpose is to provide an overall picture relating conservation laws, accounting equations, U, H, A, G, and second law equations of thermodynamics, which provide the basics for quantitatively understanding transport phenomena (and other areas of engineering as well) on a macroscopic scale, rather than to give handbook-type details and examples. General references offering further discussion of the development and application of macroscopic equations include texts in

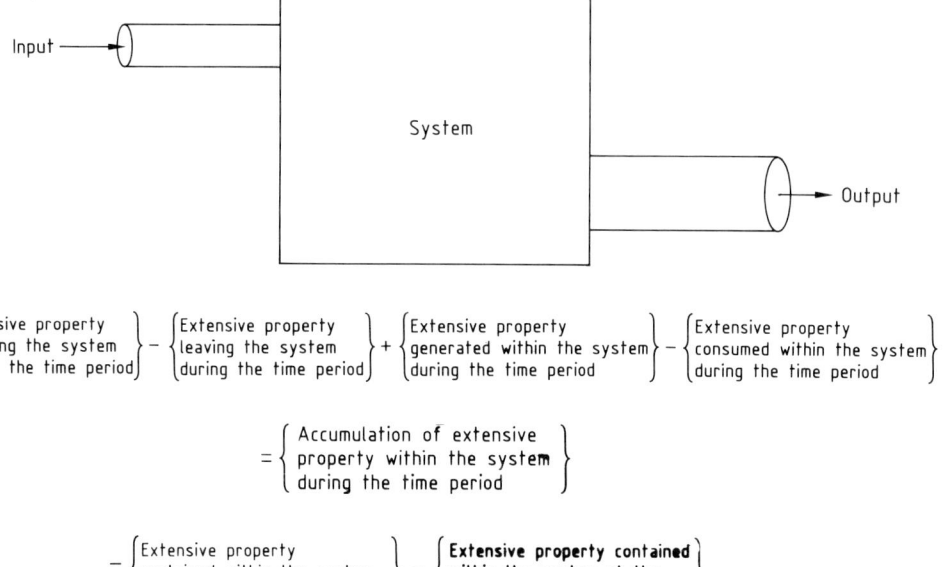

Figure 31. The concept of accounting for an extensive property—the cornerstone of transport phenomena
Input and output represent an exact exchange with the surroundings; a conserved property has no generation or consumption term.

the areas of transport phenomena [4.1]–[4.3], thermodynamics [4.4], [4.5], chemical engineering unit operations [4.6], and fluid mechanics [4.7].

4.1. Conservation of Total Mass

Total mass is conserved in the absence of nuclear conversions or relativistic effects. This means that for the system and its surroundings together the total amount of mass is a constant. Equivalently, if the amount of mass contained within the system changes, then the surroundings must change to an equal but opposite extent; there is a one-for-one exchange of mass between the system and its surroundings as mass moves across the system boundary. Allowing for multiple flow streams into and out of the system, an equation for the conservation of total mass contained within the system is

$$\sum_{\substack{\text{entering} \\ \text{streams}}} (\varrho\, u_{\text{avg}}\, A)_{\text{in}} - \sum_{\substack{\text{exiting} \\ \text{streams}}} (\varrho\, u_{\text{avg}}\, A)_{\text{out}}$$

$$= \frac{\mathrm{d}}{\mathrm{d}t} (\varrho V)_{\text{system}} \tag{4.1}$$

At each area A where mass exchanges occur between the system and surroundings, the fluid density and velocity components normal to the boundary surface u are appropriately averaged values. Likewise, the density of the system is averaged over the system volume V_{system} [4.1], [4.3], [4.8].

In addition to a conservation equation for total mass, accounting statements for individual species within the system can be written. If the system contains n species that together comprise the total mass (and could be molecular, ionic, or other species), then n accounting equations exist, one for each species. Because species are not necessarily conserved (except for the chemical elements), these accounting equations must allow for any generation or consumption of individual species through chemical reactions. Accordingly, for each of the n species within the system,

$$\sum_{\substack{\text{entering} \\ \text{streams}}} (\varrho_i\, u_{\text{avg},i}\, A)_{\text{in}} - \sum_{\substack{\text{exiting} \\ \text{streams}}} (\varrho_i\, u_{\text{avg},i}\, A)_{\text{out}} + r_i V_{\text{sys}}$$

$$= \frac{\mathrm{d}}{\mathrm{d}t} (\varrho_i V)_{\text{system}} \tag{4.2}$$

The densities and velocities are average values, as discussed above, and r_i is an average generation rate of species *i* per unit volume of the system. An equivalent accounting can be done in molar concentration units (see, e.g., Section 4.9.3).

Of these $n + 1$ equations, however, only n are independent inasmuch as the total mass must be the sum of the individual species mass (because total mass is conserved, the sum of all of the generation and consumption terms in the n species equations must be zero; no net generation or consumption of mass occurs). Note that the total mass of each species entering and leaving the system is obtained by summing over all entering and leaving streams.

4.2. Conservation of Linear Momentum

Linear momentum is conserved. Given a system, the amount of linear momentum contained therein may change only as a result of exchange of momentum between the system and its surroundings; linear momentum is not generated or consumed. Linear momentum may be exchanged as the result of mass entering or leaving the system and as the result of forces acting on the system by its surroundings. The rate of momentum exchange as mass enters or leaves the system is the momentum per unit mass u multiplied by the rate at which that mass enters or leaves the system. External forces act upon the system either by contact with the system's surfaces or as body forces, such as gravity, that act individually on each piece of mass within the system. The momentum of the system then changes according to

$$\sum \left[\frac{(u^2)_{avg}}{u_{avg}} \varrho u_{avg} A \right]_{in} - \sum \left[\frac{(u^2)_{avg}}{u_{avg}} \varrho u_{avg} A \right]_{out}$$

$$+ \sum F_{external} = \frac{dp_{system}}{dt} \quad (4.3)$$

The $(u^2)_{avg}$ term arises because the momentum flow rate, calculated as the momentum per unit mass u times the mass flow rate $\varrho u\, dA$, is integrated (averaged) over the cross section [4.1, Chap. 7], [4.2, Chap. 7]. If the velocity profile across the area is not flat, then $(u^2)_{avg} \neq (u_{avg})^2$.

If $\alpha_{mom} \equiv (u_{avg})^2/(u^2)_{avg}$, where α_{mom} is a measure of the extent to which the velocity profile is not flat, this equation becomes

$$\sum \left[\frac{\varrho (u_{avg})^2}{\alpha_{mom}} A \right]_{in} - \sum \left[\frac{\varrho (u_{avg})^2}{\alpha_{mom}} A \right]_{out}$$

$$+ \sum F_{external} = \frac{dp_{system}}{dt} \quad (4.4)$$

For fully developed laminar flow of a Newtonian fluid, α_{mom} is 3/4; for turbulent flow, it is nearly unity [4.9, p. 252]. Also p_{system} is integrated over the entire macroscopic system.

4.3. Conservation of Angular Momentum

Angular momentum is also a conserved property. The angular momentum of a system can change only by exchange with its surroundings. A piece of mass possesses angular momentum that is equal to the cross product of the position vector of that piece of mass and its momentum. Consequently, angular momentum enters and leaves with mass. Also, angular momentum enters the system as a result of external forces and is equal to the cross product of the position vector to the point of application of the force with that force:

$$r_{in} \times u_{in} (\varrho u_{avg} A)_{in} - r_{out} \times u_{out} (\varrho u_{avg} A)_{out}$$

$$+ \sum (r \times F_{ext}) = \frac{dL_{system}}{dt} \quad (4.5)$$

where L_{system} is the total angular momentum of the system. Again, the entering and leaving momenta are actually appropriate integral averages across the flow cross section, and L_{system} is integrated over the entire volume.

The conservation of angular momentum is not considered further here. It is useful, for example, for the analysis of torques on pumps and compressors for design purposes [4.3].

4.4. An Accounting of Mechanical Energy

Mechanical energy accounting can be made for finite-sized systems and can be obtained by integrating the continuum conservation law of linear momentum. It cannot be obtained from a macroscopic linear momentum law and consequently is an independent equation providing additional insight into flow situations. In this sense, the set of macroscopic equations differs from the continuum equations.

The mechanical energy equation is an accounting equation, as opposed to a conservation equation, because mechanical energy is not conserved. Conversion of mechanical energy to thermal or other forms of energy may occur so that the amount of mechanical energy in an isolated system is not necessarily constant. Nevertheless, such an accounting is extremely useful for certain problems, e.g., the design and sizing of piping systems and flow networks.

Derivation of the mechanical energy accounting equation is complex and lengthy [4.10]. The result is stated here for steady-state flow situations with one entrance and one exit. Mechanical energy is accounted for by the following steady-state equation (expressed as energy per unit mass of flow; the symbol ^ over a quantity implies "per unit mass"):

$$g(z_1 - z_2) + \frac{1}{2}\left[\frac{(u_{1,\mathrm{avg}})^2}{\alpha_{KE,1}} - \frac{(u_{2,\mathrm{avg}})^2}{\alpha_{KE,2}}\right] - \int_1^2 \frac{dp}{\varrho}$$
$$+ \hat{W}_s + \hat{W}_{es} - \hat{F} = 0 \qquad (4.6)$$

Here $\alpha_{KE} \equiv (u_{\mathrm{avg}})^3/(u^3)_{\mathrm{avg}}$ is another measure of the flatness of the velocity profile, this time with respect to kinetic energy. For fully developed laminar flow of a Newtonian fluid, α_{KE} is 1/2; for turbulent flow it is nearly unity [4.9, p. 252]. This result holds regardless of heat transfer or other thermal effects that may occur.

Equation (4.6) is referred to as the *extended Bernoulli equation* ($\rightarrow$ 5. Fluid Mechanics, pp. 5-21–5-24) and through $\hat{F}$ allows for the irreversible or dissipative conversion of mechanical energy to thermal energy in a manner analogous to that of Equation (1.80), the mechanical energy equation for a continuum. These losses are of the type encountered in flow through pipes where friction losses result in pressure drop in a straight run of horizontal pipe. The potential energy gz and kinetic energy $u^2/2\alpha_{KE}$ terms are evaluated (as averages over the cross section) at the entrance (1) and exit (2) points of the system. The shaft work $\hat{W}_s$ and the flow work due to the fluid extra stresses $\hat{W}_{es}$ act across the system boundary, whereas the friction losses occur both within the system (from viscous stresses inside the pipe) and across the system boundary (from viscous stresses at the pipe wall). The pressure term dp/ϱ must be integrated along a path through the process from the entrance to the exit, not necessarily an easy task for a compressible flow. If the flow is incompressible, then the extended Bernoulli equation becomes

$$g(z_1 - z_2) + \frac{1}{2}\left[\frac{(u_{1,\mathrm{avg}})^2}{\alpha_{KE,1}} - \frac{(u_{2,\mathrm{avg}})^2}{\alpha_{KE,2}}\right]$$
$$+ \frac{(p_1 - p_2)}{\varrho} + \hat{W}_s + \hat{W}_{es} - \hat{F} = 0 \qquad (4.7)$$

for which the pressure terms also use entrance and exit values only. Mechanical energy in the form of (1) upstream kinetic energy, (2) potential energy, (3) isotropic and extra stress flow work, and (4) shaft work is either redistributed downstream among these forms of mechanical energy, converted to work, or lost as mechanical energy by conversion to thermal energy.

4.5. Conservation of Total Energy

Total energy is conserved. Mass possesses energy in the form of kinetic, potential, and internal energy. When mass enters a system, it brings with it energy in these forms, and the rate at which it does so is calculated as the product of the specific energy in each of these forms (i.e., the energy per unit mass) and the rate of mass flow into the system.

Heat Q and work W exist at the system boundary but are not energy possessed by mass. Heat enters the system by either conduction or radiation because of a temperature difference across the system boundary. Work is energy transferred when a force acts through a distance upon elements of the system at the boundary. More properly, work may be expressed as

$$W = \int_{s_1}^{s_2} \mathbf{F} \cdot d\mathbf{s} \qquad (4.8)$$

The dot product of the force $\mathbf{F}$ with the displacement $d\mathbf{s}$ provides that the component of force in the direction of displacement produces work, whereas that which is normal to the displacement does not. The convention here is that both heat and work are positive when energy is added to the system.

With energy existing within the system as kinetic, potential, and internal energy, the conservation of total energy for a macroscopic system is expressed in rate forms as

$$\sum\left[\left(\hat{U} + \frac{u^2}{2\alpha_{KE}} + gz\right)(\varrho u_{avg} A)\right]_{in}$$

$$-\sum\left[\left(\hat{U} + \frac{u^2}{2\alpha_{KE}} + gz\right)(\varrho u_{avg} A)\right]_{out}$$

$$+ \dot{W} + \dot{Q} = \frac{d}{dt}[(\hat{U} + \widehat{KE} + \widehat{PE})m]_{sys}$$

The dots over W and Q imply work rate and heat-transfer rate, respectively, rather than a derivative rate of change of work or heat. Again, properties of the mass entering and leaving the system are averaged over the flow cross section.

Usually, the form of the equation is changed by separating the work required to move mass into and out of the system against the pressure (isotropic stress) at the boundary (expressed per unit mass p/ϱ) from all other work of the process, the "nonpressure" work $\dot{W}_{np}$. (The term p/ϱ is commonly referred to as "flow work". However, because it represents only the flow work due to isotropic stress and does not include that due to extra stress, it is referred here to as pressure work and all other forms as "nonpressure" work.) The statement is rewritten by combining the p/ϱ terms with the other inlet and outlet flow stream terms and replacing the total work term with $\dot{W}_{np}$. The p/ϱ terms are combined with the internal energy to give the enthalpy of the flow streams ($\hat{H} \equiv \hat{U} + p/\varrho$) and the resulting equation is

$$\sum\left[\left(\hat{H} + \frac{u^2}{2\alpha_{KE}} + gz\right)(\varrho u_{avg} A)\right]_{in}$$

$$-\sum\left[\left(\hat{H} + \frac{u^2}{2\alpha_{KE}} + gz\right)(\varrho u_{avg} A)\right]_{out}$$

$$+ \dot{W}_{np} + \dot{Q} = \frac{d}{dt}[(\hat{U} + \widehat{KE} + \widehat{PE})m]_{sys} \quad (4.10)$$

The "nonpressure" work includes work associated with volume changes of the system acting against external forces (a nonsteady-state term), all shaft work (compressors, pumps, turbines, electric motors or generators, etc.), other forms of energy (electrical, magnetic, etc.), plus—if important—any flow work due to extra (non-isotropic) stresses at the system boundaries. The last term is frequently neglected.

4.6. An Accounting of Thermal Energy

By subtracting the steady-state mechanical energy equation for one inlet (point 1) and one outlet (point 2) flow stream (Eq. 4.6) from the corresponding result for total energy conservation (Eq. 4.10), an accounting of thermal energy per unit mass is obtained which, although not independent from the total and mechanical energy equations, can be a useful way of viewing thermal effects:

$$\hat{H}_2 - \hat{H}_1 = \int_{p_1}^{p_2}\frac{dp}{\varrho} + \hat{Q} + \hat{F} \quad (4.11)$$

This result is true regardless of any changes in elevation, pressure, density, or flow velocity or the amount of shaft work, as long as a steady state exists through a single conduit. Furthermore, for a reversible situation ($\hat{F} = 0$), this result is readily identified as the thermodynamic relation $d\hat{H} = Td\hat{S} + dp/\varrho$, which is integrated to follow an element of the fluid through the process along the flow path in a sequence of quasi-equilibrium steps. For irreversible processes this thermodynamic relation still holds, but $\hat{Q}$ and $\hat{F}$ together correspond to the integral of $Td\hat{S}$; both heat transfer to the system and irreversibilities result in an increase in entropy (see Section 4.7).

These thermodynamic relations for $d\hat{U}$ (and, equivalently, for $d\hat{H}$, $d\hat{A}$, and $d\hat{G}$) may be viewed as thermal energy accounting equations, expressed in terms of thermodynamic state functions rather than the path functions $\hat{Q}$ and $\hat{F}$ of Equation (4.11). On the one hand, they represent relations between thermodynamic state variables regardless of the path or process used to bring about change; on the other hand, they represent thermal energy conversion for a specific process.

Both statements have intrinsic value in their own right. Given a change in enthalpy that occurs in a process and is calculated from such a thermal energy equation, the thermodynamic relations can be used to express this change in terms of changes in observable properties such as heat capacity, temperature, pressure, and volume. HOUGEN et al. [4.4, p. 508] recognize this separation of mechanical and thermal energy, and say that these state functions of a body do not depend upon "its external position or motion relative to other bodies." This result is considered further in Section 4.9.1.

The approach outlined in Sections 4.2–4.6 for deriving the macroscopic momentum, mechanical energy, total energy, and thermal energy equations is not the most common. More frequently, the extended Bernoulli equation is hypothesized by considering total energy for adiabatic, incompressible situations or by applying the thermodynamic realtions for $d\hat{U}$ or $d\hat{H}$ (e.g., in the form of Eq. 4.11) to total energy (Eq. 4.10) to obtain Equation (4.6) or (4.7). The approach given here is preferred because it is more direct and parallels the traditional approach used in continua (Chap. 1). This preference has also been expressed by WHITAKER [4.8].

4.7. The Second Law of Thermodynamics

The second law of thermodynamics places constraints on processes that can occur in macroscopic systems. As discussed for continua, the thermodynamic property of entropy quantifies the fact that certain processes are observed to occur in only one direction. In the context of a macroscopic law, the entropy contained within a macroscopic system changes through exchange with the surroundings and also through generation within the system as a result of irreversibilities. On the one hand, accounting for entropy is similar to accounting for other nonconserved properties because the entropy of the system may change as a result of generation within the system, as well as exchange with the surroundings. Unlike other accounting statements (e.g., mechanical energy, electrical energy, individual chemical species), however, entropy cannot be consumed within the system; it can only be generated. Also, unlike other accounting relations, an independent relation exists for the surroundings; the change in entropy of the surroundings due to this process arises because of exchange with the system and also as a result of generation of entropy within the surroundings due to the process. Again, this generation of entropy must be positive; it can never be negative. Consequently, when the two entropy accounting expressions for the system and surroundings are summed, the total entropy change of the system plus surroundings is necessarily positive (or, in the limit of a reversible process, equal to zero).

These statements are embodied in an accounting of entropy for a macroscopic system

$$\Sigma [(\hat{S}) \varrho u_{avg} A]_{in} - \Sigma [(\hat{S}) \varrho u_{avg} A]_{out}$$
$$+ \Sigma \left(\frac{\dot{Q}}{T}\right)_{system} + (\dot{S}_{gen})_{system} = \frac{d}{dt}[(\hat{S})m]_{system}$$

(4.12)

and its surroundings

$$\Sigma [(\hat{S}) \varrho u_{avg} A]_{out} - \Sigma [(\hat{S}) \varrho u_{avg} A]_{in}$$
$$+ \Sigma \left(\frac{\dot{Q}}{T}\right)_{surr} + (\dot{S}_{gen})_{surr}$$
$$= \frac{d}{dt}[(\hat{S})m]_{surr} \qquad (4.13)$$

and for the system and surroundings combined

$$\frac{d}{dt}[(\hat{S}m)_{system} + (\hat{S}m)_{surr}] \geq 0 \qquad (4.14)$$

In these equations, the exchange of entropy between the system and surroundings is the result of mass (possessing entropy per unit mass) transferring between the system and surroundings (values of $\hat{S}$ are averaged over the flow cross section).

Heat transfer is also a one-for-one-exchange between the system and its surroundings, provided the temperatures of the system and the surroundings at the boundary ($T_{system, bound}$ and $T_{surr, bound}$, respectively) are the same, this is normally the case (however, see the next paragraph concerning entropy generation). In this case, the rate of entropy added to the system ($\dot{S}_{in, system}$) due to a rate of heat transfer from the surroundings to the system $\dot{Q}$ is

$$\dot{S}_{in, system} = \frac{\dot{Q}}{T_{system, bound}} \qquad (4.15)$$

and the rate of entropy transfer to the surroundings ($\dot{S}_{in, surr}$) due to the same heat transfer is

$$\dot{S}_{in, surr} = -\frac{\dot{Q}}{T_{surr, bound}} \qquad (4.16)$$

If the temperatures of the system and surroundings at the boundary are the same, then these two terms cancel exactly and a true exchange of entropy occurs between the system and surroundings.

The generation of entropy in the system and surroundings is frequently difficult or impossible to quantify (although not always) but conceptually is the result of finite driving forces for mass, energy, or momentum transfer. Consequently, if temperature, mole fraction, or velocity gradients within the system or surroundings are dissipated as a result of heat, mass, or momentum transfer within the system, then these dissipation processes cause a movement toward a homogeneous condition which necessarily results in an increase in entropy. Likewise, if heat or momentum transfer at the system boundary results from finite differences in temperature or forces, then an entropy increase is associated with the heat or momentum transfer due to subsequent temperature or momentum equilibriation within the system and surroundings. For the system and surroundings of homogeneous but different temperatures (T_{system} and T_{surr}), the net total increase in entropy of the universe resulting from the two-step heat transfer and subsequent equilibration process is calculated according to

$$\frac{Q}{T_{\text{system}}} - \frac{Q}{T_{\text{surr}}} \geq 0 \tag{4.17}$$

as though a single-step process of transfer to the system (whose boundary temperature is T_{system}) and from the surroundings (with boundary temperature T_{surr}) has occurred.

Again, because entropy is not conserved, an accounting of entropy for either the system or the surroundings alone is not an adequate statement of the second law of thermodynamics. Furthermore, the entropy of a material is a function of state and consequently can be calculated (with respect to an arbitrarily defined reference value) in terms of heat capacities, equations of state, and energy changes associated with phase changes. Consequently, the entropy exchange associated with mass crossing the system boundary is well defined, and the change of entropy of the system, given its composition and change in state, is also well defined. As mentioned previously, however, the generation of entropy as a result of irreversibilities is not always as well defined. In some cases, such as the diffusive mixing of two ideal gases at equal pressure and temperature, the entropy increase can be calculated from theoretical considerations. In other cases, however, such as the rapid expansion of a gas against an external force, the entropy increase due to irreversibilities cannot be calculated because the expansion is associated with a turbulent velocity profile that is unpredictable and, because of velocity gradients, ultimately results in an increase in entropy.

4.8. Tabular Summary of the Macroscopic Equations

The macroscopic equations given in Sections 4.1–4.7 are summarized in Table 9. Also given are the number of independent scalar equations that are represented in each case and any limitations on the result.

4.9. Some Examples

4.9.1. Pressure Drops and Temperature Changes for Closed Channel Flow

Perhaps the most basic problem of macroscopic fluid mechanics is to calculate the pressure drop or pump requirement (or turbine output) associated with fluid flow through a pipe, tube, or other closed conduit. The basic design problem specifies a given amount of material that is to flow from one location to another at a given design flow rate. This requires either sufficient upstream kinetic, potential, and pressure energy or the addition of energy through a pump. Calculation of the required upstream energy or pump work is done by accounting for mechanical energy through the extended Bernoulli equation (Eq. 4.6 for a compressible fluid and Eq. 4.7 for an incompressible fluid). For example, to move water to an elevated storage tank, a pump must be supplied to overcome the change in potential energy and any frictional losses of mechanical energy due to the flow of water through the pipe. Kinetic energy changes may also be a factor, although these are usually relatively small.

One of the primary factors in such flow calculations is determining this mechanical energy loss term. This is not a calculation that can be made solely on theoretical grounds. Instead, it is handled through a dimensionless group correlation for a friction factor f from which the mechanical energy losses for flow at average velocity u_{avg} through a straight section of horizontal pipe of length L and diameter D are calculated

Table 9. Summary of the macroscopic equations

Equation name	Equation number	Number of independent macroscopic equations	Comments
Total mass	(4.1)	1	
Species mass (n species)	(4.2)	$n - 1$	
Linear momentum	(4.4)	3*	
Angular momentum	(4.5)	3*	
Mechanical energy (compressible)	(4.6)	1**	steady state, one flow stream
Mechanical energy (incompressible)	(4.7)	1**	steady state, one flow stream
Total energy	(4.10)	1	
Thermal energy	(4.11)	0	steady state, one flow stream
Entropy, system	(4.12)	1	
Entropy, surroundings	(4.13)	1	
Total entropy (second law of thermodynamics)	(4.14)	0	an inequality

* One equation for each spatial dimension, e.g., for a two-dimensional situation, only two nontrivial equations exist.
** Only one of these two equations can be used in any one situation.

according to

$$\hat{F} = 4f\left(\frac{L}{D}\right)\left(\frac{u_{avg}^2}{2}\right) \quad (4.18)$$

The dimensionless group correlation typically appears in graphic form (the original chart of MOODY [4.11] or its modifications) or in analytical form, convenient for computer use (given by CHURCHILL [4.12]). [Alternative definitions for the friction factor in the literature differ from each other by a factor of 2 or 4. Care must be taken that the dimensionless group correlation for f versus Re and the friction factor definition (e.g., Eq. 4.18) employed are consistent with each other.]

The friction factor depends on the Reynolds number of the flow through the pipe and, for turbulent flow, the roughness of the wall. Additional losses occur in valves, fittings, and pipe bends; relations for calculating these losses also are found in standard texts on fluid flow.

The effect of mechanical energy losses on the temperature of the fluid is normally quite small for liquids, although in some instances it may have to be taken into account. The thermal energy equation provides a basis for calculating this effect as discussed in Section 4.6. For a compressible, single-phase fluid the change in enthalpy is written in terms of heat capacity and thermodynamic $p-\varrho-T$ properties by using standard thermodynamic arguments:

$$\hat{H}_2 - \hat{H}_1 = \int_{T_1}^{T_2} c_p(\partial T)_{p_1} + \int_{p_1}^{p_2}(1 - T\beta)\frac{(\partial p)_{T_2}}{\varrho} \quad (4.19)$$

where the temperature integral is at constant pressure, the pressure integral is at constant temperature, and β is the isobaric coefficient of thermal expansion [$\beta \equiv -1/\varrho\,(\partial\varrho/\partial T)_p$]. Therefore, by combining Equations (4.11) and (4.19), in the absence of heat transfer and for a temperature-independent heat capacity c_p, the net temperature change is obtained:

$$T_2 - T_1 = \frac{1}{c_p}\left[\hat{F} + \int_{p_1}^{p_2}\frac{dp}{\varrho} - \int_{p_1}^{p_2}(1 - T\beta)\frac{(\partial p)_T}{\varrho}\right] \quad (4.20)$$

where $\hat{F}$ is given by Equation (4.18). If the fluid is truly incompressible, then $\beta = 0$, the two pressure integrals cancel exactly and the temperature increases in direct proportion to frictional losses. For compressible fluids, however, this heating is counteracted by the expansion of the fluid; if $p_2 < p_1$, then

$$\int_{p_1}^{p_2} \beta T/\varrho\,(\partial p)_T < 0$$

which, if large enough, can result in net cooling.

As an example, for the flow of oil through the Trans-Alaska pipeline between pumping stations, the appropriate parameters for this problem are $D = 1.19$ m, $L = 107$ km, $u_{avg}A = 3.18 \times 10^5$ m^3/d, $\beta = 8.17 \times 10^{-4}$ K^{-1}, $f = 0.00335$, and $c_p = 1.8$ kJ kg^{-1} K^{-1} [4.13], and the temperature rise is approximately

$$T_2 - T_1 \approx \frac{1}{c_p}\left[4f\left(\frac{L}{D}\right)\left(\frac{u_{avg}^2}{2}\right)(1 - \beta T_1)\right] = 2.7 \text{ K} \quad (4.21)$$

For comparison, neglecting the fluid compressibility results in a temperature increase of 3.7 K. Mechanical energy conversion to thermal energy in pumps, when present, also contributes to temperature increases.

Although the temperature increase can be fairly small for flow of a fluid over a short distance, it can accumulate and become appreciable if distances are long enough as in a transcontinental pipeline or a continuously circulating flow loop. Temperature changes accompanying flow can also contribute to errors in viscosity measurement.

4.9.2. Compressible Flow in a Tube

If the flow through a closed channel or tube is compressible, then accurate design and flow calculations can be considerably more complicated. The compressible form of the extended Bernoulli equation must be used (again this is for steady-state flow situations), and the friction factor is calculated in the same way as for incompressible flow by using the friction factor correlations. However, the pressure term is shown as an integral from the entrance of the flow channel to the exit. The difficulty of this calculation is that the density must be known as a function of pressure through the pipe and along the actual flow path of the fluid through the pipe. Knowing the conditions at the entrance and exit is not enough; conditions must be known from one end to the other, along the path taken by the fluid. Normally, these are unknown because of the flow complexity.

To handle this situation, idealized approximations to the flow are calculated as limiting or extreme cases. For example, if the flow of an ideal gas ($p_1 M_r = RT_1 \varrho_1$, where R is the ideal gas constant and M_r the molecular mass) is adiabatic and reversible (no heat transfer occurs to or from the gas, and mechanical energy losses are negligible—*isentropic flow*), the pressure and density are related by

$$\frac{p}{\varrho^\kappa} = \text{const.} = \frac{p_1}{\varrho_1^\kappa} \qquad (4.22)$$

where κ is the ratio of the heat capacities, c_p/c_v, so that

$$-\int_{p_1}^{p_2} \frac{dp}{\varrho} = \frac{RT_1 \kappa}{(\kappa - 1)} \left[1 - \left(\frac{p_2}{p_1}\right)^{(\kappa-1)/\kappa} \right] \qquad (4.23)$$

A second situation is *isothermal flow* of an ideal gas for which

$$\frac{p}{\varrho} = \frac{RT}{M_r} = \text{constant} \qquad (4.24)$$

and

$$-\int_{p_1}^{p_2} \frac{dp}{\varrho} = \frac{RT}{M_r} \ln \frac{p_1}{p_2} \qquad (4.25)$$

For the flow of a nonideal gas, the equation of state for that gas can be used instead of the ideal gas equation of state to obtain corresponding integrals.

For adiabatic, nonreversible flow, COULSON and RICHARDSON [4.14] give an approach for calculating the pressure integral and conclude that "the rate of flow of gas under adiabatic conditions is never more than 20% greater than that obtained for the same pressure difference with isothermal conditions." Furthermore, they point out that the difference becomes less for long pipes and that "for pipes of length at least one-thousand diameters the difference does not exceed about 5%." Thus, isothermal flow and adiabatic reversible flow are commonly assumed to bracket the actual observed situation, and for long pipelines, little difference exists between these two extremes.

One other aspect of compressible flow should be pointed out. For large pressure drops and high flow rates, a limit to the flow rate may be obtained even though the pressure drop is allowed to increase by lowering the downstream pressure. This limit is established by the sonic velocity of the gas in the pipe. If the gas reaches sonic velocity, any further reduction of the downstream pressure at constant pipe cross-sectional area results in no additional increase in velocity. The sonic velocity of the gas at the downstream end, based on thermodynamic principles and in terms of the density ϱ, pressure p, and specific entropy $\hat{S}$, is

$$u_{\text{sonic}} = \left[\left(\frac{\partial p}{\partial \varrho}\right)_{\hat{S}} \right]^{1/2} \qquad (4.26)$$

4.9.3. Heterogeneous Reaction in a Fluidized-Bed Reactor

Consider now the reaction of component A to product B carried out in a well-mixed fluidized-bed reactor (Fig. 32). Such a conversion might represent the catalytic cracking of heavy hydrocarbons in a refinery process. The reactor stream is fed to the fluidized bed as a gas with sufficient velocity to maintain the catalyst particles in a fluidized state. Once inside the reactor, reacting species A is converted to product B at the catalyst surface. However, before this reaction can occur, the reactant must first migrate to the surface of the catalyst pellet. Then, after reaction at the surface, the product B must migrate to the bulk fluid to be removed from the reactor in the exit flow stream. What complicates this situation is that the catalyst itself may be porous, so that not only does the reactant contact the exterior surface of the catalyst, but part of the reactant species also diffuses into the micropores before reacting. Consequently, some reactant is converted at a reduced concentration inside the pores and, therefore, at a lower rate than if it were converted at the concentration existing at the macroscopic exterior surface of the catalyst. Calculation of the effective reaction rate requires considering each of these mass-transfer and reaction rate processes.

To illustrate the concept, the macroscopic material balance of reacting species A is considered. Under the assumption of a well-mixed fluidized bed, the gas-phase concentration is the same everywhere in the reactor and equal to the concentration in the exit flow stream. Of course, this model is an approximation because concentration gradients exist near the catalyst pellets due to the heterogeneous reaction. The model assumes that enough reaction locations of sufficiently small size exist for them to be treated like a homogeneous reaction occurring uniformly throughout the system volume. An accounting of species A within the fluidized bed in terms of

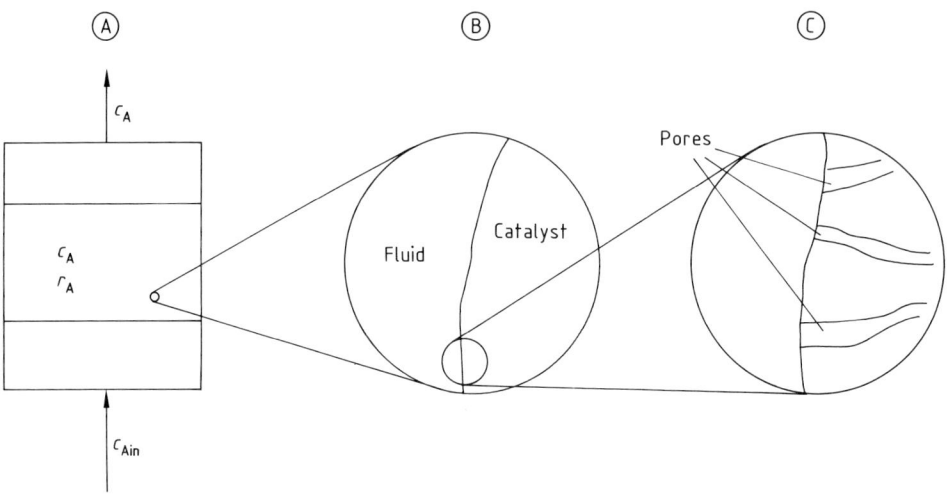

Figure 32. Modeling of the fluidized-bed reactor: macro-scale for the reactor accounting of mass (A), mesoscale for the mass-transfer rate (B), and microscale for definition and calculation of the effectiveness factor (C)
A) Macroscale: fluidized-bed region is modeled as well mixed and uniform in temperature, composition, and reaction rate;
B) Mesoscale: mass transfer occurs between the bulk fluid and the catalyst surface because of concentration differences;
C) Microscale: diffusion within the catalyst pores reduces the effective reaction rate

its concentration in the feed stream c_{Ain}, the concentration in the bulk fluid in the bed and outlet flow stream c_A, and the feed and product volumetric flow rates (F_{in} and F_{out}, respectively) is

$$c_{Ain} F_{in} - c_A F_{out} = r_A V \qquad (4.27)$$

The amount of A that enters the reactor at concentration c_{Ain} either leaves at concentration c_A or is converted to products at an effective reaction rate r_A per unit volume.

Although complicated by r_A, this result provides a clear approach to reactor design. Once an expression for this reaction rate is obtained based on transport and kinetic factors, the problem can be completed to obtain the conversion in the reactor. Then economic and design calculations can be made for the reactor in terms of fractional conversion, space time, and reactor volume and flow rates [4.15], [4.16]. The key to reactor design is to determine the overall reaction rate for a given situation. The effect of transport on this effective rate is as follows.

The transport of species from the bulk of the fluid in the fluidized bed to the catalyst surface can be estimated by using mass-transfer coefficient correlations for fluidized beds. A j factor correlation has been presented by CHU et al. [4.17] that is applicable to gas–solid and liquid–solid systems and for both fluidized-bed and fixed-bed reactors; this is presented in Table 8. Graphs of these correlations are given in [4.16], [4.17].

For porous catalysts, determination of the reaction rate at the catalyst, although primarily a problem in reaction kinetics, can also be a problem in transport phenomena because of diffusion in the catalyst pores. The chemical kinetics problem must be determined from experimental laboratory reactors. LIN provides extensive literature references to studies of a range of industrial reactions [4.16]. The transport problem is normally addressed by the use of an *effectiveness factor* η. The effectiveness factor adjusts the reaction rate to account for the fact that within the pores of a catalyst the reaction occurs at a reduced rate compared to the pellet's exterior surface. At the catalyst surface, the reaction proceeds at a rate that is dictated by the chemical kinetics and by the concentration of the reacting species at the surface. However, as the reactant becomes depleted with greater depth into the pores, the reaction rate decreases accordingly; the effectiveness of the catalyst surface within the pores is lower than that at the surface.

Calculation of an average (averaged over the entire catalyst surface, whether deep inside pores or not) effectiveness factor is based on combined diffusion and reaction inside the catalyst pores and was addressed originally by THIELE [4.18] and ARIS [4.19] and subsequently in many references on reaction and transport [4.1], [4.15], [4.20]. Such analysis leads to calculation of an average effectiveness factor for a catalyst pellet as a function of a dimensionless variable called the *Thiele modulus*. This parameter depends on the relative rates of the kinetic reaction to diffusion. If the diffusion rate into the pores is very high compared to the reaction constant, then the fact that some of the catalyst surface is available only in pores makes little difference and the effectiveness factor is close to unity. However, if the kinetic reaction rate constant is large compared to the diffusion rate, then the concentrations, and hence reaction rates, within the pores are reduced considerably, giving a substantially lower effectiveness factor.

By using the effectiveness factor η to adjust for the reaction rate at the catalyst, the reaction rate (per unit volume of catalyst particle) for a reaction that is first-order irreversible in A with rate constant k is given by

$$r_A = \eta k c_{Ai} \quad (4.28)$$

If a steady state exists at the surface of the catalyst (i.e., no accumulation of materials occurs at the catalyst surface), the transport rate of materials to the pellet equals the reaction rate at the pellet. For species A then (by using the mass-transfer coefficients of Section 3.3),

$$k_G S_p (c_A - c_{Ai}) = \eta k V_p c_{Ai} \quad (4.29)$$

where the external pellet surface area S_p and the pellet volume V_p are required because k_G is per surface area and k is per pellet volume. For a true first-order kinetic reaction, the rate constant is a function of temperature but not of species concentrations. If the reaction is not first order, an additional functionality depending on the interface concentration exists. This equation can be solved for the interface concentration in terms of the rate constants and the bulk fluid concentrations, which can then be used to express the overall reaction rate of species A in terms of the bulk phase concentration

$$r_A = K c_A \quad (4.30)$$

where

$$K = \cfrac{1}{\cfrac{1}{\eta k (V_p/S_p)} + \cfrac{1}{k_G}} \quad (4.31)$$

This shows that both the mass-transfer coefficient to the catalyst and the effectiveness factor – rate constant product act as resistances to the conversion of the reactant to the product. When the resistances are added in series, a combined rate constant is obtained.

This result leads to the notion of transport-limited versus kinetic-limited reactions. If the mass-transfer coefficient is small enough, mass transfer is the rate limiting step and reaction kinetics play no role in the overall conversion rate. Likewise, if the kinetic rate constant – effectiveness factor product is small enough (compared to the mass-transfer coefficient), it becomes the limiting part of the rate process and the mass-transfer coefficient plays no role in establishing the reaction rate. If both rate constants are of the same order, then the overall rate coefficient depends on both the kinetic and the mass-transfer rate factors.

5. References

References for Chapter 1

[1.1] R. Aris: *Vectors, Tensors, and the Basic Equations of Fluid Mechanics*, Prentice Hall, Englewood Cliffs, N.J., 1962.
[1.2] L. Brand: *Vector and Tensor Analysis*, J. Wiley & Sons, New York 1957.
[1.3] R. B. Bird, W. E. Stewart, E. N. Lightfoot: *Transport Phenomena*, J. Wiley & Sons, New York 1960.
[1.4] S. Whitaker: *Fundamental Principles of Heat Transfer*, Pergamon Press, New York 1977.
[1.5] J. C. Slattery: *Momentum, Energy and Mass Transfer in Continua*, 2nd ed., R. E. Krieger Publishing Co, Huntington, N.Y. 1981.
[1.6] L. E. Malvern: *Introduction to the Mechanics of a Continuous Medium*, Prentice Hall, Englewood Cliffs, N.J., 1969.
[1.7] G. E. Mase: *Theory and Problems of Continuous Mechanics*, McGraw-Hill, New York 1970.
[1.8] R. C. Reid, J. M. Prausnitz, T. K. Sherwood: *The Properties of Gases and Liquids*, 3rd ed., McGraw-Hill, New York 1977.
[1.9] S. Chapman, T. G. Cowling: *The Mathematical Theory of Nonuniform Gases*, Cambridge University Press, New York 1939.
[1.10] J. O. Hirschfelder, C. F. Curtiss, R. B. Bird: *Molecular Theory of Gases and Liquids*, Wiley Interscience, New York 1954.
[1.11] P. D. Neufeld, A. R. Janzen, R. A. Aziz, *J. Chem. Phys.* **57** (1972) 1100.
[1.12] R. S. Brokaw, *Ind. Eng. Chem. Process Des. Dev.* **8** (1969) 240.
[1.13] C. Truesdell: *The Elements of Continuum Mechanics*, Springer Verlag, New York 1966.
[1.14] R. Darby: *Viscoelastic Fluids*, Marcel Dekker, New York 1976.

[1.15] W. R. Schowalter: *Mechanics of Non-Newtonian Fluids*, Pergamon Press, New York 1978.

References for Chapter 2

[2.1] S. Whitaker: *Fundamental Principles of Heat Transfer*, Pergamon Press, New York 1977.
[2.2] R. B. Bird, W. E. Stewart, E. N. Lightfoot: *Transport Phenomena*, J. Wiley & Sons, New York 1960.
[2.3] A. J. Chapman: *Heat Transfer*, Macmillan Publ. Co., New York 1960.
[2.4] H. S. Carslaw, J. C. Jaeger: *Conduction of Heat in Solids*, 2nd ed., Oxford University Press, London 1959.
[2.5] L. Brand: *Vector and Tensor Analysis*, J. Wiley & Sons, New York 1947.
[2.6] J. H. Lienhard: *A Heat Transfer Textbook*, Prentice Hall, Englewood Cliffs, N.J., 1987.
[2.7] H. P. Gurney, J. Lurie, *Ind. Eng. Chem.* **15** (1923) 1170.
[2.8] H. C. Groeber, *VDI-Z.* **69** (1925) 705.
[2.9] A. Shack: *Industrial Heat Transfer*, J. Wiley & Sons, New York 1933.
[2.10] M. P. Heisler, *Trans. ASME* **69** (1947) 227–236.
[2.11] J. R. Welty, C. E. Wicks, R. E. Wilson: *Fundamentals of Momentum, Heat and Mass Transfer*, J. Wiley & Sons, New York 1976.
[2.12] L. E. Sissom, D. R. Pitts: *Elements of Transport Phenomena*, McGraw-Hill, New York 1972.
[2.13] F. W. Kroesser, S. Middleman, *Polym. Eng. Sci.* **5** (1965) 1.
[2.14] R. W. Flumerfelt, M. W. Pierick, S. L. Cooper, R. B. Bird, *Ind. Eng. Chem. Fundam.* **8** (1969) 354.
[2.15] J. Happel, H. Brenner: *Low Reynolds Number Hydrodynamics*, Prentice Hall, Englewood Cliffs, N.J., 1965.
[2.16] S. M. Selby: *Standard Mathematical Tables*, 16th ed., The Chemical Rubber Company, Cleveland 1968.
[2.17] R. B. Bird, R. C. Armstrong, O. Hassager: *Dynamics of Polymeric Liquids*, vol. 1, Fluid Mechanics, J. Wiley & Sons, New York 1977.
[2.18] A. B. Metzner, J. C. Reed, *AIChE J.* **1** (1955) 434.
[2.19] A. H. P. Skelland: *Non-Newtonian Fluid and Heat Transfer*, J. Wiley & Sons, New York 1967.
[2.20] D. W. Dodge, A. B. Metzner, *AIChE J.* **5** (1959) 189.
[2.21] S. Goldstein: *Modern Developments in Fluid Dynamics*, Oxford University Press, London 1938.
[2.22] S. Prasad: *Non-Newtonian Flow Through Constricted Geometries*, M. S. Thesis, University of Houston, Houston, TX, 1978.
[2.23] T. B. Drew: *Trans. Am. Inst. Chem. Eng.* **26** (1931) 26.
[2.24] H. L. Langhaar, *J. Appl. Mech.* **64** (1942) A-55.
[2.25] W. M. Kays, *Trans. ASME* **77** (1955) 1265.
[2.26] J. R. Sellars, M. Tribus, J. S. Klein, *Trans. ASME* **78** (1956) 441.
[2.27] E. N. Sieder, G. E. Tale, *Ind. Eng. Chem.* **28** (1936) 1429.
[2.28] H. Hausen, *Verfahrenstechnik (Berlin)* **4** (1943) 91.
[2.29] H. Schlichting: *Boundary Layer Theory*, 6th ed., McGraw-Hill, New York 1968.
[2.30] L. M. Milne-Thomson: *Theoretical Hydrodynamics*, 5th ed., Macmillan, New York 1968.
[2.31] J. Schetz: *Foundations of Boundary Layer Theory*, Prentice-Hall, Englewood Cliffs, N.J., 1984.
[2.32] E. Pohlhausen, *Z. Angew. Math. Mech.* **1** (1921) 115.
[2.33] S. W. Churchill, H. H. S. Chu, *Int. J. Heat Mass Transfer* **18** (1975) 1323.
[2.34] P. S. Klebanoff: "Characteristics of Turbulence on a Boundary Layer with Zero Pressure Gradient," NACA Report 1247, 1955.
[2.35] J. Nikuradse, *Ing. Arch.* **1** (1930) 150; *VDI-Forschungsh.* **361** (1933) 1; *Pet. Eng.* **11** (1940) 164; **11** (1940) 75; **11** (1940) 124; **11** (1940) 38; **11** (1940) 83.
[2.36] L. Prandtl, *Z. Angew. Math. Mech.* **5** (1925) 136.
[2.37] J. M. Kay, R. W. Nedderman: *An Introduction to Fluid Mechanics and Heat Transfer*, 3rd ed., Cambridge University Press, London 1974.
[2.38] H. Blasius, *VDI-Forschungsh.* **131** (1913).
[2.39] C. F. Colebrook, *J. Inst. Civ. Eng. (London)* **133** (1938–1939).
[2.40] A. A. Zhukauskas, A. B. Ambrazyavichyus: NACA Report 909, 1949.
[2.41] L. Prandtl: "Über den Reibungswiderstand strömender Luft," Reports of the Aerol. Versuchsanst. Göttingen, 3rd Series, 1927; see also: "Zur turbulenten Strömung in Rohren und Längsplatten," Reports of the Aerol. Versuchsanst. Göttingen, 4th Series, 1931.
[2.42] W. L. Friend, A. B. Metzner, *AIChE J.* **4** (1958) 393.
[2.43] S. W. Churchill, M. Bernstein, *J. Heat Transfer* **99** (1977) 300.
[2.44] S. Whitaker, *AIChE J.* **18** (1972) 361.
[2.45] S. W. Churchill, H. H. S. Chu, *Int. J. Heat Mass Transfer* **18** (1975) 1323.
[2.46] J. P. Holman: *Heat Transfer*, 6th ed., McGraw-Hill, New York 1986.
[2.47] T. Fujii, H. Imura, *Int. J. Heat Mass Transfer* **15** (1972) 755.

References for Chapter 3

[3.1] R. B. Bird, W. E. Stewart, E. N. Lightfoot: *Transport Phenomena*, J. Wiley & Sons, New York 1960.
[3.2] J. C. Slattery: *Momentum, Energy and Mass Transfer in Continua*, 2nd ed., R. E. Krieger Publishing Co, Huntington, N.Y. 1981.
[3.3] E. L. Cussler: *Diffusion Mass Transfer in Fluid Systems*, Cambridge University Press, New York 1984.
[3.4] E. L. Cussler: *Multicomponent Diffusion*, Elsevier Scientific Publishing Company, New York 1976.
[3.5] J. Crank: *The Mathematics of Diffusion*, Clarendon Press, Oxford 1956.
[3.6] R. C. Reid, J. M. Prausnitz, T. K. Sherwood: *The Properties of Gases and Liquids*, 3rd ed., McGraw-Hill, New York 1977.
[3.7] H. S. Carslaw, J. C. Jaeger: *Conduction of Heat in Solids*, Oxford University Press, London 1959.
[3.8] P. D. Neufeld, A. R. Janzen, R. A. Aziz, *J. Chem. Phys.* **57** (1972) 1100.
[3.9] R. S. Brokaw, *Ind. Eng. Chem. Process Des. Dev.* **8** (1969) 240.
[3.10] E. N. Fuller, P. D. Schettler, J. C. Giddings, *Ind. Eng. Chem.* **58** (1966) 18.
[3.11] C. R. Wilke, P. Chang, *AIChE J.* **1** (1955) 264.
[3.12] J. R. Welty, C. E. Wicks, R. E. Wilson: *Fundamentals of Momentum, Heat, and Mass Transfer*, 2nd ed., J. Wiley & Sons, New York 1976.
[3.13] J. M. Coulson, J. F. Richardson: *Chemical Engineering*, 3rd ed., vol. 1, Pergamon Press, Oxford 1977.
[3.14] A. S. Foust et al.: *Principles of Unit Operations*, 2nd ed., J. Wiley & Sons, New York 1980.

[3.15] W. L. McCabe, J. C. Smith, P. Harriott: *Unit Operations of Chemical Engineering*, 4th ed., McGraw-Hill, New York 1985.
[3.16] R. E. Treybal: *Mass-Transfer Operations*, 3rd ed., McGraw-Hill, New York 1980.
[3.17] C. O. Bennett, J. E. Myers: *Momentum, Heat, and Mass Transfer*, 2nd ed., McGraw-Hill, New York 1974.
[3.18] W. G. Whitman, *Chem. Metall. Eng.* **29** (1923) 147.
[3.19] R. Higbie, *Trans. Am. Inst. Chem. Eng.* **31** (1935) 365.
[3.20] P. V. Danckwerts, *Ind. Eng. Chem.* **43** (1951) 1460.
[3.21] P. Harriott, *Chem. Eng. Sci.* **17** (1962) 149.
[3.22] H. L. Toor, J. M. Marchello, *AIChE J.* **4** (1958) 97.
[3.23] A. L. Hines, R. N. Maddox: *Mass Transfer, Fundamentals and Applications*, Prentice Hall, Englewood Cliffs, N.J., 1985.
[3.24] W. H. Linton, T. K. Sherwood, *Chem. Eng. Prog.* **46** (1950) 258.
[3.25] E. R. Gilliland, T. K. Sherwood, *Ind. Eng. Chem.* **26** (1934) 516.
[3.26] P. Harriott, R. M. Hamilton, *Chem. Eng. Sci.* **20** (1965) 1073.
[3.27] M. J. Christian, S. P. Kezibs, *AIChE J.* **5** (1959) 61.
[3.28] C. H. Bedingfield, I. B. Drew, *Ind. Eng. Chem.* **42** (1950) 1164.
[3.29] R. L. Steinberger, R. E. Treybal, *AIChE J.* **6** (1960) 227.
[3.30] F. H. Garner, R. D. Sucking, *AIChE J.* **4** (1958) 114.
[3.31] E. J. Wilson, C. J. Geankoplis, *Ind. Eng. Chem. Fundam.* **5** (1966) 9.
[3.32] A. Sen Gupta, G. Thodos, *AIChE J.* **9** (1963) 751.
[3.33] P. N. Dwivedi, S. N. Upadhyay, *Ind. Eng. Chem. Process Des. Dev.* **16** (1977) 157.
[3.34] A. Sen Gupta, G. Thodos, *AIChE J.* **8** (1962) 608.
[3.35] J. C. Chu, J. Kalil, W. A. Wetteroth, *Chem. Eng. Prog.* **49** (1953) 141.

References for Chapter 4

[4.1] R. B. Bird, W. E. Stewart, E. N. Lightfoot: *Transport Phenomena*, J. Wiley & Sons, New York 1960.
[4.2] R. S. Brodkey, H. C. Hershey: *Transport Phenomena, A Unified Approach*, McGraw-Hill, New York 1988.
[4.3] J. R. Welty, C. E. Wicks, R. E. Wilson: *Fundamentals of Momentum, Heat, and Mass Transfer*, 2nd ed., J. Wiley & Sons, New York 1969.
[4.4] O. A. Hougen, K. M. Watson, R. A. Ragatz: *Chemical Process Principles*, part II, *Thermodynamics*, 2nd ed., J. Wiley & Sons, New York 1959.
[4.5] J. M. Smith, H. C. Van Ness: *Introduction to Chemical Engineering Thermodynamics*, 3rd ed., McGraw-Hill, New York 1975.
[4.6] W. L. McCabe, J. C. Smith, P. Harriott: *Unit Operations of Chemical Engineering*, 4th ed., McGraw-Hill, New York 1985.
[4.7] F. M. White: *Fluid Mechanics*, 2nd ed., McGraw-Hill, New York 1986.
[4.8] S. Whitaker in N. A. Peppas (ed.): *One Hundred Years of Chemical Engineering*, Kluwer Academic Publishers, Dordrecht 1989, pp. 47–109.
[4.9] J. C. Slattery: *Momentum, Energy, and Mass Transfer in Continua*, 2nd ed., R. E. Krieger Publishing Co, Huntington, N.Y. 1981.
[4.10] R. B. Bird, *Chem. Eng. Sci.* **6** (1957) 123.
[4.11] L. W. Moody, *Trans. ASME* **66** (1944) 672.
[4.12] S. W. Churchill, *Chem. Eng.* (N.Y.) **87** (1977) Nov. 7, 91.
[4.13] P. R. Hooker, W. E. Brigham, *JPT J. Pet. Technol.* **30** (1978) 747.
[4.14] J. M. Coulson, J. F. Richardson: *Chemical Engineering*, 3rd ed., vol. 1, Pergamon Press, Oxford 1977.
[4.15] O. Levenspiel: *Chemical Reaction Engineering*, J. Wiley & Sons, New York 1962.
[4.16] K.-H. Lin in R. H. Perry, C. H. Chilton (eds.): *Chemical Engineer's Handbook*, 5th ed., section 4, McGraw-Hill, New York 1973.
[4.17] J. C. Chu, J. Kalil, W. A. Wetteroth, *Chem. Eng. Prog.* **49** (1953) 141.
[4.18] E. W. Thiele, *Ind. Eng. Chem.* **31** (1939) 916.
[4.19] R. Aris, *Chem. Eng. Sci.* **6** (1957) 262.
[4.20] C. D. Holland, R. G. Anthony: *Fundamentals of Chemical Reaction Engineering*, Prentice Hall, Englewood Cliffs, N.J., 1979.

5. Fluid Mechanics

DAVID V. BOGER, Y. LEONG YEOW, Department of Chemical Engineering, The University of Melbourne, Parkville, Victoria 3052, Australia

1.	Introduction	5-2	4.1.	Classification of Fluids According to Viscosity Behavior ... 5-25
2.	Basic Equations of Fluid Mechanics	5-4	4.1.1.	Bingham Type Behavior ... 5-25
2.1.	Continuity Equation	5-4	4.1.2.	Shear-Thinning (Pseudoplastic) Fluids 5-26
2.2.	Cauchy Equations of Motion	5-7	4.1.3.	Shear-Thickening (Dilatant) Fluids ... 5-27
2.3.	Energy Transport Equation and Bernoulli Equation	5-8	4.1.4.	Fluids with Time-Dependent Viscosity (Thixotropic Fluids) ... 5-28
2.4.	Constitutive Equations and Classification of Fluids	5-8	4.2.	Fully Developed Tube Flow ... 5-28
			4.2.1.	Volumetric Flow Rate–Pressure Drop Relationship ... 5-29
3.	Newtonian Fluids	5-9	4.2.2.	Generalized Treatment ... 5-30
3.1.	Deviatoric Stress and Viscosity	5-9	4.2.3.	Velocity Distribution ... 5-31
3.2.	Navier–Stokes Equations	5-11	4.2.4.	Friction Factor–Reynolds Number Relationships ... 5-32
3.3.	Application of Navier–Stokes Equations	5-11	4.3.	Viscoelastic Fluid Mechanics ... 5-33
3.3.1.	Flow in Pipes	5-11	4.3.1.	Steady Shear Behavior of Viscoelastic Fluids ... 5-35
3.3.2.	Concentric Cylinder Flow	5-13	4.3.2.	Behavior of Viscoelastic Fluids in Oscillatory Shear Flow ... 5-36
3.3.3.	Creeping Flow Past a Sphere	5-15	4.3.3.	Examples of Constitutive Equations for Viscoelastic Fluids ... 5-37
3.4.	Some Other Important Flows	5-17	4.3.4.	Extensional Behavior of Viscoelastic Fluids ... 5-39
3.4.1.	Flow Through Granular Beds	5-17		
3.4.2.	Fluidization	5-18	4.3.5.	Accelerating and Deceleration Flow of Viscoelastic Fluids ... 5-40
3.4.3.	Gas–Liquid Flow	5-18		
3.5.	Mechanical Energy Balance for Macroscopic Systems	5-21	5.	Numerical Methods in Fluid Mechanics 5-44
3.5.1.	Fully Developed Tube Flows	5-22	5.1.	Finite Difference Method ... 5-44
3.5.2.	Accelerating and Decelerating Flows	5-22	5.2.	Finite Element Method ... 5-46
3.5.3.	The Orifice Plate and Other Flow Rate Measurement Devices	5-23	5.3.	General Remarks ... 5-49
4.	Non-Newtonian Fluids	5-24	6.	References ... 5-49

In addition to the standard symbols defined in the front matter of this volume, the following symbols are used:

A	cross-sectional area, surface area, amplitude of oscillation
C	loss coefficient
C_d	drag coefficient, discharge coefficient
d	distance
D	diameter
D_e	extrudate diameter
D_{hy}	hydraulic diameter
D_p	particle diameter
D_v	vane diameter
e	surface roughness, end correction
$\dot{e}$	rate of strain tensor
$\dot{e}_{xx}, \dot{e}_{xy}, \dot{e}_{ij}$	components of rate of strain tensor
Ev	frictional loss
Ev_{ex}	losses in valves, fittings, etc.
Ev_{fd}	losses in fully developed flows
f	friction factor
$\boldsymbol{F}$	force vector
F_d	drag force
F_i	component of force vector
G	volumetric flow rate in gas phase
G'	storage modulus
G''	loss modulus

h	elevation	λ_1	relaxation time
H	gap between plates	λ_2	retardation time
H_v	vane height	ϱ	density
i	unit vector in x direction	ϱ_p	particle density
I	identity matrix	τ	deviatoric stress tensor
j	unit vector in y direction	τ_{ij}	component of deviatoric stress tensor
J_G	superficial gas phase velocity	$\tau_{rz}, \tau_{r\theta}$	shear stress
J_L	superficial liquid phase velocity	τ_w	wall shear stress
k	thermal conductivity, parameter in the power-law equation	τ_y	yield stress
		$\tau_{1/2}$	parameter of Ellis and Meter model
$\mathbf{k}$	unit vector in z direction	Φ	interpolation function
K'	intercept in the log–log plot of τ_w versus $8V/D$	ψ_1	first normal stress coefficient
		ψ_2	second normal stress coefficient
L	length, volumetric flow rate in liquid phase	Ψ	stream function
L_e	equivalent length	$\omega_1, \omega_2, \omega$	angular velocity, vorticity
L_0	initial length	Ω	angular velocity
n	flow-behavior index of power law equation		
$\mathbf{n}$	unit outward pointing normal		
n'	slope of log–log plot of τ_w versus $8V/D$		
N_1	first normal stress		
p	isotropic pressure		
$\mathbf{q}$	heat flux vector		
Q	volumetric flow rate		
r	radial coordinate		
r_0	plug radius		
R_c, R_x, R_y	residuals		
R, R_1, R_2	radius		
Re	Reynolds number		
Re'	generalized Reynolds number for time-independent fluid		
S	path joining two points		
t	time		
T	temperature		
$\mathbf{T}$	stress tensor		
$T_{ij}, T_{xx}, T_{xy}, T_{xz}$	stress components		
$\mathbf{u}$	velocity vector		
u_f	fluidizing velocity		
$u_i, u_x, u_y, u_r, u_t, u_0$	velocity components		
U, U_0, U_∞	velocity		
V	average velocity, volume		
W	mass flow rate		
W_s	shaft work		
$\mathbf{x}$	position vector		
x_i	components of position vector		
z	elevation		
α	parameter in Meter model		
γ	shear strain		
γ_0	amplitude of sinusoidal shear strain		
$\dot\gamma$	shear rate		
$\dot\gamma_0$	amplitude of sinusoidal shear rate		
Γ, Γ_m	torque		
δ_{ij}	Kronecker delta		
ε	porosity		
$\dot\varepsilon_0$	constant extensional rate		
η	viscosity		
η_{el}	extensional viscosity		
η_{pl}	plastic viscosity		
η_0	viscosity parameter of Meter, Maxwell, and Oldroyd-B models		
η'	dynamic viscosity		
η''	component of complex viscosity		
θ	azimuthal coordinate		
λ	bulk viscosity, time constant		
λ_{el}	extensional viscosity		

1. Introduction

The traditional approach for writing an article or a text book on fluid mechanics has been to deal with the Newtonian fluid and the resulting mechanics associated with such materials. Thus the basic tools (in a sequence of events such as those illustrated in Figure 1) are treated in detail to illustrate how to solve a Newtonian fluid mechanics problem. The laws of conservation of mass (continuity equation) and conservation of momentum (Cauchy momentum equations or equations of motion) are established for a general system and in multidimensional form. From these equations it then becomes clear that the solution to any fluid mechanics problem is not possible unless a relationship is developed between the stress tensor and rate of deformation tensor, or, between the stresses and velocity gradients in the system. Such an equation is termed the constitutive equation. For Newtonian fluids the relationship between the stress tensor and the rate of deformation is linear, the proportionality constant is the viscosity. Substitution of this relationship into the Cauchy momentum equations leads to the Navier–Stokes equations. Thus the Navier–Stokes equations, in conjunction with the viscosity, form the basis for the solution of classical fluid mechanics problems. Much of the effort in classical fluid mechanics is devoted to establishing methods of solution of the Navier–Stokes equations. These methods can be analytical in nature and in recent times, of course, are numerical. While one whole section of this article is devoted to numerical methods because of their importance in modern fluid mechanics, no attention is directed towards classical analytical methods such as those used in the solution of boundary layer and slow flow

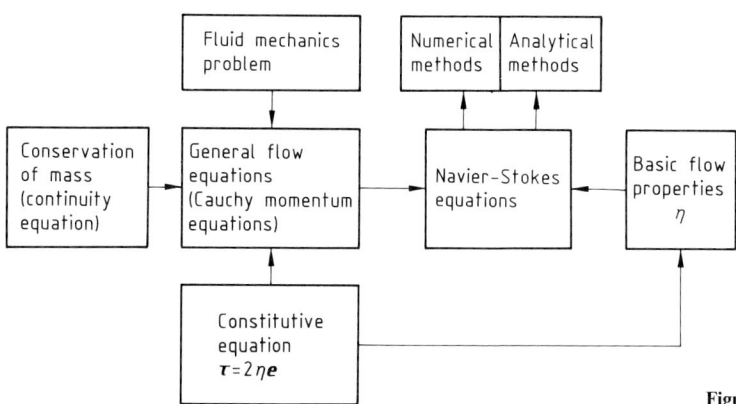

Figure 1. Newtonian fluid mechanics

(creeping flow) problems. The emphasis is directed towards the industrial chemist, process scientist, and chemical engineer who encounter a vast range of materials in today's industries where the Newtonian fluid is only one small subset of a vast array of fluid behavior which can be observed. Figure 2 illustrates the range of behavior that can be encountered in terms of the viscosity of fluids and in terms of whether materials are classified as being inelastic or viscoelastic fluids. It is clear from examination of Figure 2 that a vast range of behavior in terms of the viscosity of materials can be and is observed. An industrial chemist or process engineer will encounter low molecular mass liquids and gases (Newtonian materials) but will also be dealing with polymer solutions, polymer melts, rubbers, mineral suspensions, food products, pharmaceuticals, and energy products such as coal oil and coal water fuels, etc. In such systems the viscosity can be constant, but more likely will vary with shear rate or rate of flow. Viscosity can be a function of shear rate and time of shear; or it can be a function of shear rate, time, and thermal history and each of these viscosity behaviors can be observed in inelastic and/or viscoelastic fluids. In addition many materials will not flow until a certain stress is exceeded (e.g., toothpaste). This stress, termed yield stress, can be present in both inelastic and viscoelastic fluids. Therefore, it is important to be aware of the vast range of material behavior present in industry.

In Chapter 2 the basic equations for fluid mechanics—the continuity equation, the Cauchy equations of motion, and the energy transport and Bernoulli equations—are established. Chapter 3 deals with Newtonian fluids and Newtonian fluid mechanics where the Navier–Stokes equations are developed. In addition, the mechanical energy balance is established for macroscopic

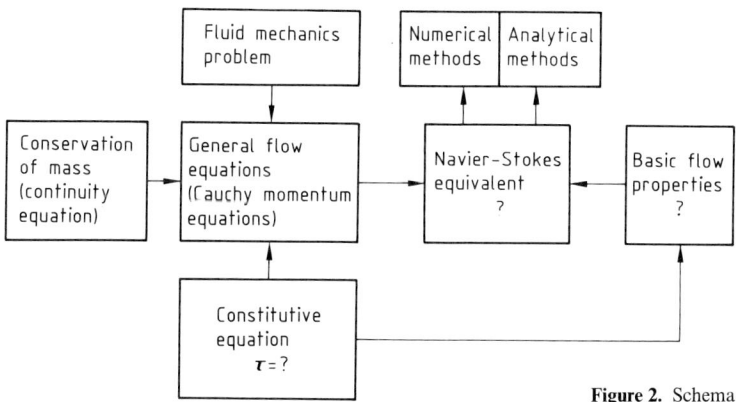

Figure 2. Schematic classification of fluid behavior

systems in order that pipe flow systems can be designed. The emphasis in the article is directed toward pipe flows because the process engineer is more concerned with flow in conduits than in other geometries. Little emphasis is placed on the details associated with turbulence because process engineers are primarily concerned with turbulent flow in pipes of Newtonian fluids and are not concerned with design of aerofoil structures, etc.

Chapter 4 concentrates on non-Newtonian fluids. In this chapter fluids are classified according to their viscosity behavior and are examined in some depth in fully developed tube flow. Viscoelastic fluid mechanics are also summarized—the steady shear and dynamic shear behavior of viscoelastic fluids are reviewed and examples of simple constitutive equations are given. The article is completed with a brief examination of numerical methods in fluid mechanics.

The emphasis of this article is shifted toward non-Newtonian fluid mechanics, the sequence of events and tools of importance are illustrated in Figure 3. Here the basic conservation of mass (continuity equation) and the conservation of momentum (Cauchy momentum equations) remain the equations of fundamental importance, but, the relevant constitutive equation in most cases is still unknown. Thus the Navier–Stokes equivalent set of equations in many cases is not known and because any number of constitutive equations are available it is not entirely clear what basic flow property information is required for the most general viscoelastic material. It is only now (1990) that fluid mechanics problems are solved for the first time for viscoelastic fluids. The solutions evolving now for such problems will be extremely important in the future. It is also of considerable importance to keep abreast of the numerical software packages which have been developed in this area particularly in regard to the processing of polymers [1], [2].

2. Basic Equations of Fluid Mechanics

The basic equations of fluid mechanics can be derived from a small number of fundamental physical laws such as conservation of mass, of momentum (Newton's second law), and of energy. Although very general statements of these laws can be written down (applicable to all substances, solids as well as fluids), in fluid mechanics these laws are formulated in terms of physical variables such as velocity u, pressure p, temperature T, and density ϱ. The resulting statements may be less general but they are more directly applicable in analyzing fluid mechanical problems.

2.1. Continuity Equation

The requirement that mass be conserved at every point in a flowing fluid imposes certain restrictions on the velocity u and density ϱ. Consider an arbitrary region in space, of volume V, bounded by surface A, through which the fluid flows (see Fig. 4). The rate of *increase in mass* within this region is given by the volume integral

$$\int_V \frac{\partial \varrho}{\partial t} \, dV \tag{1}$$

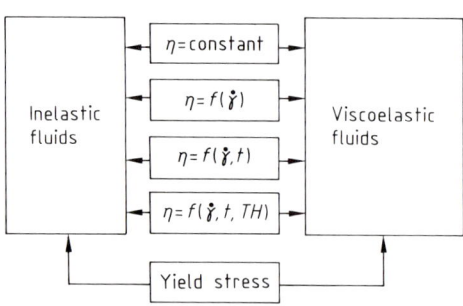

Figure 3. Non-Newtonian fluid mechanics
TH = Thermal history

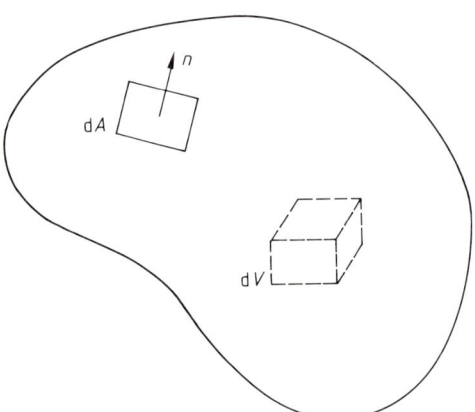

Figure 4. Fluids flowing through an arbitrary region in space of volume V and surface area A

The rate at which fluid is *leaving the region* is given by the surface integral of the outward pointing normal velocity multiplied by the fluid's density taken over the surface A

$$\int_A \varrho \mathbf{u} \cdot \mathbf{n} \, dA = \int_V \nabla \cdot (\varrho \mathbf{u}) \, dV \tag{2}$$

where $\mathbf{n}$ is the unit outward pointing normal at any point on A. The divergence theorem has been used to convert the surface integral into a volume integral. Mass conservation requires the sum of the rate of accumulation in V and the rate of outflow from V to be zero:

$$\int_V \left[\frac{\partial \varrho}{\partial t} + \nabla \cdot (\varrho \mathbf{u}) \right] dV = 0 \tag{3}$$

Since V is arbitrary, the integrand itself must vanish at every point in space:

$$\frac{\partial \varrho}{\partial t} + \nabla \cdot (\varrho \mathbf{u}) = 0 \tag{4}$$

Equation (4) is usually referred to as the *continuity equation*. It is a statement of mass conservation in terms of fluid velocity and density. Expanding the second term and regrouping, the equation can be written as

$$\frac{\partial \varrho}{\partial t} + \mathbf{u} \cdot \nabla \varrho + \varrho \nabla \cdot \mathbf{u} = 0 \tag{5}$$

or

$$\frac{D\varrho}{Dt} + \varrho \nabla \cdot \mathbf{u} = 0 \tag{6}$$

where D/Dt is used to represent the operator $\frac{\partial}{\partial t} + \mathbf{u} \cdot \nabla$. It is known as the substantial derivative operator. By following the change in the density $\varrho(\mathbf{x}, t)$ of a fluid element occupying position $\mathbf{x}$ at time t and its density $\varrho(\mathbf{x} + \delta \mathbf{x}, t + \delta t)$ at a later time $t + \delta t$ when it is occupying position $\mathbf{x} + \delta \mathbf{x}$, it can be seen that $D\varrho/Dt$ is the rate of change in density experienced by a fluid element as it moves about in space. $D\varrho/Dt$ should not be confused with $\partial \varrho/\partial t$ which is the rate of change in density at a fixed point in space. The substantial derivative operator can also be applied to the velocity $\mathbf{u}$, temperature T, or other properties associated with a fluid element with a similar physical interpretation of the result.

A number of special cases of the continuity equation are frequently encountered. For exam-ple, under steady-state conditions the equation reduces to

$$\nabla \cdot (\varrho \mathbf{u}) = 0 \tag{7}$$

Much of the subsequent analyses will be restricted to fluids that are incompressible and homogeneous. For such fluids the continuity equation takes on a particularly simple form

$$\nabla \cdot \mathbf{u} = 0 \tag{8}$$

In many practical problems the fluid flow can be regarded as approximately two-dimensional. For such problems, Equation 8 (in cartesian coordinates) takes on the form

$$\frac{\partial u_x(x, y)}{\partial x} + \frac{\partial u_y(x, y)}{\partial y} = 0 \tag{9}$$

Equation (9) and, hence, mass conservation are identically satisfied if the velocity components are given by

$$u_x = \frac{\partial \Psi(x, y)}{\partial y} \quad \text{and} \quad u_y = -\frac{\partial \Psi(x, y)}{\partial x} \tag{10}$$

where $\Psi(x, y)$, the *stream function*, is any scalar function of x and y. A simple physical interpretation can be given to Ψ. Consider a line of constant Ψ in the $x - y$ plane, along such a line

$$\delta \Psi = \frac{\partial \Psi}{\partial x} \delta x + \frac{\partial \Psi}{\partial y} \delta y = 0 \tag{11}$$

i.e.,

$$-u_y \delta x + u_x \delta y = 0$$

or

$$\frac{dy}{dx} = \frac{u_y}{u_x} \tag{12}$$

Since dy/dx is the local slope of the line, Equation (12) shows that the velocity vector is everywhere tangential to lines of constant Ψ. Such lines are referred to as *streamlines*. They give the instantaneous flow pattern of the fluid. By integrating along any path S joining streamlines where $\Psi = \Psi_A$ and $\Psi = \Psi_B$, it can be seen that

$$\Psi_B - \Psi_A = \int_S \frac{d\Psi}{dS} dS = \int u_n dS \tag{13}$$

i.e., volumetric flow crossing S; u_n is the velocity component normal to S. Equation (13) shows that the difference in stream function between

Table 1. Continuity and Navier–Stokes equations for incompressible homogeneous fluids in cartesian, cylindrical, and spherical coordinates

Cartesian	Cylindrical	Spherical

Continuity equation

Cartesian:
$$\frac{\partial u_x}{\partial x} + \frac{\partial u_y}{\partial y} + \frac{\partial u_z}{\partial z} = 0$$

Cylindrical:
$$\frac{\partial (r u_r)}{r \partial r} + \frac{1}{r}\left(\frac{\partial u_\theta}{\partial \theta}\right) + \frac{\partial u_z}{\partial z} = 0$$

Spherical:
$$\frac{1}{r^2}\frac{\partial (r^2 u_r)}{\partial r} + \frac{1}{r \sin\theta}\frac{\partial (u_\theta \sin\theta)}{\partial \theta} + \frac{1}{r \sin\theta}\frac{\partial u_\varphi}{\partial \varphi} = 0$$

Navier–Stokes equation

Cartesian:
$$\varrho\left(\frac{\partial u_x}{\partial t} + u_x\frac{\partial u_x}{\partial x} + u_y\frac{\partial u_x}{\partial y} + u_z\frac{\partial u_x}{\partial z}\right)$$
$$= -\frac{\partial p}{\partial x} + \eta\left(\frac{\partial^2 u_x}{\partial x^2} + \frac{\partial^2 u_x}{\partial y^2} + \frac{\partial^2 u_x}{\partial z^2}\right)$$

$$\varrho\left(\frac{\partial u_y}{\partial t} + u_x\frac{\partial u_y}{\partial x} + u_y\frac{\partial u_y}{\partial y} + u_z\frac{\partial u_y}{\partial z}\right)$$
$$= -\frac{\partial p}{\partial y} + \eta\left(\frac{\partial^2 u_y}{\partial x^2} + \frac{\partial^2 u_y}{\partial y^2} + \frac{\partial^2 u_y}{\partial z^2}\right)$$

$$\varrho\left(\frac{\partial u_z}{\partial t} + u_x\frac{\partial u_z}{\partial x} + u_y\frac{\partial u_z}{\partial y} + u_z\frac{\partial u_z}{\partial z}\right)$$
$$= -\frac{\partial p}{\partial z} + \eta\left(\frac{\partial^2 u_z}{\partial x^2} + \frac{\partial^2 u_z}{\partial y^2} + \frac{\partial^2 u_z}{\partial z^2}\right)$$

Cylindrical:
$$\varrho\left(\frac{\partial u_r}{\partial t} + u_r\frac{\partial u_r}{\partial r} + \frac{u_\theta}{r}\frac{\partial u_r}{\partial \theta} - \frac{u_\theta^2}{r} + u_z\frac{\partial u_r}{\partial z}\right)$$
$$= -\frac{\partial p}{\partial r} + \eta\left[\frac{\partial}{\partial r}\left(\frac{1}{r}\frac{\partial}{\partial r}(r u_r)\right) + \frac{1}{r^2}\frac{\partial^2 u_r}{\partial \theta^2} - \frac{2}{r^2}\frac{\partial u_\theta}{\partial \theta} + \frac{\partial^2 u_r}{\partial z^2}\right]$$

$$\varrho\left(\frac{\partial u_\theta}{\partial t} + u_r\frac{\partial u_\theta}{\partial r} + \frac{u_\theta}{r}\frac{\partial u_\theta}{\partial \theta} + \frac{u_r u_\theta}{r} + u_z\frac{\partial u_\theta}{\partial z}\right)$$
$$= -\frac{1}{r}\frac{\partial p}{\partial \theta} + \eta\left[\frac{\partial}{\partial r}\left(\frac{1}{r}\frac{\partial}{\partial r}(r u_\theta)\right) + \frac{1}{r^2}\frac{\partial^2 u_\theta}{\partial \theta^2} + \frac{2}{r^2}\frac{\partial u_r}{\partial \theta} + \frac{\partial^2 u_\theta}{\partial z^2}\right]$$

$$\varrho\left(\frac{\partial u_z}{\partial t} + u_r\frac{\partial u_z}{\partial r} + \frac{u_\theta}{r}\frac{\partial u_z}{\partial \theta} + u_z\frac{\partial u_z}{\partial z}\right)$$
$$= -\frac{\partial p}{\partial z} + \eta\left[\frac{1}{r}\frac{\partial}{\partial r}\left(r\frac{\partial u_z}{\partial r}\right) + \frac{1}{r^2}\frac{\partial^2 u_z}{\partial \theta^2} + \frac{\partial^2 u_z}{\partial z^2}\right]$$

Spherical:
$$\varrho\left(\frac{\partial u_r}{\partial t} + u_r\frac{\partial u_r}{\partial r} + \frac{u_\theta}{r}\frac{\partial u_r}{\partial \theta} + \frac{u_\varphi}{r \sin\theta}\frac{\partial u_r}{\partial \varphi} - \frac{u_\theta^2 + u_\varphi^2}{r}\right)$$
$$= -\frac{\partial p}{\partial r} + \eta\left[\frac{1}{r^2}\frac{\partial}{\partial r}\left(r^2\frac{\partial u_r}{\partial r}\right) - \frac{2}{r^2}u_r - \frac{2}{r^2}\frac{\partial u_\theta}{\partial \theta} - \frac{2}{r^2}u_\theta \cot\theta - \frac{2}{r^2\sin\theta}\frac{\partial u_\varphi}{\partial \varphi}\right.$$
$$\left.+ \frac{1}{r^2\sin^2\theta}\frac{\partial^2 u_r}{\partial \varphi^2}\right]$$

$$\varrho\left(\frac{\partial u_\theta}{\partial t} + u_r\frac{\partial u_\theta}{\partial r} + \frac{u_\theta}{r}\frac{\partial u_\theta}{\partial \theta} + \frac{u_\varphi}{r \sin\theta}\frac{\partial u_\theta}{\partial \varphi} + \frac{u_r u_\theta}{r} - \frac{u_\varphi^2 \cot\theta}{r}\right)$$
$$= -\frac{1}{r}\frac{\partial p}{\partial \theta} + \eta\left[\frac{1}{r^2}\frac{\partial}{\partial r}\left(r^2\frac{\partial u_\theta}{\partial r}\right) + \frac{1}{r^2\sin\theta}\frac{\partial}{\partial \theta}\left(\sin\theta\frac{\partial u_\theta}{\partial \theta}\right) + \frac{2}{r^2}\frac{\partial u_r}{\partial \theta} - \frac{u_\theta}{r^2\sin^2\theta} - \frac{2\cos\theta}{r^2\sin^2\theta}\frac{\partial u_\varphi}{\partial \varphi}\right.$$
$$\left.+ \frac{1}{r^2\sin^2\theta}\frac{\partial^2 u_\theta}{\partial \varphi^2}\right]$$

$$\varrho\left(\frac{\partial u_\varphi}{\partial t} + u_r\frac{\partial u_\varphi}{\partial r} + \frac{u_\theta}{r}\frac{\partial u_\varphi}{\partial \theta} + \frac{u_\varphi}{r \sin\theta}\frac{\partial u_\varphi}{\partial \varphi} + \frac{u_\varphi u_r}{r} + \frac{u_\theta u_\varphi}{r}\cot\theta\right)$$
$$= -\frac{1}{r\sin\theta}\frac{\partial p}{\partial \varphi} + \eta\left[\frac{1}{r^2}\frac{\partial}{\partial r}\left(r^2\frac{\partial u_\varphi}{\partial r}\right) + \frac{1}{r^2\sin\theta}\frac{\partial}{\partial \theta}\left(\sin\theta\frac{\partial u_\varphi}{\partial \theta}\right) - \frac{u_\varphi}{r^2\sin^2\theta} + \frac{2}{r^2\sin\theta}\frac{\partial u_r}{\partial \varphi} + \frac{2\cos\theta}{r^2\sin^2\theta}\frac{\partial u_\theta}{\partial \varphi}\right.$$
$$\left.+ \frac{1}{r^2\sin^2\theta}\frac{\partial^2 u_\varphi}{\partial \varphi^2}\right]$$

two streamlines is equal to the volumetric flow rate across any line joining the two streamlines.

The continuity equation and the stream function take on different forms when written in non-cartesian coordinates. Listing of the equations of fluid mechanics in a number of commonly encountered coordinates can be found in most standard textbooks on this subject, e.g., in [3]. A partial list is given in Table 1.

2.2. Cauchy Equations of Motion

Before deriving the equations of motion for fluids, it is necessary to examine the *forces* acting on a fluid element as it moves about in space. Consider again the arbitrary volume V in space (see Fig. 4). One of the external forces acting on the fluid inside V is its weight which is given by the volume integral

$$\int_V \varrho \boldsymbol{g} \, dV \tag{14}$$

where $\boldsymbol{g}$ is the acceleration due to gravity. Another external force is that exerted by the fluid outside V. This force acts across the closed surface A and can be expressed in terms of the stress tensor $\boldsymbol{T}$ on the surface. In cartesian coordinates this stress tensor has nine components $T_{xx}, T_{xy}, T_{xz}, T_{yx}, T_{yy} \ldots T_{zz}$, or more concisely T_{ij}; $i, j = 1, 2$ or 3. x_1, x_2, x_3 are identified with x, y and z, respectively. T_{ij} is the force per unit area in direction j on a surface with normal in the i direction. The force $\delta \boldsymbol{F}$ acting on an area dA which has a normal $\boldsymbol{n}$ is given by

$$\delta \boldsymbol{F} = \boldsymbol{T} \cdot \boldsymbol{n} \, dA \quad \text{or} \quad \delta F_i = \sum_{j=1}^{3} T_{ji} n_j \, dA \tag{15}$$

Equation (15) is often referred to as the *Cauchy stress principle*. For the fluids commonly encountered, the stress tensor can be taken to be symmetric, i.e., $T_{ij} = T_{ji}$ for all combinations of i and j. Thus, the external force acting through the closed surface A is given by the surface integral

$$\int_A \boldsymbol{T} \cdot \boldsymbol{n} \, dA \tag{16}$$

which is equivalent to

$$\int_V \nabla \cdot \boldsymbol{T} \, dV \tag{17}$$

where the divergence theorem is again used to convert the surface integral to a volume integral. The i component of the vector $(\nabla \cdot \boldsymbol{T})$ is

$$(\nabla \cdot \boldsymbol{T})_i = \sum_{j=1}^{3} \frac{\partial T_{ij}}{\partial x_j} \tag{18}$$

To obtain the equations of motion in terms of $\boldsymbol{T}$, $\boldsymbol{u}$, and $(\varrho \boldsymbol{g})$, it is convenient to start with a statement of conservation of momentum which asserts that: the rate of change of momentum inside the volume V and the net rate at which momentum is being convected out of the volume is equal to the sum of the external forces acting on the volume. The rate of accumulation is given by

$$\int_V \frac{\partial (\varrho \boldsymbol{u})}{\partial t} \, dV \tag{19}$$

The rate at which momentum is being convected out is given by the surface integral

$$\int_A \varrho \boldsymbol{u} (\boldsymbol{u} \cdot \boldsymbol{n}) \, dA = \int_V \nabla \cdot (\varrho \boldsymbol{u} \boldsymbol{u}) \, dV \tag{20}$$

where $\varrho \boldsymbol{u}$ is the momentum per unit volume and $\boldsymbol{u} \cdot \boldsymbol{n}$ is the volumetric flux leaving A. Conservation of momentum in V then requires

$$\int_V \frac{\partial (\varrho \boldsymbol{u})}{\partial t} \, dV + \int_V \nabla \cdot (\varrho \boldsymbol{u} \boldsymbol{u}) \, dV$$
$$= \int_V \varrho \boldsymbol{g} \, dV + \int_V \nabla \cdot \boldsymbol{T} \, dV \tag{21}$$

Since V is arbitrary, the integrands must be equal

$$\frac{\partial (\varrho \boldsymbol{u})}{\partial t} + \nabla \cdot (\varrho \boldsymbol{u} \boldsymbol{u}) = \varrho \boldsymbol{g} + \nabla \cdot \boldsymbol{T} \tag{22}$$

Upon expanding the terms on the left-hand side and regrouping, Equation (22) becomes

$$\boldsymbol{u} \left[\frac{\partial \varrho}{\partial t} + \nabla \cdot (\varrho \boldsymbol{u}) \right] + \varrho \left[\frac{\partial \boldsymbol{u}}{\partial t} + \boldsymbol{u} \cdot \nabla \boldsymbol{u} \right]$$
$$= \varrho \boldsymbol{g} + \nabla \cdot \boldsymbol{T} \tag{23}$$

From the continuity equation (Eq. 4) the first group of terms on the left-hand side is zero. The second group of terms can be identified as the product of ϱ and the substantial derivative of velocity. Therefore

$$\varrho \frac{D \boldsymbol{u}}{Dt} = \varrho \boldsymbol{g} + \nabla \cdot \boldsymbol{T} \tag{24}$$

This is the equation of motion for fluids. It is often referred to as the *Cauchy momentum equation* or equation of motion. The left-hand side is referred to as the inertia term. The first term on the right-hand side is the force per unit volume acting on a fluid element arising from gravitational body force. The second term is the force per unit volume acting on the fluid element as a result of the spatial variation of the stress tensor. The form of the Cauchy equations of motion in different coordinate systems can be found in [3].

2.3. Energy Transport Equation and Bernoulli Equation

In flow problems where a substantial variation in temperature occurs, an additional equation is needed to relate the local temperature T of the fluid to variables such as velocity, velocity gradient, and temperature gradient. Such an equation takes the general form

$$\varrho C_V \frac{DT}{Dt} = -\nabla \cdot q - T\left(\frac{\partial \varrho}{\partial T}\right)_V \nabla \cdot u$$
$$+ \text{dissipation terms}$$

where q is the heat flux and C_V is the heat capacity at constant volume. This equation simplifies to

$$\varrho C_p \frac{DT}{Dt} = k\nabla^2 T + \text{dissipation terms} \quad (25)$$

for an incompressible fluid with a heat capacity, C_p, at constant pressure and when the temperature dependence of the thermal conductivity k can be neglected. In arriving at Equation (25), Fourier's expression for heat conduction ($q = -k\nabla T$) was used to express q in terms of temperature gradient. Equation (25), the energy transport equation, is a statement of conservation of energy and is a form of the First Law of Thermodynamics. It equates the rate of energy transfer to a fluid element by thermal conduction and the rate of dissipation/conversion of kinetic energy into thermal energy to the accumulation of thermal energy within the element. The dissipation terms arise as a result of the deformation suffered by the fluid element. The derivation of the energy transport equation and the exact form of the dissipation terms are given in [3]. In the following sections only isothermal flows will be discussed. In such flows the dissipation terms are negligible and the energy transport equation is satisfied. This greatly simplifies the analysis.

In many flow fields of practical interest, a simple statement of conservation of kinetic and potential energy can be derived. This is obtained by forming the dot product between the velocity vector u, and the equations of motion (24), and integrating the result between any two points on the same streamline. The resulting equation takes the form

$$\frac{1}{2}u_1^2 + gh_1 + \frac{p_1}{\varrho} = \frac{1}{2}u_1^2 + gh_2 + \frac{p_2}{\varrho} + \text{losses} \quad (26)$$

This is one form of the well-known *Bernoulli equation*; h, u, and p are the vertical elevation, the velocity, and the isotropic pressure of the fluid respectively. Subscript 1 and 2 denote where these variables are to be evaluated on the streamline. Losses refer to the conversion of mechanical (kinetic and potential) energy into internal (thermal) energy between the two points. In deriving this equation, the fluid is taken to be incompressible and the flow to be steady. Relaxation of these assumptions will result in a more general form of the Bernoulli equation [1]. The Bernoulli equation finds many practical applications, especially in flows where the losses are small or can be estimated with reasonable accuracy. In many applications the spatial average value of the variables at point 1 and point 2 are used instead of local point values, resulting in further simplification (see Section 3.5).

2.4. Constitutive Equations and Classification of Fluids

The equations derived from the conservation laws do not contain any information about the mechanical properties of fluids. An applied force or stress acting on fluids with different properties will result in different flow patterns. Conversely, the same flow pattern will induce different stresses in different fluids. The diverse mechanical properties of fluids are described by equations that relate the local stress tensor to the local flow kinematics. Such equations are known as *rheological constitutive equations*.

One of the most important kinematic variables that appear in rheological constitutive equations is the *rate of strain tensor* e. This is defined as

$$e = \frac{1}{2}[\nabla u + (\nabla u)^T] \quad (27)$$

or in cartesian coordinates

$$e_{ij} = \frac{1}{2}\left(\frac{\partial u_i}{\partial x_j} + \frac{\partial u_j}{\partial x_i}\right) \quad (28)$$

The rate of strain tensor is a symmetric tensor. Simple physical interpretations can be given to the components of e. For example, e_{11} or e_{xx} is the rate at which a fluid element is being stretched in the x direction, e_{12} or e_{xy} is the rate at which the fluid element is being sheared in the $x-y$ plane. The form of the rate of strain and other kinematic tensors in different coordinate systems can be found in [4]. It should be mentioned that some authors, particularly those working on non-Newtonian fluids, define the rate of strain tensor without the factor of 1/2 as shown in Equation (27). This difference in the definition can lead to considerable confusion and should carefully noted.

A convenient way of classifying fluids is according to the form of their constitutive equations. For a large class of fluids, the stress tensor T is a linear function of the rate of strain tensor e. Such fluids are referred to as Newtonian fluids. If T is a non-linear function of e the fluids are said to be non-Newtonian. In chemically more complex fluids, the stress is not only a function of e, it also depends on the deformation or the entire history of deformation suffered by the fluid. Such fluids may also exhibit solidlike elastic behavior and are known as viscoelastic fluids. (Non-Newtonian and viscoelastic fluids are treated in greater detail in Chap. 4).

3. Newtonian Fluids

3.1. Deviatoric Stress and Viscosity

In a stagnant fluid, the stress at any point is the hydrostatic pressure. This is an isotropic compressive stress, i.e., the stress has the same magnitude in any direction. Such a stress can be represented by

$$T = -p I \quad (29)$$

or in component form

$$T_{ij} = -p \delta_{ij} \quad (30)$$

where I is the identity tensor

$$\begin{bmatrix} 1 & 0 & 0 \\ 0 & 1 & 0 \\ 0 & 0 & 1 \end{bmatrix}$$

and δ_{ij} is the Kronecker delta

$$\begin{aligned} \delta_{ij} &= 1 \quad i = j \\ \delta_{ij} &= 0 \quad i \neq j \end{aligned} \quad (31)$$

Unless otherwise stated, i and j take on values 1, 2, or 3 in all the equations in this article. The shear components (when i is not equal to j) of the isotropic stress are identically zero and the three normal components take on the same value p, the magnitude of the pressure; p is a function of the vertical position in the fluid. It is customary to have a negative sign before p so that for hydrostatic compressive stress, the numerical value associated with p is positive. All these are in agreement with the well known nature of hydrostatic pressure.

For a fluid in motion the stress will, in general, no longer be isotropic. At any point the stress is a combination of the hydrostatic pressure and the stresses arising from the deformation experienced by the fluid. It is convenient to decompose the stress tensor T_{ij} into an isotropic part $-p\delta_{ij}$ and a nonisotropic part τ_{ij},

$$T_{ij} = -p\delta_{ij} + \tau_{ij} \quad (32)$$

τ_{ij}, usually referred to as the deviatoric stress tensor, is a consequence of the deformation. For a fluid element that has not experienced any deformation, the deviatoric stress vanishes. In a stagnant fluid the deviatoric stress is therefore zero and the resulting isotropic stress can be identified with the hydrostatic pressure. However, because of fluid motion, the isotropic component of the stress tensor will, in general, be different from the hydrostatic pressure.

The nature of the relationship between deviatoric stress and deformation depends on the rheological constitutive equation of the fluid. The constitutive equation completely describes the deformation arising from a specified deviatoric stress and, conversely, gives the stresses needed to produce a specified deformation. For a large class of incompressible fluids the deviatoric stress τ is directly proportional to the rate of strain tensor e:

$$\tau = 2\eta e \quad \text{or} \quad \tau_{ij} = 2\eta e_{ij} \quad (33)$$

where η the constant of proportionality, is a property of the fluid and is known as the viscosity. Equation (33) in the *Newtonian constitutive equation*. It is an example of a rheological constitutive equation used to relate kinematic tensors, such as the rate of strain tensor, to the deviatoric stress. Most commonly encountered low molecular mass liquids and gases follow the Newtonian constitutive equation, they are called Newtonian fluids. Viscosity is the most important flow property of fluids. It is a strong function of temperature. The SI unit of viscosity is pascal–second. It should be pointed out that when dealing with compressible fluids, such as gases, the Newtonian constitutive equation must be modified to allow for compressibility. This is done by introducing a bulk viscosity λ into the Newtonian constitutive equation

$$\tau = \lambda \nabla \cdot \boldsymbol{u} \boldsymbol{I} + 2\eta \boldsymbol{e} \tag{34}$$

The bulk viscosity term is zero for incompressible fluids. This article is concerned almost exclusively with incompressible fluids, thus the bulk viscosity term will not appear in all subsequent equations and discussions. For further discussion on compressible flow and bulk viscosity see [5].

Specialized instruments known as viscometers have been developed for measuring viscosity. The basic principles of such instruments can be explained by examining the rate of strain and the deviatoric stress tensors in a fluid contained between two parallel plates, shown in Figure 5. The lower plate is held stationary while the upper is moving parallel to itself, with a constant velocity u_0. Let F denote the force required to maintain the motion of the upper plate, A the area of the plates, and d the spacing between the plates. In this flow, the velocity of the fluid varies linearly from zero at the lower plate to u_0 at the upper plate. With reference to the coordinates shown, it can be seen that the only nonzero components of the rate of strain tensor are

$$e_{xy} = e_{yx} = \frac{1}{2}\frac{du_x}{dy} = \frac{1}{2}\frac{u_0}{d} \tag{35}$$

where u_0/d is the velocity gradient across the plate spacing. This velocity gradient is often referred to as the *shear rate* and denoted by $\dot{\gamma}$. In this simple flow field (the steady shear flow) for Newtonian fluids, the only nonzero components of the deviatoric stress are

$$\tau_{xy} = \tau_{yx} = \frac{F}{A} \tag{36}$$

By definition, the viscosity is given by

$$\eta = \frac{\tau_{xy}}{2 e_{xy}} = \frac{F}{A} \bigg/ \left(\frac{u_0}{d}\right) \tag{37}$$

In a viscometer F and u_0 are measured, and A and d are obtained from the dimensions of the instrument. From these data the viscosity can be calculated. A detailed description of viscometric techniques is given in [6]–[8].

Typical values of viscosity (in Pa s at 20 °C or as indicated) for several fluids are given in the following list. These values show the large variation of viscosity that can be encountered in fluid mechanical problems as well as the variation of viscosity with temperature.

Steam (373 K)	1.28×10^{-5}
Carbon dioxide	1.46×10^{-5}
Air (273 K)	1.71×10^{-5}
(293 K)	1.81×10^{-5}
(373 K)	2.18×10^{-5}
(473 K)	2.58×10^{-5}
Pentane	2.34×10^{-4}
Hexane	3.26×10^{-4}
Heptane	4.09×10^{-4}
Octane	5.45×10^{-4}
Benzene	6.47×10^{-4}
Mercury	1.55×10^{-3}
Water (273 K)	1.79×10^{-3}
(283 K)	1.30×10^{-3}
(293 K)	1.00×10^{-3}
(303 K)	7.98×10^{-4}
Olive oil	8.40×10^{-2}
Castor oil	9.86×10^{-2}
Glycerol (273 K)	1.21×10^{1}
(293 K)	1.49
(298 K)	0.942
(303 K)	0.622
Low-density polyethylene at zero shear rate	
(385 K)	2.65×10^{5}
(403 K)	1.05×10^{5}
(443 K)	2.38×10^{4}
(463 K)	1.40×10^{4}

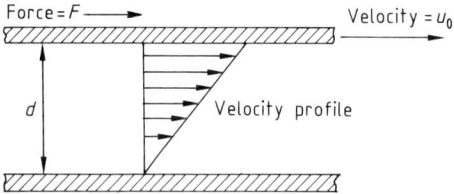

Figure 5. Shearing motion between two parallel plates

3.2. Navier–Stokes Equations

When the Newtonian constitutive equation is substituted into the Cauchy equations of motion (Eq. 24), the following equation is obtained:

$$\varrho \frac{D\mathbf{u}}{Dt} = -\nabla p + \eta \nabla^2 \mathbf{u} + \varrho \mathbf{g} \qquad (38)$$

or in cartesian component form

$$\varrho \frac{\partial u_i}{\partial t} + \sum_{j=1}^{3} \varrho u_j \frac{\partial u_i}{\partial x_j} = -\frac{\partial p}{\partial x_i} + \eta \sum_{j=1}^{3} \frac{\partial^2 u_i}{\partial x_j^2} + \varrho g_i \qquad (39)$$

The first term on the right-hand side is the isotropic pressure term and the second term is the viscous term. This equation together with the continuity equation forms a set of four equations which can be solved to give the three unknown velocity components and the isotropic pressure for any flow of a Newtonian fluid. Equation (39), known as the Navier–Stokes equation, describes the motion of Newtonian fluids. The equivalent forms in some of the more commonly encountered coordinate systems are listed in Table 1. Viscosity is a strong function of temperature in nonisothermal flow, where the variation of viscosity with temperature cannot be ignored, the Navier–Stokes equations must be modified. Thus, it now takes the form

$$\varrho \frac{\partial u_i}{\partial t} + \sum_{j=1}^{3} \varrho u_j \frac{\partial u_i}{\partial x_j}$$
$$= -\frac{\partial p}{\partial x_i} + \sum_{j=1}^{3} \frac{\partial}{\partial x_j}\left(\eta(T) \frac{\partial u_i}{\partial x_j}\right) + \varrho g_i \qquad (40)$$

This equation must now be solved simultaneously with the continuity equation and the energy transport equation for the unknowns—the three velocity components, the isotropic pressure, and the temperature. It is assumed that the variation of viscosity with temperature is a known experimentally measured quantity. Variable viscosity adds to the difficulty of solving the equations describing the flow. The complexity of these equations is such that they often do not have simple analytical solutions and computers are used to generate approximate numerical answers.

3.3. Applications of Navier–Stokes Equations

3.3.1. Flow in Pipes

Steady flow of Newtonian fluids in pipes is probably the most commonly encountered flow field. It is also one of the small number of flow fields for which the Navier–Stokes equations have a simple analytical solution. In this flow the only nonvanishing velocity component is that along the axis of the pipe. It is convenient, in this case, to write down the equations of motion in cylindrical coordinates. The axial velocity u_z is a function of the radial coordinate only and is independent of the other coordinates and time. Examination of the Navier–Stokes equations in cylindrical coordinates in Table 1 shows that the only nontrivial component is

$$0 = -\frac{\Delta p}{L} + \eta \left[\frac{1}{r}\frac{d}{dr}\left(r \frac{du_z}{dr}\right)\right] \qquad (41)$$

The driving force of this flow is the applied pressure gradient $\Delta p / L$; the effect of gravity has been left out in this equation. Equation (41) can be integrated to give

$$u_z = \frac{1}{4\eta} \frac{\Delta p}{L} r^2 + A \ln r + B \qquad (42)$$

where A and B are the constants of integration that can be determined by considering the boundary conditions. The velocity along the axis of the tube is finite; this then requires A to be zero. At the pipe wall $r = R$, it is assumed that the fluid adheres to the wall, i.e., $u_z = 0$. This is the well known no-slip boundary condition that is normally imposed at the interface between a solid wall and a flowing fluid. This boundary condition is satisfied for

$$B = -\frac{1}{4\eta} \frac{\Delta p}{L} R^2$$

Thus the resulting velocity profile in the pipe is given by

$$u_z = \frac{R^2}{4\eta}\left(-\frac{\Delta p}{L}\right)\left(1 - \frac{r^2}{R^2}\right) \qquad (43)$$

From Equation (43) the volumetric flow rate Q through the pipe can be obtained by integrating

$$Q = \int_0^R 2\pi r u_z \, dr = \frac{\pi R^4}{8\eta}\left(-\frac{\Delta p}{L}\right) \quad (44)$$

The average velocity V in the pipe is given by

$$V = \frac{R^2}{8\eta}\left(-\frac{\Delta p}{L}\right) \quad (45)$$

In terms of V, the velocity profile in the pipe takes the simple form

$$u_z = 2V\left(1 - \frac{r^2}{R^2}\right) \quad (46)$$

Thus the velocity profile is a parabola and the maximum velocity, which occurs at the centre of the pipe, is twice the average velocity. For a given pipeline, the driving force needed to attain a specified average velocity is given by

$$\frac{\Delta p}{L} = -\frac{8\eta V}{R^2} \quad (47)$$

Equation (45) is the *Hagen–Poiseuille equation*. It can be put in a dimensionless form

$$f = -\frac{D \Delta p}{\frac{4L}{\varrho V^2/2}} = \frac{16}{\varrho V D/\eta} = \frac{16}{Re} \quad (48)$$

where D is the diameter of the pipe, $Re = \varrho V D/\eta$, and f is as defined by this equation; f is the friction factor and Re is the Reynolds number. Both these quantities are dimensionless. Such dimensionless numbers are widely encountered in fluid mechanical analysis. Equation (48) shows that f is a function of Re only. It will be shown later that shear stress at the pipe wall is given by $D\Delta p/4L$, while shear rate at the pipe wall is $8 V/D$. Thus the ratio of the two measurable quantities $\Delta p D/4 L$ and $8 V/D$ gives the viscosity of the fluid. Measurement in pipe flow can be used for viscosity measurements.

For flow in pipes with $Re < 2100$, the simple relationship between f and Re in Equation (48) has been confirmed by experimental measurements. Such flows are known as laminar flows. As the Reynolds number is increased above 2100, the flow gradually loses it steady nature and becomes more and more chaotic. The expressions derived for the velocity profile, volumetric flow rate, and friction factor are no longer valid. The chaotic fluid motion associated with flows at high Reynolds number ($Re > 4000$) is known as turbulent flow. Turbulent flow is characterized by rapid and apparently random fluctuations in fluid velocity and pressure. The same definition of f and Re given in Equation (48) can be used in turbulent flow. However, the relationship between these two dimensionless numbers in turbulent flow can not be determined analytically and a number of empirical and semiempirical correlations have been developed. In the range of $4000 < Re < 10^5$, experimental data of f versus Re closely follow the empirical *Blasius equation*

$$f = 0.079 \, Re^{-1/4} \quad (49)$$

Another equation that can be used to describe the relationship between f and Re is the *von Karman–Nikuradse equation*

$$\frac{1}{\sqrt{f}} = 4.0 \log(Re\sqrt{f}) - 0.4 \quad (50)$$

Equation (50) is valid for $Re \geq 4000$ [5]. The constants 4.0 and 0.4 are determined from experiment. Analysis of turbulent flow requires special techniques and will not be discussed in this article. For further details see [9].

Equations (48)–(50) which relate the friction factor to the Reynolds number were derived for pipes with a smooth inner surface. In commercial pipes surface irregularities are invariably present depend on the material of construction and the manufacturing process. A quantitative measure of this surface irregularity is the *relative roughness* of the pipe which is defined as the ratio of the average surface roughness e to the pipe diameter D. The relative roughness has to be included as an additional factor in the correlations for friction factor. For most commercial pipes the relative roughness is much smaller than unity and surface irregularities have no noticeable effect on pressure drop as long as the flow remains *laminar*. Thus Equation (48) remains valid for rough pipes. In *turbulent flow*, the situation is quite different. The friction factor is altered significantly even for small e/D. Empirical equations have been developed to relate the friction factor to the relative roughness and the Reynolds number. One example is the *Colebrook equation*:

$$\frac{1}{\sqrt{f}} = -4.0 \log\left(\frac{e}{D} + \frac{4.67}{Re\sqrt{f}}\right) + 2.28 \quad (51)$$

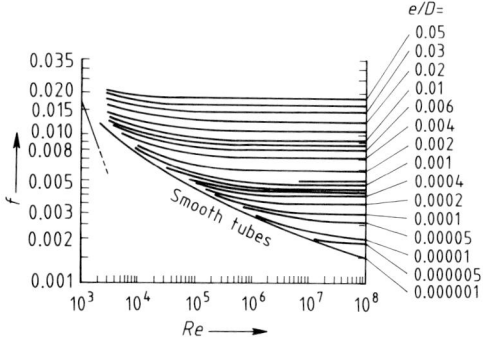

Figure 6. Friction factor for flow in circular pipes at different Reynolds number and for different relative pipe roughness (e/D) [1]

A disadvantage is that for a given Re and e/D, Equation (51) must be solved iteratively for f. Another difficulty in using the Colebrook equation is the large uncertainty associated with the relative roughness.

The friction factor is important in the sizing of pipelines. It is customary to plot f against Re in the form of a log–log plot. An example of such a plot is shown in Figure 6 (*Moody chart*). The relative roughness of the pipe is included in this plot as an independent parameter.

3.3.2. Concentric Cylinder Flow

Another exact solution of the Navier–Stokes equation is the one that describes the velocity profile and stresses generated in a fluid contained between the annular gap of two rotating concentric cylinders (see Fig. 7). This flow is of great practical interest because it is the basis of a large class of commercial viscometers. It is also of considerable theoretical importance; it is one of the few flows for which the transition (via a series of progressively more complicated flows) of the simple laminar solution to turbulent flow has been carefully observed and analyzed theoretically.

To simplify the analysis, the length of the cylinders are assumed to be long compared to their radii so that the end effects can be ignored. There is no applied pressure gradient in this flow. Fluid motion is brought about by the rotation of the cylinders. Let ω_1 and ω_2 denote the angular velocity of the outer and inner cylinders respectively. The corresponding linear velocity of the cylinders are $\omega_1 R_1$ and $\omega_2 R_2$, where R_1 and R_2 are the respective radii. The only nonvanishing velocity component, in cylindrical coordinates, is the azimuthal component u_θ. The Navier–Stokes equations reduces to

$$-\varrho\left(\frac{u_\theta^2}{r}\right) = -\frac{dp}{dr} \tag{52}$$

$$0 = \eta \frac{d}{dr}\left[\frac{1}{r}\frac{d}{dr}(r u_\theta)\right] \tag{53}$$

Equation (53) can be integrated to give u_θ in terms of the radial coordinate and the boundary conditions on the inner and outer cylinder. Using the newly obtained velocity, Equation (52) can in turn be integrated to give the pressure. The resulting expressions for u_θ and p are

$$u_\theta = \frac{1}{R_2^2 - R_1^2}\left[r(R_2^2\omega_2 - R_1^2\omega_1) + \frac{R_1^2 R_2^2}{r}(\omega_1 - \omega_2)\right]$$

$$p = p_2 + \frac{\varrho}{R_2^2 - R_1^2}\left[\frac{(R_2^2\omega_2 - R_1^2\omega_1)^2(r^2 - R_2^2)}{2}\right.$$

$$+ 2R_1^2 R_2^2(\omega_1 - \omega_2)(R_2^2\omega_2 - R_1^2\omega_1)\ln\frac{r}{R_2}$$

$$\left.+ R_1^4 R_2^4(\omega_1 - \omega_2)^2\left(\frac{1}{r^2} - \frac{1}{R_2^2}\right)\right]$$

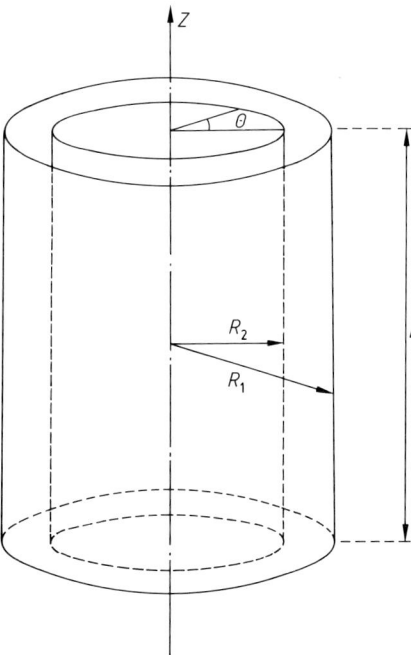

Figure 7. Flow between two concentric cylinders

where p_2 is the pressure at the inner cylinder wall. It can easily be verified that the no-slip boundary condition is satisfied at the outer and inner cylindrical walls.

In most commercial concentric cylinder viscometers, the outer cylinder is held fixed while the inner one is rotated at a steady angular velocity by an applied torque Γ. For this special case, the velocity in the annular gap is given by

$$u_\theta = \frac{1}{R_2^2 - R_1^2}\left(rR_2^2\omega_2 - \frac{R_1^2 R_2^2 \omega_2}{r}\right) \tag{54}$$

The corresponding shear rate and shear stress on the inner wall are

$$\dot{\gamma}_{r\theta_2} = \frac{2\omega_2}{1 - (R_2/R_1)^2} \tag{55}$$

$$\tau_{r\theta_2} = \frac{\Gamma}{2\pi R_2^2 L} \tag{56}$$

where L is the length of the cylinder. In viscometry, these two expressions are combined to yield an explicit expression for the viscosity of the fluid:

$$\eta = \frac{\tau_{r\theta_2}}{\dot{\gamma}_{r\theta_2}} = \frac{\Gamma}{L}\frac{1}{4\pi\omega_2}\left(\frac{1}{R_2^2} - \frac{1}{R_1^2}\right) \tag{57}$$

With Equation (57) the viscosity of the fluid can be calculated from the applied torque per unit height of the cylinders Γ/L and the measured angular velocity ω_2. For Newtonian fluids, in principle, only a single reading of the torque together with the corresponding angular velocity is needed to determine the viscosity. For non-Newtonian fluids, the situation is more complex and a series of data points are needed to determine the viscosity as a function of shear rate [6].

According to the solution of the Navier-Stokes equations, the streamlines should form a family of circles concentric with the cylinders. At low Reynolds number this is indeed the observed flow pattern. However, as the Reynolds number is increased (either by increasing the angular velocity or decreasing the viscosity) this simple flow loses its stability and gives way to a new steady flow in which the streamlines are spirals whose axes are concentric with the cylinders. A sketch of the new flow pattern, at sufficiently large Re, is shown in Figure 8. The vortices formed by the fluid are known as Taylor vortices. The transition between the two steady flows can

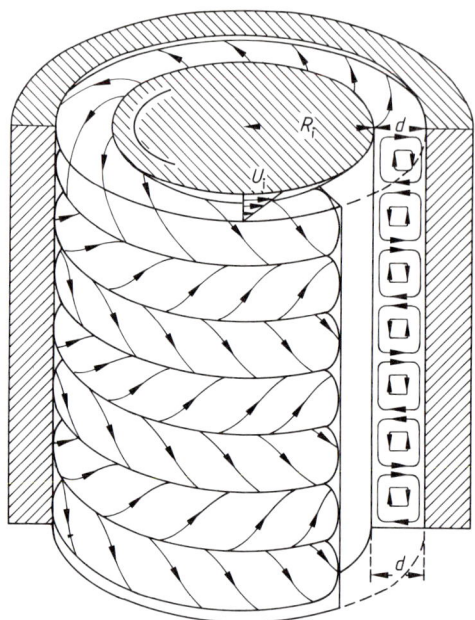

Figure 8. Taylor vortices [5] (reproduced with permission of McGraw-Hill)

be observed by plotting the torque Γ against the shear rate $\dot{\gamma}$ (see Fig. 9). The abrupt change in slope of the $\Gamma-\dot{\gamma}$ plot can be used to locate the transition point accurately. In using Equation

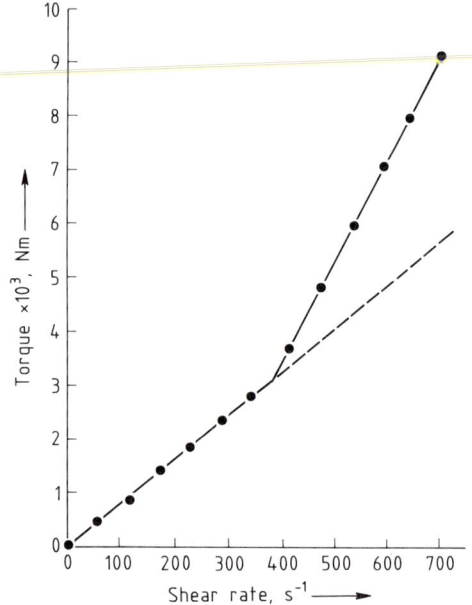

Figure 9. Change in slope in the torque–shear rate plot for flow between concentric cylinders indicating the onset of Taylor vortices [1]

(57) to obtain the viscosity from concentric viscometer measurements, it is important to ensure that all the data points are taken prior to the onset of transition flow. At even higher Reynolds number, the Taylor vortices in turn become unstable and are replaced by a series of progressively more complicated flows, which ultimately lead to turbulent flow. The detailed step by step mapping of the laminar-to-turbulent transition has been investigated experimentally and analyzed theoretically with excellent agreement between the two.

In making viscosity measurements, it is common practice to use more than one viscometer so that the experimental data covers as wide a range of shear rate as possible. This step is particularly important if the possibility exists that the fluid may not be Newtonian (i.e., it has a viscosity that is a function of shear rate). Figure 10 shows a log–log plot of shear stress against shear rate obtained using four different viscometers. The data at low shear rates were obtained using a concentric cylinder viscometer. Those at higher shear rates were obtained from the volumetric flow rate and pressure gradient data from three different capillary tubes. (Experimental techniques associated with capillary viscometer are described in Section 4.2). The fact that all the data points, from four different viscometers, fall on the same straight line with slope equal to unity means that the fluid is Newtonian over the shear rates covered and the viscosity can be read off from the intercept of the straight line with the vertical line at $\dot{\gamma} = 1.0$; η is, in this case, ca. 0.91 Pa s.

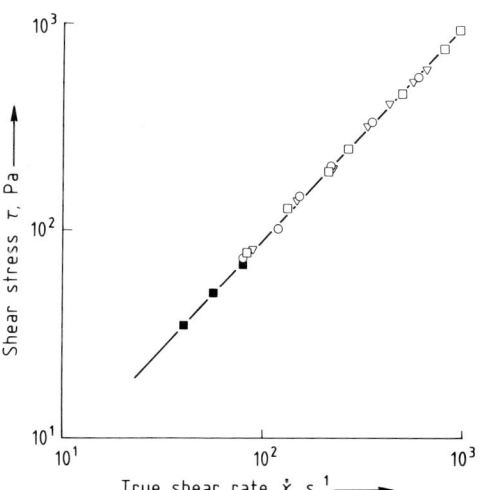

Figure 10. Log–log plot of shear stress versus shear rate for a Newtonian fluid (glycerol at 21.6 °C) obtained using four different viscometers. □ $D = 5.573$ mm, $L = 300$ mm, $L/D = 53.8$; ▽ $D = 5.569$ mm, $L = 477$ mm, $L/D = 85.7$; ○ $D = 5.595$ mm, $L = 700$ mm, $L/D = 125.1$; ■ concentric cylinder viscometer

3.3.3. Creeping Flow Past a Sphere

Steady flow of a Newtonian fluid past a sphere has been the subject of numerous investigations. This flow is of great practical importance in its own right, but also because it illustrates the steps involved in the evaluation of the drag force exerted by a flowing fluid on an object. It is also of considerable theoretical interest because of the part it played in the development of some of the modern mathematical techniques, such as singular perturbation and matched asymptotic analyses, used in solving flow problems.

The mathematical analysis is greatly simplified if the inertia terms in the Navier–Stokes equations are assumed to be small compared to the viscous terms and thus are ignored. This assumption is valid if the fluid velocity or density is small, its viscosity large, or if the sphere diameter D is small. These conditions are equivalent to requiring the Reynolds number $Re = \varrho U_0 D/\eta$, to be small; U_0 is the undisturbed velocity of the fluid at a large distance from the sphere. Flows with low Reynolds number are often referred to as *inertialess* or *creeping flows*.

In spherical coordinates, the equations for creeping flow past a sphere are

$$0 = -\frac{\partial p}{\partial r} + \eta \left[\frac{1}{r^2} \frac{\partial}{\partial r}\left(r^2 \frac{\partial u_r}{\partial r}\right) \right.$$
$$+ \frac{1}{r^2 \sin\theta} \frac{\partial}{\partial \theta}\left(\sin\theta \frac{\partial u_r}{\partial \theta}\right)$$
$$\left. - \frac{2u_r}{r^2} - \frac{2}{r^2} \frac{\partial u_\theta}{\partial \theta} - \frac{2}{r^2} u_\theta \cot\theta \right]$$

$$0 = -\frac{1}{r} \frac{\partial p}{\partial \theta} + \eta \left[\frac{1}{r^2} \frac{\partial}{\partial r}\left(r^2 \frac{\partial u_\theta}{\partial r}\right) \right.$$
$$+ \frac{1}{r^2 \sin\theta} \frac{\partial}{\partial \theta}\left(\sin\theta \frac{\partial u_\theta}{\partial \theta}\right)$$
$$\left. + \frac{2}{r^2} \frac{\partial u_r}{\partial \theta} - \frac{u_\theta}{r^2 \sin^2\theta} \right]$$
(58)

It is convenient to locate the center of the sphere at the origin of the spherical coordinates and to have the undisturbed upstream velocity of the fluid parallel to the $\theta = 0$ line (see Fig. 11). In this coordinate system, u_r and u_θ are the only two nonvanishing velocity components. The Newtonian fluid is assumed to adhere to the surface of the sphere (the no-slip boundary condition), i.e.,

$$u_r = u_\theta = 0 \quad \text{at} \quad r = D/2 \qquad (59)$$

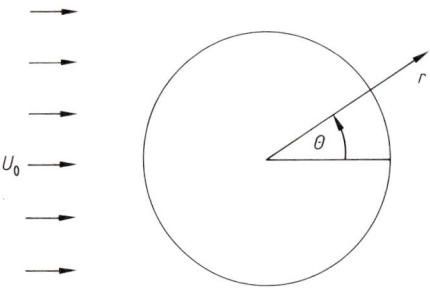

Figure 11. Definitive sketch for flow past a sphere

Far away from the sphere, u approaches the uniform velocity U_0, i.e.,

$$u_r = U_0 \cos\theta, \quad u_\theta = -U_0 \sin\theta \quad \text{as} \quad r \to \infty \quad (60)$$

In addition, the velocity components have to satisfy the continuity equation

$$\frac{1}{r^2}\frac{\partial}{\partial r}(r^2 u_r) + \frac{1}{r\sin\theta}\frac{\partial}{\partial\theta}(u_\theta \sin\theta) \quad (61)$$

Equations (58) and (61) can be solved to give the following expressions for the velocity components and pressure

$$u_r = U_0\left[1 - \frac{3}{4}\frac{D}{r} + \frac{1}{2}\left(\frac{D}{2r}\right)^3\right]\cos\theta$$

$$u_\theta = -U_0\left[1 - \frac{3}{8}\frac{D}{r} - \frac{1}{4}\left(\frac{D}{2r}\right)^3\right]\sin\theta$$

$$p = p_0 - \left(\frac{3\eta U_0}{D}\right)\left(\frac{D}{2r}\right)^2 \cos\theta \quad (62)$$

where p_0 is the uniform pressure far from the sphere. For mathematical details leading to these expressions see [1].

To calculate the *drag force* on the sphere, the isotropic pressure and deviatoric stress on the surface of the sphere must be evaluated. These are given, respectively, by

$$p = p_0 - \frac{3\eta U_0}{D}\cos\theta \quad (63)$$

$$\tau_{r\theta} = \eta\left[r\frac{\partial}{\partial r}\left(\frac{u_\theta}{r}\right) + \frac{1}{r}\frac{\partial u_r}{\partial\theta}\right]_{r=D/2} = -\frac{3\eta U_0}{D}\sin\theta$$

The other components of the deviatoric stress are zero on the surface. The drag force F_d is obtained by integrating the resulting Cauchy stress over the entire surface of the sphere. From the symmetry of the problem, it is clear that the drag force is in the same direction as the undisturbed velocity:

$$F_d = \int_{\varphi=0}^{\pi}\int_{\theta=0}^{2\pi}\left[\left(-p_0 + \frac{3\eta U_0}{D}\cos\theta\right)\cos\theta\right.$$
$$\left. + \frac{3\eta U_0}{D}\sin^2\theta\right]R^2\sin\theta\,d\theta\,d\varphi$$

$$= 3\pi\eta D U_0$$

This result is often referred to as *Stokes law*. One of the practical applications of Stokes law is the *falling ball viscometer*. Here the terminal velocity of a small sphere of known diameter that falls through a fluid is measured. When the sphere has attained its terminal velocity, the drag force on the sphere is exactly balanced by it weight, i.e.:

$$F_d = \frac{4}{3}\pi(D/2)^3\Delta\varrho\, g = 3\pi\eta D U_0 \quad (64)$$

where $\Delta\varrho$ is the difference between the density of the solid sphere and the fluid. This then yields a simple expression which allows the fluid viscosity to be calculated from measurable quantities such as the densities of the fluid and the sphere, and the terminal velocity:

$$\eta = \frac{1}{18}\frac{D^2 \Delta\varrho\, g}{U_0} \quad (65)$$

Usually the drag force F_d is expressed in terms of a dimensionless *drag coefficient* C_d obtained by dividing F_d by $\frac{1}{2}\varrho U_0^2$ and by the cross-sectional area of the sphere normal to the velocity U_0; thus the drag coefficient is

$$C_d = \frac{3\pi\eta D U_0}{\frac{1}{2}\varrho U_0^2 \pi D^2/4} = \frac{24\eta}{\varrho U_0 D} \quad (66)$$

In terms of C_d and the Reynolds number $Re = \varrho U_0 D/\eta$, Stokes law takes the form

$$C_d = 24/Re \quad (67)$$

This result was obtained for Re much smaller than unity: in this case the error incurred is acceptable for most practical purposes. For larger Re, the inertia terms can no longer be ignored and additional terms must be included in the expression for C_d. The result now takes the form of an infinite series in Re, the leading terms are

$$C_d = \frac{24}{Re}\left[1 + \frac{3}{16}Re + \frac{9}{160}(\ln Re)Re^2 + \ldots\right] \quad (68)$$

The validity of this series extends to Re of the order of unity. The mathematical analysis leading to this series is described in [2].

Expressions (67) and (68) for C_d are based on the assumption that the flow remains steady and laminar and Re is small. As Re is increased, the flow pattern around the sphere becomes more and more complicated. (See the numerically generated flow pattern around a cylinder shown in Figure 47, p. 5-46) The laminar flow eventually gives way to turbulent flow, and C_d is expected to change dramatically with Re. Experimentally measured drag coefficients as a function of Re are shown in Figure 12. The validity of Equations (67) and (68) is restricted to Re less than ca. unity with the range of validity of Equation (68) slightly larger than that of Equation (67). For Re in the range 1 to 10^3, C_d can be approximated by

$$C_d \approx 18\, Re^{-0.6} \quad (69)$$

And for $10^3 \leq Re \leq 2 \times 10^5$, C_d is approximately constant

$$C_d \approx 0.44 \quad (70)$$

For a detailed description of the flow phenomena at increasing Re, including the sudden decrease in C_d at $Re \approx 2 \times 10^5$, see [1].

3.4. Some Other Important Flows

3.4.1. Flow Through Granular Beds

The flow of fluids through granular beds occurs widely in industry and in nature, some examples are flow of water down a filter bed, seepage of oil through porous underground soil structure, and flow of chemicals in a packed bed reactor. The key variable in the study of flow through a granular bed is the volumetric flow rate per unit cross-sectional area of the bed, Q/A. This flow rate depends on the density ϱ and viscosity η of the fluid, the average size (and shape) of the particles D_p that make up the bed, the applied pressure gradient $\Delta p/L$ across the bed, and the porosity of the bed ε. The relationship between these variables can be expressed in the general form

$$\Delta p/L = f(Q/A, \varrho, \eta, \varepsilon, D_p)$$

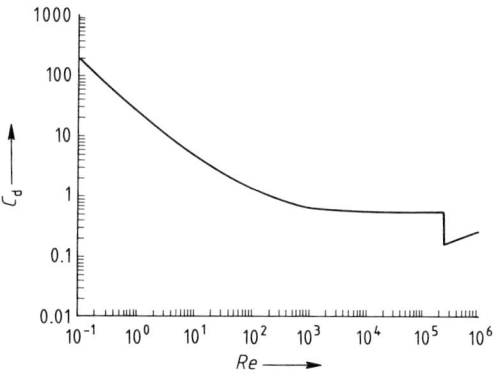

Figure 12. Log–log plot of drag coefficient on a sphere as a function of the Reynolds number

The exact form of the function in this equation is expected to be complex and cannot be determined from fundamental considerations only. It must therefore be determined by experimental measurements. A clearer understanding of the influence of each of these variables can be achieved by presenting the experimental results in a dimensionless form. For the present problem, the relevant dimensionless variables are an appropriately defined friction factor f and an appropriately defined Reynolds number:

$$f = \frac{D_p \varepsilon^3}{\varrho U_\infty^2 (1-\varepsilon)} \frac{\Delta p}{L} \quad (71)$$

$$Re = \frac{\varrho D_p U_\infty}{(1-\varepsilon)\eta} \quad (72)$$

U_∞ is the superficial velocity and is defined by $U_\infty = Q/A$. From the accumulated data in the literature, it can be shown that f and Re are reasonably correlated by the following semi-empirical expressions:

$$\begin{aligned} f &= 150/Re & \text{for} \quad Re < 10 \\ f &= 1.75 + \frac{150}{Re} & \text{for} \quad 10 < Re < 1000 \\ f &= 1.75 & \text{for} \quad 1000 < Re \end{aligned} \quad (73)$$

This set of expressions for f is known as the *Ergun relationship*. At low Re, corresponding to inertialess laminar flow through the bed, the friction factor is proportional to the reciprocal of Re, as is the case of laminar flow in a circular pipe. This appears to be a general property of laminar flow of Newtonian fluids. The Ergun relationship was based on experimental data—pressure drop in granular beds of spherical particles. For nonspherical particles, D_p must be replaced by the hydraulic diameter, which is defined by $D_{hy} = 4 \times$ volume of particle/surface area of particle wetted by fluid. For a granular bed of particles which deviate greatly from sphericity, a shape factor may have to be introduced.

3.4.2. Fluidization

Fluidization is a common operation in the process industry in which a fluid stream, usually gas, is used to ensure thorough mixing of a bed of solid particles. The gas stream is passed vertically upward through the solid bed. The pressure drop across the solid bed (or equivalently the upward drag force on the particles) increases with the gas velocity. At a sufficiently high gas velocity the drag force becomes equal to the weight of the individual particles and the particles enter a weightless state. Further increase in gas velocity is not accompanied by an increase in pressure drop (see Fig. 13), instead the bed begins to expand. The bed of solid particles now acquires many of the properties normally associated with a fluid, such as the ability to flow freely; it is said to be fluidized. The weightless particles move rapidly and randomly throughout the expanded bed, thereby ensuring thorough mixing. This is exploited, for example, in a fluidized-bed reactor where the solid particles may be the catalyst on which an exothermic chemical reaction takes place. The random motion in the fluidized state results in good particle–particle and particle–wall contact and hence in an efficient removal of the heat of reaction.

One of the key variables in the operation of a fluidized bed is the minimum superficial gas velocity u_f needed to maintain the particles in the fluidized state. This minimum velocity can be estimated using the Ergun equation (73) for the pressure drop Δp in a packed bed. The force exerted by the gas stream on a bed of solid particles of cross-sectional area A is given by $A \Delta p$. At the point of fluidization, this is equal to the weight of the bed, i.e.

$$A \Delta p = (\varrho_p - \varrho) g A H (1 - \varepsilon) \quad (74)$$

where H is the height of the bed and ε is its porosity; $(\varrho_p - \varrho)$ is the difference between the density of the solid particles and the gas; and Δp is given by the Ergun equation. Since, in most fluidized beds, the particle size is small and the Reynolds number is often < 10, the friction factor in the Ergun equation can be approximated by $150/Re$. Substituting this result into Equation (74) gives the following expression for u_f

$$u_f = \frac{(\varrho_p - \varrho) g D_p^2 \varepsilon^3}{150 \eta (1 - \varepsilon)} \quad (75)$$

The difficulty with using Equation (75) lies in the uncertainty of estimating the value of ε, the porosity of the bed at the point of fluidization. In the absence of experimental data, $\varepsilon^3/(1 - \varepsilon)$ can be approximated by 0.091. For a more accurate estimation of the minimum fluidization velocity, it is also necessary to take into account the non-spherical shape of the particles. A large number of semi-empirical correlations exist in the literature on fluidization which take into account these and additional factors [10].

3.4.3. Gas–Liquid Flow

Many industrial operations require the handling of fluid streams composed of two distinct phases. The two phases can be a gas and a liquid, two immiscible liquids, or a suspension of fine solid particles in a gas or liquid. The flow behavior of a two-phase stream is considerably more

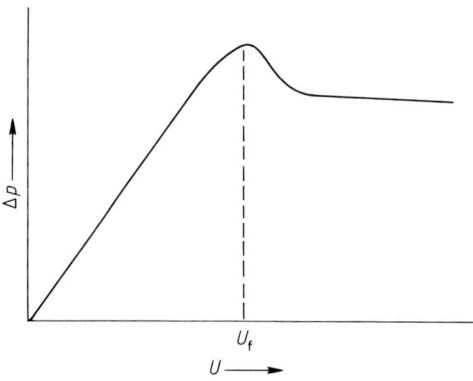

Figure 13. Variation of pressure drop Δp with flow rate u in a granular bed showing the leveling of pressure drop after the onset of fluidization (u_f)

complicated than that of a single-phase fluid. Investigation of two-phase flow is an active area of research with an extensive literature. This section provides only a brief qualitative description of gas–liquid flow in tubes, probably the most frequently encountered two-phase flow.

Flow in Vertical Tubes. One of the most distinctive features of gas–liquid flow is the large number of flow patterns that can be observed. The flow patterns in a pipe for example, depend on the velocity of the two phases, their physical properties, pipe diameter, and the geometrical arrangement of the pipe. Some of the flow patterns that can be observed in a gas–liquid stream flowing up a vertical tube are shown in Figure 14 together with the terms used to describe the various flow regimes.

In the *bubble flow regime* (Fig. 14 A), observed at very low gas flow rates, the gas phase exists in the form of many small bubbles distributed in the continuous liquid phase. As the gas rate is increased, the bubbles grow and coalesce to form large slugs. The transition from bubble flow to *slug flow* (Fig. 14 B) is gradual. The slug diameter is of the order of the tube diameter. Most of the gas phase now resides within the slugs with only a small fraction dispersed as small bubbles in the liquid bridges separating the slugs. A thin liquid film remains between the slug and the tube wall. When the gas rate is increased further this film suddenly becomes unstable and *churn flow* (Fig. 14 C) takes over. Here the flow is

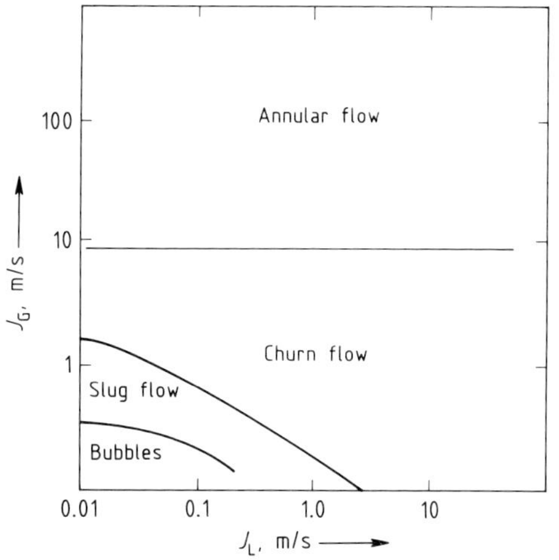

Figure 15. Map of different flow regimes for gas–liquid flow in vertical pipes [11]

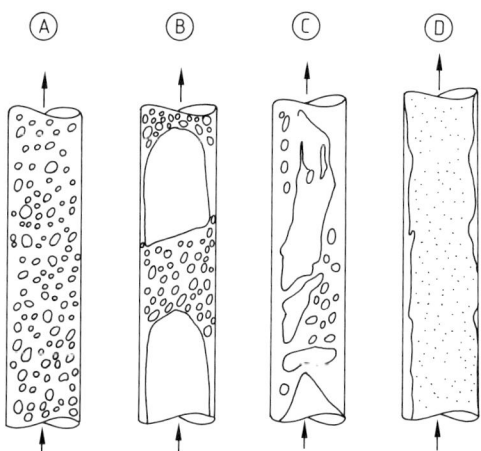

Figure 14. Flow patterns for gas–liquid flow in vertical pipes [12]
A) Bubble flow; B) Slug flow; C) Churn flow; D) Annular flow

highly chaotic and the slugs eventually lose their identity. At yet higher gas rates the flow becomes more steady. The gas phase now becomes the continuous phase occupying the central core of the tube with an *annular*, usually wavy, *liquid film* (Fig. 14 D) lining the tube wall. There is also a fine mist of liquid droplets entrained in the continuous gas phase.

One of the objectives of gas–liquid flow experimentation is to determine the conditions for the transition from one flow regime to another. Experimental observations are often summarized in the form of a *flow regime map*. Figure 15 is a typical map for gas–liquid flow in a vertical tube. The horizontal and the vertical axes are the superficial liquid velocity J_L and gas phase velocity J_G, respectively. These are defined by

$$J_L = L/(\pi D^2/4)$$
$$J_G = G/(\pi D^2/4)$$

where L is the volumetric flow rate of the liquid phase and G is that of the gas phase; D is the tube diameter. The boundaries between the various flow regimes are far from well defined. They also depend on the physical properties of the gas and liquid, and on the diameter of the tube. Like most flow regime maps reported in the literature, Figure 15 is based on data collected for air–water systems at normal temperature and pressure in a tube 25 mm in diameter. Using this map to determine the flow pattern of two-phase systems

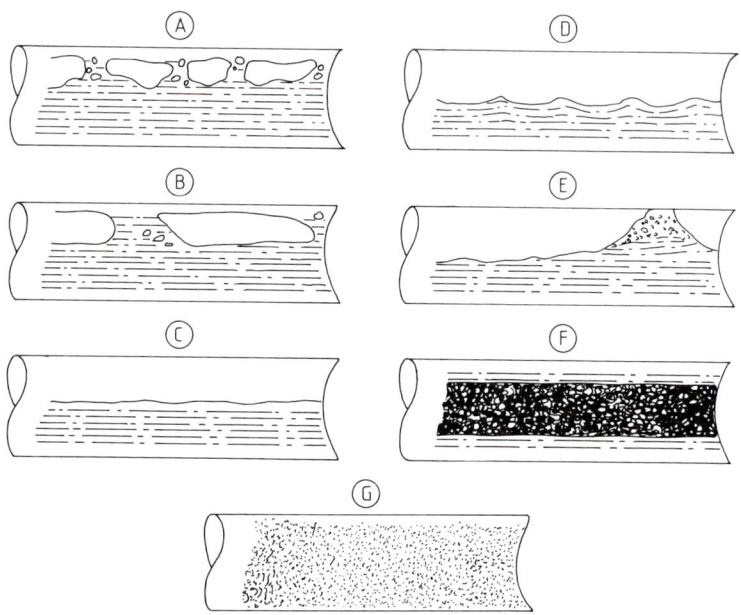

Figure 16. Flow patterns for gas–liquid flow in horizontal pipes [13]
A) Bubble flow; B) Plug flow; C) Stratified flow; D) Wavy flow; E) Slug flow; F) Annular flow; G) Spray

with vastly different properties or in much larger or smaller tubes is not recommended.

Flow in Horizontal Tubes. As expected, the observed flow patterns of a gas–liquid stream in a horizontal tube differ significantly from that in a vertical tube. In the horizontal tube the two phases are more likely to segregate because of their density difference. Figure 16 is the horizontal analogue of Figure 14. Similar terms are used to describe the observed flow patterns at increasing gas flow rates. At low gas flow rates, the gas phase exists as *bubbles* (Fig. 16A) which move along the upper part of the tube at approximately the same speed as the liquid phase. Increasing gas flow will gradually lead to *plug flow* (Fig. 16B), where alternate plugs of liquid and gas move along the upper part of the tube. In *stratified* (Fig. 16C) and in *wavy flow* (Fig. 16D) the two phases occupy different parts of the tube. In stratified flow the gas–liquid interface is relatively smooth while in wavy flow the higher gas velocity causes the interface to become wavy. When the interface becomes grossly distorted, the gas phase becomes separated into *slugs* (Fig. 16E) which move through the tube at a higher speed than the liquid phase. In *annular flow* (Fig. 16F), the liquid flows as a thin film lining the wall of the tube and the gas flows at a higher speed in the central core. The liquid film is thicker at the bottom half of the tube and is generally not smooth. A part of the liquid phase exists as droplets entrained in the gas stream. At very high gas rates, the liquid film completely disintegrates and the liquid phase is carried as a *fine spray* (Fig. 16G) by the high speed gas stream.

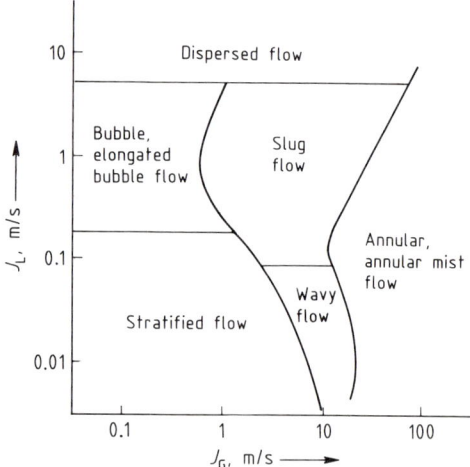

Figure 17. Map of different flow regimes for gas–liquid flow in horizontal pipes [14]

The regime map for gas–liquid flow in a horizontal tube is shown in Figure 17. The vertical axis is the superficial liquid velocity and the horizontal axis is the gas velocity. Figure 17 was constructed from data for air–water mixtures in tubes with diameter < 100 mm.

Apart from mapping the flow regimes in gas–liquid flows, considerable effort has been directed at solving the equations of motion for two-phase flow and the development of correlations that can be used in design calculations. Details of these can be found in [11]–[15].

3.5. Mechanical Energy Balance for Macroscopic Systems

The solution of pipe network flow problems for a macroscopic system such as that illustrated in Figure 18 is often of extreme importance. Once again the three basic conservation principles are required but this time expressed for a system of finite size. At steady state the statements of conservation of mass and energy for the control volume shown in Figure 18 are

$$\varrho_1 \langle u_1 \rangle A_1 = \varrho_2 \langle u_2 \rangle A_2 = W \tag{76}$$

and

$$\frac{1}{2} \Delta \frac{\langle u^3 \rangle}{\langle u \rangle} + g \Delta z + \int_{p_1}^{p_2} \frac{dp}{\varrho} + Ev + W_s = 0 \tag{77}$$

where W is the mass flow rate, ϱ the fluid density, A the cross-section of the flow, u the flow velocity, z the height above some arbitrary datum, p the pressure, g the acceleration due to gravity, W_s the shaft work, i.e., pump work, Ev the friction loss, and $\langle \rangle$ is used to denote average value. Each of the terms in Equation (77) has the unit of energy per unit mass; Δ refers to the outlet minus the inlet conditions. Since the velocity u across any flow cross-section is not in general uniform, an average velocity V defined by

$$V = \langle u \rangle = \frac{1}{A} \int_A u \, dA \tag{78}$$

and for tube flow by

$$V = \langle u \rangle = \frac{1}{\pi R^2} \int_0^R 2\pi r u \, dr \tag{79}$$

must be used in the mass and energy balances. For turbulent flow prior knowledge of the velocity distribution at a particular cross-section is of little consequence because the velocity is essentially uniform across the cross-section, i.e.,

$$\langle u \rangle = V \quad \text{and} \quad \frac{\langle u^3 \rangle}{\langle u \rangle} = \frac{V^3}{V} = V^2 \tag{80}$$

and for an incompressible fluid

$$\int_{p_1}^{p_2} \frac{dp}{\varrho} = \frac{1}{\varrho}(p_2 - p_1) \tag{81}$$

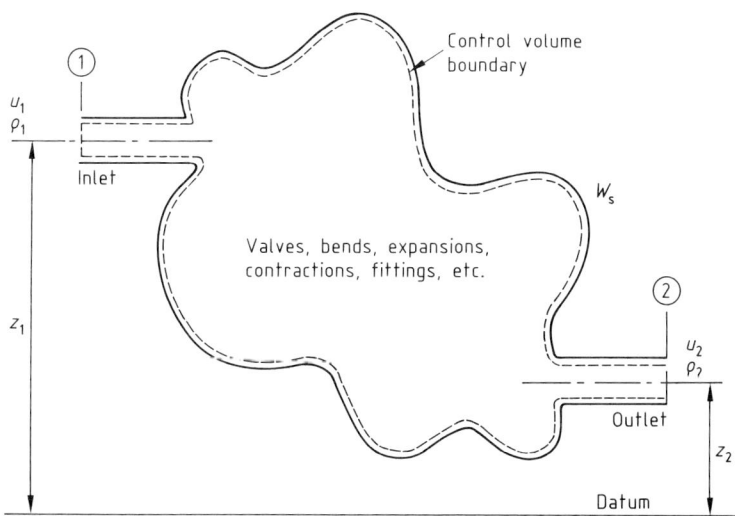

Figure 18. Generalized representation of a control volume for a flow system

Thus, for *turbulent flow* of an incompressible fluid Equation (77) reduces to:

$$\Delta\left(\frac{V^2}{2}\right) + g\Delta z + \frac{p_2 - p_1}{\varrho} + Ev + W_s = 0 \quad (82)$$

Equation (82) is sometimes referred to as the *mechanical energy balance*. If in addition no shaft work W_s crosses the system boundary, then

$$\Delta\left(\frac{V^2}{2}\right) + g\Delta z + \frac{p_2 - p_1}{\varrho} + Ev = 0 \quad (83)$$

this is identical to Equation (26), the Bernoulli equation. Both the mechanical energy balance and the Bernoulli equation are not generally applicable to all flow situations.

For isothermal flow of incompressible fluids (*laminar* and *turbulent flow*)

$$\varrho_1 \langle u_1 \rangle A_1 = \varrho_2 \langle u_2 \rangle A_2 = W \quad (76)$$

and

$$\frac{1}{2}\Delta\frac{\langle u^3 \rangle}{\langle u \rangle} + g\Delta z + \frac{p_2 - p_1}{\varrho} + Ev + W_s = 0 \quad (84)$$

The problem is to evaluate the terms in Equation (84) so that the performance of an existing flow system can be evaluated or a new flow system can be designed [16].

The major unknown in Equation (84) is the friction loss term Ev. The friction loss term is conveniently split into a contribution due to fully developed flows in long, straight sections of the pipe and an excess contribution due to losses in valves, bends, fittings, expansions, contractions, etc:

$$Ev = Ev_{fd} + Ev_{ex}$$

Ev_{ex} is determined from experimental observation whereas Ev_{fd} can in some instances be determined from fundamental principles. The evaluation of Ev_{fd} for tube flows is considered in Section 3.5.1.

3.5.1. Fully Developed Tube Flows

Consider the control volume shown in Figure 19 for fully developed tube flow. Application of Equation (84) to this control volume yields

$$Ev_{fd} = \frac{p_1 - p_2}{\varrho} \quad (85)$$

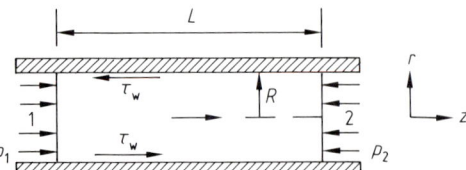

Figure 19. Control volume for fully developed tube flow

That is, the energy loss per unit mass is simply the pressure drop divided by the fluid density. Application of a *force balance* to the same control volume yields

$$\pi R^2 (p_1 - p_2) = \tau_w 2\pi R L \quad (86)$$

or

$$\tau_w = \frac{R(p_1 - p_2)}{2L} = \frac{D(p_1 - p_2)}{4L} \quad (87)$$

and at any radius r

$$\tau_{rz} = \frac{r(p_1 - p_2)}{2L} \quad (88)$$

Elimination of $(p_1 - p_2)$ from Equations (87) and (88) leads to

$$\tau_{rz} = \frac{r}{R}\tau_w \quad (89)$$

Equation (89) indicates a linear distribution of shear stress in a tube from zero at the centerline to τ_w at the wall. This relationship is valid for Newtonian and non-Newtonian fluids and for laminar and turbulent flows.

Frictional losses for fully developed flows are expressed in terms of the friction factor (Eq. 48). Combination of Equations (85), (87), and (48) leads to the *Fanning equation*:

$$Ev_{fd} = \frac{2fV^2 L}{D} \quad (90)$$

which is the basic equation for defining the frictional losses for fully developed flow in pipes. For Newtonian fluids f is given by Equation (48) for laminar flow and by Equations (49), (50), and (51) for turbulent pipe flow.

3.5.2. Accelerating and Decelerating Flows

In any flow system (in addition to fully developed flows in long straight pipes) it is essential that energy losses which occur in the accelerating and decelerating flow in valves, bends, expansions, contractions, and fittings can be estimated. For turbulent flow these losses must be evaluated from available experimental data, whereas numerical simulation results are now becoming available for some laminar flows.

As an example consider the flow through a sudden contraction in pipe diameter as is illustrated in Figure 20. Application of the mechanical energy balance (Eq. 84) between stations 1 and 2 yields

$$\frac{p_1 - p_2}{\varrho} = \frac{1}{2}\Delta\frac{\langle u_z^3 \rangle}{\langle u_z \rangle} + Ev_{1,2} \tag{91}$$

Suppose the flow is turbulent, then

$$\frac{1}{2}\Delta\frac{\langle u_z^3 \rangle}{\langle u_z \rangle} = \frac{1}{2}\Delta V^2 = \frac{V_2^2 - V_1^2}{2} \tag{92}$$

For incompressible flow, conservation of mass (Eq. 76) yields

$$V_2 = V_1\left(\frac{D_1}{D_2}\right)^2 \tag{93}$$

and then

$$\frac{1}{2}\Delta V^2 = \frac{V_2^2}{2}\left[1 - \left(\frac{D_2}{D_1}\right)^4\right] = \frac{V_2^2}{2}, \quad \text{if } D_1 \gg D_2 \tag{94}$$

Also

$$Ev_{1,2} = Ev_{fd_{1,2}} + Ev_{ex_{1,2}} \tag{95}$$

Then

$$Ev_{ex_{1,2}} = \left(\frac{p_1 - p_2}{\varrho} - \frac{V_2^2}{2}\right) - Ev_{fd_{1,2}} \tag{96}$$

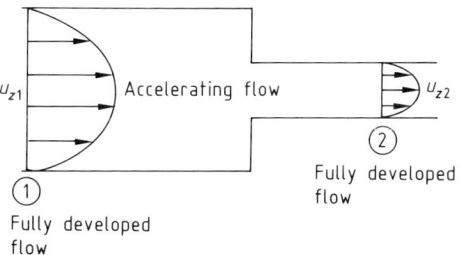

Figure 20. Schematic diagram for flow through a sudden contraction

Now $Ev_{fd_{1,2}}$ will be known at a fixed flow rate, as will V_2. Hence measurement of $p_1 - p_2$ is sufficient to define $Ev_{ex_{1,2}}$. It is found that the resulting measurements for most situations can be generalized in the form

$$\frac{Ev_{ex}}{\frac{V^2}{2}} = C \tag{97}$$

where C is the dimensionless *loss coefficient*. Loss coefficients are tabulated in many standard handbooks such as in [17]. A great deal of confusion can arise in the use of loss coefficients. Sometimes C is referred to as the *number of velocity heads*. This number times $V^2/2$ yields the excess energy loss for that particular fitting. If loss coefficients are used to estimate losses in valve and fittings, etc., then the total frictional loss for a flow system is simply the sum of the individual components:

$$Ev_{\text{total}} = \sum\frac{2fV^2L}{D} + \sum\frac{CV^2}{2} \tag{98}$$

Another and perhaps more convenient way to determine Ev_{total} for a flow system is to use the Fanning equation in the form

$$Ev_{\text{total}} = \sum\frac{2fV^2L}{D} + \sum\frac{2fV^2L_e}{D} \tag{99}$$

where L is the length of straight pipe of diameter D between fittings and L_e the approximate length of straight pipe of diameter D which would have the same energy loss as the fitting. Equivalent length data for various fittings are available in standard references, for example in [18]. A sufficient amount of data for C and L_e have been tabulated for good design of turbulent pipe flow systems. Very little data are available for laminar flows and almost no results are available for non-Newtonian fluids in both laminar and turbulent flows.

3.5.3. The Orifice Plate and Similar Flow Rate Measurement Devices

In many chemical plants a simple device known as an orifice plate, (see Fig. 21) is used to measure the volumetric flow rate of fluids in turbulent pipe flow. In its simplest form the orifice plate is a disk with usually (but not always) a circular hole through which the fluid flows. Con-

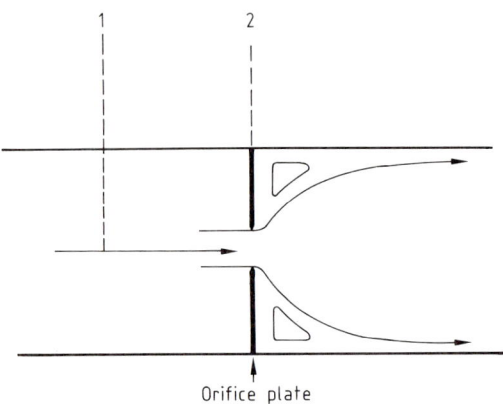

Figure 21. Flow through an orifice plate

sider station 1 upstream of the plate and station 2 at or just downstream of the plate. Station 1 is sufficiently far upstream so that the flow there is undisturbed by the presence of the orifice plate. Neglecting losses due to fluid friction, the average pressure and velocities between these two stations are related by Equation (83). In addition, from a mass balance

$$u_1 A_1 = u_2 A_2 \qquad (100)$$

Combining these two equations leads to

$$Q = u_1 A_1 = A_1 \sqrt{\frac{2 \Delta p}{\varrho \left(\frac{A_1^2}{A_2^2} - 1 \right)}} \qquad (101)$$

Equation (101) defines the volumetric flow rate Q, in the pipe in terms of the measured pressure difference between stations 1 and 2. The losses between 1 and 2 are not taken into consideration. Losses can be accounted for by the incorporation of an empirical correction factor known as the discharge coefficient C_d into the equation for volumetric flow rate

$$Q = C_d A_1 \sqrt{\frac{2 \Delta p}{\varrho \left(\frac{A_1^2}{A_2^2} - 1 \right)}} \qquad (102)$$

C_d is usually in the range 0.59–0.65. Its exact value depends on the geometry of the orifice plate, such as the size and shape of the hole and the location of the pressure measurement points. It also depends, to a lesser extent, on the Reynolds number in the pipe. C_d can either be determined by direct calibration or, in the case of an orifice plate of standard design, from tables and correlations published by organisations such as the ISO.

A number of other flow measurement devices based on the Bernoulli equation with an empirical correction for losses are also in common use. Examples are the *pitot tube* and the *venturi meter*. In each case the volumetric flow rate is obtained from the measured pressure difference using an equation similar to that for the orifice plate. The value of C_d must be determined for each device.

4. Non-Newtonian Fluids [4], [19]

For fluids such as polymer solutions, molten plastics, emulsions, suspensions and many others, the deviatoric stress tensor may not be directly proportional to the rate of strain tensor and the simple Newtonian constitutive equation does not describe the flow behavior of these fluids. For such fluids, the viscosity, defined as the ratio of shear stress to shear rate, is now a function of the rate of strain tensor e:

$$\tau = 2 \eta(e) e \qquad (103)$$

Equation (103) is the generalized Newtonian constitutive equation. The viscosity of some non-Newtonian fluids can change by several orders of magnitude as a result of shearing. Such behavior has serious practical consequences for the handling of these fluids. Apart from their variable viscosity, these complex fluids often exhibit other flow behavior—such as stress relaxation, normal stresses, and stresses that depend on strain history—which are not observed in Newtonian fluids and also cannot be accounted for by the generalized Newtonian constitutive equation. Some non-Newtonian fluids exhibit some of the characteristics of an elastic solid and thus are referred to as *viscoelastic fluids*. The simple imaginary experiment shown in Figure 22 helps to visualize the difference between inelastic and viscoelastic fluids. Imagine a fluid confined between two flat plates where a force is applied to the upper plate, resulting in the displacement of a fluid particle just below the upper plate through a distance dx. Also imagine a similar experiment for an elastic solid. The work done in displacing the fluid or solid particle a distance dx

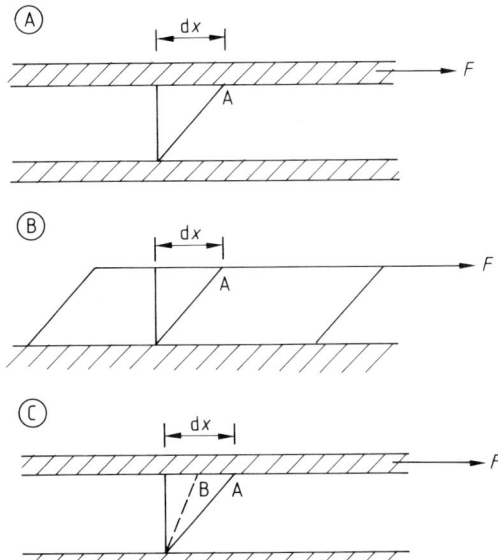

Figure 22. Inelastic and viscoelastic fluids
A) Viscous, inelastic liquid, work dissipated as heat; B) Elastic solid, work recoverable; C) Viscoelastic fluid, part of work recoverable

is then $F\,dx$. If an inelastic fluid is confined between the flat plates, on removing the force F, the fluid particle at point A will remain at point A (Fig. 22 A). All of the work done on an inelastic fluid is dissipated as heat and cannot be recovered. After removing the force acting on the ideally elastic or Hookean solid, the solid particle at point A will recoil to its original position (Fig. 22 B). All of the work done on an elastic solid is recoverable. For a viscoelastic fluid partial elastic recovery is observed. The fluid particle at point A recovers to an intermediate position B (Fig. 22 C). A portion of the work done on a viscoelastic fluid is recoverable and the remainder is dissipated as heat. In fact, the amount of recovery depends on the rate at which the displacement dx is imposed. In general, increasing the displacement rate increases the recovery. An inelastic fluid can be described sufficiently by measuring its viscosity and by determining the variation of viscosity with shear rate, time of shear, and, in some instances, thermal history. The viscosity is a steady state concept and is also applicable to viscoelastic fluids without any modification. In addition to the viscosity, however, there are other fundamental properties which must be measured to adequately characterize a viscoelastic fluid. The distinction between inelastic and viscoelastic fluids is clearly one of time scales. If manifestation of recoverable elastic deformation is observed on a time scale of interest, then the material is viscoelastic. In practical terms molecular relaxation times of $< 10^{-3}$ s can probably be ignored in all cases. Viscoelastic fluids will be dealt with in Section 4.3.

4.1. Classification of Fluids According to Viscosity Behavior

In Section 3.1 it has been shown that in one-dimensional flows the shear rate $\dot{\gamma}$ is equal to the velocity gradient for that flow. Since the viscosity is measured in one-dimensional flows (viscometric flows) the fluid behavior is classified according to the observed shear stress–shear rate behavior in one dimension. A graph of the measured shear stress as a function of shear rate is called the *flow curve*.

4.1.1. Bingham Type Behavior

Bingham type material exhibits a *yield stress*. That is, the shear stress τ must exceed a certain yield value τ_y before the fluid deforms and flows. Classical Bingham plastic behavior is illustrated in Figure 23 and can be described by

$$\tau = \left(\frac{\tau_y}{|\dot{\gamma}|} + \eta_{pl}\right)\dot{\gamma} \quad \text{for} \quad \tau \geq \tau_y \tag{104a}$$

i.e.,

$$\eta = \frac{\tau_y}{|\dot{\gamma}|} + \eta_{pl} \quad \text{for} \quad \tau \geq \tau_y \tag{104b}$$

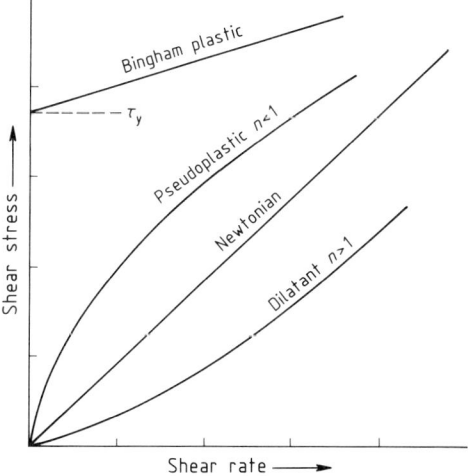

Figure 23. Typical flow curves

The stress is undetermined when $|\tau| < \tau_y$. The term $\dot\gamma/|\dot\gamma|$ is necessary to ensure that the stress takes on the same sign as the shear rate and η_{pl} is the *plastic viscosity*. The true viscosity η approaches η_{pl} only when $\eta_{pl}|\dot\gamma| \gg \tau_y$. In contrast the viscosity for the Bingham and any yield stress material becomes unbounded as $|\dot\gamma| \to 0$.

When the shear stress–shear rate function is nonlinear the data are often fit by a *Herschel–Buckley model*:

$$\tau = \left(\frac{\tau_y}{|\dot\gamma|} + k\dot\gamma^{n-1}\right)\dot\gamma \quad \text{for} \quad |\tau| \geq \tau_y \tag{105a}$$

$$\eta = \frac{\tau_y}{|\dot\gamma|} + k|\dot\gamma|^{n-1} \quad \text{for} \quad |\tau| \geq \tau_y \tag{105b}$$

The parameters k and n are obtained by plotting $\tau - \tau_y$ versus $\dot\gamma$ on logarithmic coordinates. Another equation often used to describe yield stress materials is the *Casson equation*

$$\tau = \left(\frac{\tau_y^{1/2}}{|\dot\gamma|^{1/2}} + \eta_{pl}^{1/2}\right)^2 \dot\gamma \quad \text{for} \quad |\tau| \geq \tau_y \tag{106a}$$

$$\eta = \left(\frac{\tau_y^{1/2}}{|\dot\gamma|^{1/2}} + \eta_{pl}^{1/2}\right)^2 \quad \text{for} \quad |\tau| \geq \tau_y \tag{106b}$$

Typical shear stress–shear rate data for a meat extract concentrate and for a tomato soup concentrate are presented in Figures 24 and 25, respectively. The meat extract behaves like a Bingham solid while the tomato soup concentrate exhibits Herschel–Buckley behavior.

The yield stress is associated with a three-dimensional structure that must deform elastically before flow can occur. It is commonly observed in fine particle suspension systems, especially at high particle concentrations. Common examples are toothpaste, paint, oil well drilling muds, foams, and many food and pharmaceutical products.

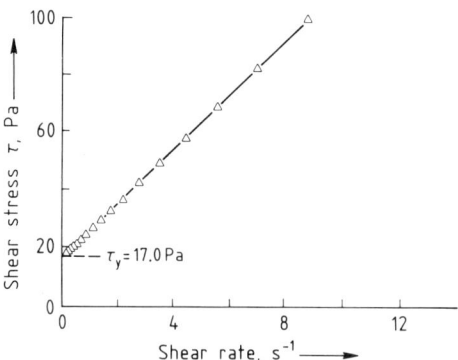

Figure 24. Shear stress–shear rate data (flow curve) for a meat extract

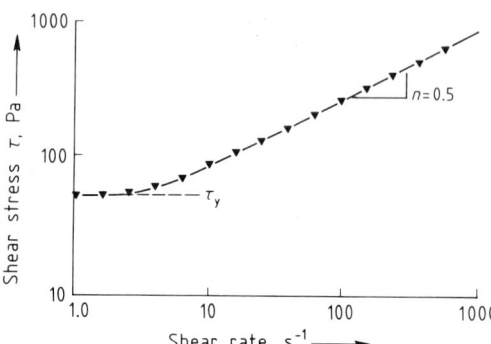

Figure 25. Shear stress–shear rate data (flow curve) for a tomato soup concentrate

4.1.2. Shear-Thinning (Pseudoplastic) Fluids

Solutions and melts of flexible macromolecules and many suspensions exhibit pseudoplastic shear stress–shear rate behavior such as that illustrated in Figure 23 and shown specifically for a polymer solution in Figure 26. Here the ratio $\tau/\dot\gamma$ (i.e., η) is a decreasing function of $\dot\gamma$. Such a fluid has a shear-thinning viscosity and is referred to as pseudoplastic. The decrease of viscosity with increasing rate of deformation can usually be attributed to the breakdown of a structure at the colloidal or molecular level. Macromolecules will become more aligned and hence less entangled and less resistant to deformation as the deformation rate is increased.

The viscosity of pseudoplastic fluids is typically plotted versus shear rate on logarithmic coordinates because the viscosity for such fluids can change by orders of magnitude over several decades of shear rate. The data for the polyacrylamide solution shown in Figure 26 are typical for a macromolecular fluid. The viscosity approaches a constant value at low shear rates called the *zero-shear viscosity* η_0 and a constant value in the limit of very high shear rates called the *infinite shear viscosity* η_∞. Experimental measurements in both regions can be difficult, particularly in the infinite shear region. Thus data are often only available between the two limits. Between the two limits the shear stress (or viscosity)

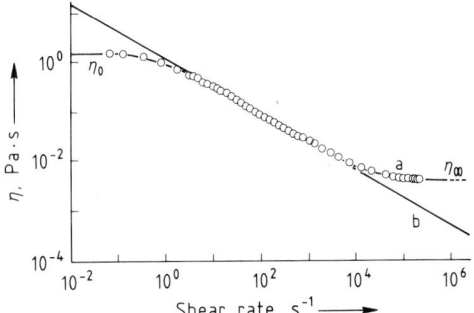

Figure 26. Viscosity as a function of shear rate for a polyacrylamide solution

is often linear with shear rate over several decades on logarithmic coordinates. In this restricted region

$$\log \tau = \log k + n \log \dot\gamma \quad (107\,\text{a})$$

$$\log \eta = \log k + (n - 1) \log \dot\gamma \quad (107\,\text{b})$$

where n is the slope in the linear region of the plot ($n = 0.41$ for the data in Fig. 26). Thus

$$\tau = k|\dot\gamma|^{n-1}\dot\gamma \quad (108\,\text{a})$$

$$\eta = k|\dot\gamma|^{n-1} \quad (108\,\text{b})$$

Equation (108) is known as the *power law* and sometimes as the Ostwald–de Waele model. This law is empirical and must fail for both high and low shear rates. The factor k is very temperature sensitive (and concentration sensitive for suspensions), but n is typically insensitive to temperature (and concentration) changes.

Many empirical and semi-empirical equations have been proposed to represent the viscosity data for shear-thinning fluids. The *Meter model*:

$$\eta = \eta_\infty + \frac{\eta_0 - \eta_\infty}{1 + \left|\dfrac{\tau}{\tau_{1/2}}\right|^{\alpha-1}} \quad (109)$$

has four parameters, $\eta_0, \eta_\infty, \tau_{1/2}$, and α with the following asymptotic behavior

$$\frac{\tau}{\tau_{1/2}} \to 0, \quad \eta \approx \eta_0 \quad (109\,\text{a})$$

$$\frac{\tau}{\tau_{1/2}} \to \infty, \quad \eta \approx \eta_\infty \quad (109\,\text{b})$$

and for

$$1 \ll \left|\frac{\tau}{\tau_{1/2}}\right|^{\alpha-1} \ll \frac{\eta_0}{\eta_\infty}$$

then

$$\eta = \frac{\eta_0 - \eta_\infty}{\tau_{1/2}^{1-\alpha}}|\tau|^{1-\alpha} \quad (109\,\text{c})$$

The Meter model exhibits power-law behavior in the intermediate region, with $\alpha = 1/n$; $\tau_{1/2}$ is the stress at which the viscosity is midway between the values of η_0 and η_∞. If η_∞ is taken to be zero, which is often a good approximation for molten polymers, then $\tau_{1/2}$ is the stress at which the viscosity drops to one-half the zero-shear value. In this case Equation (109) is known as the *Ellis model*. Line (a) through the data in Figure 26 is the fit of the Meter model to the data with $\eta_0 = 1.43$ Pa s, $\eta_\infty = 0.004$ Pa s, $\tau_{1/2} = 1.21$ Pa, and $\alpha = 2.43$. The power-law fit is line (b) with $n = 0.41$ ($= 1/\alpha$). The Ellis model follows the Meter model at low shear rates (incorporating the zero-shear viscosity behavior) and the power-law at higher shear rates.

4.1.3. Shear-Thickening (Dilatant) Fluids

Some suspensions with a high concentration of solids and with a relatively uniform particle-size distribution have a viscosity that increases with shear rate. An example for a titanium dioxide suspension is shown in Figure 27 where at the

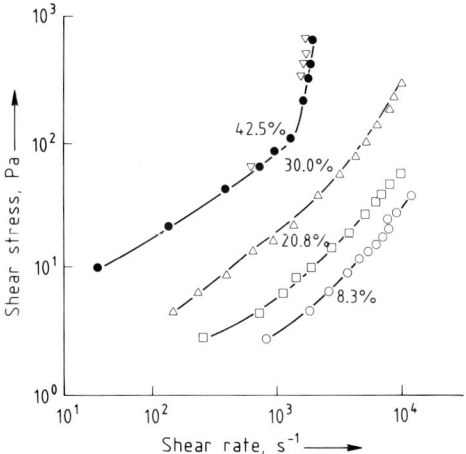

Figure 27. Shear stress–shear rate for a titanium dioxide suspension [20]

highest volume fraction of solids (42.5%) the viscosity increases with shear rate beyond a critical shear rate. The increase in stress observed at this point is related to a shear-induced structure. The phenomena is easily illustrated in the kitchen by mixing a concentrated suspension of corn starch and water. Data for shear-thickening liquids are usually fit by equations such as those for shear-thinning fluids but with $\eta_\infty = 0$ and $n > 1$ (or $\alpha < 1$). Dramatic equipment failures have been documented in the processing industries where a dilatant fluid response was not anticipated.

4.1.4. Fluids with Time-Dependent Viscosity (Thixotropic Fluids)

A special nomenclature is used in cases when the viscosity varies with the time of shear and the shear rate. A material whose apparent viscosity *decreases* with the time of shear is said to be *thixotropic* and one whose viscosity *increases* with time of shear is said to be *rheopectic*. The behavior of these materials depends not only on the time of shear but also on the past shear and thermal history. Rheopectic fluids are encountered very rarely and hence the discussion in this section will be confined to thixotropic fluids.

If the flow curve is measured in a single experiment in which the shear rate is steadily increased at a constant rate from zero to some maximum value and then decreased at the same rate to zero shear rate, a form of a hysteresis loop as is shown in Figure 28 is observed. The position of this loop, its shape, and the area within the

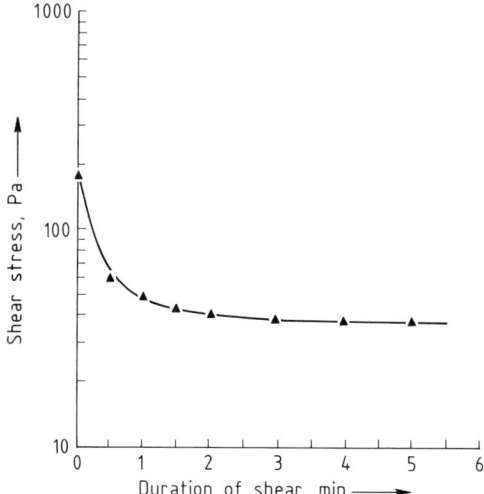

Figure 29. Shear stress versus time of shear for an Australian waxy crude oil at 10 °C and a shear rate of 458 s^{-1}

loop depends on the rate at which the shear rate is increased and decreased as well as the past thermal and shear history of the material. This type of measurement has little meaning for the solution of practical flow problems. What is required is a series of experimentally determined shear stress – shear rate – time curves which can then be converted to a series of flow curves with time of shear as the parameter. Because of the dependence of these curves on shear and thermal history it is important that the curves represent the behavior of a sample of material which has experienced a history approximately the same as that anticipated for the fluid whose flow behavior is to be analyzed.

The difference between thixotropy and pseudoplasticity is thought to be the time element in structural breakdown, which is finite and measurable for the thixotropic fluid and in contrast very small and undetectable for the pseudoplastic fluid. A practical example of time-dependent behavior is the waxy crude oils which in addition also exhibit a yield stress (see Fig. 29). It should be noted that a fluid can show time-dependent and viscoelastic behavior (for viscoelastic behavior see Section 4.3). Some liquid crystal polymers appear to be rheopectic and viscoelastic.

4.2. Fully Developed Tube Flow

The flow in tubes (pipes) of circular cross section is the flow field of the greatest interest.

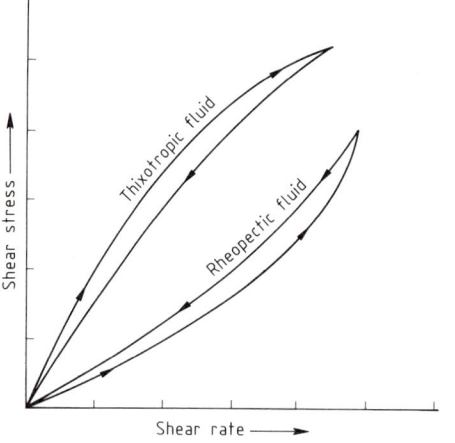

Figure 28. Time dependent fluids

For non-Newtonian fluids this is not only important because of processing applications but also because of the extreme importance of tube flow in the measurement of viscosity. Both laminar and turbulent flows will be considered. However, because of the sticky, gooey, or highly viscous nature of non-Newtonian fluids, most applications are in laminar flow. For instance a typical Reynolds number in polymer processing might be 10^{-3}. The energy required to reach turbulent flow for many non-Newtonian fluids is simply prohibitive. The treatment here is confined to fully developed tube flows, i.e., flow in long straight tubes where the velocity profile is not a function of axial position. Important results required from the discussion of Newtonian fluids include the definition of the friction factor

$$f = \frac{\tau_w}{\frac{\varrho V^2}{2}} \quad (48)$$

the general results for the shear stress in tube flow, i.e.,

$$\tau_{rz} = \frac{r \Delta p}{2L} \quad (88)$$

$$\tau_w = \frac{R \Delta p}{2L} = \frac{D \Delta p}{4L} \quad (87)$$

and

$$\tau_{rz} = \frac{r}{R} \tau_w \quad (89)$$

and Equation (44) which defines the volumetric flow rate

$$Q = \int_0^R 2\pi r u_z \, dr \quad (44)$$

The results obtained for fully developed tube flow will be valid for time-independent inelastic and viscoelastic fluids. Elastic effects are important in accelerating and decelerating flows and in time-dependent flow fields only.

4.2.1. Volumetric Flow Rate—Pressure Drop Relationship

Equation (44) can be used to obtain a general relationship between Q and τ_w for fully developed tube flows. Integration by parts yields

$$Q = \pi[u_z r^2 - \int r^2 \, du_z]_0^R \quad (110)$$

At the tube wall $(r = R)$ $u_z = 0$, then

$$Q = -\pi \int_0^R r^2 \, du_z \quad (111)$$

The shear rate in fully developed tube flow is du_z/dr and is a unique function of the shear stress τ_{rz}:

$$\frac{du_z}{dr} = -f(\tau_{rz}) \quad (112)$$

The minus sign is introduced because du_z/dr is negative (u_z decreases from a maximum value at $r = 0$ to zero at $r = R$). Then

$$du_z = -f(\tau_{rz}) \, dr \quad (113)$$

Eliminating r from Equations (111) and (113) with Equation (89) and then substituting for du_z in Equation (111) with Equation (113) yields

$$\frac{Q}{\pi R^3} = \frac{1}{\tau_w^3} \int_0^{\tau_w} \tau_{rz}^2 f(\tau_{rz}) \, d\tau_{rz} \quad (114)$$

Equation (114) is an important result which can be used in two ways: (1) *integrated* directly for a particular fluid model such as those represented by Equations (105), (106), (108), and (109) or (2) *differentiated* which leads to the definition of the wall shear rate for any fluid.

Consider the integration of Equation (114) for a power-law fluid. For tube flow of a power-law fluid

$$\tau_{rz} = -k\left(\frac{du_z}{dr}\right)^n \quad (115)$$

and

$$f(\tau_{rz}) = -\frac{du_z}{dr} = \left(\frac{\tau_{rz}}{k}\right)^{1/n} \quad (116)$$

Substitution for $f(\tau_{rz})$ in Equation (114) and integration leads to

$$\frac{Q}{\pi R^3} = \frac{n}{(3n+1)}\left(\frac{\tau_w}{k}\right)^{1/n} = \frac{n}{(3n+1)}\left(\frac{R \Delta p}{2Lk}\right)^{1/n} \quad (117)$$

For $n = 1$ Equation (117) reduces to the Hagen–Poiseuille equation derived in Section 3.3.1. The pressure drop for a pseudoplastic fluid is relatively insensitive to changes in the flow rate as compared to a Newtonian fluid. Thus flow rate measuring devices like the orifice plate that rely on the measurement of differential pressure are not very suitable for shear-thinning fluids, particularly if the flow behavior index n is low.

Equation (114) can be integrated for any of the semi-empirical equations which have been proposed to represent shear stress–shear rate data. The volumetric flow rate–wall shear stress results obtained by this integration for a number of fluids may be summarized as follows ($\tau_w = R\Delta p/2L$):

Newtonian fluid $\quad \dfrac{Q}{\pi R^3} = \dfrac{1}{4}\dfrac{\tau_w}{\eta}$

Bingham plastic

$$\frac{Q}{\pi R^3} = \frac{\tau_w}{4\eta_{pl}}\left[1 - \frac{4}{3}\frac{\tau_y}{\tau_w} + \frac{1}{3}\left(\frac{\tau_y}{\tau_w}\right)^4\right]$$

Power-law $\quad \dfrac{Q}{\pi R^3} = \dfrac{n}{3n+1}\left(\dfrac{\tau_w}{k}\right)^{1/n}$

Ellis fluid $\quad \dfrac{Q}{\pi R^3} = \dfrac{1}{4\eta_0}\tau_w + \dfrac{\tau_w^\alpha}{(\alpha+3)\eta_0\tau_{1/2}^{\alpha-1}}$

4.2.2. Generalized Treatment

If Equation (114) is multiplied by τ_w^3 and differentiated with respect to τ_w

$$\left.\frac{du_z}{dr}\right|_{\tau_w} = f(\tau_w) = \frac{1}{\pi R^3}\left[3Q + \Delta p\frac{dQ}{d\Delta p}\right] \quad (118)$$

where $f(\tau_w)$ is the wall shear rate $\dot\gamma_w$. In arriving at the above, Equation (87) was used to replace τ_w by the more directly measurable quantity Δp. Equation (118) can be rearranged to

$$\dot\gamma_w = \left(\frac{3n'+1}{4n'}\right)\left(\frac{8V}{D}\right) \quad (119)$$

where

$$n' = \frac{d\ln\tau_w}{d\ln\dfrac{8V}{D}} \quad (120)$$

and V is the average velocity in the tube $\left(V = \dfrac{Q}{\pi R^2}\right)$.

Equation (118) and the rearranged form (Eq. 119) is the *Rabinowitsch–Mooney equation*

which enables the calculation of the wall shear rate from the measurable quantities, R, Q or V, and Δp. Thus measurements in fully developed tube or capillary flow can be used to determine the shear stress (Eq. 87) and the shear rate (Eq. 119 and 120) at the tube wall. Since the shear stress–shear rate curve is a function of the material and not the apparatus used, a plot of τ_w versus $\dot\gamma_w$ is one way in which a flow curve can be determined for any time-independent fluid.

The parameter n' is the slope of the log–log plot of τ_w versus $8V/D$ and must not be confused with the flow-behavior index n in the power law. The flow-behavior index n is constant over the range of shear rates for which it is defined, whereas n' can vary with $8V/D$. Since n' in Equation (120) represents the slope of a line, the line may be represented by an equation of the form

$$\ln\tau_w = n'\ln 8V/D + \ln K' \quad (121)$$

If n' is not constant, but varies with τ_w then Equation (121) represents the tangent to the curve at a given point. On the other hand, if n' is constant, Equation (121) represents the actual relationship between τ_w and $8V/D$. When n' is constant, it is useful to write Equation (121) in the form

$$\tau_w = K'\left(\frac{8V}{D}\right)^{n'} \quad (122)$$

Comparing Equation (122) to Equation (117) shows that for a power-law fluid, $n = n'$ and

$$k = K'\left(\frac{4n}{3n+1}\right)^n \quad (122a)$$

It is not necessary to compute the slope of the experimentally determined log–log plot of τ_w versus $8V/D$ if calculations for laminar tube flow are to be made. The plot can be used directly for design or process evaluation, provided that process conditions are within the range of the experimental data. In fact, a single experimental determination of τ_w in a convenient model at a given value of V/D enables the prediction of the pressure drop in a larger pipe of the same value of V/D.

The terms n' and K' need only be determined when the τ_w versus $8V/D$ data are to be used for other laminar flow geometries or for turbulent flow. It is important to realize that n' and K', as determined from the log–log plot, are only valid for the shear rate or the $8V/D$ range in which

4.2.3. Velocity Distribution

For fully developed laminar flow, the velocity distribution in a tube for any time-independent fluid is obtained by integrating the velocity gradient from the tube wall $r = R$ to any radial position $r = r$, assuming no slip at the wall:

$$u_z(r) = \int_r^R \left(-\frac{du_z}{dr}\right) dr \qquad (123\,\text{a})$$

or

$$u_z(r) = \frac{R}{\tau_w} \int_{\tau_{rz}}^{\tau_w} f(\tau_{rz}) d\tau_{rz} \qquad (123\,\text{b})$$

For a power-law fluid where $f(\tau_{rz}) = (\tau_{rz}/k)^{1/n}$ and remembering that $\tau_{rz} = \frac{r}{R}\tau_w$ and $\tau_{rz} = r\Delta p/2L$ yields

$$u_z(r) = \frac{n+1}{R^n}\left(\frac{\Delta p}{2kL}\right)^{\frac{1}{n}}\left(\frac{n}{n+1}\right)\left[1 - \left(\frac{r}{R}\right)^{\frac{n+1}{n}}\right] \qquad (124)$$

when Equation (123 b) is integrated. In terms of the average velocity

$$V = \frac{2\pi}{\pi R^2}\int_0^R r u_z\, dr = R^{(n+1)/n}\left(\frac{\Delta p}{2kL}\right)^{\frac{1}{n}}\left(\frac{n}{3n+1}\right) \qquad (125)$$

and hence

$$u_z(r) = \left(\frac{3n+1}{n+1}\right) V\left[1 - \left(\frac{r}{R}\right)^{\frac{n+1}{n}}\right] \qquad (126)$$

Equation (126) defines the velocity distribution for *power-law fluids* in fully developed tube flow. The velocity gradient at the tube wall is

$$\left.\frac{du_z}{dr}\right|_{r=R} = \dot{\gamma}_w = \frac{3n+1}{4n}\left(\frac{8V}{D}\right) \qquad (127)$$

As expected Equation (127) is identical to Equation (119) when $n = n'$ but more importantly for a *Newtonian fluid* ($n' = n = 1$)

$$\dot{\gamma}_w = \frac{8V}{D} \qquad (128)$$

Thus $8V/D$ is the wall shear rate for a Newtonian fluid in laminar tube flow.

For a *Bingham material* similar calculations yield the following velocity distribution:

$$u_z(r) = \frac{R^2}{4\eta_{pl}}\left(\frac{\Delta p}{L}\right)\left(1 - \frac{r_0^2}{R}\right) \quad 0 \le r \le r_0 \qquad (129\,\text{a})$$

and

$$u_z(r) = \frac{R^2}{4\eta_{pl}}\frac{\Delta p}{L}\left[1 - \left(\frac{r}{R}\right)^2\right] - \frac{\tau_y R}{\eta_{pl}}\left[1 - \left(\frac{r}{R}\right)\right]$$
$$r_0 \le r \le R \qquad (129\,\text{b})$$

where $r_0 = 2L\tau_y/\Delta p$ and η_{pl} is the plastic viscosity of the fluid. Equations (129a) and (129b) show that there is a region of radius r_0, in the center of the pipe, where the fluid moves as a solid plug with a uniform velocity. The volumetric flow rate in a pipe for a Bingham material is given on p. 5-30, where τ_y is the yield stress and τ_w is the wall shear stress. This equation is only valid when τ_w is greater than τ_y. Flow in the pipe will not occur if τ_w is less than τ_y. Thus

$$\frac{\Delta p}{L} > \frac{2\tau_y}{R} \qquad (130)$$

is necessary for the initiation of flow in a pipe for yield stress materials.

The yield stress of a material can be determined by extrapolation of the basic shear stress–shear rate data to zero shear rate. However, in view of the importance of the

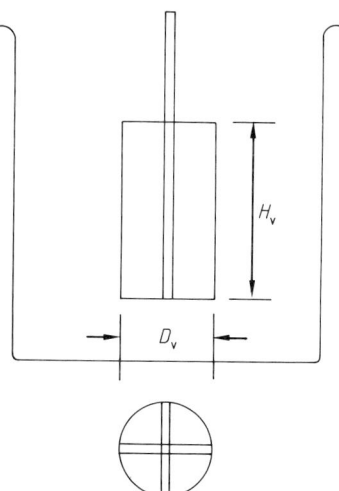

Figure 30. The vane used for yield stress measurement
D_v = vane diameter; H_v = vane height

yield stress special techniques have been developed for its measurement. In one such technique, a vane is inserted into the fluid (see Fig. 30). The minimum torque Γ_m required to initiate a rotation motion in the vane is recorded. The yield stress is then related to this torque by

$$\Gamma_m = \frac{\pi D_v^3}{2}\left(\frac{H_v}{D_v} + \frac{1}{3}\right)\tau_y \tag{131}$$

In deriving Equation (131) it is assumed that the yielding surface is the cylindrical surface and the two end circular disks traced out by the vane. For short vanes, end effects may become important and must be taken into consideration [21].

4.2.4. Friction Factor–Reynolds Number Relationships

General procedures applicable to all types of fluids are of course desirable because of the difficulty in classifying a fluid as one type or another, and because of the tendency of some fluids to change from one type to another with changing shear rate, temperature, or composition. At present generalizations are established only for time-independent fluids.

Laminar Flow. For laminar flow of Newtonian fluids in circular pipes, the friction factor satisfies the relation (see Section 3.3.1)

$$f = \frac{16}{Re} \tag{48}$$

with the Reynolds number defined as $Re = (DV\varrho)/\eta$. For the laminar flow of any time-independent fluid in a tube, the wall shear stress is given by

$$\tau_w = K'\left(\frac{8V}{D}\right)^{n'} \tag{132}$$

where, in general, K' and n' are not constant, but depend on $8V/D$. When Equation (132) is substituted into the defining equation for the friction factor one obtains

$$f = \frac{16}{Re'} \tag{133}$$

where

$$Re' = \frac{D^{n'} V^{2-n'}}{K' 8^{n'-1}} \varrho \tag{134}$$

which is a generalized Reynolds number valid for all time-independent fluids. In the special case where $n' = 1$ and $K' = \eta$, Equation (134) reduces to the Reynolds number for a Newtonian fluid. For a power-law fluid $n' = n$ and Equation (122a) is valid.

Transition Flow. So far only the laminar flow of non-Newtonian fluids in tubes has been considered and the important question, when is the flow laminar, has been avoided. For Newtonian fluids, the question is answered by calculating the Reynolds number and comparing it to the empirically determined value of 2100 below which the flow is laminar. A similar criterion has been sought for non-Newtonian fluids by METZNER and REED [22] and DODGE and METZNER [23]. METZNER and REED first considered that stable laminar flow in tubes usually ends when the friction factor becomes < 0.008, which corresponded to a generalized Reynolds number of 2100. In the later work by DODGE and METZNER, which led to the generalized $f-Re$ chart shown in Figure 31, laminar flow was found to end at generalized Reynolds numbers which increase slightly as n' decreases. At n' values of 1.0, 0.726, and 0.38, the transition region appeared to begin at Re' of about 2100, 2700 and 3100, respectively. Others have examined the flow transition question for non-Newtonian fluids but the conservative conclusion remains:

$$Re' \leq 2100 \tag{135}$$

has been accepted and is used to establish whether the flow is laminar or turbulent for a non-Newtonian fluid.

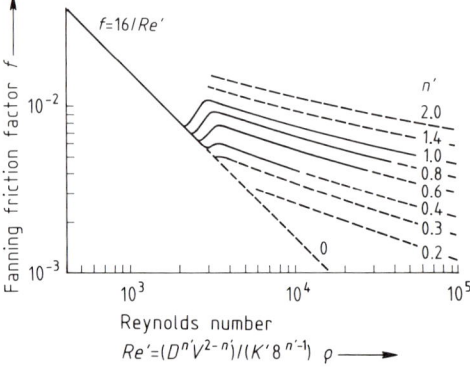

Figure 31. Friction factor design chart for non-Newtonian fluids
——— Experimental regions; – – – extrapolated regions

Turbulent Flow. The friction factor chart for time-independent inelastic fluids is reproduced in Figure 31. The friction factor is defined by

$$\sqrt{\frac{1}{f}} = \frac{4.0}{(n')^{0.75}} \log [Re' f^{1-n'/2}] - \frac{0.4}{n'^{1.2}} \qquad (136)$$

Equation (136) reduces to the von Karman–Nikuradse equation (50) for Newtonian fluids when $n' = 1$. The solid lines in Figure 31 are based on experimental results, and the dashed lines are extrapolations made using Equation (136). The constants in Equation (136) were evaluated from experimental data. If the fluid does not obey the power law with an n' and K' which vary with $8V/D$, then a trial-and-error procedure is required to calculate f, τ_w, or Δp. First a value of Δp is assumed, then the approximate values of n' and K' are found from the laminar flow curve (τ_w versus $8V/D$) at the assumed τ_w. From these parameters, Re' is calculated and f is obtained from Figure 31. If the pressure drop calculated from the friction factor does not agree with the original assumed value, another iteration is carried out. The process converges quickly.

4.3. Viscoelastic Fluid Mechanics

Materials such as molten plastics, solutions of synthetic macromolecules and many biological fluids—apart from possessing the properties normally associated with purely viscous fluids—also exhibit a range of behavior which is normally associated with elastic solids. For this reason such materials have been given the name viscoelastic fluids. In a viscoelastic fluid the stress generated by deformation is no longer a simple function of the instantaneous rate of strain. It also depends, as in solids, on the strain and, for large strains, on the strain history experienced by the fluid. As discussed earlier, in a purely viscous fluid, all the energy required to produce the deformation is dissipated as heat while in a perfectly elastic solid this energy is stored and is completely recovered when the forces acting on the solid are removed. In a viscoelastic fluid, this energy is partly dissipated and partly stored, the ratio of dissipation to storage depends on the fluid, the nature of the deformation, and the history of the deformation. Thus, depending on the flow, a viscoelastic fluid may exhibit, in varying degrees, the elastic properties of solids and the viscous properties of fluids.

It is therefore not surprising that some of the flow behaviors of viscoelastic fluids have no parallel in purely viscous fluids. A number of these viscoelastic phenomena are reproduced in Figures 32 to 35. Figure 32 shows a rod slowly rotating in a beaker filled with a viscoelastic fluid. The rotation causes the viscoelastic fluid to climb up the rod. Under similar conditions, a purely viscous fluid is undisturbed by the slow rotation and its surface remains essentially flat. Rod climbing by a viscoelastic fluid is often referred to as the *Weissenberg effect*. When a viscoelastic fluid is extruded from a capillary (see Fig. 33) the resulting fluid stream may swell up to several times the capillary diameter. This phenomenon is known as *extrudate swell*. Newtonian fluids also exhibit extrudate swell at small Reynolds number, but the increase in diameter is only 13%. Large extrudate swell is observed in polymer extrusion and clearly has to be taken into account in the design of extrusion shaping dies. Figure 34 shows a *tubeless syphon*, where a viscoelastic fluid is being sucked vertically upwards into a tube. The suction tube can be lifted to a

Figure 32. Rod-climbing by a viscoelastic fluid

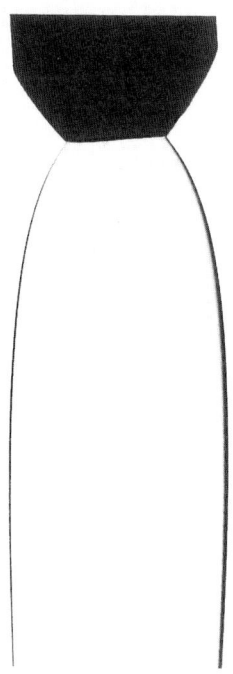

Figure 33. Extrudate swell by a viscoelastic fluid

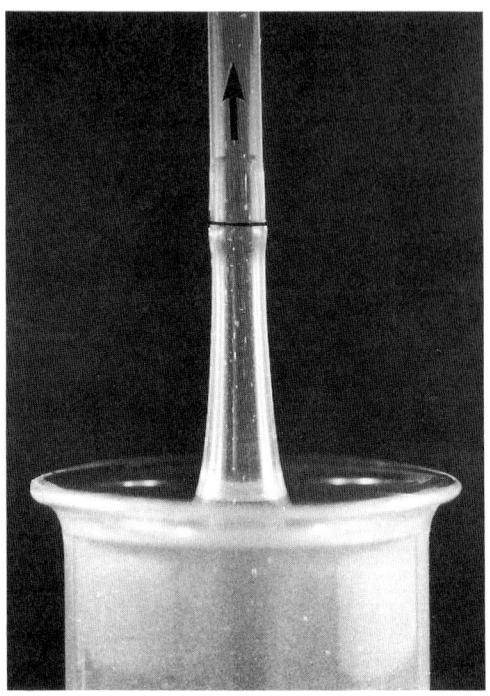

Figure 34. Viscoelastic fluid being sucked up (tubeless syphon)

large height (well over four to five tube diameters in the case shown) above the viscoelastic fluid without the syphoning action being interrupted. In comparison, for a purely viscous fluid, syphoning ceases as soon as the tube is lifted out of the fluid. The tubeless syphon illustrates the large tensile stress that can be sustained in a viscoelastic fluid. Figure 35 is a comparison of the streamline patterns formed by a Newtonian fluid and a viscoelastic fluid as they flow through a pipe contraction. Figure 35A shows a Newtonian fluid; Figures 35B–D show a viscoelastic fluid at increasing flow rate. For the Newtonian fluid, two small recirculating vortices can be ob-

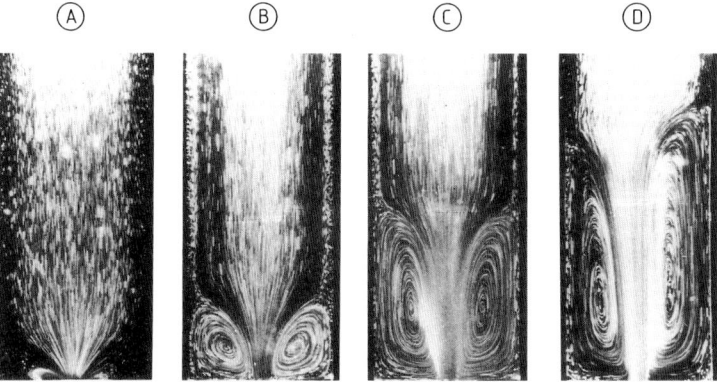

Figure 35. Comparison of the flow patterns exhibited by a Newtonian fluid and by a viscoelastic fluid A) Newtonian fluid; B)–D) Viscoelastic fluid at increasing flow rates

served at the contraction plane. The size of the Newtonian vortex, shown at low Reynolds number, is not very sensitive to changes in flow rate. For viscoelastic fluids, the situation is quite different. At low flow rates the viscoelastic vortices are approximately of the same size as those for the Newtonian fluid, but they rapidly increase in size, up to four times the size of Newtonian vortex, as the flow rate is increased. The flow then becomes unstable and the vortices begin to pulsate, growing and decreasing in size in some erratic manner. Although the streamline pattern of the Newtonian fluid can be obtained by solving the Navier–Stokes equations, attempts to understand the viscoelastic flow patterns have so far been, at most, only partially successful [24]. The most likely source of the problem here is the present inability to describe, with the necessary degree of accuracy, the complex rheological behavior of the viscoelastic fluid when it undergoes shearing and stretching deformation as it flows through the contraction.

Some of the more fundamental aspects of viscoelastic fluid mechanics will be discussed in the next section. The treatment is, however, of a more descriptive nature.

4.3.1. Steady Shear Behavior of Viscoelastic Fluids

Under steady shear the shear stress T_{xy} generated in a viscoelastic fluid is related to the shear rate by

$$T_{xy} = \eta(\dot{\gamma})\dot{\gamma}$$

where η is the shear-rate dependent viscosity of the fluid. Figure 26 is the plot of the viscosity of a solution of polyacrylamide, a typical viscoelastic fluid. Like most non-Newtonian fluids, most viscoelastic fluids exhibit shear thinning, and, over a range of shear rates, behave like a power-law fluid. However, if the normal components of the stress tensor in the x direction, T_{xx}, and in the y direction, T_{yy}, are measured, it will be found that they are, in general, not equal. This is not observed in Newtonian and generalized Newtonian fluids. The difference in T_{xx} and T_{yy} is referred to as the *first normal stress difference* N_1. Similarly, it will be found that T_{yy} and T_{zz} are unequal, and the difference $T_{yy} - T_{zz}$ is known as the *second normal stress difference*. The first and second normal stress difference are related to the shear rate by

$$T_{xx} - T_{yy} = \psi_1(\dot{\gamma})\dot{\gamma}^2$$

$$T_{yy} - T_{zz} = \psi_2(\dot{\gamma})\dot{\gamma}^2$$

The functions ψ_1 and ψ_2 are known as the *first* and *second normal stress coefficients*, respectively. They are a function of the shear rate $\dot{\gamma}$: η, ψ_1 and ψ_2 are known as the viscometric functions of the fluid which together describe the response of the viscoelastic fluid under steady shear. ψ_1 is measured in special viscometers, which measure the normal stress, as well as the shear stress, exerted by the test fluid when it is being sheared [8]. Figure 36 is a plot of the first normal stress difference exhibited by a polyacrylamide solution. For comparison, the shear stress is also included in Figure 36. It can be seen that the normal stress is very much larger than the shear stress which means that this viscoelastic fluid can be expected to behave quite differently from a purely viscous shear-thinning fluid with similar shear stress versus shear rate behavior. The second normal stress difference is usually much smaller than the first normal stress difference and is very difficult to measure accurately. It is not normally reported in the literature.

In laboratory investigations of viscoelasticity, it is very useful to be able to isolate the effects of shear thinning and fluid inertia from that of fluid elasticity so as to concentrate on the latter. For this purpose, a class of special test fluids have been developed. They are commonly referred to as *Boger fluids* or ideal elastic fluids. Typically, they are prepared by dissolving a high molecular mass solute in a highly viscous Newtonian solvent. The resulting solutions are characterized by their near constant viscosity, over a wide range of shear rates, and they exhibit all the elastic properties normally associated with viscoelastic fluids. Boger fluids are used extensively in the comparison of experimen-

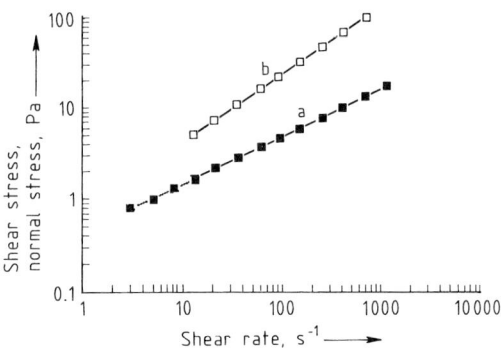

Figure 36. Normal stress versus shear rate for a viscoelastic fluid
a) Shear stress τ; b) First normal stress difference N_1

tally observed elastic response of fluids with that predicted by viscoelastic constitutive equations. They play an important role in the development and selection of such constitutive equations.

4.3.2. Behavior of Viscoelastic Fluids in Oscillatory Shear Flow

In steady shear motion the viscoelastic fluid is subjected to a constant rate of strain which does not reveal the dependence of the current stress on the strain history. Viscoelastic behavior under a time-dependent rate of strain is frequently studied experimentally by putting the fluid through an oscillatory shear flow, such as that generated when the fluid is sheared between two parallel plates: one of the plates is held stationary and the other oscillates sinusoidally with a frequency ω. The resulting shear rate and shear stress both have the same frequency ω, but there is a phase difference between them. If the oscillatory shear strain is represented by

$$\gamma = \gamma_0 \sin(\omega t) \qquad (137)$$

where γ_0 is the amplitude of the sinusoidal shear strain, then the shear rate experienced by the fluid is given by

$$\dot{\gamma} = \gamma_0 \omega \cos(\omega t) = \dot{\gamma}_0 \cos(\omega t) \qquad (138)$$

$\dot{\gamma}_0 (= \omega \gamma_0)$ is used to denote the amplitude of the sinusoidal shear rate.

It is customary to express the resulting shear stress in the form

$$T_{xy}(\omega) = \gamma_0 [G'(\omega) \sin \omega t + G''(\omega) \cos \omega t] \qquad (139)$$

The terms associated with $G'(\omega)$ and $G''(\omega)$ are the in-phase and the out-phase component of the shear stress. $G'(\omega)$ is usually referred to as the *storage modulus* of the fluid, it is related to the storage of energy. $G''(\omega)$ is the *loss modulus* of the fluid and is connected with energy dissipation. As indicated above, $G'(\omega)$ and $G''(\omega)$ are functions of ω, and for small amplitude oscillatory shear, they are independent of γ_0. For purely viscous fluids $G'(\omega)$ is zero.

$G'(\omega)$ and $G''(\omega)$ are material properties of a viscoelastic fluid. Two other material functions, related to $G'(\omega)$ and $G''(\omega)$ are also in common use. These are the complex viscosity of the viscoelastic fluid, defined by

$$\eta'(\omega) = G''(\omega)/\omega \qquad (140a)$$

and

$$\eta''(\omega) = G'(\omega)/\omega \qquad (140b)$$

The function $\eta'(\omega)$ is called the *dynamic viscosity* of the fluid; $\eta''(\omega)$ does not seem to have a generally accepted name. For a Newtonian fluid $\eta'(\omega)$ is equal to the viscosity of the fluid and $\eta''(\omega)$ is zero. It is evident that a large number of material functions are needed just to describe the behavior of viscoelastic fluids in shear. At low shear rates and low angular frequencies, the material functions for steady-shear and oscillatory shear are related to one another. The dynamic viscosity at very low frequencies is related to the steady shear viscosity at very low shear rates by

$$\lim_{\omega \to 0} \eta'(\omega) = \lim_{\dot{\gamma} \to 0} \eta(\dot{\gamma}) \qquad (141)$$

Furthermore, the storage modulus at very low frequencies is related to the first normal stress coefficient $\psi_1(\dot{\gamma})$ at very low shear rates by

$$\lim_{\omega \to 0} \left(\frac{2G'(\omega)}{\omega^2} \right) = \lim_{\dot{\gamma} \to 0} \left(\frac{N_1(\dot{\gamma})}{\dot{\gamma}^2} \right) = \lim_{\dot{\gamma} \to 0} \psi_1(\dot{\gamma}) \qquad (142)$$

These limiting relationships are very useful for checking the consistency of experimental data on viscoelastic fluids. A number of other empirical relationships have been developed to relate the steady and oscillatory data. An example of these is the Cox–Merz rule which relates $\eta(\dot{\gamma})$ to $\eta'(\omega)$ and $\eta''(\omega)$:

$$\eta(\dot{\gamma}) = [\eta'(\omega)^2 + \eta''(\omega)]^{1/2}|_{\omega = \dot{\gamma}} \qquad (143)$$

The Cox–Merz rule is often used to extend the steady shear viscosity data to large shear rates where only oscillatory measurements have been carried out. Further details and the limitations of this and other similar relationships are given in [4].

$G'(\omega)$ and $G''(\omega)$ or equivalently $\eta'(\omega)$ and $\eta''(\omega)$ are measured with special viscometers in which one of the shearing surfaces can be driven sinusoidally [8]. A typical set of oscillatory data is shown in Figure 37. These data were extracted from a comprehensive study carried out on polyethylene [25]. Oscillatory measurements are now routinely carried out by polymer manufacturers as a means of characterizing their products.

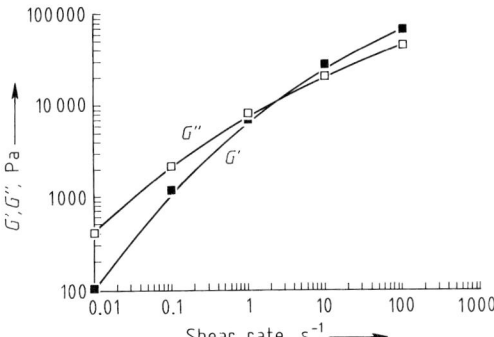

Figure 37. Storage and loss modulus of a low density polyethylene [25]
■ Storage modulus G'; □ Loss modulus G''

4.3.3. Examples of Constitutive Equations for Viscoelastic Fluids

Numerous constitutive equations have been proposed to describe the diverse relationship, observed in viscoelastic fluids, between the stresses and the various kinematic quantities used to measure the strain and the rate of strain experienced by the fluid. One of the earliest viscoelastic constitutive equations is the linear Maxwell equation. In shear flow, the stress T_{xy} and the shear rate $\dot{\gamma}$ are related by a constitutive equation of the form

$$\lambda \frac{dT_{xy}}{dt} + T_{xy} = \eta_0 \dot{\gamma} \qquad (144)$$

λ and η_0 are the two material parameters of the linear Maxwell equation where λ is the time constant of the fluid and η_0 is its viscosity. Under steady shear, the stress is not a function of time, hence the first term on the left-hand side vanishes and the linear Maxwell equation reduces to the Newtonian constitutive equation where the stress is related to the shear rate via the viscosity η_0. However, the introduction of the time constant λ means that, in general, the stress is no longer a unique function of the instantaneous shear rate. For example, during the start up of the shear flow or in oscillatory shear test, the stress is out of phase with the instantaneous shear rate.

In particular, in oscillatory shear test the oscillation of the top plate is given by $A \sin \omega t$. The velocity of the top plate is

$$u_0 = A\omega \cos \omega t$$

For most viscoelastic fluids of practical interest, experimental conditions are such that the inertia terms in the Cauchy equations of motion, Equation (24), are small and can be ignored. As a result, the velocity variation across the gap separating the plates is linear and the velocity profile is given by

$$u(y,t) = \frac{A\omega y}{H} \cos(\omega t) \qquad (145)$$

where H is the gap between plates. The shear rate is given by

$$\dot{\gamma}(t) = \frac{A\omega}{H} \cos(\omega t) = \gamma_0 \omega \cos(\omega t) \qquad (146)$$

where $\gamma_0 = A/H$ is used to denote the amplitude of the shear strain. According to the linear Maxwell constitutive equation, the stress exerted by the fluid on the oscillating top plate is given by

$$\lambda \frac{dT_{xy}}{dt} + T_{xy} = \gamma_0 \omega \cos(\omega t) \qquad (147)$$

This equation can be solved to give

$$T_{xy} = \gamma_0 \left(\frac{\eta_0 \omega \cos \omega t}{1 + \lambda^2 \omega^2} + \frac{\eta_0 \lambda \omega^2}{1 + \lambda^2 \omega^2} \sin \omega t \right)$$
$$+ \text{transient terms} \qquad (148)$$

The transient terms decay away exponentially. As oscillatory measurements are taken at large t, the transient terms can be ignored. The resulting sinusoidally fluctuating stress can be written in the form

$$T_{yx} = \gamma_0 \left(\frac{\eta_0 \omega}{1 + \lambda^2 \omega^2} \cos \omega t + \frac{\eta_0 \lambda \omega^2}{1 + \lambda^2 \omega^2} \sin \omega t \right) \qquad (149)$$

From the definition of $G'(\omega)$ and $G''(\omega)$ and of $\eta'(\omega)$ and $\eta''(\omega)$ it can then be seen that for a linear Maxwell fluid

$$G'(\omega) = \frac{\eta_0 \lambda \omega^2}{1 + \lambda^2 \omega^2} \quad G''(\omega) = \frac{\eta_0 \omega}{1 + \lambda^2 \omega^2} \qquad (150\text{a})$$

$$\eta'(\omega) = \frac{\eta_0}{1 + \lambda^2 \omega^2} \quad \eta''(\omega) = \frac{\eta_0 \lambda \omega}{1 + \lambda^2 \omega^2} \qquad (150\text{b})$$

By choosing the appropriate value for the viscosity η_0 and time constant λ, these functions can provide an adequate description of the observed behavior of real viscoelastic fluids. The linear Maxwell equation does not satisfy all the requirements of modern continuum mechanics for a theoretically sound constitutive equation. Numerous modifications of the Maxwell equations can be found in the literature. Some of the modifications were made so as to satisfy the requirements of continuum mechanics, others resulted from improved understanding of the molecular structure and dynamics of the polymer constituents of viscoelastic fluids [4], [26].

A constitutive equation that has been very popular in recent years is the *Oldroyd-B equation*. It relates the deviatoric stress tensor τ to the strain tensor γ by an equation of the form

$$\lambda_1 \tau_{(1)} + \tau = \eta_0 (\gamma_{(1)} + \lambda_2 \gamma_{(2)}) \quad (151)$$

where λ_1 and λ_2 are the relaxation and retardation time of the Oldroyd-B fluid, respectively and η_0 is, as before, the viscosity of the fluid. Subscripts (1) and (2) are used to denote the first and second convected derivatives of the stress and the strain tensor. The definition of the convected and other derivatives can be found in [4]. The introduction of these derivatives is required by the theory of constitutive equations in continuum mechanics. The derivative $\gamma_{(1)}$ is equal to the rate of strain tensor e. The Oldroyd-B equation satisfies all the requirements of continuum mechanics. In terms of the three material parameters, λ_1, λ_2, and η_0 the viscometric functions (i.e., the steady shear properties) and the oscillatory behavior of the Oldroyd-B equation can be shown to be

$$\eta = \eta_0 \quad \psi_1 = 2\eta_0(\lambda_1 - \lambda_2) \quad \psi_2 = 0 \quad (152\,\text{a})$$

$$G'(\omega) = \frac{\eta_0(\lambda_1 - \lambda_2)\omega^2}{1 + (\lambda_1 \omega)^2}$$
$$G''(\omega) = \frac{\eta_0 \omega (1 + \lambda_1 \lambda_2 \omega^2)}{1 + (\lambda_1 \omega)^2} \quad (152\,\text{b})$$

$$\eta'(\omega) = \frac{\eta_0(1 + \lambda_1 \lambda_2 \omega^2)}{1 + (\lambda_1 \omega)^2}$$
$$\eta''(\omega) = \frac{\eta_0(\lambda_1 - \lambda_2)\omega}{1 + (\lambda_1 \omega)^2} \quad (152\,\text{c})$$

As with the linear Maxwell equation, the parameters of the Oldroyd-B equations are chosen so that these functions give an approximate description of the measured viscometric and oscillatory properties of real fluids. It is interesting

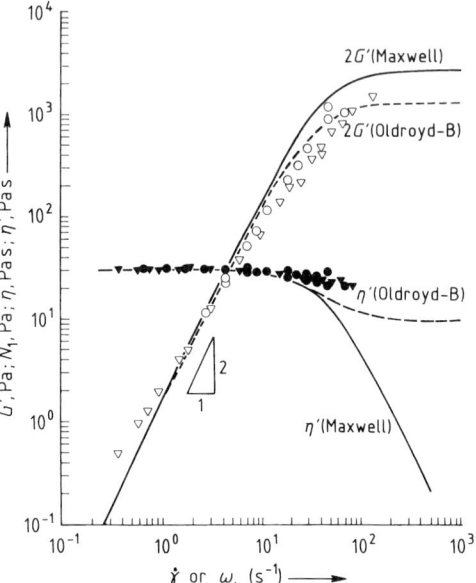

Figure 38. Comparison of the observed normal stress and storage modulus of a silicone oil against that predicted by the Maxwell and the Oldroyd-B constitutive equations, fluid: silicone oil
○ N_1, Pa; ▽ $2 G'$, Pa; ● η, Pa s; ▼ η', Pa s

to note that both linear Maxwell and the Oldroyd-B equations give a constant viscosity. This means that these equations are particularly suitable for describing the rheological behavior of viscoelastic fluids that are not shear-thinning. In Figure 38 the measured $G'(\omega)$ and $\eta'(\omega)$ for a silicone oil are plotted against ω. This fluid can be regarded as a non-shear-thinning viscoelastic fluid (Boger fluid). The expression for these two quantities according to the Maxwell and the Oldroyd-B equations are also plotted on the figure. Standard procedures have been applied to extract the fluid parameters of these two constitutive equations [27]. The Oldroyd-B equation provides a better description of the rheological behavior of this viscoelastic fluid than the simpler Maxwell equation. For comparison, the steady shear data for this silicone oil are also included in Figure 38. The experimental data confirm the limiting relationship between N_1 and $2 G'$ and between η and η' at low $\dot{\gamma}$ and ω.

The linear Maxwell and the Oldroyd-B equations are just two examples of the large number of viscoelastic constitutive equations in the literature. The construction of constitutive equations is guided by the principles of continuum mechanics and by the understanding of the dynamical properties of the macromolecules and the solvent that make up the viscoelastic fluids. For more details on these and related topics see [28], [29].

One of the recent developments in viscoelastic fluid mechanics is the use of Boger fluids to gain a physical understanding of constitutive equations. With these ideal test fluids good agreement between experimental observations and theoretical predictions could be observed for the first time. It has also been possible to relate the parameters in the constitutive equations to the physical properties of the constituents of Boger fluids.

4.3.4. Extensional Behavior of Viscoelastic Fluids

In many of the flow fields encountered in industrial processes the fluid undergoes a stretching deformation. A specific example is the deformation suffered by a fluid element when it is forced to flow through a channel or tube of gradually or abruptly changing cross-sectional area. This kind of deformation is usually referred to as extensional deformation. In extensional deformation the macromolecules in the viscoelastic fluid are being stretched and may generate large tensile stresses as a result of this. The behavior of a viscoelastic fluid in extensional deformation can be quite different from its behavior in shear deformation.

An idealized flow field that can be used to investigate the extensional behavior of a viscoelastic fluid is shown in Figure 39. The tensile stress T_{zz} applied to stretch the specimen of test fluid, in the form of a circular cylinder, is measured. The applied stress T_{zz} is varied so that the axial velocity gradient is uniform within the fluid sample and remains constant in time, i.e.,

$$e_{zz} = \frac{du_z}{dz} = \dot{\varepsilon}_0 = \frac{1}{L}\frac{dL}{dt} \qquad (153)$$

where $\dot{\varepsilon}_0$ is the constant extensional rate, which can be identified with the ratio of the instantaneous rate of increase in length of the fluid sample $dL(t)/dt$ to its instantaneous length $L(t)$. It can be assumed that the azimuthal velocity component in the test specimen is zero. The continuity equation for an incompressible fluid then requires the velocity components in the axial and radial directions to be given by

$$\begin{aligned} u_z &= \dot{\varepsilon}_0 z \\ u_r &= -\tfrac{1}{2}\dot{\varepsilon}_0 r \\ u_\theta &= 0 \end{aligned} \qquad (154)$$

The rate of strain tensor, in cylindrical coordinates, is given by

$$e = \begin{vmatrix} \dot{\varepsilon}_0 & 0 & 0 \\ 0 & -\tfrac{1}{2}\dot{\varepsilon}_0 & 0 \\ 0 & 0 & -\tfrac{1}{2}\dot{\varepsilon}_0 \end{vmatrix} \qquad (155)$$

A characteristic of extensional flow is the vanishing off-diagonal elements in the rate of strain tensor. This idealized constant-rate fluid stretching experiment, known as the steady axisymmetric extensional flow, is the extensional equivalent of steady shear flow. It is a difficult experiment to carry out and consequently reliable extensional properties of fluids are difficult to obtain. One of the difficulties becomes apparent when Equation (153) is integrated to give

$$L = L_0 e^{\dot{\varepsilon}_0 t} \qquad (156)$$

where L_0 is the initial length of the cylindrical fluid test specimen. The exponential increase in length of the test specimen makes it very difficult to sustain measurement for a period long enough for the effects of initial conditions to be ignored. Specialized rheometers have been designed to overcome this and other related problems [8]. However, good elongational flow measurements are rare.

An important extensional property of a viscoelastic fluid is its *extensional viscosity* η_{el}. In terms of the idealized extensional measurement

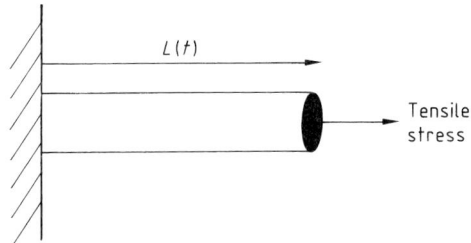

Figure 39. Schematic diagram of extensional flow

described above, η_{el} is defined by

$$\eta_{el} = \frac{T_{zz} - T_{rr}}{\dot{\varepsilon}_0} \quad (157)$$

The difference between the normal stress in the axial direction T_{zz} and that in the radial direction T_{rr} gives the deviatoric tensile stress arising from the extensional deformation. The definition of extensional viscosity is analogous to the definition of shear viscosity. Like its shear counterpart, the extensional viscosity is in general a function of the extensional rate $\dot{\varepsilon}_0$.

The extensional viscosity predicted by different constitutive equations can be found in the literature [30]. The extensional viscosity of a Newtonian fluid with constant viscosity η_0 is relatively simple to obtain. The total normal stress in the axial and the radial directions are given by

$$T_{zz} = -p + 2\eta \frac{du_z}{dz} = -p + 2\eta_0 \dot{\varepsilon}_0$$

$$T_{rr} = -p + 2\eta \frac{du_r}{dr} = -p - \eta_0 \dot{\varepsilon}_0$$

Hence

$$\eta_{el} = \frac{T_{zz} - T_{rr}}{\dot{\varepsilon}_0} = 3\eta_0 \quad (158)$$

According to this result, the extensional viscosity of a Newtonian fluid is three times its shear viscosity. This result has been verified experimentally by TROUTON in 1906 and extensional viscosity is often referred to as *Trouton viscosity* [31]. The extensional viscosity of viscoelastic fluids is, in general, much larger than three times it shear viscosity.

The extensional viscosity obtained for the Oldroyd-B equation can be shown to be

$$\eta_{el}(\dot{\varepsilon}_0) = 3\eta_0 \left[\frac{(1 - \lambda_2 \dot{\varepsilon}_0)(1 + 2\lambda_1 \dot{\varepsilon}_0)}{(1 + \lambda_1 \dot{\varepsilon}_0)(1 - 2\lambda_1 \dot{\varepsilon}_0)} \right] \quad (159)$$

At low extensional rate, the Newtonian result of $3\eta_0$ is again obtained. As the extensional rate is increased, the extensional viscosity increases rapidly. This large extensional viscosity is in agreement with the observed extensional behavior of viscoelastic fluids. However, according to the Oldroyd-B equation, the extensional viscosity grows without bound as the extensional rate approaches $\lambda_1/2$. This singularity in the relationship between extensional rate and extensional viscosity is physically unrealistic and is an indication that the Oldroyd-B equation must be modified. Comparison of the measured extensional viscosity with that predicted by a constitutive equation provides an additional test of the validity of the constitutive equation.

It is clear that fluid elasticity has greatly increased the complexity of fluid motion. There are a large number of flow phenomena which are only observed in viscoelastic fluids. Many of these have practical implications for industrial processes. For further details on the mechanics of viscoelastic fluids see [4], [19], [26].

4.3.5. Accelerating and Decelerating Flows of Viscoelastic Fluids

Fluid elasticity does not affect the energy requirements for fully-developed/laminar tube flow. However, in accelerating and decelerating flows, such as in the entrance and exit of a tube, the influence of fluid elasticity becomes quite pronounced. Entrance and exit effects are conveniently discussed with reference to the capillary rheometer, an important instrument for the measurement of fundamental flow properties of fluids.

One essential feature of the capillary viscometer is that the wall shear stress can be directly determined from the measured fully-developed flow pressure drop (see Eq. 87). However, for a laboratory-scale capillary viscometer it is more practical to measure the overall pressure drop from the upstream fluid reservoir to the exit of the tube, rather than the pressure drop in the tube itself. Schematics of a pressure-driven and a ram-driven capillary rheometer are shown in Figure 40. In the *pressure-driven* instrument, the independent variable is the shear stress, whereas in the *ram-driven instrument* the independent variable is the shear rate. Shear rates of $> 10^3 \text{ s}^{-1}$ are easily obtained in the capillary rheometer. This is one of its main advantages over conventional rotation instruments.

Assuming that the flow is laminar and that the fluid is a time-independent inelastic or viscoelastic fluid, corrections to the measured pressure drop $(p_{gas} - p_{atm})$ for pressure-driven rheometers and $(p_{app} - p_{atm})$ for ram-driven rheometers may have to be made due to the following effects:

1) Head of fluid above the tube exit
2) Kinetic energy effects
3) Tube entrance and exit losses
4) Weight of filament after exit (this will be ignored here)

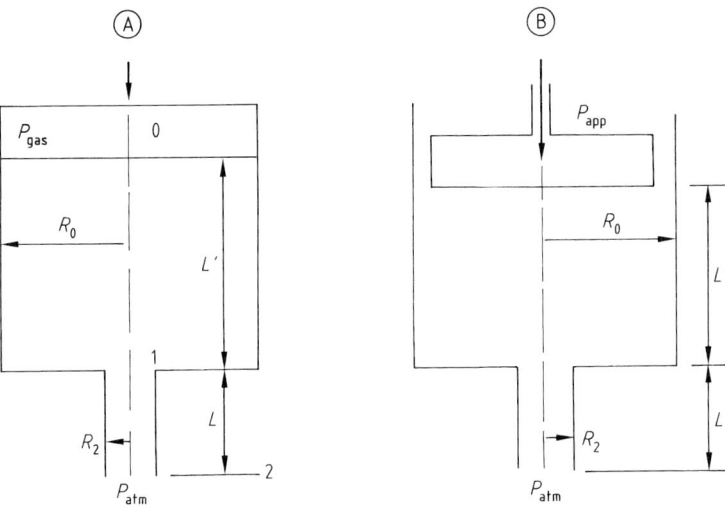

Figure 40. Schematic diagram of (A) pressure and (B) ram driven capillary rheometers

$$P_{gas} - P_{atm} = \Delta p_{fd_1} + \Delta p_{en} + \Delta p_{fd_2} + \Delta p_{ex} + \Delta KE + \Delta PE \quad (160)$$

Equation (160) is a mechanical energy balance written between surface 0 and 2 as shown in Figure 40: Δp_{fd_1} and Δp_{fd_2} are the fully-developed flow pressure drops in the reservoir and tube, respectively; Δp_{en} is the entry loss over and above the fully-developed flow loss for the flow into the tube; and Δp_{ex} is the exit loss over and above the fully-developed flow loss for flow out of the tube to the atmosphere; ΔKE and ΔPE are the kinetic and potential energy losses, respectively. Kinetic and potential energy effects can normally be neglected for polymer melts and for concentrated solutions and suspensions. In addition, the reservoir diameter is usually much greater than that of the downstream tube, so that the fully-developed pressure drop in the upstream tube or reservoir can also be neglected. Thus, Equation (160) becomes

$$P_{gas} - P_{atm} = \Delta p_{en} + \Delta p_{fd_2} + \Delta p_{ex} \quad (161)$$

where

$$\Delta p_{fd_2} = \frac{2\tau_w L}{R}$$

If high L/R capillary tubes are used for the pressure drop–flow rate measurements ($L/R \geq 200$), fundamental shear stress–shear rate data can be determined directly with the capillary rheometer because

$$\frac{2\tau_w L}{R} \gg \Delta p_{en} + \Delta p_{ex} \quad (162)$$

Therefore

$$\tau_w = \frac{R(P_{gas} - P_{atm})}{2L} \quad (163)$$

i.e., the wall shear stress can be directly determined from the measured pressure drop, and the wall shear rate is specified by the Rabinowitsch–Mooney equation

$$\dot{\gamma}_w = \left(\frac{3n' + 1}{4n'}\right)\left(\frac{8V}{D}\right) \quad (164)$$

where

$$n' = \frac{d \ln \tau_w}{d \ln \frac{8V}{D}} \quad (165)$$

However, for many materials in tubes with a high L/R ratio, the pressures required to obtain shear rates of interest are prohibitive, and methods have been derived to correct for entry and exit effects when low L/R capillary tubes are used. The method most commonly used is that first suggested by BAGLEY in 1957 [32]:

$$\Delta p_{en} + \Delta p_{ex} = \frac{2\tau_w L_e}{R} = 2\tau_w e \qquad (166)$$

where e is the dimensionless extra length of tube which defines the exit and entry losses in excess of the fully-developed flow losses.

$$p_{gas} - p_{atm} = 2\tau_w(L/R + e) \qquad (167)$$

and

$$\tau_w = \frac{p_{gas} - p_{atm}}{2(L/R + e)} \qquad (168)$$

The end correction e is determined by first plotting $(p_{gas} - p_{atm})/(2 L/R)$ versus $8 V/D$ on log–log coordinates. Different lines or curves such as those illustrated in Figure 41 will be obtained for different L/R tubes. From such a graph $p_{gas} - p_{atm}$ can be determined as a function of L/R for various values of $8 V/D$. A linear plot of $p_{gas} - p_{atm}$ versus L/R is then made for various values of $8 V/D$. The end correction e is the intercept on the abscissa of this plot (see Fig. 42). The end correction is then a known function of $8 V/D$ and hence the corrected value of τ_w can be computed as a function of $8 V/D$ from Equation (168) whereas the true shear rate can be computed with the aid of Equation (164) and the slope of a log–log plot of τ_w versus $8 V/D$ (Eq. 165).

The capillary rheometer is an important instrument for the measurement of the shear stress as a function of shear rate, particularly for molten polymers where measurements can be made at processing shear rates. It is also an instrument for determining quantitative measurement of fluid elasticity. For instance, the end correction is strongly influenced by the elasticity of the fluid. Figure 43 shows end correction re-

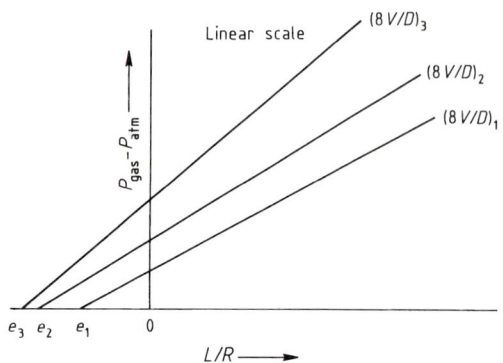

Figure 42. Determination of the end correction

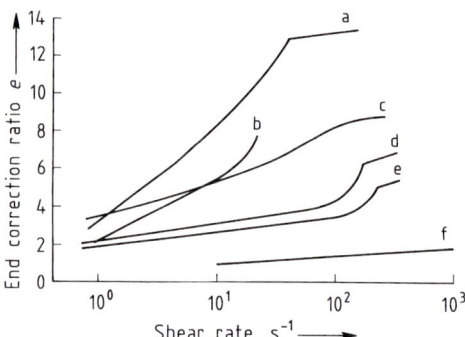

Figure 43. Typical values of the end correction as a function of shear rate [33]
a) Polystyrene; b) Poly(methyl methacrylate); C) Low-density polyethylene MFI 2.0; d) High-density polyethylene MFI 0.25; e) High-density polyethylene; f) Polyacetal

Table 2. The end correction for inelastic power-law fluids

Power-law index, n	$\Delta p_{en}/2\tau_w$	$\Delta p_{ex}/2\tau_w$	e
1.0	0.588	0.246	0.834
0.5	1.34	0.26	1.60
0.3	1.76	0.28	2.04
0.167	2.33	0.59	2.92

sults for molten polymers while Table 2 lists the end correction for inelastic power-law fluids as determined from the numerical solution of the tube entry and exit flow problem [34].

In the absence of fluid elasticity, the end correction increases as a result of the shear-thinning characteristics of the fluid, but not to the extent of the values observed for many commercial polymers which in general are shear-thinning and also highly elastic. Even higher end correc-

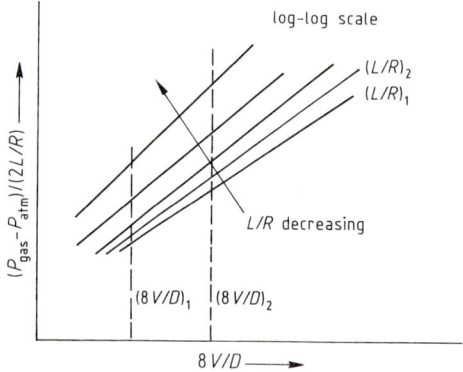

Figure 41. First plot of capillary rheometer data for determination of the end effect

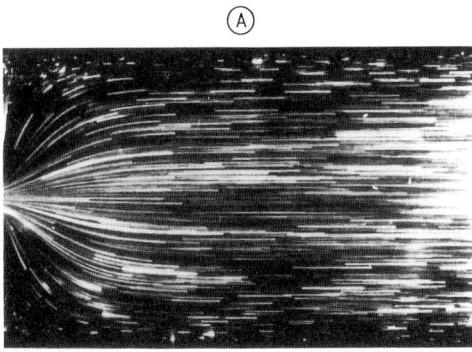

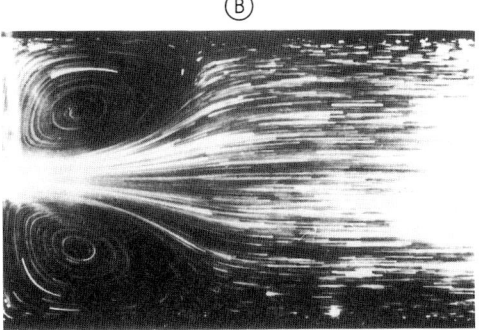

Figure 44. Comparison of entry flow patterns for a Newtonian and a non-shear thinning elastic fluid with the same viscosity
Reservoir to tube diameter ratio is 7.67
A) Newtonian; B) Non-shear thinning elastic

tion values than those illustrated for molten polymers in Figure 43 have been observed for concentrated polymer solutions. The end correction is used to differentiate between different polymers and in fact to distinguish between different grades of the same polymer (see curves d and e in Fig. 43).

Significantly different behavior is indeed observed in inlet and exit flows of inelastic and viscoelastic fluids. Figure 44 shows streamline photographs obtained for an inelastic Newtonian fluid (Fig. 44 A) and for an elastic fluid which shows no shear thinning (Fig. 44 B). Both fluids have identical viscosities. Flow is from left to right in the photographs and represents flow from the reservoir *into* the tube of a capillary rheometer. For the inelastic Newtonian fluid a small secondary flow is present in the corner of the reservoir. This cell remains essentially constant in size with increasing flow rate and ultimately disappears for Reynolds numbers > ca. 0.1 when the fluid inertia starts to become important. For the viscoelastic fluid of the same viscosity (2000 Pa s) the secondary flow is much larger. The size of the secondary flow grows with increasing flow rate for Reynolds numbers < 0.1 and continues to grow until the flow becomes unstable. For a molten polymer being extruded through a die, the flow instability results in a distorted extrudate. The flow phenomena, called melt fracture, represents an upper limit on the rate at which a molten polymer can be extruded. Similar flow instabilities are not observed in tubular inlet flows for inelastic fluids.

For a viscoelastic fluid the *exit flow* from a capillary tube also differs significantly from that of an inelastic fluid. For fully-developed flow of a viscoelastic fluid in a tube, a tension along the streamlines associated with the deviatoric normal stresses is present. When the fluid passes through the exit of the tube to the atmosphere, it will relax the tension along the streamlines by contracting in the longitudinal direction. For an incompressible fluid, this results in a lateral expansion of the fluid. This relaxation phenomena results in extrudate swell (see Fig. 33), where the diameter of the extrudate D_e is significantly greater than the internal tube diameter. For large tube length to diameter ratios the extrudate swell can be estimated as follows [35]:

$$\frac{D_e}{D} = 0.1 + \left[1 + \frac{1}{2}\left(\frac{\tau_{11} - \tau_{22}}{2\tau_{12}}\right)^2_w\right]^{1/6}$$

where the subscript w indicates that the stresses are to be evaluated at the wall shear rate for fully-developed flow. Die swell ratios of 2 or more are not unusual in the processing of molten polymers. Since the die swell depends not only on the particular polymer but also on the operating conditions such as temperature and flow rate, the industrial problems related to extrudate swell are particularly complex and challenging.

Fluid elasticity does not effect the energy requirements for fully-developed *laminar* tube flow, that is

$$f_{\text{viscoelastic}} = f_{\text{inelastic}} \tag{169}$$

The equivalent conclusion, however, is not applicable for viscoelastic fluids in *turbulent flow*. Here considerable reduction in the friction factor below the expected inelastic value is observed. This *drag reduction phenomenon*, first observed by TOMS in 1948, has received considerable attention in the literature because of its possible commercial significance. Parts per million of certain polymers dissolved in water can reduce the

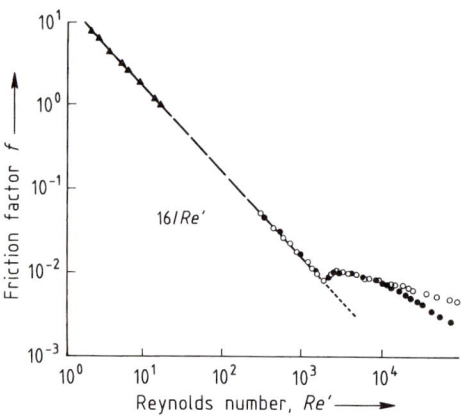

Figure 45. Friction factor–Reynolds number data for Boger fluids, distilled water, and a solution of 296 ppm by weight of polyethylene oxide in distilled water in an 8.46 mm-diameter pipe [36]. (Reproduced by permission of the American Institute of Chemical Engineers © 1975 AIChE)
▲ Boger fluid; ○ Distilled water; ● 296 ppm by weight polyethylene oxide

friction factor considerably; drag reduction by as much as 90% has been observed. Figure 45 shows some friction factor–Reynolds number data for a very dilute aqueous polymer solution. The data agrees with the laminar flow prediction but deviates from the solvent line characterization for turbulent flow for Reynolds numbers $> 10^4$.

5. Numerical Methods in Fluid Mechanics

Many of the flows of practical importance have irregular geometry and their boundary conditions are often mathematically difficult to handle. Furthermore, the fluid properties may vary rapidly with flow and thermal conditions. These complications mean that the set of equations that describe the fluid motion are unlikely to have an analytic solution. Even in those rare cases where an analytic solution can be found, it may be in a form, that is not convenient for practical use, e.g., as a slowly converging infinite series. For such flows, it is necessary to obtain an approximate numerical solution to the governing equations. Numerical methods have always been used, but their development has been greatly accelerated due to the availability of modern digital computers that perform the repetitive calculations. Computational fluid mechanics is now a well established subject capable of producing highly accurate predictions of the fluid motion at great speed. The ever increasing computing power combined with the graphical capabilities of the current generation of computers has established numerical solution of fluid mechanical problems as an extremely effective way of generating and presenting information about fluid flows. In this short section, the principles of two general numerical methods for solving the steady-state Navier–Stokes equations and the continuity equation in two dimensions will be described. Further details are given in [37]–[41].

5.1. Finite Difference Method

(→ 1. Mathematics in Chemical Engineering, pp. **1**-70 – **1**-73)

In the finite difference method, the partial derivatives in the governing equations are approximated by finite differences. To do this, the flow field of interest is divided into a regular grid as shown in Figure 46.

At any grid point (i, j), the first derivative $\partial u_x/\partial x$ in the continuity equation can be approximated by

$$\left(\frac{\partial u_x}{\partial x}\right)_{i,j} = \frac{u_{x_{i+1,j}} - u_{x_{i-1,j}}}{2\Delta x} \tag{170}$$

Subscripts are used to indicate the grid point at which the variable is to be evaluated. The error

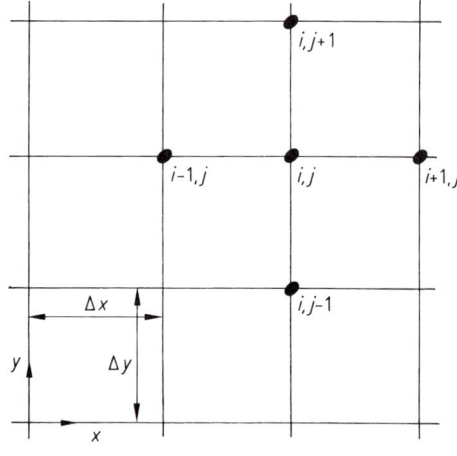

Figure 46. Finite difference grid

incurred in this approximation is dependent on the size of the grid Δx. It approaches zero as Δx approaches zero. A similar approximation can be obtained for $\partial u_y/\partial y$. It is common practice to make Δy equal to Δx. The finite difference approximation of the incompressible continuity equation at the grid point (i, j) is

$$u_{x_{i+1,j}} + u_{y_{i,j+1}} - u_{x_{i-1,j}} - u_{y_{i,j-1}} = 0 \qquad (171)$$

This is an algebraic equation for the unknowns $u_{x_{i+1,j}}, u_{x_{i-1,j}}$ etc. A similar equation can be written down for each of the grid points, $1 \leq i \leq N_i$ and $1 \leq j \leq N_j$, where N_i and N_j are the number of grid points in the x and y directions, respectively.

At the grid point (i, j), $\partial^2 u_x/\partial x^2$, a typical second derivative, can be approximated by the finite difference:

$$\left(\frac{\partial^2 u_x}{\partial x^2}\right)_{i,j} = \frac{u_{x_{i+1,j}} + u_{x_{i-1,j}} - 2u_{x_{i,j}}}{\Delta x^2} \qquad (172)$$

Replacing all the partial derivatives by their finite difference approximations, the following finite difference equivalent of the x-component of the Navier–Stokes equation is obtained:

$$\varrho\left[u_{x_{i,j}}\left(\frac{u_{x_{i+1,j}} - u_{x_{i-1,j}}}{2\Delta x}\right) + u_{y_{i,j}}\left(\frac{u_{x_{i,j+1}} - u_{x_{i,j-1}}}{2\Delta y}\right)\right]$$
$$= \frac{p_{i+1,j} - p_{i-1,j}}{2\Delta x} + \eta\left(\frac{u_{x_{i+1,j}} + u_{x_{i-1,j}} - 2u_{x_{i,j}}}{\Delta x^2}\right)$$
$$+ \eta\left(\frac{u_{x_{i,j+1}} + u_{x_{i,j-1}} - 2u_{x_{i,j}}}{\Delta y^2}\right) \qquad (173)$$

A similar finite difference equation can be written for the y-component of the Navier–Stokes equation. The finite differences have converted the differential equations describing the fluid flow to a set of simultaneous algebraic equations for the unknown velocity components and pressure at the grid points. After the incorporation of the boundary conditions, they can, in principle, be solved for the unknown variables. However, because of the complexity of the Navier–Stokes equations, the success of the finite difference method depends critically on how the finite difference equations are handled. They often have to be completely reorganized and rewritten in various forms that are numerically and computationally more manageable. For example, in this description of the finite difference method, the physical variables u_x, u_y and p appear explic-

itly. The finite difference method can be reformulated in terms of the stream function Ψ, or a combination of stream function and the vorticity ω. For two-dimensional flow ω is defined as the z component of curl of the velocity vector:

$$\omega = -\frac{\partial u_x}{\partial y} + \frac{\partial u_y}{\partial x} \qquad (174)$$

In two-dimensional flows, the stream function or stream function-vorticity formulations are, in fact, used in preference to the velocity–pressure formulation. Further details, particularly regarding the choice of numerical schemes for handling the resulting equations, can be found in the references listed.

In this introductory description of the finite difference method, the partial derivatives are approximated by central differences. Other finite difference approximation schemes can be applied. Irrespective of the finite difference scheme employed (because of the inertia terms on the left-hand side of the Navier–Stokes equations) the resulting algebraic equations are nonlinear in the unknown velocity components. These equations are not easy to solve directly. In one of the numerical procedures developed to deal with the nonlinear terms, the typical inertia term $u_x \partial u_x/\partial x$ is approximated by

$$u_x \frac{\partial u_x}{\partial x} \approx u_{x_{i,j}}^{(n-1)}\left(\frac{u_{x_{i+1,j}}^{(n)} - u_{x_{i-1,j}}^{(n)}}{2\Delta x}\right) \qquad (175)$$

Here the superscript n is used to denote the values of the unknowns at the nth iteration. At the nth iteration the values of the $(n - 1)$th iteration are already known. This way the nonlinear terms are approximated as linear ones and the finite difference equations become a set of linear algebraic equations for which many computationally efficient methods of solution are available. The calculation is terminated when the difference between two successive iterations is less than some preassigned small tolerance. To start the iterative procedure, it is necessary to supply the zeroth iteration of the unknown variables $u_{i,j}^{(0)}$, etc. In the absence of any information, these zeroth iteration values can be taken to be zeros. Convergence of this straightforward iterative scheme is not guaranteed and even if it converges, it may require a large number of iterations. Convergence, for example, becomes slower and slower as the Reynolds number of the flow is increased and may fail completely when the Reynolds number becomes too large. Many numerical pro-

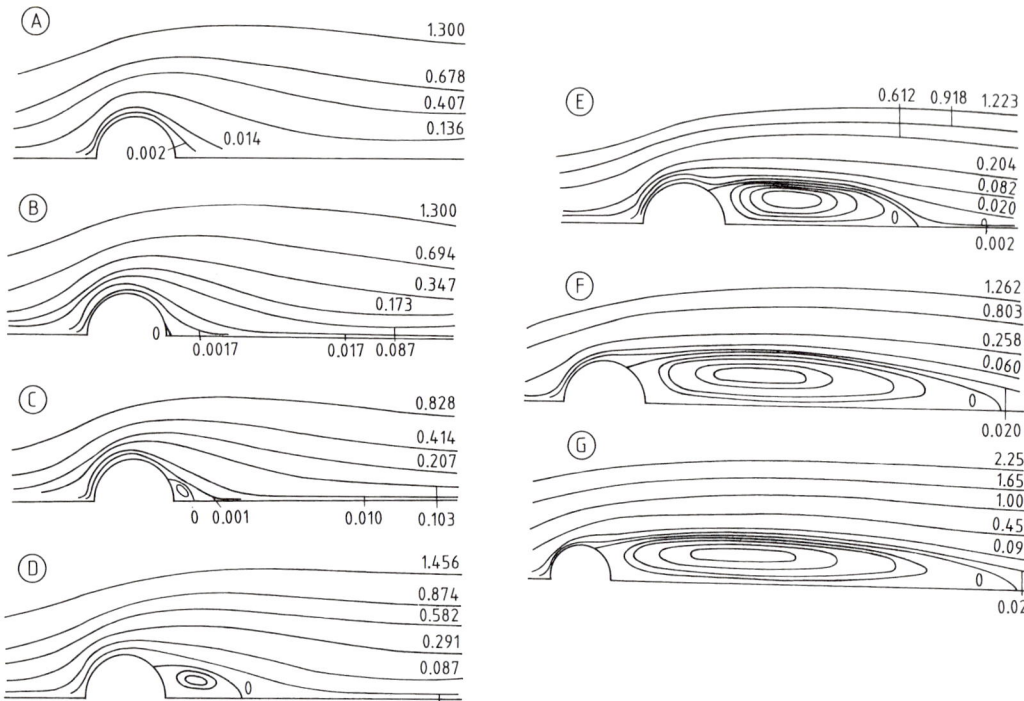

Figure 47. Streamlines around a cylinder at increasing Reynolds number obtained by finite difference computation [42] (reproduced with permission of Cambridge University Press)
A) $Re = 5$; B) $Re = 7$; C) $Re = 10$; D) $Re = 20$; E) $Re = 40$; F) $Re = 70$; G) $Re = 100$

cedures have been developed to improve the convergence performance. Iterative procedures have also been developed to deal with the nonlinearity introduced by variable fluid properties.

Figure 47 shows the streamlines of the flow of a Newtonian fluid past a cylinder at increasing Reynolds number obtained by finite difference computation using the stream function–vorticity formulation [42]. Numerical convergence at a Reynolds number as high as 100 was achieved by a specially developed iteration scheme. The effects of increasing Reynolds number show up clearly in the increasing size of the recirculating wake behind the cylinder. These numerical results are in excellent agreement with the available experimental data.

5.2. Finite Element Method

(→ 1. Mathematics in Chemical Engineering, pp. 1-76 – 1-80)

The finite element technique was originally developed by structural engineers for calculating the stresses and strains in structures of complex shapes. It has been generalized and developed as a numerical technique for solving differential equations, particularly partial differential equations. As with the finite difference method, the differential equations for the unknown variables are converted into a set of algebraic equations which can then be solved for the unknown variables at discrete points. The application of finite element computation to fluid mechanics started in the early 1960s and has since been greatly refined. A brief description of the finite element technique based on the Galerkin approach will be presented here. The mathematical principles and numerical details of the *Galerkin finite element* method, particularly the handling of boundary conditions, can be found in [37]–[41].

In finite element approximation, the flow field of interest is again subdivided by a grid into a large number of small connected domains of different shapes and sizes known as elements. (Following the general practice in finite element computation, the grid points, referred to as the nodal points, are identified by a single subscript i instead of the double subscripts i, j in finite difference.) The unknown variables, u_x, u_y, and p are approximated by series of the form

$$u_x(x, y) = \sum_{i=1}^{N} u_{xi} \Phi_i(x, y)$$

$$u_y(x, y) = \sum_{i=1}^{N} u_{yi} \Phi_i(x, y) \qquad (176)$$

$$p(x, y) = \sum_{i=1}^{N'} p_i \Psi_i(x, y)$$

where N (and N', see below) is the total number of nodal points in the grid; u_{xi}, u_{yi}, and p_i are unknown numerical coefficients to be determined; $\Phi_i(x, y)$, for $i = 1$ to N, and $\Psi_i(x, y)$ for $i = 1$ to N' are two sets of known functions defined for each of the nodal points. They are usually referred to as the interpolation functions. While one is free to choose the form of these functions, $\Phi_i(x, y)$ and $\Psi_i(x, y)$ are usually restricted to simple functions such as low-order polynomials. In many finite element computations $\Phi_i(x, y)$ are taken to be quadratic functions of x and y and $\Psi_i(x, y)$ to be linear functions of x and y. Both $\Phi_i(x, y)$ and $\Psi_i(x, y)$ take the value of unity at the nodal point i and decrease to zero at all the neighboring nodes surrounding node i. These functions are also defined to be identically zero beyond these neighboring nodes, i.e., Φ_i and Ψ_i are nonzero only in those elements which have node i as one of its nodal points. The quadratic interpolating function $\Phi_i(x, y)$ is also defined to be zero at the midpoints (the mid-side nodes) of each side of the elements that have node i as one of its nodes. The total number of nodes N for the quadratic interpolation functions includes the mid-side nodes. Thus N', the total number of nodes for the linear interpolation function, is $< N$. In order that the series representation approaches the true solution as N and N' become large, the interpolation functions have to satisfy a number of mathematical requirements. For example, all the interpolation functions have to be continuous across the common boundary of two adjacent elements.

Typical examples of the interpolating functions for quadrilateral and triangular elements are shown in Figure 48. In the finite element method, elements of different sizes and shapes can be mixed relatively easily. This greatly simplifies the discretization of irregularly shaped flow fields and is the major advantage of the finite element method over the finite difference method. When the series representations (Eq. 176) are substituted in the equations of motion and the continuity equation, the following are obtained:

$$\varrho \left[\sum_{i=1}^{N} u_{xi} \Phi_i \left(\sum_{i=1}^{N} u_{xi} \frac{\partial \Phi_i}{\partial x} \right) + \sum_{i=1}^{N} u_{yi} \Phi_i \left(\sum_{i=1}^{N} u_{xi} \frac{\partial \Phi_i}{\partial y} \right) \right]$$
$$+ \sum_{i=1}^{N'} p_i \frac{\partial \Psi_i}{\partial x} - \eta \left[\sum_{i=1}^{N} u_{xi} \frac{\partial^2 \Phi_i}{\partial x^2} + \sum_{i=1}^{N} u_{xi} \frac{\partial^2 \Phi_i}{\partial y^2} \right]$$
$$= R_x(x, y) \qquad (177\text{a})$$

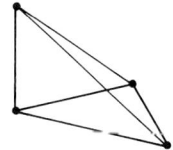
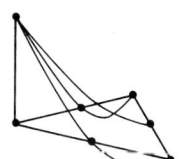

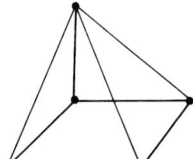
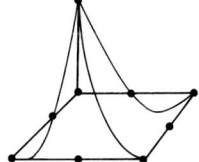
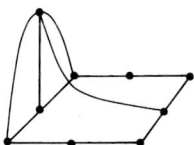

Figure 48. Typical finite element linear and quadratic interpolation functions

$$\varrho\left[\sum_{i=1}^{N} u_{xi}\Phi_i\left(\sum_{i=1}^{N} u_{yi}\frac{\partial\Phi_i}{\partial x}\right) + \sum_{i=1}^{N} u_{yi}\Phi_i\left(\sum_{i=1}^{N} u_{yi}\frac{\partial\Phi_i}{\partial y}\right)\right]$$
$$+ \sum_{i=1}^{N'} p_i\frac{\partial\Psi_i}{\partial y} - \eta\left[\sum_{i=1}^{N} u_{yi}\frac{\partial^2\Phi_i}{\partial x^2} + \sum_{i=1}^{N} u_{yi}\frac{\partial^2\Phi_i}{\partial y^2}\right]$$
$$= R_y(x, y) \tag{177b}$$

$$\sum_{i=1}^{N} u_{xi}\frac{\partial\Phi_i}{\partial x} + \sum_{i=1}^{N} u_{yi}\frac{\partial\Phi_i}{\partial y} = R_c(x, y) \tag{177c}$$

In general the series representations do not satisfy the governing equations and leave behind residuals $R_x(x, y)$, $R_y(x, y)$, and $R_c(x, y)$. These residuals are functions of the unknown coefficients u_{xi}, u_{yi}, p_i, as well as x and y. An approximate solution is obtained by finding the set of the unknown coefficients that minimize, in some sense, these residuals. In the case of Galerkin finite element method, this is done by requiring the $R_x(x, y)$ and $R_y(x, y)$ to be orthogonal to all the $\Phi_i(x, y)$ and $R_c(x, y)$ to be orthogonal to all the $\Psi_i(x, y)$ over the region of interest, i.e.,

$$\int_A R_x(x, y)\Phi_k(x, y) dA = 0 \quad k = 1 \text{ to } N \tag{178a}$$
$$\int_A R_y(x, y)\Phi_k(x, y) dA = 0 \quad k = 1 \text{ to } N \tag{178b}$$
$$\int_A R_c(x, y)\Psi_k(x, y) dA = 0 \quad k = 1 \text{ to } N' \tag{178c}$$

A is the area occupied by the flow field. After performing the integration, these become a set of algebraic equations which can be solved for the unknown coefficients u_{xi}, u_{yi}, p_i. It will be found that the number of independent algebraic equations, after taking into consideration the boundary conditions, is exactly equal to the number of unknown numerical coefficients in the series representation. From the definition of the interpolation functions, it is clear that at node j, $x = x_j$ and $y = y_j$, the only nonzero terms in the series representation are $\Phi_j(x_j, y_j)$ and $\Psi_j(x_j, y_j)$ where they take on the value of unity. Hence

$$u_x(x_j, y_j) = u_{xj}\Phi_j(x_j, y_j) = u_{xj} \tag{179a}$$
$$u_y(x_j, y_j) = u_{yj}\Phi_j(x_j, y_j) = u_{yj} \tag{179b}$$
$$p(x_j, y_j) = p_j\Psi_j(x_j, y_j) = p_j \tag{179c}$$

Thus the numerical coefficients in the series representation are also the values of the unknown velocity components and pressure at the nodal points.

A considerable amount of algebraic manipulation and numerical integration must be carried out to set up and solve these algebraic equations. Some of these steps require considerable ingenuity in computer programming. In a well designed finite element computer package most of these are taken care of by the computer requiring a minimal amount of human intervention. Because of the inertia terms and variable fluid properties, the algebraic equations are again nonlinear in the unknown coefficients. A procedure similar to that outlined above for handling the set of nonlinear algebraic equations arising from the finite difference method can also be applied here, the details for implementing the iterative procedure may, of course, be quite different.

In the finite element method u_x, u_y, and p appear explicitly in the computation. It is again possible to formulate a finite element scheme where the unknown variable is the stream function or a combination of stream function and vorticity. In the case of viscoelastic fluids, it is also common practice to have the deviatoric stress components τ_{xy}, τ_{xx}, and τ_{yy} as well as u_x, u_y, and p appearing explicitly in the finite element computation scheme.

Figure 49 shows the streamlines obtained by finite element computation for the flow in a branching channel. The finite element grid employed is shown in Figure 49A. The use of a small number of triangular elements together with rectangular elements has made the discretization of the irregularly shaped flow field a relatively simple task. A Newtonian fluid, travel-

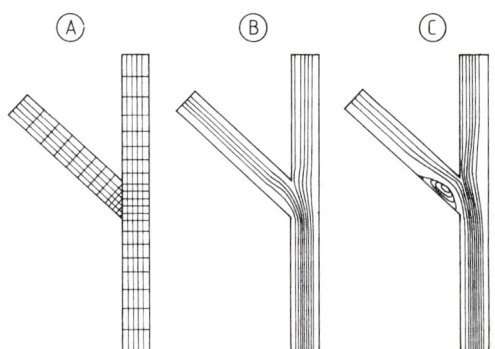

Figure 49. Finite element mesh and streamlines in a branching channel at increasing Reynolds number obtained by finite element computation
A) Finite element mesh; B) Streamlines for $Re = 10$; C) Streamlines for $Re = 100$

ing vertically upwards, enters the channel as a single stream and leaves as two separate streams, generally of unequal size. The effects of increasing Reynolds number, based on upstream channel width, now show up as a recirculating zone in the branch channel. At low Reynolds number, the flow is divided equally between the two branches (Fig. 49 B). As the Reynolds number is increased (Fig. 49 C), the flow through the straight channel increases at the expense of the branch channel.

5.3. General Remarks

Both the finite difference method and the finite element method are now routinely used to solve fluid flow problems. A detailed discussion of the relative merits of these two methods would not serve a very useful purpose and is certainly out of place here. It is, however, generally agreed that the principles of the finite difference method are easier to understand and simpler to implement on computers. In contrast the finite element method can cope with irregular flow fields very efficiently, better than the finite difference method.

The development of a computer program to solve a nontrivial flow problem numerically is a major undertaking that requires expertise not only in fluid mechanics, but also in numerical analysis, computer programming, and organization. It is not a task that can normally be carried out by a single person working in isolation. A number of commercial software packages specially designed for solving fluid flow problems are now available. Some of these packages are very well designed and simple to use. Most of them incorporate special procedures for dealing with highly nonlinear flow problems. They usually have built-in computer graphics for efficient presentation of the large amount of numerical results generated by these packages. Intelligent use of these computing tools coupled with a sound understanding of the physics and mathematics of fluid mechanics has lead rapidly to advances in the solution of flow problems that would otherwise be intractable.

6. References

[1] M. M. Denn: *Process Fluid Mechanics*, Prentice-Hall, Englewood Cliffs 1980.
[2] S. W. Churchill: *Viscous Flow the Practical Use of Theory*, Butterworths, Boston 1988.
[3] R. B. Bird, W. E. Stewart, E. N. Lightfoot: *Transport Phenomena*, J. Wiley & Sons, New York 1960.
[4] R. B. Bird, R. C. Armstrong, O. Hassager: *Dynamics of Polymeric Liquids*, 2nd ed., vol. 1, J. Wiley & Sons, New York 1987.
[5] H. Schlichting: *Boundary-Layer Theory*, 7th ed., McGraw-Hill, New York 1977.
[6] K. Walters: *Rheometry*, Chapman and Hall, London 1975.
[7] R. W. Whorlow: *Rheological Techniques*, Ellis Horwood, London 1980.
[8] J. M. Dealy: *Rheometers for Molten Plastics*, Van Nostrand, New York 1982.
[9] J. O. Hinze: *Turbulence*, 2nd ed., McGraw-Hill, New York 1975.
[10] D. Kunii, O. Levenspiel: *Fluidization Engineering*, R. E. Krieger Pub. Co., Huntington 1977.
[11] J. M. Kay, R. M. Nedderman: *Fluid Mechanics and Transfer Processes*, Cambridge University Press, Cambridge 1985.
[12] Y. Taitel, A. E. Dukler, *AIChE J.* **26** (1980) 345.
[13] G. E. Alves, *Chem. Eng. Progr.* **50** (1954) 449.
[14] J. M. Mandhane, G. H. Gregory, K. Aziz, *Int. J. Multiphase Flow* **1** (1974) 537.
[15] P. Griffith: "Two-Phase Flow," in W. M. Rohsenow, J. P. Hartnett, E. N. Ganic (eds.): *Handbook of Heat Transfer Fundamentals*, 2nd ed., McGraw-Hill, New York 1985.
[16] F. A. Holland: *Fluid Flow for Chemical Engineers*, E. Arnold, London 1973.
[17] R. H. Perry, D. W. Green: *Perry's Chemical Engineers Handbook*, 6th ed., McGraw-Hill, New York 1984.
[18] A. S. Foust et al.: *Principles of Unit Operations*, 2nd ed., J. Wiley & Sons, New York 1980.
[19] H. A. Barnes, J. F. Hutton, K. Walters: *An Introduction to Rheology*, Elsevier, Amsterdam 1989.
[20] A. B. Metzner, M. Whitlock, *Trans. Soc. Rheol.* **2** (1958) 239.
[21] Q. D. Nguyen, D. V. Boger, *J. Rheol. (N.Y.)* **29** (1985) 335.
[22] A. B. Metzner, J. C. Reed, *AIChE J.* **1** (1957) 434.
[23] D. W. Dodge, A. B. Metzner, *AIChE J.* **5** (1955) 189.
[24] D. V. Boger: "Viscoelastic Flows Through Contractions," in J. L. Lumley, M. Van Dyke, H. L. Reed (eds.): *Annual Review of Fluid Mechanics*, vol. 19, Ann. Reviews Inc., Palo Alto 1987.
[25] J. Meissner, *Pure Appl. Chem.* **42** (1975) 553.
[26] R. B. Bird, C. F. Curtiss, R. C. Armstrong, O. Hassager: *Dynamics of Polymeric Liquids*, 2nd ed., vol. 2, J. Wiley & Sons, New York 1987.
[27] G. Prilutski, R. K. Gupta, T. Sridhar, M. E. Ryan, *J. Non-Newtonian Fluid Mech.* **12** (1983) 233.
[28] M. Doi, S. F. Edwards: *The Theory of Polymer Dynamics*, Oxford University Press, Oxford 1986.
[29] R. L. Larson: *Constitutive Equations for Polymer Melts and Solutions*, Butterworths, Boston 1988.
[30] C. J. S. Petrie: *Elongational Flows*, Pitman, London 1979.
[31] F. T. Trouton, *Proc. R. Soc. London A* **77** (1906) 426.

[32] E. B. Bagley, *J. Appl. Phys.* **28** (1957) 624.
[33] J. A. Brydson: *Flow Properties of Polymer Melts*, Butterworth, London 1970.
[34] D. V. Boger, R. K. Gupta, R. I. Tanner, *J. Non-Newtonian Fluid Mech.* **4** (1978) 239.
[35] R. I. Tanner, *J. Polym. Sci. Part A-2* **8** (1970) 2067.
[36] P. S. Virk, *AIChE J.* **21** (1975). 625.
[37] T. J. Chung: *Finite Element Analysis in Fluid Dynamics*, McGraw-Hill, New York 1978.
[38] M. M. Gupta: "Numerical Methods for Viscous Flow Problems," in A. S. Mujumdar, R. A. Mashelkar (eds.): *Advances in Transport Processes*, vol. 1, Wiley, New York 1980.
[39] A. J. Baker: *Finite Element Computational Fluid Mechanics*, McGraw-Hill, New York 1983.
[40] O. C. Zienkiewicz, K. Morgan: *Finite Elements and Approximation,* Wiley & Sons, London 1983.
[41] M. J. Crochet, A. R. Davies, K. Walters: *Numerical Simulation of Non-Newtonian Flow,* Elsevier, Amsterdam 1984.
[42] S. C. R. Dennis, G. Z. Chang, *J. Fluid Mech.* **42** (1970) 471.

6. Estimation of Physical Properties

ULFERT ONKEN, Fachbereich Chemietechnik, Universität Dortmund, Dortmund, Federal Republic of Germany

HANNS-INGOLF PAUL, Bayer AG, Leverkusen, Federal Republic of Germany

1.	Introduction ... 6-3	4.2.3.	Liquid–Liquid Equilibria ... 6-30
1.1.	Scope ... 6-3	4.2.4.	Solid–Liquid Equilibria ... 6-30
1.2.	Data Sources ... 6-3	5.	**Thermodynamic Data of Chemical Reactions** ... 6-31
1.3.	Theoretical and Empirical Methods ... 6-4	5.1.	Definitions ... 6-31
2.	**Molecular and Macroscopic Properties** ... 6-5	5.1.1.	Reaction Enthalpy ... 6-31
2.1.	Intermolecular Forces ... 6-5	5.1.2.	Gibbs Energy of Reaction and Chemical Equilibrium ... 6-31
2.2.	Theorem of Corresponding States ... 6-7	5.2.	Group Contribution Methods for the Estimation of Enthalpies and Gibbs Energies of Formation ... 6-33
2.3.	Estimation of Critical Properties ... 6-7	5.3.	Applications ... 6-35
2.4.	Other Characteristic Constants of Pure Compounds ... 6-8	5.3.1.	Reaction Enthalpy ... 6-35
3.	**Thermal and Caloric Properties of Single-Phase Systems** ... 6-11	5.3.2.	Chemical Equilibrium ... 6-35
3.1.	Pressure–Volume–Temperature Behavior of Gases and Liquids ... 6-11	6.	**Transport Properties of Pure Compounds and Mixtures** ... 6-37
3.1.1.	Pure Gases ... 6-11	6.1.	Viscosity ... 6-37
3.1.2.	Pure Liquids ... 6-13	6.1.1.	Viscosity of Pure Gases ... 6-38
3.1.3.	Mixtures ... 6-16	6.1.2.	Viscosity of Gas Mixtures ... 6-40
3.1.3.1.	Gas Mixtures ... 6-16	6.1.3.	Viscosity of Pure Liquids ... 6-40
3.1.3.2.	Liquid Mixtures ... 6-17	6.1.4.	Viscosity of Liquid Mixtures ... 6-40
3.2.	Caloric Properties ... 6-17	6.2.	Thermal Conductivity ... 6-42
3.2.1.	Heat Capacity of Ideal Gases ... 6-17	6.2.1.	Thermal Conductivity of Gases at Low Pressure ... 6-42
3.2.2.	Heat Capacity of Pure Liquids ... 6-18	6.2.2.	Thermal Conductivity of Gases at High Pressure ... 6-45
3.2.3.	Heat Capacities of Liquid Mixtures ... 6-19	6.2.3.	Thermal Conductivity of Gas Mixtures ... 6-46
4.	**Phase Equilibria** ... 6-20	6.2.4.	Thermal Conductivity of Liquids ... 6-46
4.1.	Pure Compounds ... 6-20	6.2.5.	Thermal Conductivity of Liquid Mixtures ... 6-48
4.1.1.	Vapor Pressure, Boiling Point, and Melting Point ... 6-20	6.3.	Diffusion Coefficients ... 6-48
4.1.2.	Enthalpy of Vaporization ... 6-22	6.3.1.	Diffusion Coefficients of Gases at Low and Moderate Pressures ... 6-48
4.2.	Mixtures ... 6-23	6.3.2.	Diffusion Coefficients of Liquids ... 6-50
4.2.1.	Vapor–Liquid Equilibria with Liquid-Phase Activity Coefficients ... 6-24	7.	**Surface Tension of Liquids** ... 6-52
4.2.1.1.	Basic Considerations ... 6-24	7.1.	Surface Tension of Pure Liquids ... 6-53
4.2.1.2.	Data Correlation: Binary and Multicomponent Mixtures of Nonelectrolytes ... 6-25	7.2.	Surface Tension of Liquid Mixtures ... 6-53
4.2.1.3.	Prediction of Equilibrium Data ... 6-26	8.	**References** ... 6-55
4.2.2.	Vapor–Liquid and Gas–Liquid Equilibria with Equations of State ... 6-28		

Symbols

a	activity
B, B'	second virial coefficient, m³/mol or Pa
c	molar heat capacity, J mol^{-1} K^{-1}
C, C'	third virial coefficient, (m³/kmol)² or Pa²
D	diffusion coefficient, m²/s
E	energy, J; Eucken factor
f	fugacity, Pa
G	molar Gibbs energy (free enthalpy), J/mol
ΔG_f	Gibbs energy of formation, J/mol
ΔG_f^0	Gibbs energy of formation at standard pressure, J/mol
ΔG_r	Gibbs energy of reaction, J/mol
H	molar enthalpy, J/mol
ΔH_f	enthalpy of formation, J/mol
ΔH_f^0	enthalpy of formation at standard pressure, J/mol
ΔH_r	enthalpy of reaction, J/mol
ΔH^{lv}	enthalpy of vaporization, J/mol
$\Delta H_{T_b}^{lv}$	enthalpy of vaporization at boiling temperature, J/mol
ΔH^{sv}	enthalpy of sublimation, J/mol
ΔH^{trs}	enthalpy of phase transition, J/mol
I	increment
k	Boltzmann constant, J/K
K	equilibrium constant
K_p	equilibrium constant expressed in partial pressure
M	molecular mass, g/mol
n	number of moles
p	pressure, Pa (bar)
P	parachor, m³ kg$^{0.25}$ s$^{-0.5}$ mol^{-1}
Q	group interaction parameter
r	distance, m; number of components of a mixture
R	universal gas constant, J mol^{-1} K^{-1}
S	entropy, J/K
T	temperature, K, °C
V	molar volume, m³/mol
w	weight fraction
x	mole fraction in liquid phase
y	mole fraction in gas phase
z	compressibility factor; coordination number
α	relative volatility
β	compressibility, Pa^{-1}
ε	Lennard–Jones energy, J
γ	activity coefficient
η	dynamic viscosity, Pa · s
λ	thermal conductivity, W m^{-1} K^{-1}
μ	chemical potential, J/mol; dipole moment, Debye
ν	kinematic viscosity, m²/s; number of groups in a molecule; stoichiometric coefficient
ϱ	density, kg/m³
σ	Lennard–Jones length, 10^{-10} m; surface tension, N/m
ω	acentric factor
Δ	difference
χ	Stiel polar factor
Γ	group activity coefficient
φ	fugacity coefficient
Π	Poynting correction
Ω	collision integral

Subscripts

b	at normal boiling point (101.325 kPa)
c	critical property
D	diffusion
id	ideal state (gas, mixture)
i, j, k	components
m	mixture; at melting point
p	constant pressure
r	reduced quantity; reaction
sat	state of saturation
T	constant temperature
v	constant volume
0	pure component

Superscripts

C	combinatorial
E	excess
g	gas
int	internal
l	liquid
s	solid
ref	reference
R	residual
trans	translational
trs	transition
v	vapor
vap	vaporization
0	reference-state value, atmospheric pressure
∞	infinite dilution
cal	calculated value
exp	measured value

1. Introduction

1.1. Scope

The design of chemical processes requires a knowledge of the values of the properties of the individual chemical compounds and mixtures to be handled, as well as data on the chemical reaction systems involved in the process. Important types of relevant data are given in Table 1; other data may, however, also be required (e.g., for optical or electrical properties).

Although all the classes of data listed in Table 1 are relevant for the design and operation of chemical processes, only some of them are essential. Thus basic design of a process unit for liquid–liquid extraction requires data on both the selectivity and capacity of the solvent, whereas data on toxicity and inflammability are not necessary, although they are important for plant operation. The present review of estimation methods covers only those groups of data that are essential for process design. Reaction velocity data cannot, however, be included, despite the fact that they are very important. On the basis of present knowledge reaction velocities cannot be predicted with an acceptable degree of reliability and this situation is not expected to change in the near future.

Before estimating data, data tables and other sources should be consulted for measured values (see Section 1.2). If no such data can be found (as is often the case), experimental determination of missing data can be considered. For several properties (e.g., density or vapor pressure) such experiments can be performed without much expense. However, for those physical properties whose measurement requires appreciable experimental effort (e.g., phase equilibria), information from data sources is rather scarce.

Another, even more crucial problem in data retrieval is the fact that multicomponent mixtures have to be handled in most chemical processes. Since the concentrations in these mixtures are important variables that may vary both locally and with time, it is practically impossible to provide all experimental data for the required concentration ranges.

These considerations show that methods for data estimation and prediction are necessary tools for the design of chemical processes. This is true not only for planning production plants, but also for process development. Many property data are not available, especially in processes for new products. Experimental determination is often impossible due to shortage of time. In these cases qualified data estimation is necessary. Other areas in which data estimation and prediction are widely used are feasibility and comparative process studies. These activities must often be performed within a short time and with a limited financial budget.

Table 1. Data for chemical processes

Type of compounds, material, system	Type of data
Individual chemical compounds (starting and end products, intermediates)	physical properties[a] 1) thermal state properties,[a] e.g., density, vapor pressure 2) caloric properties,[a] e.g., heat capacity, latent heats 3) transport properties[a] chemical properties[b] (safety, toxicity, environmental hazards)
Mixtures, nonreacting	physical properties[a] phase equilibria[a]
Mixtures, reacting	chemical equilibrium,[a] reaction enthalpy[a] reaction velocity[a,b] catalytic activity[a,b]
Materials of construction	mechanical properties[b] corrosion[b]

[a] Data required for basic design of processes. [b] Data estimation not possible.

1.2. Data Sources

The first step in obtaining the data needed for a design problem is a search for experimental data; such data can generally be considered to be more reliable than predicted data. Moreover, most methods of data estimation require knowledge of certain property data for the respective compounds (e.g., boiling point, density, critical temperature, and critical pressure). Besides, prediction of mixture properties is more reliable when it is based on qualified experimental data for the pure components.

Available data sources range from handbooks for everyday work (D'Ans–Lax [1.1], Handbook of Chemistry and Physics [1.2]) to the comprehensive tables of Landolt–Börnstein [1.3] and the numerous specialized data collections, e.g., the Dechema Data Series [1.4] and other data books (see General References [B]–[J] and → 12. Information and Documentation, pp. **12**-72–**12**-74.

An efficient and rapidly developing way of obtaining data for physical and chemical properties and equlibria is data retrieval with computerized data banks (e.g., [1.5]–[1.7]). Prediction methods for several types of data (e.g., phase equilibria) have been incorporated into these data banks.

1.3. Theoretical and Empirical Methods

Methods for estimating property data [A 1]–[A 9] may differ in nature. They may be based on laws of physics and physical chemistry, or they may employ purely empirical rules; most methods can be classified somewhere between these two extremes.

The enthalpy of vaporization, for example, can be determined with a thermodynamic equation, i.e., the Clausius–Clapeyron equation for the dependence of vapor pressure p^{lv} on the absolute temperature T

$$\frac{dp^{lv}}{dT} = \frac{\Delta H^{lv}}{T(V^v - V^l)} \qquad (1.1)$$

where ΔH^{lv} is the enthalpy of vaporization, V^v is the volume of the vapor phase, and V^l the volume of the liquid phase. Equation (1.1) is a rigorous relation; when it is used to determine the vaporization enthalpy, vapor pressure data at several temperatures must be known as well as the densities of the vapor and liquid phases at the saturated vapor pressure p^{lv}. At low vapor pressures ($\leq$ ca. 0.1 MPa) V^l is small compared to V^v and may therefore be neglected. Furthermore, the vapor phase may be assumed to show ideal behavior. With these simplifications Equation (1.1) can be expressed as:

$$\frac{d \ln p^{lv}}{d(1/T)} = -\frac{\Delta H^{lv}}{R} \qquad (1.2)$$

where R is the gas constant. For vaporization at 0.1 MPa the deviations in ΔH^{lv} caused by neglecting V^l are usually < 1% and $\leq$ ca. 5% by assuming ideal gas behavior for the vapor phase. Real gas behavior may be accounted for by the second virial coefficient B' (see Section 3.1, Eq. 3.13):

$$p^{lv} V^v = RT(1 + B' p^{lv}) \qquad (1.3)$$

Substituting this into Equation (1.2) gives:

$$\frac{d \ln p^{lv}}{d(1/T)} = -\frac{\Delta H^{lv}}{R(1 + B' p^{lv})} \qquad (1.4)$$

In contrast to this exact procedure, enthalpies of vaporization $\Delta H^{lv}_{T_b}$ at the boiling point T_b can be estimated roughly with the simple rule of Pictet and Trouton which states that the ratio of $\Delta H^{lv}_{T_b}$ to T_b should have the same value for all liquids. As can be seen from Table 2, this simple rule is true for many compounds:

$$\Delta H^{lv}_{T_b}/T_b = \text{const.} \approx 88 \text{ J K}^{-1} \text{ mol}^{-1} \qquad (1.5)$$

This behavior is presumably related to intermolecular forces. A possible explanation is that the energy necessary for vaporizing a liquid is mainly used to separate the liquid molecules from each other. The strength of intermolecular forces should thus be more or less independent of the type of compound and molecular structure.

Associating liquids obviously do not obey the rule of Pictet and Trouton. The carboxylic acids (formic and acetic acid) show negative deviations for Trouton's constant and form dimers in both the liquid and the vapor phases. This means that in the vaporization process the effective molar mass is higher than that of the monomeric acid, resulting in a higher value of $\Delta H^{lv}_{T_b}$ and consequently of Trouton's constant. For the associating liquids in Table 2 showing positive deviations from Equation (1.5) (i.e., ammonia, water, methanol, ethanol) association is limited to the liquid phase; additional energy is required to vaporize these liquids because the bonds responsible for association have to be broken.

For liquified gases with very low boiling points (e.g., helium, hydrogen) Trouton's constant is much lower than 88 J mol^{-1} K^{-1}, whereas some of the higher boiling metals show a tendency towards higher values for this parameter. Evidently at the reference temperature T_b of Equation (1.5) conditions are not really comparable. Several modifications of Equation (1.5) have therefore been proposed of which only that proposed by KISTIAKOWSKY is mentioned here, because it does not require additional data [1.8]:

$$\frac{\Delta H^{lv}_{T_b}}{T_b} = 36.6 + R \ln T_b \text{ [J kmol}^{-1}\text{K}^{-1}\text{]} \qquad (1.6)$$

Table 2. Rule of Pictet and Trouton

Substance	$\Delta H^{lv}_{T_b}$, kJ/mol	T_b, K	$\Delta H^{lv}_{T_b}/T_b$, J mol^{-1} K^{-1}
He	0.8	4.2	19.8
Ne	1.8	27.1	65.0
Ar	6.5	87.3	74.5
H_2	9.0	20.4	44.3
N_2	5.6	77.4	72.1
CH_4	8.2	111.6	73.3
HCl	16.2	188.1	85.8
CS_2	26.8	319.3	83.8
CCl_4	30.0	349.9	85.8
$n\text{-}C_6H_{14}$	29.0	341.9	84.7
$n\text{-}C_8H_{18}$	34.6	398.5	86.8
C_6H_6	30.8	353.3	87.2
$C_2H_5OC_2H_5$	26.6	307.8	86.4
C_2H_5Cl	24.6	285.3	86.2
NH_3	23.4	239.7	97.4
H_2O	40.7	373.2	109.0
C_2H_5OH	38.7	351.6	110.1
$C_6H_5NH_2$	45.1	457.5	98.6
CH_3COOH	23.7	391.7	60.5
Hg	59.1	629.8	93.9
Na	89.0	1163.0	76.6
Zn	114.7	1180.0	97.2
Pb	179.5	2025.0	88.6
Ni	374.3	2915.0	113.8

Example. Vaporization enthalpy of diethyl ether

Boiling point [1.9]: $T_b = 34.60\,°C$ (307.75 K)
Vapor pressure [1.9] at 30.00 °C (303.15 K):
$$p^{lv} = 0.08629 \text{ MPa}$$
at 40.00 °C (313.15 K):
$$p^{lv} = 0.12280 \text{ MPa}$$

Enthalpy of vaporization at boiling point ($\Delta H^{lv}_{T_b}$, kJ/mol):
Calculated value
   Clausius–Clapeyron equation:
     simplified version (Eq. 1.2)   27.86
     with second virial coefficient
     ($B' = -0.43$ MPa^{-1} [1.10]), (Eq. 1.4)   26.65
   Pictet–Trouton rule (Eq. 1.5)   27.10
   Equation (1.6)   25.90

Experimental value [1.11]   26.70

The above example demonstrates that results from theoretical equations will be more reliable than estimations based on empirical rules provided that the theoretical relationships have not been grossly simplified. The increase in accuracy obtained by using the more exact Equation (1.4) instead of the simplified Equation (1.2) was only possible by the input of additional data. Such data are often not available, and a suboptimum result has to suffice. As with the rule of Pictet and Trouton, pitfalls may often be encountered when many other empirical and semi-empirical methods are applied to compounds with special or extreme molecular properties, for example, those that form hydrogen bonds (e.g., water, hydrogen fluoride, trichloromethane, alcohols, carboxylic acids) or that have a very low molecular mass (hydrogen, helium).

2. Molecular and Macroscopic Properties

2.1. Intermolecular Forces

Physical macroscopic properties of substances are determined by molecular properties and interactions. Molecular properties may be classified as internal or external. *Internal properties* (e.g., spectroscopic properties and chemical reactivity) are directly determined by the chemical structure of the molecule, i.e., composition, type of chemical bond, and steric structure. *External properties* are mainly relevant to molecular interaction (e.g., molecular cross section, dipole moment, polarizability) and are of more general character although they are also related to chemical structure. A strict distinction between internal and external molecular properties cannot be made. For example, optical spectra with distinct narrow lines in the gas phase may show line broadening when the spectrum is recorded from the liquid phase. This is obviously due to molecular interaction.

Thermal behavior of matter (i.e., the dependence of density on temperature and pressure, phase equilibria etc.) has its origin in interaction between molecules, caused by forces of attraction and repulsion.

Intermolecular forces include:

1) Electrostatic forces between ions, permanent dipoles and multipoles (e.g., quadrupoles), as well as ions and dipoles
2) Induction forces, e.g., dipoles caused by the influence of the electric field of permanent dipoles on the electronic cloud of nonpolar molecules (displacement of electrons)
3) Dispersion forces (so-called Van der Waals forces), i.e., forces caused by the interactions of fluctuating dipoles of nonpolar molecules
4) Chemical forces leading to the formation of associates and complexes

Molecular interactions are functions of the spatial positions of molecules relative to each other. These positions are determined by the potential energy E of the particles, which is the sum of the repulsive and attractive forces, E_{rep} and E_{att}, respectively. E_{rep} affects E positively while E_{att} contributes negatively:

$$E = E_{rep} - E_{att} \qquad (2.1)$$

In the simple case of the interaction between two spherically symmetric molecules, E is a function of the distance r between the molecules only. For the general case of nonspherical molecules E is dependent on the distance and the orientation of the molecules. The contribution of the attractive forces E_{att} to the potential energy is in general a function of the reciprocal of the sixth power of the intermolecular distance r (except in the case of ions and chemical forces):

$$E_{att} \sim \frac{1}{r^6} \qquad (2.2)$$

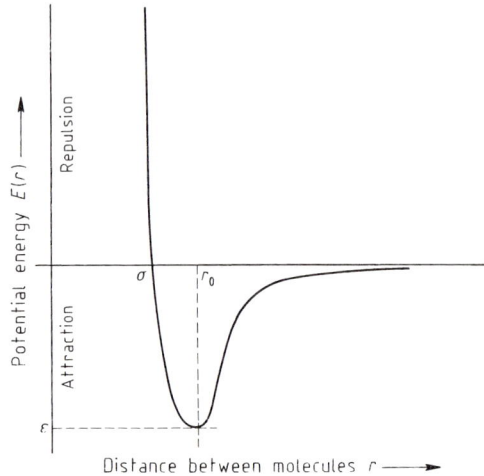

Figure 1. The Lennard–Jones 12-6 potential energy of spherical molecules

Consequently, Equation (2.2) is valid for all types of "normal" nonpolar and polar species, but not for compounds or mixtures showing association or dissociation.

Little is known about the characteristics of repulsive forces. They are, however, obviously of stronger influence at very small distances than attractive forces. A commonly used relation between E_{rep} and the distance r is

$$E \sim \frac{1}{r^{12}} \qquad (2.3)$$

The Lennard–Jones (LJ) 12–6 potential is obtained from Equations (2.1)–(2.3) and describes the dependence of $E(r)$ on the intermolecular distance

$$E(r) = 4\varepsilon \left[\left(\frac{\sigma}{r}\right)^{12} - \left(\frac{\sigma}{r}\right)^6 \right] \qquad (2.4)$$

where ε is the LJ energy and σ the LJ length; these parameters are characteristic for a given species of molecules. As shown in Figure 1, ε is the minimum of the potential energy at a distance r_0. The LJ length σ determines the distance between two molecules at which E is zero. The two characteristic parameters of the LJ potential and other two-parameter models can, in principle, be related not only to molecular data, but also to macroscopic properties (e.g., critical temperature and critical pressure). Hence it should be possible to determine the molecular parameters ε and σ from macroscopic quantities. However, in practice such calculations using statistical mechanics are rather cumbersome. Therefore empirical relations have been proposed, for example by STIEL and THODOS [2.1]:

$$\varepsilon = 65.3 \, k \, T_c \, z_c^{18/5} \qquad (2.5)$$

$$\sigma = 1.866 \times 10^{-3} \, V_c^{1/3} \, z_c^{-6/5} \; [10^{-10} \text{ m}] \qquad (2.6)$$

where k is the Boltzmann constant, T_c the critical temperature, z_c the critical compressibility factor, and V_c the critical molar volume.

Equations (2.5) and (2.6) demonstrate the analogy between the intermolecular potential and the theorem of corresponding states as a general concept for the pVT behavior of fluid phases (see Section 2.2). As a consequence this molecular approach using a two-parameter potential function is also called the microscopic corresponding states theorem. The application of both concepts is not limited to spherical molecules; they may also be used for aspherical molecules with low to moderate polarity.

Among the various two–parameter potential functions [2.2] the LJ potential is preferred because of its simplicity. Thus Equation (2.4) is used for the estimation of virial coefficients and transport properties (e.g., viscosity, diffusion coefficients). In order to account for effects of polar structures, a third parameter may be introduced, as in the Stockmayer potential [2.3]. The

Stockmayer potential is an extension of the LJ potential and takes the dipole moment into consideration.

2.2. Theorem of Corresponding States

VAN DER WAALS was the first to successfully quantify the finding that the pressure volume temperature (pVT) behavior in the gas and liquid states of many compounds is similar. He proposed an equation of state containing two parameters a and b specific for the compound considered:

$$\left(p + \frac{a}{V^2}\right)(V - b) = RT \tag{2.7}$$

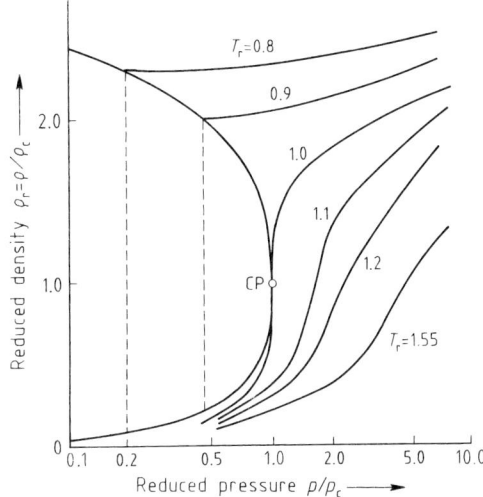

Figure 2. Pressure–density diagram of a pure substance

where p is the pressure, V the molar volume, R the gas constant, and T the absolute temperature. This equation (which is discussed in more detail in Section 3.1) describes not only the nonideal gas phase, but also the liquid-phase compressibility and the region of coexisting liquid and vapor phases including the critical point. Since the van der Waals constants a and b are related to the critical properties (denoted here by the subscript c), the introduction of dimensionless reduced variables T_r, p_r, and V_r defined by

$$T_r = \frac{T}{T_c}; \quad p_r = \frac{p}{p_c}; \quad V_r = \frac{V}{V_c} \tag{2.8}$$

yields a universal "reduced" equation of state which is in principle valid for all substances. This procedure is justified when all compounds show the same pVT behavior described in reduced variables. Figure 2 shows a pressure–density–temperature diagram in reduced variables. This theorem of corresponding states for fluid phases is conveniently expressed using the compressibility factor z defined by (cf. Section 3.1.1)

$$z = \frac{pV}{RT} \tag{2.9}$$

Hence the following universal function is obtained:

$$z = f(T_r, p_r) \tag{2.10}$$

Equation (2.10) implicitly contains the characteristics of the respective chemical compound because T_r and p_r are defined by the critical parameters T_c and p_c. This approach employs only two characteristic parameters to describe the pVT behavior of a compound. Hence satisfactory agreement between theory and experiment cannot be expected for all types of substances. However, for a large number of "normal" compounds (e.g., weakly polar, nonassociating substances, see Section 1.3) reasonable quantitative representation of pVT behavior can be achieved by applying methods based upon the corresponding states principle with two parameters [2.4]. Calculations of gas-phase properties are generally more reliable than those of the liquid state. The region close to the critical point cannot be represented with sufficient accuracy. For compounds with a very low boiling point (e.g., hydrogen, nitrogen), modified critical parameters must be used for the definitions of the reduced variables [2.5]. In order to apply corresponding states methods to polar compounds a third characteristic parameter is necessary, e.g., the acentric factor or the polar factor of Stiel (Section 2.4).

2.3. Estimation of Critical Properties

In view of the benefits of the principle of corresponding states much attention has been devoted to the critical properties of pure substances. Knowledge of these properties is important for many methods of estimating pure component properties.

Since measurement of critical volume, pressure, and temperature is difficult and tedious, estimation methods are useful tools for providing information about the critical constants. Most estimation techniques employ the principle of additivity which assumes that a physical property can be obtained from the sum of the contributions of the atoms, functional atomic groups, and bonds of the molecule considered. These group contribution or incremental methods are widely used for the prediction of physical properties.

Critical Temperature. According to the empirical Guldberg rule the ratio of the normal boiling point T_b to the critical temperature T_c is equal to 2/3:

$$T_b/T_c = 2/3 \quad (2.11)$$

Based on the more general form of this rule

$$T_c = \text{const.} \, T_b \quad (2.12)$$

various methods have been proposed for predicting critical temperatures [2.5], [2.6], e.g., the widely accepted and easy-to-handle procedure of Lydersen [2.7]. Recently, this method was improved by JOBACK who took advantage of new experimental data [2.8]. Joback's modification of Lydersen's expression is given by

$$T_c = \frac{T_b}{0.584 + 0.965 \sum s_i I_{T_c} - (\sum s_i I_{T_c})^2} \quad (2.13)$$

where s_i is the number of structural groups of type i and I_{T_c} is the Joback increment of functional groups given in Table 3. The information needed for the estimation of the critical temperature is the normal boiling point and Lydersen's or Joback's increments which are tabulated for many functional groups, e.g., [2.6].

Critical Pressure. For the prediction of critical pressure p_c JOBACK [2.8] used an equation containing the number of atoms N in the molecule and the increments I_{p_c} for the functional groups (Table 3):

$$p_c = \frac{0.1}{(0.113 + 0.0032 \, N - \sum s_i I_{p_c})^2} \, [\text{MPa}] \quad (2.14)$$

Critical Volume. The difficulties encountered in the experimental determination of the critical molar volume V_c lead to difficulties in the accurate prediction of this quantity too. Using Joback's method [2.8] only increments are needed:

$$V_c = 17.5 + \sum s_i I_{V_c} \quad (2.15)$$

where V_c is the critical volume [cm^3/mol] and I_{V_c} are Joback's increments (Table 3).

Example. Calculation of critical properties of chlorobenzene, C_6H_5Cl, with Joback's method and for comparison the results obtained with Lydersen's method

Functional groups: 5 = C− (ring), 1 = C (ring), 1−Cl; T_b = 404.9 K [2.9]; superscripts: exp = experimental value, cal = calculated value

Critical temperature: Deviation
$T_c^{\text{exp}} = 632.4$ K $T_c^{\text{cal}} - T_c^{\text{exp}}$
Joback (Eq. 2.13): $\sum s_i I_{T_c} = 0.0658$
$T_c^{\text{cal}} = 404.9/(0.584 + 0.0635$
     $- 0.00433) = 629.5$ K $- 2.9$ K
Lydersen $T_c^{\text{cal}} = 629.6$ K $- 2.8$ K

Critical pressure:
$p_c^{\text{exp}} = 4.52$ MPa $(p_c^{\text{cal}} - p_c^{\text{exp}})/p_c^{\text{exp}}$
Joback (Eq. 2.14): $\sum s_i I_{p_c} = 0.0014$
$p_c^{\text{cal}} = 0.1/[0.113 + (0.032 \times 12)$
     $- 0.0014] = 4.44$ MPa $- 1.7\%$
Lydersen $p_c^{\text{cal}} = 4.55$ MPa $+ 0.6\%$

Critical volume:
$V_c^{\text{exp}} = 308$ cm^3/mol $(V_c^{\text{cal}} - V_c^{\text{exp}})/V_c^{\text{exp}}$
Joback (Eq. 2.15): $\sum s_i I_{V_c} = 295.0$
$V_c^{\text{cal}} = 17.5 + 295.0 = 312.5$ cm^3/mol $+ 1.5\%$
Lydersen $V_c^{\text{cal}} = 310.0$ cm^3/mol $+ 0.6\%$

Discussion. Table 4 gives an overview of several methods for the estimation of critical properties together with average uncertainties of the predicted values. The uncertainty in the prediction of the critical pressure is considerably higher than that for the prediction of critical temperature; for the critical volume it is often still greater. Higher errors are also observed for molecules of larger size or complex structure because relevant experimental data are scarce. The application of group contribution methods leads to systematic errors when the properties of molecules with strongly interacting groups are estimated. In addition, these methods cannot distinguish between certain isomers, e.g., o-, m-, and p-derivatives of benzene.

A more accurate but more complicated method was proposed by AMBROSE [2.10], [2.11]. Comprehensive compilations of critical data have been published for organic compounds [2.14], [2.15] and inorganic compounds [2.16].

2.4. Other Characteristic Constants of Pure Compounds

The Acentric Factor ω. In order to extend corresponding states methods to compounds and mixtures consisting of nonspherical and po-

Table 3. Increments used in the estimation of critical temperature T_c, pressure p_c, volume V_c, normal boiling point T_b, and melting point T_m (Joback's method) [2.6]

Type of increment	I_{T_c}	I_{p_c}	I_{V_c}	I_{T_b}	I_{T_m}
Nonring increments					
$-CH_3$	0.0141	−0.0012	65	23.58	−5.10
$>CH_2$	0.0189	0	56	22.88	11.27
$>CH-$	0.0164	0.0020	41	21.74	12.64
$>C<$	0.0067	0.0043	27	18.25	46.43
$=CH_2$	0.0113	−0.0028	56	18.18	−4.32
$=CH-$	0.0129	−0.0006	46	24.96	8.73
$=C<$	0.0117	0.0011	38	24.14	11.14
$=C=$	0.0026	0.0028	36	26.15	17.78
$\equiv CH$	0.0027	−0.0008	46	9.20	−11.18
$\equiv C-$	0.0020	0.0016	37	27.38	64.32
Ring increments					
$-CH_2-$	0.0100	0.0025	48	27.15	7.75
$>CH-$	0.0122	0.0004	38	21.78	19.88
$>C<$	0.0042	0.0061	27	21.32	60.15
$=CH-$	0.0082	0.0011	41	26.73	8.13
$=C<$	0.0143	0.0008	32	31.01	37.02
Halogen increments					
$-F$	0.0111	−0.0057	27	−0.03	−15.78
$-Cl$	0.0105	−0.0049	58	38.13	13.55
$-Br$	0.0133	0.0057	71	66.86	43.43
$-I$	0.0068	−0.0034	97	93.84	41.69
Oxygen increments					
$-OH$ (alcohol)	0.0741	0.0112	28	92.88	44.45
$-OH$ (phenol)	0.0240	0.0184	−25	76.34	82.83
$-O-$ (nonring)	0.0168	0.0015	18	22.42	22.23
$-O-$ (ring)	0.0098	0.0048	13	31.22	23.05
$>C=O$ (nonring)	0.0380	0.0031	62	76.75	61.20
$>C=O$ (ring)	0.0284	0.0028	55	94.97	75.97
$O=CH-$ (aldehyde)	0.0379	0.0030	82	72.24	36.90
$-COOH$ (acid)	0.0791	0.0077	89	169.09	155.50
$-COO-$ (ester)	0.0481	0.0005	82	81.10	53.60
$=O$ (except as above)	0.0143	0.0101	36	−10.50	2.08
Nitrogen increments					
$-NH_2$	0.0243	0.0109	38	73.23	66.89
$>NH$ (nonring)	0.0295	0.0077	35	50.17	52.66
$>NH$ (ring)	0.0130	0.0114	29	52.82	101.51
$>N-$ (nonring)	0.0169	0.0074	9	11.74	48.84
$-N=$ (nonring)	0.0255	−0.0099		74.60	
$-N=$ (ring)	0.0085	0.0076	34	57.55	68.40
$-CN$	0.0496	−0.0101	91	125.66	59.89
$-NO_2$	0.0437	0.0064	91	152.54	127.24
Sulfur increments					
$-SH$	0.0031	0.0084	63	63.56	20.09
$-S-$ (nonring)	0.0119	0.0049	54	68.78	34.40
$-S-$ (ring)	0.0019	0.0051	38	52.10	79.93

lar molecules, the Pitzer acentric factor ω was defined as a third parameter:

$$\omega = -\log\left(\frac{p^{lv}}{p_c}\right)_{T_r=0.7} - 1 \quad (2.16)$$

where p^{lv} is the vapor pressure of the pure component. The Pitzer acentric factor is intended to give a measure of nonsphericity or acentricity of a molecule's potential force field. For an ideal spherical molecule it should be zero. Acentric factors for some compounds are given below [2.6]:

Compound	ω	Compound	ω
Methane	0.008	Water	0.344
Nonane	0.445	Ammonia	0.250
Cyclohexane	0.212	Hydrogen	−0.216
Benzene	0.212	Phenol	0.438
Methanol	0.556	Aniline	0.384
1-Octanol	0.587		

Table 4. Selected methods for the prediction of critical data of organic compounds

Method	Applicability	Information required	Error
Lydersen 1955 [2.7]	T_c, p_c, V_c with increments for 41 functional groups; $>$Si$<$ increments for T_c and p_c; $-$B$-$ increment for T_c	T_b for T_c M for p_c	T_c: 5–12 K p_c: 4–10% V_c: 9%
Joback 1984 [2.8]	T_c, p_c, V_c with increments for 40 functional groups	T_b for T_c	T_c: 5–10 K p_c: 6% V_c: 5–10%
Ambrose 1978 [2.10] 1979 [2.11]	T_c (52), p_c (52), V_c (35) with increments for functional groups (number of functional groups given in parentheses)	T_b for T_c M for p_c	T_c: 5–10 K p_c: 5% V_c: 3–5%
Fedors 1982 [2.12]	T_c with increments for 48 functional groups	T_b not required	T_c: 12–25 K
Forman, Thodos 1958 [2.13]	T_c and p_c from Van der Waals constants a and b, which are determined with increments for functional groups; not for aldehydes, secondary and tertiary alcohols, sulfur compounds	T_b not required	T_c: 5–12 K p_c: 3–10%

Knowing the critical temperature and pressure of a substance and its vapor pressure p^{lv} (i.e., with the Antoine coefficients A, B, and C, Eq. 4.5), the acentric factor can be calculated as follows:

$$\omega = \log\left[\frac{p_c}{\exp\left(A - \frac{B}{0.7\,T_c + C}\right)}\right] - 1 \quad (2.17)$$

If only the normal boiling point T_b is known the Equation of Edmister [2.17] gives a reasonable estimate (p_c in atmospheres):

$$\omega = \frac{3}{7}\left(\frac{\log p_c}{T_c/T_b - 1}\right) - 1 \quad (2.18)$$

The average uncertainty of this approximation is 5% [2.18].

Example. Calculation of the acentric factor for benzoic acid
$T_c = 752$ K; $p_c = 45.6$ bar; $T_b = 523$ K; $\omega^{\exp} = 0.620$ [2.5]; Antoine constants [2.6] $A = 10.5432$, $B = 4190.70$, $C = -125.2$

Substituting into Equation (2.17)

$$\omega^{\text{cal}} = \log\left[\frac{p_c}{\exp\left(10.5432 - \frac{4190.7}{(0.7 \times 752) - 125.2}\right)}\right] - 1$$

$$= 0.617$$

Relative deviation = 0.6%

With Equation (2.18)

$$\omega^{\text{cal}} = \frac{3}{7}\left(\frac{1}{(752/523) - 1}\right)\log\left(\frac{45.6}{1.013}\right) - 1 = 0.618$$

Relative deviation = 0.3%

The Stiel Polar Factor χ. This factor relates the actual reduced vapor pressure to the value calculated by the Pitzer equation (Eq. 2.16) at the reduced temperature $T_r = 0.6$:

$$\chi = \log\left(\frac{p^{lv(\exp)}}{p^{lv(\text{Pitzer})}}\right)_{T_r = 0.6} \quad (2.19)$$

It is therefore also a measure of polarity and is related to the acentric factor by

$$\chi = \log\left(\frac{p^{lv}}{p_c}\right)_{T_r = 0.6} + 1.7\,\omega + 1.552 \quad (2.20)$$

Example. Calculation of the Stiel polar factor for benzoic acid $p^{lv}(T_r = 0.6) = 0.099$ bar (Antoine Eq. 4.5 [2.5])

Substituting into Equation (2.20):

$$\chi = \log\left(\frac{0.099}{45.6}\right) + 1.7 \times 0.620 + 1.552 = -0.0573$$

Parachor. The parachor P_i is a parameter characteristic for a pure liquid compound i and depends on its surface tension σ_i. It has been defined by SUGDEN [2.19] as

$$P_i = \sigma_i^{1/4}\,\frac{M_i}{\varrho_i^l - \varrho_i^v} \quad (2.21)$$

where ϱ_i^l and ϱ_i^v are the liquid and saturated vapor densities of pure compound i, respectively. At temperatures considerably lower than the critical temperature the density of the vapor ϱ_i^v can be neglected compared to the density of the liquid ϱ_i^l, thus simplifying Equation (2.21) to

$$P_i = \sigma_i^{1/4} \frac{M_i}{\varrho_i^l} = \sigma_i^{1/4} V_i^l \qquad (2.22)$$

Table 5. Quayle's increments for the estimation of parachors P_i [cm^3 g$^{0.25}$ s$^{-0.5}$ mol^{-1}] [2.20]

Group	Increment
Carbon–hydrogen	
C	9.0
H	15.5
CH$_3$	55.5
CH$_2$ in $-$(CH$_2)_n$	
$\quad n < 12$	40.0
$\quad n > 12$	40.3
Alkyl groups	
1-Methylethyl	133.3
1-Methylpropyl	171.9
1-Methylbutyl	211.7
2-Methylpropyl	173.3
1-Ethylpropyl	209.5
1,1-Dimethylethyl	170.4
1,1-Dimethylpropyl	207.5
1,2-Dimethylpropyl	207.9
1,1,2-Trimethylpropyl	243.5
C$_6$H$_5$	189.6
Special groups	
$-$COO$-$	63.8
$-$COOH	73.8
$-$OH	29.8
$-$NH$_2$	42.5
$-$O$-$	20.0
$-$NO$_2$	74
$-$NO$_3$ (nitrate)	93
$-$CO(NH$_2$)	91.7
R$-$[$-$CO$-$]$-$R' (ketone)	
$\quad$R + R' = 2	51.3
$\quad$R + R' = 3	49.0
$\quad$R + R' = 4	47.5
$\quad$R + R' = 5	46.3
$\quad$R + R' = 6	45.3
$\quad$R + R' = 7	44.1
$-$CHO	66
O (not noted above)	20
N (not noted above)	17.5
S	49.1
P	40.5
F	26.1
Cl	55.2
Br	68.0
I	90.3
Ethylenic bonds	
Terminal	19.1
2,3-position	17.7
3,4-position	16.3
Triple bond	40.6
Ring closure	
Three-membered	12
Four-membered	6.0
Five-membered	3.0
Six-membered	0.8

The parachor is virtually constant with temperature; it is approximately an additive quantity which can be determined by summing up increments. Several such methods based on Equation (2.20) with σ_i in [mN/m] ($\equiv$ [dyn/cm]) and V_i^l in [cm^3/mol] have been proposed. The method of Quayle [2.20] is recommended, the increments for this method are given in Table 5.

3. Thermal and Caloric Properties of Single-Phase Systems

3.1. Pressure–Volume–Temperature Behavior of Gases and Liquids

3.1.1. Pure Gases

The pressure–volume–temperature (pVT) behavior can be calculated with equations of state. In this review emphasis is placed upon basic principles and easy-to-handle methods. More detailed information is given in [3.1]–[3.3].

For n moles of an ideal gas consisting of rigid molecules of zero volume and zero interaction potential energy the following equation holds

$$pV = RT \qquad (3.1)$$

where V is the molar volume.

The ideal gas law is a universal limiting law valid for all substances when the density ϱ approaches 0. Equation (3.1) can usually be used with sufficient accuracy for gases at low pressure $p \ll p_c$. The deviation of the behavior of real substances with respect to Equation (3.1) is usually expressed by the compressibility factor z

$$z = \frac{pV}{RT} \qquad (3.2)$$

The dependence of z on pressure is plotted in Figure 3 for a real fluid which obeys the principle of corresponding states. Below the critical pressure z is less than unity, i.e., the attractive forces are dominant. Above T_c and p_c, z exceeds unity due to increasing influence of the repulsive forces.

In order to describe real fluid behavior an expression must be found of the form

$$z = f(T, p) \quad \text{or} \quad z = f'(T, V) \qquad (3.3)$$

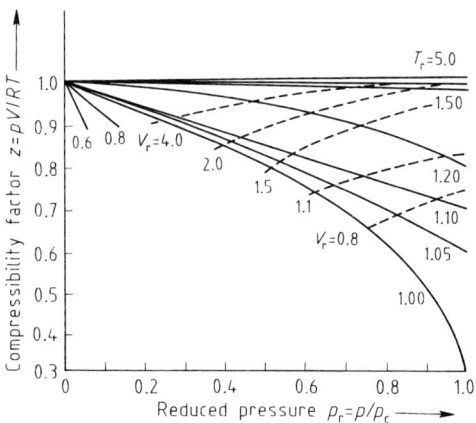

Figure 3. Generalized compressibility chart [3.4]

that is capable of describing pVT behavior for a wide variety of substances over a wide range of pressure and temperature. The three following fundamental boundary conditions must be satisfied:

$$\left(\frac{\partial p}{\partial V}\right)_{T_c} = 0 \tag{3.4}$$

$$\left(\frac{\partial^2 p}{\partial V^2}\right)_{T_c} = 0 \tag{3.5}$$

$$V \to b \quad \text{for} \quad p \to \infty \tag{3.6}$$

Cubic Equations of State. In his pioneering work of 1873 VAN DER WAALS introduced his famous equation of state which is the basic form for many other cubic (in volume) expressions:

$$\left(p + \frac{a}{V^2}\right)(V - b) = RT \tag{3.7}$$

The substance-specific constants a and b can be determined using conditions (3.4) and (3.5):

$$a = \frac{27}{64} \frac{(RT_c)^2}{p_c} \tag{3.8}$$

$$b = \frac{1}{8} \frac{RT_c}{p_c} \tag{3.9}$$

One of the most widely used two-parameter equations was suggested in 1949 by REDLICH and KWONG (RK):

$$p = \frac{RT}{V-b} - \frac{a}{T^{1/2} V(V+b)} \tag{3.10}$$

Since the parameters a and b are substance-specific constants, the equation is completely determined by the critical properties. The RK equation and its subsequent modifications are more accurate than the Van der Waals equation. One very succesful modification of the attractive term in the RK equation was recommended by SOAVE [3.5]:

$$p = \frac{RT}{V-b} - \frac{a_c[1 + M(\omega)(1 - T_r^{1/2})]^2}{V(V+b)} \tag{3.11}$$

where

$$M(\omega) = (0.48 + 1.574\,\omega - 0.176\,\omega^2) \tag{3.11a}$$

$$a_c = 0.42748\,(RT_c)^2/p_c \tag{3.11b}$$

$$b = 0.08664\,(RT_c)/p_c \tag{3.11c}$$

The pVT behavior can be predicted with a generalized equation when information on the critical properties and the acentric factor ω is available.

These equations are particularly useful for hydrocarbons. According to [3.6] the average error for this class of substances is 1–2%. Deviations increase in the vicinity of the critical point and for polar substances.

Virial Equation of State. Statistical mechanics yield an equation of state in which the compressibility factor z is given as a polynomial in reciprocal molar volume:

$$\frac{pV}{RT} = z = 1 + \frac{B}{V} + \frac{C}{V^2} + \ldots \tag{3.12}$$

In this so-called virial equation the virial coefficients $B, C \ldots$ are functions of temperature only and can be related to molecular interaction parameters. Another form of the virial equation uses pressure as the independent parameter:

$$\frac{pV}{RT} = z = 1 + B'p + C'p^2 + \ldots \tag{3.13}$$

The coefficients of the two forms of the virial Equations (3.12) and (3.13) are related, for example as follows:

$$B' = \frac{B}{RT} \tag{3.14}$$

Use of the virial equation is very convenient because of its simple form. It is usually truncated after the second term; application of this form is limited to reduced densities < 0.5. Experimental data on second and, in part, third virial coefficients of pure gases and mixtures are given in [3.7].

Calculation of Second Virial Coefficients. Many calculation schemes for the prediction of second virial coefficients have been published, e.g., [3.8]–[3.10]. The frequently used method of Tsonopoulos [3.9] will be described here briefly. The equations can be used for most nonpolar or slightly polar substances. Real fluid behavior is calculated using Pitzer's approach

$$z = 1 + B \frac{p_c}{RT_c} \frac{p_r}{T_r} = z^{(0)} + \omega z^{(1)} \qquad (3.15)$$

$$B \frac{p_c}{RT_c} = b^{(0)} + \omega b^{(1)} \qquad (3.16)$$

with

$$b^{(0)} = 0.1445 - \frac{0.33}{T_r} - \frac{0.1385}{T_r^2}$$
$$- \frac{0.0121}{T_r^3} - \frac{6.07 \times 10^{-4}}{T_r^8} \qquad (3.16a)$$

$$b^{(1)} = 0.0637 + \frac{0.331}{T_r^2} - \frac{0.423}{T_r^3} - \frac{0.008}{T_r^8} \qquad (3.16b)$$

For strongly polar compounds TSONOPOULOS suggested the following modification of Equation (3.16)

$$B \frac{p_c}{RT_c} = b^{(0)} + \omega b^{(1)} + \frac{c}{T_r^6} - \frac{d}{T_r^8} \qquad (3.17)$$

For several substances (e.g., ketones, aldehydes, ethers, nitriles) TSONOPOULOS gives recommendations for the calculation of the parameter c

$$c = -2.113 \times 10^{-4} \mu_r - 3.885 \times 10^{-21} \mu_r^8 \qquad (3.18)$$

where

$$\mu_r = \frac{10^6 \mu_p^2 p_c}{T_c^2} \qquad (3.19)$$

(p_c in MPa, T in K, and the dipole moment μ_p in Debye). The parameter d vanishes for substances that do not exhibit hydrogen bonding. For other compounds it has to be determined by experiment.

Example. Calculation of the second virial coefficient of acetone, C_3H_6O, at 310 K with Tsonopoulos's method

$M = 58.08$ g/mol, $T_c = 508.1$ K, $p_c = 4.7$ MPa, $\mu = 2.9$ Debye, $\omega = 0.304$

With $T_r = 0.610$ substitution into Equations (3.16a) and (3.16b) gives $b^{(0)} = -0.8537$, $b^{(1)} = -1.3276$.

Substituting into Equation (3.17) and neglecting the third and fourth terms yields:

$$B_{\text{nonpolar}} = -1.1303 \text{ m}^3/\text{kmol}$$

With $\mu_r = 153.107$ from Equation (3.19), $c = -33.525 \times 10^{-3}$ from Equation (3.18), and substituting into Equation (3.17):

$$B_{\text{polar}} = -1.715 \text{ m}^3/\text{kmol}$$

Experimental: $B = -1.730$ m^3/kmol [3.7]

Relative deviation = 0.9 %

Other Equations of State. The representation of pVT behavior of polar gases and liquids with two-parameter cubic equations of state is usually unsatisfactory. Accuracy can be improved significantly by applying equations of the Benedict–Webb–Rubin (BWR) type:

$$p = \frac{RT}{V} + \left(B_0 RT - A_0 - \frac{C_0}{T^2}\right) \frac{1}{V^2} + \frac{(bRT-a)}{V^3}$$
$$+ \frac{a\alpha}{V^6} + \frac{c}{T^2 V^3}(1 + \beta/V^2) \exp(-\beta/V^2) \qquad (3.20)$$

where A_0, B_0, C_0, a, b, c, α, and β are substance-specific constants. An advantageous modification is the generalized form suggested by LEE and KESLER [3.11] and its extension by PLÖCKER et al. [3.12]. For this type of equation of state the compressibility factor of a real fluid is related to the properties of a simple fluid ($\omega = 0$), e.g., to those of n-octane as a reference fluid [3.2].

A BWR extension with 20 substance-specific parameters and remarkable accuracy was suggested in the 1970s by BENDER [3.13]. In 1987 POLT reported these constants for 51 substances [3.14]. Figure 4 shows the $p\varrho T$ behavior of toluene calculated with the Bender equation.

3.1.2. Pure Liquids

Because of their importance in industrial processes, density data are available for most organic liquids. Usually at least one experimental value for ambient conditions can be found. It

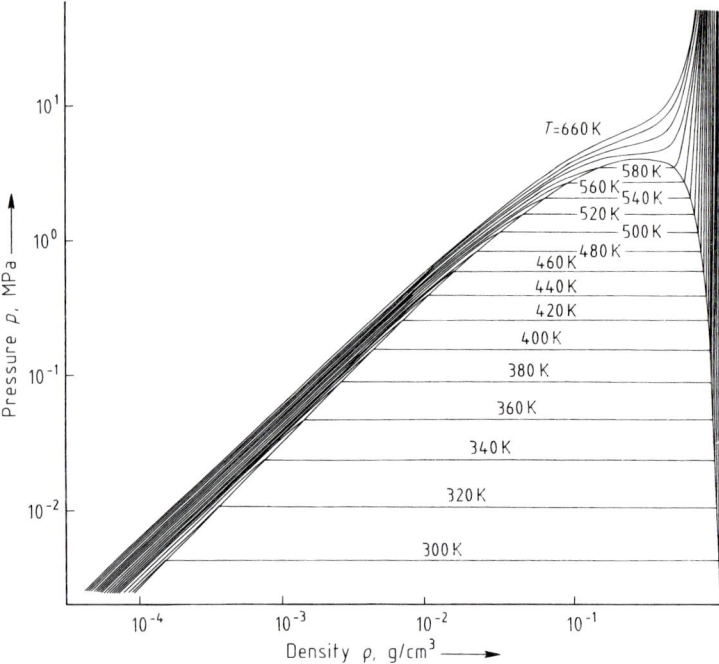

Figure 4. Pressure–density behavior of toluene calculated with the Bender equation of state [3.14]

is therefore more important to gain information about the temperature dependence of liquid densities rather than to estimate liquid densities from other properties. Nevertheless, some methods have been published for the estimation of liquid densities, e.g., [3.15].

Schroeder's Method. SCHROEDER suggested a very simple, convenient method for the estimation of saturated liquid densities at the normal boiling point [3.16]. He proposed an additive rule for the prediction of molar volumes. The number of atoms of each type of element in the molecule (i.e., carbon, hydrogen, oxygen, and nitrogen, etc.) must be counted, multiplied by a weighting factor, and summed up. Additional increments are then added to account for the molecular structure (i.e., double bonds) as shown in Table 6.

The method gives surprisingly good estimates of molar volume with an average error of ca. 3%. Exceptions are substances that exhibit hydrogen bonding or a high degree of association.

Example. Calculation of liquid molar volume V_b^l of ethyl acetate, $CH_3COOCH_2CH_3$, at the normal boiling point (101.325 kPa) using Schroeder's method

Groups: $4\,C + 2\,O + 8\,H$
Calculated: $V_b^l = 4 \times 0.007 + 2 \times 0.007 + 8 \times 0.007 = 0.098 \text{ m}^3/\text{kmol}$
Experimental [3.16]: $V_b^l = 0.0978 \text{ m}^3/\text{kmol}$
Relative deviation $= 0.2\%$

Table 6. Increments for the prediction of molar volumes of pure liquids at the normal boiling point according to Schroeder's method [3.16]

Element	Increment, m³/kmol	Bond	Increment, m³/kmol
Carbon	0.007	double bond (C=C)	0.007
Hydrogen	0.007	triple bond (C≡C)	0.014
Oxygen	0.007	ring systems	0.007
Nitrogen	0.007	(including naphthalene, anthracene)	
Bromine	0.0315		
Chlorine	0.0245		
Fluorine	0.0105		
Iodine	0.0385		
Sulfur	0.0210		

More Sophisticated Estimation Methods. Many analytical methods have been published, they are mostly based upon the principle of corresponding states. Consequently, standard density values are necessary for which the critical density can be used. A single experimental value can often be used for very accurate extrapolations with respect to temperature.

The Gunn–Yamada expression is convenient and very accurate (without any additional adjusted parameters) for predicting liquid molar volumes V^l at the state of saturation [3.18]:

$$V^l/V_{sc} = V_r^{(0)}(1 - \omega I) \qquad (3.21)$$

where I and $V_r^{(0)}$ are generalized functions and V_{sc} is a scaling parameter.
For $0.2 < T_r < 1.0$:

$$I = 0.29607 - 0.09045\, T_r - 0.04842\, T_r^2 \qquad (3.22)$$

For $0.2 < T_r < 0.8$:

$$V_r^{(0)} = 0.33593 - 0.33953\, T_r + 1.51941\, T_r^2 \\ - 2.02512\, T_r^3 + 1.11422\, T_r^4 \qquad (3.23)$$

For $0.8 < T_r < 1.0$

$$V_r^{(0)} = 1.0 + 1.3(1 - T_r)^{1/2} \log(1 - T_r) \\ - 0.50879(1 - T_r) - 0.91534(1 - T_r)^2 \qquad (3.24)$$

In terms of $V_{0.6}$ (the saturated liquid molar volume at the reduced temperature $T_r = 0.6$), V_{sc} is defined as

$$V_{sc} = \frac{V_{0.6}}{(0.3862 - 0.0866\,\omega)} \qquad (3.25)$$

If $V_{0.6}$ is not available, V_{sc} can be estimated as follows (p_c in MPa):

$$V_{sc} = R\, T_c (0.2920 - 0.0967\,\omega) \frac{0.10132}{p_c} \qquad (3.26)$$

Usually, one experimental value of V^l under ambient conditions ($T_r \approx 0.6$) is available and can be applied as a reference value V^{ref} at T_r^{ref}. In these cases the following form of the method of Gunn and Yamada is useful:

$$\frac{V^l}{V^{ref}} = \frac{V_r^{(0)}(T_r)}{V_r^{(0)}(T_r^{ref})} \cdot \frac{[1 - \omega I(T_r)]}{[1 - \omega I(T_r^{ref})]} \qquad (3.27)$$

Example. Calculation of the saturated liquid molar volume V^l of n-propylamine, $CH_3CH_2CH_2NH_2$, at 293.15 K using the method of Gunn and Yamada

$M = 59.112$ g/mol, $T_c = 497$ K, $\omega = 0.303$,
$p_c = 4.81$ MPa [3.17]

Calculated:

From Equation (3.23) $V_r^{(0)} = 0.3839$
From Equation (3.26) $V_{sc} = 0.2287$ m^3/kmol
With Equation (3.22) $I = 0.2259$
Substituting in Equation (3.21) $V^l = 0.08178$ m^3/kmol
Experimental: $V^l = 0.0824$ m^3/kmol [3.17]
Relative deviation $= 0.75\%$

The more recently published method of Hankinson and Thomson achieves higher accuracy for liquid density [3.19]. This technique employs specific constants for various groups of substances as well as three substance-specific parameters that have been tabulated for more than 400 compounds (e.g., in [3.3], [3.19]) or can be estimated. An overview of estimation methods is given in Table 7.

Correlation of Liquid Density Data. Each of the methods that use a reference value may also be used to correlate experimental liquid density data. The Rackett equation has proven to be very successful [3.20]:

$$\frac{V}{V_c} = z_c^{(1 - T_r)^{2/7}} \qquad (3.28)$$

where z_c can be used as an adjustable parameter.

Table 7. Selected methods for the determination of saturated liquid molar volumes of pure organic compounds

Method	Range of application		Information required[a]	Error, %
	Substances	Temperature		
Rackett [3.20]	nonpolar and polar	$T \geq T^{ref} - 100$ K	$T_c, z_c, \omega, (\varrho^{ref})$	$(T - T^{ref})/100$
Yen, Woods [3.21]	nonpolar and polar	$0.3 \leq T_r \leq 1$	ϱ_c, T_c, z_c, z	1
Gunn, Yamada [3.18]	nonpolar and weakly polar	$0.3 \leq T_r \leq 0.99$	$T_c, \omega, \varrho^{ref}$	$(T - T^{ref})/100$
Hankinson, Thomson [3.19]		$0.25 \leq T_r \leq 1.0$	T_c^b, V_c^b, ω^b, increments	<2
Bhirud [3.22]	nonpolar (polar)		$T_c, \omega, (\varrho^{ref})$	1

[a] Properties in parentheses are not absolutely necessary.
[b] Tabulated in [3.3], [3.19]; can also be calculated.

Pressure Dependence of Liquid Density. With the isothermal compressibility β_T

$$\beta_T = -\frac{1}{V}\left(\frac{\partial V}{\partial p}\right)_T \tag{3.29}$$

and the relation [3.23]

$$\frac{\beta^{1/7}}{V} = \frac{\beta_{sat}^{1/7}}{V_{sat}} \tag{3.30}$$

the following approximation holds:

$$V^1/V_{sat}^1 = [1 + 9\beta_{sat}(p - p_0)]^{-1/9} \tag{3.31}$$

where β_{sat} is estimated by the relation

$$\beta_{sat} = \frac{V_c}{RT_c}[(1 - 0.89\omega^{1/2})$$
$$\cdot \exp(6.9547 - 76.2853\,T_r + 191.3060\,T_r^2$$
$$- 203.5472\,T_r^3 + 82.7631\,T_r^4)] \tag{3.32}$$

This method holds for the range of reduced temperature $0.35 < T_r < 0.98$ and for reduced pressure of $p_r < 10$ with an average uncertainty of 3%.

3.1.3. Mixtures

3.1.3.1. Gas Mixtures

The constants used in the methods for the calculation of pVT behavior in gas mixtures must be related to the relevant parameters of the pure components and to the composition of the mixture. Additional correlation parameters may be necessary. The objective of this section is to provide some information on mixing rules.

Virial Equation. Statistical thermodynamics yield the following rigorous relation for the second virial coefficient B_m of a mixture of components i and j with a quadratic dependency on the molar fractions y_i and y_j:

$$B_m = \sum_i^r \sum_j^r y_i y_j B_{ij} \tag{3.33}$$

The virial coefficient $B_{ii} \equiv B_i$ of the pure component i can be calculated as described in Section 3.1. For the coefficients of the mixture B_{ij} similar methods can be employed if the critical properties of the mixture and the acentric factor ω_{ij} are known.

PRAUSNITZ discusses simple rules that yield sufficiently accurate results for practical applications [3.24]:

$$T_{cij} = (T_{ci}\,T_{cj})^{0.5}(1 - k_{ij}) \tag{3.34}$$

where k_{ij} is a specific parameter that has to be adjusted to experimental data.

$$V_{cij} = \left(\frac{V_{ci}^{1/3} + V_{cj}^{1/3}}{2}\right)^3 \tag{3.35}$$

$$Z_{cij} = \frac{Z_{ci} + Z_{cj}}{2} \tag{3.36}$$

The following relations can also be readily derived:

$$p_{cij} = \frac{Z_{cij}\,R\,T_{cij}}{V_{cij}} \tag{3.37}$$

$$\omega_{ij} = \frac{\omega_i + \omega_j}{2} \tag{3.38}$$

Pseudocritical Properties. Since many methods take advantage of the principle of corresponding states using reduced quantities for temperature, pressure, volume, and critical compressibility, knowledge of these properties in mixtures is important. If a mixture with a fixed composition is assumed to behave like a pure fluid, so-called pseudocritical constants can be introduced that serve as scaling factors.

Many attempts to calculate pseudocritical properties have been published. Some basic methods are given below (the subscript m denotes mixture):

1) Kay's rule (1936) for T_{cm} [3.25]

$$T_{cm} = \sum y_i\,T_{ci} \tag{3.39}$$

Deviations from results obtained by using more complicated methods are less than 2% [3.6].

2) p_{cm} according to Prausnitz and Gunn [3.26]

$$p_{cm} = R\,T_{cm}\frac{\sum(y_i\,z_{ci})}{\sum y_i\,V_i^g} \tag{3.40}$$

3) Joffe's rule for the acentric factor ω_m of the mixture [3.27]

$$\omega_m = \sum y_i\,\omega_i \tag{3.41}$$

Algebraic expressions of varying complexity have been suggested for the mixing rules for each equation of state applicable to mixtures. They are summarized and discussed extensively in [3.3].

3.1.3.2. Liquid Mixtures

Molar volumes V_m^l of mixtures comprising nonreacting components which do not differ greatly in structure can be calculated for low and moderate pressure ($p < p_c$) by Amagat's rule:

$$V_m^l = \sum x_i V_{0i}^l \qquad (3.42)$$

where x_i is the mole fraction of component i. Deviations for real mixtures do not generally exceed $\pm 1\%$ of V_m^l.

3.2. Caloric Properties

Enthalpy is the basic thermodynamic quantity used in energy balances for chemical processes. According to rigorous thermodynamics, the difference in molar enthalpy (H) between two states (denoted by subscripts 1 and 2) can be calculated via

$$H(T_2, p_2) - H(T_1, p_1) = \int_{T_1}^{T_2} c_p(T, p) \, dT$$
$$+ \int_{p_1}^{p_2} \left(\frac{\partial H}{\partial p}\right) dp \qquad (3.43)$$

where c_p is the molar heat capacity at constant pressure. Since heat capacity data for high pressures are scarce and the variation in H between two states is independent of the path of integration, the difference in H is preferably determined via

$$H_2 - H_1 = [H(T_1, p^0) - H(T_1, p_1)]$$
$$+ \int_{T_1}^{T_2} c_p^0(T) \, dT$$
$$- [H(T_2, p^0) - H(T_2, p_2)] \qquad (3.44)$$

where the superscript 0 indicates the reference state shown by an arrow in Figure 5. As a consequence, the temperature dependence of enthalpy is calculated at zero pressure, i.e., in the ideal gas state. The dependence on pressure is given by the so-called departure functions (e.g., $H-H^0$, $S-S^0$) that can be determined with an appropriate equation of state. For the heat capacity the rigorous thermodynamic equation is

$$c_p - c_p^0 = -T \int_{p^0}^{p} \left(\frac{\partial^2 V}{\partial T^2}\right) dp' \qquad (3.45)$$

Expressions for many equations of state as well as extensive tabulations are given in [3.3], [3.6], [3.23].

3.2.1. Heat Capacity of Ideal Gases

Classical kinetic theory leads to the simple relation $c_p^0 = 1/2\, R$ for each translational degree of freedom, a contribution of $1/2\, R$ for each axis of rotation, and of R for each vibrational degree of freedom. Neglecting vibrational degrees of freedom, a primitive estimate for c_p^0 applicable to small molecules can be obtained from the relation

$$c_p^0 = R(1 + f/2) \qquad (3.46)$$

where f is the sum of translational and rotational degrees of freedom of the molecule.

Example. Consider molecular nitrogen N_2 with five degrees of freedom (three translational, two rotational). At moderate temperatures (ca. 800 K) Equation (3.46) yields $c_p = 29.1$ J mol^{-1} K^{-1} whereas the IUPAC compilation [3.28] for N_2 reports $c_p = 29.2$ J mol^{-1} K^{-1} at 300 K and 0.1 MPa.

This approximation is not of sufficient accuracy for practical engineering purposes over a wide range of temperatures. Hence empirical estimation methods are commonly accepted for the calculation of the heat capacity of ideal gases.

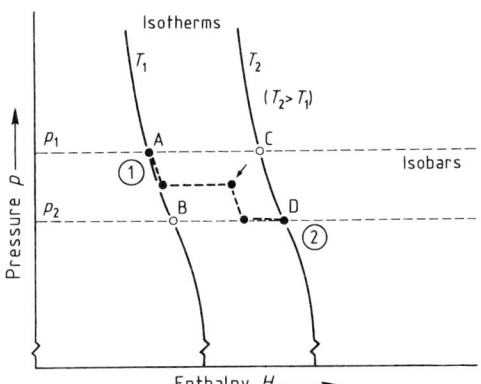

Figure 5. Integration paths for enthalpy calculation between states 1 (T_1, p_1) and 2 (T_2, p_2)
Arrow denotes the reference value referred to in the text

Group Contribution Methods. Joback's method [3.29] is a simple additive group contribution method based on a polynomial in temperature using the same structural groups as in his method for the estimation of critical properties (Section 2.3). For a molecule with r different types of groups, c_p^0 can be calculated by

$$c_p^0 = \left(\sum_{j}^{r} s_j I_a - 37.93\right) + \left(\sum_{j}^{r} s_j I_b + 0.210\right) T$$

$$+ \left(\sum_{j}^{r} s_j I_c - 3.91 \times 10^{-4}\right) T^2$$

$$+ \left(\sum_{j}^{r} s_j I_d + 2.06 \times 10^{-7}\right) T^3 \quad (3.47)$$

where s_j is the number of the group j and the increments I_a, I_b, I_c, and I_d, are component-specific parameters and are tabulated for various groups not only for the ideal-gas heat capacity but also for the enthalpy of formation ΔH_{f298}^0 and the Gibbs enthalpy of formation ΔG_{f298}^0 (see Chap. 5, Table 15). The average error in c_p^0 is usually less than 2% [3.3].

Example. Heat capacity c_p^0 of diisopropyl ether, $(CH_3)_2CHOCH(CH_3)_2$, $M = 102.177$ g/mol

Group	s_j	I_a	I_b	I_c	I_d
$-CH_3$	4	19.5	-8.08×10^{-3}	1.53×10^{-4}	-9.67×10^{-8}
$>CH-$	2	-23.0	0.204	-2.65×10^{-4}	1.20×10^{-7}
$-O-$	1	23.3	-6.32×10^{-2}	1.11×10^{-4}	-5.48×10^{-8}

Substitution into Equation (3.47):
$c_p^0(800 \text{ K}) = 313.1$ J mol^{-1} K^{-1}
Experimental value:
$c_p^0 = 311.5$ J mol^{-1} K^{-1} at 800 K [3.3]
Relative deviation = 0.5%

Benson's method [3.30] yields accurate values not only for c_p^0 but also for ΔH_{f298}^0 and S_{298}^0. Many functional groups have been incorporated into this method. Its strength lies in the fact that the neighborhood of each group is considered. For special classes of molecules and isomers additional corrections may be applied (e.g., for rings with heteroatoms).

Other methods for the estimation of molar heat capacities are listed in Table 8.

Heat Capacities of Ideal Gas Mixtures. The heat capacity of a mixture c_{pm}^0 in the ideal gas state can be calculated using the simple mixing rule

$$c_{pm}^0 = \sum x_i c_{pi}^0 \quad (3.48)$$

3.2.2. Heat Capacity of Pure Liquids

Several approaches have been proposed for determining the heat capacity of liquids, i.e., theoretical methods, corresponding states, Watson's thermodynamic cycles, or group contribution methods. These methods are discussed extensively in [3.3].

A successful group contribution method was proposed by CHUEH and SWANSON [3.36]. It can be used to calculate the heat capacities of liquids at room temperature (293.15 K) for many organic compounds. The average uncertainty is 2–3%.

MISSENARD published an incremental method for the prediction of liquid heat capacities in the temperature range 250–373 K [3.37]. This method is limited, however, to reduced tempera-

Table 8. Selected incremental methods for the calculation of ideal gas heat capacities

Method	Range of application			Error, %
	Substances	T, K	p, kPa	
Dobratz 1941 [3.31]	no triple bonds			
Rihani, Doraiswamy 1965 [3.32]	no triple bonds			2–3
Benson 1968, 1969 [3.30]		280–1100	<300	1
Thinh 1976 [3.33]	hydrocarbons only	200–1500		1–2
Yoneda 1979 [3.34]	hydrocarbons only			<2
Joback 1984 [3.29]				1–2
Harrison, Seaton 1988 [3.35]		280–1500		3

tures $T_r < 0.75$. It employs the equation

$$c_p^l(T) = \sum s_i I_i \quad (3.49)$$

and the number s_i of group i and the increments I_i (values are given in Table 9).

In analogy with the Benson method for the estimation of ideal gas heat capacities (see Section 3.2.1) LURIA and BENSON suggested a method for the prediction of the heat capacity of liquid hydrocarbons [3.38]. Their method employs a polynomial of fifth order in temperature. Effects of neighboring structural groups were considered in defining the increments.

Several estimation methods based upon the principle of corresponding states have been discussed by BONDI [3.39]. Particularly at higher temperatures these methods are superior to the group contribution methods. They require, however, additional information (see also Table 10). BONDI [3.39] proposed a modification of Rowlinson's method:

$$c_p^l - c_p^0 = R\left\{1.45 + \frac{0.45}{1 - T_r} + 0.25\,\omega\left[17.11\right.\right.$$

$$\left.\left. + 25.2\frac{(1 - T_r)^{1/3}}{T_r} + \frac{1.742}{1 - T_r}\right]\right\}$$

(3.50)

where c_p^0 is the heat capacity in the ideal gas state.

Example. Calculation of heat capacity of liquid nitrobenzene, $C_6H_5NO_2$, at 303.2 K using the method of Missenard [3.37]

$M = 123.112$ g/mol

Experimental: $c_p^l = 177.4$ J mol^{-1} K^{-1} [3.17]

Calculated:

	$I_{C_6H_5}$	I_{NO_2}
298 K	117.2	65.7
323 K	123.4	66.9
303 K	118.4	65.9

Substitution into Equation (3.49): $c_p^l = 184.3$ J mol^{-1} K^{-1}
Relative deviation = 3.9%

3.2.3. Heat Capacities of Liquid Mixtures

The heat capacity of liquid mixtures can usually be calculated as the arithmetic mean value of the individual components because deviations from observed values do not exceed 5% if the components are similar in structure and do not

Table 9. Increments of structural groups for the prediction of liquid heat capacities (Missenard's method) [3.37]

Structural group	Temperature T, K					
	248	273	298	323	348	373
$-H$	12.5	13.4	14.6	15.5	16.7	18.8
$-CH_3$	38.5	40.0	41.6	43.5	45.8	48.3
$>CH_2$	27.2	27.6	28.2	29.1	29.9	31.0
$>CH-$	20.9	23.8	24.9	25.7	26.6	28.0
$-\overset{\mid}{\underset{\mid}{C}}-$	8.4	8.4	8.4	8.4	8.4	
$-C\equiv C-$	46.0	46.0	46.0	46.0		
$-O-$	28.9	29.3	29.7	30.1	30.5	31.0
$-CO-$ (ketone)	41.8	42.7	43.5	44.4	45.2	46.0
$-OH$	27.2	33.5	43.9	52.3	61.7	71.1
$-COO-$ (ester)	56.5	57.7	59.0	61.1	63.2	64.9
$-COOH$	71.1	74.1	78.7	83.7	90.0	94.1
$-NH_2$	58.6	58.6	62.8	66.9		
$-NH-$	51.0	51.0	51.0			
$-\overset{\mid}{N}-$	8.4	8.4	8.4			
$-CN$	56.1	56.5	56.9			
$-NO_2$	64.4	64.9	65.7	66.9	68.2	
$-NH-NH-$	79.5	79.5	79.5			
C_6H_5- (phenyl)	108.8	113.0	117.2	123.4	129.7	136.0
$C_{10}H_7-$ (naphthyl)	179.9	184.1	188.3	196.6	205.0	213.0
$-F$	24.3	24.3	25.1	25.9	27.0	28.2
$-Cl$	28.9	29.3	29.7	30.1	30.8	31.4
$-Br$	35.1	35.6	36.0	36.4	37.2	38.1
$-I$	39.3	39.7	40.4	41.0		
$-S-$	37.2	37.7	38.5	39.3		

Table 10. Selected methods for the prediction of heat capacities of liquids

Method	Range of application		Information required	Error, %	Type of method
	Substances	Temperature			
Bondi–Rowlinson 1966 [3.39]	not recommended for polar substances	$0.4 \leq T_r \leq 0.96$	c_p^0, T_c, ω	5	corresponding states
Yuan–Stiel 1970 [3.40]		$0.4 \leq T_r \leq 0.95$	c_p^0, T_c, ω	5–10	corresponding states
Chueh–Swanson 1973 [3.36]		293 K		<3	group contribution
Missenard 1965 [3.37]		$248 < T < 373$ K, but $T_r < 0.75$		5	group contribution
Luria–Benson 1977 [3.38]	hydrocarbons only				group contribution

exhibit hydrogen bonding:

$$c_{pm}^l = \sum x_i c_{p_i}^l \qquad (3.51)$$

A comprehensive study of the heat capacity behavior of liquid mixtures is given in [3.41].

4. Phase Equilibria

4.1. Pure Compounds

4.1.1. Vapor Pressure, Boiling Point, and Melting Point

Vapor Pressure Correlation Functions. Many correlation functions have been published which are related to the Clausius–Clapeyron equation:

$$\frac{dp^{lv}}{dT} = \frac{\Delta H^{lv}}{T(V^v - V^l)} \qquad (4.1)$$

where ΔH^{lv} is the enthalpy of vaporization, and V^l and V^v are the molar volumes of the liquid and vapor phase, respectively. With $\Delta H^{lv} \approx$ const., $V^v = RT/p$, and $V^v \gg V^l$ (see Section 1.3), integration yields

$$\ln \frac{p_1^{lv}}{p_2^{lv}} = \frac{\Delta H^{lv}}{R}\left(\frac{1}{T_2} - \frac{1}{T_1}\right) \qquad (4.2)$$

where R is the gas constant. This relation leads to the simple *August equation* with two adjustable constants

$$\ln p^{lv} = A + \frac{B}{T} \qquad (4.3)$$

Integrating Equation (4.1) using a linear function in temperature for the enthalpy of vaporization gives the *Rankine–Kirchhoff equation*

$$\log p^{lv} = A + \frac{B}{T} + C \log T \qquad (4.4)$$

Figure 6 shows a typical vapor pressure versus temperature curve exhibiting an inflection point. Two-parameter equations such as Equation (4.3) are only able to represent a linear relationship between $\ln p^{lv}$ and $1/T$. Three-parameter equations, such as the frequently used *Antoine equation*,

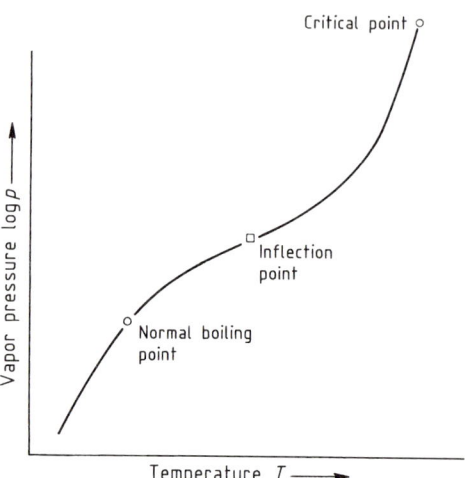

Figure 6. Vapor pressure curve of a pure liquid

$$\log p^{lv} = A - \frac{B}{T+C} \qquad (4.5)$$

can also describe positive or negative curvatures. The description of the vapor pressure of a pure component, within or close to the experimental uncertainty over a wide range of temperature, is possible using the so-called *Frost–Kalkwarf equation*:

$$\ln p^{lv} = A + \frac{B}{T} + C \ln T + \frac{D p^{lv}}{T^2} \qquad (4.6)$$

which is implicit in pressure. This relation, along with a number of other correlation forms (e.g., Wagner, Riedel, and Riedel–Plank–Miller equations), achieves the best performance for practical purposes [4.1] (not considering polynomial functions such as the Chebyshev polynomial series proposed by AMBROSE [4.2]). Experimental vapor pressure data are given in [4.3].

Estimation Methods. Methods for estimating the vapor pressures of pure substances are of low accuracy compared to the quality of measurement. For several homologous series (e.g., straight-chain alkanes) p^{lv} can, however, be predicted with considerable accuracy.

Equation (4.3) provides a rough estimate if p^{lv} data for two temperatures, e.g., the boiling point T_b and the critical point (p_c and T_c), are known.

To provide reasonable predictions of vapor pressure over a wider range of temperature the normal boiling point must be known from experiments. Using critical temperatures and pressures (which can be estimated with the methods discussed in Chap. 2), generalized functions such as the Lee–Kesler method [4.4] or the Gomez–Nieto–Thodos equation [4.5] yield vapor pressures with uncertainties of a few percent. The Gomez method is based upon the following equation:

$$\ln p_r^{lv} = \beta(T_r^{-m} - 1) + \delta(T_r^7 - 1) \quad (4.7)$$

with relations for the parameters β, m, and δ as follows:

1) Nonpolar organic and inorganic substances

$$\beta = -4.267 - \frac{221.79}{h^{2.5} \exp(0.0384 h^{2.5})}$$
$$+ \frac{3.8126}{\exp(2272.44/h^3)} + H \quad (4.8)$$

where $H = 0$ with exceptions for H_2 ($H = 0.19904$), He ($H = 0.41815$), and Ne ($H = 0.02319$).

$$m = 0.78425 \exp(0.089315 h) - 8.5217 \exp(-0.74826 h) \quad (4.9)$$

$$\delta = \frac{-\ln(p_c/1.01325) + (1 - T_{br}^{-m})\beta}{T_{br}^7 - 1} \quad (4.10)$$

h in Equations (4.8) and (4.9) is defined as

$$h = \frac{\ln(p_c/1.01325)}{(1/T_{br}) - 1} \quad (4.11)$$

2) Polar compounds (except alcohols and water)

$$\delta = 0.08594 \exp(7.462 \times 10^{-4} T_c) \quad (4.12)$$

$$m = 0.466 T_c^{0.166} \quad (4.13)$$

$$\beta = \frac{\delta(1 - T_{br}^7) + \ln(p_c/1.01325)}{(1 - T_{br}^{-m})} \quad (4.14)$$

3) Alcohols and water

$$\delta = \frac{2.464}{M \exp(9.8 \times 10^{-6} M T_c)} \quad (4.15)$$

$$m = 0.0052 M^{0.29} T_c^{0.72} \quad (4.16)$$

β from Equation (4.14)

Example. Calculation of vapor pressure of benzene, C_6H_6, with the Gomez method

$T_b = 353.2$ K, $T_c = 562.2$ K, $p_c = 4.89$ MPa [4.6]

Substituting into Equations (4.8)–(4.11):

$h = 6.5513$, $m = 1.3446$, $\delta = 0.1530$, $\beta = -4.2955$

Experimental vapor pressure from [4.7]

T, K	p^{lv}, kPa		Rel. deviation, cal − exp, %
	exp.	cal.	
288	7.784	8.01	3.0
330	46.63	46.92	0.6
400	352.4	351.7	−0.2

Boiling Point and Melting Point. Several methods have been developed for the prediction of the normal boiling point T_b, i.e., the boiling temperature at a pressure of 101.325 kPa. Some are discussed in [4.1]. A convenient method for organic compounds proposed by OGATA and TSUCHIDA [4.8] is based on the simple equation

$$T_b = py + q + 273.15 \quad [K] \quad (4.17)$$

The parameter y is related to the alkyl groups, and p and q are characteristic constants for other functional groups. This method is limited to compounds with one functional group besides the alkyl group. The authors report an accuracy of 2 to 3 K for 91 % of more than 600 substances investigated.

For the estimation of both boiling and melting temperatures of pure compounds JOBACK [4.9] developed an incremental method in which he employs the same structural groups as in his method for critical data estimation:

normal boiling point
$$T_b = 198.0 + \Sigma I_{T_b} \quad [K] \quad (4.18)$$
melting point $T_m = 122.0 + \Sigma I_{T_m} \quad [K] \quad (4.19)$

Increments I_{T_b} and I_{T_m} are given in Table 3.

Boiling points of 438 organic compounds calculated with this method showed an average error of 12.9 K [4.6]. This appears to be rather large, but considering that no experimental information is required, the T_b data are acceptable as approximate values (cf. example below). This is not, however, true for the estimation of melting points. Here a test of Joback's method on 388 organic compounds yielded an average error of 23 K; moreover, deviations of more than 100 K may occur, e.g., for benzene 171 K (calc) and 279 K (exp), for cyclohexane 168.5 K (calc) and 280 K (exp).

The reason for these discrepancies is the large variety of intermolecular structures in the crystalline solid state. In fact, almost all compounds can form an individual crystalline solid state (and often more than one such state) as a separate phase, whereas only a few liquid phases can coexist; in most multicomponent systems only one or two liquid phases are present. This means that specific information is required for describing solid phases whereas more general relations are sufficient for the treatment of liquid or gas phases. Consequently for reliable prediction of the temperature of solid–liquid phase changes specific information for the solid concerned is needed.

Example. Calculation of normal boiling point of ethyl acrylate, $CH_2=CHCOOCH_2CH_3$
Experimental: $T_b = 373.0$ K [4.6]

1) Ogata–Tsuchida method (Eq. 4.17)

	y	p	q
$CH_2=CH-COOR$		0.918	29.0
CH_3CH_2	77.1		

$T_b = 372.9$ K
Deviation $= -0.07$ K

2) Joback's method (Eq. 4.18)

$-CH_3$	$>CH_2$	$=CH-$	$=CH_2$	$-COO-$
23.58	22.88	24.96	18.18	81.10

$T_b = 368.7$ K
Deviation $= -4.3$ K

4.1.2. Enthalpy of Vaporization

Determination from Vapor Pressure Data. As has been shown in Section 1.3 the vaporization enthalpy of a pure liquid can be determined from vapor pressure data with the rigorous equation of Clausius and Clapeyron (Eq. 1.1). Its approximate version (Eq. 1.2) yields

$$\Delta H^{lv} = RT^2 \frac{d \ln p^{lv}}{dT} \quad (4.20)$$

With the Antoine Equation (4.5) for the differential quotient $d \ln p^{lv}/dT$ the following relation is obtained:

$$\Delta H^{lv} = RT \frac{B}{(T+C)^2} \quad (4.21)$$

Equation (4.21) can be readily used to calculate ΔH^{lv} with the aid of tabulated Antoine constants [4.3]. For moderate to high pressures ($p^{lv} >$ 0.1 MPa) the simplifying assumptions leading to Equation (1.2) are not justified. This means that

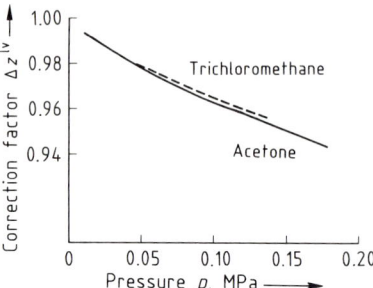

Figure 7. Influence of the Haggenmacher correction on the prediction of the heats of vaporization

the molar volumes of the liquid and the vapor must be taken into account. In Equation (1.4) this has been done only for the vapor phase.

A convenient way to consider both vapor and liquid molar volume is described by HAGGENMACHER [4.10]. He employs the difference in the compressibility factors of the vapor and the liquid:

$$\Delta z^{lv} = z^v - z^l = \frac{p}{RT}(V^v - V^l) \quad (4.22)$$

which can be approximated by the relation

$$\Delta z^{lv} = \left(1 - \frac{p_r^{lv}}{T_r^3}\right)^{1/2} \quad (4.23)$$

Use of Equations (4.22) and (4.23) allows determination of $V^v - V^l$. Introduction of this term into Equation (4.1) improves the calculation of the enthalpy of vaporization except in the region close to the critical point. Correction factors for acetone and trichloromethane are shown in Figure 7.

Other Methods. Vapor pressure data are often unavailable or of insufficient accuracy for the application of the Clausius–Clapeyron equation. Other estimation methods then have to be used for the estimation of the vaporization enthalpies. Two of these are described in Section 1.3 (Eqs. 1.5 and 1.6). VETERE [4.11] suggested the following set of correlations (ΔH^{lv}, J/mol; M, g/mol; T_b, K) [4.1]:

1) Hydrocarbons

$$\Delta H^{lv}_{T_b} = RT_b \left[7.0054 + 1.647 \log M \right.$$
$$\left. + \frac{0.7806[T_b - (263\,M)^{0.581}]^{-1.037}}{M} \right] \quad (4.24)$$

Table 11. Selected methods for the prediction of the enthalpy of vaporization ΔH^{lv} of pure substances

Method	Range of application		Information required*	Error, %
	Substances	Temperature		
Riedel 1954 [4.14]		$T = T_b$	T_b, T_c, p_c	2
Chen 1965 [4.15]		$T = T_b$	T_b, T_c, p_c	2
Vetere 1973 [4.11]		$T = T_b$	$p^{lv}(T), T_c, p_c$	2**
Antoine equation, Haggenmacher 1946 [4.10]		$0.4 \leq T_r \leq 0.6$	$p^{lv}(T), T_c, p_c, \varrho^l$	≈ 1**
Wagner 1973 [4.16]		$0.5 \leq T_r \leq 0.8$ $T^{vls} < T < 0.95\, T_c$	$p^{lv}(T), (T_c, p_c)$	≈ 1**
Kobayashi 1984 [4.17]	nonpolar, weakly polar	$T_r \leq 0.98$	T_c, ω	1.5
Watson 1943 [4.18]	nonpolar, weakly polar	$0.5 \leq T_r \leq 0.75$	$T_c, \Delta H^{lv}_{\text{ref}}$	$0.05(T - T^{\text{ref}})$
Lee–Kesler–Pitzer 1975 [4.4]	nonpolar, weakly polar	$0.5 \leq T_r \leq 0.98$	T_c, p_c, ω	≈ 1.5

* In addition to the corresponding parameters and increments. ** Depends on the accuracy of vapor pressure data.

2) Alcohols, organic acids, methylamine

$$\Delta H^{lv}_{T_b} = R T_b \left(9.7643 + 1.5748 \log T_b - \frac{3.1018\, T_b}{M} \right.$$
$$\left. + \frac{0.0176375\, T_b^2}{M} - \frac{2.57131 \times 10^{-5}\, T_b^3}{M} \right) \quad (4.25)$$

3) Esters

$$\Delta H^{lv}_{T_b} = R T_b \left(5.501 + 1.901 \log T_b + \frac{0.04852\, T_b}{M} \right.$$
$$\left. + \frac{0.53689 \times 10^{-3}\, T_b^2}{M} - \frac{0.6977 \times 10^{-6}\, T_b^3}{M} \right) \quad (4.26)$$

4) Other polar compounds

$$\Delta H^{lv}_{T_b} = R T_b \left(5.3405 + 1.8453 \log T_b + \frac{0.047109\, T_b}{M} \right.$$
$$\left. + \frac{0.52125 \times 10^{-3}\, T_b^2}{M} - \frac{0.67738 \times 10^{-6}\, T_b^3}{M} \right) \quad (4.27)$$

Corresponding states methods (e.g., Pitzer's correlation) should also be mentioned. These methods are discussed in [4.12].

A summary of estimation methods for ΔH^{lv} appears in Table 11. Figure 8 shows a comparison of different estimation methods. Experimental data can be found in [4.13].

Temperature Dependence. The temperature dependence of the vapor pressure can be represented satisfactorily using Watson's equation [4.18]:

$$\Delta H^{lv}_T = \Delta H^{lv}_{T_b} \left[\frac{1 - T_r}{1 - T_{br}} \right]^n \quad (4.28)$$

A value of 0.38 is recommended for the exponent n, but if other values of ΔH^{lv} are available this equation can be used efficiently for the correla-

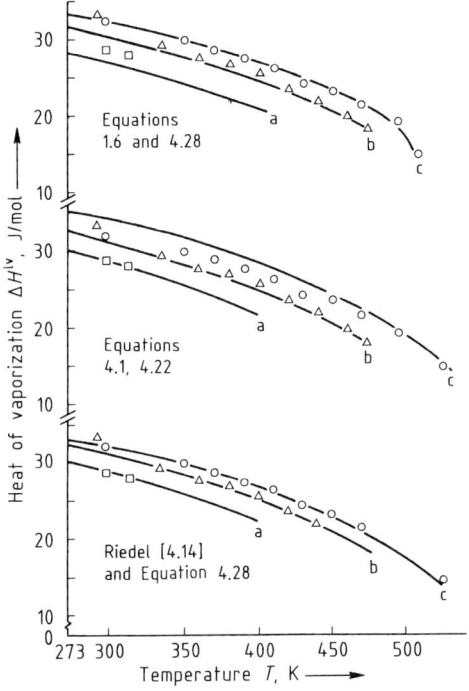

Figure 8. Comparison of estimation methods for the prediction of heat of vaporization of chloromethanes
a) Dichloromethane; b) Trichloromethane; c) Tetrachloromethane
Open symbols denote experimental data, lines denote predicted values

tion of the temperature dependence of ΔH^{lv} by adjusting the exponent n.

4.2. Mixtures

Calculation of phase equilibria is based on the thermodynamic conditions for equilibrium, i.e., equality of temperature T, pressure p, and fugacities f_i of each component i in the coexisting phases. This means that for the mathematical description of phase equilibria, expressions representing fugacities as a function of composition are required. Two types of formulations are used which differ in the state of reference. With the ideal gas as state of reference, the fugacity f_i of component i in a mixture is given by

$$f_i = x_i \varphi_i P \tag{4.29}$$

where x_i is the concentration of i (usually as mole fraction) and φ_i the fugacity coefficient (with $\varphi_i = 1$ for ideal gases).

With an arbitrary state of reference, which is very often the pure component i at its vapor pressure, the following relation is defined:

$$f_i = x_i \gamma_i f_i^0 \tag{4.30}$$

with γ_i as activity coefficient of i in the mixture and f_i^0 as fugacity of i at the state of reference.

4.2.1. Vapor–Liquid Equilibria with Liquid-Phase Activity Coefficients

4.2.1.1. Basic Considerations

For the correlation and prediction of vapor–liquid equilibria at low to moderate pressure (< ca. 1.0 MPa) Equation (4.30) is usually applied to the liquid phase. The vapor phase is either considered to be ideal or a correction for real gas behavior is applied (e.g., by estimated second virial coefficients). The reference state in the liquid phase for each component i in the mixture is the pure liquid i with a vapor pressure p_{0i}^{lv} at temperature T. With this reference state Equation (4.30) for component i in a liquid mixture is given by

$$f_i^l = x_i \gamma_i p_{0i}^{lv} \varphi_{0i} \exp\left(\frac{1}{RT}\int_{p_{0i}}^{p} V_{0i}^l \, dp\right) \tag{4.31}$$

which includes two corrections, i.e., the fugacity coefficient φ_{0i} at pressure p_{0i} accounting for deviation from ideal gas behavior and the exponential term called the Poynting factor which corrects for the difference of fugacity between p_{0i}^{lv} and system pressure p. If the liquid volume V_{0i}^l is assumed to be constant (independent of pressure), the Poynting factor Π_i is simply given by

$$\Pi_i = \exp\left(\frac{V_{0i}^l(p - p_{0i}^{lv})}{RT}\right) \tag{4.32}$$

Equality of fugacities in the vapor and liquid phases yields with Equations (4.29) and (4.31) the following general relation for vapor–liquid equilibria:

$$y_i \varphi_i p = x_i \gamma_i p_{0i}^{lv} \varphi_{0i} \Pi_i \tag{4.33}$$

with y_i and x_i as the mole fractions of i in the vapor and liquid, respectively.

At pressures up to several hundred kilopascal (several bar) the fugacity coefficients φ_i and φ_{0i} and the Poynting factor Π_i are usually not significantly different ($< 2.3 \times 10^{-3}$) from unity. This means that a simplified version of Equation (4.33) can be used for the majority of vapor–liquid equilibria without much loss in accuracy:

$$y_i p = x_i \gamma_i p_{0i}^{lv} \tag{4.34}$$

With Equation (4.34) the vapor phase is considered ideal and all nonideality effects are attributed to the liquid-phase activity coefficient γ_i. For $\gamma_i = 1$, Equation (4.34) transforms into Raoult's law for the partial pressure p_i in ideal mixtures:

$$p_i = y_i p = x_i p_{0i}^{lv} \tag{4.35}$$

The activity coefficients γ_i in liquid mixtures are directly related to an important thermodynamic quantity characterizing nonideal mixing behavior, i.e., the molar excess Gibbs energy of mixing ΔG^E which is defined as the difference in the molar Gibbs energy of mixing between real and ideal mixtures:

$$\Delta G^E = \Delta G - \Delta G^{id} \tag{4.36}$$

This quantity is related to the activity coefficients γ_i of all components i in a mixture by the following equation:

$$\Delta G^E = RT \sum x_i \ln \gamma_i \tag{4.37}$$

Conversely, the activity coefficients γ_i in a mixture can be obtained from ΔG^E when this quantity is known as a function of composition (i.e., of all x_i) according to:

$$RT \ln \gamma_i = \left(\frac{\partial n \cdot \Delta G^E}{\partial n_i}\right)_{T, p, n_j} \tag{4.38}$$

where n is the total number of moles and n_i the number of moles of component i.

4.2.1.2. Data Correlation: Binary and Multicomponent Mixtures of Nonelectrolytes

Numerous correlation methods have been published. They include flexible polynomials (e.g., Margules, von Laar, Redlich–Kister series) and other more sophisticated semitheoretical models, especially those based on the concept of local composition [Wilson, Non-Random Two-Liquid (NRTL), and UNIversal QUAsi-Chemical (UNIQUAC) equations].

Relations for some of these model equations are given in Table 12. If reliable experimental data are available, equations with two parameters per binary system are usually suitable for correlating and extrapolating available data with respect to composition and, to a limited degree, to temperature. Numerous papers have been published comparing different ΔG^E models. But no essential advantages for any of these correlation methods could be found with respect to their ability to fit data.

Table 12. Model equations for calculating the molar excess Gibbs energy of mixing ΔG^E in liquid mixtures

Model	Parameter	Equation
Margules-3 (binary systems)	A B	$\ln \gamma_1 = x_2^2 [A + 2(B - A)x_1]$ $\ln \gamma_2 = x_1^2 [B + 2(A - B)x_2]$
Wilson [4.19]	$\Lambda_{ij}, \Lambda_{ji}$ $(\Lambda_{ii} = \Lambda_{jj} = 1)$	$\dfrac{\Delta G^E}{RT} = -\sum_i^r x_i \ln\left(\sum_j^r x_j \Lambda_{ij}\right)$ $\ln \gamma_i = -\ln\left(\sum_j^r x_j \Lambda_{ij}\right) + 1 - \sum_k^r \dfrac{x_k \Lambda_{ki}}{\sum_j x_j \Lambda_{kj}}$ With $\Lambda_{ij} = \dfrac{V_{0j}^A}{V_{0i}^A} \exp\left(-\dfrac{\lambda_{ij} - \lambda_{ii}}{RT}\right)$
NRTL [4.20]	$\tau_{ij}, \tau_{ji}, \alpha_{ij} = \alpha_{ji}$	$\dfrac{\Delta G^E}{RT} = -\sum_i^r x_i \dfrac{\sum_j^r \tau_{ji} G_{ji} x_j}{\sum_k^r G_{ki} x_k}$ $\ln \gamma_i = \dfrac{\sum_i^r \tau_{ji} G_{ji} x_j}{\sum_k^r G_{ki} x_k} + \sum_j^r \dfrac{x_j G_{ij}}{\sum_k^r G_{kj} x_k}\left(\tau_{ij} - \dfrac{\sum_k^r x_k \tau_{kj} G_{kj}}{\sum_k^r G_{kj} x_k}\right)$ With $\tau_{ij} = \dfrac{g_{ij} - g_{ii}}{RT}$ $G_{ij} = \exp\{-\alpha_{ij}\tau_{ij}\}$
UNIQUAC [4.21]	τ_{ij}, τ_{ji}	$\dfrac{\Delta G^E}{RT} = \sum_i x_i \ln S_i + \dfrac{z}{2}\sum_i q_i x_i \ln \dfrac{A_i}{S_i} - \sum_i q_i x_i \ln \sum_j x_j A_j \tau_{ji}$ $\ln \gamma_i = \ln \gamma_i^C + \ln \gamma_i^R$ $\ln \gamma_i^C = 1 - S_i + \ln S_i - \dfrac{z}{2} q_i \left(1 + \dfrac{S_i}{A_i} + \ln \dfrac{S_i}{A_i}\right)$ $\ln \gamma_i^R = q_i \left(1 - \ln \dfrac{\sum_j q_j x_j \tau_{ji}}{\sum_j q_j x_j} - \sum_j \dfrac{q_j x_j \tau_{ij}}{\sum_k q_k x_k \tau_{kj}}\right)$ with $\tau_{ij} = \exp\dfrac{-\Delta u_{ij}}{T}$ $\Delta u_{ii} = \Delta u_{jj} = 0$ $z = 10$ S_i, A_i, see Equations (4.42) and (4.43)

Another important aspect has to be considered. The majority of published vapor–liquid equilibrium data are concerned with binary systems, only a few with ternary systems, and very few with systems of four and more components. In practical chemical engineering problems, however, systems with at least five or more components often have to be treated. Here the models based on the concept of local composition show the advantage that only binary parameters are required for the representation of multicomponent systems. The local composition concept was introduced by WILSON [4.19]; the essential feature of this idea is that it considers only binary interactions between neighboring molecules in the liquid mixture.

Other models using the local composition concept are the NRTL [4.20] and UNIQUAC [4.21] equations. In contrast to the Wilson equation these models are able to describe the coexistence of two liquid phases, i.e., liquid–liquid phase splitting.

4.2.1.3. Prediction of Equilibrium Data

Solution-of-Groups Concept. Measurement of vapor–liquid equilibria is difficult, time-consuming, and costly. Thus reliable and widely applicable estimation methods are of utmost interest. Numerous attempts to derive methods for the prediction of the properties of nonelectrolyte liquid mixtures using only pure component information (e.g., dipole moment or parachor) have failed. In contrast, methods based on structural groups have turned out to be quite successful. The starting point for these concepts is the assumption that a mixture is composed of structural groups and not of molecules. Consequently interactions between structural groups are considered instead of interactions between molecules. The large variety of molecular mixtures resulting from the huge number of compounds is reduced to a limited number of combinations of structural groups (e.g., $-CH_3$, $>CH_2$, $-OH$, $-CHO$, $>C=O$). This "solution-of-groups" concept was originally introduced to phase equilibrium calculations by REDLICH et al. [4.19] and WILSON and DEAL [4.23]. The first method of practical importance based on this concept was the Analytical Solution Of Groups (ASOG) method of DERR and DEAL [4.24].

ASOG Method. This method employs an athermal Flory–Huggins term to consider size effects and the Wilson equation for the interactions of the structural groups. Therefore the activity coefficient γ_i is split into two parts:

$$\ln \gamma_i = \ln \gamma_i^C + \ln \gamma_i^R \qquad (4.39)$$

where the combinatorial part $\ln \gamma_i^C$ considers size effects and the residual part $\ln \gamma_i^R$ the group interactions. Equation (4.39) is the basic relationship of the solution-of-groups concept. The residual part $\ln \gamma_i^R$ is related to the pure liquid i according to

$$\ln \gamma_i^R = \sum_k v_k^i [\ln \Gamma_k - \ln \Gamma_k^{(i)}] \qquad (4.40)$$

where v_k^i is the number of groups of type k in component i; Γ_k and $\Gamma_k^{(i)}$ are the corresponding group activity coefficients in the mixture and in the pure liquid, respectively.

UNIFAC Method. More recently FREDENSLUND et al. [4.25] derived the UNIversal Functional group Activity Coefficients (UNIFAC) method. This group contribution model applies the UNIQUAC equation [4.21] to the solution-of-groups concept (Eq. 4.39). Since this model is most widely accepted and applied, it is described here in more detail. The combinatorial part contains pure component parameters only and is the Staverman expression taken from the UNIQUAC model:

$$\ln \gamma_i^C = 1 - S_i + \ln S_i + \frac{z}{2} q_i \left(1 - \frac{S_i}{A_i} + \ln \frac{S_i}{A_i}\right) \qquad (4.41)$$

where x_i is the mole fraction. The volume fraction S_i and surface fraction A_i are defined as follows:

$$S_i = \frac{r_i}{\sum_j r_j x_j} \qquad (4.42)$$

$$A_i = \frac{q_i}{\sum_j q_j x_j} \qquad (4.43)$$

where the parameters r_i and q_i represent the relative volume and relative surface area of the component i; z is the coordination number and is usually fixed as $z = 10$. These size parameters have been tabulated for many groups [4.26] or can be calculated [4.27].

The residual part considers group–group interactions via the group activity coefficient Γ_i following the UNIQUAC model for the main groups m and n:

$$\ln \Gamma_k = Q_k \left[1 - \ln\left(\sum \Theta_m \Psi_{mk}\right) - \sum_m \frac{\Theta_m \Psi_{km}}{\sum_n \Theta_n \Psi_{nm}} \right] \quad (4.44)$$

with the area fraction Θ_m of group m defined by

$$\Theta_m = \frac{Q_m X_m}{(\sum Q_n X_n)} \quad (4.45)$$

where X_m is the group mole fraction and Q_m the group surface area. The group energy term Ψ_{nm} for the interaction between the groups m and n is defined by

$$\Psi_{nm} = \exp\left(\frac{-a_{mn}}{T}\right) \quad (4.46)$$

At present, interaction parameters a_{mn} are known for 44 main groups, some of which comprise subgroups (e.g., main group CH_2 with subgroups C, CH, CH_2, and CH_3). Several smaller molecules (e.g., water, methanol) and some molecules with special structures (e.g., pyridine, N-methylpyrrolidone) have been defined as main groups.

The interaction parameters have been determined from reliable, thermodynamically consistent experimental vapor–liquid equilibrium data [4.28]. The present state can be seen in Figure 9.

The predictive ability of the UNIFAC method is quite remarkable. Provided accurate pure-component vapor pressures are known, errors in vapor phase composition are usually ca. 1 mol%. Typical results of UNIFAC predictions are displayed in Figures 10 and 11, which demonstrate the predictive power of the UNIFAC method. In cases where only little or unreli-

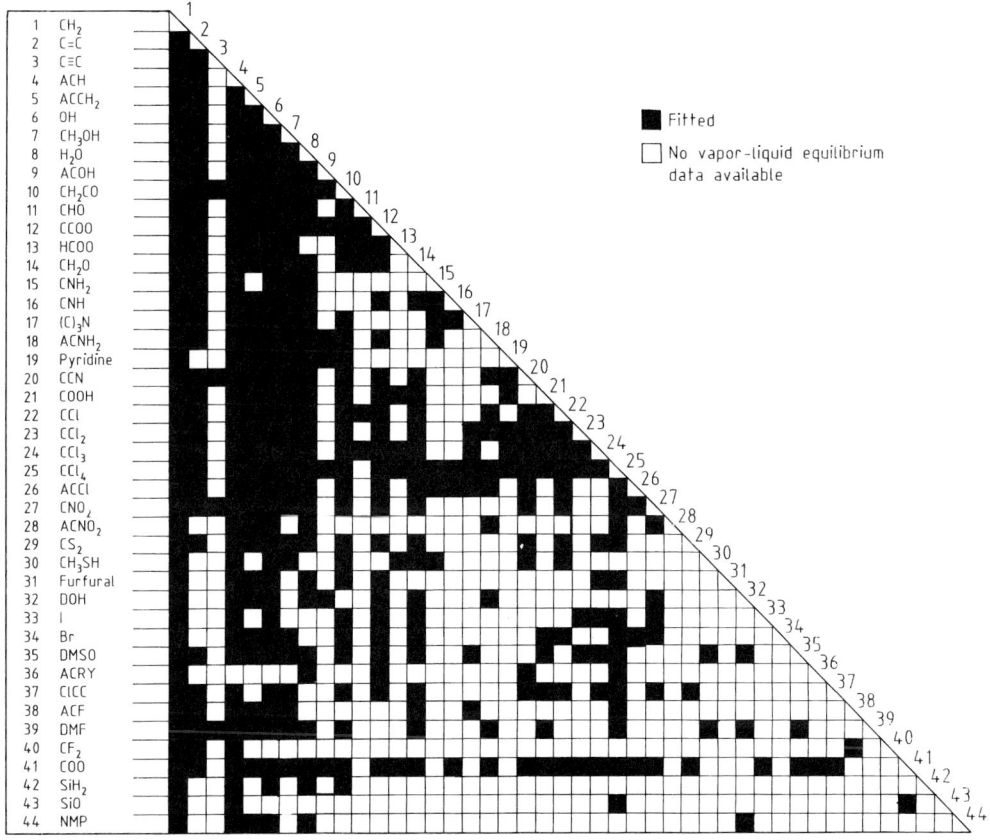

Figure 9. Status (1990) of the UNIFAC parameter matrix [4.28]
Abbreviations: A = aromatic; ACRY = acrylic group; $(C)_3N$ = tertiary amine; ClCC = Cl(C=C); DMF = dimethylformamide; DMSO = dimethyl sulfoxide; DOH = 1,2-ethanediol; NMP = N-methylpyrrolidone

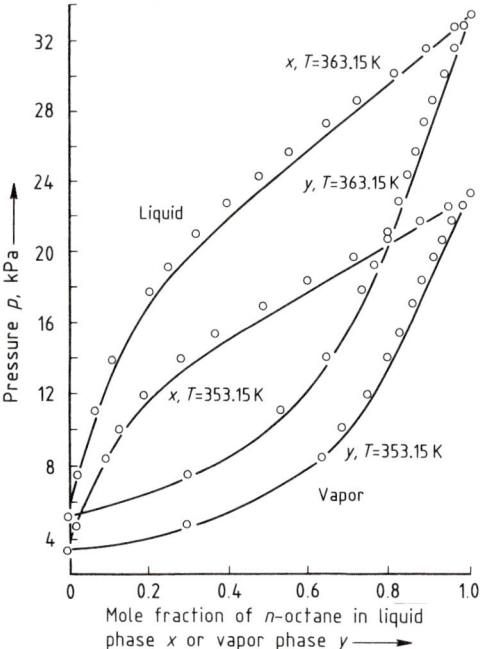

Figure 10. Results of UNIFAC predictions (lines) compared with experimental data (circles, [4.29]) for the system n-octane – benzyl chloride

able primary data are available for the determination of group interaction parameters (e.g., for combinations with aldehydes) serious deviations may be expected.

This original UNIFAC method has additional limitations:

1) Limited temperature dependence leading to unsatisfactory prediction of heat of mixing data
2) In many cases isomers cannot be distinguished from one another
3) Neighboring effects (i.e., interaction between "strong" functional groups within the same molecule) are not considered.

Much effort has been and still is devoted to overcoming these shortcomings, e.g., [4.30] – [4.32]. GMEHLING et al. [4.28], [4.33], [4.34] have included experimental data for both heats of mixing and activity coefficients at infinite dilution into their data base from which they determined temperature-dependent group interaction parameters. They showed that this modified UNIFAC method can be used to predict temperature-dependent activity coefficients at both bulk composition and high dilution and also mixing enthalpies with satisfactory accuracy.

A different approach in applying group contributions for the prediction of activity coefficients was proposed by KEHAIAN et al. [4.35]. Their method is based on Barker's lattice theory; interaction parameters have only been published for a limited number of group combinations.

For practical purposes the classical UNIFAC method with its large interaction parameter matrix provides a useful tool for the prediction of vapor – liquid equilibria of unknown systems and also for the selection of solvents in extractive and azeotropic distillation [4.36]. The current matrix of UNIFAC interaction parameters is given in [4.31].

Extensions of the UNIFAC method for polymer solutions and for gas solubilities. In order to apply the UNIFAC method to polymer solutions OISHI and PRAUSNITZ proposed the inclusion of a free volume term $\ln \gamma_i^{FV}$ in Equation (4.39) [4.37]:

$$\ln \gamma_i = \ln \gamma_i^C + \ln \gamma_i^R + \ln \gamma_i^{FV} \quad (4.47)$$

For gas – liquid equilibria the activity coefficient approach (Eq. 4.30) employing the pure liquid as reference state cannot apparently be applied when the system temperature exceeds the critical temperature of one of the components, as is the case with solubilities of so-called permanent gases (e.g., nitrogen, oxygen, methane, ethylene) in liquids at ambient temperature. A hypothetical liquid state of the gaseous component i may then be defined by extrapolating the standard fugacity f_i^o beyond the critical temperature. Using this hypothetical liquid state NOCON et al. [4.38], [4.39] obtained satisfactory predictions of solubilities of nitrogen, oxygen, carbon dioxide, methane, and ethylene in hydrocarbons and alcohols with a UNIFAC method based on Equation (4.47) (i.e., by including a free volume term).

4.2.2. Vapor – Liquid and Gas – Liquid Equilibria with Equations of State

Using the ideal gas as reference state for the fugacities (Eq. 4.29), the equilibrium condition of equal fugacities in the vapor and liquid phases yields for component i:

$$y_i \varphi_i^v = x_i \varphi_i^l \quad (4.48)$$

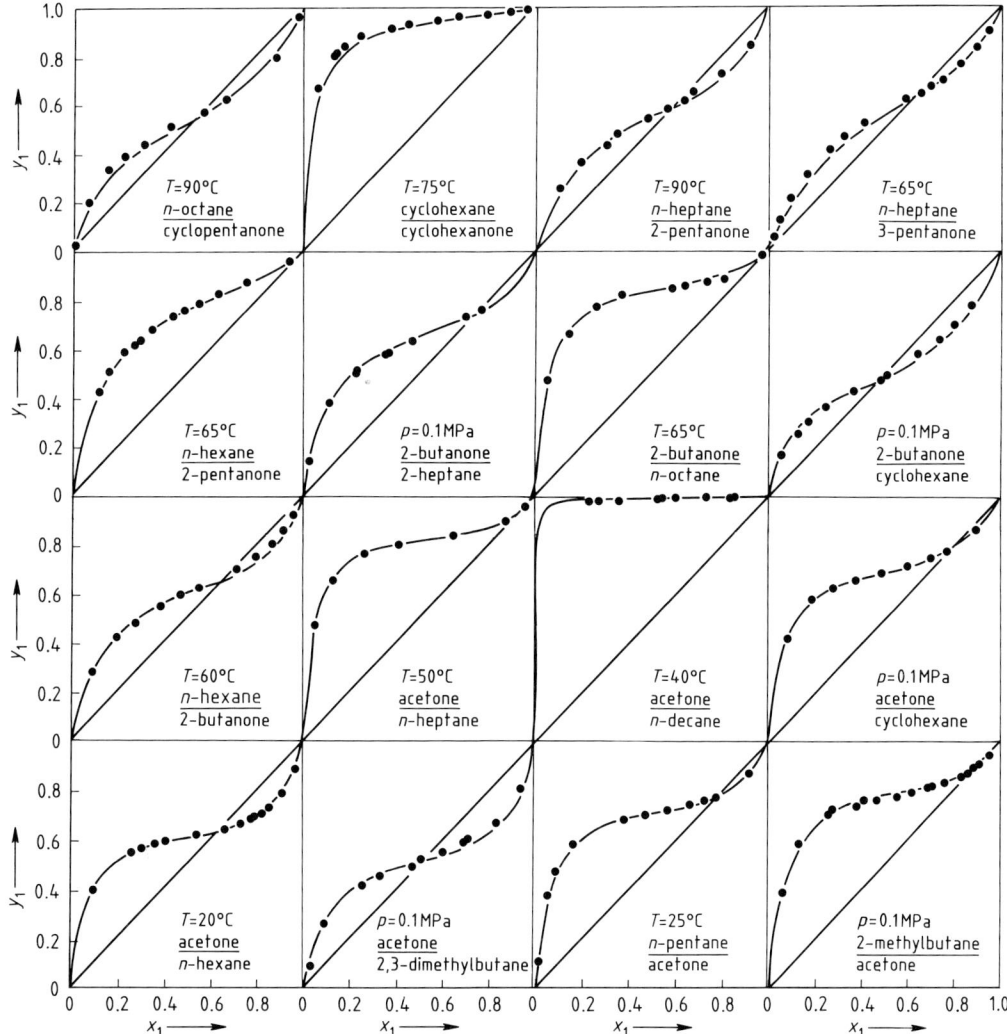

Figure 11. Results of UNIFAC predictions (curves) compared with experimental data (dots) for 16 ketone–alkane mixtures x_1 = mole fraction of underlined component in the liquid phase; y = mole fraction of underlined component in the vapor phase
Only two interaction parameters (a_{CH_2, CH_2CO} = 476.4 K, a_{CH_2CO, CH_2} = 26.76 K) have been used for all diagrams [4.31].

With this relation the same thermodynamic model can be used for both phases. The fugacity coefficient can be calculated from the rigorous relation

$$\ln \varphi_i = \frac{1}{RT} \int_0^p \left[\left(\frac{\partial V}{\partial n_i} \right)_{T, p, n_{j \neq i}} - \frac{RT}{p'} \right] dp' \qquad (4.49)$$

employing an appropriate equation of state. The advantage of this approach is that no standard-state fugacity is necessary. This is particularly useful when the mixture consists of both sub- and supercritical components. Numerous equations of state have been published but accurate prediction of phase equilibria of mixtures is not yet possible from pure-component data only. Hence, binary interaction parameters (see Section 3.1) must be obtained from experimental data for mixtures.

Modern statistical mechanics have achieved remarkable progress in the last two decades [4.40], [4.41]. Nevertheless, computation time is still considerable. Adequate force models are available only for rather simple molecules. As a consequence, the molecular methods have not yet become established for engineering calculations.

Which is the most appropriate, most powerful equation of state for the systems of industrial engineering? An extensive compilation of high-pressure vapor–liquid equilibria and gas–liquid equilibria data is presented by KNAPP et al. [4.42]. The capability of four common equations of state (Lee–Kesler–Plöcker, LKP; Benedict–Webb–Rubin–Starling, BWRS; Redlich–Kwong–Soave, RKS; and Peng–Robinson, PR) in representing experimental data was tested for a large number of binary systems. The calculations of KNAPP et al. show that there is no principal advantage of any of these equations.

The cubic equations (PR or RKS) should be used for calculations of phase equilibria when the components involved are not highly polar or do not differ too much in size. Interaction parameters must be obtained from experimental data in most cases. For systems containing strongly polar compounds, equations of state combined with density-dependent mixing rules or with association models ("chemical hypothesis") must be applied. These concepts can account for nonrandom mixing and ordering effects in the liquid phase at low temperature. For further details see [4.29], [4.46], [4.47].

A promising method for the prediction of high-pressure vapor–liquid equilibria and gas solubilities is offered by the introduction of the group contribution concept to equations of state [4.48]–[4.50].

4.2.3. Liquid–Liquid Equilibria

Phase splitting of a liquid mixture into two liquid phases occurs when a single liquid phase is thermodynamically unstable. In theory thermodynamic stability analysis may therefore be employed for the computation of liquid–liquid equilibria, but this entails high mathematical complexity. The equilibrium condition of equal fugacities for each component in the two phases is usually used instead, although this condition is only necessary, but not sufficient. Using Equation (4.30) for the fugacities f_i' and f_i'' in phases I and II with the same reference state for f_i^0 yields the following relation:

$$x_i' \gamma_i' = x_i'' \gamma_i'' \qquad (4.50)$$

The products $x_i' \gamma_i'$ and $x_i'' \gamma_i''$ are the so-called activities. Since Equation (4.50) holds for each component of a liquid–liquid system, it should be possible to predict liquid–liquid equilibria when the ΔG^E function (Eq. 4.37) and consequently the activity coefficients of the individual components in the multicomponent system are known (e.g., from vapor–liquid equilibria or from prediction methods developed for these phase equilibria, such as UNIFAC). This procedure entails several problems however:

1) The majority of vapor–liquid equilibrium data and the associated activity coefficient data have been determined at other, usually higher temperatures than are of interest for liquid–liquid equilibria.
2) The accuracy of activity coefficients γ_i for modeling and predicting liquid–liquid equilibria has to be considerably higher than for vapor–liquid equilibria, because they are much more sensitive to small errors in γ_i.
3) Systems with phase splitting often show rather high temperature dependence of activity coefficients, which are in general predicted less reliably.

Modifications of the UNIFAC method, either by creating a separate UNIFAC parameter matrix for liquid–liquid equilibrium prediction [4.43] or by including experimental liquid–liquid equilibria data into the determination of interaction parameters [4.28], [4.34], have achieved limited success. Some binary systems could be modeled satisfactorily, but only rough estimations appear to be possible for multicomponent systems.

4.2.4. Solid–Liquid Equilibria

Solid–liquid equilibria may be roughly classified into systems without miscibility of the components in their solid states and systems with formation of mixed crystals. In nonelectrolytes the first type of behavior is much more frequent. For systems of this type the following relation describes the dependence of the liquid mole fraction x_i^l when x_i^l exceeds the eutectic mole fraction at temperature T:

$$\ln (x_i^l \gamma_i^l) = -\frac{\Delta H_i^{sl}}{RT}\left(1 - \frac{T}{T_{m,0i}}\right) \qquad (4.51)$$

where ΔH_i^{sl} is the enthalpy of melting and $T_{m,0i}$ the melting point of pure component i (terms of higher order have been neglected in Eq. 4.51). The liquid-phase coefficient γ_i^l can be estimated with the UNIFAC method [4.44]. At high concentrations of i the equilibrium temperature T (i.e., the freezing point of a liquid solution or melt) can be predicted quite reliably without knowing the value of the liquid-phase activity coefficient γ_i^l, because γ_i^l generally approaches unity at these concentrations. With this assumption, rearrangement of Equation (4.51) (with some simplifications) yields the well-known Raoult equation for the lowering of the freezing point of solutions at low solute concentrations:

$$T_{m,i} - T = \frac{R\,T_{m,0i}}{\Delta H_i^{sl}}(1 - x_i) \qquad (4.52)$$

5. Thermodynamic Data of Chemical Reactions

5.1. Definitions

Thermodynamic data of chemical reactions are highly important in process design for two reasons:

1) Molar reaction enthalpies (heats of reaction) ΔH_r are required in the design of chemical reactors, since the heat effect connected with the reaction must be known to control the reaction temperature.
2) The molar Gibbs energy ΔG_r (also called the free enthalpy) of a chemical reaction is directly related to the reaction equilibrium; knowledge of ΔG_r allows prediction of equilibrium concentrations.

5.1.1. Reaction Enthalpy

Reaction enthalpies can be determined from the enthalpies of formation ΔH_f of the compounds taking part in the reaction. Thus, for the reaction

$$\alpha A + \beta B \rightleftharpoons \gamma C + \delta D \qquad (5.1)$$

the reaction enthalpy ΔH_r is given by

$$\Delta H_r = \gamma \Delta H_{fC} + \delta \Delta H_{fD} - \alpha \Delta H_{fA} - \beta \Delta H_{fB} \qquad (5.2)$$

or by the general form

$$\Delta H_r = \sum v_i \Delta H_{fi} \qquad (5.3)$$

where v_i is the stoichiometric coefficient of compound i (according to the rules of chemical thermodynamics, v_i is positive for products and negative for educts) and ΔH_{fi} is the enthalpy of formation of compound i.

For the calculation of reaction enthalpies the formation enthalpies of many compounds have been tabulated for a standard state of 1.01325 bar = 1 atm and 298.15 K [5.1]–[5.8]. They are designated as ΔH_{f298}^0 (with superscript 0 for standard pressure and subscript 298 for standard temperature, 298 K). Since 1982 0.1 MPa = 1 bar has been recommended as standard pressure, but in chemical thermodynamics and relevant data compilations 1.01325 bar is still in use. Generally the standard formation enthalpy ΔH_{f298}^0 of a compound is given for its most stable state at standard pressure and temperature. For the elements ΔH_{f298}^0 is zero by definition in their most stable state at standard conditions.

The standard reaction enthalpy ΔH_{r298}^0 for Equation (5.1) is thus given by

$$\Delta H_{r298}^0 = \gamma \Delta H_{f298,C}^0 + \delta \Delta H_{f298,D}^0 - \alpha \Delta H_{f298,A}^0 - \beta \Delta H_{f298,B}^0 \qquad (5.4)$$

or in general

$$\Delta H_{r298}^0 = \sum v_i \cdot \Delta H_{f298,i}^0 \qquad (5.5)$$

At temperature T the heat of reaction ΔH_{rT}^0 is calculated from ΔH_{r298}^0 with the aid of the molar heat capacities (ideal gas state) c_{pi}^0 of the reactants:

$$\Delta H_{rT}^0 = \Delta H_{r298}^0 + \int_{298.15}^{T} \Delta c_{pr}\, dT \qquad (5.6)$$

with $\Delta c_{pr} = \sum v_i c_{pi}$. This is in fact a heat balance. Therefore in the case of phase changes between standard-state and reaction conditions (T, p) for one or several reactants the respective phase change enthalpies ΔH^{trs} also have to be included yielding the following general expression:

$$\Delta H_{rT}^0 = \Delta H_{r298}^0 + \int_{298.15}^{T} \Delta c_{pr}\, dT + \sum v_i \Delta H_i^{trs} \qquad (5.7)$$

5.1.2. Gibbs Energy of Reaction and Chemical Equilibrium

Gibbs energies of reaction are calculated analogously to reaction enthalpies. For the reaction of Equation (5.1) the Gibbs energy ΔG_r is equal to the sum of the Gibbs energies of formation ΔG_f of the reactants multiplied by their stoichiometric coefficients:

$$\Delta G_r = \gamma \Delta G_{fC} + \delta \Delta G_{fD} - \alpha \Delta G_{fA} - \beta \Delta G_{fB} \quad (5.8)$$

The general form of Equation (5.8) is

$$\Delta G_r = \sum v_i \Delta \dot{G}_{fi} \quad (5.9)$$

In order to determine the standard Gibbs energy of reaction ΔG^0_{r298} with Equation (5.8) or (5.9) the relevant standard quantities of formation ΔG^0_{f298} have to be used.

Another way of calculating ΔG^0_{r298} uses the definition

$$\Delta G_r = \Delta H_r - T\Delta S_r \quad (5.10)$$

Application of Equation (5.10) requires ΔH^0_{r298} from Equation (5.5) and ΔS^0_{r298} obtained from standard entropies S^0_{298} according to Equation (5.11):

$$\Delta S^0_{r298} = \sum v_i S^0_{298} \quad (5.11)$$

Data for ΔG^0_{f298} and S^0_{298} can be found for many compounds in data collections, e.g., in [5.1], [5.4], [5.6], [5.7].

Conditions for equilibrium at constant pressure and temperature yield the following relation between the Gibbs energy of reaction ΔG^0_{rT} (ideal gas at standard pressure = 1.01325 bar = 101.325 kPa) and equilibrium constant K:

$$\Delta G^0_{rT} = -RT \ln K(T) \quad (5.12\,\text{a})$$

For the chemical reaction of Equation (5.1) K is defined as

$$K = \frac{f_C^\gamma f_D^\delta}{f_A^\alpha f_B^\beta} \quad (5.13)$$

K is a function of temperature only, but not of pressure. The fugacities f_i can be considered as ideal gas partial pressures of the reactants A, B, C, and D at low pressure when real gases do not show much difference from ideal gas phase behavior. The "thermodynamic" (true) equilibrium constant K is then practically equal to the equilibrium constant K_p expressed in partial pressures $p_i = y_i p$ (y_i is the mole fraction of component i in the gas phase) of the reactants as:

$$K_p = \frac{p_C^\gamma p_D^\delta}{p_A^\alpha p_B^\beta} \approx K \quad (5.14)$$

For all those reactions for which values of either ΔG^0_{f298} or ΔH^0_{f298} and S^0_{298} are available for each reactant the value of K at 25 °C (298.15 K) can easily be determined:

$$\ln K_{298.15} = -\frac{\Delta G^0_{r298}}{298.15\,R} \quad (5.12\,\text{b})$$

The dependence of K on temperature is given by the Van't Hoff Equation:

$$\frac{d \ln K}{dT} = \frac{\Delta H^0_{rT}}{RT^2} \quad (5.15)$$

Integration of Equation (5.15) with ΔH^0_{rT} as a function of temperature according to Equation (5.6) yields a general relation for calculating K at temperature T when it is known at temperature T_0 (e.g., 298.15 K):

$$\ln K(T) = \ln K_{T_0} - \frac{\Delta H^0_{rT_0}}{R}\left(\frac{1}{T} - \frac{1}{T_0}\right)$$

$$-\frac{1}{RT}\int_{T_0}^{T}\Delta c_{pr}\,dT + \frac{1}{R}\int_{T_0}^{T}\frac{\Delta c_{pr}}{T}\,dT \quad (5.16)$$

For a first approximation the integrals in Equation (5.16) can be neglected which means that the reaction enthalpy ΔH_r is considered to be independent of temperature. With K known at $T_0 = 298.15$ K this first approximation is

$$\ln K(T) = \ln K_{298.15} + \frac{\Delta H^0_{r298}}{R}\left(\frac{1}{298.15} - \frac{1}{T}\right) \quad (5.17)$$

In a second approximation the sum of the molar heat capacities Δc_{pr} is considered to be constant $[\Delta c_{pr}(T) = \overline{\Delta c_{pr}}]$:

$$\ln K(T) = \ln K_{298.15} + \frac{\Delta H^0_{r298}}{R}\left(\frac{1}{298.15} - \frac{1}{T}\right)$$

$$+\frac{\overline{\Delta c_{pr}}}{R}\left(\ln\frac{T}{298.15} + \frac{298.15}{T} - 1\right) \quad (5.18)$$

For approximate estimations Equations (5.17) and (5.18) are sufficient; accurate predictions, especially at high temperatures, require evaluation of the integrals in Equation (5.16).

Concerning the effect of pressure on equilibrium concentrations the influence of real gas behavior on K_p has to be considered. At low to moderate pressures (up to several hundred kilopascal) K_p is approximately equal to K

Table 13. Effect of pressure on equilibrium of methanol synthesis at 300 °C (CO + 2 H$_2$ ⇌ CH$_3$OH) $K = 2.316 \times 10^{-4}$

p, bar (MPa)	K_φ	K_p
10 (1)	0.96	2.41×10^{-4}
25 (2.5)	0.90	2.57×10^{-4}
50 (5)	0.80	2.90×10^{-4}
100 (10)	0.61	3.80×10^{-4}
200 (20)	0.38	6.09×10^{-4}
300 (30)	0.27	8.58×10^{-4}

(Eq. 5.14). But with higher pressure the difference between K and K_p increases and cannot be neglected at pressures > ca. 1 MPa. This effect can be predicted when the pVT behavior of the reaction mixture is known. Using the relation

$$f_i = \varphi_i y_i p = \varphi_i p_i \quad (5.19)$$

where φ_i is the fugacity coefficient for component i, Equation (5.13) is transformed as follows:

$$K = \frac{p_C^\gamma p_D^\delta}{p_A^\alpha p_B^\beta} \cdot \frac{\varphi_C^\gamma \varphi_D^\delta}{\varphi_A^\alpha \varphi_B^\beta} = K_p K_\varphi \quad (5.20)$$

The parameter K_φ represents the effect of real gas behavior on chemical equilibrium. The magnitude of this effect on high-pressure equilibria can be seen from Table 13, where data for the methanol synthesis equilibrium are given.

5.2. Group Contribution Methods for the Estimation of Enthalpies and Gibbs Energies of Formation

Especially in processes for new products, thermodynamic data (ΔH_f^0, ΔG_f^0) are often not available for all compounds and their estimation is extremely helpful.

Chemical reactions can be considered as processes in which molecular bonds are broken and others are formed. The thermodynamic functions for a chemical reaction can therefore be considered as the sum of the contributions of those chemical bonds which are altered (i.e., broken or formed) by the reaction. In a similar approach the structural groups of the reacting molecules can be used as the basis for estimating thermodynamic reaction quantities, in particular of organic substances.

Various such group contribution methods have been proposed for the estimation of standard enthalpies, Gibbs energies of formation, and standard entropies (Table 14). The method of Joback [5.9] is explained in more detail.

In the Joback method ΔH_{f298}^0, ΔG_{f298}^0, and the molar heat capacities c_p^0 are calculated via Equation (5.21), (5.22), and (3.45):

$$\Delta H_{f298}^0 = 68.29 + \sum s_i I_{Hi} \; [\text{kJ/mol}] \quad (5.21)$$

$$\Delta G_{f298}^0 = 53.88 + \sum s_i I_{Gi} \; [\text{kJ/mol}] \quad (5.22)$$

where s_i is the number of structural groups of type i and I_{Hi} and I_{Gi} are the increments of structural groups i for enthalpy and Gibbs energy of formation, respectively.

The structural groups are the same as in the Joback method for the estimation of critical data (Chap. 2). The values of the group contributions I_{Hi} for ΔH_{f298}^0; I_{Gi} for ΔG_{f298}^0; and I_a, I_b, I_c, and I_d for the coefficients of the polynomial for c_p^0 are given in Table 15.

As is evident from Equations (5.21) and (5.22), standard formation enthalpies ΔH_f^0 and standard Gibbs energy of formation ΔG_f^0 can be obtained directly for 298.15 K only. In order to calculate these quantities and the equilibrium constant K for other temperatures, the molar heat capacities c_p^0 have to be evaluated as a function of temperature with Equation (3.45); ΔH_f^0 and K can be then determined with Equations (5.6) and (5.16), respectively.

Table 14. Group contribution methods for the estimation of ΔH_f^0, ΔG_f^0, and S^0

Authors	Estimated parameters	Form in which contributions are present	Average error, kJ/mol	Secondary sources for increments; remarks
Benson [5.11], [5.12]	ΔH_f^0, S^0	I_H (298), I_S (298); $I_{c(T)}$	10	[5.10]
Franklin [5.14]	ΔH_f^0	$I_H(T)$	15	[5.18]; in kcal/mol
Joback [5.9]	ΔH_f^0, ΔG_f^0	$I_H(298)$; $I_G(298)$; $I_c = f(T)$	15	this article (Table 15); [5.10]
van Krevelen and Chermin [5.16]	ΔG_f^0	$I_G = A + BT$	15	[5.17], [5.18]; in kcal/mol
Verma and Doraiswamy [5.15]	ΔH_f^0	$I_H = A + BT$	15	[5.18]; in kcal/mol
Yoneda [5.13]	ΔH_f^0, S^0	$I_H(298)$, $I_S(298)$; $I_c = f(T)$	10	[5.10]

Table 15. Increments for the calculation of enthalpies and Gibbs energies of formation by Joback's method [5.9]

Type of increment	I_H, kJ/mol	I_G, kJ/mol	Molar heat capacities,* J mol^{-1} K^{-1}			
			Δ_a	Δ_b	Δ_c	Δ_d
Nonring increments						
$-CH_3$	-76.45	-43.96	$1.95 E+1$	$-8.08 E-3$	$1.53 E-4$	$-9.67 E-8$
$>CH_2$	-20.64	8.42	$-9.09 E-1$	$9.50 E-2$	$-5.44 E-5$	$1.19 E-8$
$>CH-$	29.89	58.36	$-2.30 E+1$	$2.04 E-1$	$-2.65 E-4$	$1.20 E-7$
$>C<$	82.23	116.02	$-6.62 E+1$	$4.27 E-1$	$-6.41 E-4$	$3.01 E-7$
$=CH_2$	-9.63	3.77	$2.36 E+1$	$-3.81 E-2$	$1.72 E-4$	$-1.03 E-7$
$=CH-$	37.97	48.53	-8.00	$1.05 E-1$	$-9.63 E-5$	$3.56 E-8$
$=C<$	83.99	92.36	$-2.81 E+1$	$2.08 E-1$	$-3.06 E-4$	$1.46 E-7$
$=C=$	142.14	136.70	$2.74 E+1$	$-5.57 E-2$	$1.01 E-4$	$-5.02 E-8$
$\equiv CH$	79.30	77.71	$2.45 E+1$	$-2.71 E-2$	$1.11 E-4$	$-6.78 E-8$
$\equiv C-$	115.51	109.82	7.87	$2.01 E-2$	$-8.33 E-6$	$1.39 E-9$
Ring increments						
$-CH_2-$	-26.80	-3.68	-6.03	$8.54 E-2$	$-8.00 E-6$	$-1.80 E-8$
$>CH-$	8.67	40.99	$-2.05 E+1$	$1.62 E-1$	$-1.60 E-4$	$6.24 E-8$
$>C<$	79.72	87.88	$-9.09 E+1$	$5.57 E-1$	$-9.00 E-4$	$4.69 E-7$
$=CH-$	2.09	11.30	-2.14	$5.74 E-2$	$-1.64 E-6$	$-1.59 E-8$
$=C<$	46.43	54.05	-8.25	$1.01 E-1$	$-1.42 E-4$	$6.78 E-8$
Halogen increments						
$-F$	-251.92	-247.19	$2.65 E+1$	$-9.13 E-2$	$1.91 E-4$	$-1.03 E-7$
$-Cl$	-71.55	-64.31	$3.33 E+1$	$-9.63 E-2$	$1.87 E-4$	$-9.96 E-8$
$-Br$	-29.48	-38.06	$2.86 E+1$	$-6.49 E-2$	$1.36 E-4$	$-7.45 E-8$
$-I$	21.06	5.74	$3.21 E+1$	$-6.41 E-2$	$1.26 E-4$	$-6.87 E-8$
Oxygen increments						
$-OH$ (alcohol)	-208.04	-189.20	$2.57 E+1$	$-6.91 E-2$	$1.77 E-4$	$-9.88 E-9$
$-OH$ (phenol)	-221.65	-197.37	-2.81	$1.11 E-1$	$-1.16 E-4$	$4.94 E-8$
$-O-$ (nonring)	-132.22	-105.00	$2.55 E+1$	$-6.32 E-2$	$1.11 E-4$	$-5.48 E-8$
$-O-$ (ring)	-138.16	-98.22	$1.22 E+1$	$-1.26 E-2$	$6.03 E-5$	$-3.86 E-8$
$>C=O$ (nonring)	-133.22	-120.50	6.45	$6.70 E-2$	$-3.57 E-5$	$2.86 E-9$
$>C=O$ (ring)	-164.50	-126.27	$3.04 E+1$	$-8.29 E-2$	$2.36 E-4$	$-1.31 E-7$
$O=CH-$ (aldehyde)	-162.03	-143.48	$3.09 E+1$	$-3.36 E-2$	$1.60 E-4$	$-9.88 E-8$
$-COOH$ (acid)	-426.72	-387.87	$2.41 E+1$	$4.27 E-2$	$8.04 E-5$	$-6.87 E-8$
$-COO-$ (ester)	-337.92	-301.95	$2.45 E+1$	$4.02 E-2$	$4.02 E-5$	$-4.52 E-8$
$=O$ (except as above)	-247.61	-250.83	6.82	$1.96 E-2$	$1.27 E-5$	$-1.78 E-8$
Nitrogen increments						
$-NH_2$	-22.02	14.07	$2.69 E+1$	$-4.12 E-2$	$1.64 E-4$	$-9.76 E-8$
$>NH$ (nonring)	53.47	89.39	-1.21	$7.62 E-2$	$-4.86 E-5$	$1.05 E-8$
$>NH$ (ring)	31.65	75.61	$1.18 E+1$	$-2.30 E-2$	$1.07 E-4$	$-6.28 E-8$
$>N-$ (nonring)	123.34	163.16	$-3.11 E+1$	$2.27 E-1$	$-3.20 E-4$	$1.46 E-7$
$-N=$ (nonring)	23.61					
$-N=$ (ring)	55.52	79.93	8.83	$-3.84 E-3$	$4.35 E-5$	$-2.60 E-8$
$=NH$	93.70	119.66	5.69	$-4.12 E-3$	$1.28 E-4$	$-8.88 E-8$
$-CN$	88.43	89.22	$3.65 E+1$	$-7.33 E-2$	$1.84 E-4$	$-1.03 E-7$
$-NO_2$	-66.57	-16.83	$2.59 E+1$	$-3.74 E-3$	$1.29 E-4$	$-8.88 E-8$
Sulfur increments						
$-SH$	-17.33	-22.99	$3.53 E+1$	$-7.58 E-2$	$1.85 E-4$	$-1.03 E-7$
$-S-$ (nonring)	41.87	33.12	$1.96 E+1$	$-5.61 E-3$	$4.02 E-5$	$-2.76 E-8$
$-S-$ (ring)	39.10	27.76	$1.67 E+1$	$4.81 E-3$	$2.77 E-5$	$-2.11 E-8$

* E = exponent of 10: $E+1 = 10^1$, $E-2 = 10^{-2}$ etc.

Deviations of ΔH^0_{f298} and ΔG^0_{f298} calculated with the Joback method from literature data are mostly below 10 kJ/mol; in a few cases, however, differences of more than 20 kJ/mol were found [5.10]. With the methods of Benson [5.11], [5.12] and of Yoneda [5.13] errors are somewhat smaller because they use larger numbers of groups which allow higher differentiation in structure. Application of these methods is, however, rather tedious. In both methods ΔG^0_{f298} is obtained via Equation (5.10), formulated for the reaction of formation for which ΔH^0_{f298} and the standard entropy S^0_{298} are estimated with group contributions. In the same way, molar heat capacities c^0_p are determined when Equations (5.6) and (5.16) are applied for ΔH^0_r and K at other temperatures.

Straightforward estimation of ΔH^0_f and ΔG^0_f at different temperatures is also possible [5.14]–[5.16]. Application of these older methods is convenient because they do not require estimation of heat capacity data. Results from these methods are, however, in kilocalories per mole.

5.3. Applications

5.3.1. Reaction Enthalpy

In the determination of reaction enthalpies with Equation (5.3) the enthalpy of formation is often not known for all reactants. The following procedure can then be adopted:

Example. Calculation of reaction enthalpy for ethoxylation of n-butanol at 390 K

Monoethers of ethylene glycol (e.g., butyl ethyl ether, n-butoxyethanol) are produced by ethoxylation of alcohols, i.e., by addition of ethylene oxide:

$$C_4H_9OH \text{ (l)} + H_2C\overset{O}{-}CH_2 \text{ (g)} \xrightarrow{\text{Cat.}}$$
$$C_4H_9OCH_2CH_2OH \text{ (l)} \quad (5.22)$$

The design of the reactor for the ethoxylation of n-butanol at 390 K is dependent on the enthalpy of this reaction (ΔH_{r390}). Determination of ΔH_{r390} consists of three steps:

1) Calculation of ΔH^0_{r298} from the standard formation enthalpies ΔH^0_{f298} with Equation (5.5)
2) Calculation of ΔH^0_{r390} with Equation (5.6)
3) Calculation of the ΔH_{r390} with n-butanol and n-butoxyethanol as liquids and ethylene oxide in the gaseous state, using Equation (5.7)

Step 1. ΔH^0_{r298} with Equation (5.5).

ΔH^0_{f298} n-butanol -274.9 kJ/mol
ΔH^0_{f298} ethylene oxide -52.67 kJ/mol
ΔH^0_{f298} n-butoxyethanol: no data available
Estimation with Joback's method [5.9]

Group i	s_i	I_{Hi}	$s_i I_{Hi}$
$-CH_3$	1	-76.45	-76.45
$-CH_2-$	5	-20.64	-103.20
$-OH$	1	-208.04	-208.04
$-O-$	1	132.22	-132.22

$\sum s_i I_{Hi} = -519.91$ kJ/mol

Equation (5.21): $\Delta H^0_{f298} = 68.29 - 519.91$
$= -451.62$ kJ/mol

Equation (5.5): $\Delta H^0_{r298} = -451.62 + 274.9 + 52.67$
$= -124.05$ kJ/mol

Step 2. ΔH^0_{r390} with Equation (5.6)

Sum of molar heat capacities:

	v_i	a	b	c	d
n-Butanol	-1	4.543	4.1782×10^{-1}	-2.242×10^{-4}	4.62×10^{-8}
Ethylene oxide	-1	-37.79	3.682×10^{-1}	-3.467×10^{-4}	20.86×10^{-8}
n-Butoxyethanol	1	28.255	5.4462×10^{-1}	-2.220×10^{-4}	1.52×10^{-8}
		61.502	-2.414×10^{-1}	3.489×10^{-4}	-23.96×10^{-8}
		$= \Delta a$	$= \Delta b$	$= \Delta c$	$= \Delta d$

$$\int_{T_0}^{T} \Delta c_{pr} dT = \int_{T_0}^{T} (a + bT + cT^2 + dT^3) dT$$
$$= \Delta a(T - T_0) + \frac{\Delta b}{2}(T^2 - T_0^2)$$
$$+ \frac{\Delta c}{2}(T^3 - T_0^3) + \frac{\Delta d}{4}(T^4 - T_0^4)$$

$$\int_{298.15}^{390} \Delta c_{pr} dT = 924.0 \text{ J mol}^{-1}$$

$\Delta H^0_{r390} = -124.05 + 0.924 = -123.13$ kJ/mol

Step 3. ΔH_{r390} with Equation (5.7)

$\sum v_i \Delta H^{trs}_i = \Delta H^{lv}_{n\text{-butoxyethanol}} - \Delta H^{lv}_{n\text{-butanol}}$
$= 48.517 - 43.781 = 4.736$ kJ/mol

$\Delta H_{r390} = -123.13 + 4.74 = -118.4$ kJ/mol

5.3.2. Chemical Equilibrium

With Equation (5.12a) the equilibrium constant K of a chemical reaction can be determined. The necessary thermodynamic data have been tabulated for many inorganic and organic compounds. However, in using results of such calculations, inaccuracies in the original data must be considered. Thus enthalpies and Gibbs energies of formation are derived from calorimetric measurements: formation enthalpies of some organic compounds are usually determined from measured heats of combustion. With combustion enthalpies of ca. ≥ 1000 kJ/mol, inaccuracies of about ± 1 kJ/mol in the final values of ΔG^0_r and ΔH^0_r have to be expected independent of the absolute value of the resulting quantities. If the Gibbs energy of formation has to be estimated for one or more of the reactants (cf. Section 5.2), the error in ΔG^0_r is ≥ 10 kJ/mol.

In Table 16 an example of the effect of errors in ΔG^0_r on K and on equilibrium concentration is given. Obviously, K values obtained in this way are too inaccurate to be used for exact process calculations without experimental verification.

Table 16. Deviations of equilibrium constant K caused by errors in ΔG_r^0 ($\Delta G_r^0 = 0.0$ kJ/mol at 500 K)

Error in ΔG_r^0, kJ/mol	K
+ 1.0 (max.)	0.786
− 1.0 (min.)	1.272
+ 10.0 (max.)	0.090
− 10.0 (min.)	11.08

For preliminary investigations and for feasibility studies, however, predicted Gibbs energies of reaction with limited accuracy are helpful. For $|\Delta G^0|$ values > 40 kJ/mol, the equilibrium is either highly favorable for the reaction (for negative ΔG_r^0) or the reaction is practically impossible (for positive ΔG_r^0).

With knowledge of ΔG_r^0 it is possible to predict whether a reaction is impossible or thermodynamically feasible, i.e., whether the reaction equilibrium may yield the desired products if equilibrium can be achieved. Equilibrium may not, however, be reached in practice. For example, the velocity of the equilibrium reaction may be so low that no detectable conversion occurs. In other cases consecutive reactions of the equilibrium reaction may be so fast that product recovery becomes practically impossible.

The application of ΔG_r^0 for the selection of processes and process conditions will now be demonstrated for hydrocarbon pyrolysis. Figure 12 shows the temperature dependence of the Gibbs energy of formation for several hydrocarbons. All of the hydrocarbons apart from acetylene show increasing instability with higher temperatures, i.e., increasing decomposition into the elements carbon (as graphite) and hydrogen. The much lower temperature dependence of ethylene and benzene means that these compounds are less stable than paraffinic hydrocarbons (e.g., C_6H_{14} and $C_{20}H_{42}$) at low temperature and more stable at high temperature. Therefore, in pyrolysis of naphtha (C_5-C_{10} paraffins) for the production of ethylene (steam cracking), temperatures of 500–900 °C are employed. Apparently it is pointless to try to develop a catalyst for the synthesis of ethylene from naphtha at lower temperatures (e.g., 300–400 °C). On the other hand naphtha pyrolysis yielding acetylene requires higher temperatures, preferably > 1150 °C; with methane as raw material even higher temperatures are necessary.

Due to increasing instability at higher temperature, reaction time has to be kept short to prevent decomposition; fast cooling of the reaction mixture after leaving the reactor is also necessary.

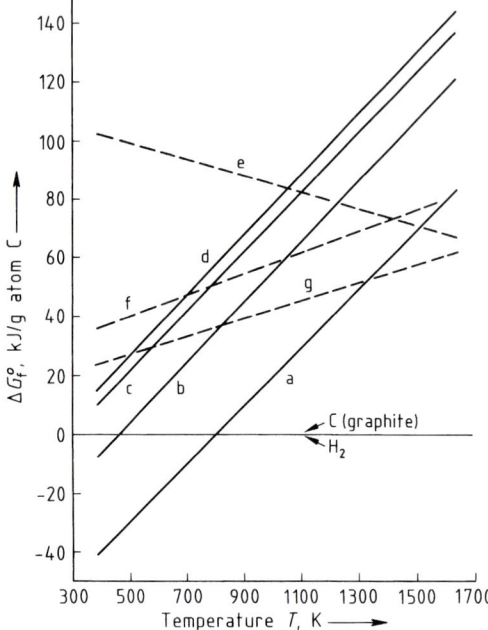

Figure 12. Temperature dependence of the Gibbs energy of formation for selected hydrocarbons
a) Methane; b) Ethane; c) Hexane; d) Eicosane (n-$C_{20}H_{42}$); e) Acetylene; f) Ethylene; g) Benzene

Example. Calculation of equilibrium constant for the reaction of 2-propanol to diisopropyl ether.

2-Propanol is produced by reacting propene with water. In a subsequent reaction diisopropyl ether is formed at ca. 150–250 °C:

$$2\,H_3C-\underset{OH}{CH}-CH_3 \rightleftharpoons \underset{CH_3}{\overset{CH_3}{HC}}-O-\underset{CH_3}{\overset{CH_3}{CH}} + H_2O$$

The loss in 2-propanol due to this reaction might be reduced by altering the reaction temperature. The temperature dependency of the equilibrium constant must therefore be determined. This can be done in three steps:

1) Determination of ΔG_{r298}^0 (with Eq. 5.8) and of the equilibrium constant K_{298} (with Eq. 5.12b)
2) First approximation for K_T at temperature T (Eq. 5.17) with ΔH_{r298}^0 by using Equation (5.5).
3) Second approximation for K_T with Equation (5.18)

Step 1. ΔG_{r298}^0 with Equation (5.8) and K_{298} with Equation (5.12b)

Sum of the standard Gibbs energies and standard enthalpies of formation; data from [5.7]

	v_i	ΔG^0_{f298}, kJ/mol	ΔH^0_{f298}, kJ/mol
2-Propanol	-2	-173.71	-272.77
Diisopropyl ether	1	-121.96	-319.03
Water	1	-228.77	-242.00
		$\Delta G^0_{r298} = -3.31$	$\Delta H^0_{r298} = -15.96$

$$\ln K_{298} = -\frac{-3310}{8.314 \times 298.15} = 1.3353$$

$$K_{298} = 3.801$$

Step 2. First approximation of $\ln K = f(T)$ with Equation (5.17)

$$\ln K(T) = 1.3353 + \frac{-15960}{8.314}\left(\frac{1}{298.15} - \frac{1}{T}\right)$$

T, °C	$\ln K(T)$	$K(T)$
0	1.9246	6.852
25	1.3353	3.801
50	0.8372	2.310
100	0.0412	1.042
150	-0.5667	0.567
200	-1.0461	0.351
250	-1.4338	0.238

Step 3. Higher accuracy for $\ln K = f(T)$ with Equation (5.18)

Example: Reaction temperature $T = 473.15$ K ($= 200$ °C), calculate $\Delta c_{pr} = \sum v_i c_{pi}$ (385.65 K) using Joback's method

	v_i	$c^0_{pi\,385}$, J mol^{-1} K^{-1}
2-Propanol	-2	106.79
Diisopropyl ether	1	191.71
Water	1	34.35
	$\overline{\Delta c_{pr}}$	$= 12.48$

Evaluation of the last term of the sum in Equation (5.18):

$$\frac{\overline{\Delta c_{pr}}}{R}\left(\ln\frac{T}{298.15} + \frac{298.15}{T} - 1\right)$$
$$= \frac{12.48}{8.314}\left(\ln\frac{473.15}{298.15} + \frac{298.15}{473.15} - 1\right) = 0.138$$

$\exp(0.138) = 1.148$; $K(473.15\,K) = 0.403\,K$

Comparison with the result from Step 2 ($K = 0.351$) shows that the contribution of the last term of Equation (5.18) is only minor.

Although formation of diisopropyl ether is less favored at higher temperature (250 °C), the equilibrium concentration of this product is still considerable. Moreover, the velocity of the sequential reaction increases with temperature. Therefore, a decrease in diisopropyl ether formation is not to be expected when the reaction temperature is increased.

In the above calculation tabulated data of ΔG^0_{f298} are used, in the following K_{298} is determined from ΔG^0_{f298} estimated with Joback's method.

Calculation of K_{298} in Step 1 using ΔG^0_{r298} determined with Joback's method

	v_i	ΔG^0_{f298}, kJ/mol
2-Propanol	-2	-164.88
Diisopropyl ether	1	-110.24
Water	1	-228.56
		$\Delta G^0_{r298} = -9.04$

$$\ln K = -\frac{-9040}{8.314 \times 298.15} = 3.6469$$

$$K_{298} = 38.4$$

The value of K obtained with Joback's method is ten times higher than that determined with tabulated ΔG^0_{f298} data.

6. Transport Properties of Pure Compounds and Mixtures

Transport properties largely determine the type and dimensions of equipment in the process industries. They are therefore of great importance in the engineering design of many processes—not just those involved in the chemical industry. Although transport properties are macroscopic nonequilibrium quantities, they may, to a certain extent, be related to static equilibrium quantities. In particular, kinetic theory has been very useful in explaining the mechanisms of heat, mass, and momentum transport in dilute gases. For a detailed description of transport phenomena, see → 4. Transport Phenomena (this volume).

From the ratios of the transport coefficients of the liquid to the gas phase given below it is evident that transport processes in the two phases differ considerably:

Density ratio	ϱ^l/ϱ^g	ca. 10^3
Viscosity ratio	η^l/η^g	10–100
Thermal conductivity ratio	λ^l/λ^g	10–100
Diffusion coefficient ratio	D^l/D^g	ca. 10^{-4}

6.1. Viscosity

The *dynamic viscosity* η of a fluid is defined as the ratio of the shear stress to velocity gradient. The SI unit of dynamic viscosity is Pa·s = N·s/m^2 but poise is also still in use where

$1\,P = 1\,g\,s^{-1}\,cm^{-1} = 10^{-1}\,N \cdot s/m^2$ and
$1\,cP = 1\,mPa \cdot s$

The *kinematic viscosity* v [m²/s] is defined by

$$v = \frac{\eta}{\varrho} \tag{6.1}$$

and is still often expressed in Stokes [cm²/s].

The following discussion is restricted to Newtonian fluids and thus includes most organic liquids and solvents but excludes many polymer solutions, melts, and slurries. Typical viscosity behavior of a pure substance (ethanol) is shown in Figure 13.

6.1.1. Viscosity of Pure Gases

Gases at Low Pressure (< 1 MPa). Kinetic theory yields the following relation for gases consisting of rigid, noninteracting, elastic spherical molecules with a Maxwellian velocity distribution:

$$\eta = C' \frac{M^{1/2}}{\sigma^2} T^{1/2} \tag{6.2}$$

where η is expressed in µPa·s and σ is the collision diameter in 10^{-10} m [Å]; M is the molecular mass in g/mol and C' a constant ($C' = 2.6693$). CHAPMAN and ENSKOG described the effect of intermolecular forces by considering the potential energy of atoms and molecules [6.1]. They found that viscosity can be expressed in the general form

$$\eta = C' \frac{M^{1/2}}{\sigma^2 \Omega_v} T^{1/2} \tag{6.3}$$

where Ω_v, the collision integral, can be calculated from an appropriate potential energy function, such as that of Lennard–Jones (see Section 2.1).

Despite the simplifying assumptions used when deriving Equation (6.3), the accuracy of the viscosity of nonpolar gases is quite acceptable (ca. 1 %).

Information on the potential energy function and the molecular parameters for more complex molecules is, however, scarce. Values for the molecular parameters and solutions for Ω_v for various compounds are tabulated in [6.2], [6.3].

Figure 14 shows viscosity data of some organic gases and vapors. In polar gases dipole–dipole interactions must be taken into account. Consequently the collision integral is calculated employing suitable potential functions, e.g., the Stockmayer potential [6.5]. The accuracy in the estimation of viscosity of polar gases by Equation (6.3) is usually better than 2%.

Corresponding States Methods. Several methods based upon Equation (6.3) and the principle of corresponding states employ the reduced viscosity η_r, defined as:

$$\eta_r = \eta \left(\frac{R T_c}{M^3 p_c^4} \right)^{1/6} \tag{6.4}$$

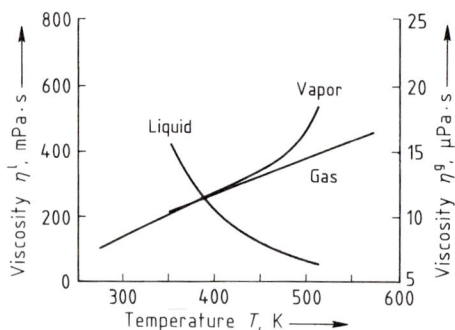

Figure 13. Viscosity behavior of ethanol [6.4]

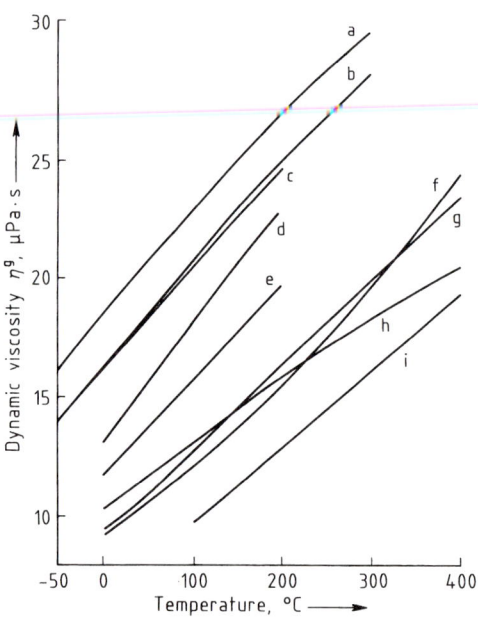

Figure 14. Viscosity data of some gases and vapors [6.4]
a) Helium; b) Carbon dioxide; c) Nitrogen; d) Hydrogen chloride; e) Hydrogen sulfide; f) Water; g) Ammonia; h) Methane; i) Hydrogen cyanide

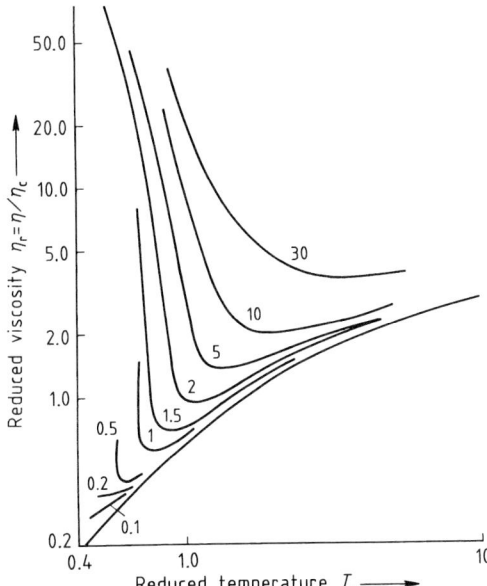

Figure 15. Generalized reduced viscosities as a function of reduced temperature [6.7]
Values on the curve denote the reduced pressure $p_r = p/p_c$.

Figure 15 shows the generalized reduced viscosities as a function of reduced temperature. In the accurate method proposed by LUCAS [6.8] the basic relation for the reduced viscosity is

$$\eta_r \beta = [0.807\, T_r^{0.618} - 0.357 \exp(-0.449\, T_r) \\ + 0.34 \exp(-4.058\, T_r) + 0.018]\, F_P^0 F_Q^0 \quad (6.5)$$

where

$$\beta = 1.76 [10^{-4}\, T_c M^{-3} p_c^{-4}]^{1/6} \quad (6.6)$$

η is in μPa·s and p_c is in MPa. The correction factors F_P^0 and F_Q^0 account for polarity and quantum effects, respectively, and can be obtained from critical data. In terms of the reduced dipole moment μ_r

$$\mu_r = 524.6 \frac{p_c}{T_c^2} \mu^2 \quad (6.7)$$

three different equations have been formulated for F_P^0:

$$F_P^0 = 1 \qquad\qquad 0 \le \mu_r < 0.022 \quad (6.7\mathrm{a})$$

$$F_P^0 = 1 + 30.55 (0.292 - z_c)^{1.72} \quad 0.022 \le \mu_r < 0.075 \quad (6.7\mathrm{b})$$

$$F_P^0 = 1 + 30.55 (0.292 - z_c)^{1.72} [0.96 + 0.1(T_r - 0.79)] \\ 0.075 \le \mu_r \quad (6.7\mathrm{c})$$

The factor F_Q^0 can be set to unity except in the case of quantum gases (He, H_2, D_2). The errors in predicting gas viscosities are usually < 3 % for nonpolar compounds and somewhat higher for polar substances. Table 17 gives an overview on important estimation methods.

Effect of Pressure on Gas Viscosity. Above ca. 1 MPa, pressure has an increasing influence on gas viscosity particularly in the vicinity of the critical point.

LUCAS has suggested an efficient method for the prediction of the viscosity of dense gases with errors usually < 5 % [6.8], [6.9]. The method correlates the reduced viscosity η/η_0 (where η_0 is the low-pressure viscosity) with pressure. The Lucas technique offers the advantage that the viscosity is expressed in terms of temperature and pressure, whereas other methods employ density as a variable and require an additional equation of state. Finally, if no experimental data are available, the low-pressure viscosity η_0 can be estimated with the same model (according to Eq. 6.5).

Example. Calculation of viscosity of gaseous carbon monoxide using the Lucas method

$M = 28.01$ g/mol, $T_c = 132.9$ K, $\mu = 0.1$ Debye, $p_c = 3.5$ MPa [6.5].

Table 17. Selected methods for estimating low-pressure viscosity of gases

Method	Range of application		Information required[a]	Error, %
	Substances	Temperature		
Kinetic theory	polar and nonpolar		$\sigma, \mu, \Omega_v, \varepsilon_0/k$	ca. 2
Chung 1984 [6.14]	not quantum gases		T_c, V_c, μ, σ	2
Bromley, Wilke 1951 [6.15]	polar compounds		T_c, V_c	ca. 5
Thodos 1955 [6.16]	not hydrogen, helium or molecular halogens	$T_r \le 2.5$	T_c, p_c, V_c, μ	ca. 4
Reichenberg 1975 [6.17]		$T_r \le 2$	T_c, p_c, V_c, μ	1^b, 3^c
Lucas 1980 [6.9]	not highly associated gases	$T_r \le 3$	T_c, p_c, z_c, μ	1^b, 3^c

[a] In addition to the corresponding parameters and increments. [b] Nonpolar. [c] Polar.

From Equation (6.6) $\beta = 7.02 \times 10^{-2}$ $(\mu Pa \cdot s)^{-1}$
Since $\mu_r = 1.04 \times 10^{-3}$ (Eq. 6.7) it follows that $F_P^0 = 1$.
Hence using Equation (6.5) $\eta^g = 15.1$ µPa·s
Experimental value $\eta^g = 15.4$ µPa·s at 250 K [6.10]
Relative deviation = -2.0%

6.1.2. Viscosity of Gas Mixtures

Extending the kinetic theory of Chapman and Enskog, viscosities of multicomponent gas mixtures can be approximated by

$$\eta_m = \frac{\sum y_i \eta_i}{\sum y_i \Phi_{ij}} \qquad (6.8)$$

WILKE [6.11] and more recently BROKAW [6.12], [6.13] have published methods for calculating the binary interaction parameter Φ_{ij}. Deviations with respect to experimental data are usually $< 5\%$ [6.2]. LUCAS has developed a calculation scheme that may be used if no pure component viscosity data are available [6.8], [6.9]. These methods are described in more detail, e.g., in [6.6].

6.1.3. Viscosity of Pure Liquids

Many methods have been published for estimating liquid viscosities. These methods can be applied in the temperature range slightly above the triple point up to reduced temperatures of about 0.75–0.8 at moderate pressures. A method for the prediction of viscosities at the melting point is given in [6.18].

No single method is clearly superior to the others. Since the method of Van Velzen et al. [6.19] is well established it is presented here. This group contribution method is based upon the following empirical relation (η in mPa·s):

$$\eta = 10^{\left[B\left(\frac{1}{T} - \frac{1}{T_0}\right)\right]} \qquad (6.9)$$

where T_0 is a fictive temperature and B is a parameter characteristic for the molecular structure of the substance concerned. It is determined using an equivalent chain length N^+

$$N^+ = N + \sum n_i I_N \qquad (6.10)$$

where N is the total number of carbon atoms, I_N the increment, and n_i the number of increment I_N. The parameter B is given by

$$B = B_a + \sum I_B \qquad (6.11)$$

For molecules with $N^+ \leq 20$ and $N^+ > 20$:

$N^+ \leq 20$: $T_0 = 28.86 + 37.439\,N^+ - 1.3547\,(N^+)^2$
$\qquad\qquad + 0.02076\,(N^+)^3$ (6.12a)

$N^+ > 20$: $T_0 = 8.164\,N^+ + 238.59$ (6.12b)

and

$N^+ \leq 20$: $B_a = 24.79 + 66.885\,N^+ - 1.3173\,(N^+)^2$
$\qquad\qquad - 0.00377\,(N^+)^3$ (6.13a)

$N^+ > 20$: $B_a = 530.59 + 13.740\,N^+$ (6.13b)

The structure parameters are given in Table 18. Since the method is only fairly accurate (errors 10–15%) it cannot easily be applied to complex molecules [6.6]. Moreover, the predictions of viscosity for the first members of homologous series often yield higher errors.

Example. Calculation of viscosity of liquid diethylamine $(CH_3CH_2)_2NH$, using the method of Van Velzen et al. [6.19]

With $N = 4$: $\qquad N^+ = 4 + 1.390 + 0.461 \times 4$
$\qquad\qquad\qquad\qquad = 7.234$
For secondary amines: $\qquad I_B = 25.39 + 8.744\,N^+$
$\qquad\qquad\qquad\qquad = 88.64$
From Equation (6.13a): $B_a = 438.27$
From Equation (6.12a): $T_0 = 236.66$ K

At 283.2 K $\eta_{cal}^l = 0.4306$ mPa·s, $\eta_{exp}^l = 0.3878$ mPa·s [6.20], rel. deviation 11.0%
At 310.8 K $\eta_{cal}^l = 0.2944$ mPa·s, $\eta_{exp}^l = 0.2732$ mPa·s [6.20], rel. deviation 7.8%

At higher temperatures ($T_r > 0.75$) the method of Letsou and Stiel [6.25] should be used. Their method is based upon the principle of corresponding states and valid in the range $0.76 < T_r < 0.98$. Only liquids at saturation pressure can be considered. The critical properties as well as the acentric factor must be known.

An overview of selected estimation methods for the viscosity of liquids is given in Table 19. In addition, viscosities of some liquids are presented as a function of temperature in Figure 16.

6.1.4. Viscosity of Liquid Mixtures

The viscosity of a liquid mixture is not usually a linear function of composition and is very difficult to predict. A first estimate of the viscosity of liquid mixtures can be obtained by

$$\ln \eta = \sum x_i \ln \eta_i \qquad (6.14)$$

Methods to date do not allow confident predictions, particularly when dealing with highly po-

Table 18. Parameters of structural groups in the method of Van Velzen et al. [6.19] for estimating viscosities of liquids

Group	I_N	I_B
n-Alkanes	0	0
Isoalkanes	$1.389 - 0.238\,N$	15.51
Saturated hydrocarbons with two methyl groups in iso position	$2.319 - 0.238\,N$	15.51
n-Alkenes	$-0.152 - 0.042\,N$	$-44.94 + 5.410\,N^+$
n-Alkadienes	$-0.304 - 0.084\,N$	$-44.94 + 5.410\,N^+$
Isoalkenes	$1.237 - 0.280\,N$	$-36.01 + 5.410\,N^+$
Isoalkadienes	$1.085 - 0.322\,N$	$-36.01 + 5.410\,N^+$
Hydrocarbon with one double bond and two methyl groups in iso position	$2.626 - 0.518\,N$	$-36.01 + 5.410\,N^+$
Hydrocarbon with two double bonds and two methyl groups in iso position	$2.474 - 0.560\,N$	$-36.01 + 5.410\,N^+$
Cyclopentanes	$0.205 + 0.069\,N$	$-45.96 + 2.224\,N^+$
	$3.971 - 0.172\,N$	$-339.67 + 23.135\,N^+$
	1.48	$-272.85 + 25.041\,N^+$
Cyclohexanes	$6.517 - 0.311\,N$	$-272.85 + 25.041\,N^+$
	0.60	$-140.04 + 13.869\,N^+$
Alkyl benzenes	$3.055 - 0.161\,N$	$-140.04 + 13.869\,N^+$
Polyphenyls	$-5.340 + 0.815\,N$	$-188.40 + 9.558\,N^+$
Alcohols		
Primary	$10.606 - 0.276\,N$	$-589.44 + 70.519\,N^+$
Secondary	$11.200 - 0.605\,N$	497.58
Tertiary	$11.200 - 0.605\,N$	928.83
Diols (correction)		557.77
Phenols (correction)	$16.17 - N$	213.68
−OH on side chain to aromatic ring (correction)	−0.16	213.68
Acids	$6.795 + 0.365\,N$	$-249.12 + 22.449\,N^+$
	10.71	$-249.12 + 22.449\,N^+$
Iso acids		$-249.12 + 22.449\,N^+$
Acids with aromatic nucleus in structure (correction)	4.81	$-188.40 + 9.558\,N^+$
Esters	$4.337 - 0.230\,N$	$-149.13 + 18.695\,N^+$
Esters with aromatic nucleus in structure (correction)	$-1.174 + 0.376\,N$	$-140.04 + 13.869\,N^+$
Ketones	$3.265 - 0.122\,N$	$-117.21 + 15.781\,N^+$
Ketones with aromatic nucleus in structure (correction)	2.70	$-760.65 + 50.478\,N^+$
Ethers	$0.298 + 0.209\,N$	$-9.39 + 2.848\,N^+$
Aromatic ethers	$11.5 - N$	$-140.04 + 13.869\,N^+$
Halogenated compounds:		
Fluoride	1.43	5.75
Chloride	3.21	−17.03
Bromide	4.39	$-101.97 + 5.954\,N^+$
Iodide	5.76	−85.32
Special configurations (corrections):		
$C(Cl)_x$	$1.91 - 1.459x$	−26.38
−CCl−CCl−	0.96	0
$-C(Br)_x-$	0.50	$81.34 - 86.850x$
−CBr−CBr−	1.60	−57.73
CF_3, in alcohols	−3.93	341.68
CF_3, in other compounds	−3.93	25.55
Aldehydes	3.38	$146.45 - 25.11\,N^+$
Aldehydes with an aromatic nucleus in structure (correction)	2.70	$-760.65 + 50.478\,N^+$
Anhydrides	$7.97 - 0.50\,N$	−33.50
Anhydrides with an aromatic nucleus in structure (correction)	2.70	$-760.65 + 50.478\,N^+$
Amides	$13.12 + 1.49\,N$	$524.63 - 20.72\,N^+$
Amides with an aromatic nucleus in structure (correction)	2.70	$-760.65 + 50.478\,N^+$
Amines:		
Primary	$3.581 + 0.325\,N$	$25.39 + 8.744\,N^+$
Primary amine in side chain of aromatic compound (correction)	−0.16	0
Secondary	$1.390 + 0.461\,N$	$25.39 + 8.744\,N^+$
Tertiary	3.27	$25.39 + 8.744\,N^+$
Primary amines with NH_2 group on aromatic nucleus	$15.04 - N$	0
Nitro compounds:		
1-nitro	$7.812 - 0.236\,N$	$-213.14 + 18.330\,N^+$
2-nitro	5.84	$-213.14 + 18.330\,N^+$
3-nitro	5.56	$-338.01 + 25.086\,N^+$
4-nitro; 5-nitro	5.36	$-338.01 + 25.086\,N^+$
Aromatic nitro-compounds	$7.812 - 0.236\,N$	$-213.14 + 18.330\,N^+$

Table 19. Selected methods for estimating the viscosity of pure liquids

Method	Range of application		Information required*	Error, %**
	Substances	Temperature		
Thomas 1946 [6.21]	no strongly polar and naphthenic compounds but sulfur-containing substances	$T \geq T_m + 5$ K $T < T_b$ $T_r \leq 0.75$	T_c, ϱ^l	20
Morris 1964 [6.22]	no sulfur-containing compounds, no highly branched substances	$T \geq T_m + 5$ K $T_r \leq 0.75$	T_c	14
Van Velzen et al. 1972 [6.19]	no sulfur-containing compounds, not for the first members of homologous series	$T_r \leq 0.75$		10–15
Orrick–Erbar 1974 [6.23]	no sulfur- or nitrogen-containing compounds	$T_r \leq 0.75$	T_c, M, ϱ^l	15
Przezdziecki–Sridhar 1985 [6.24]	no sulfur- or nitrogen-containing compounds, no alcohols	$T_r \leq 0.75$	$T_c, p_c, V_c, M, \omega, \varrho^l$	15
Letsou–Stiel 1973 [6.25]	nonpolar, weakly polar compounds at saturation	$0.76 \leq T_r \leq 0.98$	T_c, p_c, M, ω	10

* In addition to the corresponding parameters and increments. ** In some cases errors may be considerably higher.

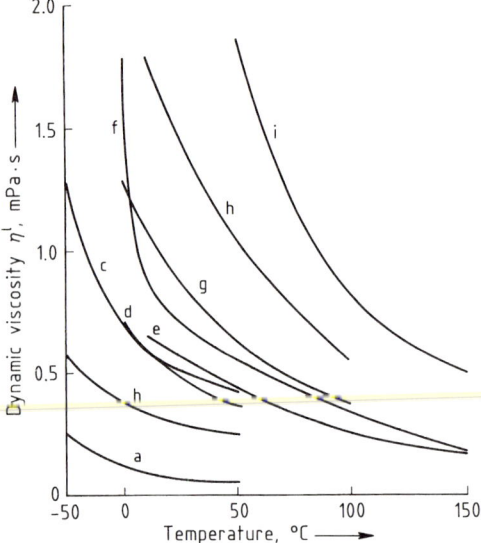

Figure 16. Dynamic viscosity η^l of some liquids [6.4]
a) Hydrogen chloride; b) Acetone; c) Trichloromethane; d) Methanol; e) Benzene; f) Water; g) Decane; h) Formic acid; i) Aniline

lar compounds or components differing greatly in size.

On the basis of Eyring's theory, MCALLISTER published a relation for the kinematic viscosity of binary and ternary mixtures which allows estimations for chemically similar compounds with average errors of 15% [6.26]. For details and recommendations see [6.27].

Viscosity data can be found, e.g., in [6.28]. The pressure–temperature dependence of viscosity for 50 pure fluids together with a comprehensive discussion of theories for dense fluids are given in [6.29].

6.2. Thermal Conductivity

The thermal conductivity λ is defined as the proportionality constant between heat flux and temperature gradient. It is usually expressed in terms of $\mathrm{W\,m^{-1}\,K^{-1}}$.

6.2.1. Thermal Conductivity of Gases at Low Pressure

This chapter discusses methods for the prediction of thermal conductivities at pressures up to 1 MPa. The lower boundary is the region of extremely low pressures ($< 10^{-2}$ kPa) where the mean free path of molecules becomes large compared to the macroscopic dimensions of the apparatus (Knudsen domain). Thermal conductivities of some organic gases and vapors are shown as a function of temperature in Figure 17. In the limiting case of noninteracting monatomic hard spheres, kinetic theory leads to the basic relation

$$\lambda^g = 2.63 \times 10^{-23} \frac{T^{1/2}}{M^{1/2} \sigma^2 \Omega_v} \qquad (6.15)$$

where T is the absolute temperature, M the molecular mass, σ the characteristic dimension of the molecule, and Ω_v the collison integral. Equation (6.15) gives the temperature dependence of λ^g at low pressure.

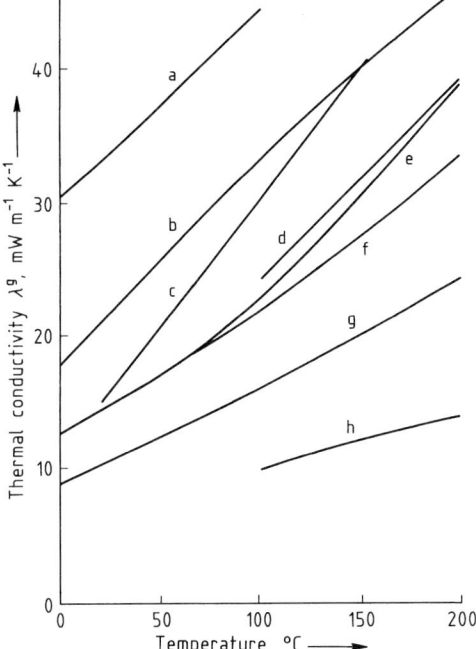

Figure 17. Thermal conductivities λ^g of some organic gases and vapors (data from [6.30])
a) Methane; b) Ethane; c) Dimethylamine; d) Water; e) Ethanol; f) Dimethyl ether; g) Chloromethane; h) Trichloromethane

The thermal conductivity λ and the viscosity η of a gas are closely related. Successful prediction methods (e.g., the Eucken model and the Mason and Moschick approach) have been proposed based on the dimensionless Eucken factor E

$$E = \frac{\lambda M}{\eta c_v} \quad (6.16)$$

where c_v is the molar heat capacity at constant volume. For further details see, e.g., [6.6].

For the estimation of the thermal conductivity of gases, Roy and Thodos [6.31] used an approach based upon the reduced thermal conductivity λ_r

$$\lambda = \frac{\lambda_r}{[210(M^3 T_c/p_c^4)^{1/6}]} \quad (6.17)$$

with the pressure p_c expressed in bar. As proposed by Eucken, the reduced thermal conductivity is separated into two terms

$$\lambda_r = \lambda_r^{\text{trans}} + \lambda_r^{\text{int}} \quad (6.18)$$

The term λ_r^{trans} considers the translational energy of the molecules and is determined by correlating dilute gas properties with temperature. The second term λ_r^{int} takes internal (i.e., rotational and vibrational) energy contributions into account and employs specific polynomial coefficients for different classes of compounds.

$$\lambda_r^{\text{trans}} = 8.757\,[\exp(0.0464\,T_r) - \exp(-0.2412\,T_r)] \quad (6.19)$$

$$\lambda_r^{\text{int}} = \beta\,[A\,T_r + B\,T_r^2 + C\,T_r^3] \quad (6.20)$$

The coefficients A, B, and C are tabulated for nine classes of organic compounds (e.g., paraffins, olefins, alcohols). The additional parameter β is determined by employing group increments I_i

$$\beta = \sum I_i \quad (6.21)$$

The method of Roy and Thodos yields estimates for both polar and nonpolar organic substances at ambient pressure. The constants for this method are given in Tables 20 and 21. The only specific data required are the molecular mass and the critical temperature. The average error is ca. 5 %.

Example. Estimation of the thermal conductivity of 2-butanol, $CH_3CH_2CH(OH)CH_3$, at 480 K by the method of Roy and Thodos

$M = 74.122$ g/mol, $T_b = 372.66$ K, $T_c = 535.95$ K, $p_c = 41.95$ bar [6.20]

$\beta = (-CH_3) + (-CH_2-) + (-CH_2-) + (-CH_2)$
$\quad + (-OH) = 0.73 + 2.0 + 3.18 + 3.68 + 4.62 = 14.21$

Table 20. Coefficients A, B, and C for the Roy and Thodos method for predicting thermal conductivities of liquids (Eq. 6.20) [6.6]

Substance	$A\,T_r + B\,T_r^2 + C\,T_r^3$
Saturated hydrocarbons*	$-0.152\,T_r + 1.191\,T_r^2 - 0.039\,T_r^3$
Olefins	$-0.255\,T_r + 1.065\,T_r^2 + 0.190\,T_r^3$
Acetylenes	$-0.068\,T_r + 1.251\,T_r^2 - 0.183\,T_r^3$
Naphthalenes and aromatics	$-0.354\,T_r + 1.501\,T_r^2 - 0.147\,T_r^3$
Alcohols	$1.000\,T_r^2$
Aldehydes, ketones, ethers, esters	$-0.082\,T_r + 1.045\,T_r^2 + 0.037\,T_r^3$
Amines and nitriles	$0.633\,T_r^2 + 0.367\,T_r^3$
Halides	$-0.107\,T_r + 1.330\,T_r^2 - 0.223\,T_r^3$
Cyclic compounds (e.g., pyridine, thiophene, ethylene oxide, dioxane, piperidine)	$-0.354\,T_r + 1.501\,T_r^2 - 0.147\,T_r^3$

* Not recommended for methane.

Table 21. Parameters for the Roy and Thodos method for predicting thermal conductivities of gases [6.6]

Type of group	ΔI	Type of group	ΔI
Paraffinic hydrocarbons		**Aromatics**	
Base group, methane	0.73	Benzene	13.2
First methyl substitution (1)	2.00	Methyl-substituted benzenes	$13.2 - n\,5.28$ [b]
Second methyl substitution (2)	3.18	$C \approx 5.21 \times 10^{-2}\,M + 1.82 \times 10^{-3}\,M^2$	$M < 120$ [c]
Third methyl substitution (3)	3.68		
Fourth and successive methyl substitutions (4)	4.56	**Ethers**	
Type of substitutions [a]		$-CH_2OH \rightarrow -CH_2-O-CH_3$	2.46
$1 \leftarrow 2 \rightarrow 1$	3.64	**Nitriles** Type of $-CN$ addition	
$1 \leftarrow 2 \rightarrow 2$	4.71	On methane	5.43
$1 \leftarrow 2 \rightarrow 3$	5.79	$CH_3-CH_3 \rightarrow CH_3-CH_2-CN$	7.12
$2 \leftarrow 2 \rightarrow 2$	5.79	$-CH=CH_2 \rightarrow -CH=CH-CN$	6.29
$1 \leftarrow 3 \rightarrow 1$	3.39		
$\downarrow$		**Acids and esters**	
1		$-CH_2-O-CH_3 \rightarrow$	0.75
$1 \leftarrow 3 \rightarrow 1$	4.50	$\quad\quad -CH_2-O-CH=O$	
$\downarrow$		$-CH_2-O-CH_2- \rightarrow$	0.31
2		$\quad\quad -CH_2-O-\overset{O}{\overset{\|}{C}}-$	
$1 \leftarrow 3 \rightarrow 1$	5.61		
$\downarrow$			
3		**Primary amines, type of substitution**	
		On methane	2.60
Olefinic, acetylenic hydrocarbons		$1 \leftarrow 1$	3.91
First double bond $\quad 1 \leftrightarrow 1$	-1.19	$1 \leftarrow 2 \rightarrow 1$	5.08
$\quad\quad 1 \leftrightarrow 2$	-0.65	$2 \leftarrow 2 \rightarrow 1$	7.85
$\quad\quad 2 \leftrightarrow 2$	-0.29	$1 \leftarrow 3 \rightarrow 1$	6.50
Second double bond $\quad 2 \leftrightarrow 1$	-0.17	$\downarrow$	
Any acetylenic bond	-0.83	1	
Halides		**Secondary amines**	
First halogen substitution on methane		$CH_3-NH_2 \rightarrow CH_3-NH-CH_3$	3.31
Fluorine	0.26	$-CH_2-NH_2 \rightarrow -CH_3-NH-CH_3$	4.40
Chlorine	1.38		
Bromine	1.56	**Tertiary amines**	
Iodine	2.70	$CH_3-NH-CH_3 \rightarrow (CH_3)_3 \equiv N$	2.59
Second and successive halogen substitutions on methane		$-\overset{H}{\overset{\|}{CH_2}}-N-CH_2- \rightarrow$	3.27
Fluorine	0.38	$\quad\quad -CH_2-\overset{CH_3}{\overset{\|}{N}}-CH_2-$	
Chlorine	2.05		
Bromine	2.81	$-CH_2-\overset{H}{\overset{\|}{N}}-CH_3-CH_2-N-(CH_3)_2$	2.94
Substitutions on ethane and higher hydrocarbons			
Fluorine	0.58	**Cyclic, ring member**	
Chlorine	2.93	$-CH_2-$	4.25
		$-CH=$	3.50
Aldehydes, ketones		$-NH-$	4.82
$-CH_2-CH_3 \rightarrow -CH_2-CHO$	1.93	$-N=$	3.50
$-CH_2-CH_2-CH_2- \rightarrow$	2.80	$-O-$	3.61
$\quad\quad -CH_2-CO-CH_2-$		$=S=$	7.01
		$C = \Sigma \Delta C - 7.83$	
Alcohols, type of $-OH$ substitution			
On methane	3.79	Type of carbon atom:	
$1 \leftarrow 1$	4.62	[a] $1 =$ primary; $2 =$ secondary;	
$2 \leftarrow 1$	4.11	$\quad 3 =$ tertiary; $4 =$ quaternary.	
$3 \leftarrow 1$	3.55	[b] $n =$ number of methyl groups.	
$4 \leftarrow 1$	3.03	[c] $M =$ molecular mass, g/mol.	
$1 \leftarrow 2 \rightarrow 1$	4.12		
Naphthenes	-1.0		

with $A = 0$, $B = 1$, $C = 0$

From Equation (6.19): $\lambda^{\text{trans}} = 2.073$
From Equation (6.20): $\lambda^{\text{int}} = 11.398$
From Equation (6.21): $\lambda = 0.0316$ W m^{-1} K^{-1}
Experimental: $\lambda = 0.0324$ W m^{-1} K^{-1} [6.32]
Relative deviation $= -0.8\%$

Other more sophisticated methods have been suggested based upon the principle of corresponding states, e.g., [6.33]. These methods employ the Eucken factor and require much more information on the properties of the substances considered, i.e., the critical properties (T_c, p_c, and V_c), the heat capacity at constant volume c_v, and the viscosity.

6.2.2. Thermal Conductivity of Gases at High Pressure

Typical $p \varrho \lambda$ behavior of fluids is displayed in Figure 18. Some methods used for gases at low pressures, particularly the corresponding states methods using the Eucken factor, have been extended to correlate high-pressure thermal conductivity data. For the region close to the critical point, however, other methods must be chosen.

The method of Chung et al. [6.35] is typical and will be briefly described here. These authors extended their equation for low-pressure thermal conductivity data:

$$\lambda^g = \frac{31.2 \, \theta \eta^0}{M} \left(\frac{1}{G_2} + B_6 y \right) + q B_7 y^2 T_r^{1/2} G_2 \quad (6.22)$$

with λ expressed in terms of the reduced density function y (V, V_c in cm^3/mol)

$$y = \frac{V_c}{6V} \quad (6.23)$$

and the parameters

$$q = 3.586 \times 10^{-3} \frac{(T_c/M)^{1/2}}{V_c^{2/3}}$$

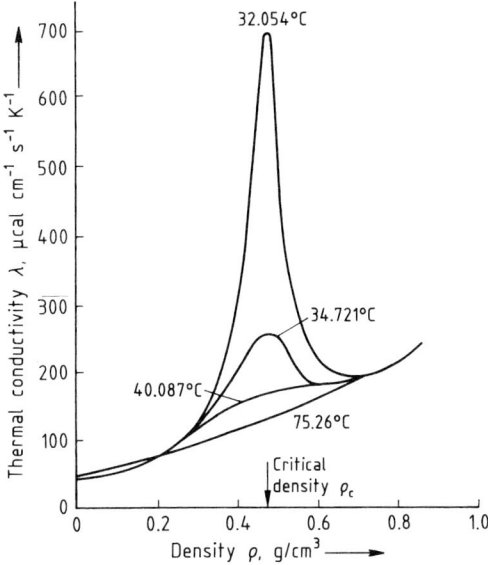

Figure 18. Isotherms of the thermal conductivity of carbon tetrachloride [6.34]

η^0 viscosity at low pressure, Pa · s
G_1, G_2 generalized functions of reduced density
$B_1 - B_7$ parameters which are functions of the acentric factor ω, the reduced dipole moment μ (Eq. 6.31), and the association factor k (Eq. 6.24)
θ function of T_r, c_v/R, and ω (Eq. 6.27)

Additional equations:

$$B_i = a_i + b_i \omega + c_i \mu_r + d_i k \quad (6.24)$$

The parameters $a_i - d_i$ are listed in Table 22.

$$G_1 = \frac{1 - 0.5 y}{(1 - y)^3} \quad (6.25)$$

$$G_2 = \frac{B_1 [1 - \exp(-B_4 y)] + B_2 G_1 \exp(B_5 y) + B_3 G_1}{y(B_1 B_4 + B_2 + B_3)} \quad (6.26)$$

Table 22. Constants (Eq. 6.24) for Chung's method [6.6, p. 522]*

i	a_i	b_i	c_i	d_i
1	2.4166 E + 0	7.4824 E − 1	− 9.1858 E − 1	1.2172 E + 2
2	− 5.0924 E − 1	− 1.5094 E + 0	− 4.9991 E + 1	6.9983 E + 1
3	6.6107 E + 0	5.6207 E + 0	6.4760 E + 1	2.7039 E + 1
4	1.4543 E + 1	− 8.9139 E + 0	− 5.6379 E + 0	7.4344 E + 1
5	7.9274 E − 1	8.2019 E − 1	− 6.9369 E − 1	6.3173 E + 0
6	− 5.8634 E + 0	1.2801 E + 1	9.5893 E + 0	6.5529 E + 1
7	9.1089 E + 1	1.2811 E + 2	− 5.4217 E + 1	5.2381 E + 2

* E = exponent of 10: E + 1 = 10^1, E + 2 = 10^2 etc.

$$\theta = 1 + \alpha \left(\frac{0.215 + 0.28288\alpha - 1.061\beta + 0.26665 Z}{0.6366 + \beta Z + 1.061\alpha\beta} \right) \quad (6.27)$$

$$\alpha = c_v/R - 1.5 \quad (6.28)$$

$$\beta = 0.7862 - 0.7109\,\omega + 1.3168\,\omega^2 \quad (6.29)$$

$$Z = 2.0 + 10.5\, T_r \quad (6.30)$$

$$\mu_r = \frac{131.3\,\mu}{(T_c V_c)^{1/2}} \quad (6.31)$$

Data for the associaton factor k are given, e.g., in [6.6, p. 396].

Since no appropriate correlation for the polar factor is available, the application of the Chung method is limited to nonpolar and a few polar substances. Errors are usually $< 8\%$ [6.6].

Example. Calculation of thermal conductivity of carbon dioxide at 473.15 K and 30 MPa

$M = 44.01$ g/mol, $\omega = 0.239$, $T_c = 304.1$ K, $V_c = 93.9$ cm^3/mol [6.6], $\mu = 0$ Debye, hence from Equation (6.31): $\mu_r = 0$, $V = 112.95$ cm^3/mol, $\varrho = 8.8537 \times 10^{-3}$ mol/cm^3, $c_v = 39.5$ J mol^{-1} K^{-1} [6.36], $\eta^0 = 23.04 \times 10^{-6}$ (at 473.15 K, 0.1 MPa).

Calculation: $G_1 = 1.4560$, $G_2 = 0.6489$, $\alpha = 3.251$, $\beta = 0.692$, $\theta = 1.8687$, $z = 18.34$
$\lambda = 0.04968$ W m^{-1} K^{-1}
Experimental: $\lambda = 0.05181$ W m^{-1} K^{-1} [6.10]
Relative deviation $= -4.1\%$

The reliable estimation of thermal conductivities in dense gases is only possible for nonpolar substances; information about the critical properties, the acentric factor, the constant-volume heat capacity c_v at low pressure, and low-pressure viscosity data are required for the prediction. The accuracy of estimation methods is usually $< 10\%$ [6.6]. An overview of estimation methods is given in Table 23.

6.2.3. Thermal Conductivity of Gas Mixtures

As a first approximation the thermal conductivity of a gas mixture can be described as a linear function of composition. Deviations increase with increasing polarity of the components and increasing difference in size and structure.

Methods for the prediction of mixture data are based upon the theoretically derived Wassiljewa equation:

$$\lambda_m = \frac{\sum y_i \lambda_i}{\sum y_j A_{ij}} \quad (6.32)$$

The binary interaction parameters A_{ij} can be determined employing the methods of Mason–Saxena or Lindsay [6.6]. Errors are usually $< 8\%$ according to [6.6], but greater deviations are possible for components differing greatly in size and polarity. Nevertheless, reasonable estimates of thermal conductivities of mixtures can be obtained from pure component data using a linear combination of λ_i in mole fraction.

6.2.4. Thermal Conductivity of Liquids

Thermal conductivities of liquids are significantly higher than those of gases due to their much higher density. Liquid thermal conductivities λ^l are usually in the range 0.1–0.17 W m^{-1} K^{-1} (Fig. 19) and are, like liquid densities, only a weak function of pressure. Comprehensive compilations of thermal conductivity data are given in [6.40], [6.41]. Since the numerical variation of λ^l values is small, estimation is comparatively easy, except for strongly polar or associating substances. Empirical methods usu-

Table 23. Selected methods for estimating the thermal conductivity of gases at low pressure

Method	Range of application		Information required*	Error, %
	Substances	Temperature and pressure		
Eucken and modifications	polar and nonpolar compounds, no associating gases		M, c_v, η	10–15
Stiel–Thodos 1964 [6.39]	no associating gases	$T_r < 10$ $10^{-4} < p_r < 0.2$	M, T_c, p_c c_v, η	11
Misic–Thodos 1961 [6.37]	hydrocarbons only		M, T_c, p_c, c_p	8
Roy–Thodos 1968, 1970 [6.31], [6.38]	preferably for polar compounds		M, T_c, p_c	8
Ely–Hanley 1983 [6.33]		low- and high-pressure version	$M, T_c, V_c, z_c, \omega, c_v$	5–7
Chung et al. 1980 [6.35]	weakly polar but preferably nonpolar compounds	$T_r < 5$	$M, T_c, \omega, c_v (\eta)$	5–7

* In addition to the corresponding model parameters and increments.

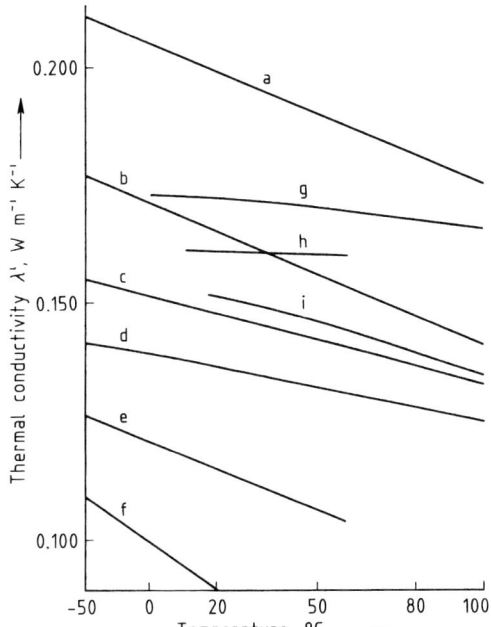

Figure 19. Thermal conductivities λ^l of some organic liquids (data from [6.30])
a) Methanol; b) Ethanol; c) 1-Butanol; d) 1-Octanol; e) Trichloromethane; f) Chlorodifluoromethane; g) Aniline; h) Pyridine; i) Nitrobenzene

ally provide estimates of considerable accuracy, often comparable to experimental uncertainty.

Sato's Method. SATO found that a simple empirical relationship exits between λ^l at the normal boiling point T_b and the molecular mass M [6.42]:

$$\lambda^l(T_b) = \frac{1.11}{M^{1/2}} \quad (6.33)$$

In order to extend this relation the equation [6.43] can be employed to give

$$\lambda^l = B[3 + 20(1 - T_r)^{2/3}] \quad (6.34)$$

where B is a substance-specific parameter [6.48]. Combination of Equations (6.33) and (6.34) yields

$$\lambda^l = \frac{1.11}{\sqrt{M}} \cdot \frac{3 + 20(1 - T_r)^{2/3}}{3 + 20(1 - T_b/T_c)^{2/3}} \quad (6.35)$$

The only information required is the critical temperature and the normal boiling point. Errors are usually < 15%, but are generally higher for strongly polar substances.

Method of Latini and Baroncini [6.44], [6.45]. This estimation method is based upon the relation

$$\lambda^l = \frac{A^*}{M^\beta} \frac{T_b^\alpha}{T_c^\gamma} \frac{(1 - T_r)^{0.38}}{T_r^{1/6}} \quad (6.36)$$

Specific data required for the application of the method are the normal boiling point T_b, the critical temperature T_c, and the molecular mass. For chlorofluorohydrocarbons the method is applicable up to $T_r < 0.9$. The exponents α, β, γ and the parameter A^* are given in Table 24 for various organic compounds. According to [6.6] the errors in estimating liquid thermal conductivity are usually < 10%.

Both estimation methods can be employed reliably up to $T_r < 0.65$. Numerous other methods have been published, but are either of limited applicability or require much more specific information without offering significant improvement in reliability and accuracy.

An overview of methods for estimating the thermal conductivities of liquids is given in Table 25.

Example. Estimation of thermal conductivity of liquid isopropylbenzene, C_9H_{12}, at 323 K. $M = 120.194$ kg/kmol, $T_b = 425.56$ K, $T_c = 631.13$ K [6.20]

Experimental: $\lambda^l = 0.1187$ W m^{-1} K^{-1} [6.45]
Calculation:
Sato's method (Eq. 6.35): $\lambda^l = 0.125$ W m^{-1} K^{-1}
Relative deviation = 5.1%
Latini's method (Eq. 6.36): $\lambda^l = 0.119$ W m^{-1} K^{-1}
Relative deviation = 0.2%

Effect of Temperature on Thermal Conductivity. At low to moderate pressures (< 1 MPa) and over a limited range of temperature, the thermal

Table 24. Constants for the Latini–Baroncini method (from [6.6])

Compounds	A^*	α	β	γ
Saturated hydrocarbons	0.0035	1.2	0.5	0.167
Olefins	0.0361	1.2	1.0	0.167
Cycloparaffins	0.0310	1.2	1.0	0.167
Aromatics	0.0346	1.2	1.0	0.167
Alcohols	0.00339	1.2	0.5	0.167
Organic acids	0.00319	1.2	0.5	0.167
Ketones	0.00383	1.2	0.5	0.167
Esters	0.0415	1.2	1.0	0.167
Ethers	0.0385	1.2	1.0	0.167
Chlorofluorohydrocarbons				
R 20, R 21, R 23	0.562	0	0.5	−0.167
Others	0.494	0	0.5	−0.167

Table 25. Selected methods for estimating the thermal conductivities of liquids

Method	Range of application		Information required	Error, %
	Substances	Temperature		
Robbins–Kingrea 1962 [6.46]	no nitrogen- or sulfur-containing compounds	$0.4 \leq T_r \leq 0.9$	$T_b, T_c, c_p, \varrho^l, \Delta H_b^{lv}$	5
Missenard 1965 [6.47]			$M, T_b, c_p^l, \varrho_b^l$	8
Sato–Riedel 1973 [6.48]	no low molecular mass hydrocarbons, no highly polar compounds	$0.5 \leq T_r \leq 0.75$	M, T_b, T_c	15
Baroncini–Latini 1978, 1984 [6.44], [6.45]	no nitrogen- or sulfur-containing compounds, no aldehydes $50 \leq M \leq 250$		M, T_b, T_c	8
Negvekar–Daubert 1987 [6.49]	no sulfur-containing compounds	$0.3 \leq T_r \leq 0.8$	T_c	6

conductivity decreases linearly with temperature

$$\lambda^l = A - BT \qquad (6.37)$$

For more accurate description over wider temperature ranges, higher order polynomial series are recommended, preferably with $(1 - T_r)$ as an independent variable.

Effect of Pressure on Thermal Conductivity. The thermal conductivity increases significantly at pressures above $p_r > 0.5$. MISSENARD suggested an empirical approach for the dimensionless thermal conductivity normalized by the thermal conductivity at low pressure (p^0) [6.50]:

$$\frac{\lambda^1(p_r)}{\lambda^1(p^0)} = 1 + C p_r^{0.7} \qquad (6.38)$$

Values of the generalized parameter $C = f(T_r, p_r)$ are reported by the author for several compounds.

6.2.5. Thermal Conductivity of Liquid Mixtures

The thermal conductivity of organic liquid mixtures λ_m can usually be approximated with sufficient accuracy by linear combination of the thermal conductivities of the pure components.

For binary systems JAMIESON et al. recommended the following relation in terms of the weight fraction w_i [6.41]:

$$\lambda_m = w_1 \lambda_1 + w_2 \lambda_2$$
$$- \alpha(\lambda_2 - \lambda_1)(1 - w_2^{1/2}) w_2 \qquad (6.39)$$

which employs an adjustable binary parameter α. If no experimental mixture data are available, α is set to unity.

For multicomponent mixtures LI suggested an equation with volume fractions [6.51]:

$$\lambda_m = 2 \sum_i^r \sum_j^r \Phi_i \Phi_j \frac{1}{(1/\lambda_i + 1/\lambda_j)} \qquad (6.40)$$

where the volume fraction Φ_i of component i is defined by

$$\Phi_i = \frac{x_i V_i}{(\sum x_j V_j)} \qquad (6.41)$$

6.3. Diffusion Coefficients

The diffusion coefficient D is defined as the proportionality factor between mass flux and the concentration gradient; it is usually given in [cm²/s]. In this discussion only molecular diffusion is considered with concentration gradients (i.e., primarily gradients in chemical potential) as driving force in systems which are free of external force fields and are homogeneous with respect to temperature and pressure. Diffusion is treated comprehensively in [6.5], [6.52].

6.3.1. Diffusion Coefficients of Gases at Low and Moderate Pressures

Molecular Theory. Statistical mechanics lead, via integration of the Boltzmann equation, to the Chapman and Enskog equation for the mutual diffusion of the components i and j:

$$D_{ij} = 3/16(2\pi k T)^{1/2} \frac{(1/M_i + 1/M_j)^{1/2}}{\pi N \sigma_{ij}^2 \Omega_D} f_D \qquad (6.42)$$

where k is the Boltzmann constant; M_i, M_j the molecular masses of components i, j respectively; N the number of molecules per unit volume; σ_{ij} a characteristic length; and f_D a correction term. Introducing the ideal gas law into this equation gives a basic relation that is valid for spherical molecules in dilute gases:

$$D_{ij} = 0.01333 \frac{T^{1.5}[1/M_i + 1/M_j]^{0.5}}{p} \frac{1}{\sigma_{ij}^2 \Omega_D} \quad (6.43)$$

This expression (p in MPa) contains the characteristic length σ_{ij} (in 10^{-10} m) and the dimensionless collison integral Ω_D, which can be determined from an appropriate intermolecular potential. For monatomic gases the potential function of Lennard and Jones may be chosen (values of σ_{ij} and Ω_D for the Lennard–Jones potential are given for some components, e.g., in [6.6]).

Employing simple combination rules, i.e.,

$$\sigma_{ij} = \frac{\sigma_i + \sigma_j}{2} \quad \text{and} \quad \varepsilon_{ij} = (\varepsilon_i \varepsilon_j)^{1/2} \quad (6.44)$$

diffusion coefficients can be determined with reasonable accuracy without considering concentration effects, even for complicated molecules. For mixtures containing highly polar compounds more sophisticated intermolecular potential functions must be applied. Diffusion coefficients of some components in air are shown as a function of temperature in Figure 20.

Method of Fuller et al. FULLER and coworkers suggested an incremental method for D_{ij} [cm²/s] based on Equation (6.43) [6.52], [6.53]:

$$D_{ij} = \frac{1.013[1/M_i + 1/M_j]^{1/2}}{10^4 p} \frac{T^{1.75}}{[(\Sigma I_D)_i^{1/3} + (\Sigma I_D)_j^{1/3}]^2} \quad (6.45)$$

with p in MPa where $(\Sigma I_D)_i$ and $(\Sigma I_D)_j$ are the diffusion volumes of components i and j, respectively which are given by FULLER et al. for simple molecules such as the noble gases, nitrogen, and oxygen [6.53]. Some values for diffusion volumes (ΣI_D) follow:

He	2.67	CO	18.0
Ne	5.98	CO_2	26.9
Ar	16.2	N_2O	35.9
Kr	24.5	NH_3	20.7
Xe	32.7	H_2O	13.1
H_2	6.12	SF_6	71.3
D_2	6.84	Cl_2	38.4
N_2	18.5	Br_2	69.0
O_2	16.3	SO_2	41.8
Air	19.7		

For other molecules (ΣI_D) is calculated by summing up the increments I_D for the elements and structural groups in the molecule. Values for these increments follow [6.53]:

C	15.9	F	14.7
H	2.31	Cl	21.0
O	6.11	Br	21.9
N	4.54	I	29.8
Ring, aromatic	−18.3	S	22.9
Ring, heterocyclic	−18.3		

The authors claim an average accuracy in the prediction of diffusion coefficients of ca. 4%.

Example. Calculation of D_{12} of ethylene–water using Fuller's method

1) C_2H_4: $M = 28.05$ g/mol; $(\Sigma I_D)_{C_2H_4} = 2C + 4H$
   $= 2 \times 15.9 + 4 \times 2.31 = 41.04$
2) H_2O: $M = 18.02$ g/mol; $(\Sigma I_D)_{H_2O} = 2H + 1O$
   $= 2 \times 2.31 + 1 \times 6.11 = 10.73$

From Equation (6.45): $D_{12} = 0.242$ cm²/s
(at 0.1 MPa, 328.4 K)

Experimental: $D_{12} = 0.236$ cm²/s
(at 328.4 K) [6.6]

Relative deviation = 2.7%

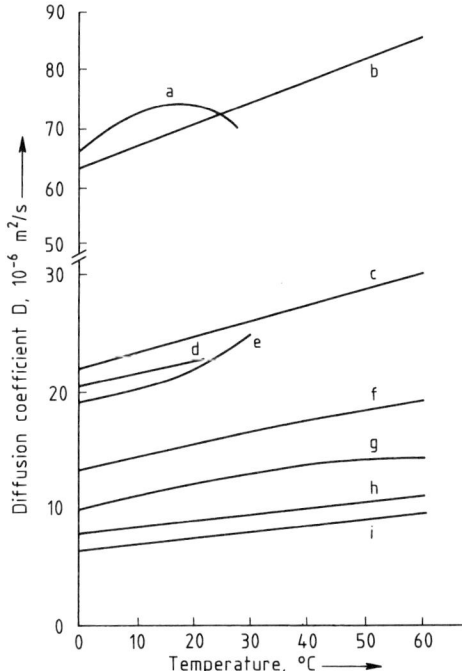

Figure 20. Diffusion coefficients of selected gases in air at ambient pressure
a) Hydrogen; b) Helium; c) Water; d) Ammonia; e) Methane; f) Methanol; g) Ethanol; h) Benzene; i) Hexane

Temperature Dependence of Binary Diffusion Coefficients. Molecular theory (Eq. 6.42) predicts the following form for the temperature dependence of binary diffusion coefficients:

$$D \sim T^{1.5} \tag{6.46}$$

[neglecting $\Omega_D(T)$]. For real systems the exponent is often about 1.8.

Effect of Pressure on Binary Diffusion Coefficients of Gases. In general, diffusion coefficients decrease with increasing pressure according to Equation (6.43). At higher pressures ($p \gg p_c$) the influence of pressure becomes more complex [6.54], [6.55] and corresponding states methods must be employed.

Effect of Concentration on Diffusion Coefficients of Gases. Binary diffusion coefficients can usually be assumed to be independent of composition. A discussion of the theory of diffusion in multicomponent gas mixtures is given, e.g., in [6.5], [6.52], [6.56]. A procedure for predicting multicomponent diffusion phenomena is given in [6.57].

6.3.2. Diffusion Coefficients of Liquids

The Stokes–Einstein equation provides the basis for many empirical correlations of diffusion coefficients in liquids. This relation describes the movement of a large hypothetical spherical particle through a solvent consisting of infinitely small molecules. With these assumptions binary diffusion coefficients are given as a function of solvent viscosity η_j and the "molecular" radius of the solute r_i:

$$D_{ij} = \frac{kT}{6\pi \eta_j r_i} \tag{6.47}$$

The effect of solvent viscosity on binary diffusion coefficients is shown in Figure 21.

Since Equation (6.47) only holds for simple systems, a variety of empirical correlations have been published. They are usually applicable to solvent–solute systems with low to moderate viscosity ($< 20 – 40$ mPa · s).

Diffusion Coefficients at Infinite Dilution. The diffusion coefficient D_{ij}^∞ for solute i at infinite dilution in solvent j is a useful reference value for the determination of binary diffusion coefficients of liquid mixtures.

The estimation technique of Hayduk and Minhas comprises several equations for different types of solutions [6.59]:

1) Nonelectrolyte systems:

$$D_{ij}^\infty = \frac{P_j^{0.5}}{P_i^{0.42}} \frac{1.55 \times 10^{-8} T^{1.29}}{V_{oj}^{0.23} \eta_j^{0.92}} \tag{6.48}$$

where V_{oj} [cm³/mol] is the liquid molar volume of solvent j at its normal boiling point, η_j its viscosity [mPa · s], and P the parachor (see Section 2.4) [cm³ g$^{0.25}$ s$^{-0.5}$ mol^{-1}]. Using the definition of the parachor as $P \equiv \sigma_i^{0.25} V_{oi}$ (Eq. 2.22 where σ_i is the surface tension of component i), Equation (6.48) can be rewritten as

$$D_{ij}^\infty = 1.55 \times 10^{-8} \frac{V_{oj}^{0.27} T^{1.29} \sigma_j^{0.125}}{V_{oi}^{0.42} \eta_j^{0.92} \sigma_i^{0.105}} \tag{6.49}$$

In many cases the surface tensions σ_i and σ_j are similar and their ratio can be set to unity, making this expression very convenient.

2) n-Paraffin solutions:

$$D_{ij}^\infty = 1.33 \times 10^{-7} \frac{T^{1.47}}{V_{oi}^{0.71} \eta_j^c} \tag{6.50}$$

where the exponent $c = (10.2/V_{oi}) - 0.791$. According to the authors errors are $< 4\%$.

3) Organic solutes in water:

$$D_{ij}^\infty = 1.25 \times 10^{-8}(V_{oi}^{-0.19} - 0.292)T^{1.52} \eta_j^\delta \tag{6.51}$$

where the parameter $\delta = (9.58/V_{oi}^l) - 1.12$.

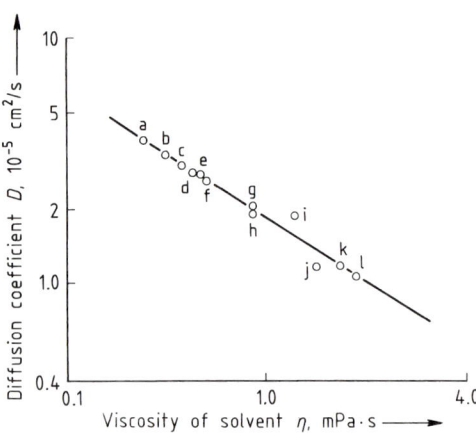

Figure 21. Diffusivity of carbon tetrachloride in various solvents [6.58]
a) Hexane; b) Heptane; c) Isooctane; d) Methanol; e) Toluene; f) Benzene; g) Carbon tetrachloride; h) Cyclohexane; i) Ethanol; j) Dioxane; k) Kerosine; l) Decalin

According to the authors the errors are usually $< 10\%$.

Example. Calculation of infinite dilution diffusion coefficients using the Hayduk–Minhas method

1) Acetone in ethyl acetate [6.20]:

	Acetone	Ethyl acetate
M, g/mol	58.08	88.106
V_{0i}^l, cm³/mol	73.52	97.83
η_i, mPa·s		0.4508
σ_i, N/m	0.02332	0.02375

From Equation (6.49): $D_{12}^\infty = 2.59 \times 10^{-5}$ cm²/s
Experimental: $D_{12}^\infty = 3.18 \times 10^{-5}$ cm²/s at 293 K [6.7]
Relative deviation $= -18.6\%$

2) Toluene in n-hexane [6.20]:

	Toluene	n-Hexane
M, g/mol	92.14	86.177
V_{0i}^l, cm³/mol	106.9	131.60
η_i, mPa·s		0.2942

Substituting $c = -0.6956$ in Equation (6.50):
$D_{12}^\infty = 4.90 \times 10^{-5}$ cm²/s
Experimental: $D_{12}^\infty = 4.12 \times 10^{-5}$ cm²/s at 298 K [6.7]
Relative deviation $= 18.9\%$

3) Aniline in water [6.20]:

	Aniline	Water
M, g/mol	93.128	18.02
V_{0i}^l, cm³/mol	77.42	
η_i, mPa·s		1.002

Substituting $\delta = -0.996$ in Equation (6.51):
$D_{12}^\infty = 1.021 \times 10^{-5}$ cm²/s
Experimental: $D_{12}^\infty = 0.92 \times 10^{-5}$ cm²/s at 293 K [6.7]
Relative deviation $= 11.0\%$

The application of the method of Hayduk et al. is somewhat more complicated for systems with associating compounds (e.g., organic acids).

Effect of Composition on Binary Diffusion Coefficients in Liquids. The dependence of binary diffusion coefficients on concentration is shown in Figure 22. In mixtures with compounds differing greatly in size and polarity the effect of concentration on diffusion coefficients cannot be neglected. Even a linear combination of diffusion coefficients at infinite dilution does not yield accurate results. The thermodynamic correction term $(\partial \ln a_i / \partial \ln x_i)$ is common to many estimation methods as in the Darken equation [6.60] for a binary mixture of components i and j

$$D_{ij} = (x_i D_{ij}^\infty + x_j D_{ji}^\infty) \frac{\partial \ln a_i}{\partial \ln x_i} \quad (6.52)$$

where x is the mole fraction and a the activity; $a_i(x_i)$ can be obtained from vapor–liquid equi-

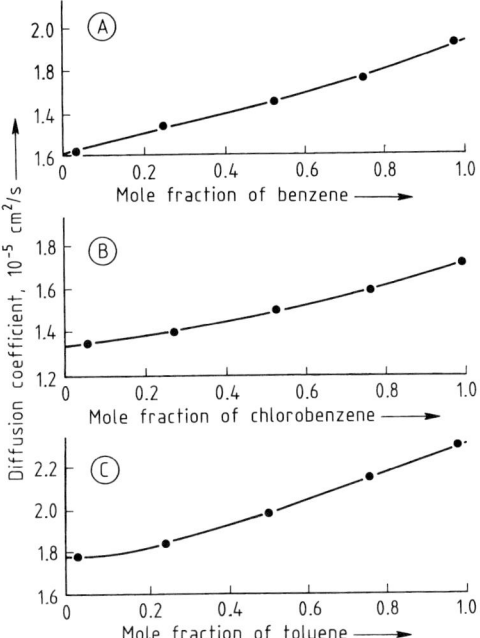

Figure 22. Effect of composition on binary diffusion coefficients [6.63]
A) Benzene–carbon tetrachloride at 298.41 K; B) Chlorobenzene–bromobenzene at 299.93 K; C) Toluene–chlorobenzene at 300.11 K

librium data. From the Gibbs–Duhem equation it follows that

$$\frac{\partial \ln a_i}{\partial \ln x_i} = \frac{\partial \ln a_j}{\partial \ln x_j} \quad (6.53)$$

Application of the NRTL equation (Section 4.2.1.2) yields for a binary mixture:

$$\frac{\partial \ln a_1}{\partial \ln x_1} = 1 - 2 x_1 x_2 \left[\frac{\tau_{21} G_{21}^2}{(x_1 + x_2 G_{21})^3} + \frac{\tau_{12} G_{12}^2}{(x_2 + x_1 G_{12})^3} \right] \quad (6.54)$$

For systems showing ideal mixing behavior the differential quotient $\partial \ln a_i / \partial \ln x_i$ is equal to zero. In view of the fact that Equation (6.53) in some cases overcorrects for nonideal behavior, RATHBURN and BABB recommended the introduction of an exponential constant f for the thermodynamic correction [6.61]:

$$D_{ij} = (x_i D_{ij}^\infty + x_j D_{ji}^\infty) \left(\frac{\partial \ln a_i}{\partial \ln x_i} \right)^f \quad (6.55)$$

Table 26. Selected methods for estimating diffusion coefficients in liquid solutions at infinite dilution

Method	Range of applicability	Information required*	Error, %**	Remarks
Wilke–Chang 1955 [6.65]	not reliable for water as solute	$M_2, \eta_2, V_1^l(T_b)$, association parameters	11 (w) 20 (o)	
Scheibel 1954 [6.66]	usually appropriate for aqueous solutions	$M_2, \eta_2, V_1^l(T_b)$	11 (w) 20 (o)	similar to Wilke–Chang
Othmer–Thakar 1953 [6.67]	preferably for aqueous solutions	V_{01}^l, η_{water}	5 (w)	
Reddy–Draiswamy 1967 [6.68]		$M_2, V_{01}^l, V_{02}^l, \eta_2$	22 (o) 25 (w)	
Lusis–Ratcliff 1968 [6.69]	preferably for alkane–alkane systems, not for aqueous solutions	$V_{01}^l, V_{02}^l, \eta_2$	29 (o)	
Tyn–Calus 1975 [6.70]	different correlations for different types of systems	$M_2, V_{01}^l, V_{02}^l, P_1, P_2, \eta_2$	9	
Nakanishi 1978 [6.71]	not for highly viscous solvents ($\eta < 20$–30 mPa·s)	$V_{01}^l, V_{02}^l, \eta_2$	14	extended version
Hayduk–Minhas 1982 [6.59]	different correlations for different types of systems	$M_2, V_{02}^l, V_{01}^l, \eta_2, P_1, P_2, (\sigma_i)$	10	
Siddiqi–Lucas 1986 [6.62]	including dissolved gases, two versions	$V_{01}^l, V_{02}^l, \eta_2$	13 (o) 20 (w)	

* Subscripts: 1 = solute; 2 = solvent. ** w = aqueous solutions; o = organic solutions.

The values for f originally proposed by RATHBURN and BABB (in parentheses) have been reevaluated by SIDDIQI and LUCAS [6.62]: $f = 0.6$ (0.6) for positive deviations from Raoult's law and $f = 0.4$ (0.3) for negative deviations.

Another correlation has been proposed by VIGNES [6.64]:

$$D_{ij} = [(D_{ij}^\infty)^{xj}(D_{ji}^\infty)^{xi}] \frac{\partial \ln a_i}{\partial \ln x_i} \quad (6.56)$$

SIDDIQI and LUCAS found that the average error was less than 5% for the prediction of 79 data sets. For systems with polar components they report deviations of 11%. Table 26 gives an overview of important estimation methods.

7. Surface Tension of Liquids

Surface and interfacial tension influence mass and heat transfer processes considerably. In this chapter only the liquid–vapor interface is discussed.

Surface tension σ is expressed in units of [N/m] or preferably in [mN/m] ($\equiv$ [dyn/cm]). Surface tensions of organic liquids vary between 20 and 50 mN/m at moderate pressure ($p \gg p_c$); experimental data can be found in the compilation of JASPER [7.1]. Water exhibits an extraordinarily high surface tension of 72 mN/m at room temperature. The variation of surface tension with temperature of some organic compounds is shown in Figure 23. Salt melts typically have surface tensions of ca. 100–200 mN/m and metal melts of 300–1000 mN/m. At low to moderate pressures ($p < p_c$) the influence of inert gases can be neglected.

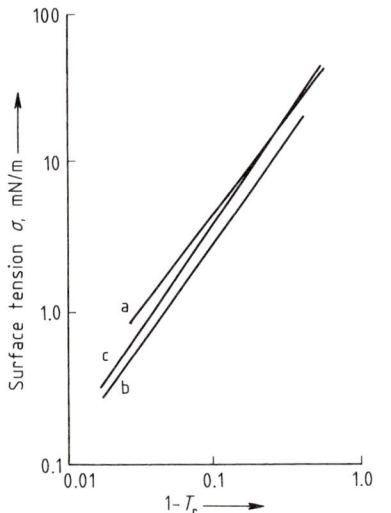

Figure 23. Variation of surface tension with temperature [7.2]
a) Acetic acid; b) Diethyl ether; c) Ethyl acetate

7.1. Surface Tension of Pure Liquids

Two basically empirical categories of methods are used for the estimation of surface tension. One type uses the parachor P and the other employs the principle of corresponding states.

Parachor Methods. The parachor (see Section 2.4) is a characteristic parameter which, according to its definition in Equation (2.21), is a function of surface tension. It can be estimated by incremental methods, e.g., the method of Quayle [7.3] (increments in Table 5). Surface tensions predicted with this method using Equation (2.21) or (2.22) usually show errors of less than 5% for nonpolar and weakly polar compounds and 5–10% for compounds with hydrogen bounds [7.4].

Example. Calculation of surface tension of acetamide, CH_3CONH_2, at 378.2 K from parachor estimated with the method of Quayle [7.3]

Structural groups: 1 (CH_3-), 1 ($-CONH_2$) (see Table 5)
$P = 55.5 + 91.7 = 147.2$ cm^3 g$^{0.25}$ s$^{-0.5}$ mol^{-1}

With $M = 59.068$ g/mol, $\varrho^l = 977.05$ kg/m^3, $p_{0i}^{lv} = 12.4$ hPa [7.5] and $\varrho^v = p_{0i}^{lv} M/(RT) = 0.023$ kg/m^3
From Equation (2.21): $\sigma = 0.0352$ N/m
Experimental: $\sigma = 0.03696$ N/m at 378.2 K [7.5]
Relative deviation $= -4.9\%$
(Deviation between Eqs. 2.21 and 2.22 = 0.01%)

Corresponding States Methods. These methods are based upon a relation proposed by VAN DER WAALS [7.6]:

$$\sigma = k\, T_c^{1/3} p_c^{2/3} (1 - T_r)^m \tag{7.1}$$

where k and m are constants, T_c is the critical temperature, p_c the critical pressure, and T_r the reduced temperature. Employing the Stiel polar factor χ (see Section 2.4), HAKIM et al. [7.7] transformed Equation (7.1) into a relation which is particularly useful for alcohols (p_c in bar, σ in N/m):

$$\sigma = p_c^{2/3} T_c^{1/3} \left(\frac{1-T_r}{0.4}\right)^m \frac{Q_p(\chi, \omega)}{1000} \tag{7.2}$$

with the functions Q_p and m in terms of the Stiel polar factor χ and the acentric factor ω

$$Q_p = 0.156 + 0.365\,\omega - 1.754\,\chi \\ - 13.57\,\chi^2 - 0.506\,\omega^2 + 1.287\,\omega\chi \tag{7.3}$$

$$m = 1.21 + 0.5385\,\omega - 14.61\,\chi - 32.07\,\chi^2 \\ - 1.656\,\omega^2 + 22.03\,\omega\chi \tag{7.4}$$

Errors are usually between 5 and 10%.

A simpler relation based on Equation (7.1) was proposed by BROCK and BIRD [7.8]; it should only be used for nonpolar or weakly polar liquids.

Example. Calculation of surface tension of 1-butanol, $CH_3CH_2CH_2CH_2OH$, at 303.2 K according to Hakim et al. [7.7]

$M = 74.122$ g/mol, $p_c = 44.13$ bar, $T_c = 563.0$ K,
$\chi = -0.07$, $\omega = 0.593$ [7.5]

From Equation (7.4): $m = 0.8981$
From Equation (7.3): $Q_p = 0.1974$
From Equation (7.2): $\sigma = 0.02314$ N/m
Experimental: $\sigma = 0.02378$ N/m [7.5]
Relative deviation $= -2.7\%$

7.2. Surface Tension of Liquid Mixtures

Organic Systems. Surface tensions of mixtures of organic compounds can be estimated by the simple mixing rule:

$$\sigma_m = \sum x_i \sigma_i \tag{7.5}$$

where x_i is the mole fraction of component i or by the more general relation

$$\sigma_m^\beta = \sum x_i \sigma_i^\beta \tag{7.6}$$

with the exponent β.

An exponent $\beta = -1$ is sometimes used:

$$\sigma_m = \left(\sum \frac{x_i}{\sigma_i}\right)^{-1} \tag{7.7}$$

Equation (2.22) can be extended for nonaqueous mixtures:

$$\sigma_m = \left[\sum P_i \left(\frac{\varrho_m^l x_i}{M}\right)\right]^4 \tag{7.8}$$

where x_i and y_i are the mole fractions of compound i in the liquid and vapor phases, respectively, and

$$\frac{1}{\varrho_m} = (\sum x_i/\varrho_i) \tag{7.9}$$

Again, the parachor of the pure compounds is employed (see Section 2.4). When experimental surface tension data for the mixture are available, P_i can also be used as a correlation parameter. The accuracy of Equation (7.8) is usually $<10\%$. If the parachor is applied as an ad-

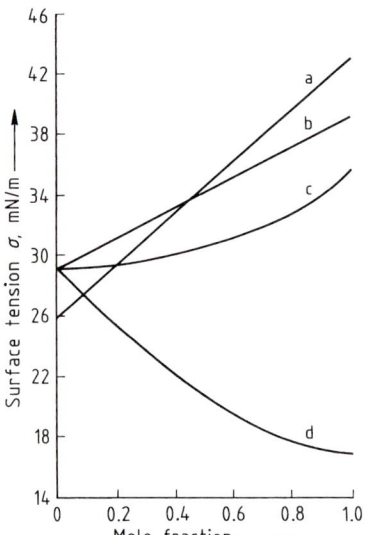

Figure 24. Surface tension of binary organic mixtures at 298 K [7.8]
A) Nitrobenzene–carbon tetrachloride; b) Acetophenone–benzene; c) Nitrobenzene–benzene; d) Diethyl ether–benzene
The mole fraction refers to the underlined component.

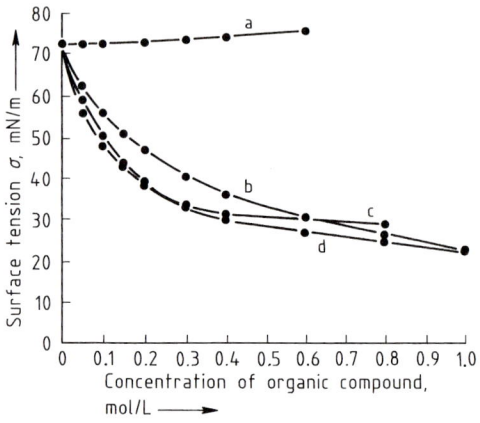

Figure 25. Surface tension of aqueous solutions of organic compounds (data from [7.9])
a) Sucrose; b) Methanol; c) Acetonitrile; d) Ethanol

justable parameter using experimental information, precision is of course much higher. The concentration dependence of the surface tension of some binary organic solutions is shown in Figure 24.

A large number of more sophisticated methods have been published based on the principle of corresponding states or on phenomenological or statistical thermodynamics. These methods are discussed in detail in [7.4].

Aqueous Systems. The surface tension of water is strongly affected by small amounts of dissolved organic compounds because hydrophobic molecules are repelled by the water molecules and are therefore enriched at the surface. The surface tension of aqueous solutions of organics therefore depends on the hydrophobicity of the organic solute with a highly nonlinear influence of solute concentration. In general, the surface tension of water is greatly decreased by hydrophobic solutes (Fig. 25).

For estimating the effect of organic solutes on the surface tension of water MEISSNER et al. [7.10] have proposed a method which is based on the following equation:

$$\sigma_m = \sigma_{0w}\left[1 - 0.411 \log\left(1 + \frac{x_i}{A}\right)\right] \quad (7.10)$$

where σ_{0w} is the surface tension of pure water, x_i the molar concentration of the organic component, and A is a substance-specific constant. Values of A ($\times 10^4$) for a number of organic substances follow:

Propionic acid	26	Ethyl propionate	3.1
1-Propanol	26	Propyl acetate	3.1
2-Propanol	26		
Methyl acetate	26	Valeric acid	1.7
		Isovaleric acid	1.7
Propylamine	19	Pentanol	1.7
Methyl ethyl ketone	19	Isopentanol	1.7
		Propyl propionate	1.0
Butyric acid	7	Caproic acid	0.75
Isobutyric acid	7	Heptanoic acid	0.17
Butanol	7	Octanoic acid	0.034
Isobutanol	7	Decanoic acid	0.0025
Propyl formate	8.5		
Ethyl acetate	8.5		
Methyl propionate	8.5		
Diethyl ketone	8.5		

The accuracy of this method is reported to be better than 15%. Another more precise but more complicated estimation method was suggested by TAMURA et al. [7.11]. This method is applicable to binary mixtures only and is discussed in detail in [7.4].

8. References

General References

Methods

[A1] S. Bretsznajder: *Prediction of Transport and Other Physical Properties of Fluids,* Pergamon Press, Oxford 1971.
[A2] R. P. Daubert, T. E. Danner: Design Institute for Physical Property Data (DIPPR)/American Institute of Chemical Engineers (AIChE): *Data Prediction Manual,* Pennsylvania State University 1983.
[A3] G. Hecht et al.: *Berechnung thermodynamischer Stoffwerte von Gasen und Flüssigkeiten,* VEB Deutscher Verlag für Grundstoffindustrie, Leipzig 1966.
[A4] C. Jochum, M. G. Hicks, J. Sunkel (eds.): *Physical Property Prediction in Organic Chemistry,* Springer Verlag, Berlin 1988.
[A5] R. C. Reid, T. K. Sherwood: *The Properties of Gases and Liquids,* 2nd ed., McGraw-Hill, New York 1966.
[A6] R. C. Reid, J. M. Prausnitz, T. K. Sherwood: *The Properties of Gases and Liquids,* 3rd ed., McGraw-Hill, New York 1977.
[A7] R. C. Reid, J. M. Prausnitz, B. E. Poling: *The Properties of Gases and Liquids,* 4th ed., McGraw-Hill, New York 1987.
[A8] Z. Sterbacek, B. Biskup, P. Tausk: *Calculation of Properties Using Corresponding-State Methods,* Elsevier, Amsterdam 1979.
[A9] S. Weiß et al. (eds.): *Verfahrenstechnische Berechnungsmethoden,* part 7: Stoffwerte, VCH Verlagsgesellschaft, Weinheim 1986.

General data books and series

[B1] D. Behrens, R. Eckermann (eds.): *Chemistry Data Series,* Dechema, Frankfurt/Main 1977.
[B2] D'Ans-Lax: *Taschenbuch für Chemiker und Physiker,* 3rd ed., Springer Verlag, Berlin [vol. 1 (1967), vol. 2 (1983), vol. 3 (1970)].
[B3] R. P. Daubert, T. E. Danner: Design Institute for Physical Property Data (DIPPR)/American Institute of Chemical Engineers (AIChE): *Data Compilation Tables of Properties of Pure Compounds,* Pennsylvania State University 1985.
[B4] Engineering Science Data Unit (ESDU): *Physical Data, Chemical Engineering, Engineering Sciences Data,* London 1959.
[B5] H. Landolt, R. Börnstein: *Zahlenwerte und Funktionen aus Naturwissenschaft und Technik,* 6th ed., Springer Verlag, Berlin 1950.
[B6] *Physical Sciences Data,* vols. 1–37, Elsevier, Amsterdam 1987.
[B7] Thermodynamics Research Centre (TRC): *Selected Values of Properties of Chemical Compounds,* Texas A & M University, College Station, Texas, 1982.
[B8] Y. S. Touloukian, C. Y. Ho (eds.): *CINDAS Data Series on Material Properties,* vols. 1– Hemisphere, Washington 1981.
[B9] Y. S. Touloukian, C. Y. Ho: *Thermophysical Properties of Matter,* vols. 1–13, IFI/Plenum Press, New York 1970.
[B10] VDI-Wärmeatlas, 5th ed., VDI-Verlag, Düsseldorf 1988.
[B11] R. C. Weast, M. J. Astle: *Handbook of Chemistry and Physics,* 70th ed., CRC Press, Boca Raton, Florida, 1989.

Universal handbooks

[C1] *Beilsteins Handbuch der Organischen Chemie,* 4th ed., Springer Verlag, Berlin 1958.
[C2] *Gmelin Handbuch der anorganischen Chemie,* 8th ed., Verlag Chemie, Weinheim 1925.

Thermodynamic Data of Pure Compounds

[D1] T. Barin, O. Knacke, O. Kubaschewski: *Thermochemical Properties of Inorganic Substances,* vol. 1, Springer Verlag, Berlin 1973, vol. 2, Verlag Stahleisen, Düsseldorf 1977.
[D2] T. Barin: *Thermochemical Data of Pure Substances,* VCH Verlagsgesellschaft, Weinheim 1989.
[D3] M. W. Chase et al.: *Joint Army Navy Air Force (JANAF) Thermochemical Tables,* 3rd ed., American Chemical Society, American Institute of Physics, Washington, D.C., 1986.
[D4] J. D. Cox, G. Pilcher: *Thermochemistry of Organic and Organometallic Compounds,* Academic Press, London 1970.
[D5] J. B. Pedley, R. D. Naylor, S. P. Kirby: *Thermochemical Data of Organic Compounds,* 2nd ed., Chapman and Hall, New York 1986.
[D6] F. D. Rossini: *Selected Values of Chemical Thermodynamic Properties,* National Bureau of Standards, Circular 500, Washington 1952.
[D7] D. R. Stull, G. C. Sinke: "Thermodynamic Properties of the Elements," in: *Advances in Chemistry Series,* vol. 18, American Chemical Society, Washington, D.C., 1956.
[D8] D. R. Stull, E. F. Westrum, G. C. Sinke: *The Chemical Thermodynamics of Organic Compounds,* Wiley, New York 1969.
[D9] A. L. Suris: *Handbook of Thermodynamic High Temperature Process Data,* Hemisphere, Washington 1987.
[D10] A. Tamir, E. Tamir, K. Stephan: *Heats of Phase Change of Pure Components and Mixtures,* Elsevier, Amsterdam 1983.
[D11] Y. S. Touloukian et al.: *Thermodynamic and Transport Data Properties of Gases, Liquids and Solids,* McGraw-Hill, New York 1959.
[D12] Y. S. Touloukian, T. Makita: *Specific Heat. Nonmetallic Liquids and Gases,* IFI/Plenum Press, New York 1970.
[D13] Y. S. Touloukian, E. H. Buyco: *Specific Heat. Nonmetallic Solids,* IFI/Plenum Press, New York 1970.
Further reference: [G4]

Thermodynamic Data of Mixtures

[E1] W. Arlt, M. E. A. Macedo, P. Rasmussen, J. M. Sorensen: *Liquid–Liquid Equilibrium Data Collection,* Chemistry Data Series, vol. 5, parts 1–4, Dechema, Frankfurt/Main 1979.
[E2] C. Christensen, J. Gmehling, P. Rasmussen, U. Weidlich, T. Holderbaum: *Heats of Mixing Data Collection,* Chemistry Data Series, vol. 3, parts 1–3, Dechema, Frankfurt/Main 1984.
[E3] H. Engels: *Phase Equilibria and Enthalpies of Electrolyte Solutions,* Chemistry Data Series, vol. 11, part 1, Dechema, Frankfurt/Main (in preparation).

[E4] J. Gmehling, U. Onken et al.: *VLE Data Collection,* vol. 1: part 1 (1977), part 1a (1981), part 1b (1988), part 2a (1985, reprint), part 2b (1987), part 2c (1982), part 2d (1982), part 2e (1988), part 3/4 (1979), part 3a/4a (in prep.), part 5 (1982), part 5a (in prep.), part 6a (1980), part 6b (1980), part 6c (1984), part 6d (in prep.), part 7 (1980), part 7a (in prep.), part 8 (1984), part 8a (in prep.), Dechema, Frankfurt/Main 1977.

[E5] J. Gmehling et al.: *Activity Coefficients at Infinite Dilution,* Chemistry Data Series, vol. 9, parts 1–3, Dechema, Frankfurt/Main 1986.

[E6] IUPAC: *Solubility Data Series,* Pergamon Press, Oxford 1979.

[E7] H. Knapp, R. Döring, L. Öllrich, U. Plöcker, J. M. Prausnitz: *VLE for Mixtures of Low-Boiling Substances,* Chemistry Data Series, vol. VI, Dechema, Frankfurt/Main 1982.

[E8] H. Knapp, M. Teller, R. Langhorst: *Solid–Liquid Equilibria Data Collection,* Chemistry Data Series, vol. 8, part 1, Dechema, Frankfurt/Main 1987.

[E9] A. Seidell, W. F. Linke: *Solubilities of Inorganic and Organic Compounds,* 4th ed., vols. 1–2, Princeton [vol. 1 (1958), vol. 2 (1965)].

[E10] H. Stephen, T. Stephen: *Solubilities of Inorganic and Organic Compounds,* Pergamon Press, Oxford 1963, reprinted 1979.

[E11] N. K. Voskresenskaya (ed.): *Handbook of Solid–Liquid Equilibria in Systems of Anhydrous Inorganic Salts,* Keter Press, Jerusalem 1970.

[E12] J. F. Zemaitis, D. M. Clark, M. Rafal, N. C. Scrivner: *Handbook of Aqueous Electrolyte Thermodynamics,* AIChE, New York 1986.

Further reference: [D10]

Critical data

[F1] D. Ambrose: "Correlation and Estimation of Vapour Liquid Critical Properties, I. Critical Temperatures of Organic Compounds," National Physical Laboratory, Teddington, *NPL Rep. Chem.* **92** (1978) corr. March 1981, Teddington, *NPL Rep. Chem.* **98** (1979).

[F2] K. H. Simmrock, R. Janowsky, A. Ohnsorge: *Critical Data of Pure Substances,* Chemistry Data Series, vol. II, part I, Dechema, Frankfurt/Main 1986.

Vapor pressure

[G1] T. Boublik, V. Fried, E. Hala: *The Vapor Pressures of Pure Substances,* 2nd ed., Elsevier, Amsterdam 1984.

[G2] S. Ohe: *Computer Aided Data Book of Vapor Pressure,* Data Book Publishing, Tokyo 1976.

[G3] I. Wichterle, J. Linek: *Antoine Vapour Pressure Constants of Pure Compounds,* Academia, Praha 1971.

[G4] B. J. Zwolinski, R. C. Wilhoit: *Handbook of Vapour Pressures and Heats of Vaporisation of Hydrocarbon and Related Compounds,* Thermodynamic Research Center, Department of Chemistry, Texas 1971.

pVT behavior

[H1] J. Cholinski, A. Szafranski et al.: *Second Virial Coefficients for Organic Compounds,* Institute of Physical Chemistry, Polish Academy of Sciences, Warszawa 1985.

[H2] J. H. Dymond, E. B. Smith: *The Virial Coefficients of Pure Gases and Mixtures: A Critical Compilation,* Clarendon Press, Oxford 1980.

Transport data

[I1] K. Stephan, T. Heckenberger: *Thermal Conductivity and Viscosity Data of Fluid Mixtures,* Chemistry Data Series, vol. 10, part 1, Dechema, Frankfurt/Main 1989.

[I2] K. Stephan, K. Lucas: *Viscosity of Pure Dense Fluids,* Plenum Press, New York 1979.

[I3] D. S. Viswanath, G. Natarjan: *Data Book on the Viscosity of Liquids,* Hemisphere, New York 1989.

Selected compounds

[J1] R. R. Dreisbach: "Physical Properties of Chemical Compounds," vols. I–III, in: *Advances in Chemistry Series,* vols. 15, 22, 29, American Chemical Society, Washington, DC [vol. I (1955), vol. II (1959), vol. I-II (1961)].

[J2] E. W. Flick: *Industrial Solvents Handbook,* 3rd ed., Noyes Data Corporation, New Jersey 1985.

[J3] A. L. Horvath: *Physical Properties of Inorganic Compounds,* Fletcher & Son, Norwich, UK, 1975.

[J4] J. A. Riddick, W. B. Bunger, T. K. Sakano: *Organic Solvents. Physical Properties and Methods of Purification,* 4th ed., vol. II, Wiley, New York 1986.

[J5] R. M. Stephenson, S. Malanowski: *Handbook of the Thermodynamics of Organic Compounds,* Elsevier, Amsterdam 1987.

[J6] Thermodynamics Research Centre (TRC): *Selected Values of Properties of Hydrocarbons and Related Compounds,* vol. 1, Texas A & M University, College Station, Texas, 1977.

[J7] Thermodynamics Research Centre (TRC): *TRC Thermodynamic Tables—Hydrocarbons,* vols. 1–11, Texas A & M University, College Station, Texas, 1942.

[J8] Thermodynamics Research Centre (TRC): *TRC Thermodynamic Tables—Non-Hydrocarbons,* vols. 1–8, Texas A & M University, College Station, Texas, 1955.

[J9] J. Timmermanns: *The Physico-Chemical Constants of Pure Organic Compounds,* Elsevier, New York [vol. 1 (1950), vol. 2 (1965)].

[J10] Y. S. Touloukian (ed.): *Thermophysical Properties of High Temperature Solid Materials,* vols. 1–6, MacMillan, New York 1967.

References for Chapter 1

[1.1] D'Ans-Lax: *Taschenbuch für Chemiker und Physiker,* 3rd ed., Springer Verlag, Berlin [vol. 1 (1967), vol. 2 (1983), vol. 3 (1970)].

[1.2] R. C. Weast, M. J. Astle: *Handbook of Chemistry and Physics,* 70th ed., CRC Press, Boca Raton, Florida, 1989.

[1.3] H. Landolt, R. Börnstein: *Zahlenwerte und Funktionen aus Naturwissenschaft und Technik,* 6th ed., Springer Verlag, Berlin 1950.

[1.4] D. Behrens, R. Eckermann (eds.): *Chemistry Data Series,* Dechema, Frankfurt/Main 1977.

[1.5] U. Onken, J. Rarey-Nies, J. Gmehling: "The Dortmund Data Bank: A Computerized System for Retrieval, Correlation, and Prediction of Thermodynamic Properties of Mixtures," *Int. J. Thermophys.* **10** (1989) 739–747.

[1.6] C. Jochum, M. G. Hicks, J. Sunkel (eds.): *Physical Property Prediction in Organic Chemistry*, Springer Verlag, Berlin 1988.

[1.7] Fachinformationszentrum (FIZ) Chemie, Mailbox 12 60 50, D-1000 Berlin 12.

Design Institute for Physical Property Data (DIPPR), STN International, 2540 Olentangy River Road, P.O.Box 2228, Columbus, Ohio, 43202, USA.

STN International, Mailbox 2465, D-7500 Karlsruhe 1, FRG.

STN International, c/o Japan Association for International Chemical Information, Gakkai Center Building, 2-4-16 Yayoi, Bunkyo-ku, Tokyo, Japan.

Physical Properties Data Service (PPDS), Institution of Chemical Engineers, 165–171 Railway Terrace, Rugby, CV21 3HQ, England.

[1.8] W. Kistiakowsky, *Z. Phys. Chem.* **107** (1923) 65.

[1.9] J. Timmermanns: *The Physico-Chemical Constants of Pure Organic Compounds*, Elsevier, New York [vol. 1 (1950), vol. 2 (1965)].

[1.10] Thermodynamics Research Centre (TRC): *Selected Values of Properties of Chemical Compounds*, Texas A & M University, College Station, Texas, 1982.

[1.11] J. A. Riddick, W. B. Bunger, T. K. Sakano: *Organic Solvents. Physical Properties and Methods of Purification*, 4th ed., vol. II, Wiley, New York 1986.

References for Chapter 2

[2.1] L. I. Stiel, G. Thodos, *Chem. Eng. Data Ser.* **7** (1962) 234.

[2.2] J. S. Rowlinson: *Liquids and Liquid Mixtures*, 2nd ed., Butterworth, London 1969.

[2.3] J. O. Hirschfelder, C. F. Curtiss, R. B. Bird: *Molecular Theory of Gases and Liquids*, 4th print, Wiley, New York 1967.

[2.4] L. C. Nelson, E. F. Obert, *Trans. ASME* **76** (1954) 1057.

[2.5] R. C. Reid, J. M. Prausnitz, T. K. Sherwood: *The Properties of Gases and Liquids*, 3rd ed., McGraw-Hill, New York 1977.

[2.6] R. C. Reid, J. M. Prausnitz, B. E. Poling: *The Properties of Gases and Liquids*, 4th ed., McGraw-Hill, New York 1987.

[2.7] A. L. Lydersen: "Estimation of Critical Properties of Organic Compounds," *Univ. Wisconsin Coll. Eng., Eng. Exp. Stn. Rept. 3*, Madison, Wisc. April 1955.

[2.8] K. G. Joback, MS thesis, Massachusetts Institute of Techn., Cambridge, Mass., 1984.

[2.9] J. Timmermanns: *The Physico-Chemical Constants of Pure Organic Compounds*, Elsevier, New York [vol. 1 (1950), vol. 2 (1965)].

[2.10] D. Ambrose: "Correlation and Estimation of Vapour Liquid Critical Properties, I. Critical Temperatures of Organic Compounds," *NPL Rep. Chem. UK Natl. Phys. Lab. Div. Chem. Stand.* **92** (1978) (corr. March 1980).

[2.11] D. Ambrose, *NPL Rep. Chem. UK Natl. Phys. Lab. Div. Chem. Stand.* **98** (1979).

[2.12] R. F. Fedors, *Chem. Eng. Commun.* **16** (1982) 149.

[2.13] J. C. Forman, G. Thodos, *AIChE J.* **4** (1958) 356; **6** (1960) 206.

[2.14] A. P. Kudchadkar, G. H. Alani, B. J. Zwolinski, *Chem. Rev.* **68** (1968) 659.

[2.15] K. H. Simmrock, R. Janowsky, A. Ohnsorge: *Critical Data of Pure Substances*, Chemistry Data Series, vol. II, part I, Dechema, Frankfurt/Main 1986.

[2.16] J. F. Mathews, *Chem. Rev.* **72** (1972) 71.

[2.17] W. C. Edmister, *Pet. Refin.* **37** (1958) no. 4, 173.

[2.18] S. Weiß et al. (eds.): *Verfahrenstechnische Berechnungsmethoden*, part 7: Stoffwerte, VCH Verlagsgesellschaft, Weinheim 1986.

[2.19] S. Sudgen, *J. Chem. Soc.* **125** (1924) 1177.

[2.20] O. R. Quayle, *Chem. Rev.* **53** (1953) 439.

References for Chapter 3

[3.1] J. S. Rowlinson: *Liquids and Liquid Mixtures*, 2nd ed., Butterworth, London 1969.

[3.2] J. M. Prausnitz, R. N. Lichtenthaler, E. Gomes de Azevedo: *Molecular Thermodynamics of Fluid-Phase Equilibria*, Prentice-Hall, Englewood Cliffs, N.Y., 1986.

[3.3] R. C. Reid, J. M. Prausnitz, B. E. Poling: *The Properties of Gases and Liquids*, 4th ed., McGraw-Hill, New York 1987.

[3.4] L. C. Nelson, E. F. Obert, *Trans. ASME* **76** (1954) 1057.

[3.5] G. Soave, *Chem. Eng. Sci.* **27** (1972) 1197.

[3.6] R. C. Reid, J. M. Prausnitz, T. K. Sherwood: *The Properties of Gases and Liquids*, 3rd ed., McGraw-Hill, New York 1977.

[3.7] J. H. Dymond, E. B. Smith: *The Virial Coefficients of Pure Gases and Mixtures: A Critical Compilation*, Clarendon Press, Oxford 1980.

[3.8] K. S. Pitzer, R. F. Curl, *J. Am. Chem. Soc.* **79** (1957) 2369.

[3.9] C. Tsonopoulos, *AIChE J.* **20** (1974) 263–272; **21** (1975) 827.

[3.10] J. G. Hayden, J. P. O'Connell, *Ind. Eng. Chem. Process Des. Dev.* **14** (1975) no. 3, 209.

[3.11] B. J. Lee, M. G. Kesler, *AIChE J.* **21** (1975) 510.

[3.12] U. Plöcker, H. Knapp, J. M. Prausnitz, *Ind. Eng. Chem. Process Des. Dev.* **17** (1978) 324.

[3.13] E. Bender: *Die Berechnung von Phasengleichgewichten mit der thermischen Zustandsgleichung*, Habilitationsschrift, Ruhr Universität, Bochum 1971.

[3.14] A. Polt, Ph. D. thesis, Universität Kaiserslautern 1987.

[3.15] J. Partington: *An Advanced Treatise on Physical Chemistry*, vol. I: Fundamental Principles: The Properties of Gases, Longmans, Green & Co., New York 1949.

[3.16] F. Schroeder in [3.15].

[3.17] J. A. Riddick, W. B. Bunger: *Organic Solvents, Physical Properties and Methods of Purification*, 3rd ed., vol. II, 1970; 4th ed., 1986, J. Wiley, New York.

[3.18] R. D. Gunn, T. Yamada, *AIChE J.* **17** (1971) 1341.

[3.19] R. W. Hankinson, G. H. Thomson, *AIChE J.* **25** (1979) 653.

[3.20] H. G. Rackett, *J. Chem. Eng. Data* **15** (1970) 514.

[3.21] L. C. Yen, S. S. Woods, *AIChE J.* **16** (1966) 95.

[3.22] V. L. Bhirud, *AIChE J.* **24** (1978) 1127.

[3.23] S. Weiß et al. (eds.): *Verfahrenstechnische Berechnungsmethoden*, part 7: Stoffwerte, VCH Verlagsgesellschaft, Weinheim 1986.

[3.24] J. M. Prausnitz: *Molecular Thermodynamics of Fluid-Phase Equilibria*, Prentice-Hall, Englewood Cliffs, N.Y., 1969.

[3.25] W. B. Kay, *Ind. Eng. Chem. Ind. Ed.* **28** (1936) 1014.

[3.26] J. M. Prausnitz, R. D. Gunn, *AIChE J.* **4** (1958) 430, 494.
[3.27] J. Joffe, *Ind. Eng. Chem. Fundam.* **10** (1971) 532.
[3.28] S. Angus et al. (eds.): *IUPAC International Thermodynamic Tables of the Fluid State: Carbon Dioxide*, Pergamon Press, Oxford 1973.
[3.29] K. G. Joback, MS thesis, Massachusetts Institute of Techn., Cambridge, Mass., 1984.
[3.30] S. W. Benson et al.: "Additive Rules for the Estimation of Thermochemical Properties," *Chem. Rev.* **69** (1969) 279; *J. Phys. Colloid Chem.* **52** (1948) 1060. S. W. Benson: *Thermochemical Kinetics*, Wiley, New York 1968.
[3.31] C. J. Dobratz, *Ind. Eng. Chem. Ind. Ed.* **33** (1941) 759.
[3.32] D. N. Rihani, L. K. Doraiswamy, *Ind. Eng. Chem. Fundam.* **4** (1965) 17.
[3.33] T.-P. Thinh, T. K. Trong, *Can. J. Chem. Eng.* **54** (1976) 344.
[3.34] Y. Yoneda, *Bull. Chem. Soc. Jpn.* **52** (1979) 1297.
[3.35] B. K. Harrison, W. H. Seaton, *Ind. Eng. Res.* **27** (1988) 1536–1540.
[3.36] C. F. Chueh, A. C. Swanson, *Chem. Eng. Prog.* **69** (1973) no. 7, 83; *Can. J. Chem. Eng.* **51** (1973) 596.
[3.37] A. Missenard, *Compte Rendue* **260** (1965) 5521.
[3.38] M. Luria, S. W. Benson, *J. Chem. Eng. Data* **22** (1977) 90.
[3.39] A. Bondi, *Ind. Eng. Chem. Fundam.* **5** (1966) 443.
[3.40] T.-F. Yuan, L. J. Stiel, *Ind. Eng. Chem. Fundam.* **9** (1970) 393.
[3.41] A. S. Teja, *J. Chem. Eng. Data* **28** (1983) 83.

References for Chapter 4

[4.1] S. Weiß et al. (eds.): *Verfahrenstechnische Berechnungsmethoden*, part 7: Stoffwerte, VCH Verlagsgesellschaft, Weinheim 1986.
[4.2] D. Ambrose, J. F. Counsell, A. J. Davenport, *J. Chem. Thermodyn.* **6** (1974) 693.
[4.3] T. Boublik, V. Fried, E. Hala: *The Vapor Pressures of Pure Substances*, 2nd ed., Elsevier, Amsterdam 1984.
[4.4] B. J. Lee, M. G. Kesler, *AIChE J.* **21** (1975) 510.
[4.5] M. Gomez-Nieto, G. Thodos, *Ind. Eng. Chem. Fundam.* **16** (1977) 254.
[4.6] R. C. Reid, J. M. Prausnitz, B. E. Poling: *The Properties of Gases and Liquids*, 4th ed., McGraw-Hill, New York 1987.
[4.7] J. A. Riddick, W. B. Bunger: *Organic Solvents, Physical Properties and Methods of Purification*, 3rd ed., vol. II, 1970; 4th ed., 1986, J. Wiley, New York.
[4.8] Y. Ogata, M. Tsuchida, *Ind. Eng. Chem.* **49** (1957) no. 3, 415.
[4.9] K. G. Joback, MS thesis, Massachusetts Institute of Techn., Cambridge, Mass., 1984.
[4.10] J. E. Haggenmacher, *J. Am. Chem. Soc.* **68** (1946) 1633.
[4.11] A. Vetere in [2.5, p. 209].
[4.12] R. C. Reid, J. M. Prausnitz, T. K. Sherwood: *The Properties of Gases and Liquids*, 3rd ed., McGraw-Hill, New York 1977.
[4.13] A. Tamir, E. Tamir, K. Stephan: *Heats of Phase Change of Pure Components and Mixtures*, Elsevier, Amsterdam 1983.
[4.14] L. Riedel, *Chem. Ing. Tech.* **26** (1954) 679.
[4.15] N. H. Chen, *J. Chem. Eng. Data* **10** (1965) 207.

[4.16] W. Wagner, *Cryogenics* **13** (1973) 470.
[4.17] R. Kobayashi, J. Magee, A. Sivaraman, *Fluid Phase Equilib.* **16** (1984) 1.
[4.18] K. M. Watson, *Ind. Eng. Chem. Ind. Ed.* **35** (1943) 398.
[4.19] G. M. Wilson, *J. Amer. Chem. Soc.* **86** (1964) 127.
[4.20] H. Renon, J. M. Prausnitz, *Ind. Eng. Chem. Process Des. Dev.* **8** (1969) 413; *AIChE J.* **15** (1969) 785.
[4.21] D. S. Abrams, J. M. Prausnitz, *AIChE J.* **21** (1975) 116.
[4.22] O. Redlich, E. L. Derr, G. Pierotti, *J. Am. Chem. Soc.* **81** (1959) 2283.
[4.23] G. M. Wilson, C. H. Deal, *Ind. Eng. Chem. Fundam.* **1** (1962) 20.
[4.24] E. L. Derr, C. H. Deal, *Ind. Chem. Eng. Symp. Ser.* **3** (1969) 40.
[4.25] A. Fredenslund, R. L. Jones, J. M. Prausnitz, *AIChE J.* **21** (1975) 1086.
[4.26] A. Fredenslund, J. Gmehling, P. Rasmussen: *Vapour Liquid Equilibria Using UNIFAC-A Group-Contribution Method*, Elsevier, Amsterdam – New York 1977.
[4.27] A. Bondi: *Physical Properties of Molecular Liquids, Crystals and Glasses*, Wiley, New York 1968.
[4.28] J. Gmehling, D. Tiegs, U. Knipp, *Fluid Phase Equilib.* **54** (1990) 147.
[4.29] D. Luedecke, J. M. Prausnitz, *Fluid Phase Equilib.* **22** (1985) 1.
[4.30] S. Skjold-Jørgensen, B. Kolbe, J. Gmehling, P. Rasmussen, *Ind. Eng. Chem. Process Des. Dev.* **18** (1979) 714.
[4.31] J. Gmehling, P. Rasmussen, A. Fredenslund, *Ind. Eng. Chem. Process Des. Dev.* **21** (1982) 118; revisions: E. A. Macedo, U. Weidlich, J. Gmehling, P. Rasmussen, *Ind. Eng. Chem. Process Des. Dev.* **22** (1983) 676; D. Tiegs, J. Gmehling, P. Rasmussen, A. Fredenslund, *Ind. Eng. Chem. Res.* **26** (1987) 159; new compilation: H. Hansen et al., *Ind. Eng. Chem. Res.* **30** (1991) (in press).
[4.32] B. Larsen, P. Rasmussen, A. Fredenslund, *Ind. Eng. Chem. Prod. Res. Dev.* **26** (1987) 2274.
[4.33] J. Gmehling, *Fluid Phase Equilib.* **30** (1986) 119.
[4.34] U. Weidlich, J. Gmehling, *Ind. Eng. Chem. Prod. Res. Dev.* **26** (1987) 1372.
[4.35] H. V. Kehaian, S. I. Sandler, *Fluid Phase Equilib.* **17** (1984) 139.
[4.36] B. Kolbe, J. Gmehling, U. Onken, *Inst. Chem. Eng. Symp. Ser.* **56** (1979); *Ber. Bunsen Ges. Phys. Chem.* **83** (1979) 1133.
[4.37] T. Oishi, J. M. Prausnitz, *Ind. Eng. Chem. Process Des. Dev.* **17** (1978) no. 3, 333.
[4.38] G. Nocon, U. Weidlich, J. Gmehling, U. Onken, *Ber. Bunsen Ges. Phys. Chem.* **87** (1983).
[4.39] G. Nocon, U. Weidlich, J. Gmehling, J. Menke, U. Onken, *Fluid Phase Equilib.* **13** (1983) 381.
[4.40] K. E. Gubbins, *Chem. Eng. Prog.* **85** (1989) 38.
[4.41] M. Luckas, K. Lucas, *Fluid Phase Equilib.* **45** (1989) 7–13.
[4.42] H. Knapp et al.: *VLE for Mixtures of Low-Boiling Substances*, Chemistry Data Series, vol. VI, Dechema, Frankfurt/Main 1982.
[4.43] F. Magnussen, P. Rasmussen, A. Fredenslund, *Ind. Eng. Chem. Process Des. Dev.* **20** (1981) 331.
[4.44] J. Gmehling, T. F. Anderson, J. M. Prausnitz, *Ind. Eng. Chem. Fundam.* **17** (1978).
[4.45] K. Tochigi, D. Tiegs, J. Gmehling, K. Kojima, *J. Chem. Eng. Jpn.* **23** (1990) 423.

[4.46] J. Vidal, *Chem. Eng. Sci.* **31** (1978) 1077.
[4.47] W. B. Whiting, J. M. Prausnitz, *Fluid Phase Equilib.* **9** (1982) 119.
[4.48] S. Skjøld-Jørgensen, *Fluid Phase Equilib.* **16** (1984) 317.
[4.49] S. Skjøld-Jørgensen, *Ind. Eng. Chem. Res.* **27** (1988) 110.
[4.50] K. Tochigi, K. Kurihara, K. Kojima, *Ind. Eng. Chem. Res.* **29** (1990) 2142.

References for Chapter 5

[5.1] H. Landolt, R. Börnstein: *Zahlenwerte und Funktionen aus Naturwissenschaft und Technik*, 6th ed., Springer Verlag, Berlin 1950.
[5.2] G. J. Janz: *Thermodynamic Properties of Organic Compounds*, Academic Press, New York 1967.
[5.3] M. W. Chase et al.: *Joint Army Navy Air Force (JANAF) Thermochemical Tables*, 3rd ed., *Journal of Physical and Chemical Reference Data, Suppl.* **14,1**, American Chemical Society, American Institute of Physics, Washington, D.C., 1986.
[5.4] J. D. Cox, G. Pilcher: *Thermochemistry of Organic and Organometallic Compounds*, Academic Press, London 1970.
[5.5] F. D. Rossini: *Selected Values of Chemical Thermodynamic Properties*, National Bureau of Standards, Circular 500, Washington 1952.
[5.6] D. R. Stull, G. C. Sinke: "Thermodynamic Properties of the Elements," in: *Advances in Chemistry Series*, vol. 18, American Chemical Society, Washington, D.C., 1956.
[5.7] D. R. Stull, E. F. Westrum, G. C. Sinke: *The Chemical Thermodynamics of Organic Compounds*, Wiley, New York 1969.
[5.8] Thermodynamics Research Centre (TRC): *Selected Values of Properties of Chemical Compounds*, Texas A & M University, College Station, Texas, 1982; *Selected Values of Properties of Hydrocarbons and Related Compounds*, vols. 1, Texas A & M University, College Station, Texas, 1977; *TRC Thermodynamic Tables – Non-Hydrocarbons*, vols. 1–8, Texas A & M University, College Station, Texas, 1955; *TRC Thermodynamic Tables – Hydrocarbons*, vols. 1–11, Texas A & M University, College Station, Texas, 1942.
[5.9] K. G. Joback: MS thesis, Massachusetts Institute of Techn., Cambridge, Mass., 1984.
[5.10] R. C. Reid, J. M. Prausnitz, B. E. Poling: *The Properties of Gases and Liquids*, 4th ed., McGraw-Hill, New York 1987.
[5.11] S. W. Benson: *Thermochemical Kinetics*, Wiley, New York 1968.
[5.12] S. W. Benson et al., *Chem. Rev.* **69** (1969) 279.
[5.13] Y. Yoneda, *Bull. Chem. Soc. Jpn.* **52** (1979) 1297.
[5.14] J. L. Franklin, *Ind. Eng. Chem.* **41** (1949) 1070; *J. Chem. Phys.* **21** (1953) 2029.
[5.15] K. K. Verma, L. K. Doraiswamy, *Ind. Eng. Chem. Fundam.* **4** (1965) 389.
[5.16] D. W. Van Krevelen, H. A. G. Chermin, *Chem. Eng. Sci.* **1** (1951) 66; *Chem. Eng. Sci.* **1** (1952) 238.
[5.17] R. C. Reid, J. M. Prausnitz, T. K. Sherwood: *The Properties of Gases and Liquids*, 3rd ed., McGraw-Hill, New York 1977.
[5.18] R. C. Reid, T. K. Sherwood: *The Properties of Gases and Liquids*, 2nd ed., McGraw-Hill, New York 1966.

References for Chapter 6

[6.1] S. Chapman, T. G. Cowling: *The Mathematical Theory of Nonuniform Gases*, Cambridge, NY, 1939.
[6.2] S. Weiß et al. (eds.): *Verfahrenstechnische Berechnungsmethoden*, part 7: Stoffwerte, VCH Verlagsgesellschaft, Weinheim 1986.
[6.3] R. C. Reid, J. M. Prausnitz, T. K. Sherwood: *The Properties of Gases and Liquids*, 3d ed., McGraw-Hill, New York 1977.
[6.4] *VDI Wärmeatlas*, 5th ed., VDI Verlag, Düsseldorf 1988.
[6.5] J. O. Hirschfelder, C. F. Curtiss, R. B. Bird: *Molecular Theory of Gases and Liquids*, 4th print, Wiley, New York 1967.
[6.6] R. C. Reid, J. M. Prausnitz, B. E. Poling: *The Properties of Gases and Liquids*, 4th ed., McGraw-Hill, New York 1987.
[6.7] O. A. Uyehara, K. M. Watson, *Natl. Pet. News* **36** (1944) R 714.
[6.8] K. Lucas: *Phase Equilibria and Fluid Properties in the Chemical Industry*, Dechema, Frankfurt 1980, p. 573.
[6.9] K. Lucas, *Chem. Ing. Tech.* **53** (1981) 959.
[6.10] *VDI-Wärmeatlas*, 4th ed., VDI-Verlag, Düsseldorf 1984.
[6.11] C. R. Wilke, *J. Chem. Phys.* **18** (1950) 517.
[6.12] R. S. Brokaw, NASA Tech. Note D-4496, Apr. 1968.
[6.13] R. S. Brokaw, *Ind. Eng. Chem. Process Des. Dev.* **8** (1969) 240.
[6.14] T.-H. Chung, L. L. Lee, K. E. Starling, *Ind. Eng. Chem. Fundam.* **23** (1984) 8.
[6.15] L. A. Bromley, C. R. Wilke, *Ind. Eng. Chem.* **43** (1951) 1641.
[6.16] G. Thodos, *AIChE J.* **1** (1955) 165; **2** (1956) 508; **3** (1957) 428.
[6.17] D. Reichenberg, *AIChE J.* **16** (1970) 854; *AIChE J.* **21** (1975) 181.
[6.18] E. N. da C. Andrade, *Endeavour* **13** (1954) 117.
[6.19] D. Van Velzen, R. L. Cardozo, H. Langenkamp, *Ind. Eng. Chem. Fundam.* **11** (1972) 20.
[6.20] J. A. Riddick, W. B. Bunger: *Organic Solvents, Physical Properties and Methods of Purification*, 3rd ed., vol. II, 1970, 4th ed., 1986, J. Wiley, New York.
[6.21] L. H. Thomas, *J. Chem. Soc.* **1946**, 573.
[6.22] P. S. Morris, MS thesis, Polytechnic Institute of Brooklyn, Brooklyn, N.Y., 1964.
[6.23] C. Orrick, J. H. Erbar in [6.6, p. 456].
[6.24] J. W. Przezdziecki, T. Sridhar, *AIChE J.* **31** (1985) 333.
[6.25] A. Letsou, L. I. Stiel, *AIChE J.* **19** (1973) 409.
[6.26] R. A. McAllister, *AIChE J.* **6** (1960) 427.
[6.27] J. B. Irving: "Viscosities of Binary Liquid Mixtures: A Survey of Mixture Equations," *NEL Rep. GB* **630** (1977); "Viscosities of Binary Liquid Mixtures: The Effectiveness of Mixture Equations," *NEL Rep. GB* **631** (1977).
[6.28] Y. S. Touloukian, C. Y. Ho: "Viscosity. Thermophysical Properties of Matter," *The TPRC Data Ser.*, vol. 2, Plenum Press, New York 1975.
[6.29] K. Stephan, K. Lucas: *Viscosity of Pure Dense Fluids*, Plenum Press, New York 1979.
[6.30] W. Blanke (ed.): *Thermophysikalische Stoffgrößen*, Springer Verlag, Berlin 1989.
[6.31] D. Roy, G. Thodos, *Ind. Eng. Chem. Fundam.* **7** (1968) 529.

[6.32] N. B. Vargaftig: *Tables on the Thermophysical Properties of Gases and Liquids*, 2nd ed., Hemisphere, Washington 1975.
[6.33] J. F. Ely, J. M. Hanley, *Ind. Eng. Chem. Fundam.* **22** (1983) 90.
[6.34] L. A. Guildner, *Proc. Natl. Acad. Sci. USA* **44** (1958) 1149.
[6.35] T. H. Chung, L. L. Lee, K. E. Starling, *Ind. Eng. Chem. Fundam.* **19** (1980) 186.
[6.36] S. Angus et al. (eds.): *IUPAC: International Thermodynamic Tables of the Fluid State: Carbon Dioxide*, Pergamon Press, Oxford 1973.
[6.37] D. Misic, G. Thodos, *AIChE J.* **7** (1961) 264; *J. Chem. Eng. Data* **9** (1963) 540.
[6.38] D. Roy, G. Thodos, *Ind. Eng. Chem. Fundam.* **9** (1970) 71.
[6.39] L. I. Stiel, G. Thodos, *AIChE J.* **7** (1961) 611; *AIChE J.* **10** (1964) 26.
[6.40] Y. S. Touloukian, C. Y. Ho: "Thermal Conductivity. Thermophysical Properties of Matter," *The TPRC Data Ser.*, vol. 3, Plenum Press, New York 1972.
[6.41] D. T. Jamieson, J. B. Irving, J. S. Tudhope: *Liquid Thermal Conductivity: A Data Survey to 1973*, H. M. Stationery Office, Edinburgh 1975.
[6.42] T. Maejima in [6.6, p. 550].
[6.43] L. Riedel, *Chem. Ing. Tech.* **21** (1949) 349; **23** (1951) 59, 321, 465.
[6.44] G. Latini, M. Pacetti, *Therm. Conduct.* **15** (1978) 245.
[6.45] C. Baroncini, G. Latini, P. Pierpaoli, *Int. J. Thermophys.* **5** (1984) no. 4, 387.
[6.46] L. A. Robbins, C. L. Kingrea, *Hydrocarbon Process. Pet. Refiner* **41** (1962) 133.
[6.47] A. Missenard, *Compte Rendue* **260** (1965) 5521.
[6.48] K. Sato in [6.6, p. 550].
[6.49] M. Negvekar, T. E. Daubert, *Ind. Eng. Chem. Prod. Res. Dev.* **26** (1987) 1362.
[6.50] A. Missenard, *Rev. Gen. Therm.* **5** (1970) no. 101, 649.
[6.51] C. C. Li, *AIChE J.* **22** (1976) 927.
[6.52] W. Jost: *Diffusion in Solids, Liquids, Gases*, Academic Press, New York 1960.
[6.53] E. N. Fuller, J. C. Giddings, *J. Gas Chromatogr.* **3** (1965) 222.
E. N. Fuller, K. Ensley, J. C. Giddings, *Ind. Eng. Chem.* **58** (1966) no. 5, 18.
[6.54] L. S. Tee, G. R. Kuether, R. C. Robinson, W. E. Stewart, *Am. Pet. Inst. Div. Refin.* May 1966.
[6.55] S. Takanishi, *J. Chem. Eng. Jpn.* **7** (1974) 417.
[6.56] E. L. Cussler: *Diffusion: Mass Transfer in Fluid Systems*, Cambridge University Press, Cambridge 1984, Chaps. 3, 7.

[6.57] D. Reinhardt, K. Dialer, *Chem. Eng. Sci.* **36** (1981) 1557.
[6.58] W. Hayduk, S. C. Cheng, *Chem. Eng. Sci.* **26** (1971) 635.
[6.59] W. Hayduk, B. S. Minhas, *Can. J. Chem. Eng.* **60** (1982) 295.
[6.60] L. S. Darken, *Trans. Am. Inst. Min. Metall. Eng.* **175** (1948) 184.
[6.61] R. E. Rathburn, A. L. Babb, *Ind. Eng. Chem. Process Des. Dev.* **5** (1966) 273.
[6.62] M. A. Siddiqi, K. Lucas, *Can. J. Chem. Eng.* **64** (1986) 839.
[6.63] H. R. Kamal, L. N. Canjar, *AIChE J.* **8** (1962) no. 3, 329.
[6.64] A. Vignes, *Ind. Eng. Chem. Fundam.* **5** (1966) 189.
[6.65] C. R. Wilke, P. Chang, *AIChE J.* **1** (1955) 264.
[6.66] E. G. Scheibel, *Ind. Eng. Chem.* **46** (1954) 2007.
[6.67] D. F. Othmer, M. S. Thakar, *Ind. Eng. Chem.* **45** (1953) 589.
[6.68] K. A. Reddy, L. K. Draiswamy, *Ind. Eng. Chem. Fundam.* **6** (1967) 77.
[6.69] M. A. Lusis, G. A. Ratcliff, *Can. J. Chem. Eng.* **46** (1968) 385.
[6.70] M. T. Tyn, W. F. Calus, *J. Chem. Eng. Data* **20** (1975) 106.
[6.71] K. Nakanishi, *Ind. Eng. Chem. Fundam.* **17** (1978) 253.

References for Chapter 7

[7.1] J. J. Jasper, *J. Phys. Chem. Ref. Data* **1** (1972) 841.
[7.2] D. B. Macleod, *Trans. Faraday Soc.* **19** (1923) 38.
[7.3] Q. R. Quayle, *Chem. Ref.* **53** (1953) 439.
[7.4] R. C. Reid, J. Prausnitz, B. E. Poling: *The Properties of Gases and Liquids*, 4th ed., McGraw-Hill, New York 1987.
[7.5] J. A. Riddick, W. B. Bunger, T. K. Sakano: *Organic Solvents. Physical Properties and Methods of Purification*, 4th ed., vol. II, Wiley, New York 1986.
[7.6] J. D. Van der Waals, *Z. Phys. Chem.* **13** (1894) 716.
[7.7] D. I. Hakim, D. Steinberg, L. I. Stiel, *Ind. Eng. Chem. Fundam.* **10** (1971) 174.
[7.8] J. R. Brock, R. B. Bird: *AIChE J.* **1** (1955) 174.
[7.9] W. Blanke (ed.): *Thermophysikalische Stoffgrößen*, Springer Verlag, Berlin 1989.
[7.10] H. P. Meissner, A. S. Michaelis, *Ind. Eng. Chem. Ind. Ed.* **41** (1949) 2782.
[7.11] M. Tamura, M. Kurata, H. Odani, *Bull. Chem. Soc. Jpn.* **28** (1955) 83.

7. Construction Materials in Chemical Industry

HUBERT GRÄFEN, Bayer AG, Leverkusen, Federal Republic of Germany

1.	Introduction	7-1
2.	Material Requirements	7-1
2.1.	Processablility and Joining	7-2
2.2.	Mechanical Stability and its Dependence on Temperature	7-2
2.3.	Corrosion Resistance	7-4
2.4.	Resistance to Wear	7-5
3.	Choice of Materials	7-5
4.	Quality Assurance Through Material Tests and Checking of Fabrication and Functioning	7-7
5.	Properties and Applications of Materials	7-8
5.1.	Steels	7-8
5.1.1.	Unalloyed and Low-Alloy Steels for Vessels and Pipelines	7-8
5.1.2.	Steels with High-Temperature Strength	7-10
5.1.3.	Heat-Resistant Steels	7-10
5.1.4.	Steels for Low Temperatures	7-11
5.1.5.	Steels Resistant to Pressurized Hydrogen	7-11
5.1.6.	Stainless Steels	7-13
5.1.6.1.	Technical Properties	7-16
5.1.6.2.	Chemical Properties	7-18
5.1.6.3.	Development State of Stainless Cr–Ni Steels	7-21
5.2.	Cast Iron	7-22
5.3.	Nickel and Nickel Alloys	7-23
5.3.1.	Nickel–Copper Alloys	7-24
5.3.2.	Nickel–Chromium Alloys	7-24
5.3.3.	Nickel–Molybdenum and Nickel–Molybdenum–Chromium Alloys	7-25
5.4.	Aluminum and Aluminum Alloys	7-26
5.5.	Copper and Copper Alloys	7-28
5.6.	Lead and Lead Alloys	7-30
5.7.	Zinc and Zinc Alloys	7-31
5.8.	Tin and Tin Alloys	7-32
5.9.	Titanium, Zirconium, Niobium, and Tantalum	7-32
5.10.	Organic Materials	7-35
5.10.1.	Selection Criteria	7-36
5.10.2.	Properties and Application Criteria	7-36
5.10.3.	Thermosetting Plastics	7-38
5.11.	Inorganic Nonmetallic Materials	7-39
5.11.1.	Glass	7-39
5.11.2.	Graphite	7-39
5.11.3.	Refractory and Acid-Resistant Bricks	7-40
5.11.4.	Engineering Ceramics	7-40
6.	References	7-41

1. Introduction

Components of chemical plant are generally subjected to thermal, chemical, and mechanical stresses. The combination of these stresses places very heavy demands on plant materials, especially with regard to corrosion. Thus even unalloyed and low-alloy steels have to meet very strict quality requirements. They must have high purity, for example, and their nonmetallic inclusions must be finely dispersed, since they affect the ability of plant to withstand corrosion cracking, such as stress corrosion cracking and damage by hydrogen. By far the most important materials are the highly alloyed stainless steels and nickel-based alloys, though aluminum, copper and their alloys, and refractory metals, and organic and inorganic materials are also important. In this article the applications of materials in process plant manufacture are described and development trends are discussed.

2. Material Requirements

In the choice of materials for production plant in the chemical industry, there are three basic considerations:

1) The processability of the material in its commercially available form (sheet, piping, profiles, etc.)
2) The ability of materials to withstand production processes. This is a complex property that includes mechanical stability and its dependence on temperature, resistance to corrosion, and possibly resistance to wear also.
3) The costs of materials, of their processing, and of the inspections of the chemical apparatus during its useful lifetime.

As so many factors have to be taken into consideration, choosing materials is not easy, especially if the plant is to be used for a recently developed chemical process. The prospect of a suitable choice is best where the material scientist, chemical engineer, plant designer, and plant engineer have worked closely together [1], [2].

2.1. Processability and Joining

Shaping (e.g., bending, rounding, and flanging), separating (e.g., cutting and machining) and joining (e.g., welding, bonding, and pipe rolling) are particularly important processes in the fabrication of chemical plant. Coating and processes that modify the properties of materials (e.g., quenching, tempering, nitriding, age hardening) are also widely used. The choice of fabrication processes depends on the properties of the material concerned, e.g., on its suitability for cold shaping or welding. Fabrication must not affect materials so drastically that their resistance to the conditions of subsequent use is significantly impaired.

Joint welding is by far the most important joining process in chemical plant fabrication, while build-up welding is used widely both for coating in original plant fabrication, and for repairs [3]–[5]. The weldability of a component depends on the weldability of the material and on the design and fabrication of the part [6]. Although each factor may be decisive on its own, the interplay of factors must never be overlooked. It is thus pointless to choose a steel with the highest possible yield strength unless one has checked that dangerous peak stress will not occur (through deficiencies of design) and that welding defects will not be caused by the use of welding techniques unsuited to the material.

Bonding is used for materials that cannot be welded or whose properties are changed excessively by welding [7], [8]. As adhesive bonds cannot be exposed to elevated temperatures, this method of joining parts has acquired little importance for chemical apparatus.

2.2. Mechanical Stability and its Dependence on Temperature [9]

Chemical apparatus in use is subjected to widely differing mechanical stresses. The possible variations of mechanical stress with time are shown schematically in Figure 1. These stresses may be monoaxial, biaxial, or triaxial.

The behavior of materials under polyaxial stress—the states of stress in plant components are almost always of this kind—is seldom known as such, because strength data are usually available only as yield strength and tensile strength for loads exerted monoaxially in a tensile test. Therefore, where a particular exposure is concerned, the material's mechanical stability must be assessed by comparing its known mechanical strength values with a stress calculated according to the appropriate theory of strength [10].

In the fabrication of chemical apparatus preference is given to materials that are easily shaped, since they react to the application of excessive force by undergoing energy-consuming changes of shape instead of simply breaking. Temperature has an important influence on the mechanical stability of materials. As the temperature rises, strength decreases, ease of shaping increases, and creep behavior becomes the main factor determining mechanical stability.

Thus, above a limit temperature, which depends on the material, the yield strength or tensile strength at the envisaged operating temperature can no longer serve as a characteristic value for calculating the stress of a chemical apparatus (Fig. 2) [9].

The strength of metals increases with decreasing temperature. Therefore, components intended for use at room temperature might be

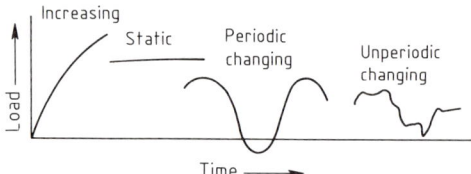

Figure 1. Forms of the time function of mechanical stress

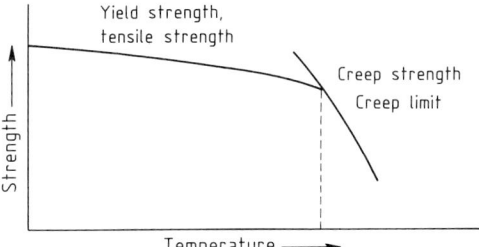

Figure 2. Dependence of characteristic strength data on temperature (schematic)

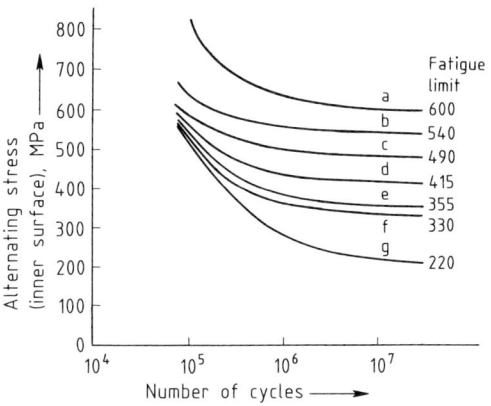

Figure 4. Influence of the surface quality on alternating stress tests on high-presure tubes (nominal bore 6 mm; nominal pressure 250 MPa) made of alloy 30CrMoV9 (material no. 1.7707; HB30 ≈ 3000 MPa; tensile strength = 900–1100 MPa)
a) Cold-worked; b) Nitrided; c) Electropolished; d) Nitrided; e) Polished; f) and g) As-delivered

expected to be more resistant to loads exerted at lower temperatures. That this is only partly true is explained by the deformation behavior of materials. The characteristic strain values of metals tend to drop suddenly as the temperature decreases, and this is accompanied by an increase in the tendency to undergo brittle fracture. The inherent risk of brittle fracture is not accounted for directly by strength calculations.

Susceptibility to brittle fracture is represented by values for elongation at rupture and reduction of area in the tensile test, and also by the transition temperature obtained from a plot of the notched-bar impact energy versus temperature (Fig 3) (see also → 10. Mechanical Properties and Testing of Metallic Materials). In choosing a material to suit a particular set of requirements, adequate safety must be ensured by taking this transition temperature into consideration, as well as the influences exerted on the material during its production and processing. In the case of high-strength materials, it is also ad-

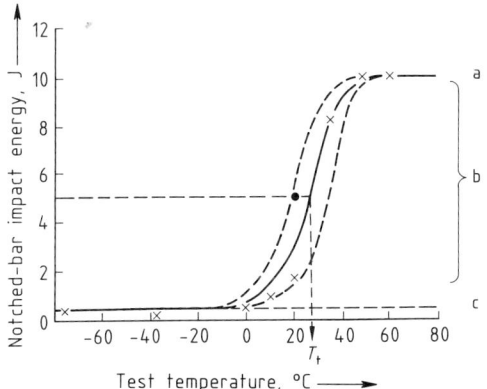

Figure 3. Transition temperature T_t of the notched-bar impact energy
a) High level, fracture on working; b) Transition range, mixed fracture; c) Low level, brittle fracture

visable to perform fracture toughness testing [11].

Where alternating stresses occur, determining suitable characteristic material values that will enable a part to be designed so that it satisfies safety requirements is often difficult, because the combinations of loads exerted in practical use cannot be accounted for fully by measuring the alternating stress endurance limit (e.g., by the Wöhler method). The shape and surface condition of the part must also be considered (Fig. 4). More realistic criteria can be obtained by determining the resistance to service conditions of components themselves. The currently valid AD (Arbeitsgemeinschaft Druckbehälter) leaflets [12] on calculations for pressure vessels or their components are based on the assumption that these vessels are normally subjected to static loads. If the pressure fluctuates, the resultant additional stresses can be taken into account according to AD leaflet S 1, in which allowance for them is made by reducing the permissible stresses (Fig. 5). Prediction of useful lifetime is particularly difficult when alternating stresses and creep stresses are superimposed.

In such cases there is still no accepted method for prediction of useful life. It would also appear unrealistic to seek "universally valid" theories and better to simulate typical stress processes in test specimens similar to plant components, thus providing the design engineer with useful life data for particular kinds of materials and prob-

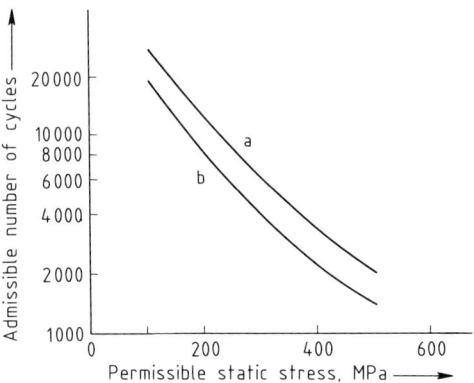

Figure 5. Admissible number of cycles for pressure vessels as a function of stress
a) Range of alternating compressive stress = 90%;
b) Range of alternating compressive stress = 100%

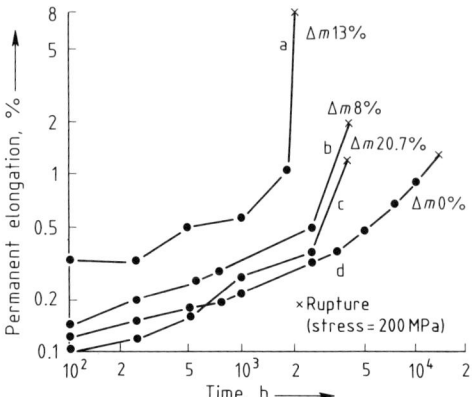

Figure 6. Creep rupture behavior of copper at 95 °C
a) 0.5 M H_2SO_4 (air saturated); b) 0.5 M H_2SO_4 (air); c) Air (precorroded in 0.5 M H_2SO_4); d) Air
Δm = Mass loss, wt%

lems. According to the available experience it is best to design plant and components so that they can withstand the stress that gives the shortest useful life.

2.3. Corrosion Resistance
(→ 8. Corrosion)

The longevity of most apparatus depends not only on mechanical loads but also on the nature and spectrum of the corrosive ambient medium. As a rule the characteristic symptom of corrosion damage is not the loss of material but an impairment of function and load-bearing capacity. Uniform corrosion, for example, may cause a considerable loss of mass before the serviceability of a part begins to suffer. Localized or selective corrosion proceeds rapidly towards the inside of a part, such that a notching effect is exerted on parts that bear mechanical loads. Corrosion of this kind may rapidly cause vessels to leak, or parts to fail through low-ductility fracture.

If the surface becomes creviced or pitted, the stress becomes nonuniform, and the operational stability of the component under alternating stresses is impaired. This also reduces the static fatigue resistance under constant load, as shown by the example in Figure 6 [13]. The static fatigue resistance of copper samples exposed simultaneously to corrosion and mechanical loading is impaired to a much greater extent that of precorroded samples that have already undergone a considerably greater loss of mass and are then stressed mechanically. This exemplifies the interaction between corrosion and creep. For iron-based and nickel-based alloys, this interrelation affects the static load fatigue resistance of parts exposed to corrosion processes at high temperatures.

Basically it may be assumed that the behavior of a part exposed to corrosive influences depends largely on the stability of the material, which, in turn, depends mainly on the film which forms at the surface. Efforts are therefore being made to find alloys that can improve the conditions for surface film formation. It is also necessary to ensure that the microstructures of alloys remain as stable as possible, so that demixing (which, for example, could impair the passivation capability of alloys) cannot occur in welding, hot forming, etc. This is particularly important because precipitation may strengthen the tendency towards local activation (local removal of the passive layer), whereby local corrosion may be initiated.

Highly localized corrosion that results from simultaneous chemical attack and mechanical stress is particularly serious. The criterion of failure here is the occurrence of low-ductility fracture. Stress corrosion cracking, fatigue cracking, and water-induced cracking are phenomena of this kind. They are particularly important with regard to serviceability, useful life, and operational reliability.

The special danger of anodic stress corrosion cracking and hydrogen-induced cracking is that they can rarely be detected while the cracks are spreading and before the apparatus starts to leak or a part breaks. Although anodic stress corrosion cracks outwardly resemble brittle fractures, the metal itself retains its ductility.

The initiation of cracking by the various stress corrosion cracking mechanisms that result from the action of a given corrodent on a given material depends very much on the mechanical stresses involved.

In the case of fatigue cracking the loss of fatigue strength for finite life and loss of fatigue strength have caused considerable difficulties in the dimensioning of parts.

2.4. Resistance to Wear
(see also → 9. Abrasion and Erosion)

Since all production processes, except chemical reactions, involve physical operations such as comminution, conveyence, and separation of phases, the wear resistance of materials is important for many types of chemical apparatus. Models of the various kinds of wear have been developed [14]; in practice, however, the conditions are complex, for not only may several kinds of wear occur simultaneously, but overlap and interaction of wear and corrosion processes can also arise.

Thus, as a rule, only practical trials reveal the actual stress conditions involved. Compared to corrosion, however, wear is less problematic because it does not cause cracklike damage, and its effects are more easily repaired, e.g., by build-up welding.

3. Choice of Materials

Designers of chemical apparatus must pay careful attention to materials as well as the purpose for which the apparatus is intended. This is important at all stages, from initial planning to the detailed drawing of the finalized design. For a selection process, however, not just the function and properties needed by the components, but also the operations by which the components are fabricated from the material, must be taken into consideration. Not only do these operations call for definite properties on the part of materials, they may also affect the behavior of components under service conditions.

The demands made on chemical plant in use are becoming increasingly strict, and the variety of fabrication processes and range of materials available are growing. These factors, together with the need to use economical fabrication techniques, have given the choice of materials a complexity that calls for systematic consideration. A basic procedure for the selection of materials for chemical plant is shown in Figure 7. Clearly the demands of the application and those of the fabrication process must be taken into account as fully as possible so that the overall requirements are correctly formulated and properly interpreted in the fixing of verifiable property data. At the same time the behavior of materials under operating conditions and those of fabrication must be known so that one can judge reliably whether or not a given material is suitable for a given purpose.

Often, these preconditions cannot be met entirely. In many cases, therefore, model experiments and serial testing are necessary as empirical aids to selection. Establishing and then interpreting the combination of mechanical and corrosion-chemical exposures, all of which are often supplemented by wear, is particularly difficult for chemical plant. Frequently, corrosion behavior is treated in the selection procedure as the decisive criterion. In fact, experience in the operation of chemical plant has shown that much of the damage that occurs to chemical apparatus arises from the use of materials that are insufficiently resistant to corrosion under practical conditions, and that damage of purely mechanical origin is generally much less frequent [15].

Therefore, technical rules exist for designing and dimensioning apparatus to meet combined mechanical and thermal stresses. The standardized calculation procedures based on strength and ductility data are intended to provide the designer with a proven basis on which to work. However, where wear processes (corrosion and abrasion) are concerned, that is not possible; here, instead, special design criteria must be worked out in each individual case.

A shortlist of materials can be drawn up from corrosion tables, which give the corrosion behavior of specific materials [16], [17]. Nevertheless, the most extensive tables can never take into consideration all the conditions of a given practical case. Almost always there is more than one corrosive agent, because chemical processes normally involve starting materials and end products, intermediates and byproducts, often with unknown properties, and solids that promote

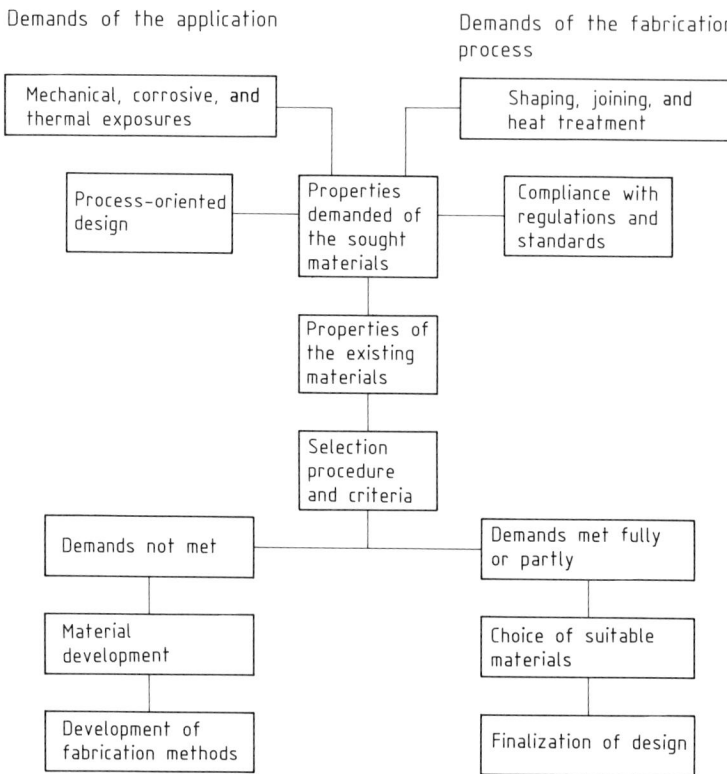

Figure 7. Basic procedure for the choice of materials for chemical apparatus

wear. Furthermore, reactions are influenced by pressure and temperature. Thus the corrosion behavior of materials cannot be given in tables with any degree of accuracy, and terms such as resistant or nonresistant may be inappropriate in an actual application. The results of laboratory tests are considerably more useful.

The most reliable predictions are those based on plant tests performed either in a plant engaged in practical production or in a pilot plant. Often, however, such tests are very laborious. Even in cases where tests have been carried out, corrosion may still occur, possibly because the stability of a material has been impaired by excessive cold forming or by excessive heat input in welding, or because the states of the material in the test plant and in the subsequently constructed chemical apparatus were not identical.

In view of this complexity, close cooperation between material scientists, chemical engineers, chemists, designers, apparatus manufacturers, and—where chemical apparatus subject to compulsory testing, e.g., pressure vessels, is concerned—the representatives of the officially recognized supervisory authorities, is particularly desirable.

Because corrosion reactions are so complex and depend on so many factors, an assessment of the corrosion behavior of a given material–corrodent system will differ greatly according to the method of investigation. False interpretations can be avoided and comparison improved if one has a good knowledge of the kinds of information provided by the chosen test method as well as knowing its limitations. It is therefore understandable that standardized corrosion tests are few in number or contain more general instructions. Frequently, they cover only the states of materials that are most favorable from the aspect of corrosion chemistry (that is to say the homogeneous structural states existing at the time of delivery), describing, for example, the testing of stainless steels for resistance to intergranular corrosion (DIN 50 914) or the testing of stainless steels in boiling nitric acid (DIN 50 921 and ASTM A 262-70). A comparison of the sensitivities of unalloyed and low-alloyed steels to intergranular stress corrosion cracking is given in DIN 50 915.

A common feature of all investigation methods is that they refer to the material, and not directly to the stresses that occur in practice. Supplementary chemical or electrochemical tests simulating the conditions that occur in practice are therefore necessary in the selection of materials. Chemical tests are performed on samples exposed to gaseous and liquid corrodents at the operating pressure and temperature, whereas in electrochemical corrosion tests, the dependence of corrosion on potential is investigated and provides information on the effects of variables which alter the potential.

The results of chemical corrosion tests and electrochemical corrosion tests may differ fundamentally. The differences arise because in chemical tests the corrosion potential may vary with time. In electrochemical tests the potential is fixed, so that, though they are more informative in some respects, their practical value may be limited in others.

The characteristic stability data needed in the choice of materials can be obtained in two ways:

1) By inserting suitable material specimens in pilot plants or in existing, but not yet entirely satisfactory, production plants,
2) By performing laboratory corrosion tests.

With regard to the corrosive medium in corrosion tests, careful attention must be paid to a number of important factors, such concentration, temperature, dissolved gases, impurities, solid matter, and rate of flow. As there are so many parameters, tests should use suitable specimens in a plant and under conditions as close as possible to those of practical use.

The specimens should be placed at several representative locations. In a distillation column, for example, they should be placed in the pit, in the vicinity of the feed point, and in the head.

The shape of a specimen used depends on the types of corrosion expected. Where general corrosion predominates, it is sufficient to use welded sheets that have the surface quality of the parts to be used in practice. If stress corrosion cracking is involved, plastically and elastically stressed specimens or tuning fork specimens are used.

Material specimens are generally attached to interior parts, such as an agitator or a thermometer protection tube. To prevent polarization of specimens through contact with plant components, and consequent falsification of measurements, the fixing screws are placed in insulating sheaths.

In addition to the exposure of test specimens, the parts of a pilot plant itself should be examined as a source of further information on corrosion behavior. For critical plant units such as heat exchangers, it has been found advisable to remove a tube from time to time, to cut it open, and to examine the inside and outside surfaces for corrosion symptoms.

If stability tests cannot be performed in pilot installations, laboratory tests under conditions as close as possible to those of practice must be carried out. They should also be carried out as a supplementary measure in cases where plant tests have not clearly revealed the time dependency of the corrosion processes, and severer conditions of attack may therefore help to complete the picture.

General guidelines on the conduct of corrosion tests are given in DIN 50 905. These guidelines should be followed as a route to reciprocally comparable results that can be transferred to the plant component in question.

For further details on chemical and electrochemical corrosion tests, see → 8. Corrosion.

4. Quality Assurance through Material Tests and Checking of Fabrication and Functioning

To ensure that a chemical apparatus will be properly fabricated and that its functioning free from disturbances, many tests must be carried out before and during fabrication, before the apparatus is put into operation, and while it is in actual use. At the stage of testing of materials when they leave the factory and arrive at the plant manufacturer's establishment it is normally fairly easy to ascertain whether or not they are of the required quality, as the required property data will have been agreed on the basis of quality standards or similar specifications. The same cannot be said of fabrication checks, which include, in particular, checking the construction for compliance with the blueprints and examining the welding work. Here the test engineer has more latitude, especially in interpreting the results of nondestructive tests. It must be emphasized that the first step in fabrication testing should be a thorough visual inspection; this gives the tester an overall impression of the care taken by the manufacturer. Checking compliance with legislation and official regulations is mainly the responsibility of the Technical Control Associations and the plant safety inspection departments of the large chemical firms. Functioning tests check the behavior of the apparatus under conditions close to those of practice. Of exceptional importance are the regular inspections of plant in use that are intended to reveal incipient damage early enough to ensure that plant shutdown can be avoided and any necessary repairs need not be undertaken in haste. Once again, visual inspection takes precedence. Much emphasis is also placed on nondestructive methods—thickness measurement, for example, and tests that reveal initial cracking [18]. Testing of plant in service also provides knowledge that may be useful in selecting materials for new chemical apparatus by revealing weaknesses of design, materials, or fabrication.

5. Properties and Applications of Materials

5.1. Steels

($\rightarrow$ Steel, treated in the A series)

Steels are still the materials most commonly used for chemical plant. Their variety of alloy compositions and the range of variation of their properties permit an exceptional degree of adaptation to practical requirements.

5.1.1. Unalloyed and Low-Alloy Steels for Vessels and Pipelines

Sheet and piping made of unalloyed and low-alloyed steels with carbon contents up to ca. 0.25 wt % are used extensively in chemical plant for vessels (pressure vessels, storage vessels, etc.) and pipelines that are not exposed to particularly severe corrosion. The main standards and designations of several typical steels are compiled in Table 1. Comparable steels with comparable properties are described in the standard specifications of other countries—see, for example, SAE, AISI, ASTM (United States), BS (Great Britain), NF (France), SS (Sweden), UNI (Italy) and NBN (Belgium). The steels are normally processed in the normalized state.

Owing to the introduction of weldable fine-grained structural steels with yield strengths of ≥ 360 MPa, boiler plate consisting of unalloyed steels, whose yield strengths are considerably lower, is used on a smaller scale than it was formerly. The use of high-strength heat-treated structural steels with yield strengths of ≥ 700 MPa [19]–[23] will presumably increase. Pipes consisting of welded steel strip will be used increasingly instead of seamless pipes [24].

Steels are suitable for fusion welding by all processes, but the welding filler metals must be suited to the base material. Specific welding guidelines must be adhered to increasingly as the alloy content and yield strength rise; these guidelines can be found by consulting either the standards or manufacturers of steels and welding filler materials. Processing guidelines for weldable fine-grained structural steels are compiled in DIN standards.

Normalized fine-grained structural steels are now very popular. They combine the excellent strength and toughness imparted by fine-grain hardening [25] with the advantages offered by modern ladle metallurgy in respect of purity and microalloying.

The recently introduced blast or shot injection of calcium compounds to reduce the sulfide content of steel (TN process) and the addition of cerium and zirconium to produce globular, finely distributed sulfide inclusions should also be

Table 1. Quality standards for unalloyed and low-alloy steels widely used in chemical plant

Standard	Short title	Typical steels	
		Short name	Material number
DIN 17 100	Steels for general structural purposes	RSt 37–2 St 44–2	1.0038 1.0044
DIN 1629	Seamless unalloyed circular steel tubes for special requirements	St 37.0 St 44.0 St 52.0	1.0254 1.0256 1.0421
DIN 17 155	Steel plates and strips for pressure purposes	H I H II 15 Mo 3	1.0345 1.0425 1.5415
DIN 17 102	Weldable, fine-grained construction steels (normalized)	StE 285 StE 355 WStE 460	1.0486 1.0562 1.8935
DIN 17 175	Seamless tubes of heat-resistant steels	St 35.8 St 45.8 15 Mo 3	1.0305 1.0405 1.5415
Stahl-Eisen-Werkstoff-blatt 087–81	Structural steels resistant to weathering	WTSt 37–2 WTSt 37–3	1.8960 1.8961

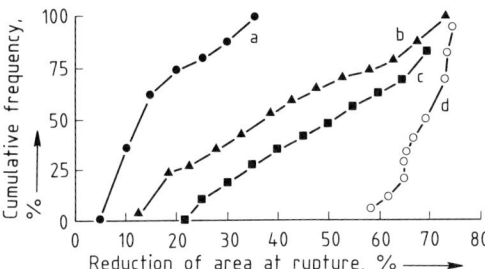

Figure 8. Influence of sulfur content on the reduction of area at rupture in the thickness direction of StE 355 steel plates
a) 40–55 mm plate, 0.015–0.054% S; b) 40 mm plate, ≤ 0.010% S; c) ≥ 50 mm plate, ≤ 0.006% S; d) 20–50 mm plate, TN treated, 0.002% S

mentioned. These processes improve purity, block segregation, weldability, and toughness.

In special cases, these desulfurization methods enable the sulfur content to be reduced to below 0.005%. Consequently, and through the influence exerted on the nature of the precipitation, the anisotropy of the steel's properties is reduced considerably, as may be seen from the percentage reductions of area at fracture of fine-grained structural steels in the thickness direction of the sheet (Fig. 8) [26]. Figure 9 [26] shows that the fatigue resistance is also improved.

Where damage occurs, the leak-before-failure behavior imparted by high toughness is exceptionally favorable because the risk of sudden failure as a result of unstable crack propagation is practically eliminated. The high purity of fine-grained steels, together with the fine dispersion of their nonmetallic inclusions, notably the sulfides, is very favorable with regard to corrosion resistance. Coarse sulfides promote the formation and growth of cracks both in hydrogen-induced crack formation and in stress corrosion cracking [27]. Upon chemical exposure to aqueous solutions, the sulfides, especially under the influence of the hydrolytic acidification that occurs in cracks, may be converted to H_2S, which is a vigorous promoter of hydrogen cracking.

It is now believed that crack growth in anodic stress corrosion cracking is also promoted by hydrogen. That would explain why steel StE 355, in particular, behaves favorably towards such crack-initiating media as alkalies, nitrates, and liquid ammonia. This steel is now a standard material for these media. For greater safety, the finished apparatus should be subjected to stress relief annealing.

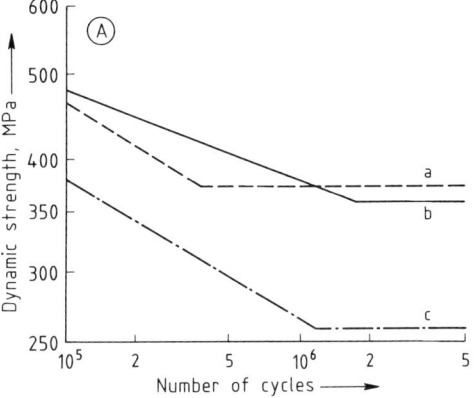

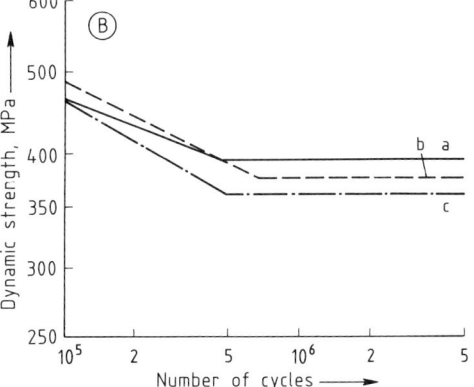

Figure 9. Influence of the TN process on the dynamic strength of steel StE 355 (stress ratio $R = 0$, thickness of plate: 50 mm)
A) Untreated; B) Calcium treated (TN process)
a) Transverse; b) Longitudinal; c) Normal

Considerable progress has been made through efforts to improve the economy of structural steels by raising their strength. Starting from steel St 52-3, it has been possible to raise the yield strength from 355 to ca. 900 MPa. The first step in this direction consisted in raising the alloy contents of normalized structural steels. But the scope for improvement in this way was limited by the fact that increasing the alloy content impairs the cold cracking resistance in welding. Agreement has therefore been reached that the yield strength of normalized steels should not exceed 500 MPa.

Considerably higher yield strengths, even at low alloy contents, are obtainable by quenching and tempering. In the case of heat-resistant structural steels, air tempering is preferred. Water quenching followed by tempering, on the other hand, is more favorable for steel needing high toughness. By far the most important representative of this group of steels in terms of quantity is StE 690, whose minimum yield strength is 690 MPa (Euronorm 137). Steels with comparable yield strengths are also used on a large scale for pressure vessels and other chemical apparatus, such as the pressure-bearing parts of multistory vessels. StE 890 has been used to an increasing extent recently [28].

5.1.2. Steels with High-Temperature Strength

High-temperature structural steels are used in the chemical industry mainly for heavily stressed parts of steam boilers (e.g., collectors, drums, piping) and for pressure vessels and tubes operating at up to ca 800 °C [29]. These materials are also very important in reactor technology. According to their maximum service temperatures that can be maintained for long periods, the following groups are distinguished:

1) For temperatures up to ca. 400 °C: unalloyed steels, in some cases melted as fine-grained structural steels, whose main strength-related property is the high-temperature yield strength. These steels are standardized in DIN 17 155 (boiler plate types), DIN 17 175 (pipes) and DIN 17 102 (fine-grained structural steels) and are suitable for all fusion welding methods. Typical steels belonging to this group are the grades HI (1.0345), St 35.8 (1.0305), and WStE 355 (1.0565).
2) For temperatures of 400–550 °C: unalloyed and low-alloy steels having good long-term, high-temperature strength-related property values. The main alloying elements are manganese (up to 1.3 wt %), chromium (up to 2.5 wt %), and molybdenum (up to 1.2 wt %), vanadium also being used occasionally (up to 0.5 wt %). These materials are standardized in DIN 17 155, DIN 17 175, DIN 17 240 (nuts and bolts), and DIN 17 245 (steel castings). These steels are suitable for fusion welding, though preheating and post-welding annealing may be necessary, details of which can be found in the standards.
3) For temperatures of 550–600 °C: martensitic steels containing up to ca. 12 wt % chromium and additions of molybdenum, vanadium, nickel, and tungsten [30]. These are standardized in DIN 17 459 (sheet, pipes, forgings) and in DIN 17 245 (steel castings). These steels are supplied in the quenched and tempered state; a typical representative is X 20 CrMoV 12 1 (1.4922). Being air-hardenable, these steels must be kept at 250 to 450 °C during welding, after which they must be cooled to ca. 120 °C and then immediately annealed.
4) Temperatures of 600–800 °C: austenitic steels containing 16–21 wt% chromium, 11–32 wt% nickel, and additions of molybdenum, tungsten, niobium, tantalum, aluminum, and other elements, e.g., X 8 CrNiNb 16 13 (material no. 1.4961), X 8 CrNiMoNb 16 16 (material no. 1.4981), and X 8 NiCrAlTi 32 21 (material no. 1.4959), which are standardized in DIN 17 459.

Table 2. Approximate temperature limits for the use of steels in a weakly oxidizing flue-gas atmosphere

Steel	Material number	Standard	Approximate temperature, °C
St 35.8	1.0305	DIN 17 175	500
St 45.8	1.0405	DIN 17 175	500
15 Mo 3	1.5415	DIN 17 155	530
13 CrMo 44	1.7335	DIN 17 155	560
10 CrMo 9 10	1.7380	DIN 17 155	590
X 20 CrMoV 12 1	1.4922	DIN 17 175	600
X 8 CrNiNb 16 13	1.4961	DIN 17 459	750
X 8 CrNiMoNb 16 16	1.4981	DIN 17 459	750

These steels are welded with filler metals of the same composition, and at the lowest possible heat input to avoid hot cracking of the weld metal.

As austenitic steels are very expensive, they are used only in the zones which reach the highest temperatures, outside of which low-alloy steels are used. Special techniques have to be used to form welds between ferritic and austenitic steels because of the diffusion processes that occur in welding and because of the differences in thermal expansion [31].

For temperatures above 800 °C only iron–chromium–nickel alloys and nonferrous alloys based on cobalt or nickel are suitable [32], [33]; see also Section 5.3.

Table 2 gives approximate figures for the highest temperatures at which a number of steels with high strength at elevated temperature can be used in a weakly oxidizing flue-gas atmosphere. Long-term heat stability values are compiled in [34]; the importance of these values in calculating the useful lives of components subjected to this exposure is discussed in [35].

5.1.3. Heat-Resistant Steels

Heat-resistant steels are those that have good strength-related property data and are distinguished by exceptional resistance to exposure for short or long periods to hot gases or combustion products at temperatures exceeding 550 °C, and which are thus resistant to scaling [36]. Rotary furnaces, cracking units, and muffle furnaces are examples of chemical plant in which these conditions arise.

Resistance to scaling is obtained mainly by alloying with chromium, but further improvements are possible if silicon and aluminum are added. The most commonly used heat-resistant steels are standardized in Stahl-Eisen-Werkstoffblatt 470-76. Distinctions are made between the ferritic steels, such as X 10 CrAl 7 (1.4713), X 10 CrAl 13 (1.4724) and X 10 CrAl 24 (1.4762); the ferritic-austenitic steel X 20 CrNiSi 25 4 (1.4828); and austenitic steels such as X 15 CrNiSi 20 12 (1.4828), X 15 CrNiSi 25 20 (1.4841) and X 10 CrNiAlTi 32 20 (1.4876).

The above-mentioned publication also gives long-term heat resistance values for periods of up to 10^5 h. The highest service temperatures listed, which extend to 1200 °C, apply to air. They may be reduced greatly by admixtures to the air, e.g., of water vapor or sulfur-containing or carburizing matter, because the reaction products do not form sufficiently thick surface layers.

Within certain temperature ranges, heat-resistant steels tend to become brittle. Ferritic steels with $\geq$ 12 wt % chromium do so between 400 and 530 °C. This so-called 475 °C embrittlement can be eliminated by annealing briefly above 600 °C. In the case of ferritic steels with $\geq$ 17 wt % chromium and austenitic steels, an intermetallic iron–chromium σ-phase, which causes severe embrittlement, is formed between 600 and 900 °C. It endangers particularly the weld interfaces.

Within the same temperature range, austenitic steels precipitate chromium carbides, which further reduce the toughness. The steel most severely affected is X 15 CrNiSi 25 20 (1.4841), which should be used only above 900 °C. The σ-phase and the carbides can be redissolved by annealing above 1000 °C, followed by quenching.

The heat-resistant steels can be welded by the usual methods, provided the guidelines for alloyed steels are followed. Suitable filler metals are listed in Stahl-Eisen-Werkstoffblatt 470-76. It should be noted that in ferritic steels with $\geq$ 12 wt % chromium, coarse grains, which can no longer be removed by heat treatment, are formed at temperatures above 900 °C.

Heat-resistant cast steel, as used in the manufacture of pipes by centrifugal casting, for example, is standardized in Stahl-Eisen-Werkstoffblatt 471-76.

5.1.4 Steels for Low Temperatures

Refrigeration is very important in the chemical industry, e.g., in the fractional distillation of hydrocarbons or storage and transportation of liquid gases. Steels for the pressure vessels must still have sufficient toughness at the lowest operating temperature. The degree of low-temperature toughness depends particularly on the steel's composition and heat treatment. Notched-bar impact energies, measured on DVM (Deutscher Verband für Materialforschung und -prüfung e. V., Berlin) longitudinal specimens, of $\geq$ 40 and $\geq$ 60 J/cm^2, for cast steel and other steels, respectively, are regarded as evidence of sufficient low-temperature toughness. Hence the lowest service temperature of a steel with good low-temperature tenacity is that at which the notch impact energy is still above this limit. Steels for low temperatures can be divided into four groups:

1) Unalloyed aluminum-killed steels for service temperatures down to -50 °C in the normalized state and -80 °C in the heat-treated state; e.g., TTSt 35 (1.1101).
2) Unalloyed and low-alloy weldable fine-grained structural steels in the normalized state for service temperatures down to -60 °C, e.g., TStE 380 (1.8910).
3) Nickel-alloyed heat-treatable steels with 1.5–9 wt % Ni for service temperatures of -100 to -190 °C, e.g., 12 Ni 19 (1.5680).
4) Austenitic chromium–nickel steels for service temperatures extending close to absolute zero; e.g., the steels specified in DIN 17 440 and DIN 17 441.

The relevant quality standards are DIN 17 173 and DIN 17 174 for groups 1 and 3 and DIN 17 102 for group 2, and Stahl-Eisen-Werkstoffblatt 685–82 for cast steel with good low-temperature toughness.

In the welding of steels for low temperatures it is necessary to use filler materials that give a weld metal whose strength and toughness are equal to those of the base material [37]–[39].

5.1.5. Steels Resistant to Pressurized Hydrogen [40], [41]

Steels for high-pressure plant in the chemical industry can divided into the following two groups:

1) Steels for components subjected to purely mechanical loads or to pressurized hydrogen, in both cases at temperatures of $\leq$ 200 °C,
2) Steels for components exposed to pressurized hydrogen at temperatures > 200 °C,

For the conditions of group 1, as present in the production of low-density polyethylene (pressures of up to about 400 MPa), mainly unalloyed and low-alloy heat-treatable steels are used; these are standardized in DIN 17200 and in the case of large forgings, in Stahl-Eisen-Werkstoffblatt 550-76. If relatively good corrosion resistance is required, stainless heat-treatable chromium and chromium–nickel steels according to DIN 17440, and occasionally hardenable stainless steels, are used instead [42]. The periodic alternation of the pressure subjects the components to a severe fatigue-inducing stress. Their shape stability therefore depends decisively not just on the choice of materials, but also on their being designed to suit the conditions of exposure as well as on satisfactory fabrication and installation [43].

Hot pressurized hydrogen, which is needed in many high-pressure syntheses (e.g., ammonia synthesis and pressure hydrogenation) dissociates on the surface of many steels, diffuses into the metal, and reacts with carbon to form methane, thus causing decarburization, embrittlement, and cracking:

$$Fe_3C + 2 H_2 \rightarrow 3 Fe + CH_4$$

This reaction between hydrogen and cementite occurs mainly at the grain boundaries, the grains thus losing their cohesion. Hydrogen and methane accumulate, giving rise to high pressure whose cleaving action leads to internal microcracks. Together with the stresses exerted on the component and the loss of mechanical strength, this results finally in pronounced brittle fracture.

The resistance of carbon steels to attack by pressurized hydrogen depends partly on the ambient conditions, but also on the microstructure, cold working, effects of welding, impurities, and heat treatment. Elements that form stable carbides, such as chromium, molybdenum, tungsten, vanadium, titanium, and niobium, can be added to the steel to prevent the reaction between hydrogen and cementite. The pressurized-hydrogen-resistant steels now in use contain up to 16 wt% Cr, and also in many cases 0.2–1.5 wt% Mo. Some types are additionally alloyed with up to 0.85 wt% V (Table 3). The low-alloy ferritic steels are suitable for fusion welding, provided that they are preheated. The chromium steels have to be kept at 250–450 °C throughout welding, and partially annealed immediately afterwards. The austenitic steel (1.4988) tends to suffer hot cracking on welding. Short-term tests give very little indication of the limits to the use of steels of this kind, because the time factor is much too important and the attack of pressurized hydrogen on steels has a significant incubation period. As an illustration, Figure 10 gives stability limits for various steels, as compiled by NELSON [44]. It can be seen that even at low hydrogen partial pressures (e.g., 2 MPa) an unalloyed steel should be used only below 300 °C. Austenitic steels with 18% Cr and ca. 9% Ni have good resistance to pressurized

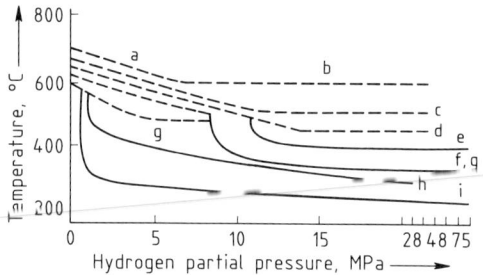

Figure 10. Resistance diagram for the attack of pressurized hydrogen on steels (Nelson diagram)
a) 1.25 Cr 0.5 Mo steel; b) 5.0 Cr 0.5 Mo steel; c) 8.0 Cr 0.5 Mo steel; d) 2.25 Cr 1.0 Mo steel; e) 2.0 Cr 0.5 Mo steel; f) 1.25 Cr 0.5 Mo steel; g) 1.0 Cr 0.5 Mo steel; h) 0.5 Mo steel; i) Mild steel

Table 3. Composition of pressurized-hydrogen-resistant steels

Steel type	Material number	Composition, wt%				
		C	Cr	Mo	V	Others
25 CrMo 4	1.7218	0.22/0.29	0.90/1.20	0.15/0.25		
16 CrMo 9 3	1.7281	0.12/0.20	2.0/2.5	0.30/0.40		
26 CrMo 7	1.7259	0.22/0.30	1.5/1.8	0.20/0.25		
24 CrMo 10	1.7273	0.20/0.28	2.3/2.6	0.20/0.30		
10 CrMo 11	1.7276	0.08/0.12	2.7/3.0	0.20/0.30		
10 CrMoV 10	1.7766	0.15/0.20	2.7/3.0	0.20/0.30	0.10/0.20	
20 CrMoV 13 5	1.7779	0.17/0.23	3.0/3.3	0.50/0.60	0.45/0.55	
X 20 CrMoV 12 1	1.4922	0.17/0.23	11.0/12.5	0.80/1.20	0.25/0.35	
X 8 CrNiMoVNb 16 13	1.4988	0.04/0.10	15.5/17.5	1.10/1.50	0.60/0.85	Ni: 12.5/14.5 N: 0.07/0.13

hydrogen; they can be exposed to hydrogen throughout the range of temperatures used in normal high-pressure processes.

5.1.6. Stainless Steels [45]

Among the highly alloyed steels, the most important group comprises those that are chemically resistant. They have chromium contents ≥ 14 wt%. As the steels of this group resist heat as well as chemicals, this group also includes a considerable number of the heat-resistant steels discussed in Section 5.1.3. Often they contain nickel in addition to chromium. The more highly alloyed corrosion-resistant (acid-resistant) steels must additionally withstand general corrosion and localized corrosion in relatively aggressive corrodents (salt solutions; acids, even at fairly high concentrations). Characteristic alloying components, apart from chromium and nickel, are molybdenum, copper, and in some cases silicon.

There are many kinds of steels and alloys, each being represented by a material number (DIN 17007) or by chemical composition (DIN 17006).

The mechanical properties of steels are determined by the microstructure, which depends on composition and heat treatment. Stainless steels are divided according to microstructure into four groups:

1) Martensitic (hardenable) steels with > 0.12 wt% C and ≤ 15 wt% Cr. They are hardened at temperatures $> 1000\,°C$ (e.g. cutting steels). After being hardened they can be improved by tempering at 500 to 600 °C—their strength thus being reduced to a desired lower level—and henceforth combine high strength with good ductility.
   In the heat treatment, a ferritic structure, with precipitated carbides of high chromium content of type $M_{23}C_6$ or M_7C_3, is formed. Fixation of chromium and the formation of chromium-depleted zones reduces the resistance to corrosion. To equalize the chromium content of the ferritic matrix again it is necessary to anneal the steel for a fairly long time at elevated temperature. Although this restores the corrosion resistance to some extent, it impairs the strength. As some of the chromium is still bound as carbide, the corrosion resistance after annealing procedure is still considerably poorer than that of the martensite.
2) Ferritic steels with body-centered cubic lattice (α-phase). Those of greatest importance are the ferritic chromium steels with ca. 17 wt% chromium. Their mechanical and technical properties, however, are unsatisfactory; when welded, they tend to become coarse-grained and brittle. Nevertheless they have the advantage of resisting stress corrosion cracking in chloride-containing media.
3) Austenitic steels with face-centered cubic lattice (γ-phase). Particularly important are the austenitic chromium–nickel steels with ca. 18 wt% Cr and 10 wt% Ni, but without addition of molybdenum. Their strength is relatively low and their ductility very good. In chloride-containing corrodents, however, they undergo transgranular stress corrosion cracking.
4) Ferritic–austenitic steels combine good mechanical properties with improved resistance to stress corrosion cracking. They are therefore used mainly under conditions which may cause fatigue corrosion and stress corrosion cracking in austenitic steels.

Pure iron has a ferritic structure. The alloying elements that may be added to iron are either ferrite-forming (chromium, molybdenum, silicon, titanium) or austenite-forming (carbon, nitrogen, nickel, manganese).

The austenitic chromium–nickel steels, with or without molybdenum, are used in the solution-heat-treated state (solution heat treatment temperature 1000–1100 °C). At solution heat treatment temperatures these steels are close to the boundary of the $\alpha + \gamma$ field in the iron–chromium–nickel phase diagram (Fig. 11). As the content of chromium increases, more nickel or nitrogen must be added to maintain the austenitic structure. If the content of molybdenum is raised, then it is also necessary to increase the content of austenitizers and/or to reduce the chromium content.

With increasing chromium content, and especially increasing molybdenum content, the tendency to segregate intermetallic compounds

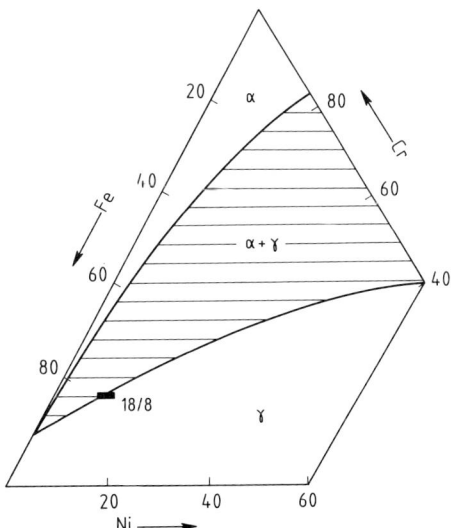

Figure 11. Ternary phase diagram of Fe–Cr–Ni (section at 1100 °C)

(σ-phase, χ-phase, Laves phase, Fe₂Mo), increases greatly (see Fig. 12). If the precipitation tendency is strong enough, intermetallic compounds may already be precipitated during the relatively short-lived heat exposure of welding. Hence the addition of molybdenum, which is desirable for reasons of corrosion chemistry, has an upper limit. In the case of austenitic chromium–nickel steels it is about 6 wt% Mo.

Addition of nitrogen retards the precipitation of the intermetallic phases and compounds considerably, as well as improving the mechanical properties by introducing nitrogen atoms into interstices in the metal lattice. Nitrogen-alloyed austenitic and ferritic–austenitic steels are therefore of industrial importance.

In ferritic–austenitic steels, the ferritizers chromium and molybdenum are enriched in the ferrite phase, whose nickel content is correspondingly depleted. The austenitic phase contains, conversely, more nickel and less chromium and molybdenum. These differences of concentration are relatively slight, however, and are only important for the use of ferritic–austenitic steels in borderline cases. The alloy composition of these steels is also chosen such that the ferritic and austenitic phases are present in the structure in approximately equal proportions and the chromium content of the austenite is not less than 17 wt%.

Annealing within certain temperature ranges may cause carbides, nitrides, and intermetallic compounds to be precipitated in the microstructure of stainless steels. Owing to the special importance of chromium and molybdenum in imparting corrosion resistance to stainless steels, the precipitation of phases and compounds that contain these elements exert strong effects. The segregation of chromium- and molybdenum-rich phases depletes the matrix in these elements. The matrix therefore loses its resistance to corrosion.

In ferritic chromium steels and austenitic chromium–nickel steels, the chromium-rich σ-phase may precipitate on annealing at 600–900 °C. For ferritic chromium steels with up to ca. 18 wt% chromium this precipitation has no importance, while in the case of molybdenum-free austenitic chromium–nickel steels it is, at the most, important only in connection with the welded material. For reasons of welding technique (prevention of hot cracking) the welded material of normal austenitic steels almost always contains some δ-ferrite, which decomposes when the steel is annealed. In a corrosion expo-

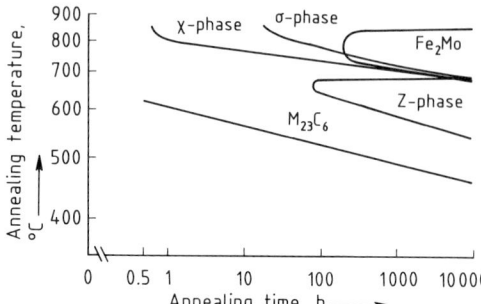

Figure 12. Precipitation fields of chromium-rich carbides $M_{23}C_6$, σ- and χ-phase, molybdenum-rich Laves phase (Fe₂Mo), and of a complex nitride (Z-phase) for the steel X 2 CrNiMo 18 14 3

sure this decomposed δ-ferrite may be selectively dissolved. Damage must be expected where a coherent δ-ferrite network is present (proportion of δ-ferrite in the structure > 10%).

The designations of the most important stainless steels for chemical plant are compiled in DIN 17 440 (Table 4).

Chemical engineering also uses a number of other special materials notable for their high resistance to pitting corrosion and stress corrosion cracking, as well as to mineral acids (Table 5).

Particularly high molybdenum contents are present in the steels listed in Table 6. The stability of the austenite is due to nickel and nitrogen.

Unlike ferritic and austenitic steels, ferritic–austenitic steels have a two-phase structure, which, in contrast to the composition of austenitic steels, is obtained by raising the contents of ferrite-stabilizing elements, such as chromium and silicon, and reducing the austenite-stabilizing nickel content (Table 7). By virtue of their high yield strength of at least 450 MPa, steels of this kind are used for components which, while exposed to corrosive media, are additionally subjected to wear (cavitation, erosion) and vibration. Being more resistant than commercial austenitic steels to stress corrosion cracking in neutral chloride solutions, they are also being used increasingly to handle aggressive cooling water. A commonly used material of this group is X 2 CrNiMoN 22 5 (1.4462), which, in addition to about 22 wt% Cr, has a nickel content of about 5.5 wt%, an Mo content of about 3 wt%, and an austenitic structure proportion of about 60% [46], [47].

A specially developed austenitic steel that contains about 4 wt% Si (X 2 CrNiSi 18 15,

Table 4. Composition of stainless steels

Short name (DIN)	US Standard (AISI)	Material number	Composition, wt%				
			C	Cr	Mo	Ni	Others
Ferritic and martensitic steels							
X 7 Cr 13	403	1.4000	<0.08	12.0–14.0			
X 7 CrAl 13	405	1.4002	<0.08	12.0–14.0			Al 0.10–0.30
X 10 Cr 13	410	1.4006	0.08–0.12	12.0–14.0			
X 15 Cr 13		1.4024	0.12–0.17	12.0–14.0			
X 20 Cr 13	420	1.4021	0.17–0.22	12.0–14.0			
X 40 Cr 13		1.4034	0.40–0.50	12.0–14.0			
X 45 CrMoV 15		1.4116	0.42–0.48	13.8–15.0	0.45–0.60		V 0.10–0.15
X 8 Cr 17	430	1.4016	<0.10	15.5–17.5			
X 8 CrTi 17	439	1.4510	<0.10	16.0–18.0			Ti ≥ 7 × % C
X 8 CrNb 17		1.4511	<0.10	16.0–18.0			Nb ≥ 12 × % C
X 6 CrMo 17	434	1.4113	<0.07	16.0–18.0	0.90–1.20		
X 12 CrMoS 17	430 F	1.4104	0.10–0.17	15.5–17.5	0.20–0.30		S 0.15–0.35
X 22 CrNi 17	431	1.4057	0.15–0.23	16.0–18.0		1.5–2.5	
Austenitic steels							
X 12 CrNiS 18 8	303	1.4305	<0.15	17.0–19.0		8.0–10.0	S 0.15–0.35
X 5 CrNi 18 9	304	1.4301	<0.07	17.0–20.0		8.5–10.0	
X 5 CrNi 19 11	305/308	1.4303	<0.07	17.0–20.0		10.5–12.0	
X 2 CrNi 18 9	304 L	1.4306	<0.03	17.0–20.0		10.0–12.5	
X 10 CrNiTi 18 9	321	1.4541	<0.10	17.0–19.0		9.0–11.5	Ti > 5 × % C
X 10 CrNiNb 18 9	347	1.4550	<0.10	17.0–19.0		9.0–11.5	Nb > 8 × % C
X 5 CrNiMo 18 10	316	1.4401	<0.07	16.5–18.5	2.0–2.5	10.5–13.5	
X 2 CrNiMo 18 10	316 L	1.4404	<0.03	16.5–18.5	2.0–2.5	11.0–14.0	
X 10 CrNiMoTi 18 10	316 Ti	1.4571	<0.10	16.5–18.5	2.0–2.5	10.5–13.5	Ti > 5 × % C
X 10 CrNiMoNb 18 10	316 Cb	1.4580	<0.10	16.5–18.5	2.0–2.5	10.5–13.5	Nb > 8 × % C
X 5 CrNiMo 18 12	316	1.4436	<0.07	16.5–18.5	2.5–3.0	11.5–14.0	
X 2 CrNiMo 18 12	316 L	1.4435	<0.03	16.5–18.5	2.5–3.0	12.5–15.0	
X 2 CrNiMo 18 16	317 L	1.4438	<0.03	17.0–19.0	3.0–4.0	15.0–17.0	
X 2 CrNiN 18 10	304 LN	1.4311	<0.03	17.0–19.0		9.0–11.5	N 0.12–0.20
X 2 CrNiMoN 18 12	316 LN	1.4406	<0.03	16.5–18.5	2.0–2.5	10.5–13.5	N 0.12–0.20
X 2 CrNiMoN 18 13	316 LN	1.4429	<0.03	16.5–18.5	2.5–3.0	12.0–14.5	N 0.14–0.22

Table 5. Composition of special alloyed stainless steels

Steel type, short name (DIN)	Material number	Composition, wt%						
		C	Si	Mn	Cr	Ni	Mo	Other
X 3 CrNiMoN 17 13 5	1.4439	≤0.04	≤1	≤2	17.5	13.5	4.5	N
X 2 NiCrAlTi 32 20	1.4558	≤0.04	≤0.7	≤1	21.5	33.5		Ti, Al
X 1 NiCrMoCu 25 20	1.4539	≤0.02	≤1	≤2	20	25	4.5	Cu
X 1 CrNiMoN 25 25 2	1.4465	≤0.02	≤0.7	≤2	25	23.5	2.3	Al, Cu
X 1 NiCrMoCuN 31 27	1.4563	≤0.02	≤0.7	≤2	27	31	3.5	Cu
NiCr21Mo	2.4858	≤0.03	≤0.5	≤1	21.5	42	3	Ti, Al

Table 6. High-alloy stainless steels containing ≥ 6 wt% molybdenum

DIN designation	ASTM designation	Composition, wt%						Trade mark
		C	Cr	Ni	Mo	Cu	N	
	S 31254	0.02	20	18	6.1	0.7	0.2	Awesta 254 SMO
1.4529		0.02	20	25	6.5	1	0.15	Nirosta 4529
			17	16	6.3	1.6	0.16	VEW A 963
		0.03	20	24	6.0			AC 6 X

1.4361) and has already undergone trials under practical conditions has remarkably good resistance to highly concentrated nitric acid and chromic acid. As Figure 13 shows, its advantages are particularly pronounced at HNO_3 concentrations above 75 %, where it is superior to stainless steels of all other types [48]. In the meantime a variant has been developed for extremely severe HNO_3 exposure (5.3 wt % Si, 17.5 wt % Ni, 0.015 wt % C) [49].

Table 7. Composition of ferritic–austenitic (duplex) steels

Name/material number	Short designation (DIN)	Typical composition, wt%						
		C	Si	Mn	Cr	Mo	Ni	N
1.4462	X 2 CrNiMoN 22 5 3	0.012	0.34	0.65	22.1	3.0	5.7	0.14
Ferralium 255	X 2 CrNiMoCu 25 5	0.019	1.4	1.4	26	3.3	5.5	0.2
Noridur 9.4460	G–X 3 CrNiMoCuN 24 6	0.03	1.5	1.5	25	3.1	6	+
A 905	X 3 CrMnNiMoN 25 4 4	0.034		5.8	26	2.3	3.7	0.36

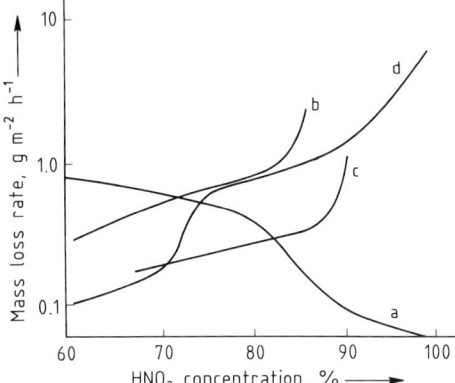

Figure 13. Area-related mass loss rate of stainless steels in boiling nitric acid (exposure time: 50–90 d)
a) X 2 CrNiSi 18 15; b) X 2 CrNi 18 9; c) X 1 CrMo 26 1; d) X 1 CrNi 25 21

Ferritic steels have also undergone special development. Because the mechanical properties of the welded joints were unsatisfactory, the commercial types used until now were rarely chosen for chemical plant. A substantial improvement may be expected from the titanium-stabilized chromium and chromium–molybdenum steels of low carbon and nitrogen content and from the nickel-alloyed chromium–molybdenum steels. Progress in the field of ferritic steels has been marked by the use of special methods for obtaining low C + N contents in their manufacture (ELI = extra low interstitial steels) [50]. Three groups can now be distinguished: 1) steels with 25–28 wt% Cr, without nickel, in some cases with an addition of Mo, and in all cases with a C + N content limited to 0.015 wt% (superferrites); 2) molybdenum-containing steels with 18 wt% Cr, which, although the proportions of the aforementioned interstitial elements have been lowered, are additionally stabilized with titanium or niobium; and 3) stabilized molybdenum- and nickel-containing ferritic steels with high Cr contents, such as X 1 CrNiMoNb 28 4 2 (Remanit 4575) and X 2 CrNiMoTi 25 4 4 (Monit).

The superior toughness of the last-mentioned steel types, as compared with that of a conventional Cr steel, is illustrated by the transition temperature of the notch impact strength.

Stabilized ferritic steels should be of particular interest in connection with the handling of corrosive media, especially chloride-containing cooling waters (e.g., river water), owing to their good corrosion properties, suitability for welding, and availability as semifinished products, as well as for economic reasons [51].

For reasons of cost, and because of certain design problems, stainless cast steel is often preferred to stainless rolled and wrought steels as a material for chemical plant. The chemical composition and mechanical properties of stainless cast steels are fairly similar to those of the rolled and wrought steels. Table 8 lists several important types, especially those used for chemical pumps.

5.1.6.1. Technical Properties

The mechanical properties of the four groups of stainless steels are shown in Table 9.

Between their melting points and room temperature, ferritic steels are pure or mainly ferritic. The production of a fine-grained structure requires forming at temperatures below 800 °C, heating beyond 800 °C, and cooling. Heating above 1000 °C (welding) enlarges the grain, thus embrittling the steel and causing precipitation of carbides. Temperatures above 1100 °C affect the structure to such an extent that the steel breaks when subsequently loaded. These steels cannot be hardened by heat treatment. Being less expensive than austenitic steels, and more resistant to chloride-containing solutions, they are used for pipelines intended for chloride-containing waters, as well as for heat exchangers and condensers.

Table 8. Cast stainless and special steels

Steel type	Designation	Material number	Composition, wt%							
			C	Si	Mn	Cr	Ni	Mo	Cu	Others
Cast ferritic stainless steel	G-X 8 CrNi 13	1.4008	≤0.09	≤1.0	≤1.0	13.0	1.5	≤0.5		
Cast martensitic stainless steel	G-X 5 CrNi 13 4	1.4313	≤0.07	≤1.0	≤1.5	13.0	4.0	≤0.7		
Cast austenitic stainless steel	G-X 6 CrNi 18 9	1.4308	≤0.07	≤2.0	≤1.5	19.0	10.0			
	G-X 6 CrNiMo 18 10	1.4408	≤0.07	≤1.5	≤1.5	19.0	11.0	2.5		
	G-X 6 CrNiMo 1713	1.4448	≤0.07	≤1.0	≤2.0	17.0	13.5	4.5		
	G-X 5 CrNiNb 18 9	1.4552	≤0.06	≤1.5	≤1.5	19.0	10.0			Nb ≥ 8 × % C
	G-X 5 CrNiMoNb 18 10	1.4581	≤0.06	≤1.5	≤1.5	19.0	11.5	2.3		Nb ≥ 8 × % C
	G-X 7 CrNiMoCuNb 18 18	1.4585	≤0.08	≤1.5	≤2.0	17.5	20.0	2.3	2.1	Nb ≥ 8 × % C
Cast special austenitic steel	G-X 7 NiCrMoCuNb 25 20	1.4500	≤0.08	≤1.5	≤2.0	20.0	25.0	3.0	2.0	Nb ≥ 8 × % C
	G-X 3 CrNiSiN 2013	9.4306	≤0.04	4.5	4.5	20.0	13.0	≤0.05		N
Cast ferritic–austenitic stainless steel	G-X 3 CrNiMoCu 24 6	9.4460	≤0.04	≤1.5	≤1.5	25.0	6.0	2.4	3.1	N

Table 9. Mechanical properties of the four main steel types

Steel type	Ferritic[a]	Martensitic[b]	Austenitic[c]	Austenitic–ferritic[d]
Yield strength, MPa	ca. 300	450–600	ca. 230[e]	ca. 500[f]
Tensile strength, MPa	450–650	600–950	ca. 600	ca. 750
Fracture strain, %	ca. 25	14–18	45	ca. 30
Notch impact energy, J[g]	30–55		ca. 150	

[a] Normal steels with 16–18 wt% Cr, annealed. [b] Quenched and annealed. [c] Solution annealed. [d] Annealed at 1100 °C and quenched. [e] 0.2% yield strength. [f] 1% yield strength; [g] DVM specimen, longitudinal.

Martensitic steels exist only in the pearlitic and martensitic forms, without intermediate stages. They are hardened by heating to 980–1100 °C, depending on carbon content, cooling in oil or air, and tempering at a temperature above 600 °C. The steels containing less than 0.4 wt% carbon serve as heat-treatable materials with good mechanical properties, while those with higher carbon contents are used as hardened steels. Certain types are resistant to very high temperatures and to pressurized water. Having good resistance to erosion and cavitation as well as high fatigue strength, they are used also for propellers and impellers.

Austenitic steels with low chromium and nickel contents are so unstable that martensite may be formed at low temperatures or through cold forming. High chromium contents favor the formation of ferritic structures. Fine granulation results from recrystallization during hot forming. The grain-growth rate is much lower than that of ferritic steel. The final heat treatment consists of heating to 1000–1100 °C and quenching in water or air. Hardening by heat treatment is not possible; the strength is improved by cold or hot–cold forming. The most conspicuous properties of these steels are a low yield strength, high resistance to fracture, high strength at elevated temperatures, resistance to pressurized water, heat resistance, permanent strain at low loads, and a tendency to creep. An important application is the fabrication of pressure vessels.

In austenitic–ferritic steels the ratio of ferrite to austenite is determined by the composition and heat treatment. It increases with the annealing temperature and rate of cooling. Attention is drawn to the risk of 475 °C embrittlement (hardness increase, loss of toughness, and loss of chemical resistance through separation into phases of high iron and chromium content [52] and σ-phase formation), as well as to the fact that, in comparison with purely austenitic steels, these steels are less sensitive to stress corrosion cracking. Their applications include pump manufacture. Special grades with high wear resistance are used for bushings.

Ferritic, austenitic, and ferritic–austenitic stainless steels can be welded by practically all the well-known methods. The Argon arc (TIG, MIG) process, resistance welding (spot or seam welding) and the recently introduced plasma arc and electron beam welding methods are particularly suitable for ferritic chromium steels. Austenitic chromium–nickel steel is recommended as the electrode material in addition to material of the same composition as the base. If, for reasons of corrosion (particularly the risk of stress corrosion cracking), either of these materials cannot be used, an austenitic filler can serve for the lower layers and a ferritic filler for the cover pass. In view of the marked grain growth tendency, as little heat as possible should be introduced. It is therefore advisable to use thin electrodes and low currents.

As scale formation may be accompanied by the formation of chromium-depleted surface zones, protective gas should be used to prevent scale formation on welding. Cleanliness of the seam, which must be without splashes and undercutting, is also important. Where the demands are particularly severe, as in the case of components subjected to vibration, the surface quality can be improved by subsequent grinding.

Unstabilized 17 wt % chromium steels that have been welded must always be annealed since they are otherwise sensitive to intergranular corrosion, even in mild corrodents. Where this treatment is impracticable, stabilized grades should be used. Although the latter benefit from heat treatment, it is not absolutely necessary. Stabilized steels and molybdenum-containing steels also exhibit lower growth tendency. Apart from imparting resistance to intergranular corrosion, additions of stabilizing elements also improve the resistance to pitting corrosion.

The principal welding methods for austenitic steels are TIG, MIG, and submerged-arc welding. Apart from using steels that have no intergranular susceptibility after welding (stabilized steels and ELC types) it is particularly important to avoid hot cracking and loss of corrosion resistance through precipitation of intermetallic phases. Appropriate recommendations can be found in the processing guidelines.

5.1.6.2. Chemical Properties

The resistance of stainless steels to corrosive media is determined by their passive behavior, which depends in turn on the alloying elements and their concentrations.

In this connection, chromium is particularly important, but the effects of nickel, molybdenum, and copper are also significant. In particular, chromium reduces the passivation current density and the residual current density of the steel in the passive state. The main effect of molybdenum is a reduction of the passivation current density and facilitation of repassivation, which promotes the transition from the active to the passive state.

In general it is considered that a corrosion rate of 0.1 mm/a ($0.1 \, \text{g m}^{-2} \text{h}^{-1}$) represents the highest degree of corrosion resistance attainable under practical conditions. Corresponding data can be found in stability tables and diagrams and in manufacturers' literature. Often, however, stability diagrams, of which Figure 14 is an example, apply only to chemically pure corrodents. In the case of sulfuric acid, for example, even relatively small amounts of oxidizing agents (NO_2^-, Fe^{3+}) permit the use of stainless steels in sulfuric acid at concentrations up to ca. 50 % and temperatures considerably higher than those shown in Figure 14. Thus figures for resistance to pure corrodents should in many cases be regarded as initial approximations. The influences of ventilation, i.e., the corrodent's oxygen content, and of the state of motion (e.g., laminar or turbulent flow) must be taken into account also.

Uniform surface corrosion is of relatively minor importance for stainless steels. Localized corrosion, however, may lead quickly to component failure even where the corrosion rate is low, and must therefore be avoided as far as possible. Normally, the susceptibility of metallic materials to localized corrosion increases with the resistance to general corrosion. In localized corrosion of stainless steels, the locally active areas are very small compared to the area of the passive surface. Therefore, the local anodic dissolution current density is very high. Damage caused to passive materials by stress corrosion cracking and fatigue corrosion, which occur with virtually no loss of mass, is particularly hazardous. The corrosion may penetrate even relatively thick components within a short time. The types of localized corrosion that are most important in stainless steels are intergranular corrosion (grain-boundary attack), pitting corrosion (pit formation), stress corrosion (corrosion cracking), and fatigue corrosion (corrosion fatigue cracking).

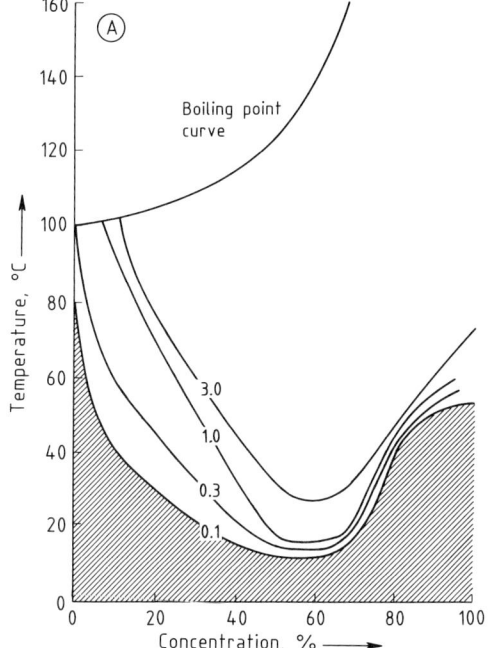

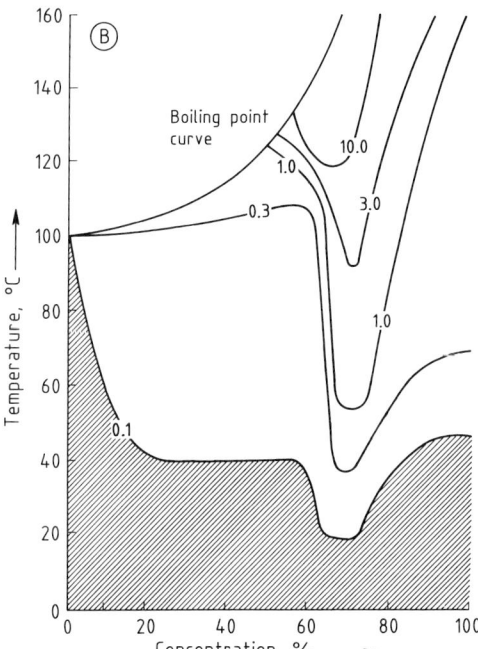

Figure 14. Isocorrosion curves of chromium–nickel–molybdenum stainless steels in sulfuric acid
A) 0.05% C, 17% Cr, 13% Ni, 5% Mo; B) 0.05% C, 20% Cr, 25% Ni, 3% Mo, 2% Cu

Intergranular corrosion is a selective process in which the areas close to the grain boundaries are corroded. In all cases it has one of the following causes:

1) Depletion of alloying elements important to chemical resistance in areas close to grain boundaries, caused by formation of precipitates, in which these elements have accumulated, at the grain boundaries. In joint welding, these precipitates may be formed in the heat-affected zones bordering on the weld seams.
2) Chemical attack on precipitates at the grain boundaries, e.g., by concentrated nitric acid on precipitates of titanium carbonitrides in titanium-stabilized steels.
3) Preferential attack by accumulations of accompanying and trace elements in the steel, e.g., phosphorus, silicon, and boron (which stimulate the anodic metal dissolution at the grain boundaries).
4) Cell formation between the matrix and precipitates at the grain boundaries.

In stainless steels, the depletion of the alloying elements chromium and molybdenum resulting from precipitation of carbides with high chromium and molybdenum contents is the most important cause of intergranular corrosion in industry.

At low temperatures the solubility of carbon in austenitic chromium–nickel steels and nickel–chromium alloys is very low, but it increases markedly as the temperatures rises. The carbon dissolved in the austenitic solid solution at high annealing temperatures may precipitate at the grain boundaries at lower annealing temperatures as a carbide of type $M_{23}C_6$ with high chromium and molybdenum content. In this way zones of depleted chromium and molybdenum content, which grow together to form coherent chromium-depleted areas as the carbide precipitation progresses, are formed round the precipitated carbides. Through the precipitation of chromium-rich carbides (75–90 wt% Cr) the chromium content in the vicinity of the grain boundaries falls below the so-called resistance limit of ca. 12 wt% chromium.

The susceptibility to intergranular corrosion is indicated by the grain-boundary corrosion fields in the plot of annealing time versus annealing temperature (Fig. 15). The shape and sizes of these fields are best described by the highest temperature that still causes susceptibility to intergranular corrosion and by the shortest annealing time leading to sensitization of grain boundaries. This shortest annealing time (represented by the "nose" of the grain-boundary corrosion field) is situated in the temperature range around 650 °C for molybdenum-free austenitic steels and in the

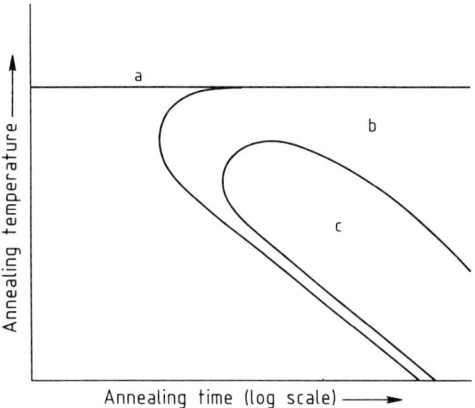

Figure 15. Diagram of grain-boundary disintegration
a) Solubility limit for carbon; b) Precipitation of chromium-rich carbides; c) Intergranular corrosion

range around 750 °C for molybdenum-containing steels.

When the temperature falls after welding, it passes through the range in which austenitic chromium–nickel steels become susceptible to intergranular corrosion. The following steps are taken to make the steel resistant to intergranular corrosion after welding:

1) The carbon content is reduced to levels at which neither temperature nor the exposure time causes sensitization (ELC steels).
2) Stabilization with additions of the carbide-forming element titanium or niobium/tantalum. The affinity of these alloying elements for carbon is greater than that of chromium. They form special carbides or carbonitrides of the type M (C, N). Due to the affinity of both elements for nitrogen, the steel's nitrogen content, as well as its carbon content, must be taken into account in calculating the stabilization ratio.

In ferritic chromium steels, the rate at which chromium-rich carbides are precipitated is very high. After annealing at temperatures above 850 °C, even very rapid cooling cannot prevent their precipitation, and the formation of a structure susceptible to intergranular corrosion. In the temperature range around 800 °C, however, the very high diffusion rates permitted by the body-centered cubic lattice of the α mixed crystal causes rapid replenishment of chromium by diffusion from the chromium-rich matrix into the chromium-depleted grain boundary areas. As a reduction of the carbon content would not eliminate intergranular susceptibility until an extremely low level had been reached (< 0.005 wt%), stabilization, even of ferritic chromium steels, by addition of niobium/tantalum, and especially titanium, has great industrial importance.

Pitting corrosion of stainless steels occurs only when they are in the passive state, and is then caused almost exclusively by chloride ions in aqueous solution. It is therefore also known as chloride corrosion. In the initiating step of pitting corrosion chloride ions are adsorbed locally by the passive layer, whereupon the passivity breaks down. The mechanism of this corrosion has not yet been fully explained.

In the pitting corrosion of stainless steels, the influence of potential is important. Pitting corrosion only occurs when the potential exceeds the pitting potential. In the presence of a corrosive medium, pitting corrosion therefore occurs only if the corrosion potential of the steel is more positive than the pitting potential.

With regard to the attacking medium, the main factors by which the pitting potential is lowered (and therefore the risk of pitting corrosion heightened) are increase of chloride concentration and temperature, and reduction of pH [53].

With regard to the material, the pitting potential can be shifted towards more positive potentials by raising the chromium content and especially the molybdenum content; the alloying elements nickel and manganese, on the other hand, have practically no influence.

The effects of chromium and molybdenum are described, and the resistances of the various alloys estimated, with the help of the pitting resistance equivalent PRE:

$$PRE = wt\% \; Cr + 3.3 \times wt\% \; Mo \; [54]$$

Over a wide range of percentages of these elements, the pitting potential depends linearly on the pitting resistance equivalent, in as far as the structure is that given by solution annealing. Deviations are attributable to homogenization of the structure by relatively high nitrogen contents.

Stress corrosion cracking, which is likewise initiated mainly by chlorides, accounts for most of the damage suffered by stainless steels. The basic prerequisites of stress corrosion cracking are as follows:

1) The action of a specific corrodent on a material susceptible to stress corrosion cracking, i.e., the presence of a critical material/corrodent system.
2) Static tensile stresses, possibly superimposed by infrequent dynamic loads. Only tensile loads are effective.

Pressure does not cause stress corrosion cracking; in fact, in many cases it is an effective means of preventing corrosion of this type.

Stress corrosion cracking of stainless steels with mainly transgranular progression of the corrosion cracks may occur in the following corrodents:

1) Chloride solutions (threshold temperature 50 °C);
2) Alkaline solutions (threshold temperature at 40–50 wt% NaOH, 100 °C). In addition, intergranular cracking occurs at high temperatures in dilute alkaline solutions.

If the steel is in the sensitized state in which it is susceptible to intergranular cracking, and the tensile stress and potential exceed critical values, intergranular stress corrosion cracking occurs also in, e.g., water at high temperatures (Fig. 16) [55].

Under conditions that cause stress corrosion cracking the following materials can be used instead of normal stainless austenitic 18 Cr 10 Ni steels: stainless ferritic chromium steels with 17–28 wt% chromium, with or without molybdenum; austenitic–ferritic steels (duplex steels), whose use is restricted in some cases to temperatures below 300 °C; austenitic steels with increased nickel and/or molybdenum content; and nickel-based alloys. Apart from molybdenum, an alloying element that decisively improves the stress corrosion cracking resistance of stainless austenitic steels is nickel (at nickel contents exceeding about 40 wt%, stress corrosion cracking no longer occurs).

Corrosion fatigue is the term used for crack damage which occurs in a component exposed simultaneously to alternating mechanical stress and a corrodent. It may be caused in all metallic materials by unspecific corrodents, differing in this respect very much from stress corrosion cracking, for whose occurrence specific corrodents are responsible. The main symptom of corrosion fatigue is the loss of endurance strength.

Fatigue cracking of stainless steels differs according to whether they are in the active or passive states. In the active state numerous cracks are formed, whereas in the passive state few cracks, or perhaps just a single precrack, are formed. Damage to stainless steel plant through fatigue cracking is, however, relatively rare [56].

5.1.6.3. Development State of Stainless Cr–Ni Steels

The corrosion resistance of stainless steels is benefitting from the attainment of particularly high degrees of purity, as given by electroslag remelting, for example. At the same time the strength-related properties are being improved without loss of ductility.

To make possible the increases in Cr and Mo content on which these improvements depend it is necessary to increase the nitrogen content. This is achieved by renitrogenizing the steel under pressure and simultaneously adding alloying elements that increase the nitrogen's solubility [57], [58], e.g., manganese.

Of two steels now undergoing trials, namely X 3 CrNiMoNoN 23 27 6 4 and X 3 CrNiMn-MoNbN 23 17 8 4, the latter is superior in respect of strength-related properties in consequence of precipitation hardening by a finely dispersed intermetallic Z-phase (0.2% yield strength at room temperature ca. 500 MPa; at 500 °C, > 300 MPa). This steel, which is without Nb and therefore free from precipitation, is superior in ductility (notch impact energy at room temperature > 300 J), though at 500 °C it still has a 0.2% yield strength of ca. 250 MPa [59].

The stress corrosion cracking behavior of steels of this type is shown in Figure 17. The steel with the highest N content is the one most resistant to boiling 35% $MgCl_2$ solution. In boiling

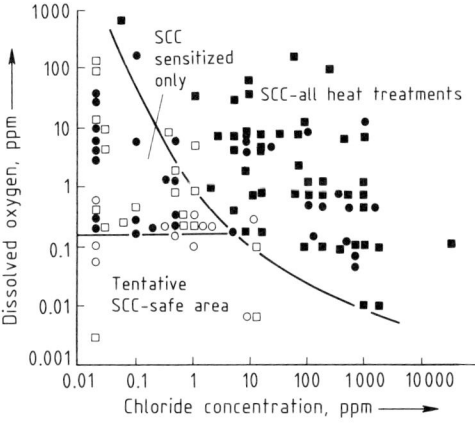

Figure 16. Concentration ranges of dissolved oxygen and chloride that may lead to stress corrosion cracking (SSC) of type 304 stainless steels in water at temperatures in the range 260–300 °C. Applied stresses in excess of yield strength and test times in excess of 1000 h, or strain rates greater than $10^{-5} s^{-1}$
● = Sensitized, SCC; ○ = Sensitized, no SCC; ■ = Annealed, SCC; □ = Annealed, no SCC

seawater all the steels achieved times to failure of more than 1000 h at loads above the 0.2% yield strength [60].

The steel X 3 CrNiMnMoNbN 23 17 5 3 (1.4565) is commercially available. With a pitting resistance equivalent of about 48 (in which the high nitrogen content is taken into account) it is more resistant to pitting corrosion than any other stainless steel (cf. Fig. 18). Only NiCr22Mo-9Nb, better known as Incoloy 625, performs comparably well.

Some mechanical properties of alloy 1.4565 are as follows:

0.2% Yield strength	≥ 500 N/mm^2
0.2% Yield strength at 300 °C	≥ 300 N/mm^2

Notch impact energy at -269 °C	≥ 200 J
Elongation	$> 50\%$ at room temperature
Reduction of area	$> 70\%$ at room temperature

As the nitrogen contents of the aforementioned steels in the molten state are below the solubility limit, these materials have good welding behavior. The filler metals used for them at present are relatively highly alloyed nickel-based materials of the Ni–Mo–Cr type.

5.2. Cast Iron [60], [61]
($\rightarrow$ Steel, treated in the A series)

The term cast iron is used for unalloyed, low-alloy, and highly alloyed iron–carbon cast materials containing ca. 2–5 wt% carbon and 0.8–3 wt% silicon, the carbon being mainly in the form of graphite. The fundamental structure of unalloyed and low-alloy cast iron is ferritic, ferritic–pearlitic, or pearlitic; that of highly-alloyed cast iron may be any of these, but also austenitic. The main alloying elements are nickel, chromium, and copper. The main quality standards are DIN 1691, for unalloyed and low-alloy lamellar graphite cast iron; DIN 1693, for unalloyed and low-alloy spheroidal graphite cast iron; and DIN 1694, for austenitic cast iron.

Mechanical properties of the various types of cast iron are listed in Table 10. The strength of gray cast iron depends on the thickness of the casting's cross section (wall thickness). Empirical values for the strength of castings are given in Figure 19.

The exceptionally low plastic deformability and impact strength of lamellar graphite cast iron in comparison with nodular graphite cast iron are explained by the graphite lamellae, which act as internal notches. Unalloyed and low-alloy lamellar graphite cast iron and nodular graphite cast iron are attacked only slightly by

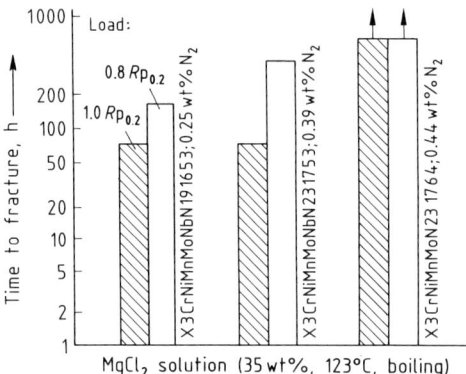

Figure 17. Stress corrosion behavior of nitrogen-containing stainless steels in boiling MgCl$_2$ solution (35 wt%, 123 °C)

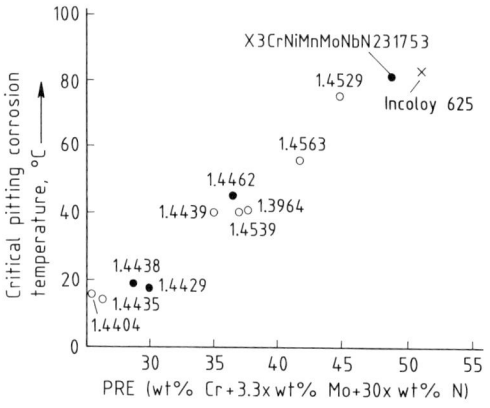

Figure 18. Pitting resistance equivalent (PRE) and critical pitting corrosion temperature for various steels and Incoloy 625 in a 10 wt% solution of FeCl$_3 \cdot$ 6 H$_2$O

Table 10. Mechanical properties of the various types of cast iron

Cast iron type	Tensile strength σ_u, MPa	Fracture strain, %	Notch impact energy, J
Unalloyed and low-alloyed laminar	100–450	< 1	
Unalloyed and low-alloyed nodular	400–800	2–22	9–14 (-20 °C)
Austenitic laminar	140–280	1–3	
Austenitic nodular	390–500	1–40	7–34

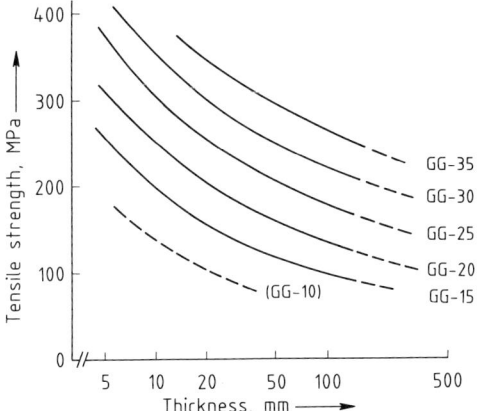

Figure 19. Mean tensile strength of gray cast iron versus the section thickness of the casting (according to DIN 1691)

pure sulfuric acid and pure phosphoric acid. Their resistance to soil corrosion and seawater attack is also good. Depending on their composition and the form of the graphite, austenitic cast materials [62], [63] withstand corrosion by alkalies and sulfuric acid. The austenitic grades that withstand moderately high and high temperatures are used for pumps, valves, cylinder linings, waste-gas pipes, and furnace parts, for example. The corrosion-resistant grades are used in the food, artificial silk, and plastics industries as well as for pipes and vessels. Mention should also be made of the thermal shock stability of GGL-(laminar)NiCr 30 3 (material no. 0.6676) and the low-temperature ductility of GGG(nodular)-NiMn 23 4 (material no. 0.7673).

A special grade that withstands hot nitric acid and hot sulfuric acid is the iron–silicon material containing 14–18 wt% silicon. Its hardness and brittleness constitute a great disadvantage because they restrict the choice of shaping techniques to casting and grinding.

Lamellar and nodular graphite cast iron can be welded hot with filler materials of the same composition, and cold with filler materials of different composition [64], [65].

5.3. Nickel and Nickel Alloys [66]–[68]

(→ Nickel, → Nickel Alloys, both treated in the A series)

Owing to their excellent corrosion resistance to various aggressive media, nickel and its alloys are of growing importance in chemical engineering. Furthermore, Ni–Cr alloys, especially those with additions of silicon and aluminum, have good high-temperature strength by virtue of their high creep strengths, and good heat-resistance to hot gases and combustion products.

The most important unalloyed nickel grades are Ni 99.6 (2.4060), nickel content $\geq$ 99.6 wt%; LC-Ni 99.6 (2.4061), nickel content $\geq$ 99.6 wt%; Ni 99.2 (2.4066), nickel content $\geq$ 99.2 wt%; LC-Ni 99 (2.4060), nickel content $\geq$ 99.0 wt%.

The nickel content (according to DIN 17 740) may include up to 1 wt% cobalt. Ni 99.6 may contain up to 0.08 wt% carbon, and Ni 99.2 up to 0.1 wt% carbon; after prolonged exposure to temperatures above about 300 °C, the carbon may be precipitated as grain boundary graphite, thus making the material brittle [69]. Frequent use is therefore made of LC (low-carbon) Ni, which has a maximum carbon content of 0.02 wt% [70], [71].

Plant consisting of nickel or nickel-plated steels is used mainly to produce and process alkalies [71], [72]. The good resistance of pure nickel to sodium hydroxide solution can be seen in Figure 20. Nickel is also used to handle acid halides [73], seawater, brackish water, dilute air-free and nonoxidizing mineral acids and salt solutions, and fatty acids. It can also be exposed to dry chlorine or hydrogen chloride at temperatures up to 535 °C.

Nickel has good high-temperature strength. Up to about 300 °C its yield strength and tensile

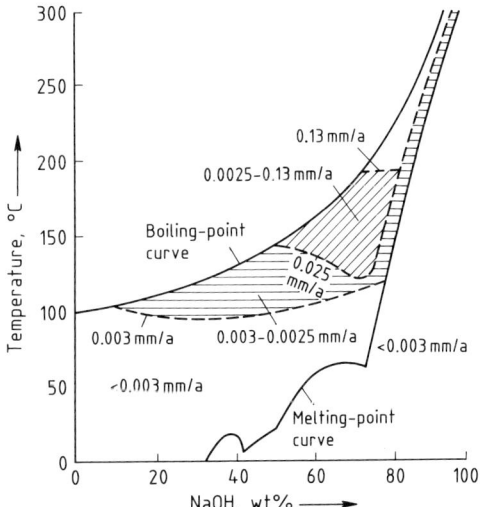

Figure 20. Corrosion resistance of pure nickel in sodium hydroxide, related to concentration and temperature [72]

strength undergo practically no change [69]. Nickel is very sensitive to sulfur and sulfur compounds [69], [70], [74]. Above 400 °C sulfur penetrates nickel along the grain boundaries and forms a eutectic mixture of nickel and nickel sulfide, embrittling the material. This process is rapid at 550–650 °C. The surface must therefore be cleaned thoroughly before welding and each annealing treatment. Sulfur-free lubricants must be used in hot and cold forming. The furnace atmosphere for heat treatment should be sulfur-free and reducing or neutral; if freedom from sulfur cannot be ensured, the furnace atmosphere should be slightly oxidizing.

All the welding methods normally used for steel, except submerged-arc welding [74], are suitable for nickel. Suitable weld filler materials are available for combinations with unalloyed or low-alloy steels, austenitic stainless steels, and nickel–copper, copper–nickel, and nickel–molybdenum alloys. Nickel parts can also be joined by brazing and soldering.

Cast nickel [70] has the characteristic properties of forged nickel, but is always slightly alloyed with carbon, silicon, and manganese to improve the flow of the molten metal and density of the castings.

5.3.1. Nickel–Copper Alloys

In chemical engineering only two alloys with more than 50 wt% nickel have gained importance, namely NiCu30Fe (2.4360, Inconel alloy 400) and NiCu30Al (2.4375), which are standardized in DIN 17 743. Both are sold as sheets, strip, pipe, rods, wires, and forgings.

The alloy NiCu30Fe can be exposed to seawater and brackish water, solutions of nonoxidizing salts, alkalies [75], cold or warm nonoxidizing acids [73], and dry bromine. It is also resistant to hydrofluoric acid [76], though stress corrosion cracking may occur in hydrofluoric acid vapors containing air, as well as to a lesser extent in the liquid phase. Stress corrosion cracking has been caused in these alloys by corrosive media containing mercury or its compounds. Where there is a risk of stress corrosion cracking, stress relief annealing should be carried out at 550 °C, and in exceptional cases at up to 650 °C, and followed by slow cooling in the furnace. As NiCu30Fe is expensive, solid sheet is often replaced in the manufacture of process plant by steel sheet plated with this alloy.

NiCu30Al is an age-hardenable alloy with similar corrosion behavior to NiCu30Fe. Because of the risk of stress corrosion cracking it should be used in the age-hardened state; it cannot be exposed to hydrofluoric acid, however.

Both materials are weldable by the usual methods; they can also be brazed. Soft soldering is less frequently used. Like pure nickel, both alloys are sensitive to sulfur and sulfur compounds above ca. 400 °C.

5.3.2. Nickel–Chromium Alloys [68]

Nickel–chromium alloys, with various iron contents and up to ca. 9 wt% molybdenum and 2 wt% copper, are normally used where the corrosion resistance of highly alloyed stainless steels is inadequate. They exhibit good resistance to acids and to pitting corrosion, crevice corrosion, and stress corrosion cracking. The types most important in chemical engineering, together with their main applications, are listed in Table 11.

Table 11. Nickel–chromium alloys

DIN designation	Material no.	U.S. trade name	Exposure tolerated
NiCr15Fe	2.4816	Inconel alloy 600	seawater, cooling water, alkalies, hot gases, combustion products
NiCr23Fe	2.5851	Inconel alloy 601	as 2.4816
NiCr29Fe	2.4642	Inconel alloy 690	alkalies
LC-NiCr15Fe	2.4817		
NiCr20CuMo	2.4660	Carpenter 20 Cb-3	sulfuric acid, phosphoric acid
NiCr21Mo	2.4858	Inconel alloy 825	seawater, cooling water, alkalies
NiCr21Mo6Cu	2.4641		sulfuric acid, phosphoric acid, seawater, cooling water
NiCr22Mo6Cu	2.4618		as 2.4641
NiCr22Mo7Cu	2.4619	Hastelloy alloy G-3	acids, mixed acids
NiCr22Mo9Nb	2.4856	Inconel alloy 625	acids, seawater, cooling water
NiCr21Fe18Mo	2.4603	Hastelloy alloy G-30	acids, HNO_3–HF pickling acid

The molybdenum-free alloy NiCr15Fe, like nickel, has good resistance to caustic alkalies, even surpassing it in this respect under oxidizing conditions in consequence of its high chromium content [77].

NiCr15Fe, NiCr23Fe and NiCr29Fe have greater high-temperature strength than nickel and can be exposed to oxidizing sulfur-free atmospheres at temperatures up to 1100 °C, and to reducing sulfur-free atmospheres at up to 1150 °C. These alloys are also used in nitriding gases (480–600 °C) and under carburizing conditions (800–950 °C). Like pure nickel and the other nickel alloys, they are sensitive to sulfur and its compounds at temperatures above ca. 400 °C; this must be taken into account where processing, e.g., by welding, is envisaged. Nickel–chromium–iron–molybdenum alloys combine the good resistance of nickel–molybdenum alloys under reducing conditions with the good resistance of nickel–chromium alloys under oxidizing conditions. This explains their good resistance to acids and salt solutions. With the low-carbon grades and those stabilized with titanium or niobium there is virtually no risk of intergranular corrosion.

Chromium–nickel casting alloys (60/40, 50/50 and 35/65) are used in chemical engineering and have good resistance to corrosion by fuel ash [78].

5.3.3. Nickel–Molybdenum and Nickel–Molybdenum–Chromium Alloys

Nickel–molybdenum and nickel–molybdenum–chromium alloys are among the chemically most resistant metallic materials. The main representatives of this group are listed in Table 12.

Nickel–molybdenum alloys are not passifiable, since they contain no chromium. Their good resistance to reducing acids is explained by the low rate of corrosion in the active state, for which molybdenum is responsible.

The alloy NiMo30 (2.4810), consisting of nickel, 26–30 wt% molybdenum and 5–7 wt%

Table 12. Nickel–molybdenum and nickel–molybdenum–chromium alloys

Material no.	Designation	US trade name
2.4617	NiMo28	Hastelloy alloy B-2
2.4610	NiMo16Cr16Ti	Hastelloy alloy C-4
2.4819	NiMo16Cr15W	Hastelloy alloy C-276
2.4608	NiCr21Mo14W	Hastelloy alloy C-22

iron withstands hydrochloric acid at all concentrations. Nickel–molybdenum alloys of this kind cannot be used under oxidizing conditions, however, because even atmospheric oxygen increases the rate of corrosion considerably. Alloying with chromium makes them passive in oxidizing corrodents, and therefore resistant to corrosion under these conditions also. In addition they retain their low rates of corrosion in the active state; therefore, like NiMo16Cr16Ti (2.4610) and NiCr21Mo14W (2.4608), for example, they can be used under reducing and oxidizing conditions. NiCr21Mo14W has proved itself under severely oxidizing conditions.

Since the corrosion resistance of nickel–molybdenum and nickel–molybdenum–chromium alloys depends on molybdenum and/or chromium, precipitation of intermetallic compounds containing these elements may make the materials susceptible to intergranular corrosion. Precipitation of molybdenum-rich compounds leads to intergranular susceptibility in the active state (in hydrochloric acid, for example), while precipitation of compounds rich in chromium causes susceptibility to intergranular corrosion and intergranular crack formation in the passive state [79].

Exact proportioning of the alloying elements, but, above all, minimization of the silicon and carbon contents (< 0.03 wt%) and additional stabilization [80], [81], give alloys that tolerate hot forming, annealing, and welding without acquiring intergranular sensitivity; this is demonstrated by the time–temperature precipitation diagram of Hastelloy alloy C-4 (Fig. 21).

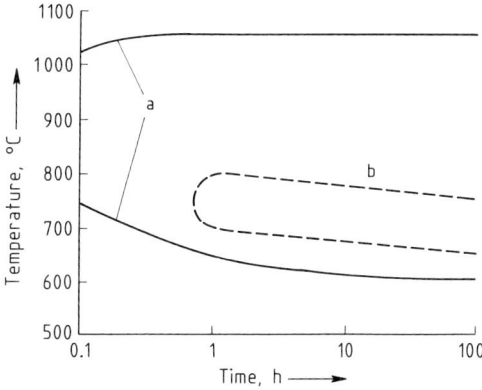

Figure 21. Time–temperature sensitization diagrams for Hastelloy alloy C-276 (a) and Hastelloy alloy C-4 (b); test for intercrystalline corrosion in accordance with ASTM G-28

Hot and cold forming of these high-nickel materials is difficult owing to their high water resistance and work-hardening tendency. They are weldable by all the commonly used methods. In the production of weld joints it is advisable, in the interest of good corrosion resistance, to start by making a thin root pass and then to weld the filler and cover passes while cooling the root side of the weld. If the design and size of a component permit heat treatment, solution annealing at 1150 °C, followed by water quenching, is recommended [79]. Steel can be explosion-cladded with any of these alloys, and they are often used for this purpose. Above 400 °C they are sensitive to sulfur and sulfur compounds, like all materials with high nickel contents; this must be taken into account in connection with annealing and welding.

5.4. Aluminum and Aluminum Alloys

(→ Aluminum, → Aluminum Alloys, treated in the A series)

The high strength and low density of aluminum make it highly suitable for lightweight structures. Its economic value in the chemical industry, however, depends more on its corrosion behavior [82], [83]. Pure and very pure aluminum according to DIN 1712 comprise the grades Al99, Al99.5, Al99.8, and Al99.99.

The good corrosion resistance of aluminum, in spite of its very negative standard electrode potential of − 1.66 V, is due to the formation of nonconducting Al_2O_3 passive films. The corrosion resistance depends on the purity of the material, which for chemical plant should not be below 99.5 wt% Al. As the protective Al_2O_3 surface layer is largely insoluble in the pH range of 4.5 – 8.8 (Fig. 22), aluminum materials have very good resistance to corrosion in approximately neutral aqueous media. The corrosion behavior and the main applications are listed in Table 13 [82].

Because the Al_2O_3 passive film is nonconducting, relatively thick corrosion-inhibiting Al_2O_3 layers can be produced by anodic oxidation. They consist of a very thin and almost nonporous dielectric underlayer (barrier film) and a fine-pored top layer (Fig. 23), which can be dyed and compacted [84]. Hard anodizing is an anodic oxidation process that gives particularly hard and wear-resistant oxide films for technical purposes [85]. Apart from aluminum itself, numerous wrought and cast alloys based on Al–Mg, Al–Mn, Al–Mg–Si, and Al–Zn–Mg are suitable for hard anodizing.

High-purity aluminum is used in chemical engineering mainly as a cladding material. Su-

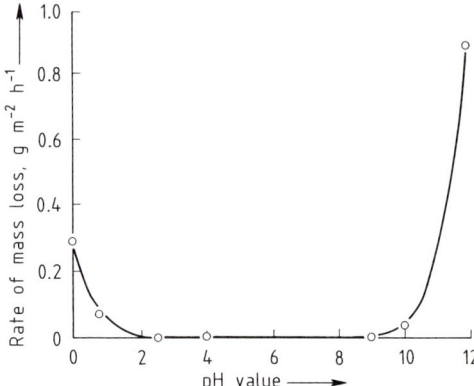

Figure 22. Influence of pH on the solubility of the Al_2O_3 film on aluminum

Table 13. Examples for the corrosion behavior of aluminum alloys in various media

Corrosive environment	Corrosion resistance* Al 99.5	AlMg	AlCuMg	Application
Ethanol water-containing dry	1 – 2 5			
Acetylene, dry	1		2 – 3	Pressure bottles
Liquid ammonia, dry	1 – 2	1 – 2		Cooling elements
Ethane	1	1		Pressure bottles
Atmosphere industrial	2 – 3	3 – 2	3	Construction, cars
marine	1 – 2	1	3 – 5	Shipbuilding
Benzene	1	1	1	Tanks, apparatus
Gasoline leaded, wet	1 3 – 6	1	1 – 3	
Distilled water	1 – 2	1 – 2		
Freon	1	1		Cooling machines
Ice	1	1	2	Cooling plant
Seawater	2 – 3	1 – 2	3 – 5	

* 1 = good resistance; 2 = resistant; 3 = low resistance; 4 = still usable; 5 = barely resistant; 6 = not resistant

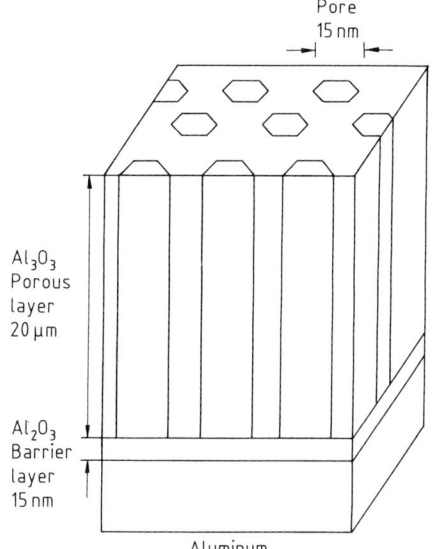

Figure 23. Structure of an oxide film on aluminum, produced by anodizing

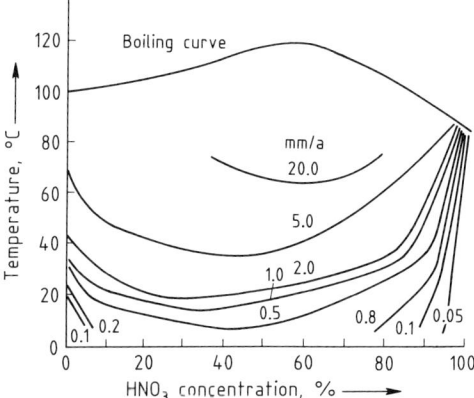

Figure 24. Corrosion behavior of Al 99.5 in nitric acid

Mn are easy to process and exhibit corrosion resistance similar to that of pure aluminum.

The main representatives of the self-hardening materials are Al–Mg alloys, whose strength is further improved by cold forming. Their corrosion resistance is approximately equal to that of pure aluminum (Al99.5), which they even surpass where seawater is concerned (cf. Table 13). AlMg3, AlMg5, AlMg2Mn0.8, and AlMg4.5Mn therefore belong to the group of aluminum materials known as seawater resistant.

However, with increasing magnesium content (especially for Mg >5 wt%), the susceptibility of Al–Mg alloys to intergranular corrosion increases through the formation of Al_8Mg_5 grain-boundary precipitates (Fig. 25). There is also a risk of susceptibility to stress corrosion cracking [87].

The stability can be raised by reducing the magnesium content and adding manganese. A typical representative of the materials obtained in this way is AlMg4.5Mn, which has good strength-related properties and corrosion resistance and is used both for purposes involving seawater contact and in chemical engineering. Additions of Mn and Cr improve the resistance to chloride-induced pitting corrosion [88].

Age-hardenable Al–Cu–Mg alloys (2.8–4.8 wt% Cu, 0.4–1.8 wt% Mg) have high strength (e.g., AlCu4Mg2: $\sigma_u = 440$ MPa) but low corrosion resistance. Apart from having little resistance to atmospheric corrosion, they are

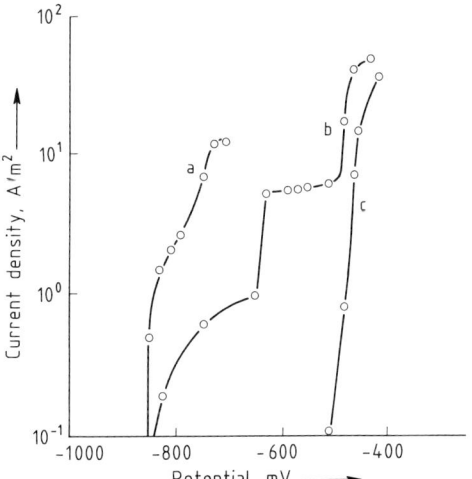

Figure 25. Breakthrough potentials for intergranular corrosion
a) Al_8Mg_5; b) AlMg7 phase with grain boundary precipitates; c) AlMg7 phase, homogenized

per-pure aluminum and AlSi 12 are outstandingly resistant to highly concentrated nitric acid (Fig. 24), even at high temperatures. They also withstand concentrated acetic acid well [86].

The main alloying elements that are used to increase the strength of aluminum are Mg, Si, Zn, Cu, and Mn. Wrought alloys are standardized in DIN 1725, Part 1, and cast alloys in DIN 1725, Part 2. Both groups include self-hardening (alloy hardening) and hardenable (precipitation hardening) materials.

Aluminum–manganese alloys are of the self-hardening type. Alloys containing 0.8–1.5 wt%

particularly susceptible to stress corrosion cracking [89].

Age-hardenable Al–Mg–Si alloys with good heat forming behavior (0.4–3.5 wt% Mg, 0.3–1.5 wt% Si) are more resistant to corrosion than Al–Cu–Mg alloys. Although relatively high silicon contents raise the strength, they reduce the resistance to intergranular corrosion by causing the formation of heterogeneous grain boundary precipitates. Therefore, where corrosion is expected, it is preferable to choose materials with low silicon contents (e.g., AlMgSi0.5). These materials, however, are not fully resistant to pitting and intergranular corrosion in media with high chloride contents.

After artificial ageing, the alloys based on Al–Zn–Mg, and especially Al–Zn–Mg–Cu, have higher strength (up to 530 MPa) than all other aluminum materials, but they are susceptible to pitting and intergranular corrosion in chloride-containing media.

Important cast aluminum alloys for mechanical and chemical plant are those based on Al–Si (5 to 20% Si), Al–Mg (3 to 10% Mg), or Al–Mg–Si or Al–Si–Mg; apart from having high strength and adequate ductility, they are distinguished by good corrosion resistance, including resistance to seawater.

Various treatments and methods are used to protect aluminum from corrosion. High-strength aluminum alloys, for example, are frequently plated with pure aluminum (Al99.5). Nonmetallic inorganic coatings such as chromate and phosphate increase the corrosion resistance and also serve as primers for paints [83].

The thickness of the natural oxide film on aluminum can be increased by boiling the metal in deionized water or by steaming it, as a result of which an AlOOH (boehmite) layer with a thickness of 1–2 μm is formed above the Al_2O_3 barrier layer [84]. Another possibility is the use of impressed current or sacrificial anodes (zinc) to reduce the potential below the values at which the risk of pitting or intergranular corrosion is critical [84].

5.5. Copper and Copper Alloys [90], [91]

(→ Copper and → Copper Alloys, both treated in the A series)

The good corrosion behavior of copper and copper-based alloys has led to their extensive use under moist atmospheric conditions and in the handling of drinking water, industrial water, and water at high temperatures. Their many outdoor applications as fresh water pipes, fittings, condensers, heat exchangers in seawater desalination plants, in chemical apparatus, and for many other purposes are explained not only by good resistance to corrosion but also by their good workability, strength-related properties, and high thermal and electrical conductivity.

In accordance with its position in the electrochemical series (standard potential $Cu/Cu^+ = 0.34$ V) copper (DIN 1787) has good resistance to corrosion, as do high-copper alloys. The good resistance to approximately neutral to alkaline aqueous media (not including water containing NH_3) results from the formation of oxide films which consist of Cu_2O or CuO (depending on the nature of the medium and the corrosion potential) and which afford good protection. In outdoor applications, including those in marine climates, copper is substantially resistant; this explains its use in the construction industry. Its main field of application, however, is plant for drinking, cooling, and industrial water [82]. Here, too, it has good resistance, though under unfavorable conditions (as when deposits are formed or the nature of the water is unsuitable) pitting corrosion cannot be entirely ruled out [92].

In inorganic and organic acids the corrosion rate depends largely on the presence of oxidizing agents. In nonoxidizing acids in the absence of oxygen and at room temperature this rate remains low; in oxidizing acids such as sulfuric acid, it increases with the oxygen content.

In aqueous ammonia solutions under conditions in which surface films are not formed (high alkalinity), copper is attacked severely. In approximately neutral solutions capable of forming surface films, especially aqueous media containing NO_2^-, some susceptibility to stress corrosion cracking, even on the part of pure copper, cannot be entirely excluded [93].

Alloying, particularly with Zn, Al, Ni, or Sn, raises the strength of copper. In the case of Cu–Zn alloys (DIN 17 660) single phase α-phase alloys (Zn ≤ 37 wt%) and (α + β)-alloys (37–46 wt% Zn) are distinguished, the former being more resistant to corrosion. The (α + β)-alloys tend to suffer preferential attack on the β-phase, which has the higher Zn content. The alloys of both types, but above all the (α + β)-alloys, suffer dezincification corrosion in chloride-containing waters (Fig. 26). Dezincification of the α-

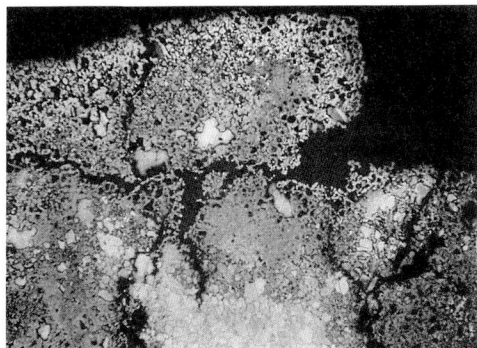

Figure 26. Dezincification of α-brass in tap water

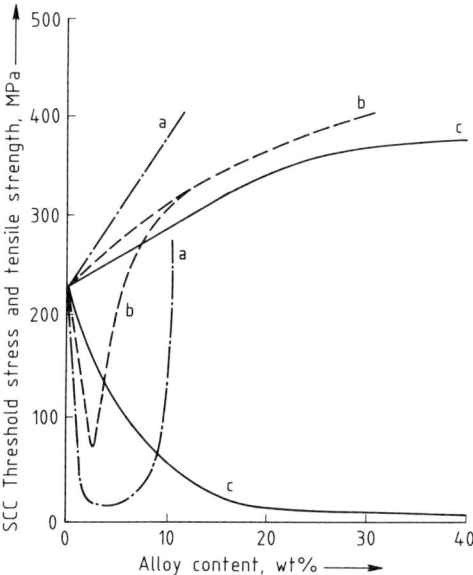

Figure 27. Influence of the content of alloying elements on ammonia-induced SCC of copper alloys
a) CuAl; b) CuNi; c) CuZn
Upper curves: tensile strength
Lower curves: SCC treshold stress

phase alloys is counteracted by adding As, Sn, Sb, and P as additional alloying elements [94]. Among the ternary alloys, CuZn28Sn and CuZn20Al (2 wt% Al) have been used successfully for many years as materials for cooler, condenser, and heat exchanger tubes in cooling waters with relatively low contamination levels. CuZn20Al is remarkable also for its good resistance to seawater and also to erosion–corrosion.

In cooling water containing NH_3 or H_2S there is a risk of stress corrosion cracking. The critical stress for ammonia-induced stress corrosion cracking in Cu–Zn materials decreases with increasing Zn content, and is only about 10 MPa for Cu70Zn30 (Fig. 27) [95]. Residual stresses in components that have not been sufficiently annealed may therefore suffice to initiate crack formation. To prevent stress corrosion cracking, stress relief annealing, followed by testing for residual stresses according to DIN 50 916, is necessary.

Of the Cu–Al alloys standardized in DIN 17 665, the homogeneous α-alloys (Al ≤ 7.8 wt%) are among those that withstand corrosion, examples being CuAl5, CuAl5As and CuAl8. In their resistance to seawater they even surpass pure copper [82]. In addition to the increased Al content, their improved resistance results from the formation of adherent, highly protective oxide films that contain Cu_2O and Al_2O_3 [96].

The two-phase (α + β)-alloys (Al > 7.8 wt%) are less resistant. The heterogeneous multiphase alloys (Al > 10 wt%), which, apart from the α-phase, also contain the γ_2-phase and (depending on the heat treatment) martensitic phases, suffer not only corrosion at relatively high rates, but also preferential dealuminization. Additions of Ni, Fe, and Mn improve the corrosion behavior of the heterogeneous alloys (e.g., CuAl11Ni, CuAl10Fe, CuAl9Mn).

Aluminum bronzes are used in process plant manufacture, where their applications include pump parts (gears) and fittings, including those of seawater pumps, and condenser tubing.

Although the homogeneous α-alloys are those most resistant to uniform corrosion and dealuminization, they show heightened susceptibility to ammonia-induced stress corrosion cracking; this applies particularly to Cu–Al materials with about 4 wt% Al (Fig. 27). The heterogeneous materials with Al > 8 wt%, on the other hand, have good resistance to stress corrosion cracking, but are sensitive to selective corrosion [95].

Technically interesting Cu–Sn materials (tin bronzes) standardized in DIN 17 662 contain up to 9 wt% and up to 14 wt% Sn, respectively, depending on whether they are wrought alloys or cast alloys [82]. Their strength depends on the tin content and degree of cold forming. With favorable working properties and resistance to alternating stresses, together with resistance to corrosion in neutral salt solutions and alkaline solutions (except those containing NH_3), they are used for screws, springs (CuSn2), pipes

(CuSn6) and chemical plant components (CuSn4, CuSn6).

Cast materials consisting of binary Cu–Sn bronzes and multicomponent alloys with additions of Pb (Sn–Pb bronzes) or Zn (gun metal) have proved themselves in marine technology, machinery, and process plant. Owing to their lubrication properties, which are outstanding in some cases, they are used for bearings, screws, worm gears, toothed wheels, pumps, and turbine blades, etc.

Copper–nickel alloys (DIN 17664 and DIN 17658) form a continous mixed-crystal series, which improves their corrosion behavior. The Cu–Ni materials of technical interest (CuNi10, CuNi20, CuNi30), which generally contain iron and manganese additions, are among the copper-based materials most resistant to corrosion. At levels of up to about 1.5 wt%, iron further improves the corrosion behavior. The good corrosion resistance of the frequently used materials CuNi10Fe and CuNi30Fe results from the formation of oxidic surface films consisting of Cu_2O on the metal side and complex corrosion products with high percentages of Fe and Ni on the solution side [97].

The high corrosion resistance of Cu–Ni–Fe materials, especially in seawater, but also in ammonia-containing media (brackish water), explains their growing importance in marine technology (seawater desalination plants) and as tube materials for condensers, coolers, and heat exchangers [82]. These alloys also withstand erosion–corrosion and cavitation well. Chloride-induced pitting corrosion occurs occasionally under unfavorable conditions. In comparison with other copper-based materials, Cu–Ni–Fe alloys with Ni contents of ≥ 10% exhibit substantially better resistance to ammonia-induced intergranular stress corrosion cracking (Fig. 27). They are immune to stress corrosion cracking in chloride-containing media [97].

As the Cu–Ni–Fe alloys of technical interest tend to form Fe–Ni-rich grain boundary precipitates at 350–650 °C [98], it is assumed that the homogenized, precipitation-free materials are those most resistant to intergranular stress corrosion cracking. Hence the homogenized state has always been preferred for component fabrication. Recent investigations into the influence of the state of precipitation on the stress corrosion cracking behavior of the alloy CuNi10Fe1.5 in ammonia-containing solutions have shown, however, that specific heat treatments are capa-

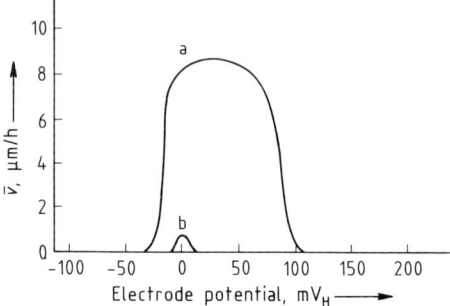

Figure 28. Mean crack velocity $\bar{v}$ of CuNi10Fe1.5 in 1 M $[(NH_4)_2SO_4 + NH_3]$ (pH 9, 50 °C) as a function of the electrode potential
a) Homogenized; b) Aged (50 h, 500 °C)

ble of producing material states considerably superior to the homogeneous states [97] (Fig. 28). It has also been demonstrated that the stress corrosion cracking behavior is determined exclusively by the hardness of the material [97].

The variety of copper-based materials is such that reliable use depends on very careful choice: the material must be exactly suited to the conditions of use.

5.6. Lead and Lead Alloys [82], [99]

(→ Lead and → Lead Alloys, both treated in the A series)

Lead owes its good corrosion resistance to its ability to form dense, firmly adherent surface films consisting of lead sulfates, carbonates, or oxides, depending on the corrosive medium [99].

In outdoor applications and approximately neutral waters the protection from corrosion arises from the formation of basic lead carbonates of low solubility $(Pb(OH)_2 \cdot 2\,PbCO_3)$, which may also contain lead sulfates. In acids and alkalies, lead normally suffers severe surface corrosion (Fig. 29). However, it is distinguished by outstanding resistance to sulfuric acid and good resistance to phosphoric acid and chromic acid; this results from the low solubility of the sulfate, phosphate, or chromate surface films [99]. Lead is therefore a useful material for fittings, pipes, pump parts, and other mechanical components and also for leadings (reaction vessels, electrolysis tanks) [82], [100].

For greater strength and finer grain, lead is alloyed with antimony (0.5–13 wt% Sb). Lead–antimony alloys can be age hardened (precipita-

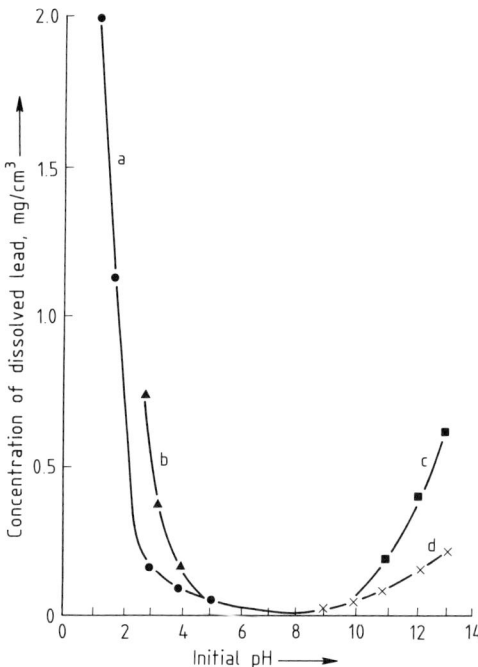

Figure 29. Dependence of the corrosion behavior of lead in aqueous solutions on pH
a) HNO$_3$; b) Acetic acid; c) Ba(OH)$_2$; d) NaOH

Table 14. Linear corrosion rate of multicomponent lead alloys in boiling sulfuric acid

Alloy	Corrosion rate, mm/a		
	50% H$_2$SO$_4$	70% H$_2$SO$_4$	80% H$_2$SO$_4$
Pb 99.9 Cu	0.48	8.23	a
Pb Cu Pd (0.06, 0.1)	0.12	0.21	0.22
Pb Cu Au (0.06, 0.1)	1.86	0.10	0.30
Pb Sb Pd (1.1, 0.1)	0.17	0.19	a
Pb Cu Sn Pd (0.05, 0.12, 0.10)	0.01	0.10	0.26
Pb Cu Sn Pd (0.10, 0.13, 0.2)	0.01	0.05	0.19
Pb Cu Sn Au (0.04, 0.05, 0.10)	0.15	0.23	1.95
Pb Ni Sn Pd (0.10, 0.10, 0.10)	0.09	0.28	2.80
Pb Te Sn Pd (0.10, 0.10, 0.10)	0.09	0.29	3.50

^{a} Dissolves.

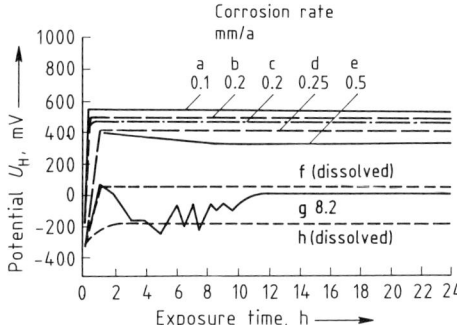

Figure 30. Potential–time curves of lead alloys in boiling 70% H$_2$SO$_4$
a) Pb Cu Sn Pd (0.05 0.12 0.1); b) Pb 99.9 Cu–0.1 Pd; c) Pb 99.985–0.1 Pd; d) Pb Cu Sn Au (0.04 0.05 0.1); e) Pb Cu Sn (0.12 0.12); f) Pb 99.985–0.1 Au; g) Pb 99.9 Cu; h) Pb 99.985

tion hardening), which gives them further strength (hard lead). These alloys are used particularly where mechanical stresses must be withstood in addition to corrosion, as in the case of accumulator plates, pumps, valves, and impellers [82]. The corrosion resistance can be further improved by small additions of As and Se.

Compared to pure lead, alloys based on Pb–Cu or Pb–Cu–Sn, and also alloys of the Pb–Cu–Sn–Pd type with extremely low contents of Cu (0.01–0.1 wt%), Sn (0.05–1.12 wt%), and Pd (0.10 wt%), have improved resistance to corrosion by hot and boiling sulfuric acid [101]; see Table 14.

As the solubilities of Cu and Pd in lead are extremely low, these alloying elements are present in the precipitated state, and fine dispersion of the particles is beneficial. The precipitates are capable of reducing the hydrogen overvoltage and accelerating oxygen reduction, thus shifting the potential of the metal into the passive range and consequently improving its corrosion resistance (Fig. 30) [101].

Accordingly, alloys consisting of high-purity lead and copper, or where particularly good corrosion resistance is needed, of lead and palladium, are used for chemical apparatus. The Pb–Cu–Sn–Pd alloys also have relatively high creep strength [101].

5.7. Zinc and Zinc Alloys
(→ Zinc Alloys, treated in the A series)

In DIN 1706 a distinction is made between high-purity zinc (Zn 99.995, Zn 99.99, Zn 99.95) and technical zinc (Zn 99.5, Zn 98.5 and Zn 97.5).

Through outdoor exposure and even mild chemical exposure (contact with drinking water, for example) zinc forms protective surface films that consist mainly of basic zinc carbonates and which adhere to the metallic zinc fairly firmly.

Zinc is severely attacked by hot water and steam. A fairly high corrosion rate can also be expected in industrial air containing SO_2 [82]. Zinc is not resistant to acids and strong alkalies.

The main use of zinc is the protection of steel from corrosion, where it is used mainly in the form of metallic coatings, but also as sacrificial anodes.

The most important zinc alloys are the die casting alloys GD-ZnAl 4 and GD-ZnAl 4-Cu 1; in plant manufacture they are of only minor importance.

5.8. Tin and Tin Alloys
(→ Tin Alloys, treated in the A series)

The purity requirements for tin as specified in DIN 1704 range from 98 to 99.90 wt% Sn.

Due to the formation of oxidic surface films of low solubility, tin has good corrosion behavior in outdoor applications and in water [82]. Tin is attacked severely by halogens, halogen compounds, and alkalies. It is substantially resistant to numerous foods and drinks. It also has good resistance to corrosion by organic acids (acetic, citric, maleic, tartaric, and lactic acid), especially in the absence of oxygen. Tin is used mainly in the food industry as a protective metallic coating on appliances and containers.

Tin–lead alloys with 30–60 wt% Sn (e.g., SnPb40, PbSn50, PbSn40), possibly with additions of Sb and Cu, are used as soft solders (DIN 1707). Tin alloys with Sn contents of 80–91 wt% are important bearing materials (sliding bearings); examples are Sn80Sb12 (5–7 wt% Cu, 11–13 wt% Sb, 1–3 wt% Pb) and Sn80Sb18 (16–20 wt% Sb, 1–3 wt% Pb).

5.9. Titanium, Zirconium, Niobium, and Tantalum [102], [103]

The refractory metals of groups 4 and 5 of the periodic table have now gained considerable importance in the fabrication of chemical plant. Having excellent passivity, which results from the formation of oxidic surface films, such materials as titanium, tantalum, niobium, and zirconium are outstandingly resistant to many media.

In VdTÜV-Werkstoffblatt 230/1 unalloyed titanium is allocated to four groups having different degrees of purity and strength (Table 15). As the forming behavior of types belonging to group IV is limited, only those belonging to groups I to III are normally used in chemical engineering. The corrosion resistance of these types can be improved by alloying with 0.15–0.25 wt% palladium [104].

In the United States, unalloyed titanium (C.P. Ti = commercially pure titanium) is classified in four groups (grades 1 to 4; see Table 16) but the percentages of permissible additives are greater than those laid down in VdTÜV-Werkstoffblatt 230/1. Two palladium-alloyed titanium types are termed grade 7 and grade 11.

The high price of Pd explains the introduction of the molybdenum- and nickel-containing

Table 15. Grades of unalloyed titanium

Material	Material number	Max. impurity content, wt%					Yield strength (1%), MPa
		Fe	O	N	C	H	
Ti I	3.7025	0.15	0.12	0.05	0.06	0.013	200
Ti II	3.7035	0.20	0.18	0.05	0.06	0.013	270
Ti III	3.7055	0.25	0.25	0.05	0.06	0.013	350
Ti IV	3.7065	0.30	0.35	0.05	0.06	0.013	410

Table 16. U.S. Classification of titanium

Type	ASTM B-265-79 grade	Composition, wt% (max.)						
		Fe	O	N	C	H	Pd	Others
C.P.	1	0.20	0.18	0.03	0.10	0.015		
C.P.	2	0.30	0.25	0.03	0.10	0.015		
C.P.	3	0.30	0.35	0.05	0.10	0.015		
C.P.	4	0.50	0.40	0.05	0.10	0.015		
C.P.-Pd	7	0.30	0.25	0.03	0.10	0.015	0.12–0.25	
C.P.-Pd	11	0.20	0.18	0.03	0.10	0.015	0.12–0.25	
C.P.-Mo + Ni	12	0.30	0.25	0.03	0.08	0.015		0.2–0.4 Mo, 0.6–0.9 Ni
Ticorex-A*		0.03	0.05	0.008	0.004	0.0015		0.05 Ru, 0.5 Ni

* Nippon Mining Co.

alloy grade 12, which, like the ruthenium- and nickel-containing alloy Ticorex A, is superior to unalloyed titanium in crevice corrosion resistance.

Finally, mention should be made of a titanium alloy containing 5% tantalum which has been developed to the production stage in Japan. It has improved resistance to nitric acid.

A Werkstoffblatt for the alloy tantalum–2.5 tungsten, in which growing interest is being shown on account of its strength, is in preparation.

The mechanical properties of titanium, zirconium and tantalum, and also a selection of their physical data, are compiled in Tables 17 and 18.

The fundamental corrosion-chemical properties of these materials for chemical plant are

1) The stability of titanium under oxidizing conditions
2) The stability of zirconium under reducing and alkaline conditions
3) The stability of tantalum under oxidizing and reducing conditions.

In the case of titanium, which is the most favorable material in this group with respect to density, particular attention must be drawn to its resistance to chlorides and oxidizing agents. For example, titanium has been used successfully for heat exchangers exposed to seawater, brackish water, nitric acid, acetic acid, chromic acid, moist chlorine, chlorine dioxide, bleaching solutions, sodium chlorate, and sodium chlorite.

The use of palladium as an alloying element extends the range of applications of titanium in acids with reducing effects, such as sulfuric acid and hydrochloric acid, and improves the resistance to crevice corrosion [104].

Corrosion resistance data for titanium, zirconium, and tantalum are compiled in Table 19. A common feature of the special metals is a pronounced sensitivity to hydrofluoric acid and fluorides, with a risk of hydrogen embrittlement in the case of extreme exposures. Titanium cannot be used in hydrochloric acid or sulfuric acid. However, its resistance to these media can be improved considerably by adding oxidizing agents. Titanium and zirconium cannot be used in fuming nitric acid, because of the possibility of pyrophoric reactions and stress corrosion cracking.

Of the refractory metals, tantalum has by far the best corrosion behavior in hot, concentrated mineral acids, apart from hydrofluoric acid. In hot, concentrated sulfuric acid its corrosion resistance is comparable with that of glass or cast ferrosilicon and is surpassed only by that of noble metals such as gold and platinum. It is also highly resistant to hydrochloric acid and phosphoric acid and has no pyrophoric tendency in

Table 17. Mechanical properties of titanium, zirconium, and tantalum

Property	Titanium 3.7035 group II	Titanium 3.7055 group III	Zirconium grade 702	Tantalum
Tensile strength, MPa	400–550	470–600	295–442	274
0.1% Yield strength, MPa	280	360	197–295	196
Creep strength 150 °C (10^5 h), MPa	150	170	176	255
250 °C (10^5 h), MPa	110	130	117	225
Elongation, %	22	18	25–35	15
Hardness HB 30, MPa	1400	1600		
Hardness HV 30, MPa			1180–1570	825–1470

Table 18. Some physical properties of titanium, zirconium, and tantalum

Property	Titanium 3.7025	Zirconium	Tantalum (Ta-ES)
Density, g/cm^3	4.5	6.53	16.6
mp, K	1975	2125	3271
Thermal conductivity, J m^{-1} s^{-1} K^{-1}	17	21	55.25
Modulus of elasticity (300 K), GPa	108	94.176	172.5

Table 19. Corrosion resistance of titanium, zirconium, and tantalum

Medium	Concentration, wt%	T, °C	Corrosion rate, mm/a Titanium	Zirconium	Tantalum
Hydrochloric acid (aerated)	5	20	<0.05	<0.05	<0.001
	15	35	2.4	<0.08	<0.001
	37	35	15.0	<0.08	<0.001
Sulfuric acid (aerated)	10	35	1.2	<0.05	<0.001
	40	35	8.5	<0.05	<0.001
Nitric acid (fuming)			inflamm. in air	inflamm. in air	resistant
	0.001	all	not resistant	not resistant	not resistant
Sodium hydroxide	10	100	<0.05	<0.05	1.0
	40	80	<0.1	<0.05	not resistant

fuming nitric acid. Tantalum is, however, attacked by oleum, even at room temperature, and also by hot alkaline solutions. Although it has high ductility, its strength can be increased considerably by alloying with 2–10 wt% W, with little effect on the corrosion resistance.

Tantalum–tungsten alloys exhibit good resistance to sulfuric acid, which may be seen in Figure 31 [105]. The alloy most favorable in this respect is Ta2.5W.

At normal pressure tantalum and its alloy with 2.5 wt% tungsten are completely resistant to hydrochloric acid at all concentrations and temperatures [106]. If, however, hydrochloric acid is handled under pressure in tantalum apparatus, damage through hydrogen absorption, and finally through hydrogen embrittlement, is possible. An autoclave insert failed in consequence of hydrogen embrittlement after being in use for only 8 h. The range of conditions under which tantalum is at risk through hydrogen embrittlement is shown in Figure 32.

Alloying tantalum with niobium gives favorably priced materials with similar resistance to corrosion that have a good chance of becoming established in the chemical industry and of playing a part similar in importance to that of tantalum itself. This applies particularly to use in nitric acid. Figure 33 shows the corrosion rates of tantalum–niobium alloys in sulfuric acid (70%) at 165 °C. As, however, the rate of corrosion increases with the niobium content, only alloys containing up to 40 wt% Nb are of technical interest. The mechanical properties of tantalum, tantalum alloys, and niobium are shown in Table 20.

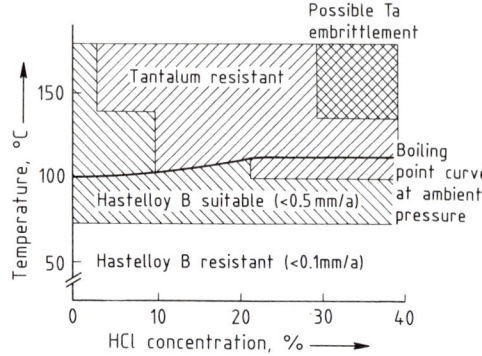

Figure 32. Comparison of the corrosion resistance of tantalum and Hastelloy alloy B in hydrochloric acid

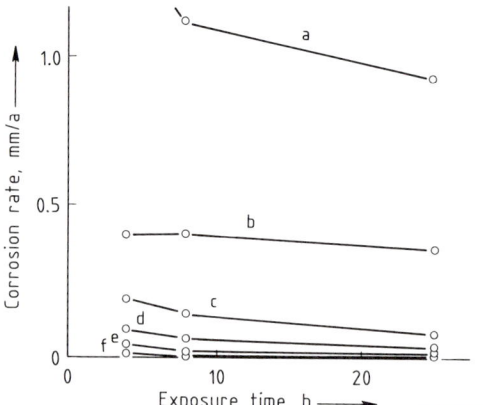

Figure 33. Corrosion rates of tantalum–niobium in boiling 70 vol% sulfuric acid at 165 °C
a) Ta–75 Nb; b) Ta–60 Nb; c) Ta–50 Nb; d) Ta–40 Nb; e) Ta–25 Nb; f) Ta

Table 20. Mechanical properties of tantalum, tantalum alloys, and niobium

Property	Ta	Ta 2.5 W	TaNb 60/40	Nb type 2
Tensile strength, MPa	225	290	276	170
0.2% Yield strength, MPa	140	205	193	105
Elongation, %	15	25	25	25

Under oxidizing conditions zirconium is often less resistant than titanium. Its resistance to nonoxidizing media nevertheless greatly exceeds that of titanium. Zirconium is thus practically passive in hot, fluoride-free sulfuric acid. Zirconium, like titanium, is resistant to many organic acids, but is severely attacked by hydrofluoric acid. Unlike the other refractory metals, zirconi-

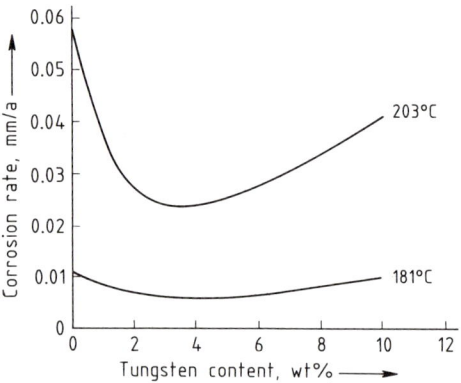

Figure 31. Corrosion rates of tantalum–tungsten alloys in sulfuric acid (96%)

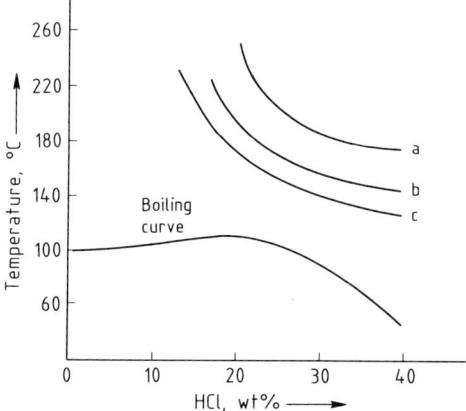

Figure 34. Corrosion of zirconium in hydrochloric acid
a) 5 mm/a; b) 0.5 mm/a; c) 0.13 mm/a

um is resistant to hot alkaline solutions and is therefore the most suitable material for these media in many cases.

Zirconium also has good resistance to corrosion by hydrochloric acid (Fig. 34), though it is vital to ensure that oxidizing agents, such as Fe(III) or Cu(II) compounds, are not present. At 30 °C and an acid concentration of 32%, as little as 5 ppm of such a compound increases the corrosion rate ten-fold (from 7.6×10^{-4} mm/a to 7.6×10^{-3} mm/a) [107], [108]. For critical corrosion conditions, heat treatment for 3 h at 750–790 °C, especially after welding, is advisable in order to maximize the resistance [109].

Refractory metals have a wide range of applications in the chemical industry. They are used for reactors, columns, agitators, pipelines, fittings, bellows, and pumps, for example. Sensing heads, such as thermo-feelers, magnetic floats and membranes, that must also function reliably under adverse conditions are protected by cladding with special metals. Heat exchangers made from special metals and consisting of coils, rods, candle-shaped elements, or tube bundles are used to cool or heat corrosive media.

The components may consist entirely of the special metal or of suitable composites. In many cases nonadherent lining or sheathing of the substrate material is sufficient; cladding is appropriate where heavy loads are exerted, good heat transmission is needed, or negative pressure exists. As the special metals tend to form intermetallic phases with steel at high temperatures (1000 °C), they are normally applied by explosion cladding, possibly with intermediate layers (e.g., copper) [110], [111].

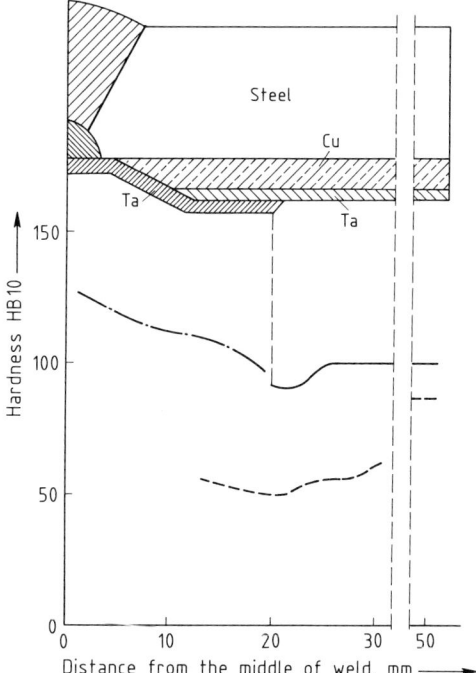

Figure 35. Design of a butt-weld for explosion-cladded sheet (mild steel/tantalum with copper intermediate layer)
–·– Hardness curve in tantalum cover strip; —— Hardness curve in tantalum cladding; – – – Hardness curve in copper intermediate layer

Although special metals have good welding behavior, their reactions with atmospheric gases must be taken into account. Hence welding is only possible under an inert gas or in high vacuum. It is essential to avoid alloying with iron-based materials in welding. Since the melting point of tantalum, for example, is roughly twice as high as that of steel, specially designed butt-welds are necessary in the processing of explosion-cladded metal sheet (cf. Fig. 35) [112].

5.10. Organic Materials [113]–[115]

To an increasing extent the growing requirements on the availability and reliability of chemically exposed plant components are necessitating the use of particularly corrosion-resistant materials. In this respect the chemical nature of organic materials gives them certain advantages over metallic and inorganic nonmetallic materials, provided the operating temperatures do not reach critical levels. In many cases they are well suited for plant that handles the media most commonly encountered at chemical works, namely hydrochloric acid, caustic soda solution, hypochlorite solution, sulfuric acid, waters, and salt solutions of all kinds.

The main types of organic materials are thermoplastics, thermosetting plastics, elastomers, glass-fiber-reinforced plastics, and foamed plastics. Composite and surface coating materials are being used increasingly too (see → 8. Corrosion).

One of the most important processing methods for plastics is welding. A survey of the welding methods is given in [116]. A comprehensive account of the design of plastic components, with detailed consideration of strength calculations and dimensioning, is contained in VDI-Taschenbuch T 21 [117].

5.10.1. Selection Criteria

The stability required of plastics varies according to whether a plastic component must simply withstand corrosion, or fulfil structural functions also. In general no risk arises if the physical properties of a coating or facing consisting of a synthetic material are impaired through the action of a medium; but the same process in a statically or dynamically stressed part may lead to its premature failure. Similarly, swelling and stress corrosion cracking may be nonhazardous, and the quality of joints relatively unimportant, in nonstructural plastics, but hazardous, or critical, in those with structural functions. The permeability of polymers is less serious in components than in coatings, especially those on metallic substrates. Thus the envisaged conditions of use must be carefully considered before materials are chosen. As important parameters, the solutions to which the components will be exposed, including all their constituents (such as solvents, amines, phenols, compounds with high vapor pressures, oxidizing agents), the maximum and long-term service temperatures, the pressures and vacuum conditions, and the static and dynamic loads, must receive special attention. Only when the various factors have been established can a list of foreseeably suitable materials be drawn up for more detailed consideration.

5.10.2. Properties and Application Criteria

Unlike the surface removal corrosion of metallic materials, the action of liquid media on organic materials leads mainly to swelling, caused by absorption of the medium. In many cases the degree of swelling of an organic material in individual media and the dependence of swelling on time and temperature as well as its possible reversibility constitute adequate criteria for the determination of suitability. The stability data published in manufacturers' and other literature are usually based on swelling tests. These data are adequate where coatings or linings, e.g., of rubber or plastics, are concerned, and in a number of other cases.

Where static or dynamically loaded parts are concerned, swelling tests alone must be considered tentative, enabling one merely to draw up a shortlist of materials for further consideration. Here it is far more important to ascertain how the strength of materials is affected by prolonged exposure to chemicals, the degree to which creep occurs, and whether or not the conditions are favorable to stress corrosion cracking. Losses of long period creep resistance caused by liquid media are preferably expressed as resistance factors. A resistance factor is the ratio of the time to failure in the test medium to the time to failure in water [118]. In some media this ratio depends on the mechanical stresses exerted on the sample; in other cases it is largely independent of them. Resistance factors for glass-reinforced plastics can be derived from the differences in creep behavior in tensile tests under various ambient conditions (as ratios between creep moduli, for example) [119].

The stress corrosion cracking of organic polymers is generally an entirely physical process, with diffusion and swelling processes and internal and/or external stresses playing a major part [120]. The term is used only where cracking caused by exposure to chemicals, like that of metals, occurs exclusively in the presence of tensile stresses. Tests are performed in various media and at various temperatures on panels or pipes subjected to constant deformation or constant stress in the long period tensile creep test [118]. Susceptibility to stress corrosion cracking is increased by exposure to wetting agents, organic acids, solvents, and oxidizing media, by shaping and jointing operations, by internal and external stresses, by prolonged exposure to high temperatures in processing, by temperature alteration, and by notching. In many cases the sensitivity of finished parts to stress corrosion cracking can be reduced by thermal after-treatment.

The thermoplastics most widely used in chemical engineering are PVC (polyvinyl chloride), PE (polyethylene), PP (polypropylene) and PB (polybutene).

Poly(vinyl chloride) has been used for decades due to its outstanding resistance to chemicals. It is harder and tougher than other thermoplastics, but at temperatures below 20 °C it has less impact strength than polyethylene (PE) and polybutene (PB). Above its softening temperature of 80 °C, poly(vinyl chloride) can be shaped by plastic deformation. The most favorable deformation range is 110–130 °C. Original shaping (e.g., by injection molding, extrusion, or welding) requires higher temperatures, however (170–200 °C). The upper service temperature limit is 60 °C, or possibly a lower temperature, depending on the chemical exposure.

Polyethylene is a partly crystalline material with high impact strength at temperatures down to ca. −100 °C. High-density polyethylene is preferred for chemical plant because it has greater strength and stiffness than the low-density types. It can be shaped thermoplastically at temperatures above 130 °C and welded at ca. 200 °C. The upper service temperature limit is 60 °C, or 80 °C if glass fibers are incorporated. The resistance to nonoxidizing media is excellent. The swelling caused by hydrocarbons leads to losses of strength that can be taken into account with the aid of resistance factors. The susceptibility to stress corrosion cracking increases with the degree of crystallization, but falls with increasing molecular mass. High molecular mass material is less easily processed, however, and its properties may suffer if it is exposed to elevated processing temperatures for long periods.

Polypropylene (PP) has properties similar to those of polyethylene, but greater impact, notched impact, and shock resistance, especially at temperatures below 0 °C. The upper service temperature limit is 80 °C, or 100 °C if glass fibers are present. Polypropylene is almost as resistant to chemicals as hard polyethylene, but more sensitive to oxidizing media and more strongly swelling in organic solvents. In common with hard polyethylene it has a certain permeability to organic media.

The applications of polypropylene include exhaust ducts, internal chimney tubing, separators, rotary filters, filter fabrics and pressure plates, hydrocyclones, and gas scrubbers.

Polybutene (PB) is one of the more recent polymeric materials that have gained acceptance in chemical engineering. Its strength properties lie between those of soft polyethylene (LD-PE) and polypropylene. Its long period creep resistance exceeds even that of polypropylene. Although polybutene is soluble in several organic solvents at elevated temperatures, its susceptibility to stress corrosion cracking is low. The upper service temperature limit is 90 °C. The main applications of polybutene are high-temperature pressure pipes, filter pressure plates, strip-wound vessels, and pumps.

Polytetrafluorethylene (PTFE) is the thermally and chemically most resistant plastic used in chemical engineering. Its service temperature range extends from −200 to 250 °C. It is chemically resistant to virtuallly all media except molten alkalies and elemental fluorine. However, fillers added to the plastic to improve its extrudability or mechanical strength may reduce its resistance to chemicals. The outstanding abhesive behavior and exceptionally favorable sliding properties deserve special mention. Applications include pipe linings, bellows, hoses, pump parts, tube bundle heat exchangers, steam release nozzles, conveyor belts, seals, stuffing box packings, and filter fabrics.

Poly(vinylidene fluoride) (PVDF) is a partly crystalline plastic of high thermal stability. It can be used at temperatures up to 100 °C, or, if glass-filled, up to about 120 °C. PVDF is resistant to most organic and inorganic acids. In alkalies, especially caustic soda solution, it tends to suffer stress corrosion cracking. Its main application is in pipelines used to convey acids.

Elastomers and ebonite based on natural or synthetic rubber [121], [82] are used mainly for surface protection (see → 8. Corrosion). Elastomers also have many applications in chemical engineering as seals, membranes, bellows, vibration dampers, hoses, and Moineau pump stators, as well as being used for space-saving folding containers for the transportation of corrosive materials by land and sea.

5.10.3. Thermosetting Plastics

The main materials in this group are phenolic, furane, epoxy, vinyl ester, and unsaturated polyester (UP) resins. They are filled with glass fibers, graphite, or quartz and processed to tamping compositions, hot molding compounds, sheet, and putties. Being brittle, they are rarely used directly for mechanically stressed parts unless reinforcing materials and fillers have been incorporated. Having high crosslink density, thermosetting plastics are insoluble and cannot be melted, and their strength and long period creep resistance are without the marked dependence on temperature exhibited by thermoplastics and elastomers. Individual parts can be joined to one another with catalytically hardening cements.

Solid parts made from glass-reinforced reaction resins are distinguished not only by chemical stability but also by relatively low weight, together with high strength, and by good emergency running behavior. They are being used increasingly for composite structures exploiting the high stability and heat resistance of glass-fiber-reinforced reactive resins with the superior chemical resistance of thermoplastic coatings and linings.

For exposure to aggressive media, parts consisting entirely of glass-reinforced plastics usually receive an at least 2.5 mm thick coating consisting mainly of resin. This protective coating, in which the proportion of glass is low, acts as a barrier to corrosion and protects the load-bearing, glass-reinforced structure from the action of destructive media. The resins used are unsaturated polyester, vinyl ester, epoxy, or furane resins.

Table 21 gives a general idea of the corrosion resistance of reaction resins used in the fabrication of apparatus and pipes. Glass-fiber-reinforced resins are used for corrosion-resistant articles of all kinds, especially pipes for waste water and waste air disposal, stacks, siphons, vats, transportation and storage containers, downpipes, filter pressure plates, and frames. The upper service temperature limits extend to about 120 °C, depending on the resin in question.

More frequently, however, parts consisting of glass-reinforced plastics are provided with linings. The lining materials most widely used at present are the thermoplastics polypropylene (PP), unplasticized polyvinyl chloride (PVC-U) and polyvinylidene fluoride (PVDF).

In the fabrication of composite structures, thermoplastic liners are normally produced as a first step and than reinforced with a glass-filled reaction resin. Thermoplastic linings consisting of sheet material are placed end to end and joined by welding. The normal welding techniques are hot gas, heated tool, and hot gas extrusion welding.

Figure 36 depicts a PVC-lined UP-GF apparatus, seen here as a demonstration exhibit with parts removed to show structural details. Another application is shown in Figure 37. This is a

Table 21. Chemical resistance of reaction resins*

Type of resin	Resistance towards			
	Acids	Alkalies	Solvents	Oxidizing agents
Polyester	+/0	+/−	0/−	+
Vinyl ester	+/0	+	0/−	+
Epoxy	0/−	+	+/0	+/−
Furane	+	+	+	0/−

* + = resistant; 0 = conditionally resistant; − = not resistant

Figure 36. PVC-lined UP-GF apparatus

Figure 37. UP-GF tanks and pipes with PVC liners, containing HCl and NaOCl solutions

storage facility for solutions containing HCl and NaOCl, consisting of UP-GF tanks and pipes with PVC liners. The design of parts consisting of glass-reinforced plastics is described in [122]. Properties, processing, and applications are described in [123]. The use of organic materials in the form of corrosion-resistant coatings and linings represents an important application in chemical engineering (see → 8. Corrosion).

5.11. Inorganic Nonmetallic Materials

5.11.1. Glass [124], [125]

Pipelines, apparatus, and even entire plants are now being produced increasingly from borosilicate glass in order to solve exceptionally critical corrosion problems. A complete range of glass parts for assembly by the user, ranging from simple pipe sections and moldings to fittings, vessels and column components, and even to complex heat exchangers and their accessories, is now commercially available.

Borosilicate glass has a very wide range of uses and can be exposed permanently to temperatures up to 200 °C. It is permanently stable to virtually all media except hydrofluoric acid and strong alkalies. Above 200 °C it is attacked fairly severely by all acids, especially phosphoric acid [126]. Other advantages include the smoothness of its surfaces, catalytic inertness, and nontoxicity.

Glass can be combined with other materials—plastics and metals, for example—thus opening up a wide range of technical possibilities. Glass apparatuses for heating purposes can be provided with elements consisting of stainless steel, highly corrosion-resistant nickel alloys, titanium, or tantalum. Pumps and valves of all kinds can be sealed with polytetrafluoroethylene components. PTFE is also used in the form of bellows to compensate for thermal expansion and to damp vibration. Important applications of borosilicate glass include distillation and rectification units, absorbers and gas scrubbers, nitric acid concentrators, acid recovery plant, cleaning units, and piping.

Although quartz glass is very resistant to thermal shock due to its extremely low coefficient of thermal expansion, its high cost limits technical use to special articles (e.g., observation windows).

The use of enamel coatings, made from vitreous enamels and partly crystalline enamels, to protect tanks, apparatus, columns, pipelines, pumps and other items from corrosion is described in → 8. Corrosion.

5.11.2. Graphite [82], [127], [128]

Graphite has been used as a special material in chemical engineering since the mid-1930s. Heat exchangers, simply cemented together from panels and strip, were the first application. Now, all structural elements, such as pipes, solid and hollow cylinders, slabs, panels, and profiles, can be produced from impregnated graphite (hard-fired carbon or electrographite).

As the various graphite production methods give porous materials, impregnation with phenol–formaldehyde or furane condensation resins by the vacuum-pressure process is necessary if impermeability to gases and liquids is required, the effect of this treatment being that all the pores are closed. The quality of the finished material depends on the properties of the chosen resin. This applies particularly to the permeability, corrosion resistance, and upper service temperature limits (which are in the region of 165 °C). For greater stability and higher service temperature limits, alternative impregnating agents, such as polytetrafluorethylene, are used. Carbon filling of the pores in graphite can be carried out by impregnating with cokable materials, which are then decomposed thermally, this treatment being repeated several times. The resulting material has good corrosion resistance and a high service temperature limit, but is less strong and more brittle than the resin-impregnated grades.

Resin-impregnated electrographite is the synthetic carbon material most widely used in chemical engineering. It is highly resistant to corrosion by organic and inorganic acids, alkalies, alcohols and many solvents, and other organic compounds. It is easily worked, has high thermal conductivity, and withstands even severe temperature shocks. As graphite is a brittle material with practically no plastic deformation, the parts must be so designed that tensile and shear stresses are avoided as far as possible.

The main applications of graphite are heat exchangers of all kinds, columns, sprinkling coolers, centrifugal pumps, valves, pipelines, and small parts of many kinds. The use of graphite in plant for production of dry hydrogen chloride and in the treatment of waste water and waste air deserves special mention.

5.11.3. Refractory and Acid-Resistant Bricks [129]–[132]

This term covers ceramic materials that are still free from deformation at temperatures exceeding 1600 °C and still do not soften at those exceeding 1800 °C. Bricks are normally shaped and fired, but are sometimes used in the form of castings. Apart from refractoriness, such properties as softening under pressure, thermal conductivity, chemical resistance, resistance to temperature alternation, abrasion resistance, shape stability, and slagging tendency are of great importance.

Refractory bricks serve to enclose spaces in which reactions take place and must last as long as possible under the given conditions. In view of the variety of these conditions they must be chosen for then intended application in accordance with their chemical and technological properties. The following classification of refractory materials is based on their main consituents: products of high silica content, alumina-based materials, basic refractory bricks, neutral refractory products, carbon bricks, and silicon carbide bricks.

For corrosion protection in difficult cases, chemical apparatus can be lined with corrosion-resistant bricks, tiles, or moldings. These are attached to the wall or to one another with mortar or putty. Sealing and insulating layers may be used additionally between the wall of the apparatus and the lining.

5.11.4. Engineering Ceramics [133], [134]

Oxide Ceramics. The most important oxide ceramic material is aluminum oxide, Al_2O_3. Beryllium oxide has applications as a moderator in nuclear reactors, and zirconium dioxide as a rocket material. In Table 22 the main mechanical and thermal properties of aluminum oxide are compared with those of hard metal. A conspicuous property of Al_2O_3 is its great hardness. The strength values are practically constant over a wide temperature range. This is because the increase in plasticity (and hence in the ability to dissipate local peak stresses) which occurs as the temperature rises compensates for the thermally induced loss of strength. On the other hand the impact strength (unnotched) of $0.2 J/cm^2$ at 20 °C, is low compared to the notched impact strength of $100 J/cm^2$ for structural steel. Thus Al_2O_3 is a brittle material.

Alumina has very good corrosion resistance, withstanding acids, molten metals, many glasses, and slag. It has advantages over metals particularly where corrosive and thermal stresses are superimposed on mechanical stresses.

One of the commonest applications of alumina ceramics is therefore the sealing of passages leading into vessels and spaces in which corrosive media are present. In the form of slide rings and mechanical seals alumina is superior to all other materials in resistance to corrosion and wear. Its use is therefore leading to the replacement of conventional stuffing boxes on rotating shafts by mechanical seals, especially where high rotation speeds occur. Aluminum oxide can be used in oxidizing and reducing atmospheres and under vacuum at working temperatures of up to 1950 °C.

Table 22. Properties of Al_2O_3 ceramic and a hard metal

Property	Al_2O_3 ceramic	WC Hard metal with 6% Co
Density, g/cm³	3.9	14.9
Bending strength, MPa	300	1900
Compression strength, MPa	2500	5000
Breaking elongation in compression test, %	<0.1	ca. 0.5
Modulus of elasticity, GPa	390	620
Vickers hardness	1750	1600
Thermal conductivity, $W m^{-1} K^{-1}$	30	80
Coefficient of thermal expansion, $10^{-6}/K$	8	5

Silicon Carbide. Silicon carbide has not only high strength (bending strength up to 650 MPa), but also a low coefficient of thermal expansion of only 4.8×10^{-6} K^{-1} and a high thermal conductivity of 42 W K^{-1} m^{-1}. These properties explain the very good resistance of silicon carbide products to changes of temperature. Since silicon carbide also has excellent resistance to corrosion, it is used as a structural material for components subjected to severe thermal and chemical loads.

Silicon carbide is used mainly as a grinding agent, e.g., for grey cast iron, nonferrous metals, glass, and ceramics. With great resistance to temperature alternation and corrosion, SiC thermocouple protection tubes are used in contact with liquid aluminum in aluminum holding furnaces. Silicon carbide products are used for the muffles of indirectly heated smelting furnaces for zinc, aluminum, copper and their alloys, and for tank linings. With good electrical conductivity at high temperatures, coupled with resistance to oxidation, SiC heating elements are used in oxidizing atmospheres at temperatures up to 1500 °C.

Silicon Nitride. The most remarkable properties of silicon nitride include good resistance to chemical attack by acids and molten nonferrous metals, resistance to temperature fluctuation (coefficient of thermal expansion 3×10^{-6} K^{-1}), and heat resistance. The bending strength of hot-pressed silicon nitride is about 700 MPa at 20 °C and still as high as 290 MPa at 1400 °C. At 1000 °C dense silicon nitride still has a 100-h long-term creep resistance of 200 MPa. In the case of the aluminum industry, materials are needed that are not wetted by the molten metal, have good resistance to temperature fluctuation, and high strength at temperatures up to ca. 800 °C. As silicon nitride meets these requirements, it is used for, among other things, thermocouple protection tubes and ascending pipes in low-pressure casting. The hardness of silicon nitride gives it high wear resistance, which, together with the low friction coefficient ($\mu = 0.1 - 0.2$), favors its use as a material for bearings and sliding contacts. Slide rings, roller bearings, and tube drawing plugs are typical applications. Silicon nitride is also used as a cutting material.

Ceramic materials are often used in the chemical industry in the form of thermally sprayed coatings that afford protection against wear and corrosion. This field of application is described in → 8. Corrosion.

6. References

[1] E. Gaube et al.: "Werkstoffe für den Apparatebau in der chemischen Technik," in: *Winnacker-Küchler*, 4th ed., vol. 1, pp. 452–503.
[2] H. Titze: *Elemente des Apparatebaues*, 2nd ed., Springer Verlag, Berlin – Heidelberg – New York 1967.
[3] K. Wellinger, F. Eichhorn, P. Gimmel: *Schweißen*, Kröner, Stuttgart 1964.
[4] O. Becken: "Grundlagen und Anwendung," in: *Handbuch des Schutzgasschweißens*, part I, Deutscher Verlag für Schweißtechnik, Krefeld 1969.
[5] J. Ruge: "Werkstoffe," in: *Handbuch der Schweißtechnik*, 2nd ed., vol. 1, Springer Verlag, Berlin – Heidelberg – New York 1980.
[6] W. Schönherr: "Über die Eignung der Werkstoffe zum Schweissen und thermischen Trennen," *Schweissen Schneiden* **23** (1971) 441–442.
[7] G. Habenicht: *Kleben, Grundlagen, Technologie, Anwendung*, 2nd ed., Springer Verlag, Berlin – Heidelberg – New York – London – Paris – Tokyo – Hongkong – Barcelona 1990.
[8] R. J. Schliekelmann, F. Mittrop: *Metallkleben – Konstruktion und Fertigung in der Praxis*, Deutscher Verlag für Schweißtechnik, Krefeld 1971.
[9] M. Pfender: "Über das Festigkeitsverhalten metallischer Werkstoffe und Konstruktionsteile," in: *Das Gesicht des Bruches metallischer Werkstoffe*, vols. 1 and 2, Allianz-Versicherungs-AG, München 1956.
[10] K. Wellinger, H. Dietmann: *Festigkeitsberechnung*, Kröner, Stuttgart 1969.
[11] D. Broek: *Elementary Engineering Fracture Mechanics*, Noordhoff International Publ., Leyden 1974.
H. Blumenauer, G. Pusch: *Technische Bruchmechanik*, VEB Deutscher Verlag für Grundstoffindustrie, Leipzig 1982 (distributed by Springer, Vienna – New York).
[12] Merkblätter der Arbeitsgemeinschaft für Druckbehälter.
[13] M. Litzkendorf: *Über den Zusammenhang zwischen Korrosion und Festigkeitsverhalten hochlegierter Chrom- und Chrom-Nickel-Stähle*, Dissertation, TH Darmstadt 1974.
[14] K. Wellinger, H. Uetz: "Gleitverschleiß, Spülverschleiß, Strahlverschleiß unter der Mitwirkung von körnigen Stoffen," *VDI Forschungsh.* **449** (1955).
[15] T. Günther *Chem. Ing. Tech.* **42** (1970) 774–780.
[16] DECHEMA: *DECHEMA-Werkstoff-Tabelle*, Frankfurt, published since 1953.
D. Behrens (ed.): *DECHEMA Corrosion Handbook*, VCH Verlagsgesellschaft, Weinheim, vol. 1 (1981) – vol. 6 (1990), to be continued.
[17] F. Ritter: *Korrosionstabellen metallischer Werkstoffe*, 4th ed., Springer Verlag, Wien 1958.
[18] H. Gräfen, K. Steiger: "Zerstörungsfreie Prüfungen bei Betrieb und Instandhaltung von Chemieanlagen," *DECHEMA Monogr.* **99** (1985) 137–157.
[19] H. Adrian, F. Brühl: "Die Entwicklung der hochfesten schweißbaren Stähle für den Stahlbau und ihre Anwendung und Verarbeitung," *Stahl Eisen* **86** (1966) 645–662, 670–672.
[20] K. Schaar: "Hochfeste Stähle im Druckbehälterbau," *Tech. Überwach. (Essen)* **7** (1966) 145–150.

[21] K. H. Piehl: "Herstellung, Eigenschaften, Verarbeitung und Anwendung von hochfesten Feinkornbaustählen," *Rheinstahl-Tech.* **7** (1969) 71–83.
[22] F. Heisterkamp, L. Meyer: "Mechanische Eigenschaften perlitarmer Baustähle," *Thyssenforschung* **3** (1971) 44–65.
[23] F. Heisterkamp, D. Lauterborn, H. Hübner: "Technologische Eigenschaften, Verarbeitbarkeit und Anwendungsmöglichkeiten perlitarmer Baustähle," *Thyssenforschung* **3** (1971) 66–76.
[24] E. Weber, W. Rudat: "Herstellung der Stahlrohre," in: *Rohrleitungen, Theorie und Praxis*, Springer Verlag, Berlin – Heidelberg – New York 1967.
[25] W. Dahl, H. Hengstenberg, H. Behrens: "Sprödbruchverhalten unlegierter und niedriglegierter Baustähle im Kerbschlagbiege- und Zugversuch in Abhängigkeit von Korngrößen, Temperatur, Vorverfestigung und Alterung," *Stahl Eisen* **88** (1968) 578.
[26] C. Straßburger: "Stahl – auch weiterhin ein moderner Werkstoff," *Thyssen Tech. Ber.* **13** (1987) no. 1, 89–102.
[27] E. Lenz, N. Wieling: "Einflußgrößen auf die Rißinitiierung und auf das Rißwachstum in niedriglegierten Feinkornbaustählen in Hochtemperaturwasser," *Vorträge der 20. Sitzung des Arbeitskreises Bruchvorgänge, Frankfurt 1988*, Deutscher Verband für Materialforschung und -prüfung, Berlin 1988, pp. 127–167.
[28] D. Uwer, H. Dißelmeyer: "Erfahrungen mit der Herstellung, Verarbeitung und Anwendung des hochfesten wasservergüteten Baustahles XAB090," *Schweissen Schneiden* **38** (1986) no. 9, 430–436.
[29] C. Florin et al.: "Über die Grundlagen der Warmfestigkeit, dargestellt an legierten Stählen," *Arch. Eisenhüttenwes.* **41** (1970) 777–787.
[30] H. Wisniowski: "Hochwarmfeste 12%-Chromstähle," *DEW Tech. Ber.* **9** (1969) 117–134.
[31] H. Wirtz: "Anwendung," in: *Verhalten der Stähle beim Schweißen*, 2nd ed., vol. 2: Deutscher Verlag für Schweißtechnik GmbH, Düsseldorf 1977.
[32] K. Bungart: "Hochtemperaturwerkstoffe auf Nickel- und Kobaltbasis," *DEW Tech. Ber.* **9** (1969) 146–162.
[33] H. Kiessler: "Normung hochwarmfester Stähle und Legierungen," *DEW Tech. Ber.* **9** (1969) 110–117.
[34] *Ergebnisse deutscher Zeitstandversuche*, Verlag Stahleisen mbH, Düsseldorf 1969.
[35] W. Ruttmann, H. Kaes: "Bedeutung der Änderung der Langzeitwerte," *Mitt. Ver. Großkesselbetr.* **48** (1968) 1–9.
[36] K. Bungardt, H. Krainer, H. Schrader: "Nichtrostende, hitzebeständige sowie hochwarmfeste Stähle und Legierungen," *Stahl Eisen* **84** (1964) 1796–1811.
[37] D. Schneider: "Schweißverbindungen an kaltzähen Stählen im Druckbehälterbau," *Tech. Überwach. (Essen)* **7** (1966) 317–321.
[38] K.-A. Ebert: "Betrachtungen zum Lichtbogenhandschweißen der kaltzähen Stähle," *Schweissen Schneiden* **18** (1966) 125–137.
[39] H. Jesper, K. Achtelik: *Kaltzähe Stähle*, publication no. 33, Nickel-Informationsbüros GmbH, Düsseldorf 1964.
[40] H. Spähn: "Druckwasserstoffangriff auf unlegierter und niedriglegierter Stähle im Temperaturbereich oberhalb 200 °C," in D. Kuron (ed.): *Wasserstoff und Korrosion*, Bonner Studien-Reihe, Verlag Irene Kuron, Bonn 1986, pp. 155–201.
[41] H. H. Buchter: *Apparate und Armaturen der Chemischen Hochdrucktechnik*, Springer Verlag, Berlin – Heidelberg – New York 1967.
[42] W. Wessling, F. Schmöning, H. E. Bock: "Aushärtbare nichtrostende Stähle," *Sie und wir*, Company brochure, Stahlwerke Südwestfalen, 1970, no. 6, pp. 41–59.
[43] H. Sigwart: "Werkstofftechnische Grundlagen der Konstruktion," in Dubbel: *Taschenbuch für den Maschinenbau*, 13th ed., vol. 1, Springer, Berlin – Heidelberg – New York 1970.
[44] American Petroleum Institute, Refining Department: *Steels for Hydrogen Service at Elevated Temperatures and Pressures in Petroleum Refineries and Petrochemical Plants*, API-Publication 941, 3rd ed., May 1983.
[45] *Nichtrostende Stähle*, 2nd ed., Verlag Stahleisen, Düsseldorf 1989.
[46] H. Wehner, H. Speckhardt: "Zum Ausscheidungs- und Korrosionsverhalten eines ferritisch-austenitischen Chrom-Nickel-Molybdän-Stahles nach kurzzeitiger Glühbehandlung unter besonderer Berücksichtigung des Schweißens," *Z. Werkstofftech.* **10** (1979) 317–332.
[47] W. Wessling, H. E. Bock, W. Fuchs: "Neuere Entwicklungen von korrosionsbeständigen Sonderstählen für das Chemiewesen," *Z. Werkstofftech.* **4** (1973) 186–195.
[48] E.-M. Horn, A. Kügler: "Entwicklung, Eigenschaften, Verarbeitung und Einsatz des hochsiliciumhaltigen Stahls X2CrNiSi1815," *Z. Werkstofftech.* **8** (1977) 362–370.
[49] G. Hochörtler, E.-M. Horn: "Austenitic Stainless Steel With Approximately 5.3% Silicon," *Met. Corros. Proc. Int. Congr. Met. Corros. 8th* **2** (1981) 1471.
[50] R. Oppenheim, H. Kiesheyer, *Thyssen Edelstahl Tech. Ber.* **8** (1982) 97–100, 111–114.
[51] K. Fäßler, H. Spähn: "Werkstoffauswahl bei wasserseitiger Beanspruchung," in: "Wärmeaustauscher," *DECHEMA Monogr.* **87** (1980) 51–72.
[52] E.-M. Horn, D. Kuron, H. Gräfen: "Lochkorrosion an passiven Legierungssystemen der Elemente Eisen, Chrom und Nickel," *Z. Werkstofftech.* **8** (1977) no. 2, 37–68.
[53] E.-M. Horn, *VDI Ber.* **235** (1975) 39–45.
[54] K. Lorenz, G. Medawar, *Thyssenforschung* **1** (1969) no. 3, 97–108.
[55] R. W. Staehle, M. O. Speidel (eds.): *Handbook on Stress Corrosion Cracking and Corrosion Fatigue*, ARPA (Advanced Research Projects Agency), Washington 1978.
[56] H. Spähn, *Z. Phys. Chem. (Leipzig)* **234** (1967) 1.
[57] M. A. Harzenmoser, P. J. Uggowitzer: "Neue aufgestickte austenitisch-rostfreie Stähle und Duplexstähle," in: *Moderne Stähle*, vol. 1, Verlag der Schweizerischen Akademie der Werkstoffwissenschaften, Zürich 1987, pp. 205–218.
[58] F. P. Pickering: "Physical Metallurgical Developments of Stainless Steels," in: *Stainless Steels 84*, The Institute of Metals, London 1985.
[59] G. Gümpel, T. Ladwein, E. Michel, F. H. Strom: "Entwicklungen bei austenitischen Stählen mit erhöhten Festigkeitseigenschaften für den Einsatz im chemischen Apparatebau," *Thyssen Edelstahl Tech. Ber.* **14** (1988) no. 1, 12–25.
[60] *Gußeisen Handbuch*, Gießerei-Verlag GmbH, Düsseldorf 1963.

[61] K. Röhrig, D. Wolters: "Gußeisen mit Lamellengraphit und karbidischen Gußeisen," in: *Legiertes Gußeisen*, vol. 1, Gießerei-Verlag GmbH, Düsseldorf 1970.
[62] *Die Ni-Resist-Gußeisenwerkstoffe*, publication no. 16, 3rd ed., International Nickel Deutschland GmbH, Düsseldorf 1968.
[63] O. Nickel: "Eigenschaften und Anwendungen austenitischer Gußeisenwerkstoffe in Erdölraffinerien und petrochemischen Werken," *Erdöl Kohle Erdgas Petrochem.* **18** (1965) 121–127, 197–205.
[64] H. Grundmann: *Schweißen von Gußeisenwerkstoffen und Stahlguß*, Deutscher Verlag für Schweißtechnik GmbH, Düsseldorf 1971.
[65] U. Draugelates: "Konstruktionsschweißen von Eisengußwerkstoffen mit Kugelgraphit, Leistungsstand und Verbindungsgüte," *VDI Ber.* **469** (1982) 95–103.
[66] K. E. Volk (ed.): *Nickel und Nickellegierungen*, Springer Verlag, Berlin – Heidelberg – New York 1970.
[67] W. Betteridge: *Nickel and Its Alloys*, MacDonald and Evans Ltd., Estover – Plymouth 1977.
[68] U. Heubner et al.: *Nickel Alloys and High-Alloy Special Stainless Steels*, VDM, Expert Verlag, Sindelfingen 1987.
[69] G. R. Pease: *Metallkundliche Vorgänge beim Schweißen von Nickel und Nickellegierungen*, publication no. 6, International Nickel Deutschland GmbH, Düsseldorf 1957.
[70] *Nickel-Eigenschaften, Verarbeitung und Verwendung*, publication no. 32, 1st ed., International Nickel Deutschland GmbH, Düsseldorf 1964.
[71] *Beständigkeit von Nickel und Nickellegierungen gegenüber Ätzalkalien*, publication no. 8, 3rd ed., International Nickel Deutschland GmbH, Düsseldorf 1963.
[72] W. Z. Friend: *Corrosion of Nickel and Nickel-Base Alloys*, J. Wiley and Sons, New York – Chichester – Brisbane – Toronto 1980.
[73] Wiggin, Huntington: *Korrosionsbeständige Legierungen*, publication no. 3385, Nickel Alloys International S.A., Brüssel 1969.
[74] *Schweißen von Nickel und seinen Legierungen*, publication no. 9, 1st ed., International Nickel Deutschland GmbH, Düsseldorf 1967.
[75] *Nickel-Kupfer-Legierungen*, publication no. 20, 2nd ed., International Nickel Deutschland GmbH, Düsseldorf 1966.
[76] *Korrosionsbeständigkeit nickelhaltiger Werkstoffe gegenüber Fluor, Fluorwasserstoff, Flußsäure und anderen Fluorverbindungen*, publication no. 59, 1st ed., International Nickel Deutschland GmbH, Düsseldorf 1970.
[77] A. J. Sedricks: *Corrosion of Stainless Steels*, J. Wiley and Sons, New York – Chichester – Brisbane – Toronto 1979.
[78] *Chrom-Nickel-Legierungen gegen Brennstoffaschen-Korrosion*, publication no. 2942 D, International Nickel Deutschland GmbH, Düsseldorf 1965.
[79] H. Gräfen, G. Böhm: "Die interkristalline Korrosion von Ni-Mo- und Ni-Mo-Cr-Legierungen," *Z. Metallkd.* **51** (1960) 245–252; *Nickel-Ber.* **19** (1961) 3–9.
[80] H. Gräfen: "Eigenschaften von Ni–Cr–Mo-Legierungen und ihre Verbesserungsmöglichkeiten," *Chem. Ing. Tech.* **35** (1963) 229–235.
[81] R. Köcher: "Neue Werkstoffe für den Chemie-Apparatebau und ihre schweißtechnische Verarbeitung," *VDI Z.* **112** (1970) 749–808.
[82] W. Schatt (ed.): *Werkstoffe des Maschinen-, Anlagen- und Apparatebaues*, VEB-Verlag, Leipzig 1983.
[83] Aluminium-Zentrale (ed.): *Aluminium-Taschenbuch*, Aluminium-Verlag, Düsseldorf 1984.
[84] W. Huppatz, in: H. Gräfen et al. (eds.): *Die Praxis des Korrosionsschutzes*, Expert Verlag, Grafenau 1981, pp. 64–81.
[85] A. Oechslin: "Hartanodisieren von Aluminiumlegierungen," *Schweiz. Alum. Rundsch.* **19** (1969) 8.
[86] F. F. Berg: *Korrosionsschaubilder*, VDI-Verlag, Düsseldorf 1965.
[87] W. Hornig, Diplomarbeit, Universität Erlangen-Nürnberg 1977.
[88] H. Holten, H. Sigurdsson, *Werkst. Korros.* **28** (1977) 475–477.
[89] E. Wendler-Kalsch, *VDI-Ber.* **365** (1980) 17–35.
[90] *Kupfer – Eigenschaften, Verarbeitung, Verwendung*, Deutsches Kupfer-Institut e.V., Berlin 1961.
[91] *Kupfer und Kupferlegierungen in Industrie und Handwerk*, Deutsches Kupfer-Institut, e.V., Berlin 1952.
[92] O. v. Franqué, *Sanit., Heiz., Klimatech.* **1984**, nos. 1 + 2, 34–40, 94–97.
[93] E. Wendler-Kalsch in: *Korrosion in Kalt- und Warmwassersystemen der Hausinstallation*, DGM, Oberursel 1984, pp. 51–70.
[94] E. E. Langenegger, F. P. A. Robinson: "The Role of Arsenic in Preventing the Dezincification of α-Brass," *Corrosion (Houston)* **25** (1969) 137–143.
[95] D. H. Thompson, A. W. Tracy, *Metals Trans.* **1949**, 100.
A. W. Blackwood, N. S. Stoloff, *ASM Trans. Q.* **62** (1969) 677.
[96] E. Wendler-Kalsch, *Z. Werkstofftech.* **13** (1982) 129–137.
[97] H. Vogel, Dissertation, Universität Erlangen-Nürnberg 1985.
[98] H. Vogel, E. Wendler-Kalsch, *Z. Metallkd.*, **75** (1984) 217–221.
[99] W. Hofmann: *Lead and Lead Alloys*, Springer Verlag, Berlin – Heidelberg – New York 1970.
[100] W. W. Krysko: "Homogen verbleiter Stahl als Konstruktionswerkstoff," *Werkst., Korros.* **15** (1964) 631–633, 726–728, 991–995; **17** (1966) 235–238.
[101] H. Gräfen, D. Kuron: "Neuentwicklung von Bleilegierungen aufgrund elektrochemischer Untersuchungen über das Korrosionsverhalten in Schwefelsäure," *Werkst., Korros.* **21** (1970) 3–16.
[102] M. Hörmann, D. Lupton, E. Heintze, E.-M. Horn: "Sondermetalle und ihre Anwendung im Chemieapparatebau," *Z. Werkstofftech.* **18** (1987) 139–147, 186–194.
[103] K. Rüdinger: "Moderne Werkstoffe, Auswahl – Prüfung – Anwendung, Übersichten über Sondergebiete der Werkstofftechnik für Studium und Praxis," *Z. Werkstofftech.* **9** (1978) 181–188.
[104] K. Rüdinger: "Technologische Eigenschaften und Korrosionsverhalten einer Titanlegierung mit 0,2 % Palladium," *Werkst., Korros.* **16** (1965) 109–115.
[105] H. Schussler: *Corrosion Data Survey on Tantalum*, Fansteel Inc., North Chicago, Ill., 1972, p. 104.
[106] H. Dickmann, E.-M. Horn, U. Gramberg: Erfahrungen mit hochschmelzenden Metallen in der Chemietechnik, VDI-W Meeting, Köln, Dec. 6–7, 1984, pp. 87–95.

[107] D. R. Knittel, R. T. Webster: "Corrosion Resistance of Zirconium and Zirconium Alloys in Inorganc Acids and Alkalies," in: E. W. Kleefisch (ed.): "Industrial Applications of Titanium and Zirconium," *ASTM Spec. Tech. Publ.* **728** (1981) 191–203.

[108] R. D. Knittel: "Zirconium," in: P. A. Schweitzer (ed.): *Corrosion and Corrosion Protection Handbook*, Marcel Dekker, New York – Basel 1983, pp. 209–210.

[109] T. Günther: "Zusammenhang zwischen Gefüge und Korrosionsverhalten bei Zirconium," *Werkst. Korros.* **30** (1979) publication no. 5, 308–321.

[110] R. Köcher: "Nichteisenmetalle im Apparate- und Anlagenbau," *Chem. Ing. Tech.* **55** (1983) no. 10, 752–762.

[111] U. Gramberg, E.-M. Horn, K.-O. Cavalar: "Explosionsplattierte Bleche in Anlagen, Erfahrungen aus der Chemietechnik," VDI-W Meeting: Explosionsplattieren – ein modernes Verfahren zur Herstellung von Hochleistungsverbundsystemen, Düsseldorf, Dec. 1, 1983, pp. 29–35.

[112] R. Köcher, *Werkst. Korros.* **28** (1977) 166–173.

[113] *Kunststoff-Handbuch* (12 vols.), Hanser, München 1963–1975.

[114] H. Domininghaus: *Die Kunststoffe und ihre Eigenschaften*, 3rd ed., VDI-Verlag, Düsseldorf 1988.

[115] W. Nestler: "Kunststoffe im Apparatebau," *Kunststoffe* **60** (1970) 719–732.

[116] Th. Hadick: "Die schweißtechnische Praxis," in: *Schweißen von Kunststoffen für Praktiker und Konstrukteur*, vol. 6, Deutscher Verlag für Schweißtechnik GmbH, Düsseldorf 1970.

[117] R. Taprogge: *Konstruieren mit Kunststoffen*, VDI-Taschenbuch 21. VDI-Verlag, Düsseldorf 1971.

[118] G. Ehrbar: "Kunststoffe im chemischen Apparatebau," *Kunststoffe* **53** (1963) 845–854.

[119] E. Dolfen: Bemessungsgrundlagen für tragende Bauelemente aus glasfaserverstärkten Kunststoffen, insbesondere aus durch Glaseidenmatten bewährten Polyesterharzen, Dissertation, TH Aachen, 1968.

[120] F. Fischer: "Spannungsrißbildung und Spannungsrißkorrosion bei Kunststoffen," *Z. Werkstofftech.* **1** (1970) no. 2, 74–83.

[121] S. Boström: *Kautschukhandbuch*, vol. 5, Verlag Berliner Union, Stuttgart 1962, pp. 340–369.

[122] R. Grünewald: "Konstruieren mit glasfaserverstärkten Kunststoffen," *Kunststoffberat. Rundsch. Tech.* **5** (1967) 356–361.

[123] H. Haferkamp: *Glasfaserverstärkte Kunststoffe*, VDI-Taschenbuch 17. VDI-Verlag, Düsseldorf 1970.

[124] S. Lohmeyer et al.: "Kontakt und Studium," in: *Werkstoff Glas*, vol. 22, Lexika-Verlag, Grafenau/Württemberg 1979.

[125] H. Scholze: *Glas*, 2nd ed., Springer Verlag, Berlin – Heidelberg – New York 1977.

[126] A. Peters: "Chemisches Verhalten und physikalische Daten von Borosilikatglas," *Haus Tech. Vortragsveröff.* **211** (1969) 6–10.

[127] F. Brandmair: "Zweckmäßige Apparatekonstruktionen aus Kunstkohle und Elektrographit und deren Einsatzgebiete," *Werkst. Korros.* **17** (1966) 10–17.

[128] H. Wurmseher: "Graphit im chemischen Apparatebau," *Tech. Mitt.* **62** (1969) 226–230.

[129] F. Harders, S. Kienow: *Feuerfestkunde*, Springer Verlag, Berlin – Göttingen – Heidelberg 1960.

[130] K. Konopicky: *Feuerfeste Baustoffe*, Verlag Stahleisen, Düsseldorf 1957.

[131] J. Agst: *Die feuerfesten Baustoffe*, 6th ed., Scimarowsky and Agst, Kapellen/Moers 1971.

[132] F. K. Falcke: *Kleines Handbuch des Säureschutzbaues*, Verlag Chemie, Weinheim 1966.

[133] H. Lahnang, H. Scholze: "Keramische Werkstoffe," in: *Keramik*, part 2: 6th ed., Springer Verlag, Berlin – Heidelberg – New York – Tokyo 1983.

[134] H.-J. Bargel et al. in H.-J. Bargel, G. Schulze (eds.): *Werkstoffkunde*, 5th ed., VDI-Verlag, Düsseldorf 1988.

8. Corrosion

HUBERT GRÄFEN, ELMAR-MANFRED HORN, HARTMUT SCHLECKER, HELMUT SCHINDLER, Bayer AG, Leverkusen, Federal Republic of Germany

1.	Introduction	8-2
2.	Basic Electrochemistry	8-2
2.1.	Electrochemical Equilibrium	8-2
2.2.	Electrochemical Kinetics	8-4
2.2.1.	Corrosion Elements	8-5
2.2.2.	Mixed Electrode and Mixed Potential	8-6
2.2.3.	Overpotential and Polarization	8-6
2.3.	Current Density – Potential Curves	8-7
2.4.	Hydrogen Overpotential and Acid Corrosion	8-9
2.5.	Oxygen Overpotential and Oxygen Corrosion	8-11
2.6.	Protective Film Formation and Passivity	8-13
3.	Types of Electrolytic Corrosion	8-14
3.1.	Uniform and Nonuniform Surface Corrosion	8-14
3.2.	Forms of Localized Corrosion without Applied Mechanical Stress	8-16
3.2.1.	Pitting Corrosion	8-16
3.2.2.	Crevice Corrosion	8-19
3.2.3.	Selective Corrosion	8-21
3.2.3.1.	Intergranular Corrosion	8-21
3.2.3.2.	Spongiosis	8-25
3.2.3.3.	Dezincification	8-25
3.2.3.4.	Line- and Layer-Type Corrosion	8-26
3.3.	Forms of Local Corrosion under Mechanical Stress	8-28
3.3.1.	Stress Corrosion Cracking	8-28
3.3.1.1.	Unalloyed and Low-Alloy Steels	8-29
3.3.1.2.	Stainless Steels	8-31
3.3.1.3.	Nonferrous Metals	8-33
3.3.1.4.	Special Types of Stress Corrosion Cracking	8-34
3.3.1.5.	Hydrogen-Induced Cracking at Low Temperatures	8-35
3.3.1.6.	Hydrogen Corrosion at Elevated Temperatures	8-37
3.3.2.	Corrosion Fatigue	8-39
3.3.3.	Differentiation between Types of Corrosion Cracking	8-42
4.	High-Temperature Corrosion	8-42
4.1.	Principles of High-Temperature Corrosion	8-42
4.1.1.	Thermodynamic Principles	8-42
4.1.2.	Kinetics of Scaling	8-43
4.2.	Oxidation of Alloys	8-45
4.3.	Sulfurization of Metals and Alloys	8-45
4.4.	Corrosion by Combustion Gases	8-46
4.5.	Effect of Ash, Deposits, and Salts	8-47
4.6.	Corrosion by Synthesis Gases	8-47
4.7.	Oxidation Behavior of Technical Alloys	8-49
4.7.1.	Steels	8-49
4.7.2.	Nickel Alloys	8-51
4.7.3.	Cobalt Alloys	8-52
5.	General Corrosion Protection Procedures	8-52
5.1.	Corrosion Protection by Coating	8-52
5.1.1.	Metal Coatings	8-53
5.1.2.	Inorganic Nonmetallic Protective Coatings	8-57
5.1.2.1.	Enamel	8-58
5.1.2.2.	Brick Lining	8-59
5.1.2.3.	Thermally Sprayed Coatings	8-60
5.1.3.	Organic Lining and Coating Materials	8-61
5.1.3.1.	Rubber Linings	8-62
5.1.3.2.	Rubber – Plastic Composite Linings	8-62
5.1.3.3.	Linings Consisting of Thermally Crosslinkable Thermosetting Plastics	8-64
5.1.3.4.	Reinforced Reaction Resin Coatings, Crosslinkable with Catalysts	8-64
5.1.3.5.	Thermoplastic Linings	8-64
5.1.3.6.	Paints and Powder Coatings	8-66
5.2.	Inhibitors	8-66
6.	Electrochemical Corrosion Protection	8-67
7.	Corrosion Testing	8-70
7.1.	Test Methods	8-70
8.	References	8-73

1. Introduction

When in use, all materials are exposed to environmental effects that can impair or even destroy their usefulness or that of components made out of them. Most of the damage in the case of metals is caused by corrosion—reactions with the environment which can lead to a measurable change in the material (cf. DIN 50 900, Part 1 [1]). The reactions are mainly electrochemical, although they can also involve chemical or metallo-physical processes.

Material changes due exclusively to mechanical effects are attributable to wear rather than to corrosion. However, a corrosive reaction with the environment can be brought about by erosion of protective coatings (erosive corrosion).

The tendency of metals to change with loss of energy into the disordered, more stable thermodynamic state represented by compounds is the reason for corrosion. However, kinetic barriers, mainly provided by the formation of protective layers, often make metals resistant to corrosion and allow them to be used in technological applications.

The types of material changes brought about by corrosion are manifold. A distinction is made between uniform and nonuniform surface corrosion, pitting, selective corrosion, and cracking. Of these, the local corrosion processes are particularly critical.

Since corrosion reactions generally occur on the metal surface, they are termed interfacial processes and can be represented by a phase diagram. A distinction is made between the metal phase, the liquid, gaseous, or solid medium, and the phase boundary (Fig. 1). The corrosion process takes place at the metal–medium phase boundary and is therefore a heterogeneous reaction in which the structure and condition of the reaction surface play a significant role. Essential points, for example, are whether the surface is uncoated, whether it is covered with an adhesive, compact or loose, porous coating, or whether its properties have been changed by machining and processing. Furthermore, the corrosive medium must be transported to the surface, and the corrosion products removed. Hence, material transport phenomena, including free convection and diffusion into surface layers, must be taken into account.

2. Basic Electrochemistry

(See also → Electrochemistry, treated in the A series)

If the corrosive medium is an electrolyte solution, the resulting corrosion reactions are electrochemical. This means that material transport in the form of metal ions and charge exchange in the form of electrons take place at the metal–solution phase boundary because of the conductivity caused in the liquid phase by mobile anions and cations and the electron conductivity of the metals.

Aqueous corrosive media are electrolyte solutions, and hence by far the greatest proportion of damage arising through corrosion is due to electrochemical processes. Since in metals these are largely identical with electrolysis processes, metal corrosion in aqueous media is best described as electrolytic corrosion. The most important definitions are set out in DIN 50 900, Part 2 [2].

2.1. Electrochemical Equilibrium

When a piece of metal is dipped into water a number of processes take place. Metal atoms leave the solid phase and enter the liquid as cations. This disturbs the equilibrium in the metal phase; the metal now contains a surplus of electrons, thus becoming electrically negative, while a positively charged ion cloud forms in the liquid. This cloud is held close to the metal sur-

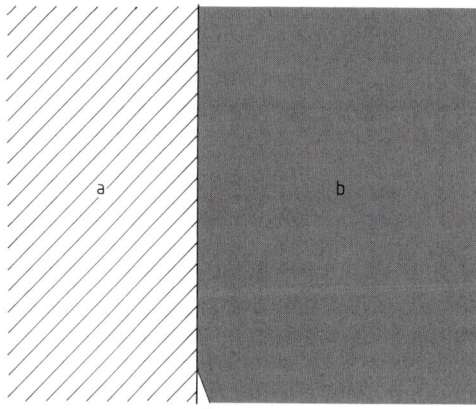

Figure 1. Corrosion as a phase interface reaction
a) Metal; b) Solid, liquid, or gaseous medium; c) Phase boundary

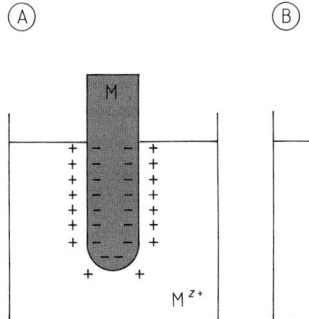 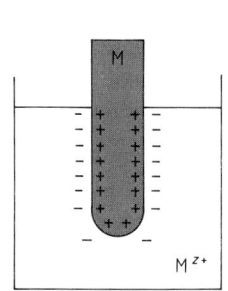

Figure 2. Formation of a potential difference between electrode and electrolyte
A) $\mu_{M^{z+}}$ (metal) $> \mu_{M^{z+}}$ (solution);
B) $\mu_{M^{z+}}$ (metal) $< \mu_{M^{z+}}$ (solution)
$\mu =$ Chemical potential

face by the attractive force of the electrons, resulting in an electrical double layer in which an electrical field, directed from the ion cloud to the metal surface, exists. The dissolution of the metal is self-inhibitory because all departing metal ions must overcome this field. With increasing concentration of the ion cloud, the exit energy released is less and less capable of supplying the energy needed to overcome the electrical field, which results in a dynamic equilibrium: the rate of metal dissolution equals the rate of metal deposition, with the exchange current density i_0. The double layer results in a potential difference between metal and solution (Fig. 2A).

If ions of the metal are already present in the liquid when the metal electrode is immersed, far fewer metal ions will go into solution. If the concentration is higher than that of the equilibrium double layer, metal ions will deposit on the metal surface and return to the solid state, and the direction of the potential difference at the phase boundary is reversed (Fig. 2B).

A potential difference also forms with an electrode that is not attacked by the electrolyte solution (e.g., platinum): through electron loss or gain, the dissolved substances are changed into an oxidized or reduced form (redox couple). An example is the following redox system:

$$Fe^{3+} + e^- \rightarrow Fe^{2+}$$

The immersed inert electrode acts as an electron donor or acceptor. Electrons are exchanged at the metal surface until an electrochemical equilibrium is reached between the electrons in the metal and in the solution.

Gas electrodes are a further important type of electrode in corrosion. These involve the formation, on the metal surface, of gases from dissolved ions, or of ions from dissolved gases. In this instance the metal has the function of consuming or providing electrons. In pure form such processes occur on metals (e.g., platinum) which themselves exhibit no reaction with electrolyte solutions. Of greatest importance in corrosion processes are hydrogen and oxygen electrodes. The reduction of hydrogen ions in acid solutions ($2H^+ + 2e^- \rightarrow H_2$) and the reduction of dissolved oxygen in approximately neutral solutions ($O_2 + 4e^- + 2H_2O \rightarrow 4OH^-$) are the most important cathodic reactions.

Standard Potentials and Electrochemical Series. The potential of the electrodes (half cells) described above is measured by connecting them by means of a solution bridge to a second electrode, whose potential is used as a reference. The potential difference between the two half cells, with zero current flow, is then measured. An electrode potential U measured in this way differs from the absolute potential $\Delta\varphi$ by an unknown constant which is dependent on the reference electrode. The potential difference arising at the point of contact of the electrolyte solutions (diffusion potential) can largely be eliminated by interposing concentrated salt solutions with ions of identical mobility. The additional potential changes caused by the instrument leads between electrode and instrument in each of the two cells can be disregarded since they are in opposing directions.

The hydrogen electrode in a given concentration (platinized platinum electrode immersed in a solution with a H^+ activity of 1 and purged with H_2 at 100 kPa) is generally used as the standard electrode; its potential is arbitrarily set at zero. The potentials of electrodes measured against the standard hydrogen electrode are designated by the symbol U_H. Figure 3 shows the determination of the electrode potential of silver. The potentials of other metal electrodes, relative to the standard hydrogen electrode, in a solution of their own salts with metal-ion activity 1 are called standard potentials.

The electrochemical series of metals is a listing of the standard potentials U_H^0 of metals determined at 25 °C. Some of the most important standard potentials are listed in Table 1.

Equilibrium electrode potentials at other concentrations are calculated from the Nernst

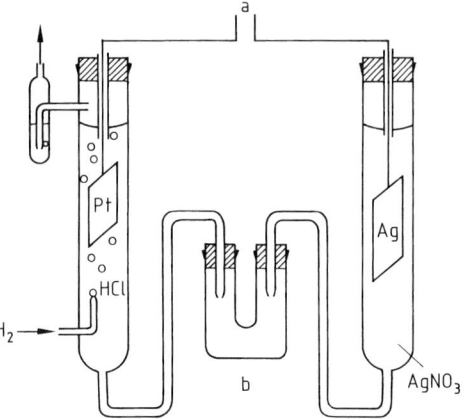

Figure 3. Determination of the electrode potential of silver
a) High-impedance voltmeter; b) Solution bridge

Table 1. Some important standard potentials

Electrode	U_H^0, V
Mg/Mg^{2+}	− 2.350
Al/Al^{3+}	− 1.660
Zn/Zn^{2+}	− 0.762
Fe/Fe^{2+}	− 0.440
Ni/Ni^{2+}	− 0.236
Sn/Sn^{2+}	− 0.141
Pb/Pb^{2+}	− 0.126
H$_2$/H$^+$	0.000
Cu/Cu^{2+}	+ 0.345
O$_2$/OH$^-$	+ 0.402
Ag/Ag$^+$	+ 0.800
Hg/Hg^{2+}	+ 0.861
Cl$_2$/Cl$^-$	+ 1.359
Au/Au^{3+}	+ 1.500

equation:

$$U_H = U_H^0 + \frac{RT}{zF} \cdot \ln a_{M^{z+}} \qquad (1)$$

where R is the gas constant, z is the valency of the metal ions, F is the Faraday constant, and $a_{M^{z+}}$ is the activity of potential-determining metal ions in the solution.

For a redox electrode the equation is

$$U_H = U_H^0 + \frac{RT}{zF} \cdot \ln \frac{a_{ox}}{a_{red}} \qquad (2)$$

The change in equilibrium potential U_H^0 in this instance depends on the activity quotient of the oxidized and reduced species in solution. The equation for a gas electrode, e.g., the hydrogen electrode, is as follows:

$$U_H = U_H^0 + \frac{RT}{zF} \ln a_{H^+} - \frac{RT}{zF} \ln p_{H_2} \qquad (3)$$

The deviations from U_H^0 are due to the activity of the hydrogen ions and the hydrogen partial pressure. If both are unity, then $U_H^0 = U_H$ and the conditions of the standard hydrogen electrode prevail.

2.2. Electrochemical Kinetics

If two half cells with differing electrode potentials are joined together, an electrochemical element (voltaic cell) arises with a current flow from the more negative to the more positive electrode, caused by the potential difference between the two electrodes (Fig. 4). Since the electrons released at the anode through metal dissolution migrate to the cathode and are consumed there (in this instance by metal deposition) the flow of current continues as long as the reaction partners remain available. The constant electrical field arising at one electrode at equilibrium, described in Section 2.1 (double layer), is not evident in this case and, therefore, cannot slow down the process. Accordingly, an electrochemical reaction can only take place when the electrode potential deviates from the equilibrium potential, whereby the rate of dissolution of the metal or reaction of

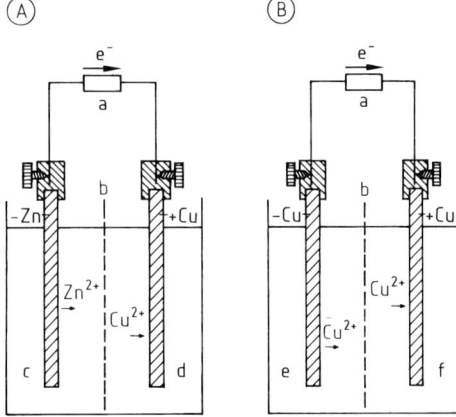

Figure 4. Electrode cells
A) Daniell cell; B) Concentration cell
a) External resistance; b) Diaphragms; c) ZnSO$_4$ solution; d) CuSO$_4$ solution; e) Dilute CuSO$_4$ solution; f) Concentrated CuSO$_4$ solution

the electrons is described by a current intensity I, which is equivalent to the respective mass conversion in accordance with Faraday's Law. For the dissolved metal mass ΔM and the transported electric charge Q in A · s the following equation applies:

$$-\Delta M = A_r Q / z F \qquad (4)$$

where A_r is the relative atomic mass of the metal, z is the ion valency, and F is the Faraday constant (96 520 A · s or 26.8 A · h). This constant gives the amount of charge required for the dissolution of 1 mol of a metal. For example, 1 A · h dissolves 1.04 g of pure iron. As a rule of thumb, the following applies for unalloyed steels:

$$0.1 \text{ mA cm}^{-2} \simeq 1 \text{ g m}^{-2} \text{ h}^{-1}$$

In contrast to electrolysis with current supply, in spontaneous corrosion with current yield, the dissolving electrode is known as the anode and its counterpart, at which, for example, cations are deposited, as the cathode. That is, the electrode at which negative charges enter the electrolyte solution is the cathode.

Voltaic cells can also be formed by combining metal–salt electrodes with metal–gas electrodes or metal–redox electrodes. The metal dissolution that occurs at the anode corresponds to the partial process that leads to material destruction in electrochemical corrosion. This also occurs when the electrolytic solution initially contains no metal ions, which is usually the case in corrosion processes, because only one element is needed, which is formed by the coupling of an electron-delivering (anode) and an electron-consuming (cathode) electrode. For this to occur, the cathode potential must be the higher of the equilibrium potentials of the individual electrode reactions. Therefore, noble metals such as gold do not corrode in aqueous solutions because, with their high anodic potentials, more positive cathode reactions are no longer available.

2.2.1. Corrosion Elements

A corroding metal–solution system forms anode and cathode areas, whose sizes range from a few lattice constants to macroscopic areas. Figure 5 shows a comparison between a galvanic and a corrosion cell. The corrosion cell, like the

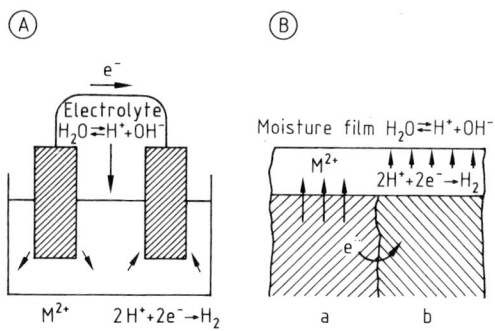

Figure 5. Comparison of a galvanic cell and a corrosion cell
A) Galvanic cell; B) Corrosion cell
a) Anode; b) Cathode
Anode process: $M \rightarrow M^{2+} + 2\,e^-$ (dissolution of metal)
Cathode process: $2\,H^+ + 2\,e^- \rightarrow H_2$ (evolution of hydrogen)

galvanic cell, can arise only when at least two electrode processes take place, one providing the electrons and the other consuming them. This corresponds to the composition of a galvanic cell consisting of at least two half cells.

Metal destruction (dissolution) takes place in the material area with the more electronegative potential (anode) since this is where the metal ions from the surface enter the solution. The electrons released in the process migrate to the electropositive cathode and are consumed there at the phase boundary by reducible substances present in the solution (e.g., cations, dissolved oxygen).

The anodic process (anode reaction) thus consists of metal dissolution (oxidation)

$$M \longrightarrow M^{2+} + 2\,e^-$$

while the cathodic process (cathode reaction) is a reduction:

(a) $2\,H^+ + 2\,e^- \longrightarrow H_2$ hydrogen ion reduction
(b) $O_2 + 2\,H_2O + 4\,e^- \longrightarrow 4\,OH^-$ oxygen reduction
(c) $\begin{cases} Cu^{2+} + 2\,e^- \longrightarrow Cu \\ Fe^{3+} + e^- \longrightarrow Fe^{2+} \end{cases}$ metal reduction

In practice, the cathode reactions (a) and (b) are particularly important because of their frequency. In the corrosion of metals in acid solutions, the reduction of hydrogen ions is the main reaction, whereas oxygen reduction is the main reaction in the corrosion of metals in neutral, aerated electrolyte solutions (e.g., salt solutions, seawater, atmosphere).

2.2.2. Mixed Electrode and Mixed Potential

A corroding metal surface represents a collection of corrosion elements, with the constancy, the spread, and the attendant macroscopic interface of the anode and cathode surfaces depending on the system. The distribution of local anodic and cathodic current densities i_a and i_c at the surface of an electrode can differ appreciably. Possibilities are $i_a = i_c$, $i_a - i_c = + i_E$ and $i_a - i_c = - i_E$. The current i_E per unit area of the corrosion element can thus be zero, positive, or negative. Only in certain borderline cases are the anode and cathode surfaces physically separated from one another (local elements). The principle of additivity (a superimposition of all electrode reactions taking place on a metal surface, see Section 2.3), first formulated by WAGNER and TRAUD [3], makes these statements comprehensible. A multiple or mixed electrode of this kind is homogeneous when i_E is zero, and is heterogeneous when, as a result of locally differing current densities, it is nonzero. The mixed potential of a homogeneous mixed electrode is identical over the entire surface, whereas a heterogeneous mixed electrode exhibits different mixed potentials.

Considering the corrosion system as a whole, the sum of the cathode currents must equal the sum of the anode currents for reasons of electroneutrality. Thus the cathode and anode reactions affect one another reciprocally, and the slowest (most inhibited) determines the overall reaction rate. In many cases the cathode reaction is the slowest (cathode-controlled corrosion). Lack of external current can only arise at a given mixed potential, the free corrosion potential.

2.2.3. Overpotential and Polarization

The corrosion behavior of metals cannot be predicted from the position of their standard potentials in the electrochemical series because the potential of an electrode changes with the current density. If an electrode in which only one electrode process takes place is termed a working electrode and the resultant potential a working potential, then the difference between working potential and the Nernst equilibrium potential is called overpotential; it is caused by reaction restraints. In general, polarization is defined as the shift in potential of working electrodes within a corrosion element. In such an element, at least two electrode reactions occur, whose overpotentials are superimposed, resulting in the polarization effect.

Transfer overpotential is the inhibition of transfer of charge carriers (ions and electrons) as a result of the electrical field on the metal surface. The potential barrier between metal and solution can only be overcome by ions of sufficiently high energy, which requires a corresponding deviation from the equilibrium potential of the electrode reaction. In reversible electrochemical reactions, anodic current dominates for $\Delta U > 0$, and cathodic current for $\Delta U < 0$. Near the equilibrium potential, the external current is the difference of the two partial current densities:

$$i = i_0 \left[\exp\left(\frac{\Delta U \cdot z F \alpha^+}{RT}\right) - \exp\left(\frac{\Delta U \cdot z F \alpha^-}{RT}\right) \right] \quad (5)$$

where i_0 is the exchange current density at the equilibrium potential of the reaction, and α is the charge transfer factor. For high overpotentials the second exponential expression is negligible and the equation simplifies to

$$i = i_0 \exp\left(\pm \frac{\Delta U \cdot z F \alpha}{RT}\right) \quad (6)$$

Taking logarithms and transforming gives the Tafel equation,

$$\log i = \log i_0 \pm \frac{\Delta U \cdot z F \alpha}{RT} \cdot \frac{1}{2.3} \quad (7)$$

$$\Delta U = \pm 2.3 \frac{RT}{zF\alpha} \log i_0 \pm 2.3 \frac{RT}{zF\alpha} \log i \quad (8)$$

which in simplified form is written:

$$\Delta U = a + b \log i \quad (9)$$

For the anodic dissolution of metals the following applies:

$$a = -2.3 \frac{RT}{\alpha z F} \log i_0 \quad \text{and} \quad b = 2.3 \frac{RT}{\alpha z F}$$

Since a and b are constants, plotting the anode and cathode overpotential curves of a simple electrode reaction in semi-logarithmic representation results in straight lines (Tafel lines) when the overpotential ΔU is plotted on the linear and the current density i on the logarithmic axis (Fig. 6). Individually, however, they cannot be directly measured; their superimposition gives a

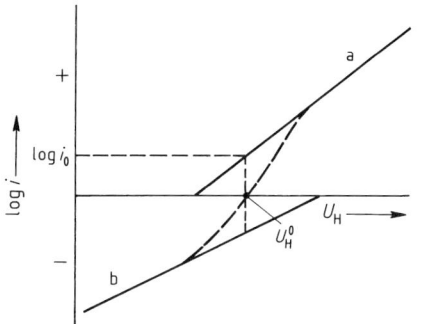

Figure 6. Overpotential curve for a simple electrode reaction
a) Overpotential curve for the anode reaction; b) Overpotential curve for the cathode reaction
U_H^0 = Equilibrium potential; i_0 = Exchange current density

sum curve, the actual overpotential curve, as shown by the dashed line in Figure 6.

The value of α generally lies between 0.45 and 0.55; as an approximation a value of 0.5 can be used. The constant b indicates the gradient of the straight lines. For $z = 2$, b is almost always 0.05 – 0.06 V (for $z = 1$, b is 0.1 – 0.12 V). Since in a number of electrode reactions the situation is complicated by the occurrence of multistep reactions, the Tafel equation has only limited applicability.

In the dissolution and deposition of most metals, the activation energy is small and hence the transfer overpotential is low. In transition metals such as iron and nickel it is much higher. Conspicuously high overpotentials indicate electrode processes in which hydrogen and oxygen are involved.

The effect of concentration c, resulting from the concentration changes at the electrode, leads to a concentration overpotential:

$$\Delta U = 2.3 \frac{RT}{nF} \log c \qquad (10)$$

At 18 °C, $2.3\, RT/F = 0.058$ V. Consequently, a concentration shift (more precisely a change in activity) by a factor of 10 results in a potential shift of 58, 29, or 19 mV, depending on the number of electrons n involved (change of valency). The concentration overpotential plays an important role where a reaction partner (e.g., oxygen) is consumed by corrosion. In this case the corrosion rate is determined by the rate of transport (by convection and diffusion) of the oxygen required for the cathodic reaction. Since for each electrode reaction the supply of species to, and removal from, the interface is essential, any barrier to such diffusion processes leads to a diffusion overpotential.

When one of the reaction steps at the metal surface is a purely chemical reaction, any inhibition of that reaction leads to a reaction overpotential. It is conceivable, for example, that an oxidizing agent needed to sustain the cathodic reaction is formed by a chemical reaction. If this is the slowest individual process the reaction overpotential will be determined by this reaction and the transport of material to the electrode.

A further barrier to corrosion reactions is provided by electrical resistance. When the anodic and cathodic reactions at the metal surface take place with locally different current densities (heterogeneous mixed electrode), resistance in the current circuits can cause a measurable drop in potential (resistance polarization). This resistance polarization is a linear function of the current. Resistance polarization frequently arises through the formation of passive films. The resulting relationship between the change in potential and the current usually no longer follows Ohm's law, but instead is subject to a logarithmic relationship.

2.3. Current Density – Potential Curves

Theoretical predictions of the overpotential (polarization) at a given current density are generally not possible, and it is impossible to determine the current densities to be expected during corrosion (a measure of the corrosion rate) from the equilibrium potentials. The relationship between potential and current density can normally only be determined experimentally.

The results of such measurements are known as current density – potential curves. They represent cumulative curves given by the superimposition of the current density – potential curves of the individual reactions. For simple electrodes with defined electrode processes, these are the overpotential curves (see Fig. 6). For metals exposed to electrolytic attack, superimposition of several overpotential curves gives the actual current density – potential curves that are of significance in corrosion testing and research. Figure 7 illustrates how two overpotential curves superimpose to form a current density – potential curve when both reactions take place at a metal

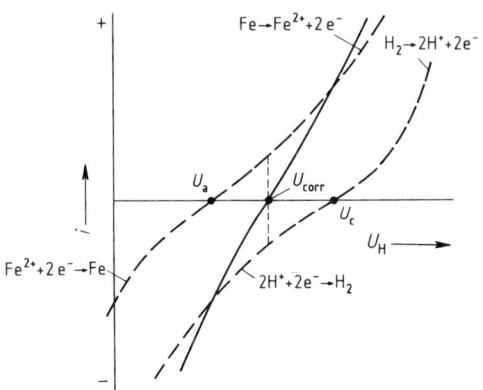
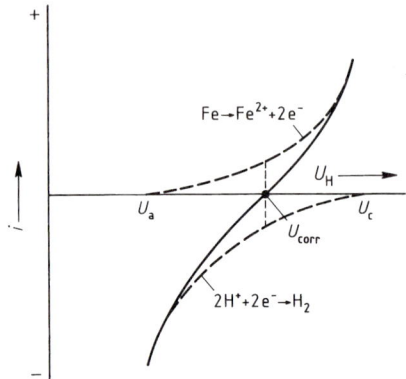

Figure 7. Superposition of two overpotential curves to form a current density–potential curve
U_a = Equilibrium potential of the overpotential curve for the anode reaction; U_c = Equilibrium potential of the overpotential curve for the cathode reaction; U_{corr} = Corrosion potential (rest potential)

Figure 8. Superposition of the anodic and cathodic partial reactions to form a current density–potential curve (simplified version compared to Fig. 7)
---- = Partial current curves; ——— = Cumulative current curve; U_{corr} = Corrosion potential

surface. The potential is plotted on the abscissa and the current density on the ordinate. This presentation corresponds to the normal practice of measuring current density at a given potential. Figure 7 shows the superposition of the overpotential curves of a hydrogen and an iron electrode. The metal surface on which the two reactions occur appears to be without external current, although current flows on the metal surface. However, these are internal processes that have no outward effect since the anodic current must be the same as the cathodic current. In the example given this means that the extent of metal dissolution must be equal to the extent of hydrogen evolution. When there is no resistance in the circuit, which can be assumed for electrolytes with good conductivity, this condition is met at the common free corrosion potential (mixed potential) U_{corr}. The U_{corr} is the intersection of the measured cumulative current–potential curve with the abscissa and is also called the rest potential when the electrode reactions are homogeneous.

Since only the cathode branch (hydrogen reduction) is of interest on the overpotential curve of the hydrogen electrode (right) and only the anode branch (metal dissolution) on the iron electrode (left), the curve can be simplified, as shown in Figure 8. Electrical resistance in the circuit of a dissolving metal electrode always leads to reduced metal dissolution.

Even though normal corrosion without external current is only shown by the point of intersection of the cumulative current density–potential curve and the abscissa (U_{corr}), the anode and cathode branches, i.e., the full course of the curve measurable with external current supply, are still of interest since they reflect all the peculiarities of the polarization curves involved. From such diagrams, therefore, conclusions can often be drawn as to the corrosion performance of metals under given environmental conditions.

The procedure for determining a current density–potential curve is to immerse the metal electrode in the corrosive medium. After an incubation period, the mixed potential corresponding to the outwardly currentless state becomes established on the metal. If a circuit is now placed on the electrode and external current is applied, the result is a shift in potential (polarization). Current density–potential curves are therefore also known as polarization curves. Depending on the direction of the external current imposed, this is termed anodic or cathodic polarization (anodic: potential becomes positive; cathodic: potential becomes negative).

Determination of Corrosion Current Density. Simple, purely transfer-related electrode reactions give cumulative current density–potential curves of the type shown by the unbroken line in Figure 9. At the points p_1 and p_2 it swings into the overpotential curves of the respective part reactions, because beyond these points, only the anodic or the cathodic reaction exists. At these points the equilibrium potentials of the reverse

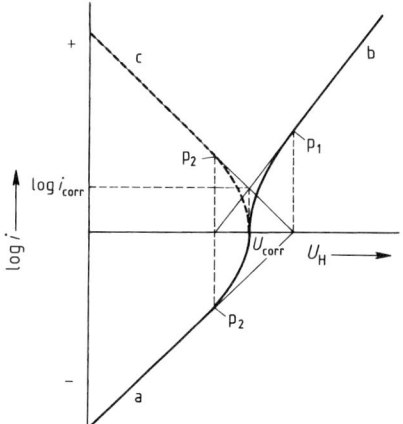

Figure 9. Determination of the corrosion current density i_{corr}
a) Overpotential curve for the cathode reaction; b) Overpotential curve for the anode reaction; c) Reflection of curve a

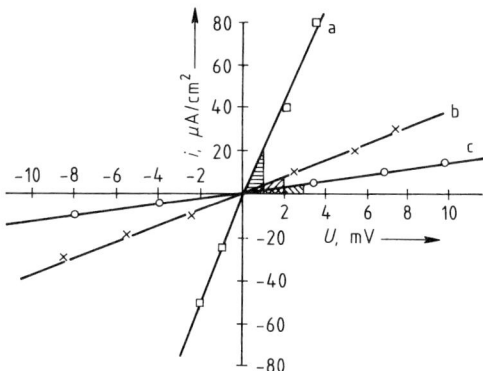

Figure 10. Determination of the corrosion rate i_{corr} from polarization resistances

$$i_{corr} = 0.02 \text{ V} \cdot \frac{\Delta i \, [\text{A/cm}^2]}{\Delta U \, [\text{V}]}$$

a) $t = 80\,°C$, $i_{corr} = 0.48$ mA/cm²; b) $t = 60\,°C$, $i_{corr} = 0.09$ mA/cm²; c) $t = 40\,°C$, $i_{corr} = 0.023$ mA/cm²

reaction are exceeded and superimposition no longer occurs. These pure overpotential curves thus form linear Tafel lines, which, after reflection of the cathode curve in the x-axis, can be made to intersect by extrapolation in the direction of the abscissa. The ordinate section at the point of intersection is then $\log i_{corr}$, i.e., the log of the corrosion current density at the free corrosion potential (rest potential), from which the corrosion rate can be calculated by Faraday's law (see Section 2.2).

The simple correlation between polarization resistance and corrosion rate at the free corrosion potential developed by WAGNER and TRAUD [3], was later extended by STERN and GEARY for practical application [4]. By this method the corrosion rate is estimated from the gradient of the current density–potential curve (polarization resistance):

$$i_{corr} = \frac{1}{2.3\left(\frac{1}{b_a} + \frac{1}{b_c}\right)} \left(\frac{di}{dU}\right)_{i=0} = B\left(\frac{di}{dU}\right)_{i=0} \quad (11)$$

where i_{corr} = corrosion current density in A/cm²; b_a, b_c = constants (gradients of the Tafel lines); and $(di/dU)_{i=0}$ = gradient of the current density–potential curve at the rest potential in A cm^{-2} V^{-1}.

This procedure is useful when quantitative determination based on Tafel lines is no longer possible; for example, when the corrosion products form deposits. The polarization resistance is measured at the rest potential (free corrosion potential) within a range of ≤ 20 mV. The constant B is often ca. 20 mV and can be determined for exact measurements.

An example of determining the corrosion rate by this method is shown in Figure 10 [5]. However, for very small corrosion currents that are independent of potential in some ranges (e.g., in passive metals), this method cannot be used to determine the corrosion rate. Nor is the method applicable when non-transfer-related part reactions occur, e.g., when the corrosion current is dependent on diffusion processes at the specimen surface. Also, the constant B can be time-dependent when dense protective films form, which further restricts the applicability of the method [6].

For local corrosion phenomena, determination of the corrosion rate is of secondary importance. In fact, in processes limited to specific potential ranges, the threshold potential (i.e., the potential above which local corrosion begins), which depends on both the corrosive medium and the alloy composition, is more important.

2.4. Hydrogen Overpotential and Acid Corrosion

The cathodic reduction of hydrogen ions

$$2\,H^+ + 2\,e^- \rightleftharpoons H_2$$

takes place in a number of complex steps. Under certain conditions, acid corrosion can induce hydrogen embrittlement when special effects influence the kinetics of hydrogen reduction, which takes place up to the stage of molecular hydrogen formation (recombination). In many metals, considerable overpotential is needed to initiate hydrogen evolution. Furthermore, the activity of metal surfaces can be impaired by contamination with impurities. This principle is applied in deliberate inhibition (e.g., use of pickling inhibitors, see Section 5.2). A slower recombination rate generally leads to a higher concentration of adsorbed hydrogen at the surface, which in turn can result in increased diffusion of atomic hydrogen into the metal. This may eventually lead to embrittlement in various metals and alloys.

In aqueous solution, the hydrogen ion from the dissociation of water ($H_2O \rightleftarrows H^+ + OH^-$) or acids (e.g., $HCl \rightleftarrows H^+ + Cl^-$) is hydrated, forming a hydronium ion:

$$H^+ + H_2O \longrightarrow H_3O^+$$

The hydronium ion (H_{aq}^+) must first diffuse through the liquid to the metal surface

$$H_{aq}^+ \longrightarrow (H_{aq}^+)_{\text{metal surface}}$$

For discharge of the hydronium ion to occur, an electron must pass through the metal–electrolyte phase boundary; the discharge reaction (Volmer reaction) is therefore known as a transfer reaction:

$$(H_{aq}^+)_{\text{metal surface}} + e^- \longrightarrow H_{ad}$$

Initially, the hydrogen remains adsorbed in atomic form on the metal surface. For the formation of molecular hydrogen, there are two possibilities: either two adsorbed hydrogen atoms combine chemically to form an adsorbed hydrogen molecule (Tafel reaction):

$$H_{ad} + H_{ad} = (H_2)_{ad}$$

or an electrochemical reaction takes place between an adsorbed hydrogen atom, a hydronium ion, and an electron (Heyrovsky reaction):

$$H_{ad} + (H_{aq}^+)_{\text{metal surface}} + e^- \longrightarrow (H_2)_{ad}$$

Thus there are two possibilities for the formation of adsorbed molecular hydrogen: two Volmer reactions and one Tafel reaction, or one Volmer reaction and one Heyrovsky reaction (Fig. 11).

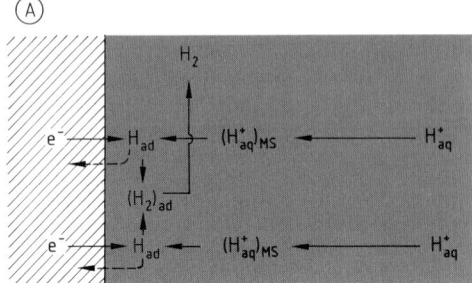
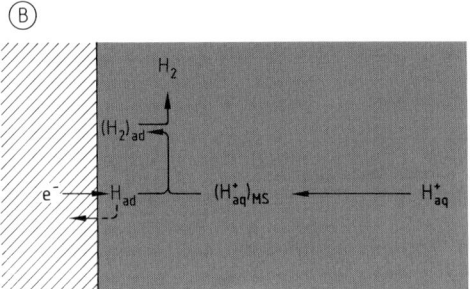

Figure 11. Reaction schematic of cathodic hydrogen formation
A) By the Volmer–Tafel mechanism; B) By the Volmer–Heyrovsky mechanism (after KAESCHE, see general references)
▨ = Metal; ☐ = Boundary layer solution
$(H_{aq}^+)_{MS}$ = Hydronium ion H_3O^+ on metal surface

The formation of adsorbed molecular hydrogen is followed by the transition of H_2 molecules into the gaseous phase:

$$(H_2)_{ad} \longrightarrow H_2$$

This reaction sometimes consists of several distinguishable individual steps. It is occasionally observed with metals which dissolve in acid, for example, that the hydrogen bubbles always occur at a few spots where bubble formation is presumably energetically favorable. In this case, the reaction consists of diffusion to those spots with subsequent desorption. For very slow hydrogen evolution, bubble formation is unnecessary because the hydrogen molecules can dissolve rapidly enough in the aqueous phase, diffuse through it to the liquid surface, and be desorbed into the ambient gaseous phase. The transition of atomic hydrogen into the metallic phase is discussed in Section 3.3.1.5.

The considerable overpotential for hydrogen evolution observed in many metals (in Hg and Pb, > 0.5 V) is due to inhibition in one of the part reactions. The slowest partial process determines the rate of the overall reaction. There is practically no inhibition in platinum, which possesses the lowest hydrogen overpotential. Other "active" metals are palladium, strontium, and rhodium.

2.5. Oxygen Overpotential and Oxygen Corrosion

Acid corrosion arises only rarely in pure form because it requires the corrosive solution to be completely free of air. However, this is not usually the case, and in acid solutions an additional electrode process often occurs:

$$O_2 + 4e^- + 2H_2O \rightleftharpoons 4OH^-$$

An important feature of a corrosion process is the reduction of the dissolved oxygen by formation of hydroxyl ions (OH^-). Since the equilibrium potential is linked through the concentration of the OH^- ions with the concentration of the hydrogen ions (or H_3O^+ ions) by the ionic product of water

$$(a_{H^+}) \cdot (a_{OH^-}) = K_w = 10^{-14} \quad (\text{at } 25\,°C),$$

the equilibrium potential is pH dependent. The standard potential is ca. 0.4 V at 25 °C, a partial pressure of 100 kPa, and an OH^- ion activity of 1 (pH 14). Equilibrium potentials under other conditions can be calculated by the Nernst equation:

$$U_H = U_H^0 - \frac{RT}{F}\ln a_{OH^-} + \frac{RT}{4F}\ln p_{O_2} \quad (12)$$

Interpretation of the equation for $p_{O_2} = 100$ kPa is shown in Figure 12, in which, for comparison, the equilibrium potential of the hydrogen electrode for $p_{H_2} = 100$ kPa is also given. It can be seen that the potential lines run parallel, and that at all pH values the equilibrium potential of the oxygen electrode is more than 1.2 V higher than that of the hydrogen electrode. This explains why the presence of oxygen often promotes electrochemical corrosion. Quite apart from the fact that a higher cathode potential normally increases the potential difference and hence the current density in local elements, anodic destruction pro-

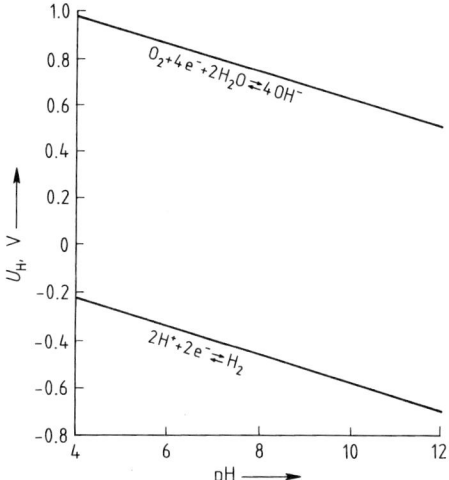

Figure 12. Equilibrium potentials for the oxygen and hydrogen electrodes

cesses are now possible at potentials that are above that of the hydrogen cathode and which cannot occur in the absence of oxygen. An example is the behavior of copper, which is only attacked by electrolytes when oxygen is present. On a bare copper surface, the loss of metal is directly proportional to the quantity of cathode-converted oxygen. In practice, however, there is usually fairly rapid formation of oxidic protective films (consisting at room temperature mostly of Cu_2O and at higher temperatures mainly of CuO), which restrict corrosion to a harmless extent. Only when the formation of these films is inhibited can the oxygen corrosion of copper progress linearly as a function of time.

The working potential of the oxygen electrode generally differs greatly from its equilibrium potential. This leads to a considerable overpotential, which in most cases is a concentration overpotential. Oxygen is consumed at the cathode, leading to depletion of oxygen in the cathode boundary layer. If the cathodic process is to continue, additional oxygen must be supplied from the electrolyte. The amount of oxygen converted electrochemically—and, therefore, the corrosion rate—depends on the rate of subsequent oxygen supply, which is diffusion controlled and is determined by:

1) The diffusion coefficient, which increases with increasing temperature
2) The difference between the oxygen concentration at the metal surface and in the corrosive

medium, which increases with increasing oxygen content of the solution
3) The thickness of the boundary layer, which depends on whether the flow is laminar or turbulent, with smaller thicknesses being associated with higher flow rates

The quantity of oxygen diffusing to the cathode per unit time and area is directly proportional to the diffusion coefficient and the difference in oxygen concentration, and inversely proportional to the thickness of the boundary layer. Therefore, an increase in temperature and a higher concentration of oxygen in the electrolyte increase the corrosion current density, whereas an increase in the boundary layer thickness reduces the current density.

The occurrence of a concentration overpotential in oxygen corrosion is clearly shown by the cathodic overpotential curve. As demonstrated in Figure 13, a diffusion-limited current density appears on the curve, which is indicated by the parallelism of the cathodic current curve and the potential axis (diffusion current density). In this region the current density can only be further increased if the concentration of available oxygen is increased. The figure also shows the behavior of copper towards corrosive solutions containing oxygen.

If the corrosion potentials in metals (mixed potentials) are within the range of the diffusion-limited current, the corrosion current density is independent of the potential. This leads to the conclusion that normally the rate of oxygen corrosion is determined solely by oxygen diffusion to the cathode and that other factors are only of minor influence. In particular, the pH of the solution in this instance does not have the importance often attached to it in practice.

The following equation shows that areas in which a mainly cathodic electrode process prevails tend to alkalize through the accumulation of hydroxyl ions:

$$O_2 + 4e^- + 2H_2O \longrightarrow 4OH^-$$

If phenolphthalein is used as an indicator for hydroxyl ions and potassium hexacyanoferrate(III) as an indicator for iron(II) ions, the anode and cathode areas can be made visible in the case of corroding iron. If, for example, an iron nail is placed in a gelatine solution mixed with these two indicators, after a while the head and point of the nail start to turn blue because of the formation of Prussian blue, whereas the middle of the nail turns red. This experiment shows that it is mainly the anodic process that takes place at the mechanically worked points, whereas the cathodic process with the formation of hydroxyl ions occurs in the middle of the nail.

This separation of the two distinct processes in oxygen corrosion can also be observed when a ring of rust forms under a saltwater droplet on an iron surface some distance from the outer edge of the droplet and not at the point of minimum oxygen supply. The dissolved iron ions then combine as they diffuse with hydroxyl ions to form sparingly soluble iron hydroxide, which turns to rust on absorption of oxygen, and deposits (Fig. 14).

The process can be written as follows:

$$Fe \longrightarrow Fe^{2+} + 2e^-$$
$$2Fe^{2+} + O \longrightarrow 2Fe^{3+} + O^{2-}$$
$$O + 2e^- \longrightarrow O^{2-}$$
$$O^{2-} + H_2O \longrightarrow 2OH^-$$
$$Fe^{2+} + 2OH^- \longrightarrow Fe(OH)_2$$
$$2Fe(OH)_2 + O + H_2O \longrightarrow 2Fe(OH)_3$$
$$Fe(OH)_3 \longrightarrow FeO(OH) + H_2O$$

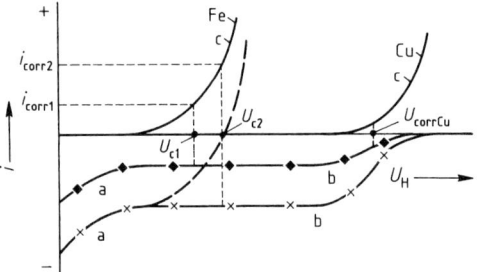

Figure 13. Current–potential curve for oxygen corrosion
——— Partial current curve; ——— Current–potential curve at oxygen concentration 2; ◆◆ Oxygen concentration 1; ✶✶ Oxygen concentration 2
a) $2H^+ + 2e^- \rightarrow H_2$; b) $O_2 + 2H_2O + 4e^- \rightarrow 4OH^-$; c) $M \rightarrow M^{2+} + 2e^-$

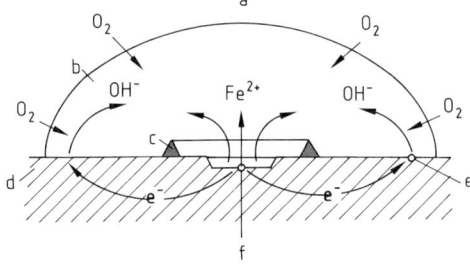

Figure 14. Mechanism of corrosion under a saltwater droplet
a) Air; b) Drop of salt solution; c) Rust ring; d) Iron; e) Cathodic oxygen reduction $\frac{1}{2}O_2 + H_2O + 2e^- \rightarrow 2OH^-$; f) Anodic metal dissolution $Fe \rightarrow Fe^{2+} + 2e^-$

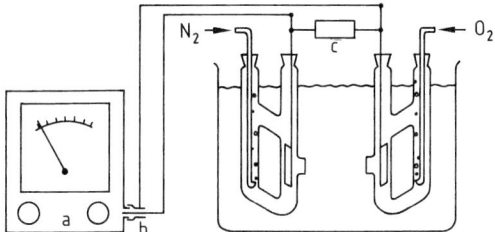

Figure 15. Basic circuit diagram of an aeration cell
a) Millivoltmeter; b) Input with inner conductor and screen; c) Short circuit resistance

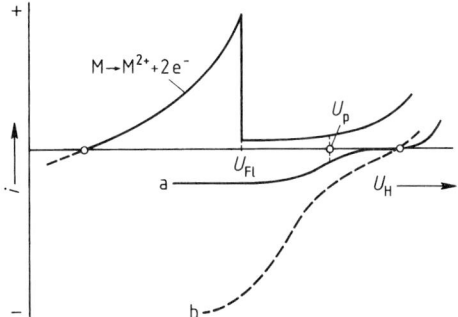

Figure 16. Current–potential curve for metal dissolution with drop in current due to passivation
a) Redox system 1: conservation of passivity; b) Redox system 2: generation of passivity
U_p = Rest potential in the passive state; U_{Fl} = Flade potential

Using a simple test arrangement, EVANS showed that different metal surface aeration in oxygen corrosion results in the formation of local elements (aeration elements) [7]. An improved model experiment is illustrated in Figure 15.

Although the nitrogen-purged iron electrode (anode) is corroded, no corrosion is found on the oxygen-purged iron electrode (cathode). The actual effect consists of inhibition of the anodic process through the formation of OH⁻ ions at the O_2 electrode. Consequently, the term aeration element is somewhat misleading, in that the element formation in this corrosion process is the result of a change in pH. This can be readily proven since element formation does not occur in sufficiently buffered solutions.

2.6. Protective Film Formation and Passivity

The corrosion reactions described above require an active, i.e., mainly bare, metal surface. In practice, however, many metals tend to form protective films which can have a major influence on the electrolytic process. Dense films provide protection and are thus referred to as protective coatings or films. A protective iron oxide film is known to form, for example, in boiler pipes. In contrast, porous rust layers provide no protection. The good resistance of aluminum, which is the result of an Al_2O_3 film forming in air, can also be produced artificially by anodic oxidation.

In addition to these macroscopic protective films, a number of metals possess optically invisible, pore-free oxide films with particularly good protective action; this gives rise to the passivity of metals. The occurrence of passivity can best be observed electrochemically. Figure 16 shows an anodic metal dissolution curve with the typical drop in current at the onset of passivity. The metal initially dissolves actively in the solution. With increasing potential the dissolution current rises until a potential (passivation potential, often called Flade potential U_{Fl}) is reached at which there is a sudden sharp drop in metal dissolution. The very low metal dissolution remains practically constant over a relatively large potential range, the passive range, before rising again at the beginning of the so-called transpassive range. To achieve the passive state, the redox system (as Fig. 16 shows) must provide a cathodic current that is at least as large as the anodic current needed to form the passive film at the passivation potential. A pore-free oxide film of ca. 10 nm thickness then forms at the electrode surface, suppressing the previous intense metal dissolution.

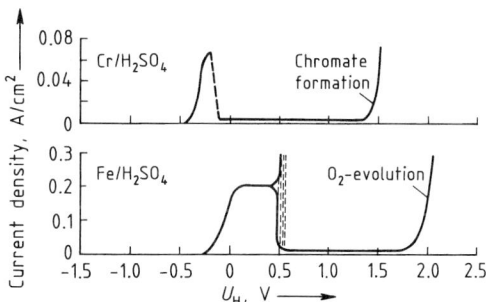

Figure 17. Passive behavior of chromium and iron in sulfuric acid

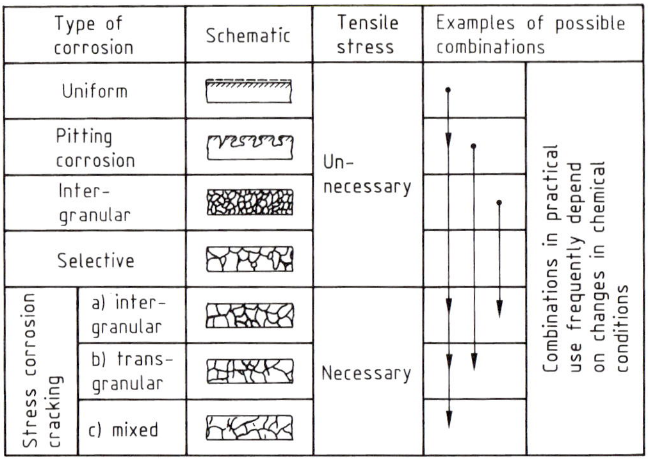

Type of corrosion		Schematic	Tensile stress	Examples of possible combinations			
Uniform			Un-necessary			Combinations in practical use frequently depend on changes in chemical conditions	
Pitting corrosion							
Inter-granular							
Selective							
Stress corrosion cracking	a) inter-granular		Necessary				
	b) trans-granular						
	c) mixed						

Figure 18. Types of corrosion

The minimal residual anodic current at the electrode in the passive range is caused by the slow dissolution and constant reformation of the passive film. The passivation potential and the dissolution in the passive range differ for each metal or alloy. Of economic significance are metals with a moderate passivation potential and minimum dissolution in the passive state. Figure 17 shows current–potential curves of iron and chromium in sulfuric acid. The passivation of chromium begins more than 500 mV before that of iron and requires a considerably lower current density. The good passivation capability of chromium can be imparted to iron by alloying with > 12 wt% chromium. This forms the basis for the development of all stainless and acid-resistant steels. Further improvements can be obtained by adding nickel, molybdenum, and copper to these alloys.

Under certain circumstances the passivation capacity of various alloys results in special corrosion phenomena. These arise mainly because the passive film exhibits local defects as a result of external influences (special corrosive media) or structure-related effects (deposits, grain boundary disturbances, segregation, etc.). This leads to localized corrosion processes, including selective corrosion, intergranular corrosion, and pitting corrosion. Under the influence of static mechanical tensile stress, various types of stress corrosion cracking can arise, while alternating mechanical stress leads to corrosion fatigue, which occurs in both fully active and passive alloy systems.

Figure 18 shows schematically the different types of corrosion together with the possible combinations which can arise in practice. Corrosion fatigue is not considered here because mechanistically it is not confined to passive alloys. It is therefore treated separately (Section 3.3.2).

3. Types of Electrolytic Corrosion

The surface of a material or component is subject to various environmental influences when in use. The mechanical, thermal, and chemical (corrosive) parameters may act individually, or two or even three factors can combine, the significance of each differing according to the type of stress. The corrosive factor, as shown in Figure 19, can be divided into chemical and electrochemical components. Electrochemical corrosion can take place with or without mechanical stress. The types of electrochemical corrosion that occur are also shown in Figure 19 and are discussed in more detail in the following sections.

3.1. Uniform and Nonuniform Surface Corrosion

DIN 50900, Part 1 [1] defines uniform surface corrosion as corrosion with practically equal mass loss over the entire surface (homogeneous mixed electrode), and nonuniform corrosion (shallow pit formation) as corrosion with locally different mass losses. The cause of nonuniform corrosion is the presence of corrosion elements (heterogeneous mixed electrode) [2]. The common feature of both types of corrosion

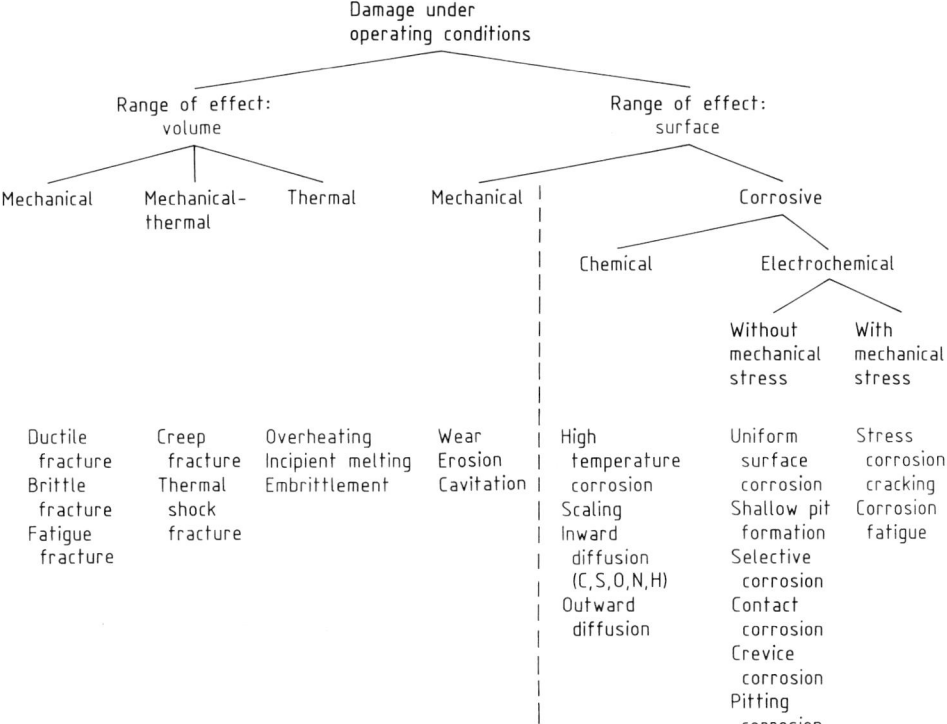

Figure 19. Overview of damage to metallic materials under operating conditions

is the geometry of the damaged area. The surface spread of these areas is generally greater than the depth. Accordingly, compared with localized types of corrosion such as pitting, these types of corrosion are limited in extent.

Uniform corrosion is without doubt the easiest form of corrosion to recognize and to control. In the ideal case, material loss runs parallel to the surface of the corroded component, the reduction in wall thickness being given by:

$$\Delta M_{tot}/\varrho \cdot A$$

where ΔM_{tot} = total mass of the dissolved metal, A = area of corroded surface, and ϱ = density of the metal.

In terms of the corrosion current densities, this implies the equality of the real local corrosion current density I_R and the mean corrosion current density i_R, ($I_R = i_R$), where the following relationship exists [8]:

$$i_R = \frac{1}{A} \int I_R \cdot dA$$

A practical example of almost uniform surface corrosion is as follows [9]: a pipe made of unalloyed steel St 35 (material no. 1.0308) used at ca. 90 °C for the transport of service water showed material erosion of the inner surface after three years in operation; a layer of corrosion product (mainly iron oxide) had formed on the inside with a practically constant thickness over the entire area. Given the operating conditions, the material corrosion had to be due to oxygen corrosion.

A further example is illustrated in Figure 20. A thermometer protection tube made of the highly corrosion resistant nickel alloy NiMo 30 (material no. 2.4810) was immersed at ca. 20 °C in an aqueous, acidic chloride solution to which sodium nitrite was added, releasing nitrogen oxides. The protection tube, consisting of a larger-diameter tube and a smaller-diameter tube extension welded together, suffered severe corrosion with the exception of the weld joint. A chromium-containing filler metal of the type nickel alloy NiMo 16 Cr (material no. 2.4812) had been used. The uniform surface corrosion of the chromium-free alloy without passivation capability NiMo 30 is due to the presence of oxidants. Under these conditions, the weld containing chromium is considerably more resistant.

In practice, uniform surface corrosion is easy to control in the majority of cases, especially because the reduction in wall thickness during operation can be monitored by nondestructive testing or by suspending a specimen of the material in the corrosive agent and monitoring the corrosion [10]. Suitable measures to combat uniform

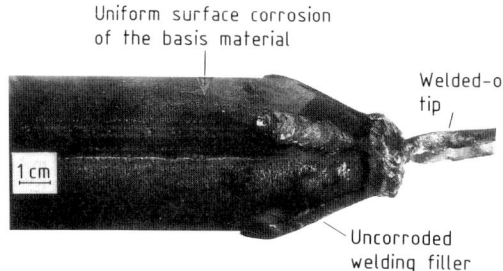

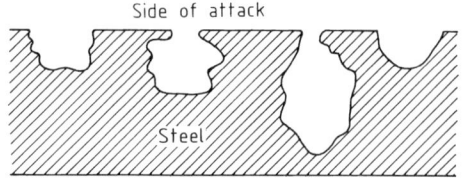

Figure 21. Shapes of pits in pitting corrosion

Figure 20. Uniform corrosion of a thermometer protection tube made of NiMo 30 (material no. 2.4810)
Welding filler: NiMo 16 Cr (material no. 2.4812)

corrosion include greater wall thicknesses, a more resistant material, and cladding or protective coatings.

In the case of nonuniform corrosion (shallow pit formation), the real corrosion current density I_R and mean corrosion current density i_R are not equal. Accordingly, nonuniform corrosion is characterized by a locally differing corrosion rate. The conditions for this may occur, for example, during the course of corrosion through a nonhomogeneous covering of the reaction product on the material.

3.2. Forms of Localized Corrosion without Applied Mechanical Stress

This general term covers pitting corrosion, crevice corrosion, and selective corrosion, including the subgroups of intergranular corrosion, spongiosis, dezincification, and line- and layer-type corrosion.

3.2.1. Pitting Corrosion

In DIN 50 900, Part 1 [1], pitting in the broadest sense is defined as the form of corrosion in which craterlike, surface-hollowing, or pinpricklike cavities occur (see Fig. 21). Outside the pitted areas practically no surface corrosion is observed. The depth of the pitted area is generally equal to or greater than its diameter. Pitting always arises when the material surface is covered with a corrosion-resistant coating that has pores or defects, the latter being there initially or arising through a local reaction with the corrosive medium. The following are regarded as being protective layers:

1) A natural oxide scale or rolling scale in unalloyed or low-alloy steels
2) Protective films (e.g., paint, bitumen coatings, or oil films)
3) Metallic coatings (e.g., zinc)
4) Salt layers formed under operating conditions, such as the lead sulfate layers that form on lead in sulfuric acid, or the copper(I) oxide or malachite layer formed on copper in drinking water
5) A film produced by the adsorption of inhibitors (e.g., chromates)
6) The passive film, usually consisting of only few molecular layers, such as found in iron, nickel, chromium, and iron- and nickel-based alloys with passivation capacity

For pitting to occur in metals coated with protective layers as in (1) to (3), there must be defects in the protective coating and an electrolyte solution acting as a corrosive medium. The resulting corrosion element consists of the cathode (intact coating) and the anode (pores in the coating). The attack on the base material in the region of the pores acts downwards. Depth progression depends on the size and activity of the cathode surface. Pitting in cases (4) and (5) in corrosive media is due to an incomplete protective layer on the material surface. The intensity of corrosion depends on the area ratio between anode (uncoated surface) and cathode (coated surface).

Pitting on materials with a passive coating, such as unalloyed or low-alloy steels, stainless and acid-resistant ferritic chromium steels, and stainless and acid-resistant austenitic chromium–nickel steels, is caused by electrolytes that contain chloride, bromide, and, in some cases, iodide ions. Since chloride ions in particular are responsible for this kind of corrosion, it is often called *chloride corrosion*. Halide ions (excluding fluoride, which does not cause pitting) are adsorbed at discrete locations of the passive coating (formation of cavity nuclei). After an induction period the passive coating is penetrated and pitting corrosion initiated.

An example of localized corrosion in the form of pits in natural protective coatings is the corrosion that arises through cracks in the rolling scale of unalloyed or low-alloy steels. If, for example, the rolling scale of a steel plate has spalled at a tiny point, and the sheet is immersed in a salt solution containing oxygen, an electric current flows between the oxide film (the cathode) and the bare metal (the anode). At the anode, iron dissolves as iron(II) ions, whereas at the cathode oxygen is reduced:

$$1/2\, O_2 + H_2O + 2\,e^- \longrightarrow 2\,OH^-$$

This oxygen reduction, which determines the flow of current and hence the corrosion of the iron at the uncoated spot, continues as long as oxygen can diffuse unhindered to the cathode (see Section 2.5). This local dissolution process can even perforate sheet and piping. This kind of corrosion can be avoided by completely removing the rolling scale.

Similarly, corrosion processes take place at pores and defects in electrically conducting coatings that are more noble than the base material. In paints, the propagation rate of cavities depends on the possibility of substance transport through the coating.

Pitting Corrosion in Stainless Steels. The conditions that lead to chloride-induced pitting corrosion in stainless steels in chemical plants are well-known and have been extensively examined for this group of materials [11].

Corrosion Parameters. Figure 22 shows the current density–potential curve of a material with passivation capacity in an initially chloride-free corrosive medium. Starting from the free corrosion potential U_{corr}, the steel passes, with increasing anode polarization after exceeding the passivation potential, through the passive range and reaches the breakthrough potential at U_b.

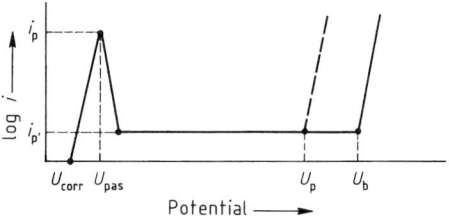

Figure 22. Current density–potential curve of a passifiable alloy, with and without pitting corrosion
U_{corr} = Corrosion potential; U_{pas} = Passivation potential; U_p = Pitting potential; U_b = breakthrough potential; i_p = Passivation current density; $i_{p'}$ = Passive current density

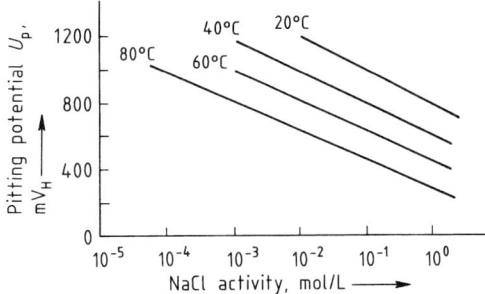

Figure 23. Pitting potential of diffusion-annealed X 3 CrNiMoN 17 13 5 (material no. 1.4439) in NaCl solution as a function of chloride ion activity and temperature

While in the active range, the alloying elements go into solution in their lowest valencies, during corrosion above U_b, iron(III) and chromium(VI) compounds are formed. If chlorides are now added to the corrosive medium, an increase in current is observed at U_p, this being due to pitting corrosion. Chlorides (and also bromides and iodides) thus decrease the passive range of the material.

The pitting potential shifts to more negative values with increasing chloride ion concentration, rising temperature, and decreasing pH; i.e., susceptibility to pitting increases. On the other hand, the susceptibility to pitting in a given corrosive medium can be decreased by higher flow rates of the medium.

The dependence of the pitting potential U_p on various factors has been explained for a number of austenitic chromium–nickel–(molybdenum) steels by using diffusion-annealed specimens [12]. It was demonstrated, for example, that the pitting potential U_p is directly proportional to the logarithm of chloride ion activity. This is shown in Figure 23 for a nitrogen-alloyed steel with high molybdenum content (X 3 CrNiMoN 17 13 5, material no. 1.4439). Note that pitting occurs, albeit at high potentials, even at chloride ion concentrations of 2 ppm. A temperature change results in a parallel shift of the characteristic curve. The temperature dependence of the pitting potential is shown in Figure 24, in which U_p is plotted against the reciprocal of absolute temperature. Figure 25 shows that an increase in the pH of the corrosive medium shifts the pitting potential U_p to more positive values, to the extent of 55–60 mV per decade of the hydrogen ion concentration. This value corresponds to the factor $2.3\, RT/nF$ for $n = 1$ at 20 °C in Equation (10), which describes

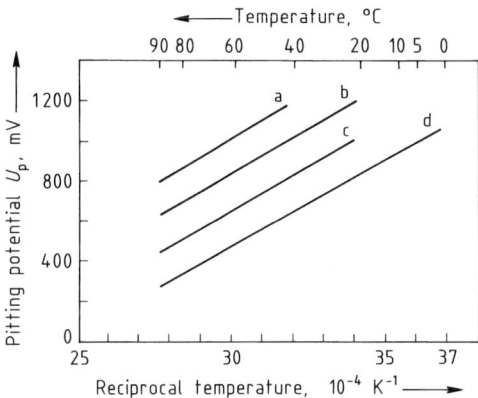

Figure 24. Pitting potentials U_p of diffusion-annealed X 3 CrNiMoN 17 13 5 (material no. 1.4439) in NaCl solutions as a function of temperature
a) 0.001 N NaCl solution; b) 0.01 N; c) 0.1 N; d) 1 N

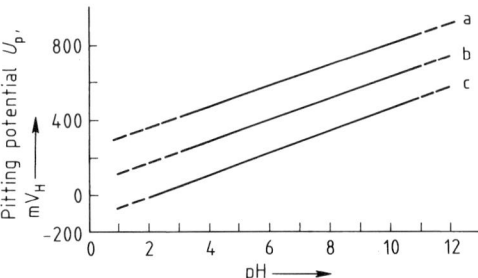

Figure 25. Pitting potentials U_p of diffusion-annealed X 3 CrNiMoN 17 13 5 (material no. 1.4439) in NaCl solutions as a function of temperature
a) 1 N, $t = 40\,°C$; b) 0.1 N, $t = 80\,°C$; c) 1 N, $t = 80\,°C$

the dependence of the overpotential on the concentration of the potential-determining ion. A change in hydrogen ion activity by a factor of 10 makes $\Delta U = 58$ mV (see Section 2.2.3):

$$\Delta U = (2.3\, RT/nF) \log a$$

Pitting in austenitic chromium–nickel steels can be avoided in many cases by lowering the chloride ion concentration, reducing the temperature, and raising the pH of the corrosive medium as high as possible. If these measures cannot be taken, or cannot be taken to an adequate degree, other materials must be chosen.

Material Parameters. For a given steel, the susceptibility to pitting decreases with increasing surface finishing. A ground surface is more susceptible to pitting than a pickled one, which in turn is more susceptible than a polished surface. A further decisive point is structural uniformity. Rolled surfaces (i.e., surfaces parallel to the rolling direction) are less prone to pitting than short-transverse-cut sections exhibiting open microsegregations. Oxide and sulfide inclusions, carbide precipitates, and intermetallic phases also increase susceptibility to pitting. Cold forming can also cause a marked increase in the incidence of pitting, although the influence of cold forming has not yet been finally established and contrary results have also been found.

Effect of Alloying Elements. For a standard chromium–nickel steel (e.g., X 6 CrNiTi 18 10, material no. 1.4541) with 17.0–19.0 wt% chromium and 9.0–11.5 wt% nickel, the resistance to pitting can be improved by increasing the contents of chromium and molybdenum, but is not affected by higher nickel concentrations. In stainless steels, 3% chromium is equivalent to ca. 1% molybdenum. This finding led to the formulation of a *pitting resistance equivalent* (PRE), which is defined as [13]:

PRE = wt% chromium + 3 × wt% molybdenum

Recently, some investigations have suggested a factor of 3.3 for molybdenum [14].

Figure 26 shows that U_p is directly proportional to the PRE. Therefore, austenitic steels with the highest possible chromium and molybdenum contents exhibit extremely high pitting resistance.

The positive effect of nitrogen on the resistance to pitting corrosion of stainless austenitic steels has recently been studied in greater detail. Examinations of austenitic materials with different Cr, Mo, and N contents indicated a PRE factor of ca. 13 for nitrogen [15]; other authors

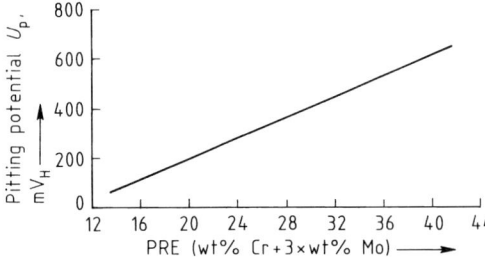

Figure 26. Pitting potential U_p of austenitic chromium–nickel–molybdenum steels in a NaCl solution as a function of the pitting resistance equivalent (PRE)

give values of 20 or even 30 [16]. However, the effectiveness of nitrogen is clearly linked to specific Cr and Mo percentages. To what extent nitrogen exercises its own pitting-inhibiting effect or whether its structure-stabilizing influence is the cause has yet to be clarified.

The high-alloy, austenitic stainless steels X 2 CrNiMoN 17 13 5 (material no. 1.4439) [17], X 1 CrNiMoN 25 25 2 (material no. 1.4465) [18], [19] and X 1 CrNiMoN 25 22 2 (material no. 1.4466) [20], which have PRE's of ca. 31, are widely used in chemical engineering [21]–[25]. Even higher PRE's are exhibited by the materials X 1 NiCrMoCu 25 20 5 (material no. 1.4539) [19], [26], X 1 NiCrMoCu 31 27 4 (material no. 1.4563) [19], X 2 CrNiMoCuN 20 18 6 [19], [27], and X 1 NiCrMoCuN 25 20 6 (material no. 1.4529) [19], [28]. These high-alloy materials have also proved successful in a wide variety of applications, such as chemical apparatus, flue gas desulfurization, and marine technology [24], [29]–[34].

Factors affecting pitting corrosion in austenitic chromium–nickel–(molybdenum) steels are listed in Table 2. An example of pitting corrosion is given in Figure 27, which shows a section of a forged T-piece made of X 5 CrNiMoTi 25 25 (material no. 1.4577) into which a thermocouple protection tube was inserted. In the gap between the steel wall and the protection tube, a sublimable, chloride-containing salt accumulated causing considerable pitting corrosion. The problem was solved by using the same material under changed plant conditions.

Pitting corrosion is not restricted to stainless steels; other passive metals tend to undergo pitting in solutions containing halides. These in-

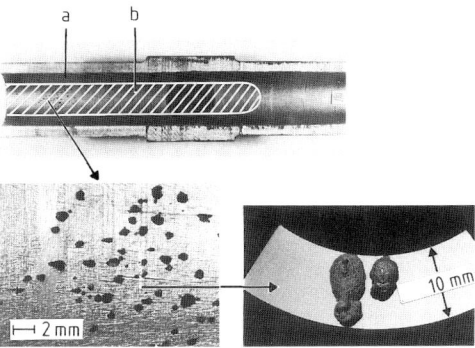

Figure 27. Pitting corrosion on a high-pressure T-piece of material no. 1.4577 at 180 °C and 13 MPa as the result of accumulation of chloride in the gap between the pipe (a) and the thermometer protection tube (b)

clude, for example, aluminum and aluminum alloys, nickel alloys with passivation capacity such as NiMo16Cr16Ti, and titanium. Nickel-based alloys are far less prone to corrosion attack by chlorides than corrosion-resistant Cr–Ni austenitic steels and are thus seldom at risk. Aluminum alloys show some susceptibility and have pitting potentials that are determined by the chloride ion activity and temperature but not by the pH. Because of the low electron conductivity of the oxide scale of aluminum, however, the corrosion rate remains low if it is not enhanced by contact corrosion with metals such as copper, iron, nickel, and their alloys, or with electrically conducting deposits with rest potentials more positive than the pitting potential of aluminum (intensification of the cathode part reaction) [35]. Under extreme conditions (e.g., in a solution with 65% $ZnCl_2$, 5% HCl, and 1% $FeCl_3$ at 100 °C) [36], titanium is prone to pitting corrosion. So far pitting has not been reported for tantalum and niobium.

3.2.2. Crevice Corrosion

Crevice corrosion is also a form of localized corrosion. It amounts to locally intensified corrosion in crevices and thus results exclusively from the design of a specific component [37], [38]. In unalloyed steels it usually occurs as nonuniform or uniform corrosion, and in stainless steels it can also occur as pitting corrosion. It can be regarded as a variation of pitting corrosion.

Although the mechanism of crevice corrosion has not yet been fully explained, the elec-

Table 2. Parameters affecting the pitting corrosion of austenitic chromium–nickel steels

Favoring pitting	Hindering pitting
Material parameters	
Rough surface	polished surface
Inhomogeneous structure	homogeneous structure
Cold forming	
Corrodent parameters	
High chloride ion concentration	low chloride ion concentration
High temperature	low temperature
Low pH value	higher pH value
Stationary attacking agent	higher flow rate
Alloying parameters	
Low chromium contents	high chromium contents
Low molybdenum contents	high molybdenum contents

trolyte in the crevice contains less oxygen than the electrolyte outside the crevice because diffusion is impeded. The vast majority of crevice corrosion phenomena can thus be explained by the formation of an aeration element (see Section 2.5). Through the lack of oxygen in the crevice, anodic regions form in which metal ions go into solution. As a result, the conditions in the crevice electrolyte usually change, this being towards greater corrosiveness (e.g., lower pH through hydrolysis). The time dependence of oxygen depletion as a function of the crack width and depth is shown in Figure 28 [39].

Whether or not crevice corrosion occurs largely depends on the crevice geometry. Crevice widths are particularly critical when the distance between the crevice-forming surfaces is less than 1 mm. Since the reactions taking place in the crevice cannot be influenced from the outside, damage due to crevice corrosion can only be avoided through the prevention of crevices.

In ferritic and austenitic stainless steels, crevice corrosion is almost always initiated by local activation. This can be induced in a crevice by oxidant depletion, if necessary supplemented by halides. The passivity then breaks down. The access of oxidants to the material surface, and hence the passivity, may also be hindered by local deposits.

Numerous incidences of damage in crevices in components are due to crevice corrosion; for example, in the critical areas between the tubes and tube sheets in heat exchangers. Other commonly found defects are crevices caused by incomplete welding, especially of root passes, which can lead to crevice corrosion under critical corrosion conditions (Fig. 29). Typical damage due to crevice corrosion on a plate heat exchanger made of X 5 CrNiMo 17 13 (material no. 1.4449) is illustrated in Figure 30. The medium was an almost saturated sodium chloride solution containing small amounts of free chlorine. The product temperature was 50–70 °C. The first plates were already porous after five weeks at the points where the grooves touch those of the neighboring sheet. The damage is due to heavy local corrosion in the region of the points of contact (crevices). Remarkably, pipes operating with the same product and made of the same material, but which do not present the conditions for crevice corrosion, have been in use for more than ten years with negligible corrosion.

The conditions for crevice corrosion are not only present when the crevice is formed by metallic materials, but also when the surface of a passive material is coated with nonconductors such as seals, packings, and product deposits, under which the material can then be prone to pitting corrosion. Corrosive media containing chloride, which can neutralize passivity, are particularly dangerous in this respect.

Figure 28. Dependence of oxygen depletion in crevice corrosion on crack width and depth
Crack width: A) 1.65 mm; B) 0.74 mm; C) 0.40 mm; D) 0.15 mm
Time: a) 6 min; b) 15 min; c) 30 min; d) 1 h; e) 1.5 h; f) 2 h

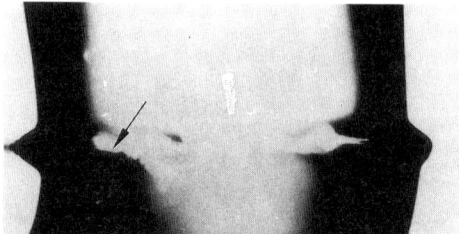
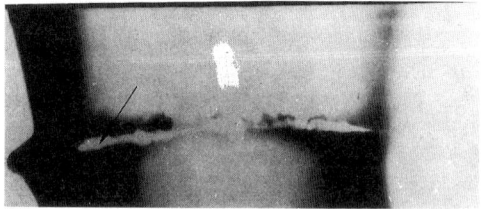

Figure 29. Crevice corrosion of a pipe (material no. 1.4571) (radiograph)

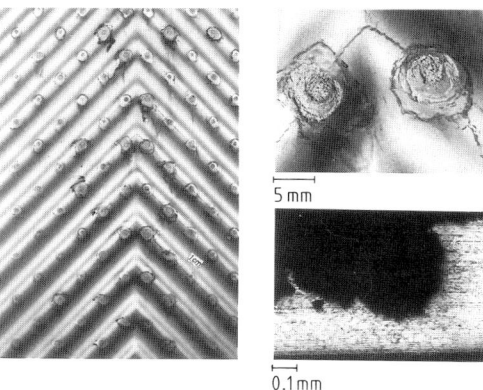

Figure 30. Crevice corrosion of a plate heat exchanger (material no. 1.4449)

An example is a horizontal copper pipe for drinking water, in which typical pitting corrosion damage has formed under deposits. The first perforations appeared after eight-months' use. The deposits consisted primarily of residual solder paste and iron oxides and were found solely in the bottom half of the pipe. After removing the deposited layers by pickling, numerous pitted spots could be seen, mainly under pimples in the protective coating. Since the cause of damage can only be improper handling of solder paste during soldering, this kind of damage, which was frequently found some years ago in the Federal Republic of Germany, ceased to exist after the changeover to brazing. The corrosion-inducing effect of iron oxides is obviously due to their impairing the formation of a dense copper(I) oxide layer. Iron oxides do not cause corrosion damage when they come into contact with previously formed, dense protective layers.

3.2.3. Selective Corrosion

Selective corrosion is defined as corrosion in which specific microstructural constituents, regions close to grain boundaries, or alloying components are preferentially attacked [1].

Intergranular corrosion, spongiosis, dezincification, and line- and layer-type corrosion are classified in this subgroup. Of particular importance is intergranular corrosion.

3.2.3.1. Intergranular Corrosion

Intergranular corrosion in electrolyte solutions arises when the resistance of zones close to the grain boundary, compared with the resistance of areas further from the grain boundary, is reduced due to microstructural changes in the grain boundary regions. This type of corrosion is particularly characteristic of passive materials, in which large differences between the corrosion rates of the matrix and the grain boundary areas are possible if the chemical composition of the grain boundary zones is changed by discontinuous precipitation processes in the supersaturated solid solution. Intergranular corrosion is characterized by the fact that grain boundary zones are corroded preferentially in a groovelike manner until individual grains are loosened and fall out of the microstructure. Given heavy intergranular corrosion, the entire structure can collapse [40]–[42].

The reduced resistance of the grain boundary regions is due to the following:

1) Depletion of alloying elements that are important for the maintenance of the corrosion resistance in the grain boundary regions through precipitation of phases containing these elements along the grain boundaries
2) Formation of microstructural components at the grain boundaries that are preferentially attacked in a corrosive medium
3) Accumulation at the grain boundaries of impurities which stimulate anodic dissolution

Of these causes, depletion of alloying elements has the greatest practical significance. This occurs when grain boundary precipitates are formed that contain one or more alloying elements in a markedly higher concentration than the matrix.

Precipitation processes of this kind are always caused by heat treatments, such as sensitizing annealing, that are inappropriate for the alloy in question. For the austenitic chromium–nickel–(molybdenum) steels used for the fabrication of chemical plant equipment, the critical temperature range is 400–800 °C. Chromium depletion through formation of chromium-rich carbides, mostly of the type $M_{23}C_6$, is the main cause of intergranular corrosion in these steels. The precipitation of chromium nitrides—of importance is only the chromium-rich nitride Cr_2N [43]—can initiate intergranular corrosion, especially in ferritic steels. Since the intermetallic phases in stainless steels contain appreciably less chromium than carbides and nitrides and their deposition is far slower, the chromium depletion related to these phases is minimal.

Molybdenum depletion at grain boundaries through carbides rich in molybdenum and through intermetallic phases can also lead to intergranular corrosion. For standard molybdenum steels, molybdenum depletion is always linked to a far greater chromium depletion and therefore plays only a minor role. In special high-alloy molybdenum-containing steels, and especially in nickel-based alloys with very high molybdenum contents, molybdenum depletion is a major cause of susceptibility to intergranular corrosion. This is particularly significant in the chromium-free alloy NiMo 30 (material no. 2.4810).

In NiMo 30, molybdenum-rich carbides and intermetallic phases of the type Ni_4Mo and Ni_3Mo precipitate readily [44]. Figure 31 shows intergranular corrosion in a welded pipe. In the meantime, NiMo 30 has been superseded by NiMo 28 (material no. 2.4617), which resists the formation of grain boundary precipitates in the weld heat-affected zone.

In the high-silicon austenitic chromium–nickel special steel X 1 CrNiSi 18 15 4, which has high resistance to concentrated nitric acid [45]–[47], susceptibility to intergranular corrosion depends on the separation of a carbide $M_{23}C_6$ and on a π-phase of type $M_{11}(C, N)_2$, containing chromium, nickel, silicon, carbon, and nitrogen [48].

The susceptibility to intergranular corrosion of a chemically resistant steel depends on the carbon content, the duration of sensitizing, the sensitizing temperature, and the temperature of prior solution annealing; it is represented in a time–temperature diagram. Curves of this type are shown in Figure 32 for an austenitic steel with ca. 18% chromium, 8% nickel, and various

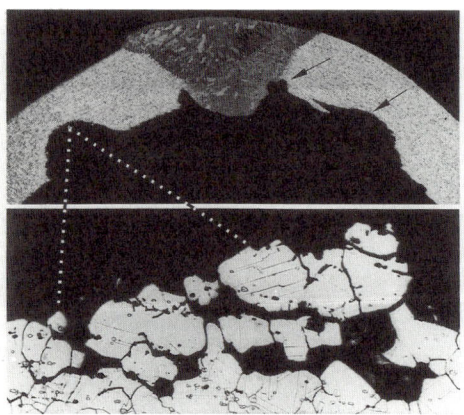

Figure 31. Intergranular corrosion in the weld heat-affected zone of a pipe (made of NiMo 30) caused by hydrochloric acid

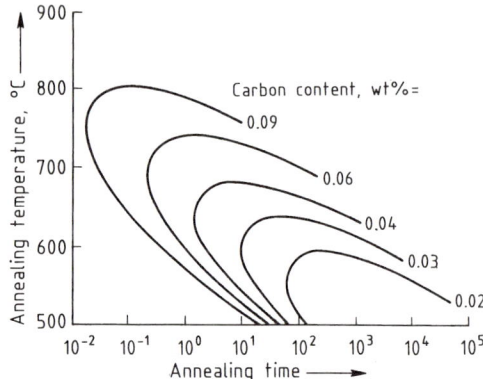

Figure 32. Time–temperature sensitization curves of an unstabilized austenic steel (18% Cr, 8% Ni) as a function of carbon content

carbon contents [49]. The curves encompass the respective area of susceptibility.

The effect of the solution annealing temperature is especially pronounced in stabilized steels (i.e., those that contain titanium and niobium for bonding of the carbon). With increasing solution annealing temperature, the curves encompassing the area of susceptibility shift to lower sensitizing temperatures and longer annealing times, as shown in Figure 33 [50]. The reason for this shift is that with increasing solution annealing temperature, the amount of carbon which is in solution and, therefore, available for precipitation in the critical temperature range, also increases. At the same time, grain coarsening occurs, reducing the grain boundary surfaces available as precipitation sites. This also leads to greater chromium depletion because the increased amount of available carbon on the grain surfaces can then bind more chromium. However, the considerably lower susceptibility of stabilized steels to grain disintegration, compared with unstabilized steels, is due to the formation of carbides, which leads to a lower concentration of dissolved carbon. Since the solubility of the carbides (TiC and NbC) decreases with decreasing temperature, proneness to intergranular corrosion may be eliminated by annealing (stabilization) these steels at a lower temperature (900–950 °C) to remove dissolved carbon by precipitation of the carbides. Another possibility of lowering the susceptibility to grain disintegration is to reduce the carbon content to below 0.03% (ELC grades).

The carbon solubility is further reduced by a higher nickel content, thus increasing the tenden-

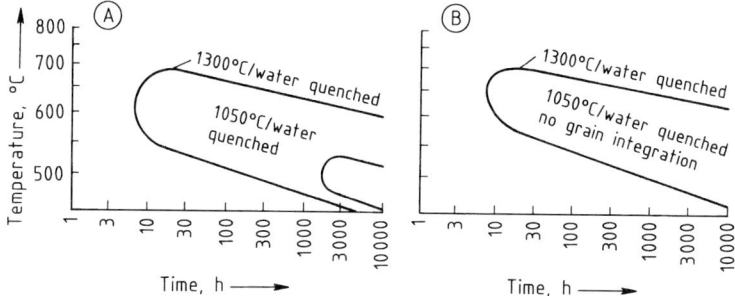

Figure 33. Time–temperature sensitization curves of stabilized steels as a function of solution annealing and sensitizing temperatures (test criterion: Strauss test)
A) Material no. 1.4541; B) Material no. 1.4550

cy to precipitate. This shifts the grain disintegration areas to shorter tempering times and higher temperatures, as demonstrated in Figure 34, which compares the grain disintegration curves of X 2 CrNi 19 11 (material no. 1.4306, 11% Ni), the titanium-stabilized grade X 10 NiCrAlTi 32 20 (material no. 1.4876, 32% Ni), and the alloy NiCr 15 Fe (> 72% Ni). The carbon content of the materials is ca. 0.02% in each case. Higher nitrogen contents also shift the beginning of precipitation to shorter times (Fig. 35). In practice, however, grain disintegration is only expected above ca. 0.2% N [43].

Of the ferritic steels, those containing ca. 17 wt% chromium are of industrial significance. Final heat treatment is performed at 750–850 °C with subsequent air or water cooling. In unstabilized grades the carbon is practically all precipitated as $M_{23}C_6$, without the formation of chromium-depleted zones. As the annealing temperature rises, the carbon solubility increases, the carbides dissolve completely or partially, and reprecipitate on cooling as $M_{23}C_6$ at the grain boundaries, giving rise to chromium-depleted grain boundary zones if the cooling rate is sufficiently high. In contrast to austenitic steels, rapid cooling from a high temperature is critical for ferritic steels; the higher mobility of the atoms in the body-centered cubic lattice causes a high precipitation rate of the carbides, and rapid post-diffusion of chromium into the depleted zones occurs. Ferritic steels with 17% Cr can therefore suffer intergranular corrosion in the high-temperature zone immediately adjacent to the weld if this is cooled so rapidly after welding that chromium depletion is not eliminated; an example is shown in Figure 36. Ferritic chromium steels can be stabilized by addition of titanium or niobium, but compared with the austenitic standard steels, a comparatively higher concentration of these elements is necessary [51].

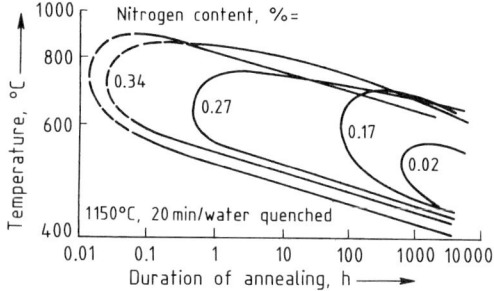

Figure 35. Time–temperature sensitization curves of unstabilized, nitrogen-containing, austenitic steels with ca. 18% Cr, 10% Ni, and 0.01% C

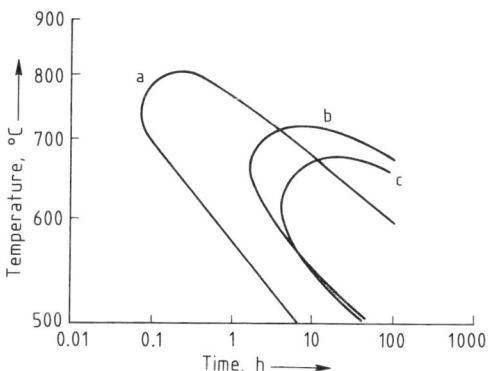

Figure 34. Time–temperature sensitization curves of austenitic steels
a) NiCr 15 Fe, 0.017% C; b) X 10 NiCrAlTi 32 20, 0.021% C, 0.35% Ti; c) X 2 CrNi 19 11, 0.020% C

Figure 36. Intergranular corrosion in the weld heat-affected zone of a 17% Cr steel

So-called superferrites also exist, such as X 1 CrNiMoNb 28 4 2 (material no. 1.4575), characterized by C + N contents ≤ 0.040%, which greatly improves the weldability of these materials.

A special form of intergranular corrosion is observed in titanium-stabilized stainless steels containing ca. 17 wt% chromium: it is caused by precipitation of titanium carbonitrides at the grain boundaries. The carbonitrides are dissolved by strongly oxidizing solutions such as nitric acid, leaving grooves (Fig. 37). In such cases preference should be given to the niobium-stabilized grade since NbC is hardly attacked by oxidizing acids [52].

Marked susceptibility to grain disintegration can also be observed in the highly corrosion-resistant nickel-based alloys, such as the chromium- and molybdenum-containing type NiMo 16 Cr (material no. 2.4812) (Fig. 38), which is, however, no longer in use. In addition to carbides, grain boundary precipitates in this material also consist of an intermetallic chromium- and molybdenum-rich intermetallic phase, which leads to depletion of chromium and molybdenum in the grain boundary zones. The critical temperature range for precipitation is very wide (500–1150 °C) [44]. More recent types of this family of alloys, such as NiMo16Cr16Ti (material no. 2.4610) and NiCr21Mo14W (material no. 2.4602), exhibit a considerably reduced susceptibility due to their lower carbon and silicon contents (both < 0.1%); thus, there is no further risk of intergranular corrosion in practical applications [53]–[57].

Intergranular corrosion can also occur in other metals and alloys in connection with grain boundary precipitation, for example, in AlMg5

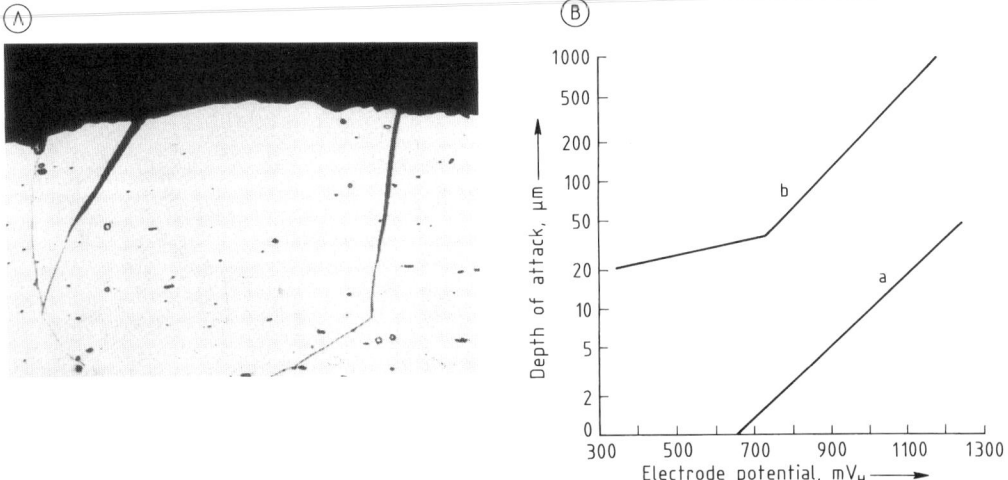

Figure 37. Intergranular corrosion of a Ti-stabilized 17% Cr steel (X 7 CrTi 17) in boiling sulfuric acid
A) Mode of attack; heat treatment 30 min, 1300 °C, water quenched; test time 24 h; U_H = 650 mV
B) Depth of attack as a function of electrode potential; heat treatment: a) 30 min, 850 °C, water quenched; b) As a) + 30 min, 1300 °C, water quenched

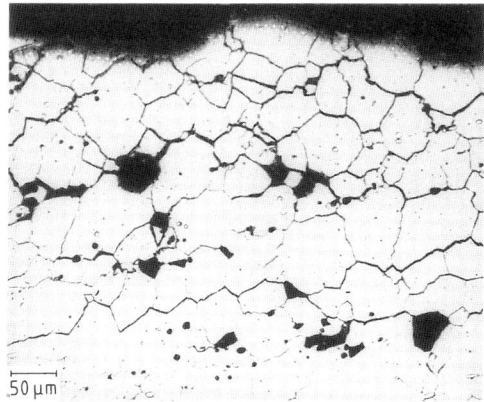

Figure 38. Intergranular corrosion of NiMo 16 Cr as a result of improper heat treatment

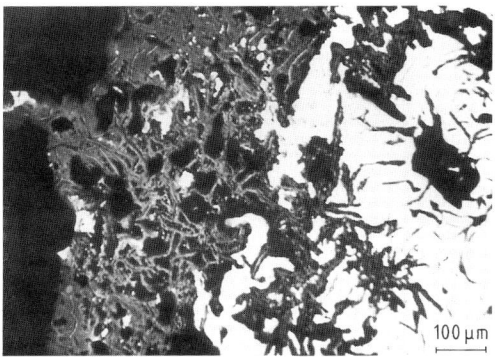

Figure 39. Spongiosis of a cast iron pump used for well water
The light areas in the picture are ferrite, the dark areas are graphite.

alloy, NiCr30Fe, or titanium (in methanol with traces of water and HCl). Solution-annealed austenitic chromium–nickel steels can suffer intergranular corrosion in strong oxidants such as chromium-containing, azeotropic, or superazeotropic nitric acid, even though no precipitation can be detected. It is assumed that phosphorus contents above 0.01 wt% and silicon contents above 0.1 wt% are responsible for this form of corrosion [40].

3.2.3.2. Spongiosis

Spongiosis, also known as graphitization, is a form of corrosion found only in gray cast iron. Because of inadequate formation of a protective layer, ferrite and pearlite are converted primarily to iron(II) oxide hydrate, whereas the graphite structure and the distribution and percentage of the phosphide eutectic remain unchanged [1], [58]. In this instance the metallic bond is lost, leading to complete loss of strength in the damaged areas of the component (Fig. 39). The remarkable feature of this process is that the outer form of the component is so well maintained that tool marks can often be seen.

Spongiosis occurs primarily in cast iron pipes laid in acid earth or in calcareous loamy soil. It can be avoided by applying adhesive, crack-free protective coatings, such as those based on tar. Damage caused by spongiosis is also observed in cast-iron feed-water preheaters and pumps for drinking water and seawater [58].

When ferritic or pearlitic cast iron comes into contact with pure water or aqueous salt solutions, spongiosis occurs. It begins without exception in the immediate vicinity of the graphite, giving rise to a porous coating consisting of graphite, carbides, sulfides, phosphides, and nitrides. A sufficiently high oxygen content in the corrosive medium then leads to a greater deposition of corrosion products on this primary coating, which thereby becomes sealed, finally forming a protective coating [59]. The advance of spongiosis is therefore to be expected when the corrosive medium has a low oxygen concentration.

3.2.3.3. Dezincification

Dezincification is the selective corrosion of copper–zinc alloys (brass) with more than 15 % zinc, in which porous copper deposits form on the corroded parts [1], [60]. Although less common, similar phenomena have also been observed in other copper alloys such as copper–aluminum (aluminum stripping), copper–tin (detinning), and copper–nickel (nickel stripping) [61]–[64].

Dezincification can occur, for example, in tap water, river water, seawater, and liquids containing chloride. The rate of this corrosion process is increased by plastic deformation of the material [65]. Dezincification can occur by two mechanisms:

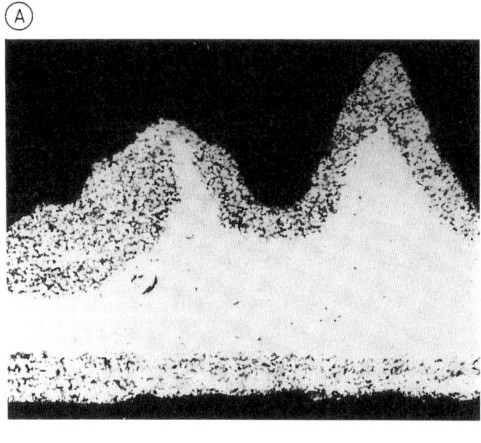

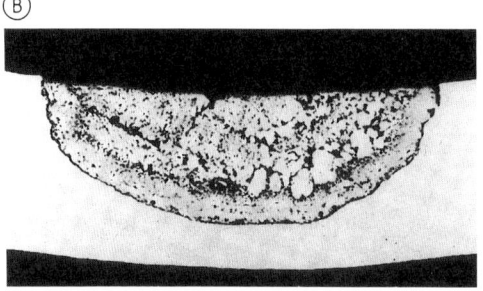

Figure 40. Types of dezincification
A) Layer dezincification of a brass water faucet (GK-CuZn 37Pb); B) Plug-type dezincification in a brass pipe (CuZn 28 Sn 1)

1) going into solution of both alloy constituents (copper and zinc) through dissolution of the solid solution and cementation of the more noble component (copper), or
2) selective dissolution of the less noble constituent from the solid solution and accumulation of the exposed atoms through surface diffusion.

It is assumed that the two mechanisms work either simultaneously or one after the other [66], [67].

According to their appearance, a distinction is made between layer dezincification and plug-type dezincification (Fig. 40). In layer dezincification the entire surface is corroded practically uniformly. This form of corrosion is observed when a component made of a copper–zinc alloy comes into contact with nobler metals in an electrolyte (e.g., seawater). Crevices and cracks such as threads and surface damage promote corrosion [60].

Far more critical is plug-type dezincification, which mainly arises in heat exchanger tubes upon deposition of sand grains, oil residues, rust, mussels, and algae with formation of aeration elements. Even thick tubes can be perforated in this way [60], [68]. Dezincification can be impeded by adding small amounts of arsenic or phosphorus.

3.2.3.4. Line- and Layer-Type Corrosion

In these forms of selective corrosion the attack occurs practically parallel to the rolling direction. Layer corrosion is observed in the aluminum alloys AlMg, AlZnMg, AlCuMg, and AlZnMgCu. The material undergoes corrosion in chloride-containing media by exfoliation parallel to the surface. This occurs if a rolling process gives rise to a laminar structure of intermetallic compounds or when lamellar zones of differing concentration were already present in the ingot [69], [70]. Figure 41 shows layer corrosion in an aluminum alloy [70].

In contrast to layer corrosion in aluminum alloys, line-type corrosion can occur in many materials, although it is also linked to forged and rolled products. The cause of line-type corrosion may be due to both the banded structure of precipitations and microsegregation, i.e., insufficient concentration compensation.

Because of the high degree of deformation, sheet products exhibit a particularly marked sandwich structure with regard to the concentration distribution of the alloying elements. If the corrosion resistance of a material depends on the concentration of the segregating alloying element, local differences in corrosion depth can

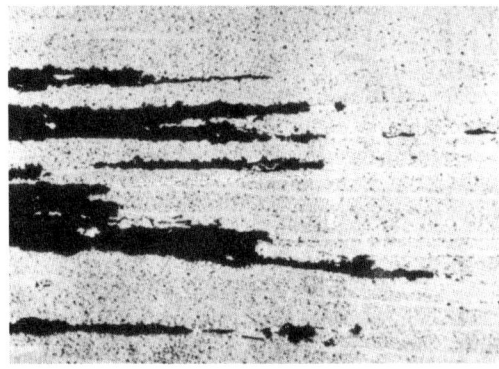

Figure 41. Layer corrosion of an Al–Zn–Mg alloy

occur if fabricated parts are subjected to stress in the fiber direction. In practice this is the case at the face of rolled-in tubes of heat exchangers, the cutting edges of sieve trays, or valves and valve seats in forged armatures [42], [71]–[75].

A good example is provided by support rigs of tube bundles made of material no. 1.4571, which after several months' exposure in a hot, weakly acidic, chloride-containing suspension had fractured because of transgranular stress corrosion cracking (see Section 3.3.1.2). One detail of the fractured fixtures, with many cracks, on which clear signs of pitting corrosion can be seen, is shown in Figure 42. Metallographic examination (see Fig. 42 B) showed that the transgranular cracks produced selective corrosion, which spread along the ferrite lines. At the point marked by an arrow in Figure 42 B it can be seen that the extended ferrite islands were not dissolved; instead, the corrosion spread along the segregated austenite rims [42].

Figure 43 shows a nozzle head made of Ni-Mo 30 (material no. 2.4810), which was manufactured from a nonuniformly annealed bar. After 200 h in operation, signs of corrosion that look like pitting appear. Figure 43 B shows line- or band-like precipitation in the fiber direction which is responsible for the pitting-like attack.

Figure 44 shows selective corrosion caused by segregation. The SEM photographs of the face of a heat exchanger tube made of X 1 CrNiSi 18 15 4 which was subjected to boiling, highly concentrated nitric acid for 60 d showed evidence of line-type corrosion. The sides of the

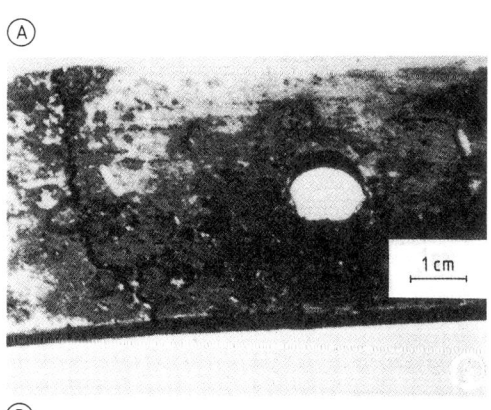

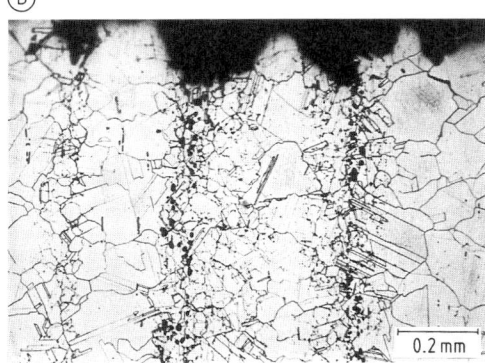

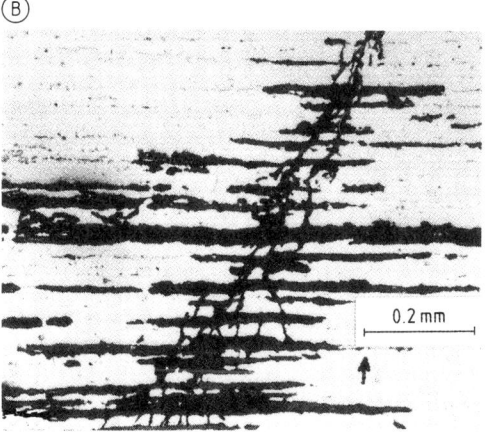

Figure 42. Transgranular stress corrosion cracking of a steam coil support
A) Macroscopic view; B) Microsection showing selective corrosion along the ferrite lines (→)

Figure 43. Selective corrosion of a NiMo30 (material no. 2.4810) nozzle head of in a region exhibiting lines of precipitates
A) Nozzle head; B) Microstructure of nozzle head

corrosion crevices are coated with several layers of the corrosion product SiO_2 [9], [48].

Selective corrosion caused by differences in concentration can be reduced by homogenizing heat treatment or by laying one weld pass over the edges of sheets and tubes.

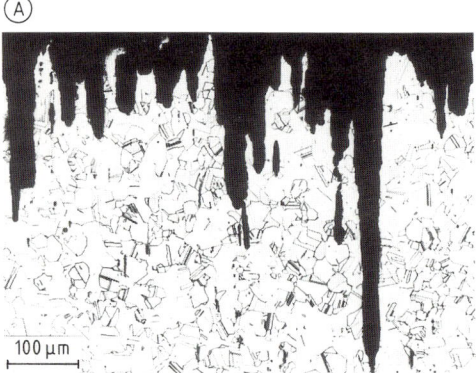

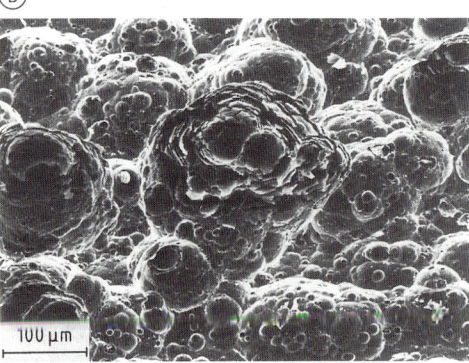

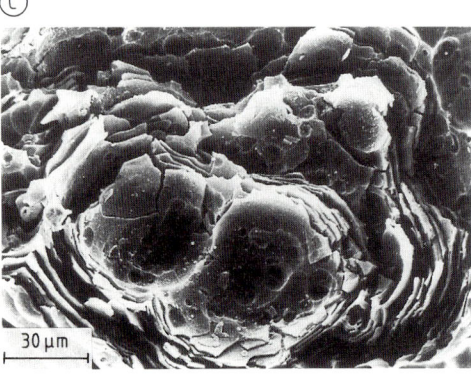

Figure 44. Selective corrosion caused by segregation on the face of a heat exchanger tube in boiling, highly concentrated HNO_3. Material: X 1 CrNiSi 18 15 4
A) Longitudinal section; B) and C) Scanning electron micrographs of the attacked face

3.3. Forms of Local Corrosion under Mechanical Stress

3.3.1. Stress Corrosion Cracking

Stress corrosion cracking is one of the most unpleasant forms of corrosion because it usually occurs unexpectedly and can very quickly lead to cracking in equipment such as reactors, vessels, and piping, and to fractures in components of all kinds. Depending on the alloy system and corrosive medium, the characteristics are inter- or transgranular brittle cracking, branching out to varying degrees into the material and eventually causing fracture of the remaining cross section. General surface corrosion arising in this way is either nonexistent or immeasurably small. Deposited corrosion products are seldom found.

Normal anodic stress corrosion cracking is caused by a combination of mechanical tensile stress and local electrolyte dissolution processes when certain conditions are met. Firstly, the corrosive medium must have a specific effect on the respective alloy, and in addition the alloy in contact with the electrolyte in this material/corrosive medium system must be prone to stress corrosion cracking. The tensile stress must also be sufficiently high. Susceptible systems, for example, are stainless austenitic steels in chloride-containing solutions or unalloyed and low-alloy steels in nitrate solutions. In contrast, unalloyed and low-alloy steels are not susceptible to stress corrosion cracking in chloride solutions.

Protective layers are a prerequisite for stress corrosion cracking to occur on the surface of a component under the influence of an electrolyte. Local crack-inducing corrosion processes are caused by selective degradation of these protective or passive layers. In this instance, the corrosive medium can cause local perforation of the layer (e.g., media containing chlorides with passive, stainless austenitic CrNi steels), or its passivity can be impaired by the influence of the prevailing tensile stress to the extent that only a locally incomplete protective layer is formed (e.g., solutions of alkali hydroxides or nitrates with unalloyed steels). However, this simplified view of the formation of local anodes in stress corrosion cracking merely explains the initial stage during the induction period; it sheds no light on the actual cracking period, for which there are now several interpretations.

The most common interpretation of the mechanism of cracking is based on a "periodic electrochemical–mechanical process". This suggests that cracking is an alternating sequence of relatively slow anodic dissolution in the crack base and sudden mechanical crack propagation. In some alloys intermittent cracking has actually been found, but in many other cases no evidence of stepwise cracking has been produced.

A further theory on the cracking mechanism is that it is a purely electrochemical process in the base of the crack (active path corrosion), the rapid crack propagation rate observed in most cases being due to greatly increased dissolution of the material as a result of plastic deformation at the crack tip with simultaneous passivation of the crack walls. Greatly increased dissolution during the plastic deformation of a metal has been demonstrated [77].

A distinction is made between anodic and hydrogen-induced cathodic stress corrosion cracking, in which atomic hydrogen generated from a cathodic process in a corrosion reaction penetrates the metal lattice and, in combination with internal and external stresses, causes transgranular cracking and in heat-treated steels preferentially intergranular cracking. This kind of stress corrosion cracking thus belongs to the category of hydrogen damage or embrittlement (see Section 3.3.1.5).

A more recent theory on the mechanism of anodic stress corrosion cracking, based partly on tests on steels, combines the electrochemical process of local metal dissolution with hydrogen embrittlement at the base of the crack caused by the atomic hydrogen forming during corrosion, which may be of major significance to crack propagation.

In addition to classic stress corrosion cracking, systems are known in which the static tensile stress is not sufficient to create cracking (e.g., carbonate–bicarbonate solutions). In such cases, critical slow strain rates are necessary to initiate stress corrosion cracking. This type of corrosion is called strain-induced or nonclassical stress corrosion [78]–[81].

3.3.1.1. Unalloyed and Low-Alloy Steels

In unalloyed and low-alloy steels, the corrosive media (alkalies, nitrates) produce passive protective layers which have a certain instability at high-angle grain boundaries and are thus decomposed under sufficient stress, initiating local corrosion. The resulting stress corrosion cracking is therefore an intergranular process in these steels (Fig. 45). It is assumed that the special activity of the grain boundaries is determined by accumulation of alloy phases. This theory is supported by a number of test results. It has been found, for example, that cementite and nitrides or carbonitrides at the grain boundaries in unalloyed steels produce a high susceptibility to stress corrosion cracking [82].

The susceptibility to stress corrosion cracking of a steel with 0.011 wt% carbon subject to different annealing conditions, measured under constant tensile stress in boiling $Ca(NO_3)_2$ solution, is shown in Table 3 [82]. The marked improvement when the carbon is kept in solution is clearly evident. Elements promoting passivity such as chromium and molybdenum clearly reduce the tendency to cracking, although in alloyed steels of this kind, the type of heat treatment also plays a major role [83].

On the basis of these observations the induction phase in intergranular cracking can be interpreted as the preferential dissolution of inhomogeneous grain boundaries with formation of deep cavities with corresponding stress peaks,

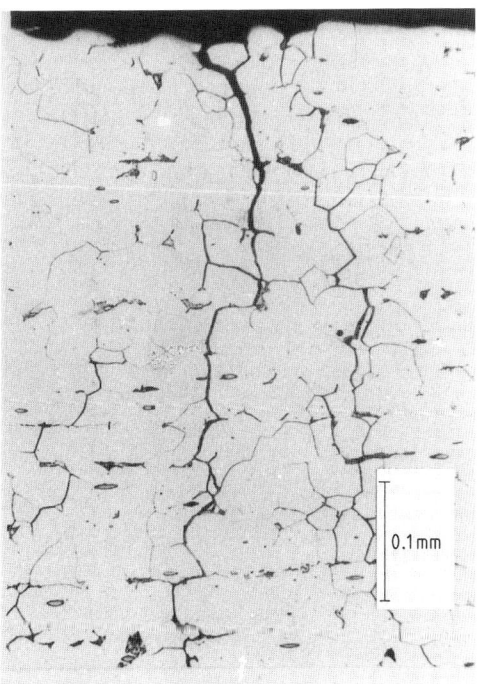

Figure 45. Intergranular stress corrosion cracking of Thomas steel (normalized) in a nitrogen-purged 20 wt% NaOH solution at 110 °C

Table 3. Influence of grain interface cementite on the stress corrosion behavior of a mild, unalloyed steel [a]

Heat treatment	Time to fracture in minutes at a tensile stress of			
	$0.3 \cdot \sigma_u$ [b]	$0.4 \cdot \sigma_u$	$0.5 \cdot \sigma_u$	$0.6 \cdot \sigma_u$
Normalized (920 °C, 20 min, air)	288	30	104	48
Normalized + 720 °C, 24 h, liq. N$_2$ (carbon held in solution)	> 15 000	> 15 000	132	54
Normalized + 730 °C, 2 h, water + 500 °C, 4 h, water (tertiary cementite precipitation)	96	66	18	12

[a] Analysis of the steel: C 0.011%, Si 0.02%, Mn < 0.005%, P 0.007%, Si < 0.005%, Ni 0.02%; N, Al, Cr, and Mo lie beneath the detection limit. [b] Ultimate tensile strength σ_u (normalized) = 213 MPa.

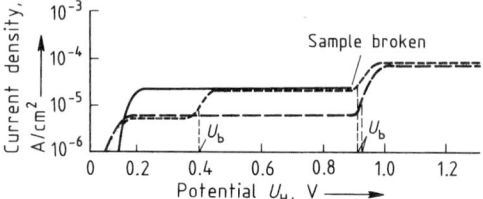

Figure 46. The influence of load on the grain boundary breakthrough potential U_b
Load: —— unloaded; ----- $0.2 \cdot \sigma_u$; —— $0.4 \cdot \sigma_u$
($\sigma_u = 438$ MPa)

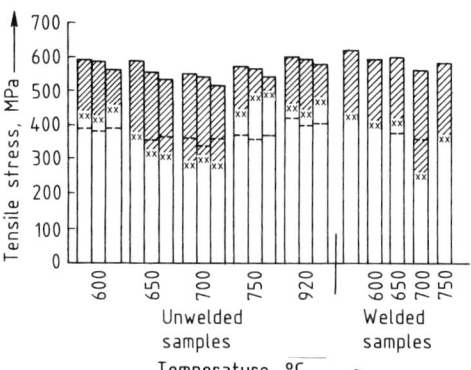

Figure 47. Threshold tensile stresses for stress corrosion cracking of the fine-grained structural steel StE 355 as a function of the annealing temperature as determined with unwelded and welded samples
☐ Tensile strength; -- Yield strength; xx Threshold stress for stress corrosion cracking; ▨ Range of stress corrosion cracking

which initiate the actual cracking. Basically, a lower stress limit for the induction of intergranular stress corrosion cracking in a given corrosive medium exists for each steel and for each structural state resulting from annealing, below which stress corrosion cracking cannot be initiated.

Current density–potential curves for stressed and unstressed tensile test specimens in crack-inducing media show that the unstressed specimen exhibits no rise in current until ca. 900 mV U_H, which indicates the start of grain boundary corrosion (Fig. 46) [84]. At 20% of the tensile strength σ_u, corrosion begins at 400 mV, i.e., a shift of 500 mV in the direction of the actual corrosion potential in the steel. At a stress of $0.4 \sigma_u$ the grain boundary breakthrough potential reaches the corrosion potential, and the steel can now rupture without external polarization through stress corrosion cracking. As it does, the critical tensile stress for dissolution is reached.

In practice, the considerable influence of the stress level can also be seen by the fact that weld seams are preferentially attacked by stress corrosion cracking on account of shrinkage stresses and the stresses caused by structural changes due to the effect of heat input during welding.

The susceptibility to stress corrosion cracking of unalloyed and low-alloy steels is thus indicated by the extent to which the grain boundary breakthrough potential shifts under tensile stress. If, for a low tendency to shift, the stress limit needed to induce cracking is greater than the yield strength, industrial application of a steel of this kind is possible given careful monitoring of the nominal stress. The structural composition formed through annealing is also of importance, as is shown in Figure 47 for the fine-grained structural steel StE 355 [85].

Stress corrosion cracking in alkali-metal hydroxide solutions shows a number of variations. Whereas in nitrate solutions the grain boundary breakthrough potential only means a restriction of the potential range of stress corrosion cracking towards the cathodic direction (there is no limit potential for cracking at the anodic side), in alkalies there is both a cathodic and an anodic limit potential, as shown in Figure 48 [86]. In this instance, stress corrosion cracking, as illustrated

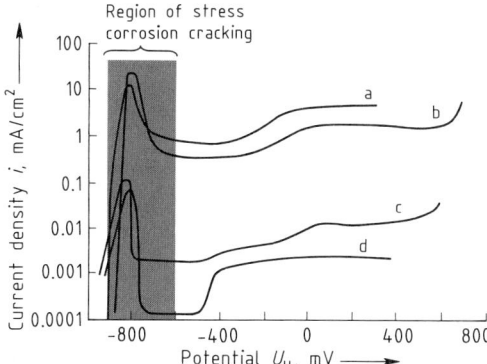

Figure 48. Current density–voltage curves of unalloyed steels in alkaline solution and position of stress corrosion cracking
a) 33% NaOH, boiling, holding time 60 s per measuring point [87]; b) 35% NaOH, boiling, holding time 30 s per measuring point [213]; c) 35% KOH, 80 °C, holding time 5 h per measuring point [86]; d) Anodic partial current–voltage curve for dissolution of iron in 4 N KOH (ca. 20%), 90 °C [76]

Figure 49. Stress corrosion cracking in an unalloyed steel storage tank for liquid ammonia

by the current density–potential curve, is restricted to the potential range corresponding to formation of a passive layer that is initially disturbed at the grain boundaries. Susceptibility disappears once complete passivation has been attained. In alkalies, initiation of cracking requires tensile stresses comparable to the yield strength. In contrast to nitrate solutions, therefore, the limit stress to initiate stress corrosion cracking in alkalies is characterized in most cases by the requirement for small amounts of permanent elongation [87].

In alloyed air-hardened steels (e.g., 13 CrMo 44), thorough tempering at 750 °C is needed after welding to eliminate the hardness structure. Only when the hardness is below 220 HV are these steels resistant to stress corrosion cracking.

In addition to the known corrosive effects of alkali-metal hydroxide and nitrate solutions, intergranular stress corrosion cracking in unalloyed and low-alloy steels in contact with ammonium carbonate [81], [88], [89] and crude methanol (methanol with a low concentration of impurities) has now also been observed [90], [91]. When this group of materials comes into contact with various other aggressive substances, stress corrosion cracking occurs, primarily with transgranular characteristics.

In the past, cases of this sort of crack damage in storage tanks and transportation containers for liquid ammonia (both at room temperature and at 0 °C) were found in increasing numbers (Fig. 49) and simulated in laboratory experiments [92]–[95]. In a few cases the formation of cracks was even observed in areas where the vapor phase had condensed, and in low-temperature storage (-33 °C) [96].

Transgranular stress corrosion cracking also occurs in solutions of $CO-CO_2-H_2O$ containing oxygen [97], [98]. Recent results also indicate that unalloyed and low-alloy steels may be sensitive to anodic stress corrosion cracking in aqueous CO_2 solutions. Note, however, that considerable hydrogen permeation has been observed under critical conditions ($t > 40$ °C, $p_{CO_2} > 1$ MPa, $\sigma > 0.5 \cdot \sigma_{ys}$); a certain amount of hydrogen-induced stress corrosion cracking in the overall cracking activity must therefore be assumed [99], [100].

3.3.1.2. Stainless Steels

Chloride-containing corrosive media can give rise to transgranular stress corrosion cracking in austenitic stainless steels (e.g., steels of composition 18% Cr, 10% Ni, or 18% Cr, 12% Ni, and 2% Mo). It occurs particularly often through the high chloride content of cooling water (usually river water) in heat exchanger tubes: in many cases in the region of salt deposits that arise through local evaporation with insufficient

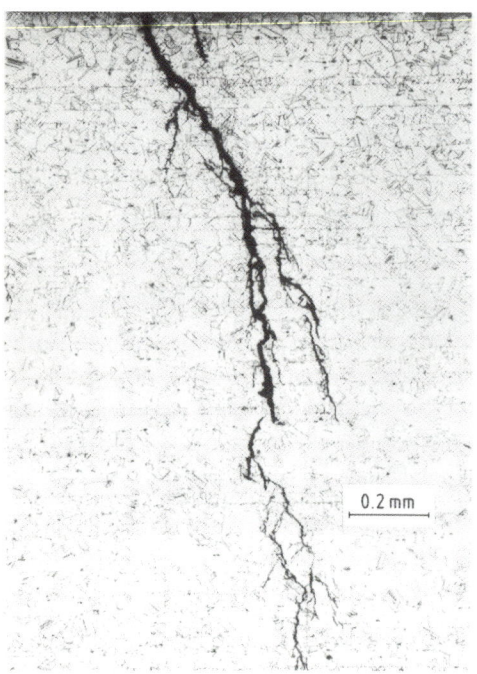

Figure 50. Stress corrosion cracking of a loop sample of X 10 CrNiMoNb 18 10 (material no. 1.4580) caused by an alkaline solution at 280–300 °C and 3 MPa

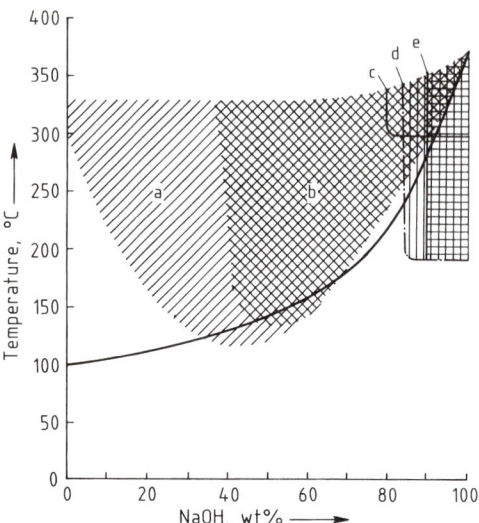

Figure 51. Stress corrosion cracking regions of various metallic materials in sodium hydroxide
a) 18/8 CrNi; b) 18/8/2 CrNiMo; c) Monel; d) Nickel; e) Inconel
—— Boiling curve at 100 kPa

water throughput or incomplete filling of the tube [101].

Alkali-metal hydroxide solutions can also cause stress corrosion cracking in austenitic stainless steels (Fig. 50), albeit only at elevated temperature, often above the boiling point of the corrosive medium (Fig. 51). Therefore, damage occurs only occasionally and primarily in pressure vessels. However, in Figure 52 lower temperature limits are given [102].

It must be assumed in transgranular stress corrosion cracking that local accumulations of foreign atoms in areas with lattice defects outside the grain boundaries, i.e., dislocations and slip lines, are responsible for local activation, which can often be recognized by the formation of etch pits in the presence of corrosive media or through a particular proneness to pitting.

Less prone to stress corrosion cracking are high-alloy austenitic steels that contain more chromium and nickel or have additional copper and molybdenum contents. Particularly resistant are nickel-based alloys with chromium and/or molybdenum, in which stress corrosion cracking occurs only under extreme stress. Commercial ferritic stainless chromium steels also show very

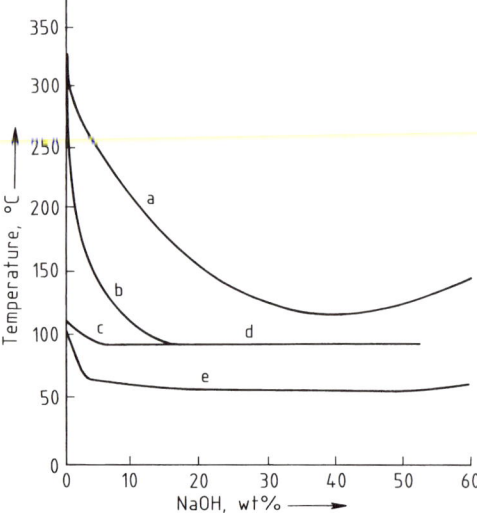

Figure 52. Stress corrosion cracking (SCC) of austenitic stainless steels in sodium hydroxide solution: temperature and concentration limits [102]
a) Rapid stress corrosion failure (one day), unsensitized; b) Delayed SCC failure, unsensitized type 304; c) Delayed SCC failure, sensitized type 304; d) Delayed SCC failure type 304, both sensitized and unsensitized (100–300 d); e) Tentative safe SCC limit

Table 4. Comparison of the yield strength and the nominal stress σ_N present with normal laying out with the limit stress necessary to initiate stress corrosion cracking for three austenitic steels

Steel		0.2% yield strength, MPa	Limit stress, MPa	$\sigma_N = \dfrac{\sigma_{0.2}}{1.5}$, MPa
Mat. no.	Designation (DIN 17006)			
1.4550	X 6 CrNiNb 18 10	180	160	120
1.4435	X 2 CrNiMo 18 14 3	200	130	132
1.4429	X 2 CrNiMoN 18 133	280	100	155

high resistance, but for other reasons (e.g., welding sensitivity) are seldom used in equipment fabrication. Ferritic–austenitic steels are readily applicable in approximately neutral halide solutions. Apart from chlorides, which are responsible for most of the damage, bromides and iodides can also cause stress corrosion cracking, although such cases are rare.

A tensile stress limit must also be exceeded for transgranular stress corrosion cracking to occur in stainless steels. The limit stresses are shown in Table 4 for three standard steels in boiling 42% MgCl$_2$ solution—the standard laboratory solution for investigating stress corrosion cracking [103]. The limit stresses are always lower than the yield stress, as the comparison with the 0.2% yield strength shows. Components are usually designed with a 150% safety margin based on the yield strength. The nominal stresses calculated for the individual steels are also listed in Table 4. The limit stress is higher than the nominal stress only for the steel with the material no. 1.4550. This does not, however, ensure safety against stress corrosion cracking, since stress peaks and local yield processes cannot be avoided in these highly ductile steels, which, for example, are not usually heat-treated after welding. It must also be expected that, as pitting begins, the required degree of stress is reached because of the excess stress at the edges of the pits. Transgranular stress corrosion cracking can occur in the presence of halides under two conditions:

1) Below the limit stress through shifting of the corrosion potential towards the pitting potential by oxidants, whereby, depending on the corrosive medium, the increase in potential need sometimes only be small, and subsequent exceeding of the limit stress through excess stress at the location of pitting
2) Above the limit stress through the occurrence of local yield processes with the formation of local corroded spots on the steel surface with high current density and due to perforation of the surface layer by slip lines as the start of cracking

If stress-relief annealing is to be carried out on stainless austenitic steels to avoid stress corrosion cracking, temperatures above 900 °C are required because of the high strength at elevated temperatures. The use of more resistant materials can therefore be regarded as a more suitable alternative.

3.3.1.3. Nonferrous Metals

In the presence of specific corrosive media, nonferrous alloys are also prone to stress corrosion cracking. Some aluminum alloys that contain Zn and Mg, for example, are sensitive to chloride solutions and seawater, and copper alloys, especially brass, are attacked by moist ammonia, causing cracking [104]. In aluminum alloys cracking is intergranular, and the formation of the cracks is related to the degree of chemical activity of the grain boundaries, as shown in Figure 53 for alloy AlMg 9 as a function of heat treatment [105].

At a tempering temperature of 180 °C, the activity of the grain boundaries is particularly high, and the life of loop-shaped specimens in the alternating immersion test in sodium chloride solution is particularly short. In Cu–Zn alloys,

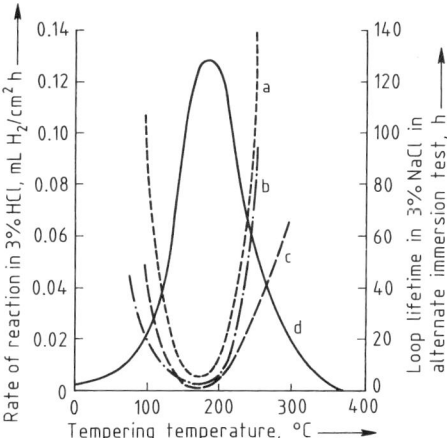

Figure 53. Stress corrosion cracking behavior (a–c) of loop samples in 3% NaCl solution (lifetime in alternate immersion test) and chemical activity (hydrogen evolution) (d) of the grain boundaries in hydrochloric acid for 1 mm aluminum alloy sheet with 9.1% Mg

cracking unter stress corrosion cracking conditions depends on the structure, intergranular cracking being prevalent in α alloys and transgranular in β alloys.

3.3.1.4. Special Types of Stress Corrosion Cracking

An uncommon type of stress corrosion, which is of importance in practice, is transgranular cracking of copper-containing stainless austenitic Cr–Ni–Mo steels in sulfuric acid. Cracking depends on the concentration and temperature of the acid and the level of mechanical stress [106]. In boiling sulfuric acid, the critical concentration range in which the transgranular stress corrosion cracking, depicted in Figure 54, occurs is 20–60%.

In addition to the iso-corrosion curves of the steel X 5 CrNiMoCuNb 18 18, Figure 55 shows the stress corrosion cracking zone. The data refer to aerated sulfuric acid as the corrosive medium. A notable feature is that cracking does not occur at all temperatures and concentrations, but only above a given corrosion rate, which under the conditions in this case is around 0.2 mm/a. In many cases copper deposits on the steel, which is due to the prior corrosion of the copper-containing solid solution.

The formation of copper-containing protective layers, which apparently exhibit local defects under the influence of tensile stress, is a prerequisite for this type of corrosion. There must therefore first be a minimum corrosion rate, as otherwise the required concentration in the solution for copper deposition cannot arise. However, if the corrosion rate is too high, a copper layer will

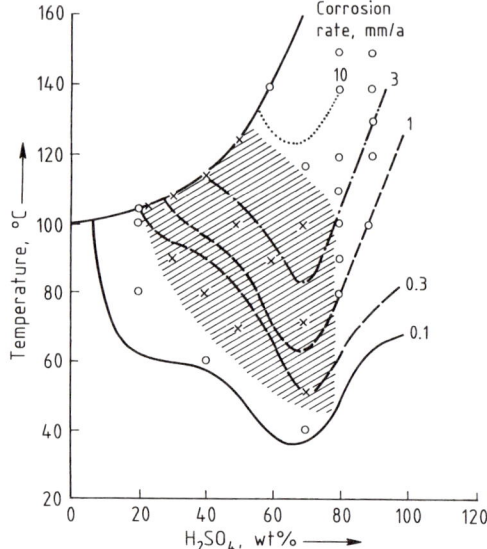

Figure 55. Iso-corrosion curves of X 5 CrNiMoNuNb 18 18 and field of SCC
▨ = Stress corrosion cracking region; o = Sample without stress corrosion cracking; × = Sample with stress corrosion cracking; ——— = Boiling curve

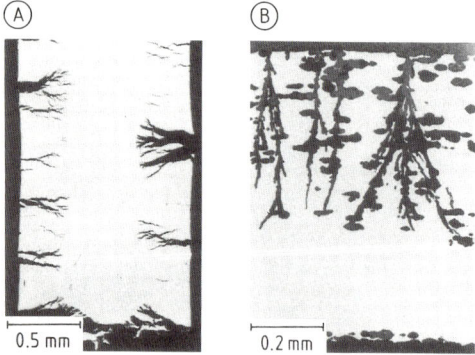

Figure 54. Transgranular stress corrosion cracking of stainless steel X 5 CrNiMoCuNb 18 18 in sulfuric acid
A) Tensile specimen, stress $0.3 \cdot \sigma_{ys}$, 50% H_2SO_4, boiling;
B) Bent-beam specimen, 30% H_2SO_4, boiling

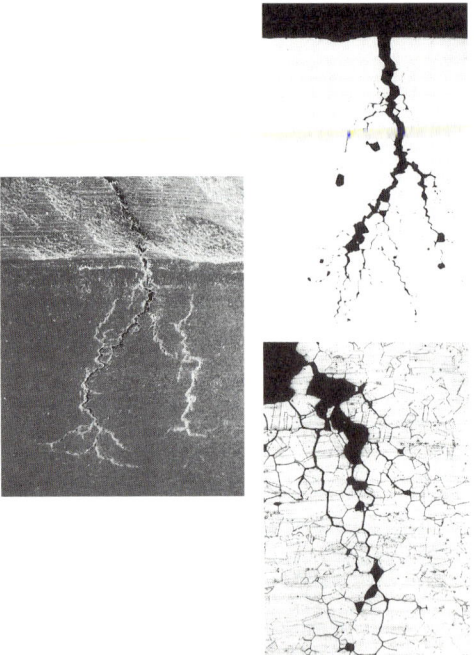

Figure 56. Intergranular stress corosion cracking of a loop sample of X 3 CrNiMoN 17 13 5 (material no. 1.4439) caused by saturated ammonium nitrate solution at 180 °C

not form to a sufficient extent and no cracking occurs. In this case, too, the susceptibility of copper-containing, austenitic stainless steels to stress corrosion cracking can be eliminated by increasing the nickel content to over 32%.

Transgranular stress corrosion cracking is induced not only in austenitic stainless steels but also in nickel alloys by aqueous alkaline solutions. Since the initiation temperature for cracking is high, often exceeding the boiling point of the alkaline solution, it occurs primarily in pressure vessels. The precise conditions are set out in Figure 51 [107]. In ammonium nitrate solutions a type of intergranular stress corrosion cracking (see Fig. 56) of high-alloy austenitic stainless steels is observed at temperatures close to the decomposition point of ammonium nitrate [108].

3.3.1.5. Hydrogen-Induced Cracking at Low Temperatures

The absorption of atomic hydrogen at temperatures below 100 °C changes the toughness and ductility of steel; the effect increases with increasing steel strength. This can be recognized by the disappearance of the distinctive yield point and the loss of reduction of area and elongation in tensile testing.

The quantity of hydrogen H_{ab} being absorbed by the metallic phase from an electrolyte solution in which hydrogen ions are discharged, corresponds only to the hydrogen gas pressure in the solution (generally 100 kPa, given undisturbed equilibrium between H_{ad} and H_2 gas):

$$H_{aq}^+ \rightarrow (H_{aq}^+)_{surface} \xrightarrow[e-]{reduction} H_{ad} \begin{array}{l} \nearrow H_{ad} + H_{aq}^+ + e^- \rightarrow H_2 \uparrow \\ \rightarrow 2\,H_{ad} \rightarrow H_2 \uparrow \\ \searrow H_{ab} \end{array}$$

Only when the steps involving recombination of atomic hydrogen are inhibited can an increased H_{ad} concentration and hence a higher value for H_{ab} occur. Some substances result in a particularly high retention of atomic hydrogen on a steel surface in corrosion reactions by inhibiting recombination. They are known as promotors or poisons, and include compounds of S, P, As, Se, Te, and Sb (e.g., thiocyanates and thiosulfates) as well as CO and cyanides. They cause a considerable increase in the concentration of hydrogen absorbed in the metal, which formally corresponds to an increased hydrogen pressure in the gas phase. This can give rise to considerable embrittlement of steel even under normal pressure.

Apart from AsH_3, COS, HCN, and CO, above all H_2S is responsible for many cases of damage, in particular in the chemical, petroleum, petrochemical, and electroplating industries.

In mild steels, blisters, cracks, and terraced fractures (caused by slag and sulfide lines) can occur under sufficiently high hydrogen uptake and static tensile stress below the yield point at low temperature (e.g., room temperature): this results from the formation of high pressure hydrogen in structural defects as shown in Figure 57 [109]. Residual stress is often sufficient to generate local plastic deformation and initiate cracking.

Cracking occurs far more readily and rapidly in intermediate-strength steels. In cold parts of high-pressure petrochemical plants, cracks are often found in shaped components made of heat-treatable steel through the attack of condensates containing H_2S. The same applies to equipment for drilling and extracting natural gas [110].

Hydrogen-induced embrittlement in high-strength steels is less dependent on the occurrence of local internal hydrogen pressure zones, because they cannot induce local plastic deformation in such steels. It depends on the interac-

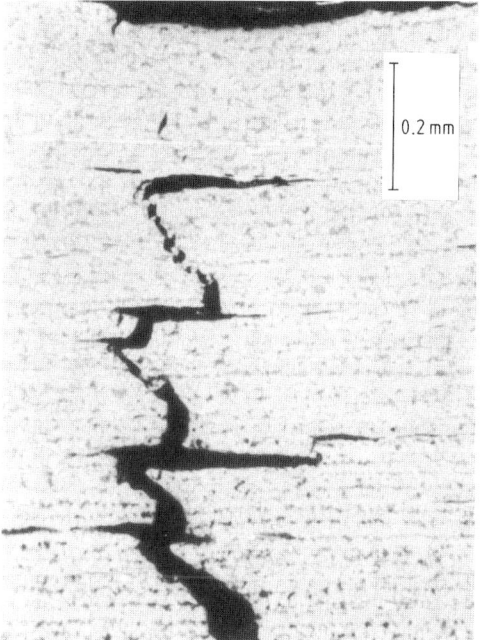

Figure 57. Terrace-shaped fractures in a natural gas pipeline of unalloyed steel (X 42 API Std 5 LX) caused by condensate containing H_2S

tion between dissolved and mobile hydrogen and stress fields, precipitation regions, and crack tips during a local plastic deformation process. The proneness to embrittlement by plastic deformation under the influence of low-pressure molecular hydrogen was reported by HOFMANN et al. [111].

With pure molecular hydrogen, atomic hydrogen can originate on places of plastic deformation (notches, scratches) by interaction between molecules and the resulting active surface of steels. The absorbed hydrogen causes crack initiation and propagation. This occurs particularly under low-cycle fatigue conditions. Damage occurs, e.g., at inner die marks in high-pressure gas cylinders for hydrogen transport that are often filled and emptied [112], [113].

A purely optical differentiation between hydrogen-induced cracks and cracks produced by anodic stress corrosion cracking is often extremely difficult. Figure 58 shows a case that occurred in a spiral-welded natural gas pipeline made of St 52.3 as a result of condensate containing H_2S and CO_2 in the region of the weld and in a neighboring stress zone. The observation that weld seam areas and cold-formed zones are particularly prone to cracking is a general one. The proneness to cracking of welded equipment in the presence of nascent hydrogen can usually be eliminated by stress-relieving annealing.

A striking feature is provided by the "fish eyes" sometimes seen in fractures, indicating that hydrogen was involved in the bursting process (see Fig. 59 A). They arise when the yield limits are exceeded at the edges of hydrogen-filled cavities through diffusion of atomic hydrogen into the surrounding area, leaving a confined, speck-shaped brittle fracture with a bright fracture surface. In this process the hydrogen dissociates at the plastically deforming cavity wall.

An example of the tremendous effect hydrogen-induced cracking can have on high-strength steel is the explosion of a light steel bottle (40 L volume) upon filling (Fig. 59). It was made of a low-alloy heat-treatable manganese steel with a yield strength of 840 MPa and a tensile strength

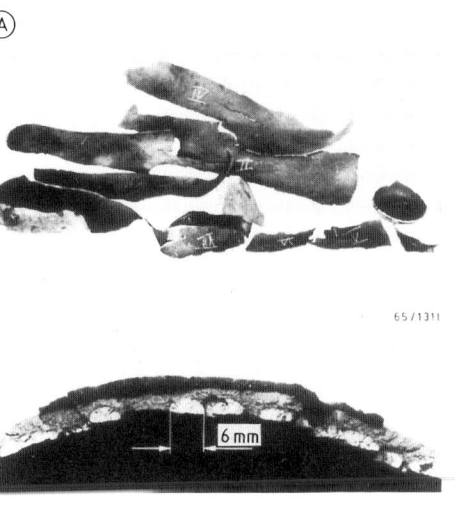

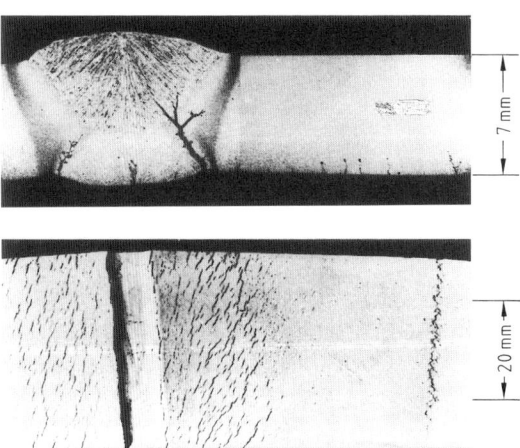

Figure 58. Hydrogen-induced crack formation in a spiral-welded natural gas pipeline (made of St 52.3) caused by condensate containing H_2S and CO_2

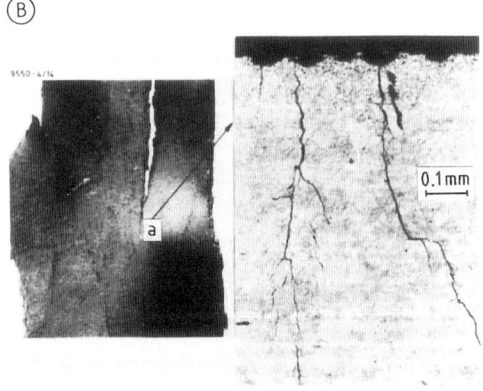

Figure 59. An exploded carbon monoxide lightweight steel bottle made of a heat-treatable manganese steel ($\sigma_{ys} = 840$ MPa)
A) General view and fracture surface; B) Fragment showing course of crack: a) Micrograph

of 950 MPa. The bottle had been in use for 2 ½ years, filled with technically pure CO up to 20 MPa [107], [114]. In addition to the presence of CO as promotor, a major part in this damage was played by the existence of a corrosive water sump in the bottle since the reaction leading to the formation of atomic hydrogen is electrolytic. Numerous cracks were found in the area covered with water.

Just how important the absolute absence of moisture films is, especially in bottles made of high-strength steel, is shown by the damage in the form of hydrogen-induced stress corrosion cracking caused after several years of being filled with pure CO_2 (8–10 MPa) to a number of fire extinguisher bottles, which burst. The initiator and source of hydrogen was the weakly acidic carbonic acid formed through the presence of water, which acted on the mostly horizontally-positioned bottles along the inner surface of the cylinder and caused cracking [114]. Ultrahigh-strength steels are so prone to hydrogen-induced stress corrosion cracking that they can suffer brittle fracture under the effect of atmospheric humidity.

If there is a danger of electrolyte-induced atomic hydrogen formation and it is not possible to eliminate the electrolyte (e.g., by drying), the use of steels as soft as possible and careful stress-relieving annealing of the components should be considered. Since austenitic Cr–Ni steels in standard supply form and after normal processing are practically unaffected by hydrogen they can be used in difficult cases.

Because of their tendency to hydride formation, the embrittlement of titanium, zirconium, and tantalum by hydrogen is particularly pronounced. The proneness to gas absorption restricts the use of these metals. They are not only attacked by nascent hydrogen, they also become brittle at elevated temperatures in gases containing nitrogen and oxygen. In addition to uniform corrosion, these metals also show hydrogen embrittlement in corrosion reactions in acidic solutions (Fig. 60). In the case of tantalum, hydrogen embrittlement can occur after long periods in use, even when the corrosion rate is practically negligible (e.g., several thousandths of a millimeter per year) [115].

3.3.1.6. Hydrogen Corrosion at Elevated Temperatures

The effect of pressurized molecular hydrogen on steel must be borne in mind in a number of high-pressure syntheses, e.g., in high-pressure hydrogenation and ammonia production. At sufficiently elevated temperatures (> 230 °C) pressurized hydrogen decarburizes carbon steels. The results of this process are embrittlement, poorer mechanical properties, grain-boundary loosening and a splitting effect at defect sites caused by reaction products. As a consequence, brittle fractures occur (Fig. 61).

Figure 62 shows decarburization and grain loosening in the microstructure of an unalloyed steel as a result of pressurized hydrogen attack. The visible changes within the corroded areas

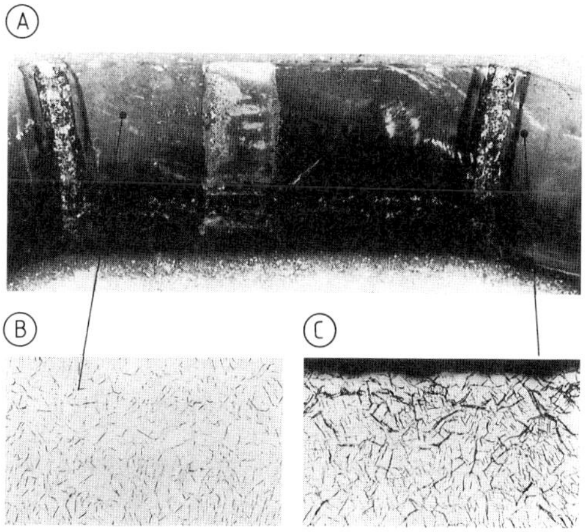

Figure 60. Corrosion of a weld of a titanium pipe
A) General view; B) and C) Embrittlement as the result of hydrogen absorption (structure exhibiting hydride needles); bending angle up to fracture 12° (A) and 4° (B)

Figure 61. Brittle fracture of a fitting (T-piece) as a result of pressurized hydrogen attack

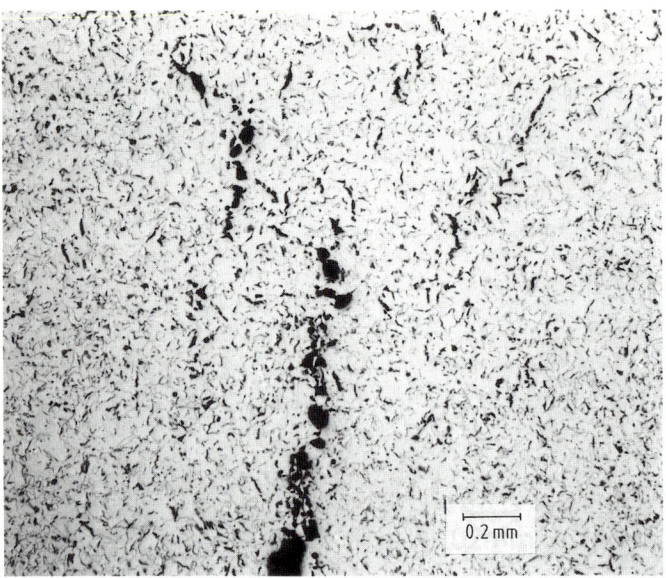

Figure 62. Damage to a St 35.8 pipe (NH_3 plant) by pressurized hydrogen

cause significant changes in mechanical properties. In the decarburized zone, the tensile strength, elongation, reduction in area, and hardness drop to very low levels as the carbon content decreases.

Even the first observations of damage of this kind in the development of ammonia synthesis showed that the effect of pressurized hydrogen on steel is both chemical and physical [116]. The physical effect presumably takes place in different stages. Since molecular hydrogen does not diffuse in the steel, the first step must be surface adsorption with dissociation at elevated temperature. In this process the metallic wall acts as a catalyst. With pure, dry pressurized hydrogen, for example, dissociation produced by yielding phenomena on the steel surface can supply critical quantities of diffusible hydrogen even at room temperature.

The next step after formation of adsorbed hydrogen is absorption of the hydrogen. The ensuing chemical effect of hydrogen on steel is marked by decarburization of pearlite and formation of hydrocarbons, usually methane:

$$Fe_3C + 2 H_2 \longrightarrow 3 Fe + CH_4$$

The reaction generally occurs in areas adjacent to the grain boundaries, with the grains losing their cohesion. In addition to methane, recombined hydrogen molecules accumulate at the loosened points producing a high pressure, which leads initially to internal microcracking. Component stress and lowered mechanical properties finally combine to produce marked brittle fracture.

Apart from the ambient conditions, the resistance of carbon steels to pressurized hydrogen attack depends on the grain size, cold forming, welding effects, impurities, and heat treatment. To prevent the reaction of hydrogen with pearlite, carbon-bonding elements such as chromium, molybdenum, tungsten, vanadium, titanium, and niobium, whose carbides are considerably more stable than cementite, can be added to the steel. These alloying elements considerably increase the resistance of steels to hydrogen attack [117].

Steels resistant to pressurized hydrogen that are now in use contain 2–6% chromium and often a further 0.2–0.6% molybdenum. Some types are additionally alloyed with vanadium up to 0.85% and tungsten up to 0.45%. Short-term testing gives little information on where the limits of application of such steels lie, especially since corrosion by pressurized hydrogen normally starts after a considerable incubation period. However, experience gained with a wide variety of installations does provide reliable indicators. Figure 63, for example, shows the resistance lim-

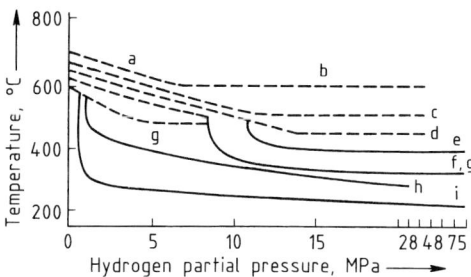

Figure 63. Resistance diagram for the attack of pressurized hydrogen on steels (Nelson diagram) [118]
a) 1.25 Cr 0.5 Mo steel; b) 5.0 Cr 0.5 Mo steel; c) 8.0 Cr 0.5 Mo steel; d) 2.25 Cr 1.0 Mo steel; e) 2.0 Cr 0.5 Mo steel; f) 1.25 Cr 0.5 Mo steel; g) 1.0 Cr 0.5 Mo steel; h) 0.5 Mo steel; i) Mild steel

its for various steels [118]. Even at low hydrogen partial pressures (e.g., 2 MPa) a carbon steel should not be used above a temperature of 300 °C; small amounts of molybdenum are sufficient to increase resistance appreciably. Austenitic steels with 18% chromium and ca. 10% nickel possess good resistance to pressurized hydrogen attack and can be used with hydrogen over the the entire temperature range of common high-pressure processes.

3.3.2. Corrosion Fatigue

If in addition to alternating mechanical stresses a component is exposed to a constant corrosive effect through the environment, the resulting damage is called corrosion fatigue. The Wöhler curve describes the fatigue behavior of a metallic material without corrosion. The corrosive influence shifts the area of fatigue strength for finite life ($< 10^7$ load cycles) to lower numbers of cycles and shows in addition, as depicted in Figure 64, no further fatigue strength [119]. Accordingly, only a fatigue strength for finite life depending on different parameters can be achieved under a load combination of this kind.

The corrosion process itself is localized by slip processes at the surface generated by the mechanical stress. The corrosive effect can also be recognized from the corrosion products that often deposit in cracks. These are found extensively at the crack beginning, the area of slower crack propagation, but are not much in evidence at the crack tip, the area of mostly rapid propagation. Extensively corroded cracks indicate a high number of cycles to failure, low load amplitude, and severe corrosion. The component finally breaks when the remaining cross section is stressed to such an extent by the continuing corrosion fatigue that it cracks through spontaneous rupture. As an example Figure 65 shows damage to a pipe elbow made of 13 CrMo 4 4 steel caused by alternating thermal stress with simultaneous exposure to steam (low frequency, long service period, marked corrosion).

In addition to the metallographic examination, microfractography can also be used to identify the cause of damage since in areas where corrosion of the rupture surface and deposition are not too extensive, striations are often found in corrosion fatigue (Fig. 66), indicating the action of alternating stress [120]. The decision as to whether pure fatigue failure or corrosion fatigue is involved can be facilitated by the appear-

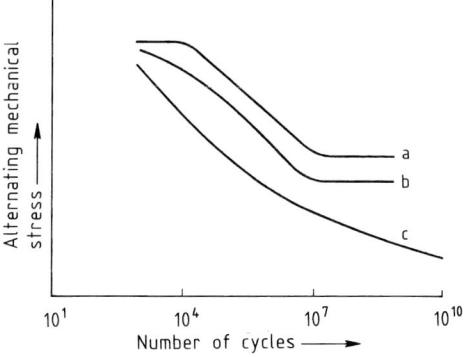

Figure 64. Wöhler curves, in air, after precorrosion, and under corrosion fatigue conditions
a) Fatigue strength in air; b) Fatigue strength after precorrosion; c) Corrosion fatigue

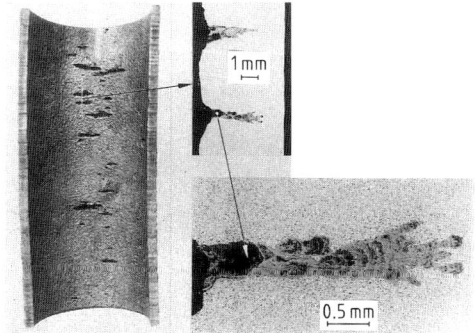

Figure 65. Cracks in a pipe elbow made of 13 CrMo 4 4 caused by alternating thermal stress and by corrosion
Conditions: 340 °C, steam, 10^5 h (11.4a)

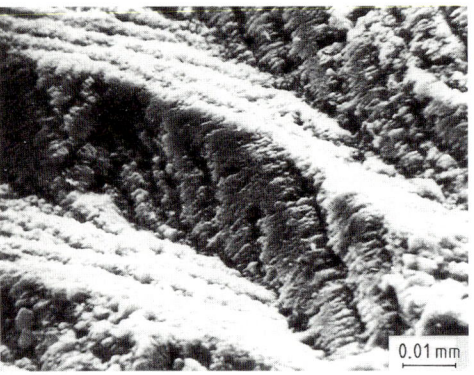

Figure 66. Fatigue striations on the fracture surface of an oil cooler pipe made of brass SoMs 76, cracked by corrosion fatigue due to river water

ance of corrosion phenomena or products on the rupture surface.

Electromechanical methods have also been employed to study the mechanism of corrosion fatigue [121]. If the potential changes occurring at the surface of the specimen during corrosion fatigue are recorded, three stages of corrosion are found for steels with an active surface.

In the *preliminary stage*, potential drops in the negative direction are observed, the cause of which can be seen as the appearance of slip lines on the specimen surface with increased activity, i.e., increased rate of reaction with the electrolyte.

In the *second stage*, these local anodes increase in number and area, forming notches with high mechanical stress and hence high dissolution, which is marked by a further denobling of the potential.

In the *final stage*, interaction of localized corrosion and increased notching finally leads to cracking and thus to the spontaneous rupture of the remaining cross section.

If corrosion fatigue of unalloyed and low-alloy steels is allowed to occur potentiostatically and the changes in current are measured as corrosion progresses, it is seen that marked current peaks only occur at the stage of rapid crack propagation [122].

In passive steels the current rises only after local activation, at which point the cracking stage begins immediately. This is because only one crack forms in the otherwise passive surface, causing extreme localization of the corrosion process and hence a correspondingly high current density at the crack base, which explains the extraordinarily rapid crack propagation. Stainless steels can of course become fully active in appropriate corrosive media, whereupon their corrosion fatigue behavior is similar to that of active steels. Whereas a single smooth break generally occurs in the passive state, without any further signs of corrosion attack [123], the indicators in corrosion fatigue in the active state are numerous cracks, a fissured break due to the converging of various crack levels, and general corrosion attack. A comparison of active and passive corrosion fatigue is shown in Figure 67 [124].

Since a frequently repeated minimum stress amplitude is needed to initiate corrosion fatigue in stainless steels in passivating solutions, it must be assumed that this load limit is connected with the mechanical stress capacity of passive layers. Only when particularly marked slip starts at one point on the surface will the layer be cracked. At that point a constantly repetitive process begins, in which a new activation process at the same spot always follows repassivation. Each new passivation process consumes metal, deepens the corrosion, and increases the stress peak until, because of the constantly rising stress, repassivation is no longer possible. The resistance of passive metallic materials under fatigue conditions in electrolytes is therefore largely dependent on three factors:

1) the extent of the load amplitude in relation to the mechanical strength of the material and the passive layer;

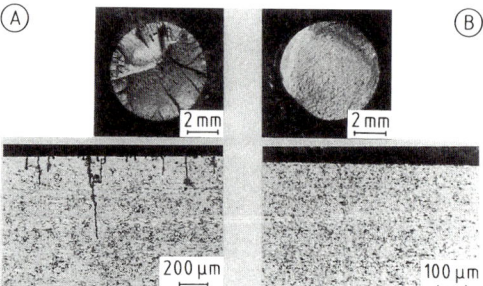

Figure 67. Active and passive corrosion fatigue
A) Active state: severe corrosive attack, fissuring of the fracture surface, several cracks; B) Passive state: no evidence of corrosive attack, smooth fracture surface, no secondary cracks

2) the passivation properties of the electrolyte; and
3) the passivation tendency of the material, as determined by its alloying elements.

The mechanical properties of the alloy determine the occurrence of slip. The higher the fatigue strength in air the greater the stress amplitude needed to initiate sufficiently extensive slip, i.e., the higher the resistance to corrosion fatigue in the passive state.

In the case of "protective layers" the mechanical strength and the adhesion of these layers and their healing capacity in corrosive media are of special significance in corrosion fatigue. For thicker layers, which themselves are subject to fatigue, a special Wöhler curve applies, whose path is determined by the fatigue strength of the layer if it is lower than that of the base material (Fig. 68). If the operating stress is below the fatigue strength of the material but above that of the protective layer, the layer cracks at the points of highest stress, and local corrosion begins. The corrosion that forms provides the starting point for subsequent cracks. Corrosion fatigue of metallic materials with protective layers is therefore preceded by a period of strain-induced corrosion, formerly known as stress-induced corrosion [125].

To increase the resistance to corrosion fatigue, the same steps can be taken that give high structural strength (good surface, no notches or grooves, avoidance of stress peaks through good geometrical design and processing). However, to these are added the increased corrosion resistance of the materials used, since greater chemical resistance plays a decisive role in achieving longer life.

An especially noteworthy feature is a process known as low-cycle corrosion fatigue, in which a stress producing constantly repetitive deformation amplitudes and corrosion cause damage after only a low alternating load cycle. The interaction of plastic alternating strain and a corrosive electrolyte solution leads to rupture, especially in components with notches or fissures, after only a

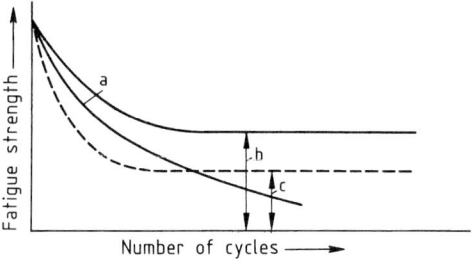

Figure 68. Cracking of a protective layer as a precondition for corrosion fatigue (strain-induced corrosion)
a) Corrosion fatigue; b) Fatigue strength in air; c) Fatigue strength determined by the protective layer

Table 5. Types of corrosion cracking [126]

Type of cracking	Stress corrosion cracking		Corrosion fatigue		Strain-induced corrosion cracking[a]
	Classical (stress-induced)	Nonclassical (strain-induced)	Active state	Passive state	
Type of stress	constant elongation (strain rate $\dot{\varepsilon} = 0$)	continuous strain ($\dot{\varepsilon} > 0$)	pulsating strain		alternating elongation (with nonstatic load)
Specimen and test method	smooth clamped specimen (constant elongation)	notched tensile test specimen (constant load)	smooth tensile specimen, pulsating or oscillating test		any form of specimen, alternating load (mechanical and chemical)
		smooth tensile test specimen (slow strain rate)			
Corrosion system	critical system parameter values		no critical system parameter values only in systems with film formation		
Characteristics of cracking	inter- or transgranular		mostly transgranular		transgranular
Number of cracks	often many cracks		many cracks	few single cracks	few cracks

[a] Formerly known as stress-induced corrosion

small number of load cycles if the corrosion is long-acting due to the low cycle frequency, especially in the tensile phase (slow change in load). Embrittlement and crack propagation also be caused under these conditions by gases (e.g., H_2) and media that show hardly any corrosive effect under static load (see Section 3.3.1.5).

3.3.3. Differentiation Between Types of Corrosion Cracking

The term stress corrosion cracking often refers to very different types of cracking under corrosive attack. A clear distinction must be made between their corrosion mechanisms. The different types of corrosion cracking are set out in Table 5. The distinction between the types of corrosion cracking is made difficult by overlapping. This overlapping is found in particular between nonclassical stress corrosion cracking and corrosion fatigue and between corrosion fatigue in the passive state and strain-induced corrosion cracking.

In the more recent literature the term *strain-induced corrosion* is commonly found, which is a generic term for nonclassical stress corrosion cracking, corrosion fatigue with low cycle frequency, and strain-induced corrosion cracking. For the purposes of a clear distinction between the different types of cracking this term should be replaced by the following:

1) strain-induced stress corrosion cracking for nonclassical stress corrosion;
2) low cycle corrosion fatigue for strain-induced fatigue corrosion with low-frequency load cycles on materials and protective layers; and
3) strain-induced corrosion cracking for corrosion with cracking of materials with protective layers as a result of layer damage on elongation change and corrosion during shutdown periods (non-stationary operation).

4. High-Temperature Corrosion

4.1. Principles of High-Temperature Corrosion

4.1.1. Thermodynamic Principles

The reaction of a metal M in normal state (100 kPa) with an oxidizing gas X_2 produces a solid reaction product MX, which usually forms a layer on the surface ($M/MX/X_2$). The free enthalpy of such a reaction is negative for practically all metal oxides, i.e., oxides are stable in atmospheres containing oxygen. Metals therefore are not stable, and oxidation or corrosion occurs always, although it is not visible in many cases because of the low reaction rate.

With rising temperature the free enthalpy approaches zero. This means that at a given temperature, equilibrium exists between metal, metal oxide, and oxygen (100 kPa pressure). All phases are stable next to one another, i.e., under the prevailing conditions the decomposition pressure of the metal oxide is equal to the oxygen pressure of the surrounding atmosphere. At lower oxygen pressures the compound decomposes into its elements.

At room temperature, therefore, practically all metals and metal alloys—with the exception of gold, whose oxide decomposition pressure is greater than the oxygen partial pressure of air—are unstable in air and form surface oxide films. At higher temperatures, all metal compounds decompose in principle. This is also the case at moderate temperatures for the oxides of silver, mercury, platinum, and palladium, whose oxygen partial pressures (decomposition pressures) between 140 °C and ca. 800 °C exceed the oxygen pressure of air. At elevated temperatures, however, this also applies to the oxides of copper, iron, and other less noble metals. For most oxides, however, the oxygen partial pressures required for dissociation are too low to be achieved experimentally, or else the decomposition temperatures are above the boiling points of the metals or metal compounds.

The decomposition pressure of a metal oxide p_{O_2}, i.e., the oxygen partial pressure of the surrounding atmosphere at which the oxide is just stable at a given temperature, can be calculated from Equation (13):

$$G_0 = RT \ln (1/p_{O_2}) = 4.574\, T \log p_{O_2} \qquad (13)$$

Equilibrium values (e.g., decomposition pressures) therefore determine the existence of compounds in metal oxidation, but the rate at which they form is a kinetic problem. Because of the formation of layers of low porosity that separate metal and gaseous phase, the reaction rates of metals vary appreciably. Corrosion resistance in metals therefore means that the oxidation rates under the conditions in which they are used in practice are low.

4.1.2. Kinetics of Scaling

When a clean metal surface is subjected to attack from an oxidizing gas X_2, e.g., oxygen, nitrogen, or chlorine, an oxidized film forms between gas and metal, provided that the reaction product is not fragile or volatile and the oxygen is not completely soluble in the structure. This film is usually pore-free and gas-tight, and separates the reaction partners M and X_2. However, the reaction does not come to a standstill at elevated temperatures. Thus, at least one of the reaction partners must diffuse through the MX film. The mechanism and the rate of diffusion of the reaction partners through the reaction products are decisive factors in the reaction of metals with gases. They are closely linked with imperfections in the MX layer.

All crystalline solids exhibit defects in and departure from the ideal lattice structure, particularly at elevated temperature. Either lattice points remain unoccupied (vacancies) or lattice elements deposit between the regular lattice points (interstitial lattice points). These point defects determine material transport in a solid. In addition, there are a number of linear and face defects (dislocations, grain boundaries, etc.), which although of importance to the mechanical properties of a solid are less significant for material transport.

When metals react with gases, the main corrosion products are ionic compounds which can be stoichiometric or nonstoichiometric. Generally, only defect ions (ion conductors) arise in stoichiometric compounds (e.g., AgCl, NaCl). Four border cases of imperfections are possible: When cation vacancies in the lattice and cations at interstitial lattice sites are found in an undisturbed anion lattice, the cations are mobile. In the opposite case the anions are mobile. In compounds with anion and cation vacancies both can migrate, as they can when an equal number of cations and anions are present at interstitial lattice sites.

Nonstoichiometric compounds show electron defects in addition to ion defects because of electroneutrality. In chemical terms this can be equated to the occurrence of ions of different valency (e.g., Fe^{3+} in FeO). Since the electrons are more mobile than ions by a factor of $10^3 - 10^6$, these compounds are practically pure electron conductors (semiconductors). They can be divided into electron-surplus conductors, and electron-deficient conductors, according to whether the electric current is transported by free electrons or by defect electrons.

The mechanism of metal oxidation can be divided into an initial phase and a phase of diffusion-controlled scaling. During the initial phase, reaction steps occur that are largely controlled by processes taking place at the phase interfaces, such as:

1) cleavage of the molecules of the oxidizing medium into atoms and chemisorption;
2) formation of a nonepitactic, polycrystalline, primary oxide film; and
3) occurrence of monocrystalline, epitactic nuclei on the metal surface and the spread of these nuclei through lateral propagation.

These steps of the first reaction phase take place rapidly at elevated temperatures and high partial pressures of the oxidizing medium with

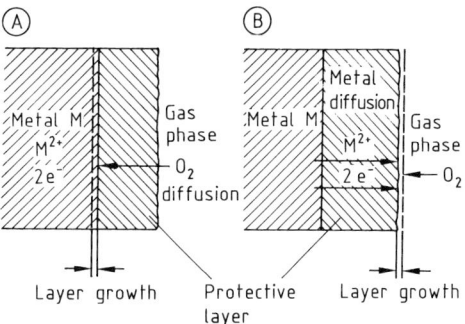

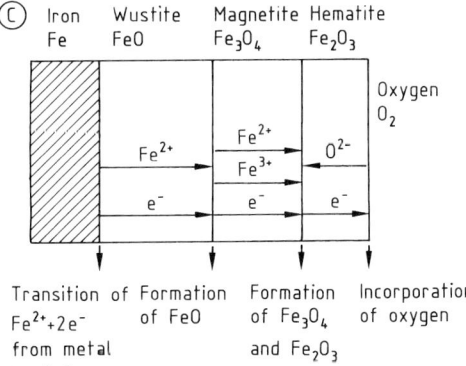

Figure 69. Transport processes and layer growth
A) Layer growth with preferential oxygen diffusion; B) Layer growth with preferential metal diffusion; C) Diffusion processes and phase interface reactions for the oxidation of iron at $> 570\,°C$

formation of protective films. After the formation of thin layers, a protective film propagates, as is observed in the scaling of Fe, Ni, and Co at moderate and elevated temperatures, in the form of a dense layer. The reaction can only then continue if at least one of the reactants can diffuse through the scale. At elevated temperatures, positive partial pressures of the scaling atmosphere, and a certain layer thickness, the phase boundary reactions are generally rapid. Therefore, transport processes such as the diffusion of oxygen to the metal or the transport of metal through the initial layer or scale, become the slowest and hence the rate determining process (Fig. 69). After layer formation the oxidation reaction thus continues by way of diffusion of the metal or nonmetal in the form of ions, electrons, or defect electrons through the film.

Time Laws. For the formation of very thin films (about 10 nm) at low temperature, logarithmic, reciprocal-logarithmic or cubic time laws can be deduced. If diffusion determines the rate of formation of thick and compact scale, a parabolic time law applies

$$\frac{dx}{dt} = \frac{k'}{x} \tag{14}$$

which in integrated form reads:

$$x^2 = 2k't \tag{15}$$

where x is the thickness of the scale, t is the reaction time, and k' is the scale constant.

If the rate is determined by phase boundary reactions a linear law applies

$$dx/dt = k \quad \text{or} \quad x = kt \tag{16}$$

where k is a constant. Knowledge of the rate-determining step is needed if the oxidation rate is to be influenced. This can be determined from detailed kinetic measurements. As the scale increases in thickness, however, diffusion in the scale will at some point become the rate-determining partial step.

The mechanism of scale formation also determines whether a crack in the film surface layer can mend or not. If metal ions and electrons migrate outwards, the oxide film propagates at the oxide–atmosphere interface and the crack quickly closes. But when oxygen ions migrate inwards the oxide film propagates at the metal–oxide interface and the crack cannot close.

The mechanical properties of the oxide film, in particular the formability and state of stress, determine the occurrence of cracks. If the formability of the film is poor the contact between metal and oxide is lost as long as ion migration of the metal is from the inside to the outside. Oxygen migration from the outside to the metal–oxide interface leads to formation of a new oxide scale, which propagates into the metal.

Wagner's Theory of Metal Oxidation. Assuming that only ions and electrons migrate in the scale and not neutral atoms, CARL WAGNER established a formula in the 1930s whereby the parabolic scale constant of a pure metal can be calculated from the free formation enthalpy of the corrosion product, the electrical conductivity of the protective layer, and the transport numbers of cations, anions, and electrons in the film [127]:

$$k = \frac{300}{96\,500\,N \cdot \varepsilon} \int_{p_{x_2}^{(i)}}^{p_{x_2}^{(a)}} \frac{1}{Z_2} (U_1 + U_2) U_3 \, x \, d\ln p_{x_2} \tag{17}$$

where:

k = rational scale constant in $\text{mol cm}^{-1}\,\text{s}^{-1}$
N = Loschmidt's number
ε = elementary charge
Z_2 = charge of the nonmetal in the protective film
U = transport number with index 1 for cation, 2 for anion and 3 for electron $(U_1 + U_2 + U_3 = 1)$
x = electrical conductivity in $\Omega^{-1}\cdot\text{cm}^{-1}$
p = partial pressure of the nonmetal X_2 with index i at the interface M/MX and index a at the interface MX/X_2.

The rate of oxidation with rate-determining diffusion in the scale is calculated from this equation. This is the maximum oxidation rate for a gas-tight protective layer. The equation applies for the oxidation, sulfurization, nitriding, and halogenation of a metal ($X_2 = O_2, S_2, N_2$, halogen). Two border cases can be differentiated with the equation. For $(U_1 + U_2) \simeq 1$ (ion conductors), k is determined by the parameter U_3, the electron transport number. For $U_3 \simeq 1$ (electron conductors), k is determined by $(U_1 + U_2)$.

4.2. Oxidation of Alloys

Whereas the oxidation rate of pure metals for the case of rate-determining diffusion of ions and electrons in the protective film can be predicted, this is not possible for the oxidation of alloys. It will probably never be possible to predict the oxidation rate of alloys because too many effects can overlap. The following effects must be considered in the oxidation of alloys:

1) the influence of the base metal,
2) composition of the alloy,
3) formation enthalpy of the oxides, sulfides, and halides in question,
4) effect of dissolved oxides of the alloy components on the disorder of protective oxide or sulfide films,
5) formation of mixed oxides,
6) formation of ternary oxides or sulfides,
7) diffusion within the alloy,
8) solubility and diffusion of oxygen in the metal phase (internal oxidation), and
9) relationship between the formation rate of different oxides.

Adding more noble metals has little influence on the oxidation of the base metal. The more noble metal will accumulate in metallic form, e.g., Cu in steel (danger of red shortness), at the alloy–scale interface. Higher contents can lead to a marked retardation of oxidation because diffusion within the alloy could become the slowest partial step.

If the oxidation rate of a metal is to be changed appreciably this can only be achieved by adding less noble metals. In this case the oxide of the alloying element is more stable than that of the base metal. The oxidation rate will be lower if the concentration is high enough for the formation of a sealed oxide film, and the diffusion rate of the ions and electrons in this oxide is lower than in the oxide of the base metal. Examples of such oxides are Cr_2O_3, Al_2O_3, and SiO_2. The metals of these oxides provide the scale resistance of heat-resistant alloys. The metals Ag, Cu, Ni, and Fe can dissolve oxygen. If these metals contain small amounts of less noble metals (Al, Cr, Si, Ti) these can be oxidized within the metal phase (internal oxidation). Internal oxidation does not occur as long as the concentration of the alloying elements is so high that a sealed oxide film of the alloying element forms on the surface.

4.3. Sulfurization of Metals and Alloys

The reaction of metals and alloys with sulfur is basically identical to oxidation with oxygen. Thus, Wagner's formulae are also applicable here. However, the sulfurization rate of most metals is far higher than the oxidation rate in air or oxygen because of the following:

1) Most sulfides show far greater disorder than the corresponding oxides. Therefore, substance transport in them is greater.
2) Some sulfides form relatively low-melting eutectics with metals. The protective effect of the film is thereby lost. Examples of this are:

Fe–FeS   mp 965 °C
FeO–FeS   mp 940 °C
Fe–FeO–FeS   mp 925 °C
Co–Co_4S_3 mp 880 °C
Ni–Ni_2S_3 mp 645 °C

The low melting point of the Ni–Ni_2S_3 eutectic is the reason for the high sensitivitiy of high-nickel materials to sulfur. The resistance to attack by sulfur can be improved by adding chromium. However, the protective effect of chromium is not as high for attack by S_2 as it is for attack by O_2 (see Figs. 70 and 71) [128], [129].

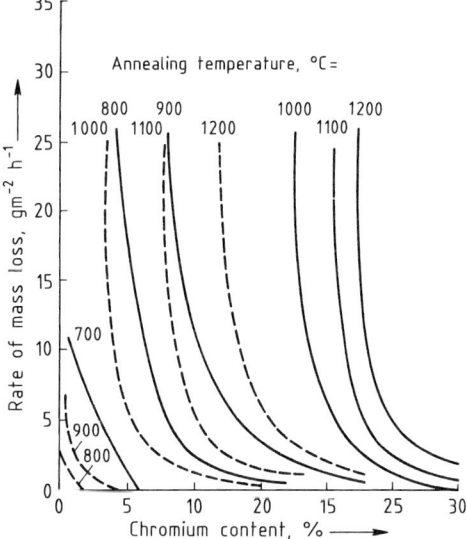

Figure 70. Influence of the silicon and chromium contents on the scaling resistance of steel during annealing in air for 120 h
—— 0.5–1.0 % Si; --- 2.0–3.0 % Si

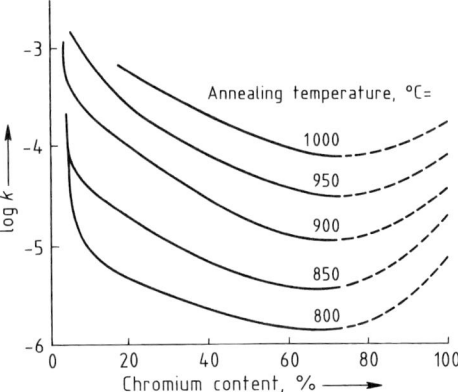

Figure 71. Dependence of the sulfurization constant k of steels on the chromium content in the temperature range 800–1000 °C

$$k = \left(\frac{\Delta m}{A}\right)^2 \cdot \frac{1}{t}$$

Δm = Mass increase, g; A = Surface area, m²; t = time, min

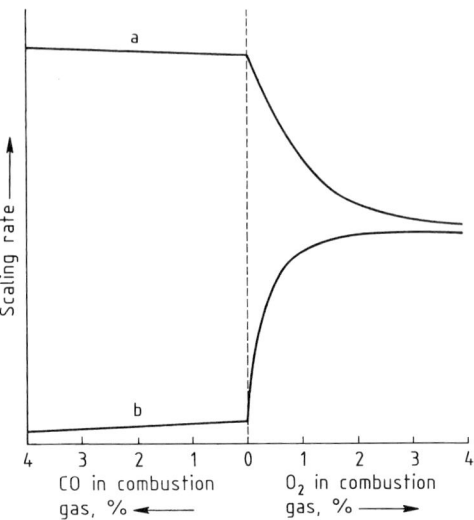

Figure 72. Scaling of soft steel at 850 °C in combustion gases from sulfur-containing and sulfur-free fuel
a) 0.15% SO_2; b) SO_2-free

4.4. Corrosion by Combustion Gases [130]

The effect of combustion gases from fossil fuels on metals depends on whether

1) the fuel has been burnt completely (air surplus) or not (air deficiency),
2) the fuel contains sulfur, and
3) the combustion gases contain ash or dust particles.

The effect on the scaling behavior of steel of air surplus and deficiency from sulfur-containing fuels and sulfur-free fuels is shown schematically in Figure 72. In sulfur-free gases, oxidation with surplus air is somewhat greater than in incompletely burnt gases. In gases containing sulfur the situation is reversed. In combustion gases containing oxygen, the sulfur content of the gas has practically no influence, whereas the scaling rate increases in incompletely burnt gases [131]. The behavior depicted in Figure 71 is observed in many cases. To understand this behavior, thermodynamic and kinetic considerations are necessary.

The isothermal section through the Fe–O–S system (Fig. 73) shows how the stability of the individual phases changes as a function of the oxygen and sulfur pressure. For a given sulfur content of a gas, the oxides are stable at high

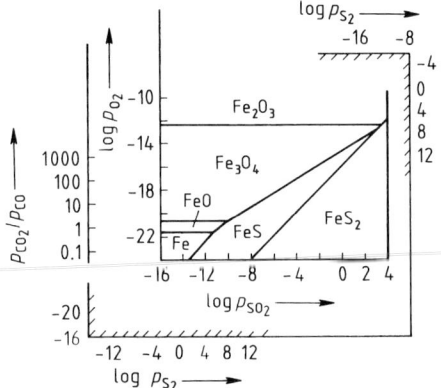

Figure 73. Isothermal section through the Fe–O–S system at 700 °C

oxygen pressure; at low oxygen pressure, however, the sulfides are stable. That is why oxides are stable in fully burnt gases containing sulfur, and sulfides are stable in incompletely burnt gases because of their very low oxygen pressure. These thermodynamic considerations alone, however, are not sufficient to avoid sulfide formation in gases containing sulfur and oxygen. Kinetic measurements show that where oxides exist, both oxide and sulfide form simultaneously if scaling does not follow a parabolic time law, i.e., if the diffusion of ions and electrons in the oxide film is not the slowest partial step [130].

4.5. Effect of Ash, Deposits, and Salts

Many corrosion problems arise through deposition of ash and dust on the metal surface. These deposits can react at elevated temperatures, even in the solid state, with the protective films of metals and alloys to form new compounds with different transport properties. However, corrosion becomes particularly intense when fusible phases form because either the deposits contain components with relatively low melting points, or low-melting eutectics arise between the deposits and protective films. A well-known phenomenon is oil ash containing vanadium (Fig. 74) [132], which in the presence of surplus oxygen creates a critical oxidation rate. Other oxides such as PbO, MoO_3, and B_2O_3 can also form low-melting eutectics. Alkali-metal sulfates are also often responsible for increased corrosion (Fig. 75) [133]. The reactions are extremely complex. Since most salts and oxides in molten state are dissociated into ions, electrochemical processes, similar to those in aqueous solutions, take place. These reactions can therefore be investigated with the same electrochemical methods as are employed to study corrosion in aqueous media (Fig. 76) [134].

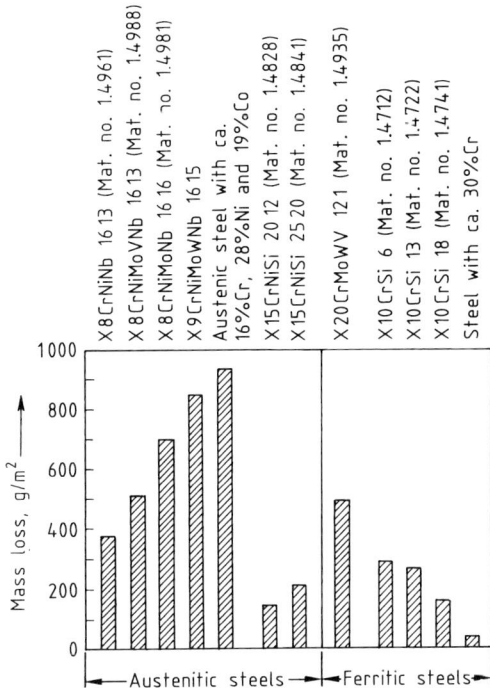

Figure 75. Corrosion rates of various high temperature and heat-resistant steels embedded in pure potassium sulfate after annealing for 700 h at 650 °C in an air–steam environment containing SO_2 (dew point 165 °C)

4.6. Corrosion by Synthesis Gases

In high-pressure synthesis the corrosive effect of hydrogen is not the only problem: there is also the problem of corrosive gas mixtures involved in the reaction. Their reaction products must also be taken into account. In ammonia synthesis, for example, there is the additional aspect of the nitriding effect of ammonia, which results in the formation of a nitride layer and embrittlement of the material beneath this layer. The behavior of various steels in an ammonia–hydrogen gas mixture is shown in Figure 77 [135].

After Armco iron, austenitic steels are least prone to corrosion. Ammonia has a nitriding and embrittling effect on unalloyed and low-alloy steels at 250 °C, and in austenitic steels corrosion begins at ca. 350 °C (Fig. 78) [136]. However, a standard 18-8 CrNi steel shows only a very shallow nitriding depth, and experience has proven that this steel can be used for years in the ammonia cycle-gas at 450 °C. After ten years the penetration depth of nitration is 1 mm at most.

Carbon monoxide is also important in that it is frequently used as synthesis gas, e.g., in methanol synthesis. Under pressure, carbon monoxide attacks unalloyed and low-alloy steels above

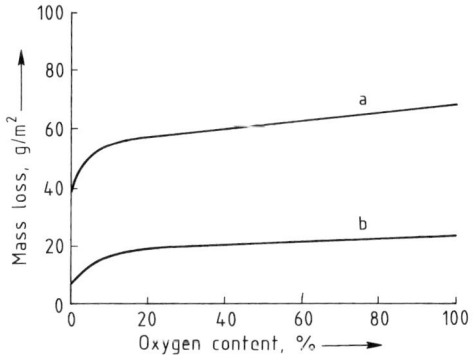

Figure 74. Dependence of the mass increase of steel X 20 CrNiSi 25 4 (material no. 1.4821) after annealing for 6 h in V_2O_5-containing oil ashes on the oxygen content of the gas
a) Synthetic oil ash (V_2O_5:Na_2O = 3), annealing temperature 800 °C; b) Natural oil ash (V_2O_5:Na_2O = 1), annealing temperature 700 °C

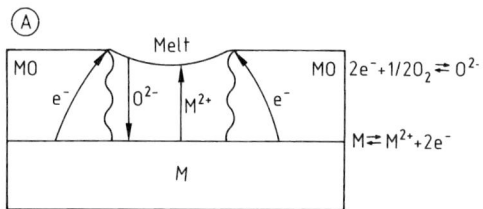

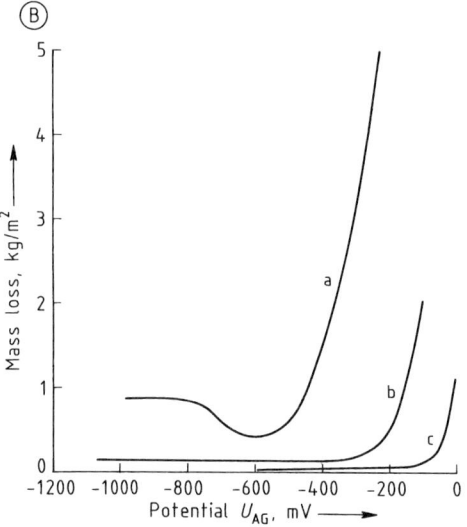

Figure 76. High-temperature corrosion in salt melts
A) Schematic of corrosion; B) Dependence of the corrosion in salt melts on the potential (at 750 °C for 23 h)
a) 10 CrMo 9 10; b) X 10 CrAl 18; c) X 10 CrAl 24

Figure 77. The effect of a H_2–NH_3 gas mixture on various steels ($t = 450$–520 °C; $p = 32.5$–90 MPa)

130–140 °C, forming iron pentacarbonyl. Above 350 °C, corrosion practically ceases again because the carbonyl becomes unstable. In high-alloy chromium and chromium–nickel steels the damage is appreciably less. Chromium steels with 30 % Cr and austenitic steels with 25 % Cr and 20 % Ni are completely stable. Figure 79 shows the corrosive effect of CO on various steels in a short-term test with pure CO–H_2 mixtures at two CO partial pressures [137]. A striking feature is the shift in the corrosion maximum of the 18-9 CrNi steel at the higher CO partial pressure. This possibly corresponds to the beginning of nickel carbonyl formation, which after a period of time leads to a chromium-enriched and, therefore, more stable surface layer.

If the iron carbonyl produced on CO attack is carried into the synthesis cycle, carbonyl decomposition can occur at elevated temperatures with the formation of active, pyrophoric iron,

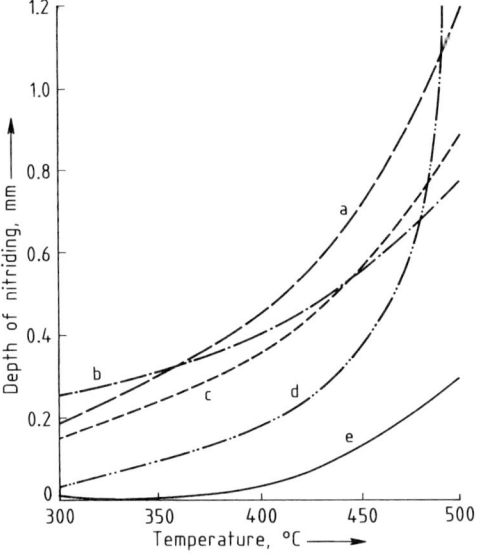

Figure 78. Depth of nitriding of various steels following nitridation in ammonia for an annealing time of 300 h
a) 13 CrMo 4 4; b) 17 CrMoV 10; c) 10 CrMo 9 10; d) X 8 Cr 17; e) X 8 CrNiNb 16 13

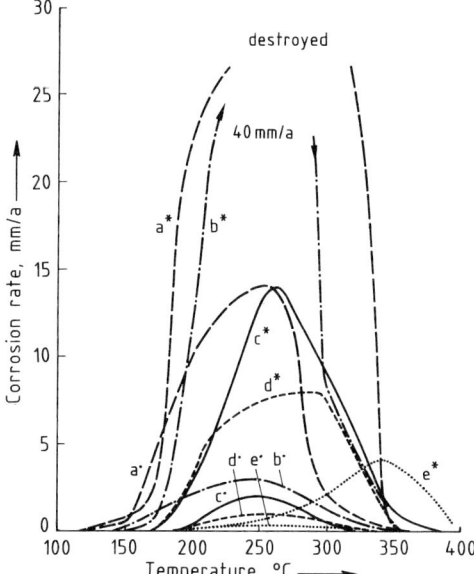

Figure 79. Corrosion behavior of various steels in CO–H$_2$ gas mixtures at two CO partial pressures
a) Boiler plate; b) 3.3 Cr/0.3 Mo; c) 13.1 Cr/0.34 Mo; d) 16.2 Cr; e) 18-9 CrNi
CO partial pressure: · = 12.5 MPa; * = 35 MPa

which in turn catalyses the reaction

$$2\,CO \longrightarrow CO_2 + C$$

Since this reaction is highly exothermic, individual apparatus parts can overheat [138]. In this way expansion has occurred in high-pressure vessels; the carbonyl produced must therefore be removed.

4.7. Oxidation Behavior of Technical Alloys

4.7.1. Steels

The scaling of steels in air has so far been investigated more thoroughly than in other corrosive media. Behavior in air is often the basis for a comparison of steels with respect to their applicability. However, since the corrosion conditions in technical gases can differ appreciably from those in air, extrapolation of parameters determined in air to the service behavior of steels is only possible to a limited extent.

The time function of oxidation is described by the equation:

$$\Delta m^n = k\,t$$

where Δm is the mass loss or increase, t the time, k the scale constant, and n a dimensionless exponent. The exponent n indicates the type of rate-determining reaction step and gives information on resistance. Values of $n \geq 2$ generally indicate resistance capacity, and values < 2 often imply lack of resistance. If n and k are known, metal losses can be calculated for any corrosion time. However, n is frequently variable because of transition from resistance to lack of resistance (breakdown), e.g., through a change in the oxide composition of alloy steels.

Unalloyed steels can be used in air up to 550 °C and low-alloy steels up to ca. 600 °C. The applicability of high-alloy steels is determined by the alloy contents, with special importance attached to Cr, Si, and Al, as demonstrated in Figures 70, 80, and 81 [139]. Water vapor and carbon dioxide in air generally worsen the scaling behavior of steels. The resistance of steels in water vapor is of particular importance in steam boilers and heat exchangers. To date it has been investigated at temperatures up to 800 °C.

Scaling in water vapor on low-alloy steels is similar to that measured in air at a temperature some 50–100 °C higher. This applies to high-al-

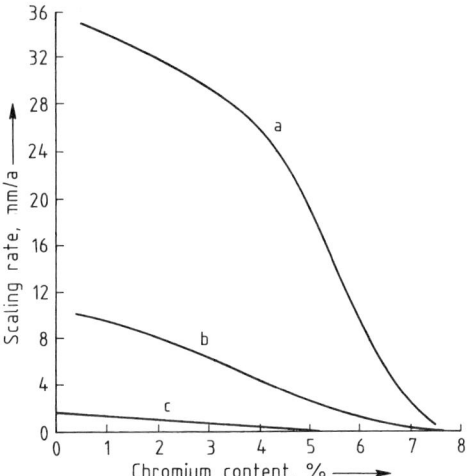

Figure 80. Influence of the chromium content of steels containing 0.15% C and 0.7–0.9% Si on the scaling resistance in air
Annealing temperature, °C: a) 800; b) 700; c) 600

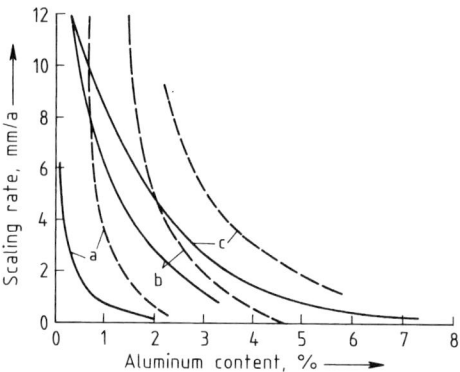

Figure 81. Influence of the Al and Cr contents on the scaling resistance of steel when annealed in air for 120 h
a) 6% Cr; b) 2% Cr; c) 0% Cr
——— 600 °C; --- 700 °C

Graphite contents in the scale lead to a loosening of the oxide film and accelerate the oxidation rate.

The addition of carbon monoxide to carbon dioxide reduces the scaling rate. The effect is particularly marked in the low-temperature region in unalloyed and low-alloy steels. Water content, on the other hand, has a negative effect. In addition to high carbon dioxide pressure, it is the cause of premature breakdown, i.e., the formation of noncovering oxide films, whereby in addition to the time of breakdown, the scaling rate after breakdown is also influenced. For dry gas ($< 5-10$ ppm H_2O) scaling follows a parabolic time law; higher water contents (see Fig. 82

loy steels even at elevated temperatures. Of significance here is the shift of breakdown towards greater scaling: the breakdown in low-alloy steels can be about 50 °C lower and in high-alloy steels up to 150 °C lower. The improved scaling behavior provided by Cr, Si, and Al corresponds to that obtained in air. A positive effect has also been found for molybdenum. In steels with higher chromium content, nickel makes a pronounced improvement.

Resistance to carbon dioxide is required of certain components in gas-cooled nuclear reactors, e.g., in heat exchanger tubing and fuel element cladding, where temperatures up to 800 °C exist. Data on high-temperature corrosion in carbon dioxide provide further indication of the performance of heat-resistant steels in exhaust gases arising from combustion of pure gases, especially when corrosion tests are run in carbon dioxide containing water vapor. Basically the same time laws apply to scaling in carbon dioxide as to scaling in air, but with carbon dioxide, graphite deposits in the scale can occur in unalloyed and low-alloy steels, and carburization in austenitic steels. The penetration of carbon into an alloy can lead to the following processes, which alter properties and reduce scaling resistance:

1) lower melting point (by ca. 350 °C in Ni–Cr alloys)
2) carbide formation with chromium depletion: cubic $Cr_{23}C_6$, triclinic Cr_7C_3, and orthorhombic Cr_3C_2; and
3) formation of a brittle, intermetallic σ-phase.

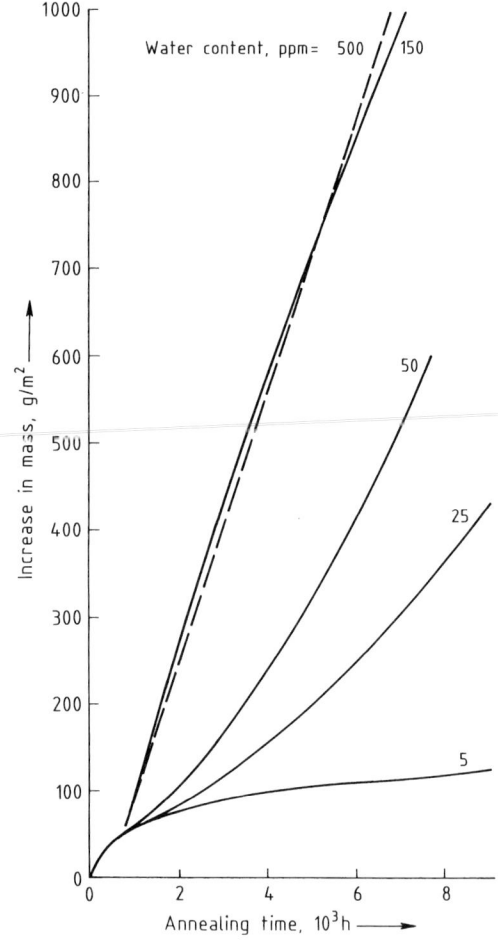

Figure 82. Scaling behavior of a steel similar to 13 CrMo 4 4 and dependence on the water content of carbon dioxide at 500 °C

[140]) give practically straight lines, with gradients that increase with increasing water content. GOODISON et al. found in low-alloy steel at 420 °C and 2.9 MPa gas pressure the following relation between water content and scaling rate after breakthrough [141]: mass increase in g/m² after 1000 h = 5 × water content in ppm. In contrast, methane and hydrogen contents delay breakthrough in unalloyed and low-alloy steels. Niobium and small amounts of rare earths (ca. 0.1 % cerium or yttrium) also have a positive effect on the scaling resistance of steels.

4.7.2. Nickel Alloys

Nickel alloys with more than 15 % Cr show good oxidation properties, which are due largely to the presence of an outer Cr_2O_3 layer. The adhesive strength of oxide films can also be greatly improved by small additions of cerium, lanthanum, and calcium (ca. 0.1 %) (Ni–Cr heat conducting alloys). The influence of temperature and atmospheric composition on the performance of Ni–Cr and Ni–Cr–Fe heat conducting alloys is shown in Figure 83 [142]. The sharp drop in life span is caused by the formation of liquid carbide eutectics through carbon absorption. Generally, higher application temperatures than in air are possible, as is shown by the curve drawn for performance in air.

Nickel has a high affinity for sulfur and combines with it to form the eutectic $Ni-Ni_3S_2$, which melts at 645 °C. However, even at ca. 100 °C below this temperature, annealing in sulfurous gases leads to penetration of sulfur into the nickel, especially along the grain boundaries (Fig. 84).

In reducing gases, intergranular damage due to sulfur (H_2S) is more pronounced than in oxidizing gases (SO_2). Generally, the resistance of

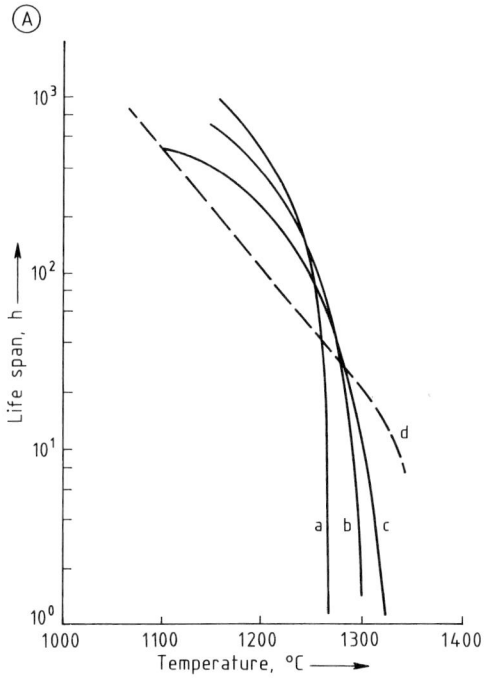
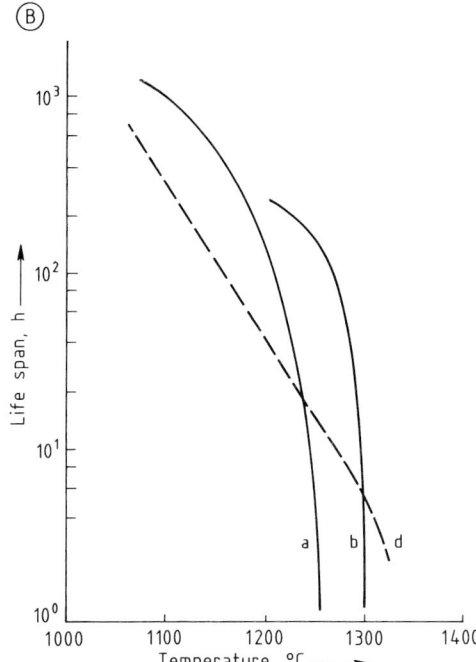

Figure 83. Dependence of the lifespan of Ni–Cr alloys on temperature in various gases
A) NiCr 80 20; B) NiCr 30 20 (Fe-containing)

Curve	Gas composition, vol %					
	H_2	CO	CO_2	CH_4	N_2	O_2
a	40	20			40	
b	20	13	7	2	58	
c	25	7	4	3	59	2
d			air			

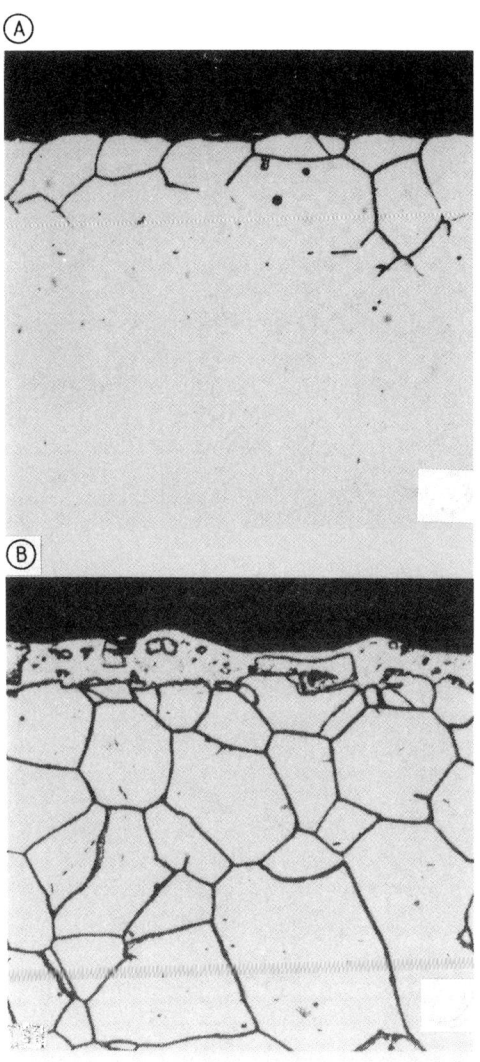

Figure 84. Cross section of a nickel sheet after annealing in nitrogen containing 0.5% sulfur dioxide
A) Annealing at 600 °C; B) Annealing at 640 °C

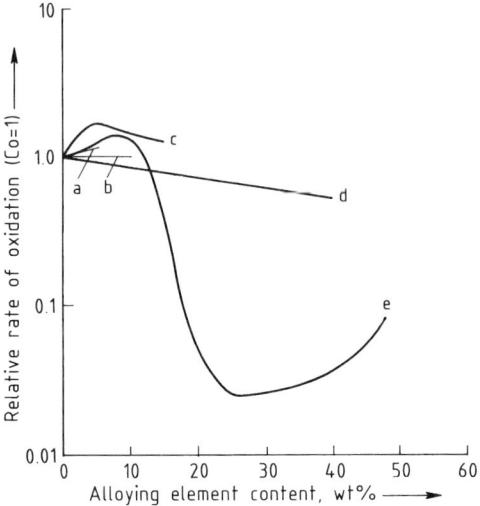

Figure 85. Influence of alloying elements on cobalt oxidation at 1000 °C
a) Al; b) Mo; c) W; d) Ni; e) Cr

nickel–chromium alloys to sulfur attack increases with increasing chromium content. Suitably good values are achieved with alloys NiCr 50 50 and NiCr 60 40.

4.7.3. Cobalt Alloys

Oxidation of Co–Cr alloys is in many ways similar to that of Ni–Cr alloys. The rate of cobalt oxidation is faster than for nickel. Under equivalent conditions CoO is even richer in cation vacancies than NiO.

Small Cr contents increase the rate of reaction, but at 20% Cr the reaction rate starts to decrease and exhibits a minimum at 25–30% Cr (see Fig. 85). The minimum value depends on the pressure. More chromium is needed to stabilize a protective film since the diffusion coefficient of chromium in cobalt is lower than for chromium in nickel. However, since the adhesion strength of the film on Co–Cr alloys is poorer than on Ni–Cr alloys—despite the identical oxidation rate of the Cr-containing Co alloys with Cr_2O_3 protective film—the practical oxidation resistance is lower. Other alloying elements, as Figure 85 shows, have little influence on scale resistance [143].

5. General Corrosion Protection Procedures

5.1. Corrosion Protection by Coating

The best possible corrosion protection can be achieved through the choice of suitable materials (see → 7. Construction Materials in Chemical Industry). Given extensive corrosion attack, this can mean compulsory use of expensive materials that may not always prove satisfactory in mechanical terms. A way out of this dilemma is provided by a composite system: the underlying function of such a system is provided by a cheap base material with good mechanical properties while the corrosion resistance is imparted by a coating material with relatively low thickness.

5.1.1. Metal Coatings

Of the various procedures for forming metal–metal composite systems, cladding is the most significant. Materials of greater thickness are combined by various processes with the base material. Compared with cladding, lining or the other coating methods are of minor importance. A distinction can be made between the following:

Cladding: roll cladding, explosive cladding, weld overlay cladding;
Lining; and
Coating: immersion, electrolytic metal deposition, diffusion deposition, hot dip metal coating, evaporation coating, and metal spray coating.

Roll Cladding. Strip and sheet metal are mainly produced by this method. Prerolled base and coating materials with clean surfaces are heated to the required temperature and rolled in contact with one another. Strength-reducing oxidation in the bonding zone can be avoided either by rolling the base metal/cladding sandwich within a hermetically sealed envelope of low-cost steel or by interleaving layers of nickel or mild steel between the base and cladding. Steel as a base metal can be clad with stainless steels, silver, nickel, copper, aluminum, and their alloys. Roll-clad semifinished products can be further processed by well-established fabrication methods, although care must be taken in hot forming and heat treatment.

The process for cladding tubes is carried out in two stages: backward impact extrusion of a multilayer solid followed by forward extrusion of the tube blank [144]. Forming with high-pressure hammers facilitates the binding process [145]. For unalloyed steel as the base material, stainless steel and titanium have so far been examined as cladding materials; interleaving sheaths such as Cu–Ni–Fe alloys are required when cladding with titanium.

Explosive Cladding. By using high-energy shock waves, composite systems can be produced from materials which for metallurgical or technological reasons can be neither roll-clad nor weld overlay plated [146]. In this process, the schematic for which is given in Figure 86, the cladding material impacts the substrate at high speed. If the collision angle reaches a critical value, a liquid metal jet, which is essential for a solid bond to form, is ejected from the contact region. Practically all metals can be combined with one another by explosive cladding. Because of the small depth of the binding zone, the mechanical properties are generally not impaired. While cladding of flat or tubular products is the state of the art, problems still arise when using this method to produce other curved composite systems.

Like roll-clad products, explosively clad metals can be further processed by all fabrication procedures, although special welding conditions may be required. The problems of heat treating systems of this kind are exemplified by the properties listed in Table 6 for a combination of fine-grained structural steel and titanium.

Joining of semifinished clad products is an important step in the production of components and equipment. This presents no particular problems in roll cladding because base material and coating can be welded seperately. However, to obtain the desired properties of the system the respective recommendations must be observed [147].

Joining explosive-clad semifinished products, on the other hand, presents a number of problems because metal combinations are often involved which cannot be combined thermally. Thus, in such cases design measures must be tak-

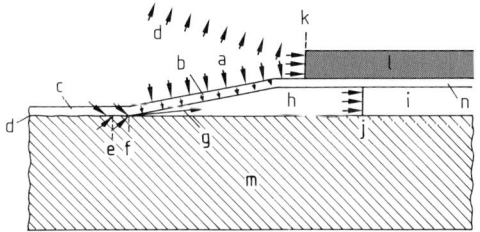

Figure 86. Schematic of explosive cladding
a) Expanding detonation products (swathes); b) Accelerated coating sheet; c) Cladded coating sheet; d) Undulating binding zone; e) Metal interface layers flowing in a laminar or turbulent manner under high pressure; f) Collision point, very high pressure (1000–10 000 MPa); g) Jet of material at high speed; h) Air in a collision state; i) Air at atmospheric pressure; j) Advancing collision front in the air; k) Detonation front in the explosive; l) Unused explosive; m) Base plate or substrate; n) Sheet to be clad

Table 6. Heat treatment of fine-grained structural steel and titanium

Treatment	Temperature, °C
WStE 460	
Hot forming	> 800
Normalizing	900–940
Recrystallization annealing	> 600
Stress-relieving annealing	530–580
Titanium	
Carbon diffusion from steel into titanium	> 550
Iron diffusion from steel into titanium	> 800
Hot forming	> 650
Recrystallization annealing	> 500
Stress-relieving annealing	400–500

en to prevent the substrate and cladding materials from mixing [148].

Weld overlay cladding is primarily economical for fairly thick plates and large forgings. On account of their low cost and only slight penetration, two processes are particularly significant: submerged arc welding [149] with strip electrodes and plasma hot-wire welding [150].

In submerged arc welding the strip electrode is fed from a coil into the flux bed and fused in the arc. The width of the weld deposit corresponds approximately to the width of the electrode, and mixing with the substrate depends on the process parameters. The procedure is preferably used to clad stainless steels. Because of burn off and mixing, the electrode must be overalloyed to compensate for the loss of important elements.

In plasma hot-wire deposition welding, an oscillating plasma jet under inert gas melts the surface of the base material. Wires of the cladding material are fed into the molten layer, melt through the passage of current, and form the cladding. The heat input in this process is small. Consequently mixing is very limited, as are thermal influences on the base material. There is practically no burn off, and the method is far more widely applicable than submerged arc welding [151], [152].

There are two nonwelding methods in which the coating material is also in molten form. In homogeneous lead coating, lead is melted with burners and applied to the steel surface. Intermediate coatings, e.g., tin solder, are needed to improve adhesion.

Centrifugally cast composite pipes are used for special applications (e.g., cracking tubes in the petrochemical industry) [153]. Initially the casting of the outer shell is produced in a rotating cast-iron mold. After a suitable cooling period, during which the inner surface of the casting must be protected against oxidation, the melt to form the inner shell of the composite pipe is poured in. Practical experience has been gained with a composite of unalloyed or low-alloy steels with high-alloy, corrosion-resistant grades.

Lining. A further possibility of protecting equipment and components against corrosion consists of lining with corrosion-resistant materials. This lining is generally attached to the base material (as long as a thick-walled self-supporting liner is not preferred) by local welding. In this case strips of the lining material are welded to the base material and the edges of the lining material welded to these strips. If the two materials cannot be welded to each other, a form-fitting bond is used, whereby the lining material is mechanically locked into conical slots in the base material. This corrosion protection measure has recently started to decline: one reason is the design problems presented by a lining subjected to alternating thermal stress.

A special lining method is the Resista Clad process, which allows thin liners (0.15–2.5 mm) to be applied during manufacture. The base and lining materials (e.g., steel, nickel alloys, titanium, tantalum) are attached to one another by resistance welding with an intermediate layer. Depending on the service conditions, the distance between the weld seams can vary from 20 to 150 mm. Apparatus such as vessels, columns, and heat exchangers can be operated in vacuum (seam distance ca. 20 mm). Depending on the coating material, service temperatures of up to 350 °C are possible [154]–[156]. Apparatus and equipment manufactured by the Resista Clad method have also been successfully used in flue gas desulfurization [157].

Coating. With regard to the protection they offer against corrosion, the methods listed on page **8**-53 for producing coatings [158] are generally only of minor importance in chemical engineering because they cannot guarantee adequate resistance under the extreme conditions encountered in some areas of the chemical industry.

Through immersion, metals such as copper or silver are deposited from an aqueous solution or molten salt onto the material to be protected.

The coating thickness is ca. 1 µm and the adhesion strength is not always satisfactory.

In electrolytic methods, deposition of metals is improved by using a constant or pulsating direct current along the principle of electrolysis [159]. Voltages of 1–10 V and current densities of 5–100 mA/cm^2 are used. Depending on the base metal, coating metals such as nickel, copper, zinc, tin, chromium, silver, and gold can be applied to the base material either directly or with intermediate layers. Nickel, for example, is deposited on steels both directly and on an intermediate copper layer [160]. A chromium coating on nickel or copper intermediate layers also offers good application possibilities.

Coating with tantalum by molten salt electrolysis [161] has gained importance in chemical engineering. This process has proved successful primarily for fairly small parts since it provides compact, nonporous coatings with good adhesion and thicknesses above 200 µm. Base materials include steels, copper, nickel, and their alloys [162], [163].

In the diffusion coating process the coating material is deposited at elevated temperature from the pulverized or gaseous state. The metal diffuses into the lattice of the base metal with formation of a high-alloy surface zone. To produce zinc or aluminum coatings (sherardizing and alitizing), the part to be coated is embedded in powders of the corresponding metal chlorides. Chromium coatings are deposited by inchromizing from gaseous or molten chromium(II) chloride.

Hot-dip metal coating is being used increasingly in automated, continuous plant, in particular for the production of semifinished products, but also for mass-produced parts. The part is immersed in a molten bath of a coating metal with a comparatively low melting point. This is a common method for coating with zinc, tin, aluminum, and lead.

Evaporation coating [164] involves the production of very pure, nonporous surface coatings by condensation of vacuum-evaporated metallic materials (physical vapor deposition, PVD). However, the extremely low coating thickness permits application of evaporation-coated parts in chemical engineering only in exceptional cases.

Metal spray coating involves wire melt spraying and powder spraying from gas, arc spraying, or plasma spraying torches. Powder spraying is important for metals (and several nonmetals) that cannot be made into wires. Common to all these processes are the melting of the coating material in the heat source as it leaves the torch, the atomization of the coating material, and the impinging and adhesion of the metal droplets on the bare metal surface of the object to be coated. As the adhesion of sprayed metal coats is entirely mechanical, the adherence depends very much on the physical nature of the substrate. A roughened and clean (shot-blasted) surface is therefore a prerequisite for good adherence.

The most important materials used for chemical plant include metals with high melting points and good resistance to corrosion, such as tantalum and molybdenum, but also titanium and nickel alloys, including hard metal alloys.

Alloys based on Ni–Cr and Co–Cr, containing borides, silicides, and carbides, and applied by powder flame spraying, are used to protect components from corrosion and wear. The inherent pore volume of a coating can be reduced to about 1 % by a subsequent densifying heat treatment [165]. Figure 87 shows the structure of a flame-sprayed and subsequently densified hard metal coating consisting of Ni–Cr–Si–B. The Si and B form fluxes by alloying and lower the melting point. Exceptional hardness (60–70 R_c) and good adherence can be obtained when W_2C is introduced into the surface of the substrate. A conveyor screw made of mild steel (St 37) protected by Ni–Cr–Si–B provides an example of the improvements in durability achievable by these techniques. The useful life in this instance was increased 12-fold. The life of an iron sludge pump casing spray-coated with a W_2C/Ni–Cr–Si–B alloy, which was subsequently densified at

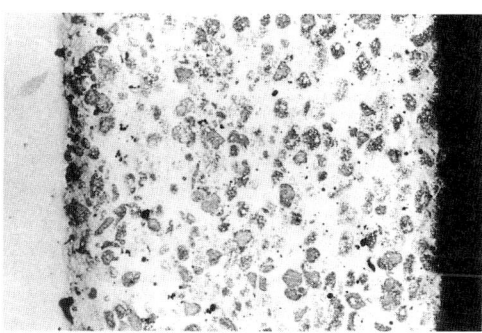

Figure 87. Microstructure perpendicular to the surface of a flame-sprayed and densified coating of a Ni–Cr–Si–B alloy

1050 °C was extended 10 times. Note that subsequent thermal densification may distort the component and alter the structure of the material.

A prerequisite for good resistance to corrosion is that the protective coating must be dense and free from pores. Flame-sprayed coatings, with their low density and numerous oxide inclusions, are therefore unsuitable in many cases for exposure to aggressive media. Plasma-sprayed coatings are denser and more homogeneous.

Plotting current density against potential curves is a quick way of investigating the electrochemical behavior of thermally sprayed coatings and gaining an initial impression of their resistance to corrosion. Figure 88 shows current density–potential curves for plasma-sprayed Ni–Ta–Cr–B alloys (containing τ-borides) in 0.1 N sulfuric acid. All three alloys show good resistance to corrosion by virtue of their chemical composition and the fact that the coatings are very dense. Figure 89 shows the microstructure of a coating of this type on unalloyed steel. Although alloys forming τ-borides are still under development, they will become increasingly important in the protection of chemical plant components [166].

The high-velocity flame spraying (Jet-Kote) process, which became commercially available only a few years ago, is similar to powder flame spraying, but differs in that the spray particles are accelerated to high velocities in the gun, with the result that the adherence and density of the coating are superior to those achieved by conventional powder flame spraying. However, some kinds of spray powder cannot be applied by this technique. Figure 90 shows the good ad-

Figure 89. Microstructure perpendicular to the surface of a plasma-sprayed Ni–Ta–Cr–B alloy containing τ-borides

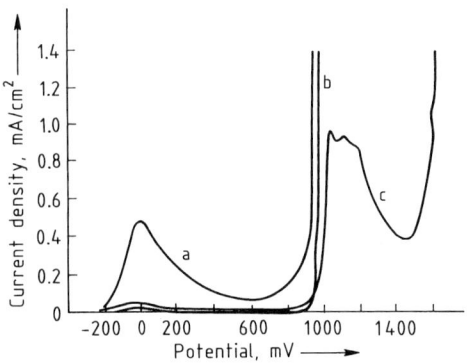

Figure 88. Current density–potential curves for various alloys containing τ-borides in 0.1 N H_2SO_4
a) Ni25Ta8CrB; b) Ni20Cr3TaB; c) Ni38Ta7CrB

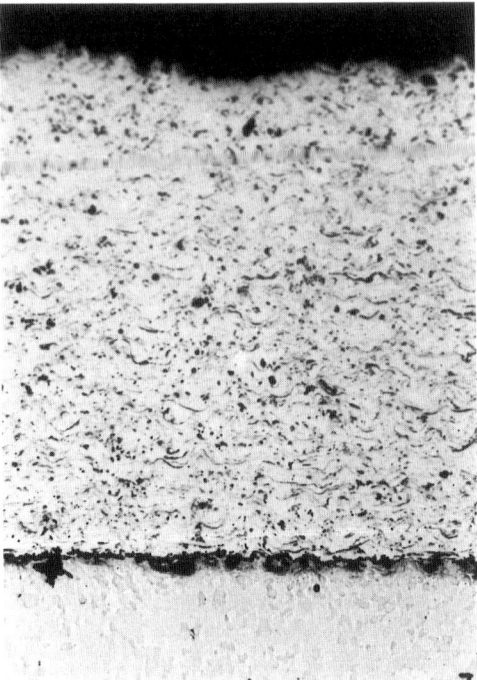

Figure 90. Hastelloy C-4 coating applied by powder flame spraying (Jet-Kote process)

Figure 91. Vacuum chamber for plasma spraying of Ti, Ta, and Nb

herence and low porosity of a Hastelloy C-4 coating applied by the Jet-Kote process.

Spray coatings consisting of metals with high melting points are applied mainly by plasma spraying, as used particularly for oxide ceramic coatings. Coat thicknesses of 0.1 to several millimeters are feasible and good adherence is obtained. The coating is relatively dense, but contains individual closed pores. Vacuum or low-pressure plasma spraying is a further development of atmospheric plasma spraying. The process is carried out in a pressure chamber (Fig. 91), which is evacuated to 20 Pa. While the spraying is in progress, argon is introduced until the pressure reaches about 5 kPa. The resulting inertization prevents the reactions between molten particles and the surrounding atmosphere that occur in atmospheric plasma spraying [167]. This modification gives the process substantial advantages over conventional plasma spraying. In particular, the method enables reactive metals such as Ti, Ta, and Nb to be applied as very dense, nonporous coatings. At present the process suffers from the limitations imposed by the dimensions of the chamber and the high cost ($1.5-2 \times 10^6$ DM).

5.1.2. Inorganic Nonmetallic Protective Coatings

Important inorganic nonmetallic coatings used for the protection of chemical plant include, above all, enamels with high silicon dioxide and low boric acid contents, brick linings based on silicates and carbon, as well as thermally sprayed coatings based on metal oxides and metal carbides (see Table 7). Metal nitrides and metal borides are also used, but less often because chemically they are less resistant. Having low thicknesses of ≤ 20 μm, coatings formed on metal and plastic substrates by chemical and physical deposition from the gas phase (CVD and PVD) are of only minor importance in corrosion

Table 7. Inorganic nonmetallic protective coatings used in chemical engineering

Material	Applications[a]
Enameling (0.8–2.2 mm)	
Vitreous or partly crystalline	Acidic and organic media up to 230 °C (340 °C[b]) (except HF and H_3PO_4)
	Alkaline media up to 60 °C
Brick lining	
Acid-resistant bricks	Neutral, acidic, and organic media up to 200 °C
Carbon bricks and electrographite, impregnated	Nonoxidizing acidic, alkaline, or organic media up to 180 °C
Carbon bricks and electrographite, unimpregnated	Nonoxidizing media up to 400 °C
Refractory bricks	Lining of reaction zones in which temperatures of up to ca. 1600 °C occur
Thermally sprayed coatings	
Metal oxides	Wear protection, neutral and acidic media up to 100 °C
Metal carbides	Wear protection, neutral and alkaline media up to 80 °C
Ceramic–metal mixtures (cermets)	Special applications and intermediate coatings
CVD and PVD coatings	Corrosion-inhibiting wear protection

[a] The indicated temperature limits depend on the agressiveness and concentration of the medium and may in certain cases be too high. If no experience is available, it is advisable to carry out corrosion tests. [b] With increased compressive prestressing, which extends the temperature range to 340 °C and enhances the resistance to thermal shock [174].

Table 8. Main properties of the constituents of the enamel/mild steel system

Property	Enamel	Boiler plate
Adhesive strength, steel/enamel, MPa	100	
Compressive strength, MPa	800–1000	2000 (60% compression)
Tensile strength, MPa	70–90	400
Coefficient of thermal expansion (20–400 °C), K^{-1}	$80-95 \times 10^{-7}$	135×10^{-7}
Compressive prestress, MPa	130	
Elastic modulus, GPa	70	210
Vickers hardness, MPa	6000	1100
Fracture strain, %	0.15–0.3	25
Yield strength, MPa		200
Impact force (DIN 51 155), N	40–80	
Coefficient of thermal conductivity, $kJ\,m^{-1}h^{-1}K^{-1}$	3.35	188
Specific heat (10–100 °C), $kJ\,m^{-1}h^{-1}K^{-1}$	0.84	0.46
Specific resistivity, $\Omega \cdot cm$	$10^{12}-10^{14}$	0.002
Dielectric strength (room temperature), kV/mm	20–30	
Density, g/cm^3	2.5	7.8
Softening point, °C	790	
Melting point, °C	960	1510
Enamel thickness, mm	1–2	

protection. The same applies to protective coatings formed by reaction of the metal surface with acidic phosphate, chromate, or oxalate solutions; these are used particularly as passivating primers for paints. Lining with tubes and other shaped parts consisting of fusion-cast basalt [168] has gained some importance as a means of protecting plant from corrosion and abrasion, and the same is true of the use of tungsten carbide platelets for screw cladding [169].

The resistance of protective materials depends mainly on whether they are based on oxides or on silicates, consist of carbon, or have been sealed by impregnation with synthetic resins (see Table 8). In general it can be assumed that the oxide and silicate coating and lining materials used for chemical plant are outstandingly resistant to acids (except hydrofluoric acid and phosphoric acid), salt solutions, and organic compounds of all kinds. As a rule they resist aqueous alkaline solutions up to 50 °C. Unimpregnated carbon materials withstand temperatures up to 400 °C, while those that are impregnated with synthetic resins are stable up to 180 °C. With silicate materials the chemical attack, which occurs almost entirely on the surface, depends greatly on the pH and water content of the attacking medium. Water-free acids, except phosphoric acid, do not attack silicate materials. The resistance of carbon materials depends on the resistance of the resin impregnating agents and putty joints.

5.1.2.1. Enamel

An enameling is a highly stable combination consisting of enamel (an inorganic oxide-based glass flux) and a metallic substrate. Known in the process industries as chemical enamel, the combination of enamel with steel has for decades been firmly established in chemical processing plants. Its stability is such that many chemical processes are dependent on it, and many products can only be manufactured economically in enameled apparatus [170]. By virtue of their structure [171]—highly stable silicate glass in a firm bond, created by fusion, with supporting steel—the traditional domain of enamel coatings is processes that involve reactions with acids at high temperatures. However, the development of enamel formulations and enameling technology have considerably extended this traditional field of application: enamel is now being used increasingly as an all-round material for exposure to neutral and alkaline media as well [172]. This applies equally to vitreous enamels and to the partly crystalline ceramic enamels. Table 8 lists the main properties of the partners in the enamel/unalloyed steel system. The stability of this system, represented as iso-corrosion curves for a corrosion rate of 0.1 mm/a, is shown in Figures 92 and 93 for characteristic acids and alkalies. Stability tests are performed at atmospheric pressure according to DIN/ISO 2743, 2744, and 2745, and under pressure according to DIN

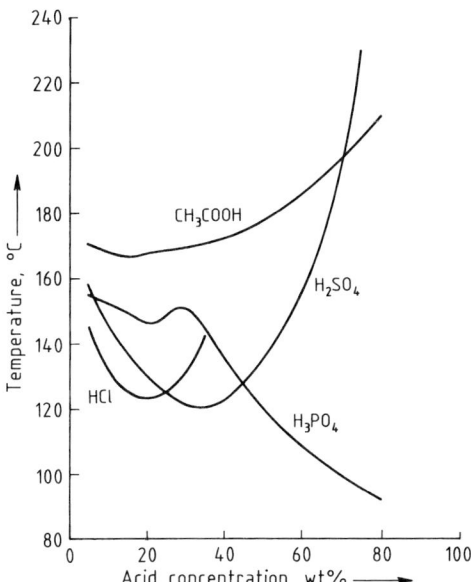

Figure 92. Iso-corrosion curves (0.1 mm/a) for chemical enamel in acids

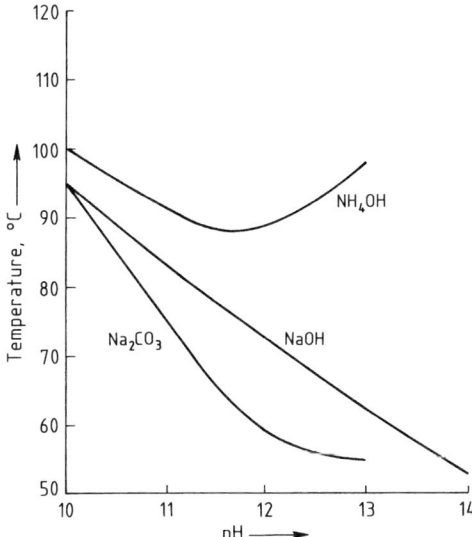

Figure 93. Iso-corrosion curves (0.1 mm/a) for chemical enamel in alkalies

51 174. The temperature dependence of the corrosion rate obeys the Arrhenius equation. A corrosion rate of 0.1 mm/a is normally the highest at which enameled apparatus can be used economically, and it should not be exceeded, especially in cases where intermediate exposure to alkalies is involved, e.g., in neutralization reactions and cleaning processes. Vitreous and ceramic enamels differ little in their resistance. If a glassy cover coating is not present, the chemical resistance of partly crystalline enamels is in some cases considerably lower than that of glassy coatings of comparable composition. Ceramic enamels are considerably superior to vitreous enamels in mechanical behavior, especially impact resistance, and to some extent also in resistance to abrasion. The acid resistance of enamels can be improved considerably by adding small amounts of SiO_2 to the acid, which has a marked inhibiting effect [173].

Special enamels are now available which, for example, are 300% more resistant to alkalies (though less resistant to acids than normal enamels), which show considerably less tendency to attract deposits because they have substantially improved surface quality (owing to an increase in surface tension), or which, because they contain refractory ingredients, can be fired on heat-resistant substrates at temperatures up to 1200 °C [174]. These high-temperature enamels give additional protection from scale formation and corrosion to alloys and metals that are intrinsically heat-resistant, thus simultaneously prolonging the useful lifetime of the plant and enabling it to withstand high temperatures. A notable improvement in the already high standard of chemical enameling is expected to result from the use of an enamel that combines the property patterns of vitreous and ceramic enamels. This product, introduced at Achema 1988 [175], has a reinforcing mechanism which is independent of the matrix material and absorbs the energy that tends to cause fractures. This "in situ" micro-reinforcement enables the mechanical stability of chemical apparatus coated with this enamel to surpass even that of plant coated with good ceramic enamels, without concessions having to be made in regard to chemical stability.

5.1.2.2. Brick Lining

Acid-resistant bricks, carbon bricks, and electrographite afford additional protection to parts of plants which, owing to the severity of the chemical or thermal exposure, are not adequately protected by coatings and linings.

Refractory bricks and other materials are used, both with and without a lining between the steel and bricks, to enclose reaction chambers.

Depending on the conditions to be withstood, the products used include basic and neutral materials with high silicon dioxide contents, carbon bricks, and silicon carbide bricks. The structure of a chemically resistant brick lining is shown schematically in Figure 94.

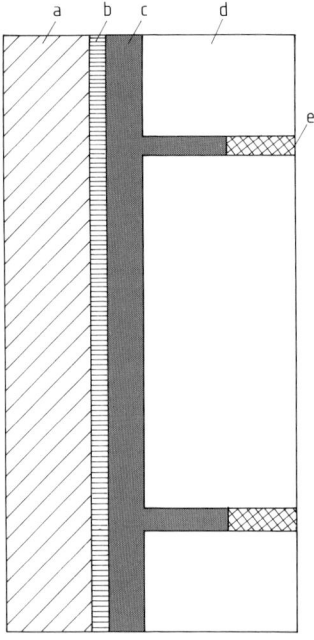

Figure 94. Schematic of a chemically resistant brick lining
a) Substrate (concrete or steel); b) Resistant sealing compound; c) Bonding cement; d) Bricks or tiles; e) Pointing cement

5.1.2.3. Thermally Sprayed Coatings

Improvements in sprayable materials and spraying techniques have already resulted in the widespread use of thermally sprayed coatings consisting of inorganic nonmetallic materials [176]. These can be applied reliably by flame spraying, detonation spraying, or plasma spraying (see also → Metals, Surface Treatment, **A16**, pp. 429–437). Metal oxides, metal carbides, or their mixtures are melted in a plasma flame at ca. 20 000 °C and sprayed onto the surfaces of the workpieces; coatings with a thickness of 0.1 mm to several millimeters can be produced in this way. Subsequent heat treatment of the coating can make it almost nonporous through the formation of a glassy phase. Most of the pores, which account for 1–20 % of the volume, are closed. Corrosion-resistant coatings can be built up without difficulty by spraying several coats on top of one another. In Table 9 the principle spraying processes are classified according to rate of application, supplementary spraying materials, and source of energy. Figure 95 shows a polished section through a wear-resistant coating on a channel entrance in a graphite exchanger [177].

Coatings of these types can be produced on carbon steels and stainless steels. Another application is shown in Figure 96. The fiber deflection roll shown here, which is coated with Al_2O_3, lasts three times longer than an otherwise identical roll with hard chromium plating.

Table 9. Classification of spraying methods

Method	Energy source	Operating temperature, K	Partial velocity, m/s	Spraying material		Capacity, kg/h
				Product form	Material	
Flame spraying	fuel gas, acetylene–oxygen	3200	50–200	wire	mostly metals	5–8 (metals)
				powder	all types	1–2 (ceramics)
Jet-Kote high-speed flame spraying	fuel gas, slow-burning	2900	350–600	powder	metals and ceramics	2–6
Detonation spraying	fuel gas	3500	650	powder	all types	4–6
Plasma spraying	electricity	3000–20 000	350–450	powder	all types	4–8 (metals) 2–4 (ceramics)
Arc spraying	electricity	4000	150	wire	metals	15–20

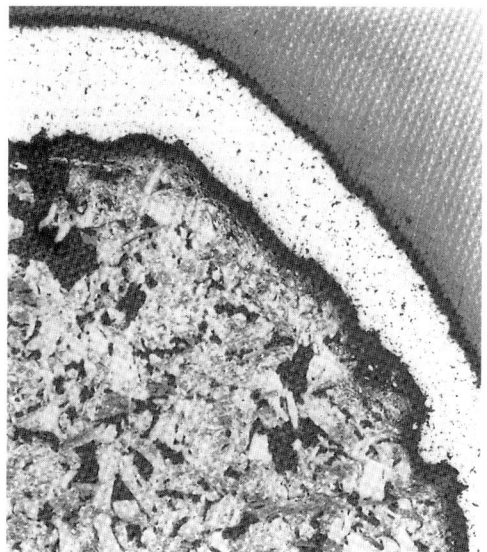

Figure 95. Polished section through a plasma-sprayed coating on a channel entrance in a graphite heat exchanger

Figure 96. Fiber deflection roll, spray coated with Al_2O_3

5.1.3. Organic Lining and Coating Materials

The range of high-molecular lining and coating materials has increased substantially, both with respect to types and grades of materials and with respect to their properties and applications. In general it can be said that modification of well-known and proven materials by copolymerization has improved important mechanical properties of linings and coatings, such as their impact resistance, while compounding with randomly distributed conductive carbon has enhanced the electrical conductivity of plastics so much—without appreciably impairing their me-

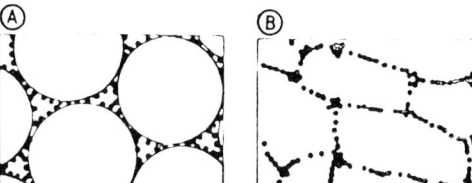

Figure 97. Schematic of materials with covered particle structure [179]. The diameter of the electrically conducting particles is considerably smaller than that of the plastic particles
A) Original structure; B) Deformed structure (e.g., after processing)

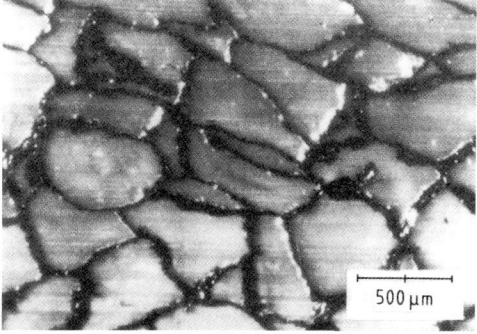

Figure 98. Section through an electrically conducting pressed sheet, consisting of high-density polyethylene filled with carbon black and having a covered particle structure

chanical and chemical stability—that the linings or coatings can now even be exposed to combustible liquids and vapors (Figs. 97–99) [178].

Table 10 gives a general idea of the thermal stability of coatings as a function of chemical exposure. Standards and guidelines concerning organic lining and coating materials can be found in DIN 28 053, 55 929; VDI 2531–2539; TRbF 401–403; and DVS 2202, 2301.

Effects of Media. Polymers, unlike metals, are subject to swelling rather than corrosion. Chloride-induced pitting corrosion, to which austenitic Cr–Ni steels are particularly susceptible, is unknown in polymeric materials. Stress corrosion cracking, however, is possible under conditions that cause the medium to attack the polymer chemically [180]. Stress corrosion cracking can only occur, however, if, at the same time, stresses resulting from manufacture, coating or lining, or the conditions of use are present.

Where polymeric materials are concerned, attention should be paid to diffusion and perme-

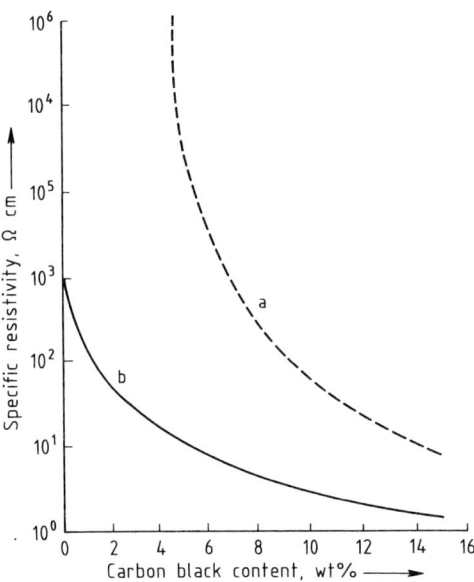

Figure 99. Specific resistivity as a function of carbon black content
a) Uniform carbon black distribution; b) Covered particle structure

ation, irrespective of stability behavior. Even if the protective material is stable, gaseous or vaporous substances may penetrate organic coatings and linings and, if corrosive, attack the substrate, thus resulting in loss of adhesion, blister formation, and failure of the coating or lining to protect the substrate. The factors which influence this process include temperature, temperature gradient, the vapor pressure of the medium, the thickness of the lining, and the chemical affinity of the lining material and of the medium to which it is exposed.

Protective Coatings and Linings. Protective coatings or linings consisting of organic materials are applied to metals, especially steel, to glass-filled UP, VE, and EP resins, and to concrete and masonry. They protect the substrate from chemical attack. Recommendations for the design and production of the items to be protected can be found in VDI-Richtlinie 2532 for metals and in VDI-Richtlinie 2533 for concrete and masonry; recommendations on the choice of coating materials and methods are given in VDI-Richtlinie 2531.

5.1.3.1. Rubber Linings

Rubber linings (see also VDI-Richtlinie 2537) consist of ebonite or soft rubber based on natural or synthetic polymers. They account for the largest group of materials used for surface protection in chemical plant. Synthetic elastomers are becoming more and more important since they have greater chemical and mechanical stability than natural materials, and less susceptibility to mechanical damage.

In recent years halogen butyl and self-vulcanizing systems based on the chloroprene rubber Baypren have become established and proved their worth, especially where they have been used for in situ treatment of large items such as storage tanks, waste water purification plant and flue gas desulfurization equipment [181]. The sulfochlorinated polyethylene Hypalon has good resistance to acidic and alkaline oxidizing agents and to chromic acid and chlorine bleaching solution.

Being more highly crosslinked, ebonite linings are even more resistant to swelling and permeation than soft rubber linings. As, however, they have to be vulcanized in workshop autoclaves, their use is generally restricted to relatively small items whose volume does not exceed about 100 m^3.

Plant components of up to about 1000 m^3 in volume can also be ebonite-lined on site if hot water or piped steam is available for vulcanization. For adequate crosslinking, the steel must be heated to 90–95 °C, which is only possible if effective exterior insulation is provided. The largest tank known to have been ebonite-lined on site has a capacity of 8000 m^3. It has a triple-ply lining, which was applied as a 5-mm thick single sheet and consists of a soft rubber adhesive layer, a diffusion-proof ebonite layer, and a corrosion-resistant soft rubber layer. Since 1982 this tank has been continuously exposed to concentrated hydrochloric acid containing chlorine and solvents without suffering damage. It was installed at the New Martinsville works of Mobay in the United States.

5.1.3.2. Rubber–Plastic Composite Linings

The stability of rubber linings, especially to organic compounds that diffuse readily in rubber and damage it severely, can be improved by the simultaneous use of plastics. In such cases a sup-

Table 10. Organic lining and coating materials used in chemical engineering

Material[a]	Application[b]			
	Acids, salt solutions	Alkaline solutions	Oxidizing agents	Organic media
Thermoplastic linings (1–4 mm)				
PVC with plasticizer	50 °C	50 °C	50 °C	
PVC without plasticizer	60 °C	60 °C	60 °C	
PE-HD	60 °C	60 °C		30 °C
PP with contact adhesive	50 °C	50 °C		30 °C
PP with fabric lamination	80 °C	80 °C		30 °C
PVDF (partly fluorinated)	100 °C		80 °C	30 °C
E/CTFE (partly fluorinated)	100 °C	100 °C	80 °C	30 °C
FEP, PFA (fully fluorinated)	120 °C	120 °C	120 °C	120 °C
PTFE (fully fluorinated)	200 °C	200 °C	200 °C	200 °C
Rubber linings (3–4 mm)				
Soft rubber based on natural and synthetic rubbers (NR, CR, IIR, NBR, CSM, FPM)	Nonoxidizing acidic to alkaline media up to 90 °C			
Ebonite based on natural and synthetic rubbers (NR, NR/IR, SBR, NBR)	Nonoxidizing acidic to alkaline media up to 110 °C			
Graphite-filled thermosetting plastic linings (4–5 mm)				
PF	120 °C			120 °C
FU	120 °C	120 °C		120 °C
EP	100 °C	120 °C		80 °C
Reinforced reaction resin coatings				
Glass fiber laminate coatings (3–0.5 mm)				
UP-GF and VE-GF	70 °C	50 °C	30 °C	30 °C
FU-GF	70 °C	70 °C		70 °C
EP-GF	70 °C	70 °C	30 °C	30 °C
Grouts and sprayed coats, reinforced with glass flakes (1–2 mm)				
UP and VE, based on bisphenol A	100 °C			
VE, based on epoxy–novolak acrylate	150 °C			
Paint and powder coatings (0.2–2 mm)				
Liquid coating materials				
Catalyst-curing resins (PUR, EP, UP, VE)	Nonoxidizing acidic to alkaline media up to 80 °C			
Thermosetting resins (PF, EP, PF/EP)	Nonoxidizing acidic to alkaline or organic media up to 120 °C			
Thermoplastic powders				
PE-HD	Nonoxidizing acidic to alkaline media up to 50 °C			
E/VAL	Nonoxidizing acidic to alkaline media up to 50 °C			
PA	Nonoxidizing neutral and alkaline or organic media up to 60 °C			
PVDF (partly fluorinated)	Acidic to neutral media up to 80 °C			
E/TFE, E/CTFE (partly fluorinated)	Acidic to neutral media up to 80 °C			
PFA, FEP (fully fluorinated)	Acidic to alkaline or organic media up to 150 °C			

[a] E/TFE Ethylene/tetrafluoroethylene, E/CTFE Ethylene/chlorotrifluoroethylene, EP Epoxide, E/VAL Ethylene/vinyl alcohol, FEP Tetrafluoroethylene/hexafluoropropylene, FU Furane, PA Polyamide, PCTFE Polychlorotrifluroethylene, PE-HD High density polyethylene, PF Phenol formaldehyde, PFA Perfluoro-alkoxyalkane, PP Polypropylene, PTFE Polytetrafluoroethylene, PUR Polyurethane, PVC Poly(vinyl chloride), PVDF Poly(vinylidene fluoride), UP Unsaturated polyester, VE Vinyl or phenacrylic ester, UP-GF Fiberglass-reinforced unsaturated polyester, VE-GF Fiberglass-reinforced vinylester, FU-GF Fiberglass-reinforced furane, EP-GF Fiberglass-reinforced epoxide, CR Chloroprene rubber, CSM Chlorosulfonyl polyethylene, FPM Vinylidene fluoride/hexafluoropropylene copolymer, IIR Isobutane–isoprene rubber, NBR Nitrile–butadiene rubber, NR Natural rubber, SBR Styrene–butadiene rubber.
[b] The indicated temperature limits depend on the aggressiveness, concentration, and penetration capability of the medium and may in individual cases be too high. If no experience is available, corrosion tests are recommended.

plementary lining consisting of a thermoplastic chosen for its resistance to organic or oxidizing compounds is applied on top of a soft rubber or ebonite lining. Where the item to be protected will be exposed to temperatures of up to 50 °C in practical use, the polypropylene or polyvinylidene chloride sheets, which are the preferred materials, can be bonded to the rubber reliably with an isocyanate crosslinking chloroprene rubber adhesive. For higher temperatures, as reached in reaction vessels, for example, a bond must be formed by vulcanizing the rubber in contact with the plastic [182] or by placing a fabric on the rubber, applying a heat-resistant epoxy resin to it, adding the thermoplastic sheet, and applying pressure by means of a vacuum [183].

5.1.3.3. Linings Consisting of Thermally Crosslinkable Thermosetting Plastics

Phenol, furan, and epoxy resins are the main representatives of this group. After addition of plasticizers, graphite, and fibers they are applied like rubber linings as 4–6 mm thick sheets and crosslinked three-dimensionally in pressure vessels. In common with rubber linings they can also be applied to items made of steel or fiberglass-filled plastics. The crosslink densities of the finished linings are so high that the material is neither soluble nor capable of melting, and its strength and creep resistance are not dependent on temperature to the same extent as those of thermoplastics. Thermosetting plastics are used mainly to line columns, flues, and other plant items that have to withstand severe exposure to chemicals and heat.

5.1.3.4. Reinforced Reaction Resin Coatings, Crosslinkable with Catalysts

Fiber-Reinforced Laminate Coatings. Fiber-reinforced reaction resins (see also VDI-Richtlinie 2536) are used mainly to coat large storage tanks on site. Generally the coating is a fiberglass-mat-reinforced laminate, but sometimes it is applied by the fiber spraying technique. The coatings are based on unsaturated polyester, vinyl ester, epoxy, and furan resins, and, like those described in Section 5.1.3.3, they are thermosetting. The structure of a laminate coating is shown in Figure 100. Plant protected in this way generally consists of steel, but in some cases of concrete. Coatings highly resistant to acids, alkalies, or even solvents can be produced, depending on the type of resin chosen. Storage tanks for concentrated chlorine bleaching solution have been protected successfully with fiberglass-reinforced unsaturated polyester resins based on bisphenol A fumarates or vinylesters based on bisphenol A acrylates. For more than 10 years fiberglass-filled vinyl esters based on epoxy–

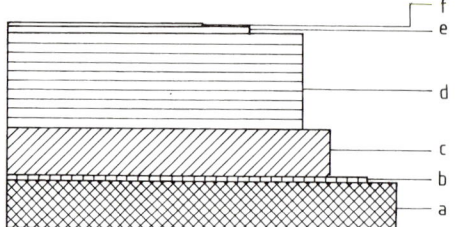

Figure 100. Structure of a laminate coating (schematic)
a) Substrate (steel or concrete); b) Primer (ca. 0.1 mm); c) Grout (1 mm); d) Glass mats (2 mm); e) Glass veil; f) Sealing coat (0.2 mm)

novolak acrylates have been protecting tanks of up to 5000 m^3 for the treatment of acidic or alkaline wastewater containing organic compounds. Fiberglass–furan (FU) laminates withstand solvents best, even at elevated temperatures. At the moment, however, the lack of ductility exhibited by FU resins restricts their application to containers of up to 100 m^3 in volume. Because of the differences between the expansion coefficients of the substrate and coating, fiberglass laminates can only be exposed to temperatures of up to 70 °C.

Grouts and Spray Coats Reinforced with Glass Flakes. Thermosetting unsaturated polyester (PE) and vinyl ester (VE) resins reinforced by glass flakes, which have a thickness of only a few microns and lie parallel to the surface of the substrate, are applied as grouts and sprayed coats, especially to protect gas ducts in flue gas desulfurization units, even in cases where condensates form [184]. Other applications of these easily applied coatings, which, when silanized flakes are used, are also relatively diffusion-proof, include the covers of large rubber-lined, brick-lined, or laminate-coated tanks. Details of their structure are given in Figure 101.

5.1.3.5. Thermoplastic Linings

These are available as semi-finished goods in the form of foils, sheets, panels, and tubes. The preferred materials are poly(vinyl chloride) with or without plasticizers (PVC-U, PVC-P), high-density polyethylene (PE-HD), polypropylene (PP), poly(vinylidene fluoride) (PVDF), ethylene chlorotrifluoroethylene (E/CTFE), tetrafluoroethylene/hexafluoropropylene (FEP), and polytetrafluoroethylene (PTFE). They are used to protect plant made of steel, concrete, or fiber-

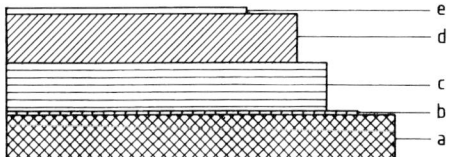

Figure 101. Structure of a grout coating reinforced with glass flakes (schematic)
a) Steel; b) Primer (ca. 0.1 mm); c) Grout (ca. 0.9 mm, 3 × 2-mm glass flakes); d) Grout (ca. 0.9 mm, 3 × 2-mm glass flakes); e) Topcoat (ca. 0.1 mm)

Figure 102. PTFE lining with welded-in sockets, before assembly

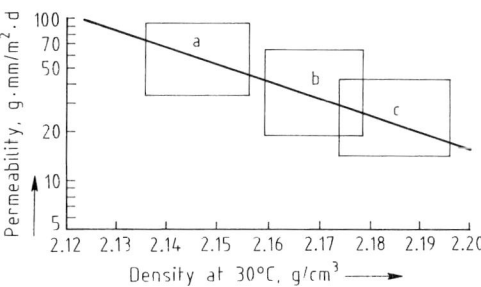

Figure 103. Permeability of PTFE in water at 137.5 °C and 0.34 MPa as a function of density
a) Ram-extruded products; b) Paste-extruded products; c) Peeled foil

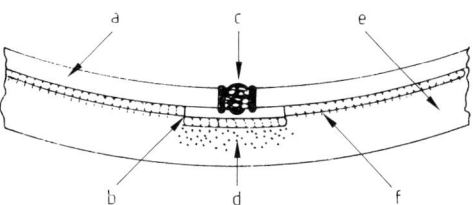

Figure 104. Structure of a fiberglass/thermoplastic laminate [186]
a) Thermoplastic; b) Thermoplastic patch strip; c) Weld; d) Carbon-filled resin target; e) FRP; f) Fiberglass

glass-reinforced unsaturated polyester, vinyl ester or epoxy resins, and their application in the manufacture of vessels and other plant is increasing rapidly. Except for PTFE, which can be processed only by the sintering technique and calls for special methods where forming and welding are required, all of these thermoplastic lining materials can be formed, welded, and bonded without difficulty. Figure 102 shows a loose-insert lining of PTFE peeled foil with welded-in sockets which merely have to be attached to the flanges of the vessel and its sockets after the insert has been introduced. If the lined apparatus is to have a long useful life, especially at high service temperatures, the apparatus must be well insulated against loss of heat and should have ventilation openings in the wall beneath the lining. Figure 103 shows how the resistance of such linings to permeation by gases and steam depends on the density of the PTFE and therefore on the production process [185].

Thermoplastic lining materials, with the exception of PTFE, are attached to the substrate with contact adhesives or with heat-resistant systems based on unsaturated polyester (UP), vinyl or phenacrylic ester (VE), or epoxide (EP), the adhesive or resin composition being applied to a glass or synthetic fiber fabric, the underside of which is fused into the pipe or panel. For service temperatures above 80 °C it is particularly desirable to use glass fabric because this improves resistance to hydrolysis. The use of knitted glass fiber is always recommended if the laminated sheet is likely to be deformed considerably while being processed, as in the fabrication of round-bottomed vessels, columns, or large pipes. The structure of a thermoplastic/fiberglass laminate is shown schematically in Figure 104. The weld joint is additionally reinforced by a fabric-reinforced cover tape. This reinforcement is particularly necessary in the case of linings that must be antistatic or electrically conductive, especially because these cannot be tested for pores and cracks with high-voltage instruments. Liners of this kind having good ductility are now available as copolymers with carbon contents of less than 5 wt%. They have superseded the brittle homopolymers in which up to 25% of carbon had to

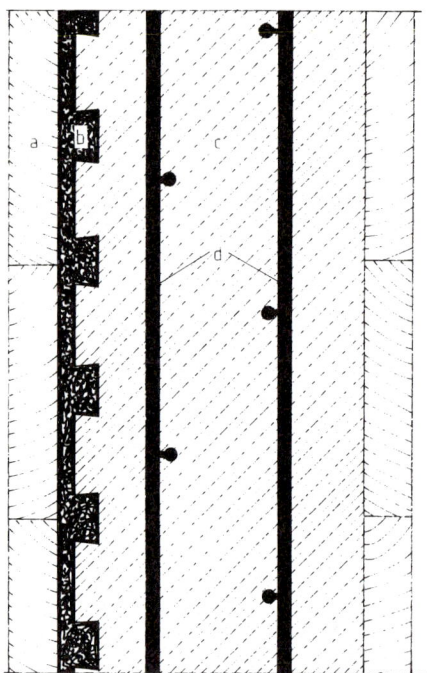

Figure 105. Structure of a Bekaplast lining [188]
a) Casing; b) Bekaplast panel; c) Concrete; d) Reinforcement

be incorporated [187]. Figure 105 shows the structure of Bekaplast, a composite consisting of a thermoplastic liner and concrete.

5.1.3.6. Paints and Powder Coatings

Liquid Coating Materials. Liquid coating materials are classified as physically drying (here the coating material gels after the solvent or dispersing agent has evaporated), thermally crosslinking, or catalytically crosslinking. Their applications depend on the polymer base, the thickness at which the coat can be applied, and—where heated plant items are concerned—on the magnitude and direction of the heat flow. In the chemical industry tube bundle heat exchangers coated on the cooling water side with stoving finishes are an important application of liquid coating materials in which they have shown that they are able to withstand the respective operating conditions continuously for up to ten years. Large storage tanks are being coated increasingly with epoxy-based high-solids formulations or with solvent-free polyurethane (PU), EP, UP, or VE resins by the low- or high-pressure process, and with or without heat, depending on the technical requirements of the system. The PU and EP resins are also modified with mineral coaltar pitch, particularly to reduce their cost and improve their stability to aqueous media [189].

Powder Coatings. Temperature sensitive plant items can be coated reliably with thermoplastics or thermosetting plastics by the techniques of fluidized-bed sintering, powder sintering, flame spraying, and electrostatic powder coating. The range of applications is similar to that of catalytically and thermally crosslinking resins. Details of surface protection by powder plastics that are melted to form coatings can be found in VDI-Richtlinie 2538.

In powder sintering, the powder is sprinkled, tipped, or flung onto the metal parts after they have been heated to 200–400 °C. Excess powder is removed and the plastic coat formed by melting is reheated until it becomes a continuous film.

In fluidized-bed sinter coating the heated substrate is dipped into a fluidized bed of the plastic powder. A 0.1–1.5 mm-thick film is formed, depending on the heat content of the part.

In electrostatic powder spraying, the plastic powder is sprayed by compressed air from a spray nozzle to which a high d.c. voltage is applied. The current gives the plastic particles an electric charge which causes most of them to land on the earthed substrate as soon as this enters the vicinity of the polymer dust cloud. The difference in polarity between the substrate and the particles and the throwing power ensures that the metal object is uniformly coated, even on surfaces that do not face the cloud. The residual powder is separated in a cyclone and reused. Films with a thickness of 50–1000 μm can be obtained, depending on the coating equipment and properties of the powder. The powder, which adheres to the metal through Coulomb forces, is melted in a stove to form a continuous film. Not only components, but also complete apparatuses, are treated in this way.

5.2. Inhibitors [190], [191]

Inhibitors are substances that reduce or eliminate the aggressiveness of a corrosive medium and are either already contained in the corrosive medium or are specifically added to it. A distinction is made between electrochemical, chemical, and physical inhibitors [192].

Electrochemical inhibitors retard or prevent the anodic and/or cathodic partial reactions, i.e., they influence the reaction at the metal/corrosive medium interface. Chemical inhibitors can react both with the material and form protective coatings and with the medium itself or its constituents and thus diminish its aggressiveness. Physical inhibitors form adsorption layers on the metal surface, which block the corrosion reaction. Inhibitors that influence the electrochemical electrode reactions are subdivided according to their mode of action and site of action in the area of the metal/medium phase boundary, the subdivision being between interface inhibitors, electrolyte film inhibitors, membrane inhibitors, and passivators.

Industrial application of inhibitors involves acidic, neutral, and alkaline solutions as well as the gas phase [193]. Physical and electrochemical inhibitors can be used for acids, whereas all three types are employed in neutral and alkaline solutions. Gas and vapor phase inhibitors (i.e., substances with high vapor pressure) have gained importance, especially in atmospheric corrosion protection in storage and transportation [194].

The following compounds are used as inhibitors in acid solutions: amines, amino-imidazolines, amino- and nitrophenols, aminotriazole, aldehydes, benzothiazol, dibenzyl sulfoxide, dithiophosphonic acids, guanidine derivatives, ureas, phosphonium salts, sulfonium salts, sulfonic acids, thioethers, thioureas, and thiocarbanoyl disulfides. Amino alcohols, aminobenzimidazole, benzoates, quinoline derivatives, cinnamates, fatty amines, polyether amines, silicates, and triazoles are used as inhibitors in neutral or weakly alkaline solutions, while for strongly alkaline solutions, aldehydes and fatty amines are used.

Inhibitors exist for the protection of metals and their alloys [195]–[199] and for the most varied of uses, including pickling acids, service water, brine, coolants, oils, heat-transfer agents, and electroplating baths [200]–[202]. Often, using a combination of several inhibitors gives a synergistic effect, whereby the protective action achieved generally far exceeds the sum of the effectivenesses of the individual components.

For the chemical industry inhibitors in open and closed water circulation systems and in oil refining are highly significant [202]. In oil refineries, for example, the cracking that may occur in heat-treatable steels through the action of weakly acidic solutions containing H_2S and H_2O can be eliminated by the addition of ammonia saturated with oxygen. The H_2S is oxidized to polysulfide, which forms a good protective coating on the steel [110].

Corrosion inhibitors for mineral oils include alkyl imidazoles, amines, aminopyridines, quaternary ammonium salts, alkyl sulfonates, borate esters, hydroxy amines, naphthenic acids, phosphate esters, and stearates.

Oxygen and carbon dioxide corrosion can cause serious economic loss in water circulation systems in which water is heated, evaporated, and condensed. Oxygen-binding additives offer good protection (hydrazine, Levoxin, sodium sulfite) and are used extensively in boiler feed water preparation. Carbon dioxide corrosion can be controlled by neutralizing and film-forming amines and by vapor phase inhibitors. In practice derivatives of phosphoric acid, organophosphates, and chelates have proved successful as inhibitors in open cooling circulation systems.

Note, however, that there are conditions under which inhibitors can give rise to detrimental local corrosion (pitting corrosion). This is the case when the amount of inhibitor is insufficient. Under these conditions only part of the surface can be covered, thus giving rise to a local element. Corrosive attack is particularly extensive at the uncovered anode areas because of increased corrosion current density, and deep cavities penetrate into the material. Similarly, if the inhibitor is too readily reduced at the cathodic areas of the metal surface, increased corrosion can result because compact protective films are not formed. Since there are no universally applicable inhibitors they must be carefully selected and examined for each specific case. In doing so, inhibition of metal dissolution is not the only point to be considered; there is also hydrogen absorption [202].

In many cases corrosimeters are successfully used to monitor inhibition.

6. Electrochemical Corrosion Protection

Potential plays an essential role in corrosion processes. Not only the corrosion rate but also the occurrence of local corrosion phenomena, such as pitting and stress corrosion cracking, depend on the potential. If the critical potential range for corrosion is known, a condition can be established through polarization in which no corrosion occurs or it is negligible. Depending on the direction of the polarization this electrochemical corrosion protection is anodic or cathodic [203].

In addition to the well-known application of cathodic corrosion protection to underground pipelines, there has been an increased use for the internal protection of containers and pipes. Initially, galvanic anodes were used to this effect, like the ones now used, for example, to protect the interiors of tankers and boilers. However, since these anodes are often subject to heavy inherent corrosion, especially with the highly aggressive media often found in the chemical industry, external current systems with insoluble anode material have now largely replaced these.

The possibilities of cathodic protection are limited in two respects. The cathodic polarization required for protection can lead to cathode corrosion in some systems. Hydrogen evolution can also cause damage such as embrittlement of the steel and debonding of a protective coating (e.g., paint). This kind of protective coating is often combined with cathodic protection in order to keep the current demand low. The attack of acidic solutions on equipment parts cannot therefore be prevented in general by a cathodic protection system. Cathodic polarization without evolution of large amounts of hydrogen is at best a possibility with copper alloys in acidic solutions [204].

For these reasons the cathodic corrosion protection of chemical plant parts is restricted to neutral solutions, service water, and alkalies. Various types of anode are available for the protection of container interiors. Platinized disk electrodes, bar anodes, and titanium basket anodes are used [205], [206]. They are now supplied ready for installation and are designed in such a way that in continuous operation they can emit 8–10 A at a current density of 6–8 A/dm^2.

Aluminum and ferrosilicon anodes are suitable for the protection of hot-water tanks [207]. The use of aluminum anodes in a hot-water apparatus by the Guldager electrolysis method [208] has additional effects: an indirect protective action on subsequent piping, since a protective film (1–1.5 mm thick) is built up by anode-formed aluminum oxide hydrates. In addition to alkalization at the cathode, a shift in the $CaCO_3 - CO_2$ equilibrium leads to partial softening of water, $CaCO_3$ being deposited at the cathode.

Corrosion protection through imposition of a passive state in metallic materials by anodic polarization has found some applications in the chemical industry.

The practical design of equipment for anodic protection requires intensive laboratory studies to determine the passivation current densities and the potential range of passivity as a function of various parameters such as temperature, concentration, and rate of flow for the respective application. The properties of the cathode material must also be studied.

The protective current must be potentiostatically controlled if the material to be protected shows rapid activation on disconnection. Intermittent operation may also be chosen, whereby the protection current connects or disconnects on dropping below or rising above a limit potential.

All common systems can be used as reference electrodes, e.g., calomel, Ag/AgCl, and Mg/MgO electrodes, which help to monitor the apparatus and control the on-off processes. Platinum, platinized materials, tantalum, and lead are used as cathode materials in acidic solutions. Nickel is suitable in alkaline solutions.

Anodic protection against acidic solutions has been used in a number of chemical processes and in the transportation and storage of liquids. Unalloyed steels can be protected in this way in salt solutions with nitrates and sulfates and in nitric and sulfuric acid [209], although there are limits imposed in sulfuric acid by temperature and concentration [210]. Stainless chromium and chromium–nickel steels are particularly suited to anodic protection. It has so far been practiced with H_2SO_4, oleum, and H_3PO_4.

In the production of sulfuric acid, including heat recovery and the reconditioning and recycling of spent acids, it is necessary to handle acids at elevated temperatures and various concentrations. Corrosion damage that considerably impairs the availability of plant has occurred in sulfuric acid coolers, for example. Damage of this kind can be prevented by anodic protection.

The commonly used austenitic stainless steels exhibit satisfactory resistance to corrosion by sulfuric acid at low concentrations ($< 20\%$), and high concentrations ($> 70\%$) below a critical temperature. If at high sulfuric acid concentrations (> 90%) the temperature exceeds ca. 70 °C, corrosion, differing in severity according to the composition of the steel concerned, occurs, and the steel may alternate between the active and passive states [211].

Anodic protection enables materials to be used under unfavorable conditions, provided they are passifiable in sulfuric acid. In the handling of sulfuric acid at concentrations of 93–99%, Cr–Ni steels (material nos. 1.4541 and 1.4571) can be used economically at temperatures up to 160 °C. This allows operation within a range of temperatures (120–160 °C) suitable for heat recovery.

The anodic protection technique now enables air coolers and tube bundles in sulfuric acid plants to be protected from corrosion reliably and economically. In 1966 anodic protection was provided in the Federal Republic of Germany for the air coolers of a sulfuric acid production plant for the first time. Since then, a combined cooler surface area exceeding 10 000 m² in air-cooled and water-cooled sulfuric acid plants has been protected in this way worldwide. The installed initial electrical direct current output of the potentiostats is > 25 kW, corresponding to an energy requirement of 2.5 W/m² for the surface needing protection [212].

Owing to the narrowness of the potential range within which stress corrosion cracking occurs in unalloyed and low-alloy steels in alkalies, anodic protection against this particular kind of corrosion can be provided by impressed current. As the current densities needed to maintain passivity are quite low, this form of protection is also economically favorable, particularly for large plant equipment that cannot be annealed. Cathodic protection, though theoretically possible, cannot be recommended, since excessive surface corrosion caused by removal of top layers from the steel surface would have to be tolerated as the alternative to stress corrosion cracking. Further reasons are that higher current densities are needed and that a power failure would immediately end the protection. A passivated surface, on the other hand, does not become active immediately, so that anodic protection can also be operated intermittently.

Anodic inhibition of stress corrosion cracking was first provided on an industrial scale in a large plant for the production of hydrogen by electrolysis of potassium hydroxide solution [86]. In November 1968, after preliminary trials on a laboratory scale, the chemical industry's

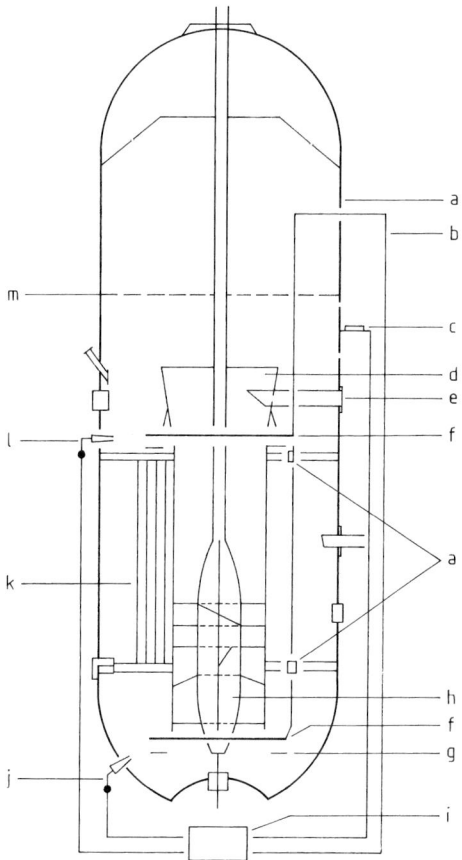

Figure 106. Anodic protection of a caustic vaporizer (content 115 m³, surface area 2400 m²) against stress corrosion cracking
a) PTFE; b) Cathode; c) Anode; d) Central tube e) Liquor entry; f) Ring electrode; g) Insulation; h) Stirrer; i) Potentiostat; j) Electrode E_2; k) Vaporizing tubes; l) Electrode E_1; m) Filling level

first anodically protected large-scale plant, a sodium hydroxide solution evaporator with a capacity of 142 t, was put into operation. In the meantime further plants have been equipped in the same way [214]. The structure of the evaporator and the arrangement of the counter electrodes and reference electrodes are shown schematically in Figure 106.

Because of its good passivity, titanium is also suited for anodic protection, e.g., in mineral and organic acids; tantalum cathodes are generally used [215].

7. Corrosion Testing [216]–[218]

In the design of chemical apparatus, apart from process technology, material selection is especially important. With careful attention paid to economy, material selection must take account of all the requirements arising out of the operating function of the chemical apparatus and the manufacturing process of the product. It is often difficult to comprehend and take account of the interaction of mechanical and chemicocorrosion stresses, which are often compounded by wear.

The resistance parameters needed to select a material can be obtained in two ways:

1) by incorporating suitable test specimens in test plants or in existing, not yet optimally designed production plants, and
2) by corrosion testing in the laboratory.

Both methods of corrosion testing are similiar in that on the corrosive medium side a number of important parameters such as concentration, temperature, dissolved gases, impurities, solids, and the flow rate must be carefully considered. Because of the many parameters involved, when material specimens are incorporated in a test or production plant, efforts should always be made to ensure that the specimens are subjected as close as possible to operating stress. The material specimen is geared to the types of corrosion expected.

Material specimens are mostly attached to internals, e.g., to an agitator or a thermometer protection tube. To avoid unwanted polarization of the specimens through contact with plant components and a subsequent incorrect measurement, the specimens are bolted with insulating sleeves. In many cases it is sufficient to wrap the bolts with PTFE tape. At low flow rates and low temperatures the specimens can be incorporated with the help of plastic binding or tape.

If corrosion tests in pilot plants are not possible, laboratory tests sufficiently close to practice should be carried out. These should also be carried out when operating tests do not clearly indicate the time function of corrosion processes and thus more intense corrosion conditions can complete the picture.

Three groups of variables determine the corrosion of metallic materials [219]:

1) material variables (chemical composition, heat treatment, and surface condition);
2) corrosive medium variables (pH, temperature, flow rate); and
3) potential.

In a laboratory test used as the basis for material selection the biggest problem is the correct choice of these variables, which is the determining factor for simulating operating conditions.

Chemical corrosion tests focus primarily on resistance to surface and selective corrosion. In general the effect of material and corrosive medium variables can be understood with these methods. The variable potential, on the other hand, is more or less undefined and can experience time changes depending on the properties of the various partial reactions involved in corrosion. The fluctuation range of the potentials found in practice cannot be taken into acount in the immersion test. Often, therefore, the results obtained in chemical corrosion testing using electrochemical methods must be further differentiated to take account of the variable potential. If the corrosion rate determined in chemical corrosion testing depends heavily on the potential, chemical testing has little to offer, but if there is only little potential dependence it is more reliable.

By means of electrochemical corrosion tests, the dependence of corrosion on the potential can be investigated and indicators obtained as to the parameters which influence the potential. There can be fundamental differences between the results of chemical and electrochemical corrosion tests. This is due to the fact that in chemical tests the potential can change with time. This is deliberately avoided in electrochemical tests, with partly increased predictability and partial restriction of the potential practical uses, e.g., in the event of strong time dependence. Dependence on the following variables is a major consideration in electrochemical corrosion tests: potential U_H (in relation to the standard hydrogen electrode), current density i (in relation to the geometric surface of the specimen), and time t.

7.1. Test Methods

Process-Related Corrosion Tests. As mentioned above, tests to determine the corrosion resistance of materials in industrial corrosive media can be performed in the laboratory, in pilot plants, and in existing production plants. With regard to the choice of locations for specimens it should be borne in mind that reactors and other

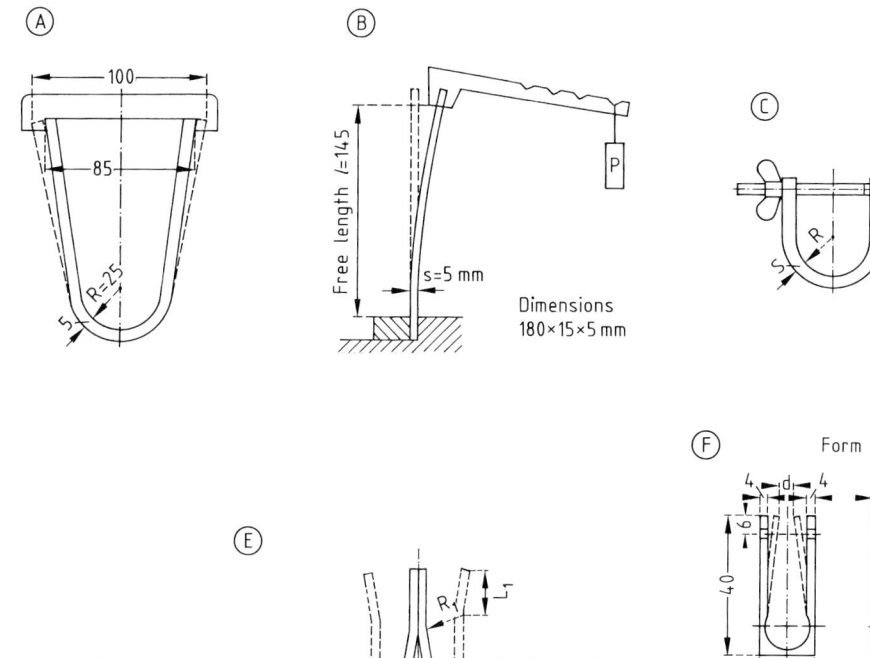

Figure 107. Stress corrosion cracking specimens
A), C), and E) Typical U-bent specimens; B) Cantilever beam specimen; D) Bent beam specimen, three point loaded; F) Tuning fork specimen

apparatus may be attacked differently by the liquid and vapor phases of the corrosive agent and at the three-phase interface of the liquid, vapor, and material. As a rule, therefore, material specimens must be exposed at each of these phases. Corrosion testing principles are described in [220]–[222].

The characteristics of individual forms of corrosion are taken into consideration by providing appropriate corrosion specimens. Welded coupons having the surface quality of the material used later in practice are sufficient for determining uniform corrosion rates and acquiring general information on the type of local corrosion. Resistance to crevice corrosion can be determined by using specimens as described in ASTM G 78 [223]. Conditions of heat transfer can be simulated by using hot-wall/cool-wall specimens under temperature-controlled conditions [221].

Sensitivity to stress corrosion cracking can be determined with elastically or plastically prestressed specimens, such as those of the Jones, U-bent, and C-ring types (Fig. 107) [224]–[226]. Specimens sensitive to stress corrosion cracking can also be produced by introducing heavy tensile stresses into the surface by means of stress-inducing grinding [227].

Laboratory corrosion tests are discussed in [220], [222]. To adjust the parameters of temperature and/or pressure it may be necessary to perform corrosion tests in an autoclave.

In pilot and production plants, the flow conditions are determined by the nature of the apparatus and process, but in laboratory tests they have to be individually chosen. To determine the effects of static or very gently moving media it is sufficient to stir the medium with an agitator. If, however, exposure of the material to flowing media is expected, special corrosion tests are essential for simulation: for example, circulation tests with pipe or channel flow and use of rotating discs or cylinders as specimens [228].

Special laboratory investigations of the resistance to stress corrosion cracking may include the use of machinery. In such cases tensile test specimens are tested under defined mechanical conditions, such as constant stress or constant strain rate [126].

Standardized Methods. Standardized corrosion tests are performed in the laboratory with standardized solutions; they are used mainly to determine general corrosion characteristics, as well as in fundamental investigations and quality control. Various methods [229]–[234], the choice of which depends on the type of material concerned, are used to determine the resistance of highly alloyed steels and nickel-based alloys to intergranular corrosion. There is also a standardized method for determining resistance to pitting and crevice corrosion [235]. The general resistance of various materials to stress corrosion cracking (SCC) can likewise be determined in test solutions. Depending on the material and corrodent, the following standards are used:

1) Stainless steels, transgranular SCC: $MgCl_2$ solution [216], [236]; NaCl solution [241]
2) Unalloyed and low-alloy steels: intergranular SCC [224]; hydrogen-induced cracking H_2S [237]
3) Cu materials in NH_3 [238]; in NH_4^+ solution [239]
4) Al materials [240], [241]

The standard methods also include those used particularly, but not exclusively, to test anticorrosion coatings in corrosive atmospheres, e.g., those containing SO_2 and Cl^-.

Field Tests. Field tests include those in which specimens are surrounded by aggressive soils, atmospheres, or waters (e.g., seawater). Atmospheric and water tests, which are performed both on unprotected and on protected (coated) materials, require special methods [242, Parts 1 and 2]. Here it is important that control specimens be tested simultaneously in order to predetermine the corrosive conditions at the testing site.

Atmospheric Corrosion Testing. It is customary to perform tests in special climate-chambers in addition to field tests. The tests are used for comparison, but are also valuable for determining the behavior of anticorrosive films and coatings. The conditions used to obtain the appropriate atmospheres, constant or alternating condensed water climates, with and without the presence of such additional substances as sulfur dioxide and salt spray, at various pH values, are specified in [243]–[249].

Electrochemical Measurement. Electrochemical methods are used mainly in the laboratory [250]–[252]. In special cases (potential measurements; corrosion meters) they can also be used in industrial plants.

Electrochemical methods, in contrast to chemical methods, provide mainly qualitative data on corrosion systems. They are very suitable for studying corrosion mechanisms and the influences of parameters, but also for investigating the value and effectiveness of active corrosion protection measures. Current density versus potential curves (see Chap. 2, Figs. 13, 17) provide information on the effects of alloy constituents and inhibitors. Quantitative data on loss of mass rates can be gained by extrapolating the Tafel lines or by measuring the polarization resistance (see Section 2.3; Figs. 9, 10). The latter method also forms the basis of a number of commercially available corrosion meters, which are used to determine corrosion rates and to investigate the effectiveness of inhibitors.

Depending on the purpose of an investigation, such tests are performed either at a given potential (potentiostatic tests) or at a potential which changes in a defined manner (potentiodynamic tests). Galvanostatic or galvanodynamic tests are also performed frequently. Besides investigation of the relationship between current density and potential, simple measurement of the (free) corrosion potential U_{corr} (open loop potential), or its dependence on time, is a useful way of studying corrosion systems, for the level at which the potential of the material settles is very important with regard to the damage suffered. Potential measurements are also fundamental, though not alone sufficient, in the investigation of contact corrosion [253], [254]. Electrochemical methods are also used successfully in the basic

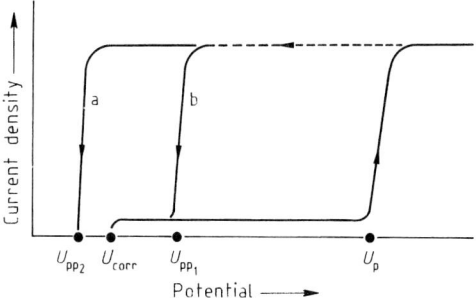

Figure 108. Determination of susceptibility to pitting corrosion (schematic potentiodynamic current density versus potential curves)
a) $U_{pp} < U_{corr}$: pitting corrosion susceptibility; b) $U_{pp} > U_{corr}$: no pitting corrosion susceptibility

investigation of local corrosion, e.g., crevice and pitting corrosion.

A special method for determining the sensitivity of stainless steels to pitting corrosion in industrial corrodents is the measurement of the pitting passivation potential U_{pp}. Here the potential, which is initially adjusted to the free corrosion potential U_{corr}, is slowly increased until, at the pitting potential U_p (see Section 3.2.1) pitting—recognizable from the steep increase of the current (Fig. 108)—is forced to occur. After the rate of pitting corosion has been stabilized by a further increase of potential the scanning direction is reversed until the current acquires values similar to those of the passive current density. The potential at which this occurs is known as the pitting passivation potential U_{pp}. Comparison of U_{pp} with U_{corr} indicates whether pitting corrosion will occur ($U_{pp} < U_{corr}$) or not ($U_{pp} > U_{corr}$) in the corrosion system concerned [255]. The main advantage of this process is that the repassivation potential U_{pp} is more easily reproduced than the pitting potential U_p, and that within wide limits it is independent of the rate of change of the potential.

8. References

General references

H. H. Uhlig: *The Corrosion Handbook*, J. Wiley & Sons, New York/Chapmann & Hall Ltd., London 1948.
F. N. Speller: *Corrosion, Causes and Prevention*, McGraw-Hill, New York – London 1951.
F. Ritter: *Korrosionstabellen metallischer Werkstoffe*, 4th ed., Springer Verlag, Wien 1958.
F. Tödt: *Korrosion und Korrosionsschutz*, 2nd ed., de Gruyter, Berlin 1961.
K. Vetter: *Elektrochemische Kinetik*, Springer Verlag, Berlin – Göttingen – Heidelberg 1961.
F. L. Laque, H. R. Copson: *Corrosion Resistance of Metals and Alloys*, 2nd ed., Reinhold Publ., New York 1963.
U. R. Evans: *Einführung in die Korrosion der Metalle*, Verlag Chemie, Weinheim 1965.
U. R. Evans: *The Corrosion and Oxidation of Metals; Scientific Principles and Practical Applications*, 1st. and 2nd suppl. vols., Edward Arnold Ltd., London 1960, 1968, 1976.
H. Kaesche: *Die Korrosion der Metalle*, 2nd ed., Springer Verlag, Berlin – Heidelberg – New York 1979.
Mannesmann-Röhren-Werke, *Lexikon der Korrosion*, vols. 1 and 2, Düsseldorf 1970.
H. E. Hömig: *Metall und Wasser. Eine Einführung in die Korrosionskunde*, Vulkan-Verlag, Essen 1971.
H. Gräfen, F. Kahl, A. Rahmel (eds.): *1. Korrosionum: Die Bedeutung der Korrosion für Planung, Bau und Betrieb von Anlagen der chemischen und petrochemischen Technik sowie in der Mineralölindustrie*, Verlag Chemie, Weinheim 1974.
B. J. Monitz, W. I. Pollock (eds.): *Process Industries Corrosion*, NACE, Houston, Texas, 1986.
R. N. Parkins (ed.): *Corrosion Processes*, Applied Science Publishers Ltd., London – New York 1982.
D. Kuron (ed.): *Wasserstoff und Korrosion*, Bonner Studien Reihe, Verlag I. Kuron, Bonn 1986.
R. Gibala, R. F. Hehemann (eds.): *Hydrogen Embrittlement and Stress Corrosion Cracking*. Papers presented at a symposium held at Case Western Reverse University on June 1 – 3, 1980, ASM, Ohio 1984.
G. Kortüm: *Lehrbuch der Elektrochemie*, 5th ed., Verlag Chemie, Weinheim 1972.
C. H. Hamann, W. Vielstich: *Elektrochemie I*, Verlag Chemie, Weinheim 1975.
Verein Deutscher Eisenhüttenleute (ed.): *Prüfung und Untersuchung der Korrosionsbeständigkeit von Stählen*, Verlag Stahleisen, Düsseldorf 1973.
A. Rahmel, W. Schwenk: *Korrosion und Korrosionsschutz von Stählen*, Verlag Chemie, Weinheim 1977.
H. Gräfen et al. in W. J. Bartz, Techn. Akademie Esslingen (ed.): *Die Praxis des Korrosionsschutzes, Kontakt und Studium*, vol. 64, Expert Verlag, Grafenau 1981.
DIN-Taschenbuch 219: *Korrosion und Korrosionsschutz, Beurteilung – Prüfung – Schutzmaßnahmen*, Beuth-Verlag, Berlin – Köln 1987.
W. Fischer (ed.): *Korrosionsschutz durch Information und Normung*, Kommentar zum DIN-Taschenbuch 219, Schriftenreihe der Arbeitsgemeinschaft Korrosion e. V., Verlag I. Kuron, Bonn 1988.
Annual Book of ASTM Standards 1988, Section 3; *Metals Test Methods and Analytical Procedures*, vol. 03.02, Wear and Erosion, Metal Corrosion, American Society for Testing and Materials, Philadelphia 1988.
W. v. Baeckmann, W. Schwenk, W. Prinz: *Handbuch des kathodischen Korrosionsschutzes*, 3rd ed., VCH Verlagsgesellschaft, Weinheim 1989.
P. J. Gellings: *Introduction to Corrosion Prevention and Control for Engineers*, Delft University Press, Delft 1976.
L. L. Shreir: *Corrosion*, 2nd ed., vols. 1 and 2, Newnes-Butterworth, London – Boston 1976.
Corrosion Data Survey Metals Section, 6th ed., NACE, Houston, Texas, 1985.
D. Behrens (ed.): *DECHEMA Corrosion Handbook*, Corrosive Agents and Their Interaction with Materials,

vols. 1–5, VCH Verlagsgesellschaft, Weinheim – New York 1987.
M. Fischer, K. Hauffe, W. Wiederholt: *Passivierende Filme und Deckschichten*, Springer Verlag, Berlin 1956.
K. Hauffe: *Oxidation von Metallen und Metallegierungen*, Springer Verlag, Berlin 1956.
O. Kubaschewski, B. E. Hopkins: *Oxidation of Metals*, Butterworth, London 1962.
J. Bernard: *L'Oxydation des Metaux*, vols. 1 and 2, Gauthier-Villars, Paris 1962 and 1964.
M. Pfeiffer, M. Thomas: *Zunderfeste Legierungen*, Springer Verlag, Berlin 1963.
P. Kofstad: *High-Temperature Oxidation of Metals*, J. Wiley, New York 1966.
J. C. Scully: *Fundamentals of Corrosion*, Pergamon Press, Oxford 1966.
Z. A. Foroulis: *High Temperature Metallic Corrosion of Sulfur and its Compounds*, The Electrochemical Society, New York 1970.
H.-J. Engell, A. Rahmel: "Verzunderung von Metallen durch Gase. 50 Jahre Tamman'sche Zunderformel," *Korrosion* **23** (1971).
A. Rahmel, H. Manenc: "Mechanische Eigenschaften und Haftung von Zunderschichten. Einfluß auf die Oxidation von Metallen," *Korrosion* **24** (1973).

Specific references

[1] DIN 50 900, Korrosion der Metalle. Begriffe, Part 1, Allgemeine Begriffe, Beuth-Verlag, Berlin–Köln 1982.
[2] DIN 50 900, Korrosion der Metalle. Begriffe, Part 2, Elektrochemische Begriffe, Beuth-Verlag, Berlin–Köln 1984.
[3] C. Wagner, W. Traud, *Z. Elektrochem. Angew. Phys. Chem.* **44** (1938) 391–402.
[4] M. Stern, A. L. Geary, *J. Electrochem. Soc.* **104** (1957) 56–63.
[5] E. Heitz, *Werkst. Korros.* **19** (1968) 773–781.
[6] E. Heitz, W. Schwenk, *Werkst. Korros.* **27** (1976) 241–245.
[7] U. R. Evans, *J. Inst. Met.* **30** (1923) 239.
[8] H. Kaesche, *Z. Metallkd.* **61** (1970) 94–101.
[9] U. Gramberg, E.-M. Horn, *VDI-Ber.* **243** (1975) 75–92.
[10] H. Gräfen, K. Gerischer, E.-M. Horn, *Z. Werkstofftech.* **4** (1973) no. 4, 169–186.
[11] E.-M. Horn, D. Kuron, H. Gräfen, *Z. Werkstofftech.* **8** (1977) 37–68.
[12] E.-M. Horn, *VDI-Ber.* **235** (1975) 39–49.
[13] K. Lorenz, G. Medawar, *Thyssenforschung* **1** (1969) no. 3, 97–108.
[14] E. Wallis: *Einfluß des Molybdängehaltes auf die Lochkorrosionsbeständigkeit hochlegierter nichtrostender Stähle und Nickel–Chrom–Eisen-Legierungen mit unterschiedlichen Gefügezuständen unter Berücksichtigung der Wirksummenrelation Massen-% Cr + 3,3 Massen-% Mo*, Dissertation, TU Clausthal 1988.
[15] M. Renner, U. Heubner, M. B. Rockel, E. Wallis, *Werkst. Korros.* **37** (1986) 183–190.
[16] G. Herbsleb, *Werkst. Korros.* **33** (1982) 334–340.
[17] DIN 17 440, issued 07. 85; VdTÜV-Werkstoffblatt 405, issued 12. 87.
[18] VdTÜV-Werkstoffblatt 486, issued 09. 87.
[19] Stahl-Eisen-Werkstoffblatt des Vereins Deutscher Eisenhüttenleute, SEW 400, issued 04. 88.
[20] VdTÜV-Werkstoffblatt 415, issued 09. 84.
[21] A. Bäumel, E.-M. Horn, G. Siebers, *Werkst. Korros.* **23** (1972) no. 11, 973–983.
[22] H. Brandis, G. Lennartz, R. Oppenheim, *DECHEMA-Monogr.* **78** (1975) 235–258.
[23] A. Diebold, H. Weingerl, *Werkst. Korros.* **25** (1974) no. 3, 175–179.
[24] J. C. Carlen, B. Kwarnbäck, *Werkst. Korros.* **25** (1974) no. 9, 653–658.
[25] W. Stewen, H. Jörgens, M. Remke, M. Litzkendorf, *Glückauf* **121** (1985) no. 13, 1030–1033.
[26] VdTÜV-Werkstoffblatt 421, issued 02. 82/03. 84/09. 86.
[27] VdTÜV-Werkstoffblatt 473, issued 06. 85.
[28] M. B. Rockel, M. Renner, *Werkst. Korros.* **35** (1984) 537–542.
[29] *Nickel Topics* **34** (1981) no. 4, 15.
[30] J. C. Crum, E. L. Hibner, R. W. Ross, Jr., *Mater. Perform.* **20** (1981) no. 2, 9–13.
[31] Sandvik Steel AB: Sandvik Sanicro 28 – References and booked orders for the process industry, Oct. 1985.
[32] Avesta Technical Information: Avesta 254 SMO – Referenzen und Anwendungsgebiete, issued 02. 1982.
[33] W. Heimann, *VDI-Ber.* **674** (1988) 41–63.
[34] R. Kirchheiner, W. Römer: VDM-Report no. 9, Oct. 1987.
[35] H. Kaesche, *Werkst. Korros.* **14** (1963) 557–574.
[36] W. R. Fischer: *Zur Lochfraßkorrosion*, Dissertation, TH Braunschweig 1964.
[37] K. Fässler, *VDI-Ber.* **235** (1975) 51–68.
[38] K. Fässler: "Gesichtspunkte der konstruktiven Gestaltung chemisch beanspruchter Apparate," in H. Gräfen, F. Kahl, A. Rahmel (eds.): *1. Korrosionum, die Bedeutung der Korrosion für Planung, Bau und Betrieb von Anlagen der chemischen und petrochemischen Technik sowie der Mineralölindustrie*, Verlag Chemie, Weinheim 1974, pp. 136–165.
[39] R. Münster: "Erscheinungsformen der Korrosion," in [38] pp. 3–29.
[40] G. Lennartz, *VDI-Ber.* **235** (1975) 169–182.
[41] G. Herbsleb, *VDI-Ber.* **243** (1975) 103–118.
[42] E.-M. Horn, T. Günther, *Chem. Ing. Tech.* **41** (1969) no. 18, 991–999.
[43] G. Grützner: *Über die interkristalline Korrosion stickstofflegierter 18/10 Chrom-Nickel-Stähle*, Dissertation, RWTH Aachen 1971.
[44] H. Gräfen, G. Böhm, *Z. Metallkd.* **51** (1960) 245–252.
[45] A. Bäumel, E.-M. Horn, H. Gräfen, *Proc. Int. Cong. Met. Corros.* 5th **1972**, 934–941.
[46] A. Kratzer, B. Pieger, H. Tischner, E.-M. Horn: "A Low Carbon Stainless Steel for Use in Nitric Acid at Higher Concentrations and Temperatures," ICMC, 9th International Congress on Metallic Corrosion, Toronto, Canada, June 3–7, 1984.
[47] E.-M. Horn, H. Kohl, *Werkst. Korros.* **37** (1986) 57–69.
[48] E.-M. Horn, A. Kügler, *Z. Werkstofftech.* **8** (1977) 362–370, 410–417.
[49] H. J. Rocha, *DEW-Tech. Ber.* **2** (1962) no. 1, 16–24.
[50] Mannesmann-Röhren-Werke: *Lexikon der Korrosion*, vol. 1, Düsseldorf 1970, p. 55.
[51] I. Class, H. Gräfen, *Werkst. Korros.* **11** (1960) no. 9, 529–547.
[52] G. Herbsleb, *Werkst. Korros.* **19** (1968) 406–412.

[53] Publication H-2007 and H-2019 of Cabot Wrought Products Division, now Haynes International, 1983/1984.
[54] E. Scheil, I. Class, H. Gräfen, DE 1 210 566, 1961, US 3 203 792, 1964.
[55] R. B. Leonard, *Corrosion* (*Houston*) **25** (1969) no. 5, 222–228.
[56] R. W. Kirchner, W. L. Silence, *Mater. Prot. Perform.* **10** (1971) no. 1, 11–15.
[57] R. W. Kirchner, F. G. Hodge, *Werkst. Korros.* **24** (1973) no. 12, 1042–1049.
[58] Kh. G. Schmitt-Thomas, G. Fenzel, *Maschinenschaden* **38** (1965) 94–97.
[59] E. Fot, *Schweiz. Arch.* **30** (1964) 329–344, 384–398.
[60] E. Kauczor, *Metall* (*Berlin*) **20** (1966) 1165–1168.
[61] R. Heidersbach, *Corrosion* (*Houston*) **24** (1968) 38–44.
[62] "Quaterly Report: Clarification of De-alloying Phenomena," *Corrosion* (*Houston*) **26** (1970) 445–447.
[63] W. B. Brooks, *Corrosion* (*Houston*) **24** (1968) 171.
[64] L. Leontaritis, E.-M. Horn, *Werkst. Korros.* **31** (1980) 179–185.
[65] P. Rothenbacher, *Corros. Sci.* **10** (1970) 391–400.
[66] H. G. Feller, *Z. Metallkd.* **58** (1967) 875–885.
[67] E. E. Langenegger, F. P. A. Robinson, *Corrosion* (*Houston*) **25** (1969) 59–66.
[68] G. Schikorr: "Häufige Korrosionsschäden an Metallen und ihre Vermeidung," in Landesgewerbeamt Baden-Württemberg (ed.): *Schriften zur Gewerbeförderung*, vol. 3, Verlag K. Wittwer, Stuttgart.
[69] F. Tödt: *Korrosion und Korrosionsschutz*, 2nd ed., De Gruyter, Berlin 1961, p. 418.
[70] D. G. Evans, P. W. Jeffrey: "Exfoliation Corrosion of AlZnMg Alloys," in R. W. Staehle, B. F. Brown, J. Kruger, A. Agrawal (eds.): *Localized Corrosion*, International Corrosion Conference Series, Nat. Assoc. of Corrosion Eng., Houston, Texas 1974, pp. 614–622.
[71] H. Weingerl, H. Straube, R. Blöch, *Werkst. Korros.* **27** (1976) 69–77.
[72] E. Schürmann, H.-J. Voss, *Arch. Eisenhüttenwes.* **48** (1977) 129–132.
[73] A. Diebold, H. Weingerl, *Werkst. Korros.* **28** (1977) 240–243.
[74] A. Bäumel, *Schweißen Schneiden* **27** (1975) 227–230.
[75] A. Bäumel, *Werkst. Korros.* **27** (1976) 687–693.
[76] W. Schwarz, W. Simon, *Ber. Bunsenges. Phys. Chem.* **67** (1963) 108–117.
[77] K. Gerischer, *Z. Metallkd.* **46** (1955) 661.
[78] R. N. Parkins: *5th Symp. Linepipe Research*, Amer. Gas Ass. Cat. No. L 30 174, Report U, Houston 1974.
[79] E. Wendler-Kalsch, *Werkst. Korros.* **31** (1980) 534–542.
[80] W. Schwenk: "Einflußgrößen bei der Korrosion von Stählen unter Rißbildung mit besonderer Berücksichtigung der mechanischen Belastungsparameter," *Mannesmann – Forschungsbericht*, no. 944, 1983.
[81] H. Diekmann et al., *Stahl Eisen* **103** (1983) no. 18, 895–901.
[82] R. Münster, H. Gräfen, *Arch. Eisenhüttenwes.* **36** (1965) 227–284.
[83] W. Rädeker, B. N. Miskra, *Werkst. Korros.* **21** (1970) 691–698.
[84] H. Gräfen, *Mitt. Ver. Großkesselbesitzer* **73** (1961) 280–289.
[85] P. Drodten, *Rheinstahl-Tech.* **10** (1972) no. 3, 97–106.
[86] H. Gräfen D. Kuron, *Arch. Eisenhüttenwes.* **36** (1965) 285–291.
[87] K. Bohnenkamp, *Arch. Eisenhüttenwes.* **39** (1968) 361–368.
[88] E. Wendler-Kalsch, *Corros. Sci.* **23** (1983) no. 6, 601–612.
[89] K. J. Kessler, E. Wendler-Kalsch, *Werkst. Korros.* **28** (1977) 78–85.
[90] K. Matsukura, K. Sato: "Effects of Metallurgical Factors of Low Carbon Steel Sheets on Stress Corrosion Cracking in Methanol Solutions," *Tetsu to Hagane* **63** (1977) no. 6, 1016–1025.
[91] E. Wendler-Kalsch: "Korrosion und Spannungsrißkorrosion niedriglegierter Stähle in Methanol, (Project-no. G 6.1/7)," *FEKKs Symposium*, Lahnstein 1987.
[92] B. E. Wilde: Stress Corrosion Cracking of ASTM A 517, Steel in Liquid Ammonia: Environmental Factors, National Association of Corrosion Engineers, vol. 37, no. 3 (1981).
[93] H. Gräfen et al., *Werkst. Korros.* **36** (1985) 203–215.
[94] L. Lunde: "Ammonia Plant Safety," *AIChE* **24** (1984) 154; *Symp. on Safety in Ammonia Plants and Related Facilities AIChE*, Denver, CO, August 29–31, 1983, report no. iFE/KR/E-83/007.
[95] K. Fäßler, H. Spähn: "Grundlagen der Spannungsrißkorrosion unlegierter Stähle in flüssigem Ammoniak – Einflußgrößen und Gegenmaßnahmen," Essen, VGB-Konf. Werkstoffe, 205–228 (1989).
[96] A. Heuser, H. Spähn, G. H. Wagner, *Materialprüfung* **31** (1989) no. 3, 73–79.
[97] M. Kowaka, S. Nagata, *Boshoku Gijutsu* **21** (1972) no. 4, 165–171.
[98] H. Gräfen, H. Schlecker: "$CO–CO_2–H_2O$," *GWF Gas Wasserfach Gas Erdgas* **126** (1985) no. 4, 195–204.
[99] G. Schmitt:, *GWF Gas Wasserfach Gas Erdgas* **122** (1981) no. 2, 49–54.
[100] G. Schmitt, H. Schlerkmann: *Met. Corros. Proc. Int. Congr. Met. Corros. 8th*, Mainz 1981, vol. 1, p. 426.
[101] H. Gräfen, *Werkst. Korros.* **23** (1972) 247–254.
[102] M. O. Speidel in A. J. Sedriks: *Corrosion of Stainless Steels*, Wiley & Sons, New York 1979, p. 173.
[103] H. Spähn, G. H. Wagner, U. Steinhoff, *TÜ* **14** (1973) no. 10, 292–299.
[104] L. Logan, *J. Res. Nat. Bur. Stand.* (*U.S*) **56** (1956) R.P. 2662, 159.
[105] A. Beerwald, H. Gröber, *Z. Aluminium* **1940**, 502–510.
K. Matthaes, *Korrosion* **9** (1958) 5–21.
[106] H. Gräfen, *Werkst. Korros.* **16** (1965) 876–879.
[107] H. Gräfen, H. Spähn, *Chem. Ing. Tech.* **39** (1967) 138–146.
[108] E.-M. Horn, H. Schlecker: unpublished results.
[109] W. Dahl, H. Stoffels, H. Hengstenberg, C. Düren, *Stahl Eisen* **87** (1967) 125–136.
[110] I. Class, *Werkst. Korros.* **6** (1955) 237–245.
[111] W. Hofmann, W. Rauls, *Arch. Eisenhüttenwes.* **32** (1961) 169–171.
W. Hofmann, W. Rauls, *Arch. Eisenhüttenwes.* **34** (1963) 925–934.
[112] G. Enterlein, M. Kesten, D. Schlegel, K. F. Windgassen, *Z. Werkstofftech.* **13** (1982) 290–297.
[113] H. Gräfen, T. Günther, *Z. Werkstofftech.* **10** (1979) no. 11, 373–390.

[114] H. Spähn, G. H. Wagner, U. Steinhoff, *TÜ* **14** (1973) no. 9, 260–264.
[115] H. Böhm, *Schweiz. Arch.* **33** (1967) 339–363.
[116] C. Bosch: Nobel lecture, 21. 5. 1932 in Stockholm, *VDI Z.* **77** (1933) 305–317; *Chem. Fabr.* **6** (1933) 127–142.
[117] I. Class, *Stahl Eisen* **80** (1960) 1117–1135.
[118] API Refining Department: *Steels for Hydrogen Service at Elevated Temperatures and Pressures in Petroleum Refineries and Petrochemical Plants*, API Publication 941, May 1983 (Copyright 1967 by G. A. Nelson).
[119] H. Spähn, *VDI-Ber.* **235** (1975) 103–115.
[120] U. Gramberg, T. Günther, H. Palla, *Werkst. Korros.* **26** (1975) 461–464.
[121] H. Spähn, *Metalloberfläche* **16** (1962) 267–272.
[122] H. Spähn, *Metalloberfläche* **16** (1962) 335–340.
[123] K. Risch, *Z. Werkstofftech.* **17** (1986) 6–17.
[124] H. Spähn, *Z. Phys. Chem. (Leipzig)* **234** (1967) 1–25.
[125] K. Wellinger, K. Lehr, *Mitt. VGB* **49** (1969) 190–201; *Tech. Wiss. Berichte MPA-Stuttgart* (1969) no. 69-02.
[126] DIN 50 922 (Oct. 1985), Korrosion und Metalle. Untersuchung der Beständigkeit von metallischen Werkstoffen gegen Spannungsrißkorrosion.
[127] C. Wagner, *Z. Phys. Chem. Abt. B* **21** (1933) 25–41; *Z. Phys. Chem. Abt. B* **32** (1936) 447–462.
[128] E. Houdremont, G. Bandel, *Arch. Eisenhüttenwes.* **11** (1937/38) 131–138.
[129] S. Mrowec, T. Welec, T. Werber, *Oxid. Met.* **1** (1969) 93–120.
[130] A. Rahmel, *VDI-Ber.* **235** (1975) 145–154.
[131] A. Preece in: *High-Temperature Steels and Alloys for Gas Turbines*, London 1952, (Spec. Rep. Iron Steel Inst. no. 43), pp. 149–152.
[132] K. Wickert, *Nickel Ber.* **24** (1966) 177–186.
[133] A. Rahmel in H. R. Johnson, D. J. Littler (eds.): *The Mechanism of Corrosion by Fuel Impurities*, Plenum Press, New York 1963, pp. 556–570.
[134] U. Jäkel, W. Schwenk, *Werkst. Korros.* **26** (1975) 521–529.
A. Rahmel, E. Tarar-Moisescu, *Werkst. Korros.* **26** (1975) 513–520.
[135] V. Cihal, "Corrosion Mechanism in Ammonia Synthesis Equipment," *First International Congress on Metallic Corrosion*, London 1962, pp. 591–596.
[136] U. Jäkel, W. Schwenk, *Werkst. Korros.* **22** (1971) 1–7.
[137] V. van Rossum, *Chem. Ing. Tech.* **25** (1953) 481–487.
[138] L. Raichle, *Chem. Ing. Tech.* **28** (1956) 203–213.
[139] Mannesmann-Röhrenwerke, Düsseldorf: *Lexikon der Korrosion*, vol. 1, pp. 133–134.
[140] C. G. Stevens, J. Board, *Br. Corros. J.* **4** (1969) 80–85.
[141] D. Goodison, R. J. Harris, P. Goldenbaum, *Brit. Corros. J.* **4** (1969) 293–300.
[142] H. Pfeiffer, G. Sommer, *Werkst. Korros.* **13** (1962) 667–677.
[143] O. Kubaschewski, B. E. Hopkins: *Oxidation of Metals*, Butterworths, London 1962.
[144] J. Möller, *Metall* **26** (1972) 820–825.
[145] J. Möller, *Ind. Anz.* **96** (1974) no. 38, 856–857.
[146] U. Richter, *Schweißtechnik (Vienna)* **7** (1975) 116–120.
[147] DIN 8553 (April 1970): Verbindungsschweißen plattierter Stähle.
[148] U. Richter, *Schweißen & Schneiden* **25** (1973) 218–220.
[149] J. Chene, P. Luginbühl, *Tech. Rundsch. Sulzer* **2** (1973) 73–78.
[150] W. Ruckdeschel, *Schweißtechnik (Zürich)* **63** (1973) 229–241.
[151] B. Bouaifi, J. Krohn, U. Draugelates, *DECHEMA-Monogr.* **103** (1986) 165–175.
[152] B. Bouaifi, I. Graf, U. Draugelates, *Prakt. Metallogr.* **25** (1988) no. 11, 543–554.
[153] D. Fuchs, H. Preisendanz, P. Schüler, *DEW-Tech. Ber.* **13** (1973) 137–142.
[154] Pfaudler-Werke AG, Data Sheet 249-1: "Resista-Clad, ein Verfahren, das Metalle dauerhaft verbindet."
[155] The Pfaudler Company, Data-Sheet DS49-300-1: "Resista-Clad, Weld-Bonded Cladding Process," 1984.
[156] The Pfaudler Company, Data-Sheet DS49-301-1: "Resista-Clad, Physical & Chemical Performance Data," 1985.
[157] M. X. Cerny, *Power Eng.* (1987) no. 8, 36–37.
[158] W. Burckhardt, *Technik* **26** (1971) 697–704; **27** (1972) 563–570, 528–531; **30** (1975) 239–245.
[159] L. L. Shreir, *Ind. Finish (London)* **8** (1955) 261–267, 389–400.
[160] L. Winkler, *Metall (Berlin)* **31** (1977) 506–509.
[161] J. Wurm, *DECHEMA-Monogr.* **76** (1974) 1486–1504, 81–89.
[162] A. W. Berger, *CAV* (1980) 82–84.
[163] K. Breitwieser, R. Engelmann, *Maschinenmarkt* **85** (1979) 393–400.
[164] W. Paatsch, *Metall (Berlin)* **30** (1976) 332–336.
[165] H. Gräfen, *VDI-Ber.* **624** (1986) 273.
[166] H. Gräfen: "Werkstoffe des Chemieanlagenbaus, Stand der Entwicklung," in *50 Jahre Werkstofftechnik an der Fakultät für Maschinenwesen*, Colloquy at the Technische Universität München (1989) pp. 38–63.
[167] K. Kreisel, E. Protogerakis, *DECHEMA-Monogr.* **103** (1986) 191–205.
[168] H. Schindler: "Nichtmetallische Werkstoffe für die Chemietechnik," *Fortschritte der Verfahrenstechnik*, vol. 15, VDI-Verlag GmbH, Düsseldorf 1977, pp. 445–459.
[169] H. Schindler: "Nichtmetallische Werkstoffe für die Chemietechnik," *Fortschritte der Verfahrenstechnik*, vol. 18, VDI-Verlag GmbH, Düsseldorf 1980, pp. 439–456.
[170] H. Gräfen, U. Gramberg, H. Schindler, *Werkst. Korros.* **30** (1979) 297–307.
[171] A. H. Dietzel: *Emaillierung*, Springer Verlag, Berlin-Heidelberg-New York 1981.
[172] R. Lorentz: "Korrosion von Chemieemail durch wäßrigneutrale Medien," *Mitteilungen des Vereins Deutscher Emailfachleute e.V.*, vol. 34/5, (1986) 65–76.
[173] H. Scharbach: "Glas- und Glaskeramikemail für den technischen Einsatz," *Swiss Chem.* **5** (1983) 55–60.
[174] Pfaudler-Werke AG, Schwetzingen, Printed publication FCP5M 12/87.
[175] Literature supplied by Pfaudler-Werke AG, Schwetzingen, at ACHEMA 1988 in Frankfurt.
[176] E. Kreisel, E. Protogerakis: "Thermisches Spritzen in Chemieapparate-Maschinenbau-Anlagen, Werkstoffe, Anwendungen," *Chem. Ing. Tech.* **59** (1987) no. 2, 118–122.
[177] Sigri GmbH, Meitingen, Printed publication 853 01 10, 1983.

[178] K. H. Möbius: "Füllstoffhaltige elektrisch leitfähige Kunststoffe," *Kunststoffe* **78** (1988) no. 1, 53–58.
[179] A. Malliaris, D. T. Turner: "Influence of Particle Size on the Electrical Resistivity of Compacted Mixtures of Polymeric and Metallic Powders," *J. Appl. Phys.* **42** (1971) no. 2, 614–618.
[180] H. Schindler, H. Gräfen, *Chem. Ing. Tech.* **57** (1985) no. 7, 597–602.
[181] H. Gräfen: "Fortschritte beim Einsatz organischer Werkstoffe im Korrosionsschutz," *VDI-Ber.* **670** (1988) 451–468.
[182] E. Oelke: "Kunststoffe in Chemieanlagen," *VDI-Gesellschaft Kunststofftechnik*, VDI-Verlag GmbH, Düsseldorf 1986, pp. 167–183.
[183] T. Hasky in: "Oberflächenschutz mit organischen Werkstoffen im Behälter-, Apparate- und Rohrleitungsbau," *VDI-Gesellschaft Werkstofftechnik*, VDI-Verlag GmbH, Düsseldorf 1980, pp. 101–129.
[184] E. Schacht: "Erfahrungen mit Gummierungen und Beschichtungen in Anlagen zur Rauchgasentschwefelung," *VDI-Ber.* **674** (1988) 263–281.
[185] A. Swozil, G. Ullmann, *Chem. Ing. Tech.* **52** (1980) no. 4, 292–298.
[186] J. E. Niesse: "Innovations in Organic Linings," *Chem. Eng. Prog.* **82** (1986) no. 6, 55–62.
[187] G. Dexheimer, H. Schindler, internal report, Bayer AG, 1989.
[188] "Bekaplast-Säureschutzverkleidung," company brochure, Steuler Industriewerke, Höhr-Grenzhausen.
[189] H. R. Rottkämer in: [6], 169–184.
[190] J. I. Bregmann: *Corrosion Inhibitors*, The Macmillan Comp., New York 1963.
[191] H. J. Rother: "Die Praxis des Korrosionsschutzes," *Corrosion Inhibitors*, The Maxmillan Comp., New York 1963, pp. 250–266.
[192] H. Fischer, *Werkst. Korros.* **23** (1972) 445–465.
[193] K. Risch, *Werkst. Korros.* **25** (1974) 727–734.
[194] E. Rabald, *Werkst. Korros.* **5** (1954) 368–392.
[195] Corrosion Inhibition, Proceedings of the Int. Conference on Corrosion Inhibition, NACE Houston 1988
[196] D. Kuron, H.-J. Rother, R. Holm, S. Storp, *Werkst. Korros.* **37** (1986) 83–93.
[197] M. N. Desai, *Werkst. Korros.* **23** (1972) 483–487.
[198] M. N. Desai, *Werkst. Korros.* **24** (1973) 707–716.
[199] J. C. Kora, S. C. Makwana, K. C. Koshel, N. K. Patel, *Werkst. Korros.* **25** (1974) 753–756.
[200] M. Brooke, *Chem. Eng.* **12** (1954) 230–234.
[201] D. Kuron, H. J. Rother, H. Gräfen, *Werkst. Korros.* **32** (1981) 409–421.
[202] G. Schmitt, B. Olbertz, *Werkst. Korros.* **29** (1978) 451–456.
G. Schmitt, B. Olbertz, *Werkst. Korros.* **35** (1984) 99–106.
G. Schmitt, B. Olbertz, K. H. Kurtz, *Werkst. Korros.* **35** (1984) 107–110.
[203] H. Gräfen, G. Herbsleb, F. Paulekat, W. Schwenk, *Werkst. Korros.* **22** (1971) 16–31.
[204] E.-M. Horn, R. Kilian, H. Stiepel, H. Gräfen, *Werkst. Korros.* **23** (1972) 967–973.
[205] J.-W. Kühn, V. Burgsdorff, H. Richter in W. v. Baeckmann, W. Schwenk (eds.): *Handbuch des kathodischen Korrosionsschutzes*, Verlag Chemie, Weinheim 1971, pp. 342–354.
[206] H. Gräfen, F. Paulekat in: W. v. Baeckmann, W. Schwenk, W. Prinz, see general references.
[207] W. von Baeckmann, *Chem. Ztg. Chem. Appar.* **87** (1963) 395–404.
[208] U. Heinzelmann in: [205], pp. 330–341.
[209] W. P. Banks, J. P. Sudbury, *Corrosion (Houston)* **19** (1963) 300–307.
[210] J. M. Stammen, *Mater. Proc.* **7** (1968) no. 12, 33–35.
[211] F. Paulekat, H. Gräfen, D. Kuron, *Werkst. Korrosion* **33** (1982) 254–262.
[212] D. Kuron, H. Gräfen, *Chem. Ing. Tech.* **60** (1988) 604–612.
[213] M. J. Humphries, R. N. Parkins, *Corros. Sci.* **7** (1967) 745–761.
[214] H. Gräfen et al., *Werkst. Korros.* **22** (1971) 16–31.
[215] J. B. Cotton, *Werkst. Korros.* **11** (1960) 152–155.
[216] Verein Deutscher Eisenhüttenleute (ed.): *Prüfung und Untersuchung der Korrosionsbeständigkeit von Stählen*, Verlag Stahleisen, Düsseldorf 1973.
[217] DIN-Taschenbuch 219, see general references.
[218] Annual Book of ASTM Standards 1988, see general references.
[219] H. Gräfen, G. Herbsleb, in: [38], pp. 55–69.
[220] DIN 50 905: Korrosion der Metalle. Korrosionsuntersuchungen Part 1 Grundsätze; Part 2 Korrosionsgrößen bei gleichmäßiger Flächenkorrosion; Part 3 Korrosionsgrößen bei ungleichmäßiger und örtlicher Korrosion ohne mechanische Belastung; Part 4 Durchführung von chemischen Korrosionsversuchen ohne mechanische Belastung in Flüssigkeiten im Laboratorium (1987).
[221] ASTM G 4-84: Standard Method for Conducting Corrosion Coupon Tests in Plant Equipment (1984).
[222] ASTM G 31-72: Standard Practice for Laboratory Immersion Corrosion Testing of Metals (1985).
[223] ASTM G 78-83: Standard Guide for Crevice Corrosion Testing of Iron-Base and Nickel-Base Stainless Alloys in Seawater and other Chloride Containing Aqueous Environments (1983).
[224] DIN 50915: Prüfung von unlegierten und niedriglegierten Stählen auf Beständigkeit gegen interkristalline Spannungsrißkorrosion (1985).
[225] ASTM G 30-79: Standard Practice for Making and Using U-Bend Stress-Corrosion Test-Specimens (1984).
[226] ASTM G 38-73: Standard Practice for Making and Using C-Ring Stress-Corrosion Test Specimens (1984).
[227] K. Risch, *Werkst. Korros.* **36** (1985) 55–63.
[228] DIN 50920: Korrosion der Metalle. Korrosionsuntersuchungen in strömenden Flüssigkeiten. Part 1 Allgemeines (1985).
[229] DIN 50914: Prüfung nichtrostender Stähle auf Beständigkeit gegen interkristalline Korrosion. Kupfersulfat-Schwefelsäure-Verfahren. Strauß-Test (1984).
[230] Stahl-Eisen-Prüfblatt 1877: Prüfung der Beständigkeit hochlegierter, korrosionsbeständiger Werkstoffe gegen interkristalline Korrosion, VDEh (1979).
[231] DIN 50921: Korrosion der Metalle, Prüfung nichtrostender austenitischer Stähle auf Beständigkeit gegen örtliche Korrosion in stark oxidierenden Säuren. Korrosionsversuch in Salpetersäure durch Messung des Massenverlustes (Prüfung nach Huey) (1984).
[232] ASTM A 262-86: Standard Practices for Detecting Susceptibility to Intergranular Attack in Austenitic Stainless Steels (1986).
[233] Euronorm 114-72.
[234] ASTM G 28-85: Standard Test Methods of Detecting Susceptibility to Intergranular Attack in Wrought, Nickel-Rich, Chromium-Bearing Alloys (1985).
[235] ASTM G 48-76: Standard Test Methods for Pitting and Crevice Corrosion Resistance of Stainless Steels

and Related Alloys by the Use of Ferric Chloride Solution (1980).
[236] ASTM G 36-87: Standard Practice for Evaluating Stress-Corrosion-Cracking Resistance of Metals and Alloys in a Boiling Magnesium Chloride Solution (1987).
[237] NACE Standard TM-01-077.
[238] DIN 50916: Prüfung von Kupferlegierungen. Spannungsrißkorrosionsversuche mit Ammoniak. Part 1 Prüfung von Rohren, Stangen und Profilen (1976); Part 2 Prüfung von Bauteilen (1985).
[239] ASTM G 37-85: Standard Test Method for Use of Mattsson's Solution of pH 7.2 to Evaluate the Stress-Corrosion Cracking Susceptibility of Copper-Zinc Alloys (1985).
[240] ASTM G 47-79: Standard Test Method for Determining Susceptibility to Stress-Corrosion Cracking of High-Strength Aluminum Alloy Products (1984).
[241] ASTM G 44-88: Standard Practice for Evaluating Stress Corrosion Cracking Resistance of Metals and Alloys by Alternate Immersion in 3.5% Sodium Chloride Solution (1988).
[242] DIN 50917: Korrosion der Metalle. Naturversuche. Freibewitterung (1979).
[243] DIN 50017: Klimate und ihre technische Anwendung. Kondenswasser-Prüfklimate (1982).
[244] DIN 50018: Korrosionsprüfungen. Beanspruchung im Kondenswasser-Wechselklima mit schwefeldioxidhaltiger Atmosphäre (1978).
[245] DIN 50021: Korrosionsprüfungen. Sprühnebelprüfungen mit verschiedenen Natriumchloridlösungen (1975).
[246] ASTM B 117-85: Standard Method of Salt Spray (Fog) Testing (1985).
[247] ASTM G 87-84: Standard Practice for Conducting Moist SO_2 Tests (1984).
[248] ASTM G 91-86: Standard Practice for Monitoring Atmospheric SO_2 Using the Sulfation Plate Technique (1986).
[249] ASTM G 85-85: Standard Practice for Modified Salt Spray (Fog) Testing (1985).
[250] DIN 50918: Korrosion der Metalle. Elektrochemische Korrosionsuntersuchungen (1978).
[251] ASTM G 3-74: Standard Practice for Conventions Applicable to Electrochemical Measurements in Corrosion Testing (1981).
[252] ASTM G 5-87: Standard Reference Test Method for Making Potentiostatic and Potentiodynamic Anodic Polarization Measurements (1987).
[253] DIN 50919: Korrosion der Metalle. Korrosionsuntersuchungen der Kontaktkorrosion in Elektrolytlösungen (1984).
[254] ASTM G 71-81: Standard Guide for Conducting and Evaluating Galvanic Corrosion Tests in Electrolytes (1986).
[255] D. Kuron, H. Gräfen, Z. Werkstofftechn. **8** (1977) 182–191.

9. Abrasion and Erosion

KLAUS SCHNEEMANN, Hüls AG, Marl, Federal Republic of Germany

1.	Introduction	9-1	5.1. Sliding Wear, Elastic Rolling Wear, and Oscillation Wear	9-18
2.	Types of Wear and Wear Mechanisms	9-2	5.2. Abrasion Wear	9-19
3.	Behavior of Materials	9-8	5.3. Damage by Particle Erosion	9-20
3.1.	Metals	9-9	5.4. Solid Particles—Free Erosion	9-22
3.2.	Plastics and Elastomers	9-13	5.5. Damage Caused by Erosion–Corrosion	9-23
3.3.	Ceramics	9-14	6. References	9-25
4.	Surface Treatment and Coatings	9-14		
5.	Practical Examples of Abrasion and Erosion Damage	9-17		

1. Introduction

Practical experience with industrial equipment, machinery, and plant has shown that components have only limited service lives. Damage and ultimate failure of the component can occur as a result of changes in the material that originate at the surface, even if the components are designed such that long-term action of the forces alone causes neither fracture nor undue deformation.

If the reactions responsible for the damage are of electrochemical or predominantly chemical nature, the term corrosion is normally used, whereas mechanical damage to the surface of the component is defined as wear. Attempts to avoid a loss of material due to wear, or at least to reduce the loss, concentrate on making the affected surface more resistant to wear. This can be achieved by mechanical, thermal, or thermochemical treatment of the surface or by applying or depositing metallic coatings. Under some circumstances the wear conditions can be changed by design measures so that the danger for the affected component surface is eliminated or reduced to a tolerable level.

With few exceptions (e.g., running-in of bearings), wear in engineering means an undesired change that causes very high costs every year; in a highly developed, industrialized country this can amount to ca. 1–2% of the gross national product [1].

Excluding the contribution from the automobile sector, the proportions occurring in the various branches of industry can be divided up approximately as shown in Table 1. From this, it can be seen that the plant construction typical of the chemical industry plays an insignificant role, and wear is correctly known as "the problem child of mechanical engineering" [2].

Wear, friction, and lubrication are described under the term tribology as the science of the study, industrial application, and modification of the phenomena and processes occurring between surfaces which are acting against each other and moving relative to one another; this includes boundary surface interactions between solids, and between solids and their gaseous or liquid surroundings. Since at least two components of a system are involved in wear, it is not a pure material characteristic, but only a system characteristic. Wear itself is generally under-

Table 1. Fixed costs and personnel costs caused by abrasive and erosive wear in the Federal Republic of Germany in 1984

Sector	Costs, 10^9 DM
Mining	1.66
Foundry industry	1.2
Quarrying	1.1
Building	1.53
Plastics industry	0.75
Power engineering	0.16
Total	6.4

stood as progressive loss of material from the surface of a solid body caused by mechanical action, i.e., contact and relative motion with a solid, liquid, or gaseous phase.

2. Types of Wear and Wear Mechanisms

The treatment of wear must take the diversity of tribological processes into account, and this requires precise analysis of the loads and of the appearance of the damage. It is usual to subdivide the large number of wear processes into types of wear and wear mechanisms, in which different mechanisms have to be allocated to one and the same type of wear.

The kinematic conditions and the types of materials involved in the wear determine the types of wear, such as sliding wear, elastic rolling wear, impact wear, and shock wear.

Oscillating wear stress, or oscillation wear, is caused by oscillating sliding and by oscillating sliding parts when rolling or even on impact. Wear mechanisms characterize the energetic and material interactions between the individual elements of a wear system. DIN 50 320 differentiates between adhesion, abrasion, surface destruction, and tribochemical reaction as the principle mechanisms (Table 2) [3].

Adhesion. If the micro-roughnesses on the "clean" surfaces of two solid bodies are in contact with each other, high surface pressures are generated locally; relative movements cause cold welding which can lead to separation of material. Material transfer, scales, and shear dimples are typical wear patterns.

Abrasion. When two bodies of different hardnesses slide against each other, abrasion causes the softer surface of the base body to be scratched by the harder counter body. This harder body can be a micro-roughness of the counter body or a wear particle that has already been removed. The penetration of the harder counter body and the relative movement produce wear grooves or furrows in the softer body, for which reason it is also called furrow wear (Fig. 1). If the softer material is plastically deformed and is displaced to the furrow edges, the process is called microplowing. Material is not removed in a single event but by the simultaneous action of many

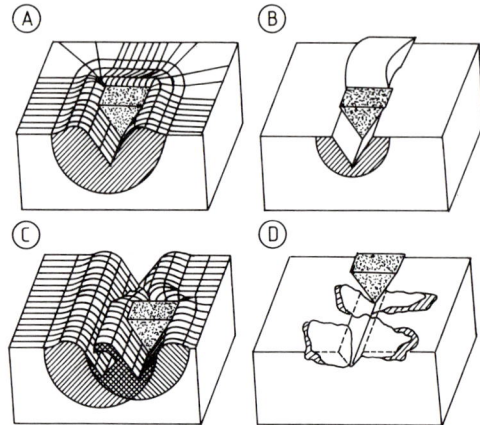

Figure 1. Mechanisms of material damage in furrow wear
A) Microplowing; B) Microcutting; C) Microfatigue; D) Microgouging

abrasive particles or by repeated furrowing by one particle [4].

More often the material is removed by cutting; so-called microchips are obtained. In ductile materials these two processes are predominant, but in relatively brittle materials micro-particles break away and microgouging occurs.

Depending on the number of components involved, a distinction is made between two-body and three-body abrasion wear. Sliding, shock, and rolling are possible types of wear, see Table 3 [5].

Oscillating mechanical loads in the surface of a solid can lead to surface destruction by material fatigue. This appears as formation and propagation of cracks, and may lead to separation of material particles.

Tribological stress in the system can lead to tribochemical reactions in addition to purely physical processes. As a result layers can appear on or between the touching surfaces. In general these wear mechanisms do not act in pure form in practice; they usually occur in combinations.

According to [4] the relative proportions of microplowing and microcutting can be determined from the profile of a wear groove (Fig. 2). The so-called f_{ab} value is defined by

$$f_{ab} = \frac{A_v - (A_1 + A_2)}{A_v}$$

so that for the two extreme cases, pure microplowing and pure microcutting, f_{ab} becomes 0 and 1, respectively.

Table 2. Classification of types of wear (based on DIN 50320)

System	Tribiological loading	Type of wear	Active mechanisms [a] (individual or combined)			
			Adhesion	Grooving	Surface destruction	Tribochem. reactions
Solid–intermediate material–solid [b]	Sliding, rolling, impact				●	○
Solid–solid	Sliding	Sliding wear	●	○	○	●
	Rolling	Rolling wear	○	○	●	○
	Oscillation	Oscillating wear (fretting)	●	●	●	●
	Impact	Impact wear	○	○	●	○
Solid–particle	Impact	Abrasive impact wear [c]		●	●	○
	Sliding [d]	Abrasive sliding wear [c]		●		○
Solid–particles–solid	Sliding	Three-body abrasive wear	○	●	●	○
	Rolling		○	●	●	○
	Impact		○	○	●	○
Solid–particle–liquid	Flow	Hydroabrasive wear		●	●	○
Solid–particle–gas	Flow	Sliding jet wear	○	●	●	○
	Flow, impact	Impact and inclined jet wear	○	●	●	○
Solid–liquid	Flow, vibration	Cavitation erosion			●	○
	Impact	Drop impingement erosion			●	○
	Flow	Liquid erosion			○	●
Solid–gas	Flow	Gas erosion				●

The middle section spanning multiple rows is labeled "Abrasion" (for solid-particle, solid-particles-solid, solid-particle-liquid, solid-particle-gas rows) and "Erosion" (for solid-liquid rows).

[a] ● Principle active mechanisms; ○ additional active mechanisms. [b] With complete separation of the two solid surfaces. [c] Two-body abrasive wear. [d] Also rolling.

Table 3. Comparison of the characteristics of abrasive sliding, abrasive impact, and three-body abrasive wear

Classification	Abrasive sliding wear		Abrasive impact wear	Three-body abrasive wear
	Two-body abrasive wear			
Symbol	1 ↑ 2	2 ↑ 1	2 ↓↓ 1	1 ↑ 3 2 / 1 ↓ 3 2 / 1 ↔ 3 2
Body 1	Wearing material surface			
Counterbody 2	Surface peaks of solid abrasive	Bound abrasive	Coarse-grained abrasive	Wearing material surface
Intermediate material 3		Free abrasive		Particulate solid
Surrounding medium			Liquid or gas	
Relative motion of body and counterbody	Sliding	Sliding	Impact, sliding	Sliding	Rolling	Impact, sliding
Motion of abrasive	Sliding	Sliding	Impact, sliding	Sliding, rolling		Sliding, rolling, impact
Relative wear intensity	High	Low	Very high	Low, high	Low, high	Very high
Practical examples	Rock drilling	Pressing of refractory brick (grinding, honing)	Discharge points in solid transport, deflector plates	Undesired processes in bearings, joints		Comminution and grinding processes
Test systems	Tool/stone	Abrasive paper test	Wear pot test	Buffing gear for mass goods	Block/ring block/wheel	Roller/roller roller/surface ball mill	Model jaw crusher

| Sliding, rolling, flow |

(additional row cells for "Relative motion of body and counterbody": Sliding, rolling, flow appears under Two-body free abrasive)

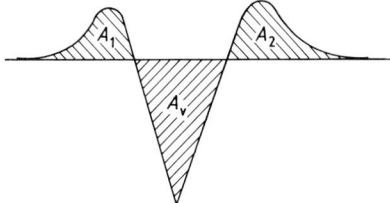

Figure 2. Profile of a wear groove

The linear wear intensity $W_{l/s}$ is defined as

$$W_{l/s} = \frac{\Delta h}{s}$$

where Δh represents the reduction of thickness of the worn component and s is the wear path.

Using the f_{ab} value allows the wear intensity to be expressed as

$$W_{l/s} = \varphi f_{ab} \cdot \frac{p}{H_{def}}$$

where p is the surface pressure, H_{def} is the hardness of the wearing material in the strain-hardened state, and φ is the form factor.

The f_{ab} value is a function of the effective surface deformation caused by the abrading particles, the ductility of the worn material, and the material's strain-hardening behavior [6]. The value decreases with increasing ductility, increasing strain-hardening capability, and decreasing surface deformation.

According to the above equation, a high wear resistance means a small f_{ab} value and a high material hardness in the strain-hardened state. According to this model, hardness alone often fails to provide a reliable assessment of the expected wear resistance.

Figure 3A shows the wear resistance $1/W_{l/s}$ as a function of hardness for various metals, and Figure 3B shows the resistance as a function of the ratio HV_{def}/f_{ab} in accordance with the above equation. For different materials with the same hardness, the wear resistance increases with increasing strain-hardening capability; see Figure 4 [4].

This assumes that in the test method the wear takes place at the so-called upper shelf of the wear characteristic. In the treatment of wear problems, WAHL was the first to draw attention to the connection between abrasive particle hardness and wear [7]. He found that at a certain value of the particle hardness the wear increases

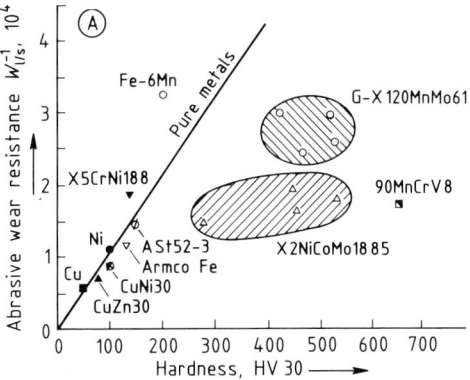

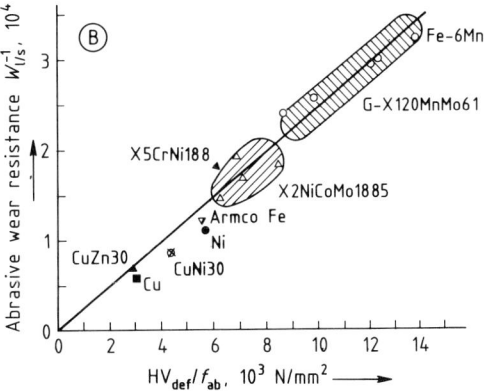

Figure 3. Dependence of the abrasive wear resistance of metallic materials on the hardness of the material (A) and on the ratio of the particle hardness to the factor f_{ab} (B), measured by the abrasive paper method, (SiC, 80 grain, $p = 3.54$ MPa)

sharply, and at higher values remains virtually constant; see Figure 5.

Note, however, that the pure furrowing described here occurs in tests with bonded particles, a condition which occurs relatively rarely in practical situations. More typical is the case where unbonded, loose particles are carried over the surface of the component, and additional rolling motion is also possible.

With increasing particle mobility, the wear caused by loose particles represents the transition from abrasive sliding wear to erosion (abrasive–erosive attack).

Erosion. Erosion phenomena are caused by flow, in which the flowing fluid itself is erosive or the fluid carries particles that are unable to follow the filaments of flow at surface irregularities. In this case a shock component is often superim-

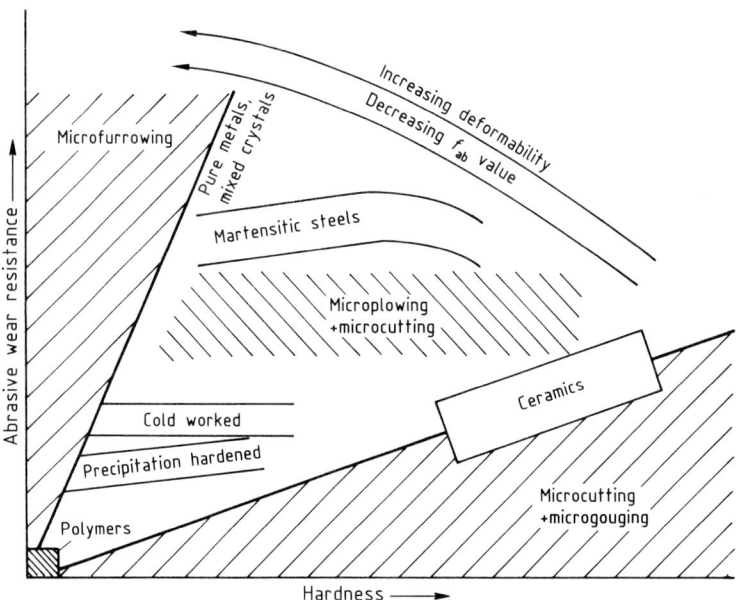

Figure 4. Schematic representation of the abrasive wear resistance in the upper shelf and the wear mechanisms plotted versus the material hardness

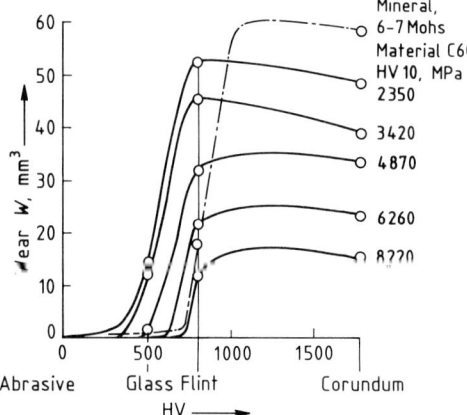

Figure 5. Lower and upper shelf wear characteristic (abrasive paper method, 80 grain, $p = 0.37$ MPa)

posed on the plowing mechanism; this causes local deformation and destruction.

If the fluid is a gas, tribochemical reactions with the solid are only possible at fairly high temperatures where the ablation mechanism can act through the processes of sublimation, vaporization, or fusion.

Pure liquid erosion occurs relatively rarely because even at high flow velocities the forces acting are hardly sufficient to cause removal of material; exceptions to this are additional corrosive elements and the involvement of cavitation.

Cavitation. Since the term cavitation is used to describe the processes occuring in the liquid and also the resulting damage to the material surface, the latter is often called cavitational erosion to avoid confusion. These processes are caused by the formation and subsequent collapse of vapor or gas bubbles in liquids. When the pressure falls, dissolved gases evolve from small gaseous nuclei, or evaporation takes place if the pressure falls locally below the vapor pressure. In regions of higher pressure in the liquid the bubbles implode. If this implosion takes place directly at the surface of the component, the cyclic, locally confined compressive stresses finally lead to microscopic fatigue phenomena. This resembles the behavior of materials under vibratory stress. If the vapor bubbles formed do not implode directly at the surface the shock intensity is attenuated by the interposed liquid layer [8]. Damage is therefore only caused by bubbles which collapse directly at or close to the surface.

If the pressure differences are caused solely by flow, the resulting process is called flow cavitation. However, vibrations of sufficiently high frequency can initiate vibrational cavitation.

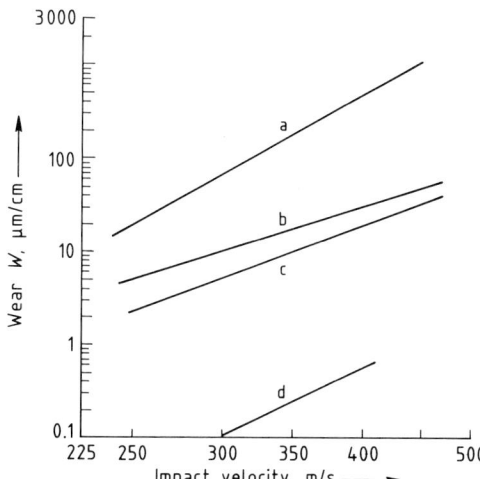

Figure 6. Dependence of drop impingement wear (wear depth per centimeter impinging liquid column) on the impact velocity for various materials
a) Glass; b) Aluminum; c) Polyurethane; d) Sintered alumina

The material damage described can take place by purely mechanical means, but in practice it is often considerably intensified by superimposed corrosion stress.

Comparable damage can occur through the repeated impact of drops or liquid jets at high velocity. After the deformability has become exhausted in tough materials, initial cracks appear; these develop into breaks and progress to deep fissuring. This wear, known as drop impingement, depends on the impact velocity (Fig. 6) and the angle of impact [9].

In chemical technology and process engineering, the most important instances of erosion are those where the fluid contains solid particles. If the fluid is a liquid this type of wear is called hydroabrasive wear or scouring wear; with gaseous fluids it is called jet wear. In both types of wear, plowing is active as the essential mechanism, and with steeper angles of impact there is an additional component causing destruction of the material. Tribochemical reactions also occur.

During hydraulic transport of solids in pipelines, the flow path lines near the surface travel essentially parallel to the surface. The erosive effect of the particles is generally slight provided that the flow remains undisturbed. Flow separation at irregularities causes turbulence, which can force the particles against the wall and produce erosion. Typical irregularities in pipelines are the protruding roots of welds. Furthermore, in turbulent flow more collisions between particles occur, leading to impacts with the enclosing walls.

During hydraulic transport the particles follow the streamlines of the fluid better than in pneumatic transport due to the smaller difference in densities. Sliding jet wear is always present in horizontal pipeline transport due to the influence of gravity.

Where changes in direction of flow occur, (e.g., at pipe bends) the particles, due to their higher density, are pressed to an increasing extent against the surface, where they have an erosive action (Fig. 7) [10].

The wear increases distinctly with increasing velocity and can be described by the empirically determined expression

$$W(\alpha) = C v^{n(\alpha)}$$

where α is the angle of impact, C is a constant that depends on the material and the abrasive, v is the velocity, and n is a velocity exponent which depends on the material, the abrasive, and the angle.

This equation is also valid for jet wear, in which material is removed by impacting and/or plowing particles in a carrier gas. Because of the large number of forms which occur in practice it is necessary to distinguish between the sliding jet, in which the particle slides over the surface of the workpiece, and the impact jet, in which the particle strikes perpendicularly; all intermediate forms of jet ($0° \leq \alpha \leq 90°$) are called inclined jets.

The basic process in jet wear is the impact of a particle against the surface at some arbitrary angle. About 90% of the kinetic energy of the particle is converted into plastic deformation

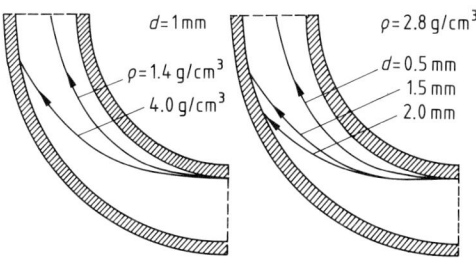

Figure 7. Path of motion of hydraulically transported particles of various densities and diameters (calculated for water at 18 °C and a flow rate of 2 m/s)

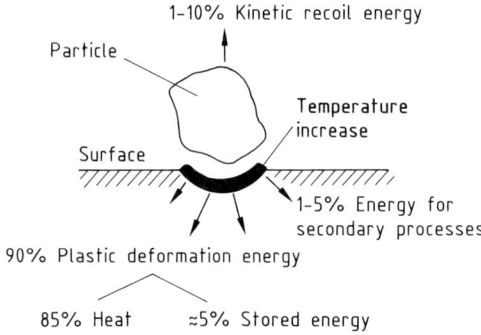

Figure 8. Energy balance for the impact, without breakage, of a hard particle on a ductile metal surface

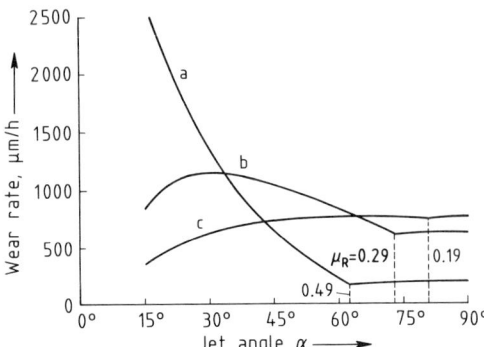

Figure 10. Influence of the jet angle on the wear rate
a) Rubber; b) St 37; c) C60H

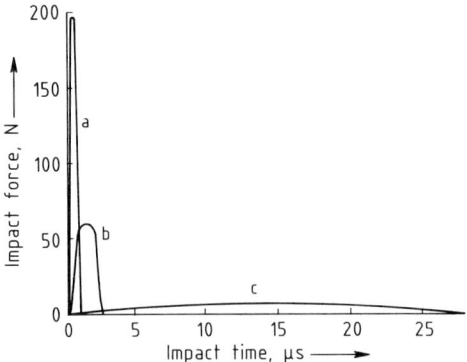

Figure 9. Variation of the calculated impact force of a sphere (diameter 1 mm, velocity 100 m/s) with time for:
a) C6OH; b) St 37; c) Rubber

and heat (Fig. 8) [11]. The main energy conversion takes place in the softer body [12].

Estimates give values of a few thousand megapascals [13] for the pressures occurring on impact and of microseconds [14] for the impact times (Fig. 9). The fatigue mechanism is dominant where the impact is predominantly perpendicular, while at shallow incident jet angles the particle has a plowing action on the surface under the influence of a normal force component [15]. These result in different requirements for the material involved:

1) Very hard materials are best for withstanding plowing processes
2) Impact processes cause little damage if the energy can be absorbed elastically or plastically.

Since hard materials normally have low ductility, wear is low at small angles, but fragments are removed easily with perpendicular impact. Soft, ductile materials exhibit higher wear under furrowing conditions. At $\alpha = 0°$, the vertical component of the force theoretically approaches zero, so after a maximum the wear should also approach zero, but this situation does not actually occur in practice (Fig. 10) [15].

Erosion–Corrosion. Under certain corrosive conditions many metals form covering layers. If these are sufficiently dense they act as protective films against corrosive removal of material. An example of this is the protective layer of iron oxide formed in unalloyed or low-alloy boiler tubes. Erosion–corrosion is understood as the combined action of mechanical surface removal and corrosion. With some soft and loose layers the shear forces obtained with pure flowing liquids at medium flow velocities are sufficient to damage the protective layer without the involvement of abrasive solid particles. Where drop impingement or cavitation are involved the mechanical removal of material is understandable. On the other hand, diffusion-controlled corrosion inhibition may be nullified in flowing fluids as a result of saturation of the boundary layer [16]. The resulting purely chemical removal of material therefore no longer corresponds to the usual term of erosion–corrosion and is now called flow-induced corrosion [17].

3. Behavior of Materials

The many outward forms assumed by wear mechanisms make it necessary to design the component to be resistant to wear, at least on the surface under attack. This can be accomplished by selecting a suitable base material or by modifying the surface by mechanical, thermal, or thermochemical treatment or by applying protective coverings of other materials [18]. Adequate wear resistance of a component is normally only one of several criteria in the requirement profile. Adequate strength, toughness, corrosion resistance, and other properties, such as ease of repair, are requirements which must also be met. Economic criteria also have to be considered. In general, theoretical considerations and laboratory tests do not lead to a quick solution, and in most cases the final suitability is proved under operational conditions. Wear problems can often be overcome by changes in design or in the process technology. Metals, plastics, ceramics, and protective coatings or layers are compared below purely from the aspect of material technology.

3.1. Metals

There is still no comprehensive correlation available between wear and hardness of materials. Depending on the wear mechanism, other properties, such as the state of the material's microstructure, also play a significant role. Because of the simplicity of the hardness test it is, however, useful to know which wear processes depend essentially on hardness alone. In any case, the loading of the material is confined to the surface region, and in abrasive wear a penetration process occurs that is similar to the hardness tests. Unlike hardness tests, an additional tangential force component also has to be taken into account in the case of wear, which initiates effects such as increase in temperature (possibly associated with phase changes), reaction layer formation, and strain hardening by plastic deformation, and can therefore lead to permanent changes in the tribological system.

It is usual to subdivide furrow wear into:

1) counterbody furrowing, in which the mineralogical grains are fixed on the surface of the counterbody (e.g., a grinding wheel), and
2) particle furrowing, in which the wear is caused by freely moving particles.

In trials both types show a similar dependence of the wear rate on the hardness of the attacking particle (lower shelf/upper shelf characteristic); see Figs. 11 [19] and 12 [20].

For steels an increase in wear resistance is observed with increasing hardness, whereby increasing the hardness by alloying elements has a greater effect than increasing the hardness by heat treatment.

With cast iron the wear resistance also tends to increase with increasing hardness. The maximum resistance values increase in the following order: gray cast iron with lamellar or spheroidal graphite, pearlitic cast iron, and martensitic cast iron (Fig. 13) [21].

In the upper shelf region and in the presence of normal abrasive particles, all pure metals exhibit a linear relationship between resistance and material hardness (Fig. 14) [22].

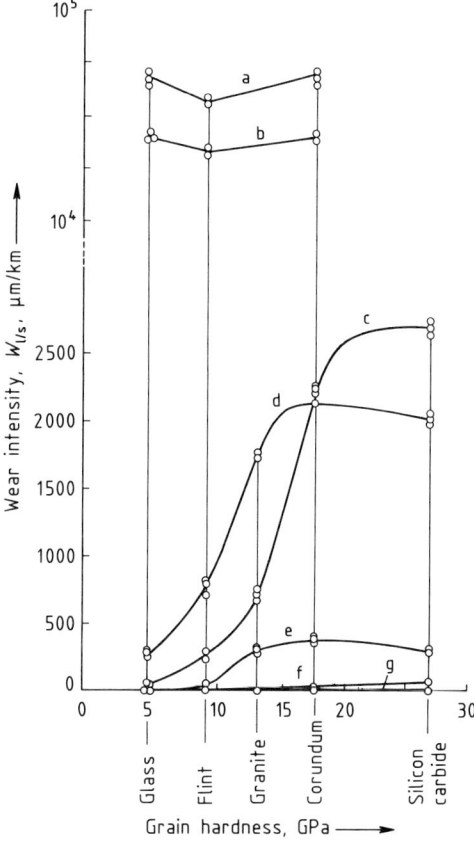

Figure 11. Dependence of wear intensity on the hardness of the abrasive particles for counterbody furrowing for various materials (abrasive paper method, $p = 0.1$ MPa, $v = 0.25$ m/s)
a) Rubber; b) Polystyrene; c) Cast basalt; d) Steel St 37; e) C60H; f) Hard metal G4; g) Hard metal H2

Jet Wear. Materials in chemical plants are usually transported through pipelines and converted in enclosing containers. Compressors and fans are used as the conveying units when gas-

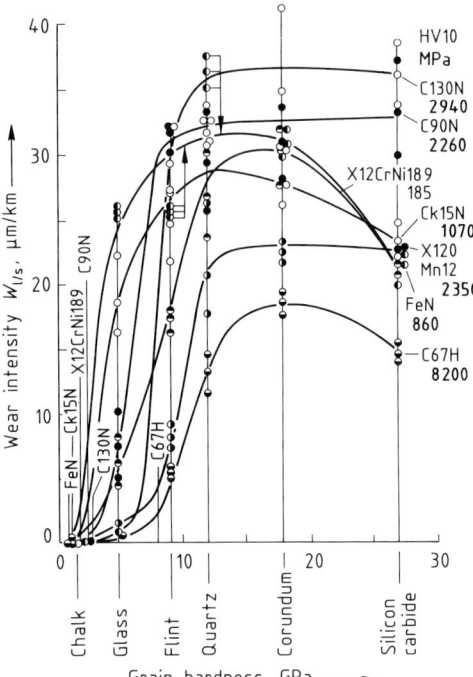

Figure 12. Dependence of wear intensity on the hardness of the abrasive particles for particle furrowing for various materials

eous materials are transported, while pumps are normally used for liquids. The resulting wear phenomena are therefore associated with the type of flow involved, and the various types of wear are covered by the term erosion.

Even at very high velocities pure gases are only capable of damaging the enclosing solid body when the temperature is high enough to lead to thermally induced reactions. Problems with gas conveying are therefore only to be expected when entrained solid particles are present. In jet wear the material is elastically or elastically/plastically deformed by impact of the particles against the surface.

Apart from particle velocity the most important factor in jet wear is the jet angle: brittle metals exhibit increasing wear with increasing angle of impact, with a maximum at 90°; whereas with tough metals, wear increases up to an angle of 15°–40°, after which the loss decreases at steeper angles. The ductility and strain hardening at the surface are thought to be responsible for this behavior (Fig. 15) [23].

The relationship between wear resistance and material hardness known from abrasion is valid only at small jet angles because of the comparable wear mechanisms. According to BITTER [24], abrasion is dominant with the sliding jet and surface destruction with the impacting jet.

For the first-named conditions WAHL [25] gives an appraisal of working characteristics (Table 4) which relates predominantly to steel and cast iron.

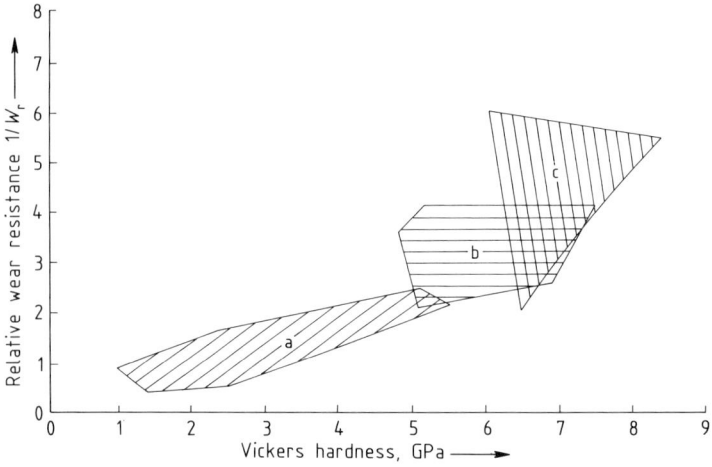

Figure 13. Wear resistance of cast iron towards particle furrowing (wear pot method, quartz sand)
a) Martensitic cast iron alloyed with Cr and Ni; b) Unalloyed and low-alloy pearlitic cast iron; c) Steels with 0.15–0.5% C and gray cast iron with lamellar or spheroidal graphite

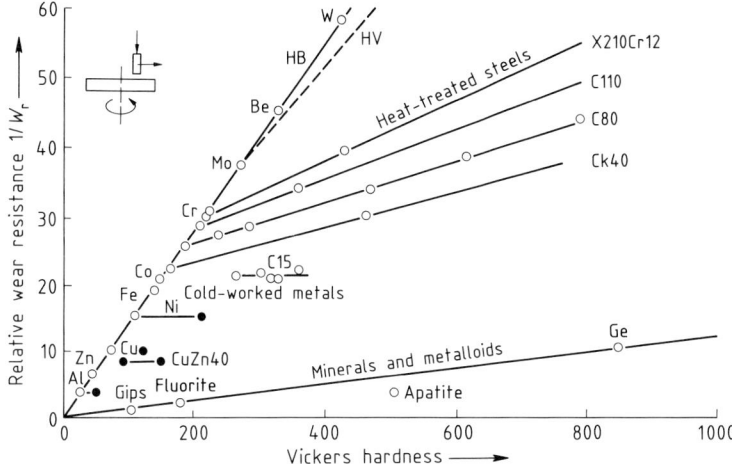

Figure 14. Linear relationship between wear resistance and hardness in the upper shelf region (abrasive paper method, corundum, 180 grain, $p = 0.94$ MPa)

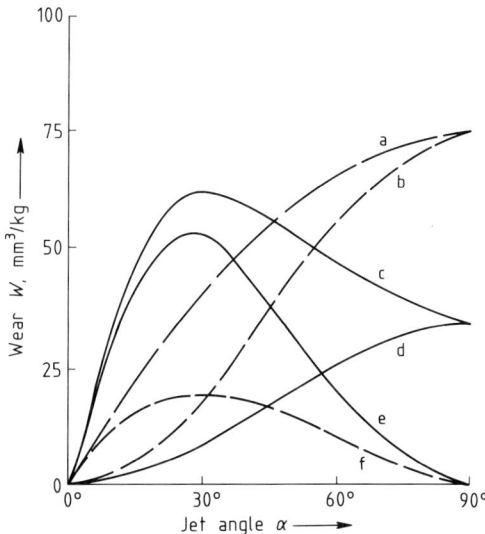

Figure 15. Impact jet and sliding jet components in jet wear
a) Experimental curve for cast iron (2.8% C, 21% Cr); b) Experimental curve for St 37; c) Impact jet component for cast iron; d) Impact jet component for St 37; e) Sliding jet component for cast iron; f) Sliding jet component for St 37

Wear with Liquids. The forces that particle-free liquids exert on metal surface are so low at the usual flow velocities of a few meters per second that they do not cause any removal of material. According to HEIL, no erosion effects could be detected with plain carbon steel, even at 60 m/s, in a test method similar to that with the rotating disc [26]. Damage caused by liquid flow alone is therefore relatively rare in practice. The situation is different if, for example, a liquid under high pressure emerges with high velocity into a space of lower pressure; it may have an erosive action on the metal at the exit point. This applies to an even greater extent if the fluid forms dense protective layers with the metal. If these layers are removed mechanically, they must be re-formed, otherwise erosion will act in combination with corrosion, which is known as erosion – corrosion. This situation occurs very frequently in practice as many metals can only be used, for example, with water because a thin, usually oxidic layer protects the metal against corrosive attack. Table 5 gives guide values for maximum permissible flow velocities for pure water and, for comparison, for the more aggressive seawater [27].

Since technical pure liquids usually contain very fine solid particles, it is often difficult in practice to differentiate between pure fluid erosion and so-called hydro-abrasive wear. As with jet wear, the number, hardness, and velocity of entrained particles determine the extent to which the protective covering layer is removed and the extent of the pure metal wear. If mechanical stressing due to entrained particles is predominant then hardness and strength are of decisive importance for metals, as shown in Figure 16 for copper alloys [28].

Table 4. Performance properties of wear resistant materials

Material	Tensile strength, MPa	Fracture strain, %	Wear resistance
Low-alloy austenitic manganese steel	600	15	very low
Austenitic manganese hard steel	550	50	medium
Unalloyed steels	420	20	extremely low
High-strength low-alloy steels	600	18	extremely low
Pearlitic steels	500–600	12	extremely low
Martensitic steels	500–600	15–20	medium
Ledeburitic steels	600	2	high
Heat-treated 66 (gray cast iron)	250	< 0.5	very low
Heat-treated 666 (nodular iron)	500	3	low
Pearlitic white cast iron	250	< 0.5	high
Martensitic white cast iron	350	< 0.5	very high
High-chromium white cast iron	500	< 0.5	very high
Co–Cr–W–C alloy	500	< 0.5	very high
Hard metal	500	< 0.5	extremely high

Table 5. Maximum permissible flow rates for pure water and seawater

Material	Flow rate, m/s	
	Pure water	Seawater
Aluminum	1.2–1.5	1.0
Copper	1.8	1.0
Copper + As	2.1	1.0
Copper + Fe	4.0	1.5
CuZn28Sn	2.0–2.4	1.5–2.0
Al bronze	ca. 3.0	ca. 2.0
CuNi10Fe	5.0	2.4
CuNi30Fe	6.0	4.5
Steel	3–6	2–5
Nickel alloys	30	15–25
Plastics	6–8	6–8

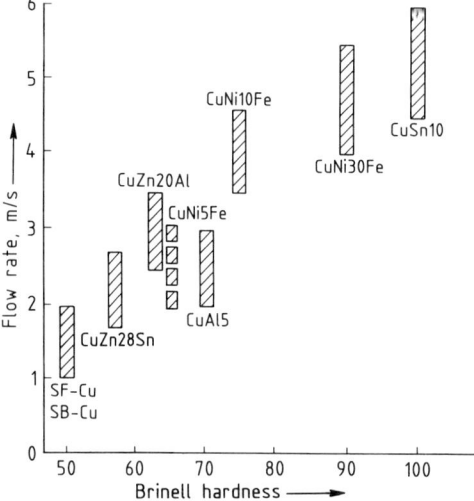

Figure 16. Maximum tolerable flow rates (water) as a function of material hardness for various copper alloys

Cavitation and Drop Impingement. According to GREIN all materials succumb to the exceptionally high mechanical stress occurring during cavitation if the intensity is sufficiently high [29]. In soft materials such as aluminum, cavitation erosion appears in the form of plastic indentations caused by single implosions. With higher strength metals the damage starts by roughening of the surface after a certain incubation time; as the destruction progresses the interconnected holes eventually form a spongy structure.

In general, the strength and the fatigue strength are the most important material parameters for resistance to cavitation; hardness is therefore often used as a measure. With similar microstructures, the metal with the higher hardness has the more favorable behavior, and where different metals have the same hardness, the resistance improves with increasing toughness. As cavitation attack takes place locally, the behavior is determined by the structural constitution to a greater extent with this type of erosion than with others. A homogeneous, fine-grained microstructure is required; soft inclusions and soft, or very brittle, grain boundaries reduce the resistance.

The following factors increase the resistance to cavition erosion:

1) High corrosion resistance
2) Homogenity
3) Ductility
4) Compressive residual stress
5) High strain-hardening capability
6) Smooth surface
7) Fine-grained structure
8) Fine-grained hard inclusions
9) High content of hard components
10) Layer structure
11) Wrought structure

while the following reduce the resistance:

1) Low corrosion resistance
2) Heterogenity
3) Brittleness
4) Tensile residual stress
5) Low strain-hardening capability
6) Rough surface
7) Coarse-grained structure
8) Coarse-grained hard inclusions
9) High content of soft components
10) Dendritic structure
11) Cast structure

The data provided by PILTZ [30] and REINGANS [31] can be used as guides for the selection of materials.

As with other types of erosion, the superposition of a corrosion process also has to be taken into account where damage by cavitation occurs. Removal of material by corrosion after destruction of protective covering layers often represents the more intensive attack, and the corrosion resistance of the material is then the dominant property.

A comparable stress occurs with drop impingement, characterized by repeated, short-time liquid impacts. The comments made above about cavitation also apply here to the material behavior.

Since residual stresses counteract the external loading of the material they can increase resistance to destruction perceptibly. In this sense, strain hardening has a similar effect to carburization and nitriding [32].

3.2. Plastics and Elastomers

The relationship between wear and surface hardness obtained for metals would predict a comparatively poor behavior for polymers. However, their special structural features give rise to properties that can play a special role in wear.

The viscoelastic deformation behavior is characterized by time-, temperature- and velocity-dependent deformation processes. Relatively low levels of hardness and strength, high plasticity, low thermal conductivity, and high thermal expansion are effects of the weak secondary bonding forces between the macromolecules and their coiled structures.

In particular, the low tendency to adhesion gives polymers their good slip characteristics with steels as the sliding partners—in the absence of additional abrasive particles—because of the

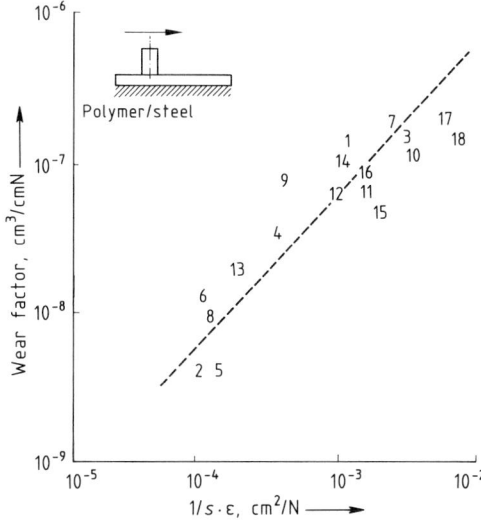

Figure 17. Relationship between wear and the reciprocal product of tensile strength S and fracture strain ε for polymers

1) Poly(methyl methacrylate); 2) Low-density polyethylene; 3) Polystyrene; 4) Polyoxymethylene; 5) Polyamide 66; 6) Polypropylene; 7) Epoxy resin; 8) Polytetrafluoroethylene; 9) PMMA–acrylonitrile copolymer; 10) Polyester; 11) Polychlorotrifluoroethylene; 12) Polycarbonate; 13) Polyamide 11; 14) ABS; 15) Poly(phenylene oxide); 16) Polysulfone; 17) Poly(vinyl chloride); 18) Poly(vinylidene chloride)

low frictional forces involved, and the slip system is characterized by additional emergency running properties. Polyamide and PTFE occupy the prime positions here as they possess good cohesive linkage properties compared with other unreinforced polymers [33], [34].

If abrasive sliding stress is present, the dependence on hardness known for metals cannot really be depicted in the same way. It has been demonstrated that polymers exhibit a good relationship between wear resistance and crack propagation energy, or even between wear and the product of tensile strength and fracture strain; see Figure 17 [35].

Due to their material properties, polymers have proved successful where streams of small particles cause impact stress in addition to sliding wear, i.e., with abrasive impact wear and with erosive attack. Although polymers generally have poor resistance to abrasive sliding attack, their ductility, especially of elastomers, leads to a behavior superior to that of metals when the impacting component is dominant (Fig. 18) [36].

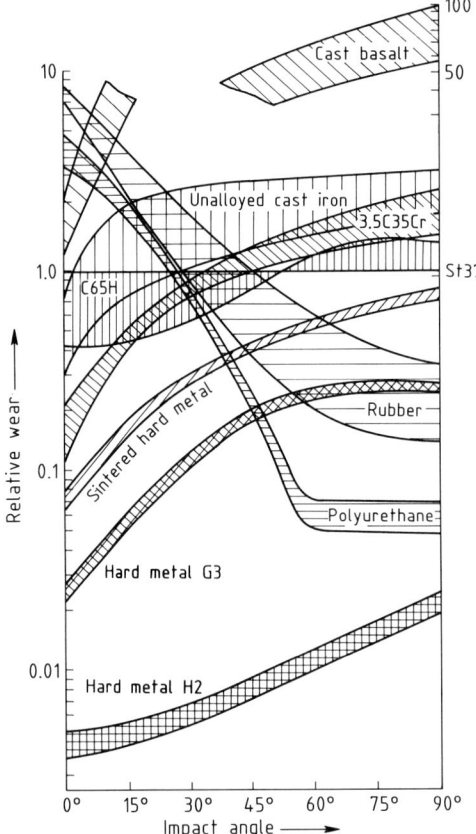

Figure 18. Wear relative to St 37 as a function of impact angle for various materials (abrasive: quartz sand, particle size 0.2–1.5 mm, HV = 1250)

very different wear behavior of these two materials (Fig. 19).

3.3. Ceramics

Ceramic materials have gained increasing importance over the last few decades; in addition to corrosion resistance and suitability for high temperatures the resistance to wear is the important property of these materials. The above-mentioned properties of the sintered body can be varied over a wide range by adjusting the powdered raw material.

The nonmetallic, mechanically resistant materials which are constituents of the ceramic materials can be subdivided into oxidic and nonoxidic materials [41]. High melting points and hardnesses are the outstanding properties, as shown in Table 7.

Alumina is the most important oxidic abrasion-resistant material. Metal carbides are in some ways superior to oxides with respect to hardness and melting point, but they are much more brittle than the oxides and are only used in isolated instances as wearing bodies. Silicon carbide is characterized by its low thermal expansion and high thermal conductivity and has proved to be more resistant to thermal shock than oxides. Zirconia is tougher than alumina; its modulus of elasticity is only about half as large and is comparable with that of steel. Zirconia is therefore very suitable for compound structures with steel. At present the applications of ceramic sintered materials in chemical plant construction are slide rings, pump parts, and slide bearings.

Their behavior therefore differs significantly depending on the angle of impact. The material becomes heated due to internal friction, which can lead to complete failure at high jet intensities.

The preferred elastomers include the polyurethanes and synthetic rubbers because of their outstanding resistance to wear. In polyurethanes, greater resistance is found in the hardness range 70–95 Shore A, whereas normal grades of rubber reach their optimum between 50 and 70 Shore A [37]. It is not possible to separate the influencing factors systematically with respect to tribological behavior because of the large number of additives, types of rubber, and applications (see Table 6) [38], [39].

If, for rubber and C 60 H steel, the amount of wear relative to St 37 steel is plotted versus the impact angle and the hardness of the jet material then, according to [40], it is possible to show the

4. Surface Treatment and Coatings

The fact that wear starts at the surface of the workpiece suggests that only the tribologically stressed surface should be strengthened instead of making the entire component out of wear-resistant material. It should be borne in mind that the base material has to fulfill other, for example, load-bearing functions as well as surface stressing, i.e., the surface which has been made wear-resistant must not interfere unduly with the component.

If the material involved can be hardened then various methods can be used to harden the sur-

Table 6. Jet wear of polymers and metals

Material	Hardness	Abrasive*	Relative wear W/W_{St37}
Steel T 80 H	590 HV	II	0.109
Polyurethane	18 Shore D	II	0.143
Poly(vinyl chloride)	5 Shore D	II	0.143
Polyurethane	34 Shore D	II	0.403
Poly(vinyl chloride)	10 Shore D	II	0.42
Rubber	17 Shore D	II	0.57
Poly(vinyl chloride)	14 Shore D	II	0.96
Steel St 37	122 HV	I	1.0
Low-pressure polyethylene	60 Shore D	I	1.06
Steel St 34	124 HV	I	1.07
Poly(vinyl chloride)	17 Shore D	II	1.12
Polyamide 6, Grilon R 50	62 Shore D	I	1.33
Polyamide 6, Grilon R 70	64 Shore D	I	1.33
Copper	99 HV	I	1.36
High-pressure polyethylene	42 Shore D	II	1.4
Low-pressure polyethylene	58 Shore D	II	1.4
Polyamide 11, Rilsan Besvo	71 Shore D	I	1.81
Low-pressure polyethylene	58 Shore D	I	2.0
Low-pressure polyethylene	60 Shore D	II	2.0
Polyamide 6, Ultramid	70 Shore D	I	2.21
Aluminum	39 HV	II	2.68
Brass	150 HV	I	2.76
Aluminum	29 HV	II	3.23
Polyamide 11	69 Shore D	I	3.31
Poly(vinyl chloride)	52 Shore D	II	4.2
Poly(vinyl chloride)	78 Shore D	II	6.3
Resitex	89 Shore D	II	8.2
Poly(vinyl chloride)	76 Shore D	II	8.5
Glass	6–7 Mohs	II	9.7
Lead	4 HV	II	10.5
Plexiglas	85 Shore D	II	10.75
Pertinax	92 Shore D	II	18.5
Epoxy resin with glass fiber	86 Shore D	II	19.5
Epoxy resin with hardener and quartz powder	84 Shore D	II	31

* Abrasive I: sand, HV = 5000 MPa, particle size ≤ 0.9 mm;
 Abrasive II: sand, HV = 7200–8100 MPa, particle size 0.3–0.5 mm

face to a certain depth by structural transformation, making it resistant to wear. These include induction, flame hardening, HF impulse, electron, and laser jet methods. In case hardening, a steel which originally cannot be hardened is made hardenable by inward diffusion of carbon, sometimes also with nitrogen; hardening is then carried out by rapid quenching. The hardness achieved is governed essentially by the contents of dissolved C and N; the usual depths of application are 0.5 to 2 mm.

Nitride layers are produced after treatment in a salt bath or in a gas atmosphere by inward diffusion, usually below 600 °C. Unlike transformation hardening, which is effected by lattice distortion as a result of embedded carbon atoms, nitriding and boriding provide the surface with an exceptional increase in hardness by the formation of an intermetallic bonding layer. These bonding layers also increase the general corrosion resistance, but they are so thin that they can break by mechanical point loading.

Hardfacing, on the other hand, is a thick layer process. By means of various welding techniques, wear resistant alloys in rod, wire, or powder form build strong metallurgical bonds in the fused state with the surface, which is also fused. Mixing should be kept as low as possible. There is an exceptionally large variety of additives available on the market and, among other things, the requirement for mechanical workability must be taken into account. Most filler metals are alloyed on an Fe-base, but Ni or Co can also be the main constituents where high-

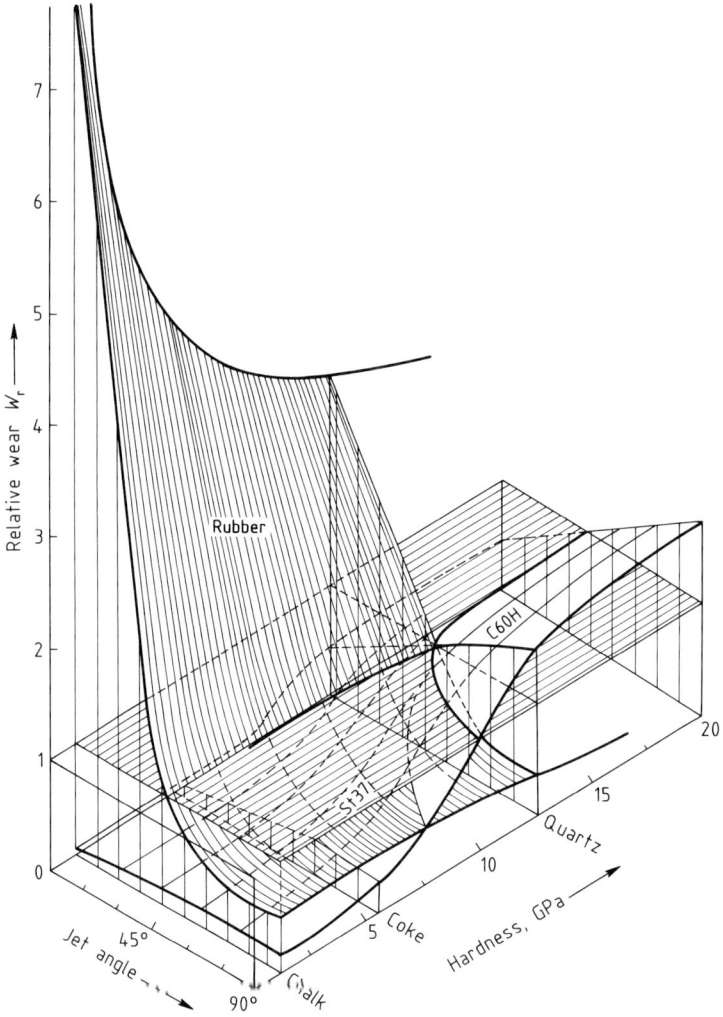

Figure 19. Dependence of jet wear on the jet angle and the hardness of the impinging mineral

temperature and/or additional corrosion resistance is required. Where flat or rotationally symmetrical surfaces have to be coated over large areas, automatic or fully-mechanized welding is much cheaper than manual welding [42].

Sheet metal plating by roll-bonded or explosion plating is not always as effective as weld plating with respect to adhesion. As the plating material must be very plastically deformable to achieve an intimate bond with the supporting material, highly wear resistant and high-strength materials are unsuitable for the purpose. Nevertheless, this method is widely used in chemical technology because to a large extent corrosion stresses are superimposed on the wear processes.

The selection of highly corrosion resistant alloys applied to mechanically heavily stressed base bodies provides satisfactory solutions to the majority of problems in the construction of vessels and pipes [43].

Thermal spraying can cause even greater adhesion problems than those of the above-mentioned coating methods. Unlike hardfacing, after the coating additive has been fused in a flame, arc, or plasma it is projected against the cold or only slightly heated base. Adhesion to the surface and the density of the layer itself can differ greatly depending on the intensity of the spraying and the possibility of reaction with the surrounding medium on the way to the base materi-

Table 7. Hard materials

Material	mp, °C	ϱ, g/cm^3	Vickers hardness HV$_{0.2}$
Oxides			
Al$_2$O$_3$	2050	3.9	2300
ZrO$_2$	2700	5.7	1100
TiO$_2$	1860	4.2	1000
Carbides			
TiC	3150	4.9	3200
TaC	3780	14.5	1790
WC	2600	15.7	2080
SiC	2180	3.2	3000
B$_4$C	2450	2.5	3500
Nitrides			
TiN	2950	5.2	2450
TaN	3090	13.8	3230
Si$_3$N$_4$	1900	3.2	1400
BN (cubic)	3000*	3.48	8000
Borides			
TiB$_2$	2900	4.4	3480
ZrB$_2$	2990	6.0	2200
Carbon			
Diamond	3750	3.5	8000–10 000

* Sublimes.

al. From a micrograph it can be seen that, depending on the process used, the layers have a particular structure caused by agglomeration of many fine globules of molten material, which become deformed on impact and leave a visible porosity in the composite material.

The most important groups of materials for wire flame spraying and arc spraying are

1) Low- and high-alloy steels
2) Bronzes, nonferrous metals such as nickel, copper, zinc, Monel metal, brass, aluminum

Powder flame spraying is a versatile method of thermal coating and also handles

1) Special steel alloys with embedded mechanically resistant materials
2) Ceramics with fairly high melting points
3) So-called self-flowing Cr–Ni–B–Si alloys which, after the actual spraying process—without any adhesion layer—are submitted in a second operation to subsequent treatment at about 1100 °C (liquid-phase sintering); this produces virtually pore-free layers with good adhesion characteristics.

Substantially higher temperatures are reached in the plasma process, so protective layers can be produced from high-melting oxides, carbides, and borides. The higher thermal energy and higher velocity produce improved homogeneity and adhesive strength.

The quality of this method is improved by the use of an inert atmosphere under reduced pressure. Individual alloying elements are not burnt off, oxide and nitride formation is suppressed, and adhesion and layer structure is improved.

Typical materials and applications for vacuum plasma spraying are:

1) M–Cr–Al–Y alloys for protection against corrosion by hot gases
2) Carbides (WC, Co–Cr$_3$C$_2$, NiCr–TiC–NbC) for protection against abrasive wear and erosion
3) Oxides and refractory metals for the formation of thick protective coatings [44]

The use of chemical vapor deposition (CVD) and physical vapor deposition (PVD) for forming coatings of carbides, borides, nitrides, and oxides has increased. The CVD method has been furthest developed for the deposition of TiC, TiN, CrC, WC, and Al$_2$O$_3$ [45]–[48].

Surface treatment and coating of metals are discussed in more detail in → 8. Corrosion, pp. **8**-53–**8**-67 and in → Metals, Surface Treatment, treated in the A series.

5. Practical Examples of Abrasion and Erosion Damage

Chemical industry with its manifold process steps involves conveying products from vessels or tanks through pipelines to other vessels where new products are produced by chemical reactions, and then to still further apparatuses where the solutions or mixture of products are separated from one another. The process steps also include mechanical processing such as comminution, kneading, and granulating. Production of large quantities requires a continuous material flow. This can be achieved most easily with fluid phases which are transported in conveying units. Erosion and abrasion are therefore the predominant types of wear in the main chemical plant components, while sliding wear, elastic rolling wear, and oscillation wear are found more in drives and machinery.

Transport velocities remain within defined limits and material flows are relatively constant, so characteristic wear data which have been determined under realistic conditions are quite readily transferable. Wear, as a form of damage, does not play a dominant part in chemical plants. Nevertheless, it does occur and the following examples give an impression of the diversity.

5.1. Sliding Wear, Elastic Rolling Wear, and Oscillation Wear

The pulsation in gases which have been compressed in reciprocating compressors must be adequately damped before the gases are transferred to heat exchangers for cooling. This damping was not sufficient in the case of oxygen compression described here, so that the baffles were continuously striking against the supporting spacer tubes; they deformed them increasingly and caused sliding wear to the tubes seated tightly against them (Fig. 20).

Instances of damage caused by sliding occur quite frequently in piston machines. In the rotary compressor shown (Roots type) for ammonia compression the damage occurred after approximately 60 000 h operation (Fig. 21).

The centrifugal pump is the most important hydraulic machine for setting a fluid in motion and for increasing the pressure of the transported medium. Its roller bearings are highly stressed parts with a limited life. This can be shortened considerably if axial shocks are transmitted from the flowing liquid as a result of large numbers of stops and starts, i.e., predominantly discontinuous operation. This is often indicated by running tracks displaced towards the edges (Fig. 22).

Predominantly transient stressing (radial impacts) occurred at the piston of an oxygen compressor where the surface of a stuffing box chamber ring showed tribochemical reaction zones as a result of heating in contact with oxygen (Fig. 23). Because of the danger of spontaneous ignition of metals in the presence of compressed oxygen any processes involving friction in such machines must be treated with great caution.

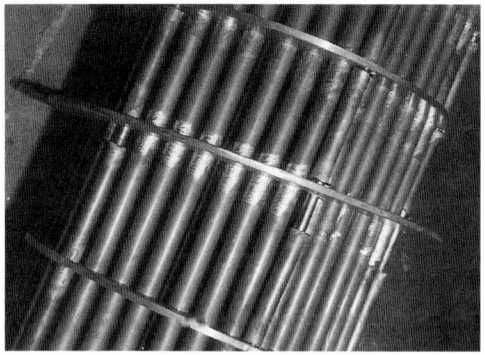

Figure 20. Sliding wear on copper tubes of a heat exchanger for oxygen

Figure 22. Surface destruction of a ball-bearing inner race as a result of intermittent axial overloading

Figure 21. Sliding wear damage on the rotary piston of a Roots compressor for ammonia

Figure 23. Tribochemical reaction and surface destruction on the brass piston of an oxygen compressor

5.2. Abrasion Wear

Where abrasive impact wear occurs in hammer mills, such as those used for comminution of raw coal in power stations, continuous replacement of plain plates of unalloyed steel is preferred to the application of sustantially more expensive materials with only moderately improved service lives.

Three-body abrasive wear occurs relatively frequently in machines used for process engineering as the materials processed often contain additives with abrasive properties. This is particularly true for the large number of barrel extruders used for kneading and milling processes. It is found, for example, in the draining of unvulcanized rubber, in which the moisture content is reduced from 60 to 10% under a pressure of 6–8 MPa. Screw elements made of 13% Cr steel withstood these conditions for several years without any signs of wear. However, when the unvulcanized rubber contains carbon black the service life of screws made of X 40 Cr 13 is only about one year, even after hard-facing with Hastelloy C (16% Cr, 16% Mo, remainder Ni), Figure 24. This carbon black can be processed, for example, in a wet comminution mill: 1.5 t/h carbon black is ground with 30 t/h water at 3000 rpm between a stator and a rotor adjusted to a gap of approximately 0.4 mm. Both parts have impact ribs with special profiling and various flow diverters. In this case the most economical solution was again to design the elements as wearing parts made of 18% Cr steel and replace them after 7000–10 000 h operation (Fig. 25).

The screw housings which enclose the screw shafts are exposed to high stresses, similar to those in the screw combs of the extruders. The product, which is processed under high pressures, stresses the surface to an exceptional extent, especially in the so-called kneading zones of worm extruders. If glass fibers or glass balls are added to the product (in this case, polypropylene), even wear-protecting layers (60% Ni, 18% Co, 13% Cr, 2% Mo, 1% C) applied by centrifugal casting have only limited service life (Fig. 26). A suitable method of coating with appropriate protective material should be selected to suit the type of stressing.

Permanent magnet pumps with no seals are being used increasingly for the sake of imission protection. The medium to be transported, in this case liquefied petroleum gas, flows through a gap of less than 1 mm around the non-contact internal rotor, which is fitted with permanent magnets. Magnetite particles (Fe_3O_4) originating from the pipeline adhere to the outer surface of the internal rotor and cause permanent sliding

Figure 25. Impact ribs (18% Cr steel) of a wet comminution mill eroded by water containing carbon black

Figure 24. Three-body abrasive wear, caused by carbon black, on a draining screw, hard-faced with Hastelloy C, used for unvulcanized rubber

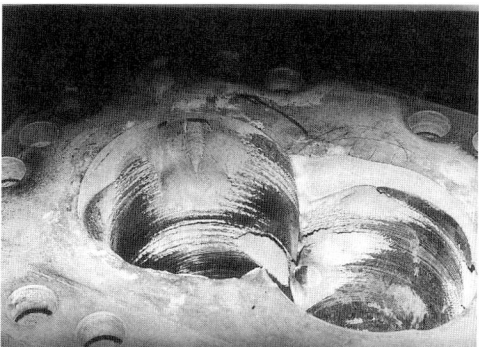

Figure 26. Abrasive wear caused by glass-fiber-containing polypropylene on the high-alloy protective coating (60% Ni, 18% Co, 2% Mo, 1% C) of a screw housing

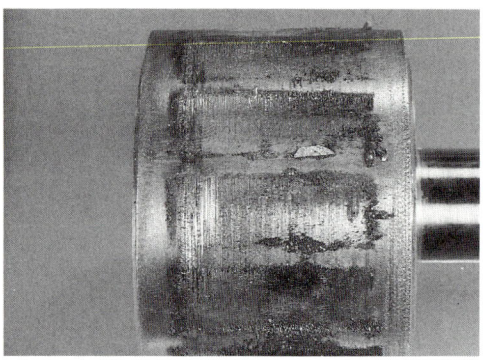

Figure 27. Abrasive wear on the inner rotor of a canned pump caused by adhering magnetite particles

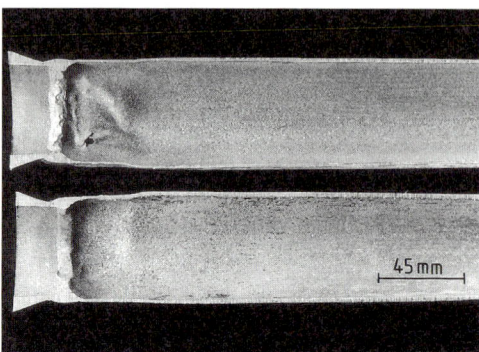

Figure 28. Hydroabrasive wear caused by a liquid containing carbon black in an unalloyed steel pipe as a result of turbulences behind a weld

abrasion at the isolation shell and the rotor, (Fig. 27); finally, this causes the entire pump system to leak.

5.3. Damage by Particle Erosion

Abrasive wear is found predominantly in conveying equipment, but in pipelines and apparatus erosive wear is more typical, and is caused by tribological stressing during the flow of fluids, usually with the involvement of solid particles. Unlike the welded joints in apparatus, the butt welds in pipelines can normally only be made from the outside. Depending on geometrical factors (e.g., edge misalignment, difference in wall thickness) and the welding method, weld seam roots are often produced which have a very considerable sag. This can cause a great reduction in the flow cross section and also, depending on the flow conditions, marked turbulences behind the roots. Figure 28 shows a section of a pipe made of St 35.8 (unalloyed steel) in which the carbon black content in a liquid mixture of organic compounds caused a breach in the wall after 15 months' operating time as a result of hydroabrasion. The wall retained its original thickness only a few centimeters behind the point of turbulence, which illustrates the effect of the angle of incidence of the particle on the surface.

High mechanical stressing occurs when dispersions of plastics are atomized at high pressure. Under some circumstances the acceleration in a cone leading to the jet hole (a few tenths of a millimeter in diameter) and passage through the hole can change the geometry of a jet in a

Figure 29. Pump cover (18-8 CrNi cast) eroded by sharp-edged catalyst particles in the liquid

very short time by erosion. This can affect hard metals and oxide ceramics as well as steel. Due to the forces of acceleration produced in centrifugal pumps, they are particularly liable to hydroabrasive wear. Corrosion-resistant pumps made of cast CrNi (1.4408) had to be replaced after only 6 months' operating time; this was due to the erosion damage shown in Figure 29 because the sharp-edged CuBi catalyst suspended in the aqueous solution had almost eroded through the pump cover.

Even a ducted-wheel pump made of G-X3 CrNiMoCu 24 6 (Noridur 9.4460) hardly lasted any longer, but the tempered and quenched cast alloy G-X170 CrMo 25 2 (Niroloy NL 262) had a service life of several years.

When solid particles are carried in a gas stream and not in a guided liquid stream, any changes in flow direction are very important as, due to the large difference in density, the solid

Figure 30. Bend in a tube furnace for ethylene production, worn by sliding jet wear (material: CrNi 30/30 alloy)

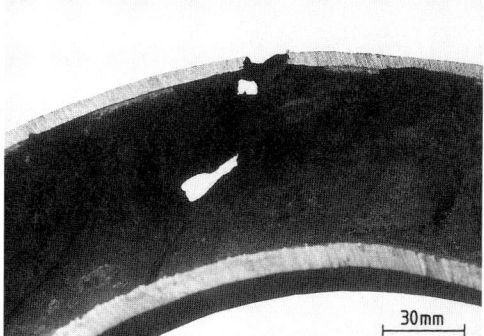

Figure 31. Tube elbow of a waste heat boiler whose protective coating has been worn away in a restricted area by inclined jet wear, which ultimately penetrated the tube

particles are pressed with high force on to the tube surface which guides the flow. The velocities of the transported gas may be very high and then, as a result of intensive sliding jet wear at localized points, lead to short term, but usually accelerated, loss of wall thickness and to rupture of the tube bends.

An example of this is provided by bends from tube furnaces for the production, for example, of ethylene. Because of the high operating temperatures of over 900 °C, the required creep strength and oxidation resistance are usually obtained by using high-alloy materials (e.g., 30% Cr, 30% Ni) and considerably increasing the wall thickness at the bends. Figure 30 shows localized penetration of this type. The wear stressing occurs during repeated decoking with superheated steam.

A comparable situation occurs in waste heat boilers that utilize the high temperature of synthesis gas (CO, H_2) produced in an oil gasification plant for generating steam. Condensing and solidifying particles of slag often form a dense and tightly adhering coating in the gas-carrying tubes, with the result that over a long period the system provides its own protection against jet wear on the heat-resistant, low-alloy steels used. When there are changes in the mode of operation, this protective action often fails locally—the gas velocity apparently plays an important role here—and the protective layer and the tube elbow are worn away in a very restricted area, as if with a milling cutter (Fig. 31).

Superimposition of flow and transient stressing by particles in the gas flow, which is characteristic of impact wear and inclined jet wear, is not found to any great extent in chemical plants

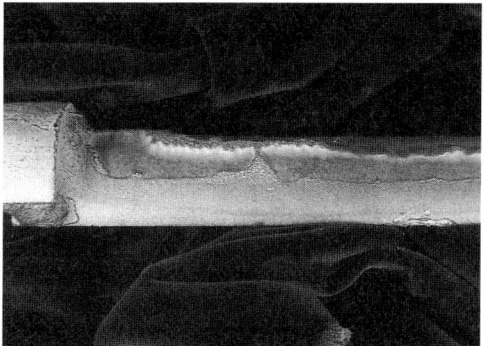

Figure 32. Plasma-coated protective shell of a steam superheater tube, damaged by impact jet wear at ca. 1000 °C

but does occur in power stations. It affects the first contact heating surfaces where the ash particles, at temperatures up to 1000 °C rising at ca. 10 m/s with the flue gas, lead to impact jet wear. These tubes, which are under high steam pressure, are therefore usually protected with heat-resistant wear-protecting half shells. In the case described, comparative tests have shown that high-alloy austenitic alloys (with 25% Cr, 20% Ni, and Si) last longer than high-chromium ferritic steels that have been plasma coated with an oxide layer (Fig. 32). It is assumed that metallurgical changes and chemical reactions caused by the high temperature combine with the purely mechanical stressing to form the complex outward appearance.

However, if the wear-induced break occurs in steam tubes, the expanding steam escaping at high velocity causes erosion on adjacent tubes. Although in this case there are again solid particles entrained in the flue gas, microscopic obser-

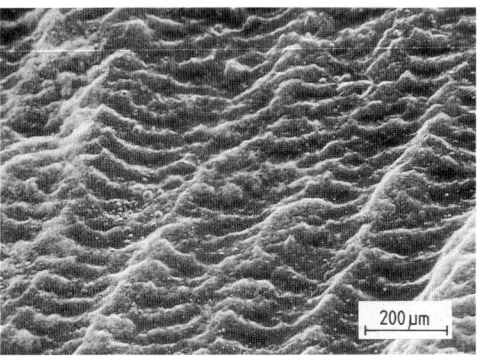

Figure 33. Scanning electron micrograph of the surface of a steam tube (material: 12% Cr steel) damaged by a steam jet

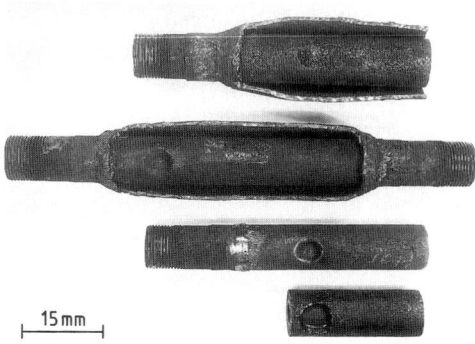

Figure 34. Steam-heated high-pressure pipe showing liquid erosion damage due to condensate carried with the steam

vation of the affected surface shows that, unlike the situation with particle jet wear, there is an ripple structure, typical of steam jet action (Fig. 33).

5.4. Solid Particles—Free Erosion

Heat transfer systems are one of the basic requirements of thermal process technology. Depending on the purpose of the heat exchange, the terms evaporator, liquefier or condenser, superheater, boiler, and many others, are also used.

If the flows of material taking part in the heat transfer are separated from one another by walls, the process is called indirect heat exchange. In general, the flow conditions and prevailing thermal conditions are sufficiently defined and maintained, so only in critical cases does mechanical damage occur by erosion alone.

Liquid erosion damage is to be expected if heating steam contains a proportion of condensate—again dependent on the angle of incidence. When compared with austenitic steels, steam-heated, high-pressure pipes made of low-alloy ferritic steel prove to be insufficiently resistant if the heating is carried out with incompletely dried steam at 2 MPa (Fig. 34).

Rotary compressors include liquid ring pumps, in which a rotor eccentric to the housing opens and closes chambers of different sizes with the aid of a liquid ring; they are used as compressors and vacuum pumps. Under some circumstances cavitation of spontaneously vaporizing liquid can lead to damage to the rotor at the narrowest gap between the rotor and the hous-

Figure 35. Cavitation damage on a cast iron pump rotor

ing, as in the example shown in Figure 35, where the rotor is made of nodular cast iron GGG 42. The different resistance to cavitational stressing can be seen clearly on the surface, where the harder iron phosphide eutectic is largely retained, while the iron matrix has been worn away (Fig. 36). There is also the possibility that a certain corrosive component has contributed to the damage, so this may be a case of cavitation corrosion.

Comparable conditions also caused the damage shown in the following examples. Damage occurred only in the region of the gas inlet of a shell-and-tube heat exchanger in which the tubes carry water in closed circuit for cooling. This was because local vaporization of the water followed immediately by collapse of the vapor bubbles led to deep holes, while immediately after the gas inlet point the existing protective layer remained completely undamaged on the unalloyed steel.

Figure 36. Micrograph of the rotor shown in Figure 35; although the iron matrix has been removed by cavitation, the more resistant iron phosphide eutectic remained unaffected

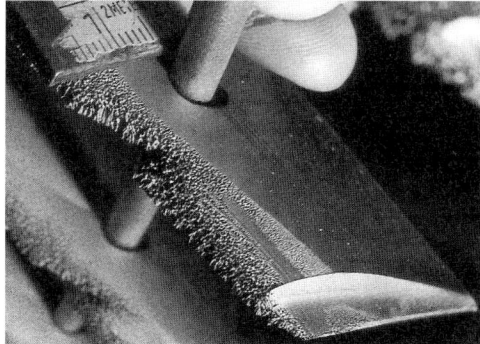

Figure 38. Impingement attack on a steam turbine blade (X 20 Cr 13)

Figure 37. Cavitation erosion in a steel (St. 35.8) heat exchanger tube due to excessive heating

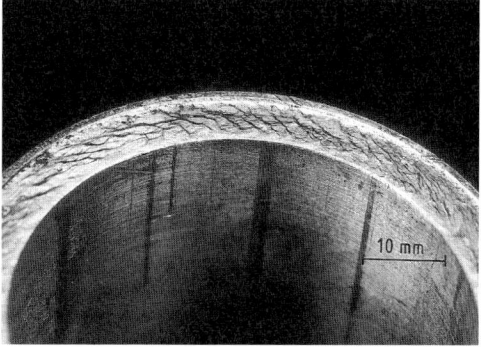

Figure 39. Surface destruction of the case (18-8 CrNi steel) of a canned motor pump

The weakening visible in the micrograph resembles that of pitting corrosion, and clearly the already broken protective layer remains electrochemically active so that a corrosive removal mechanism is superimposed. The hammering effect of the micro jet becomes obvious from Figure 37. In steam turbines even a chromium steel X 20 Cr 13 with otherwise good resistance properties is permanently damaged by impingement of droplets at a water content of approximately 8 % at room temperature (Fig. 38).

A type of drop impingement can also lead to purely mechanical damage and deformation on such parts where a liquid-filled gap, in this case between parts of a canned motor pump made of steel 1.4408, is submitted to transient stressing in the axial direction with simultaneous oscillating radial movement without contact between the metal surfaces (Fig. 39).

The reason for this unusual type of stressing was incorrect assembly in which the jointed parts had not been correctly secured.

5.5. Damage Caused by Erosion–Corrosion

The chemical resistance of many metals to electrolytes is achieved because the reaction products adhere to the surface, hindering ionic transport that controls the electrochemical dissolution process. Guide values can therefore be provided, for example, for the maximum velocity of drinking water up to which certain copper alloys, steels, and titanium retain their protective coatings, and only when the measured values are exceeded is removal of the protective layer to be expected. In many cases unalloyed steel and cooling water operating under defined flow conditions represent the most economical solution for heat exchangers in the chemical industry. Even then, the alternating actions of erosion and corrosion can cause significant damage due to the design and the associated changes in flow direction, although with smooth flow it is possible to maintain a protective covering layer for many years.

Cases of this type of damage can be clearly demonstrated in heat exchangers from a CO conversion process in which the gas preheats the recirculating water (nominal pressure 6.4 MPa) to a maximum of 220 °C: excessive water velocities through perforated impact plates, at changes in direction, and at tube openings through baffles prevented the formation of a complete protective layer; this led to progressive

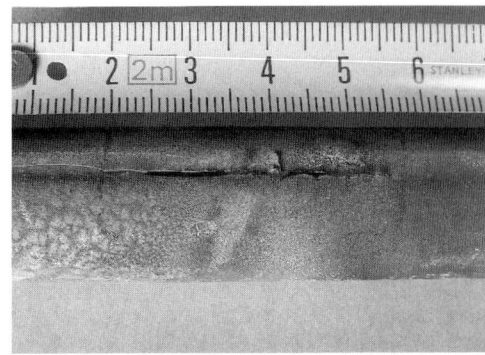

Figure 42. Erosion–corrosion caused by dripping molten acid (material: 18-8 CrNi steel)

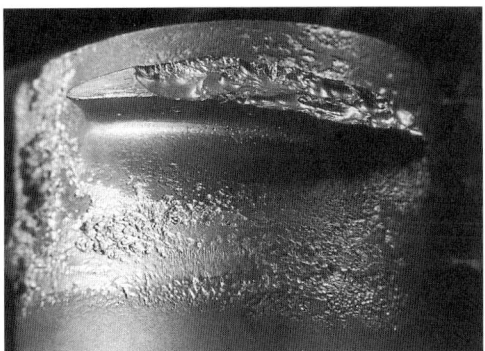

Figure 43. Superposition of erosion and corrosion in a kneading element of a rubber extruder (material: X 35 CrMo 17, ionitrided)

Figure 40. Due to the high flow velocity of the cooling water, the protective layer was removed from heat exchanger tubes (St. 35.8) and sheets (St. 37), leading to intensive erosion–corrosion

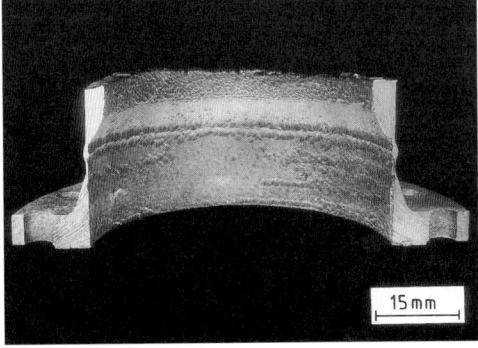

Figure 41. Erosion–corrosion due to turbulence of the liquid caused by the reduction in diameter of a pipe (nominal diameter 300 mm, 15 Mo 5)

wear through liquid erosion accelerated by electrolytic dissolution of the metal (Fig. 40). It was possible to eliminate this type of erosion on the internal surfaces of vessels completely and permanently when they were spray-coated with stainless Cr–Ni steel.

Loss of material by preferential erosion–corrosion caused by the flow conditions can be seen at a product outlet connection of a wash column made of 15 Mo 3 (DN 300). The high velocity together with the slightly corrosive medium is entirely responsible for the loss in wall thickness (Fig. 41).

Occasionally the free fall of a corrosive liquid is also sufficient to remove protective layers and promote the corrosion process, as can be seen in a melting apparatus for dodecanedioic acid (Fig. 42).

Crystalline acid with a moisture content of 10–20 % is fed through a filling hopper and melted by heating tubes made of 18-8 CrNi steel, and the vapor produced is driven off. This damage, a combination of erosion and corrosion, occurred at the region of highest temperature. The complex stress profile of the extruder screw may also be demonstrated with the example of a kneader for unvulcanized rubber, in which, as a result of a localized escape of fumaric acid and very small quantities of water, serious damage occurred within one day due to superimposed erosion and corrosion of a kneading element made of quenched and tempered ionitrided chromium steel X 35 CrMo 17 (Fig. 43).

Conclusion. In chemical plants the problems of wear play a significantly smaller role than those of errosion. Damage in the component equipment and pipelines is met relatively rarely, although abrasion and erosion can occasionally be very detrimental to the potential availability of the plant components at risk. However, it is a different question with machines and conveying units, in which exceptionally high wear stresses can sometimes result from the influence of high flow velocities and high accelerations at housing walls, and especially at sliding seal elements. The manufacturing industries make use of the most modern materials technology to control these problems. Nevertheless, design measures also have to be chosen so that components which are exposed to particular stresses and have limited service lives are designed as easily replaceable wearing parts.

6. References

[1] BMFT-Report: *Damit Rost und Verschleiß nicht Milliarden fressen*, BMFT, Bonn 1984.
[2] B. Genath: "Der Verschleiß ist das Sorgenkind im Maschinenbau," *VDI-Nachr.* **52** (1971) no. 47, 1, 6, 7.
[3] DIN 50 320: Verschleiß, Begriffe – Systemanalyse von Verschleißvorgängen, Gliederungen der Verschleißgebiete, Dec. 1979.
[4] K. H. Zum Gahr: "Grundlagen des Verschleißes," in: *Metallische und nichtmetallische Werkstoffe und ihre Verarbeitungsverfahren im Vergleich*, part III, *VDI-Ber.* **600**.3 (1987) 29 – 56.
[5] H. Uetz, K. Sommer: "Abrasiv-Gleitverschleiß," in: H. Uetz (ed.): *Abrasion und Erosion*, Carl Hanser Verlag, München – Wien 1986, pp. 108 – 157.
[6] K. H. Zum Gahr: "Einfluß des Makroaufbaus von Stahl/Polymer-Faserverbundwerkstoffen auf den Abrasivverschleiß," *Z. Werkstofftech.* **16** (1985) 297 – 305.
[7] H. Wahl: "Verschleißprobleme im Braunkohlenbergbau," *Braunkohle Wärme Energie* **3** (1951) 75 – 87.

[8] R. Schulmeister: "Zur Untersuchung der Werkstoffzerstörung durch Kavitation und Korrosion mit Ultraschall-Koppelschwingern," *Metalloberfläche* **21** (1967) no. 1, 17 – 25.
[9] H. Rieger: *Kavitation und Tropfenschlag*, Werkstofftechnische Verlagsgesellschaft, Karlsruhe 1977.
[10] H. Brauer, E. Kriegel: "Probleme des Verschleißes von Rohrleitungen beim pneumatischen und hydraulischen Feststofftransport," *Maschinenmarkt* **71** (1965) no. 68, 140 – 151.
[11] J. M. Hutchings: "Some Comments on the Theoretical Treatment of Erosive Particle Impacts," *Proc. 5th Int. Conf. on Erosion by Solid and Liquid Impact*, Cambridge 1979.
[12] G. Gommel: *Stoßuntersuchungen Stahlkugel/Stahlplatte im Zusammenhang mit Strahlmittelzertrümmerung und Strahlverschleiß*, Dissertation, TH Stuttgart 1966.
[13] J. S. Rinehard, J. Pearson: *Behavior of Metals Under Impulsive Load*, American Society of Metals, Cleveland 1954.
[14] K. Wellinger, H. Uetz: "Strahlverschleiß," *Tech. Rundsch.*, 8 (1958) 1 – 8.
[15] J. Föhl: "Strahl- und Spülverschleiß," in: K. H. Zum Gahr (ed.): *Reibung und Verschleiß*, Deutsche Gesellschaft für Metallkunde, Oberursel 1983, pp. 157 – 176.
[16] H.-G. Heitmann, W. Kastner: "Erosionskorrosion in Wasser-Dampf-Kreisläufen – Ursachen und Gegenmaßnahmen, *VGB Kraftwerkstech.* **62** (1982) 211 – 219.
[17] H. Tischner: "Korrosionserscheinungen in strömenden Medien am Beispiel von Chemiepumpen," *Chem. Ing. Tech.* **62** (1989) no. 3, 220 – 228.
[18] C. Razim, C. Düll, W. Räuchle: "Über die Beeinflussung der Bauteil-Grundeigenschaften durch Verschleißschutzschichten," *VDI-Ber.* **333** (1979) 11 – 22.
[19] K. Wellinger, H. Uetz: "Gleitverschleiß, Spülverschleiß, Strahlverschleiß unter der Wirkung von körnigen Stoffen," *VDI-Forschungsh.* **449**, ed. B (1955) no. 21.
[20] M. Y. Gürleyik: *Gleitverschleißuntersuchungen an Metallen und nichtmetallischen Hartstoffen unter Wirkung körniger Gegenstoffe*, Dissertation, TH Stuttgart 1967.
[21] F. Henke: "Niedrig- und hochlegierter verschleißfester Vergütungsstahlguß," *Gießerei-Prax.* **1975**, no. 23/24, 377 – 407.
[22] M. M. Krushchov, M. A. Babichev: "Experimental Fundaments of Abrasive Wear Theory," *Russ. Eng. J.* (*Engl. Transl.*) (1964) no. 6, pp. 43 – 48.
[23] H. Uetz, K. J. Groß: "Strahlverschleiß," in: H. Uetz (ed.): *Abrasion und Erosion*, Carl Hanser Verlag, München – Wien 1986, pp. 236 – 278.
[24] J. G. A. Bitter: "A Study of Erosion Phenomena," Part I: *Wear* **6** (1963) 5 – 21; Part II: *Wear* **6** (1963) 69 – 190.
[25] W. Wahl: "Unterschiedliche Werkstoffbewährung bei abrasiv beanspruchten Bauteilen," *VDI-Ber.* **600**.3 (1987) 245 – 286.
[26] K. Heil: *Erosionskorrosion an unlegierten Eisenwerkstoffen in schnellströmenden Wässern*, Dissertation, TH Darmstadt 1979.
[27] D. Kuron: "Korrosion durch Kühlwasser und Schutzmaßnahmen," in W. J. Bartz (ed.): *Die Praxis des Korrosionsschutzes*, Expert Verlag, Grafenau 1981.
[28] H. Sick: "Die Erosionsbeständigkeit von Kupferwerkstoffen gegenüber strömendem Wasser," *Werkst. Korros.* **23** (1972) no. 1, 12 – 18.

[29] H. Grein: "Kavitation – eine Übersicht," *Sulzer Forschungsh.* **1974**, 87–112.

[30] H. H. Piltz: *Werkstoffzerstörung durch Kavitation*, VDI-Verlag, Düsseldorf 1966.

[31] W. J. Rheingans: "Cavitation in Hydraulic Turbines," Symp. on Erosion and Cavitation, *ASTM Spec. Tech. Publ.* **307** (1962) 17–31.

[32] K. H. Habig: *Verschleiß und Härte von Werkstoffen*, Carl Hanser Verlag, München – Wien 1980, p. 215.

[33] H. Uetz, J. Wiedemeyer: *Tribologie der Polymere*, Carl Hanser Verlag, München – Wien 1985.

[34] J. Wiedemeyer: "Deutung des tribologischen Verhaltens ungeschmierter Thermoplaste auf der Basis von Modellrechnungen sowie experimentellen Ergebnissen," *Fortschr. Ber. VDI Z. Reihe 5* **96** (1985).

[35] J. K. Lancaster: "Abrasive Wear of Polymers," *Wear* **14** (1969) 223–239.

[36] K. Wellinger, H. Uetz, G. Gommel: "Verschleiß durch Wirkung von körnigen mineralischen Stoffen," *Materialprüfung* **9** (1967) no. 5, 153–160.

[37] H.-D. Ruprecht: "Elastomere-Polyurethane," in: H. Uetz (ed.): *Abrasion und Erosion*, Carl Hanser Verlag, München – Wien 1986, pp. 438–450.

[38] H. P. Lachmann: "Elastomere-Gummi," in H. Uetz (ed.): *Abrasion und Erosion*, Carl Hanser Verlag, München – Wien 1986, pp. 451–465.

[39] H. Brauer, E. Kriegel: "Untersuchungen über den Verschleiß von Kunststoffen und Metallen," *Chem. Ing. Techn.* **35** (1963) 697–707.

[40] K. Wellinger, H. Uetz: "Verschleiß durch körnige mineralische Stoffe," *Aufbereit. Tech.* **4** (1963) 193–204, 319–335.

[41] E. Dörre: "Nichtmetallische Hartstoffe," in H. Uetz (ed.): *Abrasion und Erosion*, Carl Hanser Verlag, München – Wien 1986, pp. 451–465.

[42] W. Wahl, I. Kretschmer, J. Wabnegger: "Auftragschweißen," in H. Uetz (ed.): *Abrasion und Erosion*, Carl Hanser Verlag, München – Wien 1986, pp. 374–394.

[43] H. Gräfen: "Beschichtungen in der Chemietechnik," in: Beschichtungen für Hochleistungs-Bauteile, *VDI-Ber.* **624** (1986) 273–296.

[44] H.-M. Höhle: "Thermische Spritzverfahren," in: Beschichtungen für Hochleistungs-Bauteile, *VDI-Ber.* **624** (1986) 71–83.

[45] F. Wendl: "Aktuelle Trends bei der Oberflächenbehandlung von Werkzeugen in der Kunststoffverarbeitung," *Thyssen Edelstahl Tech. Ber.* **15** (1989) no. 2, 110–125.

[46] H.-A. Mathesius: "Herstellen von verschleißfesten Schichten mit Hilfe von CVD-Verfahren," in: Beschichtungen für Hochleistungs-Bauteile, *VDI-Ber.* **624** (1986) 37–48.

[47] H. Weiß: "Elektrochemische Beschichtung und Sonderverfahren der Oberflächentechnik," *Thyssen Edelstahl Tech. Ber.* **15** (1989) no. 2, 85–114.

[48] H. Simon, M. Thoma: *Angewandte Oberflächentechnik für metallische Werkstoffe*, Carl Hanser Verlag, München – Wien 1985.

10. Mechanical Properties and Testing of Metallic Materials

ALI FATEMI, The University of Toledo, Toledo, Ohio 43606, United States

1.	Introduction ... 10-2	2.5.	Hardness Tests ... 10-11
1.1.	Concepts of Stress and Strain ... 10-2	2.5.1.	Rockwell Hardness Test ... 10-11
1.1.1.	Stress ... 10-2	2.5.2.	Brinell Hardness Test ... 10-11
1.1.2.	Strain ... 10-2	2.5.3.	Vickers Hardness Test ... 10-13
1.2.	Elastic Deformation and Properties ... 10-2	2.5.4.	Knoop Hardness Test ... 10-13
1.2.1.	Young's Modulus ... 10-3	2.5.5.	Correlation of Hardness with Strength ... 10-13
1.2.2.	Shear Modulus ... 10-3	2.6.	Impact Testing ... 10-13
1.2.3.	Poisson's Ratio ... 10-3	2.6.1.	Notched Bar Impact Test ... 10-13
1.2.4.	Relation Between Elastic Constants ... 10-4	2.6.2.	Dynamic Tear Test ... 10-14
1.3.	Plasticity and Flow ... 10-4	2.7.	Creep Test ... 10-14
2.	Mechanical Testing ... 10-4	2.8.	Stress Relaxation Tests ... 10-15
2.1.	Tensile Test ... 10-4	3.	Fatigue Loading ... 10-16
2.1.1.	Significance and Use ... 10-4	3.1.	Principles and Terminology ... 10-16
2.1.2.	Apparatus and Specimen ... 10-5	3.1.1.	Fatigue Phenomena ... 10-16
2.1.3.	Stress–Strain Diagram ... 10-5	3.1.2.	Fatigue Fracture Surfaces ... 10-17
2.1.4.	Strength Properties ... 10-5	3.1.3.	Fatigue Behavior ... 10-17
2.1.5.	Ductility ... 10-6	3.2.	Cyclic Stress-Controlled Fatigue ... 10-18
2.1.6.	Toughness and Resilience ... 10-7	3.2.1.	Stress–Life Curves ... 10-18
2.1.7.	True Stress–True Strain Diagram ... 10-7	3.2.2.	Fatigue Tests ... 10-18
2.1.8.	Strain-Rate Sensitivity ... 10-8	3.2.3.	Mean Stress Effects ... 10-19
2.1.9.	Temperature Effects ... 10-8	4.	Fracture Mechanics ... 10-19
2.2.	Compression Test ... 10-8	4.1.	Principles and Terminology ... 10-19
2.3.	Bend and Flexure ... 10-9	4.2.	Fracture Toughness ... 10-20
2.3.1.	Elastic Bending Stress and Deflection ... 10-9	4.2.1.	Fracture Toughness Testing ... 10-21
2.3.2.	Bend and Flexure Test ... 10-9	5.	References ... 10-23
2.4.	Shear and Torsion ... 10-10		
2.4.1.	Direct Shear Stress and Test ... 10-10		
2.4.2.	Torsional Shear Stress and Test ... 10-10		

In addition to the symbols defined in the front matter of this volume, the following symbols are used:

a crack length
A cross-sectional area
A_0 initial cross-sectional area
A_f final cross-sectional area
C distance to neutral surface in beam bending
d diameter
D indenter diameter
E elastic (Young's) modulus
f function
F force or load
F_{max} maximum load
G shear modulus or modulus of rigidity
h_0 initial height
h_f final height
I moment of inertia
K stress intensity factor, strength coefficient
K' strain-rate constant
K_c fracture toughness
K_{Ic} plane-strain fracture toughness
l length

l_0 initial gauge length
l_f final gauge length
Δl change in length
m strain-rate sensitivity
M bending moment
n strain-hardening exponent
N number of cycles
r radius
R stress ratio
S shear strength
t thickness, indentation depth, time
Δt time increment
T torque
v displacement
V shear force
w width
Y dimensionless parameter
γ shear strain
γ_{max} maximum shear strain
δ_{max} maximum deflection
ε normal (axial) strain
ε_e elastic normal strain
ε_{eng} engineering strain
ε_p plastic normal strain
ε_{true} true strain
$\dot{\varepsilon}$ creep rate
$\dot{\varepsilon}_{true}$ true strain rate
$\Delta \varepsilon$ strain increment
μ Poisson's ratio
σ normal (axial) stress
σ_a alternating stress or stress amplitude
σ_{eng} engineering stress
σ_f fracture stress
σ_{fat} fatigue strength for $\sigma_m = 0$
σ_m mean stress
σ_{max} maximum stress
σ_{min} minimum stress
σ_{true} true stress
σ_u ultimate tensile strength
σ_{ys} yield strength
$\Delta \sigma$ stress range
τ shear stress
τ_{ave} average shear stress
τ_{max} maximum shear stress
Φ displacement angle or angle of twist

1. Introduction

1.1. Concepts of Stress and Strain

1.1.1. Stress

Stress σ is defined as the force divided by the cross-sectional area on which the force acts

$$\sigma = \frac{F}{A} \qquad (1)$$

where F is the force and A is the area. It represents the intensity of the reaction force at any point in the body as imposed by service loads, assembly conditions, fabrication, and thermal changes.

If the imposed load is perpendicular to the area over which it acts, the resulting stress is a normal (or axial) stress σ. There are two general types of normal stress: tensile and compressive. A pulling force produces tensile normal stress, whereas a pushing force produces compressive normal stress. When the load acts parallel to the area, the resulting stress is referred to as shear stress τ.

In SI units, stress is measured in newtons per square meter or pascals.

1.1.2. Strain

When stresses are applied to a body, they cause deformation of the body and result in changes in dimensions or distortion; the body is then said to be strained. Strains resulting from changes in dimension are normal strains ε, whereas strains resulting from distortion of the body are shear strains γ. A normal strain is, by definition, the axial deformation (elongation or contraction) per unit length

$$\varepsilon = \frac{\Delta l}{l} \qquad (2)$$

where Δl is the change in length and l is the length. A normal strain resulting from elongation is called a tensile strain, and a normal strain resulting from contraction is called a compressive strain.

A change in the shape of a body is caused by the application of shearing stresses. Shearing stresses cause relative displacement of the upper and lower surfaces of the body, as shown in Figure 1. The angular change between the two perpendicular lines is defined as the shear strain γ. Strain is a nondimensional quantity, but it is frequently expressed in meters per meter or in percent.

1.2. Elastic Deformation and Properties

Elastic deformation is a temporary deformation that is fully recoverable when the load is

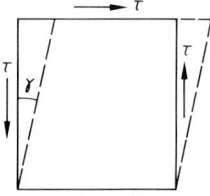

Figure 1. Distortion of a body as a result of the application of shear stress.
— undeformed; --- deformed

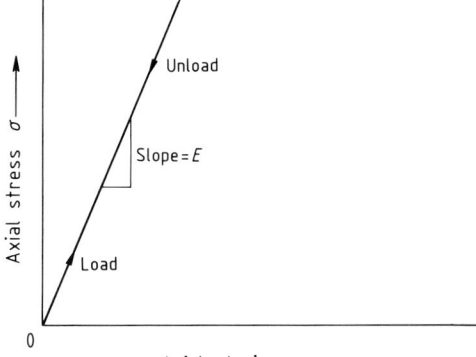

Figure 2. Elastic axial stress–strain diagram

removed. The amount of elastic deformation that a metal can undergo is small because metal atoms are displaced from their original positions during elastic deformation, but not to the extent that they take up new positions.

1.2.1. Young's Modulus

If a long steel wire is pulled by an increasing load, the length of the wire increases proportionately with the load. The relationship between stress resulting from the load ($\sigma = F/A$) and strain resulting from elongation of the wire ($\varepsilon = \Delta l/l$) is given by Hooke's law:

$$\sigma = E\varepsilon \qquad (3)$$

where the proportionality constant E is a material property, called the elastic modulus or Young's modulus. This value represents the stiffness of the material (its resistance to elastic strain), and for most typical metals its magnitude is in the range of 45–410 GPa. A stress–strain diagram representative of elastic deformation for loading and unloading is shown in Figure 2.

As expected, imposition of a compressive stress also evokes elastic behavior. However, stress–strain characteristics and Young's modulus are virtually the same for both tensile and compressive stresses in the elastic range.

1.2.2. Shear Modulus

Shear stress and strain are related through:

$$\tau = G\gamma \qquad (4)$$

where G is a material property called the shear modulus, sometimes referred to as the modulus of rigidity. The shear modulus represents the

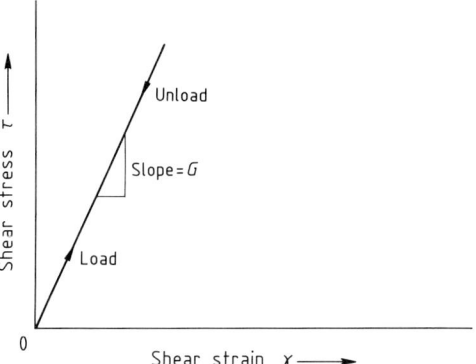

Figure 3. Elastic shear stress–strain diagram

slope of the linear elastic region of the shear stress–strain diagram, as shown in Figure 3. For many metals, G is ca. $0.4\,E$.

1.2.3. Poisson's Ratio

When a uniaxial tensile force is applied to a rod or bar, it results in extension (or contraction in the case of a compressive force). Existing simultaneously with this elongation, however, is a lateral or transverse contraction (or extension for a compressive force) in the directions normal to the applied stress. The ratio of the magnitude of the transverse change per unit length (transverse strain) to the magnitude of the longitudinal change in length per unit length (longitudinal strain) is called Poisson's ratio μ:

$$\mu = \frac{|\text{Lateral strain}|}{|\text{Longitudinal strain}|} \qquad (5)$$

Table 1. Elastic constants for some common metallic materials

Material	Young's modulus E, 10^3 MPa	Shear modulus G, 10^3 MPa	Poisson's ratio μ
Aluminum	71	26	0.34
Brass	110	41	0.34
Copper	128	48	0.34
Iron	207	80	0.29
Magnesium	45	17	0.29
Nickel	200	77	0.30
Silver	83	30	0.37
Titanium	113	43	0.32

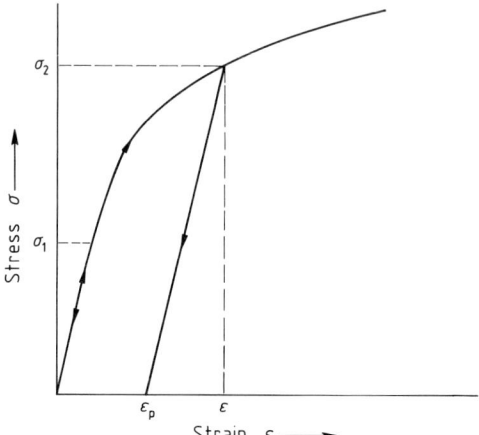

Figure 4. Plastic deformation preceded by elastic deformation

Poisson's ratio is a material elastic property and for typical metals is on the order of 0.3.

1.2.4. Relation Between Elastic Constants

The three elastic constants (E, G, and μ) in linear elasticity are related to each other by the equation:

$$G = \frac{E}{2(1+\mu)} \qquad (6)$$

Therefore, if two of the elastic constants are given or measured, the third can be calculated. Typical values of elastic constants for metallic materials are given in Table 1.

1.3. Plasticity and Flow

Many materials undergo permanent or non-recoverable deformation when stressed beyond a certain level. This is called plastic deformation; it remains in the material after the load or stress has been removed. Plastic deformation results from permanent displacement of atoms or molecules from their original positions in the lattice. This displacement is due to the breaking of bonds between atoms or molecules and the net movement of large numbers of them relative to one another.

A constant load of sufficient magnitude applied to a material results in continuously increasing deformation; this phenomenon is called flow.

Figure 4 illustrates the idea of plastic deformation. If the material is stressed to σ_1 and then released, the strain returns to zero (elastic behavior). However, when the material is stressed to level σ_2 and then released, the material recovers to the extent ($\varepsilon - \varepsilon_p$), where ε_p is the plastic strain remaining after the load or stress has been removed. Therefore, for stress levels beyond the elastic limit, the total strain is composed of elastic, ε_e, and plastic, ε_p, portions:

$$\varepsilon = \varepsilon_e + \varepsilon_p \qquad (7)$$

2. Mechanical Testing

The mechanical properties of materials determine their characteristics and behavior under applied service forces or loads. Specifically, the mechanical behavior of a material characterizes its response (or deformation) when subjected to an applied load. Important mechanical properties include stiffness, strength, ductility, and hardness. In this chapter, several tests that are used to determine these properties are discussed.

2.1. Tensile Test

2.1.1. Significance and Use

The tensile test (or pull test) provides information on the strength and ductility of materials under a uniaxial tensile stress condition. The results of this test are often used to evaluate the mechanical properties of materials and to compare materials. Details of tensile testing are given in ASTM E8M-89 [1].

2.1.2. Apparatus and Specimen

A typical setup for a tensile test is shown in Figure 5. A typical specimen has a diameter of 12.5 mm and a gauge length of 62.5 mm as shown in Figure 6 [1]. The ends of the specimen are screwed into a pair of grips, which are part of the tensile testing machine (Fig. 5). The grips are then pulled apart, often by mechanical means. The load on the specimen is recorded continuously by means of a load cell, which is often a calibrated stiff spring. Extension of the gauge section is measured by an extensometer, which is anchored to the specimen gauge section at the start of the test. Raw data are recorded in terms of load and extension.

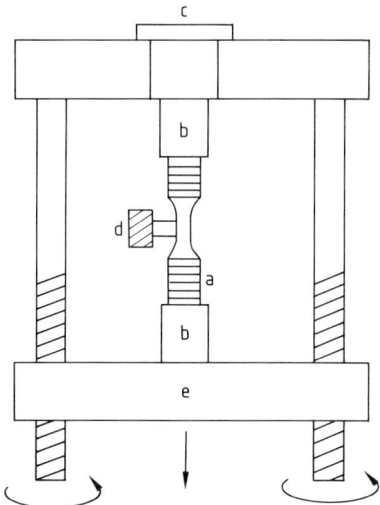

Figure 5. Schematic of a tension test setup
a) Specimen; b) Grips; c) Load cell; d) Extensometer; e) Moving crosshead

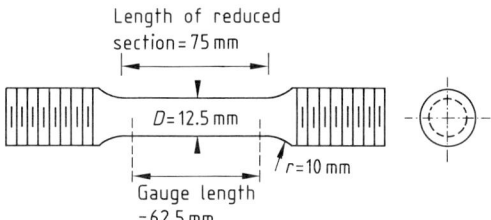

Figure 6. Typical round tension test specimen [1]

2.1.3. Stress–Strain Diagram

Load versus gauge extension data are converted to engineering stress σ_{eng} and engineering strain ε_{eng} as follows:

$$\sigma_{eng} = \frac{\text{Load}}{\text{Initial area}} = \frac{F}{A_0} \tag{8}$$

$$\varepsilon_{eng} = \frac{\text{Change in length}}{\text{Initial length}} = \frac{\Delta l}{l_0} \tag{9}$$

A typical engineering stress–strain diagram for a mild steel is shown in Figure 7, which is divided into two distinct regions of elastic deformation and plastic deformation.

Stretching of the metal in the elastic portion of the stress–strain curve (the initial linear portion) is fully recoverable when the load is removed. At higher forces the material behaves in a plastic manner. Plastic deformation is permanent; upon removal of the load, only the elastic deformation (which is often very small compared to the plastic deformation) is recovered. The plastic region is the nonlinear portion of the stress–strain diagram.

2.1.4. Strength Properties

The *proportional limit* is the limit within which Hooke's law (Eq. 3) applies; it is shown in Figure 7. Young's modulus E is determined from the slope of this linear portion of the curve and measures the stiffness of a material (see Section 1.2.1).

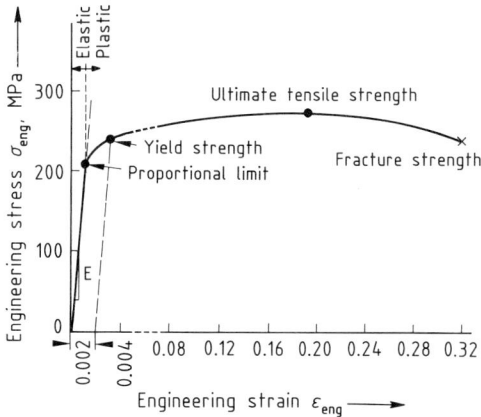

Figure 7. Engineering stress–strain diagram for a mild steel

The stress level at which plastic deformation begins is referred to as the *yield point*. At this point the material undergoes rather rapid and extensive plastic deformation with little accompanying increase in load. Since the transition from elastic to plastic behavior is usually gradual, accurate assessment of the yield point is often difficult. Therefore, the *yield strength* is generally found by an offset method. A line parallel to the initial linear elastic line is drawn from an offset point (typically at a strain of 0.002). The intersection of this line with the stress–strain curve determines the yield strength for the material, as shown in Figure 7. The yield strength of a metal is a measure of its resistance to plastic deformation and is probably the most important mechanical property used in design.

The *ultimate tensile strength* is the maximum strength reached in the engineering stress–strain curve, as shown in Figure 7. Even though the ultimate tensile strength is relatively unimportant for selection or fabrication of materials, its value is often reported in handbooks because it is easy to measure and some of the other material mechanical properties that are more difficult to measure, can be estimated from it. All deformation up to this point in the stress–strain curve is uniform throughout the specimen gauge section. However, at this maximum stress, deformation does not remain uniform, and one region starts deforming more than others (a large local decrease in cross section). This phenomenon is called *necking* and is prominent in ductile materials. Figure 8 shows a schematic representation of this local deformation.

As soon as a localized neck begins to form, a drop in the force occurs due to the decrease in the cross-sectional area at this point. With the onset of necking, the stress distribution changes from uniaxial to triaxial.

After the neck has developed, further plastic deformation is constrained to its vicinity and fracture ultimately occurs at this point. The *fracture* or *rupture strength* corresponds to the stress at fracture, as shown in Figure 7.

2.1.5. Ductility

Ductility is another important mechanical property; it measures the amount of deformation that a material has sustained at fracture. A material that undergoes very little deformation before rupture is said to be *brittle*, whereas one that undergoes a large amount of plastic deformation before rupture is said to be *ductile*. Engineering stress–strain curves for a brittle and a ductile metal are shown in Figure 9.

Ductility is expressed by two measures: percent elongation and percent reduction in area. *Percent elongation* describes the amount that the specimen stretches before fracture and is calculated from

$$\% \text{ elongation} = \frac{l_f - l_0}{l_0} \times 100 \tag{10}$$

where l_0 is the initial gauge length and l_f is the final gauge length. The *percent reduction in area* describes the amount of thinning that the specimen undergoes during the test and can be determined from the following equation:

$$\% \text{ reduction in area} = \frac{A_0 - A_f}{A_0} \times 100 \tag{11}$$

where A_0 is the initial gauge area and A_f is the final gauge area. Both percent elongation and percent reduction in area also give an index of metal quality. If defects such as porosity and inclusions are present, decreases in percent elon-

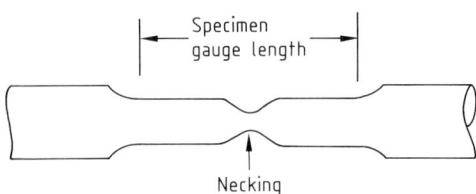

Figure 8. Necking during tension testing

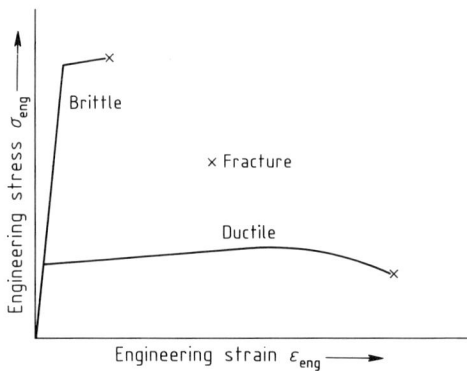

Figure 9. Engineering stress–strain curves for a ductile and a brittle material

Table 2. Tensile properties for some metallic materials [2]

Material	Treatment	Yield strength σ_{ys}, MPa	Ultimate tensile strength σ_u, MPa	Elongation, %	Reduction in area, %
Steel alloys					
1015	as-rolled	315	420	39	61
1050	as-rolled	415	725	20	40
1080	as-rolled	585	965	12	17
4340	Q + T* (205 °C)	1675	1875	10	38
4340	Q + T* (650 °C)	855	965	19	60
301	annealed plate	275	725	55	
403	annealed bar	275	515	35	
Aluminum alloys					
2024	T3	345	485	18	
2024	T6, T651	395	475	10	
7075	T6	505	570	11	
7075	T73	415	505	11	
Titanium alloys					
Ti–5Al–2.5Sn	annealed	805	860	16	40
Ti–6Al–4V	annealed	925	995	14	30

* Quenched and tempered

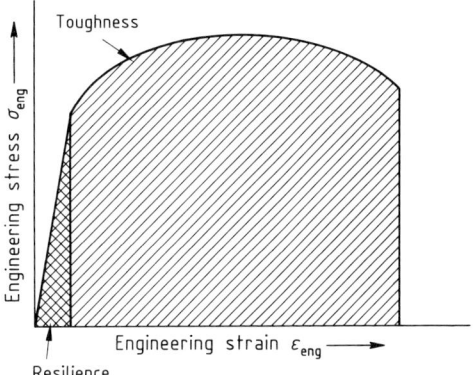

Figure 10. Schematic representation of toughness and resilience

gation and percent reduction in area are often observed.

Tensile properties for some metallic engineering materials are listed in Table 2 [2].

2.1.6. Toughness and Resilience

Toughness is a measure of the ability of a material to absorb energy before fracturing and is given by the area under the engineering stress–strain curve. Toughness is particularly important for materials subjected to dynamic or impact loads, such as automobile bumpers, shock absorbers, and airplane landing gear. Under dynamic loads, toughness is determined by means of an impact test in which a certain load is applied suddenly to a specimen (see Section 2.6).

Resilience is a measure of a material's ability to absorb energy in the elastic range. The *modulus of resilience* is given by the area under the linear portion of the uniaxial stress–strain curve. A high-resilience material has high strength and low elastic modulus. A schematic representation of toughness and resilience is shown in Figure 10.

2.1.7. True Stress–True Strain Diagram

In the engineering stress–strain diagram (Fig. 7), engineering stress decreases beyond the ultimate tensile strength, which seems to indicate that the material is becoming weaker. However, in reality the material's strength increases. The reason for the decrease of stress in the engineering stress–strain curve is that the calculated stress is based on the original cross-sectional area, whereas the area actually changes continually with plastic deformation. Calculation of the true values of stress σ_{true} and strain ε_{true} are based on the instantaneous area and length of the specimen as follows:

$$\sigma_{true} = \frac{\text{Load}}{\text{True area at the time}} = \frac{F}{A} \quad (12)$$

$$\varepsilon_{true} = \int_{l_0}^{l} \frac{dl}{l_0} = \ln\left(\frac{l}{l_0}\right) = \ln\left(\frac{A_0}{A}\right) \quad (13)$$

True and engineering stress and strain are related by the following equations:

$$\sigma_{true} = \sigma_{eng}(1 + \varepsilon_{eng}) \quad (14)$$

$$\varepsilon_{true} = \ln(1 + \varepsilon_{eng}) \quad (15)$$

The true stress–strain curve is compared to the engineering stress–strain curve in Figure 11. As can be seen from the figure, the true stress necessary to sustain increasing strain continues to increase after necking.

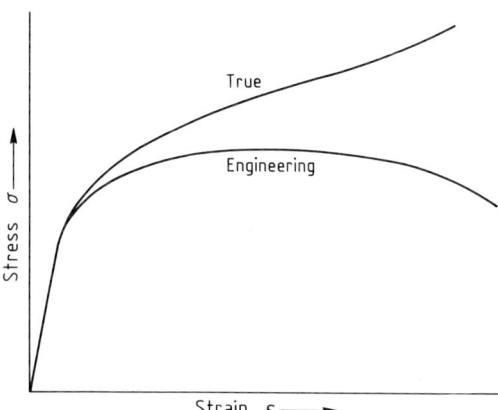

Figure 11. Comparison of true and engineering stress–strain diagrams

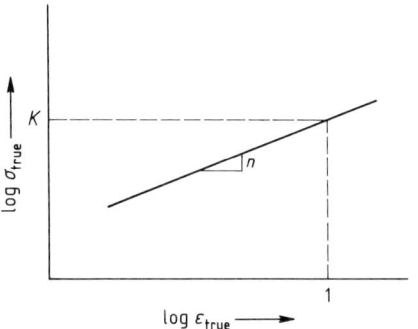

Figure 12. Plot of log σ_{true} versus log $\varepsilon_{\text{true}}$

The true stress–strain curve is important in certain instances involving high plastic deformation of ductile materials, for example, in metal forming. If log σ_{true} is plotted against log $\varepsilon_{\text{true}}$, a straight line is obtained, as shown in Figure 12. The equation of this line is given by

$$\sigma_{\text{true}} = K \varepsilon_{\text{true}}^n \tag{16}$$

where K is a material property called the *strength coefficient* and n is the *strain-hardening exponent*, which gives a measure of the material's work-hardening behavior. The higher the strain-hardening exponent, the more difficult the material is to work (deform) i.e., higher stress is necessary for a given strain.

2.1.8. Strain-Rate Sensitivity

An increase in loading rate (strain rate) usually results in an increase in flow stress of the material and is often adequately represented by the following empirical equation:

$$\sigma_{\text{true}} = K'(\dot{\varepsilon}_{\text{true}})^m \tag{17}$$

where $\dot{\varepsilon}_{\text{true}}$ is the true strain rate ($d\varepsilon_{\text{true}}/dt$), m is the *strain-rate sensitivity*, and K' is a constant. The value of m varies between zero (indicating a strain-rate-insensitive material) and unity (indicating a highly strain-rate-sensitive material). Strain-rate sensitivity generally increases with increasing temperature. The effect of strain rate on the tensile properties of a mild steel at room temperature is shown in Figure 13 [3].

2.1.9. Temperature Effects

In general, tensile properties are significantly affected by temperature. The modulus of elasticity, yield strength, and ultimate tensile strength all decrease with increasing temperature. However, the ductility of the material, as measured by percent elongation or percent reduction in area, generally increases with increasing temperature.

2.2. Compression Test

Compressive properties are used in the analysis of structures subjected to compressive loads or bending loads that result in compressive normal stresses, as well as in processes involving large compressive deformations such as metal working (e.g., rolling of steel). Mechanical properties obtained from a compression test generally include Young's modulus, yield strength, and

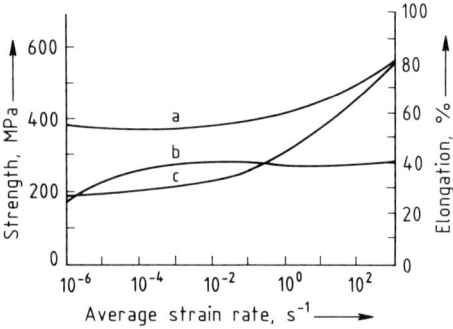

Figure 13. Effect of strain rate on tensile properties of a mild steel [3]
a) Ultimate tensile strength; b) Total elongation; c) Yield strength

compressive strength. These properties are obtained by a test conducted in a manner similar to the tensile test, except that the test specimen used is often in the form of a solid circular cylinder and the force is compressive. The resulting stress–strain behavior in the plastic region is similar to its tensile counterpart, but no necking occurs and the mode of fracture is different from that for tension.

Equations (8) and (9) are also used for calculating the compressive stress and strain. In a compression test, the specimen contracts along the load direction; therefore, the final length l_f is smaller than the initial length l_0 in Equation (9), resulting in a compressive strain that is negative. In compression testing, the specimen geometry and loading should be such that the specimen does not buckle; otherwise, failure may occur by instability rather than by crushing. Details of compression testing are given in ASTM E9-89a [4]

2.3. Bend and Flexure

2.3.1. Elastic Bending Stress and Deflection

The stress in an elastically deformed beam or bar can be calculated by using the flexure formula

$$\sigma = \frac{MC}{I} \qquad (18)$$

where M is the applied bending moment, C is the distance from the neutral plane (the plane with zero stress) to the point at which the stress is being calculated, and I is the moment of inertia for the beam cross section. The value of the bending moment M depends on the magnitude of the applied load, the length of the beam, and the location of the supports. The moment of inertia I is a geometrical property of the cross section. The maximum stress σ_{max} and deflection δ_{max} for two common beam geometries and support configurations are shown in Figure 14.

For the cantilever beam (Fig. 14 A), maximum stress occurs at the supported end of the beam, with tension on top and compression on the bottom, and maximum deflection at the free end of the beam. For the simply supported configuration (Fig. 14 B), maximum stress and deflection occur at the middle of the beam, with tension on the bottom and compression on top.

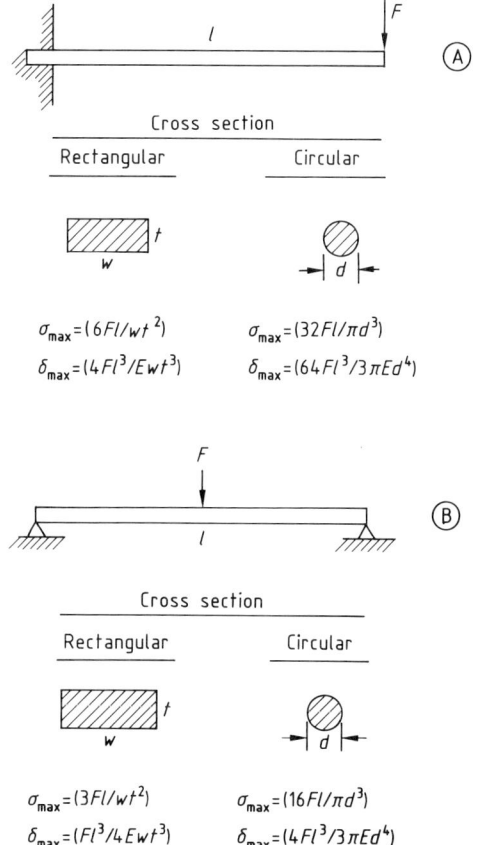

Figure 14. Maximum stress σ_{max} and maximum deflection δ_{max} in selected beams
a) Cantilever beam with end load; b) Simple supports with center load

Note that the equations given in Figure 14 apply when the load magnitude is such that only elastic deformation results.

2.3.2. Bend and Flexure Test

The bend test is a method of evaluating the ductility of a material, as evidenced by its ability to resist cracking during bending. Because the loading in such a test consists of both elastic and plastic deformation (particularly in ductile materials), the flexure and deflection equations given in Section 2.3.1 cannot be used.

In a bend test, the test specimen is a bar with either rectangular or circular cross section. Bending forces are usually applied through one of the three arrangements shown in Figure 15 [5].

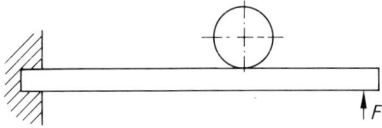

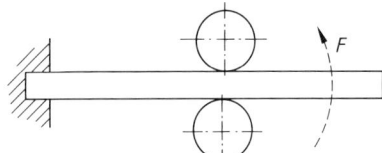

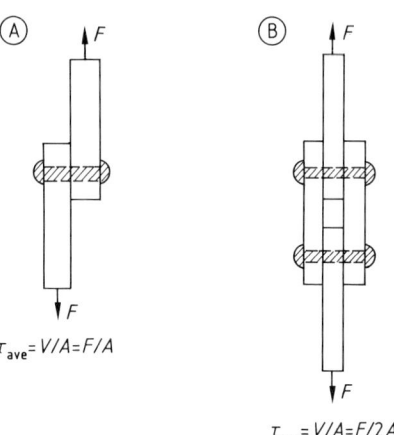

$\tau_{ave} = V/A = F/A$

$\tau_{ave} = V/A = F/2A$

Figure 16. Single- and double-shear joints
a) Single; b) Double

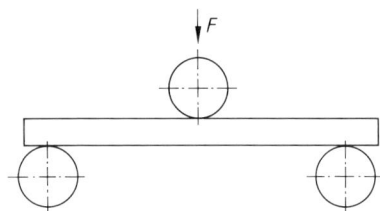

Figure 15. Bend test arrangements [5]

Bending is continued to a specified angle of bend and inside radius of curvature. If complete fracture does not occur, the convex surface of the bar is examined for cracks, and the number and size of the cracks can be used as the criterion for failure. Therefore, the bend test does not provide a quantitative method of evaluating the ductility of a material. Details of the bend test can be found in ASTM E290-87 [5]

2.4. Shear and Torsion

2.4.1. Direct Shear Stress and Test

Mechanical joints, such as rivets, bolts, and pins, often experience direct shearing stress, which is defined as the shearing force divided by the cross-sectional area of the rivet, bolt, or pin:

$$\tau_{ave} = \frac{V}{A} \quad (19)$$

where V is the shear force and A is the cross-sectional area. Equation (19) gives the average shear stress that the rivet, bolt, or pin experi-

ences. Rivets, bolts, or pins that resist shearing forces at a single cross section are called single-shear joints, and those resisting shearing forces at two cross sections are called double-shear joints, as illustrated in Figure 16. The average shearing stress for each condition is also given in Figure 16.

The shear strength of a joint is determined by a shear test in which the joint is usually subjected to double-shear loading (Fig. 16 B). The load required to fracture the joint is determined, and the shear strength is calculated from

$$S = \frac{F_{max}}{2A} = \frac{F_{max}}{2(\pi d^2/4)} = \frac{2 F_{max}}{\pi d^2} \quad (20)$$

where S is the maximum shear strength, F_{max} is the maximum load in the test, and d is the diameter of the joint.

2.4.2. Torsional Shear Stress and Test

A torque applied to a circular bar results in twisting of the bar, as shown in Figure 17. Point A on the circular bar moves to point A' when a torque T is applied. Details of shear testing are given in ASTM B 565-87 [6].

A shear strain is produced that varies linearly with the radius of the bar. The maximum shear strain γ_{max} occurs on the lateral surface and is given by

$$\gamma_{max} = \frac{\Phi d}{2l} \quad (21)$$

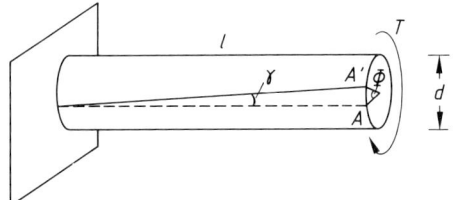

Figure 17. Torsion of a solid bar

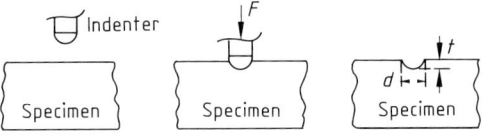

Figure 18. Hardness test

where d is the diameter, Φ is the displacement angle, and l is the length.

For elastic twisting, the displacement angle Φ, which is referred to as the angle of twist, can be calculated from

$$\Phi = \frac{32\,Tl}{\pi d^4 G} \qquad (22)$$

In elastic twisting, the shear stress also varies linearly with the radius. The maximum shear stress τ_{max} occurs on the lateral surface and is given by

$$\tau_{max} = \frac{16\,T}{\pi d^3} \qquad (23)$$

The shear stress and strain are related by the shear modulus G discussed in Section 1.2.2.

A torsional test is analogous to a tensile test in that shear stress τ is plotted against shear strain γ. The slope of the initial linear portion is the shear modulus G (Fig. 3), which is related to Young's modulus and Poisson's ratio by Equation (6), discussed in Section 1.2.4. Because of this relationship, if Young's modulus and Poisson's ratio are determined from a tensile test, a torsional test is not required to determine G.

2.5. Hardness Tests

Hardness tests are probably the most frequently performed mechanical tests because they are nondestructive and usually no special specimen need be prepared.

Hardness tests measure the resistance to penetration of the surface of a material by an indenter with a force applied to it. The hardness of materials depends on the type of binding forces between atoms, ions, or molecules and, like strength, increases with the magnitude of these forces. Because indentation occurs by plastic deformation in metals and alloys, hardness is inherently related to the material's plastic resistance and, therefore, to its yield strength and wear resistance.

A hard indenter (such as hardened steel, tungsten carbide, or diamond) of a standard shape (either round or pointed) is pressed into the surface of the material under a known force. The resulting area or depth of indentation is measured, which in turn is related to the hardness number. This process is shown schematically in Figure 18.

Depending on the material tested and the type of test, various combinations of load and indenters are used. The most common methods of hardness testing for metals are the Rockwell, Brinell, Vickers, and Knoop methods. Table 3 summarizes the type of indenters and types of impressions associated with these four common hardness tests.

2.5.1. Rockwell Hardness Test

The Rockwell hardness test is widely used; it has scales (Rockwell A, Rockwell B, etc.) for different hardness ranges. A small-diameter steel ball is used as the indenter for soft materials, and a diamond cone for hard materials. The depth of penetration is measured automatically by the testing machine and converted to a Rockwell hardness number. Therefore, each measurement requires only a few seconds. The Rockwell scale is designated by an R with appropriate scale letter as a subscript (i.e., R_B indicates Rockwell B). For each scale, hardness values range up to 100. Details of the Rockwell hardness test are given in ASTM E18-89a [8].

2.5.2. Brinell Hardness Test

In the Brinell hardness test, a single scale covers a wide range of material hardness. A hard spherical indenter (10 mm in diameter) is forced

Table 3. Hardness tests [7]

Test	Indenter	Shape of indentation Side view	Top view	Load	Formula for hardness number	
Brinell	10-mm sphere of steel or tungsten carbide			F	$\mathrm{BHN} = \dfrac{2F}{\pi D(D - \sqrt{D^2 - d^2})}$	
Vickers	diamond pyramid	136°	$d_1 \times d_1$	F	$\mathrm{VHN} = \dfrac{1.72 F}{d_1^2}$	
Knoop microhardness	diamond pyramid	$l/b = 7.11$ $b/t = 4.00$		F	$\mathrm{KHN} = \dfrac{14.2 F}{l^2}$	
Rockwell A, C, D	diamond cone	120°		60 kg, 150 kg, 100 kg	$R_A =$, $R_C =$, $R_D =$	100–500 t
B, F, G	$\tfrac{1}{16}$-inch-diameter steel sphere			100 kg, 60 kg, 150 kg	$R_B =$, $R_F =$, $R_G =$	130–500 t
E	$\tfrac{1}{8}$-inch-diameter steel sphere			100 kg	$R_E =$	

into the surface of the metal (a 500-kg load is used for softer metals such as copper and aluminum; a 3000-kg load for hard metals such as steel, iron, and alloys). The Brinell hardness number (BHN) is a function of both the magnitude of the load and the diameter of the resulting indentation. The diameter of indentation is measured with a special microscope. A schematic of the Brinell hardness test is shown in Figure 19.

The Brinell hardness number is defined as the load divided by the surface area of the indentation and is calculated from the following equation:

$$\mathrm{BHN} = \dfrac{F}{\left(\dfrac{\pi}{2}\right) D (D - \sqrt{D^2 - d^2})} = \dfrac{F}{\pi D t} \quad (24)$$

where F is the applied load in kilogram-force;

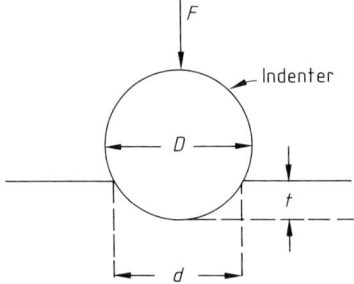

Figure 19. Schematic of the Brinell hardness test

and D is the diameter of the indenter, d the diameter of the impression, and t the indentation depth—all in millimeters. This hardness testing method is relatively insensitive to irregularities in the material surface because the impression left by the indenter is large. Details of the Brinell hardness test are given in ASTM E10-84 [9].

2.5.3. Vickers Hardness Test

The Vickers hardness test is termed a microhardness test because it forms a small indentation of microstructural dimensions. A small diamond pyramid is pressed into the surface of the specimen, and the diagonal of the square impression is measured. Based on this measurement, the Vickers hardness number (VHN) is read from a chart. Because a small indentation is made, Vickers hardness testing is often used for measurements of small regions, for example, in assessing the relative hardness of various phases in multiphase alloys. A microscope is required to obtain the measurement. Details of the Vickers hardness test can be found in ASTM E92-82 [10].

2.5.4. Knoop Hardness Test

The Knoop test is also a microhardness test, and the resulting impression requires measurement under a microscope. A skewed diamond indenter is used in this test, and the Knoop hardness number (KHN) is calculated as the force divided by the projected indentation area. The Knoop hardness scale is approximately equivalent to the Vickers hardness scale. The Knoop hardness test is often used in measuring the microhardness of small areas of metallic material microstructures and for brittle materials such as ceramics. Details of the Knoop hardness test are given in ASTM E384-84 [11].

2.5.5. Correlation of Hardness with Strength

Since hardness and tensile strength are both indicators of a metal's resistance to plastic deformation, a strong correlation exists between them. For example, a strong linear correlation exists between the Brinell hardness number and the ultimate tensile strength σ_u of steel, as follows:

$$\sigma_u = 3.45 \text{ BHN} \qquad (25)$$

where tensile strength is given in megapascals. Figure 20 shows the conversion among Brinell, Vickers, and Rockwell tests, as well as the correlation between the hardness numbers and ultimate tensile strength; conversion tables are given in ASTM E140-88 [13].

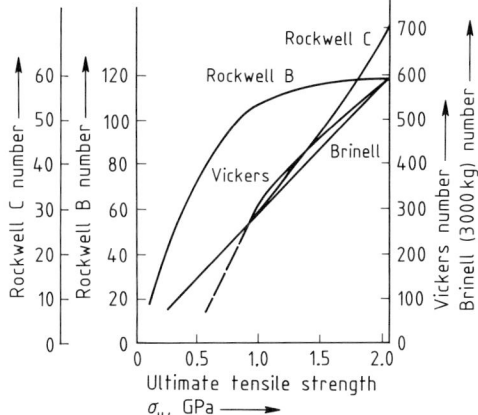

Figure 20. Correlation of hardness with tensile strength and conversion among Brinell, Vickers, and Rockwell tests [12]

2.6. Impact Testing

Impact tests measure a material's resistance to failure under an impact load. The ability of the material to resist an impact load without failure is also referred to as *toughness*. This resistance to failure by fracture is particularly important for notched components. Two common methods of impact testing are the notched bar impact test and the dynamic tear test.

2.6.1. Notched Bar Impact Test

The notched bar impact test is the most common laboratory test to measure impact energy. A Charpy V-notch specimen, shown in Figure 21, is often used. The Charpy V-notch specimen is placed in an impact testing machine, as shown in Figure 22 [14]. A heavy pendulum is released from a known height h_0. After striking and breaking the specimen, the pendulum stops at a lower, final height h_f. If the mass of the pendulum and its initial and final elevations are known, the energy absorbed by the fracture can be determined. The energy is usually expressed in joules. In general, materials that have both high strength and high ductility have large impact energies (or high toughness).

The impact energy of a material is usually very sensitive to temperature. At high temperature, the material behaves in a ductile manner, and a large absorbed energy is required for fracture. At low temperature, the material behaves in

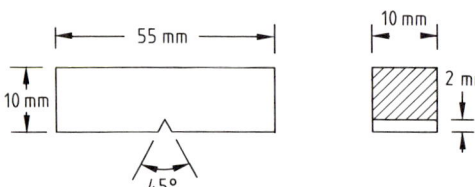

Figure 21. Charpy V-notch specimen

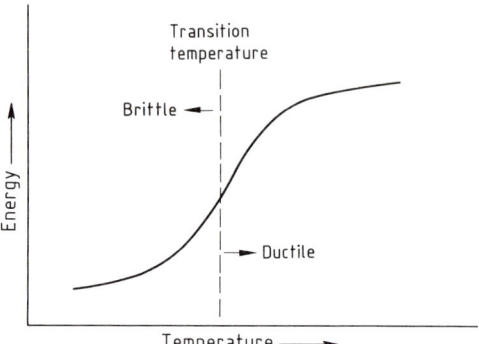

Figure 23. Effect of temperature on impact energy

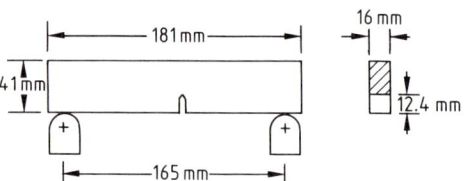

Figure 24. Dynamic tear test specimen and supports

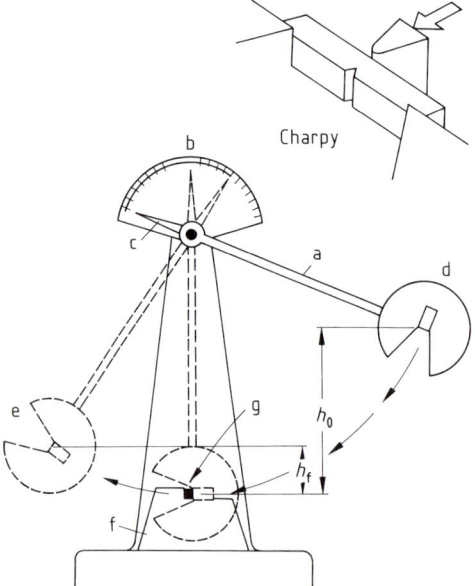

Figure 22. Standard Charpy impact test machine [14]
a) Hammer; b) Scale; c) Pointer; d) Starting position; e) End of swing; f) Anvil; g) Specimen

a brittle manner, and little deformation is observed at fracture. Therefore, the absorbed energy is low. The temperature at which the material changes behavior from ductile to brittle is called the *transition temperature*. The transition temperature can fall between ca. -100 and $+100\,°C$, depending on the alloy and the test conditions. Figure 23 shows the effect of temperature on the impact energy of a typical material. Details of the notched bar impact test can be found in ASTM E23-88 [15].

2.6.2. Dynamic Tear Test

The dynamic tear test is conducted to determine the resistance of a metallic material to rapid progressive fracture. This resistance is measured as dynamic tear energy, which is the total energy required to fracture a dynamic tear specimen.

A single-edge notched beam specimen is used as the dynamic tear specimen and is simply supported on its ends, as shown in Figure 24. The specimen is then impact loaded in the middle by a pendulum or drop weight. The total energy during separation of the specimen as it fractures is measured as the dynamic tear energy for the material tested. Details of dynamic tear testing can be found in ASTM E604-83 [16].

2.7. Creep Test

Permanent deformation of materials at elevated temperature when subjected to a constant externally applied load or stress below the yield strength is called *creep*. In metallic materials, this phenomenon usually occurs at temperatures of about 40 % of the melting point. The creep test measures the resistance of a material to deformation under a static load at high temperature. A cylindrical specimen is placed in a furnace, and a constant stress or load is applied while a constant temperature is maintained. As soon as the constant load or stress is applied the specimen un-

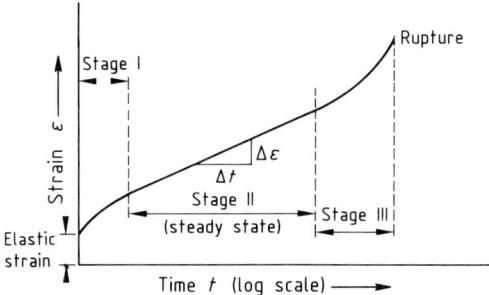

Figure 25. Typical creep curve of strain versus time at constant load or stress and constant elevated temperature

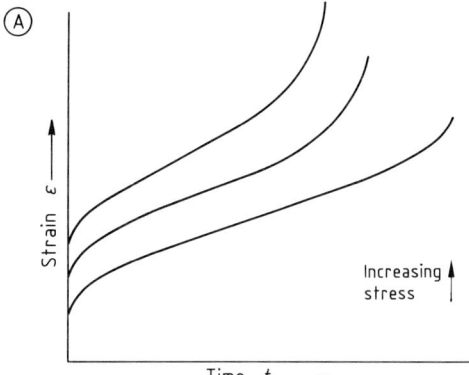

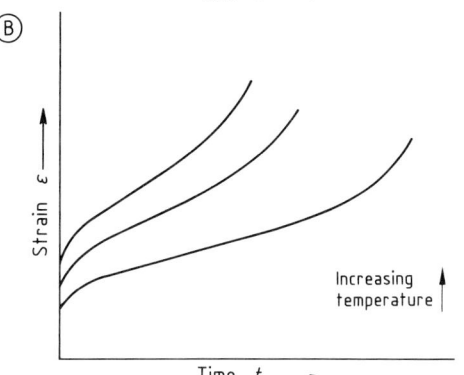

Figure 26. A) Variation of creep curve with stress; B) Variation of creep curve with temperature

dergoes a small elastic strain. After this initial instantaneous elastic deformation, the strain increases with time, as shown in Figure 25.

The creep curve has three distinct stages. Stage I is the primary stage, during which the strain rate $(d\varepsilon/dt)$ decreases, which indicates that the material is experiencing an increase in hardness. The second stage is the linear region, which indicates a constant strain rate and is referred to as the steady-state region. This stage of creep usually has the longest duration. The slope of the steady-state region of the creep curve is the creep rate $\dot{\varepsilon}$:

$$\dot{\varepsilon} = \frac{\Delta\varepsilon}{\Delta t} \qquad (26)$$

In stage III, the strain rate increases due to the increase in true stress as a result of a reduction in cross-sectional area (necking), until failure occurs. This failure is called rupture, and the time required for failure to occur is the rupture time. The creep rate of a material is a function of both the applied stress level and the test temperature. Figure 26 shows the variation of the creep curve with stress and temperature.

To predict the expected lifetime of a component for a given combination of stress and temperature, stress rupture curves are often plotted. These curves are obtained from a series of creep tests by plotting the logarithm of the stress versus the logarithm of rupture life (Fig. 27). Details of the creep test can be found in ASTM E139-83 [17].

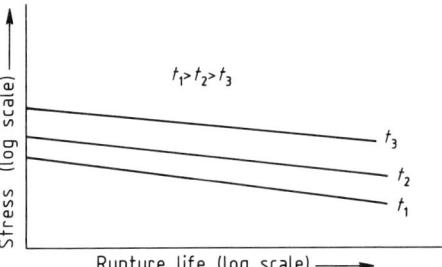

Figure 27. Stress rupture curves

2.8. Stress Relaxation Tests

Stress relaxation refers to the time-dependent decrease in stress under a constant constraint and environment. Stress relaxation test data are often necessary for the design of mechanically fastened joints such as bolted and riveted assemblies, and shrink-fit components, for which permanent tightness is of major concern.

Stress relaxation tests for a material can be conducted under tension, compression, bending, or torsion loads. In general, a specimen is sub-

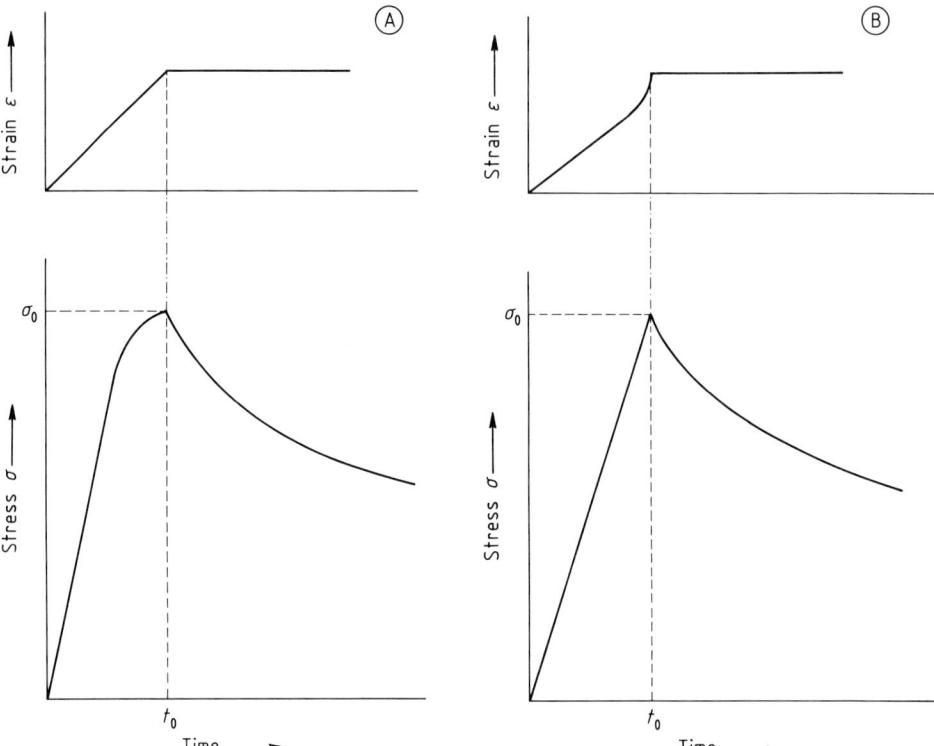

Figure 28. Stress relaxation test
A) At constant strain rate; B) At constant load rate

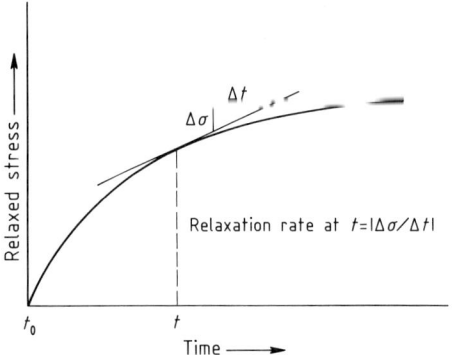

Figure 29. Typical relaxation curve

jected to an increasing load until a predetermined strain is reached. The time at which this initial strain is attained is taken as the reference (zero) time. Then, while the predetermined strain is held constant, the stress during the test is monitored continuously. This procedure under constant strain rate as well as constant load rate is shown in Figure 28 [18].

Relaxed stress, which is the initial stress minus the remaining stress at a given time, is then plotted against time (Fig. 29). The absolute value of the slope of the stress relaxation curve at a given time gives the relaxation rate. Stress relaxation testing is described in ASTM E328-86 [18].

3. Fatigue Loading

3.1. Principles and Terminology

3.1.1. Fatigue Phenomena

In many applications and service conditions, metallic components and parts are subjected to repetitive or cyclic stresses. Repeated stress usually arises in moving parts that rotate, bend, or vibrate. Under cyclic stress, the material usually fails at a stress level far below its maximum static strength (yield strength or ultimate tensile strength). Failures that occur under repeated or cyclic stress are called fatigue fractures.

3.1.2. Fatigue Fracture Surfaces

A typical fatigue failure usually originates at a point of stress concentration. Stress concentrations generally result from either geometrical discontinuities in a part (e.g., keyways, holes, and fillets) or metallurgical defects (e.g., inclusions and voids). Once a fatigue crack has nucleated, the effect of stress concentration increases and the crack grows during, and as a result of, cyclic stress. The crack growth rate is primarily a function of the magnitude, and frequency of the cyclic stress. As the crack grows, the stressed area decreases in size, which results in increased magnitude of the stress, leading ultimately to rapid fracture of the remaining area. Figure 30 is a schematic representation of this process for a typical fatigue failure in rotating bending.

As can be seen in Figure 30, the fracture surface has three distinct regions: (1) the crack nucleation site or sites, (2) the crack growth region, and (3) the final fracture region. Since the crack surfaces are pressed together repeatedly during crack growth, the crack growth region is smooth, whereas the final fracture region usually exhibits a rough surface. In addition, the crack growth region generally contains line markings referred to as "beach marks", which result from different periods of crack extension and various positions of the crack front during its growth. The size of the crack growth region, compared to the final fracture region, depends mainly on the magnitude of the cyclic stress and the material toughness. As the magnitude of stress decreases or toughness increases, the size of the crack growth region on the fracture surface increases. In contrast to the gross deformation observed in a tensile failure for a ductile material (i.e., necking; see Section 2.1.4), typical fatigue fractures do not exhibit gross plastic deformation. Therefore, fatigue failures are often characterized as brittle failures, even for ductile materials.

3.1.3. Fatigue Behavior

The applied stress history can be completely random or simple and repetitive. Examples of a random history and a simple sinusoidal history are given in Figure 31.

The applied cyclic stress can be axial, flexural (bending), or torsional. The standard stress variables in fatigue loading are shown in Figure 31 B and are defined as follows:

$$\text{Stress range} = \Delta\sigma = \sigma_{max} - \sigma_{min} \quad (27)$$

$$\text{Stress amplitude} = \sigma_a = \frac{\Delta\sigma}{2} = \frac{\sigma_{max} - \sigma_{min}}{2} \quad (28)$$

$$\text{Mean stress} = \sigma_m = \frac{\sigma_{max} + \sigma_{min}}{2} \quad (29)$$

$$\text{Stress ratio} = R = \frac{\sigma_{min}}{\sigma_{max}} \quad (30)$$

For completely reversed stressing the mean stress σ_m is zero, and $R = -1$.

Depending on the number of cycles that a part must withstand during its intended lifetime, the fatigue behavior can be classified into low-

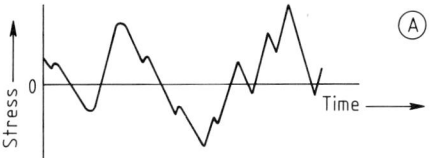

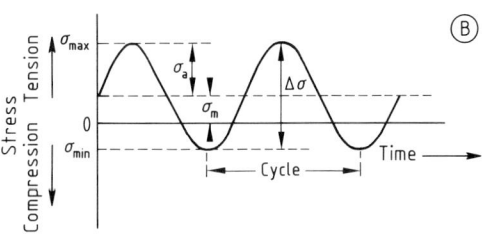

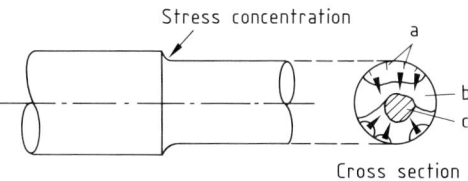

Figure 30. Schematic of a typical fatigue failure process in rotating bending
a) Crack nucleation; b) Crack growth region; c) Final fracture region

Figure 31. Cyclic variation of stress with time
A) Random history; B) Sinusoidal history

cycle or high-cycle regimes. The strain–life approach is generally used to characterize the low-cycle (typically less than 10^5 cycles) fatigue of materials, whereas the stress–life approach (see Section 3.2.1) is often used for high-cycle fatigue applications. Cyclic plasticity is an important consideration in the treatment of low-cycle fatigue, whereas strain is primarily elastic in the high-cycle fatigue regime.

The strain–life approach or strain-controlled fatigue is mainly used to evaluate the fatigue life of notched parts because significant plastic strains usually exist at the root of a notch. In this approach, the fatigue life of the notched part is related to the life of a small smooth specimen cycled to the same strain level as the material at the root of the notch. Further details of strain-controlled low-cycle fatigue testing are given in ASTM E606-80 [19].

The treatment of material fatigue behavior in this section covers only stress-controlled, constant-amplitude cyclic loading, shown in Figure 31 B. For treatment of variable amplitude or random loading and strain-controlled fatigue, see [20].

3.2. Cyclic Stress-Controlled Fatigue

3.2.1. Stress–Life Curves

Traditionally, the stress–life method has been widely used in design against fatigue. A series of tests are performed under stress-controlled conditions by using smooth polished specimens. The data from these tests are then plotted in the form of a stress–life (S–N) curve in which the stress causing failure is plotted against the number of cycles to failure. Figure 32 shows typical S–N curves for a steel and an aluminum alloy [21]. As Figure 32 indicates, the higher the stress amplitude, the smaller is the number of cycles to failure. For steel, leveling off occurs in the S–N curve with no decrease in fatigue strength as the number of cycles increases. The stress level at which the S–N curve becomes horizontal is called the *fatigue limit* or *endurance limit*. The endurance limit is the stress below which fatigue failure does not occur. Ferrous materials usually have S–N curves exhibiting an endurance limit. For steel, the fatigue limit is estimated to be about half the ultimate tensile strength. For materials such as the aluminum alloy in Figure 32, the S–N curve continues to decline and such materials do not have an endurance limit.

3.2.2. Fatigue Tests

The rotating bending fatigue test is the most commonly used small-scale test, in which the smooth polished specimen is subjected to alternating compression and tension stresses of equal magnitude while it rotates. The apparatus is depicted schematically in Figure 33 [20]. A dead weight W produces a uniform bending moment over the test section of the specimen as it rotates. The stress is then calculated from the flexure formula (see Section 2.3.1).

This type of test, however, cannot be used to simulate loading conditions with a mean stress (i.e., $R \neq -1$) because in rotating bending, the loading is completely reversed ($R = -1$). Therefore, for fatigue testing with a mean stress, an axially loaded specimen, shown in Figure 34, is often used [2]. Pulsating tension or tension–compression axial stress is provided by means of a hydraulic or mechanical actuator. A standard procedure for constant amplitude axial fatigue testing is given in ASTM E466-82 [22].

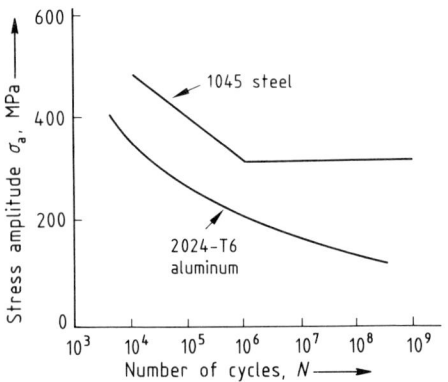

Figure 32. Typical S–N curves for a steel and an aluminum alloy [21]

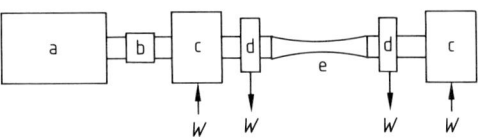

Figure 33. Schematic of rotating bending fatigue test machine [20]
a) Motor; b) Flexible coupling; c) Main bearings; d) Load bearings; e) Test piece

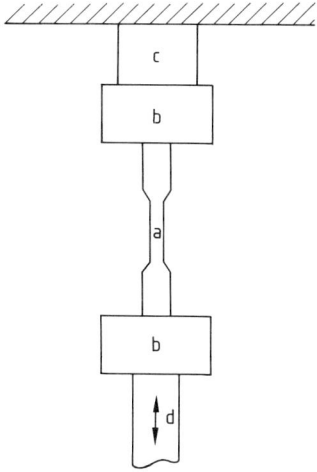

Figure 34. Axial fatigue loading [2]
a) Specimen; b) Grips; c) Load cell; d) Moving actuator

3.2.3. Mean Stress Effects

Mean stress can substantially influence a material's fatigue behavior. In general, tensile mean stresses ($\sigma_m > 0$) are detrimental to fatigue behavior, whereas compressive mean stresses ($\sigma_m < 0$) are beneficial. Typical S–N diagrams with differing mean stress levels are shown in Figure 35.

As can be seen from Figure 35, cyclic life decreases with increasing mean stress level at a given alternating stress σ_a. To account for the effect of mean stress on long-life fatigue strength, one of the following empirical relations is generally used:

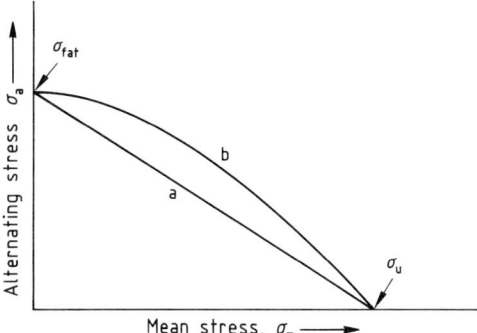

Figure 36. Relations for mean stress effects
a) Goodman; b) Gerber

Goodman: $\quad \sigma_a = \sigma_{fat}\left(1 - \dfrac{\sigma_m}{\sigma_u}\right) \quad$ (31)

Gerber: $\quad \sigma_a = \sigma_{fat}\left[1 - \left(\dfrac{\sigma_m}{\sigma_u}\right)^2\right] \quad$ (32)

where σ_{fat} and σ_a are fatigue strengths for the conditions $\sigma_m = 0$ and $\sigma_m \neq 0$, respectively. These equations are shown graphically in Figure 36. Most experimental data fall between the two lines plotted in Figure 36. Therefore, the Goodman relation is a more conservative criterion of mean stress.

4. Fracture Mechanics

4.1. Principles and Terminology

Fracture mechanics is used to study the behavior of components and structures containing cracks or other flaws. If the geometry and size of the crack or flaw are known, the fracture

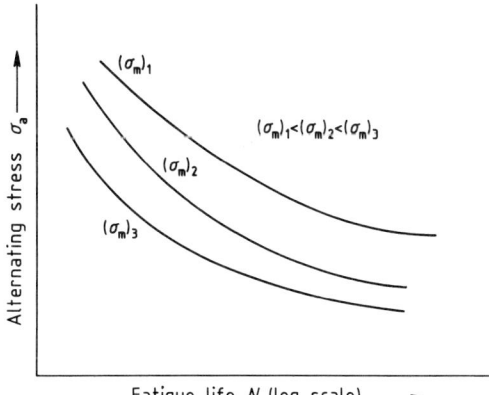

Figure 35. Effect of mean stress on fatigue life

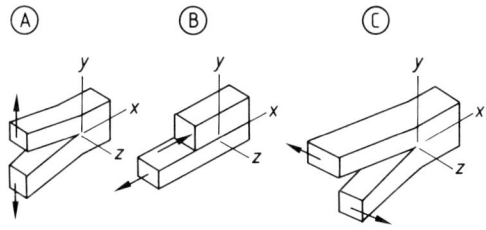

Figure 37. Crack extension modes
A) Mode I; B) Mode II; C) Mode III

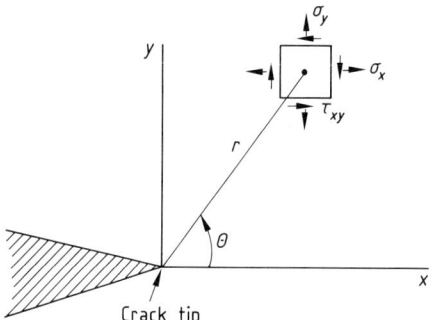

Figure 38. Stresses in the vicinity of a crack tip

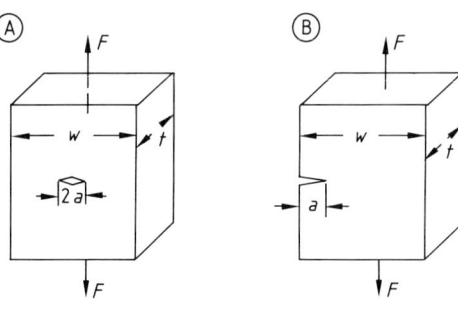

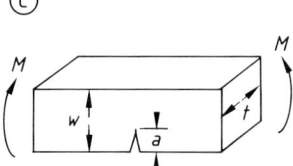

Figure 39. Several common crack configurations
A) Center-cracked plate in tension; B) Single-edge cracked plate in tension; C) Single-edge cracked beam in pure bending

strength of the part or structure can be evaluated by applying the principles of fracture mechanics.

Figure 37 shows the three modes of loading under which a crack can extend. Mode I, the opening or tensile mode, is the most common. Mode II is the sliding or shear mode, and mode III, the tearing or antiplane mode.

Stresses in the vicinity of the crack tip, shown in Figure 38, are given by

$$\sigma_y = \frac{K}{\sqrt{2\pi r}} f_1(\theta) \quad (33)$$

$$\sigma_x = \frac{K}{\sqrt{2\pi r}} f_2(\theta) \quad (34)$$

$$\tau_{xy} = \frac{K}{\sqrt{2\pi r}} f_3(\theta) \quad (35)$$

where θ and r are geometric parameters, defined in Figure 38, and $f_1(\theta)$, $f_2(\theta)$, and $f_3(\theta)$ are functions of θ.

These equations indicate that the stresses at a given point near the crack tip are dependent only on K, which is a stress field parameter, defined as the *stress intensity factor*, that depends on crack geometry, configuration, and loading. The general form of K is given as

$$K = \sigma Y \sqrt{\pi a} \quad (36)$$

where σ is the nominal applied stress, a is crack length, and Y is a dimensionless parameter that depends on specimen and crack geometry. Typical mode I stress intensity expressions for several common specimen configurations, shown in Figure 39, are as follows [20]: for a center-cracked plate in tension (Fig. 39A),

$$K_1 = \sigma\sqrt{\pi a} \cdot \sqrt{\sec(\pi a/w)}, \text{ where } \sigma = \frac{F}{wt} \quad (37)$$

for a single-edge cracked plate in tension (Fig. 39B),

$$K_1 = \sigma\sqrt{\pi a}\left[1.12 - 0.231\left(\frac{a}{w}\right) + 10.55\left(\frac{a}{w}\right)^2 - 21.72\left(\frac{a}{w}\right)^3 + 30.39\left(\frac{a}{w}\right)^4\right], \quad (38)$$

where $\sigma = \dfrac{F}{wt}$

and for a single-edge cracked beam in pure bending (Fig. 39C),

$$K_1 = \sigma\sqrt{\pi a}\left[1.12 - 1.40\left(\frac{a}{w}\right) + 7.33\left(\frac{a}{w}\right)^2 - 13.08\left(\frac{a}{w}\right)^3 + 14.0\left(\frac{a}{w}\right)^4\right], \quad (39)$$

where $\sigma = 6 M/tw^2$

The subscript I in these equations denotes mode I (opening or tensile mode) loading. Stress intensity factor expressions for other common crack geometry and loadings can be found in [2], [20].

4.2. Fracture Toughness

Fracture toughness is the critical value of the stress intensity factor and refers to the condition in which a crack extends in a rapid (unstable) manner. Fracture toughness can therefore be considered the limiting value of the stress intensity factor and is designated K_c. If the fracture toughness and the size and geometry of the crack in a material are known, the fracture stress σ_f can be calculated from

$$\sigma_f = \frac{K_c}{Y\sqrt{\pi a}} \quad (40)$$

In general, the interaction of material properties (i.e., K_c), design stress, and crack size controls the fracture condition in a cracked component. If any two of these three parameters are known, the third can be calculated from Equation (40).

Although fracture toughness can be considered a material property, its value depends on thickness. In general, a material can exhibit ductile behavior in a thin sheet, whereas the same material may fracture in a brittle manner in a thick plate. The thicker the material, the lower is the fracture toughness, as shown in Figure 40. As thickness increases, K_c approaches an asymptotic minimum value called the *plane-strain fracture toughness*, denoted by K_{Ic}. Values of K_{Ic} for several metallic materials are given in Table 4 [23].

In general, the fracture toughness of metals and alloys decreases as yield strength increases. Therefore, even though the fracture resistance of a component may be increased by choosing a higher strength material, if a crack (or flaw) exists, the fracture resistance can actually decrease. Also, the fracture toughness of metals depends on strain rate and environmental conditions such as temperature and corrosion. In general, as temperature decreases, fracture toughness usually decreases, whereas yield strength generally increases.

Table 4. Values of K_{Ic} for some metallic materials [23]

Material	K_{Ic}, MPa·m$^{1/2}$	Yield strength σ_{ys}, MPa
Steel		
4340	99	860
4340	60	1515
52100	14	2070
Aluminum		
2024	26	455
7075	24	495
7178	33	490
Titanium		
Ti–6Al–4V	115	910
Ti–6Al–4V	55	1035

4.2.1. Fracture Toughness Testing

Because K_{Ic} represents a lower limiting value of fracture toughness it can be considered an important material property. By using this property, a relationship between failure stress and crack or defect size can be obtained (Eq. 40).

Different specimen configurations can be used for K_{Ic} testing. Acceptable specimen configurations for K_{Ic} testing according to the ASTM standard are the bend, arc-shaped, compact, and disk-shaped compact specimens (Fig. 41) [24].

The crack length a is nominally equal to the thickness, which is usually half of the width w. For a valid plane-strain fracture toughness test, the specimen thickness and crack length should satisfy the following criteria:

$$\text{thickness } t \text{ and crack length } a \geq 2.5 \left(\frac{K_{Ic}}{\sigma_{ys}}\right)^2 \quad (41)$$

The test sample is usually fatigue loaded to extend the machined notch for a sharp initial crack. A clip gauge is then placed at the crack mouth in order to monitor crack displacement as the specimen is loaded. Typical load–displacement plots are shown in Figure 42 [24].

To determine a valid K_{Ic} value, the following procedure is used [24]: the secant $0F_5$ is drawn through the origin of the test plot with a slope that is 5% lower than the tangent line 0A (see Fig. 42). If the load at every point on the load–displacement that precedes F_5 is lower than F_5, then $F_5 = F_Q$. This situation is shown in Figure 42, type I. However, if a maximum load precedes F_5 and exceeds it, this maximum load is F_Q (types II and III in Fig. 42). Once F_Q has been determined according to this procedure, K_Q is calculated from it. For the compact tension specimen

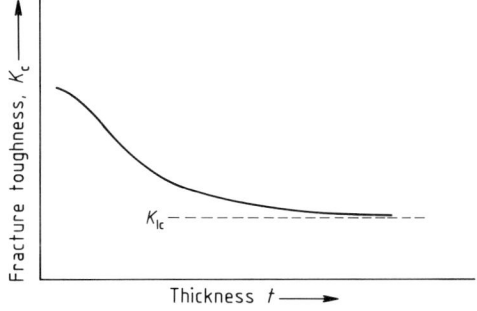

Figure 40. Variation of fracture toughness with thickness

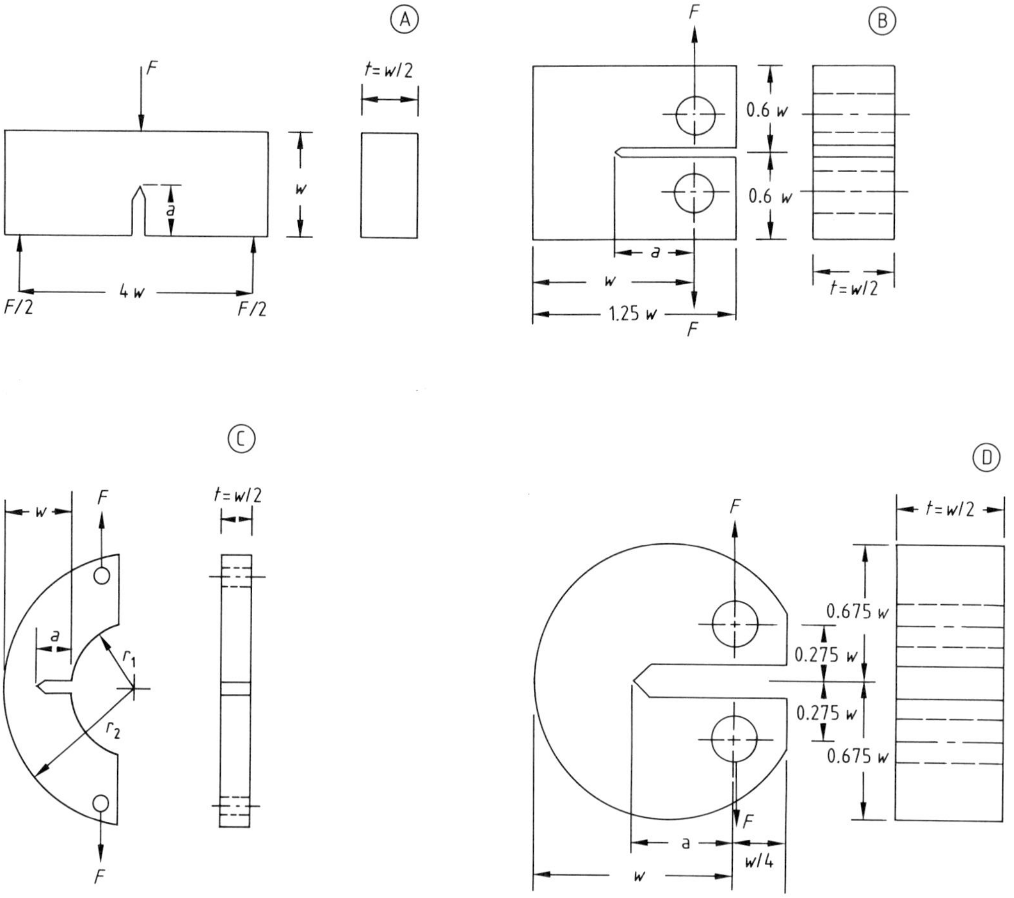

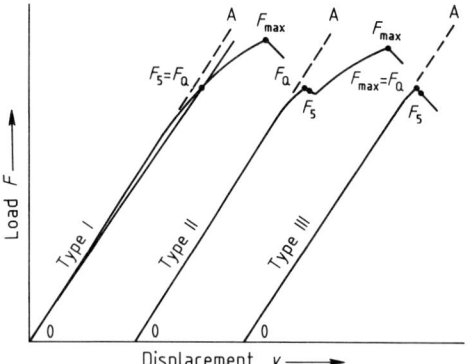

Figure 41. Configurations of K_{Ic} test specimens [24]
A) Bend specimen; B) Compact specimen; C) Arc-shaped specimen; D) Disk-shaped compact specimen

Figure 42. Load–displacement plots in a K_{Ic} test [24]

(Fig. 41 B), K_Q is given by [24]

$$K_Q = \frac{F_Q}{t\sqrt{w}} f\left(\frac{a}{w}\right) \tag{42}$$

where

$$f\left(\frac{a}{w}\right) = \frac{\left(2+\frac{a}{w}\right)\left(0.886 + 4.64\frac{a}{w} - 13.32\frac{a^2}{w^2} + 14.76\frac{a^3}{w^3} - 5.6\frac{a^4}{w^4}\right)}{\left(1-\frac{a}{w}\right)^{3/2}} \tag{43}$$

Similar K_Q expressions for other specimen geometries are given in [24]. If K_Q satisfies Equation (41), then $K_Q = K_{Ic}$. Otherwise, a thicker or more deeply cracked specimen should be used. For further details, see [24].

5. References

[1] Annual Book of ASTM Standards, vol. 3.01, 1989, Designation E8M-89: Standard Test Methods for Tension Testing of Metallic Materials.
[2] R. W. Hertzberg: *Deformation and Fracture Mechanics of Engineering Materials*, 3rd ed., Wiley Interscience, New York 1989, p. 21.
[3] M. J. Manjoine: "Influence of Rate of Strain and Temperature on Yield Stresses of Mild Steel," *J. Appl. Mech.* **66** (1944) A211-A218.
[4] In [1], Designation E9-89a: Standard Test Methods for Compression Testing of Metallic Materials at Room Temperature.
[5] In [1], Designation E290-87: Standard Test Method for Semi-Guided Bend Test for Ductility of Metallic Materials.
[6] In [1], Designation B565-87: Standard Method for Shear Testing of Aluminum and Aluminum-Alloy Rivets and Cold-Heading Wire and Rods.
[7] H. W. Hayden, W. G. Moffatt, J. Wulff: *The Structure and Properties of Materials*, vol. III, Wiley Interscience, New York 1965, p. 12.
[8] In [1], Designation E18-89a: Standard Test Methods for Rockwell Hardness and Rockwell Superficial Hardness of Metallic Materials.
[9] In [1], Designation E10-84: Standard Test Method for Brinell Hardness of Metallic Materials.
[10] In [1], Designation E92-82: Standard Test Method for Vickers Hardness of Materials.
[11] In [1], Designation E384-84: Standard Test Method for Microhardness of Materials.
[12] R. A. Flinn, P. K. Trojan: *Engineering Materials and Their Applications*, 3rd ed., Houghton Mifflin Co., 1986, p. 84.
[13] In [1], Designation E140-88: Standard Hardness Conversion Tables for Metals.
[14] H. W. Hayden, W. G. Moffatt, J. Wulff: *The Structure and Properties of Materials,* vol. III, Wiley Interscience, New York 1965, p. 13.
[15] In [1], Designation E23-88: Standard Test Methods for Notched Bar Impact Testing of Metallic Materials.
[16] In [1], Designation E604-83: Standard Test Method for Dynamic Tear Testing of Metallic Materials.
[17] In [1], Designation E139-83: Standard Practice for Conducting Creep, Creep–Rupture and Stress–Rupture Tests of Metallic Materials.
[18] In [1], Designation E328-86: Standard Methods for Stress Relaxation Tests for Materials and Structures.
[19] In [1], Designation E606-80: Standard Recommended Practice for Constant Amplitude Low-Cycle Fatigue Testing.
[20] H. O. Fuch, R. I. Stephens: *Metal Fatigue in Engineering*, Wiley Interscience, New York 1980.
[21] H. W. Hayden, W. G. Moffatt, J. Wulff: *The Structure and Properties of Materials,* vol. III, Wiley Interscience, New York 1965, p. 15.
[22] In [1], Designation E466-82: Standard Practice for Conducting Constant Amplitude Fatigue Tests of Metallic Materials.
[23] J. E. Shigley, C. R. Mischke: *Mechanical Engineering Design,* 5th ed., McGraw-Hill, New York 1989, p. 224.
[24] In [1], Designation E399-83: Standard Test Method for Plane-Strain Fracture Toughness of Metallic Materials.

11. Nondestructive Testing

KANJI ONO, Department of Materials Science and Engineering, University of California, Los Angeles, California 90024, United States

1.	Introduction ... 11-1	2.5.	Penetrant Methods ... 11-21
2.	**Methods of Nondestructive Testing** ... 11-2	2.6.	**Magnetics and Electromagnetics** ... 11-22
2.1.	**Material Identification** ... 11-2	2.6.1.	Magnetic Particle Inspection ... 11-22
2.2.	**Radiography** ... 11-4	2.6.2.	Magnetic Leakage Field Inspection . 11-23
2.2.1.	Radiation ... 11-4	2.6.3.	Other Magnetic Inspection Methods ... 11-24
2.2.2.	Attenuation ... 11-4	2.6.4.	Eddy-Current Inspection ... 11-24
2.2.3.	Imaging ... 11-6	2.6.5.	Microwave Inspection ... 11-27
2.2.4.	Recording Images ... 11-6	2.7.	**Thermal and Optical Methods** ... 11-27
2.2.5.	Screens ... 11-7	2.7.1.	Thermal Inspection ... 11-27
2.2.6.	Radiographic Quality ... 11-8	2.7.2.	Visual and Optical Inspection ... 11-28
2.2.7.	Radiographic Practices ... 11-8	2.7.3.	Optical Holography ... 11-28
2.2.8.	Radiographic Interpretation ... 11-9	2.8.	**Leak Testing** ... 11-29
2.2.9.	Other Techniques ... 11-10	2.8.1.	Pressurized Gas Leaks ... 11-29
2.2.10.	Radiation Safety ... 11-11	2.8.2.	Pressurized Liquid Leaks ... 11-30
2.3.	**Ultrasonics** ... 11-11	2.8.3.	Vacuum System Leaks ... 11-30
2.3.1.	Waves ... 11-11	3.	**Uses** ... 11-30
2.3.2.	Generation and Detection ... 11-13	3.1.	**Metallic Pressure Vessels and Piping** 11-30
2.3.3.	Inspection Methods ... 11-15	3.2.	**Composite Pressure Vessels and Piping** ... 11-31
2.3.4.	Display Methods ... 11-16		
2.3.5.	Interpretation ... 11-17	3.3.	**Weldments** ... 11-32
2.3.6.	Inspection Standards ... 11-18	3.4.	**Other Uses** ... 11-32
2.4.	**Acoustic Emission** ... 11-18	4.	**References** ... 11-33
2.4.1.	Detection ... 11-19		
2.4.2.	Material Behavior ... 11-19		
2.4.3.	Source Location ... 11-19		
2.4.4.	Acoustic Emission Monitoring ... 11-20		

1. Introduction

Nondestructive testing of materials has become an indispensable part of modern industry. Beginning with the oil and whiting method, a forerunner of the penetrant testing method used for crack detection, a variety of tests—based on numerous physical principles—have been developed. At the start, the primary task was to find flaws and discontinuities. With the advent of fracture mechanics, the measurement of flaw dimensions became more urgent. Increasing use of fiber-reinforced plastics in the past decade has posed new challenges in material characterization.

Techniques of nondestructive evaluation have made substantial advances. However, many problems still remain because the current design of components and structures demands higher efficiency and the material capability is exploited to its fullest extent. For example, a bonded joint of composite materials is a desirable design approach, yet no nondestructive testing method is available to assure the strength of the joint.

This article provides brief summaries of widely used methods of nondestructive testing, covering basic concepts, techniques in common use, and examples of industrial applications. Less well developed but promising methods are also introduced. Because the field of nondestructive testing is broad, all methods of importance to the chemical industry are included, but those applicable primarily to the aerospace, nuclear, and steel industries are covered only incidentally.

Two major problems, corrosion and cracking, confront nondestructive test engineers in the chemical and petroleum industries. Ferrous and corrosion-resistant nonferrous alloys used in vessels and piping are often subjected to chloride- and hydrogen-containing environments. Hydrogen-induced cracking and stress corrosion cracking are found frequently. Cracking of welds from residual stress or poor weld quality is another persistent problem during service. Many vessels and pipes are constructed from carbon steel, and general corrosion from long-term service must be monitored. In addition, fiber-reinforced plastic components require special attention because traditional nondestructive testing methods cannot adequately evaluate their reliability.

Nondestructive testing attempts to discover indications of discontinuities and abnormal conditions in materials and structures. If an indication is found to be beyond an acceptable limit, it is a flaw and the material (or structure) must be repaired or rejected. A very small gas hole in a weld is innocuous and can be allowed to remain. Beyond a certain size, however, it is no longer tolerated, while cracks of any size are typically rejected. Flaws and discontinuities are treated as interchangeable because the size limits are set by design considerations.

Flaws are introduced at three stages. The first is during primary manufacturing. Shrinkage cavities, inclusions, cracks, blowholes, and other faults occur in castings. Flaws in forgings may be traced to ingots that contain segregations, pipes (shrinkage cavities), inclusions and, hydrogen-induced cracks or hydrogen flakes. During forging, forging bursts (internal tears), surface flaws (such as cold shuts, laps, folds, and seams), burning, and decarburization are often the result of improper setup and control. Rolling and drawing of ingots and billets generate flaws similar to those of the forging process. In addition, lamination and chevron defects may occur. The second stage is fabrication into structures, in which welding is the most important process. Because the fusion of metals is induced, many flaws are also introduced. These include gas porosity, slag entrapment, incomplete fusion, incomplete penetration, and cracks. In addition, surface flaws such as undercut, mismatch, and underfill occur. The third stage of flaw generation is during service. Overloading and fatigue loading produce cracks, and corrosion leads to pitting, general reduction of wall thickness, and stress corrosion cracking. Exposure to hydrogen results in hydrogen-induced cracking, and fluid flow produces erosion damage. High-temperature environments in furnaces and reactors cause creep and oxidation, leading to wall thinning and cracking. The nature and origin of the flaws listed here are discussed in [1].

Detailed information on nondestructive testing can be found in introductory texts [2], training guides [3], handbooks [4], codes and standards [5]–[7], and a series of monographs and conference proceedings [8], [9]. Several journals and a newsletter are published in English [10], and publications in other languages are also available. More than a dozen national societies for nondestructive testing, such as the American Society for Nondestructive Testing (ASNT), the British Institute for Nondestructive Testing (BINDT), Deutsche Gesellschaft für Zerstörungsfreie Prüfung e. V. (DGZfP), and the Japanese Society for Nondestructive Inspection (JSNDI), exist worldwide, and the International Institute of Welding (IIW) and the International Standards Organization (ISO) are active in establishing standards. The Institut für Zerstörungsfreie Prüfverfahren (IZfP; Saarbrücken, Federal Republic of Germany) and the Electric Power Research Institute (EPRI) NDT Center are research institutes dedicated to studies of nondestructive testing.

2. Methods of Nondestructive Testing

2.1. Material Identification

To verify that an alloy is within specified composition limits, quantitative chemical analysis is the only answer. However, a system of rapid identification of metals and alloys is useful in confirming the alloy type or in sorting mixed lots of alloys. Several techniques are used singly or in combination. The basic technique is to use mass density (specific gravity) for preliminary grouping into light (Al, Mg, Ti), medium (steel, Cu, Ni, Zn), heavy (Pb, Mo, Ag), and very heavy (W, Ta, Pt) alloys. Some alloys have distinct colors, but most have a white or grayish white color.

X-Ray Fluorescent Analysis. Each element generates characteristic X rays at several specific energy levels when it is irradiated with electromagnetic radiation having a higher energy spectrum. Characteristic X-ray energies range from 1 keV for sodium to 118 keV for uranium K radiation, and from 1 keV for zinc to 17 keV for uranium L radiation. Fluorescent X rays are detected with an energy-discriminating proportional counter or solid-state detector and are electronically processed to identify the chemical composition of an alloy sample. Because of the high absorption of low-energy X rays, only the surface of the sample can be analyzed.

Radioactive isotopes are often used as the source of radiation for exciting fluorescent X rays. An isotope source of 1–100 millicuries (mCi) can be small and portable. Some sources must be replaced periodically. Regular X-ray sources or small, air-cooled X-ray tubes can also be used. For X-ray energy analysis, a solid-state detector commonly used in a scanning electron microscope (SEM) for elemental detection can be employed, but such a detector requires constant cooling with liquid nitrogen. Another limitation arises from the resolution of the X-ray energy analyzer; elements lighter than aluminum cannot be evaluated. In the electron microprobe analyzer, energetic electrons are used to generate fluorescent X rays; a high-resolution energy spectrum analyzer is employed, which is capable of detecting an element as light as boron. Small samples can be analyzed in an SEM or microprobe.

Chemical Spot Testing. Chemical spot testing utilizes chemical color reactions, which are carried out on a sample, on filter paper, or on a spot plate. A test kit consists of about 100 solutions. Carbon steels, for example, can be identified by placing a drop of 50% nitric acid on a fresh surface. After 5 min, the spot is rinsed with water; a brown spot indicates carbon steel. See [11] for a listing of solutions and identification tests.

Spark Testing. The spark testing technique classifies ferrous alloys according to their chemical composition by visual examination of the sparks thrown off when a sample is held against a high-speed grinding wheel. The test is fast and economical, yet an experienced operator can identify a variety of alloys with reasonable accuracy [1], [12].

Hardness Testing. Hardness testing is not strictly nondestructive because an indenter is forced into the surface of a test piece. The resulting indentation is, however, often sufficiently small to be left on a finished product. The indenter is a ball, cone, or pyramid made of hardened steel or diamond. Constant force is applied on the indenter, and the diameter, length, or depth of the indentation is used as a measure of material hardness. The Brinell, Rockwell, Vickers, and Knoop hardness tests are commonly employed [13] (see also → Mechanical Properties and Testing of Metallic Materials). Because each alloy has a certain hardness range, knowing the hardness aids in material identification. Hardness measurement is also a nondestructive test method for the yield and tensile strength of ductile alloys based on empirical correlations. In brittle ceramic materials, cracks emanate from the corners of a Vickers hardness indentation. The length of these cracks has been found to be related to the fracture toughness of ceramics. Thus, nearly nondestructive determination of ceramic fracture toughness can be conducted.

Electrical and Magnetic Testing. Electrical resistivity can be used to classify metals and alloys and the state of heat treatment of many alloys. One technique involves use of a four-point probe with four equally spaced tips. The probe passes a direct current (I_{dc}) between the two outermost tips and measures the potential (V_t) between two inner tips. For samples whose thickness is more than three times the distance between two tips (S_t), the electrical resistivity (ϱ_e) is

$$\varrho_e = 2\pi S_t V_t / I_{dc}$$

Another technique uses the eddy-current principle (see Section 2.6.4). A probe containing a coil(s) is placed on the surface of a sample. Changes in coil impedance are analyzed and indicated as electrical conductivity (the inverse of resistivity) values on the analyzer output device, for example, a digital readout.

The magnetic properties of alloys can be checked with a hand magnet to classify them as strongly magnetic, weakly magnetic, or nonmagnetic. Austenitic stainless steels and some nickel alloys are weakly magnetic or nonmagnetic, depending on cold working and heat treatment. The eddy-current instrument for electrical conductivity testing can also be used for quantitative evaluation of magnetic permeability.

Heated and cold electrodes are brought in contact with the surface to evaluate the thermoelectric properties of a sample [6, E 977]. This is useful in sorting alloys by comparative measurements.

Composite Phase Testing. In fiber-reinforced plastic composites, the amount of fiber reinforcement is an important material variable. Destructive testing for fiber content uses acid digestion and ignition techniques to eliminate the resin matrix. Ultrasonic velocity has been shown to depend on fiber content, although few dedicated instruments are in use. An instrument based on elastostatic principles has also been developed [32]; it can determine the fiber content and the porosity fraction from differential compressibility measurements (i.e., measuring the thickness of a sample under different static loading conditions) together with destructive calibration procedures. When the density and composition of matrix and fiber phases differ, as in the case of a resin matrix with glass fibers, X- or gamma-ray absorption can be used to determine the fiber content.

2.2. Radiography

Radiography utilizes the differential attenuation of penetrating radiation to find internal flaws such as porosity and compositional changes, as well as variations in thickness and cracks. X rays, gamma rays, and neutrons are used and are imaged by photographic films, fluoroscopic screens, and more recently, electronic imaging devices. Thickness changes of 1 – 2 % and flaw sizes of a few tenths of a millimeter can typically be detected. This method examines a large volume of material simultaneously, and most welds are radiographed routinely. However, radiation safety measures must be rigorously enforced. For details, see [4c], [14], and more than 20 ASTM standards [6].

2.2.1. Radiation

X rays are generated when high-energy electrons strike a metallic target. When the electrons are decelerated at the target, a continuous spectrum of X-ray photons is produced, having energies lower than that of the electrons. The peak X-ray intensity occurs at about one-third of the maximum photon energy. At photon energies of less than 118 keV, characteristic X rays are also produced and their energies are specified by the target element. A common choice of target material is tungsten, but copper, iron, and cobalt are also used. Tubes are used to generate X rays, with tube voltages up to 1000 kV. These consist basically of a diode with a filament for electron emission (cathode) and a target (anode), between which an accelerating potential is applied. The cathode has a focusing cup to control the size of the electron beam. The anode is fluid cooled, and the target is usually tilted to direct the X rays generated. The vacuum envelope is made partly of glass or ceramic for electrical insulation and has a thin beryllium X-ray window.

Higher energy X rays are generated by using an electron linear accelerator (a linac). Electron energies of 1 – 30 MeV can be obtained with commercial equipment, which employs microwave power to accelerate electrons in a series of cylindrical cavities. Betatrons, as well as Van de Graaff and resonant transformers, have also been utilized, but they are larger and less convenient than linacs.

Gamma rays are produced by radioisotopes such as cobalt-60 and cesium-137. Although radioisotope sources require radiation shielding and safe handling equipment, no power supply is needed. This results in portability, but the radiation intensity is low and long exposure times are common. Radioisotopes (e.g., ^{60}Co) are also a convenient source of the high-energy radiation needed to penetrate thicker test pieces.

Neutrons are produced by nuclear reactors, deuterium accelerators, or isotopes such as californium-252. Because of the capability of neutrons to detect explosives, neutron radiography is used by the military. Industrial applications are uncommon.

2.2.2. Attenuation

Both X rays and gamma rays are attenuated in a substance by absorption and scattering. At lower energy levels, an incoming photon is absorbed by ionization of an atom (photoelectric effect). This effect decreases with increasing photon energy E_p as $E_p^{-3.5}$, except when the energy corresponds to the electronic binding energy, at which point absorption increases greatly. Such increases are called absorption edges. Another absorption process, in which the photon energy

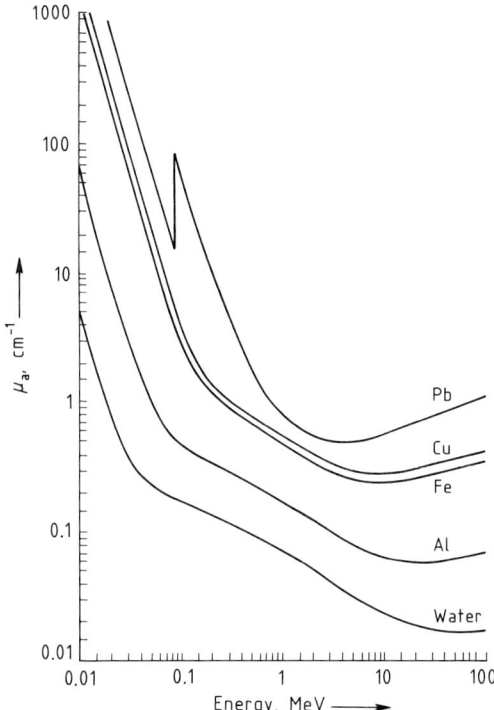

Figure 1. Absorption coefficient μ_a versus energy for various materials

is used to create an electron–positron pair, occurs above 1 MeV. Scattering of a photon by an electron occurs either coherently without changing the kinetic energy (Rayleigh scattering) or incoherently with loss of the photon energy and changes in photon direction (Compton scattering). Figure 1 shows changes in attenuation for several materials as a function of energy.

Radiation intensity I decreases exponentially with the thickness x of a uniform material:

$$I = I_0 \exp(-\mu_a x) = I_0 \exp(-N_a \sigma_a x)$$

where I_0 is the incident intensity, μ_a the linear attenuation coefficient, N_a the number of atoms per unit volume, and σ_a the atomic attenuation coefficient. When μ_a is divided by mass density ϱ, the mass attenuation coefficient (μ_a/ϱ) is obtained to express attenuation per unit areal weight. In steels and nickel or copper alloys, photoelectric absorption is dominant at a photon energy of less than 100 keV. In light metals (aluminum alloys), this range shifts to less than 40–50 keV. The atomic attenuation coefficient increases with atomic number. The exponential attenuation equation is valid when the photon energy spectrum remains unaffected by passage through the material. This condition exists when a collimator is used in front of the radiation detector to exclude scattered radiation (i.e., the so-called narrow-beam geometry). Here, all scattered photons are assumed to be absorbed. In conventional radiographs, scattered photons reach the recording medium; this condition is known as broad-beam geometry. The broad-beam attenuation coefficient is smaller than the narrow-beam attenuation coefficient and depends on the photon energy spectrum and the penetration depth. Published data on attenuation coefficients are based on the narrow-beam geometry. When estimating radiographic exposures, narrow-beam data may result in inaccuracy, especially with low-energy X rays and thick test pieces.

The radiation intensity I_B in the broad-beam geometry is expressed as

$$I_B = I_0 B \exp(-\mu_a x)$$

where B is the buildup factor, which is equal to the ratio of the total intensity (direct beam plus scattered beam) to the direct-beam intensity. The buildup factor increases with increasing thickness of the test piece, and the rate of increase of B is higher for lower energy radiation (see Fig. 2 for B in steel).

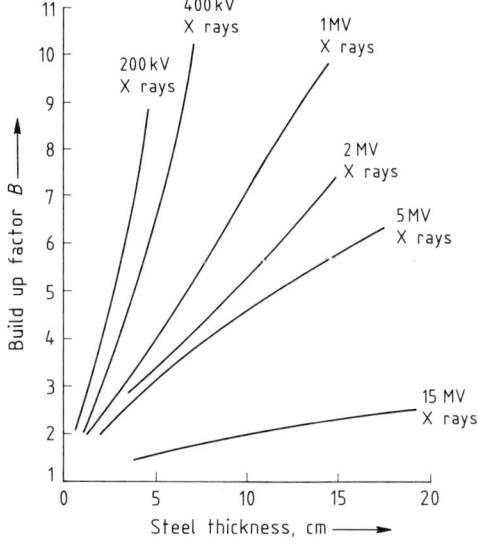

Figure 2. Buildup factor as a function of thickness

When X rays are generated at the target of an X-ray tube, the energy spectrum contains the low-energy portions. As the beam passes through the tube window, higher proportions of low-energy X rays are absorbed, hardening the spectrum. This spectrum change is quantified by equivalent filtration with a thickness of aluminum and is called inherent filtration. When harder (higher average X-ray energy) radiation is required, a filter is inserted into the beam path.

After electromagnetic radiation is emitted from the source, the beam spreads as it travels. As the distance from the source increases, the same radiation intensity is distributed over a larger area and the intensity varies inversely with the square of the distance. If the intensity is I_1 at a distance x_1, the intensity I_2 at x_2 is

$$I_2 = I_1 (x_1/x_2)^2$$

which is called the inverse-square law.

2.2.3. Imaging

Radiographic images are formed in analogy to geometric optics because of straight-line propagation of X and gamma rays. A point source of radiation is generally used. However, the X- and gamma-ray sources are finite in size, typically ranging from 0.5 to 5 mm. When an image (or shadow) of an object at the source-to-object distance x_o is formed at the source-to-image distance x_i, the image is enlarged by a factor x_i/x_o. This enlargement factor is nearly unity in conventional radiography because the film is placed immediately behind the object or test piece. By using a microfocus X-ray tube with a source size of 1–10 µm, a ten- to hundredfold magnification can be realized in microradiography (X-ray microscopy) applications. Here, a test piece is placed near the source and the film is further away.

Image distortion occurs when a two-dimensional object is imaged on a nonparallel image plane. A three-dimensional object always gives a distorted image because the enlargement factor is not constant over the object.

The finite spot size of X- and gamma-ray sources produces regions of partial shadow at the edges of an object. These are known as penumbra and result in a geometric unsharpness U_g, given by

$$U_g = F_s (x_i - x_o)/x_o$$

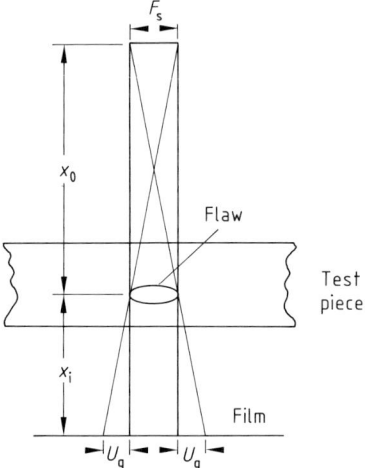

Figure 3. Geometrical unsharpness in radiography

where F_s is the source spot size (see Fig. 3). Geometric unsharpness and image distortion are reduced by using a larger source-to-object distance x_o. Increasing x_o, however, reduces the radiation intensity, and the exposure time must therefore be increased.

2.2.4. Recording Images

The most common method of recording radiographic images is the use of X-ray films. Films make a permanent record of images of large areas. These have a thin emulsion coating on both sides of a film base or a thin clear plastic support. The emulsion is typically 25 µm thick and contains a microscopic suspension of silver halide grains. Both X-ray and gamma-ray energy cause a latent image to form. A visible image is revealed by a developer that converts the exposed silver halide to metallic silver. The image is then fixed by washing off unexposed halide grains and drying. Most X-ray films are for direct exposure to X rays. Some are used with fluorescent screens and are more sensitive to visible light emitted by the screens.

The exposure of a film causes an inherent unsharpness, which results from the scattering of secondary electrons in the film emulsion. Typical values of inherent unsharpness are 0.01–0.05 mm for 100–250-kV X rays.

Darkening of a film is measured as density D by transmitting light through it. The density is defined as

$$D = \log(I_0/I_t)$$

where I_0 is the intensity of the incident light and I_t is that of transmitted light. Only 1 % of light is transmitted at a density of 2, and 0.01% at $D = 4$. Since low values of D give low sensitivity and high values make viewing difficult, standard procedures restrict D values. Valid density ranges for single and composite viewings are 1.5–4 and 2–4, respectively. For $D > 2$, a high-intensity illuminator is required. The density of an exposed film is dictated by exposure, film characteristics, and film development conditions. Exposure is the product of the radiation intensity and the time of exposure. In radiography, relative exposure is generally used, with a certain exposure condition including X-ray source, energy, filters, source-to-film distance, film type, and developing condition, being set as reference.

The characteristic curve, also known as the H and D curve after HURTER and DRIFFIELD, expresses the relation between the relative exposure applied to a film and the resulting density. The relative exposure is plotted on a logarithmic scale. Figure 4 shows examples of such curves, which are useful for determining suitable exposure conditions and comparing the speed and contrast of various film types. Speed describes how fast a particular density can be produced. In Figure 4, type 1 has the lowest speed and type 4 the highest. High-speed film tends to have coarse grains, whereas low-speed film has fine grains. Such graininess results from clustering of individual silver halide grains. Film contrast, also called film gradient, is the difference in density due to a change in exposure and is expressed by the slope G of the characteristic curve. A higher contrast gives a larger difference in density for the same variation in exposure, allowing better resolving of details, but the range of usable exposure conditions is narrow. For type 1 film in Figure 4, the log of relative exposure must be in the range 4.75–5.15 to obtain high contrast and densities of 2–4. This film has a narrow latitude.

For a given range of densities on the characteristic curve, the average slope G_{av} can be defined. A change in density ΔD for a given exposure time is related as

$$\Delta D = -0.43\, G_{av}\, \mu_a \Delta x / B$$

where μ_a is the attenuation coefficient for the radiation used, Δx is the change in thickness, and B is the buildup factor. A change in D of more than 0.006 is required for a perceptible density change, so the minimum thickness change that can be detected radiographically is expressed as

$$-(2.3 \times 0.006)\, B / \mu_a G_{av}$$

That is, higher attenuation coefficients, obtained with low-energy radiation, and steeper slopes of the characteristic curve increase radiographic sensitivity. The buildup factor, on the other hand, decreases the sensitivity due to non-image-forming scattered radiation. Both B and μ_a increase with decreasing radiation energy, but μ_a is affected much more than B.

The shape of characteristic curves is affected by the development time: film contrast increases with development time, which also gives some increase in speed. The energy spectrum of the radiation also affects the speed. The characteristic curve is shifted to the higher exposure side as the radiation energy increases from less than 100 kV to several hundred kV or more.

For a given combination of radiation source, source-to-film distance, material to be radiographed, type of film, and film development technique, a radiographic exposure chart for a particular density can be prepared. When the thickness of a test piece varies within the area of coverage, a latitude chart may be used to help determine the correct exposure. The latitude chart is similar to the exposure chart except that a band corresponding to a range of densities is given (Fig. 5).

2.2.5. Screens

Lead foil screens are commonly used in direct contact with films to produce an intensifying ac-

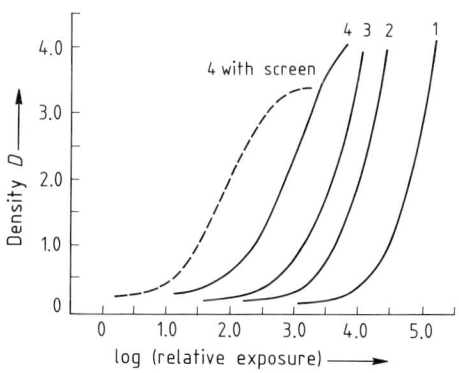

Figure 4. Characteristic curves for films

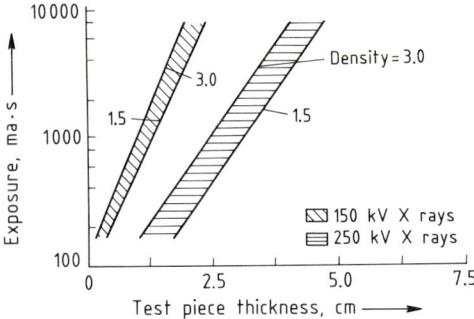

Figure 5. Latitude chart for type 2 film

tion at as low as 90 kV, depending on the thickness of the material and the foil. The intensification arises through emission of secondary electrons by the lead upon X-ray photon bombardment. Higher energy radiation gives greater intensification. A screen in front of the film also filters out scattered radiation arising from the test piece and improves radiographic quality. Lead screens, 0.13 mm thick, are generally used in front and back for radiography of 6-mm-thick (or more) steel with sources of more than 135 kV. For higher energy radiation (e.g., ^{60}Co) 0.25-mm-thick lead screens are used. The rear screen also protects the film from backscattered radiation. For this purpose, a lead sheet is used in the back of the film holder as well. Other screen materials include copper, gold, tantalum, and lead oxide.

Fluorescent screens are used only when the exposure would be extremely long without them. They are employed in combination with film optimized for high sensitivity to emitted light. The screens are coated with phosphors such as $CaWO_4$ and ZnCdS and can produce 2–25-fold intensification. However, light scattering within the phosphor gives rise to screen unsharpness, reducing image definition on the film. Typical values of screen unsharpness for commercially available screens are 0.1–0.4 mm. Another drawback of fluorescent screens is the appearance of screen mottle on the finished radiograph. This arises from statistical variations in fluorescence and appears as a larger scale graininess.

2.2.6. Radiographic Quality

Image quality is influenced by many variables that contribute to radiographic contrast and image clarity. Superior contrast is obtained by a large change in thickness Δx, lower radiation energy, less scattered radiation, exposure in the high-film-gradient region, adequate film development, and use of films with high gradients. Good image clarity results from low total unsharpness, which is defined as the square-root of the sum of the squares of geometrical, inherent, and screen unsharpnesses. Motion unsharpness also arises due to movement of the test piece. Typical values of total unsharpness with good film techniques for 100-kV X rays are 0.05 mm; 200-kV X rays, 0.10 mm; 400-kV X rays, 0.20 mm; iridium-192 gamma rays, 0.25 mm; and cobalt-60 gamma rays, 0.50 mm.

Practical indicators of image quality are known as penetrameters or image quality indicators. One ASTM design [6, E 1025] is based on rectangular or round sheets with thickness T of 0.13–1.3 mm, having holes of size $1T$, $2T$, and $4T$ ($1T$ and $2T$ for round penetrameters). A penetrameter whose thickness corresponds to 2 % of that of the test piece is generally used. Another type of penetrameter [6, E 477] employs a set of six wires of different sizes. The penetrameter material is the same as that of the test piece. The penetrameter is placed directly on the test piece, facing the source and remote from the area of inspection. Image quality is usually expressed as the size of the smallest penetrameter feature, such as hole size or wire diameter, that is clearly visible in the processed radiograph. The minimum acceptable quality level corresponds to the visibility of a $2T$ hole with a 2 % thickness penetrameter. This quality level is expressed as 2-2 T. Equivalent penetrameter sensitivity (in percent) is defined as $100\sqrt{Th/x}\sqrt{2}$, where x is the test piece thickness, T is the penetrameter thickness, and h is the hole size. Other measures of equivalent sensitivity are also used.

2.2.7. Radiographic Practices

The positions of the radiation source, the test piece, and the film are chosen to best reveal expected flaws with minimum geometrical distortion. In the case of cracklike flaws, the X- or gamma-ray beam must be oriented parallel to the flaw. The orientation matters little for void-like flaws, but the size or shape of the image will change unless the beam passes normal to the plane of the flaw.

Scattered radiation decreases radiographic sensitivity. Scattering from wall reflections and internal holes or openings can be reduced by using lead masks that cover the holes or openings and by surrounding the test piece with metallic shot or a liquid absorber. Backscatter from objects behind a film (away from the source) is eliminated by a lead screen or sheet. A collimator is placed near the source and reduces the beam size to cover only the test piece, thus eliminating spurious scattering.

Flat test pieces can best be radiographed by a normally incident beam with films below the test pieces. Curved plates can be inspected similarly, but minimum distortion is achieved with the source at the center of the radius of curvature. Inspection of a cylindrical test piece requires section-equalizing techniques with a solid cradle, liquid absorber, or shim stocks (Fig. 6). Double-film techniques with films of different speed, and multiple exposures with different radiation energies can be employed instead of these section-equalizing techniques. Overexposure of the outer portion of the radiographic image can be corrected either by exposing a film of lower sensitivity or by using lower energy radiation. Tubular sections are inspected by single- or double-wall techniques (Fig. 7). Single-wall techniques use an external radiation source and the film inside the cavity. Alternatively, the source is placed inside the cavity and the film on the outside. For the latter, a panoramic X-ray source that emits radiation 360° around its axis may be used. Single-wall techniques usually take a single image through a wall or weld and provide good radiographic sensitivity. Double-wall, double-image techniques are used when single-wall techniques are difficult to employ, e.g., for small cross sections (< 9 cm). These produce radiographs on which the images of both walls are recorded and are also applied to the inspection of diagonally opposed corner welds in rectangular box sections. Circumferential butt welds can be inspected with the offset or corona techniques (Fig. 8). For smaller diameters, two images can be exposed, but the single-image technique gives a better radiograph for large-diameter pipes.

Inspection of complex shapes requires multiple exposures with different viewing directions. Combinations of the above-mentioned techniques are chosen to reduce the image to a simple shape for easy interpretation. Welds are usually inspected to reveal flaws in weld metals and in heat-affected zones. A beam direction normal to the surface of the weld is used. For T-joints or lap joints, the beam direction is inclined to both

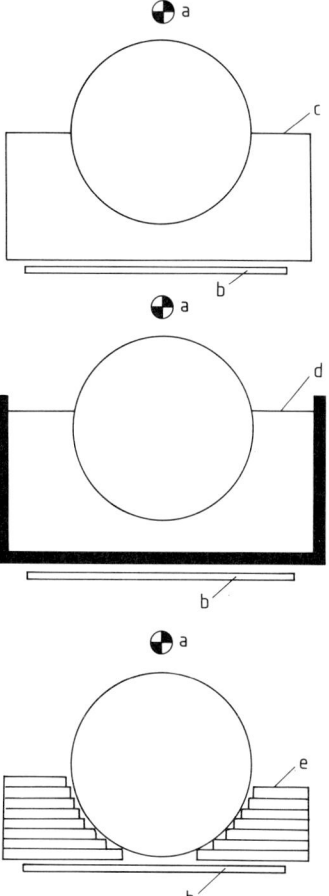

Figure 6. Three methods of equalizing radiographic density
a) Radiation source; b) Film; c) Solid cradle; d) Liquid absorber; e) Shims

sides of the weld joint. When two adjacent welds are made, two beam directions are usually required, but the images may be recorded on a single film.

2.2.8. Radiographic Interpretation

The radiographic appearance of the types of flaw usually found in castings and weldments has been compiled in volumes of reference radiographs, which are graded according to flaw size and severity. The ASTM has produced 12 volumes of reference radiographs [6]. Radiographic acceptance criteria are defined from these reference radiographs, permitting acceptable limits of

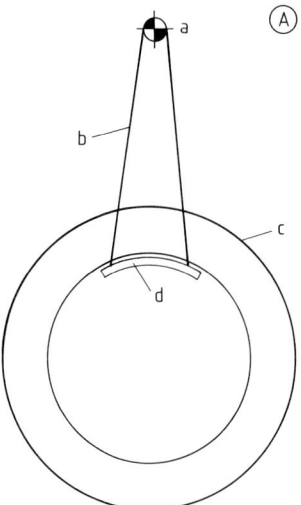

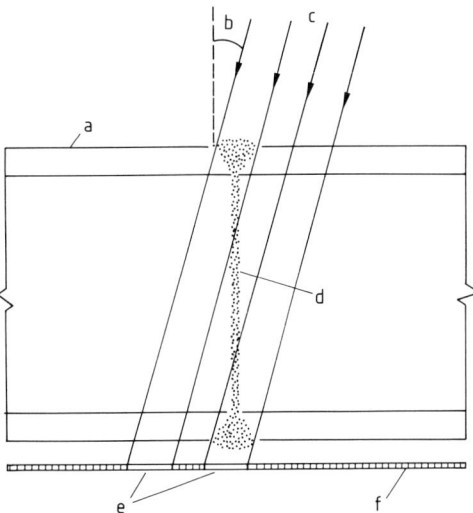

Figure 8. Corona (offset) radiographic technique
a) Pipe; b) Offset angle; c) Radiation beam; d) Weld; e) Images of weld; f) Film

2.2.9. Other Techniques

Computerized tomography, developed for medical applications, can also be used in inspection of lightweight composites. Conventional tomography equipment is expensive and as yet impractical for routine commercial use. Film-based tomography may, however, become a useful method in the future [15].

Real-time radiographic inspection techniques have advanced substantially within the past decade because of improvements in electronic image processing. However, they are five to ten times less sensitive to cracks and are used mainly in electronic component inspection and airport baggage screening [6, E 1000].

Microfocus X-ray tubes can readily provide a magnification of two to ten times because of their small spot size (ca. 10 μm). Although their low X-ray output requires long exposure times or use of an electronic image intensifier (combining the real-time fluoroscopic and image-processing equipment), a microfocus unit allows the inspection of small flaws in ceramics and composites.

Inspection of thick sections requires high-energy radiation sources. Steel sections up to 5 cm can be radiographed with medium energy X rays (250–400 kV) or iridium-192, and up to 10 cm, with 1-MV X rays or cobalt-60. Higher energy radiation improves radiographic sensitivity for

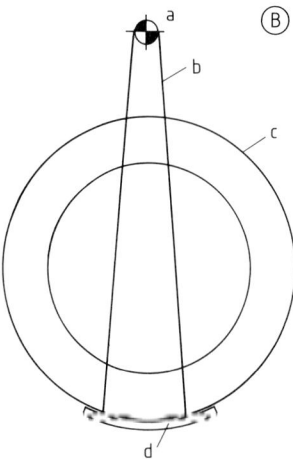

Figure 7. Radiographic inspection
A) Single-wall and B) Double-wall radiographic technique
a) Radiation source; b) Beam; c) Pipe; d) Film

flaw types, sizes, and densities. Linear-type flaws, such as cracks and incomplete fusion, are prohibited regardless of size in most criteria [5].

Of the service-induced flaws, corrosion in a flat or tubular section can be detected because the loss of material results in dark areas in the radiograph. In contrast, tight cracks from stress corrosion and hydrogen effects are difficult to detect. Similarly, short fatigue cracks are hard to locate.

thick sections, but requires specialized equipment.

2.2.10. Radiation Safety

Exposure to radiation is harmful to health and must be strictly controlled. Radiation safety is practiced by adhering to established procedures of handling radiation equipment and radioactive materials, by incorporating adequate shielding design and safety interlock systems, by thoroughly educating personnel about safe practices, by monitoring radiation levels of the working environment, and by enforcing the limit on allowable exposure. Proper licensing requirements must be maintained.

2.3. Ultrasonics

Ultrasonic testing uses high-frequency mechanical (elastic) waves in the range of 20 kHz to 50 MHz. Elastic waves are reflected and refracted at interfaces where elastic moduli and density are discontinuous. Cracks, voids, nonmetallic inclusions, and back surfaces reflect incident waves. By sending a pulse and receiving its reflection, the presence and location of flaws can be defined in analogy to radar. This is the pulse-echo technique. When laminations, disbonding, and other flaws are present, the intensity of transmitted waves is attenuated. By measuring the intensity on the back surface, the through-transmission technique detects flaws via attenuation. When the material under inspection is uniform, the time of transit of elastic waves through the material can be converted to the thickness by determining the wave velocity on a calibration sample. This time-of-flight technique is used in thickness meters. Time-of-flight and attenuation measurements on samples of known thickness are used to characterize material properties, such as grain size of metals or fiber and void content in composites.

The ability to locate and measure internal flaws, the ability to internally inspect three-dimensional objects from a single surface, and nonhazardous operation are advantages of ultrasonic testing. Disadvantages include the complexity in interpreting some results and the difficulty of inspecting parts with rough surfaces, irregular or small shapes, or near surface flaws. Operator training requirements are also higher than for most other nondestructive testing methods; see [6], [16] for further details.

2.3.1. Waves

Elastic waves travel with characteristic velocity in a gas, liquid, or solid. In gases and liquids, only compressive waves exist; i.e., the medium is alternately compressed or rarefied. Waves travel parallel to the direction of compression and rarefaction; hence, they are also called longitudinal waves. In solids, compressive waves and shear waves are present. In shear or transverse waves, elastic displacement occurs perpendicular to the direction of wave propagation. These waves are shown schematically in Figure 9 A and B. Here, the distance between two points of the same phase is the wavelength λ_u.

Most ultrasonic testing utilizes frequencies of 1 – 25 MHz. Higher frequencies allow better spatial resolution due to the shorter wavelength. However, more attenuation occurs above 10 MHz in steels as the wavelength approaches the grain size. In highly attenuating fiber-reinforced composites, frequencies of 1 MHz or lower are commonly used. In most structural steels and wrought alloys, attenuation is low, and ultrasonic testing has good penetrating power to depths of 1 – 10 m.

The velocity V_c of plane compressional waves in an isotropic medium is given by

$$V_c = \sqrt{(\lambda + 2\mu)/\varrho}$$

where λ and μ are Lamé constants and ϱ is mass density. Similarly, the shear wave velocity V_s is

$$V_s = \sqrt{\mu/\varrho}$$

Wavelength is equal to wave velocity V divided by frequency f. For steels, $V_c \approx 6$ mm/µs, and for $f = 1$ MHz, $\lambda_u \approx 6$ mm; for water, $V_c \approx 1.5$ mm/µs and $\lambda_u \approx 1.5$ mm. Table 1 lists typical values of V for various materials.

By measuring wave velocities, the elastic constants, Young's modulus E, and Poisson's ratio v, can be found:

$$E = \frac{\varrho V_s^2 [3(V_c/V_s)^2 - 4]}{(V_c/V_s)^2 - 1}$$

$$v = \frac{(V_c/V_s)^2 - 2}{2(V_c/V_s)^2 - 2}$$

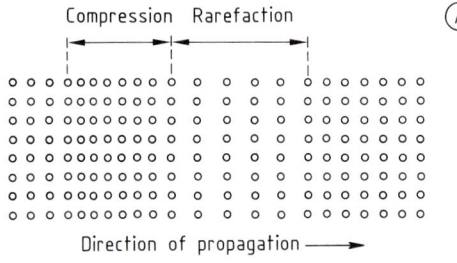
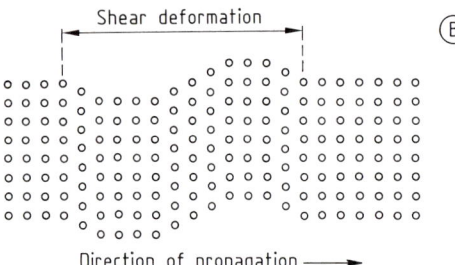
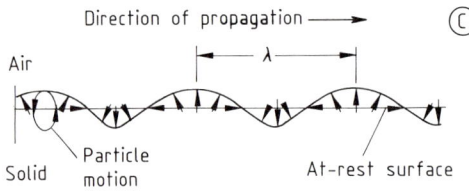
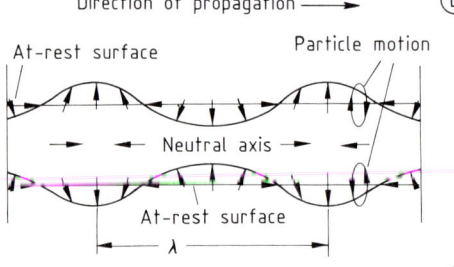
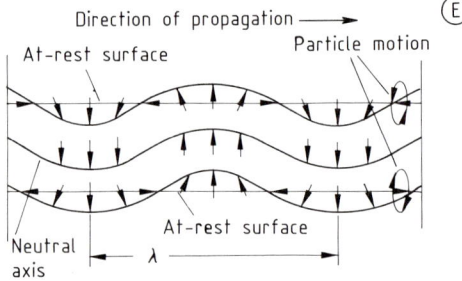

Figure 9. Ultrasonic waves
A) Compressive wave; B) Shear wave; C) Surface wave; D) Symmetrical plate wave; E) Asymmetrical plate wave

Table 1. Typical ultrasonic velocities

Material	V_c, mm/µs	V_s, mm/µs
Carbon steels	5.94	3.24
304 stainless steel	5.64	3.07
Cast iron	3.5–5.6	2.2–3.2
Inconel	5.82	3.02
Brass (70/30)	3.83	2.05
Titanium (pure)	6.10	3.12
Glass	5.77	3.43
Poly(methyl methacrylate) (Plexiglas)	2.67	1.12

Surface (or Rayleigh) waves are also used in ultrasonic testing. These travel along the surface with approximately 90 % of the shear wave velocity. The wave energy is concentrated within about one wavelength of the surface, as shown in Figure 7C. Rayleigh waves follow curved surfaces and corners, but are reflected when the ratio of the radius of curvature r to the wavelength λ_u is less than 1.6. About half the wave energy is reflected for $r/\lambda_u < 0.6$. In comparison to bulk waves (compressive and shear), Rayleigh waves are attenuated less and are well suited to inspecting parts that have complex contours.

Lamb waves (plate waves) propagate in sheets and plates in which the wavelength and thickness are of the same order of magnitude. Symmetric and asymmetric modes of propagation exist (Fig. 7D and E). These waves have different phase and group velocities, each dependent on frequency and thickness. Until recently, the application of Lamb waves to ultrasonic testing has been limited due to their complex propagation characteristics.

When ultrasonic waves are normally incident on a plane interface of two materials, 1 and 2, part of the wave energy is transmitted through the interface, and the remainder is reflected. The ratio of transmitted energy to incident wave energy, known as the transmission coefficient T_i, is expressed in terms of the acoustic impedance Z, the product of mass density and wave velocity ($Z = \varrho V_c$):

$$T_i = 4Z_1 Z_2/(Z_1 + Z_2)^2$$
$$= 4\varrho_1 \varrho_2 V_{c1} V_{c2}/(\varrho_1 V_{c1} + \varrho_2 V_{c2})^2$$

The reflection coefficient $R_i = 1 - T_i$. At an interface of water and steel, $T_i = 0.12$ and $R_i = 0.88$; that is, nearly 90 % of the wave energy is reflected.

When a plane wave is obliquely incident on an interface, both compressional and shear waves are reflected or refracted regardless of the nature of the incident wave. This is known as mode conversion, and the angles of reflection and refraction are defined by Snell's law (see Fig. 10):

$$\frac{\sin\theta_{c1}}{V_{c1}} = \frac{\sin\theta_{s1}}{V_{s1}} = \frac{\sin\theta_{c2}}{V_{c2}} = \frac{\sin\theta_{s2}}{V_{s2}}$$

For water (1) and steel (2) with an incident angle θ_{c1} of less than 45°, compressional waves in water are reflected at θ_{c1}; refracted compressional and shear waves exit from the interface at θ_{c2} and θ_{s2}. Note that no shear wave exists in liquids. The angle

$$\theta_{c1} = \sin^{-1}(V_{c1}/V_{c2}) = 14.5°$$

is called the first critical angle because θ_{c2} becomes 90° and the compressional waves in steel no longer exist for higher values of θ_{c1}. The second critical angle is at $\sin^{-1}(V_{c1}/V_{s2}) = 27.4°$, above which no shear wave is observed in steel and total reflection occurs. Between the first and second critical angles, only shear waves are refracted in steel. This is advantageous because interpretation is easier with only one wave mode, and higher intensity waves are produced for shear waves refracted at 35–60°. Partly due to geometrical considerations, 45° and 60° shear wave transmission paths are most commonly used. For thin sections, 70° shear waves are also employed.

Useful depth of inspection is determined by the attenuation of ultrasonic waves. The acoustic pressure decreases exponentially with depth, as in the case of X rays. However, the attenuation of ultrasonic waves is usually expressed in decibels per meter (dB/m). General ranges of inspection are up to 1–10 m in steels, in nickel, titanium and magnesium alloys with fine microstructures, and in all aluminum alloys; these have attenuation rates of 0.1–1 dB/m. Most cast metals have a lower inspection range of 0.1–1 m. Stainless steel weldments and fiber-reinforced composites have high attenuation, and the inspection range extends to less than 0.1 m.

In highly attenuating materials such as cast nonferrous materials and fiber-reinforced composites, the wave velocities are frequency dependent. This dispersion affects waveforms as they propagate because higher frequency contents are attenuated.

2.3.2. Generation and Detection

Ultrasonic waves are generated and detected by using a transducer element that converts electrical signals to mechanical ones and vice versa [17]. A thin disk of piezoelectric crystal or polarized ceramic is used. Quartz, lithium sulfate, and lithium niobate are used as single-crystal elements, whereas lead zirconate–titanate is the most common ceramic element. The transducer element is enclosed in a protective casing with a thin ceramic or plastic faceplate and an attenuative backing such as a tungsten–epoxy mixture (see Fig. 11 A). The backing is needed to dampen the resonance of the transducer element and to broaden the frequency response. The casing also acts as an electrical shield. A thicker faceplate (a delay tip) may be used to separate the front-surface signal from the large excitation pulse. The frequency characteristics of the packaged transducer or search unit are dictated primarily by the thickness resonance of the transducer element, which occurs when the thickness equals one-half of the wavelength, causing constructive interference of waves reflected at the surface.

To generate ultrasonic pulses with a search unit, an electronic pulse generator is used to produce a train of short pulses or tone bursts (up to ten oscillations). These are repeated 50 to 2000 times per second. A short pulse contains a wide spectrum of frequencies, while the frequency spectrum of a tone burst is centered at its base frequency. When the search unit is excited with a pulse (or tone burst), the frequency spectrum of the ultrasonic wave output is the product of the frequency spectra of the input pulse (or tone burst) and the search unit.

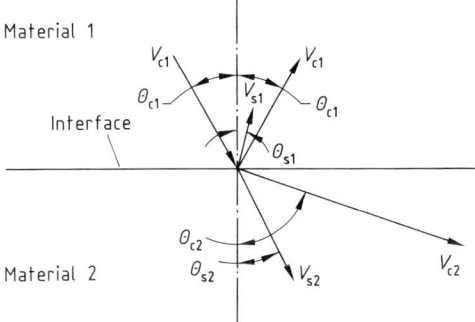

Figure 10. Mode conversion of ultrasonic waves

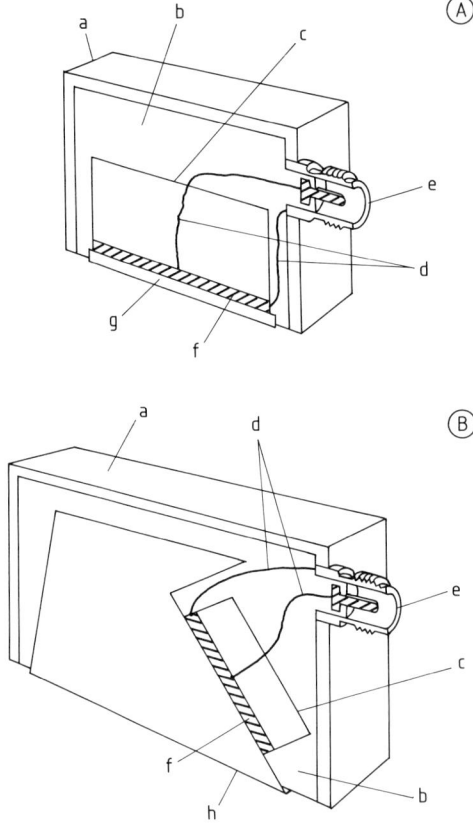

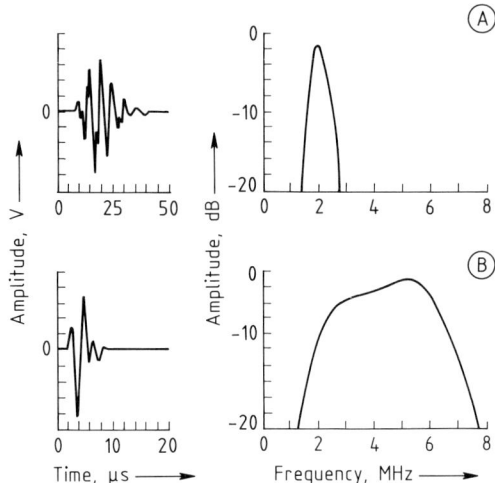

Figure 12. Ultrasonic waveforms and their frequency spectra
A) Narrow band; B) Broad band

Figure 11. Typical ultrasonic search units
A) Straight-beam unit; B) Angle-beam unit
a) Case; b) Epoxy potting; c) Tungsten-loaded backing; d) Electrical connections; e) Cable connector; f) Transducer element; g) Faceplate; h) Plastic wedge

The waveform of the output can be obtained from the frequency spectrum by inverse Fourier transform. This waveform W_o can also be expressed as the convolution of the input waveform W_i and the input response W_s:

$$W_o(t) = \int W_i(t - \tau) W_s(\tau) d\tau$$

where W_s is the characteristic waveform for an infinitely short pulse (delta-function shape). Typical waveforms and the frequency spectra of narrow-band and broad-band search units are shown in Figure 12. High peak voltages (300–500 V) are needed for a short pulse because only part of the pulse energy is used. This often leads to the failure of thin, high-frequency transducer elements. With a tone burst, the base frequency must be matched to the nominal resonance frequency of the search unit. Since the tone burst is limited in bandwidth, the peak voltage can be much lower to obtain the same ultrasonic energy output. Alternatively, a much stronger wave can be generated with a tone burst than with a short pulse for a given applied voltage.

The sound field immediately in front of the faceplate is basically a plane wave, but the intensity varies depending on the position and distance from the search unit because of diffraction effects. This region is known as the near field, which extends to a distance N_x given by

$$N_x = D_t^2/4\lambda_u$$

where D_t is the diameter of the transducer element. For flaw detection, the near field is difficult to use because of the intensity variation (near-field effect). Near-field effects in the material under inspection can be eliminated by placing a delay tip between it and the search unit, whereby the near-field effect is confined to the delay tip. The region beyond N_x is known as the far field and a spherical wave propagates as shown in Figure 13. No diffraction fringe occurs, and the wave intensity decreases with the square of the distance. The spread φ of the far-field sound beam is dictated by λ_u/D_t and is given by

$$\varphi = 2\sin^{-1}(1.22\lambda_u/D_t)$$

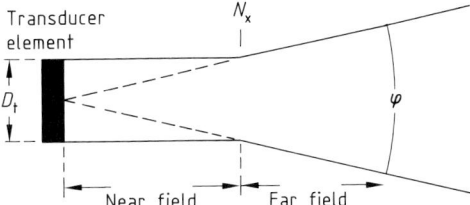

Figure 13. Near- and far-field regions

For example, $\varphi = 44°$ in steel at 1 MHz with $D_t = 19$ mm. For the same D_t at 5 MHz, φ decreases to 8.7°. A smaller value of D_t allows flaws to be more accurately located at short distances, but this advantage is lost as the beam spreads with increased penetration. A higher frequency beam spreads less for a given value of D_t or allows the use of a smaller D_t.

For special laboratory studies, ultrasonic pulses can be generated by means of the thermoelastic effect with a strong laser beam. Electromagnetic transducers allow noncontact generation and detection. They are suited for automated inspection of pipes and plates at elevated temperature or with rough surfaces, for which noncontact generation is desirable for ease of use. Commercially available units can be used up to 3 MHz. A strong magnetic field is required, and a special powering unit must be used. Nonmagnetic test pieces lower the efficiency. Limits in the frequency range of operation and lower signal-to-noise ratios are some of the disadvantages.

By using the principle of refraction, ultrasonic search units can be designed to generate angle beams. A plastic wedge is housed with a transducer element as shown in Figure 11 B. The direction of the beam is valid for a specified material. Surface wave search units are also made with the same design concept.

In the pulse-echo mode, the same search unit is used for generation and detection of ultrasonic waves. Dual-element search units are also used which house separate transducer elements for detection and generation. Two separate search units are employed for the through-transmission mode, in which one is dedicated for receiving the ultrasonic waves. Received ultrasonic waves produce low-level electrical signals. Electronic amplifiers with low noise levels (even for a wide bandwidth of 0.1 – 50 MHz) are used to provide good signal-to-noise ratios. When a single search unit is used for transmission and receiving, an input protection circuit is required and the input amplifier must be capable of rapid recovery from an overload due to the excitation pulse.

The sound beam can be focused by means of an acoustic lens attached to the front surface of the transducer. This requires a water path between the lens and the test piece. Transducers can be spherically focused to a spot or cylindrically to a line. For the latter, a rectangular transducer element is used. Spherical focusing achieves the same for pipe and tubing inspection. Both can increase near-surface resolution without increasing the transducer frequency. Focusing also reduces the effects of surface roughness and contour.

Focal distance is given in water path length. Path length in a solid is converted to an equivalent path length in water by multiplying by the ratio of sound velocities. Focal distance should then be equal to (water path) + (solid path) × (sound velocity in solid) ÷ (sound velocity in water). Search units of various focal distances (typically, 1 – 25-cm water path) are available.

2.3.3. Inspection Methods

When a search unit is placed directly over the surface of a part under test in contact inspection, a thin layer of liquid couplant is used to provide a low loss path for wave transmission. Water, oils, glycerol, greases, and resins are used as couplants. Certain soft rubbers have been specially formulated to provide dry coupling of a search unit. In immersion inspection, the part under inspection and the search unit are placed in a tank of water. Here, a water path exists in place of a couplant layer. A search unit is usually fitted with a waterproof connector and attached to a manipulator. It is usually set manually but is increasingly used under computer control.

In conventional immersion inspection, the search unit and the test piece are placed in a water-filled tank. The search unit is connected via a manipulator and an extension tube to an electromechanical scanning device (or a probe manipulator bridge). The basic scanning device provides X – Y movements of prearranged scanning patterns. Advanced units add computer control that adapts to the shape of a test piece. Turntables and roller drives allow efficient scanning of round disks and cylindrical test pieces. A rotating reflector inside a tube can be used to reflect the ultrasonic beam of an immersion

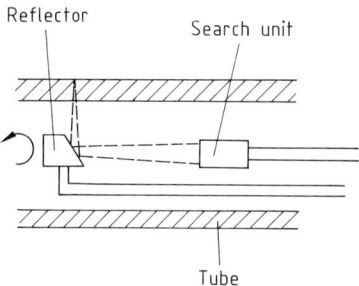

Figure 14. Ultrasonic inspection of pipes

search unit toward the tube wall, allowing the inspection of thickness changes or flaws by means of normal or shear beams (Fig. 14) [18]. Scanning is synchronized with ultrasonic data acquisition, and flaw indications are displayed at corresponding locations. An X–Y recorder has been used for such a display, but recent equipment can store the scan position and ultrasonic data from 10^7–10^8 locations in digital form. Real-time or post-test analysis of the stored data provides detailed color displays.

Wheel-type search units consist of a stationary transducer element, a liquid path inside a rubber tire, and a rolling wheel to provide continuous contact with a test piece. The transducer element can be placed for straight-beam or angle-beam inspection. The wheel unit can be stationary and a test piece moved past it, or it may be moved over a fixed test piece.

A search unit can be fitted with a squirter, a nozzle through which water streams out under pressure, which provides a column of water for a sound beam to reach the test piece. The squirter technique can be combined with robotics. This combination eliminates the size limitation of immersion testing, can be adapted to complex contours, and is suited for automatic operation. Another method, known as the bubbler method, uses a small tank with overflowing water. A search unit is placed in this tank, and a test piece is moved over it.

2.3.4. Display Methods

Results of ultrasonic testing are commonly presented in A-scan, B-scan, and C-scan and, more recently, with reconstructed images. The A-scan presentation is a plot of amplitude versus time, as shown in Figure 15. Front reflection,

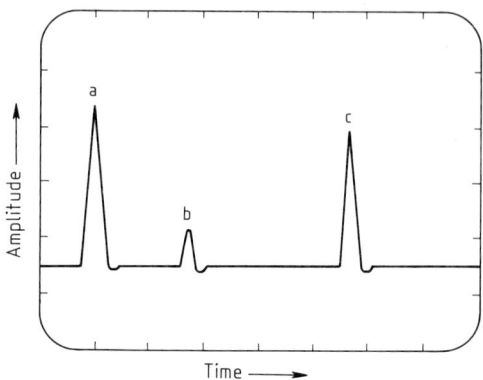

Figure 15. A-scan display for pulse-echo technique
a) Front reflection; b) Flaw echo; c) Back reflection

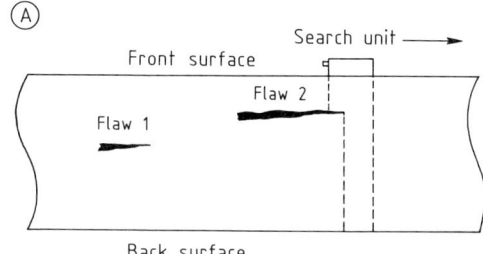

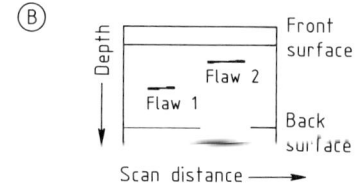

Figure 16. A) Pulse-echo technique; B) B-scan display

back reflection, and a flaw echo are indicated on the time axis, which corresponds to depth. The height of the flaw echo is related to the size of the flaw. In the absence of a flaw, an A-scan over a longer period also shows the ultrasonic attenuation of the test piece. Multiple back reflections, whose heights decrease exponentially, are detected.

The B-scan presentation is used to display the depth and length of a flaw. The transducer is moved along the front surface, as shown in Figure 16. When a flaw echo exceeds a predetermined threshold, an indication is recorded on the display along with indications of the front- and

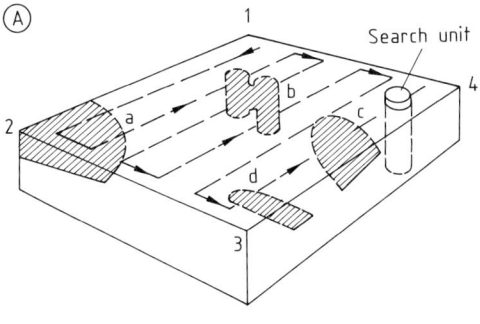

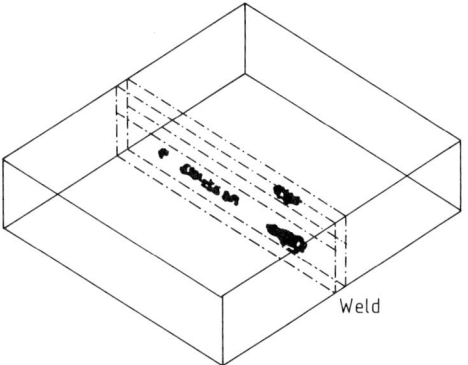

Figure 18. Reconstructed image of weld flaws

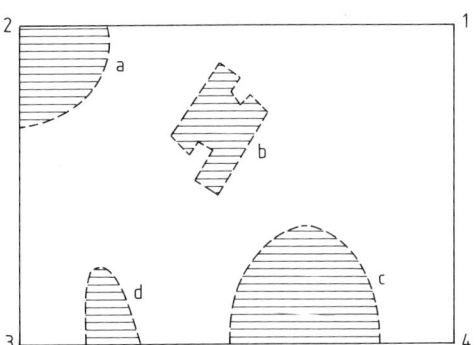

Figure 17. A) Pulse-echo technique; B) C-scan display

back-surface signals. Convenient visual observation is possible with manual scanning and a storage oscilloscope. Mechanized scanning allows more accurate determination of the position of a flaw.

The C-scan presentation is most useful in identifying defective areas in a plane view of a test piece. The transducer is moved back and forth over the front surface (Fig. 17). When a flaw echo above a threshold is present, it is recorded on an X–Y recorder, which is also scanned synchronously. The use of several electronic timing gates, each sensing the flaw echo over a specified time interval, enables scanning for flaws at predetermined depth ranges. Newer inspection systems are usually equipped with a computer that controls scanning, ultrasonic data acquisition (in digitized form), and display. These allow data storage, display of C-scan records at different depths, and a variety of image processing. For example, flaws at different depths can be displayed in different colors.

In specialized inspection systems, A-scan records are accumulated along with location data during a more complex mechanical scanning of various shapes. For tubular products, helical scanning is used, either internally or externally. For pressure vessels, remote-controlled crawlers are used. Industrial robots are used on contoured surfaces. Three-dimensional images of flawed regions can be reconstructed from these data, and different perspectives of flaw images can be obtained. A reconstructed view of flaws in a weld is shown in Figure 18. In both reconstructed imaging and C-scan displays, the use of color enhances visual perception.

2.3.5. Interpretation

Echo intensity in an A-scan display depends on the size, shape, and distance of the flaw. For disks of various sizes in water, the intensity of the echo varies as shown in Figure 19 [16]. The reduction in echo intensity results from attenuation and spreading of the beam. Compensation for this reduction is provided by a distance–amplitude correction (DAC) circuit in most ultrasonic instruments. By comparing the echo intensities of flaws with those of flat bottom holes in reference blocks, flaw sizes can be estimated. Another approach is to use the ratio of echo intensity to the back-surface reflection as an indicator of an unacceptable flaw. An echo intensity equal to 50% of the back reflection is one such rejection criterion.

The shape of the echo depends on the shape, orientation, and sound-reflecting characteristics of an interface. Smooth metal–air interfaces that are normal to the beam produce sharp echoes. Curved or rough interfaces (pores, cracks, lami-

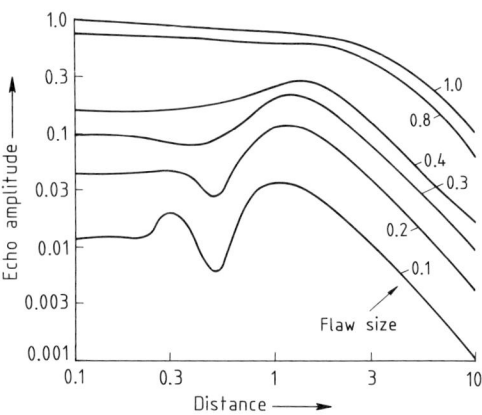

Figure 19. Distance–echo amplitude–flaw size diagram

nations) produce broadened echoes. Changing the beam orientation and the position of the search unit often provides clues to flaw shape and orientation. Loss of back reflection is another factor in flaw size estimation. This is due to the reflection of a sound beam by a flaw, which may not produce a flaw echo if the reflected beam is directed away from the search unit. In this case, the loss of back reflection indicates the presence of a flaw. Scattering from slags, inclusions, and large grain structures also results in the loss of back reflection.

The large grain structures of anisotropic materials (e.g., stainless steel or brass) produce beam scattering, which is indicated as random echoes between the front and back reflections. These are called grass and make the discrimination of small flaw echoes difficult. Spurious reflections are also produced by fillets. A choice of suitable inspection position is important for complex geometries.

2.3.6. Inspection Standards

Ultrasonic inspection is performed under specified procedures established by national bodies, such as the ASTM, DIN, and JIS. They require the use of standard reference blocks, which may be of different sizes prepared from different alloys, containing holes, slots, or notches. Two types of standard blocks are widely used. One type is the area–amplitude or distance–amplitude block, including ASTM E 127 blocks. Each of the ASTM blocks is 51 mm in diameter and has a 19-mm-deep flat-bottom hole. The size of the holes varies from 0.40 to 6.4 mm in diameter, and the distance from the surface to the bottom of the hole varies from 3.2 to 152 mm. By changing the hole size at a given distance, these blocks can be used to relate the amplitude of the flaw signal to the area of a flaw, and to check the linearity and sensitivity of a pulse-echo inspection system. For a given hole size at various depths, distance–amplitude blocks can evaluate variations of echo amplitude with distance for straight-beam inspection in a given material. These blocks must be made from the same material as the test piece in order to estimate flaw sizes at various depths. However, a set of blocks can be used as reference blocks for performance calibration of a test system.

Another reference block is known as the IIW type. It is a nearly rectangular steel plate (25 × 100 × 360 mm) with a curved edge of radius 100 mm at one end. The block has a notch, a slot, and large and small holes (50 and 1.5 mm diameter), which are used together with the curved edge to determine the sensitivity, propagation angle and beam spread of angle-beam search units. The IIW block can also be used for simple sensitivity evaluation of straight-beam units. A miniature version based on the same concept is also used.

2.4. Acoustic Emission

Acoustic emission (AE) refers to transient elastic waves generated from localized sources within a material or structure. Such transient waves originate at, for example, cracks, defective welds, and yielding regions in metals and at delaminations, fibers, and matrix fractures in composite materials. Sensitive ultrasonic sensors detect acoustic emission signals, from which the location and activity level of the source can be evaluated. Acoustic emission requires the application of a stimulus to a test piece. In pressure vessels and piping, hydrostatic pressurization is usually employed to activate acoustic emission sources. Acoustic emission has become the inspection method of choice for fiberglass-reinforced composite vessels and piping and, more recently, for metallic vessels, including industrial gas transport trailers and railroad tank cars; see [4e], [19], [20].

2.4.1. Detection

In acoustic emission testing, signals in the ultrasonic frequency range of 30 kHz to 2 MHz are usually detected. Airborne noise interferes with AE measurements at lower frequencies, and signal attenuation makes the higher frequency range difficult to use. In composite testing, the 30–150-kHz range is common, and the 100-kHz to 1-MHz range is used for metals. Both burst-type (pulselike) and continuous-type (random noise) signals are generated. Burst-type emissions arise from distinct events involving the release of elastic energy, such as crack advances, fiber fracture, and delamination. Continuous-type emissions are produced by many overlapping events and are observed from plastic deformation of metals.

Sensors used for acoustic emission detection are 40–60 dB more sensitive than those used for ultrasonic testing. One type enhances the sensitivity by using the resonance of a transducer element, which is usually a piezoelectric disk. These are resonant or narrow-band sensors and are most commonly employed in acoustic emission testing. The second type has a broadened frequency response and uses a backing material behind a transducer element; its construction is essentially identical to that of ultrasonic sensors. For specialized high-fidelity capture of signal waveforms, a conical piezoelectric element with matched backing and a capacitive sensor can be used.

Acoustic emission activity is measured by the rate of burst-type emissions (AE event count rates) and by the averaged signal intensity of continuous-type emissions (e.g., root-mean-square voltages of amplified signals). The cumulative number or rate of oscillations that cross a predetermined threshold value (called AE counts or AE count rates), some measures of signal energy, the duration of burst-type signals, and the distribution of peak amplitudes are also employed.

2.4.2. Material Behavior

Plastic deformation of most structural alloys emits acoustic emission that reaches a maximum near the yield stress and diminishes with work hardening. The composition, microstructure, and heat treatment of alloys affect acoustic emission behavior. Cold-worked metals produce hardly any acoustic emission because AE phenomena are irreversible; therefore, acoustic emission is absent until previously applied stress levels are exceeded. Some alloys (e.g., austenitic stainless steels and 2219 aluminum alloy) have inherently low acoustic emission activity during plastic deformation. Stressing of various types of composite materials produces very high levels of acoustic emission. Matrix cracking is dominant at low stress. Glass-fiber fracture and delamination generate intense signals at higher stress.

Subcritical crack growth emits acoustic emission at increasing rates as the crack nears criticality. Some ductile cracks and fatigue crack growth have lower acoustic emission activity. Such cracks can still be located by detecting secondary AE sources, such as crack face interference and crushing of corrosion and oxidation products. Decohesion and fracture of nonmetallic inclusions are also prominent sources of acoustic emission.

Other important acoustic emission sources include corrosion (gas evolution), stress corrosion (cracking), martensitic transformation, solidification (hot cracking), oxidation (oxide cracking and spalling), gas and fluid leakage, and welding processes (metal and slag cracking, transformation, and contamination). Magnetization of steels produces magnetoacoustic emission, which can be used to evaluate heat treatment and residual stress.

2.4.3. Source Location

Acoustic emission monitoring utilizes the structure under test to discover flaws. Acoustic emission signals emitted from the flaws travel through the structure to sensors mounted on its surface. When the structure is hot, a waveguide is used to mount a sensor. The flaws are located by analyzing the signals received. Two general approaches are employed in flaw location. The zone location method uses multiple sensors. The flaw is presumed to exist within the zone that belongs to the sensor receiving the strongest AE signals. This method is suitable for locating leaks that produce continuous emissions and flaws in highly attenuating media, such as fiber-reinforced plastics, but is also used in testing metal vessels. Typical results are shown in Figure 20 [21]. The discrete source location method also employs multiple sensors, but uses triangulation techniques based on the exact differences in time

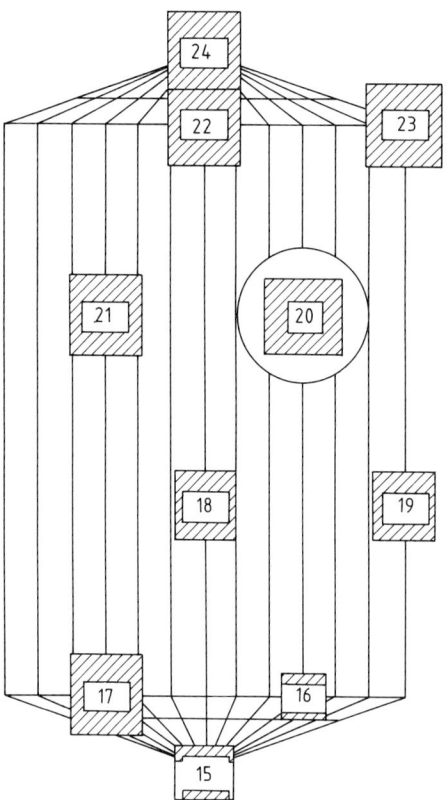

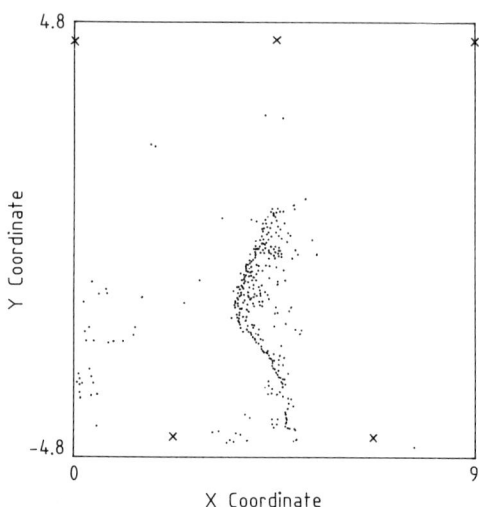

Figure 20. Zone location results from pressurization test of a hydrocracker reactor
Sensor 20 indicates a significant flaw requiring immediate inspection; sensors 17–19 and 21–24 show flaws that require follow-up evaluation

Figure 21. Discrete source location from pipe subjected to sour gas solution
X marks the sensor position; dots indicate located acoustic emissions sources on an X–Y plot

between the arrival of a given event at different sensors. It provides more accurate positions of acoustic emission sources. This approach requires an AE signal to reach at least two (linear location) or three (planar location) sensors. Attenuation of the signals must be low to avoid the use of an excessive number of sensors and signal processing channels. Most metal vessels and piping can be inspected by this approach. An example is shown in Figure 21 [22].

2.4.4. Acoustic Emission Monitoring

Acoustic emission testing of metallic pressure vessels is conducted by using one of the source location methods. The vessels are pressurized according to applicable code specifications. Sources of emissions are usually crack growth, yielding, delamination, corrosion effects (corrosion product cracking or spalling, local yielding of thinned sections), stress corrosion cracking, and embrittlement. Most weld defects also exhibit acoustic emission when stressed. When metals are ductile, AE activity is low and AE test results should be evaluated carefully. Once the metals have been embrittled by the environment or by low temperature, even early stages of stressing activate AE sources. In fact, acoustic emission is the best method for detecting hydrogen-induced cracking and stress corrosion cracking. The flaws located are usually confirmed by other nondestructive test methods. In small, thick-walled metallic vessels, the use of acoustic emission examination alone has been approved in lieu of ultrasonics or radiography [5, Sect. VIII]. Highway gas-trailer tube testing can be performed with acoustic emission instead of the formerly used hydrostatic recertification process. Many chemical and petrochemical vessels, including tank cars and tank trailers, are monitored by acoustic emission during and between service without being emptied, thus minimizing the length of plant shutdown [33].

Composite vessels and piping are monitored with acoustic emission. This is one of the most successful AE applications (see Section 3.2).

2.5. Penetrant Methods

Inspection with liquid penetrants relies on the capillary action of a surface-wetting fluid and provides a reliable method for locating minute surface flaws down to micron sizes. Surface cracks, laps, porosity, shrinkage areas, laminations, and other discontinuities can be detected as long as they are open to the surface. The shape, size, and composition of the test piece have no effect, but materials with rough or porous surfaces cannot be inspected; see [1], [4b].

Penetrant testing consists of (1) surface preparation, (2) penetration, (3) penetrant removal, (4) development, and (5) inspection; see Figure 22. Initial cleaning of the surface removes oil, grease, dirt, surface coatings, rust, scale, and other foreign objects that give rise to false indications or interfere with wetting by a penetrant. Chemical etching may be required to eliminate a cold-worked layer. A film of liquid penetrant is applied to cover the surface to be inspected. The film remains on the surface for a penetration (dwell) period specified for the penetrant used. Excess penetrant remaining on the surface is then removed, leaving penetrant that has seeped into discontinuities. For postemulsifiable penetrants, application of emulsifier is required before removal. A developing agent is applied to cover the surface in a thin coating. The developer draws penetrant out of the discontinuities and broadens the apparent width of the visible indications. The test piece is then viewed under suitable lighting, either in white light for visible-dye penetrants or ultraviolet (black) light with darkened background for fluorescent penetrants.

Fluorescent and visible-dye penetrants are made with water-washable, postemulsifiable, or solvent-removable fluids. Different levels of flaw detection sensitivity are available with any combination, but visible-dye penetrant systems are less sensitive than fluorescent ones. Water-washable penetrants contain emulsifiers and can be removed by water. This requires careful washing so that penetrants in flaws are not removed. Postemulsifiable penetrants require application of emulsifier following a penetrant dwell period. Excess penetrants become washable with water and are rinsed off. Penetrant in the flaws remains water insoluble, providing excellent retention capability. Although postemulsifiable penetrants are more expensive, they provide greater sensitivity and can reveal hairline cracks and shallow discontinuities. Solvent-removable penetrants

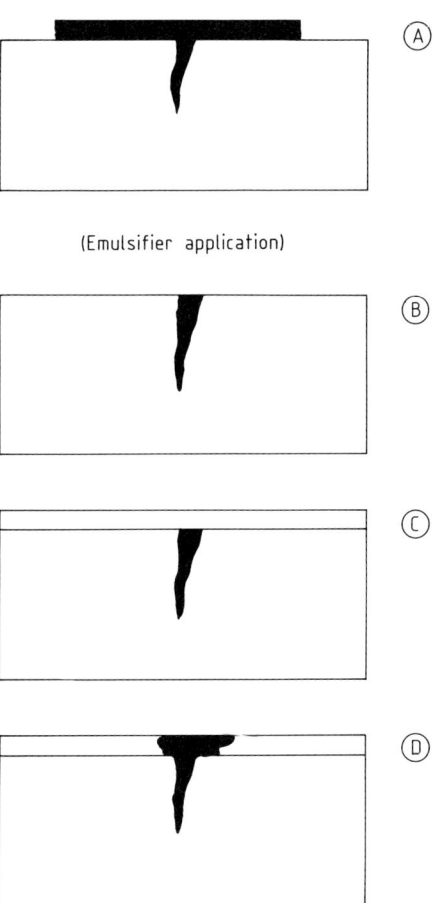

Figure 22. Steps of penetrant testing
A) Penetrant application; B) Surface Cleaning; C) Application of Developer; D) Inspection
Emulsifiers are used only for postemulsifiable penetrants.

are used for limited, special applications because the use of solvent is labor intensive and in some locations requires an extensive pollution abatement system. Minimum facilities are needed especially for solvent-removable, visible-dye penetrants, which are convenient for use in remote locations.

Reference photographs of typical flaw indications are available to establish inspection criteria [6, E433]. The smallest flaws that can be detected by penetrant inspection with a laser and a photodetector are cracks of ca. 1-µm width. Reliable visual inspection of 25-µm cracks can be made. Most inspection has been done by the human eye, but automated scanning equipment has increased inspection speed substantially for quantity inspection of identical or similarly shaped parts [23].

The quality of penetrant inspection systems can be monitored by the use of cracked-block comparators or suitable materials with known flaws. A number of comparators have been developed, including cracked chrome-plate panels, fractured glass panels, unglazed ceramic disks, and nickel–copper laminates. Quenched aluminum plates are the best known example.

Inspection of high-strength steels and titanium- or nickel-based alloys with penetrants requires special attention to the sulfur and chlorine content of all materials used. These elements should be kept below 1%. Some penetrants are shock sensitive in the presence of liquid oxygen. Special penetrants must be used if the surface will come in contact with liquid oxygen. Another limitation of penetrant systems is that most penetrants cannot be used much above ambient temperature because they may be baked onto the surface and lose efficiency. A crayon penetrant system has been developed for high-temperature use up to 270 °C.

2.6. Magnetics and Electromagnetics

2.6.1. Magnetic Particle Inspection

Magnetic particles become attached to the surface and to some subsurface discontinuities when a ferromagnetic test piece is magnetized. The particles (e.g., iron or soft ferrite), which have high permeability and low coercive force, are coated with colored or fluorescent pigments for increased visibility. They are applied as dry particles or as suspensions in water or oil. Magnetic leakage fields of a discontinuity collect the particles, indicating its location, size, and shape by color contrast or fluorescence. Nonmagnetic materials cannot be tested with this method. Such materials include austenitic stainless steels and many heat-resistant nickel alloys; see [1], [4f], [6, E 709].

In the continuous method, magnetization is applied continuously while the magnetic particles are administered. Because the high current used causes heating, it is used for only a short time during which the magnetic particles must be applied. In the residual method, a pulse magnetization is applied, after which magnetic particles are brought in contact with the test piece. The residual method can be used when the test piece has adequate remanence to retain a leakage field for attracting magnetic particles. Most structural steels fall into this category, but low-carbon steels and soft magnetic materials require the use of the continuous method.

Magnetic leakage fields are formed when a discontinuity lies normal to the applied magnetic field. The direction of magnetization must be selected to reveal expected flaws. Several methods of magnetization are used: (1) the flow of direct or alternating current through a test piece, and (2) the placement of a test piece within the magnetic field of a magnetizing coil or yoke. Because of the eddy-current effect (see Section 2.6.4), an a.c.-induced magnetic field concentrates near the outer surface and is used for surface flaw detection.

To detect a transverse or circumferential flaw (e.g., a weld crack in a butt-welded pipe joint) a longitudinal magnetic field is necessary. To produce the required field, a magnetizing coil can be wound over the pipe (Fig. 23 A), the pipe can be placed between two pole pieces of a magnet (Fig. 23 B), or the two poles of a magnetizing yoke (a C-shaped electromagnet with flexible poles) can be placed across the flaw (Fig. 23 C). The test piece can also be placed inside a stationary magnetizing coil. To reveal a longitudinal flaw (e.g., an axial crack in a welded pipe) the

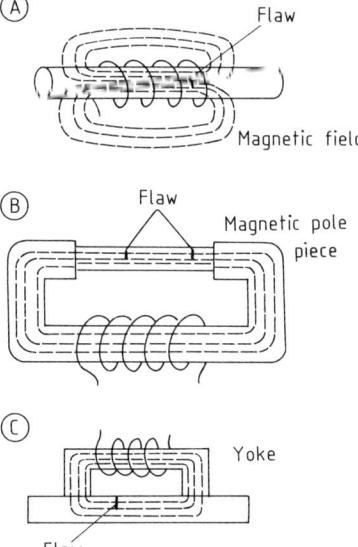

Figure 23. Longitudinal magnetization methods
A) Encircling coil; B) Between magnetic pole pieces; C) Magnetic yoke

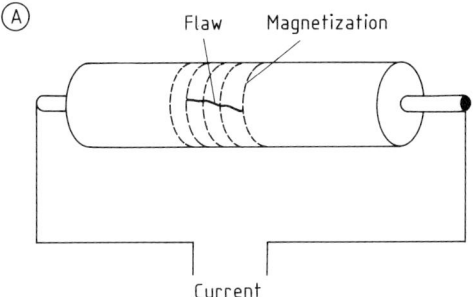

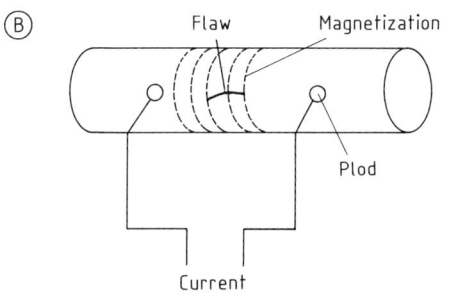

Figure 24. Circular magnetization methods
A) Current passing through a sample (head-shot); B) Current passing locally through two plods

magnetic field must exist circumferentially. This can be induced by diametrically applying a magnetizing yoke. Alternatively, a high current can be passed axially parallel to the flaw to produce a circular magnetic field (Fig. 24 A). By using two plod contacts placed along the weld, a current can be passed parallel to the crack (Fig. 24 B). For short hollow cylinders, electrical current may be passed through a central conductor.

The levels of magnetization of ferrous materials must be of the order of 1 T (10 000 gauss). Typical ferromagnetic materials have a relative permeability of more than 240. For direct, circular magnetization, the required magnetization can be achieved by 200–350 A per linear centimeter in diameter. For longitudinal magnetization of a short test piece having a length-to-diameter ratio (L/D) of 2–15, the number of required ampere-turns is $45\,000/(L/D)$.

Interpretation of magnetic particle indications requires the ability to reject irrelevant indications. Reference photographs of indications on steel castings have been compiled and can be used to establish acceptance and rejection criteria.

Variations of magnetic particle testing have been developed. The magnetic rubber method polymerizes in situ a liquid rubber containing magnetic particles. During curing, the particles concentrate in the region of a leakage field. This allows the inspection of recessed areas (e.g., bolt holes) and makes a permanent record. Magnetic printing and painting are also useful. These employ magnetic tape and paint for recording and indication.

2.6.2. Magnetic Leakage Field Inspection

Magnetic leakage fields from a surface of subsurface flaw can be detected and measured by Hall-effect probes, magnetic diodes, moving coils, and magnetometers. Leakage fields can in principle provide quantitative information on the flaw. The leakage fields have tangential and perpendicular components, and their distribution for simple flaw shapes has been calculated. Added to the leakage fields are large, but slowly varying bias magnetic fields. Most materials, however, have nonlinear magnetic properties that vary with processing conditions. Thus, estimation of flaw size and shape is difficult except by methods that use reference flaws for comparison; see [4d], [24].

Hall-effect probes are made with submillimeter dimensions and have high sensitivity. Some have a built-in amplifier and a sensing element as small as 100×25 µm. The probe size must not be much larger than the dimensions of the leakage field. Two probes are often used to cancel the effect of bias fields. Magnetic diodes work similarly to Hall-effect probes. A moving coil probe senses the rate of change of magnetic flux lines through the coil. The coil is moved uniformly over the surface of the sample. It must be small to achieve as high a sensitivity as Hall-effect probes.

A flux-gate magnetometer or a Förster probe utilizes the nonlinear magnetic hysteresis of a ferrite core, on which two coils are wound [2a, pp. 328–332]. The primary coil of the probe is excited at a carrier frequency, and the second harmonics in the secondary coil are detected. The output is sensitive to the magnetic field acting on the ferrite core and has a sensitivity hundreds of times greater than that of Hall-effect probes. Nuclear magnetic resonance magnetometers are also very sensitive. Because they use microwave frequencies, these are confined to specialized studies.

Applications of magnetic leakage fields use either continuous field excitation or residual leakage fields (fields existing after the applied magnetic field is removed). The latter require high-sensitivity probes such as flux-gate magnetometers. Magnetic leakage field methods are used extensively in the inspection of tubular shapes. Excitation fields and sensing probes can be placed on the outside or inside to search for axial and transverse flaws. For internal inspection, a drive section, a magnetizing and sensing section, and a recording or data output section are combined into a form of inspection "pig," which is universally jointed to permit movement through bent tubes. For smaller diameter tubes (< 5-cm inside diameter), permanent magnets can be employed for magnetization. Flaws such as pitting, holes, splits, hydrogen damage, and cracks in heat exchangers, pipelines, and chemical plant tubing can be inspected by using this method.

2.6.3. Other Magnetic Inspection Methods

The magnetic properties of a ferromagnetic material, such as initial permeability, saturation induction, coersive force, and remanence, are affected by its metallurgical history. Of these, permeability and coersive force are most often used for material characterization (i.e., hardness, heat treatment, alloy content, and grain structure). Applications include sorting of steel, ferrite content determination in stainless steel welds and castings, and detection of ferromagnetic impurities [25].

The magnetic Barkhausen effect and magnetoacoustic emission originate from the motion of magnetic domain boundaries [26]. These are also suited for material characterization and detection of residual stresses.

2.6.4. Eddy-Current Inspection

A changing electromagnetic field induces eddy currents in an electrical conductor. The eddy currents create a secondary electromagnetic field, which is detected through its effects on the primary exciting coil or by means of a separate sensor. In nonmagnetic materials or ferromagnetic materials that are magnetized to more than 98% saturation, the electrical conductivity of the materials is the primary property controlling the secondary field. In unsaturated magnetic materials, magnetic permeability has a larger effect on the secondary field. Moreover, the permeability of a material is sensitive to a variety of material properties, including heat treatment, cold working, composition, and magnetization level. Eddy-current testing of ferromagnetic materials has been hampered by this variation in permeability and by the difficulty of applying fields high enough to achieve saturation. This method can measure a combination of electrical conductivity and magnetic permeability, as well as any parameter or discontinuity that affects these two properties. Consequently, the detection of heat treatment and hardness changes and the sorting of dissimilar alloys can be performed. The detection of corrosion and flaws in tubular products has been an important application of this method; see [4d], [22].

Eddy-current testing can be carried out without direct contact, and signal processing is entirely electronic. Thus, an automated inspection system can achieve high inspection rates.

Because of the eddy current, externally applied electromagnetic fields are attenuated in a conductor. At the skin depth D_s, the field strength decreases to $1/e$, where

$$D_s = 1/\sqrt{\pi f \mu_m \sigma_e}$$

where f is the frequency, μ_m the magnetic permeability, and σ_e the electrical conductivity. In terms of the electrical resistivity ($\varrho_e = 1/\sigma_e$) in $\mu\Omega \cdot$ cm, and the relative permeability μ_r (which is unity for nonmagnetic materials, and 200–800 in structural steels), D_s (in mm) can be expressed as

$$D_s = 50\sqrt{\varrho_e/f\mu_r}$$

Typical values of D_s are given in Figure 25 as a function of frequency.

The basic element of an eddy-current instrument is an inductor or a coil. When its inductance is L_0 and its resistance is R, the coil impedance Z_c is given by

$$Z_c = \sqrt{(2\pi f L_0)^2 + R^2}$$

Z_c is affected by a conductor in proximity, and such changes are usually plotted in a normalized impedance plane representation (Fig. 26). The abscissa is the resistive component ($R/2\pi f L_0$), and the ordinate is the inductive component ($2\pi f L/2\pi f L_0$). When a solid rod is inserted into an encircling coil, the coil impedance Z_c varies

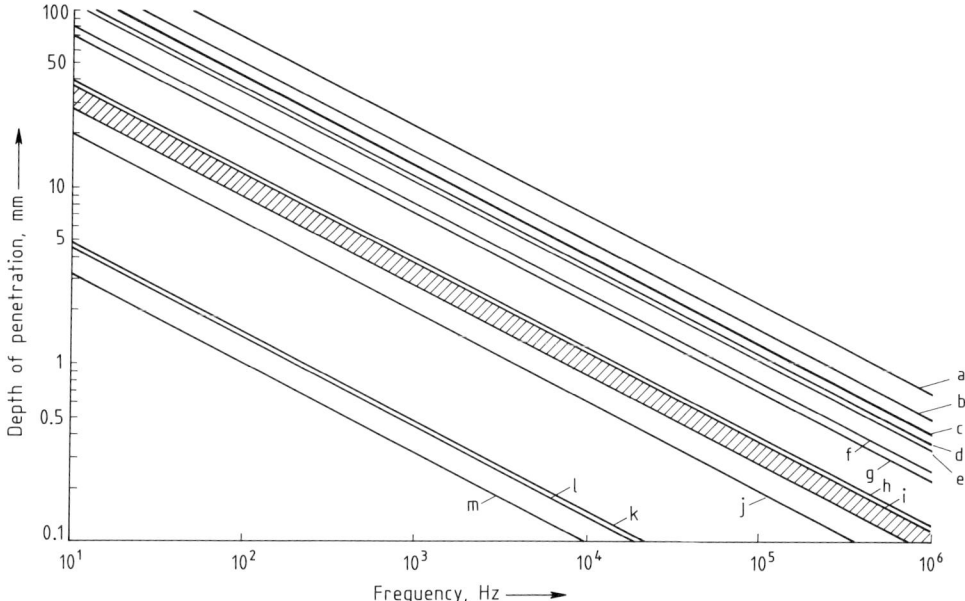

Figure 25. Skin depth versus frequency for various metallic materials
a) Hastelloy; b) Inconel; c) 300 stainless steels; d) 400 stainless steels, titanium; e) Zirconium, Monel; f) Nickel–silver (20 Ni); g) Lead; h) Brass (30 Zn); i) Aluminum alloys; j) Copper (pure); k) Carbon steel; l) Alloy steel; m) Iron (pure)

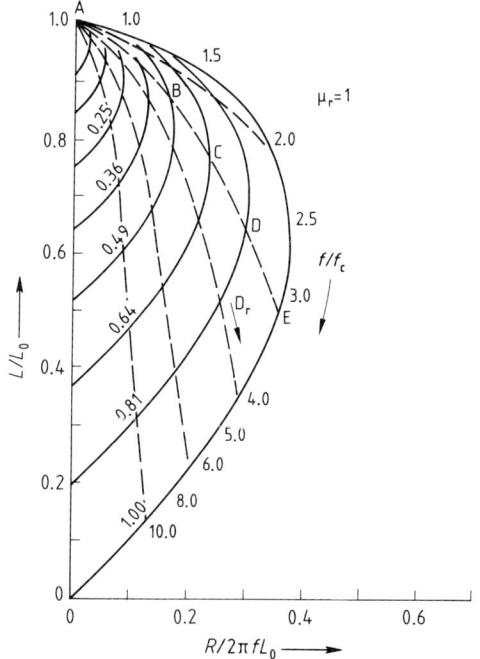

Figure 26. Normalized coil impedance plot of inductive versus resistive components for variations in frequency or conductivity (solid lines for fill factors of 1, 0.81, 0.64, ...) and diameter (dashed lines for $f/f_c = 2, 3, 4, 6,$ and 10) of a nonmagnetic solid cylinder

from A (no rod) to B (a fill factor or ratio of rod area to coil area of 0.49), C (0.64), and D (0.81) through E (1.0). Both the magnitude Z_c and the phase angle ($\tan^{-1} 2\pi f L/R$) vary with changing rod diameter D_r along the dashed lines. The value given for the curve of the unit fill factor is the ratio of the frequency f to the characteristic frequency f_c and is equal to

$$f/f_c = f\mu_r \sigma_e D_r^2 / 5066$$

where σ_e is given in $\mu\Omega^{-1}$ cm^{-1} and D_r in mm. Thus, when σ_e varies, Z_c changes along the solid curve. These variations in Z_c can be measured by detecting the voltage and current in the coil. This example gives the basis for an absolute (single) coil inspection method for variations in rod diameter. The same method can be used to inspect variations in alloy composition or heat treatment for constant-diameter rods and longitudinal flaws in such rods. These flaws lower the effective conductivity of the rod in the direction of the eddy currents. A basically identical method can inspect the diameter and thickness of tubing by means of an encircling coil.

For practical instruments, two coils are used in a bridge circuit to detect small differences. A reference rod is placed in one coil, and the other coil is used to inspect a rod or tubing. Another approach is to use one coil for excitation and the other for sensing. These coils are usually inductively coupled, but they are separated for the remote field eddy-current techniques. Differential sensing methods can also be used for the coupled-coil approach [27].

Inspection frequencies of 200 Hz to 6 MHz are used in eddy-current testing. Lower frequencies have greater penetration depth, but the sensitivity to flaws and speed of inspection decrease. For surface flaw inspection of nonferrous alloys, frequencies in the megahertz range can be used. Ferromagnetic materials require lower frequencies because of the poor penetration (high μ_r).

A single coil or a set of coils can be used as a probe to be placed over the flat surface of a conductor. This arrangement is used to measure conductivity by first calibrating the instrument with electrical conductivity standards of known values. The frequency is chosen so that the thickness of test pieces and standards is at least 2.6 times the skin depth. For conductivity measurements, the coil-to-metal distance (lift-off) is kept constant or compensated.

Changes in lift-off can be measured while the conductivity effect is kept negligible by controlling the frequency. This allows measurement of thickness for nonconductive coatings on conductive materials. The thickness of sheet materials can be measured by adjusting the frequency so that it is of the same order as the skin depth D_s.

The encircling coil technique passes a tube lengthwise through a coil energized at one or more frequencies (Fig. 27 A). The electrical impedance of the coil is modified by the proximity of the tube, the tube dimensions, electrical conductivity, saturating magnetic field, magnetic permeability, and metallurgical or mechanical discontinuities in the tube. Encircling coils are used for small-diameter tubing (< 8 cm) because the sensitivity to flaws deteriorates for larger diameter tubes.

A probe with one or more sensors in close proximity to a test piece is used in the probe technique (Fig. 27 B). The probe is generally small and examines only a limited area. For welded tubular products, only the weld is typically examined.

A coil can be placed inside a tube (Fig. 27 C). The internal coil can be concentric, the same type as an encircling coil, for exciting and sensing circumferential eddy currents. Two concentric coils

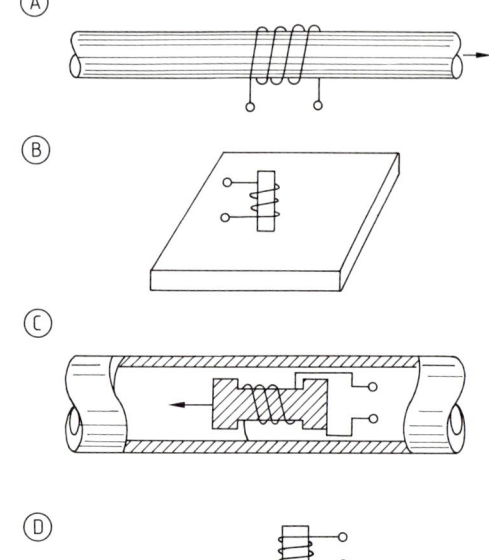

Figure 27. Examples of inspection coil arrangement
A) Encircling coil; B) Probe coil (with or without core); C) Inside coil (with or without core); D) Probe coil (rotating probe or tube)

can be used as an exciter and a sensor, each of which may be placed inside or outside. When one is outside and the other inside, this is a transmission technique and the sensor is affected only by those electromagnetic fields that have passed through the wall. The two coils can also be located inside or outside. This reflection technique is suited for detecting primarily inner or outer surface flaws.

A smaller probe can be used to scan helically the inner or outer surface of a tube. These rotating probe techniques are suitable for detecting corrosion and cracks in tubing (Fig. 27 D).

Reference standards for flaws are required to set eddy-current test conditions. Samples with actual flaws and those with drilled holes and machined notches of known size are employed. Notches are milled, filed, or electrodischarge machined on the outer or inner surface.

For magnetic materials, a saturation magnetic field is applied to reduce the permeability and its variation. For tubular products, an encircling saturation coil is employed. Permanent magnets and U-shaped electromagnets are used in other instances.

For eddy-current inspection of small surface flaws, various probe designs have been developed. These typically have a ferrite core and concentrate electromagnetic fields to a gap in the core. The gap is placed on the surface of a test piece, which increases the flaw detection sensitivity.

2.6.5. Microwave Inspection

Microwaves are electromagnetic radiation with frequencies of 300 MHz to 300 GHz. For nondestructive testing, frequencies of 1–20 GHz are generally employed. By measuring attenuation and reflection, this method can be used to analyze the moisture content of dielectric materials; to detect chemical changes, voids, and discontinuities in ceramics, polymers, and fiberglass-reinforced composites; and to gauge thickness. For conductive materials, skin depth is very low and most of the incident wave is reflected. Only reflection techniques can be used for metals. For nonconductive materials, transmission, reflection, and standing-wave techniques are used. The techniques can employ different modes of microwave generation. Frequency may be fixed, swept, or pulse modulated. Continuous waves or pulsed waves are generated [4 d].

Basic transmission techniques use a microwave generator, two antennas for transmitting and receiving, and a phase-sensitive detector or peak-amplitude detector. With the phase-sensitive detector, the in-phase and out-of-phase (quadrature) components can be determined for a fixed-frequency system. Attenuation and phase change may be used to determine the moisture content, dimensional changes, and property variation of dielectric materials. The presence of local anomalies can also be detected. By using a pulse-modulated transmission and a peak-amplitude detector, the velocity of propagation can be measured. This is useful in evaluating polymerization, esterification, distillation, vulcanization, and evaporation in polymers.

Reflection techniques use one antenna. When a metal backing exists or a reflector is placed behind a dielectric sample, property measurements similar to those made with transmission techniques can be achieved. Instruments for dielectric property determination are referred to as microwave dielectrometers. Both fixed-frequency and swept-frequency (also known as frequency-modulated, FM) reflection techniques are also used to detect delaminations (if a detectable air gap exists), inclusions, voids, and other three-dimensional anomalies in dielectric materials.

Since microwaves are reflected at an interface, standing waves are produced when parallel interfaces or reflectors exist; for example, the front and back faces of a plastic plate. The patterns of standing waves are used for distance measurements and thickness gauging.

Moisture measurements are perhaps the most widely used application of microwaves. Water has a high dielectric loss compared to most other dielectric materials, and its content can be determined by sensing dielectric constants with transmission or reflection techniques. Transmission techniques may be employed at high water levels (50–85%). Reflection measurements with a resonant cavity can achieve high sensitivity over a narrower range of moisture content.

Another use of microwaves occurs in eddy-current inspection. In contrast to the usual frequencies up to 10 MHz, a ceramic crystal resonator is used in a sensor, extending the frequency to several gigahertz. Resolution of surface flaws is greatly improved.

2.7. Thermal and Optical Methods

2.7.1. Thermal Inspection

Heat flow and temperature changes are detected to locate flaws and abnormal heat distribution. Temperature can be measured by many types of substances and devices such as heat-sensitive paints, papers, and phosphors; liquid crystals; thermocouples; thermistors; and bolometers. These have been used in limited applications because direct contact is required and only point-by-point measurements are possible with some methods. Massive changes in thermal conductivity from thermal fatigue cracking, delamination, and stress corrosion cracking may be detected by using coatings of liquid crystals and heat-sensitive substances and applying certain heat flow patterns [1].

Noncontact thermography utilizes IR radiation emitted by the surfaces of a test piece. The principle used is similar to radiation pyrometry, but modern thermographic cameras can obtain false-color images of the surface, with each color representing a narrow range of temperature having intervals as small as 0.1 °C. Infrared optics are also available to image microscopic areas. Thermographic cameras can be employed to evaluate the temperature distribution of a component in service. When an abnormal condition exists, such an examination can reveal it. Transient heat flow can also disclose an anomaly in thermal conductivity that may result from delamination and unbonding. Temperature changes on the surface can be induced by an external heat source. Higher temperatures are observed where the section thickness has been reduced by corrosion. This method is applicable to inspection of piping.

Another method of inducing a temperature differential is to apply mechanical vibration. This is known as vibrothermography [34] and can be applied to flaw detection in composite materials subjected to fatigue loading.

A scanning high-power laser can be combined with an IR detector. The unbonding of a coated layer can be revealed because the temperature of the coating at the irradiated unbonded spot rises more than that at well-bonded regions. Photoacoustic spectrometry employs modulated laser beams, irradiating mainly particulate matter. Thermal excitation leads to detectable emission of acoustic waves. This method has been used in evaluating ceramic powders and chemicals.

2.7.2. Visual and Optical Inspection

Evaluating the surface of a structure with the human eye is generally the first step in any nondestructive inspection [36]. Visual inspection is facilitated by using a magnifier, a borescope, a fiberscope, or a videoscope. Video technology has also become an integral part of this method for examination and recording of images. Borescopes and fiberscopes are used for remote viewing. Borescopes are low-magnification devices with a rigid tube support for optical lenses. Their length may be extensible. Fiberscopes are made with a bundle of flexible optical fibers and provide magnification up to 200 ×. These are available in diameters as small as 0.35 mm and are often combined with a video camera. Videoscopes employ a small image sensor, which is typically a charge-coupled device consisting of many light-sensitive elements. The image is sent electronically to a video monitor. Videoscopes provide higher resolution and greater depth of field than fiberscopes. Images from fiberscopes and videoscopes are displayed on a monitor screen and recorded on a video recorder.

Long-distance microscopes have a large working distance, up to a few hundred times that of conventional microscopes. This allows the examination of small objects 1–2 m away with 5-μm resolution.

By using a laser as the light source and scanning rapidly over a surface, transmitted, reflected or scattered light can be used to characterize the surface by imaging the flaws and discontinuities and by measuring the surface roughness. Diffraction of light scattered from certain kinds of surfaces (e.g., weave patterns) can be used to detect unacceptable surface patterns.

Optical interferometry is used to obtain surface profiles via interference fringes [38]. Noncontact displacement measurements can be made with a resolution of 10 nm. Two-dimensional surface profiles are also obtained by laser speckle reflectometry [37], which is employed mainly for measurement of roughness.

2.7.3. Optical Holography [35]

Optical holography utilizes coherent monochromatic light from a laser. First, both the amplitude and the phase of coherent light from a test piece are recorded on a film or plate along with the reference beam (Fig. 28 A). Three-dimensional images are formed from the holographic film or plate when it is irradiated with the same type of light (Fig. 28 B). The regenerated image can then be used as a three-dimensional template against which any deviations in the shape or dimensions of the test piece can be observed and measured. With no part of the optical system or test piece position changed, the test piece is irradiated with the same reference beam and an inspector views it through the hologram. Interference patterns are seen with optical fringes surrounding anomalies where the test piece has moved short distances. In a double-exposure technique, the surface of the test piece is slightly distorted between exposures, which are recorded on a single holographic film, thereby producing a holographic fringe pattern. Another method utilizes the interfering holographic image effects of vibrating test surfaces.

These methods are used to inspect brazed or adhesively bonded honeycomb-sandwich struc-

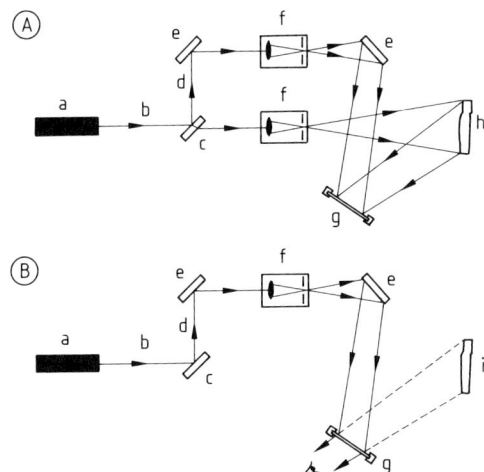

Figure 28. Basic holography setup
A) Recording hologram; B) Reconstructing image
a) Laser; b) Laser beam; c) Beam splitter; d) Reference beam; e) Mirrors; f) Spatial filters; g) Photographic plate; h) Test piece; i) Virtual image of test piece

tures, coatings, and tires for lack of bonding, delamination, and other flaws. They are also useful in vibration analysis of rotating and vibrating components.

2.8. Leak Testing

Leak testing detects the leakage of a liquid or gas from an internally pressurized vessel or the leakage of a gas into a vacuum vessel. The location of a leak, the amount of leakage, or both can be determined. A number of methods have been developed with different sensitivities and detection principles [1], [4a].

2.8.1. Pressurized Gas Leaks

Two approaches are used to detect leaks in pressurized gas systems. The amount of gas lost is determined by the measurement of weight or pressure. Both methods are insensitive, and leak location is not possible. If two vessels are available, a differential pressure gauge can be used to increase the sensitivity of the pressure-gauging method.

The other approach is to detect the escaping gas; many techniques are available for this. Acoustic methods rely on sound or ultrasound generated by turbulent flow through a leak. If the leak is large, audible sounds are emitted. Stethoscopes can be used for smaller leaks. Still smaller leaks can be detected with ultrasonic probes, which operate for airborne ultrasound in the range of 30–45 kHz. Airborne ultrasound can be detected from as far away as 30 m and works well for overhead gas pipelines. Parabolic sound collectors are useful. With low background noise, turbulent leaks on the order of $1 \text{ mPa} \cdot \text{m}^3 \text{ s}^{-1}$ can be detected. An acoustic emission sensor (30–100 kHz) can be coupled directly to the leaking vessel. By using multiple AE sensors, a leak can be located. This method is effective in leak detection especially in buried gas pipelines.

Bubble testing is applied to small vessels that can be immersed in a liquid. Oils and glycerol are more sensitive to leaks than water, and reduced pressure above the liquid is also helpful. Bubble-forming solutions are applied to the surface of larger or fixed-pressure vessels and piping. Soap solutions can be used for simple tasks, but specially formulated solutions have been developed for more consistent bubbling. Sensitivity on the order of $10^{-2} \text{ mPa} \cdot \text{m}^3 \text{ s}^{-1}$ and qualitative estimates of leak size are feasible.

Flow detection is applicable to a vessel if it can be placed in a larger enclosure with an orifice. The flow of leaking gas from the orifice is measured by using a flow meter. Leaks of $1 \text{ µPa} \cdot \text{m}^3 \text{ s}^{-1}$ can be detected.

Specific-gas detectors are used to respond to halogens, combustible and odorous gases, gases that cause chemical reactions, and certain other gases. The human nose detects very low levels of odorous gases but quickly becomes insensitive. A halogen-sensitive detector uses a diode that ionizes halogen atoms and detects the ionic current. Air is drawn into the diode through a sniffing probe. If a vessel does not contain halogens, a tracer gas containing halogen compounds can be mixed with the gas in the vessel for leak detection. Combustible gas and sulfur hexafluoride gas detectors are similar to the halogen detector. Common uses of the halogen and combustible gas detectors are for air-conditioner maintenance and in unventilated, enclosed areas where leaked gases accumulate. Leaked gas can react chemically with a coating or tape placed on the surface of a vessel to change its color. Ammonia is generally used for this method. Another type of gas detector works like a thermocouple vacuum gauge and a Pirani gauge [39]. Gases with a

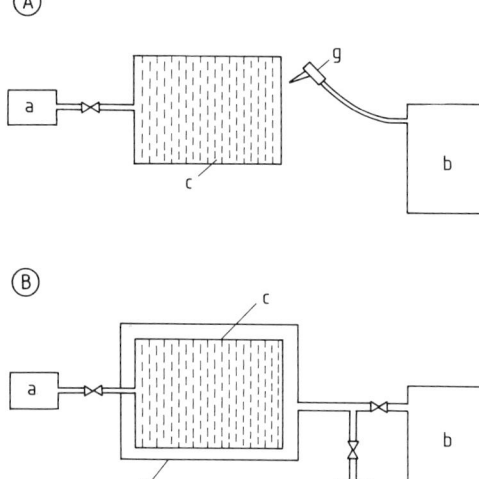

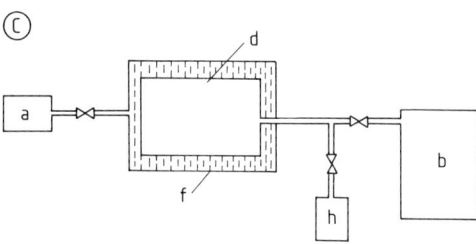

Figure 29. Helium leak testing
a) Helium source; b) Mass spectrometer; c) System under test (helium filled); d) System under test (evacuated); e) Evacuated enclosure; f) Helium-filled enclosure; g) Probe; h) Vacuum pump

different thermal conductivity (e.g., hydrogen) can be sensed relative to air by this method.

Mass spectrometers detect the distribution of atomic or molecular weights of different components of a gas. Helium mass spectrometers are most commonly used in leak detection. Either a sniffing probe is used to draw leaked gas from a helium-filled vessel (Fig. 29 A) or an evacuated enclosure may be used to collect the leakage from a helium-filled vessel (Fig. 29 B). When a vessel can be evacuated (even partially), it can be placed in an enclosure filled with helium (Fig. 29 C). This is the most sensitive method, and a sensitivity of $nPa \cdot m^3 s^{-1}$ is often achieved.

2.8.2. Pressurized Liquid Leaks

Liquid leaks are visually detectable if leak rates are high. Dye penetrants may be introduced into a vessel and procedures for penetrant testing applied on the outside of the vessel. As in the detection of chemically reactive gases, the vessel surface can be coated with pH-sensitive paint, paste or tape, and color changes can be monitored. Liquid level loss monitoring is simple and is used commonly for underground tanks. The liquid-containing tank can be pressurized and sealed, and the pressure drop monitored.

Acoustic methods are also useful when turbulent flow occurs. Acoustic emission techniques are used to detect leaks through valves, steam condensate lines, and flat-bottomed tanks. Metal and plastic pipelines are monitored for leaks by means of multiple acoustic emission sensors during hydrostatic testing.

2.8.3. Vacuum System Leaks

Vacuum systems are normally tested in the vacuum condition. Most vacuum-sealing systems are not suited for internal pressurization, and container walls may not withstand pressure reversal. Helium mass spectrometers are generally used for leak detection. Vacuum pressure gauges and halogen detectors may also be used. The vacuum system is connected to the helium detector, and a stream of helium gas is directed at a suspect area. To determine the total leak rate, the system is enclosed in a container filled with helium.

In testing high-vacuum systems, virtual leaks may be a substantial source of leakage. These involve gradual desorption of gases from surfaces or components within a vacuum system.

3. Uses

3.1. Metallic Pressure Vessels and Piping

Pressure vessels are constructed in various sizes and shapes, but invariably these are manufactured and inspected according to the American Society of Mechanical Engineers (ASME) Boiler and Pressure Vessel Code [5]. Section V of the ASME code establishes methods of nondestructive testing and covers visual, radiographic, ultrasonic, acoustic emission, magnetic particle, penetrant, and eddy-current methods.

The code specifies inspection methods and acceptance criteria for plates, forgings, and tubes, before and during fabrication, and for vessels and piping after fabrication. The most likely locations of flaws are at various weldments (see Section 3.3).

Acoustic emission testing is used to localize defective areas during initial hydrotesting. It is also employed for in-service inspection and continuous monitoring, in which process fluids are used for pressurization. Results indicate the locations of active AE sources, which are graded according to their severity. The severity of the source is classified according to the average energy of the most energetic emissions, energy release rates, number of AE events during a load hold period, number of events during a load increase, number of high-amplitude events, etc. These provide real-time indications of defective areas of the pressure vessel being tested and allow prevention of catastrophic failure of the vessel. The defective areas identified are then inspected by using other test methods. Acoustic emission testing is applied on various types of vessels, including ammonia spheres and tanks, hydroformer reactors, and catalytic cracking units [5, Section V], [28].

Surface examination for cracks, corrosion, and contamination relies mainly on visual methods, which are used for all welds not internally tested after a hydrotest. Magnetic particle and penetrant methods are also applied for crack detection. For surface inspection of vessel interiors, plant shutdown is required.

Generally, radiographic and ultrasonic methods are used for internal flaw detection. Both X- and gamma-ray sources are used in radiography. Access to the vessel interior is often difficult, which limits the use of radiography. For heavy steel sections (> 10 cm), field inspection becomes difficult because high-energy sources are needed. Ultrasonic methods, especially with shear waves, are widely used for weld inspection. Thickness measurement using the normal beam is commonly practiced for corrosion monitoring. Those areas having a high corrosion rate or history of attack are inspected thoroughly.

3.2. Composite Pressure Vessels and Piping

The performance of fiber-reinforced plastic vessels and piping has been poor, and many failures have been recorded. Apart from acoustic emission, no satisfactory test exists for determining the structural adequacy of fiber-reinforced plastic equipment. Radiography is difficult because resins have low absorption coefficients. Fiber-reinforced plastics are anisotropic, and ultrasonic attenuation is very high, making pulse-echo techniques hard to use. In addition, the various types of life-limiting flaws are quite different and some of them are not amenable to conventional ultrasonic inspection. Fiber fracture and fiber–matrix debonding are difficult to detect with ultrasonics.

The use of acoustic emission for testing fiber-reinforced plastic vessels and piping has been highly successful, and standard test procedures have been established [6, E 1067], [22], [29], [30]. Test vessels and piping are pressurized up to 150% of the maximum allowable working pressure. The procedures are designed to locate substantial flaws, which are then evaluated by other techniques such as ultrasonic testing (delamination) or visual (resin loss) and penetrant (matrix cracking) inspection. A vessel (or piping) to be tested can be new, in-service, or repaired, and different steps are specified for atmospheric, vacuum, and pressurized vessels. A vessel that has been in service must be preconditioned by reducing the operating pressure; for this, the maximum operating pressure within the previous year must be known. Acoustic emission instrumentation should have a sufficient number of channels to localize sources by using the zone location method; that is, many AE sensors are installed to completely cover the vessel with the corresponding zone marked around each sensor. Acoustic emission activities of flaws within each zone are detected, and the zone represents the approximate position of these flaws. High-frequency (100–200 kHz) sensors are used for zone location. Two or more low-frequency (25–75 kHz) sensors are used to evaluate the adequacy of coverage of the high-frequency sensors. If a low-frequency sensor detects acoustic emission whereas none of the high-frequency sensors do, the latter must be relocated. Sensors are positioned to detect structural flaws at critical sections of the test vessel, such as high-stress areas, geometrical discontinuities, nozzles, manways, repaired regions, support rings, and visible flaws.

Pressurization of a vessel during AE testing proceeds in steps, with pressure hold periods. For atmospheric vessels, the pressure is held at 50, 75, 87.5, and 100% of the test pressure. Pressure vessels are stressed with 10% increments, with depressure increments (also 10%) above

30 % of the test pressure. A test is terminated whenever a rapid increase in AE activity indicates an impending failure. Acoustic emission data from the high-frequency sensors are used for evaluation, whereas the low-frequency sensors generally detect acoustic emission from significant flaws. Detected flaws are graded according to using several criteria, including emissions during pressure hold periods, felicity ratio (the ratio of the load at the onset of significant emissions to the maximum prior pressure), total AE counts, high-amplitude events, and long-duration events. Emissions during hold indicate continuing permanent damage and lack of structural integrity. For in-service vessels, the felicity ratio criterion (when it is less than 0.95) is an important measure of previous damage. High-amplitude events indicate structural (fiber) damage, especially in new vessels. Long-duration events are characterized by measured area of the rectified signal envelope (MARSE), which is an indicator of combined signal and amplitude duration. Large MARSE values result from delamination, adhesive bond failure, and crack growth.

3.3. Weldments

Nondestructive testing methods used for completed fusion weldments include (1) visual inspection, (2) radiography, (3) ultrasonic pulse echo, (4) magnetic particle and leakage field testing, (5) liquid penetrant testing, (6) leak testing, and (7) acoustic emission testing.

For many noncritical welds, integrity is assured mainly by visual inspection to look for cracks, bead thickness, bead contour, undercut, overlap, and spatter. For critical welds, both faces and root surfaces are examined, especially for cracks, undercut, root penetration, and unfilled craters. These are tested further by radiography and ultrasonics for internal flaws.

Radiography is commonly used for detecting porosity, slag entrapment, and inclusions. A round or oval dark spot represents the image of a pore. Slag inclusions appear along the weld edge as irregular or continuous dark lines, whereas tungsten inclusions give rise to single or clustered light spots. Cracks are sometimes visible in radiographs as dark narrow irregular lines, but the lack of any radiographic image of cracks does not assure their absence. Radiography may be unable to detect incomplete fusion and incomplete penetration because of their small effects on X-ray absorption. They appear as very narrow dark lines.

Ultrasonic pulse-echo techniques are used effectively for the detection and location of planar defects such as laminations, unbonded areas, cracks, hidden surfaces, and other flaws. They are also used to reveal a lack of root penetration, porosity, and unbonded sidewalls of fusion zones. Ultrasonic testing and radiography are complementary techniques, and both are used for inspection of critical welds. When no access to the opposite side of the weldment is available, ultrasonic testing is the only option for internal flaw inspection.

Shear wave beams are generally used in weld inspection. To detect longitudinal flaws (along the weld), the search unit is moved along a zigzag scanning path either with sharp changes in direction or with right-angle changes. To detect transverse flaws in welds, the search unit is placed on the base metal surface at the edge of the weld. The sound beam is directed into the weld by angling the search unit at ca. 15°. The scanning of the search unit is parallel to the weld.

Magnetic particle testing and penetrant testing are used mainly for surface-breaking flaws (also for subsurface flaws in magnetic particle testing). These methods are relatively inexpensive but can reveal even fine cracks with clarity. In magnetic particle testing, the magnetic field is applied in two mutually perpendicular directions.

Leak testing of welded vessels uses a tracer gas under pressure or vacuum. Welds are tested for leak location or for leak rate. Acoustic emission tests for large-scale structures can narrow the areas of inspection by locating sources of acoustic emission. Typically, the structure is stressed by pressurization while AE sensors are mounted in arrays on the surface of the structure. When the applied stress exceeds the previously applied stress level, acoustic emission activities increase rapidly. The positions of such activities are identified by triangulation or the zone-location method. Often, the source of acoustic emission is localized to specific weld regions which are inspected further by ultrasonic and radiographic methods.

3.4. Other Uses

Polymeric materials are adversely affected by improper curing, inclusions, porosity, environmental degradation, machining and impact damage, and fretting. These are detected by specialized application of various nondestructive testing methods. Ultrasonic propagation characteristics are used to reveal anisotropy of elastic moduli, film and fiber orientation, structural relaxation, glass transition, degree of cross-linking, and state of cure in polymers.

Dielectric measurements are used to monitor the curing process [31]. In situ radio-frequency impedance is determined for this purpose. Microwave transmission and reflection methods are used for flaw detection (cracks, voids, inclusions) and for monitoring moisture, thickness, and cure. Optical techniques are employed to measure the state of mechanical strain in transparent polymers (photoelasticity). This is used widely in studying stress concentration effects in model components.

Thermographic techniques are useful in testing polymers because they have low thermal conductivities. Infrared spectroscopy provides information on molecular structure and has also been used to measure stress concentration effects near crack tips.

Radiography of polymers requires low-energy (ca. 20 kV) radiation because polymers have low absorbance. To enhance contrast, high atomic number penetrants are used. Test pieces are immersed in fluids such as tetrabromoethane and diiodobutane. This is only beneficial for surface-breaking flaws. Although difficult to use, neutron radiography is effective for polymers.

Paper and paperboard products require special testing methods during production. Radiation thickness gauging with beta emitters (^{85}Kr and ^{90}Sr) is used for the measurement of basis weight (areal weight). For determining ash in paper due to fillers and coating minerals, the attenuation and fluorescence of low-energy photons (4–5 keV) are used.

Various on-line optical methods are employed in determining opacity, surface roughness, color, and reflectance. Infrared absorption in the 1.95-μm region is used for moisture measurement. For heavyweight paper, moisture is measured by microwave methods.

The mechanical properties of a paper can be probed by means of ultrasonic wave propagation. An online caliper system obtains the thickness and ultrasonic propagation data and the results are correlated to the mechanical strength of the paper.

4. References

[1a] *Metals Handbook*, "Inspection and Quality Control," 8th ed., vol. 11, Amer. Soc. Metals, Metals Park, Ohio 1976.
[1b] *Metals Handbook*, "Nondestructive Evaluation and Quality Control," 9th ed., vol. 17, ASM International, Metals Park, Ohio 1989.
[2a] D. E. Bray, R. K. Stanley: *Nondestructive Evaluation*, McGraw-Hill, New York 1989.
[2b] R. Halmshaw: *Nondestructive Testing*, Arnold, London 1987.
[2c] *Nondestructive Testing*, "A Survey," NASA SP-5113, NASA, Washington D.C. 1973.
[3] *Nondestructive Testing Study Guides*, "Basic Magnetic Particle Testing, Penetrant Testing, Eddy-Current Testing and Radiography (1983)," General Dynamics, Convair Div., 1977–1983.
[4] *Nondestructive Testing Handbook*, 2nd ed., vol. 1, "Leak Testing," 1982; vol. 2, "Liquid Penetrant Tests," 1982; vol. 3, "Radiography," 1985; vol. 4, "Electromagnetic Methods," 1987; vol. 5, "Acoustic Emission Testing," 1987; vol. 6, "Magnetic Particle Testing," 1989, ASNT, Columbus, Ohio.
[5] ASME Boiler and Pressure Vessel Code, ASME, New York 1989.
[6] *1989 Annual Book of ASTM Standards*, vol. 3.03, "Nondestructive Testing," ASTM, Philadelphia 1989.
[7] *Redi-Reference Guide*, "An Issue of Materials Evaluation," *published yearly by* ASNT, Columbus, Ohio.
[8] R. S. Sharpe (ed.): *Research Techniques in Nondestructive Testing*, vol. 1 1970, vol. 2 1973, vol. 3 1977, vol. 4 1980, vol. 5 1981, vol. 6 1982, vol. 7 1984, vol. 8 1985, Academic Press, London.
[9a] D. O. Thomson, D. E. Chimenti (eds.): *Quantitative NDE*, vol. 1–7, (1982–1988) Plenum Publishing, New York.
[9b] *Proc. World Conference on Nondestructive Evaluation*, 11th, Las Vegas, Nov. 1985.
J. Boogard, G. M. van Dijk (eds.): *Non-Destructive Testing (Proc. World Conference on Nondestructive Evaluation)*, Elsevier, Amsterdam 1989.
[9c] *17th Symposium on Nondestructive Evaluation*, San Antonio, April 1989.
[9d] J. M. Farley, R. W. Nichols (eds.): *Nondestructive Testing*, vol. 1–4, Pergamon Press, Oxford 1988.
[10] Materials Evaluation; NDT International; Research in Nondestructive Evaluation; Brit. J. Nondestructive Testing; Soviet J. Nondestructive Testing; NTIAC Newsletter.
[11] F. Feigl, U. Anger: *Spot Tests in Inorganic Analysis*, 6th ed., Elsevier, Amsterdam 1972.
[12] G. Tschorn: *Spark Atlas of Steels*, Macmillan Publ. Co., New York 1963.
[13] A. R. Fee, R. Sagabache, E. L. Tabolski, *Metals Handbook*, "Hardness Testing," 9th ed., vol. 8, ASM International, Metals Park 1985, pp. 69–113.
[14] R. Halmshaw: *Industrial Radiography-Theory and Practice*, Applied Science Pub., Englewood, NJ 1982.
[15] R. A. Armistead, R. N. Yancey, *Mater. Eval.* **47** (1989) 487–491.
B. D. Hansche, *Mater. Eval.* **47** (1989) 741–745.
[16] J. Krautkrämer, H. Krautkrämer: *Ultrasonic Testing of Materials*, 3rd ed., Springer Verlag, Berlin 1983.
[17] M. G. Silk: *Ultrasonic Transducers for Nondestructive Testing*, A. Hilger, Bristol 1984.
[18] C. Broere et al., in [9d], vol. 4, pp. 2424–2432.

[19] Progress in Acoustic Emission, (a) I (1982), (b) II (1984), (c) III (1986), (d) IV (1988), Japan Soc. Non-Destructive Inspection, Tokyo.
[20] Many articles in *J. Acoustic Emission*, vols. 1–9 (1982–1990).
[21] R. K. Miller et al., *J. Acoustic Emission* **8** (1989) 25–29.
[22] R. Davies, in [19c], pp. 9–25.
[23] K. Göbbels, G. Ferrano, in [9d] vol. 4, p. 2763.
[24] W. Lord (ed.): *Electromagnetic Methods of Nondestructive Testing*, Gordon and Breach, New York 1985.
[25] D. C. Jiles, *NDT Int.* **21** (1988) 311–319.
[26] K. Ono, in [19c] pp. 200–212.
[27] V. S. Cecco, F. L. Sharp, *NDT Int.* **22** (1989) 217–221.
[28] T. J. Fowler, in [19c] pp. 150–162.
[29] *Recommended Practice for Acoustic Emission Testing of Fiberglass Reinforced Plastic Tanks/Vessels*, Soc. Plastics Industry, New York 1982.
[30] T. J. Fowler, *Chem. Proc.* (*Chicago*) March 1984, 24–27.
[31] F. I. Mopsik et al., *Mater. Eval.* **47** (1989) 448–453.
[32] C. A. Salvado, *Proc. Design and Manufacturing of Advanced Composites*, ASM International, Metals Park 1989, pp. 111–119.
[33] T. J. Fowler, J. A. Blessing, P. J. Conlisk, *AECM-3* (Third Int. Symp. Acoustic Emission from Composite Materials, Paris 1989), ASNT, Columbus 1989, pp. 16–27.
[34] S. S. Russell, E. G. Henneke, *NDT Int.* **17** (1984) 19–25.
[35] J. W. Wagner, in [16], pp. 405–431.
[36] R. C. Anderson, *Inspection of Methods: Visual Examination*, vol. 1, ASM, Metals Park, Ohio 1983.
[37] E. P. Chiang in C. P. Grover (ed.): *Optical Testing and Metrology*, vol. 661, SPIE International Soc. Optical Eng., Bellingham 1986, pp. 249–261.
[38] R. Jones, C. Wykes, *Holographic and Speckle Interferometry*, Cambridge University Press, Cambridge 1983.
[39] In [4a], pp. 346–368.

12. Information and Documentation

WENDY A. WARR, Information Services Section, ICI Pharmaceuticals, Macclesfield, United Kingdom (Chaps. 1–5, 7–11)

CLAUS SUHR, BASF Aktiengesellschaft, Ludwigshafen, Federal Republic of Germany (Chap. 6)

1.	The Scientific Journal	12-3
1.1.	The Primary Literature	12-3
1.2.	Trends in Scientific Publications	12-3
1.3.	Function of the Scientific Journal	12-4
1.4.	Quality and Prestige	12-4
1.5.	Document Delivery	12-5
1.6.	Translation	12-6
1.7.	Copyright	12-6
1.8.	Alternatives to the Conventional Journal	12-6
2.	Abstracting and Indexing Services	12-7
2.1.	Introduction	12-7
2.2.	Chemical Abstracts Service	12-8
2.2.1.	Statistics	12-8
2.2.2.	Document Analysis	12-8
2.2.3.	Indexes	12-9
2.2.4.	Chemical Nomenclature	12-9
2.2.5.	Computerization at CAS	12-9
2.2.6.	Other CAS Products and Services	12-10
2.3.	Institute for Scientific Information	12-10
2.3.1.	Science Citation Index	12-10
2.3.2.	Index Chemicus	12-11
2.3.3.	Current Chemical Reactions	12-11
2.4.	Other Abstracting and Indexing Services	12-11
2.5.	Current Awareness Services	12-11
2.6.	Future of Abstracting and Indexing Services	12-12
3.	Tertiary Literature	12-12
3.1.	Introduction	12-12
3.2.	Reviews	12-12
3.3.	Encyclopedias and Handbooks	12-13
3.3.1.	Encyclopedias	12-13
3.3.2.	The Handbuch Concept	12-13
3.3.3.	The Dictionary of Organic Compounds	12-15
3.4.	Sources Concerned with Chemical Reactions	12-16
3.5.	Other Reference Books	12-16
4.	Gray Literature	12-17
4.1.	Introduction	12-17
4.2.	Characteristics	12-17
4.3.	Organizations Specializing in Gray Literature	12-17
4.4.	Reports	12-18
4.5.	Official Publications	12-19
4.6.	Conference Proceedings	12-19
4.7.	Theses	12-19
5.	Business and Economic Information	12-20
5.1.	Introduction	12-20
5.2.	Company Information	12-20
5.3.	Products and Markets	12-22
5.3.1.	Product Information	12-22
5.3.2.	Market Information	12-23
5.4.	News Services	12-24
5.5.	Legal Information	12-26
5.6.	Economics and Finance	12-26
5.6.1.	Hard-Copy Sources	12-26
5.6.2.	Online Sources	12-27
6.	Patent Information	12-28
6.1.	Introduction	12-28
6.2.	The Volume of Patent Literature	12-28
6.3.	Origin and Significance of Patent Literature in Patent-Granting Procedures	12-28
6.4.	Content and Layout of Patent Documents	12-31
6.5.	Patent Gazettes, Patent Registers, and Other Literature	12-34
6.6.	Patent Abstracts	12-38

6.7.	Principles and Methods of Patent Information Management in the Chemical Industry	12-38	8.3.1.	Microcomputer Data Bases 12-59
			8.3.2.	Data Bases on CD-ROM 12-60
6.7.1.	Indexing and Retrieval of Bibliographic Patent Data 12-38		**8.4.**	**Records Management** 12-60
			8.4.1.	Microforms 12-60
6.7.2.	Indexing and Retrieval of the Technical Disclosure of Patent Documents 12-41		8.4.2.	Document Image Processing 12-61
			8.4.3.	Comparison of Micrographics and Optical Filing 12-62
6.7.3.	Collecting, Storing, and Making Patent Literature Available 12-42		8.4.4.	Information Retrieval 12-62
			9.	**Numeric and Factual Data Bases (Data Banks)** 12-63
6.8.	**The Role of Patent Information in Industry** 12-43		9.1.	Introduction 12-63
6.9.	**Organizational Aspects of Patent Information Management** 12-44		9.2.	Types of Data Banks and Data ... 12-64
			9.3.	Quality Control 12-64
7.	**Information Technology** 12-44		9.4.	Spectral Data Bases 12-65
			9.4.1.	Nuclear Magnetic Resonance (NMR) Spectroscopy 12-65
7.1.	Introduction 12-44			
7.2.	**Hardware** 12-45		9.4.2.	Infrared (IR) Spectral Data Bases 12-66
7.2.1.	Digital Computers 12-45			
7.2.2.	Computer Peripherals 12-45		9.4.3.	Mass Spectral Data Bases 12-66
7.2.3.	Data Storage 12-47		9.4.4.	Building Spectral Data Collections 12-67
7.2.4.	Microcomputers 12-49			
7.3.	**Software** 12-50		9.4.5.	Spectral Search Systems 12-68
7.3.1.	Systems Software and Application Software 12-50		9.5.	Crystallographic Data Bases 12-70
			9.5.1.	Molecular Sequence Data Banks . 12-71
7.3.2.	Programming Languages, Compilers, and Interpreters 12-50		9.6.	Chemical and Physical Property Data Bases 12-71
7.3.3.	Organization of Data 12-50		9.6.1.	Beilstein 12-71
7.3.4.	Information Retrieval Packages .. 12-51		9.6.2.	Other Handbooks and Encyclopedias Available Online .. 12-72
7.3.5.	Microcomputer Software 12-51			
7.3.6.	Software Engineering 12-52		9.6.3.	Data Banks on the CIS System .. 12-72
7.4.	**Telecommunications and Networks** 12-52		9.6.4.	Thermodynamics and Thermophysical Property Data Banks 12-72
7.5.	**Distributed Computing** 12-53			
			9.6.5.	Other Chemical and Physical Property Data Bases ...., 12-74
8.	**Records Management, Online Searching, and Information Retrieval** 12-53		9.7.	Toxicology, Hazard, and Environmental Data Banks 12-74
8.1.	Introduction 12-53		9.8.	Special Applications 12-76
8.2.	**Online Searching** 12-53		9.9.	The Future of Data Banks 12-76
8.2.1.	Introduction 12-53		**10.**	**Chemical Structure Handling** 12-77
8.2.2.	Equipment 12-53			
8.2.3.	Benefits and Problems 12-54		**10.1.**	**Chemical Structure Representation** 12-77
8.2.4.	Data-Base Producers and Vendors 12-55		10.1.1.	Systematic Nomenclature 12-77
			10.1.2.	Fragmentation Codes 12-77
8.2.5.	System Features and Search Strategies 12-55		10.1.3.	Linear Notations 12-78
			10.1.4.	Connection Tables 12-78
8.2.6.	Costs 12-56		**10.2.**	**Compound Registration** 12-79
8.2.7.	Gateways, Front Ends, and Microcomputer Software 12-56			
			10.3.	**Techniques of Substructure Searching** 12-79
8.2.8.	Full-Text Online Data Bases 12-57			
8.2.9.	Graphics Display 12-58		**10.4.**	**Current Research in Substructure Searching** 12-80
8.2.10.	Electronic Document Delivery ... 12-58			
8.2.11.	Videotex 12-59		10.4.1.	Generic Chemical Structure Handling 12-80
8.3.	**Public Data Bases for In-House Use** 12-59			

10.4.2.	Substructure Searching in Files of Three-Dimensional Structures	12-81	10.7.2.	Substructural Analysis and Data-Base Techniques ... 12-95
10.4.3.	Similarity Searching ... 12-83		10.7.3.	Molecular Modeling ... 12-96
10.4.4.	Use of Parallel Computer Hardware ... 12-83		10.8.	Reaction Indexing ... 12-96
			10.9.	Computer-Aided Synthesis Design . 12-98
10.5.	Operational Substructure Search Systems ... 12-84		11.	Artificial Intelligence ... 12-98
10.5.1.	History ... 12-84		11.1.	Introduction ... 12-98
10.5.2.	CAS ONLINE ... 12-85		11.2.	Machine Architectures and Neural Networks ... 12-98
10.5.3.	CIS/SANSS ... 12-86			
10.5.4.	Cambridge Structural Data Base (CSD) System ... 12-87		11.3.	Search, Problem Solving, and Planning ... 12-99
10.5.5.	DARC ... 12-87		11.4.	Theorem Proving and Logic Programming ... 12-99
10.5.6.	HTSS ... 12-89			
10.5.7.	MACCS ... 12-89		11.5.	Human–Computer Interaction .. 12-99
10.5.8.	OSAC ... 12-90		11.5.1.	Speech Simulation and Recognition ... 12-99
10.5.9.	Softron Substructure Search System (S4) ... 12-90			
10.5.10.	Proprietary Systems ... 12-90		11.5.2.	Natural Language Processing (NLP) ... 12-100
10.6.	Chemical Structure Software for Microcomputers ... 12-91		11.6.	Expert Systems ... 12-101
10.6.1.	Graphics Terminal Emulation ... 12-91		11.6.1.	Definition and Features ... 12-101
10.6.2.	Scientific Word Processing and Structure Drawing Software ... 12-91		11.6.2.	Knowledge Representation ... 12-101
10.6.3.	Structure Management Software . 12-92		11.6.3.	Knowledge Engineering ... 12-102
10.6.4.	Special Application Software ... 12-92		11.6.4.	Inference Engine ... 12-102
10.6.5.	Software for 3-D Molecular Graphics and Modeling ... 12-94		11.6.5.	Software and Hardware ... 12-102
			11.6.6.	Advantages and Disadvantages . 12-103
10.7.	Structure–Activity Relationships and Drug Design ... 12-94		11.6.7.	Applications ... 12-103
			11.7.	Hypermedia ... 12-105
10.7.1.	Statistical Approaches ... 12-95		12.	References ... 12-105

1. The Scientific Journal

1.1. The Primary Literature

The primary literature includes journal articles, patents, theses, reports, and conference papers. Patents are considered in Chapter 6. Theses, reports, conference papers, and related items belong to what is often known as the "'gray" literature (Chap. 4). The present chapter concerns original published papers in scientific journals.

To read the primary literature the scientist needs access to a library. WOLMAN has written a useful introduction to library organization for the user of chemical information [1.1]. There are several standard works on chemical information sources [1.2]–[1.8].

A number of specialized journals (e.g., *Scientometrics* and the *Journal of Documentation*) are devoted to studies of the primary and secondary literature. Books are also available about scientific publishing [1.9], [1.10]. The Primary Communications Research Center at Leicester (UK) was a center of expertise from 1976 until 1986 [1.11].

1.2. Trends in Scientific Publications

The first scientific journal, *Philosophical Transactions of the Royal Society*, was published in 1665. At the beginning of the nineteenth century there were about 100 scientific journals. According to Ulrich's International Periodicals Directory, 10 000 serials titles were published in 1951 and 71 000 in 1987. The highest growth rate occurred in the 1960s and was particularly evident for the sciences. Growth in most fields slowed in the 1980s. Chemical Abstracts now handles about 10 000 journals in chemistry or chemically related disciplines. The increase in the number of scientific papers included in Chemical

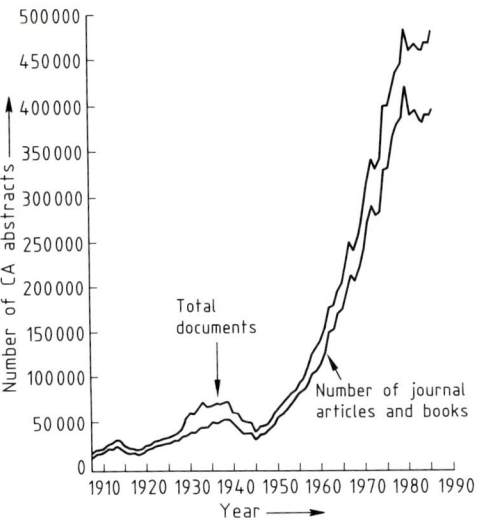

Figure 1. Number of scientific papers included in Chemical Abstracts from 1907 to 1986 (reproduced with permission from [1.12])

Table 1. Language of publication of journal literature abstracted in Chemical Abstracts, as percentage of total journal literature abstracted

Language	1961	1966	1972	1978	1983	1988
English	43.4	54.9	58.0	62.8	68.6	73.4
Russian	18.4	21.0	22.4	20.4	15.8	12.0
Japanese	6.3	3.1	3.9	4.7	4.4	4.1
German	12.3	7.1	5.5	5.0	3.5	3.3
Chinese	*	0.5	*	0.3	1.9	2.8
French	5.2	5.2	3.9	2.4	1.5	1.1
Polish	1.9	1.8	1.2	1.1	0.8	0.7
Spanish	0.6	0.5	0.6	0.7	0.5	0.5
Czech	1.9	0.9	0.6	0.5	0.4	0.3
Italian	2.4	2.1	0.8	0.6	0.5	0.3
Korean	*	*	0.2	0.2	0.3	0.3
Others	7.7	2.9	2.9	1.3	1.8	1.2

* Included in "Others" for year.

Abstracts between 1907 and 1986 is shown in Figure 1.

As the scope and volume of chemistry have increased, some journals have had to divide into new sections and many new specialized journals have appeared. The exponential growth in journal size is leveling off.

Scientific journals suffer from ever-increasing production costs and, consequently, higher subscription rates. The increasing number of journals and their increased subscription costs cause libraries to give serious consideration to which subscriptions they should discontinue.

There has also been a noticeable trend in language of publication. Germany was the leading scientific nation up to the beginning of the twentieth century but since then the United States has achieved dominance of the scientific literature. Table 1 shows the rise in the use of English since 1961.

1.3. Function of the Scientific Journal

The journal serves as a foundation for the advancement of science. Its main functions are archival storage, current awareness, quality control, and author recognition [1.13], [1.14].

Until the mid-nineteenth century, journals were archival in character, recording the text of papers presented oraly at scientific meetings. Scientists started to rely more upon journals rather than meetings for their information and journals began to develop an alerting or current awareness function.

Nowadays, journal scanning is still useful for current awareness but probably not as useful as "the invisible college"—conferences, seminars, informal visits, letters, and telephone calls. Formal papers are not written until the completion of a project, and publication and abstracting delays diminish the current awareness value of the published article. Publication delays of up to 8 months or more are inevitable because of the quality control process. Authors submit papers and the editor sends them to referees for assessment. Quality control is discussed further in Section 1.4.

1.4. Quality and Prestige

The range of journals in circulation provides a subtle system of quality control. A reviewer may reject an article for one journal but accept it for a "lesser" journal. Not all journals are "learned"; there is a place also for "news" journals such as *Chemical and Engineering News* or *Chemistry in Britain*.

Articles in learned journals are of four types: full paper, note, communication, or review. A note is shorter than a full paper and describes more limited findings. A communication is more urgent and provides a preliminary report of important results. Reviews are considered in Section 3.2.

Broadly speaking, journals with a high rejection rate (by the editor and/or the peer reviewers) have high prestige. Interdisciplinary primary journals such as *Science*, *Nature*, or *Experientia* are likely to have a particularly high rejection rate: over 90% for *Science* compared with about 45% for the *Journal of Chemical Information and Computer Sciences*.

Due to increasing costs and number of publications, librarians use various criteria for selection of journals, for example citation analysis of the primary and secondary literature, and journal-use pattern studies [1.15].

The theory of *citation analysis of the primary literature* [1.16]–[1.19] is as follows. When researchers cite a journal article, it indicates that this article has influenced them. The more frequently that a journal is cited, the more often the scientific community indicates the influence or impact of that journal. Science Citation Index (see Section 2.3.1) has a data base of literature citations and it produces an annual analysis called the *Journal of Citation Reports* [1.20]. In 1987 eight chemical journals were listed amongst the 30 most cited journals:

> Journal of Biological Chemistry
> Journal of the American Chemical Society
> Journal of Chemical Physics
> Biochimica et Biophysica Acta
> Biochemistry
> Biochemical Journal
> Journal of Organic Chemistry
> Journal of Physical Chemistry.

Dangers in rating the prestige or quality of a journal purely on its rank in such a list are discussed in [1.19]. New journals, journals that subdivide, and journals that change title are at a disadvantage in citation analysis ranking. Some important journals are often read but little cited. Journal reputation, circulation, availability, and the extent of library holdings and coverage by secondary services, all have an impact on the citation frequency.

Ranked lists of journals can also be produced according to the frequency of their *citations in the secondary literature*. The Chemical Abstracts Service Source Index [1.21] lists the most frequently cited journals in Chemical Abstracts. As in the Science Citation list, the *Journal of Biological Chemistry* is first (1988), *Biochimica et Biophysica Acta* is second, and the *Journal of the American Chemical Society* fourth. Such rankings have certain merits compared with those produced from citation analysis of the primary literature. Recently published journals have a better chance of finding a higher place as do popular journals, such as *New Scientist*.

Many of the most prestigious journals are produced by learned societies but there are also several highly regarded commercial publishers.

There is an unproven theory that higher prestige in scientific journals goes hand in hand with low readability [1.22]. There are many formulae for measuring readability [1.23], [1.24].

1.5. Document Delivery

If the article a scientist wants is not available in the library, he may request a photocopy from another library, use a commercial service to obtain a copy, or write to one of the article's authors to obtain a reprint. Many scientists have access to a library that can arrange an interlibrary loan.

The developing nations [1.25], [1.26] and the Eastern Bloc [1.27] have particular problems because their libraries do not stock many journal titles and obtaining copies of articles may be costly and difficult. Writing to an author for a reprint is the cheapest option.

Many guides or periodicals enable users to locate primary journals and check titles and publication data. In North America the first sources to consult are the Union List of Serials in Libraries of the United States and Canada, or Chemical Abstracts Service Source Index [1.21]. Other countries have their own national sources (e.g., the British Library).

The use of optical storage media, electronic mail, and telefacsimile opens the way for the development of large-scale automated document delivery systems [1.28]–[1.31]. See also Section 8.2.10.

Libraries and information centers which operate document supply services include:

> The British Library Document Supply Center
> Colorado Technical Reference Center
> Delft University of Technology Library
> Technical Research Center of Finland
> Centre National de la Recherche Scientifique (France)

The cost of these services is beyond the reach of most developing countries. A cooperative solution has been proposed for the International Information System for the Agricultural Sciences and Technology, AGRIS [1.26].

Commercial services for document delivery are operated by Chemical Abstracts Service (their Document Delivery Services, CASDDS), by the Institute for Scientific Information (The Genuine Article) and by the University Microfilms International Article Clearing House. Some private organizations also offer services (e.g., INFO from Information on Demand Inc. of Berkeley, California).

1.6. Translation

English is the dominant language of scientific publishing (see Table 1) and many publications appear totally or partially in English even though that is not the native language of the publisher's country. *Chemische Berichte* and *Angewandte Chemie* are notable examples. Nevertheless large numbers of articles are published in other languages, and scientists may require English translations. Non-English speaking scientists also require translations of articles published in English.

The scientist should first try to find an abstract either in the original journal (many journals produce abstracts in more than one language) or from an abstracting service (see Chap. 2). If the abstract looks interesting, a translation of the whole article may be needed.

Many journals, especially those in Russian, are translated into English cover to cover. The Chemical Abstracts Service Source Index [1.21] and the indexes produced by national translation centers are convenient ways of identifying and locating such journals.

In most developed countries translation centers collect translations into their native language, e.g., the National Translation Center at the John Crerar Library in Chicago, USA and the Centre National de la Recherche Scientifique, CNRS, France. The International Translation Center (ITC) in Delft, The Netherlands, is a translation clearing house which was originally set up to translate "difficult" languages into Western European ones but which now also handles translations of European languages. It prints the World Index of Scientific Translation. World Translations Index, a file produced by CNRS and ITC, is available online on DIALOG.

If a translation of an article is not available, a near equivalent may possibly be found, for example an article in another language by the same author, a review article, or a book. If, however, a ready-made translation or an equivalent cannot be found, the scientist will have to use a dictionary [1.1], ask a colleague to do a translation, or pay for a custom-made translation. The use of computers in translating is considered in Section 11.5.2.

1.7. Copyright

In many countries "writings" are protected by copyright law e.g., the Copyright, Designs and Patents Act 1988 in the United Kingdom. In the United States the Copyright Law enacted in October 1976 became effective on January 1, 1978 and will be described here as an example.

For works created on or after January 1, 1978, copyright lasts for the life of the author plus 50 years.

Copyright is not restricted to the written word. Composers, artists, and computer programmers are also protected. A "writing", unlike a patentable invention, need not be novel. The copyright owner has the exclusive right to reproduce the work, prepare derivative works, distribute copies and, in the case of musical and dramatic works and the like, to perform and display the work publicly.

There are "fair-use" exceptions to these exclusive rights. For example, copies may be taken for teaching, scholarship, and research provided certain rules are obeyed. Libraries and archives can make and distribute a single copy of any work if they do so without commercial advantage and conform to certain restrictions.

The law also covers copyright transfer. Nowadays most publishers of scientific papers require copyright transfer before publication. They can then more easily grant permissions to reproduce, make text available in full text form for online searching, and so on.

Useful summaries of the law and its impact have been published [1.32]–[1.34]. The Copyright Clearance Center operates a scheme of collection of authorization fees for photocopying copyright publications [1.35].

Copyright statute revision has always lagged behind technology developments. New guidelines are needed to protect the rights of creators and copyright owners, while allowing scientific users to harness the advantages of new technologies in electronic document replication and delivery.

1.8. Alternatives to the Conventional Journal

A tendency to conservatism exists in the domain of primary communication. The price of inappropriate innovation is high for publishers and editors. Authors wish to perpetuate the system on which their prestige depends. The majority of journals are still, therefore, printed on paper, but other methods of information transfer have been investigated [1.36].

Conventional journals can be reproduced in microform but most such journals have corresponding printed editions. Microform versions can be useful for libraries short of shelf space.

For many years there have been discussions about the value of synopsis or abstract journals

[1.37], [1.38]. The current awareness function is supposed to be served by the synopsis while the archiving function is served by the filing of microfiche or separates. The *Journal of Chemical Research* appears as a synopsis journal plus a microfiche and a miniprint version.

The production of supplementary material to published journals is much more common; experimental detail, crystallographic structure data, and computer programs, are deposited separately in a repository. The United States has had a national depository (the National Auxiliary Publications Service) since 1937. The British Library Document Supply Center has had a similar scheme since 1969. In the Soviet Union, VINITI (the All-Union Institute for Scientific and Technical Information) runs a deposition scheme. The Royal Society of Chemistry and the American Chemical Society also store supplementary material for certain journals.

Computers have also been introduced into publishing [1.39]. "Electronic publishing" refers to the use of the computer in the production of publications and also in the electronic distribution of text via computer terminals. The use of a generalized mark-up format for electronic publishing of a hard-copy journal can facilitate conversion of data into an online full-text data base [1.40]. The Chemical Journals Online (CJO) service on Scientific and Technical Information Network (STN) International contains the full text of the American Chemical Society's primary research journals (CJACS), the Royal Society of Chemistry's ten primary journals (CJRSC), J. Wiley and Son's five primary polymer journals (CJWILEY), the Association of Official Analytical Chemists' primary journal (CJAOAC), Elsevier Journals (CJELSEVIER) and VCH's international edition of *Angewandte Chemie* (CJVCH) [1.41]. Full-text data bases are considered in Section 8.2.8.

Pergamon's *Tetrahedron Computer Methodology* claims to be the first scientific journal to be simultaneously published in hard copy and in electronic form (floppy disks).

2. Abstracting and Indexing Services

2.1. Introduction

Scholars have always been concerned with the speedy identification of, location of, and access to the contents of learned journals. The need for devices to aid them increased with the explosive growth of the primary literature [2.1]. Abstracting and indexing services (sometimes known as secondary information services, or the secondary literature) are designed to help scholars keep abreast of the primary literature. It is estimated that more than 1500 abstracting–indexing publications worldwide are concerned with learned publication.

Table 2. Abstract journals in chemistry

Journal	Year published
Chemisches Zentralblatt	1830–1969
Chemical Abstracts	1907–present
British Abstracts	1926–1953
Nippon Kagaku Soran, Second Series*	1927–1974
Bulletin Signalétique	1940–1983
Referativnyi Zhurnal, Khimiya	1953–present
Current Abstracts of Chemistry and Index Chemicus	1960–present
Chemischer Informationsdienst (now ChemInform)	1970–present
Kagaku Gijutsu Bunken Sokuho, Kagaku, Kagakukogyo Hen*	1974–present

* Nippon Kagaku Soran changed to Kagaku Gijutsu Bunken Sokuho, Kagaku, Kagakukogyo Hen (Current Bibliography on Science and Technology, Chemistry and Chemical Engineering) in 1974.

The major abstracts journals in chemistry, many of which are now discontinued, are listed in Table 2. (Patent services are considered in Chap. 6.)

Various common-interest groups in secondary information have emerged in the United States including the International Council of Scientific Unions Abstracting Board (ICSUAB), the National Federation of Abstracting and Indexing Services (NFAIS), the Information Industry Association (IIA), and the Association of Information and Dissemination Centers (ASIDIC) [2.2]. In Europe, the European Association of Information Services (EUSIDIC) and the European Information Industry Association (EIIA) have more generalized interests than abstracting and indexing. A Global Alliance of Information Industry Associations (GAIIA) was set up in 1989.

Abstracts originally served an alerting function: they gave scholars a quick and easy method of keeping up with the rapidly growing literature. Abstracts are now an indispensable filtering tool in retrospective access [2.3]. The scientist who has located an article in a printed index, or from an online search, can scan an "informative" abstract to decide whether it is worth reading the original article in full. An informative abstract is defined by the American National Standards Institute (ANSI) as follows [2.4]:

"A well-prepared abstract enables readers to identify the basic content of a document quickly and accurately, to determine its relevance to their interests, and thus to decide whether they need to read the document in its entirety."

2.2. Chemical Abstracts Service
[2.5], [2.6]

The American Chemical Society started to publish *Chemical Abstracts* (CA) in 1907, to make the results of chemical research throughout the world accessible to scientists in the United States. At this time more publications appeared in German than in English, and American chemists were dissatisfied with coverage of the American literature in European abstracting journals. What began as a national organization has become the international *Chemical Abstracts Service* (CAS) based in Columbus, Ohio. The American Chemical Society gave CAS the mission of abstracting the complete world chemistry literature. The history of abstracting at CAS is bound up with the definitions of "complete", "abstract," and "chemistry" [2.7]. The motto of Editorial Operations at CAS today is "quality, comprehensiveness, and timeliness".

CAS employs nearly 1500 people, over a third of them chemists. Highly trained staff analyze and input information, and computers assist in editing and reviewing prior to publication.

Over 1.5×10^6 documents are examined every year in the fields of chemistry, chemical engineering, and related sciences.

The median currency ("up-to-dateness") of all abstracts in CA, measured from the publication date of the original article to the date the abstract was published in CA, is about 90 days. For the nearly 800 journals covered by Chemical Titles (see Section 2.2.5), the median is less than 2 months.

2.2.1. Statistics

Just under 12 000 abstracts were published by CA in 1907; in 1989, nearly 490 000 were published and the cumulative number of abstracts had risen to nearly 12.75×10^6. It took CA 30 years to publish the first 1×10^6 abstracts.

Nowadays nearly 1×10^6 abstracts are added every two years. Over 1.5×10^6 documents are examined by CAS each year, including patent documents, conference proceedings, government reports, books, and 10 000 scientific journals. The documents come from more than 150 countries and are written in more than 50 languages.

By the end of 1989 nearly 10×10^6 chemical substances were known to the CAS Registry System (see Sections 2.2.2 and 2.2.5); over 600 000 new compounds were registered in 1989.

The eleventh collective index (see Section 2.2.3) is reputedly the world's largest index with 28×10^6 entries.

2.2.2. Document Analysis

In the early days CAS used large numbers of volunteer abstractors. Nowadays more than 95 % of the abstracts are prepared in-house. The Japan Association for International Chemical Information (JAICI) abstracted and indexed over 13 000 Japanese patent documents for CAS in 1988 and the Fachinformationszentrum Chemie (FIZ Chemie) in Germany abstracted and indexed more than 4000 German-language documents. CAS document and structure analysts have expertise in a wide range of scientific disciplines and most of them have skills in languages other than English.

A document analyst starts by selecting items from the document and translating titles into English. Editorial Co-ordination Services flag key bibliographic information for data entry. This data is entered and then checked by a second keyboard operator. Each source document and an abstract preparation sheet are then sent to a document analyst, who prepares an informative English-language abstract and generates index entries and keywords. Documents are batched up and converted into computer-readable form.

At this point it is necessary to index, or "register" chemical structures into the CAS Registry System. CAS assigns a Registry Number of the form [nnnnnn-nn-n] (where n is a one-digit number) to each unique compound. In order to check whether a compound is novel its name is first matched against the Registry Nomenclature File. If no name-match is found, either the name is a new synonym for a known compound (and will need adding to the Registry Nomenclature File) or the compound is novel. A registry structure sheet is generated for compounds that are not successfully name-matched or that have multiple index names. These structure sheets are sent to an analyst who resolves the name-match hits and draws structures for new compounds.

Another analyst then edits the abstract and index work units online. Batches of data without structure sheets are released for publication; structure sheets are sent on to the structure input process.

Each structure diagram is converted into computer-readable form. The CAS Chemical Registry System converts the machine structure record into a unique connection table by means of the Morgan algorithm [2.8]. The connection table (discussed further in Section 10.1.4) can be matched against the 10×10^6 other substances in the structure file. If a match is found, the previously assigned Registry Number is retrieved. If no match is found, a new Registry Number is automatically assigned by the computer and its record is added to the file.

Registry editors then resolve inconsistencies before new structure registration sheets are sent on to nomenclature specialists. Staff assign unique and unambiguous index names to each new substance (see Section 2.2.4). These names are keyed into the computer and added to the CAS ONLINE Registry File.

At the next stage, the computer selects appropriate material for a particular publication or service and converts it into the proper format. A final quality check is made before pages of CA and its indexes are dispatched for printing and distribution. Other CAS services are produced in-house. The data from CA and its indexes are also added to STN International for online searching.

2.2.3. Indexes

Documents are abstracted and indexed according to five main subject areas and 80 sections. CA indexes were published annually until 1962 and thereafter semiannually. The CA Collective Indexes combine the contents of ten individual volume indexes into single, organized listings [2.6], [2.9]. The Eleventh Collective Index contains author, chemical substance, general subject, patent, and formula indexes; an index of ring systems; and an Index Guide. The Index Guide explains the use of the indexes and gives an extensive summary of CA nomenclature.

2.2.4. Chemical Nomenclature [2.10]–[2.12]

There are many different types of nomenclature, each having advantages and disadvantages for certain purposes. CA nomenclature is designed for use in a large printed index where each name must be unique and unambiguous [2.13].

The International Union of Pure and Applied Chemistry (IUPAC) is responsible for an international standard in chemical nomenclature [2.14]–[2.17]. One substance may, however, have several IUPAC names. CA names conform with IUPAC principles but CA has had to make additional rules to achieve uniqueness.

With the publication of the Ninth Collective Index, CA simplified the nomenclature system and made it more systematic [2.18]. The systematic nomenclature has, however, some disadvantages: names tend to be longer and relatively unfamiliar names are used for familiar compounds, e.g., benzenamine for aniline.

Fully systematic nomenclature should allow a computer to generate chemical names for chemical structures or vice versa. The introduction of computers to the CAS system in the early 1960s led to the development of well-defined translation procedures for converting CAS systematic names to atom–bond connection tables [2.19]. Computer programs were used to validate the CAS index names in early records and to convert them to a common internal form for storage [2.20]. These routines have since been extended to provide editing and verification facilities for new index names [2.21].

The University of Hull (UK) has developed microcomputer software to convert chemical names into connection tables, from which chemical structures can subsequently be displayed [2.22]–[2.24].

In 1984–1985 the Beilstein Institute developed a program, VICA, to convert German chemical names into structures on a mainframe IBM computer for the first version of the Beilstein Online data base (see Section 9.6.1) [2.25]. The program is being developed to handle English names. The Beilstein Institute is also developing a microcomputer program, AUTONOM, which will convert structures to names [2.25]. At the moment neither VICA nor AUTONOM handles stereochemistry.

2.2.5. Computerization at CAS

The world's first computer-produced periodical, Chemical Titles, was introduced by CAS in 1961 and covers ca. 750 chemically oriented journals. Each article is indexed by author and keyword with a reference to the journal in which it appears and each is subsequently reported in CA.

The bibliographic CA data base has been compiled since 1967 and is accessible for online searches (see Section 8.2.4).

Since 1965 all chemical substances mentioned in the literature have been recorded according to the CA Registry System [2.26]–[2.36]. Pre-1965 data are now being gradually added. A two-dimensional structural formula, stereochemical descriptor, all labeled atoms, and unusual valences are recorded for each registry number. In addition, the molecular formula, the

CA Index Name, and all known trivial and trade names are linked to the CA Registry Number.

The Registry File can be used to ascertain whether a substance with a certain name has already been mentioned under the same or another name. About 6000 substances are checked by CA every day.

In the mid-1980s CAS began a Registry Enhancements project to determine how the Registry System might be improved in response to user demand. Important aspects of the project are changes in the areas of polymers, biological macromolecules, stereochemistry, coordination compounds, and inorganic compounds and materials [2.37].

2.2.6. Other CAS Products and Services

The Ring Systems Handbook [2.6] which first appeared in 1984, contains information about all ring and cage systems in the CAS Chemical Registry System. The Registry Handbook – Common Names on microform lists synonymous common names and CA Registry Numbers. CAS run a search service on the CAS data base and on a biosequence data base on request.

Chemical Industry Notes, both computer-readable and in hard copy, is a weekly journal of extracts from about 100 leading industry and trade journals. The extracts are grouped into eight sections: production, pricing, sales, facilities, products and processes, corporate activities, government activities, and people.

To provide convenient international access to data bases in a wide range of scientific and technical areas, CAS has joined with FIZ Karlsruhe in the Federal Republic of Germany, and the Japan Information Center of Science and Technology (JICST), to operate the Scientific and Technical International Network known as STN International. Some of the online data bases on this service are considered in Chapter 8. CAS also support microcomputer software to help users access some of these data bases (see Section 10.6.4). CAS and FIZ Karlsruhe are also cooperating in the development of a numeric data service. Some numeric data bases are already available on STN International (see Sections 9.4.1, 9.6.1, and 9.6.4).

CAS provide support for registration of substances in, and addition of CAS Registry Numbers to, a number of non-CAS data bases, for example BioSciences Information Service's BIOSIS Previews File. CAS have also applied their expertise in software development to the automation of patent processing in the U.S. Patent and Trademark Office.

2.3. Institute for Scientific Information

The Institute for Scientific Information (ISI) is a for-profit organization, founded in 1960. It has about 650 employees worldwide but is based in Philadelphia. ISI is best known in the field of citation indexing but it also produces other important abstracting and indexing publications and services. ISI's current awareness products are covered in Section 2.5.

2.3.1. Science Citation Index

An article in a learned journal traditionally cites references to other related articles. For the purpose of citation indexing, the references are *cited articles* and the article citing them is the *citing article*. A *citation index* is an ordered list of cited articles each of which is accompanied by a list of citing articles. The citing article is identified by a source citation, the cited article by a reference citation. The index is arranged by reference citations. Any source citation may subsequently become a reference citation.

Use of a citation index enables a searcher to locate a chain or network of related articles, starting from one literature reference, without the need to use keywords, subject terms, or chemical names. Scientists can determine which subsequent papers have cited a particular reference and can thus move both backwards and forwards in time in a literature search.

Science Citation Index [2.38], [2.39] was first published in 1961; it is now published six times a year and cumulated annually. There are also five- and ten-year cumulations. The Citation Index is a series of indexes and is now available in printed form, as an online data base, and as a compact disc (see Chap. 8). It covers annually about 3500 journals and a few hundred nonjournal titles from more than 100 disciplines, including chemistry. The Citation Index lists cited documents alphabetically by name of the first author. The Source Index is alphabetically arranged by author with a separate index by organization. The Permuterm Subject Index is based on the original words in the titles of items covered. All significant words in a title are coupled together.

The theory of citation analysis is briefly discussed in Section 1.4. Another extension of citation indexing, cocitation clustering, is now also used for automatic hierarchical classification and mapping of the literature [2.39].

2.3.2. Index Chemicus

Index Chemicus (between 1970 and 1986 called Current Abstracts of Chemistry and Index Chemicus) is a weekly abstracting and indexing service which has been published since 1960. It details new organic compounds and syntheses reported in about 110 of the journals most important to organic chemists. It is claimed that over 90% of all new organic compounds appear in this small selection of journals. Index Chemicus endeavors to print an abstract less than 45 days after publication of the original article. Unfortunately patents are not covered. The strong points of Index Chemicus are currency and the easy-to-read, highly graphic abstracts that include structural diagrams and reaction flows. The ISI accession number permits ordering of the original article through ISI's The Genuine Article service (see Section 1.5). The indexes to Index Chemicus are cumulated quarterly and annually and include a permuted (rotated) molecular formula index.

2.3.3. Current Chemical Reactions

This ISI tool was first published in 1979. It prints flow diagrams of new and newly modified reactions and syntheses, with abstracts and bibliographic details, abstracted from over 120 journals and some books. About 300 000 reactions are reported annually.

Current Chemical Reactions is also available as an in-house computer-readable data base for use with Molecular Design Limited's REACCS software (see Section 10.8).

2.4. Other Abstracting and Indexing Services

The abstract journals listed in Table 2 (p. 12-7), even those now discontinued, are useful in addition to CA because they cover time periods CAS does not and they have a different scope.

ChemInform (formerly Chemischer Informationsdienst) has been produced since 1970 by FIZ Chemie and Bayer. It is an abstracting journal that mainly covers publications on new reactions and synthetic methods from about 250 major journals. It is published weekly, since 1987 fully in English.

Online data bases produced by abstracting and indexing services are described in Chapter 8. For further details of the hard-copy (and online) products, see [2.6], [2.10], [2.40]–[2.45]. A large directory of abstracting and indexing services is available [2.46]. Important services are listed below:

The Royal Society of Chemistry produces a number of current awareness bulletins (Section 2.5).

In the field of agricultural science CAB International, previously called the Commonwealth Agricultural Bureaux, provides an outstanding abstracting and indexing service.

The National Library of Medicine in the United States and Excerpta Medica in Europe are important sources of medicinal chemical information.

The Rubber and Plastics Research Association (RAPRA) produces abstracts, indexes, and an online data base.

BioSciences Information Service (BIOSIS) produces Biological Abstracts and other services.

Engineering Information produces Engineering Index and its computer-readable form COMPENDEX.

Information Service for the Physics and Engineering Communities (INSPEC) is a product of the Institution of Electrical Engineers in the United Kingdom.

The Deutsche Gesellschaft für chemisches Apparatewesen (DECHEMA) abstracts and indexes about 400 engineering journals.

Cambridge Scientific Abstracts are a large for-profit publisher of abstracts in several fields.

Chemical reactions are covered by Current Chemical Reactions (see Section 2.3.3), Methods in Organic Synthesis from the Royal Society of Chemistry (see Section 2.5), and the compendia described in Section 3.4.

2.5. Current Awareness Services

Many of the vendors mentioned earlier in this chapter also produce current awareness services. Indeed some of the hard-copy products such as Current Chemical Reactions are used by scientists for current awareness.

Selective dissemination of information (SDI) refers to those activities that help researchers stay up to date with the literature on a specific topic. A search strategy used for an online data base search can be saved and then run periodically against updated parts of the data base. Most search systems can also issue a series of commands which automatically run the strategy against each new update of one or more data bases and have printouts made on a regular basis. Such a stored strategy is known as an *SDI profile*.

The Royal Society of Chemistry produces several current awareness periodicals and data bases:

> Chemical Hazards in Industry
> Laboratory Hazards Bulletin
> Methods in Organic Synthesis
> Natural Product Updates
> Current Biotechnology Abstracts
> Chemical Engineering Abstracts
> Theoretical Chemical Engineering Abstracts
> Mass Spectrometry Bulletin
> Chemical Business Bulletins
> Chemical Business Update

The ISI weekly service, Current Contents, reproduces tables of contents of about 1000 of the most important research journals in chemistry and related sciences. A similar publication, Chemical Titles, from CAS, is described in Section 2.2.5.

Automatic Subject Citation Alert (ASCA) is an SDI service based on the ISI online data base. ASCATOPICS (renamed Research Alert, as of Feb. 1st 1990) is a series of about 350 profiles, any selection of which can be run by ISI as a service for a scientist. Patent information is, however, not included and there are no abstracts.

CA SELECTS is a very valuable current awareness service supplied by CAS. It provides complete CA abstracts and bibliographic citations, and also covers patents but is not as fast as ASCATOPICS. The number of topics in 1989 was 209.

2.6. Future of Abstracting and Indexing Services

The continuing growth of the primary literature has led to continuing expansion of secondary services. There has been a "migration" from print products to electronic services and this is likely to continue as more countries and users gain access to online data bases.

A trend towards in-house systems, not simply CD-ROM data bases (Chap. 8), is likely. In-house systems are more cost effective for regularly used data, there are fewer problems with telecommunications, and the software can be tailored better to the users.

More and more "end users" are gaining access to online information, bypassing the services of an intermediary information scientist. The intermediary is, however, usually needed for complex searches.

New technology (Chap. 7) will continue to have an impact on information services. One result is the integration of primary and secondary publishing. When journals are photocomposed, the machine-readable text can be reused in other ways, including input to secondary services (see Section 1.8). As the full text of journals becomes available online the traditional boundaries between primary and secondary publishing will blur.

3. Tertiary Literature

3.1. Introduction

Definitions of tertiary literature vary but here it is assumed to be evaluated literature based on primary and secondary sources, including reviews, handbooks, and encyclopedias. For convenience, other standard reference works in chemistry are also included.

Journals and patents nearly always contain more current information than books. Encyclopedias, even more so than books, have a problem with obsolescence. Nevertheless good books and encyclopedias are an excellent source for an overview or an evaluation.

This chapter covers a selection of the most important reviews, books and encyclopedias. There are several standard works on chemical information sources that are more comprehensive [3.1]–[3.8].

3.2. Reviews

Because of the enormous growth in chemistry a review writer cannot both cover a topic from its origins and evaluate all recent references. Reviews are therefore more suitable for updating a researcher's knowledge than for providing the educator with an overview.

Key review journals include Chemical Reviews (American Chemical Society), Annual Reports on the Progress of Chemistry (Royal Society of Chemistry), and Chemical Society Reviews (Royal Society of Chemistry). The Society of Chemical Industry produces Critical Reports on Applied Chemistry. Russian Chemical Reviews provide an English translation of Uspekhi Khimii. The Royal Society of Chemistry has produced more than 40 titles in its series Specialist Periodical Reports.

Reviews are designated in the Volume Indexes of Chemical Abstracts by the letter R. The Institute for Scientific Information's Index to Scientific Reviews is available both in hard copy and online.

3.3. Encyclopedias and Handbooks

3.3.1. Encyclopedias

Kirk–Othmer. The Encyclopedia of Chemical Technology [3.9] is commonly referred to as Kirk–Othmer after the names of the original editors. The third edition was produced between 1978 and 1984. It consists of 24 volumes, a supplement, and an index, the main work containing over 1200 articles, 9×10^6 words, 6000 tables, and 5000 figures. Chemical Abstracts Registry Numbers were included.

The articles are well set out and easy to read with emphasis on applied chemistry. Kirk–Othmer is an indispensable tool in almost every chemistry library but has a bias towards American practice. The full text of this encyclopedia is available online and on CD-ROM.

A one-volume version, Kirk–Othmer Concise Encyclopedia of Chemical Technology, was published in 1985. Subject-oriented reprint volumes (e.g., on antibiotics) are also available.

Ullmann's. The fourth edition of Ullmanns Enzyklopädie der Technischen Chemie [3.10], in 25 volumes (in German) began in 1972 and was completed in 1984. The fifth edition, under the title Ullmann's Encyclopedia of Industrial Chemistry, is being published in 36 volumes in English [3.11]. Work began in 1984 and is expected to be complete by 1996. Volumes A1–A28 contain articles about industrial chemicals, product groups, and production processes covering all branches of the chemical and allied industries. The eight B volumes form a basic knowledge series covering chemical engineering fundamentals, analytical methods, environmental protection, and plant safety. Each year a cumulative index appears.

Ullmann's Encyclopedia is international in authorship and coverage. To aid the reader, all but the shortest articles start with a table of contents and the printing is in easy-to-read columns.

The advantages and disadvantages of Ullmanns in comparison to Kirk–Othmer are discussed in [3.12] which, however, erroneously assumes that there is a significant price difference between the two.

3.3.2. The Handbuch Concept

The Handbuch is an old tradition in German chemistry. It is quite different from U.S. handbooks, being multivolume and more extensive in scope and coverage.

In the United States, Handbuch volumes tend to be regarded as obsolete, incomprehensible, and difficult to use. It does take months, or even years, to evaluate data for an accurate and reputable handbook, but coverage often goes back to the beginnings of chemistry, which is not true of, for example, Chemical Abstracts.

At least two major handbooks (Beilstein and Gmelin) are now published in English, making them accessible to a wider scientific audience. New tools are also being produced to facilitate the use of these handbooks.

The Beilstein Handbook. Beilstein's Handbook of Organic Chemistry [3.13]–[3.15] is the oldest, best-known reference work in organic chemistry. It takes its name from FRIEDRICH KONRAD BEILSTEIN who produced the first edition between 1881 and 1883. Coverage goes back to the beginning of organic chemistry (1830). Substances are included in Beilstein if they are organic compounds; if they have known, verified constitutions; if they are pure; if syntheses for them are known; and if data are available on them. Information is abstracted from journals, patents, monographs, and other publications. Since 1985 electronic abstracting methods have been employed [3.16]. The Beilstein Institute in Frankfurt, a nonprofit organization, employs over 170 permanent staff, including 120 chemists and 500 external abstractors. The aim is critical evaluation of the primary literature and the ordered presentation of verified facts and data. During evaluation, contradictions are cleared up, duplicate material and trivia are eliminated, and errors in the primary literature are corrected.

Descriptions of compounds cover constitution and configuration; natural occurrence and isolation from natural sources; preparation and manufacture; chemical and physical properties; structural and energy parameters; characterization and analysis; and salts and addition compounds.

Beilstein consists of more than 350 volumes. The original work (basic series) was divided into 27 volume numbers and the five supplementary series adopt the same classification and volume number arrangement (Table 3). The fifth supplementary series is published in English. Up to the end of 1960, 1.5×10^6 compounds had been described. Characteristic of the Handbook is the Beilstein classification system, which is explained in [3.17].

Organic compounds are distributed over the 27 volume numbers as shown in Table 4. Within

Table 3. The series of the Beilstein Handbook (4th edition)

Series	Abbreviations	Literature years covered	Spine label color	Status
Basic Series	H	Up to 1909	green	complete
Supplementary Series I	EI	1910–1919	dark red (on brown cover)	complete
Supplementary Series II	EII	1920–1929	white	complete
Supplementary Series III	EIII	1930–1949	blue	complete
Supplementary Series III/IV*	EIII/IV	1930–1959	blue/black	complete
Supplementary Series IV	EIV	1950–1959	black	complete
Supplementary Series V** (in English)	EV	1960–1979	red (on blue cover)	1984–

* Volumes 17–27 of Supplementary Series III and IV (heterocyclic compounds) are combined in a joint issue.
** The first volumes published in this series (17–27) relate to heterocyclic compounds based on a survey of user requirements.

Table 4. Contents of Beilstein volumes

Beilstein volume number	Compound class
1–4	acyclics
5–16	isocyclics
17–27	heterocyclics type and number of heteroatoms
17, 18	1 O
19	2 O, 3 O ...
20–22	1 N
23–25	2 N
26	3 N, 4 N ...
27	1 N, 1 O; 1 N, 2 O ... 2 N, 1 O; 2 N, 2 O ... further heteroatoms*

* E.g., B, Si, P but not S, Se, Te.

these classes further ordering is based on type and number of functional groups. At the top of the even-numbered pages is the designation of the compound class to which the compounds dealt with on that page belong. At the top of the odd-numbered pages are the coordinating reference, the series number, volume number, and system number. The coordinating reference indicates the page of the Basic Series to which the item would have been assigned had the substance been known then. The system number is the unit of the Beilstein System of structure classification. Values run from 1 to 4720 and are dependent on structural features.

If an item is not being covered for the first time, a back reference is given to earlier volumes and page numbers.

To locate a compound in Beilstein the user can use the molecular formula index, the subject (i.e., compound) index, or the system number. Each volume has formula and subject indexes; there are also cumulative and collective indexes. Once the compound has been located in a particular volume, the system number and coordinating reference can be used to find the same compound in the same volume number of other supplements.

A microcomputer program called SANDRA (Structure and Reference Analyzer) aids the user in locating the appropriate volume of Beilstein once he draws in the chemical structure concerned [3.18], [3.19].

Beilstein Online, an electronic version of the Handbook, is now available [3.16], [3.19]. The data base can be searched by structure, substructure, and other features (see Section 9.6.1). The online version contains more compounds than the Handbook because it also contains more current, nonevaluated data.

There are significant differences between Beilstein and Chemical Abstracts [3.19], [3.20] and a well-stocked library needs both. Beilstein covers the literature back to 1830, Chemical Abstracts only to 1907. Chemical Abstracts, however, is much more current than Beilstein. Beilstein gives immediate access to validated numeric data and facts, whereas Chemical Abstracts gives the reader bibliographic references to sources containing such facts. The Beilstein classification locates parents and derivatives on nearby pages. However, Chemical Abstracts permits subject and author searches in the hard copy and Beilstein does not.

Gmelin. The Gmelin Handbook of Inorganic and Organometallic Chemistry [3.21] is an invaluable source of information on elements, inorganic compounds, and organometallic compounds. It is named after LEOPOLD GMELIN, who published his first Handbuch from 1817 to 1819. The handbook is now produced by the Gmelin Institute of Inorganic Chemistry and Related Sciences, which is part of the Max Planck Society for the Advancement of Sciences. The Gmelin Institute is housed in the same building in Frankfurt as the Beilstein Institute and employs almost as many scientific staff.

Work on the current eighth edition was begun in 1922. This edition now comprises over 600 volumes (December 1989) with nearly 190 000 pages. About 320 000 elements, compounds, and systems were described up to 1987, and about 14 000 more are added each year. Each substance is described by information about occurrence or methods of preparation, physical and structural properties, and chemical behavior. Gmelin includes many useful numerical data, graphs, and diagrams. There is extensive coverage of applied aspects and commercial manufacturing practice. Since the mid-1970s more attention has been paid to toxicological and environmental issues, uses, and applications.

The Main Volume series started in 1924 with a reporting period beginning around the middle of the 18th century and ending at best some months before publication date of the volume under consideration. This series is almost complete. Supplement volumes, which began to appear in 1937, continue the subject. The reverse of the title page for each volume shows the latest date through which literature for that volume is evaluated. A New Supplement Series started to appear in 1970. The earlier volumes are in German, but volumes produced since 1982 are entirely in English.

The Gmelin classification is based upon 71 system numbers for the various elements or combinations of them (Table 5). A compound consisting of two or more elements is indexed in the volume(s) pertaining to the element of highest system number. Formerly, librarians arranged volumes by increasing system number. The Gmelin Institute now recommends libraries to shelve the volumes alphabetically by atomic symbol rather than by system number.

The classical way of looking a compound up is to find the system number, find the appropriate volume, and then use the Table of Contents. All volumes have contents tables in English and German. Some volumes or system numbers even have particular formula indexes. All compounds up to 1987 are included in the English-language Gmelin Formula Index. The first column in this index is the empirical molecular formula in alphabetical order of the elements; C and H are not treated separately here. The second column gives the usual empirical formula as employed in the body of the Handbook text. The third column contains the system number and the fourth column the volume and page numbers. Both the Gmelin Formula Index and the Complete Catalog, which also exists in printed form, are available online via STN International.

In 1991 the Gmelin Institute will offer a Factual Data Bank as a further online product. This data bank will also include the most up-to-date facts which are not yet published in the printed Handbook.

3.3.3. The Dictionary of Organic Compounds

The Dictionary of Organic Compounds is a successor to the dictionary first compiled by HEILBRON and BUNBURY in 1934. The fifth edition was published in five volumes in 1982 [3.22]. Annual supplements have been published since then. The work is indexed by name, molecular formula, heteroatom, and CAS Registry Number. The main work has 50 000 entries covering 150 000 compounds. Each annual supplement has a further 3000–4000 entries.

The dictionary is selective in coverage: it describes the 5% of compounds that are most widely known and used. Selection is made by industrial and academic experts.

Structural formulae, names, CAS Registry Numbers, physical, chemical, and (where available) biological properties, and literature references are listed for each compound.

The dictionary is phototypeset from a sophisticated electronic data base [3.23]. The chemical structure diagrams are currently artworked and scanned photoelectronically but work is in progress on a new methodology for linking structures and text [3.24].

The Dictionary of Organic Compounds, plus six other dictionaries published by Chapman and Hall, form the HEILBRON online data base on DIALOG. This data base contains 250 000 organic substances, with CAS Registry Numbers; structures can be displayed graphically using the DIALOGLINK software (see Section 8.2.9).

Table 5. System numbers used in the Gmelin classification

System No.	Symbol	Element
1		noble gases
2	H	hydrogen
3	O	oxygen
4	N	nitrogen
5	F	fluorine
6	Cl	chlorine
7	Br	bromine
8	I	iodine
	At	astatine
9	S	sulfur
10	Se	selenium
11	Te	tellurium
12	Po	polonium
13	B	boron
14	C	carbon
15	Si	silicon
16	P	phosphorus
17	As	arsenic
18	Sb	antimony
19	Bi	bismuth
20	Li	lithium
21	Na	sodium
22	K	potassium
23	NH_4	ammonium
24	Rb	rubidium
25	Cs	cesium
	Fr	francium
26	Be	beryllium
27	Mg	magnesium
28	Ca	calcium
29	Sr	strontium
30	Ba	barium
31	Ra	radium
32	Zn	zinc
33	Cd	cadmium
34	Hg	mercury
35	Al	aluminum
36	Ga	gallium
37	In	indium
38	Tl	thallium
39		rare earths
40	Ac	actinium
41	Ti	titanium
42	Zr	zirconium
43	Hf	hafnium
44	Th	thorium
45	Ge	germanium
46	Sn	tin
47	Pb	lead
48	V	vanadium
49	Nb	niobium
50	Ta	tantalum
51	Pa	protactinium
52	Cr	chromium

System No.	Symbol	Element
53	Mo	molybdenum
54	W	tungsten
55	U	uranium
56	Mn	manganese
57	Ni	nickel
58	Co	cobalt
59	Fe	iron
60	Cu	copper
61	Ag	silver
62	Au	gold
63	Ru	ruthenium
64	Rh	rhodium
65	Pd	palladium
66	Os	osmium
67	Ir	iridium
68	Pt	platinum
69	Tc	technetium
70	Re	rhenium
71		transuranium elements

3.4. Sources Concerned with Chemical Reactions

Houben–Weyl [3.25]. The aims of Houben–Weyl are to deal critically and comprehensively with experimental methods; to give examples with full experimental detail; to include relevant theoretical background; and to involve experts to ensure quality. Over 80 volumes have been published so far in German.

Theilheimer. WILLIAM THEILHEIMER started to produce his renowned yearbooks, Synthetic Methods in Organic Chemistry [3.26], in 1946. Since 1974 (vol. 30) the series has been derived from the Journal of Synthetic Methods (Derwent Publications). The combination of Theilheimer and the journal forms the basis of Derwent's on-line Chemical Reactions Documentation Service (CRDS) [3.27], [3.28] and of the Theilheimer data base for use in-house with the reaction indexing software REACCS (see Section 10.8).

Other Reaction Compendia. A number of multivolume works are devoted to chemical reaction information [3.29]–[3.33].

3.5. Other Reference Books

Several multivolume treatises of lesser scope than Beilstein and Gmelin have been published [3.34]–[3.40].

The Merck Index [3.41], now in its 11th edition, is of particular use to the pharmaceutical and agrochemical industries. This one-volume reference book contains short descriptions of over 10 000 biologically active compounds. Computer-assisted production methods have been used for the ninth edition onwards and the work is also available online.

The CRC Handbook of Data on Organic Compounds, HODOC II contains chemical, physical, and spectral data for 30 000 compounds [3.42]. Annual supplements are printed. The publishers are considering methods of mounting the data online.

Potentially hazardous chemicals are covered in SAX's handbook [3.43]. BRETHERICK's publications on chemical hazards are equally important [3.44], [3.45]. The most complete list of toxic effects of chemicals is the Registry of Toxic Effects of Chemical Substances (RTECS) published by the U.S. National Institute for Occupational Safety and Health (NIOSH) and also available online.

The Aldrich Chemical Company, Sadtler Research Laboratories, and the Royal Society of Chemistry all produce hard-copy libraries of spectra. Analytical data are covered in more detail in Chapter 9.

Fuller lists of reference works concerning physical properties, analytical chemistry, and safety and related data are given in [3.1]–[3.8].

4. Gray Literature

4.1. Introduction

Gray literature, otherwise termed "nonconventional", "fugitive", "informal", "ephemeral", "invisible," or "underground", is literature which is not available through normal bookselling channels and which has characteristics such as small circulation and poor bibliographic control. It includes reports, theses, conference papers, preprints, official publications, translations, house journals, trade literature, and working papers. However, not all the literature in these categories is "gray". For example, conference proceedings may be published as books, many official publications are commercially available, and many journals are translated cover to cover.

Translations have been discussed in Section 1.6 and supplementary material in Section 1.8.

4.2. Characteristics [4.1]–[4.3]

Nonconventional documents often have small print runs because they are intended for a small audience. They may be produced very rapidly, for example by in-house publication, but often have variable standards of editing, legibility, and physical presentation. The issuing organization is likely to be a university, research institution, government department, public body, or commercial agency. The documents are usually poorly publicized, have inadequate bibliographic control (e.g., author, title, producer), and bear unusual codes or identification numbers. Gray literature is not easily available to the average library. It is more likely to be produced in a local language than the conventional literature. Sometimes it comes in a format (e.g., microfilm or microfiche) unacceptable to the average user. A nonconventional document is frequently large and may be an issue of a serial appearing at irregular intervals. The cost to libraries of tracing, reproducing, and distributing such documents is very high.

Exploitation of the huge body of knowledge in the much underused gray literature would make sense. Many problems could be overcome if the producers realized the value of the information and improved its accessibility [4.1], [4.4], [4.5]:

1) Documents should be produced to better physical and bibliographic standards.
2) Producers should be less restrictive about what is released, should indicate the confidentiality of documents, and should widely publicize only that material which is for unlimited distribution.
3) Copies should be sent to appropriate secondary services and national gray literature centers; and to national depositories, copyright libraries, and specialist collections which can provide bibliographic and physical access to the documents concerned.
4) Producers should have larger print runs to meet the demand generated by more publicity.

4.3. Organizations Specializing in Gray Literature

In view of the problems described in Section 4.2, the tasks of collection, bibliographic control, and circulation of gray literature fall upon specialized organizations.

United States. One the best known centers is the National Technical Information Service

(NTIS) in the United States. This government agency has been acquiring, recording, reformatting, and promoting American research reports since the late 1960s. Over 2×10^6 reports are available from NTIS and about 70 000 new ones are added annually. They contain the results of U.S. and foreign government research and developments that have not generally been previously published elsewhere. They are available in hard copy and on microfiche and are sold worldwide directly or through appointed national agents. NTIS lists the documents fortnightly in the Government Reports Announcements and Index.

Other U.S. government agencies produce printed and online secondary services for thousands of reports every year: the Department of Energy (DOE), the National Aeronautics and Space Administration (NASA), and the Educational Resources Information Center (ERIC).

Europe. The Federal Republic of Germany has a decentralized system for the provision of literature, including gray literature [4.6]. Specialist libraries such as the Technische Informationsbibliothek (TIB) in Hannover and the Central Agricultural Library in Bonn have national responsibility for providing services in their subject areas.

The Centre de Documentation of the Centre National de la Recherche Scientifique (CNRS) in France and the British Library Document Supply Centre (BLDSC) have an interest in gray, as well as in conventional literature. The British Library has been collecting gray literature since the late 1960s and has a vast collection of reports, theses, translations, supplementary publications, local government documents, and semipublished conference proceedings [4.7], [4.8]. It publicizes its holdings in British Reports Translations and Theses, Index to Conference Proceedings Received, and Current Serials Received.

One of the most significant developments is the establishment of the System for Information on Gray Literature in Europe (SIGLE) [4.1], [4.5], [4.9]–[4.12]. Details of the literature are contained in an online data base (on BLAISE-LINE, Sunist, and STN International) produced by national centers in Belgium, France, Italy, Luxembourg, The Netherlands, the Federal Republic of Germany, and the United Kingdom. The centers belong to an association known as the European Association for Gray Literature Exploitation (EAGLE). SIGLE holds information on ca. 150 000 documents covering pure and applied sciences, and technology since 1980, and economics, social sciences, and humanities since 1984. The data base increases by 30 000 records per year.

International Agencies. The International Nuclear Information System (INIS) of the International Atomic Energy Agency (IAEA) microcopies the full text of nonconventional documents and makes them available on demand. It supports this service with the bibliographic tool INIS Atomindex.

The Food and Agriculture Organization (FAO) has a document supply service for reports and produces a hard-copy index, AGRINDEX, through the International Information System for Agricultural Sciences and Technology (AGRIS) [4.13].

Many publications of the World Health Organization (WHO) can be obtained through booksellers and standard channels but its production of internal, unpublished documents is disseminated through the bimonthly bulletin WHODOC: Index to WHO Technical Documents.

The International Agency for Research on Cancer (IARC) handles both conventional and gray literature.

The International Federation of Library Associations (IFLA) Office of International Lending and the Worldwide Network of Agricultural Libraries (AGLINET) supply documents through interlibrary loan schemes.

Commercial Organizations. Congressional Information Service in the United States indexes, abstracts, and microfilms all U.S. Congress publications. Micromedia index Canadian federal, provincial, and local government publications, as well as reports of research institutes and professional associations. Chadwyck Healey in the United Kingdom indexes non-HMSO (Her Majesty's Stationery Office) publications.

4.4. Reports [4.5], [4.9]

Bibliographic control of report literature is better in the United States than elsewhere (NTIS, NASA, and DOE, see Section 4.3).

In the Federal Republic of Germany the Fachinformationszentrum Energie, Physik, Mathematik (FIZ) produces the monthly

Forschungsberichte aus Technik und Naturwissenschaften. National centers such as CNRS and BLDSC mentioned in Section 4.3 handle reports and contribute them to SIGLE.

The European Communities Commission produces and disseminates many reports. Some of these can be obtained through one of the European Documentation Centers or from a European Communities Depository Library (see also Section 4.5).

Research reports on publicly funded research projects in Japan, as well as governmental reports concerning policy on science and technology are now included in JICST-E, a data base produced by the Japan Information Center of Science and Technology, available online on STN International. Over 4000 reports, including gray literature, are covered.

4.5. Official Publications [4.3], [4.8]

Several international organizations were mentioned in Section 4.3.

In the United States, reports from the Environmental Protection Agency (EPA), DOE, the Food and Drug Administration (FDA), the Centers for Disease Control (CDC) and others, are covered by NTIS. All aspects of U.S. government information are covered in [4.14].

Large numbers of United Kingdom Government publications appear in Chadwyck Healey's Catalog of British Official Publications not Published by HMSO. Some of this gray material is available from BLDSC.

The EUR reports of the European Communities are not generally deposited in European Documentation Centers but are microfilmed by the Commission of the European Communities and are input to the SIGLE data base.

4.6. Conference Proceedings

Conference material causes particular problems for the librarian and the research worker. It is estimated that 30–50% of conference papers are never published and those proceedings that are printed can appear up to 3 or more years after the conference has been held. Documents published before, during, and after a conference vary in terminology, contents, size, and value. Conference material is published erratically through many channels and is often not adequately refereed. It does not always report original work and much of it is not noted in abstracting journals [4.15].

The BLDSC announces its acquisitions in the Index to Conference Proceedings Received (ICP), which is also available online on BLAISE. ISI's Index to Scientific and Technical Proceedings (ISTP) is available in hard copy and online. Other useful listings include the Bibliographic Guide to Conference Publications, the Union List of Conference Proceedings in Libraries of the Federal Republic of Germany including Berlin (West) [4.16], and the Samkatalog over nyanskaffat Konferenstryck.

The Conference Papers Index (CPI) produced by Cambridge Scientific Abstracts is available online on DIALOG. Each meeting is assigned a unique number which, where available, matches that used in the hard-copy World Meetings publications [4.17]–[4.19] compiled by the World Meetings Information Center.

More than 1.4×10^6 references to meeting abstracts are included in SciSearch, the online version of ISI's Science Citation Index.

"Conference Proceeding" is a searchable designation for Chemical Abstracts online (on several host computers). Analytical Abstracts offers a keyword-indexed Conference Title field for easy access to meeting publications in analytical chemistry.

4.7. Theses [4.5]

The majority of American dissertations are abstracted in Dissertations Abstracts International published by University Microfilms International (UMI). Section B and its indexes concern science and engineering. The same publishers also produce Comprehensive Dissertations Index 1861–1972, American Doctoral Dissertations and Masters Abstracts. Many universities make their dissertations available through UMI but some have to be approached directly.

In the United Kingdom, BLDSC is the main source of most post-1970 British doctoral theses. The theses are kept on microfilm. The University of London provides a service for its own theses.

In France, theses have been recorded since 1884 in the Catalogue des Thèses et Écrits Academiques and Supplement D of Bibliographie de la France. PASCAL Explore – E99-Congrès, Rapports, Thèses (formerly Bulletin Signalétique-Section 401-Congrès, Rapports, Thèses) lists French reports and other gray literature originating in France. French scientific theses are available from CNRS.

German theses are included in the Jahresverzeichnis der Hochschulschriften der DDR, der BRD und West Berlin [4.20]. Scientific theses since 1983 (FRG) appear in Forschungsberichte aus Technik und Naturwissenschaften.

French, British, and German theses are included in SIGLE.

Canadian theses are available through the National Library of Canada in Ottawa, which has published the listing Canadian Theses since 1984.

In the Soviet Union, dissertations are only occasionally cited but are mostly published in summary form as VINITI papers.

5. Business and Economic Information

5.1. Introduction

Business information can be loosely defined as the information that is required by a business to give it competitive advantage in the marketplace. This encompasses a broad range of information on companies, markets, products, current affairs, legislation, economics, and finance, for which there are a great many printed and electronic sources.

Directories and handbooks are the "first-stop" guides to sources, e.g., [5.1], [5.2]. Essential primary publications include newspapers, trade and business periodicals, statistical publications, company annual reports, and market research reports. External agencies such as research associations, government departments, local authorities, independent information brokers, embassies, chambers of commerce, and stock exchanges are further important suppliers of business information.

Business information must be up-to-date for competitive advantage and in this respect online data bases have made a most significant contribution. Online business information has been slower to develop than the scientific and technical data bases. However, in recent years the number of business data bases has increased enormously. Financial, company, news, and market information have become higher revenue earners than the scientific and technical data bases and have an even larger growth potential.

Development in the business data-base market has been geared towards the provision of full-text data bases with frequent updating. Especially in the financial sector, information is continuously updated and available to the user as soon as it appears. The previous domination of information produced by and for the North American markets has been challenged by the emergence of European data bases. Europeans now have access to online information that is tailored to their own requirements and markets. This trend has been accelerated by the advent of the Single European Market. Details of the growing numbers of hosts and data bases available to the online user of business information are given in [5.3], [5.4].

More recent developments in optical disc technology mean that many data bases are also commercially available on CD-ROM (see Section 7.2.3). Information in the form of text, graphics, images, or data may be downloaded to a microcomputer's memory and manipulated with appropriate software. Commercially available CD-ROM's of relevance to the business information user include text of directories, newspapers, and company data. At the time of writing, the market is expanding, but CD-ROM is still at a fairly experimental stage.

5.2. Company Information

Company information can range from simple needs (e.g., the name and address of a company, or its directors) to more detailed information about financial performance or international activities (acquisitions and mergers).

A company's own annual reports and accounts provide authoritative information on financial data, turnover, trading profits before and after tax, earnings, dividends, and capital expenditure. This information originates from legal requirements necessitating companies to deposit certain documents with appropriate authorities or from the need to disclose certain financial information so that shares can be traded on the stock exchange. The quality and quantity of the information can vary from country to country. The United Kingdom, Republic of Ireland, Denmark, Greece, and Luxembourg, for example, all have centralized registration systems but companies in other countries register with a local registrar or regional chamber of commerce.

Company annual reports and company information from government agencies such as the Securities and Exchange Commission (USA) or Companies House (UK) are the most important sources of primary information. Providers of company information, including Inter-Company Comparisons (ICC), Jordans, Extel, and Dun and Bradstreet, all supply information based on these data.

Directories. In addition to the above sources, large numbers of directories and yearbooks give basic company information. Amongst the more prominent are the Kompass directories which are available for most European countries. Listed below are some key publications that provide a useful starting point for most enquiries. Further sources can be found in the Directory of Directories (Gale Research Co.), Current European Directories (CBD Research), and Current British Directories (CBD Research).

International
International Stock Exchange Official Yearbook, Macmillan
Chemical Industry Directory, Benn Business Information Services
Principal International Businesses, Dun and Bradstreet
Directory of Multinationals, Macmillan

Europe
Chemical Company Profiles: Western Europe, IPC Industrial Press
Europe's 15 000 Largest Companies, ELC International
Major Chemical and Petroleum Companies of Europe, Graham and Trotman
Major Companies of Europe, Graham and Trotman

Federal Republic of Germany
Handbuch der Deutschen Aktiengesellschaften, Verlag Hoppenstedt, gives detailed financial information on 2500 public companies
Handbuch der Gross-Unternehmen, Verlag Hoppenstedt, describes 22 000 major companies with over 100 employees or a turnover of DM 10×10^6
West German Middle-sized Companies, Verlag Hoppenstedt

Scandinavia
Major Companies of Scandinavia, Graham and Trotman, lists 1000 Scandinavian companies

United Kingdom
CRO Directory (on microfiche)
Financial Times Industrial Companies, Volume II Chemicals, Longman Group
Kelly's Business Directory, Kelly's Directories
Key British Enterprises, Dun and Bradstreet
Macmillan Top 20 000 Unquoted Companies, Macmillan
Sell's Directory of Products and Services, Sell's Publications

United States
Chemical Company Profiles, The Americas, IPC Industrial Press
Chem Sources USA, Directories Publishing Company
Directory of Chemical Products – United States, SRI International
Major Companies of the USA, Graham and Trotman
The Million Dollar Directory, Dun and Bradstreet, also available online
Moody's Manuals, Dun and Bradstreet
Standard and Poor's Register of Corporations, Directors and Executives, Standard and Poor's Corporation

India
Indian Chemical Directory, Technical Press publications

Far East
Major Companies of the Arab World, Graham and Trotman, covers 20 countries including Egypt, Iraq, Algeria
Asia's 7500 Largest Companies, ELC International

Japan
Diamonds Japan Business Directory, Diamond Lead Co.
Japan Chemical Directory, Chemical Daily Co.
Japan Company Handbook, Oriental Economist, gives detailed information on 1st and 2nd section companies

Other more specialized directories give a particular type of company information. Who Owns Whom (Dun and Bradstreet) is also available as an online data base and covers the United Kingdom, North America, Continental Europe, Australia, and the Far East. Each area has two volumes: the first lists subsidiary companies and gives their parents, the second lists parent companies and their subsidiaries. The Directory of Directors (Thomas Skinner) gives the names of directors of public and private companies. The Stock Exchange Companies (Financial Times) tables the performance of the 1000 largest listed UK companies over the past five years. It also gives details of every Unlisted Securities Market company. The Times 1000 (Times Books Ltd) ranks the 1000 largest UK companies by size and gives brief financial details; it also contains 500 European and other companies. The Register of Defunct Companies (Stock Exchange Press) is also useful and there are also many specific directories, such as Duns Guide to Healthcare Companies (Duns Marketing Services).

Many other key sources for company information are available in public libraries or by subscription. Extel Company Card Services (Extel Statistical Services) cover about 3400 quoted and unquoted UK companies and also large European, Australian, and North American companies. The cards give details of company activities, chairman's statements, balance sheets, dividend records, board members, profit and loss accounts, yields, earnings, and capital history. The service also covers Unlisted Securities and Third Market companies. The McCarthy Press Cuttings Service provides weekly press cuttings on UK, European, Australian, and North American companies. ICC Business Ratio Reports provide standard ratios for intercompany comparisons. The reports on different areas of industry, including the chemical industry, analyze up to 100 leading UK companies, giving basic data and ratios from company accounts, company addresses, directors, name of holding company, and principal activities. ICC Financial Surveys cover over 160 sectors of industry in the United Kingdom and for each company give date of accounts, turnover, total assets, current liabilities, and payments to directors. Credit Reporting Services are offered by Dun and Bradstreet, ICC, and Infocheck. They provide confidential, detailed financial analyses, and credit ratings on UK companies. Stockbrokers Reports are produced on industry sectors and individual companies.

Information Available Online and on CD-ROM. Much of the information contained in printed directories is also available online. General chemical industry data bases, such as the Chemical Business Newsbase (Royal Society of Chemistry), cover business aspects of the European chemical industry. This includes company information in the form of abstracts from annual reports, press releases, and promotional material. Specialist data bases concentrate on specific areas of company information, such as trademarks, mergers, and acquisitions.

The impetus for the development of CD-ROM products has come from the United States. Compact Disclosure (Disclosure Inc.) is the CD-ROM equivalent of its online data base and is updated monthly. Lotus Development Corporation produces Compustat data bases on CD-ROM giving investment information on U.S. companies. The information can be manipulated in-house using Lotus software. Moody's CD-ROM (Dun and Bradstreet) gives financial and business information on all U.S. Stock Exchange companies. Standard and Poors' Corporations on CD-ROM gives details of 9000 U.S. public companies. European products include European Kompass on Disc and Kompass on Disc. The former covers 300 000 European companies and is available in five languages, the latter covers 160 000 UK companies.

5.3. Products and Markets

Information on competitors' products is vital to most companies for finding detailed technical data on new products, plus sales and marketing information, or simply to find out the producer of a particular product or the product range of a particular company. On a broader scale, companies need to keep up to date with general trends in their area of the market and to have access to market research to successfully target their own research, development, and production activities.

5.3.1. Product Information

Directories. To find out which companies make a particular product, the best sources of information are directories with product indexes. A comprehensive listing of trade directories is Croner's Trade Directories of the World (Gale Research Company, 1985). The Kompass series of directories (Kompass Publishers) covers most Western and some other countries; the first volume in each country set is indexed by products and services. Key British Enterprises (Dun and Bradstreet), Kelly's Business Directory (Kelly's Directories), and Sell's Directory of Products and Services (Sell's Publications) also have product indexes and are useful sources of UK company product information. The Thomas Register of American Manufacturers (Thomas Publishing Company) is a key source for the United States. Directories can also be used for finding out which products a particular company makes, although they rarely give much detail. Some trade journals contain trade directory information and a useful guide to these is Trade Directory Information in Journals (British Library).

Trade Mark Information. The major UK source for finding out which company owns a particular trade name is UK Trade Names (Kompass Publishers). Other directories such as Sell's Directory of Products and Services (Sell's Publications) and Kelly's Business Directory (Kelly's Directories) include trade name indexes. The major U.S. source is The Trade Names Directory (Gale Research). For European and international trade mark information, local company directories with a trade name index must be used. The Kompass series is a good starting point. For Japan, Diamond's Japan Business Directory (Diamond Lead Company) has a trade mark index. Gardener's Chemical Synonyms and Trade Names is useful for chemicals. The Trade Mark Registry (State House, High Holborn, London WC1) maintains a listing of registered trade marks which can be consulted by the public for a fee. However not all UK trade marks are registered, because there is no legal compulsion for a company to do so. The Patent Office, (25 Southampton Buildings, Chancery Lane, London WC2A 1AW) produces the weekly Trade Marks Journal which advertises newly registered trade marks for the United Kingdom. The U.S. Patent and Trademark Office (The Commissioner of Patents and Trade Marks, Patent and Trademark Office, Washington D.C., 20231) can supply information for the United States.

Trade Literature. For more detailed information on a product, the company's own trade literature is useful, although it may not be easily obtainable. The British Library's Science Reference and Information Service (25 Southampton Buildings, Chancery Lane, London WC2A 1AW) keeps a large collection of trade literature, covering around 12 000 companies (mainly UK) and a range of industries. Another source of trade literature may be the appropriate trade association. The chemical industry has many such associations, the most appropriate can be found in [5.5]. Trade associations for the United States, European, and Eastern countries are given in [5.2]. The country's embassy may also be able to assist.

Press Cuttings. The McCarthy Press Cutting Service includes a classified products and services sequence in its system of cards. It covers UK quoted companies and some larger European, Australian, and North American companies. The weekly service is available on subscription, but is also held in large public libraries.

Online Sources. The user should consult the Online Business Sourcebook [5.3] or Directory of Online Data bases [5.4] for a comprehensive listing of online sources. Four useful data bases for the chemical industry are Chemical Business Newsbase (Royal Society of Chemistry), Chemical Industry Notes (Chemical Abstracts Service), European Chemical News (European Chemical News), and East European Chemical Monitor (Business International S/A). They all contain data on companies, new products, and markets.

The directory Wer Liefert Was? (Wer Liefert Was? GmbH) lists $> 1 \times 10^6$ links between products and suppliers in Europe, and is available online and on CD-ROM. It operates in five languages and its 55 000 addresses can be manipulated by the user's own word-processing software. The Thomas Register of American Manufacturers is available on CD-ROM as well as in hard copy (DIALOG and Thomas Publishing Company).

5.3.2. Market Information

Guides to Market Research Reports. Market research can be carried out in-house or contracted out to a market research company. Both options are expensive and it is well worth first finding out whether the information required already exists as a published market research report. There are several guides to market research reports:

Marketing Surveys Index, MSI (UK) Ltd (updated monthly with a very detailed subject index, covering products as well as general areas; also available online)

Market Research: a Guide to British Library Holdings, British Library (all reports listed are available to the general public)

Marketsearch, British Overseas Trade Board

FINDEX: the Directory of Market Research Reports, Studies and Surveys, U.S. National Standards Association (also available as an online data base, covering 12 000, mainly U.S., reports)

The major producers of market research reports are Inter-Company Comparisons (ICC), Mintel Publications, the Economist Intelligence Unit, Frost and Sullivan, Jordan & Sons, Arthur D. Little, and Euromonitor. Many of these concentrate on consumer and retail markets. For coverage of the chemical industry, it is necessary to turn to more specialized publishers such as Chem Systems International Ltd, SRI International, Kline, the Freedonia Group, and IAL Consultants Ltd.

A typical report contains a detailed review of the industry structure, major companies involved, market size and trends, product sales by volume, value and market share, recent developments, and future prospects. Some reports may deal solely with technology and not with markets. The following list gives typical market research reports covering the chemical industry:

Biotechnology Products, Key Note Publications (1988), covers UK
Britain's Plastics Industry, Jordan & Sons (1987), covers UK
Chemfacts: United Kingdom, Chemical Intelligence Services (1987), covers UK
Chemfacts: Polypropylene, Chemical Intelligence Services (1987), worldwide coverage
Chemfacts: Polyethylene, Chemical Intelligence Services (1987), worldwide coverage
Chemfacts: West Germany, Chemical Intelligence Services (1987), covers FRG
Custom Chemical Synthesis (1987), IAL Consultants, covers France
Dyes and Organic Pigments (1988), Freedonia Group, covers USA
Electronic Chemicals (1987), IAL Consultants, covers France
Ethylene Oxide and Derivatives (1988), Freedonia Group, covers USA
Fertiliser Outlook (1988), Freedonia Group, covers USA
UK Chemical Manufacturers and Distributors (1987), ICC Financial Surveys, covers UK

Market research periodicals include Marketing in Europe (monthly, Economist Publications), Market Research Great Britain (monthly, Euromonitor), and Marketing Trends (semi annual, AC Nielson). The last is available in British, American, German, Spanish, and Portuguese editions. These contain topical marketing information and are a useful means of keeping abreast of current trends.

Online Sources. ICC International Business Research (ICC Information Group) gives abstracts of all ICC's Keynote Market Reports. Arthur D. Little Online (Arthur D. Little) gives full text for around 80% of their publications, including industry forecasts and market reviews. The data base's coverage is mainly United States, although some is international. The Media Expenditure Analysis Ltd. (MEAL) data base gives information on advertising and expenditure in the United Kingdom.

Marketing Associations. If published sources fail to provide the necessary information then the user can commission a market research company to produce a report. Examples of marketing associations able to give advice and further information are:

Association of Market Survey Organisations Ltd
　60 Kenilworth Road Ltd
　Leamington Spa
　Warwickshire CV32 6JY

Industrial Marketing Research Association
　11 Bird Street
　Lichfield
　Staffs WS13 6PW
American Marketing Association
　22 South Riverside Plaza
　Suite 606
　Chicago IL 60606
Japan Marketing Research Association
　No. 20 Sankyo Building
　11-5 Iidabashi-3-chome
　Chiyoda-ku
　Tokyo 101
TMO Consultants
　22 rue de Quatre Septembre
　75002 Paris

5.4. News Services

Scanning newspapers and business periodicals has been the traditional way of keeping up to date with business news and developments. Periodicals such as Chemical Insight (published semimonthly by Hyde Chemical Publications) cover the chemical industry on an international basis, whilst others like Japan Chemical Week help to monitor events in specific countries. Timely access to new information on competitor activity and other market developments clearly help a business to make informed decisions, but delays in receiving foreign publications can result in news being stale by the time it is read.

The 1980s have seen the development of fulltext online data bases designed to provide rapid access to industry news. Reuter Textline (previously owned by Finsbury Data as Textline) contains abstracts and full text articles from over 1400 newspapers and journals. It covers the major UK dailies, various provincial newspapers, European and international press, and trade press. It also includes all Reuter wires from journalists and has a specialist section for chemicals. It has records from 1980 onwards and is updated daily. NEXIS (Mead Data Central International) is a full-text data base that is updated rapidly and has four main files: NEXIS Magazines, NEXIS Newspapers, NEXIS Newsletters, and NEXIS Wire Services. It is biased towards U.S. sources, but is expanding its European coverage. PTS Promt (Predicasts Inc.) covers company and government news, market information, and details of new products and processes. Coverage is international with a U.S. bias.

Profile Information (Profile Information, a subsidiary of Financial Times Information

Online Ltd) has vast files of full-text sources covering a wide range of company, business, marketing, and industry news. Sources include The Washington Post, Financial Times, Asahi News Service, BBC Summary of News Broadcasts, and TASS Newswire. Infomat (previously BIS-Infomat and now also produced by Predicasts) covers around 500 European and international newspapers and journals. It is not full text but produces informative abstracts and aims to give good coverage of the European Single Market.

Nikkei Telecom Japan News and Retrieval is a real-time data base which displays news as soon as it is filed. It gives English-language coverage of Japanese newspaper items and articles, often several hours before the newspaper is published in Japan. There are also files covering Tokyo Stock Market and Money Market figures. A broadcasting mode is available, whereby real-time news is displayed as soon as it comes into Nikkei Telecom.

Chemical Business Newsbase (Royal Society of Chemistry), covers trends and current affairs in the European chemical industry and end markets. Chemical Industry Notes (Chemical Abstracts Service) covers worldwide business news on the chemical processing industries, giving information on production, pricing, sales, products, processes, and corporate activities.

Routine monitoring of an appropriate data base allows the user to keep abreast of all developments in a certain business area in a fraction of the time it would take to scan manually all the appropriate publications. The full range of on-line news data bases can be found in [5.6]

CD-ROM is less useful for news information than online services due to delays in updating but this is an expanding area. The National Newspaper Index (Information Access Company) indexes five major U.S. newspapers. It covers four years of data and is updated monthly. Newspaper Abstracts Ondisc (University Microfilms International) gives brief abstracts for a range of U.S. newspapers. ABI-INFORM, (University Microfilms International) contains abstracts of articles from over 800 business journals. Its sister product, Business Periodicals Ondisc, gives access to the full text of the current and backfile issues of 300 business and management journals.

Some of the newspapers and periodicals of interest to the chemical industry follow. For more detailed listings for individual countries, see [5.2].

United Kingdom (general business)
Financial Times
The Guardian
The Independent
The Times
The Observer
The Accountant (monthly, Lafferty Publications Ltd. London)
Business (monthly, Business People Publications, London)
Campaign (weekly)
The Economist (weekly, Economist Newspapers Ltd, London)

United Kingdom (chemical industry)
British Journal of Pharmaceutical Practice
British Polymer Journal
Chemical Engineer
Chemistry & Industry
Chemistry in Britain
European Chemical News
Fertiliser International
Laboratory Products
Industrial Chemistry Bulletin
Pharmaceutical Business News
Practical Biotechnology
Process Engineering
Process Equipment News
Scrip World Pharmaceutical News
Speciality Chemicals

United States (general business)
Journal of Commerce
New York City Tribune
New York Post
New York Times
The Wall Street Journal
Barrons (weekly, Dow Jones)
Business Week (weekly, McGraw Hill)
Forbes (fortnightly, Forbes Inc.)
Fortune (monthly, Time Inc.)
Harvard Business Review (bimonthly)

United States (chemical industry)
American Chemical Society Lab Guide
Butane–Propane News
CEC–The Process Industry Catalog
Chemical Business
Chemical and Engineering News
Chemical Industry Product News
Chemical Marketing Reporter
Chemical Week
Industrial Chemical News
Soap, Cosmetics, Chemical Specialties

European and International coverage (chemical industry)
Chemic Anlagen & Verfahren Europe
Chemical & Engineering News
Chemical Week
Chemical Products
Chemische Industrie International
European Chemical News
European Plastics News
International Labmatic

5.5. Legal Information

Each country has its own ever-changing legal system and there are obvious problems in finding legal information for an unfamiliar territory. Interpretation of the law is very much the concern of the experts and a major consideration is the currency of the information. Many important reference works are published in loose-leaf format so that they can be constantly updated. Online information is especially valuable but the most prominent online legal data base, LEXIS (Mead Data Central), is designed for the exclusive use of lawyers.

Company law usually comes to the fore in a take over, merger, or flotation. The law affecting day-to-day business covers such items as trade descriptions, weights and measures, sale of goods, industrial relations, health and safety, planning, and environmental law.

For most industries in the United Kingdom the primary source of information on new legislation and cases is the relevant trade association or trade publications. For more detailed study the user should consult up-to-date text books or Halsbury's Laws of England (Butterworths).

The same comments with regard to trade associations apply to EEC law. Useful sources of information on European legislation are to be found in [5.7], [5.8]. Local law in each member state often has as much impact as community law.

Each state of the United States has its own public, criminal, and private law, both common law and statutory. Unifying influences include the decisions of the Supreme Courts and the lower federal courts and the requirement of the Constitution that every state shall give full faith and credit to the public acts, records, and judicial proceedings of every other state. States tend, however, to adopt uniform laws on specific issues such as sales and commercial transactions. It is even more important to obtain expert legal advice in the United States, if only because public authorities and other companies are likely to have lawyers readily available. The two main online data bases in the United States are LEXIS (Mead Data Central) and WESTLAW (West Publishing).

5.6. Economics and Finance

To do business successfully in any country, information on its economic background and current economic and financial situation is vital. Every country has its own unique economic and financial situation and information sources therefore differ. A good starting point is one of the regional or international organizations that collect and disseminate information from various sources.

5.6.1. Hard-Copy Sources

The United Nations publishes a vast amount of data and a useful guide is UNDOC Current Index: United Nations Document Index (bimonthly, United Nations, New York). Publications such as the World Trade Annual, Yearbook of National Account Statistics, Monthly Bulletin of Statistics, and The Growth of World Industry are of interest to businesses in the chemical industry.

In the United Kingdom Her Majesty's Stationery Office (HMSO) is one of the major information providers. Economic Trends, published monthly, charts statistics which give the background to current UK economic trends. Financial Statistics is a monthly compilation of key financial and monthly statistics, including government income and expenditure, public sector borrowing, banking statistics, money supply, institutional investment, company finance and liquidity, security prices, and exchange and interest rates. UK National Accounts (The "Blue Book") gives detailed estimates of production, income, and expenditure for the UK. UK Balance of Payments (The "Pink Book") covers balance of payment statistics for the preceding eleven years. Business Monitors are arranged by industry sector and give statistics on sales of UK manufactured products. They include import and export data, producer price indices, and employment figures. The Economic Commission for Europe publishes the Economic Survey of Europe and the Economic Bulletin of Europe annually. The Economic Commission for Asia and the Far East produces The Statistical Yearbook for Asia and the Far East. The Economic Commission for Africa produces Foreign Trade Statistics of Africa and African Economic Indicators.

The EEC is a prolific producer of economic information in nine languages. Reference texts, such as Europe in Figures, cover the socio-economic situation in the EEC. The Official Journal of the EEC has several series: Series A, economic trends; Series B, business and computer survey results; Series C, communications; and Series L, legislation. Data for Short-Term Economic Analysis is published eleven times a year and is a review of the main quantitative data in relation to the community and its member states, covering economy, employment, industrial production, prices, finance, and balance of payments. National Accounts ESA–Aggregates is useful for country comparisons, since it covers accounts, surveys, and statistics of the United States, Japan, and European countries, harmonizing the results in accordance with the European System of Integrated Economic Accounts (ESA). The Consumer Price Index, Money and Finance, and ECU–EMS Information are all EEC periodicals offering current financial information.

The Organization for Economic Cooperation and Development (OECD) produces useful sources of economic information on Canada, Japan, the United States, Australia, New Zealand, Yugoslavia, and Western Europe. The OECD Economic Survey Series annually analyzes developments, prospects, demand, wages, money and capital markets, balance of payments, and government policies, with an issue for each country covered. OECD Main Economic Indicators (monthly) includes records of industrial production and consumer price indices. The Chemical Industry (an annual) gives production, price, and investment data for the major branches of the chemical industry.

Other OECD publications can be accessed via its Catalog of Publications, published every two to three years and updated in between by supplements.

The International Monetary Fund (IMF) produces the fortnightly IMF Survey which details the Fund's international banking activities; its Annual Report: Exchange Arrangements, and Exchange Restrictions provides information on foreign exchange rate systems.

The Economist Intelligence Unit (EIU) provides a wider scope of economic and political information. Its Country Reports (quarterly) monitor political, economic, and business developments in over 165 countries. Country Profiles are produced annually for each country, giving a survey of political and economic background. They are a useful starting point for the user embarking on an investigation of a company or industry in an unfamiliar country. Economic Prospects provides in-depth forecasting reports for selected countries. EIU Business Updates are monthly summaries of economic news forecasts and statistics on the major OECD countries. European Trends is a quarterly analysis and discussion of developments in the EEC.

5.6.2. Online Sources

For a full listing of data bases of economic and financial information, the user should consult [5.2], [5.6].

International. Data Resources Inc. produce DRI World Forecast, DRI External Debt, DRI International Cost Forecasting, and DRI Current Economic Indicators. The last of these sources provides major financial and economic indicators for 35 countries (including 15 developing countries). It is updated monthly and dates back to 1960. The IMF produces IMF International Financial Statistics and IMF Balance of Payments. The OECD Main Economic Indicators are available online, providing a respected source of data on economic affairs of OECD member countries. For trade statistics, Tradstat (Data-Star) is believed to cover 85% of world trade.

United Kingdom. The Bank of England Databank (Bank of England) gives access to a wide range of time series on financial indicators for the United Kingdom allowing trends to be studied and predictions to be generated. The Central Statistics Office (CSO) Data Bank has a wide coverage of monthly, quarterly, and annual statistics and includes data from the hard-copy versions of UK National Accounts and UK Balance of Payments.

Europe. Probably the most important data base is Cronos–Eurostat, produced by the European Communities Statistical Office. It covers over 900 000 time series of economic data, including general statistics, national accounts, industry and services, and foreign trade. Other major data bases include DRI Europe (Data Resources Inc.), which provides comprehensive European indicators, and DRI European Forecast.

United States. Citibase (Citicorp Data Base Services) is a U.S. macroeconomic data base and contains 5000 time series, most going back to

1947. Other data bases include DRI CFS Cost and Industry (DRI Europe Ltd) which includes export and import price indices by commodity; DRI Long-Term Industry Forecast (DRI Europe Ltd) which gives ten year forecasts for U.S. industrial activity; and MMS Equity Market Analysis (MMS International) which offers analysis of fiscal and trade policy, and economic factors affecting U.S. equity prices.

Other Countries. Eastern Bloc Economic Statistics (Vienna Institute for Comparative Economic Studies) contains over 7000 time series of economic data for Bulgaria, Czechoslovakia, the former German Democratic Republic, Hungary, Poland, Romania, the Soviet Union, and Yugoslavia. It is available through Reuters. DRI Asian Forecast (DRI Europe Ltd) covers eleven countries and gives detailed projections of economic and financial conditions. Of similar use are DRI Latin American Forecast and DRI Middle East and African Forecast.

6. Patent Information

6.1. Introduction

Industrial inventions constitute an important part of industrial property and are accorded legal protection on the basis of national laws and international patent conventions. To obtain protection for an invention, a formal application has to be filed with the respective national or supranational patent office, together with a detailed description of the invention and a precise statement of the matter for which protection is claimed. In the course of the ensuing registration and granting procedure, the patent office issues one or more legal documents which will be termed *patent literature*, the information they contain will be termed *patent information*.

For a more detailed explanation of the legal aspects of industrial property protection, see → 13. Patents (this volume). The present chapter is concerned with patent literature from the aspect of information management.

Laws pertaining to intellectual property differ from country to country. This is also the case for the documents issued by the authorities administering industrial property protection (e.g., published patent applications, patent specifications, utility model descriptions, inventors' certificates).

Apart from obvious differences in language, format, and layout of the documents, the procedures for obtaining industrial property rights, their terms, and scope of protection vary from one patent system to another. This fact is very important for intellectual property specialists, but less so for information management. All these documents contain descriptions of industrial inventions. For the sake of clarity differences between these documents will be ignored; attention will be concentrated on patents which are the typical and most important kind of industrial property rights.

A patent document not only describes an industrial invention, it also defines the scope of protection applied for or granted by the authority. Specific problems concerning the management of patent information result from this two-tier function, these problems are either absent from or less severely felt in the handling of other technical literature. There is, therefore, justification for dealing with patent information management as a separate item within the framework of information technology.

6.2. The Volume of Patent Literature

Worldwide, industrial property titles are published by some 180 national and 4 supranational authorities. Their present annual output is in the range of 1×10^6 documents. The total amount of patent literature issued so far is estimated at 25×10^6 documents, representing an invaluable stock of technical knowledge.

In 1988 CAS abstracted 80 795 chemical patent documents issued by 29 patent offices [6.1]. These are *first cases* (see p. **12**-40) covering the majority of all chemical inventions published.

Almost 54% of the chemical patent documents are issued by the Japanese Patent Office. The shares of the European Patent Office and the United States Patent and Trademark Office are 10.3 and 6.9%, respectively.

6.3. Origin and Significance of Patent Literature in Patent-Granting Procedures

The publication of patent documents is intimately linked to the successive steps leading from filing a patent application to obtaining an enforceable patent. In a number of national and supranational patent systems (e.g., UK, France, FRG, European patent system, and Patent Co-operation Treaty system), the text of a patent application is laid open to public inspection by the patent office 18 months after the priority date (see Section 6.7.1) or date on which it was filed (see below), examination being deferred until later. Publication is simultaneously announced in the *patent gazette*. Figure 2 (p. **12**-30) shows the title page of a published European patent application, Figure 3 (p. **12**-31) shows the corresponding entry in the European Patent Gazette.

The publication of a patent application does not constitute a monopoly—it is just an early public warning that a monopoly has been applied for. Published patent applications are the most important documents in patent information management because they are the earliest disclosures of new inventions which are often not made public anywhere else. Since the inventions they describe have not yet been examined for patentability, they are also the broadest presentations of the inventions disclosed, i.e., they are the documents which carry the highest information content in patent literature. Consequently, published patent applications play a major role in the current awareness of research and development scientists, in patent examiners' files, and as a source of information retrieval in industry.

In most patent systems a patent is granted only after the disclosed invention has been thoroughly examined for patentability. The preeminent requirements to be met are novelty, nonobviousness, and industrial applicability.

An invention is deemed *novel* if it has not already been described before the date the patent application is filed or, in technical jargon, if no prior art exists. Any divulgation, be it in written, oral, or other form, made by the inventor or anybody else before first filing constitutes prior art (in certain patent systems the inventor is granted a short preclusive period). Even an application filed by another person before the filing date of the examined application, but not yet published by the patent office, can be cited as prior art against the "younger" patent application if it claims the same invention. Novelty is assumed to exist if no prior art has been detected by the patent examiner. If prior art comes to the knowledge of the patent office after granting, the patent might be restricted or even declared invalid. Prior art may be discovered in (technical) literature of any kind. Patent literature is the most promising source material to be searched for prior art.

Nonobviousness of an invention means that its crucial idea would not have been immediately apparent to persons skilled in the art, but embodies an "inventive step". To determine whether an invention is obvious or not, it is judged by a specialist whose technical background is the field the invention belongs to. The technical background is based on the pertinent technical literature, especially the patent literature.

If, after thorough examination, an invention meets the requirements of patentability, a patent is granted. A patent specification with the claim(s) approved by the examiner is issued (Fig. 4, p. **12**-32) and the grant is announced in the patent gazette (Fig. 5, p. **12**-34). In a number of patent systems the public is then invited to notify the patent office within a certain period of time (the *opposition period*) of any objections to the patentability of the invention. The most productive source of relevant prior art not yet considered by the patent examiner is the patent literature. Presentation of relevant prior art by an opponent may lead to partial or total revocation of the patent.

Even after the opposition period has expired a patent can be attacked in a nullity suit on the grounds of nonpatentability, which is also usually based on the patent literature.

A granted patent entitles the patentee to prohibit any other person from making use of his invention. The patentee may permit a party to use his invention under a special *licence agreement* in which the extent of use and the licence fee are stipulated. The licence fee is determined by the "strength" of the patent, i.e., how difficult it is to circumvent it.

Patents are limited by territory and time. A patent is effective only within the borders of the state that has granted it. If the patentee wants protection for his invention in other countries, he has to file patent applications with the patent offices of these countries as well. A patent only runs for a certain period of time—a maximum of 20 years in many countries. The patentee has to pay an annual fee to keep the patent in force, if he fails to do so it will expire. When a patent has expired anybody may make use of the invention.

The essential requirements of patentability are virtually the same in every patent system; however, granting procedures vary [6.2], [6.3]. Thus, not every patent office will publish patent

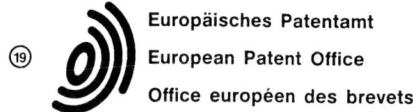

(19) **Europäisches Patentamt**
European Patent Office
Office européen des brevets

(11) Publication number: **0 327 356 A1**

(12) **EUROPEAN PATENT APPLICATION**

(21) Application number: **89301001.7**

(22) Date of filing: **02.02.89**

(51) Int. Cl.[4]: **C 07 D 301/10**
B 01 J 23/66, B 01 J 23/50

(30) Priority: **03.02.88 CN 88100400**

(43) Date of publication of application:
09.08.89 Bulletin 89/32

(84) Designated Contracting States: **DE GB NL**

(71) Applicant: **CHINA PETROCHEMICAL CORPORATION**
24 Xiaoguan Street Anwai
Beijing (CN)

(84) Designated Contracting States: **DE GB NL**

(71) Applicant: **RESEARCH INSTITUTE OF BEIJING YANSHAN PETROCHEMICAL CORPORATION**
9 Fonghuanting Road
Yanshan District Beijing (CN)

(84) Designated Contracting States: **DE**

(72) Inventor: **Jin, Jiquan Research Institute of Beijing Yanshan**
Petrochemical Corporation
Yanshan District Beijing (CN)

Jin, Guoquan Research Institute of Beijing Yanshan
Petrochemical Corporation
Yanshan District Beijing (CN)

Xu, Yong Research Institute of Beijing Yanshan
Petrochemical Corporation
Yanshan District Beijing (CN)

Shang, Liandi Tianjin Chemical Research Institute
of Ministry of China Chemical Industry
Tianjin (CN)

Luo, Guochun Tianjin Chemical Research Institute
of Ministry of China Chemical Industry
Tianjin (CN)

(74) Representative: **Knowles, Audrey Elizabeth**
624 Pershore Road
Selly Park Birmingham B29 7HG (GB)

(54) **High efficiency silver catalysts for the production of ethylene oxide via ethylene oxidation.**

(57) This invention relates to a process of preparing silver-containing catalysts and their carriers for the production of ethylene oxide via ethylene oxidation and also to the applications of said catalysts in producing ethylene oxide . A commercial trihydrated α-alumina , boehmite , carbonaceous materials , a fluxing agent , fluoride and a binder are mixed with water , kneaded andextruded to form strips which are cut and shaped. The shaped bodies are then dried , calcined and converted to α-alumina bodies i.e.carriers. This process is characterized by using trihydrated α-alumina , boehmite alumina and carbonaceous materials which have a good matching of particle sizes and proportions in preparing alumina carriers with the following pore structure:

specific surface area	0.2 - 2 m^2/g
pore volume	> 0.5 ml/g
pore radius	> 30 μ, 25-10% of total volume
	< 30 μ, 75-90% of total volume

Said alumina carriers are impregnated with silver compounds and promoters, and then dried , activated and used in ethylene oxidation for making ethylene oxide. The selectivity of the catalyst reaches from 83 to 84 percent.

Figure 2. The title page of a European patent application

(51) C07D 301/10
B01J 23/66
B01J 23/50
(25) En
(21) 89301001.7
(84) DE, GB, NL
(30) 03.02.88 CN 88100400
(54) ● Hochwirksame Silberkatalysatoren zur Herstellung von Ethylenoxid durch Ethylenoxidation.
● High efficiency silver catalysts for the production of ethylene oxide via ethylene oxidation.
● Catalyseurs à argent à haute efficacité, pour l'oxydation de l'éthylène en oxyde d'éthylène.
(71) CHINA PETROCHEMICAL CORPORATION, 24 Xiaoguan Street Anwai, Beijing, CN
(84) DE, GB, NL
(71) RESEARCH INSTITUTE OF BEIJING YANSHAN PETROCHEMICAL CORPORATION, 9 Fonghuanting Road, Yanshan District Beijing, CN
(84) DE
(72) Jin, Jiquan Research Institute of Beijing Yanshan, Yanshan District Beijing, CN
Jin, Guoquan Research Institute of Beijing Yanshan, Yanshan District Beijing, CN
Xu, Yong Research Institute of Beijing Yanshan, Yanshan District Beijing, CN
Shang, Liandi Tianjin Chemical Research Institute, Tianjin, CN
Luo, Guochun Tianjin Chemical Research Institute, Tianjin, CN
(74) Knowles, Audrey Elizabeth, 624 Pershore Road, Selly Park Birmingham B29 7HG, GB

(11) 0 327 356
A1
(26) En
(22) 02.02.89

Figure 3. Announcement of patent application (Fig. 2) in the European Patent Bulletin

applications prior to examination; in some countries (e.g., USA and USSR), the invention is not divulged before the patent is granted.

6.4. Content and Layout of Patent Documents

The layout and format of patent documents have been standardized under the guidance of the World Intellectual Property Organisation (WIPO) in Geneva [6.4].

The *title page* displays a number of bibliographic data which are each assigned a unique numerical code (INID Code, Fig. 6, p. **12**-35). The INID Code numbers greatly facilitate the identification of bibliographic data in docu-

ments written in languages the reader is not familiar with.

Important bibliographic data given on title pages are:

1) The double-digit *country code* (INID Code number 11) together with the document number (patent number) by which the document is unequivocally identified (Table 6, p. **12**-33).
2) The *classification symbols* (INID Code number 51) allocating the invention to a defined technical field. In 26 patent systems [6.5] the classification symbols are taken from the International Patent Classification (IPC, see Section 2.7.8.2).
3) The *filing date* (INID Code number 22), which determines the term of the patent.
4) The details of the *priority* or priorities claimed (INID Code numbers 31–34) identifying the country or supranational patent system in which a patent application was first filed for that invention and the filing date(s). The concept of priority is discussed further in Section 6.7.1.
5) Bibliographic details of *publications cited as prior art* in the course of examination proceedings or of publications containing prior art that have to be considered (INID Code number 56). The latter information is retrieved in a search carried out by or for the patent office.
6) An *abstract* giving a short account of the invention (INID Code number 57).

Bibliographic patent data are very important pieces of information that are extensively used in patent information management (see Section 6.7.1).

The *patent description* discloses the invention in every detail, as required by patent laws and international patent conventions. Only what is disclosed is patentable, and a patent may be revoked if the invention is not disclosed in sufficient detail for it to be reproduced by a skilled worker. The invention is also usually expounded in a number of examples supported by appropriate drawings or structure diagrams.

A *patent claim* is a concise description of the essential features of the invention. It is a legally binding definition of the product or process for which protection is sought. A patent document may contain more than one claim possibly referring to different aspects of the invention (e.g., a product and a process for manufacturing it).

In many countries patents protecting chemical substances can be obtained. These composition-of-matter claims provide the broadest possible protection because they also cover any preparation or application of that particular compound or compounds.

Claimed chemical compounds are usually expressed in terms of structural formulae. When a claim covers a whole class of chemical com-

⑲ **Europäisches Patentamt**
European Patent Office
Office européen des brevets

⑪ Publication number: **0 097 504**
B1

⑫ **EUROPEAN PATENT SPECIFICATION**

㊺ Date of publication of patent specification: **07.01.87**

(51) Int. Cl.⁴: **B 60 R 19/18**

㉑ Application number: **83303515.7**

㉒ Date of filing: **17.06.83**

(54) **Core material for automobile bumpers.**

㉚ Priority: **19.06.82 JP 105660/82**

㊸ Date of publication of application:
04.01.84 Bulletin 84/01

㊺ Publication of the grant of the patent:
07.01.87 Bulletin 87/02

(84) Designated Contracting States:
DE GB IT

(56) References cited:
DE-A-2 751 077
DE-B-1 794 025
Chemical Abstracts vol. 71, no. 13, 29 September 1969, Columbus, Ohio, USA; H.H. Lubitz "Minicel polypropylene foam", page 26, column 2, abstract no. 62035q
Chemical Abstracts vol. 82, no. 10, 10 March 1975, Columbus, Ohio, USA; D.P. HUG et al. "Polyolefin structural foam for automotive use", page 86, column 1, abstract no. 59198s
Patent Abstracts of Japan, vol. 7, no. 97, 23 April 1983, page 115C163
Patent Abstract of Japan, vol. 7, no. 100, 28 April 1983, page 9C164

(73) Proprietor: **Japan Styrene Paper Corporation**
1-1, 2-chome, Uchisaiwai-cho
Chiyoda-ku Tokyo (JP)

(73) Proprietor: **NISSAN MOTOR CO., LTD.**
No.2, Takara-cho, Kanagawa-ku
Yokohama-shi Kanagawa-ken 221 (JP)

(72) Inventor: **Adachi, Akira**
515-14, Joza
Sakura-shi Chiba-ken (JP)
Inventor: **Kubota, Takashi**
3268-249, Nishinomiya-cho
Utsunomiya-shi Tochigi-ken (JP)
Inventor: **Okada, Yukio**
Shiroyama-Danchi 2-104 4589, Ohba
Fujisawa-shi Kanagawa-ken (JP)
Inventor: **Miyazaki, Kenichi**
4-14-11, Naka-machi
Machida-shi Tokyo (JP)
Inventor: **Hagiwara, Taro**
7-36-1-518, Sagamiohno
Sagamihara-shi Kanagawa-ken (JP)

(74) Representative: **Myerscough, Philip Boyd et al**
J.A.Kemp & Co. 14, South Square Gray's Inn
London, WC1R 5EU (GB)

Note: Within nine months from the publication of the mention of the grant of the European patent, any person may give notice to the European Patent Office of opposition to the European patent granted. Notice of opposition shall be filed in a written reasoned statement. It shall not be deemed to have been filed until the opposition fee has been paid. (Art. 99(1) European patent convention).

Figure 4. A European patent specification

Table 6. List of countries and other entities issuing or registering industrial property titles, in the order of their codes*

Code	Country	Code	Country	Code	Country
AE	United Arab Emirates	FR	France	MU	Mauritius
AF	Afghanistan			MV	Maldives
AG	Antigua and Barbuda	GA	Gabon	MW	Malawi
AI	Anguilla	GB	United Kingdom	MX	Mexico
AL	Albania	GD	Grenada	MY	Malaysia
AN	Netherlands Antilles	GH	Ghana	MZ	Mozambique
AO	Angola	GI	Gibraltar		
AR	Argentina	GM	Gambia	NE	Niger
AT	Austria	GN	Guinea	NG	Nigeria
AU	Australia	GQ	Equatorial Guinea	NI	Nicaragua
		GR	Greece	NL	Netherlands
BB	Barbados	GT	Guatemala	NO	Norway
BD	Bangladesh	GW	Guinea-Bissau	NP	Nepal
BE	Belgium	GY	Guyana	NR	Nauru
BF	Burkina Faso			NZ	New Zealand
BG	Bulgaria	HK	Hong Kong		
BH	Bahrain	HN	Honduras	OM	Oman
BI	Burundi	HT	Haiti		
BJ	Benin	HU	Hungary	PA	Panama
BM	Bermuda			PE	Peru
BN	Brunei Darussalam	ID	Indonesia	PG	Papua New Guinea
BO	Bolivia	IE	Ireland	PH	Philippines
BR	Brazil	IL	Israel	PK	Pakistan
BS	Bahamas	IN	India	PL	Poland
BT	Bhutan	IQ	Iraq	PT	Portugal
BU	Burma	IR	Iran (Islamic Republic of)	PY	Paraguay
BW	Botswana	IS	Iceland		
BZ	Belize	IT	Italy	QA	Qatar
CA	Canada				
CF	Central African Republic	JM	Jamaica	RO	Romania
		JO	Jordan	RW	Rwanda
CG	Congo	JP	Japan		
				SA	Saudi Arabia
				SB	Solomon Islands
CH	Switzerland			SC	Seychelles
CI	Côte d'Ivoire	KE	Kenya	SD	Sudan
CL	Chile	KH	Democratic Kampuchea	SE	Sweden
CM	Cameroon	KI	Kiribati	SG	Singapore
CN	China	KM	Comoros	SH	Saint Helena
CO	Colombia	KN	Saint Christopher and Nevis	SL	Sierra Leone
CR	Costa Rica			SM	San Marino
CS	Czechoslovakia	KP	Democratic People's Republic of Korea	SN	Senegal
CU	Cuba			SO	Somalia
CV	Cape Verde	KR	Republic of Korea	SR	Suriname
CY	Cyprus	KW	Kuwait	ST	Sao Tome and Principe
		KY	Cayman Islands	SU	Soviet Union
DD	German Democratic Republic	LA	Laos	SV	El Salvador
DE	Germany, Federal Republic of	LB	Lebanon	SY	Syria
		LC	Saint Lucia	SZ	Swaziland
DJ	Djibouti	LI	Liechtenstein		
DK	Denmark	LK	Sri Lanka	TD	Chad
DM	Dominica	LR	Liberia	TG	Togo
DO	Dominican Republic	LS	Lesotho	TH	Thailand
DZ	Algeria	LU	Luxembourg	TN	Tunisia
		LY	Libya	TO	Tonga
EC	Ecuador			TR	Turkey
EG	Egypt	MA	Morocco	TT	Trinidad and Tobago
ES	Spain	MC	Monaco	TV	Tuvalu
ET	Ethiopia	MG	Madagascar	TW	Taiwan, Province of China
		ML	Mali	TZ	United Republic of Tanzania
FI	Finland	MN	Mongolia		
FJ	Fiji	MR	Mauritania	UG	Uganda
FK	Falkland Islands (Malvinas)	MS	Montserrat	US	United States of America
		MT	Malta	UY	Uruguay

Table 6. (Continued)

Code	Country	Code	Country	Code	Country
VA	Vatican City (Holy See)	VU	Vanuata	ZA	Republic of South Africa
VC	Saint Vincent and the Grenadines	WS	Samoa	ZM	Zambia
VE	Venezuela			ZR	Zaire
VG	British Virgin Islands	YD	Democratic Yemen	ZW	Zimbabwe
		YE	Yemen		
VN	Vietnam	YU	Yogoslavia		

* Reprinted from Patent Information and Documentation Handbook, copyright World Intellectual Property Organization, Geneva. Reproduced by permission of the publishers.

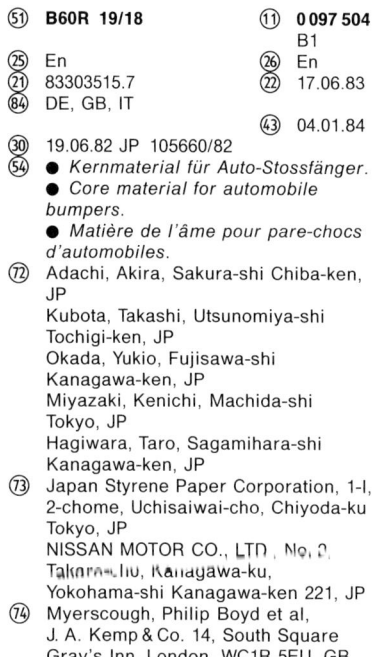

Figure 5. Announcement of a granted patent (Fig. 4) in the European Patent Bulletin

pounds with a common structural feature (substructure) this class will usually be defined by a *generic formula*, sometimes (wrongly) called Markush formula. Generic formulae pose specific problems for information retrieval (Section 6.7.2).

When a patent infringement suit comes to court, the scope of protection is determined on the basis of the patent claim(s), interpreted in the light of the description. Obviously, careful wording of patent claims is essential to obtain adequate protection. The formulation of patent claims requires experience and skill. Complex language and involved logic are frequently encountered because the value of a patent, its ability to stand up to litigation, and the economic reward for disclosing the invention to competitors depend heavily on the wording of the patent claims.

6.5. Patent Gazettes, Patent Registers, and Other Literature

In addition to the weekly patent documents, each patent office publishes a patent gazette listing the patent applications laid open to public inspection, the patents granted, and any other patent-related transactions that have occurred in that particular week. These transactions become legally valid the moment they are notified in the patent gazette; therefore, careful cognizance of the patent gazette entries is mandatory to patent information management (see Section 6.7.1).

In addition to the patent gazette, each patent office runs a register in which the gazette entries are cumulated and the current legal status is shown for every patent or application. A number of patent offices offer public inquiry services (telephone, telex, online access) [6.6].

Other patent-related literature is also relevant to patent information management. Journals published by the patent offices, professional societies, or commercial publishers report on new national laws, regulations, and patent office procedures or international conventions as well as recent court decisions having a bearing on patent matters. These publications must also be watched closely and continuously.

INID CODES AND MINIMUM REQUIRED FOR THE IDENTIFICATION OF BIBLIOGRAPHIC DATA

(10) Document identification

- *(11) Number of the document
- *(12) Plain language designation of the kind of document
- *(13) Kind of document code according to WIPO Standard ST.16
- **(19) WIPO Standard ST.3 code, or other identification, of the office publishing the document

Notes: (i) ** Minimum data element for patent documents only.

(ii) with the proviso that when data coded (11) and (13), or (19), (11) and (13), are used together and on a single line, category (10) can be used, if so desired.

(20) Domestic filing data

- *(21) Number(s) assigned to the application(s), e.g. "Numéro d'enregistrement national", "Aktenzeichen"
- *(22) Date(s) of filing application(s)
- *(23) Other date(s), including date of filing complete specification following provisional specification and date of exhibition
- (24) Date from which industrial property rights may have effect
- (25) Language in which the published application was originally filed
- (26) Language in which the application is published

(30) Priority Data

- *(31) Number(s) assigned to priority application(s)
- *(32) Date(s) of filing of priority application(s)
- *(33) WIPO Standard ST.3 Code identifying the national patent office allotting the priority application number or the organization allotting the regional priority application number; for international applications filed under the PCT, the Code "WO" is to be used
- (34) For priority filings under regional or international arrangements, the WIPO Standard ST.3 Code identifying at least one country party to the Paris Union for which the regional or international application was made

Notes: (i) With the proviso that when data coded (31), (32) and (33) are used together and on a single line, category (30) can be used, if so desired. If an ST.3 Code identifying a country for which a regional or international application was made is published, it should be identified as such using INID Code (34) and should be on a line separate from that of elements coded (31), (32) and (33) or (30).

(ii) The presentation of priority application numbers should be as recommended in WIPO Standards ST.10/C and in ST.34.

(40) Date(s) of making available to the public

- **(41) Date of making available to the public by viewing, or copying on request, an <u>unexamined</u> document, on which no grant has taken place on or before the said date
- **(42) Date of making available to the public by viewing, or copying on request, an <u>examined</u> document, on which no grant has taken place on or before the said date

*For the meaning of this asterisk, see paragraph 6 of this Standard.

0861r/PIC

Figure 6. INID codes as defined by the World Intellectual Property Organization (Standard ST. 9) Reprinted by permission of the publishers.

* Codes relate to data elements which are considered to be the minimum elements which should appear in the first page of a document and in an entry in an official gazette.

(43) Date of publication by printing or similar process of an unexamined document, on which no grant has taken place on or before the said date

(44) Date of publication by printing or similar process of an examined document, on which no grant or only a provisional grant has taken place on or before the said date.

(45) Date of publication by printing or similar process of a document on which grant has taken place on or before the said date

(46) Date of publication by printing or similar process of the claim(s) only of a document

(47) Date of making available to the public by viewing, or copying on request, a document on which grant has taken place on or before the said date

Note: ** Minimum data element for patent documents only, the minimum data requirement being met by indicating the date of making available to the public the document concerned

(50) Technical information

* (51) International Patent Classification

(52) Domestic or national classification

(53) Universal Decimal Classification

* (54) Title of the invention

(55) Keywords

(56) List of prior art documents, if separate from descriptive text

Note: Attention is drawn to WIPO Standard ST.14 in connection with the citation of references on the front page of patent documents and in search reports attached to patent documents.

(57) Abstract or claim

(58) Field of search

(60) References to other legally or procedurally related domestic patent documents including unpublished applications therefor.

*(61) Number and, if possible, filing date of the earlier application, or number of the earlier publication, or number of earlier granted patent, inventors' certificate, utility model or the like to which the present document is an addition

*(62) Number and, if possible, filing date of the earlier application from which the present document has been divided out

*(63) Number and filing date of the earlier application of which the present document is a continuation

*(64) Number of the earlier publication which is "reissued"

(65) Number of a previously published patent document concerning the same application

(66) Number and filing date of the earlier application of which the present document is a substitute, i.e., a later application filed after the abandonment of an earlier application for the same invention.

Notes: (i) Priority data should be coded in category (30).

(ii) Code (65) is intended primarily for use by countries in which the national laws require that re-publication occurs at various procedural stages under different publication numbers and these numbers differ from the basic application numbers.

Figure 6. (Continued)

(70) Identification of parties concerned with the document

    **(71)   Name(s) of applicant(s)

    (72)   Name(s) of inventor(s) if known to be such

    **(73)   Name(s) of grantee(s)

    (74)   Name(s) of attorney(s) or agent(s)

    **(75)   Name(s) of inventor(s) who is (are) also applicant(s)

    **(76)   Name(s) of inventor(s) who is (are) also applicant(s) and grantee(s)

    Notes   (i) ** For documents on which grant has taken place on or before the date of making available to the public, and gazette entries relating thereto, the minimum data requirement is met by indicating the grantee, and for other documents by indication the applicant.

            (ii) (75) and (76) are intended primarily for use by countries in which the national laws required that the inventor and applicant are normally the same. In other cases (71) or (72) or (71), (72) and (73) should generally be used.

(80) Identification of data related to International Conventions other than the Paris Convention

    (81)   Designated State(s) according to the PCT

    (83)   Information concerning the deposit of microorganisms, e.g. under the Budapest Treaty

    (84)   Designated contracting states under regional patent conventions

    (85)   Date of fulfillment of the requirements of Articles 22 and/or 39 of the PCT for introducing the national procedure according to the PCT

    (86)   Filing data of the regional or PCT application, i.e. application filing date, application number, and, optionally, the language in which the published application was originally filed

    (87)   Publication data of the regional or PCT application, i.e. publication date, publication number, and, optionally, the language in which the application is published

    (88)   Date of deferred publication of the search report

    (89)   Document number, date of filing, and country of origin of the original document according to the CMEA Agreement on Mutual Recognition of Inventors' Certificates and other Documents of Protection for Inventions

    Notes: (i) The codes (86) and (87) are intended to be used:

            - on national documents when identifying one or more of the relevant filing data or publication data of a regional or PCT application, or

            - on regional documents when identifying one or more of the relevant filing data or publication data of another regional or PCT application.

        (ii) all data in code (86) should be presented together and preferably on a single line.

       (iii) all data in code (87) should also be presented together and preferably on a single line.

Figure 6. (Continued)

6.6. Patent Abstracts

Abstracts play an important role in patent information management because they are much easier to handle, to store, and to read than the rather bulky patent documents. The abstracts must be prepared carefully and, together with the corresponding bibliographic data, must give the essential features of the inventions with at least one typical example and, if necessary, chemical formulae and drawings.

The abstracts printed on the title pages of patent documents seldom fulfil these stringent requirements because they have no legal significance, and patent offices do not put much emphasis on controlling their quality.

High-quality patent abstracts have been available for many years from commercial information services (Fig. 7). They can serve a number of purposes in patent information management: packaged suitably and distributed according to the needs of the readers they are welcome media giving up-to-date information on new technical developments. They may be grouped according to the IPC symbols shown in the abstract headings and directly used by the addressees in the form of card files for manual search purposes.

Patent searchers can use patent abstracts to assess the relevance of search results obtained in electronic retrieval systems and can modify their search strategies accordingly. The selected patent abstracts can subsequently be sent to the end users.

Patent abstracts are commercially available either printed on paper (single abstracts on cardboard or bound in bulletins) or in electronic form. At present, the most detailed and comprehensive patent abstracts are available on paper only. It is expected that they will soon appear in digitized form on optical storage media (e.g., CD-ROM).

6.7. Principles and Methods of Patent Information Management in the Chemical Industry

Because of the two-tier function of patent literature (Section 6.1), patent information management has to serve the needs of two user communities: the first is mainly interested in the industrial inventions disclosed in patent literature, and the second focusses on the legal implications of patent documents.

The first community comprises research and development scientists. They have to be informed speedily and comprehensively of new technical developments and must be supported by providing immediate access to the huge stock of technical knowledge embodied in the patent literature.

Patent specialists belonging to the second community are engaged in filing patent applications, protecting and enforcing corporate rights, and opposing impeding monopolies of other parties. They must be supplied with accurate, up-to-date information on existing patents and their legal status.

The novelty of an invention is an essential condition of its patentability (Section 6.3). Therefore, an invention can be patented only once in a given patent system (an invention can be patented in different patent systems, see Section 6.7.1). Although duplicate inventions may occur in the patent literature because in some patent systems applications are published before they have been examined, the redundancy rate is extremely low compared with other technical literature. The greatest care must therefore be taken in indexing and retrieving patent literature, because if relevant information is lost it may not be recoverable elsewhere.

In sectors of technology where research activity in competing industrial companies is high, large numbers of closely related inventions may appear in the patent literature within fairly short periods of time. In-depth indexing is required to discriminate between these inventions and to prevent the searcher from being overwhelmed by masses of irrelevant documents.

6.7.1. Indexing and Retrieval of Bibliographic Patent Data

More than one (three on average) patent application is usually filed for an invention in different patent systems. In 1883 an international agreement, the Paris Convention for the Protection of Industrial Property, was drawn up to regulate the rights of foreign patent applicants [6.7]. When a patent application has been filed in one of the member states of the Paris Convention the applicant may file applications in other member states within one year from the first filing date without his first application or any other descrip-

| 88-221149/32 A88 D15 J01 (A14 A26) AGEN 22.01.87 | A(10-E21B, 11-B5A, 12-W11A) D(4-A1A, 4-A1E, 4-B6) J(1-C3) D0040 |

88-221149/32   A88 D15 J01 (A14 A26)   AGEN 22.01.87
AGENCY OF IND SCI TECH   *DE 3801-690-A
03.07.87-JP-165285 (+ JP-011337) (04.08.88) B01d-13 B01d-53/22
C02f-01/44 C08j-03/24 C08j-05/22
Poly-ion complex polymer membrane - for sepg. water from organic substance in liq. or vapour form
C88-098649

Other Priority: 04.03.87-JP-047495 03.07.87-JP-165284

Polymer membrane for sepg. water from an organic substance by pervaporation of permeation of water vapour consists of a synthetic polymer (I) with an anionic gp. associated with a synthetic polymer (II) with a cationic gp. by an ionic bond to form a polyion complex on the surface of the membrane and/or in the membrane.

USE/ADVANTAGE
The membrane is water-resistant and durable and has a satisfactory permeation rate and high sepn. coefft. over a wide concn. range of an organic substance in soln. It is useful for sepg. water from e.g. alcohols (MeOH, EtOH, 1-PrOH, 2-PrOH, n-BuOH), ketones (acetone, MEK, ethers (THF, dioxane), organic acids (formic, acetic), aldehydes (acetaldehyde, propionaldehyde) or amines (pyridine, picoline) or gaseous mixts. contg. water and one or more of these cpds.

PREFERRED COMPOSITION
(I) is crosslinked with a crosslinking agent (pref. a polyfunctional epoxide, amine, methylolmelamine or NCO cpd.) forming a covalent bond with the polymer and is polyacrylic acid or its metal or ammonium salt.
The cationic gp. of (II) is a prim., sec., tert. or quat. amino gp. and (II) pref. is polyallylamine or polyethyleneimine; or (II) has a quat. ammonium salt gp. in its main chain and pref. is of formula (IIA), (IIB), (IIC) or (IID):

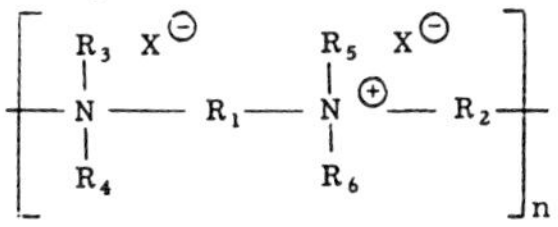

$$\left[\begin{array}{c} R_3 \; X^{\ominus} \\ | \\ -N - R_1 - N^{\oplus} - R_2 - \\ | \qquad\qquad | \\ R_4 \qquad\quad R_6 \end{array} \right]_n \quad (IIA)$$

DE 3801690-A+

$$\left[\begin{array}{c} CH_3 \; X^{\ominus} \qquad\qquad CH_3 \; X^{\ominus} \\ | \qquad\qquad\qquad\qquad | \\ -N^{\oplus} - CH_2-CH_2-N^{\oplus}-CH_2-\bigcirc-CH_2- \\ | \qquad\qquad\qquad\qquad | \\ CH_3 \qquad\qquad\qquad CH_3 \end{array} \right]_n$$

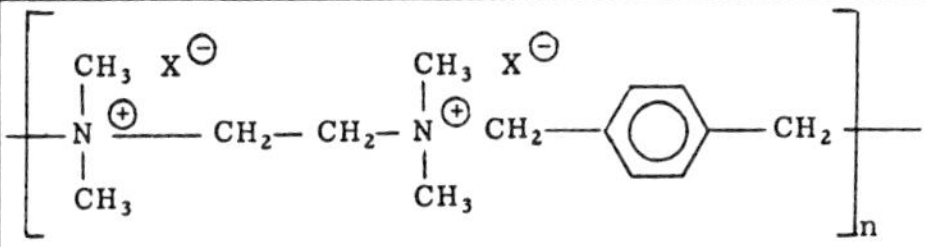

(IIB)

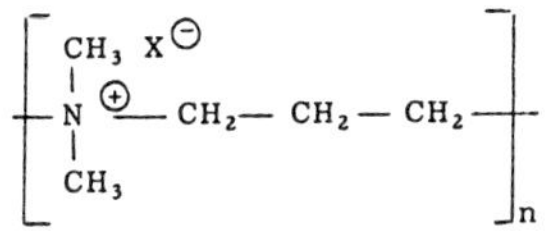

$$\left[\begin{array}{c} CH_3 \; X^{\ominus} \\ | \\ -N^{\oplus} - CH_2-CH_2-CH_2- \\ | \\ CH_3 \end{array} \right]_n \quad (IIC)$$

$$\left[\begin{array}{c} CH_3 \; X^{\ominus} \\ | \\ -N^{\oplus} - CH_2-CH-CH_2- \\ | \qquad\qquad | \\ CH_3 \qquad\quad OH \end{array} \right]_n \quad (IID)$$

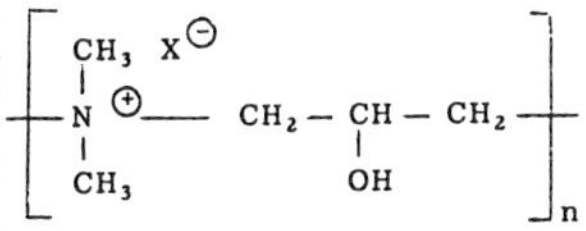

R_1 and R_2 = alkylene with $\geqslant$ 2C or a hydroxyalkylene, alicyclic or aromatic gp.;
R_{3-6} = 1-3C (hydroxy)alkyl;
X^- = a halide counter-ion.
The membrane pref. is a composite membrane with a skin of the polyion complex and a porous substrate.

PREPARATION
The membrane can be prepd. by dipping a (I) or (II) membrane, which has been rendered insol. by crosslinking, in a (II) or (I) soln. For a composite membrane, a porous polymer membrane with an insolubilised (I) coating is dipped in a (II) soln.

EXAMPLE
An aq. soln. of polyacrylic acid (viscosity 24 cP for 1% aq. soln.) was neutralised with aq. NH_3 and diluted with water to give a 0.5% polyammonium acrylate soln. This was spread on a polyethersulphone ultrafiltration membrane and dried in air to give a 0.4 μ thick soln. The composite membrane was dipped in a slightly acidic 0.5% soln. of polyallylamine hydrochloride (pH 3.5) in 1 : 1 EtOH/water for 10 min. at room temp., giving a polyion complex membrane.

DE 3801690-A+/1

88-221149/32   A 95 : 5 EtOH/water mixt. at 0.1 kg/cm² excess pressure and 70°C was supplied to the prim. (main) side of (A) the complex membrane or (B) the original polyammonium acrylate membrane. The permeation rate was (A) 0.69, (B) 0.54 kg/m².h; and the sepn. coefft. (A) 107, (B) 38. (22pp016RBHDwgNo0/0).

D0041

DE 3801690-A/2

Figure 7. A patent abstract issued as published by Derwent Publications Ltd., London
Reprinted by permission of the publishers.

tion of the invention made public during this one year period being cited as prior art against the follow-up applications. This special right is called *priority* (more specifically, *convention priority*). It is only granted when explicitly claimed, and the details of the priority claimed comprising the country (or supranational patent system), date, and number of the first-filed application are some of the most important bibliographic patent data. The grant of priority is restricted to follow-up applications whose disclosure is identical to that of the first-filed applications, and proof of this has to be provided. The applicant may cumulate the disclosures of several first-filed applications into one follow-up application for which he then claims more than one priority.

The Paris Convention is supervised by the World Intellectual Property Organisation; 100 states have so far signed the Convention [6.8]. The Paris Convention has made the filing of patent applications much more secure. Nevertheless, 3–5% of follow-up applications are filed without priorities being claimed by the applicants.

Priority cannot only be claimed for follow-up applications in different countries. In some patent systems it is also possible to claim the priority of a previous application filed in the same patent system and based on the same invention (*domestic priority*). The applicant may develop the invention further and file a new application whithin one year. Domestic priority has the same effect as the Convention priority with respect to prior art. When a domestic priority is claimed, the previous application on which it is based becomes ineffective. Domestic and Convention priorities cannot be added up to periods of more than one year.

A cluster of patents and patent applications based on the same invention is called a *patent family*. The patent documents of a patent family are published on different dates depending on the administrative procedures in the patent offices. The first issued document of a patent family is called a *first case*. First cases are not necessarily identical with the first filed applications.

Although the documents of a patent family may differ in the wording of their patent claims, their technical disclosures are essentially the same (otherwise the priorities of the first-filed applications could not be claimed). This has the advantage that it suffices to index the disclosure of one patent family member document, normally the first case. As a result, the indexing effort is reduced to about one third. Subsequent documents of the patent family are simply registered by means of their bibliographic data. The members of each patent family have a unique common entry: the application number of the patent application on which their priority is based. Therefore, patent family files (*patent data banks*) use the *priority application numbers* as access keys.

The allocation of incoming patent documents to patent families was a formidable and demanding task—priority details had to be collated and transferred to card files. Nowadays bibliographic patent data are available on electronic media [6.9] and building patent data banks is no longer a fundamental problem. Moreover, comprehensive patent data banks are made accessible to online searches by commercial patent information services.

Building up a patent data bank is by no means simple. Multiple priorities, multiple applicants, relationships to patents of addition, continuation applications, continuation-in-part applications, etc. create a complex network of cross references. Other bibliographic information (e.g., applicants' names, inventors' names, patent numbers) must be made searchable as well; at least the applicants' names and the patent numbers should function as secondary access keys.

Because corporate patent specialists use patent data as a basis for legal actions, patent data bank entries have to be carefully checked both intellectually and by elaborate computer programs. Patent data banks are important working tools for the patent information manager. Not only do they provide a reliable control mechanism for indexing activities and coverage, they are also used for looking up bibliographic patent data and patent family relationships. Thus, patent specialists who know the data of a certain patent may easily find the data of corresponding patents in other patent systems.

A patent data bank established by computer processing of priority information can only locate patents and patent applications with identical priority application numbers. It therefore fails to identify follow-up patent applications or patents for which no priority has been claimed. Nonpriority cases (in jargon: noncons) can only be detected by careful intellectual analysis of the patent documents on the basis of their technical disclosure.

Not only information on patent families is relevant to patent specialists. They often also want to know about the *legal status* of a patent or patent application, e.g., when the application was laid open to public inspection, whether a patent has already been granted, and if so when, whether a search has been applied for and if so what the result was. Such questions can be answered by a case-by-case check of the patent registers or from a file which has been regularly fed with the relevant entries selected from the patent gazettes. This is tedious and costly. Fortunately, commercial information services have recently established *patent status data banks* that can be accessed online.

The exploitation of patent data banks and patent status data banks for *competitor intelligence* has been recommended in the literature [6.10]–[6.16]. It is certainly not difficult to use bibliographic patent data to compile comprehensive statistical charts displaying, for example, the development of patent activities in specific areas of technology. Much more relevant to industrial companies, but also much more demanding, is the design of a computerized warning system that alert specialists at an early stage to new trends in technology from competitors' recent patent applications. For this a number of conditions have to be met:

1) The patent data processed must be complete and reliable.
2) The computer software has to be fine-tuned to pick up meaningful fluctuations and ignore trivial ones.
3) The output obtained by purely numerical operations has to be translated into meaningful results by critical analysis of the corresponding patent documents and their disclosures.

6.7.2. Indexing and Retrieval of the Technical Disclosure of Patent Documents

The objective of indexing is to compress the information contained in a document into a sufficiently small number of controlled descriptors which can be more easily stored and more reliably retrieved than the full text. This can be done in a variety of ways. Consistency is an important aspect of indexing [6.17], [6.18].

The methods used for indexing patent literature are in principle the same as for other technical literature and will be dealt with in Chapter 10. Only approaches that are characteristic of indexing (chemical) patent literature will be discussed here.

Classification is still the indexing method most widely employed by patent offices. In 1954 several national governments agreed upon an internationally uniform classification scheme that is now employed in 26 patent systems worldwide: the International Patent Classification (IPC) [6.19]. The IPC symbols are printed on the title pages of patent documents (INID Code number 51). Every five years the IPC is revised and adapted to recent developments in technology by a group of experts under the guidance of the World Intellectual Property Organisation. The fifth edition of the IPC (IPC^5) took effect in 1990. It divides technology into eight sections (A to H) and comprises 64 000 hierarchical classificatory units (classes, subclasses, main groups and subgroups).

The IPC Manual has nine volumes giving extensive guidance to users. The overriding problem with the IPC, which is used by many patent offices, is to keep its application consistent [6.20]. In addition, the IPC is a function-oriented classification which causes particular problems in chemistry where patent information is largely substance-oriented. Recent IPC revisions have taken this into account but fully satisfactory solutions have not yet been devised.

Although the IPC certainly has its merits as a search tool for chemical patent literature, the chemical industry prefers different indexing and retrieval methods. *Coordinate indexing* (Section 10.1.2) has been widely used for both structural and nonstructural concepts.

The development of *topological indexing* of structural concepts started in the chemical industry in the late 1950s and has made enormous progress in recent years. Topological indexing of single (fully defined) chemical structures and graphical query formulation of substructure searches in connection table representations have become commonplace (Section 10.1.4). However, a form of structural formulae exists that is characteristic of chemical patent literature and poses considerable problems to topological indexing: *generic formulae* (often, but incorrectly, called Markush formulae) [6.21]. Type I generic formulae represent finite, albeit sometimes extensive, classes of chemical compounds:

$R^1 = H, CH_3, C_2H_5$
$R^2 = H, CH_3$
$R^3 = H, -CH<^{CH_3}_{CH_3}$
$X = O, S$

Up-to-date topological indexing and retrieval systems can correctly map type I generic formulae and allow substructure searches to be done in the corresponding connection table representations.

Type II generic formulae (Fig. 8) represent virtually infinite classes of chemical compounds defined by structural features that cannot be mapped unambiguously in ordinary connection table representations. Up to very recently type II generic formulae could only be indexed by making use of fragment codes. Sophisticated topological indexing and retrieval methods for type II generic formulae are now emerging.

Commercial information services [6.22]–[6.25] have been quick to offer online topological full structure and substructure searches in single compound representations but have found it quite difficult to cope with comprehensive indexing of generic formulae and to implement reliable and generally applicable topological search algorithms for substructure searches. The fragment codes the searcher has to use instead frequently produce search results at low relevance ratios, making comprehensive patent searches elaborate and time-consuming.

6.7.3. Collecting, Storing, and Making Patent Literature Available

In each patent system patent documents are only issued by the patent office, which publishes them in complete, well-ordered series and in a fixed sequence. Thus, the patent offices' output is easy to control, there is no "gray literature".

(51) C07 C-93/14
(21) 88119667.9
(22) 25.11.88
(30) 27.11.87
19.04.88
10.05.88
(84) AT, BE, CH, DE, ES, FR, GB, GR, IT, LI, LU, NL, SE
(54) **Substituierte Alkylamin-Derivate** ◇
Substituted alkylamine derivatives ◇
Dérivés substitués d'alkylamine
(71) BANYU PHARMACEUTICAL CO., LTD., 2–3, Nihonbashi Honcho 2-chome, Chuo-ku Tokyo 103, JP
(74) Kraus, Walter, Dr., et al, Patentanwälte Kraus, Weisert & Partner Thomas-Wimmer-Ring 15, D-8000 München 22, DE
(57) A substituted alkylamine derivative represented by the general formula

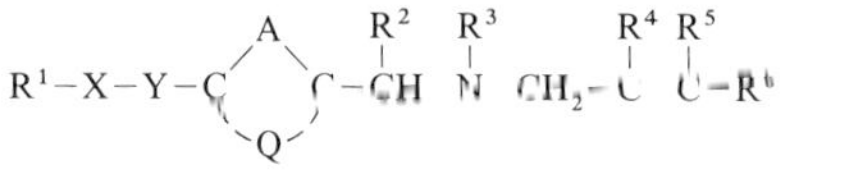

(11) EP 0 318 860 A2
(25) En (26) En
(43) 07.06.89
JP 299584/87
JP 96286/88
JP 113310/88

(72) Takezawa, Hiroshi, Hachioji-shi Tokyo, JP; Hayashi, Masahiro, Ichikawa-shi Chiba-ken, JP; Iwasawa, Yoshikazu, Naka-gun Kanagawa-ken, JP; Hosoi, Masaaki, Kawasaki-shi Kanagawa-ken, JP; Iida, Yoshiaki, Yokohama-shi Kanagawa-ken, JP; Tsuchiya, Yoshimi, Funabashi-shi Chiba-ken, JP; Horie, Masahiro, Itabashi-ku Tokyo, JP; Kamei, Toshio, Tachikawa-shi Tokyo, JP
(51) C07 C-103/82 C07 C-103/75
C07 D-307/12 C07 C-121/75
C07 D-307/42 C07 D-263/32
C07 D-277/24 C07 D-213/30
C07 D-295/08 C07 D-261/08

$$R^1-X-Y-C\underset{Q}{\overset{A}{\diagdown}}C-\overset{R^2}{\underset{|}{C}}H\ \overset{R^3}{\underset{|}{N}}\ CH_2-C\overset{R^4}{\underset{|}{}}\overset{R^5}{\underset{|}{C}}-R^6 \qquad (I)$$

wherein A represents a methine group, a nitrogen atom, an oxygen atom or a sulfur atom; Q represents a group which may contain one or two hetero atoms selected from the group consisting of nitrogen, oxygen and sulfur atoms and forms a 5- or 6-membered aromatic ring together with the adjacent carbon atoms and A (the aromatic ring may contain one substituent, or two identical or different substituents, selected from the class consisting of halogen atoms, hydroxyl groups, lower alkyl groups and lower alkoxy groups); X and Y may be identical or different and each represents an oxygen atom, a sulfur atom, a carbonyl group, a group of the formula -CHR^a- in which R^a represents a hydrogen atom or a lower alkyl group, or a group of the formula -NR^b- in which R^b represents a hydrogen atom or a lower alkyl group, or when taken together, X and Y represent a vinylene or ethynylene group; R^1 represents a lower alkenyl group which may be substituted, a cycloalkenyl group which may be substituted, a lower alkynyl group which may be substituted, an aryl group which may be substituted, or a heterocyclic group which may be substituted; R^2 represents a hydrogen atom or a lower alkyl group; R^3 represents a hydrogen atom, a lower alkyl group, a lower alkenyl group, a lower alkynyl group or a cycloalkyl group; R^4 and R^5 may be identical or different and each represents a hydrogen atom or a halogen atom; and R^6 represents an acyclic hydrocarbon group which may be substituted (which may contain 1 or 2 unsaturated bonds selected from double and triple bonds); a cycloalkyl group which may be substituted, or a phenyl group which may be substituted; provided that when either one of X and Y represents an oxygen or sulfur atom, or the group -NR^b- in which R^b is as defined above, the other represents a carbonyl group or the group -CHR^a- in which R^a is as defined above, and a nontoxic salt thereof. The said compound is useful as pharmaceuticals, particularly for the treatment and prevention of hypercholesterolemia, hyperlipemia and arteriosclerosis.

Figure 8. A type II generic formula in a patent application

Patent documents never go out of print and are always available, albeit sometimes as photocopies. There is no copyright on patent documents.

Patent documents can be purchased from patent offices or from commercial distributors on the basis of the IPC; subscriptions to individual classes can be made.

Patent information is offered on different media. *Printed documents* are easiest to deal with by readers, but most cumbersome to handle by patent information managers. Collections of patent documents on paper require considerable storage capacity and are, therefore, expensive to maintain. They have to be restricted to documents from a few patent systems and pertaining to a narrow selection of IPC classes according to the immediate needs of the users.

Reeled microfilm saves a lot of space but should be used only for dated material that is rarely accessed. *Microfiches* are somewhat handier to use. *Aperture cards,* available only from a small number of patent offices, can be sorted and copied automatically. Patent literature on *CD-ROM* marks a breakthrough. The European Patent Office started to issue its published patent applications in digitized (facsimile) form on CD-ROM ('Espace') in 1989, and the German Patent Office and the United States Patent and Trademark Office are to follow suit. The Espace material (60 CD-ROM's annually) is inexpensive and easy to handle, provided that the necessary equipment is at hand.

Although patent documents printed on paper can be arranged as classified (IPC) files for manual searches, this is only feasible for small specialized collections. Large collections of patent literature should always be employed in conjunction with information retrieval systems. Their output then leads the searcher to the position of the document in the corresponding archive (paper, microfilm, etc.), which can then simply be arranged in numerical order. Patent literature in digitized form is now opening up completely new opportunities for combining information retrieval and document display.

6.8. The Role of Patent Information in Industry [6.26]

The importance of patent literature for current awareness has already been mentioned in foregoing sections. Research and development scientists should be informed of new technical developments in two ways: they should regularly and promptly receive, on the basis of carefully tailored information profiles and in easily manageable quantities, patent abstracts (see Section 6.6) or patent documents (selected by IPC classes, see Section 6.7.2) reporting on new patent literature published in the areas they are working on. In addition, the scientists should be encouraged to read abstract bulletins covering wide, interdisciplinary areas of technology with some bearing on their work.

In contrast to the current awareness activities triggered off by publications, retrospective searches are problem-oriented [6.27]. They have to help a research scientist starting work in an unfamiliar area to gain a first idea of how to tackle the problem, to obtain a firm concept for a solution, and eventually to work out a description of his invention on which a patent application can be based.

When embarking on research in a new field, a scientist is well advised to ask for a comprehensive search that unearths all the prior art. A careful study of patent literature dating back a number of years is also worthwhile. Many ideas disclosed in documents for which the corresponding patents have already expired may have been well ahead of their time and may not have been put into practice for technical or economic reasons. Circumstances may have changed in the meantime and reconsideration of these ideas might be worthwhile. Old patent literature is not covered by most commercial online patent data bases; it is only searchable in the card files that have fortunately been kept by many industrial information departments.

As an invention assumes concrete form, additional searches on specific details might be necessary. When the text of a patent application is being formulated a further search based on the proposed patent claim(s) should be carried out in anticipation of the patent examiner's objections.

The searches considered so far are *prior art searches*: a clearly defined general idea of an invention (the most clearly defined being the patent claim) is taken as the starting point of a query, and the search is aimed at retrieving patent documents or other literature in which the idea is described either identically or more specifically. The query should be formulated sufficiently broadly to retrieve literature relevant to both novelty and obviousness. The more comprehensive the result of a prior art search is, the better are the chances of standing up to examination and opposition proceedings and of obtaining an enforceable patent.

The moment a patent application is filed the invention enters the realm of the patent specialist, who will do his best to convince the examiner of the patentability of the invention. When a patent is granted, the specialist may have to defend it in opposition proceedings or in a nullity suit. Once a strong patent has been obtained, it must be enforced against infringement and exploited by licensing to make optimal use of the monopoly granted. The patent information manager supports the patent specialist by conducting searches in the patent literature. The patent specialist is involved in warding off imminent monopolies of competitors by filing opposition cases, naturally based on prior art searches.

When a new product is manufactured and launched, it is particularly important to make sure that an existing patent is not infringed. Infringement can be very costly and might even endanger the company; for this reason a careful search has to be conducted to exclude every possibility of patent infringement. *Infringement searches* are different from prior art searches. The starting point of an infringement search is the new product or process for its manufacture and the search aims at retrieving every patent still in force that either directly claims the product or process in question or covers it in more general terms. No commercial patent information service offers retrieval software with which infringement searches can be carried out as easily as prior art searches. It is common practice to generalize the query step by step, and the degree of sophistication of the retrieval software determines how much irrelevant material has to be sifted out by the searcher. The retrieved patent documents must then be checked against the patent status data bank to find out which patents are still in force.

6.9. Organizational Aspects of Patent Information Management

The organizational position of patent information management within an industrial enterprise is determined by the two-tier role of patent literature: the legal aspect calls for allocating patent information management to the patent department, whereas the information aspect suggests its close coordination with research and development.

Patent information is much more frequently used by research personnel than by patent specialists. It is for this reason and because up-to-date information retrieval facilities relying heavily on electronic data processing are more efficiently employed to capacity when centralized that preference should be given to the integration of patent information management into overall corporate information management activities.

All industrial companies have to index their technical literature at high cost for corporate use. For many years cooperation schemes have therefore existed in information management, and especially in patent information management. Since indexing technical literature has become more and more the domain of commercial services, cooperation between industrial companies has been carried on in the form of user groups. These groups not only survey and criticize the commercial information providers' activities, but also test their new services and give competent advice on how to improve them. Examples of international patent information user groups are the Patent Documentation Group (PDG) [6.28], which has existed since 1957, and the International Association of Producers and Users of Online Patent Information (OLPI), established in 1986 [6.29]. Cooperation of these organizations is being extended to patent offices. A considerable number of national (patent) information user groups also exist.

7. Information Technology

7.1. Introduction

In the twentieth century there has been an exponential growth in the published literature. Chemical Abstracts recorded about 12 000 abstracts in 1907 but nearly 490 000 in 1989. Chemical companies have also experienced an enormous increase in the number of internal records; very large collections of chemical structures and related property data have been established in-house. It would not have been possible to manage these huge quantities of data without computers or the great advances made in information technology since the 1970s.

This brief introduction to information technology covers only those aspects of greatest relevance to online searching, and to in-house documentation and chemical structure handling systems.

7.2. Hardware [7.1], [7.2]

7.2.1. Digital Computers

A computer's memory consists of a large number of electronic switches represented as 1 (on) or 0 (off). A combination of switches, or bits, can be used to represent numbers, letters, and other data to be handled by the computer. The American Standard Code for Information Interchange (ASCII) was created to assign binary codes to data. Eight bits, forming a byte, are used to represent each character. (Large IBM machines use the extended binary coded decimal information code, EBCDIC.)

Modern computers use semiconductor memory, silicon usually being the basic material. The number of integrated circuits that can be stored in a single chip has increased rapidly from tens of transistors in 1973 (small scale integration, SSI) up to millions of transistors (ultralarge scale integration, ULSI) in 1990.

A computer consists of a central processing unit (CPU), which is an assembly of electronic circuits linked to data storage devices, input devices, and output devices.

First-generation computers (1946–1955) were based upon valve technology. They were very large and had very poor performance. Programming was laborious using machine code (binary) or an assembly level language (a symbolic version of machine language). Second-generation computers (1956–1963) used transistors and had compilers. Programs could be written more quickly and conveniently in a high-level language which the compiler converted into machine instructions. Third-generation computers (1964–1981) had integrated circuits, semiconductor memories, and time-sharing operating systems (see Section 7.3) and made large-scale use of disk storage. The current fourth generation of computing uses very large-scale integration, VLSI (hundreds of thousands of transistors on a single chip).

Research continues into fifth generation computing. The Japanese Fifth Generation Project [7.3] began in 1982. It was closely followed by the British 5-year "Alvey" program (1983) [7.4]; the European Strategic Program for Research and Development in Information Technologies (ESPRIT, 1984); and initiatives sponsored by American, French, German, and East European goverments and industry [7.5], [7.6].

Advances in microelectronics could increase processing power by a factor of 1000 [7.7], [7.8]. However, physical constraints (e.g., the speed at which electrons can move and the need to dissipate heat) will eventually prevent further large increases in the performance of any one processor. Research is, therefore, also taking place into "parallel processing", the design of systems which can perform large numbers of simultaneous operations [7.9]. Some applications of parallel processing in information science are considered in Sections 8.4.4 and 10.4.4.

Another development is reduced instruction set computer (RISC) technology which makes more efficient use of the CPU by reducing the number of different machine instructions.

Studies are also being carried out into man–machine interfaces [7.10], which could provide more sophisticated methods of interaction between computers and users (Section 11.5) and into expert systems which allow inferences to be drawn from knowledge encoded in machine-readable form (Section 11.6).

The most powerful computers are termed supercomputers. Corporate large computers with hundreds of users are referred to as mainframes. Minicomputers are rather less powerful multiuser machines. Personal computers are often referred to as microcomputers (see Section 7.2.4).

7.2.2. Computer Peripherals

Peripherals are the means of getting information into and out of computers.

Printers [7.1], [7.11], [7.12]. Paper-based input (cards and tapes) is now of historical interest only but printed output continues to be popular.

Serial printers (e.g., dot matrix, daisy wheel, thermal, ink-jet) print a single character at a time. Dot matrix printers are noisy and do not produce particularly high-quality prints but they print quickly and cheaply. Daisy wheel printers are slower and more expensive than dot matrix printers, and are also noisy, but they produce prints of much higher quality. Thermal printers are much quieter but are slow and the image degrades with time and on exposure to light. Inkjet printers (→ Imaging Technology, **A13**, pp. 588–599) can produce high-quality print at high speed but they are more expensive and less reliable than daisy wheel printers. They can print graphics and characters and are capable of high-quality color printing.

Line printers were the traditional method of producing large quantities of print from mainframe computers, printing an entire line at a time.

Laser printers can produce a whole page of high-quality print at a time; they have become commonplace despite their (comparatively) high cost.

Computer-output microfilm (COM) is considered in Section 8.4.1.

Graph Plotters [7.11]. Both drum plotters and flatbed plotters use pens on paper. They retain a market niche for special applications, especially where high quality or extensive use of color is required.

Computer Terminals. The use of teletypewriters is now rare and visual display units (VDUs) are commonplace. The VDU has a display screen, a data store, a character generator, a keyboard for data input, and a communications interface. "Intelligent" terminals have local processing capabilities. See also → Display Technology, treated in the A series.

Graphics Terminals [7.13]. In the early 1980s graphics terminals were expensive and were restricted to uses such as molecular modeling (see Section 10.7.3) and computer-aided design and manufacture (in an engineering environment). With the advent of microcomputers (see Section 7.2.4), most scientists now have access to graphics systems for handling chemical structures for producing graphs and charts, and for desktop publishing. An important factor is the resolution of the graphics display system. Low resolution is adequate for most home computing whereas computer-aided design and manufacture require very high resolution. Most chemical structure handling systems require intermediate resolution.

There are two main ways of producing graphics on a cathode ray tube (CRT) display: raster displays and vector displays. Raster displays work much like a television: the screen is scanned by a large predetermined number of horizontal lines that provide uniform coverage of the display space. Vector displays work by drawing lines on the screen. The latter are more expensive and are gradually being overtaken by raster technology, except for specialized uses (e.g., rotating molecules in three dimensions) where both interactivity and high resolution are required. (Interactivity refers to the speedy, effective communication between humans and machine when response is fast.)

Light pens are used to input two-dimensional visual data to a graphics terminal [7.1], [7.13]. One end of the pen houses a light-sensitive transistor which senses the illuminated pixels (picture elements) on the screen. The action of pointing the pen at the screen requires the user to hold an arm up for extended periods of time. *Graphics tablets* do not have this disadvantage since they act as horizontal "drawing pads" [7.1], [7.13]. A tablet is used in conjunction with a "pen" which has a microswitch in the tip. When the tip is depressed, the coordinates of the pen-down position are transmitted to the graphics screen.

Two-dimensional data are more commonly input to a microcomputer by means of a *mouse*, which is a small hand-held body with a freely turning ball underneath it [7.1], [7.13]. As the mouse is moved around, the movements of the ball are transmitted to the cursor on the VDU. Mechanical mice run on any flat surface; optical mice require a special pad with grid lines. Mice and related devices are useful not only for inputting drawings, but also for moving a cursor in selection of menu items in a menu-driven program.

Touch screens have a grid of IR beams in front of them, which can be broken by a finger or a suitable implement. They are not very precise but are useful for inexperienced users operating menu-driven software in, for example, public information systems.

Image Scanning [7.14]–[7.16]. Digital raster scanning and optical character recognition (see below) are alternatives to key punching for the conversion of a paper document into electronic form.

Image scanning uses digital facsimile transmission technology to capture a page image and convert it into a digital code that can be displayed as a visual image. However, text and numbers within an image, once stored, cannot be interpreted or processed. If the input pages are manually indexed they can be retrieved as visual images but items within each page (image) cannot be retrieved.

The images are represented electrically as a stream of very small picture elements (pels or pixels). Each pixel may exist in one of two states, black or white, represented by one bit of data. (Color and half-tone images require a more complex methodology.) The collection of individually processable pixels is referred to as a bit map (Fig. 9). Bit-mapped images occupy a considerable amount of storage space compared with text and they are usually compressed algorithmically to reduce storage requirements and increase transmission speed.

Images may also be represented by vectors. For vectorization the line structure of the image is analyzed and described in terms of its coordinates; it is specified as a set of descriptions of its component lines and shapes.

The PATDPA system which involves vectorization for further compression and efficiency is described in Section 8.2.9. Other online systems that can display images are described in the same section.

Optical Character Recognition (OCR) [7.15]. Optical character recognition equipment (Section 11.5.2), such as that of Kurzweil, reads individual printed or typed characters and converts them into digital character codes. The text can then be indexed, edited, reformatted, or used in an information retrieval system. Optical character recogniton is much more dependent on the quality of the source document than digital raster scanning. Manual checking of input material and correction of misinterpreted characters may be necessary.

Bar Codes. Bar codes represent information as black and white bars in a logical pattern. A beam of light is shone across the bars and the reflected light is converted into voltage differences that the computer can recognize.

Bar codes have been used in library applications (e.g., to register borrowed books) and in inventory systems for laboratory chemicals to record bottle movements.

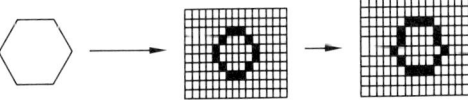

Figure 9. Creation and scale-up of a bit-map image

7.2.3. Data Storage
(→ Information Storage Materials, A series)

Textual, numeric, graphic, audio, and video data may be recorded. Comparisons of some storage media, including speeds, costs, and capacities are given in Tables 7 and 8.

Paper. The widespread implementation of office technology has not apparently decreased the the volume of paper in offices and document collections. Paper occupies a lot of space, may deteriorate in storage, and has poor security. However, paper records are necessary for some legal purposes.

Microforms [7.2]. Microform images are recorded on photographic film in roll-film format or flat, microfiche format. They may be used in computer-based information systems (see Section 8.4.1) or may be handled manually using reader–printers. They occupy about 2% of the space occupied by the original documents and therefore offer considerable savings in storage space and cost for text and graphics in records collections, archives, and libraries. However, retrieval is slow and users are resistant to the manual operations involved. Microforms are therefore best used for high-volume archiving where little retrieval is needed.

Magnetic Media (see → Information Storage Materials, **A14**, pp. 172–196). Magnetic media (tape and disks) have the advantage of being reusable: the data can be overwritten many times if required. *Magnetic tape* is portable, making transfer of data from one system to another possible, but the data has to be read serially, which is slow.

A *magnetic disk* has a film of magnetic material on a disk. The whole disk is rotated while the read–write heads move over it. Both fixed and removable disks are available. The disks allow random access, which is faster, and can store more data than a tape (0.5–1.2 Gbytes compared with 20 Mbytes for a tape).

Digital Optical Discs [7.18]. Optical media offer huge storage capacity at low cost. They are not prone to magnetic destruction and should not in theory deteriorate with repeated use since the disc is not physically touched as it spins in the reading device.

Table 7. Cost, capacity, durability, and speed of data storage media*

Medium	Capacity of one unit, Mbytes	Cost (1989), $/MByte	Removable	Durability	Speed
CD-ROM	550	0.005	yes	stable	medium
Online	unlimited	≤ 200	no		very slow
Paper	0.002	7	yes	stable	extremely slow
Tape	60	0.5–1	yes	volatile	very slow
Floppy disk	0.36–1.4	1–2	yes	volatile	medium
Hard disk	10–600	10–20	no	volatile	very fast
Bernoulli box	20–40	5	yes	volatile	fast
WORM	10–400	0.5–1.5	yes	stable	medium fast

* Reproduced with permission from a presentation given by STUART MARSON to the American Chemical Society Division of Chemical Information, Dallas, Texas, April 1989.

Table 8. Comparison of different storage systems listed according to capacity*

Storage medium	Capacity per unit, Mbits	Number of A4 pages	Density, bit/mm^2	Write/read transfer rate
One A4 page (2000 characters)	0.016	1	0.45	150 bit/s
Semiconductor memory	0.256	16	10×10^3	5 Mbit/s
Magnetic bubble memory	1	62.5	15×10^3	50 kbit/s
Magnetic disk	560	35 000	15×10^3	15 Mbit/s
Magnetic tape (dp)	720	45 000	1×10^3	10 Mbit/s
Music cassette, 60 min (analogue)	(860)	62 500	2×10^3	15 kHz
Audio disk (analogue)	(1200)	75 000	10×10^3	20 kHz
Holographic memory	10 000	630 000	1×10^6	100 Mbit/s
Compact disc	15 000	940 000	270×10^3	4.5 Mbit/s
Optical disc, 30 cm	20 000	1.3×10^6	2×10^6	10 Mbit/s
Digital optical disc, 30 cm	30 000	1.9×10^6	470×10^3	16 Mbit/s
Video tape (analogue)	(150 000)	9.4×10^6	120×10^3	8 MHz
Video play v.l.p.	(150 000)	9.4×10^6	2.7×10^6	10 MHz
Human brain	1×10^6 (long-term storage)	62.5×10^6	(10^9/cm^3)	1 bit/s (long-term memory) 50 bit/s (short-term memory)

* Reproduced with permission using data supplied by Polygram and Siemens [7.17].

Technology depends on the use of a high-powered laser to create a physical change in the surface of the disc. The laser can burn a pit in the surface, create a bubble, or change the state of the surface in a more subtle way (e.g., a change from the crystalline to the amorphous state). The pattern thus created is equivalent to a series of zeroes and ones. A low-powered laser system can read the pattern by measuring light reflectance from the surface.

Three types of optical memory are of significance in information processing: compact disc-read only memory (CD-ROM), write-once read many times (WORM) discs and erasable/rewriteable discs (→ Information Storage Materials, **A14**, pp. 196–234).

CD-ROM [7.1], [7.18]–[7.26]. Compact discs were originally designed to reproduce high-fidelity sound for use in the home. The market success of CD audio, and the similarity of production technology for CD-ROM, gave CD-ROM a head start in information processing.

The manufacturing process involves premastering, mastering, and replication. In premastering the data are converted to a master tape. The tape is read by a minicomputer and a glass master disc is cut. The disc is coated with a photoresist and a laser beam creates a pattern corresponding to the data to be encoded. An etching process then produces a series of pits and "lands" (holes or no holes). The glass master is converted to a "stamper" in a complex duplication process. The CD-ROMs are produced from the stamper by injection molding of a polycarbonate resin. Each CD-ROM is finally given reflective and protective coatings.

About 600 Mbytes of data (equivalent to 270 000 typed A4 pages) can be stored on one 12-cm (4.72 inch) CD-ROM. In 1986 it was estimated that it cost about $3000 to produce a master but only about $15 for each CD-ROM. Costs have been decreasing ever since. However, the complexity of the production process means that CD-ROM is not suitable for data that need regular updating.

Since CD-ROM is a read-only medium, it cannot be repeatedly overwritten as can a magnetic tape or disk. This, however, is an advantage for the integrity of storage of data bases and documents. CD-ROMs have slower access times than magnetic disks.

A CD-ROM can be used by only one user at any one time and is read by a "player" attached to a microcomputer. Some applications of CD-ROM are considered in Sections 8.2.10 and 8.3.2.

Variations on CD-ROM such as compact disc interactive (CD-I) [7.22] for storing text, still pictures, and audio or digital video interactive (DVI) [7.23], which can handle motion pictures, text, graphics, and audio, are not yet widely applicable in information and documentation.

WORM. This differs from CD-ROM in that the disc is supplied as a blank and the user writes in his own information. Once the disc is written, the data cannot be modified or added to. WORMs are used by reports collections and archives which have to store vast numbers of documents (see Section 8.4.2). The discs are commonly 12 inches in diameter and about 100 of them can be used together in digital "juke boxes", giving huge storage capacity.

Erasable discs are only just emerging from the laboratory and are still very expensive. They are more applicable to computer science than to electronic publishing or records management [7.27].

Other Optical Storage Media. In 1988 ICI introduced a new, inexpensive optical data storage medium called Digital Paper [7.28], [7.29] consisting of an extremely thin, flexible polymer film fabric coated with an IR-sensitive dye. Information is stored in much the same way as optical discs but at much higher densities and the material can be made into sheets like paper, stamped out as discs, or wound onto cassettes as optical tape. A 12-inch reel of 35-mm Digital Paper could store 1000 Gbytes of data (as much as 1600 CD-ROMs or 1×10^9 sheets of paper).

Drexler Technology's laser cards carry data on a device similar to a credit card [7.30], [7.31].

Semiconductor and Bubble Memories. Semiconductor-based storage devices include *random access memory* (RAM) and *read only memory* (ROM) integrated circuits.

ROM contains data which are to be permanently stored and need to be directly addressed and rapidly accessed by the CPU. The data are not lost when the machine is switched off. RAM differs from ROM in that all the data in RAM are lost once the machine is switched off.

Bubble memory is the subject of much research [7.1], [7.2]. The name derives from "bubbles" (minute cylindrical magnetic domains) that are propelled through a thin film of magnetic material. The presence of a bubble signals a one and the absence of a bubble represents a zero. Permanent storage is possible within the CPU rather than on an external device; the stored data are not lost when the power is switched off (unlike data on RAM). Bubble memories are compact, have no electromechanical parts, and (unlike disks) use very little power. Access is faster than with magnetic storage media but capacity is lower.

7.2.4. Microcomputers [7.32], [7.33]

First-generation microcomputers were relatively slow and had low-resolution graphics; many had no graphics capability. In general they were used for local data handling and were not linked as terminals to mainframes or in local area networks. With the evolution of terminal emulation software and communication boards that facilitated these links, microcomputers replaced dumb terminals as the preferred method of connecting to mainframe computers. Soon after, networking software and hardware allowed microcomputers to be grouped into local area networks. However, for computationally intensive tasks requiring high-resolution graphics (e.g., molecular modeling and computer-aided design), the personal computer had to be enhanced by special graphics circuit boards, screens, and additional coprocessors (i.e., extra boards). The scientific workstation was designed to fill this gap.

Scientific workstations are characterized by high-resolution graphics, either raster (e.g., Silicon Graphics), or vector (e.g., Apollo Domain). They offer a choice of up to 256 colors, window-oriented user environments (see Section 7.3.5), and multitasking operating systems. They often have high-bandwidth network links to other workstations or mainframes and shared computational facilities.

Third-generation microcomputers based on 32-bit microprocessors have the computational power of earlier minicomputers. (The terms 8-bit, 16-bit, 32-bit etc. refer to the size of normal machine instruction that the CPU can process). Examples are the IBM PS/2 and the Apple Macintosh II, equipped with Intel 80386 or Motorola 68020–68030 chips, respectively. These machines can process 4×10^6 instructions per second, comparable to the processing power of a DEC VAX 11/780 minicomputer of the mid-1980s. They offer high-resolution graphics at a price an order of magnitude lower than that of the typical scientific workstation of the early 1980s.

Scientific workstation manufacturers responded with newer models at lower prices to compete with top-range microcomputers. Mainframe and minicomputer manufacturers also started to enter the workstation market. The distinction between the top end of the personal computer range and the bottom end of the workstation range is thus blurred.

7.3. Software

7.3.1. Systems Software and Application Software

Software is a generic term covering the concepts, procedures, and instructions which cause computer systems to perform useful tasks. It is usually thought of in terms of individual programs and integrated collections of programs (systems or packages).

Software may be divided into two categories: *systems software* which controls the execution of other programs and utilizes hardware effectively, and *applications software* which covers programs written to satisfy a particular user need.

Systems software is generally supplied by the hardware manufacturer. It includes operating systems; assemblers, compilers, and interpreters; programs for controlling input and output devices and copying data between storage media; and utilities for sorting, merging, and editing files, and controlling program libraries.

Operating systems are programs which enable users' software to run efficiently, making use of hardware and software resources [7.34]. They include control and allocation of processing power, storage space, and input and output devices; maintaining security; and handling files, timesharing, networking.

7.3.2. Programming Languages, Compilers, and Interpreters [7.35]–[7.37]

Ultimately a computer can only operate in binary machine code. Assembly language is a more convenient, mnemonic form of machine code. Its use nowadays is restricted to those applications or part-programs where another language would not be satisfactory. Third-generation, high-level languages such as FORTRAN (for scientific applications) and COBOL (for business applications) are much easier to program and to understand. Fourth-generation languages are mentioned in Section 7.3.6.

A written program must be compiled (i.e., converted into machine instructions) before it can be executed. A compiler sits in memory and operates as a program. The data to be compiled are known as the source program and the compiled program is called an object program.

An interpreter converts individual source statements into machine code as they are needed. Interpreters are quicker and more flexible to use than compilers but a compiler produces object programs that are independent of the compiler and execute more quickly.

One of the best known microcomputer languages, BASIC, can be either compiled or interpreted. The language C is becoming popular for microcomputers because programs written in C can be more easily made to run on a variety of machines.

7.3.3. Organization of Data

Data have to be organized to maximize hardware and software performance. The disposition of the data on a mass storage device is known as the physical organization but programmers are more concerned with logical organization. Items may be filed logically in sequence (a serial file) or they may be organized with special indexes used to find relevant items. Data-base management systems (see Section 7.3.4) can be used to provide complicated logical arrangements of data.

The nature of the physical organization may have a bearing on optimizing the software efficiency. For example, random access to items on a CD-ROM is slower than access to data on a magnetic disk, but it is then quite efficient for reading a number of items serially. Information retrieval software for CD-ROMs must be written bearing these factors in mind [7.20], [7.38].

7.3.4. Information Retrieval Packages

Data-base management systems and text retrieval systems are important software packages for information retrieval (see also Section 8.4.4).

Data-base management sytems are best suited to handling structured, numerical, and factual data, and short items of text, particularly if rapid updating is required [7.1], [7.39]–[7.41]. They allow different users different views of common data. Frequently they offer a high-level query language (e.g., SQL), fourth-generation programming languages, applications generators, and report generators (see Section 7.3.6). A data dictionary and data definition language allow careful definition of data elements and their relationships. The logical structure of the data may be hierarchical, networked, or relational. The systems are described as relational if they allow the interrelation of data elements in answering complex queries.

Text retrieval systems are best suited for large volumes of unstructured text [7.41]–[7.44]. The file structures are simpler than with database management systems but text retrieval systems offer specialized features such as proximity searching and thesaurus control (see Section 8.4.4).

The two types of information retrieval system tend to overlap, as data-base management systems add text-handling capabilities and text retrieval packages acquire data-base management facilities.

7.3.5. Microcomputer Software [7.32], [7.33]

Spreadsheets. Spreadsheet programs consist of a matrix of boxes or "cells" into which the user enters data or formulae. All spreadsheets print out data exactly as they are recorded but some also allow the production of graphs, bar charts, pie charts, etc. These programs are particularly useful for modeling "what-if" operations by changing variables and inspecting the results.

Data-Base Management. Data-base managers allow storage of data and text, and retrieval in many formats. Data bases can be created; data can be entered on data entry screens, reviewed using query commands, and reported using report generators.

Word Processing. Word processing software is used to compose text on a video display unit instead of on a typewriter. Corrections are easy to make: words, paragraphs, and blocks of text can be deleted and inserted, and spelling checker modules can be used. The final document can be printed in a number of formats.

Data Communications. This software allows communication with other computers, for example those built into laboratory instruments or the remote host computers holding scientific data bases [7.33], [7.45]. The advantages of microcomputers for uploading and downloading information in online searching of public data bases are considered in Section 8.2.7.

Graphics. Graphics software covers a wide range from simple data plotting and screen image-making programs to computer-aided design and other high-resolution design systems. Specialized software for handling chemical structures is considered in Chapter 10.

Drawing programs allow graphs and charts to be generated from data; painting programs allow manipulation of each pixel to generate images. Drawing programs are needed for analysis and communication of scientific data; painting programs are used in creative scientific communications (e.g., audiovisual aids for presentations).

Integrated Software. Integrated programs (e.g., Lotus' Symphony) combine two or more of the major program types described above.

Application environments are application programs that run under the microcomputer's operating system but provide a more advanced user interface and additional features such as concurrent operation and coresident programs. Applications programs can then be written to include code that interacts with the application environment. An example is X-windows, written at the Massachusetts Institute of Technology and adapted by various vendors. With Microsoft Windows for the IBM-PC, for example, two or more application programs can be run and information can be exchanged between applications. Windows is a graphics-oriented interface and makes the IBM-PC user interface more similar to that of the Apple Macintosh, which is characterized by the use of windows, icons, mice, and pointing devices (WIMPs).

GEM, from Digital Research, is another visually oriented user interface. It is a single-tasking environment and also uses WIMPs.

7.3.6. Software Engineering

The production of application programs is highly labor intensive. Developments in software always lag behind those in hardware and this can lead to a bottleneck. One solution is increased availability of packaged software. Another is the use of fourth-generation languages which allow the programmer to write instructions in something akin to natural language leaving the computer to generate code automatically.

The emergence of the new discipline of software engineering has encouraged improvements in programming practice, such as structured (modular) programming [7.46], [7.47]. Computer-aided software engineering (CASE) aims to cover the whole spectrum of software development from the technical definitions of the system to the way it is managed. The tools and methods involved include graphics, "front-end" design tools, fourth-generation languages, and object-oriented technology. Object-oriented programming systems are a way of developing packaged software that draws heavily from common experience and the manner in which real-world objects relate to each other.

7.4. Telecommunications and Networks
[7.2], [7.32], [7.48]

The function of digital networks is to interconnect computers and other devices so that data can be transferred. Online searching of remote data bases, electronic mail, and other telecommunications services are all dependent upon the transmission of digital data.

Simple data networks that connect a few pieces of equipment which are close together may simply involve lengths of cable. Access to distant computers means using the public telephone network or a private telephone line leased from a telecommunications company.

Telephone systems were originally developed for transmitting sound and are not ideal for transmission of data. The efficiency of data transmission is measured in terms of "bandwidth" (measured in Hertz), which is roughly equivalent to the maximum rate of data transmission (measured in bits per second).

Most telephone networks depend upon analogue signals whereas data consists of digital (on–off) pulses. Connection of a terminal or microcomputer to a distant computer therefore requires a modem (modulator–demodulator) to convert the digital data into analogue signals, and vice versa.

Many countries are now implementing new networks for transmitting digitized speech and data. The United Kingdom's Integrated Services Digital Network (ISDN) will gradually replace the Public Switched Telephone Network (PSTN).

In busy long-distance public networks it is important to transmit the maximum amount of data along each data line. One technique used for this is multiplexing. Data from more than one sender are interleaved for transmission by one multiplexer and unscrambled at the receiving end by another multiplexer.

Local area networks are designed to link equipment within a much smaller geographical framework. Ring networks involve a ring of cabling with each computer device connected separately into the ring. Broadcast networks (e.g., Ethernet, developed by Xerox Corporation) use coaxial cable as the transmission medium and employ separate transmitter–receivers with repeaters to boost the signals. Star and mesh networks are also in use.

International services such as the British IPSS (International Packet Switching Service) and PSS, the European Euronet, the American Tymnet and Telenet, the French Transpac, and the Canadian Datapac networks, use packet switching. This technique employs computers to control data flow and does not provide a dedicated physical path between the sender and the recipient. Messages are broken up into segments which are sent separately, maximizing bandwidth usage. The rate at which data are sent need not be fixed and devices with different data rates can communicate with each other.

Data rates on the public telephone network are limited to 48 kbits/s (often less) but local area networks can transmit data much faster (10 Mbits/s).

Wide-band digital networks rely on new technologies such as satellite links and fiber optics.

The rules governing the flow of data are called protocols. Unfortunately there is more than one common protocol. A first step towards a general-purpose network standard, allowing connection of any device to a network, is the Open System Interconnection (OSI) reference model put forward by the International Standards Organization (ISO).

7.5. Distributed Computing

Increasing standardization of user interfaces and the user friendliness of software implementing menus, icons, and pointing devices have facilitated moving from one application to another. Consistent user interface, connectivity, integration, and open architecture are important for the distributed computing environment of the early 1990s. The term *connectivity* addresses the communications and other issues relating to the interfacing of and transfer of data between applications running on personal computers and mainframes. *Integration* implies a higher level of connectivity in which the relative personal computer and mainframe applications have been designed to work closely together. *Open architecture* is an approach to software design which allows and encourages integration of software components from different producers by the use of well-documented, modular interfaces to system components.

Personal computers are becoming increasingly powerful and new operating systems offering multitasking and multiple users are enhancing their capabilities still further. Applications which involve access to a centralized data base or require sharing of information between users, are becoming more efficient as consistent user interfaces appear and standards for integration and open architecture are implemented.

8. Records Management, Online Searching, and Information Retrieval

8.1. Introduction

Both internal (corporate) and external (public) data bases can be accessed online. To the information scientist or research chemist, the distinction between internal and external data bases is somewhat artificial. Data from an external data base can also be downloaded to make an in-house data base. In some cases the same software can be used for both the external data base and the in-house one. This chapter is concerned with information retrieval from personal, corporate, and public data bases. Chemical structure information is discussed in Chapter 10.

8.2. Online Searching

8.2.1. Introduction

In an online retrieval system a user can directly interrogate a machine-readable data base on a remote computer [8.1]–[8.6].

Online bibliographic data bases first arose in the late 1960s as byproducts of primary printed publications. As computer typesetting was introduced, more material for publication became available in machine-readable form. The creation of machine-readable data bases by the secondary information services coincided with the development of long-distance telecommunications networks, such as Tymnet and Telenet in the United States. The online industry started when two organizations with spare computing capacity, Lockheed [8.7] and SDC [8.8], provided the software and necessary computing facilities to enable the data bases to be stored and searched interactively via telecommunications networks. Nowadays many data bases are published only in machine-readable form, with no print equivalent.

Most online data bases are bibliographic in nature. The author, title, and source of the document are indexed by the data-base producer, usually with additional controlled descriptors or keywords, and possibly with an abstract. Full-text data bases are, however, also available. In addition to bibliographic data, these carry the full text of documents, footnotes, cited references, and captions for graphics or figures. At present graphics are usually omitted.

Data bases that carry factual and numeric data are described in Chapter 9. Systems permitting the retrieval of chemical structural information are described in Chapter 10.

8.2.2. Equipment

Online searching involves two-way (interactive) communication between the user and a remote computer. The user will most likely have a keyboard plus cathode ray tube display for inputting queries and receiving the output. The input device may be a "dumb" terminal, an "intelligent" terminal, or a microcomputer. For accessing systems that permit input and display of graphics, a graphics terminal or suitable microcomputer is needed. In most cases a modem converts the digital signals of the computer into the analogue signals carried by the telephone line. A modem is not required if the input device is connected directly to the remote computer or is linked to a network that allows external access. Telecommunications are discussed in Section 7.4.

The advantages of using a microcomputer are covered in Section 8.2.7.

8.2.3. Benefits and Problems

Online information systems offer rapid and convenient access to a multitude of references, facts, and chemical structures. Online information is almost always more up-to-date than that available in print. Another advantage is the number of access points. The indexes to a printed volume limit the number of ways in which information can be accessed. Moreover, in searching online the user can combine terms using Boolean logic, for example:

```
A or B or C or D
A and B and C and D
(A and B) or (C and D) not (E and F)
```

Keywords, authors, formulae, patent numbers, chemical names, and many other fields, can be searched in one or more data bases. In some systems one query can be simultaneously submitted to more than one data base. "Cross-file" searching amongst certain data bases is also possible (i.e., carry a cross file of reference numbers selected in one query and search for records bearing those reference numbers in another data base or system).

Considerable expertise is needed to carry out all but the simplest of online searches. For many years almost all searches were carried out by an information scientist or skilled intermediary. In recent years there has been an increasing tendency for users (scientists, lawyers, or managers) to do the search themselves. End user searching has many advantages [8.9]. End users have immediate access to information relating to their problems. Their flow of ideas is not interrupted and searching can be performed interactively (e.g., by modifying the original query). Use of subject expertise in the search formulation is increased. The end user may be encouraged to use systems and to do searches when the intermediary is not available. Users become more skilled at explaining their requirements if they use an intermediary on occasions. End-user searching allows an information unit to concentrate its expertise on more complex problems. In organizations with few (or no) skilled intermediaries, end users are obliged to do their own searching, or to pay an information center or broker.

The average online system is not very user-friendly [8.10]. Even if the user knows which data base to consult and the host, the logon procedure is laborious. The host's command languages and error messages are not user-friendly. Indeed proper use of such software requires extensive training and voluminous documentation. Each vendor supplies separate contracts and passwords. Customer service is not available around the clock. Document delivery mechanisms are cumbersome. Data-base names vary from system to system. (In 1985 there were 17 names for the various CAS files on six different computer systems.) Commands also vary from system to system. Attempts to standardize on one command language [8.11] have made little impact on the average user. Online Inc. of Weston, Connecticut produce an International Command Chart comparing the command languages of various systems.

End users take longer than an intermediary to run a search [8.12]. Some systems designed for end users do not have all of the searching capabilities of the traditional systems. End users have difficulty interpreting output and revising an ineffectual search strategy; they are also unaware of or do not use search aid tools such as thesauri. Some users do not understand Boolean logic; many do not use very complex search strategies. End users are infrequent users and therefore forget commands and search protocol.

A reasonable compromise in large commercial organizations is for end users to do the "quick-and-easy" searches while experts perform complex searches, or searches on which major financial decisions might depend.

The development of tools to facilitate end user searching is considered in Section 8.2.7 and Chapter 10.

8.2.4. Data-Base Producers and Vendors

The company (or system) which has the hardware and software that makes a group of data bases available is known as a *host* or *vendor*. Some American sources refer to hosts as *data banks*, whereas many European sources use the word data bank for a data base of factual and numerical information. A company which constructs a data base is known as a *data-base producer*. Some data-base producers are also the vendors of their own (and other) data bases.

About 4500 online data bases are now available (1990) on about 600 different hosts [8.13], [8.14]. Directories are available in hard copy and online. Major hosts are:

Dialog Information Services, Palo Alto, California
Maxwell Online, McLean, Virginia (incorporating the older services BRS, SDC/ORBIT, and Pergamon Infoline)
STN International, Columbus, Ohio (see Section 2.2.6)
Questel (Télésystèmes), Paris
Mead Data Central of Dayton, Ohio
European Space Agency Information Retrieval Service (ESA/IRS), Frascati, Italy
DataStar, Berne, Switzerland
Deutsches Institut für Medizinische Dokumentation und Information (DIMDI), Köln, FRG.

The preeminent chemical data base is that produced by CAS (see Chap. 2). Chemical structure searching is possible on STN and Questel. Versions of the CAS data base are available on various hosts but only STN International offers searchable abstracts online and pre-1967 registry data.

If vendors do not offer chemical structure searching, compounds have to be located by chemical names, registry numbers, or molecular formulae. Often the vendor supplies a chemical dictionary file where the user may find the preferred name or a registry number for a compound. The preferred name, or the reference number, can then be used for access to data in another file.

Several chemical hazards bibliographic data bases are available [8.15].

SCISEARCH, the online version of Science Citation Index (see Section 2.3.1) is available on more than one host. It is not a bibliographic data base in the same sense as others listed here, but it is useful to include it here. SYNGE has reported on its use as a complement to Chemical Abstracts [8.16].

In the patent information area, Derwent Publications' World Patents Index (WPI) and WPI/L (where "L" stands for "latest") are of great importance. Searching generic structures is possible (see Sections 6.7 and 10.4.1). IFI Plenum's CLAIMS files cover U.S. patents and have chemical data back to 1950. APIPAT and APILIT are produced by the American Petroleum Institute. The Chemical Abstracts data base also covers patents. MARPAT is STN International's service for searching generic structures from patents (see Section 10.4.1). The French patent office, INPI, has data on Questel. INPADOC is the most comprehensive collection of worldwide patent literature and it builds up and reports the various Patent Family Collections. Mead Data Central offers the full text of U.S. patents as LEXPAT (this is not simply a bibliographic data base, but is mentioned for completeness).

Derwent Publications' Chemical Reactions Document Service (CRDS) is a rather complex system based on various codes (see Section 3.4).

Other data bases are mentioned in Chapters 2 and 3. For a fuller listing of data bases in chemistry and related disciplines, see [8.13], [8.14], [8.17]–[8.20].

Data banks (i.e., files of factual and numeric data) of interest to chemists are considered in Chapter 9. Beilstein Online (see Sections 3.3.2 and 9.6.1) is one of the most important data banks.

8.2.5. System Features and Search Strategies

A checklist of the functions and capabilities of the systems of many major vendors has been compiled [8.21]:

Access to system
Data bases mounted
Data base selection
Treatment of records
Searching capabilities
Printing offline
Limits on capacity
Saving searches
Multiple file and cross-file searching
Searching assistance while online
Vocabulary assistance during a search
Simplified searching options
Information on costs and system usage
Online ordering capability
Documentation available from vendor
Training available from vendor
Vendor charges
Electronic messaging
Software package enhancements

Online bibliographic data bases are based on the *inverted file structure*. Typically, an inverted file structure consists of several related files. The "main" or "record" file stores the bibliographic records themselves. These records may be maintained in any order, usually the order in which they are acquired. To search a large data base for all the occurrences of one term would be very time-consuming if this file were treated as a "direct file" and searched sequentially from beginning to end. To speed up the search process, and facilitate more complex searches for combinations of terms, an index file is created. The index is usually stored and used in a two-step operation.

The main index stores a list, typically in alphabetical order, of all the terms in the records file together with the number of times that term occurs in the file. A secondary index stores the record numbers where each term is found. When a search is initiated, the software begins by searching the main index file and if the search term is found reports the number of occurrences (postings) to the user. If the user wishes to display the results, the secondary index (postings) file is accessed to obtain the record numbers in question. Only at this stage does the software actually go to the records file and pull out the correct text identified by the record number.

Searching using an inverted data-base structure is therefore very fast and efficient because in the first instance only the main index need be searched. Data-base vendors have introduced their own refinements to this basic structure.

In addition to the use of Boolean operators, word stem searching (truncation: e.g., CRYST* to find crystal, crystallography, crystallographic, etc.), proximity operators, and other facilities may be available.

Formulation of a search is not simple. A strategy should be developed on paper before costs are incurred formulating the search online [8.19]. Detailed examples of search formulation for the Chemical Abstracts data base are given in [8.18], [8.22]. A complete search may require the use of more than one data base.

The complementary nature of the SCISEARCH and Chemical Abstracts data bases is described in [8.16].

8.2.6. Costs

Data-base producers receive royalties from online vendors and in some cases charge customers subscriptions. Often, online searchers receive a discount on online costs if they subscribe to the hard-copy version of a machine-readable data base.

Online vendors used to base their charging heavily on the number of minutes for which the user was online (connect charges). However, due to development of offline search formulation and rapid searching using fast modems, vendors introduced other charging systems. Charges for opening a data base, for each search term used, for each hit displayed, or for each offline print ordered are all used [8.23].

8.2.7. Gateways, Front Ends, and Microcomputer Software [8.24]–[8.31]

Definitions. The problems described in Section 8.2.3 have been addressed by software packages, loosely called gateways, front ends, and intelligent interfaces, many of them operating on the user's microcomputer. The microcomputer has the advantage of allowing the user to do online searching on the same "terminal" used for other tasks. The user thus has an ever-increasing volume of storage space, can make use of specialized searching and utility software, and can keep statistics on performed searches.

Gateways are systems that allow the user to switch easily from one host computer to another, often with a simplification of invoicing procedures, and sometimes permitting the use of just one command language for more than one host system. Terminal emulation software (sometimes also called a communications package) [8.29], [8.30], [8.32]–[8.34] is required so that the remote computer regards the user's personal computer as a terminal and can communicate with it. Sophisticated communications software offers additional features such as automated logging-on and the ability to capture data output from the remote computer on disk (downloading).

Front ends [8.10], [8.34]–[8.38] offer many more features, including some at the "back end" of the search:

1) Access to several hosts
2) Data base selection
3) Automatic dialing and logon, including the pursuit of an alternative if the chosen route is unobtainable, and the handling of error messages
4) Offline search formulation and storage of profiles
5) Presearch editing and uploading (allowing interaction and uploading of selected statements)
6) Online running of the search when required, including facilities to cope with error messages

7) Help features: there should be a different question-and-answer dialogue, or menu-driven interface, for novices from that needed by expert users
8) Viewing and printing of results
9) Transfer to a different data base
10) Downloading
11) Logoff
12) Reformatting and post-processing of the search results, with more powerful features than those available in word-processing and other utility software

Search Formulation. The user interface is usually menu-driven or works in question-and-answer mode. Menus are essential for the naïve and frustrating for the experienced: a good system offers both user and expert modes. Natural language interfaces and other artificial intelligence techniques to facilitate searching for the novice (e.g., TOME Searcher) are described in Section 11.5.2. Verity's TOPIC [8.39]–[8.41] is a concept-based retrieval system for use in distributed computing environments. The Intelligent Test Management System is an expert system from Information Access Systems [8.40]. For a historical review of earlier systems, see [8.42].

Uploading [8.31]. Formulation of a search offline saves money (by eliminating connect time) and reduces the stress on the user (the "taximeter syndrome": the urgency of formulating the search quickly, and perhaps carelessly, because money is being spent by the minute). The user can then log on to a remote computer and "upload" the query.

Storage of search profiles and uploading is useful in the following cases [8.31]:

1) For the creation, rapid transmission, and re-use of "hedges" (i.e., the set of terms that defines a common or regularly used search concept)
2) For a search to be used against several data bases
3) For a search that is run daily or weekly
4) For a "boiler-plate" search, where only a few terms, amongst many, are changed at each execution of the search
5) For a "canned" search (e.g., one which is stored to run at 3 a.m. at cheap telecommunications rates)

Downloading [8.34], [8.43]. The major benefits of using a microcomputer for online searching were first thought to be automatic logon and offline search strategy preparation. Improved storage technology, faster telecommunications, and the availability of suitable software made downloading onto disks possible. It is both faster and cheaper to omit the printer, and multiple copies of the search output may be made from the disk. However, the most important benefits of downloading are the possibilities of manipulating the data and producing in-house data bases or bulletins. Word-processing or data-base management software can be used for this but specialized packages (see Section 8.3.1) offer advantages [8.25], [8.34], [8.44]–[8.46].

Other Search Aids. Some vendors offer simplified systems on the host computer for the benefit of end users. Examples are BRS BRKTHRU on Maxwell Online and DIALOG's Knowledge Index.

"Gateway services" such as Infotap's Intelligent Information and Istel's Infosearch [8.47] also enable the searcher to access data bases on different hosts via one logon procedure and password. Some gateway services, such as EASYNET [8.48], [8.49], also offer help in choosing a data base and formulating a search strategy.

8.2.8. Full-Text Online Data Bases [8.50]

Full-text data bases comprise the complete texts of articles, books, newspapers, encyclopedias and so on, available in a machine-readable form in which every word in the entire text (except designated stopwords such as "the" and "and") can be searched.

In 1988 DIALOG offered the entire text of 355 periodicals [8.51]. Mead Data Central's NEXIS system offers the full text of many magazines and newspapers [8.50]. Mead also offers the text of U.S. patents in LEXPAT. STN International's Chemical Journals Online (CJO) [8.52] is described in Section 1.8. The full text of the Kirk–Othmer Encyclopedia of Chemical Technology [8.53] (see Section 3.3.1) is available online [8.54].

Most full-text data bases are searched with the same sort of software as is used for searching bibliographic data bases. This is rather unsatisfactory because with full-text data bases the context of the search terms is very important. In a search for (A AND B) false drops will occur if the terms A and B are separated by ten pages of text. Use of Boolean operators plus term truncation and synonym searching is not sufficient. Proximity operators (e.g., DIALOG's proximity operator "S") are needed to ensure that search terms occur in the same chapter, paragraph, or sentence.

A further enhancement to full-text searching would be the use of a term count feature. Information retrieval systems usually report the total number of occurrences (or postings) of a term in the whole data base. The greater the number of occurrences, the greater the relevance of the document.

Synonym listings, ranking, term weighting, and document clustering [8.55] are other techniques that can enhance retrieval [8.56]. These are discussed in Section 8.4.4.

Specialized techniques are needed for the construction of full-text data bases [8.57], [8.58]. The use of a generalized markup format for electronic publishing of a hard-copy journal can facilitate conversion of the data into an online full-text data base.

Large backfiles of journals are not available and end users are not yet avid searchers. Unfortunately graphics, mathematical symbols, and tabular material are not usually mounted online because of the expense. This reduces the usefulness of scientific journals online. More advanced electronic publishing systems will facilitate the inclusion of graphics in full-text journals online [8.59].

8.2.9. Graphics Display

On STN International PATGRAPH is the patents graphics subfile of the German patent data base PATDPA. It contains patent drawings, chemical structures, and complex mathematical formulae published in addition to the abstract of German first patent publications.

An international facsimile standard called "Group 4" has been established by the Comité Consultatif Internationale Télégraphique et Téléphonique (CCITT). Graphic data for PAT-DPA are scanned in this format and converted to outline vector representation (see Section 7.2.2) using a system called SCORE (Scan Conversion for Outline Representation of Images) developed by Imagin, Karlsruhe, FRG [8.60]–[8.64]. The transmission of text, patent drawings, images, and chemical structures via telecommunications networks is possible since text and vectorized graphics are transmitted as ASCII data (see Section 7.2.1). Transmission of vectorized data is much more effective than transmission of the original digitized graphics: the compression factor is more than 60%. Vectorization and the appropriate software (STN Express, see Section 10.6.4) facilitate the display and printing of graphics on a wide range of graphics screens and printers.

The DIALOG data bases TRADEMARK-SCAN-FEDERAL and HEILBRON contain graphics which can be displayed (but not searched) with DIALOGLINK software [8.65], [8.66]. Macintosh microcomputer users can print them using DIALOG Image Catcher. TRADE-MARKSCAN-FEDERAL contains all active trademark applications and registrations filed in the United States Patent and Trademark Office. Trademark images have been digitized and compressed. HEILBRON is the online version of the Dictionary of Organic Compounds, DOC5 (see Section 3.3.3) [8.67]. It contains scanned chemical structures.

A more recent development in DOC5 is the "OCR-ing" of chemical graphics [8.68] (for a description of optical character recognition, OCR, see Section 11.5.2). From Chapman and Hall's Dictionary of Organic Compounds, 55 000 chemical structure diagrams have been scanned by a new program, Recog, to produce parameterized diagrams which can be converted by a further program, Constr, into a substructure-searchable data base. This technology opens up new possibilities for producing structure-searchable data bases from hard copy and is rather different from the usual concept of drawing structures and publishing with the same structure drawing package (see Chap. 10). IBM are also studying OCR of chemical structures.

8.2.10. Electronic Document Delivery
[8.69]–[8.72]

Improvements in document delivery were alluded to in Section 1.5. Electronic mail is a comparatively recent innovation as far as online searching is concerned. Several hosts have had some form of electronic mailbox system for online document ordering for some time. It was not until 1988, however, that the integration of electronic messaging, facsimile (FAX), and telex with searching facilities became available. A customer can now immediately download a search into his mailbox and deliver the result to the requestor's mailbox or convert it into a fax or telex.

The British Library project, QUARTET, is researching electronic document delivery [8.69], [8.70]. They envisage a "just in time" concept for information. The user pays only for what is needed. Users find articles they want or have the bibliography checked by online searching, request information from the British Library by electronic mail, and receive documents by Group 4 FAX over ISDN. The Group 4 FAX standard has been incorporated into the new international standards that allow compound documents to be transmitted between various devices. QUARTET makes use of the ADONIS (Article Delivery Over Network Information Systems) CD-ROM data base of articles in biomedical journals [8.71].

8.2.11. Videotex [8.73]–[8.79]

Videotex is synonymous with *viewdata* (the latter term is usually used in the United Kingdom). Both terms are generic and include any interactive systems for transmitting text or graphics stored in computer data bases via telecommunications networks for display on television screens. Color is also part of the standard. The classic example of viewdata in the United Kingdom is the PRESTEL system.

Videotex should not be confused with *teletext*. Information from teletext systems is also displayed on a television screen but forms part of the regular broadcast transmission. Teletext is noninteractive. Two examples of teletext systems are Ceefax and Oracle in the United Kingdom.

Videotex systems can be accessed using special terminals, modified television sets, or microcomputers with videotex software. Standard asynchronous communications software cannot be used.

The approach to data bases is different from that for bibliographic systems because videotex systems are designed for use by the general public. Emphasis is therefore on ease of use and this is reflected in both the data-base structures and information retrieval techniques. Videotex systems should be simple to use, low cost, and designed for a large number of users.

Data is stored in tree structures. The nodes of the tree are called pages, each page being identified by a unique page number. The page numbers are dictated by the position of the page in the tree. Each page consists of one or more frames. The frame numbers are distinguished by a letter following the page number.

Videotex systems are menu-driven. The user can browse through increasingly specific frames of information or access a page directly. Page numbers may be located through the printed indexes or index pages online. Good menu design and up-to-date indexes are essential.

The use of tree structures in a menu-driven system is simple but limits information retrieval facilities. Some information providers have therefore developed better types of index plus limited keyword searching.

Videotex systems can also be set up in-house or by using a bureau. Videotex has proved very useful in the travel industry, stock control, sales records, pharmaceutical wholesaling and distributing, clinical trial management, adverse drug reaction reporting, and in communications with sales representatives in the field.

European videotex systems which were created in the 1970s and early 1980s were set up on a national basis. Each country had its own policy for technical standards, networks, terminal distribution, subsidy to information providers and tarification. The general objective, however, of all systems was to provide a medium as cheap and simple as the telephone for distributing quickly changing information (e.g., public transport timetables and directories).

France, with its distribution policy of free or cheap terminals, had the largest installed European base of 5.2×10^6 terminals by February 1990. The Federal Republic of Germany and the United Kingdom had only 163 000 and 155 000 terminals, respectively.

8.3. Public Data Bases For In-House Use

8.3.1. Microcomputer Data Bases

In-house data bases can be purchased (on tape, floppy disk, or CD-ROM) or can be constructed in-house from downloaded data, copyright permitting.

Downloading is time-consuming and expensive. Hit files from various searches and hosts need to be merged and duplicates have to be detected and removed. The search software for the in-house data base may be different from the hosts' information retrieval packages. Display features differ. File inversions on the microcomputer may well tie it up for hours. However, as storage capacity, the speed of PCs, and software improve, many more users will start constructing data bases from downloaded data.

Much software is now available for downloading data and making personal data bases. Examples are Pro-Cite with Biblio-Links, Reference Manager with Capture, Bib/Search, and In-

Magic with Headform [8.25], [8.34], [8.44]–[8.46]. Many of the major text retrieval packages such as STATUS, CAIRS, and BASIS are available in microcomputer versions [8.80].

Some information providers sell data on floppy disks, e.g., Current Contents from ISI (see Section 2.5). The greater capacity of CD-ROM has, however, made it a more suitable medium for the supply of information for use in-house.

8.3.2. Data Bases on CD-ROM

CD-ROM as a storage medium is discussed in Section 7.2.3. Its large capacity and the fact that it is a read-only medium make it suitable for supplying full-text public data bases for in-house use. It also has the advantage that the full "text" can contain graphics.

There has been much controversy as to whether CD-ROMs can compete financially with online data bases [8.81]. In searching a CD-ROM the user does not have to worry about telecommunications problems or connect charges. The retrieval software is more user-friendly than that of online systems (it is often hypertext-like, see Section 11.7), although it may not be as fast or powerful. Unfortunately CD-ROMs cannot be updated as fast as online data bases, consequently CD-ROM is most suitable where currency is not essential but data base access is very frequent. Enormous data bases, such as Chemical Abstracts, are too large to be supplied on CD-ROM because too many discs would be needed.

The uses of CD-ROM are described in [8.82]–[8.85]. Directories of available data bases have also been published [8.13], [8.86]–[8.88].

DIALOG offer a number of their online data bases on CD-ROM as DIALOG OnDisc Products. The National Library of Medicine's MEDLINE is one of these. CSA Compact Cambridge (a Cambridge Scientific Abstracts service) also offer MEDLINE, as do SilverPlatter of Wellesley MA. Compact Cambridge produce many CD-ROMs for the medical community, including Drug Information Source, produced in collaboration with the American Society of Hospital Pharmacists.

The EMBASE data base of Excerpta Medica in Amsterdam is marketed on CD-ROM by SilverPlatter.

The MICROMEDEX TOMES Plus System (Microinfo of Alton, UK) offers the Registry of Toxic Effects of Chemical Substances (RTECS) and other health and safety data. OSH-ROM (SilverPlatter) contains three health and safety data bases, including RTECS.

BIOSIS's Biological Abstracts on CD-ROM are available through SilverPlatter.

CD-ROMs of interest to the agriculture industry include the Royal Society of Chemistry's Pesticides Disc, produced in collaboration with Maxwell Communication Corporation; the UN Food and Agriculture Organization's AGRIS Data, marketed by SilverPlatter; and CAB Abstracts, from CAB International, also marketed by SilverPlatter.

ISI sell a CD-ROM version of Science Citation Index.

The ChemLink Fine Chemicals Data Base (text only), produced by Chemron of San Antonio, TX, lists sources for 137 000 chemicals. The CD-ROM version is sold by Microinfo. (An online version is up on DIALOG.) The Aldrich Chemical Company produces its catalog on CD-ROM.

The European Patent Office's ESPACE system, with software written by JOUVE, has text and graphics for recent European and PCT patents on multiple CD-ROMs. Chadwyck Healey offer the text of U.S. patents since 1973 but there are no graphics.

Whitakers Books in Print and Bowkers Books in Print are both available on CD-ROM. IBM and Oxford University Press collaborated in the production of the Oxford English dictionary on CD-ROM. A more recent product is Molecular Structures in Biology, a reference tool for exploring protein structures [8.89]. The software supplied allows both molecular graphics options and special hypertext facilities.

Chapman and Hall and Maxwell Communication Corporation are collaborating to produce a substructure-searchable data base of natural products on CD-ROM. No CD-ROM product yet offers substructure search.

8.4. Records Management [8.90]–[8.92]

Most organizations have large amounts of proprietary data (reports, correspondence, laboratory notebooks, etc.). Policies and procedures need to be established for managing such records: determining what should be stored and how, establishing retention times, assigning security categories, controlling archives, and adhering to standards required by regulatory authorities.

8.4.1. Microforms [8.93], [8.94]

There are two principle microform products, flat microfiche and roll microfilm.

Roll microfilm, usually 16 mm, is the preferred medium for purely archival microfilming; 3000–10 000 documents can be stored per film. Roll film can be optically marked (with a "blip") or coded to provide a degree of automated retrieval. If a computerized index to the documents is held they do not have to be sorted prior to filming.

Microfiche is an alternative to film but is more expensive to produce. Fiche is preferred to film if a number of copies are to be made and higher quality is needed. Standard microfiche carry 98 frames but 60–420 frames are also used.

An updatable version of microfiche is the *microfilm jacket*, where strips of 16-mm microfilm are inserted into transparent pockets on a standard fiche. Updating such systems is labor-intensive but can save retrieval time if the contents of a single file are not scattered across multiple films.

Computer output can be linked directly to the microfilming process, thus eliminating the need for an intermediate paper copy. *Computer output microfilm* (COM) may be generated in-house, or the data may be spun onto a magnetic tape which is sent to an outside bureau for COM production, usually on fiche rather than film.

Major vendors in the micrographics area are Kodak, Bell and Howell, 3M, and Agfa.

As a storage medium, microforms offer great space saving (98% compared to paper) and stability (with an archival life of up to 100 years). They can be easily and speedily produced, duplicated, and transported. Microfilm can be generated directly from computer files, can be accessed by computerized indexes (with some difficulty), and copes with a variety of textual and graphical material.

Many users dislike reading from film rather than paper but the main problem is the slow, two-step retrieval of information. Identification of the appropriate roll or fiche and the location of the relevant frame can be automated (computer-assisted retrieval, CAR). For roll film attention has been focused on automatically locating the frame after manual selection of the roll. With fiche more attention has been paid to automated access to the relevant fiche with manual location of the desired frame. Computers allow automatic prefiling and random retrieval on fields such as account number or date, rather than roll and frame numbers. CAR systems are usually dependent on key entry for indexing purposes but many organizations deploy equipment with integral bar code readers that can automatically index and capture the document image in one pass.

Kodak's KAR, Agfa's ADMIS (Agfa Document Management Information System), and Bell and Howell's DP1000 are well-known CAR systems. In Kodak's KIMS the retrieved microfilmed image is scanned electronically and converted into a digital bit stream for transmission to a remote terminal. The image can be viewed or processed further.

8.4.2. Document Image Processing
[8.93], [8.95]–[8.97]

Document Image Processing (DIP) is the technology of storing and retrieving documents from WORM (Write Once Read Many) optical discs (see Section 7.2.3). These systems are also called optical filing systems. A 12-inch disk can hold up to 2 Gbytes of data (ca. 50 000 A4 pages).

Basic DIP systems provide fast access and retrieval (10–30 s), concurrent access by several users, indexing and availability for retrieval of new documents within hours of being received, enormous storage space savings compared with paper, and remote access by a PC over a dial-up telephone line with automatic delivery by facsimile transmission.

The constituent technologies include scanners, optical storage (discs, drives, and "jukebox" selectors), OCR, indexing and text retrieval software, and facsimile [8.98]–[8.102].

Documents are scanned using a facsimile-type scanner. The images scanned in raster form are indexed, compressed, and stored on an optical disc as a digital bit map. Retrieval consists of carrying out a search using an index system, and passing the request via a document manager unit (which may be connected also to a temporary magnetic storage unit) to the optical disc. The retrieved image is displayed on a high-resolution screen. Documents can be output to a laser printer.

Vendors of DIP include Wang (WIIS), FileNet (FileNet and British Olivetti), CACL (Intelligent Archive), Ingenium Software (Archea), Philips (Megadoc), Callhaven (MARS), Kodak (KIMS), IBM (ImagePlus), Document Systems Limited (InfoPlus), Image System Integrators (Image System), and Xionics (DIP-X). The micrographics companies Kodak, 3M, and Bell and Howell are involved in both CAR and DIP. With Kodak's KIMS 4000 system, users can scan documents and store them on optical disc, then download the index data to an IBM PS/2 or IBM mainframe computer. Multiple users can access a document index data base and search for documents.

One use for DIP in the pharmaceutical industry is the filing of a New Drugs Application (NDA). Unfortunately the legal status of optical disc files has not been established and companies are obliged to retain paper copies of all the data they file optically. Microfilm has legal status in most countries.

The European Patent Office (EPO) has scanned a huge backfile of patents into raster image format and would also like to accept all new patent applications in electronic form. The text must be held in character-coded form and the image data in raster or vector form. Patent applications can be filed on floppy disk or paper. The EPO uses a modified version of ISO 8879 (Standard Generalized Mark-up Language, SGML) to mark-up patents for entry into full-text data bases [8.103], [8.104]. The text can be edited and printed as required.

There is thus a dual system. For backfile conversion CCITT Group 4 facsimile standards were followed (see Section 8.2.9) and the patents were stored as images. This means that they are searchable only by the keywords used to index them and not by content. The patent text cannot be edited. For current needs, the patent text was captured in character-coded form and marked-up for entry into full-text data bases. It can thus be searched, edited, and printed.

The storage of the average A4 page on an imaging system occupies ca. 30 times more space than that required by normal character codes. As most business documents consist of a large amount of text, total image capture is a wasteful method.

New, standard document architectures will eventually allow image documents to be broken down into image and coded components for storage and distribution, and rebuilt in the workstation. Line loadings and storage space requirements will then be very much less and the current need to separate image system networks from data networks may disappear. Cheaper hardware with greater functionality will allow separation of documents into the required architectural components. Higher speed lines will aid distribution of documents.

Two developing international document architecture standards are Office Document Architecture (ODA) and Office Document Interchange Format (ODIF) under the Open System Interconnection (OSI) model. Computing companies have their own standards: for example IBM's Document Content Architecture (DCA)/Document Interchange Format (DIF), and DEC's Compound Document Architecture (CDA)/Digital Document Interchange Format (DDIF).

8.4.3. Comparison of Micrographics and Optical Filing [8.105]

Optical filing allows huge storage capacity on a single disc, with jukeboxes allowing access to up to 200 discs. Large collections of documents and drawings can be held and retrieved automatically in seconds. Once character-coded data can be integrated, there will be further advantages. Retrieval from microforms is slower and less convenient.

Micrographics is a well-established technology whereas DIP is much newer, lacking in some standards, and very expensive. Microfilm is stable for up to 100 years: a life of 10 years is claimed for optical discs. The legal status of optically stored documents is in doubt.

An image on an optical disc can be accessed by any user connected to the system. Few micrographics systems allow remote access by multiple users over telecommunciation lines. Multiple users are commonly accommodated by multiple microfiche copies.

All the user of an DIP system requires to access the images is a single workstation; there is no need for both a computer terminal and an image terminal as with CAR systems. Finally, DIP systems offer image enhancement facilities for poor quality documents.

8.4.4. Information Retrieval [8.25], [8.34], [8.56], [8.80], [8.106]

Types of Data Base. A data base differs from a conventional file in that the user expects to be able to access any piece of data in more than one way.

Multi-indexed data bases have the inverted file structure described in Section 8.2.5. The records are stored in one file and accessed by any key included in the main index. This type of data base is very easy for the user to create and maintain. Entries are made in the main file and the system automatically does the indexing. Only records of the same type can be stored.

A *hierarchic data base structure* can store records of different types and works on the principle of sets of information and the relationships between them. A data base of sales orders, for example, may contain records relating to orders, products, parts of products, and customers. Customers will order parts, products may have many parts, orders will have dates, etc. There may be one-to-one, one-to-many, or many-to-one relationships between the various items. In a hierarchic data base all the records are linked together by means of pointers so that relationship can be identified. Setting up the data structures is complicated.

A *relational data base* has very complicated data structures. The principle is again based on the relationship between records but the data base can relate records from different files. Records are linked and cross-referenced by the contents of fields that are common to the different files.

Data-Base Software. Simple filing systems are designed to emulate card index systems and information retrieval is usually only possible using one or two keys designated when the system is set up.

Data-base management systems and text retrieval software were compared in Section 7.3.4. Text retrieval software is designed to cope with large volumes of unstructured text. It is generally easier to use than a data-base management system but, because it is concerned more with relevance and recall than with precision, it has fuller retrieval facilities. These are similar to those found in major online systems, e.g., Boolean searching, term truncation, field searching, and proximity searching.

In contrast to a data-base management system, records are usually stored sequentially. The inverted index model is likely to be used. Report generation and sorting facilities are usually available.

Unlike a data-base management system, concurrency is low, i.e, it is not usually possible to use the same data for two different activities at the same time.

Many text retrieval packages are on the market [8.80]. Some are used by major online vendors, e.g., DataStar uses BASIS, which is marketed by Information Dimensions. Other well-known packages are TRIP (Paralog), ASSASSIN (Associated Knowledge Systems), and BRS/Search (BRS/Search).

There are also many information retrieval packages for use on microcomputers [8.25], [8.34], [8.80].

Future Developments. Work carried out in the 1970s and 1980s has been aimed at producing a new type of document retrieval system which is both more effective, in terms of the relevancy of information retrieved, and more efficient in terms of human effort.

Manual indexing is time-consuming, expensive, and prone to error: automatic indexing could offer advantages [8.107]–[8.111]. It is possible to identify potential indexing terms (ignoring stop words such as "and" and "the") in the machine-readable titles and abstracts of the documents to be searched. These content words are then passed to a stemming algorithm which finds the word root. For example, "absorbed" and "absorption" would both reduce to "absorb".

In work pioneered by SALTON [8.112]–[8.114] a document in a collection is described in terms of weighted vectors, a set of numbers representing importance-weighted concepts. Queries are assigned vectors in the same way and the collection is searched by matching document and query vectors. Retrieved documents are reported in rank order, queries are modified (relevance feedback) and the search is repeated until the user is satisfied. SPARCK–JONES [8.107], [8.108], [8.115], CLEVERDON [8.116] and WILLETT [8.56] have also published extensively in this area.

The problems involved in formulating a search using Boolean logic can be avoided if a best match or nearest neighbor search algorithm is used to rank the documents in a collection in order of decreasing similarity with the query [8.56], [8.116]. Similarity is measured using the weights assigned to search terms. The documents output as hits can be inspected in order of decreasing similarity, giving the user control over the size of output.

WILLETT has studied the use of parallel processing (see Section 7.2.1) in document clustering [8.55], [8.117]. He has also compared statistically-based techniques of information retrieval with knowledge-based ones. Expert systems and WILLETT's comparison are considered in Section 11.6.7.

The recent development of hypertext is detailed in Section 11.7.

9. Numeric and Factual Data Bases (Data Banks)

9.1. Introduction

This chapter describes publicly available collections of numeric, textual, and factual data relating to chemical compounds. In Europe, such data bases are known as data banks. Chemistry is particularly well served by the bibliographic and chemical structure data bases (Chaps. 8 and 10). The compilation of numeric and factual data bases is equally important because they allow systematic analysis of large numbers of related structures and associated data; however, the development of these data banks has been less well coordinated. A large number and variety of data banks are available online to users. Data can cover several different properties of a chemical compound, but most data banks assemble facts relating to one category of data (e.g., NMR spectra) in such a way that the user does not need to access the original information source. For a data bank to be of value, the data must be correct and validated.

9.2. Types of Data Banks and Data

Data banks in chemistry may be broadly classified as follows:

1) Spectral data
2) X-ray crystallographic data
3) Chemical and physical properties
4) Toxicology and biological properties
5) Environmental and hazard data
6) Data relating to legislation

Many data banks contain data on more than one of these categories. Compilation of a large, high-quality data bank demands motivation, justification, funding, coordination, and organization. The impetus behind the creation of a data bank has tended to come either as a result of legislation or in response to a need by a group of users. Because of the limited space available in the primary literature, it is not always possible to include detailed experimental results, and data banks have been created as a depository for complete experimental results.

Chemical data banks contain a variety of data types: structural, textual, numerical, and graphical. Software for handling bibliographic data has been in use since the early days of machine-readable data base preparation, but chemical structures require specialized software. Much research has been devoted to this topic [9.1]–[9.8], with the result that the software for storage, retrieval, and manipulation of chemical structural information is probably more advanced than that available for handling most other data types.

Textual data is used in nearly all data banks, e.g., for recording chemical names and synonyms, or for chemical hazards and action to be taken in cases of emergency. Emergency information must be up-to-date and accurate.

Numerical data, coordinates, and codes present different problems. The storing of numerical data is easier than text, but facts should be validated and correct. Specialized search techniques are needed. For example, the user may wish to search for a range of values and combine the data search with a chemical structural search. Three-dimensional structure coordinates require graphics hardware and software.

An integrated approach to numeric, textual, bibliographic, and graphical information is required so that data can be transferred between files, or data banks created that incorporate all three types of information. Integrated systems for storage, manipulation, retrieval, and printing of structures and text are less well advanced, but are required for hard-copy versions of data banks, such as DOC5 [9.9].

9.3. Quality Control

The input of accurate, validated data is of paramount importance in the creation of any data bank, but particularly so where the data is used to assess the risk to humans and the environment [9.10]. Data may be used for making major environmental policy decisions at national and international levels, or for calculations and predictions.

Just because data is available online, it is not necessarily accurate because it may not have been evaluated. Errors already in a data bank are difficult to remove. They are best prevented at source. Transcription errors can be avoided by taking the data directly from the primary source. Few organizations can afford the high-cost, labor-intensive methodology used for evaluating data in the Beilstein Handbook [9.11]. Regrettably, high accuracy was offset by poor currency, and a new philosophy has been adopted for the online data base still with high quality control standards [9.12] (see Section 9.6.1).

The Standard Reference Data (SRD) program at the National Institute of Standards and Technology (NIST) improves data quality by supporting ongoing data centers; funding short-term data evaluation projects; and cooperating with other groups in the government and private sectors [9.13].

The Cambridge Crystallographic Database (see Section 9.5) also contains highly accurate, quality-controlled data. Bond lengths are calculated from the published cell and atomic coordinates and are compared with the bond lengths quoted by the author [9.14].

Data evaluation and quality control are more difficult in the field of toxicology. The Registry of Toxic Effects of Chemical Substances (RTECS) data bank (Section 9.7), compiled and updated by the National Institute for Occupational Safety and Health (NIOSH), only uses data from refereed journals.

Data accuracy has been studied in the development of the Pesticides Properties Data Base at the USDA Agricultural Research Service (ARS) [9.15]. Few bibliographic abstracts contained the required data, and few hard-copy data sources contained literature citations. A consistent and objective data evaluation scheme is being developed, together with a thorough report of experimental procedures and conditions.

Some of the factors chemists need to consider in determining the reliability of nonevaluated data are listed in [9.16]. The quality of spectral data bases is discussed by SHELLEY who is concerned about errors, omissions, lack of diversity of structures, and the problems of integrating spectral and chemical structural search systems [9.17].

The two main commercially-available mass spectral data bases, the Wiley Registry of Mass Spectral Data and the Mass Spectrometry Data Center data base produced by NIST–EPA, both use a Quality Index (QI) algorithm [9.18] (see Section 9.4.3).

9.4. Spectral Data Bases

Although there are over 1×10^7 known chemical compounds, most spectral data bases cover less than 100 000 compounds. The number of hard-copy compilations of spectral data is 20–30 times larger than the number of data bases [9.19]. Details of the main analytical data bases are given below. The software used to search the data bases is described in Section 9.4.5. Structure elucidation from spectra is covered in Section 11.6.7.

9.4.1. Nuclear Magnetic Resonance (NMR) Spectroscopy

Carbon-13 NMR Data Base on the Chemical Information System (CIS). This data base [9.20] was originally constructed by the Royal Dutch Chemical Society and currently contains only 11 700 spectra. The CIS [9.21], [9.22] is a collection of online data banks with the same chemical substructure search facility, the Structure and Nomenclature Search System (SANSS, see Section 9.6.3).

Carbon-13 NMR Data Base of the Fachinformationszentrum (FIZ). This data base has been available online on STN International since December 1987. It contained about 68 000 spectra of 60 000 compounds in 1988. The data base and retrieval software are related to those of the SPECINFO system (Section 9.4.5).

The sources of the data are journals, spectral catalogs, and unpublished spectra from BASF, ICI, and the Gesellschaft für Biotechnologische Forschung.

Data include chemical shifts, experimental conditions, literature references, molecular formulae, structures, coupling constants, and relaxation times. The data-base-specific software was developed at BASF with the following retrieval options:

1) Search of reference spectra by entering the lines of a measured spectrum
2) Search of reference spectra by entering a name fragment of a chemical compound
3) Search of spectra via full or partial molecular formula
4) Search of similar spectra
5) Search for structure fragments based on a measured spectrum
6) Search of compounds and their spectra with a defined structure or substructure
7) Estimation of chemical shifts for a defined structure

Bruker Spectroscopic Data Base is only available to users of Bruker spectrometers. It contains 19 000 carbon-13 NMR spectra and 900 proton NMR spectra.

Sadtler Laboratories Data Base. In 1989 this data base contained 28 000 carbon-13 NMR spectra which will soon be in full digital format, linked to a chemical substructure search system (see Section 9.4.5).

Collection of the National Chemical Laboratory for Industry (NCLI). This is part of the Japanese integrated online Spectral Data Base System [9.23]. It lists 6000 proton NMR spectra and 5700 carbon-13 NMR spectra. Software is provided for the user to look up or predict a spectrum, or match an unknown spectrum.

Other NMR data bases include the Varian data base. Tsukuba University (Japan) produces a collection of carbon-13 NMR spectra of polymers on CD-ROM.

9.4.2. Infrared (IR) Spectral Data Bases

Many large collections of IR spectra were built up when the spectrometers were prism and grating instruments. Fourier transform infrared (FT-IR) spectrometers, which generate digitized spectra, are now commonplace and new collections of data are being constructed.

Aldrich–Nicolet Digital FT-IR Data Base and the Sigma–Nicolet Biochemical Library. The Aldrich–Nicolet collection contained 10 600 compounds in 1987, and the Sigma–Nicolet library contained 10 400 compounds. The data bases are designed for use on personal computers. Software is supplied for matching peak intensities and locations from an unknown spectrum [9.24].

Sadtler Research Laboratories. This is the largest commercially available collection of IR spectra with about 60 000 spectra largely from prism and grating spectrometers. FT-IR spectra are being added.

Coblentz Society Spectra. The Coblentz Society has digitized 4400 of the spectra from the collection of 10 500 compiled and evaluated in collaboration with the Joint Committee on Atomic and Molecular Physical Data (JCAMP). The data base will be available for use with personal computers.

Clearinghouse for Digital IR Spectra. This project was initiated at the University of California–Riverside in October 1986 for the construction of a data base of digitized FT-IR spectra. It involves an automated algorithm for evaluating the spectra [9.25].

IR Data Committee of Japan (IRDC). This computer file has ca. 19 000 peak wavenumbers and intensities with search software requiring entry of wavenumbers and intensities in order of decreasing intensity. Spectra retrieved are listed in order of the probability of being a correct match. The possibility of fully digitizing the spectra has been discussed.

NCLI Japan. The Spectral Database System mentioned in Section 9.4.1 also includes 22 500 IR spectra [9.23]. All the spectra were determined at NCLI under carefully controlled conditions. The IR spectra were transferred to the data base directly, in digital form, from an FT-IR instrument. The data base is available online in Japan.

Other IR Spectral Data Bases. Some 3300 vapor phase spectra from the Environmental Protection Agency (EPA) laboratories are available through instrument manufacturers. Coded IR data on 145 000 compounds, compiled by the American Society for Testing and Materials (ASTM) are available from Chemir Laboratories and Sadtler Research Laboratories and online on the Canadian Scientific Numeric Data System.

9.4.3. Mass Spectral Data Bases

Mass spectral data bases are almost entirely devoted to electron impact (EI) spectra.

The National Institute of Standards and Technology, Environmental Protection Agency, Mass Spectrometry Data Center Data Base (NIST/EPA/MSDC Data Base). NIST used to be called the National Bureau of Standards (NBS) so the data base is still often referred to as NBS/EPA/MSDC.

The data base was originally constructed by S. HELLER of EPA and G. MILNE of the National Institutes of Health (NIH) [9.20],[9.26]–[9.29]. The Mass Spectrometry Data Center of Nottingham later became involved. In November 1988 the data base consisted of 50 000 spectra, each one corresponding to a unique chemical compound.

Quality control of the chemical structural information [9.30] is achieved by cooperation with CAS in the allocation of CAS Registry Numbers of systematic names. The spectra are evaluated using a Quality Index (QI) algorithm [9.18] based on that developed by MCLAFFERTY [9.31]. The NIST data base uses the following quality factors, each with a value between zero and one.

QF1 electron voltage
QF2 peaks at mass to charge ratio (m/z) above that corresponding to the molecular mass
QF3 illogical neutral losses
QF4 isotopic abundance accuracy
QF5 number of peaks in a spectrum
QF6 lower mass limit of the spectrum
QF7 sample purity
QF8 calibration date
QF9 similarity index of calibration mass spectrum

To obtain the Quality Index for the spectrum, all the quality factors are multiplied, then the product is multiplied by 1000. The Quality Index is calculated for each compound to be entered into the data base. If a new spectrum is discovered for an old compound, the new spectrum displaces the old one if it has a higher QI.

The data base is available online on CIS as MSSS (Section 9.4.5). It is distributed on tape without search software and in a PC version with software. The data base can be searched by identification number, CAS Registry Number, chemical name, molecular formula, molecular mass, and abundances of ten major peaks.

Wiley Registry of Mass Spectral Data. This collection is maintained by F. MCLAFFERTY at Cornell University. The data base, containing 123 704 spectra of 108 173 compounds, is available from J. Wiley and Sons on magnetic tape or CD-ROM. The magnetic tape version is distributed without search software, but software for matching unknown spectra is available free from Cornell University [9.32]–[9.36]. The data base is available online on CIS (Section 9.4.5).

In contrast to the NIST/EPA/MSDC data base, the Wiley data base has more spectra than compounds. A Quality Index algorithm is used, but is based on only seven Quality Factors (QF1–QF6 and a seventh called the source of the spectrum) [9.30].

The Eight Peak Index. This collection of partial spectra (65 000 eight-peak spectra for 52 332 compounds, indexed by molecular mass, chemical formula, and most abundant ions) is available on tape from the Royal Society of Chemistry.

NCLI (Japan). The Spectral Database System (see Sections 9.4.1 and 9.4.2) includes 10 000 mass spectra [9.23].

Japan Information Center for Science and Technology (JICST). The JICST runs an online mass spectral data base system searchable by name, formula, CAS Registry Number, and peaks. It is based on the NIST/EPA/MSDC data base, with 6000 additional spectra from the Mass Spectrometry Society of Japan.

MASS–LIB is a software package for automatic evaluation of low-resolution mass spectra. It was developed by the Max-Planck Institut für Kohlenforschung (FRG) and is sold by MSP Friedli and Co. (Switzerland) for use on VAX/VMS computer systems. It implements both the Wiley and NIST spectral data bases, allows input and maintenance of the user's own spectral libraries, and offers graphical representation of chemical structures.

9.4.4. Building Spectral Data Collections

Spectral data-base systems tend to specialize in just one particular technique. However, structure elucidation and other work need a multi-spectroscopy data bank. HELLER illustrates the lack of overlap of data banks with the diagram shown in Figure 10 [9.30].

Collections of data, such as those of Nicolet or NCLI, put together in a single laboratory by systematically determining spectra of compounds, are likely to be of high quality but database building is expensive and relatively slow. Collections put together through donations of spectra from laboratories can be built relatively quickly and inexpensively, but the data may be inaccurate. Collection of data from the literature (e.g., the Wiley registry) can quickly increase data-base size but many spectra are likely to be incomplete.

Most spectral data bases were started as a common need of a group of users. The most use-

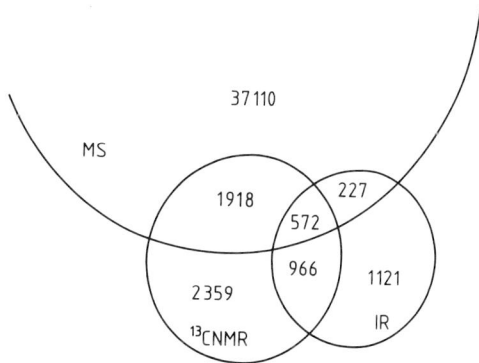

Figure 10. Lack of overlap in spectral data bases (reproduced with permission of Ellis Horwood [9.30])

ful data bases have been created as a result of extensive organization and coordination and, most importantly, with adequate funding. The size and relevance of a data bank can be improved in a large industrial organization by integrating the sample-management and reference data-base systems, thus allowing internal data to be added to external data.

9.4.5. Spectral Search Systems

Until recently, most spectral data bases did not contain chemical structural data and did not have facilities for substructure searching. The CIS data bases with the SANSS software were an exception (see Sections 9.4.1 and 9.6.3). The software developed by BREMSER at BASF (see Section 9.4.1) offers a form of structure searching based on codes.

Sophisticated chemical structure handling systems that have been linked to spectral data systems include Molecular Design Limited's MACCS and the software used for CAS ONLINE data bases and FIZ's C13NMR on STN International [9.1], [9.7], [9.37], [9.38]. Two important microcomputer software packages are the Chemist's Personal Software Series (CPSS) from Molecular Design Limited [9.6], [9.7], [9.39], [9.40] and PSIDOM from Hampden Data Services [9.6], [9.7], [9.39], [9.41] (see Chap. 10).

In addition to searches for specific terms or compounds, and for spectra with certain peaks, there is a need for library or full-spectrum search, i.e., a best-match algorithm which finds those spectra that resemble most closely the spectrum of an unknown compound. Two well-known algorithms for whole spectrum searches of mass spectra are the Biemann KB, or forward search algorithm [9.42] and the McLafferty, Probability-Based Matching (PBM) algorithm [9.32], [9.33], [9.43].

When FT-IR digital spectra with accurate wavelength registration became available, full spectral search methods for IR spectra were also developed [9.24], [9.44]–[9.51].

Spectrum search of the carbon-13 NMR data base on CIS uses an algorithm developed by CLERC and coworkers [9.52]. The SPECINFO and CSEARCH systems also offer search software.

The following examples illustrate systems where chemical substructure searching is linked to a spectral data system. Systems for handling chemical structures are described in Chapter 10.

Spectral Data Bases on CIS. The CIS offers five spectroscopy data banks:

1) CNMR   Carbon-13 NMR Spectral Search System
2) IRSS   Infrared Search System
3) MSSS   Mass Spectral Search System
4) NMRLIT Literature Search System
5) WMSSS   Wiley Mass Spectral Search System

Chemical structures are searched using the Structure and Nomenclature Search System, SANSS [9.53]. Further information on CIS is given in Section 9.6.3.

CNMR permits search by chemical shift, and analysis and display are available.

IRSS permits retrieval of known spectra and analysis of unknowns. The data base includes approximately 3000 spectra taken from the EPA collection plus contributions from the Boris Kidric Institute in Yugoslavia. A graphics package allows display of spectra on graphics terminals. MSSS contains the NIST/EPA/MSDC mass spectra of over 42 000 compounds which can be searched on the basis of peak and intensity requirements, as well as by Biemann and probability-based matching (PBM) techniques. Searches by molecular mass, molecular formula, and partial formula are also permitted. WMSSS parallels this system, but uses the Wiley data base.

NMRLIT permits searching the index to Nuclear Magnetic Resonance Literature Abstracts and Index (Preston Publications, Inc.). Currently there are over 43 000 references covering abstracts published from 1964 through December 1984. Searches may be on subject, nucleus, author, or general reference.

SPECINFO is a structure-related, multidimensional spectroscopic data base system for VAX computers that was written by BREMSER and coworkers at BASF [9.54]–[9.57]. Since January 1990 it has been marketed by Chemical Concepts of Weinheim (FRG). Not only is SPECINFO multidimensional (NMR, IR, and mass spectra), but it also incorporates chemical structure handling software. The carbon-13 NMR data base and similar software are available online on STN International. The data available in SPECINFO are shown in Table 9.

The in-house system uses color graphics displays. Full connection tables are stored for each compound but structure searching is done by HOSE (Hierarchically Ordered Spherical Description of Environment) codes or HORD (Hierarchically Ordered Ring Description) ring

Table 9. Data available on SPECINFO*

Type of spectra	Number of spectra and structures
C-13 NMR	100 000
P-31 NMR	2 000
O-17 NMR	800
N-15 NMR	1 000
H-1 NMR	20 000
Mass spectra	3 000
Mass spectra (NIST)	40 000
IR	16 500

* Wiley mass spectra are not incorporated.

codes stored in inverted files. Structure queries are entered by naming various fragments and giving numbers for ring positions. The input is textual but the structure is displayed graphically.

The data bases can be searched for a variety of single parameters (e.g., molecular formulae, NMR coupling constants, mass spectral masses), CAS ring index codes, and compound names. Spectral data can be displayed both graphically and as a listing.

Input of spectral data is somewhat laborious at present. Registration of additional spectra has recently been made available. Automatic search of incoming data is now a facility.

The only NMR search available at present is a sequential line search but SAHO (Spectral Appearance in Hierarchical Order) and similar search features (available on STN International) will eventually be added. SPECINFO incorporates useful NMR spectral prediction and coupling constant prediction software. There are plans to link SPECINFO with MASS-LIB.

MASS-LIB. MASS-LIB is a set of programs developed by HENNEBERG of the Max Planck Institut für Kohlenforschung (FRG) for the evaluation of low-resolution mass spectra [9.58]. It is sold by MSP Friedli and Co. (Switzerland) for use on VAX/VMS computer systems. From 1st January 1991, Chemical Concepts has the right to sell MASS-LIB. It implements both Wiley and NIST mass spectral data bases and allows users to maintain their own libraries. Structures can be displayed but not searched. Compared with SPECINFO, the strength of MASS-LIB lies in its SISCOM (Search for Identical and Similar Components) library search and in its larger data base. Its storage algorithm allows almost the whole spectrum to be used for comparison without an unacceptable overhead in terms of speed of search.

MACCS-II CSEARCH Implementation. CSEARCH is a system written by W. ROBIEN and coworkers of the University of Vienna [9.59]. Its data base contains 15 772 chemical structures and carbon-13 NMR spectra.

Molecular Design Limited has adapted the data to the MACCS-II environment (see Section 10.5.7) to allow search of peaks and structures; simple and complex spectral displays; and output of structures and spectra in tabular form. Users may also register their own data and enter the mapping graphically.

Structure and substructure search use standard MACCS-II capabilities including the "Power Search Module" (see Section 10.5.7).

Mapping when registering data or studying search output can be done on a simple display of structure, tabulated shifts and multiplicities, and a spectral representation: "clicking the mouse" on an atom indicates the relevant peak and vice versa. In complex displays, the list of shift values of the spectrum can be scrolled and the user can "zoom in" on part of the spectrum. Full spectra can be visually compared.

Fluorine-19 NMR Data Base. Fraser Williams (Scientific Systems) Ltd. has written special software to go with its fluorine-19 NMR data base. When structures and spectral data are displayed together, the relevant atoms are highlighted by dark "blobs". The data base plus PC-SABRE search software are available for use on an IBM PC [9.6].

Sadtler PC Spectral Search Libraries. Sadtler's PC SEARCH software operates on IBM PC/AT and PS/2 computers. Full spectrum search (IR and carbon-13 NMR) is possible and the hits are listed in order of closeness of matching. Two IR spectra can be displayed in different colors on one screen and the relevant chemical structures shown. A difference spectrum can be obtained.

The software operates under the application environment Microsoft Windows. The chemical structures for the 59 000 compounds for which IR spectra are available have been registered in a PsiBase data base [9.6], [9.7], [9.39], [9.41]. Substructure searching is therefore possible and a hit list from a PsiBase search can be taken into PC SEARCH for spectral search and display.

There will also be a 28 000 compound carbon-13 NMR data base and possibly a UV–visible spectral data base in the same system.

9.5. Crystallographic Data Bases

More than 100000 full three-dimensional (3-D) crystal structure analyses have been reported, of which over 50% have been published since 1980 [9.60]. Because of the complexity of the atomic coordinate data, they cannot all be recorded in the primary literature. Molecular graphics data and theoretical chemistry data (e.g., molecular orbitals) are particularly well suited to data bank storage. Although printed handbooks of crystallographic data are available, online data banks have become the main source of three-dimensional data and provide the information in a comprehensive, versatile, and readily accessible form. Sophisticated software enables structures to be retrieved, manipulated, and compared.

The Cambridge Structural Database (CSD). The Cambridge Crystallographic Data Center (CCDC) was set up in 1965 to compile a computerized data base on 3-D molecular structures. The CSD is the world's largest data base of evaluated experimental results, it contains over 73 000 compounds with the addition of about 8500 per year (1990) [9.1], [9.60]–[9.62]. Each structure in the CSD has an entry in three separate files: the bibliographic information (BIB), the chemical connectivity (CONN), and the numeric data (DATA); 68% of the information occurs in the DATA file. The files are linked by a reference code.

All information is thoroughly checked and evaluated before being incorporated. Numeric data are also subjected to computerized internal consistency checking (see Section 9.3). The percentage of entries still in error in the CSD is 1%, compared with ca. 15% in the literature [9.63].

The data base is searched using a Search, Retrieve, Analysis, and Display (SRAD) program [9.64]. Version 4 of the CSD system (see Section 10.5.4), released in 1989, uses high quality graphics for query input and for the display of "hits", and the data base is augmented by two-dimensional structural diagrams for 90% of the compounds. The search output has been improved by highlighting the substructural search fragment.

The Protein Data Bank. The Protein Data Bank, produced by the Brookhaven National Laboratory, is supported by the U.S. National Science Foundation and NIH. It was established in 1971 as a computerized archive for three-dimensional structures of biological macromolecules [9.60], [9.65].

Over 1200 biological macromolecules have been crystallized, of which about 400 structures have been determined. In 1987, PDB held about 350 coordinate entries, with bibliographic references for most of the remaining structures. The format adopted by the PDB has been accepted as a standard for the interchange of atomic coordinates [9.66].

NBS Crystal Data. This data base is produced with the support of the NIST Office of Standard Reference Data. There are about 125 000 entries covering crystallographic, chemical, and physical data, with over 7000 annual additions. Each data entry is critically evaluated at NIST before being incorporated. The data base is available online through CRYSTDAT by the cooperation of the Canadian Institute for Scientific and Technical Information (CISTI), the Canadian Scientific Numeric Database Service (CAN/SND), and the NBS Crystal Data Center (USA). Although full structure searching is not possible, lattice-formula search techniques [9.67] enable data for unknown compounds to be matched with entries in the data base and the compound can be identified.

CAN/SND also provide access to the Cambridge Crystal Structure Database, the Protein Data Bank, the Inorganic Crystal Structure Database, and the NRCC Metals Crystallographic Data File also described in this Section.

Other Crystallographic Data Bases [9.60]. The *Inorganic Crystal Structure Database (ICSD)* is produced by the Gmelin Institute in cooperation with the University of Bonn [9.68]. It contains 29 000 records (1990) of inorganic crystal structures and can be searched via STN International.

The *NRCC Metals Crystallographic Data File (CRYSTMET)* contains critically evaluated crystallographic and bibliographic data for metallic phases, determined by diffraction methods. The file is maintained by the Canada Institute for Scientific and Technical Information (CISTI) through the Scientific Numeric Database office (CAN/SND). The data base contains about 11 000 entries for compounds where the composition is clearly defined and the space group and unit cell have been determined.

The creation of a *Biological Macromolecule Crystallization Data Base* is supported by NIST. To be included, the data must have been published and the macromolecules must have at least one crystal form for which the unit cell and space group have been established by X-ray diffraction analysis.

The *Powder Diffraction File* is distributed and maintained by the Joint Committee on Powder Diffraction Standards (JCPDS)-International Center for Diffraction Data, Pennsylvania, and is a collection of single-phase X-ray powder diffraction patterns. It contains over 44 000 patterns, with the addition of about 2000 per year.

The *Crystal Data Identification File (CDIF)* is owned by the International Center for Diffraction Data (USA) and contains crystal class and unit-cell information for about 60 000 structures. All cell data are reduced to a standard form, and software allows searching on the basis of cell dimensions.

9.5.1. Molecular Sequence Data Banks

A chapter on chemical data banks would be incomplete without reference to the growing number of protein sequence data banks. Over 10 000 protein sequences have been determined so far [9.69]. To cope with the task of assembling evaluated data, CODATA recommended in 1984 that three protein data bases should cooperate to form the Protein Identification Resource (PIR) International Association of Protein Sequence Data Collection Centers [9.70]. The three data bases are the PIR at the National Biomedical Research Foundation (PIR–NBRF) in the United States, the Martinsreid Institute for Protein Sequences (MIPS) in the Federal Republic of Germany, and the Japanese International Protein Information Databank (JIPID).

The National Library of Medicine (NLM) is also cooperating internationally to collect molecular sequence data; 13 data banks register molecular sequences deposited with them by researchers [9.71].

Chemical Abstracts Service are setting up two data bases (CASSEQ–Nucleic Acid and CASSEQ–Protein) to hold all the sequences in their abstracted references [9.72].

9.6. Chemical and Physical Property Data Bases

Until the advent of computer data banks, the chemical and physical properties of compounds were to be found in one of the encyclopedias or handbooks described in Chapter 3. Many of these are now available online, the most comprehensive being the Beilstein handbook, available as Beilstein Online (Section 9.6.1) [9.11], [9.12], [9.73], [9.74]. Many other data banks containing chemical and physical properties of compounds are offered online but most are independent of one another and searches cannot be linked. The Chemical Information System (CIS), however, is an integrated collection of over 30 data banks [9.75], [9.76], linked by a central index file, the SANSS file [9.53] (Section 9.6.3).

9.6.1. Beilstein

The Beilstein Handbook is the world's largest compilation of evaluated data on organic chemistry (see Section 3.3.2). Beilstein Online, first released in 1988, provides more up-to-date information than the printed version, and also enables searches to be carried out in English. It is available on STN International and DIALOG, and will also be released on ORBIT (Maxwell Online).

The data base contains two types of records [9.12]:

1) Organic substances cited in the Beilstein Handbook, Basic Series and Supplementary Series I–IV (1830–1960). These records constitute the *Full File*.
2) Organic substances from the Beilstein collection of excerpts from the primary literature (1960–1980) that have not yet been critically reviewed and are being published as Supplementary Series V. These records constitute the *Short File*.

Information in the Full File has been checked for errors and redundancies, but the Short File is created from data abstracted directly from the primary literature without checking. Data in the Short File will be added to the Full File when it has been checked.

For each compound, information is stored in the structural and factual files. Structure searching is covered in Chapter 10. The data in the factual file can be divided into seven parts:

1) Identifiers: molecular formulae and registry numbers for identification and search of the compounds
2) Structure-related data: information on the purity of the compound and alternative structure representations

3) Preparative data: yield, solvents, temperature, pressure, etc.
4) Physical properties: structure and energy parameters; physical properties of the pure compound; physical properties of multicomponent systems
5) Chemical behavior: reactions with other chemicals
6) Physiological behavior and applications: toxicity, biological functions, ecological data
7) Characterization of derivatives and salts

There are over 500 fields, of which more than 300 are searchable. The user can search for structures and substructures using graphical input or string search; for numerical terms using Boolean logic; for keywords; or for other key fields such as molecular formula, CAS Registry Number, or Beilstein Registry Number.

9.6.2. Other Handbooks and Encyclopedias Available Online

Several of the reference works described in Chapter 3 are available online [9.77] and cover chemical and physical properties. The HEILBRON data base, available on DIALOG since 1986, corresponds to the Dictionary of Organic Compounds, Fifth Edition (DOC5) and six related dictionaries (Section 3.3.3) [9.9]. HEILBRON contains physical and chemical data plus bibliographic references to 250 000 substances.

Other reference works available as online data banks include:

1) The Kirk–Othmer Encyclopedia of Chemical Technology, available via DIALOG and BRS (Maxwell Online).
2) The Merck Index, available on DIALOG, BRS, and CIS.
3) The Formula Index (GFI) and the Complete Catalog from the Gmelin Handbook of Inorganic and Organometallic Chemistry, accessible via STN International. The Gmelin Institute is also preparing a Factual Data Bank.
4) HODOC, produced by CRC Press and containing information from the second edition of the Chemical Rubber Company's Handbook of Data on Organic Compounds (HODOC II), available online via STN International.

9.6.3. Data Banks on the CIS System

The Chemical Information System (CIS) is a collection of about 30 data banks developed originally by the U.S. government agencies NIH and EPA, and now marketed by CIS Inc., a subsidiary of Fein–Marquart Associates. The independent data banks are linked through a central file, the Structure and Nomenclature Search System (SANSS). Chemical structure searching is discussed in Chapter 10.

Over 350 000 chemical substances are searchable by full structure, substructure, or by name or partial name. Structure query entry can be facilitated with the front-end package, SuperStructure [9.6]. Most of the data banks can be searched using a common command language Text Data Retrieval System (TDRS), which also allows transfer of data from one data bank to another. The compounds are identified and cross-referenced by their CAS Registry Numbers.

The information on the data banks can be classified into:

1) Spectral data banks (Section 9.4.5)
2) Chemical and physical property data banks (Table 10)
3) Toxicology, environmental, and hazard data (Section 9.7)
4) Pharmaceutical data banks
5) Analysis and modeling data banks.

The pharmaceutical data bases are: Drug Information Fulltext (DIF), which contains the complete contents of two reference volumes published by the American Society of Hospital Pharmacists; The Merck Index Online (see Section 9.6.2); Physicians' Desk Reference Online, which contains the full text for prescription, over-the-counter, and ophthalmological drugs.

The analysis and modeling data banks are: ARTHUR, which is used in the formulation and evaluation of models for incompletely understood data sets; CHEMLAB (Chemical Modeling Laboratory), which provides capabilities for three-dimensional conformational analysis, molecular orbital calculations, and estimation of many chemical properties; MLAB/CLAB (Mathematical Modeling System including Cluster Analysis Laboratory), which is an interactive system for mathematical modeling, cluster analysis, and pattern recognition of numeric data.

9.6.4. Thermodynamics and Thermophysical Property Data Banks

Of all the chemical properties, thermodynamics is the most important for many scientists [9.78]. It is particularly important that thermodynamics data bases contain accurate, validated information, as the searcher is likely to calculate other properties from retrieved data.

STN International Data Bases. There are three thermodynamic data bases on STN International: DIPPR, JANAF, and NBSTHERMO. STN International provides access to a number of numeric data bases [9.77], [9.79]. Several of these have already been described: the carbon-13 NMR data base (Section 9.4.1), ICSD (Section 9.5), Beilstein Online (Section 9.6.1) and HODOC (Section 9.6.2). Numeriguide, produced by the American Chemical Society, is a compilation of the lists of properties in each of the STN numeric files.

DIPPR contains physical property data, textual information, and CAS Registry Numbers for over 1000 commercially important chemical compounds. Data are compiled and evaluated by the Design Institute for Physical Property Data (DIPPR) of the American Institute of Chemical Engineers. DIPPR is also available on Numerica.

JANAF contains the Joint Army, Navy, and Air Force Thermochemical Tables and is produced by the U.S. Department of Commerce, NIST. There are 1100 records of critically evaluated chemical thermodynamic properties of inorganic substances.

NBSTHERMO is also produced by NIST and corresponds to the data in the NBS Tables of Chemical Thermodynamic Properties. It contains over 8200 records of critically evaluated chemical thermodynamic properties of inorganic and organic substances.

Numerica Data Bases. Technical Database Services, Inc. of New York also provide thermodynamic data bases through their Numerica service.

Data Files from the Thermodynamics Research Center (TRC). The TRC at Texas A&M University has created two numerical data bases of thermodynamic properties [9.80]: the TRC Selected Data File, which contains selected values of properties published in the TRC Thermodynamic Tables, and the TRC Source Data File, which contains properties of pure organic compounds that have been published in the open literature.

The TRC Thermophysical Property Datafile 1: Vapor Pressure is available online through Numerica. It contains about 18 000 records of experimental data for 3500 chemicals. Vapor pressure and boiling points for over 5500 chemicals can be calculated interactively using critically evaluated Antoine coefficients.

The Physical Property Data Service (PPDS) is an integrated system which provides thermodynamic and transport property data for pure components and mixtures. It is maintained at the National Engineering Laboratory and the Institute of Chemical Engineers (UK) and is available online through Numerica. Supplementary programs allow the users to incorporate their own data.

Table 10. CIS data banks of physical and chemical properties

Name of data bank	Producer	Number of substances covered	Type of data available
CESARS (Chemical Evaluation Search and Retrieval System)	Office of Materials Control, State of Michigan	194	physical and chemical properties, carcinogenicity and toxicity, environmental fate, hazard and handling information
CHRIS (Chemical Hazard Response Information System)	U.S. Coastguard	1 156	labeling, physical and chemical properties, health hazards, fire hazards, chemical reactivity, water pollution
ENVIROFATE (Environmental Fate)	EPA Office of Toxic Substances, Syracuse Research Corporation	450	physical and chemical properties, environmental transformation rates
ISHOW (Information System for Hazardous Organics in Water)	EPA Environmental Research Laboratory, University of Minnesota, EPA Office of Toxic Substances	5 400	melting points, boiling points, partition coefficients, acid dissociation constants, water solubility, vapor pressures
OHM/TADS (Oil and Hazardous Materials/Technical Assistance Data System)	EPA	1 400	physical, chemical, biological, toxicological, and commercial data
SOLUB (aqueous solubility data base)	University of Arizona College of Pharmacy	2 600	evaluated data on aqueous solubility
THERMO (thermodynamics data base)	NIST	15 000*	evaluated thermodynamic data

* Entries.

DECHEMA Thermophysical Property Data Bank (DETHERM). The Deutsche Gesellschaft für Chemisches Apparatewesen, Chemische Technik and Biotechnologie e.V. (DECHEMA) produces a number of reference works, including the DECHEMA Chemistry Data Series [9.81].

The DETHERM data bank is produced in conjunction with FIZ and is available online via Numerica. It includes three systems:

1) DETHERM-SDR (Data Retrieval System),
2) DETHERM-SDC (Data Calculation System), and
3) DETHERM-SDS (Data Synthesis System).

DETHERM-SDR is a compilation of physical property data on 3000 components abstracted from the literature. DETHERM-SDC is used to compute properties and phase equilibria from the stored data. DETHERM-SDS can be used to create new pure-component data for subsequent use with the SDC programs.

Other Thermodynamic and Thermophysical Property Data Banks. The THERMO data base on CIS is included in Table 10. The data base F*A*C*T (Facility for the Analysis of Chemical Thermodynamics) uses some of the same data sources as THERMO, although it only contains data on 2400 substances. The user can calculate thermodynamic equilibria or phase diagram boundaries.

Thermodata (France) produces a bibliographic data base Thermodoc and three thermodynamic data banks: Thermocomp, Thermalloy, and Thermosalt.

Thermodynamic properties of organic substances can be calculated using data from the data base EPIC (Estimate of Properties for Industrial Chemistry) available online through CIGL Inc. in Belgium.

Online data bases of vapor–liquid equilibrium data [9.76] are CHEMTRAN and the Dortmund VLE Data Bank (ChemShare Corp.), and the Physical Property Data Service (PPDS) produced by the Institute of Chemical Engineers (UK) and available through Numerica.

9.6.5. Other Chemical and Physical Property Data Bases

The Pesticides Properties Database is being developed by the USDA Agricultural Research Service (ARS) to provide essential data on the chemical and physical properties of pesticides [9.15]. All the data obtained from the literature are critically evaluated. The data base will help in regulatory assessment of pesticides and in developing models for determining the contamination of drinking water supplies by pesticides entering the soil.

Codura Publications Inc. produce Poly-Probe, a data base of physical property data on plastics.

9.7. Toxicology, Hazard, and Environmental Data Banks

Toxicology data are very important [9.82], but there are more problems associated with their use and creation than with most other types of data bank. It is difficult to use a toxicology data bank without a good knowledge of the subject and some experience in the use of information systems. There are problems with evaluation and quality control of data. Checks for internal consistency are difficult and repeating the measurements is not normally feasible.

CIS Toxicology Data Banks. The Registry of Toxic Effects of Chemical Substances (RTECS) is compiled by NIOSH and is the largest chemical toxicology data bank, containing information on 90 000 substances. Data cover acute and chronic toxicity measurements; primary skin and eye irritant data; carcinogen, mutagen, and tumorigen data. Negative findings are not recorded and the data are not evaluated. Thus, RTECS cannot necessarily be used as a substitute for the original source. However, the linking of chemical substructure searching with data searching allows correlation of structure and toxic effects. The data bank is available on CIS, and as part of the CHEM-BANK CD-ROM from SilverPlatter.

CIS contains several other data banks that include toxicological, environmental, and hazard information [9.75]; CESARS, CHRIS, ENVIROFATE, ISHOW and OHM/TADS are listed in Table 10 (p. 12-73). The six remaining toxicological data banks are shown in Table 11. In addition, CIS provide a bibliographic data base on GastroIntestinal Absorption (GIABS) and an index to unpublished health and safety studies submitted under the Toxic Substances Control Act (TSCATS).

Chemical Hazards Data Banks. Most of the CIS data banks relating to hazards have already been covered: RTECS, OHM/TADS, CHRIS and CESARS (Tables 10 and 11). CIS also provide the HAZINF data bank (Hazardous Chemicals Information and Disposal Guide). This contains hazard ratings, chemical and physical properties, hazardous reactions, fire hazards, physiological properties, health hazards, and waste and spillage disposal instructions for about 220 chemicals or classes of chemicals. The data base was created by the Department of Chemistry, University of Alberta.

A comparison of the coverage of chemical hazard data banks is given in [9.83]. The Hazardous Substances Data Bank (HSDB), produced by the National Library of Medicine, Toxicology Information Program, was found to have the widest subject coverage and is a good source for human toxicity information. Although it only contains approximately 6500 compounds, much of the data is evaluated [9.82].

The National Chemical Emergency Center at Harwell (UK) produces the microcomputer system CHEMDATA, which provides information for identifying hazards and procedures for dealing with chemical products in emergencies. Safety measures for 18 000 substances are available. CHEMDATA is one of the seven sources of information on the Dangerous Goods CD-ROM from Springer Verlag. Also available on the CD-ROM are:

1) Hommel Handbook of Dangerous Goods
2) The Operation Files for chemical incidents of the Swiss Fire Brigades
3) The Fluka Catalog
4) The Swiss Toxic Substances List, containing 76 000 substances
5) SUVA, the Swiss Accident Insurance Institution, which contains the safety codes of liquids and gases
6) VCI, the handbook of firms in the chemical industry (Econ-Verlag, FRG)

The SilverPlatter CD-ROM, CHEM-BANK, contains information on hazardous chemicals (the RTECS, CHRIS, and OHM/TADS data banks). The data bank TOSCA (Toxic Substances Control Act) Initial Inventory from the EPA is to be added.

The Environmental Chemicals Data and Information Network (ECDIN). The ECDIN data bank is produced at the Joint Research Center of the Commission of the European Communities at Ispra in Italy [9.84]. It contains data related to actual or potential chemical pollutants. The

Table 11. CIS toxicological data banks

Name of data bank	Producer	Number of compounds covered	Type of data available
AQUIRE (Aquatic Information Retrieval)	EPA Environmental Research Lab, EPA Office of Toxic Substances	4 000	acute, chronic, bioaccumulative, sublethal effects data from experiments performed on freshwater and saltwater organisms
CCRIS (Chemical Carcinogenesis Research Information System)	National Cancer Institute	1 000	assay results and test conditions for carcinogenicity, mutagenicity, tumor promotion, and cocarcinogenicity
CTCP (Clinical Toxicology of Commercial Products)	5th edition of publication by Gleason, Hodge, Gosselin and Smith	20 000*	toxicity data for about 20 000 common commercial products, including symptom and treatment data
DERMAL (Dermal Absorption)	EPA Office of Toxic Substances	655	data on toxic effects, absorption, distribution, metabolism and excretion related to dermal application of chemical substances
GENETOX (Genetic Toxicity)	National Institute of Environmental Health, Oak Ridge National Laboratory, Environmental National Laboratory	2 600	mutagenicity information tested against 38 biological systems
PHYTOTOX (Plant Toxicity)	University of Oklahoma, Dept. of Botany and Microbiology; EPA	1 000	biological effects of application of chemicals to plants
RTECS (Registry of Toxic Effects of Chemical Substances)	NIOSH	90 000	acute and chronic toxicity data

* Commercial product.

broad spectrum of parameters and properties supports the evaluation of the risks connected with the use of the substances. Links are established between various types of data, enabling the data bank to be used for correlation analyses. Public access is via the DC Host Center in Copenhagen and the data are managed under the ADABAS system. Chemical structure handling uses a modified form of the CROSSBOW system (Section 10.1.3), but chemical names and synonyms are the most common way of retrieving structures.

One of the ECDIN files is the European Inventory of Existing Chemical Substances (EINECS) which gives the chemical substances that were on the EEC market between January 1971 and September 1981. All these substances have been registered with CAS and their Registry Numbers are available on ECDIN. Substances not in the EINECS inventory must be notified before they can be marketed in the EEC.

Data Banks Produced as a Result of Legislation. Legislation has been the impetus behind the production of a number of data banks (e.g., EINECS). The CIS data bank TSCAPP was compiled as a result of the Toxic Substances Control Act which required the establishment of an inventory of chemicals in U.S. commerce from 1975 through 1977. The TSCAPP file contains plant and production data for over 55 000 chemicals on the Toxic Substances Control Act Inventory. The TSCATS file on CIS (Toxic Substances Control Act Test Submissions) contains about 15 000 records to over 1200 compounds.

Four other CIS files were created as a result of legislation:

1) *Federal Register Search Sytem (FRSS)* provides a cross-reference to all citations to a chemical or class of chemicals cited in the Federal Register from 1 January 1977 to November 1983, but is no longer being updated.
2) *Suspect Chemicals (SUSPECT)* contains regulatory and advisory data for ca. 4000 potentially hazardous chemicals, including explosive limits. The information is taken from U.S. government regulations, National Fire Protection Hazard Ratings (NFPA), and Threshold Limit Values and Biological Exposure Indices (ACGIH).
3) *Chemical Activity Status Report (CASR)* lists chemicals that are studied by the EPA in the course of regulatory or scientific research activities.
4) *Industry File Index System (IFIS)* provides summaries of EPA regulations on particular industries and chemical substances.

Material Safety Data Sheets (MSDSs) are required to accompany the transfer of chemicals from one company to another. CIS provides access to more than 1600 MSDSs from J.T. Baker Co. and more than 1400 MSDSs from Mallinckrodt Inc. The MSDSs are formatted according to the OSHA guidelines, and contain information on safe handling, storage, and disposal.

Sigma – Aldrich Chemical Co. produce a CD-ROM of 30000 MSDSs for about 15 000 compounds [9.85].

9.8. Special Applications

The ChemQuest data base on ORBIT (Maxwell Online) is created from the catalogs of 54 chemical suppliers. This fine chemicals directory can be searched by chemical name, trade name, molecular formula, CAS Registry Number, or by structure or substructure. Compounds can be ordered online.

A fine chemicals data base is available as the CD-ROM ChemLink data base from Chemron Inc. [9.85] and online on DIALOG. Catalogs from only 31 suppliers are included, but it is cheaper to use than ChemQuest.

9.9. The Future of Data Banks

In the field of chemistry, future progress should be directed towards the improvement of data quality and better integration of existing systems. Several barriers to building chemical data bases have been identified [9.19]; scientists have little experience with data-base management systems; most data bases are built for individual use and are hard to adapt to more general use; users have not been involved enough in the design of data bases; very few ways for accessing data bases are available to the chemical community at large; funding for data-base building rarely is available. Another problem is the lack of skill and interest to generate and use these data bases [9.20].

Most of the data banks described in this chapter have developed independently without standardization. Work on standards is being carried out by the National Standard Reference Data System which was established in the United States in 1963 to coordinate production and dissemination of critically evaluated data in the physical sciences. IUPAC is concerned with collecting and compiling numeric data sources; the Committee on Data for Science and Technology

(CODATA) has been working on standards since 1966 [9.86]. World Data Centers (WDC) have existed since the 1950s [9.87]. More international cooperation is needed in the development of on-line data banks if the vast amount of data being produced is to be made available in a usable form. The use of CD-ROM as a storage medium has great potential in the dissemination of data banks.

10. Chemical Structure Handling

10.1. Chemical Structure Representation
[10.1]–[10.4]

10.1.1. Systematic Nomenclature

The chemist's preferred method of communicating structural information is the two-dimensional (2-D) chemical structure.

In all chemical information systems there is a need to reduce the structure to a "name" that can be stored and manipulated by a computer. Systematic nomenclature (see Section 2.2.4) in its usual sense is not useful here for many reasons. Systematic names are often long and complicated. When a new area of chemistry is opened up the appropriate nomenclature is not available for months or even years. Systematic rules are often inconsistent or ill-defined. In IUPAC nomenclature there is inconsistency in the retention of trivial names and there may be more than one acceptable name for a single compound. In a useful computerized system any one compound must have only one name and any name must be convertible to only one structure. CAS has systematized nomenclature sufficiently to give a unique name for most compounds but has had to make its own rules. Trivial names are still in common usage for some applications.

Thus, while systematic nomenclature does have advantages, for example, in printed indexes, the chemical information specialist has turned to other methods for the internal representation of structures in the computer.

10.1.2. Fragmentation Codes

A fragmentation code is a collection of predetermined small substructures (e.g. six-membered ring, carbonyl group) each represented by a number, or sometimes a combination of letters and numbers. In the early days each fragment could be represented by a hole in a punch card. Thus a hydroxy group might be represented by punching a hole in the third row of the seventieth column of an 80-column punch card, represented on paper as 70/3. A punch card with appropriate holes was generated for each compound and the structure drawn by hand on the card. Mechanical card-sorting equipment (or even simple rods) were then used to find all the cards with a hole 70/3 (or any other required holes) to perform a search for hydroxy compounds.

Nowadays the codes are automatically derived from a form of computer representation and searched by a computer program.

There are two types of fragmentation code: fixed and open-ended. A *fixed fragmentation code* consists of a fixed number of fragments, and is not easily modified if new features need to be included. If a new fragment is invented, all previously processed structures will need recoding.

The fragments of an *open-ended code* are constructed from a structure representation by a computer algorithm according to a particular set of rules. Consequently, an open-ended code can be more responsive than a fixed code to the changing characteristics of the compound collection. Also, more precise retrieval from the file may be possible because the code offers a much greater variety of search terms.

Derwent Publications manually encodes structures into two fixed fragmentation codes: the Ring Code [10.5] and the New Chemical Code [10.6] for use with Derwent's RingDoc and World Patents Index Services, respectively.

An important and sophisticated, open-ended fragmentation code is GREMAS (Generic REtrieval by MAgnetic tape Search) [10.7]–[10.9] developed by the Internationale Dokumentationsgesellschaft für Chemie (IDC), a consortium of seven German companies. This code involves fragment terms which are constructed according to a number of broad structural classifications, each of which can be subdivided using a three-level hierarchy. These classifications can be used to characterize fragment occurrences at varying levels of specificity and the way in which terms can be combined, thus reducing the ambiguity of the representation and increasing the efficiency of searching.

Fragmentation codes are inevitably an ambiguous method of structure representation. They describe the various parts of a molecule but

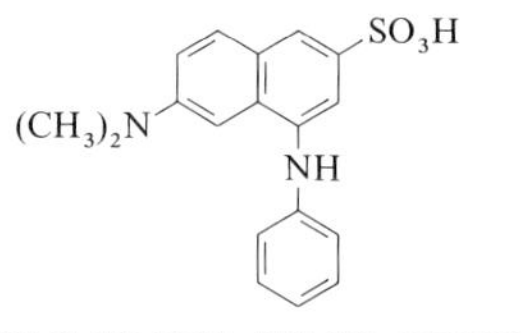

IUPAC: B6₂N(C)₂4SO₂Q8: *IO*N/B6
WLN: L66J BMR&DSWQ IN1&1

Figure 11. Comparison of Wiswesser Line Notation and Dyson/IUPAC Notations

not how those parts are linked together (i. e., the full topology). Now that other methods of structure representation are available allowing the description of the full topology, fragmentation codes are used mainly for the initial screening stage in substructure searching (see Section 10.3).

10.1.3. Linear Notations

Linear notations are a compact way of representing a chemical structure by strings of alphanumeric characters. Two notations, Wiswesser Line Notation (WLN) [10.10]–[10.12] and Dyson/IUPAC [10.13] are compared in Figure 11.

WLN was the best known of the notations in the chemical industry for many years and was used by ISI for its Index Chemicus Registry System [10.14]. With a few minor exceptions, WLN is an unambiguous representation: it fully represents the constitution and connectivity of a molecular structure. Moreover, if the encoding rules are followed precisely, the notations are unique (only one linear string is possible for any compound), which makes WLN appropriate for registration (see Section 10.2) and the production of printed listings.

The ICI CROSSBOW system (Computerized Retrieval of Organic Structures Based on Wiswesser) used not only WLN but also fragment codes and connection tables (see Section 10.1.4) based on WLN [10.15]–[10.18]. CROSSBOW was widely used worldwide until the early 1980s.

WLN has some of the disadvantages of systematic nomenclature; it is not suitable for handling tautomerism, mesomerism, and stereochemistry. One of its principle advantages, compactness, has disappeared with the advent of low-cost, high-density storage. Its use has therefore given way to the use of connection tables (see Section 10.1.4). Many companies have converted their WLN files to connection table files using programs such as DARING [10.19].

A notation still in use by a select group is SMILES (Simplified Molecular Input Line Entry System) [10.20]. Its use with software supplied by Daylight Chemical Information Systems is described in Section 10.4.2.

10.1.4. Connection Tables [10.3], [10.4], [10.21]

A connection table contains a list of all of the non-hydrogen atoms within a structure, together with bond information that describes the exact manner in which the individual atoms are linked together. Hydrogen atoms are not included since their presence or absence can be deduced from the bond orders and atomic types.

Many different types of connection table are used. In a *redundant* connection table each bond appears twice (Fig. 12); in a *nonredundant* connection table each linkage is listed only once (Fig. 13). In *bond-implicit* connection tables (e.g., the CROSSBOW system), the bonds are not defined explicitly but are included in the "nodes" which characterize each atom. Figures 12 and 13 both show *bond-explicit* connection tables. They take up more storage space than bond-implicit tables but the information can be interpreted more easily by a computer program.

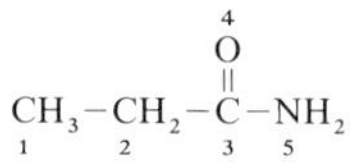

$$CH_3 - CH_2 - \overset{\overset{4}{O}}{\underset{3}{C}} - NH_2$$
$$1 2 3 5$$

Atom number	Atomic symbol	Bond connection	Attached atom number	Bond connection	Attached atom	Bond connection	Attached atom
1	C	1	2				
2	C	1	1	1	3		
3	C	1	2	2	4	1	5
4	O	2	3				
5	N	1	3				

Figure 12. A redundant connection table

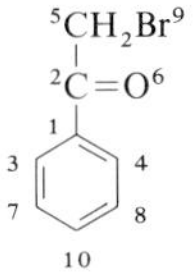

```
         ⁵CH₂Br⁹
          |
         ²C=O⁶
          |
          1
       3 / \ 4
        |   |
       7 \ / 8
          10
```

Node no.	Atom	Connected to	Bond type
1	C		
2	C	1	−1
3	C	1	*5
4	C	1	*5
5	C	2	−1
6	O	2	−2
7	C	3	*5
8	C	4	*5
9	Br	5	−1
10	C	7	*5
Ring closure		8–10	*5

1 = single bond   5 = alternating bond
2 = double bond   * = cyclic bond
− = non-cyclic bond

Figure 13. A nonredundant connection table (reproduced with permission from an internal publication of Chemical Abstracts Service)

Connection tables are unambiguous since they give a complete description of the topology of the compound which they represent. However, they are generally not unique since there are very many different ways of numbering the atoms. There are algorithms to produce a unique connection table from a randomly numbered version. This process is usually known as "canonicalization", although the correct mathematical term is "canonization". The Morgan algorithm used by Chemical Abstracts Service (CAS) was mentioned in Section 2.2.2.

In most modern chemical information systems structures are entered graphically and the connection table is derived by the computer. Fragmentation codes can also be derived algorithmically from a connection table.

The great diversity of connection table formats has led to moves to establish standard connection table formats for the exchange of chemical structure data. The Standard Molecular Data (SMD) file format is one of these [10.22], [10.23].

10.2. Compound Registration

An essential function of a chemical information system is the ability to search a chemical structures file for the presence of a given compound, perhaps represented in the file in a different but equivalent form. Compound registration is the procedure by which new compounds and accompanying data are added to a structure file and associated data files, respectively. A structure file in which a single and unique record of each compound is maintained is known as a registry file. A unique identifier is usually assigned to each structure record in a registry file.

The CA Registry File was described in detail in Section 2.2.2. The CA Registry Number is used not only in CAS publications and services but also in data banks, primary journals, handbooks, and compound lists.

Registration of a new compound involves matching a canonical representation [usually a connection table, although a unique WLN (Section 10.1.3) was used in older systems] against others already in the registry system. A modification of the Morgan algorithm is used in many systems [10.24], [10.25].

Registration may take place batchwise or on-line. In a *batch* registration system a batched, holding file of new records is periodically sorted and matched, usually against a serial file structure. In an *online* registration system, direct access to the structure file is required every time a record needs matching, but file update is then immediate, i.e., the new structure is available for searching within seconds of registration.

Direct matching of connection tables is inefficient for large files and various methods have been employed to reduce the match to a small part of the file. A simple method is isomer sorting (i.e., partitioning using molecular formula) [10.26], [10.27]. Another quick method involves generating so-called "hash codes" for every structure record [10.28]–[10.30].

10.3. Techniques of Substructure Searching [10.2], [10.4], [10.31]

Chemical structure diagrams can be regarded as topological graphs in which the atoms are nodes and the bonds are edges. The problem of substructure searching is that of matching two graphs, one for the query structure and one for the file structure. This type of mapping is called *subgraph isomorphism*. A "brute force" approach involves trying all possible permutations. In the worst case situation the time taken to match a query structure against those on file increases exponentially with the number of nodes. The subgraph isomorphism problem belongs to a class of mathematical problems known as NP-complete.

Because matching is so time-consuming much research on substructure searching has focused on the development of efficient and effective screening methodologies, i.e., eliminating large numbers of structures that could not possibly be matched so that only a small part of the registry file needs to be submitted to the atom-by-atom matching procedure.

Screening Systems. Molecular formula offers an obvious screening possibility: for example, if palladium compounds are not required there is no point in doing atom-by-atom search on connection tables containing palladium atoms. Unfortunately most organic compounds contain carbon, hydrogen, and a few other elements, for which molecular formula screens are of no use. Thus screening systems based on algorithmically generated structural fragments have been devised. Atom-centered, bond-centered, and ring descriptor fragments have been investigated [10.32]–[10.36]. The BASIC (Basel Information Center) group of companies in Switzerland has compiled a dictionary of fragments [10.37] which formed the basis of the original CAS ONLINE substructure search system on STN International [10.38].

The DARC (Description, Acquisition, Retrieval and Correlation) system [10.39] offered commercially by Questel also uses "limited environment" screens but these are open-ended rather than controlled by a dictionary. They are discussed in more detail in Section 10.5.5.

Atom-by-Atom Search. Once screening has been completed (typically reducing the file to 1% of its original size), the second, atom-by-atom stage has to be carried out for those molecules which matched the query at the first stage. Most systems use "backtracking" [10.40] which involves matching one pair of atoms, then trying to establish matches for their neighbors, and so on. If at any stage a match cannot be established, the last unsuccessful match is "undone" and another one is tried. Partitioning, or set reduction techniques have also been tried [10.41]–[10.43]. These reduce the number of structure atoms which can be mapped to each of the atoms in the query and the number of combinations which need to be tested.

"Relaxation" is a technique which has been used more recently [10.44]. It is thought that parallel computer hardware (see Sections 7.2.1 and 10.4.4) is needed for this technique to be implemented efficiently [10.45].

Tree-Structured Inverted Files. The computationally intensive part of the substructure search can be reduced by preprocessing the data base to produce a tree-structured set of inverted files. Data-base building becomes slower but searching is more rapid. Two systems which use this technique are the Chemical Information System (CIS) originally developed by the National Institutes of Health (NIH) and the Environmental Protection Agency (EPA) [10.46]–[10.51] and the Hierarchical Tree Substructure Search (HTSS) developed in Hungary [10.52]. The CIS system offers three search methods: a Ring Probe (based on ring systems present in the structure), a Fragment Probe (based on augmented atom fragments), and an atom-by-atom search. HTSS extends the tree search file in the data-base preprocessing in such a way that substructure search does not need an atom-by-atom stage.

10.4. Current Research in Substructure Searching

10.4.1. Generic Chemical Structure Handling [10.53],[10.54]

Generic chemical structures, also known as Markush structures, are most frequently found in chemical patents as a means of expressing families of substances (see Section 6.7.2). They pose special problems. A single generic structure may cover millions of individual substances. The structure diagram may contain complexities such as variable bond orders and attachment positions, and optional rings. The generic description may be a mixture of text and graphics and the text may contain generic expressions such as "aryl", and non-structural concepts such as "electron-withdrawing group". All these factors have meant that chemical patents were best handled by fragmentation codes until recently. Examples are the Derwent chemical fragmentation code (see Section 10.1.2) and IDC's GREMAS code (see Section 10.1.2 and p. **12**-81).

The Sheffield University Project. Research on generic structure handling has been carried out at the University of Sheffield (UK) since 1979 [10.55]–[10.57]. An input representation for generic structures (GENSAL) has been developed [10.58]. It has a context-free grammar similar to that of modern programming languages and can describe a large proportion of the generic structures found in patents. The GENSAL Interpreter program [10.59] checks that the GENSAL statements are syntactically and semantically correct, and then generates an internal representation of the structure, the Extended Connection Table Representation (ECTR) [10.60]. Generic nomenclature terms are intrepreted by a chemical structure grammar, TOPOGRAM [10.61]. This can be used both to generate atom-centered and bond-centered screens for the generic radicals and to determine whether a substituent group in the query structure belongs to the class of radicals described by a generic radical term in the file structures. It can also determine common membership between generic radical terms.

A two-stage search process has been developed. The first stage uses algorithmically generated screens [10.62]. Atom-by-atom searching of generic chemical structures is not possible. A relaxation algorithm is used, which, due to the fact that it is inherently parallel in nature, offers the potential for implementation on a parallel processing machine [10.62].

A reduced chemical graph approach has also been developed to describe and match the gross structural features of substances. Some of the nodes are collapsed together, forming a smaller, simpler graph which is quicker to search [10.63]–[10.65].

Graphics-Based Systems [10.54]. As well as providing access to the CAS Registry File for its member companies, IDC runs a patent service and recodes patents from Derwent's Chemical Patents Index using the GREMAS code. The GREMAS system has a particularly good capacity for handling generic concepts in storage and retrieval and gives high precision in searching compared with other fragmentation codes [10.7]. However, it is difficult to learn and expensive to run. IDC is now developing a system allowing automatic generation of GREMAS codes from generic structures input using Sheffield University's GENSAL and Hampden Data Services (HDS) graphics structure input program PsiGen (see Section 10.6.3).

Hoechst, one of IDC's members, has developed a program GREDIA to generate a GREMAS search strategy from graphical input of a query [10.66].

Derwent Publications, Questel, and the French Patent Office, INPI (Institut National de la Propriété Industrielle) have collaborated in a project to enhance the DARC software (see Section 10.5.5) to permit generic data-base searching [10.67]. The new Markush DARC system will replace Derwent's fragmentation-code-based system. The backfile of fragments is already searchable using a microcomputer program called TOPFRAG, which allows the searcher to draw a generic structure using HDS's PsiGen techniques. TOPFRAG converts the structure into Derwent codes and formulates the online search strategy so that the user need not learn the rules of the coding system.

The efficiency of Derwent's Chemical Code, IDC's GREMAS system, and Markush DARC has recently been compared [10.9].

CAS introduced MARPAT in 1990 for searching a data base of generic chemical structures [10.64], [10.68].

10.4.2. Substructure Searching in Files of Three-Dimensional Structures

Three-dimensional (3-D) data are obtained mainly from X-ray and neutron diffraction methods. The 3-D coordinates, and cell and symmetry information, can be used to calculate a variety of intra- and intermolecular geometrical parameters. There are now well over 100 000 crystal structures in the public domain [10.69]. The major data bases are the Protein Data Bank, the Cambridge Structural Database (CSD), the Inorganic Crystal Structure Database, and the Metals Crystallographic Data File (see Section 9.5). Quantum mechanics and molecular mechanics programs may be used to calculate 3-D data, but the above data bases contain experimentally measured values.

The topography of a chemical structure (i.e., the arrangement of the atoms and bonds in 3-D space) can be represented in a connection table by replacing the interatomic connections with interatomic distances. A facility for 3-D substructure searching is of particular interest in computer-aided drug design. The task is often referred to as *pharmacophoric pattern matching*, where a pharmacophoric pattern is the geometrical arrangement of atoms in 3-D space responsible for some observed biological activity [10.70]. The first 3-D program was Molpat [10.71]. Lederle Laboratories have an in-house system, 3DSEARCH [10.72].

Research at Sheffield University. Research has been carried out at Sheffield University into techniques for searching a data base of 3-D chemical structures, analogous to those used for 2-D structures [10.73]. As with 2-D substructure searching, an initial and rapid search is used to eliminate large numbers of molecules followed by a more computationally demanding search (this time geometric) to determine whether the precise query pattern is present. The screen search is based on screens which correspond to pairs of atoms together with a distance range [10.74]. A range of algorithms has been studied for the geometric search [10.75] and multiprocessor systems are being studied to increase the efficiency of 3-D substructure searching (see Section 10.4.4).

A related problem to 3-D pharmacophoric pattern searching is the identification of a pattern responsible for some biological activity. Work previously done on the identification of maximal common substructures (MCS) for 2-D structures (see Section 10.8) has been extended to 3-D structures [10.76], [10.77]. Thus the largest common substructure in a set of molecules possessing the activity of interest is likely to contain the pharmacophoric pattern. Algorithms for the identification of an MCS are very computationally demanding and may perhaps be made more efficient by the use of parallel processing (see Section 10.4.4).

Cambridge Crystallographic Data Center. The Cambridge Crystallographic Data Center's GSTAT program locates a fragment, calculates user-specified geometry, and selects fragments on the basis of limiting values supplied for any of the derived parameters. Even then, statistical analyses may be required to answer a 3-D query completely. Integration of 3-D searching with the 1-D and 2-D capabilities of the CSD program QUEST is being effected via a 1:1 graph matching of chemical and crystallographic connection tables, to be followed by a careful generation of 3-D screening mechanisms. An improved statistics package is also being developed.

Work on 2-D and 3-D similarity searching has also been carried out.

MACCS-3D. Molecular Design Limited (MDL), the vendors of MACCS (see Section 10.5.7) have recently released MACCS-3D for the organized storage, searching, and retrieval of static 3-D molecular models and model-related data, including atom and atom-pair data.

Models in MACCS-3D are MACCS-II structures for which additional 3-D information (e.g., Cartesian coordinates and partial atomic charges) is stored. A 2-D structure may have 3-D models associated with it, each corresponding to a different conformation or shape. Each model is associated with a single structure and has its own model-related data that may be general to the model or specific to individual atoms or atom pairs.

MACCS-3D allows exact-match, geometric, and submodel searching of 3-D models with geometric constraints specified to a certain degree of tolerance. Thus the user can retrieve molecules not only with a specific chemical structure, but with a specific conformation as well.

The user sets constraints (distances, angles, dihedral angles, and spheres of exclusion) on user-defined geometric objects (points, lines, planes, centroids, and normal vectors).

The multicolored display in MACCS-3D allows the user to compare queries with retrieved conformations by superimposing the query on each model retrieved.

A number of data bases are supplied by MDL for use with the software. Tripos' CONCORD program (developed by PEARLMAN and coworkers at the University of Texas) [10.78] has been used to create 3-D versions of MACCS 2-D data bases.

ChemDBS-3D. Chemical Design Ltd. have developed ChemDBS-3D, which operates in close integration with their other Chem-X molecular modeling modules.

ChemDBS-3D searches 3-D structure data bases for compounds matching a modeled pharmacophore in some low-energy conformation. It uses a 3-D distance screening method which enables all low-energy conformers to be searched but requires only one set of coordinates to be stored. The screening rate is 100 000 compounds per second. The search query (spatial distribution of centers) is perceived automatically from a modeled pharmacophore. Centers include user-definable classes of atoms with some specific property (e.g., hydrogen donor). Classical substructure search and property searches are also available. The results of screening and searching are written in nonconformationally dependent answer sets. Answer sets may be combined using arithmetic and logical operators, used as the basis for further searches, or exported to other systems. Conformations that match the original query are regenerated and stored in the results data base.

Aladdin. Daylight Chemical Information Systems offers the Aladdin system originally developed at Abbott Laboratories [10.79]. Aladdin searches data bases of 3-D structures to find compounds that meet substructural and geometric criteria. In addition, searches can be based on the ability of compounds to fit into predefined binding sites. Geometric objects (points, lines, and planes) are constructed from the coordinates of atoms in appropriate substructure environments using the GCL language. Any object can be generated from previously defined objects, for example, a point as the mean of several previously defined points.

Search queries in Aladdin specify the ranges of distances, angles defined by three points, torsion (dihedral) angles, and plane angles that the geometric objects must match to constitute hits. Output files include details of each test, hard copies of structures with hit atoms identified, and data that can be used to generate molecular graphics of retrieved structures. Molecular models can be viewed as stereo images.

Aladdin is one of a number of modules that work within the framework provided by Daylight's chemical information system, Daymenus, which originated with the MedChem project at Pomona College, Claremont, California. Daymenus includes a data base that supports 3-D data, stores associated chemical and biological information, and permits substructure/similarity searching. Daymenus also provides interfaces to CONCORD and other molecular modeling software.

10.4.3. Similarity Searching

Registration involves exact-match retrieval; substructure search involves partial-match retrieval. Another possibility is a best-match or nearest-neighbor retrieval algorithm, where those structures in the file most similar to an input target structure are identified. Usually, the fragment screens that characterize the target structure are compared with the screens for each of the structures in the data base in order to determine the similarity; the structures are then sorted in order of decreasing similarity [10.80], [10.81]. This system allows chemists to browse in a data base without having to specify a substructure search precisely; it also gives them control over the volume of output [10.82].

Similarity measures can be used not just for matching a single target structure against a data base, but also for the pairwise comparison of all of the molecules in a data base to generate a clustering of them. Clustering methods can be used to group substructure search output and to select compounds for biological testing in drug design programs [10.83], [10.84].

10.4.4. Use of Parallel Computer Hardware

A well-known example of the use of parallel computer hardware to increase search speeds is provided by the CAS ONLINE system (see Section 10.5.2) where the early version of the Registry file was partitioned across thirteen pairs of minicomputers. The screens were stored as sequential lists on half of the pairs of computers and hits from a screen search were transferred to the other halves of the pairs for atom-by-atom searching. The hardware architecture made up for slowness in the search software, and screens and new compounds could be easily added to the system. The software, however, was not transportable.

In the newer "search engine" system implemented in 1989 [10.85], eleven advanced minicomputers are used, two of which serve as search managers, while nine search portions of the file in parallel. The main difference is that the screens are now stored as inverted lists. Search is now much faster (< 1 min as opposed to 5 min) and the system is transportable to single-processor environments.

The Beilstein Institute and Softron have developed a different multiprocessor architecture for substructure searching [10.86].

Four possible architectures are recognized for parallel computers:

1) Single instruction stream, single data stream (SISD)
2) Single instruction stream, multiple data stream (SIMD)

3) Multiple instruction stream, single data stream (MISD)
4) Multiple instruction stream, multiple data stream (MIMD).

MISD is unrealizable in practice. SISD represents the serial architecture of most computers to date, in which individual instructions are executed in sequence on one data type at a time.

An SIMD computer, the Distributed Array Processor (DAP), has been used at Sheffield University in research in nearest neighbor (similarity) searching in document data bases [10.87], [10.88] and in clustering of chemical structures [10.89], [10.90]. Cluster analysis is a multivariate, statistical technique that allows automatic production of classifications, usually by calculating all of the interobject similarities, dissimilarities, or distances. The use of clustering in drug-screening programs and in postprocessing of substructure search output is mentioned in Section 10.4.3.

Researchers in Sheffield have also used the DAP in searching techniques for 3-D macromolecules from the Protein Data Bank [10.90]. Their system, Protein Online Substructure Searching – Ullman Method (POSSUM), allows substructure searches to be carried out for user-defined motifs (i.e., patterns of secondary structure elements in 3-D space) [10.91].

POSSUM makes use of the fact that the common helix and strand secondary structure elements are approximately linear repeating structures and that such an element can be described by a vector drawn along its linear axis. The set of vectors corresponding to the secondary structure elements in a protein or a query motif can then be used to describe the structure of that protein in 3-D space, with the relative orientation of the helix and strand elements being defined by the interline angles and distances. Proteins and motifs may be regarded as labeled graphs, with the nodes of the graph corresponding to the linear representation of the helices. The performance of the DAP in this system is discussed in [10.90].

The parallelism in an SIMD system, such as the DAP, arises from the fact that different processors can execute a program on their own local data. This is also true of a MIMD system but, in addition, the different processors can execute their own programs simultaneously and communicate between themselves as required. Early reports simulated the use of MIMD processing for substructure searching [10.45], [10.92] and in maximal common substructure algorithms [10.93].

More recently researchers at Sheffield University have used the INMOS transputer for the implementation of chemical substructure searching [10.87], [10.90], [10.94]. This transputer is a microprocessor which can be linked to others in a network, and programmed using a special language, occam. A multiprocessing system can only be used efficiently if the task to be carried out can be broken down into subtasks which can be allocated to various processors. In the Sheffield research, a "farm" or "pool" of processors is used. The subgraph isomorphism process has been implemented in two different ways: data base parallelism and algorithm parallelism. The former attempts to increase the speed of searching by distributing the data base across the nodes of the farm; the latter attempts to increase the speed of individual searches.

10.5. Operational Substructure Search Systems

10.5.1. History

The ICI CROSSBOW system [10.15]–[10.18] was the first substructure search system for in-house data bases to become established worldwide. It used not only WLN but also fragment codes and connection tables derived from WLN (see Sections 10.1.3 and 10.1.4). The CROSSBOW connection table was bond-implicit [10.21] as opposed to the bond-explicit connection table used by CAS.

The earliest system which allowed substructure searching involving chemical connectivity input and structure display, was the NIH/EPA CIS (see Section 10.5.3). Its Structure and Nomenclature Search System (SANSS) eventually allowed access to a large range of public data bases. The system allowed access from teletype terminals and was not graphics-based.

The CA Registry file became substructure-searchable online in the early 1980s, first by means of the DARC (Description, Acquisition, Retrieval and Correlation) system (see Section 10.5.5) and, soon afterwards, in the CAS ONLINE Service (see Section 10.5.2). These were the first online graphics substructure search systems.

Chemical reaction systems such as LHASA and SECS (Section 11.6.7) had long used graphics [10.95] but it was some time before the first in-house, proprietary system appeared. This was Upjohn's Compound Information System, COUSIN [10.96], [10.97].

Neither COUSIN nor the CAS ONLINE Messenger software was portable or commercially available. From the early 1980s there was a big demand for user-friendly interactive access to in-house chemical structure data bases. The market leaders became MACCS (Section 10.5.7) marketed by Molecular Design Ltd (MDL). DARC in-house and OSAC (Organic Structures Accessed by Computer) from the ex-Leeds University team ORAC Ltd (see Section 10.5.8) appeared later.

In graphics systems of this type the user draws a 2-D structural diagram representing his substructure query. The system automatically performs the search, usually by algorithmically generating fragment codes, doing a screen search, and then doing an atom-by-atom search.

Many organizations acquiring these new systems already had structures files on chemical typewriters or encoded in WLN and wanted to generate MACCS data bases or DARC connection tables automatically. Interconversion thus became a theme of the mid-1980s. ELDER's DARING program has been widely used for this purpose [10.19]. Further software is required to convert the DARING connection tables to MACCS or other versions, and to generate structure coordinates needed for the graphics display [10.19].

Once many in-house, chemical structure data bases had been built, users began to realize that it was more efficient to use commercially available structure handling software for structures alone (or structures and a minimal amount of related property data) and to take advantage of data-base management systems to handle property data [10.98], [10.99]. The vendors of chemical structure handling systems now allow for interfaces to data base management systems.

By the mid-1980s the advent of the microcomputer had started to make a big impact on the world of chemical information [10.100], [10.101]. The problem was no longer the lack of systems and data bases but rather the proliferation of systems which could not be linked in a seamless manner. Commercial and technical factors cause severe limitations on users [10.102] both as regards drawing chemical structures and accessing files with these structures [10.102].

10.5.2. CAS ONLINE [10.38], [10.103]

The CA Registry system, which contained about 10×10^6 compounds by 1990, has been described in Sections 2.2.2. and 2.2.5.

A number of files under the generic title CAS ONLINE are available online on STN International (see Section 2.2.6). The system software, Messenger, includes chemical substructure search, text retrieval and numeric data searching facilities.

Chemical structures and Registry Numbers are held in the CA Registry file. There are four ways of searching the structures:

1) EXA (or exact) search retrieves the input structure and its stereoisomers, homopolymers, ions, radicals and isotopically labeled compounds.
2) FAM (or family) search retrieves the same structures as EXA, plus multicomponent compounds, copolymers, addition compounds, mixtures, and salts.
3) For SSS a range of possible substituents and bonds can be used in the input structure and a substructure search is performed.
4) CSS (closed substructure search) is akin to a family search with variable groups allowed. (The concept of closing substitution is discussed later.)

Structures can be input textually using alphanumeric characters or graphically using a graphics terminal or personal computer. A number of graphics front-end packages can be used (see Sections 8.2.7 and 10.6.4). STN Express is recommended and is marketed by STN. Detailed examples of search strategies have been published [10.104]–[10.106]. STN also supply microcomputer tutorial programs under the name STN Mentor [10.107]. A sample search over 5% of the file can be run at no cost (other than connect charges). This projects whether the search will succeed or whether it is too wide.

The search can be limited by using a Registry Number range, adding screens, choosing search terms (e.g., to exclude polymers), and using hydrogen counts. Every position in the molecule is considered free for substitution unless otherwise specified. Thus the exact or minimum number of hydrogen atoms desired at each position must be specified with HCOUNT.

Certain generic groups can be specified (e.g., CY for any cyclic group). Variable point of attachment is also possible.

The CA Registry System automatically identifies substances with several possible chemically equivalent representations such as tautomers or aromatic rings [10.108]. Certain explicit single and double bonds in such circumstances are replaced by normalized bonds, and the associated migrating tautomeric hydrogen is associated with groups of atoms rather than single atoms. Since the single–double bond patterns and specific migrating group locations have all been replaced by normalized data, all forms of the tautomeric structure lead to the same Registry structure record. Unfortunately some users find it hard to understand the CAS conventions, since they are not quite the same as text-book definitions. Keto–enol tautomerism, for example, is not recognized. STN Express takes care of this problem, making single bonds single or normalized and double bonds double or normalized, where a tautomeric or aromatic system is possible.

At the moment stereochemical discrimination in search is not possible but the addition of true stereochemistry to the Registry file is planned in the early 1990s.

The output from a CA Registry file search includes Registry Number, CA index name, molecular formula, up to 50 synonyms, structure diagram, and full bibliographic information for the 10 most recent references about the substance. The total number of references for the substance in the CA file and an indication of further references in the CAOLD file are also given.

Once a Registry file search is complete, it is possible to switch to the CA file for a further search and for retrieval of the corresponding bibliographic information. The CA file is the only online file to contain CA abstracts and is not licensed to other hosts. The abstract text is searchable and displayable from 1970 onwards and older abstracts will be gradually added.

Information prior to 1967 is supplied as abstract numbers in CAOLD, where most Registry Numbers for compounds cited between 1957 and 1967, and retrieved in the Registry file, can be searched.

There are three other files in the CAS ONLINE family: MARPAT for generic structures [10.64], [10.68]; CASREACT, a reaction index (Section 10.8); and CA previews, where information is initially entered prior to abstracting and indexing.

The CAS ONLINE search system has screening and atom-by-atom search stages. The screen set was originally developed by the BASIC group of companies in Switzerland [10.37], [10.38] based on earlier work done at CAS and Sheffield University. CAS added further generic screens and CAS ONLINE uses 12 highly posted screen types. The original CAS ONLINE search software was not as sophisticated as some, but speed was achieved by parallel processing. The newer "search engine" system is faster still (see Section 10.4.4).

Comparisons of CAS ONLINE with other systems have appeared in the literature [10.109]–[10.114].

A number of companies, (e.g., Kodak [10.115]) maintain their corporate data bases as private registries under the CAS ONLINE system.

In 1989 CAS and Questel agreed to explore ways of collaborating and linking CAS ONLINE and DARC services. They also acquired minority shareholdings in Hampden Data Services who produce microcomputer software (including STN Express) for CAS, Questel, Derwent, and other vendors.

10.5.3. CIS/SANSS [10.46]–[10.51]

This system was originally designed for teletype terminals but graphics input is now possible using a PC-based front-end SuperStructure (see Section 10.6.4). SANSS is the substructure searching module for a large number of data banks on CIS (see Sections 9.4.5 and 9.6.3). CIS gives direct links to toxicological and spectral data rather than to bibliographies.

The structures are held as CAS connection tables. Tautomeric and aromatic systems have normalized bonds. The system contains about 350 000 structures, with molecular formulae, names, and synonyms. Since this file is much smaller than the CA Registry file, wider, more generalized substructure searches can be carried out without fear of an excessive output of hits. Moreover all the compounds are "useful" in the sense that they occur in data banks of commercial or environmental importance.

Searches can be limited to the contents of particular data banks. The searcher may employ name fragments (user-defined), molecular formula fragments (including ranges of acceptable element counts), and molecular mass ranges, as well as a controlled vocabulary of functional group and ring system codes to generate a small set of compounds on which to perform an exact substructure search.

If a ring skeleton is known but the number and/or positions of heteroatoms in the ring and/or points of substitution is undefined, a generic search at the desired specificity level can be run.

Atom-centered fragments of a particular type can also be searched. Sets from any of these operations may be combined with Boolean logical operators.

By using a variety of simple textual commands, the user can create the substructural fragments (user-defined type) for an atom-by-atom substructure search of a previously formed set of compounds.

It is not possible to block sites on the substructure with hydrogen to prevent substitution at that atom. Variable atoms but not groups can be specified. Stereochemistry is not handled.

The output record for each compound contains names, molecular formula, structure, and a list of all the collections in which the compound appears. Bibliographic information is not part of the system but can be located by downloading retrieved Registry Numbers and transferring them to another online retrieval system for use as search terms.

A nested-tree structure-searching system is used which operates on the data base of connection tables and uses an inverted file organization to provide rapid search response (see Section 10.3).

Comparisons of SANSS and other systems have been published [10.111], [10.114], [10.116], [10.117].

An adaptation of SANSS has been incorporated into the Drug Information System of the National Cancer Institute in the United States [10.118]–[10.120].

10.5.4. Cambridge Structural Data Base (CSD) System

The data base supplied by the Cambridge Crystallographic Data Center (Section 9.5) is one of only a handful of sources [10.69] of experimental 3-D structural data and a unique source of such data for organocarbon compounds. It is thus an invaluable tool in computer-aided drug design and in the analysis of geometrical parameters to obtain new insights into structure and bonding.

The data base is supplied complete with searching software. The core program QUEST in Version 3 allows users to search bibliographic chemical connectivity and selected numerical data within a single query. Tests of individual fields are combined via logical operators to form a complete question. Version 4 (released in 1989) has a graphics input and output package which replaces the alphanumeric coding of the search question. Queries are formulated using a fully interactive, menu-driven interface. Substructure, textual, and numeric searches may all be specified.

Menus allow generation of a substructure which may be edited, saved for future use, or combined with further bibliographic or numerical questions. The complete query is then passed to the data-base search software.

Mapping of the crystal connectivity onto the chemical connectivity is promised in Version 5 of CSD. Version 4 is available in both machine-independent and VAX/VMS specific hardware versions. It operates on all Tektronix terminals. It can also be used on certain workstations and under the UNIX operating system.

The Cambridge Crystallographic Data Center is working on 3-D substructure and similarity searching to be integrated with the 1-D and 2-D capabilities of the QUEST program.

Comparison of CSD with CAS ONLINE and MACCS. It is instructive to compare CSD with the leading online and in-house systems, CAS ONLINE (plus STN Express, Section 10.5.2) and MACCS (Section 10.5.7), respectively. Molecular Design Limited produce a microcomputer package, ChemBase, but it is not truly a front end to MACCS in the way that STN Express works with CAS ONLINE, so MACCS alone is used for comparison here. However, ChemBase has many features similar to STN Express. MACCS and CSD are both minicomputer/mainframe based products. STN Express has the advantage of being able to use advanced microcomputer features such as drop-down menus. For simplicity CAS ONLINE and its STN Express front end are called STN in the following.

The CSD provides over 60 system templates (predrawn structures which the user can select and modify). In CSD and STN the templates are displayed graphically and selected using the mouse, crosswires, or other pointing device. In MACCS (at least until 1991) the user has to know the file name of each template and select by name.

All three systems allow users to save their own templates. They all permit freehand drawing and a multiplicity of ways for constructing structures from all the atoms of the periodic table. STN and CSD offer "Feldmann ring notation" (e.g., 66U6D5 specifies the steroid skeleton).). CSD offers more structure-drawing facilities than MACCS but is not as user-friendly as STN.

CSD draws structures on a grid system, producing "neat" structures with consistent bond lengths. STN allows "rubber-banding" of bonds so that atoms can be placed where desired. MACCS has neither a grid nor "rubber-banding".

STN and MACCS allow a roughly drawn structure to be "cleaned". STN allows the cleaning to be "undone". CSD and STN allow a variable point of attachment for a group, MACCS does not. STN has a wide range of "short-cuts" (e.g., SO_3H may be selected from a menu and the system will interpret this into the full topology of the group for structure storage and search). CSD has a similar "groups menu". MACCS has no equivalent. STN allows a rectangular window to be drawn around a part of any structure so that operations may be performed in that window alone. CSD's window can be irregular in shape. MACCS has no windowing facility.

Once a search is complete in MACCS the hits are made into a list. Creation of data-base subsets for offspring searching in CSD is somewhat clumsy and wastes disk space. However, Boolean logic is used in search statements in CSD whereas MACCS only allows Boolean logic for hit lists. Subsetting in STN is a very recent feature (mid-1990).

CSD incorporates similarity searching software. Two different coefficients can be used, the output is ranked, and the coefficient value is displayed. There is no similarity search in standard MACCS, but an additional module can be purchased. STN does not offer similarity searching.

The CSD data base contains over 73 000 structural diagrams. In the search output the substructural search fragment is highlighted for rapid visual interpretation; this feature is not offered in MACCS or STN.

An additional CSD module, BUILDER, will soon (1991) be offered to allow users to build their own data bases. MACCS is already a data-base-building package and data bases are commercially available for use with it. At the moment STN is for use only with public online data bases.

10.5.5. DARC [10.39], [10.121]–[10.126]

The DARC system was conceived by DUBOIS of the University of Paris in 1963. It is now marketed by Questel, a subsidiary of Télésystèmes.

Generic DARC is used to access four online data bases included in the CA Registry: EURECAS, POLYCAS, MINICAS, and UPCAS. EURECAS is the main file, excluding polymers, POLYCAS contains polymers, MINICAS is a sample file (1% of EURECAS), and UPCAS contains the most recent monthly update. Ab-

Volume B1   p. **12**-87

         left column, 2nd paragraph, 3rd line.

in Version 3 allows users to search bibliographic data,

stracts and pre-1965 data are not available; they are mounted only on STN. Chemical Abstracts connection tables are converted into DARC code for searching online on the Questel host.

Text structure input of a substructure query is particularly easy but graphics entry is normally used. This can be done with an online, interactive menu system, or offline using DARC CHEM-LINK (a PC-based program with menus identical to those used online) or using STN Express (see Section 10.6.4).

A parent molecule is first built, then graph modifiers (up to 20 different groups or atoms) are specified. Boolean logic is allowed. Offspring searching can be carried out as follows. A structure is drawn and submitted to the first-stage search. Further searches can be carried out on the hits from this fragment search without incurring additional fees. Hits can be viewed after a first-stage search, whereas on STN the hits at the first fragment stage cannot be displayed.

The search can be limited by isolating rings (e.g., benzene is acceptable but not naphthalene) and by specifying the minimum and maximum number of rings, non-hydrogen atoms, and components.

If an atom is open to substitution, a "free site" must be specified. In CAS ONLINE it is assumed that substitution is allowed unless hydrogen atoms are specified.

Once a substructure search has been carried out, structures can be displayed and the Registry Numbers can be transferred to search bibliographic files on Questel Plus (the CA bibliographic file, the Merck Index, the Janssen catalog, and the Eurosynthèses data base of dangerous chemicals, EECDNG). Cross-file searching with INPI patents files and Derwent's WPI/WPIL is also possible. This is particularly useful, compared with CAS ONLINE, because WPI/WPIL are not available on STN. A memory command can be used to sort and rank on any desired field in a set of answers. Unfortunately a hit file from a Questel Plus data base or a WPI/WPIL search cannot be used to generate a query for searching EURECAS. On STN it will soon be possible to take the Registry Numbers out of CA file hit output and use them to search the CA Registry file.

The DARC screening system is based on an open-ended fragment code—Fragments Reduced to an Environment which is Limited (FRELs). Two concentric layers of atoms around a focus are described and topological fragments of up to eight atoms in a row are then made. The focus can be an atom or a bond: the more highly connected the focus, the more discriminating the FREL. A tree structure with growing specificity is created from the fragments. The screens are file dependent and optimal for the file. An advantage is gained by adding a few extra screens for items such as ring information. The nature of the FRELs means that small molecules are handled better in DARC than in CAS ONLINE and system limits are less likely to be exceeded. Very large data bases (millions of structures) can be handled on a single processor machine. Search times for the CA Registry on CAS ONLINE and on Questel DARC are comparable.

Comparisons of DARC with other systems have been published [10.109]–[10.114], [10.116], [10.117].

Markush DARC. Whereas Generic DARC allows generic queries to be put to a data base of specific structures, Markush DARC allows generic queries of a generic structure data base [10.8]. Data bases searchable using Markush DARC are WPIM (Derwent Chemical Patents Index since 1987) and MPHARM (an INPI file of U.S. and European pharmaceutical patents since 1978 and French pharmaceutical patents since 1961). The bibliographies of WPIM are in WPIL and those for MPHARM in PHARM, also mounted by Questel. MPHARM and PHARM form the INPI data base PHARM-SEARCH.

Since generic structural expressions cannot be simply represented by connection tables, Markush DARC has descriptors, known as superatoms, to represent types of structural elements (e.g., HEA stands for heteroaryl). Since specific compounds are not coded in the data base (which covers the scope of the invention, not all specific possibilities), there is no link between generic structures containing superatoms and the specific compounds to which they correspond. Thus a search for a substructure containing a superatom retrieves only structures containing the corresponding superatom.

A full structure, substructure, or generic substructure query can be phrased on- or offline; the search is performed; structures are displayed with highlighting of the relevant part of each structure; and bibliographic files can then be cross-searched.

Markush DARC is a collaborative development of Questel, INPI, and Derwent Publications. Questel are also collaborating with the European Patent Office (EPO) in the EPO Query system, EPOQUE.

DARC-In-house. In addition to online searching systems, Questel markets software for in-house use. DARC In-house is similar to MACCS in its functions (Section 10.5.7) and handles stereochemistry. Query input and structure searching are similar to those in Generic DARC; hash codes are used for full-structure search. Features such as report generation and structure registration are not applicable to online DARC. The DARC Communication Modules [10.127], [10.128] allow links to data-base management systems [10.129] and links are possible to microcomputer software.

The DARC "tool kit" is a set of stand-alone modules for editing structures and queries; searching on structure, substructure, and molecular formula; displaying and printing data; and handling queries and sets of answers. Communication modules allow links to data-base management systems, text retrieval software, and other applications.

Links between DARC In-house and the reaction indexing system ORAC (see Section 10.8) have been demonstrated.

10.5.6. HTSS

HTSS was developed in Hungary [10.52], [10.130]. It is now marketed by ORAC in Leeds (UK) and is the substructure search software used for accessing the Beilstein data base on Maxwell Online.

An HTSS data base has a hierarchic tree structure, which is created from the connection tables of the individual chemical structures by classification of all the atoms in the data base according to the number of nearest neighbors, atom type, and other classifiers. The subtree concept enables the search to be restricted to portions of the data base. If the required chemical structure is in the data base, then the tree of this structure is embedded in the tree of all the molecules which form the data base. The substructure search involves a "tree-walk" from root to leaves via the branches of the relevant tree which corresponds to the query substructure. A substructure is found if a leaf can be reached for each atom in the structure during the tree-walk; if not, the structures in the data base most similar to those requested by the query are found, enabling simultaneous retrieval of related structures. The HTSS approach integrates the screening and atom-by-atom searching steps used in conventional substructure searching systems, thus eliminating the time-consuming atom-by-atom search.

The time taken to search the tree does not increase linearly with the size of the data base, as is the case with most other systems. HTSS should therefore be suitable for use with very large data bases. HICKS and JOCHUM have shown that it is faster than DARC in many cases but they only vouch for its performance on files of up to 600 000 structures [10.113]. Walking across the tree sometimes needs a large number of disk accesses, especially when there are many free sites on the query structure.

Another possible disadvantage is the amount of time needed to generate the tree, although structures can be added to the data base without having to rebuild the whole tree.

An advantage of HTSS is that it can be implemented in almost identical forms on machines ranging from small microcomputers [10.131] to large mainframes.

10.5.7. MACCS

MACCS is the Molecular ACCess System marketed since 1979 by Molecular Design Limited (MDL) for integrated chemical information management on mainframe and minicomputers [10.106], [10.132]–[10.140]. The current version (MACCS-II) is the most popular of the systems for handling in-house chemical data bases and is available for a variety of makes of computer and graphic devices. It is an interactive, menu-driven graphics system, with facilities for graphical input, storage, retrieval, display, and printing of molecular information.

Structures input graphically are checked for valence errors and can be registered into corporate data bases. Templates are available (by file name, not by structure) to aid the graphics construction of diagrams. All the atoms of the periodic table, charges, and isotopes can be handled. Six-membered rings with alternating double and single bonds are recognized as aromatic. An additional "substance module" or "Mod S" handles more complex aromatic systems and has sophisticated features for mixtures and polymers. In basic MACCS tautomers and stereoisomers are checked before a compound can be registered.

Substructure search queries are entered using another menu in which a range of permissible atoms or bonds can be specified. A selection of atoms can be excluded as substituents at any position. All positions are assumed to be open to substitution unless hydrogen atoms are specified. Stereochemical discrimination in search is possible.

After a search, structures can be viewed one at a time on the screen. Data are displayed on different screens but software is available to print forms containing both structures and data for each record on a data base.

Generic queries (i.e., those where a group R is attached to the query skeleton) can be handled with the additional "Power Search Module" which also offers similarity searching.

The system is based on connection tables. Input structures are converted to a unique, nonvariant SEMA name (Stereochemically Extended Morgan Algorithm) [10.141]. Full-structure search uses hash coding of the SEMA name. The substructure search system uses about 1000 fragments (called keys) in a set similar to that of CAS, but smaller. The screens are held as inverted lists. Atom-by-atom search follows key search. According to [10.113] the system is limited to files of fewer than 350 000 compounds.

MDL sell a range of data bases for use with MACCS (e.g., Prous Science Publishers' Drug Data Report and The Fine Chemicals Directory). Data-base interface modules can be purchased to allow links to data-base management systems and the Customization Module allows the user to automate routine procedures, create customized menus with graphics buttons, and optimize the system interface to the local environment.

MACCS-3D offers 3-D substructure search (see Section 10.4.2). MDL also sell a reaction indexing package REACCS (see Section 10.8) with molecule handling features similar to those of MACCS. A set of microcomputer programs, CPSS (Chemist's Personal Software Series), is also available and includes ChemBase whose features are similar to those of MACCS (see Section 10.6.3). ChemBase is more user-friendly than MACCS and has some extra structure-handling features. However, this means that the two systems are not totally compatible, although a common file (the MOL or SD file) can be exchanged between the two. ChemBase is designed for use only with small personal data bases.

MDL also sell software for molecular modeling (CHEMLAB-II) and structure–activity relationships (ADAPT) [10.142], [10.143].

10.5.8. OSAC

Organic Structures Accessed by Computer (OSAC) is marketed by ORAC of Leeds (UK) and bears similarities to their reaction indexing system ORAC (see Section 10.8). MACCS, DARC, and OSAC are compared in [10.144].

10.5.9. Softron Substructure Search System (S4)

This system was developed by the Beilstein Institut and Softron of Gräfelfing (FRG). All molecules are encoded by using each atom in turn as the starting atom. For a molecule of n atoms there are n connection tables. The codes, which are very compact, are sorted, stored, and indexed. With this highly redundant representation, all hits can be retrieved with just a few sequential reads, carried out on a part of the file determined by the index. The need for many time-consuming random accesses is eliminated and atom-by-atom search is often unnecessary. Stereochemical and tautomeric search are not yet available.

S4 is claimed to perform faster than any other available system and can handle very large data bases (millions of compounds) [10.113]. It is also very suitable for searching CD-ROM data bases, where random access must be minimized because access is slow [10.145].

S4 is a full structure and substructure searching module. The complete system as implemented by Beilstein also requires the MOLKICK structure display and query editor interface (see Section 10.6.4). S4 is used by the Beilstein Institute in-house and is operated by DIALOG for their Beilstein online data base. It is also available commercially as an in-house system.

10.5.10. Proprietary Systems

Upjohn's COUSIN system (see Section 10.5.1) was developed in-house before user-friendly, interactive, graphics-based systems such as MACCS and DARC were available. Most companies choose to integrate a commercial structure handling package (and usually also a data-base management system) into their in-house integrated systems, rather than writing their own substructure search system [10.99]. Exceptions are, however, Bayer's ReSy (Research System), Pfizer's SOCRATES [10.146], and Philip Morris's MIMS [10.147]. The BASIC and IDC companies have had access to Chemical Abstracts and patents data in-house for many years. The development of graphics front ends for IDC's GREMAS system was described in Section 10.4.1.

Sandoz is developing an end-user interface comprising a directory, a graphics module, aids for formulating property searches, and a communication module [10.148]. The in-house directory contains Beilstein, CA, and Fine Chemicals Directory structures in a hash-coded form. Sandoz corporate data are also available. Full structure and limited generic structure searching are possible but not substructure search. The graphics module is DARC. The communication module must give access to STN in a simplified, cost-effective form acceptable to Sandoz end users. "Hedges" such as "preparation" or "toxicology" have been set up for them. The provision of a uniform, consistent, STN/in-house substructure search facility is Sandoz's next priority. The problem of obtaining a universal interface to corporate and public data concerns most of the chemical and pharmaceutical industry [10.102].

10.6. Chemical Structure Software for Microcomputers [10.149]

10.6.1. Graphics Terminal Emulation

Graphics communication software, also known as graphics terminal emulation software, was the first area in which PCs made an impact on chemical information. Until such packages became available the only way of using services such as CAS ONLINE or DARC in graphics mode was to use very expensive graphics terminals. PCs supplied a cheaper alternative and allowed new possibilities such as offline query formulation and downloading (see Section 8.2.7).

10.6.2. Scientific Word Processing and Structure Drawing Software [10.149]–[10.151]

A scientific manuscript may contain not only text but also mathematical equations, diagrams, and chemical structures. Until recently only the text could be word-processed. Chemical structures and diagrams had to be manually pasted into the final document. Scientific word processors now allow the editing of equations, chemical structures, and text.

There are three methods of creating diagrams:

1) The character- or font-based method
2) The mark-up method
3) A graphics interface

The majority of packages with full-function word processing use the first method. Atoms and bond types are represented as single characters or fonts and the user builds a structure one character at a time. Some of these packages are described as WYSIWYG (What You See is What You Get) because the image on the screen corresponds very closely to the final print. The mark-up approach uses in-line codes that describe the drawing as a set of mnemonic instructions. The advantage of this is that machine-independent ASCII files can be written. However, the packages cannot be described as WYSIWYG. In the graphics WYSIWGG approach, a true graphics interface is used and free-hand drawing is possible together with selection of items from menus using a mouse or other pointing device.

With many scientific word processors a different application is used for each image type (equations, structures etc.). Thus, a true compound document that can be manipulated with one editor is not produced. Very few of the chemical structure drawing packages "understand chemistry", store connection tables, and offer substructure searching. Those that do are considered in Section 10.6.3.

ChemDraw from Cambridge Scientific Computing is probably the most popular structure-drawing package on the grounds of ease of use and print quality. It is available only for Macintosh computers. Various systems (e.g., CAS ONLINE, DARC, Daylight, and ORAC) offer ChemDraw links of one sort or another.

Graphics from external programs or scientific instruments can be incorporated into a document in the form of bit-mapped images (see Sections 7.2.2 and 11.5.2). This method has two disadvantages: the images require large amounts of computer storage space and they cannot be edited because they can only be changed bit by bit (see Fig. 9, p. **12**-47). However, for hard-copy information a bit-mapped image is the only option. For electronically available information, a second option is a format known as a graphics metafile. Metafile images can be rotated and scaled-up or scaled-down, but they are not truly computable. The ideal method of importing graphics data is by means of a standard file format. Several instrument manufacturers are trying to agree upon such a standard.

10.6.3. Structure Management Software
[10.149], [10.152]

Structure management software differs from structure drawing software in that exact structure and substructure searches are possible because connection tables are stored. Some of the programs also store and retrieve related text and numeric data. Structure management software allows users to build personal data bases of structures and related data. Compounds can be input graphically and registered into a data base. Records can be retrieved by structure or substructure search and hits may be displayed and printed.

ChemBase [10.149], [10.152]–[10.154]. ChemBase is one of the four modules of MDL's CPSS software for IBM-PCs. The others are ChemText, for structure drawing and scientific word processing; ChemTalk Plus, a PC-based communications package (see Section 10.6.4); and ChemHost, a mainframe or minicomputer program allowing corporate MACCS and REACCS data bases to be linked to ChemBase ones. ChemBase uses the same molecule editor as ChemText and presentation-quality graphics output are possible from both.

An initial data base of structures and reactions is provided as well as a large collection of graphically displayed templates. A mouse is required for graphics input. A variety of bond types are available including in-the-plane and out-of-plane bonds for designating stereochemistry. Valence checking takes place as the structure is drawn. Molecular formula and molecular mass are automatically generated. Substructure search is rather slow and has no screening stage, so ChemBase is suitable for data bases of only a few thousand structures. ChemBase is the only structure management system to handle reaction data bases and reaction-specific searches.

PsiBase [10.149], [10.152], [10.155], [10.156]. PsiBase (marketed by Hampden Data Services) incorporates a module, PsiGen, for creating, editing, and displaying chemical structure diagrams using drop-down menus and a mouse. PsiBase, including software for structure management and substructure searching, is available for both IBM and Macintosh microcomputers.

With PsiGen the user can draw structures freehand, build them from graphically displayable templates, or construct them using Feldmann notation. Commands can be entered by either menu buttons or keyboard commands. A variety of bond types are available including in-the-plane and out-of-plane bonds for designating stereochemistry. Parts of a structure can be enclosed in a rectangular "window" and operated upon independently from the rest of the structure. Valency checking takes place as structures are drawn. The molecular formula of a registered molecule is calculated.

Substructure searching is substantially faster than in ChemBase. Chemical reactions may be drawn, but not searched.

PsiGen is the structure drawing interface adopted by CAS for STN Express (see Sections 10.5.2 and 10.6.4), by Derwent Publications for TOPFRAG and TORC (see Section 10.6.4), and by Sadtler Laboratories for their spectral search system (see Section 9.4.5).

Other Structure Management Systems. HTSS (see Section 10.5.6) is available for use on microcomputers. PC-SABRE/PICASSO, from Fraser Williams (Scientific Systems) provides facilities for input, registration, storage, search, retrieval, and display of chemical structures and associated data. It also permits presentation-quality graphics output.

10.6.4. Special Application Software

Conversion of Connection Tables to Structures. ARGOS (Automatically Represents Graphics of Chemical Structures) is a program sold by Springer Verlag to convert ASCII connection table files into structures and to display them including stereochemistry, charges, and radicals. Fraser Williams (Scientific Systems) sell a program, PC-REWARD, which generates 2-D structure diagram coordinates from connection tables such as those produced by their program PC-DARING (see Section 10.1.3). The diagrams can be transferred to the PC-SABRE search package (see Section 10.6.3).

Property Prediction. TopKat from Health Designs Inc. is a menu-driven software package for the prediction of toxicity of structure input using the SMILES notation (see Section 10.1.3). It predicts rat oral LD_{50}, mutagenicity (Ames test), carcinogenicity, teratogenicity, and rabbit skin and eye irritation.

CompuDrug of Budapest and Austin, Texas sell Metabolexpert to predict metabolites of potential drugs. A related program, HazardExpert estimates health hazard effects in seven different biological systems, using the TSCA data base. Both programs allow the user to enter a chemical structure graphically.

Front Ends [10.157], [10.158]. Chemical structure searching presents a need for customized front ends, allowing the scientist to make use of the 2-D chemical structure diagram. When a user draws a query in the CAS ONLINE or DARC systems, information commands are sent through the telecommunications network to the central computer, which in turn interprets the commands and creates the query image for display at the user's terminal [10.101].

There are tremendous advantages to be gained if the diagram can be created at the local computer. This reduces the load on the central computer, reduces telecommunications traffic, and reduces the cost to the user. The stress caused by trying to remember all the commands that are required to create the diagram while the costs are ticking away (taxi-meter syndrome) is also reduced. Using an offline query negotiation, the user can think about the query and ensure that it is right before logging onto the online host.

In some of the packages which came on the market in the late 1980s, the writer of the front-end software (e.g., STN Express and DARC Chemlink) and the online host have collaborated closely, so that the best features of both systems can be used. The host must define standard formats for queries and also provide the appropriate software "hooks" in its retrieval system. However, products of this type tend to be exclusive: the front end is ideally used for only one host.

Other products (e.g., ChemTalk Plus) have been written without the collaboration of an online host, and are therefore prone to malfunction should the online system change minor features of its file structure or command language. These products depend on conversion of the input graphics structure to the text string used by the host. They cannot produce some of the "finer" features available in a collaborative venture.

ChemConnection is a Macintosh desk accessory written by SoftShell International Limited. Structures drawn with the ChemIntosh interface can be placed on the clipboard and converted to the text structure input string required for searching the STN Registry file. Uploading requires a communications program such as Versa Term Pro.

ChemTalk Plus is a module of MDL's CPSS software (see Section 10.6.3) for IBM-PCs. It provides terminal emulation facilities for online interactive access to MDL mini/mainframe software and (in conjunction with ChemHost) allows file transfer, structure and data searching, and downloading between MACCS and ChemBase. It also provides facilities for accessing public services such as CAS ONLINE and Beilstein Online. Offline query preparation, automatic logon, uploading, query execution, and display of results are permitted. For Beilstein Online (but not CAS ONLINE), the Chemists' Access System (a development of ChemTalk Plus) allows downloading of structures to make a substructure-searchable ChemBase data base.

DARC CHEMLINK is an IBM-PC-based package written to look as much as possible like DARC's standard chemical structure entry. It is exclusively for use with chemical structures on Télésystème's Questel.

MOLKICK [10.158], [10.159] is a package written by Softron for the Beilstein Institute. It uploads structures drawn with Beilstein–Softron graphics for searches of the CAS ONLINE Registry file on STN or Télésystèmes Questel, or for searching Beilstein Online on STN or DIALOG. Structures are converted into the CAS text structure input string or into a ROSDAL (Representation of Structure Diagram Arranged Linearly) string [10.158]. A function key allows rapid transfer in and out of MOLKICK and the online system. The program is quick, is memory-resident, and can be used with more than one terminal emulation package.

SANDRA (Structure and Reference Analyzer) is an IBM-PC program which takes the chemical structure of a compound and directs the user to where to find references to that compound in the Beilstein Handbook. The following information is output: the series, volume, and subvolume numbers; Hauptwerk page numbers (H-page); system number or range; degree of unsaturation; and carbon number (see Section 3.3.2).

STN Express is front-end software that provides access to STN International structure data bases (e.g., the CA Registry file) [10.106]. It is also a front end to Télésystèmes Questel. Features of the program include guided search, offline chemical structure query formulation, and offline search and strategy formulation. Help is supplied with tautomerism and other details. The program has special search-and-display features, data capture, and automatic logon. The guided-search feature allows the user to input a search query through a series of menus and then have it automatically processed in the STN data base. Offline chemical structure building is menu-driven and requires the use of a mouse. The structure input software is compatible with PsiGen (Section 10.6.3). The program provides predefined search strategies for general subjects (e.g., toxicology) that take advantage of data bases provided by STN. Graphics are fully integrated with text and can be captured in transcript files and printed. Versions are available for both Macintosh and IBM-PC.

SuperStructure is the Fein Marquart/Fraser Williams graphics front end for CIS (see Section 10.5.3). It evolved from the microcomputer-based structure input program written for the National Cancer Institute's Drug Information System [10.119], [10.120].

TOPFRAG converts chemical structures drawn on the PC to the correct Derwent chemical fragmentation codes, and compiles them into fully time-ranged search strategies for direct input to Derwent data bases on ORBIT, DIALOG, or Questel. Automatic conversions include tautomer perception and allocation of ring index numbers. The program allows the selection, through a menu-driven interface, of nonstructural concepts such as activities or uses. It also allows the user to edit input strategies. TOPFRAG handles generic structures and uses the PsiGen structure-drawing interface (see Section 10.6.3).

An equivalent program, TORC, (To Ring-Code) also uses PsiGen, but generates codes for searching Derwent Publications' data base of the pharmaceutical literature, Ringdoc.

Miscellaneous. Fraser Williams (Scientific Systems) sell a program, Markout, which accepts a series of structures, calculates a generic skeleton structure, and tabulates the various R groups on that structure in a matrix of R groups and data. Sorting on various fields is possible and the matrix can be used in structure–activity studies.

CASKit from DH Limited (Santa Cruz, California) can be used to capture graphics files from CAS ONLINE and process them into a form suitable for use with CPSS software.

10.6.5. Software for 3-D Molecular Graphics and Modeling

Three-dimensional drawing programs may serve as front ends to software on more powerful machines, but are basically limited to displaying and manipulating structures. Molecular modeling programs can calculate and predict a variety of physicochemical features of the drawn molecule. They may also create files compatible with modeling systems for use on more powerful machines. Modeling programs include an energy minimization function, through molecular orbital and/or molecular mechanics calculations.

The structure input features of both 3-D drawing and modeling programs are often less sophisticated than those offered by structure drawing and structure management programs (see Sections 10.6.2 and 10.6.3). However they offer more display options (e.g., stick, ball-and-stick, and space-filling displays). Some programs can also generate an ORTEP (Oak Ridge thermal ellipsoid plot) display. Rotation of the 3-D structure is also a feature. Hard-copy output is usually feasible.

Alchemy (Tripos Associates) is a well-known example for molecular modeling on an IBM-PC. It can be linked to the CAS Registry file via STN Express. CONCORD (see Section 10.4.2) has been used to generate 3-D coordinates for more than 4×10^6 organic substances in the CAS Registry. A 3-D molecular model can be uploaded from Alchemy through STN Express so that a search can be conducted in the Registry file. Literature references are thus found. The 3-D coordinates of interesting substances can then be downloaded through STN Express into Alchemy, ready for modeling and manipulation.

10.7. Structure–Activity Relationships and Drug Design [10.160]–[10.163]

Quantitative structure–activity relationship (QSAR) techniques involve the correlation of physical, biological, and chemical data with variables characterizing the molecular structures in a data set. Most of the applications have been in pharmaceutical and agrochemical research because of the costs involved in synthesis and testing of thousands of compounds. The methods are, however, applicable in other areas, such as environmental chemistry [10.164]. In an agrochemical or pharmaceutical synthetic program, SAR methods may be used for lead optimization or lead generation. Lead optimization attempts to optimize activity within a given series of compounds by systematic modification of compounds previously identified as active. Lead generation attempts to identify classes of interesting compounds that might be active.

Various property descriptors have been used: hydrophobicity, octanol–water partition coefficients, pK_a, electronic descriptors (Hammett sigma constants, dipole moments, molar refractivities, ionization potentials). These are correlated with 2-D substructural fragments (e.g., augmented atoms, topological torsions) or 3-D substructural fragments (e.g., pharmacophores).

Techniques may be classified as statistical approaches (Section 10.7.1); substructural analysis and data-base techniques (Section 10.7.2); and molecular modeling approaches (Section 10.7.3). Molecular modeling and quantum mechanics can only be used for small numbers of structures; statistical methods can handle hundreds of compounds; substructural analysis can handle thousands.

10.7.1. Statistical Approaches

Free–Wilson Additivity Model [10.165]. This approach does not use physicochemical parameters but assumes that each constituent in a given location makes an additive and constant contribution to the overall activity of the molecule. A series of equations can then be written expressing the biological activity of the molecules in a set in terms of a constant activity plus contributions from each substituent. The equations are solved by regression analysis. The modification used by FUJITA and BAN overcomes the problem of the need for multiple substitution sites [10.166].

Hansch Analysis [10.167], [10.168]. This involves the correlation of physicochemical properties with the observed biological activities within congeneric series of compounds, mainly using multiple regression analysis. HANSCH combined several properties into a multiparameter model in which the biological response is modeled in terms of electronic, steric, dispersion, and hydrophobic components. HAMMETT and TAFT's earlier work [10.169] formed the basis of the electronic and steric contributions; the dispersion component is represented by the molar refractivity and the most popular hydrophobic parameter is the octanol–water partition coefficient. The success of the model has led to the inclusion of many other parameters [10.169]–[10.172]. However, the method is usually limited to structural variations within a given congeneric series where there is a high probability that the same mode of action pertains for all of the compounds.

Pattern Recognition. Pattern recognition methods were first used in chemistry for structure elucidation in the field of spectral information [10.173] but this led to studies involving predictions of biological activity [10.143], [10.174].

An example of a pattern recognition technique is a linear learning machine, where compounds are considered as points in multidimensional space, each dimension of which corresponds to one of the structural features used in the characterization of the compounds in the data set. The machine tries to identify a hyperplane dividing the space so that all of the active compounds lie on one side of the plane. Other pattern recognition techniques have been used [10.143], [10.160]. An interactive program, ADAPT, based on pattern recognition techniques [10.142], [10.143] can be used in conjunction with MDL's MACCS software (see Section 10.5.7).

Cluster Analysis. Unlike pattern recognition, which categorizes compounds into predefined classes, cluster analysis attempts to detect groups present in multivariate data sets in order to identify classes. Clustering in chemical information is described in [10.83].

10.7.2. Substructural Analysis and Data-Base Techniques

The Free–Wilson approach (see Section 10.7.1) forms the basis of the lead generation technique known as substructural analysis in which a given substructure is assumed to make a constant and additive contribution to the overall activity of a molecule, irrespective of the type of molecule in which it occurs. Structural features which can be derived automatically from computer-readable structure representations, and used in QSAR and property prediction studies are described in [10.175].

The first approach to substructural analysis used features derived from a fragmentation code [10.176]. The approach has been extended and refined and is referred to as the statistical–heuristic method [10.177]–[10.179]. It has been used in the selection of compounds for antitumor screening at the National Cancer Institute. Compounds are ranked in order of decreasing probability of activity.

A multiple regression analysis technique uses both WLN and connection tables as representations from which structural features can be automatically derived [10.180]–[10.186].

The use of data-base techniques such as 3-D searching was described in Section 10.4.2.

10.7.3. Molecular Modeling
[10.187]–[10.192]

Physical models of molecules are of limited value, especially for large molecules. They may take a long time to build and can be easily damaged. They are also difficult to manipulate and represent a static molecule without indication of its energy. Molecular graphics is a much more useful aid in visualizing 3-D structure. It combines interactive computer graphics with computational procedures such as molecular mechanics or molecular orbital calculations [10.193], [10.194].

A number of molecular modeling systems are commercially available (e.g., Chem-X from Chemical Design, QUANTA from Polygen, and SYBYL from Tripos Associates). They enable the user to manipulate objects in real-time on the screen, rotate them, translate them, and scale them to any size.

The 3-D shape and structure of a molecule are important to its biological activity and molecular graphics techniques are increasingly being used in the field of drug design. They allow 3-D interactions to be viewed between a potential drug molecule and its biological receptor site.

If crystal structure data for a potential new drug are not available, the 3-D structures can be produced from calculations based on standard bond lengths, bond angles, and dihedral angles. The 2-D structure is drawn and input to a molecular mechanics energy minimization program. A distance geometry approach and quantum mechanical calculations are other methodologies for obtaining preferred conformations.

Molecular graphics systems can be used to study interactions as a substrate molecule approaches the site of biological action. Using a technique known as plane clipping, an image can be moved in a direction perpendicular to the plane of the screen so that the molecular surface can be viewed from points both inside and outside the molecule. Techniques for enhancing 3-D perspective include the shading and color-highlighting of atoms and bonds, variation in the intensity of illumination (known as depth cueing), and the removal of "hidden" lines.

Molecular graphics can be used to identify pharmacophores (see Section 10.4.2).

10.8. Reaction Indexing [10.195]

Two major problems need to be overcome before retrieval systems for reactions become as well-established as systems for storage and retrieval of individual molecules. The first is the very wide range of query types that has to be handled. These include reactions where both reactant and product are fully specified; reactions giving rise to a specified structure or substructure; reactions of a specified structure or substructure; and a substructural transformation in which reactant and product substructures are specified.

The second problem is the organization of the large volumes of data that need to be stored to answer other types of query. Apart from topological descriptions of any chemical structures, a system must hold experimental conditions (temperature, pressure, yield, etc), citation details, and a reaction analysis (i.e., some description of the change that has taken place). This last is the most important, and most difficult element.

VLADUTZ suggested that reactions could be indexed by substructural changes, and introduced the concept of a reaction site [10.196]. This consists of all the bonds altered during the reaction, including any heteroatoms directly connected to an atom in the reaction site and any atom connected by a multiple bond to the site.

The automatic identification of such reaction sites involves comparing connection tables to identify common features on two sides of a reaction equation [10.195], [10.197], [10.198].

Operational reaction indexing systems such as GREMAS (Section 10.1.2) and CRDS (Section 3.4) have been considered elsewhere. The present section relates to graphics-based systems for storage and retrieval of chemical reaction information.

CASREACT [10.199], [10.200]. CASREACT is one of the CAS ONLINE data bases available on STN International (Section 10.5.2). Indexing for the service began in January 1985. At the moment only reactions in ca. 100 journals covered in CA's Organic Section are included in the data base, patents are not included. By 1988 the data base size had grown to 625 000 single-step reactions; ca. 170 000 are added annually. The system is document-based rather than reaction-based. The total number of matching reactions in an answer set of two or more documents cannot be identified.

Substructure searching is done in the CA Registry file. The whole file is not searched because the subset of compounds relevant to CAS-REACT can be retrieved by use of a screening fragment. Once the Registry file search is complete the hits are crossed over into CASREACT and associated with a "rôle" (e.g., "reactant").

The display of hit reactions consists of the reaction "map" (A + B → C + D for example), followed by a reaction scheme of chemical structures and Registry Numbers for the reactants, products, reagents, catalysts, and solvents. Abbreviated names are given for common catalysts, reagents, and solvents; a full CAS Index Name is given for unusual molecules. Only the Registry Numbers are searchable, not the names or abbreviated names. A Note (NTE) statement comments on hazards, unusual conditions (e.g., photolysis), and failed reactions. The final part of the display is the bibliography, which is not searchable in CASREACT but can be searched by crossing into CA File.

A comparison of CASREACT and REACCS has been published [10.201].

Organic Reactions Accessed by Computer (ORAC) [10.202]–[10.205]. ORAC is a graphics-based reaction indexing system written at Leeds University and sold by ORAC Ltd of Leeds. The related system for handling chemical structure data bases, OSAC, was described in Section 10.5.8. Searchable features in ORAC include reactants, products, intermediates, reagents, yield, reaction conditions, author, substructure, and reactions. The reaction search allows retrieval on the basis of reaction site or functional group changes. Keywords are also used for searching and retrieval because they can convey concepts which are difficult to express in terms of structural features (e.g., mechanism, selectivity, or reagent types).

Up to 32 "boxes" (data bases) can be handled. These include in-house data bases and the ORAC data base of over 116 000 reactions (1990 figure) from the current literature and key tertiary publications. They can be searched separately or in any combination.

There is a similarity searching feature. A Host Language Interface allows links to other systems both for synthesis planning and for QSAR and modeling [10.206].

REACCS [10.106], [10.204], [10.207]–[10.209]. REACCS is MDL's Reaction Access system and is related to MACCS (see Section 10.5.7). It is an interactive graphical data-base management system for storing and retrieving chemical reaction information. Reactants, products, catalysts, reagents, and solvents are substructure-searchable, with graphics input and output of the diagrams. Reaction centers are perceived. Variations on a given reaction are grouped together. Thus reactions are not duplicated if they differ only in yield or experimental conditions. Stereochemistry can be stored and searched. Multistep reactions are indexed by making manual entries for each individual step and the overall step. Similarity searching is offered.

One of the reasons for REACCS' prominence is the availability of several large and useful data bases with the system. These include a Current Literature File (CLF), the Theilheimer data base, and the Journal of Synthetic Methods data (see Section 3.4), a REACCS-readable version of Organic Synthesis, METALYSIS (a data base of metal-mediated chemistry), and CHIRAS (a data base of asymmetric synthesis).

Other Reaction Indexing Systems. SYNLIB [10.204], [10.205], [10.210] differs from REACCS and ORAC in that it could be classed as a knowledge-based system. It was designed to facilitate browsing in an evaluated collection of representative reactions. Users can also build their own in-house data bases and modify the supplied knowledge base. Comparisons of SYNLIB, ORAC, and REACCS have been published [10.204], [10.205].

Pfizer have written their own system, CONTRAST [10.211], as part of their in-house system SOCRATES [10.146].

10.9. Computer-Aided Synthesis Design

Programs for computer-aided synthesis design may be knowledge-based or depend on physical parameters. In the latter structural transformations are represented in a generalized manner by manipulating an abstract model of atoms and electrons.

LHASA, SECS, and CASP (see Section 11.6.7) are well-known knowledge-based systems. Other examples are CHIRON (Chiral Synthon) [10.212], and SYNCHEM [10.213], [10.214].

Systems based on physical parameters are EROS, IGOR, SYNGEN, and CAMEO. EROS (Elaboration of Reactions for Organic Synthesis) [10.215] uses DUGUNDJI and UGI's mathematical theory of constitutional chemistry [10.216].

IGOR (Interactive Generation of Organic Reactions) is based on the same model [10.217]. It has been used to identify novel reaction types [10.218].

SYNGEN aims to find the shortest and most economical synthetic routes to a given target structure from available starting materials [10.219], [10.220]. A rigorous mathematical form is used for describing reactions. Structures are represented numerically, enabling all possible reactions to be considered.

CAMEO (Computer Assisted Mechanistic Evaluation of Organic Reactions) is based on mechanistic analyses and operates in a synthetic, rather than retrosynthetic direction. It is an interactive program designed to predict the products of organic reactions from specified starting materials and conditions.

The AIPHOS system combines both approaches to synthesis planning [10.221].

11. Artificial Intelligence

11.1. Introduction

The goal of artificial intelligence (AI) is to construct computer programs which exhibit behavior that would be called "intelligent" when observed in humans. These programs seek to emulate human responses, for example, to solve problems, to learn from experience, to understand language and to interpret visual signs [11.1]–[11.9].

The roots of AI go back to 1936 when ALAN TURING first theorized about teaching computers to perform tasks based on logic. In 1950 he proposed what has come to be called the Turing test [11.10]. If a machine could fool its human partner into believing it to be human by participating in a wide range of conversations, then it could be concluded that the machine had passed the Turing test.

AI will be of even greater significance in the future because of its links with fifth-generation computers (see Chap. 7) [11.11], [11.12].

DAVIES defines the six classical branches of AI as machine architectures, expert systems (Section 11.6), natural language processing (Section 11.5.2), robotics, speech simulation and recognition (Section 11.5.1), and vision [11.13]. Robotics and computer vision (attempting to make machines identify and track objects they "see", and make decisions about them) are of little interest to the information profession at present. Optical character recognition (see Sections 7.2.2 and 11.5.2) is not usually regarded as part of vision technology.

CERCONE and MCCALLA include as AI subareas: search, problem solving, and planning (Section 11.3); theorem proving and logic programming (Section 11.4); knowledge representation (Section 11.6.2); learning; computer-aided instruction; game playing; automatic programming; and AI tools [11.7].

Machine learning strives to develop methods for automating the acquisition of new information, new skills, and new ways of organizing existing information [11.14]. Automatic programming (getting the computer to program itself) is now considered to be a formal area of computer science separate from AI itself. Computer-aided instruction [11.15], game playing, and miscellaneous AI tools are beyond the scope of this article.

11.2. Machine Architectures and Neural Networks

Machine architecture work deals with fundamentally different designs for computers, e.g., parallel processing (Sections 7.2.1 and 10.4.4). Most new types of hardware require sophisticated new software.

Neural networks represent a new way of organizing the computation process and are often implemented by simulation rather than by building specialist hardware [11.16]. They are data processing systems that learn by example. Design is inspired by the way in which the brain works with a large number of single processing elements operating in parallel. Neural networks are most powerful when applied to problems whose solution requires knowledge that is difficult to specify. Possible applications are in image recognition, computer vision, speech processing, robotics, and knowledge processing.

11.3. Search, Problem Solving, and Planning

There are two basic ingredients in intelligent behavior: possession of knowledge and search. The organization of knowledge is considered in Section 11.6.2. Search refers to the ability to create a space of possibilities large enough to contain the solution to a problem, and then searching for that solution.

Attacking a problem involves moving from a starting state to a goal. In chess the starting state is the initial board layout and the goal is checkmate. The number of possible states from the starting state to the goal defines the state space, which for chess is about 1×10^{120}. Each move produces a branch path that grows towards a potential goal. As the number of possible pathways increases, a combinatorial explosion takes place. Combinatorial explosion can limit the capability of an intelligent program so other problem-solving techniques, such as problem reduction and heuristic search [11.7], have been developed.

11.4. Theorem Proving and Logic Programming

Theorem proving is the process of making logical deductions from a noncontradictory set of axioms specified in first-order logic or predicate calculus. The process can be automated using a method called resolution [11.7].

Many simple problem-solving tasks can be formulated and solved using a theorem-proving approach, but this has not been widely applicable because of the problems of combinatorial explosion. However, theorem proving is at the heart of the more recent development of logic programming. The AI language PROLOG (PROgramming in LOGic), developed in 1972 [11.17] is based on a resolution theorem prover (see Section 11.6.5).

11.5. Human–Computer Interaction
[11.15], [11.18].

The user interface (the way information is entered or output) was discussed in Chapter 7. The use of user-hostile command languages is gradually giving way to user-friendly methods such as the selection of items from menus using pointing devices and touch screens. None of these are AI techniques. Much AI research has, however, been done on human–computer interaction (HCI), sometimes also called man–machine interaction (MMI), particularly in computational linguistics.

11.5.1. Speech Simulation and Recognition

Many of the issues of natural language understanding overlap those of speech understanding (see Section 11.5.2).

Speech as an output medium is well established. Early speech synthesizers digitized a speech signal for subsequent reproduction but more advanced systems generate the signal as required. Such synthesis methods are more flexible in operation and less demanding of computer storage, while producing more acceptable, natural speech output. Speech synthesis chips have been employed in toys, car and aircraft alarm systems, and automatic telephone answering.

The use of speech as an input medium is a more challenging problem [11.19]. Minimal acoustic differences between many words, background noise, and pronunciations cause problems. A Kurzweil machine can currently recognize a vocabulary of about 5000 words, when spoken as discrete units and not run together as in rapid human speech. A small vocabulary is sufficient for certain applications such as bibliographic retrieval [11.20], password or telephone number identification, air traffic control, and voice activation of domestic appliances or robots.

Connected speech recognition remains an unsolved problem. An example of the state-of-the-art is the HEARSAY-II system developed at Carnegie Mellon University [11.21].

11.5.2. Natural Language Processing (NLP)

Natural Language Interfaces. Rigid interfaces, protocols, command languages, and syntax requirements are serious barriers to the widespread use of computers. The user would prefer to express his requirements in his own language in a form such as "Give me a good 1988 review article about the treatment of asthma".

Unfortunately natural language is complex and ambiguous. A computer can easily compare character strings but it cannot detect the equivalence of "I hold a book" and "The volume is in my hands" [11.22].

Functional, relatively reliable, NLP systems are now available for several question-and-answer environments with a small possible vocabulary and a small number of queries and responses. An example is MYCIN, an early system for medical diagnosis [11.9], [11.22], [11.23]. The expert data base searching system TOME searcher [11.24]–[11.26] has a natural language interface for a domain-specific knowledge base in the field of information technology.

Understanding natural language involves three levels of interpretation: syntactic, semantic, and pragmatic [11.7], [11.27]. *Syntactic processes* parse sentences to clarify the grammatical relationships between words in a sentence. *Semantics* is concerned with assigning meaning to the various syntactic constituents. Semantic analysis converts the output of syntactic analysis into an extended predicate calculus, or a semantic network with quantification, and resolves ambiguities arising from the multiple meanings of words. *Pragmatics* attempts to relate individual sentences to one another and to their context.

Machine Translation [11.28]–[11.31] Machine translation has had a checkered history over the last 40 years and many translators still regard translation programs as "toys" of the academic researcher. The translation of technical manuals demands few cultural and stylistic skills and only a limited vocabulary may be required. Computer systems can be customized to handle such material reasonably well. Up to 80% of a human translator's time is spent in terminological research and document production. There is, therefore, considerable scope for "tools" such as electronic dictionaries linked to word processing software. The vendors of machine translation systems hope that the computer can relieve humans of the boring and repetitive elements of their tasks.

The Commission of the European Communities' machine translation system SYSTRAN has been operational since 1976 and is available for several European language pairs. Translations are processed at a rate of up to 400 000 words per hour. The EUROTRA project has been running since 1982 and is designed for the nine official languages of the EEC (72 language pairs).

Most Japanese electronics companies are developing or selling machine translation systems, mainly to translate English into Japanese. NEC has a prototype system that translates between Chinese, English, Japanese, Korean, Spanish and Thai using a "Pivot Method" where an intermediate language is used as a go-between. Fujitsu has Japanese-to-English and Japanese-to-German systems for business contracts, technical documents, and manuals. Researchers at ATR Telephony Research Laboratories, in cooperation with the Center for Machine Translation at Carnegie Mellon University, are applying neural network principles to speech recognition and automatic translation systems. They hope to mount a real-time, telephone translation system in Japanese–English/English–Japanese.

Optical Character Recognition (OCR). Progress is being made in computer recognition of written input (OCR). Sophisticated OCR machines (e.g., the Kurzweil) can be trained to recognize almost any print face. "Intelligent" optical character recognition systems are referred to as ICR, rather than OCR.

Scanning is done through charge coupled devices (CCDs) that transform light reflected from the image into signals that are a function of the light intensity. The image is then scanned and broken down into minute picture elements (pixels) that are sent to a computer where the digital files are created. The files are then processed by software for a particular application. The image and application files can be compressed and decompressed to more efficient sizes.

Several approaches can be used to translate the pixel bits to the multibit bytes used to define word characters. Feature extraction is based on the principle that each character has distinct features, regardless of font or spacing considerations. The software analyzes the scanned character, builds a features list, and then determines which character has most or all of these features. Some more advanced systems use AI techniques and can learn the characteristics of a new font. Some systems even attempt to read handwriting.

Neural network technology promises further breakthroughs in the reading area.

An example of the practical use of OCR in chemical information is the way CAS converted their older indexes into machine-readable form (see Section 2.2.5).

11.6. Expert Systems [11.8], [11.32]–[11.40]

11.6.1. Definition and Features

An expert system handles real-world, complex problems requiring an expert's interpretation. It solves these problems using a computer model of human reasoning [11.36]. Expert systems are the most important component of research in intelligent knowledge-based systems (IKBS), which are in turn a component of fifth-generation systems (see Chap. 7) [11.11], [11.12].

Artificial intelligence and expert systems differ from traditional computing in three ways:

1) They work with symbols rather than numbers.
2) They reason with heuristics ("rules of thumb") rather than algorithms (precisely defined instructions and decisions). The heuristic method of problem solving involves trying potential solutions, evaluating the result, and then modifying the procedure.
3) AI and expert systems work with interpretative rather than compiled languages, allowing the expression of concepts difficult to encode in traditional languages. Problems expressed in AI languages are transformed directly into machine actions during run time. However, compiled programs execute much more rapidly and efficiently than interpreted ones.

Expertise is acquired and codified during interviews with experts in the relevant problem area (domain). It is codified as a collection of facts and rules that constitute the *knowledge base*. Some systems are able to handle uncertain information using numerical values to denote the degree of confidence, credibility, or plausibility. The knowledge base is controlled and operated by the *inference engine*. The inference engine is separate from the knowledge base, thus allowing the knowledge base to be updated and extended without interfering with the overall control structure. In many expert systems interaction with the system occurs through a *user interface*, not directly with the knowledge base. The system should be interactive in that it can ask the user questions when the information it has is incomplete or conflicting.

11.6.2. Knowledge Representation
[11.4], [11.32]

For a program to exhibit intelligence it must have access to large amounts of knowledge and it must know how to manipulate and use that knowledge [11.37].

Knowledge must be represented inside an expert system so that it can be used effectively, modified, and augmented. Representation must be accessible and flexible. The expert system must also be able to explain its actions to humans.

A number of different knowledge representation "paradigms" have emerged. One of the earliest represented knowledge in *semantic networks* where facts are stored at nodes and relationships between the facts are represented by arcs. One of the most common relationships in a semantic network is the so-called "ISA" link which allows facts (e.g., dogs have tails) to be attached to classes of objects (e.g., dogs) and then inherited by specific objects in the class (e.g., Fido, Rover).

Each node in a semantic net has to be meticulously specified. Generic objects are not handled efficiently; difficulties are encountered in interconnecting object attributes and in handling unspecified attributes or default conditions. Expert systems developers therefore started to study *frames* for the storage of declarative knowledge. Here data are presented in a hierarchy of facts where lower orders automatically inherit all the characteristics of higher orders. Thus a specific breed of dog must exhibit all the features of dogs in general. Frames are composed of slots. For example, a frame for a chemical could contain slots for formula, structure, and physical properties. Slots may be further subdivided into facets. Slots may also contain hypotheses that relate to the expert system's functioning, rules about program situations, or pointers to other frames. Frames allow an expert system to retain more information about a specific item than it must explicitly express each time. Quantities of diverse information can be conveniently handled and extra slots can be added for new data.

First-order predicate calculus represents information by means of formulae. Function, variable, and constant symbols are set off with parentheses, brackets, and commas, respectively, to form statements. True or false propositions express relationships between specific and generic objects. More complex expressions can be written using connectives such as logical AND, logical OR, "implies", and "there exists".

The knowledge base not only stores declarative knowledge. It also tells how the data can be manipulated to solve the problem (procedural knowledge). The most common method of data manipulation involves production rules of the type: IF (condition) THEN (action).

The condition and the action often have several components, e.g., IF the unknown is a bird AND the unknown cannot fly AND the unknown has black and white feathers THEN it is an ostrich with probability 0.8. Predicate logic and other techniques can also be used to represent procedural knowledge.

11.6.3. Knowledge Engineering

Domain knowledge has to be elicited from the human expert by a "knowledge engineer" using interviewing and observation techniques. The knowledge engineer has to structure the knowledge so that it can be properly represented by the computer system. As knowledge engineering is slow, methods are being been sought to automate knowledge acquisition or allow systems to learn from examples presented in a natural way.

Computerized tools have been designed to help the knowledge engineer (see Section 11.6.5). Many personal-computer-based expert system shells and some larger, more complex expert system building environments incorporate more reasoning strategies, and have choices of representation language, graphics, and links to other languages.

11.6.4 Inference Engine [11.34], [11.41], [11.42]

The inference engine (control structure) is the central program which manipulates the data in the knowledge base to reach conclusions. The engine may approach a problem by beginning with either hypotheses or facts. A *goal-directed expert system* uses a reverse-chaining or backward-chaining algorithm. The program starts with a limited set of possible hypotheses and attempts to prove the validity of each one by examining all the factors. The knowledge base is searched to find a rule which concludes the initial hypothesis. The IF clauses from this rule then become the hypotheses for the next search level. The process continues until all of the remaining IF clauses are known to be true (in which case the hypothesis is true) or until no more rules apply (and the hypothesis is false).

A *data-driven expert system* uses a forward-chaining mechanism. The program begins with a list of facts that are known to be true. Each rule in the knowledge base is tested to see if all of its IF clauses are contained in the list of known facts. When such a rule is found, the system adds the THEN clauses from the rule to the list of known facts. All the rules in the knowledge base are scanned repetitively until no new facts can be concluded.

If the rules are formulated with weights or confidence factors to their conclusions, the system can produce an answer which is neither true nor false. Multiple answers are possible with degrees of importance or predictability.

The performance of an expert system may be increased by using heuristic rules to eliminate possible but unlikely solutions. The added knowledge is called meta-knowledge; in knowledge bases which use production rules, meta-knowledge is incorporated as meta-rules. These rules instruct the system how to choose which rule to use when more than one is relevant. For example, in an organic synthesis expert system a meta-rule might be IF multiple reactions have the same product THEN use the reaction with the highest yield.

11.6.5. Software and Hardware

Successful AI programs can be written in traditional computer languages but specialized AI languages are much more common in expert systems.

LISP (from LISt Processing) was developed in 1960 and is the most commonly used AI language in the United States [11.32]. It is used to define and manipulate irreducible objects associated with an alphanumeric label, called an atom. Atoms may be assigned values which are combinations of atoms and operators arranged in a list structure. Atoms are stored in memory and are located by means of pointers. Lists are represented as a collection of memory cells whose contents are pointers to other memory cells. Manipulation of lists involves manipulation of their pointers and application of some simple logic processes. LISP and LISP-like languages are typified by their interactive nature, the emphasis on symbolic expressions, and a tree-oriented approach to data structures.

PROLOG is used by most Europeans and Japanese. It is a higher level language than LISP, allowing a fact to be made in one statement where LISP would require many more programming steps. However LISP is more flexible and has more sophisticated programming aids. Advocates of PROLOG argue that it is more readable, provides a more suitable natural language interface, and can execute in a parallel fashion that improves speed of execution. It supports the easy construction and use of relational data bases.

PROLOG [11.12], [11.17], [11.43] is based on predicate calculus (see Section 11.6.2) which allows the user to program relationships and qualities or attributes.

Both compiler and interpreter forms (see Section 7.3.2) of LISP and PROLOG exist.

Expert systems can be run on traditional computer hardware but specialized minicomputers, usually LISP-oriented, give better results. PROLOG adapters for such systems are available. Microcomputer versions of both LISP and PROLOG are available and LISP compilers are offered for some general-purpose hardware.

The most common and cheapest method of developing a small expert system is by using a commercially available PC-based "shell". This is an empty knowledge base with its own representation language and inference engine. The user can concentrate on the development of the knowledge base, rather than the programming procedures. Shells have put the software in the hands of the expert rather than the knowledge engineer (see Section 11.6.3).

11.6.6. Advantages and Disadvantages

A well-written expert system performs consistently (although "mindlessly") for 24 h a day in an hostile environment. It can also release experts from tedious tasks for more important work. Many simultaneous users can be accommodated by an expert system.

Knowledge used by experts is generally learned through experience and is not written down. An expert system captures knowledge that could otherwise be lost. An average user may, however, be too trusting of the results output by the computer and may fail to ask for an explanation.

Characteristically, an expert system covers only a narrow domain. As the domain increases, the number of contingencies that can be covered decreases. Building and maintaining a large knowledge base is difficult. Even for a narrow domain, constructing the knowledge base is slow and laborious. Another problem is speed of execution.

Further problems and trends are discussed in [11.34].

11.6.7. Applications

Medical Diagnosis. The best known expert system in the field of medical diagnostics is MYCIN [11.9], [11.23] which advises on the treatment of bacterial infections, using backward-chaining techniques. It is written in LISP and has three modules: a consultation program (the user interface), an explanation program (the inference engine), and the knowledge acquisition program (the knowledge base and a maintenance interface). The physician enters data to record the patient's history and symptoms. If information is missing, MYCIN will request more data or infer it from the knowledge base. MYCIN then reports (and explains) its conclusions and suggests an appropriate treatment.

Librarianship and Information Retrieval. The online public access catalog (OPAC) should be amenable to the expert system approach [11.34], [11.40], [11.44]–[11.46]. An OPAC offers a user-friendly interface to locate documents in library catalogs. Visitors to the U.S. Library Corporation can use an intelligent CD-ROM-based OPAC, the Bibliofile Intelligent Catalog, which uses both sound and graphics [11.47].

Online searching is also an area ripe for expert systems developments. The contents of a knowledge base for information retrieval from an online data base must be divided between system knowledge and subject knowledge, e.g., the command language and retrieval techniques as opposed to the choice of search terms and synonyms.

Some experimental information systems for online searching, are reviewed in [11.48], methodologies are discussed in [11.24]. The PLEXUS system [11.49], [11.50] links an intelligent search interface to a conventional data base, whereas the GENERIS system [11.51] incorporates the conventional structure of a relational data base but can also represent semantic relationships between items in the tables.

The TOME Searcher front end [11.24]–[11.26] allows people without experience to make use of data bases. It has semantic networks based on thesauri. The user can train the system to learn new terminology. The system runs on an IBM personal computer and has four components: a set of menu-driven functions which help the user set up search parameters; a natural language interface for search queries; a step-by-step question and answer module to refine the automatically generated search strategy; and a communications module to call up the search service, upload the search, and download the results. ESA, STN, DIALOG, and Maxwell Online can be accessed.

WILLETT and coworkers [11.52] have compared the knowledge-based approach to reference retrieval (exemplified by PLEXUS) with statistically based techniques (exemplified by Sheffield University's INSTRUCT program [11.53]–[11.55]).

INSTRUCT supports facilities for natural language query processing, best-match and cluster-based searching, user-initiated query expansion based on string similarity or term co-occurrence data, automatic relevance feedback based on probabilistic term weighting, and a browsing capability. An operational version for interactive browsing and ranking has been implemented in an industrial environment [11.56].

Ease-of-use and cognitive and behavioral aspects of retrieval are discussed in [11.57]–[11.60].

Recent publications have discussed the application of expert systems to library and information science work [11.15], [11.39], [11.40], [11.46], [11.61], [11.62].

Computer-Aided Synthesis Design. The first attempt to codify and organize organic syntheses was the Logic and Heuristics Applied to Synthetic Analysis (LHASA) program [11.63], [11.64]. The goal is a target molecule and the system works backwards through a synthesis tree by a method called retrosynthetic analysis, tracing possible precursors from a "transform library" of reactions. In Europe, LHASA UK handles the use of the program and maintenance of the transform library.

The Simulation and Evaluation of Chemical Syntheses (SECS) project [11.65]–[11.67] was initiated in 1969 to focus on stereochemistry, stereoelectronic effects, and other aspects not considered by LHASA. SECS also uses retrosynthetic analysis and a transform library. A consortium of seven major German and Swiss companies, the computer-assisted synthesis planning (CASP) project, has worked with SECS.

For other synthesis planning programs with different strategies see Section 10.9 [11.68]–[11.76].

Analytical Chemistry and Structure Elucidation. The interpretation of spectra has proved a fruitful area for the application of expert systems theory [11.32], [11.77]–[11.84]. The earliest venture was the DENDRAL (**dendr**itic **al**gorithm) project which used mass and carbon-13 NMR spectra [11.85]–[11.87].

BREMSER's work [11.88]–[11.91] is of interest because of the variety of spectral techniques, his substructure searching technology, and the commercial availability of the software (see Chap. 9). His IDIOTS (Infrared Spectra Documentation and Interpretation Operating with Transcripts and Structures) [11.88] system is, however, not commercially available.

CHEMICS (Combined Handling of Elucidation Methods for Interpretable Chemical Structures) uses IR, and carbon-13 and proton NMR spectra [11.92]–[11.94].

PAIRS (Program for the Analysis of Infrared Spectra) does not employ a data base of IR spectra but attempts to parallel the reasoning used in interpreting IR spectra [11.95]–[11.97].

MCLAFFERTY and coworkers developed STIRS (Self-Training, Interpretive, and Retrieval System) to interpret unknown mass spectra [11.98].

CASE (Computer Assisted Structure Elucidation) [11.99] follows a new strategy [11.100]. It handles more than one type of NMR spectra and an IR spectrum interpreter is under development.

Other systems have been reported [11.101]–[11.104].

Other Applications. Other applications of expert systems are listed below:

MACSYMA, a symbolic manipulation system that functions as a mathematical aid.

R1, an expert system that aids in the configuration of Digital Equipment Corporation VAX Computer systems.

PROSPECTOR, an expert geologist system that seeks commercially exploitable mineral deposits.

MOLGEN, an expert system for designing experiments in molecular biology. The Imperial Cancer Relief

Fund has a PROLOG system for protein topology [11.105].

The DOCENT expert system deals with macromolecules [11.106], [11.107]. It is a molecular modeling application which represents macromolecules by "generalized cylinders".

WIZARD is a symbolically based conformational analysis program which reasons about intra- and intermolecular forces [11.108]–[11.110].

The AIMB program (Analogy and Intelligence in Model Building) is a symbolic, nonnumerical approach to model building and conformational analysis [11.111], [11.112].

Metabolexpert predicts metabolic pathways by logic programming [11.113].

For other examples of systems in the molecular graphics area see [11.41], [11.114], [11.115]. There are also industrial applications of expert systems in chemistry [11.41], [11.115] and environmental systems [11.116].

11.7. Hypermedia [11.117]–[11.122]

The concept of an "intelligent data base" in a knowledge-based integrated information system, involves object orientation (see Section 7.3.6), expert systems, and hypermedia. The history of hypermedia goes back to 1945 and an article published by VANNEVAR BUSH [11.123]. BUSH had a vision of a device called Memex [11.123], [11.124], an electronic desk that could access the text of linking files in seconds. Inspired by BUSH's ideas TED NELSON coined the word "hypertext" in the 1960s, for a type of nonsequential reading and writing that links different text nodes [11.125]. Implementation of the theory was not possible until Office Workstations Ltd. International's Guide software and Apple Computer's HyperCard software were introduced in 1987.

A hypertext system allows the user to link pieces of information, creating trails through the associated materials. The idea is akin to the use of footnotes, references, and "See Also's" in a printed tool. The reader can "browse and hunt" in an electronic book. In early hypertext systems all links were preprogrammed. Interactive links in newer systems give the user more freedom.

A hypertext system may be envisaged as consisting of "cards" containing text or other information and connected by directed links. In most systems a card may have several outgoing links. Users navigate through the system by following the links. They may also backtrack by following the links they have used in the reverse direction.

"Landmarks" are especially prominent cards, for example, because they are directly accessible from many other cards. A hypertext system has two bidirectional navigational dimensions: a linear dimension used to move back and forward among the text pages of a given section; and a nonlinear dimension used for hypertext jumps.

Essential features of a hypertext system are an intuitive graphical user interface (mouse, buttons, icons, pull-down menus, consistency, Chap. 7), cross-hierarchical links, an object-oriented environment and scripting language, and a flexible format for information, not necessitating a formal data-base structure.

Hypermedia is an extension of the concept of hypertext and implies incorporation of color, graphics, images, sound, and animation (video).

Some problems still require resolution. Setting up a hypertext data base can be a very time-consuming proposition. The user may tend to feel lost while navigating the system. Text searching is limited and nontextual information (e.g., video) cannot be searched; a multiuser environment is not possible; there are high data storage requirements; and data storage and compression/decompression standards need establishing. The large storage space requirements for graphic and video have led to the association of hypertext systems with optical disc storage (see Section 7.2.3) and some CD-ROM data-base systems have hypertext-related retrieval software (see Chap. 8).

12. References

References for Chapter 1

[1.1] Y. Wolman: *Chemical Information. A Practical Guide to Utilization*, Wiley-Interscience, Chichester 1988.
[1.2] K. Subramanyam: *Scientific and Technical Information Resources*, Marcel Dekker, New York 1981.
[1.3] R. T. Bottle: *Use of the Chemical Literature*, 3rd ed., Butterworths, London 1979.
[1.4] R. E. Maizell: *How to Find Chemical Information A Guide for Practicing Chemists, Educators, and Students*, 2nd ed., John Wiley, New York 1987.
[1.5] A. Antony: *Guide to Basic Information Sources in Chemistry*, J. Wiley & Sons, Halsted Press, New York 1979.
[1.6] M. G. Mellon: *Chemical Publications, their Nature and Use*, 5th ed., McGraw-Hill, New York 1982.
[1.7] H. Skolnik: *The Literature Matrix of Chemistry*, Wiley-Interscience, New York 1982.
[1.8] M. Mücke: *Die Chemische Literatur. Ihre Erschließung und Benutzung*, VCH Verlagsgesellschaft, Weinheim 1982.
[1.9] A. J. Meadows (ed.): *The Scientific Journal*, Aslib, London 1979.
[1.10] A. J. Meadows (ed.): *The Growth of Science Publishing in Europe*, Elsevier, Amsterdam 1980.
[1.11] A. J. Meadows, *Libr. Review* **37** (1988) 7–16.

[1.12] H. Schulz: *From CA to CAS ONLINE*, VCH Verlagsgesellschaft, Weinheim 1988.
[1.13] J. M. Ziman: *Public Knowledge: The Social Dimension of Science*, Cambridge University Press, London 1968.
[1.14] J. R. Ravetz: *Scientific Knowledge and its Social Problems*, Penguin, Harmondsworth 1973.
[1.15] S. M. Dhawan, S. K. Phull, S. P. Jain, *J. Doc.* **36** (1980) no. 1, 24–41.
[1.16] E. Garfield: *Citation Indexing – Its Theory and Application in Science, Technology, and Humanities*, Wiley-Interscience, New York 1979.
[1.17] E. Garfield, *Science (Washington, D.C.)* **178** (1972) 471–479.
[1.18] R. Todorov, W. Glänzel, *J. Inf. Sci.* **14** (1988) 47–56.
[1.19] M. H. MacRoberts, B. R. MacRoberts, *J. Am. Soc. Inf. Sci.* **40** (1989) no. 5, 342–349.
[1.20] E. Garfield (ed.): *SCI Journal Citation Reports*, "A Bibliometric Analysis of Science Journals in the ISI Data Base," Institute for Scientific Information, Philadelphia 1987 (published annually).
[1.21] Chemical Abstracts Service Source Index (CASSI), *1907–1989 Cumulative Index*, Chemical Abstracts Service, Columbus, Ohio 1989.
[1.22] J. Hartley, M. Trueman, A. J. Meadows, *J. Inf. Sci.* **14** (1988) 69–75.
[1.23] R. F. Flesch, *J. Appl. Psychol.* **32** (1948) 221–233.
[1.24] G. R. Klare, *Reading Research Quarterly* **10** (1974–1975) 62–101.
[1.25] C. R. H. Inman, *J. Inf. Sci.* **6** (1983) 159–164.
[1.26] E. K. Samaha, *Information Development*, **3** (1987) no. 2, 103–107.
[1.27] H. H. Budzier, *Zentralbl. Bibliothekswesen* **100** (1986) no. 3, 93–101.
[1.28] W. Tuck, D. Archer, M.-C. Hayet, C. McKnight: *Project Quartet, LIR Report 76*, British Library, London 1990.
[1.29] W. Tuck, *Netlink* **5** (1989) no. 1, 5–8.
[1.30] R. M. Campbell, B. T. Stern, *Microcomputers for Information Management* **4** (1987) no. 2, 87–107.
[1.31] F. A. Mastroddi, J. Page in: *Electronic Publishing: State of the Art Report*, Pergamon Infotech, Maidenhead 1987, p. 37.
[1.32] B. F. Polansky, B. H. Weil, *J. Chem. Inf. Comput. Sci.* **25** (1985) 153–159.
[1.33] B. H. Weil, B. F. Polansky, *J. Chem. Inf. Comput. Sci.* **24** (1984) 43–50.
[1.34] B. F. Polansky in J. S. Dodd (ed.): *The ACS Style Guide*, American Chemical Society, Washington DC 1986, p. 137.
[1.35] D. P. Waite, *J. Chem. Inf. Comput. Sci.* **22** (1982) 63–66.
[1.36] P. J. Hills (ed.): *Trends in Information Transfer*, Francis Pinter, London 1982.
[1.37] L. C. Cross, *Aslib Proc.* **26** (1974) no. 11, 425–429.
[1.38] A. A. Manten, *J. Inf. Sci.* **1** (1980) 293–296.
[1.39] R. G. Lerner et al., *Annu. Rev. Inf. Sci. Technol.* **18** (1983) 127–149.
[1.40] D. P. Martinsen, R. A. Love, L. R. Garson, *Online (Weston Conn.)* **13** (1989) no. 2, 121–133.
[1.41] J. A. Hearty, *Information Services and Use* **8** (1988) 93–105.

References for Chapter 2

[2.1] M. Cooper, *J. Am. Soc. Inf. Sci.* **33** (1982) no. 3, 152–156.
[2.2] L. W. Granick, *J. Am. Soc. Inf. Sci.* **33** (1982) no. 3, 175–182.
[2.3] R. J. Rowlett, *J. Chem. Inf. Comput. Sci.* **25** (1985) no. 3, 159–163.
[2.4] B. H. Weil, *J. Am. Soc. Inf. Sci.* **21** (1970) 351–357.
[2.5] H. Schulz: *From CA to CAS ONLINE*, VCH Verlagsgesellschaft, Weinheim 1988.
[2.6] R. E. Maizell: *How to Find Chemical Information. A Guide for Practicing Chemists, Educators, and Students*, 2nd ed., John Wiley, New York 1987.
[2.7] D. B. Baker, J. W. Horiszny, W. V. Metanomski, *J. Chem. Inf. Comput. Sci.* **20** (1980) 193–201.
[2.8] H. L. Morgan, *J. Chem. Doc.* **5** (1965) 107–113.
[2.9] D. F. Zaye, W. V. Metanomski, A. J. Beach, *J. Chem. Inf. Comput. Sci.* **25** (1985) 392–399.
[2.10] M. Mücke: *Die Chemische Literatur. Ihre Erschließung und Benutzung*, VCH Verlagsgesellschaft, Weinheim 1982.
[2.11] R. S. Cahn, O. C. Dermer: *Introduction to Chemical Nomenclature*, 5th ed., Butterworths, London 1979.
[2.12] P. Fresenius: *Organic Chemical Nomenclature. Introduction to the Basic Principles*, Ellis Horwood, Chichester 1989.
[2.13] M. G. Robiette in R. Lees, A. Smith (ed.): *Chemical Nomenclature Usage*, Ellis Horwood, Chichester 1983, p. 74.
[2.14] *Nomenclature of Organic Chemistry, Sections A, B, C, D, E, F and H, (The Blue Book)*, Pergamon Press, Oxford 1979.
[2.15] *Nomenclature of Inorganic Chemistry, (The Red Book)*, 2nd ed., Butterworths, London 1971. [3rd ed. Blackwells Scientific, due January 1990].
[2.16] *Compendium of Analytical Nomenclature, (The Orange Book)*, 2nd ed., Blackwells Scientific, Oxford 1987.
[2.17] *Biochemical Nomenclature and Related Documents, (The Compendium)*, Biochemical Society, London 1978.
[2.18] N. Donaldson et al., *J. Chem. Doc.* **14** (1974) 3–14.
[2.19] G. G. Vander Stouw, I. Naznitsky, J. E. Rush, *J. Chem. Doc.* **7** (1967) 165–169.
[2.20] G. G. Vander Stouw, P. M. Elliott, A. C. Isenberg, *J. Chem. Doc.* **14** (1974) 185–193.
[2.21] G. G. Vander Stouw, *J. Chem. Inf. Comput. Sci.* **15** (1975) 232–236.
[2.22] D. I. Cooke-Fox, G. H. Kirby, J. D. Rayner, *J. Chem. Inf. Comput. Sci.* **29** (1989) 101–105.
[2.23] D. I. Cooke-Fox, G. H. Kirby, J. D. Rayner, *J. Chem. Inf. Comput. Sci.* **29** (1989) 106–112.
[2.24] D. I. Cooke-Fox, G. H. Kirby, J. D. Rayner, *J. Chem. Inf. Comput. Sci.* **29** (1989) 112–118.
[2.25] L. Goebels in J. Gasteiger (ed.): *Software-Entwicklung in der Chemie 2*, Springer-Verlag, Berlin 1988, p. 57.
[2.26] P. G. Dittmar, R. E. Stobaugh, C. E. Watson, *J. Chem. Inf. Comput. Sci.* **20** (1980) 111–121.
[2.27] R. G. Freeland, S. A. Funk, L. J. O'Korn, G. A. Wilson, *J. Chem. Inf. Comput. Sci.* **19** (1979) 94–97.
[2.28] J. E. Blackwood, P. S. Elliott, R. E. Stobaugh, C. E. Watson, *J. Chem. Inf. Comput. Sci.* **17** (1977) 3–8.
[2.29] G. G. Vander Stouw, C. Gustafson, J. D. Rule, C. E. Watson, *J. Chem. Inf. Comput. Sci.* **16** (1976) 213–218.
[2.30] A. Zamora, D. L. Dayton, *J. Chem. Inf. Comput. Sci.* **16** (1976) 219–222.
[2.31] R. Stobaugh, *J. Chem. Inf. Comput. Sci.* **20** (1980) 76–82.

[2.32] J. Mockus, R. E. Stobaugh, *J. Chem. Inf. Comput. Sci.* **20** (1980) 18–22.
[2.33] J. P. Moosemiller, A. W. Ryan, R. E. Stobaugh, *J. Chem. Inf. Comput. Sci.* **20** (1980) 83–88.
[2.34] A. W. Ryan, R. E. Stobaugh, *J. Chem. Inf. Comput. Sci,* **22** (1982) 22–28.
[2.35] K. A. Hamill, R. D. Nelson, G. G. Vander Stouw, R. E. Stobaugh, *J. Chem. Inf. Comput. Sci.* **28** (1988) 175–179.
[2.36] R. E. Stobaugh, *J. Chem. Inf. Comput. Sci.* **28** (1988) 180–187.
[2.37] G. G. Vander Stouw in W. A. Warr (ed.): *Chemical Structures, The International Language of Chemistry,* Springer Verlag, Berlin 1988, p. 211.
[2.38] E. Garfield, *Science (Washington, D.C.)* **144** (1964) 649–654.
[2.39] E. Garfield, *J. Chem. Inf. Comput. Sci.* **25** (1985) 170–174.
[2.40] Y. Wolman: *Chemical Information. A Practical Guide to Utilization,* Wiley-Interscience, Chichester 1988.
[2.41] K. Subramanyam: *Scientific and Technical Information Resources,* Marcel Dekker, New York 1981.
[2.42] R. T. Bottle: *Use of the Chemical Literature,* 3rd ed., Butterworths, London 1979.
[2.43] A. Antony: *Guide to Basic Information Sources in Chemistry,* J. Wiley & Sons, Halsted Press, New York 1979.
[2.44] M. G. Mellon: *Chemical Publications, their Nature and Use,* 5th ed., McGraw-Hill, New York 1982.
[2.45] H. Skolnik: *The Literature Matrix of Chemistry,* Wiley-Interscience, New York 1982.
[2.46] J. Schmittroth (ed.): *Abstracting and Indexing Services Directory,* Gale Research, Detroit 1982–1983.

References for Chapter 3

[3.1] Y. Wolman: *Chemical Information. A Practical Guide to Utilization,* Wiley-Interscience, Chichester 1988.
[3.2] K. Subramanyam: *Scientific and Technical Information Resources,* Marcel Dekker, New York 1981.
[3.3] R. T. Bottle: *Use of the Chemical Literature,* 3rd ed., Butterworths, London 1979.
[3.4] R. E. Maizell: *How to Find Chemical Information. A Guide for Practicing Chemists, Educators, and Students,* 2nd ed., John Wiley, New York 1987.
[3.5] A. Antony: *Guide to Basic Information Sources in Chemistry,* J. Wiley & Sons, Halsted Press, New York 1979.
[3.6] M. G. Mellon: *Chemical Publications, their Nature and Use,* 5th ed., McGraw-Hill, New York 1982.
[3.7] H. Skolnik: *The Literature Matrix of Chemistry,* Wiley-Interscience, New York 1982.
[3.8] M. Mücke: *Die Chemische Literatur. Ihre Erschließung und Benutzung,* VCH Verlagsgesellschaft, Weinheim 1982.
[3.9] *Kirk–Othmer,* 3rd ed., 1978–1984.
[3.10] *Ullmann,* 4th ed., 1972–1984.
[3.11] *Ullmann,* 5th ed., 1985 to date.
[3.12] J. Matley, *Chem. Eng. Int. Ed.* **93** (1986) no. 8, 95–97.
[3.13] *Beilstein,* 1918 to date (now published in English).
[3.14] R. Luckenbach, R. Ecker, J. Sunkel, *Angew. Chem. Int. Ed. Engl.* **20** (1981) 841–849.
[3.15] R. Luckenbach, *J. Chem. Inf. Comput. Sci.* **21** (1982) 82–83.
[3.16] C. Jochum in W. A. Warr (ed.): *Chemical Structures, The International Language of Chemistry,* Springer Verlag, Berlin 1988, p. 187.
[3.17] *How to Use Beilstein,* Springer Verlag, Berlin 1984.
[3.18] A. J. Lawson in W. A. Warr (ed.): "Graphics for Chemical Structures: Integration with Text and Data," *ACS Symp. Ser.* **341** (1987) 80.
[3.19] S. R. Heller, *Database* **10** (1987) no. 4, 47–52.
[3.20] H. O. House, *J. Chem. Inf. Comput. Sci.* **24** (1984) 277.
[3.21] *Gmelin,* 1922 to date.
[3.22] J. Buckingham (ed.): *Dictionary of Organic Compounds,* 5th ed., Chapman and Hall, London 1982 and annual supplements thereafter.
[3.23] J. Buckingham, *CHEMTECH* **15** (1985) no. 11, 674–679.
[3.24] P. Hyams, *Inf. World Rev.* **1989**, July, 18.
[3.25] *Houben-Weyl,* 4th ed., 1952 to date.
[3.26] W. Theilheimer: *Synthetic Methods of Organic Chemistry,* Karger, Basel 1946 to date.
[3.27] A. F. Finch, *J. Chem. Inf. Comput. Sci.* **26** (1986) no. 1, 17–22.
[3.28] A. F. Finch in P. Willett (ed.): *Modern Approaches to Chemical Reaction Searching,* Gower, Aldershot 1986, p. 36.
[3.29] *Organic Syntheses,* 2nd ed., Wiley, New York 1941 to date.
[3.30] W. Dauben (ed.): *Organic Reactions,* Wiley, New York, 1942 to date.
[3.31] L. G. Wade (ed.): *Compendium of Organic Synthetic Methods,* Wiley, New York 1977 to date (Volumes 1 and 2 were edited by I. T. Harrison and S. Harrison).
[3.32] M. Fieser (ed.): *Fieser and Fieser's Reagents for Organic Synthesis,* Wiley, New York 1967 to date.
[3.33] *Inorganic Syntheses,* Wiley, New York 1939 to date.
[3.34] S. Coffey or M. F. Ansell (ed.): *Rodd's Chemistry of Carbon Compounds,* 2nd ed., Elsevier, Amsterdam 1964 to date.
[3.35] D. Barton, W. D. Ollis (eds.): *Comprehensive Organic Chemistry,* Pergamon Press, Oxford 1979.
[3.36] G. Wilkinson (ed.): *Comprehensive Organometallic Chemistry,* Pergamon Press, Oxford 1982.
[3.37] *The Chemistry of Heterocyclic Compounds – A Series of Monographs,* Wiley, New York 1970 to date.
[3.38] A. R. Katritzky, C. W. Rees (eds.): *Comprehensive Heterocyclic Chemistry,* Pergamon Press, Oxford 1984.
[3.39] G. Wilkinson (ed.): *Comprehensive Coordination Chemistry,* Pergamon Press, Oxford 1987.
[3.40] J. C. Bailar, H. J. Emeleus, R. Nyholm, A. F. Trotman, (eds.): *Comprehensive Inorganic Chemistry,* Pergamon Press, Oxford 1973.
[3.41] S. Budavari (ed.): *The Merck Index,* 11th ed., Merck and Company, Rahway, New Jersey 1989.
[3.42] R. C. Weast, J. G. Grasselli (eds.): *CRC Handbook of Data on Organic Compounds,* 2nd ed., CRC Press, Boca Raton, Florida 1988.
[3.43] N. I. Sax, R. J. Lewis (eds.): *Dangerous Properties of Industrial Materials,* 7th ed., Van Nostrand Reinhold, New York 1989.
[3.44] L. Bretherick: *Handbook of Reactive Chemical Hazards,* 3rd ed., Butterworths, London 1985.
[3.45] L. Bretherick (ed.): *Hazards in the Chemical Laboratory,* Royal Society of Chemistry, London 1986.

References for Chapter 4

[4.1] D. N. Wood, *IFLA J.* **10** (1984) no. 3, 278–282.
[4.2] N. W. Posnett, W. J. Baulkwill, *J. Inf. Sci.* **5** (1982) 121–130.
[4.3] V. Alberani, *Orv. Konyvtaros* **28** (1988) no. 4, 341–350.
[4.4] N. W. Posnett, *Libr. Acquisitions: Practice Theory* **8** (1984) 275–285.
[4.5] J. P. Chillag in D. F. Shaw (ed.): *Information Sources in Physics*, Butterworths, London 1985, p. 355.
[4.6] C. Hasemann, *Z. Bibliothekswesen Bibliographie* **33** (1986) no. 6, 417–427.
[4.7] D. N. Wood, *Aslib Proc.* **34** (1982) no. 11/12, 459–465.
[4.8] J. P. Chillag in D. W. Bromley, A. M. Allott (eds.): *British Librarianship and Information Work 1981–1985*, vol. 2: "Special Libraries, Materials and Processes," Library Association, London 1988, p. 95.
[4.9] V. Alberani, A. Pagamonci (eds.): "Letteratura Grigia," *Boll. Inf. Assoc. Ital. Biblioteche* **27** (1987) no. 3/4, 305–498.
[4.10] J. M. Gibb, M. Maurice, *Aslib Proc.* **34** (1982) no. 11/12, 493–497.
[4.11] C. Hasemann, *ABI-Tech.* **5** (1985) no. 4, 261–265.
[4.12] C. Salmon, L. van Simaeys, *Cah. Doc.* **3** (1980) 53–56.
[4.13] E. K. Samaha, *Inf. Dev.* **3** (1987) no. 2, 103–107.
[4.14] J. S. Robinson: *Tapping the Government Grapevine: The User-Friendly Guide to US Government Information Services*, Oryx Press, Phoenix 1988.
[4.15] H. Ogawa et al., *J. Am. Soc. Inf. Sci.* **40** (1989) no. 5, 350–355.
[4.16] E. Kohl, M. Ockenfeld: *Konferenzinformation: Hinweise zur Ermittlung und Beschaffung von Terminen und Vorträgen chemierelevanter Veranstaltungen*, Arbeitsgruppe Informationswissenschaft in der Chemie an der Universität Frankfurt am Main (AIC), Frankfurt 1981.
[4.17] *World Meetings: United States and Canada*, MacMillan, New York 1963, quarterly.
[4.18] *World Meetings: Outside United States and Canada*, MacMillan, New York 1968, quarterly.
[4.19] *World Meetings: Medicine*, MacMillan, New York 1978, quarterly.
[4.20] M. Ockenfeld: *Dissertationen als Informationsquellen. Hinweise zu ihrer Ermittlung, Beschaffung, Gestaltung und Verbreitung im Fach Chemie*, Arbeitsgruppe Informationswissenschaft in der Chemie an der Universität Frankfurt am Main (AIC), Frankfurt 1981.

References for Chapter 5

[5.1] *Croner's A–Z of Business Information*, Croner, Kingston-upon-Thames, UK 1989.
[5.2] S. Ball: *Directory of International Sources of Business Information*, Pitman, London 1989.
[5.3] P. and A. Foster (eds.): *The Online Business Sourcebook*, Headland Press, Headland, UK 1989.
[5.4] *Directory of Online Databases*, Cuandra Associates, Santa Monica, CA, issued quarterly.
[5.5] P. Millard: *Trade Associations and Professional Bodies of the UK*, Pergamon Press, Oxford 1969.
[5.6] P. Foster, A. Foster: *Online Business Sourcebook*, Headland Press, Headland, UK 1989.
[5.7] D. Lasok, J. W. Bridge: *Introduction to the Law and Institutions of the European Communities*, Butterworth, Seven Oaks, UK 1982.
[5.8] J. Jeffries: *Guide to the Official Publications of the European Communities*, Mansell, London 1981.

References for Chapter 6
General

[6.1] *Chemical Abstracts Service Statistical Summary 1907–1988*, Chemical Abstracts Service, Columbus, Ohio, CAS 1433, May 1989.
[6.2] B. M. Rimmer: *Guide to Official Industrial Property Publications*, The British Library, Science Reference Library, London 1985.
[6.3] J. Schade: *Patent-Tabelle*, 7th. ed, Heymanns Verlag, Köln–Berlin–Bonn–München 1990.
[6.4] "Standards and Recommendations Concerning Patent Documentation," in: *Patent Information and Documentation Handbook*, part 3, World Intellectual Property Organization, Geneva 1981.
[6.5] Reference [6.4], part 4.
[6.6] B. M. Rimmer: *Patent Information and Documentation in Western Europe. An Inventory of Services Available to the Public*, 3rd ed., Saur Verlag, München – London – New York – Paris 1989.
[6.7] *The Paris Convention for the Protection of Industrial Property from 1883 to 1983*, International Bureau of Intellectual Property, Geneva 1983.
[6.8] The German Democratic Republic, one of the member states of the Paris Convention, ceased to exit as of October 3, 1990.
[6.9] W. Pilch, W. Wratschko: "INPADOC: A Computerized Patent Documentation System," *J. Chem. Inf. Comput. Sci.* **18** (1978) 69.
[6.10] M. Julius et al.: "A Very Early Warning System for the Rapid Identification and Transfer of New Technology," *J. Am. Soc. Inf. Sci.* **28** (1977) 170.
[6.11] H. M. Allcock, J. W. Lotz: "Patent Intelligence and Technology – Gleaning Pseudoproprietary Information from Publicly Available Data," *J. Chem. Inf. Comput. Sci.* **18** (1978) 65.
[6.12] H. Mlodczik: "Patent Literature – a Tool for Forecasting in the Pharmaceutical Industry," *World Patent Inf.* **1** (1979) 219.
[6.13] H. Fendt: Blick in die Zukunft: Strategische Patentanalyse," *Wirtschaftswoche* no. 29, July 15, 1983.
[6.14] C. Oppenheim: "A Microcomputer Program for the Statistical Analysis of Patent Databases," *World Patent Inf.* **5** (1983) 209.
[6.15] S. M. Kaback: "Patent Statistics and Other Games," *World Patent Inf.* **6** (1984) 80.
[6.16] L. M. Fuld: *Competitor Intelligence*, J. Wiley & Sons, New York – Chichester – Brisbane – Toronto – Singapore 1985, p. 309.
[6.17] S. M. Kaback: "What's in a Patent? Information! But can I find it?" *J. Chem. Inf. Comput. Sci.* **24** (1984) 159.
[6.18] S. M. Kaback: "Access all the Information in Patents," *CHEMTECH* **15** (1985), 146.
[6.19] P. Claus: "The International Patent Classification," *World Patent Inf.* **2** (1980) 13.
[6.20] T. S. Eisenschitz: "Accuracy of Information Transfer Through Patent Classifications," *World Patent Inf.* **4** (1982) 18.
[6.21] J. M. Barnard (ed.): *Computer Handling of Generic Structures*, proceedings of a conference organized by

the Chemical Structure Association at the University of Sheffield, England, 26–29 March 1984.
[6.22] S. M. Kaback: "Retrieving Patent Information Online," *Online (Weston, Conn.)* **2** (1978) 16.
[6.23] E. S. Simmons, F. C. Rosenthal: "Patent Data bases: A Survey," *World Patent Inf.* **7** (1985) 33.
[6.24] C. S. Kulp: "Patent Databases. A survey of what is available from DIALOG, QUESTEL, SDC, PERGAMON and INPADOC, *Database* **1984**, 56.
[6.25] J. A. Silk: "Present and Future Prospects for Structural Searching of the Journal and Patent Literature," *J. Chem. Inf. Comput. Sci.* **19** (1979) 195.
[6.26] J. Stevenson: "The Use of Patent Information in Industry," *World Patent Inf.* **4** (1982) 164.
[6.27] E. S. Simmons: "The Paradox of Patentability Searching," *J. Chem. Inf. Comput. Sci.* **25** (1985) 379.
[6.28] P. Ochsenbein: "The Patent Documentation Group (PDG)," *World Patent Information* **9** (1987) 92.
[6.29] Meeting Report, *World Patent Information* **11** (1989) 48.

References for Chapter 7

[7.1] P. F. Burton, J. H. Petrie: *The Librarian's Guide to Microcomputers for Information Management*, Van Nostrand Reinhold (UK), Wokingham 1986.
[7.2] P. Zorkoczy: *Information Technology An Introduction*, Pitman, London 1985.
[7.3] P. C. Treleven, I. G. Lima, *Computer* **15** (1982) no. 8, 79–88.
[7.4] Her Majesty's Stationery Office (HMSO): *A Programme for Advanced Information Technology: the Report of the Alvey Committee*, HMSO, London 1982.
[7.5] P. Bishop: *Fifth Generation Computers*, Ellis Horwood, Chichester 1986.
[7.6] P. Salenieks: *Computing: The Next Generation*, Ellis Horwood, Chichester 1988.
[7.7] A. Peled, *Sci. Am.* **257** (1987) no. 4, 35–42.
[7.8] J. D. Meindl, *Sci. Am.* **257** (1987) no. 4, 54–62.
[7.9] G. C. Fox, P. C. Messina, *Sci. Am.* **257** (1987) no. 4, 44–52.
[7.10] J. D. Foley, *Sci. Am.* **257** (1987) no. 4, 82–90.
[7.11] D. Bawden, *J. Inf. Sci.* **11** (1985) 1–8.
[7.12] R. C. Rouse, *Program* **23** (1989) no. 3, 269–275.
[7.13] W. T. Wipke in W. A. Warr (ed.): "Graphics for Chemical Structures Integration with Text and Data," *ACS Symp. Ser.* **341** (1987).
[7.14] A. E. Cawkell, *The Electronic Library* **7** (1989) no. 1, 24–28.
[7.15] A. E. Cawkell, *The Electronic Library* **7** (1989) no. 2, 106–110.
[7.16] A. E. Cawkell, *The Electronic Library* **7** (1989) no. 3, 180–184.
[7.17] *Monitor* **42** (1984) 8.
[7.18] M. Rivett, *J. Inf. Sci.* **13** (1987) 25–34.
[7.19] D. H. Davies, *J. Am. Soc. Inf. Sci.* **39** (1988) no. 1, 34–42.
[7.20] E. M. Cichocki, S. M. Ziemer, *J. Am. Soc. Inf. Sci.* **39** (1988) no. 1, 43–46.
[7.21] T. Hendley, *Inf. Media Technology* **19** (1986) no. 3, 103–106.
[7.22] M. S. White, *World Patent Information* **8** (1986) no. 3, 177–181.
[7.23] L. B. Glass, *Byte* **14** (1989) no. 5, 283–289.
[7.24] S. Lambert, S. Ropiequet: *CD-ROM: The New Papyrus*, Microsoft Press, Redmond, WA 1986.
[7.25] C. Oppenheim (ed.): *CD-ROM Fundamentals to Applications*, Butterworth, London 1988.
[7.26] C. Sherman: *The CD-ROM Handbook*, McGraw-Hill, New York 1988.
[7.27] M. H. Kryder, *Sci. Am.* **257** (1987) no. 4, 72–81.
[7.28] *Electronic and Optical Publishing Review* **8** (1988) no. 2, 102–103.
[7.29] D. Pountain, *Byte* **14** (1989) no. 2, 274–280.
[7.30] G. A. Pierce, *Proc. Int. Soc. Optical Engineering* **899** (1988) 31–33.
[7.31] R. B. Barnes, F. J. Sukernick, *J. Information and Image Management* **19** (1986) no. 10, 34–38.
[7.32] G. I. Ouchi: *Personal Computers for Scientists*, American Chemical Society, Washington, DC 1986.
[7.33] R. Alberico: *Microcomputers for the Online Searcher*, Meckler, Westport, CT 1987.
[7.34] P. J. Denning, R. L. Brown, *Sci. Am.* **251** (1984) 80–89.
[7.35] N. Wirth, *Sci. Am.* **251** (1984) 48–57.
[7.36] L. G. Tesler, *Sci. Am.* **251** (1984) 58–66.
[7.37] M. Lesk, *Annu. Rev. Inf. Sci. Technol.* **19** (1984) 97–128.
[7.38] T. Oren, G. A. Kildall, *IEEE Spectrum* **4** (1986) 49–54.
[7.39] L. A. Kurtz, *Program* **18** (1984) 1–15.
[7.40] F. D. Gault in J. R. Humble, V. E. Hampel (eds.): *Data Base Management in Science and Technology*, Elsevier, Amsterdam 1984.
[7.41] J. H. Ashford, *Program* **18** (1984) 16–45.
[7.42] R. Kimberley (ed.): *Text Retrieval A Directory of Software*, 3rd ed., Gower, Aldershot 1990.
[7.43] P. F. Burton, H. Gates, *Program* **19** (1985) 1–19.
[7.44] J. H. Ashford, *Program* **18** (1984) 124–146.
[7.45] W. A. Warr in D. E. Meyer, W. A. Warr, R. A. Love (eds.): *Chemical Structure Software for Personal Computers*, American Chemical Society, Washington, DC, 1988, p.37.
[7.46] B. W. Boehm, *Computer* **20** (1987) no. 9, 43–57.
[7.47] I. Somerville: *Software Engineering*, Addison-Wesley, Wokingham 1985.
[7.48] D. M. Jennings et al., *Science (Washington, D.C.)* **231** (1986) 943–950.

References for Chapter 8

[8.1] R. J. Hartley, E. M. Keen, J. A. Large, L. A. Tedd: *Online Searching: Principles and Practice*, Bowker-Saur, London 1990.
[8.2] J. Convey, C. Bingley: *Online Information Retrieval. An Introductory Manual to Principles and Practice*, 3rd ed., Library Association, London 1989.
[8.3] G. Turpie: *Going Online, 1988*, Aslib, London 1988.
[8.4] C. H. Fenichel, T. H. Hogan: *Online Searching: A Primer*, Learned Information, Marlton, NJ, 1989.
[8.5] H. Stack: *Online Searching Made Simple*, PJB Publications, London 1988.
[8.6] *Online Searching in Science and Technology: An Introductory Guide to Equipment and Search Techniques*, British Library Science Reference and Information Service, London 1988.
[8.7] R. K. Summit in A. Kent, H. Lancourt (eds.): *Encyclopedia of Library and Information Science*, vol. 7, Marcel Dekker, New York 1972.
[8.8] C. A. Cuadra, *J. Chem. Inf. Comput. Sci.* **15** (1975) 48–51.
[8.9] W. A. Warr, A. R. Haygarth Jackson, *J. Chem. Inf. Comput. Sci.* **28** (1988) 68–72.
[8.10] D. T. Hawkins, L. R. Levy, *Online (Weston, Conn.)* **9** (1985) no. 6, 30–36.

[8.11] M. Morrison, *Online (Weston, Conn.)* **13** (1989) no. 4, 46–52.
[8.12] R. V. Janke, *Online (Weston, Conn.)* **7** (1983) no. 5, 12–29.
[8.13] K. Y. Marcaccio (ed.): *Computer-Readable Data bases: A Directory and Data Sourcebook*, Gale Research, Detroit 1989.
[8.14] *Directory of Online Databases*, Cuadra Associates, Santa Monica, CA, issued quarterly.
[8.15] D. M. Cipra, C. F. Damron, *Database* **8** (1985) no. 2, 23–30.
[8.16] R. L. M. Synge, *J. Chem. Inf. Comput. Sci.* **30** (1990) no. 1, 33–35.
[8.17] D. T. Hawkins, *Database* **8** (1985) no. 2, 31–41.
[8.18] Y. Wolman: *Chemical Information. A Practical Guide to Utilization*, Wiley-Interscience, Chichester 1988.
[8.19] R. E. Maizell: *How to Find Chemical Information A Guide for Practicing Chemists, Educators, and Students*, 2nd ed., John Wiley, New York 1987.
[8.20] M. Mücke: *Die Chemische Literatur. Ihre Erschließung und Benutzung*, VCH Verlagsgesellschaft, Weinheim 1982.
[8.21] A. B. Piternick, *Online Rev.* **13** (1989) no. 6, 457–476.
[8.22] H. Schulz: *From CA to CAS ONLINE*, VCH Verlagsgesellschaft, Weinheim 1988.
[8.23] J. Witiak, *Database* **11** (1988) no. 2, 95–96.
[8.24] K. D. Lehmann, H. Strohl-Goebel (eds.): *The Application of Microcomputers in Information, Documentation and Libraries*, Elsevier North-Holland, Amsterdam 1987.
[8.25] P. Leggate, H. Dyer, *The Electronic Library* **4** (1986) no. 1, 38–49.
[8.26] S. J. Kolner, *Online (Weston, Conn.)* **9** (1985) no. 1, 37–42.
[8.27] S. J. Kolner, *Online (Weston, Conn.)* **9** (1985) no. 2, 39–46.
[8.28] S. J. Kolner, *Online (Weston, Conn.)* **9** (1985) no. 3, 44–50.
[8.29] S. J. Kolner, *Online (Weston, Conn.)* **9** (1985) no. 4, 27–34.
[8.30] S. J. Kolner, *Online (Weston, Conn.)* **9** (1985) no. 6, 42–50.
[8.31] S. J. Kolner, *Online (Weston, Conn.)* **10** (1986) no. 4, 32–36.
[8.32] P. F. Burton, J. H. Petrie: *The Librarian's Guide to Microcomputers for Information Management*, Van Nostrand Reinhold (UK), Wokingham 1986.
[8.33] P. Nieuwenhuysen, *Online Rev.* **11** (1987) no. 6, 363–367.
[8.34] P. Nieuwenhuysen, *The Electronic Library* **6** (1988) no. 3, 168–172.
[8.35] L. R. Levy, D. T. Hawkins, *Online (Weston, Conn.)* **10** (1986) no. 1, 33–40.
[8.36] D. T. Hawkins, L. R. Levy, *Online (Weston, Conn.)* **10** (1986) no. 3, 49–58.
[8.37] W. A. Warr, *Database* **10** (1987) no. 3, 122–128.
[8.38] R. Walsh, *Aslib Information* **16** (1988) nos. 11/12, 282–283.
[8.39] S. Cisler, *Online (Weston, Conn.)* **12** (1988) no. 6, 99–102.
[8.40] *Monitor* **91** (1988) 5–8.
[8.41] C. Oppenheim, *Advanced Information Report*, March 1990, 7–9.
[8.42] C. A. Kehoe, *Online Rev.* **9** (1985) no. 6, 489–505.
[8.43] *The Electronic Library* **4** (1986) no. 1, 30–33.

[8.44] T. A. Hanson, *Aslib Proc.* **41** (1989) no. 9, 267–274.
[8.45] N. Hoyle, *Database* **10** (1987) no. 1, 73–78.
[8.46] A. N. Grosch, *Online Rev.* **12** (1988) no. 6, 375–386.
[8.47] C. Rodwell, P. Clayton, *Advanced Information Report* **10** (1988) no. 8, 9–11.
[8.48] J. K. Pemberton, *Online (Weston, Conn.)* **10** (1986) no. 3, 17–24.
[8.49] C. Tenopir, *Library J.* **1986**, 48–49.
[8.50] C. Tenopir, *Annu. Rev. Inf. Sci. Technol.* **19** (1984) 215–246.
[8.51] R. Summit, A. Lee, *Serials Rev.* **3** (1988) 7–10.
[8.52] J. A. Hearty, *Information Services and Use* **8** (1988) 93–105.
[8.53] *Kirk–Othmer*, 3rd ed., 1978–1984.
[8.54] E. W. Johnson, M. P. Kutz, *Electronic Publishing and Bookselling* **2** (1984) no. 1, 17–19.
[8.55] P. Willett, *Information Processing and Management* **24** (1988) 577–597.
[8.56] P. Willett (ed.): *Document Retrieval Systems*, Taylor Graham, London 1988.
[8.57] R. G. Lerner et al., *Annu. Rev. Inf. Sci. Technol.* **18** (1983) 127–149.
[8.58] D. P. Martinsen, R. A. Love, L. R. Garson, *Online (Weston Conn.)* **13** (1989) no. 2, 121–133.
[8.59] J. A. Hearty, *Information Services and Use* **8** (1988) 93–105.
[8.60] G. Tittlbach: "State of the Art Report," in: *Electronic Publishing*, Pergamon, Oxford 1987, p. 91.
[8.61] G. Tittlbach, *Nachr. Dok.* **37** (1986) nos. 4–5, 198–204.
[8.62] G. Tittlbach, *J. Chem. Inf. Comput. Sci.* **26** (1986) no. 1, 13–17.
[8.63] W. Detemple, *Online Rev.* **13** (1989) no. 2, 155–160.
[8.64] W. Niedermeyr in W. A. Warr (ed.): *Graphics for Chemical Structures. Integration with Text and Data*, ACS Symposium Series 341, American Chemical Society, Washington, D.C., 1987, p. 143.
[8.65] N. J. Thompson, *Online (Weston, Conn.)* **13** (1989) no. 3, 15–26.
[8.66] T. B. Chadwick, *Online (Weston, Conn.)* **13** (1989) no. 3, 28–30.
[8.67] J. Buckingham (ed.): *Dictionary of Organic Compounds*, 5th ed., Chapman and Hall, London 1982 and annual supplements thereafter.
[8.68] P. Hyams, *Information World Review*, July 1989, 18.
[8.69] W. Tuck, D. Archer, M.-C. Hayet, C. McKnight: *Project Quartet, LIR Report 76*, British Library, London 1990.
[8.70] W. Tuck, *Netlink* **5** (1989) no. 1, 5–8.
[8.71] R. M. Campbell, B. T. Stern, *Microcomputers for Information Management* **4** (1987) no. 2, 87–107.
[8.72] F. A. Mastroddi, J. Page in: *Electronic Publishing: State of the Art Report*, Pergamon Infotech, Maidenhead 1987, p. 37.
[8.73] A. Buscain, *Aslib Proc.* **37** (1985) nos. 6/7, 249–256.
[8.74] D. D. Baird, *Aslib Proc.* **37** (1985) nos. 6/7, 257–265.
[8.75] A. J. Metcalf, *Aslib Proc.* **37** (1985) nos. 6/7, 267–271.
[8.76] C. H. Jacobs, *Aslib Proc.* **37** (1985) nos. 6/7, 273–276.
[8.77] P. N. Hunter, *Aslib Proc.* **37** (1985) nos. 6/7, 277–280.

[8.78] W. R. Tuck, *Aslib Proc.* **38** (1985) no. 3, 85–92.
[8.79] R. Veith, *Annu. Rev. Inf. Sci. Technol.* **18** (1983) 3–28.
[8.80] R. Kimberley (ed.): *Text Retrieval A Directory of Software*, 3rd ed., Gower, Aldershot 1990.
[8.81] M. M. K. Hlava (ed.): Bulletin of the American Society for Information Science, Oct/Nov 1987, 14–27.
[8.82] C. Oppenheim (ed.): *CD-ROM Fundamentals to Applications*, Butterworth, London 1988.
[8.83] C. Sherman: *The CD-ROM Handbook*, McGraw-Hill, New York 1988.
[8.84] S. E. Arnold, L. Rosen: *Managing the New Electronic Information Products*, Riverside Data, Sudbury, MA, 1989.
[8.85] J. P. Roth (ed.): *CD-ROM Applications and Markets*, Meckler, Westport, CT 1988.
[8.86] J. Mitchell, J. Harrison (eds.): *The CD-ROM Directory 1990*, TFPL Publishing, London 1989.
[8.87] N. Desmaris (ed.): *CD-ROMs in Print 1990: An International Guide*, Meckler, Westport, CT, 1989.
[8.88] S. Oberlin, J. Cox: *The Microsoft CD-ROM Yearbook (1989–1990)*, Microsoft Press, Redmond, WA 1989.
[8.89] J. M. Burridge, *Biochem. Soc. Trans.* **17** (1989) 840–841.
[8.90] I. A. Penn et al.: *Records Management Handbook*, AIIM Publications, Silver Spring, MD 1989.
[8.91] P. Emmerson (ed.): *How to Manage Your Records: A Guide to Effective Practice*, ICSA, Cambridge 1989.
[8.92] S. James: *Records Management: An Introduction*, TFPL Publishing, London 1989.
[8.93] 1989 International Micrographics Source Book Including Related Imaging Technologies, Microfilm Publishing, Larchmont, NY 1989.
[8.94] R. J. Focarelli et al.: *The Microform Connection*, AIIM Publications, Silver Spring, MD, 1989.
[8.95] A. Shiel: *Optical Disk Storage and Document Image Processing: A Guide and Directory*, 2nd ed., Cimtech, Hatfield 1990.
[8.96] J. P. Roth, B. A. Berg: *Software for Optical Storage*, AIIM Publications, Silver Spring, MD, 1989.
[8.97] W. Saffady: *Optical Storage Technology 1989: A State of the Art Review*, Meckler, Westport, CT, 1989.
[8.98] A. E. Cawkell, *The Electronic Library* **7** (1989) no. 1, 24–28.
[8.99] A. E. Cawkell, *The Electronic Library* **7** (1989) no. 2, 106–110.
[8.100] A. E. Cawkell, *The Electronic Library* **7** (1989) no. 3, 180–184.
[8.101] A. E. Cawkell, *The Electronic Library* **7** (1989) no. 4, 248–250.
[8.102] A. E. Cawkell, *The Electronic Library* **7** (1989) no. 5, 317–323.
[8.103] R. G. Lerner et al., *Annu. Rev. Inf. Sci. Technol.* **18** (1983) 127–149.
[8.104] D. P. Martinsen, R. A. Love, L. R. Garson, *Online (Weston Conn.)* **13** (1989) no. 2, 121–133.
[8.105] W. Saffady: *Optical Disks versus Micrographics as Document Storage and Retrieval Technologies*, Meckler, Westport, Connecticut 1988.
[8.106] G. Salton, M. J. McGill: *Introduction to Modern Information Retrieval*, McGraw-Hill Computer Science Series, New York 1983.
[8.107] K. Sparck-Jones, *Information Processing and Management* **24** (1988) no. 6, 703–711.
[8.108] K. Sparck-Jones, J. I. Tait, *J. Doc.* **40** (1984) no. 1, 50–66.
[8.109] I. G. Hendry, P. Willett, F. E. Wood, *Program* **20** (1986) 245–263.
[8.110] I. G. Hendry, P. Willett, F. E. Wood, *Program* **20** (1986) 382–393.
[8.111] S. J. Wade, P. Willett, *Program* **22** (1988) 44–61.
[8.112] G. Salton, *Commun. ACM* **29** (1986) 648–656.
[8.113] G. Salton: *Automatic Text Processing: The Transformation, Analysis and Retrieval of Information*, Addison-Wesley, Reading 1989.
[8.114] G. Salton, C. Buckley, *J. Am Soc. Inf. Sci.* **41**, no. 4 (1990) 288–297.
[8.115] K. Sparck-Jones, *J. Inf. Sci.* **1** (1980) 325–332.
[8.116] C. W. Cleverdon, *Information Services and Uses* **4** (1984) 37–47.
[8.117] P. Willett, *J. Inf. Sci.* **15** (1989) nos. 4/5, 223–236.

References for Chapter 9

[9.1] J. E. Ash et al.: *Communication, Storage and Retrieval of Chemical Information*, Ellis Horwood, Chichester 1985.
[9.2] J. M. Barnard in C. Citroen, J. M. Griffith (eds.): *Perspectives in Information Management*, vol. 1, Butterworths, Guildford 1989.
[9.3] R. E. Maizell: *How to Find Chemical Information*, 2nd ed., Wiley, New York 1987, pp. 152–200.
[9.4] W. A. Warr (ed.): *Graphics for Chemical Structures: Integration with Text and Data*, ACS Symposium Series 341, American Chemical Society, Washington, DC, 1987.
[9.5] W. A. Warr (ed.): *Chemical Structures, The International Language of Chemistry*, Springer Verlag, Berlin 1988.
[9.6] D. E. Meyer, W. A. Warr, R. A. Love (eds.): *Chemical Structure Software for Personal Computers*, American Chemical Society, Washington, DC, 1988.
[9.7] W. A. Warr in W. A. Warr (ed.): *Chemical Structure Information Systems: Interfaces, Communication and Standards*, ACS Symposium Series 400, American Chemical Society, Washington, DC, 1989, p. 1.
[9.8] J. Figueras in B. W. Rossiter (ed.): *Physical Methods of Chemistry*, vol. 1: "Components of Scientific Instruments and Applications of Computers to Chemical Research," Wiley, New York 1986, p. 687.
[9.9] J. Buckingham (ed.): *Dictionary of Organic Compounds*, 5th ed., Chapman and Hall, London 1982 and annual supplements thereafter.
[9.10] H. P. Kollig, *Toxicol. Environ. Chem.* **17** (1988) 287–311.
[9.11] R. Luckenbach, R. Ecker, J. Sunkel, *Angew. Chem. Int. Ed. Engl.* **20** (1981) 841–849.
[9.12] C. Jochum in W. A. Warr (ed.): *Chemical Structures. The International Language of Chemistry*, Springer Verlag, Berlin 1988, p. 187.
[9.13] D. R. Lide, *Science (Washington, D. C)* **212** (1981) 1343–1349.
[9.14] F. H. Allen et al., *J. Appl. Crystallogr.* **7** (1974) 73–78.
[9.15] S. R. Heller, K. Scott, D. W. Bigwood, *J. Chem. Inf. Comput. Sci.* **29** (1989) 159–162.
[9.16] R. E. Maizell: *How to Find Chemical Information*, 2nd ed., Wiley, New York 1987, pp. 327–330.
[9.17] C. A. Shelley in J. Zupan (ed.): *Computer-Supported Spectroscopic Databases*, Ellis Horwood, Chichester 1986, p. 6.

[9.18] G. W. A. Milne et al., *Org. Mass Spectrom.* **17** (1982) no. 11, 547–552.
[9.19] J. R. Rumble, D. R. Lide, *J. Chem. Inf. Comput. Sci.* **25** (1985) 231–235.
[9.20] S. R. Heller, *J. Chem. Inf. Comput. Sci.* **25** (1985) 224–231.
[9.21] S. R. Heller, G. W. A. Milne, R. J. Feldmann, *Science (Washington, D. C)* **195** (1977) 253–259.
[9.22] G. W. A. Milne, C. L. Fisk, S. R. Heller, R. Potenzone, *Science (Washington, D. C)* **215** (1982) 371–375.
[9.23] O. Yamamoto et al., *Anal. Sci.* **4** (1988) no. 3, 233–239.
[9.24] S. R. Lowry, D. A. Huppler, C. R. Anderson, *J. Chem. Inf. Comput. Sci.* **25** (1985) 235–241.
[9.25] P. R. Griffiths, C. L. Wilkins, *Appl. Spectrosc.* **42** (1988) no. 4, 538–545.
[9.26] S. R. Heller, *Anal. Chem.* **44** (1972) 1951–1961.
[9.27] S. R. Heller, R. J. Feldmann, H. M. Fales, G. W. A. Milne, *J. Chem. Doc.* **13** (1973) 130–133.
[9.28] R. S. Heller, G. W. A. Milne, R. J. Feldmann, S. R. Heller, *J. Chem. Inf. Comput. Sci.* **16** (1976) 176–178.
[9.29] S. R. Heller, R. S. Heller, D. P. Martinsen, *Adv. Mass Spectrom.* **8B** (1980) 1578–1581.
[9.30] S. R. Heller in J. Zupan (ed.): *Computer-Supported Spectroscopic Databases*, Ellis Horwood, Chichester 1986, p. 118.
[9.31] D. D. Speck, R. Venkataraghavan, F. W. McLafferty, *Org. Mass Spectrom.* **13** (1978) no. 4, 209–213.
[9.32] F. W. McLafferty, R. H. Hertel, R. D. Villwock, *Org. Mass Spectrom.* **9** (1974) no. 7, 690–702.
[9.33] G. M. Pesyna, R. Venkataraghavan, H. E. Dayringer, F. W. McLafferty, *Anal. Chem.* **48** (1976) 1362–1368.
[9.34] I. K. Mun, R. Venkataraghavan, F. W. McLafferty, *Anal. Chem.* **49** (1977) 1723–1726.
[9.35] B. L. Atwater (Fell), R. Venkataraghavan, F. W. McLafferty, *Anal. Chem.* **51** (1979) 1945–1949.
[9.36] F. W. McLafferty et al., *Int. J. Mass Spectrom. Ion Phys.* **47** (1983) 317–319.
[9.37] S. Anderson, *J. Mol. Graphics* **2** (1984) 83–90.
[9.38] N. A. Farmer, M. P. O'Hara, *Database* **3** (1980) 10–25.
[9.39] D. E. Meyer in W. A. Warr (ed.): *Chemical Structures. The International Language of Chemistry*, Springer Verlag, Berlin 1988, p. 251.
[9.40] D. del Rey in W. A. Warr (ed.): *Graphics for Chemical Structures: Integration with Text and Data*, ACS Symposium Series 341, American Chemical Society, Washington, D.C. 1987, p. 48.
[9.41] W. G. Town, *Chem. Br.* **25** (1989) no. 11, 1118–1120.
[9.42] H. S. Hertz, R. A. Hites, K. Biemann, *Anal. Chem.* **43** (1971) 681–691.
[9.43] B. L. Atwater, D. B. Stauffer, F. W. McLafferty, D. W. Peterson, *Anal. Chem.* **57** (1985) 899–903.
[9.44] A. Hanna, J. C. Marshall, T. L. Isenhour, *J. Chromatogr. Sci.* **17** (1979) 434–440.
[9.45] M. D. Erickson, *Appl. Spectrosc.* **35** (1981) 181–184.
[9.46] C. W. Small, G. T. Rasmussen, T. L. Isenhour, *Appl. Spectrosc.* **33** (1979) 444–450.
[9.47] M. F. Delaney, P. C. Uden, *Anal. Chem.* **51** (1979) 1242–1243.
[9.48] J. A. de Haseth, L. V. Azarraga, *Anal. Chem.* **53** (1981) 2292–2295.
[9.49] G. W. Milne, S. R. Heller, *J. Chem. Inf. Comput. Sci.* **20** (1980) 204–211.
[9.50] S. R. Lowry, D. A. Huppler, *Anal. Chem.* **53** (1981) 889–893.
[9.51] S. R. Lowry, D. A. Huppler, *Anal. Chem.* **55** (1983) 1288–1291.
[9.52] J. T. Clerc, R. Schwarzenbach, J. J. Meili, H. Koenitzer, *Org. Magn. Reson.* **8** (1976) no. 1, 11–16.
[9.53] A. B. Wagner, *Online Rev.* **10** (1986) no. 3, 173–183.
[9.54] M. Passlack, W. Bremser in J. Zupan (ed.): *Computer-Supported Spectroscopic Databases*, Ellis Horwood, Chichester 1986, p. 92.
[9.55] W. Bremser, *Angew. Chem. Int. Ed. Engl.* **27** (1988) 247–260.
[9.56] W. Bremser, R. Neudert, *Eur. Spectrosc. News* **75** (1987) 10–27.
[9.57] R. Neudert, W. Bremser, H. Wagner, *Org. Mass Spectrom.* **22** (1987) 321–329.
[9.58] H. Damen, D. Henneberg, B. Weimann, *Anal. Chim. Acta* **103** (1978) 289–302.
[9.59] H. Kalchhauser, W. Robien, *J. Chem. Inf. Comput. Sci.* **25** (1985) 103–108.
[9.60] *Crystallographic Databases*, International Union of Crystallography, Chester 1987.
[9.61] O. Kennard et al., *Chem. Br.* **11** (1975) 213–216.
[9.62] C. Kratky, W. Robien, *Österr. Chem. Z.* **89** (1988) no. 3, 58–62.
[9.63] S. Bellard in: *Crystallographic Databases*, International Union of Crystallography, Chester 1987, p. 39.
[9.64] F. H. Allen et al., *Acta Crystallogr. Sect. B* **B35** (1979) 2331–2339.
[9.65] E. E. Abola, F. C. Bernstein, T. F. Koetzle, in P. J. Glaeser (ed.): *The Role of Data in Scientific Progress*, Elsevier, New York 1985.
[9.66] I. D. Brown, *Acta Crystallogr. Sect. A* **A41** (1985) 399.
[9.67] A. D. Mighell, V. L. Himes, *Acta Crystallogr. Sect. A* **A42** (1986) 101–105.
[9.68] G. Bergerhoff, R. Hundt, R. Sievers, I. D. Brown, *J. Chem. Inf. Comput. Sci.* **23** (1983) 66–69.
[9.69] M. J. E. Sternberg, S. A. Islam, *Biochem. Soc. Trans.* **17** (1989) 845–847.
[9.70] H. W. Mewes, A. Elzanowski, D. G. George, *Biochem. Soc. Trans.* **17** (1989) 843–845.
[9.71] *Blaise Newsletter* No. 93 (1988) 19–20.
[9.72] CODATA Conference on Scientific and Technical Data in a New Era, 26–29 Sept. 1988, Karlsruhe, FRG.
[9.73] R. E. Buntrock, *Database* **13** (1990) no. 3, 99–100.
[9.74] S. R. Heller (ed.): *The Beilstein Online Database: Implementation, Content, and Retrieval*, ACS Symposium Series 436, American Chemical Society, Washington 1990.
[9.75] D. J. Huddart, *Aslib Proc.* **40** (1988) no. 5, 133–137.
[9.76] D. T. Hawkins, *Database* **8** (1985) no. 2, 31–41.
[9.77] F. C. Allan, W. R. Ferrell, *Database* **12** (1989) no. 3, 50–58.
[9.78] C. Dutheuil, *Newsidic* **91** (1988) 9–12.
[9.79] *Online Rev.* **14** (1990) no. 1, 54.
[9.80] *The Electronic Library* **6** (1988) no. 5, 381.
[9.81] *DECHEMA Chemistry Data Series*, Deutsche Gesellschaft für Chemisches Apparatewesen, Chemische Technik und Biotechnologie e.V., Frankfurt, 10 volumes.
[9.82] D. Bawden, *Aslib Proc.* **40** (1988) no. 3, 79–85.
[9.83] F. E. Wood, A. T. Berrie, H. R. Plampin, M. L. Wilkinson-Tough, *J. Inf. Sci.* **15** (1989) 269–276.
[9.84] O. Norager in W. A. Warr (ed.): *Chemical Structures: The International Language of Chemistry*, Springer Verlag, Berlin 1988.

[9.85] S. R. Heller, *J. Chem. Inf. Comput. Sci.* **29** (1989) no. 2, 135–136.
[9.86] Y. Wolman: *Chemical Information. A Practical Guide to Utilization*, Wiley, Chichester 1983, Chap. 6.
[9.87] S. V. Meschel, *Online Rev.* **8** (1984) no. 1, 77–101.

References for Chapter 10

[10.1] W. A. Warr in R. Lees, A. Smith (eds.): *Chemical Nomenclature Usage*, Ellis Horwood, Chichester 1983, p.124.
[10.2] P. Willett, *J. Chemom.* **1** (1987) 139–155.
[10.3] J. Ash et al., (eds.): *Communication, Storage and Retrieval of Chemical Information*, Ellis Horwood, Chichester 1985, Chap. 5, p. 128.
[10.4] J. Figueras in B. W. Rossiter (ed.): *Physical Methods of Chemistry*, vol. 1: "Components of Scientific Instruments and Applications of Computers to Chemical Research," Wiley, New York 1986, p. 687.
[10.5] D. Bawden, T. Devon, *Database* **3** (1980) no. 3, 29–39.
[10.6] S. M. Kaback, *J. Chem. Inf. Comput. Sci.* **20** (1980) 1–6.
[10.7] R. Fugmann in J. E. Ash, E. Hyde (eds.): *Chemical Information Systems*, Ellis Horwood, Chichester 1975, p. 195.
[10.8] C. Suhr, E. von Harsdorf, W. Dethlefsen in J. M. Barnard (ed.): *Computer Handling of Generic Chemical Structures*, Gower, Aldershot 1984, p.10.
[10.9] U. Schoch-Grübler, *Online Rev.* **14** (1990) no. 2, 95–108.
[10.10] P. A. Baker, G. Palmer, P. W. L. Nichols in J. E. Ash, E. Hyde (eds.): *Chemical Information Systems*, Ellis Horwood, Chichester 1975, p.97.
[10.11] W. J. Wiswesser, *J. Chem. Inf. Comput. Sci.* **22** (1982) 88–93.
[10.12] G. Palmer, *Chem. Br.* **6** (1970) 422.
[10.13] G. M. Dyson in J. E. Ash, E. Hyde (eds.): *Chemical Information Systems*, Ellis Horwood, Chichester 1975, p.130.
[10.14] C. E. Granito, M. D. Rosenberg, *J. Chem. Doc.* **11** (1971) 251–256. E. Garfield, M. Sim, *Pure Appl. Chem.* **49** (1977) 1803.
[10.15] D. R. Eakin in J. E. Ash, E. Hyde (eds.): *Chemical Information Systems*, Ellis Horwood, Chichester 1975, p.227.
[10.16] D. R. Eakin, E. Hyde, G. Palmer, *Pestic. Sci.* **5** (1974) 319–326.
[10.17] E. E. Townsley, W. A. Warr in W. J. Howe, M. M. Milne, A. F. Pennell (eds.): *Retrieval of Medicinal Chemical Information*, ACS Symposium Series 84, American Chemical Society, Washington, D.C., 1978, p.73.
[10.18] W. A. Warr in: *Proceedings of the 5th International Online Information Meeting*, Learned Information, Oxford 1981, p.391.
[10.19] W. A. Warr, *J. Mol. Graphics* **4** (1986) 165–169.
[10.20] D. J. Weininger, *J. Chem. Inf. Comput. Sci.* **28** (1988) 31–36.
[10.21] J. E. Ash in J. E. Ash, E. Hyde (eds.). *Chemical Information Systems*, Ellis Horwood, Chichester 1975, p.156.
[10.22] H. Bebak et al., *J. Chem. Inf. Comput. Sci.* **29** (1989) 1–5.
[10.23] J. M. Barnard, *J. Chem. Inf. Comput. Sci.* **30** (1990) 81–96.
[10.24] G. Moreau, *Nouv. J. Chim.* **4** (1980) 17–22.
[10.25] W. T. Wipke, T. M. Dyott, *J. Am. Chem. Soc.* **96** (1974) 4825–4834.
[10.26] M. F. Lynch, J. Orton, W. G. Town, *J. Chem. Soc. C* **1969**, 1732–1736.
[10.27] J. H. R. Bragg, M. F. Lynch, W. G. Town, *J. Chem. Doc.* **10** (1970) 125–128.
[10.28] R. G. Freeland, S. A. Funk, L. J. O'Korn, G. A. Wilson, *J. Chem. Inf. Comput. Sci.* **19** (1979) 94–97.
[10.29] W. T. Wipke, S. K. Krishnan, G. I. Ouchi, *J. Chem. Inf. Comput. Sci.* **18** (1978) 32–37.
[10.30] D. Bawden et al., *J. Chem. Inf. Comput. Sci.* **21** (1981) 83–86.
[10.31] J. M. Barnard in W. A. Warr (ed.): *Chemical Structures, The International Language of Chemistry*, Springer Verlag, Berlin 1988, p.113.
[10.32] M. F. Lynch in J. E. Ash, E. Hyde (eds.): *Chemical Information Systems*, Ellis Horwood, Chichester 1975, p.177.
[10.33] J. E. Crowe, M. F. Lynch, W. G. Town, *J. Chem. Soc. C* **1970**, 990–996.
[10.34] G. W. Adamson, M. F. Lynch, W. G. Town, *J. Chem. Soc. C* **1971**, 3702–3706.
[10.35] G. W. Adamson et al., *J. Chem. Doc.* **13** (1973) 153–157.
[10.36] G. W. Adamson, S. E. Creasey, J. P. Eakins, M. F. Lynch, *J. Chem. Soc. Perkin Trans.* **1973**, 2071–2076.
[10.37] W. Graf, H. K. Kaindl, H. Kniess, R. Warszawski, *J. Chem. Inf. Comput.* **22** (1982) 177–181.
[10.38] P. G. Dittmar et al., *J. Chem. Inf. Comput. Sci.* **23** (1983) 93–102.
[10.39] R. Attias, *J. Chem. Inf. Comput. Sci.* **23** (1983) 102–108.
[10.40] L. C. Ray, R. A. Kirsch, *Science (Washington, D. C.)* **126** (1957) 814–819.
[10.41] S. H. Unger, *Commun. ACM* **7** (1964) 26–34.
[10.42] E. H. Sussenguth, *J. Chem. Doc.* **5** (1965) 36–43.
[10.43] J. Figueras, *J. Chem. Doc.* **12** (1972) 237–244.
[10.44] A. von Scholley, *J. Chem. Inf. Comput. Sci.* **24** (1984) 235–241.
[10.45] V. J. Gillet et al., *J. Chem. Inf. Comput. Sci.* **26** (1986) 118–126.
[10.46] S. R. Heller, *J. Chem. Inf. Comput. Sci.* **25** (1985) 224–231.
[10.47] G. W. A. Milne, C. L. Fisk, S. R. Heller, R. Potenzone, *Science (Washington D.C.)* **215** (1982) 371–375.
[10.48] G. W. Milne, S. R. Heller, *J. Chem. Inf. Comput. Sci.* **20** (1980) 204–211.
[10.49] S. R. Heller, G. W. A. Milne, R. J. Feldmann, *Science (Washington, D. C.)* **195** (1977) 253–259.
[10.50] R. J. Feldmann, *J. Chem. Inf. Comput. Sci.* **17** (1977) 157–163.
[10.51] G. W. A. Milne et al., *J. Chem. Inf. Comput. Sci.* **18** (1978) no. 4, 181–185.
[10.52] M. Z. Nagy, S. Kozics, T. Veszpremi, P. Bruck in W. A. Warr (ed.): *Chemical Structures, The International Language of Chemistry*, Springer Verlag, Berlin 1988, p.127.
[10.53] J. M. Barnard (ed.): *Computer Handling of Generic Chemical Structures*, Gower, Aldershot 1984.
[10.54] J. M. Barnard, *Database* **10** (1987) no. 3, 27–34.
[10.55] M. F. Lynch, *World Patent Inf.* **8** (1986) 85–91.

[10.56] M. F. Lynch, J. M. Barnard, S. M. Welford, *J. Chem. Inf. Comput. Sci.* **25** (1985) 264–270.

[10.57] G. Downs, V. Gillet, J. Holliday, M. F. Lynch in W. A. Warr (ed.): *Chemical Structures, The International Language of Chemistry*, Springer Verlag, Berlin 1988, p.151.

[10.58] J. M. Barnard, *J. Chem. Inf. Comput. Sci.* **21** (1981) 151–161.

[10.59] J. M. Barnard, M. F. Lynch, S. M. Welford, *J. Chem. Inf. Comput. Sci.* **24** (1984) 66–71.

[10.60] J. M. Barnard, M. F. Lynch, S. M. Welford, *J. Chem. Inf. Comput. Sci.* **22** (1982) 160–164.

[10.61] S. M. Welford, M. F. Lynch, J. M. Barnard, *J. Chem. Inf. Comput. Sci.* **21** (1981) 161–168.

[10.62] S. M. Welford, M. F. Lynch, J. M. Barnard, *J. Chem. Inf. Comput. Sci.* **24** (1984) 66–71.

[10.63] V. J. Gillet et al., *J. Chem. Inf. Comput. Sci.* **27** (1987) 126–137.

[10.64] American Chemical Society, US 4 642 762, 1987 (W. Fisanick).

[10.65] T. Nakayama, Y. Fujiwara, *J. Chem. Inf. Comput. Sci.* **23** (1983) 80–87.

[10.66] C. Fricke, I. Nickelsen, R. Fugmann, J. Sander, *Tetrahedron Comput. Methodol.* **2** (1989) no. 3, 167–175.

[10.67] K. E. Shenton, P. Norton, E. A. Fearns in W. A. Warr (ed.): *Chemical Structures, The International Language of Chemistry*, Springer Verlag, Berlin 1988, p.169.

[10.68] W. Fisanick, *J. Chem. Inf. Comput. Sci.* **30** (1990) 145–154.

[10.69] F. H. Allen, M. F. Lynch, *Chem. Br.* **25** (1989) no. 11, 1101–1104, 1108.

[10.70] P. Gund, *Prog. Mol. Subcell. Biol.* **5** (1977) 117–143.

[10.71] Y. C. Martin, M. G. Bures, P. Willett in K. B. Lipkowitz, D. B. Boyd (eds.): *Reviews in Computational Chemistry*, VCH, New York 1990, p. 213–256.

[10.72] R. P. Sheridan et al., *J. Chem. Inf. Comput. Sci.* **29** (1989) 255–260.

[10.73] A. T. Brint, E. Mitchell, P. Willett in W. A. Warr (ed.): *Chemical Structures, The International Language of Chemistry*, Springer Verlag, Berlin 1988, p.131.

[10.74] S. E. Jakes, P. Willett, *J. Mol. Graphics* **4** (1986) 12–20.

[10.75] A. T. Brint, P. Willett, *J. Mol. Graphics* **5** (1987) 49–56.

[10.76] A. T. Brint, P. Willett, *J. Chem. Inf. Comput. Sci.* **27** (1987) 152–158.

[10.77] C. W. Crandall, D. H. Smith, *J. Chem. Inf. Comput. Sci.* **23** (1983) 186–197.

[10.78] A. Rusinko III et al., *J. Chem. Inf. Comput. Sci.* **29** (1989) 251–255.

[10.79] J. H. Van Drie, D. Weininger, Y.C. Martin, *J. Comput.-Aided Mol. Des.* **3** (1989) no. 3, 225–251.

[10.80] R. E. Carhart, D. H. Smith, R. Venkataraghavan, *J. Chem. Inf. Comput. Sci.* **25** (1985) 64–73.

[10.81] P. Willett, V. Winterman, D. Bawden, *J. Chem. Inf. Comput. Sci.* **26** (1986) 36–41.

[10.82] D. Bawden in W. A. Warr (ed.): *Chemical Structures, The International Language of Chemistry*, Springer Verlag 1988, p. 145.

[10.83] P. Willett: *Similarity and Clustering in Chemical Information Systems*, Research Studies Press, Letchworth 1987.

[10.84] P. Willett, V. Winterman, D. Bawden, *J. Chem. Inf. Comput. Sci.* **26** (1986) 109–118.

[10.85] N. Farmer et al. in W. A. Warr (ed.): *Chemical Structures, The International Language of Chemistry*, Springer Verlag, Berlin 1988, p. 283.

[10.86] C. Jochum, T. Worbs in W. A. Warr (ed.): *Chemical Structures, The International Language of Chemistry*, Springer Verlag, Berlin 1988, p. 279.

[10.87] P. Willett, *J. Inf. Sci.* **15** (1989) nos. 4/5, 223–236.

[10.88] C. A. Pogue, E. M. Rasmussen, P. Willett, *Parallel Computing* **8** (1988) 399–407.

[10.89] E. M. Rasmussen, G. M. Downs, P. Willett, *J. Comput. Chem.* **9** (1988) no. 4, 378–386.

[10.90] H. M. Grindley et al. in H. R. Collier (ed.): *Chemical Information, Information in Chemistry, Pharmacology and Patents*, Springer Verlag, Berlin 1989, p. 253.

[10.91] P. J. Artymiuk, D. W. Rice, E. M. Mitchell, P. Willett, *J. Inf. Sci.* **15** (1989) 287–298.

[10.92] W. T. Wipke, D. Rogers, *J. Chem. Inf. Comput. Sci.* **24** (1984) 255–262.

[10.93] A. T. Brint, P. Willett, *J. Mol. Graphics* **5** (1987) 200–207.

[10.94] A. T. Brint et al., *Parallel Computing* **8** (1988) 295–300.

[10.95] W. T. Wipke in W. A. Warr (ed.): "Graphics for Chemical Structures Integration with Text and Data," *ACS Symp. Ser.* **341** (1987).

[10.96] W. J. Howe, T. R. Hagadone, *J. Chem. Inf. Comput. Sci.* **22** (1982) 8–15.

[10.97] W. J. Howe, T. R. Hagadone, *J. Chem. Inf. Comput. Sci.* **22** (1982) 182–186.

[10.98] D. Bawden, *Chem. Br.* **25** (1989) no. 11, 1107–1108.

[10.99] T. R. Hagadone in W. A. Warr (ed.): *Chemical Structures, The International Language of Chemistry*, Springer Verlag, Berlin 1988, p. 23.

[10.100] W. G. Town, *Chem. Br.* **25** (1989) no. 11, 1118–1120.

[10.101] W. G. Town in W. A. Warr (ed.): *Chemical Structures, The International Language of Chemistry*, Springer Verlag, Berlin 1988, p. 243.

[10.102] W. A. Warr (ed.): *Chemical Structure Information Systems: Interfaces, Communication and Standards*, ACS Symposium Series 400, American Chemical Society, Washington, DC, 1989.

[10.103] N. A. Farmer, M. P. O'Hara, *Database* **3** (1980) 10–25.

[10.104] Y. Wolman: *Chemical Information. A Practical Guide to Utilization*, Wiley-Interscience, Chichester 1988.

[10.105] H. R. Pichler in H. Schulz (ed.): *From CA to CAS Online*, VCH, Weinheim 1988, p. 146.

[10.106] S. V. Kasparek: *Computer Graphics and Chemical Structures*, Wiley-Interscience, New York 1990.

[10.107] R. Buntrock, *Database* **11** (1988) no. 1, 87–88.

[10.108] J. Mockus, R. E. Stobaugh, *J. Chem. Inf. Comput. Sci.* **20** (1980) 18–22.

[10.109] H. R. Pichler, *Chem. Labor Betr.* **34** (1983) no. 5, 188–196.

[10.110] U. Jordis, O. Oberhauser, *Österr. Chem. Z.* **1982**, 311–314.

[10.111] S. R. Heller in: *Proceedings of the 7th International Online Information Meeting*, Learned Information, Oxford 1983, p. 81.

[10.112] A. Meurling, *Database* **13** (1990) no. 1, 54–63.

[10.113] M. G. Hicks, C. Jochum, *J. Chem. Inf. Comput. Sci.* **30** (1990) 191–199.

[10.114] J. Ash et al., (eds.): *Communication, Storage and Retrieval of Chemical Information*, Ellis Horwood, Chichester 1985, Chap. 7, p. 182.

[10.115] A. P. Lurie in W. A. Warr (ed.): *Chemical Structures, The International Language of Chemistry*, Springer Verlag, Berlin 1988, p. 77.

[10.116] R. E. Maizell: *How to Find Chemical Information A Guide for Practicing Chemists, Educators, and Students*, 2nd ed., John Wiley, New York 1987.

[10.117] A. B. Wagner, *Online Rev.* **10** (1986) no. 3, 173–183.

[10.118] G. W. A. Milne in: *Proceedings of the 7th International Online Meeting*, Learned Information, Oxford 1983, p. 99.

[10.119] G. W. A. Milne in W. A. Warr (ed.): *Graphics for Chemical Structures: Integration with Text and Data*, ACS Symposium Series 341, American Chemical Society, Washington, DC, 1987, p. 102.

[10.120] J. R. McDaniel, A. E. Fein in W. A. Warr (ed.): *Graphics for Chemical Structures: Integration with Text and Data*, ACS Symposium Series 341, American Chemical Society, Washington, DC, 1987, p. 62.

[10.121] J. E. Dubois, H. Viellard, *Bull. Soc. Chim. Fr.* **1968**, 900–919.

[10.122] J. E. Dubois, J. P. Anselmini, M. Chastrette, F. Hennequin, *Bull. Soc. Chim. Fr.* **1969**, 2439–2448.

[10.123] J. E. Dubois, D. Laurent, *Bull. Soc. Chim. Fr.* **1969**, 2449–2455.

[10.124] J. E. Dubois, H. Viellard, *Bull. Soc. Chim. Fr.* **1971**, 839–848.

[10.125] J. E. Dubois, *J. Chem. Doc.* **13** (1973) 8–13.

[10.126] G. Bauer in: *Proceedings of the 5th International Online Information Meeting*, Learned Information, Oxford 1981, p. 377.

[10.127] J.-P. Gay, G. Auneveux, F. Chabernaud in W. A. Warr (ed.): *Chemical Structure Information Systems, Interfaces, Communication and Standards*, ACS Symposium Series 400, American Chemical Society, Washington, DC, 1989, p. 89.

[10.128] J.-P. Gay, H. Alardo in H. R. Collier (ed.): *Chemical Information: Information in Chemistry, Pharmacology and Patents*, Springer Verlag, Berlin 1989, p. 221.

[10.129] A. J. C. M. de Jong, A. M. C. Deibel in W. A. Warr (ed.): *Chemical Structures, The International Language of Chemistry*, Springer Verlag 1988, p. 45.

[10.130] P. Bruck, M. Z. Nagy, S. Kozics in: *Proceedings of the 11th International Online Information Meeting*, Learned Information, Oxford 1987, p. 41.

[10.131] S. R. Heller in: *Proceedings of the 11th International Online Information Meeting*, Learned Information, Oxford 1987, p. 25.

[10.132] S. Anderson, *J. Mol. Graphics* **2** (1984) 83–90.

[10.133] G. W. Adamson, J. M. Bird, G. Palmer, W. A. Warr, *J. Chem. Inf. Comput. Sci.* **25** (1985) 90–92.

[10.134] G. W. Adamson, J. M. Bird, G. Palmer, W. A. Warr, *J. Mol. Graphics* **4** (1986) 165–169.

[10.135] S. Barcza, L. A. Kelly, S. S. Wahrman, R. E. Kirschenbaum, *J. Chem. Inf. Comput. Sci.* **25** (1985) 55–59.

[10.136] S. Barcza, H. W. Mah, M. H. Myers, S. S. Wahrman, *J. Chem. Inf. Comput. Sci.* **26** (1986) 198–204.

[10.137] T. M. Johns in W. A. Warr (ed.): *Graphics for Chemical Structures: Integration with Text and Data*, ACS Symposium Series 341, American Chemical Society, Washington, DC 1987, p. 18.

[10.138] A. P. Lurie in W. A. Warr (ed.): *Chemical Structures, The International Language of Chemistry*, Springer Verlag, Berlin 1988, p. 77.

[10.139] W. A. Warr in: *Proceedings of the 7th International Online Information Meeting*, Learned Information, Oxford 1983, p. 91.

[10.140] T. Legatt, A. Saltzman, *Drug Inf. J.* **20** (1986) 51–56.

[10.141] W. T. Wipke, T. M. Dyott, *J. Am. Chem. Soc.* **96** (1974) 4834–4842.

[10.142] A. J. Stuper, P. C. Jurs, *J. Chem. Inf. Comput. Sci.* **16** (1976) 99–105.

[10.143] A. J. Stuper, W. E. Brugger, P. C. Jurs: *Computer-Assisted Studies of Chemical Structure and Biological Function*, Wiley, New York 1979.

[10.144] D. Magrill in W. A. Warr (ed.): *Chemical Structures, The International Language of Chemistry*, Springer Verlag, Berlin 1988, p. 53.

[10.145] L. Domokos, C. Jochum, H. Maier in H. R. Collier (ed.): *Chemical Information: Information in Chemistry, Pharmacology and Patents*, Springer Verlag, Berlin 1989, p. 191.

[10.146] D. Bawden et al. in W. A. Warr (ed.): *Chemical Structures, The International Language of Chemistry*, Springer Verlag, Berlin 1988, p. 63.

[10.147] J. Kao, V. Day, L. Watt, *J. Chem. Inf. Comput. Sci.* **25** (1985) no. 2, 129–135.

[10.148] H. K. Kaindl in H. R. Collier (ed.): *Chemical Information: Information in Chemistry, Pharmacology and Patents*, Springer Verlag, Berlin 1989, p. 63.

[10.149] D. E. Meyer, W. A. Warr, R. A. Love (eds.): *Chemical Structure Software for Personal Computers*, American Chemical Society, Washington, DC, 1988.

[10.150] J. F. Barstow, D. del Rey, J. S. Laufer, *Am. Lab. (Fairfield, Conn.)* **20** (1988) no. 7, 82–85.

[10.151] C. K. Gerson, R. A. Love, *Anal. Chem.* **59** (1987) no. 17, 1031A–1048A.

[10.152] D. E. Meyer in W. A. Warr (ed.): *Chemical Structures: The International Language of Chemistry*, Springer Verlag, Berlin 1988, p. 251.

[10.153] D. del Rey in W. A. Warr (ed.): *Graphics for Chemical Structures: Integration with Text and Data*, ACS Symposium Series 341, American Chemical Society, Washington, DC, 1987, p. 48.

[10.154] C. Seiter, P. Cohan, *Int. Lab.* **17** (1987) no. 7, 62–67.

[10.155] W. G. Town, *Chem. Br.* **25** (1989) no. 11, 1118–1120.

[10.156] W. G. Town in: *Proceedings of the 11th International Online Information Meeting*, Learned Information, Oxford 1987, p. 33.

[10.157] W. A. Warr, M. P. Wilkins, *Online (Weston, Conn.)* **14** (1990) no. 3, 50–54.

[10.158] J. M. Barnard, C. J. Jochum, S. M. Welford in W. A. Warr (ed.): *Chemical Structure Information Systems, Interfaces, Communication, and Standards*, ACS Symposium Series 400, American Chemical Society, Washington, DC, 1989, p. 76.

[10.159] S. R. Heller, *Database* **10** (1987) no. 4, 47–52.

[10.160] R. Franke: *Theoretical Drug Design Methods*, Elsevier, Amsterdam 1984.

[10.161] J. G. Topliss (ed.): *Quantitative Structure-Activity Relationships of Drugs*, Academic Press, New York 1983.
[10.162] Y. C. Martin, *J. Med. Chem.* **24** (1981) 229–237.
[10.163] J. L. Fauchere (ed.): *QSAR: Quantitative Structure-Activity Relationships in Drug Design*, Alan R. Liss, New York 1989.
[10.164] S. Borman, *Chem. Eng. News*, Feb. 19, 1990, 20–23.
[10.165] S. M. Free, J. W. Wilson, *J. Med. Chem.* **7** (1964) 395–399.
[10.166] T. Fujita, T. Ban, *J. Med. Chem.* **14** (1971) 148–152.
[10.167] C. Hansch, *Acc. Chem. Res.* **2** (1969) 232–239.
[10.168] S. H. Unger: *Consequences of the Hansch Paradigm for the Pharmaceutical Industry*, Academic Press, New York 1980.
[10.169] C. Hansch, *Drug Dev. Res.* **1** (1981) 267–309.
[10.170] C. Hansch, A. J. Leo: *Substituent Constants for Correlation Analysis in Chemistry and Biology*, Wiley Interscience, New York 1979.
[10.171] Y. C. Martin: *Quantitative Drug Design. A Critical Introduction*, Marcel Dekker, New York 1978.
[10.172] Y. C. Martin in E. J. Ariens (ed.): *Drug Design*, Academic Press, New York 1979, p. 8.
[10.173] P. C. Jurs, T. L. Isenhour: *Chemical Applications of Pattern Recognition*, Wiley, New York 1975.
[10.174] P. C. Jurs, T. R. Stouch, M. Czerwinski, J. N. Narvaez, *J. Chem. Inf. Comput. Sci.* **25** (1985) 296–308.
[10.175] D. Bawden, *J. Chem. Inf. Comput. Sci.* **23** (1983) 14–22.
[10.176] R. D. Cramer, G. Redl, C. E. Berkoff, *J. Med. Chem.* **17** (1974) 533–535.
[10.177] L. Hodes, G. F. Hazard, R. I. Geran, S. Richman, *J. Med. Chem.* **20** (1977) 469–475.
[10.178] L. Hodes, *J. Chem. Inf. Comput. Sci.* **21** (1981) 128–132.
[10.179] L. Hodes, *J. Chem. Inf. Comput. Sci.* **21** (1981) 132–136.
[10.180] G. W. Adamson, D. Bawden, *J. Chem. Inf. Comput. Sci.* **15** (1975) 215–220.
[10.181] G. W. Adamson, D. Bawden, *J. Chem. Inf. Comput. Sci.* **16** (1976) 161–165.
[10.182] G. W. Adamson, D. Bawden, *J. Chem. Inf. Comput. Sci.* **17** (1977) 164–171.
[10.183] G. W. Adamson, D. Bawden, *J. Chem. Inf. Comput. Sci.* **20** (1980) 97–100.
[10.184] G. W. Adamson, D. Bawden, D. T. Saggers, *Pestic. Sci.* **15** (1984) 31–39.
[10.185] G. W. Adamson, J. A. Bush, *Nature (London)* **248** (1974) 406–407.
[10.186] G. W. Adamson, J. A. Bush, *J. Chem. Soc. Perkin Trans.* **1** (1976) 168–172.
[10.187] R. Langridge, T. E. Ferrin, I. D. Kuntz, M. L. Connolly, *Science (Washington, D. C.)* **211** (1981) 661–666.
[10.188] C. Humblet, G. R. Marshall, *Drug Dev. Res.* **1** (1981) 409–434.
[10.189] P. Gund, J. D. Andose, J. B. Rhodes, G. M. Smith, *Science (Washington, D. C.)* **208** (1980) 1425–1431.
[10.190] A. J. Hopfinger, *J. Med. Chem.* **28** (1985) 1133–1139.
[10.191] N. C. Cohen, *Adv. Drug Res.* **14** (1985) 42–145.
[10.192] S. H. Unger, *Drug Inf. J.* **21** (1987) no. 3, 267–275.
[10.193] G. W. A. Milne, J. S. Driscoll, V. E. Marquez in H. R. Collier (ed.): *Chemical Information: Information in Chemistry, Pharmacology and Patents*, Springer Verlag, Berlin 1989, p. 19.
[10.194] J. G. Vinter, M. Harris, *Chem. Br.* **25** (1989) no. 11, 1111–1116.
[10.195] P. Willett: *Modern Approaches to Chemical Reaction Searching*, Gower, Aldershot 1988.
[10.196] G. E. Vladutz, *Inf. Storage Retr.* **1** (1963) 117–146.
[10.197] P. Willett, *J. Chem. Inf. Comput. Sci.* **20** (1980) 93–96.
[10.198] J. J. McGregor, P. Willett, *J. Chem. Inf. Comput. Sci.* **21** (1981) 137–140.
[10.199] P. E. Blower, R. C. Dana in P. Willett (ed.): *Modern Approaches to Chemical Reaction Searching*, Gower, Aldershot 1986, p. 146.
[10.200] P. E. Blower et al. in W. A. Warr (ed.): *Chemical Structures, The International Language of Chemistry*, Springer Verlag, Berlin 1988, p. 399.
[10.201] R. E. Buntrock, *Database* **11** (1988) no. 6, 124–127.
[10.202] A. P. Johnson in W. A. Warr (ed.): *Chemical Structures, The International Language of Chemistry*, Springer Verlag, Berlin 1988, p. 297.
[10.203] A. P. Johnson, A. P. Cook in P. Willett (ed.): *Modern Approaches to Chemical Reaction Searching*, Gower, Aldershot 1986, p. 184.
[10.204] E. Zass, S. Muller, *Chimia* **40** (1986) no. 2, 38–50.
[10.205] J. H. Borkent, F. Oukes, J. H. Noordik, *J. Chem. Inf. Comput. Sci.* **28** (1988) 148–150.
[10.206] A. P. Johnson et al. in W. A. Warr (ed.): *Chemical Structure Information Systems: Interfaces, Communication and Standards*, ACS Symposium Series 400, American Chemical Society, Washington, 1989, p. 50.
[10.207] T. E. Moock, J. G. Nourse, D. Grier, W. D. Hounshell in W. A. Warr (ed.): *Chemical Structures, The International Language of Chemistry*, Springer Verlag, Berlin 1988, p. 303.
[10.208] G. Grethe, D. del Rey, J. G. Jacobson, M. Van Duyne in W. A. Warr (ed.): *Chemical Structures, The International Language of Chemistry*, Springer Verlag, Berlin 1988, p. 315.
[10.209] W. T. Wipke et al. in P. Willett (ed.): *Modern Approaches to Chemical Reaction Searching*, Gower, Aldershot 1988, p. 92.
[10.210] D. F. Chodosh in P. Willett (ed.): *Modern Approaches to Chemical Reaction Searching*, Gower, Aldershot 1986, p. 118.
[10.211] D. Bawden, S. Wood in P. Willett (ed.): *Modern Approaches to Chemical Reaction Searching*, Gower, Aldershot 1986, p. 78.
[10.212] S. Hanessian, F. Major, S. Leger, *New Methods Drug. Res.* **1** (1985) 201–224.
[10.213] H. Gelernter et al., *Science (Washington, D. C.)* **197** (1977) 1041–1049.
[10.214] K. K. Agarwal, D. L. Larsen, H. Gelernter, *Comput. Chem.* **2** (1978) 75–84.
[10.215] J. Gasteiger et al., *Top. Curr. Chem.* **137** (1987) 19–73.
[10.216] J. Dugundji, I. Ugi, *Top. Curr. Chem.* **39** (1973) 19–64.
[10.217] J. Bauer, R. Herges, E. Fontain, I. Ugi, *Chimia* **39** (1985) 43–53.
[10.218] R. Herges in W. A. Warr (ed.): *Chemical Structures, The International Language of Chemistry*, Springer Verlag, Berlin 1988, p. 385.
[10.219] J. B. Hendrickson, *Acc. Chem. Res.* **19** (1986) 274–281.

[10.220] J. B. Hendrickson, A. G. Toczko, *J. Chem. Inf. Comput. Sci.* **29** (1989) 137–145.

[10.221] K. Funatsu, S. I. Sasaki, *Tetrahedron Comput. Methodol.* **1** (1988) 27–37.

References for Chapter 11

[11.1] A. Barr, E. A. Feigenbaum (eds.): *The Handbook of Artificial Intelligence*, vol. I, William Kaufmann, Los Altos, CA, 1981.

[11.2] A. Barr, E. A. Feigenbaum (eds.): *The Handbook of Artificial Intelligence*, vol. II, William Kaufmann, Los Altos, CA, 1982.

[11.3] P. R. Cohen, E. A. Feigenbaum (eds.): *The Handbook of Artificial Intelligence*, vol. III, William Kaufmann, Los Altos, CA, 1982.

[11.4] E. Rich: *Artificial Intelligence*, McGraw-Hill, New York 1983.

[11.5] P. H. Winston: *Artificial Intelligence*, 2nd ed., Addison-Wesley, Reading 1984.

[11.6] L. F. Lunin, L. C. Smith, *J. Am. Soc. Inf. Sci.* **35** (1984) no. 5, 278–279.

[11.7] N. Cercone, G. McCalla, *J. Am. Soc. Inf. Sci.* **35** (1984) no. 5, 280–290.

[11.8] A. Bonnet: *Artificial Intelligence: Promise and Performance*, Prentice Hall, Hemel Hempstead 1985.

[11.9] S. C. Shapiro (ed.): *Encyclopedia of Artificial Intelligence*, Wiley-Interscience, Chichester 1987.

[11.10] A. M. Turing, *Mind* **59** (1950) no. 236, 433–460.

[11.11] E. A. Feigenbaum, P. McCorduck: *The Fifth Generation: Artificial Intelligence and Japan's Computer Challenge to the World*, Addison-Wesley, Reading 1983.

[11.12] N. K. Herther, *Microcomputers for Information Management* **3** (1986) no. 1, 31–45.

[11.13] P. Davies, *Advanced Information Report*, March 1990, 1–4.

[11.14] P. Langley, J. G. Carbonell, *J. Am. Soc. Inf. Sci.* **35** (1984) no. 5, 306–316.

[11.15] P. Davies: *Artificial Intelligence: Potential for Application in the Information Industry*, EUSIDIC Research Report (EUSIDIC, Calne, Wiltshire, UK) 1988, Knowledge Software Management, 63 College Piece, Mortimer, Reading, UK. An updated version is in the press, Learned Information, Medford NJ.

[11.16] R. C. Johnson, C. Brown: *Cognizers: Neural Networks and Machines That Think*, Wiley, Chichester 1988.

[11.17] A. Colmerauer, H. Kanoui, R. Pasero, Ph. Roussel: "Un Système de Communication Homme-machine en Français," Res. Rep. Groupe Intelligence Artificielle, Faculté des Sciences de Luminy, Marseilles 1973.

[11.18] J. D. Foley, *Sci. Am.* **257** (1987) no. 4, 82–90.

[11.19] R. Bisiani, *Ann. N. Y. Acad. Sci.* **405** (1983) 39–47.

[11.20] F. J. Smith, R. J. Linggard, *Lect. Notes Comp. Sci.* **146** (1983) 275–288.

[11.21] L. D. Erman, F. Hayes-Roth, V. R. Lesser, D. R. Reddy, *Comput. Surveys* **12** (1980) 213–253.

[11.22] D. T. Hawkins, *Online (Weston, Conn.)* **11** (1987) no. 5, 91–98.

[11.23] E. H. Shortliffe: *Computer-Based Medical Consultations: MYCIN*, American Elsevier, New York 1976.

B. G. Buchanan, E. H. Shortliffe (eds.): *Rule-Based Expert Systems: The MYCIN Experiments of the Stanford Heuristic Programming Project*, Addison-Wesley, Reading 1984.

[11.24] D. Gross, *Online Rev.* **12** (1988) no. 5, 283–289.

[11.25] B. Vickery, A. Vickery, *J. Inf. Sci.* **16** (1990) 65–70.

[11.26] A. Vickery, *Aslib Information* **17** (1989) nos. 11/12, 271–274.

[11.27] R. Grishman, *J. Am. Soc. Inf. Sci.* **35** (1984) no. 5, 291–296.

[11.28] *La Traduction Assistée par Ordinateur*, Observatoire des Industries de la Langue, Paris 1989.

[11.29] Proceedings of the Congress on Machine Translation held in Munich, 16–18 August 1989, Deutsche Gesellschaft für Dokumentation, Frankfurt.

[11.30] P. Mayorcas (ed.): *Translating and the Computer 10*, Aslib, London 1989.

[11.31] M. Nagao: *Machine Translation: How Far Can it Go?*, Oxford University Press, Oxford 1989.

[11.32] R. E. Dessy, *Anal. Chem.* **56** (1984) 1200A–1212A.

[11.33] J. K. Kastner, S. J. Hong, *Eur. J. Operational Res.* **18** (1984) no. 3, 285–292.

[11.34] N. S. Yaghmai, J. A. Maxin, *J. Am. Soc. Inf. Sci.* **35** (1984) no. 15, 297–305.

[11.35] R. Forsyth: *Expert Systems: Principles and Case Studies*, Chapman and Hall, London 1984.

[11.36] S. M. Weiss, C. A. Kulikowski: *A Practical Guide to Designing Expert Systems*, Rowman and Allanheld, Lanham, MD, 1984.

[11.37] P. C. Jurs: *Computer Software Applications in Chemistry*, John Wiley and Sons, New York 1986, Chap. 15, p. 212.

[11.38] A. Hart: *Expert Systems: An Introduction for Managers*, Kogan Page, London 1988.

[11.39] M. O'Neill, *Aslib Proc.* **41** (1989) no. 4, 163–168.

[11.40] R. Alberico, M. Micco: *Expert Systems for Reference and Information Retrieval*, Meckler, London 1990.

[11.41] B. A. Hohne, T. H. Pierce (eds.): *Expert System Applications in Chemistry*, ACS Symposium Series 408, American Chemical Society, Washington, DC, 1989.

[11.42] B. M. Carrington, *Database* **13** (1990) no. 2, 47–50.

[11.43] W. F. Clocksin, C. S. Mellish: *Programming in PROLOG*, Springer-Verlag, Berlin 1981.

[11.44] C. Paice, *Aslib Proc.* **38** (1986) no. 10, 343–353.

[11.45] N. N. Mitev, S. Walker in: *Advances in Intelligent Retrieval: Informatics 8*, Aslib, London 1986, p. 215.

[11.46] A. Kemp: *Computer-Based Knowledge Retrieval*, Aslib, London 1988.

[11.47] N. Harrison, B. Murphy, *Library HiTech.* **19** (1987) no. 5, 77–80.

[11.48] D. T. Hawkins, *Online (Weston, Conn.)* **12** (1988) no. 1, 31–43.

[11.49] A. Vickery, H. M Brooks, B. Robinson, *J. Doc.* **43** (1987) no. 1, 1–23.

[11.50] A. Vickery, H. M. Brooks, *Information Processing and Management* **23** (1987) 99–117.

[11.51] J. Hares, M. Thomas, *Computing*, June 1988, 14–17.

[11.52] S. Wade et al., *Online Rev.* **12** (1988) no. 2, 91–108.

[11.53] I. G. Hendry, P. Willett, F. E. Wood, *Program* **20** (1986) 245–263.
[11.54] I. G. Hendry, P. Willett, F. E. Wood, *Program* **20** (1986) 382–393.
[11.55] S. J. Wade, P. Willett, *Program* **22** (1988) 44–61.
[11.56] S. J. Wade, P. Willett, D. Bawden, *J. Inf. Sci.* **15** (1989) 249–260.
[11.57] N. J. Belkin, H. M. Brooks, P. J. Daniels, *Int. J. of Man-Machine Studies* **27** (1987) no. 2, 127–144.
[11.58] N. J. Belkin, W. B. Croft, *Annu. Rev. Inf. Sci. Technol.* **22** (1987) 109–146.
[11.59] H. M. Brooks, N. J. Belkin, P. J. Daniels in: *Advances in Intelligent Retrieval: Informatics 8*, Aslib, London 1986, p. 191.
[11.60] H. M. Brooks, P. J. Daniels, N. J. Belkin, *Journal of Information and Image Management* **12** (1986) nos. 1/2, 37–44.
[11.61] I. Wormell (ed.): *Knowledge Engineering: Expert Systems and Information Retrieval*, Taylor Graham, London 1987.
[11.62] M. O'Neill, A. Morris, *The Electronic Library* **7** (1989) no. 5, 295–300.
[11.63] E. J. Corey, A. K. Long, S. D. Rubenstein, *Science (Washington, D. C.)* **228** (1985) 408–418.
[11.64] D. A. Pensak, E. J. Corey in W. T. Wipke, W. J. Howe (eds.): *Computer-Assisted Organic Synthesis*, ACS Symposium Series 61, American Chemical Society, Washington, DC, 1977, p. 1.
[11.65] W. T. Wipke et al. in W. T. Wipke, W. J. Howe (eds.): *Computer-Assisted Organic Synthesis*, ACS Symposium Series 61, American Chemical Society Washington, DC, 1977, p. 97.
[11.66] W. T. Wipke, G. I. Ouchi, S. Krishnan, *Artificial Intelligence* **11** (1978) 173–193.
[11.67] P. Gund et al., *J. Chem. Inf. Comput. Sci.* **20** (1980) 88–93.
[11.68] M. Bersohn, A. Esack, *Chem. Rev.* **76** (1976) 269–282.
[11.69] J. Gasteiger, M. G. Hutchings, P. Low, H. Saller in T. H. Pierce, B. A. Hohne (eds.): *Artificial Intelligence Applications in Chemistry*, ACS Symposium Series 306, American Chemical Society, Washington, DC, 1986, p. 258.
[11.70] J. Gasteiger, P. Röse, H. Saller, *J. Mol. Graphics* **6** (1988) no. 2, 87–97.
[11.71] W. T. Wipke, W. J. Howe (eds.): *Computer-Assisted Organic Synthesis*, ACS Symposium Series 61, American Chemical Society, Washington, DC, 1977.
[11.72] W. T. Wipke, D. P. Dolata in T. H. Pierce, B. A. Hohne (eds.): *Artificial Intelligence Applications in Chemistry*, ACS Symposium Series 306, American Chemical Society, Washington, DC, 1986, p. 188.
[11.73] A. J. Gushurst, W. L. Jorgensen, *J. Org. Chem.* **53** (1988) no. 15, 3397–3408.
[11.74] H. Gelernter et al., *Science (Washington, D. C.)* **197** (1977) 1041–1049.
[11.75] K. Funatsu, S. Sasaki, *Tetrahedron Comput. Methodol.* **1** (1988) no. 1, 39–51.
[11.76] J. B. Hendrickson et al. in B. A. Hohne, T. H. Pierce (eds.): *Expert System Applications in Chemistry*, ACS Symposium Series 408, American Chemical Society, Washington 1989, p. 62.
[11.77] D. H. Smith (ed.): *Computer-Assisted Structure Elucidation*, ACS Symposium Series 54, American Chemical Society, Washington, DC, 1977.
[11.78] W. Bremser et al.: *Carbon-13 NMR Spectral Data*, 3rd ed., Verlag Chemie, Weinheim 1981.
[11.79] R. E. Dessy, *Anal. Chem.* **56** (1984) 1312A-1332A.
[11.80] N. A. B. Gray: *Computer-Assisted Structure Elucidation*, John Wiley, New York 1986.
[11.81] N. A. B. Gray, *Anal. Chim. Acta* **210** (1988) 9–32.
[11.82] J. Zupan (ed.): *Computer-Supported Spectroscopic Databases*, Ellis–Horwood, Chichester 1986.
[11.83] T. P. Bridge, M. H. Williams, A. F. Fell, *Chem. Br.*, Nov. 1987, 1085–1088.
[11.84] P. C. Jurs: *Computer Software Applications in Chemistry*, John Wiley, New York 1986, Chap. 16, p. 219.
[11.85] B. G. Buchanan, E. A. Feigenbaum, *Artificial Intelligence* **11** (1978) 5–24.
[11.86] R. K. Lindsay, B. G. Buchanan, E. A. Feigenbaum, J. Lederberg: *Applications of Artificial Intelligence for Organic Chemistry; the DENDRAL Project*, McGraw-Hill, New York 1980.
[11.87] N. A. B. Gray, *Prog. Nucl. Magn. Res. Spectrosc.* **15** (1982) 201–248.
[11.88] M. Passlack, W. Bremser in J. Zupan (ed.): *Computer-Supported Spectroscopic Databases*, Ellis–Horwood, Chichester 1986, p. 92.
[11.89] W. Bremser, R. Neudert, *Eur. Spectrosc. News* **75** (1987) 10–27.
[11.90] R. Neudert, W. Bremser, H. Wagner, *Org. Mass Spectrom.* **22** (1987) 321–329.
[11.91] W. Bremser, *Angew. Chem. Int. Ed. Engl.* **27** (1988) 247–260.
[11.92] S. Sasaki, Y. Kudo, *J. Chem. Inf. Comput. Sci.* **25** (1985) 252–257.
[11.93] K. Funatsu, N. Miyabayashi, S. Sasaki, *J. Chem. Inf. Comput. Sci.* **28** (1988) 18–28.
[11.94] K. Funatsu, Y. Susuta, S. Sasaki, *J. Chem. Inf. Comput. Sci.* **29** (1989) 6–17.
[11.95] H. B. Woodruff, *Trends Anal. Chem.* **3** (1984) 72–75.
[11.96] H. B. Woodruff, S. A. Tomellini, G. M. Smith in T. H. Pierce, B. A. Hohne (eds.): *Artificial Intelligence Applications in Chemistry*, ACS Symposium Series 306, American Chemical Society, Washington, DC, 1986, p. 312.
[11.97] L. S. Ying, S. P. Levine, S. A. Tomellini, S. R. Lowry, *Anal. Chim. Acta* **210** (1988) 51–62.
[11.98] F. W. McLafferty, D. B. Stauffer, *J. Chem. Inf. Comput. Sci.* **25** (1985) 245–252.
[11.99] C. A. Shelley, M. E. Munk, *Anal. Chim. Acta* **133** (1981) 507–516.
[11.100] M. E. Munk, M. Farkas, A. Lipkis, B. D. Christie, *Mikrochim. Acta* **1986** II (1987) 199–215.
[11.101] L. A. Gribov, *Anal. Chim. Acta* **122** (1980) 249–256.
[11.102] J. Zupan, M. Novic, S. Bohanec, M. Razinger, *Anal. Chim. Acta* **200** (1987) 333–345.
[11.103] H. J. Luinge, J. H. Van der Maas, *Anal. Chim. Acta* **223** (1989) 135–147.
[11.104] H. Kalchhauser, W. Robien, *J. Chem. Inf. Comput. Sci.* **25** (1985) 103–108.
[11.105] C. J. Rawlings, *Biochem. Soc. Trans.* **17** (1989) 851–855.
[11.106] C. Trindle, *J. Mol. Graphics* **6** (1988) no. 2, 67–73.
[11.107] C. Trindle in B. A. Hohne, T. H. Pierce (eds.): *Expert System Applications in Chemistry*, ACS Symposium Series 408, American Chemical Society, Washington, DC, 1989, p. 92.

[11.108] D. P. Dolata, A. R. Leach, K. Prout, *Journal of Computer-Aided Molecular Design* **1** (1987) 73–85.

[11.109] D. P. Dolata, R. E. Carter, *J. Chem. Inf. Comput. Sci.* **27** (1987) 36–47.

[11.110] D. P. Dolata, A. R. Leach, K. Prout in W. G. Richards (ed.): *Computer-Aided Molecular Design*, IBC Technical Services, London 1989, p. 67.

[11.111] W. T. Wipke, M. A. Hahn, *Tetrahedron Comput. Methodol.* **1** (1988) 141–167.

[11.112] M. A. Hahn, W. T. Wipke in W. A. Warr (ed.): *Chemical Structures, The International Language of Chemistry*, Springer Verlag, Berlin 1988, p. 269.

[11.113] F. Darvas, *J. Mol. Graphics* **6** (1988) no. 2, 80–86.

[11.114] T. Koschmann et al., *J. Mol. Graphics* **6** (1988) no. 2, 74–79.

[11.115] T. H. Pierce, B. A. Hohne (eds.): *Artificial Intelligence Applications in Chemistry*, ACS Symposium Series 306, American Chemical Society, Washington, DC, 1986.

[11.116] J. M. Hushon (ed.): *Expert Systems for Environmental Applications*, ACS Symposium Series 431, American Chemical Society, Washington, DC, 1990.

[11.117] C.-C. Chen, *Microcomputers for Information Management* **6** (1989) no. 2, 77–97.

[11.118] C.-C. Chen, *Microcomputers for Information Management* **6** (1989) no. 2, 135–145.

[11.119] L. Davenport, B. Cronin, *J. Inf. Sci.* **15** (1989) 369–372.

[11.120] C. Franklin, *Online (Weston, Conn.)* **13** (1989) no. 3, 37–49.

[11.121] *Commun. ACM* **31** (1988), special issue on hypertext.

[11.122] *J. Am. Soc. Inf. Sci.* **40** (1989) no. 3, special volume on hypertext.

[11.123] V. Bush, *The Atlantic Monthly*, July 1945, 101–108.

[11.124] V. Bush in: *Science is Not Enough*, Apollo Editions, New York 1969, p. 75.

[11.125] T. H. Nelson: *Literary Machines*, Tempus Press (Microsoft), Tell City, IN, 1987.

13. Patents

WERNER HAUF, Patentstelle für die Deutsche Forschung der Fraunhofer Gesellschaft zur Förderung der Angewandten Forschung, München, Federal Republic of Germany

1. Basic Concepts of Industrial Property Rights 13-2
1.1. Novelty, Inventive Step, Industrial Application 13-2
1.2. Priority, Paris Convention 13-3
1.3. European Patent Convention (EPC) 13-4
1.4. Community Patent Convention (CPC) 13-4
1.5. International Patent Classification (IPC) 13-5
1.6. Patent Cooperation Treaty (PCT) 13-5
1.7. Other Industrial Property Rights 13-5
1.8. Rights to Inventions 13-7
1.9. Licensing, Group Exemption of License Contracts 13-7
2. European Patent Granting Procedure 13-9
2.1. Procedure up to Grant at the European Patent Office 13-9
2.1.1. Filing 13-9
2.1.2. Search 13-9
2.1.3. Examination 13-10
2.1.4. Refusal or Grant 13-10
2.1.5. Publication 13-10
2.1.6. Opposition Procedure 13-10
2.1.7. Default of Time Limits 13-11
2.2. European Patent Application 13-11
2.2.1. Content of the Description (Art. 78, EPC) 13-11
2.2.2. Form and Content of Claims 13-12
2.2.3. Patent Claims in Differing Categories 13-12
3. Examples for the Wording of Claims 13-13
3.1. Pharmaceutical Invention 13-13
3.1.1. Subject Matter 13-13
3.1.2. Opposition Procedure 13-14
3.2. Chemical Invention 13-14
3.2.1. Subject Matter 13-14
3.2.2. Opposition Procedure 13-15
3.2.3. Wording of the Claims 13-15
3.3. Computer-Related Invention 13-16
3.3.1. Subject Matter 13-16
3.3.2. Opposition Procedure 13-16
4. U.S. Procedure up to Grant 13-17
4.1. Differing Terms and Practice 13-18
4.1.1. Filing of the Application, Preclusive Period of Novelty 13-18
4.1.2. Interference Process, First Inventor 13-18
4.1.3. Disclosure, Best Form of Embodiment, Enablement 13-18
4.1.4. Patent Misuse 13-19
4.1.5. Fraud on the Patent Office 13-19
4.2. Conditions for Patenting, Patent Categories, Patent Claims 13-19
4.2.1. Novelty 13-19
4.2.2. Inventive Step 13-19
4.2.3. Patents for Plants, Microorganisms, Animals 13-19
4.2.4. Medicaments, Pharmaceuticals, and Cosmetics 13-20
4.2.5. Patent Categories, Patent Claims 13-20
4.3. Granting Procedure at the U.S. Patent Office 13-20
4.3.1. Examination, Grant, and Appeal 13-21
4.3.2. Divisional Applications, Continuation, Continuation in Part 13-21
4.3.3. Reexamination; Appeal Before the Supreme Court and the District Courts 13-21
4.4. Design Patents 13-21
5. Japanese Patent Granting Procedure 13-22
6. Excerpt from the EPC 13-22
7. References 13-23

This article applies especially to those working in the field of research and development and is intended to draw their attention to significant conditions for legally securing their results. However, the article in its present scope cannot cover all aspects of the protection of industrial property. Important questions on the patent authorities' procedure are reserved for specialists (patent attorneys) and are not dealt with in this article. The applications of property rights should be handled by these experts, especially in cases involving critical restriction of the novelty of an invention with respect to the state of the art.

For the sake of exemplification, the European Patent Law according to the European Patent Convention (EPC, October 7, 1977) and the patent granting procedure before the European Patent Office (EPO) are dealt with as being representative for national patent law and the partially similar granting procedures in the individual countries. Selected examples from the Technical Board of Appeal of the EPO are explanatory only and are not selected according to their significance for the further development of patent law.

1. Basic Concepts of Industrial Property Rights

New technical developments and research results are exploitable particularly if the owner thereby obtains an advantage in the market over his competitor. This advantage may be secret know-how, or frequently, secret know-how combined with protective rights such as patents, trademarks, semiconductor protective rights, design patents, and copyrights for software. Secret know-how in this context means research results and results of developments as well as empirical know-how kept secret from competitors. Know-how that has become accessible to the public, particularly to specialists (i.e., prepublished know-how) may be used by everybody if the owner has not taken particular precautions for its protection. This is in contrast to the generally better known copyright, which arises with creation of the individual artistic work and protects it for the author without registration. Because this protection is insufficient for the technical field, the above-mentioned protective rights requiring official registration were created.

A person working in research and development should examine his research results, acquired know-how, and empirical knowledge regularly to see whether it would be advisable to apply for protective rights for any part of them, either provisionally, in order to forestall possible parallel developments, or in the longer term if this is of significance for future marketing.

1.1. Novelty, Inventive Step, Industrial Application

In order to be patentable or to be eligible for utility model protection, technical inventions must be:

1) novel, i.e., they are not part of the public, generally accessible state of the art (public domain) (Art. 54, EPC),
2) inventive, i.e., a person skilled in the art should not be in a position to deduce them from the state of the art in an obvious manner (Art. 56, EPC), and
3) industrially applicable (Art. 57, EPC).

Novelty of an invention is primarily established if it on the whole is not part of the state of the art and is not described in a publication. The state of the art is everything which has been made accessible to the public by written or oral description, by use, or in any other way before the date of filing of the European patent application (Art. 54/3, EPC). Unpublished patent applications of the EPO (Art. 54/4, EPC) with an earlier filing date are also part of the state of the art, but only with effect for the countries designated in the patent applications (contracting states of the EPC, cf. Chap. 2).

The publication by the inventor (applicant) of his own research and development results is particularly detrimental to novelty because in most cases the publication is largely identical with the invention. The U.S. patent right, the Canadian patent right, and the German utility models, for example, provide a period of grace, as formerly existed in national patent laws, as regards the inventor's own publication (1 year in the United States, 2 years in Canada). Information given in seminars, committees with a restricted number of not directly known participants without confidential character, leaflets, and posters exhibited at trade fairs, may be sufficient to diminish or destroy novelty. In any event, the published material that partially or entirely anticipates the invention can no longer be protected in a patent granting procedure.

For the case of obvious unlawful disclosure to the disadvantage of the applicant (inventor), a term of six months is provided in which the entitled applicant may make a subsequent European application (Art. 55a, EPC). Since the term is limited and the burden of proof lies with the inventor, special care should be exercised when handling patentable results.

The reverse case: if the intention of the inventor is to prevent a third party, to whom his invention is presumed to be known, from obtaining a patent and obstructing the continuation of his work by means of the patent, rapid publication is advisable. However, this has the effect that everybody is free to use the invention industrially.

Inventive Activity. The most usual case in practice is that the individual features of the invention in part or in their entirety are known from several publications. These publications are frequently not known to the applicant (inventor) on the date of filing but only become evident later, for example, during the granting procedure. In the granting procedure, it must be determined whether a person skilled in the art could arrive at the invention simply by combining the individual, previously known features, without having to take an inventive step.

The scope of the patent application then has to be restricted against the known state of the art (i.e., the characterizing part in an individual two-part claim after "characterised in that . . .", cf. Chap. 3).

In some cases the characterizing part consists of a combination of features of the invention which may be known partially or in their entirety. It represents a new invention if its effect is not known and this effect exceeds the total of the individual characteristics (combination claim). However, the invention (combination of the features known per se) is patentable if it represents an enhancement of the state of the art and is nonobvious to a person skilled in the art.

Indications of an inventive action may be the solution of a long-existing problem or need (technical progress) but also serendipity, the overcoming of technical prejudice, or small improvements leading to a particular economic success in the case of mass-produced goods.

In contrast, equivalent measures, changes in the dimensions (scale changes), craftsmanship, and kinematic reversals have to be classified as foreseeable developments which the average person skilled in the art may obtain by means of routine experiments. This does not represent an inventive action.

Methods for surgical and therapeutic treatment of human and animal bodies and diagnostic methods carried out on human and animal bodies (Art. 52/4, Art. 57, EPC) are not applicable industrially and hence are not patentable. Substances and compositions, known per se and used for the first time in such procedures are excluded and hence patentable (pharmaceuticals, 1st and 2nd medical indications).

Plant varieties and animals, and essentially biological processes for breeding plants and animals, e.g., by crossing or selection (Art. 53 B, EPC), are not patentable. Biological (sexual) reproduction itself is expressly excluded from patentability. In some countries, specific plants and vegetable products may be protected under the Plant Varieties Protection Law.

Microbiological processes and products obtained from them (e.g., alcohol obtained by fermentation processes and/or the yeast used) are an important exception. The microorganism should be deposited at an authorized official depository if it is not reproducibly described or if it cannot be determined from the microbial process.

Discoveries, scientific theories (e.g., the solution of differential equations or algorithms), esthetic form creations, plans, rules, games, and computer programs (software) are not considered to be inventions (Art. 52/2, Art. 52/3, EPC). If data processing programs are integrated in a clearly technical process and processed directly with a technical effect the data processing programs may be indirectly patentable, i.e., with the technical process and the technical device (control systems). Control circuits consisting of a number of function groups have a technical character; however, software does not become technical if it is merely stored in an electronic memory.

1.2. Priority, Paris Convention

Patents are effective for each individual country in which they are granted in the form of a monopoly that, according to the national patent law, concedes the owner the exclusive right to the industrial use of the patented subject matter or the patented process. Achieving far-reaching territorial protection necessitates applying for several national patents, e.g., in countries in which an important competitor has its headquarters or in which the invention should be protected for an important market. A European producer with a competitor residing in Japan primarily will prefer patent protection in the United States as joint market, rather than patent protection for the Japanese market, although this would additionally allow him to have the production prohibited in Japan. Although production can be shifted to a patent-free country, this is not a problem in certain technical fields. However, obtaining patents in a great number of industrial countries is financially prohibitive. In the pharmaceutical field, numerous subsequent applications are usual for securing the markets.

As early as 1883 major industrialized countries assembled in the Paris Convention (PC) for the Protection of Industrial Property. At present there are more than 80 member states. The most important regulation is the granting of a one-year priority term for subsequent applications in a Contracting State of the Convention with the same date of filing and content as the national patent application, which the national Patent Office confirms to the foreign Patent Office in the form of a priority document.

This also holds true for claiming the priority of several national patent applications or utility models. A further regulation refers to officially recognized international trade fairs in which inventions are exhibited for the first time. A priority document for the content of the exhibited invention may be issued at, for example, the Hannover trade fair, in an office established for this purpose by the German Patent Office. A patent application having the same content must subsequently be filed within a time limit, which differs from country to country.

The high cost of subsequent applications suggests the idea of a "world patent", i.e., obtaining a once-filed and -granted patent having validity in all industrialized countries. Before the advent of the EPC patent, applications in 14 languages with official examination in nine countries (the rest were registered patents without examination) were necessary in 22 west-European states to achieve covering protection. Considerable differences as regards juridical security and the extent of a monopoly, like the patent, were unavoidable in that case.

1.3. European Patent Convention (EPC)

In the meantime 15 contracting states have ratified this convention for the joint granting of European Patents, among them all EEC states. The Luxembourg Government Conferences of 1969–1972 were followed by the signing in Munich in 1973 of a convention for creating a common European patent right (law) in the form of the EPC, which exists in the individual contracting states in addition to the national patent law. Thus it is still possible to apply for and obtain individual national patents in the respective countries or, as an alternative, at the European Patent Office in Munich, with branches in The Hague and Berlin, that is valid in the contracting states indicated by the applicant. When the European patent is granted, it is separated into a bundle of national patents corresponding to the designated countries and based essentially on the respective national patent law; i.e., infringement and nullity suits (Art. 64, Art. 138/1, EPC) are decided upon before national courts according to national law.

The contracting states had to undergo a necessary alignment of their national patent laws; this has significance especially in the pharmaceutical, agricultural, and horticultural sectors. An important uniform regulation is that products produced according to patented processes for which partially national patent prohibitions existed are protected as well. The lifetime of the patents is uniformly limited to 20 years (Art. 63, EPC). The cost of a European patent approximately amounts to the cost of three to four national patents.

1.4. Community Patent Convention (CPC)

A joint patent is planned for the entire EEC, similar to the U.S. patent which is valid for all states of the United States. The convention only comes into force when all member states of the EEC have ratified it (introduction is aimed for 1991). The Community Patent Convention may, however, under certain circumstances, come into force in 1993 but only with effect for a majority of the member states that have ratified the convention by then. For this an additional decision of all member states is required (1991, Luxembourg Conference 1989).

The community patent will offer uniform protection throughout the EEC and can be transferred and declared invalid only uniformly. In addition to the currently existing Board of Appeal of the EPC, which corresponds partially to the national Patent Courts, a central European Court of Appeal will be established (COPAC, Community Patent Appeal Court). This court will make the final decisions in lawsuits concerning the infringement and validity of common patents and will ensure that the EPC and the CPC are applied according to the same standards. The common patent will be granted and administered by the European Patent Office.

The EEC will then have:

1) community patents having uniform protection (supranational) throughout the EEC (excluding the above-mentioned temporary reservations),
2) European patents with protection in the contracting states, and
3) national patents with protection in the member state in which they are applied for.

1.5. International Patent Classification (IPC)

The classification of patent literature into technical fields was standardized internationally in 1971. It includes the fields: A, human necessities; B, performing operations; C, chemistry and metallurgy; D, textiles and paper; E, fixed constructions; F, mechanical engineering, lighting, heating, weapons, blasting; G, physics; H, electricity.

1.6. Patent Cooperation Treaty (PCT)

At present 35 states belong to the international convention concluded at the Washington Conference in 1970. It provides for a collective application administered by the World Intellectual Property Organization (WIPO) in Geneva. Individual patent application authorities were not established, but national patent authorities such as the National Patent Offices or the European Patent Office are specially designated offices at which a collective application may be filed in the prescribed official language (national language).

The PCT procedure offers the applicant several advantages, especially in case of subsequent applications that claim the priority of one or more (national) primary applications:

1) The PCT patent application may be filed on the last day before the expiration of priority (Art. 8, PCT) in the own (official) language with the Office of the state in which the applicant has his residence (Art. 11, PCT). Time-consuming translations are needed only when apportioning the PCT application and starting the granting procedure in the individual countries.
2) The International Search Report on the state of the art is issued with a term (three months) and, if the application was filed with the EPO, corresponds to the European Search Report.
3) Contracting states may be named as designated countries, whereby Europe (i.e., an application with the European Patent Office), is treated as a country. When the fee of designation for ten countries is paid, all countries of the International Patent Cooperation are regarded as designated.
4) There is also the possibility to start a provisional international examination with confidential character.

The cost of the PCT granting procedure on the whole is higher than the sum of the individually carried out national granting procedures. However, under certain circumstances this is compensated by the time gained after the priority date, because the decision upon the number of states, or whether the PCT application should be prosecuted at all, need only be made after the international search report has been issued or, even later, after the conclusion of the international examination. Only then do the considerable costs for the individual national procedures become due.

1.7. Other Industrial Property Rights

In addition to the patents for the protection of technical inventions, other types of protective rights exist. They are exemplified here by those in the Federal Republic of Germany.

Utility models (Utility Model Law, version of January 1, 1987) are protective rights for inventions that are restricted to tools and articles of daily use (parts). Hence, only an object, but not a process such as a manufacturing or working process, may be the subject matter of the application. In future the utility model protection is to be extended to chemical substances and pharmaceuticals (substance protection).

The requirements the object must meet as subject of an application are as follows: (1) new design, device, and control system, (2) be based on an inventive step, and (3) industrial applicability.

Utility models are registered rights without examination. They are examined only upon a substantiated request for cancellation by a third party. The requirements for novelty and inventive step are lower than for patents. Public use abroad is not detrimental to novelty, and there is a period of grace (preclusive period) of 6 months with respect to one's own publications. The priority (date of application) of the applicant's own patent application may be claimed up to two months after grant or final settlement of an opposition procedure. The period of protection, with extension, is six years after registration in the register of utility models (a maximum of eight years after the date of priority of the application).

Semiconductor Protection Law (1986). This law concerns the protection of the topographic type of electronic semiconductor products. It is a registered right and protects the three-dimensional structure of microchips, parts of this structure, representation for the production of the topography against copying and exploitation (masks), but not microchips produced through analysis of protected microchips (reverse engineering). The topography (similar to the case of the copyright) must exhibit original elements (individuality, similar to the case of the copyright).

The period of protection is ten years after the first nonconfidential commercial exploitation and the application at the German Patent Office, which is subsequently possible within two years. Anyone may file a substantiated request for cancellation. Not only may claims for damages be made against infringers, as is the case in the patent and utility model laws, they may also be prosecuted.

Design patents are registered rights that protect the esthetic individuality of industrial patterns and models (designs). The inventor or his client is directly entitled to the design patent. Although industrial copying is forbidden, the making of individual copies for noncommercial purposes is allowed. The design application, for deposition with the German Patent Office, includes a graphic or photographic representation of the design and the classes of goods to which the design is to be assigned (cf. trademarks). The period of protection is five years and may be extended to a maximum of twenty years on payment of fees. In case of mass-produced goods (e.g., household appliances) the design patent offers the possibility of obtaining industrial protection for the esthetic image, even if valid patents cannot be obtained due to the extensive state of the art.

Copyright for Software. Data processing programs as such are not patentable. There are differences as regards judicature of the EPO and national judicature concerning the restriction of the patent prohibition according to Art. 53/2c, EPC. Like works of art, software enjoys copyright through the creation of the work as such. The German copyright law has been adapted thereto. In addition, further protection for software can be achieved by trademarks and by action against unfair competition (shrink and wrap license).

Trademarks designate the origin of products or services and protect their marketing. In contrast to patents they do not offer any right of monopoly with a time limit; they lapse in case of nonusage or dwindling distinctiveness of the product. They can be sold only together with the business. The trademark protects those who have achieved recognition for their products in the market by continuous expenditure for advertising and quality standards, as the inventor is to be rewarded with the right of monopoly with a time limit for publishing his secret know-how and obtaining an advantage as regards marketing by the publication of the patent specification and the enlargement of the state of the art combined therewith.

The registration of trademarks and service marks is carried out after examination within classes of goods or classes of services for a period of ten years, with the possibility of obtaining an extension for a further ten years. A European trademark convention is to be prepared. According to the Paris Convention, subsequent applica-

1.8. Rights to Inventions

The inventor or his successor in title is entitled to the rights of the invention, which implies the right of patent application at home and abroad (Art. 60, EPC). Most inventions applied for are those of employees, whose substantive right passes over to the employer. The individual right to be named as the inventor remains with the employee. The transfer of rights is regulated differently in each country.

In the Federal Republic of Germany the transfer of rights between the employer and the employee is determined by the Employees Invention Law (Arbeitnehmererfindungsgesetz); this transfer has to be carried out in each individual case, and rights cannot be uniformly transferred (e.g., in a working contract covering future inventions). The employee is obliged to report service inventions (i.e., inventions made during his professional work for the company); this also includes so-called free inventions (i.e., inventions made in the employee's off-duty time), which should be indicated as such in the report to the employer. Service inventions are inventions made during the time of employment that are based mainly on experience of the company (or the public administration as employer) and the duties of the inventor, even if they are filed within a certain period of time after the termination of employment. Within a term of four months after the written report the employer may wholly or partly transfer rights to himself by unilateral declaration in view of the inventor (unlimited or limited claim). He is then obliged to apply for protective rights and to establish the remuneration of the inventor. Unused rights for foreign applications have to be released or paid for by the employer before the expiration of the priority year. The minimum remuneration to be paid on exploitation of the invention, in addition to the inventor's salary, is based on the theoretically obtainable amount (purchase, license) a free inventor could achieve through exploitation of the invention. The additional know-how already remunerated by the salary as well as the position of the inventor in the company, the support given by the company, and the means made available have to be taken into consideration as reducing the value of the invention. According to these guidelines, a proportionality factor is calculated, which is on average 10–25 % of the value of the invention, determined by comparison (license analogy) with the fictitious license contract of a free inventor. So-called supply inventions made in the field of research and development and not exploited commercially give the employer a supposed "benefit" concerning his marketing position if the fees for maintenance of the patent have been paid for a long period of time. As a rule, the inventor is remunerated or the rights to the invention are returned to him (release). For inventions made by employees of the public services, participation in the marketing profits, which was previously agreed upon or fixed, may be arranged instead of the claim. There is an exceptional regulation for professors and research assistants at universities. They are free to exploit their inventions; if the university has made particular means available it may participate in the marketing profit of the inventor in order to cover the cost of those means. However, if research projects are funded by a nonuniversity sponsor, the reserved rights of exploitation for the sponsor have to be taken into account.

1.9. Licensing, Group Exemption of License Contracts

From the day of their publication, European patents and European patent applications have the same rights as national patents and patent applications (Art. 64, Art. 67, EPC) and are subject to the same financial regulations as national patents and national patent applications of the respective (designated) state (Art. 74): they may be licensed (Art. 73, EPC), charged with rights, or transferred (Art. 71, 72, EPC). Infringement suits are decided upon before the national courts according to national law, as is the examination of the legal validity frequently combined therewith (nullity suit); cf. Section 1.4.

The infringement suit will be suspended, if necessary, until the nullity suit started by the infringer has been settled. The sole use by the patentee and the right to prevent third parties from this use covers the production, offering for sale, marketing, and use of articles protected by a patent and the use or offering for sale of protected processes, including articles produced by means of these processes.

In license contracts the following rights can be granted:

1) exclusive licenses with the right to grant sublicenses,
2) nonexclusive licenses, and
3) sole licenses, i.e., an exclusive right with the use reserved for the licensor himself.

"Mixed contracts" are found most frequently, either in the form of know-how contracts, in which secret know-how is licensed, and the patents are essentially supplementary, or in the form of patent–license contracts in which the protective scope (Art. 69, EPC) of the licensed patents is sufficient as subject matter of the contract and additional know-how only serves for carrying out the technical teaching.

If the invention is of a general nature (i.e., having several potential applications), several exclusive license contracts with several licensees for different fields of application (differing technical fields) may be made. Further items in addition to the type of license and grant of a license, return services (royalty per unit, cost sum), minimum royalty, settlement of account, notice of termination, future cooperation, and treatment of future protective rights, are emphasized here as being significant for licensing research results, with further adaptive development by the licensee: the guarantee for the functional capacity of the licensed object does not yet cover its "operativeness" or the complete suitability for use of a marketable product. The licensee has to take over the legal responsibility for the product in the production; this responsibility may be transferred to the licensor in case of a principally substandard subject matter of the license (extended recourse). It is therefore advisable to agree upon a limitation of liability based on the actual state of development (functional model, functional proof in test series, prototype, pilot production, testing in practice), this being described as precisely as possible in the preamble, if results from research and development (e.g., basic research), especially from private persons, are licensed.

The right of monopoly for the marketing of a patented product is restricted by national and European antitrust regulations intended to secure unobstructed circulation of goods and full competition (e.g., FRG; law against restrictions of competition, antitrust law) for the licensee by prohibiting restriction of marketing areas, quantitative restriction, and restricting purchasing obligations and the like. In international trade throughout Europe, Article 85 (prohibition standard) and Article 85/3 (possibility of release) of the EEC contract are particularly extensive standards for securing free flow of goods in the EEC; they are, however, somewhat contradictory to the monopoly of the patent right.

Article 85 of the EEC Contract:
1) Incompatible with and prohibited are all regulations between ... being suitable to obstruct or effect free flow of goods between member states and aim to obstruct, restrict or falsify competition within the Common Market, especially ... (control of production, marketing, technical development, investments; division of markets or resources; differing treatment of trading partners with disadvantage of competition; special payments of the licensee surmounting the subject matter of the contract.
2) The agreements or conclusions prohibited by this article are invalid.
3) The conditions of paragraph 1 may be declared inapplicable, on agreements ... conclusions ..., agreed behaviour ..., which ... contribute to the promotion of technical or economic progress, ..."

In the past this led to the practice that such license contracts have to be submitted to and approved by the EEC Commission in order to examine a possible market-dominating position of the license partners.

The Exemption Regulations of groups of patent and know-how license contracts issued by the EEC Commission determine what may be contained, must be contained, and may not be contained in licensing agreements (white, black, and exempted lists, respectively).

For example, in patent license contracts it is not permitted to make agreements which automatically extend the lifetime of the contract beyond the lifetime of the essential patent, e.g., to the lifetime of a younger patent later assigned to the license contract. The licensor may not forbid nullity suits by the licensee.

The lifetime of know-how contracts is restricted to ten years and the essential secret know-how has to be identified. The know-how receiver may not be obliged:

1) to accept quality instructions, goods, or services from the know-how donor as far as this is not necessary for the technically faultless use of the know-how (reported technology),
2) to supply particular groups of purchasers only or use specific sales outlets only, or
3) to accept restrictions when setting prices and discounts.

2. European Patent Granting Procedure

2.1. Procedure up to Grant at the European Patent Office

Only a small portion of the European patent applications filed at present are first applications (about 5%), the rest are based on the priority of one or more national patent applications. As previously explained, in the PCT procedure an application practice that utilizes the time delay of the priority year, has been developed.

The European Patent Granting Procedure takes the following course: (1) Search, (2) Examination, Granting, or Rejection, (3) possible Opposition Procedure, (4) possible Appeal Procedure. The departments (Directorate General DG) formed at the European Patent Office correspond to

DG 1: Search (branches in The Hague and Berlin)
DG 2: Examination and Opposition
DG 3: Appeals
DG 4: Administration
DG 5: Legal Affairs and International Relations

2.1.1. Filing

Filing offices for patent applications are the EPO and its branches and the national Patent Authorities of the contracting states. The official languages are German, English, and French. Filing in a national language of the contracting states is admissible, but three months after filing or 13 months after the date of filing of the earliest application whose priority has been claimed, a translation in one of the above-mentioned official languages must be filed. This becomes the language of proceedings before the EPO. Further requests may be worded in the national language; however, within a term of one month the applicant must file a translation in the selected language of proceedings (Art. 14, EPC).

The EPO acknowledges receipt of the European patent application, which is directly filed or received through the national authorities. In some of the contracting states, filing via the national authorities (Patent Office) is obligatory if the applicant resides in the contracting state and no national first application has been filed. The national filing is obligatory if the subject matter of the application has significance for national security and a secret patent application is to be expected; this is ordered by the national Patent Authority. There are no secret patent applications or secret patents in the European procedure.

Fees (with a term or the day of application) which have to be payed to the EPO without demand are: application fee (1 month, extended term 2 months with surcharge), search fee (1 month, extended term 2 months with surcharge), fee for the 11th claim and each following claim (1 month), designation fee (12 months, extension 1 month after the day of application of the earliest priority application). Should the payments not arrive in time the application is considered irrevocably withdrawn (loss of priority). The applicant has to prove that he ordered the payment 10 days before the expiration of the term.

The following contracting states at present may be designated in patent applications: Belgium (BE), Federal Republic of Germany (DE), France (FR), Italy (IT), Luxembourg (LU), The Netherlands (NL), Austria (AT), Sweden (SE), Spain (ES), Switzerland with Liechtenstein (CH), Portugal (PT), United Kingdom (GB), Denmark (DK), Greece (GR). A later designation of further states is not possible; however, further prosecution of the application in individual contracting states may be given up, but only after the grant of the patent.

First the patent application is subjected to an examination on filing (Art. 90). A date of filing is granted if it fulfills the minimum requirements of a patent application. These are the identity of the applicant, the specification with one or more patent claims, and the designation of at least one contracting state. The completeness of the application papers is examined in the formal examination (Art. 91) and, if necessary, the applicant is given the opportunity to correct deficiencies in the application (e.g., designation of inventors, abstract, priority document). Formal drawings filed later, e.g., with reference to the patent claims, would change the original contents of the disclosure: the applicant only has the choice of doing without (formal drawings and reference numerals) or to defer the day of application to the day of the later filing of the formal drawings.

2.1.2. Search

The European Search Report of the state of the art relevant to the invention is issued for the patent application according to the patent claims, with due regard of the specification and the formal drawings, and mailed to the applicant with a list and characterization of the literature (e.g., "X": having special significance as individual document; "Y": having special significance in connection with other documents; "A": technological background, etc.) as well as copies thereof, and notice of publication in the official gazette is given. It is left to the applicant's discretion to adapt and refile the specification and the claims immediately after receiving the search report; examination has to be requested and the examination fee payed in any case six months after receipt of the European search report.

Should the patent application contain two (or more) inventions, the search report is issued for one (partial) invention only. By paying a further search fee (with a term ranging from two to six weeks) the applicant may obtain a search report for the second invention and request a further granting procedure thereof. The unity of the invention is decided upon in the subsequent official examination proceedings and, if necessary, the search fee refunded should the decision of the search department not have been correct.

2.1.3. Examination

In the official communication, the applicant is requested to agree upon an allowable version of the original patent claims, restricted with respect to state of the art traced in the search report. An oral hearing for discussing the patentable version may be appointed on request or officially. However, in contrast to an oral hearing in the national procedure, the result is binding.

In some national procedures the (suspended) examination is not carried out; in Germany and Japan the request for examination may be forwarded within seven years after application. Prior art is searched for during the examination procedure.

2.1.4. Refusal or Grant

The EPO informs the applicant by means of a preliminary notification of the grant of the patent after termination of the examintion procedure, the version of which the applicant can either oppose or (tacitly) consent to. The granting fee (printing fee) has to be payed within three months and the translation of the patent claims into both other official languages prepared. However, the legally valid grant will become effective only with its being mentioned in the European Patent Bulletin. At the same time the European patent specification with certificate is issued.

The European patent now is apportioned into national patents; most of the contracting states request a translation into the national language, and different terms have to be complied with. It is left to the applicant's discretion to further prosecute national patents in a restricted number of the designated countries.

2.1.5. Publication

The European patent application is published, together with the search report, 18 months after application or the earliest day of priority, independently from the examination procedure. This publication is in general effected before the grant of the patent and at the same time gives the possibility to inspect the official files, for which previously the consent of the applicant was necessary.

2.1.6. Opposition Procedure

Everybody has the possibility within a period of time of nine months to lodge opposition substantiated with new material against the grant of the patent by the EPO. Grounds of opposition can be based only on (see also Chap. 6):

Art. 52—patentable inventions,
Art. 53—exceptions regarding patentability,
Art. 54—novelty,
Art. 55—nonprejudicial disclosure,
Art. 56—inventive step
Art. 57—industrial applicability,
Art. 100—the invention is not disclosed sufficiently clearly and completely for it to be carried out by a person skilled in the art, or
—the subject matter extends beyond the content of the application as filed.

The opposition is a part of the granting procedure and is examined by the examination department of the EPO.

An appeal may be lodged within two months against the decision of the opposition department concerning maintenance or revocation of the European patent. This appeal must be substantiated within four months. The appeal is presented to the Technical or Legal Board of Appeal. These Boards may present questions having principal significance or as regards interpretation by the EPO, respectively, to the enlarged Board of Appeal for decision (e.g., 2nd medical indication).

2.1.7. Default of Time Limits

Defaults of term which are not remediable are: paying the application fee, search fee, priority term, request for reestablishment in rights, payment of the designation fee, and requests for examination. For all defaults of the terms set by the EPO, reestablishment in rights may be requested within a term of two months upon payment of a fee. The applicant and his representative have to meet increased requirements concerning diligence.

2.2. European Patent Application

Special care should be taken when wording the description of the invention or the specification of the patent application to obtain a legally valid protective right having a sufficient scope that cannot be circumvented. It is advantageous to trace the state of the art, which is extensive in many fields, by (partially) searching the patent literature for substantiating the invention, especially as regards the field of research and development.

It is often worthwhile to emphasize and protect important special features of the invention instead of taking fundamental scientific ideas to form the subject matter of the application. The aim is to obtain an industrial patent to prevent competitors from imitation or infringement.

It may not be advisable to apply for a patent if:

1) the illegal use of a protected production process cannot be proven from features of the resulting product,
2) a process for the determination of measurement techniques is directed at the user in the laboratory and not at the producer of the measurement devices, and
3) if the disclosure of a patented "recipe" enables everybody to produce a protected product, but the applicant cannot prove an infringement of his patent and is thus unable to forbid it.

The implementing regulations of the EPC contain the requirements for the request for grant (form for patent applications, Rule 26) and the basic drafting of the patent application (Rule 27) (cf. information brochure of the EPO *The Way to the European Patent*). The specification of the application must disclose the invention such that a person skilled in the art should be able to reproduce the invention without unreasonably long experimentation and without his having to take inventive steps or make conclusions (Art. 83, EPC).

The protective scope of the patent is determined by the patent claims (Art. 69) and as regards interpretation, the specification and formal drawings are taken into consideration.

2.2.1. Content of the Description (Art. 78, EPC)

Rule 27 EPC:
(1) In the specification
a) the title of the invention mentioned in the request for grant of a European patent shall be indicated in the beginning;
b) the technical field the invention refers to shall be indicated;
c) the prior state of the art has to be indicated as far as according to the applicant's knowledge it is considered useful regarding comprehension of the invention, producing the European Search Report, and the examination; literature from which the state of the art may be derived shall be indicated as well;
d) the invention as characterized in the patent claims shall be represented in such a way that according thereto, the technical problem underlying the invention, even if it is not expressively mentioned, and its solution may be comprehended; moreover, advantageous effects of the invention, if necessary, shall be indicated by referring to the prior art;
e) formal drawings, if there are any, shall be provided with short legends;
f) at least one way for reproducing the claimed invention shall be indicated in detail; this shall be done, where it is appropriate, by giving examples and, if necessary, by referring to drawings;

g) shall be indicated expressively, if it cannot obviously be derived from the specification or the type of the invention, in which way the subject matter of the invention is applicable industrially.
(2) The specification shall be filed in the way and sequence indicated in paragraph 1, as far as due to the type of invention a deviating form or sequence shall cause better comprehension or more concise representation.

The formal drawings may not contain any explanations except concise essential details (section A–B, vapor, water, etc.) but only reference numerals concerning the text (Rule 32). If a microorganism is part of the disclosure it is to be deposited on the day of filing in a culture collection specified by the EPO (Rule 28).

Attached drawings do not have any individual disclosure contents. The same is true for the abstract which as a summary of the complete application shows the technical field and the technical problem underlying the invention as well as its solution by the invention and facilitates the research.

2.2.2. Form and Content of Claims

Rule 29: (1) The claims shall define the matter for which protection is sought in terms of the technical features of the invention. Wherever appropriate claims shall contain:
(a) a statement indicating the designation of the subject matter of the invention and those technical features which are necessary for the definition of the claimed subject matter but which, in combination, are part of the prior art;
(b) a characterising portion—preceded by the expression "characterised in that" or "characterised by" stating the technical features which, in combination with the features stated in subparagraph (a), it is desired to protect.
(2) Subject to Article 82, a European patent application may contain two or more independent claims in the same category (product, process, apparatus or use) where it is not appropriate, having regard to the subject matter of the application, to cover this subject matter by a single claim.
(3) Any claim stating the essential features of an invention may be followed by one or more claims concerning particular embodiments of that invention.

(4) Any claim which includes all the features of any other claim (dependent claim) shall contain, if possible at the beginning, a reference to the other claim and then state the additional features which it is desired to protect. A dependent claim shall also be admissible where the claim it directly refers to is itself a dependent claim. All dependent claims referring back to a single previous claim, and all dependent claims referring back to several previous claims shall be grouped together to the extent and in the most appropriate way possible.
(5) The number of the claims shall be reasonable in consideration of the nature of the invention claimed. If there are several claims, they shall be numbered consecutively in arabic numerals.
(6) Claims shall not, except where absolutely necessary, rely, in respect of the technical features of the invention, on references to the description or drawings. In particular, they shall not rely on such references as: "as described in part ... of the description", or "as illustrated in figure ... of the drawings".
(7) If the European patent application contains drawings, the technical features mentioned in the claims shall preferably, if the intelligibility of the claim can thereby be increased, be followed by reference signs relating to these features and placed between parentheses. These reference signs shall not be construed as limiting the claim.

2.2.3. Patent Claims in Differing Categories

Rule 30:
Article 82 shall be construed as permitting in particular that one and the same European patent application may include:
(a) in addition to an independent claim for a product, an independent claim for a process specially adapted for the manufacture of the product, and an independent claim for a use of the product; or
(b) in addition to an independent claim for a process, an independent claim for an apparatus or means specifically designed for carrying out the process; or
(c) in addition to an independent claim for a product, an independent claim for a process specially adapted for the manufacture of the product, and an independent claim for an apparatus or means specifically designed for carrying out the process.

3. Examples for the Wording of Claims

The following examples concerning the wording of claims are taken from the official gazette of the EPO according to two points of view:

- examples of pharmaceutical, chemical, and computer inventions,
- examples illustrating section II, EPC, material patent right (I. patentability) with the following articles (excerpt):

Art. 52 – patentable inventions,
Art. 53 – exceptions from patentability,
Art. 54 – nonprejudicial disclosure,
Art. 56 – inventive step,
Art. 57 – industrial applicability,
Art. 123 – amendments.

3.1. Pharmaceutical Invention

(EPO, Official Journal 6/1984), pp. 265–271)

3.1.1. Subject Matter

Simethicone Antacid Tablet (European Patent Application)

Abstract: A tablet containing at least two separate and discrete volume portions one of which contains simethicone and the other of which contains antacid. A barrier separates the two volume portions to maintain the simethicone out of contact with the antacid and to prevent migration of the simethicone from its volume portion of the tablet into the volume portion containing the antacid, and vice versa. The simethicone is physically combined with the other ingredients of the tablet in such a manner that the simethicone is available relatively immediately for antifoaming action, and its availability does not depend upon the breakdown of a maxtrix.

The application contains the following one independent (main) claim and eight dependent product claims (Art. 82, Rule 30). These claims correspond to a U.S. patent application (open claim, cf. Section 3.2.5), whose priority was used to file the European patent application in question:

1) A tablet containing simethicone and an antacid, said tablet *comprising*: a first volume portion containing said simethicone; a second volume portion containing said antacid; each of said first and second volume portions being separate and discrete from the other volume portion; and barrier means between said first and second volume portions for maintaining the simethicone in said first volume portion out of contact with the antacid in the second volume portion and for preventing migration of ingredients from one volume portion to another; said simethicone being exterior of any matrix formed by any of the other ingredients in said tablet, the availability of simethicone for antifoaming action being independent of the breakdown of any such matrix.

2) A tablet as recited in Claim 1 and *comprising*: two layers each constituting one of said first and second volume portions; said barrier means being sandwiched between said two layers.

3) A tablet as recited in Claim 2, wherein: the layer constituting said first volume portion comprises simethicone and a solid carrier composed of simethicone adsorbing material.

4) A tablet as recited in Claim 1 and *comprising*: an inner core constituting said first volume portion; and an outer layer constituting said second volume portion and encompassing said inner core; said barrier means being disposed between said inner core and said outer layer and encompassing said inner core.

5) A tablet as recited in Claim 4, wherein said inner core *comprises* simethicone and a solid carrier composed of simethicone adsorbing material.

6) A tablet as recited in Claim 4, wherein: said inner core comprises liquid simethicone; and said barrier means *comprises* a container for said liquid simethicone.

7) A tablet as recited in Claim 6, wherein: said barrier means is a soft, chewable gelatin capsule shell.

8) A tablet as recited in Claim 1 and *comprising*: an inner core constituting said second volume portion and containing said antacid; and an outer layer constituting said first volume portion and containing said simethicone; said barrier means being disposed between said inner core and said outer layer and encompassing said inner core.

9) A tablet as recited in Claim 8, wherein: said outer layer comprises simethicone and a solid carrier composed of simethicone adsorbing material.

The invention represents a so-called problem invention. According to the EPO Guidelines for Examination the discovery of an as yet unrecognized problem may, in certain circumstances, give rise to patentable subject matter. In retrospect, the inventive solution seems to be trivial and obvious in itself, having stated the problem or the desired effect of the problem clearly enough This example of a galenic form, a tablet, contains two active substances: an antacid ingredient as a gastrointestinal agent such as e.g., magnesium hydroxide or aluminum hydroxide, and silicone oil (simethicone), acting as an antiflatulent. It was generally known that the antacid component impedes or even prevents the release of the silicone oil, strongly absorbed by the antacid agent, when both components are mixed as the active substance. Thus the antifoaming function is reduced in the gastrointestinal tract. Tablets with barriers to separate various ingredients are generally known in the prior art. Other prior art teaches suspending or entraining microscopic particles of simethicone within a matrix of sorbitol or glycerol–corn syrup, then mixing with antacid and tableting. The release of simethicone depends on the breakdown of the matrix in the stomach, which is relatively slow, and occurs only by dissolution if there is no digestion of the matrix in the stomach.

For these reasons a simpler production method of a mixture is actually preferred, with simethicone reversibly absorbed on a large amount of filler material (lactose, sorbit, saccharose) and the antacid agent in a two layer tablet. Even in this case, the release of simethicone turned out to be retarded or prevented after a given shelf time.

3.1.2. Opposition Procedure

It appeared to the inventor that simethicone migrates, against all expectations, from the absorbed state into the solid antacid layer and becomes absorbed therein. The inventive solution of the problem is to provide a barrier in the tablet to prevent this interaction.

Against the refusal of the examiner, who objected that there is no surprising effect demonstrating an inventive step to solve a well-known problem, the applicant lodged an appeal at the Technical Board of Appeal of the EPO. The Board raised objections against the patentability of the claims in a communication to the appellant. After an oral hearing the following independent (main) claim among the new claims was submitted.

> 1) A tablet containing simethicone and an antacid, said tablet comprising: a first volume portion containing said simethicone and a solid carrier composed of simethicone adsorbing material; a second volume portion containing said antacid; each of said first and second volume portions being separate and discrete from the other volume portion; said simethicone being exterior of any matrix formed by any of the other ingredients in said tablet, the availability of simethicone being independent of the breakdown of any such matrix *characterised in that* there are barrier means between said first and second volume portions for maintaining the simethicone in said first volume portion out of contact with the antacid in the second volume portion and for preventing migration of ingredients from one volume portion to another.

The Board stated in the reasons for the decision that the trend in the art was to avoid barriers, which are cumbersome to manufacture. It was preferred to combine simethicone with a large excess of a carrier intended to prevent migration and absorption by the antacid. The granules of such material are tableted in admixture with granules of antacid, although the possibility of forming continuous layers in a single tablet was expressly mentioned in the prior art. In view of the fact that this was the object of a particular state-of-the-art document (patent) and an effective separation had not been achieved, the modified tablets claimed in the present application are therefore novel and show an improved performance.

The question regarding the inventive step, in relation to the modification of the layered tablet of the state of the art as suggested by the present applicants, is not whether a person skilled in the art could have inserted a barrier between the layers but whether he would have done so in expectation of some improvement or advantage. Since the tablet (of the cited document) was, on the face of it and from what was assumed in view of its commercialization, a satisfactory answer to the problem of undesirable migration, the addition of a barrier would have appeared superfluous, wasteful, and devoid of any technical effect. In view of the recognition that a barrier has, after all, a substantial effect, the result was not predictable and the claimed modification involves an inventive step on this basis.

Decision of the Technical Board of Appeal:

- the decision of the Examining Division is set aside
- the case is remitted to the first instance with the order to grant an European patent on the basis of the following documents: (1) description, (2) claims, (3) drawings (stemming from the original European patent application and material submitted to the court during the appeal).

International Classification: A 61 K 9/24

A	Section A—Human Necessities
A 61	Subsection: Medical and Veterinary Science, Hygiene
A 61 K	Preparation for Medical, Dental, or Toilet Purposes
A 61 K 9/00	Medicinal Preparations by Special Physical Form
A 61 K 9/24	Layered or Laminated Unitary Dosage Forms.

3.2. Chemical Invention

(EPO, Official Journal 7/1985, pp. 209–216)

3.2.1. Subject Matter

This example represents an invention by purposeful selection from a domain or multitude known per se (in the prior art), a so-called selection invention, e.g.,:

1) inventions concerning chemical processes, this may be a distinct quantity, range, percentage of mass of a reactant or a distinct range of factors i.e., temperature, pressure, pH value;
2) inventions concerning (chemical) products, i.e., a new compound taken from a group of compounds which is generally known to a person skilled in the art. However the requirements for patenting (novelty and inventive step) must be proven. To illustrate this, the decisions of the Technical Board of Appeal (EPO) may be cited in some interesting cases, beginning with the present one:

Thiochloroformates. The sub-range is novel not by virtue of an effect which occurs only within it; but this effect permits the inference that what is involved is not an arbitrarily chosen specimen from the prior art but another invention (purposeful selection). (Diastomere, *Off. J. EPO* 8/1982, pp. 296–303). In the case of one of a number of chemical substances described by its structural formula by a prior publication, that substance's particular stereospecific configuration (threo form)—though not explicitly mentioned—is anticipated if it proves to be the inevitable but undetected result of one of a number of processes adequately described in a prior publication by indication of the starting compound and the process. In such cases, novelty by selection cannot be claimed, since none of the possible combinations of all the listed starting compounds and process variants introduce a new element—indispensable for substance selection—that would result in a true and not just "identical" modification of the starting substances.

The subject matter of the European patent application concerns a process for the preparation of thiochloroformates. These compounds are usually prepared by reaction of mercaptans with phosgene in the presence of a catalyst. In the prior art a citation describes a process of this kind in which catalytic amounts of a carboxylic acid amide (2 mol% relative to the starting mercaptan) are used. Although this process permits yields of up to 90% of the theoretical maximum, it is not possible to reprocess the reaction deposits by distillation without previously removing the catalyst by washing with water or aqueous hydrochloric acid, causing decomposition and discoloration of the distillate or sublimation of solids. The (technical) problem was to improve the yield of the process of preparation of thiochloroformates by phosgenating mercaptans and at the same time to simplify the reprocessing of the reaction mixture.

A European patent was granted with a main process claim, defining a sub-range from 0.02–0.2 mol% for the catalyst (in total 5 claims cited below).

3.2.2. Opposition Procedure

An opponent filed opposition against the European patent, which was rejected. Then the opponent appealed at the Technical Board of Appeal of the EPO, stating that the catalytic amounts of amide used had no lower limit and therefore included the amount claimed in the contested patent. According to the examples in the citation the concentration of the catalyst lies between 2 and 13 mol%, yields of between 44 and 90% being achieved according to the type of catalyst and the reprocessing technique. The Board had to establish whether the process according to the patent under consideration involves an inventive step, stating:

A fairly broad range of numbers delimited by minimum and maximum values (in this case 0–100 mol%) does not necessarily represent a disclosure, ruling out the selection of a sub-range, of all the numerical values between these minimum and maximum values if the sub-range selected is narrow (in this case, 0.02–0.2 mol%) and sufficiently far removed from the known range illustrated by means of examples (in this case, 2–13 mol%).

The teaching of the citation was that the yields indicated in each example and the high yield obtained in certain cases indicate that catalytic amounts of a carboxylic acid amide as catalyst are to be used from 2 mol% upwards, it being a matter of common knowledge that larger amounts of catalyst produce higher yields. A person skilled in the art would have tended to increase the concentration of catalyst rather than reduce it to solve the given problem.

The proposal by the patent proprietor, whereby the amount of catalyst is to be purposely reduced in spite of the conventional teaching, was therefore not obvious to a person skilled in the art. The Board decided that the appeal is dismissed, summarizing that the teaching according to Claim 1 is new and involves an inventive step.

3.2.3. Wording of the Claims

1) A process for the preparation of thiochloroformates by reaction of mercaptans with phosgene in the presence of at least one carboxylic acid amide and/or urea derivative as catalyst, characterized in using the catalyst in an amount of from 0.02 to 0.2, preferably 0.05 to 0.1 mol%, relative to the starting mercaptan.
2) The process as claimed in Claim 1, characterized in using as mercaptan compounds of the formula R–SH

in which R is alkyl, having preferably 1–18, especially 6 carbon atoms; cycloalkyl, having preferably 5–6, especially 6 carbon atoms; alkenyl, having preferably 3–8, especially 3–4 carbon atoms; aryl, preferably phenyl or naphthyl, especially phenyl; aralkyl preferably benzyl; a heterocyclic radical, preferably thienyl or furyl (optionally substituted by inert substituents on the SH group), whereby as inert substituents the following groups may be used: halogen, preferably F, Cl, Br, especially Cl; alkoxy, preferably C_1–C_4 alkoxy, especially methoxy; aryloxy, preferably phenoxy; carboalkoxy, preferably carbo(C_1–C_4)-alkoxy, especially carbomethoxy; and—except in the case that R = alkyl—also alkyl, preferably having 1–4 carbon atoms, especially methyl.

3) The process as claimed in Claims 1 and 2, characterized in using as mercaptans C_1–C_8 alkylmercaptans, phenylmercaptan, or tolylmercaptans.
4) The process as claimed in Claims 1 and 2, characterized, in using as catalysts compounds of the formula $R^1CONR^2R^3$ in which R^1 = H, alkyl having preferably from 1 to 4 carbon atoms, phenyl or the group NR^2R^3 where R^2 and R^3 independently from each other, are H, alkyl preferably having from 1 to 4 carbon atoms, phenyl, with the proviso that at least one of the radicals R^1, R^2, and R^3 = H when $R^1 \neq NR^2R^3$ and at least one of the radicals R^2 and R^3 = H when $R^1 \neq NR^2R^3$ and two of the radicals R^1, R^2, R^3 together, that either $R^1 + R^2$ or R^3, or $R^2 + R^3$, may furthermore be alkylene, preferably having from 3 to 6 carbon atoms.
5) The process as claimed in Claims 1 to 4, characterized in using as catalysts dimethyl formamide, tetramethyl urea, and/or N-methylpyrrolidone.

The main independent claim refers to a new feature of the process, the dependent claims refer to a variety of catalysts, known per se in this context.

International Classification: C 07 C 154/00

C	Section C—Chemistry and Metallurgy
C 07	Subsection: Organic Chemistry
C 07 C	Acyclic and carbocyclic compounds
C 07 C 154/00	Thiocarbonic acid; derivatives thereof
C 07 C 154/02	Dithiocarbonic acid derivatives, e.g., xanthates

3.3. Computer-Related Invention

(EPO, Official Journal 1/1987, pp. 14–23)

Mathematical Method/Computer Program

3.3.1. Subject Matter

The European patent application of an applicant from the United States, claiming the priority of his U.S. application was refused by the Examining Division of the EPO, which stated that the method claims were related to a mathematical method which is not patentable by virtue of Article 52(2), 52(3), EPC. The dependent method claims did not additionally represent the technical features, as required in Rule 29 (Section 2.2.2). In summary: the normal implementation of mathematical methods by a program on a known computer could not be regarded as an invention.

The applicants intended to improve the digital image processing, achieving greater efficiency in digital filtering techniques (object of the invention). A familiar example of digital image processing is the restoration or enhancement of images taken by satellites. The two-dimensional digital image may be considered as an array of dots (pixels), with each pixel being assigned values representing the brightness, luminance, and color of the system. This pixel array may be viewed mathematically as a matrix of data for the purposes of digital image processing (digital filtering). These techniques are well known and use mathematical operations such as Fourier transform filtering. Conventional filtering methods require a great amount of computation. The present invention provides a processing method which substantially reduces the number of computations.

However the Examining Division of the EPO argued further that the disclosure relates to special purpose hardware (a block diagram for an electrical circuit is shown) which is to be put into practice by the person skilled in the art of designing circuitry. This circuitry can perform the specific operations which are precisely defined by mathematical expressions. A filter designer is able to "reduce" a mathematically specified filter to its circuit form. Thus the mathematics are merely a shorthand by which a technical function is to be described, and not the totality of the invention. In the claims the process steps might be said to be defined in terms of a novel algorithm.

3.3.2. Opposition Procedure

The applicants lodged an appeal against the refusal, considering that only an algorithm per se might be excluded by Article 52(2), EPC; a process carried out in accordance with an algorithm is clearly not excluded, being no different in principle from any other sort of technical definition of a process. What should determine patentability is the substance of what is being claimed, not its manner of definition.

The Board remitted the case to the Examining Division for further prosecution. The decision of the Board is summarized in the following Head Notes and the citation of the Official Journal:

1) Even if the idea underlying an invention may be considered to reside in a mathematical method a claim directed to a technical process in which the method is used does not seek protection for the mathematical method as such.
2) A computer of known type set up to operate according to a new program cannot be considered as forming part of the state of the art as defined by Art. 54(2), EPC.
3) A claim directed to a technical process, which process is carried out under the control of a program (whether by means of hardware or software), cannot be regarded as relating to a computer program as such.
4) A claim which can be considered as being directed to a computer set up to operate in accordance with a specified program (whether by means of hardware or software) for controlling or carrying out a technical process cannot be regarded as relating to a computer program as such.

"Clearly a method for obtaining and/or reproducing an image of a physical object or even an image of a simulated object (as in computer-aided design/computer-aided manufacturing (CAD/CAM) systems) may be used, e.g., in investigating properties of the object or designing an industrial article and is therefore susceptible of industrial application. Similarly, a method for enhancing or restoring such an image, without adding to its informational content, has to be considered as susceptible of industrial application within the meaning of Article 57, EPC."

The applicant, having directed the characterizing part of the independent claim of his application mainly to mathematical features (steps of the method), presented amended claims of method and apparatus for the further prosecution within the scope of the disclosure. These claims now relate to the digital processing of images in the form of a two-dimensional data array:

1) A method of digitally processing images in the form of two-dimensional data array having elements arranged in rows and columns in which an operator matrix of a size substantially smaller than the size of the data array is convolved with the data array, including sequentially scanning the elements of the data array with the operator matrix, characterized in that the method includes repeated cycles of sequentially scanning the entire data array with a small generating kernel operator matrix to generate a convolved array and then replacing the data array as a new data array; the small generating kernel remaining the same for any single scan of the entire data array and also comprising at least a multiplicity of elements, nevertheless being of a size substantially smaller than is required of a conventional operator matrix in which the operator matrix is convolved with data array only once, and the cycle being repeated for each previous new data array by selecting the small generating kernel operator matrices and the number of cycles according to conventional error minimisation techniques until the last new data array generated is substantially the required convolution of the original data array with the conventional operator matrix (independent method claim).

Main classification and supplementary classification of the search report:

International Classification: G 06 F 15/20, H 04 N 7/12

G	Section G – Physics
G 06	Computing, Calculating, Counting
G 06 F	Digital Computers in which at least part of the computation is effected electrically; Arrangements for Handling Digital Data
G 06 F 15/00	Data-processing Equipment characterized by the combination of functions covered by group 7/00 (Selection of materials for use in image receiving members, i.e., for reversal by physical contact; Manufacture thereof) and at least one other main group.
G 06 F 15/20	.... characterized by a Specific Application so far as the Design or Construction of the Computing Part is Concerned.
H	Section H – Electricity
H 04	Electrical Communication Technique
H 04 N	Telephonic Communication
H 04 N 7/00	Television Systems
H 04 N 7/12	System in which the Picture Signal is transmitted via one channel or a plurality of parallel channels, the bandwidth of each channel being less than the bandwidth of the Picture Signal.

4. U.S. Procedure up to Grant

The legal source for the patent granting procedure is the U.S. patent law, version 1952 (35 USC United States Code) having received decisive supplements from anglo-saxon legal practice (case law, common law): common terms like e.g., double patenting, patent misuse, fraud of the PTO (Patent and Trademark Office) originate from general jurisdiction.

Starting from the European patent granting procedure described in Chapters 1–3, the differences as regards the U.S. patent granting procedure are dealt with in particular.

4.1. Differing Terms and Practice

4.1.1. Filing of the Application, Preclusive Period of Novelty

The right of filing the application lies exclusively with the inventor, even if the rights to the invention have already been transferred (e.g., to the employer). Therefore, declarations of inventors are filed simultaneously with the U.S. patent application. A team of inventors may own an invention jointly and file an application; however, separate applications have to be provided if, for example, one inventor has invented the product and another the process.

The U.S. patent law provides for a period of grace of novelty of one year. The priority date of an invention is not necessarily identical with the date of filing with the U.S. Patent Office.

4.1.2. Interference Process, First Inventor

The development of an invention according to the American legal practice (§ 102, 35 USC) covers the following steps:

1) conception of invention;
2) actual reduction to practice—realizing a physical form;
3) constructive reduction to practice—industrial utility in the form of a patent application;
4) diligence—realization of the invention before the patent application, i.e., between "actual reduction" and "constructive reduction to practice". This has some significance when determining the first inventor. The first inventor is determined in a formal interference procedure.

The often lengthy interference procedure is suggested by the Examiner and started by the Commissioner of the PTO at the Board for Interference Procedures. According to § 135 35 USC the interference procedure has to be carried out in the case of colliding patent applications (i.e., two or more applications describing the same invention either in whole or in part) or in case of a collision between a patent application and a recently granted patent in order to determine the first inventor. Once the first inventor has been determined, colliding patent claims are awarded to the first inventor.

Foreign applicants, having conceived their inventions outside of the United States, have only the priority of their original application for proving a date of invention before the date of filing of the U.S. application, because, according to § 104, 199, USC, only acts carried out in the national territory of the United States are admissible as evidence.

4.1.3. Disclosure, Best Form of Embodiment, Enablement

The wording of the claims of a U.S. application in comparison to a European application is often performed in greater detail to meet the requirements of industrial practicability of the invention by the average expert (person skilled in the art). In the national (European) procedures, proof of industrial nonpracticability is sufficient for starting a nullity suit by a third party. However, the industrial practicability by the average expert is partly ascertained in the U.S. examination procedure (enablement of the person skilled in the art).

A (chemical) product is not patentable if the steps of its production that are not described in the application cannot be simply taken from the general state of the art. In such a case, it is not only necessary to sufficiently disclose the details of the production but also to indicate the best mode of embodiment as well (methods of carrying out the invention), which the applicant would probably withhold as secret know-how in some cases. In infringement procedures, the absence of the best mode of embodiment may be decisive for the survival of the patent. The extent of the patent claims have to correspond to the complete practicability of a machine or process (enablement) and corresponds to the protective scope of the patent (§ 112, 35 USC). It is recommendable to revise European patent applications, which nearly always are kept quite general, before filing them with the U.S. Patent Office and to word the claims (increased number) as well as the specification in a more detailed manner. Supplements to application that become necessary during the examination proceedings, e.g., for clarifying or defining technical terms used in the description, can be obtained only with difficulty and at high costs by expert opinions or (authorized) declaration of the inventor.

4.1.4. Patent Misuse

According to American law, the monopoly of expressly granted rights to the inventor is combined with the heightened duty to not broaden them inadmissibly. Often the courts consider certain conditions of license contracts, such as dividing up markets, obligations of delivery, fixing of prices, and conditions regarding license fees, as patent misuse. In such cases the patentee may be divested of the right of enforceability, until the misuse is removed (purge of misuse).

4.1.5. Fraud on the Patent Office

Dishonest behavior of the patentee during the patent granting procedure for the grant of a monopoly is regarded in the same light. In particular, in a later infringement suit this will be an obstacle for enforcing the disputed patent.

According to the rules of the U.S. Patent Office, the patentee has the duty to deal with the PTO with the utmost of candor and honesty, i.e., to provide the entire disclosure of his invention with "good faith and duty". Thus it is binding to disclose essential material having significance for the novelty of the invention to be patented and which has been indicated in parallel granting procedures in other countries. The filing term is three months after filing the application or promptly after publication, respectively.

Representing the invention (e.g., experimental results) advertently or inadvertently in an incomplete or unsatisfactory manner may be considered as such misuse and punished, causing the loss of the patent rights if thereby wrong conclusions will be drawn.

Statements given during the examination proceedings in view of the U.S. Patent Office are given as attested declarations or affidavits, e.g., regarding the date of the conception of an invention, for refuting a publication detrimental to novelty that appeared between the development of the invention and the date of application for the patent and being traced during the examination proceedings. Such affidavits are rare in the case of foreign patentees, because only the actions taken for developing the invention on American territory or their becoming known to a limited number of persons skilled in the art through the inventor are admissible as proof (§ 104, U.S. Patent Law).

4.2. Conditions for Patenting, Patent Categories, Patent Claims

(35 USC Part II Patentability of Inventions and Grant of Patents, Chapters 10, 15)

4.2.1. Novelty
(§ 102 Novelty, Loss of Right to Patent)

Printed publications published more than a year before the U.S. application, granted national and foreign patents (as well as utility models), circulation, and public pre-use in the United States are generally detrimental to novelty. When a foreign applicant misses the term of priority for a patent application outside the United States, he may still file a U.S. patent application having the same contents due to the term of the period of grace for novelty of one year. However, this will not be possible if the original application has already been patented abroad. An official publication of the original application by a patent office is not detrimental to novelty.

4.2.2. Inventive Step
(§ 103 Conditions for Patentability, Nonobvious Subject Matter)

The inventive achievement (inventive step in the European Proceedings) corresponds to the definition of the "not being obvious to the person skilled in the art" in the U.S. proceedings.

4.2.3. Patents for Plants, Microorganisms, and Animals

Plants are patentable in the United States (§ 161 Patents for Plants) as are bred species, mutants, and hybrids, but not naturally occurring or propagated species. There is no protective law for species in the United States as exists in some European countries.

Microorganisms selected or obtained by genetic engineering have to be regarded patentable after the decision Diamond v. Chakrabarty (447 US 303 1990) of the Supreme Court (legal subject matters of inventions "should include everything moving on earth and made by man"). The microorganisms have to be deposited before or on the date of application. Following the decision Allen (2 USPQ2d 1425, 1987 polyploid oysters) nonhuman polycellular living organisms occurring in nature, including animals, also count as patentable subject matter (Manual of Patent Examining Procedure: 2105 – Declaration of the Commissioner of USPTO).

In April 1989 the U.S. Patent Office granted a patent (US 4 736 866) to the University of Harvard for a genetically changed nonhuman mammal, in this case a genetically changed mouse (Oncomouse) that is prone to breast cancer. The European application 8 530 449 was rejected with the decision dated 14th of July 1989 by the Examining Division of the EPO (according to Art. 53 b, Art. 83).

4.2.4. Medicaments, Pharmaceuticals, and Cosmetics

Novel compounds having unexpected characteristics and made from starting materials that are known per se are patentable. The starting materials, the method of production, and the mode of use have to be sufficiently disclosed in the application.

Since 1984 (Drug Price Competition and Patent Term Restoration Act 1984) the lifetime of the patents for human medicaments and some other products may be prolonged for half of the testing time during the time of the pending patent application up to a maximum of 5 years. This only applies to one patent for any active ingredient if the premarketing review process by the Food and Drug Administration (FDA) has been delayed.

4.2.5. Patent Categories, Patent Claims

In § 100 b (definitions) "process" is taken to mean a process, the way, as well as also a novel use of processes, machines, devices, compositions, or substances, being known per se.

The preamble (first part) and the characterizing part are mostly separated from each other by the words "comprising" (open-ended claim) or "consisting of" (close-ended claim).

The so-called open-ended claim defines the species of an invention and does not contain features of the protected invention. The wording of the first part follow the elements (features) which constitute the invention or the steps of an inventive process, irrelevant of being new or belonging to the state of the art. Both parts may be separated by the word "comprising". Such a claim is interpreted in practice e.g., in the case of an infringement that adding further elements are of no importance for the invention, when the named elements exist.

In contrast to the open claim is the separation of both parts by the words "consisting of", which in practice means that the invention consists of the named elements (features) only: a combination consisting of A, B, C, and D reduces the extent of the claim to these elements only. Adding a further element E, for example, does not constitute an infringement of the patent.

In wording the claim of an improvement, a "Jepson-claim" is frequently used. Similar to the European form of wording claims, the known features are represented in the first part and the new features in the characterizing part. To interpret the scope of the patent the whole claim has to be considered in regard to the features of the invention.

In comparison to national European procedures of grant it is unusual to characterize the general inventive idea, the wording of the claims rather should contain structural data. Functional features in combination claims are allowed in a claim together with the word "means". "Product by process" claims are possible, "product" and "apparatus" may be claimed in one invention, however, "product" and "process" may be claimed only if they are dependent on each other and cannot be represented separately. Claims should be worded with decreasing protective range starting with the broadest claim (main claim, generic claim). Claims being referred back several times are admissible, however, such a claim may not refer back to exactly such a one. As already explained, the wording of the patent claims corresponds to the protective scope.

4.3. Granting Procedure at the U.S. Patent Office

The USPTO (U.S. Patent and Trademark Office) examines patent applications chronologically according to their filing dates. Requests to accelerate the proceedings for certain reasons (marketing to be started, facts regarding infringement, age of the inventor, environmental inventions) are possible. In exceptional cases it is possible to file an application in another language than English, however, an authorized and exact translation thereof must be filed within two months. Patent applications filed in an incomplete state may be completed within six months and are then given the date of completion as filing date.

4.3.1. Examination, Grant, and Appeal

The examination at the U.S. Patent Office, when issuing an official action, covers

1) the formal part (form, declaration of inventors, formal drawings, completeness of the disclosure, uniformity) with objections, and
2) the subject matter by means of the traced state of the art, the examination of the patent claims (§ 102, 103, patentability) with possible rejection of individual claims or all claims.

Only rejections may later be appealed against before the Court of Customs and Patent Appeal (CCPA). Other statements of the Examiner to be revised can be made only as petition to the Commissioner of the U.S. Patent and Trademark Office.

In the case of lacking disclosure of the invention, the rejection of all patent claims has to be expected since they are not supported in the specification. In the case of a rejection due to lack of inventive step (§ 103) the examiner shows in the official action the difference between the citation and the individual claims. In the reply, the objections made in the official letter have to be dealt with individually and the reply has to be filed within a given term of minimally 30 days and maximally six months (usually three months). A later introduction of "new material", namely words, supplements, deletions, supplements to the formal drawings, is very difficult and is restricted to a request to the USPTO made in the reply. In contrast to European procedures it is not usual and inadmissible to file a new specification in which, e.g., the state of the art cited by the Patent Office has been considered. New patent claims suggested by the examiner have to be added to the previously filed claims and numbered subsequently or deleted and amended. After a final rejection, amendments are no longer possible or only in a very restricted manner (amendment after rejection, Rule 116). In case of grant the examiner issues a notice of allowance with the action of grant; the base issue fee has to be paid. The patent is granted for a duration of 17 years from the date of grant. After 3, 7, and 11 years, maintenance fees have to be paid; small enterprises, private inventors, universities, and nonprofit organizations only have to pay the generally reduced patent fees (50%).

4.3.2. Divisional Applications, Continuation, Continuation in Part

The patent claims rejected by the examiner during the granting procedure may be prosecuted in different ways in the form of an additional application. If the original application contains two (or more) inventions a *divisional* application has to be filed.

If the rejected set of claims contain a joint, generic main claim covering the invention entirely, the part of the claims divided out may be continued in an additional application (*continuation*).

If the set of excluded claims is combined with new matter, of course, provided that there is a main claim covering the new material, a *continuation in part* is advisable.

In both cases the later application is referred back to the original application with the request that it is covered by the invention and the original disclosure. The case may occur that the examiner, pointing to the inadmissible double patenting, requests to prosecute two patent applications in one application of the same applicant if he is of the opinion that only one invention exists.

4.3.3. Reexamination; Appeal Before the Supreme Court and District Courts

Since 1980 the examination procedure may be resumed before the Patent Office, even on request of third parties, instead of a nullity suit before the Court if e.g., sufficient material detrimental to novelty has been traced in the meantime (reexamination).

Appeals must be directed to the Board of Appeals and Interference but, in contrast to the European procedures, also to the Court of Appeals for the Federal Circuit or the District Court of Columbia. The Federal Courts are responsible for patent infringements.

4.4. Design Patents

In the United States, design patents have the character of patents with a wording of claims, in contrast to e.g., European design patents. The priority of American Design Patent Applications may possibly be claimed for applications in some European countries.

5. Japanese Patent Granting Procedure

The legal basis is the patent law from 1985. In addition to patents with a lifetime of maximum 2 years starting with the date of application utility models may also be filed. The procedure is adapted to the European patent granting procedure with postponed examination (7 years). The Japanese Patent Office accepts versions of claims in applications having a foreign priority worded according to European or U.S. rules. Compulsory licenses may be granted after three years on nonused patents.

6. Excerpt from the EPC

Article 100 Grounds for opposition

Opposition may only be filed on the grounds that:
(a) the subject-matter of the European patent is not patentable within the terms of Articles 52 to 57;
(b) the European patent does not disclose the invention in a manner sufficiently clear and complete for it to be carried out by a person skilled in the art;
(c) the subject-matter of the European patent extends beyond the content of the application as filed, or, if the patent was granted on a divisional application or on a new application filed in accordance with Article 61, beyond the content of the earlier application as filed.

Article 52 Patentable inventions

(1) European patents shall be granted for any inventions which are susceptible of industrial application, which are new and which involve an inventive step.
(2) The following in particular shall not be regarded as inventions within the meaning of paragraph 1:
(a) discoveries, scientific theories and mathematical methods;
(b) aesthetic creations;
(c) schemes, rules and methods for performing mental acts, playing games or doing business, and programs for computers;
(d) presentations of information.
(3) The provisions of paragraph 2 shall exclude patentability of the subject-matter or activities referred to in that provision only to the extent to which a European patent application or European patent relates to such subject-matter or activities as such.
(4) Methods for treatment of the human or animal body by surgery or therapy and diagnostic methods practised on the human or animal body shall not be regarded as inventions which are susceptible of industrial application within the meaning of paragraph 1. This provision shall not apply to products, in particular substances or compositions, for use in any of these methods.

Article 53 Exceptions to patentability

European patents shall not be granted in respect of:
(a) inventions the publication or exploitation of which would be contrary to "ordre public" or morality, provided that the exploitation shall not be deemed to be so contrary merely because it is prohibited by law or regulation in some or all of the Contracting States;
(b) plant or animal varieties or essentially biological processes for the production of plants or animals; this provision does not apply to microbiological processes or the products thereof.

Article 54 Novelty

(1) An invention shall be considered to be new if it does not form part of the state of the art.
(2) The state of the art shall be held to comprise everything made available to the public by means of a written or oral description, by use, or in any other way, before the date of filing of the European patent application.
(3) Additionally, the content of European patent applications as filed, of which the dates of filing are prior to the date referred to in paragraph 2 and which were published under Article 93 on or after that date, shall be considered as comprised in the state of the art.
(4) Paragraph 3 shall be applied only in so far as a Contracting State designated in respect of the later application, was also designated in respect of the earlier application as published.
(5) The provisions of paragraphs 1 to 4 shall not exclude the patentability of any substance or composition, comprised in the state of the art, for use in a method referred to in Article 52, paragraph 4, provided that its use for any methods referred to in that paragraph is not comprised in the state of the art.

Article 55 Non-prejudicial disclosures

(1) For the application of Article 54 a disclosure of the invention shall not be taken into consideration if it occurred no earlier than six months preceding the filing of the European patent application and if it was due to, or in consequence of:
(a) an evident abuse in relation to the applicant or his legal predecessor, or
(b) the fact that the applicant or his legal predecessor has displayed the invention at an official, or officially recognised, international exhibition falling within the terms of the Convention on international exhibitions signed at Paris on 22 November 1928 and last revised on 30 November 1972.
(2) In the case of paragraph 1(b), paragraph 1 shall apply only if the applicant states, when filing the European patent application, that the invention has been so displayed and files a supporting certificate within the period and under the conditions laid down in the Implementing Regulations.

Article 56 Inventive step

An invention shall be considered as involving an inventive step if, having regard to the state of the art, it is not obvious to a person skilled in the art. If the state of the art also includes documents within the meaning of Article 54, paragraph 3, these documents are not to be considered in deciding whether there has been an inventive step.

Article 57 Industrial application

An invention shall be considered as susceptible of industrial application if it can be made or used in any kind of industry, including agriculture.

7. References

As an introduction to industrial property rights, a broad variety of international literature is available. The patent offices mostly edit guidelines and brochures for the information of the inventor about important facts concerning patent applications, utility models, and the granting procedure.

[1] K. Haertel (ed.): *European Patent Convention*, Carl Heymanns Verlag, Cologne 1980.
[2] European Patent Office (ed.): *How to get an European Patent, Guide for Applicants*, WILA-Verlag für Wirtschaftswerbung, Munich.
[3] American Intellectual Property Law Association, Arlington, Virginia: *An Overview of Intellectual Property; How to Protect and Benefit from Your Ideas; What is a patent?*

European Patent Law

[4] B. White et al. (eds.): *Encyclopedia of U.K. and European Patent Law* 1977.
Chartered Institute of Patent Lawyers (ed.): *European Patents Handbook*, vols. 1–3.
[5] P. Methely: *Le droit européen des brevets d'invention*, 1978.
[6] F. Panel: *La protection des inventions en droit européen des brevets*, 1977.
[7] P. Rolf: *EPÜ, GPÜ, PCT, Leitfaden der Internationalen Patentverträge*, 1977.

U.S. Patent Law

[8] R. Calvert: *The Encyclopaedia of Patent Practice and Invention Management*, 1964.
[9] D. S. Chisum: *Patents. A Treatise on the Law of Patentability, Validity and Infringement*, vols. 1–7, 1979.
[10] P. Goldstein: *Commentary of the New Patent Act 1954, Introduction to Sec. 35 U.S.C.A.*
[11] A. R. Miller, M. H. Davis: *Intellectual Property Patents, Trademarks and Copyright*, 1983.
[12] P. Rosenberg: *Patent Law Fundamentals*, 2nd ed. 1980 (2 vols.).

International Manuals

[13] *Manual for the Handling of Applications for Patents, Designs and Trademarks Throughout the World*, Octrooibureau les en Stigter, Amsterdam.

FEB 11 1991